From Spinors to Supersymmetry

Supersymmetry is an extension of the successful Standard Model of particle physics; it relies on the principle that fermions and bosons are related by a symmetry, leading to an elegant predictive structure for quantum field theory. This textbook provides a comprehensive and pedagogical introduction to supersymmetry and spinor techniques in quantum field theory. By using the two-component spinor formalism for fermions, the authors provide many examples of practical calculations relevant for collider physics signatures, anomalies, and radiative corrections. They present in detail the component-field and superspace formulations of supersymmetry and explore related concepts, including the theory of extended Higgs sectors, models of grand unification, and the origin of neutrino masses. Numerous exercises are provided at the end of each chapter. Aimed at graduate students and researchers, this volume provides a clear and unified treatment of theoretical concepts that are at the frontiers of high-energy particle physics.

Herbi K. Dreiner is Professor of Physics at the University of Bonn. He received his PhD from the University of Wisconsin, and also worked at the Deutsches Elektronen-Synchrotron (DESY), the University of Oxford, ETH Zürich, and the Rutherford Appleton Laboratory. He is the author of over 100 scientific papers on supersymmetry and has received several teaching prizes. He is a founder of the Bonn Physics Show, for which he received the European Physical Society High Energy Physics (EPS-HEP) Outreach Prize.

Howard E. Haber is Distinguished Professor of Physics at the University of California, Santa Cruz. He received his PhD from the University of Michigan and is a Fellow of the American Physical Society. He is a co-author of *The Higgs Hunter's Guide*, published by Perseus Books in 1990. He won an Alexander von Humboldt Research Award in 2009 and was a co-recipient of the 2017 American Physical Society J. J. Sakurai Prize for Theoretical Particle Physics.

Stephen P. Martin is Professor of Physics at Northern Illinois University, where he received the university's highest award for teaching excellence. He gained his PhD from the University of California, Santa Barbara and is a Fellow of the American Physical Society. He has authored over 100 papers in high-energy physics.

From Spinors to Supersymmetry

HERBI K. DREINER
University of Bonn

HOWARD E. HABER
University of California, Santa Cruz

STEPHEN P. MARTIN
Northern Illinois University

CAMBRIDGE
UNIVERSITY PRESS

CAMBRIDGE
UNIVERSITY PRESS

Shaftesbury Road, Cambridge CB2 8EA, United Kingdom

One Liberty Plaza, 20th Floor, New York, NY 10006, USA

477 Williamstown Road, Port Melbourne, VIC 3207, Australia

314–321, 3rd Floor, Plot 3, Splendor Forum, Jasola District Centre, New Delhi – 110025, India

103 Penang Road, #05–06/07, Visioncrest Commercial, Singapore 238467

Cambridge University Press is part of Cambridge University Press & Assessment, a department of the University of Cambridge.

We share the University's mission to contribute to society through the pursuit of education, learning and research at the highest international levels of excellence.

www.cambridge.org
Information on this title: www.cambridge.org/9780521800884

DOI: 10.1017/9781139049740

First published 2023

A catalogue record for this publication is available from the British Library.

A Cataloging-in-Publication data record for this book is available from the Library of Congress

ISBN 978-0-521-80088-4 Hardback

Additional resources for this publication at www.cambridge.org/spinors-to-susy

Contents

Part III Realistic Supersymmetric Models

Part IV Sample Calculations in the Standard Model and Its Supersymmetric Extension

Part V The Appendices

Preface

The discovery of the Higgs scalar boson in 2012 has put the Standard Model of particle physics on a firm footing. We now have a technically complete theoretical description of all collider phenomena that are known as of this writing. However, there are compelling reasons to believe that the Standard Model of particle physics must be extended. In our view, the foremost of these is that the squared-mass parameter of the Higgs field is quadratically sensitive, through radiative corrections, to every other larger mass scale to which it couples, directly or indirectly. This quadratic sensitivity is an indication of the so-called hierarchy problem.

The classical gravitational interaction lies outside the Standard Model. Using the fundamental constants $\hbar$, c, and Newton's gravitational constant G_N, one can construct a quantity with the units of energy called the Planck scale,

$$M_P c^2 \equiv \left(\frac{\hbar c^5}{G_N}\right)^{1/2} \simeq 1.221 \times 10^{19}\ \text{GeV}.$$

At the Planck energy scale, the quantum mechanical aspects of gravity can no longer be neglected. In particular, the gravitational potential energy Φ of a particle of mass m, evaluated at its Compton wavelength, $r_c = \hbar/(mc)$, must satisfy

$$|\Phi| \sim \frac{G_N m^2}{r_c} = \frac{G_N m^3 c}{\hbar} \lesssim 2mc^2$$

to avoid particle–antiparticle pair creation by the gravitational field. Since the creation of particle–antiparticle pairs is an inherently quantum mechanical phenomenon, quantum gravitational effects cannot be ignored if $m \gtrsim M_P$ [up to $\mathcal{O}(1)$ constants].[1] Thus, the Standard Model cannot be a theory of fundamental particles and their interactions at energy scales of order the Planck scale and above.

Consequently, the Standard Model must break down at some energy scale Λ that is bounded from above by the Planck scale. However, there are strong hints that Λ may in fact lie below $M_P c^2$. The quantization of weak hypercharge, the way that the fermion representations of the Standard Model fit into SU(5) and SO(10) multiplets, and the convergence of running gauge couplings (via renormalization group evolution) all hint at some sort of full or partial unification of forces, the scale of which (if it exists) must be very high to evade proton decay and other bounds. Of course, this might be just a coincidence, and the hierarchy problem should not be viewed as hinging on the existence of unification. Other affirmative,

[1] Note that for $m = M_P$, the Schwarzschild radius $r_s \equiv 2G_N m/c^2 \sim r_c$, which also suggests that the quantum mechanical nature of gravity cannot be neglected at mass scales above M_P.

and perhaps stronger, hints of the presence of mass scales far above the electroweak scale but below M_P include: (i) the presence of nonzero neutrino masses, which are most naturally explained with the seesaw mechanism; (ii) the puzzle of the origin of baryogenesis, which cannot be explained in the Standard Model alone because of the lack of sufficient CP violation; (iii) the solution of the strong CP problem, which can be explained by axions with a very high Peccei–Quinn breaking scale; and, (iv) the fact that many independent observations in astrophysics and cosmology point to the existence of dark matter.

Supersymmetry is the principle that fermions and bosons are related by a symmetry. Remarkably, this simple statement leads ineluctably to a beautiful and predictive structure for quantum field theory. Of course, beauty is a subjective criterion, but historically it has been acknowledged as a useful guide to progress, as attested to by Einstein, Dirac, and many others. From a practical point of view, supersymmetry provides a way of tremendously reducing the hierarchy problem associated with the Higgs field, and nontrivially it also provides avenues of approach to the other outstanding new physics puzzles noted above, especially the dark matter and unification ideas. We have written this book to convey the essential features of supersymmetry as an extension of the Standard Model, but it is another purpose of ours to also provide a toolbox that is useful not only within supersymmetry but in other aspects of particle physics at the high-energy frontier.

A fundamental observation about physics at the electroweak scale is that it is chiral; the left-handed and right-handed components of fermions are logically distinct objects that have different gauge transformation properties. This points to the use of the two-component spinor formalism, which by construction treats left-handed and right-handed fermions separately from the start. Furthermore, within the context of supersymmetric field theories, two-component spinors enter naturally, due to the spinorial nature of the symmetry generators themselves, and the holomorphic structure of the superpotential. Despite this, textbooks on quantum field theory usually present calculations, such as cross sections, decay rates, anomalies, and radiative corrections, in the four-component spinor language. Parity-conserving theories such as QED and QCD are well suited to the four-component spinor methods. There is also a certain perceived advantage to familiarity. However, as we progress to phenomena at and above the scale of electroweak symmetry breaking, it seems increasingly natural to employ two-component spinor notation, in harmony with the transformation properties dictated by the physics.

One often encounters the misconception that even though the two-component fermion language may be better for devising many theories, including supersymmetry, it is somehow inherently ill-suited or unwieldy for practical calculations of physical observables. One of the goals of this book is to demonstrate that two-component fermion notation is just as useful for analyzing quantum field theories as it is for formulating them. The two-component spinor formalism employed here applies equally well to Dirac fermions such as the Standard Model quarks and charged leptons and to Majorana fermions such as the neutrinos of the Standard Model or the neutralinos predicted by supersymmetry.

We have therefore organized the material of this book with two primary goals in mind. The first goal is to present a comprehensive treatment of the two-component spinor formalism as applied in quantum field theory in 3+1 spacetime dimensions, and its applications in the Standard Model and extensions thereof. This is the focus of Part I, which comprises the first six chapters. In Part I, we have also reserved one chapter for providing a detailed translation between the two-component and four-component fermion formalisms, so that the reader can make connections with the vast majority of textbooks that employ the latter in their development of quantum field theory and the Standard Model of particle physics. The second goal is to provide a pedagogical introduction to the construction and application of supersymmetric theories in establishing realistic theories of fundamental particles and interactions beyond the Standard Model. This is accomplished in Parts II and III of this book. Finally, we provide in Part IV sample calculations in the Standard Model and its supersymmetric extension in some detail using the two-component spinor formalism. These calculations, which encompass decay processes, scattering processes, and radiative corrections, illustrate the power of these methods in practical calculations of experimental observables. Additional details of our toolbox of techniques are collected in Part V, consisting of 11 separate appendices.

There is a huge literature on spinors and supersymmetry. Many references to the original literature that are not explicitly cited in this book can be found in reviews and summer school lecture notes by the authors that appeared in Refs. [1–3]. The mathematical treatment of spinors and their applications to physics can be found in numerous textbooks (e.g., see [B1–B15]). Textbook treatments of supersymmetry and supergravity are also abundant (e.g., see [B16–B45]) along with many reviews and summer school lecture notes (e.g., see Refs. [2–27]). Additional references that trace the origins of supersymmetry can be found in [B46].[2]

We hope that this book serves as a useful addition to the literature cited above. In particular, we have strived to provide a level of detail not typically found in other textbooks. We have also attempted to address many questions that students often pose when first encountering this material that are typically ignored in the standard treatments. Numerous exercises are provided at the end of each chapter (and at the end of most of the appendices). These exercises serve a number of different purposes. Some of the exercises require rather straightforward calculations that allow the student to gain a better understanding of the material. Other exercises represent the treatment of additional topics that would have been included in the main text if space permitted. In these cases, we have often quoted the final result (which the student is encouraged to verify), which may be employed elsewhere in the book. Finally, some of the exercises are more substantial and would constitute a good term project for a special topics course. In this way, we believe that this book can provide a good basis for an advanced class in particle theory (where the instructor focuses on some subset of the book's chapters for a one-semester course), or provide a useful resource for self-study.

[2] Books are listed in the Bibliography and are indicated by the prefix B. Review articles, lecture notes, and journal articles are independently numbered and appear separately in the References.

Prerequisites for this book include a basic knowledge of quantum field theory (e.g., see [B47–B74]), gauge theories (e.g., see [B75–B81]), and the Standard Model of particle physics and beyond (e.g., see [B82–B96]). Some review of this material is also provided in Part I of this book. Important mathematical methods employed in this book are treated in the appendices, including Lie groups and algebras (e.g., see [B97–B125]) and matrix algebra (e.g., see [B126–B129]).

Although we have endeavored to produce a book that is free of errors, it is very likely that typographical errors, misprints, and various inconsistencies have escaped our attention. We will therefore maintain updated information on any errors that we discover after the initial publication, which can be found on the home page of our book:

`www.cambridge.org/spinors-to-susy`

Of course, the authors would be most grateful to readers who detect any errors in our book and provide us with the relevant details. We will be happy to list the errata along with the appropriate citations on the home page of this book.

We conclude this preface by acknowledging that the Large Hadron Collider (LHC), operating at center-of-mass energies up to 13 TeV, has not seen any evidence of supersymmetry as of this writing, confounding the hopes of many. Several comments are in order here. First, the fact that the Higgs boson mass has turned out to be 125 GeV (rather than, say, 115 GeV, as seemed plausible when the LHC turned on) hints at considerably larger masses for the supersymmetric partners of the Standard Model particles than one might have guessed previously. This is because of the way that the superpartner masses feed logarithmically into the Higgs boson mass in radiative corrections, as explained in Sections 13.8 and 19.8. From this point of view, it is hardly surprising that direct evidence for supersymmetry has not been found within the presently explored mass ranges. While this seemingly leads to a little hierarchy problem associated with the ratio of the superpartner masses to the electroweak symmetry-breaking scale, that hierarchy remains minuscule compared to the problem of the large hierarchy between the electroweak and Planck scales in non-supersymmetric extensions of the Standard Model.

Furthermore, the existence of a 125 GeV Higgs boson is quite compatible with supersymmetric particles somewhat above the TeV scale, as had been noted long before the LHC turned on. Many other competitor theories either predicted no Higgs boson at all, or predicted a much heavier Higgs boson, or simply did not dare to make any firm prediction about it whatsoever. Supersymmetry compares favorably in this regard. Finally, the lack of any LHC evidence for other new physics means that many of the alternatives to supersymmetry that have been proposed to address the hierarchy problem are either eliminated or highly disfavored, and in our view none of these approaches can be said to be in better shape than supersymmetry.

Therefore, while we of course make no guarantees, for it would be foolish to do so, we remain optimistic that supersymmetry is an essential feature of the fundamental laws of physics, and we look forward to this being clarified as the great adventure of exploration continues. We hope that this book will be useful along the way.

Acknowledgments

This book is the culmination of a project that has lasted 20 years. The conception of this book project was strongly encouraged by Robert N. Cahn, who provided us with the initial connection with Cambridge University Press. We are especially grateful to Simon Capelin for guiding us for many years during this effort and for his patience and understanding over numerous missed deadlines. Although we had hoped to have a finished project in time for his retirement celebration, a few unforeseen events followed by the Covid-19 pandemic ultimately delayed the final delivery of the manuscript. We would like to thank Simon's successor, Vince Higgs, for his continued confidence and support through the latter stages of our project, Sarah Armstrong and Clare Dennison for guiding us through the final steps to publication, John King for his superb copy-editing skills, and Suresh Kumar and the team of LaTeX experts at Cambridge University Press for their assistance on a number of typographical issues.

Our book incorporates some material that was originally published in a chapter by Stephen P. Martin, entitled "A Supersymmetry Primer," which appeared in *Perspectives on Supersymmetry*, edited by Gordon L. Kane (World Scientific, Singapore, 1998), pp. 1–98, and subsequently updated in *Perspectives on Supersymmetry II*, edited by Gordon L. Kane (World Scientific, Singapore, 2010), pp. 1–153. In addition, some of the exercises in our book first appeared in lecture notes by Howard E. Haber and Laurel Stephenson Haskins, entitled "Supersymmetric Theory and Models," which was published in *Anticipating the Next Discoveries in Particle Physics*, Proceedings of the 2016 Theoretical Advanced Study Institute in Elementary Particle Physics (TASI-2016), edited by R. Essig and I. Low (World Scientific, Singapore, 2018), pp. 355–499. Lastly, we have made substantial use of the review article by Herbi K. Dreiner, Howard E. Haber, and Stephen P. Martin, "Two-component spinor techniques and Feynman rules for quantum field theory and supersymmetry," Physics Reports **494**, 1–196 (2010), which was published by Elsevier. We are grateful to World Scientific and Elsevier for granting permission for the reuse of these materials in this book.

There are many colleagues, whose collaboration and/or counsel were critical in assembling the material for this book. We are especially pleased to acknowledge Steven Abel, Ben Allanach, Wolfgang Altmannshofer, Nima Arkani-Hamed, Howard Baer, Jonathan Bagger, Thomas Banks, Riccardo Barbieri, R. Michael Barnett, Prudhvi Bhattiprolu, Jon Butterworth, Marcela Carena, J. Alberto Casas, Zackaria Chacko, Ali Chamseddine, Piotr Chankowski, Kiwoon Choi, Seong Youl Choi, Andrew Cohen, Timothy Cohen, Nathaniel Craig, Sacha Davidson, Sally

Dawson, Athanasios Dedes, André de Gouvêa, Francesco D'Eramo, Jordy de Vries, Marco Diaz, Savas Dimopoulos, Michael Dine, Lance Dixon, Abdelhak Djouadi, Patrick Draper, Manuel Drees, John Ellis, Henriette Elvang, Jonathan Feng, Patrick Fox, Jürgen Fuchs, Tony Gherghetta, Gian Giudice, Rohini Godbole, Stefania Gori, Sebastian Grab, Yuval Grossman, John Gunion, Lawrence Hall, Tao Han, Sven Heinemeyer, Ralf Hempfling, María Herrero, Christopher Hill, D.R. Timothy Jones, Jan Kalinowski, Gordon Kane, Shinya Kanemura, Jong-Soo Kim, Christopher Kolda, Michael Krämer, Graham Kribs, Paul Langacker, Heather Logan, Ian Low, Christoph Luhn, Markus Luty, Joseph Lykken, Víctor Martín Lozano, John Mason, Stephen Mrenna, Hitoshi Murayama, Ann Nelson, Hans Peter Nilles, Yosef Nir, Keith Olive, Deva O'Neil, Toby Opferkuch, Siannah Peñaranda, Michael Peskin, Stefan Pokorski, Alex Pomarol, Stefano Profumo, Mariano Quirós, Stuart Raby, Pierre Ramond, Lisa Randall, Riccardo Rattazzi, Peter Richardson, Janusz Rosiek, Graham Ross, Subir Sarkar, Martin Schmaltz, Christoph Schweigert, Nausheen Shah, João Silva, Pietro Slavich, Mark Srednicki, Laurel Stephenson Haskins, Raman Sundrum, Timothy Tait, Xerxes Tata, Jamie Tattersall, John Terning, Jesse Thaler, Scott Thomas, Marc Thormeier, Lorenzo Ubaldi, Carlos Wagner, Zeren Simon Wang, Georg Weiglein, James Wells, and Peter Zerwas.

In addition, we are grateful to many of our students and postdocs who have read earlier drafts of this book and have made numerous suggestions for improvement: Annika Buchholz, Eric Carlson, Joseph Connell, Florian Domingo, Janis Dücker, Nicolas Fernandez, Julian Günther, John Kehayias, Dominik Köhler, Saurabh Nangia, Hiren Patel, and Martin Schürmann.

The three authors of this book would like to gratefully acknowledge the Deutsche Forschungsgemeinschaft and Bundesministerium für Bildung und Forschung, the US Department of Energy, and the National Science Foundation for providing longstanding support through research grants that fostered our investigations of quantum field theory, physics of the Standard Model and beyond, and supersymmetric theory and phenomena. The research performed under these grants has deepened our understanding of many of the topics treated in this book and provided us with an opportunity to present some of the results of our explorations in a pedagogical setting.

Finally, for their work on promoting and nurturing national and international scientific collaborations, we wish to acknowledge and celebrate the Alexander von Humboldt Foundation and the Aspen Center for Physics (the latter located just 10 miles down the road from the picturesque Maroon Lake that adorns the cover of this book). These two institutions were instrumental in the research lives of the three authors of this book, and both played a significant role in providing the collaborative opportunities that allowed the authors to carry out this work.

We dedicate this book to our respective spouses, Heike, Marjorie, and Jeanette, who are probably even more enthusiastic than we are that this book project has finally come to a conclusion.

Acronyms and Abbreviations

1PI	one-particle irreducible
2HDM	two-Higgs doublet model
AMSB	anomaly-mediated supersymmetry breaking
ATLAS	A Toroidal LHC Apparatus general-purpose detector
BCH	Baker–Campbell–Hausdorff
BPS	Bogomol'nyi–Prasad–Sommerfield
BR	branching ratio
BRST	Becchi–Rouet–Stora–Tyutin
$\mathbb{C}$	set of complex numbers
CKM	Cabibbo–Kobayashi–Maskawa
CM	center-of-momentum
CMS	Compact Muon Solenoid general-purpose detector
CMSSM	Constrained Minimal Supersymmetric Standard Model
CPT	charge conjugation/parity/time reversal
DGS	discrete gauge symmetry
DM	dark matter
$\overline{\text{DR}}$	modified minimal subtraction scheme using DRED
DRED	dimensional reduction
DREG	dimensional regularization
EM	electromagnetic
FCNC	flavor-changing neutral current
FI	Fayet–Iliopoulos
FT	Fourier transform
GBP	generalized baryon parity
GF	gauge-fixed
GIM	Glashow–Iliopoulos–Maiani
GLP	generalized lepton parity
GMP	generalized matter parity
GMSB	gauge-mediated supersymmetry breaking
GUT	grand unified theory
GWP	Glashow–Weinberg–Paschos
HM	"HiggsMass" scheme
IR	infrared
irrep	irreducible representation
LEP	Large Electron-Positron Collider
LHC	Large Hadron Collider

Lips	Lorentz-invariant phase space
LSP	lightest supersymmetric particle
LSZ	Lehmann–Symanzik–Zimmermann
MHV	maximally helicity-violating
$\overline{\text{MS}}$	modified minimal subtraction scheme using DREG
MSSM	Minimal Supersymmetric Standard Model
mSUGRA	minimal supergravity
NLSP	next-to-lightest supersymmetric particle
NMSSM	Next-to-Minimal Supersymmetric Standard Model
NSVZ	Novikov–Shifman–Vainshtein–Zakharov
NUHM	nonuniversal Higgs mass model
OS	on-shell
PMNS	Pontecorvo–Maki–Nakagawa–Sakata
PMSB	Planck-scale-mediated supersymmetry breaking
pMSSM	phenomenological MSSM
PQ	Peccei–Quinn
PS	Pati–Salam
QCD	quantum chromodynamics
QED	quantum electrodynamics
Q4S	"quasi-4-dimensional" space
$\mathbb{R}$	set of real numbers
RG	renormalization group
RGE	renormalization group equation
RPV	R-parity-violating
SE-MSSM	seesaw-extended MSSM
SM	Standard Model
SQED	superpartner contribution of SUSY-QED
SUSY	supersymmetry (or supersymmetric)
SUSYGUT	supersymmetric grand unified theory
SUSY-QED	supersymmetric QED
SUSY-QCD	supersymmetric QCD
SVD	singular value decomposition
TASI	Theoretical Advanced Study Institute in Particle Physics
UV	ultraviolet
VEV	vacuum expectation value
WB	Wess–Bagger
XMSB	extra-dimensional-mediated supersymmetry breaking
YM	Yang–Mills
$\mathbb{Z}$	set of integers

PART I

SPIN-1/2 FERMIONS IN QUANTUM FIELD THEORY, THE STANDARD MODEL, AND BEYOND

1 Two-Component Formalism for Spin-1/2 Fermions

In this chapter, we examine the incorporation of spin-1/2 fermions into quantum field theory. Underlying the relativistic theory of quantized fields is special relativity and the invariance of the Lagrangian under the Poincaré group, which comprise Lorentz transformations and spacetime translations (e.g., see [B130–B132]).

1.1 The Lorentz Group and Its Lie Algebra

Under an *active* Lorentz transformation, $\Lambda^\mu{}_\nu$, the spacetime coordinates, x^μ, transform as $x'^\mu = \Lambda^\mu{}_\nu x^\nu$, where there is an implicit sum over the repeated index ν (see Appendix A.4 for our conventions for spacetime notation). One can regard $\Lambda^\mu{}_\nu$ as the matrix element located in row μ and column ν of the 4×4 matrix Λ. The condition that $g_{\mu\nu}x^\mu x^\nu$ is invariant under Lorentz transformations implies that

$$\Lambda^\mu{}_\nu g_{\mu\rho} \Lambda^\rho{}_\lambda = g_{\lambda\nu} \,. \tag{1.1}$$

The set of all Λ that satisfy eq. (1.1) forms a Lie group O(1,3).

Equation (1.1) implies that Λ has the following two properties: (i) $\det \Lambda = \pm 1$ and (ii) $|\Lambda^0{}_0| \geq 1$. Thus, Lorentz transformations fall into four disconnected classes, which can be denoted by a pair of signs: $\big(\text{sgn}(\det\Lambda)\,,\,\text{sgn}(\Lambda^0{}_0)\big)$. The proper orthochronous Lorentz transformations correspond to $(+,+)$ and are continuously connected to the identity. The group of such transformations forms a subgroup of O(1,3), which we shall denote by $\text{SO}_+(1,3)$.[1] If $\Lambda \in \text{SO}_+(1,3)$, we can generate all elements of the other classes of Lorentz transformations by introducing the space-inversion matrix, $\Lambda_P = \text{diag}(1,-1,-1,-1)$, the time-inversion matrix $\Lambda_T = \text{diag}(-1,1,1,1)$ and the spacetime-inversion matrix $\Lambda_P\Lambda_T = -\mathbb{1}_{4\times 4}$. Then, $\{\Lambda\,,\,\Lambda_P\Lambda\,,\,\Lambda_T\Lambda\,,\,\Lambda_P\Lambda_T\Lambda\}$ spans the full Lorentz group.

Infinitesimal Lorentz transformations must be proper and orthochronous, since these are continuously connected to the identity. These transformations are best studied by exploring the properties of the SO(1,3) Lie algebra, denoted henceforth by $\mathfrak{so}(1,3)$. Later (see Section 1.7), we shall examine the implications of the improper Lorentz transformations: space inversion and time inversion.

The most general proper orthochronous Lorentz transformation matrix Λ is characterized by a rotation angle θ about an axis $\hat{\boldsymbol{n}}$ $[\vec{\boldsymbol{\theta}} \equiv \theta\hat{\boldsymbol{n}}]$ and a boost vector

[1] In this notation, the S (which stands for "special") corresponds to the condition $\det \Lambda = 1$ and the subscript + corresponds to $\Lambda^0{}_0 \geq +1$.

$\vec{\zeta} \equiv \hat{v} \tanh^{-1} \beta$ [where $\hat{v} \equiv \vec{v}/|\vec{v}|$ is the unit velocity vector and $\beta \equiv |\vec{v}|/c$].[2] The 4×4 matrix Λ is given by[3]

$$\Lambda = \exp\left(-\tfrac{1}{2}i\theta_{\rho\sigma}s^{\rho\sigma}\right) = \exp\left(-i\vec{\theta}\cdot\vec{s} - i\vec{\zeta}\cdot\vec{k}\right) = \exp\begin{pmatrix} 0 & \zeta^1 & \zeta^2 & \zeta^3 \\ \zeta^1 & 0 & -\theta^3 & \theta^2 \\ \zeta^2 & \theta^3 & 0 & -\theta^1 \\ \zeta^3 & -\theta^2 & \theta^1 & 0 \end{pmatrix}, \tag{1.2}$$

where $\theta_{\rho\sigma} = -\theta_{\sigma\rho}$, and the $s^{\rho\sigma} = -s^{\sigma\rho}$ are six independent 4×4 antisymmetric matrices that satisfy the commutation relations of the Lie algebra $\mathfrak{so}(1,3) \simeq \mathfrak{sl}(2,\mathbb{C})_{\mathbb{R}}$,[4]

$$[s^{\alpha\beta}, s^{\rho\sigma}] = i(g^{\beta\rho}\, s^{\alpha\sigma} - g^{\alpha\rho}\, s^{\beta\sigma} - g^{\beta\sigma}\, s^{\alpha\rho} + g^{\alpha\sigma}\, s^{\beta\rho}). \tag{1.3}$$

The matrix elements of $s^{\rho\sigma}$ are given explicitly as follows (e.g., see [B133]):

$$(s^{\rho\sigma})^{\mu}{}_{\nu} = i(\delta^{\sigma}_{\nu}\, g^{\mu\rho} - \delta^{\rho}_{\nu}\, g^{\mu\sigma})\,. \tag{1.4}$$

In eq. (1.2), we have also defined $\theta^i \equiv \frac{1}{2}\epsilon^{ijk}\theta^{jk}$, $\zeta^i \equiv \theta^{i0} = -\theta^{0i}$, $s^i \equiv \frac{1}{2}\epsilon^{ijk}s^{jk}$, and $k^i \equiv s^{0i} = -s^{i0}$, after noting that $\theta_{ij} = \theta^{ij}$ and $\theta_{0i} = -\theta^{0i}$. Here, the indices $i,j,k \in \{1,2,3\}$ and $\epsilon^{123} = +1$.

In light of eqs. (1.2) and (1.4), an infinitesimal orthochronous Lorentz transformation is given by

$$\Lambda^{\mu}{}_{\nu} \simeq \delta^{\mu}_{\nu} - \tfrac{1}{2}i\theta_{\rho\sigma}(s^{\rho\sigma})^{\mu}{}_{\nu} = \delta^{\mu}_{\nu} + \tfrac{1}{2}\theta_{\rho\sigma}(\delta^{\sigma}_{\nu}\, g^{\mu\rho} - \delta^{\rho}_{\nu}\, g^{\mu\sigma}) = \delta^{\mu}_{\nu} + \tfrac{1}{2}(\theta^{\mu}{}_{\nu} - \theta_{\nu}{}^{\mu})\,. \tag{1.5}$$

Since $\theta^{\mu}{}_{\nu} = g^{\mu\rho}\theta_{\rho\nu} = -g^{\mu\rho}\theta_{\nu\rho} = -\theta_{\nu}{}^{\mu}$, it follows that

$$\Lambda^{\mu}{}_{\nu} \simeq \delta^{\mu}_{\nu} + \theta^{\mu}{}_{\nu} \simeq (\mathbb{1}_{4\times 4} - i\vec{\theta}\cdot\vec{s} - i\vec{\zeta}\cdot\vec{k})^{\mu}{}_{\nu}\,, \tag{1.6}$$

where $\mathbb{1}_{4\times 4}$ is the 4×4 identity matrix. That is, the s^i generate infinitesimal three-dimensional rotations in space and the k^i generate infinitesimal Lorentz boosts.

The spin-0 and (massive) spin-1 fields transform under a general proper orthochronous Lorentz transformation as[5]

$$\phi'(x') = \phi(x)\,, \qquad \text{spin } 0\,, \tag{1.7}$$

$$A'^{\mu}(x') = \Lambda^{\mu}{}_{\nu}A^{\nu}(x)\,, \qquad \text{spin } 1\,. \tag{1.8}$$

For a field of spin s, the general transformation law reads

$$\psi'_{\alpha}(x') = \exp\left(-\tfrac{1}{2}i\theta_{\mu\nu}S^{\mu\nu}\right)_{\alpha}{}^{\beta}\, \psi_{\beta}(x)\,, \tag{1.9}$$

[2] Henceforth, we work in units where $\hbar = c = 1$ (see Appendix A.3).

[3] All symmetry transformations in this chapter are defined from the active point of view. For a passive Lorentz transformation, where the coordinate frame is transformed and the four-vectors are held fixed, simply replace $\{\vec{\theta}, \vec{\zeta}\}$ with $\{-\vec{\theta}, -\vec{\zeta}\}$.

[4] If $\{\boldsymbol{t_1}, \boldsymbol{t_2}, \boldsymbol{t_3}\}$ is a basis for the complex Lie algebra $\mathfrak{sl}(2,\mathbb{C})$, then the notation $\mathfrak{sl}(2,\mathbb{C})_{\mathbb{R}}$ corresponds to the realification of $\mathfrak{sl}(2,\mathbb{C})$, which yields a real Lie algebra consisting of real linear combinations of the six generators, $\{\boldsymbol{t_1}, i\boldsymbol{t_1}, \boldsymbol{t_2}, i\boldsymbol{t_2}, \boldsymbol{t_3}, i\boldsymbol{t_3}\}$.

[5] Equations (1.7) and (1.8) can also be written as $\phi'(x) = \phi(\Lambda^{-1}x)$ and $A'^{\mu}(x) = \Lambda^{\mu}{}_{\nu}A^{\nu}(\Lambda^{-1}x)$. The transformation law for a massless spin-1 gauge field is more complicated and has the form $A'^{\mu}(x) + \partial^{\mu}\Omega(x,\Lambda) = \Lambda^{\mu}{}_{\nu}A^{\nu}(\Lambda^{-1}x)$, as indicated in eq. (5.9.31) of [B72], since $A'^{\mu}(x)$ and the gauge-transformed $A'^{\mu}(x) + \partial^{\mu}\Omega(x,\Lambda)$ are physically equivalent.

where the $S^{\mu\nu}$ are (finite-dimensional) irreducible matrix representations of the Lie algebra of the Lorentz group, and α and β label the components of the matrix representation space. The dimension of this space is related to the spin of the particle. In particular, $S^{\mu\nu}$ is an antisymmetric tensor, $S^{\mu\nu} = -S^{\nu\mu}$, that satisfies the commutation relations of $\mathfrak{so}(1,3)$ [eq. (1.3)]. Different irreducible finite-dimensional representations of $\mathfrak{so}(1,3)$ correspond to particles of different spin. In analogy with s^i and k^i defined below eq. (1.4), we identify the following pieces of $S^{\mu\nu}$:

$$S^i \equiv \tfrac{1}{2}\epsilon^{ijk}S^{jk}\,, \qquad K^i \equiv S^{0i}\,, \tag{1.10}$$

where $i, j, k \in \{1,2,3\}$. Using eq. (1.3), it follows that S^i and K^i satisfy the commutation relations

$$[S^i, S^j] = i\epsilon^{ijk}S^k\,, \tag{1.11}$$

$$[S^i, K^j] = i\epsilon^{ijk}K^k\,, \tag{1.12}$$

$$[K^i, K^j] = -i\epsilon^{ijk}S^k\,. \tag{1.13}$$

It is convenient to define the following linear combinations of the generators:

$$\vec{S}_+ \equiv \tfrac{1}{2}(\vec{S} + i\vec{K})\,, \tag{1.14}$$

$$\vec{S}_- \equiv \tfrac{1}{2}(\vec{S} - i\vec{K})\,. \tag{1.15}$$

Then, eqs. (1.11)–(1.13) decouple and yield two independent $\mathfrak{su}(2)$ Lie algebras:

$$[S_+^i, S_+^j] = i\epsilon^{ijk}S_+^k\,, \tag{1.16}$$

$$[S_-^i, S_-^j] = i\epsilon^{ijk}S_-^k\,, \tag{1.17}$$

$$[S_+^i, S_-^j] = 0\,. \tag{1.18}$$

The finite-dimensional irreducible representations of $\mathfrak{su}(2)$ are well known: these are the $(2s+1)\times(2s+1)$ representation matrices corresponding to spin s, where $s = 0, \frac{1}{2}, 1, \frac{3}{2}, \ldots$ (whose matrix elements appear in most textbooks on quantum mechanics; e.g., see [B134]). Hence, the irreducible representations of the Lorentz group can be characterized by two numbers (s_+, s_-), where $s_\pm$ is nonnegative and either an integer or a half-integer. The eigenvalues of $\vec{S}_\pm^{\,2}$ are given by $s_\pm(s_\pm+1)$. The dimension of the representation corresponding to (s_+, s_-) is $(2s_+ +1)(2s_- +1)$.

Using eqs. (1.9) and (1.10), an infinitesimal Lorentz transformation is given by

$$M \equiv \exp\left(-\tfrac{1}{2}i\theta_{\mu\nu}S^{\mu\nu}\right) \simeq \mathbb{1} - i\vec{\theta}\boldsymbol{\cdot}\vec{S} - i\vec{\zeta}\boldsymbol{\cdot}\vec{K}\,, \tag{1.19}$$

where θ^i and ζ^i are defined following eq. (1.2) and $\mathbb{1}$ is the identity. The simplest (trivial) representation is the one-dimensional $(0,0)$ representation, which corresponds to a spin-0 scalar field. In this representation, $\vec{S} = \vec{K} = 0$ and we recover from eq. (1.9) the transformation law given in eq. (1.7). The spin-1 transformation law [see eq. (1.8)] corresponds to the four-dimensional $(\frac{1}{2}, \frac{1}{2})$ representation. However, in a quantum field theory of massive spin-1 fields, only three of the four degrees of freedom are physical. Moreover, in gauge theories of massless spin-1 fields, gauge invariance introduces an additional constraint and only two degrees of freedom are physical. This is described in detail in [B72], to which we refer the reader.

1.2 The Poincaré Group and Its Lie Algebra

The Poincaré group is a semidirect product of the group of spacetime translations and the Lorentz group (e.g., see [B112, B132]). In particular, under a Poincaré transformation, the spacetime coordinates transform as $x'^{\mu} = \Lambda^{\mu}{}_{\nu}x^{\nu} + a^{\mu}$, where Λ satisfies eq. (1.1) and a^{μ} is a constant four-vector. In examining the behavior of the fields under a Poincaré transformation, we shall assume that Λ is a proper orthochronous Lorentz transformation [see eq. (1.2)].

Under a Poincaré transformation, a field of spin s transforms according to eq. (1.9) after identifying x' with the Poincaré transformed spacetime coordinate. It is convenient to rewrite this transformation law as follows:

$$\psi'_{\alpha}(x) = \exp(-\tfrac{1}{2}i\theta_{\mu\nu}S^{\mu\nu})_{\alpha}{}^{\beta}\, \psi_{\beta}\left(\Lambda^{-1}(x-a)\right)\,, \tag{1.20}$$

where we have used $x = \Lambda^{-1}(x'-a)$ and redefined the dummy variable x' by removing the prime. Under an infinitesimal Poincaré transformation, we expand eq. (1.20) about $\Lambda = \mathbb{1}_{4\times 4}$ and $a = 0$ to obtain[6]

$$\psi'_{\alpha}(x) \simeq \left[\mathbb{1} + ia_{\mu}P^{\mu} - \tfrac{1}{2}i\theta_{\mu\nu}(L^{\mu\nu} + S^{\mu\nu})\right]_{\alpha}{}^{\beta}\, \psi_{\beta}(x)\,, \tag{1.21}$$

where P^{μ} and $L^{\mu\nu}$ are the differential operators[7]

$$P^{\mu} \equiv i\partial^{\mu}\,, \qquad L^{\mu\nu} \equiv i(x^{\mu}\partial^{\nu} - x^{\nu}\partial^{\mu})\,, \tag{1.22}$$

and the total angular momentum tensor is $J^{\mu\nu} \equiv L^{\mu\nu} + S^{\mu\nu}$. The four generators of spacetime translations (P^{μ}) and the six generators of Lorentz transformations ($J^{\mu\nu}$, $\mu < \nu$) satisfy the following commutation relations of the Poincaré algebra:[8]

$$[P^{\mu}, P^{\nu}] = 0\,, \tag{1.23}$$

$$[J^{\mu\nu}, P^{\rho}] = i(g^{\nu\rho}P^{\mu} - g^{\mu\rho}P^{\nu})\,, \tag{1.24}$$

$$[J^{\mu\nu}, J^{\rho\sigma}] = i(g^{\nu\rho}J^{\mu\sigma} - g^{\mu\rho}J^{\nu\sigma} - g^{\nu\sigma}J^{\mu\rho} + g^{\mu\sigma}J^{\nu\rho})\,. \tag{1.25}$$

It is convenient to introduce the Pauli–Lubański vector w^{μ}:

$$w^{\mu} \equiv -\tfrac{1}{2}\epsilon^{\mu\nu\rho\sigma}J_{\nu\rho}P_{\sigma} = (\vec{\boldsymbol{J}}\cdot\vec{\boldsymbol{P}}\,;\, P^0\vec{\boldsymbol{J}} + \vec{\boldsymbol{\mathcal{K}}}\times\vec{\boldsymbol{P}})\,, \tag{1.26}$$

in a convention where $\epsilon^{0123} = 1$, where $J^i \equiv \frac{1}{2}\epsilon^{ijk}J^{jk}$ and $\mathcal{K}^i \equiv J^{0i}$. It follows that $w_{\mu}P^{\mu} = 0$. The Poincaré algebra possesses two independent Casimir operators $P^2 \equiv P_{\mu}P^{\mu}$ and $w^2 \equiv w_{\mu}w^{\mu}$, which commute with the generators P^{μ} and $J^{\mu\nu}$.

The representations of the Poincaré group, which correspond to particle states of nonnegative energy P^0 with definite mass and spin, can be labeled by the eigenvalues of the Casimir operators, P^2 and w^2, when acting on the physical states.

[6] The operators $\mathbb{1}$, P^{μ}, and $L^{\mu\nu}$ include an implicit factor of $\delta_{\alpha}{}^{\beta}$, whereas the spin operator $S^{\mu\nu}$ depends nontrivially on α and β (except for the case of spin 0, where $S = 0$).

[7] We recognize P^{μ} and $L^{\mu\nu}$ as the quantum mechanical four-vector momentum operator and tensor orbital angular momentum operator, respectively, in the coordinate representation.

[8] As demonstrated in Exercise 1.2, eq. (1.24) is a consequence of the transformation law of the four-vector P^{μ} under Lorentz transformations.

The eigenvalue of P^2 is m^2, where m is the mass. To see the physical interpretation of w^2, we first consider the case of $m \neq 0$. In this case, we are free to evaluate w^2 in the particle rest frame (since w^2 is a Lorentz scalar). In this frame, $w^\mu = (0\,;\, m\vec{S})$, where S^i is defined in eq. (1.10). Hence, $w^2 = -m^2\vec{S}^{\,2}$, with eigenvalues $-m^2 s(s+1)$ [where $s = 0, \frac{1}{2}, 1, \ldots$]. We conclude that massive (positive-energy) states can be labeled by (m, s), where m is the mass and s is the spin of the state.

If $m = 0$, the previous analysis is not valid, since we cannot evaluate w^2 in the rest frame. Nevertheless, if we take the $m \to 0$ limit, it follows from the results above that either $w^2 = 0$ or the corresponding states have infinite spin. We reject the second possibility (which does not appear to be realized in Nature) and assume that $w^2 = 0$. Thus, we must solve the equations $w^2 = P^2 = w_\mu P^\mu = 0$. It is simplest to choose a frame in which $P = (P^0; 0, 0, P^0)$ where $P^0 > 0$. In this frame, it is easy to show that $w = (w^0; 0, 0, w^0)$. That is, in any Lorentz frame,

$$w^\mu = hP^\mu\,, \tag{1.27}$$

where h is called the helicity operator. In particular,

$$[h\,,\, P^\mu] = [h\,,\, J^{\mu\nu}] = 0\,, \tag{1.28}$$

which means that the eigenvalues of h can be used to label states of the irreducible massless representations of the Poincaré algebra. From eq. (1.27), we derive[9]

$$h = \frac{w^0}{P^0} = \frac{\vec{J}\cdot\vec{P}}{P^0} = \frac{\vec{S}\cdot\vec{P}}{P^0}\,. \tag{1.29}$$

Eigenvalues of h are called the helicity (and are denoted by λ). In particular, note that for massless states, the eigenvalue of $\vec{S}\cdot\vec{P}/P^0$ is equal to that of $\vec{S}\cdot\hat{P}$ (where $\hat{P} \equiv \vec{P}/|\vec{P}|$). Moreover, $\vec{S}\cdot\hat{P}$ corresponds to the projection of the spin along its direction of motion with a spectrum consisting of $\lambda = 0, \pm\frac{1}{2}, \pm 1, \ldots$, where $\lambda \to -\lambda$ under a CPT transformation. Thus, in any quantum field theory realization of massless particles, both $\pm|\lambda|$ helicity states must appear in the theory. It is common to refer to a massless (positive-energy) state of helicity $|\lambda|$ as having spin-$|\lambda|$.

1.3 Spin-1/2 Representation of the Lorentz Group

We first focus on the simplest nontrivial irreducible representations of the Lorentz algebra: the two-dimensional representations $(\frac{1}{2}, 0)$ and $(0, \frac{1}{2})$. The corresponding two-dimensional representations of the Lorentz generators are explicitly given by

$$(\tfrac{1}{2}, 0):\quad \vec{S}_+ = \tfrac{1}{2}(\vec{S} + i\vec{K}) = \tfrac{1}{2}\vec{\sigma}\,, \qquad \vec{S}_- = \tfrac{1}{2}(\vec{S} - i\vec{K}) = 0\,, \tag{1.30}$$

which corresponds to $\vec{S} = \vec{\sigma}/2$ and $\vec{K} = -i\vec{\sigma}/2$, and

$$(0, \tfrac{1}{2}):\quad \vec{S}_+ = \tfrac{1}{2}(\vec{S} + i\vec{K}) = 0\,, \qquad \vec{S}_- = \tfrac{1}{2}(\vec{S} - i\vec{K}) = \tfrac{1}{2}\vec{\sigma}\,. \tag{1.31}$$

[9] The three-vector orbital angular momentum operator is given by $L^i \equiv \frac{1}{2}\epsilon^{ijk}L^{jk}$ [see eq. (1.22)]. Hence, $\vec{L} = \vec{x} \times \vec{P}$ and it follows that $\vec{J}\cdot\vec{P} = (\vec{L} + \vec{S})\cdot\vec{P} = (\vec{x} \times \vec{P} + \vec{S})\cdot\vec{P} = \vec{S}\cdot\vec{P}$.

Hence, we can identify $\vec{S} = \vec{\sigma}/2$ and $\vec{K} = i\vec{\sigma}/2$. Here, $\vec{\sigma} = (\sigma^1, \sigma^2, \sigma^3)$ are the usual Pauli spin matrices,

$$\sigma^1 = \begin{pmatrix} 0 & 1 \\ 1 & 0 \end{pmatrix}, \qquad \sigma^2 = \begin{pmatrix} 0 & -i \\ i & 0 \end{pmatrix}, \qquad \sigma^3 = \begin{pmatrix} 1 & 0 \\ 0 & -1 \end{pmatrix}. \tag{1.32}$$

It is convenient to define a fourth Pauli matrix,

$$\sigma^0 = \mathbb{1}_{2\times 2}, \tag{1.33}$$

where $\mathbb{1}_{2\times 2}$ is the 2×2 identity matrix. We can then define the four Pauli matrices in a unified notation, $\sigma^\mu = (\mathbb{1}_{2\times 2}\,; \vec{\sigma})$.[10]

Consider the infinitesimal Lorentz transformation in the $(\frac{1}{2}, 0)$ representation. Inserting the $(\frac{1}{2}, 0)$ generators [eq. (1.30)] into eq. (1.19) yields

$$M = \exp(-\tfrac{1}{2}i\vec{\theta}\cdot\vec{\sigma} - \tfrac{1}{2}\vec{\zeta}\cdot\vec{\sigma}) \simeq \mathbb{1}_{2\times 2} - \tfrac{1}{2}i\vec{\theta}\cdot\vec{\sigma} - \tfrac{1}{2}\vec{\zeta}\cdot\vec{\sigma}. \tag{1.34}$$

A two-component $(\frac{1}{2}, 0)$ spinor field is denoted by $\chi_\alpha(x)$, and transforms as

$$\chi'_\alpha(x') = M_\alpha{}^\beta \chi_\beta(x), \qquad \alpha, \beta \in \{1, 2\}. \tag{1.35}$$

By definition, M carries undotted spinor indices, as indicated by $M_\alpha{}^\beta$. In our conventions for the location of the spinor indices, we sum implicitly over a repeated index pair in which one index is lowered and one index is raised.

If M is a matrix representation of SL(n,$\mathbb{C}$), then M^*, $(M^{-1})^\mathsf{T}$ and $(M^{-1})^\dagger$ are also matrix representations of the same dimension. For $n > 2$, all four representations are inequivalent. For SL(2,$\mathbb{C}$), there are at most two distinct matrix representations corresponding to a given dimension: (j_1, j_2) and (j_2, j_1). Using eq. (1.34) and the following property of Pauli matrices,

$$\sigma^2 \vec{\sigma} (\sigma^2)^\mathsf{T} = \vec{\sigma}^\mathsf{T}, \tag{1.36}$$

where the transpose of the σ-matrices are $\vec{\sigma}^\mathsf{T} = (\sigma^1, -\sigma^2, \sigma^3)$, it follows that M and $(M^{-1})^\mathsf{T}$ are related by

$$(M^{-1})^\mathsf{T} = i\sigma^2 M (i\sigma^2)^\mathsf{T}. \tag{1.37}$$

Since $(i\sigma^2)^\mathsf{T} = (i\sigma^2)^{-1}$, the matrices M and $(M^{-1})^\mathsf{T}$ are related by a similarity transformation, corresponding to a unitary change in basis. Hence, M and $(M^{-1})^\mathsf{T}$ are equivalent representations.[11]

It is convenient to introduce the two-component spinor field $\chi^\alpha(x)$, which under the contragredient representation $(M^{-1})^\mathsf{T}$ transforms as

$$\chi'^\alpha(x') = [(M^{-1})^\mathsf{T}]^\alpha{}_\beta \chi^\beta(x) = [i\sigma^2 M (i\sigma^2)^\mathsf{T}]^\alpha{}_\beta \chi^\beta(x). \tag{1.38}$$

[10] The beginning student often misinterprets the symbol σ^2 to mean the square of σ. In the notation of eqs. (1.32) and (1.33), the superscripts are analogous to contravariant indices. Although the σ^μ do not transform under Lorentz transformations, we will see shortly that one can create contravariant four-vectors by suitably employing σ^μ [see eq. (1.102)].

[11] This corresponds to the well-known result that the **2** and $\overline{\mathbf{2}}$ representations of SU(2) are equivalent.

This motivates the definitions

$$\epsilon^{\alpha\beta} \equiv i\sigma^2 = \begin{pmatrix} 0 & 1 \\ -1 & 0 \end{pmatrix}, \qquad \epsilon_{\alpha\beta} \equiv (i\sigma^2)^{-1} = \begin{pmatrix} 0 & -1 \\ 1 & 0 \end{pmatrix}. \tag{1.39}$$

In particular, the nonzero components of the epsilon symbols are

$$\epsilon^{12} = -\epsilon^{21} = \epsilon_{21} = -\epsilon_{12} = 1. \tag{1.40}$$

We also introduce the two-index symmetric Kronecker delta symbol,

$$\delta_1^1 = \delta_2^2 = 1, \qquad \delta_2^1 = \delta_2^1 = 0. \tag{1.41}$$

The epsilon symbols satisfy

$$\epsilon_{\alpha\beta}\epsilon^{\rho\tau} = -\delta_\alpha{}^\rho\delta_\beta{}^\tau + \delta_\alpha{}^\tau\delta_\beta{}^\rho, \tag{1.42}$$

from which it follows that[12]

$$\epsilon_{\alpha\beta}\epsilon^{\beta\gamma} = \delta_\alpha{}^\gamma, \qquad \epsilon^{\gamma\beta}\epsilon_{\beta\alpha} = \delta^\gamma{}_\alpha. \tag{1.43}$$

Finally, the following equation, often called the Schouten identity, is noteworthy:

$$\epsilon_{\alpha\beta}\epsilon_{\gamma\delta} + \epsilon_{\alpha\gamma}\epsilon_{\delta\beta} + \epsilon_{\alpha\delta}\epsilon_{\beta\gamma} = 0. \tag{1.44}$$

Equations (1.35) and (1.38) imply that

$$\chi^\alpha = \epsilon^{\alpha\beta}\chi_\beta, \qquad \chi_\alpha = \epsilon_{\alpha\beta}\chi^\beta. \tag{1.45}$$

That is, the epsilon symbols can be used to raise or lower a spinor index. In particular, in raising or lowering an index of a spinor quantity, *adjacent* spinor indices are summed over when multiplied on the *left* by the appropriate epsilon symbol. As noted below eq. (1.37), M and $(M^{-1})^\mathsf{T}$ are equivalent representations. Hence, χ_α and χ^α are equally good candidates for the $(\frac{1}{2}, 0)$ representation.

Consider next the infinitesimal Lorentz transformation in the $(0, \frac{1}{2})$ representation [eqs. (1.19) and (1.31)]:

$$(M^{-1})^\dagger = \exp(-\tfrac{1}{2}i\vec{\theta}\cdot\vec{\sigma} + \tfrac{1}{2}\vec{\zeta}\cdot\vec{\sigma}) \simeq \mathbb{1}_{2\times 2} - \tfrac{1}{2}i\vec{\theta}\cdot\vec{\sigma} + \tfrac{1}{2}\vec{\zeta}\cdot\vec{\sigma}, \tag{1.46}$$

after taking the inverse of the hermitian conjugate of eq. (1.34). A two-component $(0, \frac{1}{2})$ spinor field is denoted by $\eta^{\dagger\dot{\alpha}}(x)$, and transforms as

$$\eta'^{\dagger\dot{\alpha}}(x') = (M^{-1})^{\dagger\,\dot{\alpha}}{}_{\dot{\beta}}\,\eta^{\dagger\dot{\beta}}(x), \qquad \dot{\alpha}, \dot{\beta} \in \{\dot{1}, \dot{2}\}. \tag{1.47}$$

By definition, $(M^{-1})^\dagger$ carries dotted spinor indices, as indicated by $(M^{-1})^{\dagger\,\dot{\alpha}}{}_{\dot{\beta}}$. Here, the "dotted" indices have been introduced to distinguish the $(0, \frac{1}{2})$ representation from the $(\frac{1}{2}, 0)$ representation.

The equivalent description of this representation is obtained via the conjugate representation M^*. Taking the complex conjugate of eq. (1.37), it follows that M^*

[12] In light of eq. (1.41), the distinction between $\delta_\alpha{}^\gamma$ and $\delta^\gamma{}_\alpha$ is somewhat pedantic. Nevertheless, it is useful to keep track of this distinction when reinterpreting such equations in terms of matrix multiplication.

is related by a similarity transformation to $(M^{-1})^\dagger$. The two-component spinor field $\eta^\dagger_{\dot\alpha}(x)$ under the conjugate representation transforms as

$$\eta'^\dagger_{\dot\alpha}(x') = (M^*)_{\dot\alpha}{}^{\dot\beta}\eta^\dagger_{\dot\beta}(x), \tag{1.48}$$

where

$$\eta^\dagger_{\dot\alpha} = \epsilon_{\dot\alpha\dot\beta}\eta^{\dagger\,\dot\beta}, \qquad \eta^{\dagger\,\dot\alpha} = \epsilon^{\dot\alpha\dot\beta}\eta^\dagger_{\dot\beta}, \tag{1.49}$$

and

$$\epsilon^{\dot\alpha\dot\beta} \equiv (\epsilon^{\alpha\beta})^* = \begin{pmatrix} 0 & 1 \\ -1 & 0 \end{pmatrix}, \qquad \epsilon_{\dot\alpha\dot\beta} \equiv (\epsilon_{\alpha\beta})^* = \begin{pmatrix} 0 & -1 \\ 1 & 0 \end{pmatrix}. \tag{1.50}$$

Hence, $\epsilon^{\dot\alpha\dot\beta}$ and $\epsilon^{\alpha\beta}$ are numerically equal, but the dotted and undotted indices transform differently under Lorentz transformations [see eqs. (1.35) and (1.47)] and must always be kept distinct. Likewise, we define the Kronecker delta symbol with dotted indices, $\delta^{\dot\beta}_{\dot\alpha} \equiv (\delta^\beta_\alpha)^*$.

The dotted epsilon tensor satisfies

$$\epsilon_{\dot\alpha\dot\beta}\epsilon^{\dot\rho\dot\tau} = -\delta_{\dot\alpha}{}^{\dot\rho}\delta_{\dot\beta}{}^{\dot\tau} + \delta_{\dot\alpha}{}^{\dot\tau}\delta_{\dot\beta}{}^{\dot\rho}, \tag{1.51}$$

from which it follows that

$$\epsilon_{\dot\alpha\dot\beta}\epsilon^{\dot\beta\dot\gamma} = \delta_{\dot\alpha}{}^{\dot\gamma}, \qquad \epsilon^{\dot\gamma\dot\beta}\epsilon_{\dot\beta\dot\alpha} = \delta^{\dot\gamma}{}_{\dot\alpha}. \tag{1.52}$$

Likewise, the Schouten identity analogous to that of eq. (1.44) also holds:

$$\epsilon_{\dot\alpha\dot\beta}\epsilon_{\dot\gamma\dot\delta} + \epsilon_{\dot\alpha\dot\gamma}\epsilon_{\dot\delta\dot\beta} + \epsilon_{\dot\alpha\dot\delta}\epsilon_{\dot\beta\dot\gamma} = 0. \tag{1.53}$$

Note that the $(\frac{1}{2}, 0)$ and $(0, \frac{1}{2})$ representations are related by conjugation. That is, if ψ_α is a $(\frac{1}{2}, 0)$ fermion, then $(\psi_\alpha)^\dagger$ transforms as a $(0, \frac{1}{2})$ fermion. In this context, the conjugation symbol (†) denotes complex conjugation for classical fields or hermitian conjugation for quantum field operators. In particular, we shall identify[13]

$$\psi^\dagger_{\dot\alpha} \equiv (\psi_\alpha)^\dagger. \tag{1.54}$$

This means that we can, and will, describe all fermion degrees of freedom using only fields defined as left-handed $(\frac{1}{2}, 0)$ fermions ψ_α, and their conjugates. In combining spinors to make Lorentz tensors (as in Section 1.4), it is useful to regard $\psi^\dagger_{\dot\alpha}$ as a row vector, and ψ_α as a column vector. Likewise, it follows that $\psi_\alpha = (\psi^\dagger_{\dot\alpha})^\dagger$. The Lorentz transformation property of $\eta^\dagger_{\dot\alpha}$ then follows from eq. (1.48), which can be rewritten as $[\eta_\alpha(x)]^\dagger \to [\eta_\beta(x)]^\dagger (M^\dagger)^{\dot\beta}{}_{\dot\alpha}$, where $(M^\dagger)^{\dot\beta}{}_{\dot\alpha} = (M^*)_{\dot\alpha}{}^{\dot\beta}$ reflects the definition of the hermitian adjoint matrix as the complex conjugate transpose of the matrix.

Spinors labeled with one undotted or one dotted index are sometimes called spinors of rank one [or more precisely, spinors of rank (1,0) or (0,1), respectively].

[13] In this book, the dotted-index notation is used in association with the dagger to denote conjugation, as specified in eq. (1.54). In contrast, many references in the supersymmetry literature employ the bar notation made popular by Wess and Bagger [B40], where $\overline{\psi}_{\dot\alpha} \equiv \psi^\dagger_{\dot\alpha} \equiv (\psi_\alpha)^\dagger$.

One can also define spinors of higher rank, which possess more than one spinor index, with Lorentz transformation properties that depend on the number of undotted and dotted spinor indices. In particular, for a spinor of rank (m, n) denoted by $S_{\alpha_1\alpha_2\cdots\alpha_m\dot{\beta}_1\dot{\beta}_2\cdots\dot{\beta}_n}$, each lowered undotted α-index transforms separately according to $M_{\alpha_i'}{}^{\alpha_i}$ and each lowered dotted $\dot{\beta}$-index transforms according to $(M^*)_{\dot{\beta}_i'}{}^{\dot{\beta}_i}$.

The epsilon symbols can also be used to raise or lower undotted or dotted indices of spinors of higher rank. For example, it is natural to define

$$A^{\gamma\delta} \equiv \epsilon^{\gamma\alpha}\epsilon^{\delta\beta}A_{\alpha\beta}\,, \qquad\qquad A_{\gamma\delta} \equiv \epsilon_{\gamma\alpha}\epsilon_{\delta\beta}A^{\alpha\beta}\,. \tag{1.55}$$

In the special case where $A^{\alpha\beta} = \psi^\alpha\xi^\beta$ is a product of rank-one spinors, eq. (1.55) is not just natural but necessary, as it follows directly from eq. (1.45). However, in other cases there can be a different sign associated (by convention) with raising and lowering spinor indices, because of the antisymmetry of the epsilon symbols (in contrast to the symmetry of the spacetime metric used to raise and lower spacetime indices). This sign convention can be defined independently for distinct higher-rank spinors (even in the case where the higher-rank spinors possess the same index structure). Indeed, as a consequence of our epsilon symbol conventions of eq. (1.39), the epsilon symbols themselves satisfy

$$\epsilon^{\gamma\delta} = -\epsilon^{\gamma\alpha}\epsilon^{\delta\beta}\epsilon_{\alpha\beta}\,, \qquad\qquad \epsilon_{\gamma\delta} = -\epsilon_{\gamma\alpha}\epsilon_{\delta\beta}\epsilon^{\alpha\beta}\,, \tag{1.56}$$

in contrast to eq. (1.55). The above results (and similar ones with dotted indices) show that some care is required, since the extra overall minus signs of eq. (1.56) in comparison to eq. (1.55) might otherwise have been unexpected. This reflects an awkwardness imposed by the epsilon symbol conventions of eq. (1.39), rather than an inconsistency. When employing the conventions used in this book, one must be aware of this sign issue when using the epsilon symbols to explicitly raise or lower two or more spinor indices of higher-rank spinors. Fortunately, such manipulations are quite rare in practical calculations.

We will encounter two types of spinor quantities in this book. First, we shall examine spinor fields that are used to construct Lagrangians for field theoretic models. These spinors are anticommuting objects. For example, the product of two anticommuting spinors, $\psi^\alpha\psi^\beta$ is antisymmetric with respect to the interchange of the spinor indices α and β. Hence, this product of spinors must be proportional to $\epsilon^{\alpha\beta}$. Similar conclusions hold for the corresponding spinor products with lowered undotted indices and with raised and lowered dotted indices, respectively. Thus,

$$\psi^\alpha\psi^\beta = -\tfrac{1}{2}\epsilon^{\alpha\beta}\psi\psi\,, \qquad\qquad \psi_\alpha\psi_\beta = \tfrac{1}{2}\epsilon_{\alpha\beta}\psi\psi\,, \tag{1.57}$$

$$\psi^{\dagger\,\dot{\alpha}}\psi^{\dagger\,\dot{\beta}} = \tfrac{1}{2}\epsilon^{\dot{\alpha}\dot{\beta}}\psi^\dagger\psi^\dagger\,, \qquad\qquad \psi^\dagger_{\dot{\alpha}}\psi^\dagger_{\dot{\beta}} = -\tfrac{1}{2}\epsilon_{\dot{\alpha}\dot{\beta}}\psi^\dagger\psi^\dagger\,, \tag{1.58}$$

where $\psi\psi \equiv \psi^\gamma\psi_\gamma$ and $\psi^\dagger\psi^\dagger \equiv \psi^\dagger_{\dot{\gamma}}\psi^{\dagger\,\dot{\gamma}}$. Note that the minus signs above can be understood to be a consequence of the extra minus sign that arises when the indices of the epsilon symbol are lowered or raised [see eqs. (1.55) and (1.56)].

A free anticommuting fermion field can be expanded in terms of sums of plane waves multiplied by spinor wave functions times the appropriate creation or annihilation operator. These wave functions are *commuting* two-component spinors that satisfy the free-field Dirac equation, whereas the creation and annihilation operators satisfy the usual canonical anticommutation relations. The properties of commuting two-component spinor wave functions will be examined in detail in Chapter 2.

1.4 Bilinear Covariants of Two-Component Spinors

It is well known that one can construct Lorentz scalar, vector, and tensor quantities that are bilinear in the fermion fields. Subsequently, these vector and tensor quantities can be further combined to make new Lorentz scalars. Such scalars can be used in the construction of Lagrangians for theories of fermion fields. The Lagrangian of a relativistic quantum field theory is Lorentz invariant. That is, the Lagrangian must transform as a Lorentz scalar.

The simplest quantum field theory of fermions consists of a single $(\frac{1}{2}, 0)$ fermion field, which transforms under a Lorentz transformation via

$$\xi_\alpha \to M_\alpha{}^\beta \xi_\beta \,. \tag{1.59}$$

In order to construct a Lorentz-invariant bilinear of the two-component spinor fields (which is a Lorentz scalar), we first note a basic property of the Lorentz transformation matrix M. From the definition of the determinant,

$$\epsilon^{\alpha\beta} M_\alpha{}^\rho M_\beta{}^\sigma = \epsilon^{\rho\sigma} \det M = \epsilon^{\rho\sigma} \,, \tag{1.60}$$

where we have used $\det M = 1$ at the last step. It then follows that the bilinear combination

$$\xi\xi \equiv \xi^\alpha \xi_\alpha = \epsilon^{\alpha\beta} \xi_\beta \xi_\alpha \tag{1.61}$$

is invariant under Lorentz transformations. Similarly, given a single $(0, \frac{1}{2})$ fermion fields, the bilinear combination

$$\xi^\dagger \xi^\dagger \equiv \xi^\dagger_{\dot\alpha} \xi^{\dagger\,\dot\alpha} = \epsilon_{\dot\alpha\dot\beta} \xi^{\dagger\,\dot\beta} \xi^{\dagger\,\dot\alpha} \tag{1.62}$$

is also Lorentz invariant. Note carefully the placement of the spinor indices. We have adopted a convention that the heights of indices must be consistent in the sense that lowered indices must always be contracted with raised indices. Moreover, after an appropriate arrangement of the spinors [using (anti)commutativity properties if necessary], the undotted and dotted indices are contracted with the placement of indices as follows:

$${}^{\alpha}{}_{\alpha} \qquad \text{and} \qquad {}_{\dot\alpha}{}^{\dot\alpha} \,. \tag{1.63}$$

Equation (1.63) is often described by saying that undotted indices are contracted by going from the northwest to the southeast, whereas dotted indices are contracted by going from the southwest to the northeast. Once this convention is adopted, the spinor indices can be omitted without ambiguity as shown in eqs. (1.61) and (1.62).

More generally, one can consider a theory of a multiplet of fermions fields. Since the corresponding Lagrangian is bilinear in the fermion fields, it is sufficient to consider two $(\frac{1}{2},0)$ fermion fields and their Lorentz transformation properties,

$$\xi_\alpha \to M_\alpha{}^\beta \xi_\beta\,, \qquad \eta_\alpha \to M_\alpha{}^\beta \eta_\beta\,. \tag{1.64}$$

In light of eq. (1.60), it follows that the bilinear combination

$$\chi\eta \equiv \chi^\alpha \eta_\alpha = \epsilon^{\alpha\beta}\chi_\beta\eta_\alpha \tag{1.65}$$

is invariant under Lorentz transformations. Similarly, given two $(0,\frac{1}{2})$ fermion fields, one can construct a Lorentz-invariant bilinear quantity,

$$\chi^\dagger\eta^\dagger \equiv \chi^\dagger_{\dot\alpha}\eta^{\dagger\dot\alpha} = \epsilon_{\dot\alpha\dot\beta}\chi^{\dagger\dot\beta}\eta^{\dagger\dot\alpha}\,. \tag{1.66}$$

Having adopted the convention indicated in eq. (1.63), the spinor indices can be omitted without ambiguity as indicated in eqs. (1.65) and (1.66).

We consider next the behavior of eqs. (1.65) and (1.66) under hermitian conjugation (for quantum field operators) or complex conjugation (for classical fields) [see eq. (1.54)]. Conjugation of a spinor product reverses the order of the spinors. Hence, it follows that

$$(\chi\eta)^\dagger = (\chi^\alpha\eta_\alpha)^\dagger = \eta^\dagger_{\dot\alpha}\chi^{\dagger\dot\alpha} = \eta^\dagger\chi^\dagger\,, \tag{1.67}$$

which holds for both commuting and anticommuting two-component spinors. Finally, we note that for *anticommuting* two-component fermion fields,

$$\chi\eta = \epsilon^{\alpha\beta}\chi_\beta\eta_\alpha = -\epsilon^{\alpha\beta}\eta_\alpha\chi_\beta = \epsilon^{\beta\alpha}\eta_\alpha\chi_\beta = \eta\chi\,, \tag{1.68}$$

where we have used $\epsilon^{\alpha\beta} = -\epsilon^{\beta\alpha}$ and relabeled the dummy indices. Likewise,

$$\chi^\dagger\eta^\dagger = \epsilon_{\dot\alpha\dot\beta}\chi^{\dagger\dot\beta}\eta^{\dagger\dot\alpha} = -\epsilon_{\dot\alpha\dot\beta}\eta^{\dagger\dot\alpha}\chi^{\dagger\dot\beta} = \epsilon_{\dot\beta\dot\alpha}\eta^{\dagger\dot\alpha}\chi^{\dagger\dot\beta} = \eta^\dagger\chi^\dagger\,. \tag{1.69}$$

Combining eqs. (1.67) and (1.69) yields

$$(\chi\eta)^\dagger = \chi^\dagger\eta^\dagger\,. \tag{1.70}$$

It follows that the quantity $\chi\eta + \chi^\dagger\eta^\dagger$ is Lorentz invariant and hermitian.

In order to construct fermion bilinears that transform as Lorentz vectors and tensors, we must employ the sigma matrices that have been introduced in eqs. (1.32) and (1.33):

$$\sigma^\mu = (\mathbb{1}_{2\times2}\,;\,\vec{\boldsymbol\sigma})\,, \qquad \overline{\sigma}^\mu = (\mathbb{1}_{2\times2}\,;\,-\vec{\boldsymbol\sigma})\,, \tag{1.71}$$

where $\mu \in \{0,1,2,3\}$. Note that these sigma matrices have been defined with an upper (contravariant) index. They are related to the sigma matrices with lower (covariant) indices in the usual way:

$$\sigma_\mu = g_{\mu\nu}\sigma^\nu = (\mathbb{1}_{2\times2}\,;\,-\vec{\boldsymbol\sigma})\,, \qquad \overline{\sigma}_\mu = g_{\mu\nu}\overline{\sigma}^\nu = (\mathbb{1}_{2\times2}\,;\,\vec{\boldsymbol\sigma})\,. \tag{1.72}$$

The sigma matrices possess a specific spinor index structure: $\sigma_{\mu\alpha\dot{\beta}}$ and $\overline{\sigma}_{\mu}^{\dot{\alpha}\beta}$. To verify this assertion, we first note that

$$p^\mu \sigma_\mu = p^0 \mathbb{1}_{2\times 2} - \vec{\boldsymbol{p}}\boldsymbol{\cdot}\vec{\boldsymbol{\sigma}} = \begin{pmatrix} p^0 - p^3 & -p^1 + ip^2 \\ -p^1 - ip^2 & p^0 + p^3 \end{pmatrix} \tag{1.73}$$

is a hermitian 2×2 matrix. Moreover, any hermitian 2×2 matrix can be written in the form of eq. (1.73). Since $Mp^\mu \sigma_\mu M^\dagger$ is also hermitian, there exists a p'^μ such that

$$p'^\mu \sigma_\mu = M p^\mu \sigma_\mu M^\dagger \,. \tag{1.74}$$

Using $\det(p^\mu \sigma_\mu) = (p^0)^2 - |\vec{p}|^2$ and $\det M = 1$, it follows that

$$(p'^0)^2 - |\vec{\boldsymbol{p}}\,'|^2 = (p^0)^2 - |\vec{\boldsymbol{p}}|^2 \,, \tag{1.75}$$

and hence p^μ and p'^μ are related by a Lorentz transformation Λ. More significantly, Λ is determined precisely by the same parameters $\vec{\boldsymbol{\theta}}$ and $\vec{\boldsymbol{\zeta}}$ that determine M. Specifically, one can show that the following is an identity:

$$M \sigma_\mu M^\dagger = \sigma_\nu \Lambda^\nu{}_\mu \,, \tag{1.76}$$

where the infinitesimal form for M is given by eq. (1.34) and Λ is the 4×4 Lorentz transformation matrix given by eq. (1.2). Equations (1.74) and (1.76) imply that $p'^\nu = \Lambda^\nu{}_\mu p^\mu$. Thus, eq. (1.76) provides a mapping[14] from the set of 2×2 complex matrices of unit determinant (M) to the set of proper orthochronous Lorentz transformations (Λ). Similar conclusions can be drawn by considering $p^\mu \overline{\sigma}_\mu$. In this case, we find

$$p'^\mu \overline{\sigma}_\mu = (M^{-1})^\dagger p^\mu \overline{\sigma}_\mu M^{-1} \,, \tag{1.77}$$

and $p'^\nu = \Lambda^\nu{}_\mu p^\mu$ as a consequence of

$$(M^{-1})^\dagger \overline{\sigma}_\mu M^{-1} = \overline{\sigma}_\nu \Lambda^\nu{}_\mu \,. \tag{1.78}$$

The index structure of σ_μ and $\overline{\sigma}_\mu$ can be deduced from eqs. (1.76) and (1.78) by requiring these equations to be Lorentz covariant with respect to both spacetime and spinor indices. Using the index structure of the Lorentz transformation matrices M [eq. (1.35)] and $M^\dagger$ [the inverse of eq. (1.47)], it follows that the spinor index structure of eq. (1.76) is given by

$$\Lambda^\nu{}_\mu \sigma_{\nu\alpha\dot{\alpha}} = M_\alpha{}^\beta \sigma_{\mu\beta\dot{\beta}} (M^\dagger)^{\dot{\beta}}{}_{\dot{\alpha}} \,. \tag{1.79}$$

Likewise, using the index structure of the Lorentz transformation matrices $(M^{-1})^\dagger$ [eq. (1.47)] and M^{-1} [the inverse of eq. (1.35)], it follows that the spinor index structure of eq. (1.78) is given by

$$\Lambda^\nu{}_\mu \overline{\sigma}_\nu^{\dot{\alpha}\alpha} = [(M^{-1})^\dagger]^{\dot{\alpha}}{}_{\dot{\beta}}\, \overline{\sigma}_\mu^{\dot{\beta}\beta} \, (M^{-1})_\beta{}^\alpha \,. \tag{1.80}$$

[14] In fact, eq. (1.76) implies that both M and $-M$ map onto the same Lorentz transformation Λ. Since M is in general an SL(2,$\mathbb{C}$) matrix, and Λ is an $\mathrm{SO}_+(1,3)$ matrix, the mapping specified by eq. (1.76) implies that $\mathrm{SO}_+(1,3) \cong \mathrm{SL}(2,\mathbb{C})/\mathbb{Z}_2$.

These results confirm the spinor index structure of σ_μ and $\overline{\sigma}_\mu$,

$$\sigma_{\mu\alpha\dot{\alpha}} \quad \text{and} \quad \overline{\sigma}_\mu^{\dot{\alpha}\alpha} . \tag{1.81}$$

Both σ^μ and $\overline{\sigma}^\mu$ are hermitian matrices, i.e., $(\sigma^\mu)^\dagger = \sigma^\mu$ and $(\overline{\sigma}^\mu)^\dagger = \overline{\sigma}^\mu$, or equivalently

$$(\sigma^\mu_{\alpha\dot{\beta}})^* = \sigma^\mu_{\beta\dot{\alpha}} , \qquad (\overline{\sigma}^{\mu\dot{\alpha}\beta})^* = \overline{\sigma}^{\mu\dot{\beta}\alpha} . \tag{1.82}$$

Some useful relations between σ^μ and $\overline{\sigma}^\mu$ are

$$\sigma^\mu_{\alpha\dot{\alpha}} = \epsilon_{\alpha\beta}\epsilon_{\dot{\alpha}\dot{\beta}}\overline{\sigma}^{\mu\dot{\beta}\beta} , \qquad \overline{\sigma}^{\mu\dot{\alpha}\alpha} = \epsilon^{\alpha\beta}\epsilon^{\dot{\alpha}\dot{\beta}}\sigma^\mu_{\beta\dot{\beta}} , \tag{1.83}$$

$$\epsilon^{\alpha\beta}\sigma^\mu_{\beta\dot{\alpha}} = \epsilon_{\dot{\alpha}\dot{\beta}}\overline{\sigma}^{\mu\dot{\beta}\alpha} , \qquad \epsilon^{\dot{\alpha}\dot{\beta}}\sigma^\mu_{\alpha\dot{\beta}} = \epsilon_{\alpha\beta}\overline{\sigma}^{\mu\dot{\alpha}\beta} . \tag{1.84}$$

Finally, we note the following identities, which can be used to systematically simplify expressions involving products of σ and $\overline{\sigma}$ matrices:

$$\sigma^\mu_{\alpha\dot{\alpha}}\overline{\sigma}_\mu^{\dot{\beta}\beta} = 2\delta_\alpha{}^\beta\delta^{\dot{\beta}}{}_{\dot{\alpha}} , \tag{1.85}$$

$$\sigma^\mu_{\alpha\dot{\alpha}}\sigma_{\mu\beta\dot{\beta}} = 2\epsilon_{\alpha\beta}\epsilon_{\dot{\alpha}\dot{\beta}} , \tag{1.86}$$

$$\overline{\sigma}^{\mu\dot{\alpha}\alpha}\overline{\sigma}_\mu^{\dot{\beta}\beta} = 2\epsilon^{\alpha\beta}\epsilon^{\dot{\alpha}\dot{\beta}} , \tag{1.87}$$

$$[\sigma^\mu\overline{\sigma}^\nu + \sigma^\nu\overline{\sigma}^\mu]_\alpha{}^\beta = 2g^{\mu\nu}\delta_\alpha{}^\beta , \tag{1.88}$$

$$[\overline{\sigma}^\mu\sigma^\nu + \overline{\sigma}^\nu\sigma^\mu]^{\dot{\alpha}}{}_{\dot{\beta}} = 2g^{\mu\nu}\delta^{\dot{\alpha}}{}_{\dot{\beta}} , \tag{1.89}$$

$$\sigma^\mu\overline{\sigma}^\nu\sigma^\rho = g^{\mu\nu}\sigma^\rho - g^{\mu\rho}\sigma^\nu + g^{\nu\rho}\sigma^\mu + i\epsilon^{\mu\nu\rho\kappa}\sigma_\kappa , \tag{1.90}$$

$$\overline{\sigma}^\mu\sigma^\nu\overline{\sigma}^\rho = g^{\mu\nu}\overline{\sigma}^\rho - g^{\mu\rho}\overline{\sigma}^\nu + g^{\nu\rho}\overline{\sigma}^\mu - i\epsilon^{\mu\nu\rho\kappa}\overline{\sigma}_\kappa , \tag{1.91}$$

in a convention where $\epsilon^{0123} = 1$. A more comprehensive set of identities is provided in Appendix B.

Computations of cross sections and decay rates generally require traces of alternating products of σ and $\overline{\sigma}$ matrices. For example,

$$\mathrm{Tr}[\sigma^\mu\overline{\sigma}^\nu] = \mathrm{Tr}[\overline{\sigma}^\mu\sigma^\nu] = 2g^{\mu\nu} , \tag{1.92}$$

$$\mathrm{Tr}[\sigma^\mu\overline{\sigma}^\nu\sigma^\rho\overline{\sigma}^\kappa] = 2\left(g^{\mu\nu}g^{\rho\kappa} - g^{\mu\rho}g^{\nu\kappa} + g^{\mu\kappa}g^{\nu\rho} + i\epsilon^{\mu\nu\rho\kappa}\right) , \tag{1.93}$$

$$\mathrm{Tr}[\overline{\sigma}^\mu\sigma^\nu\overline{\sigma}^\rho\sigma^\kappa] = 2\left(g^{\mu\nu}g^{\rho\kappa} - g^{\mu\rho}g^{\nu\kappa} + g^{\mu\kappa}g^{\nu\rho} - i\epsilon^{\mu\nu\rho\kappa}\right) . \tag{1.94}$$

Traces involving a larger even number of σ and $\overline{\sigma}$ matrices can be systematically obtained from eqs. (1.92)–(1.94) by repeated use of eqs. (1.88) and (1.89) and the cyclic property of the trace. Traces involving an odd number of σ and $\overline{\sigma}$ matrices cannot arise, since there is no way to connect the spinor indices consistently.

In addition to manipulating expressions containing anticommuting fermion quantum fields, we often must deal with products of *commuting* spinor wave functions that arise when evaluating the Feynman rules. In the following expressions we denote the generic spinor by z_i. In the various identities listed below, an extra minus sign arises when interchanging the order of two anticommuting fermion fields of a given spinor index height. It is convenient to introduce the notation

$$(-1)^A \equiv \begin{cases} +1\,, & \text{commuting spinors,} \\ -1\,, & \text{anticommuting spinors.} \end{cases} \tag{1.95}$$

The following identities hold for the z_i:

$$z_1 z_2 = -(-1)^A z_2 z_1\,, \tag{1.96}$$
$$z_1^\dagger z_2^\dagger = -(-1)^A z_2^\dagger z_1^\dagger\,, \tag{1.97}$$
$$z_1 \sigma^\mu z_2^\dagger = (-1)^A z_2^\dagger \overline{\sigma}^\mu z_1\,, \tag{1.98}$$
$$z_1 \sigma^\mu \overline{\sigma}^\nu z_2 = -(-1)^A z_2 \sigma^\nu \overline{\sigma}^\mu z_1\,, \tag{1.99}$$
$$z_1^\dagger \overline{\sigma}^\mu \sigma^\nu z_2^\dagger = -(-1)^A z_2^\dagger \overline{\sigma}^\nu \sigma^\mu z_1^\dagger\,, \tag{1.100}$$
$$z_1^\dagger \overline{\sigma}^\mu \sigma^\rho \overline{\sigma}^\nu z_2 = (-1)^A z_2 \sigma^\nu \overline{\sigma}^\rho \sigma^\mu z_1^\dagger\,, \tag{1.101}$$

and so on. Equations (1.96) and (1.97) for the case of anticommuting two-component spinors have already been derived in eqs. (1.68) and (1.69). The derivation of the other results above can be obtained similarly.

We may now employ the following spinor product to obtain a bilinear covariant that transforms as a Lorentz four-vector:

$$\chi \sigma^\mu \eta^\dagger \equiv \chi^\alpha \sigma^\mu_{\alpha\dot{\beta}} \eta^{\dagger\,\dot{\beta}}\,. \tag{1.102}$$

Note that the convention of eq. (1.63) is respected and allows us to suppress the spinor indices. Moreover, the structure of the spinor and Lorentz indices guarantees that $\chi\sigma^\mu\eta^\dagger$ transforms as a Lorentz four-vector. Nevertheless, an explicit proof is instructive. Using the transformation laws for χ^α and $\eta^{\dagger\,\dot{\beta}}$ [eqs. (1.38) and (1.47), respectively], it follows that

$$\chi'(x')\sigma^\mu \eta'^{\,\dagger}(x') = \chi^\alpha(x)[M^{-1}\sigma^\mu (M^{-1})^\dagger]_{\alpha\dot{\alpha}} \eta^{\dagger\,\dot{\alpha}}(x) = \Lambda^\mu{}_\nu \chi(x)\sigma^\nu \eta^\dagger(x)\,, \tag{1.103}$$

which is the proper transformation law for a four-vector field. The last step of eq. (1.103) was obtained using eq. (1.300) [which is an alternative form of eq. (1.76)].

Likewise, we may define the spinor product

$$\eta^\dagger \overline{\sigma}^\mu \chi \equiv \eta^\dagger_{\dot{\alpha}} \overline{\sigma}^{\mu\dot{\alpha}\alpha} \chi_\alpha\,. \tag{1.104}$$

Indeed, $\eta^\dagger\overline{\sigma}^\mu\chi$ also transforms as a Lorentz four-vector. As above, by using the transformation laws of χ_α and $\eta^\dagger_{\dot{\beta}}$ [eqs. (1.35) and (1.48), respectively], one can verify that $\eta^\dagger\overline{\sigma}^\mu\chi$ possesses the transformation law of a four-vector field, as a consequence of eq. (1.299) [or its alternative form, eq. (1.78)].

We consider next the behavior of eqs. (1.102) and (1.104) under conjugation. Using the hermiticity properties of the sigma matrices, it follows that

$$(\chi\sigma^\mu\eta^\dagger)^\dagger = \eta\sigma^\mu\chi^\dagger\,, \tag{1.105}$$
$$(\chi^\dagger\overline{\sigma}^\mu\eta)^\dagger = \eta^\dagger\overline{\sigma}^\mu\chi\,, \tag{1.106}$$

which hold for both commuting and anticommuting two-component spinors.

The two four-vector quantities introduced in eqs. (1.102) and (1.104) are not independent. In particular, one may rewrite eq. (1.104) in terms of σ^μ by using the following identity for anticommuting two-component spinor fields [see eq. (1.98)]:

$$\eta^\dagger \overline{\sigma}^\mu \chi = -\chi \sigma^\mu \eta^\dagger \,. \tag{1.107}$$

Combining eqs. (1.105) and (1.107) yields

$$(\chi \sigma^\mu \eta^\dagger)^\dagger = -\chi^\dagger \overline{\sigma}^\mu \eta \,. \tag{1.108}$$

One can generalize the construction of a Lorentz four-vector in terms of a spinor product as follows. We shall introduce a bispinor, $V_{\alpha\dot{\beta}}$, which is a spinor of rank (1, 1). There is a one-to-one correspondence between $V_{\alpha\dot{\beta}}$ and the associated Lorentz four-vector V^μ defined by[15]

$$V^\mu \equiv \tfrac{1}{2} \overline{\sigma}^{\mu \dot{\beta}\alpha} V_{\alpha\dot{\beta}} \,. \tag{1.109}$$

Using eq. (1.85), we can determine $V_{\alpha\dot{\beta}}$ in terms of V^μ:

$$V_{\alpha\dot{\beta}} = V^\mu \sigma_{\mu\alpha\dot{\beta}} \,. \tag{1.110}$$

In particular, if V^μ is a real four-vector, then $V_{\alpha\dot{\beta}}$ is hermitian (and vice versa). To clarify this last remark, consider the bispinor $V_{\alpha\dot{\beta}}$ regarded as a 2×2 matrix. Then,

$$(V^{\mathsf{T}})_{\alpha\dot{\beta}} \equiv V_{\beta\dot{\alpha}} \,, \qquad (V^*)_{\alpha\dot{\beta}} \equiv (V_{\alpha\dot{\beta}})^* \,, \qquad (V^\dagger)_{\alpha\dot{\beta}} \equiv (V_{\beta\dot{\alpha}})^* . \tag{1.111}$$

In taking the transpose of $V_{\alpha\dot{\beta}}$, the rows and columns of V are interchanged without altering the fact that the first spinor index is undotted and the second spinor index is dotted. A hermitian bispinor satisfies $V = V^\dagger$, or equivalently $V_{\alpha\dot{\beta}} = (V_{\beta\dot{\alpha}})^*$.

Next we consider bilinear covariants that transform as Lorentz second-rank tensors. Such quantities are simply obtained by employing a product of a σ and $\overline{\sigma}$ matrix inside the spinor product. Using the hermiticity properties of the sigma matrices, it follows that

$$(\chi \sigma^\mu \overline{\sigma}^\nu \eta)^\dagger = \eta^\dagger \overline{\sigma}^\nu \sigma^\mu \chi^\dagger \,, \tag{1.112}$$

which is valid for both commuting and anticommuting two-component spinors. Moreover, the following identity is applicable in the case of anticommuting two-component spinors:

$$\eta \sigma^\mu \overline{\sigma}^\nu \chi = \chi \sigma^\nu \overline{\sigma}^\mu \eta \,. \tag{1.113}$$

Combining eqs. (1.112) and (1.113) yields

$$\eta \sigma^\mu \overline{\sigma}^\nu \chi = \chi \sigma^\nu \overline{\sigma}^\mu \eta = (\chi^\dagger \overline{\sigma}^\nu \sigma^\mu \eta^\dagger)^\dagger = (\eta^\dagger \overline{\sigma}^\mu \sigma^\nu \chi^\dagger)^\dagger \,. \tag{1.114}$$

In light of eqs. (1.88) and (1.89), it is sufficient to focus on bilinear covariants

[15] More generally, there is a one-to-one correspondence between a spinor of rank (n, n) and a rank-n contravariant Lorentz tensor (e.g., see Ref. [28]). Details are left for the reader.

that transform as antisymmetric second-rank tensors. Using the matrices σ^μ and $\overline{\sigma}^\mu$ we first define[16]

$$(\sigma^{\mu\nu})_\alpha{}^\beta \equiv \frac{i}{4}\left(\sigma^\mu_{\alpha\dot{\alpha}}\overline{\sigma}^{\nu\dot{\alpha}\beta} - \sigma^\nu_{\alpha\dot{\alpha}}\overline{\sigma}^{\mu\dot{\alpha}\beta}\right), \tag{1.115}$$

$$(\overline{\sigma}^{\mu\nu})^{\dot{\alpha}}{}_{\dot{\beta}} \equiv \frac{i}{4}\left(\overline{\sigma}^{\mu\dot{\alpha}\alpha}\sigma^\nu_{\alpha\dot{\beta}} - \overline{\sigma}^{\nu\dot{\alpha}\alpha}\sigma^\mu_{\alpha\dot{\beta}}\right). \tag{1.116}$$

Equivalently, we can use eqs. (1.88) and (1.89) to write

$$(\sigma^\mu\overline{\sigma}^\nu)_\alpha{}^\beta = g^{\mu\nu}\delta_\alpha{}^\beta - 2i(\sigma^{\mu\nu})_\alpha{}^\beta, \tag{1.117}$$

$$(\overline{\sigma}^\mu\sigma^\nu)^{\dot{\alpha}}{}_{\dot{\beta}} = g^{\mu\nu}\delta^{\dot{\alpha}}{}_{\dot{\beta}} - 2i(\overline{\sigma}^{\mu\nu})^{\dot{\alpha}}{}_{\dot{\beta}}. \tag{1.118}$$

The components of $\sigma^{\mu\nu}$ and $\overline{\sigma}^{\mu\nu}$ are easily evaluated:

$$\sigma^{ij} = \overline{\sigma}^{ij} = \tfrac{1}{2}\epsilon^{ijk}\sigma^k, \tag{1.119}$$

$$\sigma^{i0} = -\sigma^{0i} = -\overline{\sigma}^{i0} = \overline{\sigma}^{0i} = \tfrac{1}{2}i\sigma^i. \tag{1.120}$$

The hermiticity of $\vec{\boldsymbol{\sigma}}$ implies that $(\sigma^{\mu\nu})^\dagger = \overline{\sigma}^{\mu\nu}$, or equivalently

$$(\sigma^{\mu\nu}{}_\alpha{}^\beta)^* = \overline{\sigma}^{\mu\nu\dot{\beta}}{}_{\dot{\alpha}}. \tag{1.121}$$

The matrices $\sigma^{\mu\nu}$ and $\overline{\sigma}^{\mu\nu}$ satisfy self-duality relations:

$$\sigma^{\mu\nu} = -\tfrac{1}{2}i\epsilon^{\mu\nu\rho\kappa}\sigma_{\rho\kappa}, \qquad \overline{\sigma}^{\mu\nu} = \tfrac{1}{2}i\epsilon^{\mu\nu\rho\kappa}\overline{\sigma}_{\rho\kappa}. \tag{1.122}$$

A number of useful properties involving $\sigma^{\mu\nu}$ and $\overline{\sigma}^{\mu\nu}$ can be derived. For example, eq. (1.42) implies that

$$(\sigma^{\mu\nu})_\alpha{}^\beta = \epsilon_{\alpha\tau}\epsilon^{\beta\gamma}(\sigma^{\mu\nu})_\gamma{}^\tau, \qquad (\overline{\sigma}^{\mu\nu})^{\dot{\alpha}}{}_{\dot{\beta}} = \epsilon^{\dot{\alpha}\dot{\tau}}\epsilon_{\dot{\beta}\dot{\gamma}}(\overline{\sigma}^{\mu\nu})^{\dot{\gamma}}{}_{\dot{\tau}}, \tag{1.123}$$

$$\epsilon^{\tau\alpha}(\sigma^{\mu\nu})_\alpha{}^\beta = \epsilon^{\beta\gamma}(\sigma^{\mu\nu})_\gamma{}^\tau, \qquad \epsilon_{\dot{\tau}\dot{\alpha}}(\overline{\sigma}^{\mu\nu})^{\dot{\alpha}}{}_{\dot{\beta}} = \epsilon_{\dot{\beta}\dot{\gamma}}(\overline{\sigma}^{\mu\nu})^{\dot{\gamma}}{}_{\dot{\tau}}, \tag{1.124}$$

$$\epsilon_{\gamma\beta}(\sigma^{\mu\nu})_\alpha{}^\beta = \epsilon_{\alpha\tau}(\sigma^{\mu\nu})_\gamma{}^\tau, \qquad \epsilon^{\dot{\gamma}\dot{\beta}}(\overline{\sigma}^{\mu\nu})^{\dot{\alpha}}{}_{\dot{\beta}} = \epsilon^{\dot{\alpha}\dot{\tau}}(\overline{\sigma}^{\mu\nu})^{\dot{\gamma}}{}_{\dot{\tau}}. \tag{1.125}$$

Numerous identities involving $\sigma^{\mu\nu}$ and $\overline{\sigma}^{\mu\nu}$ can be derived. The most useful identities are listed in Appendix B.

Previously [see eqs. (1.19), (1.30), and (1.31)], we noted that the Lorentz transformation matrices for the $(\frac{1}{2},0)$ and $(0,\frac{1}{2})$ representations, respectively, are given by

$$\exp\left(-\tfrac{1}{2}i\theta_{\mu\nu}S^{\mu\nu}\right) \simeq \begin{cases} \mathbb{1}_{2\times2} - \frac{1}{2}i\vec{\boldsymbol{\theta}}\cdot\vec{\boldsymbol{\sigma}} - \frac{1}{2}\vec{\boldsymbol{\zeta}}\cdot\vec{\boldsymbol{\sigma}}, & \text{for } (\frac{1}{2},0), \\ \mathbb{1}_{2\times2} - \frac{1}{2}i\vec{\boldsymbol{\theta}}\cdot\vec{\boldsymbol{\sigma}} + \frac{1}{2}\vec{\boldsymbol{\zeta}}\cdot\vec{\boldsymbol{\sigma}}, & \text{for } (0,\frac{1}{2}). \end{cases} \tag{1.126}$$

From this we deduce that

$$S^{\mu\nu} = \begin{cases} \sigma^{\mu\nu}, & \text{for the } (\frac{1}{2},0) \text{ representation}, \\ \overline{\sigma}^{\mu\nu}, & \text{for the } (0,\frac{1}{2}) \text{ representation}. \end{cases} \tag{1.127}$$

[16] The reader is cautioned that $\sigma^{\mu\nu}$ and $\overline{\sigma}^{\mu\nu}$ are sometimes defined in the literature without the factor of i in eqs. (1.115) and (1.116) (as in [B135]), or with an overall factor of $\frac{1}{2}i$ (as in [11]) instead of $\frac{1}{4}i$.

Table 1.1 Independent bilinear covariants involving a pair of two-component spinor fields.

bilinear covariants	number of components
$\chi\eta$	1
$\chi^\dagger\eta^\dagger$	1
$\chi\sigma^\mu\eta^\dagger$	4
$\chi^\dagger\overline{\sigma}^\mu\eta$	4
$\chi\sigma^{\mu\nu}\eta$	3
$\chi^\dagger\overline{\sigma}^{\mu\nu}\eta^\dagger$	3

The corresponding exponentiated forms of eqs. (1.34) and (1.46) are, respectively,

$$M = \exp\left(-\tfrac{1}{2}i\theta_{\mu\nu}\sigma^{\mu\nu}\right), \tag{1.128}$$

$$(M^{-1})^\dagger = \exp\left(-\tfrac{1}{2}i\theta_{\mu\nu}\overline{\sigma}^{\mu\nu}\right). \tag{1.129}$$

From eq. (1.127), it follows that $S^i = \frac{1}{2}\epsilon^{ijk}S^{jk} = \frac{1}{2}\sigma^i$ for both $(\frac{1}{2},0)$ and $(0,\frac{1}{2})$ representations. Applying the discussion below eq. (1.26), we find that $w^2 = -m^2\vec{\boldsymbol{S}}^2$ with eigenvalues $-\frac{3}{4}m^2$ as expected for a spin-1/2 fermion.

We may now construct two bilinear covariants that transform as second-rank antisymmetric tensors:

$$\chi\sigma^{\mu\nu}\eta \equiv \chi^\alpha(\sigma^{\mu\nu})_\alpha{}^\beta\eta_\beta\,, \tag{1.130}$$

$$\chi^\dagger\overline{\sigma}^{\mu\nu}\eta^\dagger \equiv \chi^\dagger_{\dot\alpha}(\overline{\sigma}^{\mu\nu})^{\dot\alpha}{}_{\dot\beta}\,\eta^{\dagger\dot\beta}. \tag{1.131}$$

Again, we note that the convention of eq. (1.63) is respected and allows us to suppress the spinor indices. Under conjugation, eq. (1.121) yields

$$(\chi\sigma^{\mu\nu}\eta)^\dagger = \eta^\dagger\overline{\sigma}^{\mu\nu}\chi^\dagger\,, \tag{1.132}$$

which is valid for both commuting and anticommuting two-component spinors. Next, we use eqs. (1.99) and (1.100) to derive the following identity for anticommuting two-component spinor fields:

$$\chi\sigma^{\mu\nu}\eta = -\eta\sigma^{\mu\nu}\chi\,, \qquad \chi^\dagger\overline{\sigma}^{\mu\nu}\eta^\dagger = -\eta^\dagger\overline{\sigma}^{\mu\nu}\chi^\dagger\,. \tag{1.133}$$

The complete list of independent bilinear covariants is given in Table 1.1. To show that the list is complete, we note that an arbitrary 2×2 complex matrix can be written as a complex linear combination of four linearly independent 2×2 matrices, which may be chosen, e.g., to be $\sigma^\mu_{\alpha\dot\beta}$ $(\mu = 0,1,2,3)$. However, in constructing the bilinear covariants, both undotted and dotted two-component spinors are employed. Thus, the relevant space is the 16-dimensional product space of two independent spaces of 2×2 matrices, spanned by the 16 matrices

$$\{\delta_\alpha{}^\beta\,,\quad \delta^{\dot\alpha}{}_{\dot\beta}\,,\quad \sigma^\mu_{\alpha\dot\beta}\,,\quad \overline{\sigma}^{\mu\dot\alpha\beta}\,,\quad \sigma^{\mu\nu}{}_\alpha{}^\beta\,,\quad \overline{\sigma}^{\mu\nu\dot\alpha}{}_{\dot\beta}\}. \tag{1.134}$$

Inside the 16-dimensional product space there exist 4-dimensional subspaces spanned by $\{\delta_\alpha{}^\beta\,,\ \sigma^{\mu\nu}{}_\alpha{}^\beta\}$ and $\{\delta^{\dot\alpha}{}_{\dot\beta}\,,\ \overline{\sigma}^{\mu\nu\dot\alpha}{}_{\dot\beta}\}$, respectively.

Here, it is important to note that $\sigma^{\mu\nu}{}_\alpha{}^\beta$ and $\overline{\sigma}^{\mu\nu\dot\alpha}{}_{\dot\beta}$ each possess only three independent components due to the self-duality relations of eq. (1.122). Hence, the total number of independent components of the bilinear covariants is indeed equal to 16, as indicated in Table 1.1. One consequence of these considerations is that products of three or more σ or $\overline{\sigma}$ matrices can always be expressed as a linear combination of the matrices given in eq. (1.134). Examples can be found in Appendix B.

In a theory of a single anticommuting two-component spinor field, there are only six independent components of the nonvanishing bilinear covariants. To obtain this result, let us set $\chi = \eta \equiv \xi$ in the bilinear covariants listed in Table 1.1. Using eqs. (1.107) and (1.133), it follows that $\xi^\dagger\overline{\sigma}^\mu\xi = -\xi\sigma^\mu\xi^\dagger$ and $\xi\sigma^{\mu\nu}\xi = \xi^\dagger\overline{\sigma}^{\mu\nu}\xi^\dagger = 0$. One can also derive the latter result as follows. Using eqs. (1.43), (1.68), and (1.130),

$$\xi\sigma^{\mu\nu}\xi\ \epsilon_{\beta\gamma}(\sigma^{\mu\nu})_\alpha{}^\beta\,\xi^\alpha\xi^\gamma = -\tfrac{1}{2}\epsilon_{\beta\gamma}\epsilon^{\alpha\gamma}(\sigma^{\mu\nu})_\alpha{}^\beta\,\xi\xi = \tfrac{1}{2}\xi\xi\,\mathrm{Tr}\ \sigma^{\mu\nu} = 0\,. \tag{1.135}$$

The derivation of $\xi^\dagger\overline{\sigma}^{\mu\nu}\xi^\dagger = 0$ is similar. One consequence of this result is that the magnetic moment of a real Majorana fermion must vanish.

Finally, it is noteworthy that products of two bilinear covariants can often be simplified by using Fierz identities. For example, denoting a generic (commuting or anticommuting) two-component spinor by z_i, the identities

$$(z_1z_2)(z_3z_4) = -(z_1z_3)(z_4z_2) - (z_1z_4)(z_2z_3)\,, \tag{1.136}$$
$$(z_1^\dagger z_2^\dagger)(z_3^\dagger z_4^\dagger) = -(z_1^\dagger z_3^\dagger)(z_4^\dagger z_2^\dagger) - (z_1^\dagger z_4^\dagger)(z_2^\dagger z_3^\dagger)\,, \tag{1.137}$$

are a consequence of the Schouten identities [eqs. (1.44) and (1.53)]. Similarly, one can employ eqs. (1.85)–(1.87) to derive

$$(z_1\sigma^\mu z_2^\dagger)(z_3^\dagger\overline{\sigma}_\mu z_4) = -2(z_1z_4)(z_2^\dagger z_3^\dagger)\,, \tag{1.138}$$
$$(z_1^\dagger\overline{\sigma}^\mu z_2)(z_3^\dagger\overline{\sigma}_\mu z_4) = 2(z_1^\dagger z_3^\dagger)(z_4z_2)\,, \tag{1.139}$$
$$(z_1\sigma^\mu z_2^\dagger)(z_3\sigma_\mu z_4^\dagger) = 2(z_1z_3)(z_4^\dagger z_2^\dagger)\,. \tag{1.140}$$

A more comprehensive list of Fierz identities is provided in Appendix B.3.

1.5 Lagrangians for Free Spin-1/2 Fermions

We now have the ingredients to construct a Lorentz-invariant hermitian Lagrangian, $\mathscr{L}$, for free fermions.[17] First, consider a theory of a single two-component fermion ξ. In light of Section 1.4, the following quantity is a Lorentz scalar:

$$i\xi^\dagger_{\dot\alpha}\,\overline{\sigma}^{\mu\dot\alpha\beta}\partial_\mu\xi_\beta \equiv i\xi^\dagger\,\overline{\sigma}^\mu\partial_\mu\xi. \tag{1.141}$$

[17] More accurately, $\mathscr{L}$ is the Lagrangian density, although the word "density" is typically dropped.

The factor of i is inserted since the hermitian quantity

$$\tfrac{1}{2}i\,\xi^\dagger\,\overline{\sigma}^\mu \overleftrightarrow{\partial}_\mu\xi \equiv \tfrac{1}{2}i\left[\xi^\dagger\,\overline{\sigma}^\mu(\partial_\mu\xi) - (\partial_\mu\xi^\dagger)\overline{\sigma}^\mu\xi\right] \tag{1.142}$$

differs from $i\xi^\dagger\,\overline{\sigma}^\mu\partial_\mu\xi$ by a total divergence. Hence, eq. (1.141) is a candidate for a kinetic energy term in the Lagrangian. A second hermitian Lorentz scalar,

$$m\xi\xi + m^*\xi^\dagger\xi^\dagger\,, \tag{1.143}$$

is a candidate for a mass term in the Lagrangian for a theory of fermions.

Thus, the Lagrangian for a free two-component fermion is given by

$$\mathscr{L} = i\xi^\dagger\,\overline{\sigma}^\mu\partial_\mu\xi - \tfrac{1}{2}(m\xi\xi + m^*\xi^\dagger\,\xi^\dagger)\,. \tag{1.144}$$

Without loss of generality, we can absorb the phase of the parameter m into the definition of the field ξ. Henceforth, we shall always work in a convention in which m is real and nonnegative. The corresponding field equations are

$$i\overline{\sigma}^\mu\partial_\mu\xi - m\xi^\dagger = 0\,, \tag{1.145}$$

$$i\sigma^\mu\partial_\mu\xi^\dagger - m\xi = 0\,, \tag{1.146}$$

where eq. (1.146) is the conjugate of eq. (1.145). This is a theory of a (neutral) Majorana spin-1/2 fermion.

Using the Noether procedure, one can determine the canonical energy–momentum tensor $T^\mu{}_\nu$ and the canonical angular momentum density $M^\mu{}_{\nu\lambda}$ (e.g., see Ref. [29]), which satisfy conservation laws $\partial_\mu T^\mu{}_\nu = 0$ and $\partial_\mu M^\mu{}_{\nu\lambda} = 0$, after applying the field equations. In particular, for a theory with N two-component $(\frac{1}{2},0)$ spinor fields $\xi_{\alpha j}$ and the corresponding two-component $(0,\frac{1}{2})$ spinor fields $\xi^{\dagger\dot\alpha j}$ (where $j\in\{1,2,\ldots,N\}$),

$$T^\mu{}_\nu = \frac{\partial\mathscr{L}}{\partial(\partial_\mu\xi_{\alpha j})}\partial_\nu\xi_{\alpha j} + \frac{\partial\mathscr{L}}{\partial(\partial_\mu\xi^{\dagger\dot\alpha j})}\partial_\nu\xi^{\dagger\dot\alpha j} - \delta^\mu_\nu\mathscr{L}\,, \tag{1.147}$$

$$M^\mu{}_{\nu\lambda} = x_\nu T^\mu{}_\lambda - x_\lambda T^\mu{}_\nu - i\frac{\partial\mathscr{L}}{\partial(\partial_\mu\xi_{\alpha j})}(\sigma_{\nu\lambda})_\alpha{}^\beta\xi_{\beta j} - i\frac{\partial\mathscr{L}}{\partial(\partial_\mu\xi^{\dagger\dot\alpha j})}(\overline{\sigma}_{\nu\lambda})^{\dot\alpha}{}_{\dot\beta}\xi^{\dagger\dot\beta j}\,, \tag{1.148}$$

where there is an implicit sum over the repeated index j. In the case of $N=1$, where $\mathscr{L}$ is given by eq. (1.144), we have

$$\frac{\partial\mathscr{L}}{\partial(\partial_\mu\xi_\alpha)} = i\xi^\dagger_{\dot\beta}\overline{\sigma}^{\mu\dot\beta\alpha}\,, \qquad \frac{\partial\mathscr{L}}{\partial(\partial_\mu\xi^{\dagger\dot\alpha})} = 0\,. \tag{1.149}$$

The corresponding time-independent "charges" are the momentum four-vector and the total angular momentum tensor, respectively:

$$P_\nu = \int d^3x\,T^0{}_\nu\,, \qquad J_{\nu\lambda} = \int d^3x\,M^0{}_{\nu\lambda}\,. \tag{1.150}$$

Using the canonical anticommutation relations of the quantum fields $\xi(\vec{x},t)$ and $\xi^\dagger(\vec{x},t)$, it is straightforward to show that P_ν and $J_{\nu\lambda}$ satisfy the commutation relations of the Poincaré algebra eqs. (1.23)–(1.25). Moreover, P_ν and $J_{\nu\lambda}$, defined in eq. (1.150), are *hermitian* operators[18] that act on the quantum Hilbert space (see Exercise 1.11). Hence, $\exp(ia_\mu P^\mu)$ and $\exp(-\frac{1}{2}i\theta_{\mu\nu}J^{\mu\nu})$ serve as *unitary* representations of the Poincaré group.

Next, consider a theory of two free two-component fermion fields, ξ_1 and ξ_2. The corresponding Lagrangian can be written in the following form:[19]

$$\mathscr{L} = i\xi_1^\dagger\bar{\sigma}^\mu\partial_\mu\xi_1 + i\xi_2^\dagger\bar{\sigma}^\mu\partial_\mu\xi_2 - \tfrac{1}{2}m_1(\xi_1\xi_1 + \xi_1^\dagger\xi_1^\dagger) - \tfrac{1}{2}m_2(\xi_2\xi_2 + \xi_2^\dagger\xi_2^\dagger)\,, \tag{1.151}$$

where m_1 and m_2 are real and nonnegative mass parameters. A special case arises when $m_1 = m_2 = m$. In this case one can perform complex field redefinitions

$$\chi = \frac{1}{\sqrt{2}}(\xi_1 + i\xi_2)\,, \qquad \eta = \frac{1}{\sqrt{2}}(\xi_1 - i\xi_2)\,, \tag{1.152}$$

to obtain

$$\mathscr{L} = i\chi^\dagger\,\bar{\sigma}^\mu\partial_\mu\chi + i\eta^\dagger\,\bar{\sigma}^\mu\partial_\mu\eta - m(\chi\eta + \chi^\dagger\,\eta^\dagger)\,. \tag{1.153}$$

The corresponding field equations for χ and η are

$$i\bar{\sigma}^\mu\partial_\mu\chi - m\eta^\dagger = 0\,, \qquad i\bar{\sigma}^\mu\partial_\mu\eta - m\chi^\dagger = 0\,, \tag{1.154}$$

$$i\sigma^\mu\partial_\mu\eta^\dagger - m\chi = 0\,, \qquad i\sigma^\mu\partial_\mu\chi^\dagger - m\eta = 0\,. \tag{1.155}$$

The canonical energy–momentum tensor and angular momentum density are obtained from eqs. (1.147) and (1.148), respectively. One can again verify that the hermitian operators P_ν and $J_{\nu\lambda}$ defined by eq. (1.150) satisfy the commutation relations of the Poincaré algebra.

The theory of two free two-component fermion fields of equal mass [eq. (1.151) with $m_1 = m_2$] possesses a global internal O(2) symmetry,[20]

$$\xi_i \to O_{ij}\xi_j\,, \quad \text{where } O^{\mathsf{T}}O = \mathbb{1}_{2\times 2}\,. \tag{1.156}$$

Corresponding to this symmetry is a hermitian conserved Noether current,

$$J^\mu = i(\xi_1^\dagger\bar{\sigma}^\mu\xi_2 - \xi_2^\dagger\bar{\sigma}^\mu\xi_1)\,, \tag{1.157}$$

with a corresponding conserved charge, $Q = \int J^0 d^3x$. In the ξ_i-basis, the Noether current is off-diagonal. However, after converting to the χ–η basis via eq. (1.152), the current is diagonal:

$$J^\mu = \chi^\dagger\bar{\sigma}^\mu\chi - \eta^\dagger\bar{\sigma}^\mu\eta\,, \tag{1.158}$$

which provides the motivation for this basis choice. In particular, we can now identify χ and η as fields of definite and opposite charge Q. In particular, eq. (1.153) is

[18] In contrast, the finite-dimensional irreducible matrix representations of the Lorentz group generators $J^i = \frac{1}{2}\epsilon^{ijk}J^{jk}$ and $K^i = J^{0i}$ are hermitian and antihermitian, respectively.

[19] The theory specified by eq. (1.144) is in a form in which the mass matrix is diagonal. In Section 1.6, we will generalize to the case of n two-component fields ($n \geq 2$) and discuss the general procedure of fermion mass matrix diagonalization.

[20] The kinetic energy term alone is invariant under a larger global U(2) symmetry. However, when $m \neq 0$, the global symmetry of the free-fermion field theory is O(2).

invariant under the Noether U(1) global symmetry transformations, $\chi \to e^{i\theta}\chi$ and $\eta \to e^{-i\theta}\eta$. Together, χ and η constitute a single Dirac spin-1/2 fermion.

The U(1) symmetry transformations identified above correspond to the SO(2) symmetry transformations specified in eq. (1.156) with $\det O = 1$. The Lagrangian given in eq. (1.153) is also invariant under the discrete $\mathbb{Z}_2$ symmetry transformation that interchanges the fermion fields χ and η. Indeed, the semidirect product $\mathrm{SO}(2)\rtimes\mathbb{Z}_2$ is isomorphic to the O(2) symmetry group identified in eq. (1.156).[21]

Finally, we introduce the concept of *off-shell* degrees of freedom and contrast this with *on-shell* or physical degrees of freedom.[22] Given a particle with four-momentum p, real particle propagation must satisfy $p^2 = m^2$ (the mass-shell condition), whereas for virtual particles (e.g., particle exchange inside Feynman diagrams), the four-momentum p and the mass m are independent. The field operator corresponding to real particle propagation satisfies the field equations, whereas the field operator of a virtual particle is not constrained by the field equations.

For scalar (spin-0) fields, there is no distinction between on-shell and off-shell degrees of freedom. The on-shell scalar field satisfies the Klein–Gordon equation, but this does not alter the fact that a real (complex) scalar field consists of one (two) degree(s) of freedom. For a nonzero spin-s field, the number of on-shell degrees of freedom is less than the number of off-shell degrees of freedom. Although all the components of an on-shell spin-s field satisfy the Klein–Gordon equation, the spin-s field equations provide additional constraints that reduce the number of degrees of freedom originally present. It is instructive to demonstrate how the number of off-shell degrees of freedom is reduced in the case of free-fermion field theory.

A theory of a neutral self-conjugate fermion is described in terms of a single complex two-component fermion field $\xi_\alpha(x)$. This corresponds initially to four off-shell degrees of freedom, since ξ_α and $\xi^\dagger_{\dot\alpha} \equiv (\xi_\alpha)^\dagger$ are independent degrees of freedom. But, when the field equations are imposed [see eq. (1.145)], $\xi^\dagger(x)$ is determined in terms of $\xi(x)$. Thus, the number of physical on-shell degrees of freedom for a neutral self-conjugate fermion is equal to two.

A theory of a charged fermion is described in terms of a pair of mass-degenerate two-component fermions. We shall work in a basis where the charged fermion is represented by a pair of two-component fermion fields χ and η. The number of off-shell degrees of freedom for a charged fermion is eight. Applying the field equations determines $\chi^\dagger$ in terms of η and determines $\eta^\dagger$ in terms of χ [see eqs. (1.154) and (1.155)]. Thus, the number of physical on-shell degrees of freedom for a charged fermion is equal to four.

Although we have illustrated the counting of degrees of freedom for free-fermion field theory, the same counting applies in an interacting theory.

21 In the ξ_i-basis, one can identify $\mathbb{Z}_2 = \{\boldsymbol{I}_2\,,\,\sigma^3\}$, where $\boldsymbol{I}_2$ is the 2×2 identity matrix and $\sigma^3 = \mathrm{diag}(1\,,\,-1)$. Note that $\mathrm{O}(2)= \{hk\,|\,h\in\mathrm{SO}(2)\,,\,k\in\mathbb{Z}_2\}$, where $\mathrm{SO}(2)\cap\mathbb{Z}_2 = \boldsymbol{I}_2$ and SO(2) is a normal subgroup of O(2), which imply that $\mathrm{O}(2)\cong\mathrm{SO}(2)\rtimes\mathbb{Z}_2$. However, $\mathbb{Z}_2$ is *not* a normal subgroup of O(2) since $R\sigma^3R^{-1}\notin\{\boldsymbol{I}_2\,,\,\sigma^3\}$ for rotations $R\in\mathrm{SO}(2)$ by an angle $\theta\neq\frac{1}{2}n\pi$ (for integer n). Hence, the conditions for a direct product structure are not satisfied.

22 Here, off-shell is shorthand for "off mass shell" and on-shell is shorthand for "on mass shell." These two terms distinguish between virtual and real particle propagation, respectively.

1.6 The Fermion Mass Matrix and Its Diagonalization

We now generalize the discussion of Section 1.5 and consider a collection of n free anticommuting two-component spin-1/2 fields, $\hat{\xi}_{\alpha i}(x)$, which transform as $(\frac{1}{2},0)$ fields under the Lorentz group. Here, α is the spinor index, and i is a flavor index ($i \in \{1,2,\ldots,n\}$) that labels the distinct fields of the collection. The free-field Lagrangian is given by

$$\mathscr{L} = i\hat{\xi}^{\dagger i}\bar{\sigma}^{\mu}\partial_{\mu}\hat{\xi}_i - \tfrac{1}{2}M^{ij}\hat{\xi}_i\hat{\xi}_j - \tfrac{1}{2}M_{ij}\hat{\xi}^{\dagger i}\hat{\xi}^{\dagger j}\,, \tag{1.159}$$

where

$$M_{ij} \equiv (M^{ij})^*. \tag{1.160}$$

Note that M is a complex symmetric matrix, since the product of anticommuting two-component fields satisfies $\hat{\xi}_i\hat{\xi}_j = \hat{\xi}_j\hat{\xi}_i$ [with the spinor contraction rule according to eq. (1.63)].

In eq. (1.159), we have employed the U(n)-covariant tensor calculus [B103, B122] for *flavor tensors* labeled by the flavor indices i and j. Each left-handed $(\frac{1}{2},0)$ fermion always has an index with the opposite height of the corresponding right-handed $(0,\frac{1}{2})$ fermion. Raised indices can only be contracted with lowered indices and vice versa. Flipping the heights of all flavor indices of an object corresponds to complex conjugation, as in eq. (1.160). In particular, we generalize eq. (1.54) as follows:[23]

$$\psi^{\dagger\, i}_{\dot{\alpha}} \equiv (\psi_{\alpha i})^{\dagger}\,. \tag{1.161}$$

If $M = 0$, then the free-field Lagrangian is invariant under a global U(n) symmetry. That is, for a unitary matrix U, with matrix elements $U_i{}^j$, and its hermitian conjugate defined by

$$(U^{\dagger})_i{}^j = (U_j{}^i)^* \equiv U^j{}_i\,, \tag{1.162}$$

with $U_i{}^k(U^{\dagger})_k{}^j = \delta_i^j$, the massless free-field Lagrangian is invariant under the transformations

$$\hat{\xi}_i \longrightarrow U_i{}^j\hat{\xi}_j\,, \qquad\qquad \hat{\xi}^{\dagger i} \longrightarrow U^i{}_j\hat{\xi}^{\dagger j}\,. \tag{1.163}$$

For $M \neq 0$, eq. (1.159) remains formally invariant under the global U(n) symmetry if M acts as a spurion field (e.g., see [B51, B60, B65]) with the appropriate tensorial transformation law, $M^{ij} \longrightarrow U^i{}_k U^j{}_\ell M^{k\ell}$.

Expressions consisting of flavor vectors and second-rank flavor tensors have natural interpretations as products of vectors and matrices. As a result, the flavor indices

[23] In the case at hand, we have more specifically chosen all of the left-handed fermions to have lowered flavor indices, which implies that all of the right-handed fermions have raised flavor indices. However, in cases where a subset of left-handed fermions transform according to some representation R of a (global) symmetry and a different subset of left-handed fermions transform according to the conjugate representation $\overline{R}$, it is often more convenient to employ a raised flavor index for the latter subset of left-handed fields.

can be suppressed, and the resulting expressions can be written in an index-free matrix notation. To accomplish this, one must first assign a particular flavor-index structure to the matrices that will appear in the index-free expression. For example, given the second-rank flavor tensors introduced above, we define the matrix elements of M to be M^{ij} and the matrix elements of U to be $U_i{}^j$. Note that $(U^\dagger)_i{}^j$ has the same flavor-index structure as U.

As a simple example, in an index-free notation eq. (1.163) reads: $\hat{\xi} \longrightarrow U\hat{\xi}$ and $\hat{\xi}^\dagger \longrightarrow \hat{\xi}^\dagger U^\dagger$. A slightly more complicated example is exhibited below:

$$U^i{}_k M^{k\ell} = (U^\dagger)_k{}^i M^{k\ell} = (U^* M)^{i\ell}, \tag{1.164}$$

where we have used $(U^\dagger)^\mathsf{T} = U^*$ in obtaining the final result. That is, in matrix notation with suppressed indices, $U^i{}_k M^{k\ell}$ corresponds to the matrix U^*M. Thus, in an index-free notation, the tensorial transformation law for the spurion field M is given by $M \longrightarrow U^* M U^\dagger$.

In general, the matrix M is not diagonal, so we cannot directly identify $\hat{\xi}_i(x)$ as a field of definite mass. If the Lagrangian were also to include interaction terms, we would call $\hat{\xi}_i(x)$ an *interaction* eigenstate field.[24] One can diagonalize the mass matrix and rewrite the Lagrangian in terms of mass eigenstates $\xi_{\alpha i}$ that exhibit real nonnegative masses m_i, respectively. To do this, we introduce the *mass* eigenstates:

$$\hat{\xi}_i = \Omega_i{}^j \xi_j\,, \qquad \hat{\xi}^{\dagger i} = \Omega^i{}_j \xi^{\dagger j}\,, \tag{1.165}$$

where Ω is a unitary $n \times n$ matrix [to be determined by eq. (1.167) below], and $\Omega^i{}_j \equiv (\Omega_i{}^j)^*$. Under this transformation, the kinetic energy term is unchanged, $\xi^{\dagger i}\overline{\sigma}^\mu \partial_\mu \xi_i = \hat{\xi}^{\dagger i}\overline{\sigma}^\mu \partial_\mu \hat{\xi}_i$, but the mass term is transformed:

$$M^{ij}\hat{\xi}_i\hat{\xi}_j = M^{ij}\Omega_i{}^k\Omega_j{}^\ell \xi_k\xi_\ell = \sum_k m_k \xi_k \xi_k\,, \tag{1.166}$$

where the m_k are real and nonnegative, and $M^{ij}\Omega_i{}^k\Omega_j{}^\ell = m_k\delta^{k\ell}$ (no sum over k). Equivalently, in matrix notation with suppressed indices, $\hat{\xi} = \Omega\xi$ and

$$\Omega^\mathsf{T} M \Omega = \boldsymbol{m} \equiv \mathrm{diag}(m_1, m_2, \ldots). \tag{1.167}$$

In general, the m_i are *not* the eigenvalues of M. Rather, they are the *singular values* of the matrix M, which are defined to be the nonnegative square roots of the eigenvalues of $M^\dagger M$. See Appendix G for further details.

One can prove that a unitary matrix Ω satisfying eq. (1.167) always exists. This result corresponds to the Takagi diagonalization of a general complex symmetric matrix, which is discussed in more detail in Appendix G.11. That is, for any complex symmetric matrix M there exists a unitary matrix Ω such that $\Omega^\mathsf{T} M \Omega$ is diagonal with nonnegative entries given by the positive square roots of the eigenvalues of $M^\dagger M$ (or $MM^\dagger$). To compute the values of the (nonnegative real) diagonal elements of $\boldsymbol{m}$, note that eq. (1.167) implies that

$$\Omega^\dagger M^\dagger M \Omega = \boldsymbol{m}^\dagger \boldsymbol{m} = \boldsymbol{m}^2\,. \tag{1.168}$$

[24] In general, we shall use "hatted" fields to represent interaction eigenstate fields and "unhatted" fields to describe states of definite mass.

The matrix $M^\dagger M$ is hermitian and thus it can be diagonalized by a unitary matrix. Hence, the diagonal elements of $\boldsymbol{m}$ are the nonnegative square roots of the corresponding eigenvalues of $M^\dagger M$, as asserted above. However, if $M^\dagger M$ has degenerate eigenvalues, then eq. (1.168) *cannot* be employed to determine the unitary matrix Ω that satisfies eq. (1.167). A more general technique for determining Ω that works in all cases is given in Appendix G.2.[25] If the singular values of M are nondegenerate, then Ω is unique up to permutations of the columns of Ω and a possible overall multiplication of any column of Ω by -1. The nonuniqueness of Ω in the case of degenerate singular values is discussed briefly below eq. (G.13).

Thus, in terms of the mass eigenstates, the general form for the free-field Lagrangian of n flavors of two-component spinors is given by

$$\mathscr{L} = i\xi^{\dagger i}\overline{\sigma}^\mu\partial_\mu\xi_i - \tfrac{1}{2}\sum_i m_i(\xi_i\xi_i + \xi^{\dagger i}\xi^{\dagger i})\,. \tag{1.169}$$

If all the masses are nondegenerate, then this Lagrangian does not exhibit any nontrivial global flavor symmetries. However, if some of the masses are degenerate, global flavor symmetries will exist (some of which may still be respected once interactions are included). For example, if all masses are degenerate, then the model exhibits an O(n) global symmetry corresponding to $\xi_i \to O_{ij}\xi_j$ with $O^{\mathsf{T}}O = \boldsymbol{I}_n$.

The case where some of the spin-1/2 fermion fields correspond to massive Dirac fermions carrying a conserved charge merits special attention. If χ_α is a charged massive field, then there must be an associated independent two-component spinor field η_α of equal mass with the opposite charge. Although the mass diagonalization procedure described above is always applicable, it is often more convenient to employ mass eigenstates that are also eigenstates of the charge operator. In the case of two fields of opposite charge, this means writing the corresponding Lagrangian in the form given by eq. (1.153) [whereas the form given by eq. (1.151) results from the diagonalization procedure described above].

Consider a collection of such free anticommuting charged massive spin-1/2 fields, which can be represented by pairs of two-component interaction eigenstate fields $\hat{\chi}_{\alpha i}(x)$, $\hat{\eta}^i_\alpha(x)$. These fields transform in (possibly reducible) representations of the unbroken symmetry group that are conjugates of each other. This accounts for the opposite flavor-index heights of $\hat{\chi}_i$ and $\hat{\eta}^i$ (see footnote 23). The free-field Lagrangian is given by

$$\mathscr{L} = i\hat{\chi}^{\dagger i}\overline{\sigma}^\mu\partial_\mu\hat{\chi}_i + i\hat{\eta}^\dagger_i\overline{\sigma}^\mu\partial_\mu\hat{\eta}^i - M^i{}_j\hat{\chi}_i\hat{\eta}^j - M_i{}^j\hat{\chi}^{\dagger i}\hat{\eta}^\dagger_j\,, \tag{1.170}$$

where M is an arbitrary complex matrix with matrix elements $M^i{}_j$, and

$$M_i{}^j \equiv (M^i{}_j)^*\,. \tag{1.171}$$

If $M = 0$, then the free-field Lagrangian is invariant under a global U(n)×U(n)

[25] For example, it is possible to have a nondiagonal matrix M such that $M^\dagger M$ is proportional to the identity matrix. In this case, Ω drops out completely from eq. (1.168), and must be determined directly from eq. (1.167).

symmetry. That is, for a pair of unitary matrices U_L and U_R, with matrix elements given respectively by $(U_L)_i{}^j$ and $(U_R)^i{}_j$, and the corresponding hermitian conjugates defined by

$$(U_L^\dagger)_j{}^i = [(U_L)_i{}^j]^* \equiv (U_L)^i{}_j\,, \qquad (U_R^\dagger)^j{}_i = [(U_R)^i{}_j]^* \equiv (U_R)_i{}^j\,, \quad (1.172)$$

the massless free-field Lagrangian is invariant under the transformations

$$\hat{\chi}_i \longrightarrow (U_L)_i{}^j\hat{\chi}_j\,, \quad \hat{\chi}^{\dagger i} \longrightarrow (U_L)^i{}_j\hat{\chi}^{\dagger j}\,, \quad \hat{\eta}^i \longrightarrow (U_R)^i{}_j\hat{\eta}^j\,, \quad \hat{\eta}^\dagger_i \longrightarrow (U_R)_i{}^j\hat{\eta}^\dagger_j\,. \quad (1.173)$$

For $M \neq 0$, eq. (1.170) remains formally invariant under the $\mathrm{U}(n)\times\mathrm{U}(n)$ symmetry if M acts as a spurion field, which transforms as $M^i{}_j \to (U_L)^i{}_k(U_R)_j{}^\ell M^k{}_\ell$ (or equivalently, $M \longrightarrow U_L^* M U_R^\dagger$, in an index-free matrix notation with suppressed flavor indices).

In order to diagonalize the fermion mass matrix, we introduce the mass eigenstates χ_i and η^i and unitary matrices L and R, with matrix elements given respectively by $L_i{}^k$ and $R^i{}_k$, such that

$$\hat{\chi}_i = L_i{}^k\chi_k\,, \qquad \hat{\eta}^i = R^i{}_k\eta^k\,, \quad (1.174)$$

and demand that $M^i{}_jL_i{}^kR^j{}_\ell = m_k\delta^k_\ell$ (no sum over k), where the m_k are real and nonnegative. Equivalently, if we employ matrix notation with suppressed indices, then $\hat{\chi} = L\chi$, $\hat{\eta} = R\eta$, and

$$L^{\mathsf{T}}MR = \boldsymbol{m} = \mathrm{diag}(m_1, m_2, \ldots) \quad (1.175)$$

is a diagonal matrix whose diagonal elements m_i are real and nonnegative. The singular value decomposition of linear algebra, discussed more fully in Appendix G.3, states that for any complex matrix M, unitary matrices L and R exist such that eq. (1.175) is satisfied. It then follows that

$$L^{\mathsf{T}}(MM^\dagger)L^* = \boldsymbol{m}\boldsymbol{m}^\dagger = \boldsymbol{m}^2\,, \quad (1.176)$$

$$R^\dagger(M^\dagger M)R = \boldsymbol{m}^\dagger\boldsymbol{m} = \boldsymbol{m}^2\,. \quad (1.177)$$

That is, since $MM^\dagger$ and $M^\dagger M$ are both hermitian,[26] they can be diagonalized by unitary matrices. The diagonal elements of $\boldsymbol{m}$ are therefore the nonnegative square roots of the corresponding eigenvalues of $MM^\dagger$ (or $M^\dagger M$). In terms of the mass eigenstates,

$$\mathscr{L} = i\chi^{\dagger i}\bar{\sigma}^\mu\partial_\mu\chi_i + i\eta^\dagger_i\bar{\sigma}^\mu\partial_\mu\eta^i - m_i(\chi_i\eta^i + \chi^{\dagger i}\eta^\dagger_i)\,. \quad (1.178)$$

The mass matrix now consists of 2×2 blocks $\left(\begin{smallmatrix} 0 & m_i \\ m_i & 0\end{smallmatrix}\right)$ along the diagonal. More importantly, the squared-mass matrix is diagonal with doubly degenerate entries m_i^2 that will appear in the denominators of the propagators of the theory. For $m_i \neq 0$, each χ_i–η^i pair describes a charged Dirac fermion consisting of four on-shell real

[26] Although $MM^\dagger$ and $M^\dagger M$ are not equal in general, these two matrices possess the same real nonnegative eigenvalues.

degrees of freedom.[27] In addition, eq. (1.178) yields an even number of massless two-component fermions.

Given an arbitrary collection of two-component left-handed $(\frac{1}{2},0)$ fermions, the distinction between Majorana and Dirac fermions depends on whether the Lagrangian is invariant under a global (or local) continuous symmetry group G, and the corresponding multiplet structure of the fermion fields. If no such continuous symmetry exists, then the fermion mass eigenstates will consist of Majorana fermions. If the Lagrangian is invariant under a symmetry group G, then the collection of two-component fermions will break up into a sum of multiplets that transform irreducibly under G. As described in Appendix H, a representation R can be either a real, pseudoreal, or complex representation of G. If a multiplet transforms under a real representation of G, then the corresponding fermion mass eigenstates are Majorana fermions.[28] If a multiplet transforms under a complex representation of G, then the corresponding fermion mass eigenstates are Dirac fermions. In particular [as noted above eq. (1.170)], if the χ_i transform under the representation R, then the η^i transform under the conjugate representation $\overline{R}$.

The case where a multiplet of two-component left-handed fermions transform under a pseudoreal representation of G has not been explicitly treated above. The simplest example of this kind is a model of $2n$ multiplets (or "flavors") of two-component SU(2)-doublet[29] fermions, $\hat{\psi}_{ia}$ (where $i \in \{1,2,\ldots,2n\}$ labels the flavor index and a labels the SU(2) doublet index). The free-field Lagrangian is given by

$$\mathscr{L} = i\hat{\psi}^{\dagger ia}\overline{\sigma}^{\mu}\partial_{\mu}\hat{\psi}_{ia} - \tfrac{1}{2}\left(M^{ij}\epsilon^{ab}\hat{\psi}_{ia}\hat{\psi}_{jb} + \text{h.c.}\right)\,, \tag{1.179}$$

where ϵ^{ab} is the antisymmetric SU(2)-invariant tensor defined in Appendix A.8 such that $\epsilon^{12} = -\epsilon^{21} = +1$. As $\epsilon^{ab}\hat{\psi}_{ia}\hat{\psi}_{jb}$ is antisymmetric under the interchange of flavor indices i and j, it follows that M is a complex antisymmetric matrix. To identify the fermion mass eigenstates ψ_{ja}, we introduce a unitary matrix U (with matrix elements $U_i{}^j$) such that $\hat{\psi}_{ia} = U_i{}^j\psi_{ja}$ and demand that

$$U^{\mathsf{T}}MU = \boldsymbol{N} \equiv \text{diag}\left\{\begin{pmatrix} 0 & m_1 \\ -m_1 & 0\end{pmatrix}, \begin{pmatrix} 0 & m_2 \\ -m_2 & 0\end{pmatrix}, \ldots, \begin{pmatrix} 0 & m_n \\ -m_n & 0\end{pmatrix}\right\}, \tag{1.180}$$

where $\boldsymbol{N}$ is written in block-diagonal form consisting of 2×2 matrix blocks appearing along the diagonal, and the m_j are real and nonnegative. Equation (1.180) corresponds to the reduction of a complex antisymmetric matrix to its real normal

[27] Of course, one could always choose instead to treat the Dirac fermions in a basis where the mass matrix is diagonal [as in eq. (1.169)], where the corresponding two-component fermion fields are not eigenstates of the charge operator. Inverting eq. (1.152) for each Dirac fermion yields $\xi_{2i-1} = (\chi_i + \eta^i)/\sqrt{2}$ and $\xi_{2i} = i(\eta_i - \chi^i)/\sqrt{2}$. However, it is rarely, if ever, convenient to do so; practical calculations only require that the squared-mass matrix $M^{\dagger}M$ is diagonal, and it is of course more convenient to employ fields that carry well-defined charges.

[28] This is a slight generalization of the more restrictive definition that requires Majorana fermions to transform trivially under the group G. The gluinos, which transform under the (real) adjoint representation of the color SU(3) group, are Majorana fermions according to our more general definition.

[29] The doublet representation of SU(2) is pseudoreal, as noted above eq. (H.10).

form, which is discussed in more detail in Appendix G.4. In order to compute the m_k, we first note that

$$U^\dagger M^\dagger M U = \text{diag}(m_1^2\,,\, m_1^2\,,\, m_2^2\,,\, m_2^2\,,\, \ldots\,,\, m_n^2\,,\, m_n^2)\,. \tag{1.181}$$

Hence, the m_j are the nonnegative square roots of the corresponding eigenvalues of $M^\dagger M$. Since the dimension of the doublet representation of SU(2) provides an additional degeneracy factor of 2, eq. (1.181) implies that the mass spectrum consists of $2n$ pairs of mass-degenerate two-component fermions, which are equivalent to $2n$ Dirac fermions. In particular,

$$\mathscr{L} = \sum_{i=1}^{2n} i\psi^{\dagger ia}\overline{\sigma}^\mu \partial_\mu \psi_{ia} - \sum_{i=1}^{n} \left(m_i \epsilon^{ab}\, \psi_{2i-1,\,a}\psi_{2i,\,b} + \text{h.c.}\right)\,. \tag{1.182}$$

In the general case of a pseudoreal representation R (of dimension d_R), the SU(2)-invariant epsilon tensor is replaced by a more general $d_R \times d_R$ unitary antisymmetric matrix C [defined in eq. (H.10)]. Thus, the analysis above can be repeated virtually unchanged. By defining

$$\chi_{ia} \equiv \psi_{2i-1,\,a}\,, \qquad \eta^{ia} \equiv C^{ab}\psi_{2i,\,b}\,, \qquad i \in \{1,2,\ldots,n\}\,; \quad a \in \{1,2,\ldots,d_R\}\,, \tag{1.183}$$

with an implicit sum over the repeated index b, the resulting Lagrangian given by

$$\mathscr{L} = \sum_{i=1}^{n} i\chi^{\dagger ia}\overline{\sigma}^\mu \partial_\mu \chi_{ia} + i\eta^\dagger_{ia}\overline{\sigma}^\mu\partial_\mu\eta^{ia} - m_i\left(\chi_{ia}\eta^{ia} + \chi^{\dagger ia}\eta^\dagger_{ia}\right) \tag{1.184}$$

describes a free field theory of nd_R Dirac fermions [see eq. (1.178)]. Therefore, if a multiplet of two-component left-handed fermions transforms under a pseudoreal representation of G, then the corresponding fermion mass eigenstates are Dirac fermions. If eq. (1.179) contains an odd number of pseudoreal fermion multiplets, then the (antisymmetric) mass matrix M is odd-dimensional and thus has an odd number of zero eigenvalues [according to eq. (G.21)]. But as d_R must be even, it follows that the pseudoreal fermion multiplet contains an even number of massless two-component fermions.

In conclusion, the mass diagonalization procedure of an arbitrary field theory of fermions yields (in general) a set of massless two-component fermions, a set of massive neutral Majorana fermions [as in eq. (1.169)], and a set of massive charged Dirac fermions [as in eq. (1.178)]. Hence, in the most general theory of spin-1/2 fields, the Lagrangian can be written in terms of two-component $(\frac{1}{2},0)$ fermion fields $\hat{\psi}_i(x)$, which consists of neutral fermions $\hat{\xi}_i$, and charged fermion pairs $\hat{\chi}_i$ and $\hat{\eta}_i$. The mass-eigenstate basis is achieved by a unitary rotation $U_i{}^j$ on the flavor indices. In matrix form,

$$\hat{\psi} \equiv \begin{pmatrix} \hat{\xi} \\ \hat{\chi} \\ \hat{\eta} \end{pmatrix} = U\psi \equiv \begin{pmatrix} \Omega & 0 & 0 \\ 0 & L & 0 \\ 0 & 0 & R \end{pmatrix} \begin{pmatrix} \xi \\ \chi \\ \eta \end{pmatrix}, \tag{1.185}$$

where Ω, L, and R are constructed as described above. The result of the mass diagonalization procedure in a general theory therefore always consists of a collection of Majorana fermions as in eq. (1.169), plus a collection of Dirac fermions as in eq. (1.178).

1.7 Discrete Spacetime and Internal Symmetries

The Lorentz bilinear covariant quantities given in Table 1.1 transform as a scalar, vector, or second-rank tensor under a proper orthochronous Lorentz transformation. But it also proves useful to study the behavior of the bilinear covariants under an improper Lorentz transformation. To accomplish this, it is sufficient to study the behavior of the bilinear covariants under parity (P) and under time reversal (T). In addition, for bilinear covariants that depend on a pair of two-component spinor fields, a discrete transformation that can be interpreted as charge conjugation (C) is also relevant. These symmetries act on the Hilbert space of states via the unitary operator $\mathcal{P}$, the antiunitarity operator $\mathcal{T}$, and the unitary operator $\mathcal{C}$, respectively. The second-quantized fermion fields, which are operators that act on the Hilbert space, transform under P, T, and C via similarity transformations: $\xi \to \mathcal{P}\xi\mathcal{P}^{-1}$, $\xi \to \mathcal{T}\xi\mathcal{T}^{-1}$, and $\xi \to \mathcal{C}\xi\mathcal{C}^{-1}$.

The transformation laws for the two-component spin-1/2 fermions under P and T cannot be determined a priori from the transformation laws under proper Lorentz transformations. This can be understood by noting that the two-component spinors transform under SL(2,$\mathbb{C}$), which is the universal covering group of $SO_+(1,3)$, the group of proper orthochronous Lorentz transformations.[30] In particular, SL(2,$\mathbb{C}$) does not contain space inversion or time inversion. As a result, there is some freedom in defining the action of P and T on the two-component fermion fields. We shall fix this freedom by demanding that the quantum theory of free (noninteracting) fermion fields is invariant under the P and T discrete symmetry transformations. That is, $\mathscr{L}^P(\vec{x},t) = \mathscr{L}(-\vec{x},t)$ and $\mathscr{L}^T(\vec{x},t) = \mathscr{L}(\vec{x},-t)$, which ensures that the corresponding free-field action is invariant under P and T, respectively.

We expect that under a parity transformation a $(\frac{1}{2},0)$ fermion is transformed into a $(0,\frac{1}{2})$ fermion and vice versa. Thus, it is convenient to define a hermitian matrix P such that[31]

$$\mathcal{P}\xi_\alpha(x)\mathcal{P}^{-1} \equiv i\eta_P P_{\alpha\dot\beta}\xi^{\dagger\dot\beta}(x_P)\,, \quad \mathcal{P}\xi^\dagger_{\dot\alpha}(x)\mathcal{P}^{-1} \equiv -i\eta_P^* P_{\beta\dot\alpha}\xi^\beta(x_P)\,, \tag{1.186}$$

where $x_P \equiv (t\,;\,-\vec{x})$ and η_P is initially an arbitrary complex phase. Note that the

[30] As Lie groups, $SO_+(1,3) \cong SL(2,\mathbb{C})/\mathbb{Z}_2$, corresponding to a double covering of $SO_+(1,3)$ by SL(2,$\mathbb{C}$). For example, the SL(2,$\mathbb{C}$) matrices $\mathbb{1}_{2\times2}$ and $-\mathbb{1}_{2\times2}$ correspond to the identity element $\mathbb{1}_{4\times4} \in SO_+(1,3)$. For fermions, this is significant, since the fermion wave function changes by an overall minus sign under a 360° rotation.

[31] The phases η_P, η_T, and η_C defined in this section should not be confused with the two-component spinor field $\eta(x)$.

hermiticity of P implies that $(P^*)_{\alpha\dot{\beta}} = P_{\beta\dot{\alpha}}$. By imposing $\mathscr{L}^P(\vec{x},t) = \mathscr{L}(-\vec{x},t)$, where $\mathscr{L}$ is the free-fermion Lagrangian [eq. (1.144)], the matrix P is determined and the possible values of η_P are restricted. Here, we follow the standard convention where the mass term is of the form $\mathscr{L}_m = -\frac{1}{2}m(\xi\xi + \xi^\dagger\xi^\dagger)$ with m real and nonnegative.[32] The invariance of the kinetic energy term [eq. (1.141)] determines $P_{\alpha\dot{\beta}} = \sigma^0_{\alpha\dot{\beta}}$ (up to an overall sign that can be absorbed into the definition of η_P). To derive this result, we note that any hermitian 2×2 matrix can be written as $P = a_\mu \sigma^\mu$. Invariance of the kinetic energy term then implies that $a_0 = \pm 1$ and $a_i = 0$. It is convenient to separate out an explicit factor of i in the definition of η_P, as we have done in eq. (1.186). Then, the invariance of the mass term requires $\eta_P = \eta_P^*$ or $\eta_P = \pm 1$, as shown in Section 1.8.

Time reversal transforms a $(\frac{1}{2},0)$ fermion into itself [and likewise for the $(0,\frac{1}{2})$ fermion]. Thus, it is convenient to define a hermitian matrix T such that

$$\mathcal{T}\xi_\alpha(x)\mathcal{T}^{-1} = \eta_T T_{\beta\dot{\alpha}}\xi^\beta(x_T)\,, \qquad \mathcal{T}\xi^\dagger_{\dot{\alpha}}(x)\mathcal{T}^{-1} = \eta_T^* T_{\alpha\dot{\beta}}\xi^{\dagger\dot{\beta}}(x_T)\,, \tag{1.187}$$

where $x_T \equiv (-t\,;\,\vec{x})$ and η_T is initially an arbitrary complex phase. Here, it should be noted that $\mathcal{T}\xi_\alpha(\vec{x},t)\mathcal{T}^{-1}$ transforms as $\xi^\dagger_{\dot{\alpha}}$ and $\mathcal{T}\xi^\dagger_{\dot{\alpha}}(\vec{x},t)\mathcal{T}^{-1}$ transforms as ξ_α, which explains the spinor index structure of eq. (1.187). This behavior is a consequence of the fact that $\mathcal{T}$ is an antiunitary operator that satisfies $\mathcal{T}z\mathcal{T}^{-1} = z^*$ for any complex number z. By imposing $\mathscr{L}^T(\vec{x},t) = \mathscr{L}(\vec{x},-t)$, where $\mathscr{L}$ is the free-fermion Lagrangian [eq. (1.144)], the matrix T is determined and the possible values of η_T are restricted. Invariance of the kinetic energy term [eq. (1.141)] determines $T_{\alpha\dot{\beta}} = \sigma^0_{\alpha\dot{\beta}}$ (up to an overall sign that can be absorbed into the definition of η_T). Invariance of the mass term requires $\eta_T = \eta_T^*$ or $\eta_T = \pm 1$, as shown in Section 1.9.

Finally, we examine charge conjugation. In a free-fermion theory, charge conjugation simply interchanges particles and antiparticles, so $\mathcal{C}^2 = 1$. In a theory of a single neutral two-component Majorana fermion (which is its own antiparticle), $\mathcal{C}\xi_\alpha\mathcal{C}^{-1} = \eta_C\xi_\alpha$. Then, $\mathcal{C}^2 = 1$ implies that $\eta_C^2 = 1$, and we conclude that $\eta_C = \pm 1$. In the free-fermion theory, charge conjugation is trivial (and one can choose $\eta_C = 1$). When interactions are included, the charge-conjugation symmetry may uniquely fix the sign of η_C. Charge conjugation is less trivial in a theory of charged fermions. Consider a theory of a pair of two-component fermion fields of equal (nonzero) mass. The corresponding free-fermion Lagrangian is given by eq. (1.151) with $m_1 = m_2$. As before, we assume that the phases of the fields have been chosen such that m_1 and m_2 are real and positive. In this case, the Lagrangian exhibits a global O(2) symmetry, $\xi_i \to C_{ij}\xi_j$, where $C \in \mathrm{O}(2)$. Corresponding to this symmetry is a hermitian conserved Noether current [eq. (1.157)] with a corresponding conserved charge Q. Under the action of charge conjugation, the eigenvalues of Q change sign, i.e., $\mathcal{C}Q\mathcal{C}^{-1} = -Q$. Hence it follows that

$$\mathcal{C}J^\mu\mathcal{C}^{-1} = -J^\mu\,. \tag{1.188}$$

We can use eq. (1.188) to fix the form of the charge-conjugation matrix C. To

[32] As previously noted, if $m = |m|e^{i\theta}$, then θ can be removed by appropriately rephasing the field ξ.

accomplish this, consider the transformation law of the fields under charge conjugation:

$$\mathcal{C}\xi_i(x)\mathcal{C}^{-1} = C_{ij}\xi_j(x)\,, \qquad \mathcal{C}\xi_i^\dagger(x)\mathcal{C}^{-1} = C_{ij}\xi_j^\dagger(x)\,, \tag{1.189}$$

where C is a real orthogonal 2×2 matrix. Using eq. (1.188) and the explicit form of J^μ [eq. (1.157)], it follows that $\det C = -1$. The most general transformation of this type is given by

$$\begin{pmatrix} \xi_1^C \\ \xi_2^C \end{pmatrix} = \begin{pmatrix} 1 & 0 \\ 0 & -1 \end{pmatrix} \begin{pmatrix} \cos\theta & \sin\theta \\ -\sin\theta & \cos\theta \end{pmatrix} \begin{pmatrix} \xi_1 \\ \xi_2 \end{pmatrix} . \tag{1.190}$$

However, one can always redefine the fermion fields to absorb the angle θ (while leaving the free-field Lagrangian unchanged). Thus, without loss of generality one can choose $C = \eta_C \sigma^3$, where $\eta_C = \pm 1$. Explicitly,

$$\mathcal{C}\xi_1(x)\mathcal{C}^{-1} = \eta_C\xi_1(x)\,, \qquad \mathcal{C}\xi_2(x)\mathcal{C}^{-1} = -\eta_C\xi_2(x)\,. \tag{1.191}$$

It is often more convenient to employ two-component fermion fields of definite (and opposite) charge, denoted by χ and η. More precisely, for a fermion of charge $+1$ (in arbitrary units), the fields χ and $\eta^\dagger$ correspond to charge $+1$ fields and the fields $\chi^\dagger$ and η correspond to charge -1 fields. The fields χ and η can be expressed in terms of ξ_1 and ξ_2 using eq. (1.152). Acting on χ and η, the charge-conjugation transformation [eq. (1.191)] is given by

$$\mathcal{C}\chi(x)\mathcal{C}^{-1} = \eta_C\eta(x)\,, \qquad \mathcal{C}\eta(x)\mathcal{C}^{-1} = \eta_C\chi(x)\,. \tag{1.192}$$

Free field theories are always separately invariant under P, T, and C. When interactions are included, this may no longer be the case. To test whether an interacting field theory is invariant under one or more of the discrete symmetries, one must exhibit some choice of the phases (η_P, η_T, and η_C) of the fields for which the Lagrangian of the interacting theory is invariant. However, even if none of the discrete symmetries is separately conserved, a deep theorem of quantum field theory asserts that the combined discrete symmetry of CPT must be conserved by all relativistic local (free or interacting) Lorentz-invariant quantum field theories (e.g., see [B72]). The CPT invariance of a quantum field theory can be checked by showing that the phases η_P, η_T, and η_C can always be chosen in a way that is consistent with CPT invariance. In particular, as shown in Ref. [30], the CPT invariance of the theory is consistent with the choice of phases such that

$$\eta_{CPT} \equiv \eta_C\eta_P\eta_T = 1\,, \tag{1.193}$$

for all two-component fermion fields, independently of the particle species. In particular, for a charged fermion represented by a pair of two-component fermion fields of opposite charge, χ and η,

$$\mathcal{CPT}\chi(x)(\mathcal{CPT})^{-1} = -i\chi^\dagger(-x)\,, \quad \mathcal{CPT}\eta(x)(\mathcal{CPT})^{-1} = -i\eta^\dagger(-x)\,. \tag{1.194}$$

We now examine P, T, and C transformations in more detail.

1.8 Parity Transformation of Two-Component Spinors

Under the parity transformation (or more precisely, the space-inversion transformation), $x_P^\mu = (\Lambda_P)^\mu{}_\nu x^\nu = (t\,;\,-\vec{x})$, where

$$\Lambda_P = \begin{pmatrix} 1 & 0 & 0 & 0 \\ 0 & -1 & 0 & 0 \\ 0 & 0 & -1 & 0 \\ 0 & 0 & 0 & -1 \end{pmatrix}. \tag{1.195}$$

Consider a theory of a single two-component fermion field $\xi(x)$. Under parity the two-component spinor transforms as

$$\mathcal{P}\xi_\alpha(x)\mathcal{P}^{-1} \equiv \xi^P_\alpha(x) = i\eta_P \sigma^0_{\alpha\dot\beta}\xi^{\dagger\dot\beta}(x_P)\,, \tag{1.196}$$

$$\mathcal{P}\xi^\alpha(x)\mathcal{P}^{-1} \equiv \xi^{P\alpha}(x) = -i\eta_P \overline{\sigma}^{0\dot\beta\alpha}\xi^\dagger_{\dot\beta}(x_P)\,, \tag{1.197}$$

$$\mathcal{P}\xi^\dagger_{\dot\alpha}(x)\mathcal{P}^{-1} \equiv \xi^{P\dagger}_{\dot\alpha}(x) = -i\eta_P^* \sigma^0_{\beta\dot\alpha}\xi^\beta(x_P)\,, \tag{1.198}$$

$$\mathcal{P}\xi^{\dagger\dot\alpha}(x)\mathcal{P}^{-1} \equiv \xi^{P\dagger\dot\alpha}(x) = i\eta_P^* \overline{\sigma}^{0\dot\alpha\beta}\xi_\beta(x_P)\,, \tag{1.199}$$

where $|\eta_P| = 1$. Equation (1.197) is obtained from eq. (1.196) by raising the indices [see eq. (1.45)] and using eq. (1.84). Equations (1.198) and (1.199) are obtained from eqs. (1.196) and (1.197) by hermitian conjugation, respectively. Note that when the parity transformation is applied twice, we obtain

$$(\xi^P)^P = -\xi\,, \qquad (\xi^{\dagger P})^P = -\xi^\dagger\,. \tag{1.200}$$

This is consistent with the more general result that, for any neutral self-conjugate spin-s field, $\mathcal{P}^2 = (-1)^{2s}$.

If one redefines $\xi \to e^{i\theta}\xi$ [and $\xi^\dagger \to e^{-i\theta}\xi^\dagger$], then $\eta_P \to e^{2i\theta}\eta_P$. If the mass of the fermion is nonzero,[33] we may restrict the choice of the phase η_P by establishing a convention in which the mass parameter m in the Lagrangian of eq. (1.144) is taken to be real and positive. Then, $\eta_P = \pm 1$ as a consequence of the parity invariance of $\xi\xi + \xi^\dagger\xi^\dagger$, as we now demonstrate.[34]

To show that the Lagrangian of eq. (1.144) is invariant under parity, we investigate the transformation properties of the scalar bilinear covariants. For $\xi_1\xi_2$ and $\xi_1^\dagger\xi_2^\dagger$ we obtain

$$\xi_1^P(x)\xi_2^P(x) = \eta_{P1}\eta_{P2}\xi_1^\dagger(x_P)\xi_2^\dagger(x_P)\,, \tag{1.201}$$

$$\xi_1^{P\dagger}(x)\xi_2^{P\dagger}(x) = (\eta_{P1}\eta_{P2})^*\xi_1(x_P)\xi_2(x_P)\,, \tag{1.202}$$

[33] For massless fermions, there are no additional restrictions in the choice of phases that enter the discrete symmetry transformation laws. However, in order to have a continuous $m \to 0$ limit, we will choose these phases according to the restrictions (if any) of the massive theory.

[34] The standard phase convention in the literature (and in textbooks that treat such things carefully; e.g., see [B136]) absorbs the factor of i into η_P in eqs. (1.196)–(1.199). In this convention, $\eta_P = \pm i$. In this chapter, we have chosen the simpler procedure of making the factor of i explicit so that in our convention $\eta_P = \pm 1$.

where we have used $\sigma^0_{\gamma\dot{\alpha}}\overline{\sigma}^{0\dot{\alpha}\beta} = \delta^\beta_\gamma$. As promised, $\xi\xi + \xi^\dagger\xi^\dagger$ transforms as a scalar for $\eta_P = \pm 1$. In this case, the hermitian linear combination $i(\xi\xi - \xi^\dagger\xi^\dagger)$ transforms as a pseudoscalar.

The Lorentz vector bilinear covariants transform as

$$\xi_1^{P\dagger}(x)\overline{\sigma}^\mu\xi_2^P(x) = \eta^*_{P1}\eta_{P2}(\Lambda_P)^\mu{}_\nu\xi_1(x_P)\sigma^\nu\xi_2^\dagger(x_P)\,, \tag{1.203}$$

$$\xi_1^P(x)\sigma^\mu\xi_2^{P\dagger}(x) = \eta_{P1}\eta^*_{P2}(\Lambda_P)^\mu{}_\nu\xi_1^\dagger(x_P)\overline{\sigma}^\nu\xi_2(x_P)\,, \tag{1.204}$$

where Λ_P is given in eq. (1.195).[35] In deriving these results, we have made use of[36]

$$\overline{\sigma}^{0\dot{\beta}\alpha}\,\sigma^\mu_{\alpha\dot{\gamma}}\,\overline{\sigma}^{0\dot{\gamma}\beta} = (\Lambda_P)^\mu{}_\nu\,\overline{\sigma}^{\nu\dot{\beta}\beta}\,, \tag{1.205}$$

$$\sigma^0_{\beta\dot{\alpha}}\,\overline{\sigma}^{\mu\dot{\alpha}\gamma}\,\sigma^0_{\gamma\dot{\beta}} = (\Lambda_P)^\mu{}_\nu\,\sigma^\nu_{\beta\dot{\beta}}\,. \tag{1.206}$$

The kinetic energy term of the action [see eq. (1.141)] is therefore a scalar under parity[37] (independently of the choice of η_P). This is easily deduced from eq. (1.203) by noting that $\partial_\mu = (\Lambda_P)^\nu{}_\mu\partial^P_\nu$ (where $\partial^P_\nu \equiv \partial/\partial x^\nu_P$ is the parity-transformed derivative).

For the Lorentz tensor bilinear covariants, we have

$$\xi_1^P(x)\sigma^{\mu\nu}\xi_2^P(x) = \eta_{P1}\eta_{P2}(\Lambda_P)^\mu{}_\rho(\Lambda_P)^\nu{}_\tau\xi_1^\dagger(x_P)\overline{\sigma}^{\rho\tau}\xi_2^\dagger(x_P)\,, \tag{1.207}$$

$$\xi_1^{P\dagger}(x)\overline{\sigma}^{\mu\nu}\xi_2^{P\dagger}(x) = \eta^*_{P1}\eta^*_{P2}(\Lambda_P)^\mu{}_\rho(\Lambda_P)^\nu{}_\tau\xi_1(x_P)\sigma^{\rho\tau}\xi_2(x_P)\,. \tag{1.208}$$

In deriving these results, we have made use of

$$\overline{\sigma}^{0\dot{\beta}\alpha}\,\sigma^{\mu\nu}{}_\alpha{}^\gamma\,\sigma^0_{\gamma\dot{\alpha}} = (\Lambda_P)^\mu{}_\rho(\Lambda_P)^\nu{}_\tau\,\overline{\sigma}^{\rho\tau\dot{\beta}}{}_{\dot{\alpha}}\,, \tag{1.209}$$

$$\sigma^0_{\beta\dot{\alpha}}\,\overline{\sigma}^{\mu\nu\dot{\alpha}}{}_{\dot{\gamma}}\,\overline{\sigma}^{0\dot{\gamma}\alpha} = (\Lambda_P)^\mu{}_\rho(\Lambda_P)^\nu{}_\tau\,\sigma^{\rho\tau}{}_\beta{}^\alpha\,. \tag{1.210}$$

In theories with multiple two-component fermion fields with no global symmetries, the parity properties of the kth fermion are given by eqs. (1.196)–(1.199), with $(\eta_P)_k = \pm 1$ (in the convention where all mass parameters are real). When interactions are included, then the theory is parity-conserving if there is some choice of the $(\eta_P)_k$ such that the action is invariant under parity. If there is an internal global symmetry ($\xi_i \to U_{ij}\xi_j$) under which the Lagrangian is invariant, then there are relations among the parity transformations of the fermion fields. Here, we examine the simplest case of two mass-degenerate fermion fields. As shown in Section 1.7, a theory with a pair of mass-degenerate two-component fermion fields possesses a conserved charge Q [see eq. (1.157)]. This implies that one can choose linear combinations of the fermions that are eigenstates of Q. In particular, χ and η [with the

[35] Numerically, $(\Lambda_P)^\mu{}_\nu = g_{\mu\nu}$, and many books therefore employ $g_{\mu\mu}$ (no sum over μ) in the parity transformation of the bilinear covariants. We prefer the more accurate notation above.

[36] Roughly speaking, the factor of $(\Lambda_P)^\mu{}_\nu$ converts σ^μ into $\overline{\sigma}^\mu$, while the spinor index structure is preserved by the multiplication of the appropriate factors of the identity matrix (either σ^0 or $\overline{\sigma}^0$).

[37] Under parity, the Lagrangian satisfies $\mathcal{P}\mathscr{L}(x)\mathcal{P}^{-1} = \mathscr{L}(x_P)$. Integrating this result to get the action $S \equiv \int \mathscr{L}\, d^4x$ yields $\mathcal{P}S\mathcal{P}^{-1} = S$ as desired.

properties noted in eq. (1.192)] are states of definite (and opposite) charge. The transformation of the fields χ, η, $\chi^\dagger$, and $\eta^\dagger$ under parity are given by

$$\mathcal{P}\chi_\alpha(x)\mathcal{P}^{-1} \equiv \chi^P_\alpha(x) = i\eta_P\sigma^0_{\alpha\dot\beta}\eta^{\dagger\dot\beta}(x_P)\,, \tag{1.211}$$

$$\mathcal{P}\eta_\alpha(x)\mathcal{P}^{-1} \equiv \eta^P_\alpha(x) = i\eta^*_P\sigma^0_{\alpha\dot\beta}\chi^{\dagger\dot\beta}(x_P)\,, \tag{1.212}$$

$$\mathcal{P}\chi^{\dagger\dot\alpha}(x)\mathcal{P}^{-1} \equiv \chi^{P\dagger\dot\alpha}(x) = i\eta^*_P\overline{\sigma}^{0\dot\alpha\beta}\eta_\beta(x_P)\,, \tag{1.213}$$

$$\mathcal{P}\eta^{\dagger\dot\alpha}(x)\mathcal{P}^{-1} \equiv \eta^{P\dagger\dot\alpha}(x) = i\eta_P\overline{\sigma}^{0\dot\alpha\beta}\chi_\beta(x_P)\,. \tag{1.214}$$

That is, the phases that appear in the parity transformation laws of two-component fermion fields of opposite charge are complex conjugates of each other. This ensures that the mass term of the free Lagrangian, $m(\chi\eta + \chi^\dagger\eta^\dagger)$ with m real, is parity invariant. If we return to the case of a single uncharged two-component fermion by taking $\chi = \eta \equiv \xi$ in the transformation laws above [eqs. (1.211) and (1.212)], then consistency of the χ and η transformation laws implies that $\eta_P = \eta^*_P$ as previously noted. In the case of the charged fermion pair, the phase η_P is not constrained. However, we can always choose to work in another basis. For example, if we work in terms of the fields ξ_1 and ξ_2 [eq. (1.152)], then it follows that

$$\mathcal{P}\xi_{1\alpha}(x)\mathcal{P}^{-1} = (\mathrm{Re}\,\eta_P)i\sigma^0_{\alpha\dot\beta}\xi_1^{\dagger\dot\beta}(x_P) - (\mathrm{Im}\,\eta_P)i\sigma^0_{\alpha\dot\beta}\xi_2^{\dagger\dot\beta}(x_P)\,, \tag{1.215}$$

$$\mathcal{P}\xi_{2\alpha}(x)\mathcal{P}^{-1} = (\mathrm{Im}\,\eta_P)i\sigma^0_{\alpha\dot\beta}\xi_1^{\dagger\dot\beta}(x_P) + (\mathrm{Re}\,\eta_P)i\sigma^0_{\alpha\dot\beta}\xi_2^{\dagger\dot\beta}(x_P)\,. \tag{1.216}$$

If we choose $\eta_P = \eta^*_P$, then ξ_1 and ξ_2 have simple transformation properties under parity:

$$\mathcal{P}\xi_{i\alpha}(x)\mathcal{P}^{-1} = i\eta_P\sigma^0_{\alpha\dot\beta}\xi_i^{\dagger\dot\beta}(x_P)\,, \qquad i = 1, 2\,. \tag{1.217}$$

That is, the fields ξ_1 and ξ_2 obey the parity transformation laws of a single two-component fermion field [eqs. (1.196)–(1.199)], with the same phase η_P in each case. Had one chosen $\eta_P \neq \eta^*_P$ in eqs. (1.211) and (1.212), then the parity transformation laws of ξ_1 and ξ_2 would have been more complicated [eqs. (1.215) and (1.216)]. However in this case, one is free to make a further SO(2) rotation to transform ξ_1 and ξ_2 into new fields that exhibit the simpler parity transformation laws of eq. (1.217).

We noted previously that $\mathcal{P}^2 = -1$ for a neutral self-conjugate spin-1/2 fermion. However, for a charged fermion, eqs. (1.211) and (1.212) imply that

$$\mathcal{P}^2\chi_\alpha(x)(\mathcal{P}^2)^{-1} = -\eta_P^2\chi_\alpha(x)\,, \tag{1.218}$$

$$\mathcal{P}^2\eta^{\dagger\dot\alpha}(x)(\mathcal{P}^2)^{-1} = -\eta_P^2\eta^{\dagger\dot\alpha}(x)\,. \tag{1.219}$$

That is, $\mathcal{P}^2 = -\eta_P^2$ when acting on χ and $\eta^\dagger$ (these are fields with the same value of the conserved charge). Note that unless $\eta_P = \pm 1$ or $\pm i$, $\mathcal{P}^2$ is not an element of the Lorentz group! Some authors insist that $\mathcal{P}^2 = \pm 1$ when applied to charged fermions, although there is nothing inconsistent with a more general phase. In fact, for any choice of η_P, one can always redefine $\mathcal{P}$ (by multiplication by an appropriate gauge transformation) to have any desired behavior. Nevertheless, there is some motivation for choosing $\eta_P = \pm 1$ so that $\mathcal{P}^2 = -1$ as in the case of the neutral self-conjugate spin-1/2 fermion.

One can now work out the parity properties of bilinear covariants constructed out of the fields of a charged fermion pair. The results are listed in Table C.2.

1.9 Time-Reversal of Two-Component Spinors

Under the time-reversal transformation (or more precisely, the time-inversion transformation), $x_T^\mu = (\Lambda_T)^\mu{}_\nu x^\nu = (-t\,;\,\vec{x})$, where

$$\Lambda_T = \begin{pmatrix} -1 & 0 & 0 & 0 \\ 0 & 1 & 0 & 0 \\ 0 & 0 & 1 & 0 \\ 0 & 0 & 0 & 1 \end{pmatrix} . \tag{1.220}$$

Since the time-reversal transformation is antiunitary, it requires careful consideration. Time reversal is treated differently in the first-quantized and the second-quantized theories.[38] Here, we shall only consider the second-quantized version of time reversal, which is governed by an antiunitary operator $\mathcal{T}$ that acts on the Hilbert space. Consider a theory of a single two-component fermion field $\xi(x)$. Under time reversal, the two-component spinor transforms as[39]

$$\mathcal{T}\xi_\alpha(x)\mathcal{T}^{-1} \equiv \xi^{t\dagger}_{\dot\alpha}(x) = \eta_T \sigma^0_{\beta\dot\alpha}\xi^\beta(x_T)\,, \tag{1.221}$$

$$\mathcal{T}\xi^\alpha(x)\mathcal{T}^{-1} \equiv \xi^{t\dagger\dot\alpha}(x) = -\eta_T \overline{\sigma}^{0\dot\alpha\beta}\xi_\beta(x_T)\,, \tag{1.222}$$

$$\mathcal{T}\xi^\dagger_{\dot\alpha}(x)\mathcal{T}^{-1} \equiv \xi^t_\alpha(x) = \eta_T^* \sigma^0_{\alpha\dot\beta}\xi^{\dagger\dot\beta}(x_T)\,, \tag{1.223}$$

$$\mathcal{T}\xi^{\dagger\dot\alpha}(x)\mathcal{T}^{-1} \equiv \xi^{t\alpha}(x) = -\eta_T^* \overline{\sigma}^{0\dot\beta\alpha}\xi^\dagger_{\dot\beta}(x_T)\,, \tag{1.224}$$

where $|\eta_T| = 1$. The notation requires some explanation. Due to the antiunitary nature of the operator $\mathcal{T}$, the two-component spinor quantity $\mathcal{T}\xi_\alpha(x)\mathcal{T}^{-1}$ transforms as the complex conjugate of a $(\frac{1}{2}, 0)$ spinor with a lower spinor index. Thus, we shall denote this quantity by $\xi^{t\dagger}_{\dot\alpha}(x)$. Equation (1.222) is obtained from eq. (1.221) by first raising the indices [eq. (1.49)] and then using eq. (1.84). Equations (1.223) and (1.224) are obtained from eqs. (1.221) and (1.222) by hermitian conjugation, respectively.

The time-reversal operator also satisfies the following property:

$$\mathcal{T} z \mathcal{T}^{-1} = z^*\,, \tag{1.225}$$

[38] The wave functions of a first-quantized theory are replaced by quantum field operators in the corresponding second-quantized theory. In contrast to time reversal, the first-quantized and second-quantized treatments of parity and charge conjugation are identical. See [B135] for a more detailed discussion.

[39] The choice of the superscript t in eqs. (1.221)–(1.224) instead of T has been made in order to avoid confusion with the symbol for the matrix transpose.

for any complex number z. Thus, when time reversal is applied twice in succession, one can use $\mathcal{T}\eta_T\mathcal{T}^{-1} = \eta_T^*$ and $|\eta_T| = 1$ to obtain

$$(\xi^t)^t = -\xi\,, \qquad (\xi^{\dagger\,t})^t = -\xi^\dagger\,. \tag{1.226}$$

This is consistent with the more general result that

$$\mathcal{T}^2 = (-1)^{2s} \tag{1.227}$$

for any (charged or neutral) spin-s quantum field.

If one redefines $\xi \to e^{i\theta}\xi$ [and $\xi^\dagger \to e^{-i\theta}\xi^\dagger$], then $\eta_T \to e^{-2i\theta}\eta_T$. As before, we may restrict the choice of the phase η_T by choosing the convention in which the mass parameter m in the Lagrangian of eq. (1.144) is taken to be real and nonnegative. Then, as shown below, $\eta_T = \pm 1$ as a consequence of the invariance of $\xi\xi + \xi^\dagger\xi^\dagger$ under time reversal.

To show that the Lagrangian of eq. (1.144) is invariant under time reversal, we investigate the transformation properties of the scalar bilinear covariants. For $\xi_1\xi_2$ and $\xi_1^\dagger\xi_2^\dagger$ we obtain

$$\mathcal{T}\xi_1(x)\xi_2(x)\mathcal{T}^{-1} = \mathcal{T}\xi_1^\alpha\mathcal{T}^{-1}\mathcal{T}\xi_{2\alpha}\mathcal{T}^{-1} = \xi_1^{t\dagger\dot{\alpha}}(x)\xi_{2\dot{\alpha}}^{t\dagger}(x) = \eta_{T1}\eta_{T2}\xi_1(x_T)\xi_2(x_T) \tag{1.228}$$

and

$$\mathcal{T}\xi_1^\dagger(x)\xi_2^\dagger(x)\mathcal{T}^{-1} = (\eta_{T1}\eta_{T2})^*\xi_1^\dagger(x_T)\xi_2^\dagger(x_T)\,. \tag{1.229}$$

Thus, $\xi\xi + \xi^\dagger\xi^\dagger$ is even under time reversal if $\eta_T = \pm 1$. In this case, the hermitian linear combination $i(\xi\xi - \xi^\dagger\xi^\dagger)$ is odd under time reversal (after noting that $\mathcal{T}i\mathcal{T}^{-1} = -i$).

The Lorentz vector bilinear covariants transform as

$$\mathcal{T}\xi_{1\dot{\alpha}}^\dagger\overline{\sigma}^{\mu\dot{\alpha}\alpha}\xi_{2\alpha}\mathcal{T}^{-1} = (\mathcal{T}\xi_{1\dot{\alpha}}^\dagger\mathcal{T}^{-1})(\mathcal{T}\overline{\sigma}^{\mu\dot{\alpha}\alpha}\mathcal{T}^{-1})(\mathcal{T}\xi_{2\alpha}\mathcal{T}^{-1})\,, \tag{1.230}$$

$$\mathcal{T}\xi_1^\alpha\sigma^\mu_{\alpha\dot{\alpha}}\xi_2^{\dagger\dot{\alpha}}\mathcal{T}^{-1} = (\mathcal{T}\xi_1^\alpha\mathcal{T}^{-1})(\mathcal{T}\sigma^\mu_{\alpha\dot{\alpha}}\mathcal{T}^{-1})(\mathcal{T}\xi_2^{\dot{\alpha}}\mathcal{T}^{-1})\,. \tag{1.231}$$

In light of eqs. (1.221)–(1.224), in order to have the spinor indices properly matched in eqs. (1.230) and (1.231), one must make use of the following results:

$$\mathcal{T}\sigma^\mu_{\alpha\dot{\beta}}\mathcal{T}^{-1} = \sigma^\mu_{\beta\dot{\alpha}}\,, \qquad \mathcal{T}\overline{\sigma}^{\mu\dot{\alpha}\beta}\mathcal{T}^{-1} = \overline{\sigma}^{\mu\dot{\beta}\alpha}\,, \tag{1.232}$$

which follow from $\mathcal{T}\vec{\boldsymbol{\sigma}}\mathcal{T}^{-1} = \vec{\boldsymbol{\sigma}}^* = \vec{\boldsymbol{\sigma}}^{\mathsf{T}}$. We then obtain

$$\begin{aligned}
\mathcal{T}\xi_1^\dagger(x)\overline{\sigma}^\mu\xi_2(x)\mathcal{T}^{-1} &= \eta_{T1}^*\eta_{T2}\xi_1^{\dagger\dot{\gamma}}(x_T)\sigma^0_{\alpha\dot{\gamma}}(\overline{\sigma}^{\mu\beta\alpha})\sigma^0_{\delta\dot{\beta}}\xi_2^\delta(x_T) \\
&= -\eta_{T1}^*\eta_{T2}\xi_2^\delta(x_T)\sigma^0_{\delta\dot{\beta}}(\overline{\sigma}^{\mu\beta\alpha})\sigma^0_{\alpha\dot{\gamma}}\xi_1^{\dagger\dot{\gamma}}(x_T) \\
&= \eta_{T1}^*\eta_{T2}(\Lambda_T)^\mu{}_\nu\xi_2(x_T)\sigma^\nu\xi_1^\dagger(x_T) \\
&= -\eta_{T1}^*\eta_{T2}(\Lambda_T)^\mu{}_\nu\xi_1^\dagger(x_T)\overline{\sigma}^\nu\xi_2(x_T)\,.
\end{aligned} \tag{1.233}$$

The minus sign in the second step arises because of the anticommutation of the spinors. In the third step, we have made use of

$$\overline{\sigma}^{0\dot{\beta}\alpha}\,\sigma^{\mu}_{\alpha\dot{\gamma}}\,\overline{\sigma}^{0\dot{\gamma}\beta} = -(\Lambda_T)^{\mu}{}_{\nu}\,\overline{\sigma}^{\nu\dot{\beta}\beta}\,, \tag{1.234}$$

$$\sigma^{0}_{\beta\dot{\alpha}}\,\overline{\sigma}^{\mu\dot{\alpha}\gamma}\,\sigma^{0}_{\gamma\dot{\beta}} = -(\Lambda_T)^{\mu}{}_{\nu}\,\sigma^{\nu}_{\beta\dot{\beta}}\,, \tag{1.235}$$

which follow from eqs. (1.205) and (1.206) after noting that $(\Lambda_T)^{\mu}{}_{\nu} = -(\Lambda_P)^{\mu}{}_{\nu}$. Finally, applying eq. (1.107) yields the final result given in eq. (1.233). Similarly,

$$\mathcal{T}\xi_1(x)\sigma^{\mu}\xi_2^{\dagger}(x)\mathcal{T}^{-1} = -\eta_{T1}\eta^{*}_{T2}(\Lambda_T)^{\mu}{}_{\nu}\xi_1(x_T)\sigma^{\nu}\xi_2^{\dagger}(x_T)\,. \tag{1.236}$$

The kinetic energy term of the action [see eq. (1.141)] is therefore a scalar under time reversal (independently of the choice of η_T). This is easily deduced from eq. (1.233) by noting that $\partial_\mu = (\Lambda_T)^{\nu}{}_{\mu}\partial^t_{\nu}$ (where $\partial^t_\nu = \partial/\partial x^{\nu}_T$ is the time-reversal-transformed derivative) and remembering to complex conjugate the factor of i.

For the Lorentz tensor bilinear covariants, we examine

$$\mathcal{T}\xi_1^{\alpha}(\sigma^{\mu\nu})_{\alpha}{}^{\beta}\xi_{2\beta}\mathcal{T}^{-1} = (\mathcal{T}\xi_1^{\alpha}\mathcal{T}^{-1})(\mathcal{T}(\sigma^{\mu\nu})_{\alpha}{}^{\beta}\mathcal{T}^{-1})(\mathcal{T}\xi_{2\beta}\mathcal{T}^{-1})\,, \tag{1.237}$$

and an analogous equation involving $\overline{\sigma}^{\mu\nu}$. We then employ results similar to those of eq. (1.232):

$$\mathcal{T}(\sigma^{\mu\nu})_{\alpha}{}^{\beta}\mathcal{T}^{-1} = (\overline{\sigma}^{\mu\nu})^{\dot{\beta}}{}_{\dot{\alpha}}\,, \qquad \mathcal{T}(\overline{\sigma}^{\mu\nu})^{\dot{\alpha}}{}_{\dot{\beta}}\mathcal{T}^{-1} = (\sigma^{\mu\nu})_{\beta}{}^{\alpha}\,. \tag{1.238}$$

The rest of the derivation is straightforward, and the end result is

$$\mathcal{T}\xi_1(x)\sigma^{\mu\nu}\xi_2(x)\mathcal{T}^{-1} = -\eta_{T1}\eta_{T2}(\Lambda_T)^{\mu}{}_{\rho}(\Lambda_T)^{\nu}{}_{\tau}\xi_1(x_T)\sigma^{\rho\tau}\xi_2(x_T)\,, \tag{1.239}$$

$$\mathcal{T}\xi_1^{\dagger}(x)\overline{\sigma}^{\mu\nu}\,\xi_2^{\dagger}(x)\mathcal{T}^{-1} = -\eta^{*}_{T1}\eta^{*}_{T2}(\Lambda_T)^{\mu}{}_{\rho}(\Lambda_T)^{\nu}{}_{\tau}\xi_1^{\dagger}(x_T)\overline{\sigma}^{\rho\tau}\,\xi_2^{\dagger}(x_T)\,, \tag{1.240}$$

where we have used eq. (1.133) and the following results:

$$\overline{\sigma}^{0\dot{\beta}\alpha}\,\sigma^{\mu\nu}{}_{\alpha}{}^{\gamma}\,\sigma^{0}_{\gamma\dot{\alpha}} = (\Lambda_T)^{\mu}{}_{\rho}(\Lambda_T)^{\nu}{}_{\tau}\,\overline{\sigma}^{\rho\tau\dot{\beta}}{}_{\dot{\alpha}}\,, \tag{1.241}$$

$$\sigma^{0}_{\beta\dot{\alpha}}\,\overline{\sigma}^{\mu\nu\dot{\alpha}}{}_{\dot{\gamma}}\,\overline{\sigma}^{0\dot{\gamma}\alpha} = (\Lambda_T)^{\mu}{}_{\rho}(\Lambda_T)^{\nu}{}_{\tau}\,\sigma^{\rho\tau}{}_{\beta}{}^{\alpha}\,, \tag{1.242}$$

which follow immediately from eqs. (1.209) and (1.210).

In theories with multiple two-component fermion fields with no global symmetries, the time-reversal properties of each fermion are given by eqs. (1.221)–(1.224), with $\eta_T = \pm 1$ (in the convention where all mass parameters are real). When interactions are included, if there is some choice of the η_T such that the action is invariant under time reversal, then the theory is time reversal invariant. If there is an internal global symmetry ($\xi_i \to U_{ij}\xi_j$) under which the Lagrangian is invariant, then there are relations among the time-reversal transformations of the fermion fields. Here, we again examine the simplest case of two mass-degenerate fermion fields, ξ_1 and ξ_2, and employ the linear combinations χ and η [eq. (1.152)] corresponding to the fields of definite charge. The transformations of the fields χ, η, $\chi^{\dagger}$, and $\eta^{\dagger}$ under time reversal are given by

$$\mathcal{T}\chi_\alpha(x)\mathcal{T}^{-1} \equiv \chi^{t\dagger}_{\dot{\alpha}}(x) = \eta_T \sigma^0_{\beta\dot{\alpha}}\chi^\beta(x_T)\,, \tag{1.243}$$
$$\mathcal{T}\eta_\alpha(x)\mathcal{T}^{-1} \equiv \eta^{t\dagger}_{\dot{\alpha}}(x) = \eta^*_T \sigma^0_{\beta\dot{\alpha}}\eta^\beta(x_T)\,, \tag{1.244}$$
$$\mathcal{T}\chi^{\dagger\dot{\alpha}}(x)\mathcal{T}^{-1} \equiv \chi^{t\alpha}(x) = -\eta^*_T \overline{\sigma}^{0\dot{\beta}\alpha}\chi^\dagger_{\dot{\beta}}(x_T)\,, \tag{1.245}$$
$$\mathcal{T}\eta^{\dagger\dot{\alpha}}(x)\mathcal{T}^{-1} \equiv \eta^{t\alpha}(x) = -\eta_T \overline{\sigma}^{0\dot{\beta}\alpha}\eta^\dagger_{\dot{\beta}}(x_T)\,. \tag{1.246}$$

That is, the phases that appear in the time-reversal transformation laws of two-component fermion fields of opposite charge are complex conjugates of each other. If we apply the time-reversal operator twice,

$$\mathcal{T}^2\chi_\alpha(x)(\mathcal{T}^2)^{-1} = -\chi_\alpha(x)\,, \qquad \mathcal{T}^2\eta_\alpha(x)(\mathcal{T}^2)^{-1} = -\eta_\alpha(x)\,, \tag{1.247}$$
$$\mathcal{T}^2\chi^{\dagger\dot{\alpha}}(x)(\mathcal{T}^2)^{-1} = -\chi^{\dagger\dot{\alpha}}(x)\,, \qquad \mathcal{T}^2\eta^{\dagger\dot{\alpha}}(x)(\mathcal{T}^2)^{-1} = -\eta^{\dagger\dot{\alpha}}(x)\,, \tag{1.248}$$

and so $\mathcal{T}^2 = -1$ for both neutral and charged spin-1/2 fermions [as noted in eq. (1.227)]. In contrast to the case of $\mathcal{P}^2$ considered in Section 1.8, the phase η_T drops out in the computation of $\mathcal{T}^2$ due to the antilinearity of the time-reversal operator.

Notice that if we return to the case of a single uncharged two-component fermion by taking $\chi = \eta \equiv \xi$ in the transformation laws above [eqs. (1.243) and (1.244)], then consistency of the χ and η transformation laws implies that $\eta_T = \eta^*_T$ as previously noted. In the case of the charged fermion pair, the phase η_T is not constrained. However, we can always choose to work in another basis. For example, if we work in terms of the fields ξ_1 and ξ_2 [eq. (1.152)], then it follows that

$$\mathcal{T}\xi_{1\alpha}(x)\mathcal{T}^{-1} = (\mathrm{Re}\,\eta_T)\sigma^0_{\beta\dot{\alpha}}\xi^\beta_1(x_T) - (\mathrm{Im}\,\eta_T)\sigma^0_{\beta\dot{\alpha}}\xi^\beta_2(x_T)\,, \tag{1.249}$$
$$\mathcal{T}\xi_{2\alpha}(x)\mathcal{T}^{-1} = -(\mathrm{Im}\,\eta_T)\sigma^0_{\beta\dot{\alpha}}\xi^\beta_1(x_T) - (\mathrm{Re}\,\eta_T)\sigma^0_{\beta\dot{\alpha}}\xi^\beta_2(x_T)\,. \tag{1.250}$$

If we choose $\eta_T = \eta^*_T$, then ξ_1 and ξ_2 have simple transformation properties under time reversal:

$$\mathcal{T}\xi_{1\alpha}(x)\mathcal{T}^{-1} = \eta_T\sigma^0_{\beta\dot{\alpha}}\xi^\beta_1(x_T)\,, \tag{1.251}$$
$$\mathcal{T}\xi_{2\alpha}(x)\mathcal{T}^{-1} = -\eta_T\sigma^0_{\beta\dot{\alpha}}\xi^\beta_2(x_T)\,. \tag{1.252}$$

That is, the fields ξ_1 and ξ_2 obey the standard time-reversal transformation laws of a single two-component fermion field [eqs. (1.221)–(1.224)], but with *opposite* sign phase factors η_T in each case. Had one chosen $\eta_T \neq \eta^*_T$ in eqs. (1.243) and (1.244), then the time-reversal transformation laws of ξ_1 and ξ_2 would have been more complicated [eqs. (1.249) and (1.250)]. But, as before, one is free to make a further SO(2) rotation to transform ξ_1 and ξ_2 into new fields that do exhibit the simpler time-reversal transformation laws [eq. (1.217)].

The time-reversal properties of bilinear covariants constructed out of the fields of a charged fermion pair are exhibited in Table C.2.

1.10 Charge Conjugation of Two-Component Spinors

Charge conjugation was introduced in Section 1.7. The charge-conjugation operator is a discrete operator that interchanges particles and their C-conjugates. Here, conjugation refers to some conserved charge operator Q, where $\mathcal{C}Q\mathcal{C}^{-1} = -Q$. The conjugate of the conjugate field is the original field, so that $\mathcal{C}^2 = 1$. For a single two-component fermion field,

$$\mathcal{C}\xi_\alpha(x)\mathcal{C}^{-1} \equiv \xi^C_\alpha(x) = \eta_C\xi_\alpha(x)\,, \tag{1.253}$$

$$\mathcal{C}\xi^\dagger_{\dot\alpha}(x)\mathcal{C}^{-1} \equiv \xi^{C\dagger}_{\dot\alpha}(x) = \eta^*_C\xi^\dagger_{\dot\alpha}(x)\,, \tag{1.254}$$

where $|\eta_C| = 1$. In this case, no conserved charge exists, so that charge conjugation is trivial. In particular, as noted in Section 1.7, the invariance of the mass term $\xi\xi + \xi^\dagger\xi^\dagger$ implies that $\eta^*_C = \eta_C$, and hence $\eta_C = \pm 1$.[40] The behavior of the bilinear covariants under C is simple:

$$\xi_1^{C(\dagger)}\mathcal{O}\,\xi_2^{C(\dagger)} = \eta_{C1}\eta_{C2}\xi_1^{(\dagger)}\mathcal{O}\,\xi_2^{(\dagger)}\,, \tag{1.255}$$

for any $\mathcal{O} = \mathbb{1}_{2\times 2}, \sigma_\mu, \overline{\sigma}_\mu, \sigma_{\mu\nu}$, or $\overline{\sigma}_{\mu\nu}$, where the superscript daggers inside the parentheses are either present or absent depending on the choice of $\mathcal{O}$.

In theories with multiple two-component fermion fields with no global symmetries, the charge-conjugation properties of each fermion are given by eqs. (1.253) and (1.254), with $\eta_C = \pm 1$. When interactions are included, if there is some choice of the η_C such that the action is invariant under charge conjugation, then the theory is charge conjugation invariant. If there is an internal global symmetry ($\xi_i \to U_i{}^j\xi_j$) under which the Lagrangian is invariant, then there are relations among the charge-conjugation transformations of the fermion fields. Here, we consider the simplest case of two mass-degenerate fermion fields, ξ_1 and ξ_2, and identify the conserved charge as $Q = \int J^0 d^3x$, where the conserved current J^μ is given in eq. (1.157). In terms of the linear combinations χ and η [eq. (1.152)] corresponding to the fields of definite charge, we define the charge-conjugation transformations as follows:

$$\mathcal{C}\chi_\alpha(x)\mathcal{C}^{-1} \equiv \chi^C_\alpha(x) = \eta_C\eta_\alpha(x)\,, \qquad \mathcal{C}\chi^\dagger_{\dot\alpha}(x)\mathcal{C}^{-1} \equiv \chi^{C\dagger}_{\dot\alpha}(x) = \eta^*_C\eta^\dagger_{\dot\alpha}(x)\,, \tag{1.256}$$

$$\mathcal{C}\eta_\alpha(x)\mathcal{C}^{-1} \equiv \eta^C_\alpha(x) = \eta^*_C\chi_\alpha(x)\,, \qquad \mathcal{C}\eta^\dagger_{\dot\alpha}(x)\mathcal{C}^{-1} \equiv \eta^{C\dagger}_{\dot\alpha}(x) = \eta_C\chi^\dagger_{\dot\alpha}(x)\,. \tag{1.257}$$

Notice that if we return to the case of a single uncharged two-component fermion by taking $\chi = \eta \equiv \xi$ in the transformation laws above [eqs. (1.256) and (1.257)], then consistency of the χ and η transformation laws implies that $\eta_C = \eta^*_C$ as previously noted. In the case of the charged fermion pair, the phase η_C is not constrained, i.e., $\mathcal{C}^2 = 1$ independently of the value of the phase η_C. However, we can always choose to work in another basis. For example, if we work in terms of the fields ξ_1 and ξ_2

[40] In a theory with just one fermion field, no physical quantity can depend on the sign of η_C since the fermion field must appear quadratically in the Lagrangian. So, without loss of generality, one can take $\eta_C = 1$.

[eq. (1.152)], then it follows that

$$\mathcal{C}\xi_1(x)\mathcal{C}^{-1} = (\mathrm{Re}\,\eta_C)\xi_1(x) + (\mathrm{Im}\,\eta_C)\xi_2(x)\,, \tag{1.258}$$

$$\mathcal{C}\xi_2(x)\mathcal{C}^{-1} = (\mathrm{Im}\,\eta_C)\xi_1(x) - (\mathrm{Re}\,\eta_C)\xi_2(x)\,. \tag{1.259}$$

If we choose $\eta_C = \eta_C^*$, then ξ_1 and ξ_2 have simple transformation properties under charge conjugation:

$$\mathcal{C}\xi_1(x)\mathcal{C}^{-1} = \eta_C\xi_1(x)\,, \qquad \mathcal{C}\xi_2(x)\mathcal{C}^{-1} = -\eta_C\xi_2(x)\,. \tag{1.260}$$

That is, the fields ξ_1 and ξ_2 obey the standard charge-conjugation transformation laws of a single two-component fermion field [eqs. (1.253) and (1.254)], but with *opposite* sign phase factors η_C in each case. Had one chosen $\eta_C \neq \eta_C^*$ in eqs. (1.256) and (1.257), then the charge-conjugation transformation laws of ξ_1 and ξ_2 would have been more complicated [eqs. (1.258) and (1.259)]. But once again, one is free to make a further SO(2) rotation to transform ξ_1 and ξ_2 into new fields that do exhibit the simpler charge-conjugation transformation laws [eq. (1.260)].

One can now work out the charge-conjugation properties of bilinear covariants constructed out of the fields of a charged fermion pair. The results are listed in Table C.2.

1.11 CP and CPT Conjugation of Two-Component Spinors

Given the results of the previous three sections, it is a simple matter to work out the effects of CP and CPT transformations on the two-component spinors and the bilinear covariants. Here, we focus on the case of two mass-degenerate fermion fields, ξ_1 and ξ_2, with the associated conserved charge J^μ given in eq. (1.157). In terms of the linear combinations χ and η [eq. (1.152)] corresponding to the fields of definite charge, the CP transformations of the two-component spinors are defined by first applying $\mathcal{P}$ and then applying $\mathcal{C}$ as follows:

$$\mathcal{CP}\,\chi_\alpha(x)(\mathcal{CP})^{-1} = i\eta_{CP}\sigma^0_{\alpha\dot\beta}\chi^{\dagger\dot\beta}(x_P)\,, \tag{1.261}$$

$$\mathcal{CP}\,\eta_\alpha(x)(\mathcal{CP})^{-1} = i\eta^*_{CP}\sigma^0_{\alpha\dot\beta}\eta^{\dagger\dot\beta}(x_P)\,, \tag{1.262}$$

where $\eta_{CP} \equiv \eta_C\eta_P$. Again, observe that if we return to the case of a single uncharged two-component fermion by taking $\chi = \eta \equiv \xi$ in the transformation laws above [eqs. (1.261) and (1.262)], then consistency of the χ and η transformation laws implies that $\eta_{CP} = \eta^*_{CP}$ as expected. In the case of the charged fermion pair, the phase η_{CP} is not constrained. However, if we work in terms of the fields ξ_1 and ξ_2 [eq. (1.152)] and choose $\eta_{CP} = \pm 1$ (as required for uncharged two-component fields), then as before it follows that the ξ_i have simple transformation properties:

$$\mathcal{CP}\xi_{1\alpha}(x)(\mathcal{CP})^{-1} = i\eta_{CP}\sigma^0_{\alpha\dot{\beta}}\xi_1^{\dagger\dot{\beta}}(x_P)\,, \tag{1.263}$$

$$\mathcal{CP}\xi_{2\alpha}(x)(\mathcal{CP})^{-1} = -i\eta_{CP}\sigma^0_{\alpha\dot{\beta}}\xi_2^{\dagger\dot{\beta}}(x_P)\,. \tag{1.264}$$

That is, the fields ξ_1 and ξ_2 obey the standard CP transformation laws of a single two-component fermion field [eqs. (1.261) and (1.262)], but with *opposite* sign phase factors η_{CP} in each case. This is the origin of the oft-quoted statement in the literature that a theory of two mass-degenerate Majorana fermions of opposite CP quantum numbers can be combined into a single Dirac fermion.

In general, when acting on charged fermion fields, $\mathcal{CP} \neq \mathcal{PC}$ (e.g., see [B133]). A simple computation yields $\mathcal{PC} = (\eta_P^*)^2\mathcal{CP}$ when acting on the χ and $\eta^\dagger$ fields.[41] However, for self-conjugate spin-1/2 fermion fields, $\mathcal{PC} = \mathcal{CP}$ since $\eta_P = \pm 1$. This provides another motivation for the choice of $\eta^*_{CP} = \eta_{CP}$ above in the case of charged fermion fields. One can also examine the behavior of the fermion fields under the other possible products of $\mathcal{P}$, $\mathcal{T}$, and $\mathcal{C}$. Again, the order of the operators can be significant. For example, $\mathcal{TC} = (\eta_C^*\eta_T^*)^2\mathcal{CT}$ and $\mathcal{TP} = -(\eta_P^*)^2\mathcal{PT}$ when acting on the χ and $\eta^\dagger$ fields, due to the antilinearity of the time-reversal operator. For self-conjugate spin-1/2 fermion fields, $\mathcal{TC} = \mathcal{CT}$ and $\mathcal{TP} = -\mathcal{PT}$ since η_P, η_T, and η_C are real (and equal to either ± 1). One is also free to employ these phase conventions in the case of charged fermion fields.

Finally, we consider the effect of a CPT transformation.[42] The corresponding transformation laws under CPT conjugation take on even simpler forms:

$$\mathcal{CPT}\chi_\alpha(x)(\mathcal{CPT})^{-1} = -i\eta_{CPT}\chi^\dagger_{\dot{\alpha}}(-x)\,, \tag{1.265}$$

$$\mathcal{CPT}\eta_\alpha(x)(\mathcal{CPT})^{-1} = -i\eta^*_{CPT}\eta^\dagger_{\dot{\alpha}}(-x)\,, \tag{1.266}$$

where $\eta_{CPT} \equiv \eta_C\eta_P\eta_T$ and $-x \equiv (-t\,;\,-\vec{x})$. Again, we observe that, when $\chi = \eta \equiv \xi$, consistency of the CPT transformation laws implies that for self-conjugate neutral fermion fields, $\eta_{CPT} = \pm 1$. Note that when $\mathcal{CPT}$ is applied twice,[43]

$$(\mathcal{CPT})^2\chi_\alpha(x)[(\mathcal{CPT})^2]^{-1} = -(\eta^*_{CPT})^2\chi_\alpha(x)\,, \tag{1.267}$$

$$(\mathcal{CPT})^2\eta_\alpha(x)[(\mathcal{CPT})^2]^{-1} = -(\eta_{CPT})^2\eta_\alpha(x)\,. \tag{1.268}$$

Consequently, for self-conjugate neutral spin-1/2 fermion fields, it follows that $(\mathcal{CPT})^2 = -1$. As before, if we also choose $\eta_{CPT} = \pm 1$ in the case of charged fields χ and η, then one obtains simple CPT-transformation laws when the charged fermion fields are rewritten in terms of ξ_1 and ξ_2 [eq. (1.152)]:

$$\mathcal{CPT}\xi_{i\alpha}(x)(\mathcal{CPT})^{-1} = -i\eta_{CPT}\xi^\dagger_{i\dot{\alpha}}(-x)\,, \qquad i = 1,2\,. \tag{1.269}$$

[41] When acting on the $\chi^\dagger$ and η fields (whose conserved charge is opposite to that of χ and $\eta^\dagger$), one must employ the complex conjugate of the phases above.

[42] In principle, there are six possible orderings of $\mathcal{P}$, $\mathcal{T}$, and $\mathcal{C}$, where each ordering results in a product that differs by at most a phase factor that is easily determined from the results previously obtained above [B133].

[43] Using the results quoted above, when applied to the spin-1/2 fermions fields χ and $\eta^\dagger$ (see footnote 41), $(\mathcal{CPT})^2 = (\eta^*_{CPT})^2\mathcal{C}^2(\mathcal{PT})^2 = -(\eta^*_{CPT})^2$, where we have noted that $\mathcal{C}^2 = 1$ and $(\mathcal{PT})^2 = -1$ independently of the phase choices.

As previously noted [see eq. (1.193)], one is free to choose a phase convention such that, for any two-component fermion field, $\eta_{CPT} \equiv \eta_C \eta_P \eta_T = +1$.

More generally, one can show that, for an arbitrary interacting Lorentz-invariant local quantum field theory, CPT invariance is consistent with eq. (1.193) for any particle (bosonic or fermionic). Moreover, for a spin-s boson represented by a complex tensor quantum field $\phi_{\mu_1\mu_2\cdots\mu_s}(x)$ [B72],

$$\mathcal{CPT}\,\phi_{\mu_1\mu_2\cdots\mu_s}(x)(\mathcal{CPT})^{-1} = (-1)^s\,\phi^\dagger_{\mu_1\mu_2\cdots\mu_s}(-x)\,, \tag{1.270}$$

after imposing eq. (1.193) on the suitably defined P, T, and C phases of the bosonic fields (e.g., see Ref. [30]).

Consider the bilinear covariants of the form $\xi_i^{(\dagger)}\mathcal{O}\xi_j^{(\dagger)}$ where $\mathcal{O} = \mathbb{1}_{2\times 2}, \sigma_\mu, \overline{\sigma}_\mu, \sigma_{\mu\nu}$, or $\overline{\sigma}_{\mu\nu}$, where the superscript daggers inside the parentheses are either present or absent depending on the choice of $\mathcal{O}$. These can be used to construct hermitian bilinear quantities, denoted by B, B^μ, and $B^{\mu\nu}$, that are either scalar, vectors, or second-rank antisymmetric tensors with respect to proper Lorentz transformations, respectively. Then, starting from eq. (1.269) and using eqs. (1.232) and (1.238) and the anticommutativity of the fermion fields, it follows that

$$\mathcal{CPT}\,B(x)\,(\mathcal{CPT})^{-1} = B(-x)\,, \tag{1.271}$$

$$\mathcal{CPT}\,B^\mu(x)\,(\mathcal{CPT})^{-1} = -B^\mu(-x)\,, \tag{1.272}$$

$$\mathcal{CPT}\,B^{\mu\nu}(x)\,(\mathcal{CPT})^{-1} = B^{\mu\nu}(-x)\,, \tag{1.273}$$

provided that the phase η_{CPT} is chosen to be the same for *every* two-component fermion field (as noted above). Thus, we see that all Lorentz tensors of the same (integer) rank behave the same way under a CPT transformation [see eq. (1.270)]. This ensures that the Lagrangian for any local quantum field theory (which is a hermitian Lorentz scalar) is CPT invariant.

Exercises

1.1 (a) Consider an infinitesimal (proper orthochronous) Lorentz transformation parameterized by the rotation angle vector $\vec{\boldsymbol{\theta}}$ and the boost vector $\vec{\boldsymbol{\zeta}}$. For an infinitesimal boost, $\vec{\boldsymbol{\zeta}} \equiv \hat{\boldsymbol{v}}\tanh^{-1}\beta \simeq \beta\hat{\boldsymbol{v}} \equiv \vec{\boldsymbol{\beta}}$, since $\beta \ll 1$. Using eq. (1.6), show that the actions of the infinitesimal boosts and rotations on the space-time coordinates are

$$\text{Rotations:}\qquad \begin{cases} \vec{\boldsymbol{x}} \to \vec{\boldsymbol{x}}' \simeq \vec{\boldsymbol{x}} + (\vec{\boldsymbol{\theta}} \times \vec{\boldsymbol{x}})\,, \\ t \to t' \simeq t\,, \end{cases} \tag{1.274}$$

$$\text{Boosts:}\qquad \begin{cases} \vec{\boldsymbol{x}} \to \vec{\boldsymbol{x}}' \simeq \vec{\boldsymbol{x}} + \vec{\boldsymbol{\beta}}t\,, \\ t \to t' \simeq t + \vec{\boldsymbol{\beta}}\cdot\vec{\boldsymbol{x}}\,, \end{cases} \tag{1.275}$$

with analogous transformations for any contravariant four-vector.

(b) The 4×4 boost matrices $\vec{\boldsymbol{k}} = (k^1\,,\,k^2\,,\,k^3)$ are given explicitly by $(k^i)^\mu{}_\nu = i(\delta^i_\nu\, g^{\mu 0} - \delta^0_\nu\, g^{\mu i})$ [cf. eq. (1.4)]. Prove that the Lorentz transformation corresponding to a pure boost in the direction $\hat{\boldsymbol{v}}$ is given by

$$\Lambda \equiv \exp[-i\vec{\boldsymbol{\zeta}}\boldsymbol{\cdot}\vec{\boldsymbol{k}}] = \mathbb{1}_{4\times 4} - i\hat{\boldsymbol{v}}\boldsymbol{\cdot}\vec{\boldsymbol{k}}\sinh\zeta - (\hat{\boldsymbol{v}}\boldsymbol{\cdot}\vec{\boldsymbol{k}})^2[\cosh\zeta - 1]\,, \tag{1.276}$$

where $\hat{\boldsymbol{v}} = \vec{\boldsymbol{\zeta}}/\zeta$ and $\zeta \equiv |\vec{\boldsymbol{\zeta}}|$. Consider the rest frame of a particle of mass m. Using the Lorentz transformation matrix Λ to boost to a frame where the particle four-momentum is given by $(E\,;\,\vec{\boldsymbol{p}})$, show that $E = m\cosh\zeta$ and $\vec{\boldsymbol{p}} = \hat{\boldsymbol{v}}\, m\sinh\zeta$, where $\vec{\boldsymbol{v}}$ is the velocity of the boosted frame (and $\hat{\boldsymbol{v}} \equiv \vec{\boldsymbol{v}}/|\vec{\boldsymbol{v}}|$).

(c) Consider a boost parameterized by $\vec{\boldsymbol{\zeta}}$ that takes $\vec{\boldsymbol{x}} \to \vec{\boldsymbol{x}}'$. Using the results of part (b), obtain an explicit expression for $\vec{\boldsymbol{x}}'$ in terms of $\vec{\boldsymbol{x}}$. Show that by taking $\vec{\boldsymbol{x}} = 0$, one can use the relation $\vec{\boldsymbol{x}}' = \vec{\boldsymbol{v}}t'$ to determine the velocity $\vec{\boldsymbol{v}}$ of the boosted frame. Then, verify that $\vec{\boldsymbol{\zeta}} = \hat{\boldsymbol{v}}\tanh^{-1}\beta$.

(d) Prove that the exponentiation of the infinitesimal Lorentz transformation of part (a) coincides with the correct expression for the finite Lorentz transformation only if one replaces $\vec{\boldsymbol{\beta}}$ with $\vec{\boldsymbol{\zeta}}$ in eq. (1.275) before performing the exponentiation. Explain.

1.2 The Pauli–Lubański vector is defined by

$$w^\mu \equiv -\tfrac{1}{2}\epsilon^{\mu\nu\rho\sigma} J_{\nu\rho} P_\sigma\,, \tag{1.277}$$

in a convention where $\epsilon^{0123} = 1$. Derive the following properties:

(a) $w_\mu P^\mu = 0$ (1.278)

(b) $[P^\mu\,,\,w^\nu] = 0$ (1.279)

(c) $[J^{\mu\nu}\,,\,w^\rho] = i(g^{\nu\rho}w^\mu - g^{\mu\rho}w^\nu)$ (1.280)

(d) $[w^\mu\,,\,w^\nu] = i\epsilon^{\mu\nu\rho\sigma}w_\rho P_\sigma$ (1.281)

(e) $w^\rho w_\rho = -\tfrac{1}{2}J_{\mu\nu}J^{\mu\nu}P_\sigma P^\sigma + J_{\mu\sigma}J^{\nu\sigma}P^\mu P_\nu\,.$ (1.282)

Note that eq. (1.280) confirms that w^μ is a Lorentz four-vector [cf. eq. (1.285)].

(f) Using eqs. (1.279) and (1.280), show that $w^2 \equiv w^\rho w_\rho$ is a Casimir operator of the Poincaré algebra by verifying that $[w^2\,,\,P^\mu] = [w^2\,,\,J^{\mu\nu}] = 0$.

1.3 An nth-rank Lorentz tensor operator, $T^{\mu_1\mu_2\cdots\mu_n}$, which acts on a Hilbert space $\mathcal{H}$, satisfies

$$\exp\left(\tfrac{1}{2}i\theta_{\rho\tau}J^{\rho\tau}\right)T^{\mu_1\mu_2\cdots\mu_n}\exp\left(-\tfrac{1}{2}i\theta_{\kappa\sigma}J^{\kappa\sigma}\right) = \Lambda^{\mu_1}{}_{\nu_1}\Lambda^{\mu_2}{}_{\nu_2}\cdots\Lambda^{\mu_n}{}_{\nu_n}T^{\nu_1\nu_2\cdots\nu_n}\,, \tag{1.283}$$

where the $J^{\rho\tau} = -J^{\tau\rho}$ (which are operators that act on $\mathcal{H}$) represent the generators of the Lorentz group, and the $\theta_{\rho\tau} = -\theta_{\rho\tau}$ parameterize the 4×4 Lorentz transformation matrix $\Lambda^\mu{}_\nu$ [see eqs. (1.2) and (1.4)].

(a) Expanding eq. (1.283) to first order in the $\theta_{\rho\tau}$, derive the commutation relations

$$[J^{\rho\tau}, T^{\mu_1\mu_2\cdots\mu_n}] = -i\big(g^{\rho\mu_1}T^{\tau\mu_2\cdots\mu_n} - g^{\tau\mu_1}T^{\rho\mu_2\cdots\mu_n} + g^{\rho\mu_2}T^{\mu_1\tau\cdots\mu_n} - g^{\tau\mu_2}T^{\mu_1\rho\cdots\mu_n} + \cdots + g^{\rho\mu_n}T^{\mu_1\mu_2\cdots\tau} - g^{\tau\mu_n}T^{\mu_1\mu_2\cdots\rho}\big). \tag{1.284}$$

In particular, show that any Lorentz four-vector operator V^μ and second-rank Lorentz tensor operator $T^{\mu\nu}$ must satisfy

$$[J^{\rho\tau}, V^\mu] = i(g^{\tau\mu}V^\rho - g^{\rho\mu}V^\tau), \tag{1.285}$$

$$[J^{\rho\tau}, T^{\mu\nu}] = i(g^{\tau\mu}T^{\rho\nu} - g^{\rho\mu}T^{\tau\nu} + g^{\tau\nu}T^{\mu\rho} - g^{\rho\nu}T^{\mu\tau}). \tag{1.286}$$

(b) Using the results of part (a), show that the commutation relations of the Poincaré algebra imply that P^μ is a Lorentz four-vector operator and $J^{\mu\nu}$ is a second-rank antisymmetric Lorentz tensor operator.

1.4 A spinor operator of rank (m,n), denoted by $Q^{\dot\alpha_1\dot\alpha_2\cdots\dot\alpha_n}_{\alpha_1\alpha_2\cdots\alpha_m}$, which acts on a Hilbert space $\mathcal{H}$, satisfies

$$\exp\left(\tfrac{1}{2}i\theta_{\rho\tau}J^{\rho\tau}\right) Q^{\dot\alpha_1\dot\alpha_2\cdots\dot\alpha_n}_{\alpha_1\alpha_2\cdots\alpha_m} \exp\left(-\tfrac{1}{2}i\theta_{\kappa\sigma}J^{\kappa\sigma}\right) = M_{\alpha_1}{}^{\beta_1}M_{\alpha_2}{}^{\beta_2}\cdots M_{\alpha_m}{}^{\beta_m}\overline{M}^{\dot\alpha_1}{}_{\dot\beta_1}\overline{M}^{\dot\alpha_2}{}_{\dot\beta_2}\cdots\overline{M}^{\dot\alpha_n}{}_{\dot\beta_n}Q^{\dot\beta_1\dot\beta_2\cdots\dot\beta_n}_{\beta_1\beta_2\cdots\beta_m}, \tag{1.287}$$

where [see eqs. (1.128) and (1.129)]

$$M_\alpha{}^\beta \equiv \exp\left(-\tfrac{1}{2}i\theta_{\mu\nu}\sigma^{\mu\nu}\right)_\alpha{}^\beta, \tag{1.288}$$

$$\overline{M}^{\dot\alpha}{}_{\dot\beta} \equiv (M^{-1})^{\dagger\,\dot\alpha}{}_{\dot\beta} \equiv \exp\left(-\tfrac{1}{2}i\theta_{\mu\nu}\overline{\sigma}^{\mu\nu}\right)^{\dot\alpha}{}_{\dot\beta}. \tag{1.289}$$

(a) Expanding eq. (1.287) to first order in the $\theta_{\rho\tau}$, derive the commutation relations

$$\begin{aligned}[Q^{\dot\alpha_1\dot\alpha_2\cdots\dot\alpha_n}_{\alpha_1\alpha_2\cdots\alpha_m}, J^{\mu\nu}] = {}& \sigma^{\mu\nu}{}_{\alpha_1}{}^{\beta_1}Q^{\dot\alpha_1\dot\alpha_2\cdots\dot\alpha_n}_{\beta_1\alpha_2\cdots\alpha_m} + \sigma^{\mu\nu}{}_{\alpha_2}{}^{\beta_2}Q^{\dot\alpha_1\dot\alpha_2\cdots\dot\alpha_n}_{\alpha_1\beta_2\cdots\alpha_m} + \cdots \\ & + \sigma^{\mu\nu}{}_{\alpha_m}{}^{\beta_m}Q^{\dot\alpha_1\dot\alpha_2\cdots\dot\alpha_n}_{\alpha_1\alpha_2\cdots\beta_m} \\ & + \overline{\sigma}^{\mu\nu\dot\alpha_1}{}_{\dot\beta_1}Q^{\dot\beta_1\dot\alpha_2\cdots\dot\alpha_n}_{\alpha_1\alpha_2\cdots\alpha_m} + \overline{\sigma}^{\mu\nu\dot\alpha_2}{}_{\dot\beta_2}Q^{\dot\alpha_1\dot\beta_2\cdots\dot\alpha_n}_{\alpha_1\alpha_2\cdots\alpha_m} + \cdots \\ & + \overline{\sigma}^{\mu\nu\dot\alpha_n}{}_{\dot\beta_n}Q^{\dot\alpha_1\dot\alpha_2\cdots\dot\beta_n}_{\alpha_1\alpha_2\cdots\alpha_m}. \end{aligned} \tag{1.290}$$

(b) Using the results of part (a), the commutation relations of spinor operators Q_α and $Q^{\dagger\dot\alpha}$, which transform as the $(\frac{1}{2},0)$ and $(0,\frac{1}{2})$ representations of the Poincaré algebra respectively, are given by

$$[Q_\alpha, J^{\mu\nu}] = \sigma^{\mu\nu}{}_\alpha{}^\beta Q_\beta, \qquad [Q^{\dagger\dot\alpha}, J^{\mu\nu}] = \overline{\sigma}^{\mu\nu\dot\alpha}{}_{\dot\beta}Q^{\dagger\dot\beta}. \tag{1.291}$$

One can also derive similar results for spinors with raised undotted indices and lowered dotted indices. For example, show that

$$[Q^\alpha, J^{\mu\nu}] = -Q^\beta\sigma^{\mu\nu}{}_\beta{}^\alpha, \qquad [Q^\dagger_{\dot\alpha}, J^{\mu\nu}] = -Q^\dagger_{\dot\beta}\overline{\sigma}^{\mu\nu\dot\beta}{}_{\dot\alpha}. \tag{1.292}$$

(c) Consider the generators P^μ and $J^{\mu\nu}$ of the Poincaré algebra. It is convenient to define

$$P_{\alpha\dot{\beta}} \equiv P_\mu \sigma^\mu_{\alpha\dot{\beta}}\,, \qquad J_\alpha{}^\beta \equiv J_{\mu\nu}(\sigma^{\mu\nu})_\alpha{}^\beta\,, \qquad \overline{J}^{\dot{\alpha}}{}_{\dot{\beta}} \equiv J_{\mu\nu}(\overline{\sigma}^{\mu\nu})^{\dot{\alpha}}{}_{\dot{\beta}}\,, \tag{1.293}$$

where α and β [$\dot{\alpha}$ and $\dot{\beta}$] are undotted [dotted] two-component spinor indices. Show that the above relations can be inverted:

$$P^\mu = \tfrac{1}{2}\overline{\sigma}^{\mu\dot{\alpha}\beta} P_{\beta\dot{\alpha}}\,, \qquad J^{\mu\nu} = \tfrac{1}{2}\left[(\sigma^{\mu\nu})_\alpha{}^\beta J_\beta{}^\alpha + (\overline{\sigma}^{\mu\nu})^{\dot{\alpha}}{}_{\dot{\beta}} \overline{J}^{\dot{\beta}}{}_{\dot{\alpha}}\right]. \tag{1.294}$$

In light of the above results, show that P^μ transforms as a $(\frac{1}{2},\frac{1}{2})$ representation and $J^{\mu\nu}$ transforms as a $(1,0)\oplus(0,1)$ representation of the Poincaré algebra.

1.5 Consider a proper orthochronous Lorentz transformation that is a pure boost in the direction $\hat{\boldsymbol{v}}$.

(a) Prove the following identity:

$$\exp\left(\tfrac{1}{2}\vec{\boldsymbol{\zeta}}\boldsymbol{\cdot}\vec{\boldsymbol{\sigma}}\right) = \boldsymbol{I}_2 \cosh(\zeta/2) + \hat{\boldsymbol{v}}\boldsymbol{\cdot}\vec{\boldsymbol{\sigma}}\sinh(\zeta/2)\,, \tag{1.295}$$

where $\boldsymbol{I}_2$ is the 2×2 identity matrix, $\hat{\boldsymbol{v}} = \vec{\boldsymbol{\zeta}}/\zeta$, and $\zeta \equiv |\vec{\boldsymbol{\zeta}}|$.

(b) Using the result of parts (a) and recalling the expressions for E and $\vec{\boldsymbol{p}}$ from Exercise 1.1(b), show that, for a pure boost (where $\zeta^i = \theta^{i0} = -\theta^{0i}$ and $\theta^{ij} = 0$), the Lorentz transformations for the $(\frac{1}{2},0)$ and $(0,\frac{1}{2})$ representations, respectively, are given by

$$\exp\left(-\tfrac{1}{2}i\theta_{\mu\nu}S^{\mu\nu}\right) = \begin{cases} \exp\left(-\tfrac{1}{2}\vec{\boldsymbol{\zeta}}\boldsymbol{\cdot}\vec{\boldsymbol{\sigma}}\right) = \sqrt{\dfrac{p\cdot\sigma}{m}}\,, & \text{for } (\tfrac{1}{2},0)\,, \\[2ex] \exp\left(\tfrac{1}{2}\vec{\boldsymbol{\zeta}}\boldsymbol{\cdot}\vec{\boldsymbol{\sigma}}\right) = \sqrt{\dfrac{p\cdot\overline{\sigma}}{m}}\,, & \text{for } (0,\tfrac{1}{2})\,, \end{cases} \tag{1.296}$$

where $\sqrt{p\cdot\sigma}$ and $\sqrt{p\cdot\overline{\sigma}}$ are explicitly given in eq. (A.73).

1.6 (a) Derive the following two identities:

$$\sigma^{\mu\nu}\sigma^\rho - \sigma^\rho\overline{\sigma}^{\mu\nu} = i(g^{\nu\rho}\sigma^\mu - g^{\mu\rho}\sigma^\nu)\,, \tag{1.297}$$

$$\overline{\sigma}^{\mu\nu}\overline{\sigma}^\rho - \overline{\sigma}^\rho\sigma^{\mu\nu} = i(g^{\nu\rho}\overline{\sigma}^\mu - g^{\mu\rho}\overline{\sigma}^\nu)\,. \tag{1.298}$$

(b) Using the infinitesimal forms for Λ, M, and $(M^{-1})^\dagger$ given in eqs. (A.67)–(A.69), prove the following two results:

$$M^\dagger\overline{\sigma}^\mu M = \Lambda^\mu{}_\nu\,\overline{\sigma}^\nu\,, \tag{1.299}$$

$$M^{-1}\sigma^\mu(M^{-1})^\dagger = \Lambda^\mu{}_\nu\,\sigma^\nu\,. \tag{1.300}$$

Show that eqs. (1.299) and (1.300) are alternative versions of eqs. (1.78) and (1.76), respectively.

(c) Equations (1.299) and (1.300) are also valid for the exponential forms of M and Λ. Starting from eq. (1.299) with $M = \exp(-\frac{1}{2}i\vec{\theta}\cdot\vec{\sigma} - \frac{1}{2}\vec{\zeta}\cdot\vec{\sigma})$, show that

$$\Lambda^\mu{}_\nu = \tfrac{1}{2}\,\mathrm{Tr}\big(M^\dagger\overline{\sigma}^\mu M\sigma_\nu\big)\,. \tag{1.301}$$

Obtain an explicit expression for the 4×4 matrix Λ in terms of $\vec{\theta}$, $\vec{\zeta}$.

(d) Obtain an explicit expression for Λ using eq. (1.2), and verify that it is equivalent to the result obtained in part (c). (*HINT*: See Ref. [31].)

1.7 Suppose that χ and η are two-component commuting or anticommuting spinors. Prove the following hermiticity conditions:

$$(\chi\Sigma\eta)^\dagger = \eta^\dagger\Sigma_r\chi^\dagger\,, \qquad (\chi\Sigma\eta^\dagger)^\dagger = \eta\Sigma_r\chi^\dagger\,, \qquad (\chi^\dagger\Sigma\eta)^\dagger = \eta^\dagger\Sigma_r\chi\,, \tag{1.302}$$

where Σ stands for any sequence of alternating σ and $\overline{\sigma}$ matrices, and Σ_r is obtained from Σ by reversing the order of all of the σ and $\overline{\sigma}$ matrices.

1.8 (a) Prove the following Schouten identities:

$$\epsilon_{\alpha\beta}\epsilon_{\gamma\delta} + \epsilon_{\alpha\gamma}\epsilon_{\delta\beta} + \epsilon_{\alpha\delta}\epsilon_{\beta\gamma} = 0\,, \tag{1.303}$$

$$\epsilon_{\dot\alpha\dot\beta}\epsilon_{\dot\gamma\dot\delta} + \epsilon_{\dot\alpha\dot\gamma}\epsilon_{\dot\delta\dot\beta} + \epsilon_{\dot\alpha\dot\delta}\epsilon_{\dot\beta\dot\gamma} = 0\,. \tag{1.304}$$

HINT: If a quantity $T_{\alpha\beta\gamma\delta}$ is totally antisymmetric with respect to the interchange of any pair of two-component spinor indices, then show that $T_{\alpha\beta\gamma\delta} = 0$.

(b) Suppose that ξ, χ, and η are two-component commuting or anticommuting spinors. Derive the following identity:

$$\xi_\alpha(\chi\eta) = -\chi_\alpha(\eta\xi) - \eta_\alpha(\xi\chi). \tag{1.305}$$

1.9 Using the notation of eq. (1.95), derive the following identities:

$$z_1\sigma^{\mu\nu}z_2 = (-1)^A z_2\sigma^{\mu\nu}z_1\,, \tag{1.306}$$

$$z_1^\dagger\overline{\sigma}^{\mu\nu}z_2^\dagger = (-1)^A z_2^\dagger\overline{\sigma}^{\mu\nu}z_1^\dagger\,, \tag{1.307}$$

where z_1 and z_2 are two-component spinors.

1.10 Show that the free-field Dirac equations given in eq. (1.154) are form invariant under a proper orthochronous Lorentz transformation, $x'^\mu = \Lambda^\mu{}_\nu x^\nu$. That is, by writing $\chi'_\alpha(x') = M_\alpha{}^\beta\chi_\beta(x)$ and $\eta'^{\dagger\dot\alpha}(x') = (M^{-1})^{\dagger\dot\alpha}{}_{\dot\beta}\eta^{\dagger\dot\beta}(x)$, and using eq. (A.62) to obtain $\partial_\mu = \Lambda^\nu{}_\mu\partial'_\nu$, show that χ'_α and $\eta'^{\dagger\dot\alpha}$ satisfy the Dirac equations in the transformed reference frame if eq. (1.299) is satisfied.

1.11 Consider the Lagrangian for a free two-component fermion given by eq. (1.144).

(a) Using the Noether procedure, show that the invariance of the action under time translations, space translations, and Lorentz transformations implies that the four-momentum operator $P^\mu = (P^0\,;\,\vec{\boldsymbol{P}})$ and the angular momentum operator $J^{\mu\nu}$ defined in eq. (1.150) are given by

$$P^0 = \int d^3x\, \left\{\xi^\dagger(x)\, i\vec{\boldsymbol{\sigma}}\boldsymbol{\cdot}\vec{\boldsymbol{\nabla}}\xi(x) + \tfrac{1}{2}m\left[\xi(x)\xi(x) + \xi^\dagger(x)\xi^\dagger(x)\right]\right\}, \quad (1.308)$$

$$\vec{\boldsymbol{P}} = -\int d^3x\, \xi^\dagger_{\dot\alpha}(x)\, i\overline{\sigma}^{0\dot\alpha\beta}\vec{\boldsymbol{\nabla}}\xi_\beta(x)\,, \quad (1.309)$$

$$J^{\mu\nu} = \int d^3x \left\{\xi^\dagger_{\dot\alpha}(x)\left[i(x^\mu\partial^\nu - x^\nu\partial^\mu)\overline{\sigma}^{0\dot\alpha\beta} + (\overline{\sigma}^0\sigma^{\mu\nu})^{0\dot\alpha\beta}\right]\xi_\beta(x)\right.$$
$$\left. + \tfrac{1}{2}m(x^\mu g^{0\nu} - x^\nu g^{0\mu})\left[\xi(x)\xi(x) - \xi^\dagger(x)\xi^\dagger(x)\right]\right\}, \quad (1.310)$$

where $\xi(x)$ and $\xi^\dagger(x)$ satisfy the field equations [see eqs. (1.145) and (1.146)].
Defining $J^i = \frac{1}{2}\epsilon^{ijk}J^{jk}$ and $\mathcal{K}^i = J^{0i}$, show that

$$\vec{\boldsymbol{J}} = \int d^3x\, \xi^\dagger(\vec{\boldsymbol{x}},t)\left[-i\vec{\boldsymbol{x}}\times\vec{\boldsymbol{\nabla}} + \tfrac{1}{2}\vec{\boldsymbol{\sigma}}\right]\xi(\vec{\boldsymbol{x}},t)\,, \quad (1.311)$$

$$\vec{\boldsymbol{\mathcal{K}}} = -i\int d^3x \left\{\xi^\dagger(\vec{\boldsymbol{x}},t)\left[\left(t\vec{\boldsymbol{\nabla}} + \vec{\boldsymbol{x}}\,\partial_0\right) + \tfrac{1}{2}\vec{\boldsymbol{\sigma}}\right]\xi(\vec{\boldsymbol{x}},t)\right.$$
$$\left. - \tfrac{1}{2}im\vec{\boldsymbol{x}}\left[\xi(\vec{\boldsymbol{x}},t)\xi(\vec{\boldsymbol{x}},t) - \xi^\dagger(\vec{\boldsymbol{x}},t)\xi^\dagger(\vec{\boldsymbol{x}},t)\right]\right\}. \quad (1.312)$$

(b) Prove that P^0, $\vec{\boldsymbol{P}}$, and $\vec{\boldsymbol{J}}$ are hermitian operators.

(c) Show that $\vec{\boldsymbol{\mathcal{K}}}$ given in eq. (1.312) can be re-expressed as

$$\vec{\boldsymbol{\mathcal{K}}} = t\vec{\boldsymbol{P}} - \tfrac{1}{2}i\int d^3x\, \vec{\boldsymbol{x}}\,\xi^\dagger(\vec{\boldsymbol{x}},t)\,\overleftrightarrow{\partial}_0\xi(\vec{\boldsymbol{x}},t)\,, \quad (1.313)$$

where

$$\xi^\dagger(\vec{\boldsymbol{x}},t)\,\overleftrightarrow{\partial}_0\xi(\vec{\boldsymbol{x}},t) \equiv \xi^\dagger(\vec{\boldsymbol{x}},t)\frac{\partial\xi(\vec{\boldsymbol{x}},t)}{\partial t} - \frac{\partial\xi^\dagger(\vec{\boldsymbol{x}},t)}{\partial t}\xi(\vec{\boldsymbol{x}},t)\,. \quad (1.314)$$

HINT: To derive eq. (1.313), it is convenient to make use of the following identity:

$$\frac{i}{2}\int d^3x\,\partial_k\left[x^i\,\xi^\dagger_{\dot\alpha}(\vec{\boldsymbol{x}},t)\overline{\sigma}^{k\dot\alpha\beta}\xi_\beta(\vec{\boldsymbol{x}},t)\right] = 0\,, \quad (1.315)$$

which is valid under the assumption that the field $\xi_\beta(\vec{\boldsymbol{x}},t)$ and its hermitian conjugate vanish at spacial infinity. Further simplification can be achieved by employing the field equations [eqs. (1.145) and (1.146)].

(d) Using eqs. (1.308), (1.309), (1.311), and (1.313) and imposing the field equations for ξ and $\xi^\dagger$, verify explicitly that

$$\frac{dP^0}{dt} = \frac{d\vec{\boldsymbol{P}}}{dt} = \frac{d\vec{\boldsymbol{J}}}{dt} = \frac{d\vec{\boldsymbol{\mathcal{K}}}}{dt} = 0\,. \quad (1.316)$$

(e) In contrast to the finite-dimensional representations of the K^i [defined in eq. (1.10)] which are antihermitian, show that $\vec{\mathcal{K}}$ given by eq. (1.313) is a hermitian operator.

(f) Generalize the above results to the Dirac Lagrangian given in eq. (1.153).

1.12 (a) Show that there is no solution for M in eq. (1.299) for $\Lambda = \Lambda_P$, where Λ_P is given by eq. (1.195).

(b) Show that the free-field Dirac equations in two-component spinor notation given in eqs. (1.154) and (1.155) are form invariant under a parity transformation, $x'^\mu = (\Lambda_P)^\mu{}_\nu x^\nu$. That is, by writing $\chi'_\alpha(x') = iP_{\alpha\dot{\beta}}\eta^{\dagger\,\dot{\beta}}(x)$ and $\eta'^{\dagger\,\dot{\alpha}}(x') = i\overline{P}^{\dot{\alpha}\beta}\chi_\beta(x)$, where $\overline{P}^{\dot{\alpha}\beta} \equiv \epsilon^{\beta\alpha}\epsilon^{\dot{\alpha}\dot{\beta}}(P^\dagger)_{\alpha\dot{\beta}}$, show that χ' and $\eta'^\dagger$ satisfy the Dirac equation in the space-reflected reference frame if

$$\overline{P}^{-1}\overline{\sigma}^\mu P = (\Lambda_P)^\mu{}_\nu\sigma^\nu\,, \qquad P^{-1}\sigma^\mu\overline{P} = (\Lambda_P)^\mu{}_\nu\overline{\sigma}^\nu\,. \tag{1.317}$$

Determine P and $\overline{P}$ (up to an overall phase).

1.13 (a) Show that there is no solution for M in eq. (1.299) for $\Lambda = \Lambda_T$, where Λ_T is given by eq. (1.220).

(b) Starting from eqs. (1.244) and (1.245), show that

$$\mathcal{T}\chi^\dagger_{\dot{\alpha}}(x)\mathcal{T}^{-1} \equiv \chi^t_{\dot{\alpha}}(x) = \eta^*_T\sigma^0_{\alpha\dot{\beta}}\chi^{\dagger\dot{\beta}}(x_T)\,, \tag{1.318}$$

$$\mathcal{T}\eta^\alpha(x)\mathcal{T}^{-1} \equiv \eta^{t\dagger\dot{\alpha}}(x) = -\eta^*_T\overline{\sigma}^{0\dot{\alpha}\beta}\eta_\beta(x_T)\,. \tag{1.319}$$

(c) Show that the free-field Dirac equations in two-component spinor notation given in eqs. (1.154) and (1.155) are form invariant under a time reversal transformation, $x'^\mu = (\Lambda_T)^\mu{}_\nu x^\nu$. That is, by writing $\chi'_\alpha(x') = T_{\alpha\dot{\beta}}\chi^{\dagger\dot{\beta}}(x)$ and $\eta'^{\dagger\dot{\alpha}}(x') = -\overline{T}^{\dot{\alpha}\beta}\eta_\beta(x)$, where $\overline{T}^{\dot{\alpha}\beta} \equiv \epsilon^{\beta\alpha}\epsilon^{\dot{\alpha}\dot{\beta}}(T^\dagger)_{\alpha\dot{\beta}}$, show that χ' and $\eta'^\dagger$ satisfy the Dirac equation in the time-reversed reference frame if

$$\overline{T}^{-1}\overline{\sigma}^\mu T = -(\Lambda_T)^\mu{}_\nu\sigma^\nu\,, \qquad T^{-1}\sigma^\mu\overline{T} = -(\Lambda_T)^\mu{}_\nu\overline{\sigma}^\nu\,. \tag{1.320}$$

Determine T and $\overline{T}$ (up to an overall phase).

1.14 Consider the Lagrangian of two free two-component fermion fields, ξ_1 and ξ_2:

$$\mathscr{L} = i\xi_1^\dagger\overline{\sigma}^\mu\partial_\mu\xi_1 + i\xi_2^\dagger\overline{\sigma}^\mu\partial_\mu\xi_2 - m(\xi_1\xi_2 + \xi_1^\dagger\xi_2^\dagger) - \tfrac{1}{2}M(\xi_2\xi_2 + \xi_2^\dagger\xi_2^\dagger)\,. \tag{1.321}$$

(a) Determine the mass eigenstates and find the corresponding masses of the two fermions.

(b) In the seesaw mechanism for light neutrino masses, one assumes that $m \ll M$. From the results of part (a), find the mass of the lightest fermion (keeping terms of order m^2/M). Determine the numerical value of M, assuming $m = m_\tau = 1.777$ GeV and $m_{\nu_\tau} \simeq 5\times 10^{-2}$ eV.

1.15 Consider the following Lagrangian of n free two-component fermion fields, ξ_i:

$$\mathscr{L} = iK_i{}^j\,\hat{\xi}^{\dagger i}\,\bar{\sigma}^\mu\partial_\mu\hat{\xi}_j - \tfrac{1}{2}M^{ij}\hat{\xi}_i\hat{\xi}_j - \tfrac{1}{2}M_{ij}\hat{\xi}^{\dagger i}\hat{\xi}^{\dagger j}\,. \tag{1.322}$$

(a) What are the constraints on $K_i{}^j$, assuming that the Lagrangian is hermitian?

(b) Define a new set of fields ξ_i in terms of the $\hat{\xi}_i$, such that the kinetic energy term is canonical, i.e., $K_i{}^j = \delta_i{}^j$). Re-express the Lagrangian in terms of the new fields ξ_i and show that the resulting expression takes the form given in eq. (1.159).

(c) How does the presence of $K_i{}^j$ affect the mass diagonalization procedure of Section 1.6?

1.16 Prove that

$$\mathcal{CPT}\,\xi_1(x)\xi_2(x)\,(\mathcal{CPT})^{-1} = \xi_1^\dagger(-x)\xi_2^\dagger(-x)\,. \tag{1.323}$$

HINT: Pay attention to the location of the suppressed spinor indices.

1.17 Given a Dirac fermion field, which consists of two anticommuting two-component fermion fields of opposite charge, χ and η, consider the hermitian bilinear covariant, $\eta\chi + \chi^\dagger\eta^\dagger$.

(a) Show that

$$\mathcal{C}\,(\eta\chi + \chi^\dagger\eta^\dagger)\,\mathcal{C}^{-1} = \eta\chi + \chi^\dagger\eta^\dagger\,. \tag{1.324}$$

That is, $\eta\chi + \chi^\dagger\eta^\dagger$ is C-even.

(b) Show that if χ and η had been *commuting* two-component fermion fields, then $\mathcal{C}\,(\eta\chi + \chi^\dagger\eta^\dagger)\,\mathcal{C}^{-1} = -(\eta\chi + \chi^\dagger\eta^\dagger)$. In this case, one would have concluded that $\eta\chi + \chi^\dagger\eta^\dagger$ is C-odd.

(c) Repeat parts (a) and (b) by examining the behavior of the hermitian bilinear covariant, $\chi^\dagger\sigma^\mu\chi + \eta\sigma^\mu\eta^\dagger$ under charge conjugation. Treat the cases of $\mu = 0$ and $\mu = 1, 2, 3$ separately. What is the physical significance of your result for $\mu = 0$ (in the case of *anticommuting* fermion fields)?

1.18 Consider a multiplet of n neutral two-component fermion fields, $\hat{\xi}_k$, governed by the free-field Lagrangian given in eq. (1.159).

(a) Under a CP transformation,

$$\mathcal{CP}\,\hat{\xi}_{k\alpha}(x)(\mathcal{CP})^{-1} = i\hat{\eta}_{CP}\sigma^0_{\alpha\dot{\beta}}\hat{\xi}_k^{\dagger\dot{\beta}}(x_P)\,, \tag{1.325}$$

where $\hat{\eta}_{CP}$ is either ± 1, independently of the value of k. Show that the free-field Lagrangian [eq. (1.159)] is CP invariant if and only if the neutral fermion mass matrix M is a real symmetric matrix.

(b) A real symmetric matrix is diagonalized by a real orthogonal matrix Z:

$$Z^{\mathsf{T}} M Z = \mathrm{diag}(\varepsilon_1 m_1\,, \varepsilon_2 m_2\,, \ldots\,, \varepsilon_n m_n)\,, \tag{1.326}$$

where the m_k are real and nonnegative and the $\varepsilon_k m_k$ are the real eigenvalues of M with corresponding signs $\varepsilon_k = \pm 1$. (Conventionally, we assign $\varepsilon_k = 1$ if the corresponding $m_k = 0$.) We identify the fermion mass-eigenstate fields ξ_k by introducing a unitary matrix Ω such that

$$\hat{\xi}_j = \Omega_j{}^k \xi_k\,. \tag{1.327}$$

Show that the Takagi diagonalization of M is achieved by

$$\Omega_j{}^k = \varepsilon_k^{1/2} Z_{jk}\,, \qquad \text{no sum over } k\,, \tag{1.328}$$

where

$$\varepsilon_k^{1/2} = \begin{cases} 1\,, & \text{if } \varepsilon_k m_k \geq 0, \\ i\,, & \text{if } \varepsilon_k m_k < 0. \end{cases} \tag{1.329}$$

(c) Using eqs. (1.325) and (1.327), evaluate $\mathcal{CP}\,\xi_{k\alpha}(x)(\mathcal{CP})^{-1}$. Show that if the neutral fermion mass matrix M is a real symmetric matrix, then

$$\mathcal{CP}\,\xi_{k\alpha}(x)(\mathcal{CP})^{-1} = i\varepsilon_k \tilde{\eta}_{CP} \sigma^0_{\alpha\dot{\beta}} \xi_k^{\dagger\dot{\beta}}(x_P)\,, \tag{1.330}$$

where the ε_k are the signs defined in eq. (1.326), and there is no sum over the repeated index k. That is, neutral fermion fields corresponding to positive and negative eigenvalues of the real symmetric neutral fermion mass matrix have CP quantum numbers of opposite sign.

1.19 (a) Consider a free complex scalar-field Lagrangian, $\mathscr{L} = \partial^\mu \Phi^\dagger \partial_\mu \Phi - m^2 \Phi^\dagger \Phi$. Show that the free-field action is separately invariant under P, T, and C, respectively, where the corresponding transformation laws for Φ are

$$\mathcal{P}\,\Phi(\vec{x},t)\mathcal{P}^{-1} = \eta_P \Phi(-\vec{x},t)\,, \tag{1.331}$$

$$\mathcal{T}\,\Phi(\vec{x},t)\mathcal{T}^{-1} = \eta_T \Phi(\vec{x},-t)\,, \tag{1.332}$$

$$\mathcal{C}\,\Phi(\vec{x},t)\mathcal{C}^{-1} = \eta_C \Phi^\dagger(\vec{x},t)\,, \tag{1.333}$$

and η_P, η_T, and η_C are arbitrary complex phases. In a convention where $\eta_{CPT} \equiv \eta_C \eta_P \eta_T = 1$, verify that $\mathcal{CPT}\,\Phi(x)(\mathcal{CPT})^{-1} = \Phi^\dagger(-x)$ [cf. eq. (1.270)].

(b) Repeat part (a) in the case of a free complex massive vector field V_μ. Determine the transformation laws for V_0 and V_i ($i = 1,2,3$) that replace eqs. (1.331)–(1.333) and show that the corresponding free-field action is separately invariant under P, T, and C, respectively [30]. In a convention where $\eta_{CPT} = 1$, show that $\mathcal{CPT}\,V_\mu(x)(\mathcal{CPT})^{-1} = -V_\mu^\dagger(-x)$ in agreement with eq. (1.270).

2 Feynman Rules for Spin-1/2 Fermions

In this chapter, we devise a set of Feynman rules to describe matrix elements of processes involving spin-1/2 fermions. The rules are developed for two-component fermions and are then applied to tree-level decay and scattering processes and the fermion self-energy functions in the one-loop approximation.

2.1 Fermion Creation and Annihilation Operators

We begin by describing the properties of a free neutral massive anticommuting spin-1/2 field, denoted $\xi_\alpha(x)$, which transforms as $(\frac{1}{2}, 0)$ under the Lorentz group. The field ξ_α therefore describes a Majorana fermion. The free-field Lagrangian was given in eq. (1.144), which we repeat here for the convenience of the reader:

$$\mathscr{L} = i\xi^\dagger \overline{\sigma}^\mu \partial_\mu \xi - \tfrac{1}{2}m(\xi\xi + \xi^\dagger\xi^\dagger) . \tag{2.1}$$

On-shell, ξ satisfies the free-field Dirac equation,

$$i\overline{\sigma}^{\mu\dot{\alpha}\beta}\partial_\mu \xi_\beta = m\xi^{\dagger\dot{\alpha}} . \tag{2.2}$$

By virtue of the field equations, the field $\xi_\alpha(x)$ can be expanded in a Fourier series,

$$\xi_\alpha(x) = \sum_s \int \frac{d^3\boldsymbol{p}}{(2\pi)^{3/2}(2E_{\boldsymbol{p}})^{1/2}} \left[x_\alpha(\vec{\boldsymbol{p}}, s)a(\vec{\boldsymbol{p}}, s)e^{-ip\cdot x} + y_\alpha(\vec{\boldsymbol{p}}, s)a^\dagger(\vec{\boldsymbol{p}}, s)e^{ip\cdot x}\right] , \tag{2.3}$$

where $E_{\boldsymbol{p}} \equiv (|\vec{\boldsymbol{p}}|^2 + m^2)^{1/2}$, and the creation and annihilation operators $a^\dagger$ and a satisfy anticommutation relations:

$$\{a(\vec{\boldsymbol{p}}, s), a^\dagger(\vec{\boldsymbol{p}}\,', s')\} = \delta^3(\vec{\boldsymbol{p}} - \vec{\boldsymbol{p}}\,')\delta_{ss'} , \tag{2.4}$$

and all other anticommutators vanish. Applying eq. (1.145) to eq. (2.3), we find that the x_α and y_α satisfy momentum space Dirac equations. We shall write these equations out explicitly and study the properties of these two-component spinor wave functions in Section 2.2.

The hermitian conjugate of eq. (2.3) yields the Fourier series expansion of $\xi^\dagger_{\dot\alpha}(x)$:

$$\xi^\dagger_{\dot\alpha}(x) = \sum_s \int \frac{d^3\boldsymbol{p}}{(2\pi)^{3/2}(2E_{\boldsymbol{p}})^{1/2}} \left[x^\dagger_{\dot\alpha}(\vec{\boldsymbol{p}},s) a^\dagger(\vec{\boldsymbol{p}},s) e^{ip\cdot x} + y^\dagger_{\dot\alpha}(\vec{\boldsymbol{p}},s) a(\vec{\boldsymbol{p}},s) e^{-ip\cdot x} \right] . \tag{2.5}$$

We shall employ covariant normalization of the one-particle states. That is, we act with one creation operator on the vacuum with the following convention:

$$|\vec{\boldsymbol{p}},s\rangle \equiv (2\pi)^{3/2}(2E_{\boldsymbol{p}})^{1/2} a^\dagger(\vec{\boldsymbol{p}},s)\,|0\rangle \,, \tag{2.6}$$

so that

$$\langle \vec{\boldsymbol{p}},s|\vec{\boldsymbol{p}}\,',s'\rangle = (2\pi)^3(2E_{\boldsymbol{p}})\delta^3(\vec{\boldsymbol{p}}-\vec{\boldsymbol{p}}\,')\delta_{ss'} \,. \tag{2.7}$$

Therefore,

$$\langle 0|\,\xi_\alpha(x)\,|\vec{\boldsymbol{p}},s\rangle = x_\alpha(\vec{\boldsymbol{p}},s)e^{-ip\cdot x} \,, \qquad \langle 0|\,\xi^\dagger_{\dot\alpha}(x)\,|\vec{\boldsymbol{p}},s\rangle = y^\dagger_{\dot\alpha}(\vec{\boldsymbol{p}},s)e^{-ip\cdot x} \,, \tag{2.8}$$

$$\langle \vec{\boldsymbol{p}},s|\,\xi_\alpha(x)\,|0\rangle = y_\alpha(\vec{\boldsymbol{p}},s)e^{ip\cdot x} \,, \qquad \langle \vec{\boldsymbol{p}},s|\,\xi^\dagger_{\dot\alpha}(x)\,|0\rangle = x^\dagger_{\dot\alpha}(\vec{\boldsymbol{p}},s)e^{ip\cdot x} \,. \tag{2.9}$$

It should be emphasized that $\xi_\alpha(x)$ is an anticommuting spinor field, whereas x_α and y_α are *commuting* two-component spinor wave functions. The anticommuting properties of the fields are carried by the creation and annihilation operators $a^\dagger(\vec{\boldsymbol{p}},s)$ and $a(\vec{\boldsymbol{p}},s)$.

2.2 Properties of Two-Component Spinor Wave Functions

In this section, we explore in some detail the properties of the two-component spinor wave functions $x_\alpha(\vec{\boldsymbol{p}},s)$ and $y_\alpha(\vec{\boldsymbol{p}},s)$. Applying eq. (2.2) to eq. (2.3), the two-component spinor wave functions, x_α, and y_α, satisfy momentum space Dirac equations. These conditions can be written down in a number of equivalent ways:

$$(p\cdot\overline{\sigma})^{\dot\alpha\beta}x_\beta = my^{\dagger\dot\alpha} \,, \qquad (p\cdot\sigma)_{\alpha\dot\beta}y^{\dagger\dot\beta} = mx_\alpha \,, \tag{2.10}$$

$$(p\cdot\sigma)_{\alpha\dot\beta}x^{\dagger\dot\beta} = -my_\alpha \,, \qquad (p\cdot\overline{\sigma})^{\dot\alpha\beta}y_\beta = -mx^{\dagger\dot\alpha} \,, \tag{2.11}$$

$$x^\alpha(p\cdot\sigma)_{\alpha\dot\beta} = -my^\dagger_{\dot\beta} \,, \qquad y^\dagger_{\dot\alpha}(p\cdot\overline{\sigma})^{\dot\alpha\beta} = -mx^\beta \,, \tag{2.12}$$

$$x^\dagger_{\dot\alpha}(p\cdot\overline{\sigma})^{\dot\alpha\beta} = my^\beta \,, \qquad y^\alpha(p\cdot\sigma)_{\alpha\dot\beta} = mx^\dagger_{\dot\beta} \,. \tag{2.13}$$

Using the identities $[(p\cdot\sigma)(p\cdot\overline{\sigma})]_\alpha{}^\beta = p^2\,\delta_\alpha{}^\beta$ and $[(p\cdot\overline{\sigma})(p\cdot\sigma)]^{\dot\alpha}{}_{\dot\beta} = p^2\,\delta^{\dot\alpha}{}_{\dot\beta}$, one can check that both x_α and y_α must satisfy the mass-shell condition, $p^2 = m^2$ (or equivalently $p^0 = E_{\boldsymbol{p}}$). We will later see that eqs. (2.10)–(2.13) are often useful for simplifying matrix elements.

The quantum number s labels the spin or helicity of the spin-1/2 fermion. We shall examine two approaches for constructing the spin-1/2 states. In the first approach, we consider the particle in its rest frame and quantize the spin along a fixed axis specified by the unit vector $\hat{\boldsymbol{s}} \equiv (\sin\theta\cos\phi\,,\, \sin\theta\sin\phi\,,\, \cos\theta)$ with polar angle θ and azimuthal angle ϕ with respect to a fixed z-axis.[1] The corresponding spin states will be called fixed-axis spin states. The relevant basis of two-component spinors χ_s are eigenstates of $\frac{1}{2}\vec{\boldsymbol{\sigma}}\cdot\hat{\boldsymbol{s}}$, i.e.,

$$\tfrac{1}{2}\vec{\boldsymbol{\sigma}}\cdot\hat{\boldsymbol{s}}\,\chi_s = s\chi_s\,, \qquad s = \pm\tfrac{1}{2}\,. \tag{2.14}$$

It is convenient to normalize χ_s such that

$$\chi_s^\dagger \chi_{s'} = \delta_{ss'}\,. \tag{2.15}$$

This fixes χ_s up to an overall phase, which can be chosen by convention:

$$\chi_{1/2}(\hat{\boldsymbol{s}}) = e^{i\zeta}\begin{pmatrix} \cos\dfrac{\theta}{2} \\ e^{i\phi}\sin\dfrac{\theta}{2}\end{pmatrix}, \qquad \chi_{-1/2}(\hat{\boldsymbol{s}}) = e^{-i\zeta}\begin{pmatrix} -e^{-i\phi}\sin\dfrac{\theta}{2} \\ \cos\dfrac{\theta}{2}\end{pmatrix}. \tag{2.16}$$

A detailed examination of the properties of the two-component spinors χ_s along with a discussion of the choice of phase conventions for ζ can be found in Appendix E.

The fixed-axis spin states described above are not very convenient for particles in relativistic motion. Moreover, these states cannot be employed for massless particles since no rest frame exists. Thus, a second approach is to consider helicity states and the corresponding basis of two-component helicity spinors χ_λ that are eigenstates of $\frac{1}{2}\vec{\boldsymbol{\sigma}}\cdot\hat{\boldsymbol{p}}$, i.e.,

$$\tfrac{1}{2}\vec{\boldsymbol{\sigma}}\cdot\hat{\boldsymbol{p}}\,\chi_\lambda = \lambda\chi_\lambda\,, \qquad \lambda = \pm\tfrac{1}{2}\,. \tag{2.17}$$

Here $\hat{\boldsymbol{p}}$ is the unit vector in the direction of the three-momentum, with polar angle θ and azimuthal angle ϕ with respect to a fixed z-axis. That is, the two-component helicity spinors can be obtained from the fixed-axis spinors by replacing $\hat{\boldsymbol{s}}$ by $\hat{\boldsymbol{p}}$ and identifying θ and ϕ as the polar and azimuthal angles of $\hat{\boldsymbol{p}}$.

For fermions of mass $m \neq 0$, it is possible to define the spin four-vector S^μ, which is specified in the rest frame by $(0; \hat{\boldsymbol{s}})$. The unit three-vector $\hat{\boldsymbol{s}}$ corresponds to the axis of spin quantization in the case of fixed-axis spin states. In an arbitrary reference frame, the spin four-vector satisfies $S\cdot p = 0$ and $S\cdot S = -1$. After boosting from the rest frame to a frame in which $p^\mu = (E\,,\,\vec{\boldsymbol{p}})$ [see eq. (A.83)], one finds

$$S^\mu = \left(\frac{\vec{\boldsymbol{p}}\cdot\hat{\boldsymbol{s}}}{m}\,;\,\hat{\boldsymbol{s}} + \frac{(\vec{\boldsymbol{p}}\cdot\hat{\boldsymbol{s}})\,\vec{\boldsymbol{p}}}{m(E+m)}\right). \tag{2.18}$$

Where necessary, we shall write $S^\mu(\hat{\boldsymbol{s}})$ to emphasize the dependence of S^μ on $\hat{\boldsymbol{s}}$.

The spin four-vector for helicity states is defined by taking $\hat{\boldsymbol{s}} = \hat{\boldsymbol{p}}$. Equation (2.18) then reduces to

$$S^\mu = \frac{1}{m}\left(|\vec{\boldsymbol{p}}|\,;\,E\hat{\boldsymbol{p}}\right). \tag{2.19}$$

[1] In the literature, it is a common practice to choose $\hat{\boldsymbol{s}} = \hat{\boldsymbol{z}}$. However, in order to be somewhat more general, we shall not assume this convention here.

In the nonrelativistic limit, the spin four-vector for helicity states is $S^\mu \simeq (0\,;\hat{\boldsymbol{p}})$, as expected.[2] In the high-energy limit ($E \gg m$), $S^\mu = p^\mu/m + \mathcal{O}(m/E)$. For a massless fermion, the spin four-vector does not exist (as there is no rest frame). Nevertheless, one can obtain consistent results by working with massive helicity states and taking the $m \to 0$ limit at the end of the computation. In this case, one can simply use $S^\mu = p^\mu/m + \mathcal{O}(m/E)$; in practical computations the final result will be well defined in the zero mass limit. In contrast, for massive fermions at rest, the helicity state does not exist without reference to some particular boost direction, as noted in footnote 2.

It is convenient to introduce

$$\sqrt{p\cdot\sigma} \equiv \frac{(E_{\boldsymbol{p}}+m)\,\mathbb{1}_{2\times 2} - \vec{\boldsymbol{\sigma}}\cdot\vec{\boldsymbol{p}}}{\sqrt{2(E_{\boldsymbol{p}}+m)}}\,, \tag{2.20}$$

$$\sqrt{p\cdot\overline{\sigma}} \equiv \frac{(E_{\boldsymbol{p}}+m)\,\mathbb{1}_{2\times 2} + \vec{\boldsymbol{\sigma}}\cdot\vec{\boldsymbol{p}}}{\sqrt{2(E_{\boldsymbol{p}}+m)}}\,. \tag{2.21}$$

These matrix square roots are defined to be the unique nonnegative definite hermitian matrices (i.e., with nonnegative eigenvalues) whose squares are equal to the nonnegative definite hermitian matrices $p\cdot\sigma$ and $p\cdot\overline{\sigma}$, respectively.[3] Using eqs. (A.84) and (A.85) with $S_R^\mu = (0\,;\hat{\boldsymbol{s}})$, two important formulae are obtained:

$$\sqrt{p\cdot\sigma}\,S\cdot\overline{\sigma}\,\sqrt{p\cdot\sigma} = m\,\vec{\boldsymbol{\sigma}}\cdot\hat{\boldsymbol{s}}\,, \tag{2.22}$$

$$\sqrt{p\cdot\overline{\sigma}}\,S\cdot\sigma\,\sqrt{p\cdot\overline{\sigma}} = -m\,\vec{\boldsymbol{\sigma}}\cdot\hat{\boldsymbol{s}}\,. \tag{2.23}$$

These results can also be derived directly by employing the explicit form for the spin vector S^μ [eq. (2.18)].

The two-component spinor wave functions x and y can now be given explicitly in terms of the χ_s, whose explicit form is given in eq. (2.16). First, we note that eq. (2.10) when evaluated in the rest frame yields $x_1 = y^{\dagger 1}$ and $x_2 = y^{\dagger 2}$. That is, as column vectors, $x_\alpha(\vec{\boldsymbol{p}}=0) = y^{\dagger\dot{\alpha}}(\vec{\boldsymbol{p}}=0)$ can be expressed in general as some linear combination of the χ_s (where $s = \pm\frac{1}{2}$). Hence, we may choose $x_\alpha(\vec{\boldsymbol{p}}=0,s) = y^{\dagger\dot{\alpha}}(\vec{\boldsymbol{p}}=0,s) = \sqrt{m}\chi_s$, where the factor of $\sqrt{m}$ reflects the standard relativistic normalization of the rest-frame spin states. These wave functions can be boosted to an arbitrary frame using eq. (1.296). The resulting undotted spinor wave functions are given by

$$x_\alpha(\vec{\boldsymbol{p}},s) = \sqrt{p\cdot\sigma}\,\chi_s\,, \qquad x^\alpha(\vec{\boldsymbol{p}},s) = -2s\chi^\dagger_{-s}\sqrt{p\cdot\overline{\sigma}}\,, \tag{2.24}$$

$$y_\alpha(\vec{\boldsymbol{p}},s) = 2s\sqrt{p\cdot\sigma}\,\chi_{-s}\,, \qquad y^\alpha(\vec{\boldsymbol{p}},s) = \chi^\dagger_s\sqrt{p\cdot\overline{\sigma}}\,, \tag{2.25}$$

[2] Strictly speaking, $\hat{\boldsymbol{p}}$ is not defined in the rest frame. In practice, helicity states are defined in some moving frame with momentum $\vec{\boldsymbol{p}}$. The rest frame is achieved by boosting in the direction of $-\vec{\boldsymbol{p}}$.

[3] Note that $p\cdot\sigma$ and $p\cdot\overline{\sigma}$ are nonnegative matrices due to the implicit mass-shell condition satisfied by p^μ.

and the dotted spinor wave functions are given by

$$x^{\dagger\dot{\alpha}}(\vec{p},s) = -2s\sqrt{p\cdot\overline{\sigma}}\,\chi_{-s}\,, \qquad x^{\dagger}_{\dot{\alpha}}(\vec{p},s) = \chi^{\dagger}_{s}\sqrt{p\cdot\sigma}\,, \tag{2.26}$$

$$y^{\dagger\dot{\alpha}}(\vec{p},s) = \sqrt{p\cdot\overline{\sigma}}\,\chi_{s}\,, \qquad y^{\dagger}_{\dot{\alpha}}(\vec{p},s) = 2s\chi^{\dagger}_{-s}\sqrt{p\cdot\sigma}\,. \tag{2.27}$$

Note that eqs. (2.24)–(2.27) imply that the x and y spinors are related:

$$y(\vec{p},s) = 2sx(\vec{p},-s)\,, \qquad y^{\dagger}(\vec{p},s) = 2sx^{\dagger}(\vec{p},-s)\,. \tag{2.28}$$

The phase choices in eqs. (2.24)–(2.27) are consistent with those employed for four-component spinor wave functions (see Appendix E). We again emphasize that in eqs. (2.24)–(2.27), one may either choose χ_s to be an eigenstate of $\vec{\sigma}\cdot\hat{s}$, where the spin is measured in the rest frame along the quantization axis $\hat{s}$, or choose χ_s to be an eigenstate of $\vec{\sigma}\cdot\hat{p}$ (in this case we shall write $s=\lambda$), which yields the helicity spinor wave functions.

The following equations can now be derived:

$$(S\cdot\overline{\sigma})^{\dot{\alpha}\beta}x_{\beta}(\vec{p},s) = 2sy^{\dagger\dot{\alpha}}(\vec{p},s)\,, \qquad (S\cdot\sigma)_{\alpha\dot{\beta}}y^{\dagger\dot{\beta}}(\vec{p},s) = -2sx_{\alpha}(\vec{p},s)\,, \tag{2.29}$$

$$(S\cdot\sigma)_{\alpha\dot{\beta}}x^{\dagger\dot{\beta}}(\vec{p},s) = -2sy_{\alpha}(\vec{p},s)\,, \qquad (S\cdot\overline{\sigma})^{\dot{\alpha}\beta}y_{\beta}(\vec{p},s) = 2sx^{\dagger\dot{\alpha}}(\vec{p},s)\,, \tag{2.30}$$

$$x^{\alpha}(\vec{p},s)(S\cdot\sigma)_{\alpha\dot{\beta}} = -2sy^{\dagger}_{\dot{\beta}}(\vec{p},s)\,, \qquad y^{\dagger}_{\dot{\alpha}}(\vec{p},s)(S\cdot\overline{\sigma})^{\dot{\alpha}\beta} = 2sx^{\beta}(\vec{p},s)\,, \tag{2.31}$$

$$x^{\dagger}_{\dot{\alpha}}(\vec{p},s)(S\cdot\overline{\sigma})^{\dot{\alpha}\beta} = 2sy^{\beta}(\vec{p},s)\,, \qquad y^{\alpha}(\vec{p},s)(S\cdot\sigma)_{\alpha\dot{\beta}} = -2sx^{\dagger}_{\dot{\beta}}(\vec{p},s)\,. \tag{2.32}$$

For example, using eqs. (2.22) and (2.23) and the definitions above for $x_{\alpha}(\vec{p},s)$ and $y^{\dagger\dot{\alpha}}(\vec{p},s)$, we find (suppressing spinor indices)

$$\sqrt{p\cdot\sigma}\,S\cdot\overline{\sigma}\,x(\vec{p},s) = \sqrt{p\cdot\sigma}\,S\cdot\overline{\sigma}\,\sqrt{p\cdot\sigma}\,\chi_s = m\vec{\sigma}\cdot\hat{s}\,\chi_s = 2sm\,\chi_s\,. \tag{2.33}$$

Multiplying both sides of eq. (2.33) by $\sqrt{p\cdot\overline{\sigma}}$ and noting that $\sqrt{p\cdot\overline{\sigma}}\sqrt{p\cdot\sigma}=m$, we end up with

$$S\cdot\overline{\sigma}\,x(\vec{p},s) = 2s\sqrt{p\cdot\overline{\sigma}}\,\chi_s = 2sy^{\dagger}(\vec{p},s)\,. \tag{2.34}$$

All the results of eqs. (2.29)–(2.32) can be derived in this manner.

The consistency of eqs. (2.29)–(2.32) can also be checked as follows. First, each of these equations yields

$$(S\cdot\sigma)_{\alpha\dot{\alpha}}(S\cdot\overline{\sigma})^{\dot{\alpha}\beta} = -\delta_{\alpha}^{\beta}\,, \qquad (S\cdot\overline{\sigma})^{\dot{\alpha}\alpha}(S\cdot\sigma)_{\alpha\dot{\beta}} = -\delta^{\dot{\alpha}}_{\dot{\beta}}\,, \tag{2.35}$$

after noting that $4s^2=1$ (for $s=\pm\frac{1}{2}$). From eqs. (1.88) and (1.89) it follows that $S\cdot S=-1$, as required. Second, if one applies

$$(p\cdot\sigma\,S\cdot\overline{\sigma} + S\cdot\sigma\,p\cdot\overline{\sigma})_{\alpha}{}^{\beta} = 2p\cdot S\,\delta_{\alpha}{}^{\beta}\,, \tag{2.36}$$

$$(p\cdot\overline{\sigma}\,S\cdot\sigma + S\cdot\overline{\sigma}\,p\cdot\sigma)^{\dot{\alpha}}{}_{\dot{\beta}} = 2p\cdot S\,\delta^{\dot{\alpha}}{}_{\dot{\beta}}\,, \tag{2.37}$$

to eqs. (2.10)–(2.13) and eqs. (2.29)–(2.32), it follows that $p\cdot S=0$.

One can combine the results of eqs. (2.10)–(2.13) and eqs. (2.29)–(2.32) as follows:

$$p^\mu - 2smS^\mu)\overline{\sigma}_\mu^{\dot\alpha\beta}x_\beta(\vec{p},s) = 0\,, \qquad (p_\mu - 2smS_\mu)\sigma^\mu_{\alpha\dot\beta}x^{\dagger\dot\beta}(\vec{p},s) = 0\,, \tag{2.38}$$

$$(p^\mu + 2smS^\mu)\overline{\sigma}_\mu^{\dot\alpha\beta}y_\beta(\vec{p},s) = 0\,, \qquad (p_\mu + 2smS_\mu)\sigma^\mu_{\alpha\dot\beta}y^{\dagger\dot\beta}(\vec{p},s) = 0\,, \tag{2.39}$$

$$x^\alpha(\vec{p},s)\sigma^\mu_{\alpha\dot\beta}(p_\mu - 2smS_\mu) = 0\,, \qquad x^\dagger_{\dot\alpha}(\vec{p},s)\overline{\sigma}_\mu^{\dot\alpha\beta}(p^\mu - 2smS^\mu) = 0\,, \tag{2.40}$$

$$y^\alpha(\vec{p},s)\sigma^\mu_{\alpha\dot\beta}(p_\mu + 2smS_\mu) = 0\,, \qquad y^\dagger_{\dot\alpha}(\vec{p},s)\overline{\sigma}_\mu^{\dot\alpha\beta}(p^\mu + 2smS^\mu) = 0\,. \tag{2.41}$$

Equations (2.24)–(2.41) also apply to the helicity wave functions $x(\vec{p},\lambda)$ and $y(\vec{p},\lambda)$ simply by replacing s with λ and $S^\mu(\hat{s})$ [eq. (2.18)] with $S^\mu(\hat{p})$ [eq. (2.19)].

The above results are applicable only for massive fermions (where the spin four-vector S^μ exists). We may treat the case of massless fermions directly by employing helicity spinors in eqs. (2.24)–(2.27). Putting $E = |\vec{p}|$ and $m = 0$, we easily obtain

$$x_\alpha(\vec{p},\lambda) = \sqrt{2E}\,(\tfrac{1}{2} - \lambda)\,\chi_\lambda\,, \qquad x^\alpha(\vec{p},\lambda) = \sqrt{2E}\,(\tfrac{1}{2} - \lambda)\chi^\dagger_{-\lambda}\,, \tag{2.42}$$

$$y_\alpha(\vec{p},\lambda) = \sqrt{2E}\,(\tfrac{1}{2} + \lambda)\,\chi_{-\lambda}\,, \qquad y^\alpha(\vec{p},\lambda) = \sqrt{2E}\,(\tfrac{1}{2} + \lambda)\chi^\dagger_\lambda\,, \tag{2.43}$$

or, equivalently,

$$x^{\dagger\dot\alpha}(\vec{p},\lambda) = \sqrt{2E}\,(\tfrac{1}{2} - \lambda)\,\chi_{-\lambda}\,, \qquad x^\dagger_{\dot\alpha}(\vec{p},\lambda) = \sqrt{2E}\,(\tfrac{1}{2} - \lambda)\chi^\dagger_\lambda\,, \tag{2.44}$$

$$y^{\dagger\dot\alpha}(\vec{p},\lambda) = \sqrt{2E}\,(\tfrac{1}{2} + \lambda)\,\chi_\lambda\,, \qquad y^\dagger_{\dot\alpha}(\vec{p},\lambda) = \sqrt{2E}\,(\tfrac{1}{2} + \lambda)\chi^\dagger_{-\lambda}\,. \tag{2.45}$$

It follows that

$$(\tfrac{1}{2} + \lambda)\,x(\vec{p},\lambda) = 0\,, \qquad (\tfrac{1}{2} + \lambda)\,x^\dagger(\vec{p},\lambda) = 0\,, \tag{2.46}$$

$$(\tfrac{1}{2} - \lambda)\,y(\vec{p},\lambda) = 0\,, \qquad (\tfrac{1}{2} - \lambda)\,y^\dagger(\vec{p},\lambda) = 0\,. \tag{2.47}$$

The significance of eqs. (2.46) and (2.47) is clear: for massless fermions, only one helicity component of x and y is nonzero.[4] These results are the basis for the spinor helicity method, which is described in more detail in Appendix F.

Equations (2.46)–(2.47) can also be derived by carefully taking the $m \to 0$ limit of eqs. (2.38) and (2.39) applied to the helicity wave functions $x(\vec{p},\lambda)$ and $y(\vec{p},\lambda)$. We then replace mS^μ with p^μ, which is the leading term in the limit of $E \gg m$. Using the results of eqs. (2.10) and (2.11) and dividing out by an overall factor of m (before finally taking the $m \to 0$ limit) reproduces eqs. (2.46) and (2.47).

Noting that $\frac{1}{2}(1 + 2s\,\vec{\sigma}\cdot\hat{s})\chi_{s'} = \frac{1}{2}(1 + 4ss')\chi_{s'} = \delta_{ss'}\chi_{s'}$ (since $s,\, s' = \pm\frac{1}{2}$), one can now obtain the spin projection matrices,

$$\chi_s\chi_s^\dagger = \tfrac{1}{2}\,(1 + 2s\,\vec{\sigma}\cdot\hat{s})\sum_{s'}\chi_{s'}\chi_{s'}^\dagger = \tfrac{1}{2}\,(1 + 2s\,\vec{\sigma}\cdot\hat{s})\,, \tag{2.48}$$

where, at the second step, we have employed the completeness relation given in eq. (E.20). Making use of eq. (2.22) for $\vec{\sigma}\cdot\hat{s}$, it follows that

$$\chi_s\chi_s^\dagger = \tfrac{1}{2}\left(1 + \frac{2s}{m}\sqrt{p\cdot\sigma}\,S\cdot\overline{\sigma}\,\sqrt{p\cdot\sigma}\right). \tag{2.49}$$

[4] Applying this result to massless neutrinos yields left-handed neutrinos ($\lambda = -1/2$) and right-handed antineutrinos ($\lambda = +1/2$).

Hence, with both spinor indices in the lowered position,

$$x(\vec{p},s)x^\dagger(\vec{p},s) = \sqrt{p\cdot\sigma}\,\chi_s\chi_s^\dagger\,\sqrt{p\cdot\overline{\sigma}} = \tfrac{1}{2}\sqrt{p\cdot\sigma}\left[1+\frac{2s}{m}\sqrt{p\cdot\overline{\sigma}}S\cdot\overline{\sigma}\sqrt{p\cdot\sigma}\right]\sqrt{p\cdot\sigma}$$
$$= \tfrac{1}{2}\left[p\cdot\sigma+\frac{2s}{m}p\cdot\sigma S\cdot\overline{\sigma}p\cdot\sigma\right] = \tfrac{1}{2}\left[p\cdot\sigma-2smS\cdot\sigma\right]. \qquad (2.50)$$

In the final step above, we simplified the product of three dot products by noting that $p\cdot S=0$ implies that $S\cdot\overline{\sigma}\,p\cdot\sigma=-p\cdot\overline{\sigma}\,S\cdot\sigma$. The other spin projection formulae for massive fermions can be similarly derived and are given below:

$$x_\alpha(\vec{p},s)x^\dagger_{\dot\beta}(\vec{p},s) = \tfrac{1}{2}(p_\mu-2smS_\mu)\sigma^\mu_{\alpha\dot\beta}\,, \qquad (2.51)$$
$$y^{\dagger\dot\alpha}(\vec{p},s)y^\beta(\vec{p},s) = \tfrac{1}{2}(p^\mu+2smS^\mu)\overline{\sigma}_\mu^{\dot\alpha\beta}, \qquad (2.52)$$
$$x_\alpha(\vec{p},s)y^\beta(\vec{p},s) = \tfrac{1}{2}\left(m\delta_\alpha{}^\beta-2s[S\cdot\sigma\,p\cdot\overline{\sigma}]_\alpha{}^\beta\right), \qquad (2.53)$$
$$y^{\dagger\dot\alpha}(\vec{p},s)x^\dagger_{\dot\beta}(\vec{p},s) = \tfrac{1}{2}\left(m\delta^{\dot\alpha}{}_{\dot\beta}+2s[S\cdot\overline{\sigma}\,p\cdot\sigma]^{\dot\alpha}{}_{\dot\beta}\right). \qquad (2.54)$$

The hermitian conjugates of the above results yield an equivalent set of formulae:

$$x^{\dagger\dot\alpha}(\vec{p},s)x^\beta(\vec{p},s) = \tfrac{1}{2}(p^\mu-2smS^\mu)\overline{\sigma}_\mu^{\dot\alpha\beta}\,, \qquad (2.55)$$
$$y_\alpha(\vec{p},s)y^\dagger_{\dot\beta}(\vec{p},s) = \tfrac{1}{2}(p_\mu+2smS_\mu)\sigma^\mu_{\alpha\dot\beta}\,, \qquad (2.56)$$
$$y_\alpha(\vec{p},s)x^\beta(\vec{p},s) = -\tfrac{1}{2}\left(m\delta_\alpha{}^\beta+2s[S\cdot\sigma\,p\cdot\overline{\sigma}]_\alpha{}^\beta\right), \qquad (2.57)$$
$$x^{\dagger\dot\alpha}(\vec{p},s)y^\dagger_{\dot\beta}(\vec{p},s) = -\tfrac{1}{2}\left(m\delta^{\dot\alpha}{}_{\dot\beta}-2s[S\cdot\overline{\sigma}\,p\cdot\sigma]^{\dot\alpha}{}_{\dot\beta}\right). \qquad (2.58)$$

For the case of massless spin-1/2 fermions, we must use helicity spinor wave functions. The corresponding massless projection operators can be obtained directly from the two-component spinor wave functions given in eqs. (2.42)–(2.45):

$$x_\alpha(\vec{p},\lambda)x^\dagger_{\dot\beta}(\vec{p},\lambda) = (\tfrac{1}{2}-\lambda)p\cdot\sigma_{\alpha\dot\beta}\,, \qquad x^{\dagger\dot\alpha}(\vec{p},\lambda)x^\beta(\vec{p},\lambda) = (\tfrac{1}{2}-\lambda)p\cdot\overline{\sigma}^{\dot\alpha\beta}, \qquad (2.59)$$
$$y^{\dagger\dot\alpha}(\vec{p},\lambda)y^\beta(\vec{p},\lambda) = (\tfrac{1}{2}+\lambda)p\cdot\overline{\sigma}^{\dot\alpha\beta}\,, \qquad y_\alpha(\vec{p},\lambda)y^\dagger_{\dot\beta}(\vec{p},\lambda) = (\tfrac{1}{2}+\lambda)p\cdot\sigma_{\alpha\dot\beta}, \qquad (2.60)$$
$$x_\alpha(\vec{p},\lambda)y^\beta(\vec{p},\lambda) = 0\,, \qquad y_\alpha(\vec{p},\lambda)x^\beta(\vec{p},\lambda) = 0, \qquad (2.61)$$
$$y^{\dagger\dot\alpha}(\vec{p},\lambda)x^\dagger_{\dot\beta}(\vec{p},\lambda) = 0\,, \qquad x^{\dagger\dot\alpha}(\vec{p},\lambda)y^\dagger_{\dot\beta}(\vec{p},\lambda) = 0. \qquad (2.62)$$

As a check, one can verify that the above results follow from eqs. (2.51)–(2.58) by replacing s with λ, setting $mS^\mu=p^\mu$, and taking the $m\to 0$ limit at the end of the computation.

Having listed the projection operators for definite spin projection or helicity, we may now sum over spins to derive the spin-sum identities. These arise when computing squared matrix elements for unpolarized scattering and decay. There are only four basic identities, but for convenience we list each of them with the two-index height permutations that can occur in squared amplitudes by following the rules given in this chapter. The results can be derived by inspection of the spin projection operators, since summing over $s = \pm\frac{1}{2}$ simply removes all terms linear in the spin four-vector S^μ. Thus, we obtain

$$\sum_s x_\alpha(\vec{p},s)x^\dagger_{\dot\beta}(\vec{p},s) = p\cdot\sigma_{\alpha\dot\beta}\,, \qquad \sum_s x^{\dagger\dot\alpha}(\vec{p},s)x^\beta(\vec{p},s) = p\cdot\overline{\sigma}^{\dot\alpha\beta}\,, \tag{2.63}$$

$$\sum_s y^{\dagger\dot\alpha}(\vec{p},s)y^\beta(\vec{p},s) = p\cdot\overline{\sigma}^{\dot\alpha\beta}\,, \qquad \sum_s y_\alpha(\vec{p},s)y^\dagger_{\dot\beta}(\vec{p},s) = p\cdot\sigma_{\alpha\dot\beta}\,, \tag{2.64}$$

$$\sum_s x_\alpha(\vec{p},s)y^\beta(\vec{p},s) = m\delta_\alpha{}^\beta\,, \qquad \sum_s y_\alpha(\vec{p},s)x^\beta(\vec{p},s) = -m\delta_\alpha{}^\beta\,, \tag{2.65}$$

$$\sum_s y^{\dagger\dot\alpha}(\vec{p},s)x^\dagger_{\dot\beta}(\vec{p},s) = m\delta^{\dot\alpha}{}_{\dot\beta}\,, \qquad \sum_s x^{\dagger\dot\alpha}(\vec{p},s)y^\dagger_{\dot\beta}(\vec{p},s) = -m\delta^{\dot\alpha}{}_{\dot\beta}\,. \tag{2.66}$$

These results are applicable both to spin sums and to helicity sums, and hold for both massive and massless spin-1/2 fermions.

One can generalize the above massive and massless projection operators by considering products of two-component spinor wave functions, where the spin or helicity of each spinor can be different. These are the Bouchiat–Michel formulae, which are derived in Appendix E.5.

2.3 Charged Two-Component Fermion Fields

Consider a collection of free anticommuting two-component spin-1/2 fields, $\xi_{\alpha i}(x)$, which transform as $(\frac{1}{2},0)$ fields under the Lorentz group, where i is a flavor index. Each $\xi_{\alpha i}(x)$ can now be expanded in a Fourier series [see eq. (2.3)]:

$$\xi_{\alpha i}(x) = \sum_s \int \frac{d^3\vec{p}}{(2\pi)^{3/2}(2E_{\boldsymbol{p}i})^{1/2}}\left[x_\alpha(\vec{p},s)a_i(\vec{p},s)e^{-ip\cdot x} + y_\alpha(\vec{p},s)a_i^\dagger(\vec{p},s)e^{ip\cdot x}\right]\,, \tag{2.67}$$

where $E_{\boldsymbol{p}i} \equiv (|\vec{p}|^2 + m_i^2)^{1/2}$, and the creation and annihilation operators $a_i^\dagger$ and a_i satisfy anticommutation relations:

$$\{a_i(\vec{p},s), a_j^\dagger(\vec{p}\,',s')\} = \delta^3(\vec{p}-\vec{p}\,')\delta_{ss'}\delta_{ij}\,. \tag{2.68}$$

We employ covariant normalization of the one-particle states, i.e., we act with one creation operator on the vacuum with the following convention:

$$|\vec{p},s,i\rangle \equiv (2\pi)^{3/2}(2E_{\boldsymbol{p}i})^{1/2}a_i^\dagger(\vec{p},s)\,|0\rangle\,, \tag{2.69}$$

so that $\langle\vec{p},s,i\,|\,\vec{p}\,',s',j\rangle = (2\pi)^3(2E_{\boldsymbol{p}i})\delta^3(\vec{p}-\vec{p}\,')\delta_{ij}\delta_{ss'}$.

In the case of two mass-degenerate massive fermion fields, i.e., $m_1 = m_2 \neq 0$, eq. (1.169) possesses a global internal O(2) flavor symmetry, $\xi_i \to \mathcal{O}_i{}^j\xi_j$ ($i \in \{1,2\}$), where $\mathcal{O}^\mathsf{T}\mathcal{O} = \mathbb{1}_{2\times2}$. Corresponding to this symmetry is a conserved hermitian Noether current:

$$J^\mu = i(\xi^{\dagger 1}\overline{\sigma}^\mu\xi_2 - \xi^{\dagger 2}\overline{\sigma}^\mu\xi_1)\,, \tag{2.70}$$

with a corresponding conserved charge, $Q = \int J^0 d^3x$. In the ξ_1–ξ_2 basis, the Noether current is off-diagonal. It is thus convenient to define a new basis of fields:

$$\chi \equiv \frac{1}{\sqrt{2}}(\xi_1 + i\xi_2)\,, \qquad \eta \equiv \frac{1}{\sqrt{2}}(\xi_1 - i\xi_2)\,. \tag{2.71}$$

With respect to the χ–η basis, the Noether current is diagonal:

$$J^\mu = \chi^\dagger \bar{\sigma}^\mu \chi - \eta^\dagger \bar{\sigma}^\mu \eta\,. \tag{2.72}$$

That is, the fermions χ and η are eigenstates of the charge operator Q with corresponding eigenvalues ± 1. In terms of the fermion fields of definite charge, the free-field fermion Lagrangian is given by[5]

$$\mathscr{L} = i\chi^\dagger \bar{\sigma}^\mu \partial_\mu \chi + i\eta^\dagger \bar{\sigma}^\mu \partial_\mu \eta - m(\chi\eta + \chi^\dagger \eta^\dagger)\,. \tag{2.73}$$

On-shell, χ and η satisfy the free-field Dirac equations:

$$i\bar{\sigma}^\mu \partial_\mu \chi - m\eta^\dagger = 0\,, \qquad i\bar{\sigma}^\mu \partial_\mu \eta - m\chi^\dagger = 0\,. \tag{2.74}$$

In the χ–η basis, the global internal SO(2) symmetry (which is the maximal subgroup of O(2) that is continuously connected to the identity) is realized as the U(1) symmetry $\chi \to e^{i\theta}\chi$ and $\eta \to e^{-i\theta}\eta$, where θ is the rotation angle that defines the SO(2) rotation matrix.

Together, χ and $\eta^\dagger$ constitute a single Dirac fermion. In the case of multiple flavors of Dirac fermions,

$$\chi_{\alpha i}(x) = \sum_s \int \frac{d^3\vec{p}}{(2\pi)^{3/2}(2E_{pi})^{1/2}} \left[x_\alpha(\vec{p}, s) a_i(\vec{p}, s) e^{-ip\cdot x} + y_\alpha(\vec{p}, s) b_i^\dagger(\vec{p}, s) e^{ip\cdot x} \right], \tag{2.75}$$

$$\eta^i_\alpha(x) = \sum_s \int \frac{d^3\vec{p}}{(2\pi)^{3/2}(2E_{pi})^{1/2}} \left[x_\alpha(\vec{p}, s) b^i(\vec{p}, s) e^{-ip\cdot x} + y_\alpha(\vec{p}, s) a^{\dagger i}(\vec{p}, s) e^{ip\cdot x} \right], \tag{2.76}$$

where $E_{pi} \equiv (|\vec{p}|^2 + m_i^2)^{1/2}$. The relative heights of the flavor indices (lower for χ and upper for η) is a consequence of the U(1) charge of η that is the negative of the corresponding U(1) charge of χ. The creation and annihilation operators $a^{\dagger i}$, $b_i^\dagger$, a_i, and b^i satisfy anticommutation relations:

$$\{a_i(\vec{p}, s), a^{\dagger j}(\vec{p}\,', s')\} = \{b^j(\vec{p}, s), b_i^\dagger(\vec{p}\,', s')\} = \delta^3(\vec{p} - \vec{p}\,')\delta_{s,s'}\delta_i^j\,, \tag{2.77}$$

and all other anticommutators vanish. We now must distinguish between two types of one-particle states, which we can call fermion (F) and antifermion ($\bar{F}$):

$$|\vec{p}, s, i; F\rangle \equiv (2\pi)^{3/2}(2E_{pi})^{1/2} a^{\dagger i}(\vec{p}, s)\,|0\rangle\,, \tag{2.78}$$

$$|\vec{p}, s, i; \bar{F}\rangle \equiv (2\pi)^{3/2}(2E_{pi})^{1/2} b_i^\dagger(\vec{p}, s)\,|0\rangle\,. \tag{2.79}$$

[5] Although the fermion mass matrix is not diagonal in the χ–η basis, this is not an obstacle to the subsequent analysis, as one only needs a diagonal *squared*-mass matrix, $M^\dagger M$, to ensure that the denominators of propagators are diagonal. Equation (2.71) provides the explicit Takagi diagonalization of the Dirac fermion matrix $\left(\begin{smallmatrix}0 & 1\\ 1 & 0\end{smallmatrix}\right)$. See Appendix G.3 for the mathematical interpretation of this special case.

Note that both $\eta(x)$ and $\chi^\dagger(x)$ can create $|\vec{p}, s; F\rangle$ from the vacuum, while $\eta^\dagger(x)$ and $\chi(x)$ can create $|\vec{p}, s; \overline{F}\rangle$. The one-particle wave functions are given by

$$\langle 0|\, \chi_\alpha(x)\, |\vec{p}, s; F\rangle = x_\alpha(\vec{p}, s)e^{-ip\cdot x}\,, \qquad \langle 0|\, \eta^\dagger_{\dot\alpha}(x)\, |\vec{p}, s; F\rangle = y^\dagger_{\dot\alpha}(\vec{p}, s)e^{-ip\cdot x}\,, \quad (2.80)$$

$$\langle F; \vec{p}, s|\, \eta_\alpha(x)\, |0\rangle = y_\alpha(\vec{p}, s)e^{ip\cdot x}\,, \qquad \langle F; \vec{p}, s|\, \chi^\dagger_{\dot\alpha}(x)\, |0\rangle = x^\dagger_{\dot\alpha}(\vec{p}, s)e^{ip\cdot x}\,, \quad (2.81)$$

$$\langle 0|\, \eta_\alpha(x)\, |\vec{p}, s; \overline{F}\rangle = x_\alpha(\vec{p}, s)e^{-ip\cdot x}\,, \qquad \langle 0|\, \chi^\dagger_{\dot\alpha}(x)\, |\vec{p}, s; \overline{F}\rangle = y^\dagger_{\dot\alpha}(\vec{p}, s)e^{-ip\cdot x}\,, \quad (2.82)$$

$$\langle \overline{F}; \vec{p}, s|\, \chi_\alpha(x)\, |0\rangle = y_\alpha(\vec{p}, s)e^{ip\cdot x}\,, \qquad \langle \overline{F}; \vec{p}, s|\, \eta^\dagger_{\dot\alpha}(x)\, |0\rangle = x^\dagger_{\dot\alpha}(\vec{p}, s)e^{ip\cdot x}\,, \quad (2.83)$$

and the eight other single-particle matrix elements vanish.

2.4 Feynman Rules for External Fermion Lines

Let us consider a general Feynman diagram in which fermions of definite mass (i.e., the so-called fermion mass eigenstates) can appear as initial and final states. The rules for assigning two-component external-state spinors of momentum $\vec{p}$ and spin label s are as follows:

- For an initial-state (incoming) left-handed $(\frac{1}{2}, 0)$ fermion: $x(\vec{p}, s)$
- For an initial-state (incoming) right-handed $(0, \frac{1}{2})$ fermion: $y^\dagger(\vec{p}, s)$
- For a final-state (outgoing) left-handed $(\frac{1}{2}, 0)$ fermion: $x^\dagger(\vec{p}, s)$
- For a final-state (outgoing) right-handed $(0, \frac{1}{2})$ fermion: $y(\vec{p}, s)$

These rules are summarized in the mnemonic diagram shown in Fig. 2.1.

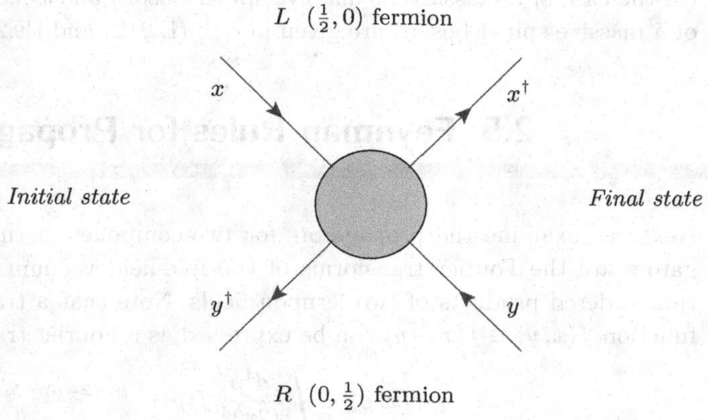

Fig. 2.1 The external wave function spinors should be assigned as indicated here, for initial-state and final-state left-handed $(\frac{1}{2}, 0)$ and right-handed $(0, \frac{1}{2})$ fermions.

In general, the two-component external-state fermion wave functions are distinguished by their Lorentz group transformation properties, rather than by their particle or antiparticle status as in four-component Feynman rules. This helps to explain why two-component notation is especially convenient for (i) theories with Majorana particles, in which there is no fundamental distinction between particles and antiparticles, (ii) theories like the Standard Model and its supersymmetric extension where the left- and right-handed fermions transform under different representations of the gauge group, and (iii) problems with polarized particle beams.

In contrast to four-component Feynman rules (treated in Chapter 3), the direction of the arrow does *not* correspond to the flow of charge or fermion number. The two-component Feynman rules for external fermion lines simply correspond to the formulae for the one-particle wave functions in eqs. (2.8) and (2.9) [with the convention that $|\vec{\boldsymbol{p}}, s\rangle$ is an initial-state fermion and $\langle\vec{\boldsymbol{p}}, s|$ is a final-state fermion]. In particular, the arrows indicate the spinor index structure, with fields of undotted indices flowing *into* any vertex and fields of dotted indices flowing *out* of any vertex.

The rules above apply to any mass-eigenstate two-component fermion external wave functions. It is noteworthy that the same rules apply for the two-component fermions governed by the Lagrangians of eq. (1.169) [Majorana] and eqs. (1.178) or (1.184) [Dirac].

The corresponding rules for an external boson of momentum $\vec{\boldsymbol{k}}$ are well known (see, e.g., [B64]):

- For an initial-state (incoming) or final-state (outgoing) spin-0 boson : 1
- For an initial-state (incoming) spin-1 boson : $\varepsilon^\mu(\vec{\boldsymbol{k}}\,, \lambda)$
- For a final-state (outgoing) spin-1 boson : $\varepsilon^\mu(\vec{\boldsymbol{k}}\,, \lambda)^\star$

The explicit forms of the spin-1 polarization vectors $\varepsilon^\mu(\vec{\boldsymbol{k}}\,, \lambda)$ for helicity $\lambda = \pm 1$ (in the case of a massless or massive spin-1 boson) and helicity $\lambda = 0$ (in the case of a massive spin-1 boson) are given in eqs. (E.212) and (E.214), respectively.

2.5 Feynman Rules for Propagators

Next, we examine the propagators for two-component fermions. Fermion propagators are the Fourier transforms of the free-field vacuum expectation values of time-ordered products of two fermion fields. Note that a translationally invariant function $f(x, y) \equiv f(x - y)$ can be expressed as a Fourier transform:

$$f(x,y) = \int \frac{d^4p}{(2\pi)^4}\, \hat{f}(p)\, e^{-ip\cdot(x-y)}\,. \tag{2.84}$$

It is convenient to introduce the notation $f(x, y)_{\rm FT} \equiv \hat{f}(p)$. For example,

$$\langle 0|\, T\xi_\alpha(x)\xi^\dagger_{\dot\beta}(y)\, |0\rangle_{\rm FT} \equiv \int d^4w\, \langle 0|\, T\xi_\alpha(x)\xi^\dagger_{\dot\beta}(y)\, |0\rangle\, e^{ip\cdot w}\,, \qquad w \equiv x - y\,, \tag{2.85}$$

where the translationally invariant expectation values such as $\langle 0|\, T\xi_\alpha(x)\xi^\dagger_{\dot\beta}(y)\,|0\rangle$ are functions of the coordinate difference $w \equiv x - y$.

We first consider the action for a single massive neutral two-component fermion $\xi_\alpha(x)$ governed by the Lagrangian given by eq. (2.1), coupled to an anticommuting two-component fermionic source term $J_\alpha(x)$:

$$S = \int d^4x\,(\mathscr{L} + J\xi + \xi^\dagger J^\dagger)$$
$$= \int d^4x \left\{ \tfrac{1}{2}\left[i\xi^\dagger\overline{\sigma}^\mu\partial_\mu\xi + i\xi\sigma^\mu\partial_\mu\xi^\dagger - m(\xi\xi + \xi^\dagger\xi^\dagger)\right] + J\xi + \xi^\dagger J^\dagger \right\}, \tag{2.86}$$

where we have split the kinetic energy term symmetrically into two terms. The generating functional is given by

$$W[J, J^\dagger] = N \int \mathcal{D}\xi\,\mathcal{D}\xi^\dagger\, e^{iS[\xi,\,\xi^\dagger,\,J,\,J^\dagger]}, \tag{2.87}$$

where N is a normalization factor chosen such that $W[0,0] = 1$, and $\mathcal{D}\xi\,\mathcal{D}\xi^\dagger$ is the integration measure. It is convenient to Fourier transform the fields $\xi(x)$, $\xi^\dagger(x)$ and sources $J(x)$, $J^\dagger(x)$ in eq. (2.87), and rewrite the action in terms of the corresponding Fourier coefficients $\hat\xi(p)$, $\hat\xi^\dagger(p)$, $\hat J(p)$, and $\hat J^\dagger(p)$:

$$\xi_\alpha(x) = \int \frac{d^4p}{(2\pi)^4} e^{-ip\cdot x}\hat\xi_\alpha(p)\,, \qquad \xi^\dagger_{\dot\alpha}(x) = \int \frac{d^4p}{(2\pi)^4} e^{ip\cdot x}\hat\xi^\dagger_{\dot\alpha}(p)\,, \tag{2.88}$$

$$J_\alpha(x) = \int \frac{d^4p}{(2\pi)^4} e^{-ip\cdot x}\hat J_\alpha(p)\,, \qquad J^\dagger_{\dot\alpha}(x) = \int \frac{d^4p}{(2\pi)^4} e^{ip\cdot x}\hat J^\dagger_{\dot\alpha}(p)\,. \tag{2.89}$$

Furthermore, we introduce the integral representation of the delta function:

$$\delta^{(4)}(x - x') = \int \frac{d^4p}{(2\pi)^4} e^{-ip\cdot(x-x')}\,. \tag{2.90}$$

In order to rewrite eq. (2.87) in a more convenient matrix form, we introduce the following definitions:

$$\Omega(p) \equiv \begin{pmatrix} \hat\xi^{\dagger\dot\alpha}(-p) \\ \hat\xi_\alpha(p) \end{pmatrix}, \qquad X(p) \equiv \begin{pmatrix} \hat J_\alpha(p) \\ \hat J^{\dagger\dot\alpha}(-p) \end{pmatrix},$$

$$\mathcal{M}(p) \equiv \begin{pmatrix} p\cdot\sigma_{\alpha\dot\beta} & -m\,\delta_\alpha{}^\beta \\ -m\,\delta^{\dot\alpha}{}_{\dot\beta} & p\cdot\overline{\sigma}^{\dot\alpha\beta} \end{pmatrix}. \tag{2.91}$$

Note that $\mathcal{M}$ is a hermitian matrix. We can then rewrite the action [eq. (2.86)] in the following matrix form [after using eqs. (1.96) and (1.97) to write the product of the spinor field and the source in a symmetrical fashion]:

$$S = \frac{1}{2}\int \frac{d^4p}{(2\pi)^4}\left(\Omega^\dagger\mathcal{M}\Omega + \Omega^\dagger X + X^\dagger\Omega\right). \tag{2.92}$$

The linear term in the field Ω can be removed by a field redefinition,

$$\Omega' = \Omega + \mathcal{M}^{-1}X\,. \tag{2.93}$$

In terms of Ω', the action now takes the convenient form

$$S = \frac{1}{2}\int \frac{d^4p}{(2\pi)^4}\left(\Omega'^\dagger \mathcal{M}\Omega' - X^\dagger \mathcal{M}^{-1}X\right), \tag{2.94}$$

where the inverse of the matrix $\mathcal{M}$ is given by

$$\mathcal{M}^{-1} = \frac{1}{p^2 - m^2 + i\varepsilon}\begin{pmatrix} p\cdot\overline{\sigma}^{\dot{\alpha}\beta} & m\,\delta^{\dot{\alpha}}{}_{\dot{\beta}} \\ m\,\delta_\alpha{}^\beta & p\cdot\sigma_{\alpha\dot{\beta}}\end{pmatrix}, \tag{2.95}$$

where the $m^2 \to m^2 - i\varepsilon$ prescription has been introduced to ensure the convergence of the path integral [eq. (2.87) with S given by eq. (2.94)].

The Jacobian of the field transformation exhibited in eq. (2.93) is unity. Hence, one can insert the new action, eq. (2.94), in the generating functional, eq. (2.87), to obtain (after dropping the primes on the two-component fermion fields)

$$\begin{aligned} W[\hat{J},\hat{J}^\dagger] &= N\int \mathcal{D}\xi\,\mathcal{D}\xi^\dagger \exp\left\{\frac{i}{2}\int\frac{d^4p}{(2\pi)^4}\left(\Omega^\dagger\mathcal{M}\Omega - X^\dagger\mathcal{M}^{-1}X\right)\right\} \\ &= N\left[\int \mathcal{D}\xi\,\mathcal{D}\xi^\dagger\,\exp\left\{\frac{i}{2}\Omega^\dagger\mathcal{M}\Omega\right\}\right]\exp\left\{-\frac{i}{2}\int\frac{d^4p}{(2\pi)^4}X^\dagger\mathcal{M}^{-1}X\right\} \\ &= \exp\left\{-\frac{i}{2}\int\frac{d^4p}{(2\pi)^4}X^\dagger\mathcal{M}^{-1}X\right\}, \end{aligned} \tag{2.96}$$

where we have defined the normalization constant N such that $W[0,0] = 1$. Inserting the explicit forms for X and $\mathcal{M}$ into eq. (2.96), we obtain

$$\begin{aligned} W[\hat{J},\hat{J}^\dagger] = \exp\Bigg\{-\frac{1}{2}\int\frac{d^4p}{(2\pi)^4}\Bigg(&\hat{J}^\alpha(-p)\frac{ip\cdot\sigma_{\alpha\dot{\beta}}}{p^2-m^2+i\varepsilon}\hat{J}^{\dagger\dot{\beta}}(-p) \\ &+\hat{J}^\dagger_{\dot{\alpha}}(p)\frac{ip\cdot\overline{\sigma}^{\dot{\alpha}\beta}}{p^2-m^2+i\varepsilon}\hat{J}_\beta(p) + \hat{J}^\alpha(-p)\frac{im\delta_\alpha{}^\beta}{p^2-m^2+i\varepsilon}\hat{J}_\beta(p) \\ &+\hat{J}^\dagger_{\dot{\alpha}}(p)\frac{im\delta^{\dot{\alpha}}{}_{\dot{\beta}}}{p^2-m^2+i\varepsilon}\hat{J}^{\dagger\dot{\beta}}(-p)\Bigg)\Bigg\}. \end{aligned} \tag{2.97}$$

Using eq. (1.98), it is convenient to rewrite the first two terms of the integrand on the right-hand side of eq. (2.97) in two different ways:

$$\begin{aligned} &\frac{1}{2}\int\frac{d^4p}{(2\pi)^4}\left[\hat{J}^\alpha(-p)\frac{ip\cdot\sigma_{\alpha\dot{\beta}}}{p^2-m^2+i\varepsilon}\hat{J}^{\dagger\dot{\beta}}(-p) + \hat{J}^\dagger_{\dot{\alpha}}(p)\frac{ip\cdot\overline{\sigma}^{\dot{\alpha}\beta}}{p^2-m^2+i\varepsilon}\hat{J}_\beta(p)\right] \\ &\qquad = \int\frac{d^4p}{(2\pi)^4}\,\hat{J}^\alpha(-p)\frac{ip\cdot\sigma_{\alpha\dot{\beta}}}{p^2-m^2+i\varepsilon}\hat{J}^{\dagger\dot{\beta}}(-p) \\ &\qquad = \int\frac{d^4p}{(2\pi)^4}\,\hat{J}^\dagger_{\dot{\alpha}}(p)\frac{ip\cdot\overline{\sigma}^{\dot{\alpha}\beta}}{p^2-m^2+i\varepsilon}\hat{J}_\beta(p)\,, \end{aligned} \tag{2.98}$$

where we have changed integration variables from p to $-p$ in relating the two terms above. The vacuum expectation value of the time-ordered product of two spinor fields in configuration space is obtained by taking two functional derivatives of the generating functional with respect to the sources J and $J^\dagger$ and then setting $J = J^\dagger = 0$ at the end of the computation (e.g., see [B64]). For example,

$$\begin{aligned}\langle 0|T\xi_\alpha(x_1)\xi^\dagger_{\dot\beta}(x_2)|0\rangle &= \left(-i\frac{\overrightarrow{\delta}}{\delta J^\alpha(x_1)}\right) W[J,J^\dagger]\left(-i\frac{\overleftarrow{\delta}}{\delta J^{\dagger\dot\beta}(x_2)}\right)\Bigg|_{J=J^\dagger=0} \\ &= N\int \mathcal{D}\xi\,\mathcal{D}\xi^\dagger\;\xi_\alpha(x_1)\xi^\dagger_{\dot\beta}(x_2)\exp\left(i\int d^4x\,\mathcal{L}\right), \qquad (2.99)\end{aligned}$$

where the functional derivatives act in the indicated direction (which ensures that no extra minus signs are generated due to the anticommutativity properties of the sources and their functional derivatives). To obtain the two-point functions involving the product of two spinor fields with different combinations of dotted and undotted spinors, it may be more convenient to write $J\xi = \xi J$ and/or $\xi^\dagger J^\dagger = J^\dagger\xi^\dagger$ in eq. (2.87). One can then easily verify the following expressions for the four possible two-point functions:

$$\langle 0|T\xi_\alpha(x_1)\xi^\dagger_{\dot\beta}(x_2)|0\rangle = \left(-i\frac{\overrightarrow{\delta}}{\delta J^\alpha(x_1)}\right) W[J,J^\dagger]\left(-i\frac{\overleftarrow{\delta}}{\delta J^{\dagger\dot\beta}(x_2)}\right)\Bigg|_{J=J^\dagger=0}, \qquad (2.100)$$

$$\langle 0|T\xi^{\dagger\dot\alpha}(x_1)\xi^{\beta}(x_2)|0\rangle = \left(-i\frac{\overrightarrow{\delta}}{\delta J^\dagger_{\dot\alpha}(x_1)}\right) W[J,J^\dagger]\left(-i\frac{\overleftarrow{\delta}}{\delta J_{\beta}(x_2)}\right)\Bigg|_{J=J^\dagger=0}, \qquad (2.101)$$

$$\langle 0|T\xi^{\dagger\dot\alpha}(x_1)\xi^\dagger_{\dot\beta}(x_2)|0\rangle = \left(-i\frac{\overrightarrow{\delta}}{\delta J^\dagger_{\dot\alpha}(x_1)}\right) W[J,J^\dagger]\left(-i\frac{\overleftarrow{\delta}}{\delta J^{\dagger\dot\beta}(x_2)}\right)\Bigg|_{J=J^\dagger=0}, \qquad (2.102)$$

$$\langle 0|T\xi_\alpha(x_1)\xi^{\beta}(x_2)|0\rangle = \left(-i\frac{\overrightarrow{\delta}}{\delta J^\alpha(x_1)}\right) W[J,J^\dagger]\left(-i\frac{\overleftarrow{\delta}}{\delta J_{\beta}(x_2)}\right)\Bigg|_{J=J^\dagger=0}. \qquad (2.103)$$

As an example, we provide details for the evaluation of eq. (2.100). Making use of eqs. (2.97) and (2.98), we obtain

$$\langle 0|T\xi_\alpha(x_1)\xi^\dagger_{\dot\beta}(x_2)|0\rangle = \frac{\overrightarrow{\delta}}{\delta J^\alpha(x_1)}\left(\int\frac{d^4p}{(2\pi)^4}\,\hat{J}^\alpha(-p)\frac{ip\cdot\sigma_{\alpha\dot\beta}}{p^2-m^2+i\varepsilon}\hat{J}^{\dagger\dot\beta}(-p)\right)\frac{\overleftarrow{\delta}}{\delta J^{\dagger\dot\beta}(x_2)}. \qquad (2.104)$$

The chain rule for functional differentiation and the inverse Fourier transforms of eq. (2.89) yield

$$\frac{\delta}{\delta J^\alpha(x_1)} = \int d^4p_1\,\frac{\delta\hat{J}^\beta(-p_1)}{\delta J^\alpha(x_1)}\,\frac{\delta}{\delta\hat{J}^\beta(-p_1)} = \int d^4p_1\,e^{-ip_1\cdot x_1}\frac{\delta}{\delta\hat{J}^\alpha(-p_1)}, \qquad (2.105)$$

$$\frac{\delta}{\delta J^{\dagger\dot\beta}(x_2)} = \int d^4p_2\,\frac{\delta\hat{J}^{\dagger\dot\alpha}(-p_2)}{\delta J^{\dagger\dot\beta}(x_2)}\,\frac{\delta}{\delta\hat{J}^{\dagger\dot\alpha}(-p_2)} = \int d^4p_2\,e^{ip_2\cdot x_2}\frac{\delta}{\delta\hat{J}^{\dagger\dot\beta}(-p_2)}. \qquad (2.106)$$

Applying eqs. (2.105) and (2.106) to eq. (2.104), we obtain

$$\langle 0|T\xi_\alpha(x_1)\xi^\dagger_{\dot\beta}(x_2)|0\rangle = \int \frac{d^4p}{(2\pi)^4}\, e^{-ip\cdot(x_1-x_2)}\, \frac{ip\cdot\sigma_{\alpha\dot\beta}}{p^2-m^2+i\varepsilon}\,. \tag{2.107}$$

Likewise, the same methods can be employed to obtain the vacuum expectation values of products of two undaggered or products of two daggered two-component spinor fields.

One can also apply canonical methods for deriving the fermion propagators. These can be obtained by inserting the free-field expansion of the two-component fermion field and evaluating the spin sums using the formulae given in eqs. (2.63)–(2.66). Of course, the end result is the same. We quote the relevant results below.

For the case of a single neutral two-component fermion field ξ of mass m,

$$\langle 0|\, T\xi_\alpha(x)\xi^\dagger_{\dot\beta}(y)\,|0\rangle_{\rm FT} = \frac{i}{p^2-m^2+i\varepsilon}\sum_s x_\alpha(\vec{\boldsymbol p},s)x^\dagger_{\dot\beta}(\vec{\boldsymbol p},s) = \frac{i}{p^2-m^2+i\varepsilon}\,p\cdot\sigma_{\alpha\dot\beta}\,, \tag{2.108}$$

$$\langle 0|\, T\xi^{\dagger\dot\alpha}(x)\xi^{\beta}(y)\,|0\rangle_{\rm FT} = \frac{i}{p^2-m^2+i\varepsilon}\sum_s y^{\dagger\dot\alpha}(\vec{\boldsymbol p},s)y^{\beta}(\vec{\boldsymbol p},s) = \frac{i}{p^2-m^2+i\varepsilon}\,p\cdot\overline{\sigma}^{\dot\alpha\beta}\,, \tag{2.109}$$

$$\langle 0|\, T\xi^{\dagger\dot\alpha}(x)\xi^\dagger_{\dot\beta}(y)\,|0\rangle_{\rm FT} = \frac{i}{p^2-m^2+i\varepsilon}\sum_s y^{\dagger\dot\alpha}(\vec{\boldsymbol p},s)x^\dagger_{\dot\beta}(\vec{\boldsymbol p},s) = \frac{i}{p^2-m^2+i\varepsilon}\,m\delta^{\dot\alpha}{}_{\dot\beta}\,, \tag{2.110}$$

$$\langle 0|\, T\xi_\alpha(x)\xi^{\beta}(y)\,|0\rangle_{\rm FT} = \frac{i}{p^2-m^2+i\varepsilon}\sum_s x_\alpha(\vec{\boldsymbol p},s)y^{\beta}(\vec{\boldsymbol p},s) = \frac{i}{p^2-m^2+i\varepsilon}\,m\delta_\alpha{}^{\beta}\,. \tag{2.111}$$

These results have an obvious diagrammatic representation, as shown in Fig. 2.2. Note that the direction of the momentum flow p^μ here is determined by the creation operator that appears in the evaluation of the free-field propagator. Arrows on fermion lines always run away from dotted indices at a vertex and toward undotted indices at a vertex.

There are two types of two-component fermion propagators. The first type preserves the direction of arrows, so it has one dotted and one undotted index. For this type of propagator, it is convenient to establish a convention where p^μ in the diagram is defined to be the momentum flowing in the direction of the arrow on the fermion propagator. With this convention, the two rules above for propagators of the first type can be summarized by a single rule, as shown in Fig. 2.3. Here the choice of the σ or the $\overline{\sigma}$ version of the rule is uniquely determined by the height of the indices on the vertex to which the propagator is connected.[6] These heights should always be chosen so that they are contracted as in eq. (1.63). It should be

[6] The second form of the rule in Fig. 2.3 arises when one flips diagram (b) of Fig. 2.2 around by a 180° rotation (about an axis perpendicular to the plane of the diagram), and then relabels $\dot\alpha \to \dot\beta$ and $\beta \to \alpha$.

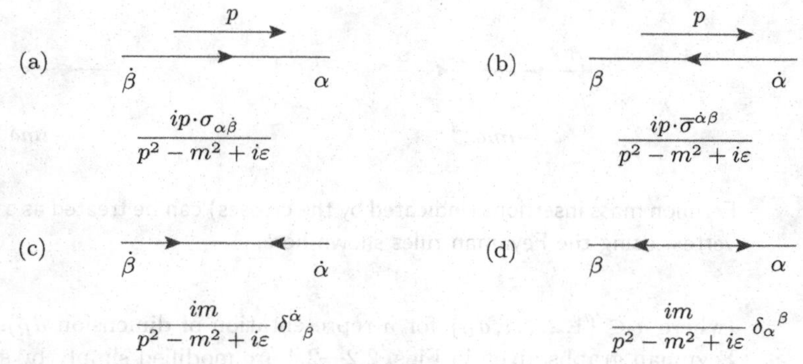
(a)
p
$\dot\beta$
α
$\dfrac{ip\cdot\sigma_{\alpha\dot\beta}}{p^2-m^2+i\varepsilon}$
(b)
p
β
$\dot\alpha$
$\dfrac{ip\cdot\overline{\sigma}^{\dot\alpha\beta}}{p^2-m^2+i\varepsilon}$
(c)
$\dot\beta$
$\dot\alpha$
$\dfrac{im}{p^2-m^2+i\varepsilon}\,\delta^{\dot\alpha}{}_{\dot\beta}$
(d)
β
α
$\dfrac{im}{p^2-m^2+i\varepsilon}\,\delta_{\alpha}{}^{\beta}$

Fig. 2.2 Feynman rules for propagator lines of a neutral two-component spin-1/2 fermion.

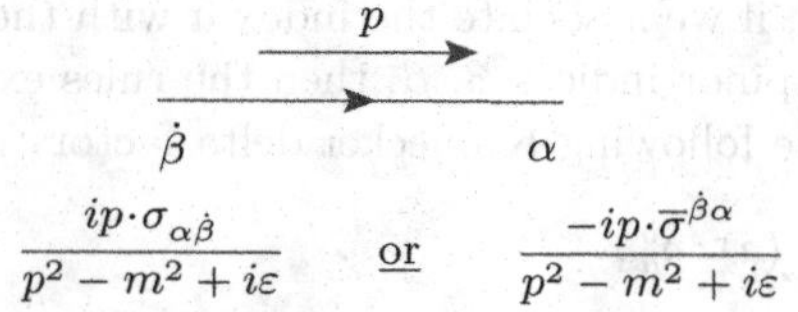
p
$\dot\beta$
α
$\dfrac{ip\cdot\sigma_{\alpha\dot\beta}}{p^2-m^2+i\varepsilon}$ or $\dfrac{-ip\cdot\overline{\sigma}^{\dot\beta\alpha}}{p^2-m^2+i\varepsilon}$

Fig. 2.3 This single rule summarizes the rules of Fig. 2.2(a) and (b).

noted that in diagrams (a) and (b) of Fig. 2.2 as drawn, the indices on the σ and $\overline{\sigma}$ read from right to left. In particular, the Feynman rules for the propagator can be employed with the spinor indices suppressed, provided that the arrow-preserving propagator lines are traversed in the direction parallel [antiparallel] to the arrowed line segment for the $\overline{\sigma}$ [σ] version of the rule, respectively.

The second type of propagator shown in diagrams (c) and (d) of Fig. 2.2 does not preserve the direction of arrows, and corresponds to an odd number of mass insertions. The indices on $\delta^{\dot\alpha}{}_{\dot\beta}$ and $\delta_{\alpha}{}^{\beta}$ are staggered as shown to indicate that $\dot\alpha$ and α are to be contracted with expressions to the left, while $\dot\beta$ and β are to be contracted with expressions to the right, in accordance with eq. (1.63). As in Fig. 2.3, alternative and equivalent versions of the rules corresponding to diagrams (c) and (d) of Fig. 2.2 can be given for which the indices on the Kronecker deltas are staggered as $\delta^{\dot\beta}{}_{\dot\alpha}$ and $\delta_{\beta}{}^{\alpha}$. These versions correspond to flipping the two respective diagrams by 180° and relabeling the indices $\dot\alpha \to \dot\beta$ and $\beta \to \alpha$.

Starting with massless fermion propagators, one can also derive the massive fermion propagators by employing mass insertions as interaction vertices, as shown in Fig. 2.4. By summing up an infinite chain of such mass insertions between massless fermion propagators, one can reproduce the massive fermion propagators of both types.

The above results for the propagator of a Majorana fermion can be generalized to a multiplet of mass-eigenstate Majorana fermions, $\xi_{\alpha a}(x)$ [such as a color octet of gluinos], which transforms as a real representation R of a (gauge or flavor) group G

$$-im\delta_{\alpha}{}^{\beta} \qquad\qquad -im\delta^{\dot{\alpha}}{}_{\dot{\beta}}$$

Fig. 2.4 Fermion mass insertions (indicated by the crosses) can be treated as a type of interaction vertex, using the Feynman rules shown here.

(where $a \in \{1, 2, \ldots, d_R\}$ for a representation of dimension d_R). In this case, the Feynman graphs given in Figs. 2.2—2.4 are modified simply by specifying a group index a and b at either end of the propagator line. The corresponding Feynman rules then include an additional Kronecker delta factor in the group indices. In particular, if we associate the index a with the spinor indices α, $\dot{\alpha}$ and the index b with the spinor indices β, $\dot{\beta}$, then the rules exhibited in Fig. 2.2(a) and (b) would include the following Kronecker delta factors:

$$(a)\ \delta_a^b\,, \qquad\qquad (b)\ \delta_b^a\,, \tag{2.112}$$

and the factors of m in the rules in Fig. 2.2(c) and (d) would be replaced by

$$(c)\ \delta_c^a m^{cd}\delta_d^b = m_a\delta^{ab}\,, \qquad\qquad (d)\ \delta_a^c m_{cd}\delta_b^d = m_a\delta_{ab} \tag{2.113}$$

(with no sum over the repeated index a), where m^{cd} and $m_{cd} \equiv m^{cd}$ are diagonal matrices with real nonnegative diagonal elements m_c. We have introduced the separate symbol m_{cd} in order to maintain the convention that two repeated group indices are summed when one index is raised and one is lowered. If the Lagrangian is invariant under the symmetry group G, then a multiplet of Majorana fermions corresponding to an irreducible representation R has a common mass $m = m_a$.

It is convenient to treat separately the case of charged massive fermions. Consider a charged Dirac fermion of mass m, which is described by a pair of two-component fields $\chi(x)$ and $\eta(x)$, whose free-field Lagrangian is given by eq. (2.73). We shall work in a basis of fields where the action, including external anticommuting sources, is given by

$$S[\chi, \chi^\dagger, \eta, \eta^\dagger, J_\chi, J_\chi^\dagger, J_\eta, J_\eta^\dagger] = \int d^4x \Big[i\chi^\dagger\bar{\sigma}^\mu\partial_\mu\chi + i\eta^\dagger\bar{\sigma}^\mu\partial_\mu\eta - m(\chi\eta + \chi^\dagger\eta^\dagger) + J_\chi\chi + \chi^\dagger J_\chi^\dagger + J_\eta\eta + \eta^\dagger J_\eta^\dagger \Big]. \tag{2.114}$$

Following the techniques employed in deriving the neutral fermion propagators, we introduce Fourier coefficients for all the fields and sources and define

$$\Omega_c(p) \equiv \begin{pmatrix} \hat{\eta}^{\dagger\dot{\alpha}}(-p) \\ \hat{\chi}_\alpha(p) \end{pmatrix}, \qquad X_c(p) \equiv \begin{pmatrix} \hat{J}_{\eta\alpha}(p) \\ \hat{J}_\chi^{\dagger\dot{\alpha}}(-p) \end{pmatrix}. \tag{2.115}$$

The action functional, eq. (2.114), can then rewritten in matrix form as before (but

(a) $\quad \chi \xrightarrow{\quad p \quad} \chi$
$\quad\quad\quad \dot\beta \quad\quad\quad\quad \alpha$

(b) $\quad \eta \xrightarrow{\quad p \quad} \eta$
$\quad\quad\quad \dot\beta \quad\quad\quad\quad \alpha$

$$\frac{ip\cdot\sigma_{\alpha\dot\beta}}{p^2-m^2+i\varepsilon} \quad \text{or} \quad \frac{-ip\cdot\overline{\sigma}^{\dot\beta\alpha}}{p^2-m^2+i\varepsilon} \quad\quad\quad \frac{ip\cdot\sigma_{\alpha\dot\beta}}{p^2-m^2+i\varepsilon} \quad \text{or} \quad \frac{-ip\cdot\overline{\sigma}^{\dot\beta\alpha}}{p^2-m^2+i\varepsilon}$$

(c) $\quad \chi \xrightarrow{\quad\quad} \leftarrow \eta$
$\quad\quad\quad \dot\beta \quad\quad\quad\quad \dot\alpha$

(d) $\quad \chi \leftarrow \xrightarrow{\quad\quad} \eta$
$\quad\quad\quad \beta \quad\quad\quad\quad \alpha$

$$\frac{im}{p^2-m^2+i\varepsilon}\delta^{\dot\alpha}{}_{\dot\beta} \quad\quad\quad\quad\quad\quad \frac{im}{p^2-m^2+i\varepsilon}\delta_\alpha{}^\beta$$

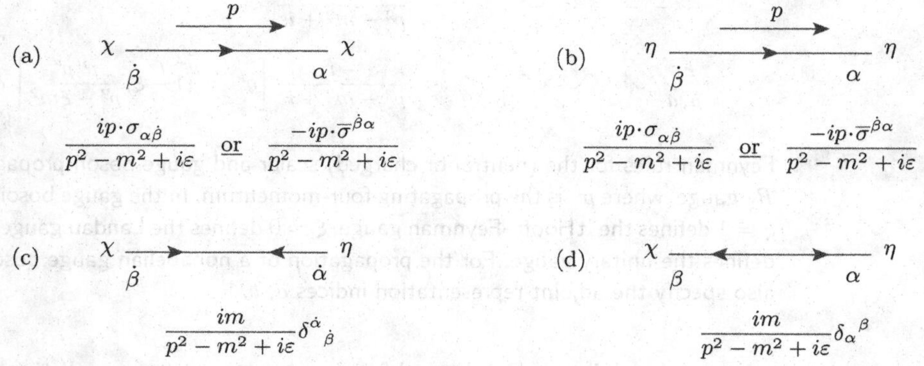

Fig. 2.5 Feynman rules for propagator lines of a pair of charged two-component fermions with a Dirac mass m. As in Fig. 2.3, the direction of the momentum is taken to flow from the dotted to the undotted index in diagrams (a) and (b).

with no overall factor of $1/2$):

$$S = \int \frac{d^4p}{(2\pi)^4} \left(\Omega_c^\dagger\mathcal{M}\Omega_c + \Omega_c^\dagger X_c + X_c^\dagger\Omega_c\right), \tag{2.116}$$

where $\mathcal{M}$ is again given by eq. (2.91). The remaining calculation proceeds as before with few modifications, and yields the Dirac two-component fermion free-field propagators.

As in the derivation of eqs. (2.108)–(2.111), one can also apply canonical methods for deriving the Dirac fermion propagators. These can be obtained by employing the free-field expansions given by eqs. (2.75) and (2.76), and the appropriate spin-sum formulae given in eqs. (2.63)–(2.66). We quote the relevant results below.

$$\langle 0|\, T\chi_\alpha(x)\chi_{\dot\beta}^\dagger(y)\,|0\rangle_{\mathrm{FT}} = \langle 0|\, T\eta_\alpha(x)\eta_{\dot\beta}^\dagger(y)\,|0\rangle_{\mathrm{FT}} = \frac{i}{p^2-m^2+i\varepsilon}\, p\cdot\sigma_{\alpha\dot\beta}, \tag{2.117}$$

$$\langle 0|\, T\chi^{\dagger\dot\alpha}(x)\chi^\beta(y)\,|0\rangle_{\mathrm{FT}} = \langle 0|\, T\eta^{\dagger\dot\alpha}(x)\eta^\beta(y)\,|0\rangle_{\mathrm{FT}} = \frac{i}{p^2-m^2+i\varepsilon}\, p\cdot\overline{\sigma}^{\dot\alpha\beta}, \tag{2.118}$$

$$\langle 0|\, T\chi_\alpha(x)\eta^\beta(y)\,|0\rangle_{\mathrm{FT}} = \langle 0|\, T\eta_\alpha(x)\chi^\beta(y)\,|0\rangle_{\mathrm{FT}} = \frac{i}{p^2-m^2+i\varepsilon}\, m\,\delta_\alpha{}^\beta, \tag{2.119}$$

$$\langle 0|\, T\chi^{\dagger\dot\alpha}(x)\eta_{\dot\beta}^\dagger(y)\,|0\rangle_{\mathrm{FT}} = \langle 0|\, T\eta^{\dagger\dot\alpha}(x)\chi_{\dot\beta}^\dagger(y)\,|0\rangle_{\mathrm{FT}} = \frac{i}{p^2-m^2+i\varepsilon}\, m\,\delta^{\dot\alpha}{}_{\dot\beta}. \tag{2.120}$$

For all other combinations of fermion bilinears, the corresponding two-point functions vanish. These results again have a simple diagrammatic representation, as shown in Fig. 2.5. Note that, for Dirac fermions, the propagators with opposing arrows (proportional to a mass) necessarily change the identity (χ or η) of the two-component fermion, while the single-arrow propagators are diagonal in the fields. In

$$\frac{i}{p^2 - m^2 + i\varepsilon}$$

$$\frac{-i}{p^2 - m^2 + i\varepsilon}\left[g^{\mu\nu} - (1-\xi)\frac{p^\mu p^\nu}{p^2 - \xi m^2}\right]\delta^{ab}$$

Fig. 2.6 Feynman rules for the (neutral or charged) scalar and gauge boson propagators, in the R_ξ gauge, where p^μ is the propagating four-momentum. In the gauge boson propagator, $\xi = 1$ defines the 'tHooft–Feynman gauge, $\xi = 0$ defines the Landau gauge, and $\xi \to \infty$ defines the unitary gauge. For the propagation of a nonabelian gauge boson, one must also specify the adjoint representation indices a, b.

processes involving such a charged fermion, one must of course distinguish between the χ and η fields.

The above results for the propagator of a Dirac fermion can be generalized to a multiplet of mass-eigenstate Dirac fermions, $\chi_{\alpha i}$, η^i_α, which transform under a (gauge or flavor) group G. In this case, the Feynman graphs given in Fig. 2.5 are modified simply by specifying a group index i and j at either end of the propagator line. The corresponding Feynman rules then include an additional Kronecker delta factor in the group indices. In particular, if we associate the group index i with the spinor indices α, $\dot\alpha$ and the index j with the spinor indices β, $\dot\beta$, then the rules exhibited in Fig. 2.5(a) and (b) include the following Kronecker delta factors:

$$(a)\ \delta_i^j\,, \qquad\qquad (b)\ \delta^i_j\,, \qquad (2.121)$$

and the factors of m in the rules exhibited in Fig. 2.5(c) and (d) are replaced by

$$(c)\ \delta_i^\ell m_\ell{}^n \delta_n^j = m_i \delta_i^j\,, \qquad\qquad (d)\ \delta^i_\ell m^\ell{}_n \delta^n_j = m_i \delta^i_j\,, \qquad (2.122)$$

where $m^\ell{}_n$ and $m_\ell{}^n \equiv m^\ell{}_n$ are diagonal matrices with real nonnegative diagonal elements m_ℓ, and there is no sum over the repeated index i. (Here, we have introduced the separate symbol $m_\ell{}^n$ in order to maintain the convention that two repeated group indices are summed when one index is raised and one index is lowered.) As before, if the Lagrangian is invariant under the symmetry group G, then an irreducible multiplet of Dirac fermions has a common mass $m = m_i$.

Finally, we exhibit in Fig. 2.6 the Feynman rules for the (neutral or charged) scalar boson propagator and for the gauge boson propagator in the R_ξ gauge, with gauge parameter ξ, which will be treated in more detail in Chapter 4.

2.6 Feynman Rules for Fermion Interactions

We next discuss the interaction vertices for fermions with bosons. Renormalizable Lorentz-invariant interactions involving fermions must consist of bilinears in the fermion fields, which transform as a Lorentz scalar or vector, coupled to the appropriate bosonic scalar or vector field to make an overall Lorentz scalar quantity.

Let us write all of the two-component left-handed $(\frac{1}{2},0)$ fermions of the theory as $\hat{\psi}_i$, where i runs over all of the gauge group representation and flavor degrees of freedom. In general, the $(\frac{1}{2},0)$-fermion fields $\hat{\psi}_i$ consist of Majorana fermions $\hat{\xi}_i$, and Dirac fermion pairs $\hat{\chi}_i$ and $\hat{\eta}^i$ after mass terms (both explicit and coming from spontaneous symmetry breaking) are taken into account. Likewise, consider a multiplet of scalar fields $\hat{\phi}_I$, where I runs over all of the gauge group representation and flavor degrees of freedom. In general, the scalar fields $\hat{\phi}_I$ consist of real scalar fields[7] $\hat{\varphi}_I$ and pairs of complex scalar fields $\hat{\Phi}_I$ and $\hat{\Phi}^{\dagger I} \equiv (\hat{\Phi}_I)^\dagger$, where the dagger denotes complex conjugation for classical fields and hermitian conjugation for quantum field operators. In matrix form,

$$\hat{\psi} \equiv \begin{pmatrix} \hat{\xi} \\ \hat{\chi} \\ \hat{\eta} \end{pmatrix}, \qquad \hat{\phi} \equiv \begin{pmatrix} \hat{\varphi} \\ \hat{\Phi} \\ \hat{\Phi}^\dagger \end{pmatrix}. \tag{2.123}$$

By dividing up the fermions into Majorana and Dirac fermions and the spin-0 fields into real and complex scalars, we are assuming implicitly that some of the indices I and i correspond to states of a definite (global) U(1) charge (denoted in the following by q_I and q_i, respectively).

The most general set of Yukawa interactions of the scalar fields with a pair of fermion fields is then given by

$$\mathscr{L}_{\rm int} = -\tfrac{1}{2}\hat{Y}^{Ijk}\hat{\phi}_I\hat{\psi}_j\hat{\psi}_k - \tfrac{1}{2}\hat{Y}_{Ijk}\hat{\phi}^{\dagger I}\hat{\psi}^{\dagger j}\hat{\psi}^{\dagger k}\,, \tag{2.124}$$

where $\hat{Y}_{Ijk} = (\hat{Y}^{Ijk})^*$ is symmetric under the interchange of j and k. We have suppressed the spinor indices here; the product of two-component spinors is always performed according to the index convention indicated in eq. (1.63). The Yukawa Lagrangian [eq. (2.124)] must be invariant under

$$\begin{aligned} &\hat{\xi}_i \to \hat{\xi}_i\,, \qquad \hat{\chi}_i \to e^{iq_i\theta}\hat{\chi}_i\,, \qquad \hat{\eta}^i \to e^{-iq_i\theta}\hat{\eta}^i\,, \qquad \hat{\varphi}_i \to \hat{\varphi}_i\,, \\ &\hat{\Phi}_I \to e^{iq_I\theta}\hat{\Phi}_I, \qquad \hat{\Phi}^{\dagger I} \to e^{-iq_I\theta}\hat{\Phi}^{\dagger I}\,, \end{aligned} \tag{2.125}$$

where the q_i are the U(1) charges of the corresponding Dirac fermions and the q_I are the U(1) charges of the corresponding complex scalars. Consequently, the form of the $\hat{Y}^{Ijk}$ is constrained:

$$\hat{Y}^{Ijk} = 0\,, \quad \text{unless} \quad q_I + q_j + q_k = 0\,. \tag{2.126}$$

Of course, any other conserved symmetries will impose additional selection rules on the Yukawa couplings $\hat{Y}^{Ijk}$.

The hatted fields are the interaction eigenstate fields. However, the computation of matrix elements for physical processes is more conveniently done in terms of

[7] Classical real scalar fields are promoted to hermitian field operators in quantum field theory. Nevertheless, in this book we shall employ the nomenclature of classical fields by referring to real and complex fields instead of hermitian and nonhermitian fields.

the propagating mass-eigenstate fields. In Chapter 1, we showed that the mass-eigenstate charged and neutral two-component fermion fields can be identified by performing a singular value decomposition and a Takagi diagonalization of the corresponding fermion mass matrices, respectively. The mass-eigenstate fields ψ are related to the interaction eigenstate fields $\hat{\psi}$ by a unitary rotation $U_i{}^j$ on the flavor indices. In matrix form,

$$\hat{\psi} \equiv \begin{pmatrix} \hat{\xi}_i \\ \hat{\chi}_i \\ \hat{\eta}^i \end{pmatrix} = U\psi \equiv \begin{pmatrix} \Omega_i{}^j & 0 & 0 \\ 0 & L_i{}^j & 0 \\ 0 & 0 & R^i{}_j \end{pmatrix} \begin{pmatrix} \xi_j \\ \chi_j \\ \eta^j \end{pmatrix}, \tag{2.127}$$

where Ω, L, and R are constructed as described previously in Chapter 1 [see eqs. (1.167) and (1.175)].

In contrast, the mass-eigenstate scalar fields are identified by performing a unitary similarity transformation of the corresponding scalar squared-mass matrix. Consider a collection of free commuting real spin-0 fields, $\hat{\varphi}_i(x)$, where the flavor index i labels the distinct scalar fields of the collection. The free-field Lagrangian is given by[8]

$$\mathscr{L} = \tfrac{1}{2}\partial^\mu \hat{\varphi}_i \partial_\mu \hat{\varphi}_i - \tfrac{1}{2} M^2_{ij} \hat{\varphi}_i \hat{\varphi}_j \,, \tag{2.128}$$

where M^2 is a real symmetric matrix. We diagonalize the scalar squared-mass matrix by introducing mass eigenstates φ_i and the real orthogonal matrix Q such that $\hat{\varphi}_i = Q_{ij}\varphi_j$, with $M^2_{ij} Q_{ik} Q_{j\ell} = m_k^2 \delta_{k\ell}$ (no sum over k). In matrix form,

$$Q^{\mathsf{T}} M^2 Q = \boldsymbol{m}^2 = \mathrm{diag}(m_1^2, m_2^2, \ldots) \,, \tag{2.129}$$

where the squared-mass eigenvalues m_k^2 are real.[9] This is the standard diagonalization problem for a real symmetric matrix.

Next, consider a collection of free commuting complex spin-0 fields, $\hat{\Phi}_i(x)$. For complex fields, we follow the conventions for flavor indices specified in Appendix A.2 [e.g., $\hat{\Phi}^{\dagger i} \equiv (\hat{\Phi}_i)^\dagger$]. The free-field Lagrangian is given by

$$\mathscr{L} = \partial^\mu \hat{\Phi}^{\dagger i} \partial_\mu \hat{\Phi}_i - (M^2)_j{}^i \hat{\Phi}^{\dagger j} \hat{\Phi}_i \,, \tag{2.130}$$

where M^2 is a hermitian matrix [i.e., $(M^2)_j{}^i = (M^2)^i{}_j$ in the notation of eq. (1.172)]. We diagonalize the scalar squared-mass matrix by introducing mass eigenstates Φ_i and the unitary matrix W such that $\hat{\Phi}_i = W_i{}^k \Phi_k$ (and $\hat{\Phi}^{\dagger i} = W^i{}_k \Phi^{\dagger k}$), with $(M^2)^i{}_j W_i{}^k W^j{}_\ell = m_k^2 \delta_\ell^k$ (no sum over k). In matrix form,

$$W^\dagger M^2 W = \boldsymbol{m}^2 = \mathrm{diag}(m_1^2, m_2^2, \ldots) \,, \tag{2.131}$$

where the squared-mass eigenvalues m_k^2 are real (see footnote 9). This is the standard diagonalization problem for a hermitian matrix.

[8] Since the scalar fields are real, there is no need to distinguish between raised and lowered flavor indices.

[9] If the vacuum corresponds to a local minimum (or flat direction) of the scalar potential, then the squared-mass eigenvalues of M^2 are real *and* nonnegative.

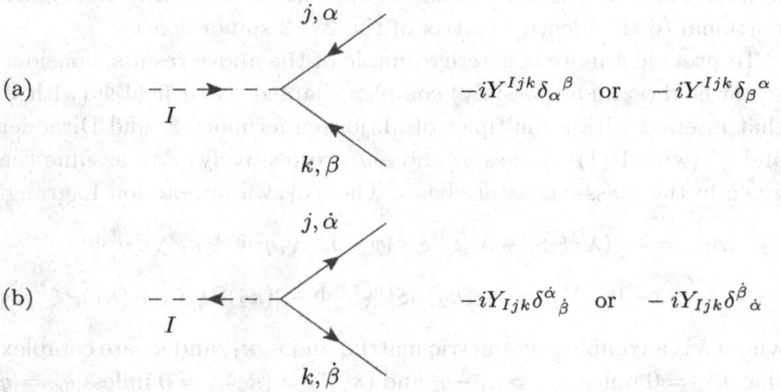

Fig. 2.7 Feynman rules for Yukawa couplings of scalars to two-component fermions in a general field theory. The choice of which rule to use depends on how the vertex connects to the rest of the amplitude. When indices are suppressed, the spinor index part is always proportional to the identity matrix.

Thus, the scalar mass-eigenstate fields ϕ are related to the interaction eigenstate fields $\hat\phi$ by a unitary rotation $V_I{}^J$ on the flavor indices. In matrix form,

$$\hat\phi \equiv \begin{pmatrix} \hat\varphi_I \\ \hat\Phi_I \\ \hat\Phi^{\dagger I} \end{pmatrix} = V\phi \equiv \begin{pmatrix} Q_I{}^J & 0 & 0 \\ 0 & W_I{}^J & 0 \\ 0 & 0 & W^I{}_J \end{pmatrix} \begin{pmatrix} \varphi_J \\ \Phi_J \\ \Phi^{\dagger J} \end{pmatrix}, \tag{2.132}$$

where $W^I{}_J = (W_I{}^J)^*$, and Q and W are constructed according to eqs. (2.129) and (2.131).

Thus, we may rewrite eq. (2.124) in terms of mass-eigenstate fields:

$$\mathscr{L}_{\text{int}} = -\tfrac{1}{2}Y^{Ijk}\phi_I\psi_j\psi_k - \tfrac{1}{2}Y_{Ijk}\phi^{\dagger I}\psi^{\dagger j}\psi^{\dagger k}, \tag{2.133}$$

where

$$Y^{Ijk} = Y^{Ikj} \equiv V_J{}^I U_m{}^j U_n{}^k \hat{Y}^{Jmn}. \tag{2.134}$$

Note that eq. (2.126) implies that $Y^{Ijk} = 0$ unless $q_I + q_j + q_k = 0$. The corresponding Feynman rules that arise from the Yukawa interaction Lagrangian are shown in Fig. 2.7. If the scalar ϕ_I is complex, then one can associate an arrow with the flow of analyticity, which would point into the vertex in (a) and out of the vertex in (b). That is, the arrow on the scalar line keeps track of the height of the scalar flavor index entering or leaving the vertex.

In Fig. 2.7, two versions are given for each Feynman rule. The choice of which rule to use is dictated by the height of the indices on the fermion lines that connect to the vertex. These heights should always be chosen so that they are contracted as in eq. (1.63). However, when all spinor indices are suppressed, the scalar–fermion–

fermion rules will have an identical appearance for both cases, since they are proportional to the identity matrix of the 2×2 spinor space.

To provide a more concrete example of the above results, consider a real neutral scalar field ϕ and a (possibly) complex charged scalar field Φ (with U(1) charge q_Φ) that interact with a multiplet of Majorana fermions ξ_i and Dirac fermion pairs χ_j and η^j (with U(1) charges q_j and $-q_j$, respectively). We assume that all fields are given in the mass-eigenstate basis. The Yukawa interaction Lagrangian is given by

$$\begin{aligned}\mathscr{L}_{\text{int}} = & -\tfrac{1}{2}(\lambda^{ij}\xi_i\xi_j + \lambda_{ij}\xi^{\dagger i}\xi^{\dagger j})\phi - \kappa^i{}_j\chi_i\eta^j\Phi + \kappa_i{}^j\chi^{\dagger i}\eta_j^\dagger\Phi^\dagger \\ & - [(\kappa_1)^i{}_j\xi_i\eta^j + (\kappa_2)_{ij}\xi^{\dagger i}\chi^{\dagger j}]\Phi - [(\kappa_2)^{ij}\xi_i\chi_j + (\kappa_1)_i{}^j\xi^{\dagger i}\eta_j^\dagger]\Phi^\dagger ,\end{aligned} \quad (2.135)$$

where λ is a complex symmetric matrix, and κ, κ_1, and κ_2 are complex matrices such that $\kappa^i{}_j = 0$ unless $q_\Phi = q_j - q_i$ and $(\kappa_1)^i{}_j = (\kappa_2)_{ij} = 0$ unless $q_\Phi = q_j$ [flavor-index conventions are specified in eqs. (1.160) and (1.171)]. The corresponding Feynman rules of Fig. 2.7(a) are obtained by identifying $Y^{Iij} = \lambda^{ij}$, $\kappa^i{}_j$, $(\kappa_1)^i{}_j$, and $(\kappa_2)^{ij}$ for the undotted fermion vertices $\phi\xi_i\xi_j$, $\Phi\chi_i\eta^j$, $\Phi\xi_i\eta^j$, and $\Phi^\dagger\xi_i\chi_j$, respectively.[10] The corresponding Feynman rules of Fig. 2.7(b) for the dotted fermion vertices are governed by the complex-conjugated Yukawa couplings, $Y_{Ijk} \equiv (Y^{Ijk})^*$.

The renormalizable interactions of vector bosons with fermions and scalars arise from gauge interactions. As shown in Section 4.3, these interaction terms of the Lagrangian derive from the respective kinetic energy terms of the fermions and scalars when the derivative is promoted to the covariant derivative:

$$(\nabla_\mu)_i{}^j \equiv \delta_i{}^j\partial_\mu + ig_a A_\mu^a (\boldsymbol{T^a})_i{}^j , \quad (2.136)$$

where the index a labels the real (interaction eigenstate) vector bosons A_μ^a. There is an implicit sum over the thrice-repeated index a, which runs over the adjoint representation of the gauge group.[11] The $(\boldsymbol{T^a})_i{}^j$ are hermitian representation matrices of the generators of the Lie algebra of the gauge group acting on the left-handed fermions (for further details, see Appendix H). For a U(1) gauge group, the $\boldsymbol{T^a}$ are replaced by real numbers corresponding to the U(1) charges of the left-handed $(\frac{1}{2}, 0)$ fermions. The gauge group G can be expressed (in general) as a direct product of simple groups and one or more copies of the U(1) group. Each of the simple groups and U(1) groups contained in the direct product is associated with gauge coupling constant g_a. In particular, the generators $\boldsymbol{T^a}$ separate out into distinct classes, each of which is identified with one of the a simple groups or U(1) groups of the direct product, where $g_a = g_b$ if $\boldsymbol{T^a}$ and $\boldsymbol{T^b}$ are in the same class. If G is simple, then $g_a = g$ for all a.

In the gauge-interaction basis for the left-handed $(\frac{1}{2}, 0)$ two-component fermions,

[10] For the $\Phi^\dagger\xi_i\chi_j$ vertex, we should reverse the direction of the arrow on the scalar line in Fig. 2.7(a) [and likewise for the corresponding hermitian-conjugated vertex of Fig. 2.7(b)], in which case all arrows on the charged scalar and fermion lines would represent the direction of flow of the conserved U(1) charge.

[11] Since the adjoint representation is a real representation, the height of the adjoint index a is not significant. The choice of a subscript or superscript adjoint index is based solely on typographical considerations.

the corresponding interaction Lagrangian is given by

$$\mathscr{L}_{\rm int} = -g_a A^a_\mu \hat{\psi}^{\dagger i}\,\bar{\sigma}^\mu (\boldsymbol{T^a})_i{}^j \hat{\psi}_j\,. \tag{2.137}$$

In the case of spontaneously broken gauge theories, one must diagonalize the vector boson squared-mass matrix. The form of eq. (2.137) still applies where A^a_μ are gauge boson fields of definite mass, although in this case for a fixed value of a, $g_a\boldsymbol{T^a}$ [which multiplies A^a_μ in eq. (2.137)] is some linear combination of the original $g_a\boldsymbol{T^a}$ of the unbroken theory. That is, the hermitian matrix gauge field $(A_\mu)_i{}^j \equiv A^a_\mu(\boldsymbol{T^a})_i{}^j$ appearing in eq. (2.137) can always be re-expressed in terms of the *physical* mass-eigenstate gauge boson fields.

In terms of the physical gauge boson fields, $A^a_\mu\boldsymbol{T^a}$ consists of a sum over real neutral gauge fields multiplied by hermitian generators, and complex charged gauge fields multiplied by nonhermitian generators. For example, in the electroweak Standard Model, $G = \mathrm{SU}(2)_L\times\mathrm{U}(1)_Y$ with gauge bosons and generators W^a_μ and $\boldsymbol{T^a} = \frac{1}{2}\tau^a$ for SU(2) and B_μ and $\boldsymbol{Y}$ for U(1), where the τ^a are the usual Pauli matrices. After diagonalizing the gauge boson squared-mass matrix [see eq. (4.148)],

$$gW^a_\mu\boldsymbol{T^a} + g'B_\mu\boldsymbol{Y} = \frac{g}{\sqrt{2}}(W^+_\mu\boldsymbol{T^+} + W^-_\mu\boldsymbol{T^-}) + \frac{g}{c_W}(\boldsymbol{T^3} - \boldsymbol{Q}s^2_W)\,Z_\mu + e\boldsymbol{Q}A_\mu\,, \tag{2.138}$$

where $c_W \equiv \cos\theta_W$, $s_W \equiv \sin\theta_W$, $\boldsymbol{Q} = \boldsymbol{T^3} + \boldsymbol{Y}$ is the generator of the unbroken $\mathrm{U}(1)_{\rm EM}$, $\boldsymbol{T^\pm} \equiv \boldsymbol{T^1} \pm i\boldsymbol{T^2}$, and $e = gs_W = g'c_W$. Here, the W^a_μ and B_μ are gauge fields of the unbroken theory. The massive gauge boson charge eigenstate fields of the broken theory consist of massive charged gauge bosons, $W^\pm \equiv (W^1 \mp iW^2)/\sqrt{2}$, a neutral massive gauge boson, $Z \equiv W^3c_W - Bs_W$, and the massless photon, $A \equiv W^3s_W + Bc_W$.

If an unbroken (global or local) U(1) symmetry exists, then the physical gauge bosons will be eigenstates of the conserved U(1) charge. If the U(1) symmetry group is orthogonal to the gauge group under which the A^a_μ transform, then all the gauge bosons are neutral with respect to the U(1) charge. For example, in the case of the interaction of a gluon with a pair of Majorana fermion gluinos, the gluon is a gauge boson that transforms under the SU(3) color group, which is orthogonal to the conserved $\mathrm{U}(1)_{\rm EM}$. That is, gluinos are color-octet, electrically neutral fermions. In contrast, in the case of the interaction of a Z with a pair of Majorana neutralinos, $\mathrm{U}(1)_{\rm EM}$ is not orthogonal to the electroweak $\mathrm{SU}(2)_L\times\mathrm{U}(1)_Y$ gauge group. Nevertheless, the Z gauge boson interactions of the neutralinos are allowed as they conserve electric charge.

To obtain the desired Feynman rule, we rewrite eq. (2.137) in terms of mass-eigenstate fermion fields. The resulting interaction Lagrangian can be rewritten as

$$\mathscr{L}_{\rm int} = -A^a_\mu \psi^{\dagger i}\,\bar{\sigma}^\mu (G^a)_i{}^j \psi_j\,, \tag{2.139}$$

where the A^a_μ are the mass-eigenstate gauge fields (of definite U(1) charge, if relevant), and

$$(G^a)_i{}^j = g_a U^k{}_i (\boldsymbol{T^a})_k{}^m U_m{}^j\,, \tag{2.140}$$

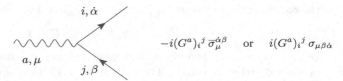

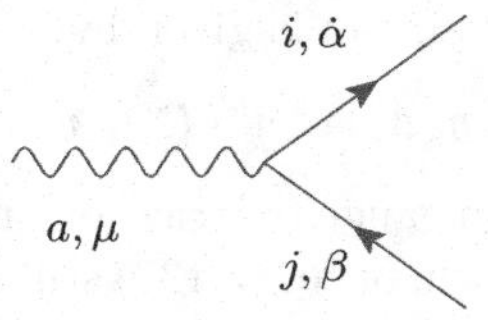

$$-i(G^a)_i{}^j\,\overline{\sigma}_\mu^{\dot{\alpha}\beta} \quad \text{or} \quad i(G^a)_i{}^j\,\sigma_{\mu\beta\dot{\alpha}}$$

Fig. 2.8 The Feynman rules for two-component fermion interactions with gauge bosons. The choice of which rule to use depends on how the vertex connects to the rest of the amplitude. The G^a are defined in eq. (2.140). The index a runs over both neutral and charged (mass-eigenstate) gauge bosons, consistent with charge conservation at the vertex.

or in matrix form, $G^a = g_a U^\dagger \boldsymbol{T^a} U$ (no sum over a). For values of a corresponding to the neutral gauge fields, the G^a are hermitian matrices. The corresponding Feynman rule is shown in Fig. 2.8.

First, consider the gauge interactions of neutral Majorana fermions. The Majorana fermions consist of left-handed $(\frac{1}{2}, 0)$ interaction eigenstate fermions $\hat{\xi}_i$ that transform under a real representation of the gauge group. After converting from the interaction eigenstates $\hat{\xi}_i$ to the mass eigenstates ξ_i using eq. (1.165), the Lagrangian for the gauge interactions of Majorana fermions is given by

$$\mathscr{L}_{\text{int}} = -A_\mu^a \xi^{\dagger i}\,\overline{\sigma}^\mu (G^a)_i{}^j \xi_j\,, \tag{2.141}$$

where the A_μ^a are *neutral* (real) mass-eigenstate gauge fields, and

$$(G^a)_i{}^j = g_a \Omega^k{}_i (\boldsymbol{T^a})_k{}^m \Omega_m{}^j\,, \tag{2.142}$$

or in matrix form, $G^a = g_a \Omega^\dagger \boldsymbol{T^a} \Omega$ (no sum over a). Note that the G^a are hermitian matrices. The corresponding Feynman rule takes the same form as the generalized rule shown in Fig. 2.8, with a restricted to values corresponding to the neutral mass-eigenstate gauge bosons.

Next, consider the gauge interactions of charged Dirac fermions. A Dirac fermion consists of pairs of left-handed $(\frac{1}{2}, 0)$ interaction eigenstate fermions $\hat{\chi}_i$ and $\hat{\eta}^i$ that transform as conjugate representations of the gauge group (hence the opposite flavor-index heights). The fermion mass matrix couples χ-type and η-type fields as in eq. (1.170). In the coupling to the interaction eigenstate gauge fields, if the $(\boldsymbol{T^a})_i{}^j$ are matrix elements of the hermitian representation matrices of the generators acting on the $\hat{\chi}_i$, then the $\hat{\eta}^i$ transform in the complex conjugate representation with the corresponding generator matrices $-(\boldsymbol{T^a})^* = -(\boldsymbol{T^a})^{\mathsf{T}}$, i.e., with matrix elements $-(\boldsymbol{T^a})_j{}^i$. Hence, the Lagrangian for the gauge interactions of Dirac fermions can be written in the form

$$\mathscr{L}_{\text{int}} = -g_a A_a^\mu \hat{\chi}^{\dagger i}\,\overline{\sigma}_\mu (\boldsymbol{T^a})_i{}^j \hat{\chi}_j + g_a A_a^\mu \hat{\eta}_i^\dagger\,\overline{\sigma}_\mu (\boldsymbol{T^a})_j{}^i \hat{\eta}^j\,. \tag{2.143}$$

We can now rewrite eq. (2.143) in terms of mass-eigenstate fermion fields using eq. (1.174), and express the hermitian matrix gauge field $A^\mu \equiv A_a^\mu \boldsymbol{T^a}$ in terms

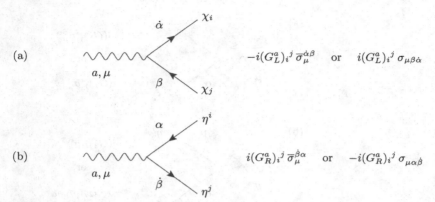

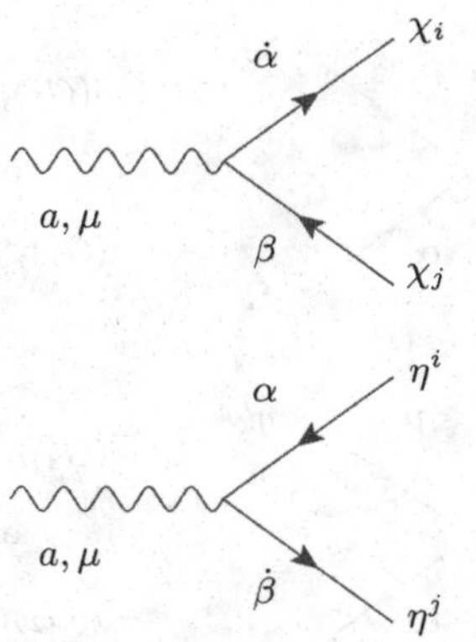

(a) $-i(G^a_L)_i{}^j\,\overline{\sigma}^{\dot\alpha\beta}_\mu$ or $i(G^a_L)_i{}^j\,\sigma_{\mu\beta\dot\alpha}$

(b) $i(G^a_R)_i{}^j\,\overline{\sigma}^{\dot\beta\alpha}_\mu$ or $-i(G^a_R)_i{}^j\,\sigma_{\mu\alpha\dot\beta}$

Fig. 2.9 The Feynman rules for the interaction of a gauge boson and a pair of Dirac fermions (each formed by χ and η of the appropriate flavor index). The fermion lines are labeled by the corresponding two-component left-handed $(\frac{1}{2},0)$ fermion fields. The matrices G^a_L and G^a_R depend on the group generators for the representation carried by the χ_i according to eq. (2.145). The index a runs over both neutral and charged (mass-eigenstate) gauge bosons, consistent with charge conservation at the vertex.

of mass-eigenstate gauge fields (of definite U(1) charge, if relevant). The resulting interaction Lagrangian is then given by

$$\mathscr{L}_{\rm int} = -A^\mu_a\left[\chi^{\dagger i}\,\overline{\sigma}_\mu(G^a_L)_i{}^j\chi_j - \eta^\dagger_i\,\overline{\sigma}_\mu(G^a_R)_j{}^i\eta^j\right], \tag{2.144}$$

where $A^\mu_a G^a_L$ and $A^\mu_a G^a_R$ are hermitian matrix-valued gauge fields, with

$$(G^a_L)_i{}^j = g_a L^k{}_i(\boldsymbol{T^a})_k{}^m L_m{}^j\,, \qquad (G^a_R)_j{}^i = g_a R^m{}_j(\boldsymbol{T^a})_m{}^k R_k{}^i\,. \tag{2.145}$$

In matrix form, $G^a_L = g_a L^\dagger \boldsymbol{T^a} L$ and $G^a_R = g_a R^\dagger \boldsymbol{T^a} R$ (no sum over a). For values of a corresponding to neutral gauge fields, G^a_L and G^a_R are hermitian matrices. The corresponding Feynman rules for the gauge interactions of Dirac fermions are shown in Fig. 2.9. Note that χ_i with its arrow pointing out of the vertex and η^i with its arrow pointing into the vertex collectively represent the same Dirac fermion.

Finally, consider the interaction of a charged vector boson W with a fermion pair consisting of one Majorana and one Dirac fermion. The U(1) charge of the W is denoted by q_W. As before, we represent the Majorana fermion by the two-component field ξ_i and the Dirac fermion by the pair of two-component fields χ_j and η^j (with U(1) charges q_j and $-q_j$, respectively). All fields are assumed to be in the mass-eigenstate basis. The interaction Lagrangian is given by[12]

$$\mathscr{L}_{\rm int} = -W_\mu[(G_1)_j{}^i\chi^{\dagger j}\overline{\sigma}^\mu\xi_i - (G_2)_{ij}\xi^{\dagger i}\overline{\sigma}^\mu\eta^j] - W^\dagger_\mu[(G_1)^j{}_i\,\xi^{\dagger i}\overline{\sigma}^\mu\chi_j - (G_2)^{ij}\eta^\dagger_j\overline{\sigma}^\mu\xi_i]\,, \tag{2.146}$$

where G_1 and G_2 are arbitrary complex matrices, with $(G_1)^i{}_j \equiv [(G_1)_i{}^j]^*$ and

[12] The sign in front of G_2 is conventionally chosen to match the sign of the term proportional to G^a_R in eq. (2.144).

(a)

$$-i(G_1)_j{}^i\overline{\sigma}_\mu^{\dot\beta\alpha}$$

or

$$i(G_1)_j{}^i\sigma_{\mu\alpha\dot\beta}$$

(b)

$$i(G_2)_{ij}\overline{\sigma}_\mu^{\dot\alpha\beta}$$

or

$$-i(G_2)_{ij}\sigma_{\mu\beta\dot\alpha}$$

(c)

$$-(G_1)^j{}_i\overline{\sigma}_\mu^{\dot\alpha\beta}$$

or

$$(G_1)^j{}_i\sigma_{\mu\beta\dot\alpha}$$

(d)

$$i(G_2)^{ij}\overline{\sigma}_\mu^{\dot\beta\alpha}$$

or

$$-i(G_2)^{ij}\sigma_{\mu\alpha\dot\beta}$$

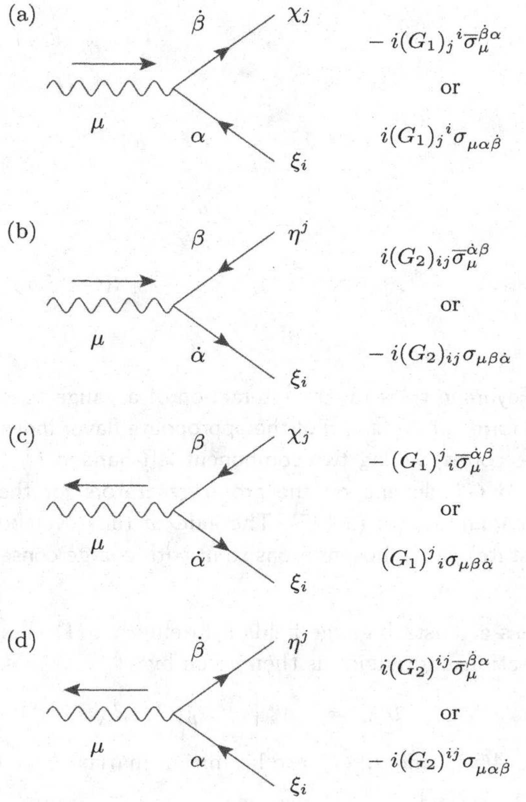

Fig. 2.10   The Feynman rules for the interactions of a charged vector boson (with U(1) charge q_W) with a fermion pair consisting of one Majorana fermion ξ_i and one Dirac fermion formed by χ_j and η^j (with corresponding U(1) charges q_j and $-q_j$). The fermion lines are labeled by the corresponding two-component left-handed $(\frac{1}{2},0)$ fermion fields. The matrix couplings G_1 and G_2 are defined in eq. (2.146). Note that $(G_1)_j{}^i = (G_2)_{ij} = 0$ unless $q_W = q_j$. The arrows indicate the direction of flow of the U(1) charges of the fermion and boson fields.

$(G_2)^{ij} \equiv [(G_2)_{ij}]^*$, such that $(G_1)_j{}^i = (G_2)_{ij} = 0$ unless $q_W = q_j$. The interactions of eq. (2.146) yield the Feynman rules exhibited in Fig. 2.10. Note that rules (c) and (d) are the complex conjugates of rules (a) and (b), respectively, corresponding to a reversal of the flow of the U(1) charge through the interaction vertex.

In Figs. 2.8–2.10, two versions are given for each of the boson–fermion–fermion Feynman rules. The correct version to use depends in a unique way on the heights of indices used to connect each fermion line to the rest of the diagram. For example, the way of writing the vector–fermion–fermion interaction rule depends on whether we used $\psi^{\dagger i}\overline{\sigma}^\mu\psi_j$, or its equivalent form $-\psi_j\sigma^\mu\psi^{\dagger i}$, in eq. (2.137). Note the different heights of the undotted and dotted spinor indices that adorn σ^μ and $\overline{\sigma}^\mu$. The choice of which rule to use is thus dictated by the height of the indices on the lines that connect to the vertex. These heights should always be chosen so that they are contracted as in eq. (1.63).

2.7 General Structure and Rules for Feynman Graphs

When computing an amplitude for a given process, all possible diagrams should be drawn that conform with the rules for external wave functions, propagators, and interactions, respectively. Starting from any external wave function spinor (or from any vertex on a fermion loop), factors corresponding to each propagator and vertex should be written down from left to right, following the line until it ends at another external-state wave function (or at the original point on the fermion loop). If one starts a fermion line at an x or y external-state spinor, it should have a raised undotted index in accordance with eq. (1.63). Or, if one starts with an $x^\dagger$ or $y^\dagger$, it should have a lowered dotted spinor index. Then, all spinor indices should always be contracted as in eq. (1.63). If one ends with an x or y external-state spinor, it will have a lowered undotted index, while if one ends with an $x^\dagger$ or $y^\dagger$ spinor, it will have a raised dotted index. For arrow-preserving fermion propagators and gauge vertices, the preceding determines whether the σ or $\overline{\sigma}$ rule should be used.

With only a little practice, one can write down amplitudes immediately with all spinor indices suppressed. In particular, the following two rules must be satisfied:

- For any scattering matrix amplitude, factors of σ and $\overline{\sigma}$ must alternate. If one or more factors of σ and/or $\overline{\sigma}$ are present, then x and y must be followed [preceded] by a σ [$\overline{\sigma}$], and $x^\dagger$ and $y^\dagger$ must be followed [preceded] by a $\overline{\sigma}$ [σ].

These requirements automatically dictate whether the σ or $\overline{\sigma}$ version of the rule for arrow-preserving fermion propagators and gauge vertices are employed in any tree-level Feynman diagram. In loop diagrams, we must add one further requirement that governs the order of the σ and $\overline{\sigma}$ factors as one traverses around the loop:

- Arrow-preserving propagator lines must be traversed in a direction parallel [antiparallel] to the arrowed line segment for the $\overline{\sigma}$ [σ] version of the propagator rule.

The latter rule is simply a consequence of the order of the spinor indices in Fig. 2.3.

For fermion lines that are not closed loops, the second rule is automatically realized provided that the requirements of the first rule above are satisfied. However, for closed fermion loops, one must use the correct fermion propagator corresponding to the direction around the loop one has chosen to follow in writing down the spinor trace with suppressed indices. For example, having employed a σ [$\overline{\sigma}$] rule at one vertex attached to the loop, one must then traverse the loop from that vertex point in a direction parallel [antiparallel] to the arrow-preserving propagator lines in the loop. Indeed, this rule is crucial for obtaining the correct sign for the triangle anomaly calculation in Chapter 5.

Symmetry factors for identical particles are implemented in the usual way. Fermi–Dirac statistics are implemented by the following two additional rules:

- Each closed fermion loop gets a factor of -1.
- A relative minus sign is imposed between terms contributing to a given amplitude whenever the ordering of external-state spinors (written left-to-right in a formula) differs by an odd permutation.

Amplitudes generated according to these rules will contain objects of the form

$$a = z_1 \Sigma z_2 \,, \tag{2.147}$$

where z_1 and z_2 are commuting external spinor wave functions, x, $x^\dagger$, y, or $y^\dagger$, and Σ is a sequence of alternating σ and $\overline{\sigma}$ matrices. The complex conjugate of this quantity is obtained by applying the results of eqs. (B.51)–(B.56), and is given by

$$a^* = z_2^\dagger \Sigma_r z_1^\dagger \,, \tag{2.148}$$

where Σ_r is obtained from Σ by reversing the order of all the σ and $\overline{\sigma}$ matrices and using the same rule for suppressed spinor indices. We emphasize that, in principle, it does not matter in what direction a diagram is traversed while applying the rules. However, for each diagram one must include a sign that depends on the ordering of the external fermions. This sign can be fixed by first choosing some canonical ordering of the external fermions. Then for any graph that contributes to the process of interest, the corresponding sign is positive (negative) if the ordering of external fermions is an even (odd) permutation with respect to the canonical ordering. If one chooses a different canonical ordering, then the resulting amplitude changes by an overall phase (is unchanged) if this ordering is an odd (even) permutation of the original canonical ordering.[13] This is consistent with the fact that the S-matrix element is only defined up to an overall sign, which is not physically observable.[14]

[13] For a process with exactly two external fermions, it is convenient to apply the Feynman rules by starting from the same fermion external state in all diagrams. That way, all terms in the amplitude have the same canonical ordering of fermions and there are no additional minus signs between diagrams. However, if there are four or more external fermions, it often happens that there is no way to choose the same ordering of external-state spinors for all graphs when the amplitude is written down. Then the relative signs between different graphs must be chosen according to the relative sign of the permutation of the corresponding external fermion spinors. This guarantees that the total amplitude is antisymmetric under the interchange of any pair of external fermions.

[14] Recall that the S-matrix element S_{fi} is related to the invariant matrix element $\mathcal{M}_{fi}$ by $S_{fi} = \delta_{fi} + (2\pi)^4 \delta^{(4)}(p_f - p_i)\, i\mathcal{M}_{fi}$, where p_f (p_i) is the total four-momentum of the final (initial) state. If $f \neq i$ (i.e., the final and initial states are distinct), then $\delta_{fi} = 0$ in which case the invariant matrix element is only defined up to an overall (unphysical) sign. However, if $f = i$, the most convenient choice for the canonical ordering of external fermions is the one that yields $\langle f | i \rangle = \delta_{fi}$ (with no extra minus sign), which then fixes the absolute sign of the invariant matrix element.

Different graphs contributing to the same process will often have different external-state wave function spinors, with different arrow directions, for the same external fermion. Furthermore, there are no arbitrary choices to be made for arrow directions, as there are in some four-component Feynman rules for Majorana fermions (as discussed in Chapter 3). Instead, one must add together *all* Feynman graphs that obey the rules.

2.8 Examples of Feynman Diagrams and Amplitudes

Some simple examples based on the interaction vertices of Section 2.6 will help clarify the rules of Section 2.7. In the tree-level Feynman graphs of this section, we shall label all two-component fermion lines by their corresponding left-handed $(\frac{1}{2}, 0)$ fields. (A slightly different labeling convention is introduced in Section 17.1.)

2.8.1 Tree-Level Decays

Scalar Boson Decay to Fermions

Let us first consider a theory with a multiplet of uncharged, massive $(\frac{1}{2}, 0)$ fermions ξ_i, and a real scalar ϕ, with interaction

$$\mathscr{L}_{\text{int}} = -\tfrac{1}{2}\left(\lambda^{ij}\xi_i\xi_j + \lambda_{ij}\xi^{\dagger i}\xi^{\dagger j}\right)\phi\,, \tag{2.149}$$

where $\lambda_{ij} \equiv (\lambda^{ij})^*$ and $\lambda^{ij} = \lambda^{ji}$. Consider the decay $\phi \to \xi_i(\vec{\boldsymbol{p}}_1, s_1)\xi_j(\vec{\boldsymbol{p}}_2, s_2)$ [for a fixed choice of i and j], where by $\xi_i(\vec{\boldsymbol{p}}, s)$ we mean the one-particle state given by eq. (2.69).

Two diagrams contribute to this process, as shown in Fig. 2.11. The matrix element is

$$\begin{aligned} i\mathcal{M} &= y_i^{\alpha}(\vec{\boldsymbol{p}}_1, s_1)(-i\lambda^{ij}\delta_{\alpha}{}^{\beta})y_{\beta j}(\vec{\boldsymbol{p}}_2, s_2) + x^{\dagger i}_{\dot\alpha}(\vec{\boldsymbol{p}}_1, s_1)(-i\lambda_{ij}\delta^{\dot\alpha}{}_{\dot\beta})x^{\dagger\dot\beta j}(\vec{\boldsymbol{p}}_2, s_2) \\ &= -i\lambda^{ij}y_i(\vec{\boldsymbol{p}}_1, s_1)y_j(\vec{\boldsymbol{p}}_2, s_2) - i\lambda_{ij}x^{\dagger i}(\vec{\boldsymbol{p}}_1, s_1)x^{\dagger j}(\vec{\boldsymbol{p}}_2, s_2) \end{aligned} \tag{2.150}$$

for fixed flavor indices i and j (no sum over the repeated flavor indices). The

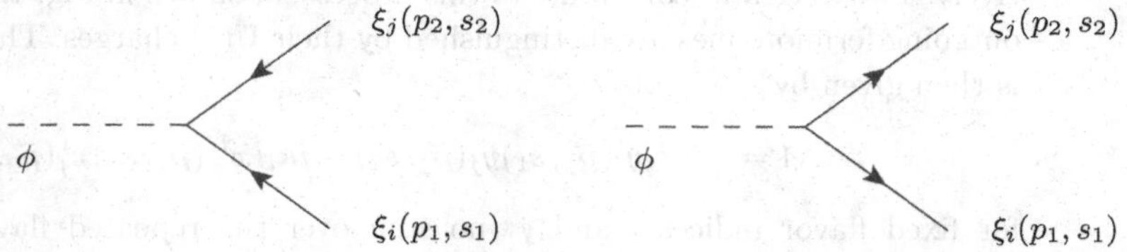

Fig. 2.11 The two tree-level Feynman diagrams contributing to the decay of a neutral scalar into a pair of Majorana fermions.

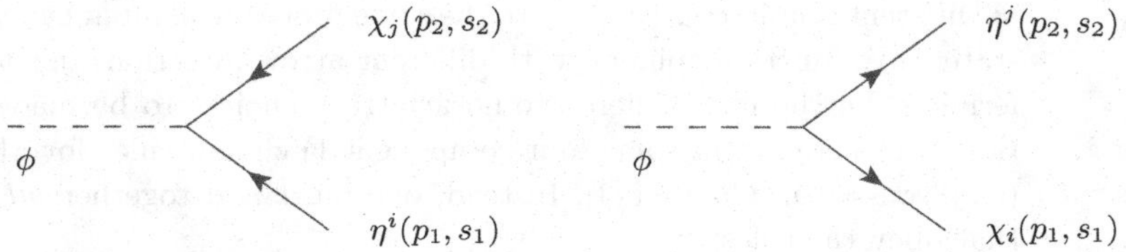

Fig. 2.12 The two tree-level Feynman diagrams contributing to the decay of a neutral scalar into a pair of Dirac fermions. The χ_i–η^i and χ_j–η^j pairs, each with oppositely directed arrows, comprise Dirac fermion states with flavor indices i and j, respectively.

second line could be written down directly by recalling that the sum over suppressed spinor indices is taken according to eq. (1.63). Note that if we reverse the ordering for the external fermions, the overall sign of the amplitude changes sign. This is easily checked, since for the commuting spinor wave functions (x and y), the spinor products in eq. (2.150) change sign when the order is reversed [see eqs. (1.96) and (1.97)]. This overall sign is not significant and depends on the order used in constructing the two-particle state. One could even make the choice of starting the first diagram from fermion 1 and the second diagram from fermion 2:

$$i\mathcal{M} = -i\lambda^{ij} y_i(\vec{p}_1, s_1) y_j(\vec{p}_2, s_2) - (-1) i\lambda_{ij} x^{\dagger j}(\vec{p}_2, s_2) x^{\dagger i}(\vec{p}_1, s_1) . \tag{2.151}$$

Here, the first term establishes the canonical ordering of fermions, i.e., (12). Hence, the contribution from the second diagram includes a relative minus sign inside the parentheses in eq. (2.151). Indeed, eqs. (2.150) and (2.151) are equivalent. In the computation of the total decay rate for the case of $i = j$, one must multiply the integral over the total phase space by 1/2 to account for the identical particles.

Next, we consider a theory of a massive neutral scalar boson that couples to a multiplet of Dirac fermions. We denote the corresponding two-component fields by χ_i and η^i. For simplicity, we take all the U(1) charges of the χ_i to be equal (and opposite to the charges of the η^i). The corresponding U(1)-invariant interaction is

$$\mathscr{L}_{\text{int}} = -(\kappa^i{}_j \chi_i \eta^j + \kappa_i{}^j \chi^{\dagger i} \eta^\dagger_j)\phi , \tag{2.152}$$

where $\kappa_i{}^j = (\kappa^i{}_j)^*$. Consider the decay $\phi \to f_i(\vec{p}_1, s_1)\bar{f}^j(\vec{p}_2, s_2)$, where by $f(\vec{p}, s)$ and $\bar{f}(\vec{p}, s)$ we mean the one-particle states given by eqs. (2.78) and (2.79), respectively. Two diagrams contribute to this process, as shown in Fig. 2.12. Note that the outgoing fermion lines are distinguished by their U(1) charges. The matrix element is then given by

$$i\mathcal{M} = -i\kappa^j{}_i y^i(\vec{p}_1, s_1) y_j(\vec{p}_2, s_2) - i\kappa_i{}^j x^{\dagger i}(\vec{p}_1, s_1) x^\dagger_j(\vec{p}_2, s_2) \tag{2.153}$$

for fixed flavor indices i and j (no sum over the repeated flavor indices). The heights of the flavor indices of the y and $x^\dagger$ spinor wave functions in eq. (2.153) [which differ from those appearing in eq. (2.150)] are conventional and chosen for convenience. Indeed, in subsequent equations below, we will typically suppress these flavor indices when there is no ambiguity in associating a spinor wave function with the corresponding flavor of the external state.

Note that the matrix element for $\phi \to f_i(\vec{p}_1, s_1)\bar{f}^j(\vec{p}_2, s_2)$ is identical to that of $\phi \to \xi_i(\vec{p}_1, s_1)\xi_j(\vec{p}_2, s_2)$ after replacing λ^{ij} with $\kappa^i{}_j$. However, for fixed $i = j$, the rate for scalar boson decay to $f_i\bar{f}^i$ is twice that of $\xi_i\xi_i$ due to the final-state identical particles in the latter case, as noted above. One also arrives at the same conclusion if one treats a single Dirac fermion as a pair of mass-degenerate two-component fields ξ_1 and ξ_2 [see eq. (2.71)]. Due to the U(1) symmetry, the scalar Yukawa interactions are diagonal in the ξ_1–ξ_2 basis, so the rate for scalar decay into the Dirac fermion pair is equal to the incoherent sum of the rate for decay into $\xi_1\xi_1$ and $\xi_2\xi_2$.

Vector Boson Decay to Fermions

Consider next the decay of a massive neutral vector boson A_μ into a pair of Majorana fermions, $A_\mu \to \xi_i(\vec{p}_1, s_1)\xi_j(\vec{p}_2, s_2)$, following from the interaction

$$\mathscr{L}_{\text{int}} = -G_i{}^j A^\mu \xi^{\dagger i}\bar{\sigma}_\mu \xi_j\,, \tag{2.154}$$

where G is a hermitian coupling matrix. Two diagrams contribute to this process, as shown in Fig. 2.13. We start from the fermion with momentum p_1, spin vector s_1, and flavor i and end at the fermion with momentum p_2, spin vector s_2, and flavor j, using the rules of Fig. 2.8. The resulting amplitude for the decay is

$$i\mathcal{M} = \varepsilon^\mu \left[-iG_i{}^j x^\dagger(\vec{p}_1, s_1)\bar{\sigma}_\mu y(\vec{p}_2, s_2) + iG_j{}^i y(\vec{p}_1, s_1)\sigma_\mu x^\dagger(\vec{p}_2, s_2)\right]\,, \tag{2.155}$$

where ε^μ is the vector boson polarization vector, and the flavor indices of the y and $x^\dagger$ spinor wave functions have been suppressed. Note that we have used the $\bar{\sigma}$ version of the vector–fermion–fermion rule (see Fig. 2.8) for the first diagram of Fig. 2.13 and the σ version for the second diagram of Fig. 2.13, as dictated by the implicit spinor indices (which have also been suppressed).

However, we could have chosen to evaluate the second diagram of Fig. 2.13 using the $\bar{\sigma}$ version of the vector–fermion–fermion rule by starting from the fermion with momentum p_2 and spin vector s_2. In that case, the term $iG_j{}^i y(\vec{p}_1, s_1)\sigma_\mu x^\dagger(\vec{p}_2, s_2)$

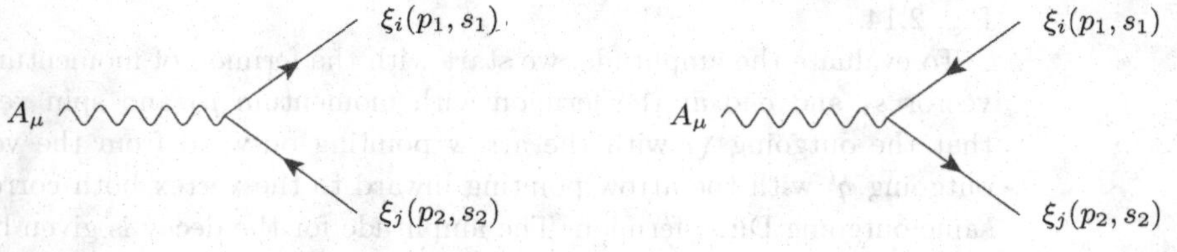

Fig. 2.13 The two tree-level Feynman diagrams contributing to the decay of a massive neutral vector boson A_μ into a pair of Majorana fermions.

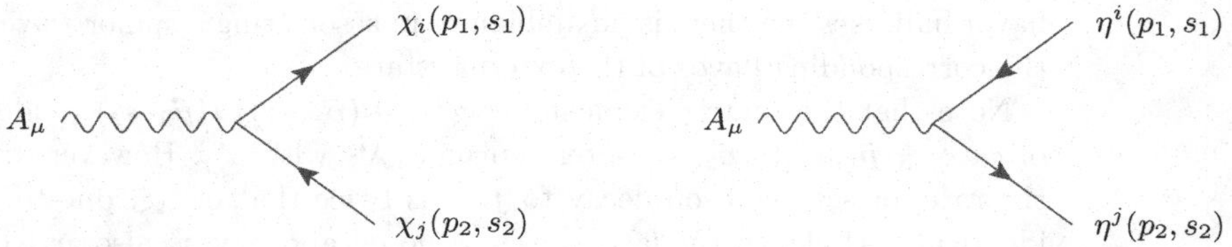

Fig. 2.14 The two tree-level Feynman diagrams contributing to the decay of a massive neutral vector boson A_μ into a pair of Dirac fermions. The χ_i–η^i and χ_j–η^j pairs, each with oppositely directed arrows, comprise Dirac fermion states with flavor indices i and j, respectively.

in eq. (2.155) is replaced by

$$(-1)[-iG_j{}^i x^\dagger(\vec{p}_2, s_2)\overline{\sigma}_\mu y(\vec{p}_1, s_1)]\,. \qquad (2.156)$$

In eq. (2.156), the factor of $-iG_j{}^i$ arises from the use of the $\overline{\sigma}$ version of the vector–fermion–fermion rule, and the overall factor of -1 appears because the order of the fermion wave functions has been reversed; i.e., (21) is an odd permutation of (12). This is in accord with the ordering rule stated at the end of Section 2.7. Thus, the resulting amplitude for the decay of the vector boson into the pair of Majorana fermions now takes the form

$$i\mathcal{M} = \varepsilon^\mu \left[-iG_i{}^j x^\dagger(\vec{p}_1, s_1)\overline{\sigma}_\mu y(\vec{p}_2, s_2) + iG_j{}^i x^\dagger(\vec{p}_2, s_2)\overline{\sigma}_\mu y(\vec{p}_1, s_1)\right]\,, \qquad (2.157)$$

which coincides with eq. (2.155) after using $y\sigma^\mu x^\dagger = x^\dagger\overline{\sigma}^\mu y$ [see eq. (1.98) with commuting spinors]. Equation (2.157) explicitly exhibits the property that the amplitude is antisymmetric under the interchange of the two external identical fermions. Again, the absolute sign of the total amplitude is not significant and depends on the choice of ordering of the outgoing states.

Next, we consider the decay of a massive neutral vector boson into a pair of Dirac fermions. Each Dirac fermion is described by the two-component fields χ_i and η^i, which possess equal and opposite U(1) charges, respectively. The corresponding interaction Lagrangian is given by

$$\mathscr{L}_{\rm int} = -A^\mu[(G_L)_i{}^j\,\chi^{\dagger i}\overline{\sigma}_\mu\chi_j - (G_R)_j{}^i\,\eta_i^\dagger\overline{\sigma}_\mu\eta^j\,]\,, \qquad (2.158)$$

where G_L and G_R are hermitian. There are two contributing graphs, as shown in Fig. 2.14.

To evaluate the amplitude, we start with the fermion of momentum p_1 and spin vector s_1 and end at the fermion with momentum p_2 and spin vector s_2. Note that the outgoing χ_i with the arrow pointing outward from the vertex and the outgoing η^i with the arrow pointing inward to the vertex both correspond to the same outgoing Dirac fermion. The amplitude for the decay is given by

$$\begin{aligned} i\mathcal{M} &= \varepsilon^\mu \left[-i(G_L)_i{}^j x^\dagger(\vec{p}_1, s_1)\overline{\sigma}_\mu y(\vec{p}_2, s_2) - i(G_R)_i{}^j y(\vec{p}_1, s_1)\sigma_\mu x^\dagger(\vec{p}_2, s_2)\right] \\ &= \varepsilon^\mu \left[-i(G_L)_i{}^j x^\dagger(\vec{p}_1, s_1)\overline{\sigma}_\mu y(\vec{p}_2, s_2) - i(G_R)_i{}^j x^\dagger(\vec{p}_2, s_2)\overline{\sigma}_\mu y(\vec{p}_1, s_1)\right]. \qquad (2.159) \end{aligned}$$

As in the case of the decay to a pair of Majorana fermions, we have exhibited a second form for the amplitude in eq. (2.159) in which the $\overline{\sigma}$ version of the vertex Feynman rule has been employed in both diagrams. Of course, the resulting amplitude must be the same in each method (up to a possible overall sign of the total amplitude that is not determined).

2.8.2 Tree-Level Annihilation Processes

In this section, we evaluate the rate for fermion pair annihilation into a scalar boson. We shall consider the $2 \to 1$ annihilation processes $\xi(\vec{p}_1, s_1)\xi(\vec{p}_2, s_2) \to \phi$ and $f(\vec{p}_1, s_1)\bar{f}(\vec{p}_2, s_2) \to \phi$, respectively. The corresponding amplitudes are given by eqs. (2.150) and (2.153) with $y \to x$ and $x^\dagger \to y^\dagger$ (for simplicity, we neglect flavor). In the computation of the cross sections, there is no extra factor required to account for the case of identical particles in the initial state. That is, the cross section for $f(\vec{p}_1, s_1)\bar{f}(\vec{p}_2, s_2) \to \phi$ is equal to the cross section for $\xi(\vec{p}_1, s_1)\xi(\vec{p}_2, s_2) \to \phi$ after replacing λ with κ.

This may at first seem puzzling given that a Dirac fermion can be represented by a pair of mass-degenerate two-component fields χ_1 and χ_2. But recall the standard procedure for the calculation of decay rates and cross sections in field theory: *average over unobserved degrees of freedom of the initial state and sum over unobserved degrees of freedom of the final state.* This mantra is well known for dealing with spin and color degrees of freedom, but it is also applicable to degrees of freedom associated with global internal symmetries. Thus, the cross section for the annihilation of a Dirac fermion pair into a neutral scalar boson can be obtained by computing the *average* of the cross sections for $\xi_1(\vec{p}_1, s_1)\xi_1(\vec{p}_2, s_2) \to \phi$ and $\xi_2(\vec{p}_1, s_1)\xi_2(\vec{p}_2, s_2) \to \phi$. Since the annihilation cross sections for $\xi_1\xi_1$ and $\xi_2\xi_2$ are equal, we confirm the annihilation cross section for the Dirac fermion pair obtained above in the χ–η basis. Since the latter is conceptually simpler, subsequent computations involving Dirac fermions will be performed in the χ–η basis.

The annihilation rate of fermions enters the analysis of the event flux due to the annihilation of dark matter in the halo of our galaxy. Let us compare the rates in the case that the dark matter is either a Majorana or a Dirac fermion. Suppose the annihilation involves two fermions whose number densities are n_1 and n_2, respectively. Then the observer on Earth who integrates along the line of sight to the annihilation events that are detected sees a flux of events proportional to

$$\frac{dN_{\text{events}}}{dA\,dt} \sim \int n_1 n_2 \, \langle \sigma_{\text{ann}} v_{\text{rel}} \rangle \, d\ell \,, \tag{2.160}$$

where σ_{ann} is the annihilation cross section, v_{rel} is the relative velocity of the annihilating particles [see eq. (D.123)], and $\langle \cdots \rangle$ refers to a thermal average over the velocity distribution of dark matter particles in the halo. We now compare the case of the annihilation of a single species of Majorana particles and the annihilation of a Dirac fermion–antifermion pair (assumed to have the same mass and couplings). We assume that the number density of Dirac fermions and antifermions and the corresponding number density of Majorana fermions are all the same (and denoted by n). At the beginning of this subsection, we showed that $\sigma_{\rm ann}$ is the same for the annihilation of a single species of Majorana and Dirac fermions. For the Dirac case, $n_1 n_2 = n^2$. For the Majorana case, because the Majorana fermions are identical particles, given N initial-state fermions in a volume V, there are $N(N-1)/2$ possible scatterings. In the thermodynamic limit where $N, V \to \infty$ at fixed $n \equiv N/V$, we conclude that $n_1 n_2 = \frac{1}{2}n^2$ for a single species of annihilating Majorana fermions. Hence the event flux rate for the annihilation of a Dirac fermion–antifermion pair is double that of a single species of Majorana fermions.[15] The extra factor of 1/2 can also be understood by noting that in the case of annihilating dark matter particles (in the large N limit), all possible scattering axes occur and are implicitly integrated over. But integrating over 4π steradians double-counts the annihilation of identical particles (in the same way it does in the computation of the decay rate of a scalar into identical fermions, discussed above). Hence, one must include a factor of $\frac{1}{2}$ in this case by replacing $n_1 n_2 = n^2$ by $\frac{1}{2}n^2$ in eq. (2.160).

The relic abundance of primordial dark matter particles in the universe is inversely proportional to $\langle \sigma_{\rm ann} v_{\rm rel} \rangle$ (e.g., see [B137]). By similar arguments to the ones just presented, it follows that the relic abundance of a single species of Majorana fermions would be twice that of a single species of Dirac fermions.

2.8.3 Tree-Level Scattering Processes

The next level of complexity consists of diagrams that involve fermion propagators. In this section, we examine a variety of tree-level $2 \to 2$ scattering processes. In the examples that follow, we shall ignore the flavor index and consider scattering processes that involve a single flavor of Majorana or Dirac fermion.

Two-Body Scattering of a Boson and a Neutral Fermion

For our first example of this type, consider the tree-level matrix element for the scattering of a neutral scalar and a two-component neutral massive fermion ($\phi\xi \to \phi\xi$), with the interaction Lagrangian given above in eq. (2.149). Using the corresponding Feynman rules, there are eight contributing diagrams. Four are depicted in Fig. 2.15; there are another four diagrams (not shown) where the initial-state and final-state scalars are crossed (i.e., the initial-state scalar is attached to the same vertex as the final-state fermion).

We shall write down the amplitudes for the four diagrams shown in Fig. 2.15, starting with the final-state fermion line and moving toward the initial-state fermion

[15] This is also consistent with the interpretation of a Dirac fermion as a pair of mass-degenerate Majorana fermions.

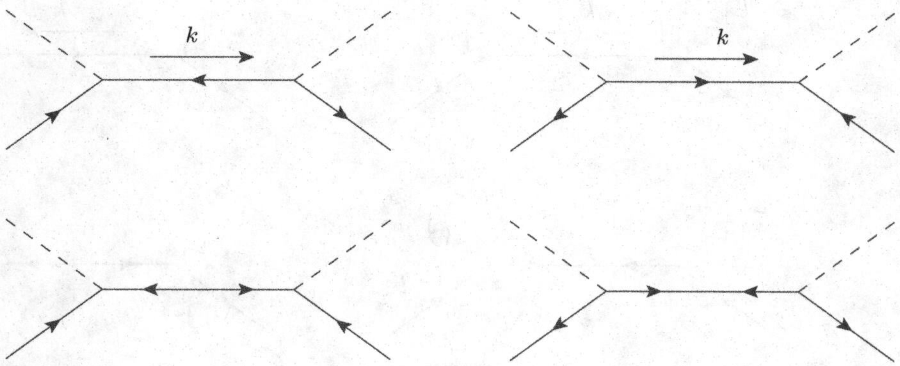

Fig. 2.15 Tree-level Feynman diagrams contributing to the elastic scattering of a neutral scalar and a Majorana fermion. There are four more diagrams, obtained from these by crossing the initial and final scalar lines.

line. We then obtain

$$i\mathcal{M} = \frac{i}{k^2 - m_\xi^2}\Big\{(-i\lambda)(-i\lambda^*)\left[x^\dagger(\vec{p}_2, s_2)\,\overline{\sigma}\!\cdot\! k\, x(\vec{p}_1, s_1) + y(\vec{p}_2, s_2)\,\sigma\!\cdot\! k\, y^\dagger(\vec{p}_1, s_1)\right]$$
$$+ m_\xi\left[(-i\lambda)^2 y(\vec{p}_2, s_2)x(\vec{p}_1, s_1) + (-i\lambda^*)^2 x^\dagger(\vec{p}_2, s_2)y^\dagger(\vec{p}_1, s_1)\right]\Big\} + (\text{crossed})\,, \tag{2.161}$$

where k^μ is the sum of the two incoming (or outgoing) four-momenta, and (p_1, s_1), (p_2, s_2) are the four-momenta and spins of the incoming and outgoing fermions, respectively. The notation "(crossed)" refers to the contribution to the amplitude from diagrams that have the initial and final scalars interchanged. Note that we could have evaluated the diagrams above by starting with the initial vertex and moving toward the final vertex. It is easy to check that the resulting amplitude is the negative of the one obtained in eq. (2.161); the overall sign change simply corresponds to swapping the order of the two fermions and has no physical consequence. The overall minus sign is a consequence of eqs. (1.96)–(1.98) and the minus sign difference between the two ways of evaluating the propagator that preserves the arrow direction.

Next, we compute the tree-level matrix element for the scattering of a neutral vector boson and a neutral massive two-component fermion ξ with the interaction Lagrangian of eq. (2.154). Again there are eight diagrams: the four diagrams depicted in Fig. 2.16 plus another four (not shown) where the initial-state and final-state vector bosons are crossed.

Starting with the final-state fermion line and moving toward the initial state, we

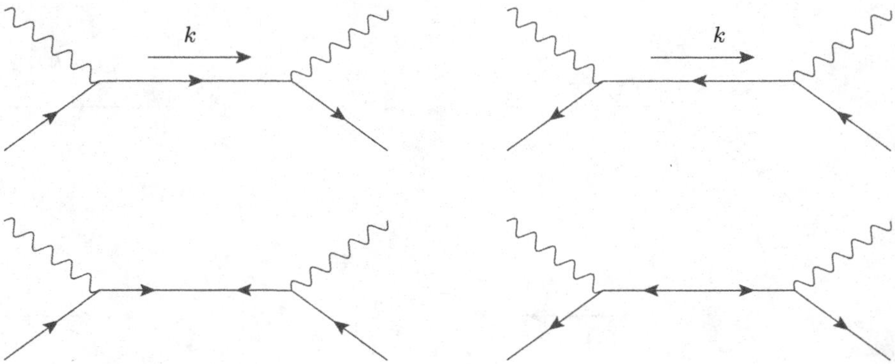

Fig. 2.16 Tree-level Feynman diagrams contributing to the elastic scattering of a neutral vector boson and a Majorana fermion. There are four more diagrams, obtained from these by crossing the initial and final scalar lines.

obtain

$$\begin{aligned} i\mathcal{M} = \frac{i}{k^2 - m_\xi^2}\bigg\{ &(-iG)^2 x^\dagger(\vec{\boldsymbol{p}}_2, s_2)\,\overline{\sigma}\cdot\varepsilon_2^*\,\sigma\cdot k\,\overline{\sigma}\cdot\varepsilon_1\, x(\vec{\boldsymbol{p}}_1, s_1) \\ &+ (iG)^2 y(\vec{\boldsymbol{p}}_2, s_2)\,\sigma\cdot\varepsilon_2^*\,\overline{\sigma}\cdot k\,\sigma\cdot\varepsilon_1 y^\dagger(\vec{\boldsymbol{p}}_1, s_1) \\ &+ (-iG)(iG) m_\xi y(\vec{\boldsymbol{p}}_2, s_2)\,\sigma\cdot\varepsilon_2^*\,\overline{\sigma}\cdot\varepsilon_1\, x(\vec{\boldsymbol{p}}_1, s_1) \\ &+ (iG)(-iG) m_\xi x^\dagger(\vec{\boldsymbol{p}}_2, s_2)\,\overline{\sigma}\cdot\varepsilon_2^*\,\sigma\cdot\varepsilon_1\, y^\dagger(\vec{\boldsymbol{p}}_1, s_1)\bigg\} + (\text{crossed}), \end{aligned} \tag{2.162}$$

where ε_1 and ε_2 are the initial and final vector boson polarization four-vectors, respectively. As before, k^μ is the sum of the two incoming (or outgoing) four-momenta, (p_1, s_1), (p_2, s_2) are the four-momenta and spins of the incoming and outgoing fermions, respectively, and "(crossed)" indicates that the terms from diagrams in which the initial and final vector bosons are interchanged. Alternatively, if one starts with an initial-state fermion and moves toward the final state, the resulting amplitude is the negative of the one obtained in eq. (2.162), as expected.

Two-Body Scattering of a Boson and a Charged Fermion

We first consider the scattering of a Dirac fermion with a neutral scalar. We denote the Dirac mass of the fermion by m_D. The left-handed fields χ and η have opposite charges (which we take to be $Q = +1$ and -1, respectively) and interact with the scalar ϕ according to

$$\mathscr{L}_{\text{int}} = -\phi[\kappa\chi\eta + \kappa^*\chi^\dagger\eta^\dagger], \tag{2.163}$$

where κ is a coupling parameter. Then, for the elastic scattering of the $Q = +1$ fermion and a scalar, the diagrams of Fig. 2.17 contribute at tree level, plus an-

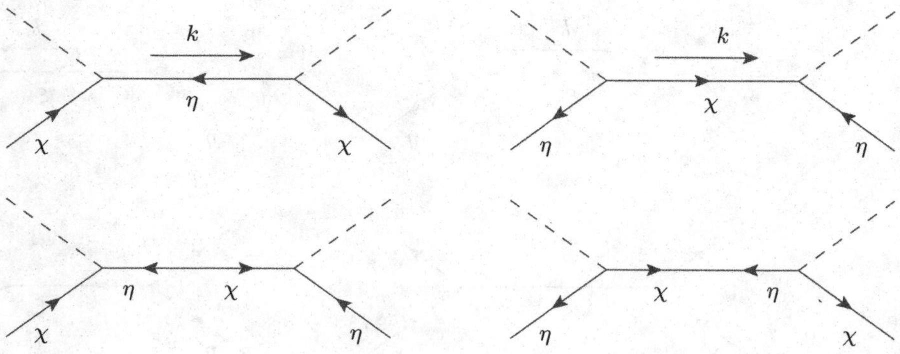

Fig. 2.17 Tree-level Feynman diagrams contributing to the elastic scattering of a neutral scalar and a charged fermion. There are four more diagrams, obtained from these by crossing the initial and final scalar lines.

other four diagrams (not shown) where the initial-state and final-state scalars are crossed. Now, these diagrams match precisely those of Fig. 2.15. Thus, applying the Feynman rules yields the same matrix element, eq. (2.161), previously obtained for the scattering of a neutral scalar and neutral two-component fermion, with the replacement of λ with κ and m_ξ with m_D.

We next examine the scattering of a Dirac fermion and a charged scalar, where both the scalar and the fermion have the same absolute value of the charge. As above, we denote the charged $Q = \pm 1$ fermion by the pair of two-component fermions χ and η and the (intermediate state) neutral two-component fermion by ξ. The charged $Q = \pm 1$ scalar is represented by the complex scalar field Φ and its hermitian conjugate. The interaction Lagrangian takes the form

$$\mathscr{L}_{\rm int} = -\Phi[\kappa_1 \eta\xi + \kappa_2^* \chi^\dagger \xi^\dagger] - \Phi^\dagger[\kappa_2 \chi\xi + \kappa_1^* \eta^\dagger \xi^\dagger] . \tag{2.164}$$

Consider the scattering of an initial boson–fermion state into its charge-conjugated final state via the exchange of a neutral fermion. The relevant diagrams are shown in Fig. 2.18, plus the corresponding diagrams with the initial and final scalars crossed. We define the four-momentum k to be the sum of the two initial-state four-momenta, as shown in Fig. 2.18. The derivation of the amplitude is similar to the ones given previously, and we end up with

$$\begin{aligned} i\mathcal{M} = \frac{-i}{k^2 - m_\xi^2} \Big\{ & \kappa_1^* \kappa_2 [x^\dagger(\vec{\boldsymbol{p}}_2, s_2)\, \overline{\sigma}\cdot k\, x(\vec{\boldsymbol{p}}_1, s_1) + y(\vec{\boldsymbol{p}}_2, s_2)\, \sigma\cdot k\, y^\dagger(\vec{\boldsymbol{p}}_1, s_1)] \\ & + m_\xi \left[\kappa_2^2 y(\vec{\boldsymbol{p}}_2, s_2) x(\vec{\boldsymbol{p}}_1, s_1) + (\kappa_1^*)^2 x^\dagger(\vec{\boldsymbol{p}}_2, s_2) y^\dagger(\vec{\boldsymbol{p}}_1, s_1)\right] \Big\} + \text{(crossed)}. \end{aligned} \tag{2.165}$$

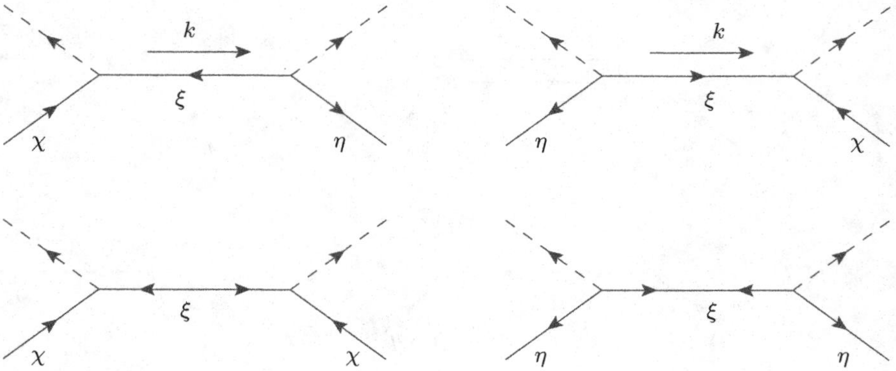

Fig. 2.18 Tree-level Feynman diagrams contributing to the scattering of an initial charged scalar and a charged fermion into its charge-conjugated final state. The unlabeled intermediate state is a neutral fermion. There are four more diagrams, obtained from these by crossing the initial and final scalar lines.

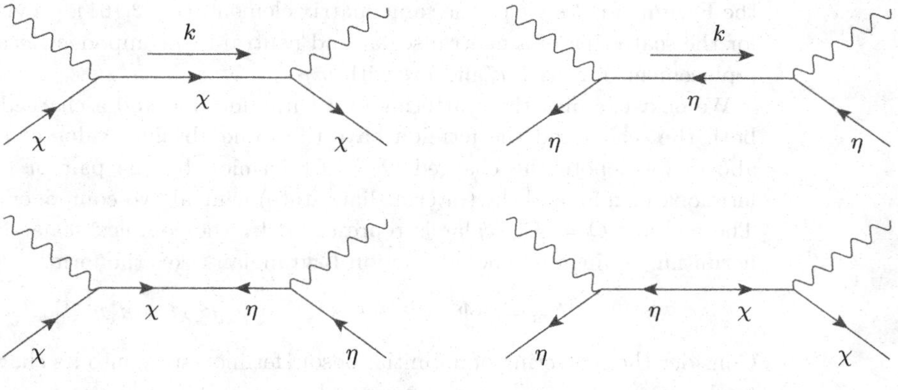

Fig. 2.19 Tree-level Feynman diagrams contributing to the elastic scattering of a neutral vector boson and a Dirac fermion. There are four more diagrams, obtained from these by crossing the initial and final vector lines.

The scattering of a charged fermion and a neutral spin-1 vector boson can be similarly treated. For example, consider the amplitude for the elastic scattering of a charged fermion of mass m_D and a neutral vector boson. Again taking the interactions as given in eq. (2.158), the relevant diagrams are those shown in Fig. 2.19, plus four diagrams (not shown) obtained from these by crossing the initial-state and final-state vector bosons. Applying the Feynman rules of Fig. 2.9, we obtain the following matrix element:

$$
\begin{aligned}
i\mathcal{M} = \frac{-i}{k^2 - m_D^2} \Big\{ & G_L^2 x^\dagger(\vec{\boldsymbol{p}}_2, s_2)\, \overline{\sigma}\cdot\varepsilon_2^*\, \sigma\cdot k\, \overline{\sigma}\cdot\varepsilon_1\, x(\vec{\boldsymbol{p}}_1, s_1) \\
& + G_R^2 y(\vec{\boldsymbol{p}}_2, s_2)\, \sigma\cdot\varepsilon_2^*\, \overline{\sigma}\cdot k\, \sigma\cdot\varepsilon_1 y^\dagger(\vec{\boldsymbol{p}}_1, s_1) \\
& + m_D G_L G_R y(\vec{\boldsymbol{p}}_2, s_2)\, \sigma\cdot\varepsilon_2^*\, \overline{\sigma}\cdot\varepsilon_1\, x(\vec{\boldsymbol{p}}_1, s_1) \\
& + m_D G_L G_R x^\dagger(\vec{\boldsymbol{p}}_2, s_2)\, \overline{\sigma}\cdot\varepsilon_2^*\, \sigma\cdot\varepsilon_1\, y^\dagger(\vec{\boldsymbol{p}}_1, s_1) \Big\} + (\text{crossed})\,, \qquad (2.166)
\end{aligned}
$$

where the assignments of momenta and spins are as before.

Two-Body Fermion–Fermion scattering

Finally, let us work out an example with four external-state fermions. Consider the case of elastic scattering of two identical Majorana fermions due to scalar exchange, governed by the interaction of eq. (2.149). The diagrams for scattering initial fermions labeled $1, 2$ into final-state fermions labeled $3, 4$ are shown in Fig. 2.20. The resulting invariant matrix element is

$$
\begin{aligned}
i\mathcal{M} = (-1)\frac{-i}{s - m_\phi^2} \Big\{ & \lambda^2 (x_1 x_2)(y_3 y_4) + (\lambda^*)^2 (y_1^\dagger y_2^\dagger)(x_3^\dagger x_4^\dagger) \\
& + |\lambda|^2 \left[(x_1 x_2)(x_3^\dagger x_4^\dagger) + (y_1^\dagger y_2^\dagger)(y_3 y_4) \right] \Big\} \\
+ \frac{-i}{t - m_\phi^2} \Big\{ & \lambda^2 (y_3 x_1)(y_4 x_2) + (\lambda^*)^2 (x_3^\dagger y_1^\dagger)(x_4^\dagger y_2^\dagger) \\
& + |\lambda|^2 \left[(x_3^\dagger y_1^\dagger)(y_4 x_2) + (y_3 x_1)(x_4^\dagger y_2^\dagger) \right] \Big\} \\
+ (-1)\frac{-i}{u - m_\phi^2} \Big\{ & \lambda^2 (y_4 x_1)(y_3 x_2) + (\lambda^*)^2 (x_4^\dagger y_1^\dagger)(x_3^\dagger y_2^\dagger) \\
& + |\lambda|^2 \left[(x_4^\dagger y_1^\dagger)(y_3 x_2) + (y_4 x_1)(x_3^\dagger y_2^\dagger) \right] \Big\}\,, \qquad (2.167)
\end{aligned}
$$

where $x_i \equiv x(\vec{\boldsymbol{p}}_i, s_i)$, $y_i \equiv y(\vec{\boldsymbol{p}}_i, s_i)$, $s = (p_1 + p_2)^2$, $t = (p_1 - p_3)^2$, $u = (p_1 - p_4)^2$, and m_ϕ is the mass of the exchanged scalar. We have chosen the canonical ordering of external fermions to be 3142 (corresponding to the t-channel contribution). For elastic scattering, this choice of canonical ordering guarantees that if no scattering occurs then the S-matrix is just equal to the identity with no extraneous minus sign (as explained in footnote 14). The relative minus signs between the t-channel diagram and the s- and u-channel diagrams [shown in parentheses in eq. (2.167)] are obtained by observing that both 1234 and 4132 are odd permutations of 3142. If we had crossed the initial-state fermion lines instead of the final-state fermion then the same relative signs for the u-channel diagrams would have been obtained.

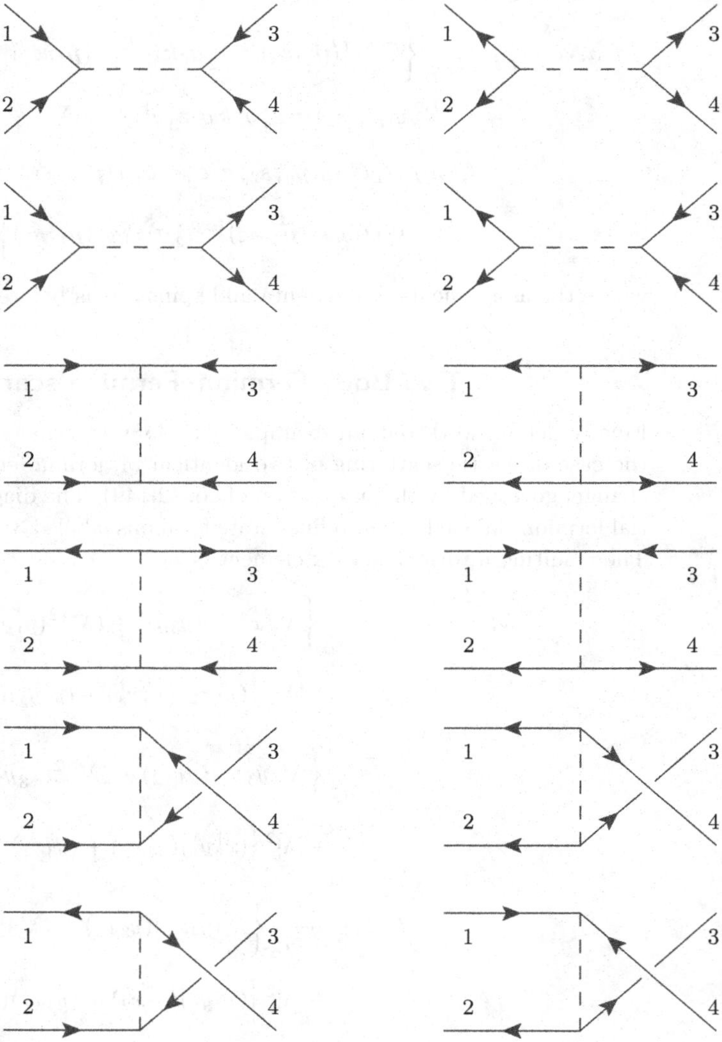

Fig. 2.20 Tree-level Feynman diagrams contributing to the elastic scattering of identical Majorana fermions via scalar exchange in the s-channel (top two rows), t-channel (middle two rows), and u-channel (bottom two rows).

Equation (2.167) can be factorized with respect to the scalar line:

$$\mathcal{M} = \frac{(\lambda x_1 x_2 + \lambda^* y_1^\dagger y_2^\dagger)(\lambda y_3 y_4 + \lambda^* x_3^\dagger x_4^\dagger)}{s - m_\phi^2} - \frac{(\lambda y_3 x_1 + \lambda^* x_3^\dagger y_1^\dagger)(\lambda y_4 x_2 + \lambda^* x_4^\dagger y_2^\dagger)}{t - m_\phi^2} + \frac{(\lambda y_4 x_1 + \lambda^* x_4^\dagger y_1^\dagger)(\lambda y_3 x_2 + \lambda^* x_3^\dagger y_2^\dagger)}{u - m_\phi^2}\,. \tag{2.168}$$

This is a common feature of Feynman graphs with a virtual boson.

This example also illustrates that, in contrast to the four-component fermion formalism, the two-component fermion Feynman rules typically yield many more diagrams, but the contribution of each of the diagrams is correspondingly simpler. Thus, the complexity of the final result is similar in both formalisms.

2.8.4 Nonrelativistic Potential Due to Scalar or Pseudoscalar Exchange

Consider two distinguishable fermions and a scalar–fermion–fermion Yukawa interaction given by eq. (2.135). We can compute the force law that the fermions experience due to exchange of a spinless boson. That is, we shall derive the Yukawa potential as a function of the separation distance of the two fermions in the static limit.

To carry out this computation, we compute the invariant matrix element for two-body fermion–fermion elastic scattering in the nonrelativistic limit. The relevant diagrams are shown in Fig. 2.20. As our two fermions are distinguishable, only the t-channel graphs (shown in the middle row of Fig. 2.20) are relevant. As a result, the matrix element for the elastic scattering of two Majorana fermions is given by the t-channel contribution of eq. (2.168),

$$i\mathcal{M} = \frac{i}{m_\phi^2 - t}(\lambda y_3 x_1 + \lambda^* x_3^\dagger y_1^\dagger)(\lambda y_4 x_2 + \lambda^* x_4^\dagger y_2^\dagger). \tag{2.169}$$

The choice of the overall sign is fixed by the canonical ordering of the external fermions.[16] Although the two fermions are distinguishable, we have assumed for simplicity that their (complex) Yukawa coupling strengths are the same and given by λ. For the scattering of two distinguishable Dirac fermions, the resulting expression for the scattering amplitude is identical to eq. (2.169), with λ replaced by the appropriate complex Yukawa coupling κ.

We denote the masses of the distinguishable fermions by m_1 and m_2. In the nonrelativistic limit, $p_1 \simeq (m_1\,;\,\vec{\boldsymbol{p}}_1)$ and $p_3 \simeq (m_1\,;\,\vec{\boldsymbol{p}}_3)$, so that

$$m_\phi^2 - t \simeq |\vec{\boldsymbol{p}}_1 - \vec{\boldsymbol{p}}_3|^2 + m_\phi^2 \equiv |\vec{\boldsymbol{q}}|^2 + m_\phi^2\,, \tag{2.170}$$

where $\vec{\boldsymbol{q}} \equiv \vec{\boldsymbol{p}}_3 - \vec{\boldsymbol{p}}_1 = \vec{\boldsymbol{p}}_2 - \vec{\boldsymbol{p}}_4$ is the momentum-transfer three-vector. Two separate cases will be considered.

In the first case, λ is a real coupling. This corresponds to the exchange of a $J^{\rm PC} = 0^{++}$ scalar. Using the nonrelativistic forms of eqs. (E.62) and (E.68) for the spinor bilinears, it is only necessary to keep the leading term. It then follows that

$$i\mathcal{M} = \frac{4i|\lambda|^2 m_1 m_2}{|\vec{\boldsymbol{q}}|^2 + m_\phi^2}\delta_{s_1 s_3}\delta_{s_2 s_4}\,. \tag{2.171}$$

In the second case, λ is purely imaginary, and we will write $\lambda = i|\lambda|$ (the overall

16 As noted below eq. (2.167), the canonical ordering of the external fermions in two-body elastic scattering is determined by the requirement that $\langle f|i\rangle = +1$ for $f = i$ (see footnote 14).

sign is not significant). This corresponds to the exchange of a $J^{\rm PC} = 0^{-+}$ pseudoscalar. Again, we use the nonrelativistic forms of eqs. (E.62) and (E.68) for the spinor bilinears. However, in this case the leading term cancels and we must retain the $\mathcal{O}(|\vec{p}|/m)$ terms appearing in the nonrelativistic limit of the spinor bilinears. In this case, we find

$$i\mathcal{M} = \frac{i|\lambda|^2}{|\vec{q}|^2 + m_\phi^2}(\vec{q}\cdot\hat{s}^a\tau^a_{s_3 s_1})\,(\vec{q}\cdot\hat{s}^b\tau^b_{s_4 s_2})\,. \tag{2.172}$$

We choose the spin quantization axis to lie along the z-direction. That is, according to eq. (E.26), we choose $(\hat{s}^1\,,\,\hat{s}^2\,,\,\hat{s}^3) = (\hat{x}\,,\,\hat{y}\,,\,\hat{z})$, in which case one can rewrite eq. (2.172) in the more traditional way,

$$i\mathcal{M} = \frac{i|\lambda|^2}{|\vec{q}|^2 + m_\phi^2}(\vec{q}\cdot\vec{\sigma}_{s_3 s_1})\,(\vec{q}\cdot\vec{\sigma}_{s_4 s_2})\,, \tag{2.173}$$

where $\vec{\sigma} \equiv \hat{x}\tau^1 + \hat{y}\tau^2 + \hat{z}\tau^3$ are the usual spin-1/2 Pauli matrices and the subscripted spin labels on $\vec{\sigma}$ should be interpreted in the same way as indicated below eq. (E.29). Thus, pseudoscalar exchange yields a spin-dependent force law.

The nonrelativistic potential that arises from the t-channel scalar or pseudoscalar exchange is obtained by comparing the relativistic scattering amplitude $\mathcal{M}$ with the Born approximation for scattering off a potential $V(\vec{x})$ in nonrelativistic quantum mechanics. Taking into account the difference between the conventions for the normalization of relativistic and nonrelativistic single-particle states, one finds that the static potential is given by

$$V(\vec{x}) = -\frac{1}{4m_1 m_2}\int \frac{d^3q}{(2\pi)^3}\,\mathcal{M}(\vec{q})e^{i\vec{q}\cdot\vec{x}}\,, \tag{2.174}$$

in a convention where the invariant amplitude is defined as in footnote 14. Inserting the scattering amplitude for scalar (S) exchange, one obtains the well-known attractive spin-independent Yukawa potential,

$$V_{\rm S}(\vec{x}) = -\frac{|\lambda|^2}{4\pi r}\,e^{-m_\phi r}\,\delta_{s_1 s_3}\delta_{s_2 s_4}\,, \tag{2.175}$$

where $r \equiv |\vec{x}|$. For the case of pseudoscalar (PS) exchange, one can easily evaluate the integral in eq. (2.174) by writing $q_j q_k e^{i\vec{q}\cdot\vec{x}} = -\nabla_j\nabla_k e^{i\vec{q}\cdot\vec{x}}$. The end result is

$$\begin{aligned} V_{\rm PS}(\vec{x}) &= \frac{|\lambda|^2}{16\pi m_1 m_2}(\vec{\sigma}_{s_3 s_1}\cdot\vec{\nabla})(\vec{\sigma}_{s_4 s_2}\cdot\vec{\nabla})\,\frac{e^{-m_\phi r}}{r} \\ &= \frac{|\lambda|^2 m_\phi^2}{16\pi m_1 m_2}\Biggl\{\left[-\frac{4\pi}{3m_\phi^2}\delta^{(3)}(\vec{x}) + \frac{e^{-m_\phi r}}{r}\right]\vec{\sigma}_{s_3 s_1}\cdot\vec{\sigma}_{s_4 s_2} \\ &\quad + \left[\frac{1}{(m_\phi r)^2} + \frac{1}{(m_\phi r)} + \frac{1}{3}\right]\left[\frac{3\,(\vec{\sigma}_{s_3 s_1}\cdot\vec{x})(\vec{\sigma}_{s_4 s_2}\cdot\vec{x})}{r^2} - \vec{\sigma}_{s_3 s_1}\cdot\vec{\sigma}_{s_4 s_2}\right]\frac{e^{-m_\phi r}}{r}\Biggr\}, \end{aligned} \tag{2.176}$$

after employing eq. (2.409).

2.9 Self-Energy of Scalar and Vector Bosons

In order to go beyond the tree-level approximation, one must learn how to treat the self-energy functions of external states in order to identify the physical pole masses and the corresponding wave function renormalization constants defined in the on-shell renormalization scheme. These quantities are then used in the treatment of external particles in scattering and decay processes via the Lehmann–Symanzik–Zimmermann (LSZ) eduction formalism.[17]

Before tackling the more involved analysis required in the treatment of external fermion states, we begin by considering the self-energy functions for boson fields, taking into account the possibilities of loop-induced mixing and absorptive parts corresponding to decays to intermediate states. After first treating scalar fields in detail, we then examine the new features that arise in the case of vector gauge boson fields.

2.9.1 Self-Energy of Scalar Fields

Consider a theory with real scalar degrees of freedom $\hat{\phi}_i$ labeled by a flavor index $i =\in \{1, 2, \ldots, N\}$. The theory is assumed to contain arbitrary interactions, which we will not need to refer to explicitly. The tree-level scalar squared-mass matrix is diagonalized and the corresponding scalar mass eigenstates ϕ_i are identified. With respect to the basis of scalar mass eigenstates, the diagonal squared-mass matrix of the scalars is denoted by

$$\boldsymbol{m}^2_{ij} = m_j^2 \delta_{ij} \qquad \text{(no sum over the repeated index } j), \tag{2.177}$$

where m_j^2 is the tree-level squared mass of the jth scalar.

The full, loop-corrected Feynman propagators with four-momentum p^μ are defined by the Fourier transforms of vacuum expectation values of time-ordered products of bilinears of the fully interacting unrenormalized scalar fields. We define

$$\langle 0 | \, T\phi_i(x)\phi_j(y) \, | 0 \rangle_{\text{FT}} = i\boldsymbol{A}_{ij}(s) \, , \tag{2.178}$$

where $s \equiv p^2$ and the matrix $\boldsymbol{A}$ satisfies a symmetry condition, $\boldsymbol{A}^{\mathsf{T}} = \boldsymbol{A}$, and the reality condition $\boldsymbol{A}^\star = \boldsymbol{A}$. Here, we have introduced the star symbol to mean that a quantity $\boldsymbol{Q}^\star$ is obtained from $\boldsymbol{Q}$ by taking the complex conjugate of all Lagrangian parameters appearing in its calculation, but not taking the complex conjugates of Euclideanized loop integral functions, whose imaginary (absorptive) parts correspond to fermion decay widths to multiparticle intermediate states. That is, the dispersive part of $\boldsymbol{A}$ is a real symmetric matrix and the absorptive part of $\boldsymbol{A}$ is a pure imaginary symmetric matrix.

The diagrammatic representations of the full propagator is displayed in Fig. 2.21, where $\boldsymbol{A}_{ij}$, defined in eq. (2.178), is an $N \times N$ matrix function. The full propagator

[17] For a detailed treatment of the LSZ formalism, see, e.g., Chapter 7 of [B66].

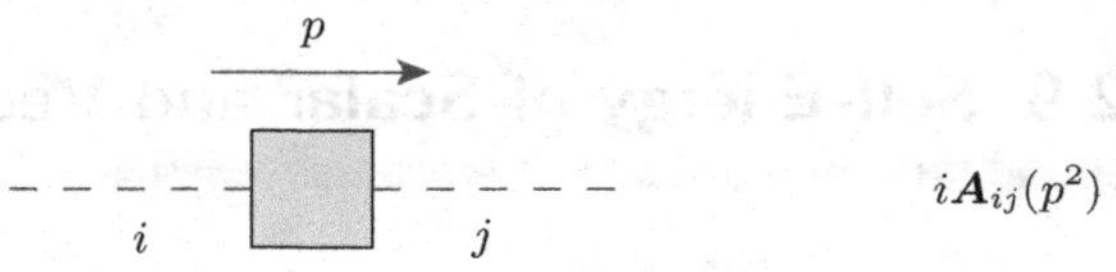

Fig. 2.21 The full, loop-corrected propagator for scalars is associated with the matrix function $\boldsymbol{A}(p^2)_{ij}$. The shaded box represents the sum of all connected Feynman diagrams, with external legs included. Although the four-momentum p flows from left to right, $\boldsymbol{A}(p^2)_{ij}$ is unchanged if the direction of the momentum is reversed.

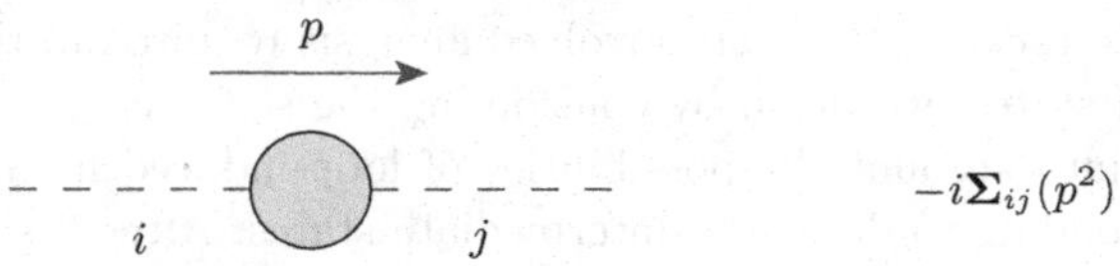

Fig. 2.22 The self-energy functions for scalars are associated with the function $\boldsymbol{\Sigma}_{ij}(p^2)$. The shaded circle represents the sum of all one-particle irreducible, connected Feynman diagrams, and the external legs are amputated. Although the four-momentum p flows from left to right, $\boldsymbol{\Sigma}(p^2)_{ij}$ is unchanged if the direction of the momentum is reversed.

function is given by

$$\boldsymbol{A}_{ij}(s) = \delta_{ij}/(s - m_i^2) + \cdots , \tag{2.179}$$

with no sum on i. The $\boldsymbol{A}_{ij}$ are functions of the Lorentz-invariant quantity $s \equiv p^2$, and of the masses and couplings of the theory.

The computation of the full propagators can be organized, as usual in quantum field theory, in terms of one-particle irreducible (1PI) self-energy functions. These are formally defined to be the sum of Feynman diagrams, to all orders in perturbation theory (with the corresponding tree-level graph *excluded*), that cannot be split into two separate diagrams by cutting through one internal line. Diagrammatically, the 1PI self-energy functions are defined in Fig. 2.22. As in the case of the full loop-corrected propagator, $\boldsymbol{\Sigma}$ satisfies a symmetry condition, $\boldsymbol{\Sigma}^{\mathsf{T}} = \boldsymbol{\Sigma}$, and the reality condition $\boldsymbol{\Sigma}^{\star} = \boldsymbol{\Sigma}$, where the star symbol was defined below eq. (2.178).

To compute the full propagator, consider the following diagrammatic identity:

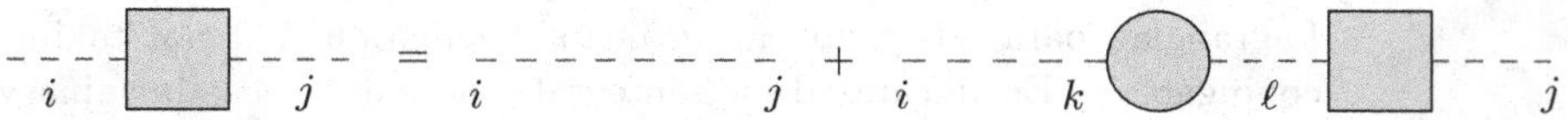

(2.180)

The diagrammatic equation above can be expressed algebraically as

$$F = T + TSF , \tag{2.181}$$

where F is the full loop-corrected propagator, T is the tree-level propagator, and

S is the 1PI self-energy function. To solve for F, we multiply eq. (2.181) on the left by T^{-1} and on the right by F^{-1} to obtain $T^{-1} = F^{-1} + S$. Thus, it follows that[18]

$$F = [T^{-1} - S]^{-1} . \tag{2.182}$$

That is,

$$\boldsymbol{A}^{-1}(s) = s\mathbb{1} - \boldsymbol{m^2} - \boldsymbol{\Sigma}(s) , \tag{2.183}$$

where $\mathbb{1}$ is the $N \times N$ identity matrix and $\boldsymbol{m^2}$ is the tree-level diagonal squared-mass matrix of the scalars [see eq. (2.177)].

The pole mass can be found most easily by considering the rest frame of the (off-shell) scalar, in which the space components of p^μ vanish. Setting $p^\mu = (\sqrt{s}\,;\,\boldsymbol{0})$, we search for values of s where the inverse of the full propagator has a zero eigenvalue. This is equivalent to setting the determinant of the inverse of the full propagator to zero. The end result is that the poles of the full propagator (which are in general complex), which are of the form

$$s_{\text{pole},n} \equiv M_n^2 - i\Gamma_n M_n \tag{2.184}$$

for $n = 1, 2, \ldots, N$, are formally the solutions to the nonlinear equation

$$\det\left[s\mathbb{1} - \boldsymbol{m^2} - \boldsymbol{\Sigma}(s)\right] = 0 . \tag{2.185}$$

The quantity Γ_n in eq. (2.184) is positive and is interpreted as the total decay rate of the (unstable) particle of mass M_n.

Some care is required in using eq. (2.185), since the squared pole mass always has a *nonpositive* imaginary part, while the loop integrals used to find the self-energy functions are complex functions of a real variable s that is given an infinitesimal *positive* imaginary part. Therefore, eq. (2.185) should be solved iteratively by first expanding the matrix self-energy function $\boldsymbol{\Sigma}$ in a series in s about either $m_n^2 + i\varepsilon$ or $M_n^2 + i\varepsilon$. The complex quantities $s_{\text{pole},n}$, which can be identified as the complex squared pole masses, are renormalization group and gauge invariant physical observables.

Corresponding to the pole $s_{\text{pole},n}$ of the full propagator $i\boldsymbol{A}(s)$, one can identify the residue at the pole as follows. Recall that $\boldsymbol{A}(s)$ was defined in eq. (2.178) in terms of the unrenormalized scalar fields. To define the renormalized real scalar fields (denoted with a subscript r), we introduce a matrix of wave function renormalization constants,

$$\phi_{r,i} = (Z_\phi^{-1/2})_{ij}\phi_j , \tag{2.186}$$

where there is an implicit sum over the flavor index j. The unrenormalized full

[18] Alternatively, one can solve eq. (2.181) by iteration and sum the resulting geometric series,

$$F = T + TS(T + TS(T + TS(\cdots))) = T + TST + TSTST + \cdots = T[1 + ST + (ST)^2 + \cdots]$$
$$= T[1 - ST]^{-1} = (T^{-1})^{-1}[1 - ST]^{-1} = [(1 - ST)T^{-1}]^{-1} = [T^{-1} - S]^{-1} .$$

loop-corrected propagator with four-momentum p^μ is defined in eq. (2.178). The corresponding renormalized full loop-corrected propagator is

$$i\boldsymbol{A}_r(s)_{ij} \equiv \langle 0|\, T\phi_{r,i}(x)\phi_{r,j}(y)\,|0\rangle_{\rm FT} = \left[Z_\phi^{-1/2} i\boldsymbol{A}(s)(Z_\phi^{-1/2})^{\mathsf{T}}\right]_{ij} . \quad (2.187)$$

That is, the renormalized loop-corrected propagator matrix is related to the corresponding unrenormalized loop-corrected propagator matrix by

$$\boldsymbol{A}_r(s) = Z_\phi^{-1/2}\boldsymbol{A}(s)(Z_\phi^{-1/2})^{\mathsf{T}} . \quad (2.188)$$

In particular, in the limit of $s \to s_{{\rm pole},n}$,

$$\boldsymbol{A}_r(s)_{ij} = \frac{\delta_{in}\delta_{nj}}{s - s_{{\rm pole},n}} + \boldsymbol{A}_{r,0}(s_{{\rm pole},n})_{ij} + \mathcal{O}(s - s_{{\rm pole},n}) , \quad \text{for all } n = 1, 2, \ldots, N, \quad (2.189)$$

where $\boldsymbol{A}_{r,0}(s)_{ij}$ has a finite nonzero limit as $s \to s_{{\rm pole},n}$. Hence, one can identify

$$\boldsymbol{A}_{ij}(s) \sim \frac{(Z_\phi^{1/2})_{in}(Z_\phi^{1/2})_{jn}}{s - s_{{\rm pole},n}} \quad (\text{as } s \to s_{{\rm pole},n}). \quad (2.190)$$

That is, $(Z_\phi)_{nn}$ is the residue of $\boldsymbol{A}(s)_{nn}$ at the pole $s = s_{{\rm pole},n}$.

The matrix of wave function renormalization constants are of the form

$$(Z_\phi)_{ij} = \delta_{ij} + (\delta Z_\phi)_{ij} , \quad (2.191)$$

where the contributions to $(\delta Z_\phi)_{ij}$ begin at one-loop order. As a result, at one-loop accuracy we can treat the diagonal elements and the off-diagonal elements separately.

First, we determine the complex pole masses $m_{p,n}$ by using eq. (2.185) to perturbatively solve for $s_{pole,n} = m_{p,n}^2$. Note that the argument of the determinant in eq. (2.185) is of the form $A_j\delta_{ij} + B_{ij}$, where B_{ij} is of one-loop order. Thus, the first-order perturbative solutions to $\det(A_j\delta_{ij} - B_{ij}) = 0$ are $A_j = B_{jj}$. Hence,

$$m_{p,n}^2 = m_n^2 + \boldsymbol{\Sigma}_{nn}(m_n^2) , \quad (2.192)$$

for each n. In light of eq. (2.184), it follows that, at one-loop accuracy,

$$M_n^2 = m_n^2 + {\rm Re}\,\boldsymbol{\Sigma}_{nn}(m_n^2) , \quad (2.193)$$

$$\Gamma_n = -\,{\rm Im}\,\boldsymbol{\Sigma}_{nn}(m_n^2)/m_n . \quad (2.194)$$

The squared-mass counterterm is defined by

$$\delta m_n^2 \equiv M_n^2 - m_n^2 . \quad (2.195)$$

In light of eq. (2.193),

$$\delta m_n^2 = {\rm Re}\,\boldsymbol{\Sigma}_{nn}(m_n^2) . \quad (2.196)$$

Note that, after taking the square root of eq. (2.193), the pole mass at one-loop accuracy is given by

$$M_n = m_n + \frac{1}{2m_n}\,{\rm Re}\,\boldsymbol{\Sigma}_{nn}(m_n^2) . \quad (2.197)$$

Next, we shall determine the elements of the matrix Z_ϕ by employing on-mass-shell renormalization conditions. These quantities can be determined from either eq. (2.189) or eq. (2.190). Our interest in Z_ϕ is primarily associated with the computation of scattering matrix elements. If one evaluates the contributing Feynman diagrams based on the bare Lagrangian, then the LSZ reduction formulae instruct us to replace any self-energy insertion on an external leg with the appropriate factors of $(Z_\phi^{1/2})_{ij}$, as discussed in Section 2.11. Alternatively, one can employ the counterterm approach in which the unperturbed Lagrangian is expressed in terms of renormalized parameters and fields, in which case extra Feynman rules appear for counterterms that render the scattering matrix element finite. In this approach, the wave function renormalization constants appear explicitly in the counterterm Feynman rules.

Strictly speaking, each on-shell external particle should correspond to an asymptotic state of the theory, which means that the corresponding external scalar should be stable. This implies that the absorptive part of $\boldsymbol{\Sigma}_{nn}(m_n^2)$ must vanish, which yields $\Gamma_n = 0$ for all $n = 1, 2, \ldots, N$. In this case, it is straightforward to determine the on-mass-shell conditions that specify the $(Z_\phi)_{ij}$.

The diagonal elements of Z_ϕ are fixed by eq. (2.190):

$$(Z_\phi)_{nn} = \lim_{p^2 \to m_{p,n}^2} (p^2 - m_{p,n}^2)\boldsymbol{A}_{nn}(p^2)\,. \tag{2.198}$$

In the one-loop approximation, $(\boldsymbol{A}^{-1})_{nn} = (\boldsymbol{A}_{nn})^{-1}$. Hence, it follows that

$$\begin{aligned}\left[\boldsymbol{A}(p^2)_{nn}\right]^{-1} &= p^2 - m_n^2 - \boldsymbol{\Sigma}_{nn}(p^2) \\ &= p^2 - m_n^2 - \boldsymbol{\Sigma}_{nn}(m_n^2) - (p^2 - m_{p,n}^2)\boldsymbol{\Sigma}'_{nn}(m_n^2) + \mathcal{O}((p^2 - m_{p,n}^2)^2) \\ &= (p^2 - m_{p,n}^2)\left[1 - \boldsymbol{\Sigma}'_{nn}(m_n^2)\right] + \mathcal{O}((p^2 - m_{p,n}^2)^2)\,,\end{aligned} \tag{2.199}$$

after employing eq. (2.192), where

$$\boldsymbol{\Sigma}'_{nn}(m_n^2) \equiv \left.\frac{d\boldsymbol{\Sigma}(p^2)_{nn}}{dp^2}\right|_{p^2=m_n^2}\,. \tag{2.200}$$

At one-loop accuracy, the arguments of $\boldsymbol{\Sigma}$ and $\boldsymbol{\Sigma}'$ can be taken to be either m_n^2 or $m_{p,n}^2$, since the difference of these two quantities is of one-loop order. Hence,

$$(Z_\phi)_{nn} = 1 + (\delta Z_\phi)_{nn} = \left[1 - \boldsymbol{\Sigma}'_{nn}(m_n^2)\right]^{-1}. \tag{2.201}$$

Dropping terms beyond one-loop order, it follows that

$$(\delta Z_\phi)_{nn} = \boldsymbol{\Sigma}'_{nn}(m_n^2)\,. \tag{2.202}$$

Returning to eq. (2.188), we can rewrite this equation, to one-loop accuracy, as

$$\begin{aligned}\boldsymbol{A}_r(p^2) &= \left[(Z_\phi^{1/2})^{\mathsf{T}}(p^2\mathbb{1} - \boldsymbol{m^2} - \boldsymbol{\Sigma}(p^2))Z_\phi^{1/2}\right]^{-1} \\ &= \left[p^2\mathbb{1} - \boldsymbol{m^2} + \tfrac{1}{2}p^2(\delta Z_\phi + \delta Z_\phi^{\mathsf{T}}) - \tfrac{1}{2}(\delta Z_\phi^{\mathsf{T}}\boldsymbol{m^2} + \boldsymbol{m^2}\delta Z_\phi) - \boldsymbol{\Sigma}(p^2)\right]^{-1}.\end{aligned} \tag{2.203}$$

We now define the subtracted self-energy function $\widehat{\boldsymbol{\Sigma}}(p^2)$ such that

$$\boldsymbol{A}_r(p^2) = \left[p^2\mathbb{1} - \boldsymbol{m^2} - \boldsymbol{\delta m^2} - \widehat{\boldsymbol{\Sigma}}(p^2)\right]^{-1}, \tag{2.204}$$

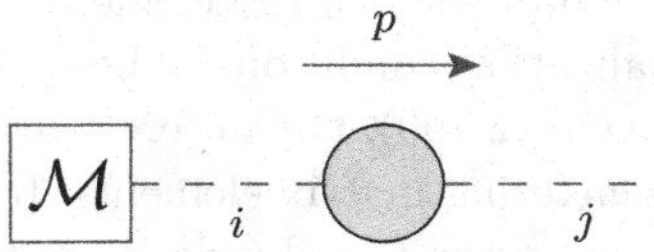

Fig. 2.23 The insertion of a subtracted self-energy function on an outgoing on-shell external line with flavor index j vanishes in the case of off-diagonal mixing with flavor $i \neq j$. The white box inscribed with an $\mathcal{M}$ represents the rest of the Feynman diagram (which may contain additional external fields) and the gray circle represents the subtracted self-energy function. The four-momentum p flows from left to right.

where $\boldsymbol{m^2}$ and $\boldsymbol{\delta m^2}$ are diagonal matrices with diagonal elements m_n^2 and δm_n^2, respectively. Comparing eqs. (2.203) and (2.204), it follows that

$$\widehat{\boldsymbol{\Sigma}}(p^2) = \boldsymbol{\Sigma}(p^2) - \boldsymbol{\delta m^2} - \tfrac{1}{2}p^2(\delta Z_\phi + \delta Z_\phi^{\mathsf{T}}) + \tfrac{1}{2}(\delta Z_\phi^{\mathsf{T}}\boldsymbol{m^2} + \boldsymbol{m^2}\delta Z_\phi). \quad (2.205)$$

We can recover eq. (2.202) by noting that the condition

$$\left.\frac{d\widehat{\boldsymbol{\Sigma}}_{nn}(p^2)}{dp^2}\right|_{p^2=m_n^2} = 0 \quad (2.206)$$

yields the one-loop expression for $(\delta Z_\phi)_{nn}$. Moreover, in the absence of an absorptive contribution to the self-energy functions, it also follows that $\widehat{\boldsymbol{\Sigma}}_{nn}(m_n^2) = 0$.

Given two unequal flavors, $i \neq j$, let us assume that the corresponding masses are nondegenerate, $m_i \neq m_j$. Equation (2.189) can be used to derive the conditions that fix the values of $(\delta Z_\phi)_{ij}$ for $i \neq j$, as shown in Exercise 2.15. These conditions have a natural diagrammatic interpretation. In particular, for any Feynman diagram, the insertion of a flavor off-diagonal subtracted self-energy function on an on-shell external scalar line vanishes. This on-shell renormalization condition ensures that there is no flavor mixing in the corrections associated with external on-shell scalars.

Consider a scattering or decay process involving an outgoing on-shell scalar of flavor j. One can insert a subtracted self-energy function on the on-shell external line, as shown in Fig. 2.23. It follows that the on-shell renormalization condition to remove flavor mixing associated with external on-shell states requires that

$$\mathcal{M}_i \widehat{\boldsymbol{\Sigma}}_{ij}(m_j^2) = 0\,, \quad (2.207)$$

where $i\mathcal{M}$ represents the contribution to the S-matrix element of the rest of the Feynman diagram, as shown in Fig. 2.23, and there is an implicit sum over $i \neq j$. Thus, we shall demand that $\widehat{\boldsymbol{\Sigma}}_{ij}(m_j^2) = 0$ for $i \neq j$. It then follows from eq. (2.205) that, in the one-loop approximation,

$$(\delta Z_\phi)_{ij} = \frac{2\boldsymbol{\Sigma}_{ij}(m_j^2)}{m_j^2 - m_i^2}\,, \qquad \text{for } i \neq j. \quad (2.208)$$

One can repeat this analysis for the insertion of the subtracted self-energy function on an ingoing on-shell external line of flavor i. In this case, $\widehat{\boldsymbol{\Sigma}}_{ij}(m_i^2) = 0$ for $j \neq i$ and one obtains

$$(\delta Z_\phi)_{ji} = \frac{2\boldsymbol{\Sigma}_{ij}(m_i^2)}{m_i^2 - m_j^2}\,, \qquad \text{for } j \neq i. \tag{2.209}$$

Note that $(\delta Z_\phi)_{ij}$ does not possess any particular symmetry under the interchange of i and j, since

$$(\delta Z_\phi)_{ij} + (\delta Z_\phi)_{ji} = \frac{2\left[\boldsymbol{\Sigma}_{ij}(m_j^2) - \boldsymbol{\Sigma}_{ij}(m_i^2)\right]}{m_j^2 - m_i^2}\,, \qquad \text{for } i \neq j. \tag{2.210}$$

The expression for $(\delta Z)_{ij}$, for $i \neq j$, is singular in the limit of $m_i = m_j$. However, in light of eq. (2.210), the symmetric part of $(\delta Z)_{ij}$ is well defined in the equal-mass limit. Indeed, one can show that only the symmetric part of $(\delta Z)_{ij}$ occurs in the renormalization of physical observables. For example, if two masses, say m_1 and m_2, are degenerate, then one can introduce a charged scalar field as discussed below eq. (2.212).

In summary, the on-shell conditions that determine the matrix elements of Z_ϕ are

$$\widehat{\boldsymbol{\Sigma}}_{ij}(m_i^2) = \widehat{\boldsymbol{\Sigma}}_{ij}(m_j^2) = \left.\frac{d\widehat{\boldsymbol{\Sigma}}_{nn}(p^2)}{dp^2}\right|_{p^2=m_n^2} = 0\,, \quad \text{for all } i,j = 1,2,\ldots,N. \tag{2.211}$$

If the scalar squared-mass matrix possesses a doubly degenerate eigenvalue, then it is convenient to define a complex scalar field

$$\Phi^\pm(x) = \frac{\phi_1(x) \pm i\phi_2(x)}{\sqrt{2}}\,, \tag{2.212}$$

where ϕ_1 and ϕ_2 are the mass-degenerate real scalar fields. In this case, there exists a global U(1) symmetry corresponding to an SO(2) rotation of the real fields ϕ_1 and ϕ_2 or a U(1) rephasing of the complex field, $\Phi^\pm \to e^{\pm iq\theta}\Phi^\pm$, where q is the charge of the scalar. Although the formalism above is adequate to treat such cases, it is convenient to separately treat the case of a collection of n scalars with the same nonzero charge. In this case, scalars of charge q will be denoted by Φ_i with the flavor index in the lowered position, and scalars of charge $-q$ will be denoted by $\Phi^i \equiv (\Phi_i)^\dagger$ with the flavor index in the raised position.

For a collection of n charged scalar fields, we define

$$\langle 0|\, T\Phi_i(x)\Phi^{\dagger j}(y)\,|0\rangle_{\text{FT}} = i\boldsymbol{A}_i{}^j(s)\,, \tag{2.213}$$

$$\langle 0|\, T\Phi^{\dagger i}(x)\Phi_j(y)\,|0\rangle_{\text{FT}} = i(\boldsymbol{A}^\mathsf{T})^i{}_j(s)\,, \tag{2.214}$$

where $s \equiv p^2$ and the matrix $\boldsymbol{A}$ satisfies the hermiticity condition $\left[\boldsymbol{A}^\mathsf{T}\right]^\star = \boldsymbol{A}$, and the star symbol is being employed as discussed below eq. (2.178). That is, the dispersive part of $\boldsymbol{A}$ is hermitian and the absorptive part of $\boldsymbol{A}$ is antihermitian.

The diagrammatic representations of the full propagator are displayed in Fig. 2.24, where $\boldsymbol{A}_i{}^j$, defined in eq. (2.213), is an $N \times N$ matrix function. Note that $\boldsymbol{A}_i{}^j$ represents the creation of flavor j from the vacuum at spacetime point y, followed by

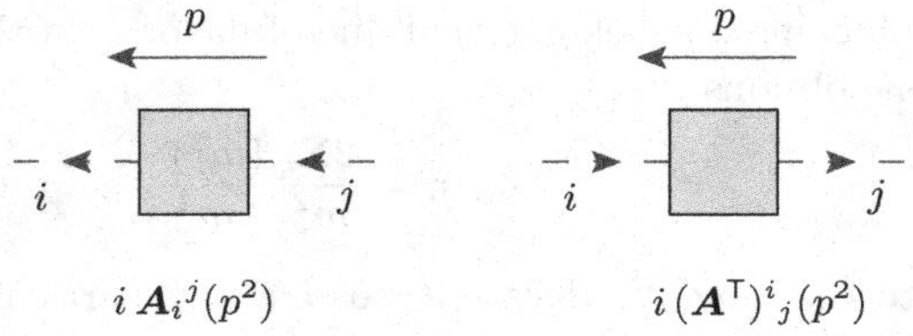

Fig. 2.24 The full, loop-corrected propagators for charged scalars are associated with functions $\boldsymbol{A}(p^2)_i{}^j$ [and its matrix transpose], as shown. The shaded boxes represent the sum of all connected Feynman diagrams, with external legs included. The four-momentum p flows from right to left.

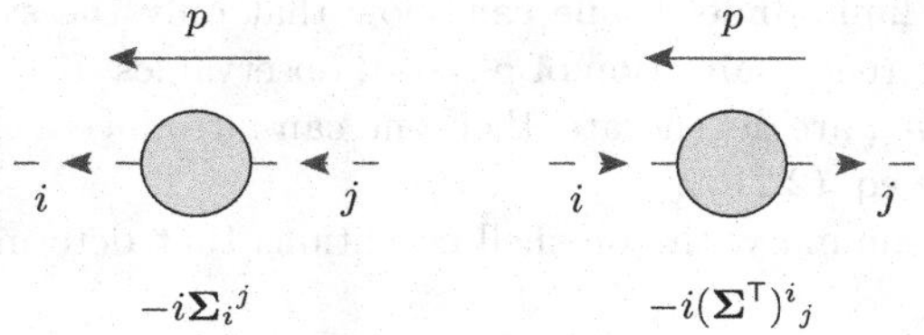

Fig. 2.25 The self-energy functions for the charged scalars are associated with functions $\boldsymbol{\Sigma}(s)_i{}^j$ [and its matrix transpose], as shown. The shaded circles represent the sum of all one-particle irreducible, connected Feynman diagrams, and the external legs are amputated. The four-momentum p flows from right to left.

the annihilation of flavor i at spacetime point x into the vacuum, as indicated by the directions of the arrows on the dashed lines that represent the charged scalar field (which also indicate the direction of charge flow). As previously indicated in the caption to Fig. 2.21, the full propagator function is unchanged if the direction of momentum flow is reversed.

Again, we organize the computation of the full propagators in terms of the 1PI self-energy functions. Diagrammatically, the 1PI self-energy functions are defined in Fig. 2.25. As in the case of the full loop-corrected propagator, $\boldsymbol{\Sigma}$ satisfies the hermiticity condition, $[\boldsymbol{\Sigma}^{\mathsf{T}}]^\star = \boldsymbol{\Sigma}$. In the treatment below, we shall assume that the scalars are stable, in which case the absorptive parts of the self-energy functions are zero. However, it is sometimes desirable to treat unstable particles as external states. There are a number of subtleties that arise in doing so, which will be glossed over in the following.

To compute the full propagator, consider the following diagrammatic identity:

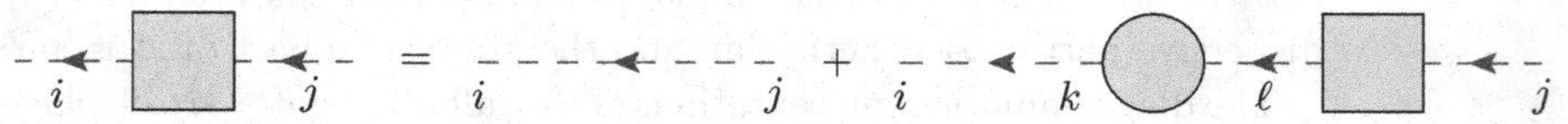

(2.215)

and a similar diagrammatic identity where the arrows point in the opposite direction. We can combine the two diagrammatic identities into a single matrix equation, which is represented diagrammatically by

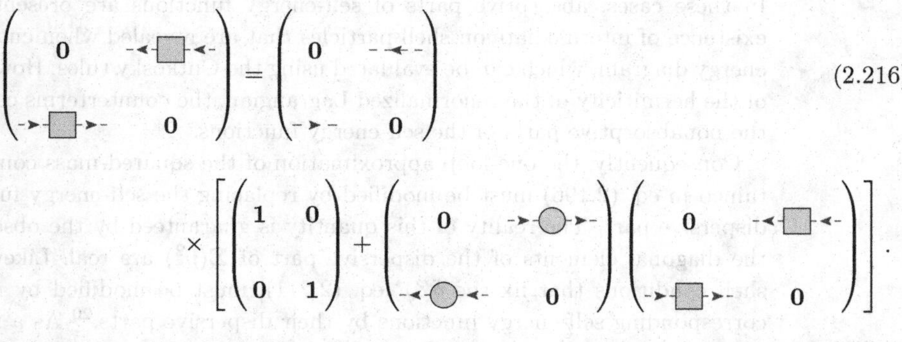

$$\begin{pmatrix} 0 & \text{-\!◄\!■\!◄\!-} \\ \text{-\!►\!■\!►\!-} & 0 \end{pmatrix} = \begin{pmatrix} 0 & \text{-\!◄\!-} \\ \text{-\!►\!-} & 0 \end{pmatrix} \tag{2.216}$$

$$\times \left[\begin{pmatrix} 1 & 0 \\ 0 & 1 \end{pmatrix} + \begin{pmatrix} 0 & \text{-\!►\!●\!►\!-} \\ \text{-\!◄\!●\!◄\!-} & 0 \end{pmatrix} \begin{pmatrix} 0 & \text{-\!◄\!■\!◄\!-} \\ \text{-\!►\!■\!►\!-} & 0 \end{pmatrix} \right].$$

Indeed, one can verify that the above diagrammatic equation yields eq. (2.215) and a second identity for the corresponding full loop-corrected propagators with the arrows reversed. Note that we have chosen the flavor labels in the appropriate left-to-right order to permit the interpretation of eq. (2.216) as a matrix equation.

The solution to eq. (2.216) is diagrammatically represented by

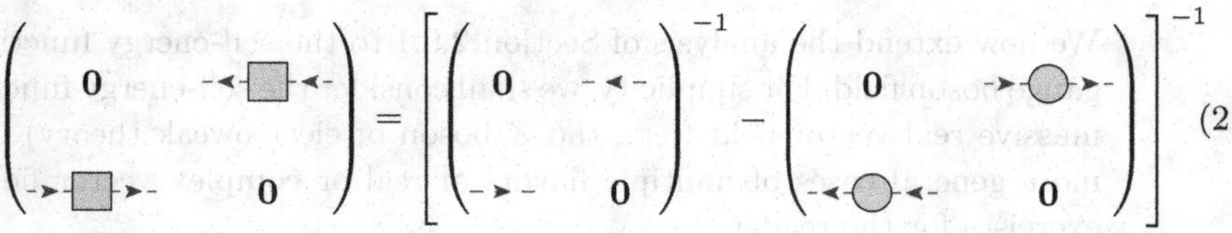

$$\begin{pmatrix} 0 & \text{-\!◄\!■\!◄\!-} \\ \text{-\!►\!■\!►\!-} & 0 \end{pmatrix} = \left[\begin{pmatrix} 0 & \text{-\!◄\!-} \\ \text{-\!►\!-} & 0 \end{pmatrix}^{-1} - \begin{pmatrix} 0 & \text{-\!►\!●\!►\!-} \\ \text{-\!◄\!●\!◄\!-} & 0 \end{pmatrix} \right]^{-1} \tag{2.217}$$

That is, we obtain a $2n \times 2n$ matrix equation for the full propagator functions,

$$\begin{pmatrix} 0 & i\boldsymbol{A} \\ i\boldsymbol{A}^{\mathsf{T}} & 0 \end{pmatrix} = \begin{pmatrix} 0 & -i(s\mathbb{1} - \boldsymbol{m}^2 - \boldsymbol{\Sigma}^{\mathsf{T}}) \\ -i(s\mathbb{1} - \boldsymbol{m}^2 - \boldsymbol{\Sigma}) & 0 \end{pmatrix}^{-1}, \tag{2.218}$$

which is equivalent to

$$\boldsymbol{A}^{-1}(s) = s\mathbb{1} - \boldsymbol{m}^2 - \boldsymbol{\Sigma}(s). \tag{2.219}$$

Thus, the only difference between the cases of a collection of real scalar fields and a collection of complex charged scalar fields is that the reality condition satisfied by the full propagator and self-energy functions in the former case is replaced by a hermiticity condition in the latter case. In particular, in the one-loop approximation,

$$(\delta Z_\phi)_k{}^k = \left. \frac{d\Sigma_k{}^k(p^2)}{dp^2} \right|_{p^2 = m_k^2}, \qquad (\delta Z_\phi)_i{}^j = \frac{2\Sigma_i{}^j(m_j^2)}{m_j^2 - m_i^2}, \quad \text{for } i \neq j. \tag{2.220}$$

In the case of two mass-degenerate charged scalars, the antihermitian part of the off-diagonal element of $(\delta Z)_i{}^j$ is singular, whereas the hermitian part is well defined. One can show that only the hermitian part of $(\delta Z)_i{}^j$ occurs in the renormalization of physical observables [32].

One can modify the various formulae obtained above and pretend that it is proper to treat the matrix elements for scattering and/or decay of unstable particles using the same Feynman diagram techniques that are applied to true asymptotic states.[19] In these cases, absorptive parts of self-energy functions are present due to the existence of intermediate on-shell particles that are revealed when cutting the self-energy diagram, which can be evaluated using the Cutkosky rules. However, in light of the hermiticity of the renormalized Lagrangian, the counterterms can only affect the nonabsorptive parts of the self-energy functions.

Consequently, the one-loop approximation of the squared-mass counterterm obtained in eq. (2.196) must be modified by replacing the self-energy function by its dispersive part. The reality of this quantity is guaranteed by the observation that the diagonal elements of the dispersive part of $\mathbf{\Sigma}(p^2)$ are real. Likewise, the on-shell conditions that fix the δZ_ϕ [eq. (2.211)] must be modified by replacing the corresponding self-energy functions by their dispersive parts.[20] As a result, in the one-loop approximation, we can evaluate the matrix δZ_ϕ of a collection of scalar fields by using the results obtained in this section but replacing the self-energy functions by their corresponding dispersive parts.

2.9.2 Self-Energy of Vector Gauge Boson Fields

We now extend the analysis of Section 2.9.1 to the self-energy function of a vector gauge boson field. For simplicity, we shall consider the self-energy function of a single massive real vector field (e.g., the Z boson of electroweak theory). Extensions to more general cases of multiple flavors of real or complex vector fields are left as exercises for the reader.[21]

The full, loop-corrected Feynman propagators with four-momentum p^μ are defined by the Fourier transforms of vacuum expectation values of time-ordered products of bilinears of the fully interacting unrenormalized real massive vector fields, denoted below by V_μ. We define the two-point connected Green function,

$$\langle 0|\, TV_\mu(x)V_\nu(y)\,|0\rangle_{\rm FT} = -i\mathscr{D}_{\mu\nu}(p)\,. \tag{2.221}$$

The tree-level propagator of the boson of mass m in the R_ξ gauge, which we denote below by

$$-i\mathscr{D}_{\mu\nu}(p) \equiv \frac{-i}{s-m^2}\left[g_{\mu\nu} - (1-\xi)\frac{p_\mu p_\nu}{s-\xi m^2}\right], \tag{2.222}$$

where $s \equiv p^2$, was previously exhibited in Fig. 2.6.

19 Ultimately, any results obtained in this matter will necessarily be approximate in Nature. A rigorous computation would require a treatment of the production of the unstable particle from the scattering of stable particles, in which case all initial and final states are true asymptotic states of the theory.

20 In the notation of [B76] and Ref. [33], one replaces the self-energy function $\mathbf{\Sigma}$ with $\widetilde{\rm Re}\,\mathbf{\Sigma}$, where $\widetilde{\rm Re}$ takes the real part of the loop integral but not of any complex Lagrangian parameter appearing in $\mathbf{\Sigma}$. However, this procedure has its drawbacks, as discussed in Ref. [34].

21 In electroweak theory, one must allow for Z–γ mixing at one loop, which requires an extension of the analysis presented here to the case of two flavors of real vector boson fields.

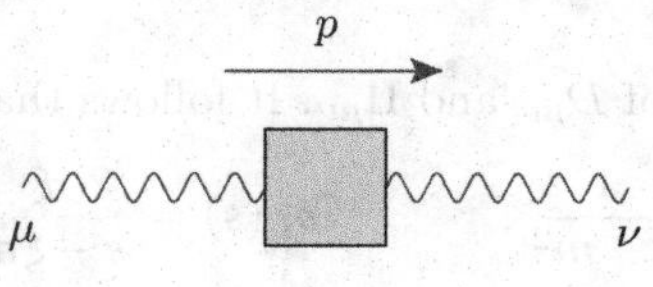

$-i\mathscr{D}_{\mu\nu}(p)$

Fig. 2.26 The full, loop-corrected propagator for real vector bosons is associated with the function $\mathscr{D}_{\mu\nu}(p)$, as shown. The shaded box represents the sum of all connected Feynman diagrams, with external legs included. Although the four-momentum p flows from left to right, $\mathscr{D}_{\mu\nu}(p)$ is unchanged if the direction of the momentum is reversed.

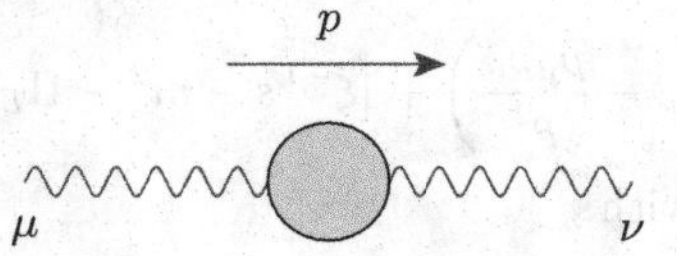

$i\Pi_{\mu\nu}(p) \equiv i[g_{\mu\nu}A_{VV}(p^2) + p_\mu p_\nu B_{VV}(p^2)]$

Fig. 2.27 The self-energy function for neutral vector bosons is associated with the function $\Pi_{\mu\nu}(p))$, as shown. The shaded circle represents the sum of all one-particle irreducible, connected Feynman diagrams, and the external legs are amputated. Although the four-momentum p flows from left to right, $\Pi_{\mu\nu}(p)$ is unchanged if the direction of the momentum is reversed.

The diagrammatic representation of the full propagator is displayed in Fig. 2.26, where $\mathscr{D}_{\mu\nu}(p)$ is defined in eq. (2.221). The computation of the full propagator can be organized, as usual in quantum field theory, in terms of 1PI self-energy functions. Diagrammatically, the 1PI self-energy function is defined in Fig. 2.27.

To compute the full propagator, consider the following diagrammatic identity:

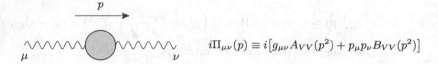

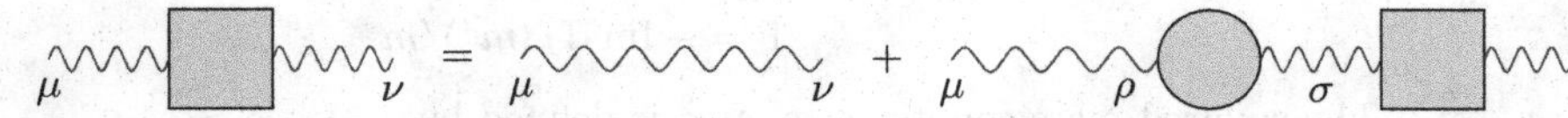

(2.223)

The diagrammatic equation above can be expressed algebraically as

$$-i\mathscr{D}_{\mu\nu} = -iD_{\mu\nu} + (-iD_{\mu\rho})i\Pi^{\rho\sigma}(-i\mathscr{D}_{\sigma\nu}), \tag{2.224}$$

where $i\Pi_{\mu\nu}(p)$ is the 1PI self-energy function exhibited in Fig. 2.27. Multiplying eq. (2.224) on the left by $(D^{-1})_\alpha{}^\mu$ and on the right by $(\mathscr{D}^{-1})^\nu{}_\beta$ yields

$$\mathscr{D}^{-1}_{\alpha\beta}(p) = D^{-1}_{\alpha\beta}(p) - \Pi_{\alpha\beta}(p). \tag{2.225}$$

In order to compute the inverses above, it is convenient to expand the corresponding second-rank tensors into their transverse and longitudinal parts, e.g.,

$$\mathscr{D}_{\mu\nu}(p) = \mathscr{D}_T(s)\left(g_{\mu\nu} - \frac{p_\mu p_\nu}{p^2}\right) + \mathscr{D}_L(s)\frac{p_\mu p_\nu}{p^2}, \tag{2.226}$$

and a similar decomposition of $D_{\mu\nu}$ and $\Pi_{\mu\nu}$. It follows that

$$\mathscr{D}_T(s) = \frac{1}{s-m^2}\,, \qquad \mathscr{D}_L(s) = \frac{\xi}{s-\xi m^2}\,. \tag{2.227}$$

One can now determine the inverse of $\mathscr{D}_{\mu\nu}(p)$ by computing the inverses of the scalar functions $\mathscr{D}_L$ and $\mathscr{D}_T$:

$$\mathscr{D}^{-1}_{\mu\nu}(p) = \mathscr{D}^{-1}_T(s)\left(g_{\mu\nu} - \frac{p_\mu p_\nu}{p^2}\right) + \mathscr{D}^{-1}_L(s)\frac{p_\mu p_\nu}{p^2}\,. \tag{2.228}$$

Hence, eq. (2.225) yields

$$\mathscr{D}^{-1}_{\mu\nu}(p) = \left[s - m^2 - \Pi_T(s)\right]\left(g_{\mu\nu} - \frac{p_\mu p_\nu}{p^2}\right) + \left[\xi^{-1}s - m^2 - \Pi_L(s)\right]\frac{p_\mu p_\nu}{p^2}\,. \tag{2.229}$$

Inverting this result, we end up with

$$\mathscr{D}_{\mu\nu}(p) = \frac{1}{s - m^2 - \Pi_T(s)}\left(g_{\mu\nu} - \frac{p_\mu p_\nu}{p^2}\right) + \frac{1}{\xi^{-1}s - m^2 - \Pi_L(s)}\,\frac{p_\mu p_\nu}{p^2}\,. \tag{2.230}$$

The physical mass of the vector boson corresponds to the pole of the transverse part of the full, loop-corrected Feynman propagator. As before, we consider the rest frame of the (off-shell) vector boson, in which the space components of p^μ vanish. Setting $p^\mu = (\sqrt{s}\,;\,\mathbf{0})$, we search for values of s that satisfy

$$s = m^2 + \Pi_T(s)\,. \tag{2.231}$$

The solution to this equation (which is in general complex), is of the form

$$s_{\rm pole} \equiv M^2 - i\Gamma M\,. \tag{2.232}$$

It follows that, at one-loop accuracy,

$$M^2 = m^2 + {\rm Re}\,\Pi_T(m^2)\,, \tag{2.233}$$

$$\Gamma = -\,{\rm Im}\,\Pi_T(m^2)/m\,. \tag{2.234}$$

The squared-mass counterterm, δm^2 is defined by

$$\delta m^2 \equiv M^2 - m^2\,. \tag{2.235}$$

In light of eq. (2.233),

$$\delta m^2 = {\rm Re}\,\Pi_T(m^2)\,. \tag{2.236}$$

Corresponding to the pole $s_{\rm pole}$ of the transverse part of the full propagator, $\mathscr{D}_T$, one can identify the residue at the pole following the analysis already presented in the case of scalar fields. Hence, one can identify

$$\mathscr{D}_T(s) \sim \frac{Z_V}{s - s_{\rm pole}} \qquad (\text{as } s \to s_{\rm pole}). \tag{2.237}$$

Hence, at one-loop accuracy following the derivation of eq. (2.202), the wave function renormalization constant of the vector field is $Z_V = 1 + \delta Z_V$, where

$$\delta Z_V = \Pi'_T(m^2)\,, \tag{2.238}$$

and $\Pi_T'(m^2) \equiv (d\Pi_T(s)/ds)_{s=m^2}$.

In the literature, it is common practice to write

$$i\Pi_{\mu\nu}(p) = i\big[A_{VV}(s)g_{\mu\nu} + B_{VV}(s)p_\mu p_\nu\big]\,. \tag{2.239}$$

That is,

$$A_{VV}(s) = \Pi_T(s)\,, \qquad B_{VV}(s) = \frac{\Pi_L(s) - \Pi_T(s)}{s}\,. \tag{2.240}$$

Thus, δm^2 and δZ_V are fixed by $A_{VV}(s)$ and its derivative evaluated at $s = m^2$, respectively. The extension to the case of multiple flavors of gauge bosons (real and/or complex) is now straightforward, where the role of $\boldsymbol{\Sigma}(s)$ in the case of scalar fields is now played by a suitably generalized version of $A_{VV}(s)$.

2.10 Self-Energy of Two-Component Fermions

Having introduced the formalism of the one-loop self-energy function for bosons in Section 2.9, we can now extend the formalism to the treatment of spin-1/2 fermions. In this section, we discuss the self-energy functions of a multiplet of two-component fermion fields consisting of Majorana and/or Dirac fermions, taking into account the possibilities of loop-induced mixing and absorptive parts corresponding to decays to intermediate states. The implications for two-component fermion field wave function renormalization are subsequently treated in Section 2.10.2.

2.10.1 Self-Energy Functions and Pole Masses of Two-Component Fermions

Consider a theory with left-handed fermion degrees of freedom $\hat{\psi}_i$ labeled by a flavor index $i \in \{1, 2, \ldots, N\}$. Associated with each $\hat{\psi}_i$ is a right-handed fermion $\hat{\psi}^{\dagger i}$, where the flavor labels are treated as described below eq. (1.160). The theory is assumed to contain arbitrary interactions, which we will not need to refer to explicitly. We diagonalize the fermion mass matrix and identify the fermion mass eigenstates ψ_i as indicated in eq. (2.127). In general, the mass eigenstates consist of Majorana fermions ξ_k ($k = 1, \ldots, N-2n$) and Dirac fermion pairs χ_ℓ and η_ℓ ($\ell = 1, \ldots, n$).[22] With respect to this basis, the symmetric $N \times N$ tree-level fermion mass matrix, $\boldsymbol{m}^{ij}$, is made up of diagonal elements m_k and 2×2 blocks $\left(\begin{smallmatrix} 0 & m_\ell \\ m_\ell & 0 \end{smallmatrix}\right)$ along the diagonal, where the m_k and m_ℓ are real and nonnegative. Since $\boldsymbol{m}^{ij}$ is real, the height of the flavor indices is not significant. Nevertheless, it is useful to define $\overline{\boldsymbol{m}}_{ij} \equiv \boldsymbol{m}^{ij}$ in order to maintain the convention that two repeated flavor

[22] In order to have a unified description, we shall take the flavor index of all left-handed fields (including η_k) in the lowered position [in contrast to the convention adopted in eq. (2.76)] when considering a collection of two-component fermion fields that contains both Majorana and Dirac fermions.

indices are summed when one index is raised and the other is lowered.[23] In matrix notation,

$$\overline{\boldsymbol{m}} = \boldsymbol{m}^{\dagger} . \tag{2.241}$$

Moreover, $\overline{\boldsymbol{m}}_{ik}\boldsymbol{m}^{kj} = \boldsymbol{m}^{ik}\overline{\boldsymbol{m}}_{kj} = m_i^2\delta_i^j$ is a diagonal matrix.

The full, loop-corrected Feynman propagators with four-momentum p^μ are defined by the Fourier transforms of vacuum expectation values of time-ordered products of bilinears of the fully interacting two-component fermion fields [see eqs. (2.84) and (2.85)]. Following eqs. (2.108)–(2.111), we define $s \equiv p^2$ and

$$\langle 0|\, T\psi_{\alpha i}(x)\psi^{\dagger j}_{\dot\beta}(y)\,|0\rangle_{\rm FT} = ip\!\cdot\!\sigma_{\alpha\dot\beta}\, \boldsymbol{C}_i{}^j(s)\,, \tag{2.242}$$

$$\langle 0|\, T\psi^{\dagger\dot\alpha i}(x)\psi^{\beta}_{j}(y)\,|0\rangle_{\rm FT} = ip\!\cdot\!\overline{\sigma}^{\dot\alpha\beta}\, (\boldsymbol{C}^{\mathsf{T}})^i{}_j(s)\,, \tag{2.243}$$

$$\langle 0|\, T\psi^{\dagger\dot\alpha i}(x)\psi^{\dagger j}_{\dot\beta}(y)\,|0\rangle_{\rm FT} = i\delta^{\dot\alpha}{}_{\dot\beta}\, \boldsymbol{D}^{ij}(s)\,, \tag{2.244}$$

$$\langle 0|\, T\psi_{\alpha i}(x)\psi^{\beta}_{j}(y)\,|0\rangle_{\rm FT} = i\delta_\alpha{}^\beta\, \overline{\boldsymbol{D}}_{ij}(s)\,, \tag{2.245}$$

where $\boldsymbol{C}_i{}^j$, $\boldsymbol{D}^{ij}$, and $\overline{\boldsymbol{D}}_{ij}$ defined above are each $N \times N$ matrix functions and

$$(\boldsymbol{C}^{\mathsf{T}})^i{}_j \equiv \boldsymbol{C}_j{}^i\,. \tag{2.246}$$

In light of eq. (1.83), one can derive eq. (2.243) from eq. (2.242) by first writing

$$\psi^{\dagger\dot\alpha i}(x)\psi^{\beta}_{j}(y) = -\epsilon^{\beta\alpha}\,\epsilon^{\dot\alpha\dot\beta}\psi_{\alpha j}(y)\psi^{\dagger i}_{\dot\beta}(x)\,, \tag{2.247}$$

where the minus sign arises because of the anticommutativity of the fields; the interchange of x and y (after FT) simply changes p^μ to $-p^\mu$.

In general, $\boldsymbol{D}$ and $\overline{\boldsymbol{D}}$ are complex symmetric matrices, and the full, loop-corrected propagator matrices satisfy hermiticity conditions,

$$[\boldsymbol{C}^{\mathsf{T}}]^\star = \boldsymbol{C}\,, \qquad \overline{\boldsymbol{D}} = \boldsymbol{D}^\star\,, \tag{2.248}$$

where the star symbol is being employed as discussed below eq. (2.178). That is, the dispersive part of $\boldsymbol{C}$ is hermitian and the absorptive part of $\boldsymbol{C}$ is antihermitian. The diagrammatic representations of the full propagators are displayed in Fig. 2.28.

The second diagram of Fig. 2.28, when flipped by 180° about the vertical axis that bisects the diagram, is equivalent to the first diagram of Fig. 2.28 (with $p \to -p$, $\alpha \to \beta$, $\dot\beta \to \dot\alpha$, and $i \leftrightarrow j$). In analogy with Fig. 2.3, the first two diagrammatic rules of Fig. 2.28 can be replaced by a single rule shown in Fig. 2.29, after using eq. (2.246) to rewrite the second version of the rule in terms of $\boldsymbol{C}^{\mathsf{T}}$. Indeed, by using the $\overline{\sigma}$ version of the rule shown in Fig. 2.29 and flipping the corresponding diagram by 180° as described above, one reproduces the rule of the second diagram of Fig. 2.28.

In what follows, we prefer to keep the first two rules of Fig. 2.28 as separate entities. This will permit us to conveniently assemble the four diagrams of Fig. 2.28

[23] We will soon be suppressing the indices, so it is convenient to employ the bar on $\overline{\boldsymbol{m}}_{ij}$ to indicate its lowered index structure.

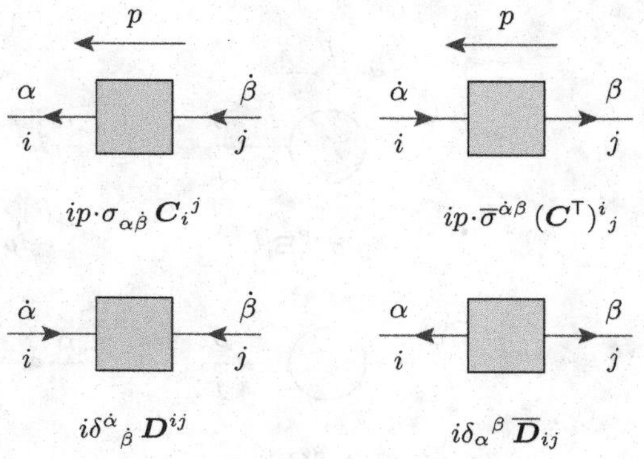

Fig. 2.28 The full, loop-corrected propagators for two-component fermions are associated with functions $\boldsymbol{C}(p^2)_i{}^j$ [and its matrix transpose], $\boldsymbol{D}(p^2)^{ij}$, and $\overline{\boldsymbol{D}}(p^2)_{ij}$, as shown. The shaded boxes represent the sum of all connected Feynman diagrams, with external legs included. The four-momentum p flows from right to left.

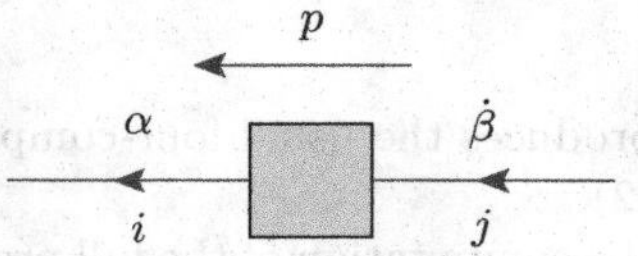

$ip{\cdot}\sigma_{\alpha\dot{\beta}}\, \boldsymbol{C}_i{}^j$ or $-ip{\cdot}\overline{\sigma}^{\dot{\beta}\alpha}\, (\boldsymbol{C}^{\mathsf{T}})^j{}_i$

Fig. 2.29 The first two diagrammatic rules of Fig. 2.28 can be summarized by a single diagram. Here, the choice of the σ or $\overline{\sigma}$ version of the rule is uniquely determined by the height of the spinor indices on the vertex to which the full loop-corrected propagator is connected (see Fig. 2.3 and the accompanying text).

into a 2×2 block matrix of two-component propagators. Moreover, in contrast to the full propagator of scalar fields, the full propagators for two-component fermion fields proportional to $\boldsymbol{C}$ and $\boldsymbol{C}^{\mathsf{T}}$ depend on the direction of momentum flow. By choosing the momentum flow in the two-component propagators from right to left, the left-to-right orderings of the spinor labels of the diagrams coincide with the ordering of spinor indices that appear in the corresponding algebraic representations. Thus, we can multiply diagrams together and interpret them as the product of the respective algebraic quantities taken from left to right in the normal fashion.

Given the tree-level propagators of Fig. 2.2, the full propagator functions are

$$\boldsymbol{C}_i{}^j = \delta_i{}^j/(s - m_i^2) + \cdots \tag{2.249}$$

$$\boldsymbol{D}^{ij} = \boldsymbol{m}^{ij}/(s - m_i^2) + \cdots \tag{2.250}$$

$$\overline{\boldsymbol{D}}_{ij} = \overline{\boldsymbol{m}}_{ij}/(s - m_i^2) + \cdots, \tag{2.251}$$

with no sum on i in each case. They are functions of the external momentum invariant s and the masses and couplings of the theory. Inserting the leading terms [eqs. (2.249)–(2.251)] into Fig. 2.28 and organizing the result in a 2×2 block matrix

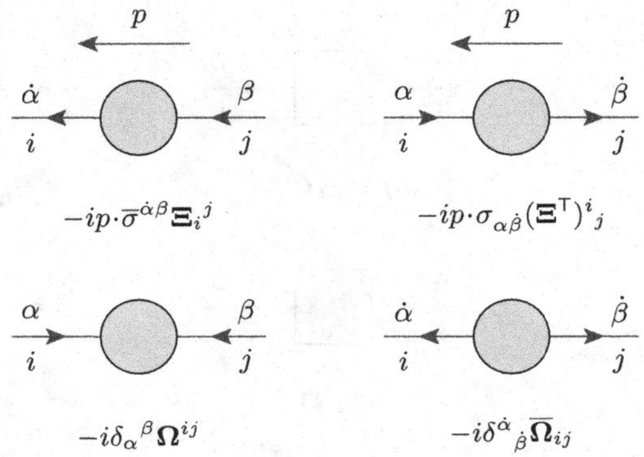

Fig. 2.30 The self-energy functions for two-component fermions are associated with functions $\Xi(s)_i{}^j$ [and its matrix transpose], $\Omega(s)^{ij}$, and $\overline{\Omega}(s)_{ij}$, as shown. The shaded circles represent the sum of all one-particle irreducible, connected Feynman diagrams, and the external legs are amputated. The four-momentum p flows from right to left.

of two-component propagators reproduces the usual four-component fermion tree-level propagator given in eq. (3.192).

As discussed below eq. (2.179), the computation of the full propagators can be organized in terms of 1PI self-energy functions. Diagrammatically, the 1PI self-energy functions are defined in Fig. 2.30. As in the case of the full loop-corrected propagators, Ω and $\overline{\Omega}$ are complex symmetric matrices, and the self-energy functions satisfy hermiticity conditions,

$$[\Xi^{\mathsf{T}}]^\star = \Xi\,, \qquad\qquad \overline{\Omega} = \Omega^\star\,, \tag{2.252}$$

where the star symbol was defined below eq. (2.178) and $(\Xi^{\mathsf{T}})^i{}_j \equiv \Xi_j{}^i$.

We illustrate the computation of the full propagator by considering first the following diagrammatic identity (with momentum p flowing from right to left):

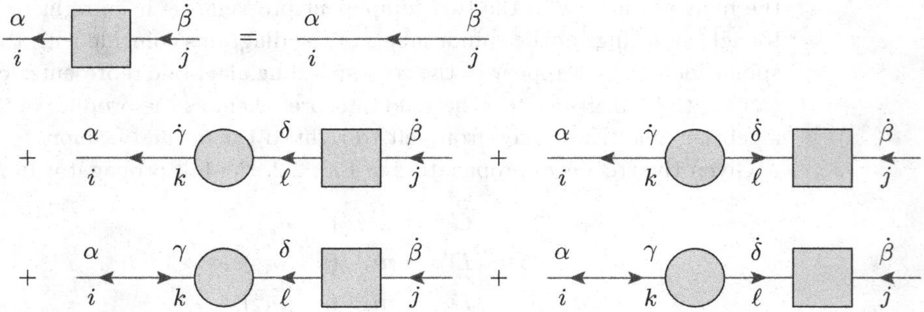

(2.253)

Similar diagrammatic identities can be constructed for the three other full loop-corrected propagators of Fig. 2.28. The resulting four equations are of the form,

$$F = T + TSF, \tag{2.254}$$

where F is the matrix of full loop-corrected propagators, T is the matrix of tree-level propagators, and S is the matrix of self-energy functions. Expressing eq. (2.254) in terms of diagrams,

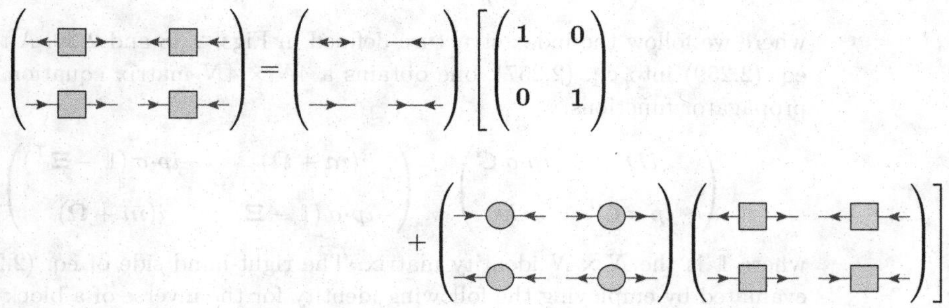

$$\begin{pmatrix} \boxed{} & \boxed{} \\ \boxed{} & \boxed{} \end{pmatrix} = \begin{pmatrix} & \\ & \end{pmatrix} \left[\begin{pmatrix} 1 & 0 \\ 0 & 1 \end{pmatrix} \right.$$

$$\left. + \begin{pmatrix} \bigcirc & \bigcirc \\ \bigcirc & \bigcirc \end{pmatrix} \begin{pmatrix} \boxed{} & \boxed{} \\ \boxed{} & \boxed{} \end{pmatrix} \right] \tag{2.255}$$

which, when expanded out, yields eq. (2.253) and the corresponding identities for the three other full loop-corrected propagators of Fig. 2.28. Note that we have chosen the labeling and momentum flow in Figs. 2.28 and 2.30 such that the spinor and flavor labels of the diagrams appear in the appropriate left-to-right order to permit the interpretation of eq. (2.255) as a matrix equation.

We have already encountered eq. (2.254) in our analysis of the self-energy function of scalar fields [see eq. (2.181)]. The solution to this equation was given in eq. (2.182), which we repeat below:

$$F = [T^{-1} - S]^{-1}. \tag{2.256}$$

This equation can be represented diagrammatically as follows:

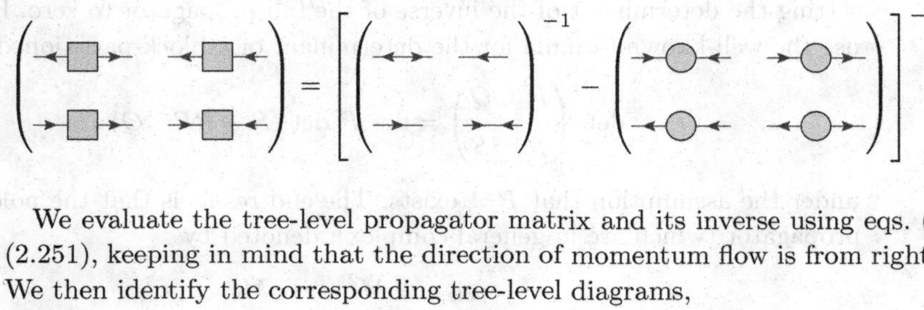

$$\begin{pmatrix} \boxed{} & \boxed{} \\ \boxed{} & \boxed{} \end{pmatrix} = \left[\begin{pmatrix} & \\ & \end{pmatrix}^{-1} - \begin{pmatrix} \bigcirc & \bigcirc \\ \bigcirc & \bigcirc \end{pmatrix} \right]^{-1} \tag{2.257}$$

We evaluate the tree-level propagator matrix and its inverse using eqs. (2.249)–(2.251), keeping in mind that the direction of momentum flow is from right to left. We then identify the corresponding tree-level diagrams,

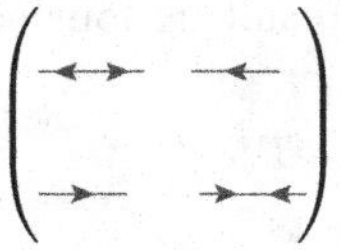

$$= \frac{1}{s - m_i^2} \begin{pmatrix} i\,\overline{m}_{ij}\,\delta_\alpha{}^\beta & ip\cdot\sigma_{\alpha\dot\beta}\,\delta_i{}^j \\ ip\cdot\overline{\sigma}^{\dot\alpha\beta}\,\delta^i{}_j & i\,m^{ij}\,\delta^{\dot\alpha}{}_{\dot\beta} \end{pmatrix}, \tag{2.258}$$

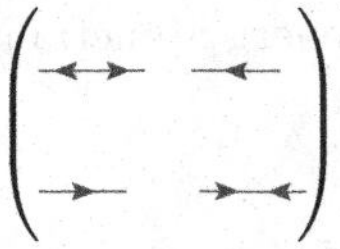

$$^{-1} = \begin{pmatrix} i\,m^{ij}\,\delta_\alpha{}^\beta & -ip\cdot\sigma_{\alpha\dot\beta}\,\delta^i{}_j \\ -ip\cdot\overline{\sigma}^{\dot\alpha\beta}\,\delta_i{}^j & i\,\overline{m}_{ij}\,\delta^{\dot\alpha}{}_{\dot\beta} \end{pmatrix}, \tag{2.259}$$

where we follow the index structure defined in Figs. 2.28 and 2.30. After inserting eq. (2.259) into eq. (2.257), one obtains a $4N \times 4N$ matrix equation for the full propagator functions:

$$\begin{pmatrix} i\overline{\boldsymbol{D}} & ip\cdot\sigma\,\boldsymbol{C} \\ ip\cdot\overline{\sigma}\,\boldsymbol{C}^{\mathsf{T}} & i\boldsymbol{D} \end{pmatrix} = \begin{pmatrix} i(\boldsymbol{m} + \boldsymbol{\Omega}) & -ip\cdot\sigma\,(\mathbb{1} - \boldsymbol{\Xi}^{\mathsf{T}}) \\ -ip\cdot\overline{\sigma}\,(\mathbb{1} - \boldsymbol{\Xi}) & i(\overline{\boldsymbol{m}} + \overline{\boldsymbol{\Omega}}) \end{pmatrix}^{-1}, \tag{2.260}$$

where $\mathbb{1}$ is the $N \times N$ identity matrix. The right-hand side of eq. (2.260) can be evaluated by employing the following identity for the inverse of a block-partitioned matrix (e.g., see [B128]):

$$\begin{pmatrix} P & Q \\ R & S \end{pmatrix}^{-1} = \begin{pmatrix} (P - QS^{-1}R)^{-1} & (R - SQ^{-1}P)^{-1} \\ (Q - PR^{-1}S)^{-1} & (S - RP^{-1}Q)^{-1} \end{pmatrix}, \tag{2.261}$$

under the assumption that all inverses appearing in eq. (2.261) exist. Applying this result to eq. (2.260), we obtain

$$\boldsymbol{C}^{-1} = s(\mathbb{1} - \boldsymbol{\Xi}) - (\overline{\boldsymbol{m}} + \overline{\boldsymbol{\Omega}})(\mathbb{1} - \boldsymbol{\Xi}^{\mathsf{T}})^{-1}(\boldsymbol{m} + \boldsymbol{\Omega}), \tag{2.262}$$

$$\boldsymbol{D}^{-1} = s(\mathbb{1} - \boldsymbol{\Xi})(\boldsymbol{m} + \boldsymbol{\Omega})^{-1}(\mathbb{1} - \boldsymbol{\Xi}^{\mathsf{T}}) - (\overline{\boldsymbol{m}} + \overline{\boldsymbol{\Omega}}), \tag{2.263}$$

$$\overline{\boldsymbol{D}}^{-1} = s(\mathbb{1} - \boldsymbol{\Xi}^{\mathsf{T}})(\overline{\boldsymbol{m}} + \overline{\boldsymbol{\Omega}})^{-1}(\mathbb{1} - \boldsymbol{\Xi}) - (\boldsymbol{m} + \boldsymbol{\Omega}). \tag{2.264}$$

Note that eq. (2.264) is consistent with eq. (2.263) as $\boldsymbol{\Xi}^\star = \boldsymbol{\Xi}^{\mathsf{T}}$.

The pole mass can be found most easily by considering the rest frame of the (off-shell) fermion, in which the space components of p^μ vanish. This reduces the spinor index dependence to a triviality. Setting $p^\mu = (\sqrt{s}\,;\,\boldsymbol{0})$, we search for values of s where the inverse of the full propagator has a zero eigenvalue. This is equivalent to setting the determinant of the inverse of the full propagator to zero. Here we shall use the well-known formula for the determinant of a block-partitioned matrix,

$$\det \begin{pmatrix} P & Q \\ R & S \end{pmatrix} = \det P \det\,(S - RP^{-1}Q), \tag{2.265}$$

under the assumption that P^{-1} exists. The end result is that the poles of the full propagator (which are in general complex), denoted by

$$s_{\text{pole},k} \equiv M_k^2 - i\Gamma_k M_k, \tag{2.266}$$

for $k = 1, 2, \ldots, N$, are formally the solutions to the nonlinear equation[24]

$$\det\left[s\mathbb{1} - (\mathbb{1} - \boldsymbol{\Xi}^{\mathsf{T}})^{-1}(\boldsymbol{m} + \boldsymbol{\Omega})(\mathbb{1} - \boldsymbol{\Xi})^{-1}(\overline{\boldsymbol{m}} + \overline{\boldsymbol{\Omega}})\right] = 0\,. \qquad (2.267)$$

As discussed below eq. (2.185), we shall iteratively solve eq. (2.267) by first expanding the self-energy function matrices $\boldsymbol{\Xi}$, $\boldsymbol{\Omega}$, and $\overline{\boldsymbol{\Omega}}$ in a series in s about either $m_k^2 + i\epsilon$ or $M_k^2 + i\epsilon$.

The results obtained above can be applied to an arbitrary collection of fermions (Majorana and/or Dirac). However, it is convenient to treat separately the case where all fermions are Dirac fermions of the same U(1) charge, consisting of pairs of two-component fields χ_i and η^i. In this case, we revert to the convention for flavor indices established in eqs. (2.75) and (2.76). As discussed in Section 1.6, the Dirac fermion mass eigenstates are defined in eq. (1.174) and are determined by the singular value decomposition of the Dirac fermion mass matrix. With respect to the mass basis, we denote the diagonal Dirac fermion mass matrix by

$$\boldsymbol{M}_i{}^j \equiv M_i \delta_i{}^j\,, \qquad (2.268)$$

where the M_i are real and nonnegative.

At tree level, there are four propagators for each pair of χ and η fields, as shown in Fig. 2.5. The corresponding full, loop-corrected Feynman propagators with four-momentum p^μ are defined by the Fourier transforms of vacuum expectation values of time-ordered products of bilinears of the fully interacting two-component fermion fields. Following eqs. (2.242)–(2.245), we define $s \equiv p^2$ and

$$\langle 0|\, T\chi_{\alpha i}(x)\chi^{\dagger j}_{\dot\beta}(y)\,|0\rangle_{\rm FT} = ip\cdot\sigma_{\alpha\dot\beta}\, \boldsymbol{S}_{Ri}{}^j(s)\,, \qquad (2.269)$$

$$\langle 0|\, T\eta^{\dagger\dot\alpha}_i(x)\eta^{\beta j}(y)\,|0\rangle_{\rm FT} = ip\cdot\overline{\sigma}^{\dot\alpha\beta}\, (\boldsymbol{S}_{\boldsymbol{L}}^{\mathsf{T}})_i{}^j(s)\,, \qquad (2.270)$$

$$\langle 0|\, T\eta^{\dagger\dot\alpha}_i(x)\chi^{\dagger j}_{\dot\beta}(y)\,|0\rangle_{\rm FT} = i\delta^{\dot\alpha}{}_{\dot\beta}\, \boldsymbol{S}_{Di}{}^j(s)\,, \qquad (2.271)$$

$$\langle 0|\, T\chi_{\alpha i}(x)\eta^{\beta j}(y)\,|0\rangle_{\rm FT} = i\delta_\alpha{}^\beta\, (\overline{\boldsymbol{S}}{}_{\boldsymbol{D}}^{\mathsf{T}})_i{}^j(s)\,. \qquad (2.272)$$

The full, loop-corrected propagators are exhibited in Fig. 2.31. The naming and sign conventions employed for the full, loop-corrected Dirac fermion propagator functions in Fig. 2.31 derive from the corresponding functions used in the more traditional four-component treatment presented in Chapter 3 [see eq. (3.226)].

The full, loop-corrected propagators are complex matrices that satisfy the hermiticity conditions

$$[\boldsymbol{S}_{\boldsymbol{R}}^{\mathsf{T}}]^\star = \boldsymbol{S}_{\boldsymbol{R}}\,, \qquad [\boldsymbol{S}_{\boldsymbol{L}}^{\mathsf{T}}]^\star = \boldsymbol{S}_{\boldsymbol{L}}\,, \qquad \overline{\boldsymbol{S}}_{\boldsymbol{D}} = \boldsymbol{S}_{\boldsymbol{D}}^\star\,, \qquad (2.273)$$

where the star symbol was defined below eq. (2.178). In contrast to the general case of an arbitrary collection of fermions treated earlier, $\boldsymbol{S_R}$ and $\boldsymbol{S_L}$ are unrelated and $\boldsymbol{S_D}$ is a complex matrix (not necessarily symmetric).

[24] The determinant of the inverse of the full propagator [the inverse of eq. (2.260)] is equal to eq. (2.267) multiplied by $\det\left[-(\mathbb{1} - \boldsymbol{\Xi})(\mathbb{1} - \boldsymbol{\Xi}^{\mathsf{T}})\right]$. We assume that the latter does not vanish. This must be true perturbatively since the eigenvalues of $\boldsymbol{\Xi}$ are one-loop (or higher) quantities, which one assumes cannot be as large as 1.

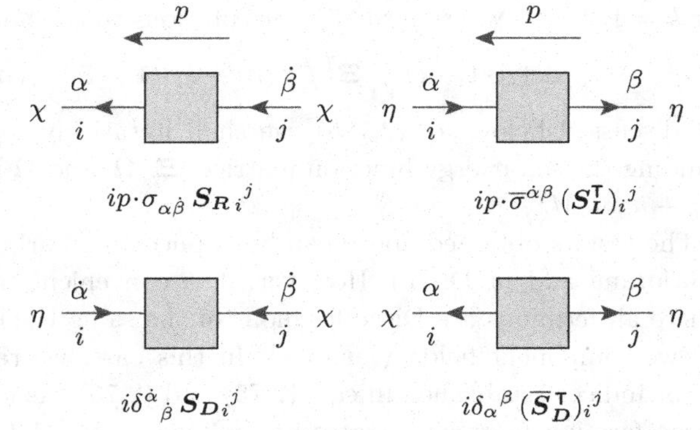

Fig. 2.31 The full, loop-corrected propagators for Dirac fermions, represented by pairs of two-component (oppositely charged) fermion fields χ_i and η^i, are associated with functions $S_R(s)_i{}^j$, $S_L^{\mathsf{T}}(s)_i{}^j$, $S_D(s)_i{}^j$, and $\overline{S}_D^{\mathsf{T}}(s)_i{}^j$, as shown. The shaded boxes represent the sum of all connected Feynman diagrams, with external legs included. The four-momentum p and the charge of χ flow from right to left.

Instead of working in a χ–η basis for the two-component Dirac fermion fields, one can Takagi diagonalize the fermion mass matrix. In the new ψ-basis,

$$\psi_{2i-1} = \frac{1}{\sqrt{2}}\left(\chi_i + \eta^i\right), \qquad \psi_{2i} = \frac{1}{i\sqrt{2}}\left(\chi_i - \eta^i\right), \qquad \text{for } i = 1, 2, \ldots, n, \tag{2.274}$$

and the loop-corrected propagators of Fig. 2.28 are applicable. It is easy to check that the number of independent functions is the same in both methods for treating Dirac fermions. In particular, the loop-corrected propagator functions in the ψ-basis are given in terms of the corresponding functions in the χ–η basis by[25]

$$C = \begin{pmatrix} S_R & 0 \\ 0 & S_L \end{pmatrix}, \qquad D = \begin{pmatrix} 0 & S_D^{\mathsf{T}} \\ S_D & 0 \end{pmatrix}, \qquad \overline{D} = \begin{pmatrix} 0 & \overline{S}_D^{\mathsf{T}} \\ \overline{S}_D & 0 \end{pmatrix}. \tag{2.275}$$

Note that C, D, and $\overline{D}$ do not inherit the flavor-index structure of S_R, S_L^{T}, S_D, and $\overline{S}_D^{\mathsf{T}}$, in light of flavor-index conventions employed in eq. (2.274).

We similarly introduce the 1PI self-energy matrix functions for the Dirac fermions in the χ–η basis, where the corresponding self-energy functions are defined in Fig. 2.32. As before, the naming and sign conventions employed for the Dirac fermion self-energy functions above derive from the corresponding functions used in the more traditional four-component treatment of Chapter 3 [see eq. (3.227)].

As in the case of the full loop-corrected propagators, the self-energy functions satisfy $\boldsymbol{\Sigma_R}$ satisfy hermiticity conditions,

$$[\Sigma_L^{\mathsf{T}}]^\star = \Sigma_L\,, \qquad [\Sigma_R^{\mathsf{T}}]^\star = \Sigma_R\,, \qquad \overline{\Sigma}_D = \Sigma_D^\star\,, \tag{2.276}$$

[25] The simple forms of C in eq. (2.275) and Ξ in eq. (2.277) motivate our definitions of S_L and Σ_R with the transpose as indicated in Figs. 2.31 and 2.32, respectively.

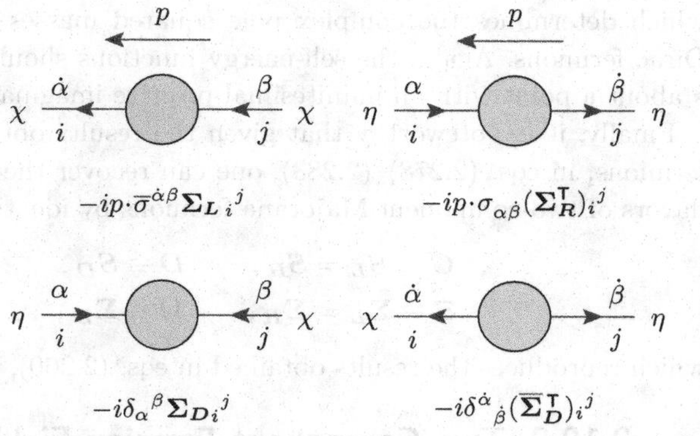

Fig. 2.32 The self-energy functions for two-component Dirac fermions, represented by pairs of two-component (oppositely charged) fermion fields χ_i and η^i, are associated with functions $\boldsymbol{\Sigma_L}(s)_i{}^j$, $\boldsymbol{\Sigma}^{\mathsf{T}}_{\boldsymbol{R}}(s)_i{}^j$, $\boldsymbol{\Sigma_D}(s)_i{}^j$, and $\overline{\boldsymbol{\Sigma}}{}^{\mathsf{T}}_{\boldsymbol{D}}(s)_i{}^j$, as shown. The shaded circles represent the sum of all one-particle irreducible, connected Feynman diagrams, and the external legs are amputated. The four-momentum p flows from right to left.

where the star symbol was defined below eq. (2.178). Moreover, $\boldsymbol{\Sigma_L}$ and $\boldsymbol{\Sigma_R}$ are unrelated and $\boldsymbol{\Sigma_D}$ is a complex matrix (not necessarily symmetric). The self-energy functions in the ψ-basis are given in terms of the corresponding functions in the χ–η basis by

$$\boldsymbol{\Xi} = \begin{pmatrix} \boldsymbol{\Sigma_L} & 0 \\ 0 & \boldsymbol{\Sigma_R} \end{pmatrix}, \qquad \boldsymbol{\Omega} = \begin{pmatrix} 0 & \boldsymbol{\Sigma}^{\mathsf{T}}_{\boldsymbol{D}} \\ \boldsymbol{\Sigma_D} & 0 \end{pmatrix}, \qquad \overline{\boldsymbol{\Omega}} = \begin{pmatrix} 0 & \overline{\boldsymbol{\Sigma}}{}^{\mathsf{T}}_{\boldsymbol{D}} \\ \overline{\boldsymbol{\Sigma}}_{\boldsymbol{D}} & 0 \end{pmatrix}. \tag{2.277}$$

In the case of Dirac fermions fields, eq. (2.257) still holds in the χ–η basis, which yields

$$\begin{pmatrix} i\overline{\boldsymbol{S}}{}^{\mathsf{T}}_{\boldsymbol{D}} & ip{\cdot}\sigma\,\boldsymbol{S_R} \\ ip{\cdot}\overline{\sigma}\,\boldsymbol{S}^{\mathsf{T}}_{\boldsymbol{L}} & i\boldsymbol{S_D} \end{pmatrix} = \begin{pmatrix} i(\boldsymbol{M}+\boldsymbol{\Sigma_D}) & -ip{\cdot}\sigma\,(\mathbb{1}-\boldsymbol{\Sigma}^{\mathsf{T}}_{\boldsymbol{R}}) \\ -ip{\cdot}\overline{\sigma}\,(\mathbb{1}-\boldsymbol{\Sigma_L}) & i(\boldsymbol{M}+\overline{\boldsymbol{\Sigma}}{}^{\mathsf{T}}_{\boldsymbol{D}}) \end{pmatrix}^{-1}. \tag{2.278}$$

Using eq. (2.261), it follows that

$$\boldsymbol{S_L}^{-1} = s(\mathbb{1}-\boldsymbol{\Sigma_R}) - (\boldsymbol{M}+\overline{\boldsymbol{\Sigma}}_{\boldsymbol{D}})(\mathbb{1}-\boldsymbol{\Sigma}^{\mathsf{T}}_{\boldsymbol{L}})^{-1}(\boldsymbol{M}+\boldsymbol{\Sigma}^{\mathsf{T}}_{\boldsymbol{D}}), \tag{2.279}$$

$$\boldsymbol{S_R}^{-1} = s(\mathbb{1}-\boldsymbol{\Sigma_L}) - (\boldsymbol{M}+\overline{\boldsymbol{\Sigma}}{}^{\mathsf{T}}_{\boldsymbol{D}})(\mathbb{1}-\boldsymbol{\Sigma}^{\mathsf{T}}_{\boldsymbol{R}})^{-1}(\boldsymbol{M}+\boldsymbol{\Sigma_D}), \tag{2.280}$$

$$\boldsymbol{S_D}^{-1} = s(\mathbb{1}-\boldsymbol{\Sigma_L})(\boldsymbol{M}+\boldsymbol{\Sigma_D})^{-1}(\mathbb{1}-\boldsymbol{\Sigma}^{\mathsf{T}}_{\boldsymbol{R}}) - (\boldsymbol{M}+\overline{\boldsymbol{\Sigma}}{}^{\mathsf{T}}_{\boldsymbol{D}}), \tag{2.281}$$

$$\overline{\boldsymbol{S}}_{\boldsymbol{D}}{}^{-1} = s(\mathbb{1}-\boldsymbol{\Sigma}^{\mathsf{T}}_{\boldsymbol{L}})(\boldsymbol{M}+\overline{\boldsymbol{\Sigma}}_{\boldsymbol{D}})^{-1}(\mathbb{1}-\boldsymbol{\Sigma_R}) - (\boldsymbol{M}+\boldsymbol{\Sigma}^{\mathsf{T}}_{\boldsymbol{D}}). \tag{2.282}$$

Note that eq. (2.282) is consistent with eq. (2.281) as $\boldsymbol{\Sigma}^{\star}_{\boldsymbol{L,R}} = \boldsymbol{\Sigma}^{\mathsf{T}}_{\boldsymbol{L,R}}$.

The pole mass is now easily computed using the technique previously outlined. In particular, eq. (2.267) becomes

$$\det\left[s\mathbb{1} - (\mathbb{1}-\boldsymbol{\Sigma}^{\mathsf{T}}_{\boldsymbol{R}})^{-1}(\boldsymbol{M}+\boldsymbol{\Sigma_D})(\mathbb{1}-\boldsymbol{\Sigma_L})^{-1}(\boldsymbol{M}+\overline{\boldsymbol{\Sigma}}{}^{\mathsf{T}}_{\boldsymbol{D}})\right] = 0, \tag{2.283}$$

which determines the complex pole squared masses, $s_{\rm pole}$, of the corresponding Dirac fermions. Again, the self-energy functions should be expanded in a series in s about a point with an infinitesimal positive imaginary part.

Finally, it is noteworthy that given the results obtained for N flavors of Dirac fermions, in eqs. (2.278)–(2.283), one can recover the corresponding results for N flavors of two-component Majorana fermions by identifying

$$\boldsymbol{C} \equiv \boldsymbol{S_L} = \boldsymbol{S_R}\,, \qquad \boldsymbol{D} \equiv \boldsymbol{S_D}\,, \qquad \overline{\boldsymbol{D}} \equiv \overline{\boldsymbol{S}}_{\boldsymbol{D}}\,, \tag{2.284}$$

$$\boldsymbol{\Xi} \equiv \boldsymbol{\Sigma_L} = \boldsymbol{\Sigma_R}\,, \qquad \boldsymbol{\Omega} \equiv \boldsymbol{\Sigma_D}\,, \qquad \overline{\boldsymbol{\Omega}} \equiv \overline{\boldsymbol{\Sigma}}_{\boldsymbol{D}}\,, \tag{2.285}$$

which reproduces the results obtained in eqs. (2.260), (2.262)–(2.264), and (2.267).

2.10.2 Two-Component Fermion Field Renormalization

Using the results obtained in the previous subsection, we shall now write down explicit expressions for the wave function renormalization constants at one-loop accuracy. We shall perform the calculation for the case of n two-component Dirac fermion fields, $\{\chi_i, \eta^i\}$, all of the same U(1) charge. Subsequently, we will obtain the corresponding result for the case of N two-component Majorana fermion fields by employing eqs. (2.284) and (2.285).

We begin by defining renormalized fields (denoted with a subscript r),

$$\chi_{r,i} = (Z_L^{-1/2})_i{}^j \chi_j\,, \qquad \eta^\dagger_{r,i} = (Z_R^{-1/2})_i{}^j \eta^\dagger_j\,, \tag{2.286}$$

where there is an implicit sum over the flavor index j. It then follows that

$$\chi_r^{\dagger\, i} = \chi^{\dagger\, j}(Z_L^{-1/2\dagger})_j{}^i\,, \qquad \eta_r^i = \eta^j (Z_R^{-1/2\dagger})_j{}^i\,. \tag{2.287}$$

The unrenormalized full loop-corrected propagators with four-momentum p^μ are defined in eqs. (2.269)–(2.272). The corresponding renormalized full loop-corrected propagators are

$$\begin{pmatrix} \langle 0\,|\,T\chi_r(x)\eta_r(y)\,|\,0\rangle_{\rm FT} & \langle 0\,|\,T\chi_r(x)\chi_r^\dagger(y)\,|\,0\rangle_{\rm FT} \\ \langle 0\,|\,T\eta_r^\dagger(x)\eta_r(y)\,|\,0\rangle_{\rm FT} & \langle 0\,|\,T\eta_r^\dagger(x)\chi_r^\dagger(y)\,|\,0\rangle_{\rm FT} \end{pmatrix}$$

$$= \begin{pmatrix} Z_L^{-1/2} i\overline{\boldsymbol{S}}{}^{\sf T}_{\boldsymbol{D}}(p^2) Z_R^{-1/2\dagger} & Z_L^{-1/2} ip\!\cdot\!\sigma \boldsymbol{S_R}(p^2) Z_L^{-1/2\dagger} \\ Z_R^{-1/2} ip\!\cdot\!\overline{\sigma} \boldsymbol{S}^{\sf T}_{\boldsymbol{L}}(p^2) Z_R^{-1/2\dagger} & Z_R^{-1/2} i\boldsymbol{S_D}(p^2) Z_L^{-1/2\dagger} \end{pmatrix}, \tag{2.288}$$

where two-component spinor indices and flavor indices have been suppressed. Denoting the renormalized two-point full loop-corrected propagators with a subscript r, it follows that

$$\boldsymbol{S_D}(p^2) = Z_R^{1/2} \boldsymbol{S_{D}}_r(p^2) Z_L^{1/2\dagger}\,, \qquad \overline{\boldsymbol{S}}{}^{\sf T}_{\boldsymbol{D}}(p^2) = Z_L^{1/2} \overline{\boldsymbol{S}}{}^{\sf T}_{\boldsymbol{D}r}(p^2) Z_R^{1/2\dagger}\,, \tag{2.289}$$

$$\boldsymbol{S_R}(p^2) = Z_L^{1/2} \boldsymbol{S_{R}}_r(p^2) Z_L^{1/2\dagger}\,, \qquad \boldsymbol{S}^{\sf T}_{\boldsymbol{L}}(p^2) = Z_R^{1/2} \boldsymbol{S}^{\sf T}_{\boldsymbol{L}r}(p^2) Z_R^{1/2\dagger}\,. \tag{2.290}$$

The wave function renormalization constants are of the form

$$(Z_L)_i{}^j = \delta_i{}^j + (\delta Z_L)_i{}^j\,, \qquad (Z_R)_i{}^j = \delta_i{}^j + (\delta Z_R)_i{}^j\,, \tag{2.291}$$

where the contributions to $(\delta Z_{L,R})_i{}^j$ begin at one-loop order, and similarly for the corresponding adjoints. As a result, at one-loop accuracy we can treat the diagonal elements and the off-diagonal elements separately. We begin by determining the complex pole masses $M_{p,k}$ by solving eq. (2.283) perturbatively for $s_{pole,k} = M^2_{p,k}$. To one-loop accuracy, we must solve

$$\det(s\mathbb{1} - \boldsymbol{M^2} - \boldsymbol{M}\boldsymbol{\Sigma}_{\boldsymbol{D}}^{\mathsf{T}} - \boldsymbol{\Sigma}_{\boldsymbol{D}}\boldsymbol{M} - \boldsymbol{\Sigma}_{\boldsymbol{R}}^{\mathsf{T}}\boldsymbol{M^2} - \boldsymbol{M}\boldsymbol{\Sigma}_{\boldsymbol{L}}\boldsymbol{M}) = 0\,, \tag{2.292}$$

where $\boldsymbol{M}_i{}^j = M_i\delta_i{}^j$ (no sum over repeated indices of the same height), and M_i is the bare mass of the ith Dirac fermion. Note that the argument of the determinant is of the form $A_j\delta^i{}_j + B^i{}_j$, where B is of one-loop order. Thus, the first-order perturbative solution to $\det(A_j\delta^i{}_j - B^i{}_j) = 0$ is $A_j = B^j{}_j$. Hence,

$$M^2_{p,k} = M^2_k\big[1 + \boldsymbol{\Sigma_L}(M^2_k)_k{}^k + \boldsymbol{\Sigma_R}(M^2_k)_k{}^k + \boldsymbol{\Sigma_D}(M^2_k)_k{}^k/M_k + \overline{\boldsymbol{\Sigma}}_{\boldsymbol{D}}(M^2_k)_k{}^k/M_k\big]\,. \tag{2.293}$$

Taking the square root, the pole mass at one-loop accuracy is given by

$$M_{p,k} = M_k + \tfrac{1}{2}\big[M_k\boldsymbol{\Sigma_L}(M^2_k)_k{}^k + M_k\boldsymbol{\Sigma_R}(M^2_k)_k{}^k + \boldsymbol{\Sigma_D}(M^2_k)_k{}^k + \overline{\boldsymbol{\Sigma}}_{\boldsymbol{D}}(M^2_k)_k{}^k\big]. \tag{2.294}$$

The pole masses $M_{p,k}$ are complex:

$$M_{p,k} \equiv M_{r,k} - \tfrac{1}{2}i\Gamma_k\,, \tag{2.295}$$

where $M_{r,k} \equiv \operatorname{Re} M_{p,k}$ and Γ_k is the width of the fermion (which is nonzero if a decay channel is kinematically allowed). We define the mass counterterm δM_k by

$$\delta M_k = M_{r,k} - M_k\,, \tag{2.296}$$

which is a quantity of one-loop order. Hence,

$$\delta M_k = \tfrac{1}{2}\operatorname{Re}\big[M_k\boldsymbol{\Sigma_L}(M^2_k)_k{}^k + M_k\boldsymbol{\Sigma_R}(M^2_k)_k{}^k + \boldsymbol{\Sigma_D}(M^2_k)_k{}^k + \overline{\boldsymbol{\Sigma}}_{\boldsymbol{D}}(M^2_k)_k{}^k\big], \tag{2.297}$$

$$\Gamma_k = -\operatorname{Im}\big[M_k\boldsymbol{\Sigma_L}(M^2_k)_k{}^k + M_k\boldsymbol{\Sigma_R}(M^2_k)_k{}^k + \boldsymbol{\Sigma_D}(M^2_k)_k{}^k + \overline{\boldsymbol{\Sigma}}_{\boldsymbol{D}}(M^2_k)_k{}^k\big]. \tag{2.298}$$

We now examine the on-mass-shell conditions that will be used to derive the matrix elements of δZ_L and δZ_R. Strictly speaking, each on-shell external particle should correspond to an asymptotic state, which means that the corresponding external fermion should be stable. This implies that the absorptive part of the self-energy functions evaluated on-shell must vanish. However, following the discussion in Section 2.9.1 (see footnote 20), in cases of unstable external fermions, we shall interpret the self-energy functions appearing below as corresponding to their dispersive parts.

First, we shall determine the diagonal elements of δZ_L and δZ_R. These elements are fixed by the following conditions:

$$\lim_{p^2\to M^2_{p,k}} (p^2 - M^2_{p,k})\overline{\boldsymbol{S}}_{\boldsymbol{D}r}(p^2)_k{}^k = \lim_{p^2\to M^2_{p,k}} (p^2 - M^2_{p,k})\boldsymbol{S}_{\boldsymbol{D}r}(p^2)_k{}^k = M_{p,k}\,, \tag{2.299}$$

$$\lim_{p^2\to M^2_{p,k}} (p^2 - M^2_{p,k})\boldsymbol{S}_{\boldsymbol{R}r}(p^2)_k{}^k = \lim_{p^2\to M^2_{p,k}} (p^2 - M^2_{p,k})\boldsymbol{S}_{\boldsymbol{L}r}(p^2)_k{}^k = 1\,. \tag{2.300}$$

Using eqs. (2.289) and (2.290), the conditions above can be rewritten in terms of the unrenormalized two-point functions:

$$\lim_{p^2\to M^2_{p,k}} (p^2 - M^2_{p,k})\overline{\boldsymbol{S}}_{\boldsymbol{D}}(p^2)_k{}^k = (Z_L^{1/2})_k{}^k M_{p,k}(Z_R^{1/2\dagger})_k{}^k\,, \tag{2.301}$$

$$\lim_{p^2\to M^2_{p,k}} (p^2 - M^2_{p,k})\boldsymbol{S}_{\boldsymbol{D}}(p^2)_k{}^k = (Z_R^{1/2})_k{}^k M_{p,k}(Z_L^{1/2\dagger})_k{}^k\,, \tag{2.302}$$

$$\lim_{p^2\to M^2_{p,k}} (p^2 - M^2_{p,k})\boldsymbol{S}_{\boldsymbol{R}}(p^2)_k{}^k = (Z_L^{1/2})_k{}^k (Z_L^{1/2\dagger})_k{}^k\,, \tag{2.303}$$

$$\lim_{p^2\to M^2_{p,k}} (p^2 - M^2_{p,k})\boldsymbol{S}_{\boldsymbol{L}}(p^2)_k{}^k = (Z_R^{1/2})_k{}^k (Z_R^{1/2\dagger})_k{}^k\,. \tag{2.304}$$

We now employ the results of eqs. (2.279)–(2.282). The key observation is that in the one-loop approximation, $(\boldsymbol{S}^{-1})_k{}^k = (\boldsymbol{S}_k{}^k)^{-1}$ for $\boldsymbol{S} = \boldsymbol{S_L}$, $\boldsymbol{S_R}$, $\boldsymbol{S_D}$, and $\overline{\boldsymbol{S}}_{\boldsymbol{D}}$. Hence, at one-loop accuracy, it follows that

$$\begin{aligned}
[\boldsymbol{S_L}(p^2)_k{}^k]^{-1} &= p^2\big[1 - \boldsymbol{\Sigma_R}(p^2)^k_k\big] - M_k^2\big[1 + \overline{\boldsymbol{\Sigma}}_{\boldsymbol{D}}(p^2)_k{}^k/M_k\big] \\
&\quad \times \big[1 + \boldsymbol{\Sigma_L}(p^2)_k{}^k\big]\big[1 + \boldsymbol{\Sigma_D}(p^2)_k{}^k/M_k\big]\,, &(2.305)\\
[\boldsymbol{S_R}(p^2)_k{}^k]^{-1} &= p^2\big[1 - \boldsymbol{\Sigma_L}(p^2)_k{}^k\big] - M_k^2\big[1 + \overline{\boldsymbol{\Sigma}}_{\boldsymbol{D}}(p^2)_k{}^k/M_k\big] \\
&\quad \times \big[1 + \boldsymbol{\Sigma_L}(p^2)_k{}^k\big]\big[1 + \boldsymbol{\Sigma_D}(p^2)_k{}^k/M_k\big]\,, &(2.306)\\
M_k[\boldsymbol{S_D}(p^2)_k{}^k]^{-1} &= p^2\big[1 - \boldsymbol{\Sigma_L}(p^2)_k{}^k\big]\big[1 - \boldsymbol{\Sigma_D}(p^2)_k{}^k/M_k\big]\big[1 - \boldsymbol{\Sigma_R}(p^2)_k{}^k\big] \\
&\quad - M_k^2\big[1 + \overline{\boldsymbol{\Sigma}}_{\boldsymbol{D}}(p^2)_k{}^k/M_k\big]\,, &(2.307)\\
M_k[\overline{\boldsymbol{S}}_{\boldsymbol{D}}(p^2)_k{}^k]^{-1} &= p^2\big[1 - \boldsymbol{\Sigma_L}(p^2)_k{}^k\big]\big[1 - \overline{\boldsymbol{\Sigma}}_{\boldsymbol{D}}(p^2)_k{}^k/M_k\big]\big[1 - \boldsymbol{\Sigma_R}(p^2)_k{}^k\big] \\
&\quad - M_k^2\big[1 + \boldsymbol{\Sigma_D}(p^2)_k{}^k/M_k\big]\,. &(2.308)
\end{aligned}$$

After dropping terms that are two-loop order and higher, it follows that:

$$\begin{aligned}
[\boldsymbol{S_L}(p^2)_k{}^k]^{-1} &= \big[1 - \boldsymbol{\Sigma_R}(M_k^2)_k{}^k\big]\big[p^2 - M_k^2(1 + \boldsymbol{\Sigma_L}(M_k^2)_k{}^k + \boldsymbol{\Sigma_R}(M_k^2)_k{}^k) \\
&\qquad + M_k(\boldsymbol{\Sigma_D}(M_k^2)_k{}^k + \overline{\boldsymbol{\Sigma}}_{\boldsymbol{D}}(M_k^2)_k{}^k)\big] \\
&= \big[1 - \boldsymbol{\Sigma_R} M_k^2)_k{}^k\big]\big[p^2 - M_k^2(1 + \boldsymbol{\Sigma_L}(M_k^2)_k{}^k + \boldsymbol{\Sigma_R}(M_k^2)_k{}^k) \\
&\qquad - M_k(\boldsymbol{\Sigma_D}(M^2)_k{}^k + \overline{\boldsymbol{\Sigma}}_{\boldsymbol{D}}(M_k^2)_k{}^k) \\
&\qquad - M_k^2(p^2 - M_k^2)(\boldsymbol{\Sigma}'_{\boldsymbol{L}}(M_k^2)_k{}^k + \boldsymbol{\Sigma}'_{\boldsymbol{R}}(M_k^2)_k{}^k) \\
&\qquad - M_k(p^2 - M_k^2)(\boldsymbol{\Sigma}'_{\boldsymbol{D}}(M_k^2)_k{}^k + \overline{\boldsymbol{\Sigma}}'_{\boldsymbol{D}}(M_k^2)_k{}^k)\big] \\
&= \big[1 - \boldsymbol{\Sigma_R}(M_k^2)_k{}^k\big](p^2 - M^2_{p,k})\big[1 - \boldsymbol{\mathcal{D}}_k\big] \\
&= (p^2 - M^2_{p,k})\big[1 - \boldsymbol{\Sigma_R}(M_k^2)_k{}^k - \boldsymbol{\mathcal{D}}_k\big] + \mathcal{O}\big((p^2 - M^2_{p,k})^2\big)\,, &(2.309)
\end{aligned}$$

where

$$\boldsymbol{\mathcal{D}}_k \equiv M_k^2\big[\boldsymbol{\Sigma}'_{\boldsymbol{L}}(M_k^2)_k{}^k + \boldsymbol{\Sigma}'_{\boldsymbol{R}}(M_k^2)_k{}^k\big] + M_k\big[\boldsymbol{\Sigma}'_{\boldsymbol{D}}(M_k^2)_k{}^k + \overline{\boldsymbol{\Sigma}}'_{\boldsymbol{D}}(M_k^2)_k{}^k\big]\,, \tag{2.310}$$

and

$$\boldsymbol{\Sigma}'(M_k^2)_k{}^k \equiv \left.\frac{d\boldsymbol{\Sigma}(p^2)_k{}^k}{dp^2}\right|_{p^2=M_k^2}\,. \tag{2.311}$$

After dropping terms of $\mathcal{O}\big((p^2-M_{p,k}^2)^2\big)$, eqs. (2.305)–(2.308) yield (at one-loop accuracy)

$$\big[\boldsymbol{S_L}(p^2)_k{}^k\big]^{-1} = (p^2-M_{p,k}^2)\big[1-\boldsymbol{\Sigma_R}(M_k^2)_k{}^k-\boldsymbol{\mathcal{D}}_k\big], \tag{2.312}$$

$$\big[\boldsymbol{S_R}(p^2)_k{}^k\big]^{-1} = (p^2-M_{p,k}^2)\big[1-\boldsymbol{\Sigma_L}(M_k^2)_k{}^k-\boldsymbol{\mathcal{D}}_k\big], \tag{2.313}$$

$$M_k\big[\boldsymbol{S_D}(p^2)_k{}^k\big]^{-1} = (p^2-M_{p,k}^2)\big[1-\boldsymbol{\Sigma_L}(M_k^2)_k{}^k-\boldsymbol{\Sigma_R}(M_k^2)_k{}^k - \boldsymbol{\mathcal{D}}_k-\boldsymbol{\Sigma_D}{}_k{}^k(M_k^2)/M_k\big], \tag{2.314}$$

$$M_k\big[\overline{\boldsymbol{S}}_{\boldsymbol{D}}(p^2)_k{}^k\big]^{-1} = (p^2-M_{p,k}^2)\big[1-\boldsymbol{\Sigma_L}(M_k^2)_k{}^k-\boldsymbol{\Sigma_R}(M_k^2)_k{}^k - \boldsymbol{\mathcal{D}}_k-\overline{\boldsymbol{\Sigma}}_{\boldsymbol{D}}(M_k^2)_k{}^k/M_k\big]. \tag{2.315}$$

We can now employ eqs. (2.301)–(2.304) to determine the diagonal elements of the wave function renormalization constants. Employing the notation of eq. (2.291), it then follows that

$$\tfrac{1}{2}(\delta Z_L+\delta Z_L^\dagger)_k{}^k = \boldsymbol{\Sigma_L}(M_k^2)_k{}^k+\boldsymbol{\mathcal{D}}_k\,, \tag{2.316}$$

$$\tfrac{1}{2}(\delta Z_R+\delta Z_R^\dagger)_k{}^k = \boldsymbol{\Sigma_R}(M_k^2)_k{}^k+\boldsymbol{\mathcal{D}}_k\,, \tag{2.317}$$

$$\tfrac{1}{2}(\delta Z_L+\delta Z_R^\dagger)_k{}^k = \boldsymbol{\Sigma_L}(M_k^2)_k{}^k+\boldsymbol{\Sigma_R}(M_k^2)_k{}^k+\boldsymbol{\mathcal{D}}_k+\frac{\overline{\boldsymbol{\Sigma}}_{\boldsymbol{D}}(M_k^2)_k{}^k+M_k-M_{p,k}}{M_k}\,, \tag{2.318}$$

$$\tfrac{1}{2}(\delta Z_L^\dagger+\delta Z_R)_k{}^k = \boldsymbol{\Sigma_L}(M_k^2)_k{}^k+\boldsymbol{\Sigma_R}(M_k^2)_k{}^k+\boldsymbol{\mathcal{D}}_k+\frac{\boldsymbol{\Sigma_D}(M_k^2)_k{}^k+M_k-M_{p,k}}{M_k}\,. \tag{2.319}$$

By making use of eq. (2.294), the last two equations above can be rewritten as

$$\tfrac{1}{2}(\delta Z_L+\delta Z_R^\dagger)_k{}^k = \frac{1}{2}\left[\boldsymbol{\Sigma_L}(M_k^2)_k{}^k+\boldsymbol{\Sigma_R}(M_k^2)_k{}^k-\frac{\boldsymbol{\Sigma_D}(M_k^2)_k{}^k-\overline{\boldsymbol{\Sigma}}_{\boldsymbol{D}}(M_k^2)_k{}^k}{M_k}\right]+\boldsymbol{\mathcal{D}}_k, \tag{2.320}$$

$$\tfrac{1}{2}(\delta Z_L^\dagger+\delta Z_R)_k{}^k = \frac{1}{2}\left[\boldsymbol{\Sigma_L}(M_k^2)_k{}^k+\boldsymbol{\Sigma_R}(M_k^2)_k{}^k+\frac{\boldsymbol{\Sigma_D}(M_k^2)_k{}^k-\overline{\boldsymbol{\Sigma}}_{\boldsymbol{D}}(M_k^2)_k{}^k}{M_k}\right]+\boldsymbol{\mathcal{D}}_k. \tag{2.321}$$

There is some freedom in solving for the diagonal elements of δZ_L, δZ_R, $\delta Z_L^\dagger$, and $\delta Z_R^\dagger$. In particular, the most general solution is consistent with the following equation:

$$\tfrac{1}{2}(\delta Z_L+\delta Z_R)_k{}^k = \tfrac{1}{2}(\delta Z_L^\dagger+\delta Z_R^\dagger)_k{}^k + C \tag{2.322}$$

for any choice of the constant C. We can set this constant to zero by appropriately exploiting the symmetry of eqs. (2.301)–(2.304) under the rephasing

$$(Z_{L,R}^{1/2})_k{}^k \to e^{i\theta_k}(Z_{L,R}^{1/2})_k{}^k\,, \qquad (Z_{L,R}^{\dagger 1/2})_k{}^k \to e^{-i\theta_k}(Z_{L,R}^{\dagger 1/2})_k{}^k\,. \tag{2.323}$$

The end result is

$$(\delta Z_L)_k{}^k = \boldsymbol{\Sigma_L}(M_k^2)_k{}^k - \frac{\boldsymbol{\Sigma_D}(M_k^2)_k{}^k - \overline{\boldsymbol{\Sigma}}_{\boldsymbol{D}}(M_k^2)_k{}^k}{2M_k} + \boldsymbol{\mathcal{D}}_k\,, \tag{2.324}$$

$$(\delta Z_R)_k{}^k = \boldsymbol{\Sigma_R}(M_k^2)_k{}^k + \frac{\boldsymbol{\Sigma_D}(M_k^2)_k{}^k - \overline{\boldsymbol{\Sigma}}_{\boldsymbol{D}}(M_k^2)_k{}^k}{2M_k} + \boldsymbol{\mathcal{D}}_k\,, \tag{2.325}$$

$$(\delta Z_L^\dagger)_k{}^k = \boldsymbol{\Sigma_L}(M_k^2)_k{}^k + \frac{\boldsymbol{\Sigma_D}(M_k^2)_k{}^k - \overline{\boldsymbol{\Sigma}}_{\boldsymbol{D}}(M_k^2)_k{}^k}{2M_k} + \boldsymbol{\mathcal{D}}_k\,, \tag{2.326}$$

$$(\delta Z_R^\dagger)_k{}^k = \boldsymbol{\Sigma_R}(M_k^2)_k{}^k - \frac{\boldsymbol{\Sigma_D}(M_k^2)_k{}^k - \overline{\boldsymbol{\Sigma}}_{\boldsymbol{D}}(M_k^2)_k{}^k}{2M_k} + \boldsymbol{\mathcal{D}}_k\,. \tag{2.327}$$

Note that eqs. (2.326) and (2.327) are consistent with eqs. (2.324) and (2.325) in light of the hermiticity conditions given in eq. (2.276).

Finally, we shall determine the off-diagonal elements of δZ_L and δZ_R. Returning to eqs. (2.289) and (2.290), we can rewrite these equations in matrix form. One can then employ eq. (2.278) to evaluate the matrix of renormalized full loop-corrected propagators to obtain

$$\begin{aligned}
&\begin{pmatrix} \overline{\boldsymbol{S}}{}^{\mathsf{T}}_{\boldsymbol{D}r}(p^2) & p\cdot\sigma\boldsymbol{S}_{\boldsymbol{R}r}(p^2) \\ p\cdot\overline{\sigma}\boldsymbol{S}^{\mathsf{T}}_{\boldsymbol{L}r}(p^2) & \boldsymbol{S}_{\boldsymbol{D}r}(p^2) \end{pmatrix} \\
&= \begin{pmatrix} Z_L^{-1/2} & 0 \\ 0 & Z_R^{-1/2} \end{pmatrix} \begin{pmatrix} \overline{\boldsymbol{S}}{}^{\mathsf{T}}_{\boldsymbol{D}}(p^2) & p\cdot\sigma\boldsymbol{S}_{\boldsymbol{R}}(p^2) \\ p\cdot\overline{\sigma}\boldsymbol{S}^{\mathsf{T}}_{\boldsymbol{L}}(p^2) & \boldsymbol{S}_{\boldsymbol{D}}(p^2) \end{pmatrix} \begin{pmatrix} Z_R^{-1/2\dagger} & 0 \\ 0 & Z_L^{-1/2\dagger} \end{pmatrix} \\
&= \begin{pmatrix} Z_L^{-1/2} & 0 \\ 0 & Z_R^{-1/2} \end{pmatrix} \begin{pmatrix} -\boldsymbol{M} - \boldsymbol{\Sigma_D}(p^2) & p\cdot\sigma[\mathbb{1} - \boldsymbol{\Sigma}^{\mathsf{T}}_{\boldsymbol{R}}(p^2)] \\ p\cdot\overline{\sigma}[\mathbb{1} - \boldsymbol{\Sigma_L}(p^2)] & -\boldsymbol{M} - \overline{\boldsymbol{\Sigma}}{}^{\mathsf{T}}_{\boldsymbol{D}}(p^2) \end{pmatrix}^{-1} \begin{pmatrix} Z_R^{-1/2\dagger} & 0 \\ 0 & Z_L^{-1/2\dagger} \end{pmatrix} \\
&= \begin{pmatrix} -Z_R^{1/2\dagger}[\boldsymbol{M} + \boldsymbol{\Sigma_D}(p^2)]Z_L^{1/2} & Z_R^{1/2\dagger}[\mathbb{1} - \boldsymbol{\Sigma}^{\mathsf{T}}_{\boldsymbol{R}}(p^2)]Z_R^{1/2}p\cdot\sigma \\ Z_L^{1/2\dagger}[\mathbb{1} - \boldsymbol{\Sigma_L}(p^2)]Z_L^{1/2}p\cdot\overline{\sigma} & -Z_L^{1/2\dagger}[\boldsymbol{M} + \overline{\boldsymbol{\Sigma}}{}^{\mathsf{T}}_{\boldsymbol{D}}(p^2)]Z_R^{1/2} \end{pmatrix}^{-1}.
\end{aligned} \tag{2.328}$$

At one-loop accuracy, we end up with

$$\begin{pmatrix} -[\boldsymbol{M} + \tfrac{1}{2}(\delta Z_R^\dagger \boldsymbol{M} + \boldsymbol{M}\delta Z_L) + \boldsymbol{\Sigma_D}(p^2)] & p\cdot\sigma[\mathbb{1} + \tfrac{1}{2}(\delta Z_R + \delta Z_R^\dagger) - \boldsymbol{\Sigma}^{\mathsf{T}}_{\boldsymbol{R}}(p^2)] \\ p\cdot\overline{\sigma}[\mathbb{1} + \tfrac{1}{2}(\delta Z_L + \delta Z_L^\dagger) - \boldsymbol{\Sigma_L}(p^2)] & -[\boldsymbol{M} + \tfrac{1}{2}(\delta Z_L^\dagger \boldsymbol{M} + \boldsymbol{M}\delta Z_R) + \overline{\boldsymbol{\Sigma}}{}^{\mathsf{T}}_{\boldsymbol{D}}(p^2)] \end{pmatrix}^{-1}. \tag{2.329}$$

We now define the subtracted self-energy functions (indicated with hats over the corresponding $\boldsymbol{\Sigma}$ functions) such that[26]

$$\begin{pmatrix} \overline{\boldsymbol{S}}{}^{\mathsf{T}}_{\boldsymbol{D}r}(p^2) & p\cdot\sigma\boldsymbol{S}_{\boldsymbol{R}r}(p^2) \\ p\cdot\overline{\sigma}\boldsymbol{S}^{\mathsf{T}}_{\boldsymbol{L}r}(p^2) & \boldsymbol{S}_{\boldsymbol{D}r}(p^2) \end{pmatrix} = \begin{pmatrix} -\boldsymbol{M} - \widehat{\boldsymbol{\Sigma}}_{\boldsymbol{D}}(p^2) - \boldsymbol{\delta M} & p\cdot\sigma[\mathbb{1} - \widehat{\boldsymbol{\Sigma}}{}^{\mathsf{T}}_{\boldsymbol{R}}(p^2)] \\ p\cdot\overline{\sigma}[\mathbb{1} - \widehat{\boldsymbol{\Sigma}}_{\boldsymbol{L}}(p^2)] & -\boldsymbol{M} - \widehat{\overline{\boldsymbol{\Sigma}}}{}^{\mathsf{T}}_{\boldsymbol{D}}(p^2) - \boldsymbol{\delta M} \end{pmatrix}^{-1}, \tag{2.330}$$

[26] Ultraviolet divergences in the $\boldsymbol{\Sigma}(p^2)$ will be canceled by divergences in $\boldsymbol{\delta M}$, δZ, and $\delta Z^\dagger$, which will then yield finite subtracted self-energy functions.

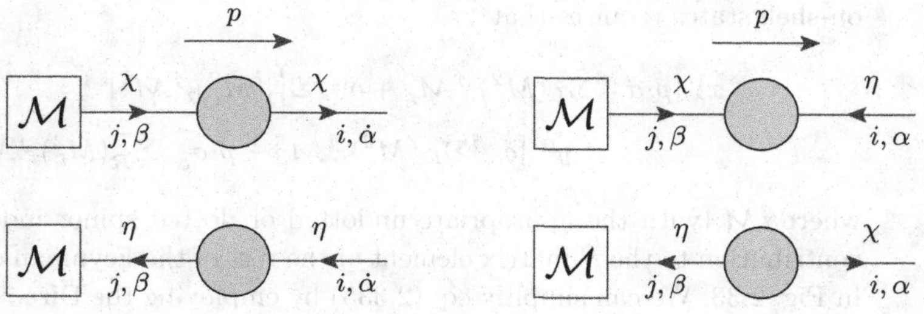

Fig. 2.33 The insertion of a subtracted self-energy function on an on-shell external line vanishes when summed over all possible configurations for the external Dirac fermion (represented by the two-component fields χ_i and η^i). Each box inscribed with an $\mathcal{M}$ represents the rest of the Feynman diagram (which may contain additional external fields) and each gray circle represents the subtracted self-energy function. In all four diagrams shown above, the four-momentum p flows from left to right.

where $\boldsymbol{\delta M}$ is a diagonal matrix with diagonal elements δM_k [see eq. (2.296)]. It then follows that, at one-loop accuracy,

$$\widehat{\boldsymbol{\Sigma}}_{\boldsymbol{L}}(p^2) = \boldsymbol{\Sigma}_{\boldsymbol{L}}(p^2) - \tfrac{1}{2}\left[\delta Z_L + \delta Z_L^\dagger\right], \tag{2.331}$$

$$\widehat{\boldsymbol{\Sigma}}^{\mathsf{T}}_{\boldsymbol{R}}(p^2) = \boldsymbol{\Sigma}^{\mathsf{T}}_{\boldsymbol{R}}(p^2) - \tfrac{1}{2}\left[\delta Z_R + \delta Z_R^\dagger\right], \tag{2.332}$$

$$\widehat{\boldsymbol{\Sigma}}_{\boldsymbol{D}}(p^2) = \boldsymbol{\Sigma}_{\boldsymbol{D}}(p^2) - \boldsymbol{\delta M} + \tfrac{1}{2}\left[\boldsymbol{M}\delta Z_L + \delta Z_R^\dagger \boldsymbol{M}\right], \tag{2.333}$$

$$\widehat{\overline{\boldsymbol{\Sigma}}}^{\mathsf{T}}_{\boldsymbol{D}}(p^2) = \overline{\boldsymbol{\Sigma}}^{\mathsf{T}}_{\boldsymbol{D}}(p^2) - \boldsymbol{\delta M} + \tfrac{1}{2}\left[\boldsymbol{M}\delta Z_R + \delta Z_L^\dagger \boldsymbol{M}\right]. \tag{2.334}$$

In order to compute $(\delta Z^L)_i{}^j$ and $(\delta Z^R)_i{}^j$ for $i \neq j$, we shall demand that, for any Feynman diagram, the insertion of a subtracted self-energy function on an on-shell external line vanishes when summed over all possible configurations for the external two-component fermion fields. This on-shell renormalization condition ensures that there is no flavor mixing in the corrections associated with external on-shell two-component fermions.

Consider a scattering or decay process with an on-shell outgoing Dirac fermion that is diagrammatically represented by either a χ field (denoted by a line whose arrow points out of the Feynman diagram) or by an $\eta^\dagger$ field (denoted by a line whose arrow points into the Feynman diagram). One can insert a subtracted self-energy function on the on-shell external line in four possible ways, as shown in Fig. 2.33.

Employing the Feynman rules exhibited in Fig. 2.1 for the external wave function spinors and the definition of the self-energy functions displayed in Fig. 2.32 (which also apply to the corresponding subtracted self-energy functions), it follows that the on-shell renormalization condition to remove flavor mixing associated with external on-shell states requires that[27]

$$x^{\dagger i}_{\dot{\alpha}}\big[p\cdot\bar{\sigma}^{\dot{\alpha}\beta}\widehat{\boldsymbol{\Sigma}}_{\boldsymbol{L}}(M_i^2)_i{}^j\mathcal{M}_\beta+\delta^{\dot{\alpha}}{}_{\dot{\beta}}\widehat{\boldsymbol{\Sigma}}^{\mathsf{T}}_{\boldsymbol{D}}(M_i^2)_i{}^j\mathcal{M}^{\dot{\beta}}\big]$$
$$+\,y^{\alpha i}\big[\delta_\alpha{}^\beta\widehat{\boldsymbol{\Sigma}}_{\boldsymbol{D}}(M_i^2)_i{}^j\mathcal{M}_\beta+p\cdot\sigma_{\alpha\dot{\beta}}\widehat{\boldsymbol{\Sigma}}^{\mathsf{T}}_{\boldsymbol{R}}(M_i^2)_i{}^j\mathcal{M}^{\dot{\beta}}\big]=0\,, \qquad (2.335)$$

where $i\mathcal{M}$ (with the appropriate undotted or dotted spinor index) represents the contribution to the S-matrix element of the rest of the Feynman diagram, as shown in Fig. 2.33. We can simplify eq. (2.335) by employing the Dirac equations given in eqs. (2.10)–(2.13), which yields

$$y^{\beta i}\mathcal{M}_\beta\big[M_i\widehat{\boldsymbol{\Sigma}}_{\boldsymbol{L}}(M_i^2)_i{}^j+\widehat{\boldsymbol{\Sigma}}_{\boldsymbol{D}}(M_i^2)_i{}^j\big]+x^{\dagger i}_{\dot{\beta}}\mathcal{M}^{\dot{\beta}}\big[\widehat{\boldsymbol{\Sigma}}^{\mathsf{T}}_{\boldsymbol{D}}(M_i^2)_i{}^j+M_i\widehat{\boldsymbol{\Sigma}}^{\mathsf{T}}_{\boldsymbol{R}}(M_i^2)_i{}^j\big]=0. \qquad (2.336)$$

In order for this equation to be satisfied, the coefficients of $y^{\beta i}\mathcal{M}_\beta$ and $x^{\dagger i}_{\dot{\beta}}\mathcal{M}^{\dot{\beta}}$ must separately vanish. Similarly, a computation that analyzes a scattering or decay process with an on-shell ingoing Dirac fermion yields

$$\mathcal{M}_{\dot{\alpha}}y^{\dagger\dot{\alpha}}_j\big[M_j\widehat{\boldsymbol{\Sigma}}_{\boldsymbol{L}}(M_j^2)_i{}^j+\widehat{\boldsymbol{\Sigma}}^{\mathsf{T}}_{\boldsymbol{D}}(M_j^2)_i{}^j\big]+\mathcal{M}^{\dot{\alpha}}x_{\alpha j}\big[\widehat{\boldsymbol{\Sigma}}_{\boldsymbol{D}}(M_j^2)_i{}^j+M_j\widehat{\boldsymbol{\Sigma}}^{\mathsf{T}}_{\boldsymbol{R}}(M_j^2)_i{}^j\big]=0, \qquad (2.337)$$

and it follows that the coefficients of $\mathcal{M}_{\dot{\alpha}}y^{\dot{\alpha}i}$ and $\mathcal{M}^\alpha x^{\dagger i}_{\dot{\alpha}}$ must separately vanish.

Hence, the following equations are obtained for $i\neq j$:

$$(\boldsymbol{M}\widehat{\boldsymbol{\Sigma}}_{\boldsymbol{L}}(M_i^2)+\widehat{\boldsymbol{\Sigma}}_{\boldsymbol{D}}(M_i^2))_i{}^j=0\,, \qquad (2.338)$$
$$(\boldsymbol{M}\widehat{\boldsymbol{\Sigma}}^{\mathsf{T}}_{\boldsymbol{R}}(M_i^2)+\widehat{\boldsymbol{\Sigma}}^{\mathsf{T}}_{\boldsymbol{D}}(M_i^2))_i{}^j=0\,, \qquad (2.339)$$
$$(\widehat{\boldsymbol{\Sigma}}_{\boldsymbol{L}}(M_j^2)\boldsymbol{M}+\widehat{\boldsymbol{\Sigma}}^{\mathsf{T}}_{\boldsymbol{D}}(M_j^2))_i{}^j=0\,, \qquad (2.340)$$
$$(\widehat{\boldsymbol{\Sigma}}^{\mathsf{T}}_{\boldsymbol{R}}(M_j^2)\boldsymbol{M}+\widehat{\boldsymbol{\Sigma}}_{\boldsymbol{D}}(M_j^2))_i{}^j=0\,. \qquad (2.341)$$

Using the results of eqs. (2.331)–(2.334) and assuming that $i\neq j$, we obtain, at one-loop accuracy,

$$\tfrac{1}{2}(\boldsymbol{M}\delta Z_L^\dagger-\delta Z_R^\dagger\boldsymbol{M})_i{}^j=(\boldsymbol{M}\boldsymbol{\Sigma}_{\boldsymbol{L}}(M_i^2)+\boldsymbol{\Sigma}_{\boldsymbol{D}}(M_i^2))_i{}^j, \qquad (2.342)$$
$$\tfrac{1}{2}(\boldsymbol{M}\delta Z_R^\dagger-\delta Z_L^\dagger\boldsymbol{M})_i{}^j=(\boldsymbol{M}\boldsymbol{\Sigma}^{\mathsf{T}}_{\boldsymbol{R}}(M_i^2)+\overline{\boldsymbol{\Sigma}}^{\mathsf{T}}_{\boldsymbol{D}}(M_i^2))_i{}^j, \qquad (2.343)$$
$$\tfrac{1}{2}(\delta Z_L\boldsymbol{M}-\boldsymbol{M}\delta Z_R)_i{}^j=(\boldsymbol{\Sigma}_{\boldsymbol{L}}(M_j^2)\boldsymbol{M}+\overline{\boldsymbol{\Sigma}}^{\mathsf{T}}_{\boldsymbol{D}}(M_j^2))_i{}^j, \qquad (2.344)$$
$$\tfrac{1}{2}(\delta Z_R\boldsymbol{M}-\boldsymbol{M}\delta Z_L)_i{}^j=(\boldsymbol{\Sigma}^{\mathsf{T}}_{\boldsymbol{R}}(M_j^2)\boldsymbol{M}+\boldsymbol{\Sigma}_{\boldsymbol{D}}(M_j^2))_i{}^j. \qquad (2.345)$$

[27] We could have considered a scattering or decay process with an on-shell outgoing Dirac antifermion that is diagrammatically represented by either an η field (denoted by a line whose arrow points out of the Feynman diagram) or by a $\chi^\dagger$ field (denoted by a line whose arrow points into the Feynman diagram). In this case, we would simply interchange the χ and η fields in Fig. 2.33. After employing $\mathcal{M}$ corresponding to the external antifermion state, the resulting amplitude would be of the same form as the one exhibited in eq. (2.335).

The solutions to these equations are unique:

$$(\delta Z_L)_i{}^j = \frac{2\big[\boldsymbol{\Sigma_L}(M_j^2)\boldsymbol{M^2} + \boldsymbol{M}\boldsymbol{\Sigma}^{\mathsf{T}}_{\boldsymbol{R}}(M_j^2)\boldsymbol{M} + \boldsymbol{M}\boldsymbol{\Sigma_D}(M_j^2) + \overline{\boldsymbol{\Sigma}}^{\mathsf{T}}_{\boldsymbol{D}}(M_j^2)\boldsymbol{M}\big]_i{}^j}{M_j^2 - M_i^2}, \tag{2.346}$$

$$(\delta Z_R)_i{}^j = \frac{2\big[\boldsymbol{M}\boldsymbol{\Sigma_L}(M_j^2)\boldsymbol{M} + \boldsymbol{\Sigma}^{\mathsf{T}}_{\boldsymbol{R}}(M_j^2)\boldsymbol{M^2} + \boldsymbol{\Sigma_D}(M_j^2)\boldsymbol{M} + \boldsymbol{M}\overline{\boldsymbol{\Sigma}}^{\mathsf{T}}_{\boldsymbol{D}}(M_j^2)\big]_i{}^j}{M_j^2 - M_i^2}, \tag{2.347}$$

$$(\delta Z_L^\dagger)_i{}^j = \frac{2\big[\boldsymbol{M^2}\boldsymbol{\Sigma_L}(M_i^2) + \boldsymbol{M}\boldsymbol{\Sigma}^{\mathsf{T}}_{\boldsymbol{R}}(M_i^2)\boldsymbol{M} + \boldsymbol{M}\boldsymbol{\Sigma_D}(M_i^2) + \overline{\boldsymbol{\Sigma}}^{\mathsf{T}}_{\boldsymbol{D}}(M_i^2)\boldsymbol{M}\big]_i{}^j}{M_i^2 - M_j^2}, \tag{2.348}$$

$$(\delta Z_R^\dagger)_i{}^j = \frac{2\big[\boldsymbol{M}\boldsymbol{\Sigma_L}(M_i^2)\boldsymbol{M} + \boldsymbol{M^2}\boldsymbol{\Sigma}^{\mathsf{T}}_{\boldsymbol{R}}(M_i^2) + \boldsymbol{\Sigma_D}(M_i^2)\boldsymbol{M} + \boldsymbol{M}\overline{\boldsymbol{\Sigma}}^{\mathsf{T}}_{\boldsymbol{D}}(M_i^2)\big]_i{}^j}{M_i^2 - M_j^2}, \tag{2.349}$$

assuming that $i \neq j$. Note that eqs. (2.348) and (2.349) are consistent with the results of eqs. (2.346) and (2.347) in light of the hermiticity conditions given in eq. (2.276).

We end this section by obtaining formulae for the wave function renormalization constants for the case of N flavors of two-component Majorana fermion fields, ξ_i. In this case, there is only one matrix of wave function renormalization constants, $(Z_M)_i{}^j$ and its hermitian conjugate, which relate the bare and renormalized fields,

$$\xi_{r,i} = (Z_M^{-1/2})_i{}^j \xi_j\,, \qquad \xi_r^{\dagger i} = \xi^{\dagger j}(Z_M^{-1/2\dagger})_j{}^i\,. \tag{2.350}$$

To obtain explicit expressions for δZ_M and $\delta Z_M^\dagger$, we can use the results obtained previously in the case of n flavors of Dirac fermions by identifying the two-component fermion fields, $\xi = \chi = \eta$, and the corresponding wave function renormalization constants, $(Z_M)_i{}^j = (Z_L)_i{}^j = \big[(Z_R)_i{}^j\big]^*$. In addition, we may employ eqs. (2.284) and (2.285) to identify the full loop-corrected propagator and self-energy functions of the two-component Majorana fermion fields. The only subtlety involves the index structure of the corresponding matrices, which do not inherit the corresponding index structures employed in the case of Dirac fermions (due to the relation $\xi_i = \chi_i = \eta^i$ used above which mixes lower and upper indices). As in Section 2.10.1, we can introduce the diagonal mass matrices $\boldsymbol{m}^{ij}$ and $\overline{\boldsymbol{m}}_{ij}$, which are numerically equal with diagonal elements m_j, in order to obtain equations that treat the flavor-index heights consistently.

For example, the pole mass at one-loop accuracy is obtained by solving

$$\det(s\mathbb{1} - \boldsymbol{m}\overline{\boldsymbol{m}} - \boldsymbol{m}\overline{\boldsymbol{\Omega}} - \boldsymbol{\Omega}\overline{\boldsymbol{m}} - \boldsymbol{\Xi}^{\mathsf{T}}\boldsymbol{m}\overline{\boldsymbol{m}} - \boldsymbol{m}\overline{\boldsymbol{m}}\boldsymbol{\Xi}) = 0\,, \tag{2.351}$$

which is consistent with the flavor-index conventions employed in Section 2.10.1. However, in formulae that do not involve sums over flavor indices, it is notationally simpler to allow for different flavor-index heights. In particular, the results for the mass counterterm and width obtained by solving eq. (2.351) are

$$\delta m_j = \mathrm{Re}\big[m_j\, \boldsymbol{\Xi}(m_j^2)_j{}^j + \tfrac{1}{2}\boldsymbol{\Omega}(m_j^2)^{jj} + \tfrac{1}{2}\overline{\boldsymbol{\Omega}}(m_j^2)_{jj}\big]\,, \tag{2.352}$$

$$\Gamma_j = -2\,\mathrm{Im}\big[m_j\, \boldsymbol{\Xi}(m_j^2)_j{}^j + \tfrac{1}{2}\boldsymbol{\Omega}(m_j^2)^{jj} + \tfrac{1}{2}\overline{\boldsymbol{\Omega}}(m_j^2)_{jj}\big]\,. \tag{2.353}$$

Likewise, the diagonal elements of δZ_M and $\delta Z_M^\dagger$ are given by

$$(\delta Z_M)_j{}^j = \boldsymbol{\Xi}(m_j^2)_j{}^j - \frac{\boldsymbol{\Omega}(m_j^2)^{jj} - \overline{\boldsymbol{\Omega}}(m_j^2)_{jj}}{2m_j} + \boldsymbol{Q}_j\,, \tag{2.354}$$

$$(\delta Z_M^\dagger)_j{}^j = \boldsymbol{\Xi}(m_j^2)_j{}^j + \frac{\boldsymbol{\Omega}(m_j^2)^{jj} - \overline{\boldsymbol{\Omega}}(m_j^2)_{jj}}{2m_j} + \boldsymbol{Q}_j\,, \tag{2.355}$$

where

$$\boldsymbol{Q}_j \equiv 2m_j^2\, \boldsymbol{\Xi}'(m_j^2)_j{}^j + m_j\big[\boldsymbol{\Omega}'(m_j^2)^{jj} + \overline{\boldsymbol{\Omega}}'(m_j^2)_{jj}\big]\,, \tag{2.356}$$

and the prime indicates differentiation with respect to p^2 which is then evaluated at $p^2 = m_j^2$.

The off-diagonal elements of δZ_M and its hermitian conjugate can easily be written in a form that respects the flavor-index height conventions:

$$(\delta Z_M)_i{}^j = \frac{2\big[\boldsymbol{\Xi}(m_j^2)\overline{\boldsymbol{m}}\boldsymbol{m} + \overline{\boldsymbol{m}}\,\boldsymbol{\Xi}^{\mathsf{T}}(m_j^2)\boldsymbol{m} + \overline{\boldsymbol{m}}\,\boldsymbol{\Omega}(m_j^2) + \overline{\boldsymbol{\Omega}}(m_j^2)\boldsymbol{m}\big]_i{}^j}{m_j^2 - m_i^2}\,, \tag{2.357}$$

$$(\delta Z_M^\dagger)_i{}^j = \frac{2\big[\overline{\boldsymbol{m}}\boldsymbol{m}\boldsymbol{\Xi}(m_i^2) + \overline{\boldsymbol{m}}\,\boldsymbol{\Xi}^{\mathsf{T}}(m_i^2)\boldsymbol{m} + \overline{\boldsymbol{m}}\boldsymbol{\Omega}(m_i^2) + \overline{\boldsymbol{\Omega}}(m_i^2)\boldsymbol{m}\big]_i{}^j}{m_i^2 - m_j^2}\,, \tag{2.358}$$

assuming that $i \neq j$. Recalling that $\boldsymbol{\Omega}$ and $\overline{\boldsymbol{\Omega}}$ are complex symmetric matrices, and the hermiticity conditions given in eqs. (2.241) and (2.252), it follows that eqs. (2.355) and (2.358) are consistent with eqs. (2.354) and (2.357), respectively.

2.11 Feynman Rules for External Fermion Lines Revisited

Consider the S-matrix elements for a scattering or decay process at higher orders in perturbation theory beyond the tree level, which are computed by evaluating the contributing connected Feynman diagrams by employing Feynman rules based on the tree-level Lagrangian prior to renormalization. The LSZ reduction formulae instruct us to replace any flavor-diagonal or flavor-off-diagonal self-energy insertion on the external lines with the corresponding diagonal or off-diagonal element of the matrix wave function renormalization constant. Hence, the Feynman rules for assigning two-component external-state spinors of momentum $\vec{\boldsymbol{p}}$, spin label s, and flavor label j specified in Section 2.4 should be modified as follows:

- For an initial χ state (incoming) left-handed $(\frac{1}{2},0)$ fermion: $(Z_L^{1/2})_i{}^j x_j(\vec{\boldsymbol{p}},s)$
- For an initial η state (incoming) left-handed $(\frac{1}{2},0)$ fermion: $(Z_R^{1/2\,\dagger})_j{}^i x^j(\vec{\boldsymbol{p}},s)$
- For an initial $\chi^\dagger$ state (incoming) right-handed $(0,\frac{1}{2})$ fermion: $(Z_L^{1/2\,\dagger})_j{}^i y^{\dagger j}(\vec{\boldsymbol{p}},s)$
- For an initial $\eta^\dagger$ state (incoming) right-handed $(0,\frac{1}{2})$ fermion: $(Z_R^{1/2})_i{}^j y_j^\dagger(\vec{\boldsymbol{p}},s)$
- For a final χ state (outgoing) left-handed $(\frac{1}{2},0)$ fermion: $(Z_L^{1/2\,\dagger})_j{}^i x^{\dagger j}(\vec{\boldsymbol{p}},s)$
- For a final η state (outgoing) left-handed $(\frac{1}{2},0)$ fermion: $(Z_R^{1/2})_i{}^j x_j^\dagger(\vec{\boldsymbol{p}},s)$
- For a final $\chi^\dagger$ state (outgoing) right-handed $(0,\frac{1}{2})$ fermion: $(Z_L^{1/2})_i{}^j y_j(\vec{\boldsymbol{p}},s)$
- For a final $\eta^\dagger$ state (outgoing) right-handed $(0,\frac{1}{2})$ fermion: $(Z_R^{1/2\,\dagger})_j{}^i y^j(\vec{\boldsymbol{p}},s)$

(2.359)

where Z_L and Z_R are the corresponding matrix wave function renormalization constants that were derived by imposing on-shell renormalization conditions in Section 2.10.2. No sum over j is implied since j refers to the fixed flavor index of the external fermion state. The index i is the flavor index of the two-component fermion line that attaches to the rest of the Feynman diagram, and there is an implicit sum over i corresponding to a sum over Feynman diagrams. Note that the heights of the flavor indices attached to the external spinor wave functions are governed by the convention established by the Fourier decomposition of the two-component fermion fields given in eqs. (2.75) and (2.76), where the flavor index i appears in the lowered position in $\chi_{\alpha i}(x)$ and $\eta_i^{\dagger\dot{\alpha}}$ and in the raised position in η_α^i and $\chi^{\dagger\dot{\alpha}i}(x)$, respectively.

In the case of two-component external-state Majorana spinors, there is only one independent complex matrix wave function renormalization constant, Z_M. The rules for assigning two-component external-state Majorana spinors of momentum $\vec{\boldsymbol{p}}$, spin label s, and flavor label j are as follows:

- For an initial ξ state (incoming) left-handed $(\frac{1}{2},0)$ fermion: $(Z_M^{1/2})_i{}^j x_j(\vec{\boldsymbol{p}},s)$
- For an initial $\xi^\dagger$ state (incoming) right-handed $(0,\frac{1}{2})$ fermion: $(Z_M^{1/2\,\dagger})_j{}^i y^{\dagger j}(\vec{\boldsymbol{p}},s)$
- For a final ξ state (outgoing) left-handed $(\frac{1}{2},0)$ fermion: $(Z_M^{1/2\,\dagger})_j{}^i x^{\dagger j}(\vec{\boldsymbol{p}},s)$
- For a final $\xi^\dagger$ state (outgoing) right-handed $(0,\frac{1}{2})$ fermion: $(Z_M^{1/2})_i{}^j y_j(\vec{\boldsymbol{p}},s)$

(2.360)

The heights of the flavor indices attached to the external spinor wave functions are governed by the convention established by the Fourier decomposition of the two-component fermion field given in eq. (2.67), where the flavor index i appears in the lowered position in $\xi_{\alpha i}(x)$ and in the raised position in $\xi^{\dagger\dot{\alpha}i}(x)$, respectively.

Similarly, the corresponding rules for an external neutral (real) boson of momentum $\vec{\boldsymbol{k}}$ are as follows:

- For an external-state (incoming or outgoing) spin-0 boson : $Z_\phi^{1/2}$
- For an initial-state (incoming) spin-1 boson : $Z_V^{1/2}\varepsilon^\mu(\vec{\boldsymbol{k}}\,,\,\lambda)$
- For a final-state (outgoing) spin-1 boson : $Z_V^{1/2}\varepsilon^\mu(\vec{\boldsymbol{k}}\,,\,\lambda)^\star$ (2.361)

where Z_ϕ and Z_V are the corresponding wave function renormalization constants for a spin-0 and spin-1 field, respectively, obtained by applying on-shell renormalization conditions, as discussed in Section 2.9.

In the case of multiple flavors of neutral or charged scalars, one can employ the matrix of wave function renormalization constants obtained in Section 2.9.1, analogously to eq. (2.360) [without the spinor wave functions, of course]. Details are left for the reader.

For tree-level calculations, the matrix wave function renormalization constants $Z_i{}^j$ can be replaced by the Kronecker delta, δ_i^j. In the one-loop approximation,

$$(Z^{1/2})_i{}^j = \delta_i^j + \tfrac{1}{2}(\delta Z)_i{}^j\,. \tag{2.362}$$

We provide two examples by revisiting the computation of the decay of a massive neutral scalar boson into a pair of Majorana or Dirac fermions. The tree-level decay amplitude for the decay of a neutral scalar into a pair of Majorana fermions was previously given in eq. (2.150). Using the rules exhibited in eq. (2.360), the decay matrix element in the one-loop approximation is given by

$$\begin{aligned}
i\mathcal{M} = &-i\lambda^{k\ell}y_i(\vec{\boldsymbol{p}}_1, s_1)\big[(1+\tfrac{1}{2}\delta Z_\phi)\delta_k^i\delta_\ell^j + \tfrac{1}{2}(\delta Z_M)_k{}^i\delta_\ell^j + \tfrac{1}{2}(\delta Z_M^{1/2})_\ell{}^j\delta_k^i\big]y_j(\vec{\boldsymbol{p}}_2, s_2)\\
&-i\lambda_{k\ell}x^{\dagger i}(\vec{\boldsymbol{p}}_1, s_1)\big[(1+\tfrac{1}{2}\delta Z_\phi)\delta_i^k\delta_j^\ell + \tfrac{1}{2}(\delta Z_M^\dagger)_i{}^k\delta_j^\ell + \tfrac{1}{2}(\delta Z_M^\dagger)_j{}^\ell\delta_i^k\big]x^{\dagger j}(\vec{\boldsymbol{p}}_2, s_2)\\
&+i\mathcal{M}_{loop}
\end{aligned} \tag{2.363}$$

for fixed flavors i and j, and an implied sum over the repeated indices k and ℓ, where $i\mathcal{M}_{loop}$ is the sum of the two diagrams contributing to the one-loop vertex correction shown in Fig. 2.34.

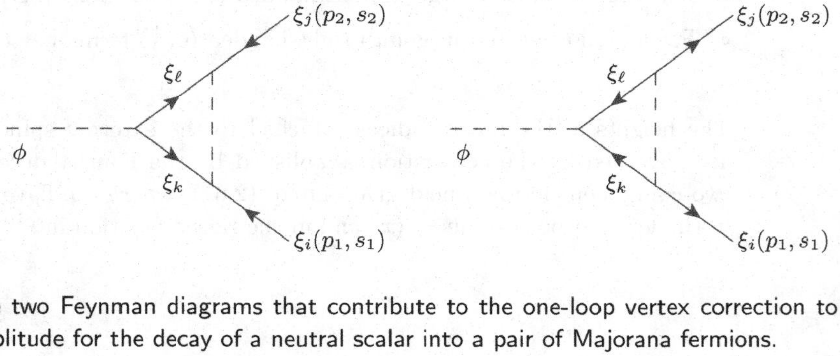

Fig. 2.34 The two Feynman diagrams that contribute to the one-loop vertex correction to the amplitude for the decay of a neutral scalar into a pair of Majorana fermions.

Likewise, the tree-level decay amplitude for the decay of a neutral scalar into a pair of Dirac fermions was previously given in eq. (2.153). Using the rules exhibited in eq. (2.359), the decay matrix element in the one-loop approximation is given by

$$\begin{aligned} i\mathcal{M} = &- i\kappa^{\ell}{}_{k}\, y^{i}(\vec{\boldsymbol{p}}_1, s_1)\big[(1+\tfrac{1}{2}\delta Z_\phi)\delta_i^k\delta_\ell^j + \tfrac{1}{2}(\delta Z_R^\dagger)_i{}^k\delta_\ell^j + \tfrac{1}{2}(\delta Z_L)_{\ell}{}^{j}\delta_i^k\big] y_j(\vec{\boldsymbol{p}}_2, s_2) \\ &- i\kappa_k{}^{\ell}\, x^{\dagger i}(\vec{\boldsymbol{p}}_1, s_1)\big[(1+\tfrac{1}{2}\delta Z_\phi)\delta_i^k\delta_\ell^j + \tfrac{1}{2}(\delta Z_L^\dagger)_i{}^k\delta_\ell^j + \tfrac{1}{2}(\delta Z_R)_{\ell}{}^{j}\delta_i^k\big] x_j^\dagger(\vec{\boldsymbol{p}}_2, s_2) \\ &+ i\mathcal{M}_{loop} \end{aligned} \qquad (2.364)$$

for fixed flavors i and j, and an implied sum over the repeated indices k and ℓ, where $i\mathcal{M}_{loop}$ is the sum of the two diagrams (not shown here) that contribute to the one-loop vertex correction. In Exercise 2.17, the reader will provide the relevant diagrams analogous to those in Fig. 2.34 and obtain the one-loop corrected formula for the decay rate of a neutral scalar boson into a Dirac fermion–antifermion pair.

Any ultraviolet divergences appearing in the matrix elements above can be absorbed into the definition of the renormalized couplings and masses.

Exercises

2.1 The Gordon identities can be used to simplify bilinear products of the form $z_1(\vec{\boldsymbol{p}}\,')\sigma^{\mu\nu}P_\nu z_2(\vec{\boldsymbol{p}})$ and $z_1(\vec{\boldsymbol{p}}\,')\sigma^{\mu\nu}q_\nu z_2(\vec{\boldsymbol{p}})$, where $P \equiv p + p'$ and $q \equiv p' - p$, m is the common fermion mass, and z_1 and z_2 are commuting two-component spinor wave functions (either x or y) that satisfy the on-shell conditions given in eqs. (2.10)–(2.13). Similar identities can be used to simplify expressions of the form $z_1^\dagger(\vec{\boldsymbol{p}}\,')\overline{\sigma}^{\mu\nu}P_\nu z_2^\dagger(\vec{\boldsymbol{p}})$ and $z_1^\dagger(\vec{\boldsymbol{p}}\,')\overline{\sigma}^{\mu\nu}q_\nu z_2^\dagger(\vec{\boldsymbol{p}})$. All together, there are 16 different Gordon identities.

(a) Derive the first set of four Gordon identities,

$$2iy(\vec{\boldsymbol{p}}\,')\sigma^{\mu\nu}P_\nu x(\vec{\boldsymbol{p}}) + q^\mu y(\vec{\boldsymbol{p}}\,')x(\vec{\boldsymbol{p}}) = m\big[x^\dagger(\vec{\boldsymbol{p}}\,')\overline{\sigma}^\mu x(\vec{\boldsymbol{p}}) - y(\vec{\boldsymbol{p}}\,')\sigma^\mu y^\dagger(\vec{\boldsymbol{p}})\big]\,, \qquad (2.365)$$

$$2ix^\dagger(\vec{\boldsymbol{p}}\,')\overline{\sigma}^{\mu\nu}P_\nu y^\dagger(\vec{\boldsymbol{p}}) + q^\mu x^\dagger(\vec{\boldsymbol{p}}\,')y^\dagger(\vec{\boldsymbol{p}}) = -m\big[x^\dagger(\vec{\boldsymbol{p}}\,')\overline{\sigma}^\mu x(\vec{\boldsymbol{p}}) - y(\vec{\boldsymbol{p}}\,')\sigma^\mu y^\dagger(\vec{\boldsymbol{p}})\big]\,, \qquad (2.366)$$

$$2iy(\vec{\boldsymbol{p}}\,')\sigma^{\mu\nu}q_\nu x(\vec{\boldsymbol{p}}) + P^\mu y(\vec{\boldsymbol{p}}\,')x(\vec{\boldsymbol{p}}) = m\big[x^\dagger(\vec{\boldsymbol{p}}\,')\overline{\sigma}^\mu x(\vec{\boldsymbol{p}}) + y(\vec{\boldsymbol{p}}\,')\sigma^\mu y^\dagger(\vec{\boldsymbol{p}})\big]\,, \qquad (2.367)$$

$$2ix^\dagger(\vec{\boldsymbol{p}}\,')\overline{\sigma}^{\mu\nu}q_\nu y^\dagger(\vec{\boldsymbol{p}}) + P^\mu x^\dagger(\vec{\boldsymbol{p}}\,')y^\dagger(\vec{\boldsymbol{p}}) = m\big[x^\dagger(\vec{\boldsymbol{p}}\,')\overline{\sigma}^\mu x(\vec{\boldsymbol{p}}) + y(\vec{\boldsymbol{p}}\,')\sigma^\mu y^\dagger(\vec{\boldsymbol{p}})\big]\,. \qquad (2.368)$$

HINT: Starting with the first term in each of the four Gordon identities above and using the definitions of $\sigma^{\mu\nu}$ and $\overline{\sigma}^{\mu\nu}$, simplify the corresponding expressions by employing the σ-matrix identities of eqs. (1.88) and (1.89) and making use of the on-shell conditions given by eqs. (2.10)–(2.13).

(b) Show that one can obtain a second set of Gordon identities by interchanging $x \leftrightarrow y$, $x^\dagger \leftrightarrow y^\dagger$, and $m \to -m$ in eqs. (2.365)–(2.368). Check these results by making use of eqs. (B.58)–(B.60), (B.64), and (B.65) in part (a).

(c) Derive the third set of four Gordon identities,

$$2ix(\vec{p}\,')\sigma^{\mu\nu}P_\nu x(\vec{p}) + q^\mu x(\vec{p}\,')x(\vec{p}) = -m\big[y^\dagger(\vec{p}\,')\overline{\sigma}^\mu x(\vec{p}) + x(\vec{p}\,')\sigma^\mu y^\dagger(\vec{p})\big]\,, \tag{2.369}$$

$$2ix^\dagger(\vec{p}\,')\overline{\sigma}^{\mu\nu}P_\nu x^\dagger(\vec{p}) + q^\mu x^\dagger(\vec{p}\,')x^\dagger(\vec{p}) = m\big[x^\dagger(\vec{p}\,')\overline{\sigma}^\mu y(\vec{p}) + y(\vec{p}\,')\sigma^\mu x^\dagger(\vec{p})\big]\,, \tag{2.370}$$

$$2ix(\vec{p}\,')\sigma^{\mu\nu}q_\nu x(\vec{p}) + P^\mu x(\vec{p}\,')x(\vec{p}) = -m\big[y^\dagger(\vec{p}\,')\overline{\sigma}^\mu x(\vec{p}) - x(\vec{p}\,')\sigma^\mu y^\dagger(\vec{p})\big]\,, \tag{2.371}$$

$$2ix^\dagger(\vec{p}\,')\overline{\sigma}^{\mu\nu}q_\nu x^\dagger(\vec{p}) + P^\mu x^\dagger(\vec{p}\,')x^\dagger(\vec{p}) = -m\big[x^\dagger(\vec{p}\,')\overline{\sigma}^\mu y(\vec{p}) - y(\vec{p}\,')\sigma^\mu x^\dagger(\vec{p})\big]\,. \tag{2.372}$$

(d) Show that one can obtain a fourth set of Gordon identities by interchanging $x \leftrightarrow y$, $x^\dagger \leftrightarrow y^\dagger$, and $m \to -m$ in eqs. (2.369)–(2.372).

2.2 The following results are quite useful in analyzing the properties of the two-component spinor fields under space inversion.

(a) Show that the x-spinor and y-spinor wave functions satisfy the relations

$$x_\alpha(\vec{p}, s) = \sigma^0_{\alpha\dot{\beta}}\, y^{\dagger\dot{\beta}}(-\vec{p}, s)\,, \tag{2.373}$$

$$y_\alpha(\vec{p}, s) = -\sigma^0_{\alpha\dot{\beta}}\, x^{\dagger\dot{\beta}}(-\vec{p}, s)\,, \tag{2.374}$$

$$x^{\dagger\dot{\alpha}}(\vec{p}, s) = -\overline{\sigma}^{0\dot{\alpha}\beta} y_\beta(-\vec{p}, s)\,, \tag{2.375}$$

$$y^{\dagger\dot{\alpha}}(\vec{p}, s) = \overline{\sigma}^{0\dot{\alpha}\beta} x_\beta(-\vec{p}, s)\,, \tag{2.376}$$

where $s = \pm\frac{1}{2}$ is the spin quantum number obtained by quantizing the spin along a fixed axis in the rest frame. Derive four additional formulae by raising the undotted spinor index in eqs. (2.373) and (2.374) and lowering the dotted spinor index in eqs. (2.375) and (2.376).

(b) Derive the corresponding formulae for the helicity spinor wave functions,

$$x_\alpha(\vec{p}, \lambda) = -i\sigma^0_{\alpha\dot{\beta}}\, y^{\dagger\dot{\beta}}(-\vec{p}, -\lambda)\,, \tag{2.377}$$

$$y_\alpha(\vec{p}, \lambda) = -i\sigma^0_{\alpha\dot{\beta}}\, x^{\dagger\dot{\beta}}(-\vec{p}, -\lambda)\,, \tag{2.378}$$

$$x^{\dagger\dot{\alpha}}(\vec{p}, \lambda) = -i\overline{\sigma}^{0\dot{\alpha}\beta}\, y_\beta(-\vec{p}, -\lambda)\,, \tag{2.379}$$

$$y^{\dagger\dot{\alpha}}(\vec{p}, \lambda) = -i\overline{\sigma}^{0\dot{\alpha}\beta}\, x_\beta(-\vec{p}, -\lambda)\,, \tag{2.380}$$

using the conventions of Appendix E.3 to define $x(-\vec{p}, \lambda)$ and $y(-\vec{p}, \lambda)$. Derive four additional formulae by raising the undotted spinor index in eqs. (2.377) and (2.378) and lowering the dotted spinor index in eqs. (2.379) and (2.380).

2.3 The following results are quite useful in analyzing the properties of the two-component spinor fields under time reversal.

(a) Show that the x-spinor and y-spinor wave functions satisfy the relations,

$$x^{\beta}(\vec{p}, s) = -2s x^{\dagger}_{\dot{\alpha}}(-\vec{p}, -s)\, \overline{\sigma}^{0\dot{\alpha}\beta}\,, \tag{2.381}$$

$$y^{\beta}(\vec{p}, s) = -2s y^{\dagger}_{\dot{\alpha}}(-\vec{p}, -s)\, \overline{\sigma}^{0\dot{\alpha}\beta}\,, \tag{2.382}$$

$$x^{\dagger}_{\dot{\beta}}(\vec{p}, s) = 2s x^{\alpha}(-\vec{p}, -s)\, \sigma^{0}_{\alpha\dot{\beta}}\,, \tag{2.383}$$

$$y^{\dagger}_{\dot{\beta}}(\vec{p}, s) = 2s y^{\alpha}(-\vec{p}, -s)\, \sigma^{0}_{\alpha\dot{\beta}}\,. \tag{2.384}$$

Derive four additional formulae by lowering the undotted spinor index in eqs. (2.381) and (2.382) and raising the dotted spinor index in eqs. (2.383) and (2.384).

(b) Derive the corresponding formulae for the helicity spinor wave functions,

$$x^{\beta}(\vec{p}, \lambda) = -2i\lambda\, x^{\dagger}_{\dot{\beta}}(-\vec{p}, \lambda)\, \overline{\sigma}^{0\dot{\alpha}\beta}\,, \tag{2.385}$$

$$y^{\beta}(\vec{p}, \lambda) = 2i\lambda\, y^{\dagger}_{\dot{\beta}}(-\vec{p}, \lambda)\, \overline{\sigma}^{0\dot{\alpha}\beta}\,, \tag{2.386}$$

$$x^{\dagger}_{\dot{\beta}}(\vec{p}, \lambda) = -2i\lambda\, x^{\alpha}(-\vec{p}, \lambda)\, \sigma^{0}_{\alpha\dot{\beta}}\,, \tag{2.387}$$

$$y^{\dagger}_{\dot{\beta}}(\vec{p}, \lambda) = 2i\lambda\, y^{\alpha}(-\vec{p}, \lambda)\, \sigma^{0}_{\alpha\dot{\beta}}\,, \tag{2.388}$$

employing the conventions of Appendix E.3 to define $x(-\vec{p}, \lambda)$ and $y(-\vec{p}, \lambda)$. Derive four additional formulae by lowering the undotted spinor index in eqs. (2.385) and (2.386) and raising the dotted spinor index in eqs. (2.387) and (2.388).

2.4 Consider a free Dirac fermion field which is represented by the two-component spinors χ and η. The expansions of these fields in terms of creation and annihilation operators are given by eqs. (2.75) and (2.76). The action of the unitary operators $\mathcal{P}$, $\mathcal{T}$, and $\mathcal{C}$, which represent space inversion, time reversal, and charge conjugation, acting on the two-component fermion field operators were derived in Chapter 1.

(a) Consider the action of $\mathcal{P}$ on the annihilation operators. Examine first the case of a Dirac fermion field, assuming that the plane-wave expansion of the Dirac field [eqs. (2.75) and (2.76)] employs fixed-axis spinor wave functions [see eq. (2.14)]. Show that

$$\mathcal{P} a(\vec{p}, s)\mathcal{P}^{-1} = i\eta_P\, a(-\vec{p}, s)\,, \tag{2.389}$$

$$\mathcal{P} b(\vec{p}, s)\mathcal{P}^{-1} = i\eta_P^{*}\, b(-\vec{p}, s)\,, \tag{2.390}$$

where the choice of phase η_P is defined in eqs. (1.211) and (1.212). Note that the factor of i has been extracted from η_P as a matter of convenience. In the case of a neutral Majorana fermion field [eq. (2.3)], show that eq. (2.389) still applies, but with η_P restricted to be ± 1.

(b) Repeat part (a) under the assumption that the plane-wave expansion of the Dirac field employs helicity spinor wave functions [eq. (2.17)]. Show that

$$\mathcal{P}a(\vec{\boldsymbol{p}},\lambda)\mathcal{P}^{-1} = -\eta_P a(-\vec{\boldsymbol{p}},-\lambda)\,, \tag{2.391}$$

$$\mathcal{P}b(\vec{\boldsymbol{p}},\lambda)\mathcal{P}^{-1} = -\eta_P^* b(-\vec{\boldsymbol{p}},-\lambda)\,, \tag{2.392}$$

employing the conventions of Appendix E.3 to define $x(-\vec{\boldsymbol{p}},\lambda)$ and $y(-\vec{\boldsymbol{p}},\lambda)$.

(c) Consider the action of $\mathcal{T}$ on the annihilation operators. In the case of a Dirac fermion field, show that

$$\mathcal{T}a(\vec{\boldsymbol{p}},s)\mathcal{T}^{-1} = 2s\eta_T a(-\vec{\boldsymbol{p}},-s)\,, \tag{2.393}$$

$$\mathcal{T}b(\vec{\boldsymbol{p}},s)\mathcal{T}^{-1} = 2s\eta_T^* b(-\vec{\boldsymbol{p}},-s)\,, \tag{2.394}$$

where the choice of phase η_P is defined in eqs. (1.247) and (1.248). In the case of a neutral Majorana fermion field, show that eq. (2.393) still applies, but with η_T restricted to be ± 1.

(d) Repeat part (c) under the assumption that the plane-wave expansion of the Dirac field employs helicity spinor wave functions. Show that

$$\mathcal{T}a(\vec{\boldsymbol{p}},\lambda)\mathcal{T}^{-1} = 2i\lambda\eta_T a(-\vec{\boldsymbol{p}},\lambda)\,, \tag{2.395}$$

$$\mathcal{T}b(\vec{\boldsymbol{p}},\lambda)\mathcal{T}^{-1} = 2i\lambda\eta_T^* b(-\vec{\boldsymbol{p}},\lambda)\,. \tag{2.396}$$

(e) Consider the action of $\mathcal{C}$ on the annihilation operators. In the case of a Dirac fermion field, show that

$$\mathcal{C}a(\vec{\boldsymbol{p}},s)\mathcal{C}^{-1} = \eta_C b(\vec{\boldsymbol{p}},s)\,, \tag{2.397}$$

$$\mathcal{C}b(\vec{\boldsymbol{p}},s)\mathcal{C}^{-1} = \eta_C^* a(\vec{\boldsymbol{p}},s)\,, \tag{2.398}$$

where the choice of phase η_C is defined in eqs. (1.256) and (1.257). Verify that eqs. (2.397) and (2.398) are valid if the spin quantum number s refers to either fixed-axis spin states or to helicity states. In the case of a neutral Majorana fermion field, show that

$$\mathcal{C}a(\vec{\boldsymbol{p}},s)\mathcal{C}^{-1} = \eta_C a(\vec{\boldsymbol{p}},s)\,, \tag{2.399}$$

where η_C is restricted to be ± 1.

(f) Determine the action of $\mathcal{CP}$ and $\mathcal{CPT}$ on the annihilation operators of a Dirac and a Majorana fermion field, respectively. Treat separately the cases of fixed-axis spin states and helicity states.

2.5 Consider a multiplet of n neutral two-component fermion fields, $\hat{\xi}_k$, governed by the free-field Lagrangian given in eq. (1.159). As noted in Exercise 1.18, if the neutral fermion mass matrix M is a real symmetric matrix, then eq. (1.159) describes a CP-invariant theory. Denoting the sign of the kth eigenvalue of M by ε_k, you showed in part (c) of Exercise 1.18 that

$$\mathcal{CP}\,\xi_{k\alpha}(x)(\mathcal{CP})^{-1} = i\varepsilon_k\hat{\eta}_{CP}\sigma^0_{\alpha\dot{\beta}}\xi_k^{\dagger\dot{\beta}}(x_P)\,, \tag{2.400}$$

where $\hat{\eta}_{CP}$ is a fixed sign that is independent of k and the $\xi_k(x)$ are the mass-eigenstate fermion fields obtained after performing a Takagi diagonalization of M. That is, the effective CP phase factor of the field $\xi_k(x)$ is $(\eta_{CP})_k \equiv \varepsilon_k\hat{\eta}_{CP}$.

(a) Consider a mass-eigenstate neutral fermion with nonzero mass in its rest frame, with spin quantum number s (obtained by quantizing the spin along a fixed axis):

$$|s_k\rangle \equiv a^\dagger(\vec{\boldsymbol{p}}_k = 0, s_k)\,|0\rangle\,, \tag{2.401}$$

where the flavor index k indicates the kth neutral fermion of the multiplet. Assuming that the vacuum state is CP-invariant, i.e., $\mathcal{CP}\,|0\rangle = |0\rangle$, use the results of Exercise 2.4 to show that $|s_k\rangle$ is an eigenstate of $\mathcal{CP}$:

$$\mathcal{CP}\,|s_k\rangle = -i\varepsilon_k\hat{\eta}_{CP}\,|s_k\rangle\,. \tag{2.402}$$

This justifies the assertion of Exercise 1.18 that neutral fermions corresponding to positive and negative eigenvalues of the real symmetric neutral fermion mass matrix have CP quantum numbers of opposite sign.

(b) Show that a Dirac fermion is equivalent to two mass-degenerate Majorana fermions with CP quantum numbers of opposite sign.

(c) If the neutral fermion is massless, then one must employ one-particle helicity states, $|\vec{\boldsymbol{p}}, \lambda\rangle$. In this case, show that

$$\mathcal{CP}\,|\vec{\boldsymbol{p}}, \lambda\rangle = -i\eta_{CP}\,|-\vec{\boldsymbol{p}}, -\lambda\rangle\,. \tag{2.403}$$

Hence, massless neutral fermions can never be eigenstates of CP. Nevertheless, in light of eq. (2.403), it is conventional to denote the CP quantum number of such a state by $-i\eta_{CP}$.

2.6 In eq. (2.17), we introduced the eigenstates of the helicity operator. If $\hat{\boldsymbol{p}}$ is a unit vector with polar angle θ and azimuthal angle ϕ with respect to a fixed z-axis, then one can define the following two-component spinors:

$$\zeta_{1/2}(\hat{\boldsymbol{p}}) = \begin{pmatrix} \cos\frac{\theta}{2} \\ e^{i\phi}\sin\frac{\theta}{2} \end{pmatrix}, \qquad \zeta_{-1/2}(\hat{\boldsymbol{p}}) = \begin{pmatrix} -e^{-i\phi}\sin\frac{\theta}{2} \\ \cos\frac{\theta}{2} \end{pmatrix}. \tag{2.404}$$

Derive the following explicit formula for $\zeta_\lambda(\hat{\boldsymbol{p}})$:

$$\zeta_\lambda(\hat{\boldsymbol{p}}) = \exp\left(-i\theta\hat{\boldsymbol{n}}\cdot\vec{\boldsymbol{\sigma}}/2\right)\,\zeta_\lambda(\hat{\boldsymbol{z}})\,, \tag{2.405}$$

where $\hat{\boldsymbol{n}} = (-\sin\phi\,, \cos\phi\,, 0)$.

HINT: Obtain $\zeta_\lambda(\hat{\boldsymbol{p}})$ from $\zeta_\lambda(\hat{\boldsymbol{z}})$ by employing the spin-1/2 rotation operator corresponding to a rotation from the $\hat{\boldsymbol{z}}$-direction to the direction of $\hat{\boldsymbol{p}}$.

2.7 Suppose that $x_\alpha(\vec{\boldsymbol{p}}, s)$ and $y_\alpha(\vec{\boldsymbol{p}}, s)$ satisfy eqs. (2.10)–(2.13) and (2.29)–(2.32).

(a) Using the identity $(p\cdot\overline{\sigma})(p\cdot\sigma) = (p\cdot\sigma)(p\cdot\overline{\sigma}) = p^2$, show that both x_α and y_α must satisfy the mass-shell condition, $p^2 = m^2$ (or equivalently, $p^0 = E_{\boldsymbol{p}}$).

(b) Show that eqs. (2.29)–(2.32) imply that

$$(S\cdot\sigma)_{\alpha\dot{\alpha}}(S\cdot\overline{\sigma})^{\dot{\alpha}\beta} = -\delta_\alpha{}^\beta\,, \qquad (S\cdot\overline{\sigma})^{\dot{\alpha}\alpha}(S\cdot\sigma)_{\alpha\dot{\beta}} = -\delta^{\dot{\alpha}}{}_{\dot{\beta}}\,, \tag{2.406}$$

and conclude that $S\cdot S = -1$.

(c) Using the identities given by eqs. (2.36) and (2.37), show that $p\cdot S = 0$ when acting on the two-component spinor wave functions, x_α and y_α.

2.8 Starting from eqs. (2.102) and (2.103), derive the following two expressions:

$$\langle 0|\, T\xi^{\dagger\dot{\alpha}}(x)\xi^\dagger_{\dot{\beta}}(y)\,|0\rangle = \int \frac{d^4p}{(2\pi)^4}\, e^{-ip\cdot(x_1-x_2)}\frac{im\delta^{\dot{\alpha}}{}_{\dot{\beta}}}{p^2-m^2+i\varepsilon}\,, \tag{2.407}$$

$$\langle 0|\, T\xi_\alpha(x)\xi^\beta(y)\,|0\rangle = \int \frac{d^4p}{(2\pi)^4}\, e^{-ip\cdot(x_1-x_2)}\frac{im\delta_\alpha{}^\beta}{p^2-m^2+i\varepsilon}\,. \tag{2.408}$$

2.9 Derive the massive fermion propagators of Fig. 2.2 by summing up an infinite chain of mass insertions between massless fermion propagators using the rules for mass insertions exhibited in Fig. 2.4.

2.10 (a) Compute the tree-level amplitude for the decay of a charged scalar boson to a fermion pair consisting of one Majorana fermion and one Dirac fermion, due to the interactions given in eq. (2.135).

(b) Compute the tree-level amplitude for the decay of a charged vector boson to a fermion pair consisting of one Majorana fermion and one Dirac fermion, due to the interactions given in eq. (2.146).

2.11 (a) Compute the tree-level amplitude for the scattering of a charged scalar boson and a Majorana fermion via an s-channel exchange of a Dirac fermion, due to the interactions given in eq. (2.135).

(b) Compute the tree-level amplitude for the scattering of a charged vector boson and a Majorana fermion via an s-channel exchange of a Dirac fermion, due to the interactions given in eq. (2.146).

2.12 Derive the identity

$$\nabla_i \nabla_j \left(\frac{1}{r}\right) = \frac{3x_i x_j - r^2\delta_{ij}}{r^5} - \frac{4\pi}{3}\delta_{ij}\,\delta^{(3)}(\vec{\boldsymbol{x}}), \tag{2.409}$$

where $r \equiv |\vec{\boldsymbol{x}}|$. This identity was used in obtaining the result given in eq. (2.176). Some caveats associated with this identity can be found in Ref. [35].

2.13 In QED, the bare mass of the photon vanishes. The Ward identities of QED (which are a consequence of gauge invariance of the theory) imply that $\Pi_{\mu\nu}(p)$ is purely transverse, which in the notation of eq. (2.226) yields $\Pi_L(s) = 0$.

(a) Show that the longitudinal part of the full loop-corrected photon propagator is given by its tree-level expression (and thus exhibits no corrections at higher orders in perturbation theory).

(b) Compute $\Pi_T(s)$ at one-loop order in QED due to an electron–positron loop. Show that $\Pi_T(0) = 0$ and conclude that the photon remains massless in the one-loop approximation. Compute the wave function renormalization constant of the photon, and show that $Z_V = [1 - \Pi'_L(0)]^{-1}$. Evaluate $\Pi'_L(0)$ at one-loop order using dimensional regularization of the ultraviolet singularity. Treat the electron–positron loop using two-component fermion Feynman rules.

(c) To conclude that the photon is massless to all orders in perturbation theory, one must argue that $\Pi_{\mu\nu}(p)$ has the form

$$\Pi_{\mu\nu}(p) = (p^2 g_{\mu\nu} - p_\mu p_\nu)F(p^2), \tag{2.410}$$

which implies that $\Pi_T(s) = sF(s)$. In QED, $F(s)$ is nonsingular as $s \to 0$, which then yields $\Pi_T(0) = 0$ to all orders in perturbation theory. Under what circumstances would you expect $F(s)$ to be singular as $s \to 0$ (which would allow for a radiatively generated photon mass in a manner consistent with the gauge invariance of the theory)?

2.14 Consider the special case of a parity-conserving vectorlike theory of Dirac fermions (such as QED or QCD). In this case, show that the following relations hold among the loop-corrected propagator functions and self-energy functions, respectively:

$$\boldsymbol{S_R} = \boldsymbol{S_L^{\mathsf{T}}}, \qquad \boldsymbol{S_D} = \boldsymbol{\overline{S}_D^{\mathsf{T}}}, \qquad \boldsymbol{\Sigma_L} = \boldsymbol{\Sigma_R^{\mathsf{T}}}, \qquad \boldsymbol{\Sigma_D} = \boldsymbol{\overline{\Sigma}_D^{\mathsf{T}}}. \tag{2.411}$$

2.15 In the on-mass-shell limit, the renormalized scalar propagator is given by eq. (2.189). Note that in the same limit the renormalized inverse propagator is not singular and thus can be written as

$$\boldsymbol{A}_r^{-1}(s)_{ij} = \boldsymbol{A}_{r,0}^{-1}(s_{\text{pole},n})_{ij} + (s - s_{\text{pole},n})\boldsymbol{A}_{r,1}^{-1}(s_{\text{pole},n})_{ij} + \mathcal{O}\big((s - s_{\text{pole},n})^2\big). \tag{2.412}$$

(a) Show that

$$A^{-1}_{r,0}(s_{\text{pole},n})_{in} = 0\,, \quad \text{for } i = 1,2,\ldots,N, \tag{2.413}$$

$$A^{-1}_{r,0}(s_{\text{pole},n})_{nj} = 0\,, \quad \text{for } j = 1,2,\ldots,N, \tag{2.414}$$

$$A^{-1}_{r,1}(s_{\text{pole},n})_{nn} = 1\,. \tag{2.415}$$

(b) By imposing $(\boldsymbol{A}\boldsymbol{A}^{-1})_{ij} = (\boldsymbol{A}^{-1}\boldsymbol{A})_{ij} = \delta_{ij}$, show that

$$A_{r,0}(s_{\text{pole},n})_{ik}A^{-1}_{r,0}(s_{\text{pole},n})_{kj} = A^{-1}_{r,0}(s_{\text{pole},n})_{ik}A^{-1}_{r,0}(s_{\text{pole},n})_{kj} = \delta_{ij}\,, \tag{2.416}$$

$$A^{-1}_{r,1}(s_{\text{pole},n})_{nj} + A_{r,0}(s_{\text{pole},n})_{nk}A^{-1}_{r,0}(s_{\text{pole},n})_{kj} = 0\,, \tag{2.417}$$

$$A^{-1}_{r,1}(s_{\text{pole},n})_{in} + A^{-1}_{r,0}(s_{\text{pole},n})_{ik}A^{-1}_{r,0}(s_{\text{pole},n})_{kn} = 0\,. \tag{2.418}$$

(c) Using the results of parts (a) and (b), show that

$$[\boldsymbol{A}^{-1}(s_{\text{pole},n})]_{in} = [\boldsymbol{A}^{-1}(s_{\text{pole},n})]_{nj} = 0\,, \quad \text{for all } i,\, j = 1,2,\ldots,N, \tag{2.419}$$

and

$$\left.\frac{d\boldsymbol{A}_r(s)_{nn}}{dp^2}\right|_{s=s_{\text{pole},n}} = 1\,. \tag{2.420}$$

Show that the above results are equivalent to the conditions exhibited in eq. (2.211).

2.16 Explicit expressions for the wave function renormalization constants $(\delta Z_L)_i{}^j$ and $(\delta Z_R)_i{}^j$ were obtained in Section 2.10.2. Suppose that two of the Dirac fermions are mass-degenerate, say, $M_i = M_j$.

(a) Show that the off-diagonal elements of the hermitian parts of $(\delta Z_L)_i{}^j$ and $(\delta Z_R)_i{}^j$, namely $\frac{1}{2}(\delta Z_L + \delta Z_L^\dagger)_i{}^j$ and $\frac{1}{2}(\delta Z_R + \delta Z_R^\dagger)_i{}^j$, are finite in the limit of $M_i \to M_j$.

(b) Show that by setting $i = j$ in the finite expressions obtained in part (a), one recovers the results of eqs. (2.316) and (2.317).

(c) Show that the off-diagonal elements of the antihermitian parts of $(\delta Z_L)_i{}^j$ and $(\delta Z_R)_i{}^j$, namely $\frac{1}{2}(\delta Z_L - \delta Z_L^\dagger)_i{}^j$ and $\frac{1}{2}(\delta Z_R - \delta Z_R^\dagger)_i{}^j$, are singular in the limit of $M_i \to M_j$. However, these off-diagonal elements do not occur in the renormalization of physical observables. Try to justify this last remark.

(d) Verify that there are $4n^2$ renormalization conditions that define the on-shell renormalization scheme employed in Section 2.10.2. Check that this number is equal to the number of counterterms [δM_j, $(\delta Z_{L,R})_i{}^j$, and $(\delta Z^\dagger_{L,R})_i{}^j$], after taking into account the rephasing freedom noted in eq. (2.323). How are these results modified in the case of N flavors of Majorana fermions?

2.17 Consider the interaction Lagrangian of eq. (2.152) that describes the interaction of a neutral scalar with a two-component Dirac fermion (i.e., no flavor indices needed).

(a) Compute the wave function renormalization constant of the neutral scalar, Z_ϕ, in the one-loop approximation due to a fermion loop. Likewise, evaluate the wave function renormalization constants of the Dirac fermion, δZ_L and δZ_R, in the one-loop approximation due to virtual scalar exchange.

(b) Draw the two Feynman diagrams (analogous to those of Fig. 2.34) that contribute to the one-loop vertex corrections of the decay of a neutral scalar into a Dirac fermion–antifermion pair. Evaluate these diagrams explicitly, using dimensional regularization in $d = 4 - 2\epsilon$ spacetime dimensions.

(c) Using eq. (2.364) and the results of parts (a) and (b), compute the sum of the tree-level and one-loop contributions to the amplitude for the decay of a neutral scalar into a fermion–antifermion pair. Obtain a formula for the decay rate that is accurate to $\mathcal{O}(\kappa_r^3)$, where the renormalized coupling κ_r has the following form:

$$\kappa_r = \kappa\big[1 + C\kappa(4\pi)^\epsilon\Gamma(\epsilon)\big]. \tag{2.421}$$

By choosing C appropriately, show that the decay rate obtained above is finite in the limit of $\epsilon \to 0$. This choice corresponds to a modified minimal subtraction ($\overline{\text{MS}}$) scheme for defining the renormalized coupling κ_r.

2.18 In Section 2.11, we introduced a modified treatment of the Feynman rules for external two-component fermion lines (referred to as Procedure B below). Show that this procedure is equivalent to Procedure A, in which only self-energy insertions on the external on-shell fermion lines that are diagonal in flavor are omitted.

(a) Procedure A: self-energy insertions on the external on-shell fermion lines that are diagonal in flavor are omitted. This means that if there is a self-energy insertion on an external on-shell fermion line that mixes flavor i and flavor $j \neq i$, we must include this diagram in our computation. The latter is evaluated using the standard Feynman rules. In particular, if i is the flavor index of the external fermion of mass m_i, then the propagator of the internal fermion of mass m_j will include a factor proportional to $(m_i^2 - m_j^2)^{-1}$.

(b) Procedure B: both diagonal and off-diagonal self-energy insertions on the external on-shell fermion lines are omitted, as described in Section 2.11. For a fixed external fermion of flavor i one must then sum over all values of j (including $j = i$). Show that the contribution of the sum over j of all terms with $j \neq i$ precisely replaces the explicit evaluation of the sum of diagrams with an off-diagonal self-energy insertion using the standard Feynman rules as indicated in part (a) above.

3 From Two-Component to Four-Component Spinors

To motivate the introduction of four component spinors for spin-1/2 particles, it is instructive to compare real and complex scalar fields that describe spin-0 particles. Consider a free scalar field theory with two real, mass-degenerate scalar fields $\phi_1(x)$ and $\phi_2(x)$. The corresponding Lagrangian is given by

$$\mathscr{L} = \tfrac{1}{2}(\partial_\mu \phi_1)(\partial^\mu \phi_1) + \tfrac{1}{2}(\partial_\mu \phi_2)(\partial^\mu \phi_2) - \tfrac{1}{2}m^2(\phi_1^2 + \phi_2^2)\,, \tag{3.1}$$

where m^2 is assumed to be positive. In this case, we can rewrite the theory in terms of a single complex scalar field,

$$\Phi = \frac{1}{\sqrt{2}}(\phi_1 + i\phi_2)\,, \qquad \Phi^\dagger = \frac{1}{\sqrt{2}}(\phi_1 - i\phi_2)\,. \tag{3.2}$$

In terms of the complex field, eq. (3.1) takes the following form:

$$\mathscr{L} = (\partial_\mu \Phi^\dagger)(\partial^\mu \Phi) - m^2 \Phi^\dagger \Phi\,. \tag{3.3}$$

Note that eq. (3.1) exhibits a global O(2) symmetry, $\phi_i \to C_{ij}\phi_j$. Corresponding to this symmetry is a conserved Noether current,

$$J^\mu = \phi_1 \partial^\mu \phi_2 - \phi_2 \partial^\mu \phi_1\,, \tag{3.4}$$

with a corresponding conserved Noether charge, $Q = \int J^0 d^3x$. The complex field $\Phi(x)$ creates and annihilates particles of definite charge Q. Comparing these results to eqs. (1.151)–(1.153) with $m_1 = m_2$, we see that in a theory of mass-degenerate spin-1/2 fermions, the ξ_i are the fermionic analogues of the real scalar fields ϕ_i. The χ and η are fermion fields of definite (and opposite) charge. Consequently, we can combine χ and $\eta^\dagger$ into one complex four-component spinor field Ψ, which is the fermionic analogue of the complex scalar field Φ.

3.1 Four-Component Spinors and Dirac Gamma Matrices

We now spell out in more detail the four-component Dirac fermion notation. A four-component Dirac spinor field, $\Psi(x)$, is made up of two mass-degenerate two-component spinor fields, $\chi_\alpha(x)$ and $\eta_\alpha(x)$, of opposite U(1) charge as follows:

$$\Psi(x) \equiv \begin{pmatrix} \chi_\alpha(x) \\ \eta^{\dagger\dot{\alpha}}(x) \end{pmatrix}\,. \tag{3.5}$$

The field equation satisfied by $\Psi(x)$ is obtained from eq. (1.154). This is the free-field Dirac equation:

$$(i\gamma^\mu \partial_\mu - m)\Psi = 0\,, \tag{3.6}$$

where the 4×4 Dirac gamma matrices, γ^μ, can be expressed in 2×2 block form as follows:

$$\gamma^\mu = \begin{pmatrix} 0 & \sigma^\mu_{\alpha\dot{\beta}} \\ \overline{\sigma}^{\mu\dot{\alpha}\beta} & 0 \end{pmatrix}. \tag{3.7}$$

Using eqs. (1.88) and (1.89), one immediately obtains the anticommutation relation that serves as the defining property of the Dirac matrices:

$$\{\gamma^\mu, \gamma^\nu\} = 2g^{\mu\nu}\mathbb{1}\,, \tag{3.8}$$

where $\mathbb{1}$ is the 4×4 identity matrix,

$$\mathbb{1} = \begin{pmatrix} \delta_\alpha{}^\beta & 0 \\ 0 & \delta^{\dot{\alpha}}{}_{\dot{\beta}} \end{pmatrix}. \tag{3.9}$$

There are many other representations for the Dirac gamma matrices that satisfy eq. (3.8). The explicit form given in eq. (3.7) is called the *chiral* representation, which corresponds to a basis choice in which γ_5 is diagonal:

$$\gamma_5 \equiv i\gamma^0\gamma^1\gamma^2\gamma^3 = -\frac{i}{4!}\,\epsilon^{\mu\nu\rho\tau}\gamma_\mu\gamma_\nu\gamma_\rho\gamma_\tau = \begin{pmatrix} -\delta_\alpha{}^\beta & 0 \\ 0 & \delta^{\dot{\alpha}}{}_{\dot{\beta}} \end{pmatrix}, \tag{3.10}$$

in a convention where $\epsilon^{0123} = +1$ [see eq. (A.20)]. Note that, independently of the representation of the Dirac gamma matrices,

$$(\gamma_5)^2 = \mathbb{1}\,, \qquad \{\gamma^\mu\,,\,\gamma_5\} = 0\,, \tag{3.11}$$

as a consequence of eq. (3.8).

It is convenient to define chiral projections operators:

$$P_L \equiv \tfrac{1}{2}(\mathbb{1} - \gamma_5) = \begin{pmatrix} \delta_\alpha{}^\beta & 0 \\ 0 & 0 \end{pmatrix}, \qquad P_R \equiv \tfrac{1}{2}(\mathbb{1} + \gamma_5) = \begin{pmatrix} 0 & 0 \\ 0 & \delta^{\dot{\alpha}}{}_{\dot{\beta}} \end{pmatrix}. \tag{3.12}$$

The (left- and right-handed) *Weyl spinor* fields, $\Psi_L(x)$ and $\Psi_R(x)$, are the four-component spinor eigenstates of γ_5 with corresponding eigenvalues -1 and $+1$, respectively (i.e., $\gamma_5\Psi_{L,R} = \mp\Psi_{L,R}$). More explicitly,

$$\Psi_L(x) \equiv P_L\Psi(x) = \begin{pmatrix} \chi_\alpha(x) \\ 0 \end{pmatrix}, \qquad \Psi_R(x) \equiv P_R\Psi(x) = \begin{pmatrix} 0 \\ \eta^{\dagger\dot{\alpha}}(x) \end{pmatrix}. \tag{3.13}$$

The generators of the Lorentz group in the $(\frac{1}{2}, 0) \oplus (0, \frac{1}{2})$ representation are[1]

$$\tfrac{1}{2}\Sigma^{\mu\nu} \equiv \frac{i}{4}[\gamma^\mu, \gamma^\nu] = \begin{pmatrix} \sigma^{\mu\nu}{}_\alpha{}^\beta & 0 \\ 0 & \overline{\sigma}^{\mu\nu\dot{\alpha}}{}_{\dot{\beta}} \end{pmatrix}. \tag{3.14}$$

[1] In most textbooks, $\Sigma^{\mu\nu}$ is called $\sigma^{\mu\nu}$. In this book, we use the symbol $\Sigma^{\mu\nu}$ so that there is no confusion with $\sigma^{\mu\nu}$ defined in eq. (1.115).

The duality conditions satisfied by $\sigma^{\mu\nu}$ and $\overline{\sigma}^{\mu\nu}$ [eq. (1.122)] imply that

$$\gamma_5\Sigma^{\mu\nu} = \tfrac{1}{2}i\epsilon^{\mu\nu\rho\kappa}\Sigma_{\rho\kappa}\,. \tag{3.15}$$

Equation (3.14) yields the spin-1/2 angular momentum matrices $\{\frac{1}{2}\Sigma^i\}$, $i = 1,2,3$, where

$$\Sigma^i \equiv \tfrac{1}{2}\epsilon^{ijk}\Sigma_{jk} = \gamma^0\gamma^i\gamma_5 = \begin{pmatrix} -(\sigma^0\overline{\sigma}^i)_\alpha{}^\beta & 0 \\ 0 & (\overline{\sigma}^0\sigma^i)^{\dot{\alpha}}{}_{\dot{\beta}} \end{pmatrix}. \tag{3.16}$$

The following gamma matrix identities are noteworthy:

$$\gamma^\mu\gamma^\nu\gamma^\rho = g^{\mu\nu}\gamma^\rho - g^{\mu\rho}\gamma^\nu + g^{\nu\rho}\gamma^\mu + i\epsilon^{\mu\nu\rho\kappa}\gamma_\kappa\gamma_5\,, \tag{3.17}$$

$$\Sigma^{\mu\nu}\gamma^\rho = i(g^{\nu\rho}\gamma^\mu - g^{\mu\rho}\gamma^\nu - \epsilon^{\mu\nu\rho\kappa}\gamma_\kappa\gamma_5)\,, \tag{3.18}$$

$$\gamma^\rho\Sigma^{\mu\nu} = i(g^{\mu\rho}\gamma^\nu - g^{\nu\rho}\gamma^\mu - \epsilon^{\mu\nu\rho\kappa}\gamma_\kappa\gamma_5)\,, \tag{3.19}$$

$$\begin{aligned}\Sigma^{\mu\nu}\Sigma^{\rho\kappa} &= i(\Sigma^{\rho\mu}g^{\kappa\nu} - \Sigma^{\rho\nu}g^{\kappa\mu} + \Sigma^{\kappa\nu}g^{\rho\mu} - \Sigma^{\kappa\mu}g^{\rho\nu}) \\ &\quad + g^{\mu\rho}g^{\nu\kappa} - g^{\nu\rho}g^{\mu\kappa} + i\epsilon^{\mu\nu\rho\kappa}\gamma_5\,.\end{aligned} \tag{3.20}$$

In addition, some useful gamma matrix trace identities are recorded below:

$$\operatorname{Tr}(\gamma^\mu\gamma^\nu) = 4g^{\mu\nu}, \tag{3.21}$$

$$\operatorname{Tr}(\gamma^\mu\gamma^\nu\gamma^\rho\gamma^\kappa) = 4(g^{\mu\nu}g^{\rho\kappa} - g^{\mu\rho}g^{\nu\kappa} + g^{\mu\kappa}g^{\nu\rho})\,, \tag{3.22}$$

$$\operatorname{Tr}(\Sigma^{\mu\nu}\Sigma^{\rho\kappa}) = 4(g^{\mu\rho}g^{\nu\kappa} - g^{\nu\rho}g^{\mu\kappa})\,, \tag{3.23}$$

$$\operatorname{Tr}\Sigma^{\mu\nu} = 0\,, \tag{3.24}$$

$$\operatorname{Tr}\gamma_5 = \operatorname{Tr}(\gamma_5\gamma^\mu) = \operatorname{Tr}(\gamma_5\gamma^\mu\gamma^\nu) = \operatorname{Tr}(\gamma_5\gamma^\mu\gamma^\nu\gamma^\rho) = 0\,, \tag{3.25}$$

$$\operatorname{Tr}(\gamma_5\gamma^\mu\gamma^\nu\gamma^\rho\gamma^\kappa) = -4i\epsilon^{\mu\nu\rho\kappa}\,, \tag{3.26}$$

$$\operatorname{Tr}(\gamma^{\mu_1}\gamma^{\mu_2}\cdots\gamma^{\mu_{2n-1}}\gamma^{\mu_{2n}}) = \operatorname{Tr}(\gamma^{\mu_{2n}}\gamma^{\mu_{2n-1}}\cdots\gamma^{\mu_2}\gamma^{\mu_1})\,, \tag{3.27}$$

$$\operatorname{Tr}(\gamma^{\mu_1}\gamma^{\mu_2}\cdots\gamma^{\mu_{2n-1}}) = 0\,, \qquad \text{for any positive integer } n\,. \tag{3.28}$$

Given a four-component spinor Ψ, we introduce four related spinors: the Dirac adjoint spinor $\overline{\Psi}$, the space-reflected spinor Ψ^P, the time-reversed spinor Ψ^t, and the charge-conjugated spinor Ψ^C, which are respectively given by

$$\overline{\Psi}(x) \equiv \Psi^\dagger(x)A = \left(\eta^\alpha(x)\,,\, \chi^\dagger_{\dot{\alpha}}(x)\right)\,, \tag{3.29}$$

$$\Psi^P(x) \equiv i\gamma^0\Psi(x_P) = \begin{pmatrix} i\sigma^0_{\alpha\dot{\beta}}\eta^{\dagger\dot{\beta}}(x_P) \\ i\overline{\sigma}^{0\dot{\alpha}\beta}\chi_\beta(x_P)\end{pmatrix}, \tag{3.30}$$

$$\Psi^t(x) \equiv -\gamma^0B^{-1}\overline{\Psi}^{\mathsf{T}}(x_T) = -\gamma^0B^{-1}A^{\mathsf{T}}\Psi^*(x_T) = \begin{pmatrix} \sigma^0_{\alpha\dot{\beta}}\chi^{\dagger\dot{\beta}}(x_T) \\ -\overline{\sigma}^{0\dot{\alpha}\beta}\eta_\beta(x_T)\end{pmatrix}, \tag{3.31}$$

$$\Psi^C(x) \equiv C\overline{\Psi}^{\mathsf{T}}(x) = D\Psi^*(x) = \begin{pmatrix} \eta_\alpha(x) \\ \chi^{\dagger\dot{\alpha}}(x)\end{pmatrix}, \tag{3.32}$$

where $x_P \equiv (t; -\vec{x})$, $x_T \equiv (-t; \vec{x})$, and

$$D \equiv CA^{\mathsf{T}}\,. \tag{3.33}$$

It is convenient to introduce a notation for left- and right-handed charge-conjugated spinors (which are also Weyl spinors) following the conventions of [B89]:[2]

$$\Psi_L^C(x) \equiv P_L \Psi^C(x) = C\overline{\Psi}_R^{\mathsf{T}}(x) = [\Psi_R(x)]^C\,, \tag{3.34}$$
$$\Psi_R^C(x) \equiv P_R \Psi^C(x) = C\overline{\Psi}_L^{\mathsf{T}}(x) = [\Psi_L(x)]^C\,. \tag{3.35}$$

Since we are employing the chiral representation for the Dirac gamma matrices, it follows that, in this representation, A, B, C, and D are explicitly given by

$$A = \begin{pmatrix} 0 & \delta^{\dot\alpha}{}_{\dot\beta} \\ \delta_\alpha{}^\beta & 0 \end{pmatrix}, \qquad C = \begin{pmatrix} \epsilon_{\alpha\beta} & 0 \\ 0 & \epsilon^{\dot\alpha\dot\beta} \end{pmatrix},$$

$$B = \begin{pmatrix} \epsilon^{\alpha\beta} & 0 \\ 0 & -\epsilon_{\dot\alpha\dot\beta} \end{pmatrix}, \qquad D = \begin{pmatrix} 0 & \epsilon_{\alpha\beta} \\ \epsilon^{\dot\alpha\dot\beta} & 0 \end{pmatrix}. \tag{3.36}$$

Note the *numerical* equalities:[3]

$$A = \gamma^0\,, \qquad B = \gamma^1\gamma^3\,, \qquad C = i\gamma^0\gamma^2\,, \qquad D = -i\gamma^2\,. \tag{3.37}$$

However, these numerical identifications do not respect the structure of the undotted and dotted spinor indices specified in eq. (3.36). In translating between two-component and four-component spinor notation, eq. (3.36) should always be used. However, in practical four-component spinor calculations, there is no harm in employing the numerical values for A, B, C, and D exhibited in eq. (3.37). The matrices A, B, C, and D can be defined independently of the gamma matrix representation. In general, these matrices must satisfy

$$A\gamma^\mu A^{-1} = \gamma^{\mu\dagger}\,, \qquad B\gamma^\mu B^{-1} = \gamma^{\mu\mathsf{T}}\,, \tag{3.38}$$
$$C^{-1}\gamma^\mu C = -\gamma^{\mu\mathsf{T}}\,, \qquad D^{-1}\gamma_\mu D = -\gamma_\mu^*\,. \tag{3.39}$$

We now impose the following additional conditions:[4]

$$\Psi = A^{-1}\overline{\Psi}^\dagger\,, \qquad (\Psi^t)^t = -\Psi\,, \qquad (\Psi^C)^C = \Psi\,. \tag{3.40}$$

The first of these conditions together with eq. (3.29) is equivalent to the statement that $\overline{\Psi}\Psi$ is hermitian. The second condition corresponds to the result previously noted, that when the time-reversal operator is applied twice to a fermion field it yields $\mathcal{T}^2 = -1$. Finally, the third condition corresponds to the statement that the charge-conjugation operator applied twice is equal to the identity operator. Using eqs. (3.38)–(3.40) and the defining property of the gamma matrices [eq. (3.8)], one

[2] The reader is warned that the opposite convention is often employed in the literature (e.g., see [B138, B139]) where Ψ_L^C is a right-handed Weyl spinor and Ψ_R^C is a left-handed Weyl spinor.

[3] These identifications have been obtained specifically in the chiral representation of the Dirac matrices. As shown in Appendix A, eq. (3.37) also holds in the Dirac representation but not in the Majorana representation.

[4] The defining relations [eqs. (3.38) and (3.39)] and the conditions given in eq. (3.40) do not uniquely fix the values of the matrices A, B, C, and D [see eq. (A.119)]. Nevertheless, the remaining freedom in defining these matrices has no effect on any of the results in this section.

can show that the matrices A, B, C, and D must satisfy the following relations independently of the gamma matrix representation:

$$A^\dagger = A\,, \qquad B^{\mathsf{T}} = -B\,, \qquad C^{\mathsf{T}} = -C\,, \tag{3.41}$$

$$CB = -\gamma_5\,, \qquad BA^{-1} = -(AB^{-1})^*\,, \qquad (AC)^{-1} = (AC)^*\,, \tag{3.42}$$

$$D^*D = DD^* = \mathbb{1}\,. \tag{3.43}$$

Using the above results, the following results are easily derived:

$$A\Gamma A^{-1} = \eta_\Gamma^A \Gamma^\dagger\,, \qquad \eta_\Gamma^A = \begin{cases} +1\,, & \text{for } \Gamma = \mathbb{1}\,,\ \gamma^\mu\,,\ \gamma^\mu\gamma_5\,,\ \Sigma^{\mu\nu}\,, \\ -1\,, & \text{for } \Gamma = \gamma_5\,,\ \Sigma^{\mu\nu}\gamma_5\,, \end{cases} \tag{3.44}$$

$$B\Gamma B^{-1} = \eta_\Gamma^B \Gamma^{\mathsf{T}}\,, \qquad \eta_\Gamma^B = \begin{cases} +1\,, & \text{for } \Gamma = \mathbb{1}\,,\ \gamma_5\,,\ \gamma^\mu\,, \\ -1\,, & \text{for } \Gamma = \gamma^\mu\gamma_5\,,\ \Sigma^{\mu\nu}\,,\ \Sigma^{\mu\nu}\gamma_5\,, \end{cases} \tag{3.45}$$

$$C^{-1}\Gamma C = \eta_\Gamma^C \Gamma^{\mathsf{T}}\,, \qquad \eta_\Gamma^C = \begin{cases} +1\,, & \text{for } \Gamma = \mathbb{1}\,,\ \gamma_5\,,\ \gamma^\mu\gamma_5\,, \\ -1\,, & \text{for } \Gamma = \gamma^\mu\,,\ \Sigma^{\mu\nu}\,,\ \Sigma^{\mu\nu}\gamma_5\,, \end{cases} \tag{3.46}$$

$$D^{-1}\Gamma D = \eta_\Gamma^D \Gamma^*\,, \qquad \eta_\Gamma^D = \begin{cases} +1\,, & \text{for } \Gamma = \mathbb{1}\,,\ \gamma^\mu\gamma_5\,,\ \Sigma^{\mu\nu}\gamma_5\,, \\ -1\,, & \text{for } \Gamma = \gamma^\mu\,,\ \gamma_5\,,\ \Sigma^{\mu\nu}\,. \end{cases} \tag{3.47}$$

Employing eqs. (3.11) and (3.42), one can write $\gamma_5\Gamma\gamma_5 = (CB)^{-1}\Gamma CB$. It then follows that

$$\gamma_5\Gamma\gamma_5 = \eta_\Gamma^B\eta_\Gamma^C\Gamma\,, \qquad \eta_\Gamma^B\eta_\Gamma^C = \begin{cases} +1\,, & \text{for } \Gamma = \mathbb{1}\,,\ \gamma_5\,,\ \Sigma^{\mu\nu}\,,\ \Sigma^{\mu\nu}\gamma_5\,, \\ -1\,, & \text{for } \Gamma = \gamma^\mu\,,\ \gamma^\mu\gamma_5\,. \end{cases} \tag{3.48}$$

One can also describe self-conjugate fermions in four-component spinor notation. These are the Majorana fermion fields, which are the analogues of the real scalar fields. That is, for a complex scalar field, one can reduce the number of degrees of freedom by a factor of 2 by imposing the reality condition $\phi^* = \phi$. For a Dirac fermion Ψ, one can likewise reduce the number of degrees of freedom by a factor of 2 by imposing the condition $\Psi^C = \Psi$.[5] Thus, we define a Majorana four-component spinor by imposing the Majorana condition,

$$\Psi_M = \Psi_M^C = D\Psi_M^*\,. \tag{3.49}$$

Explicitly, this condition in the chiral representation implies that Ψ_M takes the following form:

$$\Psi_M(x) \equiv \begin{pmatrix} \chi_\alpha(x) \\ \chi^{\dagger\dot{\alpha}}(x) \end{pmatrix}. \tag{3.50}$$

[5] Note that $\Psi^C = \Psi$ is a Lorentz covariant condition since it sets equal two quantities with the same Lorentz transformation properties [see eq. (3.57)]. The same is not true in general for the condition $\Psi^* = \Psi$, except in special representations of the gamma matrices (such as the Majorana representation).

Multiple species of fermions are indicated with a flavor index such as i and j. Dirac fermions are constructed from two-component fields of opposite charge, χ_i and η^i (hence the opposite flavor-index heights). Thus, we establish the following conventions for the flavor indices of four-component Dirac fermions:

$$\Psi_i(x) \equiv \begin{pmatrix} \chi_{\alpha i}(x) \\ \eta_i^{\dagger\dot{\alpha}}(x) \end{pmatrix}, \qquad \overline{\Psi}^i(x) = \left(\eta^{\alpha i}(x)\,,\; \chi^{\dagger i}_{\dot{\alpha}}(x)\right), \qquad \Psi^{C\,i}(x) \equiv \begin{pmatrix} \eta^i_\alpha(x) \\ \chi^{\dagger\dot{\alpha}i}(x) \end{pmatrix}. \tag{3.51}$$

Note that $\chi^{\dagger i} = (\chi_i)^\dagger$ and $\eta_i^\dagger \equiv (\eta^i)^\dagger$ following the conventions established in Section 1.6. Raised flavor indices can only be contracted with lowered flavor indices and vice versa. In contrast, Majorana fermions are neutral so there is no a priori distinction between raised and lowered flavor indices. That is,

$$\Psi_{Mi}(x) = \Psi^i_M(x) = \Psi^C_{Mi}(x) = \Psi^{C\,i}_M(x) \equiv \begin{pmatrix} \xi_{\alpha i}(x) \\ \xi^{\dagger\dot{\alpha}i}(x) \end{pmatrix},$$

$$\overline{\Psi}_{Mi}(x) = \overline{\Psi}^i_M(x) \equiv \left(\xi^\alpha_i(x)\,,\; \xi^{\dagger i}_{\dot{\alpha}}(x)\right). \tag{3.52}$$

In this case, the contraction of two repeated flavor indices is allowed in all cases, irrespective of the heights of the two indices. In the convention adopted in Section 1.6, in which all neutral left-handed $(\frac{1}{2}, 0)$ [right-handed $(0, \frac{1}{2})$] fermions have lowered [raised] flavor indices, the height of the flavor index of a four-component Majorana fermion field is meaningful when multiplied by a left-handed or right-handed projection operator. Thus, the height of the flavor index for Majorana fermions can be consistently chosen according to one of the following four cases:

$$P_L\Psi_{Mi}\,, \qquad \overline{\Psi}_{Mi}P_L\,, \qquad P_R\Psi^i_M\,, \qquad \overline{\Psi}^i_M P_R\,. \tag{3.53}$$

3.2 Four-Component Spinor Indices

The Lorentz transformation properties of the four-component spinor field can be determined from the corresponding transformation laws of the two-component spinor fields in the $(\frac{1}{2}, 0)$ and $(0, \frac{1}{2})$ representations [see eqs. (1.34) and (1.46), respectively]. The 4×4 representation matrices in the $(\frac{1}{2}, 0)\oplus(0, \frac{1}{2})$ representation of the Lorentz group are thus given by

$$\mathbb{M} = \begin{pmatrix} M & 0 \\ 0 & (M^{-1})^\dagger \end{pmatrix} = \exp\left(-\frac{i}{4}\theta_{\mu\nu}\Sigma^{\mu\nu}\right) \simeq \mathbb{1} - \tfrac{1}{4}i\theta_{\mu\nu}\Sigma^{\mu\nu}\,, \tag{3.54}$$

where the infinitesimal forms of M and $(M^{-1})^\dagger$ are given in eqs. (A.68) and (A.69). Two useful identities that follow from eqs. (3.44), (3.46), and (3.54) are

$$A\,\mathbb{M}A^{-1} = (\mathbb{M}^{-1})^\dagger\,, \tag{3.55}$$

$$C^{-1}\mathbb{M}C = (\mathbb{M}^{-1})^{\mathsf{T}}\,. \tag{3.56}$$

Therefore, under a Lorentz transformation, $x'^{\mu} = \Lambda^{\mu}{}_{\nu} x^{\nu}$,

$$\Psi'(x') = \mathbb{M}\Psi(x)\,, \qquad \overline{\Psi}'(x') = \overline{\Psi}(x)\mathbb{M}^{-1}\,. \tag{3.57}$$

The transformation law for $\overline{\Psi}$ is easily derived from that of Ψ by making use of eq. (3.55). It then follows immediately that $\overline{\Psi}(x)\Psi(x)$ is a Lorentz scalar. Similarly, one can prove that the other bilinear covariants transform like Lorentz tensors of definite rank. This requires one further result,[6]

$$\mathbb{M}^{-1}\gamma^{\mu}\mathbb{M} = \Lambda^{\mu}{}_{\nu}\gamma^{\nu}\,, \tag{3.58}$$

which is the four-component version of eqs. (1.299) and (1.300). It then follows that $\overline{\Psi}(x)\gamma^{\mu}\Psi(x)$ transforms as a Lorentz four-vector, $\overline{\Psi}(x)\Sigma^{\mu\nu}\Psi(x)$ transforms as an antisymmetric second-rank tensor, etc.

The four-component Dirac or Majorana spinor, Ψ_a, is assigned a lowered spinor index a, and is defined in terms of two-component spinors by eq. (3.5) or eq. (3.50), respectively. Four-component spinor indices, which will be chosen in general from the beginning of the lower case Roman alphabet, $a, b, c, \ldots$, can assume integer values $1, 2, 3, 4$. Under a Lorentz transformation, Ψ_a transforms as

$$\Psi_a \to \mathbb{M}_a{}^b\,\Psi_b\,. \tag{3.59}$$

In analogy with the conventions for two-component spinor indices, we sum implicitly over a pair of repeated indices consisting of a raised and a lowered spinor index. The transformation law for the Dirac conjugate spinor (often called the Dirac adjoint spinor), $\overline{\Psi} = \Psi^{\dagger}A$, is obtained from eq. (3.59) after employing $A^{\dagger} = A$ and eq. (3.55):

$$\overline{\Psi}^{\,a} \to \overline{\Psi}^{\,b}\,(\mathbb{M}^{-1})_b{}^a\,. \tag{3.60}$$

In particular, $\overline{\Psi}\Psi \equiv \overline{\Psi}^{\,a}\Psi_a$ is a Lorentz scalar, which justifies the assignment of a raised spinor index for the Dirac conjugate spinor $\overline{\Psi}^{\,a}$.

It is convenient to introduce *barred* four-component spinor indices in the transformation laws of the hermitian-conjugated four-component spinors,

$$\Psi^{\dagger}_{\dot{a}} \to \Psi^{\dagger}_{\dot{b}}\,(\mathbb{M}^{\dagger})^{\dot{b}}{}_{\dot{a}}\,, \tag{3.61}$$

$$\overline{\Psi}^{\dagger\dot{a}} \to [(\mathbb{M}^{-1})^{\dagger}]^{\dot{a}}{}_{\dot{b}}\,\overline{\Psi}^{\dagger\dot{b}}\,, \tag{3.62}$$

where there is an implicit sum over the repeated lowered and raised barred spinor indices, and

$$\Psi^{\dagger}_{\dot{a}} \equiv (\Psi_a)^{\dagger}\,, \qquad \overline{\Psi}^{\dagger\dot{a}} \equiv (\overline{\Psi}^{\,a})^{\dagger}\,. \tag{3.63}$$

The spinor index structure of the Dirac conjugation matrix A is then fixed by noting that the Dirac conjugate spinor, $\overline{\Psi}^{\,b} \equiv \Psi^{\dagger}_{\dot{a}}A^{\dot{a}b}$, has a raised unbarred spinor index, whereas the hermitian-conjugated spinor has a lowered barred spinor index.

[6] Equation (3.58) can be proved directly by inserting the infinitesimal forms for Λ and $\mathbb{M}$ given by eqs. (1.6) and (3.54) and making use of the commutator $[\Sigma^{\mu\nu}, \gamma^{\rho}] = 2i(g^{\nu\rho}\gamma^{\mu} - g^{\mu\rho}\gamma^{\nu})$, derived from eqs. (3.18) and (3.19).

The charge-conjugation matrix can be used to raise and lower four-component spinor indices, which we shall employ in defining the spinors Ψ^a, $\Psi^{\dagger\bar{a}}$, $\overline{\Psi}_a$, and $\overline{\Psi}^{\dagger}_{\bar{a}}$:

$$\Psi_a = C_{ab}\Psi^b\,, \qquad \Psi^a = (C^{-1})^{ab}\Psi_b\,, \tag{3.64}$$

$$\Psi^{\dagger}_{\bar{a}} = C_{\bar{a}\bar{b}}\Psi^{\dagger\bar{b}}\,, \qquad \Psi^{\dagger\bar{a}} = (C^{-1})^{\bar{a}\bar{b}}\Psi^{\dagger}_{\bar{b}}\,, \tag{3.65}$$

where

$$C_{\bar{a}\bar{b}} \equiv (C_{ab})^*\,, \qquad (C^{-1})^{\bar{a}\bar{b}} \equiv [(C^{-1})^{ab}]^*\,. \tag{3.66}$$

Note that, in contrast to the epsilon symbols of the two-component spinor formalism, we prefer to explicitly exhibit the inverse symbols in $(C^{-1})^{ab}$ and $(C^{-1})^{\bar{a}\bar{b}}$.

Equations (3.64) and (3.65) also apply to $\overline{\Psi}^{\,a}$, $\overline{\Psi}_a$, and their hermitian conjugates. In particular, one can identify the Dirac conjugate spinor with a lowered spinor index ($\overline{\Psi}_a$) as the charge-conjugated spinor, $\Psi^C \equiv C\overline{\Psi}^{\mathsf{T}}$, and the Dirac spinor with a raised spinor index (Ψ^a) as the Dirac conjugate of the charge-conjugated spinor, $\overline{\Psi^C} = -\Psi^{\mathsf{T}} C^{-1}$. That is,[7]

$$\Psi^C_a \equiv \overline{\Psi}_a = C_{ab}\overline{\Psi}^{\,b}\,, \qquad \overline{\Psi^C}^{\,a} = \Psi^a = (C^{-1})^{ab}\Psi_b\,. \tag{3.67}$$

The rules for raising and lowering spinor indices are consistent with the Lorentz transformation properties of eqs. (3.59)–(3.62), as a consequence of eq. (3.56). In particular, the condition for a self-conjugate four-component (Majorana) spinor, $\Psi_a \equiv \Psi^C_a = \overline{\Psi}_a$, is Lorentz covariant.

Using eqs. (3.29), (3.64), and (3.65), it then follows that

$$\overline{\Psi}_a = (A^{-1})_{a\bar{b}}\Psi^{\dagger\bar{b}}\,, \qquad \overline{\Psi}^{\,a} = \Psi^{\dagger}_{\bar{b}}A^{\bar{b}a}\,, \tag{3.68}$$

$$\Psi^{\dagger}_{\bar{a}} = \overline{\Psi}^{\,b}(A^{-1})_{b\bar{a}}\,, \qquad \Psi^{\dagger\bar{a}} = A^{\bar{a}b}\,\overline{\Psi}_b\,, \tag{3.69}$$

after making use of the properties of the matrices A and C given in eqs. (3.41) and (3.42). One can check that eqs. (3.68) and (3.69) are consistent with the Lorentz transformation properties of eqs. (3.59)–(3.62), as a consequence of eq. (3.55).

In addition to the Lorentz scalar $\overline{\Psi}\Psi \equiv \overline{\Psi}^{\,a}\Psi_a$, one can construct two additional independent Lorentz scalar quantities:

$$-\Psi^{\mathsf{T}}C^{-1}\,\Psi \equiv -\Psi_a(C^{-1})^{ab}\Psi_b = \Psi^a\Psi_a\,, \tag{3.70}$$

and the hermitian conjugate of eq. (3.70),

$$\overline{\Psi}\,C\,\overline{\Psi}^{\mathsf{T}} \equiv \overline{\Psi}^{\,a}C_{ab}\overline{\Psi}^{\,b} = \overline{\Psi}^{\,a}\overline{\Psi}_a = \Psi^{\dagger}_{\bar{a}}\Psi^{\dagger\bar{a}} = (\Psi^a\Psi_a)^{\dagger}\,, \tag{3.71}$$

after using C^{-1} and C to raise and lower the appropriate spinor indices, respectively. The penultimate equality in eq. (3.71) is a consequence of eq. (3.69). Note that a fourth possible Lorentz scalar is not independent:

$$\Psi^a\overline{\Psi}_a = (C^{-1})^{ab}C_{ac}\Psi_b\overline{\Psi}^{\,c} = -\Psi_b\overline{\Psi}^{\,b} = \overline{\Psi}^{\,b}\Psi_b\,, \tag{3.72}$$

where we have used $C^{\mathsf{T}} = -C$ and the anticommutativity of the spinors.

[7] For a Dirac spinor field defined in eq. (3.5), $\overline{\Psi}_a(x) = \Psi^C_a(x)$ is given in terms of two-component spinors by eq. (3.32), and $\Psi^a(x) = \overline{\Psi^C}^{\,a}(x) = \big(\chi^\alpha(x)\,,\,\eta^{\dagger}_{\dot{\alpha}}(x)\big)$.

The Lorentz invariance of $\overline{\Psi}{}^a\Psi_a$, $\Psi^a\Psi_a$, and $\Psi^\dagger_{\bar{a}}\Psi^{\dagger\bar{a}} = \overline{\Psi}{}^a\overline{\Psi}_a$ is manifest and demonstrates the power of the four-component spinor index notation developed above. Moreover, after invoking eq. (3.67), we note that [analogous to eq. (1.63)] *descending* contracted unbarred spinor indices and *ascending* contracted barred spinor indices can be suppressed in spinor-index-contracted products. For example,

$$\overline{\Psi}{}^a\Psi_a \equiv \overline{\Psi}\Psi = \overline{\Psi^C}\Psi^C\,, \tag{3.73}$$

after employing eq. (3.72), and

$$\Psi^a\Psi_a = \overline{\Psi^C}{}^a\Psi_a \equiv \overline{\Psi^C}\Psi\,, \qquad \overline{\Psi}{}^a\overline{\Psi}_a = \overline{\Psi}{}^a\Psi^C_a \equiv \overline{\Psi}\Psi^C. \tag{3.74}$$

The charge-conjugated spinor can also be written as $\Psi^C_a \equiv D_a{}^{\bar{c}}\Psi^\dagger_{\bar{c}}$ [see eq. (3.32)]. The spinor index structure of D (and its inverse) derives from

$$D_a{}^{\bar{c}} \equiv C_{ab}(A^{\mathsf{T}})^{b\bar{c}} = C_{ab}A^{\bar{c}b}\,, \qquad (D^{-1})_{\bar{a}}{}^c \equiv (C^*A)_{\bar{a}}{}^c = C_{\bar{a}\bar{b}}A^{\bar{b}c}\,, \tag{3.75}$$

where we have used $D^{-1} = D^*$. Combining the results of eqs. (3.64), (3.65), (3.68), and (3.69) then yields

$$\overline{\Psi}_a = D_a{}^{\bar{c}}\,\Psi^\dagger_{\bar{c}}\,, \qquad \Psi^\dagger_{\bar{a}} = (D^{-1})_{\bar{a}}{}^c\,\overline{\Psi}_c\,, \tag{3.76}$$

$$\overline{\Psi}{}^a = -\Psi^{\dagger\bar{c}}(D^{-1})_{\bar{c}}{}^a\,, \qquad \Psi^{\dagger\bar{a}} = -\overline{\Psi}{}^c\,D_c{}^{\bar{a}}\,. \tag{3.77}$$

In summary, a four-component spinor Ψ_a and its charge-conjugated spinor Ψ^C_a possess a lowered unbarred spinor index, whereas the corresponding Dirac conjugates, $\overline{\Psi}{}^a$ and $\overline{\Psi^C}{}^a$, possess a raised unbarred spinor index. The corresponding hermitian-conjugated spinors exhibit barred spinor indices (with the height of each spinor index unchanged). Following eqs. (3.64) and (3.65), one can also lower or raise a four-component unbarred or barred spinor index by multiplying by the appropriate matrix C, C^{-1}, C^*, or $(C^{-1})^*$, respectively.

The identity matrix, the gamma matrices, and their products are denoted collectively by Γ. The spinor index structure of these matrices and their inverses is given by

$$\delta_a^b\,,\ \Gamma_a{}^b\,,\ (\Gamma^{-1})_a{}^b\,, \tag{3.78}$$

where the δ_a^b are the matrix elements of the identity matrix $\mathbb{1}$. In this case, the rows are labeled by the lowered index and the columns are labeled by the raised index. As previously noted, the quantities $\overline{\Psi}{}^a\,\Gamma_a{}^b\,\Psi_b$, $\Psi^a\,\Gamma_a{}^b\,\Psi_b$, and $\overline{\Psi}{}^a\,\Gamma_a{}^b\,\overline{\Psi}{}^b$ transform as Lorentz tensors, whose rank is equal to the number of (suppressed) spacetime indices of Γ.

For the matrices A, B, C, D and their inverses, the spinor index structure is given by

$$A^{\bar{a}b}\,,\ (A^{-1})_{a\bar{b}}\,,\ B^{ab}\,,\ (B^{-1})_{ab}\,, C_{ab}\,,\ (C^{-1})^{ab}\,, D_a{}^{\bar{b}}\,,\ (D^{-1})_{\bar{a}}{}^b\,. \tag{3.79}$$

The corresponding complex-conjugated matrices exhibit the analogous spinor index structure with unbarred spinor indices changed to barred spinor indices and vice versa. Matrix transposition interchanges rows and columns. For example,

$$(\Gamma^{\mathsf{T}})^a{}_b \equiv \Gamma_b{}^a\,, \qquad (A^{\mathsf{T}})^{a\bar{b}} \equiv A^{\bar{b}a}\,, \qquad (C^{\mathsf{T}})_{ab} = C_{ba}\,, \qquad (D^{\mathsf{T}})^{\bar{a}}{}_b \equiv D_b{}^{\bar{a}}\,. \tag{3.80}$$

Hermitian conjugation is complex conjugation followed by matrix transposition. For example,

$$(\Gamma^\dagger)^{\bar{a}}{}_{\bar{b}} \equiv (\Gamma_b{}^a)^*\,, \qquad (A^\dagger)^{\bar{a}b} \equiv (A^{\bar{b}a})^*\,, \qquad (C^\dagger)_{\bar{a}\bar{b}} = (C_{ba})^*\,, \qquad (D^\dagger)^a{}_{\bar{b}} \equiv (D_b{}^{\bar{a}})^*. \tag{3.81}$$

It is now straightforward to identify the index structure of equations that involve four-component spinors. For example, specifying the four-component spinor indices of eq. (3.56) yields

$$(C^{-1})^{ab}\mathbb{M}_b{}^c C_{cd} = [(\mathbb{M}^{-1})^{\mathsf{T}}]^a{}_d \equiv (\mathbb{M}^{-1})_d{}^a\,. \tag{3.82}$$

3.3 Bilinear Covariants and Their P, T, and C Properties

The Dirac bilinear covariants are quadratic in the Dirac spinor field and transform irreducibly as Lorentz tensors. These are easily constructed from the corresponding quantities that are quadratic in the two-component fermion fields. To construct a translation table between the two-component form and the four-component forms for the bilinear covariants, we first introduce two Dirac spinor fields:

$$\Psi_i(x) \equiv \begin{pmatrix} \chi_i(x) \\ \eta_i^\dagger(x) \end{pmatrix}, \qquad \Psi_j(x) \equiv \begin{pmatrix} \chi_j(x) \\ \eta_j^\dagger(x) \end{pmatrix}, \tag{3.83}$$

where spinor indices have been suppressed. The following results, which apply to both commuting and anticommuting fermion fields, are then obtained:[8]

$$\overline{\Psi}{}^i P_L \Psi_j = \eta^i \chi_j\,, \qquad \overline{\Psi}{}^i P_R \Psi_j = \chi^{\dagger\, i} \eta_j^\dagger\,, \tag{3.84}$$

$$\overline{\Psi}{}^i \gamma^\mu P_L \Psi_j = \chi^{\dagger\, i} \overline{\sigma}^\mu \chi_j\,, \qquad \overline{\Psi}{}^i \gamma^\mu P_R \Psi_j = \eta^i \sigma^\mu \eta_j^\dagger\,, \tag{3.85}$$

$$\overline{\Psi}{}^i \Sigma^{\mu\nu} P_L \Psi_j = 2\,\eta^i \sigma^{\mu\nu} \chi_j\,, \qquad \overline{\Psi}{}^i \Sigma^{\mu\nu} P_R \Psi_j = 2\,\chi^{\dagger\, i} \overline{\sigma}^{\mu\nu} \eta_j^\dagger\,. \tag{3.86}$$

The above results can then be used to express the standard four-component spinor bilinear covariants in terms of two-component spinor bilinears:

$$\overline{\Psi}{}^i \Psi_j = \eta^i \chi_j + \chi^{\dagger\, i} \eta_j^\dagger \tag{3.87}$$

$$\overline{\Psi}{}^i \gamma_5 \Psi_j = -\eta^i \chi_j + \chi^{\dagger\, i} \eta_j^\dagger \tag{3.88}$$

$$\overline{\Psi}{}^i \gamma^\mu \Psi_j = \chi^{\dagger\, i} \overline{\sigma}^\mu \chi_j + \eta^i \sigma^\mu \eta_j^\dagger \tag{3.89}$$

$$\overline{\Psi}{}^i \gamma^\mu \gamma_5 \Psi_j = -\chi^{\dagger\, i} \overline{\sigma}^\mu \chi_j + \eta^i \sigma^\mu \eta_j^\dagger \tag{3.90}$$

$$\overline{\Psi}{}^i \Sigma^{\mu\nu} \Psi_j = 2(\eta^i \sigma^{\mu\nu} \chi_j + \chi^{\dagger\, i} \overline{\sigma}^{\mu\nu} \eta_j^\dagger) \tag{3.91}$$

$$\overline{\Psi}{}^i \Sigma^{\mu\nu} \gamma_5 \Psi_j = -2(\eta^i \sigma^{\mu\nu} \chi_j - \chi^{\dagger\, i} \overline{\sigma}^{\mu\nu} \eta_j^\dagger)\,. \tag{3.92}$$

[8] For anticommuting spinors, it is often useful to apply eq. (1.98) to rewrite $\eta^i \sigma^\mu \eta_j^\dagger = -\eta_j^\dagger \overline{\sigma}^\mu \eta^i$.

Table 3.1 A translation table relating the Dirac bilinear covariants written in terms of four-component Dirac spinor fields to the corresponding quantities expressed in terms of two-component spinor fields using the notation of eq. (3.83). These results apply to both commuting and anticommuting spinors.

$\overline{\Psi}{}^i P_L \Psi_j = \eta^i \chi_j$	$\overline{\Psi^C_i} P_L \Psi^{C\,j} = \chi_i \eta^j$
$\overline{\Psi}{}^i P_R \Psi_j = \chi^{\dagger\, i} \eta^\dagger_j$	$\overline{\Psi^C_i} P_R \Psi^{C\,j} = \eta^\dagger_i \chi^{\dagger\, j}$
$\overline{\Psi^C_i} P_L \Psi_j = \chi_i \chi_j$	$\overline{\Psi}{}^i P_L \Psi^{C\,j} = \eta^i \eta^j$
$\overline{\Psi}{}^i P_R \Psi^{C\,j} = \chi^{\dagger\, i} \chi^{\dagger\, j}$	$\overline{\Psi^C_i} P_R \Psi_j = \eta^\dagger_i \eta^\dagger_j$
$\overline{\Psi}{}^i \gamma^\mu P_L \Psi_j = \chi^{\dagger\, i} \overline{\sigma}^\mu \chi_j$	$\overline{\Psi^C_i} \gamma^\mu P_L \Psi^{C\,j} = \eta^\dagger_i \overline{\sigma}^\mu \eta^j$
$\overline{\Psi^C_i} \gamma^\mu P_R \Psi^{C\,j} = \chi_i \sigma^\mu \chi^{\dagger\, j}$	$\overline{\Psi}{}^i \gamma^\mu P_R \Psi_j = \eta^i \sigma^\mu \eta^\dagger_j$
$\overline{\Psi}{}^i \Sigma^{\mu\nu} P_L \Psi_j = 2\eta^i \sigma^{\mu\nu} \chi_j$	$\overline{\Psi^C_i} \Sigma^{\mu\nu} P_L \Psi^{C\,j} = 2\chi_i \sigma^{\mu\nu} \eta^j$
$\overline{\Psi}{}^i \Sigma^{\mu\nu} P_R \Psi_j = 2\chi^{\dagger\, i} \overline{\sigma}^{\mu\nu} \eta^\dagger_j$	$\overline{\Psi^C_i} \Sigma^{\mu\nu} P_R \Psi^{C\,j} = 2\eta^\dagger_i \overline{\sigma}^{\mu\nu} \chi^{\dagger\, j}$

Additional identities can be derived that involve the charge-conjugated four-component Dirac fermion fields. For example, using $C^{\mathsf{T}} = -C$ and $\overline{\Psi^C} = -\Psi^{\mathsf{T}} C^{-1}$, one can prove that

$$\overline{\Psi^C_i}\, \Gamma\, \Psi^{C\,j} = -(-1)^A \overline{\Psi}{}^j C\, \Gamma^{\mathsf{T}}\, C^{-1} \Psi_i = -(-1)^A \eta^C_\Gamma \overline{\Psi}{}^j\, \Gamma\, \Psi_i\,, \tag{3.93}$$

where the sign η^C_Γ is given in eq. (3.46). The factor of $(-1)^A = \pm 1$, where the upper [lower] sign is taken for commuting [anticommuting] fermion fields, respectively, arises at the second step above after reversing the order of the terms by matrix transposition. Identities involving just one charge-conjugated four-component field can also be easily obtained. For example,

$$\overline{\Psi^C_i} P_L \Psi_j = -\Psi^{\mathsf{T}}_i C^{-1} P_L \Psi_j = -\epsilon^{\alpha\beta} \chi_{\alpha i} \chi_{\beta j} = \chi_i \chi_j\,. \tag{3.94}$$

If one replaces Ψ_k with $\Psi^{C\,k}$ in eqs. (3.84)–(3.92), then in the corresponding two-component expression one simply interchanges $\chi_k \leftrightarrow \eta^k$ and $\chi^{\dagger\, k} \leftrightarrow \eta^\dagger_k$.

The above results are conveniently summarized in Table 3.1. In the case of anticommuting spinors, it is often useful to apply eq. (1.107) to eqs. (3.85), (3.89), and (3.90), in which case one may alternatively write $\overline{\Psi}{}^i \gamma^\mu P_R \Psi_j = -\eta^\dagger_j \overline{\sigma}^\mu \eta^i$, etc.

We expect a total of 16 independent components among the bilinear covariants, corresponding to the 16-dimensional space of 4×4 matrices. From eq. (3.15), we see that $\Sigma^{\mu\nu}$ and $\Sigma^{\mu\nu}\gamma_5$ are not independent. That is, the 16-dimensional space is spanned by $\mathbb{1}$, γ_5, γ^μ, $\gamma^\mu\gamma_5$, and $\Sigma^{\mu\nu}$, which matches the counting presented at the end of Section 1.4.

We shall always work in a convention where $A^\dagger = A$. It then follows that

$$(\overline{\Psi}{}^i \Gamma \Psi_j)^\dagger = \overline{\Psi}{}^j A^{-1} \Gamma A \Psi_i = \eta^A_\Gamma \overline{\Psi}{}^j \Gamma \Psi_i\,, \tag{3.95}$$

after employing eqs. (3.38) and (3.44). When $i = j$, the bilinear covariants $\overline{\Psi}\Gamma\Psi$ are either hermitian or antihermitian. However it is more convenient to redefine

$\Gamma \to i\Gamma$, when necessary, to ensure that all the bilinear covariants (with $i = j$) are hermitian. This can be achieved by modifying the list of possible choices for Γ as follows:

$$\Gamma = \mathbb{1}\,,\, i\gamma_5\,,\, \gamma^\mu\,,\, \gamma^\mu\gamma_5\,,\, \Sigma^{\mu\nu}\,,\, i\Sigma^{\mu\nu}\gamma_5\,. \tag{3.96}$$

The set of choices for Γ exhibited in eq. (3.96) corresponds to setting $\eta^A_\Gamma = 1$ in eq. (3.95), which in light of eq. (3.44) yields

$$A\Gamma A^{-1} = \Gamma^\dagger\,. \tag{3.97}$$

That is, the bilinear covariants $\overline{\Psi}\Gamma\Psi$ are hermitian if Γ satisfies eq. (3.97).

The results of Table 3.1 and eqs. (3.84)–(3.92) also apply to four-component Majorana spinors, Ψ_{Mi}, by setting $\chi_i = \eta^i \equiv \xi_i$, and $\chi^{\dagger\, i} = \eta^\dagger_i \equiv \xi^{\dagger\, i}$. This implements the Majorana condition, $\Psi_{Mi} = D\Psi^*_{Mi}$, and imposes additional restrictions on the Majorana bilinear covariants. For example, eqs. (3.46) and (3.93) imply that *anticommuting* Majorana four-component fermions satisfy[9]

$$\overline{\Psi}_{Mi}\Psi_{Mj} = \overline{\Psi}_{Mj}\Psi_{Mi}\,, \qquad \overline{\Psi}_{Mi}\gamma_5\Psi_{Mj} = \overline{\Psi}_{Mj}\gamma_5\Psi_{Mi}\,, \tag{3.98}$$

$$\overline{\Psi}_{Mi}\gamma^\mu\Psi_{Mj} = -\overline{\Psi}_{Mj}\gamma^\mu\Psi_{Mi}\,, \qquad \overline{\Psi}_{Mi}\gamma^\mu\gamma_5\Psi_{Mj} = \overline{\Psi}_{Mj}\gamma^\mu\gamma_5\Psi_{Mi}\,, \tag{3.99}$$

$$\overline{\Psi}_{Mi}\Sigma^{\mu\nu}\Psi_{Mj} = -\overline{\Psi}_{Mj}\Sigma^{\mu\nu}\Psi_{Mi}\,, \qquad \overline{\Psi}_{Mi}\Sigma^{\mu\nu}\gamma_5\Psi_{Mj} = -\overline{\Psi}_{Mj}\Sigma^{\mu\nu}\gamma_5\Psi_{Mi}\,. \tag{3.100}$$

One additional useful result is

$$\overline{\Psi}{}^i_M\gamma^\mu P_L\Psi_{Mj} = -\overline{\Psi}_{Mj}\gamma^\mu P_R\Psi^i_M\,, \tag{3.101}$$

which follows immediately from eq. (3.99). Note that in eq. (3.101) the heights of the flavor indices follow the convention established in eq. (3.53).

By setting $i = j$, it follows that

$$\overline{\Psi}_M\gamma^\mu\Psi_M = \overline{\Psi}_M\Sigma^{\mu\nu}\Psi_M = \overline{\Psi}_M\Sigma^{\mu\nu}\gamma_5\Psi_M = 0\,, \tag{3.102}$$

leaving only six independent components of the nonvanishing bilinear covariants: $\overline{\Psi}_M\Psi_M$, $\overline{\Psi}_M\gamma_5\Psi_M$, and $\overline{\Psi}_M\gamma_\mu\gamma_5\Psi_M$. This result matches the counting presented above eq. (1.135).

Due to the Majorana condition, the bilinear covariants of eqs. (3.87)–(3.92) are either hermitian or antihermitian even if $i \neq j$. Indeed, given that $A^\dagger = A$,

$$\begin{aligned}(\overline{\Psi}_{Mi}\Gamma\Psi_{Mj})^\dagger &= \overline{\Psi}_{Mj}A^{-1}\Gamma^\dagger A\Psi_{Mi} = -\Psi^{\mathsf{T}}_{Mj}C^{-1}A^{-1}\Gamma^\dagger AC\overline{\Psi}{}^{\mathsf{T}}_{Mi}\\ &= \overline{\Psi}_{Mi}C(A\Gamma A^{-1})^*C^{-1}\Psi_{Mj} = \eta^A_\Gamma\eta^C_\Gamma\,\overline{\Psi}_{Mi}\Gamma\Psi_{Mj}\,,\end{aligned} \tag{3.103}$$

where we have made use of eqs. (3.41) and (3.49) and the anticommutativity of the fermion fields. However, as previously noted, one can always redefine $\Gamma \to i\Gamma$, when necessary, such that all the bilinear covariants are hermitian. This can be achieved by modifying the list of possible choices for Γ as follows:

$$\Gamma = \mathbb{1}\,,\, i\gamma_5\,,\, i\gamma^\mu\,,\, \gamma^\mu\gamma_5\,,\, i\Sigma^{\mu\nu}\,,\, \Sigma^{\mu\nu}\gamma_5\,. \tag{3.104}$$

[9] Here, one is free to choose all flavor indices to be in the lowered position [see eq. (3.52)].

The set of choices for Γ exhibited in eq. (3.104) corresponds to setting $\eta^A_\Gamma \eta^C_\Gamma = 1$ in eq. (3.103), which in light of eqs. (3.44) and (3.46) yields

$$\Gamma^\dagger = AC\,\Gamma^{\mathsf{T}}(AC)^{-1}\,. \tag{3.105}$$

That is, the Majorana bilinear covariants $\overline{\Psi}_{Mi}\Gamma\Psi_{Mj}$ are hermitian if Γ satisfies eq. (3.105). In the case of $i=j$, a comparison of eqs. (3.96) and (3.104) implies that the Majorana bilinear covariant, $\overline{\Psi}_M\Gamma\Psi_M$, is both hermitian and antihermitian if $\Gamma = \gamma^\mu$, $\Sigma^{\mu\nu}$ or $\Sigma^{\mu\nu}\gamma_5$. Hence for these three choices of Γ, the corresponding Majorana bilinear covariants must be zero, in agreement with eq. (3.102).

The behavior of the four-component fermion field $\Psi(x)$ under P, T, and C can be obtained from the corresponding behavior of the two-component fermion fields (written in the basis of the fields of definite charge). Using the results of eqs. (1.211)–(1.214) for P, eqs. (1.243)–(1.246) for T and eqs. (1.256) and (1.257) for C, it follows that the (second-quantized versions of the) corresponding transformation laws of the four-component spinor field operators are given by[10]

$$\mathcal{P}\Psi(x)\mathcal{P}^{-1} = i\eta_P\gamma^0\Psi(x_P) = \eta_P\Psi^P(x)\,, \tag{3.106}$$

$$\mathcal{T}\Psi(x)\mathcal{T}^{-1} = \eta_T(\gamma^0A^{-1})^{\mathsf{T}}B\Psi(x_T) = \eta_T\Psi^{t*}(x)\,, \tag{3.107}$$

$$\mathcal{C}\Psi(x)\mathcal{C}^{-1} = \eta_C C\overline{\Psi}^{\mathsf{T}}(x) = \eta_C D\Psi^*(x) = \eta_C\Psi^C(x)\,, \tag{3.108}$$

after making use of eqs. (3.30)–(3.33), (3.41), and (3.42). For convenience, we also list the corresponding transformation laws for the Dirac adjoint field:

$$\mathcal{P}\overline{\Psi}(x)\mathcal{P}^{-1} = -i\eta_P^*\overline{\Psi}(x_P)\gamma^0 = \eta_P^*\overline{\Psi^P}(x)\,, \tag{3.109}$$

$$\mathcal{T}\overline{\Psi}(x)\mathcal{T}^{-1} = \eta_T^*\overline{\Psi}(x_T)B^{-1}(A^\dagger\gamma^0)^{\mathsf{T}} = \eta_T^*\overline{\Psi^{t*}}(x)\,, \tag{3.110}$$

$$\mathcal{C}\overline{\Psi}(x)\mathcal{C}^{-1} = -\eta_C^*\Psi^{\mathsf{T}}(x)C^{-1} = \eta_C^*\overline{\Psi^C}(x)\,. \tag{3.111}$$

These results apply to both Dirac and Majorana four-component fermion fields. For the transformation of Dirac fields, the phases η_P, η_T, and η_C are arbitrary. When applied to Majorana fields, the Majorana relation, $\Psi_M = D\Psi_M^*$, imposes constraints on these phases. In particular, it is straightforward to show that the transformation of $\Psi^C(x)$ under P, T, and C is given by eqs. (3.106)–(3.111) with the replacement of the phases η_P, η_T, and η_C by their complex conjugates. Hence, for Majorana fields, which satisfy $\Psi_M^C = \Psi_M$, it follows that the phases η_P, η_T, and η_C are real (and equal to ±1).

The form of the transformation law under CPT is particularly simple. After some algebraic manipulation,

$$\begin{aligned}\mathcal{CPT}\,\Psi(x)(\mathcal{CPT})^{-1} &= \eta_T(\gamma^0A^{-1})^{\mathsf{T}}B\,\mathcal{CP}\Psi(x_T)(\mathcal{CP})^{-1}\\ &= i\eta_{CPT}(\gamma^0A^{-1})^{\mathsf{T}}B\gamma^0CA^{\mathsf{T}}\Psi^*(-x)\\ &= i\eta_{CPT}(ACBA^{-1})^{\mathsf{T}}\Psi^*(-x)\,,\end{aligned} \tag{3.112}$$

[10] Once again, we remind the reader that we have deviated from the standard textbook conventions (e.g., see [B84, B133, B135]) in eq. (3.106) by separating out the factor of i from the phase η_P. As a result, in our notation $\eta_P = \pm1$ for a Majorana fermion, whereas in the standard convention $\eta_P = \pm i$ after the factor of i is absorbed into the definition of η_P.

where as before $\eta_{CPT} \equiv \eta_C\eta_P\eta_T$. Using that $CB = -\gamma_5$ [see eq. (3.42)] and making use of eq. (3.38), we end up with

$$\mathcal{CPT}\,\Psi(x)(\mathcal{CPT})^{-1} = i\eta_{CPT}[\gamma_5\Psi(-x)]^*\,, \tag{3.113}$$

$$\mathcal{CPT}\,\overline{\Psi}(x)(\mathcal{CPT})^{-1} = -i\eta^*_{CPT}[A^\dagger\gamma_5\Psi(-x)]^\mathsf{T}\,. \tag{3.114}$$

It is now straightforward to work out the properties of the bilinear covariants under P, T, C, CP, and CPT. First, we introduce some convenient notation to present the relevant results. We shall denote any tensor that is transformed by the space-inversion matrix, $\Lambda_P = \text{diag}(1, -1, -1, -1)$ with a tilde. For example,

$$\tilde{x}^\mu \equiv x^\mu_P = (\Lambda_P)^\mu{}_\nu x^\nu = (x^0\,;\,-\vec{\boldsymbol{x}})\,. \tag{3.115}$$

Since $(\Lambda_P)^\mu{}_\nu = \text{diag}(1, -1, -1, -1)$ is numerically equal to $g_{\mu\nu}$, it follows that $\tilde{x}^\mu$ is numerically equal to x_μ. Indeed, in the expressions that we present below, some books write the corresponding space-inverted Lorentz tensors by lowering all Lorenz indices. However, we prefer to employ the tilde notation to maintain the Lorentz covariance of the corresponding expressions given below. As an example, we shall use eq. (3.8) to write

$$\tilde{\gamma}^\mu = \gamma^0\gamma^\mu\gamma^0\,, \tag{3.116}$$

since $\tilde{\gamma}^\mu = (\gamma^0\,;\,-\gamma^i)$. It then follows that

$$\gamma^0\Gamma\gamma^0 = \eta^0_\Gamma\widetilde{\Gamma}\,, \qquad \eta^0_\Gamma = \begin{cases} +1\,, & \text{for } \Gamma = \mathbb{1}\,,\ \gamma^\mu\,,\ \Sigma^{\mu\nu}\,, \\ -1\,, & \text{for } \Gamma = \gamma_5\,,\ \gamma^\mu\gamma_5\,,\ \Sigma^{\mu\nu}\gamma_5\,. \end{cases} \tag{3.117}$$

Hence, we obtain

$$\mathcal{P}\overline{\Psi}^i\Gamma\mathcal{P}^{-1}\,\Psi_j(x) = \eta^*_{Pi}\eta_{Pj}\overline{\Psi}^i(\tilde{x})\gamma^0\Gamma\gamma^0\Psi(\tilde{x}) = \eta^*_{Pi}\eta_{Pj}\eta^0_\Gamma\overline{\Psi}^i(\tilde{x})\widetilde{\Gamma}\Psi_j(\tilde{x})\,. \tag{3.118}$$

To compute the time-reversed bilinear covariants, note that $\Lambda_T = \text{diag}(-1, 1, 1, 1)$, so that $x_T = -\tilde{x}$. Moreover, $\mathcal{T}$ is an antiunitary operator, which implies that $\mathcal{T}\Gamma\mathcal{T}^{-1} = \Gamma^*$. It is convenient to impose the condition given in eq. (3.97), in which case the possible choices for Γ are given in eq. (3.96). Hence, using $A^\dagger = A$ and making use of eqs. (3.38) and (3.97), it follows that

$$\begin{aligned}\mathcal{T}\,\overline{\Psi}^i(x)\Gamma\Psi_j(x)\,\mathcal{T}^{-1} &= \eta^*_{Ti}\eta_{Tj}\,\overline{\Psi}^i(-\tilde{x})B^{-1}(A\gamma^0\Gamma\gamma^0A^{-1})^*B\Psi_j(-\tilde{x}) \\ &= \eta^*_{Ti}\eta_{Tj}\eta^0_\Gamma\overline{\Psi}^i(-\tilde{x})B^{-1}(A\widetilde{\Gamma}A^{-1})^*B\Psi_j(-\tilde{x}) \\ &= \eta^*_{Ti}\eta_{Tj}\eta^0_\Gamma\overline{\Psi}^i(-\tilde{x})B^{-1}\widetilde{\Gamma}^\mathsf{T}B\Psi_j(-\tilde{x}) \\ &= \eta^*_{Ti}\eta_{Tj}\eta^0_\Gamma\eta^B_\Gamma\overline{\Psi}^i(-\tilde{x})\widetilde{\Gamma}\Psi_j(-\tilde{x})\,,\end{aligned} \tag{3.119}$$

where

$$\eta^0_\Gamma\eta^B_\Gamma = \begin{cases} +1\,, & \text{for } \Gamma = \mathbb{1}\,,\ \gamma^\mu\,,\ \gamma^\mu\gamma_5\,,\ i\Sigma^{\mu\nu}\gamma_5\,, \\ -1\,, & \text{for } \Gamma = i\gamma_5\,,\ \Sigma^{\mu\nu}\,. \end{cases} \tag{3.120}$$

To compute the charge-conjugated bilinear covariants, we shall make use of the anticommutativity of the fermion fields.[11] Then, after employing eqs. (3.29), (3.33), and (3.39), one obtains

$$\begin{aligned} \mathcal{C}\,\overline{\Psi}^i(x)\Gamma\Psi_j(x)\,\mathcal{C}^{-1} &= -\eta^*_{Ci}\eta_{Cj}\Psi^{\mathsf{T}}_i(x)C^{-1}\Gamma D\Psi^{*j}(x) \\ &= \eta^*_{Ci}\eta_{Cj}\Psi^{\dagger j}(x)(C^{-1}\Gamma C A^{\mathsf{T}})^{\mathsf{T}}\Psi_i(x) \\ &= \eta^*_{Ci}\eta_{Cj}\eta^C_\Gamma\overline{\Psi}^j(x)\Gamma\Psi_i(x)\,. \end{aligned} \quad (3.121)$$

An extra minus sign was generated by taking the transpose, which requires an interchange of the two fermion fields (resulting in a change in the order of the flavor indices i and j). Note that the charge-conjugation operation also interchanges the heights of the two flavor indices.

It is now straightforward to derive

$$\mathcal{CP}\,\overline{\Psi}^i(x)\Gamma\Psi_j(x)\,\mathcal{CP}^{-1} = \eta^*_{CPi}\eta_{CPj}\eta^0_\Gamma\eta^C_\Gamma\overline{\Psi}^j(\tilde{x})\widetilde{\Gamma}\Psi_i(\tilde{x})\,, \quad (3.122)$$

$$\mathcal{CPT}\,\overline{\Psi}^i(x)\Gamma\Psi_j(x)\,\mathcal{CPT}^{-1} = \eta^B_\Gamma\eta^C_\Gamma\overline{\Psi}^j(-x)\Gamma\Psi_i(-x)\,, \quad (3.123)$$

where $\eta_{CPi} \equiv \eta_{Ci}\eta_{Pi}$, under the assumption that Γ satisfies eq. (3.97). Employing eq. (3.48) in eq. (3.123) yields the alternative form[12]

$$\mathcal{CPT}\,\overline{\Psi}^i(x)\Gamma\Psi_j(x)\,(\mathcal{CPT})^{-1} = \overline{\Psi}^j(-x)\gamma_5\Gamma\gamma_5\Psi_i(-x)\,. \quad (3.124)$$

As noted in eq. (1.193), we have adopted a convention where $\eta_{CPT} = 1$, independently of the fermion species. Hence, there is no remaining CPT phase dependence in eqs. (3.123) and (3.124).

The P, T, and C transformation properties of the bilinear covariants obtained above are also applicable in the case where the bilinear covariants are constructed using four-component Majorana fermion fields. However, due to the Majorana condition [eq. (3.49)], the Majorana bilinear covariants can be hermitian even in the case of $i \neq j$. As noted below eq. (3.103), $\overline{\Psi}_{Mi}\Gamma\Psi_{Mj}$ is hermitian if Γ satisfies eq. (3.105). Thus, if we impose eq. (3.105) in the computations above, eqs. (3.119) and (3.123) are modified:

$$\mathcal{T}\,\overline{\Psi}_{Mi}(x)\Gamma\Psi_{Mj}(x)\,\mathcal{T}^{-1} = \eta^*_{Ti}\eta_{Tj}\eta^0_\Gamma\eta^B_\Gamma\eta^C_\Gamma\overline{\Psi}_{Mi}(-\tilde{x})\widetilde{\Gamma}\Psi_{Mj}(-\tilde{x})\,, \quad (3.125)$$

$$\mathcal{CPT}\,\overline{\Psi}_{Mi}(x)\Gamma\Psi_{Mj}(x)\,(\mathcal{CPT})^{-1} = \eta^B_\Gamma\overline{\Psi}_{Mj}(-x)\Gamma\Psi_{Mi}(-x)\,, \quad (3.126)$$

where the choices for Γ are listed in eq. (3.104). Using $\overline{\Psi}_{Mj}\Gamma\Psi_{Mi} = \eta^C_\Gamma\overline{\Psi}_{Mi}\Gamma\Psi_{Mj}$ [see eqs. (3.98)–(3.100)] and employing eq. (3.48), we can rewrite eq. (3.126) as

$$\mathcal{CPT}\,\overline{\Psi}_{Mi}(x)\Gamma\Psi_{Mj}(x)\,(\mathcal{CPT})^{-1} = \overline{\Psi}_{Mi}(-x)\gamma_5\Gamma\gamma_5\Psi_{Mj}(-x)\,. \quad (3.127)$$

Note carefully the order of the flavor indices i and j in eq. (3.127). Indeed, the CPT properties of the hermitian bilinear covariants, eq. (3.124) with $i = j$ and eq. (3.127), are consistent with the results of eqs. (1.271)–(1.273).

11 Strictly speaking, the bilinear covariants are always assumed to be normal-ordered expressions. As a result, it is legitimate to treat the fermion fields as anticommuting (that is, by ignoring the delta function that appears in the equal-time anticommutation relations).

12 One can also derive eq. (3.123) more directly, starting from eqs. (3.113) and (3.114). After imposing eq. (3.97) and using the anticommutativity of the fermion fields, one again obtains eq. (3.124).

3.4 Lagrangians for Free Four-Component Fermions

The free-field Lagrangian in four-component spinor notation can be obtained from the corresponding two-component fermion Lagrangian by employing the relevant identities for the bilinear covariants given in eqs. (3.84)–(3.92). First, consider a collection of free anticommuting four-component Majorana fields, Ψ_{Mi}, where i is a flavor index. Each four-component Majorana field satisfies the Majorana condition specified by eq. (3.49). The free-field Lagrangian (in terms of mass-eigenstate fields) may be obtained from eq. (1.169) by converting to four-component spinor notation using eqs. (3.87) and (3.89) with $\chi = \eta \equiv \xi$, which yields

$$\mathscr{L} = \tfrac{1}{2}i\overline{\Psi}_{Mi}\gamma^\mu\partial_\mu\Psi_{Mi} - \tfrac{1}{2}m_i\overline{\Psi}_{Mi}\Psi_{Mi}\,, \tag{3.128}$$

where the sum over i is implicit. The corresponding free-field equation for Ψ_{Mi} is the Dirac equation and its conjugate,

$$(i\gamma^\mu\partial_\mu - m)\Psi_{Mi} = 0\,, \qquad \overline{\Psi}_{Mi}(i\gamma^\mu\overleftarrow{\partial}_\mu + m) = 0\,. \tag{3.129}$$

For simplicity, we focus on a theory of a single four-component Majorana fermion field, $\Psi_M(x) = \Psi^C_M(x)$. One can rewrite the free-field Majorana fermion Lagrangian in terms of a Weyl fermion field, $\Psi_L(x) \equiv P_L\Psi(x)$, where $\Psi(x)$ is a four-component fermion field whose lower two components (in the chiral representation) are irrelevant for the present discussion. The Majorana and Weyl fields are related by

$$\Psi_M(x) = \Psi_L(x) + \Psi^C_R(x)\,, \tag{3.130}$$

where $\Psi^C_R(x)$ is defined in eq. (3.35). The corresponding Dirac conjugate field is given by $\overline{\Psi}_M(x) = \overline{\Psi}_L(x) + \overline{\Psi^C_R}(x)$, where

$$\overline{\Psi}_L(x) \equiv [P_L\Psi(x)]^\dagger A = \overline{\Psi}(x)P_R\,, \tag{3.131}$$

$$\overline{\Psi^C_R}(x) \equiv \overline{\Psi^C}(x)P_L = -\Psi^{\mathsf{T}}(x)C^{-1}P_L = -\Psi^{\mathsf{T}}_L(x)C^{-1}\,. \tag{3.132}$$

Using the identity[13]

$$\overline{\Psi^C_R}\gamma^\mu\partial_\mu\Psi^C_R = -\Psi^{\mathsf{T}}C^{-1}P_L\gamma^\mu\partial_\mu P_R C\overline{\Psi}^{\mathsf{T}} = \overline{\Psi}_L\gamma^\mu\partial_\mu\Psi_L + \text{total divergence}\,, \tag{3.133}$$

the Lagrangian for a single Majorana field can be written in terms of a Weyl field:

$$\mathscr{L} = i\overline{\Psi}_L\gamma^\mu\partial_\mu\Psi_L + \tfrac{1}{2}m\left(\Psi^{\mathsf{T}}_L C^{-1}\Psi_L - \overline{\Psi}_L C\overline{\Psi}^{\mathsf{T}}_L\right)\,. \tag{3.134}$$

In deriving eq. (3.134), we have employed the identity

$$(\Psi^{\mathsf{T}}C^{-1}\Psi)^\dagger = -\overline{\Psi}A^{-1}C^{-1\,*}A^{-1\,*}\overline{\Psi}^{\mathsf{T}} = -\overline{\Psi}\,C\,\overline{\Psi}^{\mathsf{T}}\,, \tag{3.135}$$

[13] In deriving eq. (3.133), we have used eq. (3.46) and the anticommutativity of the spinor fields. The total divergence can be dropped from the Lagrangian, as it does not contribute to the field equations.

which follows from the properties of the matrices A and C given in eqs. (3.41) and (3.42). The corresponding free-field equation is

$$i\gamma^\mu \partial_\mu \Psi_L = mC\overline{\Psi}_L^{\mathsf{T}}, \tag{3.136}$$

where we have used $(\overline{\Psi}_L C)^{\mathsf{T}} = -C\overline{\Psi}_L^{\mathsf{T}}$ and the anticommutativity of Ψ_L, $\overline{\Psi}_L$. The generalization of eqs. (3.130)–(3.134) to the case of a multiplet of four-component Majorana fields is straightforward and is left as an exercise for the reader.

Of course, one could have chosen instead to rewrite the four-component Majorana fermion Lagrangian in terms of a single Weyl fermion, $\Psi_R(x) \equiv P_R\Psi(x)$, in which case the upper two components (in the chiral representation) of $\Psi(x)$ are not relevant. In this case, the Majorana and Weyl fields are related by[14]

$$\Psi_M(x) = \Psi_R(x) + \Psi_L^C(x), \tag{3.137}$$

where $\Psi_L^C(x)$ is defined in eq. (3.34). The corresponding Dirac conjugate field is given by $\overline{\Psi}_M(x) = \overline{\Psi}_R(x) + \overline{\Psi_L^C}(x)$, where

$$\overline{\Psi}_R(x) \equiv [P_R\Psi(x)]^\dagger A = \overline{\Psi}(x)P_L, \tag{3.138}$$

$$\overline{\Psi_L^C}(x) \equiv \overline{\Psi^C}(x)P_R = -\Psi^{\mathsf{T}}(x)C^{-1}P_R = -\Psi_R^{\mathsf{T}}(x)C^{-1}. \tag{3.139}$$

The corresponding Weyl fermion Lagrangian is given by eq. (3.134) with L replaced by R.

Thus, a Majorana fermion can be represented either by a four-component self-conjugate field $\Psi_M(x)$ or by a single Weyl field [either $\Psi_L(x)$ or $\Psi_R(x)$]. Both descriptions are unitarily equivalent. That is, one can construct a unitary similarity transformation that connects a Majorana field operator and a Weyl field operator (and vice versa). Of course, this is hardly a surprise in the two-component spinor formalism, where both the Majorana and Weyl forms of the Lagrangian correspond to the same field theory of a single two-component spinor field $\xi_\alpha(x)$.

For $m\neq 0$, the Weyl Lagrangian given by eq. (3.134) possesses no global symmetry, and hence no conserved charge. In contrast, for $m = 0$ the Weyl Lagrangian exhibits a U(1) chiral symmetry. In a theory of massless neutrinos, the U(1) chiral charge of the neutrino is correlated with its lepton number L, and one is free to use either the Majorana or Weyl description. In the former, the neutrino is a neutral self-conjugate fermion, which is not an eigenstate of L. In the latter, $\Psi_L(x)$ corresponds to the left-handed neutrino field and $\Psi_R^C(x)$ corresponds to the right-handed antineutrino field, which are eigenstates of L with opposite sign lepton numbers. No experimental observable can distinguish between these two descriptions.

We next consider a collection of free anticommuting four-component Dirac fields, Ψ_i. The free-field Lagrangian (in terms of mass-eigenstate fields) may be obtained from eq. (1.178) by converting to four-component spinor notation. We then obtain the standard textbook result:

$$\mathscr{L} = i\overline{\Psi}^i\gamma^\mu\partial_\mu\Psi_i - m_i\overline{\Psi}^i\Psi_i. \tag{3.140}$$

[14] If Ψ is an unconstrained four-component spinor, then Ψ_L and Ψ_R are independent Weyl fields, in which case $\Psi_L + \Psi_R^C$ and $\Psi_R + \Psi_L^C$ are independent self-conjugate fields.

By writing $\Psi = \Psi_L + \Psi_R$, we see that the Lagrangian for a single Dirac field can be written in terms of two Weyl fields:

$$\mathscr{L} = i\overline{\Psi}_L\gamma^\mu\partial_\mu\Psi_L + i\overline{\Psi}_R\gamma^\mu\partial_\mu\Psi_R - m\left(\overline{\Psi}_L\Psi_R + \overline{\Psi}_R\Psi_L\right) . \tag{3.141}$$

The corresponding free-field equations are

$$i\gamma^\mu\partial_\mu\Psi_L = m\Psi_R \,, \qquad\qquad i\gamma^\mu\partial_\mu\Psi_R = m\Psi_L \,. \tag{3.142}$$

Summing these two equations yields the Dirac equation, $(i\gamma^\mu\partial_\mu - m)\Psi = 0$. The corresponding field equations are again the free-field Dirac equation [eq. (3.6)] and its conjugate.

In Section 1.5, we learned that a charged Dirac fermion was equivalent to a pair of mass-degenerate Majorana fermions. This identification carries over to four-component notation as follows. Given the four-component Dirac spinor $\Psi(x)$, we construct a pair of four-component Majorana fermions:

$$\Psi_{M1} = \frac{1}{\sqrt{2}}\left(\Psi + \Psi^C\right) , \qquad\qquad \Psi_{M2} = \frac{-i}{\sqrt{2}}\left(\Psi - \Psi^C\right) . \tag{3.143}$$

From the definition of the charge-conjugated spinor [eq. (3.32)], it follows that $(i\Psi)^C = -i\Psi^C$. Thus we see that $\Psi^C_{Mi} = \Psi_{Mi}$ (for $i = 1, 2$) are self-conjugate four-component spinors.[15] One can rewrite the Lagrangian for a single Dirac field [eq. (3.140)] in terms of the Majorana fields Ψ_{Mi}:

$$\mathscr{L} = \tfrac{1}{2}i(\overline{\Psi}_{M1}\gamma^\mu\partial_\mu\Psi_{M1} + \overline{\Psi}_{M2}\gamma^\mu\partial_\mu\Psi_{M2}) - \tfrac{1}{2}m(\overline{\Psi}_{M1}\,\Psi_{M1} + \overline{\Psi}_{M2}\,\Psi_{M2}) . \tag{3.144}$$

In deriving eq. (3.144), we have used eqs. (3.98) and (3.99) and dropped a term that is a total divergence (which does not contribute to the action).

The complex four-component Dirac field, which represents a charged spin-1/2 particle, possesses eight off-shell degrees of freedom. In particular, Ψ and $\overline{\Psi}$ are initially independent degrees of freedom. After imposing the field equations [eq. (3.6)], the four complex components are no longer independent, resulting in four physical on-shell degrees of freedom. As noted above, the complex four-component Dirac field can also be employed to describe a neutral self-conjugate spin-1/2 Majorana fermion. In this case, one must impose an additional constraint: $\Psi^C(x) = \Psi(x)$. This Majorana constraint reduces the number of degrees of freedom of the Dirac fermion by a factor of 2, resulting in four off-shell degrees of freedom before imposing the field equations and two physical on-shell degrees of freedom after imposing the field equations. Of course, this counting of degrees of freedom reproduces the result of the analysis based on two-component fermion fields given in Section 1.5.

Models of N free four-component fermions, with nondiagonal mass matrices, are easily constructed. As a pedagogical example in which both Dirac and Majorana mass terms are present, we perform the diagonalization of the neutrino mass matrix

[15] Moreover, if $\mathcal{C}\Psi\mathcal{C}^{-1} = \eta_C\Psi^C$, with the phase η_C chosen real, then $\mathcal{C}\Psi_{M1}\mathcal{C}^{-1} = \eta_C\Psi_{M1}$ and $\mathcal{C}\Psi_{M2}\mathcal{C}^{-1} = -\eta_C\Psi_{M2}$ [see eq. (1.263) for the corresponding result in two-component notation].

in a one-generation seesaw model[16] using the four-component spinor formalism. We first introduce a four-component anticommuting neutrino field ν_D, and the corresponding Weyl fields,

$$\nu_L \equiv P_L \nu_D\,, \qquad \nu_L^C \equiv P_L \nu_D^C\,, \qquad \nu_R \equiv P_R \nu_D\,, \quad \text{and} \quad \nu_R^C \equiv P_R \nu_D^C\,. \tag{3.145}$$

Note that eqs. (3.34) and (3.132) imply that the anticommuting Weyl fermion fields satisfy

$$\overline{\nu_R^C}\nu_L^C = \overline{\nu_R}\nu_L\,, \qquad\qquad \overline{\nu_L^C}\nu_R^C = \overline{\nu_L}\nu_R\,. \tag{3.146}$$

A Dirac mass term for the neutrinos in the one-generation seesaw model couples ν_L and ν_L^C (and by hermiticity of the Lagrangian, ν_R^C and ν_R), and can be written equivalently as

$$\begin{aligned} m_D(\nu_L^{\mathsf{T}} C^{-1}\nu_L^C + \nu_R^{\mathsf{T}} C^{-1}\nu_R^C) &= -m_D(\overline{\nu_R^C}\nu_L^C + \overline{\nu_L^C}\nu_R^C) \\ &= -m_D(\overline{\nu_R}\nu_L + \overline{\nu_L}\nu_R) = -m_D\overline{\nu}_D\nu_D\,, \end{aligned} \tag{3.147}$$

after making use of eq. (3.146). The Majorana mass term for the neutrinos in the one-generation seesaw model couples ν_L^C to itself (and by hermiticity of the Lagrangian, ν_R to itself), and can be written equivalently as

$$\tfrac{1}{2}M(\nu_L^{C\,\mathsf{T}} C^{-1}\nu_L^C + \nu_R^{\mathsf{T}} C^{-1}\nu_R) = -\tfrac{1}{2}M(\overline{\nu_R}\nu_L^C + \overline{\nu_L^C}\nu_R)\,. \tag{3.148}$$

We shall define the phases of the neutrino fields such that the parameters m_D and M are real and nonnegative.

Thus, the mass terms of the one-generation neutrino seesaw Lagrangian, given in eq. (6.19) in terms of two-component fermion fields, translates in four-component spinor notation to

$$\begin{aligned} \mathscr{L}_{\text{mass}} &= -\tfrac{1}{2}m_D(\overline{\nu_L}\nu_R + \overline{\nu_R}\nu_L + \overline{\nu_L^C}\nu_R^C + \overline{\nu_R^C}\nu_L^C) - \tfrac{1}{2}M(\overline{\nu_R}\nu_L^C + \overline{\nu_L^C}\nu_R) \\ &= -\tfrac{1}{2}\begin{pmatrix}\overline{\nu_R^C} & \overline{\nu_R}\end{pmatrix}\begin{pmatrix}0 & m_D \\ m_D & M\end{pmatrix}\begin{pmatrix}\nu_L \\ \nu_L^C\end{pmatrix} - \tfrac{1}{2}\begin{pmatrix}\overline{\nu_L} & \overline{\nu_L^C}\end{pmatrix}\begin{pmatrix}0 & m_D \\ m_D & M\end{pmatrix}\begin{pmatrix}\nu_R^C \\ \nu_R\end{pmatrix}, \end{aligned} \tag{3.149}$$

where we have used eq. (3.146) to write the first line of eq. (3.149) in a symmetrical fashion. Finally, we employ eqs. (3.34) and (3.132) to obtain

$$\mathscr{L}_{\text{mass}} = \tfrac{1}{2}\begin{pmatrix}\nu_L^{\mathsf{T}} & \nu_L^{C\,\mathsf{T}}\end{pmatrix} C^{-1}\begin{pmatrix}0 & m_D \\ m_D & M\end{pmatrix}\begin{pmatrix}\nu_L \\ \nu_L^C\end{pmatrix} + \text{h.c.} \tag{3.150}$$

Note that if $M = 0$, then one can write $\mathscr{L}_{\text{mass}} = -m_D\overline{\nu}_D\nu_D$ and identify ν_D as a four-component massive Dirac neutrino.

The Takagi diagonalization of the neutrino mass matrix yields two mass eigenstates, which we designate by ν_ℓ and ν_h, where ℓ and h stand for *light* and *heavy*,

[16] In Chapter 6, the seesaw model of neutrino masses is introduced using the two-component spinor formalism.

respectively. The mass-eigenstate Weyl neutrino fields are related to the interaction eigenstate Weyl neutrino fields via

$$\begin{pmatrix} \nu_L \\ \nu_L^C \end{pmatrix} = \mathcal{U} \begin{pmatrix} P_L \nu_\ell \\ P_L \nu_h^C \end{pmatrix}, \tag{3.151}$$

where $\mathcal{U}$ is a 2×2 unitary matrix that is chosen such that

$$\mathcal{U}^{\mathsf{T}} \begin{pmatrix} 0 & m_D \\ m_D & M \end{pmatrix} \mathcal{U} = \begin{pmatrix} m_{\nu_\ell} & 0 \\ 0 & m_{\nu_h} \end{pmatrix}. \tag{3.152}$$

For $M \neq 0$, the neutrino mass eigenstates are *not* Dirac fermions. In the seesaw limit of $M \gg m_D$, the corresponding neutrino masses are $m_{\nu_\ell} \simeq m_D^2/M$ and $m_{\nu_h} \simeq M + m_D^2/M$, with $m_{\nu_\ell} \ll m_{\nu_h}$. In terms of the mass eigenstates, the neutrino mass Lagrangian is

$$\mathscr{L}_{\text{mass}} = \tfrac{1}{2}\left[m_{\nu_\ell} \nu_\ell^{\mathsf{T}} C^{-1} P_L \nu_\ell + m_{\nu_h} \nu_h^{C\,\mathsf{T}} C^{-1} P_L \nu_h^C\right] + \text{h.c.}, \tag{3.153}$$

after using eq. (3.46). We now define four-component self-conjugate Majorana neutrino fields, denoted by Ψ_ℓ and Ψ_h respectively, according to eqs. (3.130) and (3.137),

$$\Psi_\ell \equiv P_L \nu_\ell + P_R C \overline{\nu}_\ell^{\mathsf{T}}, \qquad \overline{\Psi}_\ell \equiv \overline{\nu}_\ell P_R - \nu_\ell^{\mathsf{T}} C^{-1} P_L, \tag{3.154}$$

$$\Psi_h \equiv P_R \nu_h + P_L C \overline{\nu}_h^{\mathsf{T}}, \qquad \overline{\Psi}_h \equiv \overline{\nu}_h P_L - \nu_h^{\mathsf{T}} C^{-1} P_R. \tag{3.155}$$

Then, eq. (3.153) reduces to the expected form:

$$\mathscr{L}_{\text{mass}} = -\tfrac{1}{2}\left[m_{\nu_\ell} \overline{\Psi}_\ell \Psi_\ell + m_{\nu_h} \overline{\Psi}_h \Psi_h\right]. \tag{3.156}$$

A comparison with the analysis of the neutrino mass matrix given in Chapter 6 exhibits the power and the simplicity of the two-component spinor formalism, as compared to the rather awkward four-component spinor analysis presented above.

3.5 Properties of Four-Component Spinor Wave Functions

In four-dimensional Minkowski space, the free four-component Majorana field can be expanded in a Fourier series; each positive [negative] frequency mode is multiplied by a *commuting* spinor wave function $u(\vec{p}, s)$ [$v(\vec{p}, s)$] as in eq. (2.67),[17]

$$\Psi_{Mi}(x) = \sum_s \int \frac{d^3\vec{p}}{(2\pi)^{3/2}(2E_{pi})^{1/2}} \left[u(\vec{p}, s) a_i(\vec{p}, s) e^{-ip\cdot x} + v(\vec{p}, s) a_i^\dagger(\vec{p}, s) e^{ip\cdot x}\right], \tag{3.157}$$

[17] Some subtleties arise in the choice of relative phases of the creation and annihilation operators, which are related to the C, CP, and CPT transformation properties of the Majorana field. For further details, see Ref. [36].

where $E_{\boldsymbol{p}i} \equiv (|\vec{\boldsymbol{p}}|^2 + m_i^2)^{1/2}$, and the creation operators $a_i^\dagger$ and the annihilation operators a_i satisfy anticommutation relations:

$$\{a_i(\vec{\boldsymbol{p}}, s), a_j^\dagger(\vec{\boldsymbol{p}}\,', s')\} = \delta^3(\vec{\boldsymbol{p}} - \vec{\boldsymbol{p}}')\delta_{ss'}\delta_{ij}\,, \tag{3.158}$$

with all other anticommutation relations vanishing. We employ covariant normalization of the one-particle states given by eq. (2.69). It then follows that

$$\langle 0|\, \Psi_M(x)\, |\vec{\boldsymbol{p}}, s\rangle = u(\vec{\boldsymbol{p}}, s)e^{-ip\cdot x}\,, \qquad \langle 0|\, \overline{\Psi}_M(x)\, |\vec{\boldsymbol{p}}, s\rangle = \bar{v}(\vec{\boldsymbol{p}}, s)e^{-ip\cdot x}\,, \tag{3.159}$$

$$\langle \vec{\boldsymbol{p}}, s|\, \overline{\Psi}_M(x)\, |0\rangle = \bar{u}(\vec{\boldsymbol{p}}, s)e^{ip\cdot x}\,, \qquad \langle \vec{\boldsymbol{p}}, s|\, \Psi_M(x)\, |0\rangle = v(\vec{\boldsymbol{p}}, s)e^{ip\cdot x}\,. \tag{3.160}$$

These results are the four-component spinor versions of eqs. (2.8) and (2.9).

Likewise, the free Dirac field can be expanded in a Fourier series,

$$\Psi_i(x) = \sum_s \int \frac{d^3\vec{\boldsymbol{p}}}{(2\pi)^{3/2}(2E_{\boldsymbol{p}i})^{1/2}} \left[u(\vec{\boldsymbol{p}}, s)a_i(\vec{\boldsymbol{p}}, s)e^{-ip\cdot x} + v(\vec{\boldsymbol{p}}, s)b_i^\dagger(\vec{\boldsymbol{p}}, s)e^{ip\cdot x}\right], \tag{3.161}$$

where the creation operators $a_i^\dagger$ and $b_i^\dagger$ and the annihilation operators a_i and b_i satisfy anticommutation relations:

$$\{a_i(\vec{\boldsymbol{p}}, s), a_j^\dagger(\vec{\boldsymbol{p}}\,', s')\} = \delta^3(\vec{\boldsymbol{p}} - \vec{\boldsymbol{p}}')\delta_{ss'}\delta_{ij}\,, \tag{3.162}$$

$$\{b_i(\vec{\boldsymbol{p}}, s), b_j^\dagger(\vec{\boldsymbol{p}}\,', s')\} = \delta^3(\vec{\boldsymbol{p}} - \vec{\boldsymbol{p}}')\delta_{ss'}\delta_{ij}\,, \tag{3.163}$$

with all other anticommutation relations vanishing. We employ covariant normalization of the fermion (F) and antifermion ($\overline{F}$) one-particle states given by eqs. (2.78) and (2.79). It then follows that

$$\langle 0|\, \Psi(x)\, |\vec{\boldsymbol{p}}, s; F\rangle = u(\vec{\boldsymbol{p}}, s)e^{-ip\cdot x}\,, \qquad \langle 0|\, \overline{\Psi}(x)\, |\vec{\boldsymbol{p}}, s; \overline{F}\rangle = \bar{v}(\vec{\boldsymbol{p}}, s)e^{-ip\cdot x}\,, \tag{3.164}$$

$$\langle \vec{\boldsymbol{p}}, s; F|\, \overline{\Psi}(x)\, |0\rangle = \bar{u}(\vec{\boldsymbol{p}}, s)e^{ip\cdot x}\,, \qquad \langle \vec{\boldsymbol{p}}, s; \overline{F}|\, \Psi(x)\, |0\rangle = v(\vec{\boldsymbol{p}}, s)e^{ip\cdot x}\,, \tag{3.165}$$

and the four other single-particle matrix elements vanish. These results are the four-component spinor versions of eqs. (2.80)–(2.83). The Fourier expansion of the charge-conjugated free Dirac field $\Psi_i^C(x) = C\overline{\Psi}_i^{\mathsf{T}}(x)$ is given by

$$\Psi_i^C(x) = \sum_s \int \frac{d^3\vec{\boldsymbol{p}}}{(2\pi)^{3/2}(2E_{\boldsymbol{p}i})^{1/2}} \left[u(\vec{\boldsymbol{p}}, s)b_i(\vec{\boldsymbol{p}}, s)e^{-ip\cdot x} + v(\vec{\boldsymbol{p}}, s)a_i^\dagger(\vec{\boldsymbol{p}}, s)e^{ip\cdot x}\right], \tag{3.166}$$

where we have used eq. (3.169). That is, the charge-conjugation transformation interchanges the annihilation and creation operators, $a_i \leftrightarrow b_i$ and $a_i^\dagger \leftrightarrow b_i^\dagger$. Thus, if $\Psi^C(x) = \Psi(x)$, then we must identify $a = b$ and $a^\dagger = b^\dagger$, corresponding to the free Majorana field given in eq. (3.157).

The two-component spinor momentum space wave functions are related to the traditional four-component spinor wave functions according to

$$u(\vec{p}, s) = \begin{pmatrix} x_\alpha(\vec{p}, s) \\ y^{\dagger\dot{\alpha}}(\vec{p}, s) \end{pmatrix}, \qquad \bar{u}(\vec{p}, s) = \left(y^\alpha(\vec{p}, s),\, x^\dagger_{\dot{\alpha}}(\vec{p}, s)\right), \tag{3.167}$$

$$v(\vec{p}, s) = \begin{pmatrix} y_\alpha(\vec{p}, s) \\ x^{\dagger\dot{\alpha}}(\vec{p}, s) \end{pmatrix}, \qquad \bar{v}(\vec{p}, s) = \left(x^\alpha(\vec{p}, s),\, y^\dagger_{\dot{\alpha}}(\vec{p}, s)\right), \tag{3.168}$$

where the u and v spinors are related by

$$v(\vec{p}, s) = C\bar{u}(\vec{p}, s)^\mathsf{T}, \qquad u(\vec{p}, s) = C\bar{v}(\vec{p}, s)^\mathsf{T}, \tag{3.169}$$

$$\bar{v}(\vec{p}, s) = -u(\vec{p}, s)^\mathsf{T} C^{-1}, \qquad \bar{u}(\vec{p}, s) = -v(\vec{p}, s)^\mathsf{T} C^{-1}. \tag{3.170}$$

The spin quantum number takes on values $s = \pm\frac{1}{2}$, and refers either to the component of the spin as measured in the rest frame with respect to a fixed axis or to the helicity (as discussed in Section 2.2 and Appendix E). Note that the u and v spinors also satisfy

$$v(\vec{p}, s) = -2s\gamma_5 u(\vec{p}, -s), \qquad u(\vec{p}, s) = 2s\gamma_5 v(\vec{p}, -s), \tag{3.171}$$

which follows from eq. (2.28). Explicit forms for the four-component spinor wave functions in the chiral representation can be obtained using eqs. (2.24)–(2.27), where $\chi_s(\hat{s})$ is given in eq. (2.404). For helicity spinors, further simplifications result by employing eqs. (E.79)–(E.82).

One can check that u and v satisfy the Dirac equations[18]

$$(\not{p} - m)\, u(\vec{p}, s) = (\not{p} + m)\, v(\vec{p}, s) = 0, \tag{3.172}$$

$$\bar{u}(\vec{p}, s)\, (\not{p} - m) = \bar{v}(\vec{p}, s)\, (\not{p} + m) = 0, \tag{3.173}$$

corresponding to eqs. (2.10)–(2.13), and

$$(2s\gamma_5 \not{S} - 1)\, u(\vec{p}, s) = (2s\gamma_5 \not{S} - 1)\, v(\vec{p}, s) = 0, \tag{3.174}$$

$$\bar{u}(\vec{p}, s)\, (2s\gamma_5 \not{S} - 1) = \bar{v}(\vec{p}, s)\, (2s\gamma_5 \not{S} - 1) = 0, \tag{3.175}$$

corresponding to eqs. (2.29)–(2.32), where the spin vector S^μ is defined in eq. (2.18). For massive fermions, eqs. (2.51)–(2.54) correspond to the spin projection operators

$$u(\vec{p}, s)\bar{u}(\vec{p}, s) = \tfrac{1}{2}(1 + 2s\gamma_5 \not{S})\, (\not{p} + m), \tag{3.176}$$

$$v(\vec{p}, s)\bar{v}(\vec{p}, s) = \tfrac{1}{2}(1 + 2s\gamma_5 \not{S})\, (\not{p} - m). \tag{3.177}$$

To apply the above formulae to the massless case we must employ helicity states, where s is replaced by the helicity quantum number λ, and S^μ is defined by eq. (2.19). In particular, in the $m \to 0$ limit, $S^\mu = p^\mu/m + \mathcal{O}(m/E)$. Inserting this result in eqs. (3.174) and (3.175) and using the Dirac equations, it follows that the massless helicity spinors are eigenstates of γ_5:

$$\gamma_5 u(\vec{p}, \lambda) = 2\lambda u(\vec{p}, \lambda), \qquad \gamma_5 v(\vec{p}, \lambda) = -2\lambda v(\vec{p}, \lambda). \tag{3.178}$$

[18] We use the standard Feynman slash notation: $\not{p} \equiv \gamma_\mu p^\mu$ and $\not{S} \equiv \gamma_\mu S^\mu$.

Combining these results with eq. (3.171) [with s replaced by λ] yields

$$v(p,\lambda) = -2\lambda\gamma_5 u(p,-\lambda) = u(p,-\lambda)\,, \qquad \lambda = \pm\tfrac{1}{2}\,, \tag{3.179}$$

and we see that the massless u and v spinors of opposite helicity are the same.

Applying the above $m \to 0$ limiting procedure to eqs. (3.176) and (3.177) and using the mass-shell condition ($\not p \not p = p^2 = m^2$), one obtains the massless helicity projection operators corresponding to eqs. (2.59)–(2.62):

$$u(\vec{\boldsymbol p},\lambda)\bar u(\vec{\boldsymbol p},\lambda) = \tfrac{1}{2}(1+2\lambda\gamma_5)\not p\,, \tag{3.180}$$

$$v(\vec{\boldsymbol p},\lambda)\bar v(\vec{\boldsymbol p},\lambda) = \tfrac{1}{2}(1-2\lambda\gamma_5)\not p\,. \tag{3.181}$$

Summing over the spin degree of freedom, we obtain the spin-sum identities corresponding to eqs. (2.63)–(2.66):

$$\sum_s u(\vec{\boldsymbol p},s)\bar u(\vec{\boldsymbol p},s) = \not p + m\,, \qquad \sum_s v(\vec{\boldsymbol p},s)\bar v(\vec{\boldsymbol p},s) = \not p - m\,, \tag{3.182}$$

$$\sum_s u(\vec{\boldsymbol p},s)v^{\mathsf T}(\vec{\boldsymbol p},s) = (\not p + m)C^{\mathsf T}\,, \qquad \sum_s \bar u^{\mathsf T}(\vec{\boldsymbol p},s)\bar v(\vec{\boldsymbol p},s) = C^{-1}(\not p - m)\,, \tag{3.183}$$

$$\sum_s \bar v^{\mathsf T}(\vec{\boldsymbol p},s)\bar u(\vec{\boldsymbol p},s) = C^{-1}(\not p + m)\,, \qquad \sum_s v(\vec{\boldsymbol p},s)u^{\mathsf T}(\vec{\boldsymbol p},s) = (\not p - m)C^{\mathsf T}\,, \tag{3.184}$$

which are valid for both the massive case and the massless $m \to 0$ limit.

As previously noted, the results for the bilinear covariants obtained in eqs. (3.84)–(3.92) can also be applied to expressions involving the commuting spinor wave functions. Various relations among the possible bilinear covariants can be established by using eqs. (3.169) and (3.170). As an example,

$$\bar u(\vec{\boldsymbol p}_1,s_1)\Gamma v(\vec{\boldsymbol p}_2,s_2) = -v(\vec{\boldsymbol p}_1,s_1)^{\mathsf T}C^{-1}\Gamma C\bar u(\vec{\boldsymbol p}_2,s_2)^{\mathsf T} = -\eta^C_\Gamma \bar u(\vec{\boldsymbol p}_2,s_2)\Gamma v(\vec{\boldsymbol p}_1,s_1)\,, \tag{3.185}$$

$$\bar u(\vec{\boldsymbol p}_1,s_1)\Gamma u(\vec{\boldsymbol p}_2,s_2) = -v(\vec{\boldsymbol p}_1,s_1)^{\mathsf T}C^{-1}\Gamma C\bar v(\vec{\boldsymbol p}_2,s_2)^{\mathsf T} = -\eta^C_\Gamma \bar v(\vec{\boldsymbol p}_2,s_2)\Gamma v(\vec{\boldsymbol p}_1,s_1)\,, \tag{3.186}$$

for $\Gamma = \mathbb{1}\,, \gamma_5\,, \gamma^\mu\,, \gamma^\mu\gamma_5\,, \Sigma^{\mu\nu}\,, \Sigma^{\mu\nu}\gamma_5$, where the sign η^C_Γ [defined in eq. (3.46)] arises after taking the transpose and applying eq. (3.46). In particular, the (commuting) u and v spinors satisfy the following relations:

$$\bar u(\vec{\boldsymbol p}_1,s_1)P_L v(\vec{\boldsymbol p}_2,s_2) = -\bar u(\vec{\boldsymbol p}_2,s_2)P_L v(\vec{\boldsymbol p}_1,s_1)\,, \tag{3.187}$$

$$\bar u(\vec{\boldsymbol p}_1,s_1)P_R v(\vec{\boldsymbol p}_2,s_2) = -\bar u(\vec{\boldsymbol p}_2,s_2)P_R v(\vec{\boldsymbol p}_1,s_1)\,, \tag{3.188}$$

$$\bar u(\vec{\boldsymbol p}_1,s_1)\gamma^\mu P_L v(\vec{\boldsymbol p}_2,s_2) = \bar u(\vec{\boldsymbol p}_2,s_2)\gamma^\mu P_R v(\vec{\boldsymbol p}_1,s_1)\,, \tag{3.189}$$

$$\bar u(\vec{\boldsymbol p}_1,s_1)\gamma^\mu P_R v(\vec{\boldsymbol p}_2,s_2) = \bar u(\vec{\boldsymbol p}_2,s_2)\gamma^\mu P_L v(\vec{\boldsymbol p}_1,s_1)\,, \tag{3.190}$$

and four similar relations obtained by interchanging $v(\vec{\boldsymbol p}_2,s_2) \leftrightarrow u(\vec{\boldsymbol p}_2,s_2)$.

3.6 Feynman Rules for Four-Component Fermions

In this section, we shall establish a set of Feynman rules for four-component fermions that treat both Dirac and Majorana fermions on the same footing. These rules generalize the standard Feynman rules for four-component Dirac fermions found in most quantum field theory textbooks. Two advantages of the rules presented here are: (i) no factors of the charge-conjugation matrix C are required for fermion interaction vertices and propagators, and (ii) the *relative* sign between different diagrams corresponding to the same physical process is simply determined. Our rules have been obtained by translating our two-component fermion Feynman rules into the four-component spinor language. The resulting Feynman rules for four-component Majorana fermions are equivalent to the set of rules obtained in Ref. [37].

Consider first the Feynman rule for the four-component fermion propagator. Virtual Dirac fermion lines can either correspond to Ψ or Ψ^C. Here, there is no ambiguity in the propagator Feynman rule, since for free Dirac fermion fields,

$$\langle 0|\, T[\Psi_a(x)\overline{\Psi}{}^b(y)]\, |0\rangle = \langle 0|\, T[\Psi^C_a(x)\overline{\Psi^C}{}^b(y)]\, |0\rangle\ , \tag{3.191}$$

so that the Feynman rules for the Ψ and Ψ^C propagators exhibited in Fig. 3.1 are identical. The same rule also applies to a four-component Majorana fermion.[19]

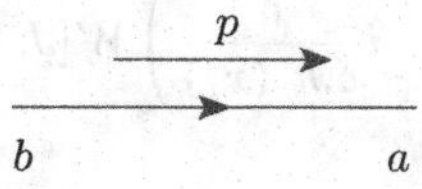

$$\frac{i(\not{p}+m)_a{}^b}{p^2-m^2+i\epsilon}$$

Fig. 3.1 Feynman rule for the propagator of a four-component fermion with mass m. The same rule applies to a Majorana, Dirac, and charge-conjugated Dirac fermion. The four-component spinor labels a and b are specified.

Using eq. (3.7), the four-component fermion propagator Feynman rule can be expressed as a partitioned matrix of 2×2 blocks,

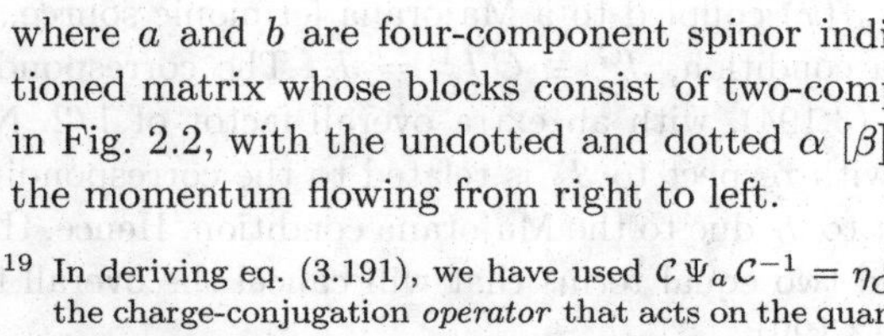

$$= \frac{i}{p^2-m^2+i\epsilon}\begin{pmatrix} m\,\delta_\alpha{}^\beta & p\cdot\sigma_{\alpha\dot{\beta}} \\ p\cdot\overline{\sigma}^{\dot{\alpha}\beta} & m\,\delta^{\dot{\alpha}}{}_{\dot{\beta}}\end{pmatrix} \tag{3.192}$$

where a and b are four-component spinor indices. That is eq. (3.192) is a partitioned matrix whose blocks consist of two-component fermion propagators defined in Fig. 2.2, with the undotted and dotted α [β] indices on the left [right] and with the momentum flowing from right to left.

19 In deriving eq. (3.191), we have used $\mathcal{C}\Psi_a\mathcal{C}^{-1} = \eta_C\Psi^C_a$ and $\mathcal{C}\overline{\Psi}{}^a\mathcal{C}^{-1} = \eta^*_C\overline{\Psi^C}{}^a$, where $\mathcal{C}$ is the charge-conjugation *operator* that acts on the quantum Hilbert space and η_C is a convention-dependent phase factor [B133, B136]. Note that $\mathcal{C}$ is a unitary operator and $\mathcal{C}\,|0\rangle = |0\rangle$ in the free-field vacuum.

The derivation of the four-component Dirac fermion propagator is treated in most modern textbooks of quantum field theory. Consider a single massive Dirac fermion $\Psi(x)$ coupled to an anticommuting four-component Dirac fermionic source term

$$J_\psi(x) \equiv \begin{pmatrix} J_{\eta\alpha}(x) \\ J_\chi^{\dagger\dot{\alpha}}(x) \end{pmatrix} . \tag{3.193}$$

The corresponding action [eq. (2.86)] in four-component notation is given by

$$S = \int d^4x \, (\mathscr{L} + \overline{J}_\psi \Psi + \overline{\Psi} J_\psi) = \int d^4x \left[\overline{\Psi}(i\partial\!\!\!/ - m)\Psi + \overline{J}_\psi \Psi + \overline{\Psi}\, J_\psi \right] . \tag{3.194}$$

Introducing the momentum space Fourier coefficients:

$$\Psi(x) = \int \frac{d^4p}{(2\pi)^4} e^{-ip\cdot x} \widehat{\Psi}(p) , \qquad J_\psi(x) = \int \frac{d^4p}{(2\pi)^4} e^{-ip\cdot x} \widehat{J}_\psi(p) , \tag{3.195}$$

we can identify the following four-component quantities with matrices of two-component quantities given in eqs. (2.91) and (2.115):

$$\widehat{\Psi}(p) = A^{-1}\Omega_c(p) , \qquad \widehat{J}_\psi(p) = X_c(p) , \qquad p\!\!\!/ - m = \mathcal{M}(p)A , \tag{3.196}$$

where A is the Dirac conjugation matrix defined in eq. (3.29). Using the results of Section 2.5, one easily derives:

$$\langle 0|T(\Psi(x_1)\overline{\Psi}(x_2))|0\rangle = \left(-i \frac{\overrightarrow{\delta}}{\delta \overline{J}_\psi(x_1)} \right) W[J, \overline{J}] \left(-i \frac{\overleftarrow{\delta}}{\delta J_\psi(x_2)} \right) \Bigg|_{J_\psi = \overline{J}_\psi = 0} \tag{3.197}$$

where

$$\begin{aligned} W[J_\psi \, , \overline{J}_\psi] &= N \int \mathcal{D}\Psi \, \mathcal{D}\overline{\Psi} \, e^{iS[\Psi, \overline{\Psi}, J_\psi, \overline{J}_\psi]} \\ &= \exp\left\{ -i \int \frac{d^4p}{(2\pi)^4} \widehat{\overline{J}}_\psi(p) \frac{p\!\!\!/ + m}{p^2 - m^2 + i\varepsilon} \widehat{J}_\psi(p) \right\} , \end{aligned} \tag{3.198}$$

and the $m^2 \to m^2 - i\varepsilon$ prescription has been introduced to ensure the convergence of $W[J_\psi \, , \overline{J}_\psi]$ with S given by eq. (3.194). Using the analogues of eqs. (2.105) and (2.106), we end up with the expected result

$$\langle 0|T(\Psi(x_1)\overline{\Psi}(x_2))|0\rangle = \int \frac{d^4p}{(2\pi)^4} \, e^{-ip\cdot(x_1 - x_2)} \, \frac{i(p\!\!\!/ + m)}{p^2 - m^2 + i\varepsilon} . \tag{3.199}$$

The analogous computation can be carried out for a single four-component Majorana fermion field $\Psi_M(x)$ coupled to a Majorana fermionic source, $J_\xi(x)$, which satisfies the Majorana condition, $J_\xi^C \equiv C\overline{J}_\xi^{\mathsf{T}} = J_\xi$. The corresponding action is similar to that of eq. (3.194), with an extra overall factor of 1/2. Note that the functional derivative with respect to $\overline{J}_\xi$ is related to the corresponding functional derivative with respect to J_ξ due to the Majorana condition. Hence, the calculation of eq. (3.198) will yield two equal terms that will cancel the overall factor of 1/2, resulting again in eq. (3.199).

We next examine the various interactions involving four-component fermions.

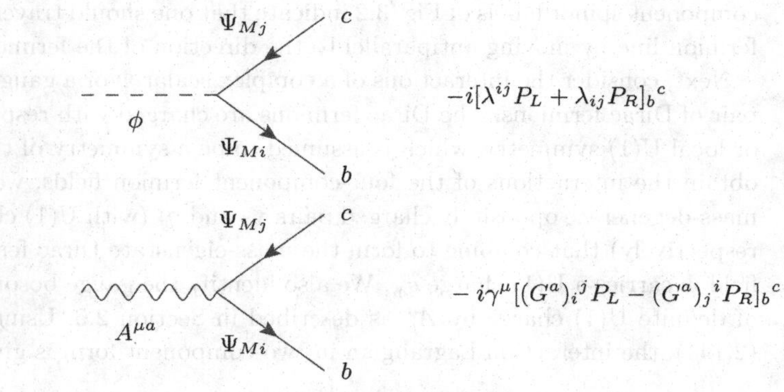

Fig. 3.2 Feynman rules for neutral scalar and gauge boson interactions with a pair of four-component Majorana fermions (labeled by four-component spinor indices b and c). The G^a are defined in eq. (2.142). The index a runs over the neutral (mass-eigenstate) gauge bosons.

First, we consider the interactions of a neutral scalar ϕ or a gauge boson A^a_μ with a pair of Majorana fermions. To obtain the interactions of the four-component fermion fields, we first identify the neutral two-component fermion mass-eigenstate neutral fields ξ_i. Using eqs. (2.135) and (2.141), the interaction Lagrangian in two-component form is given by

$$\mathscr{L}_{\text{int}} = -\tfrac{1}{2}(\lambda^{ij}\xi_i\xi_j + \lambda_{ij}\xi^{\dagger i}\xi^{\dagger j})\phi - (G^a)_i{}^j\,\xi^{\dagger i}\bar{\sigma}^\mu\xi_j A^a_\mu\,, \tag{3.200}$$

where λ is a complex symmetric matrix with $\lambda^{ij} \equiv \lambda^*_{ij}$ [see eq. (1.160)], the A^a_μ are the mass-eigenstate gauge fields, and the corresponding hermitian matrices G^a are defined in eq. (2.142). Converting this result into four-component notation yields

$$\mathscr{L}_{\text{int}} = -\tfrac{1}{2}(\lambda^{ij}\overline{\Psi}_{Mi}P_L\Psi_{Mj} + \lambda_{ij}\overline{\Psi}{}^i_M P_R\Psi^j_M)\phi - (G^a)_i{}^j\overline{\Psi}{}^i_M\gamma^\mu P_L\Psi_{Mj}A^a_\mu\,, \tag{3.201}$$

where the Ψ_{Mj} are a set of (neutral) Majorana four-component fermions. We can use eq. (3.101) to rewrite the term proportional to $(G^a)_i{}^j$ in eq. (3.201) as follows:

$$\begin{aligned}(G^a)_i{}^j\overline{\Psi}{}^i_M\gamma^\mu P_L\Psi_{Mj} &= \tfrac{1}{2}(G^a)_i{}^j\left[\overline{\Psi}{}^i_M\gamma^\mu P_L\Psi_{Mj} - \overline{\Psi}_{Mj}\gamma^\mu P_R\Psi^i_M\right]\\ &= \tfrac{1}{2}\overline{\Psi}_{Mi}\gamma^\mu\left[(G^a)_i{}^j P_L - (G^a)_j{}^i P_R\right]\Psi_{Mj}\,.\end{aligned} \tag{3.202}$$

In the last step above, the flavor indices of the four-component Majorana fermion fields have been lowered, as the heights of these indices can be arbitrarily chosen.

The corresponding four-component fermion Feynman rules are shown in Fig. 3.2. The Majorana fermion is neutral under all conserved charges (and thus equal to its own antiparticle). Hence, an arrow on a Majorana fermion line simply reflects the structure of the interaction Lagrangian; i.e., $\overline{\Psi}_M$ [Ψ_M] is represented by an arrow pointing out of [into] the vertex. These arrows are then used for determining the placement of the u and v spinors in an invariant amplitude. In particular, the four-component spinor labels of Fig. 3.2 indicate that one should traverse any continuous fermion line by moving antiparallel to the direction of the fermion arrows.

Next, consider the interactions of a complex scalar Φ or a gauge boson A^a_μ with a pair of Dirac fermions. The Dirac fermions are charged with respect to some global or local U(1) symmetry, which is assumed to be a symmetry of the Lagrangian. To obtain the interactions of the four-component fermion fields, we first identify the mass-degenerate oppositely charged pairs χ_j and η_j (with U(1) charges q_j and $-q_j$, respectively) that combine to form the mass-eigenstate Dirac fermions. The scalar field Φ carries a U(1) charge q_Φ. We also identify the gauge boson mass eigenstates of definite U(1) charge by A^a_μ as described in Section 2.6. Using eqs. (2.135) and (2.144), the interaction Lagrangian in two-component form is given by

$$\mathscr{L}_{\rm int} = -\kappa^i{}_j \chi_i \eta^j \Phi - \kappa_i{}^j \chi^{\dagger i} \eta^\dagger_j \Phi^\dagger - \left[(G^a_L)_i{}^j \chi^{\dagger i} \overline{\sigma}^\mu \chi_j - (G^a_R)_j{}^i \eta^\dagger_i \overline{\sigma}^\mu \eta^j \right] A^a_\mu \,, \quad (3.203)$$

where $\kappa_i{}^j \equiv (\kappa^i{}_j)^*$ [see eq. (1.171)] and κ is an arbitrary complex matrix coupling, subject to the conditions that $\kappa^i{}_j = 0$ unless $q_\Phi = q_j - q_i$. For the gauge boson couplings, we follow the notation of eq. (2.145). In particular, $A^a_\mu G^a_L$ and $A^a_\mu G^a_R$ are hermitian matrix-valued gauge fields, which when summed over a can contain both neutral and charged [with respect to U(1)] mass-eigenstate gauge boson fields. Converting to four-component notation yields

$$\begin{aligned} \mathscr{L}_{\rm int} = &-\kappa^i{}_j \overline{\Psi}{}^j P_L \Psi_i \Phi - \kappa_i{}^j \overline{\Psi}{}^i P_R \Psi_j \Phi^\dagger \\ &- \left[(G^a_L)_i{}^j \overline{\Psi}{}^i \gamma^\mu P_L \Psi_j + (G^a_R)_i{}^j \overline{\Psi}{}^i \gamma^\mu P_R \Psi_j \right] A^a_\mu \,, \end{aligned} \quad (3.204)$$

where the Ψ_j are a set of Dirac four-component fermions. If Φ is a real scalar field, then we shall write $\phi \equiv \Phi = \Phi^\dagger$. The corresponding four-component fermion Feynman rules are exhibited in Fig. 3.3. The rules involving the charge-conjugated Dirac fields have been obtained by using eq. (3.93). Note that the arrows on the charged scalar and Dirac fermion lines depict the flow of the conserved U(1) charge.

Finally, the interaction of a charged scalar boson Φ (with U(1) charge q_Φ) or a charged vector boson W (with U(1) charge q_W) with a fermion pair consisting of one Majorana and one Dirac fermion is examined. We denote the neutral fermion mass-eigenstate fields by ξ_i and pairs of oppositely charged fermion mass-eigenstate fields by χ_j and η^j (with U(1) charges q_j and $-q_j$, respectively). Using eqs. (2.135) and (2.146), the interaction Lagrangian is given by

$$\begin{aligned} \mathscr{L}_{\rm int} = &-[(\kappa_1)^i{}_j \xi_i \eta^j + (\kappa_2)_{ij} \xi^{\dagger i} \chi^{\dagger j}]\Phi - [(\kappa_2)^{ij} \xi_i \chi_j + (\kappa_1)_i{}^j \xi^{\dagger i}_i \eta^\dagger_j]\Phi^\dagger \\ &-[(G_1)_j{}^i \chi^{\dagger j} \overline{\sigma}^\mu \xi_i - (G_2)_{ij} \xi^{\dagger i} \overline{\sigma}^\mu \eta^j] W_\mu - [(G_1)^j{}_i \xi^{\dagger i} \overline{\sigma}^\mu \chi_j - (G_2)^{ij} \eta^\dagger_j \overline{\sigma}^\mu \xi_i] W^\dagger_\mu \,, \end{aligned} \quad (3.205)$$

where κ_1, κ_2, G_1, and G_2 are arbitrary complex coupling matrices, subject to the conditions that $(\kappa_1)^i{}_j = (\kappa_2)_{ij} = 0$ unless $q_\Phi = q_j$, and $(G_1)_j{}^i = (G_2)_{ij} = 0$ unless $q_W = q_j$. Converting to four-component spinor notation yields

$$\begin{aligned} \mathscr{L}_{\rm int} = &- \left[(\kappa_1)^i{}_j \overline{\Psi}{}^j P_L \Psi_{Mi} + (\kappa_2)_{ij} \overline{\Psi}{}^j P_R \Psi^i_M \right] \Phi \\ &- \left[(G_1)_j{}^i \overline{\Psi}{}^j \gamma^\mu P_L \Psi_{Mi} + (G_2)_{ij} \overline{\Psi}{}^j \gamma^\mu P_R \Psi^i_M \right] W_\mu + \text{h.c.} \end{aligned} \quad (3.206)$$

The corresponding four-component fermion Feynman rules are exhibited in Fig. 3.4.

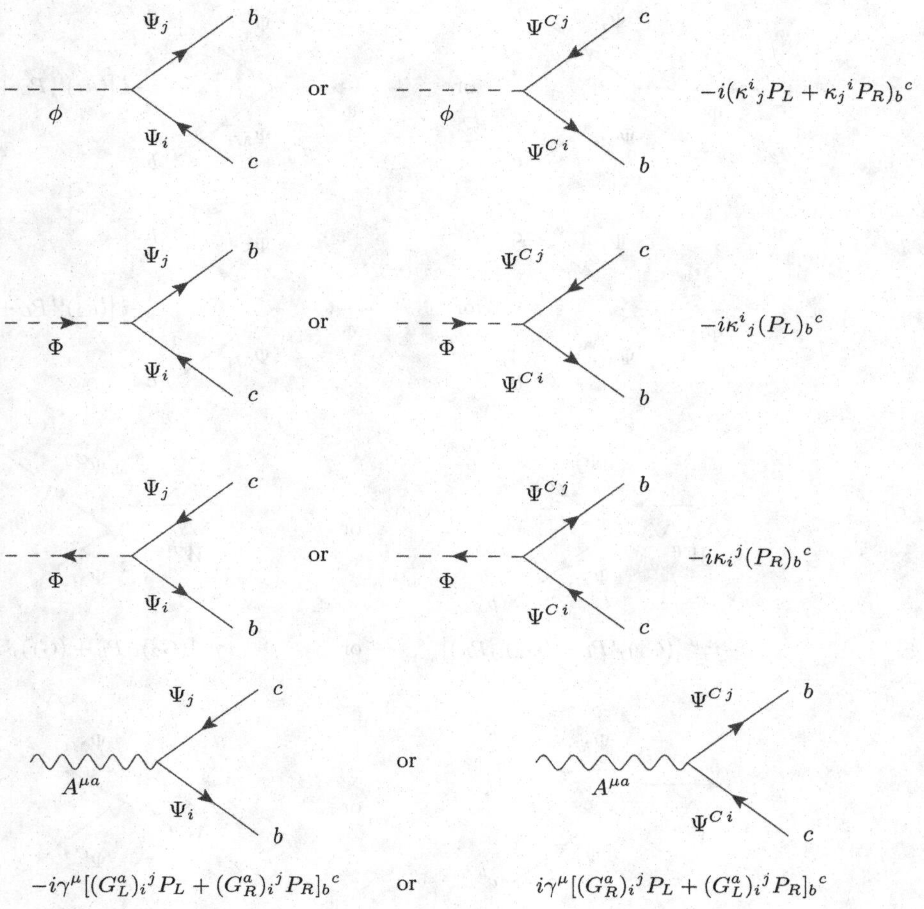

Fig. 3.3 Feynman rules for neutral scalar (ϕ), charged scalar (Φ), and gauge boson ($A^{\mu a}$) interactions with a pair of four-component Dirac fermions (labeled by four-component spinor indices b and c). In each case, one has two choices for the corresponding Feynman rule: one involving Ψ and one involving the oppositely charged Ψ^C (with the arrows of the corresponding Ψ and Ψ^C lines pointing in opposite directions). The arrows indicate the direction of flow of the U(1) charges of the Dirac fermion and charged scalar fields. The index a runs over both neutral and charged (mass-eigenstate) gauge bosons, consistent with charge conservation at the vertex.

There is an equivalent form for the interactions given by eqs. (3.203) and (3.206) where $\mathscr{L}_{\text{int}}$ is written in terms of charge-conjugated Dirac fields [after making use of eq. (3.93)]. The Feynman rules involving Dirac fermions can take two possible forms, as shown in Figs. 3.3 and 3.4.

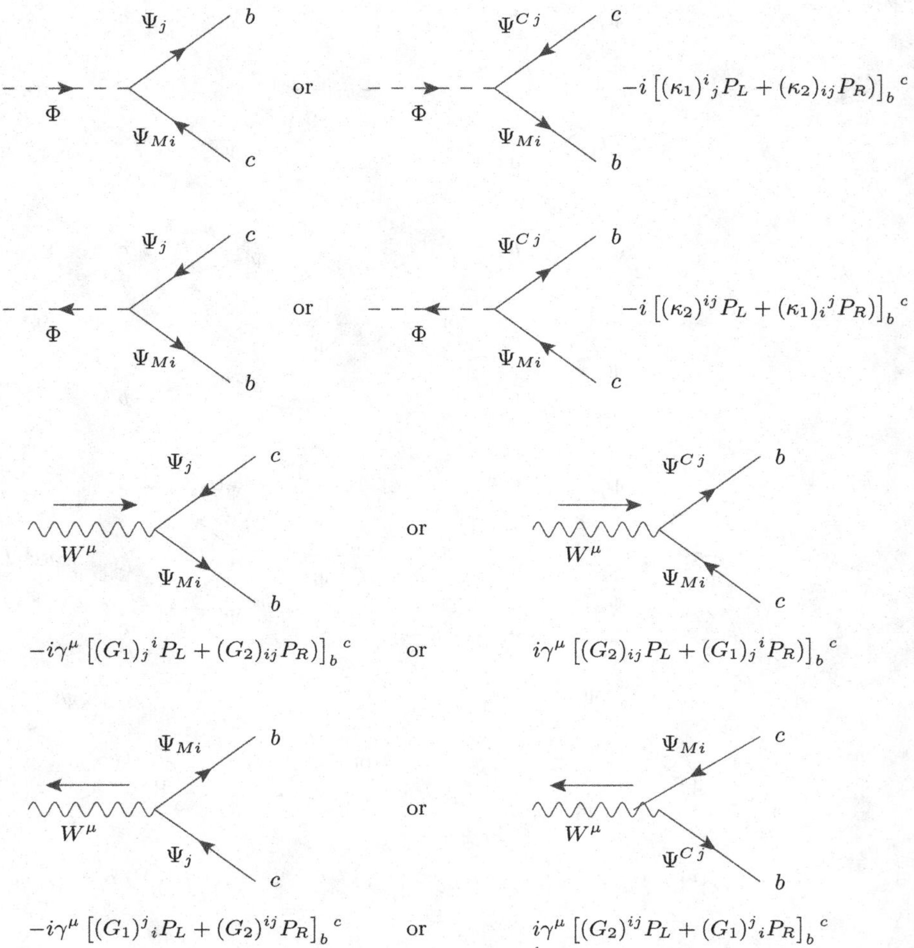

Fig. 3.4 Feynman rules for charged scalar and vector boson interactions with a fermion pair consisting of one Majorana and one Dirac four-component fermion (labeled by four-component spinor indices b and c). In each case, one has two choices for the corresponding Feynman rule: one involving Ψ and one involving the oppositely charged Ψ^C (with the arrows of the Ψ and Ψ^C lines pointing in opposite directions). The arrows of the Dirac fermion and charged bosons indicate the direction of flow of the corresponding U(1) charges. That is, the charge of the boson (either Φ or W above) must coincide with the charge of Ψ_j. The arrows of the Majorana fermions satisfy the requirement that the fermion line arrow directions flow continuously through the vertex.

3.7 Feynman Rules for External Fermion Lines

The direction of an arrow on a Dirac fermion line indicates the direction of the fermion charge flow (whereas the arrow on the Majorana fermion line is unconnected to charge flow). However, we are free to choose either a Ψ or Ψ^C line to represent a Dirac fermion at any place in a given Feynman graph.[20] For any decay or scattering process, a suitable choice of either the Ψ-rule or the Ψ^C-rule at each vertex (the choice can be different at different vertices), will guarantee that the arrow directions on fermion lines flow continuously through the Feynman diagram. Then, to evaluate an invariant amplitude, one should traverse *any* continuous fermion line (either Ψ or Ψ^C) by moving antiparallel to the direction of the fermion arrows, as indicated by the order of the four-component spinor labels in the Feynman rules of Figs. 3.3 and 3.4. Examples will be provided in Section 3.8.

For a given process, there may be a number of distinct choices for the arrow directions on the Majorana fermion lines, which may depend on whether one represents a given Dirac fermion by Ψ or Ψ^C. However, different choices do *not* lead to independent Feynman diagrams.[21] When computing an invariant amplitude, one first writes down the relevant Feynman diagrams with no arrows on any Majorana fermion line. The number of distinct graphs contributing to the process is then determined. Finally, one makes some choice for how to distribute the arrows on the Majorana fermion lines and how to label Dirac fermion lines (either as the field Ψ or its charge conjugate Ψ^C) in a manner consistent with the rules of Figs. 3.2 and 3.4. The end result for the invariant amplitude (apart from an overall unobservable phase) does not depend on the choices made for the direction of the fermion arrows.

Using the above procedure, the Feynman rules for the external fermion wave functions are the same for Dirac and Majorana fermions:

- $u(\vec{p}, s)$: incoming Ψ [or Ψ^C] with momentum $\vec{p}$ parallel to the arrow direction,
- $\bar{u}(\vec{p}, s)$: outgoing Ψ [or Ψ^C] with momentum $\vec{p}$ parallel to the arrow direction,
- $v(\vec{p}, s)$: outgoing Ψ [or Ψ^C] with momentum $\vec{p}$ antiparallel to the arrow direction,
- $\bar{v}(\vec{p}, s)$: incoming Ψ [or Ψ^C] with momentum $\vec{p}$ antiparallel to the arrow direction.

The proof that the above rules for external wave functions apply unambiguously to Majorana fermions is straightforward. Simply insert the plane-wave expansion of the Majorana field given by eq. (3.157) into eq. (3.201), and evaluate matrix elements for, e.g., the decay of a scalar or vector particle into a pair of Majorana fermions.

[20] Since the charge of Ψ^C is opposite in sign to the charge of Ψ, the corresponding arrow directions of the Ψ and Ψ^C lines must point in opposite directions.

[21] In contrast, the two-component Feynman rules developed in Section 2.4 require that two vertices differing by the direction of the arrows on the two-component fermion lines must both be included in the calculation of the matrix element.

3.8 Examples of Feynman Diagrams and Amplitudes

We now reconsider the matrix elements for scalar and vector particle decays into fermion pairs and $2 \to 2$ elastic scattering of a fermion off a scalar and vector boson, respectively. We shall compute the matrix elements using the Feynman rules for the interaction vertices given in Section 3.6, and check that the results agree with the ones obtained by two-component spinor methods developed in Chapter 2.

3.8.1 Tree-Level Decay Processes

Consider first the decay of a neutral scalar boson ϕ into a pair of Majorana fermions, $\phi \to \Psi_{Mi}(\vec{\boldsymbol{p}}_1, s_1)\Psi_{Mj}(\vec{\boldsymbol{p}}_2, s_2)$, of flavor i and j, respectively. Here, $\Psi_{Mi}(\vec{\boldsymbol{p}}, s)$ denotes the one-particle state given by eq. (2.69). The decay matrix element is given by

$$i\mathcal{M} = -i\bar{u}^i(\vec{\boldsymbol{p}}_1, s_1)(\lambda^{ij}P_L + \lambda_{ij}P_R)v_j(\vec{\boldsymbol{p}}_2, s_2)\,, \tag{3.207}$$

for fixed flavor indices i and j (no sum over the repeated flavor indices). The heights of the flavor indices are awkward when four-component Majorana fermions are involved. In particular, the height of the flavor index of the four-component spinor wave functions of a Majorana fermion matches only one of the two flavor indices of the two-component spinor wave functions that comprise the u and v spinors, as follows:

$$u_i(\vec{\boldsymbol{p}}, s) = \begin{pmatrix} x_{\alpha i}(\vec{\boldsymbol{p}}, s) \\ y^{\dagger\dot{\alpha}i}(\vec{\boldsymbol{p}}, s) \end{pmatrix}, \qquad \bar{u}^i(\vec{\boldsymbol{p}}, s) = (y_i^{\alpha}(\vec{\boldsymbol{p}}, s),\; x^{\dagger i}_{\dot{\alpha}}(\vec{\boldsymbol{p}}, s))\,, \tag{3.208}$$

$$v_j(\vec{\boldsymbol{p}}, s) = \begin{pmatrix} y_{\alpha j}(\vec{\boldsymbol{p}}, s) \\ x^{\dagger\dot{\alpha}j}(\vec{\boldsymbol{p}}, s) \end{pmatrix}, \qquad \bar{v}^j(\vec{\boldsymbol{p}}, s) = (x_j^{\alpha}(\vec{\boldsymbol{p}}, s),\; y^{\dagger j}_{\dot{\alpha}}(\vec{\boldsymbol{p}}, s))\,. \tag{3.209}$$

One can check that the result obtained in eq. (3.207) matches with that of eq. (2.150), which was derived using two-component spinor techniques. Note that if one had chosen to switch the two final states (equivalent to switching the directions of the Majorana fermion arrows), then the resulting matrix element would simply exhibit an overall sign change [due to the results of eqs. (3.187) and (3.188)]. This overall sign change is a consequence of the Fermi–Dirac statistics, and corresponds to changing which order one uses to construct the two-particle final state.

Consider next the decay of a (neutral or charged) scalar boson Φ into a pair of Dirac fermions, $\Phi \to F_i(\vec{\boldsymbol{p}}_1, s_1)\overline{F}^j(\vec{\boldsymbol{p}}_2, s_2)$, where by $F(\vec{\boldsymbol{p}}, s)$ and $\overline{F}(\vec{\boldsymbol{p}}, s)$ we mean the one-particle states given by eqs. (2.78) and (2.79). The matrix element for the decay is given by

$$i\mathcal{M} = -i\bar{u}^i(\vec{\boldsymbol{p}}_1, s_1)(\kappa^j{}_iP_L + \kappa_i{}^jP_R)v_j(\vec{\boldsymbol{p}}_2, s_2)\,, \tag{3.210}$$

for fixed flavor indices i and j (no sum over the repeated flavor indices). The heights of the flavor indices are more natural when four-component Dirac fermions are involved, since they are correlated with the U(1) charge. In particular, the heights of the flavor index of the four-component spinor wave functions of a Dirac fermion match the corresponding heights of the flavor index of the two-component fermion wave functions that comprise the u and v spinors. That is, we must replace eqs. (3.208) and (3.209) with

$$u_i(\vec{p},s) = \begin{pmatrix} x_{\alpha i}(\vec{p},s) \\ y_i^{\dagger\dot{\alpha}}(\vec{p},s) \end{pmatrix}, \qquad \bar{u}^i(\vec{p},s) = \left(y^{\alpha i}(\vec{p},s),\, x^{\dagger i}_{\dot{\alpha}}(\vec{p},s)\right), \tag{3.211}$$

$$v_j(\vec{p},s) = \begin{pmatrix} y_{\alpha j}(\vec{p},s) \\ x_j^{\dagger\dot{\alpha}}(\vec{p},s) \end{pmatrix}, \qquad \bar{v}^j(\vec{p},s) = \left(x^{\alpha j}(\vec{p},s),\, y^{\dagger j}_{\dot{\alpha}}(\vec{p},s)\right). \tag{3.212}$$

The result obtained in eq. (3.210) coincides with that of eq. (2.153), which was derived using two-component spinor techniques.

For the decay of a neutral vector boson (denoted by A_μ) into a pair of Majorana fermions, $A_\mu \to \Psi_{Mi}(\vec{p}_1,s_1)\Psi_{Mj}(\vec{p}_2,s_2)$, we use the Feynman rules of Fig. 3.2 to obtain

$$i\mathcal{M} = -i\bar{u}(\vec{p}_1,s_1)\gamma_\mu\left[G_i{}^j P_L - G_j{}^i P_R\right] v(\vec{p}_2,s_2)\varepsilon^\mu\,, \tag{3.213}$$

where ε^μ is the vector boson polarization vector, and the flavor indices of the $\bar{u}$-spinor and v-spinor wave functions have been suppressed. The above result is equivalent to eq. (2.155), which was derived using two-component spinor techniques. Again, we note that if one had chosen to switch the two final states (equivalent to switching the directions of the Majorana fermion arrows), then the resulting matrix element would simply exhibit an overall sign change [due to the results of eqs. (3.189) and (3.190)].

For $i = j$, eq. (3.213) simplifies to

$$i\mathcal{M} = iG\bar{u}(\vec{p}_1,s_1)\gamma^\mu\gamma_5 v(\vec{p}_2,s_2)\varepsilon_\mu\,, \tag{3.214}$$

where $G \equiv G_i{}^i$. The absence of a vector coupling of the vector boson to a pair of identical Majorana fermions is a consequence of the identity $\overline{\Psi}_M\gamma^\mu\Psi_M = 0$ noted below eq. (3.100).

For the decay of a (neutral or charged) vector particle A_μ into a fermion pair consisting of a Dirac fermion and antifermion, $A_\mu \to F_i(\vec{p}_1,s_1)\overline{F}{}^j(\vec{p}_2,s_2)$, the matrix element is given by

$$i\mathcal{M} = -i\bar{u}(\vec{p}_1,s_1)\gamma^\mu\left[(G_L)_i{}^j P_L + (G_R)_i{}^j P_R\right] v(\vec{p}_2,s_2)\varepsilon_\mu\,, \tag{3.215}$$

which matches the result of eq. (2.159).

Finally, we consider the decay of a charged boson to a fermion pair consisting of one Dirac fermion and one Majorana fermion. Using the Feynman rules of Fig. 3.4, we see that we have a choice of two rules for each decay process. As an example, consider the decay $W \to \Psi_{Mi}(\vec{p}_1,s_1)F_j(\vec{p}_2,s_2)$. If we apply the $W\Psi_M\Psi$ Feynman rule of Fig. 3.4, we obtain

$$i\mathcal{M} = -i\bar{u}(\vec{p}_2,s_2)\left[(G_1)_j{}^i P_L + (G_2)_{ij} P_R\right] v(\vec{p}_1,s_1)\,. \tag{3.216}$$

If we apply the corresponding $W\Psi_M\Psi^C$ Feynman rule, we obtain the negative of eq. (3.216) with $P_L \leftrightarrow P_R$ and $(\vec{p}_1, s_1) \leftrightarrow (\vec{p}_2, s_2)$. Using eqs. (3.189) and (3.190), the resulting amplitude is the negative of eq. (3.216), as expected since the order of the spinor wave functions in the two computations is reversed. A similar conclusion is obtained for the decay $\Phi \to \Psi_{Mi}F_j$.

3.8.2 Tree-Level Scattering Processes

Consider the elastic scattering of a Majorana fermion and a neutral scalar. We shall examine two equivalent ways for computing the amplitude. Following the rules previously stated, there are two possible choices for the direction of arrows on the Majorana fermion lines. Thus, we may evaluate either one of the two diagrams shown in Fig. 3.5 plus one of two diagrams from a second set of diagrams (not shown) where the initial-state and final-state scalars are crossed.

Evaluating the first diagram shown in Fig. 3.5, the scattering matrix element for $\phi\Psi_M \to \phi\Psi_M$ is given by

$$\begin{aligned} i\mathcal{M} &= \frac{-i}{s-m^2}\bar{u}(\vec{p}_2, s_2)(\lambda P_L + \lambda^* P_R)(\not{p} + m)(\lambda P_L + \lambda^* P_R)u(\vec{p}_1, s_1) + (\text{crossed}) \\ &= \frac{-i}{s-m^2}\bar{u}(\vec{p}_2, s_2)\left[|\lambda|^2\not{p} + (\lambda^2 P_L + (\lambda^*)^2 P_R)\, m\right] u(\vec{p}_1, s_1) + (\text{crossed}), \end{aligned} \tag{3.217}$$

where m is the Majorana fermion mass and $\sqrt{s}$ is the center-of-mass energy. Using eqs. (3.7) and (3.167), one recovers the results of eq. (2.161). Had we chosen to evaluate the second diagram of Fig. 3.5 instead, the resulting amplitude would have been given by

$$i\mathcal{M} = \frac{-i}{s-m^2}\bar{v}(\vec{p}_1, s_1)\left[-|\lambda|^2\not{p} + (\lambda^2 P_L + (\lambda^*)^2 P_R)\, m\right] v(\vec{p}_2, s_2) + (\text{crossed}). \tag{3.218}$$

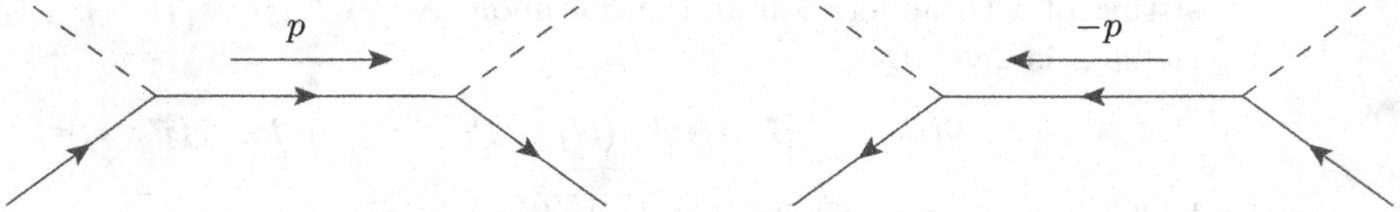

Fig. 3.5 Tree-level Feynman diagrams contributing to the elastic scattering of a neutral scalar and a Majorana fermion. Either diagram shown above (but not both) can be used to compute the scattering matrix element using four-component spinor Feynman rules. Two additional diagrams in which the initial-state and final-state scalars are crossed are not shown.

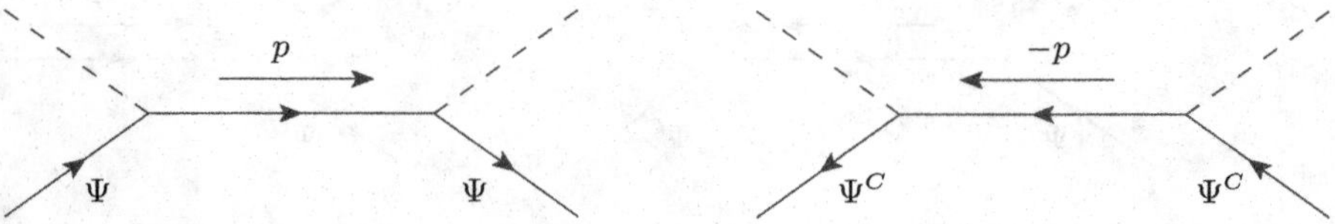

Fig. 3.6 Tree-level Feynman diagrams contributing to the elastic scattering of a charged scalar and a Majorana fermion. Similar considerations noted in the caption to Fig. 3.5 also apply here.

Using eq. (3.186), it follows that

$$\bar{v}(\vec{p}_1, s_1)v(\vec{p}_2, s_2) = -\bar{u}(\vec{p}_2, s_2)u(\vec{p}_1, s_1)\,, \tag{3.219}$$

$$\bar{v}(\vec{p}_1, s_1)\gamma^\mu v(\vec{p}_2, s_2) = \bar{u}(\vec{p}_2, s_2)\gamma^\mu u(\vec{p}_1, s_1)\,. \tag{3.220}$$

Hence, the amplitude computed in eq. (3.218) is just the negative of eq. (3.217). This is expected, since the order of spinor wave functions in eq. (3.218) is an odd permutation of the order of spinor wave functions in eq. (3.217). As in the two-component fermion Feynman rules, the overall sign of the amplitude is arbitrary, but the relative signs of any pair of diagrams is unambiguous. This relative sign is positive [negative] if the permutation of the order of spinor wave functions of one diagram relative to the other diagram is even [odd].

Next, we consider the elastic scattering of a charged fermion and a neutral scalar. Again, we examine two equivalent ways for computing the amplitude. Following our rules, there are two possible choices for the directions of the fermion line arrows, depending on whether we represent the fermion by Ψ or Ψ^C. Thus, we may evaluate either one of the two diagrams shown in Fig. 3.6, plus one of two diagrams from a second set of diagrams (not shown) where the initial-state and final-state scalars are crossed. Evaluating the first diagram shown in Fig. 3.6, the matrix element for $\phi F \to \phi F$ is given by eq. (3.217), with λ replaced by κ. Had we chosen to evaluate the second diagram instead, the resulting amplitude would have been given by eq. (3.218), with λ replaced by κ. Thus, the discussion above in the case of neutral fermion scattering processes also applies to charged fermion scattering processes.

In processes that only involve vertices with two Dirac fields, it is never necessary to use charge-conjugated Dirac fermion lines. In contrast, consider the following process that involves a vertex with one Dirac and one Majorana fermion. Specifically, we examine the scattering of a Dirac fermion and a charged scalar into its charge-conjugated final state, via the exchange of a Majorana fermion: $\Phi^\dagger F \to \Phi \overline{F}$. If one attempts to draw the relevant Feynman diagram employing Dirac fermion lines but with no charge-conjugated Dirac fermion lines, one finds that there is no possible choice of arrow direction for the Majorana fermion that is consistent with the vertex rules of Fig. 3.4. The resolution is simple: one can choose the incoming

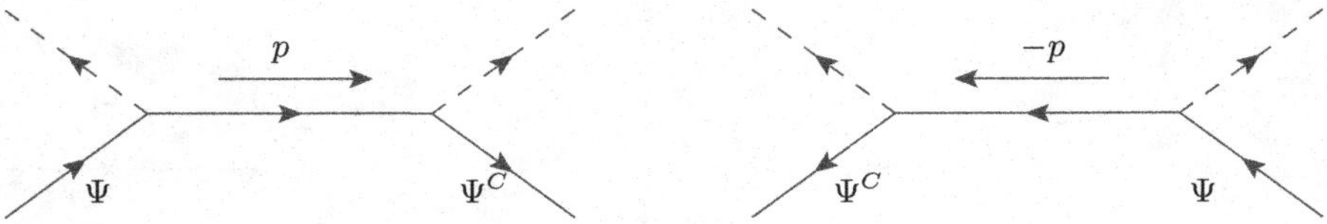

Fig. 3.7 Tree-level Feynman diagrams contributing to the elastic scattering of a charged scalar and a Dirac fermion. Similar considerations noted in the caption to Fig. 3.5 also apply here.

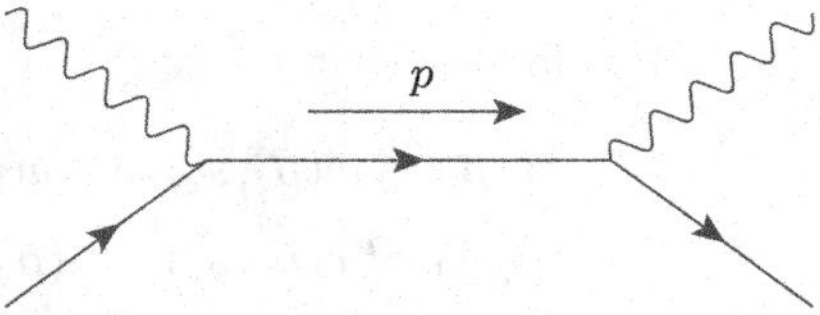

Fig. 3.8 Tree-level Feynman diagrams contributing to the elastic scattering of a neutral vector boson and a Majorana fermion. A second diagram in which the initial-state and final-state vector bosons are crossed is not shown.

line to be Ψ and the outgoing line to be Ψ^C or vice versa. Thus, we may evaluate either one of the two diagrams shown in Fig. 3.7, plus one of two diagrams from a second set of diagrams (not shown) where the initial-state and final-state scalars are crossed. Evaluating the first diagram shown in Fig. 3.7, the matrix element for $\phi F \to \phi F$ is given by

$$\begin{aligned} i\mathcal{M} &= \frac{-i}{s-m^2}\,\bar{u}(\vec{\boldsymbol{p}}_2, s_2)(\kappa_2 P_L + \kappa_1^* P_R)(\not{p} + m)(\kappa_2 P_L + \kappa_1^* P_R)u(\vec{\boldsymbol{p}}_1, s_1) + (\text{crossed}) \\ &= \frac{-i}{s-m^2}\,\bar{u}(\vec{\boldsymbol{p}}_2, s_2)\left[\kappa_1^*\kappa_2\not{p} + \left(\kappa_2^2 P_L + (\kappa_1^*)^2 P_R\right)m\right]u(\vec{\boldsymbol{p}}_1, s_1) + (\text{crossed})\,, \end{aligned} \tag{3.221}$$

where m is the Majorana fermion mass. This result is equivalent to eq. (2.165) obtained via the two-component spinor methods. Had we chosen to evaluate the second diagram instead, the resulting amplitude [after using eqs. (3.219) and (3.220)] would have been found to be the negative of eq. (3.221), as expected. As before, the relative sign between diagrams for the same process is unambiguous.

In the case of elastic scattering of a Majorana fermion of mass m and a neutral vector boson, the two contributing diagrams include the diagram shown in Fig. 3.8 plus a second diagram (not shown) where the initial-state and final-state vector bosons are crossed. Using the Feynman rules of Fig. 3.2, the Feynman rule for the $A_\mu \overline{\Psi}_M \Psi_M$ vertex is given by $iG\gamma^\mu\gamma_5$. Hence, the corresponding matrix element is

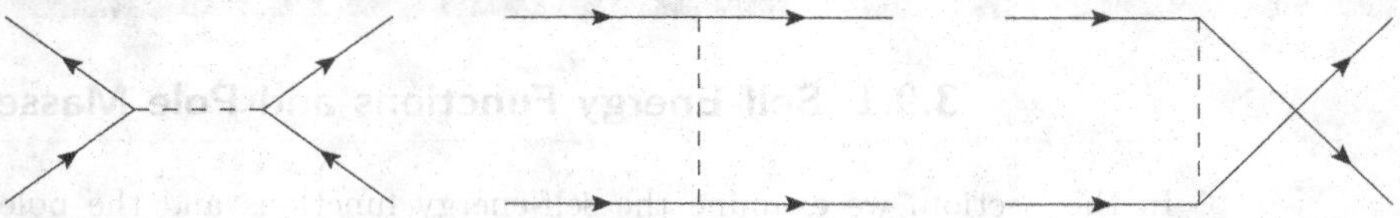

Fig. 3.9 Tree-level Feynman diagrams contributing to the elastic scattering of two identical Majorana fermions via scalar particle exchange.

given by

$$i\mathcal{M} = \frac{-iG^2}{s-m^2}\bar{u}(\vec{p}_2, s_2)\,\gamma\cdot\varepsilon_2^*\,(\not{p}-m)\,\gamma\cdot\varepsilon_1 u(\vec{p}_1, s_1) + (\text{crossed})\,, \tag{3.222}$$

where we have used $\gamma^\nu\gamma_5(\not{p}+m)\gamma^\mu\gamma_5 = \gamma^\nu(\not{p}-m)\gamma^\mu$. Using eqs. (3.7) and (3.167), one easily recovers the results of eq. (2.162).

The scattering of a Dirac fermion of mass m and a neutral vector boson can be similarly treated. The relevant Feynman graphs are the same as in the previous case, and the matrix element is given by

$$\begin{aligned} i\mathcal{M} &= \frac{-i}{s-m^2}\bar{u}(\vec{p}_2, s_2)\,\gamma\cdot\varepsilon_2^*\,(G_L P_L + G_R P_R)(\not{p}+m)\,\gamma\cdot\varepsilon_1\,(G_L P_L + G_R P_R)u(\vec{p}_1, s_1) \\ &\qquad + (\text{crossed}) \\ &= \frac{-i}{s-m^2}\bar{u}(\vec{p}_2, s_2)\,\gamma\cdot\varepsilon_2^*\left[(G_L^2 P_L + G_R^2 P_R)\not{p} + G_L G_R m\right]\,\gamma\cdot\varepsilon_1 u(\vec{p}_1, s_1) \\ &\qquad + (\text{crossed}). \end{aligned} \tag{3.223}$$

One can easily check that this result coincides with that of eq. (2.166).

Finally, we examine the elastic scattering of two identical Majorana fermions via scalar exchange. The three contributing diagrams are shown in Fig. 3.9, and the corresponding matrix element is given by

$$\begin{aligned} i\mathcal{M} &= \frac{-i}{s-m_\phi^2}\left[\bar{v}_1(\lambda P_L + \lambda^* P_R)u_2\,\bar{u}_3(\lambda P_L + \lambda^* P_R)v_4\right] \\ &\quad + (-1)\frac{-i}{t-m_\phi^2}\left[\bar{u}_3(\lambda P_L + \lambda^* P_R)u_1\,\bar{u}_4(\lambda P_L + \lambda^* P_R)u_2\right] \\ &\quad + \frac{-i}{u-m_\phi^2}\left[\bar{u}_4(\lambda P_L + \lambda^* P_R)u_1\,\bar{u}_3(\lambda P_L + \lambda^* P_R)u_2\right]\,, \end{aligned} \tag{3.224}$$

where $u_i \equiv u(\vec{p}_i, s_i)$, $v_j \equiv v(\vec{p}_j, s_j)$ and m_ϕ is the exchanged scalar mass. The relative minus sign of the t-channel graph relative to the s and u-channel graphs is obtained by noting that 3142 [4132] is an odd [even] permutation of 1234. Using eqs. (3.12), (3.167), and (3.168), the results of eq. (2.167) are recovered.

3.9 Self-Energy of Four-Component Fermions

3.9.1 Self-Energy Functions and Pole Masses

In this section, we examine the self-energy functions and the pole masses for a set of four-component fermions. We first consider four-component Dirac fermion fields Ψ_{ai}, where a is the four-component spinor index and i is the flavor index. The full, loop-corrected Feynman propagator with four-momentum p^μ is defined by the Fourier transform of vacuum expectation values of time-ordered products of bilinears of the fully interacting four-component fermion fields, using the notation for the Fourier transform introduced in eqs. (2.84) and (2.85),

$$\langle 0| \, T\Psi_{ai}(x)\overline{\Psi}^{bj}(y) \, |0\rangle_{\rm FT} = i(\boldsymbol{S}_a{}^b)_i{}^j(p)\,, \tag{3.225}$$

where

$$\boldsymbol{S}(p) \equiv \not{p}\left[P_L \boldsymbol{S}_{\boldsymbol{L}}^{\mathsf{T}}(p^2) + P_R \boldsymbol{S}_{\boldsymbol{R}}(p^2)\right] + P_L \overline{\boldsymbol{S}}{}_{\boldsymbol{D}}^{\mathsf{T}}(p^2) + P_R \boldsymbol{S}_{\boldsymbol{D}}(p^2)\,, \tag{3.226}$$

and the four-component spinor indices a and b and the flavor indices i and j have been suppressed. As in Section 2.10.1, we shall organize the computation of the full propagator in terms of the 1PI self-energy function,[22]

$$\boldsymbol{\Sigma}(p) \equiv \not{p}\left[P_L \boldsymbol{\Sigma}_{\boldsymbol{L}}(p^2) + P_R \boldsymbol{\Sigma}_{\boldsymbol{R}}^{\mathsf{T}}(p^2)\right] + P_L \boldsymbol{\Sigma}_{\boldsymbol{D}}(p^2) + P_R \overline{\boldsymbol{\Sigma}}{}_{\boldsymbol{D}}^{\mathsf{T}}(p^2)\,. \tag{3.227}$$

Diagrammatically, $i\boldsymbol{S}$ and $-i\boldsymbol{\Sigma}$ are shown in Fig. 3.10.

Fig. 3.10 The full, loop-corrected propagator for four-component Dirac fermions, $i(\boldsymbol{S}_a{}^b)_i{}^j(p)$, is denoted by the shaded box, which represents the sum of all connected Feynman diagrams, with external legs included. The self-energy function for four-component Dirac fermions, $-i(\boldsymbol{\Sigma}_a{}^b)_i{}^j(p)$, is denoted by the shaded circle, which represents the sum of all one-particle irreducible, connected Feynman diagrams with the external legs amputated. In both cases, The four-momentum p flows from right to left.

The hermiticity of the effective action implies that $\boldsymbol{S}$ and $\boldsymbol{\Sigma}$ satisfy hermiticity conditions given by

[22] Our notation in eq. (3.227) differs from that of Refs. [32, 38], as we employ $\boldsymbol{\Sigma}_{\boldsymbol{R}}^{\mathsf{T}}$ instead of $\boldsymbol{\Sigma}_{\boldsymbol{R}}$. Our motivation for this choice is that in the case of Majorana fermions [see eq. (3.239)], we simply have $\boldsymbol{\Sigma}_{\boldsymbol{L}} = \boldsymbol{\Sigma}_{\boldsymbol{R}}$, without an extra transpose (or conjugation). We have also chosen to employ $\boldsymbol{S}_{\boldsymbol{L}}^{\mathsf{T}}$ in eq. (3.226) for similar reasons.

$$[\boldsymbol{S}^{\mathsf{T}}]^{\star} = A\boldsymbol{S}A^{-1}\,, \qquad [\boldsymbol{\Sigma}^{\mathsf{T}}]^{\star} = A\boldsymbol{\Sigma}A^{-1}\,, \tag{3.228}$$

where A is the Dirac conjugation matrix [see eq. (3.29)] and the star symbol is defined below eq. (2.248). Applying eq. (3.228) to eqs. (3.226) and (3.227) then yields the following conditions for the complex matrix functions:

$$[\boldsymbol{S}_L^{\mathsf{T}}]^{\star} = \boldsymbol{S}_L\,, \qquad [\boldsymbol{S}_R^{\mathsf{T}}]^{\star} = \boldsymbol{S}_R\,, \qquad \overline{\boldsymbol{S}}_D = \boldsymbol{S}_D^{\star}\,, \tag{3.229}$$

$$[\boldsymbol{\Sigma}_L^{\mathsf{T}}]^{\star} = \boldsymbol{\Sigma}_L\,, \qquad [\boldsymbol{\Sigma}_R^{\mathsf{T}}]^{\star} = \boldsymbol{\Sigma}_R\,, \qquad \overline{\boldsymbol{\Sigma}}_D = \boldsymbol{\Sigma}_D^{\star}\,, \tag{3.230}$$

in agreement with eqs. (2.273) and (2.276).

Starting at tree level and comparing with Fig. 3.1, the full propagator function is given by

$$\boldsymbol{S}_i{}^j(p) = \frac{\not{p} + m}{p^2 - m_i^2}\,\delta_i^j + \cdots\,, \tag{3.231}$$

with no sum over i implied. The full loop-corrected propagator can be expressed diagrammatically in terms of the 1PI self-energy function:

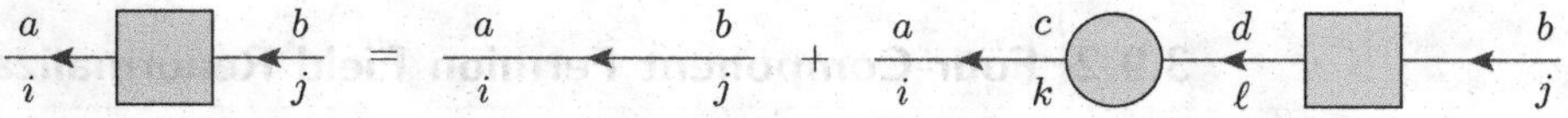

(3.232)

The algebraic representation of eq. (3.232) can be written as [see eqs. (2.181) and (2.182)]:

$$\boldsymbol{S} = \boldsymbol{T} + \boldsymbol{T}\boldsymbol{\Sigma}\boldsymbol{S} = (\boldsymbol{T}^{-1} - \boldsymbol{\Sigma})^{-1}\,, \tag{3.233}$$

where $\boldsymbol{T}_i{}^j \equiv (\not{p} + m)\delta_i^j/(p^2 - m_i^2)$ is the tree-level contribution to $\boldsymbol{S}$ given in eq. (3.231). By writing the expressions for $\boldsymbol{S}$ and $\boldsymbol{\Sigma}$ given in eqs. (3.226) and (3.227) and $\boldsymbol{T}$ in block matrix form using eq. (3.7), one can verify that eq. (3.233) is equivalent to eq. (2.278). Consequently, the complex pole masses of the corresponding Dirac fermions are again determined from eq. (2.283).

In the special case of a parity-conserving vectorlike theory of Dirac fermions (such as QED or QCD), the pseudoscalar and pseudovector parts of $\boldsymbol{S}(p)$ and $\boldsymbol{\Sigma}(p)$ must be absent. Thus, the following relations must hold among the loop-corrected propagator functions and self-energy functions, respectively:

$$\boldsymbol{S}_R = \boldsymbol{S}_L^{\mathsf{T}}\,, \qquad \boldsymbol{S}_D = [\boldsymbol{S}_D^{\mathsf{T}}]^{\star}\,, \tag{3.234}$$

$$\boldsymbol{\Sigma}_L = \boldsymbol{\Sigma}_R^{\mathsf{T}}\,, \qquad \boldsymbol{\Sigma}_D = [\boldsymbol{\Sigma}_D^{\mathsf{T}}]^{\star}\,, \tag{3.235}$$

in agreement with eq. (2.411) [after making use of eq. (2.276)].

In the case of a set of four-component Majorana fermion fields, we can still use the results of eqs. (3.226)–(3.233). However, one obtains additional constraints on the full propagator and self-energy matrix functions due to the Majorana condition $\Psi_{Mi} = C\overline{\Psi}^{\mathsf{T}}_{Mi}$. Inserting this result into eq. (3.225), and making use of the anticommutativity of the fermion fields, one easily derives:

$$\langle 0| T\Psi_{Mai}(x)\overline{\Psi}^{b}_{Mj}(y)\,|0\rangle_{\rm FT} = C_{ae}\,\langle 0| T\Psi_{Mdi}(x)\overline{\Psi}^{e}_{Mj}(y)\,|0\rangle_{\rm FT}\,(C^{-1})^{db}\,. \tag{3.236}$$

Consequently,

$$C\boldsymbol{S}^{\mathsf{T}}C^{-1} = \boldsymbol{S}\,, \qquad\qquad C\boldsymbol{\Sigma}^{\mathsf{T}}C^{-1} = \boldsymbol{\Sigma}\,. \tag{3.237}$$

Inserting the expressions for $\boldsymbol{S}$ and $\boldsymbol{\Sigma}$ [eqs. (3.226) and (3.227)] and using the result of eq. (3.93), it follows that

$$\boldsymbol{S_L} = \boldsymbol{S_R}\,, \qquad \boldsymbol{S_D} = \boldsymbol{S}^{\mathsf{T}}_{\boldsymbol{D}}\,, \qquad \overline{\boldsymbol{S}}_{\boldsymbol{D}} = \overline{\boldsymbol{S}}^{\mathsf{T}}_{\boldsymbol{D}}\,, \tag{3.238}$$

$$\boldsymbol{\Sigma_L} = \boldsymbol{\Sigma_R}\,, \qquad \boldsymbol{\Sigma_D} = \boldsymbol{\Sigma}^{\mathsf{T}}_{\boldsymbol{D}}\,, \qquad \overline{\boldsymbol{\Sigma}}_{\boldsymbol{D}} = \overline{\boldsymbol{\Sigma}}^{\mathsf{T}}_{\boldsymbol{D}}\,. \tag{3.239}$$

As expected, with these constraints the form of eq. (2.278) matches precisely with the form of eq. (2.260), corresponding to the equation for the full propagator functions for a theory of generic two-component fermion fields. In the notation of Section 2.10.1, we can therefore identify: $\boldsymbol{C} \equiv \boldsymbol{S_L} = \boldsymbol{S_R}$, $\boldsymbol{D} \equiv \boldsymbol{S_D}$, $\boldsymbol{\Xi} \equiv \boldsymbol{\Sigma_L} = \boldsymbol{\Sigma_R}$, $\boldsymbol{\Omega} \equiv \boldsymbol{\Sigma_D}$ and $\overline{\boldsymbol{\Omega}} \equiv \overline{\boldsymbol{\Sigma}}_{\boldsymbol{D}}$, as previously noted in eqs. (2.284) and (2.285).

3.9.2 Four-Component Fermion Field Renormalization

Using the results obtained in the previous section, we shall now write down explicit expressions for the wave function renormalization constants at one-loop accuracy. We shall first perform the calculation for the case of n four-component Dirac fermion fields, Ψ_i.

We begin by defining renormalized fields (denoted with a subscript r),

$$\Psi_{r,i} = (Z^{-1/2})_i{}^j\Psi_j\,, \qquad\qquad \overline{\Psi}^i_r = \overline{\Psi}^j(\overline{Z}^{-1/2})_j{}^i\,, \tag{3.240}$$

where there is an implicit sum over the flavor index j and

$$\overline{Z}^{-1/2} \equiv A^{-1}Z^{-1/2\dagger}A\,, \tag{3.241}$$

where A is defined in eq. (3.29).

The unrenormalized full loop-corrected propagator with four-momentum p^μ is defined in eq. (3.225). The corresponding renormalized full loop-corrected propagator (denoted with a subscript r) is

$$i\boldsymbol{S}_r(p) = \langle 0| T\Psi_r(x)\overline{\Psi}_r(y)\,|0\rangle_{\rm FT} = Z^{-1/2}i\boldsymbol{S}(p)\overline{Z}^{-1/2}\,, \tag{3.242}$$

where the four-component spinor indices and flavor indices have been suppressed. In light of eq. (3.233), the unrenormalized full loop-corrected inverse propagator is related to the unrenormalized self-energy function is given by

$$\boldsymbol{S}^{-1}(p) = \not p\mathbb{1} - \boldsymbol{m} - \boldsymbol{\Sigma}(p)\,, \tag{3.243}$$

where $\boldsymbol{m}$ is a diagonal matrix with diagonal elements m_j and $\mathbb{1}$ is the $n\times n$ identity matrix.

The wave function renormalization constants are of the form

$$(Z)_i{}^j = \delta_i{}^j + (\delta Z)_i{}^j\,, \qquad (\overline{Z})_i{}^j = \delta_i{}^j + (\delta\overline{Z})_i{}^j\,, \tag{3.244}$$

where the contributions to $(\delta Z)_i{}^j$ and $(\delta\overline{Z})_i{}^j$ begin at one-loop order. Thus, to one-loop accuracy, it follows that

$$\boldsymbol{S}_r^{-1}(p) = \not{p}\mathbb{1} - \boldsymbol{m} - \boldsymbol{\Sigma}(p) + \tfrac{1}{2}\delta\overline{Z}(\not{p}\mathbb{1} - \boldsymbol{m}) - \tfrac{1}{2}(\not{p}\mathbb{1} - \boldsymbol{m})\delta Z\,. \tag{3.245}$$

In general, Z and $\overline{Z}$ can be decomposed into left-handed and right-handed contributions,

$$Z = P_L Z_L + P_R Z_R\,, \qquad \overline{Z} = A^{-1}Z^\dagger A = P_R Z_L^\dagger + P_L Z_R^\dagger\,, \tag{3.246}$$

after making use of eq. (3.44). In a parity-conserving theory, $Z_L = Z_R$, but in a parity-violating theory Z_L and Z_R are unequal. It then follows that

$$(\Psi_L)_r = Z_L^{-1/2}\Psi_L\,, \qquad (\Psi_R)_r = Z_R^{-1/2}\Psi_R\,, \tag{3.247}$$

$$(\overline{\Psi}_L)_r = \overline{\Psi}_L\overline{Z}^{-1/2}\,, \qquad (\overline{\Psi}_R)_r = \overline{\Psi}_R\overline{Z}^{-1/2}\,, \tag{3.248}$$

which are equivalent to eqs. (2.286) and (2.287). In light of eq. (3.244), it follows that

$$\delta Z = P_L\delta Z_L + P_R\delta Z_R\,, \qquad \delta\overline{Z} = P_R\delta Z_L^\dagger + P_L\delta Z_R^\dagger\,. \tag{3.249}$$

Our first task is to evaluate the pole mass and width. As noted below eq. (3.233), the complex pole masses of the Dirac fermions are determined from eq. (2.283). In particular, to one-loop accuracy, we must solve eq. (2.292). As shown below eq. (2.292), in the one-loop approximation the pole masses are determined by the poles of the diagonal elements of $\boldsymbol{S}(p)$.

Inserting eq. (3.227) in the expression for the diagonal elements of $\boldsymbol{\Sigma}^{-1}(p)$ yields

$$\boldsymbol{S}^{-1}(p)_j{}^j = \not{p}(aP_L + bP_R) - m_j(cP_L + dP_R)\,, \tag{3.250}$$

where

$$a = 1 - \boldsymbol{\Sigma_L}(p^2)_j{}^j\,, \qquad b = 1 - \boldsymbol{\Sigma_R}(p^2)_j{}^j\,, \tag{3.251}$$

$$c = 1 + \boldsymbol{\Sigma_D}(p^2)_j{}^j/m_j\,, \qquad d = 1 + \overline{\boldsymbol{\Sigma}}\boldsymbol{_D}(p^2)_j{}^j/m_j\,. \tag{3.252}$$

At one-loop accuracy, $\boldsymbol{S}^{-1}(p)_j{}^j = \left[\boldsymbol{S}(p)_j{}^j\right]^{-1}$. Hence, inverting eq. (3.250) yields

$$\boldsymbol{S}(p)_j{}^j = \frac{1}{\not{p}(aP_L + bP_R) - m_j(cP_L + dP_R)} = \frac{\not{p}(aP_L + bP_R) + m_j(dP_L + cP_R)}{abp^2 - cdm_j^2}\,. \tag{3.253}$$

The pole mass $m_{p,j}$ is defined by the condition that $S(p)_j{}^j$ has a pole in the limit of $p^2 \to m_{p,j}^2$, i.e., $abp^2 - cdm_j^2 = 0$ as $p^2 \to m_{p,j}^2$. Hence, using eqs. (3.251) and (3.252), if follows that, to one-loop accuracy,

$$m_{p,j} = m_j + \tfrac{1}{2}\left[m_j\boldsymbol{\Sigma_L}(m_j^2)_j{}^j + m_j\boldsymbol{\Sigma_R}(m_j^2)_j{}^j + \boldsymbol{\Sigma_D}(m_j^2)_j{}^j + \overline{\boldsymbol{\Sigma}}\boldsymbol{_D}(m_j^2)_j{}^j\right], \tag{3.254}$$

which recovers the result obtained in eq. (2.294).

The pole masses $m_{p,j}$ are complex:

$$m_{p,j} \equiv m_{r,j} - \tfrac{1}{2} i\Gamma_j \,, \tag{3.255}$$

where $m_{r,j} \equiv \operatorname{Re} m_{p,j}$ and Γ_j is the width of the fermion (which is nonzero if a decay channel is kinematically allowed). We define the mass counterterm δm_j by

$$\delta m_j = m_{r,j} - m_j \,, \tag{3.256}$$

which is a quantity of one-loop order. Hence,

$$\delta m_j = \tfrac{1}{2} \operatorname{Re}\left[m_j \boldsymbol{\Sigma_L}(m_j^2)_j{}^j + m_j \boldsymbol{\Sigma_R}(m_j^2)_j{}^j + \boldsymbol{\Sigma_D}(m_j^2)_j{}^j + \overline{\boldsymbol{\Sigma}}_{\boldsymbol{D}}(m_j^2)_j{}^j\right], \tag{3.257}$$

$$\Gamma_j = -\operatorname{Im}\left[m_j \boldsymbol{\Sigma_L}(m_j^2)_j{}^j + m_j \boldsymbol{\Sigma_R}(m_j^2)_j{}^j + \boldsymbol{\Sigma_D}(m_j^2)_j{}^j + \overline{\boldsymbol{\Sigma}}_{\boldsymbol{D}}(m_j^2)_j{}^j\right]. \tag{3.258}$$

In the following, we shall treat the diagonal elements and the off-diagonal elements of δZ and $\delta\overline{Z}$ separately, under the assumption that the absorptive part of the self-energy functions evaluated on mass shell vanish. Since $i\boldsymbol{S}(p)_j{}^j$ is the inverse propagator of the jth Dirac fermion, we can identify the residue at the pole as the wave function renormalization constant. That is [see eq. (3.242)],

$$iS(p)\Big|_{p^2\to m_{p,j}^2} = (Z^{1/2})_j{}^j \frac{i}{\not{p} - m_{p,j}} (\overline{Z}^{1/2})_j{}^j \,. \tag{3.259}$$

At one-loop accuracy, we rewrite eq. (3.259) as

$$\lim_{p^2\to m_{p,j}^2} \boldsymbol{S}(p)_j{}^j = (1 + \tfrac{1}{2}\delta Z_j{}^j) \left(\frac{\not{p} + m_{p,j}}{p^2 - m_{p,j}^2}\right) (1 + \tfrac{1}{2}\delta\overline{Z}_j{}^j) \,. \tag{3.260}$$

Employing eq. (3.254) in the above equations, we obtain, at one-loop accuracy,

$$\begin{aligned} \lim_{p^2\to m_{p,j}^2} (p^2 - m_{p,j}^2)\boldsymbol{S}(p)_j{}^j &= \not{p}\left[1 + \tfrac{1}{2}(\delta Z_L + \delta Z_L^\dagger)_j{}^j P_R + \tfrac{1}{2}(\delta Z_R + \delta Z_R^\dagger)_j{}^j P_L\right] \\ &\quad + m\Bigg\{1 + \tfrac{1}{2}(\delta Z_L + \delta Z_R^\dagger)_j{}^j P_L + \tfrac{1}{2}(\delta Z_L^\dagger + \delta Z_R)_j{}^j P_R \\ &\quad + \frac{1}{2}\left(\boldsymbol{\Sigma_L}(m_j^2)_j{}^j + \boldsymbol{\Sigma_R}(m_j^2)_j{}^j + \frac{\boldsymbol{\Sigma_D}(m_j^2)_j{}^j + \overline{\boldsymbol{\Sigma}}_{\boldsymbol{D}}(m_j^2)_j{}^j}{m_j}\right)\Bigg\}. \end{aligned} \tag{3.261}$$

We compare this with eq. (3.253) as follows. First we expand eqs. (3.251) and (3.252) around $p^2 = m_j^2$:

$$a = 1 - \boldsymbol{\Sigma_L}(m_j^2)_j{}^j - (p^2 - m_j^2)\boldsymbol{\Sigma_L}'(m_j^2)_j{}^j \,, \tag{3.262}$$

$$b = 1 - \boldsymbol{\Sigma_R}(m_j^2)_j{}^j - (p^2 - m_j^2)\boldsymbol{\Sigma}'_{\boldsymbol{R}}(m_j^2)_j{}^j \,, \tag{3.263}$$

$$c = 1 + \boldsymbol{\Sigma_D}(m_j^2)_j{}^j/m_j + (p^2 - m_j^2)\boldsymbol{\Sigma}'_{\boldsymbol{D}}(m_j^2)_j{}^j/m_j \,, \tag{3.264}$$

$$d = 1 + \overline{\boldsymbol{\Sigma}}_{\boldsymbol{D}}(m_j^2)_j{}^j/m_j + (p^2 - m_j^2)\overline{\boldsymbol{\Sigma}}'_{\boldsymbol{D}}(m_j^2)_j{}^j/m_j \,, \tag{3.265}$$

where

$$\boldsymbol{\Sigma}'_{\boldsymbol{a}}(m_j^2)_j{}^j \equiv \left.\frac{d\boldsymbol{\Sigma}_{\boldsymbol{a}}(p^2)_j{}^j}{dp^2}\right|_{p^2=m_j^2}, \quad \text{for } a = L, R, D, \overline{D}. \tag{3.266}$$

It then follows that, at one-loop accuracy,

$$\begin{aligned}
abp^2 - cdm_j^2 &= p^2\Big[1 - \{\boldsymbol{\Sigma}_{\boldsymbol{L}}(m_j^2) + \boldsymbol{\Sigma}_{\boldsymbol{R}}(m_j^2)\}_j{}^j - (p^2 - m_j^2)\{\boldsymbol{\Sigma}'_{\boldsymbol{L}}(m_j^2) + \boldsymbol{\Sigma}'_{\boldsymbol{R}}(m_j^2)\}_j{}^j\Big] \\
&\quad - m_j\Big[m_j + \{\boldsymbol{\Sigma}_{\boldsymbol{D}}(m_j^2) + \overline{\boldsymbol{\Sigma}}_{\boldsymbol{D}}(m_j^2)\}_j{}^j + (p^2 - m_j^2)\{\boldsymbol{\Sigma}'_{\boldsymbol{D}}(m_j^2) + \overline{\boldsymbol{\Sigma}}'_{\boldsymbol{D}}(m_j^2)\}_j{}^j\Big] \\
&= (p^2 - m_j^2)\Big[1 - \boldsymbol{\Sigma}_{\boldsymbol{L}}(m_j^2)_j{}^j - \boldsymbol{\Sigma}_{\boldsymbol{R}}(m_j^2)_j{}^j - m_j^2\left[\boldsymbol{\Sigma}'_{\boldsymbol{L}}(m_j^2)_j{}^j + \boldsymbol{\Sigma}'_{\boldsymbol{R}}(m_j^2)_j{}^j\right] \\
&\qquad - m_j\left[\boldsymbol{\Sigma}'_{\boldsymbol{D}}(m_j^2)_j{}^j + \overline{\boldsymbol{\Sigma}}'_{\boldsymbol{D}}(m_j^2)_j{}^j\right]\Big] + \mathcal{O}\big((p^2 - m_j^2)^2\big) \\
&\quad - m_j^2\left[\boldsymbol{\Sigma}_{\boldsymbol{L}}(m_j^2)_j{}^j + \boldsymbol{\Sigma}_{\boldsymbol{R}}(m_j^2)_j{}^j + \frac{\boldsymbol{\Sigma}_{\boldsymbol{D}}(m_j^2)_j{}^j + \overline{\boldsymbol{\Sigma}}_{\boldsymbol{D}}(m_j^2)_j{}^j}{m_j}\right] \\
&= (p^2 - m_j^2)\Big[1 - \boldsymbol{\Sigma}_{\boldsymbol{L}}(m_j^2)_j{}^j - \boldsymbol{\Sigma}_{\boldsymbol{R}}(m_j^2)_j{}^j - m_j^2\left[\boldsymbol{\Sigma}'_{\boldsymbol{L}}(m_j^2)_j{}^j + \boldsymbol{\Sigma}'_{\boldsymbol{R}}(m_j^2)_j{}^j\right] \\
&\qquad - m_j\left[\boldsymbol{\Sigma}'_{\boldsymbol{D}}(m_j^2)_j{}^j + \overline{\boldsymbol{\Sigma}}'_{\boldsymbol{D}}(m_j^2)_j{}^j\right]\Big] + m_j^2 - m_{p,j}^2 + \mathcal{O}\big((p^2 - m_j^2)^2\big) \\
&= (p^2 - m_{p,j}^2)\Big[1 - \boldsymbol{\Sigma}_{\boldsymbol{L}}(m_j^2)_j{}^j - \boldsymbol{\Sigma}_{\boldsymbol{R}}(m_j^2)_j{}^j - m_j^2\left[\boldsymbol{\Sigma}'_{\boldsymbol{L}}(m_j^2)_j{}^j + \boldsymbol{\Sigma}'_{\boldsymbol{R}}(m_j^2)_j{}^j\right] \\
&\qquad - m_j\left[\boldsymbol{\Sigma}'_{\boldsymbol{D}}(m_j^2)_j{}^j + \overline{\boldsymbol{\Sigma}}'_{\boldsymbol{D}}(m_j^2)_j{}^j\right]\Big] + \mathcal{O}\big((p^2 - m_{p,j}^2)^2\big),
\end{aligned} \tag{3.267}$$

after using the square of eq. (3.254) in the penultimate step above.

Inserting this result back into eq. (3.253), we obtain

$$\begin{aligned}
&\lim_{p^2\to m_{p,j}^2} (p^2 - m_{p,j}^2)\boldsymbol{S}(p) \\
&= \Big[\not{p}\left[1 - \boldsymbol{\Sigma}_{\boldsymbol{L}}(m_j^2)_j{}^j P_L - \boldsymbol{\Sigma}_{\boldsymbol{R}}(m_j^2)_j{}^j P_R\right] + m_j + \overline{\boldsymbol{\Sigma}}_{\boldsymbol{D}}(m_j^2)_j{}^j P_L + \boldsymbol{\Sigma}_{\boldsymbol{D}}(m_j^2)_j{}^j P_R\Big] \\
&\quad \times\Big[1 + \boldsymbol{\Sigma}_{\boldsymbol{L}}(m_j^2)_j{}^j + \boldsymbol{\Sigma}_{\boldsymbol{R}}(m_j^2)_j{}^j + m_j^2\left[\boldsymbol{\Sigma}'_{\boldsymbol{L}}(m_j^2)_j{}^j + \boldsymbol{\Sigma}'_{\boldsymbol{R}}(m_j^2)_j{}^j\right] \\
&\qquad + m_j\left[\boldsymbol{\Sigma}'_{\boldsymbol{D}}(m_j^2)_j{}^j + \overline{\boldsymbol{\Sigma}}'_{\boldsymbol{D}}(m_j^2)_j{}^j\right]\Big] \\
&= \not{p}\Big\{1 + \boldsymbol{\Sigma}_{\boldsymbol{L}}(m_j^2)_j{}^j P_R + \boldsymbol{\Sigma}_{\boldsymbol{R}}(m_j^2)_j{}^j P_L + m_j^2\left[\boldsymbol{\Sigma}'_{\boldsymbol{L}}(m_j^2)_j{}^j + \boldsymbol{\Sigma}'_{\boldsymbol{R}}(m_j^2)_j{}^j\right] \\
&\qquad + m_j\left[\boldsymbol{\Sigma}'_{\boldsymbol{D}}(m_j^2)_j{}^j + \overline{\boldsymbol{\Sigma}}'_{\boldsymbol{D}}(m_j^2)_j{}^j\right]\Big\} \\
&\quad + m_j\Big\{1 + \boldsymbol{\Sigma}_{\boldsymbol{L}}(m_j^2)_j{}^j + \boldsymbol{\Sigma}_{\boldsymbol{R}}(m_j^2)_j{}^j + \left[\overline{\boldsymbol{\Sigma}}_{\boldsymbol{D}}(m_j^2)_j{}^j P_L + \boldsymbol{\Sigma}_{\boldsymbol{D}}(m_j^2)_j{}^j P_R\right]/m_j \\
&\qquad + m_j^2\left[\boldsymbol{\Sigma}'_{\boldsymbol{L}}(m_j^2)_j{}^j + \boldsymbol{\Sigma}'_{\boldsymbol{R}}(m_j^2)_j{}^j\right] + m_j\left[\boldsymbol{\Sigma}'_{\boldsymbol{D}}(m_j^2)_j{}^j + \overline{\boldsymbol{\Sigma}}'_{\boldsymbol{D}}(m^2)_j{}^j\right]\Big\}.
\end{aligned} \tag{3.268}$$

Comparing eqs. (3.261) and (3.268), it follows that

$$\tfrac{1}{2}(\delta Z_L + \delta Z_L^\dagger)_j{}^j = \boldsymbol{\Sigma_L}(m_j^2)_j{}^j + \boldsymbol{\mathcal{D}}_j\,, \tag{3.269}$$

$$\tfrac{1}{2}(\delta Z_R + \delta Z_R^\dagger)_j{}^j = \boldsymbol{\Sigma_R}(m_j^2)_j{}^j + \boldsymbol{\mathcal{D}}_j\,, \tag{3.270}$$

$$\tfrac{1}{2}(\delta Z_L + \delta Z_R^\dagger)_j{}^j = \boldsymbol{\Sigma_L}(m_j^2)_j{}^j + \boldsymbol{\Sigma_R}(m_j^2)_j{}^j + \boldsymbol{\mathcal{D}}_j + \frac{\overline{\boldsymbol{\Sigma}}_{\boldsymbol{D}}(m_j^2)_j{}^j + m_j - m_{p,j}}{m_j}\,, \tag{3.271}$$

$$\tfrac{1}{2}(\delta Z_L^\dagger + \delta Z_R)_j{}^j = \boldsymbol{\Sigma_L}(m_j^2)_j{}^j + \boldsymbol{\Sigma_R}(m_j^2)_j{}^j + \boldsymbol{\mathcal{D}}_j + \frac{\boldsymbol{\Sigma_D}(m_j^2)_j{}^j + m_j - m_{p,j}}{m_j}\,, \tag{3.272}$$

where

$$\boldsymbol{\mathcal{D}}_j \equiv m_j^2\left[\boldsymbol{\Sigma}'_{\boldsymbol{L}}(m_j^2)_j{}^j + \boldsymbol{\Sigma}'_{\boldsymbol{R}}(m_j^2)_j{}^j\right] + m_j\left[\boldsymbol{\Sigma}'_{\boldsymbol{D}}(m_j^2)_j{}^j + \overline{\boldsymbol{\Sigma}}'_{\boldsymbol{D}}(m_j^2)_j{}^j\right], \tag{3.273}$$

which reproduces the results of eqs. (2.316)–(2.319). Thus, following the analysis given in Section 2.10.2, and exploiting the freedom to rephase $Z_L^{1/2}$ and $Z_R^{1/2}$, we end up with eqs. (2.324)–(2.327), which we record here for the convenience of the reader:

$$(\delta Z_L)_j{}^j = \boldsymbol{\Sigma_L}(M_j^2)_j{}^j - \frac{\boldsymbol{\Sigma_D}(M_j^2)_j{}^j - \overline{\boldsymbol{\Sigma}}_{\boldsymbol{D}}(M_j^2)_j{}^j}{2M_j} + \boldsymbol{\mathcal{D}}_j\,, \tag{3.274}$$

$$(\delta Z_R)_j{}^j = \boldsymbol{\Sigma_R}(M_j^2)_j{}^j + \frac{\boldsymbol{\Sigma_D}(M_j^2)_j{}^j - \overline{\boldsymbol{\Sigma}}_{\boldsymbol{D}}(M_j^2)_j{}^j}{2M_j} + \boldsymbol{\mathcal{D}}_j\,, \tag{3.275}$$

$$(\delta Z_L^\dagger)_j{}^j = \boldsymbol{\Sigma_L}(M_j^2)_j{}^j + \frac{\boldsymbol{\Sigma_D}(M_j^2)_j{}^j - \overline{\boldsymbol{\Sigma}}_{\boldsymbol{D}}(M_j^2)_j{}^j}{2M_j} + \boldsymbol{\mathcal{D}}_j\,, \tag{3.276}$$

$$(\delta Z_R^\dagger)_j{}^j = \boldsymbol{\Sigma_R}(M_j^2)_j{}^j - \frac{\boldsymbol{\Sigma_D}(M_j^2)_j{}^j - \overline{\boldsymbol{\Sigma}}_{\boldsymbol{D}}(M_j^2)_j{}^j}{2M_j} + \boldsymbol{\mathcal{D}}_j\,. \tag{3.277}$$

Next, we determine the off-diagonal elements of δZ_L and δZ_R. It is convenient to introduce the (finite) subtracted self-energy function (denoted by $\widehat{\boldsymbol{\Sigma}}$) such that

$$\boldsymbol{S}_r^{-1}(p) = \not{p}\mathbb{1} - \boldsymbol{m_r} - \widehat{\boldsymbol{\Sigma}}(p)\,, \tag{3.278}$$

where $\boldsymbol{m_r} = \boldsymbol{m} + \boldsymbol{\delta m}$ is a diagonal matrix with diagonal elements $m_{r,j}$. Using eqs. (3.243) and (3.256), it follows that

$$\widehat{\boldsymbol{\Sigma}}_i{}^j(p) = \boldsymbol{\Sigma}_i{}^j(p) - \delta m_j \delta_i{}^j - \tfrac{1}{2}\delta\overline{Z}_i{}^j(\not{p} - m_j) - \tfrac{1}{2}(\not{p} - m_i)\delta Z_i{}^j\,, \tag{3.279}$$

where there is no sum over the repeated indices. The on-shell conditions that determine the off-diagonal wave function renormalization constants are chosen such that fermion mixing does not occur on external fermion lines of the Feynman diagram of a scattering or decay process. That is,

$$\widehat{\boldsymbol{\Sigma}}_i{}^j(p)u_j(p) = 0\,, \qquad \text{for } p^2 \to m_j^2 \text{ and } i \neq j, \tag{3.280}$$

$$\bar{u}^i(p)\widehat{\boldsymbol{\Sigma}}_i{}^j(p) = 0\,, \qquad \text{for } p^2 \to m_i^2 \text{ and } i \neq j, \tag{3.281}$$

$$\widehat{\boldsymbol{\Sigma}}_i{}^j(p)v_j(-p) = 0\,, \qquad \text{for } p^2 \to m_j^2 \text{ and } i \neq j, \tag{3.282}$$

$$\bar{v}^i(-p)\widehat{\boldsymbol{\Sigma}}_i{}^j(p) = 0\,, \qquad \text{for } p^2 \to m_i^2 \text{ and } i \neq j. \tag{3.283}$$

Employing eqs. (3.227) and (3.279) in eqs. (3.280) and (3.281), it follows that, for $i \neq j$,

$$\begin{aligned}\{\not{p}[\boldsymbol{\Sigma_L}(m_j^2)P_L + \boldsymbol{\Sigma}^{\mathsf{T}}_{\boldsymbol{R}}(m_j^2)P_R] + \boldsymbol{\Sigma_D}(m_j^2)P_L + \overline{\boldsymbol{\Sigma}}{}^{\mathsf{T}}_{\boldsymbol{D}}(m_j^2)P_R& \\ - \tfrac{1}{2}(\not{p} - m_i)\delta Z\}_i{}^j u_j(p) = 0\,,&\end{aligned} \tag{3.284}$$

$$\begin{aligned}\bar{u}^i(p)\{\not{p}[\boldsymbol{\Sigma_L}(m_i^2)P_L + \boldsymbol{\Sigma}^{\mathsf{T}}_{\boldsymbol{R}}(m_i^2)P_R] + \boldsymbol{\Sigma}^{\boldsymbol{D}}(m_i^2)P_L + \overline{\boldsymbol{\Sigma}}{}^{\mathsf{T}}_{\boldsymbol{D}}(m_i^2)P_R& \\ - \tfrac{1}{2}\delta\overline{Z}(\not{p} - m_j)\}_i{}^j = 0\,,&\end{aligned} \tag{3.285}$$

after making use of the Dirac equation [see eqs. (3.172) and (3.173)]. Using eq. (3.249) and the Dirac equation once more, we end up with

$$\begin{aligned}&m_j\{[\boldsymbol{\Sigma_L}(m_j^2)_i{}^j - \tfrac{1}{2}(\delta Z_L)_i{}^j]P_R + [\boldsymbol{\Sigma}^{\mathsf{T}}_{\boldsymbol{R}}(m_j^2)_i{}^j - \tfrac{1}{2}(\delta Z_R)_i{}^j]P_L]\} \\ &\quad + \boldsymbol{\Sigma_D}(m_j^2)_i{}^jP_L + \overline{\boldsymbol{\Sigma}}{}^{\mathsf{T}}_{\boldsymbol{D}}(m_j^2)_i{}^jP_R + \tfrac{1}{2}m_i[(\delta Z_L)_i{}^jP_L + (\delta Z_R)_i{}^jP_R] = 0\,,\end{aligned} \tag{3.286}$$

$$\begin{aligned}&m_i\{[\boldsymbol{\Sigma_L}(m_i^2)_i{}^j - \tfrac{1}{2}(\delta Z_L^\dagger)_i{}^j]P_L + [\boldsymbol{\Sigma}^{T}_{\boldsymbol{R}}(m_i^2)_i{}^j - \tfrac{1}{2}(\delta Z_R^\dagger)_i{}^j]P_R]\} \\ &\quad + \boldsymbol{\Sigma_D}(m_i^2)_i{}^jP_L + \overline{\boldsymbol{\Sigma}}{}^{\mathsf{T}}_{\boldsymbol{D}}(m_i^2)_i{}^jP_R + \tfrac{1}{2}m_j[(\delta Z_R^\dagger)_i{}^jP_L + (\delta Z_L^\dagger)_i{}^jP_R] = 0\,.\end{aligned} \tag{3.287}$$

These equations are satisfied if the coefficients of P_L and P_R, respectively, vanish. Thus we obtain four independent equations,[23]

$$\tfrac{1}{2}(\delta Z_L\boldsymbol{m} - \boldsymbol{m}\delta Z_R)_i{}^j = (\boldsymbol{\Sigma_L}(m_j^2)\boldsymbol{m} + \overline{\boldsymbol{\Sigma}}{}^{\mathsf{T}}_{\boldsymbol{D}}(m_j^2))_i{}^j\,, \tag{3.288}$$

$$\tfrac{1}{2}(\delta Z_R\boldsymbol{m} - \boldsymbol{m}\delta Z_L)_i{}^j = (\boldsymbol{\Sigma}^{\mathsf{T}}_{\boldsymbol{R}}(m_j^2)\boldsymbol{m} + \boldsymbol{\Sigma_D}(m_j^2))_i{}^j\,, \tag{3.289}$$

$$\tfrac{1}{2}(\boldsymbol{m}\delta Z_L^\dagger - \delta Z_R^\dagger\boldsymbol{m})_i{}^j = (\boldsymbol{m}\boldsymbol{\Sigma_L}(m_i^2) + \boldsymbol{\Sigma_D}(m_i^2))_i{}^j\,, \tag{3.290}$$

$$\tfrac{1}{2}(\boldsymbol{m}\delta Z_R^\dagger - \delta Z_L^\dagger\boldsymbol{m})_i{}^j = (\boldsymbol{m}\boldsymbol{\Sigma}^{\mathsf{T}}_{\boldsymbol{R}}(m_i^2) + \overline{\boldsymbol{\Sigma}}{}^{\mathsf{T}}_{\boldsymbol{D}}(m_i^2))_i{}^j\,, \tag{3.291}$$

thereby reproducing the results of eqs. (2.342)–(2.345). The solutions to these equations are unique, as noted in Section 2.10.2. For the convenience of the reader, we record the final expressions for the off-diagonal elements of δZ_L, δZ_R, and their

[23] Note that a similar computation that employs eqs. (3.227) and (3.279) in eqs. (3.282) and (3.283) also yields eqs. (3.288)–(3.291).

hermitian adjoints below:

$$(\delta Z_L)_i{}^j = \frac{2\big[\boldsymbol{\Sigma_L}(m_j^2)\boldsymbol{m^2} + \boldsymbol{m}\boldsymbol{\Sigma}_{\boldsymbol{R}}^{\mathsf{T}}(m_j^2)\boldsymbol{m} + \boldsymbol{m}\boldsymbol{\Sigma_D}(m_j^2) + \overline{\boldsymbol{\Sigma}}_{\boldsymbol{D}}^{\mathsf{T}}(m_j^2)\boldsymbol{m}\big]_i{}^j}{m_j^2 - m_i^2}\,, \tag{3.292}$$

$$(\delta Z_R)_i{}^j = \frac{2\big[\boldsymbol{m}\boldsymbol{\Sigma_L}(m_j^2)\boldsymbol{m} + \boldsymbol{\Sigma}_{\boldsymbol{R}}^{\mathsf{T}}(m_j^2)\boldsymbol{m^2} + \boldsymbol{\Sigma_D}(m_j^2)\boldsymbol{m} + \boldsymbol{m}\overline{\boldsymbol{\Sigma}}_{\boldsymbol{D}}^{\mathsf{T}}(m_j^2)\big]_i{}^j}{m_j^2 - m_i^2}\,, \tag{3.293}$$

$$(\delta Z_L^\dagger)_i{}^j = \frac{2\big[\boldsymbol{m^2}\boldsymbol{\Sigma_L}(m_i^2) + \boldsymbol{m}\boldsymbol{\Sigma}_{\boldsymbol{R}}^{\mathsf{T}}(m_i^2)\boldsymbol{m} + \boldsymbol{m}\boldsymbol{\Sigma_D}(m_i^2) + \overline{\boldsymbol{\Sigma}}_{\boldsymbol{D}}^{\mathsf{T}}(m_i^2)\boldsymbol{m}\big]_i{}^j}{m_i^2 - m_j^2}\,, \tag{3.294}$$

$$(\delta Z_R^\dagger)_i{}^j = \frac{2\big[\boldsymbol{m}\boldsymbol{\Sigma_L}(m_i^2)\boldsymbol{m} + \boldsymbol{m^2}\boldsymbol{\Sigma}_{\boldsymbol{R}}^{\mathsf{T}}(m_i^2) + \boldsymbol{\Sigma_D}(m_i^2)\boldsymbol{m} + \boldsymbol{m}\overline{\boldsymbol{\Sigma}}_{\boldsymbol{D}}^{\mathsf{T}}(m_i^2)\big]_i{}^j}{m_i^2 - m_j^2}\,, \tag{3.295}$$

for $i \neq j$.

In the case of N four-component Majorana fields, the wave function renormalization constants are again introduced as in eq. (3.240), where Ψ_i and $\Psi_{r,i}$ satisfy the corresponding Majorana conditions, $\Psi_i = C\overline{\Psi}_i^{\mathsf{T}}$ and $\Psi_{r,i} = C\overline{\Psi}_{r,i}^{\mathsf{T}}$. Imposing the Majorana condition on eq. (3.240) implies that

$$Z = C\overline{Z}^{\mathsf{T}}C^{-1}\,. \tag{3.296}$$

Using eq. (3.246), it then follows that

$$Z_L P_L + Z_R P_R = C(Z_L^\dagger P_R + Z_R^\dagger P_L)^{\mathsf{T}} C^{-1} = Z_L^* P_R + Z_R^* P_L\,, \tag{3.297}$$

after making use of eq. (3.46). Multiplying the above equation by either P_L or P_R yields the same conclusion,

$$Z_M \equiv Z_L = Z_R^*\,, \tag{3.298}$$

which defines the Majorana fermion wave function renormalization constant, Z_M.

To obtain the complex pole mass and the diagonal and off-diagonal elements of δZ_M and $\delta Z_M^\dagger$, we simply identify $\boldsymbol{\Xi} \equiv \boldsymbol{\Sigma_L} = \boldsymbol{\Sigma_R}$, and $\boldsymbol{\Omega} \equiv \boldsymbol{\Sigma_D}$, as previously noted below eq. (3.239). Applying these identifications to the computation above, we recover the results obtained at the end of Section 2.10.2, which we provide again below for the reader's convenience.

The mass counterterm and width at one-loop accuracy are given by

$$\delta m_j = \operatorname{Re}\big[m_j\,\boldsymbol{\Xi}(m_j^2)_j{}^j + \tfrac{1}{2}\boldsymbol{\Omega}(m_j^2)^{jj} + \tfrac{1}{2}\overline{\boldsymbol{\Omega}}(m_j^2)_{jj}\big]\,, \tag{3.299}$$

$$\Gamma_j = -2\operatorname{Im}\big[m_j\,\boldsymbol{\Xi}(m_j^2)_j{}^j + \tfrac{1}{2}\boldsymbol{\Omega}(m_j^2)^{jj} + \tfrac{1}{2}\overline{\boldsymbol{\Omega}}(m_j^2)_{jj}\big]\,. \tag{3.300}$$

Likewise, the diagonal elements of δZ_M and $\delta Z_M^\dagger$ are given by

$$(\delta Z_M)_j{}^j = \boldsymbol{\Xi}(m_j^2)_j{}^j - \frac{\boldsymbol{\Omega}(m_j^2)^{jj} - \overline{\boldsymbol{\Omega}}(m_j^2)_{jj}}{2m_j} + \boldsymbol{Q}_j \,, \tag{3.301}$$

$$(\delta Z_M^\dagger)_j{}^j = \boldsymbol{\Xi}(m_j^2)_j{}^j + \frac{\boldsymbol{\Omega}(m_j^2)^{jj} - \overline{\boldsymbol{\Omega}}(m_j^2)_{jj}}{2m_j} + \boldsymbol{Q}_j \,, \tag{3.302}$$

where

$$\boldsymbol{Q}_j \equiv 2m_j^2 \boldsymbol{\Xi}'(m_j^2)_j{}^j + m_j \left[\boldsymbol{\Omega}'(m_j^2)^{jj} + \overline{\boldsymbol{\Omega}}'(m_j^2)_{jj}\right], \tag{3.303}$$

and the prime indicates differentiation with respect to p^2 which is then evaluated at $p^2 = m_j^2$. Finally, the off-diagonal elements of δZ_M and its hermitian conjugate are given by

$$(\delta Z_M)_i{}^j = \frac{2\left[\boldsymbol{\Xi}(m_j^2)\overline{\boldsymbol{m}}\boldsymbol{m} + \overline{\boldsymbol{m}}\,\boldsymbol{\Xi}^{\mathsf{T}}(m_j^2)\boldsymbol{m} + \overline{\boldsymbol{m}}\,\boldsymbol{\Omega}(m_j^2) + \overline{\boldsymbol{\Omega}}(m_j^2)\boldsymbol{m}\right]_i{}^j}{m_j^2 - m_i^2}\,, \quad (\text{for } i \neq j)\,, \tag{3.304}$$

$$(\delta Z_M^\dagger)_i{}^j = \frac{2\left[\overline{\boldsymbol{m}}\boldsymbol{m}\boldsymbol{\Xi}(m_i^2) + \overline{\boldsymbol{m}}\,\boldsymbol{\Xi}^{\mathsf{T}}(m_i^2)\boldsymbol{m} + \overline{\boldsymbol{m}}\boldsymbol{\Omega}(m_i^2) + \overline{\boldsymbol{\Omega}}(m_i^2)\boldsymbol{m}\right]_i{}^j}{m_i^2 - m_j^2}\,, \quad (\text{for } i \neq j)\,, \tag{3.305}$$

where $\overline{\boldsymbol{m}} = \boldsymbol{m}^\dagger$ [see eq. (2.241)].

The expressions obtained above for δm, δZ_L, δZ_R, and δZ_m coincide with the results previously obtained in the literature (e.g., see Refs. [32, 38]).

In Section 3.10, we shall apply the expressions for the matrix wave function renormalization constants, obtained by imposing on-mass-shell renormalization conditions, to external fermion states that appear in scattering or decay processes. Strictly speaking, this application is rigorously valid only when the external fermion states are stable, and thus correspond to asymptotic states of the theory. It is for this reason that the derivations in this section were presented under the assumption that the absorptive parts of the self-energy functions vanish when evaluated on mass shell.

Nevertheless, it is sometimes desirable to treat unstable particles as external states by pretending that it is proper to treat the matrix elements for scattering and/or decay of unstable particles using the same Feynman diagram techniques that are applied to true asymptotic states. Following the discussion at the end of Section 2.9.1, in cases of unstable external fermions one can employ the formulae obtained in this section for δZ_L, δZ_R, and δZ_M (at one-loop accuracy) by replacing the corresponding self-energy functions by their dispersive parts. For a more detailed discussion of the many subtleties involved in this procedure, see Refs. [34, 39].

3.10 Feynman Rules for External Fermion Lines Revisited

Consider the S-matrix elements for a scattering or decay process at higher orders in perturbation theory beyond the tree level, which are computed by evaluating the contributing connected Feynman diagrams by employing Feynman rules based on the tree-level Lagrangian prior to renormalization. As noted in Section 2.11, the LSZ reduction formulae instruct us to replace any flavor-diagonal or flavor-off-diagonal self-energy insertion on the external lines with the corresponding diagonal or off-diagonal matrix elements of the wave function renormalization constants. Then, the Feynman rules for assigning four-component external-state spinors of momentum $\vec{\boldsymbol{p}}$, spin label s, and flavor label j specified in Section 3.7 are modified as follows:

- $(Z^{1/2})_i{}^j u_j(\vec{\boldsymbol{p}}, s)$: incoming Ψ [or Ψ^C] with flavor j and momentum $\vec{\boldsymbol{p}}$ parallel to the arrow direction,
- $\bar{u}^j(\vec{\boldsymbol{p}}, s)(\overline{Z}^{1/2})_j{}^i$: outgoing Ψ [or Ψ^C] with flavor j and momentum $\vec{\boldsymbol{p}}$ parallel to the arrow direction,
- $(Z^{1/2})_i{}^j v_j(\vec{\boldsymbol{p}}, s)$: outgoing Ψ [or Ψ^C] with flavor j and momentum $\vec{\boldsymbol{p}}$ antiparallel to the arrow direction,
- $\bar{v}^j(\vec{\boldsymbol{p}}, s)(\overline{Z}^{1/2})_j{}^i$: incoming Ψ [or Ψ^C] with flavor j and momentum $\vec{\boldsymbol{p}}$ antiparallel to the arrow direction,

(3.306)

where Z and $\overline{Z}$, which are fixed by imposing on-mass-shell renormalization conditions, can be decomposed into left- and right-handed parts following eq. (3.246). No sum over j is implied since j refers to the fixed flavor index of the external fermion state. The index i is the flavor index of the four-component fermion line that attaches to the rest of the Feynman diagram, and there will be an implicit sum over i corresponding to a sum over Feynman diagrams.

The rules exhibited in eq. (3.306) apply both to four-component Majorana and Dirac fermions; in the case of Majorana fermions, Z_L and Z_R are related by eq. (3.298). However, the conventions implicit in defining the heights of the flavor indices differ in their applications to the four-component spinor wave functions of Majorana and Dirac fermions. In particular, one should employ eqs. (3.208) and (3.209) for Majorana fermions and eqs. (3.211) and (3.212) for Dirac fermions.

For tree-level calculations, the matrix wave function renormalization constants $Z_i{}^j$ and $\overline{Z}_j{}^i$ can be replaced by the corresponding Kronecker delta. In the one-loop approximation,

$$(Z^{1/2})_i{}^j = \delta_i^j + \tfrac{1}{2}(\delta Z)_i{}^j\,, \qquad (\overline{Z}^{1/2})_j{}^i = \delta_j^i + \tfrac{1}{2}(\delta\overline{Z})_j{}^i\,. \tag{3.307}$$

We provide two examples by revisiting the computation of the decay of a massive neutral scalar boson into a pair of Majorana or Dirac fermions. The tree-level decay amplitude for the decay of a neutral scalar into a pair of Majorana fermions was

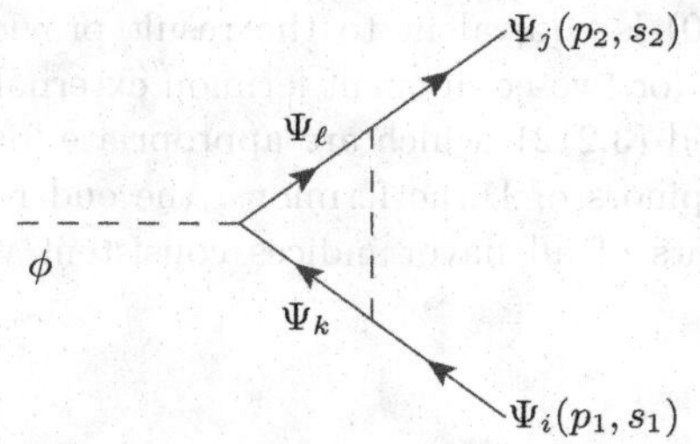

Fig. 3.11 The Feynman diagram that contributes to the one-loop vertex correction to the amplitude for the decay of a neutral scalar into a fermion–antifermion pair. The diagram above applies both in the cases of final-state Majorana and Dirac fermions.

previously given in eq. (3.207). Using the rules exhibited in eqs. (2.361) and (3.306) and employing eq. (3.246) subject to the constraint given by eq. (3.298), the decay matrix element in the one-loop approximation is given by

$$\begin{aligned} i\mathcal{M} = &-i\bar{u}^i(\vec{p}_1, s_1)(\lambda^{ij}P_L + \lambda_{ij}P_R)v_j(\vec{p}_2, s_2)\left[1 + \tfrac{1}{2}\delta Z_\phi\right] \\ &- \tfrac{1}{2}i\bar{u}^i(\vec{p}_1, s_1)\left[(\delta Z_M^\dagger P_R + \delta Z_M^\mathsf{T} P_L\right]_i{}^k(\lambda^{kj}P_L + \lambda_{kj}P_R)v_j(\vec{p}_2, s_2) \\ &- \tfrac{1}{2}i\bar{u}^i(\vec{p}_1, s_1)(\lambda^{i\ell}P_L + \lambda_{i\ell}P_R)\left[\delta Z_M P_L + \delta Z_M^* P_R\right]_\ell{}^j v_j(\vec{p}_2, s_2) \\ &+ i\mathcal{M}_{loop}\,, \end{aligned} \tag{3.308}$$

for fixed flavors i and j, and an implied sum over the repeated indices k and ℓ, where $i\mathcal{M}_{loop}$ is the one-loop vertex correction shown in Fig. 3.11. We can check that eq. (3.308) is equivalent to the result previously obtained using the corresponding rules for two-component fermion external states. Here, we must employ eqs. (3.208) and (3.209) and note that

$$(\delta Z_M^\mathsf{T})_i{}^k = (\delta Z_M)_k{}^i\,, \qquad (\delta Z_M^*)_\ell{}^j = (\delta Z_M^\dagger)_j{}^\ell\,. \tag{3.309}$$

Indeed, the end result coincides with eq. (2.363), with the heights of all flavor indices consistent with the conventions previously established.

Likewise, the tree-level decay amplitude for the decay of a neutral scalar into a pair of Dirac fermions was previously given in eq. (3.210). Using the rules exhibited in eqs. (2.361) and (3.306), the decay matrix element in the one-loop approximation is given by

$$\begin{aligned} i\mathcal{M} = &-i\bar{u}^i(\vec{p}_1, s_1)(\kappa^j{}_i P_L + \kappa_i{}^j P_R)v_j(\vec{p}_2, s_2)\left[1 + \tfrac{1}{2}\delta Z_\phi\right] \\ &- \tfrac{1}{2}i\bar{u}^i(\vec{p}_1, s_1)\left[(\delta Z_L^\dagger P_R + \delta Z_L^\dagger P_R\right]_i{}^k(\kappa^j{}_k P_L + \kappa_k{}^j P_R)v_j(\vec{p}_2, s_2) \\ &- \tfrac{1}{2}i\bar{u}^i(\vec{p}_1, s_1)(\kappa^\ell{}_i P_L + \kappa_i{}^\ell P_R)\left[\delta Z_L P_L + \delta Z_R P_R\right]_\ell{}^j v_j(\vec{p}_2, s_2) \\ &+ i\mathcal{M}_{loop}\,, \end{aligned} \tag{3.310}$$

for fixed flavors i and j, and an implied sum over the repeated indices k and ℓ, where $i\mathcal{M}_{loop}$ is the one-loop vertex correction shown in Fig. 3.11. We can check that eq. (3.310) is equivalent to the result previously obtained using the corresponding rules for two-component fermion external states. Indeed, after employing eqs. (3.211) and (3.212), which are appropriate for specifying the flavor indices of the u and v spinors of Dirac fermions, the end result coincides with eq. (3.308), with the heights of all flavor indices consistent with the conventions previously established.

Exercises

3.1 Given any representation of the Dirac gamma matrices, γ^μ [and corresponding four-component Dirac spinor $\Psi(x)$], a new representation $\widetilde{\gamma}^\mu \equiv \mathcal{S}\gamma^\mu\mathcal{S}^{-1}$ also satisfies eq. (3.8), where $\mathcal{S}$ is any nonsingular matrix. The Dirac spinor with respect to the new representation is given by $\widetilde{\Psi}(x) = \mathcal{S}\Psi(x)$. Discuss how the freedom to choose an arbitrary nonsingular matrix $\mathcal{S}$ is manifested in the two-component spinor formalism.

3.2 (a) Show that if Q is a 4×4 matrix that commutes with γ^μ (for $\mu = 0, 1, 2, 3$) and R is a 4×4 matrix that anticommutes with the γ^μ, then $Q = c_1\boldsymbol{I}_4$ and $R = c_2\gamma_5$, for some complex numbers c_1 and c_2.

(b) From the defining relations for the matrices A, B, and C [eqs. (3.38) and (3.39)], show that the matrices $A^{-1}A^\dagger$, $B^{-1}B^T$, $C^{-1}C^T$, $(BA^{-1})(BA^{-1})^*$, and $AC(AC)^*$ all commute with γ^μ, and that the matrix CB anticommutes with the γ^μ.

(c) Using parts (a) and (b), show that $B^T = -B$ and $C^T = -C$ in all gamma matrix representations.

(d) By imposing $(\Psi^C)^C = \Psi$ and $(\Psi^t)^t = -\Psi$, derive

$$AB^{-1} = -(AB^{-1})^*\,, \qquad (AC)^{-1} = (AC)^*\,. \tag{3.311}$$

Finally, impose the condition $\Psi = A^{-1}\overline{\Psi}^\dagger$ and conclude that $A^\dagger = A$. Verify that the choice of A, B, and C is still not unique, as specified in eq. (A.119).

3.3 Starting with the chiral representation of the Dirac gamma matrices, transform to the Dirac representation while keeping track of the two-component spinor indices. Show that in the Dirac representation, the four-component spinor is given by

$$\Psi = \frac{1}{\sqrt{2}}\begin{pmatrix} \xi_\alpha + \sigma^0_{\alpha\dot{\beta}}\bar{\eta}^{\dot{\beta}} \\ \bar{\eta}^{\dot{\alpha}} - \bar{\sigma}^{0\dot{\alpha}\beta}\xi_\beta \end{pmatrix}, \tag{3.312}$$

and the Dirac gamma matrices are given by

$$\gamma^0 = \begin{pmatrix} \delta_\alpha{}^\beta & 0 \\ 0 & -\delta^{\dot\alpha}{}_{\dot\beta} \end{pmatrix}, \quad \gamma^i = \begin{pmatrix} 0 & \sigma^i_{\alpha\dot\beta} \\ \overline{\sigma}^{i\dot\alpha\beta} & 0 \end{pmatrix}, \quad \gamma_5 = \begin{pmatrix} 0 & \sigma^0_{\alpha\dot\beta} \\ \overline{\sigma}^{0\dot\alpha\beta} & 0 \end{pmatrix}. \tag{3.313}$$

Evaluate the matrices A, B, and C in the Dirac representation and show that they are given by

$$A = \begin{pmatrix} \overline{\sigma}^{0\dot\alpha\beta} & 0 \\ 0 & -\sigma^0_{\alpha\dot\beta} \end{pmatrix}, \quad B = \begin{pmatrix} \epsilon^{\alpha\beta} & 0 \\ 0 & -\epsilon_{\dot\alpha\dot\beta} \end{pmatrix}, \quad C = \begin{pmatrix} 0 & -\epsilon_{\alpha\gamma}\overline{\sigma}^{0\dot\beta\gamma} \\ \epsilon^{\dot\alpha\dot\gamma}\sigma^0_{\beta\dot\gamma} & 0 \end{pmatrix}. \tag{3.314}$$

3.4 Starting with the Dirac representation of the gamma matrices (introduced in the previous exercise), transform to the Majorana representation while keeping track of the two-component spinor indices. Show that, in the Majorana representation, the four-component spinor is given by

$$\Psi = \frac{1}{\sqrt{2}} \begin{pmatrix} \xi_\alpha + \sigma^2_{\alpha\dot\beta}\overline{\eta}^{\dot\beta} \\ -\overline{\eta}^{\dot\alpha} - \overline{\sigma}^{2\dot\alpha\beta}\xi_\beta \end{pmatrix}, \tag{3.315}$$

and the gamma matrices are given by

$$\gamma^0 = \begin{pmatrix} 0 & \sigma^2 \\ -\overline{\sigma}^2 & 0 \end{pmatrix}, \quad \gamma^1 = 2i \begin{pmatrix} \sigma^{12} & 0 \\ 0 & \overline{\sigma}^{12} \end{pmatrix}, \quad \gamma^2 = \begin{pmatrix} 0 & -\sigma^2 \\ -\overline{\sigma}^2 & 0 \end{pmatrix},$$

$$\gamma^3 = -2i \begin{pmatrix} \sigma^{23} & 0 \\ 0 & \overline{\sigma}^{23} \end{pmatrix}, \quad \gamma_5 = 2i \begin{pmatrix} \sigma^{02} & 0 \\ 0 & \overline{\sigma}^{02} \end{pmatrix}, \tag{3.316}$$

where the spinor labels in eq. (3.316) have been suppressed.

Evaluate the matrices A, B, and C in the Majorana representation and show that they are given by

$$A = -2i \begin{pmatrix} 0 & \overline{\sigma}^{02\dot\alpha}{}_{\dot\beta} \\ -\sigma^{02}{}_\alpha{}^\beta & 0 \end{pmatrix}, \qquad B = \begin{pmatrix} 0 & \epsilon^{\alpha\gamma}\sigma^2_{\gamma\dot\beta} \\ -\epsilon_{\dot\alpha\dot\gamma}\overline{\sigma}^{0\dot\gamma\beta} & 0 \end{pmatrix},$$

$$C = -i \begin{pmatrix} 0 & \epsilon_{\alpha\gamma}\overline{\sigma}^{0\dot\beta\gamma} \\ -\epsilon^{\dot\alpha\dot\gamma}\sigma^0_{\beta\dot\gamma} & 0 \end{pmatrix}. \tag{3.317}$$

The factor of $-i$ in C is conventional and has been chosen so that the numerical value of CA^T is equal to the identity matrix. Verify this last result (and display the correct two-component spinor label structure).

3.5 (a) Translate the results of Exercise 1.12(b) to four-component spinor notation. Show that eq. (1.317) is equivalent to

$$S_P^{-1}\gamma^\mu S_P = (\Lambda_P)^\mu{}_\nu\gamma^\nu, \tag{3.318}$$

where

$$S_P \equiv \begin{pmatrix} 0 & P \\ \overline{P} & 0 \end{pmatrix}. \tag{3.319}$$

Verify eq. (3.318) directly in the four-component language by noting that $\Psi^P(x) = iS_P\Psi(x)$, where $S_P \equiv \gamma^0$ [see eq. (3.30)].

(b) The time-reversed four-component spinor defined in eq. (3.31) has the form $\Psi^t(x) = S\Psi^*(x_T)$, where $S \equiv -\gamma^0 B^{-1} A^{\mathsf{T}}$. Show that S satisfies the following condition:

$$S^{-1}\gamma^\mu S = -(\Lambda_T)^\mu{}_\nu \gamma^{\nu *}. \tag{3.320}$$

(c) Translate the results of Exercise 1.13(b) to four-component spinor notation. Verify that eq. (1.320) is equivalent to

$$S_T^{-1}\gamma^\mu S_T = -(\Lambda_T)^\mu{}_\nu \gamma^{\nu\dagger}, \tag{3.321}$$

where

$$S_T \equiv \begin{pmatrix} T & 0 \\ 0 & \overline{T} \end{pmatrix}. \tag{3.322}$$

Finally, in light of eqs. (3.107) and (3.110), show that $S = S_T B^*$ and use this result and that of eq. (3.322) to rederive eq. (3.320).

3.6 (a) The Racah time-reversed spinor is defined by $\Psi'(x') = [\Psi^C(x)]^t$, where $x'^\mu = (\Lambda_T)^\mu{}_\nu x^\nu$. Without assuming the chiral representation, show that $[\Psi^C(x)]^t = S_{CT}\Psi(x)$, where $S_{CT} = \gamma^0\gamma_5$. Then, verify that

$$S_{CT}^{-1}\,\gamma^\mu\, S_{CT} = (\Lambda_T)^\mu{}_\nu \gamma^\nu. \tag{3.323}$$

This is the direct analogue of the parity transformation given in eq. (3.318).

(b) What is the two-component spinor version of Racah time reflection?

3.7 (a) Explain what is wrong with the following computation:

$$\begin{aligned}
\mathcal{CPT}\Psi(x)(\mathcal{CPT})^{-1} &= \mathcal{CP}[\eta_T(\gamma^0 A^{-1})^{\mathsf{T}} B\,\Psi(x_T)](\mathcal{CP})^{-1} \\
&= \mathcal{C}[i\eta_P\eta_T\gamma^0(\gamma^0 A^{-1})^{\mathsf{T}} B\Psi(-x)]\mathcal{C}^{-1} \\
&= \eta_C C A^{\mathsf{T}}[i\eta_P\eta_T\gamma^0(\gamma^0 A^{-1})^{\mathsf{T}} B\Psi(-x)]^*, \\
&= -i\eta_C\eta_P^*\eta_T^* C A^{\mathsf{T}}\gamma^{0*}(\gamma^0 A^{-1})^\dagger B^*\Psi^*(-x),
\end{aligned}$$

where Ψ is a four-component Dirac field. In particular, by writing out the various matrices in the chiral representation and keeping the two-component spinor labels explicit, show that the last line above does not make any sense. Compare with eq. (3.112) and explain why the latter provides the correct method of computation.

(b) The field Ψ transforms under charge conjugation according to eq. (3.108). Compare the following two computations:

$$\mathcal{C}i\Psi\mathcal{C}^{-1} = (\mathcal{C}\,i\,\mathcal{C}^{-1})(\mathcal{C}\Psi\mathcal{C}^{-1}) = i\eta_C CA^{\mathsf{T}}\Psi^*\,, \tag{3.324}$$

$$\mathcal{C}i\Psi\mathcal{C}^{-1} = \eta_C CA^{\mathsf{T}}(i\Psi)^* = -i\eta_C CA^{\mathsf{T}}\Psi^*\,. \tag{3.325}$$

Which computation is correct? Explain.

3.8 (a) Generalize eqs. (3.130)–(3.134) to a multiplet of Majorana fermions.

(b) In Section 1.6, a field theory of n free two-component fermion fields is constructed with an arbitrary mass matrix M^{ij}. Convert the results of this section to four-component spinor notation. Write down the most general model of n four-component free-fermion fields and discuss the mass diagonalization procedure. Consider separately the cases of a multiplet of Majorana fermions and a multiplet of Dirac fermions.

3.9 Using the results of Exercise 2.1, one can obtain the four-component spinor versions of the Gordon identities after employing eqs. (3.167) and (3.168).

(a) Show that Exercise 2.1(a) yields the following four Gordon identities:

$$\bar{u}(\vec{\boldsymbol{p}}\,')\big[P^\mu + i\Sigma^{\mu\nu}q_\nu\big]u(\vec{\boldsymbol{p}}) = 2m\bar{u}(\vec{\boldsymbol{p}}\,')\gamma^\mu u(\vec{\boldsymbol{p}})\,, \tag{3.326}$$

$$\bar{u}(\vec{\boldsymbol{p}}\,')\big[q^\mu + i\Sigma^{\mu\nu}P_\nu\big]\gamma_5 u(\vec{\boldsymbol{p}}) = 2m\bar{u}(\vec{\boldsymbol{p}}\,')\gamma^\mu\gamma_5 u(\vec{\boldsymbol{p}})\,, \tag{3.327}$$

$$\bar{u}(\vec{\boldsymbol{p}}\,')\big[q^\mu + i\Sigma^{\mu\nu}P_\nu\big]u(\vec{\boldsymbol{p}}) = 0\,, \tag{3.328}$$

$$\bar{u}(\vec{\boldsymbol{p}}\,')\big[P^\mu + i\Sigma^{\mu\nu}q_\nu\big]\gamma_5 u(\vec{\boldsymbol{p}}) = 0\,, \tag{3.329}$$

where m is the common fermion mass, $P \equiv p + p'$, and $q \equiv p' - p$.

(b) Show that a second set of four Gordon identities can be obtained from eqs. (3.326)–(3.329) by replacing $u \to v$, $\bar{u} \to \bar{v}$, and $m \to -m$.

HINT: Use eq. (3.186) in rewriting the u spinors in terms of the v spinors in eqs. (3.326)–(3.329).

(c) Show that parts (c) and (d) of Exercise 2.1 yields the following four Gordon identities:

$$\bar{u}(\vec{\boldsymbol{p}}\,')\big[q^\mu + i\Sigma^{\mu\nu}P_\nu\big]v(\vec{\boldsymbol{p}}) = 2m\bar{u}(\vec{\boldsymbol{p}}\,')\gamma^\mu v(\vec{\boldsymbol{p}})\,, \tag{3.330}$$

$$\bar{u}(\vec{\boldsymbol{p}}\,')\big[P^\mu + i\Sigma^{\mu\nu}q_\nu\big]\gamma_5 v(\vec{\boldsymbol{p}}) = 2m\bar{u}(\vec{\boldsymbol{p}}\,')\gamma^\mu\gamma_5 v(\vec{\boldsymbol{p}})\,, \tag{3.331}$$

$$\bar{u}(\vec{\boldsymbol{p}}\,')\big[P^\mu + i\Sigma^{\mu\nu}q_\nu\big]v(\vec{\boldsymbol{p}}) = 0\,, \tag{3.332}$$

$$\bar{u}(\vec{\boldsymbol{p}}\,')\big[q^\mu + i\Sigma^{\mu\nu}P_\nu\big]\gamma_5 v(\vec{\boldsymbol{p}}) = 0\,. \tag{3.333}$$

(d) In light of the hermiticity relations of eq. (3.44), show that a fourth set of four Gordon identities can be obtained from eqs. (3.330)–(3.333) by replacing $v \to u$, $\bar{u} \to \bar{v}$, and $m \to -m$.

3.10 (a) The helicity operator acting on four-component spinors is defined by

$$h = \tfrac{1}{2}\vec{\boldsymbol{\Sigma}}\boldsymbol{\cdot}\widehat{\boldsymbol{p}}\,, \qquad \text{where } \widehat{\boldsymbol{p}} \equiv \vec{\boldsymbol{p}}/|\vec{\boldsymbol{p}}|, \tag{3.334}$$

and $\Sigma^i \equiv \gamma^0\gamma^i\gamma_5$ [see eq. (3.16)]. Using eqs. (3.167), (3.168), and the properties of the two-component helicity spinor wave functions x and y, show that

$$hu(\vec{\boldsymbol{p}},\lambda) = \lambda u(\vec{\boldsymbol{p}},\lambda)\,, \qquad hv(\vec{\boldsymbol{p}},\lambda) = -\lambda v(\vec{\boldsymbol{p}},\lambda)\,, \tag{3.335}$$

for both massive and massless spinors, where the helicity $\lambda = \pm\frac{1}{2}$.

(b) In light of part (a), show that on-shell massive helicity spinors satisfy

$$(\gamma_5 \not{S} - 2h)u(\vec{\boldsymbol{p}},\lambda) = 0\,, \qquad (\gamma_5 \not{S} + 2h)v(\vec{\boldsymbol{p}},\lambda) = 0\,, \tag{3.336}$$

where $\not{S} \equiv \gamma_\mu S^\mu$ and $S^\mu = \big(|\vec{\boldsymbol{p}}|/m\,;\, E\widehat{\boldsymbol{p}}/m\big)$. Using eqs. (3.335) and (3.336), verify that eq. (3.174) [with s replaced by λ] is satisfied.

(c) Show that, in the massless limit,

$$\lim_{m\to 0}\,(\not{S} - 1)u(\vec{\boldsymbol{p}},\lambda) = 0\,, \qquad \lim_{m\to 0}\,(\not{S} + 1)v(\vec{\boldsymbol{p}},\lambda) = 0\,. \tag{3.337}$$

Using eq. (3.336), verify that a massless helicity spinor is an eigenstates of γ_5 and an eigenstate of $2h$ with the same eigenvalue [see eqs. (3.178) and (3.335)].

3.11 The following results are useful in analyzing the properties of four-component spinor fields under space inversion.

(a) Show that the u-spinor and v-spinor wave functions satisfy the relations

$$u(\vec{\boldsymbol{p}},s) = \gamma^0 u(-\vec{\boldsymbol{p}},s)\,, \qquad v(\vec{\boldsymbol{p}},s) = -\gamma^0 v(-\vec{\boldsymbol{p}},s)\,, \tag{3.338}$$

where $s = \pm\frac{1}{2}$ is the spin quantum number obtained by quantizing the spin along a fixed axis in the rest frame.

(b) Derive the corresponding formulae for the helicity spinor wave functions

$$u(\vec{\boldsymbol{p}},\lambda) = -iPu(-\vec{\boldsymbol{p}},-\lambda), \qquad v(\vec{\boldsymbol{p}},\lambda) = -iPv(-\vec{\boldsymbol{p}},-\lambda), \tag{3.339}$$

employing the conventions of Appendix E.3 to define $u(-\vec{\boldsymbol{p}},\lambda)$ and $v(-\vec{\boldsymbol{p}},\lambda)$.

3.12 The following results are useful in analyzing the properties of four-component spinor fields under time reversal.

(a) Show that the u-spinor and v-spinor wave functions satisfy the relations

$$u^*(\vec{\boldsymbol{p}},s) = 2s(\gamma^0 A^{-1})^{\mathsf{T}} Bu(-\vec{\boldsymbol{p}},-s)\,, \tag{3.340}$$

$$v^*(\vec{\boldsymbol{p}},s) = 2s(\gamma^0 A^{-1})^{\mathsf{T}} Bv(-\vec{\boldsymbol{p}},-s)\,. \tag{3.341}$$

In the chiral representation of the Dirac gamma matrices, we may employ the numerical equalities given by eq. (3.37) to obtain $(\gamma^0 A^{-1})^{\mathsf{T}} B = \gamma^1\gamma^3$.

(b) Derive the corresponding formulae for the helicity spinor wave functions

$$u^*(\vec{p},\lambda) = -2i\lambda(\gamma^0 A^{-1})^{\mathsf{T}} B u(-\vec{p},\lambda)\,, \qquad (3.342)$$

$$v^*(\vec{p},\lambda) = 2i(\gamma^0 A^{-1})^{\mathsf{T}} B v(-\vec{p},\lambda)\,, \qquad (3.343)$$

employing the conventions of Appendix E.3 to define $u(-\vec{p},\lambda)$ and $v(-\vec{p},\lambda)$.

3.13 (a) The expansion of a free four-component (charged) Dirac field in terms of creation and annihilation operators is given by eq. (3.161). Employing this result along with the results of Exercises 3.11 and 3.12, determine the action of the operators $\mathcal{P}$, $\mathcal{T}$, and $\mathcal{C}$ on the annihilation operators a and b. Consider separately the cases of fixed-axis spin states and helicity states. Verify the results previously obtained in eqs. (2.389)–(2.398).

(b) The expansion of a free four-component (neutral) Majorana field in terms of creation and annihilation operators is given by eq. (3.157). Repeat the computation of part (a) and show that the results of eqs. (2.389), (2.391), (2.393), and (2.395) still apply, whereas the action of $\mathcal{C}$ on the annihilation operator a is given by eq. (2.399).

(c) Determine the action of $\mathcal{CP}$ and $\mathcal{CPT}$ on the annihilation operators of a Majorana and Dirac fermion field, respectively. Treat separately the cases of fixed-axis spin states and helicity states.

3.14 Consider an alternative plane-wave expansion of a neutral Majorana field:

$$\Psi_{Mi}(x) = \sum_s \int \frac{d^3\boldsymbol{p}}{(2\pi)^{3/2}(2E_{\boldsymbol{p}})^{1/2}} \left[u(\vec{p},s)a_i(\vec{p},s)e^{-ip\cdot x} + \zeta_i v(\vec{p},s)a_i^\dagger(\vec{p},s)e^{ip\cdot x}\right], \qquad (3.344)$$

where i is a flavor index and the ζ_i are complex phases called the *creation phase factors*. Although no physical observable can depend on the ζ_i, it is sometimes convenient to choose creation phase factors that differ from unity [36].

(a) Show that the Majorana condition [eq. (3.49)] must be modified to

$$\Psi_{Mi} = \zeta_i \Psi^C_{Mi}\,, \qquad (3.345)$$

where there is no sum over the repeated index i. As a result, show that a four-component Majorana fermion field $\Psi_{Mi}(x)$ can be expressed in terms of the Weyl fermion field, $\Psi_{Li}(x)$ as follows:

$$\Psi_{Mi}(x) = \Psi_{Li}(x) + \zeta_i C\overline{\Psi}^{\mathsf{T}}_{Li}(x)\,. \qquad (3.346)$$

(b) Employing eq. (3.344) in place of eq. (3.157), how are the results of eqs. (2.389), (2.391), (2.393), (2.395), and (2.399) modified?

(c) Determine the corresponding restrictions on the phases η_P, η_T, and η_C when the ζ_i are arbitrary phases.

(d) How is the mass diagonalization procedure for a multiplet of four-component interaction eigenstate Majorana fermions (previously obtained in Exercise 3.8) modified if the Majorana condition is given by eq. (3.345)?

(e) One can express the four-component mass-eigenstate Majorana fermions by multiplying the interaction eigenstate fields by a unitary matrix U that is determined by the mass diagonalization procedure obtained in part (d). Show that the matrix U depends on the creation phases, ζ_i.

(f) In a CP-conserving theory of Majorana fermions, show that a choice of the ζ_i exists such that U is a real orthogonal matrix.

3.15 Compute the tree-level amplitude for the scattering of a charged vector boson and a four-component Majorana fermion via an s-channel exchange of a Dirac fermion, due to the interactions given in eq. (3.206).

3.16 Consider the interaction Lagrangian of eq. (3.204) that describes the interaction of a neutral scalar with one flavor of four-component Dirac fermions (i.e., no flavor indices needed).

(a) Compute the wave function renormalization constant of the neutral scalar, Z_ϕ, in the one-loop approximation due to a fermion loop. Likewise, evaluate the wave function renormalization constants of the Dirac fermion, δZ_L and δZ_R, in the one-loop approximation due to virtual scalar exchange.

(b) The Feynman diagram corresponding to the one-loop vertex correction to the decay of a neutral scalar into a pair of four-component Dirac fermions is exhibited in Fig. 3.11. Evaluate this diagram explicitly, using dimensional regularization in $d = 4 - 2\epsilon$ spacetime dimensions.

(c) Using eq. (3.310) and the results of parts (a) and (b), compute the sum of the tree-level and one-loop contributions to the amplitude for the decay of a neutral scalar into a fermion–antifermion pair. Obtain a formula for the decay rate that is accurate to $\mathcal{O}(\kappa_r^3)$, where κ_r has the following form:

$$\kappa_r = \kappa\big[1 + C\kappa(4\pi)^\epsilon\Gamma(\epsilon)\big]\,. \tag{3.347}$$

By choosing C appropriately, show that the decay rate obtained above is finite in the limit of $\epsilon \to 0$. This choice corresponds to a modified minimal subtraction ($\overline{\rm MS}$) scheme for defining the renormalized coupling κ_r.

(d) Repeat the computations of parts (a)–(c) by using the method of counterterms approach to renormalization. Explain why the result obtained for the decay rate is independent of the renormalization condition used to fix the wave function renormalization constants. Verify this observation by employing an $\overline{\rm MS}$ renormalization condition to determine the wave function renormalization constants. How should the rules of eq. (3.306) be modified in this case?

4 Gauge Theories and the Standard Model

In Chapter 1, we focused on quantum field theories of free fermions. In order to construct renormalizable interacting quantum field theories, we must introduce additional fields. The requirement of renormalizability imposes two constraints. First, the couplings in the interaction Lagrangian must have nonnegative mass dimension. Thus, no perturbatively renormalizable interacting theory that consists only of spin-1/2 fields exists, since the simplest interaction term is a four-fermion interaction coupling whose mass dimension is -2. Second, the most general renormalizable interacting theories consist of scalar, fermion, and spin-1 boson fields, where the latter are either abelian (massless or massive) vector fields or nonabelian gauge fields. This exhausts all possible renormalizable interacting field theories.

The Standard Model is a spontaneously broken nonabelian gauge theory containing elementary scalars, fermions, and spin-1 gauge bosons. Typically, one refers to the spin-0 and spin-1/2 fields (which are either neutral or charged with respect to the underlying gauge group) as *matter fields*, whereas the spin-1 gauge bosons are called *gauge fields*. In this chapter we review the ingredients for constructing nonabelian (Yang–Mills) gauge theories and their breaking via the dynamics of self-interacting scalar fields. The Standard Model of fundamental particles and interactions is then exhibited, and some of its properties are described.

4.1 Abelian Gauge Field Theory

Consider quantum electrodynamics (QED), which is an abelian gauge theory that describes the interactions of electrons, positrons, and photons. The Lagrangian of QED is given by

$$\mathscr{L} = -\tfrac{1}{4}F^{\mu\nu}F_{\mu\nu} + i\chi^\dagger\,\overline{\sigma}^\mu\nabla_\mu\chi + i\eta^\dagger\,\overline{\sigma}^\mu\nabla_\mu\eta - m(\chi\eta + \chi^\dagger\,\eta^\dagger)\,, \tag{4.1}$$

where the electromagnetic field strength tensor $F_{\mu\nu}$ is defined in terms of the gauge field A_μ as

$$F_{\mu\nu} \equiv \partial_\mu A_\nu - \partial_\nu A_\mu\,, \tag{4.2}$$

and the covariant derivative ∇_μ is defined as

$$\nabla_\mu\psi \equiv (\partial_\mu + ieq_\psi A_\mu)\psi\,, \tag{4.3}$$

where $\psi = \chi$ or η with $q_\chi = -1$ and $q_\eta = +1$. Note that we have written eq. (4.1) in terms of the two-component charged fermion fields χ and η. The identification of these fields with the electron and positron (with corresponding electric charges q_ψ in units of $e > 0$) is consistent with the conventions for fermion and antifermion names and fields given in Table 17.1.

The QED Lagrangian consists of a sum of the kinetic energy term for the gauge (photon) field, $-\frac{1}{4}F_{\mu\nu}F^{\mu\nu}$, and the Dirac Lagrangian with the ordinary derivative ∂_μ replaced by a covariant derivative ∇_μ. This Lagrangian is invariant under a local U(1) gauge transformation parameterized by a function $\Lambda(x)$:

$$A_\mu(x) \to A_\mu(x) + \partial_\mu \Lambda(x)\,, \tag{4.4}$$

$$\chi(x) \to \exp[ie\Lambda(x)]\chi(x)\,, \tag{4.5}$$

$$\eta(x) \to \exp[-ie\Lambda(x)]\eta(x)\,. \tag{4.6}$$

The two-component fermion Feynman rules for electrons interacting with photons, which can be deduced from Fig. 2.9, are explicitly exhibited in Fig. 17.3. Note that the corresponding four-component Feynman rule deduced from Fig. 3.3 yields the well-known rule for the $e^+e^-\gamma$ vertex of QED (see Exercise 4.1).

One can extend the theory above by including charged scalars among the possible matter fields. For example, a complex scalar field $\Phi(x)$ of definite U(1) charge q_Φ will transform under the local U(1) gauge transformation:

$$\Phi(x) \to \exp[-ieq_\Phi \Lambda(x)]\Phi(x)\,. \tag{4.7}$$

A gauge invariant Lagrangian can be obtained from the free-field scalar Lagrangian of eq. (3.3) by replacing $\partial_\mu \Phi(x)$ with $\nabla_\mu \Phi(x) \equiv (\partial_\mu + ieq_\Phi)\Phi(x)$. One may also add gauge-invariant Yukawa interactions of the form

$$\mathscr{L}_{\rm Y} = -y^{ijk}\Phi_i \psi_j \psi_k + \text{h.c.}\,, \tag{4.8}$$

where the Φ_i consist of neutral and/or charged scalar fields and the ψ_i consist of neutral Majorana (ξ) and/or charged Dirac pairs (χ and η) of two-component fermion fields. The complex Yukawa coupling y^{ijk} vanishes unless the condition $q_{\Phi_i} + q_{\psi_j} + q_{\psi_k} = 0$ is satisfied, as required by gauge invariance.

4.2 Nonabelian Gauge Groups and Their Lie Algebras

Abelian gauge field theory can be generalized by replacing the abelian U(1) gauge group of QED with a nonabelian gauge group G. We again consider possible matter fields: multiplets of scalar fields $\Phi_i(x)$ and two-component fermion fields $\psi_i(x)$ [or equivalently, four-component fermions $\Psi_i(x)$] that are either neutral or charged with respect to G.

The symmetry group G employed in gauge theories is (generically) a direct product of a finite number of compact connected simple Lie groups and U(1) groups. The direct product of simple Lie groups [i.e., with no U(1) factors] is called a *semisimple* Lie group. Given some matter field, which we generically designate by $\phi_i(x)$, the gauge transformation under which the Lagrangian is invariant, is given by

$$\phi_i(x) \to U_i{}^j(g)\phi_j(x)\,, \qquad i,j \in \{1,2,\ldots,d_R\}\,, \tag{4.9}$$

where g is an element of G (that is, g is a specific gauge transformation) and $U(g)$ is a (possibly reducible) d_R-dimensional unitary representation R of the group G. One is always free to redefine the fields via $\phi(x) \to V\phi(x)$, where V is any fixed unitary matrix (independent of the choice of g). The gauge transformation law for the redefined $\phi(x)$ now has $U(g)$ replaced by $V^{-1}U(g)V$. If $U(g)$ is a reducible representation, then it is possible to find a V such that the $U(g)$ for *all* group elements g assume a block-diagonal form. Otherwise, the representation $U(g)$ is irreducible, which implies that the corresponding multiplets transform only among themselves. The matter fields of the gauge theory generally form a reducible representation, which can subsequently be decomposed into their irreducible components.

The *local* gauge transformation $U(g)$ is also a function of spacetime position, x^μ. Explicitly, any group element that is continuously connected to the identity takes the following form:[1]

$$U(g(x)) = \exp[-ig_a\Lambda^a(x)\boldsymbol{T^a}]\,, \tag{4.10}$$

where there is an implicit sum over the thrice-repeated index $a \in \{1,2,\ldots,d_G\}$. The $\boldsymbol{T^a}$ are a set of d_G linearly independent hermitian matrices[2] called generators of the Lie group, and the corresponding $\Lambda^a(x)$ are arbitrary x-dependent functions. The constants g_a (which are the analogues of e of the abelian theory) are called the gauge couplings. As previously noted below eq. (2.136), there is a separate coupling g_a for each simple group or U(1) group contained in the direct product group G. Thus, the generators $\boldsymbol{T^a}$ separate out into distinct classes, each of which is associated with a different simple group or U(1) group.[3] In particular, $g_a = g_b$ if $\boldsymbol{T^a}$ and $\boldsymbol{T^b}$ are in the same class. If G is simple, then $g_a = g$ for all a.

The number of linearly independent generators, d_G, of the Lie group G, which is also called the dimension of G, depends only on the abstract definition of the group (and not on the choice of representation). Moreover, the commutator of two generators is a linear combination of generators,

$$[\boldsymbol{T^a}, \boldsymbol{T^b}] = if_c^{ab}\boldsymbol{T^c}\,, \tag{4.11}$$

where the f_c^{ab} are called the structure constants of the Lie group. The linearly independent generators will always be chosen such that the f_c^{ab} are real numbers.

In studying the structure of gauge field theories, nearly all the information of

[1] Admittedly, the notation used here is not ideal. The abstract element g of the Lie group G is unrelated to the gauge coupling g_a, which is also called g when G is a simple group. However, when g appears in subsequent equations, its meaning should be obvious from the context.

[2] The condition of linear independence means that $\sum_a c^a\boldsymbol{T^a} = 0$ implies that $c^a = 0$ for all a.

[3] Each U(1) generator is diagonal, with the choice of normalization a matter of convention. However, if G is embedded in a simple group G_s, then the normalization of the U(1) generator of G is determined (see Section 6.3.4).

interest can be ascertained by focusing on infinitesimal gauge transformations. In a given representation R, with Lie group generators represented by d_R-dimensional matrices $\boldsymbol{T^a}$, any matrix representation of a group element $U(g(x))$ that is continuously connected to the identity element can be expanded about the $d_R \times d_R$ identity matrix:

$$U(g(x)) \simeq \boldsymbol{I}_{d_R} - ig_a\Lambda^a(x)\boldsymbol{T^a} . \tag{4.12}$$

Then, the infinitesimal gauge transformation corresponding to eq. (4.9) is given by $\phi_i(x) \to \phi_i(x) + \delta\phi_i(x)$, where

$$\delta\phi_i(x) = -ig_a\Lambda^a(x)(\boldsymbol{T^a})_i{}^j\phi_j(x) . \tag{4.13}$$

From the generators $\boldsymbol{T^a}$, one can reconstruct the group elements $U(g(x))$, so it is sufficient to focus on the infinitesimal group transformations. The group generators $\boldsymbol{T^a}$ span a real vector space, whose general element is $c^a\boldsymbol{T^a}$, where the c^a are real numbers.[4] One can formally define a "vector product" of any two elements of the vector space as the commutator of the two vectors. For example, using eq. (4.11), it is clear that the vector product of any two vectors $[c^a\boldsymbol{T^a}, d^b\boldsymbol{T^b}]$ is a real linear combination of the generators, which is also an element of the vector space. Consequently, this vector space is also an algebra, called a Lie algebra. Henceforth, the Lie algebra corresponding to the Lie group G will be designated by $\mathfrak{g}$. An abstract Lie algebra is characterized by one additional axiom,

$$[\boldsymbol{T^a}, [\boldsymbol{T^b}, \boldsymbol{T^c}]] + [\boldsymbol{T^b}, [\boldsymbol{T^c}, \boldsymbol{T^a}]] + [\boldsymbol{T^c}, [\boldsymbol{T^a}, \boldsymbol{T^b}]] = 0 , \tag{4.14}$$

called the Jacobi identity. Indeed, eq. (4.14) is satisfied by any three elements of an abstract Lie algebra (as well as being satisfied in general by any three matrices).

If the symmetry group G is a direct product of simple compact Lie groups and U(1) groups, then its Lie algebra $\mathfrak{g}$ is a direct sum of a finite number of simple compact Lie algebras and $\mathfrak{u}(1)$ Lie algebras.[5] A direct sum of simple Lie algebras [i.e., with no $\mathfrak{u}(1)$ factors] is called a semisimple Lie algebra. If $\boldsymbol{T^a}$ and $\boldsymbol{T^b}$ belong to different classes (i.e., different factors of the direct sum), then $[\boldsymbol{T^a}, \boldsymbol{T^b}] = 0$. Equivalently, $f^{ab}_c = 0$ if $\boldsymbol{T^a}$, $\boldsymbol{T^b}$, and $\boldsymbol{T^c}$ do not all belong to the same class. Moreover, $f^{ab}_c = 0$ if $\boldsymbol{T^a}$, $\boldsymbol{T^b}$, or $\boldsymbol{T^c}$ is a U(1) generator.

In the next section, we will see that the gauge fields transform under the adjoint representation of the global symmetry group. The explicit matrix elements of the adjoint representation generators are given by

$$(\boldsymbol{T^a})^b{}_c = -if^{ab}_c \qquad a, b, c \in \{1, 2, \ldots, d_G\} . \tag{4.15}$$

Thus, the dimension of the adjoint representation matrices coincides with the num-

[4] With the $\boldsymbol{T^a}$ hermitian, we require the c^a to be real so that $U(g)$ is unitary. Then, the $\boldsymbol{T^a}$ span a *real* Lie algebra. Mathematicians consider the elements of the real Lie algebra to be $ic^a\boldsymbol{T^a}$, with antihermitian generators $i\boldsymbol{T^a}$. Note that for real Lie algebras, the representation matrices for $\boldsymbol{T^a}$ (or $i\boldsymbol{T^a}$) may be complex or quaternionic.

[5] The Lie algebra $\mathfrak{u}(1)$ is equivalent to the vector space of real numbers. The vector product is the commutator, which vanishes for any pair of $\mathfrak{u}(1)$ elements. Thus, $\mathfrak{u}(1)$ is an *abelian* Lie algebra.

ber of generators, d_G. Hence, we shall often refer to the indices a, b, c as adjoint indices.

The choice of basis vectors (or generators) $\boldsymbol{T}^a$ is arbitrary. Moreover, the values of the structure constants f_c^{ab} depend on this choice of basis. Nevertheless, there is a canonical choice that we now adopt. The generators in an irreducible representation R of a simple Lie algebra $\mathfrak{g}$ are chosen such that[6]

$$\mathrm{Tr}(\boldsymbol{T}^a \boldsymbol{T}^b) = T_R \delta^{ab} , \tag{4.16}$$

where $T_R \equiv I_2(R)$ is called the second-order Dynkin index and depends on the irreducible representation R (see Appendix H.3). In this basis, the f_c^{ab} are completely antisymmetric under the interchange of any pair of indices. Moreover, having chosen this basis, there is no distinction between upper and lower adjoint indices. Henceforth, the real structure constants will be denoted by f^{abc} which is a totally antisymmetric third-rank tensor. For example, the structure constants of $\mathfrak{su}(2)$ are $f^{abc} = \epsilon^{abc}$. It then follows from eq. (4.15) that in the adjoint representation the $i\boldsymbol{T}^a$ are real antisymmetric matrices. The representation matrices of the corresponding Lie group elements [eq. (4.10)] are therefore real and orthogonal. That is, the adjoint representation provides a *real* representation of the Lie group and Lie algebra.

Once T_R is determined for any nontrivial irreducible representation, then the value of T_R for any other irreducible representation is fixed. Corresponding to each simple compact Lie algebra, one can identify one particular irreducible representation, called the *fundamental* (or *defining*) representation, with corresponding generators denoted by $\boldsymbol{t}^a$. The most common examples are listed in Table 4.1; e.g., the fundamental representation of the generators of $\mathfrak{su}(2)$ are $\boldsymbol{t}^a = \frac{1}{2}\sigma^a$, where the σ^a are the Pauli matrices. For the fundamental representation ($R = F$) of $\mathfrak{su}(n)$, the conventional value for the second-order Dynkin index is taken to be

$$T_F = \tfrac{1}{2} , \qquad \text{for } \mathfrak{g} = \mathfrak{su}(n). \tag{4.17}$$

The second-order Dynkin indices for the fundamental representations of $\mathfrak{so}(n)$ and $\mathfrak{sp}(n)$ are given in eqs. (H.171)–(H.173).

For $\mathfrak{g} = \mathfrak{su}(n)$ $[n \geq 3]$ and $\mathfrak{so}(4k+2)$ $[k \geq 1]$, the corresponding fundamental representations F are complex. The complex conjugate of F, denoted by $\overline{F}$, is called the *antifundamental representation*, with hermitian generators $-(\boldsymbol{t}^a)^* = -(\boldsymbol{t}^a)^{\mathsf{T}}$. In light of eq. (H.37), the corresponding second-order Dynkin index is $T_{\overline{F}} = T_F$.

For the record, we mention two other properties of Lie algebras that will be useful in this book. First, given any semisimple Lie algebra $\mathfrak{g}$ and a corresponding irreducible representation of antihermitian generators $i\boldsymbol{T}^a$, one can always find an equivalent representation $V^{-1} i\boldsymbol{T}^a V$ for some unitary matrix V. There exists some choice of V (not necessarily unique) that maximizes the number of simultaneous diagonal generators. This maximal number r_G, called the *rank* of $\mathfrak{g}$, is independent of the choice of representation, and is a property of the abstract Lie algebra. The ranks of the classical compact real Lie algebras are given in Table 4.1.

[6] More generally, in an arbitrary basis, $\mathrm{Tr}(\boldsymbol{T}^a \boldsymbol{T}^b) = T_R g^{ab}$, where g^{ab} is the Killing form (which can be used to raise and lower adjoint indices). For further details, see Appendix H.4.

Table 4.1 Simple real compact Lie algebras, $\mathfrak{g}$, of dimension d_G and rank r_G. (The Lie algebra $\mathfrak{so}(4)$ is semisimple.) The floor function $\lfloor x \rfloor$, for $x \in \mathbb{R}$, is equal to the greatest integer less than or equal to x. The fundamental (or defining) representation of $\mathfrak{g}$ refers to arbitrary linear combinations $ic^a t^a$, where the c^a are real and the t^a are the corresponding generators of $\mathfrak{g}$.

$\mathfrak{g}$	d_G	r_G	fundamental representation it^a
$\mathfrak{so}(n)$	$\frac{1}{2}n(n-1)$	$\lfloor \frac{1}{2}n \rfloor$	$n \times n$ real antisymmetric
$\mathfrak{su}(n)$	$n^2 - 1$	$n - 1$	$n \times n$ traceless complex antihermitian
$\mathfrak{sp}(n)$	$n(2n+1)$	n	$n \times n$ quaternionic antihermitian

Second, a Casimir operator is defined to be an operator that commutes with all the generators $\boldsymbol{T^a}$ of G. One can prove that a simple Lie algebra of rank r_G possesses r_G independent Casimir operators. The most important of these is the quadratic Casimir operator, which is defined in Appendix H.3 by

$$(\boldsymbol{T}^2)_i{}^j \equiv (\boldsymbol{T^a})_i{}^k (\boldsymbol{T^a})_k{}^j = C_R \delta_i{}^j \, . \tag{4.18}$$

Any operator that commutes with all the generators in an *irreducible* representation must be a multiple of the identity [see eqs. (H.60) and (H.61)]. The coefficient of $\delta_i{}^j$ depends on the irreducible representation R and is denoted by C_R. By multiplying eq. (4.18) by $\delta_j{}^i$, one derives an important theorem:

$$T_R d_G = C_R d_R \, , \tag{4.19}$$

where d_R is the dimension of the representation R. Note that the dimension of the adjoint representation ($R = A$) is $d_A = d_G$, so that eq. (4.19) yields $C_A = T_A$. As an example, for SU(n), Table 4.1 gives $d_G = n^2 - 1$ and $d_F = n$. Using eqs. (4.17) and (4.19), it follows that $C_F = (n^2-1)/(2n)$. From an explicit representation of the f^{abc} for SU(n), one can also derive $C_A = T_A = n$.

4.3 Nonabelian Gauge Field Theory

In order to construct a nonabelian gauge theory, we follow the recipe presented in the case of the abelian gauge theory. Namely, we introduce a gauge field A_μ and a covariant derivative ∇_μ. By replacing ∂_μ with ∇_μ in the kinetic energy terms of the matter fields and introducing an appropriate transformation law for A_μ, the resulting matter kinetic energy terms are invariant under local gauge transformations.

As an example, consider a scalar field theory with the Lagrangian

$$\mathscr{L} = (\partial_\mu \Phi^{\dagger i})(\partial^\mu \Phi_i) - V(\Phi, \Phi^\dagger) \, . \tag{4.20}$$

In eq. (4.20), the scalar potential V is invariant under gauge transformations, $\Phi_i(x) \to U_i{}^j(g)\Phi_j(x)$; that is,

$$V(U\Phi, (U\Phi)^\dagger) = V(\Phi, \Phi^\dagger)\,. \tag{4.21}$$

Although $\mathscr{L}$ is invariant under *global* symmetry transformations, the kinetic energy term is not invariant under *local* gauge transformations due to the presence of the derivative. In particular, under local gauge transformations,

$$\partial_\mu\Phi \to \partial_\mu(U\Phi) = U\partial_\mu\Phi + (\partial_\mu U)\Phi\,. \tag{4.22}$$

This motivates the introduction of the covariant derivative acting on a matter field that transforms according to some representation R of the symmetry group G, which was previously given in eq. (2.136) and is repeated here for the convenience of the reader:

$$(\nabla_\mu)_i{}^j = \delta_i{}^j\partial_\mu + ig_a A_\mu^a(\boldsymbol{T^a})_i{}^j\,, \qquad a \in \{1, 2, \ldots, d_G\}\,, \tag{4.23}$$

where the flavor indices $i, j \in \{1, 2, \ldots, d_R\}$. In light of the remarks below eq. (4.10), if the Lie algebra $\mathfrak{g}$ is simple, then $g_a = g$. Otherwise $g_a = g_b$ if and only if $\boldsymbol{T^a}$ and $\boldsymbol{T^b}$ belong to the same simple Lie algebra or $\mathfrak{u}(1)$ factor in the direct sum decomposition of the Lie algebra $\mathfrak{g}$.

By introducing a suitable transformation law for $A_\mu^a(x)$, one can arrange $\nabla_\mu\Phi$ to transform under a local gauge transformation as $\nabla_\mu\Phi \to U\nabla_\mu\Phi$, in which case,

$$\mathscr{L} = (\nabla_\mu\Phi)^{\dagger i}(\nabla^\mu\Phi)_i - V(\Phi, \Phi^\dagger) \tag{4.24}$$

is invariant under local gauge transformations.

The transformation law for A_μ^a is most easily expressed for the matrix-valued gauge field[7]

$$A_\mu \equiv g_a A_\mu^a \boldsymbol{T^a}\,. \tag{4.25}$$

Under local gauge transformations, the matrix-valued gauge field transforms as

$$A_\mu \to UA_\mu U^{-1} - iU(\partial_\mu U^{-1})\,. \tag{4.26}$$

Using eq. (4.26), one quickly shows that $\nabla_\mu\Phi$ transforms as expected:

$$\begin{aligned}\nabla_\mu\Phi \equiv (\partial_\mu + iA_\mu)\Phi &\to \left[\partial_\mu + iUA_\mu U^{-1} + U(\partial_\mu U^{-1})\right]U\Phi \\ &= U\nabla_\mu\Phi + \left[\partial_\mu U + U(\partial_\mu U^{-1})U\right]\Phi \\ &= U\nabla_\mu\Phi\,.\end{aligned} \tag{4.27}$$

In the last step, we noted that

$$\partial_\mu U + U(\partial_\mu U^{-1})U = \left[(\partial_\mu U)U^{-1} + U(\partial_\mu U^{-1})\right]U = \left[\partial_\mu(UU^{-1})\right]U = 0\,, \tag{4.28}$$

after employing $UU^{-1} = I$.

7 The matrix-valued gauge field that one employs in the covariant derivative depends on the representation of the matter fields on which the covariant derivative acts.

It is also useful to exhibit the infinitesimal version of the transformation law, by taking the infinitesimal form for U [eq. (4.12)]. This allows us to express the transformation law, eq. (4.26), in infinitesimal form, $A_\mu^a \to A_\mu^a + \delta A_\mu^a$, where

$$\delta A_\mu^a = g_a f^{abc} \Lambda^b A_\mu^c + \partial_\mu \Lambda^a \equiv \nabla_\mu^{ab} \Lambda^b \,, \tag{4.29}$$

and there is no implicit sum over a, since the free index a appears on both sides of eq. (4.29). The covariant derivative operator applied to fields in the adjoint representation,[8]

$$\nabla_\mu^{ab} \equiv \delta^{ab} \partial_\mu + g_a f^{abc} A_\mu^c \,, \tag{4.30}$$

has been obtained by employing eq. (4.15) for the $\boldsymbol{T^a}$ in eq. (4.23). Note that in an abelian gauge theory, $f^{abc} = 0$, in which case $\nabla_\mu = \partial_\mu$ when acting on an adjoint field. Equation (4.29) then reduces to $\delta A_\mu = \partial_\mu \Lambda$ in agreement with eq. (4.4).

To complete the construction of the nonabelian gauge field theory, we must introduce a gauge invariant kinetic energy term for the gauge fields. To motivate the definition of the gauge field strength tensor, we consider $[\nabla_\mu, \nabla_\nu]$ acting on Φ:

$$\begin{aligned} [\nabla_\mu, \nabla_\nu]\Phi &= [\partial_\mu + iA_\mu, \partial_\nu + iA_\nu]\Phi \\ &= i\{\partial_\mu A_\nu - \partial_\nu A_\mu + i[A_\mu, A_\nu]\}\Phi \,. \end{aligned} \tag{4.31}$$

The matrix-valued gauge field strength tensor $F_{\mu\nu} \equiv g_a F_{\mu\nu}^a \boldsymbol{T^a}$ is then defined by

$$F_{\mu\nu} \equiv -i[\nabla_\mu, \nabla_\nu] = \partial_\mu A_\nu - \partial_\nu A_\mu + i[A_\mu, A_\nu] \,. \tag{4.32}$$

Using eqs. (4.11) and (4.31), it follows that

$$F_{\mu\nu}^a = \partial_\mu A_\nu^a - \partial_\nu A_\mu^a - g_a f^{abc} A_\mu^b A_\nu^c \,. \tag{4.33}$$

Under a local gauge transformation, $\Phi \to U\Phi$ and $\nabla_\mu \Phi \to U \nabla_\mu \Phi$, which implies that $[\nabla_\mu, \nabla_\nu]\,\Phi \to U[\nabla_\mu, \nabla_\nu]\Phi$. Consequently, one obtains the transformation law $F_{\mu\nu}\Phi \to U F_{\mu\nu}\Phi = (U F_{\mu\nu} U^{-1}) U\Phi$. That is,

$$F_{\mu\nu} \to U F_{\mu\nu} U^{-1} \,, \tag{4.34}$$

which is the transformation law for an adjoint field. The infinitesimal form of eq. (4.34) is $F_{\mu\nu} \to F_{\mu\nu} + \delta F_{\mu\nu}$, where

$$\delta F_{\mu\nu}^a = g_a f^{abc} \Lambda^b F_{\mu\nu}^c \,. \tag{4.35}$$

Note that in an abelian gauge theory, $U F_{\mu\nu} U^{-1} = U U^{-1} F_{\mu\nu} = F_{\mu\nu}$ so that $F^{\mu\nu}$ is gauge invariant (i.e., neutral under the gauge group). For a nonabelian gauge group, $F_{\mu\nu}$ transforms nontrivially; i.e., it carries a nontrivial gauge charge.

We can now construct a gauge-invariant kinetic energy term for the gauge fields. Each simple gauge group contained in the direct product group G with corresponding gauge coupling constant $g_a = g$ contributes:

$$\mathscr{L}_{\text{gauge}} \supset -\frac{1}{4g^2 T_R} \operatorname{Tr}(F_{\mu\nu} F^{\mu\nu}) \,. \tag{4.36}$$

[8] In particular, with respect to global symmetry transformations, the gauge fields A_μ^a transform under the adjoint representation of the gauge group.

Using eq. (4.34), we see that $\mathscr{L}_{\text{gauge}}$ is gauge invariant due to the invariance of the trace under cyclic permutation of its arguments. In light of

$$\operatorname{Tr}(F_{\mu\nu}F^{\mu\nu}) = g^2 F^a_{\mu\nu} F^{\mu\nu b} \operatorname{Tr}(\boldsymbol{T}^a \boldsymbol{T}^b) = g^2 T_R F^a_{\mu\nu} F^{\mu\nu a}\,, \tag{4.37}$$

the form for $\mathscr{L}_{\text{gauge}}$ does not depend on the representation R. Adding up the contributions from all the simple groups and U(1) groups contained in the direct product group G, we end up with

$$\begin{aligned}\mathscr{L}_{\text{gauge}} &= -\tfrac{1}{4} F^a_{\mu\nu} F^{\mu\nu a} \\ &= -\tfrac{1}{4}(\partial_\mu A^a_\nu - \partial_\nu A^a_\mu - g_a f^{abc} A^b_\mu A^c_\nu)(\partial^\mu A^{\nu a} - \partial^\nu A^{\mu a} - g_a f^{ade} A^\mu_d A^\nu_e)\,,\end{aligned} \tag{4.38}$$

where there is an implicit sum over repeated indices. In contrast to abelian gauge theory, the gauge kinetic energy term of a nonabelian gauge theory generates three-point and four-point self-interactions among the gauge fields.

To summarize, if the matter Lagrangian, $\mathscr{L}_{\text{matter}}$, is invariant under a group G of global symmetry transformations, then

$$\mathscr{L} = \mathscr{L}_{\text{gauge}} + \mathscr{L}_{\text{matter}}(\partial_\mu \to \nabla_\mu) \tag{4.39}$$

is invariant under a group G of local gauge transformations. The terms in $\mathscr{L}_{\text{matter}}$ that contain explicit derivatives consist of a sum of kinetic energy terms of the various scalar and fermion matter multiplets, each of which transforms under some irreducible representation of the gauge group. In these terms, we replace the ordinary derivative with $\nabla_\mu = \partial_\mu + i g_a A^a_\mu \boldsymbol{T}^a$ and use the matrix representation $\boldsymbol{T}^a$ appropriate for each of the matter field multiplets.

Note that there is no mass term for the gauge field, since the term

$$\mathscr{L}_{\text{mass}} = \tfrac{1}{2} m^2 A^a_\mu A^{\mu a} \tag{4.40}$$

would violate the local gauge invariance. This is a tree-level result; in the next section we will discuss whether this result persists to all orders in perturbation theory.

4.4 Feynman Rules for Gauge Theories

The Feynman rules for the self-interactions of the gauge fields and for the interactions with matter are simple to obtain. The triple and quartic gauge boson self-couplings follow from the form of the gauge kinetic energy term [eq. (4.38)]. The interactions of the gauge bosons with matter are derived from the matter kinetic energy terms. For example, after replacing the ordinary derivatives with covariant derivatives in the scalar field kinetic energy terms, the gauge field dependence of ∇_μ generates cubic and quartic terms that are linear and quadratic in A_μ. A similar replacement in the fermion field kinetic energy terms yields interactions between the fermions and gauge bosons that are linear in A_μ, as exhibited by the Feynman rules for two-component [four-component] fermions previously given in Figs. 2.8—2.10 [Figs. 3.2—3.4].

However, an apparent problem is encountered when one tries to obtain the Feynman rule for the gauge boson propagator. In general the rule for the tree-level propagator is obtained by inverting the operator that appears in the part of the Lagrangian that is quadratic in the fields. In the case of gauge fields, this is the kinetic energy term, which we can rewrite as follows:

$$
\begin{aligned}
-\tfrac{1}{4}F^a_{\mu\nu}F^{\mu\nu a} &\supset -\tfrac{1}{4}(\partial_\mu A^a_\nu - \partial_\nu A^a_\mu)(\partial^\mu A^{\nu a} - \partial^\nu A^{\mu a}) \\
&= \tfrac{1}{2}A^a_\mu(g^{\mu\nu}\Box - \partial^\mu\partial^\nu)A^a_\nu \; + \text{ total derivative}\,, \qquad (4.41)
\end{aligned}
$$

where $\Box \equiv \partial^\rho\partial_\rho$. Note that the total divergence does not contribute to the action. A more significant observation is that $(g^{\mu\nu}\Box - \partial^\mu\partial^\nu)\partial_\nu = 0$, which implies that $g^{\mu\nu}\Box - \partial^\mu\partial^\nu$ has a zero eigenvalue and therefore is not an invertible operator.

The solution to this problem is to add the so-called gauge-fixing term and the Faddeev–Popov ghost fields (anticommuting adjoint fields ω^a and ω^{*a}). The justification of this procedure can be found in most quantum field theory textbooks and is most easily explained using path integral techniques. Here, we take a more practical view and simply note that the following Yang–Mills (YM) Lagrangian,

$$
\mathscr{L}_{\rm YM} = -\tfrac{1}{4}F^a_{\mu\nu}F^{\mu\nu a} - \frac{1}{2\xi}(\partial_\mu A^\mu_a)^2 + \partial^\mu\omega^*_a\nabla^{ab}_\mu\omega_b\,, \qquad (4.42)
$$

is invariant under a Becchi–Rouet–Stora–Tyutin (BRST) extended gauge symmetry, whose infinitesimal transformation laws are given by

$$
\delta A^a_\mu = \epsilon\nabla^{ab}_\mu\omega_b\,, \qquad (4.43)
$$

$$
\delta\omega_a = \tfrac{1}{2}\epsilon g_a f^{abc}\omega_b\omega_c\,, \qquad (4.44)
$$

$$
\delta\omega^*_a = -\frac{1}{\xi}\epsilon\partial_\mu A^\mu_a\,, \qquad (4.45)
$$

where ϵ is an infinitesimal anticommuting parameter. Note that the gauge transformation function $\Lambda^a(x)$ in eq. (4.29) has now been promoted to a field $\omega^a(x)$, whose transformation law is given above. Equation (4.42) generates new interaction terms involving the gauge fields and Faddeev–Popov ghosts. The Faddeev–Popov ghosts can therefore appear inside loops of Feynman diagrams. One can show that scattering amplitudes that involve only physical particles as external states satisfy unitarity. Hence the theory based on eq. (4.42) is consistent.

In particular, due to the addition of the gauge-fixing term, $\mathscr{L}_{\rm GF} = -(\partial_\mu A^\mu_a)^2/(2\xi)$, the terms of the Lagrangian that are quadratic in the gauge fields have changed, and the propagator can now be defined. Converting to momentum space, the Feynman rule for a nonabelian gauge boson propagator is exhibited in Fig. 4.1. The Feynman gauge ($\xi = 1$) and the Landau gauge ($\xi = 0$) are two of the more common gauge choices used in practical computations. Any physical quantity must be independent of ξ, which provides a good check of Feynman diagram computations in which internal gauge bosons propagate. The above considerations also apply to

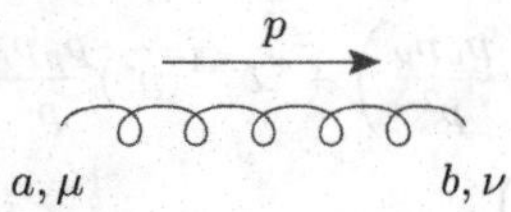

$$-iD^{ab}_{\mu\nu}(p) \equiv \frac{-i\delta^{ab}}{p^2 + i\varepsilon}\left[g_{\mu\nu} - (1-\xi)\frac{p_\mu p_\nu}{p^2}\right]$$

Fig. 4.1 Feynman rule for the propagator of a massless nonabelian gauge boson.

abelian gauge theories such as QED. In this case, one can introduce ghost fields to identify the extended BRST symmetry. However, within the class of gauge-fixing terms considered here, the ghost fields are noninteracting (since the photon does not carry any U(1) charge) and hence the ghosts can be dropped. The photon propagator still takes the form of Fig. 4.1, but with the factor of δ^{ab} removed.

Finally, consider the following question. Although the gauge boson is massless at tree level, can a nonzero mass be generated via radiative corrections? To answer this question, one must compute the gauge boson two-point function (which corresponds to the radiatively corrected inverse propagator) and check to see whether the pole of the propagator at $p^2 = 0$ is shifted. Summing up the geometric series yields an implicit equation for the fully radiatively corrected propagator [see eq. (2.224)]:

$$-i\mathscr{D}^{ab}_{\mu\nu}(p) = -i\mathscr{D}_{\mu\nu}(p)\delta^{ab}\,, \tag{4.46}$$

where

$$-i\mathscr{D}_{\mu\nu}(p) = -iD_{\mu\nu}(p) + [-iD_{\mu\lambda}(p)]i\Pi^{\lambda\rho}(p)[-i\mathscr{D}_{\rho\nu}(p)]\,. \tag{4.47}$$

In obtaining eq. (4.47), $-iD^{ab}_{\mu\nu}(p) = -iD_{\mu\nu}(p)\delta^{ab}$ is the tree-level gauge field propagator and the self-energy function, $i\Pi^{ab}_{\mu\nu}(p) = i\Pi_{\mu\nu}(p)\delta^{ab}$, is the sum over all 1PI diagrams (excluding the tree-level inverse propagator). It then follows that

$$\mathscr{D}^{-1}_{\mu\nu}(p) = D^{-1}_{\mu\nu}(p) - \Pi_{\mu\nu}(p)\,. \tag{4.48}$$

The Ward identity of the theory (a consequence of gauge invariance) implies that $p_\mu\Pi^{\mu\nu}(p) = p_\nu\Pi^{\mu\nu}(p) = 0$. Consequently, one can write

$$i\Pi^{ab}_{\mu\nu}(p) = -i(p^2 g_{\mu\nu} - p_\mu p_\nu)\Pi(p^2)\delta^{ab}\,. \tag{4.49}$$

Comparing eq. (4.49) with the decomposition into its transverse and longitudinal parts given in eq. (2.226), it follows that

$$\Pi_T(p^2) = -p^2\Pi(p^2)\,, \qquad \Pi_L(p^2) = 0\,. \tag{4.50}$$

It is convenient to decompose $\mathscr{D}_{\mu\nu}(p)$ into its transverse and longitudinal parts:

$$\mathscr{D}_{\mu\nu}(p) = \mathscr{D}_T(p^2)\left(g_{\mu\nu} - \frac{p_\mu p_\nu}{p^2}\right) + \mathscr{D}_L(p^2)\frac{p_\mu p_\nu}{p^2}\,. \tag{4.51}$$

Similarly, the decompositions of the inverses $\mathscr{D}^{-1}_{\mu\nu}(p)$ and $D^{-1}_{\mu\nu}(p)$ into their transverse and longitudinal parts can be obtained as in eq. (2.228):

$$\mathcal{D}^{-1}_{\mu\nu}(p) = \mathcal{D}_T^{-1}(p^2)\left(g_{\mu\nu} - \frac{p_\mu p_\nu}{p^2}\right) + \mathcal{D}_L^{-1}(p^2)\frac{p_\mu p_\nu}{p^2}\,, \tag{4.52}$$

$$D^{-1}_{\mu\nu}(p) = p^2 g_{\mu\nu} - \left(1 - \frac{1}{\xi}\right) p_\mu p_\nu\,. \tag{4.53}$$

Since $\Pi_{\mu\nu}(p)$ is transverse, one can use eq. (4.48) to obtain

$$\mathcal{D}_L^{-1}(p^2) = D_L^{-1}(p^2) = \frac{p^2}{\xi}\,, \qquad \mathcal{D}_T^{-1}(p^2) = p^2\left[1 + \Pi(p^2)\right]. \tag{4.54}$$

Hence, eq. (4.51) yields[9]

$$-i\mathcal{D}^{\mu\nu}(p) = \frac{-i}{p^2[1+\Pi(p^2)]}\left(g^{\mu\nu} - \frac{p^\mu p^\nu}{p^2}\right) - \frac{i\xi p^\mu p^\nu}{p^4}\,, \tag{4.55}$$

which reproduces eq. (2.230) after setting $m = 0$ and employing eq. (4.50). Thus, the pole at $p^2 = 0$ is not shifted. That is, the gauge boson mass remains zero to all orders in perturbation theory.

This elegant argument has a loophole. Namely, if $\Pi(p^2)$ develops a pole at $p^2 = 0$ with residue $-m_v^2$, then the pole of $\mathcal{D}^{\mu\nu}(p)$ shifts away from zero. In particular, $\Pi(p^2) \simeq -m_v^2/p^2$ as $p^2 \to 0$ implies that $\mathcal{D}(p^2) \simeq -i/(p^2 - m_v^2)$. This requires some nontrivial dynamics to generate a massless intermediate state in $\Pi_{\mu\nu}(p)$. Such a massless state is called a Goldstone boson. The Standard Model employs the dynamics of elementary scalar fields in order to generate the Goldstone modes. We thus turn our attention to the vector boson mass generation mechanism of the Standard Model.

4.5 Spontaneously Broken Gauge Theories

Consider a nonabelian gauge theory with scalar (and fermion) matter, given by eq. (4.39). The corresponding scalar potential function must be gauge invariant [eq. (4.21)]. In order to identify the physical scalar degrees of freedom of this model, one must minimize the scalar potential $V(\phi)$ and determine the corresponding values of the scalar fields ϕ_i at the potential minimum. These are the scalar vacuum expectation values, $v_i \equiv \langle 0|\phi_i|0\rangle$. That is, the v_i are determined by the scalar potential minimum condition

$$\left.\frac{\partial V}{\partial \phi_i}\right|_{\phi_i = v_i} = 0\,. \tag{4.56}$$

Expanding each scalar field about its vacuum expectation value (VEV) yields the tree-level scalar masses and self-couplings. However, in general the scalar fields are charged under the global symmetry group G, in which case a nonzero VEV would be incompatible with the global symmetry.

9 In eq. (4.55), we have dropped the explicit $+i\varepsilon$ that is associated with the pole of the propagator.

4.5.1 Goldstone's Theorem

Given a theory of scalars and fermions (with no gauge fields) governed by a Lagrangian that is invariant under a continuous global symmetry group, G, that is spontaneously broken, Goldstone proved the following theorem.

Theorem 4.1. *If the Lagrangian is invariant under a continuous global symmetry group G (of dimension d_G), but the vacuum state of the theory is not invariant under all G-transformations, then the theory exhibits spontaneous symmetry breaking. In this case, if the vacuum is invariant under all H-transformations, where H is a subgroup of G (of dimension d_H), then we say that the gauge group G is spontaneously broken down to H. The physical spectrum will then contain n massless scalar excitations (called Goldstone bosons), where $n \equiv d_G - d_H$.*

Goldstone's Theorem can be proved independently of perturbation theory (e.g., see [B73, B81]). However, the theorem can be easily established in the tree-level approximation. Let $\phi_i(x)$ be a multiplet of $n \equiv d_R$ real scalar fields. The scalar Lagrangian

$$\mathscr{L} = \tfrac{1}{2}(\partial_\mu \phi_i)(\partial^\mu \phi_i) - V(\phi) \tag{4.57}$$

is assumed to be invariant under a compact symmetry group G, under which the scalar fields transform as $\phi_i \to \mathcal{Q}_i{}^j \phi_j$, where $\mathcal{Q}$ is a real representation R of G. Recall that all real representations of a compact group are equivalent (via a similarity transformation) to an orthogonal representation. Thus, without loss of generality, we may take $\mathcal{Q}$ to be an orthogonal $n \times n$ matrix. The corresponding infinitesimal transformation law [see eq. (4.13)] is given by

$$\delta\phi_i = -ig_a \Lambda^a (\boldsymbol{T}^a)_i{}^j \phi_j\,, \tag{4.58}$$

where g_a and Λ^a are real and the $\boldsymbol{T}^a$ are imaginary antisymmetric matrices. One can check that the scalar kinetic term is automatically invariant under $\mathrm{O}(n)$ transformations. The scalar potential, which is not invariant in general under the full $\mathrm{O}(n)$ group, is invariant under G [which is a subgroup of $\mathrm{O}(n)$] if the scalar potential $V(\phi)$ satisfies the following condition:

$$V(\phi + \delta\phi) \simeq V(\phi) + \frac{\partial V}{\partial \phi_i}\delta\phi_i = V(\phi)\,. \tag{4.59}$$

The first (approximate) equality above is simply a Taylor expansion to first order in the field variation, while the second equality imposes the invariance assumption. Inserting the result for $\delta\phi_i(x)$ from eq. (4.58), it follows that

$$\frac{\partial V}{\partial \phi_i}(\boldsymbol{T}^a)_i{}^j \phi_j = 0\,. \tag{4.60}$$

Suppose that the scalar potential minimum, $v_i \equiv \langle 0|\phi_i|0\rangle$, is not invariant under the global symmetry group. That is,

$$\mathcal{Q}_i{}^j v_j \simeq \left[\delta_i{}^j - ig_a\Lambda^a(\boldsymbol{T}^a)_i{}^j\right] v_j \neq v_i\,. \tag{4.61}$$

It follows that there exists at least one a such that $(\boldsymbol{T}^a v)_i \neq 0$, and we conclude that the global symmetry is spontaneously broken. In general, there is a residual symmetry group H whose Lie algebra $\mathfrak{h}$ is spanned by the maximal number of linearly independent elements of the Lie algebra $\mathfrak{g}$ that annihilate the vacuum expectation value v. That is, we choose a new basis of generators for the Lie algebra $\mathfrak{g}$ (which we denote by $\widetilde{\boldsymbol{T}}^a$), such that

$$(\widetilde{\boldsymbol{T}}^a v)_i = 0\,, \qquad a = 1, 2, \ldots, d_H\,, \tag{4.62}$$

$$(\widetilde{\boldsymbol{T}}^a v)_i \neq 0\,, \qquad a = d_H + 1, d_H + 2, \ldots, d_G\,, \tag{4.63}$$

where d_H is the dimension of the maximal unbroken subgroup H. Equations (4.62) and (4.63) define the unbroken and broken generators, respectively. We then say that the symmetry group G is spontaneously broken down to the group H.

Next, we shift the field by its vacuum expectation value,

$$\phi_i \equiv v_i + \varphi_i\,, \tag{4.64}$$

and express the scalar Lagrangian in terms of the φ_i,

$$\mathscr{L} = \tfrac{1}{2}(\partial_\mu \varphi_i)(\partial^\mu \varphi_i) - \tfrac{1}{2}\mathcal{M}^2_{ij}\varphi_i\varphi_j + \mathcal{O}(\varphi^3)\,, \tag{4.65}$$

where we have imposed the scalar potential minimum condition [see eq. (4.56)], and

$$\mathcal{M}^2_{ij} \equiv \left(\frac{\partial^2 V}{\partial\phi_i\partial\phi_j}\right)_{\phi_i = v_i}\,. \tag{4.66}$$

The terms cubic and higher in the φ_i do not concern us here. After differentiating eq. (4.60) with respect to ϕ_k, one can set $\phi_k = v_k$ and invoke eq. (4.56) to obtain

$$\mathcal{M}^2_{ki}(\widetilde{\boldsymbol{T}}^a v)_i = 0\,. \tag{4.67}$$

Equation (4.67) is trivially satisfied if $\widetilde{T}^a$ is an unbroken generator [see eq. (4.63)]. In contrast, for each broken generator, $(\widetilde{\boldsymbol{T}}^a v)_i$ is an eigenvector of $\mathcal{M}^2$ with zero eigenvalue. The corresponding eigenstates can be used to identify the linear combinations of scalar fields φ_i that are massless. These are the Goldstone fields,

$$G^a \propto i\varphi_i(\widetilde{\boldsymbol{T}}^a)_i{}^j v_j\,. \tag{4.68}$$

Clearly, there are $d_G - d_H$ independent Goldstone modes, corresponding to the number of broken generators.

4.5.2 Massive Gauge Bosons

If a spontaneously broken global symmetry is promoted to a local symmetry, then a remarkable mechanism, called the Higgs mechanism, takes place. The Goldstone bosons disappear from the spectrum, and the formerly massless gauge bosons become massive. Roughly speaking, the Goldstone bosons become the longitudinal degrees of freedom of the massive gauge bosons. This result is easily demonstrated in a generalization of the tree-level analysis given previously. If the scalar sector specified in eq. (4.57) is coupled to gauge fields, then one must replace the ordinary derivative with a covariant derivative in the scalar kinetic energy term:

$$\mathscr{L}_{\rm KE} = \tfrac{1}{2}[\partial_\mu\phi_i + ig_a A^a_\mu(\boldsymbol{T^a})_i{}^j\phi_j][\partial^\mu\phi_i + ig_b A^{\mu b}(\boldsymbol{T^b})_i{}^k\phi_k]\,. \tag{4.69}$$

As above, the $\phi_i(x)$ are real fields and the $\boldsymbol{T^a}$ are purely imaginary antisymmetric matrix generators corresponding to the representation of the scalar multiplet. It is convenient to define real antisymmetric matrices[10]

$$\boldsymbol{L_a} \equiv ig_a\boldsymbol{T^a}\,. \tag{4.70}$$

If we expand the ϕ_i around their VEVs [eq. (4.64)], then eq. (4.69) yields a term quadratic in the gauge fields:

$$\mathscr{L}_{\rm mass} = \tfrac{1}{2}M^2_{ab}A^a_\mu A^{\mu b}\,, \tag{4.71}$$

where the gauge boson squared-mass matrix is given by

$$M^2_{ab} = (\boldsymbol{L_a}v, \boldsymbol{L_b}v)\,. \tag{4.72}$$

Here, we have employed a convenient notation where $(x, y) \equiv \sum_i x_i y_i$.

If $\boldsymbol{L_a}v \neq 0$ for at least one a, then the gauge symmetry is broken and at least one of the gauge bosons acquires mass. The gauge boson squared-mass matrix is real symmetric, so it can be diagonalized with an orthogonal similarity transformation:

$$\mathcal{O}M^2\mathcal{O}^{\sf T} = {\rm diag}\,(0, 0, \ldots, 0,\; m_1^2, m_2^2, \ldots)\,. \tag{4.73}$$

The corresponding gauge boson mass eigenstates are

$$\widetilde{A}^a_\mu \equiv \mathcal{O}_{ab}A^b_\mu\,. \tag{4.74}$$

Indeed, one can easily check that $M^2_{ab}A^a_\mu A^{\mu b} = \sum_a m_a^2 \widetilde{A}^a_\mu \widetilde{A}^{\mu a}$. Likewise, we may define a new basis for the Lie algebra:

$$\widetilde{\boldsymbol{L}}_a \equiv \mathcal{O}_{ab}\boldsymbol{L_b}\,. \tag{4.75}$$

It then follows that

$$(\mathcal{O}M^2\mathcal{O}^{\sf T})_{ab} = (\widetilde{\boldsymbol{L}}_a v, \widetilde{\boldsymbol{L}}_b v) = m_a^2\delta_{ab}\,. \tag{4.76}$$

In particular $m_a^2 = 0$ when $\widetilde{\boldsymbol{L}}_a v = 0$ for $a = 1, \ldots, d_H$ (corresponding to the unbroken generators) and $m_a^2 \neq 0$ when $\widetilde{\boldsymbol{L}}_a v \neq 0$ for $a = d_H + 1, \ldots, d_G$ (corresponding to the broken generators). That is, there are $d_G - d_H$ massive gauge bosons.[11]

It is instructive to revisit the scalar Lagrangian

$$\mathscr{L} = \tfrac{1}{2}\left[\partial_\mu\phi_i + (\boldsymbol{L_a})_i{}^j A^a_\mu\phi_j\right]\left[\partial^\mu\phi_i + (\boldsymbol{L_b})_i{}^k A^{\mu b}\phi_k\right] - V(\phi)\,. \tag{4.77}$$

The covariant derivative $\nabla_\mu \equiv \partial_\mu + \boldsymbol{L_a}A^a_\mu$ can also be written in terms of gauge boson mass eigenstates $\widetilde{A}^a_\mu$ and the new generators $\widetilde{\boldsymbol{L}}_a$ by using the identity

$$\widetilde{\boldsymbol{L}}_a\widetilde{A}^a_\mu = \mathcal{O}_{ac}\mathcal{O}_{ab}\boldsymbol{L_c}A^b_\mu = \boldsymbol{L_c}A^c_\mu\,. \tag{4.78}$$

[10] There is no significance to the height of the adjoint index a (see footnote 11 in Section 2.6).

[11] The number of massive gauge bosons, or equivalently the number of broken generators, is equal to the dimension of the coset space G/H.

Expanding the scalar fields in eq. (4.77) around their VEVs [eq. (4.64)], and identifying the gauge boson mass eigenstates (while not displaying terms cubic or higher in the fields and terms involving the physical scalars), we find

$$\begin{aligned}\mathscr{L} &= \tfrac{1}{2}(\partial_\mu\varphi_i)(\partial^\mu\varphi_i) + \tfrac{1}{2}m_a^2\widetilde{A}^a_\mu\widetilde{A}^{\mu a} + \tfrac{1}{2}(\widetilde{\boldsymbol{L}}_a\widetilde{A}^a_\mu v, \partial^\mu\varphi) + \tfrac{1}{2}(\partial_\mu\varphi, \boldsymbol{L}_a\widetilde{A}^{\mu a}v) + \cdots \\ &= \tfrac{1}{2}(\partial_\mu G^b)(\partial^\mu G^b) + \tfrac{1}{2}m_a^2\widetilde{A}^a_\mu\widetilde{A}^{\mu a} + m_a\widetilde{A}^a_\mu\partial^\mu G_a + \cdots\,, \end{aligned} \tag{4.79}$$

where there is an implicit sum over the thrice-repeated index a and

$$G_a = \frac{1}{m_a}(\widetilde{\boldsymbol{L}}_a v, \varphi) \qquad [\text{no sum over } a]\,. \tag{4.80}$$

We recognize G_a as the Goldstone bosons that appeared in the scalar theory with a spontaneously broken global symmetry [see eq. (4.67) and the discussion that follows]. Here, the normalization of the Goldstone field has been chosen so that G^a possesses a canonically normalized kinetic energy term.

The coupling $m_a A^a_\mu \partial^\mu G_a$ plays an integral role in the vector boson mass generation mechanism by providing a new interaction vertex:

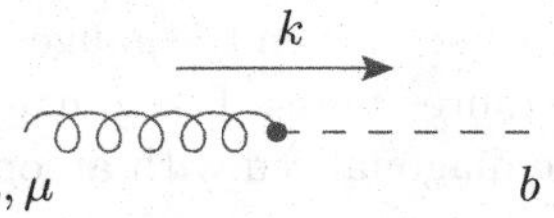

$m_a k^\mu \delta^{ab}$

We can then evaluate the contribution of the Goldstone boson propagator to $i\Pi^{\mu\nu}(k)$:

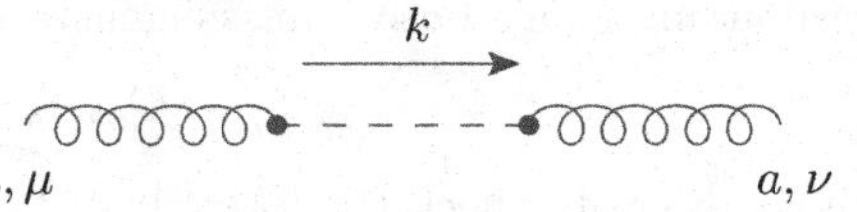

$$i\Pi^{\mu\nu}(k) = m_a^2 k^\mu(-k^\nu)\frac{i}{k^2} + \cdots = -im_a^2\frac{k^\mu k^\nu}{k^2} + \cdots\,, \tag{4.81}$$

where terms that are finite as $k^2 \to 0$ are indicated by ellipses. That is, the Feynman diagram above is the only source for a pole at $k^2 = 0$ at this order in perturbation theory. But gauge invariance requires

$$i\Pi^{\mu\nu}(k) = i(k^\mu k^\nu - k^2 g^{\mu\nu})\Pi(k^2)\,. \tag{4.82}$$

It therefore follows that

$$\Pi(k^2) \simeq -\frac{m_a^2}{k^2}\,, \tag{4.83}$$

and (up to an overall wave function renormalization that we absorb into the definition of the renormalized $\mathscr{D}$)

$$\mathscr{D}(k^2) = \frac{i}{k^2[1+\Pi(k^2)]} = \frac{i}{k^2 - m_a^2}\,. \tag{4.84}$$

We say that the gauge boson "eats" or absorbs the corresponding Goldstone boson and thereby acquires mass via the Higgs mechanism.

4.5.3 The Unitary and R_ξ Gauges

The spontaneously broken nonabelian gauge theory Lagrangian contains Goldstone boson fields. However, as we shall now demonstrate, the Goldstone bosons are gauge artifacts that can be removed by a gauge transformation. Consider the transformation law for the shifted scalar field, $\varphi_i(x) \equiv \phi_i(x) - v_i$. Promoting eq. (4.58) to a local gauge transformation, where $\Lambda^a = \Lambda^a(x)$, and noting that $\delta v_i = 0$,

$$\delta\varphi_i = -\Lambda(\varphi_i + v_i)\,, \tag{4.85}$$

where $\Lambda \equiv ig_a \boldsymbol{T}^a \Lambda^a \equiv \boldsymbol{L}_a \Lambda^a$. If we define $\widetilde{\Lambda}^a = \mathcal{O}_{ab}\Lambda^b$, then we can also write $\Lambda = \widetilde{\boldsymbol{L}}_a \widetilde{\Lambda}^a$. Then, under a gauge transformation, the Goldstone field [eq. (4.80)] transforms as $G_a \to G_a + \delta G_a$, where

$$\begin{aligned}
\delta G_a &= \frac{1}{m_a}(\widetilde{\boldsymbol{L}}_a v, \delta\varphi) = -\frac{1}{m_a}(\widetilde{\boldsymbol{L}}_a v, \Lambda\varphi + \Lambda v) \\
&= -\frac{1}{m_a}(\widetilde{\boldsymbol{L}}_a v, \Lambda\varphi) - \frac{1}{m_a}(\widetilde{\boldsymbol{L}}_a v, \widetilde{\boldsymbol{L}}_b v)\widetilde{\Lambda}^b \\
&= -\frac{1}{m_a}(\widetilde{\boldsymbol{L}}_a v, \Lambda\varphi) - m_a \widetilde{\Lambda}_a\,,
\end{aligned} \tag{4.86}$$

after using eq. (4.76) for the gauge boson masses. Note that the second term in the last line above, $m_a\widetilde{\Lambda}_a$, is an inhomogeneous term independent of φ. As a result, one can always find a $\Lambda_a(x)$ such that $G_a + \delta G_a = 0$. That is, a gauge transformation exists[12] in which the Goldstone field is completely gauged away to zero. The resulting gauge is called the unitary gauge.

So far, we have not mentioned the gauge-fixing term and Faddeev–Popov ghosts. In spontaneously broken nonabelian gauge theories, the R_ξ gauge turns out to be particularly useful.[13] The R_ξ gauge is defined by the following gauge-fixing term:

$$\begin{aligned}
\mathscr{L}_{\rm GF} &= -\frac{1}{2\xi}\left[\partial^\mu \widetilde{A}^a_\mu - \xi\,(\widetilde{\boldsymbol{L}}_a v\,,\,\varphi)\right]^2 \\
&= -\frac{1}{2\xi}(\partial^\mu \widetilde{A}^a_\mu)^2 + m_a G_a(\partial^\mu \widetilde{A}^a_\mu) - \frac{\xi m_a^2}{2} G_a G_a\,,
\end{aligned} \tag{4.87}$$

after using eq. (4.80). At this point, we notice that the term $m_a G_a(\partial^\mu \widetilde{A}^a_\mu)$ of eq. (4.87) combines with $m_a \widetilde{A}^a_\mu \partial^\mu G_a$ of eq. (4.79) to yield

$$m_a[G_a\partial^\mu \widetilde{A}^a_\mu + \widetilde{A}^a_\mu \partial^\mu G_a] = m_a \partial^\mu(G_a \widetilde{A}^a_\mu)\,, \tag{4.88}$$

which is a total divergence that can be dropped from the Lagrangian. One must

[12] In a general nonabelian gauge theory, the gauge transformation that eliminates the Goldstone fields $G_a(x)$ [for *all* values of x] cannot be written in an explicit form. Nevertheless, as shown in Section 8.5 of [B119], one can prove that such a gauge transformation must exist. See also Section 19.6 of [B73].

[13] The R stands for renormalizable, and ξ is the gauge-fixing parameter. In the R_ξ gauges, the theory is manifestly renormalizable, although unitarity is not manifest and must be separately proved. The opposite is true for the unitary gauge.

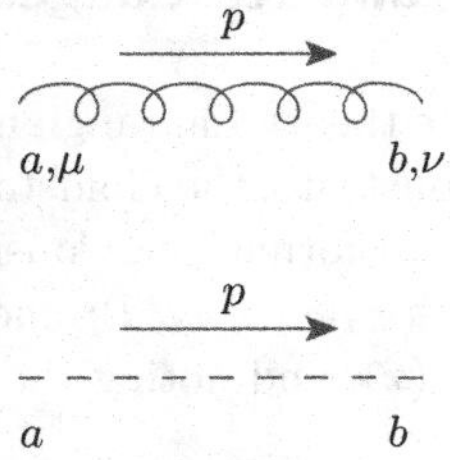

$$\frac{-i\delta^{ab}}{p^2 - m_a^2 + i\varepsilon}\left[g_{\mu\nu} - (1-\xi)\frac{p_\mu p_\nu}{p^2 - \xi m_a^2}\right]$$

$$\frac{i\delta^{ab}}{p^2 - \xi m_a^2 + i\varepsilon}$$

Fig. 4.2 Feynman rules for the propagator of the massive nonabelian gauge boson and Goldstone boson in the R_ξ gauge.

also include Faddeev–Popov ghosts (e.g., see [B75]):

$$\mathscr{L}_{\text{FP}} = \partial^\mu \omega_a^* \nabla_\mu^{ab} \omega_b - \xi \omega_a^* M_{ab}^2 \omega_b - \xi g_a g_b \omega_a^* \omega_b (\varphi\,,\, \boldsymbol{T}^b \boldsymbol{T}^a v)\,, \tag{4.89}$$

where ∇_μ^{ab} is defined in eq. (4.30) and M_{ab}^2 is the gauge boson squared-mass matrix [eq. (4.72)].

In the R_ξ gauge, the Feynman rule for the massless gauge boson shown in Fig. 4.1 remains valid. The corresponding Feynman rules for the massive gauge boson and the Goldstone boson propagators are exhibited in Fig. 4.2. Note that in the R_ξ gauge, the Goldstone bosons have acquired a squared mass equal to ξm_a^2, where m_a^2 is the mass of the associated vector boson. This is an artifact of the gauge choice. Nevertheless, for a consistent computation in the R_ξ gauge, both the Goldstone bosons and the Faddeev–Popov ghosts must be included as possible internal lines in Feynman graphs. As noted previously, any physical quantity must ultimately be independent of ξ.

As in the unbroken gauge theory, the two most useful gauges are $\xi = 1$ (now called the 't Hooft–Feynman gauge) and $\xi = 0$ (the Landau gauge). It is noteworthy that in the Landau gauge the Goldstone bosons are massless. Finally, it is instructive to consider the limit of $\xi \to \infty$. This corresponds to the unitary gauge, since the (unphysical) Goldstone bosons, whose masses have become infinitely large, are decoupled. Moreover, in this limit the massive gauge boson propagator given in Fig. 4.2 reduces to the expression shown in Fig. 4.3. The fact that the unitary gauge is a limiting case of the R_ξ gauge played a critical role in the proof that spontaneously broken nonabelian gauge theories are unitary and renormalizable [40].

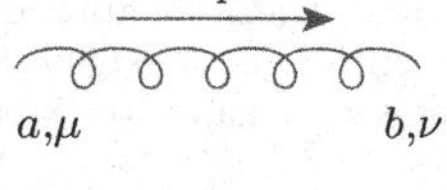

$$\frac{-i\delta^{ab}}{p^2 - m_a^2 + i\varepsilon}\left[g_{\mu\nu} - \frac{p_\mu p_\nu}{m_a^2}\right]$$

Fig. 4.3 Feynman rules for the massive nonabelian gauge boson propagator in the unitary gauge.

4.5.4 The Physical Higgs Bosons

An important check of the formalism is the counting of bosonic degrees of freedom. Assume that the multiplet of scalar fields transforms according to some d_R-dimensional real representation R under the transformation group G (which has dimension d_G). Prior to spontaneous symmetry breaking, the theory contains d_R scalar degrees of freedom and $2d_G$ vector boson degrees of freedom. The latter consists of d_G massless gauge bosons (one for each possible value of the adjoint index a), with each massless gauge boson contributing two degrees of freedom corresponding to the two possible transverse helicities.

After spontaneous symmetry breaking of G to a subgroup H (which has dimension d_H), there are $d_G - d_H$ Goldstone bosons which are unphysical (and can be removed from the spectrum by going to the unitary gauge). This leaves

$$d_R - d_G + d_H \text{ physical scalar degrees of freedom}, \tag{4.90}$$

which correspond to the physical Higgs bosons of the theory. We have also found that there are d_H massless gauge bosons (one for each unbroken generator) and $d_G - d_H$ massive gauge bosons (one for each broken generator). But, for massive gauge bosons which possess a longitudinal helicity state, we must count three degrees of freedom. Thus, we end up with

$$3d_G - d_H \text{ vector boson degrees of freedom.} \tag{4.91}$$

Adding the two yields a total of $d_R + 2d_G$ bosonic degrees of freedom, in agreement with our previous counting.

It is instructive to check that the physical Higgs bosons cannot be removed by a gauge transformation. We divide the scalars into two classes: (i) the Goldstone bosons G_a, $a = d_H + 1, d_H + 2, \ldots, d_G$ [see eq. (4.80)] and (ii) the scalar states orthogonal to G_a. These are the Higgs bosons:

$$\widetilde{H}_k = c_j^{(k)} \varphi_j \,, \tag{4.92}$$

where $k = 1, 2, \ldots, d_R - d_G + d_H$. The $c_j^{(k)}$ are real numbers that satisfy the orthogonality conditions

$$\sum_j c_j^{(k)} (\widetilde{\boldsymbol{L}}_a v)_j = 0 \,, \qquad \sum_j c_j^{(k)} c_j^{(\ell)} = \delta_{k\ell} \,. \tag{4.93}$$

Under a gauge transformation [eq. (4.85)],

$$\delta \widetilde{H}_k = \delta(c_j^{(k)} \varphi_j) = c_j^{(k)} \delta\varphi_j = -c_j^{(k)} (\Lambda\varphi + \Lambda v)_j = -c_j^{(k)} (\Lambda\varphi)_j \,, \tag{4.94}$$

where we have noted that $c_j^{(k)} (\Lambda v)_j = c_j^{(k)} \widetilde{\Lambda}^a (\widetilde{\boldsymbol{L}}_a v)_j = 0$ [after invoking eq. (4.93)]. Thus the transformation law for $\widetilde{H}_k$ is homogeneous in the scalar fields, and one cannot remove the Higgs boson field by a gauge transformation.

The states $\widetilde{H}_k$ are in general not mass eigenstates. We may write down the physical Higgs mass matrix by employing the $\{G_a, \widetilde{H}_k\}$ basis for the scalar fields.

This is accomplished by noting that the orthogonality relations of eq. (4.93) and eq. (4.76) yield

$$\varphi_j = \frac{(\widetilde{\boldsymbol{L}}_a v)_j}{m_a} G_a + c_j^{(k)} \widetilde{H}_k \,, \tag{4.95}$$

where the sum over the repeated indices a and k respectively is implied. Finally, we note that the scalar boson squared-mass matrix, given in eq. (4.66), is real symmetric and satisfies $(\mathcal{M}^2)_{ij}(\widetilde{\boldsymbol{L}}_a v)_j = 0$ [eq. (4.67)]. Thus, the nonderivative quadratic terms in the scalar part of the Lagrangian are given by

$$\mathscr{L}_{\text{scalar mass}} = -\tfrac{1}{2}(\mathcal{M}_H^2)_{k\ell}\widetilde{H}_k\widetilde{H}_\ell \,, \tag{4.96}$$

where the physical Higgs squared-mass matrix is given by

$$(\mathcal{M}_H^2)_{k\ell} = c_i^{(k)} c_j^{(\ell)} \mathcal{M}_{ij}^2 \,. \tag{4.97}$$

Diagonalizing $\mathcal{M}_H^2$ yields the Higgs boson mass eigenstates, H_k, and the corresponding eigenvalues are the physical Higgs boson squared masses.

4.6 Complex Representations of Scalar Fields

We now have nearly all the ingredients necessary to construct the Standard Model. However, in our treatment of spontaneously broken nonabelian gauge fields, we considered real scalar fields that transform under a real representation of the symmetry group. In contrast, the Standard Model employs scalars that transform under a complex representation of the symmetry group. In this section, we provide the necessary formalism that will allow us to directly treat the complex case.

Let $\Phi_i(x)$ be a set of $n \equiv d_R$ complex scalar fields. The scalar Lagrangian

$$\mathscr{L} = (\partial_\mu \Phi^{\dagger i})(\partial^\mu \Phi_i) - V(\Phi_i, \Phi^{\dagger i}) \tag{4.98}$$

is assumed to be invariant under a compact symmetry group G, under which the scalar fields transform as

$$\Phi_i \to \mathcal{U}_i{}^j \Phi_j \,, \qquad\qquad \Phi^{\dagger\, i} \to \Phi^{\dagger\, j}(\mathcal{U}^\dagger)_j{}^i \,, \tag{4.99}$$

where $\mathcal{U}$ is a complex representation[14] of G. Recall that all complex representations of a compact group are equivalent (via a similarity transformation) to a unitary representation. Thus, without loss of generality, we may take $\mathcal{U}$ to be a unitary $n \times n$ matrix. Explicitly,

$$\mathcal{U} = \exp[-ig_a \Lambda^a \mathcal{T}^a] \,, \tag{4.100}$$

[14] In this context, we call a representation complex if and only if it is not equivalent (by similarity transformation) to some real representation. This is somewhat broader than the conventional group-theoretical definition.

where the generators $\mathcal{T}^a$ are $n \times n$ hermitian matrices. The corresponding infinitesimal transformation law is

$$\delta\Phi_i(x) = -ig_a\Lambda^a(\mathcal{T}^a)_i{}^j\Phi_j(x)\,, \tag{4.101}$$

$$\delta\Phi^{\dagger\,i}(x) = +ig_a\Phi^{\dagger\,j}(x)\Lambda^a(\mathcal{T}^a)_j{}^i\,, \tag{4.102}$$

where the g_a and Λ^a are real. Note that G is generically a subgroup of U(n). Although the scalar kinetic energy term is invariant under all U(n) transformations, the scalar potential is only invariant under the subgroup of U(n) transformations that reside in G.

There are $2n$ independent scalar degrees of freedom, corresponding to the fields Φ_i and $\Phi^{\dagger\,i}$. We can also express these degrees of freedom in terms of $2n$ real scalar fields consisting of ϕ_{Aj} and ϕ_{Bj} ($j = 1, 2, \ldots, n$) defined by

$$\Phi_j = \frac{1}{\sqrt{2}}(\phi_{Aj} + i\phi_{Bj})\,, \qquad \Phi^{\dagger\,j} = \frac{1}{\sqrt{2}}(\phi_{Aj} - i\phi_{Bj})\,. \tag{4.103}$$

It is straightforward to compute the group transformation laws for the real fields ϕ_{Aj} and ϕ_{Bj}. These are conveniently expressed by introducing a $2n$-dimensional scalar multiplet,

$$\phi(x) = \begin{pmatrix} \phi_A(x) \\ \phi_B(x) \end{pmatrix}. \tag{4.104}$$

That is, $\phi_{Aj}(x) = \phi_j(x)$ and $\phi_{Bj}(x) = \phi_{j+n}(x)$. Then the infinitesimal form of the group transformation law for $\phi(x)$ is given by $\phi_k(x) \to \phi_k(x) + \delta\phi_k(x)$ for $k = 1, 2, \ldots, 2n$, where

$$\delta\phi_k(x) = -ig\Lambda^a(\boldsymbol{T}^a)_k{}^\ell\phi_\ell(x)\,, \tag{4.105}$$

and

$$i\boldsymbol{T}^a = \begin{pmatrix} -\operatorname{Im}\mathcal{T}^a & -\operatorname{Re}\mathcal{T}^a \\ \operatorname{Re}\mathcal{T}^a & -\operatorname{Im}\mathcal{T}^a \end{pmatrix}. \tag{4.106}$$

Note that $\operatorname{Re}\mathcal{T}^a$ is symmetric and $\operatorname{Im}\mathcal{T}^a$ is antisymmetric as a consequence of the hermiticity of the $\mathcal{T}^a$. Thus, $i\boldsymbol{T}^a$ is a real antisymmetric $2n \times 2n$ matrix, which when exponentiated yields a real orthogonal $2n$-dimensional representation of G. Consequently, using the real representation for $i\boldsymbol{T}^a$ [eq. (4.106)], we may immediately apply the formalism of Section 4.5 that was established for the case of a real representation, and obtain the corresponding results for the case of a complex representation.

We can also apply the above analysis to a nonabelian gauge theory based on the compact group G coupled to a multiplet of scalar fields. We assume that the scalars transform according to a (possibly reducible) d_R-dimensional complex unitary representation R of G. We proceed to transcribe some of the results of Section 4.5 to the present case. The VEV of the complex scalar field is assumed to be $\langle\Phi_i(x)\rangle \equiv \nu_i$.

Consequently, in the real representation,

$$v \equiv \langle \phi(x) \rangle = \frac{1}{\sqrt{2}} \begin{pmatrix} \nu + \nu^* \\ -i(\nu - \nu^*) \end{pmatrix} . \tag{4.107}$$

We now shift the complex fields by their VEVs:

$$\Phi_i \equiv \nu_i + \overline{\Phi}_i \,, \qquad \Phi^{\dagger i} \equiv \nu_i^* + \overline{\Phi}^{\dagger i} . \tag{4.108}$$

Using the real scalar field basis, the gauge boson mass matrix is given by eq. (4.72), which can be rewritten as

$$M^2_{ab} = g_a g_b (\boldsymbol{T}^a \boldsymbol{T}^b)_{jk} v_j v_k \,, \tag{4.109}$$

after noting that the $\boldsymbol{T}^a$ are antisymmetric. Plugging in eqs. (4.106) and (4.107), we obtain the corresponding result with respect to the complex scalar field basis [B73]:

$$M^2_{ab} = 2 g_a g_b \, \mathrm{Re}(\nu^\dagger \mathcal{T}^a \mathcal{T}^b \nu) = g_a g_b \nu^\dagger (\mathcal{T}^a \mathcal{T}^b + \mathcal{T}^b \mathcal{T}^a) \nu \,, \tag{4.110}$$

where the last step above follows from the hermiticity of the $\mathcal{T}^a$. As before, M^2 is a real symmetric matrix that can be diagonalized by an orthogonal transformation $\mathcal{O} M^2 \mathcal{O}^\mathsf{T}$ [eq. (4.73)]. The corresponding gauge boson mass eigenstates are given by eq. (4.74).

It is again convenient to introduce the antihermitian generators, $\mathcal{L}_a \equiv i g_a \mathcal{T}^a$, and then define a new basis for the Lie algebra:

$$\widetilde{\mathcal{L}}_a \equiv \mathcal{O}_{ab} \mathcal{L}_b \,. \tag{4.111}$$

It follows that

$$(\mathcal{O} M^2 \mathcal{O}^\mathsf{T})_{ab} = (\widetilde{\mathcal{L}}_a \nu)^\dagger (\widetilde{\mathcal{L}}_b \nu) + (\widetilde{\mathcal{L}}_b \nu)^\dagger (\widetilde{\mathcal{L}}_a \nu) = m_a^2 \delta_{ab} \,. \tag{4.112}$$

Hence, one can easily identify the unbroken and broken generators:

$$(\widetilde{\mathcal{T}}^a \nu)_j = 0 \,, \qquad a = 1, 2, \ldots, d_H \,, \tag{4.113}$$

$$(\widetilde{\mathcal{T}}^a \nu)_j \neq 0 \,, \qquad a = d_H + 1, d_H + 2, \ldots, d_G \,. \tag{4.114}$$

Consider next the scalar squared-mass matrix and the identification of the Goldstone bosons. Although these quantities can be obtained from first principles (see Exercise 4.2), our derivations here are based on results already obtained in Section 4.5.1, where we employ the connection between the real and complex basis outlined above. To identify the Goldstone bosons, we insert eqs. (4.104), (4.106), and (4.107) into eq. (4.80). Rewriting ϕ_A and ϕ_B in terms of the shifted complex fields $\overline{\Phi}$ and $\overline{\Phi}^\dagger$, we obtain $d_G - d_H$ Goldstone boson fields:

$$G_a = \frac{1}{m_a} \left[\overline{\Phi}^\dagger \widetilde{\mathcal{L}}_a \nu + (\widetilde{\mathcal{L}}_a \nu)^\dagger \overline{\Phi} \right] , \tag{4.115}$$

where $a = d_H + 1, d_H + 2, \ldots, d_G$. The scalar squared-mass matrix is obtained from eq. (4.66):

$$\mathscr{L}_{\text{scalar mass}} = -\tfrac{1}{2} \mathcal{M}^2_{ij} \phi_i \phi_j = -\frac{1}{2} \begin{pmatrix} \Phi_k & \Phi^{\dagger \ell} \end{pmatrix} \mathcal{M}^2 \begin{pmatrix} \Phi^{\dagger m} \\ \Phi_n \end{pmatrix} . \tag{4.116}$$

To determine the matrix $\mathcal{M}^2$, it is convenient to introduce the unitary matrix W,

$$W = \frac{1}{\sqrt{2}}\begin{pmatrix} I & -iI \\ I & iI \end{pmatrix}, \qquad W\phi = \begin{pmatrix} \Phi^\dagger \\ \Phi \end{pmatrix}, \tag{4.117}$$

and $\phi^{\mathsf{T}} W^{-1} = (\Phi \;\; \Phi^\dagger)$, where I is the $d_R \times d_R$ identity matrix. Finally, we use the chain rule to obtain

$$\mathcal{M}^2_{ij} = \left(\frac{\partial^2 V}{\partial\phi_i \partial\phi_j}\right)_{\phi_i = v_i} = W^{-1}\mathcal{M}^2 W, \tag{4.118}$$

and

$$\mathcal{M}^2 = \begin{pmatrix} \dfrac{\partial^2 V}{\partial\Phi_k \partial\Phi^{\dagger\, m}} & \dfrac{\partial^2 V}{\partial\Phi_k \partial\Phi_n} \\ \dfrac{\partial^2 V}{\partial\Phi^{\dagger\, \ell}\partial\Phi^{\dagger\, m}} & \dfrac{\partial^2 V}{\partial\Phi^{\dagger\, \ell}\partial\Phi_n} \end{pmatrix}_{\Phi_i = \nu_i}, \tag{4.119}$$

with $k, \ell, m, n \in \{1, 2, \ldots, d_R\}$. Indeed, one can check (see Exercise 4.2) that

$$\mathcal{M}^2 \begin{pmatrix} (\widetilde{\mathcal{L}}_a \nu)^{*m} \\ (\widetilde{\mathcal{L}}_a \nu)_n \end{pmatrix} = 0, \tag{4.120}$$

which confirms the identification of G^a [eq. (4.115)] as the Goldstone bosons.

Finally, we identify the physical Higgs bosons. In this case, the counting of bosonic degrees of freedom of Section 4.5.4 applies if we interpret d_R as the number of *complex* degrees of freedom, which is equivalent to $2d_R$ real degrees of freedom. Following the results of Section 4.5.4, we define the real Higgs fields,

$$\widetilde{H}_k \equiv c_j^{(k)} \overline{\Phi}^{\dagger\, j} + [c^{(k)*}]^j \overline{\Phi}_j, \tag{4.121}$$

where $k = 1, 2, \ldots, 2d_R - d_G + d_H$ and the $c_j^{(k)}$ are complex numbers that satisfy orthogonality relations,

$$(\widetilde{\mathcal{L}}_a \nu)_j \, [c^{(k)*}]^j + (\widetilde{\mathcal{L}}_a \nu)^{*j} c_j^{(k)} = 0, \tag{4.122}$$

$$[c^{(k)*}]^j c_j^{(\ell)} + [c^{(\ell)*}]^j c_j^{(k)} = \delta_{k\ell}, \tag{4.123}$$

and the repeated index j in eqs. (4.121)–(4.123) is implicitly summed over. These equations together with eq. (4.112)) can be used to derive the orthonormality of the scalar states $\{G_a, \widetilde{H}_k\}$. One can now solve for the shifted complex fields $\overline{\Phi}$ and $\overline{\Phi}^\dagger$ in terms of G_a and $\widetilde{H}_k$:

$$\overline{\Phi}_j = \frac{(\widetilde{\mathcal{L}}_a \nu)_j}{m_a} G_a + \sum_k c_j^{(k)} \widetilde{H}_k. \tag{4.124}$$

Inserting this result into eq. (4.116) and using eq. (4.120), we end up with

$$\mathcal{L}_{\text{scalar mass}} = -\tfrac{1}{2}(\mathcal{M}_H^2)_{pq}\, \widetilde{H}_p \widetilde{H}_q, \tag{4.125}$$

where the physical Higgs squared-mass matrix is given by

$$(\mathcal{M}_H^2)_{pq} = \left\{ c_k^{(p)} [c^{(q)*}]^m \frac{\partial^2 V}{\partial \Phi_k \partial \Phi^{\dagger\, m}} + c_k^{(p)} c_n^{(q)} \frac{\partial^2 V}{\partial \Phi_k \partial \Phi_n} \right.$$
$$\left. + [c^{(p)*}]^\ell [c^{(q)*}]^m \frac{\partial^2 V}{\partial \Phi^{\dagger\,\ell} \partial \Phi^{\dagger\, m}} + [c^{(p)*}]^\ell c_n^{(q)} \frac{\partial^2 V}{\partial \Phi^{\dagger\,\ell} \partial \Phi_n} \right\}_{\Phi_i = \nu_i} . \tag{4.126}$$

Note that $\mathcal{M}_H^2$ is a real symmetric matrix. Diagonalizing $\mathcal{M}_H^2$ yields the Higgs boson mass eigenstates, H_k, and the corresponding eigenvalues are the physical Higgs boson squared masses.

4.7 The Standard Model of Particle Physics

The Standard Model (SM) is a spontaneously broken nonabelian gauge theory based on the Lie algebra of the symmetry group $\mathrm{SU}(3)_C \times \mathrm{SU}(2)_L \times \mathrm{U}(1)_Y$. The color $\mathrm{SU}(3)_C$ group is unbroken, so we put it aside and focus on the spontaneous breaking of $\mathrm{SU}(2)_L \times \mathrm{U}(1)_Y$ to $\mathrm{U}(1)_{\mathrm{EM}}$. Here, we have distinguished between the hypercharge gauge group $\mathrm{U}(1)_Y$ which is broken and the electromagnetic gauge group $\mathrm{U}(1)_{\mathrm{EM}}$ which is unbroken. The breaking is accomplished by introducing an $\mathrm{SU}(2)_L$ complex doublet of scalar fields with hypercharge $Y = \frac{1}{2}$,

$$\Phi = \begin{pmatrix} \Phi^+ \\ \Phi^0 \end{pmatrix} . \tag{4.127}$$

The $\mathrm{SU}(2)_L \times \mathrm{U}(1)_Y$ covariant derivative acting on Φ is given by

$$\nabla_\mu \Phi_i = \partial_\mu \Phi_i + W_\mu^k (\mathcal{L}_k)_i{}^j \Phi_j + B_\mu (\mathcal{L}_4)_i{}^j \Phi_j \,, \tag{4.128}$$

where $i, j \in \{1, 2\}$; $k \in \{1, 2, 3\}$; and

$$(\mathcal{L}_k)_i{}^j = \tfrac{1}{2} i g (\tau^k)_i{}^j \,, \qquad (\mathcal{L}_4)_i{}^j = i g' Y \delta_i{}^j \,. \tag{4.129}$$

Here, we have introduced the $\mathrm{SU}(2)_L$ gauge fields W_μ^k, the respective gauge couplings g and g', the hypercharge-U(1) gauge field B_μ, and the $\mathrm{SU}(2)_L \times \mathrm{U}(1)_Y$ generators $\mathcal{L}_a = i g_a \mathcal{T}^a$ acting on Φ (where $a \in \{1, \ldots, 4\}$ and $\mathcal{T}^a = \{\frac{1}{2}\vec{\tau}, \frac{1}{2}\boldsymbol{I}_2\}$ with $g_{1,2,3} = g$ and $g_4 = g'$). The τ^k are the usual Pauli matrices[15] and the hypercharge operator Y is normalized such that $Y\Phi = +\frac{1}{2}\Phi$.[16]

We begin by focusing on the bosonic sector of the Standard Model. The dynamics of the scalar field are governed by the scalar potential, whose form is constrained by renormalizability and $\mathrm{SU}(2)_L \times \mathrm{U}(1)_Y$ gauge invariance:

$$V(\Phi, \Phi^\dagger) = -m^2 (\Phi^\dagger \Phi) + \lambda (\Phi^\dagger \Phi)^2 \,, \tag{4.130}$$

[15] Here, $\vec{\tau} = \vec{\sigma}$. We use a different symbol here to distinguish the τ^a from the σ^a that appear in the formalism of two-component spin-1/2 fermions.

[16] Another common normalization in the literature is $(\mathcal{L}_4)_i{}^j = \frac{1}{2} i g' Y \delta_i{}^j$, in which case the doublet of scalar fields possesses hypercharge one.

where m^2 and λ are real positive parameters. Minimizing the scalar potential yields a local minimum at $\Phi^\dagger\Phi = m^2/(2\lambda)$. Thus, the scalar doublet acquires a vacuum expectation value. It is always possible to perform an $\mathrm{SU}(2)_L$ gauge transformation to bring $\langle\Phi\rangle$ into the following form:

$$\nu \equiv \langle\Phi\rangle = \frac{1}{\sqrt{2}}\begin{pmatrix}0\\ v\end{pmatrix}, \tag{4.131}$$

where $v = m/\sqrt{\lambda}$ is real.

Using eq. (4.110), the 4×4 squared-mass matrix of the gauge bosons is easily computed. After noting that $\tau^i\tau^j + \tau^j\tau^i = 2\delta^{ij}$ and $\nu^\dagger\nu = v^2/2$, the squared-mass matrix is seen to be block-diagonal, where one of the 2×2 blocks is proportional to $\boldsymbol{I}_2$ with respect to the $\{W^1, W^2\}$ basis. In particular, we find two mass-degenerate states, W^1 and W^2, with $m_W^2 = \frac{1}{4}g^2v^2$, which determines the value of the VEV v:

$$v = \frac{2m_W}{g} = (\sqrt{2}G_F)^{-1/2} \simeq 246 \text{ GeV}, \tag{4.132}$$

where G_F is the Fermi constant of weak interactions.

The second 2×2 block is nondiagonal and is given by

$$M^2 = \frac{v^2}{4}\begin{pmatrix} g^2 & -gg' \\ -gg' & g'^2 \end{pmatrix} \tag{4.133}$$

with respect to the $\{W^3, B\}$ basis. This matrix is easily diagonalized by

$$\mathcal{O} = \begin{pmatrix} \cos\theta_W & -\sin\theta_W \\ \sin\theta_W & \cos\theta_W \end{pmatrix}, \tag{4.134}$$

where the weak mixing angle, θ_W, is defined by

$$\sin\theta_W = \frac{g'}{\sqrt{g^2 + g'^2}}. \tag{4.135}$$

From eq. (4.111), we deduce that $\widetilde{\mathcal{L}}_k = \mathcal{L}_k$ for $k = 1, 2$ and

$$\widetilde{\mathcal{L}}_3 = \mathcal{L}_3\cos\theta_W - \mathcal{L}_4\sin\theta_W = \frac{ig}{\cos\theta_W}\left[\mathcal{T}^3 - Q\sin^2\theta_W\right], \tag{4.136}$$

$$\widetilde{\mathcal{L}}_4 = \mathcal{L}_3\sin\theta_W + \mathcal{L}_4\cos\theta_W = ieQ, \tag{4.137}$$

where the QED coupling constant e is identified as

$$e = g\sin\theta_W = g'\cos\theta_W = \frac{gg'}{\sqrt{g^2 + g'^2}}, \tag{4.138}$$

and

$$Q = \mathcal{T}^3 + Y. \tag{4.139}$$

Note that eqs. (4.136), (4.137), and (4.139) are representation-independent, and thus can be applied in any representation. These results imply that $\widetilde{\mathcal{L}}_a\nu \neq 0$ for $a = 1, 2, 3$, while $\widetilde{\mathcal{L}}_4\nu = 0$. That is, $\widetilde{\mathcal{L}}_4$ is the unbroken generator, which we have identified as ieQ, where Q is the $\mathrm{U}(1)_{\mathrm{EM}}$ generator. Indeed, $\mathrm{SU}(2)_L\times\mathrm{U}(1)_Y$ is spontaneously broken down to $\mathrm{U}(1)_{\mathrm{EM}}$.

The gauge boson mass eigenstates are given by eq. (4.74). Explicitly, these are denoted by W^1_μ, W^2_μ, Z_μ, and A_μ, where

$$Z_\mu = W^3_\mu \cos\theta_W - B_\mu \sin\theta_W \,, \tag{4.140}$$

$$A_\mu = W^3_\mu \sin\theta_W + B_\mu \cos\theta_W \,. \tag{4.141}$$

Note that it follows from eq. (4.133) that $\det M^2 = 0$ and $\operatorname{Tr} M^2 = \frac{1}{4}v^2(g^2+g'^2)$. Thus, the field A_μ is massless and is identified with the photon, while the Z boson is massive, with $m_Z^2 = \frac{1}{4}v^2(g^2+g'^2)$.

The electric charge of the gauge bosons can be determined by computing the eigenvalues of the charge operator when applied to the gauge boson field. Let $\widehat{Q}$ be the charge operator that acts on the Hilbert space of quantum fields. Then, for a multiplet of gauge fields A^a_μ,

$$(\widehat{Q}A_\mu)^a = Q_{ab}A^b_\mu \,, \tag{4.142}$$

where Q_{ab} is the representation of $\widehat{Q}$ in the adjoint representation of the gauge group. The adjoint representation of the $\mathrm{SU}(2)_L\times\mathrm{U}(1)_Y$ Lie algebra consists of a direct sum of the three-dimensional adjoint representation of SU(2), given by $(T^k)_{ij} = -i\epsilon_{ijk}$, and the trivial one-dimensional representation of U(1), given by $T^4 = 0$. Thus, in the adjoint representation, Q is a 4×4 matrix. Using eq. (4.139), the explicit form for the adjoint representation matrix Q is

$$Q = \begin{pmatrix} T^3 & 0 \\ 0 & 0 \end{pmatrix} = \begin{pmatrix} 0 & -i & 0 & 0 \\ i & 0 & 0 & 0 \\ 0 & 0 & 0 & 0 \\ 0 & 0 & 0 & 0 \end{pmatrix} . \tag{4.143}$$

In eq. (4.143), Q is defined relative to the $\{W^1, W^2, Z, A\}$ basis of vector fields. But Q as exhibited in eq. (4.143) is not diagonal. The form of Q implies that Z and A are neutral under Q, whereas W^1 and W^2 are not eigenstates of $\widehat{Q}$. However, it is simple to diagonalize Q by a simple basis change. We introduce

$$W^\pm_\mu \equiv \frac{1}{\sqrt{2}}\left(W^1_\mu \mp iW^2_\mu\right) . \tag{4.144}$$

With respect to the new $\{W^+, W^-, Z, A\}$ basis,

$$Q = \begin{pmatrix} ST^3S^{-1} & 0 \\ 0 & 0 \end{pmatrix} = \begin{pmatrix} 1 & 0 & 0 & 0 \\ 0 & -1 & 0 & 0 \\ 0 & 0 & 0 & 0 \\ 0 & 0 & 0 & 0 \end{pmatrix} , \tag{4.145}$$

where $W^{(j)}_\mu = S^{(j)}{}_k W^k_\mu$ ($j=\pm$ and $k=1,2$) with

$$S = \frac{1}{\sqrt{2}}\begin{pmatrix} 1 & -i \\ 1 & i \end{pmatrix} . \tag{4.146}$$

Therefore, $\widehat{Q}W^{\pm}_{\mu} = \pm W^{\pm}_{\mu}$. That is, $W^{\pm}$ are the positively and negatively charged W bosons, respectively, with $m_W^2 = \frac{1}{4}g^2v^2$ [as noted above eq. (4.133)]. Having fixed the Z mass by experiment, the value of the W mass is now constrained by the model. At tree level, the results above imply that $m_W = m_Z \cos\theta_W$. This is often rewritten as

$$\rho \equiv \frac{m_W^2}{m_Z^2 \cos^2\theta_W} = 1\,. \tag{4.147}$$

The result $\rho = 1$ is a consequence of a "custodial" SU(2) symmetry that is implicit in the choice of the Higgs sector.[17] In theories with more complicated Higgs sectors (that involve different multiplets), the value of ρ is generally a free parameter of the model. The fact that the electroweak data is consistent with $\rho \simeq 1$ is a strong clue as to the fundamental nature of the electroweak symmetry-breaking dynamics.

Having identified the gauge boson mass eigenstates, it is useful to re-express the covariant derivative operator $\nabla_\mu = \partial_\mu + \mathcal{L}_a A^a_\mu = \partial_\mu + \widetilde{\mathcal{L}}_a \widetilde{A}^a_\mu$ in terms of these fields:

$$\nabla_\mu = \partial_\mu + \frac{ig}{\sqrt{2}}\left(\mathcal{T}^+ W^+_\mu + \mathcal{T}^- W^-_\mu\right) + \frac{ig}{\cos\theta_W}\left(\mathcal{T}^3 - Q\sin^2\theta_W\right) Z_\mu + ieQA_\mu\,, \tag{4.148}$$

where $\mathcal{T}^{\pm} \equiv \mathcal{T}^1 \pm i\mathcal{T}^2$.

We turn next to the scalar sector. First, we note that Φ is a complex $Y = +\frac{1}{2}$, SU(2)$_L$-doublet of fields, so we will need to introduce the corresponding antiparticle multiplet, which is a complex $Y = -\frac{1}{2}$, SU(2)$_L$-doublet of fields. However, in contrast to $\Phi^\dagger$, it is convenient to define the antiparticle multiplet of fields to transform under SU(2)$_L$ in the same way as the multiplet of fields Φ. Hence, we shall define the antiparticle multiplet by

$$\widetilde{\Phi}_i = \epsilon_{ij}\Phi^{\dagger j} = \begin{pmatrix} (\Phi^0)^\dagger \\ -\Phi^- \end{pmatrix}, \tag{4.149}$$

where $\Phi^- \equiv (\Phi^+)^\dagger$ and (see Appendix A.8)

$$\epsilon = i\tau^2 = \begin{pmatrix} 0 & 1 \\ -1 & 0 \end{pmatrix}. \tag{4.150}$$

The proof that $\widetilde{\Phi}$ has the desired SU(2) transformation law, $\widetilde{\Phi} \to \widetilde{\Phi} + \delta\widetilde{\Phi}$, where

$$\delta\widetilde{\Phi} = -\frac{i}{2}\vec{\theta}\cdot\vec{\tau}\,\widetilde{\Phi}\,, \tag{4.151}$$

relies on the identity among the Pauli matrices given by eq. (1.36). This result is not unexpected given that the two-dimensional representation of SU(2) is equivalent to its complex conjugate representation.

[17] The SM Higgs scalar potential [eq. (4.130)] exhibits a global SO(4) $\cong$ SU(2)$\times$SU(2) symmetry [see Exercise 4.3(a)]. One of the SU(2) factors is identified with the SU(2)$_L$ gauge group. The second SU(2) is the so-called custodial SU(2) group, of which only a subgroup U(1)$_Y$ is gauged. The custodial SU(2) is responsible for the tree-level relation $\rho = 1$. Since the custodial SU(2) is not an exact global symmetry of the SM gauge boson and fermion sectors, there exist finite radiative corrections to the quantity $\delta\rho \equiv \rho - 1$ (e.g., see [B79, B87, B89]).

The electric charge of the scalar bosons can be determined by computing the eigenvalues of the charge operator $\widehat{Q}$ when applied to the scalar field. For scalar fields in the doublet representation of $\mathrm{SU}(2)_L$,

$$(\widehat{Q}\Phi)_i = Q_i{}^j\,\Phi_j\,, \tag{4.152}$$

where $Q_i{}^j$ is the representation of $\widehat{Q}$ in the fundamental representation of the gauge group. In this case,

$$Q = \tfrac{1}{2}(\tau^3 + \boldsymbol{I}_2) = \begin{pmatrix} 1 & 0 \\ 0 & 0 \end{pmatrix}. \tag{4.153}$$

As expected, $\widehat{Q}\Phi^+ = +\Phi^+$ and $\widehat{Q}\Phi^0 = 0$. For scalar fields in the $Y = -\frac{1}{2}$ complex-conjugated doublet representation of $\mathrm{SU}(2)_L{\times}\mathrm{U}(1)_Y$, we have $\mathcal{T}^a = \{\frac{1}{2}\vec{\boldsymbol{\tau}}^*\,,\,-\frac{1}{2}\boldsymbol{I}_2\}$. Consequently, if we denote the charge operator in this representation by Q^*, then

$$(\widehat{Q}\widetilde{\Phi})_i = (Q^*)_i{}^j\widetilde{\Phi}_j\,, \quad \text{where} \quad Q^* = \tfrac{1}{2}(\tau^{3\,*} - \boldsymbol{I}_2) = \begin{pmatrix} 0 & 0 \\ 0 & -1 \end{pmatrix}. \tag{4.154}$$

Thus, $\widehat{Q}\Phi^- = -\Phi^-$ and $\widehat{Q}(\Phi^0)^\dagger = 0$, as expected.

Equation (4.115) provides an explicit formula for the Goldstone bosons. Applying this formula to the electroweak theory is straightforward, and we find

$$G_1 = \sqrt{2}\,\mathrm{Im}\,\Phi^+\,, \qquad G_2 = \sqrt{2}\,\mathrm{Re}\,\Phi^+\,, \qquad G_3 = -\sqrt{2}\,\mathrm{Im}\,\Phi^0\,. \tag{4.155}$$

The physical Higgs field must be orthonormal to the G_a, so in this case it is trivial to deduce that

$$H = \sqrt{2}\,\mathrm{Re}\,\Phi^0 - v\,. \tag{4.156}$$

Thus, the complex scalar doublet takes the form

$$\Phi = \frac{1}{\sqrt{2}}\begin{pmatrix} G_2 + iG_1 \\ v + H - iG_3 \end{pmatrix} \equiv \begin{pmatrix} G^+ \\ \frac{1}{\sqrt{2}}\left[v + H + iG^0\right] \end{pmatrix}, \tag{4.157}$$

which defines the Goldstone boson fields of definite charge: $G^\pm$ and G^0 (where $G^- = [G^+]^\dagger$). Finally, the Higgs mass is determined from eq. (4.126). In this case, there is only one physical Higgs field, so that $c_1 = 0$ and $c_2 = 1/\sqrt{2}$. Thus,

$$m_H^2 = \left\{ \frac{\partial^2 V}{\partial\Phi^0\partial\Phi^{0\dagger}} + \frac{1}{2}\frac{\partial^2 V}{\partial\Phi^0\partial\Phi^0} + \frac{1}{2}\frac{\partial^2 V}{\partial\Phi^{0\dagger}\partial\Phi^{0\dagger}} \right\}_{\Phi_i=\nu_i}. \tag{4.158}$$

Plugging in $\nu_1 = 0$ and $\nu_2 = v/\sqrt{2}$ and using the potential minimum condition $m^2 = \lambda v^2$ [see text following eq. (4.131)], we end up with

$$m_H^2 = 2\lambda v^2\,. \tag{4.159}$$

Contributions to the Goldstone boson masses derived from eq. (4.130) vanish exactly when the scalar potential minimum conditions are imposed. However, in the R_ξ gauge, the squared masses of the neutral and charged Goldstone fields receive contributions from the gauge-fixing term [see eq. (4.87)], which yield $m_G^2 = \xi m_Z^2$ and $m_{G^\pm}^2 = \xi m_W^2$, respectively.

The self-interactions of the scalar fields can be obtained by inserting eq. (4.157) into eq. (4.130). The end result is

$$\mathscr{L}_S = -\lambda\left[vH^3 + \tfrac{1}{4}\lambda H^4 + \left(\tfrac{1}{2}G^2 + G^+G^-\right)(2vH + H^2) + \left(\tfrac{1}{2}G^2 + G^+G^-\right)^2\right]. \quad (4.160)$$

The Goldstone boson couplings are relevant in calculations performed in the R_ξ gauge, since the Goldstone bosons appear as propagating fields when appearing as internal lines in a Feynman diagram.

The interactions of the gauge bosons with the Higgs bosons originate from the scalar kinetic energy term

$$\mathscr{L}_{\text{kinetic}} = (\nabla^\mu \Phi^\dagger)(\nabla_\mu \Phi)\,, \quad (4.161)$$

where the covariant derivative is given by eq. (4.148) with generators $\mathcal{T}^k = \frac{1}{2}\tau^k$ and Φ is given by eq. (4.157). In the unitary gauge, $G^\pm = G^0 = 0$, and we obtain

$$\mathscr{L}_{\text{kinetic}} = \tfrac{1}{2}(\partial^\mu H)(\partial_\mu H) + \tfrac{1}{4}g^2(v^2 + 2vH + H^2)W^{\mu +}W^-_\mu + \frac{g^2}{8\cos^2\theta_W}(v^2 + 2vH + H^2)Z^\mu Z_\mu\,. \quad (4.162)$$

As expected, eq. (4.162) yields $m_W^2 = m_Z^2\cos^2\theta_W = \frac{1}{4}g^2v^2$, as well as cubic and quartic interactions of the Higgs boson with a pair of gauge bosons.

The fermion sector of the Standard Model consists of three generations of quarks and leptons, which are represented by the two-component fermion fields listed in Table 4.2, where Y is the weak hypercharge, T^3 is the third component of the weak isospin, and $Q = T^3 + Y$ is the electric charge. After $\mathrm{SU}(2)_L \times \mathrm{U}(1)_Y$ breaking, the quark and charged lepton fields gain mass in such a way that the above two-component fields combine to make up four-component Dirac fermions:

$$U_i = \begin{pmatrix} u_i \\ \bar{u}_i^\dagger \end{pmatrix}, \qquad D_i = \begin{pmatrix} d_i \\ \bar{d}_i^\dagger \end{pmatrix}, \qquad E_i = \begin{pmatrix} e_i \\ \bar{e}_i^\dagger \end{pmatrix}, \quad (4.163)$$

where $i \in \{1, 2, 3\}$ labels the generation. The neutrinos ν_i remain massless. An extension of the Standard Model to include neutrino masses will be treated in Section 6.1.

Note that u, $\bar{u}$, d, $\bar{d}$, e, and $\bar{e}$ are two-component fields, whereas the usual four-component quark and charged lepton fields are denoted by capital letters U, D, and E. Consider a generic four-component field expressed in terms of the corresponding two-component fields, $F = (f \;\; \bar{f}^\dagger)^\mathsf{T}$. The electroweak quantum numbers of f are denoted by T_f^3, Y_f, and Q_f, whereas the corresponding quantum numbers for $\bar{f}$ are $T_{\bar{f}}^3 = 0$ and $Q_{\bar{f}} = Y_{\bar{f}} = -Q_f$. Thus we have the correspondence to our general notation [see eq. (3.5)]:

$$f \longleftrightarrow \chi, \qquad \bar{f} \longleftrightarrow \eta\,. \quad (4.164)$$

In four-component spinor notation [see eq. (3.13)], $F_L = (f \;\; 0)^\mathsf{T}$ and $F_R = (0 \;\; \bar{f}^\dagger)^\mathsf{T}$, which establishes the one-to-one correspondences

$$f \longleftrightarrow F_L\,, \qquad \bar{f}^\dagger \longleftrightarrow F_R\,. \quad (4.165)$$

Table 4.2 Standard Model fermions and their $\mathrm{SU}(3)_C \times \mathrm{SU}(2)_L \times \mathrm{U}(1)_Y$ quantum numbers. The generation indices run over $i \in \{1, 2, 3\}$. Color indices for the quarks are suppressed. The bars on the two-component antifermion fields are part of their names, and do not denote some form of complex conjugation.

two-component fermion fields		$\mathrm{SU}(3)_C$	$\mathrm{SU}(2)_L$	Y	T^3	$Q = T^3 + Y$
$Q_i \equiv$	u_i	triplet	doublet	$\frac{1}{6}$	$\frac{1}{2}$	$\frac{2}{3}$
	d_i	triplet		$\frac{1}{6}$	$-\frac{1}{2}$	$-\frac{1}{3}$
	$\bar{u}^i$	antitriplet	singlet	$-\frac{2}{3}$	0	$-\frac{2}{3}$
	$\bar{d}^i$	antitriplet	singlet	$\frac{1}{3}$	0	$\frac{1}{3}$
$L_i \equiv$	ν_i	singlet	doublet	$-\frac{1}{2}$	$\frac{1}{2}$	0
	e_i	singlet		$-\frac{1}{2}$	$-\frac{1}{2}$	-1
	$\bar{e}^i$	singlet	singlet	1	0	1

However, we shall continue to employ the two-component spinor field names when identifying the fermions of the Standard Model (and extensions thereof).

The color interactions of the quarks specified in quantum chromodynamics (QCD) are governed by the following interaction Lagrangian:

$$\mathscr{L}_{\text{int}} = -g_s A^\mu_a q^{\dagger mi}\,\overline{\sigma}_\mu (\boldsymbol{T}^a)_m{}^n q_{ni} + g_s A^\mu_a \bar{q}^\dagger_{ni}\,\overline{\sigma}_\mu (\boldsymbol{T}^a)_m{}^n \bar{q}^{mi}\,, \tag{4.166}$$

summed over the generations i, where q is a (mass-eigenstate) quark field, m and n are $\mathrm{SU}(3)_C$ color triplet indices, A^μ_a is the gluon field, g_s is the strong QCD coupling constant, and $\boldsymbol{T}^a$ are the color generators in the triplet representation of $\mathrm{SU}(3)_C$.

The electroweak interactions of the quarks and leptons are governed by the following interaction Lagrangian:

$$\begin{aligned}
\mathscr{L}_{\text{int}} = &-\frac{g}{\sqrt{2}}\left[\left(\widehat{u}^{\dagger i}\overline{\sigma}^\mu \widehat{d}_i + \widehat{\nu}^{\dagger i}\overline{\sigma}^\mu \widehat{e}_i\right) W^+_\mu + \left(\widehat{d}^{\,\dagger i}\overline{\sigma}^\mu \widehat{u}_i + \widehat{e}^{\,\dagger i}\overline{\sigma}^\mu \widehat{\nu}_i\right) W^-_\mu\right] \\
&- \frac{g}{c_W}\sum_{f=u,d,\nu,e}\left\{\left(T^3_f - s^2_W Q_f\right)\widehat{f}^{\,\dagger i}\overline{\sigma}^\mu \widehat{f}_i + s^2_W Q_f \widehat{\bar{f}}{}^\dagger_i \overline{\sigma}^\mu \widehat{\bar{f}}{}^i\right\} Z_\mu \\
&- e\sum_{f=u,d,e} Q_f\left(\widehat{f}^{\,\dagger i}\overline{\sigma}^\mu \widehat{f}_i - \widehat{\bar{f}}{}^\dagger_i \overline{\sigma}^\mu \widehat{\bar{f}}{}^i\right) A_\mu\,,
\end{aligned} \tag{4.167}$$

where $s_W \equiv \sin\theta_W$, $c_W \equiv \cos\theta_W$, the hatted fields indicate fermion interaction eigenstates, and i labels the generations. Before performing any practical computations, one must first identify the quark and lepton mass matrices and then convert from fermion interaction eigenstates to fermion mass eigenstates.

In the electroweak theory, the fermion mass matrices originate from the Higgs–fermion Yukawa interactions. The Higgs field of the Standard Model is a complex $\mathrm{SU}(2)_L$ doublet of hypercharge $Y = \frac{1}{2}$, denoted by $\boldsymbol{\Phi}_a$, where the $\mathrm{SU}(2)_L$ index $a \in \{1, 2\}$ is defined such that $\boldsymbol{\Phi}_1 \equiv \Phi^+$ and $\boldsymbol{\Phi}_2 \equiv \Phi^0$ [see eq. (4.127)]. Here, the superscripts + and 0 refer to the electric charge of the Higgs field, $Q = T^3 + Y$, with $Y = \frac{1}{2}$ and $T^3 = \pm\frac{1}{2}$. Since $\boldsymbol{\Phi}_a$ is a complex field, its complex conjugate $\boldsymbol{\Phi}^{\dagger\, a} \equiv (\Phi^-\,,\ (\Phi^0)^\dagger)$ is an $\mathrm{SU}(2)_L$ doublet scalar field with hypercharge $Y = -\frac{1}{2}$, where $\Phi^- \equiv (\Phi^+)^\dagger$. The $\mathrm{SU}(2)_L\times\mathrm{U}(1)_Y$ gauge invariant Yukawa interactions of the quarks and leptons with the Higgs field are then given by

$$\mathscr{L}_{\mathrm{Y}} = \epsilon^{ab}(\boldsymbol{y_u})^i{}_j \boldsymbol{\Phi}_a \widehat{\boldsymbol{Q}}_{bi}\widehat{\bar{u}}^j - (\boldsymbol{y_d})^i{}_j \boldsymbol{\Phi}^{\dagger\, a}\widehat{\boldsymbol{Q}}_{ai}\widehat{\bar{d}}^j - (\boldsymbol{y_e})^i{}_j \boldsymbol{\Phi}^{\dagger\, a}\widehat{\boldsymbol{L}}_{ai}\widehat{\bar{e}}^j + \text{h.c.}, \tag{4.168}$$

where ϵ^{ab} is the invariant tensor of SU(2) defined in Appendix A.8, and the flavor index placement follows that of eq. (2.135). One can rewrite eq. (4.168) more explicitly in terms of the quark and lepton fields exhibited in Table 4.2:[18]

$$\begin{aligned}-\mathscr{L}_{\mathrm{Y}} = (\boldsymbol{y_u})^i{}_j \left[\Phi^0\widehat{u}_i\widehat{\bar{u}}^j - \Phi^+\widehat{d}_i\widehat{\bar{u}}^j\right] + (\boldsymbol{y_d})^i{}_j \left[\Phi^-\widehat{u}_i\widehat{\bar{d}}^j + (\Phi^0)^\dagger\widehat{d}_i\widehat{\bar{d}}^j\right] \\ + (\boldsymbol{y_e})^i{}_j \left[\Phi^-\widehat{\nu}_i\widehat{\bar{e}}^j + (\Phi^0)^\dagger\,\widehat{e}_i\widehat{\bar{e}}^j\right] + \text{h.c.}\end{aligned} \tag{4.169}$$

The Higgs fields can be written in terms of the physical Higgs scalar H and the Goldstone bosons $G^0, G^\pm$ as

$$\Phi^0 = \frac{1}{\sqrt{2}}(v + H + iG^0)\,, \tag{4.170}$$

$$\Phi^+ = G^+ = (\Phi^-)^\dagger = (G^-)^\dagger, \tag{4.171}$$

where $v = 2m_W/g \simeq 246$ GeV. In the unitary gauge, the Goldstone bosons become infinitely heavy and decouple. We identify the quark and lepton mass matrices by setting $\Phi^0 = v/\sqrt{2}$ and $\Phi^+ = \Phi^- = 0$ in eq. (4.169):

$$(\boldsymbol{M_U})^i{}_j = \frac{v}{\sqrt{2}}(\boldsymbol{y_u})^i{}_j\,, \quad (\boldsymbol{M_D})^i{}_j = \frac{v}{\sqrt{2}}(\boldsymbol{y_d})^i{}_j\,, \quad (\boldsymbol{M_E})^i{}_j = \frac{v}{\sqrt{2}}(\boldsymbol{y_e})^i{}_j\,. \tag{4.172}$$

The neutrinos remain massless as previously indicated.

To diagonalize the quark and lepton mass matrices, we introduce four unitary matrices for the quark mass diagonalization, L_u, L_d, R_u, and R_d, and two unitary matrices for the lepton mass diagonalization, L_e and R_e [see eq. (1.174)], such that

$$\widehat{u}_i = (L_u)_i{}^j u_j\,, \qquad \widehat{d}_i = (L_d)_i{}^j d_j\,, \qquad \widehat{\bar{u}}^i = (R_u)^i{}_j \bar{u}^j\,, \qquad \widehat{\bar{d}}^i = (R_d)^i{}_j \bar{d}^j\,, \tag{4.173}$$

$$\widehat{e}_i = (L_e)_i{}^j e_j\,, \qquad \widehat{\bar{e}}^i = (R_e)^i{}_j \bar{e}^j\,, \tag{4.174}$$

where the unhatted fields u, d, $\bar{u}$, and $\bar{d}$ are the corresponding quark mass eigenstates and ν, e, and $\bar{e}$ are the corresponding lepton mass eigenstates. The fermion mass diagonalization procedure consists of the singular value decomposition of the

[18] In four-component spinor notation, the Higgs–quark Yukawa Lagrangian can be written as $-\mathscr{L}_{\mathrm{Y}} = (\boldsymbol{y_u})_i{}^j\left[(\Phi^0)^\dagger\overline{\widehat{U}}{}^i_L\widehat{U}_{jR} - \Phi^-\,\overline{\widehat{D}}{}^i_L\widehat{U}_{jR}\right] + (\boldsymbol{y_d})_i{}^j\left[\Phi^+\overline{\widehat{U}}{}^i_L\widehat{D}_{jR} + \Phi^0\,\overline{\widehat{D}}{}^i_L\widehat{D}_{jR}\right] + \text{h.c.}$, where $\widehat{U}$ and $\widehat{D}$ are the four-component up- and down-type quark spinors. In light of eq. (A.7), the Yukawa coupling matrices $(\boldsymbol{y_f})_i{}^j = [(\boldsymbol{y_f})^i{}_j]^*$ (which are also employed in [B89]) are the complex conjugates of the Yukawa coupling matrices that appear in eq. (4.169).

quark and lepton mass matrices:

$$L_u^{\mathsf T} \boldsymbol{M_U} R_u = \mathrm{diag}(m_u\,,\, m_c\,,\, m_t)\,, \tag{4.175}$$
$$L_d^{\mathsf T} \boldsymbol{M_D} R_d = \mathrm{diag}(m_d\,,\, m_s\,,\, m_b)\,, \tag{4.176}$$
$$L_e^{\mathsf T} \boldsymbol{M_E} R_e = \mathrm{diag}(m_e\,,\, m_\mu\,,\, m_\tau)\,, \tag{4.177}$$

where the diagonalized masses are real and nonnegative. Since the neutrinos are massless, we are free to define the physical neutrino fields, ν_i, as the weak $\mathrm{SU}(2)_L$ partners of the corresponding charged lepton mass-eigenstate fields. That is,

$$\widehat{\nu}_i = (L_e)_i{}^j \nu_j\,. \tag{4.178}$$

Writing out the couplings of the mass-eigenstate quarks and leptons to the gauge bosons and Higgs bosons, we first consider the charged-current interactions of the quarks and leptons. Using eq. (4.173), it follows that $\widehat{u}^{\dagger i}\overline{\sigma}^\mu \widehat{d}_i = \boldsymbol{K}_i{}^j u^{\dagger i}\overline{\sigma}^\mu d_j$, where

$$\boldsymbol{K} = L_u^\dagger L_d \tag{4.179}$$

is the unitary Cabibbo–Kobayashi–Maskawa (CKM) matrix.[19] The corresponding leptonic CKM matrix is the unit matrix due to eq. (4.178). Hence, the charged-current interactions take the form

$$\mathscr{L}_{\rm int} = -\frac{g}{\sqrt{2}}\left[\boldsymbol{K}_i{}^j u^{\dagger i}\overline{\sigma}^\mu d_j W_\mu^+ + (\boldsymbol{K}^\dagger)_i{}^j d^{\dagger i}\overline{\sigma}^\mu u_j W_\mu^- + \nu^{\dagger i}\overline{\sigma}^\mu e_i W_\mu^+ + e^{\dagger i}\overline{\sigma}^\mu \nu_i W_\mu^-\right]\,, \tag{4.180}$$

where $[\boldsymbol{K}^\dagger]_i{}^j \equiv [\boldsymbol{K}_j{}^i]^*$. Note that in the Standard Model, $\bar{u}$, $\bar{d}$, and $\bar{e}$ do not couple to the $W^\pm$. Rewriting eq. (4.180) using the four-component spinor notation employed in Chapter 3, it follows that the $W^\pm$ gauge bosons couple only to left-handed quarks and leptons [see Exercise 4.6(c)], which is the origin of the subscript L attached to the weak isospin $\mathrm{SU}(2)_L$ gauge group.

To obtain the neutral current interactions, we insert eqs. (4.173)–(4.178) into eq. (4.167). All factors of the *unitary* matrices L_f and R_f ($f = u, d, e$) cancel out, and the resulting interactions are flavor-diagonal. That is, we may simply remove the hats from the quark and lepton fields that couple to the Z and photon fields in eq. (4.167). This is the well-known Glashow–Iliopoulos–Maiani (GIM) mechanism for the flavor-conserving neutral currents.

The diagonalization of the fermion mass matrices is equivalent to diagonalizing the Yukawa couplings [see eqs. (4.172) and (4.175)–(4.177)]. Thus, we define[20]

$$y_{fi} = \frac{\sqrt{2} m_{fi}}{v}\,, \qquad f = u, d, e\,. \tag{4.181}$$

It is convenient to rewrite eqs. (4.175)–(4.177) as follows:

$$(L_f)_k{}^j (\boldsymbol{y_f})^k{}_m (R_f)^m{}_i = y_{fi}\delta_i^j\,, \qquad f = u, d, e\,, \tag{4.182}$$

[19] The CKM matrix elements V_{ij} defined in Ref. [41] are easily related to the matrix elements of $\boldsymbol{K}$. For example, $V_{tb} = \boldsymbol{K}_3{}^3$ and $V_{us} = \boldsymbol{K}_1{}^2$, etc.

[20] Boldfaced symbols are used for the nondiagonal Yukawa matrices, while non-boldfaced symbols y_{fi} are used for the diagonalized Yukawa couplings, where i labels the fermion generation. Note that the y_{fi} are real and nonnegative as a result of the sign choices employed in eq. (4.168).

with no sum over the repeated index i. Using the unitarity of L_f ($f = u,\ d$), eq. (4.182) is equivalent to the following convenient form:

$$(\boldsymbol{y_f} R_f)^k{}_i = y_{fi}(L_f^\dagger)_i{}^k\,. \tag{4.183}$$

Inserting eqs. (4.173), (4.174), and (4.178) into eq. (4.169) and making use of eqs. (4.181) and (4.183), the resulting Higgs–fermion Lagrangian is flavor-diagonal:

$$\mathscr{L}_{\rm int} = -\frac{1}{v} H \sum_i \left[m_{ui} u_i \bar{u}^i + m_{di} d_i \bar{d}^i + m_{ei} e_i \bar{e}^i \right] + \text{h.c.} \tag{4.184}$$

4.8 Parameter Count of the Standard Model

The SM Lagrangian appears to contain many parameters. However, not all these parameters are physical. One is always free to redefine the SM fields in an arbitrary manner. By suitable redefinitions, one can remove some of the apparent parameter freedom and identify the true physical independent parametric degrees of freedom.

To illustrate the procedure (following Ref. [42]), we first make a list of the parameters of the Standard Model. The gauge sector consists of three real gauge couplings (g_s, g, and g') and the QCD vacuum angle ($\Theta_{\rm QCD}$). The Higgs sector consists of one Higgs squared-mass parameter and one Higgs self-coupling (m^2 and λ). Traditionally, one trades in the latter two real parameters for the VEV ($v = 246$ GeV) and the physical Higgs mass. The fermion sector consists of three Yukawa coupling matrices $\boldsymbol{y_u}$, $\boldsymbol{y_d}$, and $\boldsymbol{y_e}$. Initially, $\boldsymbol{y_u}$, $\boldsymbol{y_d}$, and $\boldsymbol{y_e}$ are arbitrary complex 3×3 matrices, which in total depend on 27 real and 27 imaginary degrees of freedom.

However, most of these degrees of freedom associated with the Yukawa coupling matrices are unphysical. In particular, in the limit where $\boldsymbol{y_u} = \boldsymbol{y_d} = \boldsymbol{y_e} = 0$, the Standard Model possesses a global $\mathrm{U(3)}^5 \equiv \mathrm{U(3)} \times \mathrm{U(3)} \times \mathrm{U(3)} \times \mathrm{U(3)} \times \mathrm{U(3)}$ symmetry corresponding to three generations of the five $\mathrm{SU(3)}_C{\times}\mathrm{SU(2)}_L{\times}\mathrm{U(1)}_Y$ two-component spinor multiplets: (ν_i, e_i), $\bar{e}^i$, (u_i, d_i), $\bar{u}^i$, $\bar{d}^i$, where i is the generation label. Thus, one can make global $\mathrm{U(3)}^5$ rotations on the fermion fields of the Standard Model to absorb the unphysical degrees of freedom of $\boldsymbol{y_u}$, $\boldsymbol{y_d}$, and $\boldsymbol{y_e}$. A U(3) matrix can be parameterized by three real angles and six phases, so that with the most general $\mathrm{U(3)}^5$ rotation, we can apparently remove 15 real angles and 30 phases from $\boldsymbol{y_u}$, $\boldsymbol{y_d}$, and $\boldsymbol{y_e}$. However, the $\mathrm{U(3)}^5$ rotations include four exact U(1) global symmetries of the Standard Model, namely B and the three separate lepton numbers L_e, L_μ, and L_τ. Hence, one can only remove 26 phases from $\boldsymbol{y_u}$, $\boldsymbol{y_d}$, and $\boldsymbol{y_e}$. This leaves 12 real parameters (corresponding to six quark masses, three lepton masses,[21] and three CKM mixing angles) and one CP-violating imaginary degree

[21] The neutrinos in the Standard Model are automatically massless and are not counted as independent degrees of freedom in the SM parameter count.

of freedom (the phase of the CKM matrix).[22] Thus, in total one finds that the Standard Model possesses 19 independent parameters, of which 13 are associated with the flavor sector.

Exercises

4.1 Rewrite the QED Lagrangian given in eq. (4.1) in terms of the four-component spinor electron field Ψ of charge $-e$. Obtain the well-known four-component Feynman rule of QED, which identifies $ie\gamma^\mu$ with the $e^+e^-\gamma$ vertex.

4.2 Consider a scalar field theory consisting of a multiplet of n identical complex scalars, $\Phi_i(x)$. Suppose that the scalar potential, $V(\Phi, \Phi^\dagger)$, is invariant under U(n) transformations, $\Phi \to U\Phi$, where $U = \exp[-ig\Lambda^a\mathcal{T}^a]$.

(a) Show that V must satisfy

$$(\mathcal{T}^a)_i{}^j\Phi_j\frac{\partial V}{\partial\Phi_i} - (\mathcal{T}^a)_j{}^i\Phi^{\dagger\, j}\frac{\partial V}{\partial\Phi^{\dagger\, i}} = 0\,. \tag{4.185}$$

(b) Obtain two equations by first taking the derivative of eq. (4.185) with respect to Φ_k, and then taking the derivative of eq. (4.185) with respect to $\Phi^{\dagger\,\ell}$. Show that the resulting two equations, when evaluated at $\Phi_k = \langle\Phi_k\rangle \equiv v_k/\sqrt{2}$, are equivalent to eq. (4.120). Conclude that the Goldstone boson G_a is given by eq. (4.115). In particular, check that the normalization of the Goldstone boson state given in eq. (4.115) is required for a properly normalized kinetic energy term for G_a.

(c) Expand $V(\Phi, \Phi^\dagger)$ around the scalar field VEVs up to and including terms quadratic in the fields. Show that the quadratic terms are given by eqs. (4.116) and (4.119).

(d) Couple this scalar field theory to an SU(n) Yang–Mills theory. After replacing ordinary derivatives with covariant derivatives in the scalar kinetic energy term, write out the full scalar Lagrangian in the complex scalar field basis. Writing $\Phi_k = \overline{\Phi}_k + v_k/\sqrt{2}$, evaluate all terms up to and including terms quadratic in the fields. Show that one directly obtains the vector boson squared-mass matrix given in eq. (4.110).

(e) Write down the R_ξ gauge-fixing term in the complex scalar field basis. Show that the resulting couplings of the gauge field and Goldstone boson fields coincide with those previously obtained in eqs. (4.79) and (4.87).

[22] In addition, $\Theta_{\rm QCD}$ and the argument of the determinant of the quark mass matrix combine to yield a second physical CP-violating phase, $\overline{\Theta} \equiv \Theta_{\rm QCD} + \arg\det M_Q$. The strong CP problem of the Standard Model (which is reviewed in Refs. [43, 44]) consists of the observation that $|\overline{\Theta}| \lesssim 10^{-10}$, based on the experimental upper bound on the neutron electric dipole moment, whereas the (naive) expectation is $\overline{\Theta} \sim \mathcal{O}(1)$.

4.3 Consider a scalar field theory consisting of n identical complex fields Φ_i, with a Lagrangian

$$\mathscr{L} = (\partial_\mu \Phi^{\dagger i})(\partial^\mu \Phi_i) - V(\Phi^\dagger \Phi)\,, \tag{4.186}$$

where the potential function V is a function of $\Phi^{\dagger i}\Phi_i$. Such a theory is invariant under the U(n) transformation $\Phi \to U\Phi$, where U is an $n \times n$ unitary matrix.

(a) Rewrite the Lagrangian in terms of real fields ϕ_{Ai} and ϕ_{Bi} defined by

$$\Phi_j = \frac{1}{\sqrt{2}}(\phi_{Aj} + i\phi_{Bj})\,, \qquad \Phi^{\dagger j} = \frac{1}{\sqrt{2}}(\phi_{Aj} - i\phi_{Bj})\,, \tag{4.187}$$

and introduce the $2n$-dimensional real scalar field

$$\phi(x) = \begin{pmatrix} \phi_A(x) \\ \phi_B(x) \end{pmatrix}. \tag{4.188}$$

Show that the Lagrangian is actually invariant under a larger symmetry group O($2n$), corresponding to the transformation $\phi \to \mathcal{O}\phi$, where $\mathcal{O}$ is a $2n \times 2n$ orthogonal matrix.

(b) Working in the complex basis, show that the Lagrangian [eq. (4.186)] is invariant under the transformation

$$\Phi_i \to U_i{}^j \Phi_j + \Phi^{\dagger j}(V^\dagger)_j{}^i\,, \tag{4.189}$$

where U and V are complex $n \times n$ matrices, provided that the following two conditions are satisfied:

$$(i) \quad (U^\dagger U + V^\dagger V)_i{}^j = \delta_i{}^j\,, \tag{4.190}$$

$$(ii) \quad V^{\mathsf{T}} U \text{ is an antisymmetric matrix}. \tag{4.191}$$

(c) Show that the $2n \times 2n$ matrix

$$\mathcal{Q} = \begin{pmatrix} \mathrm{Re}(U+V) & -\mathrm{Im}(U+V) \\ \mathrm{Im}(U-V) & \mathrm{Re}(U-V) \end{pmatrix} \tag{4.192}$$

is an orthogonal matrix if U and V satisfy eqs. (4.190) and (4.191). Prove that any $2n \times 2n$ orthogonal matrix can be written in the form of eq. (4.192) by verifying that $\mathcal{Q}$ is determined by $n(2n-1)$ independent parameters.

(d) Define $2n \times 2n$ matrices obtained from $\mathcal{Q}$ either by setting $V = 0$ or by setting $U = 0$:

$$\mathcal{Q}_U = \begin{pmatrix} \mathrm{Re}\,U & -\mathrm{Im}\,U \\ \mathrm{Im}\,U & \mathrm{Re}\,U \end{pmatrix}, \qquad \mathcal{Q}_V = \begin{pmatrix} \mathrm{Re}\,V & -\mathrm{Im}\,V \\ -\mathrm{Im}\,V & -\mathrm{Re}\,V \end{pmatrix}. \tag{4.193}$$

Use the results of parts (b) and (c) to conclude that if U [V] is a unitary $n \times n$ matrix, then $\mathcal{Q}_U$ [$\mathcal{Q}_V$] provides an explicit embedding of the subgroup U(n)

in O$(2n)$. By writing $\mathcal{Q}_U = \exp[-ig\Lambda^a \boldsymbol{T}^a]$ and $U = \exp[-ig\Lambda^a \mathcal{T}^a]$, show that $\boldsymbol{T}^a$ is given by eq. (4.106) in terms of the $\mathcal{T}^a$.

(e) Define the unitary matrix

$$A = \begin{pmatrix} \boldsymbol{I}_n & -i\boldsymbol{I}_n \\ i\boldsymbol{I}_n & \boldsymbol{I}_n \end{pmatrix}, \tag{4.194}$$

where $\boldsymbol{I}_n$ is the $n \times n$ identity matrix. Consider a real orthogonal $2n \times 2n$ matrix R that satisfies

$$R^{\mathsf{T}} A R = A\,. \tag{4.195}$$

Using an infinitesimal analysis (where $R \simeq \boldsymbol{I} + Z$ and Z is an infinitesimal real antisymmetric $2n \times 2n$ matrix), prove that R provides an $\mathbf{n} \oplus \overline{\mathbf{n}}$ reducible representation of U(n). Verify that both $\mathcal{Q}_U$ and $\mathcal{Q}_V$ satisfy the constraint given by eq. (4.195).

(f) The natural embedding of the SO$(2n)$ subgroup in U$(2n)$ can be constructed by considering the matrix

$$\mathcal{U} = \begin{pmatrix} U & V^* \\ V & U^* \end{pmatrix}, \tag{4.196}$$

where U, V satisfy eqs. (4.190) and (4.191). Verify that $\mathcal{U}$ is a unitary matrix.

4.4 Repeat parts (a)–(d) of Exercise 4.3, under the assumption that $\Phi(x)$ is a quaternionic-valued field.

(a) Show that an analogous analysis yields an explicit embedding of the subgroup Sp(n) in U$(2n)$ [which subsequently can be embedded in O$(4n)$].

(b) How does the fundamental representation of O$(4n)$ decompose with respect to the Sp(n) subgroup? How does the fundamental representation of U$(2n)$ decompose with respect to the Sp(n) subgroup?

4.5 Consider the spontaneous breaking of a gauge group G down to U(1). The unbroken generator $Q = c_a \boldsymbol{T}^a$ is some real linear combination of the generators of G, and g_a is the gauge coupling corresponding to the generator $\boldsymbol{T}^a$.

(a) Prove that $x_b \equiv c_b/g_b$ is an (unnormalized) eigenvector of the vector boson squared-mass matrix, M^2_{ab}, with zero eigenvalue.

(b) Suppose that A_μ is the (canonically normalized) massless gauge field that corresponds to the generator Q. Show that the covariant derivative can be expressed in the form $\nabla_\mu = \partial_\mu + ieQA_\mu + \cdots$, where we have omitted terms corresponding to the other gauge bosons, and

$$e = \left[\sum_a (c_a/g_a)^2\right]^{-1/2}. \tag{4.197}$$

HINT: The vector boson mass matrix is diagonalized by an orthogonal transformation $\mathcal{O}M^2\mathcal{O}^{\mathsf{T}}$ according to eq. (4.73). The rows of the matrix $\mathcal{O}$ are constructed from the *orthonormal* eigenvectors of M^2.

(c) Evaluate $Q = c_a \boldsymbol{T}^a$ in the adjoint representation (where the generators of G are $(\boldsymbol{T}^a)^{bc} = -if^{abc}$). Show that $Q^{bc}x_c = 0$, where x_c is defined in part (a). What is the physical interpretation of this result?

(d) Prove that the commutator $[Q, M^2] = 0$, where Q is the unbroken U(1) generator in the adjoint representation and M^2 is the gauge boson squared-mass matrix. Conclude that one can always choose the eigenstates of the gauge boson squared-mass matrix to be states of definite unbroken U(1) charge.

(e) Apply the results of part (b) to the breaking of the $\mathrm{SU}(2)_L\times\mathrm{U}(1)_Y$ symmetry to $\mathrm{U}(1)_{\mathrm{EM}}$ in the Standard Model of electroweak interactions. Show that eq. (4.197) reproduces eq. (4.138).

4.6 (a) The interaction of the Higgs boson with the gauge bosons in eq. (4.162) was obtained in the unitary gauge. Compute the interactions of the Higgs and Goldstone bosons with the gauge bosons in the R_ξ gauge. For example, show that the terms that are cubic in the fields are given by

$$\begin{aligned}\mathscr{L}_{\rm int} &= (em_W A^\mu - g m_Z \sin^2\theta_W Z^\mu)(W_\mu^+ G^- + W_\mu^- G^+) \\ &\quad + \frac{g}{2\cos\theta_W} q_{k1} Z^\mu G \overleftrightarrow{\partial}_\mu H + \left[eA^\mu + \frac{g}{\cos\theta_W}(\tfrac{1}{2} - \sin^2\theta_W) Z^\mu \right] iG^+ \overleftrightarrow{\partial}_\mu G^- \\ &\quad - \tfrac{1}{2} g \left[iW_\mu^+ G^- \overleftrightarrow{\partial}^\mu (H + iG) + \text{h.c.} \right]. \end{aligned} \tag{4.198}$$

(b) Derive the interaction Lagrangian for the charged and neutral Goldstone bosons and the fermions of the Standard Model and verify that

$$\begin{aligned}\mathscr{L}_{\rm int} &= Y_{ui}[\boldsymbol{K}]_i{}^j d_j \bar{u}^i G^+ - Y_{di}[\boldsymbol{K}^\dagger]_i{}^j u_j \bar{d}^i G^- - Y_{ei}\nu_i \bar{e}^i G^- \\ &\quad + \frac{i}{\sqrt{2}}\left[Y_{di} d_i \bar{d}^i - Y_{ui} u_i \bar{u}^i + Y_{ei} e_i \bar{e}^i \right] G^0 + \text{h.c.}, \end{aligned} \tag{4.199}$$

where $\boldsymbol{K}$ is the CKM matrix and Y_{fi} (for $f = u, d, e$) is given by eq. (4.181).

(c) Rewrite the interaction Lagrangians given in eqs. (4.180) and (4.199) in terms of four-component fermion fields and show that the $W^\pm$ and $G^\pm$ couple exclusively to left-handed quarks and leptons.

4.7 The generators of $\mathrm{SU}(2)_L\times\mathrm{U}(1)_Y$ in the fundamental (defining) representation can be expressed as 4×4 matrices by employing the real representation given in eq. (4.106).

(a) Using the explicit form for this real representation, derive the gauge boson squared-mass matrix. Diagonalize the mass matrix and explicitly identify the real antisymmetric matrix generators $\widetilde{\boldsymbol{L}}_a$.

(b) Use eq. (4.80) to obtain the Goldstone bosons and identify the physical Higgs boson.

(c) Using the real representation, write down an explicit form for $\nabla_\mu \Phi$ as a four-dimensional column vector. Then, evaluate eq. (4.161) in the unitary gauge and verify the results of eq. (4.162).

4.8 In Section 4.7, we examined in detail the structure of a spontaneously broken $\mathrm{SU}(2)_L \times \mathrm{U}(1)_Y$ gauge theory, in which the symmetry breaking was due to the VEV of a $Y = \frac{1}{2}$, SU(2) doublet of complex scalar fields. In this exercise, we will replace this multiplet of scalar fields with a different representation.

(a) Consider a spontaneously broken $\mathrm{SU}(2)_L \times \mathrm{U}(1)_Y$ gauge theory with a $Y = 0$, SU(2) triplet of *real* scalar fields. Assume that the electrically neutral member of the scalar triplet acquires a vacuum expectation value. After symmetry breaking, identify the subgroup that remains unbroken. Compute the vector boson masses and the physical Higgs scalar masses in this model, and determine the corresponding ρ parameter [eq. (4.147)]. Deduce the Feynman rules for the three-point interactions among the Higgs and vector bosons.

HINT: Since the triplet of scalar fields corresponds to the adjoint representation of SU(2), the corresponding SU(2) generators that act on the triplet of scalar fields can be chosen to be $(\boldsymbol{T^a})_{bc} = -i\epsilon_{abc}$. The hypercharge operator annihilates the $Y = 0$ fields. Define $\boldsymbol{L}_a = ig_a \boldsymbol{T^a}$, and follow the methods outlined in Section 4.5.4.

(b) Consider a spontaneously broken $\mathrm{SU}(2)_L \times \mathrm{U}(1)_Y$ gauge theory with a $Y = 1$, SU(2) triplet of *complex* scalar fields. Again, assume that the electrically neutral member of the scalar triplet acquires a vacuum expectation value. After symmetry breaking, identify the subgroup that remains unbroken. Compute the vector boson masses, the corresponding ρ parameter, and the physical Higgs scalar masses in this model.

4.9 The ρ parameter, defined in eq. (4.147), is experimentally observed to be $\rho \simeq 1$. Consider a collection of scalar field multiplets of weak isospin T and hypercharge Y. In a multiplet with $Y \leq T$, one of the scalar field components is electric charge neutral. Assume that the neutral components acquire nonzero VEVs, denoted by $v_{T,Y}$. Show that to achieve a tree-level value of $\rho = 1$, the following equation must be satisfied:

$$\sum_{T,Y\neq 0} \left[T(T+1) - 3Y^2\right]|v_{T,Y}|^2 + \tfrac{1}{2}\sum_T T(T+1)|v_{T,0}|^2 = 0\,. \qquad (4.200)$$

Note that eq. (4.200) is satisfied, independently of the value of the scalar field VEVs, if the scalar multiplets consist of $Y = \frac{1}{2}$, $\mathrm{SU}(2)_L$ doublets and $Y = 0$, $\mathrm{SU}(2)_L$ singlets.

5 Anomalies

5.1 Anomalous Chiral Symmetries

Consider a collection of N two-component left-handed $(\frac{1}{2}, 0)$ fermions. The corresponding free-field Lagrangian is invariant under a global $\mathrm{U}(N)$ symmetry. When a mass term and interactions are added to the theory, the global $\mathrm{U}(N)$ symmetry is broken down to a discrete $\mathbb{Z}_2$ symmetry that reflects the fact that any term in the Lagrangian must contain an even number of fermion fields. However, in special cases (for example if there are degeneracies among the masses or if some of the masses vanish), a continuous subgroup of the global $\mathrm{U}(N)$ symmetry survives. Likewise, one can consider a collection of N Dirac fermions, each of which can be represented by a pair of two-component spinor fields. The corresponding free-field Lagrangian is invariant under a global $\mathrm{U}(N)\times\mathrm{U}(N)$ symmetry. When mass terms and interactions are included, this global symmetry is broken. The typical residual symmetry is a global vectorlike $\mathrm{U}(1)$ symmetry (such as lepton number or baryon number). Again, in the presence of mass degeneracies and/or vanishing masses, a larger subgroup of $\mathrm{U}(N)\times\mathrm{U}(N)$ may survive.

The global symmetries that survive after masses and interactions are included are *classical* symmetries. Employing the Noether procedure, each global symmetry is associated with a conserved classical current; the corresponding charges satisfy the Lie algebra of the global symmetry. However, the classical conservation law may not survive quantum corrections. In this case, we say that the global symmetry possesses an anomaly.

5.1.1 The Triangle Anomaly

The anomaly in chiral symmetries for fermions arises from the triangle diagram involving three currents carrying vector indices.[1] Since the anomaly is independent of the fermion masses, we simplify the computation by setting all fermion masses to zero. In four-component notation, the treatment of the anomaly requires care because of the difficulty in defining a consistent and unambiguous γ_5 and the Levi-Civita epsilon tensor in dimensional regularization. The same subtleties arise in two-component language, of course, but in a slightly different form since γ_5 does not appear explicitly.

[1] The discussion here parallels that given in Section 22.3 of [B73].

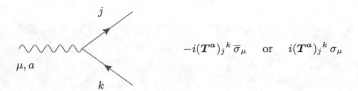

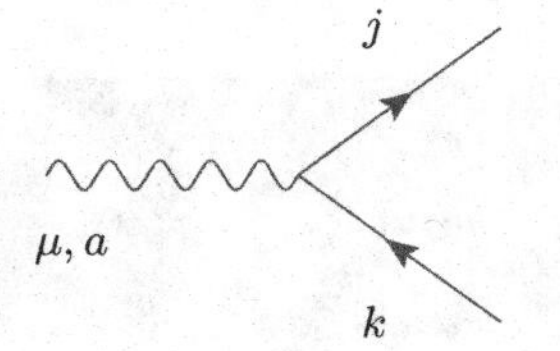

$$-i(\boldsymbol{T}^a)_j{}^k\,\overline{\sigma}_\mu \quad \text{or} \quad i(\boldsymbol{T}^a)_j{}^k\,\sigma_\mu$$

Fig. 5.1 Feynman rule for the coupling of a current with vector index μ, corresponding to the symmetry generator $\boldsymbol{T}^a$, acting on $(\frac{1}{2},0)$ [left-handed] fermions. Spinor indices are suppressed.

We shall assemble all the $(\frac{1}{2},0)$ [left-handed] two-component fermion fields of the theory into a multiplet ψ_j. For example, the fermions of the Standard Model are: $\psi_j = (\ell_k\,,\,\bar{\ell}_k\,,\,\nu_k\,,\,q_{in}\,,\,\bar{q}_{in})$, where $k = 1,2,3$ and $i = 1,2,\ldots,6$ are flavor labels and $n = 1,2,3$ are color labels. The two-component spinor indices are suppressed here. Let the symmetry generators be given by hermitian matrices $\boldsymbol{T}^a$, so that the ψ_j transform as

$$\delta\psi_j = -i\theta^a(\boldsymbol{T}^a)_j{}^k\psi_k, \tag{5.1}$$

for infinitesimal parameters θ^a. The matrices $\boldsymbol{T}^a$ form a representation R of the generators of the Lie algebra of the symmetry group. In general R will be reducible, in which case the $\boldsymbol{T}^a$ have a block–diagonal structure, where each block separately transforms (irreducibly) the corresponding field of ψ_j according to its symmetry transformation properties. Some or all of these symmetries may be gauged. The Feynman rule for the corresponding currents is the same as for external gauge bosons, as in Fig. 2.8 (but without the gauge couplings), and is shown in Fig. 5.1.

In Fig. 5.2, we exhibit the two Feynman diagrams that contribute at one loop to the three-point function of the symmetry currents. Applying the $\overline{\sigma}$ version of the Feynman rule for the currents given in Fig. 5.1, and employing the Feynman rules of Fig. 2.2 (with $m = 0$) for the propagators [traversing the loop in the direction dictated by the second rule of Section 2.7], the sum of the two triangle diagrams shown in Fig. 5.2 can be evaluated.

The resulting sum of loop integrals is

$$\begin{aligned} i\Gamma^{abc}_{\mu\nu\rho} = (-1)\int\frac{d^4k}{(2\pi)^4}\mathrm{Tr}\Bigg\{&(-i\overline{\sigma}_\mu\boldsymbol{T}^a)\frac{i(k-p+r)\cdot\sigma}{(k-p+r)^2}(-i\overline{\sigma}_\nu\boldsymbol{T}^b)\frac{i(k+r)\cdot\sigma}{(k+r)^2}\\ &\times(-i\overline{\sigma}_\rho\boldsymbol{T}^c)\frac{i(k+q+r)\cdot\sigma}{(k+q+r)^2}\\ &+(-i\overline{\sigma}_\mu\boldsymbol{T}^a)\frac{i(k-q+\ell)\cdot\sigma}{(k-q+\ell)^2}(-i\overline{\sigma}_\rho\boldsymbol{T}^c)\frac{i(k+\ell)\cdot\sigma}{(k+\ell)^2}(-i\overline{\sigma}_\nu\boldsymbol{T}^b)\frac{i(k+p+\ell)\cdot\sigma}{(k+p+\ell)^2}\Bigg\}, \end{aligned} \tag{5.2}$$

where the overall factor of (-1) is due to the presence of a closed fermion loop. The trace is taken over fermion flavor/group and spinor indices, both of which are

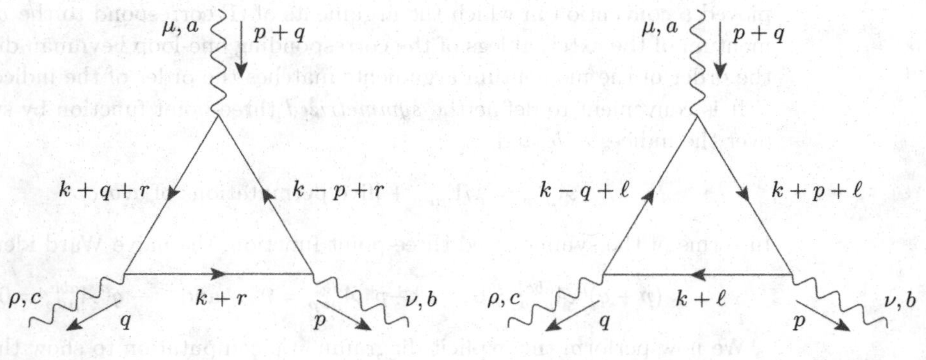

Fig. 5.2 Triangle Feynman diagrams leading to the chiral fermion anomaly. Fermion spinor and flavor indices are suppressed. The fermion momenta, as labeled, flow in the arrow directions.

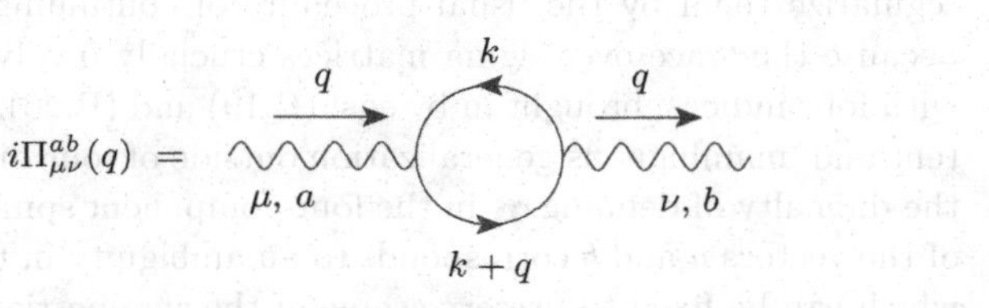

Fig. 5.3 The one-loop contribution to the current–current two-point function. The fermion momenta, as labeled, flow along the corresponding arrow directions.

suppressed. Because the individual integrals are linearly divergent, one must allow for arbitrary constant four-vectors r^μ and ℓ^μ as offsets for the loop momentum when defining the loop integrations for the two diagrams (e.g., see Ref. [45]).

The persistence of the symmetry in the quantum theory for the currents labeled by μ, a and ν, b and ρ, c implies the naive Ward identities:[2]

$$(p+q)^\mu \, i\Gamma^{abc}_{\mu\nu\rho}(-p-q,p,q) = f^{abd}\Pi^{dc}_{\nu\rho}(q) + f^{acd}\Pi^{bd}_{\nu\rho}(p)\,, \tag{5.3}$$

$$-p^\nu \, i\Gamma^{abc}_{\mu\nu\rho}(-p-q,p,q) = f^{bcd}\Pi^{da}_{\rho\mu}(p+q) + f^{bad}\Pi^{cd}_{\rho\mu}(q)\,, \tag{5.4}$$

$$-q^\rho \, i\Gamma^{abc}_{\mu\nu\rho}(-p-q,p,q) = f^{cad}\Pi^{db}_{\mu\nu}(p) + f^{cbd}\Pi^{ad}_{\mu\nu}(p+q)\,, \tag{5.5}$$

where $i\Pi^{ab}_{\mu\nu}(p)$ is the one-loop current–current two-point function shown in Fig. 5.3.

By Lorentz covariance, $\Pi^{ab}_{\mu\nu}(p)$ is a rank-two symmetric tensor that is an even function of the four-momentum p [see eq. (5.41)]. In eqs. (5.3)–(5.5), we have em-

[2] The derivation of the Ward identities is most easily achieved by first writing the three-point Green function in position space as a vacuum expectation value of the time-ordered product of three currents. After taking the divergence (with respect to the position of any one of the three currents) of the time-ordered product and using the fact that the currents are conserved ($\partial_\mu j^{a\mu} = 0$), the surviving terms can be evaluated using the equal-time commutation relations, $\delta(x^0 - y^0)[j^{a0}(x), j^{b\nu}(y)] = if^{abc}j^{c\nu}(x)\delta^4(x-y)$. Fourier-transforming the result yields the terms on the right-hand side of eqs. (5.3)–(5.5). See [B50] and Ref. [46] for further details.

ployed a convention in which the arguments of $i\Gamma$ correspond to the *outgoing* momentum of the external legs of the corresponding one-loop Feynman diagrams, and the order of the momentum arguments matches the order of the indices.

It is convenient to define the *symmetrized* three-point function by symmetrizing over the indices a, b, and c:

$$\mathcal{A}^{abc}_{\mu\nu\rho} = \tfrac{1}{6} i\Gamma^{abc}_{\mu\nu\rho} + [\text{five permutations of } a, b, c]. \tag{5.6}$$

In terms of the symmetrized three-point function, the naive Ward identities imply

$$(p+q)^\mu \mathcal{A}^{abc}_{\mu\nu\rho} = 0\,, \qquad -p^\nu \mathcal{A}^{abc}_{\mu\nu\rho} = 0\,, \quad \text{and} \quad -q^\rho \mathcal{A}^{abc}_{\mu\nu\rho} = 0\,. \tag{5.7}$$

We now perform the explicit diagrammatic computation to show that the naive Ward identities exhibited in eq. (5.7) are violated due to a quantum anomaly. Although the symmetrized three-point function is ultraviolet finite, the individual loop momentum integrals are divergent, and must be defined with care. We do not regularize them by the usual procedure of continuing to $d = 4 - 2\epsilon$ dimensions, because the trace over sigma matrices crucially involves the antisymmetric tensor with four indices, brought in by eqs. (B.19) and (B.20), for which there is no consistent and unambiguous generalization outside of four dimensions. (This is related to the difficulty of defining γ_5 in the four-component spinor formalism.) The existence of the vectors a and b corresponds to an ambiguity in the regularization procedure, which can be fixed to preserve some of the symmetries, as we will see below.

Starting from eq. (5.2), it follows from eq. (H.41) that the symmetrized three-point function is proportional to the group theory factor,

$$D^{abc} = \tfrac{1}{2}\mathrm{Tr}[\{\boldsymbol{T^a}, \boldsymbol{T^b}\}\boldsymbol{T^c}]\,, \tag{5.8}$$

where the numerical values of the D^{abc} depend on the representation R. As discussed in Appendix H.3, D^{abc} vanishes for all simple Lie groups, with the exception of SU(N) for $N \geq 3$. The D^{abc} are also nonvanishing in general for any non-semisimple compact Lie group when at least one of the group indices corresponds to U(1).

First, consider the result for $(p+q)^\mu \mathcal{A}^{abc}_{\mu\nu\rho}$. This can be simplified by rewriting

$$(p+q)^\mu = (k+q+r)^\mu - (k-p+r)^\mu\,, \tag{5.9}$$

$$(p+q)^\mu = (k+p+\ell)^\mu - (k-q+\ell)^\mu\,, \tag{5.10}$$

in the first and second diagram terms, respectively, and then applying the formulae

$$v{\cdot}\sigma\, v{\cdot}\overline{\sigma} = v^2\,, \qquad v{\cdot}\overline{\sigma}\, v{\cdot}\sigma = v^2\,, \tag{5.11}$$

which follow from eqs. (B.109) and (B.110). After rearranging the terms using the cyclic property of the trace, we obtain

$$\begin{aligned}(p+q)^\mu \mathcal{A}^{abc}_{\mu\nu\rho} &= -D^{abc}\,\mathrm{Tr}[\sigma_\kappa\overline{\sigma}_\nu\sigma_\lambda\overline{\sigma}_\rho]\, X^{\kappa\lambda}, \\ &= -2D^{abc}\left[X_{\nu\rho} + X_{\rho\nu} - g_{\nu\rho}X_\lambda{}^\lambda + i\epsilon_{\kappa\nu\lambda\rho}X^{\kappa\lambda}\right]\,, \end{aligned} \tag{5.12}$$

after applying eq. (B.19), where $\epsilon_{0123} = -1$ in the conventions of Appendix A.4.

The integral $X^{\kappa\lambda}$ is given by

$$X^{\kappa\lambda} = \int \frac{d^4k}{(2\pi)^4} \left[\frac{(k-p+r)^\kappa}{(k-p+r)^2} \frac{(k+r)^\lambda}{(k+r)^2} - \frac{(k+q+r)^\kappa}{(k+q+r)^2} \frac{(k+r)^\lambda}{(k+r)^2} + \frac{(k+\ell)^\kappa}{(k+\ell)^2} \frac{(k-q+\ell)^\lambda}{(k-q+\ell)^2} - \frac{(k+\ell)^\kappa}{(k+\ell)^2} \frac{(k+p+\ell)^\lambda}{(k+p+\ell)^2} \right]. \quad (5.13)$$

Naively, this integral appears to vanish, because the first term is equal to the negative of the fourth term after a momentum shift $k \to k-p+r-\ell$, and the second term is equal to the negative of the third term after $k \to k+q+r-\ell$. However, these momentum shifts are not valid for the individually divergent integrals. Instead, $X^{\kappa\lambda}$ can be evaluated by a Wick rotation to Euclidean space, followed by isolating the terms that contribute for large k^2 and are responsible for the integral not vanishing, and then employing the divergence (Gauss') theorem in four dimensions to rewrite $X^{\kappa\lambda}$ as an angular integral over a three-sphere with radius tending to infinity. This integral is initially evaluated at large but finite Euclidean k, with the limit $k \to \infty$ taken at the end of the computation. For example, consider a smooth function $f(k)$ of the four-momentum k with the property that the integral

$$\int d^4k f(k) \quad (5.14)$$

is at worst quadratically divergent. We define the even and odd parts of $f(k)$, respectively, by

$$f_e(k) \equiv \tfrac{1}{2}\left[f(k) + f(-k)\right], \qquad f_o(k) \equiv \tfrac{1}{2}\left[f(k) - f(-k)\right]. \quad (5.15)$$

It then follows that

$$\int \frac{d^4k}{(2\pi)^4} \left[f(k+r) - f(k)\right] = \frac{i}{(2\pi)^4} \left[2\pi^2 r_\mu \lim_{k\to\infty} k^\mu k^2 f_o(k) + \pi^2 r_\mu r_\nu \lim_{k\to\infty} k^\mu k^2 \frac{\partial}{\partial k_\nu} f_e(k) \right] \quad (5.16)$$

has a finite limit.[3] In deriving this result, we have expanded $f(k+r)$ in a Taylor series and have followed the procedure outlined above eq. (5.14). Note that the angular integration removes the even parts of $f(k)$ and $\partial f/\partial k^\nu \equiv 2k_\nu\, \partial f/\partial k^2$ from the right-hand side of eq. (5.16). The "limits" in eq. (5.16) actually correspond to an average over the three-sphere at large Euclidean k, and thus should be interpreted as

$$\lim_{k\to\infty} \frac{k^\mu k^\nu}{k^2} = \tfrac{1}{4} g^{\mu\nu}, \quad (5.17)$$

$$\lim_{k\to\infty} \frac{k^\mu k^\nu k^\rho k^\lambda}{(k^2)^2} = \frac{1}{24}\left(g^{\mu\nu} g^{\rho\lambda} + g^{\mu\rho} g^{\nu\lambda} + g^{\mu\lambda} g^{\nu\rho}\right). \quad (5.18)$$

[3] If eq. (5.14) is linearly divergent, then the second term on the right-hand side of eq. (5.16) is zero. If eq. (5.14) is logarithmically divergent or finite, then the right-hand side of eq. (5.16) vanishes.

For example, if

$$f(k)=\frac{(k-p+r)^{\kappa}(k+r)^{\lambda}}{(k-p+r)^{2}(k+r)^{2}}\,, \tag{5.19}$$

then in evaluating eq. (5.16) it is sufficient to write

$$\begin{aligned} f_o(k) &\simeq \tfrac{1}{2}(k-p+r)^{\kappa}(k+r)^{\lambda}\left[\frac{1}{(k^2)^2}+\frac{2k\cdot(p-2r)}{(k^2)^3}\right]-(k\to -k) \\ &\simeq \frac{k^{\kappa}r^{\lambda}-k^{\lambda}(p-r)^{\kappa}}{(k^2)^2}+\frac{2k^{\kappa}k^{\lambda}\,k\cdot(p-2r)}{(k^2)^3}\,, \end{aligned} \tag{5.20}$$

where we have dropped terms that do not contribute to eq. (5.16) in the limit of $k\to\infty$. Similarly,

$$\frac{\partial f_e}{\partial k_\nu}\simeq\frac{g^{\kappa\nu}k^{\lambda}+g^{\lambda\nu}k^{\kappa}}{(k^2)^2}-\frac{4k^{\kappa}k^{\lambda}k^{\nu}}{(k^2)^3}\,. \tag{5.21}$$

The evaluation of $X^{\kappa\lambda}$ is now straightforward [after using eqs. (5.17) and (5.18)]:

$$X^{\kappa\lambda}=\frac{i}{96\pi^2}\Big[g^{\kappa\lambda}(p+q)\cdot(r+\ell)+(r-2\ell)^{\kappa}(p+q)^{\lambda}+(p+q)^{\kappa}(\ell-2r)^{\lambda}\Big]\,. \tag{5.22}$$

Hence, eq. (5.12) yields the result for the anomaly in the current labeled by μ, a:

$$\begin{aligned}(p+q)^{\mu}\mathcal{A}^{abc}_{\mu\nu\rho}=\frac{i}{48\pi^2}D^{abc}\Big[&(p+q)_\nu(r+\ell)_\rho+(r+\ell)_\nu(p+q)_\rho \\ &+g_{\nu\rho}(p+q)\cdot(r+\ell)-3i\epsilon_{\nu\rho\kappa\lambda}(p+q)^{\kappa}(r-\ell)^{\lambda}\Big]. \end{aligned} \tag{5.23}$$

Repeating all of the steps starting with eq. (5.9), we similarly obtain[4]

$$\begin{aligned}-p^{\nu}\mathcal{A}^{abc}_{\mu\nu\rho}=-\frac{i}{48\pi^2}D^{abc}\Big[&p_\rho(r+\ell)_\mu+p_\mu(r+\ell)_\rho+g_{\mu\rho}p\cdot(r+\ell) \\ &-3i\epsilon_{\rho\mu\kappa\lambda}p^{\kappa}(r-\ell+2q)^{\lambda}\Big]\,, \end{aligned} \tag{5.24}$$

$$\begin{aligned}-q^{\rho}\mathcal{A}^{abc}_{\mu\nu\rho}=-\frac{i}{48\pi^2}D^{abc}\Big[&q_\mu(r+\ell)_\nu+q_\nu(r+\ell)_\mu+g_{\mu\nu}q\cdot(r+\ell) \\ &-3i\epsilon_{\mu\nu\kappa\lambda}q^{\kappa}(r-\ell-2p)^{\lambda}\Big]\,. \end{aligned} \tag{5.25}$$

Assuming that D^{abc} is nonvanishing, nonchiral anomalies (i.e., terms not proportional to the Levi-Civita epsilon tensor) will arise for all three of the currents unless we choose the arbitrary constant vectors r and ℓ such that

$$r+\ell=0\,. \tag{5.26}$$

[4] Alternatively, one can simply note that eq. (5.24) follows from eq. (5.23) by making the replacements $\mu\to\nu$, $\nu\to\rho$, $\rho\to\mu$, $r\to r+q$, $\ell\to\ell-q$, $p\to q$, and $q\to -p-q$, while eq. (5.25) follows from eq. (5.23) by making the replacements $\mu\to\rho$, $\nu\to\mu$, $\rho\to\nu$, $r\to r-p$, $\ell\to\ell+p$, $p\to -p-q$, and $q\to p$.

Using eq. (5.26), the following three anomalous Ward identities are obtained:

$$(p+q)^\mu \mathcal{A}^{abc}_{\mu\nu\rho} = \frac{1}{8\pi^2} D^{abc} \epsilon_{\nu\rho\kappa\lambda}(p+q)^\kappa r^\lambda, \tag{5.27}$$

$$-p^\nu \mathcal{A}^{abc}_{\mu\nu\rho} = -\frac{1}{8\pi^2} D^{abc} \epsilon_{\rho\mu\kappa\lambda} p^\kappa (r+q)^\lambda, \tag{5.28}$$

$$-q^\rho \mathcal{A}^{abc}_{\mu\nu\rho} = -\frac{1}{8\pi^2} D^{abc} \epsilon_{\mu\nu\kappa\lambda} q^\kappa (r-p)^\lambda. \tag{5.29}$$

If D^{abc} is nonvanishing, it is not possible to avoid an anomaly simultaneously in all three symmetries, but one can still arrange for two of the symmetries to be nonanomalous. If one wants an anomaly to arise only in the current labeled by μ, a (for example, if the symmetries labeled by b, c are gauged), one must now choose $r = p - q$. The standard result follows:

$$(p+q)^\mu \mathcal{A}^{abc}_{\mu\nu\rho} = -\frac{1}{4\pi^2} D^{abc} \epsilon_{\nu\rho\kappa\lambda} p^\kappa q^\lambda, \tag{5.30}$$

$$-p^\nu \mathcal{A}^{abc}_{\mu\nu\rho} = 0, \tag{5.31}$$

$$-q^\rho \mathcal{A}^{abc}_{\mu\nu\rho} = 0. \tag{5.32}$$

That is, one cannot gauge all three symmetries labeled by a, b, c unless $D^{abc} = 0$.

If all three currents are identical, then by Bose symmetry the anomalies of the three currents must coincide. This can be achieved by choosing $r = \frac{1}{3}(p-q)$, in which case

$$(p+q)^\mu \mathcal{A}^{abc}_{\mu\nu\rho} = -\frac{1}{12\pi^2} D^{abc} \epsilon_{\nu\rho\kappa\lambda} p^\kappa q^\lambda, \tag{5.33}$$

$$-p^\nu \mathcal{A}^{abc}_{\mu\nu\rho} = -\frac{1}{12\pi^2} D^{abc} \epsilon_{\rho\mu\kappa\lambda} p^\kappa q^\lambda, \tag{5.34}$$

$$-q^\rho \mathcal{A}^{abc}_{\mu\nu\rho} = -\frac{1}{12\pi^2} D^{abc} \epsilon_{\mu\nu\kappa\lambda} p^\kappa q^\lambda. \tag{5.35}$$

Returning briefly to the original naive Ward identities given in eqs. (5.3)–(5.5), the analysis above shows that these identities must be modified by an additional additive contribution given by the right-hand side of eqs. (5.27)–(5.29). In particular, there is no anomalous contribution proportional to f^{abc}. This can be checked explicitly by a diagrammatic computation of the two-point and three-point functions that appear in eqs. (5.3) and (5.5). We use eqs. (H.45) and (H.46) to write

$$\mathrm{Tr}(\boldsymbol{T}^a \boldsymbol{T}^b \boldsymbol{T}^c) = D^{abc}(R) + \tfrac{1}{2} i I_2(R) f^{abc}, \tag{5.36}$$

where $I_2(R)$ is the second-order Dynkin index defined in eq. (H.32) and R is the representation of the generators $\boldsymbol{T}^a$. For example, inserting this result in eq. (5.2), it follows that

$$(p+q)^\mu\, i\Gamma^{abc}_{\mu\nu\rho} = -\left[D^{abc} X^{\kappa\lambda} + \frac{i}{2} I_2(R) f^{abc} Y^{\kappa\lambda} \right] \mathrm{Tr}[\sigma_\kappa \overline{\sigma}_\nu \sigma_\lambda \overline{\sigma}_\rho], \tag{5.37}$$

where the integral $Y^{\kappa\lambda}$ is given by[5]

$$Y^{\kappa\lambda} = \int \frac{d^4k}{(2\pi)^4} \left[\frac{(k-p)^\kappa\, k^\lambda}{(k-p)^2\, k^2} - \frac{(k+q)^\kappa\, k^\lambda}{(k+q)^2\, k^2} - \frac{k^\kappa\, (k-q)^\lambda}{k^2\, (k-q)^2} + \frac{k^\kappa\, (k+p)^\lambda}{k^2\, (k+p)^2} \right]. \quad (5.38)$$

By letting $k \to -k$ in the third and fourth terms in the integrand of eq. (5.38), we see that $Y^{\kappa\lambda} = Y^{\lambda\kappa}$, and hence, by eq. (B.19),

$$-\frac{i}{2} I_2(R) f^{abc}\, Y^{\kappa\lambda} \mathrm{Tr}[\sigma_\kappa \overline{\sigma}_\nu \sigma_\lambda \overline{\sigma}_\rho] = -i I_2(R) f^{abc} \left[2Y_{\nu\rho} - g_{\nu\rho} Y_\lambda{}^\lambda \right]. \quad (5.39)$$

Since no Levi-Civita epsilon tensor appears, we can evaluate this integral in $d \neq 4$ dimensions using the standard techniques of dimensional regularization.

One can check that this result matches the diagrammatic calculation of the right-hand side of eq. (5.27). In particular, Fig. 5.3 yields

$$i\Pi^{ab}_{\mu\nu}(q) = (-1) \int \frac{d^4k}{(2\pi)^4} \mathrm{Tr} \left[(-i\overline{\sigma}_\mu \boldsymbol{T}^a) \frac{ik\cdot\sigma}{k^2} (-i\overline{\sigma}_\nu \boldsymbol{T}^b) \frac{i(k+q)\cdot\sigma}{(k+q)^2} \right]$$

$$= -I_2(R) \delta^{ab}\, \mathrm{Tr}(\overline{\sigma}_\mu \sigma_\rho \overline{\sigma}_\nu \sigma_\lambda) \int \frac{d^4k}{(2\pi)^4} \frac{k^\rho (k+q)^\lambda}{k^2 (k+q)^2}, \quad (5.40)$$

where we have used eq. (H.32). Lorentz covariance implies that

$$i\Pi^{ab}_{\mu\nu}(q) = \delta^{ab} \left[C_1(q^2) g_{\mu\nu} + C_2(q^2) q_\mu q_\nu \right], \quad (5.41)$$

where C_1 and C_2 are scalar functions of q^2. It follows that $\Pi^{ab}_{\mu\nu}(q) = \Pi^{ab}_{\mu\nu}(-q)$ and $\Pi^{ab}_{\mu\nu}(q) = \Pi^{ab}_{\nu\mu}(q)$. Consequently, we can write

$$\Pi^{ab}_{\mu\nu}(q) = \frac{i}{2} I_2(R) \delta^{ab}\, \mathrm{Tr}(\overline{\sigma}_\mu \sigma_\rho \overline{\sigma}_\nu \sigma_\lambda + \overline{\sigma}_\nu \sigma_\rho \overline{\sigma}_\mu \sigma_\lambda) \int \frac{d^4k}{(2\pi)^4} \frac{k^\rho (k+q)^\lambda}{k^2 (k+q)^2}, \quad (5.42)$$

and so no Levi-Civita epsilon tensor appears in the evaluation of the trace. As above, we are now free to evaluate the integral in $d \neq 4$ dimensions. Comparing eqs. (5.37) and (5.38) to eq. (5.42), and using eq. (5.27), the end result is

$$(p+q)^\mu\, i\Gamma^{abc}_{\mu\nu\rho}(-p-q, p, q) = I_2(R) f^{abc} \left[\Pi_{\nu\rho}(q) - \Pi_{\nu\rho}(p) \right]$$
$$+ \frac{1}{8\pi^2} D^{abc}(R) \epsilon_{\nu\rho\kappa\lambda} (p+q)^\kappa r^\lambda, \quad (5.43)$$

where we have written $\Pi^{ab}_{\nu\rho} \equiv I_2(R) \delta^{ab} \Pi_{\nu\rho}$. Indeed the terms on the right-hand side proportional to f^{abc} match those of the naive Ward identity given in eq. (5.3). As previously asserted, the anomaly only resides in the contributions to the Ward identity proportional to D^{abc}.

In writing down eq. (5.2), we chose to use the rules with $\overline{\sigma}$ matrices for the current vertices and σ matrices for the massless fermion propagators. If we had chosen the opposite prescription (i.e., σ matrices for the current vertices and $\overline{\sigma}$ matrices for

[5] Here $Y^{\kappa\lambda}$ is obtained from $X^{\kappa\lambda}$ by setting $r = \ell = 0$, since we can use dimensional regularization for this part of the computation, as explained below eq. (5.38).

the massless fermion propagators), then the order of the factors inside the trace of eq. (5.2) would have been reversed.[6] Instead of eq. (5.12), we would have obtained

$$(p+q)^\mu \mathcal{A}^{abc}_{\mu\nu\rho} = -D^{abc}\,\mathrm{Tr}[\overline{\sigma}_\kappa\sigma_\nu\overline{\sigma}_\lambda\sigma_\rho]\,\bar{X}^{\kappa\lambda}$$
$$= -2D^{abc}\left[\bar{X}_{\nu\rho} + \bar{X}_{\rho\nu} - g_{\nu\rho}\bar{X}_\lambda{}^\lambda - i\epsilon_{\kappa\nu\lambda\rho}\bar{X}^{\kappa\lambda}\right], \qquad (5.44)$$

after applying eq. (B.20), where $\bar{X}^{\kappa\lambda} = X^{\lambda\kappa}$. Inserting this result into eq. (5.44), we immediately reproduce the result of eq. (5.12), as expected.

5.1.2 The Axial Anomaly in Massless QED

It is instructive to examine the case of massless QED. The terms of the Lagrangian involving the electron fields are given by

$$\mathscr{L} \supset i\chi^\dagger\overline{\sigma}^\mu\nabla_\mu\chi + i\eta^\dagger\overline{\sigma}^\mu\nabla_\mu\eta, \qquad (5.45)$$

where $\nabla_\mu \equiv \partial_\mu + iQA_\mu$ is the covariant derivative and Q is the charge operator. Here, we identify χ as the two-component (left-handed) electron field and η as the two-component (left-handed) positron field. The corresponding eigenvalues of the charge operator are $Q\chi = -e\chi$ and $Q\eta = +e\eta$ (where $e > 0$ is the electromagnetic gauge coupling constant, or equivalently the electric charge of the positron).

At the classical level, the massless QED Lagrangian [eq. (5.45)] is invariant under a $\mathrm{U}(1)_V\times\mathrm{U}(1)_A$ global symmetry. Under a $\mathrm{U}(1)_V\times\mathrm{U}(1)_A$ transformation specified by the infinitesimal parameters θ_V and θ_A,

$$\mathrm{U}(1)_V: \qquad \delta\chi = ie\theta_V\chi, \qquad \delta\eta = -ie\theta_V\eta, \qquad (5.46)$$

$$\mathrm{U}(1)_A: \qquad \delta\chi = i\theta_A\chi, \qquad \delta\eta = i\theta_A\eta. \qquad (5.47)$$

We can combine these equations into a two-dimensional matrix equation,

$$\delta\psi_j = -i\theta_a(\boldsymbol{T_a})_j{}^k\psi_k, \qquad \text{where} \qquad \psi = \begin{pmatrix}\chi\\ \eta\end{pmatrix}, \qquad (5.48)$$

and the index a takes on two values, $a = V, A$. It follows that the $\mathrm{U}(1)_V\times\mathrm{U}(1)_A$ generators are given by

$$T_V = e\begin{pmatrix}-1 & 0\\ 0 & 1\end{pmatrix}, \qquad \text{for} \quad \mathrm{U}(1)_V, \qquad (5.49)$$

$$T_A = \begin{pmatrix}-1 & 0\\ 0 & -1\end{pmatrix}, \qquad \text{for} \quad \mathrm{U}(1)_A. \qquad (5.50)$$

The classically conserved Noether currents corresponding to the $\mathrm{U}(1)_V\times\mathrm{U}(1)_A$ global symmetry are the vector and axial-vector currents

$$J^\mu_V = -e(\chi^\dagger\overline{\sigma}^\mu\chi - \eta^\dagger\overline{\sigma}^\mu\eta), \qquad J^\mu_A = -\chi^\dagger\overline{\sigma}^\mu\chi - \eta^\dagger\overline{\sigma}^\mu\eta. \qquad (5.51)$$

[6] The arrowed fermion lines in the loop must be traversed in the direction parallel [antiparallel] to the arrow directions when the $\overline{\sigma}$ [σ] versions of the propagator rule are employed, as indicated in the second rule of Section 2.7 [and in the discussion that follows]. This rule determines the order of the factors inside the spinor trace.

Note that the interaction Lagrangian for massless QED is $\mathscr{L}_{\text{int}} = -J^\mu_V A_\mu$, which accounts for the factor of e in the definition of the vector current. The axial-vector current does not couple to the photon field; hence no coupling constant is included in its definition

Since the $U(1)_V$ symmetry is gauged, we demand that this symmetry should be anomaly-free. Thus, we make use of eqs. (5.30)–(5.32), where we identify the index pair μ, a with the axial-vector current and the index pairs ν, b and ρ, c with the vector current. Thus, we compute

$$D^{AVV} = \mathrm{Tr}\,(T_A T_V T_V) = -2e^2\,. \tag{5.52}$$

Moreover, for an abelian symmetry group, $f^{abc} = 0$. Hence, using eq. (5.30) [which also applies in this case to the unsymmetrized three-point function], the U(1) axial-vector anomaly equation reads

$$(p+q)^\mu\, i\Gamma^{AVV}_{\mu\nu\rho} = \frac{e^2}{2\pi^2}\epsilon_{\nu\rho\kappa\lambda}p^\kappa q^\lambda\,, \tag{5.53}$$

in agreement with the well-known result.[7]

We now convert eq. (5.53) into an operator equation as follows. Consider the process of two-photon production by an axial-vector current source. First, we note that $\partial_\mu J^\mu_A(x) = i[P^\mu, J_{A\mu}(x)]$, where P^μ is the momentum operator. Hence,

$$\langle p\,,\, q\,|\,\partial_\mu J^\mu_A(0)\,|\,0\rangle = i\,\langle p\,,\, q\,|\,[P^\mu\,,\, J_{A\mu}(0)]\,|\,0\rangle = i(p+q)^\mu\,\langle p\,,\, q\,|\,J_{A\mu}(0)|\,0\rangle\,. \tag{5.54}$$

We identify the S-matrix amplitude for the two-photon production as

$$i\Gamma^{AVV}_{\mu\nu\rho}\,\varepsilon^{\nu\,*}(p)\varepsilon^{\rho\,*}(q) = \langle p\,,\, q\,|\,-iJ_{A\mu}(0)|\,0\rangle\,, \tag{5.55}$$

where $\varepsilon(p)$ and $\varepsilon(q)$ are the polarization vectors for the final-state photons. Note that the factor of $-i$ on the right-hand side of eq. (5.55) has been inserted to be consistent with the Feynman rule for the axial-vector current insertion given in Fig. 5.1. Thus, using eqs. (5.53)–(5.55), we end up with

$$\begin{aligned}\langle p\,,\, q\,|\,\partial_\mu J^\mu_A(0)\,|\,0\rangle &= -\frac{e^2}{2\pi^2}\epsilon_{\nu\rho\kappa\lambda}\varepsilon^{\nu\,*}(p)\varepsilon^{\rho\,*}(q)p^\kappa q^\lambda \\ &= -\frac{e^2}{16\pi^2}\langle p\,,\, q\,|\,\epsilon_{\kappa\nu\lambda\rho}F^{\kappa\nu}F^{\lambda\rho}(0)\,|0\rangle\,,\end{aligned} \tag{5.56}$$

where $\epsilon_{\kappa\nu\lambda\rho}F^{\kappa\nu}F^{\lambda\rho} = 4\epsilon_{\kappa\nu\lambda\rho}(\partial^\kappa A^\nu)(\partial^\lambda A^\rho)$ has been used to eliminate the photon fields in favor of a product of electromagnetic field strength tensors. In deriving eq. (5.56), an additional factor of two arises due to two possible contractions of the photon fields with the external states. We thus obtain the operator form for the

[7] This result was first obtained by Adler [47]. In comparing eq. (5.53) with Adler's result, note that the normalization of the triangle amplitude in Ref. [47] differs by a factor of $(2\pi)^4$ and the opposite sign convention for ϵ_{0123} is employed.

axial-vector anomaly:[8]

$$\partial_\mu J^\mu_A = -\frac{e^2}{8\pi^2} F^{\lambda\rho}\widetilde{F}_{\lambda\rho}\,, \tag{5.57}$$

where the dual electromagnetic field strength tensor is defined by $\widetilde{F}_{\lambda\rho} \equiv \frac{1}{2}\epsilon_{\kappa\nu\lambda\rho}F^{\kappa\nu}$.

It is noteworthy that the right-hand side of eq. (5.57) is a total derivative:

$$F^{\lambda\rho}\widetilde{F}_{\lambda\rho} = \epsilon^{\kappa\nu\lambda\rho}\partial_\kappa(A_\nu F_{\lambda\rho})\,. \tag{5.58}$$

As a result, one can define a conserved current,

$$\hat{J}^\mu_A \equiv J^\mu_A + \frac{e^2}{8\pi^2}\epsilon^{\mu\nu\rho\lambda}A_\nu F_{\rho\lambda}\,, \tag{5.59}$$

which satisfies $\partial_\mu \hat{J}^\mu_A = 0$. However, $\hat{J}^\mu_A$ is manifestly gauge-dependent. Consequently, the physical consequences of the axial anomaly (such as those discussed Section 5.1.3) cannot be removed by redefining the axial-vector current as in eq. (5.59).

5.1.3 Physical Consequences of the Axial Anomaly

Consider the anomaly in the VVA triangle consisting of two photons (i.e., two vector currents) and an isovector axial-vector current constructed from quark fields,

$$J^{\mu a}_A = -q^{\dagger\, i}\overline{\sigma}^\mu(\tfrac{1}{2}\tau^a)_i{}^j q_j + \bar{q}^i\sigma^\mu(\tfrac{1}{2}\tau^a)_i{}^j \bar{q}^\dagger_j\,, \tag{5.60}$$

where i and j are SU(2) isospin labels, $\frac{1}{2}\tau^a$ are the isospin generators, and $q_1 = u$ and $q_2 = d$ (corresponding to the up and down quark fields, respectively). The axial-vector current can create a pion with four-momentum p from the vacuum:

$$\langle 0|\, J^{\mu a}_A(0)\,|\pi^b(p)\rangle = i f_\pi \delta^{ab} p^\mu\,, \tag{5.61}$$

where $f_\pi \simeq 93$ MeV is the pion decay constant. In particular, $J^{\mu 3}_A$ creates a neutral pion. The amplitude for the decay of a neutral pion to two photons can be computed by evaluating the VVA triangle graph. In the absence of the anomaly, the amplitude would vanish in the chiral limit (where the quark masses are neglected). Taking the axial anomaly into account [e.g., employing eq. (5.57) and inserting the relevant group-theoretic vertex factors],

$$\partial_\mu J^{\mu 3}_A = -\frac{e^2 N_c}{8\pi^2} F^{\lambda\rho}\widetilde{F}_{\lambda\rho}\,\mathrm{Tr}(\tfrac{1}{2}\tau^3 Q^2)\,, \tag{5.62}$$

where $Q = \mathrm{diag}(\frac{2}{3}\,,\, -\frac{1}{3})$ is the quark electric charge matrix, $F^{\lambda\rho}$ is the photon field strength tensor, and a color factor of $N_c = 3$ arises by accounting for the implicit color factors at each vertex.[9] Evaluating the trace in eq. (5.62) and writing

[8] In the literature, eq. (5.57) often occurs with the opposite sign due to a sign convention for the Levi-Civita tensor that is opposite to the one employed in this book. Here, we have reproduced the form given in [B64].

[9] Since the photon–quark coupling is diagonal in color, evaluating the triangle graph simply sums over the three possible colors circulating in the loop.

$\alpha \equiv e^2/(4\pi)$, we end up with

$$\partial_\mu J_A^{\mu 3} = -\frac{\alpha N_c}{12\pi} F^{\lambda\rho} \widetilde{F}_{\lambda\rho} \,. \tag{5.63}$$

Thus, due to the presence of the anomaly, the inverse lifetime of the neutral pion (which is dominated by the $\gamma\gamma$ final state) is given by

$$\Gamma(\pi^0 \to \gamma\gamma) = \frac{\alpha^2 N_c^2 m_\pi^3}{576\pi^3 f_\pi^2} \,, \tag{5.64}$$

in good agreement with observation.

5.1.4 Anomalies in the Baryon Number and Lepton Number Currents

As a final example, we examine the anomalous baryon number and lepton number currents in the theory of electroweak interactions For simplicity of notation, we consider a one-generation model. The baryon number current is a vector current given by

$$J_B^\mu = \tfrac{1}{3}\left[u^\dagger \bar{\sigma}^\mu u + d^\dagger \bar{\sigma}^\mu d - \bar{u}^\dagger \bar{\sigma}^\mu \bar{u} - \bar{d}^\dagger \bar{\sigma}^\mu \bar{d}\right] \,, \tag{5.65}$$

following the particle naming conventions of Table 17.1. Consider the process of gauge boson pair production by a baryon number current source. It is convenient to work in the interaction basis of gauge fields, $\{W^{\mu a}\,,\, B^\mu\}$, where $W^{\mu a}$ is an $\mathrm{SU}(2)_L$-triplet of gauge fields and B^μ is a $\mathrm{U}(1)_Y$ hypercharge gauge field. We consider triangle diagrams where one generation of quarks runs in the loop. The external vertices consist of the baryon number current source and the two gauge bosons.

The generators corresponding to the $\mathrm{SU}(2)_L$ gauge boson vertices are given in block-diagonal form by

$$\boldsymbol{T^b} = g \,\mathrm{diag}\left(\frac{\tau^b}{2} \otimes \mathbb{1}_{3\times 3}\,,\; 0\,,\, 0\right) \,, \tag{5.66}$$

where the τ^b are the Pauli matrices, $\mathbb{1}_{3\times 3}$ is the identity matrix in color space, and $\otimes$ is the Kronecker product.[10] We have included a factor of the weak $\mathrm{SU}(2)_L$ coupling g in the definition of $\boldsymbol{T^b}$, since the Feynman rule given by Fig. 5.1 does not explicitly include the gauge coupling. Likewise, the generators corresponding to the $\mathrm{U}(1)_Y$ gauge boson vertices are given in block-diagonal form by (see Table 4.2):

$$\boldsymbol{Y} = g' \,\mathrm{diag}\left(\tfrac{1}{6}\mathbb{1}_{2\times 2} \otimes \mathbb{1}_{3\times 3}\,,\; -\tfrac{2}{3}\mathbb{1}_{3\times 3}\,,\; \tfrac{1}{3}\mathbb{1}_{3\times 3}\right) \,, \tag{5.67}$$

where $\mathbb{1}_{2\times 2}$ is the identity matrix in weak isospin space, and g' is the $\mathrm{U}(1)_Y$ hypercharge gauge coupling. Finally, the generator corresponding to the baryon number current source is given in block-diagonal form by

$$\boldsymbol{B} = \tfrac{1}{3}\mathrm{diag}\left(\mathbb{1}_{2\times 2} \otimes \mathbb{1}_{3\times 3}\,,\; -\mathbb{1}_{3\times 3}\,,\; -\mathbb{1}_{3\times 3}\right) \,. \tag{5.68}$$

[10] The Kronecker product of an $n \times n$ matrix and an $m \times m$ matrix is an $nm \times nm$ matrix. In addition, the following two properties of the Kronecker product are noteworthy (e.g., see [B126]): (i) $(A \otimes B)(C \otimes D) = AC \otimes BD$, and (ii) $\mathrm{Tr}(A \otimes B) = \mathrm{Tr}\, A \;\mathrm{Tr}\, B$.

Consider first the production of two $SU(2)_L$-triplet gauge fields. We put $\boldsymbol{T}^a = \boldsymbol{B}$ and associate the indices b and c with the $SU(2)_L$-triplet gauge bosons. A simple calculation yields

$$D^{Bbc} = g^2 \operatorname{Tr}(\boldsymbol{B}\boldsymbol{T}^b\boldsymbol{T}^c) = \tfrac{1}{2}g^2\delta^{bc}\,, \tag{5.69}$$

where the superscript index B refers to the baryon number current. Since the gauged weak $SU(2)_L$ and hypercharge $U(1)_Y$ currents must be anomaly-free for the mathematical consistency of the electroweak theory, it follows that eqs. (5.30)–(5.32) apply. That is, the symmetrized amplitude for the production of $SU(2)_L$ gauge boson pairs by a baryon number source is anomalous:

$$(p+q)^\mu \mathcal{A}^{Bbc}_{\mu\nu\rho} = -\frac{g^2}{8\pi^2}\delta^{bc}\epsilon_{\nu\rho\kappa\lambda}p^\kappa q^\lambda\,. \tag{5.70}$$

Next, consider the production of two $U(1)_Y$ hypercharge gauge fields. A simple calculation yields

$$D^{BYY} = g'^2 \operatorname{Tr}(\boldsymbol{B}\boldsymbol{Y}^2) = -\tfrac{1}{2}g'^2\,. \tag{5.71}$$

That is, the symmetrized amplitude for the production of $U(1)_Y$ gauge boson pairs by a baryon number source is anomalous:

$$(p+q)^\mu \mathcal{A}^{BYY}_{\mu\nu\rho} = \frac{g'^2}{8\pi^2}\epsilon_{\nu\rho\kappa\lambda}p^\kappa q^\lambda\,. \tag{5.72}$$

Finally, the symmetrized amplitude for the associated production of an $SU(2)_L$-triplet and $U(1)_Y$ hypercharge gauge field exhibits no anomaly as the corresponding $D^{BYc} = gg' \operatorname{Tr}(\boldsymbol{B}\boldsymbol{Y}\boldsymbol{T}^c) = 0$.

The symmetrized amplitudes of the triangle diagrams involving a baryon number current source and a pair of $SU(2)_L$ or $U(1)_Y$ gauge bosons are anomalous. Since the baryon number current is a vector current, we conclude that the source of the anomaly is a VVA triangle diagram in which one of the gauge boson currents is vector (V) and the other gauge boson current is axial vector (A). Nevertheless, the gauge boson axial-vector current must be conserved, as noted above. Hence, the baryon number vector current must be anomalous. In eqs. (5.53)–(5.56), we showed how to derive the operator form of the anomaly equation from the anomalous nonconservation of the symmetrized triangle amplitude. Following the same set of steps starting with eqs. (5.70) and (5.72), one obtains the anomalous nonconservation of the baryon number vector current, in a model with N_g quark generations:

$$\partial_\mu J^\mu_B = \frac{g^2 N_g}{32\pi^2}W^{\lambda\rho b}\widetilde{W}^b_{\lambda\rho} - \frac{g'^2 N_g}{32\pi^2}B^{\lambda\rho}\widetilde{B}_{\lambda\rho}\,, \tag{5.73}$$

where $B_{\lambda\rho}$ and

$$W^b_{\lambda\rho} = \partial_\lambda W^b_\rho - \partial_\rho W^b_\lambda - g\epsilon^{bca}W^c_\lambda W^a_\rho \tag{5.74}$$

are the field strength tensors for the hypercharge $U(1)_Y$ gauge boson and $SU(2)_L$

gauge boson fields, respectively. Note that, for the $\mathrm{SU}(2)_L$ gauge fields W^a_μ,

$$\begin{aligned} W^{\lambda\rho b}\,\widetilde{W}^b_{\lambda\rho} &= 2\epsilon^{\kappa\nu\lambda\rho}\left[(\partial_\kappa W^b_\nu)(\partial_\lambda W^b_\rho) - g\epsilon^{abc}(\partial_\kappa W^a_\nu)W^b_\lambda W^c_\rho\right] \\ &= 2\epsilon^{\kappa\nu\lambda\rho}\partial_\kappa\left[W^b_\nu(\partial_\lambda W^b_\rho) - \tfrac{1}{3}g\epsilon^{abc}W^a_\nu W^b_\lambda W^c_\rho\right]. \end{aligned} \tag{5.75}$$

Strictly speaking, the triangle graphs yield only the terms on the right-hand side of eq. (5.73) that are quadratic in the gauge fields. To obtain the corresponding terms that are cubic in the gauge terms, one must compute the anomalies that arise from VVVA and VAAA box diagrams.

For completeness, we re-express the anomalous nonconservation of the baryon number current in terms of the mass-eigenstate $\mathrm{SU}(2)_L\times\mathrm{U}(1)_Y$ gauge fields:

$$\partial_\mu J^\mu_B = \frac{g^2 N_g}{16\pi^2}W^{\lambda\rho +}\widetilde{W}^-_{\lambda\rho} - \frac{g^2 N_g}{32\pi^2 c^2_W}Z^{\lambda\rho}\widetilde{Z}_{\lambda\rho} - \frac{egN_g}{32\pi^2 c_W}\left[Z^{\lambda\rho}\widetilde{F}_{\lambda\rho} + F^{\lambda\rho}\widetilde{Z}_{\lambda\rho}\right], \tag{5.76}$$

where $c_W \equiv \cos\theta_W$, and $W^\pm_{\lambda\rho}$, $Z_{\lambda\rho}$ and $F_{\lambda\rho}$ are the $W^\pm$, Z and the electromagnetic field strength tensors, respectively. The anomaly in the baryon current in electroweak theory exhibited in eq. (5.76) [or equivalently in eq. (5.73)] has physical consequences. Indeed, it is an essential ingredient in electroweak baryogenesis, in which the baryon–antibaryon asymmetry is generated during the electroweak phase transition in the early universe.

By a similar analysis, one can also compute the anomalous nonconservation of the lepton number current,

$$J^\mu_L = \ell^\dagger\overline{\sigma}^\mu\ell + \nu^\dagger\overline{\sigma}^\mu\nu - \bar{\ell}^\dagger\overline{\sigma}^\mu\bar{\ell}, \tag{5.77}$$

due to triangle diagrams with N_g generations of leptons running in the loop. In the one-generation calculation, the relevant generators are

$$\boldsymbol{T^b} = g\,\mathrm{diag}\left(\frac{\tau^b}{2}\,,\,0\right), \qquad \boldsymbol{Y} = g'\,\mathrm{diag}\left(-\tfrac{1}{2}\mathbb{1}_{2\times 2}\,,\,1\right), \qquad \boldsymbol{L} = \mathrm{diag}\left(\mathbb{1}_{2\times 2}\,,\,-1\right). \tag{5.78}$$

Thus, we end up with

$$D^{Lbc} = \tfrac{1}{2}g^2\delta^{bc}, \qquad D^{LYY} = -\tfrac{1}{2}g'^2, \qquad D^{LYc} = 0. \tag{5.79}$$

Thus, in the Standard Model with N_g generations of quarks and leptons,

$$\partial_\mu J^\mu_L = \partial_\mu J^\mu_B. \tag{5.80}$$

Hence, the $B-L$ current is conserved. However, $B-L$ is *not* anomaly-free, because the lepton number current exhibited in eq. (5.77) has both vector and axial-vector pieces. In particular, the symmetrized amplitude of the triangle diagrams with three lepton number current sources is anomalous. To avoid this anomaly, one can add a right-handed neutrino to the Standard Model. In this case the leptonic current, $J^\mu_L = \ell^\dagger\overline{\sigma}^\mu\ell + \nu^\dagger\overline{\sigma}^\mu\nu - \bar{\ell}^\dagger\overline{\sigma}^\mu\bar{\ell} - \bar{\nu}^\dagger\overline{\sigma}^\mu\bar{\nu}$, is a vector current, eq. (5.80) still holds, and the $B-L$ current is conserved and anomaly-free.

5.2 Gauge Anomalies and Their Cancellation

In Section 5.1, we focused on anomalies that arise in the conservation laws associated with global symmetries. Indeed, quantum field theories that contain anomalous global symmetries are mathematically consistent quantum theories. Moreover, the presence of an anomalous global symmetry can have profound observable consequences, as previously noted.

In contrast to an anomalous global symmetry, a quantum gauge field theory that possesses an anomalous gauge current is mathematically inconsistent. For example, certain Ward identities for Green functions involving gauge fields are no longer valid in the presence of gauge anomalies. However, such Ward identities are crucial for proving the renormalizability of the gauge field theory. In addition, anomalous gauge theories exhibit violations of unitarity in scattering processes, which are generated by higher-order Feynman diagrams that contain anomalous subdiagrams (such as anomalous triangle graphs). Consequently, an anomalous gauge theory cannot be valid to arbitrarily high energies. There must exist a cutoff Λ, above which unitarity is violated. At energies above Λ, the theory must be superseded by a more fundamental theory in which the gauge anomalies are absent.

To avoid this undesirable behavior, we shall demand that our quantum field theories must be totally free of gauge anomalies. This can happen in two different ways. Either there are no gauge anomalies whatsoever, or if present in individual diagrams the gauge anomalies are exactly canceled when all Feynman diagrams are taken into account. In this section, we will examine both possibilities.

5.2.1 Anomaly-Free Gauge Groups

In this subsection, we consider a gauge theory based on a compact semisimple Lie group G. For the present discussion, only the properties of the corresponding Lie algebra $\mathfrak{g}$ are relevant. Since semisimple Lie algebras are direct sums of simple Lie algebras, we first consider all possible compact simple Lie algebras (see Appendix H.1). These include the following Lie algebras of the classical groups:

$$\begin{aligned}&\mathfrak{su}(n)\ [\text{for } n \geq 2], \quad \mathfrak{so}(2n)\ [\text{for } n \geq 4], \quad \mathfrak{so}(2n+1)\ [\text{for } n \geq 3],\\ &\mathfrak{sp}(n)\ [\text{for } n \geq 2]\,.\end{aligned} \tag{5.81}$$

In addition, there are five exceptional Lie groups, G_2, F_4, E_6, E_7, and E_8, whose corresponding Lie algebras will be denoted by the same symbols.

The results of Section 5.1.1 imply that the triangle anomaly vanishes if

$$D^{abc}(R) \equiv \tfrac{1}{2}\mathrm{Tr}[\{\boldsymbol{T}^a_{\boldsymbol{R}}, \boldsymbol{T}^b_{\boldsymbol{R}}\}\boldsymbol{T}^c_{\boldsymbol{R}}] = 0\,, \tag{5.82}$$

where the $\boldsymbol{T}^a_{\boldsymbol{R}}$ are the generators of the Lie algebra $\mathfrak{g}$ in the irreducible representation R, corresponding to the representation of the fermions that appear as internal lines in the triangle diagram. In Appendix H.3, it is shown that $D^{abc}(R) = I_3(R)d^{abc}$, where $I_3(R)$ is the cubic index of the representation R and d^{abc} is obtained by evaluating D^{abc} in the fundamental representation, up to an overall constant that is fixed by adopting the convention in which the generators of the fundamental representation are normalized according to $\text{Tr}(\boldsymbol{T}^a_{\boldsymbol{F}}\boldsymbol{T}^b_{\boldsymbol{F}}) = I_2(F)\delta^{ab}$ and the cubic index in the fundamental representation is given by $I_3(F) = \frac{1}{2}I_2(F)$. This motivates the definition of the anomaly coefficient of the representation R,

$$A(R) \equiv \frac{2I_3(R)}{I_2(F)}, \tag{5.83}$$

which has the property that $A(F) = 1$. Following eq. (H.40), we shall employ the conventional choice of $I_2(F) = \frac{1}{2}$. If a nonvanishing d^{abc} exists, then one can construct a cubic Casimir operator, $(\boldsymbol{T}^3_{\boldsymbol{R}})_i{}^j \equiv d^{abc}(\boldsymbol{T}^a_{\boldsymbol{R}}\boldsymbol{T}^b_{\boldsymbol{R}}\boldsymbol{T}^c_{\boldsymbol{R}})_i{}^j = C_{3R}\delta_i{}^j$, that commutes with all the elements of the Lie algebra $\mathfrak{g}$. Note that C_{3R} is proportional to the cubic index [see eq. (H.53)].

The anomaly coefficients of the direct sum and the tensor product of two representations R_1 and R_2 [see eq. (H.49)] are respectively given by

$$A(R_1 \oplus R_2) = A(R_1) + A(R_2), \tag{5.84}$$
$$A(R_1 \otimes R_2) = d_{R_1}A(R_2) + d_{R_2}A(R_1). \tag{5.85}$$

In the special case of $R_1 = R_2 = R$, one can consider the symmetric and antisymmetric parts of the direct product representation $R \otimes R$, denoted respectively by $(R \otimes R)_S$ and $(R \otimes R)_A$. Then, one can show that[11]

$$A[(R \otimes R)_S] = (d_R + 4)A(R), \qquad A[(R \otimes R)_A] = (d_R - 4)A(R), \tag{5.86}$$

where d_R is the dimension of the representation R. In addition, the anomaly coefficient of a complex conjugate representation $\overline{R}$ is simply related to the anomaly coefficient of the representation R [see eq. (H.47)]:

$$A(\overline{R}) = -A(R). \tag{5.87}$$

In particular, the anomaly of a self-conjugate representation vanishes. For example, since the adjoint representation of any compact Lie algebra is a real representation, it follows that its anomaly coefficient vanishes:

$$A(R = A) = 0. \tag{5.88}$$

One can prove that the representations of the following compact simple Lie algebras are always real or pseudoreal:

$$\begin{aligned}&\mathfrak{su}(2), \quad \mathfrak{so}(4n) \ [\text{for } n \geq 2], \quad \mathfrak{so}(2n+1) \ [\text{for } n \geq 3],\\ &\mathfrak{sp}(n) \ [\text{for } n \geq 2], \mathrm{G}_2, \mathrm{F}_4, \mathrm{E}_6, \mathrm{E}_7, \mathrm{E}_8.\end{aligned} \tag{5.89}$$

In light of eq. (5.87), it immediately follows that the anomaly coefficients vanish for these Lie algebras, and the corresponding gauge theories are anomaly-free. Although the Lie algebras $\mathfrak{so}(4n+2)$ for $n \geq 2$ and E_6 possess complex representations, they

[11] See Section 8.6 of [B121] and Exercise H.4 in Appendix H.

do not possess a cubic Casimir operator. Hence the anomaly coefficients vanish for these Lie algebras as well, and the corresponding gauge theories are also anomaly-free. This leaves $\mathfrak{su}(n)$ [for $n \geq 3$] as the only possible compact simple Lie algebras with nonvanishing anomaly coefficients.

Semisimple Lie algebras are direct sums of simple Lie algebras. Consider the compact semisimple Lie algebra $\mathfrak{g}_1 \oplus \mathfrak{g}_2$. The anomaly calculation leads to the consideration of triangle diagrams with external gauge bosons of the corresponding gauge groups G_1 and G_2. The fermions that run inside the loop are representations of $\mathfrak{g}_1 \oplus \mathfrak{g}_2$ and thus can be separately charged under both gauge groups. The mixed triangle graphs in which external gauge bosons from both gauge groups appear must vanish, since they will be proportional to a $\operatorname{Tr} \boldsymbol{T}^a_{\boldsymbol{R}} = 0$ (where $\boldsymbol{T}^a_{\boldsymbol{R}}$ is the generator corresponding to the gauge group whose gauge boson appears once as an external leg of the triangle). Thus, the only triangle graphs that can survive are those in which all external gauge bosons belong to either gauge group G_1 or G_2. That is, if both $\mathfrak{g}_1$ and $\mathfrak{g}_2$ are anomaly-free then $\mathfrak{g}_1 \oplus \mathfrak{g}_2$ is anomaly-free. Conversely, if either $\mathfrak{g}_1$ or $\mathfrak{g}_2$ is the Lie algebra $\mathfrak{su}(n)$ [for $n \geq 3$], then $\mathfrak{g}_1 \oplus \mathfrak{g}_2$ may yield a nonvanishing gauge anomaly, depending on the choice of representation of the fermion that can circulate inside the triangle diagram.

The most general compact Lie algebra is a direct sum of the form

$$\mathfrak{g} \oplus \underbrace{\mathfrak{u}(1) \oplus \cdots \oplus \mathfrak{u}(1)}_{k \text{ times}}, \tag{5.90}$$

where $\mathfrak{g}$ is a compact semisimple Lie algebra (which can be expressed as a direct sum of compact simple Lie algebras), and $\mathfrak{u}(1)$ is the Lie algebra of U(1), where k is a nonnegative integer. The Lie algebra $\mathfrak{u}(1)$ possesses a single generator (denoted by Q) that commutes with all the generators of $\mathfrak{g}$. First, in a U(1) gauge theory, it is possible that $\operatorname{Tr} Q^3 \neq 0$, in which case the corresponding triangle diagram with three external U(1) gauge fields is nonvanishing. Such a gauge theory would be anomalous. Next, in the case of the Lie algebra $\mathfrak{g} \oplus \mathfrak{u}(1)$ where $\mathfrak{g}$ is simple, three classes of triangle diagrams may be nonvanishing. In particular, if $\mathfrak{g} = \mathfrak{su}(n)$ [for $n \geq 3$], then a triangle diagram with three external G-gauge bosons can be nonvanishing. Likewise, a triangle with three external U(1) gauge bosons can be nonvanishing if $\operatorname{Tr} Q^3 \neq 0$. Finally, the mixed triangle diagram with one external U(1) gauge boson and two external G-gauge bosons can be nonvanishing if $\operatorname{Tr} Q \neq 0$.[12] This last result follows after employing $\operatorname{Tr}(\boldsymbol{T}^a_{\boldsymbol{R}} \boldsymbol{T}^b_{\boldsymbol{R}}) = I_2(R)\delta^{ab}$ [see eq. (H.32)]. Finally, generalizing to the case of $\mathfrak{g} \oplus \mathfrak{u}(1)$, where $\mathfrak{g}$ is semisimple, is straightforward.

To summarize, we have identified all possible gauge theories that are anomalous due to the nonvanishing of the triangle diagram with three external gauge bosons. Among the compact simple Lie algebras, only $\mathfrak{su}(n)$ [for $n \geq 3$] with fermions in an irreducible complex representation can possess a nonvanishing anomaly coefficient. Finally, gauge groups with compact Lie algebras that contain either one or more $\mathfrak{u}(1)$ factors and/or an $\mathfrak{su}(n)$ factor for $n \geq 3$ are potentially anomalous.

12 Note that the triangle graph with two external U(1) gauge bosons and one external G-gauge boson must vanish since $\operatorname{Tr} \boldsymbol{T}^a_{\boldsymbol{R}} = 0$.

5.2.2 Cancellation of Gauge Anomalies

A mathematically consistent gauge theory must be anomaly-free. Thus, in the potentially anomalous cases identified above, one can formulate a consistent gauge theory if the anomalies are canceled. This can be achieved if the two-component fermions that are charged under the gauge group are either members of a real representation of the Lie algebra or members of a complex *reducible* representation of the Lie algebra suitably chosen such that the anomaly coefficient of the reducible representation is zero. We shall exhibit both mechanisms below.

First, consider a U(1) gauge theory with one two-component fermion of charge $q = 1$ (in units of the positron charge). This theory is anomalous since $\operatorname{Tr} Q^3 = 1$. If one adds a second two-component fermion of charge -1, then $\operatorname{Tr} Q^3 = 0$ and the corresponding gauge theory is anomaly-free. Of course, we recognize this theory as QED, which possesses a charge-conjugation symmetry $\mathcal{C}$. The one-photon state is an eigenstate of $\mathcal{C}$ with eigenvalue -1. Hence, in this theory, Green functions with an odd number of external photon fields must vanish (this is known as Furry's Theorem). Thus, the triangle diagram with three external photons vanishes by Furry's Theorem, so that this theory is trivially free of gauge anomalies. Perhaps a less trivial example is a theory with three two-component fermions with charges $q_i = 1$, 1, and $-\sqrt[3]{2}$, which is a chiral gauge theory that satisfies

$$\operatorname{Tr} Q^3 = \sum_i q_i^3 = 0\,. \tag{5.91}$$

Here, the triangle diagram with three external U(1) gauge bosons is proportional to $\operatorname{Tr} Q^3 = 0$, which vanishes. Thus the potential gauge anomalies have canceled.

It has been argued that a consistent gauge theory in the presence of gravity must be free of both gauge and gravitational anomalies [48]. By considering the triangle diagram with two external gravitons and one U(1) gauge boson with generator Q, an additional constraint arises: $\operatorname{Tr} Q = 0$ (since gravitons couple universally to all fermions), which ensures that the mixed gauge–gravitational anomaly vanishes.[13] With this additional constraint in mind, consider as an example a U(1) gauge theory with five two-component fermions with charges $q_i = 1, 5, -7, -8$, and 9, respectively. It is straightforward to check that, in this case, $\operatorname{Tr} Q = \operatorname{Tr} Q^3 = 0$. Thus, such a chiral gauge theory is free of gauge anomalies and of the mixed gauge–gravitational anomaly. Additional examples can be found in Refs. [49–51].

The last example given above seems rather artificial. There appears to be no particular reason why the theory with five two-component fermions with the charges as specified above should be special. That $\operatorname{Tr} Q = \operatorname{Tr} Q^3 = 0$ appears to be a numerical accident. However, there are other examples of anomaly-free gauge theories, in which the cancellation mechanism can be understood as the consequence of a deeper principle. The classic example where this occurs is the Standard Model. In this case, the $\mathrm{SU}(3)_C\times\mathrm{SU}(2)_L\times\mathrm{U}(1)_Y$ gauge theory is anomalous for a generic

[13] Note that in nonabelian gauge theories, the mixed gauge–gravitational anomaly automatically vanishes since $\operatorname{Tr} \boldsymbol{T}^a = 0$.

choice of fermionic matter. However, in the Standard Model, the two-component fermions comprise a reducible representation of the Lie algebra $\mathfrak{su}(3)\oplus\mathfrak{su}(2)\oplus\mathfrak{u}(1)$. In particular, the gauge quantum numbers of the two-component fermions of one generation of quarks and leptons, $q=(u,d)$, $\bar{u}$, $\bar{d}$, $\ell=(\nu,e)$, $\bar{e}$ are given by

$$(\mathbf{3},\mathbf{2},\tfrac{1}{6})\oplus(\overline{\mathbf{3}},\mathbf{1},-\tfrac{2}{3})\oplus(\overline{\mathbf{3}},\mathbf{1},\tfrac{1}{3})\oplus(\mathbf{1},\mathbf{2},-\tfrac{1}{2})\oplus(\mathbf{1},\mathbf{1},1). \tag{5.92}$$

Let us verify that the Standard Model with complete generations is free of gauge anomalies. We consider triangle diagrams with three external gauge bosons. Since $\mathfrak{su}(2)$ has no complex representations, the triangle diagrams with three external SU(2) gauge bosons vanish after summing over all the internal fermions as specified by the reducible representation given in eq. (5.92).

Consider next the triangle diagram with three external SU(3) gauge bosons (the gluons). Since there are two color triplets and two color antitriplets (the color singlets do not contribute), the SU(3) anomaly coefficients add up to zero in light of eq. (5.87). Next, triangle diagrams with exactly one external SU(3) or SU(2) gauge boson vanish exactly since $\operatorname{Tr}\boldsymbol{T}^a=0$, where $\boldsymbol{T}^a$ is the generator of $\mathfrak{su}(3)$ or $\mathfrak{su}(2)$, respectively. Next, the triangle diagram with one external hypercharge gauge boson and two SU(3) gauge bosons vanishes, since[14]

$$\operatorname{Tr}(\boldsymbol{T}^a\boldsymbol{T}^b Y)=\tfrac{1}{2}\delta^{ab}\left[2\left(\tfrac{1}{6}\right)-\tfrac{2}{3}+\tfrac{1}{3}\right]=0\,. \tag{5.93}$$

Likewise, the triangle diagram with one external hypercharge gauge boson and two SU(2) gauge bosons vanishes since, e.g.,

$$\operatorname{Tr}(\boldsymbol{T_3}^2\boldsymbol{Y})=\left(\tfrac{1}{2}\right)^2\left[3\left(\tfrac{1}{6}\right)-\tfrac{1}{2}\right]=0\,. \tag{5.94}$$

Finally, the triangle diagrams with three external hypercharge (Y) gauge bosons also vanish because

$$\operatorname{Tr}\boldsymbol{Y}^3=6\left(\tfrac{1}{6}\right)^3+3\left(-\tfrac{2}{3}\right)^3+3\left(\tfrac{1}{3}\right)^3+2\left(-\tfrac{1}{2}\right)^3+1(1)^3=0\,. \tag{5.95}$$

Moreover, the mixed gauge–gravitational anomalies cancel after noting that

$$\operatorname{Tr}\boldsymbol{Y}=6\left(\tfrac{1}{6}\right)+3\left(-\tfrac{2}{3}\right)+3\left(\tfrac{1}{3}\right)+2\left(-\tfrac{1}{2}\right)+1(1)=0\,. \tag{5.96}$$

Thus, we have confirmed that the SU(3)×SU(2)×U(1) gauge theory with the two-component fermions in the reducible representation given by eq. (5.92) is free of gauge and mixed gauge–gravitational anomalies.

It may appear that the cancellation of gauge and mixed gauge–gravitational anomalies by virtue of eqs. (5.93)–(5.96) is a numerical accident. However, in this case, there is a deeper principle at play. Consider a grand unified gauge theory with gauge group SO(10) and fermions that transform under an irreducible 16-dimensional spinor representation of the $\mathfrak{so}(10)$ Lie algebra. The gauge bosons transform under the 45-dimensional adjoint representation of $\mathfrak{so}(10)$. One can decompose any irreducible representation of $\mathfrak{so}(10)$ into a reducible representation of

[14] If $\boldsymbol{T}^a$ is the generator of the color triplet, then $-(\boldsymbol{T}^a)^{\mathsf{T}}$ is the generator of the color antitriplet. It then follows that $\operatorname{Tr}(\boldsymbol{T}^a\boldsymbol{T}^b)=\operatorname{Tr}\{(-\boldsymbol{T}^{a\mathsf{T}})(-\boldsymbol{T}^{b\mathsf{T}})\}=\tfrac{1}{2}\delta^{ab}$.

the subalgebra $\mathfrak{su}(3) \oplus \mathfrak{su}(2) \oplus \mathfrak{u}(1)$. Using the tables given in Appendix H.5.3, we can establish the following branching rules:

$$\mathbf{45} \longrightarrow (\mathbf{8},\mathbf{1},0) + (\mathbf{3},\mathbf{1},0) + (\mathbf{1},\mathbf{1},0) + (\mathbf{3},\mathbf{2},-\tfrac{5}{6}) + (\overline{\mathbf{3}},\mathbf{2},\tfrac{5}{6}) + \cdots\,, \tag{5.97}$$

$$\mathbf{16} \longrightarrow (\mathbf{3},\mathbf{2},\tfrac{1}{6}) + (\overline{\mathbf{3}},\mathbf{1},-\tfrac{2}{3}) + (\overline{\mathbf{3}},\mathbf{1},\tfrac{1}{3}) + (\mathbf{1},\mathbf{2},-\tfrac{1}{2}) + (\mathbf{1},\mathbf{1},1) + (\mathbf{1},\mathbf{1},0)\,. \tag{5.98}$$

We recognize that the adjoint representation of $\mathfrak{so}(10)$ contains within it states with the quantum numbers of the Standard Model gauge bosons. Moreover, the spinor representation of $\mathfrak{so}(10)$ contains within it states with the quantum numbers of one generation of two-component fermions of the Standard Model [see eq. (5.92)] plus one additional two-component fermion that is neutral with respect to the Standard Model gauge group. Given that an SO(10) gauge theory is anomaly-free as previously indicated, and noting that a gauge-singlet two-component fermion contributes nothing to a triangle diagram with external gauge bosons, it immediately follows that an $\mathrm{SU}(3)_C{\times}\mathrm{SU}(2)_L{\times}\mathrm{U}(1)_Y$ gauge theory with generation fermions that transform as the reducible representation given in eq. (5.92) must be anomaly-free [i.e., eqs. (5.93)–(5.95) are satisfied]. Moreover, the hypercharge $\boldsymbol{Y}$ has been promoted to a generator of $\mathfrak{so}(10)$ which must be traceless. Removing the $(\mathbf{1},\mathbf{1},0)$ two-component fermion[15] from the decomposition of the 16-dimensional representation of $\mathfrak{so}(10)$ shown in eq. (5.98) yields one generation of two-component fermions of the Standard Model that satisfies eq. (5.96). In light of the above observations, there is strong motivation to consider the possibility of grand unified models, in which the Standard Model gauge group is ultimately unified into an anomaly-free compact simple gauge group. This will be addressed further in Section 6.3.

In Chapter 13, we will examine the minimal supersymmetric extension of the Standard Model (MSSM). In this model, new fermions are introduced that are superpartners of the gauge and Higgs bosons. Thus, the corresponding reducible representation specified by eq. (5.92) under which the fermions transform will have to be enlarged. The enlarged representation will include gauginos and higgsinos, which transform the same way as the gauge bosons and Higgs bosons, respectively. Since the gauginos transform as an adjoint representation of the gauge group, the contributions of the gauginos to the triangle diagrams with three external gauge bosons vanish exactly in light of eq. (5.88). In contrast, the higgsinos present a potential problem. Since the Higgs boson transforms as $(\mathbf{1},\mathbf{2},\frac{1}{2})$ under $\mathrm{SU}(3)_C{\times}\mathrm{SU}(2)_L{\times}\mathrm{U}(1)_Y$, if one were to add a two-component fermion state with these quantum numbers to the Standard Model, one would find

$$\operatorname{Tr}\boldsymbol{Y} = (\operatorname{Tr}\boldsymbol{Y})_{\rm SM} + 2\left(\tfrac{1}{2}\right) = 1\,, \tag{5.99}$$

$$\operatorname{Tr}\boldsymbol{Y}^3 = (\operatorname{Tr}\boldsymbol{Y}^3)_{\rm SM} + 2\left(\tfrac{1}{2}\right)^3 = \tfrac{1}{4}\,. \tag{5.100}$$

Thus, such a gauge theory exhibits both a gauge anomaly and a mixed gauge–

[15] It is tempting to interpret this $\mathrm{SU}(3)_C{\times}\mathrm{SU}(2)_L{\times}\mathrm{U}(1)_Y$ gauge neutral state as a right-handed neutrino. Indeed, such a state appears in the seesaw-extended Standard Model which has been proposed to explain the existence of nonzero neutrino masses and mixing. Further details of this model can be found in Section 6.1.

gravitational anomaly. The MSSM avoids this problem by introducing a second Higgs doublet to the Standard Model of opposite hypercharge. Hence, the fermion mass spectrum contains a second higgsino state that transforms as $(\mathbf{1},\mathbf{2},-\frac{1}{2})$. In this case,

$$\mathrm{Tr}\,\boldsymbol{Y} = (\mathrm{Tr}\,\boldsymbol{Y})_{\mathrm{SM}} + 2\left(\tfrac{1}{2}\right) + 2\left(-\tfrac{1}{2}\right) = 0\,, \tag{5.101}$$

$$\mathrm{Tr}\,\boldsymbol{Y}^3 = (\mathrm{Tr}\,\boldsymbol{Y}^3)_{\mathrm{SM}} + 2\left(\tfrac{1}{2}\right)^3 + 2\left(-\tfrac{1}{2}\right)^3 = 0\,, \tag{5.102}$$

and the cancellation of gauge (and mixed gravitational-gauge) anomalies is restored. Indeed, this is not surprising, since the two higgsino states now constitute a real $(\mathbf{1},\mathbf{2},\frac{1}{2})\oplus(\mathbf{1},\mathbf{2},-\frac{1}{2})$ reducible representation of $\mathrm{SU}(3)_C\times\mathrm{SU}(2)_L\times\mathrm{U}(1)_Y$.

5.3 Discrete Gauge Anomalies

So far we have discussed both global and local continuous symmetries and their anomalies. Here we consider the anomalies of discrete symmetries. This might seem surprising. However, a discrete symmetry can arise as a remnant of a spontaneously broken gauge symmetry. The anomaly cancellation conditions for the gauge symmetry lead to meaningful constraints on the remnant discrete symmetry, which we discuss here, and also in Section 16.4 for the case of the MSSM.

5.3.1 Discrete Gauge Symmetries

When constructing models beyond the SM, new discrete symmetries are often employed to restrict the allowed set of operators. Global symmetries can be violated by quantum gravity effects, while gauge symmetries should be immune to black hole evaporation or wormhole tunneling (e.g., see Ref. [52]). Thus if we observe a (global) discrete symmetry at low energy in the laboratory, we would expect it to be protected from quantum gravity effects, only if at the fundamental level it originates from a spontaneously broken gauge symmetry.

In order to see how a discrete symmetry may arise dynamically, let us consider a $\mathrm{U}(1)_X$ gauge symmetry and the two-component fermion fields ψ_i, with $\mathrm{U}(1)_X$ charges X_i. Under a gauge transformation, $\psi_i \to e^{i\alpha(x)X_i}\psi_i$, where the index i labels the particle species. We further introduce a complex scalar field Φ with positive integer $\mathrm{U}(1)_X$ charge $X_\Phi = N$.[16] The Lagrangian can contain gauge invariant Yukawa and nonrenormalizable interactions of these fields. For example,[17]

$$-\mathcal{L}_{\mathrm{int}} \supset Y^{ij}\Phi\psi_i\psi_j + \frac{c}{M_\Phi^K}\Phi^k\prod_i \psi_i^{k_i} + \mathrm{h.c.}, \tag{5.103}$$

where the Y^{ij} are a set of dimensionless Yukawa couplings, M_Φ is a mass scale,

[16] As long as all charges are rational, we can rescale the $\mathrm{U}(1)_X$ generator such that all charges are integers.

[17] To avoid a cluttered notation we have included only one exemplary nonrenormalizable term.

and c is a dimensionless constant. In eq. (5.103), the positive integers k and k_i denote the power of the respective field in the operator. Note that $\sum_i k_i$ is an even integer (since the ψ_i are fermions), which implies that the dimension of the nonrenormalizable interaction in eq. (5.103), $K \equiv k + \frac{3}{2}\sum k_i - 4$, is an integer. The $\mathrm{U}(1)_X$ charges for these gauge invariant terms satisfy

$$N + X_i + X_j = 0\,, \quad \text{and} \quad kN + \sum k_i X_i = 0\,. \tag{5.104}$$

We next consider the $\mathrm{U}(1)_X$ symmetry to be spontaneously broken by a vacuum expectation value for the Φ field at the energy scale M_Φ, with $M_\Phi \gg m_W$. We can integrate out the heavy Φ field, and at low energy the terms in eq. (5.103) become a set of mass terms and an effective interaction for which the charges now satisfy

$$X_i + X_j = 0 \mod N\,, \qquad \sum_i k_i X_i = 0 \mod N\,. \tag{5.105}$$

Thus, from the low energy point of view only the charges modulo N are relevant. The low energy operators are now invariant under the discrete transformation of the fields, $\psi_i \to e^{2\pi i X_i/N}\psi_i$. That is, we have gone from a continuous $\mathrm{U}(1)_X$ gauge symmetry to a discrete $\mathbb{Z}_N$ symmetry at low energy. Discrete symmetries arising by such a mechanism are called *discrete gauge symmetries* (DGS).

5.3.2 Discrete Anomaly Cancellation Conditions

We next determine the constraints on the low energy discrete symmetry,[18] which arise from the anomaly cancellation conditions of the original $\mathrm{U}(1)_X$. We consider the SM gauge group $\mathrm{SU}(3)_C\times\mathrm{SU}(2)_L\times\mathrm{U}(1)_Y$ extended by a $\mathrm{U}(1)_X$ gauge symmetry, which is spontaneously broken to a discrete symmetry $\mathbb{Z}_N$ at the scale M_X. We demand that $\mathrm{U}(1)_X$ be free of gauge anomalies and free of mixed gauge–gravitational anomalies [see eq. (5.82)]:

$$D^{abc}_{AA'A''} = \mathrm{Tr}[\{\boldsymbol{T}^a_{\boldsymbol{A}}, \boldsymbol{T}^b_{\boldsymbol{A'}}\}\boldsymbol{T}^c_{\boldsymbol{A''}}] = 0\,. \tag{5.106}$$

In eq. (5.106), $\boldsymbol{T}^a_{\boldsymbol{A}}$, $\boldsymbol{T}^b_{\boldsymbol{A'}}$, and $\boldsymbol{T}^c_{\boldsymbol{A''}}$ are generators of the groups that appear in the extended SM gauge group (or the identity matrix in the case of the gravitational interaction), with $A, A', A'' \in \{C, W, Y, X, G\}$, corresponding to the gauge groups $\mathrm{SU}(3)_C$, $\mathrm{SU}(2)_L$, $\mathrm{U}(1)_Y$, and $\mathrm{U}(1)_X$, respectively, and G corresponding to the gravitational interaction. As in footnote 16, one can rescale the $\mathrm{U}(1)_X$ generator such that the $\mathrm{U}(1)_X$ charges of the SM fermion fields are integers. The discrete integer $\mathbb{Z}_N$ charges, q_i, can be related to the integer $\mathrm{U}(1)_X$ charges, X_i, via a modulo N shift:

$$X_i = q_i + m_i N, \qquad q_i,\, m_i \in \mathbb{Z}. \tag{5.107}$$

For simplicity, for the SM fermions, we here only work with generation-independent charges q_i. For convenience, we do not require the q_i to lie in the interval $[0, N-1]$. We have introduced arbitrary integers m_i to re-express the mod N condition.

[18] As shown in Ref. [53], all valid constraints on such symmetries can be understood in the low-energy theory in terms of instantons.

Beyond the SM fermions, the extended model may contain fermions that receive their mass at the high-scale M_X. We distinguish between Dirac and Majorana fermions. For a Dirac mass term we require two 2-component fermion fields, with $\mathrm{U}(1)_X$ charges X^i_{D1} and X^i_{D2}. For the mass term to be $\mathbb{Z}_N$ invariant, we must have

$$X^j_{D1} + X^j_{D2} = p_j N, \qquad p_j \in \mathbb{Z}\,. \tag{5.108}$$

Here we have again expressed the mod N conditions in terms of an arbitrary set of integers p_j. From a low energy perspective, X^j_{D1} and X^j_{D2} are unknown and are not necessarily integer valued.[19] However, they must pair up to an integer multiple of N. In contrast, a Majorana fermion pairs with itself to form a mass term. For its $\mathrm{U}(1)_X$ charge $X^{j'}_M$ we have the condition

$$2X^{j'}_M = p'_{j'} N, \qquad p'_{j'} \in \mathbb{Z}\,. \tag{5.109}$$

Note that the $X^{j'}_M$ are either half-integer or integer valued, i.e., $2X^{j'}_M$ must be an integer. Thus for N odd, the $p_{j'}$ in eq. (5.109) must be even. Even though their charges are unknown, the heavy fields may contribute to the anomaly conditions.

5.3.3 The Linear Anomaly Constraints

In order to determine the constraints on the $\mathbb{Z}_N$ charges, consider first the mixed anomaly condition where only one of the generators in eq. (5.106) is from the gauge group $\mathrm{U}(1)_X$, which yields constraints that are linear in the $\mathrm{U}(1)_X$ charges. Starting with $D_{CCX} = 0$, we replace all $\mathrm{U}(1)_X$ charges through their shifted expressions, as in eq. (5.107), and employ eq. (5.108) for the Dirac fields. We then obtain

$$\sum_{i=3,\bar{3}} q_i = -N \left[\sum_{i=3,\bar{3}} m_i + \sum_{i=3,\bar{3}} p_j \right]\,. \tag{5.110}$$

There are no Majorana fermion contributions with $\mathrm{SU}(3)_C$ charges, and for simplicity we have assumed that the relevant fermions are only in the fundamental (**3**) or antifundamental ($\bar{\mathbf{3}}$) representation of $\mathrm{SU}(3)_C$. One can in principle also include fermionic fields in nontrivial $SU(3)$ representations. Similarly for $\mathrm{SU}(2)_L$, the anomaly cancellation condition $D_{WWX} = 0$ yields

$$\sum_{i=2} q_i = -N \left[\sum_{i=2} m_i + \sum_{i=2} p_j \right], \tag{5.111}$$

again with a restriction (for simplicity) to the fundamental representation. Note that the Majorana fermions do not contribute. We defer the case D_{YYX} to the next subsection. As a third linear condition we obtain for the gravity case, $D_{GGX} = 0$,

$$\sum_i q_i = -N \left[\sum_i m_i + \sum_j p_j + \tfrac{1}{2} \sum_{j'} p'_{j'} \right]. \tag{5.112}$$

[19] Because the $\mathrm{U}(1)_X$ charges of the new fermions are unknown, one does not know how to rescale the definition of the $\mathrm{U}(1)_X$ generator to render all $\mathrm{U}(1)_X$ charges integer valued.

The sums on the right-hand side above run over all the fermion fields in the theory, including the heavy Dirac and Majorana fermions.

The heavy fermion content of the theory is unknown. Thus from the low energy point of view, the square brackets in eqs. (5.110) and (5.111) can take on arbitrary integer values. Hence, we can rewrite these two conditions as

$$\sum_{i=\mathbf{3},\bar{\mathbf{3}}} q_i = 0 \mod N\,, \qquad \sum_{i=\mathbf{2}} q_i = 0 \mod N\,. \tag{5.113}$$

The square bracket in eq. (5.112) is more involved. The first two sums can again lead to arbitrary integers. If N is odd, then the $p'_{j'}$ are even, and the third sum is also an arbitrary integer. Thus for N odd we have

$$\sum_i q_i = 0 \mod N\,, \quad \text{for } N \text{ odd}\,. \tag{5.114}$$

However, if N is even, i.e., $N' \equiv N/2$ is an integer, then the $p'_{j'}$ are arbitrary integers. One can then write

$$\sum_i q_i = -N' \left[2\sum_i m_i + 2\sum_j p_j + \sum_{j'} p'_{j'} \right]\,, \tag{5.115}$$

and $\sum_i q_i = 0 \bmod N'$. Overall, we therefore have

$$\sum_i q_i = \begin{cases} 0 \mod N, & \text{for } N \text{ odd,} \\ 0 \mod N', & \text{for } N \text{ even and } N' \equiv N/2\,. \end{cases} \tag{5.116}$$

Equations (5.113) and (5.116) provide a necessary set of conditions on the low energy fields of any discrete symmetry to be discrete gauge anomaly-free.

5.3.4 The Purely Abelian Anomalies

We next consider the anomaly cancellation conditions for abelian gauge groups. It is convenient to first rescale the $\mathrm{U}(1)_Y$ generator of the Standard Model by a factor of 6 such that the $\mathrm{U}(1)_Y$ hypercharges of all SM fermions are integers. The new hypercharges are given by

$$(Y_Q,\, Y_{\bar{u}},\, Y_{\bar{d}},\, Y_L,\, Y_{\bar{e}}) = (1,\, -4,\, 2,\, -3,\, 6)\,. \tag{5.117}$$

Employing the shift given by eq. (5.107) and using eq. (5.108), the condition $D_{YYX} = 0$ yields

$$\sum_i Y_i^2 q_i = -N \left[\sum_i Y_i^2 m_i + \sum_j (Y^j_{D1})^2 p_j \right]\,. \tag{5.118}$$

The sum over i runs over all light fields with nontrivial $\mathrm{U}(1)_Y$ *and* $\mathrm{U}(1)_X$ charges. The sum over j runs over all heavy Dirac fermions with nontrivial $\mathrm{U}(1)_Y$ *and* $\mathrm{U}(1)_X$ charges, and we have used $Y^j_{D1} = -Y^j_{D2}$ and $Y^j_M = 0$. In light of eq. (5.117), the Y_i^2 are integers and thus the left-hand side of eq. (5.118) is an integer. Unless we consider a specific model, the Y_{D1} are the unknown hypercharges of the heavy Dirac fermions. In particular, they need not be integers, even after the rescaling of the weak hypercharge specified in eq. (5.117). In general, the bracketed expression on the right-hand side of eq. (5.118) is not an integer. Thus, eq. (5.118) poses no constraint from the low energy point of view. If we construct an explicit ultraviolet complete model with given $\mathrm{U}(1)_Y$ and $\mathrm{U}(1)_X$ charges, then this equation can impose a significant constraint.

Similarly, for $D_{YXX} = 0$ and $D_{XXX} = 0$ we do not obtain new model-independent constraints on the low energy fermions.

5.3.5 Example: Standard Model with Massive Neutrinos

As an example we consider the Standard Model below the scale of electroweak breaking but with neutrino masses. We investigate the two cases of Dirac and Majorana masses separately. We assume the SM gauge group has been extended by a $\mathrm{U}(1)_X$ which is spontaneously broken at a high energy scale, with a remnant discrete $\mathbb{Z}_N$ symmetry. Besides the anomaly constraints, we shall impose $\mathbb{Z}_N$ invariance on the terms that appear in the Lagrangian. We assume $\mathrm{SU}(2)_L \times \mathrm{U}(1)_Y \to \mathrm{U}(1)_{\mathrm{EM}}$ has been broken spontaneously and the relevant mass terms have been diagonalized.

Dirac Neutrino Masses

To obtain Dirac neutrino masses, we add three singlet neutrinos $\bar{\nu}_i$ to the SM field content. We introduce a notation for the discrete $\mathbb{Z}_N$ charges of the fermions:

$$\vec{q}_i \equiv (q_{Q_i}, q_{\bar{u}_i}, q_{\bar{d}_i}, q_{L_i}, q_{\bar{e}_i}, q_{\bar{\nu}_i}) , \tag{5.119}$$

where $i \in \{1, 2, 3\}$ is the generation index. The requirement that all fermion mass terms are $\mathbb{Z}_N$ invariant is expressed in terms of the $\mathbb{Z}_N$ charges as

$$\begin{aligned} q_{Q_i} + q_{\bar{d}_i} &= 0 \mod N , \qquad & q_{Q_i} + q_{\bar{u}_i} &= 0 \mod N , \\ q_{L_i} + q_{\bar{e}_i} &= 0 \mod N , \qquad & q_{L_i} + q_{\bar{\nu}_i} &= 0 \mod N . \end{aligned} \tag{5.120}$$

Solving these 12 equations, we are left with 6 independent charges, which we choose as q_{Q_i}, q_{L_i}. The discrete charge vector is then

$$\vec{q}_i = (q_{Q_i}, -q_{Q_i}, -q_{Q_i}, q_{L_i}, -q_{L_i}, -q_{L_i}) . \tag{5.121}$$

Thus any low energy effective discrete symmetry allowing for the Dirac neutrino masses can be expressed in terms of just six integers: q_{Q_i}, q_{L_i}. However, the quarks mix via the CKM matrix [see eq. (4.179)], so they must have equal quantum numbers across generations, and similarly for the leptons and PMNS mixing [see eq. (6.6)]. We drop the generation index and have two distinct charges q_Q, q_L:

$$\vec{q} \equiv (q_Q, -q_Q, -q_Q, q_L, -q_L, -q_L) . \tag{5.122}$$

We consider the previously listed linear anomaly cancellation conditions in terms of the charges q_Q, q_L. The right-hand side of eq. (5.110) is zero in the model under

Table 5.1 Values of N satisfying the conditions eqs. (5.128) and (5.127) for $p \leq 12$.

p	1	2	3	4	5	6	7	8	9	10	11	12
allowed N			9	12	15	9,18	21	12,24	27	15,30	33	9,12,18,36

consideration. However, this imposes no additional constraint since the left-hand side of eq. (5.110) reads[20]

$$3(2 \cdot 3q_Q + 3q_{\bar{u}} + 3q_{\bar{d}}) = 3(6q_Q - 6q_Q) = 0\,. \tag{5.123}$$

This expression vanishes identically and is thus satisfied for all choices of q_Q, q_L. Likewise, eq. (5.112) yields

$$3(6q_Q + 3q_{\bar{u}} + 3q_{\bar{d}} + 2q_L + q_{\bar{e}} + q_{\bar{\nu}}) = 3(6q_Q - 6q_Q + 2q_L - 2q_L) = 0\,, \tag{5.124}$$

which is also automatically satisfied. In contrast, eq. (5.111) yields a nontrivial constraint,

$$3(3q_Q + q_L) = 0 \mod N\,. \tag{5.125}$$

As a practical application, it would be desirable to employ the $\mathbb{Z}_N$ symmetry to prohibit dangerous low energy baryon-number-violating operators, which can lead to rapid proton decay. For example, the baryon number violation operators of lowest dimension are the dimension-six operators $QQQL$, $\bar{u}\bar{u}\bar{d}\bar{e}$, and $\bar{u}\bar{d}\bar{d}\bar{\nu}$ [see eq. (6.310)]. In terms of q_Q and q_L all three operators lead to the same constraint,

$$3q_Q + q_L \neq 0 \mod N\,. \tag{5.126}$$

By defining $p \equiv (3q_Q + q_L) \bmod N$, one can rewrite the conditions of eq. (5.125) and eq. (5.126) as

$$3p = 0 \mod N\,, \qquad p \neq 0 \mod N\,. \tag{5.127}$$

If we furthermore require the absence of neutron–antineutron oscillations (which are severely constrained by experiment), one must prohibit the $\Delta B = 2$ operator $\bar{u}\bar{d}\bar{d}\bar{u}\bar{d}\bar{d}$, which leads to the additional constraint

$$6q_Q \neq 0 \mod N\,. \tag{5.128}$$

In Table 5.1 we have summarized the solutions to these two conditions for $p \leq 12$. The case of $p = 0$ is excluded by the condition specified in eq. (5.126). We have not listed solutions with $N = 3$ or 6, since they necessarily violate the condition given by eq. (5.128), for all q_Q, q_L. If we include this condition, then the smallest possible discrete symmetries are for $N = 9$, and the allowed values are $p = 3, 6$. A minimal set of discrete charges q_Q, q_L is

$$(q_Q, q_L)_p \in \{(1,0)_3; (1,3)_6; (2,0)_6; (2,6)_3; (4,0)_3; (4,3)_6\}, \quad N = 9\,. \tag{5.129}$$

[20] Here we have included $\mathrm{SU}(2)_L$ doublet, $\mathrm{SU}(3)_C$, and generation factors, and in the last step we have made use of the relations of eq. (5.120).

We have exhibited the value of p as a subscript in eq. (5.129). Some charge combinations are related by an overall charge rescaling. However, such symmetries are not always valid. For example, as shown in eq. (16.153), an equivalent $\mathbb{Z}_N$ discrete symmetry is *always* obtained by multiplying the charges by $(N-1)$ and computing mod N. We have omitted these equivalent symmetries. The next possible value is $N = 12$ with the solutions

$$(q_Q, q_L)_p \in \{(1,1)_4; (1,5)_8; (3,7)_4; (3,11)_8; (5,1)_4; (5,5)_8\}, \quad N = 12\,. \qquad (5.130)$$

The listed $\mathbb{Z}_{9,12}$ discrete symmetries give the desired Dirac neutrino mass term and prohibit dangerous $\Delta B = 1$ and $\Delta B = 2$ operators.

Majorana Neutrino Masses

If we wish to include Majorana terms and incorporate the seesaw model we must allow for the lepton-number-violating dimension-five operator [see eq. (6.1)], as well as explicit Majorana mass terms:

$$\mathscr{L}_{\rm seesaw} \supset \frac{\mathbf{F}^{ij}}{2\Lambda}[(\epsilon^{ab}\boldsymbol{\Phi}_a \boldsymbol{L}_{bi})(\epsilon^{cd}\boldsymbol{\Phi}_c \boldsymbol{L}_{dj})] + \tfrac{1}{2}M_i(\bar{\nu}_i\bar{\nu}_i + \bar{\nu}_i^\dagger \bar{\nu}_i^\dagger)\,. \qquad (5.131)$$

After electroweak symmetry breaking the first operator leads to Majorana masses for the $\mathrm{SU}(2)_L$ doublet neutrinos. The second corresponds to (heavy) Majorana masses for the singlet neutrinos. Both sets of terms are allowed by the discrete $\mathbb{Z}_N$ symmetry in the generation-independent case if

$$2q_L = 0 \text{ mod } N\,. \qquad (5.132)$$

This is a highly restrictive constraint. It excludes all the $N = 12$ solutions of eq. (5.129), and for $N = 9$ only the three solutions of eq. (5.129) with $q_L = 0$ are allowed. We thus have only a small set of $\mathbb{Z}_9$ solutions that incorporate the seesaw mechanism. Note that if only Dirac mass terms are present as discussed previously, then all the terms in eq. (5.131) must be explicitly forbidden. This is achieved for $2q_L \neq 0 \text{ mod } N$, which are the complement solutions, i.e., all the $N = 12$ solutions, as well as the $N = 9$ solutions with $q_L \neq 0$.

In models with no singlet neutrinos, the SM neutrinos obtain their mass only via the dimension-five term of eq. (5.131). In this case the gravitation anomaly cancellation condition is modified to

$$3q_L = 0 \text{ mod } N\,. \qquad (5.133)$$

The only solutions that are consistent with eqs. (5.132) and (5.133) must satisfy $q_L = 0$ mod N. In this latter case, $p = 3q_Q$ and eqs. (5.128) and (5.127) reduce to

$$3q_Q \neq 0 \text{ mod } N, \qquad 6q_Q \neq 0 \text{ mod } N\,, \quad \text{and} \quad 9q_Q = 0 \text{ mod } N\,. \qquad (5.134)$$

Possible solutions are obtained, for example, for $N = 9$ and $q_Q = 1, 2, 4, 5, 7, 8$.

We see that the discrete gauge anomaly conditions can lead to meaningful constraints on extensions of the Standard Model with an additional discrete symmetry and possible additional new fields.

Exercises

5.1 The U(1) axial anomaly in massless QED was obtained in eq. (5.53) by evaluating the one-loop VVA triangle diagram with two external vector currents and one external axial-vector current. Show that if one evaluates the one-loop AAA triangle diagram with three external axial-vector currents, then

$$(p+q)^\mu\, i\Gamma^{AAA}_{\mu\nu\rho} = -p^\nu\, i\Gamma^{AAA}_{\mu\nu\rho} = -q^\rho\, i\Gamma^{AAA}_{\mu\nu\rho} = \frac{1}{6\pi^2}\epsilon_{\nu\rho\kappa\lambda}p^\kappa q^\lambda\,. \quad (5.135)$$

That is, apart from the overall factor of e^2 in eq. (5.53), the AAA triangle anomaly is 1/3 of the VVA triangle anomaly.

5.2 Consider the one-loop VVA triangle diagram with two external vector currents (which couple with coupling strength g to a nonabelian gauge field A^a_μ) and one external singlet axial-vector current. In this case, one would replace $\boldsymbol{T^a}$ with the identity matrix in eq. (5.8) to obtain $D^{abc} = I_2(R)\delta^{bc}$, where $I_2(R)$ is the second-order Dynkin index defined in eq. (H.32) and R is the representation of the generators $\boldsymbol{T^a}$.

(a) In light of eq. (5.30), show that

$$\partial_\mu J^\mu_A = -\frac{1}{8\pi^2}\,\mathrm{Tr}\, F^{\lambda\rho}\widetilde{F}_{\lambda\rho}\,, \quad (5.136)$$

where $F_{\mu\nu} \equiv gF^a_{\mu\nu}\boldsymbol{T^a}$ [see eq. (4.33)] and $\widetilde{F}_{\lambda\rho} \equiv \frac{1}{2}\epsilon_{\kappa\nu\lambda\rho}F^{\kappa\nu}$. The existence of the singlet anomaly has been used to explain why the η' meson is significantly heavier than the pion [54].

(b) Verify that the right-hand side of eq. (5.136) is a total divergence,

$$\mathrm{Tr}\, F^{\lambda\rho}\widetilde{F}_{\lambda\rho} = 2\epsilon^{\kappa\nu\lambda\rho}\partial_\kappa\ \mathrm{Tr}\big(A_\nu\partial_\lambda A_\rho + \tfrac{2}{3}iA_\nu A_\lambda A_\rho\big)\,, \quad (5.137)$$

where the nonabelian matrix gauge field $A_\mu \equiv gA^a_\mu\boldsymbol{T^a}$. Show that eqs. (5.136) and (5.137) imply the existence of an anomalous VVVA box diagram.

5.3 In Section 5.1.2, we examined the axial anomaly in massless QED. An analysis of the relevant one-loop triangle diagrams contributing to the three-point Green function with an external axial-vector current and two gravitons (e.g., see [B50] and Ref. [55]) yields a contribution to $\partial_\mu J^\mu_A$ [see eq. (5.57)]:

$$\partial_\mu J^\mu_A = \frac{1}{384\pi^2}\epsilon^{\nu\lambda\rho\tau}R_{\alpha\beta\nu\lambda}R^{\alpha\beta}{}_{\rho\tau}\,, \quad (5.138)$$

where $R_{\alpha\beta\nu\lambda}$ is the Riemann–Christoffel curvature tensor.

(a) Show that the condition for the absence of the mixed U(1) gauge–gravitational anomaly is $\mathrm{Tr}\, Q = 0$, where $\mathrm{Tr}\, Q$ sums over all U(1) charges.

(b) Show that there are no nonzero contributions to $\partial_\mu J^\mu_A$ from diagrams with one external gauge boson and one graviton, by proving that all terms involving one gauge field strength tensor $F^{\mu\nu}$ and one curvature tensor $R_{\alpha\beta\nu\lambda}$ that could appear on the right-hand side of eq. (5.138) vanish due to symmetry properties of $F^{\mu\nu}$ and $R_{\alpha\beta\nu\lambda}$. Conclude that there is no mixed gauge–gravitational anomaly arising from a three-point Green function with two external gauge bosons and one graviton.

(c) Show that the results of part (b) can also be understood by using the properties of the interaction vertices of the relevant triangle diagram under the charge-conjugation symmetry C. (An analogous argument shows that the triangle graph with three external photons in QED vanishes due to Furry's Theorem.)

5.4 Consider a one generation version of the Standard Model with two-component fermions $(u\,,\,d)$, $\bar{u}$, $\bar{d}$, $(\nu\,,\,e)$, and $\bar{e}$ given in Table 4.2. Suppose we were to assign these states $\mathrm{U}(1)_Y$ hypercharges a, b, c, d, and e.

(a) Show that imposing the cancellation of the Standard Model gauge anomalies and the cancellation of the mixed gauge–gravitational anomaly yields the following four equations [B73]:

$$2a + b + c = 0\,, \tag{5.139}$$

$$3a + d = 0\,, \tag{5.140}$$

$$6a^3 + 3(b^3 + c^3) + 2d^3 + e^3 = 0\,, \tag{5.141}$$

$$6a + 3(b + c) + 2d + e = 0\,. \tag{5.142}$$

(b) Since b and c appear symmetrically in eqs. (5.139)–(5.142), it follows that if $b \neq c$ then any solution that interchanges b and c will yield a second solution. Moreover, note that the overall normalization of the hypercharge is a matter of convention, since one can always absorb an overall normalization factor into the definition of the hypercharge gauge coupling g'. Thus, if $e \neq 0$, then one can set $e = 1$ without loss of generality. Show that in this case, there are only two solutions to eqs. (5.139)–(5.142). One solution corresponds to the Standard Model hypercharge assignments, $a = \frac{1}{6}$, $b = -\frac{2}{3}$, $c = \frac{1}{3}$, $d = -\frac{1}{2}$, and $e = 1$, and a second solution interchanges b and c.

(c) Show that if $e = 0$ then one can choose $b = 1$ without loss of generality, in which case there is a third solution, $b = -c = 1$ and $a = c = d = e = 0$.

(d) Consider an $\mathrm{SU}(3)_C\times\mathrm{SU}(2)_L\times\mathrm{U}(1)_Y\times\mathrm{U}(1)'$ gauge theory coupled to one generation of Standard Model fermions. To avoid gauge and mixed gauge–gravitational anomalies involving the $\mathrm{SU}(3)_C$, $\mathrm{SU}(2)_L$, and $\mathrm{U}(1)'$ gauge groups, one can use the results of part (c) to assign the $\mathrm{U}(1)'$ charges. Show that the $\mathrm{SU}(3)_C\times\mathrm{SU}(2)_L\times\mathrm{U}(1)_Y\times\mathrm{U}(1)'$ gauge theory is anomalous.

(e) Now add a two-component right-handed neutrino, $\bar{\nu}$, to the one generation version of the Standard Model. Assign a $U(1)_Y$ hypercharge f to $\bar{\nu}$ and repeat the calculation of parts (a)–(c). Find all the solutions such that the corresponding $U(1)_Y$ charges yield an anomaly-free gauge theory.

(f) Show that one solution obtained in part (e) will yield the Standard Model hypercharge assignments along with $f = 0$. Check that a second solution corresponds to $a = \frac{1}{3}$, $b = c = -\frac{1}{3}$, $d = -1$, and $e = f = 1$. Identify this U(1) symmetry with a well-known global symmetry of the Standard Model.

5.5 The constraint on the $U(1)_X$ charges obtained by imposing $D_{YYX} = 0$ was given in eqs. (5.117) and (5.118). In this exercise, we consider the implications of imposing the remaining conditions $D_{YXX} = 0$ and $D_{XXX} = 0$.

(a) Using eq. (5.107), the condition $D_{YXX} = 0$ is equivalent to

$$\sum_i Y_i(q_i + m_i N)^2 = 0\,. \tag{5.143}$$

Rewrite this equation such that only $\sum_i Y_i q_i^2$ remains on the left-hand side and argue that the right-hand side need not be integer. Be mindful to include possible heavy fermions in the sum eq. (5.143) and separate them out.

(b) Perform the analogous computation for $D_{XXX} = 0$. The answer does not explicitly depend on the $X_M^{j'}$. Why? Then show that it also does not give a model-independent constraint, due to the heavy fermions.

5.6 Consider a Lagrangian that is invariant under the $\mathbb{Z}_N$ symmetries denoted by $(q_Q, q_L)_p$ as in eq. (5.129).

(a) Show that the Lagrangian is also invariant under the $\mathbb{Z}_N$ symmetry with charges rescaled by an integer $m < N$ (see Ref. [56]),

$$(q'_Q, q'_L)_{p'} = (mq_Q, mq_L)_{mp}\,. \tag{5.144}$$

Furthermore, show that if the original symmetry is anomaly-free in the sense of eq. (5.125), then the rescaled symmetry is also anomaly-free.

(b) Assume that the original charges $(q_Q, q_L)_p$ satisfy the two inequalities exhibited in eqs. (5.126) and (5.128). Give an example where the rescaled charges do not satisfy these conditions, in which case the original and rescaled symmetries are inequivalent.

(c) Verify that if $m = N - 1$, then the new symmetry is anomaly-free *and* necessarily satisfies eqs. (5.126) and (5.128). In this sense, the two symmetries $(q_Q, q_L)_p$ and $\big((N-1)q_Q, (N-1)q_L\big)_{(N-1)p}$ are equivalent. Show that if you rescale again by the same factor, you recover the original symmetry. Rescaled symmetries with $m = N - 1$ have been dropped in eqs. (5.129) and (5.130).

6 Extending the Standard Model

6.1 The Seesaw-Extended Standard Model

In the Standard Model of particle physics presented in Section 4.7, the three neutrinos are exactly massless. However, the experimental observation of neutrino mixing strongly suggests that at least two of the three neutrinos are massive. To accommodate massive neutrinos, one must slightly extend the Standard Model (e.g., see [B138–B142]). The simplest approach is to introduce an $\mathrm{SU}(2)_L\times\mathrm{U}(1)_Y$ gauge invariant dimension-five operator and a new energy scale Λ:

$$\begin{aligned}\mathscr{L}_5 &= -\frac{\widehat{\boldsymbol{F}}^{ij}}{2\Lambda}(\epsilon^{ab}\boldsymbol{\Phi}_a\widehat{\boldsymbol{L}}_{bi})(\epsilon^{cd}\boldsymbol{\Phi}_c\widehat{\boldsymbol{L}}_{dj}) + \text{h.c.}\\ &= -\frac{\widehat{\boldsymbol{F}}^{ij}}{2\Lambda}(\Phi^0\widehat{\nu}_i - \Phi^+\widehat{\ell}_i)(\Phi^0\widehat{\nu}_j - \Phi^+\widehat{\ell}_j) + \text{h.c.}\,, \end{aligned} \tag{6.1}$$

where $\widehat{\boldsymbol{F}}^{ij}$ are generalized Yukawa couplings (with $\widehat{\boldsymbol{F}}^{ij} = \widehat{\boldsymbol{F}}^{ji}$), the hatted fields indicate two-component fermion interaction eigenstates (with spinor indices suppressed), and i, j label the three generations. After electroweak symmetry breaking, the neutral component of the doublet Higgs field acquires a vacuum expectation value, and a Majorana mass matrix for the neutrinos is generated.

The diagonalization of the charged lepton mass matrix is unmodified from the treatment given in Section 4.7, where the unhatted mass-eigenstate charged lepton fields are given by eq. (4.174), and L_ℓ and R_ℓ satisfy eq. (4.177). However, the unhatted neutrino field introduced in eq. (4.178) is *not* a neutrino mass-eigenstate field when the effect of the dimension-five Lagrangian, eq. (6.1), is taken into account. To avoid confusion, we replace the unhatted neutrino fields of eq. (4.178) with new neutrino fields $\breve{\nu}_j \equiv (\nu_e, \nu_\mu, \nu_\tau)$. That is, we define

$$\widehat{\nu}_i = (L_\ell)_i{}^j\breve{\nu}_j\,. \tag{6.2}$$

We then rewrite eq. (6.1) in terms of the charged lepton mass-eigenstate field and the new neutrino field defined by eq. (6.2):

$$\mathscr{L}_5 = -\frac{\boldsymbol{F}^{ij}}{2\Lambda}(\Phi^0\breve{\nu}_i - \Phi^+\ell_i)(\Phi^0\breve{\nu}_j - \Phi^+\ell_j) + \text{h.c.}\,, \tag{6.3}$$

where $\boldsymbol{F} \equiv L_\ell^{\mathsf{T}}\widehat{\boldsymbol{F}}L_\ell$ is a complex symmetric matrix. After setting $\Phi^0 = v/\sqrt{2}$ and $\Phi^+ = \Phi^- = 0$, we identify the 3×3 complex symmetric effective light neutrino mass matrix, $\boldsymbol{M}_{\boldsymbol{\nu\ell}}$, by

$$-\mathscr{L}_{m_\nu} = \tfrac{1}{2}(\boldsymbol{M}_{\boldsymbol{\nu\ell}})^{ij}\check{\nu}_i\check{\nu}_j + \text{h.c.}\,, \tag{6.4}$$

where

$$\boldsymbol{M}_{\boldsymbol{\nu\ell}} = \frac{v^2}{2\Lambda}\boldsymbol{F}\,. \tag{6.5}$$

The current bounds on light neutrino masses [41] suggest that $v^2/\Lambda \lesssim 1$ eV, or equivalently, $\Lambda \gtrsim 10^{13}$ GeV.

The physical neutrino mass-eigenstate fields can be identified by introducing the unitary Pontecorvo–Maki–Nakagawa–Sakata (PMNS) matrix, $\boldsymbol{U}$, such that[1]

$$(\check{\nu}_\ell)_i = \boldsymbol{U}_i{}^j(\nu_\ell)_j\,, \tag{6.6}$$

where the unhatted $(\nu_\ell)_j$ fields [$j = 1, 2, 3$] denote the physical (mass-eigenstate) Majorana neutrino fields. $\boldsymbol{U}$ is determined by the Takagi diagonalization of $\boldsymbol{M}_{\boldsymbol{\nu\ell}}$ (see Appendix G.2):

$$\boldsymbol{D} \equiv \boldsymbol{U}^{\mathsf{T}}\boldsymbol{M}_{\boldsymbol{\nu\ell}}\boldsymbol{U} = \text{diag}(m_{\nu_\ell 1}\,,\, m_{\nu_\ell 2}\,,\, m_{\nu_\ell 3})\,, \tag{6.7}$$

where the singular values of $\boldsymbol{M}_{\boldsymbol{\nu\ell}}$, denoted by $m_{\nu_\ell j}$, are the physical neutrino masses.

The interaction Lagrangian of the neutrino mass eigenstates can now be determined. The charged-current neutrino interactions are given by [see eq. (4.180)]

$$\begin{aligned}\mathscr{L}_{\text{int}} &= -\frac{g}{\sqrt{2}}\left[\check{\nu}^{\dagger i}\bar{\sigma}^\mu \ell_i W^+_\mu + \ell^{\dagger i}\bar{\sigma}^\mu\check{\nu}_i W^-_\mu\right] \\ &= -\frac{g}{\sqrt{2}}\left[(\boldsymbol{U}^\dagger)_j{}^i \nu_\ell^{\dagger j}\bar{\sigma}^\mu \ell_i W^+_\mu + \boldsymbol{U}_i{}^j \ell^{\dagger i}\bar{\sigma}^\mu \nu_{\ell j} W^-_\mu\right]\,,\end{aligned} \tag{6.8}$$

where we have used eq. (6.6) to express the interaction Lagrangian in terms of the neutrino mass-eigenstate fields. The neutral current neutrino interactions are flavor-diagonal (which follows from the unitarity of $\boldsymbol{U}$), and are thus equivalent to those of the Standard Model. Finally, the couplings of the neutrinos to the Higgs and Goldstone fields arise from eq. (6.3) and from the term in eq. (4.169) proportional to $\boldsymbol{Y}_\ell$. Neglecting terms of $\mathcal{O}(m_\nu^2/v^2)$, one obtains

$$\begin{aligned}\mathscr{L}_{\text{int}} &= \frac{\sqrt{2}}{v}\sum_{i,j}\left[(m_{\nu_\ell})_j(\boldsymbol{U}^\dagger)_j{}^i(\nu_\ell)_j\,\ell_i\,G^+ - (m_\ell)_i\boldsymbol{U}_i{}^j(\nu_\ell)_j\,\bar{\ell}^i\,G^- + \text{h.c.}\right] \\ &\quad - \frac{1}{v}\sum_j (m_{\nu_\ell})_j\left[(\nu_\ell)_j(\nu_\ell)_j(H + iG^0) + \text{h.c.}\right]\,.\end{aligned} \tag{6.9}$$

The dimension-five Lagrangian, eq. (6.1), is generated by new physics beyond the Standard Model at the scale Λ. One possible realization of eq. (6.1) employs the seesaw mechanism, where a superheavy gauge-singlet neutrino with mass of order $\Lambda \gg v$ yields a light neutrino mass scale of $\mathcal{O}(v^2/\Lambda)$ and a very small mixing angle of $\mathcal{O}(v/\Lambda)$ via the Takagi diagonalization of the neutrino mass matrix.

[1] In the literature, the PMNS matrix is often defined as $\boldsymbol{V} \equiv \boldsymbol{U}^*$ (e.g., see [B92, B141]). In this book, we shall adopt $\boldsymbol{U}$ as the PMNS matrix.

The Type-I seesaw extension of the Standard Model[2] introduces SM gauge-singlet two-component neutrino fields $\bar{\nu}^I$ ($I = 1, 2, \ldots, n$), along with the most general renormalizable couplings of the $\bar{\nu}^I$ to the SM fields,

$$\mathscr{L}_{\text{seesaw}} = \epsilon^{ab}(\widehat{\boldsymbol{y}}_{\boldsymbol{\nu}})^i{}_J \boldsymbol{\Phi}_a \widehat{\boldsymbol{L}}_{bi} \widehat{\bar{\nu}}^J - \tfrac{1}{2}\widehat{\boldsymbol{M}}_{IJ}\widehat{\bar{\nu}}^I\widehat{\bar{\nu}}^J + \text{h.c.}\,, \tag{6.10}$$

where the Yukawa coupling proportional to $\widehat{\boldsymbol{y}}_{\boldsymbol{\nu}}$ is the leptonic analogue of the Higgs–quark Yukawa coupling proportional to $\widehat{\boldsymbol{y}}_{\boldsymbol{u}}$ [see eq. (4.168)]. In eq. (6.10), we have distinguished the flavor labels of three generations of SM neutrino and charged lepton fields (denoted by lower case Roman letters $i, j, \ldots$) and the flavor labels of singlet neutrino fields (denoted by upper case Roman letters I, $J, \ldots$). Note that $\widehat{\boldsymbol{y}}_{\boldsymbol{\nu}}$ is a $3 \times n$ matrix and $\widehat{\boldsymbol{M}}$ is a symmetric $n \times n$ matrix, where n is the number of singlet neutrino flavors.[3] In general, we shall not specify the value of n, which may differ from the number of SM lepton generations.

In the seesaw mechanism, the singular values of $\widehat{\boldsymbol{M}}$ are assumed to be significantly larger than the scale of electroweak symmetry breaking, which implies that[4]

$$\|\widehat{\boldsymbol{M}}\| \gg v\,. \tag{6.11}$$

If $\Lambda \sim \|\widehat{\boldsymbol{M}}\| \gg v$ then a dimension-five operator of the form given by eq. (6.1) is generated in the effective theory at energy scales below Λ.[5] In this limit, we may neglect the kinetic energy term of the gauge-singlet neutrino fields. Using the Lagrange field equations, we may solve for $\widehat{\bar{\nu}}^I$. Inserting the solution back into eq. (6.10) then yields eq. (6.1), with $\boldsymbol{F}/\Lambda$ given by

$$\widehat{\boldsymbol{F}}^{ij}/\Lambda = -(\widehat{\boldsymbol{y}}_{\boldsymbol{\nu}})^i{}_K(\widehat{\boldsymbol{y}}_{\boldsymbol{\nu}})^j{}_N(\widehat{\boldsymbol{M}}^{-1})^{KN}\,. \tag{6.12}$$

Using the definition of the SU(2)$_L$ doublet lepton field given in Table 4.2, one can rewrite eq. (6.10) more explicitly as

$$\mathscr{L}_{\text{seesaw}} = -(\widehat{\boldsymbol{y}}_{\boldsymbol{\nu}})^i{}_J\left[\Phi^0\widehat{\nu}_i\widehat{\bar{\nu}}^J - \Phi^+\widehat{\ell}_i\widehat{\bar{\nu}}^J\right] - \tfrac{1}{2}\widehat{\boldsymbol{M}}_{IJ}\widehat{\bar{\nu}}^I\widehat{\bar{\nu}}^J + \text{h.c.} \tag{6.13}$$

To analyze the physical consequences of the seesaw Lagrangian, we first express eq. (6.13) in terms of the unhatted mass-eigenstate charged lepton fields [given in eq. (4.174)], and the light neutrino fields $\breve{\nu}_i$ introduced in eq. (6.2). It is also convenient to introduce new gauge-singlet neutrino fields $\breve{\bar{\nu}}^J$ by defining

$$\widehat{\bar{\nu}}^I = N^I{}_J\breve{\bar{\nu}}^J\,, \tag{6.14}$$

where N is the unitary matrix that Takagi diagonalizes the complex symmetric matrix $\widehat{\boldsymbol{M}}$. That is,

$$\boldsymbol{M} \equiv N^{\mathsf{T}}\widehat{\boldsymbol{M}}N = \text{diag}(M_1\,,\, M_2\,, \ldots\,,\, M_n)\,, \tag{6.15}$$

[2] Alternative seesaw models, known in the literature as Type II and Type III (details can be found in [B141]), are not considered in this book.

[3] Generically, the matrices $\widehat{\boldsymbol{y}}_{\boldsymbol{\nu}}$ and $\widehat{\boldsymbol{M}}$ are complex matrices.

[4] The Euclidean matrix norm is defined by $\|\widehat{M}\| \equiv \left[\text{Tr}(\widehat{M}^\dagger\widehat{M})\right]^{1/2}$.

[5] More precisely, one can define Λ to be the minimum of the absolute value of the eigenvalues of $\widehat{\boldsymbol{M}}$. Typically, the eigenvalues of $\widehat{\boldsymbol{M}}$ are of the same order of magnitude, in which case $\|\widehat{\boldsymbol{M}}\| \sim \mathcal{O}(\Lambda)$.

where the M_I are the singular values of $\widehat{\boldsymbol{M}}$. In terms of the mass-eigenstate charged lepton fields ℓ_i and the neutrino fields $\breve{\nu}_i$ and $\breve{\nu}^I$, the seesaw Lagrangian [eq. (6.13)] is then given by

$$\mathscr{L}_{\text{seesaw}} = -(\boldsymbol{y_\nu})^i{}_J \left[\Phi^0 \breve{\nu}_i \breve{\nu}^J - \Phi^+ \ell_i \breve{\nu}^J\right] - \tfrac{1}{2}\boldsymbol{M}_{IJ}\breve{\nu}^I\breve{\nu}^J + \text{h.c.}\,, \tag{6.16}$$

where

$$\boldsymbol{y_\nu} \equiv L_\ell^{\mathsf{T}} \widehat{\boldsymbol{y}}_{\boldsymbol{\nu}} N\,. \tag{6.17}$$

As above, in the limit of $\Lambda \sim \|\widehat{\boldsymbol{M}}\| = \|\boldsymbol{M}\| \gg v$, it is also possible to directly generate the effective dimension-five operator [eq. (6.3)] in terms of the mass-eigenstate charged lepton fields and the new neutrino fields $\breve{\nu}_j$. We then identify the corresponding coefficient, $\boldsymbol{F}/\Lambda$, as

$$\boldsymbol{F}^{ij}/\Lambda = -(\boldsymbol{y_\nu})^i{}_K(\boldsymbol{y_\nu})^j{}_N(\boldsymbol{M}^{-1})^{KN}\,. \tag{6.18}$$

Recalling that $\boldsymbol{F} = L_\ell^{\mathsf{T}}\widehat{\boldsymbol{F}}L_\ell$, one can check that eq. (6.18) indeed follows from eqs. (6.12), (6.15), and (6.17).

To identify the neutrino mass matrix, we set $\Phi^0 = v/\sqrt{2}$ and $\Phi^+ = \Phi^- = 0$ in eq. (6.16):

$$-\mathscr{L}_{m_\nu} = \tfrac{1}{2}\begin{pmatrix}\breve{\nu}_i & \breve{\nu}^J\end{pmatrix}\mathcal{M}_\nu\begin{pmatrix}\breve{\nu}_k \\ \breve{\nu}^M\end{pmatrix} + \text{h.c.} \tag{6.19}$$

The neutrino mass matrix $\mathcal{M}_\nu$ is a $(3+n)\times(3+n)$ complex symmetric matrix given in block form by

$$\mathcal{M}_\nu \equiv \begin{pmatrix} \mathbb{0} & \boldsymbol{M_D} \\ \boldsymbol{M_D^{\mathsf{T}}} & \boldsymbol{M}\end{pmatrix}, \tag{6.20}$$

where $\mathbb{0}$ is the 3×3 zero matrix, $\boldsymbol{M}$ is the diagonal matrix defined in eq. (6.15), and $\boldsymbol{M_D}$ is a $3\times n$ complex matrix (called the Dirac neutrino mass matrix),

$$(\boldsymbol{M_D})^i{}_J \equiv \frac{v}{\sqrt{2}}(\boldsymbol{y_\nu})^i{}_J\,. \tag{6.21}$$

Note that if $n=3$ and $\boldsymbol{M}=\mathbb{0}$, then $\boldsymbol{M_D}$ is a 3×3 matrix that is simply the leptonic analogue of the up-type quark mass matrix $\boldsymbol{M_u}$. In this case, we would perform a singular value decomposition of $\boldsymbol{M_D}$ and identify the unhatted neutrino mass-eigenstate fields, which can be assembled into three generations of four-component Dirac neutrinos,

$$N_i = \begin{pmatrix}\nu_i \\ \bar{\nu}_i^\dagger\end{pmatrix}, \qquad i = 1,2,3\,. \tag{6.22}$$

In the seesaw model (with n not specified), we assume that $\|\boldsymbol{M}\| \gg \|\boldsymbol{M_D}\|$. In this case, the neutrino mass matrix can be perturbatively Takagi block-diagonalized as follows. Introduce the $(3+n)\times(3+n)$ (approximate) unitary matrix

$$\mathcal{U} = \begin{pmatrix} \mathbb{1}_{3\times 3} - \tfrac{1}{2}\boldsymbol{M_D^*}\boldsymbol{M}^{-2}\boldsymbol{M_D^{\mathsf{T}}} & \boldsymbol{M_D^*}\boldsymbol{M}^{-1} \\ -\boldsymbol{M}^{-1}\boldsymbol{M_D^{\mathsf{T}}} & \mathbb{1}_{n\times n} - \tfrac{1}{2}\boldsymbol{M}^{-1}\boldsymbol{M_D^{\mathsf{T}}}\boldsymbol{M_D^*}\boldsymbol{M}^{-1}\end{pmatrix}, \tag{6.23}$$

where $\mathbb{1}$ is the identity matrix (whose dimension is explicitly specified above). We define transformed [light (ℓ) and heavy (h)] neutrino states $(\breve{\nu}_\ell)_i$ and $(\breve{\nu}_h)^j$ by

$$\begin{pmatrix} \breve{\nu}_i \\ \breve{\nu}^J \end{pmatrix} = \mathcal{U} \begin{pmatrix} (\breve{\nu}_\ell)_k \\ (\breve{\nu}_h)^M \end{pmatrix} . \tag{6.24}$$

By straightforward matrix multiplication, one can verify that, to second-order accuracy in perturbation theory,

$$\mathcal{U}^{\mathsf{T}} \mathcal{M}_\nu \mathcal{U} \simeq \begin{pmatrix} -\boldsymbol{M_D M^{-1} M_D^{\mathsf{T}}} & \mathbb{O} \\ \mathbb{O}^{\mathsf{T}} & \boldsymbol{M} + \tfrac{1}{2}(\boldsymbol{M^{-1} M_D^\dagger M_D} + \boldsymbol{M_D^{\mathsf{T}} M_D^* M^{-1}}) \end{pmatrix} , \tag{6.25}$$

where $\mathbb{O}$ is the $3 \times n$ zero matrix.

We now can identify an effective 3×3 complex symmetric mass matrix $\boldsymbol{M_{\nu_\ell}}$ for the three light neutrinos as the upper left-hand block of eq. (6.25),

$$\boldsymbol{M_{\nu_\ell}} \simeq -\boldsymbol{M_D M^{-1} M_D^{\mathsf{T}}} , \tag{6.26}$$

where corrections of $\mathcal{O}(v^4/\Lambda^3)$ have been neglected. Using eqs. (6.18) and (6.21), we see that the light neutrino mass matrix obtained in eq. (6.5) has been correctly reproduced to leading order in v^2/Λ^2.

The physical light neutrino mass-eigenstate fields and their masses are identified by eqs. (6.6) and (6.7). At energy scales below the heavy neutrino mass scale, $\Lambda \equiv \|\boldsymbol{M}\|$, and we can set $\breve{\nu}_h = 0$. Neglecting corrections of $\mathcal{O}(v^2/\Lambda^2)$, eqs. (6.21)–(6.26) imply that[6]

$$\breve{\nu}_i \simeq \boldsymbol{U}_i{}^j (\nu_\ell)_j , \tag{6.27}$$

$$(\boldsymbol{y_\nu})^i{}_J \breve{\nu}^J \simeq \frac{\sqrt{2}}{v} \boldsymbol{M_{\nu_\ell} U}^{ik} (\nu_\ell)_k = \frac{\sqrt{2}}{v} \sum_k (\boldsymbol{U}^\dagger)_k{}^i (m_{\nu_\ell})_k (\nu_\ell)_k , \tag{6.28}$$

where we have employed eqs. (1.162) and (6.7) in the last step above. By using eqs. (6.27) and (6.28) to express the seesaw Lagrangian in terms of the light neutrino mass-eigenstate fields, one can verify that the resulting interactions of the light neutrinos (and charged leptons) to gauge bosons, the Higgs boson, and the Goldstone bosons reproduce the results of eqs. (6.8) and (6.9) at leading order in v^2/Λ^2.

For completeness, we examine the effective $n \times n$ complex symmetric mass matrix of the heavy neutrino states, $\boldsymbol{M_{\nu_h}}$, which is identified as the lower right-hand block in eq. (6.25):

$$\boldsymbol{M_{\nu_h}} \simeq \boldsymbol{M} + \tfrac{1}{2}(\boldsymbol{M^{-1} M_D^\dagger M_D} + \boldsymbol{M_D^{\mathsf{T}} M_D^* M^{-1}}) . \tag{6.29}$$

Although $\boldsymbol{M}$ is diagonal by definition [see eq. (6.15)], the right-hand side of eq. (6.29) is no longer diagonal due to the second-order perturbative correction. However, we

[6] Strictly speaking, eq. (6.28) should be written as

$$(\boldsymbol{y_\nu})^i{}_J \breve{\nu}^J \simeq \frac{\sqrt{2}}{v} \sum_{k,n} (\boldsymbol{U}^\dagger)_n{}^i \, \delta^{nk} (m_{\nu_\ell})_k (\nu_\ell)_k$$

to maintain covariance in the flavor indices.

do not have to perform another Takagi diagonalization, since the off-diagonal elements of the lower right-hand block only affect the physical (diagonal) masses at higher order in perturbation theory. Thus, we identify the physical heavy neutrino mass eigenstates to leading order by the unhatted fields,

$$\bar{\nu}_h^J \simeq \breve{\nu}_h^J\,, \tag{6.30}$$

with masses

$$m_{\nu_{hJ}} \simeq M_J\left(1+\frac{1}{M_J^2}\sum_i |(\boldsymbol{M_D})^i{}_J|^2\right), \tag{6.31}$$

where the M_J are the diagonal elements of $\boldsymbol{M}$ (and no sum over the repeated index J is implied). That is, the masses of the heavy neutrinos are simply given by $m_{\nu_{hJ}} \simeq M_J$, up to corrections that are of the same order as the light neutrino masses.

The interactions of the heavy neutrinos can be likewise obtained. The only unsuppressed interactions are heavy neutrino couplings to the Higgs boson and Goldstone bosons that are proportional to the Dirac neutrino mass matrix:

$$\mathscr{L}_{\rm int} = -\frac{1}{v}(\boldsymbol{U}^{\mathsf{T}}\boldsymbol{M_D})^k{}_J\bar{\nu}_h^J(\nu_\ell)_k(H+iG^0) + \frac{\sqrt{2}}{v}(\boldsymbol{M_D})^i{}_J\ell_i\bar{\nu}_h^J G^+ + \text{h.c.} \tag{6.32}$$

All other couplings of the heavy neutrinos to the $W^\pm$ and Z bosons (and additional contributions to the couplings of the heavy neutrinos to the Higgs boson and Goldstone bosons) are suppressed by (at least) a factor of $\mathcal{O}(v/\Lambda)$.

6.2 The Two-Higgs Doublet Model (2HDM)

Given that there are multiple generations of quarks and leptons, it is reasonable to entertain the possibility that the Higgs sector of electroweak theory is also non-minimal [B143]. For example, one can add additional $Y=\frac{1}{2}$, $\mathrm{SU}(2)_L$ doublet scalars and $\mathrm{SU}(2)_L\times\mathrm{U}(1)_Y$ singlet scalars while maintaining the SM tree-level prediction of $\rho = 1$ (see Exercise 4.9).

The two-Higgs doublet extension of the Standard Model (2HDM) [57] was initially introduced to provide a possible new source of CP violation mediated by neutral scalars. Subsequently, 2HDM studies focused on the new scalar degrees of freedom, which include a charged Higgs scalar pair, $H^\pm$, a neutral CP-odd Higgs scalar in the case of a CP-conserving scalar potential, and neutral scalars of indefinite CP in the case of a CP-violating scalar potential (explicit violation) or vacuum (spontaneous violation). These features yield new phenomenological signals for the production and decay of fundamental spin-0 particles.

Later in this book, the main motivation for the 2HDM arises in the context of the minimal supersymmetric extension of the Standard Model (MSSM). In Sections 7.2 and 13.2, it is shown that the Higgs sector of the MSSM is a 2HDM that is highly constrained by supersymmetry. Two different scenarios then emerge, depending on the magnitude of the mass scale of supersymmetry breaking M_S relative to the mass scale of new heavy degrees of freedom $M_{\rm 2HDM}$ that arise in the 2HDM. The absence of experimental evidence for supersymmetric particles suggests that $M_S \gtrsim 1$ TeV. If $M_{\rm 2HDM} \sim M_S$, then at energy scales below M_S the effective low-energy theory is precisely the Standard Model (with one Higgs doublet field and one physical neutral CP-even scalar state). On the other hand, if $M_{\rm 2HDM} \ll M_S$, then at energy scales below M_S (but above $M_{\rm 2HDM}$) the effective low-energy theory is a general 2HDM whose scalar potential includes terms that vanish in the supersymmetric limit.

The most general version of the 2HDM Lagrangian (which contains all possible renormalizable terms of dimension four or less allowed by the electroweak gauge invariance) possesses Higgs–fermion Yukawa interaction terms that generically yield tree-level Higgs-mediated flavor-changing neutral current (FCNC) phenomena that are incompatible with experimental observations. Such terms are absent in the MSSM, although supersymmetry-breaking effects do generate FCNC interactions at the loop-level, which can be consistent with experimental constraints. In non-supersymmetric versions of the 2HDM, one can also naturally avoid FCNCs by imposing certain discrete symmetries on the 2HDM Lagrangian, as discussed in Section 6.2.4. These symmetries reduce the parameter freedom of the 2HDM and automatically eliminate the dangerous tree-level FCNC interactions. Nevertheless, it is instructive to explore the structure of the most general 2HDM, since constrained versions of the 2HDM can then be examined as limiting cases.[7]

6.2.1 The Unconstrained 2HDM

The scalar fields of the 2HDM are complex $\mathrm{SU}(2)_L$ doublet, $Y = \frac{1}{2}$ fields, denoted by Φ_1 and Φ_2. Under the conditions of renormalizability and $\mathrm{SU}(2)_L\times\mathrm{U}(1)_Y$ gauge invariance, the most general scalar potential is given by

$$\begin{aligned}\mathcal{V} &= m_{11}^2\Phi_1^\dagger\Phi_1 + m_{22}^2\Phi_2^\dagger\Phi_2 - [m_{12}^2\Phi_1^\dagger\Phi_2 + \text{h.c.}]\\ &\quad + \tfrac{1}{2}\lambda_1(\Phi_1^\dagger\Phi_1)^2 + \tfrac{1}{2}\lambda_2(\Phi_2^\dagger\Phi_2)^2 + \lambda_3(\Phi_1^\dagger\Phi_1)(\Phi_2^\dagger\Phi_2) + \lambda_4(\Phi_1^\dagger\Phi_2)(\Phi_2^\dagger\Phi_1)\\ &\quad + \left\{\tfrac{1}{2}\lambda_5(\Phi_1^\dagger\Phi_2)^2 + \left[\lambda_6(\Phi_1^\dagger\Phi_1) + \lambda_7(\Phi_2^\dagger\Phi_2)\right]\Phi_1^\dagger\Phi_2 + \text{h.c.}\right\},\end{aligned} \tag{6.33}$$

where m_{11}^2, m_{22}^2, λ_1, λ_2, λ_3, and λ_4 are real parameters and m_{12}^2, λ_5, λ_6, and λ_7 are potentially complex parameters. Minimizing the scalar potential yields vacuum expectation values

$$\langle\Phi_1\rangle = \frac{1}{\sqrt{2}}\begin{pmatrix}0\\ v_1\end{pmatrix}, \qquad \langle\Phi_2\rangle = \frac{1}{\sqrt{2}}\begin{pmatrix}u\\ v_2\end{pmatrix}, \tag{6.34}$$

where v_1 can be chosen real and nonnegative and v_2 and u are potentially complex. Here, we have employed the $\mathrm{SU}(2)_L\times\mathrm{U}(1)_Y$ gauge freedom to bring $\langle\Phi_1\rangle$ to the form displayed in eq. (6.34). If $u \neq 0$, then $\mathrm{SU}(2)_L\times\mathrm{U}(1)_Y$ is completely broken

[7] The following subsections follow closely the 2HDM formalism developed in Refs. [58, 59].

and there is no residual $\mathrm{U}(1)_{\mathrm{EM}}$ gauge invariance remaining. The latter can occur over some fraction of the 2HDM parameter space governed by the coefficients of the scalar potential. To avoid such a possibility, we shall assume that the coefficients of the scalar potential have been chosen such that the global potential minimum breaks the $\mathrm{SU}(2)_L{\times}\mathrm{U}(1)_Y$ gauge symmetry while leaving an unbroken $\mathrm{U}(1)_{\mathrm{EM}}$, in which case $u = 0$ and $|v_1|^2 + |v_2|^2 \neq 0$. The corresponding potential minimum conditions can be obtained by replacing $\Phi_i^0 \to \Phi_i^0 + v_i/\sqrt{2}$ in the scalar potential and setting the coefficients of the terms linear in the neutral scalar fields to zero. Hence, the two vacuum expectation values v_1 and v_2, in the convention where v_1 is real and nonnegative and v_2 is potentially complex, are determined by

$$m_{11}^2 v_1 = m_{12}^2 v_2 - \tfrac{1}{2}\big(\lambda_1 v_1^2 + \lambda_{34}|v_2|^2 + \lambda_5 v_2^2\big)v_1 - \big(\lambda_6 v_2 + \tfrac{1}{2}\lambda_6^* v_2^*\big)\, v_1^2 - \tfrac{1}{2}\lambda_7 v_2^2 v_2^*\,, \tag{6.35}$$

$$m_{22}^2 v_2^* = m_{12}^2 v_1 - \tfrac{1}{2}\big(\lambda_2|v_2|^2 + \lambda_{34}v_1^2\big)v_2^* - \tfrac{1}{2}\lambda_5 v_1^2 v_2 - \tfrac{1}{2}\lambda_6 v_1^3 - \big(\lambda_7 v_2 + \tfrac{1}{2}\lambda_7^* v_2^*\big)\, v_1 v_2^*, \tag{6.36}$$

where

$$\lambda_{34} \equiv \lambda_3 + \lambda_4\,. \tag{6.37}$$

The normalization of the vacuum expectation values is fixed by the Fermi constant [see eq. (4.132)], which yields

$$v \equiv \big(|v_1|^2 + |v_2|^2\big)^{1/2} \simeq 246~\mathrm{GeV}\,. \tag{6.38}$$

It is possible that eqs. (6.35) and (6.36) yield more than one solution corresponding to a local minimum of the scalar potential. In the following, we assume that v_1 and v_2 are the vacuum expectation values corresponding to the global minimum of the scalar potential. We introduce the notation

$$v_1 \equiv v c_\beta\,, \qquad v_2 \equiv e^{i\xi} v s_\beta\,, \tag{6.39}$$

where $c_\beta \equiv \cos\beta$, $s_\beta \equiv \sin\beta$, and $0 \leq \xi < 2\pi$. Since v_1 has been taken to be real and nonnegative, we can adopt the convention where $0 \leq \beta \leq \frac{1}{2}\pi$.

In the most general 2HDM, the fields Φ_1 and Φ_2 are indistinguishable. Thus, it is always possible to define new scalar fields

$$\Phi_i' = U_i{}^j \Phi_j\,, \tag{6.40}$$

for $i, j \in \{1, 2\}$ (with the sum over the repeated index j implied), where U is a 2×2 unitary matrix, without modifying any prediction of the model. Choosing the two independent scalar fields of the 2HDM is called choosing a particular scalar field basis. Note that the transformation from the Φ-basis to the Φ'-basis leaves the kinetic energy terms unchanged. The scalar potential in the Φ'-basis has the same form as eq. (6.33) but with modified coefficients. Consequently, the coefficients that parameterize the scalar potential in eq. (6.33) are not physical observables.

To obtain a scalar potential that is more closely related to physical observables, we shall introduce the *Higgs basis* as follows. Starting from a generic Φ-basis, the

Higgs basis fields $\mathcal{H}_1$ and $\mathcal{H}_2$ are defined by the linear combinations of Φ_1 and Φ_2 such that $\langle\mathcal{H}_1^0\rangle = v/\sqrt{2}$ and $\langle\mathcal{H}_2^0\rangle = 0$. That is,

$$\mathcal{H}_1 = \begin{pmatrix}\mathcal{H}_1^+\\ \mathcal{H}_1^0\end{pmatrix} \equiv c_\beta\Phi_1 + s_\beta e^{-i\xi}\Phi_2\,, \qquad \mathcal{H}_2 = \begin{pmatrix}\mathcal{H}_2^+\\ \mathcal{H}_2^0\end{pmatrix} = e^{i\eta}\left(-s_\beta e^{i\xi}\Phi_1 + c_\beta\Phi_2\right), \tag{6.41}$$

where the complex phase factor $e^{i\eta}$ accounts for the nonuniqueness of the Higgs basis, since one is always free to rephase the Higgs basis field whose vacuum expectation value vanishes. In particular, $e^{i\eta}$ is a pseudoinvariant quantity that is rephased under a unitary basis transformation, $\Phi_i \to U_i{}^j\Phi_j$, as

$$e^{i\eta} \to (\det\, U)^{-1}e^{i\eta}\,, \tag{6.42}$$

where $\det U$ is a complex number of unit modulus.

It is convenient to introduce the following notation. We define a complex vector, $\widehat{v} = (\widehat{v}_1, \widehat{v}_2)$, of unit norm such that

$$\langle\Phi_i\rangle = \frac{v}{\sqrt{2}}\begin{pmatrix}0\\ \widehat{v}_i\end{pmatrix}, \qquad v \simeq 246~\text{GeV}, \quad \text{for } i = 1, 2, \tag{6.43}$$

in the Φ-basis. The complex conjugate of $\widehat{v}_i$ will be denoted with a raised index, $\widehat{v}^{\,i} \equiv (\widehat{v}_i)^*$. A second unit vector $\widehat{w}$ can be defined that is orthogonal to $\widehat{v}$:

$$\widehat{w}_j \equiv \widehat{v}^{\,i}\epsilon_{ij}\,, \tag{6.44}$$

where $\epsilon_{12} = -\epsilon_{21} = +1$ [see eq. (A.54)]. The complex conjugate of $\widehat{w}_i$ will be denoted with a raised index, $\widehat{w}^{\,i} \equiv (\widehat{w}_i)^*$. Note that $\widehat{v}$ and $\widehat{w}$ are orthogonal due to the vanishing of the complex dot product, $\widehat{v}^{\,j}\widehat{w}_j = \widehat{v}^{\,j}\widehat{v}^{\,i}\epsilon_{ij} = 0$. Under a unitary basis transformation $\Phi_i \to U_i{}^j\Phi_j$, the unit vectors $\widehat{v}$ and $\widehat{w}$ transform as

$$\widehat{v}_i \to U_i{}^j\,\widehat{v}_j, \quad \text{which implies that} \quad \widehat{w}_i \to (\det U)^{-1}U_i{}^j\,\widehat{w}_j. \tag{6.45}$$

Using the notation introduced above, the expressions for the Higgs basis fields given in eq. (6.41) are

$$\mathcal{H}_1 \equiv \widehat{v}^{\,i}\Phi_i\,, \qquad \mathcal{H}_2 \equiv e^{i\eta}\widehat{w}^{\,i}\Phi_i\,. \tag{6.46}$$

In light of eqs. (6.42) and (6.45), it follows that both $\mathcal{H}_1$ and $\mathcal{H}_2$ are invariant fields with respect to unitary basis transformations. From eq. (6.46), we obtain

$$\Phi_i = \mathcal{H}_1\widehat{v}_i + e^{-i\eta}\mathcal{H}_2\widehat{w}_i\,, \qquad \text{for } i = 1, 2. \tag{6.47}$$

In the Higgs basis, the scalar potential is given by

$$\begin{aligned}\mathcal{V} = {} & Y_1\mathcal{H}_1^\dagger\mathcal{H}_1 + Y_2\mathcal{H}_2^\dagger\mathcal{H}_2 + [Y_3e^{-i\eta}\mathcal{H}_1^\dagger\mathcal{H}_2 + \text{h.c.}]\\ & + \tfrac{1}{2}Z_1(\mathcal{H}_1^\dagger\mathcal{H}_1)^2 + \tfrac{1}{2}Z_2(\mathcal{H}_2^\dagger\mathcal{H}_2)^2 + Z_3(\mathcal{H}_1^\dagger\mathcal{H}_1)(\mathcal{H}_2^\dagger\mathcal{H}_2) + Z_4(\mathcal{H}_1^\dagger\mathcal{H}_2)(\mathcal{H}_2^\dagger\mathcal{H}_1)\\ & + \left\{\tfrac{1}{2}Z_5e^{-2i\eta}(\mathcal{H}_1^\dagger\mathcal{H}_2)^2 + \left[Z_6e^{-i\eta}\mathcal{H}_1^\dagger\mathcal{H}_1 + Z_7e^{-i\eta}\mathcal{H}_2^\dagger\mathcal{H}_2\right]\mathcal{H}_1^\dagger\mathcal{H}_2 + \text{h.c.}\right\}.\end{aligned} \tag{6.48}$$

The coefficients of the quadratic and quartic terms of the scalar potential exhibited in eq. (6.48) are independent of the initial choice of the Φ-basis. It then follows that Y_1, Y_2, and $Z_1, \ldots, Z_4$ are real and uniquely defined, whereas Y_3, Z_5, Z_6, and Z_7 are complex pseudoinvariant quantities that are rephased under $\Phi_i \to U_i{}^j \Phi_j$ as

$$[Y_3, Z_6, Z_7] \to (\det U)^{-1}[Y_3, Z_6, Z_7] \quad \text{and} \quad Z_5 \to (\det U)^{-2} Z_5 \,. \tag{6.49}$$

The minimization of the scalar potential in the Higgs basis yields

$$Y_1 = -\tfrac{1}{2} Z_1 v^2 \,, \qquad Y_3 = -\tfrac{1}{2} Z_6 v^2 \,. \tag{6.50}$$

This leaves 11 free parameters: 1 VEV (v), 8 real parameters, Y_2, $Z_{1,2,3,4}$, $|Z_{5,6,7}|$, and two relative phases. In assessing the phenomenological properties of the 2HDM, we shall typically assume that the dimensionless parameters Z_i do not exceed values of $\mathcal{O}(1)$, well within the perturbative domain of the model.

Because Φ_1 and Φ_2 are indistinguishable fields, there is no physical significance to the original basis of scalar fields, $\{\Phi_1 \,, \Phi_2\}$, and likewise no physical significance to v_1 and v_2 separately [beyond the constraint imposed by eq. (6.38)]. However if additional symmetries are present, such as discrete symmetries or supersymmetry, which distinguish between Φ_1 and Φ_2, then the original basis choice acquires physical significance. In this case, a parameter such as $\tan\beta \equiv |v_2|/|v_1|$, which has no physical significance in the general 2HDM, is promoted to a physical observable.

Continuing with our analysis of the general 2HDM, we next compute the tree-level masses of the scalar states. The two complex doublets of the 2HDM provide eight scalar degrees of freedom. The doublet of scalar fields in the Higgs basis can be parameterized as follows:

$$\mathcal{H}_1 = \begin{pmatrix} G^+ \\ \frac{1}{\sqrt{2}}\left(v + \varphi_1^0 + iG\right) \end{pmatrix}, \qquad \mathcal{H}_2 = \begin{pmatrix} \mathcal{H}_2^+ \\ \frac{1}{\sqrt{2}}\left(\varphi_2^0 + ia^0\right) \end{pmatrix}, \tag{6.51}$$

and the corresponding hermitian-conjugated fields are likewise defined. We identify $G^\pm$ as a charged Goldstone boson pair and G as the CP-odd neutral Goldstone boson. In particular, the identification of $G = \sqrt{2}\,\mathrm{Im}\,\mathcal{H}_1^0$ follows from the fact that we have defined the Higgs basis such that $\langle \mathcal{H}_1^0 \rangle$ is real and nonnegative. Of the remaining fields, φ_1^0 is a CP-even neutral scalar field, φ_2^0 and a^0 are states of indefinite CP quantum numbers (since CP violation in the scalar sector can arise through the interactions of $\mathcal{H}_2$), and $H^\pm$ is the physical charged Higgs boson pair. If the scalar sector is CP-violating, then φ_1^0, φ_2^0, and a^0 all mix to produce three physical neutral Higgs mass eigenstates of indefinite CP quantum numbers, henceforth denoted as h_k, $(k = 1, 2, 3)$.

To determine the Higgs mass eigenstates, one must examine the terms of the scalar potential that are quadratic in the scalar fields (after minimizing the scalar potential and defining shifted scalar fields with zero vacuum expectation values). We shall carry out this procedure in the Higgs basis by inserting $\mathcal{H}_1^0 \to \mathcal{H}_1^0 + v/\sqrt{2}$ into eq. (6.48) and using the scalar potential minimum conditions [see eq. (6.50)] to eliminate Y_1 and Y_3. One can easily check that no quadratic terms involving the Goldstone boson fields survive (as expected, since the Goldstone bosons are massless). This confirms our identification of the Goldstone fields in eq. (6.51).

The three remaining neutral fields mix, and the resulting neutral scalar squared-mass matrix in the φ_1^0–φ_2^0–a^0 basis is

$$\mathcal{M}^2 = v^2 \begin{pmatrix} Z_1 & \mathrm{Re}(Z_6\, e^{-i\eta}) & -\mathrm{Im}(Z_6\, e^{-i\eta}) \\ \mathrm{Re}(Z_6 e^{-i\eta}) & \mathrm{Re}(Z_5\, e^{-2i\eta}) + A^2/v^2 & -\tfrac{1}{2}\mathrm{Im}(Z_5\, e^{-2i\eta}) \\ -\mathrm{Im}(Z_6\, e^{-i\eta}) & -\tfrac{1}{2}\mathrm{Im}(Z_5\, e^{-2i\eta}) & A^2/v^2 \end{pmatrix}, \qquad (6.52)$$

where A^2 is the auxiliary quantity

$$A^2 \equiv Y_2 + \tfrac{1}{2}[Z_3 + Z_4 - \mathrm{Re}(Z_5 e^{-2i\eta})]v^2\,. \qquad (6.53)$$

The squared masses of the neutral scalars are the eigenvalues of the matrix $\mathcal{M}^2$. It is easy to check that all matrix elements of $\mathcal{M}^2$ are invariant with respect to unitary basis transformations, $\Phi_i \to U_i{}^j\Phi_j$, in light of eqs. (6.42) and (6.49). The squared masses of the physical neutral scalars, m_k^2 ($k = 1, 2, 3$) with no implied mass ordering, are the eigenvalues of $\mathcal{M}^2$, which are independent of the choice of η.

A necessary condition for the stability of the electroweak symmetry-breaking vacuum is that $\mathcal{M}^2$ is a positive definite matrix. For example, all the diagonal elements of $\mathcal{M}^2$ must be positive.[8] It then follows that the electroweak symmetry-breaking vacuum constitutes a global minimum. This is most easily seen by computing the vacuum energy density at the minimum, $\mathcal{V}_{\min}$, by plugging $\mathcal{H}_1 = v/\sqrt{2}$ and $\mathcal{H}_2 = 0$ into eq. (6.48). Then, using eq. (6.50) to eliminate Y_1 in favor of Z_1, we end up with

$$V_{\min} = -\tfrac{1}{8}Z_1 v^4 < 0\,. \qquad (6.54)$$

Since $V_{\min} = 0$ for the electroweak conserving vacuum (where the Higgs field vacuum expectation values vanish), we conclude that the broken electroweak vacuum is a global minimum.[9]

The real symmetric squared-mass matrix $\mathcal{M}^2$ can be diagonalized by a real orthogonal transformation of unit determinant,

$$R\mathcal{M}^2R^{\mathsf{T}} = \mathcal{M}_D^2 \equiv \mathrm{diag}\,(m_1^2\,,\, m_2^2\,,\, m_3^2)\,, \qquad (6.55)$$

where $RR^{\mathsf{T}} = I$, $\det R = 1$, and the m_k^2 are the eigenvalues of $\mathcal{M}^2$. A convenient form for R is

$$R = R_{12}R_{13}R_{23} = \begin{pmatrix} c_{12} & -s_{12} & 0 \\ s_{12} & c_{12} & 0 \\ 0 & 0 & 1 \end{pmatrix} \begin{pmatrix} c_{13} & 0 & -s_{13} \\ 0 & 1 & 0 \\ s_{13} & 0 & c_{13} \end{pmatrix} \begin{pmatrix} 1 & 0 & 0 \\ 0 & c_{23} & -s_{23} \\ 0 & s_{23} & c_{23} \end{pmatrix}$$

$$= \begin{pmatrix} c_{13}c_{12} & -s_{12}c_{23} - c_{12}s_{13}s_{23} & -c_{12}s_{13}c_{23} + s_{12}s_{23} \\ c_{13}s_{12} & c_{12}c_{23} - s_{12}s_{13}s_{23} & -s_{12}s_{13}c_{23} - c_{12}s_{23} \\ s_{13} & c_{13}s_{23} & c_{13}c_{23} \end{pmatrix}, \qquad (6.56)$$

where $c_{ij} \equiv \cos\theta_{ij}$ and $s_{ij} \equiv \sin\theta_{ij}$. Indeed, the angles θ_{12}, θ_{13}, and θ_{23} defined

[8] A more complete list of stability conditions will be given in eqs. (6.59)–(6.62).

[9] The possibility exists that, starting from eq. (6.33), one finds two different solutions to the scalar potential minimum conditions given by eq. (6.36). In this case the global minimum corresponds to the solution with the largest value of $Z_1 v^4$.

above are all invariant quantities since they are obtained by diagonalizing $\mathcal{M}^2$, whose matrix elements are independent of the choice of scalar field basis.

The neutral physical Higgs mass eigenstates are denoted by h_1, h_2, and h_3:

$$\begin{pmatrix} h_1 \\ h_2 \\ h_3 \end{pmatrix} = R \begin{pmatrix} \varphi_1^0 \\ \varphi_2^0 \\ a^0 \end{pmatrix} = RW \begin{pmatrix} \sqrt{2}\,\mathrm{Re}\,\mathcal{H}_1^0 - v \\ \mathcal{H}_2^0 \\ \mathcal{H}_2^{0\,\dagger} \end{pmatrix}, \tag{6.57}$$

which defines the unitary matrix W. The matrix RW diagonalizes the transformed squared-mass matrix, $W^\dagger \mathcal{M}^2 W$. Explicitly, we have

$$W^\dagger \mathcal{M}^2 W = v^2 \begin{pmatrix} Z_1 & \frac{1}{\sqrt{2}} Z_6 e^{-i\eta} & \frac{1}{\sqrt{2}} Z_6^* e^{i\eta} \\ \frac{1}{\sqrt{2}} Z_6^* e^{i\eta} & \frac{1}{2}(Z_3+Z_4) + Y_2/v^2 & \frac{1}{2} Z_5^* e^{2i\eta} \\ \frac{1}{\sqrt{2}} Z_6 e^{-i\eta} & \frac{1}{2} Z_5 e^{-2i\eta} & \frac{1}{2}(Z_3+Z_4)+Y_2/v^2 \end{pmatrix}. \tag{6.58}$$

The stability of the electroweak symmetry-breaking vacuum implies that $W^\dagger \mathcal{M}^2 W$ is positive definite. Hence the principal minors of $W^\dagger \mathcal{M}^2 W$ must all be positive. The positivity of the diagonal elements and the 2×2 principal minors then yields the following inequalities:

$$Z_1 > 0\,, \tag{6.59}$$

$$Z_1\big(Z_3 + Z_4 + 2Y_2/v^2\big) > |Z_6|^2\,, \tag{6.60}$$

$$Z_3 + Z_4 + 2Y_2/v^2 > |Z_5|\,. \tag{6.61}$$

Finally, the positivity of $\det \mathcal{M}^2$ leads to the constraint

$$\mathrm{Re}(Z_5^* Z_6) > \tfrac{1}{2} Z_1 \big[|Z_5|^2 - (Z_3 + Z_4 + 2Y_2/v^2)^2\big] + |Z_6|^2\big(Z_3 + Z_4 + 2Y_2/v^2\big)\,. \tag{6.62}$$

Note that the sign of the right-hand side of eq. (6.62) can be negative, so $Z_5 = 0$ and/or $Z_6 = 0$ do not in general violate the above constraints.

A straightforward calculation yields

$$RW = \begin{pmatrix} q_{11} & \frac{1}{\sqrt{2}} q_{12}^* \, e^{i\theta_{23}} & \frac{1}{\sqrt{2}} q_{12} \, e^{-i\theta_{23}} \\ q_{21} & \frac{1}{\sqrt{2}} q_{22}^* \, e^{i\theta_{23}} & \frac{1}{\sqrt{2}} q_{22} \, e^{-i\theta_{23}} \\ q_{31} & \frac{1}{\sqrt{2}} q_{32}^* \, e^{i\theta_{23}} & \frac{1}{\sqrt{2}} q_{32} \, e^{-i\theta_{23}} \end{pmatrix}, \tag{6.63}$$

where the $q_{k\ell}$ are listed in Table 6.1. The following relation satisfied by the q_{jk} is notable:

$$q_{j2}^* q_{k2} = \delta_{jk} - q_{j1} q_{k1} + i\epsilon_{jk\ell} q_{\ell 1}\,, \tag{6.64}$$

where there is an implicit sum over ℓ. Setting $j = k$ then yields

$$q_{k1}^2 + |q_{k2}|^2 = 1\,. \tag{6.65}$$

One can also derive simple sum rules that are satisfied by the q_{kj}:

$$\sum_{k=1}^{3} q_{k1}^2 = \frac{1}{2} \sum_{k=1}^{3} |q_{k2}|^2 = 1\,, \qquad \sum_{k=1}^{3} q_{k2}^2 = \sum_{k=1}^{3} q_{k1} q_{k2} = 0\,. \tag{6.66}$$

Table 6.1 Invariant combinations of neutral Higgs mixing angles θ_{12} and θ_{13}, where $c_{ij} \equiv \cos\theta_{ij}$ and $s_{ij} \equiv \sin\theta_{ij}$.

k	q_{k1}	q_{k2}
1	$c_{12}c_{13}$	$-s_{12} - ic_{12}s_{13}$
2	$s_{12}c_{13}$	$c_{12} - is_{12}s_{13}$
3	s_{13}	ic_{13}

Employing eq. (6.57), it follows that

$$h_k = q_{k1}\left(\sqrt{2}\,\mathrm{Re}\,\mathcal{H}_1^0 - v\right) + \frac{1}{\sqrt{2}}\left(q_{k2}^*\mathcal{H}_2^0 e^{i\theta_{23}} + \text{h.c.}\right). \tag{6.67}$$

Equations (6.46) and (6.51) also yield expressions for the massless neutral and charged Goldstone fields,[10]

$$G^0 = \widehat{v}^i\Phi_i^0\,, \qquad G^+ = \widehat{v}^i\Phi_i^+\,, \tag{6.68}$$

and the charged Higgs field, $\mathcal{H}^+ = e^{i\eta}\widehat{w}^i\Phi_i^+$. Nevertheless, one is always free to rephase the charged Higgs field without affecting any observable of the model. It is convenient to define the positively charged Higgs field as follows:

$$H^+ \equiv e^{i\theta_{23}}\mathcal{H}_2^+\,. \tag{6.69}$$

Note that this rephasing is conventional and does not alter the fact that H^+ is an invariant field with respect to scalar field basis transformations. The squared mass of the charged Higgs field is given by

$$m_{H^\pm}^2 = Y_2 + \tfrac{1}{2}Z_3 v^2\,. \tag{6.70}$$

One can invert eq. (6.67) and include the charged scalars to obtain

$$\mathcal{H}_1 = \begin{pmatrix} G^+ \\ \dfrac{1}{\sqrt{2}}\left(v + iG + \displaystyle\sum_{k=1}^{3} q_{k1}h_k\right)\end{pmatrix}, \qquad e^{i\theta_{23}}\mathcal{H}_2 = \begin{pmatrix} H^+ \\ \dfrac{1}{\sqrt{2}}\displaystyle\sum_{k=1}^{3} q_{k2}h_k\end{pmatrix}. \tag{6.71}$$

Although θ_{23} is an invariant parameter, it has no physical significance since it can be eliminated by rephasing $\mathcal{H}_2 \to e^{-i\theta_{23}}\mathcal{H}_2$. Thus, without loss of generality, we henceforth set $\theta_{23} = 0$.

The neutral Higgs mass eigenstates can then be expressed in terms of the Φ-basis fields by making use of eq. (6.46):

$$h_k = \frac{1}{\sqrt{2}}\left[\overline{\Phi}^{0i}(q_{k1}\widehat{v}_i + q_{k2}\widehat{w}_i e^{-i\eta}) + (q_{k1}\widehat{v}^i + q_{k2}^*\widehat{w}^i e^{i\eta})\Phi_i^0\right], \tag{6.72}$$

[10] In the R_ξ gauge, the squared masses of the neutral and charged Goldstone fields are modified by the gauge-fixing term [see eq. (4.87)], such that $m_G^2 = \xi m_Z^2$ and $m_{G^\pm}^2 = \xi m_W^2$, respectively.

where the shifted neutral fields are defined by $\overline{\Phi}_i^0 \equiv \Phi_i^0 - v\,\widehat{v}_i/\sqrt{2}$ and $\overline{\Phi}^{0i} \equiv (\overline{\Phi}_i^0)^\dagger$. The remaining freedom to define the overall sign of h_k is associated with the convention adopted for the domains of the mixing angles θ_{ij}. By convention, we choose the invariant mixing angles θ_{12} and θ_{13}, which are defined modulo π, to lie in the region

$$-\tfrac{1}{2}\pi \leq \theta_{12}\,,\, \theta_{13} \leq \tfrac{1}{2}\pi\,. \tag{6.73}$$

These angles are then uniquely determined once the mass ordering of the three neutral states h_k is fixed.[11] Furthermore, we set $\theta_{23} = 0$ as indicated below eq. (6.71). Finally, one can invert eq. (6.72) and include the charged scalars to obtain

$$\Phi_i = \begin{pmatrix} G^+\widehat{v}_i + H^+ e^{-i\eta}\widehat{w}_i \\ \dfrac{v}{\sqrt{2}}\widehat{v}_i + \dfrac{1}{\sqrt{2}}\left(iG + \displaystyle\sum_{k=1}^{3}\left(q_{k1}\widehat{v}_i + q_{k2}e^{-i\eta}\widehat{w}_i\right)h_k\right)\end{pmatrix}. \tag{6.74}$$

Note that by setting $\widehat{v} = (1 \quad 0)$ and $\widehat{w} = (0 \quad 1)$ [see Exercise 6.5(a)], one recovers the Higgs basis fields given in eq. (6.71).

Having set $\theta_{23} = 0$, eq. (6.56) yields a diagonalizing matrix $R = R_{12}R_{13}$ that depends only on the invariant angles θ_{12} and θ_{13}:

$$R = \begin{pmatrix} c_{12}c_{13} & -s_{12} & -c_{12}s_{13} \\ c_{13}s_{12} & c_{12} & -s_{12}s_{13} \\ s_{13} & 0 & c_{13}\end{pmatrix} = \begin{pmatrix} q_{11} & \text{Re }q_{12} & \text{Im }q_{12} \\ q_{21} & \text{Re }q_{22} & \text{Im }q_{22} \\ q_{31} & \text{Re }q_{32} & \text{Im }q_{32}\end{pmatrix}. \tag{6.75}$$

To obtain explicit expressions for the neutral Higgs boson squared masses, one must solve a cubic characteristic equation, which yields the eigenvalues of $\mathcal{M}^2$. The resulting expressions are unwieldy and impractical. Nevertheless, one can derive useful sum rules by employing eqs. (6.55) and (6.75). It then follows that

$$Z_1 = \frac{1}{v^2}\sum_{k=1}^{3} m_k^2 (q_{k1})^2\,, \tag{6.76}$$

$$Z_4 = \frac{1}{v^2}\left[\sum_{k=1}^{3} m_k^2 |q_{k2}|^2 - 2m_{H^\pm}^2\right], \tag{6.77}$$

$$Z_5 e^{-2i\eta} = \frac{1}{v^2}\sum_{k=1}^{3} m_k^2 (q_{k2}^*)^2\,, \tag{6.78}$$

$$Z_6 e^{-i\eta} = \frac{1}{v^2}\sum_{k=1}^{3} m_k^2\, q_{k1} q_{k2}^*\,, \tag{6.79}$$

after making use of eq. (6.70). Another noteworthy relation is the trace condition,

$$\text{Tr}\,\mathcal{M}^2 = \sum_{k=1}^{3} m_k^2 = 2Y^2 + (Z_1 + Z_3 + Z_4)v^2\,. \tag{6.80}$$

[11] Since θ_{12} and θ_{13} are defined modulo π, the boundary value points $\pm\frac{1}{2}\pi$ are physically equivalent.

Employing the results of Table 6.1, it follows that eqs. (6.76)–(6.79) yield

$$Z_1 v^2 = m_1^2 c_{12}^2 c_{13}^2 + m_2^2 s_{12}^2 c_{13}^2 + m_3^2 s_{13}^2 \,, \tag{6.81}$$

$$Z_4 v^2 + 2m_{H^\pm}^2 = m_1^2(s_{12}^2 + c_{12}^2 s_{13}^2) + m_2^2(c_{12}^2 + s_{12}^2 s_{13}^2) + m_3^2 c_{13}^2 \,, \tag{6.82}$$

$$\mathrm{Re}(Z_5\, e^{-2i\eta})\, v^2 = m_1^2(s_{12}^2 - c_{12}^2 s_{13}^2) + m_2^2(c_{12}^2 - s_{12}^2 s_{13}^2) - m_3^2 c_{13}^2 \,, \tag{6.83}$$

$$\mathrm{Im}(Z_5\, e^{-2i\eta})\, v^2 = 2 s_{12} c_{12} s_{13}(m_2^2 - m_1^2) \,, \tag{6.84}$$

$$\mathrm{Re}(Z_6\, e^{-i\eta})\, v^2 = s_{12} c_{12} c_{13}(m_2^2 - m_1^2) \,, \tag{6.85}$$

$$\mathrm{Im}(Z_6\, e^{-i\eta})\, v^2 = s_{13} c_{13}(c_{12}^2 m_1^2 + s_{12}^2 m_2^2 - m_3^2) \,. \tag{6.86}$$

Equations (6.81), (6.84), (6.85), and (6.86) can be rewritten in an equivalent form that will be particularly useful when we introduce the Higgs alignment limit in Section 6.2.3:

$$Z_1 v^2 - m_1^2 = (m_2^2 - m_1^2) c_{13}^2 s_{12}^2 + (m_3^2 - m_1^2) s_{13}^2 \,, \tag{6.87}$$

$$\mathrm{Im}(Z_5\, e^{-2i\eta}) = \frac{2 s_{13}}{c_{13}}\, \mathrm{Re}(Z_6\, e^{-i\eta}) \,, \tag{6.88}$$

$$s_{12} c_{12} c_{13} = \frac{\mathrm{Re}(Z_6\, e^{-i\eta}) v^2}{m_2^2 - m_1^2} \,, \tag{6.89}$$

$$s_{13} c_{13} = \frac{\mathrm{Im}(Z_6\, e^{-i\eta}) v^2}{m_1^2 - m_3^2 + s_{12}^2(m_2^2 - m_1^2)} \,. \tag{6.90}$$

6.2.2 Tree-Level Higgs Boson Couplings in the General 2HDM

The interactions of the scalar bosons and the vector bosons arise from the kinetic energy term of the scalars, whose form is independent of the choice of scalar field basis. Thus, employing the Higgs basis fields,

$$\mathscr{L}_{\rm kinetic} = (\nabla^\mu \mathcal{H}_1)^\dagger (\nabla_\mu \mathcal{H}_1) + (\nabla^\mu \mathcal{H}_2)^\dagger (\nabla_\mu \mathcal{H}_2) \,, \tag{6.91}$$

where the covariant derivative ∇_μ was defined in eq. (4.128). It then follows that

$$\nabla_\mu \mathcal{H}_i = \begin{pmatrix} \partial_\mu \mathcal{H}_i^+ + \left[\dfrac{ig}{c_W}\left(\tfrac{1}{2} - s_W^2\right) Z_\mu + ieA_\mu\right] \mathcal{H}_i^+ + \dfrac{ig}{\sqrt{2}} W_\mu^+ \mathcal{H}_i^0 \\ \partial_\mu \mathcal{H}_i^0 - \dfrac{ig}{2c_W} Z_\mu \mathcal{H}_i^0 + \dfrac{ig}{\sqrt{2}} W_\mu^- \mathcal{H}_i^+ \end{pmatrix} , \tag{6.92}$$

where $s_W \equiv \sin\theta_W$ and $c_W \equiv \cos\theta_W$.

Inserting eq. (6.71) into eq. (6.91) yields the interactions of the Higgs bosons and the vector bosons $V = W^\pm$ and/or Z:

$$\mathscr{L}_{VVH} = \left(gm_W W_\mu^+ W^{\mu -} + \frac{g}{2c_W} m_Z Z_\mu Z^\mu\right) q_{k1} h_k$$
$$+ e m_W A^\mu (W_\mu^+ G^- + W_\mu^- G^+) - g m_Z s_W^2 Z^\mu (W_\mu^+ G^- + W_\mu^- G^+) , \quad (6.93)$$

$$\mathscr{L}_{VHH} = -\frac{g}{4c_W}\, \epsilon_{jk\ell} q_{\ell 1} Z^\mu h_j \overleftrightarrow{\partial}_\mu h_k + \frac{g}{2c_W} q_{k1} Z^\mu G \overleftrightarrow{\partial}_\mu h_k$$
$$- \tfrac{1}{2} g \left\{ i W_\mu^+ \left[q_{k1} G^- \overleftrightarrow{\partial}^\mu h_k + q_{k2}\, H^- \overleftrightarrow{\partial}^\mu h_k \right] + \text{h.c.} \right\}$$
$$+ \tfrac{1}{2} g \left(W_\mu^+ G^- \overleftrightarrow{\partial}^\mu G + W_\mu^- G^+ \overleftrightarrow{\partial}^\mu G \right)$$
$$+ \left[i e A^\mu + \frac{ig}{c_W} \left(\tfrac{1}{2} - s_W^2\right) Z^\mu \right] (G^+ \overleftrightarrow{\partial}_\mu G^- + H^+ \overleftrightarrow{\partial}_\mu H^-) , \quad (6.94)$$

$$\mathscr{L}_{VVHH} = \left[\tfrac{1}{4} g^2 W_\mu^+ W^{\mu -} + \frac{g^2}{8c_W^2} Z_\mu Z^\mu \right] (GG + h_k h_k)$$
$$+ \left[\tfrac{1}{2} g^2 W_\mu^+ W^{\mu -} + e^2 A_\mu A^\mu + \frac{g^2}{c_W^2} \left(\tfrac{1}{2} - s_W^2\right)^2 Z_\mu Z^\mu \right.$$
$$\left. + \frac{2ge}{c_W} \left(\tfrac{1}{2} - s_W^2\right) A_\mu Z^\mu \right] (G^+ G^- + H^+ H^-)$$
$$+ \left\{ \left(\tfrac{1}{2} e g A^\mu W_\mu^+ - \frac{g^2 s_W^2}{2c_W} Z^\mu W_\mu^+ \right) \left[(q_{k1} G^- + q_{k2}\, H^-) h_k + i G^- G \right] + \text{h.c.} \right\} , \quad (6.95)$$

where the sum over pairs of repeated indices $j, k \in \{1, 2, 3\}$ is implied.

The scalar self-interactions are obtained from eq. (6.48), by expressing $\mathcal{H}_1$ and $\mathcal{H}_2$ in terms of the scalar mass eigenstates [using eq. (6.71)] and changing the overall sign to obtain the interaction Lagrangian. It is convenient to define

$$Z_{34} \equiv Z_3 + Z_4 , \qquad \overline{Z}_5 \equiv Z_5 e^{-2i\eta} , \qquad \overline{Z}_6 \equiv Z_6 e^{-i\eta} , \qquad \overline{Z}_7 \equiv Z_7 e^{-i\eta} . \quad (6.96)$$

The cubic scalar self-interactions are given by

$$\mathscr{L}_{3h} = -\tfrac{1}{2} v\, h_j h_k h_\ell \Big[q_{j1} q_{k1} q_{\ell 1} Z_1 + q_{j2} q_{k2}^* q_{\ell 1} Z_{34} + q_{j1} \operatorname{Re}(q_{k2} q_{\ell 2} \overline{Z}_5)$$
$$+ 3 q_{j1} q_{k1} \operatorname{Re}(q_{\ell 2} \overline{Z}_6) + \operatorname{Re}(q_{j2}^* q_{k2} q_{\ell 2} \overline{Z}_7) \Big] - v\, h_k H^+ H^- \Big[q_{k1} Z_3 + \operatorname{Re}(q_{k2}\, \overline{Z}_7) \Big]$$
$$- \frac{1}{2v} \Big\{ h_j h_k G\, \epsilon_{jk\ell} q_{\ell 1} (m_j^2 - m_k^2) + h_j (GG + 2G^+ G^-) q_{j1} m_j^2$$
$$+ 2 \left[h_j G^+ H^- q_{j2} (m_j^2 - m_{H^\pm}^2) + \text{h.c.} \right] \Big\} , \quad (6.97)$$

where $m_j \equiv m_{h_j}$ and there is an implicit sum over repeated indices $i, j, \ell \in \{1, 2, 3\}$.

In obtaining the couplings to Goldstone bosons (which are relevant for computations that employ the R_ξ gauge), we have made use of eqs. (6.64), (6.78), and (6.79). The quartic scalar self-interactions (excluding the Goldstone boson fields) are given by

$$\begin{aligned}\mathscr{L}_{4h} = -\tfrac{1}{8}h_j h_k h_l h_m \Big[& q_{j1}q_{k1}q_{\ell 1}q_{m1}Z_1 + q_{j2}q_{k2}q^*_{\ell 2}q^*_{m2}Z_2 + 2q_{j1}q_{k1}q_{\ell 2}q^*_{m2}Z_{34} \\ & + 2q_{j1}q_{k1}\,\mathrm{Re}(q_{\ell 2}q_{m2}\overline{Z}_5) + 4q_{j1}q_{k1}q_{\ell 1}\,\mathrm{Re}(q_{m2}\overline{Z}_6) + 4q_{j1}\,\mathrm{Re}(q_{k2}q_{\ell 2}q^*_{m2}\overline{Z}_7)\Big] \\ & - \tfrac{1}{2}h_j h_k H^+H^-\Big[q_{j2}q^*_{k2}Z_2 + q_{j1}q_{k1}Z_3 + 2q_{j1}\,\mathrm{Re}(q_{k2}\overline{Z}_7)\Big] - \tfrac{1}{2}Z_2 H^+H^-H^+H^-,\end{aligned} \tag{6.98}$$

where there is an implicit sum over repeated indices $i, j, k, \ell \in \{1,2,3\}$. The omitted quartic couplings that involve the Goldstone bosons are given in eq. (6.338).

Next, consider the Higgs–fermion Yukawa couplings. We can generalize eq. (4.169) by summing over contributions from the two Higgs fields Φ_i:[12]

$$\begin{aligned}-\mathscr{L}_{\rm Y} = (\widehat{\boldsymbol{y}}_{\boldsymbol{u}}^{\,i})^m{}_n\Big[\Phi_i^0\widehat{u}_m\widehat{\bar{u}}^n - \Phi_i^+\widehat{d}_m\widehat{\bar{u}}^n\Big] + (\widehat{\boldsymbol{y}}_{\boldsymbol{d}i})^m{}_n\Big[(\Phi^-)^i\,\widehat{u}_m\widehat{\bar{d}}^n + \Phi^{0i}\widehat{d}_m\widehat{\bar{d}}^n\Big] \\ + (\widehat{\boldsymbol{y}}_{\boldsymbol{e}i})^m{}_n\Big[(\Phi^-)^i\,\widehat{\nu}_m\widehat{\bar{e}}^n + \Phi^{0i}\widehat{e}_m\widehat{\bar{e}}^n\Big] + \text{h.c.}\,,\end{aligned} \tag{6.99}$$

where the hatted two-component spinor fields correspond to the fermion interaction eigenstates, and m, n are flavor labels.[13] In eq. (6.99), we follow our conventions where $\Phi^i \equiv (\Phi_i)^\dagger$. It is also convenient to define

$$\widehat{\boldsymbol{y}}_{\boldsymbol{f}}^{\,i} \equiv (\widehat{\boldsymbol{y}}_{\boldsymbol{f}i})^\dagger\,, \qquad \text{for } f = u, d, e. \tag{6.100}$$

Under a unitary transformation of the scalar field basis, $\Phi_i \to U_i{}^j\Phi_j$, the Yukawa coupling matrices transform covariantly: $\widehat{\boldsymbol{y}}_{\boldsymbol{f}i} \to U_i{}^j\widehat{\boldsymbol{y}}_{\boldsymbol{f}j}$. Thus, we can construct invariant matrix Yukawa couplings $\widehat{\boldsymbol{\kappa}}^{\boldsymbol{F}}$ and $\widehat{\boldsymbol{\rho}}^{\boldsymbol{F}}$ (where $F = U, D, E$) as follows:

$$\widehat{\boldsymbol{\kappa}}^{\boldsymbol{F}} \equiv \widehat{v}^j\,\widehat{\boldsymbol{y}}_{\boldsymbol{f}j}\,, \qquad \widehat{\boldsymbol{\rho}}^{\boldsymbol{F}} \equiv e^{i\eta}\widehat{w}^j\,\widehat{\boldsymbol{y}}_{\boldsymbol{f}j}\,. \tag{6.101}$$

By including the factor of $e^{i\eta}$ in the definition above, it follows that $\widehat{\boldsymbol{\kappa}}^{\boldsymbol{F}}$ and $\widehat{\boldsymbol{\rho}}^{\boldsymbol{F}}$ are invariant under a unitary transformation of the scalar field basis, in light of eqs. (6.42) and (6.45). Inserting eq. (6.47) into eq. (6.99), we end up with

$$\begin{aligned}-\mathscr{L}_{\rm Y} = (\widehat{\boldsymbol{\kappa}}^{\boldsymbol{U}})^{\dagger\, m}{}_n\Big[\mathcal{H}_1^0\widehat{u}_m\widehat{\bar{u}}^n - \mathcal{H}_1^+\widehat{d}_m\widehat{\bar{u}}^n\Big] + (\widehat{\boldsymbol{\rho}}^{\boldsymbol{U}})^{\dagger\, m}{}_n\Big[\mathcal{H}_2^0\widehat{u}_m\widehat{\bar{u}}^n - \mathcal{H}_2^+\widehat{d}_m\widehat{\bar{u}}^n\Big] \\ + (\widehat{\boldsymbol{\kappa}}^{\boldsymbol{D}})^m{}_n\Big[\mathcal{H}_1^-\widehat{u}_m\widehat{\bar{d}}^n + \mathcal{H}_1^{0\,\dagger}\widehat{d}_m\widehat{\bar{d}}^n\Big] + (\widehat{\boldsymbol{\rho}}^{\boldsymbol{D}})^m{}_n\Big[\mathcal{H}_2^-\widehat{u}_m\widehat{\bar{d}}^n + \mathcal{H}_2^{0\,\dagger}\widehat{d}_m\widehat{\bar{d}}^n\Big] \\ + (\widehat{\boldsymbol{\kappa}}^{\boldsymbol{E}})^m{}_n\Big[\mathcal{H}_1^-\widehat{\nu}_m\widehat{\bar{e}}^n + \mathcal{H}_1^{0\,\dagger}\widehat{e}_m\widehat{\bar{e}}^n\Big] + (\widehat{\boldsymbol{\rho}}^{\boldsymbol{E}})^m{}_n\Big[\mathcal{H}_2^-\widehat{\nu}_m\widehat{\bar{e}}^n + \mathcal{H}_2^{0\,\dagger}\widehat{e}_m\widehat{\bar{e}}^n\Big] \\ + \text{h.c.}\end{aligned} \tag{6.102}$$

[12] The Yukawa coupling matrices defined in eqs. (6.99)–(6.101) are the matrix transposes of the corresponding matrices introduced in Refs. [58, 59] (cf. footnote 18 in Section 4.7).

[13] In contrast to the notation used in Sections 4.7 and 13.2, in this section we shall employ hatted Yukawa coupling matrices to indicate the couplings of interaction-eigenstate fermion fields in order to explicitly distinguish these quantities from the corresponding Yukawa couplings of mass-eigenstate fermion fields.

The fermion mass matrices can be identified by setting $\mathcal{H}_1^0 = \mathcal{H}_1^{0\dagger} = v/\sqrt{2}$ and $\mathcal{H}_1^+ = \mathcal{H}_1^- = \mathcal{H}_2^+ = \mathcal{H}_2^- = \mathcal{H}_2^0 = \mathcal{H}_2^{0\dagger} = 0$ in eq. (6.102):

$$(\widehat{\boldsymbol{M}}_{\boldsymbol{U}})^m{}_n = \frac{v}{\sqrt{2}}(\widehat{\boldsymbol{\kappa}}^{\boldsymbol{U}})^{\dagger\, m}{}_n\,, \qquad (\widehat{\boldsymbol{M}}_{\boldsymbol{F}})^m{}_n = \frac{v}{\sqrt{2}}(\widehat{\boldsymbol{\kappa}}^{\boldsymbol{F}})^m{}_n\,, \quad \text{for } F = D,\, E. \tag{6.103}$$

To diagonalize the quark and lepton mass matrices, we follow the analysis given in eqs. (4.173)–(4.179). That is, we introduce four unitary matrices for the quark mass diagonalization, L_u, L_d, R_u, and R_d, and two unitary matrices for the lepton mass diagonalization, L_e and R_e, such that

$$\widehat{u}_m = (L_u)_m{}^n u_n\,, \qquad \widehat{d}_m = (L_d)_m{}^n d_n\,, \qquad \widehat{\bar{u}}^m = (R_u)^m{}_n \bar{u}^n\,, \qquad \widehat{\bar{d}}^m = (R_d)^m{}_n \bar{d}^n, \tag{6.104}$$

$$\widehat{\nu}_i = (L_e)_m{}^n \nu_j\,, \qquad \widehat{e}_m = (L_e)_m{}^n e_n\,, \qquad \widehat{\bar{e}}^m = (R_e)^m{}_n \bar{e}^n\,, \tag{6.105}$$

where the unhatted fields u, d, $\bar{u}$, and $\bar{d}$ are the corresponding quark mass eigenstates and ν, e, and $\bar{e}$ are the corresponding lepton mass eigenstates. Since the neutrinos are massless, we are free to define the physical neutrino fields, ν_m, as the SU(2) partners of the corresponding charged lepton mass-eigenstate fields. The fermion mass diagonalization procedure consists of the singular value decomposition of the quark and lepton mass matrices:

$$L_u^{\mathsf{T}}\, \widehat{\boldsymbol{M}}_{\boldsymbol{U}}\, R_u \equiv \boldsymbol{M}_{\boldsymbol{U}} = \text{diag}(m_u\,,\, m_c\,,\, m_t)\,, \tag{6.106}$$

$$L_d^{\mathsf{T}}\, \widehat{\boldsymbol{M}}_{\boldsymbol{D}}\, R_d \equiv \boldsymbol{M}_{\boldsymbol{D}} = \text{diag}(m_d\,,\, m_s\,,\, m_b)\,, \tag{6.107}$$

$$L_e^{\mathsf{T}}\, \widehat{\boldsymbol{M}}_{\boldsymbol{E}}\, R_e \equiv \boldsymbol{M}_{\boldsymbol{E}} = \text{diag}(m_e\,,\, m_\mu\,,\, m_\tau)\,, \tag{6.108}$$

where the diagonalized masses are real and nonnegative. Since no right-handed neutrino field has been introduced so far, the neutrinos are exactly massless. One can generate neutrino masses by constructing a seesaw-extended 2HDM following the analysis of Section 6.1. The details of this construction are left to the reader.

To write out the corresponding Higgs–fermion Yukawa interactions, it is convenient to define

$$\boldsymbol{\kappa}^{\boldsymbol{U}} \equiv L_u^{\mathsf{T}}\, \widehat{\boldsymbol{\kappa}}^{\boldsymbol{U}\dagger} R_u = \frac{\sqrt{2}}{v}\boldsymbol{M}_{\boldsymbol{U}}\,, \tag{6.109}$$

$$\boldsymbol{\kappa}^{\boldsymbol{D}} \equiv L_d^{\mathsf{T}}\, \widehat{\boldsymbol{\kappa}}^{\boldsymbol{D}} R_d = \frac{\sqrt{2}}{v}\boldsymbol{M}_{\boldsymbol{D}}\,, \tag{6.110}$$

$$\boldsymbol{\kappa}^{\boldsymbol{E}} \equiv L_e^{\mathsf{T}}\, \widehat{\boldsymbol{\kappa}}^{\boldsymbol{E}} R_e = \frac{\sqrt{2}}{v}\boldsymbol{M}_{\boldsymbol{E}}\,, \tag{6.111}$$

which are diagonal matrices with positive entries by construction, and

$$\boldsymbol{\rho}^{\boldsymbol{U}\dagger} \equiv L_u^{\mathsf{T}}\, \widehat{\boldsymbol{\rho}}^{\boldsymbol{U}\dagger} R_u\,, \tag{6.112}$$

$$\boldsymbol{\rho}^{\boldsymbol{D}} \equiv L_d^{\mathsf{T}}\, \widehat{\boldsymbol{\rho}}^{\boldsymbol{D}} R_d\,, \tag{6.113}$$

$$\boldsymbol{\rho}^{\boldsymbol{E}} \equiv L_e^{\mathsf{T}}\, \widehat{\boldsymbol{\rho}}^{\boldsymbol{E}} R_e\,, \tag{6.114}$$

which are arbitrary complex coupling matrices that are independent of the fermion masses.

We can then rewrite eq. (6.102) in terms of the $\boldsymbol{\kappa}$ and $\boldsymbol{\rho}$ matrices and the CKM matrix $\boldsymbol{K} \equiv L_u^\dagger L_d$ introduced in eq. (4.179). In this derivation, it is especially useful to make use of eqs. (6.109)–(6.111) to obtain

$$(\widehat{\boldsymbol{\kappa}}^{U\dagger})^m{}_n (R_u)^n{}_\ell = (\boldsymbol{\kappa}^U)^k{}_\ell (L_u)^{\dagger m}_k \,, \qquad (\widehat{\boldsymbol{\kappa}}^{D})^m{}_n (R_d)^n{}_\ell = (\boldsymbol{\kappa}^D)^k{}_\ell (L_d)^{\dagger m}_k \,. \tag{6.115}$$

It then follows that

$$(\widehat{\boldsymbol{\kappa}}^{U\dagger})^m{}_n (L_d)_m{}^k (R_u)^n{}_\ell = (\boldsymbol{\kappa}^U)^p{}_\ell \, \boldsymbol{K}_p{}^k \,, \tag{6.116}$$

$$(\widehat{\boldsymbol{\kappa}}^{D})^m{}_n (L_u)_m{}^k (R_d)^n{}_\ell = (\boldsymbol{\kappa}^D)^p{}_\ell \, (\boldsymbol{K}^\dagger)_p{}^k \,. \tag{6.117}$$

Finally, in light of eq. (6.71), we define

$$H_2 \equiv e^{i\theta_{23}} \mathcal{H}_2 \,. \tag{6.118}$$

The resulting Yukawa interactions is given by

$$\begin{aligned} -\mathscr{L}_{\rm Y} = {} & u_k \bar{u}^\ell \left(\boldsymbol{\kappa}^U \mathcal{H}_1^0 + \boldsymbol{\rho}^{U\dagger} H_2^0\right)^k{}_\ell - \boldsymbol{K}_k{}^m d_m \bar{u}^\ell \left(\boldsymbol{\kappa}^U \mathcal{H}_1^+ + \boldsymbol{\rho}^{U\dagger} H_2^+\right)^k{}_\ell \\ & + d_k \bar{d}^\ell \left(\boldsymbol{\kappa}^D \mathcal{H}_1^{0\,\dagger} + \boldsymbol{\rho}^D H_2^{0\,\dagger}\right)^k{}_\ell + \boldsymbol{K}^{\dagger\, m}_k u_m \bar{d}^\ell \left(\boldsymbol{\kappa}^D \mathcal{H}_1^- + \boldsymbol{\rho}^D H_2^-\right)^k{}_\ell \\ & + e_k \bar{e}^\ell \left(\boldsymbol{\kappa}^E \mathcal{H}_1^{0\,\dagger} + \boldsymbol{\rho}^E H_2^{0\,\dagger}\right)^k{}_\ell + \nu_k \bar{e}^\ell \left(\boldsymbol{\kappa}^E \mathcal{H}_1^- + \boldsymbol{\rho}^E H_2^-\right)^k{}_\ell + \text{h.c.} \end{aligned} \tag{6.119}$$

We now insert the expressions for $\mathcal{H}_1$ and $\mathcal{H}_2$ given in eq. (6.71) into eq. (6.119). Regarding u, d, e, and ν as row vectors and $\bar{u}$, $\bar{d}$, and $\bar{e}$ as column vectors, and employing eq. (6.118), the interactions of the physical Higgs scalars, h_k ($k = 1, 2, 3$), with the quarks and leptons are given by

$$\begin{aligned} -\mathscr{L}_{\rm Y} = {} & u \left[\frac{\boldsymbol{M_U}}{v} q_{k1} + \frac{1}{\sqrt{2}} \boldsymbol{\rho}^{U\dagger} q_{k2}\right] \bar{u}\, h_k + d \left[\frac{\boldsymbol{M_D}}{v} q_{k1} + \frac{1}{\sqrt{2}} \boldsymbol{\rho}^{D} q^*_{k2}\right] \bar{d}\, h_k \\ & + e \left[\frac{\boldsymbol{M_E}}{v} q_{k1} + \frac{1}{\sqrt{2}} \boldsymbol{\rho}^{E} q^*_{k2}\right] \bar{e}\, h_k - d \boldsymbol{K}^{\mathsf{T}} \boldsymbol{\rho}^{U\dagger} \bar{u}\, H^+ \\ & + u \boldsymbol{K}^* \boldsymbol{\rho}^D \bar{d}\, H^- + \nu \boldsymbol{\rho}^E \bar{e}\, H^- + \text{h.c.}\,, \end{aligned} \tag{6.120}$$

where $\boldsymbol{M_{U,D,E}}$ are the *diagonal* quark and charged lepton mass matrices, respectively, and flavor indices have been suppressed. In the general 2HDM, the matrices $\boldsymbol{\rho^F}$ are complex and nondiagonal. Thus, eq. (6.120) generically yields FCNC processes mediated at tree level by the exchange of the h_k. A phenomenologically acceptable model must provide an explanation for the approximate diagonality of the $\boldsymbol{\rho^F}$ matrices. This issue will be addressed further in Section 6.2.4.

It is convenient to rewrite the Higgs–fermion Yukawa couplings in terms of two 3×3 hermitian matrices $\boldsymbol{\rho^F_R}$ and $\boldsymbol{\rho^F_I}$ (for $F = U, D, E$):

$$\boldsymbol{\rho^F} = \frac{\sqrt{2}}{v} \boldsymbol{M_F^{1/2}} (\boldsymbol{\rho^F_R} + i \boldsymbol{\rho^F_I}) \boldsymbol{M_F^{1/2}} \,. \tag{6.121}$$

Including the factors of $\boldsymbol{M_F}^{1/2}$ in eq. (6.121) is convenient as it yields a simpler form for the Yukawa couplings:

$$\begin{aligned}
\mathscr{L}_{\rm Y} = &-\frac{1}{v}\Big\{u\boldsymbol{M}_{\boldsymbol{U}}^{\boldsymbol{1/2}}\big[q_{k1}\mathbb{1}+(\boldsymbol{\rho}_{\boldsymbol{R}}^{\boldsymbol{U}}-i\boldsymbol{\rho}_{\boldsymbol{I}}^{\boldsymbol{U}})q_{k2}\big]\boldsymbol{M}_{\boldsymbol{U}}^{\boldsymbol{1/2}}\big]\bar{u}\\
&+d\boldsymbol{M}_{\boldsymbol{D}}^{\boldsymbol{1/2}}\big[q_{k1}\mathbb{1}+(\boldsymbol{\rho}_{\boldsymbol{R}}^{\boldsymbol{D}}+i\boldsymbol{\rho}_{\boldsymbol{I}}^{\boldsymbol{D}})q_{k2}^*\big]\boldsymbol{M}_{\boldsymbol{D}}^{\boldsymbol{1/2}}\big]\bar{d}\\
&+e\boldsymbol{M}_{\boldsymbol{E}}^{\boldsymbol{1/2}}\big[q_{k1}\mathbb{1}+(\boldsymbol{\rho}_{\boldsymbol{R}}^{\boldsymbol{E}}+i\boldsymbol{\rho}_{\boldsymbol{I}}^{\boldsymbol{E}})q_{k2}^*\big]\boldsymbol{M}_{\boldsymbol{E}}^{\boldsymbol{1/2}}\big]\bar{e}+\text{h.c.}\Big\}h_k\\
&-\frac{\sqrt{2}}{v}\Big\{\big[u\boldsymbol{K}^*\boldsymbol{M}_{\boldsymbol{D}}^{\boldsymbol{1/2}}(\boldsymbol{\rho}_{\boldsymbol{R}}^{\boldsymbol{D}}+i\boldsymbol{\rho}_{\boldsymbol{I}}^{\boldsymbol{D}})\boldsymbol{M}_{\boldsymbol{D}}^{\boldsymbol{1/2}}\bar{d}+\nu\boldsymbol{M}_{\boldsymbol{E}}^{\boldsymbol{1/2}}\,(\boldsymbol{\rho}_{\boldsymbol{R}}^{\boldsymbol{E}}+i\boldsymbol{\rho}_{\boldsymbol{I}}^{\boldsymbol{E}})\boldsymbol{M}_{\boldsymbol{E}}^{\boldsymbol{1/2}}\bar{e}\big]H^-\\
&-d\boldsymbol{K}^{\mathsf{T}}\boldsymbol{M}_{\boldsymbol{U}}^{\boldsymbol{1/2}}(\boldsymbol{\rho}_{\boldsymbol{R}}^{\boldsymbol{U}}-i\boldsymbol{\rho}_{\boldsymbol{I}}^{\boldsymbol{U}})\boldsymbol{M}_{\boldsymbol{U}}^{\boldsymbol{1/2}}\bar{u}H^+ +\text{h.c.}\Big\}\,, \qquad (6.122)
\end{aligned}$$

where $\mathbb{1}$ is the 3×3 identity matrix.[14]

In addition to the flavor-nondiagonal neutral Higgs boson Yukawa interactions, the general 2HDM exhibits new CP-violating phenomena. There are two different sources for CP violation: the first source is associated with the scalar potential and the second source is associated with the Higgs–fermion Yukawa couplings. Explicit CP violation in the scalar potential can arise if no scalar field basis exists in which the potentially complex parameters m_{12}^2, λ_5, λ_6, and λ_7 of eq. (6.33) are simultaneously real. The general conditions that must be satisfied such that the 2HDM scalar potential is explicitly CP-conserving are rather complicated and will not be given here. Even if CP is explicitly conserved by the scalar potential, it is possible that CP is spontaneously broken by the vacuum. In this case, no basis exists in which all the scalar potential parameters and the two vacuum expectation values v_1 and v_2 are simultaneously real.

It is straightforward to show that the 2HDM scalar potential and the vacuum are CP-conserving if there exists a rephasing of the Higgs basis field $\mathcal{H}_2$ (i.e., a choice of η) such that the potentially complex parameters of the Higgs basis, Z_5, Z_6, and Z_7, are simultaneously real.[15] Equivalently, the 2HDM scalar potential and the vacuum are CP-conserving if and only if

$$\text{Im}(Z_5^* Z_6^2) = \text{Im}(Z_5^* Z_7^2) = \text{Im}(Z_6^* Z_7) = 0\,, \qquad (6.123)$$

which are invariant conditions with respect to the rephasing of the Higgs basis field $\mathcal{H}_2$. To prove this assertion, we first note that if Z_5, Z_6, and Z_7 are simultaneously real, then in the Higgs basis all scalar potential parameters and the VEVs, $v_1 = v$ and $v_2 = 0$, are simultaneously real. Conversely, if some basis exists in which all the scalar potential parameters and the two vacuum expectation values v_1 and v_2 are simultaneously real, then eq. (6.46) with $\eta = 0$ provides the transformation to the Higgs basis in which Z_5, Z_6, and Z_7 are simultaneously real.

The second source of CP violation can arise via the neutral Higgs–fermion Yukawa

[14] One can also evaluate the interaction of the fermions with the Goldstone bosons $G^\pm$ and G^0. The resulting expression coincides with the result of the Standard Model given in eq. (4.199).

[15] Note that if Z_6 is real then Y_3 is real in light of the scalar potential minimum conditions [see eq. (6.50)].

couplings. Even if the 2HDM scalar potential and vacuum are CP-conserving, in which case the neutral scalars h_k can be assigned definite CP quantum numbers, the neutral Higgs–fermion Yukawa interactions specified in eq. (6.120) are CP-violating unless $\boldsymbol{\rho}^U$, $\boldsymbol{\rho}^D$, and $\boldsymbol{\rho}^E$ are real matrices (e.g., see [B84] and Ref. [60]).

6.2.3 The Higgs Alignment Limit of the 2HDM

The tree-level couplings of the neutral field,

$$\varphi \equiv \sqrt{2}\,\mathrm{Re}\,\mathcal{H}_1^0 - v\,, \tag{6.124}$$

which resides in the scalar doublet $\mathcal{H}_1$ of the Higgs basis, are precisely those of the neutral Higgs field of the Standard Model. However, in a general 2HDM, the field φ is not a scalar mass eigenstate due to its mixing with the neutral scalar states that reside in $\mathcal{H}_2$. This implies that in general none of the neutral Higgs bosons of the 2HDM behave exactly like the Higgs boson of the Standard Model (SM).

Nevertheless, regions of the parameter space exist in which one of the neutral Higgs bosons is approximately SM-like. In this parameter regime, one of the neutral mass-eigenstate Higgs fields is approximately aligned with φ. The *alignment limit* of an extended Higgs sector (also called the Higgs alignment limit) is defined to be the region of parameter space in which one of the neutral mass-eigenstate Higgs fields is aligned in field space with the Higgs vacuum expectation value. Given that the Higgs boson observed at the CERN Large Hadron Collider (LHC) behaves as predicted by the Standard Model (to within the current experimental accuracy of the Higgs boson measurements), it follows that the parameter space of any extended Higgs sector must be close to the Higgs alignment limit.

We now examine the implications of the approximate Higgs alignment limit on the 2HDM parameter space. We shall assume that the neutral Higgs boson h_1 (with mass $m_1 = 125$ GeV) is SM-like and is to be identified with the Higgs boson observed at the LHC.[16] Starting from the neutral scalar squared-mass matrix given in eq. (6.52), an approximate Higgs alignment is realized in two scenarios:

1. $Y_2 \gg v^2$, with all quartic scalar coupling parameters $Z_i \lesssim \mathcal{O}(1)$ held fixed (this is the *Higgs decoupling limit*),
2. $|Z_6| \ll 1$, with or without decoupling.

In both cases, we can identify $h_1 \simeq \varphi$, up to a small admixture of $\sqrt{2}\,\mathrm{Re}\,\mathcal{H}_2^0$ and $\sqrt{2}\,\mathrm{Im}\,\mathcal{H}_2^0$, which implies that h_1 is SM-like.

In the decoupling regime of scenario 1, the scalars h_2, h_3, and $H^\pm$ are significantly heavier than h_1, with a characteristic squared mass of $\mathcal{O}(Y_2)$. Hence, at energy scales below $(Y_2)^{1/2}$, one can integrate out all the heavy scalar states, resulting in an effective theory that can be identified as the Standard Model with a single Higgs doublet field. Corrections to the SM Higgs properties in this effective theory scale as v^2/Y_2. Above the energy scale $(Y_2)^{1/2}$, the heavy non-SM-like

[16] In particular, we shall not presume any particular mass ordering for the three neutral Higgs scalars h_1, h_2, and h_3.

scalars reside and are approximately mass-degenerate. More precisely, the squared-mass differences between two of the heavy scalar states are of $\mathcal{O}(v^2)$, which implies that the corresponding mass splittings among the heavy scalars are of $\mathcal{O}(v^2/\sqrt{Y_2})$.

In scenario 2, approximate Higgs alignment is realized with or without decoupling, depending on whether the condition of scenario 1 is or is not satisfied. In the latter scenario, all scalar squared masses are of $\mathcal{O}(v^2)$ or less. Finally, *exact* Higgs alignment corresponds to the $Y_2 \to \infty$ limit in scenario 1 or $Z_6 = 0$ in scenario 2.

The conditions for approximate Higgs alignment can also be ascertained by requiring that the tree-level properties of $h_1 \simeq \phi$ approximate those of the SM Higgs boson, which is denoted below by $h_{\rm SM}$. Using eq. (6.93), it follows that

$$\mathcal{R}_{VV} \equiv \frac{g_{h_1VV}}{g_{h_{\rm SM}VV}} = q_{11} = c_{12}c_{13}\,, \qquad \text{where } V = W \text{ or } Z\,. \tag{6.125}$$

Due to eq. (6.73), the approximate Higgs alignment limit, $\mathcal{R}_{VV} \simeq 1$, corresponds to

$$|s_{12}|\,,\ |s_{13}| \ll 1\,. \tag{6.126}$$

In particular, eqs. (6.89) and (6.90) yield

$$|s_{12}| \simeq \left|\frac{\operatorname{Re}(Z_6 e^{-i\eta})v^2}{m_2^2 - m_1^2}\right| \ll 1\,, \tag{6.127}$$

$$|s_{13}| \simeq \left|\frac{\operatorname{Im}(Z_6 e^{-i\eta})v^2}{m_3^2 - m_1^2}\right| \ll 1\,. \tag{6.128}$$

One additional small quantity characterizes the approximate Higgs alignment limit,

$$\left|\operatorname{Im}(Z_5 e^{-2i\eta})\right| \simeq \left|\frac{2(m_2^2 - m_1^2)s_{12}s_{13}}{v^2}\right| \simeq \left|\frac{\operatorname{Im}(Z_6^2 e^{-2i\eta})v^2}{m_3^2 - m_1^2}\right| \ll 1\,, \tag{6.129}$$

as a consequence of eq. (6.88).

Finally, in the approximate Higgs alignment limit the following mass relations can be derived from eqs. (6.81)–(6.86):

$$m_1^2 \simeq v^2\big[Z_1 - s_{12}\operatorname{Re}(Z_6 e^{-i\eta}) + s_{13}\operatorname{Im}(Z_6 e^{-i\eta})\big]\,, \tag{6.130}$$

$$m_2^2 - m_3^2 \simeq v^2\big[\operatorname{Re}(Z_5 e^{-2i\eta}) + s_{12}\operatorname{Re}(Z_6 e^{-i\eta}) + s_{13}\operatorname{Im}(Z_6 e^{-i\eta})\big]\,, \tag{6.131}$$

$$m_2^2 - m_{H^\pm}^2 \simeq \tfrac{1}{2}v^2\big[Z_4 + \operatorname{Re}(Z_5 e^{-2i\eta}) + 2s_{12}\operatorname{Re}(Z_6 e^{-i\eta})\big]\,, \tag{6.132}$$

where the terms that are first order in the deviation from exact Higgs alignment are exhibited. In the decoupling limit, $m_1^2 \simeq Z_1 v^2 \ll m_2^2,\ m_3^2,\ m_{H^\pm}^2$, which guarantees that eqs. (6.127)–(6.129) are satisfied. In addition, $m_2^2 - m_3^2 \simeq m_2^2 - m_{H^\pm}^2 = \mathcal{O}(v^2)$. That is, the mass splittings among the heavy Higgs states are of order v^2/m_3. If the Higgs alignment limit is achieved via $|Z_6| \ll 1$, then eqs. (6.127)–(6.129) are still satisfied,[17] and we again obtain $m_1^2 \simeq Z_1 v^2$. However, in contrast to the decoupling limit, there is no requirement that h_2, h_3, and $H^\pm$ are significantly heavier than h_1.

In light of eqs. (6.127)–(6.129), it follows that $Z_6 = 0$ corresponds to exact Higgs

[17] More precisely, we require that $|Z_6| \ll \Delta m_{j1}^2/v^2$, where $\Delta m_{j1}^2 \equiv m_j^2 - m_1^2$ for $j = 2, 3$.

alignment where $s_{12} = s_{13} = \mathrm{Im}(Z_5 e^{-2i\eta}) = 0$. To understand the origin of the last condition, consider the neutral scalar squared mass given in eq. (6.52) in the limit of $Z_6 = 0$:

$$\mathcal{M}_0^2 = v^2 \begin{pmatrix} Z_1 & 0 & 0 \\ 0 & \frac{1}{2}Z_{345} + Y_2/v^2 & -\frac{1}{2}\,\mathrm{Im}(Z_5 e^{-2i\eta}) \\ 0 & -\frac{1}{2}\,\mathrm{Im}(Z_5 e^{-2i\eta}) & \frac{1}{2}Z_{345} - \mathrm{Re}(Z_5 e^{-2i\eta})] + Y_2/v^2 \end{pmatrix}, \quad (6.133)$$

where

$$Z_{345} \equiv Z_3 + Z_4 + \mathrm{Re}(Z_5 e^{-2i\eta}). \quad (6.134)$$

To diagonalize $\mathcal{M}_0^2$, we may take $R_{12} = R_{13} = \mathbb{1}_{3\times 3}$ in eq. (6.56), in which case eq. (6.55) reduces to

$$R_{23}\mathcal{M}_0^2 R_{23}^{\mathsf{T}} = \mathrm{diag}(m_1^2, m_2^2, m_3^2)\,. \quad (6.135)$$

Employing the $Z_6 = 0$ limit of eq. (6.52), the diagonality of $R_{23}\mathcal{M}_0^2 R_{23}^{\mathsf{T}}$ yields $\mathrm{Im}(Z_5 e^{-2i\eta}) = 0$.

Note that exact Higgs alignment is achieved when $Y_3 = Z_6 = 0$, in which case the potentially complex parameters of the scalar potential in the Higgs basis are Z_5 and Z_7. Since one is always free to rephase the Higgs basis field $\mathcal{H}_2$, one can assume that Z_5 is real without loss of generality. Thus, the only remaining potentially complex parameter of the scalar potential is Z_7. Since the neutral scalar squared-mass matrix is independent of Z_7, it follows that Higgs boson–gauge boson interactions are separately CP-conserving, and the tree-level scalar mass eigenstates are states of definite CP with squared masses

$$m_h^2 = Z_1 v^2\,, \qquad m_{H,A}^2 = Y_2 + \tfrac{1}{2}(Z_3 + Z_4 \pm Z_5)\,, \quad (6.136)$$

where h and H are CP-even scalars and A is a CP-odd scalar. The only potential source of CP violation resides in the Higgs self-interactions in the case of $\mathrm{Im}(Z_5^* Z_7^2) \neq 0$.

Moreover, the tree-level couplings of h_1 coincide with those of the SM Higgs boson. This is easily checked by setting $c_{12} = c_{13}$ in Table 6.1, which yields

$$q_{11} = q_{22} = -iq_{23} = 1\,, \qquad q_{21} = q_{31} = q_{12} = 0 \qquad \text{[exact Higgs alignment]}. \quad (6.137)$$

Inserting these results into the interaction Lagrangians obtained in Section 6.2.2,

$$\mathscr{L}_{\rm int} \supset \left\{ g m_W W_\mu^+ W^{\mu -} + \frac{g}{2c_W} m_Z Z_\mu Z^\mu - \frac{1}{v}\left(\sum_{f=u,d,e} f^{\mathsf{T}} \boldsymbol{M_F} \bar{f} + \text{h.c.} \right) \right\} h_1$$
$$+ \left[\tfrac{1}{4} g^2 W_\mu^+ W^{\mu -} + \frac{g^2}{8c_W^2} Z_\mu Z^\mu \right] (h_1)^2 - \tfrac{1}{2} Z_1 v (h_1)^3 - \tfrac{1}{8} Z_1 (h_1)^4\,, \quad (6.138)$$

which coincides with the Higgs boson interaction Lagrangian of the Standard Model if we identify $Z_1 = 2\lambda$ [see eqs. (4.160), (4.162), and (4.184)].

6.2.4 Special Forms for the Higgs–Fermion Yukawa Interactions

Generically, there is no reason why the matrices $\boldsymbol{\rho}^F_{\boldsymbol{R}}$ and $\boldsymbol{\rho}^F_{\boldsymbol{I}}$ (for $F = U, D, L$), which appear in the Higgs–fermion Yukawa interactions [see eq. (6.122)], should be approximately diagonal, as required by the absence of large FCNCs (mediated by tree-level neutral Higgs boson exchange) in Nature. Indeed, the diagonal structure of $\boldsymbol{\rho}^F_{\boldsymbol{R}}$ and $\boldsymbol{\rho}^F_{\boldsymbol{I}}$ is not stable with respect to radiative corrections in the general 2HDM, so that imposing such a condition requires an artificial fine-tuning of parameters. However, there exist special forms for the Higgs–fermion Yukawa interactions where the matrices $\boldsymbol{\rho}^F_{\boldsymbol{R}}$ and $\boldsymbol{\rho}^F_{\boldsymbol{I}}$ are automatically diagonal, due to the presence of a symmetry that guarantees that the diagonal structure is radiatively stable. In this case, tree-level Higgs-mediated FCNCs are "naturally" absent, as required by phenomenological considerations.

In a general extended Higgs model, Glashow, Weinberg, and Paschos (GWP) showed that tree-level Higgs-mediated FCNCs are absent if, for some choice of basis of the scalar fields, at most one Higgs multiplet is responsible for providing mass for quarks or leptons of a given electric charge [61, 62]. The GWP conditions can be imposed by a symmetry principle, which guarantees that the absence of tree-level Higgs-mediated FCNCs is natural. By an appropriate choice of symmetry transformation laws for the fermions and the Higgs scalars, the resulting Higgs–fermion Yukawa interactions take on the required form in some scalar field basis. In this scalar field basis, the symmetry also restricts the form of the Higgs scalar potential. Indeed, the scalar field basis in which the symmetry is manifestly realized has physical significance, and the parameters of the scalar potential and the Yukawa interactions in this basis are directly related to physical observables.

More generally, consider the Higgs–fermion Yukawa interactions of the 2HDM in the Φ-basis. Using eqs. (6.104) and (6.105) to rewrite eq. (6.99) in terms of fermion mass eigenstates, we obtain

$$\begin{aligned} -\mathscr{L}_{\rm Y} = {} & (\boldsymbol{y_u}^i)^k{}_n\big[\Phi_i^0 u_k\bar{u}^n - \Phi_i^+ \boldsymbol{K}_k{}^m d_m\bar{u}^n\big] + (\boldsymbol{y_{d\,i}})^k{}_n\big[(\Phi^-)^i(\boldsymbol{K}^\dagger)_k{}^m u_m\bar{d}^n + \Phi^{0i} d_m\bar{d}^n\big] \\ & + (\boldsymbol{y_{e\,i}})^k{}_n\left[(\Phi^-)^i\nu_k\bar{e}^n + \Phi^{0i}e_k\bar{e}^n\right] + \text{h.c.}\,, \end{aligned} \tag{6.139}$$

following the index height conventions of eq. (6.100), where there is an implicit sum over $i \in \{1,2\}$, and

$$\boldsymbol{y}^\dagger_{\boldsymbol{u}} \equiv L_u^{\mathsf{T}}\,\widehat{\boldsymbol{y}}^\dagger_{\boldsymbol{u}} R_u\,, \tag{6.140}$$

$$\boldsymbol{y_d} \equiv L_d^{\mathsf{T}}\,\widehat{\boldsymbol{y}}_{\boldsymbol{d}} R_d\,, \tag{6.141}$$

$$\boldsymbol{y_e} \equiv L_e^{\mathsf{T}}\,\widehat{\boldsymbol{y}}_{\boldsymbol{e}} R_e\,. \tag{6.142}$$

In light of eqs. (6.109)–(6.114), it follows that the basis-independent Yukawa couplings are given by

$$\boldsymbol{\kappa}^F = \frac{\sqrt{2}\boldsymbol{M_F}}{v} = \widehat{v}^i\boldsymbol{y_{f\,i}}\,, \qquad \boldsymbol{\rho}^F = e^{i\eta}\widehat{w}^i\boldsymbol{y_{f\,i}}\,. \tag{6.143}$$

Table 6.2 Four possible $\mathbb{Z}_2$ charge assignments for scalar and fermion fields. The $\mathbb{Z}_2$ symmetry is employed to constrain the Higgs–fermion Yukawa couplings, thereby implementing the GWP conditions for the (natural) absence of tree-level Higgs-mediated FCNCs.

	Φ_1	Φ_2	$\bar{u}$	$\bar{d}$	$\bar{e}$	u, d, ν, e
Type I	+	−	−	−	−	+
Type II	+	−	−	+	+	+
Type X	+	−	−	−	+	+
Type Y	+	−	−	+	−	+

Inverting eq. (6.143) yields

$$\boldsymbol{y_f}_i = \boldsymbol{\kappa^F} \widehat{v}_i + e^{-i\eta} \boldsymbol{\rho^F} \widehat{w}_i \,. \tag{6.144}$$

The GWP conditions can be implemented in four inequivalent ways [63, 64]:

1. Type-I Yukawa couplings: $\boldsymbol{y_u^1} = \boldsymbol{y_d^1} = \boldsymbol{y_e^1} = 0$,
2. Type-II Yukawa couplings: $\boldsymbol{y_u^1} = \boldsymbol{y_d^2} = \boldsymbol{y_e^2} = 0$,
3. Type-X Yukawa couplings: $\boldsymbol{y_u^1} = \boldsymbol{y_d^1} = \boldsymbol{y_e^2} = 0$,
4. Type-Y Yukawa couplings: $\boldsymbol{y_u^1} = \boldsymbol{y_d^2} = \boldsymbol{y_e^1} = 0$.

The four types of Yukawa couplings can be implemented by a $\mathbb{Z}_2$ symmetry, as shown in Table 6.2. The imposition of this discrete symmetry also restricts the form of the Higgs scalar potential given in eq. (6.33) by setting $m_{12}^2 = \lambda_6 = \lambda_7 = 0$. If, in addition, $\lambda_5 = 0$, then the $\mathbb{Z}_2$ symmetry is promoted to a continuous Peccei–Quinn $\mathrm{U}(1)_{\mathrm{PQ}}$ symmetry [65], where the scalar fields transform as $\Phi_1 \to e^{-i\theta}\Phi_1$ and $\Phi_2 \to e^{i\theta}\Phi_2$.[18] Note that the Peccei–Quinn symmetry will typically be broken by the Higgs VEVs, resulting in an (undesirable) exactly massless Goldstone boson.

However, these symmetry conditions are stronger than they need to be. It is sufficient to require that the symmetry is only respected by the dimension-four terms of the Lagrangian. This weaker condition allows for $m_{12}^2 \neq 0$, corresponding to a softly broken $\mathbb{Z}_2$ [or $\mathrm{U}(1)_{\mathrm{PQ}}$] symmetry. In this case, tree-level neutral Higgs-mediated FCNCs will still be naturally absent, although these effects can be generated via radiative corrections. Of course, FCNCs can also be generated at the loop level mediated by charged Higgs-boson exchange. Ensuring that such processes are not in conflict with experimental data will place some interesting constraints on the 2HDM parameter space (e.g., see Ref. [66]).

If one of the four possible GWP conditions is imposed, then $\boldsymbol{\rho^F}$ is proportional to $\boldsymbol{\kappa^F}$ (for $F = U, D, L$); i.e., the neutral Higgs boson–fermion interactions are flavor-diagonal. For simplicity, we restrict the following discussion to the Type-I

[18] For example, Type-II Yukawa couplings are obtained if the Yukawa Lagrangian is invariant under the $\mathrm{U}(1)_{\mathrm{PQ}}$ symmetry transformations: $\Phi_1 \to e^{-i\theta}\Phi_1$, $\Phi_2 \to e^{i\theta}\Phi_2$, $(u\,,\,d) \to e^{-i\theta}(u\,,\,d)$, $(\nu\,,\,e) \to e^{-i\theta}(\nu\,,\,e)$, $\bar{u} \to \bar{u}$, $\bar{d} \to \bar{d}$, and $\bar{e} \to \bar{e}$, where $0 \leq \theta < 2\pi$.

and Type-II Higgs–quark Yukawa couplings. First, we note that the Type-I and Type-II conditions can be expressed in a basis-independent form as follows:

$$\text{Type I:}\quad \epsilon^{ij}\,\boldsymbol{y}_{\boldsymbol{d}i}\,\boldsymbol{y}_{\boldsymbol{u}j} = \epsilon_{ij}\,\boldsymbol{y}_{\boldsymbol{d}}^{i}\boldsymbol{y}_{\boldsymbol{u}}^{j} = 0\,, \tag{6.145}$$

$$\text{Type II:}\quad \delta_i^j\,\boldsymbol{y}_{\boldsymbol{d}}^{i}\boldsymbol{y}_{\boldsymbol{u}j} = \delta_j^i\,\boldsymbol{y}_{\boldsymbol{d}i}\,\boldsymbol{y}_{\boldsymbol{u}}^{j} = 0\,. \tag{6.146}$$

Employing eq. (6.144) yields the invariant conditions

$$\text{Type I:}\quad \boldsymbol{\kappa}^{\boldsymbol{D}}\boldsymbol{\rho}^{\boldsymbol{U}} - \boldsymbol{\rho}^{\boldsymbol{D}}\boldsymbol{\kappa}^{\boldsymbol{U}} = 0\,, \tag{6.147}$$

$$\text{Type II:}\quad \boldsymbol{\kappa}^{\boldsymbol{D}}\boldsymbol{\kappa}^{\boldsymbol{U}\dagger} + \boldsymbol{\rho}^{\boldsymbol{D}}\boldsymbol{\rho}^{\boldsymbol{U}\dagger} = 0\,. \tag{6.148}$$

The softly broken $\mathbb{Z}_2$ symmetry, with the $\mathbb{Z}_2$ charge assignments specified in Table 6.2, is manifestly realized in the Φ-basis, where $\widehat{v} = (\cos\beta\,,\, e^{i\xi}\sin\beta)$ and $\widehat{w} = (-e^{-i\xi}\sin\beta\,,\,\cos\beta)$ [see eqs. (6.43) and (6.44)]. Using eq. (6.144) and the corresponding GWP conditions, the matrices $\boldsymbol{\rho_U}$ and $\boldsymbol{\rho_D}$ are diagonal and given by

$$\text{Type I:}\quad \boldsymbol{\rho}^{\boldsymbol{U}} = \frac{e^{i(\xi+\eta)}\sqrt{2}\boldsymbol{M_U}\cot\beta}{v}\,,\qquad \boldsymbol{\rho}^{\boldsymbol{D}} = \frac{e^{i(\xi+\eta)}\sqrt{2}\boldsymbol{M_D}\cot\beta}{v}\,, \tag{6.149}$$

$$\text{Type II:}\quad \boldsymbol{\rho}^{\boldsymbol{U}} = \frac{e^{i(\xi+\eta)}\sqrt{2}\boldsymbol{M_U}\cot\beta}{v}\,,\qquad \boldsymbol{\rho}^{\boldsymbol{D}} = -\frac{e^{i(\xi+\eta)}\sqrt{2}\boldsymbol{M_D}\tan\beta}{v}\,. \tag{6.150}$$

Note that the invariant Type-I and Type-II Yukawa couplings satisfy eqs. (6.147) and (6.148), as advertised. Moreover, the $\boldsymbol{\rho}^{\boldsymbol{Q}}$ are diagonal matrices, which implies that the neutral Higgs–fermion couplings specified in eq. (6.122) are flavor-diagonal.

The matrices $\boldsymbol{\rho}^{\boldsymbol{Q}}$ $(Q = U, D)$ are basis-invariant quantities, which are given in terms of the Φ-basis parameters by eqs. (6.149) and (6.150). It is instructive to examine the Φ'-basis, which is related to the Φ-basis via $\Phi' = U\Phi$, where

$$U = \begin{pmatrix} 0 & e^{-i\xi} \\ e^{i\zeta} & 0 \end{pmatrix}. \tag{6.151}$$

In the Φ'-basis, the roles of Φ_1 and Φ_2 are interchanged with respect to the Φ-basis. Thus, the softly broken $\mathbb{Z}_2$ symmetry is also manifestly realized in the Φ'-basis, where the $\mathbb{Z}_2$ charges of Φ_1 and Φ_2 specified in Table 6.2 are interchanged. Likewise, the GWP conditions specified below eq. (6.144) are also modified by interchange superscripts 1 and 2, and we shall refer to the corresponding Yukawa couplings in the Φ'-basis as Type I$'$, Type II$'$, etc. In particular, in light of

$$\begin{pmatrix} s_\beta \\ c_\beta e^{i\zeta} \end{pmatrix} = U\begin{pmatrix} c_\beta \\ s_\beta e^{i\xi} \end{pmatrix}, \tag{6.152}$$

we conclude that $\beta' = \frac{1}{2}\pi - \beta$ and $\xi' = \zeta$. Moreover, due to the pseudoinvariant nature of $e^{i\eta}$ [see eq. (6.42)], $e^{i\eta'} = (\det U)^{-1}e^{i\eta}$. Using $\det U = -e^{i(\zeta-\xi)}$, it follows that $e^{i(\xi'+\eta')} = -e^{i(\xi+\eta)}$. Thus, with respect to the parameters of the Φ'-basis, the results of eqs. (6.149) and (6.150) are modified by interchanging $\tan\beta \leftrightarrow \cot\beta$ and multiplying the resulting expressions by -1. But these are precisely the results for $\boldsymbol{\rho_U}$ and $\boldsymbol{\rho_D}$ that would have been obtained by employing the Type-I$'$ and Type-II$'$ Yukawa couplings. Thus, the Type-I$'$ and Type-II$'$ Yukawa couplings are physically identical to the Type-I and Type-II Yukawa couplings, respectively, since the former are simply related to the latter by a scalar field basis transformation in which Φ_1 and Φ_2 are interchanged.

6.2.5 The CP-Conserving 2HDM

Significant simplifications occur if one imposes CP conservation on the 2HDM scalar potential and vacuum. If CP is *explicitly* conserved, then there exists a basis for the scalar fields in which all the parameters of the scalar potential are simultaneously real. Such a basis (if it exists) is called a *real basis.* That is, we shall assume that the scalar potential is given by eq. (6.33) in which all the m_{ij}^2 and λ_k are real. We shall continue to assume that the vacuum preserves $\mathrm{U(1)_{EM}}$, in which case the minimum of the scalar potential yields the following VEVs:

$$\langle\Phi_1\rangle = \frac{v}{\sqrt{2}}\begin{pmatrix}0\\ c_\beta\end{pmatrix}, \qquad \langle\Phi_2\rangle = \frac{1}{\sqrt{2}}\begin{pmatrix}0\\ s_\beta e^{i\xi}\end{pmatrix}, \tag{6.153}$$

where $0 \leq \beta \leq \frac{1}{2}\pi$ and $0 \leq \xi < 2\pi$. The potential minimum conditions are given by eqs. (6.35) and (6.36), with $v_1 = vc_\beta$, $v_2 = vs_\beta e^{i\xi}$, and m_{12}^2, λ_5, λ_6, and λ_7 taken real. Evaluating the imaginary part of eq. (6.35), one finds two solutions for ξ (under the assumption that $\lambda_5, v_1, |v_2| \neq 0$). Either $\sin\xi = 0$ or

$$\cos\xi = \frac{m_{12}^2 - \frac{1}{2}\lambda_6 v_1^2 - \frac{1}{2}\lambda_7 |v_2|^2}{\lambda_5 v_1 |v_2|}. \tag{6.154}$$

If the right-hand side of eq. (6.154) lies outside of the range between -1 and $+1$, then only the solution $\sin\xi = 0$ remains, and the vacuum is CP-conserving. Otherwise, there is a potential for spontaneous CP violation.[19]

In this subsection, we shall assume that the 2HDM scalar potential and vacuum are both CP-conserving, in which case one can always define a real Higgs basis,

$$\mathcal{H}_1 \equiv \Phi_1 c_\beta + \Phi_2 s_\beta e^{-i\xi}, \qquad e^{-i\eta}\mathcal{H}_2 \equiv -\Phi_1 s_\beta e^{i\xi} + \Phi_2 c_\beta, \tag{6.155}$$

where $\langle\mathcal{H}_1^0\rangle = v/\sqrt{2}$ and $\langle\mathcal{H}_2^0\rangle = 0$. Since the Φ-basis is a real basis by assumption, it follows that the Higgs basis scalar potential parameters Y_1, Y_2, Y_3, and $Z_1, \ldots, Z_7$ exhibited in eq. (6.48) are all real, and $\sin\xi = \sin\eta = 0$. That is,

$$\xi, \eta = 0 \bmod \pi. \tag{6.156}$$

In particular, Y_3, Z_6, Z_7, and $\varepsilon \equiv e^{i\eta} = \cos\eta = \pm 1$ are pseudoinvariant quantities that transform under a real orthogonal basis transformation, $\Phi_i \to \mathcal{R}_i{}^j\Phi_j$, as [cf. eq. (6.49)]

$$[Y_3, Z_6, Z_7, \varepsilon] \to \det\mathcal{R}\,[Y_3, Z_6, Z_7, \varepsilon], \tag{6.157}$$

where $\det\mathcal{R} = \pm 1$. The real Higgs basis is uniquely defined up to an overall sign

[19] If eq. (6.154) is satisfied for $0 \leq \cos\xi < 1$, one must verify that this solution corresponds to a global minimum. Assuming that this is the case, then there exists a real basis in which v_2 is not real. This is a necessary condition for spontaneous CP violation. One still must check that it is not possible to transform to another real basis in which v_2 is real in order to conclude that the vacuum spontaneously breaks CP.

redefinition, $\mathcal{H}_2 \to -\mathcal{H}_2$, which preserves the reality of the Higgs basis and the condition $\langle \mathcal{H}_2 \rangle = 0$.[20]

Using eq. (6.52), the neutral Higgs squared-mass matrix in a real Higgs basis is given by

$$\mathcal{M}^2 = \begin{pmatrix} Z_1 v^2 & \varepsilon Z_6 v^2 & 0 \\ \varepsilon Z_6 v^2 & Y_2 + \frac{1}{2}(Z_3 + Z_4 + Z_5)v^2 & 0 \\ 0 & 0 & Y_2 + \frac{1}{2}(Z_3 + Z_4 - Z_5)v^2 \end{pmatrix} . \quad (6.158)$$

If $Z_6 = 0$ then $\mathcal{M}$ is diagonal and the rotation matrix defined in eq. (6.56) is the identity matrix. In this special case, the scalar h_1 is CP-even with a squared mass $m_{h_1}^2 = Z_1 v^2$ and its couplings are precisely those of the SM Higgs boson. The existence of an $H^+H^-h_2$ coupling and the absence of an $H^+H^-h_3$ coupling (assuming $Z_7 \neq 0$) suggest that h_2 is CP-even and h_3 is CP-odd. These assignments can be confirmed by examining the coupling of h_2 and h_3 to fermions (under the assumption that $\boldsymbol{\rho} \neq 0$). Finally, in the case of $Z_6 = Z_7 = 0$, the CP quantum numbers of h_2 and h_3 (which are relatively odd) can be individually determined only if $\boldsymbol{\rho} \neq 0$.[21]

Using eqs. (6.55) and (6.56) to diagonalize the neutral scalar squared-mass matrix exhibited in eq. (6.158), we see that only one nontrivial mixing angle θ_{12} is required, since $\theta_{13} = \theta_{23} = 0$. The scalar mass eigenstates are identified as two neutral CP-even scalars h_1 and h_2 and a CP-odd scalar h_3 [see eq. (6.67)]:

$$h_1 = (\sqrt{2}\,\mathrm{Re}\,\mathcal{H}_1^0 - v)\cos\theta_{12} - \sqrt{2}\,\mathrm{Re}\,\mathcal{H}_2^0 \sin\theta_{12} \,, \quad (6.159)$$

$$h_2 = (\sqrt{2}\,\mathrm{Re}\,\mathcal{H}_1^0 - v)\sin\theta_{12} + \sqrt{2}\,\mathrm{Re}\,\mathcal{H}_2^0 \cos\theta_{12} \,, \quad (6.160)$$

$$h_3 = \sqrt{2}\,\mathrm{Im}\,\mathcal{H}_2^0 \,, \quad (6.161)$$

with corresponding masses $m_i \equiv m_{h_i}$. The eigenvalues of eq. (6.158) correspond to the squared masses of two neutral CP-even scalars, h_1 and h_2,

$$m_{1,2}^2 = \tfrac{1}{2}\left\{ Y_2 + (Z_1 + \tfrac{1}{2}Z_{345})v^2 \pm \sqrt{\left[Y_2 - (Z_1 - \tfrac{1}{2}Z_{345})v^2\right]^2 + 4Z_6^2 v^4} \right\} , \quad (6.162)$$

and the squared mass of the CP-odd scalar h_3,

$$m_3^2 = Y_2 + \tfrac{1}{2}(Z_3 + Z_4 - Z_5)v^2 = m_{H^\pm}^2 + \tfrac{1}{2}(Z_4 - Z_5)v^2 \,, \quad (6.163)$$

with no mass ordering of h_1, h_2, h_3 implied. Moreover, eqs. (6.81) and (6.85) yield

$$\sin^2\theta_{12} = \frac{Z_1 v^2 - m_1^2}{m_2^2 - m_1^2} \,, \quad (6.164)$$

$$\sin\theta_{12}\cos\theta_{12} = \frac{\varepsilon Z_6 v^2}{m_2^2 - m_1^2} \,, \quad (6.165)$$

[20] In the special case of $Z_6 = Z_7 = 0$, eq. (6.156) is modified to $\eta = 0 \bmod \frac{1}{2}\pi$, which permits a class of unitary basis transformations that are not real orthogonal while preserving the reality of the scalar field basis.

[21] The 2HDM Lagrangian with $Z_6 = Z_7 = \boldsymbol{\rho} = 0$ possesses an exact $\mathbb{Z}_2$ symmetry that is preserved by the vacuum. Under the $\mathbb{Z}_2$ symmetry transformation, $H^\pm$, h_2, and h_3 (which reside in the Higgs basis field $\mathcal{H}_2$) are $\mathbb{Z}_2$-odd, whereas h_1, the gauge bosons, and fermions are $\mathbb{Z}_2$-even.

where $|\theta_{12}| \leq \frac{1}{2}\pi$ as indicated in eq. (6.73).

In the literature, when the mass ordering of the two CP-even scalars is known, then h refers to the lighter of the two CP-even scalars and H refers to the heavier of the two CP-even scalars, with corresponding masses $m_h < m_H$. It is also traditional to denote the CP-odd scalar by A with mass $m_A \equiv m_3$. Even without the explicit computation of m_1^2 and m_2^2 given in eq. (6.162), one can deduce bounds on the scalar masses by employing the following well-known theorem in linear algebra (e.g., see Corollary 10.4.7 in [B126]). Given an $n \times n$ hermitian matrix M with diagonal elements d_i and eigenvalues m_i such that $d_{\rm min}$ is the minimal diagonal entry, $d_{\rm max}$ is the maximal diagonal entry, $m_{\rm min}$ is the minimal eigenvalue, and $m_{\rm max}$ is the maximal eigenvalue, then

$$m_{\rm min} \leq d_{\rm min} \leq d_{\rm max} \leq m_{\rm max}\,. \tag{6.166}$$

It then follows that

$$m_h^2 \leq \min\{Z_1 v^2\,,\, m_A^2 + Z_5 v^2\}\,, \qquad m_H^2 \geq \max\{Z_1 v^2\,,\, m_A^2 + Z_5 v^2\}. \tag{6.167}$$

Indeed, it is a simple matter to check that these two bounds are satisfied by eq. (6.162).

It is instructive to repeat the above analysis starting in the real Φ-basis. Since $\sin\xi = 0$ [see eq. (6.156)], it follows that $s_\beta e^{\pm i\xi} = s_\beta \cos\xi = \pm s_\beta$. Thus, it is convenient to redefine

$$s_\beta \to s_\beta\, \text{sgn}(\cos\xi)\,. \tag{6.168}$$

With this modification, the range of the parameter β is extended to $-\frac{1}{2}\pi \leq \beta \leq \frac{1}{2}\pi$. After diagonalizing the 2×2 CP-even scalar squared-mass matrix (obtained in the real Φ-basis), the resulting diagonalizing angle is denoted by α. The corresponding neutral scalar mass eigenstates are then defined as follows (see [B143]):

$$h = -(\sqrt{2}\,\text{Re}\,\Phi_1^0 - v c_\beta)\sin\alpha + (\sqrt{2}\,\text{Re}\,\Phi_2^0 - v s_\beta)\cos\alpha\,, \tag{6.169}$$

$$H = (\sqrt{2}\,\text{Re}\,\Phi_1^0 - v c_\beta)\cos\alpha + (\sqrt{2}\,\text{Re}\,\Phi_2^0 - v s_\beta)\sin\alpha\,, \tag{6.170}$$

$$A = -\sqrt{2}\big[\text{Im}\,\Phi_1^0 s_\beta - \text{Im}\,\Phi_2^0 c_\beta\big]\,, \tag{6.171}$$

where h and H are CP-even scalars (with $m_h < m_H$) and A is a CP-odd scalar. Note that without further model assumptions there is no physical significance to the choice of the Φ-basis. Consequently, the angles α and β are basis-dependent quantities with no separate physical significance.

We can now make contact with the notation of eqs. (6.159)–(6.161). As previously noted, by restricting to the class of real Higgs bases, $\varepsilon \equiv e^{i\eta} = \pm 1$. Moreover, ε changes sign under $\mathcal{H}_2 \to -\mathcal{H}_2$ [see eq. (6.157)]. It is convenient to choose[22]

$$\varepsilon \equiv e^{i\eta} = \begin{cases} \text{sgn}\, Z_6\,, & \text{if } Z_6 \neq 0, \\ \text{sgn}\, Z_7\,, & \text{if } Z_6 = 0 \text{ and } Z_7 \neq 0. \end{cases} \tag{6.172}$$

[22] If $Z_6 = Z_7 = 0$, then the sign of Z_5 is no longer invariant with respect to transformations that preserve the real Higgs basis (since the sign of Z_5 changes under $\mathcal{H}_2 \to \pm i\mathcal{H}_2$). In this case, it would be more appropriate to define $\varepsilon \equiv e^{2i\eta} = \text{sgn}\, Z_5$.

Table 6.3 Invariant neutral Higgs mixing angles in the CP-conserving 2HDM, where h_1 corresponds to either the lighter (h) or heavier (H) of the two CP-even scalars, and ε is defined by eq. (6.172) in a real Higgs basis.

k	q_{k1} $(h_1 = h)$	q_{k2} $(h_1 = h)$	q_{k1} $(h_1 = H)$	q_{k2} $(h_1 = H)$
1	$s_{\beta-\alpha}$	$\varepsilon\, c_{\beta-\alpha}$	$c_{\beta-\alpha}$	$-\varepsilon\, s_{\beta-\alpha}$
2	$-\varepsilon\, c_{\beta-\alpha}$	$s_{\beta-\alpha}$	$\varepsilon\, s_{\beta-\alpha}$	$c_{\beta-\alpha}$
3	0	i	0	i

First, suppose that h_1 is the lighter of the two CP-even Higgs scalars, which implies the following identification of scalar states:

$$h_1 = h\,, \qquad h_2 = -\varepsilon\, H\,, \qquad h_3 = \varepsilon\, A\,, \tag{6.173}$$

and

$$\cos\theta_{12} = s_{\beta-\alpha}\,, \qquad \sin\theta_{12} = -\varepsilon\, c_{\beta-\alpha}\,, \tag{6.174}$$

where $s_{\beta-\alpha} \equiv \sin(\beta-\alpha)$ and $c_{\beta-\alpha} \equiv \cos(\beta-\alpha)$. Using the results of Table 6.1, the invariant quantities q_{kj} for the CP-conserving 2HDM in the case of $m_1 < m_2$ are exhibited in Table 6.3.

Equivalently, one can identify

$$\beta - \alpha = \varepsilon\, \theta_{12} + \tfrac{1}{2}\pi\,, \tag{6.175}$$

where $0 \leq \beta - \alpha \leq \pi$ [in light of eq. (6.73)]. Employing eqs. (6.164) and (6.165),

$$Z_1 v^2 = m_h^2 s_{\beta-\alpha}^2 + m_H^2 c_{\beta-\alpha}^2\,, \tag{6.176}$$

$$s_{\beta-\alpha} c_{\beta-\alpha} = -\frac{Z_6 v^2}{m_H^2 - m_h^2}\,. \tag{6.177}$$

Using eqs. (6.176) and (6.177), one can then derive an explicit expression for $c_{\beta-\alpha}$,

$$c_{\beta-\alpha} = -\varepsilon\sqrt{\frac{Z_1 v^2 - m_h^2}{m_H^2 - m_h^2}} = \frac{-Z_6 v^2}{\sqrt{(m_H^2 - m_h^2)(m_H^2 - Z_1 v^2)}}\,, \tag{6.178}$$

after employing eq. (6.172). Hence,

$$s_{\beta-\alpha} > 0\,, \qquad \varepsilon\, c_{\beta-\alpha} < 0\,, \qquad \text{if } m_1 < m_2. \tag{6.179}$$

Since θ_{12} is a basis-independent quantity, it follows that $s_{\beta-\alpha}$ is an invariant under real basis transformations, whereas $c_{\beta-\alpha}$ is a pseudoinvariant that changes sign under $\mathcal{H}_2 \to -\mathcal{H}_2$.

In the case where the SM-like Higgs boson is the lighter of the two CP-even Higgs bosons (i.e., $h_1 = h$), the conditions governing the Higgs alignment limit obtained in eqs. (6.127)–(6.132) yield

$$|c_{\beta-\alpha}| \simeq \frac{|Z_6|v^2}{m_H^2 - m_h^2} \ll 1\,, \tag{6.180}$$

$$m_h^2 \simeq v^2(Z_1 + Z_6 c_{\beta-\alpha})\,, \tag{6.181}$$

$$m_H^2 - m_A^2 \simeq v^2(Z_5 - Z_6 c_{\beta-\alpha})\,, \tag{6.182}$$

$$m_H^2 - m_{H^\pm}^2 \simeq \tfrac{1}{2}v^2(Z_4 + Z_5 - 2Z_6 c_{\beta-\alpha})\,. \tag{6.183}$$

We can then identify $m_h = 125$ GeV, corresponding to the observed SM-like Higgs boson.

Second, suppose that h_2 is the lighter of the two CP-even Higgs scalars, which implies the following identification of scalar states:

$$h_1 = H\,, \qquad h_2 = \varepsilon\, h\,, \qquad h_3 = \varepsilon\, A\,. \tag{6.184}$$

In contrast to eq. (6.174), we now identify

$$\cos\theta_{12} = c_{\beta-\alpha}\,, \qquad \sin\theta_{12} = \varepsilon\, s_{\beta-\alpha}\,. \tag{6.185}$$

Using the results of Table 6.1, the invariant quantities q_{kj} for the CP-conserving 2HDM in the case of $m_2 < m_1$ are again exhibited in Table 6.3.

Equivalently, one can identify

$$\beta - \alpha = \varepsilon\,\theta_{12}\,, \tag{6.186}$$

where $-\frac{1}{2}\pi \leq \beta - \alpha \leq \frac{1}{2}\pi$ [in light of eq. (6.73)]. Equations (6.176) and (6.177) remain valid and can be employed in deriving the following explicit expression for $s_{\beta-\alpha}$:

$$s_{\beta-\alpha} = -\varepsilon\sqrt{\frac{m_H^2 - Z_1 v^2}{m_H^2 - m_h^2}} = \frac{-Z_6 v^2}{\sqrt{(m_H^2 - m_h^2)(Z_1 v^2 - m_h^2)}}\,. \tag{6.187}$$

Hence,

$$c_{\beta-\alpha} > 0\,, \qquad \varepsilon\, s_{\beta-\alpha} < 0\,, \qquad \text{if } m_2 < m_1. \tag{6.188}$$

In this case $c_{\beta-\alpha}$ is an invariant under real basis transformations, whereas $s_{\beta-\alpha}$ is a pseudoinvariant that changes sign under $\mathcal{H}_2 \to -\mathcal{H}_2$.

Since the SM-like Higgs boson is now the heavier of the two CP-even Higgs bosons (i.e., $h_1 = H$), the conditions governing the Higgs alignment limit yield

$$|s_{\beta-\alpha}| \simeq \frac{|Z_6|v^2}{m_H^2 - m_h^2} \ll 1\,, \tag{6.189}$$

$$m_H^2 \simeq v^2(Z_1 - Z_6 s_{\beta-\alpha})\,, \tag{6.190}$$

$$m_h^2 - m_A^2 \simeq v^2(Z_5 + Z_6 s_{\beta-\alpha})\,, \tag{6.191}$$

$$m_h^2 - m_{H^\pm}^2 \simeq \tfrac{1}{2}v^2(Z_4 + Z_5 + 2Z_6 s_{\beta-\alpha})\,. \tag{6.192}$$

It follows that all Higgs boson squared masses are of $\mathcal{O}(v^2)$, corresponding to an approximate Higgs alignment limit without decoupling. In this case, one can identify $m_H = 125$ GeV, corresponding to the observed SM-like Higgs boson.[23]

Finally, we consider the standard conventions of the CP-conserving 2HDM literature, where the charged Higgs boson mass eigenstates are given by

$$H^\pm = -\Phi_1^\pm s_\beta + \Phi_2^\pm c_\beta \,. \tag{6.193}$$

Due to the factor of $e^{-i\eta}$ multiplying the charged Higgs field in eq. (6.74), we must modify our definition of $H^\pm$ in the CP-conserving 2HDM as follows:

$$H^\pm \to \varepsilon\, H^\pm \,, \tag{6.194}$$

where ε is defined in eq. (6.172).

Using Table 6.3, the bosonic couplings of the CP-conserving 2HDM can now be determined from eqs. (6.93)–(6.98) after making use of eq. (6.194), the results of Table 6.3, and the 2HDM definitions for the neutral scalar mass-eigenstate fields [either eq. (6.173) or eq. (6.184)]. For example, the couplings of the Higgs/Goldstone bosons to the gauge bosons that involve one or two neutral scalars are given by

$$\mathscr{L}_{VVH} = \left(g m_W W_\mu^+ W^{\mu\,-} + \frac{g}{2c_W} m_Z Z_\mu Z^\mu \right) \left[s_{\beta-\alpha} h + c_{\beta-\alpha} H \right] , \tag{6.195}$$

$$\begin{aligned}
\mathscr{L}_{VHH} = {} & \frac{g}{2c_W} A Z^\mu \overleftrightarrow{\partial}_\mu (c_{\beta-\alpha} h - s_{\beta-\alpha} H) + \frac{g}{2c_W} Z^\mu G \overleftrightarrow{\partial}_\mu (s_{\beta-\alpha} h + c_{\beta-\alpha} H) \\
& - \tfrac{1}{2} g \Big\{ i W_\mu^+ \Big[G^- \overleftrightarrow{\partial}^\mu (s_{\beta-\alpha} h + c_{\beta-\alpha} H) \\
& \qquad + H^- \overleftrightarrow{\partial}^\mu (c_{\beta-\alpha} h - s_{\beta-\alpha} H + iA) \Big] + \text{h.c.} \Big\} ,
\end{aligned} \tag{6.196}$$

$$\begin{aligned}
\mathscr{L}_{VVHH} = {} & \left[\tfrac{1}{4} g^2 W_\mu^+ W^{\mu\,-} + \frac{g^2}{8c_W^2} Z_\mu Z^\mu \right] (hh + HH + AA) \\
& + \Big\{ \Big(\tfrac{1}{2} e g A^\mu W_\mu^+ - \frac{g^2 s_W^2}{2c_W} Z^\mu W_\mu^+ \Big) \Big[G^- (s_{\beta-\alpha} h + c_{\beta-\alpha} H) \\
& \qquad + H^- (c_{\beta-\alpha} h - s_{\beta-\alpha} H + iA) \Big] + \text{h.c.} \Big\} .
\end{aligned} \tag{6.197}$$

The remaining VVH, VHH, and $VVHH$ couplings can be found in eqs. (6.93)–(6.95). The cubic and quartic scalar self-couplings can be obtained in a similar matter from eqs. (6.97) and (6.98) and are left as exercises for the reader.

Likewise, the Yukawa couplings of h, H, A and $H^\pm$ obtained from eq. (6.122) depend only on $\beta-\alpha$ and the matrices $\boldsymbol{M_F}$, $\varepsilon\boldsymbol{\rho_R^F}$, and $\varepsilon\boldsymbol{\rho_I^F}$, after employing eq. (6.194) and either eq. (6.173) or eq. (6.184). In order to naturally obtain flavor-diagonal neutral Higgs–fermion Yukawa couplings, we shall again impose the GWP conditions that were introduced in Section 6.2.4. In the CP-conserving 2HDM, we impose

[23] Such a scenario is disfavored by current LHC data but is not yet completely ruled out.

Table 6.4 After imposing the $\mathbb{Z}_2$ symmetries exhibited in Table 6.2, the 3×3 matrices ρ_R^F are proportional to the identity matrix, $\mathbb{1}$, and $\rho_I^F = 0$, for $F = U, D, E$ [see eq. (6.121)] . The ρ_R^F in the four types of Yukawa interactions of the CP-conserving 2HDM are given below, where ε is defined by eq. (6.172) in a real Higgs basis.

	ρ_R^U	ρ_R^D	ρ_R^E	$\rho_I^U, \rho_I^D, \rho_I^E$
Type I	$\varepsilon \mathbb{1} \cot\beta$	$\varepsilon \mathbb{1} \cot\beta$	$\varepsilon \mathbb{1} \cot\beta$	0
Type II	$\varepsilon \mathbb{1} \cot\beta$	$-\varepsilon \mathbb{1} \tan\beta$	$-\varepsilon \mathbb{1} \tan\beta$	0
Type X	$\varepsilon \mathbb{1} \cot\beta$	$\varepsilon \mathbb{1} \cot\beta$	$-\varepsilon \mathbb{1} \tan\beta$	0
Type Y	$\varepsilon \mathbb{1} \cot\beta$	$-\varepsilon \mathbb{1} \tan\beta$	$\varepsilon \mathbb{1} \cot\beta$	0

the GWP conditions in a real Φ-basis. Then, eqs. (6.149) and (6.150) yield

$$\text{Type I:} \quad \boldsymbol{\rho^F} = \frac{\sqrt{2}\boldsymbol{M_F}\,\varepsilon \cot\beta}{v}, \tag{6.198}$$

$$\text{Type II:} \quad \boldsymbol{\rho^U} = \frac{\sqrt{2}\boldsymbol{M_U}\,\varepsilon \cot\beta}{v}, \qquad \boldsymbol{\rho^{D,E}} = -\frac{\sqrt{2}\boldsymbol{M_{D,E}}\,\varepsilon \tan\beta}{v}, \tag{6.199}$$

$$\text{Type X:} \quad \boldsymbol{\rho^Q} = \frac{\sqrt{2}\boldsymbol{M_Q}\,\varepsilon \cot\beta}{v}, \qquad \boldsymbol{\rho^E} = -\frac{\sqrt{2}\boldsymbol{M_E}\,\varepsilon \tan\beta}{v}, \tag{6.200}$$

$$\text{Type Y:} \quad \boldsymbol{\rho^{U,E}} = \frac{\sqrt{2}\boldsymbol{M_{U,E}}\,\varepsilon \cot\beta}{v}, \qquad \boldsymbol{\rho^D} = -\frac{\sqrt{2}\boldsymbol{M_D}\,\varepsilon \tan\beta}{v}, \tag{6.201}$$

for $F = U, D, E$ and $Q = U, D$, after redefining s_β as indicated in eq. (6.168).[24] Due to the redefined s_β, it follows that $\beta \to -\beta$ when $\mathcal{H}_2 \to -\mathcal{H}_2$. Thus, the matrices $\boldsymbol{\rho^F}$ are invariant with respect to real orthogonal basis transformations, as expected. If we now introduce the hermitian matrices $\boldsymbol{\rho_R^F}$ and $\boldsymbol{\rho_I^F}$ following eq. (6.121), then eqs. (6.198)–(6.201) yield the results displayed in Table 6.4.

The Φ-basis (which gives meaning to the parameter $\tan\beta$) becomes well defined once the GWP conditions are imposed by implementing one of the $\mathbb{Z}_2$ symmetries specified in Table 6.2. When the $\mathbb{Z}_2$ symmetry is applied to the Higgs Lagrangian, the scalar potential parameters are constrained such that $m_{12}^2 = \lambda_6 = \lambda_7 = 0$. However, as discussed in Section 6.2.4, it is sufficient to employ a softly broken $\mathbb{Z}_2$ symmetry which permits a nonzero value of the soft-breaking parameter m_{12}^2. Of course, the dimension-four terms of the Lagrangian still respect the $\mathbb{Z}_2$ symmetry.

Consequently, the scalar field basis where $\lambda_6 = \lambda_7 = 0$ (i.e., the basis where the softly broken $\mathbb{Z}_2$ symmetry is manifestly realized) defines the Φ-basis up to the

[24] In the CP-conserving 2HDM literature, instead of redefining s_β as in eq. (6.168), the condition $\cos\xi = 1$ is conventionally imposed in the real Φ-basis, which removes the remaining freedom to redefine the real Higgs basis via $\mathcal{H}_2 \to -\mathcal{H}_2$. Hence, in this convention, $0 \le \beta \le \frac{1}{2}\pi$ and both $\tan\beta$ and α are physical parameters of the model, with $\beta - \alpha$ defined modulo π. The values of α and β then fix ε via eqs. (6.179) or (6.188).

interchange of Φ_1 and Φ_2. The remaining two-fold ambiguity is resolved by the choice of $\mathbb{Z}_2$ charges of the scalar and fermion fields.[25]

We now use the results of Table 6.4 in eq. (6.122) to obtain the Type-I, II, X, and Y Yukawa couplings of the CP-conserving 2HDM. By employing the conventional 2HDM definitions for the mass-eigenstate fields [either eq. (6.173) or eq. (6.184)] and the results of Table 6.3, we end up with the Type-I Yukawa couplings,

$$\begin{aligned}\mathscr{L}_I = &-\frac{h}{v}(s_{\beta-\alpha}+c_{\beta-\alpha}\cot\beta)\big(u\boldsymbol{M_U}\bar{u}+d\boldsymbol{M_D}\bar{d}+e\boldsymbol{M_E}\bar{e}+\text{h.c.}\big) \\ &-\frac{H}{v}(c_{\beta-\alpha}-s_{\beta-\alpha}\cot\beta)\big(u\boldsymbol{M_U}\bar{u}+d\boldsymbol{M_D}\bar{d}+e\boldsymbol{M_E}\bar{e}+\text{h.c.}\big) \\ &-\frac{A}{v}\cot\beta\big(iu\boldsymbol{M_U}\bar{u}-id\boldsymbol{M_D}\bar{d}-ie\boldsymbol{M_E}\bar{e}+\text{h.c.}\big) \\ &+\frac{\sqrt{2}}{v}\cot\beta\Big\{H^+ d\boldsymbol{K}^{\mathsf{T}}\boldsymbol{M_U}\bar{u}-H^-\big(u\boldsymbol{K}^*\boldsymbol{M_D}\bar{d}+\nu\boldsymbol{M_E}\bar{e}\big)+\text{h.c.}\Big\},\end{aligned} \tag{6.202}$$

and the Type-II Yukawa couplings,

$$\begin{aligned}\mathscr{L}_{II} = &-\frac{h}{v}\Big\{(s_{\beta-\alpha}+c_{\beta-\alpha}\cot\beta)\big(u\boldsymbol{M_U}\bar{u}+\text{h.c.}\big) \\ &\qquad+(s_{\beta-\alpha}-c_{\beta-\alpha}\tan\beta)\big(d\boldsymbol{M_D}\bar{d}+e\boldsymbol{M_E}\bar{e}+\text{h.c.}\big)\Big\} \\ &-\frac{H}{v}\Big\{(c_{\beta-\alpha}-s_{\beta-\alpha}\cot\beta)\big(u\boldsymbol{M_U}\bar{u}+\text{h.c.}\big) \\ &\qquad+(c_{\beta-\alpha}+s_{\beta-\alpha}\tan\beta)\big(d\boldsymbol{M_D}\bar{d}+e\boldsymbol{M_E}\bar{e}+\text{h.c.}\big)\Big\} \\ &-\frac{A}{v}\Big\{i\cot\beta\, u\boldsymbol{M_U}\bar{u}+i\tan\beta\big(d\boldsymbol{M_D}\bar{d}+e\boldsymbol{M_E}\bar{e}\big)+\text{h.c.}\Big\} \\ &+\frac{\sqrt{2}}{v}\Big\{\cot\beta\, H^+ d\boldsymbol{K}^{\mathsf{T}}\boldsymbol{M_U}\bar{u}+\tan\beta\, H^-\big(u\boldsymbol{K}^*\boldsymbol{M_D}\bar{d}+\nu\boldsymbol{M_E}\bar{e}\big)+\text{h.c.}\Big\}.\end{aligned} \tag{6.203}$$

The Type-X and Type-Y Yukawa couplings are similarly obtained by an appropriate modification of the Higgs–lepton Yukawa couplings above. Note that all factors of ε have conveniently canceled out of eqs. (6.202) and (6.203). Thus, following footnote 24, one can now adopt the more common CP-conserving 2HDM convention where $0 \leq \beta \leq \frac{1}{2}\pi$.

In the CP-conserving 2HDM literature, the following trigonometric identities are often used to simplify the neutral Higgs boson couplings:

$$\frac{\cos\alpha}{\sin\beta} = s_{\beta-\alpha}+c_{\beta-\alpha}\cot\beta\,, \qquad -\frac{\sin\alpha}{\cos\beta} = s_{\beta-\alpha}-c_{\beta-\alpha}\tan\beta\,, \tag{6.204}$$

$$\frac{\sin\alpha}{\sin\beta} = c_{\beta-\alpha}-s_{\beta-\alpha}\cot\beta\,, \qquad \frac{\cos\alpha}{\cos\beta} = c_{\beta-\alpha}+s_{\beta-\alpha}\tan\beta\,. \tag{6.205}$$

[25] For example, had we reversed the signs of the $\mathbb{Z}_2$ charges of Φ_1 and Φ_2 given in Table 6.2, then the roles of $\tan\beta$ and $\cot\beta$ would have been interchanged. See eq. (6.152) and the discussion that follows.

One advantage of the expressions on the right-hand sides of eqs. (6.204) and (6.205) is that the Higgs alignment limit of the Yukawa couplings are more easily obtained by setting $c_{\beta-\alpha}=0$ if h is SM-like or $s_{\beta-\alpha}=0$ if H is SM-like.

As noted above, the Φ-basis becomes meaningful once the GWP conditions are imposed by implementing one of the (softly broken) $\mathbb{Z}_2$ symmetries specified in Table 6.2. Thus, it is convenient to analyze the softly broken $\mathbb{Z}_2$ symmetric CP-conserving 2HDM directly in the real Φ-basis where $\lambda_6=\lambda_7=0$ and m_{12}^2 and λ_5 are real. We briefly summarize these results below.

The scalar potential minimum conditions [see eqs. (6.35) and (6.36)] yield

$$m_{11}^2 = m_{12}^2\tan\beta - \tfrac{1}{2}v^2\left(\lambda_1 c_\beta^2 + \lambda_{345}s_\beta^2\right), \tag{6.206}$$

$$m_{22}^2 = m_{12}^2\cot\beta - \tfrac{1}{2}v^2\left(\lambda_2 s_\beta^2 + \lambda_{345}c_\beta^2\right), \tag{6.207}$$

where $\lambda_{345}\equiv\lambda_3+\lambda_4+\lambda_5$. The squared masses for the CP-odd and charged Higgs mass eigenstates are

$$m_A^2 = \frac{2m_{12}^2}{s_{2\beta}} - \lambda_5 v^2, \tag{6.208}$$

$$m_{H^\pm}^2 = m_A^2 + \tfrac{1}{2}v^2(\lambda_5-\lambda_4). \tag{6.209}$$

The two neutral CP-even Higgs states mix according to the following squared-mass matrix:

$$\mathcal{M}^2 \equiv \begin{pmatrix} \lambda_1 v^2 c_\beta^2 + (m_A^2+\lambda_5 v^2)s_\beta^2 & [\lambda_{345}v^2 - (m_A^2+\lambda_5 v^2)]s_\beta c_\beta \\ [\lambda_{345}v^2 - (m_A^2+\lambda_5 v^2)]s_\beta c_\beta & \lambda_2 v^2 s_\beta^2 + (m_A^2+\lambda_5 v^2)c_\beta^2 \end{pmatrix}. \tag{6.210}$$

Defining the CP-even scalar mass eigenstates as in eqs. (6.169) and (6.170), the masses and mixing angle α are found from the diagonalization equation

$$\begin{aligned}\begin{pmatrix} m_H^2 & 0 \\ 0 & m_h^2\end{pmatrix} &= \begin{pmatrix} c_\alpha & s_\alpha \\ -s_\alpha & c_\alpha\end{pmatrix}\begin{pmatrix} \mathcal{M}_{11}^2 & \mathcal{M}_{12}^2 \\ \mathcal{M}_{12}^2 & \mathcal{M}_{22}^2\end{pmatrix}\begin{pmatrix} c_\alpha & -s_\alpha \\ s_\alpha & c_\alpha\end{pmatrix} \\ &= \begin{pmatrix} \mathcal{M}_{11}^2 c_\alpha^2 + 2\mathcal{M}_{12}^2 c_\alpha s_\alpha + \mathcal{M}_{22}^2 s_\alpha^2 & \mathcal{M}_{12}^2(c_\alpha^2 - s_\alpha^2) + (\mathcal{M}_{22}^2 - \mathcal{M}_{11}^2)s_\alpha c_\alpha \\ \mathcal{M}_{12}^2(c_\alpha^2 - s_\alpha^2) + (\mathcal{M}_{22}^2 - \mathcal{M}_{11}^2)s_\alpha c_\alpha & \mathcal{M}_{11}^2 s_\alpha^2 - 2\mathcal{M}_{12}^2 c_\alpha s_\alpha + \mathcal{M}_{22}^2 c_\alpha^2\end{pmatrix},\end{aligned} \tag{6.211}$$

where $s_\alpha\equiv\sin\alpha$ and $c_\alpha\equiv\cos\alpha$. Note that the two equations, Tr $\mathcal{M}^2 = m_H^2+m_h^2$ and det $\mathcal{M}^2 = m_H^2 m_h^2$, yield the following result:

$$|\mathcal{M}_{12}^2| = \sqrt{(m_H^2-\mathcal{M}_{11}^2)(\mathcal{M}_{11}^2-m_h^2)} = \sqrt{(\mathcal{M}_{22}^2-m_h^2)(\mathcal{M}_{11}^2-m_h^2)}\,. \tag{6.212}$$

Explicitly, the squared masses of the neutral CP-even Higgs bosons are given by

$$m_{H,h}^2 = \tfrac{1}{2}\left[\mathcal{M}_{11}^2 + \mathcal{M}_{22}^2 \pm \Delta\right], \tag{6.213}$$

where $m_h \le m_H$ and the nonnegative quantity Δ is defined by

$$\Delta \equiv \sqrt{(\mathcal{M}_{11}^2-\mathcal{M}_{22}^2)^2 + 4(\mathcal{M}_{12}^2)^2}\,. \tag{6.214}$$

The mixing angle α, which is defined modulo π, is evaluated by setting the off-diagonal elements of the CP-even scalar squared-mass matrix given in eq. (6.211) to zero. Hence, under the assumption that $\Delta \neq 0$,[26] it follows that

$$\cos 2\alpha = \frac{\mathcal{M}_{11}^2 - \mathcal{M}_{22}^2}{\Delta}, \qquad \sin 2\alpha = \frac{2\mathcal{M}_{12}^2}{\Delta}. \tag{6.215}$$

Instead of fixing the range of $\beta - \alpha$ as discussed below eqs. (6.175) and (6.186), a more common convention of the CP-conserving 2HDM literature consists of fixing the range of the mixing angle α such that $|\alpha| \leq \frac{1}{2}\pi$. In this latter convention, $\cos\alpha$ is nonnegative and is given by

$$\cos\alpha = \sqrt{\frac{\Delta + \mathcal{M}_{11}^2 - \mathcal{M}_{22}^2}{2\Delta}} = \sqrt{\frac{\mathcal{M}_{11}^2 - m_h^2}{m_H^2 - m_h^2}}, \tag{6.216}$$

and the sign of $\sin\alpha$ is given by the sign of $\mathcal{M}_{12}^2$. Explicitly, we have

$$\sin\alpha = \operatorname{sgn}(\mathcal{M}_{12}^2)\sqrt{\frac{\Delta - \mathcal{M}_{11}^2 + \mathcal{M}_{22}^2}{2\Delta}} = \operatorname{sgn}(\mathcal{M}_{12}^2)\sqrt{\frac{m_H^2 - \mathcal{M}_{11}^2}{m_H^2 - m_h^2}}. \tag{6.217}$$

We leave it as an exercise for the reader to check that the squared masses of the scalar mass eigenstates obtained above are consistent with the corresponding results given in terms of the Higgs basis parameters in eqs. (6.162) and (6.163).

6.2.6 The Tree-Level MSSM Higgs Sector

The tree-level Higgs sector of the MSSM is a CP-conserving Type-II 2HDM, with a scalar potential with quartic terms constrained by supersymmetry. It is convenient to adopt the notation of Section 13.8 and define[27]

$$H_{di} \equiv \epsilon_{ij}\Phi_1^j = \left((\Phi_1^0)^\dagger\,,\, -\Phi_1^-\right), \qquad H_{ui} = \Phi_{2i} = \left(\Phi_2^+\,,\, \Phi_2^0\right), \tag{6.218}$$

where i and j are SU(2) indices, $\Phi_1^j \equiv (\Phi_{1j})^\dagger$, and the SU(2) invariant tensor ϵ_{ij} is defined in eq. (A.54). Then the MSSM scalar Higgs potential is given by

$$\begin{aligned}\mathcal{V} = {} & m_d^2 H_d^\dagger H_d + m_u^2 H_u^\dagger H_u + (m_{ud}^2 \epsilon^{ij} H_{ui} H_{dj} + \text{h.c.}) \\ & + \tfrac{1}{8}(g^2 + g'^2)(H_u^\dagger H_u - H_d^\dagger H_d)^2 + \tfrac{1}{2} g^2 |H_d^\dagger H_u|^2 ,\end{aligned} \tag{6.219}$$

where

$$\epsilon^{ij} H_{ui} H_{dj} = H_u^+ H_d^- - H_u^0 H_d^0 = -\Phi_1^\dagger \Phi_2 , \tag{6.220}$$

after using $\epsilon^{ij}\epsilon_{jk} = -\delta_k^i$ [see eq. (A.54)], and

$$m_d^2 \equiv |\mu|^2 + m_{H_d}^2, \qquad m_u^2 \equiv |\mu|^2 + m_{H_u}^2, \qquad m_{ud}^2 \equiv b, \tag{6.221}$$

[26] Here, we are assuming that $m_h \neq m_H$. The case of $\Delta = 0$ (where $m_h = m_H$) is singular; in this case, the angle α is undefined since any two linearly independent combinations of h and H can serve as the physical neutral CP-even scalar states.

[27] The notation H_d, H_u (also called H_1, H_2 or $\overline{H}$, H in the literature) has the virtue of making it easy to remember which Higgs field VEVs give masses to which types of quarks and leptons.

in the notation of eq. (13.77). In light of eqs. (6.33) and (6.37), the quartic Higgs couplings in eq. (6.219) are related to the gauge couplings g and g':

$$\lambda_1 = \lambda_2 = -\lambda_{34} = \tfrac{1}{4}(g^2 + g'^2)\,, \quad \lambda_4 = -\tfrac{1}{2}g^2\,, \quad \lambda_5 = \lambda_6 = \lambda_7 = 0\,. \tag{6.222}$$

The Φ-basis, where eq. (6.222) is satisfied, corresponds to the scalar field basis in which the supersymmetry of the dimension-four terms of the scalar potential is manifestly realized. The supersymmetry is softly broken by the scalar squared-mass parameters, $m^2_{H_d}$, $m^2_{H_u}$, and b.

Note that m^2_{ud}, the only potentially complex parameter that appears in eq. (6.219), can be chosen real by an appropriate rephasing of the Higgs doublet fields. In this real scalar field basis, the minimum of the Higgs scalar potential is

$$\langle H_d^0\rangle = \frac{v_d}{\sqrt{2}} = \frac{v\cos\beta}{\sqrt{2}}\,, \qquad \langle H_u^0\rangle = \frac{v_u}{\sqrt{2}} = \frac{v\sin\beta}{\sqrt{2}}\,, \tag{6.223}$$

where v_d and v_u are real, with $v \equiv (v_d^2 + v_u^2)^{1/2} \simeq 246$ GeV. Consequently, the tree-level MSSM Higgs scalar potential and vacuum are CP-conserving.

We shall exploit the freedom to redefine $H_d \to -H_d$ and/or $H_u \to -H_u$ such that v_d and v_u are nonnegative. In this case, the parameter

$$\tan\beta \equiv \frac{v_u}{v_d} \tag{6.224}$$

is nonnegative and $0 \leq \beta \leq \frac{1}{2}\pi$. One can now transform to a real Higgs basis, in which the scalar potential parameters of the Higgs basis, Z_5, Z_6, and Z_7 are simultaneously real. It is straightforward to show that

$$\begin{aligned}
&Y_1 = -\tfrac{1}{2}Z_1 v^2\,, && Y_2 = m_A^2 + \tfrac{1}{8}(g^2 + g'^2)v^2\cos^2 2\beta\,,\\
&Y_3 = -\tfrac{1}{2}Z_6 v^2\,, && Z_1 = Z_2 = \tfrac{1}{4}(g^2 + g'^2)\cos^2 2\beta\,,\\
&Z_3 = Z_5 + \tfrac{1}{4}(g^2 - g'^2)\,, && Z_4 = Z_5 - \tfrac{1}{2}g^2\,,\\
&Z_5 = \tfrac{1}{4}(g^2 + g'^2)\sin^2 2\beta\,, && Z_7 = -Z_6 = \tfrac{1}{4}(g^2 + g'^2)\sin 2\beta\cos 2\beta\,.
\end{aligned} \tag{6.225}$$

Using these results, the properties of the tree-level MSSM Higgs sector can be derived using the results of Section 6.2.5. For example, eq. (6.167) immediately yields the following bounds on the tree-level values of the squared masses of the CP-even Higgs bosons of the MSSM:

$$m_h^2 \leq \min\{m_Z^2\cos^2 2\beta\,,\, m_A^2 + m_Z^2\sin^2 2\beta\}\,, \tag{6.226}$$

$$m_H^2 \geq \max\{m_Z^2\cos^2 2\beta\,,\, m_A^2 + m_Z^2\sin^2 2\beta\}\,. \tag{6.227}$$

Equation (6.226) is a striking result, as it implies that $m_h \leq m_Z$, in conflict with the observed Higgs boson mass of 125 GeV. We will see in Sections 13.8 and 19.8 that the radiative corrections to eq. (6.226) are significant in the MSSM, and parameter regimes of the MSSM exist in which the upper bound on the lightest CP-even Higgs boson mass can be raised to a value above 125 GeV, thereby restoring the consistency with the observed Higgs boson data.

The tree-level properties of the MSSM Higgs sector can be obtained using the results of Sections 6.2.2 and 6.2.5. In light of eq. (6.218), we can make use of eqs. (6.193) and (6.171) to identify the charged scalar boson pair,

$$H^{\pm} = H_d^{\pm} \sin\beta + H_u^{\pm} \cos\beta\,, \tag{6.228}$$

and the CP-odd neutral scalar,

$$A = \sqrt{2}\left(\operatorname{Im} H_d^0 \sin\beta + \operatorname{Im} H_u^0 \cos\beta\right)\,. \tag{6.229}$$

Likewise, it follows from eqs. (6.210) and (6.218) that the two CP-even neutral scalar mass eigenstates are determined by diagonalizing the neutral CP-even scalar squared-mass matrix with respect to the basis $\{\sqrt{2}\operatorname{Re} H_d^0 - v_d\,,\ \sqrt{2}\operatorname{Re} H_u^0 - v_u\}$:

$$\mathcal{M}^2 = \begin{pmatrix} m_A^2 \sin^2\beta + m_Z^2 \cos^2\beta & -(m_A^2 + m_Z^2)\sin\beta\cos\beta \\ -(m_A^2 + m_Z^2)\sin\beta\cos\beta & m_A^2 \cos^2\beta + m_Z^2 \sin^2\beta \end{pmatrix}\,. \tag{6.230}$$

The neutral CP-even scalar mass eigenstates are defined by eqs. (6.169) and (6.170):

$$h = -(\sqrt{2}\operatorname{Re} H_d^0 - v_d)\sin\alpha + (\sqrt{2}\operatorname{Re} H_u^0 - v_u)\cos\alpha\,, \tag{6.231}$$

$$H = (\sqrt{2}\operatorname{Re} H_d^0 - v_d)\cos\alpha + (\sqrt{2}\operatorname{Re} H_u^0 - v_u)\sin\alpha\,. \tag{6.232}$$

All scalar masses and couplings can be expressed in terms of two parameters, usually chosen to be m_A and $\tan\beta$. Using eqs. (6.208), (6.209), and (6.222), the masses of the neutral CP-odd and charged Higgs bosons are given by

$$m_A^2 = \frac{2m_{ud}^2}{\sin 2\beta} = m_d^2 + m_u^2\,, \tag{6.233}$$

after using eqs. (6.206) and (6.207), and

$$m_{H^{\pm}}^2 = m_A^2 + m_W^2\,. \tag{6.234}$$

The squared masses of the CP-even Higgs bosons h and H are eigenvalues of $\mathcal{M}^2$. The trace and determinant of $\mathcal{M}^2$ yield

$$m_h^2 + m_H^2 = m_A^2 + m_Z^2\,, \qquad m_h^2 m_H^2 = m_A^2 m_Z^2 \cos^2 2\beta\,, \tag{6.235}$$

where the CP-even Higgs masses are given by

$$m_{H,h}^2 = \tfrac{1}{2}\left(m_A^2 + m_Z^2 \pm \sqrt{(m_A^2 + m_Z^2)^2 - 4m_Z^2 m_A^2 \cos^2 2\beta}\right)\,. \tag{6.236}$$

In the convention where $0 \le \beta \le \frac{1}{2}\pi$, it is standard practice to choose the mixing angle α to lie in the range $|\alpha| \le \frac{1}{2}\pi$. However, because the off-diagonal element of $\mathcal{M}^2$ is negative [see eq. (6.230)], it follows that $-\frac{1}{2}\pi \le \alpha \le 0$. Using eqs. (6.216), (6.217), and (6.230), one obtains the following formulae for the mixing angle α as a function of m_A and β:

$$\cos\alpha = \sqrt{\frac{m_A^2 \sin^2\beta + m_Z^2 \cos^2\beta - m_h^2}{m_H^2 - m_h^2}}\,, \tag{6.237}$$

$$\sin\alpha = -\sqrt{\frac{m_H^2 - m_A^2 \sin^2\beta - m_Z^2 \cos^2\beta}{m_H^2 - m_h^2}}\,. \tag{6.238}$$

The tree-level couplings of the Higgs bosons with the gauge bosons and the Higgs self-couplings depend only the combination $\beta - \alpha$. These couplings are obtained from eqs. (6.93)–(6.98) after making use of eq. (6.173) and Table 6.3. Given the parameter ranges of α and β, it follows that $0 \leq \beta - \alpha \leq \pi$. Hence, we may employ eq. (6.178) to obtain

$$\cos(\beta - \alpha) = \frac{m_Z^2 \sin 2\beta \cos 2\beta}{\sqrt{(m_H^2 - m_h^2)(m_H^2 - m_Z^2 \cos^2 2\beta)}}\,, \tag{6.239}$$

$$\sin(\beta - \alpha) = \sqrt{\frac{m_H^2 - m_Z^2 \cos^2 2\beta}{m_H^2 - m_h^2}}\,. \tag{6.240}$$

One can check that eqs. (6.239) and (6.240) are consistent (i.e., the sum of the squares of these two equations is equal to 1) in light of eq. (6.235).

The MSSM Higgs sector employs Type-II Higgs–fermion Yukawa couplings as a consequence of supersymmetry rather than a $\mathbb{Z}_2$ symmetry. Nevertheless, the dimension-four terms of the tree-level MSSM Higgs Lagrangian respect the $\mathbb{Z}_2$ symmetry defined by the Type-II $\mathbb{Z}_2$ charges as specified in Table 6.2.[28] Hence, the tree-level MSSM Higgs–fermion Yukawa couplings are given by eq. (6.203).

The Higgs alignment limit can be achieved either in the decoupling limit where $m_A \gg m_Z$ or in the limit of $|Z_6| \ll 1$. Note that eq. (6.225) implies that $Z_6 = 0$ when $\beta = 0, \frac{1}{4}\pi$, or $\frac{1}{2}\pi$. The first and third possibilities correspond to $\tan\beta = 0$ or $\tan\beta = \infty$, respectively, neither of which yield realistic Higgs–fermion Yukawa couplings. The second possibility corresponds to $\tan\beta = 1$ and yields a tree-level value of $m_h = 0$, which is also untenable. Hence, in the tree-level approximation, the MSSM Higgs sector yields a viable SM-like Higgs boson only in the decoupling limit.[29]

In the decoupling limit where $m_A \gg m_Z$, the parameter $\cos(\beta - \alpha)$ and the tree-level neutral CP-even Higgs masses are given by the following expressions:

$$\cos(\beta - \alpha) \simeq \frac{m_Z^2 \sin 4\beta}{2m_A^2}\,, \tag{6.241}$$

$$m_h^2 \simeq m_Z^2 \cos^2 2\beta - \frac{m_Z^4 \sin^2 4\beta}{4m_A^2}\,, \tag{6.242}$$

$$m_H^2 \simeq m_A^2 + m_Z^2 \sin^2 2\beta\,, \tag{6.243}$$

which can also be obtained from eqs. (6.180)–(6.182). Using the above results along with eq. (6.234), it follows that $\cos(\beta - \alpha) \simeq 0$, up to corrections of $\mathcal{O}(m_Z^2/m_A^2)$, and $m_A \simeq m_H \simeq m_{H^\pm}$, up to corrections of $\mathcal{O}(m_Z^2/m_A)$. Indeed, this is the Higgs alignment limit of the MSSM Higgs sector, which is achieved via the decoupling of the heavy scalar states.

[28] In the MSSM, this $\mathbb{Z}_2$ symmetry is softly broken due to the nonzero parameter m_{ud}^2 in the scalar potential [eq. (6.219)].

[29] The Higgs alignment limit without decoupling in the MSSM Higgs sector can be achieved when radiative corrections are taken into account, although this parameter regime is disfavored by present experimental constraints.

6.3 Grand Unification and Unification of Couplings

The structure of the Standard Model looks rather haphazard. The gauge group, $SU(3)_C \times SU(2)_L \times U(1)_Y$, is the product of three groups and the gauge charges of the quark and lepton fields listed in Table 4.2 appear to be arbitrary. The cancellation of anomalies (discussed in Chapter 5) seems rather miraculous. Although it is often said that the electromagnetic and weak interactions are unified in the Standard Model, the electroweak interactions are governed by two independent couplings g and g'. Indeed, there is no relation in the Standard Model between the two electroweak gauge couplings and the strong coupling g_s.

The idea of grand unification of the strong and electroweak interactions is compelling.[30] In the simplest models of this type, the grand unified theory (GUT) is based on a simple gauge group G that is first spontaneously broken down to $SU(3)_C \times SU(2)_L \times U(1)_Y$ at a very large GUT energy scale, $M_{\rm GUT}$. Subsequently, electroweak symmetry breaking occurs at the scale characterized by $v = 246$ GeV. The large hierarchy of mass scales, $v \ll M_{\rm GUT}$, is not explained (and is enforced by a fine-tuning of the GUT parameters).

We first sketch out the basic ideas of grand unification by choosing $G = SU(5)$, which is the unique simple group of minimal rank that can unify the SM gauge interactions. We will also mention some advantages of extending the GUT gauge group to SO(10). We will then focus on one of the spectacular predictions of grand unification, namely the unification of couplings. Unfortunately, in the simplest grand unified extensions of the Standard Model, the unification of couplings is not consistent with the present-day experimental measurements of the three gauge couplings g_s, g, and g'. Nevertheless, it is quite remarkable that if supersymmetry is present at the TeV scale, then the unification of couplings in the simplest grand unified extensions seems to be realized! This observation is often quoted as one of the motivating factors for considering the supersymmetric extension of the Standard Model.

Another consequence of GUT models that unify the strong and electroweak interactions is that baryon number (B) and lepton number (L) are no longer symmetries of the theory at the perturbative level.[31] Consequently, in models of this type, the proton is unstable. The simplest SU(5) and SO(10) grand unified extensions of the Standard Model predict a proton lifetime of order 10^{30} years, which is ruled out by experimental searches by a few orders of magnitude. Supersymmetric grand unified models predict somewhat longer proton lifetimes, which may or may not be experimentally ruled out, depending on the details of the GUT model.

For the convenience of the reader, we have assembled in Appendix H.5.3 a number of useful tables of selected irreducible representations, tensor products, and branching rules for Lie groups that are most often used in constructing models of grand unification. More extensive tables can be found in [B116] and in Refs. [68–70].

30 See [B30, B32, B36, B93, B95] and Ref. [67] for further elaborations and a guide to the literature.

31 Recall that in the Standard Model, B and L are accidental global symmetries of the theory that are broken nonperturbatively due to anomalies, as discussed in Chapter 5.

6.3.1 SU(5) Grand Unification

We first discuss some group theory necessary for the construction of the SU(5) GUT model, first proposed by Georgi and Glashow [71]. The formalism developed in Chapter 4 can be employed in constructing an SU(5) gauge theory and the spontaneous breaking of SU(5) down to SU(3)×SU(2)×U(1). The symmetry breaking is accomplished by introducing scalar fields in the adjoint representation of SU(5).

The Lie group SU(5) has rank four, which means that the maximal number of simultaneously diagonalizable generators is four. These generators comprise the Cartan subalgebra of $\mathfrak{su}(5)$, the Lie algebra of SU(5). Eigenvalues of these diagonal generators correspond to the quantum numbers that are used to distinguish among the particle states of the model. The generators of $\mathfrak{su}(5)$ will be denoted abstractly by $\boldsymbol{T^A}$, where $A \in \{1, 2, \dots, 24\}$. In the fundamental representation, the generators of SU(5) are represented by traceless hermitian 5×5 matrices, which we denote as $\boldsymbol{t^A}$. It is convenient to exhibit explicitly 12 of the 24 generators $\boldsymbol{t^A}$ in block-diagonal form:[32]

$$\boldsymbol{t^a} = \left(\begin{array}{c|c} \frac{1}{2}\lambda^a & \mathbb{O} \\ \hline \mathbb{O} & \mathbb{O} \end{array}\right), \quad \text{for } a \in \{1, 2, \dots, 8\}, \tag{6.244}$$

$$\boldsymbol{t^{i+20}} = \left(\begin{array}{c|c} \mathbb{O} & \mathbb{O} \\ \hline \mathbb{O} & \frac{1}{2}\tau^i \end{array}\right), \quad \text{for } i \in \{1, 2, 3\}, \tag{6.245}$$

$$\boldsymbol{t^{24}} = \sqrt{\frac{3}{5}}\, \mathrm{diag}\left(-\tfrac{1}{3}, -\tfrac{1}{3}, -\tfrac{1}{3}, \tfrac{1}{2}, \tfrac{1}{2}\right), \tag{6.246}$$

where the λ^a are the Gell-Mann matrices of $\mathfrak{su}(3)$ and the τ^i are the Pauli matrices. The generators above are conventionally normalized such that

$$\mathrm{Tr}(\boldsymbol{t^A t^B}) = \tfrac{1}{2}\delta^{AB}. \tag{6.247}$$

The elements of the Cartan subalgebra of $\mathfrak{su}(5)$ are $\{\boldsymbol{t^3}, \boldsymbol{t^8}, \boldsymbol{t^{23}}, \boldsymbol{t^{24}}\}$. Moreover, one can explicitly identify the $\mathfrak{su}(3) \oplus \mathfrak{su}(2) \oplus \mathfrak{u}(1)$ subalgebra of $\mathfrak{su}(5)$, which is spanned by the 12 generators explicitly listed in eqs. (6.244)–(6.246).

Note that in the fundamental representation, an element of SU(5) is represented by $U = \exp(-i\theta^a \boldsymbol{t^A})$. Then, $U^* = \exp(i\theta^a \boldsymbol{t^{A\,*}})$, so that we can identify the generators of the $\overline{\mathbf{5}}$ representation of SU(5),

$$\boldsymbol{t^{A\,*}} = -\boldsymbol{t^{A\mathsf{T}}}. \tag{6.248}$$

The gauge bosons of SU(5) transform under the 24-dimensional adjoint representation of SU(5). The SU(5) gauge boson fields are conveniently expressed as a 5×5

[32] For completeness, we note that the generators $\boldsymbol{t^{j+8}}$ (for $j \in \{1, 2, \dots, 12\}$) are hermitian matrices with two nonzero entries. For odd [even] j, the nontrivial entry $\frac{1}{2}$ $[-\frac{1}{2}i]$ appears as the 14, 15, 24, 25, 34, 35 element of the matrix generator, respectively, and the respective complex conjugate appears in the corresponding transposed position.

traceless hermitian matrix of fields, $A_\mu(x) \equiv \sqrt{2}A_\mu^A(x)\boldsymbol{t^A}$, where the A_μ^A are real gauge boson fields.[33] Likewise, we can define a 5×5 traceless hermitian matrix of scalar adjoint fields,

$$\Phi(x) \equiv \sqrt{2}\,\Phi^A(x)\boldsymbol{t^A}\,, \tag{6.249}$$

where the Φ^A are real scalar fields. Using eq. (6.247), it follows that

$$\Phi^A(x) = \sqrt{2}\,\mathrm{Tr}\big[\boldsymbol{t^A}\Phi(x)\big]\,. \tag{6.250}$$

The indices of the matrix scalar field Φ are denoted by $\Phi_i{}^j$, where $i, j \in \{1, 2, \ldots, 5\}$. Under an SU(5) transformation U, these fields transform as

$$\Phi_i{}^j \to U_i{}^k U^j{}_\ell \Phi_k{}^\ell\,, \tag{6.251}$$

where $U^j{}_\ell \equiv (U_j{}^\ell)^* = (U^\dagger)_\ell{}^j$. In matrix form, eq. (6.251) is equivalent to

$$\Phi \to U\Phi U^\dagger\,. \tag{6.252}$$

In the limit where U is close to the identity, $U \simeq \mathbb{1} - i\theta^A\boldsymbol{t^A}$. The infinitesimal version of eq. (6.251) is given by $\Phi \to \Phi + \delta\Phi$, where

$$\delta\Phi = i\theta^A\big[\Phi, \boldsymbol{t^A}\big]\,. \tag{6.253}$$

By an appropriate choice of the range of parameters of the scalar potential, the minimum of the scalar potential is achieved for

$$\langle\Phi\rangle = c\,\boldsymbol{t^{24}}\,, \tag{6.254}$$

where c is a real constant (see Exercises 6.15 and 6.16). Note that $\boldsymbol{t^{24}}$ commutes with precisely 12 of the 24 generators of $\mathfrak{su}(5)$, namely the 12 generators listed in eqs. (6.244)–(6.246). Hence, $\delta\langle\Phi\rangle = i\theta^A[\langle\Phi\rangle, \boldsymbol{t^A}] = 0$ when the index A in eq. (6.253) is restricted to lie in the subalgebra of $\mathfrak{su}(5)$ spanned by the 12 generators listed in eqs. (6.244)–(6.246). It follows that these generators are unbroken and we can conclude that the gauge group SU(5) is spontaneously broken down to SU(3)×SU(2)×U(1). Note that the rank of the group is unchanged, which is a general feature of adjoint breaking [see Exercise 6.14(d)].

The fermion fields of the Standard Model comprise a reducible $\overline{\mathbf{5}} \oplus \mathbf{10}$ representation of $\mathfrak{su}(5)$. To see how this works, note that we have already identified the elements of the $\mathfrak{su}(3) \oplus \mathfrak{su}(2) \oplus \mathfrak{u}(1)$ subalgebra of $\mathfrak{su}(5)$, as the real linear combination of the 12 generators $\{\boldsymbol{T^1}, \ldots, \boldsymbol{T^8}, \boldsymbol{T^{21}}, \boldsymbol{T^{22}}, \boldsymbol{T^{23}}, \boldsymbol{T^{24}}\}$. The two-component fermion field that transforms as a $\overline{\mathbf{5}}$ representation of SU(5) is denoted by ψ^i ($i \in \{1, 2, \ldots, 5\}$). Its transformation law under SU(5) is given by $\psi^i \to U^i{}_j\psi^j$, where $U^i{}_j = \big[(\exp(-i\theta^a\boldsymbol{t^a})_i{}^j\big]^*$. If we let the index i run over 1,2,3, then ψ^i has the transformation properties of an SU(3) antitriplet. If we let the index i run over 4,5, then ψ^i has the transformation properties of an SU(2) doublet [keeping in mind that for SU(2), $\overline{\mathbf{2}} \cong \mathbf{2}$]. The U(1) hypercharge quantum numbers, up to an overall

[33] Here, we deviate from eq. (4.25) by omitting the gauge coupling prefactor. We also include a factor of $\sqrt{2}$ in the definition of the matrix gauge field A_μ so that $\mathrm{Tr}(A^\mu A_\mu) = A^{\mu A}A_\mu^A$ after employing eq. (6.247).

constant which is a matter of convention, is fixed by the requirement that the U(1) generator $\boldsymbol{t^{24}}$ is traceless. Hence, we have demonstrated the following branching rule, for the SU(3)×SU(2)×U(1) decomposition of $\overline{\mathbf{5}}$:

$$\overline{\mathbf{5}} \longrightarrow (\overline{\mathbf{3}},\mathbf{1},\tfrac{1}{3}) + (\mathbf{1},\mathbf{2},-\tfrac{1}{2})\,. \tag{6.255}$$

Note that $\operatorname{Tr} Y = 3\left(\frac{1}{3}\right) + 2\left(-\frac{1}{2}\right) = 0$ as required. Indeed, we can identify explicitly the generator of hypercharge Y in the defining $\mathbf{5}$ representation via

$$Y = \sqrt{\frac{5}{3}}\,\boldsymbol{t^{24}}\,. \tag{6.256}$$

In the $\overline{\mathbf{5}}$ representation, the generators of $\mathfrak{su}(5)$ are given by $-\boldsymbol{t^{a\mathsf{T}}}$ [see eq. (6.248)], implying that $Y(\overline{\mathbf{5}}) = -Y(\mathbf{5})$, which yields the hypercharge quantum numbers indicated in eq. (6.255).

The electric charge operator in the $\mathbf{5}$ representation is given by

$$Q = \boldsymbol{t^{23}} + \sqrt{\frac{5}{3}}\,\boldsymbol{t^{24}} = \operatorname{diag}\left(-\tfrac{1}{3}, -\tfrac{1}{3}, -\tfrac{1}{3}, 1, 0\right). \tag{6.257}$$

Moreover, $Q(\overline{\mathbf{5}}) = -Q(\mathbf{5})$. Higher-dimensional representations can be generated by repeated direct products of the fundamental representation, and the corresponding electric charge quantum numbers are additive. As a result, the electric charges of particles residing in SU(5) multiplets are quantized in units of $\frac{1}{3}e$. More generally, the quantization of electric charge of the fundamental particles is a prediction of grand unified models with a semisimple grand unified group. This is in contrast to the Standard Model, where electric charge quantization in units of $\frac{1}{3}e$ is not explained and the quantization of the U(1) hypercharges appears to be accidental.

The two-component fermion field that transforms as a $\mathbf{10}$ representation of SU(5) is denoted by ψ_{ij} (where $i,j \in \{1,2,\ldots,5\}$) and is antisymmetric under the interchange of the indices i and j. Indeed, an antisymmetric 5×5 matrix possesses 10 degrees of freedom. Equivalently, the $\mathbf{10}$ of SU(5) can be defined as the antisymmetric part of the direct product representation, $(\mathbf{5}\otimes\mathbf{5})_A$. The corresponding transformation law is $\psi_{ij} \to U_i{}^k U_j{}^\ell \psi_{k\ell}$. If one restricts, $i,j \in \{1,2,3\}$, then ψ_{ij} has the transformation properties of $(\mathbf{3}\otimes\mathbf{3})_A \cong \overline{\mathbf{3}}$ of SU(3). If one restricts, $i,j \in \{4,5\}$, then ψ_{ij} has the transformation properties of $(\mathbf{2}\otimes\mathbf{2})_A \cong \mathbf{1}$ of SU(2). Finally, letting $i \in \{1,2,3\}$ and $j \in \{4,5\}$ or the reverse yields a $(\mathbf{3},\mathbf{2})$ representation of SU(3)×SU(2). The U(1) hypercharge quantum numbers are additive. Hence, employing the complex conjugate of eq. (6.255), the corresponding hypercharges in $(\mathbf{5}\otimes\mathbf{5})_A$ are $-\frac{2}{3}$, 1, and $\frac{1}{6}$, respectively. That is, we have obtained the following branching rule, for the SU(3)×SU(2)×U(1) decomposition of $\mathbf{10}$:

$$\mathbf{10} \longrightarrow (\overline{\mathbf{3}},\mathbf{1},-\tfrac{2}{3}) + (\mathbf{1},\mathbf{1},1) + (\mathbf{3},\mathbf{2},\tfrac{1}{6})\,. \tag{6.258}$$

Combining the results of eqs. (6.255) and (6.258), it follows that

$$\overline{\mathbf{5}}\oplus\mathbf{10} \longrightarrow (\mathbf{3},\mathbf{2},\tfrac{1}{6}) + (\overline{\mathbf{3}},\mathbf{1},-\tfrac{2}{3}) + (\overline{\mathbf{3}},\mathbf{1},\tfrac{1}{3}) + (\mathbf{1},\mathbf{2},-\tfrac{1}{2}) + (\mathbf{1},\mathbf{1},1)\,. \tag{6.259}$$

In light of eq. (5.92), one generation of the Standard Model fermions fits precisely into a reducible $\overline{\mathbf{5}}\oplus\mathbf{10}$ representation of SU(5). Explicitly,[34]

$$\psi^i=\begin{pmatrix} \bar{d}^1 \\ \bar{d}^2 \\ \bar{d}^3 \\ e \\ -\nu \end{pmatrix}, \qquad \psi_{ij}=\frac{1}{\sqrt{2}}\begin{pmatrix} 0 & \bar{u}^3 & -\bar{u}^2 & u_1 & d_1 \\ -\bar{u}^3 & 0 & \bar{u}^1 & u_2 & d_2 \\ \bar{u}^2 & -\bar{u}^1 & 0 & u_3 & d_3 \\ -u_1 & -u_2 & -u_3 & 0 & \bar{e} \\ -d_1 & -d_2 & -d_3 & -\bar{e} & 0 \end{pmatrix}, \tag{6.260}$$

where ψ^i transforms as a $\overline{\mathbf{5}}$ and $\psi_{ij}=-\psi_{ji}$ transforms as a $\mathbf{10}$ under SU(5). Moreover, as shown in Section 5.2.2, an SU(3)×SU(2)×U(1) gauge theory with the fermions in a reducible representation as specified in eq. (6.259) is anomaly-free. Indeed, this result can also be understood from the SU(5) point of view. One can show that an SU(5) gauge theory with the fermions in a $\overline{\mathbf{5}}\oplus\mathbf{10}$ representation is anomaly-free as follows. The anomaly coefficient for a representation of a compact simple Lie group is given by a ratio of cubic indices,

$$A(R)=\frac{I_3(R)}{I_3(F)}, \tag{6.261}$$

where the cubic index is defined by $D^{abc}(R)=I_3(R)d_{abc}$ [see eqs. (5.82) and (5.83)]. In particular, since F corresponds to the five-dimensional fundamental representation of SU(5), it follows that $A(\mathbf{5})=1$ and $A(\overline{\mathbf{5}})=-1$ [see eq. (5.87)]. Noting that $\mathbf{10}=(\mathbf{5}\otimes\mathbf{5})_A$, one can employ eq. (5.86) to obtain $A(\mathbf{10})=1$.[35] Hence, eq. (5.84) yields

$$A(\overline{\mathbf{5}}\oplus\mathbf{10})=0\,. \tag{6.262}$$

To summarize, the SU(5) grand unified theory consists of fermions in a $\overline{\mathbf{5}}\oplus\mathbf{10}$ representation, gauge bosons in the 24-dimensional adjoint representation, and Higgs scalars in the 24-dimensional adjoint representation. It is rather straightforward to write down a scalar potential for the scalars that yields a vacuum expectation value $\langle\Phi\rangle=c\boldsymbol{t^{24}}$ (see Exercises 6.15 and 6.16), which spontaneously breaks the SU(5) gauge symmetry down to SU(3)×SU(2)×U(1). By assumption, $c\sim\mathcal{O}(M_{\rm GUT})$, where $M_{\rm GUT}$ is the scale of SU(5)-breaking. To be consistent with the present experimental nonobservation of proton decay, $M_{\rm GUT}\gtrsim 10^{16}$ GeV. Given that this scale lies somewhat below the Planck scale (where quantum gravitational effects can no longer be ignored), grand unification is a plausible extension of the Standard Model, which unifies the three fundamental forces of particle physics. One of the most important predictions of grand unified theories is the potential for the unification of gauge couplings, which implies that the value of the strong coupling α_s

[34] The last two components of ψ^i correspond to $\epsilon^{ij}(\psi_L)_j$ where $\psi_L=(\nu,e)$ and the weak isospin labels are $i,j\in\{1,2\}$. The upper 3×3 matrix block of ψ_{ij} corresponds to $\epsilon_{ijk}(\psi_U)^k$, where $\psi_U=(\bar{u}^1,\bar{u}^2,\bar{u}^3)$ and the SU(3) color labels $i,j,k\in\{1,2,3\}$, whereas the lower 2×2 block of ψ_{ij} is $\epsilon_{ij}\bar{e}$.

[35] One can also derive this result as follows. Since $I_3(R)$ can be evaluated for any choice of indices a, b, and c such that $d^{abc}\neq 0$, one may compute $A(\mathbf{10})$ by restricting $a,\,b,\,c\in\{1,2,3\}$. Using the $\mathfrak{su}(3)$ subalgebra of $\mathfrak{su}(5)$ [see eq. (6.244)] to express $A(\mathbf{10})$ in terms of anomaly coefficients of $\mathfrak{su}(3)$, it follows that $A(\mathbf{10})_{\mathfrak{su}(5)}=A(\overline{\mathbf{3}})_{\mathfrak{su}(3)}+2A(\mathbf{3})_{\mathfrak{su}(3)}=-1+2=1$. For another derivation based on index sum rules, see Appendix H.5.2 [see eq. (H.201)].

evaluated at the electroweak scale can be predicted given the measured values of $\alpha_{\rm EM}$ and $\sin^2\theta_W$. This prediction will be discussed further in Section 6.3.5.

To determine the quantum numbers of the gauge bosons that live in the adjoint representation of SU(5), it is sufficient to note that **24** corresponds to the traceless part of $\mathbf{5}\otimes\overline{\mathbf{5}}$. This result can be understood by writing an adjoint field schematically as $A_i{}^j = \psi_i \otimes \psi^j - \delta_i^j \psi_k \otimes \psi^k$, where there is an implicit sum over k. Note that $\mathrm{Tr}\, A = 0$ by construction. We can now use the tensor product of eq. (6.255) and its complex conjugate to obtain

$$\mathbf{24} \longrightarrow (\mathbf{8},\mathbf{1},0) + (\mathbf{1},\mathbf{3},0) + (\mathbf{1},\mathbf{1},0) + (\mathbf{3},\mathbf{2},-\tfrac{5}{6}) + (\overline{\mathbf{3}},\mathbf{2},\tfrac{5}{6}), \qquad (6.263)$$

after removing the SU(5) singlet that corresponds to the trace of $\mathbf{5}\otimes\overline{\mathbf{5}}$. We recognize $(\mathbf{8},\mathbf{1},0)+(\mathbf{1},\mathbf{3},0)+(\mathbf{1},\mathbf{1},0)$ as the $\mathrm{SU}(3)_C\times\mathrm{SU}(2)_L\times\mathrm{U}(1)_Y$ quantum numbers of the massless gauge bosons, G^a, W^i, and B. The remaining 12 gauge bosons, corresponding to $(\mathbf{3},\mathbf{2},-\frac{5}{6})+(\overline{\mathbf{3}},\mathbf{2},\frac{5}{6})$, are superheavy, with masses of $\mathcal{O}(M_{\rm GUT})$ generated by the Higgs mechanism. The doublet of colored gauge fields that transform as $(\overline{\mathbf{3}},\mathbf{2},\frac{5}{6})$ is called $(X_k\,, Y_k)$ (where k is the color index), whereas their antiparticle counterparts, $(\overline{X}^k,\overline{Y}^k)$, transform as $(\mathbf{3},\mathbf{2},-\frac{5}{6})$. More explicitly, the matrix adjoint gauge field, $A_\mu \equiv \sqrt{2}\, A_\mu^A \boldsymbol{t}^{\boldsymbol{A}}$, is a 5×5 hermitian matrix of fields given in block matrix form by

$$A_\mu \equiv \sqrt{2}\, A_\mu^A \boldsymbol{t}^{\boldsymbol{A}} = \left(\begin{array}{ccc|cc} & & & \overline{X}^1 & \overline{Y}^1 \\ & \dfrac{\lambda^a G^a_\mu}{\sqrt{2}} - \dfrac{2B}{\sqrt{30}}\mathbb{1}_{3\times 3} & & \overline{X}^2 & \overline{Y}^2 \\ & & & \overline{X}^3 & \overline{Y}^3 \\ \hline X_1 & X_2 & X_3 & \dfrac{W^3}{\sqrt{2}} + \dfrac{3B}{\sqrt{30}} & W^+ \\ Y_1 & Y_2 & Y_3 & W^- & -\dfrac{W^3}{\sqrt{2}} + \dfrac{3B}{\sqrt{30}} \end{array}\right), \qquad (6.264)$$

where $W^\pm = (W^1 \mp iW^2)/\sqrt{2}$.

The Higgs field Φ is also an SU(5) adjoint field. After SU(5) symmetry breaking, 12 of the 24 adjoint scalars correspond to the Goldstone bosons responsible for generating mass for the X and Y gauge bosons. The remaining 12 Higgs scalars are also superheavy, with masses of $\mathcal{O}(M_{\rm GUT})$. One can check that in the coupling of the gauge bosons X and Y, baryon number (B) and lepton number (L) are separately violated, although $B-L$ is preserved.

At this stage, $\mathrm{SU}(3)_C\times\mathrm{SU}(2)_L\times\mathrm{U}(1)_Y$ remains unbroken. Thus, in the final step, we shall introduce a $\mathbf{5}\oplus\overline{\mathbf{5}}$ of Higgs scalars, denoted by the complex scalar field ϕ_i (where $i\in\{1,2,\ldots,5\}$) and its complex conjugate, $\phi^i \equiv (\phi_i)^\dagger$, respectively. The branching rule for **5** with respect to $\mathrm{SU}(3)_C\times\mathrm{SU}(2)_L\times\mathrm{U}(1)_Y$ is the complex conjugate of eq. (6.255),

$$\mathbf{5} \longrightarrow (\mathbf{3},\mathbf{1},-\tfrac{1}{3}) + (\mathbf{1},\mathbf{2},\tfrac{1}{2})\,. \qquad (6.265)$$

We recognize $(\mathbf{1},\mathbf{2},\frac{1}{2})$ as having the quantum numbers of the Higgs boson of the Standard Model. It is possible to write down a scalar potential such that the electrically neutral Higgs field of $(\mathbf{1},\mathbf{2},\frac{1}{2})$ acquires a vacuum expectation value,

$\langle\phi_i\rangle = v\delta_{i5}/\sqrt{2}$, where $v = 246$ GeV governs the scale of electroweak symmetry breaking. The end result is that the SU(3)×SU(2)×U(1) gauge symmetry is broken down to SU(3)×U(1)$_{\rm EM}$ as described in Section 4.7.

The $\mathbf{5}\oplus\overline{\mathbf{5}}$ multiplet of Higgs scalars also contains colored Higgs scalar states, ϕ_k, where k labels the quantum number in the fundamental representation of SU(3). For a phenomenologically viable model, the colored Higgs scalars must have masses of $\mathcal{O}(M_{\rm GUT})$ in order to avoid scalar-mediated baryon-number-violating processes. If one considers the most general quartic potential $V(\Phi,\phi)$, consisting of multiplets of scalars in the **24** and $\mathbf{5}\oplus\overline{\mathbf{5}}$ representations of SU(5), then there will be a region of the parameter space in which the SU(5) gauge symmetry is broken down to SU(3)×U(1)$_{\rm EM}$. However, for generic choices of the parameters, one would expect $v\sim\mathcal{O}(M_{\rm GUT})$. To achieve $v\ll M_{\rm GUT}$ and to achieve a SM Higgs boson whose mass is much lighter than its colored partner requires a very precise fine-tuning of the scalar potential parameters. These fine-tunings are symptoms of the gauge hierarchy problem and the doublet–triplet splitting problem, respectively. The doublet–triplet splitting problem can sometimes be avoided by employing either a larger gauge group beyond SU(5) and/or an enlarged Higgs sector with appropriately chosen representations.[36] The gauge hierarchy problem is more profound, and provides one of the main motivations for supersymmetry, as described in more detail in Chapter 7.

The masses of the fermions (excluding the neutrino) arise due to the Yukawa couplings of ϕ with the fermions [which constitute a $\overline{\mathbf{5}}\oplus\mathbf{10}$ representation of SU(5)]. Consulting Appendix H.5.3, we note that $\overline{\mathbf{5}}\otimes\mathbf{10}=\mathbf{5}\oplus\mathbf{45}$ and $\mathbf{10}\otimes\mathbf{10}=\overline{\mathbf{5}}\oplus\overline{\mathbf{45}}\oplus\overline{\mathbf{50}}$. This means that both $\overline{\mathbf{5}}\otimes\overline{\mathbf{5}}\otimes\mathbf{10}$ and $\mathbf{5}\otimes\mathbf{10}\otimes\mathbf{10}$ contain SU(5) singlets.[37] Hence, one can write two SU(5) gauge invariant Yukawa couplings,

$$-\mathscr{L}_{\rm Yuk} = y_1\psi^i\psi_{ij}\phi^{\dagger j} + \tfrac{1}{8}y_2\epsilon^{ijk\ell m}\psi_{ij}\psi_{k\ell}\phi_m + \text{h.c.}\,, \tag{6.266}$$

where the generation labels of the two-component fermion fields have been suppressed.

In order to account for neutrino masses, we shall extend the fermion content of the SU(5) model by introducing a 16th fermion state ψ that transforms as an SU(5) singlet, which corresponds to $(\mathbf{1},\mathbf{1},0)$ under SU(3)$_C$×SU(2)$_L$×U(1)$_Y$. In this case, the seesaw-extended Standard Model can be incorporated into the grand unified SU(5) theory by adding the following terms to eq. (6.266):

$$-\delta\mathscr{L}_{\rm Yuk} = y_0\psi\psi^i\phi_i + M_N\psi\psi + \text{h.c.} \tag{6.267}$$

After electroweak symmetry breaking, one obtains the neutrino mass matrix of the seesaw-extended Standard Model discussed in Section 6.1.

Note that the Yukawa couplings of the color Higgs triplet ϕ_k (for $k\in\{1,2,3\}$)

[36] Two possible strategies for solving the doublet–triplet problem, which employ the so-called "missing doublet" mechanism and the "missing VEV" mechanism, respectively, will be discussed in the context of supersymmetric GUT models in Sections 16.3.1 and 16.3.2. These mechanisms can also be employed in non-supersymmetric GUT models.

[37] In contrast, neither $\mathbf{5}\otimes\overline{\mathbf{5}}\otimes\overline{\mathbf{5}}$ nor $\overline{\mathbf{5}}\otimes\overline{\mathbf{5}}\otimes\overline{\mathbf{5}}$ contains an SU(5) singlet.

exhibited in eqs. (6.266) and (6.267) separately violate B and L. Nevertheless, if $M_N = 0$, then the model conserves $B - L$. In order to demonstrate that $B - L$ is conserved, we shall introduce a global $\mathrm{U}(1)_X$ symmetry, denoted below by X,[38] with charges for the fermion fields and the Higgs field ϕ:

$$X(\psi) = -5\,, \qquad X(\psi^i) = 3\,, \quad X(\psi_{ij}) = -1\,, \qquad X(\phi_i) = -X(\phi^i) = 2\,. \quad (6.268)$$

In addition, the adjoint scalar Higgs fields and the adjoint gauge fields are neutral under X. It follows that the only term in eqs. (6.266) and (6.267) that violates the global $\mathrm{U}(1)_X$ symmetry is the term proportional to M_N. In the following discussion, we shall take $M_N = 0$.

In the next subsection, we will discover the origin of this $\mathrm{U}(1)_X$ symmetry. One important property of $\mathrm{U}(1)_X$ is that this symmetry can be gauged (although in the context of the SU(5) theory, it is simply a global symmetry). To verify this assertion, one must show that there are no anomalies in the conservation of the $\mathrm{U}(1)_X$ current. Thus, we shall consider triangle diagrams whose external lines consist of either SU(5) gauge bosons or a $\mathrm{U}(1)_X$ symmetry current. The internal lines of the triangle diagrams consist of the fermions that transform under the reducible representation $\mathbf{1}\oplus\overline{\mathbf{5}}\oplus\mathbf{10}$. We already know that the SU(5) gauge anomalies vanish if the fermion content consists of $\overline{\mathbf{5}}\oplus\mathbf{10}$. In addition, in the triangle diagram with two external SU(5) gauge boson and one $\mathrm{U}(1)_X$, a potential anomaly that would spoil the conservation of the $\mathrm{U}(1)_X$ current is canceled in light of

$$\mathrm{Tr}(\boldsymbol{T}^a\boldsymbol{T}^b X) = \tfrac{1}{2}\delta^{ab}\left[3I_2(\overline{\mathbf{5}}) - I_2(\mathbf{10})\right] = 0\,, \qquad (6.269)$$

where $I_2(R)$ is the index of the irreducible representation R as defined in eq. (H.32) in a convention where the index of the fundamental representation of SU(5) is given by $I_2(\mathbf{5}) = \frac{1}{2}$ [see eq. (H.40)]. In obtaining eq. (6.269), we have employed eqs. (H.36) and (H.37) to obtain $I_2(\overline{\mathbf{5}}) = \frac{1}{2}$ and $I_2(\mathbf{10}) = \frac{3}{2}$. Finally, in order to cancel the anomaly arising from the triangle diagram in which all three external lines correspond to a $\mathrm{U}(1)_X$ symmetry current, the addition of the SU(5) singlet fermion is critical. In this case,

$$\mathrm{Tr}\,X^3 = (-5)^3 + 5(3)^3 + 10(-1)^3 = 0\,. \qquad (6.270)$$

Note that we also have $\mathrm{Tr}\ X = 0$, which would eliminate a potential gravitational–gauge anomaly arising from the triangle diagram with two gravitons and a $\mathrm{U}(1)_X$ gauge boson.

Hence, we conclude that the conservation of the $\mathrm{U}(1)_X$ current is not spoiled by anomalies in the case where the fermions constitute a $\mathbf{1}\oplus\overline{\mathbf{5}}\oplus\mathbf{10}$ representation of SU(5). The significance of X is that it is related to $B - L$ as follows:

$$B - L = \tfrac{1}{5}\,(4Y - X)\,, \qquad (6.271)$$

where Y is the (gauged) hypercharge generator. To check this assertion, consider

[38] The reader should be able to distinguish between various U(1) charges such as X and Y and the SU(5) gauge bosons, X_k and Y_k, which are completely unrelated.

the fermions in the $\overline{\mathbf{5}} \oplus \mathbf{10}$. In light of eqs. (6.255), (6.258), and (6.268), we obtain

$$[B-L](\bar{u},\, \bar{d},\, u,\, d,\, e,\, \nu,\, \bar{e},\, \bar{\nu}) = (-\tfrac{1}{3},\, -\tfrac{1}{3},\, \tfrac{1}{3},\, \tfrac{1}{3},\, -1,\, -1,\, 1,\, 1). \tag{6.272}$$

In addition, if we write $\phi = (\phi_k, \phi_i)$ where k is the SU(3) color index and i is the SU(2) weak isospin index, then

$$[B-L](\phi_k,\, \phi_i) = (-\tfrac{2}{3},\, 0)\,. \tag{6.273}$$

Moreover, the X_k and Y_k gauge bosons are neutral under X while possessing hypercharge $\frac{5}{6}$. Hence, these gauge bosons possess $B-L = \frac{2}{3}$. Indeed, the X_k, Y_k, and ϕ_k fields mediate processes that violate B and L separately while conserving $B-L$. Of course, if one wishes to employ the seesaw mechanism to provide masses for the neutrinos by adding a $\Delta L = 2$ violating mass term proportional to M_N as in eq. (6.267), then X and $B-L$ are no longer conserved.

6.3.2 Grand Unification Beyond SU(5)

There is some motivation for considering unification groups with rank larger than four. For example, one generation of fermions of the Standard Model appears as a reducible $\overline{\mathbf{5}} \oplus \mathbf{10}$ representation of SU(5). It may be more elegant to consider the possibility of a single irreducible representation that can account for one generation of fermions. Indeed in SO(10), there exists a 16-dimensional irreducible spinor representation, denoted by **16**, with the following branching rule with respect to SU(5)$\times$U(1)$_X$:

$$\mathbf{16} \longrightarrow (\overline{\mathbf{5}}, 3) + (\mathbf{10}, -1) + (\mathbf{1}, -5)\,. \tag{6.274}$$

We recognize this decomposition as describing one generation of Standard Model fermions including a Standard Model gauge-singlet neutrino state $\bar{\nu}$. Moreover, the global U(1)$_X$ symmetry identified in eq. (6.268) is now a gauged U(1) subgroup of SO(10).

One can now introduce a set of scalar SO(10) multiplets and a corresponding scalar potential such that the scalar potential minimum conditions lead to the spontaneous breaking of SO(10) down to SU(3)$\times$U(1)$_{\rm EM}$. There are many choices for the symmetry-breaking patterns. A complete discussion of the allowed possibilities is beyond the scope of this section. However, let us mention one possible approach. One can employ a scalar multiplet that transforms as the **16**. In light of eq. (6.274), by choosing a scalar potential that generates a vacuum expectation value for the $(\mathbf{1}, -5)$ component of the **16**, it is clear that U(1)$_X$ is broken, leaving an unbroken SU(5). Now, one can introduce a multiplet of scalar fields that transform under the adjoint representation, **45**, of SO(10), which possesses the following branching rule with respect to SU(5)$\times$U(1)$_X$:

$$\mathbf{45} \longrightarrow (\mathbf{24}, 0) + (\mathbf{10}, 4) + (\overline{\mathbf{10}}, -4) + (\mathbf{1}, 0)\,. \tag{6.275}$$

By choosing a scalar potential that yields a vacuum expectation value for the SU(5) adjoint **24** in the direction indicated by eq. (6.254), one can reproduce the symmetry breaking of SU(5) down to $SU(3)_C \times SU(2)_L \times U(1)_Y$ as previously discussed. In the final step, one can introduce a multiplet of scalar fields that transform under the **10** of SO(10), which possesses the following branching rule with respect to $SU(5)\times U(1)_X$:

$$\mathbf{10} \longrightarrow (\mathbf{5}, 2) + (\overline{\mathbf{5}}, -2) . \tag{6.276}$$

By choosing a scalar potential that yields a vacuum expectation value for the neutral components of the **5** and $\overline{\mathbf{5}}$, one can achieve the final symmetry breaking down to $SU(3)\times U(1)_{EM}$. Indeed, the $\mathbf{5} \oplus \overline{\mathbf{5}}$ of SU(5) contains two Higgs doublet fields, and thus can provide a grand unification of the 2HDM-extended Standard Model.

Since one generation of Standard Model fermions resides in a **16**, any scalar that couples to a pair of fermions must reside in the direct product of $\mathbf{16} \otimes \mathbf{16}$,

$$\mathbf{16} \otimes \mathbf{16} = \mathbf{10} \oplus \mathbf{120} \oplus \mathbf{126} . \tag{6.277}$$

Thus, one can employ **10**, **120**, and/or $\mathbf{126} \oplus \overline{\mathbf{126}}$ multiplets of Higgs scalars to construct Yukawa couplings.[39] Indeed, the branching rule of the **10** with respect to $SU(5)\times U(1)_X$, given in eq. (6.276), corresponds to the Higgs multiplet ϕ_i and its complex conjugate ϕ^i employed in the Yukawa couplings of the SU(5) model. If the electrically neutral components of **10** acquire vacuum expectation values, then both the gauged $U(1)_X$ and $U(1)_Y$ symmetries are spontaneously broken, although $B - L$ remains unbroken. Thus, in order to break the $B - L$ (as required by the seesaw mechanism), one must employ scalars that transform under the $\mathbf{126} \oplus \overline{\mathbf{126}}$ of SO(10). The branching rule of the **126** with respect to $SU(5)\times U(1)_X$ is

$$\mathbf{126} \longrightarrow (\mathbf{1}, -10) + (\overline{\mathbf{5}}, -2) + (\mathbf{10}, -6) + (\overline{\mathbf{15}}, 6) + (\mathbf{45}, 2) + (\overline{\mathbf{50}}, -2) . \tag{6.278}$$

Thus, a term in the Yukawa couplings of the form $\psi_{16}\psi_{16}\phi^\dagger_{126}$ + h.c. can produce a mass term $m_N\psi\psi$, where ψ transforms under $SU(5)\times U(1)_X$ as $(\mathbf{1}, -5)$ if the $(\mathbf{1}, -10)$ component of **126** acquires a vacuum expectation value of $\mathcal{O}(M_N)$.

Noting that the symmetric part of eq. (6.277) is $(\mathbf{16}\otimes\mathbf{16})_S = \mathbf{10}\oplus\mathbf{126}$, it follows that the corresponding Yukawa coupling matrices are symmetric with respect to the interchange of generation indices. In contrast, the antisymmetric part of eq. (6.277) is $(\mathbf{16}\otimes\mathbf{16})_A = \mathbf{120}$, which implies that employing a **120** multiplet of scalars yields antisymmetric Yukawa coupling matrices.

Beyond SO(10), one can consider the possibility of grand unification based on the exceptional Lie group E_6. This group shares many of the good features of SU(5) and SO(10), one of which is that it possesses complex representations that are suitable for unifying chiral fermions. One generation of Standard Model fermions fits into the 27-dimensional fundamental representation, with the following branching rule with respect to $SO(10)\times U(1)$:

$$\mathbf{27} \longrightarrow (\mathbf{16}, 1) + (\mathbf{10}, -2) + (\mathbf{1}, 4) . \tag{6.279}$$

In addition to one generation of quarks and leptons, we see that **27** contains an

[39] Note that **10** and **120** are real representations, whereas **16** and **126** are complex representations of SO(10).

additional right-handed neutrino field [which transforms as a singlet under SO(10)] and new vectorlike down-type quarks and vectorlike $SU(2)_L$-doublet leptons, which transform as a **10** of SO(10) or equivalently as a $\mathbf{5} \oplus \overline{\mathbf{5}}$ with respect to SU(5). The gauge bosons of the E_6 GUT reside in the 78-dimensional adjoint representation, with the following branching rule with respect to SO(10)×U(1):

$$\mathbf{78} \longrightarrow (\mathbf{45}, 0) + (\mathbf{1}, 0) + (\mathbf{16}, -3) + (\overline{\mathbf{16}}, 3)\,. \tag{6.280}$$

Additional aspects of SO(10) and E_6 GUT models are discussed in Section 16.3.2.

6.3.3 Unification of Gauge Couplings

Grand unification theory (GUT) predicts the unification of gauge couplings at some very high energy scale [72]. The running of the couplings is dictated by the particle content of the effective theory that resides below the GUT scale. However, attempts to embed the Standard Model in an SU(5) or SO(10) unified theory do not quite succeed, as shown in Fig. 6.1. In particular, the three running gauge couplings (the strong QCD coupling g_s and the electroweak gauge couplings g and g') do not meet at a point, as exhibited by the dashed lines in Fig. 6.1. In contrast, in the case of the MSSM with supersymmetric particle masses of order 1 TeV, the renormalization group evolution is modified above the supersymmetry-breaking scale. In this case, unification of gauge couplings can be (approximately) achieved as illustrated by the lower and upper solid lines in Fig. 6.1.

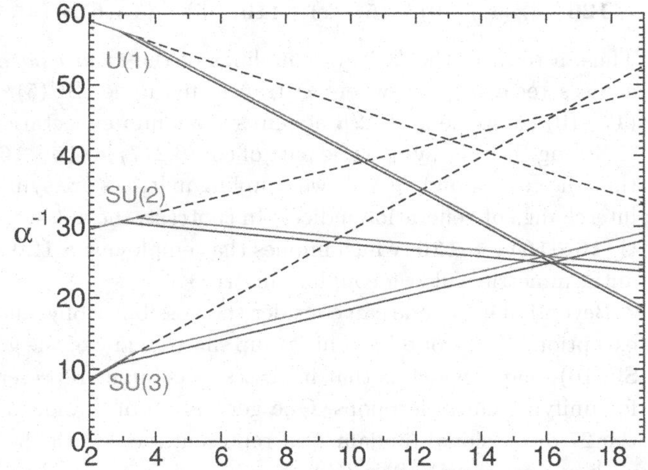

Fig. 6.1 Renormalization group evolution of the inverse gauge couplings $\alpha_a^{-1}(Q)$ in the Standard Model (dashed lines) and the MSSM (solid lines). In the MSSM case, $\alpha_3(m_Z)$ is varied between 0.120 and 0.117, and the supersymmetric particle mass thresholds are between 750 GeV and 2.5 TeV, for the lower and upper solid lines, respectively. Two-loop effects are included.

A quantitative assessment of the success of gauge coupling unification can be performed as follows. Since the electroweak gauge couplings g and g' are very well measured, first focus on these two couplings. For a given low-energy effective theory (below the GUT scale), we use the renormalization group equations to determine the couplings g and g' as a function of the energy scale. We then define $M_{\rm GUT}$ to be the scale at which these two couplings meet.

We now assume that the unification of the three gauge couplings, g_s, g, and g' occurs at $M_{\rm GUT}$. Using the renormalization group evolution of the gauge couplings [see eq. (6.288)], we can now run g_s down to the electroweak scale and compare with the experimentally measured value.

6.3.4 Normalization of the $U(1)_Y$ Coupling

In electroweak theory, the overall normalization of the $U(1)_Y$ coupling is a matter of convention. But, if the GUT gauge group is simple, then the relative normalization of the $U(1)_Y$ and $SU(2)_L$ gauge couplings is fixed. We denote the $SU(3)_C\times SU(2)_L\times U(1)_Y$ gauge couplings using the proper GUT normalization by g_3, g_2, and g_1 respectively. One can identify $g_3 \equiv g_s$ (the QCD coupling) and $g_2 = g$ (the weak isospin coupling). Our remaining task is to relate g_1 with g' (the weak hypercharge coupling). To do so, we begin by considering the covariant derivative,

$$\nabla_\mu = \partial_\mu + i\sum_a g_a \boldsymbol{T^a} A^a_\mu \,. \tag{6.281}$$

If the gauge group is a direct product group, then different sets of generators $\boldsymbol{T^a}$ are associated with the different group factors, and we must use the appropriate g_a depending on which generator it multiplies. In particular, for $SU(2)_L\times U(1)_Y$ (below the GUT scale),

$$g_a \boldsymbol{T^a} A^a_\mu \supset g T^3 W^3_\mu + g' \frac{Y}{2} B_\mu \,. \tag{6.282}$$

Above the GUT scale, the corresponding terms of the covariant derivative are

$$g_a \boldsymbol{T^a} A^a_\mu \supset g_U (T^3 W^3_\mu + T^0 B_\mu) \,, \tag{6.283}$$

where g_U is the gauge coupling of the unifying GUT group and T^0 is the properly normalized hypercharge generator. In particular, the generators $\boldsymbol{T^a}$ of the GUT group satisfy eq. (4.16), which we repeat here for the reader's convenience:

$$\mathrm{Tr}(\boldsymbol{T^a T^b}) = T(R)\delta^{ab} \,, \tag{6.284}$$

where $T(R)$ is the second-order Dynkin index of the irreducible representation R. We now set the two covariant derivatives above equal at the GUT scale:

$$g_U (T^3 W^3_\mu + T^0 B_\mu) = g T^3 W^3_\mu + g' Y B_\mu \,. \tag{6.285}$$

Noting that $g_U = g_3 = g_2 = g_1$ at the GUT scale, it follows that $g_2 = g$ and $g_1 T^0 = g' Y$. Since $T(R)$ only depends on the representation R, eq. (6.284) yields

Table 6.5 The T_3 and Y quantum numbers of the two-component fermion fields that make up one generation of SM fermions. The contributions to the traces are exhibited in the last two columns. Note that the color factors of 3 that arise when tracing over the (suppressed) color index have been included.

two-component fields	T_3	Y	$\mathrm{Tr}(T^3)^2$	$\mathrm{Tr}\,Y^2$
ψ_{Q1}	$\frac{1}{2}$	$\frac{1}{6}$	$3(\frac{1}{4})$	$3(\frac{1}{36})$
ψ_{Q2}	$-\frac{1}{2}$	$\frac{1}{6}$	$3(\frac{1}{4})$	$3(\frac{1}{36})$
ψ_U	0	$-\frac{2}{3}$	3(0)	$3(\frac{4}{9})$
ψ_D	0	$\frac{1}{3}$	3(0)	$3(\frac{1}{9})$
ψ_{L1}	$\frac{1}{2}$	$-\frac{1}{2}$	$\frac{1}{4}$	$\frac{1}{4}$
ψ_{L2}	$-\frac{1}{2}$	$-\frac{1}{2}$	$\frac{1}{4}$	$\frac{1}{4}$
ψ_E	0	1	0	1

$\mathrm{Tr}(T^3)^2 = \mathrm{Tr}(T^0)^2$. Thus,

$$g_1^2 = g'^2 \frac{\mathrm{Tr}\,Y^2}{\mathrm{Tr}(T^3)^2}\,. \tag{6.286}$$

The traces are evaluated by summing over one generation of fermions, under the assumption that it is made up of complete irreducible representations of the GUT group.[40] The relevant quantum numbers are provided in Table 6.5. Using these results, we simply add up the last two columns. These results include the appropriate color factor of 3 when tracing over the suppressed color index. Hence, we obtain $\mathrm{Tr}(T^3)^2 = 2$ and $\mathrm{Tr}\,Y^2 = \frac{10}{3}$, and eq. (6.286) yields

$$g_1^2 = \tfrac{5}{3}g'^2\,. \tag{6.287}$$

Note that in the fundamental representation of SU(5), one can identify $T^0 = \boldsymbol{t^{24}}$ and $Y = \sqrt{5/3}\,\boldsymbol{t^{24}}$ [see eq. (6.256)]. Using $g_1 T^0 = g'Y$, it follows that $g_1 = \sqrt{5/3}\,g'$ which is in agreement with eq. (6.287).

6.3.5 Gauge Coupling Running

We now examine the running of the gauge couplings in the one-loop approximation, where the gauge couplings g_i obey the differential equation

$$\frac{dg_i^2}{dt} = \frac{b_i g_i^4}{8\pi^2}\,, \qquad \text{for } i = 1, 2, 3, \tag{6.288}$$

[40] This assumption is valid in both the SU(5) and SO(10) grand unified models discussed above.

where $t = \ln Q$ (Q is the renormalization scale) and the b_i are given by

$$b_i = \tfrac{2}{3}\sum_j T(R_j^{(i)})m(R_j^{(i)}) + \tfrac{1}{6}\sum_J c_J T(R_J^{(i)})m(R_J^{(i)}) - \tfrac{11}{3}C_A(G^{(i)}), \tag{6.289}$$

where the indices j and J are employed for two-component fermions and scalars, respectively. We have assumed that the low-energy gauge group is a direct product group, $G \equiv \prod_i G^{(i)}$ and the jth two-component fermion multiplet and the Jth scalar multiplet transform irreducibly under G as $(R_j^{(1)}, R_j^{(2)}, \ldots)$ and $(R_J^{(1)}, R_J^{(2)}, \ldots)$. The multiplicity factors in eq. (6.289) are given by $m(R^{(i)}) = \prod_{k\neq i} d(R^{(k)})$, where $d(R^{(k)})$ is the dimension of the irreducible representation $R^{(k)}$, and $c_J = 1$ $[c_J = 2]$ for real [complex] scalars. Finally, $T(R^{(i)})$ and $C_A(G^{(i)})$ are group theory factors corresponding to the second-order Dynkin index of the irreducible representation $R^{(i)}$ of $G^{(i)}$ and the eigenvalue of the Casimir operator in the adjoint representation of $G^{(i)}$, respectively (see Appendix H.3).

For example, if $n \geq 2$ then $T(R = \mathbf{n}) = T(R = \overline{\mathbf{n}}) = \frac{1}{2}$, $T(R = \mathbf{n^2 - 1}) = N$ and $T(R = \mathbf{1}) = 0$ for the second-order Dynkin indices of the fundamental, anti-fundamental, adjoint, and singlet representations of $G^{(i)} = \mathrm{SU}(n)$, respectively, and $C_A(\mathrm{SU}(n)) = n$. For $\mathrm{U}(1)_Y$, we have $C_A = 0$ and $T(R) = \frac{3}{5}Y^2$, where we have employed the properly normalized U(1) hypercharge generator, $\sqrt{3/5}\,Y$, obtained in Section 6.3.4.

The solution to eq. (6.288) is

$$\frac{1}{g_i^2(m_Z)} = \frac{1}{g_U^2} - \frac{b_i}{8\pi^2}\ln\left(\frac{m_Z}{M_{\rm GUT}}\right), \tag{6.290}$$

where $M_{\rm GUT}$ is the GUT scale at which the three gauge couplings unify. Using eq. (6.290), the following two equations are obtained:

$$\sin^2\theta_W(m_Z) = \frac{g'^2(m_Z)}{g^2(m_Z) + g'^2(m_Z)} = \frac{\frac{3}{5}g_1^2(m_Z)}{g^2(m_Z) + \frac{3}{5}g_1^2(m_Z)}$$

$$= \frac{3}{8} - \frac{5}{16\pi}\alpha(m_Z)(b_1 - b_2)\ln\left(\frac{M_{\rm GUT}}{m_Z}\right), \tag{6.291}$$

$$\ln\left(\frac{M_{\rm GUT}}{m_Z}\right) = \frac{16\pi}{5b_1 + 3b_2 - 8b_3}\left(\frac{3}{8\alpha(m_Z)} - \frac{1}{\alpha_s(m_Z)}\right), \tag{6.292}$$

where $e = g\sin\theta_W$, $\alpha \equiv e^2/4\pi$, and $\alpha_s \equiv g_s^2/4\pi$.

It is convenient to introduce the parameter

$$x \equiv \frac{1}{5}\left(\frac{b_2 - b_3}{b_1 - b_2}\right). \tag{6.293}$$

Then, eqs. (6.291) and (6.292) yield

$$\sin^2\theta_W(m_Z) = \frac{1}{1+8x}\left[3x + \frac{\alpha(m_Z)}{\alpha_s(m_Z)}\right]. \tag{6.294}$$

Once we know the value of x, we can use eq. (6.294) to determine $\alpha_s(m_Z)$ given the values of $\sin^2\theta_W$ and α, evaluated at m_Z:

$$\alpha_s(m_Z) = \frac{\alpha(m_Z)}{(1+8x)\sin^2\theta_W(m_Z) - 3x}\,. \tag{6.295}$$

Inserting this last result back into eq. (6.292) yields

$$\ln\left(\frac{M_{\rm GUT}}{m_Z}\right) = \frac{2\pi(1+8x)\big[3 - 8\sin^2\theta_W(m_Z)\big]}{\alpha(m_Z)\big[5b_1 + 3b_2 - 8b_3\big]}\,. \tag{6.296}$$

Successful unification in the one-loop approximation is achieved if eq. (6.295) is satisfied by the experimental data [41]:

$$\alpha^{-1}(m_Z) = 127.951 \pm 0.009\,, \tag{6.297}$$

$$\sin^2\theta_W(m_Z) = 0.23121 \pm 0.00004\,, \tag{6.298}$$

$$\alpha_s(m_Z) = 0.1179 \pm 0.0009\,, \tag{6.299}$$

where the above quantities are defined in the five-flavor $\overline{\rm MS}$ renormalization scheme. If eq. (6.295) is approximately satisfied, then $M_{\rm GUT}$ can be evaluated by using either eq. (6.292) or eq. (6.296). We shall then require that $m_Z \ll M_{\rm GUT} < M_P$, where the Planck scale is defined by $M_P \equiv (\hbar c/G_N)^{1/2} \simeq 1.221 \times 10^{19}$ GeV and G_N is Newton's gravitational constant.

Defining the unified coupling strength by $\alpha_U \equiv g_U^2/(4\pi)$, one can use the above results to obtain

$$\alpha_U^{-1} = \frac{(5b_1+3b_2)\alpha_s^{-1}(m_Z) - 3b_3\alpha^{-1}(m_Z)}{5b_1 + 3b_2 - 8b_3}\,. \tag{6.300}$$

One can now check the reliability of the one-loop approximation employed above, which depends on the assumption that the running couplings are in the perturbative regime for all values of scales in the range $m_Z < Q < M_{\rm GUT}$.

For a more precise determination of the conditions for gauge coupling unification, one needs to include higher-order radiative corrections. The corresponding results for the running of the gauge couplings in the two-loop approximation are known but are much more complicated, since one must also take into account the effect of the Yukawa couplings (the top-quark Yukawa coupling provides the dominant effect, although in models where the bottom-quark Yukawa coupling is enhanced, its effects are also relevant). Numerically, the two-loop effects can result in $\mathcal{O}(10\%)$ shifts in the prediction of α_s given the known experimental values of $\alpha(m_Z)$ and $\sin^2\theta_W(m_Z)$.

Threshold effects can also modify the analysis presented above. At the GUT scale, the X and Y bosons and the color triplet scalar Higgs bosons all have masses that are roughly of $\mathcal{O}(M_{\rm GUT})$. But the mass splittings among these particles can be significant, and should be taken into account in the above analysis. Likewise, in models that introduce new particles near the electroweak scale, the mass splittings among these particles can also modify the analysis above. These are the high-energy and low-energy threshold effects, respectively, which can also result in $\mathcal{O}(10\%)$ shifts in the prediction of α_s. The high-energy threshold effects are especially troublesome in that they are highly model-dependent and thus introduce an irreducible systematic error to the analysis of the unification of couplings.

For our present purposes, we shall only consider the one-loop analysis with no threshold corrections. In this case, the success or failure of gauge coupling unification depends primarily on the value of the parameter x defined in eq. (6.293). Using eq. (6.295), it follows that

$$x = \frac{\sin^2\theta_W(m_Z) - \alpha(m_Z)\alpha_s^{-1}(m_Z)}{3 - 8\sin^2\theta_W(m_Z)} \simeq 0.1438\,, \tag{6.301}$$

after using the central values of the parameters given in eqs. (6.297) and (6.299).

The value of x that is predicted by the grand unification theory is determined from the values of the b_i, which are given by eq. (6.289). Assuming N_g generations of the quarks and leptons and N_h complex Higgs doublets (with hypercharge $\frac{1}{2}$), eq. (6.289) yields

$$b_3 = \tfrac{4}{3}N_g - 11\,, \tag{6.302}$$
$$b_2 = \tfrac{1}{6}N_h + \tfrac{4}{3}N_g - \tfrac{22}{3}\,, \tag{6.303}$$
$$b_1 = \tfrac{1}{10}N_h + \tfrac{4}{3}N_g\,. \tag{6.304}$$

Note that x is independent of N_g.[41] For the SM, we have $N_g = 3$ and $N_h = 1$. It follows that $b_3 = -7$, $b_2 = -\frac{19}{6}$, and $b_1 = \frac{41}{10}$. Consequently,

$$x = \frac{23}{218} = 0.1055\,. \tag{6.305}$$

The result is significantly lower than the experimentally observed value given in eq. (6.301). Thus, gauge coupling unification fails in SU(5) and SO(10) grand unification, under the assumption that the grand unification group is spontaneously broken down to $SU(3)_C \times SU(2)_L \times U(1)_Y$ at the GUT scale, with subsequent breaking to $SU(3) \times U(1)_{EM}$ at the electroweak scale.[42]

In SO(10) GUT models, the possible symmetry-breaking patterns are much richer than in SU(5) models. Thus, one can potentially achieve gauge coupling unification by employing one of the symmetry-breaking steps at an intermediate scale, M_I, between M_{GUT} and the electroweak scale. By introducing an extra parameter (namely M_I), it is possible to achieve gauge coupling unification, although in this framework one loses the prediction for $\alpha_s(m_Z)$ in terms of other low-energy parameters.

Perhaps a more convincing scenario for gauge coupling unification arises in the case of a supersymmetric extension of the Standard Model. In this scenario, the particle content of the theory at the electroweak scale is modified, which alters the running of the three gauge couplings. For example, in the MSSM, the particle content of the Standard Model (assuming N_h doublets of Higgs scalars) is extended to

41 This is an artifact of the one-loop approximation. Including higher-order corrections, one finds a weak dependence on N_g in the analysis of the unification of couplings.

42 One can also check that eqs. (6.292) and (6.296) imply that the SU(3) and U(1) couplings are equal at 7×10^{14} GeV, whereas the SU(2) and U(1) couplings are equal at 10^{13} GeV, which is yet another indication of the failure of gauge coupling unification in the Standard Model.

include spin-1/2 partners for all spin-1 gauge fields and spin-0 Higgs fields and spin-0 partners for all spin-1/2 fermion fields. Consequently, after employing eq. (6.289), the SM results quoted in eqs. (6.302)–(6.304) are modified in the MSSM as follows:

$$b_3 = 2N_g - 9\,, \tag{6.306}$$

$$b_2 = \tfrac{1}{2}N_h + 2N_g - 6\,, \tag{6.307}$$

$$b_1 = \tfrac{3}{10}N_h + 2N_g\,. \tag{6.308}$$

Once again, we note that x is independent of N_g. In the MSSM, we have $N_g = 3$ and $N_h = 2$. It follows that $b_3 = -3$, $b_2 = 1$, and $b_1 = \frac{33}{5}$, and consequently $x = \frac{1}{7}$. This result is remarkably close to the experimental value of x given in eq. (6.301). Moreover, either eq. (6.292) or eq. (6.296) yields $M_{\rm GUT} \simeq 2 \times 10^{16}$ GeV, a result that is comfortably below the Planck scale and yet large enough to explain the experimental absence of observed proton decay. Finally, by employing the MSSM values of b_1, b_2, and b_3, eq. (6.300) yields

$$\alpha_U^{-1} = \tfrac{3}{5}\left[\alpha_s^{-1}(m_Z) + \tfrac{1}{4}\alpha^{-1}(m_Z)\right] \simeq 24.25\,, \tag{6.309}$$

which is comfortably in the perturbative regime.[43] Indeed, gauge coupling unification in the MSSM provides one of the key motivations for the existence of supersymmetric partners of the SM particles at a mass scale that is close to the electroweak scale.

6.3.6 Proton Decay in Grand Unified Theories

In the Standard Model, B-violating and L-violating decays such as $p \to e^+\pi^0$ are forbidden since baryon number is an accidental symmetry of the renormalizable SM Lagrangian. Indeed, it is quite remarkable that, if one takes the fermion field content of the Standard Model and constructs the most general Lorentz invariant, $\mathrm{SU}(3)_C\times\mathrm{SU}(2)_L\times\mathrm{U}(1)_Y$-invariant Lagrangian, consisting of all possible couplings of nonnegative mass dimension (as required for renormalizability), the result will automatically be invariant under a global $\mathrm{U}(1)_{\rm B}\times\mathrm{U}(1)_{\rm L}$ symmetry (i.e., baryon and lepton number is conserved).[44] However, we have already seen in Section 6.1 that a gauge-invariant dimension-five operator involving SM fields can violate L. Likewise, a gauge-invariant dimension-six operator involving SM fields can violate both B and L (while conserving $B - L$). One such example is

$$\mathscr{L}_{\rm BLV} = \frac{g_U^2}{2M_{\rm GUT}^2}\left(\bar{d}^\dagger\bar{\sigma}^\mu e - \bar{e}^\dagger\bar{\sigma}^\mu d\right)\left(\bar{u}^\dagger\bar{\sigma}_\mu u\right) + \text{h.c.} \tag{6.310}$$

[43] Gauge coupling unification is less successful when two-loop corrections to the running of the gauge couplings in the MSSM are included. However, threshold corrections (which are highly model-dependent) can offset the effects of the two-loop corrections, thereby restoring the successful gauge coupling unification obtained in the one-loop approximation.

[44] Here, we are strictly making a perturbative statement. The results of Section 5.1.4 imply that anomalies in the baryon number and lepton number currents result in a nonperturbative violation of B and L (while maintaining $B - L$ conservation), although the practical effects of such violations are negligibly small and unobservable.

Such operators if present can mediate proton decay. In particular, the specific operator exhibited above can be responsible for the process $u + d \to \bar{u} + e^+$, which violates B and L by one unit while conserving $B - L$. If we add a spectator u quark, this process can be realized as $p \to \pi^0 e^+$.

The operator shown in eq. (6.310) arises both in SU(5) and in SO(10) grand unified theories. Focusing on the SU(5) GUT model presented earlier, it is straightforward to derive the interactions of the X and Y superheavy gauge bosons with the fermions:

$$\mathscr{L} = \frac{g_U}{\sqrt{2}}\Big\{\overline{X}^k_\mu\big[d^\dagger_k\bar{\sigma}^\mu\bar{e} - e^\dagger\bar{\sigma}^\mu\bar{d}_k + \epsilon_{kij}\bar{u}^{\dagger i}\bar{\sigma}^\mu u^j\big] + \overline{Y}^k_\mu\big[\nu^\dagger\bar{\sigma}^\mu\bar{d}_k - u^\dagger_k\bar{\sigma}^\mu\bar{e} + \epsilon_{kij}\bar{u}^{\dagger i}\bar{\sigma}^\mu d^j\big] + \text{h.c.}\Big\}. \tag{6.311}$$

Integrating out the superheavy X and Y bosons then generates the dimension-six operator shown in eq. (6.310).

One can obtain a crude estimate for the proton lifetime by employing the operator given in eq. (6.310) and using dimensional analysis. In particular, we can identify $M_{\rm GUT} \sim M_X \sim M_Y$. Dropping $\mathcal{O}(1)$ constants, one obtains

$$\tau_p \sim \frac{M^4_{\rm GUT}}{\alpha^2_U m^5_p}. \tag{6.312}$$

Taking $M_{\rm GUT} \sim 10^{16}$ GeV, $\alpha_U^{-1} \sim 25$, and a proton mass of $m_p \sim 1$ GeV, one obtains the following proton lifetime:

$$\tau_p \sim 6.25 \times 10^{66}\ \text{GeV}^{-1}(6.528 \times 10^{-25}\ \text{GeV s}) \lesssim 4 \times 10^{42}\ \text{s} \sim 10^{35}\ \text{years}, \tag{6.313}$$

which is about one order of magnitude longer than the most stringent current experimental bound. Indeed, the nonobservation of proton decay ruled out most non-supersymmetric SU(5) GUT models (and many non-supersymmetric SO(10) GUT models), which tended to favor somewhat lower values of $M_{\rm GUT} \lesssim 10^{15}$ GeV. Of course, a more detailed treatment of the proton lifetime would take into account the branching ratios into a variety of possible final states and would model the composite nature of the decaying proton and the final-state meson in terms of their quark constituents.

B-violating and L-violating dimension-six operators can also be generated by superheavy color Higgs triplet exchange. Thus, the nonobservation of proton decay also imposes constraints on the properties of the color Higgs triplets. Although these states often are somewhat lighter than the X and Y gauge bosons, their couplings to fermions are typically mass-suppressed due to Yukawa couplings proportional to light fermion masses. As a result, their corresponding contributions to proton decay amplitudes are often subdominant.

6.3.7 Fermion Mass Relations in Grand Unified Theories

The Higgs sector is the most model-dependent aspect of the GUT models. For example, there is a plethora of choices for the symmetry-breaking patterns in SO(10) models, depending on the choice of scalar multiplets. Moreover, some of the most minimal choices for Higgs fields that couple to fermions lead to fermion mass relations that are incompatible with the observed data, suggesting the need for a more complicated collection of Higgs scalar multiplets.

As an example, we return to the minimal SU(5) GUT, in which a multiplet of Higgs scalars that transform as a $\mathbf{5} \oplus \overline{\mathbf{5}}$ representation under SU(5) is introduced, denoted by the scalar fields ϕ_i and $\phi^i \equiv (\phi_i)^\dagger$, respectively. These Higgs scalars can couple to the fermions that reside in the $\overline{\mathbf{5}} \oplus \mathbf{10}$ representation of SU(5). The form of the SU(5) gauge invariant Yukawa couplings, which were given in eq. (6.266), are reproduced below, where the generation labels of the two-component spinor fields (denoted by p and q below) are made explicit:

$$-\mathscr{L}_{\text{Yuk},5} = (\boldsymbol{y_1})_{pq}(\psi^i)_p(\psi_{ij})_q\phi^{\dagger j} + \tfrac{1}{8}(\boldsymbol{y_2})_{pq}\epsilon^{ijk\ell m}(\psi_{ij})_p(\psi_{k\ell})_q\phi_m + \text{h.c.} \quad (6.314)$$

In general, $\boldsymbol{y_1}$ is a 3×3 complex matrix and $\boldsymbol{y_2}$ is a 3×3 complex symmetric matrix.[45] After spontaneous electroweak symmetry breaking, $\langle\phi_i\rangle = v_5\delta_{i5}/\sqrt{2}$ and fermions mass terms are generated:

$$-\mathscr{L}_{\text{mass},5} = \frac{v_5}{\sqrt{2}}\left\{(\boldsymbol{y_{d,5}})_{pq}(\bar{d}^i)_p(d_i)_q + (\boldsymbol{y_{\ell,5}})_{pq}\bar{\ell}_p\ell_q + (\boldsymbol{y_{u,5}})_{pq}(\bar{u}^i)_p(u_i)_q\right\} + \text{h.c.}, \quad (6.315)$$

where

$$\boldsymbol{y}_{d,5} = \boldsymbol{y}^{\mathsf{T}}_{\boldsymbol{\ell},5} = \boldsymbol{y_1}\,, \qquad \boldsymbol{y}_{u,5} = \boldsymbol{y}^{\mathsf{T}}_{\boldsymbol{u},5} = \boldsymbol{y_2}\,. \quad (6.316)$$

These SU(5)-invariant Yukawa coupling relations are valid at the GUT scale and thus serve as initial conditions for renormalization group evolution. The corresponding fermion mass matrix relations at the electroweak scale are

$$\boldsymbol{M_d} = \frac{v_5}{\sqrt{2}}\boldsymbol{y}_d(m_Z)\,, \qquad \boldsymbol{M_e} = \frac{v_5}{\sqrt{2}}\boldsymbol{y}^{\mathsf{T}}_{\boldsymbol{\ell}}(m_Z)\,, \qquad \boldsymbol{M_u} = \boldsymbol{M}^{\mathsf{T}}_{\boldsymbol{u}} = \frac{v_5}{\sqrt{2}}\boldsymbol{y}_u(m_Z). \quad (6.317)$$

One can now extend the running below m_Z down to the corresponding fermion mass scale. After diagonalizing the corresponding fermion mass matrices, one finds that the would-be mass relation $m_b = m_\tau$ at the GUT scale is modified by the running of the Yukawa couplings. The end result, when expressed in terms of the physical (pole) masses, yields roughly $m_b \sim 3m_\tau$, which is close to the experimental observations. Unfortunately, the same analysis applied to the first two generation would yield $m_s \sim 3m_\mu$ and $m_d \sim 3m_e$ which are far from the experimentally observed numbers. Indeed, by taking ratios, the dependence of the renormalization scale mostly cancels, and one would conclude, e.g., that $m_b/m_s \simeq m_\tau/m_\mu$, which again is in conflict with experimental observation.

[45] Noting that the symmetric combination $(\mathbf{10} \otimes \mathbf{10})_S = \overline{\mathbf{5}} \oplus \overline{\mathbf{50}}$, it follows that $\boldsymbol{y}_{\boldsymbol{2}}^{\mathsf{T}} = \boldsymbol{y_2}$.

Hence, the assumption that $\mathbf{5}\oplus\overline{\mathbf{5}}$ is the only Higgs scalar multiplet that couples to the fermions must be rejected. In light of the products of SU(5) representations previously noted, $\overline{\mathbf{5}}\otimes\mathbf{10}=\mathbf{5}\oplus\mathbf{45}$ and $\mathbf{10}\otimes\mathbf{10}=\overline{\mathbf{5}}\oplus\overline{\mathbf{45}}\oplus\overline{\mathbf{50}}$, one can also consider employing, e.g., the Higgs scalar multiplet, $\mathbf{45}\oplus\overline{\mathbf{45}}$, represented by the scalar fields ϕ^i_{jk} and $\phi^{\dagger jk}_i\equiv(\phi^i_{jk})^\dagger$, respectively, where

$$\phi^i_{jk}=-\phi^i_{kj}\,,\qquad \delta^j_i\,\phi^i_{jk}=0\,. \tag{6.318}$$

One can then write down two additional SU(5) gauge invariant Yukawa couplings:

$$-\mathscr{L}_{\rm Yuk,45}=\tfrac{1}{2}(\boldsymbol{y_3})_{pq}(\psi^i)_p(\psi_{jk})_q\phi^{\dagger jk}_i+\tfrac{1}{8}(\boldsymbol{y_4})_{pq}\epsilon^{ijk\ell m}(\psi_{ij})_p(\psi_{kn})_q\phi^n_{\ell m}+\text{h.c.} \tag{6.319}$$

In general, $\boldsymbol{y_3}$ is a 3×3 complex matrix and $\boldsymbol{y_4}$ is a 3×3 complex antisymmetric matrix.[46] The most general SU(3)×U(1)$_{\rm EM}$-preserving VEV of ϕ^i_{jk} is given by

$$\langle\phi^i_{jk}\rangle=v_{45}\left[\delta^i_j\delta_{k5}-\delta^i_k\delta_{j5}-4\delta^{i4}(\delta_{j4}\delta_{k5}-\delta_{k4}\delta_{j5})\right]. \tag{6.320}$$

Note that $\langle\phi^i_{jk}\rangle$ satisfies the two conditions specified in eq. (6.318) as required. That is, the only nonvanishing vacuum expectation values are $\langle\phi^4_{45}\rangle=-\langle\phi^4_{54}\rangle=-3v_{45}$ and the diagonal elements of $\langle\phi^i_{j5}\rangle=-\langle\phi^i_{5j}\rangle=v_{45}\delta^i_j$ for $i,j\in\{1,2,3\}$. After spontaneous electroweak symmetry breaking, the following fermions mass terms are generated:

$$-\mathscr{L}_{\rm mass,45}=v_{45}\left\{(\boldsymbol{y_{d,45}})_{pq}(\bar{d}^i)_p(d_i)_q+(\boldsymbol{y_{\ell,45}})_{pq}\bar{\ell}_p\ell_q+(\boldsymbol{y_{u,45}})_{pq}(\bar{u}^i)_p(u_i)_q\right\}+\text{h.c.}, \tag{6.321}$$

where

$$\boldsymbol{y_{d,45}}=-\tfrac{1}{3}\boldsymbol{y}^{\mathsf{T}}_{\boldsymbol{\ell},\mathbf{45}}=\boldsymbol{y_3}\,,\qquad \boldsymbol{y_{u,45}}=-\boldsymbol{y}^{\mathsf{T}}_{\boldsymbol{u},\mathbf{45}}=-3\boldsymbol{y_4}\,. \tag{6.322}$$

As before, these SU(5)-invariant Yukawa coupling relations are valid at the GUT scale and thus serve as initial conditions for renormalization group evolution. After diagonalizing the corresponding fermion mass matrices (via singular value decomposition, in which the physical fermion masses are nonnegative), one finds that the would-be mass relation $m_\mu=3m_s$ at the GUT scale is modified by the running of the Yukawa couplings. The end result, when expressed in terms of the physical (pole) masses, yields roughly $m_\mu\sim m_s$, which is approximately satisfied by the experimentally observed values. However, the corresponding first-generation and third-generation mass relations deduced from eq. (6.322) are in conflict with the observed down-type quark and charged lepton masses.

In summary, employing a $\mathbf{5}\oplus\overline{\mathbf{5}}$ multiplet of scalars in the Yukawa couplings yields mass relations between down-type quarks and charged leptons that reproduce the observed third-generation masses but is in conflict with the first two generation masses. In contrast, employing a $\mathbf{45}\oplus\overline{\mathbf{45}}$ multiplet of scalars alone in the Yukawa couplings yields mass relations between down-type quarks and charged leptons that reproduce the observed second-generation masses but are in conflict with the first- and third-generation masses. If one includes both the $\mathbf{5}\oplus\overline{\mathbf{5}}$ and $\mathbf{45}\oplus\overline{\mathbf{45}}$ of scalars

[46] Noting that the antisymmetric combination, $(\mathbf{10}\otimes\mathbf{10})_A=\overline{\mathbf{45}}$, it followings that $\boldsymbol{y}^{\mathsf{T}}_{\boldsymbol{4}}=-\boldsymbol{y_4}$.

with no additional conditions, then no relation exists between the Yukawa coupling matrices, $\boldsymbol{y_\ell} \equiv \boldsymbol{y_{\ell,5}} + \boldsymbol{y_{\ell,45}}$ and $\boldsymbol{y_d} = \boldsymbol{y_{d,5}} + \boldsymbol{y_{d,45}}$. However, Georgi and Jarlskog have shown [73] that one can impose discrete symmetries on the Yukawa coupling Lagrangian in such a way that after diagonalization of $\boldsymbol{y_\ell}$ and $\boldsymbol{y_d}$, one obtains mass relations $m_\tau = m_b$, $m_\mu = 3m_s$, and $m_e = \frac{1}{3}m_d$ at the GUT scale (see Exercise 6.19). Renormalization group evolution reduces the corresponding charged lepton mass in each relation by a factor of approximately 3 in each case, resulting in a mass spectrum of down-type fermions that is approximately consistent with the observed experimental mass values.

Note that in models with a $\mathbf{45}\oplus\overline{\mathbf{45}}$ multiplet of scalars, new colored scalar states exist, as is evident from the branching rule with respect to SU(3)×SU(2)×U(1):

$$\mathbf{45} \longrightarrow (\mathbf{1},\mathbf{2},\tfrac{1}{2}) + (\mathbf{3},\mathbf{1},-\tfrac{1}{3}) + (\mathbf{3},\mathbf{3},-\tfrac{1}{3}) + (\overline{\mathbf{3}},\mathbf{1},\tfrac{4}{3}) + (\overline{\mathbf{3}},\mathbf{2},-\tfrac{7}{6}) \\ + (\overline{\mathbf{6}},\mathbf{1},-\tfrac{1}{3}) + (\mathbf{8},\mathbf{2},\tfrac{1}{2})\,. \tag{6.323}$$

For a phenomenologically realistic model, the colored scalars should be superheavy with masses on the order of the GUT scale. In practice, this can be realized by an appropriate choice of the scalar potential.

In SU(5) GUT models, there is no relation between up- and down-type fermion masses since a *reducible* $\overline{\mathbf{5}}\oplus\mathbf{10}$ representation of SU(5) comprises one generation of fermions. In contrast, in SO(10) GUT models, an *irreducible* **16** representation of SO(10) comprises one generation of fermions. Indeed, unified SO(10) models with a low-energy scalar sector corresponding to the 2HDM have been proposed where $m_t = m_b = m_\tau$ at the GUT scale. In such models, one finds that $m_t/m_b \sim \tan\beta$ at the GUT scale, so that the mass hierarchy between the top and bottom quark is correlated with a large value of $\tan\beta$. The reader is referred to the literature for further details (e.g., see [B36, B93, B95]).

Exercises

6.1 Consider the mass terms of the seesaw Lagrangian given by eq. (6.10) under the assumption that $\|\widehat{\boldsymbol{M}}\| \gg v$. Note that $\widehat{\boldsymbol{M}} = \widehat{\boldsymbol{M}}^{\mathsf{T}}$.

(a) Explain why integrating out the heavy neutrino field $\widehat{\overline{\nu}}^K$ can be carried out (at leading order in $v/\|\widehat{\boldsymbol{M}}\|$) by neglecting its kinetic energy term and then solving for $\widehat{\overline{\nu}}^K$ in terms of the light neutrino field $\widehat{\nu}_j$ using the Lagrange field equations,

$$\frac{\delta\mathscr{L}_{\text{seesaw}}}{\delta\widehat{\overline{\nu}}^K} = 0\,. \tag{6.324}$$

(b) Show that by carrying out the procedure outlined in part (a),

$$\widehat{\overline{\nu}}^K = \epsilon^{cd}(\widehat{\boldsymbol{y}}_\nu)^j{}_L\boldsymbol{\Phi}_c\widehat{\boldsymbol{L}}_{dj}(\widehat{\boldsymbol{M}}^{-1})^{KL}\,. \tag{6.325}$$

(c) Plugging this result back into eq. (6.10), verify that you recover eq. (6.1), where $\widehat{\boldsymbol{F}}/\Lambda$ is given by eq. (6.12).

HINT: In obtaining the result of part (c), you should find that the second term on the right-hand side of eq. (6.10) is half the size of the first term and contributes with the opposite sign.

6.2 (a) The PMNS matrix is a 3×3 unitary matrix that is parameterized by three angles and six phases. How many of these phases are physical?

(b) If the Standard Model is extended to include the dimension-five operator given in eq. (6.4), determine the total number of free parameters that govern this extended Standard Model. How many additional parameters are there beyond those identified in Section 4.8?

6.3 Suppose all the properties of the three generations of light neutrinos are known (these include the three neutrino masses and the mixing angles and physical phases associated with the PMNS mixing matrix that you identified in Exercise 6.2). In addition, assume that the ultraviolet completion of eq. (6.4) is the seesaw-extended Standard Model with three generations of superheavy neutrinos (i.e., with $\|\boldsymbol{M}\| \gg v$), and that some future civilization can also determine the masses of the heavy neutrinos, $m_{\nu_h J}$ (for $J = 1, 2, 3$).

(a) Without further measurements of the couplings of the heavy neutrino Yukawa couplings, show that $\boldsymbol{y_\nu}$ is determined only up to an arbitrary complex orthogonal matrix R,

$$\boldsymbol{y_\nu} = \frac{i\sqrt{2}}{v} U^* \sqrt{\boldsymbol{D}}\, R \sqrt{\boldsymbol{M}}\,, \tag{6.326}$$

where the elements of the diagonal matrix $\boldsymbol{D}$, defined in eq. (6.7), are the masses of the light neutrinos, and the elements of the diagonal matrix $\boldsymbol{M}$, defined in eq. (6.15), are the approximate masses of the heavy neutrinos.

HINT: Using eqs. (6.21) and (6.26), show that eq. (6.7) can be recast into the form $RR^{\mathsf{T}} = \mathbb{1}_{3\times 3}$, and identify the complex orthogonal matrix R. Equation (6.326) is the Casas–Ibarra parameterization, first established in Ref. [74].

(b) Determine the total number of parameters that govern the seesaw-extended Standard Model discussed in Section 6.1 with three generations of heavy neutrinos, thereby extending the results of Section 4.8. Compare your answer to Refs. [75, 76].

6.4 Construct the neutrino mass matrix in the seesaw-extended 2HDM. Determine the unsuppressed interactions of the heavy neutrinos to the Higgs boson and the Goldstone bosons [the generalization of eq. (6.32)].

6.5 The 2HDM scalar potential in the Φ-basis is given by eq. (6.33). Assume that the scalar potential minimum is $\mathrm{U(1)_{EM}}$-preserving [i.e., $u = 0$ in eq. (6.34)] and choose a convention in which $v_1 = v\cos\beta$ and $v_2 = e^{i\xi}v\sin\beta$.

(a) Specifying the unit vectors $\widehat{v}$ and $\widehat{w}$ given in eqs. (6.43) and (6.44) is equivalent to fixing the scalar field basis. Show that if $\widehat{v} = (1 \quad 0)$ and $\widehat{w} = (0 \quad 1)$, then the corresponding Φ-basis scalar fields are given by $\Phi_1 = \mathcal{H}_1$ and $\Phi_2 = e^{-i\eta}\mathcal{H}_2$. Plugging these results into eq. (6.33), conclude that the scalar potential in the Higgs basis is given by eq. (6.48).

(b) Show that the coefficients of the quadratic terms of the scalar potential in the Higgs basis are given by

$$Y_1 = m_{11}^2 c_\beta^2 + m_{22}^2 s_\beta^2 - \mathrm{Re}(m_{12}^2 e^{i\xi})s_{2\beta}\,, \tag{6.327}$$

$$Y_2 = m_{11}^2 s_\beta^2 + m_{22}^2 c_\beta^2 + \mathrm{Re}(m_{12}^2 e^{i\xi})s_{2\beta}\,, \tag{6.328}$$

$$Y_3 e^{i\xi} = \tfrac{1}{2}(m_{22}^2 - m_{11}^2)s_{2\beta} - \mathrm{Re}(m_{12}^2 e^{i\xi})c_{2\beta} - i\,\mathrm{Im}(m_{12}^2 e^{i\xi})\,, \tag{6.329}$$

where $c_\beta \equiv \cos\beta$, $s_\beta \equiv \sin\beta$, $c_{2\beta} \equiv \cos 2\beta$, and $s_{2\beta} \equiv \sin 2\beta$. Note that in eq. (6.329), factors involving the phase η have canceled in light of part (a).

(c) Invert the result of parts (b) above and obtain expressions for m_{11}^2, m_{22}^2, and m_{12}^2 in terms of Y_1, Y_2, and Y_3.

(d) Show that the coefficients of the quartic terms of the scalar potential in the Higgs basis are given by

$$Z_1 = \lambda_1 c_\beta^4 + \lambda_2 s_\beta^4 + \tfrac{1}{2}\lambda_{345} s_{2\beta}^2 + 2s_{2\beta}\left[c_\beta^2\,\mathrm{Re}(\lambda_6 e^{i\xi}) + s_\beta^2\,\mathrm{Re}(\lambda_7 e^{i\xi})\right], \tag{6.330}$$

$$Z_2 = \lambda_1 s_\beta^4 + \lambda_2 c_\beta^4 + \tfrac{1}{2}\lambda_{345} s_{2\beta}^2 - 2s_{2\beta}\left[s_\beta^2\,\mathrm{Re}(\lambda_6 e^{i\xi}) + c_\beta^2\,\mathrm{Re}(\lambda_7 e^{i\xi})\right], \tag{6.331}$$

$$Z_3 = \tfrac{1}{4}s_{2\beta}^2\left[\lambda_1 + \lambda_2 - 2\lambda_{345}\right] + \lambda_3 - s_{2\beta}c_{2\beta}\,\mathrm{Re}[(\lambda_6 - \lambda_7)e^{i\xi}], \tag{6.332}$$

$$Z_4 = \tfrac{1}{4}s_{2\beta}^2\left[\lambda_1 + \lambda_2 - 2\lambda_{345}\right] + \lambda_4 - s_{2\beta}c_{2\beta}\,\mathrm{Re}[(\lambda_6 - \lambda_7)e^{i\xi}], \tag{6.333}$$

$$\begin{aligned} Z_5 e^{2i\xi} &= \tfrac{1}{4}s_{2\beta}^2\left[\lambda_1 + \lambda_2 - 2\lambda_{345}\right] + \mathrm{Re}(\lambda_5 e^{2i\xi}) + ic_{2\beta}\,\mathrm{Im}(\lambda_5 e^{2i\xi}) \\ &\quad - s_{2\beta}c_{2\beta}\,\mathrm{Re}[(\lambda_6 - \lambda_7)e^{i\xi}] - is_{2\beta}\,\mathrm{Im}[(\lambda_6 - \lambda_7)e^{i\xi}], \end{aligned} \tag{6.334}$$

$$\begin{aligned} Z_6 e^{i\xi} &= -\tfrac{1}{2}s_{2\beta}\left[\lambda_1 c_\beta^2 - \lambda_2 s_\beta^2 - \lambda_{345}c_{2\beta} - i\,\mathrm{Im}(\lambda_5 e^{2i\xi})\right] + c_\beta c_{3\beta}\,\mathrm{Re}(\lambda_6 e^{i\xi}) \\ &\quad + s_\beta s_{3\beta}\,\mathrm{Re}(\lambda_7 e^{i\xi}) + ic_\beta^2\,\mathrm{Im}(\lambda_6 e^{i\xi}) + is_\beta^2\,\mathrm{Im}(\lambda_7 e^{i\xi}), \end{aligned} \tag{6.335}$$

$$\begin{aligned} Z_7 e^{i\xi} &= -\tfrac{1}{2}s_{2\beta}\left[\lambda_1 s_\beta^2 - \lambda_2 c_\beta^2 + \lambda_{345}c_{2\beta} + i\,\mathrm{Im}(\lambda_5 e^{2i\xi})\right] + s_\beta s_{3\beta}\,\mathrm{Re}(\lambda_6 e^{i\xi}) \\ &\quad + c_\beta c_{3\beta}\,\mathrm{Re}(\lambda_7 e^{i\xi}) + is_\beta^2\,\mathrm{Im}(\lambda_6 e^{i\xi}) + ic_\beta^2\,\mathrm{Im}(\lambda_7 e^{i\xi}), \end{aligned} \tag{6.336}$$

where $s_{3\beta} \equiv \sin 3\beta$ and $c_{3\beta} \equiv \cos 3\beta$, and

$$\lambda_{345} \equiv \lambda_3 + \lambda_4 + \mathrm{Re}(\lambda_5 e^{2i\xi})\,. \tag{6.337}$$

(e) Invert the results of part (d) above and obtain expressions for the λ_i in terms of the Z_i.

6.6 In eq. (6.98), terms involving the Goldstone bosons were omitted. Using the notation of eq. (6.96), derive the following quartic self-interaction terms:

$$\begin{aligned}
\mathscr{L}_{4G} = &-\tfrac{1}{8}Z_1G^4 - \tfrac{1}{2}G^3h_j\,\mathrm{Im}(q_{j2}\overline{Z}_6) - \tfrac{1}{2}G^2(Z_1G^+G^- + Z_3H^+H^-) \\
&- \tfrac{1}{2}G^+G^-\big[Z_1G^+G^- + 2(Z_3+Z_4)H^+H^-\big] - G^+G^-(\overline{Z}_6G^-H^+ + \mathrm{h.c.}) \\
&- \tfrac{1}{2}(\overline{Z}_5G^-G^-H^+H^+ + \mathrm{h.c.}) - H^+H^-(\overline{Z}_7G^-H^+ + \mathrm{h.c.}) \\
&+ h_jG\big[\mathrm{Im}(q_{j2}\overline{Z}_6)G^+G^- + \mathrm{Im}(q_{j2}\overline{Z}_7)H^+H^-\big] \\
&- \tfrac{1}{2}G^2(\overline{Z}_6G^-H^+ + \mathrm{h.c.}) - \tfrac{1}{2}Gh_j\big[i(q^*_{j2}Z_4 - q_{j2}\overline{Z}_5)G^-H^+ + \mathrm{h.c.}\big] \\
&- \tfrac{1}{6}Gh_jh_kh_\ell\big[q_{j1}\,\mathrm{Re}(q_{k2}q_{\ell 2}\overline{Z}_5) + q_{k1}\,\mathrm{Re}(q_{j2}q_{\ell 2}\overline{Z}_5) + q_{\ell 1}\,\mathrm{Re}(q_{j2}q_{k2}\overline{Z}_5) \\
&\qquad - q_{j1}q_{k1}\,\mathrm{Re}(q_{\ell 2}\overline{Z}_6) + q_{j1}q_{\ell 1}\,\mathrm{Re}(q_{k2}\overline{Z}_6) + q_{k1}q_{\ell 1}\,\mathrm{Re}(q_{j2}\overline{Z}_6) \\
&\qquad + \mathrm{Re}(q^*_{j2}q_{k2}q_{\ell 2}\overline{Z}_7) + \mathrm{Re}(q_{j2}q^*_{k2}q_{\ell 2}\overline{Z}_7) + \mathrm{Re}(q_{j2}q_{k2}q^*_{\ell 2}\overline{Z}_7)\big] \\
&- \tfrac{1}{4}G^2h_jh_k\big[q_{j1}q_{k1}Z_1 + \mathrm{Re}(q_{j2}q^*_{k2})(Z_3+Z_4) - \mathrm{Re}(q_{j2}q_{k2}\overline{Z}_5) \\
&\qquad + q_{j1}\,\mathrm{Re}(q_{k2}\overline{Z}_6) + q_{k1}\,\mathrm{Re}(q_{j2}\overline{Z}_6)\big] \\
&- \tfrac{1}{2}h_jh_kG^+G^-\big[q_{j1}q_{k1}Z_1 + \mathrm{Re}(q_{j2}q^*_{k2})Z_3 + q_{j1}\,\mathrm{Re}(q_{k2}\overline{Z}_6) + q_{k1}\,\mathrm{Re}(q_{j2}\overline{Z}_6)\big] \\
&- \tfrac{1}{4}h_jh_k\big\{\big[(q_{j1}q^*_{k2} + q_{k1}q^*_{j2})Z_4 + (q_{j1}q_{k2} + q_{k1}q_{j2})\overline{Z}_5 + 2q_{j1}q_{k1}\overline{Z}_6 \\
&\qquad + 2\,\mathrm{Re}(q_{j2}q^*_{k2})\overline{Z}_7\big]G^-H^+ + \mathrm{h.c.}\big\},
\end{aligned} \tag{6.338}$$

where there is an implicit sum over repeated indices j, k, $\ell \in \{1,2,3\}$.

6.7 Consider the general 2HDM with neutral Higgs masses $m_j \equiv m_{h_j}$ and invariant mixing angles θ_{12} and θ_{13}. The invariant quantities q_{k1} and q_{k2} are defined in Table 6.1. Derive the following relations:

$$\mathrm{Im}(q^2_{k2}\overline{Z}_5) + 2q_{k1}\,\mathrm{Im}\,(q_{k2}\overline{Z}_6) = 0\,, \tag{6.339}$$

$$v^2\big[q_{k1}Z_1 + \mathrm{Re}(q_{k2}\overline{Z}_6)\big] = m^2_k q_{k1}\,, \tag{6.340}$$

$$v^2(q^*_{k2}Z_4 + q_{k2}\overline{Z}_5 + 2q_{k1}\overline{Z}_6) = 2q^*_{k2}(m^2_k - m^2_{H^\pm})\,, \tag{6.341}$$

where $\overline{Z}_5 \equiv Z_5e^{-2i\eta}$ and $\overline{Z}_6 \equiv Z_6e^{-i\eta}$. Likewise, show that

$$v^2\big[\mathrm{Im}(q_{j2}q_{k2}\overline{Z}_5) + q_{j1}\,\mathrm{Im}\,(q_{k2}\overline{Z}_6) + q_{k1}\,\mathrm{Im}\,(q_{j2}\overline{Z}_6)\big] = (m^2_j - m^2_k)\epsilon_{jk\ell}q_{\ell 1}\,, \tag{6.342}$$

where there is an implicit sum over the repeated index $\ell \in \{1,2,3\}$. These relations have been used in the derivation of the Goldstone boson interactions exhibited in eq. (6.97).

6.8 Consider a general 2HDM where eq. (6.123) is satisfied, so that the tree-level neutral scalar mass-eigenstates, h, H, and A are states of definite CP.

(a) Rewrite the Yukawa couplings given by eq. (6.122) using four-component spinor notation. Show that the tree-level couplings of h, H, and A to fermions are CP-violating if and only if $\boldsymbol{\rho}^F_I \neq 0$ (for at least one choice of F=U, D, E).

(b) Show that if eq. (6.123) holds true and $\rho_I^U = \rho_I^D = \rho_I^E = 0$, then the only source of CP violation in eq. (6.122) is the nonzero phase of the CKM matrix.

6.9 In the CP-conserving 2HDM, there exists a real Higgs basis in which all the dimensionless coefficients of the scalar potential, Z_i, are real.

(a) Derive the following results for the Higgs basis parameters, Z_1, Z_4, Z_5, and Z_6, in terms of the Higgs masses and $\beta - \alpha$:

$$Z_1 v^2 = m_h^2 s_{\beta-\alpha}^2 + m_H^2 c_{\beta-\alpha}^2 , \qquad (6.343)$$

$$Z_4 v^2 = m_h^2 c_{\beta-\alpha}^2 + m_H^2 s_{\beta-\alpha}^2 + m_A^2 - 2m_{H^\pm}^2 , \qquad (6.344)$$

$$Z_5 v^2 = m_h^2 c_{\beta-\alpha}^2 + m_H^2 s_{\beta-\alpha}^2 - m_A^2 , \qquad (6.345)$$

$$Z_6 v^2 = (m_h^2 - m_H^2) s_{\beta-\alpha} c_{\beta-\alpha} . \qquad (6.346)$$

(b) Using the results of part (a), derive the approximate Higgs alignment limit expressions for the two cases where either h or H is SM-like, obtained in eqs. (6.180)–(6.183) or eqs. (6.189)–(6.192), respectively.

6.10 Consider a 2HDM with a softly broken $\mathbb{Z}_2$ symmetry that imposes the condition $\lambda_6 = \lambda_7 = 0$ in the Φ-basis.

(a) In the Φ-basis, we denote $\langle \Phi_1^0 \rangle = v_1/\sqrt{2}$ and $\langle \Phi_2^0 \rangle = v_2 e^{i\xi}/\sqrt{2}$, where v_1 and v_2 are nonnegative and $\tan\beta \equiv v_2/v_1$. Show that if $\lambda_6 = \lambda_7 = 0$ in the Φ-basis, then [59]

$$\tfrac{1}{2}(Z_1 - Z_2)\sin 2\beta + \mathrm{Re}(Z_{67} e^{i\xi})\cos 2\beta + i\,\mathrm{Im}(Z_{67} e^{i\xi}) = 0 . \qquad (6.347)$$

(b) Show that if $\lambda_6 = \lambda_7 = 0$ in the Φ-basis, then the parameters of the scalar potential in the Higgs basis satisfy

$$(Z_1 - Z_2)(Z_{34} Z_{67} - Z_1 Z_7 - Z_2 Z_6 + Z_5 Z_{67}^*) - 2 Z_{67}\big(|Z_6|^2 - |Z_7|^2\big) = 0 , \qquad (6.348)$$

where $Z_{ij} \equiv Z_i + Z_j$. In the case of $Z_1 = Z_2$, $Z_5 \neq 0$, and $Z_{67} \neq 0$, eq. (6.348) must be supplemented with a second condition, $\mathrm{Im}(Z_5^* Z_{67}^2) = 0$.

HINT: you should make use of the results of part (d) of Exercise 6.5.

(c) If $\lambda_6 = \lambda_7 = 0$ in the Φ-basis and if $Z_{67} \neq 0$, show that

$$\tan 2\beta = (-1)^n \frac{2|Z_{67}|}{Z_2 - Z_1} , \qquad \xi = n\pi - \arg Z_{67} , \qquad (6.349)$$

where n is any integer. Explain the significance of the two-fold ambiguity in the value of $\tan 2\beta$.

(d) Show that the formula for $\tan 2\beta$ obtained in part (c) still holds if $Z_{67} = 0$ and $Z_1 \neq Z_2$. How are the results of part (c) modified in the special case $Z_1 = Z_2$ and $Z_{67} = 0$?

6.11 Consider the most general CP-conserving 2HDM in a real Φ-basis. If λ_6 and λ_7 are nonzero, how should the results of eqs. (6.206)–(6.210) be modified?

6.12 In the CP-conserving 2HDM with a softly broken $\mathbb{Z}_2$ symmetry, the squared masses of the CP-even neutral scalars are determined from eq. (6.211).

(a) Show that in the decoupling limit (where $m_A \gg m_h$),

$$m_h^2 = \mathcal{M}_{11}^2 \cos^2\beta + \mathcal{M}_{22}^2 \sin^2\beta + \mathcal{M}_{12}^2 \sin 2\beta + \mathcal{O}\left(\frac{v^4}{m_A^2}\right). \tag{6.350}$$

(b) Using the tree-level MSSM Higgs sector expression for $\mathcal{M}^2$, show that

$$m_h^2 = m_Z^2 \cos^2 2\beta + \mathcal{O}\left(\frac{m_Z^4}{m_A^2}\right), \tag{6.351}$$

in agreement with the result of eq. (6.242).

(c) Show that in the decoupling limit of the tree-level MSSM Higgs sector,

$$1 + \cot\alpha \cot\beta = -\frac{2m_Z^2}{m_A^2}\cos 2\beta + \mathcal{O}\left(\frac{m_Z^4}{m_A^4}\right), \tag{6.352}$$

as a consequence of eq. (6.241).

6.13 Consider the embedding of SU(3)×SU(2)×U(1) in SU(5) that is represented by the following mapping of $(h_3, h_2, e^{i\theta}) \in$ SU(3)×SU(2)×U(1) into SU(5):

$$(h_3, h_2, e^{i\theta}) \longmapsto \begin{pmatrix} e^{-2i\theta}h_3 & 0 \\ 0 & e^{3i\theta}h_2 \end{pmatrix}, \tag{6.353}$$

where h_3 is a 3×3 matrix in the fundamental representation of SU(3) and h_2 is a 2×2 representation in the fundamental representation of SU(2). Note that the determinant of the 5×5 matrix on the right-hand side of eq. (6.353) is unity as required for an element of the fundamental representation of SU(5).

(a) In light of eq. (6.353), show that elements of SU(3)×SU(2)×U(1) of the form $(e^{2in\pi/3}\mathbb{1}_{3\times 3}, e^{-in\pi}\mathbb{1}_{2\times 2}, e^{in\pi/3})$, where n is an integer, are mapped to the identity of SU(5). Explain why this means that the mapping displayed in eq. (6.353) provides an embedding of SU(3)×SU(2)×U(1)/$\mathbb{Z}_6$ in SU(5).

NOTE: The result obtained in part (a) above is equivalent to the statement that S(U(3) × U(2)) = SU(3)×SU(2)×U(1)/$\mathbb{Z}_6$ is a subgroup of SU(5), where

$$\mathrm{S}(\mathrm{U}(m) \times \mathrm{U}(n)) = \{(A,B) \in U(m) \times U(n) \,|\, \det A \cdot \det B = 1\}\,. \tag{6.354}$$

(b) In order to embed the Standard Model gauge group inside SU(5), the Standard Model gauge group must be SU(3)×SU(2)×U(1)/$\mathbb{Z}_6$ and not SU(3)×SU(2)×U(1). Under the assumption that all the particles of the Standard Model have been discovered, verify that the Standard Model gauge group is $\mathrm{SU(3)\times SU(2)\times U(1)}/\mathbb{Z}_6$. That is, identify the discrete $\mathbb{Z}_6$ subgroup of $\mathrm{SU(3)\times SU(2)\times U(1)}$ that acts trivially on all Standard Model particles.

HINT: It is convenient to redefine the hypercharge U(1) generator to be $6Y$ (where Y is the usual hypercharge of electroweak physics) so that the minimum positive hypercharge of the Standard Model multiplets is +1. Consider the element $\omega \in \mathrm{SU(3)\times SU(2)\times U(1)}$ acting on states of the Standard Model, where $\omega \equiv (e^{2i\pi/3}\mathbb{1}_{3\times 3}, -\mathbb{1}_{2\times 2}, e^{i\pi/3})$ in the defining representation of $\mathrm{SU(3)\times SU(2)\times U(1)}$. In particular, when acting on the fields $q=(u,d)$, which transform as $(\mathbf{3},\mathbf{2},6Y\!=\!1)$ under $\mathrm{SU(3)\times SU(2)\times U(1)}$, it follows that

$$\omega(q) = (e^{2i\pi/3})(-1)(e^{i\pi/3})q = q\,. \tag{6.355}$$

That is, ω acts trivially on q. To apply ω to other states of the Standard Model, note that the action of ω on a color antitriplet [octet or singlet] is to multiply the state by $e^{-2i\pi/3}$ [+1]. Likewise, the action of ω on an SU(2) doublet [triplet or singlet] is to multiply the state by -1 [+1]. Finally, the action of ω on a state of hypercharge Y is to multiply the state by $(e^{i\pi/3})^{6Y}$. For example, when acting on the multiplet of fields $\ell=(\nu,e)$, which transforms as $(\mathbf{1},\mathbf{2},6Y\!=\!-3)$ under $\mathrm{SU(3)\times SU(2)\times U(1)}$, it follows that

$$\omega(\ell) = (+1)(-1)(e^{-i\pi})\ell = \ell\,. \tag{6.356}$$

That is, ω acts trivially on ℓ. Verify that ω acts trivially on all fermionic and bosonic states of the Standard Model. Noting that ω^n is the identity operator when n is an integer multiple of 6, conclude that $\omega \in \mathbb{Z}_6$. Further details can be found in [B78].

(c) Neglecting the color degrees of freedom, show that the electroweak gauge group is $\mathrm{U(2)} \cong \mathrm{SU(2)\times U(1)}/\mathbb{Z}_2$. Identify explicitly the $\mathbb{Z}_2$ subgroup of $\mathrm{SU(2)\times U(1)}$ that acts trivially on all fermionic and bosonic states of the electroweak theory.

(d) After electroweak symmetry breaking, show that the Standard Model gauge group, $\mathrm{SU(3)\times SU(2)\times U(1)}/\mathbb{Z}_6$, is broken down to $\mathrm{SU(3)\times U(1)_{EM}}/\mathbb{Z}_3$, which is isomorphic to the group U(3). Identify explicitly the $\mathbb{Z}_3$ subgroup of $\mathrm{SU(3)\times U(1)_{EM}}$ that acts trivially on all fermionic and bosonic states of the Standard Model.

(e) More generally, show that $\mathrm{U}(n) \cong \mathrm{S}\big(\mathrm{U}(n)\times \mathrm{U}(1)\big) \cong \mathrm{SU}(n)\times\mathrm{U}(1)/\mathbb{Z}_n$ for all positive integers n [see eq. (6.354)].

6.14 Consider the bosonic sector of an SU(5) grand unified model consisting of a multiplet of real gauge boson fields A_μ^A and a multiplet of real scalar fields Φ^A, both of which transform under the 24-dimensional adjoint representation of SU(5). Following eq. (6.249), define a 5×5 hermitian traceless matrix of scalar adjoint fields, $\Phi_i{}^j(x) \equiv \sqrt{2}\Phi^A(x)(\boldsymbol{t^A})_i{}^j$. After minimizing the scalar potential of the adjoint scalar fields, one finds nonzero vacuum expectation values, $\langle\Phi\rangle_i{}^j$, which spontaneously breaks SU(5) down to a subgroup H.

(a) Using eq. (6.251), show that the unbroken generators of the fundamental representation of SU(5) commute with $\langle\Phi\rangle$. That is, $[\boldsymbol{t^A}, \langle\Phi\rangle] = 0$.

(b) Since $\langle\Phi\rangle$ is hermitian, it can be diagonalized by a unitary matrix. This implies that one can perform a field redefinition following eq. (6.252) such that $\langle\Phi\rangle$ is a real diagonal traceless matrix. Thus, without loss of generality, one may assume that $\langle\Phi\rangle$ is a linear combination of the four diagonal generators of SU(5), $\boldsymbol{t}^3$, $\boldsymbol{t}^8$, $\boldsymbol{t}^{23}$, and $\boldsymbol{t}^{24}$ [see eqs. (6.244)–(6.246)]. Using the result of part (a), show that the rank of the unbroken subgroup H is equal to four.

(c) As noted below eq. (6.254), if $\langle\Phi\rangle = c\boldsymbol{t}^{24}$, then $H = \mathrm{SU}(3)\times\mathrm{SU}(2)\times\mathrm{U}(1)$. Identify the explicit form of $\langle\Phi\rangle$ that would yield $H = \mathrm{SU}(4)\times\mathrm{U}(1)$. In this latter case, identify explicitly the unbroken U(1) subgroup.

(d) Generalize the results of parts (a) and (b) to show that if the symmetry group G is broken down to H due to the vacuum expectation values of one multiplet of real adjoint scalar fields, then the rank of H is equal to the rank of G. That is, adjoint breaking does not reduce the rank of the gauge group.

6.15 Consider a multiplet of real scalar fields, Φ^a, that transform under the adjoint representation of SU(n). As in the previous exercise, define a traceless hermitian matrix of scalar adjoint fields, $\Phi_i{}^j(x) \equiv \sqrt{2}\Phi^A(x)(\boldsymbol{t^A})_i{}^j$. Assume that the scalar potential has the form

$$V = -\tfrac{1}{2}m^2\Phi_i{}^j\Phi_j{}^i + \tfrac{1}{4}\lambda_1(\Phi_i{}^j\Phi_j{}^i)^2 + \tfrac{1}{4}\lambda_2\Phi_i{}^j\Phi_j{}^k\Phi_k{}^\ell\Phi_\ell{}^i\,, \qquad (6.357)$$

where we have assumed for simplicity that V is invariant under $\Phi \to -\Phi$. As per usual, there is an implicit sum over pairs of repeated indices.

(a) As in the previous exercise, assume without loss of generality that $\langle\Phi\rangle$ is a real diagonal traceless $n \times n$ matrix. Determine the conditions on the parameters of the scalar potential such that $\langle\Phi\rangle \neq 0$. Show that the scalar potential admits extrema only if at most two eigenvalues of $\langle\Phi\rangle$ are different.

(b) If $\langle\Phi\rangle \neq 0$, SU($n$) is broken down to a subgroup H whose rank, $n-1$, is equal to that of SU(n). The symmetry-breaking pattern will depend on the parameters of the scalar potential. Show that the possible unbroken symmetry group H is either

$$H = \mathrm{SU}(\ell)\times\mathrm{SU}(n-\ell)\times\mathrm{U}(1)\,, \quad \text{where } \ell = \begin{cases} \tfrac{1}{2}n\,, & \text{for } n \text{ even},\\ \tfrac{1}{2}(n+1)\,, & \text{for } n \text{ odd},\end{cases}$$

or

$$H = \mathrm{SU}(n-1)\times\mathrm{U}(1)\,. \qquad (6.358)$$

What are the explicit forms of $\langle\Phi\rangle$ corresponding to the various cases exhibited above? (See, e.g., Refs. [77, 78].)

(c) Identify the unbroken U(1) subgroup in each of the cases of part (b).

(d) If the discrete symmetry $\Phi \to -\Phi$ is not imposed, then a cubic term in the scalar potential proportional to $\Phi_i{}^j\Phi_j{}^k\Phi_k{}^i$ can be added. How would the results of parts (a) and (b) change in this more general case?

6.16 Consider the scalar sector of an SU(5) grand unified model consisting of an adjoint matrix Higgs field, $\Phi \equiv \sqrt{2}\,\Phi^A \boldsymbol{t}^{\boldsymbol{A}}$ and a Higgs field in the fundamental representation of SU(5), denoted by ϕ_i. A renormalizable SU(5) invariant scalar potential that respects separate discrete symmetries, $\Phi \to -\Phi$ and $\phi_i \to -\phi_i$, has the following form (e.g., see Ref. [79]):

$$V = -\tfrac{1}{2}m^2\,\mathrm{Tr}(\Phi^2) + \tfrac{1}{4}\lambda_1(\mathrm{Tr}\,\Phi^2)^2 + \tfrac{1}{4}\lambda_2\,\mathrm{Tr}\,\Phi^4 - \tfrac{1}{2}\mu^2\phi^{\dagger i}\phi_i + \tfrac{1}{4}\lambda(\phi^{\dagger i}\phi_i)^2$$
$$+\,\alpha\phi^{\dagger i}\phi_i\,\mathrm{Tr}\,\Phi^2 + \beta\phi^{\dagger i}\Phi_i{}^j\Phi_j{}^k\phi_k\,, \qquad (6.359)$$

where $\phi^{\dagger i} \equiv (\phi_i)^\dagger$. The symmetry breaking, $\mathrm{SU}(5) \to \mathrm{SU}(3)_C\times\mathrm{SU}(2)_L\times\mathrm{U}(1)_Y \to \mathrm{SU}(3)_C\times\mathrm{U}(1)_{\mathrm{EM}}$, is a result of the following nonzero expectation values:

$$\langle\phi_i\rangle = v\,\delta_{i5}/\sqrt{2}\,, \qquad \langle\Phi\rangle = V(\boldsymbol{t}^{\boldsymbol{24}} + \epsilon\,\boldsymbol{t}^{\boldsymbol{23}})\,, \qquad (6.360)$$

where $\boldsymbol{t}^{\boldsymbol{23}}$ and $\boldsymbol{t}^{\boldsymbol{24}}$ are given by eqs. (6.245) and (6.246), respectively.

(a) By minimizing the scalar potential, obtain three equations for v, V, and ϵ in terms of the scalar potential parameters.

(b) Consider the approximation in which the $\mathrm{SU}(2)_L\times\mathrm{U}(1)_Y$ breaking is neglected, by setting $\alpha = \beta = 0$ and $v = \epsilon = 0$. Determine the conditions on the scalar potential parameters such that the $\mathrm{SU}(3)_C\times\mathrm{SU}(2)_L\times\mathrm{U}(1)_Y$ vacuum is an absolute minimum of the scalar potential. Obtain formulae for the masses of the X and Y gauge bosons and show that they are of $\mathcal{O}(V)$, which corresponds to the scale of grand unification. Identify the physical Higgs scalars and determine their masses.

(c) Include the $\mathrm{SU}(2)_L\times\mathrm{U}(1)_Y$ breaking by taking $v \neq 0$. Suppose that $\alpha = \beta = 0$. Using the result of part (a), show that $\epsilon = 0$. Obtain formulae for the electroweak gauge bosons and show that they are of $\mathcal{O}(v)$. By assumption, $v \ll V$. How are the result of part (b) modified? Show that the assumption that $\alpha = \beta = 0$ is untenable by noting that, in the approximation under consideration, there exist massless physical colored scalars that can mediate proton decay.

(d) Using the results of part (a), show that if α and β are nonzero, then there exists a range of the scalar potential parameters in which the $\mathrm{SU}(3)_C\times\mathrm{U}(1)_{\mathrm{EM}}$ vacuum is an absolute minimum of the scalar potential. Show that it is possible to tune the parameters α and β such that $\epsilon \sim \mathcal{O}(\beta v^2/V^2) \ll 1$, the masses of the physical colored Higgs triplets are all of $\mathcal{O}(V)$, and at least one uncolored Higgs boson is electrically neutral with a mass of $\mathcal{O}(v)$. What can you say about the masses of the other uncolored physical Higgs scalars of the model?

6.17 In this exercise, we will employ a very simple method for computing the Dynkin index, $I_2(R)$, of a representation R of $\mathfrak{su}(5)$, which is defined by

$$\mathrm{Tr}(\boldsymbol{T}^a_{\boldsymbol{R}}\boldsymbol{T}^b_{\boldsymbol{R}}) = I_2(R)\delta_{ab}\,, \tag{6.361}$$

where the $\boldsymbol{T}^a_{\boldsymbol{R}}$ are the matrix generators of $\mathfrak{su}(5)$ in the representation R. Conventionally, $I_2(F) = \frac{1}{2}$ for the fundamental representation of $\mathfrak{su}(n)$. The computation of $I_2(R)$ makes use of the branching rules for the $\mathfrak{su}(3) \oplus \mathfrak{su}(2) \oplus \mathfrak{u}(1)$ decomposition of the irreducible representations of $\mathfrak{su}(5)$. It is then sufficient to examine the Dynkin indices of the $\mathfrak{su}(2)$ multiplets that reside in the corresponding representation of $\mathfrak{su}(5)$.

(a) Using eq. (H.39), show that for an irreducible representation of $\mathfrak{su}(2)$ of dimension $2j+1$ ($j = 0, \frac{1}{2}, 1, \frac{3}{2}, \ldots$), the Dynkin index of the corresponding $\mathfrak{su}(2)$ representation is given by $\frac{1}{3}j(j+1)(2j+1)$.

(b) The branching rule for the five-dimensional representation of $\mathfrak{su}(5)$ is given in eq. (6.265). It then follows that

$$I_2(\mathbf{5}) = 3I_2(\mathbf{1}) + I_2(\mathbf{2}) = \tfrac{1}{2}\,, \tag{6.362}$$

after evaluating $I_2(\mathbf{5})$ in terms of the Dynkin indices of $\mathfrak{su}(2)$ given in part (a). Note the factor of 3 above, which corresponds to the fact that $(\mathbf{3}, \mathbf{1}, -\frac{1}{3})$ contains three $\mathfrak{su}(2)$ singlets. Of course the result of eq. (6.362) is expected, given that the $\mathbf{5}$ is the fundamental representation of $\mathfrak{su}(5)$. Using the technique introduced above, show that $I_2(\mathbf{10}) = \frac{3}{2}$ in light of eq. (6.258).

(c) Using eq. (6.263), show that $I_2(\mathbf{24}) = 5$.

(d) The irreducible $\mathfrak{su}(5)$ representations contained in the product of two fundamental representations is $\mathbf{5}\otimes\mathbf{5} = \mathbf{15}\oplus\mathbf{10}$. Using eqs. (6.265) and (6.258), remove the multiplets contained in the $\mathbf{10}$ to obtain the branching rule for the $\mathfrak{su}(3) \oplus \mathfrak{su}(2) \oplus \mathfrak{u}(1)$ decomposition of the $\mathbf{15}$,

$$\mathbf{15} \to (\mathbf{6}, \mathbf{1}, -\tfrac{2}{3}) + (\mathbf{1}, \mathbf{3}, 1) + (\mathbf{3}, \mathbf{2}, \tfrac{1}{6})\,. \tag{6.363}$$

Using eq. (6.363), show that $I_2(\mathbf{15}) = \frac{7}{2}$.

6.18 In an SU(5) GUT with Yukawa couplings of three generations of quarks and leptons to $\mathbf{5}\oplus\overline{\mathbf{5}}$ and $\mathbf{45}\oplus\overline{\mathbf{45}}$ multiplets of scalars, the up-type quarks couple to the Higgs scalars via the following terms of the Yukawa Lagrangian:

$$\mathscr{L}_{\text{Yuk}} \supset \tfrac{1}{8}(\boldsymbol{y_2})_{pq}\epsilon^{ijk\ell m}(\psi_{ij})_p(\psi_{k\ell})_q\phi_m + \tfrac{1}{8}(\boldsymbol{y_4})_{pq}\epsilon^{ijk\ell m}(\psi_{ij})_p(\psi_{kn})_q\phi^n_{\ell m} + \text{h.c.}\,, \tag{6.364}$$

where p and q are fermion generation indices.

(a) Recall that $\psi_{ji} = -\psi_{ij}$, $\phi^n_{m\ell} = -\phi^n_{\ell m}$, and $\phi^\ell_{\ell m} = 0$ (after summing over the repeated index ℓ). Using these properties along with eq. (1.68) for the product of two-component fermion fields, show that $\boldsymbol{y_2}$ is a symmetric complex matrix and $\boldsymbol{y_4}$ is a complex antisymmetric matrix. That is,

$$\boldsymbol{y}_2^{\mathsf{T}} = \boldsymbol{y_2}\,, \qquad \boldsymbol{y}_4^{\mathsf{T}} = -\boldsymbol{y_4}\,. \tag{6.365}$$

(b) In an SO(10) GUT, the Yukawa couplings corresponding to eq. (6.364) arise from the couplings of two up-type quark fermion fields with scalars that reside in the SO(10) multiplets, **10**, **120**, and $\mathbf{126}\oplus\overline{\mathbf{126}}$. The Yukawa coupling matrix that couples two up-type quark fermion fields with the **10** and/or $\mathbf{126}\oplus\overline{\mathbf{126}}$ multiplets of scalars is symmetric, whereas the Yukawa coupling matrix that couples two up-type quark fermion fields with a **120** multiplet of scalars is antisymmetric. Show that these considerations provide another explanation for the result of eq. (6.365).

6.19 Consider an SU(5) GUT with Yukawa couplings of three generations of quarks and leptons to $\mathbf{5}\oplus\overline{\mathbf{5}}$ and $\mathbf{45}\oplus\overline{\mathbf{45}}$ multiplets of scalars. The corresponding Yukawa Lagrangian is a sum of eqs. (6.314) and (6.319). Assume that the down-type fermion Yukawa matrices have the following structure:

$$\frac{v_5}{\sqrt{2}}\boldsymbol{y_1} = \begin{pmatrix} 0 & a & 0 \\ a & 0 & 0 \\ 0 & 0 & c \end{pmatrix}, \qquad v_{45}\boldsymbol{y_3} = \begin{pmatrix} 0 & 0 & 0 \\ 0 & b & 0 \\ 0 & 0 & 0 \end{pmatrix}, \tag{6.366}$$

under the assumption that a, b, and c are real and $c \gg b \gg a$. This structure was first proposed by Georgi and Jarlskog in Ref. [73].

(a) Exhibit the down-type quark and charged lepton mass matrices obtained at the GUT scale.

(b) Show that the following approximate GUT scale mass relations are satisfied:

$$m_\tau \simeq m_b\,, \qquad m_\mu \simeq 3m_s\,, \qquad m_e \simeq \tfrac{1}{3}m_d\,. \tag{6.367}$$

(c) Using renormalization group evolution, predict the values of the of the down-type quark masses in terms of the corresponding lepton masses. You may use $\overline{\text{MS}}$ running masses to obtain your predicted results.

(d) Assuming that quark mixing arises primarily from the down-type fermion sector, show that the predicted value of the Cabibbo angle is approximately $\tan^2\theta_c \simeq \sqrt{m_d/m_s}$ and compare with the experimental observation. Does this model also fix the entire CKM matrix?

(e) In an SO(10) GUT, one can employ **10**, **120**, and $\mathbf{126}\oplus\overline{\mathbf{126}}$ multiplets of scalars in constructing the Yukawa couplings. Discuss the possible GUT scale fermion mass relations that can arise in such models.

PART II

CONSTRUCTING SUPERSYMMETRIC THEORIES

7 Introduction to Supersymmetry

7.1 Motivation: The Hierarchy Problem

Despite the successes of the Standard Model, it seems very likely that new physics is present at the TeV scale. The mere fact that the ratio M_P/m_W is so huge is already a powerful clue to the character of physics beyond the Standard Model, because of the infamous "hierarchy problem" (e.g., see Refs. [8, 80]). This is not really a difficulty with the Standard Model itself, but rather a disturbing sensitivity of the Higgs potential to new physics in almost any imaginable extension of the Standard Model. The electrically neutral part of the Standard Model Higgs field is a complex scalar H with a classical potential given by

$$V = m_H^2(H^\dagger H) + \lambda(H^\dagger H)^2 \,. \tag{7.1}$$

The Standard Model requires a nonvanishing VEV for H at the minimum of the potential. This occurs if $\lambda > 0$ and $m_H^2 < 0$, resulting in a tree-level estimate for the Higgs field VEV of $\langle H \rangle = \sqrt{-m_H^2/2\lambda}$. Experimentally, $\langle H \rangle \equiv v/\sqrt{2}$, where $v \simeq 246$ GeV, based on measurements of the properties of the weak interactions. The discovery of the Higgs boson with a mass near 125 GeV in 2012 implies that $\lambda = 0.126$ and $m_H^2 = -(93 \text{ GeV})^2$, under the assumption that the Standard Model is correct as an effective field theory [B51, B65]. (Here, λ and m_H^2 are running $\overline{\text{MS}}$ parameters evaluated at a renormalization scale equal to the top-quark mass, and include the effects of two-loop corrections.) The problem is that m_H^2 receives enormous quantum corrections from the virtual effects of every particle or other phenomenon that couples, directly or indirectly, to the Higgs field.

For example, in Fig. 7.1(a) we have a correction to m_H^2 from a loop containing a Dirac fermion f with mass m_f. If the Higgs field couples to f with a term in the Lagrangian $-\lambda_f H \overline{f} f$, then the Feynman diagram in Fig. 7.1(a) yields a correction

$$\Delta m_H^2 = -\frac{|\lambda_f|^2}{8\pi^2}\Lambda_{\text{UV}}^2 + \cdots \,. \tag{7.2}$$

Here Λ_{UV} is an ultraviolet (UV) momentum cutoff used to regulate the loop integral; it should be interpreted as (roughly) the energy scale at which new physics enters to alter the high-energy behavior of the theory. The ellipses represent terms proportional to m_f^2, which grow at most logarithmically with Λ_{UV} (and actually differ for the real and imaginary parts of H). Each of the leptons and quarks of the Standard Model can play the role of f; for quarks, eq. (7.2) should be multiplied

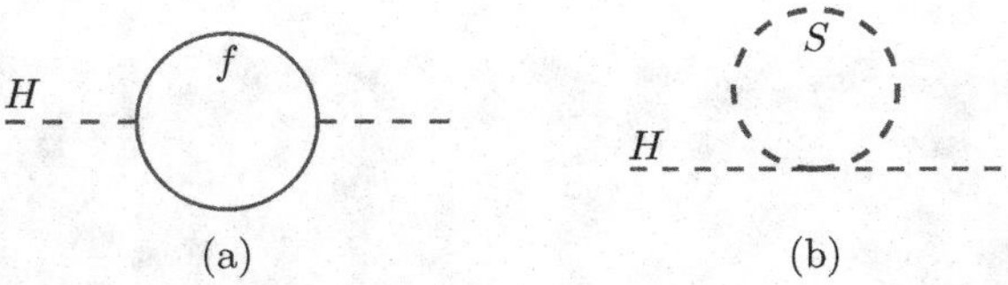

Fig. 7.1 One-loop quantum corrections to the Higgs squared mass parameter m_H^2, due to the effects of (a) Dirac fermion f and (b) scalar S loop contributions.

by 3 to account for color. The largest correction comes when f is the top quark with $\lambda_f \approx 0.94$. The problem is that if $\Lambda_{\rm UV}$ is of order M_P, say, then this quantum correction to m_H^2 is some 30 orders of magnitude larger than the aimed-for value of $m_H^2 \sim -(93 \text{ GeV})^2$. This is only directly a problem for corrections to the squared mass of the Higgs scalar boson, because quantum corrections to fermion and gauge boson masses do not have the quadratic sensitivity to $\Lambda_{\rm UV}$ found in eq. (7.2). However, the quarks and leptons and the electroweak gauge bosons Z, $W^\pm$ of the Standard Model all owe their masses to $\langle H \rangle$, so that the entire mass spectrum of the Standard Model is directly or indirectly sensitive to the cutoff $\Lambda_{\rm UV}$.

One could imagine that the solution is to simply pick a $\Lambda_{\rm UV}$ that is not too large. But then one still must concoct some new physics at the scale $\Lambda_{\rm UV}$ that not only alters the propagators in the loop, but actually cuts off the loop integral. This is not easy to do in a theory whose Lagrangian does not contain more than two derivatives, and higher-derivative theories generally suffer from a failure of either unitarity or causality (e.g., see Ref. [81]). In string theories, loop integrals are nevertheless cut off at high Euclidean momentum p by factors $\exp(-p^2/\Lambda_{\rm UV}^2)$. However, then $\Lambda_{\rm UV}$ is a string scale that is usually[1] thought to be not very far below M_P.

Furthermore, there are contributions similar to eq. (7.2) from the virtual effects of any heavy particles that might exist, and these involve the masses of the heavy particles (or other high physical mass scales), not just the cutoff. It cannot be overemphasized that merely choosing a regulator with no quadratic divergences does *not* address the hierarchy problem. The problem is not really the quadratic divergences, but rather the quadratic sensitivity to high mass scales. The latter are correlated with quadratic divergences for some, but not all, choices of ultraviolet regulator. The absence of quadratic divergences is a necessary, but not sufficient, criterion for avoiding the hierarchy problem.[2]

For example, suppose there exists a heavy complex scalar particle S with mass m_S that couples to the Higgs with a Lagrangian term $-\lambda_S (H^\dagger H)|S|^2$. Then the Feynman diagram in Fig. 7.1(b) gives a correction

$$\Delta m_H^2 = \frac{\lambda_S}{16\pi^2}\left[\Lambda_{\rm UV}^2 - 2m_S^2 \ln(\Lambda_{\rm UV}/m_S) + \cdots\right]. \tag{7.3}$$

[1] Some recent attacks on the hierarchy problem are based on the proposition that the ultimate cutoff scale is actually close to the electroweak scale, rather than the apparent Planck scale [82].

[2] See [B51] and Ref. [83] for a discussion of the hierarchy problem in the context of effective field theory.

Fig. 7.2 Two-loop corrections to the Higgs squared mass due to a heavy fermion.

If one rejects the possibility of a physical interpretation for $\Lambda_{\rm UV}$ and uses dimensional regularization on the loop integral instead of a momentum cutoff, then there will be no $\Lambda^2_{\rm UV}$ piece. However, even then the term proportional to m_S^2 cannot be eliminated without the physically unjustifiable tuning of a counterterm specifically for that purpose. This illustrates that m_H^2 is sensitive to the masses of the *heaviest* particles that H couples to; if m_S is very large, its effects on the Standard Model do not decouple, but instead make it very difficult to understand why m_H^2 is so small.

This problem arises even if there is no direct coupling between the Standard Model Higgs boson and the unknown heavy particles. For example, suppose there exists a heavy fermion F that, unlike the quarks and leptons of the Standard Model, has vectorlike quantum numbers and therefore gets a large mass m_F without coupling to the Higgs field. [In other words, an arbitrarily large mass term of the form $m_F\overline{F}F$ is not forbidden by any symmetry, including $SU(2)_L$.] In that case, no diagram like Fig. 7.1(a) exists for F. Nevertheless there will be a correction to m_H^2 as long as F shares some gauge interactions with the Standard Model Higgs field; these may be the familiar electroweak interactions, or some unknown gauge forces that are broken at a very high energy scale inaccessible to experiment. In either case, the two-loop Feynman diagrams in Fig. 7.2 yield a correction

$$\Delta m_H^2 = C_H T_F \left(\frac{g^2}{16\pi^2}\right)^2 \left[a\Lambda^2_{\rm UV} + 24 m_F^2 \ln(\Lambda_{\rm UV}/m_F) + \cdots\right], \tag{7.4}$$

where C_H and T_F are group theory factors[3] of order 1, and g is the appropriate gauge coupling. The coefficient a depends on the method used to cut off the momentum integrals. It does not arise at all if one uses dimensional regularization, but the m_F^2 contribution is always present with the given coefficient. The numerical factor $(g^2/16\pi^2)^2$ may be quite small (of order 10^{-5} for electroweak interactions), but the important point is that these contributions to Δm_H^2 are sensitive both to the largest masses and to the ultraviolet cutoff in the theory, presumably of order M_P. The "natural" squared mass of a fundamental Higgs scalar, including quantum corrections, therefore seems to be more like M_P^2 than the experimentally favored value! Even very indirect contributions from Feynman diagrams with three or more loops can give unacceptably large contributions to Δm_H^2. The argument above applies

[3] Specifically, C_H is the quadratic Casimir invariant of H, and T_F is the Dynkin index of F in a normalization such that $T_F = 1$ for a Dirac fermion (or two Weyl fermions) in a fundamental representation of $SU(n)$. See Appendix H.3 for further details.

not just for heavy particles, but for arbitrary high-scale physical phenomena such as condensates or additional compactified dimensions.

It could be that the Higgs boson field is not fundamental, but rather is the result of a composite field or collective phenomenon. Such ideas are certainly still worth exploring, although they typically present difficulties in their simplest forms. In particular, so far the 125 GeV Higgs boson does appear to have properties consistent with a fundamental scalar field. Or, it could be that the ultimate ultraviolet cutoff scale, and therefore the mass scales of all presently undiscovered particles and condensates, are much lower than the Planck scale. But, if the Higgs boson is a fundamental particle, and there really is physics far above the electroweak scale, then we have two remaining options: either we must make the rather bizarre assumption that *none* of the high-mass particles or condensates couples (even indirectly or extremely weakly) to the Higgs scalar field, or else some striking cancellation is needed between the various contributions to Δm_H^2.

It is instructive to contrast the behavior of eqs. (7.2) and (7.3) with the one-loop correction to the mass of a fermion. As an example, consider the mass of the electron in quantum electrodynamics. The one-loop correction appears in a seminal paper by Weisskopf published in 1939 [84]:

$$\Delta m_e = \frac{3\alpha m_e}{2\pi} \ln\left(\frac{\Lambda}{m}\right) + \text{finite terms}, \tag{7.5}$$

where α is the fine structure constant and Λ is an ultraviolet cutoff. This was a remarkable result, since the quantum theory of a relativistic electron produces a linear divergent contribution to the self-energy of the electron, $\Delta m_e \sim \alpha\Lambda$, due to the classical Coulomb self-energy, and a quadratically divergent contribution, $\Delta m_e \sim \alpha\Lambda^2/m_e$, due to the existence of electromagnetic field strength vacuum fluctuations, which are purely quantum mechanical effects. However, the leading quadratic and subleading linear divergences are exactly canceled in QED (due to new contributions beyond the relativistic single electron theory that arise after including the effects of the positrons), a result first obtained by Weisskopf in 1934 [85].[4]

In the 1939 paper, Weisskopf also investigated the one-loop self-energy of a boson by considering a theory of scalar particles coupled to photons with coupling e. He demonstrated that the quadratic divergence of the self-energy of the scalar is not canceled. His commentary on this result provides the foundation for what we now refer to as the hierarchy problem. Note that if one sets the electron self-energy in QED equal to $m_e c^2$ with a finite cutoff $\Lambda \equiv a^{-1}$, where a is the critical length scale, then one obtains $a \sim m_e^{-1}\exp(-1/\alpha)$, which is roughly 10^{-58} times smaller than the classical electron radius, $r_c \equiv \alpha/m$. In contrast, Weisskopf writes: "the situation is, however, entirely different for a particle with Bose statistics." Setting the scalar

[4] In the original version of the 1934 paper, Weisskopf made a computational error and the leading quadratic divergence did not cancel. The error was pointed out by Furry, whom Weisskopf acknowledges in an erratum where the cancellation of the quadratic divergence in QED was first announced. Both the original version of the 1934 paper and its erratum are in German. The English translations appear on pp. 157–168 of [B144].

self-energy to mc^2 (where m is the scalar mass) yields a much larger critical length scale, $a \sim (\alpha m^2)^{-1/2}$, as compared to the critical length scale of QED. Weisskopf concludes [84]:

> This may indicate that a theory of particles obeying Bose statistics must involve new features at this critical length [$a \sim (\alpha m^2)^{-1/2}$], or at energies corresponding to this length; whereas a theory of particles obeying the exclusion principle is probably consistent down to much smaller lengths or up to much higher energies.

The systematic cancellation of the quadratic divergent contributions to Δm_e and Δm_H^2 can only be brought about by the type of conspiracy that is better known to physicists as a symmetry. In the case of QED, where the leading quadratic and subleading linear divergences are canceled, eq. (7.5) implies that $\Delta m_e \to 0$ as $m_e \to 0$. This behavior is consistent with the invariance of the massless QED Lagrangian under a chiral transformation, $\Psi \to \exp(i\theta\gamma_5)\Psi$ (using four-component spinor notation). As for the quadratic divergence in the self-energy of the scalar, a comparison of eqs. (7.2) and (7.3) strongly suggests that the new symmetry ought to relate fermions and bosons, because of the relative minus sign between fermion-loop and boson-loop contributions to Δm_H^2. (Note that λ_S must be positive if the scalar potential is to be bounded from below.) If each of the quarks and leptons of the Standard Model is accompanied by two complex scalars with $\lambda_S = |\lambda_f|^2$, then the $\Lambda_{\rm UV}^2$ contributions of Figs. 7.1(a) and 7.1(b) will neatly cancel. Clearly, more restrictions on the theory will be necessary to ensure that this success persists to higher orders, so that, for example, the contributions in Fig. 7.2 and eq. (7.4) from a very heavy fermion are canceled by the two-loop effects of some very heavy bosons. Fortunately, the cancellation of all such contributions to scalar masses is not only possible, but is actually unavoidable, once we merely assume that there exists a symmetry relating fermions and bosons, called a *supersymmetry.*

7.2 Enter Supersymmetry

A supersymmetry (SUSY) transformation turns a bosonic state into a fermionic state, and vice versa. The operator Q that generates such transformations must be an anticommuting spinor, with

$$Q|\text{boson}\rangle = |\text{fermion}\rangle; \qquad Q|\text{fermion}\rangle = |\text{boson}\rangle. \tag{7.6}$$

Spinors are intrinsically complex objects, so $Q^\dagger$ (the hermitian conjugate of Q) is also a symmetry generator. Because Q and $Q^\dagger$ are fermionic operators, they carry spin angular momentum 1/2, so it is clear that supersymmetry must be a space-time symmetry. The possible forms for such symmetries in an interacting quantum field theory are highly restricted by the Haag–Łopuszański–Sohnius extension of the Coleman–Mandula Theorem. For realistic theories that, like the Standard Model, have chiral fermions (i.e., fermions whose left- and right-handed pieces transform differently under the gauge group) and thus the possibility of parity-violating interactions, this theorem implies that the generators Q and $Q^\dagger$ must satisfy an algebra of anticommutation and commutation relations with the schematic form

$$\{Q, Q^\dagger\} = P^\mu, \tag{7.7}$$

$$\{Q, Q\} = \{Q^\dagger, Q^\dagger\} = 0, \tag{7.8}$$

$$[P^\mu, Q] = [P^\mu, Q^\dagger] = 0, \tag{7.9}$$

where P^μ is the four-momentum generator of spacetime translations. Here we have ruthlessly suppressed the spinor indices on Q and $Q^\dagger$. In Chapter 8, we will derive the precise version of eqs. (7.7)–(7.9) with indices restored. A more comprehensive treatment of the supersymmetry algebra will be provided in Chapter 9. Here, we simply note that the appearance of P^μ on the right-hand side of eq. (7.7) is unsurprising, since it transforms under Lorentz boosts and rotations as a spin-1 object while Q and $Q^\dagger$ on the left-hand side each transform as spin-1/2 objects.

The single-particle states of a supersymmetric theory fall into irreducible representations of the supersymmetry algebra, called *supermultiplets*. Each supermultiplet contains both fermion and boson states, which are commonly known as *superpartners* of each other. By definition, if $|\Omega\rangle$ and $|\Omega'\rangle$ are members of the same supermultiplet, then $|\Omega'\rangle$ is proportional to some combination of Q and $Q^\dagger$ operators acting on $|\Omega\rangle$, up to a spacetime translation or rotation. The squared-mass operator P^2 commutes with the operators Q, $Q^\dagger$, and with all spacetime rotation and translation operators, so it follows immediately that particles that inhabit the same irreducible supermultiplet must have equal eigenvalues of P^2, and therefore equal masses.

The supersymmetry generators $Q, Q^\dagger$ also commute with the generators of gauge transformations. Therefore, particles in the same supermultiplet must also be in the same representation of the gauge group, and so must have the same electric charges, weak isospin, and color degrees of freedom.

Each supermultiplet contains equal numbers of boson (n_B) and fermion (n_F) degrees of freedom and therefore satisfies

$$n_B = n_F\,. \tag{7.10}$$

This is a consequence of the supersymmetry algebra, as shown in Section 9.4.2. The simplest possibility for a supermultiplet consistent with eq. (7.10) has a single Weyl fermion (with two spin helicity states, so $n_F = 2$) and two real scalars (each with $n_B = 1$). It is natural to assemble the two real scalar degrees of freedom into a complex scalar field; as we will see below, this provides for convenient formulation of the supersymmetry algebra, Feynman rules, supersymmetry violating effects, etc. This combination of a two-component Weyl fermion and a complex scalar field is called a *chiral* or *matter* or *scalar* supermultiplet.

The next-simplest possibility for a supermultiplet contains a spin-1 vector boson. If the theory is to be renormalizable, this must be a gauge boson that is massless, at least before the gauge symmetry is spontaneously broken. A massless spin-1 boson has two helicity states, so the number of bosonic degrees of freedom is $n_B = 2$. Its superpartner is therefore a massless spin-1/2 Weyl fermion, again with two helicity states, so $n_F = 2$. (If one tried instead to use a massless spin-3/2 fermion, the theory would not be renormalizable.) Gauge bosons must transform as the adjoint representation of the gauge group, so their fermionic partners, called *gauginos*, must also. Because the adjoint representation of a gauge group is always its own conjugate, the gaugino fermions must have the same gauge transformation properties for left-handed and for right-handed components. Such a combination of spin-1/2 gauginos and spin-1 gauge bosons is called a *gauge* or *vector* supermultiplet.

If we include gravity, then the spin-2 graviton (with two helicity states, so $n_B = 2$) has a spin-3/2 superpartner called the gravitino. The gravitino would be massless if supersymmetry were unbroken, and so has $n_F = 2$ helicity states.

There are other possible combinations of particles with spins that can satisfy eq. (7.10). However, these are always reducible to combinations[5] of chiral and gauge supermultiplets if they have renormalizable interactions, except in certain theories with "extended" supersymmetry. Theories with extended supersymmetry have more than one distinct copy of the supersymmetry generators $Q, Q^\dagger$. A brief introduction to the extended supersymmetry algebra and the corresponding supermultiplets is provided in Section 9.5. Such models are mathematically amusing, but evidently do not have any phenomenological prospects. The reason is that extended supersymmetry in four-dimensional field theories cannot allow for chiral fermions or parity violation, as observed in the Standard Model. So we will not discuss such possibilities further, although extended supersymmetry in higher-dimensional field theories might describe the real world if the extra dimensions are compactified in an appropriate way, and extended supersymmetry in four dimensions provides interesting toy models. The ordinary, nonextended, phenomenologically viable type of supersymmetric model is sometimes called $N = 1$ supersymmetry, with N referring to the number of supersymmetries (the number of distinct copies of $Q, Q^\dagger$).

In a supersymmetric extension of the Standard Model, each of the known fundamental particles is therefore in either a chiral or gauge supermultiplet, and must have a superpartner with spin differing by 1/2 unit. The first step in understanding the exciting phenomenological consequences of this prediction is to decide exactly how the known particles fit into supermultiplets, and to give them appropriate names. A crucial observation here is that only chiral supermultiplets can contain fermions whose left-handed parts transform differently under the gauge group than their right-handed parts. All of the Standard Model fermions (the known quarks and leptons) have this property, so they must be members of chiral supermultiplets.[6]

The names for the spin-0 partners of the quarks and leptons are constructed

[5] For example, if a gauge symmetry spontaneously breaks without breaking supersymmetry, then a massless vector supermultiplet will "eat" a chiral supermultiplet, resulting in a massive vector supermultiplet with physical degrees of freedom consisting of a massive vector ($n_B = 3$), a massive Dirac fermion formed from the gaugino and the chiral fermion ($n_F = 4$), and a real scalar ($n_B = 1$).

[6] In particular, one cannot attempt to make a spin-1/2 neutrino be the superpartner of the spin-1 photon; the neutrino is in a doublet, and the photon is neutral, under weak isospin.

by prepending an "s", which is short for scalar. So, generically they are called *squarks* and *sleptons* (short for "scalar quark" and "scalar lepton"), or collectively *sfermions*. The left-handed and right-handed pieces of the quarks and leptons are separate two-component Weyl fermions with different gauge transformation properties in the Standard Model, so each must have its own complex scalar partner. The symbols for the squarks and sleptons have a tilde ($\widetilde{\ }$) used to denote the superpartner of a Standard Model particle. For example, the superpartners of the left-handed and right-handed parts of the electron Dirac field are called left- and right-handed *selectrons*, and are denoted $\widetilde{e}_L$ and $\widetilde{e}_R$. It is important to keep in mind that the "handedness" here does not refer to the helicity of the selectrons (they are spin-0 particles) but to that of their superpartners. A similar nomenclature applies for the *smuons* and *staus*: $\widetilde{\mu}_L$, $\widetilde{\mu}_R$, $\widetilde{\tau}_L$, $\widetilde{\tau}_R$. The Standard Model neutrinos (neglecting their very small masses) are always left-handed, so the *sneutrinos* are denoted generically by $\widetilde{\nu}$, with a possible subscript indicating which lepton flavor they carry: $\widetilde{\nu}_e$, $\widetilde{\nu}_\mu$, $\widetilde{\nu}_\tau$. Finally, a complete list of the squarks is $\widetilde{q}_L$, $\widetilde{q}_R$ with $q = u, d, s, c, b, t$. For example, the *sbottoms*, $\widetilde{b}_L$, $\widetilde{b}_R$, and the *stops*, $\widetilde{t}_L$, $\widetilde{t}_R$ comprise the third generation of squarks. The gauge interactions of each of these squark and slepton fields are the same as for the corresponding Standard Model fermions; for instance, the left-handed squarks $\widetilde{u}_L$ and $\widetilde{d}_L$ couple to the W boson while $\widetilde{u}_R$ and $\widetilde{d}_R$ do not.

The Higgs scalar must reside in a chiral supermultiplet, since it has spin 0. Actually, it turns out that just one chiral supermultiplet is not enough. One reason for this is that if there were only one Higgs chiral supermultiplet, the electroweak gauge symmetry would suffer a gauge anomaly, and would be inconsistent as a quantum theory. As we have already observed at the end of Section 5.2.2, this can be avoided if there are two Higgs supermultiplets, one with each of $Y = \pm 1/2$, so that the total contribution to the anomaly traces from the two fermionic members of the Higgs chiral supermultiplets vanishes by cancellation. As we will see in Section 13.2, both of these are also necessary for another completely different reason: because of the structure of supersymmetric theories, only a $Y = 1/2$ Higgs chiral supermultiplet can have the Yukawa couplings necessary to give masses to charge $+2/3$ up-type quarks (up, charm, top), and only a $Y = -1/2$ Higgs chiral supermultiplet can have the Yukawa couplings necessary to give masses to charge $-1/3$ down-type quarks (down, strange, bottom) and to the charged leptons.

The $\mathrm{SU}(2)_L$-doublet complex scalar fields with $Y = 1/2$ and $Y = -1/2$ are denoted by H_u and H_d, respectively. The weak isospin components of H_u with $T_3 = (1/2, -1/2)$ have electric charges 1, 0, respectively, and are denoted (H_u^+, H_u^0). Similarly, the $\mathrm{SU}(2)_L$-doublet complex scalar H_d has $T_3 = (1/2, -1/2)$ components (H_d^0, H_d^-). The neutral scalar that corresponds to the Standard Model Higgs boson is in a linear combination of H_u^0 and H_d^0 (to be discussed further in Section 13.8). The generic nomenclature for a spin-1/2 superpartner is to append "-ino" to the name of the Standard Model particle, so the fermionic partners of the Higgs scalars are called higgsinos. They are denoted by $\widetilde{H}_u$, $\widetilde{H}_d$ for the $\mathrm{SU}(2)_L$-doublet left-handed Weyl spinor fields, with weak isospin components $\widetilde{H}_u^+$, $\widetilde{H}_u^0$ and $\widetilde{H}_d^0$, $\widetilde{H}_d^-$.

We have now found all of the chiral supermultiplets of a minimal phenomenologi-

Table 7.1 Chiral supermultiplets in the Minimal Supersymmetric Standard Model. The spin-0 fields are complex scalars, and the spin-1/2 fields are left-handed two-component Weyl fermions.

names		spin-0	spin-1/2	$SU(3)_C$, $SU(2)_L$, $U(1)_Y$
squarks, quarks	Q	$(\widetilde{u}_L\ \widetilde{d}_L)$	$(u\ d)$	$(\mathbf{3},\ \mathbf{2}\ ,\ \frac{1}{6})$
(×3 generations)	$\overline{u}$	$\widetilde{u}_R^\dagger$	$\overline{u}$	$(\overline{\mathbf{3}},\ \mathbf{1},\ -\frac{2}{3})$
	$\overline{d}$	$\widetilde{d}_R^\dagger$	$\overline{d}$	$(\overline{\mathbf{3}},\ \mathbf{1},\ \frac{1}{3})$
sleptons, leptons	L	$(\widetilde{\nu}\ \widetilde{e}_L)$	$(\nu\ e)$	$(\mathbf{1},\ \mathbf{2}\ ,\ -\frac{1}{2})$
(×3 generations)	$\overline{e}$	$\widetilde{e}_R^\dagger$	$\overline{e}$	$(\mathbf{1},\ \mathbf{1},\ 1)$
Higgs, higgsinos	H_u	$(H_u^+\ H_u^0)$	$(\widetilde{H}_u^+\ \widetilde{H}_u^0)$	$(\mathbf{1},\ \mathbf{2}\ ,\ +\frac{1}{2})$
	H_d	$(H_d^0\ H_d^-)$	$(\widetilde{H}_d^0\ \widetilde{H}_d^-)$	$(\mathbf{1},\ \mathbf{2}\ ,\ -\frac{1}{2})$

cally viable supersymmetric extension of the Standard Model. They are summarized in Table 7.1, classified according to their transformation properties under the Standard Model gauge group $SU(3)_C \times SU(2)_L \times\ U(1)_Y$, which combines left-handed u, d and ν, e degrees of freedom into $SU(2)_L$ doublets. Here we follow a standard convention that all chiral supermultiplets are defined in terms of left-handed Weyl spinors, so that the *conjugates* of the right-handed quarks and leptons (and their superpartners) appear in Table 7.1. Recall, for example, that $\bar{u}$ is a left-handed Weyl fermion, and is the conjugate of the right-handed up quark. Equivalently, it follows from the identifications of eq. (4.165) that $\widetilde{f}_L$ is the scalar superpartner of the left-handed fermion F_L, and $\widetilde{f}_R$ is the scalar superpartner of the right-handed fermion F_R. Note that the hypercharge Y is normalized as in eq. (4.139).

This protocol for defining chiral supermultiplets turns out to be very useful for constructing supersymmetric Lagrangians, as we will see in Chapter 8. It is also useful to have a symbol for each of the chiral supermultiplets as a whole, as indicated in the second column of Table 7.1. For example, Q stands for the $SU(2)_L$-doublet chiral supermultiplet containing $\widetilde{u}_L, u$ (with weak isospin component $T_3 = +1/2$), and $\widetilde{d}_L, d$ (with $T_3 = -1/2$), while $\overline{u}$ stands for the $SU(2)_L$-singlet supermultiplet containing $\widetilde{u}_R^*, \bar{u}$. There are three generations for each of the quark and lepton supermultiplets; the first-generation representatives are listed in Table 7.1. In this book, a generation index $i \in \{1, 2, 3\}$ will be affixed to the chiral supermultiplet names $(Q_i, \overline{u}_i, \ldots)$ when needed, e.g., $(\overline{e}_1, \overline{e}_2, \overline{e}_3) = (\overline{e}, \overline{\mu}, \overline{\tau})$. The bar on $\overline{u}$, $\overline{d}$, $\overline{e}$ fields is part of the name and does not denote any kind of conjugation.

The Higgs chiral supermultiplet H_d (containing H_d^0, H_d^-, $\widetilde{H}_d^0$, $\widetilde{H}_d^-$) has exactly the same Standard Model gauge quantum numbers as the left-handed sleptons and leptons L_i, for example ($\widetilde{\nu}$, $\widetilde{e}_L$, ν, e). Naively, one might suppose that we could have been more economical in our assignment by taking a neutrino and a Higgs scalar to be superpartners, instead of putting them in separate supermultiplets. This would amount to a proposal that the Higgs boson and a sneutrino should be

Table 7.2 Gauge supermultiplets in the Minimal Supersymmetric Standard Model.

names	spin-1/2	spin-1	$SU(3)_C$, $SU(2)_L$, $U(1)_Y$
gluino, gluon	$\widetilde{g}$	g	(**8**, **1** , 0)
winos, W bosons	$\widetilde{W}^{\pm}$ $\widetilde{W}^0$	$W^{\pm}$ W^0	(**1**, **3** , 0)
bino, B boson	$\widetilde{B}^0$	B^0	(**1**, **1** , 0)

the same particle. This attempt played a key role in some of the first attempts to connect supersymmetry to phenomenology, but it is now known to not work. Even ignoring the anomaly cancellation problem mentioned above, many insoluble phenomenological problems would result, including lepton number nonconservation and a mass for at least one of the neutrinos, in gross violation of experimental bounds. Thus, all of the superpartners of Standard Model particles are new particles that cannot be identified with some other Standard Model state.

The vector bosons of the Standard Model clearly must reside in gauge supermultiplets. Their fermionic superpartners are generically referred to as gauginos. The $SU(3)_C$ color gauge interactions of QCD are mediated by the gluon, whose spin-1/2 color-octet supersymmetric partner is the gluino. A tilde is again used to denote the supersymmetric partner of a Standard Model state, so the symbols for the gluon and gluino are g and $\widetilde{g}$ respectively. The electroweak gauge symmetry $SU(2)_L \times U(1)_Y$ is associated with spin-1 gauge bosons W^+, W^0, W^-, and B^0, with spin-1/2 superpartners $\widetilde{W}^+, \widetilde{W}^0, \widetilde{W}^-$, and $\widetilde{B}^0$, called *winos* and *bino*. After electroweak symmetry breaking, the W^0, B^0 gauge eigenstates mix to give mass eigenstates Z and γ. The corresponding gaugino mixtures of $\widetilde{W}^0$ and $\widetilde{B}^0$ are called *zino* ($\widetilde{Z}$) and *photino* ($\widetilde{\gamma}$); if supersymmetry were unbroken, they would be mass eigenstates with masses m_Z and 0. Table 7.2 summarizes the gauge supermultiplets of the minimal supersymmetric extension of the Standard Model.

The chiral and gauge supermultiplets in Tables 7.1 and 7.2 make up the particle content of the Minimal Supersymmetric Standard Model (MSSM). The most obvious and interesting feature of this theory is that none of the superpartners of the Standard Model particles has been discovered as of this writing. If supersymmetry were unbroken, then there would have to be selectrons $\widetilde{e}_L$ and $\widetilde{e}_R$ with masses exactly equal to $m_e \simeq 0.511$ MeV. A similar statement applies to each of the other sfermions, and there would also have to be a massless gluino and photino. These particles would have been extraordinarily easy to detect long ago. Clearly supersymmetry is a broken symmetry in the vacuum state chosen by Nature.

An important clue as to the nature of supersymmetry breaking can be obtained by returning to the motivation provided by the hierarchy problem. Supersymmetry forced us to introduce two complex scalar fields for each Standard Model Dirac fermion, which is just what is needed to enable a cancellation of the quadratically sensitive ($\Lambda^2_{\rm UV}$) pieces of eqs. (7.2) and (7.3). This sort of cancellation also requires that the associated dimensionless couplings should be related (for example $\lambda_S = |\lambda_f|^2$). The necessary relationships between couplings indeed occur in unbroken supersymmetry, as we will see in Chapter 8. In fact, unbroken supersymmetry guarantees that quadratic divergences in scalar squared masses, and therefore the quadratic sensitivity to high mass scales, must vanish to all orders in perturbation theory.[7] Now, if broken supersymmetry is still to provide a solution to the hierarchy problem even in the presence of supersymmetry breaking, then the relationships between dimensionless couplings that hold in an unbroken supersymmetric theory must be maintained. Otherwise, there would be quadratically divergent radiative corrections to the Higgs scalar masses of the form

$$\Delta m_H^2 = \frac{1}{8\pi^2}(\lambda_S - |\lambda_f|^2)\Lambda^2_{\rm UV} + \cdots \tag{7.11}$$

We are therefore led to consider "soft" supersymmetry breaking. This means that the effective Lagrangian of the MSSM can be written in the form

$$\mathscr{L} = \mathscr{L}_{\rm SUSY} + \mathscr{L}_{\rm soft}, \tag{7.12}$$

where $\mathscr{L}_{\rm SUSY}$ contains all of the gauge and Yukawa interactions and preserves supersymmetry invariance, and $\mathscr{L}_{\rm soft}$ violates supersymmetry but contains only mass terms and couplings with *positive* mass dimension. Without further justification, soft supersymmetry breaking might seem like a rather arbitrary requirement. Fortunately, we will see in Chapter 12 that theoretical models for supersymmetry breaking do indeed yield effective Lagrangians with just such terms for $\mathscr{L}_{\rm soft}$. If the largest mass scale associated with the soft terms is denoted $m_{\rm soft}$, then the additional nonsupersymmetric corrections to the Higgs scalar squared mass must vanish in the $m_{\rm soft} \to 0$ limit, so by dimensional analysis they cannot be proportional to $\Lambda^2_{\rm UV}$. More generally, these models maintain the cancellation of quadratically divergent terms in the radiative corrections of all scalar masses, to all orders in perturbation theory. The corrections also cannot go like $\Delta m_H^2 \sim m_{\rm soft}\Lambda_{\rm UV}$, because in general the loop momentum integrals always diverge either quadratically or logarithmically, not linearly, as $\Lambda_{\rm UV} \to \infty$. So they must be of the form

$$\Delta m_H^2 = m^2_{\rm soft}\left[\frac{\lambda}{16\pi^2}\ln(\Lambda_{\rm UV}/m_{\rm soft}) + \cdots\right]. \tag{7.13}$$

Here λ is schematic for various dimensionless couplings, and the ellipses stand both for terms that are independent of $\Lambda_{\rm UV}$ and for higher loop corrections (which depend on $\Lambda_{\rm UV}$ through powers of logarithms).

Because the mass splittings between the known Standard Model particles and their superpartners are just determined by the parameters $m_{\rm soft}$ appearing in $\mathscr{L}_{\rm soft}$,

[7] A simple way to understand this is to recall that unbroken supersymmetry requires the degeneracy of scalar and fermion masses. Radiative corrections to fermion masses are known to diverge at most logarithmically in any renormalizable field theory [see eq. (7.5)], so the same must be true for scalar masses in unbroken supersymmetry.

eq. (7.13) suggests that the superpartner masses should not be too huge.[8] Otherwise, we would lose our successful cure for the hierarchy problem since the $m^2_{\rm soft}$ corrections to the Higgs scalar (mass)^2 would be unnaturally large compared to the square of the electroweak breaking scale of 246 GeV. The top and bottom squarks and the winos and bino give especially large contributions to $\Delta m^2_{H_u}$ and $\Delta m^2_{H_d}$, but the gluino mass and all the other squark and slepton masses also feed in indirectly, through radiative corrections to the top and bottom squark masses. Furthermore, in most viable models of supersymmetry breaking that are not unduly contrived, the superpartner masses do not differ from each other by more than about an order of magnitude. Using $\Lambda_{\rm UV} \sim M_P$ and $\lambda \sim 1$ in eq. (7.13), one estimates that $m_{\rm soft}$, and therefore the masses of at least the lightest few superpartners, should probably not be much greater than the TeV scale, in order for the MSSM scalar potential to provide a Higgs VEV resulting in $m_W \simeq 80.4$ GeV and $m_Z \simeq 91.2$ GeV without miraculous cancellations.

However, it is useful to keep in mind that the hierarchy problem was *not* the historical motivation for the development of supersymmetry in the early 1970s. The supersymmetry algebra and supersymmetric field theories were originally concocted independently in various disguises bearing little resemblance to the MSSM. It is quite impressive that a theory that was developed for quite different reasons, including purely aesthetic ones, can later be found to provide a solution for the hierarchy problem.

One might also wonder whether there is any good reason why all of the superpartners of the Standard Model particles should be heavy enough to have avoided discovery so far. Indeed, there is a good reason! All of the particles in the MSSM that have been detected so far (except for the Higgs boson) have something in common: they would necessarily be massless in the absence of electroweak symmetry breaking. In particular, the masses of the $W^{\pm}, Z$ bosons and all quarks and leptons are equal to dimensionless coupling constants times the Higgs VEV, $v \sim 246$ GeV, while the photon and gluon are required to be massless by electromagnetic and QCD gauge invariance. Conversely, all of the undiscovered particles in the MSSM have exactly the opposite property; each of them can have a Lagrangian mass term in the absence of electroweak symmetry breaking. For the squarks, sleptons, and Higgs scalars this follows from a general property of complex scalar fields that a mass term $m^2|\phi|^2$ is always allowed by all gauge symmetries. For the higgsinos and gauginos, it follows from the fact that they are fermions in a real representation of the gauge group. So, from the point of view of the MSSM, the discovery of the top quark in 1995 marked a quite natural milestone: the already-discovered particles are precisely those that had to be light, based on the principle of electroweak gauge symmetry. There is a single exception: it has long been known that at least one neutral Higgs scalar boson had to be lighter than about 135 GeV if the minimal version of supersymmetry is correct, for reasons to be discussed in Section 13.8. The

[8] This is obviously fuzzy and subjective. Nevertheless, such subjective criteria can be useful and even essential, at least on a personal level, for making choices about what research directions to pursue, given finite time and money.

Higgs boson discovered in 2012 (with mass 125 GeV) is presumably this particle, and the fact that it was not much heavier can be counted as a successful prediction of supersymmetry.

A very important feature of the MSSM is that the superpartners listed in Tables 7.1 and 7.2 are not necessarily the mass eigenstates of the theory. This is because after electroweak symmetry-breaking and supersymmetry-breaking effects are included, there can be mixing between the electroweak gauginos and the higgsinos, and within the various sets of squarks and sleptons and Higgs scalars that have the same electric charge. The lone exception is the gluino, which is a color-octet fermion and therefore does not have the appropriate quantum numbers to mix with any other particle. The masses and mixings of the superpartners are obviously of paramount importance to experimental searches. It is perhaps slightly less obvious that these phenomenological issues are all quite directly related to one central question that is also the focus of much of the theoretical work in supersymmetry: "How is supersymmetry broken?" The reason for this is that most of what we do not already know about the MSSM has to do with $\mathscr{L}_{\text{soft}}$.

The structure of supersymmetric Lagrangians allows very little arbitrariness, as we will see in Chapter 8. In fact, all of the dimensionless couplings and all but one mass term in the supersymmetric part of the MSSM Lagrangian correspond directly to parameters in the ordinary Standard Model that have already been measured by experiment. For example, we will find out that the supersymmetric coupling of a gluino to a squark and a quark is determined by the QCD coupling constant α_s. In contrast, the supersymmetry-breaking part of the Lagrangian contains many unknown parameters and, apparently, a considerable amount of arbitrariness. Each of the mass splittings between Standard Model particles and their superpartners corresponds to terms in the MSSM Lagrangian that are purely supersymmetry-breaking in their origin and effect. These soft supersymmetry-breaking terms can also introduce a large number of mixing angles and CP-violating phases not found in the Standard Model. Fortunately, as we will see in Section 13.6, there is already strong evidence that the supersymmetry-breaking terms in the MSSM are actually not arbitrary at all. Furthermore, the additional parameters will be measured and constrained as the superpartners are detected. From a theoretical perspective, the challenge is to explain all of these parameters with a predictive model for supersymmetry breaking.

7.3 Historical Analogies

As a parable (following Murayama in Ref. [86]), consider the mass of the electron as it might have been discussed before the apotheosis of relativistic quantum field theory. If the e^- is considered to be pointlike, then its classical electrostatic energy is infinite. This divergence can be regulated by modeling the electron of charge $-e$ instead as a solid sphere of uniform charge density and radius R. Then, the Coulomb self-energy of the electron (in rationalized Gaussian units) is given by

$$\Delta E_{\text{Coulomb}} = \frac{3e^2}{20\pi R}. \tag{7.14}$$

The resulting physical mass of the electron is related to its bare mass by

$$m_{e,\text{phys}} = m_{e,\text{bare}} + (1\,\text{MeV}/c^2)\left(\frac{0.9 \times 10^{-15}\,\text{meters}}{R}\right). \tag{7.15}$$

Now, $m_{e,\text{phys}}$ is 0.511 MeV, so to avoid having to "tune" $m_{e,\text{bare}}$ to 1% accuracy, one would have to have $R \gtrsim 10^{-17}$ meters. However, any structure of the electron is certainly smaller than this. This presents a fine-tuning problem for the electron's mass in classical electrostatics.

Today, we know that the actual resolution is that the classical Coulomb self-energy is supplemented by another contribution that is intrinsically relativistic and quantum mechanical. The total self-energy calculation to first order in perturbation theory can be depicted in "old-fashioned" Feynman diagram notation as in Fig. 7.3.

Fig. 7.3 Old-fashioned Feynman diagrams for the electron self-energy, at first order in perturbation theory. In the left diagram, a virtual electron propagates forward in time, while in the right diagram, a virtual positron contributes, canceling the most severe part of the Coulomb self-energy divergence.

The left diagram includes the linear divergent classical Coulomb self-energy (which behaves as $1/R$ as $R \to 0$) and a quadratic divergent contribution due to electromagnetic field vacuum fluctuations. The right diagram involving the intermediate state virtual positron is a new feature introduced by the inclusion of antiparticles in the relativistic quantum mechanical theory of the electron.[9] This is precisely the computation that Weisskopf performed in his 1934 paper [85], as described in Section 7.1. The leading quadratic divergence and the subleading linear divergence are canceled, with a net remaining contribution that is only logarithmic, and multiplicative [cf. eq. (7.5)]:

$$m_{e,\text{phys}} = m_{e,\text{bare}}\left[1 + \frac{3\alpha}{2\pi}\ln\left(\frac{\hbar}{m_e c R}\right) + \cdots\right]. \tag{7.16}$$

This solves the apparent fine-tuning problem of the electron's physical mass, because even if $R \sim 10^{-35}$ meters, of order the Planck length [$\ell_P \equiv (\hbar G_N/c^3)^{1/2}$], the resulting correction to the electron's mass is only 18%.

[9] In modern Feynman diagram notation, the two diagrams shown in Fig. 7.3 would be combined into a single one, incorporating both effects. See Exercise 19.2, where the one-loop shift of the electron mass in QED is computed using modern techniques [see eq. (19.340)].

The presence of a positron, with the same mass and interaction strength as the electron, is no accident. It is guaranteed by the relativistic invariance of QED. Thus, the moral of the story is that a symmetry provides for the existence of a partner particle for the electron, which cancels the dangerously huge contribution to its mass. This story is in direct analogy to the argument for supersymmetry, with the electron playing the role of the Standard Model particles and the positron playing the role of the superpartners.

The precise degeneracy of the electron and the positron is due to the fact that Poincaré invariance is exact and unbroken. Another historical analogy is provided by the approximate flavor symmetry of the u, d, s quark model. In the early 1960s, Gell-Mann and Ne'eman independently realized that the known baryon resonances could be classified into irreducible representations of flavor SU(3). By the summer of 1962, the nine lightest $J^P = \frac{3}{2}^+$ baryons had all been discovered. In modern terminology, these are the Δ^-, Δ^0, Δ^+, Δ^{++} with isospin $I = 3/2$ and strangeness $S = 0$ and masses near 1232 MeV, the Σ^-, Σ^0, Σ^+ with isospin $I = 1$ and strangeness $S = -1$ and masses near 1385 MeV, and the Ξ^- and Ξ^0 with isospin $I = 1/2$ and strangeness $S = -2$ and masses near 1530 MeV. Gell-Mann and Ne'eman realized that the experimental absence of such a resonance with $S = 1$ meant that these particles must all live within the **10** (decuplet) representation of flavor SU(3), and not a larger representation such as the **15** or the **27**. They were therefore able to predict that there should exist a baryon with charge $Q = -1$, isospin $I = 0$, strangeness $S = -3$, and mass then extrapolated with the available data to be about 1685 MeV (now known to be closer to 1672 MeV), as illustrated in Fig. 7.4. Due to its strangeness assignment, the new particle could only have weak decays.

Charge → $\quad \Delta^-_{1232} \quad \Delta^0_{1232} \quad \Delta^+_{1234} \quad \Delta^{++}_{1234}$

$-$(Strangeness) $\Big\downarrow \qquad\qquad\quad \Sigma^-_{1387} \quad \Sigma^0_{1384} \quad \Sigma^+_{1383}$

$\qquad\qquad\qquad\qquad\qquad \Xi^-_{1535} \quad \Xi^0_{1532}$

$\qquad\qquad\qquad\qquad\qquad\quad \Omega^-_{1672}$

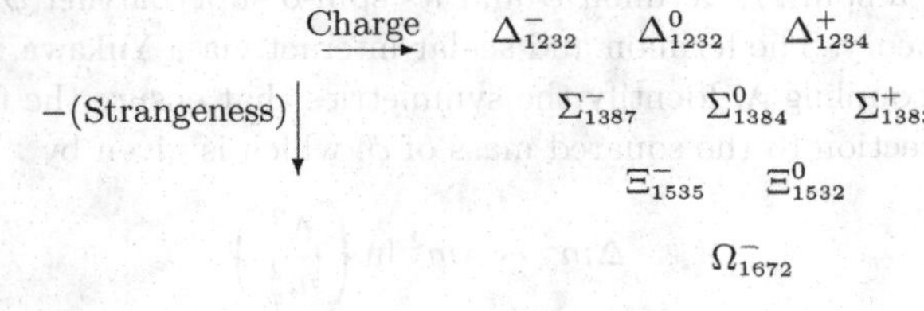

Fig. 7.4   The flavor SU(3) decuplet of the lightest $J^P = \frac{3}{2}^+$ baryons.

This partner particle, the Ω^-, was subsequently discovered in 1964 through a single event in which it decayed to $\Xi^0 \pi^-$ in a bubble chamber at Brookhaven. The theory of flavor symmetries of quarks, extended to an even more broken SU(4) with the discovery of the charm quark, has of course gone on to many more successes. The moral is that even a broken symmetry (as supersymmetry is understood to be) can be useful to not only classify particles and understand their properties, but to make striking predictions. The key is to understand not only the underlying symmetry but the patterns inherent in the symmetry breaking.

Exercises

7.1 In light of eq. (11.210), verify that the computation of Δm_e obtained using dimensional regularization in Exercise 19.2 [see eq. (19.340)] yields Weisskopf's result quoted in eq. (7.5).

7.2 (a) Consider the photon in quantum electrodynamics. Since the photon obeys Bose statistics, one might have thought, following Weisskopf's analysis of Ref. [84], that the one-loop correction to the squared mass of the photon should be quadratically divergent. However, in QED one finds that $\Delta m_\gamma^2 = 0$. Explain the origin of the cancellation of the quadratic and the subleading linear and logarithmic divergences.

(b) The quantum field theory of massive electrodynamics employs the Proca Lagrangian for a massive abelian gauge field A_μ,

$$\mathscr{L} = \mathscr{L}_{\rm QED} + \tfrac{1}{2} m_A^2 A_\mu A^\mu \,, \tag{7.17}$$

and yields a consistent renormalizable theory. How does the one-loop correction to the squared mass of the massive photon, Δm_A^2, depend on the ultraviolet cutoff Λ?

(c) In the electroweak Standard Model, determine the behavior of the one-loop squared-mass shifts, Δm_W^2, Δm_Z^2, and Δm_γ^2 as a function of the ultraviolet cutoff Λ.

7.3 Consider a spin-1/2 fermion ψ and its spin-0 superpartner ϕ in a supersymmetric theory. The fermion and scalar interact via a Yukawa interaction with Yukawa coupling λ. Identify the symmetries that ensure the form of the one-loop correction to the squared mass of ϕ, which is given by

$$\Delta m_\phi^2 \sim \lambda m_\phi^2 \ln\left(\frac{\Lambda^2}{m_\phi^2}\right) . \tag{7.18}$$

HINT: Supersymmetry enforces the relation $m_\psi = m_\phi$ to all orders in perturbation theory. Determine the form of the one-loop correction to the fermion mass, Δm_ψ, and relate this result to Δm_ϕ^2.

7.4 Consider the minimal supersymmetric extension of the seesaw-extended Standard Model described in Section 6.1. How should Table 7.1 be modified in this case?

8 Supersymmetric Lagrangians

In this chapter we will describe the construction of supersymmetric Lagrangians. Our aim is to arrive at a recipe that will allow us to write down the allowed interactions and mass terms of a general supersymmetric theory, so that later we can apply the results to the special case of the MSSM. However, we will not use the superfield language in this chapter, which is often more elegant and efficient for many purposes, but requires a more specialized machinery and might seem rather cabalistic at first. Our approach is therefore intended to be rather complementary to the superfield derivations given in Chapter 10. We begin by considering the simplest example of a supersymmetric theory in four dimensions: the free Wess–Zumino model.

8.1 A Free Chiral Supermultiplet

The minimum fermion content of any theory in four dimensions consists of a single left-handed two-component Weyl fermion ψ. Since this is an intrinsically complex object, it seems sensible to choose as its superpartner a complex scalar field ϕ. The simplest action we can write down for these fields just consists of kinetic energy terms for each:

$$S = \int d^4x \left(\mathscr{L}_{\text{scalar}} + \mathscr{L}_{\text{fermion}}\right), \tag{8.1}$$

$$\mathscr{L}_{\text{scalar}} = \partial^\mu \phi^\dagger \partial_\mu \phi, \qquad \mathscr{L}_{\text{fermion}} = i\psi^\dagger \overline{\sigma}^\mu \partial_\mu \psi. \tag{8.2}$$

This is the massless, noninteracting version of the *Wess–Zumino model* [87, 88]. It corresponds to a single chiral supermultiplet as discussed in Section 7.2.

A supersymmetry transformation should turn the scalar boson field ϕ into something involving the fermion field ψ_α. The simplest possibility for the transformation of the scalar field is

$$\delta_\epsilon \phi = \epsilon\psi; \qquad \delta_\epsilon \phi^\dagger = \epsilon^\dagger \psi^\dagger, \tag{8.3}$$

where ϵ^α is an infinitesimal, anticommuting, two-component Weyl fermion object that parameterizes the supersymmetry transformation.[1] Until Section 14.2, we will

[1] The two-component spinor formalism is ideally suited for treating supersymmetric theories and this formalism will be employed in the development of supersymmetric field theories in this book. However, one can also employ (albeit more awkwardly) the four-component spinor formalism, which can be found in a number of textbook treatments of supersymmetry.

be discussing global supersymmetry, which means that ϵ^α is a constant, satisfying $\partial_\mu \epsilon^\alpha = 0$. Since ψ has dimensions of [mass]$^{3/2}$ and ϕ has dimensions of [mass], it must be that ϵ has dimensions of [mass]$^{-1/2}$. Using eq. (8.3), we find that the scalar part of the Lagrangian transforms as

$$\delta_\epsilon \mathcal{L}_{\text{scalar}} = \epsilon \partial^\mu \psi \, \partial_\mu \phi^\dagger + \epsilon^\dagger \partial^\mu \psi^\dagger \, \partial_\mu \phi. \tag{8.4}$$

We would like for this to be canceled by $\delta_\epsilon \mathcal{L}_{\text{fermion}}$, at least up to a total derivative, so that the action will be invariant under the supersymmetry transformation. Comparing eq. (8.4) with $\mathcal{L}_{\text{fermion}}$, we see that for this to have any chance of happening, $\delta_\epsilon \psi$ should be linear in $\epsilon^\dagger$ and in ϕ and contain one spacetime derivative. Up to a multiplicative constant, there is only one possibility to try:

$$\delta_\epsilon \psi_\alpha = -i(\sigma^\mu \epsilon^\dagger)_\alpha \, \partial_\mu \phi, \qquad \delta_\epsilon \psi^\dagger_{\dot\alpha} = i(\epsilon \sigma^\mu)_{\dot\alpha} \, \partial_\mu \phi^\dagger. \tag{8.5}$$

With this guess, one immediately obtains

$$\delta_\epsilon \mathcal{L}_{\text{fermion}} = -\epsilon \, \sigma^\mu \overline{\sigma}^\nu \partial_\nu \psi \, \partial_\mu \phi^\dagger + \psi^\dagger \overline{\sigma}^\nu \sigma^\mu \epsilon^\dagger \, \partial_\mu \partial_\nu \phi \, . \tag{8.6}$$

This can be put in a slightly more useful form by employing the Pauli matrix identities eqs. (1.88) and (1.89) and using the fact that partial derivatives commute ($\partial_\mu \partial_\nu = \partial_\nu \partial_\mu$). Equation (8.6) then becomes

$$\begin{aligned} \delta_\epsilon \mathcal{L}_{\text{fermion}} = & -\epsilon \, \partial^\mu \psi \, \partial_\mu \phi^\dagger - \epsilon^\dagger \partial^\mu \psi^\dagger \, \partial_\mu \phi \\ & - \partial_\mu \left(\epsilon \sigma^\nu \overline{\sigma}^\mu \psi \, \partial_\nu \phi^\dagger - \epsilon \psi \, \partial^\mu \phi^\dagger - \epsilon^\dagger \psi^\dagger \, \partial^\mu \phi \right) . \end{aligned} \tag{8.7}$$

The first two terms here just cancel against $\delta_\epsilon \mathcal{L}_{\text{scalar}}$, while the remaining contribution is a total derivative. So we arrive at

$$\delta_\epsilon S = \int d^4x \; (\delta_\epsilon \mathcal{L}_{\text{scalar}} + \delta_\epsilon \mathcal{L}_{\text{fermion}}) = 0, \tag{8.8}$$

justifying our guess of the numerical multiplicative factor made in eq. (8.5).

We are not quite finished in demonstrating that the theory described by eq. (8.1) is supersymmetric. We must also show that the supersymmetry algebra closes; in other words, that the commutator of two supersymmetry transformations parameterized by spinors ϵ_1 and ϵ_2 is another symmetry of the theory. Using eq. (8.5)) in eq. (8.3), one finds

$$(\delta_{\epsilon_2} \delta_{\epsilon_1} - \delta_{\epsilon_1} \delta_{\epsilon_2}) \phi \equiv \delta_{\epsilon_2}(\delta_{\epsilon_1} \phi) - \delta_{\epsilon_1}(\delta_{\epsilon_2} \phi) = i(-\epsilon_1 \sigma^\mu \epsilon_2^\dagger + \epsilon_2 \sigma^\mu \epsilon_1^\dagger) \, \partial_\mu \phi. \tag{8.9}$$

This is a remarkable result; in words, we have found that the commutator of two supersymmetry transformations gives us back the derivative of the original field. In the Heisenberg picture of quantum mechanics, $i\partial_\mu$ corresponds to the generator of spacetime translations P_μ, so eq. (8.9) implies the form of the supersymmetry algebra that was foreshadowed in eq. (7.7). We will make this statement more explicit before the end of this chapter.

All of this will be for nothing if we do not find the same result for the fermion ψ, however. Using eq. (8.3) in eq. (8.5), we find

$$(\delta_{\epsilon_2} \delta_{\epsilon_1} - \delta_{\epsilon_1} \delta_{\epsilon_2}) \psi_\alpha = -i(\sigma^\mu \epsilon_1^\dagger)_\alpha \, \epsilon_2 \partial_\mu \psi + i(\sigma^\mu \epsilon_2^\dagger)_\alpha \, \epsilon_1 \partial_\mu \psi. \tag{8.10}$$

We can put this into a more useful form by applying the Fierz identity, eq. (1.305), with $\chi = \sigma^\mu \epsilon_1^\dagger$, $\xi = \epsilon_2$, $\eta = \partial_\mu \psi$, and again with $\chi = \sigma^\mu \epsilon_2^\dagger$, $\xi = \epsilon_1$, $\eta = \partial_\mu \psi$, followed in each case by an application of the identity of eq. (1.107). The result is

$$(\delta_{\epsilon_2}\delta_{\epsilon_1} - \delta_{\epsilon_1}\delta_{\epsilon_2})\psi_\alpha = i(-\epsilon_1\sigma^\mu\epsilon_2^\dagger + \epsilon_2\sigma^\mu\epsilon_1^\dagger)\,\partial_\mu\psi_\alpha + i\epsilon_{1\alpha}\,\epsilon_2^\dagger\overline{\sigma}^\mu\partial_\mu\psi - i\epsilon_{2\alpha}\,\epsilon_1^\dagger\overline{\sigma}^\mu\partial_\mu\psi. \tag{8.11}$$

The last two terms in (8.11) vanish on-shell; that is, if the equation of motion $\overline{\sigma}^\mu\partial_\mu\psi = 0$ following from the action is enforced. The remaining piece is exactly the same spacetime translation that we found for the scalar field.

The fact that the supersymmetry algebra only closes on-shell (when the classical equations of motion are satisfied) might be somewhat worrisome, since we would like the symmetry to hold even quantum mechanically. This can be fixed by a trick. We invent a new complex scalar field F, which does not have a kinetic term. Such fields are called *auxiliary*, and they are really just book-keeping devices that allow the symmetry algebra to close off-shell. The Lagrangian for F and its complex conjugate is simply

$$\mathscr{L}_{\text{auxiliary}} = F^\dagger F\,. \tag{8.12}$$

The dimensions of F are $[\text{mass}]^2$, unlike an ordinary scalar field, which has dimensions of [mass]. Equation (8.12) implies the not-very-exciting equations of motion $F = F^\dagger = 0$. However, we can use the auxiliary fields to our advantage by including them in the supersymmetry transformation rules. In view of eq. (8.11), a plausible thing to do is to make F transform into a multiple of the equation of motion for ψ:

$$\delta_\epsilon F = -i\epsilon^\dagger\overline{\sigma}^\mu\partial_\mu\psi, \qquad \delta_\epsilon F^\dagger = i\partial_\mu\psi^\dagger\overline{\sigma}^\mu\epsilon. \tag{8.13}$$

Once again we have chosen the overall factor on the right-hand sides by virtue of foresight. Now the auxiliary part of the Lagrangian transforms as

$$\delta\mathscr{L}_{\text{auxiliary}} = -i\epsilon^\dagger\overline{\sigma}^\mu\partial_\mu\psi\, F^\dagger + i\partial_\mu\psi^\dagger\overline{\sigma}^\mu\epsilon\, F, \tag{8.14}$$

which vanishes on-shell, but not for arbitrary off-shell field configurations. Now, by adding an extra term to the transformation law for ψ and $\psi^\dagger$,

$$\delta\psi_\alpha = -i(\sigma^\mu\epsilon^\dagger)_\alpha\,\partial_\mu\phi + \epsilon_\alpha F, \qquad \delta\psi^\dagger_{\dot\alpha} = i(\epsilon\sigma^\mu)_{\dot\alpha}\,\partial_\mu\phi^\dagger + \epsilon^\dagger_{\dot\alpha}F^\dagger, \tag{8.15}$$

one obtains an additional contribution to $\delta_\epsilon\mathscr{L}_{\text{fermion}}$, which cancels with $\delta_\epsilon\mathscr{L}_{\text{auxiliary}}$ up to a total derivative term. So our "modified" theory with $\mathscr{L} = \mathscr{L}_{\text{scalar}} + \mathscr{L}_{\text{fermion}} + \mathscr{L}_{\text{auxiliary}}$ is still invariant under supersymmetry transformations. Proceeding as before, one now obtains, for each of the fields $X = \phi, \phi^\dagger, \psi, \psi^\dagger, F, F^\dagger$,

$$(\delta_{\epsilon_2}\delta_{\epsilon_1} - \delta_{\epsilon_1}\delta_{\epsilon_2})X = i(-\epsilon_1\sigma^\mu\epsilon_2^\dagger + \epsilon_2\sigma^\mu\epsilon_1^\dagger)\,\partial_\mu X \tag{8.16}$$

using eqs. (8.3), (8.13), and (8.15), but without resorting to any of the equations of motion. So we have succeeded in showing that supersymmetry is a valid symmetry of the Lagrangian off-shell.

In retrospect, one can see why we needed to introduce the auxiliary field F in order to get the supersymmetry algebra to work off-shell. On-shell, the complex

Table 8.1 Counting of real degrees of freedom in the Wess–Zumino model.

	ϕ	ψ	F
on-shell ($n_B = n_F = 2$)	2	2	0
off-shell ($n_B = n_F = 4$)	2	4	2

scalar field ϕ has two real propagating degrees of freedom, which match with the two spin polarization states of ψ. Off-shell, the Weyl fermion ψ is a complex two-component object, so it has four real degrees of freedom. (Going on-shell eliminates half of the propagating degrees of freedom for ψ, because the Lagrangian is linear in time derivatives, so that the canonical momenta can be re-expressed in terms of the configuration variables without time derivatives and are not independent phase space coordinates.) To match the numbers of bosonic and fermionic degrees of freedom off-shell as well as on-shell, we had to introduce two real scalar degrees of freedom in the complex field F, which are eliminated when one goes on-shell. This counting is summarized in Table 8.1. The auxiliary field formulation is especially useful when discussing spontaneous supersymmetry breaking (see Chapter 12).

Invariance of the action under a symmetry transformation always implies the existence of a conserved current, and supersymmetry is no exception. The *supercurrent* J^μ_α is an anticommuting four-vector. It also carries a spinor index, as befits the current associated with a symmetry with fermionic generators. By the usual Noether procedure, one finds, for the supercurrent (and its hermitian conjugate) in terms of the variations of the fields $X = \phi, \phi^\dagger, \psi, \psi^\dagger, F, F^\dagger$,

$$\epsilon J^\mu + \epsilon^\dagger J^{\dagger\mu} \equiv \sum_X \delta X \, \frac{\delta \mathscr{L}}{\delta(\partial_\mu X)} - K^\mu, \tag{8.17}$$

where K^μ is an object whose divergence is the variation of the Lagrangian under the supersymmetry transformation, $\delta\mathscr{L} = \partial_\mu K^\mu$. Note that K^μ is not unique; one can always replace K^μ by $K^\mu + k^\mu$, where k^μ is any vector satisfying $\partial_\mu k^\mu = 0$, for example $k^\mu = \partial^\mu\partial_\nu a^\nu - \partial_\nu\partial^\nu a^\mu$ for any four-vector a^μ. A little work reveals that, up to the ambiguity just mentioned,

$$J^\mu_\alpha = (\sigma^\nu\overline{\sigma}^\mu\psi)_\alpha \, \partial_\nu\phi^\dagger, \qquad J^{\dagger\mu}_{\dot\alpha} = (\psi^\dagger\overline{\sigma}^\mu\sigma^\nu)_{\dot\alpha} \, \partial_\nu\phi. \tag{8.18}$$

The supercurrent and its hermitian conjugate are separately conserved:

$$\partial_\mu J^\mu_\alpha = 0; \qquad \partial_\mu J^{\dagger\mu}_{\dot\alpha} = 0, \tag{8.19}$$

as can be verified by use of the equations of motion. From these currents one constructs the conserved charges

$$Q_\alpha = \sqrt{2}\int d^3x \, J^0_\alpha, \qquad Q^\dagger_{\dot\alpha} = \sqrt{2}\int d^3x \, J^{\dagger 0}_{\dot\alpha}, \tag{8.20}$$

which are the generators of supersymmetry transformations. (The factor of $\sqrt{2}$ normalization is included to agree with an arbitrary historical convention.) As quantum mechanical operators, they satisfy

$$[\epsilon Q + \epsilon^\dagger Q^\dagger, X] = -i\sqrt{2}\,\delta X \tag{8.21}$$

for any field X, up to terms that vanish on-shell. This can be verified explicitly by using the canonical equal-time commutation and anticommutation relations,

$$[\phi(\vec{x}), \pi(\vec{y})] = [\phi^\dagger(\vec{x}), \pi^\dagger(\vec{y})] = i\delta^{(3)}(\vec{x} - \vec{y}), \tag{8.22}$$

$$\{\psi_\alpha(\vec{x}), \psi^\dagger_{\dot{\alpha}}(\vec{y})\} = (\sigma^0)_{\alpha\dot{\alpha}}\,\delta^{(3)}(\vec{x} - \vec{y}), \tag{8.23}$$

derived from the free-field theory Lagrangian eq. (8.1). Here $\pi = \partial_0\phi^\dagger$ and $\pi^\dagger = \partial_0\phi$ are the momenta conjugate to ϕ and $\phi^\dagger$, respectively.

Using eq. (8.21), the content of eq. (8.16) can be expressed in terms of canonical commutators as

$$\Big[\epsilon_2 Q + \epsilon_2^\dagger Q^\dagger,\ [\epsilon_1 Q + \epsilon_1^\dagger Q^\dagger,\ X]\Big] - \Big[\epsilon_1 Q + \epsilon_1^\dagger Q^\dagger,\ [\epsilon_2 Q + \epsilon_2^\dagger Q^\dagger,\ X]\Big]$$
$$= 2(\epsilon_1\sigma^\mu\epsilon_2^\dagger - \epsilon_2\sigma^\mu\epsilon_1^\dagger)\, i\partial_\mu X, \tag{8.24}$$

up to terms that vanish on-shell. The spacetime momentum operator is $P^\mu = (H, \vec{P})$, where H is the Hamiltonian and $\vec{P}$ is the three-momentum operator, given in terms of the canonical fields by

$$H = \int d^3x \left[\pi^\dagger\pi + (\vec{\nabla}\phi^\dagger)\cdot(\vec{\nabla}\phi) + i\psi^\dagger\vec{\sigma}\cdot\vec{\nabla}\psi\right], \tag{8.25}$$

$$\vec{P} = -\int d^3x \left(\pi\vec{\nabla}\phi + \pi^\dagger\vec{\nabla}\phi^\dagger + i\psi^\dagger\overline{\sigma}^0\vec{\nabla}\psi\right). \tag{8.26}$$

It generates spacetime translations on the fields X according to

$$[X, P^\mu] = i\partial^\mu X. \tag{8.27}$$

Rearranging the terms in eq. (8.24) using the Jacobi identity, we therefore have

$$\Big[[\epsilon_2 Q + \epsilon_2^\dagger Q^\dagger,\ \epsilon_1 Q + \epsilon_1^\dagger Q^\dagger],\ X\Big] = 2(-\epsilon_1\sigma_\mu\epsilon_2^\dagger + \epsilon_2\sigma_\mu\epsilon_1^\dagger)\,[P^\mu, X], \tag{8.28}$$

for any X, up to terms that vanish on-shell, so it must be that

$$[\epsilon_2 Q + \epsilon_2^\dagger Q^\dagger,\ \epsilon_1 Q + \epsilon_1^\dagger Q^\dagger] = 2(-\epsilon_1\sigma_\mu\epsilon_2^\dagger + \epsilon_2\sigma_\mu\epsilon_1^\dagger)\,P^\mu. \tag{8.29}$$

Now by expanding out eq. (8.29), one obtains the precise form of the supersymmetry algebra relations:

$$\{Q_\alpha, Q^\dagger_{\dot{\alpha}}\} = 2\sigma^\mu_{\alpha\dot{\alpha}} P_\mu, \tag{8.30}$$

$$\{Q_\alpha, Q_\beta\} = 0, \qquad \{Q^\dagger_{\dot{\alpha}}, Q^\dagger_{\dot{\beta}}\} = 0, \tag{8.31}$$

as promised in Section 7.2. [The commutator in eq. (8.29) turns into anticommutators in eqs. (8.30) and (8.31) in the process of extracting the anticommuting spinors ϵ_1 and ϵ_2.] The results

$$[Q_\alpha, P^\mu] = 0, \qquad [Q^\dagger_{\dot{\alpha}}, P^\mu] = 0 \tag{8.32}$$

follow immediately from eq. (8.27) and the fact that the supersymmetry transformations are global (independent of position in spacetime). This demonstration of the supersymmetry algebra in terms of the canonical generators Q and $Q^\dagger$ requires the use of the Hamiltonian equations of motion, but the symmetry itself is valid off-shell at the level of the Lagrangian, as we have already shown.

8.2 Interactions of Chiral Supermultiplets

In a realistic theory like the MSSM, there are many chiral supermultiplets, with both gauge and non-gauge interactions. In this section, our task is to construct the most general possible theory of masses and non-gauge interactions for particles that live in chiral supermultiplets. In the MSSM these are the quarks, squarks, leptons, sleptons, Higgs scalars, and higgsino fermions. We will find that the form of the non-gauge couplings, including mass terms, is highly restricted by the requirement that the action is invariant under supersymmetry transformations. (Gauge interactions will be dealt with in the following sections.)

Our starting point is the Lagrangian for a collection of free chiral supermultiplets labeled by an index i, which runs over all gauge and flavor degrees of freedom. Since we will want to construct an interacting theory with supersymmetry closing off-shell, each supermultiplet contains a complex scalar ϕ_i and a left-handed Weyl fermion ψ_i as physical degrees of freedom, plus a complex auxiliary field F_i, which does not propagate. The results of the previous section tell us that the free part of the Lagrangian is

$$\mathscr{L}_{\text{free}} = \partial^\mu \phi^{\dagger i} \partial_\mu \phi_i + i\psi^{\dagger i} \overline{\sigma}^\mu \partial_\mu \psi_i + F^{\dagger i} F_i, \tag{8.33}$$

where we sum over repeated indices i (not to be confused with the suppressed spinor indices), with the convention that fields ϕ_i and ψ_i always carry lowered indices, while their conjugates always carry raised indices. It is invariant under the supersymmetry transformation:

$$\delta_\epsilon \phi_i = \epsilon \psi_i\,, \qquad \delta_\epsilon \phi^{\dagger i} = \epsilon^\dagger \psi^{\dagger i}\,, \tag{8.34}$$

$$\delta_\epsilon (\psi_i)_\alpha = -i(\sigma^\mu \epsilon^\dagger)_\alpha\, \partial_\mu \phi_i + \epsilon_\alpha F_i\,, \qquad \delta_\epsilon (\psi^{\dagger i})_{\dot\alpha} = i(\epsilon \sigma^\mu)_{\dot\alpha}\, \partial_\mu \phi^{\dagger i} + \epsilon^\dagger_{\dot\alpha} F^{\dagger i}\,, \tag{8.35}$$

$$\delta_\epsilon F_i = -i\epsilon^\dagger \overline{\sigma}^\mu \partial_\mu \psi_i\,, \qquad \delta_\epsilon F^{\dagger i} = i\partial_\mu \psi^{\dagger i} \overline{\sigma}^\mu \epsilon\,. \tag{8.36}$$

We will now find the most general set of renormalizable interactions for these fields that is consistent with supersymmetry. We do this working in the field theory before integrating out the auxiliary fields. To begin, note that in order to be renormalizable by power counting, each term must have field content with total mass dimension ≤ 4. So, the only candidate terms are

$$\mathscr{L}_{\text{int}} = \left(-\tfrac{1}{2} W^{ij} \psi_i \psi_j + W^i F_i + x^{ij} F_i F_j\right) + \text{h.c.} - U, \tag{8.37}$$

where W^{ij}, W^i, x^{ij}, and U are polynomials in the scalar fields $\phi_i, \phi^{\dagger i}$, with degrees 1,

2, 0, and 4, respectively. [Terms of the form $F^{\dagger i}F_j$ are already included in eq. (8.33), with the coefficient fixed by the transformation rules (8.34)–(8.36).]

We must now require that $\mathscr{L}_{\text{int}}$ is invariant under the supersymmetry transformations, since $\mathscr{L}_{\text{free}}$ was already invariant by itself. This immediately requires that the candidate term $U(\phi_i, \phi^{\dagger i})$ must vanish. If there were such a term, then under a supersymmetry transformation eq. (8.34) it would transform into another function of the scalar fields only, multiplied by $\epsilon\psi_i$ or $\epsilon^\dagger\psi^{\dagger i}$, and with no spacetime derivatives or F_i, $F^{\dagger i}$ fields. It is easy to see from eqs. (8.34)–(8.37) that nothing of this form can possibly be canceled by the supersymmetry transformation of any other term in the Lagrangian. Similarly, the dimensionless coupling x^{ij} must be zero, because its supersymmetry transformation likewise cannot possibly be canceled by any other term. So, we are left with

$$\mathscr{L}_{\text{int}} = \left(-\tfrac{1}{2}W^{ij}\psi_i\psi_j + W^iF_i\right) + \text{h.c.} \tag{8.38}$$

as the only possibilities. At this point, we are not assuming that W^{ij} and W^i are related to each other in any way. However, soon we will find out that they *are* related, which is why we have chosen to use the same letter for them. Notice that eq. (1.68) tells us that W^{ij} is symmetric under $i \leftrightarrow j$.

It is easiest to divide the variation of $\mathscr{L}_{\text{int}}$ into several parts, which must cancel separately. First, we consider the part that contains four spinors:

$$\delta_\epsilon\mathscr{L}_{\text{int}}|_{4-\text{spinor}} = \left[-\frac{1}{2}\frac{\partial W^{ij}}{\partial\phi_k}(\epsilon\psi_k)(\psi_i\psi_j) - \frac{1}{2}\frac{\partial W^{ij}}{\partial\phi^{\dagger k}}(\epsilon^\dagger\psi^{\dagger k})(\psi_i\psi_j)\right] + \text{h.c.} \tag{8.39}$$

The term proportional to $(\epsilon\psi_k)(\psi_i\psi_j)$ cannot cancel against any other term. Fortunately, however, the Fierz identity eq. (1.305) implies

$$(\epsilon\psi_i)(\psi_j\psi_k) + (\epsilon\psi_j)(\psi_k\psi_i) + (\epsilon\psi_k)(\psi_i\psi_j) = 0, \tag{8.40}$$

so this contribution to $\delta_\epsilon\mathscr{L}_{\text{int}}$ vanishes identically if and only if $\partial W^{ij}/\partial\phi_k$ is totally symmetric under interchange of i, j, k. There is no such identity available for the term proportional to $(\epsilon^\dagger\psi^{\dagger k})(\psi_i\psi_j)$. Since that term cannot cancel with any other, requiring it to be absent just tells us that W^{ij} cannot contain $\phi^{\dagger k}$. In other words, W^{ij} is analytic (or holomorphic) in the complex fields ϕ_k.

Combining what we have learned so far, we can write

$$W^{ij} = M^{ij} + y^{ijk}\phi_k, \tag{8.41}$$

where M^{ij} is a symmetric mass matrix for the fermion fields, and y^{ijk} is a Yukawa coupling of a scalar ϕ_k and two fermions $\psi_i\psi_j$ that must be totally symmetric under interchange of i, j, k. It is therefore possible, and it turns out to be convenient, to write

$$W^{ij} = \frac{\partial^2 W}{\partial\phi_i\partial\phi_j}, \tag{8.42}$$

where we have introduced a useful object,

$$W = \tfrac{1}{2}M^{ij}\phi_i\phi_j + \tfrac{1}{6}y^{ijk}\phi_i\phi_j\phi_k, \tag{8.43}$$

called the *superpotential*. This is not a scalar potential in the ordinary sense; in fact, it is not even real. It is instead an analytic function of the scalar fields ϕ_i treated as complex variables.

Continuing on our vaunted quest, we next consider the parts of $\delta_\epsilon \mathcal{L}_{\text{int}}$ that contain a spacetime derivative:

$$\delta_\epsilon \mathcal{L}_{\text{int}}|_\partial = \left(iW^{ij}\partial_\mu\phi_j\,\psi_i\sigma^\mu\epsilon^\dagger + iW^i\,\partial_\mu\psi_i\sigma^\mu\epsilon^\dagger\right) + \text{h.c.} \tag{8.44}$$

Here we have used the identity eq. (1.107) on the second term, which came from $(\delta_\epsilon F_i)W^i$. Now we can use eq. (8.42) to observe that

$$W^{ij}\partial_\mu\phi_j = \partial_\mu\left(\frac{\partial W}{\partial\phi_i}\right). \tag{8.45}$$

Therefore, eq. (8.44) will be a total derivative if

$$W^i = \frac{\partial W}{\partial\phi_i} = M^{ij}\phi_j + \tfrac{1}{2}y^{ijk}\phi_j\phi_k\,, \tag{8.46}$$

which explains why we chose its name as we did. The remaining terms in $\delta_\epsilon\mathcal{L}_{\text{int}}$ are all linear in F_i or $F^{\dagger i}$, and it is easy to show that they cancel, given the results for W^i and W^{ij} that we have already found.

Actually, we can include a linear term in the superpotential without disturbing the validity of the previous discussion at all:

$$W = L^i\phi_i + \tfrac{1}{2}M^{ij}\phi_i\phi_j + \tfrac{1}{6}y^{ijk}\phi_i\phi_j\phi_k. \tag{8.47}$$

Here L^i are parameters with dimensions of $[\text{mass}]^2$, which affect only the scalar potential part of the Lagrangian. Such linear terms are only allowed when ϕ_i is a gauge singlet, and there are no such gauge-singlet chiral supermultiplets in the MSSM with minimal field content. We will therefore omit this term from the remaining discussion of this section. However, this type of term does play an important role in the discussion of spontaneous supersymmetry breaking, as we will see later.

To recap, we have found that the most general non-gauge interactions for chiral supermultiplets are determined by a single holomorphic function of the complex scalar fields, the superpotential W. The auxiliary fields F_i and $F^{\dagger i}$ can be eliminated using their classical equations of motion. The part of $\mathcal{L}_{\text{free}} + \mathcal{L}_{\text{int}}$ that contains the auxiliary fields is $F_iF^{\dagger i} + W^iF_i + W_i^\dagger F^{\dagger i}$, leading to the equations of motion

$$F_i = -W_i^\dagger\,, \qquad F^{\dagger i} = -W^i\,. \tag{8.48}$$

Thus the auxiliary fields are expressible algebraically (without any derivatives) in terms of the scalar fields.

After making the replacement[2] eq. (8.48) in $\mathcal{L}_{\text{free}} + \mathcal{L}_{\text{int}}$, we obtain the Lagrangian,

$$\mathcal{L} = \partial^\mu\phi^{\dagger i}\partial_\mu\phi_i + i\psi^{\dagger i}\overline{\sigma}^\mu\partial_\mu\psi_i - \tfrac{1}{2}\left(W^{ij}\psi_i\psi_j + W_{ij}^\dagger\psi^{\dagger i}\psi^{\dagger j}\right) - W^iW_i^\dagger. \tag{8.49}$$

[2] Since F_i and $F^{\dagger i}$ appear only quadratically in the action, the result of instead doing a functional integral over them at the quantum level has precisely the same effect.

Now that the nonpropagating fields $F_i, F^{\dagger i}$ have been eliminated, it follows from eq. (8.49) that the scalar potential for the theory is just given in terms of the superpotential by

$$\begin{aligned} V(\phi,\phi^\dagger) &= W^k W_k^\dagger = F^{\dagger k} F_k \\ &= M^*_{ik} M^{kj} \phi^{\dagger i}\phi_j + \tfrac{1}{2} M^{in} y^*_{jkn} \phi_i \phi^{\dagger j}\phi^{\dagger k} \\ &\quad + \tfrac{1}{2} M^*_{in} y^{jkn} \phi^{\dagger i}\phi_j\phi_k + \tfrac{1}{4} y^{ijn} y^*_{kln} \phi_i\phi_j\phi^{\dagger k}\phi^{\dagger l}\,. \end{aligned} \tag{8.50}$$

This scalar potential is automatically bounded from below; in fact, since it is a sum of squares of absolute values (of the W^k), it is always nonnegative. If we substitute the general form for the superpotential eq. (8.43) into eq. (8.49), we obtain, for the full Lagrangian,

$$\begin{aligned} \mathscr{L} &= \partial^\mu \phi^{\dagger i} \partial_\mu \phi_i - V(\phi,\phi^\dagger) + i\psi^{\dagger i}\bar\sigma^\mu \partial_\mu \psi_i - \tfrac{1}{2} M^{ij}\psi_i\psi_j - \tfrac{1}{2} M^*_{ij} \psi^{\dagger i}\psi^{\dagger j} \\ &\quad - \tfrac{1}{2} y^{ijk}\phi_i\psi_j\psi_k - \tfrac{1}{2} y^*_{ijk}\phi^{\dagger i}\psi^{\dagger j}\psi^{\dagger k}. \end{aligned} \tag{8.51}$$

Now we can compare the masses of the fermions and scalars by looking at the linearized equations of motion:

$$\partial^\mu\partial_\mu \phi_i = -M^*_{ik} M^{kj}\phi_j + \cdots , \tag{8.52}$$

$$i\bar\sigma^\mu \partial_\mu \psi_i = M^*_{ij}\psi^{\dagger j} + \cdots , \qquad i\sigma^\mu\partial_\mu \psi^{\dagger i} = M^{ij}\psi_j + \cdots . \tag{8.53}$$

One can eliminate ψ in terms of $\psi^\dagger$ and vice versa in eq. (8.53), obtaining [after use of the identities eqs. (1.88) and (1.89)]

$$\partial^\mu\partial_\mu \psi_i = -M^*_{ik} M^{kj}\psi_j + \cdots , \qquad \partial^\mu\partial_\mu \psi^{\dagger j} = -\psi^{\dagger i} M^*_{ik} M^{kj} + \cdots . \tag{8.54}$$

Therefore, the fermions and the bosons satisfy the same wave equation with exactly the same squared-mass matrix with real nonnegative eigenvalues, namely $(M^2)_i{}^j = M^*_{ik}M^{kj}$. It follows that diagonalizing this matrix by redefining the fields with a unitary matrix gives a collection of chiral supermultiplets, each of which contains a mass-degenerate complex scalar and Weyl fermion, in agreement with the general argument in Chapter 7.

8.3 Supersymmetric Gauge Theories

The propagating degrees of freedom in a gauge supermultiplet are a massless gauge boson field A^a_μ and a two-component Weyl fermion gaugino λ^a. The index a here runs over the adjoint representation of the gauge group [$a \in \{1,\ldots,8\}$ for $\mathrm{SU}(3)_C$ color gluons and gluinos; $a \in \{1,2,3\}$ for $\mathrm{SU}(2)_L$ weak isospin; $a = 1$ for $\mathrm{U}(1)_Y$ weak hypercharge]. The gauge transformations of the vector supermultiplet fields are

$$A^a_\mu \to A^a_\mu + \partial_\mu \Lambda^a - g f^{abc} A^b_\mu \Lambda^c , \tag{8.55}$$

$$\lambda^a \to \lambda^a - g f^{abc}\lambda^b\Lambda^c , \tag{8.56}$$

Table 8.2 Counting of real degrees of freedom for each gauge supermultiplet.

	A_μ	λ	D
on-shell ($n_B = n_F = 2$)	2	2	0
off-shell ($n_B = n_F = 4$)	3	4	1

where Λ^a is an infinitesimal gauge transformation parameter, g is the gauge coupling, and f^{abc} are the totally antisymmetric structure constants that define the gauge group. The special case of an abelian group is obtained by just setting $f^{abc} = 0$; the corresponding gaugino is a gauge singlet in that case. The conventions are such that for QED, $A^\mu = (V, \vec{\boldsymbol{A}})$ where V and $\vec{\boldsymbol{A}}$ are the usual electric potential and vector potential, with electric and magnetic fields given by $\vec{\boldsymbol{E}} = -\vec{\boldsymbol{\nabla}} V - \partial_0 \vec{\boldsymbol{A}}$ and $\vec{\boldsymbol{B}} = \vec{\boldsymbol{\nabla}} \times \vec{\boldsymbol{A}}$.

The on-shell degrees of freedom for A^a_μ and λ^a_α amount to two bosonic and two fermionic helicity states (for each a), as required by supersymmetry. However, off-shell λ^a_α consists of two complex, or four real, fermionic degrees of freedom, while A^a_μ only has three real bosonic degrees of freedom; one degree of freedom is removed by the inhomogeneous gauge transformation eq. (8.55). So, we will need one real bosonic auxiliary field, traditionally called D^a, in order for supersymmetry to be consistent off-shell. This field also transforms as an adjoint of the gauge group [i.e., like eq. (8.56) with λ^a replaced by D^a] and satisfies $(D^a)^\dagger = D^a$. Like the chiral auxiliary fields F_i, the gauge auxiliary field D^a has dimensions of [mass]^2 and no kinetic term, so it can be eliminated on-shell using its algebraic equation of motion. The counting of degrees of freedom is summarized in Table 8.2.

Therefore, the Lagrangian for a gauge supermultiplet ought to be

$$\mathscr{L}_{\text{gauge}} = -\tfrac{1}{4} F^a_{\mu\nu} F^{\mu\nu a} + i\lambda^{\dagger a} \bar{\sigma}^\mu \nabla_\mu \lambda^a + \tfrac{1}{2} D^a D^a, \tag{8.57}$$

where

$$F^a_{\mu\nu} = \partial_\mu A^a_\nu - \partial_\nu A^a_\mu - g f^{abc} A^b_\mu A^c_\nu \tag{8.58}$$

is the usual Yang–Mills field strength, and

$$\nabla_\mu \lambda^a = \partial_\mu \lambda^a - g f^{abc} A^b_\mu \lambda^c \tag{8.59}$$

is the covariant derivative of the gaugino field. To check that eq. (8.57) is really supersymmetric, one must specify the supersymmetry transformations of the fields. The forms of these follow from the requirements that they should be linear in the infinitesimal parameters $\epsilon, \epsilon^\dagger$ with dimensions of [mass]$^{-1/2}$, that $\delta_\epsilon A^a_\mu$ is real, and that $\delta_\epsilon D^a$ should be real and proportional to the field equations for the gaugino, in analogy with the role of the auxiliary field F in the chiral supermultiplet case. Thus one can guess, up to multiplicative factors, that[3]

$$\delta_\epsilon A^a_\mu = -\frac{1}{\sqrt{2}}\left(\epsilon^\dagger \overline{\sigma}_\mu \lambda^a + \lambda^{\dagger a}\overline{\sigma}_\mu \epsilon\right), \tag{8.60}$$

$$\delta_\epsilon \lambda^a_\alpha = -\frac{i}{2\sqrt{2}}(\sigma^\mu \overline{\sigma}^\nu \epsilon)_\alpha\, F^a_{\mu\nu} + \frac{1}{\sqrt{2}}\epsilon_\alpha\, D^a, \tag{8.61}$$

$$\delta_\epsilon D^a = \frac{i}{\sqrt{2}}\left(-\epsilon^\dagger \overline{\sigma}^\mu \nabla_\mu \lambda^a + \nabla_\mu \lambda^{\dagger a}\overline{\sigma}^\mu \epsilon\right). \tag{8.62}$$

The factors of $\sqrt{2}$ are chosen so that the action obtained by integrating $\mathscr{L}_{\text{gauge}}$ is indeed invariant, and the phase of λ^a is chosen for future convenience in treating the MSSM.

It is now a little bit tedious, but straightforward, to also check that

$$(\delta_{\epsilon_2}\delta_{\epsilon_1} - \delta_{\epsilon_1}\delta_{\epsilon_2})X = i(-\epsilon_1 \sigma^\mu \epsilon_2^\dagger + \epsilon_2 \sigma^\mu \epsilon_1^\dagger)\nabla_\mu X \tag{8.63}$$

for X equal to any of the gauge-covariant fields $F^a_{\mu\nu}$, λ^a, $\lambda^{\dagger a}$, D^a, as well as for arbitrary covariant derivatives acting on them. This ensures that the supersymmetry algebra of eqs. (8.30)–(8.31) is realized on gauge-invariant combinations of fields in gauge supermultiplets, as they were on the chiral supermultiplets [see eq. (8.16)]. This check requires the use of identities of eqs. (1.85), (1.90), and (1.114). If we had not included the auxiliary field D^a, then the supersymmetry algebra of eq. (8.63) would hold only after using the equations of motion for λ^a and $\lambda^{\dagger a}$. The auxiliary field satisfies a trivial equation of motion, $D^a = 0$, but this is modified if one couples the gauge supermultiplets to chiral supermultiplets, as we now do.

8.4 Gauge Interactions for Chiral Supermultiplets

Now we are ready to consider a general Lagrangian for a supersymmetric theory with both chiral and gauge supermultiplets. Suppose that the chiral supermultiplets transform under the gauge group in a representation with hermitian matrices $(\boldsymbol{T^a})_i{}^j$ satisfying $[\boldsymbol{T^a}, \boldsymbol{T^b}] = if^{abc}\boldsymbol{T^c}$. Since supersymmetry and gauge transformations commute, the scalar, fermion, and auxiliary fields must be in the same representation of the gauge group. Under an infinitesimal gauge transformation, the fields of the chiral supermultiplet transform as

$$X_i \to X_i - ig\Lambda^a(\boldsymbol{T^a}X)_i \tag{8.64}$$

[3] The supersymmetry transformations eqs. (8.60)–(8.62) are nonlinear for nonabelian gauge symmetries, since there are gauge fields in the covariant derivatives acting on the gaugino fields and in the field strength $F^a_{\mu\nu}$. By adding even more auxiliary fields besides D^a, one can make the supersymmetry transformations linear in the fields; this is easiest to see in the superfield language, as we will see later. The version here, in which those extra auxiliary fields have been removed by supergauge transformations, is called the Wess–Zumino gauge [89]. Alternatively, the supersymmetry transformations become linear if one eliminates all of the auxiliary fields and imposes the (noncovariant) light-cone gauge-fixing prescription; e.g., see Ref. [90].

for $X_i = \phi_i, \psi_i, F_i$. To construct a gauge-invariant Lagrangian, we now need to replace the ordinary derivatives in eq. (8.33) with covariant derivatives:

$$\nabla_\mu \phi_i = \partial_\mu \phi_i + igA^a_\mu (\boldsymbol{T^a}\phi)_i \,, \tag{8.65}$$
$$\nabla_\mu \phi^{\dagger i} = \partial_\mu \phi^{\dagger i} - igA^a_\mu (\phi^\dagger \boldsymbol{T^a})^i \,, \tag{8.66}$$
$$\nabla_\mu \psi_i = \partial_\mu \psi_i + igA^a_\mu (\boldsymbol{T^a}\psi)_i \,. \tag{8.67}$$

Naively, this simple procedure achieves the goal of coupling the vector bosons in the gauge supermultiplet to the scalars and fermions in the chiral supermultiplets. However, we also have to consider whether there are any other interactions allowed by gauge invariance and involving the gaugino and D^a fields, which might have to be included to make a supersymmetric Lagrangian. Since A^a_μ couples to ϕ_i and ψ_i, it makes sense that λ^a and D^a should as well.

In fact, there are three such possible interaction terms that are renormalizable (of field mass dimension ≤ 4), namely

$$(\phi^\dagger \boldsymbol{T^a}\psi)\lambda^a, \qquad \lambda^{\dagger a}(\psi^\dagger \boldsymbol{T^a}\phi), \qquad \text{and} \qquad (\phi^\dagger \boldsymbol{T^a}\phi)D^a. \tag{8.68}$$

Now one can add these terms, with unknown dimensionless coupling coefficients, to the Lagrangians for the chiral and gauge supermultiplets, and demand that the whole mess be real and invariant under supersymmetry, up to a total derivative. Not surprisingly, this is possible only if the supersymmetry transformation laws for the matter fields are modified to include gauge-covariant rather than ordinary derivatives. Also, it is necessary to include one strategically chosen extra term in $\delta_\epsilon F_i$, so

$$\delta_\epsilon \phi_i = \epsilon\psi_i \,, \tag{8.69}$$
$$\delta_\epsilon \psi_{i\alpha} = -i(\sigma^\mu \epsilon^\dagger)_\alpha \, \nabla_\mu \phi_i + \epsilon_\alpha F_i \,, \tag{8.70}$$
$$\delta_\epsilon F_i = -i\epsilon^\dagger \overline{\sigma}^\mu \nabla_\mu \psi_i \; + \; \sqrt{2} g (\boldsymbol{T^a}\phi)_i \, \epsilon^\dagger \lambda^{\dagger a} \,. \tag{8.71}$$

One can now fix the coefficients for the terms in eq. (8.68), with the result that the full Lagrangian for a renormalizable supersymmetric theory is

$$\mathcal{L} = \mathcal{L}_{\text{chiral}} + \mathcal{L}_{\text{gauge}} - \sqrt{2}g(\phi^\dagger \boldsymbol{T^a}\psi)\lambda^a - \sqrt{2}g\lambda^{\dagger a}(\psi^\dagger \boldsymbol{T^a}\phi) + g(\phi^\dagger \boldsymbol{T^a}\phi)D^a. \tag{8.72}$$

Here $\mathcal{L}_{\text{chiral}}$ means the chiral supermultiplet Lagrangian found in Section 8.2 [see eq. (8.49) or (8.51)], but with ordinary derivatives replaced everywhere by gauge-covariant derivatives, and $\mathcal{L}_{\text{gauge}}$ was given in eq. (8.57). To prove that eq. (8.72) is invariant under the supersymmetry transformations, one must use the identity

$$W^i (\boldsymbol{T^a}\phi)_i = 0. \tag{8.73}$$

This is precisely the condition that must be satisfied anyway in order for the superpotential, and thus $\mathcal{L}_{\text{chiral}}$, to be gauge invariant, since the left side is proportional to the gauge variation of W.

The last three terms in eq. (8.72) consist of interactions whose strengths are fixed to be gauge couplings by the requirements of supersymmetry, even though they are not gauge interactions from the point of view of an ordinary field theory. The first two terms are a direct coupling of gauginos to matter fields, corresponding to the "supersymmetrization" of the usual gauge boson couplings to matter fields. The last term combines with the $\frac{1}{2}D^a D^a$ term in $\mathscr{L}_{\text{gauge}}$ to provide an equation of motion,

$$D^a = -g(\phi^\dagger \boldsymbol{T^a} \phi). \tag{8.74}$$

Thus, like the auxiliary fields F_i and $F^{\dagger i}$, the D^a are expressible purely algebraically in terms of the scalar fields. Replacing the auxiliary fields in eq. (8.72) using eq. (8.74), one finds that the complete scalar potential is (recall that $\mathscr{L}$ contains $-V$)

$$V(\phi, \phi^\dagger) = F^{\dagger i} F_i + \tfrac{1}{2} \sum_a D^a D^a = W_i^\dagger W^i + \tfrac{1}{2} \sum_a g_a^2 (\phi^\dagger \boldsymbol{T^a} \phi)^2. \tag{8.75}$$

The two types of terms in this expression are called "F-term" and "D-term" contributions, respectively. In the second term in eq. (8.75), we have now written an explicit sum $\sum_a$ to cover the case that the gauge group has several distinct factors with different gauge couplings g_a. [For instance, in the MSSM the three gauge groups, $\mathrm{SU}(3)_C$, $\mathrm{SU}(2)_L$, and $\mathrm{U}(1)_Y$, have different gauge couplings g_3, g, and g'.] Since $V(\phi, \phi^\dagger)$ is a sum of squares, it is always greater than or equal to zero for every field configuration. It is an interesting and unique feature of supersymmetric theories that the scalar potential is completely determined by the *other* interactions in the theory. The F-terms are fixed by Yukawa couplings and fermion mass terms, and the D-terms are fixed by the gauge interactions.

Using Noether's procedure [see eq. (8.17)], one finds the conserved supercurrent,

$$\begin{aligned} J^\mu_\alpha = {} & (\sigma^\nu \overline{\sigma}^\mu \psi_i)_\alpha \nabla_\nu \phi^{\dagger i} + i(\sigma^\mu \psi^{\dagger i})_\alpha W_i^\dagger \\ & + \frac{1}{2\sqrt{2}} (\sigma^\nu \overline{\sigma}^\rho \sigma^\mu \lambda^{\dagger a})_\alpha F^a_{\nu\rho} + \frac{i}{\sqrt{2}} g_a \phi^\dagger \boldsymbol{T^a} \phi \, (\sigma^\mu \lambda^{\dagger a})_\alpha , \end{aligned} \tag{8.76}$$

generalizing the expression given in eq. (8.18) for the Wess–Zumino model. This result will be useful when we discuss certain aspects of spontaneous supersymmetry breaking later.

8.5 Summary: How to Build a Supersymmetric Model

In a renormalizable supersymmetric field theory, the interactions and masses of all particles are determined just by their gauge transformation properties and by the superpotential W. By construction, we found that W had to be an analytic function of the complex scalar fields ϕ_i, which are always defined to transform under supersymmetry into left-handed Weyl fermions. In an equivalent language, to be developed later, W is said to be a function of chiral *superfields*. A superfield is a single object that contains as components all of the bosonic, fermionic, and auxiliary fields within the corresponding supermultiplet, for example $\Phi_i \supset (\phi_i, \psi_i, F_i)$. (This is analogous to the way in which one often describes a weak isospin doublet or color triplet by a multicomponent field.) The gauge quantum numbers and the mass dimension of a chiral superfield are the same as those of its scalar component. In the superfield formulation, one writes, instead of eq. (8.47),

$$W = L^i \Phi_i + \tfrac{1}{2} M^{ij} \Phi_i \Phi_j + \tfrac{1}{6} y^{ijk} \Phi_i \Phi_j \Phi_k, \tag{8.77}$$

which implies exactly the same physics. The specification of the superpotential is really a code for the terms that it implies in the Lagrangian, so the reader may feel free to think of the superpotential either as a function of the scalar fields ϕ_i or as the same function of the superfields Φ_i. The derivation of all of our preceding results can be obtained somewhat more elegantly using superfield methods, which have the advantage of making invariance under supersymmetry transformations manifest by defining the Lagrangian in terms of integrals over a "superspace" with fermionic as well as ordinary commuting coordinates. In this chapter, we have avoided this extra layer of notation on purpose, in favor of the more pedestrian, but more familiar and accessible, component-field approach. The latter is at least more appropriate for making contact with phenomenology in a universe with supersymmetry breaking. Superfield methods will be discussed in the next chapter.

Given the supermultiplet content of the theory, the form of the superpotential is restricted by the requirement of gauge invariance [see eq. (8.73)]. In any given theory, only a subset of the parameters L^i, M^{ij}, and y^{ijk} are allowed to be nonzero. The parameter L^i is only allowed if Φ_i is a gauge singlet. (There are no such chiral supermultiplets in the MSSM with the minimal field content.) The entries of the mass matrix M^{ij} can only be nonzero for i and j such that the supermultiplets Φ_i and Φ_j transform under the gauge group in representations that are conjugates of each other. (In the MSSM there is only one such term, as we will see.) Likewise, the Yukawa couplings y^{ijk} can only be nonzero when Φ_i, Φ_j, and Φ_k transform in representations that can combine to form a singlet.

The interactions implied by the superpotential eq. (8.77) (with $L^i = 0$) were listed in eqs. (8.50) and (8.51), and are shown[4] in Figs. 8.1 and 8.2. Those in Fig. 8.1 are all determined by the dimensionless parameters y^{ijk}. The Yukawa interaction in Fig. 8.1(a) corresponds to the next-to-last term in eq. (8.51). For each particular Yukawa coupling of $\phi_i \psi_j \psi_k$ with strength y^{ijk}, there must be equal couplings of $\phi_j \psi_i \psi_k$ and $\phi_k \psi_i \psi_j$, since y^{ijk} is completely symmetric under interchange of any two of its indices, as shown in Section 8.2. The arrows on the fermion and scalar lines point in the direction of propagation of ϕ and ψ and opposite the direction of propagation of $\phi^\dagger$ and $\psi^\dagger$. Thus there is also a vertex corresponding to the one in Fig. 8.1(a) but with all arrows reversed, corresponding to the complex conjugate

[4] Here, the auxiliary fields have been eliminated using their equations of motion ("integrated out"). One could instead give Feynman rules that include the auxiliary fields, or directly in terms of superfields on superspace, although this is usually less useful in practical phenomenological applications.

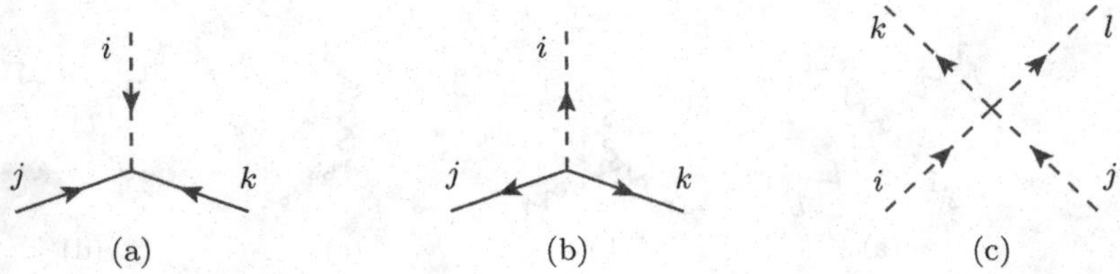

Fig. 8.1 The dimensionless non-gauge interaction vertices in a supersymmetric theory: (a) scalar–fermion–fermion Yukawa interaction y^{ijk}, (b) the complex conjugate interaction y^*_{ijk}, and (c) quartic scalar interaction $y^{ijn}y^*_{kln}$.

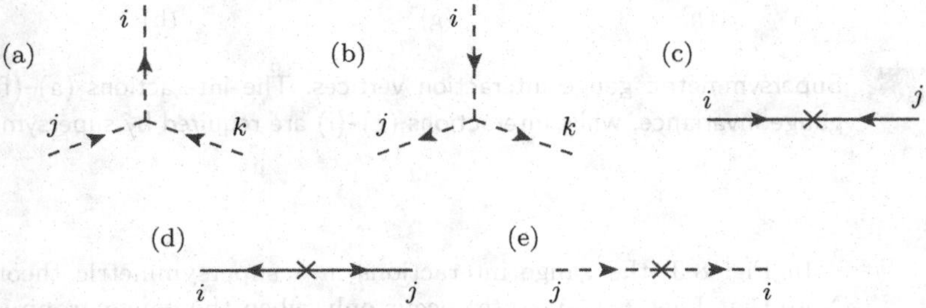

Fig. 8.2 Dimensionful supersymmetric couplings: (a) (scalar)^3 interaction vertex $M^*_{in}y^{jkn}$ and (b) the conjugate interaction $M^{in}y^*_{jkn}$, (c) fermion mass term M^{ij} and (d) conjugate fermion mass term M^*_{ij}, and (e) scalar squared-mass term $M^*_{ik}M^{kj}$.

[the last term in eq. (8.51)]. It is shown in Fig. 8.1(b). There is also a dimensionless coupling for $\phi_i\phi_j\phi^{\dagger k}\phi^{\dagger l}$, with strength $y^{ijn}y^*_{kln}$, as required by supersymmetry [see the last term in eq. (8.50)]. The relationship between the Yukawa interactions in Figs. 8.1(a),(b) and the scalar interaction of Fig. 8.1(c) is exactly of the special type needed to cancel the quadratic divergences in quantum corrections to scalar masses, as discussed in Chapter 7 [see Fig. 7.1 and eq. (7.11)].

Figure 8.2 shows the only interactions corresponding to renormalizable and supersymmetric vertices with coupling dimensions of [mass] and [mass]^2. First, there are (scalar)^3 couplings in Figs. 8.2(a),(b), which are entirely determined by the superpotential mass parameters M^{ij} and Yukawa couplings y^{ijk}, as indicated by the second and third terms in eq. (8.50). The propagators of the fermions and scalars in the theory are constructed in the usual way using the fermion mass M^{ij} and scalar squared mass $M^*_{ik}M^{kj}$. The fermion mass terms M^{ij} and M_{ij} each lead to a chirality-changing insertion in the fermion propagator; note the directions of the arrows in Fig. 8.2(c),(d). There is no such arrow-reversal for a scalar propagator in a theory with exact supersymmetry; as depicted in Fig. 8.2(e), if one treats the scalar squared-mass term as an insertion in the propagator, the arrow direction is preserved.

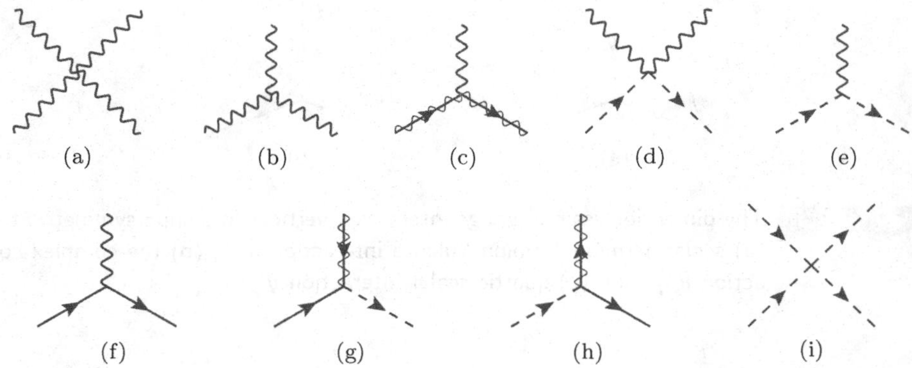

Fig. 8.3 Supersymmetric gauge interaction vertices. The interactions (a)–(f) are dictated by gauge invariance, while interactions (g)–(i) are required by supersymmetry.

In Fig. 8.3, the gauge interactions in a supersymmetric theory are exhibited. Note that Figs. 8.3(a),(b),(c) occur only when the gauge group is nonabelian, for example for $SU(3)_C$ color and $SU(2)_L$ weak isospin in the MSSM. Figs 8.3(a) and 8.3(b) are the interactions of gauge bosons, which derive from the first term in eq. (8.57). In the MSSM these are exactly the same as the well-known QCD gluon and electroweak gauge boson vertices of the Standard Model. (We do not show the interactions of ghost fields, which are necessary only for consistent loop amplitudes.) Figs. 8.3(c),(d),(e),(f) are just the standard interactions between gauge bosons and fermion and scalar fields that must occur in any gauge theory because of the form of the covariant derivative; they come from eqs. (8.59) and (8.65)–(8.67) inserted in the kinetic part of the Lagrangian. Figure 8.3(c) shows the coupling of a gaugino to a gauge boson; the gaugino line in a Feynman diagram is traditionally drawn as a solid fermion line superimposed on a wavy line. In Fig. 8.3(g) we have the coupling of a gaugino to a chiral fermion and a complex scalar [the first term in the second line of eq. (8.72)]. One can think of this as the "supersymmetrization" of Figs. 8.3(e) or 8.3(f); any of these three vertices may be obtained from any other (up to a factor of $\sqrt{2}$) by replacing two of the particles by their supersymmetric partners. There is also an interaction in Fig. 8.3(h) which is just like Fig. 8.3(g) but with all arrows reversed, corresponding to the complex conjugate term in the Lagrangian [the second term in the second line in eq. (8.72)]. Finally in Fig. 8.3(i) we have a scalar quartic interaction vertex [the last term in eq. (8.75)], which is also determined by the gauge coupling.

The results of this section can be used as a recipe for constructing the supersymmetric interactions for any model. In the case of the MSSM, we already know the gauge group, particle content and the gauge transformation properties, so it only remains to decide on the superpotential. This we will do in Section 13.2.

8.6 Soft Supersymmetry-Breaking Interactions

A realistic phenomenological model must contain supersymmetry breaking. From a theoretical perspective, we expect that supersymmetry, if it exists at all, should be an exact symmetry that is broken spontaneously. In other words, the underlying model should have a Lagrangian that is invariant under supersymmetry, but a vacuum state that is not. In this way, supersymmetry is hidden at low energies in a manner analogous to the electroweak symmetry of the Standard Model.

Many models of spontaneous symmetry breaking have indeed been proposed and we will mention the basic ideas of some of them later. These always involve extending the MSSM to include new particles and interactions at very high mass scales, and there is presently no clear consensus on exactly how this should be done. However, from a practical point of view, it is extremely useful to simply parameterize our ignorance of these issues by just introducing extra terms that break supersymmetry explicitly in the effective MSSM Lagrangian. As was argued in Chapter 7, the supersymmetry-breaking couplings should be soft (of positive mass dimension) in order to be able to naturally maintain a hierarchy between the electroweak scale and the Planck (or any other very large) mass scale. This means in particular that dimensionless supersymmetry-breaking couplings should be absent.

The possible soft supersymmetry-breaking terms in the Lagrangian of a general theory are[5]

$$\mathscr{L}_{\text{soft}} = -\left(\tfrac{1}{2}M_a\,\lambda^a\lambda^a + \tfrac{1}{6}a^{ijk}\phi_i\phi_j\phi_k + \tfrac{1}{2}b^{ij}\phi_i\phi_j + t^i\phi_i + \text{h.c.}\right) - (m^2)^i_j\phi^{j\dagger}\phi_i\,, \tag{8.78}$$

$$\mathscr{L}_{\text{maybe soft}} = -\tfrac{1}{2}c_i^{jk}\phi^{\dagger i}\phi_j\phi_k + \text{h.c.} \tag{8.79}$$

They consist of gaugino masses M_a for each gauge group, scalar squared-mass terms $(m^2)^j_i$ and b^{ij}, and (scalar)^3 couplings a^{ijk} and c_i^{jk}, and "tadpole" couplings t^i. The tadpole can only occur if ϕ_i is a gauge singlet, and so is absent from the MSSM. One might wonder why we have not included possible soft mass terms for the chiral supermultiplet fermions, like $\mathscr{L} = -\frac{1}{2}m^{ij}\psi_i\psi_j + \text{h.c.}$. The answer is that including such terms would be redundant; they can always be absorbed into a redefinition of the superpotential masses and the terms $(m^2)^i_j$ and c_i^{jk}.

A softly broken supersymmetric theory with $\mathscr{L}_{\text{soft}}$ as given by eq. (8.78) is indeed free of quadratic divergences in quantum corrections to scalar masses, to all orders in perturbation theory. For example, consider the one-loop effective potential for a gauge theory coupled to matter,

$$V_{\text{eff}}(\phi) = V_{\text{scalar}}(\phi) + V^{(1)}(\phi)\,. \tag{8.80}$$

If we regulate the divergence of the one-loop correction by a momentum cutoff Λ,

[5] A common notation in the literature defines $A^{ijk} \equiv a^{ijk}y^{ijk}$ and $B^{ij} \equiv b^{ij}M^{ij}$, where y^{ijk} and M^{ij} appear in the superpotential [see eq. (8.77)]. We shall not adopt this notation in this book (unless specifically indicated), for reasons explained in footnote 3 in Section 13.7.

then

$$V^{(1)}(\phi) = \frac{\Lambda^2}{32\pi^2}\,\mathrm{STr}\left\{M_n^2(\phi)\right\} + \frac{1}{64\pi^2}\,\mathrm{STr}\left\{M_n^4(\phi)\left[\ln\left(\frac{M_n^2(\phi)}{\Lambda^2}\right) - \frac{1}{2}\right]\right\}, \quad (8.81)$$

where $M_n^2(\phi)$ are the relevant field-dependent squared-mass eigenvalues, labeled by n, for spin 0, 1/2, and 1. "Field-dependent" means that the scalar vacuum expectation values are replaced by the corresponding scalar fields, ϕ. In eq. (8.81), we have introduced the *supertrace* of a function of the squared-mass eigenvalues, defined in general by a weighted sum over all squared-mass eigenstates with spin J:

$$\mathrm{STr}\left\{f(M^2)\right\} \equiv \sum_J (-1)^J(2J+1)\mathrm{Tr}[f(M_J^2)], \quad (8.82)$$

As we will find in Section 12.1 [see the discussion surrounding eq. (12.17)], in the presence of spontaneously broken supersymmetry (assuming that all U(1) generators are traceless), we have $\mathrm{STr}\, M^2 = 0$, in which case the quadratic divergences [i.e., the terms proportional to Λ^2 in eq. (8.81)] cancel exactly. For the case of explicit supersymmetry breaking, Girardello and Grisaru showed in Ref. [91] that if only the soft supersymmetry-breaking terms in eq. (8.78) are present, then $\mathrm{STr}\, M^2(\phi)$ is a constant independent of the scalar fields, ϕ. Such terms shift the vacuum energy, but in the context of quantum field theory neglecting gravity, they have no observable effect. In contrast, hard supersymmetry-breaking terms (for example, dimensionless couplings that break supersymmetry) will generate quadratically divergent terms in $V^{(1)}$ that are scalar-field-dependent. This is a signal that some of the parameters of the low-energy effective theory are quadratically sensitive to ultraviolet physics.

The situation is slightly more subtle if one includes the nonanalytic (scalar)^3 couplings in $\mathscr{L}_{\text{maybe soft}}$. If any of the chiral supermultiplets in the theory are singlets under all gauge symmetries, then nonzero c_i^{jk} terms can lead to quadratic divergences, despite the fact that they are formally soft according to their mass dimension. (The same is true for fermion mass terms, since as remarked above they can always be exchanged for a redefinition of the superpotential masses and c_i^{jk} and $(m^2)^i_{\ j}$.) Now, this constraint need not apply to the minimal supersymmetric extension of the Standard Model, which does not contain any gauge-singlet chiral supermultiplets. Nevertheless, the possibility of c_i^{jk} terms is nearly always neglected. The real reason for this is that it is difficult to construct models of spontaneous supersymmetry breaking in which the c_i^{jk} are not negligibly small, as explained in Section 14.3 below.

In the special case of a theory that has chiral supermultiplets that are singlets or in the adjoint representation of a simple factor of the gauge group, then there are also possible soft supersymmetry-breaking Dirac mass terms between the corresponding fermions ψ_a and the gauginos:

$$\mathscr{L} = -M^a_{\text{Dirac}}\lambda^a\psi_a + \text{h.c.} \quad (8.83)$$

This is not relevant for the MSSM with minimal field content, which does not have

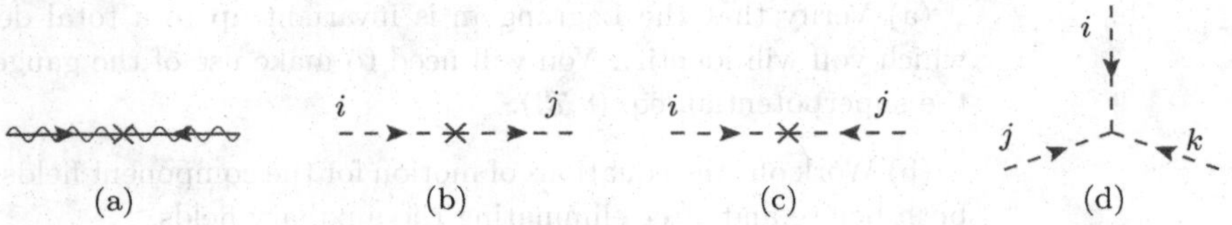

Fig. 8.4 Soft supersymmetry-breaking terms: (a) gaugino mass M_a, (b) nonanalytic scalar squared mass $(m^2)^i_j$, (c) analytic scalar squared mass b^{ij}, and (d) scalar cubic coupling a^{ijk}.

adjoint representation chiral supermultiplets. Therefore, eq. (8.78) is usually taken to be the general form of the soft supersymmetry-breaking Lagrangian.

The terms in $\mathscr{L}_{\text{soft}}$ clearly do break supersymmetry, because they involve only scalars and gauginos and not their respective superpartners. In fact, the soft terms in $\mathscr{L}_{\text{soft}}$ are capable of giving masses to all of the scalars and gauginos in a theory, even if the gauge bosons and fermions in chiral supermultiplets are massless (or relatively light). The gaugino masses M_a are always allowed by gauge symmetry. The $(m^2)^i_j$ terms are allowed for i, j such that ϕ_i, ϕ^{j*} transform in complex conjugate representations of each other under all gauge symmetries; in particular this is true of course when $i = j$, so every scalar is eligible to get a mass in this way if supersymmetry is broken. The remaining soft terms may or may not be allowed by the symmetries. The a^{ijk}, b^{ij}, and t^i terms have the same form as the y^{ijk}, M^{ij}, and L^i terms in the superpotential [compare eq. (8.78) to eq. (8.47) or eq. (8.77)], so they will each be allowed by gauge invariance if and only if a corresponding superpotential term is allowed.

The Feynman diagram interactions corresponding to the allowed non-tadpole soft terms in eq. (8.78) are shown in Fig. 8.4. For each of the interactions in Figs. 8.4(a),(c),(d) there is another with all arrows reversed, corresponding to the complex conjugate term in the Lagrangian.

Exercises

8.1 Consider the free Wess–Zumino model of Section 8.1 with supersymmetry transformations given by eqs. (8.3), (8.13), and (8.15).

(a) Verify that the total Lagrangian given by the sum of eqs. (8.2) and (8.12) is invariant up to a total derivative term, which you will identify.

(b) Check that eq. (8.16) holds for each of $X = \phi, \phi^\dagger, \psi, \psi^\dagger, F, F^\dagger$.

8.2 Consider the supersymmetric Yang–Mills Lagrangian given by eq. (8.72), with supersymmetry transformations given by eqs. (8.60)–(8.62) and (8.65)–(8.67).

(a) Verify that the Lagrangian is invariant up to a total derivative term, which you will identify. You will need to make use of the gauge invariance of the superpotential, eq. (8.73).

(b) Work out the equations of motion for the component fields in this theory, both before and after eliminating the auxiliary fields.

(c) Apply Noether's procedure of eq. (8.17) to obtain the supercurrent in eq. (8.76). Using the equations of motion for the fields, show that $\partial_\mu J^\mu_\alpha = 0$.

(d) Check that eq. (8.63) holds for each of $X = F^a_{\mu\nu}, \lambda^a, \lambda^{\dagger a}, D^a$.

8.3 Consider a theory with superpotential $W = \frac{1}{2}M^{ij}\Phi_i\Phi_j + \frac{1}{6}y^{ijk}\Phi_i\Phi_j\Phi_k$, and a supersymmetry-breaking part of the Lagrangian including a fermion mass term: $\mathcal{L}_{\psi\psi} = -\left(\frac{1}{2}m^{ij}\psi_i\psi_j + \text{h.c.}\right)$. In the text it was claimed that such a fermion mass term can be eliminated by absorbing it into a redefinition of the superpotential mass M^{ij} and the scalar interactions $(m^2)^i_j$ and c^{jk}_i. Work out the details by identifying the parameter redefinitions.

8.4 The infinitesimal gauge parameter Λ^a possesses an adjoint index a, where $a \in \{1, 2, \ldots, d_G\}$ and the dimension of the gauge group G (which is equal to the number of generators of G) is denoted by d_G. Since Λ is a function of the spacetime coordinate x, we can employ the covariant derivative in the adjoint representation to write

$$(\nabla_\mu\Lambda)^a = \partial_\mu\Lambda^a - gf^{abc}A^b_\mu\Lambda^c\,. \tag{8.84}$$

(a) Define $A_\mu \equiv A^a_\mu \boldsymbol{T^a}$ and $\Lambda \equiv \Lambda^a\boldsymbol{T^a}$, where there is an implicit sum over the repeated index a and the $\boldsymbol{T^a}$ are the generators of G. Show that

$$\nabla_\mu\Lambda = \partial_\mu\Lambda + ig[A_\mu, \Lambda]\,. \tag{8.85}$$

(b) Under a gauge transformation, $A_\mu \to A_\mu + \delta_\Lambda A_\mu$, verify that

$$\delta_\Lambda A_\mu = \nabla_\mu\Lambda\,. \tag{8.86}$$

(c) Under an infinitesimal supersymmetric transformation, the fields of the gauge supermultiplet transform as specified in eqs. (8.60)–(8.62). In light of eq. (8.58) verify that the Yang–Mills field strength transforms under an infinitesimal supersymmetric transformation as follows:

$$\delta_\epsilon F^a_{\mu\nu} = \frac{1}{\sqrt{2}}\left[\epsilon^\dagger\overline{\sigma}_\mu\nabla_\nu\lambda^a + \nabla_\nu\lambda^{\dagger a}\overline{\sigma}_\mu\epsilon\right] - (\mu \leftrightarrow \nu)\,. \tag{8.87}$$

9 The Supersymmetric Algebra

9.1 Extension of the Poincaré Algebra

Quantum fields possess definite transformation properties under the Lorentz and Poincaré groups. In the 1960s, with the discovery of new global internal symmetry groups such as the flavor SU(3) group of the quark model (based on the three quarks u, d, and s that were known at that time), the following question was considered. Since the quarks were spin-1/2 particles and the spin degree of freedom transformed under rotations which are elements of the Lorentz group, was it possible to transform a quark field, which possesses both spin and flavor degrees of freedom, under a larger group that contains both the Lorentz group and the flavor group? If spin is treated nonrelativistically, then such a construction is indeed possible. This is the quark model based on the group SU(6), which contains both the SU(3) flavor group and SU(2) [which governs the rotations of spin-1/2 particles]. However, no relativistic generalization of SU(6) was ever constructed.

The nonexistence of a relativistic extension of SU(6) was no accident. Coleman and Mandula [92] proved a very powerful no-go theorem that showed that, under the assumption that all fields interact with each other (directly or indirectly) in scattering processes, the only possible symmetry incorporating Poincaré transformations and a global internal symmetry group of transformations must be a trivial tensor product of the two symmetry groups. However, there is an implicit assumption underlying the Coleman–Mandula Theorem, which requires that the symmetry generators are closed under commutation (i.e., the commutator of any two elements of the symmetry algebra must also be an element of the symmetry algebra).

A remarkable loophole to the Coleman–Mandula Theorem was discovered by Haag, Łopuszański, and Sohnius [93], who realized that one could extend the Poincaré algebra by adding symmetry generators that obey anticommutation relations. In particular, the supersymmetry algebra specified in eqs. (8.30) and (8.31) is a nontrivial extension of the Poincaré spacetime that contains fermionic generators Q_α and $Q^\dagger_{\dot\alpha}$ whose anticommutator is proportional to the Poincaré translation generator P^μ. Hence, the supersymmetric algebra does not fall within the purview of the Coleman–Mandula Theorem.

In this chapter, we examine in detail the structure of the supersymmetric algebra. The unitary representations of the supersymmetric algebra yield supermultiplets that contain both fermions and bosons of equal mass. An excellent introduction to these topics can be found in [B29] and in Ref. [11].

9.2 The $N = 1$ Supersymmetry Algebra

We begin with the generators of the Poincaré algebra, $\{P^\mu\,,\,J^{\mu\nu}\}$, introduced in Chapter 1. These generators satisfy the commutation relations given in eqs. (1.23)–(1.25), which are reproduced below:

$$[P^\mu, P^\nu] = 0\,, \tag{9.1}$$

$$[J^{\mu\nu}, P^\lambda] = i(g^{\nu\lambda}P^\mu - g^{\mu\lambda}P^\nu)\,, \tag{9.2}$$

$$[J^{\mu\nu}, J^{\rho\lambda}] = i(g^{\nu\rho}\,J^{\mu\lambda} - g^{\mu\rho}\,J^{\nu\lambda} - g^{\nu\lambda}\,J^{\mu\rho} + g^{\mu\lambda}\,J^{\nu\rho})\,, \tag{9.3}$$

where $J^{\mu\nu} = -J^{\nu\mu}$.

The generators have specific transformation properties under the Poincaré algebra that are determined by the commutation relation with $J^{\mu\nu}$. Using the results of Exercise 1.4(c), it follows that the four-vector P^μ transforms as a $(\frac{1}{2}, \frac{1}{2})$ representation and the antisymmetric tensor $J^{\mu\nu}$ transforms as a $(1,0)\oplus(0,1)$ representation of the Poincaré algebra. The Coleman–Mandula Theorem implies that no nontrivial extension of the Poincaré algebra exists which contains additional generators that transform as a (p,q) representation of the Poincaré algebra, where $p+q$ is even (corresponding to bosonic generators that carry integer spin).

The supersymmetry algebra (also called the super-Poincaré algebra) is a nontrivial extension of the Poincaré algebra that includes fermionic generators. Haag, Łopuszański, and Sohnius showed that the only possible extension of the Poincaré algebra involves the addition of new fermionic generators that transform either as a $(\frac{1}{2},0)$ or $(0,\frac{1}{2})$ representation of the Poincaré algebra. These two-component spinor generators will be denoted by Q^i_α and its hermitian conjugate by $Q^\dagger_{\dot\alpha i} \equiv (Q^i_\alpha)^\dagger$, respectively, where $i \in \{1,2,\ldots,N\}$. In this section, we focus on the case of $N=1$, in which case the subscript i can be dropped.

We therefore begin by examining the structure of the $N=1$ supersymmetric algebra, which is obtained by adding one $(\frac{1}{2},0)$ and one $(0,\frac{1}{2})$ generator to the Poincaré algebra. Since Q_α and $Q^\dagger_{\dot\alpha}$ have no explicit dependence on the spacetime coordinate, it follows that

$$\exp(-ia_\mu P^\mu)\,Q_\alpha \exp(ia_\nu P^\nu) = Q_\alpha\,, \tag{9.4}$$

$$\exp(-ia_\mu P^\mu)\,Q^\dagger_{\dot\alpha} \exp(ia_\nu P^\nu) = Q^\dagger_{\dot\alpha}\,, \tag{9.5}$$

where the a_μ are real parameters. Working to first order in a_μ, it follows that the two-component spinor generators must commute with the translation generator P^μ:

$$[Q_\alpha\,,\,P^\mu] = [Q^\dagger_{\dot\alpha}\,,\,P^\mu] = 0\,. \tag{9.6}$$

The commutation relations given in eq. (9.6) can also be deduced by employing the following algebraic argument. Using the known transformation properties of Q_α, $Q^\dagger_{\dot\alpha}$, and P^μ under the Poincaré algebra, it follows that $[Q_\alpha\,,\,P^\mu]$ must consist of generators whose transformation properties are consistent with the tensor product

$$(\tfrac{1}{2},0) \otimes (\tfrac{1}{2},\tfrac{1}{2}) = (1,\tfrac{1}{2}) \oplus (0,\tfrac{1}{2}), \tag{9.7}$$

under the Poincaré algebra. But according to the Haag–Łopuszański–Sohnius Theorem, there are no $(1, \frac{1}{2})$ generators. This argument still leaves open the possibility that $[Q_\alpha\,, P^\mu] \propto \sigma^\mu_{\alpha\dot\beta} Q^{\dagger\dot\beta}$. However, it can be shown (see Exercise 9.1) that the proportionality constant must be zero.

As shown in Exercise 1.4(b), the transformation properties of Q_α and $Q^\dagger_{\dot\alpha}$ under the Poincaré algebra govern their commutation relations with $J^{\mu\nu}$:

$$[Q_\alpha\,, J^{\mu\nu}] = (\sigma^{\mu\nu})_\alpha{}^\beta Q_\beta\,, \qquad [Q^\dagger_{\dot\alpha}\,, J^{\mu\nu}] = -Q^\dagger_{\dot\beta}(\overline{\sigma}^{\mu\nu})^{\dot\beta}{}_{\dot\alpha}\,. \tag{9.8}$$

The Coleman–Mandula Theorem implies that one cannot obtain a consistent algebraic structure by postulating commutation relations for the Q_α and $Q^\dagger_{\dot\alpha}$. However, by declaring Q_α and $Q^\dagger_{\dot\alpha}$ to be *fermionic* generators, one can postulate *anticommutation* relations for Q_α and $Q^\dagger_{\dot\alpha}$ such that the generators $\{P^\mu\,, J^{\mu\nu}\,, Q_\alpha\,, Q^\dagger_{\dot\alpha}\}$ form a closed algebraic system. We therefore consider the three possible anticommutation relations, along with their transformation properties with respect to the Poincaré algebra:

$$\{Q_\alpha\,, Q_\beta\} \qquad (\tfrac{1}{2},0) \otimes (\tfrac{1}{2},0) = (1,0) \oplus (0,0)\,, \tag{9.9}$$

$$\{Q^\dagger_{\dot\alpha}\,, Q^\dagger_{\dot\beta}\} \qquad (0,\tfrac{1}{2}) \otimes (0,\tfrac{1}{2}) = (0,1) \oplus (0,0)\,, \tag{9.10}$$

$$\{Q_\alpha\,, Q^\dagger_{\dot\beta}\} \qquad (\tfrac{1}{2},0) \otimes (0,\tfrac{1}{2}) = (\tfrac{1}{2},\tfrac{1}{2})\,. \tag{9.11}$$

Using the results of Exercise 1.4, eqs. (9.9) and (9.10) imply that

$$\{Q_\alpha\,, Q^\beta\} = s(\sigma^{\mu\nu})_\alpha{}^\beta J_{\mu\nu} + k\delta_\alpha{}^\beta \mathbb{1}\,, \tag{9.12}$$

$$\{Q^{\dagger\dot\alpha}\,, Q^\dagger_{\dot\beta}\} = s^*(\overline{\sigma}^{\mu\nu})^{\dot\alpha}{}_{\dot\beta} J_{\mu\nu} + k^*\delta^{\dot\alpha}{}_{\dot\beta}\mathbb{1}\,, \tag{9.13}$$

where $\mathbb{1}$ is the unit operator, s and k are complex numbers, and eq. (9.13) is the hermitian conjugate of eq. (9.12). Note that we have raised and/or lowered some of the spinor indices for convenience. Since $[Q_\alpha, P^\lambda] = [Q^\dagger_{\dot\alpha}, P^\lambda] = 0$ and $[J_{\mu\nu}, P^\lambda] \neq 0$, it follows that $s = 0$. If we now lower all spinor indices, eqs. (9.12) and (9.13) with $s = 0$ yield

$$\{Q_\alpha\,, Q_\beta\} = k\epsilon_{\beta\alpha}\mathbb{1}\,, \qquad \{Q^\dagger_{\dot\alpha}\,, Q^\dagger_{\dot\beta}\} = k^*\epsilon^{\dot\beta\dot\alpha}\mathbb{1}\,, \tag{9.14}$$

and we conclude that $k = 0$ since the left-hand sides of the above equation are symmetric under the interchange of spinor indices whereas the right-hand sides are antisymmetric. Hence,

$$\{Q_\alpha\,, Q_\beta\} = \{Q^\dagger_{\dot\alpha}\,, Q^\dagger_{\dot\beta}\} = 0\,. \tag{9.15}$$

Equation (9.11) implies that the remaining anticommutation relation must be of the form

$$\{Q_\alpha\,, Q^\dagger_{\dot\beta}\} = t\sigma^\mu_{\alpha\dot\beta} P_\mu\,, \tag{9.16}$$

where t is a complex number. Multiplying eq. (9.16) by $\overline{\sigma}^{\nu\dot{\beta}\alpha}$ and taking the trace using $\text{Tr}(\sigma^\mu\overline{\sigma}^\nu) = 2g^{\mu\nu}$, it follows that

$$\overline{\sigma}_\mu^{\dot{\beta}\alpha}\{Q_\alpha\,,\,Q^\dagger_{\dot{\beta}}\} = 2tP_\mu\,. \tag{9.17}$$

In particular, for $\mu = 0$, eq. (9.17) yields

$$2tP^0 = Q_1Q_1^\dagger + Q_1^\dagger Q_1 + Q_2Q_2^\dagger + Q_2^\dagger Q_2\,. \tag{9.18}$$

Since $P^0 \geq m$ for physical states of mass m and the right-hand side of eq. (9.18) is semipositive definite, it follows that t must be real and positive.[1] One can rescale the definition of the fermionic generators Q and $Q^\dagger$ such that $t = 2$. In this convention,

$$\{Q_\alpha\,,\,Q^\dagger_{\dot{\beta}}\} = 2\sigma^\mu_{\alpha\dot{\beta}}P_\mu\,. \tag{9.19}$$

To summarize, the $N = 1$ supersymmetry algebra is spanned by the generators $\{P^\mu\,,\,J^{\mu\nu}\,,\,Q_\alpha\,,\,Q^\dagger_{\dot{\alpha}}\}$, which satisfy eqs. (9.1)–(9.3) and

$$\big[Q_\alpha\,,\,P^\mu\big] = \big[Q^\dagger_{\dot{\alpha}}\,,\,P^\mu\big] = 0\,, \tag{9.20}$$

$$\big[Q_\alpha\,,\,J^{\mu\nu}\big] = (\sigma^{\mu\nu})_\alpha{}^\beta Q_\beta\,, \tag{9.21}$$

$$\big[Q^\dagger_{\dot{\alpha}}\,,\,J^{\mu\nu}\big] = -Q^\dagger_{\dot{\beta}}(\overline{\sigma}^{\mu\nu})^{\dot{\beta}}{}_{\dot{\alpha}}\,, \tag{9.22}$$

$$\{Q_\alpha\,,\,Q_\beta\} = \{Q^\dagger_{\dot{\alpha}}\,,\,Q^\dagger_{\dot{\beta}}\} = 0\,, \tag{9.23}$$

$$\{Q_\alpha\,,\,Q^\dagger_{\dot{\beta}}\} = 2\sigma^\mu_{\alpha\dot{\beta}}P_\mu\,. \tag{9.24}$$

Note that eqs. (9.20)–(9.24) are unchanged under the U(1) phase transformation

$$Q_\alpha \to e^{-i\chi}Q_\alpha\,, \qquad Q^\dagger_{\dot{\alpha}} \to e^{i\chi}Q^\dagger_{\dot{\alpha}}\,, \tag{9.25}$$

whereas the generators P^μ and $J^{\mu\nu}$ are not transformed. One can therefore extend the $N = 1$ supersymmetry algebra by adding a bosonic generator R such that

$$e^{i\chi R}Q_\alpha e^{-i\chi R} = e^{-i\chi}Q_\alpha\,, \tag{9.26}$$

$$e^{i\chi R}Q^\dagger_{\dot{\alpha}}e^{-i\chi R} = e^{i\chi}Q^\dagger_{\dot{\alpha}}\,. \tag{9.27}$$

Expanding out to first order in χ, one easily derives the commutation relations

$$\big[R\,,\,Q_\alpha\big] = -Q_\alpha\,, \tag{9.28}$$

$$\big[R\,,\,Q^\dagger_{\dot{\alpha}}\big] = Q^\dagger_{\dot{\alpha}}\,. \tag{9.29}$$

We say that the generator Q_α has an R charge of -1. Since P^μ and $J^{\mu\nu}$ are uncharged under the $\text{U}(1)_R$ transformation, it follows that

$$\big[R\,,\,P^\mu\big] = \big[R\,,\,J^{\mu\nu}\big] = 0\,. \tag{9.30}$$

Thus, eqs. (9.1)–(9.3) and eqs. (9.20)–(9.30) define the maximally extended $N = 1$ supersymmetry algebra, which includes an additional continuous $\text{U}(1)_R$ symmetry.

[1] We reject the possibility of $t = 0$, in which case $Q = Q^\dagger = 0$ and the supersymmetric algebra reduces to the Poincaré algebra.

9.3 Representations of the $N = 1$ Supersymmetry Algebra

In Section 1.2, we identified the two Casimir operators of the Poincaré algebra, $P^2 \equiv P^\mu P_\mu$ and $w^2 \equiv w^\mu w_\mu$, where $w^\mu = -\frac{1}{2}\epsilon^{\mu\nu\rho\lambda}J_{\nu\rho}P_\lambda$ is the Pauli–Lubański vector. The unitary representations of the Poincaré algebra can be labeled by the eigenvalues of P^2 and w^2 when acting on physical states. For a timelike four-vector P^μ, it is convenient to work in the rest frame where $P^\mu = (m\,;\vec{\mathbf{0}})$, in which case $w^\mu = (0\,; m\vec{\boldsymbol{S}})$ and $w^2 = -m^2\vec{\boldsymbol{S}}^2$ with eigenvalues $-m^2 s(s+1)$, where $s = 0, \frac{1}{2}, 1, \ldots$. Thus, the massive representations are labeled by (m, s). For a fixed value of m, the corresponding spin-s representations are $(2s+1)$-dimensional. For a lightlike four-vector P^μ, it is convenient to work in a standard reference frame where $P^\mu = P^0(1\,;0\,,0\,,1)$ and $P^0 > 0$. In this frame, $w^\mu = hP^\mu$, where the helicity operator is given by $h = \vec{\boldsymbol{S}}\boldsymbol{\cdot}\hat{\boldsymbol{P}}$ [see eq. (1.29)]. Consequently, the possible eigenvalues of h are $\lambda = 0, \pm\frac{1}{2}, \pm 1, \ldots$. Under a discrete CPT transformation, λ changes sign. Hence, the massless positive-energy representations of the Poincaré algebra are specified by $|\lambda|$. For the case of $\lambda = 0$, the corresponding representation is one-dimensional. For any nonzero choice for λ, the corresponding representation is two-dimensional and reducible, as both $\pm|\lambda|$ helicity states must appear.

The unitary representations of the $N = 1$ supersymmetry algebra can be determined by using similar techniques. First, we identify the Casimir operators, which commute with all the supersymmetry algebra generators, $\{P^\mu\,, J^{\mu\nu}\,, Q_\alpha\,, Q^{\dagger\dot{\alpha}}\}$. It is clear that P^2 is a Casimir operator, since Q_α and $Q^{\dagger\dot{\alpha}}$ commute with P^μ. However, w^2 is *not* a Casimir operator of the supersymmetry algebra. To establish this result, it is straightforward to use the (anti)commutation relations of the supersymmetry algebra to prove that

$$[w^\mu\,, Q_\alpha] = i(\sigma^{\mu\nu})_\alpha{}^\beta Q_\beta P_\nu\,, \qquad [w^\mu\,, Q^\dagger_{\dot{\alpha}}] = i(\overline{\sigma}^{\mu\nu})^{\dot{\beta}}{}_{\dot{\alpha}} Q^\dagger_{\dot{\beta}} P_\nu\,. \tag{9.31}$$

Using these results, it is straightforward to derive

$$[w^2\,, Q_\alpha] = 2i\sigma^{\mu\nu}{}_\alpha{}^\beta Q_\beta w_\mu P_\nu - \tfrac{3}{4}P^2 Q_\alpha\,, \tag{9.32}$$

$$[w^2\,, Q^\dagger_{\dot{\alpha}}] = 2i\overline{\sigma}^{\mu\nu\dot{\beta}}{}_{\dot{\alpha}} Q^\dagger_{\dot{\beta}} w_\mu P_\nu - \tfrac{3}{4}P^2 Q^\dagger_{\dot{\alpha}}\,. \tag{9.33}$$

Thus, w^2 does not commute with the fermionic generators of the supersymmetric algebra. One consequence of this result is that the representations of the supersymmetry algebra consist of supermultiplets that consist of equal mass particles of different spins.

In order to deduce the possible spins that make up the irreducible supermultiplets, we shall identify a second Casimir operator of the $N = 1$ supersymmetry algebra. We begin by defining

$$B^\mu \equiv w^\mu + \tfrac{1}{4}Q^\dagger\overline{\sigma}^\mu Q\,. \tag{9.34}$$

Using eqs. (9.23), (9.24), and (9.31), one can easily derive

$$[B^\mu\,,\,Q_\alpha] = -\tfrac{1}{2}P^\mu Q_\alpha\,, \qquad\qquad [B^\mu\,,\,Q^\dagger_{\dot\alpha}] = \tfrac{1}{2}P^\mu Q^\dagger_{\dot\alpha}\,. \tag{9.35}$$

The four-vector operator B^μ possesses some of the properties of the Pauli–Lubański vector w^μ (see Exercises 1.2 and 9.4). In particular,

$$[B^\mu\,,\,B^\nu] = i\epsilon^{\mu\nu\rho\lambda}B_\rho P_\lambda\,. \tag{9.36}$$

One may be tempted to conjecture that $B^2 \equiv B_\mu B^\mu$ is a Casimir operator of the supersymmetric algebra. However, $[B^2\,,\,Q_\alpha] \neq 0$, so we must look further. The structure of eq. (9.35) suggests that we define

$$C^{\mu\nu} \equiv B^\mu P^\nu - B^\nu P^\mu\,. \tag{9.37}$$

It then follows that

$$[C^{\mu\nu}\,,\,Q_\alpha] = [C^{\mu\nu}\,,\,Q^\dagger_{\dot\alpha}] = [C^{\mu\nu}\,,\,P^\lambda] = 0\,, \tag{9.38}$$

where the first two commutators vanish as a consequence of eq. (9.35) and the last commutator vanishes as a consequence of eq. (9.182). Moreover, eqs. (9.2) and (9.183) imply that P^μ and B^μ are Lorentz four-vectors, in which case $C^{\mu\nu}$ is a second-rank Lorentz tensor. Hence,

$$C^2 \equiv C_{\mu\nu}C^{\mu\nu} = 2[B^2P^2 - (B\!\cdot\! P)^2] \tag{9.39}$$

satisfies

$$[C^2\,,\,P^\mu] = [C^2\,,\,J^{\mu\nu}] = [C^2\,,\,Q_\alpha] = [C^2,\,Q^\dagger_{\dot\alpha}] = 0\,. \tag{9.40}$$

We conclude that P^2 and C^2 are the two Casimir operators of the $N = 1$ supersymmetry algebra.

9.3.1 Massive $N = 1$ Supermultiplets

The representations of the $N = 1$ supersymmetry algebra can be labeled by the eigenvalues of P^2 and C^2 when acting on the physical states.[2] The eigenvalue of P^2 is m^2, where m is the mass. To see the physical interpretation of C^2, we first consider the case of $m \neq 0$. In this case, we are free to evaluate the Lorentz scalar C^2 in the particle rest frame. In this frame, $B^\mu = (\tfrac{1}{4}Q^\dagger\bar\sigma^0 Q\,;\, mS^i + \tfrac{1}{4}Q^\dagger\bar\sigma^i Q)$ where S^i is defined in eq. (1.10), and

$$C^2 = 2\left[B^2P^2 - (B\!\cdot\! P)^2\right] = 2m^2\left[B^2 - B_0^2\right] = -2m^2 B^iB^i\,, \tag{9.41}$$

where $B^iB^i \equiv |\vec{\boldsymbol{B}}|^2$. Moreover, if we define the rest-frame operator,

$$\mathcal{J}^i \equiv \frac{1}{m}B^i = S^i + \frac{1}{4m}Q^\dagger\bar\sigma^i Q\,, \tag{9.42}$$

[2] As in the case of the Poincaré algebra, we restrict our consideration to states of nonnegative energy P^0.

then it follows from eq. (9.36) that

$$[\mathcal{J}^i\,,\,\mathcal{J}^j] = i\epsilon^{ijk}\mathcal{J}^k\,. \tag{9.43}$$

Since the eigenvalues of $\mathcal{J}^i\mathcal{J}^i$ are $j(j+1)$ for $j = 0, \frac{1}{2}, 1, \frac{3}{2}, \ldots$, it follows that the eigenvalues of $C^2 = -2m^4\mathcal{J}^i\mathcal{J}^i$ are $-2m^4 j(j+1)$. We conclude that for timelike P^μ, the representations of the $N = 1$ supersymmetry algebra are labeled by (m, j), where j is called the *superspin* of the supermultiplet.

To construct the states of the supermultiplet, we note that in the rest frame the anticommutators given in eqs. (9.23) and (9.24) simplify to

$$\{Q_1\,,\,Q_1^\dagger\} = \{Q_2\,,\,Q_2^\dagger\} = 2m\,, \tag{9.44}$$

$$\{Q_1\,,\,Q_1\} = \{Q_2\,,\,Q_2\} = \{Q_1\,,\,Q_2\} = 0\,, \tag{9.45}$$

$$\{Q_1^\dagger\,,\,Q_1^\dagger\} = \{Q_2^\dagger\,,\,Q_2^\dagger\} = \{Q_1^\dagger\,,\,Q_2^\dagger\} = 0\,. \tag{9.46}$$

All states in a supermultiplet with superspin j are simultaneous eigenstates of P^2, $\mathcal{J}^i\mathcal{J}^i$, and $\mathcal{J}^3$ with eigenvalues m^2, $j(j+1)$, and j_3, respectively, where the possible values of j_3 are $-j, -j+1, \ldots, j-1, j$. For a fixed value of the superspin j, there exists a distinguished state of the supermultiplet, denoted by $|\Omega\rangle$, that satisfies[3]

$$Q_\beta\,|\Omega\rangle = 0\,, \qquad S_+\,|\Omega\rangle = 0\,, \tag{9.48}$$

where $S_\pm \equiv S^1 \pm iS^2$. To verify that such a state must exist, let us assume the contrary. Suppose that a simultaneous eigenstate of P^2, $\mathcal{J}^i\mathcal{J}^i$, and $\mathcal{J}^3$, denoted by $|\Psi\rangle$, is not annihilated by Q_β. In the rest frame, eq. (9.35) yields

$$[\mathcal{J}^i, Q_\beta] = [\mathcal{J}^i\,,\,Q^\dagger_{\dot\beta}] = 0\,, \tag{9.49}$$

so it follows that $Q_\beta\,|\Psi\rangle$ is also a simultaneous eigenstate of P^2, $\mathcal{J}^i\mathcal{J}^i$, and $\mathcal{J}^3$. By assumption, $Q_\beta\,|\Psi\rangle$ is not annihilated by Q_α, so we conclude that $Q_\alpha Q_\beta\,|\Psi\rangle$ is also a simultaneous eigenstate of P^2, $\mathcal{J}^i\mathcal{J}^i$, and $\mathcal{J}^3$. But we now arrive at a contradiction, since eq. (9.45) yields

$$Q_\gamma\,(Q_\alpha Q_\beta\,|\Psi\rangle) = 0\,. \tag{9.50}$$

Consequently, there must be at least one state of the supermultiplet that satisfies eq. (9.48). Using eqs. (9.42) and (9.48), it follows that

$$\mathcal{J}^i\,|\Omega\rangle = S^i\,|\Omega\rangle\,. \tag{9.51}$$

Since $S_+\,|\Omega\rangle = 0$, it follows that $|\Omega\rangle$ is also a simultaneous eigenstate of $\vec{S}^2$ and S^3 with corresponding eigenvalues $j(j+1)$ and j. Moreover, this state must be unique under the assumption that the superspin-j supermultiplet is an *irreducible* representation of the $N = 1$ supersymmetry algebra.

[3] Recall that if $|s, m_s\rangle$ are eigenstates of $\vec{S}^2$ and S^3 with corresponding eigenvalues $s(s+1)$ and m_s respectively, then

$$S_\pm\,|s, m_s\rangle = \sqrt{(s \mp m_s)(s \pm m_s + 1)}\,|s, m_s \pm 1\rangle\,. \tag{9.47}$$

Note that eq. (9.31) when evaluated in the rest frame yields

$$[S^i\,,\,Q_\alpha] = i\sigma^{i0}{}_\alpha{}^\beta Q_\beta\,, \qquad\qquad [S^i\,,\,Q^\dagger_{\dot\alpha}] = i\overline{\sigma}^{i0\dot\beta}{}_{\dot\alpha}Q^\dagger_{\dot\beta}\,. \tag{9.52}$$

Hence, one can define additional states of the supermultiplet,

$$|\Omega(j_3)\rangle \equiv (S_-)^{j-j_3}\,|\Omega\rangle\,, \qquad\qquad \text{for}\;\; j_3 = -j, -j+1, \ldots, j-1, j\,, \tag{9.53}$$

all of which satisfy

$$Q_\alpha\,|\Omega(j_3)\rangle = 0\,, \tag{9.54}$$

as a result of eq. (9.52). As before, $\mathcal{J}^i\,|\Omega(j_3)\rangle = S^i\,|\Omega(j_3)\rangle$ as a consequence of eqs. (9.42) and (9.54). It follows that $|\Omega(j_3)\rangle$ is also a simultaneous eigenstate of $\vec{\boldsymbol{S}}^2$ and S^3 with corresponding eigenvalues $j(j+1)$ and j_3. That is,

$$|\Omega(j_3)\rangle = |j, j_3\rangle\,, \tag{9.55}$$

where the rest-frame spin and its projection along the z-axis are explicitly indicated.

Starting from $|\Omega(j_3)\rangle = |j, j_3\rangle$, one can construct the remaining states of the massive supermultiplet by considering the series of states for each possible value of j_3, $|\Omega(j_3)\rangle\,,\; Q^\dagger_{\dot\alpha}\,|\Omega(j_3)\rangle\,,\; Q^\dagger_{\dot\alpha}Q^\dagger_{\dot\beta}\,|\Omega(j_3)\rangle\,, \ldots$. This series of states terminates due to eq. (9.45) and only four independent states survive (for a fixed value of j_3):

$$|\Omega(j_3)\rangle\,, \quad Q^\dagger_{\dot 1}\,|\Omega(j_3)\rangle\,, \quad Q^\dagger_{\dot 2}\,|\Omega(j_3)\rangle\,, \quad Q^\dagger_{\dot 1}Q^\dagger_{\dot 2}\,|\Omega(j_3)\rangle\,. \tag{9.56}$$

All the states of eq. (9.56) are mass-degenerate (with mass $m \neq 0$). The spins of these states can be determined by applying the operators $\vec{\boldsymbol{S}}^2$ and S^3. By virtue of eq. (9.55), we already know that $|\Omega(j_3)\rangle$ is a spin-j state with S^3-eigenvalue j_3. Next, one can use eq. (9.52) to derive

$$[S^i\,,\,Q^\dagger_{\dot\alpha}Q^\dagger_{\dot\beta}] = iQ^\dagger_{\dot\gamma}\,[\overline{\sigma}^{i0\dot\gamma}{}_{\dot\alpha}Q^\dagger_{\dot\beta} - \overline{\sigma}^{i0\dot\gamma}{}_{\dot\beta}Q^\dagger_{\dot\alpha}]\,, \tag{9.57}$$

$$[\vec{\boldsymbol{S}}^2\,,\,Q^\dagger_{\dot\alpha}Q^\dagger_{\dot\beta}] = 2iQ^\dagger_{\dot\gamma}\,[\overline{\sigma}^{i0\dot\gamma}{}_{\dot\alpha}Q^\dagger_{\dot\beta} - \overline{\sigma}^{i0\dot\gamma}{}_{\dot\beta}Q^\dagger_{\dot\alpha}]\,S^i\,. \tag{9.58}$$

It immediately follows that

$$[S^i\,,\,Q^\dagger_{\dot 1}Q^\dagger_{\dot 2}] = iQ^\dagger_{\dot 1}Q^\dagger_{\dot 2}\,\mathrm{Tr}\,\sigma^{i0} = 0\,, \tag{9.59}$$

$$[\vec{\boldsymbol{S}}^2\,,\,Q^\dagger_{\dot 1}Q^\dagger_{\dot 2}] = 2iQ^\dagger_{\dot 1}Q^\dagger_{\dot 2}S^i\,\mathrm{Tr}\,\sigma^{i0} = 0\,. \tag{9.60}$$

Applying eqs. (9.59) and (9.60) to the state $|\Omega(j_3)\rangle$, it follows that $Q^\dagger_{\dot 1}Q^\dagger_{\dot 2}\,|\Omega(j_3)\rangle$ is also a spin-j state with S^3-eigenvalue j_3. This result is easily understood. Noting that we can write

$$Q^\dagger_{\dot 1}Q^\dagger_{\dot 2} = \tfrac{1}{2}\epsilon^{\dot\alpha\dot\beta}Q^\dagger_{\dot\alpha}Q^\dagger_{\dot\beta}\,, \tag{9.61}$$

it follows that $Q^\dagger_{\dot 1}Q^\dagger_{\dot 2}$ is a *scalar* operator. This is consistent with the fact that the antisymmetric part of the tensor product of two SU(2) spinor representations is an SU(2) singlet. Thus, $Q^\dagger_{\dot 1}Q^\dagger_{\dot 2}\,|\Omega(j_3)\rangle$ and $|\Omega(j_3)\rangle$ possess the same eigenvalues with respect to $\vec{\boldsymbol{S}}^2$ and S^3.

To determine the properties of $Q^\dagger_{\dot 1}\,|\Omega(j_3)\rangle$ and $Q^\dagger_{\dot 2}\,|\Omega(j_3)\rangle$, we first note that Q_α

is a spinor operator [see eq. (1.291)] that imparts spin 1/2 to any state it acts on [see eq. (1.291)]. Moreover, eq. (9.52) yields

$$S^3 Q_1^\dagger \,|\Omega(j_3)\rangle = (j_3 + \tfrac{1}{2}) Q_1^\dagger \,|\Omega(j_3)\rangle\,, \quad S^3 Q_2^\dagger \,|\Omega(j_3)\rangle = (j_3 - \tfrac{1}{2}) Q_2^\dagger \,|\Omega(j_3)\rangle\,. \tag{9.62}$$

Hence, one can employ the standard results from the theory of angular momentum addition in quantum mechanics, which relate the tensor product basis to the total angular momentum basis. In particular,

$$|j\,,\, m\rangle = \sum_{m_1, m_2} |j_1\,,\, m_1\rangle \otimes |j_2\,,\, m_2\rangle\, \langle j_1\; j_2\,;\, m_1\; m_2\,|\, j\; m\rangle\,, \tag{9.63}$$

where $\langle j_1\; j_2\,;\, m_1\; m_2\,|\, j\; m\rangle$ are the Clebsch–Gordan coefficients. We employ the Condon–Shortley phase convention in which the Clebsch–Gordan coefficients are real (e.g., see [B122, B134]). In the present application, we require

$$|\tfrac{1}{2}\,,\, \tfrac{1}{2}\rangle \otimes |j\,,\, m - \tfrac{1}{2}\rangle = \sqrt{\frac{j + \frac{1}{2} + m}{2j+1}}\, |j + \tfrac{1}{2}\,,\, m\rangle - \sqrt{\frac{j + \frac{1}{2} - m}{2j+1}}\, |j - \tfrac{1}{2}\,,\, m\rangle\,, \tag{9.64}$$

$$|\tfrac{1}{2}\,,\, -\tfrac{1}{2}\rangle \otimes |j\,,\, m + \tfrac{1}{2}\rangle = \sqrt{\frac{j + \frac{1}{2} - m}{2j+1}}\, |j + \tfrac{1}{2}\,,\, m\rangle + \sqrt{\frac{j + \frac{1}{2} + m}{2j+1}}\, |j - \tfrac{1}{2}\,,\, m\rangle\,. \tag{9.65}$$

Equations (9.55), (9.62), (9.59), and (9.60) imply that (see Exercise 9.5)[4]

$$|\Omega(j_3)\rangle = |j\,,\, j_3\rangle\,, \tag{9.66}$$

$$Q_1^\dagger \,|\Omega(j_3)\rangle = \sqrt{\frac{j + j_3 + 1}{2j+1}}\, |j + \tfrac{1}{2}\,,\, j_3 + \tfrac{1}{2}\rangle - \sqrt{\frac{j - j_3}{2j+1}}\, |j - \tfrac{1}{2}\,,\, j_3 + \tfrac{1}{2}\rangle\,, \tag{9.67}$$

$$Q_2^\dagger \,|\Omega(j_3)\rangle = \sqrt{\frac{j - j_3 + 1}{2j+1}}\, |j + \tfrac{1}{2}\,,\, j_3 - \tfrac{1}{2}\rangle + \sqrt{\frac{j + j_3}{2j+1}}\, |j - \tfrac{1}{2}\,,\, j_3 - \tfrac{1}{2}\rangle\,. \tag{9.68}$$

$$Q_1^\dagger Q_2^\dagger \,|\Omega(j_3)\rangle = |j\,,\, j_3\rangle\,. \tag{9.69}$$

In particular, if $j_3 \neq j$ then eqs. (9.67) and (9.68) imply that $Q_1^\dagger \,|\Omega(j_3)\rangle$ and $Q_2^\dagger \,|\Omega(j_3)\rangle$ are orthogonal linear combinations of spin-$(j \pm \frac{1}{2})$ states [although these states are eigenstates of S^3 as shown in eq. (9.62)]. If $j_3 = \pm j$ then $Q_1^\dagger \,|\Omega(j)\rangle$ and $Q_2^\dagger \,|\Omega(-j)\rangle$ are states of spin $(j + \frac{1}{2})$, since both these states are eigenstates of $\vec{S}^2$ and S^3 with eigenvalues $(j + \frac{1}{2})(j + \frac{3}{2})$ and $\pm(j + \frac{1}{2})$, respectively.

The simplest example is $j = j_3 = 0$, in which case the massive supermultiplet is made up of two states of spin 0 and two states of spin 1/2. The two spin-0 states can be combined into a single complex scalar state, and the two spin-1/2 states can

[4] The fact that there are no relative phases among the four states in eqs. (9.66)–(9.69) is a consequence of the Condon–Shortley phase convention adopted here.

Table 9.1 States of an $N=1$ massive supermultiplet of superspin j. An interpretation is provided for $j=s$ and $j=s+\frac{1}{2}$ where s is a nonnegative integer. The number of bosonic and the number of fermionic degrees of freedom of the supermultiplet coincide and are equal to $2(2j+1)$.

spin	degrees of freedom	interpretation ($j=s$)	interpretation ($j=s+\frac{1}{2}$)
j	$2(2j+1)$	complex spin-s boson	"complex" spin-$(s+\frac{1}{2})$ fermion
$j+\frac{1}{2}$	$2j+2$	spin-$(s+\frac{1}{2})$ fermion	real spin-$(s+1)$ boson
$j-\frac{1}{2}$	$2j$	spin-$(s-\frac{1}{2})$ fermion	real spin-s boson

be identified as the two spin components of a two-component Majorana fermion. This is the well-known *chiral supermultiplet* of $N=1$ supersymmetry. The states of an $N=1$ massive supermultiplet of superspin j are exhibited in Table 9.1.

In summary, there are $4(2j+1)$ mass-degenerate states in a massive supermultiplet of superspin j, which are explicitly given above by eqs. (9.66)–(9.69), for $j_3=-j,-j+1,\ldots,j-1,j$. In general, a massive supermultiplet of superspin j is made up of $2(2j+1)$ states of spin j, $2j+2$ states of spin $(j+\frac{1}{2})$, and $2j$ states of spin $(j-\frac{1}{2})$. The extra two states for the case of spin $(2j+1)$ arise when $j_3=\pm j$, in which cases $Q_1^\dagger\,|\Omega(j)\rangle$ and $Q_2^\dagger\,|\Omega(-j)\rangle$ are pure states of spin $(j+\frac{1}{2})$ as previously noted. Note that the number of bosonic and the number of fermionic degrees of freedom of the massive supermultiplet coincide and are equal to $2(2j+1)$. For example, the massive supermultiplet of superspin 0 is the chiral supermultiplet mentioned above consisting of a complex scalar and a Majorana fermion. In this case the $j-\frac{1}{2}$ row of Table 9.1 is not relevant. The massive supermultiplet of superspin 1/2 consists of a (real) spin-1 boson, a (real) spin-0 boson, and two mass-degenerate Majorana fermions, which can be combined into a single Dirac fermion (called a *complex* fermion in Table 9.1).

9.3.2 Massless $N=1$ Supermultiplets

We now examine the case of zero-mass positive-energy states, where $P^2=0$ and $P^0>0$. If one multiplies eq. (9.24) by $P^\rho P^\lambda \overline{\sigma}_\rho^{\dot{\gamma}\alpha}\overline{\sigma}_\lambda^{\dot{\beta}\tau}$, one can easily derive the anticommutation relation

$$\{P^\rho\overline{\sigma}_\rho^{\dot{\gamma}\alpha}Q_\alpha\,,\,P^\lambda Q^\dagger_{\dot{\beta}}\overline{\sigma}_\lambda^{\dot{\beta}\tau}\}=2P^2P^\mu\overline{\sigma}_\mu^{\dot{\gamma}\tau}\,. \tag{9.70}$$

Thus, for $P^2=0$ we have

$$\langle\Psi|\,\{P^\rho\overline{\sigma}_\rho^{\dot{\gamma}\alpha}Q_\alpha\,,\,P^\lambda Q^\dagger_{\dot{\beta}}\overline{\sigma}_\lambda^{\dot{\beta}\tau}\}\,|\Psi\rangle=0\,, \tag{9.71}$$

for any state $|\Psi\rangle$. In the space of one-particle states, only positively normed states exist. Noting that $(P^\mu\overline{\sigma}_\mu^{\dot{\alpha}\beta}Q_\beta)^\dagger=P^\mu Q^\dagger_{\dot{\beta}}\overline{\sigma}_\mu^{\dot{\beta}\alpha}$, eq. (9.71) implies that, as operators on the space of one-particle states,

$$P^\rho \overline{\sigma}_\rho^{\dot{\gamma}\alpha} Q_\alpha = P^\lambda Q^\dagger_{\dot{\beta}} \overline{\sigma}_\lambda^{\dot{\beta}\tau} = 0\,, \qquad \text{for } P^2 = 0\,. \tag{9.72}$$

Using this result, one can evaluate the Casimir operator C^2, defined in eq. (9.39), in the case of $P^2 = 0$. In particular, using eqs. (1.278) and (9.72), we obtain

$$C^2 = -2(B\cdot P)^2 = -\tfrac{1}{8}(Q^\dagger_{\dot{\alpha}} \overline{\sigma}_\mu^{\dot{\alpha}\beta} Q_\beta P^\mu)^2 = 0\,. \tag{9.73}$$

The same conclusion can be obtained by choosing the standard reference frame for lightlike four-vectors, where $P^\mu = P^0(1\,;\,0\,,\,0\,,\,1)$. In this reference frame, the anticommutators given in eqs. (9.23) and (9.24) simplify to

$$\{Q_1\,,\,Q_1^\dagger\} = 0\,, \qquad \{Q_2\,,\,Q_2^\dagger\} = 4P_0\,, \tag{9.74}$$

$$\{Q_1\,,\,Q_1\} = \{Q_2\,,\,Q_2\} = \{Q_1\,,\,Q_2\} = 0\,, \tag{9.75}$$

$$\{Q_1^\dagger\,,\,Q_1^\dagger\} = \{Q_2^\dagger\,,\,Q_2^\dagger\} = \{Q_1^\dagger\,,\,Q_2^\dagger\} = 0\,. \tag{9.76}$$

Hence,

$$C^2 = -2(B\cdot P)^2 = -\tfrac{1}{2}P_0^2(Q_1^\dagger Q_1)^2 = \tfrac{1}{2}P_0^2 Q_1^\dagger Q_1^\dagger Q_1 Q_1 = 0\,. \tag{9.77}$$

Equation (9.72) implies a number of other operator identities when acting on the space of one-particle states. Using eq. (9.24), one easily derives

$$[Q^\alpha Q_\alpha\,,\,Q^\dagger_{\dot{\beta}}] = 4P_\mu \sigma^\mu_{\alpha\dot{\beta}} Q^\alpha\,, \qquad [Q^\dagger_{\dot{\alpha}} Q^{\dagger\,\dot{\alpha}}\,,\,Q_\beta] = -4P_\mu \sigma^\mu_{\alpha\dot{\beta}} Q^{\dagger\,\dot{\beta}}\,. \tag{9.78}$$

Applying eq. (9.72) then yields

$$[Q^\alpha Q_\alpha\,,\,Q^\dagger_{\dot{\beta}}] = [Q^\dagger_{\dot{\alpha}} Q^{\dagger\,\dot{\alpha}}\,,\,Q_\beta] = 0\,, \qquad \text{for } P^2 = 0\,. \tag{9.79}$$

Then, for any one-particle state $|\Psi\rangle$, eqs. (9.23), (9.24), and (9.79) yield

$$\begin{aligned} P_\mu \sigma^\mu_{\alpha\dot{\alpha}} Q^\beta Q_\beta\,|\Psi\rangle &= \tfrac{1}{2}\{Q_\alpha\,,\,Q^\dagger_{\dot{\alpha}}\} Q^\beta Q_\beta\,|\Psi\rangle = \tfrac{1}{2} Q_\alpha Q^\dagger_{\dot{\alpha}} Q^\beta Q_\beta\,|\Psi\rangle \\ &= \tfrac{1}{2} Q_\alpha [Q^\dagger_{\dot{\alpha}}\,,\,Q^\beta Q_\beta]\,|\Psi\rangle = 0\,. \end{aligned} \tag{9.80}$$

A similar computation of $P_\mu \sigma^\mu_{\alpha\dot{\alpha}} Q^\dagger_{\dot{\beta}} Q^{\dagger\,\dot{\beta}}$ allows us to conclude that

$$P_\mu Q^\beta Q_\beta\,|\Psi\rangle = P_\mu Q^\dagger_{\dot{\beta}} Q^{\dagger\,\dot{\beta}}\,|\Psi\rangle = 0\,, \qquad \text{for } P^2 = 0\,, \tag{9.81}$$

after multiplying through by $\overline{\sigma}_\nu^{\dot{\alpha}\alpha}$ and evaluating the resulting trace. As we are only interested in positive-energy states, we conclude that, as operators on the space of one-particle states,

$$Q^\beta Q_\beta = Q^\dagger_{\dot{\beta}} Q^{\dagger\,\dot{\beta}} = 0\,, \qquad \text{for } P^2 = 0 \text{ and } P^0 > 0\,, \tag{9.82}$$

In order to identify the massless supermultiplets of one-particle states, it is convenient to define

$$L^\mu \equiv \tfrac{1}{2}(w^\mu + B^\mu) = w^\mu + \tfrac{1}{8} Q^\dagger \overline{\sigma}^\mu Q\,. \tag{9.83}$$

Note that eqs. (9.20) and (1.279) imply that

$$[P^\mu\,,\,L^\nu] = 0\,. \tag{9.84}$$

Using eqs. (9.23), (9.24), and (9.31), one can easily derive

$$[L^\mu\,,\,Q_\alpha] = -\tfrac{1}{4}(\sigma^\mu\overline{\sigma}^\nu)_\alpha{}^\beta Q_\beta P_\nu\,, \qquad [L^\mu\,,\,Q^\dagger_{\dot\alpha}] = \tfrac{1}{4}(\overline{\sigma}^\nu\sigma^\mu)^{\dot\beta}{}_{\dot\alpha} Q^\dagger_{\dot\beta} P_\nu\,. \tag{9.85}$$

A straightforward computation then gives

$$[L^\mu\,,\,L^\nu] = i\epsilon^{\mu\nu\rho\lambda}(L_\rho + \tfrac{1}{16}Q^\dagger\overline{\sigma}_\rho Q)P_\lambda\,. \tag{9.86}$$

When $P^2 = 0$, we impose the results of eq. (9.72) to obtain

$$P^\mu L_\mu = [L^\mu\,,\,Q_\alpha] = [L^\mu\,,\,Q^\dagger_{\dot\alpha}] = 0\,, \qquad \text{for } P^2 = 0\,. \tag{9.87}$$

Moreover, if we employ the identity

$$\epsilon^{\mu\nu\rho\lambda}\overline{\sigma}_\rho = \tfrac{1}{2}i(\overline{\sigma}^\nu\sigma^\mu\overline{\sigma}^\lambda - \overline{\sigma}^\lambda\sigma^\mu\overline{\sigma}^\nu)\,, \tag{9.88}$$

it then follows from eq. (9.72) that

$$\epsilon^{\mu\nu\rho\lambda}Q^\dagger\overline{\sigma}_\rho Q P_\lambda = 0\,, \qquad \text{for } P^2 = 0\,. \tag{9.89}$$

Hence, in the massless case, eq. (9.86) simplifies to

$$[L^\mu\,,\,L^\nu] = i\epsilon^{\mu\nu\rho\lambda}L_\rho P_\lambda\,, \qquad \text{for } P^2 = 0\,. \tag{9.90}$$

Finally, we evaluate $L^\mu L_\mu$ for the positive-energy massless one-particle states. As in the analysis of the Poincaré algebra, $w^\mu w_\mu = \lim_{m\to 0}(-m^2\vec{\boldsymbol{S}}^2) = 0$. Using eq. (9.89), it follows that

$$w^\mu Q^\dagger\overline{\sigma}_\mu Q = -\tfrac{1}{2}\epsilon^{\mu\nu\rho\lambda}J_{\nu\rho}P_\lambda Q^\dagger\overline{\sigma}_\mu Q = 0\,. \tag{9.91}$$

Using $\overline{\sigma}^{\mu\dot\alpha\alpha}\overline{\sigma}^{\dot\beta\beta}_\mu = 2\epsilon^{\alpha\beta}\epsilon^{\dot\alpha\dot\beta}$, we obtain

$$\begin{aligned}(Q^\dagger\overline{\sigma}^\mu Q)(Q^\dagger\overline{\sigma}_\mu Q) &= 2\epsilon^{\dot\alpha\dot\gamma}\epsilon^{\beta\tau}Q^\dagger_{\dot\alpha}Q_\beta Q^\dagger_{\dot\gamma}Q_\tau = 2\epsilon^{\dot\alpha\dot\gamma}\epsilon^{\beta\tau}Q^\dagger_{\dot\alpha}[2P_\mu\sigma^\mu_{\beta\dot\gamma} - Q^\dagger_{\dot\gamma}Q_\beta]Q_\tau \\ &= 2(Q^\dagger_{\dot\alpha}Q^{\dagger\dot\alpha})(Q^\beta Q_\beta) - 4P^\mu Q^\dagger\overline{\sigma}_\mu Q = 0\,,\end{aligned} \tag{9.92}$$

after applying the operator identities given in eqs. (9.72) and (9.82). Hence,

$$L^\mu L_\mu = 0\,, \qquad \text{for } P^2 = 0 \text{ and } P^0 > 0\,. \tag{9.93}$$

When $P^2 = 0$ and $P^0 > 0$, the properties of L^μ [given by eqs. (9.84), (9.87), (9.90), and (9.93)] match precisely the properties of the Pauli–Lubański vector (see Exercise 1.2). Thus, we must solve the equations $L^2 = P^2 = L_\mu P^\mu = 0$. It is simplest to choose a frame in which $P^\mu = P^0(1\,;0\,,0\,,1)$ and $P^0 > 0$. In this frame, it immediately follows that $L^\mu = L^0(1\,;0\,,0\,,1)$. That is, in any Lorentz frame,

$$L^\mu = \mathcal{K}P^\mu\,, \tag{9.94}$$

where $\mathcal{K} \equiv L^0/P^0$ is called the superhelicity operator. More explicitly, in a frame where $P^\mu = P^0(1\,;0\,,0\,,1)$,

$$\mathcal{K} = h + \frac{1}{8P^0}\left(Q_1^\dagger Q_1 + Q_2^\dagger Q_2\right)\,, \tag{9.95}$$

where $h \equiv w^0/P^0 = \vec{\boldsymbol{S}}\cdot\hat{\boldsymbol{P}}$ is the usual helicity operator acting on massless one-particle states. By virtue of eqs. (9.20) and (9.87), it follows that

$$[\mathcal{K}\,,\,P^\mu] = [\mathcal{K}\,,\,Q_\alpha] = [\mathcal{K}\,,\,Q^\dagger_{\dot\alpha}] = 0\,. \tag{9.96}$$

Hence, the states of the massless supermultiplet are eigenstates of $\mathcal{K}$, with possible eigenvalues $\kappa = 0, \pm\frac{1}{2}, \pm 1, \pm\frac{3}{2}, \ldots$. In contrast, h does not commute with Q_α and $Q^\dagger_{\dot\alpha}$. Thus, the different states of the massless supermultiplet will have different helicities.

For a fixed value of the superhelicity κ, there exists a distinguished state of the supermultiplet, denoted by $|\Omega\rangle$, that satisfies

$$Q_\beta|\Omega\rangle = 0\,, \qquad \mathcal{K}\,|\Omega\rangle = \kappa\,|\Omega\rangle\,. \tag{9.97}$$

To verify that such a state must exist, let us assume the contrary. Suppose that a state of the massless supermultiplet, denoted by $|\Psi\rangle$, exists that is not annihilated by Q_β. Due to eq. (9.96), it follows that $Q_\beta\,|\Psi\rangle$ must also be a state of the massless supermultiplet. Arguing as we did below eq. (9.49), we again arrive at a contradiction. Consequently, there must be (at least) one state of the massless supermultiplet that satisfies eq. (9.97). Moreover, this state must be unique under the assumption that the massless supermultiplet with superhelicity κ is an *irreducible* representation of the $N = 1$ supersymmetry algebra.

The states of the massless supermultiplet are obtained by considering the series

$$|\Omega\rangle\,,\; Q^\dagger_{\dot\alpha}\,|\Omega\rangle\,,\; Q^\dagger_{\dot\beta}Q^\dagger_{\dot\alpha}\,|\Omega\rangle\,. \tag{9.98}$$

However, $Q^\dagger_{\dot\beta}Q^\dagger_{\dot\alpha}\,|\Omega\rangle = 0$ as a consequence of eq. (9.82), and $P^\lambda Q^\dagger_{\dot\beta}\overline{\sigma}_\lambda^{\dot\beta\tau}\,|\Omega\rangle = 0$ as a consequence of eq. (9.72). Thus, in contrast to the massive supermultiplet, the massless supermultiplet contains only two states. These two states are eigenvalues of the helicity operator h. To determine the corresponding helicities, we shall employ the standard reference frame where $P^\mu = P^0(1\,;\,0\,,\,0\,,\,1)$. Since eq. (9.72) yields $Q_1 = Q^\dagger_1 = 0$, it follows that the massless $N = 1$ supermultiplet consists of the two states $|\Omega\rangle$ and $Q^\dagger_2\,|\Omega\rangle$. Using eqs. (9.95) and (9.97), the helicities of these two states can be determined:

$$h\,|\Omega\rangle = \left[\mathcal{K} - \frac{1}{8P^0}\left(Q^\dagger_1 Q_1 + Q^\dagger_2 Q_2\right)\right]|\Omega\rangle = \kappa\,|\Omega\rangle\,, \tag{9.99}$$

$$\begin{aligned} hQ^\dagger_2\,|\Omega\rangle &= \left[\mathcal{K} - \frac{1}{8P^0}\left(Q^\dagger_1 Q_1 + Q^\dagger_2 Q_2\right)\right]Q^\dagger_2\,|\Omega\rangle \\ &= \left[\kappa Q^\dagger_2 - \frac{1}{8P^0}Q^\dagger_2\left(2P_\mu\sigma^\mu_{22} - Q^\dagger_2 Q_2\right) - \frac{1}{8P^0}Q^\dagger_1\left(2P_\mu\sigma^\mu_{12} - Q^\dagger_2 Q_1\right)\right]|\Omega\rangle \\ &= \left[\kappa - \tfrac{1}{4}(\sigma^0_{22} - \sigma^3_{22})\right]Q^\dagger_2\,|\Omega\rangle \\ &= (\kappa - \tfrac{1}{2})Q^\dagger_2\,|\Omega\rangle\,. \end{aligned} \tag{9.100}$$

Indeed, the superhelicity κ is the maximal helicity of the massless $N = 1$ super-

Table 9.2 States of an $N = 1$ massless supermultiplet of superhelicity κ and the corresponding CPT conjugates which inhabit an $N = 1$ massless supermultiplet of superhelicity $-\kappa + \frac{1}{2}$. An interpretation is provided for $\kappa = s$ and $\kappa = s - \frac{1}{2}$, where s is a positive integer. In the special case of $\kappa = \frac{1}{2}$, the scalar boson of the supermultiplet is complex, whereas for $\kappa = 1, \frac{3}{2}, 2, \ldots$ the bosonic member of the supermultiplet is real with nonzero spin. In all cases, the number of bosonic and the number of fermionic degrees of freedom coincide and are equal to 2.

helicities	degrees of freedom	interpretation ($\kappa = s$)	interpretation ($\kappa = s - \frac{1}{2}$)
$\kappa\,, -\kappa$	2	spin-s boson	spin-$(s - \frac{1}{2})$ fermion
$\kappa - \frac{1}{2}\,, -\kappa + \frac{1}{2}$	2	spin-$(s - \frac{1}{2})$ fermion	spin-$(s - 1)$ boson

multiplet. Thus, an irreducible $N = 1$ massless supermultiplet with superhelicity κ consists of two massless states with helicity κ and $\kappa - \frac{1}{2}$, respectively.[5]

Any quantum field theory realization of supersymmetry respects a discrete CPT symmetry. Since the helicity changes sign under a discrete CPT transformation, it follows that any irreducible massless supermultiplet with superhelicity κ must be accompanied by an irreducible massless supermultiplet with superhelicity $-\kappa + \frac{1}{2}$. These results are summarized in Table 9.2. For example, if $\kappa = \frac{1}{2}$, then the resulting massless supermultiplet contains two states of helicity 0 and two states of helicity $\pm\frac{1}{2}$. This is the massless limit of a chiral supermultiplet, which comprises a massless complex scalar and a massless Majorana fermion. If $\kappa = 1$, then the resulting massless supermultiplet consists of two states of helicity $\pm\frac{1}{2}$ and two states of helicity ± 1. This is a *gauge supermultiplet* (e.g., the photino and the photon of supersymmetric QED), which comprises a massless Majorana fermion and a massless spin-1 particle. The case of $\kappa = 2$ corresponds to a massless spin-3/2 and a massless spin-2 particle, which is realized in supergravity by the supermultiplet that contains the gravitino and the graviton. In all cases, the number of bosonic and the number of fermionic degrees of freedom coincide and are equal to two.

9.4 Consequences of Super-Poincaré Invariance

A Poincaré invariant quantum field theory respects the Poincaré algebra generated by $\{P^\mu\,, J^{\mu\nu}\}$, which satisfy commutation relations given by eqs. (9.1)–(9.3). One of the basic postulates of Poincaré-invariant quantum field theory states that a

[5] In the literature, it is more common to define $L^\mu = (\mathcal{K} + \frac{1}{2})P^\mu$, in which case the helicities of the massless $N = 1$ supermultiplet are $\kappa + \frac{1}{2}$ and κ (e.g., see [B20, B37]). In our opinion, the definition of the superhelicity operator given in eq. (9.94) is cleaner.

translation-invariant, Lorentz-invariant vacuum $|0\rangle$ exists such that [B66]

$$P^\mu |0\rangle = 0\,, \qquad J^{\mu\nu} |0\rangle = 0\,. \tag{9.101}$$

In particular, $\langle 0| P^\mu |0\rangle = 0$. Indeed, if $\langle 0| P^\mu |0\rangle \neq 0$, then the vacuum would not be invariant under Lorentz transformations. This is easily proven by taking the vacuum expectation value of [see eq. (1.283)],

$$\exp\left(\tfrac{1}{2}i\theta_{\rho\tau}J^{\rho\tau}\right) P^\mu \exp\left(-\tfrac{1}{2}i\theta_{\rho\tau}J^{\rho\tau}\right) = \Lambda^\mu{}_\nu P^\nu\,, \tag{9.102}$$

where the $\theta_{\rho\tau} = -\theta_{\rho\tau}$ parameterize the 4×4 Lorentz transformation matrix $\Lambda^\mu{}_\nu$ [see eqs. (1.2) and (1.4)]. Using $J^{\mu\nu}|0\rangle = 0$, it follows that

$$\langle 0| P^\mu |0\rangle = \Lambda^\mu{}_\nu \langle 0| P^\nu |0\rangle\,, \tag{9.103}$$

which holds for all Lorentz transformations Λ. Thus, it follows that $\langle 0| P^\mu |0\rangle = 0$.

A super-Poincaré invariant quantum field theory respects the supersymmetry algebra generated by $\{P^\mu\,,\, J^{\mu\nu}\,,\, Q_\alpha\,,\, Q^{\dagger\dot\alpha}\}$, which satisfy the commutation relations of the Poincaré algebra, and in addition satisfy the (anti)commutation relations given by eqs. (9.20)–(9.24). Two important consequences can be established:

1. *The vanishing of the vacuum energy is a necessary and sufficient condition for the existence of a global supersymmetric vacuum.*

2. *In a theory governed by a supersymmetric action, for a fixed nonzero P_μ the number of bosonic and the number of fermionic degrees of freedom coincide.*

We address these two results in the next two subsections.

9.4.1 The Vacuum Energy of a Globally Supersymmetric Theory

In this subsection, we prove that the vanishing of the vacuum energy is a necessary and sufficient condition for the existence of a global supersymmetric vacuum. A formal proof of this statement is obtained as follows. Consider the anticommutation relations of the fermionic generators of the supersymmetric algebra,

$$\{Q_\alpha\,,\, Q^\dagger_{\dot\beta}\} = 2\sigma^\mu_{\alpha\dot\beta} P_\mu\,. \tag{9.104}$$

Following the derivation of eq. (9.18),

$$P^0 = \tfrac{1}{4}\left[Q_1Q_1^\dagger + Q_1^\dagger Q_1 + Q_2Q_2^\dagger + Q_2^\dagger Q_2\right]. \tag{9.105}$$

Since the right-hand side of eq. (9.18) is positive semidefinite (and neither Q nor $Q^\dagger$ is the zero operator), it follows that

$$\langle 0 | P^0 | 0\rangle = 0 \quad\Longleftrightarrow\quad Q_\alpha |0\rangle = 0\,. \tag{9.106}$$

In particular, $Q_\alpha|0\rangle = 0$ implies that the vacuum is supersymmetric, in the same way that $P^\mu|0\rangle = J^{\mu\nu}|0\rangle = 0$ imply that the vacuum is translation-invariant and

Lorentz-invariant.[6] More generally, $Q_\alpha |0\rangle = 0$ implies that the vacuum expectation value of the energy–momentum tensor is zero [94].

However, eq. (9.106) is troubling for two separate reasons. First, suppose that the action of the theory is invariant under supersymmetric transformations, but the vacuum is not preserved by supersymmetry. Then, $Q_\alpha |0\rangle \neq 0$ and one says that supersymmetry is spontaneously broken. In this case, eq. (9.105) implies that $\langle 0| P^0 |0\rangle > 0$, which contradicts eq. (9.101). Thus, it appears that the spontaneous breaking of supersymmetry is not possible without breaking Lorentz invariance. Perhaps a more fundamental objection is that the concept of the vacuum energy is usually considered to be unphysical in nongravitational theories, as it is commonly asserted that only energy differences are physical. Thus, it seems to be a matter of convention to choose the vacuum energy such that $\langle 0| P^0 |0\rangle = 0$.

To overcome the objections raised above, we re-examine the concept of the vacuum energy in relativistic (nongravitational) quantum field theory. By the Noether procedure, one can derive a canonical energy–momentum tensor, $T^{(c)}_{\mu\nu}$, which is conserved, $\partial^\mu T^{(c)}_{\mu\nu} = 0$.[7] One can then formally compute the vacuum energy density by summing over the vacuum Feynman diagrams of the theory. By Lorentz covariance,

$$\langle 0| T^{(c)}_{\mu\nu} |0\rangle = \mathcal{E} g_{\mu\nu} \,, \tag{9.107}$$

where $\mathcal{E}$ is typically ultraviolet divergent. Since the Hamiltonian density is identified as $\mathscr{H} = T_{00}$, it follows that $\mathcal{E}$ is the vacuum energy density. However, one is always free to define a new subtracted energy–momentum tensor,

$$T_{\mu\nu} \equiv T^{(c)}_{\mu\nu} - \mathcal{E} g_{\mu\nu} \,, \tag{9.108}$$

which is a Lorentz-covariant expression.[8] By construction, $\partial^\mu T_{\mu\nu} = 0$ and

$$\langle 0| T_{\mu\nu} |0\rangle = 0 \,. \tag{9.109}$$

The energy–momentum tensor $T_{\mu\nu}$ plays a distinguished role in relativistic quantum field theory, since it can be used to construct the generators of spacetime translations

$$P_\mu = \int d^3x \, T_\mu{}^0 \,, \tag{9.110}$$

which satisfy $\langle 0| P_\mu |0\rangle = 0$. Indeed, P defined by eq. (9.110) is a four-vector with respect to Lorentz transformations. Likewise, one can construct a distinguished angular momentum tensor density $M_{\mu\nu\lambda}$ that can be used to construct the generators of Lorentz transformations

$$J_{\mu\nu} = \int d^3x \, M_{\mu\nu}{}^0 \,, \tag{9.111}$$

6 Equivalently, $\langle 0| \{Q_\alpha \,, Q^\dagger_{\dot\beta}\} |0\rangle = 0$, by covariance with respect to the supersymmetry algebra, since there are no spinor quantities with one undotted and one dotted index that can appear on the right-hand side of the equation. Hence, $Q_\alpha |0\rangle = 0$, which then yields $\langle 0| P^0 |0\rangle = 0$.

7 The arguments given here do not depend on whether one employs the canonical energy–momentum tensor or the improved symmetrized energy–momentum tensor.

8 For example, in the quantum theory of free fields, the vacuum energy is set to zero by defining the Hamiltonian density to be normal-ordered.

which satisfy $\langle 0|\, J_{\mu\nu}\, |0\rangle = 0$.

However, in a supersymmetric theory, another choice of the energy–momentum tensor is natural. The fermionic generators Q_α and $Q^{\dagger\dot\alpha}$ of the supersymmetric algebra are time-independent (conserved) quantities that are obtained by integrating the zeroth component of the supercurrents:

$$Q_\alpha = \int d^3x\, J^0_\alpha\,, \qquad Q^{\dagger\dot\alpha} = \int d^3x\, J^{\dagger\dot\alpha\, 0}\,. \tag{9.112}$$

In a theory governed by a supersymmetric Lagrangian, the supercurrents J^μ_α and $J^{\dagger\dot\alpha\,\mu}$ are related by supersymmetry to an energy–momentum tensor, denoted by $T^{(\rm SUSY)}_{\mu\nu}$. Then, the proper interpretation of eq. (9.104) is

$$\{Q_\alpha\,, Q^\dagger_{\dot\beta}\} = 2\sigma^\mu_{\alpha\dot\beta} \int d^3x\, T^{(\rm SUSY)}{}_\mu{}^0\,. \tag{9.113}$$

One can then rewrite the above anticommutation relation as

$$\{Q_\alpha\,, Q^\dagger_{\dot\beta}\} = 2\sigma^\mu_{\alpha\dot\beta} P_\mu + 2E_0\sigma^0_{\alpha\dot\beta}\,, \tag{9.114}$$

where P_μ is defined by eq. (9.110) and

$$E_0 \equiv \int d^3x\, \langle 0|\, T^{(\rm SUSY)}{}_0{}^0\, |0\rangle\,. \tag{9.115}$$

If $E_0 = 0$ (which corresponds to $T^{(\rm SUSY)}_{\mu\nu} = T_{\mu\nu}$), then we recover the standard supersymmetry algebra, and the vacuum is supersymmetric. If $E_0 \neq 0$, then eq. (9.114) is consistent with $\langle 0|\, P^\mu\, |0\rangle = 0$ (which is required by the Lorentz-invariant vacuum) and with $Q_\alpha\, |0\rangle \neq 0$. In particular, E_0 serves as an order parameter for broken supersymmetry.

Note that $E_0 \geq 0$, since eq. (9.114) implies that

$$E_0 = \tfrac14 \langle 0|\, Q_1 Q_1^\dagger + Q_1^\dagger Q_1 + Q_2 Q_2^\dagger + Q_2^\dagger Q_2\, |0\rangle \geq 0\,. \tag{9.116}$$

In supersymmetric theories, it is common to call E_0 the vacuum energy. Thus, if supersymmetry is spontaneously broken, then this definition of the vacuum energy is not compatible with usual conventions of quantum field theory in which the vacuum energy is defined to be zero.

Although the conclusions obtained above are correct, the derivation of eq. (9.114) is still somewhat formal. Indeed, if the vacuum breaks supersymmetry, then the integrals in eq. (9.112) do not converge when integrated over an infinite volume (this is an infrared divergence), so strictly speaking the fermionic generators Q_α and $Q^{\dagger\dot\alpha}$ are undefined.[9] Nevertheless, the supercurrents are conserved, as expected in a supersymmetric theory with no *explicit* supersymmetry breaking. In Section 12.1, we will demonstrate that, given a supersymmetric Lagrangian, if the vacuum breaks supersymmetry then a massless Goldstone fermion exists in the spectrum. The

[9] Moreover, given a nonzero value for $\langle 0|\, T^{(\rm SUSY)}{}_0{}^0\, |0\rangle$, which is a constant by translational invariance, one sees that E_0 defined in eq. (9.115) also diverges in the infinite volume limit.

long-range forces mediated by this massless particle are responsible for the nonconvergence of the integrals in eq. (9.112). Equivalently, in a spontaneously broken globally supersymmetric theory, applying Q_α to the vacuum creates a zero-momentum massless fermionic state, which is a state of infinite norm [B39].

9.4.2 Equality of Bosonic and Fermionic Degrees of Freedom in Supersymmetry

In a theory governed by a supersymmetric action, for a fixed nonzero P_μ the number of bosonic and the number of fermionic degrees of freedom coincide [see eq. (7.10)]. A formal proof of this result employs an operator introduced by Witten in Ref. [95], denoted by $(-1)^F$, with the following properties:

$$(-1)^F |B\rangle = |B\rangle \,, \qquad (-1)^F |F\rangle = -|F\rangle \,, \tag{9.117}$$

where $|B\rangle$ is a bosonic state and $|F\rangle$ is a fermionic state. Moreover, observe that the application of Q_α or $Q^\dagger_{\dot\alpha}$ to a physical state changes that state by adding half a unit of spin. An explicit example of this behavior can be seen in eqs. (9.67) and (9.68). We can summarize this behavior in the following schematic equations:

$$Q_\alpha |B\rangle = |F\rangle \,, \qquad Q_\alpha |F\rangle = |B\rangle \,, \tag{9.118}$$

and similarly for the application of $Q^\dagger_{\dot\alpha}$.

Employing the notation introduced above, note that

$$Q_\alpha(-1)^F |F\rangle = -Q_\alpha |F\rangle = -|B\rangle \,, \tag{9.119}$$

$$(-1)^F Q_\alpha |F\rangle = (-1)^F |B\rangle = |B\rangle \,, \tag{9.120}$$

and similarly for the application of $Q^\dagger_{\dot\alpha}$. It follows that Q_α [and $Q^\dagger_{\dot\alpha}$] anticommute with $(-1)^F$:

$$\{Q_\alpha \,, (-1)^F\} = \{Q^\dagger_{\dot\alpha} \,, (-1)^F\} = 0 \,. \tag{9.121}$$

Using eq. (9.121), we can evaluate the following trace over physical states of fixed momenta p^μ:

$$\begin{aligned}
\mathrm{Tr}\left[(-1)^F\{Q_\alpha \,, Q^\dagger_{\dot\beta}\}\right] &= \mathrm{Tr}\left[(-1)^F(Q_\alpha Q^\dagger_{\dot\beta} + Q^\dagger_{\dot\beta} Q_\alpha)\right] \\
&= \mathrm{Tr}\left[-Q_\alpha(-1)^F Q^\dagger_{\dot\beta} + (-1)^F Q^\dagger_{\dot\beta} Q_\alpha\right] \\
&= \mathrm{Tr}\left[-Q^\dagger_{\dot\beta} Q_\alpha(-1)^F + Q^\dagger_{\dot\beta} Q_\alpha(-1)^F\right] \\
&= 0 \,,
\end{aligned} \tag{9.122}$$

after a cyclic permutation within the trace at the penultimate step. Using eq. (9.24), we conclude that

$$\mathrm{Tr}\big[(-1)^F P^\mu\big] = 0 \,. \tag{9.123}$$

When P^μ is applied to a physical state of definite four-momentum, it yields the corresponding eigenvalue p^μ. For a fixed nonzero p^μ,

$$\mathrm{Tr}(-1)^F = \sum_{\{r\}} \langle p^\mu, \{r\}|\,(-1)^F\,|p^\mu, \{r\}\rangle = N_B(p^\mu) - N_F(p^\mu) = 0\,, \quad (9.124)$$

where $\{r\}$ indicates all other quantum numbers of the physical state. Thus, the number of bosonic (N_B) and the number of fermionic (N_F) degrees of freedom coincide.

The only case where the equality of bosonic and fermionic degrees of freedom can break down is when $P^\mu = 0$, corresponding to the vacuum state of the supersymmetric theory. Consequently, if we extend the trace over the four-momenta of all physical states of the Hilbert space, *including* $p^\mu = 0$, then we are left with

$$\mathrm{Tr}(-1)^F = N_B(p^\mu = 0) - N_F(p^\mu = 0)\,, \quad (9.125)$$

which may or may not be equal to zero. This quantity is an integer called the Witten index, which provides a key diagnostic for determining the nature of the vacuum. Note that if $\mathrm{Tr}(-1)^F \neq 0$, then eq. (9.125) implies that a zero-energy vacuum state exists. In light of the results of Section 9.4.1, it follows that supersymmetry is unbroken. For example, Witten showed that for an SU(N) supersymmetric Yang–Mills theory, $\mathrm{Tr}(-1)^F = N$, implying that a supersymmetric vacuum state exists [95]. In contrast, if $\mathrm{Tr}(-1)^F = 0$ then the existence of supersymmetry-conserving vacuum state cannot be decided without further information. In particular, if $\mathrm{Tr}(-1)^F = 0$ then either $N_B(p^\mu = 0) = N_F(p^\mu = 0) = 0$, in which case no zero-energy vacuum state exists, corresponding to spontaneously broken supersymmetry, or $N_B(p^\mu = 0) = N_F(p^\mu = 0) > 0$, in which case supersymmetry is unbroken.

9.5 Extended Supersymmetry

In Section 9.2, we noted a theorem due to Haag, Łopuszański, and Sohnius which states that the only possible extension of the Poincaré algebra involves the addition of new fermionic generators that transform either as a $(\frac{1}{2}, 0)$ or $(0, \frac{1}{2})$ representation of the Poincaré algebra, denoted by Q^i_α and its hermitian conjugate $Q^\dagger_{\dot\alpha i} \equiv (Q^i_\alpha)^\dagger$, respectively, where $i \in \{1, 2, \ldots, N\}$. Until now, we have examined in detail the case of $N = 1$. In this section, we briefly look at extended supersymmetric theories in which $N > 1$. Further details can be found in [B24] and in Refs. [11, 19].

The anticommutators of the fermionic generators can be deduced using the same techniques as the ones given in Section 9.2. Due to the extra degree of freedom specified by the "flavor" index i, the most general form for the anticommutators is given by (see Exercise 9.10)[10]

$$\{Q^i_\alpha\,,\,Q^\dagger_{\dot\beta j}\} = 2P_\mu\sigma^\mu_{\alpha\dot\beta}\delta^i_j\,, \tag{9.126}$$

$$\{Q^i_\alpha\,,\,Q^j_\beta\} = \epsilon_{\alpha\beta}Z^{ij}\,, \tag{9.127}$$

$$\{Q^\dagger_{\dot\alpha i}\,,\,Q^\dagger_{\dot\beta j}\} = -\epsilon_{\dot\alpha\dot\beta}Z_{ij}\,, \tag{9.128}$$

where $Z_{ij} \equiv (Z^{ji})^*$. The *central charges* Z^{ij} commute with all the generators of the supersymmetry algebra. Moreover, the structure of the anticommutation relations, eqs. (9.127) and (9.128), imply that the central charges are complex Lorentz scalars which satisfy

$$Z^{ij} = -Z^{ji} = -(Z_{ij})^*\,. \tag{9.129}$$

One is always free to redefine the fermionic generators by $Q^i_\alpha \to U^i{}_jQ^j$ and $Q^\dagger_{\dot\alpha i} \to (U^*)_i{}^jQ^\dagger_{\dot\beta j}$, where U is a unitary matrix. This redefinition leaves eq. (9.126) unchanged, whereas the central charges are transformed, $Z \to Z' \equiv UZU^{\mathsf{T}}$. Using eq. (G.21), one can choose the matrix V such that

$$Z' = \text{diag}\left\{\begin{pmatrix} 0 & Z_1 \\ -Z_1 & 0\end{pmatrix},\begin{pmatrix} 0 & Z_2 \\ -Z_2 & 0\end{pmatrix},\ldots,\begin{pmatrix} 0 & Z_p \\ -Z_p & 0\end{pmatrix},\mathbb{O}_{N-2p}\right\}, \tag{9.130}$$

where Z' is written in block-diagonal form with 2×2 matrices appearing along the diagonal, followed by an $(N-2p)\times(N-2p)$ block of zeros (denoted by $\mathbb{O}_{N-2p}$), and the Z_j are real and positive. Note that there are at most $\lfloor\frac{1}{2}N\rfloor$ nonzero Z_i.[11]

If $Z_{ij}=0$, then eqs. (9.126)–(9.128) are unchanged under

$$Q^i_\alpha \to U^i{}_jQ^j\,, \qquad Q^\dagger_{\dot\alpha i} \to (U^*)_i{}^jQ^\dagger_{\dot\beta j}\,, \qquad \text{where } U \equiv e^{-i\theta^a S_a}\,, \tag{9.131}$$

and the S_a are $N\times N$ hermitian matrices. Hence, the extended supersymmetry algebra can be further enlarged by adding a set of bosonic hermitian generators R_a such that

$$e^{i\theta^a R_a}Q^i_\alpha e^{-i\theta^a R_a} = (e^{-i\theta^a S_a})^i{}_jQ^j\,, \tag{9.132}$$

$$e^{i\theta^a R_a}Q^\dagger_{\dot\alpha i}e^{-i\theta^a R_a} = (e^{i\theta^a S^*_a})_i{}^jQ^\dagger_{\dot\alpha j}\,. \tag{9.133}$$

Expanding out to first order in the θ^a, it follows that

$$[Q^i_\alpha\,,\,R_a] = (S_a)^i{}_jQ^j_\alpha\,, \tag{9.134}$$

$$[Q^\dagger_{\dot\alpha i}\,,\,R_a] = -(S^*_a)_i{}^jQ^\dagger_{\dot\alpha j}\,. \tag{9.135}$$

That is, the fermionic generators Q^i_α and $Q^\dagger_{\dot\alpha i}$ are charged under a $\mathrm{U}(N)_R$ symmetry,

[10] To understand the motivation for the extra minus sign in eq. (9.128) [which is needed due to the definition of Z_{ij}], consider the case of $N=2$ where $Z^{ij}=Z\epsilon^{ij}$ and $Z_{ij}=Z^*\epsilon_{ij}$. Here we employ a convention where the antisymmetric tensor ϵ is defined such that $\epsilon^{12}=-\epsilon^{21}=\epsilon_{21}=-\epsilon_{12}=1$ [in analogy with eq. (1.39)], which is consistent with the definition $Z_{ij}\equiv(Z^{ji})^*$.

[11] Here, $\lfloor\frac{1}{2}N\rfloor$ is the greatest integer less than or equal to $\frac{1}{2}N$.

whose Lie algebra determines the commutation relations

$$[R_a\,,\,R_b] = if_{abc}R_c\,. \tag{9.136}$$

The generators R_a commute with P^μ and $J^{\mu\nu}$, as the Poincaré generators are uncharged with respect to the $\mathrm{U}(N)_R$-symmetry. That is, $\mathrm{U}(N)$ is the largest allowed R-symmetry group of N-extended supersymmetry.

If $Z^{ij} \neq 0$, then eqs. (9.127) and (9.128) are unchanged under eq. (9.131) if

$$UZU^{\mathsf{T}} = Z\,. \tag{9.137}$$

In this case, the largest possible R-symmetry group is necessarily a proper subgroup of $\mathrm{U}(N)$. As an example, if N is even and $Z_1 = Z_2 = \cdots = Z_{N/2}$ [see eq. (9.130)], then eq. (9.137) reduces to

$$UJU^{\mathsf{T}} = J\,, \qquad \text{where } J \equiv \mathrm{diag}\left\{\left(\begin{smallmatrix} 0 & 1 \\ -1 & 0\end{smallmatrix}\right), \left(\begin{smallmatrix} 0 & 1 \\ -1 & 0\end{smallmatrix}\right), \ldots, \left(\begin{smallmatrix} 0 & 1 \\ -1 & 0\end{smallmatrix}\right)\right\}. \tag{9.138}$$

This condition defines the unitary symplectic group $\mathrm{Sp}(\tfrac{1}{2}N)$. In general, the Z_{ij} are linear combinations of the R-symmetry group generators:

$$Z^{ij} = c^{aij}R_a\,. \tag{9.139}$$

Equation (9.137) then yields

$$(S_a)^i{}_k c^{bkj} = -(S^*_a)_k{}^j c^{bik}\,. \tag{9.140}$$

The representations of the extended supersymmetry algebra can be determined following the methods employed in Section 9.3. Consider first the case of an irreducible massless supermultiplet. Equation (9.72) generalizes to

$$P^\rho \overline{\sigma}_\rho^{\dot{\gamma}\alpha} Q^i_\alpha = P^\lambda Q^\dagger_{\dot{\beta} i}\overline{\sigma}_\lambda^{\dot{\beta}\tau} = 0\,, \qquad \text{for } P^2 = 0\,. \tag{9.141}$$

We choose the standard reference frame where $P^\mu = P^0(1\,;0\,,0\,,1)$. In this reference frame, eq. (9.141) reduces to $Q^i_1 = Q^\dagger_{\dot{1}i} = 0$, which immediately implies that the central charges $Z^{ij} = 0$. It is convenient to rescale the nonzero fermionic generators:

$$a_i \equiv \frac{1}{2\sqrt{P_0}}Q^i_2\,, \qquad a^\dagger_i \equiv \frac{1}{2\sqrt{P_0}}Q^\dagger_{\dot{2}i}\,. \tag{9.142}$$

In this case, the anticommutators given in eqs. (9.126)–(9.128) corresponding to $\alpha = \dot{\alpha} = 2$ simplify to

$$\{a_i\,,\,a^\dagger_j\} = \delta_{ij}\,, \qquad \{a_i\,,\,a_j\} = \{a^\dagger_i\,,\,a^\dagger_j\} = 0\,, \tag{9.143}$$

which we recognize as a Clifford algebra. Following a similar procedure to the one presented in Section 9.3, we can define the state $|\Omega\rangle$ (called the Clifford vacuum) by $a_i\,|\Omega\rangle = 0$. All the states of the massless supermultiplet are obtained by considering

$$|\Omega\rangle\,,\; a^\dagger_i\,|\Omega\rangle\,,\; a^\dagger_i a^\dagger_j\,|\Omega\rangle\,,\;\ldots\,,\; a^\dagger_1 a^\dagger_2 \cdots a^\dagger_N\,|\Omega\rangle\,, \tag{9.144}$$

where the indices $i, j \in \{1, 2, \ldots, N\}$. Since the $a^\dagger_i$ anticommute, it follows that the number of independent states of the form $a^\dagger_{i_1} a^\dagger_{i_2} \cdots a^\dagger_{i_k} |\Omega\rangle$ (corresponding to the possible distinct values of the indices $i_1, i_2, \ldots, i_k$) is given by

$$\binom{N}{k} \equiv \frac{N!}{k!(N-k)!} \,. \tag{9.145}$$

The superhelicity is defined as in eq. (9.94) using the generalization of eq. (9.83), $L^\mu \equiv w^\mu + \frac{1}{8} Q^{\dagger i} \bar{\sigma}^\mu Q_i$. An argument analogous to eq. (9.100) shows that every time a creation operator $a^\dagger_i$ is applied, the helicity is lowered by half a unit. Thus, if $h\,|\Omega\rangle = \lambda\,|\Omega\rangle$, where λ is the maximal helicity of the supermultiplet, then the helicity of the states listed in eq. (9.144) are given by

$$\lambda\,,\, \lambda - \tfrac{1}{2}\,,\, \lambda - 1\,,\, \ldots\,,\, \lambda - \tfrac{1}{2}N\,, \tag{9.146}$$

with corresponding multiplicities

$$\binom{N}{0}, \binom{N}{1}, \binom{N}{2}, \ldots, \binom{N}{N} \,. \tag{9.147}$$

The total number of states is

$$\sum_{k=0}^{N} \binom{N}{k} = 2^N \,, \tag{9.148}$$

of which 2^{N-1} are fermionic and 2^{N-1} are bosonic. If $\lambda = \frac{1}{4}N$, then the helicities of the states of the irreducible massless supermultiplet are $\lambda, \lambda - \frac{1}{2}, \ldots, -\lambda + \frac{1}{2}, -\lambda$, in which case the supermultiplet is CPT self-conjugate. Otherwise, one must explicitly add the corresponding CPT-conjugate states by adjoining a second irreducible massless supermultiplet whose helicities are given by $\frac{1}{2}N - \lambda, \ldots, 1 - \lambda, \frac{1}{2} - \lambda, -\lambda$.

We have already seen in Section 9.3 that the massless states of the irreducible $N = 1$ supermultiplet with maximal helicity $\lambda = \frac{1}{2}$ consist of a scalar state of helicity zero and a spin-1/2 fermion state of helicity $\frac{1}{2}$. To this, we must add a second irreducible supermultiplet consisting of a scalar state of helicity zero and a spin-1/2 fermion state of helicity $-\frac{1}{2}$. In principle, these two irreducible massless supermultiplets can transform under different representations with respect to some global or local symmetry. If this is the case, then the interactions of the spin-1/2 fermions are chiral.

In contrast, if $N > 1$, the interactions of the spin-1/2 particles are necessarily vectorlike. For example, consider the simplest case of $N = 2$ and $\lambda = \frac{1}{2}$. The resulting $N = 2$ massless *hypermultiplet* consists of states with helicities $\frac{1}{2}, 0, 0, -\frac{1}{2}$. In this case, the helicity $\pm\frac{1}{2}$ fermions appear in the *same* supermultiplet, and hence must transform in the same way with respect to any global or local symmetry, which implies that the interactions of the spin-1/2 particles are vectorlike. In particular, the $N = 2$ massless hypermultiplet can be decomposed into two mass-degenerate $N = 1$ chiral supermultiplets of opposite chirality. Next, we consider the case of $N = 2$ and $\lambda = 1$. The resulting $N = 2$ massless vector supermultiplet consists of states with helicities $1, \frac{1}{2}, \frac{1}{2}, 0$, in which case we must append the corresponding CPT-conjugate states. Indeed, in this case the $\pm\frac{1}{2}$ helicity states appear in different supermultiplets. However, each of these two supermultiplets (which contain a massless spin-1 particle) is a gauge supermultiplet. Since massless gauge bosons must appear in the adjoint representation of the corresponding gauge group (which is necessarily a real representation), it follows that the interactions of the fermion member of the gauge supermultiplet are vectorlike. As these conclusions apply to all $N > 1$ massless supermultiplets, the interactions of the spin-1/2 fermions of an extended supersymmetric theory must always be vectorlike. Since the weak interactions of Standard Model fermions are chiral, it follows that any viable supersymmetric extension of the Standard Model must employ $N = 1$ supersymmetry.

Nevertheless, extended supersymmetric models do possess some remarkable properties and thus are of theoretical interest. For (nongravitational) renormalizable quantum field theories, the maximal allowed helicity is $\lambda = 1$. This condition restricts $N \leq 4$. The case of $\lambda = 1$ and $N = 4$ corresponds to $N = 4$ super-Yang–Mills theory, which is known to be a finite quantum field theory. For consistent gravitational theories, the maximum allowed helicity is $\lambda = 2$. This condition restricts $N \leq 8$. The case of $\lambda = 2$ and $N = 8$ corresponds to $N = 8$ supergravity, which is the maximally supersymmetric gravitational theory.

The representations of the extended supersymmetry algebra for massive supermultiplets is more involved since in this case the central charges can be nonzero. Here, we summarize some of the main features. It is convenient to work in the rest frame, where $P^\mu = (m\,;\vec{\mathbf{0}})$. In the absence of central charges, one can simply generalize the analysis of Section 9.3 by defining annihilation and creation operators,

$$a_{\alpha i} \equiv \frac{1}{\sqrt{2m}} Q^i_\alpha \,, \qquad\qquad a^\dagger_{\dot\alpha i} \equiv \frac{1}{\sqrt{2m}} Q^\dagger_{\dot\alpha i} \,. \tag{9.149}$$

Equation (9.44) now generalizes to

$$\{a_{1i}\,, a^\dagger_{1j}\} = \delta_{ij}\,, \qquad\qquad \{a_{2i}\,, a^\dagger_{2j}\} = \delta_{ij}\,, \tag{9.150}$$

with all other anticommutators vanishing. We define the Clifford vacuum $|\Omega\rangle$ by

$$a_{\alpha i}\,|\Omega\rangle = 0\,, \qquad \text{where } i \in \{1, 2, \ldots, N\}\,, \tag{9.151}$$

and then construct the states of the massive supermultiplet by considering

$$|\Omega\rangle\,,\ a^\dagger_{\alpha i}\,|\Omega\rangle\,,\ a^\dagger_{\alpha i} a^\dagger_{\beta j}\,|\Omega\rangle\,,\ \ldots\,,\ a^\dagger_{11} a^\dagger_{21} a^\dagger_{12} a^\dagger_{22} \cdots a^\dagger_{1N} a^\dagger_{2N}\,|\Omega\rangle\,, \tag{9.152}$$

with corresponding multiplicities

$$\binom{2N}{0},\ \binom{2N}{1},\ \binom{2N}{2},\ \ldots,\ \binom{2N}{2N}. \tag{9.153}$$

The total number of states is

$$\sum_{k=0}^{2N} \binom{2N}{k} = 2^{2N}\,, \tag{9.154}$$

of which 2^{2N-1} are fermionic and 2^{2N-1} are bosonic. If the spin of $|\Omega\rangle$ is j, then the spin of $a^\dagger_{11}a^\dagger_{21}a^\dagger_{12}a^\dagger_{22}\cdots a^\dagger_{1N}a^\dagger_{2N}\,|\Omega\rangle$ is also j, since the latter state is completely antisymmetric under the interchange of any two creation operators. Thus, the maximal spin of the supermultiplet is $j+\tfrac{1}{2}N$, which corresponds to the appropriate linear combination of states obtained by multiplying $|\Omega\rangle$ by the product of N creation operators. Since $|\Omega\rangle$ is $(2j+1)$-fold degenerate (corresponding to the possible values of $j_3=-j,-j+1,\ldots,j-1,j$), it follows that the total number of states of the massive supermultiplet is $(2j+1)2^{2N}$.

The presence of nonzero central charges modifies the above analysis. Details can be found in [B38] and in Ref. [19]. We define $\lfloor\tfrac{1}{2}N\rfloor$ real nonnegative central charges Z_i according to eq. (9.130). Consistent solutions exist if and only if $m\geq\tfrac{1}{2}Z_i$ for all $i\in\{1,2,\ldots,[\tfrac{1}{2}N]\}$.[12] In this case, modified creation and annihilation operators can be defined that satisfy a Clifford algebra. One finds that this algebra contains $2N-2k$ nonzero anticommutators, where k of the Z_i are equal to $2m$. In particular, there exist $2N-2k$ creation operators that can be applied to the Clifford vacuum to produce the states of the supermultiplet. Hence, the total number of states in the massive supermultiplet is $(2j+1)2^{2(N-k)}$. The massive supermultiplet is called a long supermultiplet if $k=0$, a short supermultiplet if $0<k<\tfrac{1}{2}N$, and an ultrashort supermultiplet if $k=\tfrac{1}{2}N$. In the latter case, the number of states of a $j=0$ ultrashort supermultiplet is equal to the number of states in a massless supermultiplet with the same value of N.

We now illustrate these remarks for the case of $N=2$ and $Z_{ij}\neq 0$. In this case, there is only one central charge Z, which can be chosen positive by rephasing the fermionic generators Q^i_α and $Q^\dagger_{\dot\alpha i}$ $(i=1,2)$. That is,

$$Z^{ij}=Z\epsilon^{ij}\,,\qquad Z_{ij}=Z\epsilon_{ij}\,,\qquad \text{where } Z>0\,, \tag{9.155}$$

and the antisymmetric tensor ϵ is defined such that $\epsilon^{12}=-\epsilon^{21}=-\epsilon_{12}=\epsilon_{21}=1$. In the rest frame where $P^\mu=(m\,;\vec{0})$, eqs. (9.126)–(9.128) simplify to

$$\{Q^i_\alpha\,,\,Q^\dagger_{\dot\beta j}\}=2m\sigma^0_{\alpha\dot\beta}\delta^i_j\,,\qquad \{Q^i_\alpha\,,\,Q^j_\beta\}=Z\epsilon_{\alpha\beta}\epsilon^{ij}\,,\qquad \{Q^\dagger_{\dot\alpha i}\,,\,Q^\dagger_{\dot\beta j}\}=-Z\epsilon_{\dot\alpha\dot\beta}\epsilon_{ij}\,. \tag{9.156}$$

One can now define creation and annihilation operators,

$$a_\alpha\equiv\frac{1}{\sqrt{2}}\left(Q^1_\alpha+\epsilon_{\alpha\beta}\overline{\sigma}^{0\dot\beta\beta}Q^\dagger_{\dot\beta 2}\right)\,, \tag{9.157}$$

$$a^\dagger_{\dot\alpha}\equiv\frac{1}{\sqrt{2}}\left(Q_{\dot\alpha 1}+\epsilon_{\dot\alpha\dot\beta}\overline{\sigma}^{0\dot\beta\beta}Q^2_\beta\right)\,, \tag{9.158}$$

$$b_\alpha\equiv\frac{1}{\sqrt{2}}\left(Q^1_\alpha-\epsilon_{\alpha\beta}\overline{\sigma}^{0\dot\beta\beta}Q^\dagger_{\dot\beta 2}\right)\,, \tag{9.159}$$

$$b^\dagger_{\dot\alpha}\equiv\frac{1}{\sqrt{2}}\left(Q_{\dot\alpha 1}-\epsilon_{\dot\alpha\dot\beta}\overline{\sigma}^{0\dot\beta\beta}Q^2_\beta\right)\,, \tag{9.160}$$

12 In particular, the central charges must vanish for the massless supermultiplets, as previously noted.

which satisfy the anticommutation relations,

$$\{a_\alpha\,,\,a^\dagger_{\dot\beta}\} = (2m+Z)\sigma^0_{\alpha\dot\beta}\,, \tag{9.161}$$

$$\{b_\alpha\,,\,b^\dagger_{\dot\beta}\} = (2m-Z)\sigma^0_{\alpha\dot\beta}\,, \tag{9.162}$$

with all other anticommutators vanishing. Since $Z>0$ by assumption, and in the Hilbert space of positively normed states, $\langle\Psi|\,\{b_\alpha\,,\,b^\dagger_{\dot\beta}\}\,|\Psi\rangle > 0$, it follows that

$$m \geq \tfrac{1}{2}Z\,. \tag{9.163}$$

In the case of $0<Z<2m$, one can define the Clifford vacuum by the conditions $a_\alpha\,|\Omega\rangle = b_\alpha\,|\Omega\rangle = 0$ and construct states of the $N=2$ massive supermultiplet,

$$\begin{aligned}&|\Omega\rangle\,,\; a^\dagger_{\dot\alpha}\,|\Omega\rangle\,,\; b^\dagger_{\dot\beta}\,|\Omega\rangle\,,\; a^\dagger_{\dot\alpha}a^\dagger_{\dot\beta}\,|\Omega\rangle\,,\; b^\dagger_{\dot\alpha}b^\dagger_{\dot\beta}\,|\Omega\rangle\,,\; a^\dagger_{\dot\alpha}b^\dagger_{\dot\beta}\,|\Omega\rangle\,,\\ &a^\dagger_{\dot\alpha}a^\dagger_{\dot\beta}b^\dagger_{\dot\gamma}\,|\Omega\rangle\,,\; b^\dagger_{\dot\alpha}b^\dagger_{\dot\beta}a^\dagger_{\dot\gamma}\,|\Omega\rangle\,,\; a^\dagger_{\dot\alpha}a^\dagger_{\dot\beta}b^\dagger_{\dot\gamma}b^\dagger_{\dot\delta}\,|\Omega\rangle\,,\end{aligned} \tag{9.164}$$

which yields $16(2j+1)$ states. For example the $j=0$ supermultiplet contains five real scalars, four Majorana spin-1/2 fermions, and one real vector boson. This is an example of a long supermultiplet, with the same particle content as the corresponding supermultiplet with vanishing central charge.

For $Z=2m$, the inequality eq. (9.163) is saturated, and only the nonvanishing anticommutation relations given by eq. (9.161) are relevant. In this case, the Clifford vacuum is defined by $a_\alpha\,|\Omega\rangle = 0$ and the corresponding states of the $N=2$ massive supermultiplet are

$$|\Omega\rangle\,,\; a^\dagger_{\dot\alpha}\,|\Omega\rangle\,,\; a^\dagger_{\dot\alpha}a^\dagger_{\dot\beta}\,|\Omega\rangle\,, \tag{9.165}$$

which yields $4(2j+1)$ states. For example, the $j=0$ supermultiplet contains a complex scalar and a Majorana fermion, which is the same spin content as the $N=2$ massless hypermultiplet and that of the $N=1$ massive chiral supermultiplet. This is an example of an ultrashort supermultiplet; the corresponding states are called Bogomol'nyi–Prasad–Sommerfield (BPS) saturated states.

Exercises

9.1 Using the known transformation properties of the generators Q_α, $Q^{\dagger\,\dot\alpha}$, and P^μ under the Poincaré algebra, it follows that $[Q_\alpha\,,\,P^\mu]$ must transform as the tensor product, $(\frac{1}{2},0)\otimes(\frac{1}{2},\frac{1}{2}) = (1,\frac{1}{2})\oplus(0,\frac{1}{2})$, under the Poincaré algebra. But according the Haag–Łopuszański–Sohnius Theorem, there are no $(1,\frac{1}{2})$ generators. Hence, we can write

$$[Q_\alpha\,,\,P^\mu] = c\,\sigma^\mu_{\alpha\dot\beta}Q^{\dagger\,\dot\beta}\,, \tag{9.166}$$

for some complex number c.

(a) Prove that

$$[Q^{\dagger\,\dot\alpha}\,,\,P^\mu] = c^*\,\overline{\sigma}^{\mu\,\dot\alpha\beta}Q_\beta\,, \tag{9.167}$$

$$[[Q_\alpha\,,\,P^\mu]\,,\,P^\nu] = |c|^2(\sigma^\mu\overline{\sigma}^\nu)_\alpha{}^\beta Q_\beta\,. \tag{9.168}$$

(b) Using $[P^\mu\,,\,P^\nu]=0$ and the Jacobi identity, prove that $c=0$.

(c) Using a similar argument, prove that $[Q^{\dagger\dot\alpha}\,,\,P^\mu]=0$.

9.2 Define a four-component Majorana spinor fermionic generator,

$$Q_M = \begin{pmatrix} Q_\alpha \\ Q^{\dagger\dot\alpha}\end{pmatrix}\,. \tag{9.169}$$

The Dirac adjoint is $\overline{Q}_M = Q^\dagger A$, where A is the Dirac conjugation matrix [see eq. (3.29)].

(a) Show that the supersymmetry algebra is specified by eqs. (9.1)–(9.3) and

$$[P^\mu\,,\,Q_M] = [P^\mu\,,\,\overline{Q}_M] = 0\,, \tag{9.170}$$

$$[J^{\mu\nu}\,,\,Q_M] = -\tfrac{1}{2}\Sigma^{\mu\nu}Q_M\,, \tag{9.171}$$

$$[J^{\mu\nu}\,,\,\overline{Q}_M] = \tfrac{1}{2}\overline{Q}_M\Sigma^{\mu\nu}\,, \tag{9.172}$$

$$\{Q_M\,,\,Q_M\} = \{\overline{Q}_M\,,\,\overline{Q}_M\} = 0\,, \tag{9.173}$$

$$\{Q_M\,,\,\overline{Q}_M\} = 2\gamma^\mu P_\mu\,, \tag{9.174}$$

$$[Q_M\,,\,R] = \gamma_5 Q_M\,, \tag{9.175}$$

$$[\overline{Q}_M\,,\,R] = -\overline{Q}_M\gamma_5\,. \tag{9.176}$$

(b) Since Q_M is a Majorana spinor, it satisfies $Q_M = C\overline{Q}_M^{\mathsf T}$ [see eqs. (3.32) and (3.49)]. Show that an alternative form of eq. (9.174) is

$$\{Q_{Ma}\,,\,Q_{Mb}\} = -2(\gamma^\mu C)_{ab}P_\mu\,, \tag{9.177}$$

where a and b are four-component spinor indices (see Section 3.2).

9.3 (a) Derive the following commutation relations:

$$[w^\mu\,,\,Q_\alpha] = i(\sigma^{\mu\nu})_\alpha{}^\beta Q_\beta P_\nu\,, \qquad [w^\mu\,,\,Q^\dagger_{\dot\alpha}] = i(\overline{\sigma}^{\mu\nu})^{\dot\beta}{}_{\dot\alpha}Q^\dagger_{\dot\beta}P_\nu\,. \tag{9.178}$$

where $w^\mu = -\tfrac{1}{2}\epsilon^{\mu\nu\rho\lambda}J_{\nu\rho}P_\lambda$ is the Pauli–Lubański vector, and Q and $Q^\dagger$ are the fermionic generators of the $N=1$ supersymmetry algebra.

(b) Using eq. (9.178), derive the following commutation relations:

$$[w^2\,,\,Q_\alpha] = 2i(\sigma^{\mu\nu})_\alpha{}^\beta Q_\beta w_\mu P_\nu - \tfrac{3}{4}P^2 Q_\alpha\,, \tag{9.179}$$

$$[w^2\,,\,Q^\dagger_{\dot\alpha}] = 2i(\bar\sigma^{\mu\nu})^{\dot\beta}{}_{\dot\alpha} Q^\dagger_{\dot\beta} w_\mu P_\nu - \tfrac{3}{4}P^2 Q^\dagger_{\dot\alpha}\,. \tag{9.180}$$

9.4 Define the supersymmetric generalization of the Pauli–Lubański vector,

$$B^\mu \equiv w^\mu + \tfrac{1}{4}Q^\dagger \bar\sigma^\mu Q\,. \tag{9.181}$$

Unlike the Pauli–Lubański vector, $B_\mu P^\mu \neq 0$. However, other properties of the Pauli–Lubański vector are also satisfied by B^μ (see Exercise 1.2). Derive the following properties:

(a) $[P^\mu\,,\,B^\nu] = 0\,.$ (9.182)

(b) $[J^{\mu\nu}\,,\,B^\lambda] = i(g^{\nu\lambda}B^\mu - g^{\mu\lambda}B^\nu)\,.$ (9.183)

Note that eq. (9.183) confirms that B^μ is a Lorentz four-vector [cf. eq. (1.285)].

(c) $[B^\mu\,,\,B^\nu] = i\epsilon^{\mu\nu\rho\lambda}B_\rho P_\lambda\,.$ (9.184)

(d) The vector B^μ is not unique. Show that $B^\mu + kP^\mu$ (for any real constant k) also satisfies properties (a), (b), and (c).[13] For example, another common choice is to define

$$B'^\mu \equiv w^\mu - \tfrac{1}{8}\bar\sigma^{\mu\dot\alpha\beta}[Q_\beta\,,\,Q^\dagger_{\dot\alpha}]\,. \tag{9.185}$$

Show that $B'^\mu = B^\mu - \tfrac{1}{2}P^\mu$.

9.5 There are no relative phases among eqs. (9.66)–(9.68). This can be verified by performing the following computations.

(a) Employing eq. (9.47), compute $S_\pm Q_\alpha\,|\Omega(j_3)\rangle$ in two different ways. In the first method, use eq. (9.52). In the second method, use eqs. (9.67) and (9.68). Show that the two computations are consistent.

(b) Show that in the rest frame, eq. (9.33) reduces to

$$[\vec{S}^2\,,\,Q^\dagger_{\dot\alpha}] = 2i(\bar\sigma^{i0})^{\dot\beta}{}_{\dot\alpha} S^i Q^\dagger_{\dot\beta} - \tfrac{3}{4}Q^\dagger_{\dot\alpha}\,. \tag{9.186}$$

Using this result, show that

$$\vec{S}^2 Q^\dagger_1\,|\Omega(j_3)\rangle = \left[j(j+1) - \tfrac{1}{4} + j_3\right] Q^\dagger_1\,|\Omega(j_3)\rangle + S_+ Q^\dagger_2\,|\Omega(j_3)\rangle\,, \tag{9.187}$$

$$\vec{S}^2 Q^\dagger_2\,|\Omega(j_3)\rangle = \left[j(j+1) - \tfrac{1}{4} - j_3\right] Q^\dagger_2\,|\Omega(j_3)\rangle + S_- Q^\dagger_1\,|\Omega(j_3)\rangle\,, \tag{9.188}$$

where $S_\pm = S^1 \pm iS^2$.

[13] This nonuniqueness is not significant since the second-rank tensor $C^{\mu\nu}$ defined in eq. (9.37) is unchanged under $B^\mu \to B^\mu + kP^\mu$.

(c) Evaluate the right-hand sides of eqs. (9.187) and (9.188) by employing the results of eqs. (9.67) and (9.68). Show that one obtains the same result by directly applying the operator $\vec{S}^2$ to eqs. (9.67) and (9.68), respectively.

9.6 Define $N^\mu \equiv Q^\dagger \overline{\sigma}^\mu Q$. Derive the following commutation relations:

(a) $\bigl[N^\mu\,,\,Q_\alpha\bigr] = -2(\sigma^\nu\overline{\sigma}^\mu)_\alpha{}^\beta Q_\beta P_\nu = -2Q_\alpha P^\mu - 4i(\sigma^{\mu\nu})_\alpha{}^\beta Q_\beta P_\nu$.

(b) $\bigl[N^\mu\,,\,Q^\dagger_{\dot\alpha}\bigr] = 2(\overline{\sigma}^\mu\sigma^\nu)^{\dot\beta}{}_{\dot\alpha} Q^\dagger_{\dot\beta} P_\nu = 2Q^\dagger_{\dot\alpha} P^\mu - 4i(\overline{\sigma}^{\mu\nu})^{\dot\beta}{}_{\dot\alpha} Q^\dagger_{\dot\beta} P_\nu$.

(c) $\bigl[w^\mu\,,\,N^\nu\bigr] = i\epsilon^{\mu\nu\rho\lambda} N_\rho P_\lambda$.

(d) $\bigl[N^\mu\,,\,N^\nu\bigr] = -4i\epsilon^{\mu\nu\rho\lambda} N_\rho P_\lambda$.

(e) Use the above results to derive eqs. (9.35) and (9.36). Likewise, derive eqs. (9.85) and (9.86).

9.7 Consider the state $|\Omega\rangle$ of an irreducible $N = 1$ massive supermultiplet with superspin j that satisfies $Q_\alpha\,|\Omega\rangle = 0$. The state $|\Omega\rangle$ is a simultaneous eigenstate of $\vec{S}^2$ and S^3 with corresponding eigenvalues $j(j+1)$ and j_3, where the possible values of j_3 are $-j,\,-j+1,\ldots j-1,\,j$.

(a) Show that an irreducible massive supermultiplet contains a unique state $|\Psi\rangle$ that satisfies $Q^\dagger_{\dot\alpha}\,|\Psi\rangle = 0$. Verify that $|\Psi\rangle$ is a simultaneous eigenstate of $\vec{S}^2$ and S^3 with corresponding eigenvalues $j(j+1)$ and j_3.

(b) Under a discrete parity transformation, eqs. (1.196) and (1.198) yield

$$\mathcal{P} Q_\alpha \mathcal{P}^{-1} = i\eta_P \sigma^0_{\alpha\dot\beta} Q^{\dagger\dot\beta}\,, \qquad \mathcal{P} Q^\dagger_{\dot\alpha} \mathcal{P}^{-1} = -i\eta_P^* \sigma^0_{\beta\dot\alpha} Q^\beta\,, \tag{9.189}$$

where $\eta_P = \pm 1$. Using the results of part (a), show that $\mathcal{P}\,|\Omega\rangle \propto |\Psi\rangle$ and $\mathcal{P}|\Psi\rangle \propto |\Omega\rangle$.

HINT: Noting that $\mathcal{P} Q_\alpha \mathcal{P}^{-1}\mathcal{P}\,|\Omega\rangle = 0$, show that $\mathcal{P}\,|\Omega\rangle$ is annihilated by $Q^\dagger_{\dot\alpha}$.

(c) If $j = 0$, then $|\Omega\rangle$ and $|\Psi\rangle$ are spin-0 states. Since $\mathcal{P}^2 = \mathbb{1}$ when acting on bosonic state, show that one can choose a phase convention in which $\mathcal{P}\,|\Omega\rangle = |\Psi\rangle$ and $\mathcal{P}\,|\Psi\rangle = |\Omega\rangle$. In this convention, verify that $|\Omega\rangle \pm |\Psi\rangle$ are eigenstates of parity with eigenvalues ± 1, respectively.

(d) If $j = \frac{1}{2}$, prove that the spin-0 state $Q^\dagger_{\dot 2}\,|\Omega(\tfrac{1}{2})\rangle - Q^\dagger_{\dot 1}\,|\Omega(-\tfrac{1}{2})\rangle$ is an eigenstate of parity with eigenvalue -1.

HINT: Recall that $\mathcal{P}^2 = -\mathbb{1}$ when acting on fermionic states. Show that $|\Omega\rangle$ and $|\Psi\rangle$ can be combined into a Dirac fermion. Using the results of part (b), show that one can choose a phase convention in which $\mathcal{P}\,|\Omega\rangle = i\,|\Psi\rangle$ and $\mathcal{P}\,|\Psi\rangle = i\,|\Omega\rangle$.

9.8 Consider the state $|\Omega\rangle$ of an irreducible $N = 1$ massless supermultiplet that satisfies $Q_\alpha |\Omega\rangle = 0$ and $\mathcal{K}|\Omega\rangle = \kappa|\Omega\rangle$. As shown in eq. (9.99), the state $|\Omega\rangle$ is an eigenstate of the helicity operator h with eigenvalue κ.

(a) Show that an irreducible massless supermultiplet contains a unique state $|\Psi\rangle$ that satisfies $Q^\dagger_{\dot\alpha}|\Psi\rangle = 0$ and $\mathcal{K}|\Psi\rangle = \kappa|\Psi\rangle$.

(b) Justify the following computation:

$$\begin{aligned} w^\mu Q^\dagger_{\dot\alpha}|\Omega\rangle = hP^\mu Q^\dagger_{\dot\alpha}|\Omega\rangle &= \Big[Q^\dagger_{\dot\alpha} w^\mu + i\overline{\sigma}^{\mu\nu\dot\beta}{}_{\dot\alpha} Q^\dagger_{\dot\beta} P_\nu\Big]|\Omega\rangle \\ &= \Big[Q^\dagger_{\dot\alpha} hP^\mu - \tfrac{1}{2}(g^{\mu\nu} - \overline{\sigma}^\nu\sigma^\mu)^{\dot\beta}{}_{\dot\alpha} Q^\dagger_{\dot\beta} P_\nu\Big]|\Omega\rangle \\ &= \Big[P^\mu Q^\dagger_{\dot\alpha} h - \tfrac{1}{2}P^\mu Q^\dagger_{\dot\alpha}\Big]|\Omega\rangle \\ &= (\kappa - \tfrac{1}{2})P^\mu Q^\dagger_{\dot\alpha}|\Omega\rangle\,. \end{aligned} \tag{9.190}$$

(c) Naively, the result of part (b) suggests that the massless supermultiplet contains two states (corresponding to $\dot\alpha = 1, 2$) with helicity $\kappa - \frac{1}{2}$. Show that the state $|\Psi\rangle$, defined in part (a), is an eigenstate of h with eigenvalue $\kappa - \frac{1}{2}$, and $|\Psi\rangle$ is unique if the massless supermultiplet is irreducible.

9.9 (a) Show that the massive $j = \frac{1}{2}$ supermultiplet corresponds to a real vector field, a real scalar field, and a Dirac fermion field.

(b) Show that a massless supermultiplet with $\kappa = 2$ and its CPT-conjugates corresponds to a massless spin-3/2 and a massless spin-2 particle, which is realized in supergravity by the gravitino and the graviton.

9.10 Using Lorentz invariance, the most general form for the commutation relations of the fermionic generators of extended supersymmetry with the momentum operator is

$$[Q^i_\alpha\,, P^\mu] = X^{ij}\sigma^\mu{}_{\alpha\dot\beta} Q^{\dagger\dot\beta}_j\,, \qquad [Q^\dagger_{\dot\alpha i}\,, P^\mu] = X_{ij}\overline{\sigma}^{\mu\dot\alpha\beta} Q^j_\beta\,, \tag{9.191}$$

where X^{ij} is a complex Lorentz scalar and $X_{ij} \equiv (X^{ij})^*$. Likewise, the most general forms for the anticommutators of the fermionic generators are

$$\{Q^i_\alpha\,, Q^j_\beta\} = \epsilon_{\alpha\beta} Z^{ij} + \epsilon_{\beta\gamma} Y^{ij} (\sigma_{\mu\nu})_\alpha{}^\gamma J^{\mu\nu}\,, \tag{9.192}$$

$$\{Q^\dagger_{\dot\alpha i}\,, Q^\dagger_{\dot\beta j}\} = \epsilon_{\dot\alpha\dot\beta} Z_{ij} + \epsilon_{\dot\beta\dot\gamma} Y_{ij} (\overline{\sigma}_{\mu\nu})^{\dot\alpha}{}_{\dot\gamma} J^{\mu\nu}\,, \tag{9.193}$$

where Y^{ij} and Z^{ij} are complex Lorentz scalars, $Y_{ij} \equiv (Y^{ij})^*$ and $Z_{ij} \equiv (Z^{ij})^*$. Note that the anticommutators above imply that $Z^{ij} = -Z^{ji}$ and $Y^{ij} = Y^{ji}$ (prove this!).

(a) Prove that $X^{ij} = 0$ by considering the implications of the Jacobi identity involving the Q^i_α and P^μ. You will need to make use of two separate Jacobi identities in which Q^i_α appears either once or twice, respectively.

HINT: When employing the second Jacobi identity, you should employ anti-commutators in the appropriate places, namely,

$$[\{Q^i_\alpha\,,\,Q^j_\beta\}\,,\,P^\mu] + \{[P^\mu\,,\,Q^i_\alpha]\,,\,Q^j_\beta\} - \{[Q^j_\beta\,,\,P^\mu]\,,\,Q^i_\alpha\} = 0\,. \qquad (9.194)$$

Use eq. (9.192) and simplify the resulting expression by multiplying through by $\epsilon^{\alpha\beta}$.

(b) Using the fact that $X^{ij} = 0$ as established in part (a), show that $Y^{ij} = 0$ by examining the symmetric part of eq. (9.194).

(c) By further manipulating the relevant Jacobi identities, show that

$$[Z^{ij}\,,\,P^\mu] = [Z^{ij}\,,\,Q_\alpha] = [Z^{ij}\,,\,Q_\beta] = 0\,. \qquad (9.195)$$

Since the Z^{ij} are Lorentz scalars, it follows that $[Z^{ij}\,,\,J^{\mu\nu}] = 0$. Conclude that the Z^{ij} commute with all the fermionic and bosonic generators of the extended supersymmetry algebra.

(d) The central charges can be expanded as a linear combination of R-symmetry generators [see eq. (9.139)],

$$Z^{ij} = c^{aij} R_a\,. \qquad (9.196)$$

Show that [see eqs. (9.134) and (9.135)]

$$[Z^{ij}\,,\,R_a] = (S_a)^i{}_k Z^{kj} + (S_a)^j{}_k Z^{ik}\,, \qquad (9.197)$$

$$\left[Z^{ij}\,,\,Z^{k\ell}\right] = c^{ak\ell}(S_a)^i{}_m Z^{mj} + c^{ak\ell}(S_a)^j{}_m Z^{im}\,. \qquad (9.198)$$

By manipulating the relevant Jacobi identities, prove that the right-hand sides of eqs. (9.197) and (9.198) must vanish exactly. That is,

$$[Z^{ij}\,,\,R_a] = [Z^{ij}\,,\,Z^{k\ell}] = 0\,. \qquad (9.199)$$

(e) Construct the Casimir operator of the extended supersymmetry algebra that generalizes C^2 defined in eq. (9.39).

9.11 Consider an $N = 4$ supersymmetric Yang–Mills theory, whose spectrum consists of a CPT self-conjugate irreducible $N = 4$ massless supermultiplet with maximal helicity $\lambda = 1$. The maximal R-symmetry group is $\mathrm{U}(4)_R$.

(a) Show that the particle content of the theory consists of a (gauge) vector boson, three complex scalars, and two Dirac fermions.

(b) Verify that the particle content of an irreducible $N = 4$ massless supermultiplet is equivalent to the particle content of one $N = 1$ vector supermultiplet and three $N = 1$ chiral supermultiplets. Likewise, show that the same particle content arises from one $N = 2$ vector supermultiplet and one complex $N = 2$ hypermultiplet.

(c) Determine the transformation properties of the helicity states of an irreducible $N = 4$ massless supermultiplet with respect to the SU(4) subgroup of $U(4)_R$.

(d) Consider an irreducible $N = 3$ massless supermultiplet with maximal helicity 1. Since this supermultiplet is not CPT self-conjugate, we must add a second irreducible $N = 3$ massless supermultiplet with maximal helicity $\frac{1}{2}$. Show that the resulting particle spectrum is precisely the same as that of a CPT self-conjugate irreducible $N = 4$ massless supermultiplet with maximal helicity 1.

9.12 Consider an $N = 2$ massive supermultiplet with mass m and superspin $j = 0$.

(a) In the absence of a central charge, determine the spins of the states of the supermultiplet. Determine the transformation properties of the states with respect to the SU(2) subgroup of the R-symmetry group $U(2)_R$.

(b) If the central charge is nonvanishing, show that $SU(2)_R$ is the maximal R-symmetry group.

(c) Assuming that the central charge Z is nonvanishing but the inequality $Z < 2m$ is not saturated, show that the results of part (a) are not modified.

(d) Assuming that $Z = 2m$, determine the spins of the states of the supermultiplet. Determine the transformation properties of the states with respect to $SU(2)_R$.

(e) Repeat this exercise in the case of $j = \frac{1}{2}$. In particular, show that the number of states of the ultrashort massive supermultiplet agrees with the number of states of the $N = 2$ massless vector supermultiplet (after including the CPT-conjugate states). Moreover, if the massless vector boson of the latter acquires mass via the Higgs mechanism (by "eating" a scalar degree of freedom of the supermultiplet), show that the spins of the states of resulting spectrum coincide with that of the $N = 2$ ultrashort massive $j = \frac{1}{2}$ supermultiplet.

10 Superfields

10.1 Supercoordinates, Superspace and Superfields

Supersymmetry can be given a geometric interpretation using superspace, a manifold obtained by adding four fermionic coordinates to the usual bosonic spacetime coordinates t, x, y, z. Points in superspace are labeled by coordinates

$$x^\mu,\ \theta^\alpha,\ \theta^\dagger_{\dot\alpha}. \tag{10.1}$$

Here θ^α and $\theta^\dagger_{\dot\alpha}$ are constant complex anticommuting two-component spinors with dimension $[\text{mass}]^{-1/2}$. In the superspace formulation, the component fields of a supermultiplet are united into a single superfield, a function of these superspace coordinates. We will see below that infinitesimal translations in superspace coincide with the global supersymmetry transformations that we have already found in component-field language. Superspace thus allows an elegant and manifestly invariant definition of supersymmetric field theories.

Differentiation and integration on spaces with anticommuting coordinates are defined by analogy with ordinary commuting variables. Consider first, as a warm-up example, a single anticommuting variable η (carrying no spinor indices). Because $\eta^2 = 0$, a power series expansion in η always terminates, so a general function is linear in η:

$$f(\eta) = f_0 + \eta f_1. \tag{10.2}$$

Here f_0 and f_1 may be functions of other commuting or anticommuting variables, but not η. One of them will be anticommuting (Grassmann-odd), and the other is commuting (Grassmann-even). Then define

$$\frac{df}{d\eta} = f_1. \tag{10.3}$$

The differential operator $\frac{d}{d\eta}$ anticommutes with every Grassmann-odd object, so that if η' is distinct from η but also anticommuting, then

$$\frac{d(\eta'\eta)}{d\eta} = -\frac{d(\eta\eta')}{d\eta} = -\eta'. \tag{10.4}$$

To define an integration operation with respect to η, take

$$\int d\eta = 0, \qquad \int d\eta\, \eta = 1, \tag{10.5}$$

and impose linearity. This defines the Berezin integral for Grassmann variables, and gives

$$\int d\eta\, f(\eta) = f_1. \tag{10.6}$$

Comparing eqs. (10.3) and (10.6) shows the peculiar fact that, for an anticommuting variable, differentiation and integration are the same thing. The definition eq. (10.5) is motivated by the fact that it implies translation invariance,

$$\int d\eta\, f(\eta+\eta') = \int d\eta\, f(\eta), \tag{10.7}$$

and the integration by parts formula

$$\int d\eta\, \frac{df}{d\eta} = 0, \tag{10.8}$$

in analogy with the fundamental theorem of the calculus for ordinary commuting variables. The anticommuting Dirac delta function has the defining property

$$\int d\eta\, \delta(\eta-\eta')\, f(\eta) = f(\eta'), \tag{10.9}$$

which leads to

$$\delta(\eta-\eta') = \eta - \eta'. \tag{10.10}$$

For superspace with coordinates $x^\mu, \theta^\alpha, \theta^\dagger_{\dot\alpha}$, any superfield can be expanded in a power series in the anticommuting variables, with components that are functions of x^μ. Since there are two independent components of θ^α and likewise for $\theta^\dagger_{\dot\alpha}$, the expansion always terminates, with each term containing at most two θs and two $\theta^\dagger$s. A general superfield is therefore

$$\begin{aligned} S(x,\theta,\theta^\dagger) = {} & a + \theta\xi + \theta^\dagger\chi^\dagger + \theta\theta b + \theta^\dagger\theta^\dagger c + \theta\sigma^\mu\theta^\dagger v_\mu \\ & + (\theta^\dagger\theta^\dagger)(\theta\eta) + (\theta\theta)(\theta^\dagger\zeta^\dagger) + (\theta\theta)(\theta^\dagger\theta^\dagger)d. \end{aligned} \tag{10.11}$$

To see that there are no other independent contributions, note the identities

$$\theta_\alpha\theta_\beta = \tfrac{1}{2}\epsilon_{\alpha\beta}\theta\theta, \qquad \theta^\dagger_{\dot\alpha}\theta^\dagger_{\dot\beta} = \tfrac{1}{2}\epsilon_{\dot\beta\dot\alpha}\theta^\dagger\theta^\dagger, \qquad \theta_\alpha\theta^\dagger_{\dot\beta} = \tfrac{1}{2}\sigma^\mu_{\alpha\dot\beta}(\theta\sigma_\mu\theta^\dagger), \tag{10.12}$$

derived from eqs. (1.40) and (1.85). These can be used to rewrite any term in the forms given in eq. (10.11). Some other identities involving the anticommuting coordinates that are useful in checking results below are

$$(\theta\xi)(\theta\chi) = -\tfrac{1}{2}(\theta\theta)(\xi\chi), \qquad (\theta^\dagger\xi^\dagger)(\theta^\dagger\chi^\dagger) = -\tfrac{1}{2}(\theta^\dagger\theta^\dagger)(\xi^\dagger\chi^\dagger), \tag{10.13}$$

$$(\theta\xi)(\theta^\dagger\chi^\dagger) = \tfrac{1}{2}(\theta\sigma^\mu\theta^\dagger)(\xi\sigma_\mu\chi^\dagger), \tag{10.14}$$

$$\theta^\dagger\overline{\sigma}^\mu\theta = -\theta\sigma^\mu\theta^\dagger = (\theta^\dagger\overline{\sigma}^\mu\theta)^\dagger, \tag{10.15}$$

$$\theta\sigma^\mu\overline{\sigma}^\nu\theta = \eta^{\mu\nu}\theta\theta, \qquad \theta^\dagger\overline{\sigma}^\mu\sigma^\nu\theta^\dagger = \eta^{\mu\nu}\theta^\dagger\theta^\dagger. \tag{10.16}$$

These follow from the two-component spinor identities already given in Chapter 1.

The general superfield S could be either commuting or anticommuting, and could carry additional Lorentz vector or spinor indices. For simplicity, let us assume for the rest of this section that it is Grassmann-even and carries no other indices. Then, without further restrictions, the components of the general superfield S are 8 bosonic fields a, b, c, d, and v_μ, and 4 two-component fermionic fields $\xi, \chi^\dagger, \eta, \zeta^\dagger$. All of these are complex functions of x^μ. The numbers of bosons and fermions do agree (8 complex, or 16 real, degrees of freedom for each), but there are too many of them to match either the chiral or vector supermultiplets encountered in the previous section. This means that the general superfield is a reducible representation of supersymmetry. In Sections 10.4 and 10.5, we will see how chiral and vector superfields are obtained by imposing constraints on the general case eq. (10.11).

One can also define derivatives with respect to the anticommuting coordinates θ and $\theta^\dagger$. It is convenient to introduce the following notation:

$$\partial_\alpha \equiv \frac{\partial}{\partial\theta^\alpha}, \qquad \partial^{\dagger\dot\alpha} \equiv \frac{\partial}{\partial\theta^\dagger_{\dot\alpha}}. \tag{10.17}$$

The derivatives with respect to θ and $\theta^\dagger$ are defined in the obvious way:

$$\partial_\alpha\theta^\beta \equiv \frac{\partial}{\partial\theta^\alpha}(\theta^\beta) = \delta_\alpha^\beta, \qquad \partial_\alpha\theta^\dagger_{\dot\beta} = 0, \tag{10.18}$$

$$\partial^{\dagger\dot\alpha}\theta^\dagger_{\dot\beta} \equiv \frac{\partial}{\partial\theta^\dagger_{\dot\alpha}}(\theta^\dagger_{\dot\beta}) = \delta^{\dot\alpha}_{\dot\beta}, \qquad \partial^{\dagger\dot\alpha}\theta^\beta = 0. \tag{10.19}$$

Derivatives with respect to θ and $\theta^\dagger$ satisfy the graded Leibniz rules, which consist of the usual product rules for derivatives but with a minus sign for anticommuting through a Grassmann-odd object:

$$\partial_\alpha(fg) = (\partial_\alpha f)g + (-1)^f f(\partial_\alpha g), \tag{10.20}$$

$$\partial^{\dagger\dot\alpha}(fg) = (\partial^{\dagger\dot\alpha} f)g + (-1)^f f(\partial^{\dagger\dot\alpha} g), \tag{10.21}$$

where $(-1)^f = -1$ if f is Grassmann-odd, and $(-1)^f = +1$ if f is Grassmann-even. For example,

$$\partial_\alpha(\theta\theta) = \partial_\alpha\left(\epsilon_{\gamma\beta}\theta^\gamma\theta^\beta\right) = \epsilon_{\gamma\beta}(\delta^\gamma_\alpha\theta^\beta - \delta^\beta_\alpha\theta^\gamma) = 2\theta_\alpha, \tag{10.22}$$

$$\partial^{\dagger\dot\alpha}(\theta^\dagger\theta^\dagger) = \partial^{\dagger\dot\alpha}\left(\epsilon^{\dot\gamma\dot\beta}\theta^\dagger_{\dot\gamma}\theta^\dagger_{\dot\beta}\right) = \epsilon^{\dot\gamma\dot\beta}(\delta^{\dot\alpha}_{\dot\gamma}\theta^\dagger_{\dot\beta} - \delta^{\dot\alpha}_{\dot\beta}\theta^\dagger_{\dot\gamma}) = 2\theta^{\dagger\dot\alpha}. \tag{10.23}$$

Similarly, for anticommuting spinors ψ and $\psi^\dagger$,

$$\partial_\alpha(\psi\theta) = \partial_\alpha(\theta\psi) = \partial_\alpha(\theta^\beta\psi_\beta) = \delta_\alpha^\beta\psi_\beta = \psi_\alpha, \tag{10.24}$$

$$\partial^{\dagger\dot\alpha}(\psi^\dagger\theta^\dagger) = \partial^{\dagger\dot\alpha}(\theta^\dagger\psi^\dagger) = \partial^{\dagger\dot\alpha}(\theta^\dagger_{\dot\beta}\psi^{\dagger\dot\beta}) = \delta^{\dot\alpha}_{\dot\beta}\psi^{\dagger\dot\beta} = \psi^{\dagger\dot\alpha}. \tag{10.25}$$

Likewise, one conventionally defines

$$\partial^\alpha \equiv \frac{\partial}{\partial\theta_\alpha}, \qquad \partial^\dagger_{\dot\alpha} \equiv \frac{\partial}{\partial\theta^{\dagger\dot\alpha}}, \tag{10.26}$$

which satisfy

$$\partial^\alpha\theta_\beta \equiv \frac{\partial}{\partial\theta_\alpha}(\theta_\beta) = \delta^\alpha_\beta, \qquad \partial^\alpha\theta^\dagger_{\dot\beta} = 0, \tag{10.27}$$

$$\partial^\dagger_{\dot\alpha}\theta^{\dagger\dot\beta} \equiv \frac{\partial}{\partial\theta^{\dagger\dot\alpha}}(\theta^\dagger_{\dot\beta}) = \delta^{\dot\beta}_{\dot\alpha}, \qquad \partial^\dagger_{\dot\alpha}\theta^\beta = 0. \tag{10.28}$$

The graded Leibniz rule also applies to ∂^α and $\partial^\dagger_{\dot\alpha}$. Consequently,

$$\partial^\alpha(\theta\theta) = \partial^\alpha(\epsilon^{\beta\gamma}\theta_\gamma\theta_\beta) = \epsilon^{\beta\gamma}(\delta^\alpha_\gamma\theta_\beta - \delta^\alpha_\beta\theta_\gamma) = -2\theta^\alpha\,, \tag{10.29}$$

$$\partial^\dagger_{\dot\alpha}(\theta^\dagger\theta^\dagger) = \partial^\dagger_{\dot\alpha}(\epsilon_{\dot\beta\dot\gamma}\theta^{\dagger\dot\gamma}\theta^{\dagger\dot\beta}) = \epsilon^{\dot\beta\dot\gamma}(\delta^{\dot\gamma}_{\dot\alpha}\theta^{\dagger\dot\beta} - \delta^{\dot\beta}_{\dot\alpha}\theta^{\dagger\dot\gamma}) = -2\theta^\dagger_{\dot\alpha}\,. \tag{10.30}$$

Similarly, for anticommuting spinors ψ and $\psi^\dagger$,

$$\partial^\alpha(\psi\theta) = \partial^\alpha(\psi^\beta\theta_\beta) = -\delta^\alpha_\beta\psi^\beta = -\psi^\alpha\,, \tag{10.31}$$

$$\partial^\dagger_{\dot\alpha}(\psi^\dagger\theta^\dagger) = \partial^\dagger_{\dot\alpha}(\psi^\dagger_{\dot\beta}\theta^{\dagger\dot\beta}) = -\delta^{\dot\beta}_{\dot\alpha}\psi^\dagger_{\dot\beta} = -\psi^\dagger_{\dot\alpha}\,. \tag{10.32}$$

Note the unexpected minus sign when relating the derivatives of eqs. (10.17) and (10.26):

$$\partial^\alpha = -\epsilon^{\alpha\beta}\partial_\beta\,, \qquad \partial^\dagger_{\dot\alpha} = -\epsilon_{\dot\alpha\dot\beta}\partial^{\dagger\dot\beta}\,. \tag{10.33}$$

This is one case where the rules for raising a spinor index given in eqs. (1.45) and (1.49) do *not* apply. For example, starting from eqs. (10.24) and (10.25) and employing eq. (10.33), one can rederive eqs. (10.31) and (10.32):

$$\partial^\alpha(\psi\theta) = -\epsilon^{\alpha\beta}\partial_\alpha(\psi\theta) = -\epsilon^{\alpha\beta}\psi_\beta = -\psi^\alpha\,, \tag{10.34}$$

$$\partial^\dagger_{\dot\alpha}(\psi^\dagger\theta^\dagger) = -\epsilon_{\dot\alpha\dot\beta}\partial^{\dagger\dot\beta}(\psi^\dagger\theta^\dagger) = -\epsilon_{\dot\alpha\dot\beta}\psi^{\dagger\dot\beta} = -\psi^{\dagger\dot\alpha}\,. \tag{10.35}$$

To integrate over superspace, we define

$$d^2\theta \equiv -\tfrac{1}{4}d\theta^\alpha d\theta^\beta\epsilon_{\alpha\beta}, \qquad d^2\theta^\dagger \equiv -\tfrac{1}{4}d\theta^\dagger_{\dot\alpha}d\theta^\dagger_{\dot\beta}\epsilon^{\dot\alpha\dot\beta}, \tag{10.36}$$

so that, using eq. (10.5),

$$\int d^2\theta\,\theta\theta = 1, \qquad \int d^2\theta^\dagger\,\theta^\dagger\theta^\dagger = 1. \tag{10.37}$$

Integration of a general superfield therefore just picks out the relevant coefficients of $\theta\theta$ and/or $\theta^\dagger\theta^\dagger$ in eq. (10.11):

$$\int d^2\theta\, S(x,\theta,\theta^\dagger) = b(x) + \theta^\dagger\zeta^\dagger(x) + \theta^\dagger\theta^\dagger d(x), \tag{10.38}$$

$$\int d^2\theta^\dagger\, S(x,\theta,\theta^\dagger) = c(x) + \theta\eta(x) + \theta\theta d(x), \tag{10.39}$$

$$\int d^2\theta\, d^2\theta^\dagger\, S(x,\theta,\theta^\dagger) = d(x). \tag{10.40}$$

The Dirac delta functions with respect to integrations $d^2\theta$ and $d^2\theta^\dagger$ are

$$\delta^{(2)}(\theta - \theta') = (\theta - \theta')(\theta - \theta'), \tag{10.41}$$

$$\delta^{(2)}(\theta^\dagger - \theta'^\dagger) = (\theta^\dagger - \theta'^\dagger)(\theta^\dagger - \theta'^\dagger), \tag{10.42}$$

so that

$$\int d^2\theta\, \delta^{(2)}(\theta)\, S(x,\theta,\theta^\dagger) = S(x,0,\theta^\dagger) \;=\; a(x) + \theta^\dagger\chi^\dagger(x) + \theta^\dagger\theta^\dagger c(x), \qquad (10.43)$$

$$\int d^2\theta^\dagger\, \delta^{(2)}(\theta^\dagger)\, S(x,\theta,\theta^\dagger) = S(x,\theta,0) \;=\; a(x) + \theta\xi(x) + \theta\theta b(x), \qquad (10.44)$$

$$\int d^2\theta\, d^2\theta^\dagger\, \delta^{(2)}(\theta)\delta^{(2)}(\theta^\dagger)\, S(x,\theta,\theta^\dagger) = S(x,0,0) \;=\; a(x). \qquad (10.45)$$

The integrals of total derivatives with respect to the fermionic coordinates vanish:

$$\int d^2\theta \frac{\partial}{\partial\theta^\alpha}(\text{anything}) = 0, \qquad \int d^2\theta^\dagger \frac{\partial}{\partial\theta^\dagger_{\dot\alpha}}(\text{anything}) = 0, \qquad (10.46)$$

just as in eq. (10.8). This allows for integration by parts.

10.2 Supersymmetry Transformations the Superspace Way

To formulate supersymmetry transformations in terms of superspace, define the following differential operators that act on superfields:

$$\widehat{Q}_\alpha = i\partial_\alpha - (\sigma^\mu\theta^\dagger)_\alpha\partial_\mu, \qquad \widehat{Q}^\alpha = -i\partial^\alpha + (\theta^\dagger\overline{\sigma}^\mu)^\alpha\partial_\mu, \qquad (10.47)$$

$$\widehat{Q}^{\dagger\dot\alpha} = i\partial^{\dagger\dot\alpha} - (\overline{\sigma}^\mu\theta)^{\dot\alpha}\partial_\mu, \qquad \widehat{Q}^\dagger_{\dot\alpha} = -i\partial^\dagger_{\dot\alpha} + (\theta\sigma^\mu)_{\dot\alpha}\partial_\mu. \qquad (10.48)$$

These operators obey the graded Leibniz rules [see eqs. (10.20) and (10.21)]. Note that, in light of eq. (10.33), it follows that

$$\widehat{Q}^\alpha = \epsilon^{\alpha\beta}\widehat{Q}_\beta\,, \qquad \widehat{Q}^\dagger_{\dot\alpha} = \epsilon_{\dot\alpha\dot\beta}\widehat{Q}^{\dagger\dot\beta}. \qquad (10.49)$$

Then the supersymmetry transformation parametrized by infinitesimal ϵ, $\epsilon^\dagger$ for any superfield S is given by

$$\sqrt{2}\,\delta_\epsilon S = -i(\epsilon\widehat{Q} + \epsilon^\dagger\widehat{Q}^\dagger)S = \Big(\epsilon^\alpha\partial_\alpha + \epsilon^\dagger_{\dot\alpha}\partial^{\dagger\dot\alpha} + i\big[\epsilon\sigma^\mu\theta^\dagger + \epsilon^\dagger\overline{\sigma}^\mu\theta\big]\partial_\mu\Big)S\,. \qquad (10.50)$$

Since $\widehat{Q}$, $\widehat{Q}^\dagger$ are linear differential operators, the product or linear combination of any superfields satisfying eq. (10.50) is again a superfield with the same transformation law.

It is convenient to introduce the following notation:

$$\widehat{\delta}_\epsilon S \equiv S(x^\mu + i\epsilon\sigma^\mu\theta^\dagger + i\epsilon^\dagger\overline{\sigma}^\mu\theta,\, \theta+\epsilon,\, \theta^\dagger+\epsilon^\dagger) - S(x^\mu,\, \theta,\, \theta^\dagger)\,, \qquad (10.51)$$

which can be viewed as an infinitesimal translation in superspace,

$$\theta^\alpha \to \theta^\alpha + \epsilon^\alpha, \qquad (10.52)$$

$$\theta^\dagger_{\dot\alpha} \to \theta^\dagger_{\dot\alpha} + \epsilon^\dagger_{\dot\alpha}, \qquad (10.53)$$

$$x^\mu \to x^\mu + i\epsilon\sigma^\mu\theta^\dagger + i\epsilon^\dagger\overline{\sigma}^\mu\theta. \qquad (10.54)$$

Expanding to first order in ϵ and $\epsilon^\dagger$ and comparing with eq. (10.50), we can identify

$$\widehat{\delta}_\epsilon = \sqrt{2}\,\delta_\epsilon\,. \tag{10.55}$$

Although we originally employed δ_ϵ in Section 8.1 to avoid $\sqrt{2}$ factors in the supersymmetry transformations of the Wess–Zumino model, in this chapter it is convenient to employ $\widehat{\delta}_\epsilon$, which has the more natural definition via eq. (10.51).

It is instructive and useful to work out the supersymmetry transformations of all of the component fields of the general superfield eq. (10.11). They are

$$\widehat{\delta}_\epsilon a = \epsilon\xi + \epsilon^\dagger\chi^\dagger, \tag{10.56}$$
$$\widehat{\delta}_\epsilon \xi_\alpha = 2\epsilon_\alpha b + (\sigma^\mu\epsilon^\dagger)_\alpha(v_\mu - i\partial_\mu a), \tag{10.57}$$
$$\widehat{\delta}_\epsilon \chi^{\dagger\dot\alpha} = 2\epsilon^{\dagger\dot\alpha} c - (\overline{\sigma}^\mu\epsilon)^{\dot\alpha}(v_\mu + i\partial_\mu a), \tag{10.58}$$
$$\widehat{\delta}_\epsilon b = \epsilon^\dagger\zeta^\dagger - \tfrac{1}{2}i\epsilon^\dagger\overline{\sigma}^\mu\partial_\mu\xi, \tag{10.59}$$
$$\widehat{\delta}_\epsilon c = \epsilon\eta - \tfrac{1}{2}i\epsilon\sigma^\mu\partial_\mu\chi^\dagger, \tag{10.60}$$
$$\widehat{\delta}_\epsilon v^\mu = \epsilon\sigma^\mu\zeta^\dagger - \epsilon^\dagger\overline{\sigma}^\mu\eta - \tfrac{1}{2}i\epsilon\sigma^\nu\overline{\sigma}^\mu\partial_\nu\xi + \tfrac{1}{2}i\epsilon^\dagger\overline{\sigma}^\nu\sigma^\mu\partial_\nu\chi^\dagger, \tag{10.61}$$
$$\widehat{\delta}_\epsilon \eta_\alpha = 2\epsilon_\alpha d - i(\sigma^\mu\epsilon^\dagger)_\alpha\partial_\mu c + \tfrac{1}{2}i(\sigma^\nu\overline{\sigma}^\mu\epsilon)_\alpha\partial_\mu v_\nu, \tag{10.62}$$
$$\widehat{\delta}_\epsilon \zeta^{\dagger\dot\alpha} = 2\epsilon^{\dagger\dot\alpha} d - i(\overline{\sigma}^\mu\epsilon)^{\dot\alpha}\partial_\mu b - \tfrac{1}{2}i(\overline{\sigma}^\nu\sigma^\mu\epsilon^\dagger)^{\dot\alpha}\partial_\mu v_\nu, \tag{10.63}$$
$$\widehat{\delta}_\epsilon d = -\tfrac{1}{2}i\epsilon^\dagger\overline{\sigma}^\mu\partial_\mu\eta - \tfrac{1}{2}i\epsilon\sigma^\mu\partial_\mu\zeta^\dagger. \tag{10.64}$$

Note that since the terms on the right-hand sides all have exactly one ϵ or one $\epsilon^\dagger$, boson fields are always transformed into fermions and vice versa.

It is probably not obvious yet that the supersymmetry transformations as just defined coincide with those found in Section 8.5. This will become clear below when we discuss the specific form of chiral and vector superfields and the Lagrangians that govern their dynamics. Meanwhile, we can compute the anticommutators of $\widehat{Q}$, $\widehat{Q}^\dagger$ from eqs. (10.47) and (10.48), with the results

$$\left\{\widehat{Q}_\alpha,\, \widehat{Q}^\dagger_{\dot\beta}\right\} = 2i\sigma^\mu_{\alpha\dot\beta}\partial_\mu = 2\sigma^\mu_{\alpha\dot\beta}\widehat{P}_\mu, \tag{10.65}$$
$$\left\{\widehat{Q}_\alpha,\, \widehat{Q}_\beta\right\} = 0, \qquad \left\{\widehat{Q}^\dagger_{\dot\alpha},\, \widehat{Q}^\dagger_{\dot\beta}\right\} = 0. \tag{10.66}$$

Here, the differential operator generating spacetime translations is

$$\widehat{P}_\mu = i\partial_\mu. \tag{10.67}$$

Equations (10.65)–(10.66) have the same form as the supersymmetry algebra given in eqs. (9.15) and (9.19).

It is important to keep in mind the conceptual distinction between the unhatted objects $Q_\alpha, Q^\dagger_{\dot\alpha}, P^\mu$ appearing in Section 8.1, which are operators acting on the Hilbert space of quantum states, and the corresponding hatted objects $\widehat{Q}_\alpha, \widehat{Q}^\dagger_{\dot\alpha}, \widehat{P}^\mu$, which are differential operators acting on functions in superspace. For any superfield quantum mechanical operator X in the Heisenberg picture, the two kinds of operations are related by

$$[X,\, \epsilon Q + \epsilon^\dagger Q^\dagger] = (\epsilon\widehat{Q} + \epsilon^\dagger \widehat{Q}^\dagger)X, \tag{10.68}$$

$$[X,\, P_\mu] = \widehat{P}_\mu X. \tag{10.69}$$

10.3 Chiral Covariant Derivatives

To construct Lagrangians in superspace, we will later want to use derivatives with respect to the anticommuting coordinates, just as ordinary Lagrangians are built using spacetime derivatives ∂_μ. We will also use such derivatives to impose constraints on the general superfield in a way consistent with the supersymmetry transformations. However, ∂_α is not appropriate for this purpose, because it is not supersymmetric covariant:

$$\widehat{\delta}_\epsilon(\partial_\alpha S) \neq \partial_\alpha(\widehat{\delta}_\epsilon S)\,, \tag{10.70}$$

and similarly for $\partial^{\dagger\dot{\alpha}}$. This means that derivatives of a superfield with respect to θ_α or $\theta^\dagger_{\dot{\alpha}}$ are not superfields; they do not transform the correct way. To fix this, it is useful to define the chiral covariant derivatives

$$D_\alpha = \partial_\alpha - i(\sigma^\mu\theta^\dagger)_\alpha\partial_\mu, \qquad D^\alpha = \epsilon^{\alpha\beta}D_\beta = -\partial^\alpha + i(\theta^\dagger\overline{\sigma}^\mu)^\alpha\partial_\mu. \tag{10.71}$$

For a Grassmann-even superfield S, one can then define the antichiral covariant derivative to obey

$$\overline{D}_{\dot{\alpha}}S^\dagger \equiv (D_\alpha S)^\dagger, \tag{10.72}$$

which implies

$$\overline{D}^{\dot{\alpha}} = \partial^{\dagger\dot{\alpha}} - i(\overline{\sigma}^\mu\theta)^{\dot{\alpha}}\partial_\mu, \qquad \overline{D}_{\dot{\alpha}} = \epsilon_{\dot{\alpha}\dot{\beta}}\overline{D}^{\dot{\beta}} = -\partial^\dagger_{\dot{\alpha}} + i(\theta\sigma^\mu)_{\dot{\alpha}}\partial_\mu. \tag{10.73}$$

The following differential operators will also be useful later in this chapter:

$$D^2 = D^\alpha D_\alpha = -\partial^\alpha\partial_\alpha + 2i(\partial^\alpha\sigma^\mu_{\alpha\dot{\beta}}\theta^{\dagger\dot{\beta}})\partial_\mu + \theta^\dagger\theta^\dagger\,\Box\,, \tag{10.74}$$

$$\overline{D}^2 = \overline{D}_{\dot{\alpha}}\overline{D}^{\dot{\alpha}} = -\partial^\dagger_{\dot{\alpha}}\partial^{\dagger\dot{\alpha}} + 2i(\theta^\alpha\sigma^\mu_{\alpha\dot{\beta}}\partial^{\dagger\dot{\beta}})\partial_\mu + \theta\theta\,\Box\,, \tag{10.75}$$

where $\Box \equiv \partial_\mu\partial^\mu$.

One may now check that

$$\{\widehat{Q}_\alpha,\, D_\beta\} = \{\widehat{Q}^\dagger_{\dot{\alpha}},\, D_\beta\} = \{\widehat{Q}_\alpha,\, \overline{D}_{\dot{\beta}}\} = \{\widehat{Q}^\dagger_{\dot{\alpha}},\, \overline{D}_{\dot{\beta}}\} = 0. \tag{10.76}$$

Using the definition eq. (10.51), it follows that

$$\widehat{\delta}_\epsilon\,(D_\alpha S) = D_\alpha\left(\widehat{\delta}_\epsilon S\right), \qquad \widehat{\delta}_\epsilon\,(\overline{D}_{\dot{\alpha}} S) = \overline{D}_{\dot{\alpha}}\left(\widehat{\delta}_\epsilon S\right). \tag{10.77}$$

Thus the derivatives D_α and $\overline{D}_{\dot{\alpha}}$ are indeed supersymmetric covariant; acting on superfields, they return superfields. This crucial property makes them useful both for defining constraints on superfields in a covariant way, and for defining superspace Lagrangians involving anticommuting spinor coordinate derivatives. These derivatives are linear differential operators, obeying the graded Leibniz rules.

The chiral and antichiral covariant derivatives also satisfy the useful anticommutation identities:

$$\{D_\alpha, \overline{D}_{\dot\beta}\} = 2i\sigma^\mu_{\alpha\dot\beta}\partial_\mu, \tag{10.78}$$

$$\{D_\alpha, D_\beta\} = 0, \qquad \{\overline{D}_{\dot\alpha}, \overline{D}_{\dot\beta}\} = 0. \tag{10.79}$$

This has exactly the same form as the supersymmetry algebra in eqs. (10.65) and (10.66), but $D, \overline{D}$ should not be confused with the differential operators for supersymmetry transformations, $\widehat{Q}, \widehat{Q}^\dagger$. The operators $D, \overline{D}$ do not represent a second supersymmetry.

The reader might be wondering why we use an overline notation for $\overline{D}$, but a dagger for $\widehat{Q}^\dagger$. The reason is that the dagger and the overline denote different kinds of conjugation. The dagger on $\widehat{Q}^\dagger$ represents hermitian conjugation in the same sense that $\widehat{P}_\mu = i\partial_\mu$ is a hermitian differential operator on an inner product space, but the overline on $\overline{D}$ represents complex conjugation in the same sense that ∂_μ is a real differential operator, with $(\partial_\mu f)^* = \partial_\mu f^*$, where f is a complex function of x^μ (and f^* is the complex conjugate of f). Recall that if we define the inner product on the space of functions by

$$\langle g|f\rangle = \int d^4x\, g^*(x) f(x), \tag{10.80}$$

then, using integration by parts,

$$\langle g|\widehat{P} f\rangle = \langle f|\widehat{P} g\rangle^*. \tag{10.81}$$

Similarly, the dagger on the differential operator $\widehat{Q}^\dagger$ denotes hermitian conjugation with respect to the inner product defined by integration of complex superfunctions over superspace. To see this, define, for any two superfunctions $F(x, \theta, \theta^\dagger)$ and $G(x, \theta, \theta^\dagger)$, the inner product

$$\langle G|F\rangle = \int d^4x \int d^2\theta \int d^2\theta^\dagger\, G^* F. \tag{10.82}$$

Now one finds, by integration by parts over superspace, that with the definitions in eqs. (10.47) and (10.48),[1]

$$\langle G|\widehat{Q}^\dagger_{\dot\alpha} F\rangle = \langle F|\widehat{Q}_\alpha G\rangle^*. \tag{10.83}$$

In contrast, the definition of $\overline{D}$ is analogous to the equation $(\partial_\mu f)^* = \partial_\mu f^*$ for functions on ordinary spacetime in light of eqs. (10.72) and (10.73). In this sense, ∂_μ is a real differential operator, and similarly $\overline{D}_{\dot\alpha}$ is the conjugate of D_α. This is more than just notation; if we defined $D^\dagger_{\dot\alpha}$ from D_α in a way analogous to eq. (10.83), then one can check that it would not be equal to $\overline{D}_{\dot\alpha}$ as defined above.

[1] Note that the dagger on the operator $Q^\dagger_{\dot\alpha}$ (without the hat), defined in eq. (8.20), represents yet another sort of hermitian conjugation, in the quantum field theory Hilbert space sense.

It is useful to note that, from eq. (10.46),

$$\int d^2\theta\, D_\alpha(\text{anything}) \qquad \text{and} \qquad \int d^2\theta^\dagger\, \overline{D}_{\dot\alpha}(\text{anything}) \tag{10.84}$$

are each total derivatives with respect to x^μ. This enables integration by parts in superspace of Lagrangian terms with respect to either D_α or $\overline{D}_{\dot\alpha}$. Another useful fact is that acting three consecutive times with either of D_α or $\overline{D}_{\dot\alpha}$ always produces a vanishing result:

$$D_\alpha D_\beta D_\gamma(\text{anything}) = 0 \qquad \text{and} \qquad \overline{D}_{\dot\alpha}\overline{D}_{\dot\beta}\overline{D}_{\dot\gamma}(\text{anything}) = 0. \tag{10.85}$$

This follows from eq. (10.79), and is true essentially because the spinor indices on the anticommuting derivatives can only have two values.

10.4 Chiral Superfields

To describe a chiral supermultiplet, consider the superfield $\Phi(x,\theta,\theta^\dagger)$ obtained by imposing the constraint

$$\overline{D}_{\dot\alpha}\Phi = 0. \tag{10.86}$$

A superfield satisfying this constraint is said to be a chiral (or left-chiral) superfield, and its conjugate $\Phi^\dagger$ is called antichiral (or right-chiral) and satisfies[2]

$$D_\alpha\Phi^\dagger = 0. \tag{10.87}$$

These constraints are consistent with the transformation rule for general superfields because of eq. (10.77).

To solve the constraint eq. (10.86) in general, it is convenient to define

$$y^\mu \equiv x^\mu - i\theta\sigma^\mu\theta^\dagger, \tag{10.88}$$

and change coordinates on superspace to the set y^μ, θ^α, $\theta^\dagger_{\dot\alpha}$. In terms of these variables, the chiral covariant derivatives have the representation

$$D_\alpha = \partial_\alpha - 2i(\sigma^\mu\theta^\dagger)_\alpha\frac{\partial}{\partial y^\mu}, \qquad D^\alpha = -\partial^\alpha + 2i(\theta^\dagger\overline{\sigma}^\mu)^\alpha\frac{\partial}{\partial y^\mu}, \tag{10.89}$$

$$\overline{D}^{\dot\alpha} = \partial^{\dagger\dot\alpha}, \qquad \overline{D}_{\dot\alpha} = -\partial^\dagger_{\dot\alpha}. \tag{10.90}$$

Equation (10.90) makes it clear that the chiral superfield constraint eq. (10.86) is solved by any function of y^μ and θ only, as long as it is not a function of $\theta^\dagger$. Therefore, one can expand

$$\Phi = \phi(y) + \sqrt{2}\,\theta\psi(y) + \theta\theta F(y), \tag{10.91}$$

$$\Phi^\dagger = \phi^\dagger(y^\dagger) + \sqrt{2}\,\theta^\dagger\psi^\dagger(y^\dagger) + \theta^\dagger\theta^\dagger F^\dagger(y^\dagger), \tag{10.92}$$

[2] As remarked previously when defining a conjugate field, the dagger denotes complex conjugation for classical superfields and hermitian conjugation for quantum superfield operators.

where the factors of $\sqrt{2}$ are conventional and

$$y^{\mu\dagger} \equiv x^\mu + i\theta\sigma^\mu\theta^\dagger . \tag{10.93}$$

The chiral covariant derivatives in terms of the coordinates $(y^\dagger, \theta, \theta^\dagger)$ are also sometimes useful:

$$D_\alpha = \partial_\alpha , \qquad D^\alpha = -\partial^\alpha , \tag{10.94}$$

$$\overline{D}^{\dot\alpha} = \partial^{\dagger\dot\alpha} - 2i(\overline{\sigma}^\mu\theta)^{\dot\alpha}\frac{\partial}{\partial y^{\mu\dagger}} , \qquad \overline{D}_{\dot\alpha} = -\partial^\dagger_{\dot\alpha} + 2i(\theta\sigma^\mu)_{\dot\alpha}\frac{\partial}{\partial y^{\mu\dagger}} . \tag{10.95}$$

According to eq. (10.91), the chiral superfield-independent degrees of freedom are a complex scalar ϕ, a two-component fermion ψ, and an auxiliary field F, just as found in Section 8.1. If Φ is a free fundamental chiral superfield, then assigning it dimension $[\text{mass}]^1$ gives the canonical mass dimensions to the component fields, because θ and $\theta^\dagger$ have dimension $[\text{mass}]^{-1/2}$. Rewriting the chiral superfields in terms of the original coordinates $x, \theta, \theta^\dagger$, by expanding in a power series in the anticommuting coordinates, gives

$$\begin{aligned}\Phi = {} & \phi(x) - i\theta\sigma^\mu\theta^\dagger\partial_\mu\phi(x) - \tfrac{1}{4}\theta\theta\theta^\dagger\theta^\dagger\partial_\mu\partial^\mu\phi(x) + \sqrt{2}\theta\psi(x) \\ & - \frac{i}{\sqrt{2}}\theta\theta\theta^\dagger\overline{\sigma}^\mu\partial_\mu\psi(x) + \theta\theta F(x),\end{aligned} \tag{10.96}$$

$$\begin{aligned}\Phi^\dagger = {} & \phi^\dagger(x) + i\theta\sigma^\mu\theta^\dagger\partial_\mu\phi^\dagger(x) - \tfrac{1}{4}\theta\theta\theta^\dagger\theta^\dagger\partial_\mu\partial^\mu\phi^\dagger(x) + \sqrt{2}\theta^\dagger\psi^\dagger(x) \\ & - \frac{i}{\sqrt{2}}\theta^\dagger\theta^\dagger\theta\sigma^\mu\partial_\mu\psi^\dagger(x) + \theta^\dagger\theta^\dagger F^\dagger(x).\end{aligned} \tag{10.97}$$

Depending on the situation, eqs. (10.91) and (10.92) are sometimes a more convenient representation than eqs. (10.96) and (10.97).

By comparing the general superfield case eq. (10.11) to eq. (10.96), we see that the latter can be obtained from the former by identifying component fields:

$$a = \phi, \qquad \xi_\alpha = \sqrt{2}\,\psi_\alpha, \qquad \chi^{\dagger\dot\alpha} = 0, \qquad b = F, \tag{10.98}$$

$$c = 0, \qquad v_\mu = -i\partial_\mu\phi, \qquad \eta_\alpha = 0, \tag{10.99}$$

$$\zeta^{\dagger\dot\alpha} = -\frac{i}{\sqrt{2}}(\overline{\sigma}^\mu\partial_\mu\psi)^{\dot\alpha}, \qquad d = -\tfrac{1}{4}\partial_\mu\partial^\mu\phi. \tag{10.100}$$

It is now straightforward to obtain the supersymmetry transformation laws for the component fields of Φ, either by using $\widehat{\delta}_\epsilon\Phi = -i(\epsilon\widehat{Q} + \epsilon^\dagger\widehat{Q}^\dagger)\Phi$ or by plugging eqs. (10.98)–(10.100) into the results for a general superfield, eqs. (10.56)–(10.64). The end results are

$$\widehat{\delta}_\epsilon\phi = \sqrt{2}\,\epsilon\psi, \tag{10.101}$$

$$\widehat{\delta}_\epsilon\psi_\alpha = -i\sqrt{2}(\sigma^\mu\epsilon^\dagger)_\alpha\partial_\mu\phi + \sqrt{2}\,\epsilon_\alpha F, \tag{10.102}$$

$$\widehat{\delta}_\epsilon F = -i\sqrt{2}\epsilon^\dagger\overline{\sigma}^\mu\partial_\mu\psi, \tag{10.103}$$

in agreement with eqs. (8.3), (8.13), (8.15), after making use of eq. (10.55).

One way to construct a chiral or antichiral superfield is

$$\Phi = \overline{D}^2 S \equiv \overline{D}_{\dot\alpha}\overline{D}^{\dot\alpha} S, \qquad \Phi^\dagger = D^2 S^\dagger \equiv D^\alpha D_\alpha S^\dagger, \tag{10.104}$$

where S is any general superfield. The fact that these are chiral and antichiral, respectively, follows immediately from eq. (10.85). The converse is also true: for every chiral superfield Φ, one can find a superfield S such that eq. (10.104) is true.

Another way to build a chiral superfield is as a function $W(\Phi_i)$ of other chiral superfields Φ_i but not antichiral superfields. That is, W is holomorphic in chiral superfields treated as complex variables. This fact follows immediately from the linearity and product rule properties of the differential operator $\overline{D}_{\dot\alpha}$ appearing in the constraint eq. (10.86). It will be useful for constructing superspace Lagrangians.

10.5 Vector Superfields

A vector (or real) superfield V is obtained by imposing the constraint $V = V^\dagger$. This is equivalent to imposing the following constraints on the components of the general superfield eq. (10.11):

$$a = a^\dagger, \quad \chi^\dagger = \xi^\dagger, \quad c = b^\dagger, \quad v_\mu = v_\mu^\dagger, \quad \zeta^\dagger = \eta^\dagger, \quad d = d^\dagger. \tag{10.105}$$

It is also convenient and traditional to redefine

$$\eta_\alpha = \lambda_\alpha - \tfrac{1}{2} i(\sigma^\mu \partial_\mu \xi^\dagger)_\alpha, \qquad v_\mu = A_\mu, \qquad d = \tfrac{1}{2}D - \tfrac{1}{4}\partial_\mu\partial^\mu a. \tag{10.106}$$

The component expansion of the vector superfield is then

$$\begin{aligned} V(x,\theta,\theta^\dagger) = {} & a + \theta\xi + \theta^\dagger\xi^\dagger + \theta\theta b + \theta^\dagger\theta^\dagger b^\dagger + \theta\sigma^\mu\theta^\dagger A_\mu + \theta^\dagger\theta^\dagger\theta(\lambda - \tfrac{1}{2}i\sigma^\mu\partial_\mu\xi^\dagger) \\ & + \theta\theta\theta^\dagger(\lambda^\dagger - \tfrac{1}{2}i\overline{\sigma}^\mu\partial_\mu\xi) + \theta\theta\theta^\dagger\theta^\dagger(\tfrac{1}{2}D - \tfrac{1}{4}\partial_\mu\partial^\mu a). \end{aligned} \tag{10.107}$$

The supersymmetry transformations of these components can be obtained either from $\widehat\delta_\epsilon V = -i(\epsilon\widehat{Q} + \epsilon^\dagger\widehat{Q}^\dagger)V$ or by plugging eqs. (10.105)–(10.106) into the results for a general superfield, eqs. (10.56)–(10.64). The results are

$$\widehat\delta_\epsilon a = \epsilon\xi + \epsilon^\dagger\xi^\dagger, \tag{10.108}$$

$$\widehat\delta_\epsilon \xi_\alpha = 2\epsilon_\alpha b + (\sigma^\mu\epsilon^\dagger)_\alpha(A_\mu - i\partial_\mu a), \tag{10.109}$$

$$\widehat\delta_\epsilon b = \epsilon^\dagger\lambda^\dagger - i\epsilon^\dagger\overline{\sigma}^\mu\partial_\mu\xi, \tag{10.110}$$

$$\widehat\delta_\epsilon A^\mu = -i\epsilon\partial^\mu\xi + i\epsilon^\dagger\partial^\mu\xi^\dagger + \epsilon\sigma^\mu\lambda^\dagger - \epsilon^\dagger\overline{\sigma}^\mu\lambda, \tag{10.111}$$

$$\widehat\delta_\epsilon \lambda_\alpha = \epsilon_\alpha D - \tfrac{1}{2}i(\sigma^\mu\overline{\sigma}^\nu\epsilon)_\alpha(\partial_\mu A_\nu - \partial_\nu A_\mu), \tag{10.112}$$

$$\widehat\delta_\epsilon D = -i\epsilon\sigma^\mu\partial_\mu\lambda^\dagger - i\epsilon^\dagger\overline{\sigma}^\mu\partial_\mu\lambda. \tag{10.113}$$

A superfield cannot be both chiral and real at the same time, unless it is identically constant (i.e., independent of x^μ, θ, and $\theta^\dagger$). This follows from eqs. (10.98), (10.100), and (10.105). However, if Φ is a chiral superfield, then $\Phi+\Phi^\dagger$ and $i(\Phi-\Phi^\dagger)$ and $\Phi\Phi^\dagger$ are all real (vector) superfields.

As the notation employed in eq. (10.107) suggests, a vector superfield that is used to represent a gauge supermultiplet contains gauge boson, gaugino, and gauge auxiliary fields A^μ, λ, D as components. (Such a vector superfield V must be dimensionless in order for the component fields to have the canonical mass dimensions.) However, there are other component fields in V that did not appear in Sections 8.3 and 8.4. They are: a real scalar a, a two-component fermion ξ, and a complex scalar b, with mass dimensions respectively 0, 1/2, and 1. These are additional auxiliary fields that can be "supergauged" away. To see this, consider the "supergauge" transformation of a vector superfield V for a U(1) gauge symmetry,

$$V \to V + i(\Omega - \Omega^\dagger), \tag{10.114}$$

where $\Omega = \phi + \sqrt{2}\theta\psi + \theta\theta F + \cdots$ is a chiral superfield gauge transformation parameter. In components, this transformation is

$$a \to a + i(\phi - \phi^\dagger), \tag{10.115}$$
$$\xi_\alpha \to \xi_\alpha + i\sqrt{2}\psi_\alpha, \tag{10.116}$$
$$b \to b + iF, \tag{10.117}$$
$$A_\mu \to A_\mu + \partial_\mu(\phi + \phi^\dagger), \tag{10.118}$$
$$\lambda_\alpha \to \lambda_\alpha, \tag{10.119}$$
$$D \to D. \tag{10.120}$$

Equation (10.118) shows that eq. (10.114) provides the vector boson field with the usual gauge transformation, with parameter $\Lambda = 2\,\mathrm{Re}\,\phi$. By requiring the gauge transformation to take a supersymmetric form, it follows that independent choices of $\mathrm{Im}\,\phi$, ψ_α, and F can also change a, ξ_α, and b arbitrarily. Thus the supergauge transformation eq. (10.114) has ordinary gauge transformations as a special case.

In particular, supergauge transformations can eliminate the auxiliary fields a, ξ_α, and b completely. A superspace Lagrangian for a vector superfield must be invariant under the supergauge transformation eq. (10.114) in the abelian case, or a suitable generalization given below for the nonabelian case. After making a supergauge transformation to eliminate a, ξ, and b, the vector superfield is said to be in Wess–Zumino gauge, and is simply given by

$$V_{\text{WZ gauge}} = \theta\sigma^\mu\theta^\dagger A_\mu + \theta^\dagger\theta^\dagger\theta\lambda + \theta\theta\theta^\dagger\lambda^\dagger + \tfrac{1}{2}\theta\theta\theta^\dagger\theta^\dagger D. \tag{10.121}$$

The restriction of the vector superfield to Wess–Zumino gauge is not consistent with the linear superspace version of supersymmetry transformations. This is because $\widehat{\delta}_\epsilon(V_{\text{WZ gauge}})$ contains $-\theta^\dagger\overline{\sigma}^\mu\epsilon A_\mu + \theta\sigma^\mu\epsilon^\dagger A_\mu + \theta\theta\epsilon^\dagger\lambda^\dagger + \theta^\dagger\theta^\dagger\epsilon\lambda$, and so the supersymmetry transformation of the Wess–Zumino gauge vector superfield is not in Wess–Zumino gauge. However, a supergauge transformation can always restore $\widehat{\delta}_\epsilon(V_{\text{WZ gauge}})$ to Wess–Zumino gauge. Adopting Wess–Zumino gauge is equivalent to partially fixing the supergauge, while still maintaining the full freedom to do ordinary gauge transformations.

10.6 How to Make a Lagrangian in Superspace

So far, we have been concerned with the structural features of fields in superspace. We now turn to the dynamical issue of how to construct manifestly supersymmetric actions. A key observation is that the integral of a general superfield over all of superspace is automatically invariant:

$$\widehat{\delta}_\epsilon A = 0, \quad \text{for} \quad A = \int d^4x \int d^2\theta\, d^2\theta^\dagger\, S(x,\theta,\theta^\dagger). \qquad (10.122)$$

This follows immediately from the fact that $\widehat{Q}$ and $\widehat{Q}^\dagger$ as defined in eqs. (10.47), (10.48) are sums of total derivatives with respect to the superspace coordinates $x^\mu, \theta, \theta^\dagger$, so that $(\epsilon\widehat{Q} + \epsilon^\dagger\widehat{Q}^\dagger)S$ vanishes upon integration. As a check, eq. (10.64) shows that the $\theta\theta\theta^\dagger\theta^\dagger$ component of a superfield transforms into a total spacetime derivative.

Therefore, the action governing the dynamics of a theory can have contributions of the form of eq. (10.122), with reality of the action demanding that S is some real (vector) superfield V. From eq. (10.51), we see that the principle of global supersymmetric invariance is embodied in the requirement that the action should be an integral over superspace which is unchanged under rigid translations of the superspace coordinates. To obtain the Lagrangian $\mathscr{L}(x)$, one integrates over only the fermionic coordinates. This is often written in the notation

$$[V]_D \equiv \int d^2\theta\, d^2\theta^\dagger\, V(x,\theta,\theta^\dagger) = V(x,\theta,\theta^\dagger)\Big|_{\theta\theta\theta^\dagger\theta^\dagger} = \tfrac{1}{2}D - \tfrac{1}{4}\partial_\mu\partial^\mu a, \qquad (10.123)$$

using eq. (10.40) and the form of V in eq. (10.107) for the last equality. This is referred to as a D-term contribution to the Lagrangian (note that the $\partial_\mu\partial^\mu a$ part will vanish upon integration $\int d^4x$).

Another type of contribution to the action can be inferred from the fact that the F-term of a chiral superfield also transforms into a total derivative under a supersymmetry transformation [see eq. (10.103)]. This implies that one can have a contribution to the Lagrangian of the form

$$[\Phi]_F \equiv \Phi\Big|_{\theta\theta} = \int d^2\theta\, \Phi\Big|_{\theta^\dagger=0} = \int d^2\theta\, d^2\theta^\dagger\, \delta^{(2)}(\theta^\dagger)\,\Phi = F, \qquad (10.124)$$

using the form of Φ in eq. (10.96) for the last equality. This satisfies

$$\widehat{\delta}_\epsilon\left(\int d^4x[\Phi]_F\right) = 0. \qquad (10.125)$$

The F-term of a chiral superfield is complex in general, but the action must be real, which can be ensured if this type of contribution to the Lagrangian is accompanied by its complex conjugate:

$$[\Phi]_F + \text{h.c.} = \int d^2\theta\, d^2\theta^\dagger \left[\delta^{(2)}(\theta^\dagger)\,\Phi + \delta^{(2)}(\theta)\,\Phi^\dagger\right]. \qquad (10.126)$$

Note that the identification of the F-term component of a chiral superfield is the same in the $(x^\mu, \theta, \theta^\dagger)$ and $(y^\mu, \theta, \theta^\dagger)$ coordinates, in the sense that, in both cases, one simply isolates the $\theta\theta$ component. This follows because the difference between x^μ and y^μ is higher-order in $\theta^\dagger$. It is a useful trick, because many calculations involving chiral superfields are simpler to carry out in terms of y^μ.

Another possible try would be to take the D-term of a chiral superfield. However, this is a waste of time, because

$$[\Phi]_D = \int d^2\theta\, d^2\theta^\dagger\, \Phi = \Phi\Big|_{\theta\theta\theta^\dagger\theta^\dagger} = \tfrac{1}{4}\partial_\mu\partial^\mu\phi, \tag{10.127}$$

where the last equality follows from eq. (10.96), and ϕ is the scalar component of Φ. Equation (10.127) is a total derivative, so adding it (and its complex conjugate) to the Lagrangian has no effect.

Therefore, the two ways of making a supersymmetric Lagrangian are to take the D-term component of a real superfield, and to take the F-term component of a chiral superfield, plus the complex conjugate. When building a Lagrangian, the real superfield V used in eq. (10.123) and the chiral superfield Φ used in eq. (10.126) are usually composites, built out of more fundamental superfields. However, contributions from fundamental fields V and Φ are allowed, when V is the vector superfield for an abelian gauge symmetry and when Φ is a singlet under all symmetries.

It is always possible to rewrite a D-term contribution to a Lagrangian as an F-term contribution, by the trick of noticing that

$$\overline{D}^2(\theta^\dagger\theta^\dagger) = D^2(\theta\theta) = -4, \tag{10.128}$$

and using the fact that $\delta^{(2)}(\theta^\dagger) = \theta^\dagger\theta^\dagger$ from eq. (10.42). Thus, by integrating by parts twice with respect to $\theta^\dagger$,

$$\begin{aligned}[V]_D &= -\frac{1}{4}\int d^2\theta d^2\theta^\dagger\, V\, \overline{D}^2(\theta^\dagger\theta^\dagger) = -\frac{1}{4}\int d^2\theta d^2\theta^\dagger\, \delta^{(2)}(\theta^\dagger)\, \overline{D}^2 V + \cdots \\ &= -\tfrac{1}{4}\big[\overline{D}^2 V\big]_F + \cdots.\end{aligned} \tag{10.129}$$

The $\cdots$ indicates total derivatives with respect to x^μ, coming from the two integrations by parts. As noted in Section 10.4, $\overline{D}^2 V$ is always a chiral superfield. If V is real, then the imaginary part of eq. (10.129) is a total derivative, and the result can be rewritten as $-\frac{1}{8}\big[\overline{D}^2 V\big]_F + \text{h.c.}$

10.7 Superspace Lagrangians for Chiral Supermultiplets

In Section 10.4, we verified that the chiral superfield components have the same supersymmetry transformations as the Wess–Zumino model fields. We now have the tools to complete the demonstration of equivalence by reconstructing the Lagrangian in superspace language. Consider the composite superfield obtained by multiplying an antichiral superfield and a chiral superfield:

$$\begin{aligned}\Phi^{\dagger i}\Phi_j = {}& \phi^{\dagger i}\phi_j + \sqrt{2}\theta\psi_j\phi^{\dagger i} + \sqrt{2}\theta^\dagger\psi^{\dagger i}\phi_j + \theta\theta\phi^{\dagger i}F_j + \theta^\dagger\theta^\dagger\phi_j F^{\dagger i} \\ &+ \theta\sigma^\mu\theta^\dagger\left[-i\phi^{\dagger i}\partial_\mu\phi_j + i\phi_j\partial_\mu\phi^{\dagger i} - \psi^{\dagger i}\overline{\sigma}_\mu\psi_j\right] \\ &+ \frac{i}{\sqrt{2}}\theta\theta\theta^\dagger\overline{\sigma}^\mu(\psi_j\partial_\mu\phi^{\dagger i} - \partial_\mu\psi_j\phi^{\dagger i}) + \sqrt{2}\theta\theta\theta^\dagger\psi^{\dagger i}F_j \\ &+ \frac{i}{\sqrt{2}}\theta^\dagger\theta^\dagger\theta\sigma^\mu(\psi^{\dagger i}\partial_\mu\phi_j - \partial_\mu\psi^{\dagger i}\phi_j) + \sqrt{2}\theta^\dagger\theta^\dagger\theta\psi_j F^{\dagger i} \\ &+ \theta\theta\theta^\dagger\theta^\dagger\Big[F^{\dagger i}F_j + \tfrac{1}{2}\partial^\mu\phi^{\dagger i}\partial_\mu\phi_j - \tfrac{1}{4}\phi^{\dagger i}\partial^\mu\partial_\mu\phi_j - \tfrac{1}{4}\phi_j\partial^\mu\partial_\mu\phi^{\dagger i} \\ &\qquad + \tfrac{1}{2}i\psi^{\dagger i}\overline{\sigma}^\mu\partial_\mu\psi_j + \tfrac{1}{2}i\psi_j\sigma^\mu\partial_\mu\psi^{\dagger i}\Big],\end{aligned} \tag{10.130}$$

where all fields are evaluated as functions of x^μ (not y^μ or $y^{\mu\dagger}$). For $i = j$, eq. (10.130) is a real (vector) superfield, and the massless free-field Lagrangian for each chiral superfield is just obtained by taking the $\theta\theta\theta^\dagger\theta^\dagger$ component:

$$[\Phi^\dagger\Phi]_D = \int d^2\theta\, d^2\theta^\dagger\, \Phi^\dagger\Phi \;=\; \partial^\mu\phi^\dagger\partial_\mu\phi + i\psi^\dagger\overline{\sigma}^\mu\partial_\mu\psi + F^\dagger F + \cdots . \tag{10.131}$$

The $\cdots$ indicates a total derivative part, which may be dropped since this is destined to be integrated $\int d^4x$. Equation (10.131) is exactly the Lagrangian obtained in Section 8.1 for the massless free Wess–Zumino model.

To obtain the superpotential interaction and mass terms, recall that products of chiral superfields are also superfields. For example,

$$\Phi_i\Phi_j = \phi_i\phi_j + \sqrt{2}\theta(\psi_i\phi_j + \psi_j\phi_i) + \theta\theta(\phi_i F_j + \phi_j F_i - \psi_i\psi_j), \tag{10.132}$$

$$\begin{aligned}\Phi_i\Phi_j\Phi_k = {}& \phi_i\phi_j\phi_k + \sqrt{2}\theta(\psi_i\phi_j\phi_k + \psi_j\phi_i\phi_k + \psi_k\phi_i\phi_j) \\ &+ \theta\theta(\phi_i\phi_j F_k + \phi_i\phi_k F_j + \phi_j\phi_k F_i - \psi_i\psi_j\phi_k - \psi_i\psi_k\phi_j - \psi_j\psi_k\phi_i),\end{aligned} \tag{10.133}$$

where the presentation has been simplified by taking the component fields on the right sides to be functions of y^μ as given in eq. (10.88). More generally, any holomorphic function of chiral superfields is a chiral superfield. So, one may form a complete Lagrangian as

$$\mathscr{L}(x) = [\Phi^{\dagger i}\Phi_i]_D + \left([W(\Phi_i)]_F + \text{h.c.}\right), \tag{10.134}$$

where $W(\Phi_i)$ can be any holomorphic function of the chiral superfields (but not antichiral superfields) taken as complex variables, and coincides with the superpotential $W(\phi_i)$ that was treated in Section 8.2 as a function of the scalar components. For $W = \frac{1}{2}M^{ij}\Phi_i\Phi_j + \frac{1}{6}y^{ijk}\Phi_i\Phi_j\Phi_k$, the result of eq. (10.134) is exactly the same as eq. (8.51), after writing in component form using eqs. (10.131), (10.132), and (10.133) and integrating out the auxiliary fields.

It is instructive to obtain the superfield equations of motion from the Lagrangian eq. (10.134). The quickest way to do this is to first use the remarks at the very end of Section 10.6 to rewrite the Lagrangian as

$$\mathscr{L}(x) = \int d^2\theta\,[-\tfrac{1}{4}\overline{D}^2\Phi^{\dagger i}\Phi_i + W(\Phi_i)] \,+\, \int d^2\theta^\dagger\,[W(\Phi_i)]^\dagger . \tag{10.135}$$

Now, varying with respect to Φ_i immediately gives the superfield equation of motion,

$$-\tfrac{1}{4}\overline{D}^2\Phi^{\dagger i} + \frac{\partial W}{\partial \Phi_i} = 0, \qquad (10.136)$$

and its complex conjugate,

$$-\tfrac{1}{4}D^2\Phi_i + \frac{\partial W^\dagger}{\partial \Phi^{\dagger i}} = 0. \qquad (10.137)$$

These are equivalent to the component-level equations of motion, as can be found from the Lagrangian in Section 8.2. To verify this, it is easiest to write eq. (10.136) in the coordinate system $(y^\mu, \theta, \theta^\dagger)$, in which the first term has the simple form

$$-\frac{1}{4}\overline{D}^2\Phi^{\dagger i} = F^\dagger(y) - i\sqrt{2}\theta\sigma^\mu\partial_\mu\psi^{\dagger i}(y) - \theta\theta\partial_\mu\partial^\mu\phi^{\dagger i}(y). \qquad (10.138)$$

Since this is a chiral (not antichiral) superfield, it is simpler to write the components as functions of y^μ as shown, not $y^{\mu\dagger}$, even though the left-hand side involves $\Phi^\dagger$.

For an alternative method, consider a Lagrangian V on the full superspace, so that the action is

$$A = \int d^4x \int d^2\theta\, d^2\theta^\dagger\, V, \qquad (10.139)$$

with $V(S_i, D_\alpha S_i, \overline{D}_{\dot\alpha} S_i)$ a function of general dynamical superfields S_i and their chiral and antichiral first derivatives. Then the superfield equations of motion obtained by variation of the action are

$$0 = \frac{\partial V}{\partial S_i} - D_\alpha\left(\frac{\partial V}{\partial(D_\alpha S_i)}\right) - \overline{D}_{\dot\alpha}\left(\frac{\partial V}{\partial(\overline{D}_{\dot\alpha} S_i)}\right). \qquad (10.140)$$

In the case of the Lagrangian for chiral superfields, eq. (10.134), Lagrange multipliers $\Gamma^{\dagger i\dot\alpha}$ and Γ_i^α can be introduced to enforce the chiral and antichiral superfield constraints on Φ_i and $\Phi^{\dagger i}$, respectively. The Lagrangian on superspace is then given by

$$V = \Gamma^{\dagger i\dot\alpha}\overline{D}_{\dot\alpha}\Phi_i + \Gamma_i^\alpha D_\alpha\Phi^{\dagger i} + \Phi^{\dagger i}\Phi_i + \delta^{(2)}(\theta^\dagger)W(\Phi_i) + \delta^{(2)}(\theta)[W(\Phi_i)]^\dagger. \qquad (10.141)$$

Varying with respect to the Lagrange multipliers gives the constraints $\overline{D}_{\dot\alpha}\Phi_i = 0$ and $D_\alpha\Phi^{\dagger i} = 0$. Applying eq. (10.140) to the superfields Φ_i and $\Phi^{\dagger i}$ leads to equations of motion

$$0 = \Phi^{\dagger i} + \delta^{(2)}(\theta^\dagger)\frac{\partial W}{\partial\Phi_i} - \overline{D}_{\dot\alpha}\Gamma^{\dagger i\dot\alpha}, \qquad (10.142)$$

$$0 = \Phi_i + \delta^{(2)}(\theta)\frac{\partial W^\dagger}{\partial\Phi^{\dagger i}} - D_\alpha\Gamma_i^\alpha. \qquad (10.143)$$

Now, acting on these equations with $-\tfrac{1}{4}\overline{D}^2$ and $-\tfrac{1}{4}D^2$ respectively, and applying eqs. (10.41), (10.42), and (10.128), one again obtains eqs. (10.136) and (10.137).

10.8 Superspace Lagrangians for Abelian Gauge Theory

Now consider the superspace Lagrangian for a gauge theory, treating the U(1) case first for simplicity. The nonabelian case will be considered in the next section.

The vector superfield $V(x,\theta,\theta^\dagger)$ of eq. (10.107) contains the gauge potential A^μ. Define corresponding gauge-invariant abelian field-strength superfields by

$$\mathcal{W}_\alpha = -\tfrac{1}{4}\overline{D}^2 D_\alpha V, \qquad \mathcal{W}^\dagger_{\dot\alpha} = -\tfrac{1}{4}D^2\overline{D}_{\dot\alpha}V. \tag{10.144}$$

These are respectively chiral and antichiral by construction [see eq. (10.104)], and are examples of superfields that carry spinor indices and are anticommuting. They carry dimension $[\text{mass}]^{3/2}$. To see that $\mathcal{W}_\alpha$ is gauge invariant, note that, under a supergauge transformation of the form eq. (10.114),

$$\begin{aligned}\mathcal{W}_\alpha \to -\tfrac{1}{4}\overline{D}^2 D_\alpha[V+i(\Omega-\Omega^\dagger)] &= \mathcal{W}_\alpha - \tfrac{1}{4}i\overline{D}^2 D_\alpha\Omega = \mathcal{W}_\alpha + \tfrac{1}{4}i\overline{D}^{\dot\beta}\{\overline{D}_{\dot\beta}, D_\alpha\}\Omega \\ &= \mathcal{W}_\alpha - \tfrac{1}{2}\sigma^\mu_{\alpha\dot\beta}\partial_\mu\overline{D}^{\dot\beta}\Omega = \mathcal{W}_\alpha\,.\end{aligned} \tag{10.145}$$

The first equality follows from eq. (10.87) because $\Omega^\dagger$ is antichiral, the second and fourth equalities from eq. (10.86) because Ω is chiral, and the third from eq. (10.78).

To see how the component fields fit into $\mathcal{W}_\alpha$, it is convenient to temporarily specialize to Wess–Zumino gauge as in eq. (10.121), and then convert to the coordinates $(y^\mu,\theta,\theta^\dagger)$ as defined in eq. (10.88), with the result

$$V(y^\mu,\theta,\theta^\dagger) = \theta\sigma^\mu\theta^\dagger A_\mu(y) + \theta^\dagger\theta^\dagger\theta\lambda(y) + \theta\theta\theta^\dagger\lambda^\dagger(y) + \tfrac{1}{2}\theta\theta\theta^\dagger\theta^\dagger\,[D(y) + i\partial_\mu A^\mu(y)]\,. \tag{10.146}$$

Now application of eqs. (10.89), (10.90) yields

$$\mathcal{W}_\alpha(y,\theta,\theta^\dagger) = \lambda_\alpha + \theta_\alpha D - \tfrac{1}{2}i(\sigma^\mu\overline{\sigma}^\nu\theta)_\alpha F_{\mu\nu} + i\theta\theta(\sigma^\mu\partial_\mu\lambda^\dagger)_\alpha, \tag{10.147}$$

$$\mathcal{W}^{\dagger\dot\alpha}(y^\dagger,\theta,\theta^\dagger) = \lambda^{\dagger\dot\alpha} + \theta^{\dagger\dot\alpha}D + \tfrac{1}{2}i(\overline{\sigma}^\mu\sigma^\nu\theta^\dagger)^{\dot\alpha}F_{\mu\nu} + i\theta^\dagger\theta^\dagger(\overline{\sigma}^\mu\partial_\mu\lambda)^{\dot\alpha}, \tag{10.148}$$

where all fields are understood to be functions of y^μ and $y^{\mu\dagger}$ respectively, and

$$F_{\mu\nu} = \partial_\mu A_\nu - \partial_\nu A_\mu \tag{10.149}$$

is the ordinary component field strength. Although it was convenient to derive eqs. (10.147) and (10.148) in Wess–Zumino gauge, they must be true in general, because $\mathcal{W}_\alpha$ and $\mathcal{W}^{\dagger\dot\alpha}$ are supergauge invariant.

It is straightforward to obtain the supersymmetry transformation laws for the component fields of $\mathcal{W}_\alpha$, by using $\widehat{\delta}_\epsilon\mathcal{W}_\alpha = -i(\epsilon\widehat{Q} + \epsilon^\dagger\widehat{Q}^\dagger)\mathcal{W}_\alpha$. The end result is

$$\widehat{\delta}_\epsilon\lambda_\alpha = \epsilon_\alpha D - \tfrac{1}{2}i(\sigma^\mu\overline{\sigma}^\nu)_\alpha{}^\beta\epsilon_\beta F_{\mu\nu}\,, \tag{10.150}$$

$$\widehat{\delta}_\epsilon F_{\mu\nu} = \partial_\mu(\epsilon\sigma_\nu\lambda^\dagger + \lambda\sigma_\nu\epsilon^\dagger) - \partial_\nu(\epsilon\sigma_\mu\lambda^\dagger + \lambda\sigma_\mu\epsilon^\dagger)\,, \tag{10.151}$$

$$\widehat{\delta}_\epsilon D = -i\partial_\mu(\epsilon\sigma^\mu\lambda^\dagger - \lambda\sigma^\mu\epsilon^\dagger)\,. \tag{10.152}$$

The supersymmetric transformation laws for $\{\lambda, \lambda^\dagger, F_{\mu\nu}, D\}$ exhibited above involve only gauge invariant fields and do not require a particular choice of gauge.[3]

Equation (10.147) implies

$$[\mathcal{W}^\alpha \mathcal{W}_\alpha]_F = D^2 + 2i\lambda\sigma^\mu\partial_\mu\lambda^\dagger - \tfrac{1}{2}F^{\mu\nu}F_{\mu\nu} + \tfrac{1}{4}i\epsilon^{\mu\nu\rho\sigma}F_{\mu\nu}F_{\rho\sigma}, \qquad (10.153)$$

where now all fields on the right side are functions of x^μ. Integrating, and eliminating total derivative parts, one obtains the action

$$\int d^4x\,\mathscr{L} = \int d^4x\,\tfrac{1}{4}[\mathcal{W}^\alpha \mathcal{W}_\alpha]_F + \text{h.c.} = \int d^4x\,\left[\tfrac{1}{2}D^2 + i\lambda^\dagger\overline{\sigma}^\mu\partial_\mu\lambda - \tfrac{1}{4}F^{\mu\nu}F_{\mu\nu}\right], \qquad (10.154)$$

in agreement with eq. (8.57). Additionally, the integral of the D-term component of V itself is invariant under both supersymmetry [see eq. (10.113)] and supergauge [see eq. (10.120)] transformations. Thus, one can include a Fayet–Iliopoulos term

$$\mathscr{L}_{\rm FI} = -2\kappa[V]_D = -\kappa D, \qquad (10.155)$$

again dropping a total derivative. This type of term can play a role in spontaneous supersymmetry breaking, as we will discuss in Section 12.2.

It is also possible to write eq. (10.153) as a D-term rather than an F-term. Since $\mathcal{W}^\alpha$ is a chiral superfield with $\overline{D}_{\dot\beta}\mathcal{W}^\alpha = 0$, one can use eq. (10.144) to write

$$\mathcal{W}^\alpha \mathcal{W}_\alpha = -\tfrac{1}{4}\overline{D}^2(\mathcal{W}^\alpha D_\alpha V). \qquad (10.156)$$

Therefore, using eq. (10.129), the Lagrangian for A^μ, λ, and D can be rewritten as

$$\mathscr{L}(x) = \int d^2\theta\, d^2\theta^\dagger \left[\frac{1}{4}\left(\mathcal{W}^\alpha D_\alpha V + \mathcal{W}^\dagger_{\dot\alpha}\overline{D}^{\dot\alpha}V\right) - 2\kappa V\right]. \qquad (10.157)$$

It is instructive to count the degrees of freedom in the irreducible supermultiplet, $\{\lambda,\, \lambda^\dagger,\, F_{\mu\nu},\, D\}$. On-shell, there are two real fermionic degrees of freedom associated with the massless gaugino, after imposing the Lagrange field equations,[4]

$$i\overline{\sigma}^{\mu\dot\alpha\beta}\partial_\mu\lambda_\beta = 0\,. \qquad (10.158)$$

This matches the two real bosonic degrees of freedom corresponding to the two transverse polarizations of the massless gauge boson.

To count the off-shell bosonic degrees of freedom, one must take into account the Bianchi identity,[5]

$$\epsilon^{\mu\nu\rho\sigma}\partial_\nu F_{\rho\sigma} = 0\,, \qquad (10.159)$$

[3] In contrast, the generalization of eqs. (10.150)–(10.152) to nonabelian gauge theories shown in eqs. (8.60)–(8.62) holds only in the case of the Wess–Zumino gauge.

[4] Starting with two complex (or equivalently four real) degrees of freedom for the two-component gaugino field λ, eq. (10.158) relates the spinor components λ_1 and λ_2, thereby reducing the number of real degrees of freedom from four to two.

[5] Although it appears that the Bianchi identity yields four constraints, since the spacetime index μ is a free index, in fact only three constraints are independent. This is because one of the four constraints is redundant due to the identity $\epsilon^{\mu\nu\rho\sigma}\partial_\mu\partial_\nu F_{\rho\sigma} = 0$, which is automatically satisfied as a result of the antisymmetry of the Levi-Civita epsilon tensor.

which is satisfied independently of the field equations. This identity reduces the number of real degrees of freedom in the real antisymmetric tensor $F_{\mu\nu}$ from six to three. Adding in the one real degree of freedom associated with D, we end up with a total of four real bosonic degrees of freedom, which matches the four real off-shell fermionic degrees of freedom corresponding to λ and $\lambda^\dagger$.

Next consider the coupling of the abelian gauge field to a set of chiral superfields Φ_i carrying U(1) charges q_i. Supergauge transformations, as in eqs. (10.114)–(10.120), are parameterized by a non-dynamical chiral superfield Ω:

$$\Phi_i \to e^{-2igq_i\Omega}\Phi_i, \qquad \Phi^{\dagger i} \to e^{2igq_i\Omega^\dagger}\Phi^{\dagger i}, \tag{10.160}$$

where g is the gauge coupling. In the special case where Ω is just a real function $\phi(x)$ [independent of $\theta, \theta^\dagger$], eq. (10.160) reproduces the usual infinitesimal gauge transformation, $X_i \to X_i - igq_i\Lambda X_i$ (for $X_i = \phi_i, \psi_i, F_i$), with $\Lambda(x) = 2\phi(x)$ [see eq. (8.64)]. The kinetic term from eq. (10.131) involves the superfield $\Phi^{\dagger i}\Phi_i$, which is not supergauge invariant:

$$\Phi^{\dagger i}\Phi_i \to e^{-2igq_i(\Omega-\Omega^\dagger)}\Phi^{\dagger i}\Phi_i. \tag{10.161}$$

To remedy this, we modify the chiral superfield kinetic term in the Lagrangian to

$$\left[\Phi^{\dagger i}e^{2gq_iV}\Phi_i\right]_D. \tag{10.162}$$

The gauge transformation of the e^{2gq_iV} factor, found from eq. (10.114), exactly cancels that of eq. (10.161).

The presence of an exponential of V in the Lagrangian is possible because V is dimensionless. It might appear to be dangerous, because normally such a non-polynomial term would be nonrenormalizable. However, the gauge dependence of V comes to the rescue, as the higher-order terms can be supergauged away. In particular, evaluating e^{2gq_iV} in the Wess–Zumino gauge, the power series expansion of the exponential is simple and terminates, because

$$V^2 = \tfrac{1}{2}\theta\theta\theta^\dagger\theta^\dagger A_\mu A^\mu, \qquad V^n = 0 \qquad (n \geq 3), \tag{10.163}$$

so that

$$e^{2gq_iV} = 1 + 2gq_i(\theta\sigma^\mu\theta^\dagger A_\mu + \theta^\dagger\theta^\dagger\theta\lambda + \theta\theta\theta^\dagger\lambda^\dagger + \tfrac{1}{2}\theta\theta\theta^\dagger\theta^\dagger D) + g^2q_i^2\theta\theta\theta^\dagger\theta^\dagger A_\mu A^\mu. \tag{10.164}$$

It follows that, in the Wess–Zumino gauge and up to total derivative terms,

$$\begin{aligned}\left[\Phi^{\dagger i}e^{2gq_iV}\Phi_i\right]_D = {} & F^{\dagger i}F_i + \nabla_\mu\phi^{\dagger i}\nabla^\mu\phi_i + i\psi^{\dagger i}\overline{\sigma}^\mu\nabla_\mu\psi_i \\ & - \sqrt{2}gq_i(\phi^{\dagger i}\psi_i\lambda + \lambda^\dagger\psi^{\dagger i}\phi_i) + gq_i\phi^{\dagger i}\phi_i D,\end{aligned} \tag{10.165}$$

where ∇_μ is the gauge-covariant spacetime derivative:

$$\nabla_\mu\phi_i = \partial_\mu\phi_i + igq_iA_\mu\phi_i, \qquad \nabla_\mu\phi^{\dagger i} = \partial_\mu\phi^{\dagger i} - igq_iA_\mu\phi^{\dagger i}, \tag{10.166}$$

$$\nabla_\mu\psi_i = \partial_\mu\psi_i + igq_iA_\mu\psi_i. \tag{10.167}$$

Equation (10.165) agrees with the specialization of eq. (8.72) to the abelian case.

In summary, the superspace Lagrangian,

$$\mathcal{L} = \left[\Phi^{\dagger i} e^{2gq_i V}\Phi_i\right]_D + ([W(\Phi_i)]_F + \text{h.c.}) + \tfrac{1}{4}\left([\mathcal{W}^\alpha \mathcal{W}_\alpha]_F + \text{h.c.}\right) - 2\kappa[V]_D\,, \tag{10.168}$$

reproduces the component-form Lagrangian found in Section 8.4 in the special case of matter fields coupled to each other and to a U(1) gauge symmetry, plus a Fayet–Iliopoulos parameter κ.

10.9 Superspace Lagrangians for Nonabelian Gauge Theories

Consider a gauge group G that is (generically) a direct product of compact connected simple Lie groups and U(1) groups (see Section 4.2), which is realized on chiral superfields Φ_i in a representation R with matrix generators $(\boldsymbol{T^a})_i{}^j$,

$$\Phi_i \to \left(e^{-2ig_a\Omega^a \boldsymbol{T^a}}\right)_i{}^j\Phi_j, \qquad \Phi^{\dagger i} \to \Phi^{\dagger j}\left(e^{2ig_a\Omega^{a*}\boldsymbol{T^a}}\right)_j{}^i, \tag{10.169}$$

where there is an implicit sum over the thrice-repeated index a. The gauge couplings for the irreducible components of the Lie algebra are g_a. As in the abelian case, the supergauge transformation parameters are chiral superfields Ω^a. For each Lie algebra generator, there is a vector superfield V^a, which contains the vector gaugino boson and gaugino. The Lagrangian then contains a supergauge-invariant term

$$\mathcal{L} = \left[\Phi^{\dagger i}(e^{2g_a \boldsymbol{T^a} V^a})_i{}^j\Phi_j\right]_D. \tag{10.170}$$

It is convenient to define matrix-valued vector and gauge parameter superfields in the representation R,

$$V_i{}^j \equiv 2g_a(\boldsymbol{T^a})_i{}^j V^a, \qquad \Omega_i{}^j \equiv 2g_a(\boldsymbol{T^a})_i{}^j\Omega^a, \tag{10.171}$$

so that one can write

$$\Phi_i \to \left(e^{-i\Omega}\right)_i{}^j\Phi_j, \qquad \Phi^{\dagger i} \to \Phi^{\dagger j}\left(e^{i\Omega^\dagger}\right)_j{}^i, \tag{10.172}$$

and

$$\mathcal{L} = \left[\Phi^{\dagger i}(e^V)_i{}^j\Phi_j\right]_D. \tag{10.173}$$

For this to be supergauge invariant, the nonabelian gauge transformation rule for the vector superfields must be[6]

$$e^V \to e^{V'} = e^{-i\Omega^\dagger} e^V e^{i\Omega}. \tag{10.174}$$

In general, one cannot explicitly express V' in terms of V in closed form. Nevertheless, it is instructive to consider the case of an infinitesimal gauge transformation,

[6] In eq. (10.174), chiral supermultiplet representation indices $i, j, \ldots$ are suppressed; V and Ω with no indices stand for the matrices defined in eq. (10.171).

$V' = V + \delta V$, in which case the right-hand side of eq. (10.174) can be expanded keeping terms linear in Ω, $\Omega^\dagger$. Using eq. (H.22), it follows that, to linear order in δV and Ω,

$$e^{V+\delta V} = e^V\big[1 + f(\mathrm{ad}_V)(\delta V)\big] = e^V - i\Omega^\dagger e^V + ie^V\Omega\,, \qquad (10.175)$$

where the adjoint operator ad_V is defined in eq. (H.11) and

$$f(z) \equiv \frac{1-e^{-z}}{z} = \sum_{n=0}^{\infty} \frac{(-1)^n}{(n+1)!}\, z^n \qquad (10.176)$$

is defined by its Taylor series expansion. Multiplying eq. (10.175) on the left by e^{-V} and employing eq. (H.1) yields

$$2\sinh\!\big(\tfrac{1}{2}\,\mathrm{ad}_V\big)(\delta V) = i\,\mathrm{ad}_V\big[\sinh\!\big(\tfrac{1}{2}\,\mathrm{ad}_V\big)(\Omega + \Omega^\dagger) + \cosh\!\big(\tfrac{1}{2}\,\mathrm{ad}_V\big)(\Omega - \Omega^\dagger)\big]. \qquad (10.177)$$

Thus, it follows that

$$\delta V = \tfrac{1}{2} i\,\mathrm{ad}_V\big[\Omega + \Omega^\dagger + \coth\!\big(\tfrac{1}{2}\,\mathrm{ad}_V\big)(\Omega - \Omega^\dagger)\big]. \qquad (10.178)$$

Finally, we use the power series expansion of the hyperbolic cotangent function,

$$z\coth z = 1 + \sum_{k=1}^{\infty} \frac{2^{2k} B_{2k}}{(2k)!}\, z^{2k}\,, \qquad (10.179)$$

where the B_{2k} are the Bernoulli numbers. Hence, making use of eq. (H.12), one finds

$$\delta V = i(\Omega - \Omega^\dagger) + \tfrac{1}{2} i\,[V,\, \Omega + \Omega^\dagger] + i\sum_{k=1}^{\infty} \frac{B_{2k}}{(2k)!}\, \underbrace{[V, [\cdots [V, [V, \Omega - \Omega^\dagger]] \cdots]]}_{2k \text{ times}}, \qquad (10.180)$$

with $2k$ nested commutators. Note that eq. (10.180) is equivalent to

$$V^a \to V^a i(\Omega^a - \Omega^{a*}) - g_a f^{abc} V^b(\Omega^c + \Omega^{c*}) - \tfrac{1}{3} i g_a^2 f^{abc} f^{cde} V^b V^d (\Omega^e - \Omega^{e*}) + \cdots\,, \qquad (10.181)$$

where eqs. (4.11) and (10.171) have been used, and there is no implicit sum over the free index a. This supergauge transformation includes ordinary gauge transformations as the special case $\Omega^{a*} = \Omega^a$.

Because the second term on the right side of eq. (10.181) is independent of V^a, one can always do a supergauge transformation to Wess–Zumino gauge by choosing $\Omega^a - \Omega^{a*}$ appropriately, just as in the abelian case, so that

$$\big(V^a\big)_{\text{WZgauge}} = \theta\sigma^\mu\theta^\dagger A^a_\mu + \theta^\dagger\theta^\dagger\theta\lambda^a + \theta\theta\theta^\dagger\lambda^{\dagger a} + \tfrac{1}{2}\theta\theta\theta^\dagger\theta^\dagger D^a. \qquad (10.182)$$

After fixing the supergauge to Wess–Zumino gauge, one still has the freedom to perform ordinary gauge transformations. In the Wess–Zumino gauge, the Lagrangian contribution eq. (10.173) is polynomial, in agreement with what was found in component language in Section 8.4:

$$\big[\Phi^{\dagger i}\big(e^V\big)_i{}^j\Phi_j\big]_D = F^{\dagger i}F_i + \nabla_\mu\phi^{\dagger i}\nabla^\mu\phi_i + i\psi^{\dagger i}\bar\sigma^\mu\nabla_\mu\psi_i - \sqrt{2}g_a(\phi^\dagger \boldsymbol{T}^a\psi)\lambda^a - \sqrt{2}g_a\lambda^{\dagger a}(\psi^\dagger\boldsymbol{T}^a\phi) + g_a(\phi^\dagger\boldsymbol{T}^a\phi)D^a, \qquad (10.183)$$

where ∇_μ is the gauge-covariant derivative defined in eqs. (8.65)–(8.67).

To make kinetic terms and self-interactions for the vector supermultiplets in the nonabelian case, define a field-strength chiral superfield

$$\mathcal{W}_\alpha = -\tfrac{1}{4}\overline{D}^2\left(e^{-V} D_\alpha e^V\right), \tag{10.184}$$

generalizing the abelian case. Using eq. (10.174), one can show that it transforms under supergauge transformations as

$$\mathcal{W}_\alpha \to e^{-i\Omega}\mathcal{W}_\alpha e^{i\Omega}. \tag{10.185}$$

(The proof makes use of the fact that Ω is chiral and $\Omega^\dagger$ is antichiral, so that $\overline{D}_{\dot\alpha}\Omega = 0$ and $D_\alpha\Omega^\dagger = 0$.) This implies that $\mathrm{Tr}[\mathcal{W}^\alpha\mathcal{W}_\alpha]$ is a supergauge-invariant chiral superfield. The factors in parentheses in eq. (10.184) can be expanded as

$$e^{-V}D_\alpha e^V = D_\alpha V - \tfrac{1}{2}[V, D_\alpha V] + \tfrac{1}{6}\,[V,[V,D_\alpha V]] + \cdots, \tag{10.186}$$

where again the commutators apply in the matrix sense, and only the first two terms contribute in Wess–Zumino gauge.

The field-strength chiral superfield $\mathcal{W}_\alpha$ defined in eq. (10.184) is matrix-valued in the representation R. One can recover an adjoint representation field-strength superfield $\mathcal{W}^a_\alpha$ from the matrix-valued one by writing

$$\mathcal{W}_\alpha \equiv 2g_a \boldsymbol{T}^a \mathcal{W}^a_\alpha, \tag{10.187}$$

leading to

$$\mathcal{W}^a_\alpha = -\tfrac{1}{4}\overline{D}^2\left(D_\alpha V^a - i g_a f^{abc} V^b D_\alpha V^c + \cdots\right). \tag{10.188}$$

The terms shown explicitly are enough to evaluate this in components in Wess–Zumino gauge, with the result

$$(\mathcal{W}^a_\alpha)_{\text{WZgauge}} = \lambda^a_\alpha + \theta_\alpha D^a - \tfrac{1}{2}i(\sigma^\mu\overline{\sigma}^\nu\theta)_\alpha F^a_{\mu\nu} + i\theta\theta(\sigma^\mu\nabla_\mu\lambda^{\dagger a})_\alpha, \tag{10.189}$$

where $F^a_{\mu\nu}$ is the nonabelian field strength of eq. (8.58) and ∇_μ is the usual gauge covariant derivative from eq. (8.59).

We can now construct the kinetic terms and self-interactions for the gauge supermultiplet. Each simple Lie group contained in the direct product group G with corresponding gauge coupling constant $g_a = g$ and Dynkin index T_R contributes

$$\frac{1}{4g^2 T_R}\mathrm{Tr}[\mathcal{W}^\alpha\mathcal{W}_\alpha]_F = [\mathcal{W}^{a\alpha}\mathcal{W}^a_\alpha]_F, \tag{10.190}$$

which is invariant under both supersymmetry and supergauge transformations, and is independent of the representation R of the generators. Equation (10.190) is most easily evaluated in Wess–Zumino gauge using eq. (10.189), yielding

$$[\mathcal{W}^{a\alpha}\mathcal{W}^a_\alpha]_F = D^a D^a + 2i\lambda^a\sigma^\mu\nabla_\mu\lambda^{\dagger a} - \tfrac{1}{2}F^{a\mu\nu}F^a_{\mu\nu} + \tfrac{1}{4}i\epsilon^{\mu\nu\rho\sigma}F^a_{\mu\nu}F^a_{\rho\sigma}. \tag{10.191}$$

Since eq. (10.191) is supergauge invariant, the same expression is valid even outside of Wess–Zumino gauge.

We can now write the general renormalizable Lagrangian for a supersymmetric nonabelian gauge theory (including superpotential interactions for the chiral supermultiplets when allowed by gauge invariance):

$$\mathscr{L} = \left(\frac{1}{4} - i\frac{g_a^2\Theta_a}{32\pi^2}\right)[\mathcal{W}^{a\alpha}\mathcal{W}^a_\alpha]_F + \text{h.c.} + \left[\Phi^{\dagger i}(e^{2g_a T^a V^a})_i{}^j\Phi_j\right]_D + ([W(\Phi_i)]_F + \text{h.c.}). \tag{10.192}$$

This introduces and defines Θ_a, a CP-violating parameter, whose effect is to include a total derivative term in the Lagrangian:

$$\mathscr{L}_{\Theta_a} = \frac{g_a^2\Theta_a}{64\pi^2}\epsilon^{\mu\nu\rho\sigma}F^a_{\mu\nu}F^a_{\rho\sigma}. \tag{10.193}$$

In the nonabelian case, this can have physical effects due to topologically nontrivial field configurations (instantons). For a globally nontrivial gauge configuration with integer winding number n, one has

$$\int d^4x\, \epsilon^{\mu\nu\rho\sigma}F^a_{\mu\nu}F^a_{\rho\sigma} = \frac{64\pi^2 n}{g_a^2} \tag{10.194}$$

for a simple gauge group, so that the contribution to the path integral is

$$\exp\left\{i\int d^4x\,\mathscr{L}_{\Theta_a}\right\} = e^{in\Theta_a}. \tag{10.195}$$

Note that for nonabelian gauge groups, a Fayet–Iliopoulos term $-2\kappa[V^a]_D$ is not allowed, because it is not a gauge singlet.

When the superfields are restricted to the Wess–Zumino gauge, the supersymmetry transformations are not realized linearly in superspace, but the Lagrangian is polynomial. The nonpolynomial form of the superspace Lagrangian is thus seen to be a supergauge artifact. Within Wess–Zumino gauge, supersymmetry transformations are still realized, but nonlinearly, as we found in Sections 8.3 and 8.4.

The gauge coupling g_a and CP-violating angle Θ_a can be combined into a single holomorphic coupling,

$$\tau_a = \frac{1}{g_a^2} - i\frac{\Theta_a}{8\pi^2}. \tag{10.196}$$

Then, with redefined vector and field-strength superfields

$$\widehat{V}^a \equiv g_a V^a, \tag{10.197}$$

$$\widehat{\mathcal{W}}^a_\alpha \equiv g_a\mathcal{W}^a_\alpha = -\frac{1}{4}\overline{D}^2\left(D_\alpha\widehat{V}^a - if^{abc}\widehat{V}^b D_\alpha\widehat{V}^c + \cdots\right), \tag{10.198}$$

the gauge part of the Lagrangian is written as

$$\mathscr{L} = \tfrac{1}{4}\left[\tau_a\widehat{\mathcal{W}}^{a\alpha}\widehat{\mathcal{W}}^a_\alpha\right]_F + \text{h.c.} + \left[\Phi^{\dagger i}(e^{2T^a\widehat{V}^a})_i{}^j\Phi_j\right]_D. \tag{10.199}$$

An advantage of this normalization convention is that, when written in terms of $\widehat{V}^a$, the only appearance of the gauge coupling and Θ_a is in the τ_a in eq. (10.199). It is then sometimes useful to treat the complex holomorphic coupling τ_a as a chiral superfield with an expectation value for its scalar component. An expectation value for the F-term component of τ_a will give gaugino masses; this is sometimes a useful way to implement the effects of explicit soft supersymmetry breaking.

10.10 Nonrenormalizable Supersymmetric Lagrangians

So far, we have discussed renormalizable supersymmetric Lagrangians. However, integrating out the effects of heavy states will generally lead to nonrenormalizable interactions in the low-energy effective description. Furthermore, when any realistic supersymmetric theory is extended to include gravity, the resulting supergravity theory is nonrenormalizable as a quantum field theory. Fortunately, the nonrenormalizable interactions can be neglected for most phenomenological purposes, because they involve couplings of negative mass dimension, proportional to powers of $1/M_P$ (or perhaps $1/\Lambda_{\text{UV}}$, where Λ_{UV} is some other cutoff scale associated with new physics). This means that their effects at energy scales E ordinarily accessible to experiment are typically suppressed by powers of E/M_P (or E/Λ_{UV}). For energies $E \lesssim 1$ TeV, the consequences of nonrenormalizable interactions are therefore usually far too small to be interesting.

Still, there are several reasons why one may need to include nonrenormalizable contributions to supersymmetric Lagrangians. First, some very rare processes (like proton decay) might only be described using an effective MSSM Lagrangian that includes nonrenormalizable terms. Second, one may be interested in understanding physics at very high energy scales where the suppression associated with nonrenormalizable terms is not enough to stop them from being important. For example, this could be the case in the study of the very early universe, or in understanding how additional gauge symmetries get broken. Third, the nonrenormalizable interactions may play a crucial role in understanding how supersymmetry breaking is transmitted to the MSSM. Finally, it is sometimes useful to treat strongly coupled supersymmetric gauge theories using nonrenormalizable effective Lagrangians, in the same way that chiral effective Lagrangians are used to study hadron physics in QCD. Unfortunately, we will not be able to treat these subjects in a systematic way. Instead, we will merely sketch a few key elements of nonrenormalizable supersymmetric Lagrangians. More detailed treatments and pointers to the literature may be found in [B20, B39, B40, B69] and in Ref. [25].

A nonrenormalizable gauge-invariant theory involving chiral and vector superfields can be constructed as

$$\mathscr{L} = \Big[K(\Phi_i,\, \widetilde{\Phi}^{\dagger j})\Big]_D + \left(\Big[\tfrac{1}{4} f_{ab}(\Phi_i)\widehat{\mathcal{W}}^{a\alpha}\widehat{\mathcal{W}}^b_\alpha + W(\Phi_i)\Big]_F + \text{h.c.}\right), \qquad (10.200)$$

where, in order to preserve supergauge invariance, we define

$$\widetilde{\Phi}^{\dagger j} \equiv \left(\Phi^\dagger e^V\right)^j, \qquad (10.201)$$

with $V = 2g_a \boldsymbol{T^a} V^a = 2\boldsymbol{T^a}\widehat{V}^a$ as above, and the hatted normalization of the field-strength superfields indicated in eq. (10.198) has been used. Here, $(T^a)_i{}^j$ are the generators for the gauge group acting on the chiral superfields Φ_i. Equation (10.200) depends on couplings encoded in three functions of the superfields:

- The superpotential W, which we have already encountered in the special case of renormalizable supersymmetric Lagrangians. More generally, it can be an arbitrary holomorphic function of the chiral superfields treated as complex variables, and must be invariant under the gauge symmetries of the theory, and has dimension [mass]^3.
- The *Kähler potential* K. Unlike the superpotential, the Kähler potential is a function of both chiral and antichiral superfields, and includes the vector superfield in such a way as to be supergauge invariant. It is real, and has dimension [mass]^2. In the special case of renormalizable theories, we did not have to discuss the Kähler potential explicitly, because at tree level it is always just $K = \Phi_i \widetilde{\Phi}^{i*}$. Any additive part of K that is holomorphic in the chiral superfields (or in the antichiral superfields) does not contribute to the action, since the D-term of a chiral superfield is a total derivative on spacetime.
- The *gauge kinetic function* $f_{ab}(\Phi_i)$. Like the superpotential, the gauge kinetic function is itself a chiral superfield, and is a holomorphic function of the chiral superfields treated as complex variables. It is dimensionless and symmetric under interchange of its two indices a, b, which run over the adjoint representations of the simple and abelian component gauge groups of the model. For the nonabelian components of the gauge group, it is always proportional to δ_{ab}, but if there are two or more abelian components, then the gauge invariance of the field-strength superfield [see eq. (10.145)] allows kinetic mixing so that f_{ab} is not proportional to δ_{ab} in general. In the special case of renormalizable supersymmetric Lagrangians at tree level, it is independent of the chiral superfields, and equal to $f_{ab} = \delta_{ab}\left[1/g_a^2 - i\Theta_a/(8\pi^2)\right]$ (for fewer than two abelian components in the gauge group). More generally, it also encodes the nonrenormalizable couplings of the gauge supermultiplets to the chiral supermultiplets.

Let us see how the requirement of supergauge invariance of the Kähler potential works for nonquadratic terms. For simplicity and to avoid a proliferation of indices, let us consider only the example of a cubic term[7]

$$K_c = c_j^{k\ell}(\Phi^\dagger e^V)^j \Phi_k \Phi_\ell. \tag{10.202}$$

Here $c_j^{k\ell}$ is a coupling that must be an invariant symbol of the gauge group. By definition, this means that for any real numbers ω^a, with $\omega \equiv \omega^a \boldsymbol{T^a}$, one must have

$$c_j^{k\ell} = \left(e^{-i\omega}\right)_{k'}{}^{k}\left(e^{-i\omega}\right)_{\ell'}{}^{\ell}\left(e^{i\omega}\right)_j{}^{j'} c_{j'}^{k'\ell'}. \tag{10.203}$$

Now, if this is true for real numbers ω^a, then it is also true for chiral superfields Ω^a, since the matrix $\Omega \equiv \Omega^a \boldsymbol{T^a}$ commutes with itself.

[7] A real Lagrangian must also include the complex conjugate of eq. (10.202), but that is an irrelevant complication for the discussion that follows.

Under a supergauge transformation parameterized by chiral superfields Ω^a,

$$\Phi \to e^{-i\Omega}\Phi, \tag{10.204}$$

$$\Phi^\dagger \to \Phi^\dagger e^{i\Omega^\dagger}, \tag{10.205}$$

$$e^V \to e^{-i\Omega^\dagger} e^V e^{i\Omega}, \tag{10.206}$$

with $\Omega \equiv \Omega^a \boldsymbol{T}^a$, which yield the supergauge transformations

$$(\Phi^\dagger e^V) \to (\Phi^\dagger e^V) e^{i\Omega}, \tag{10.207}$$

$$(e^V \Phi) \to e^{-i\Omega^\dagger}(e^V\Phi). \tag{10.208}$$

Hence, the supergauge transformation of K_c is given by

$$K_c \to c_{j'}^{k'\ell'} \left(\Phi^\dagger e^V e^{i\Omega}\right)^{j'} \left(e^{-i\Omega}\Phi\right)_{k'} \left(e^{-i\Omega}\Phi\right)_{\ell'}$$

$$= \left[\left(e^{-i\Omega}\right)_{k'}{}^{k} \left(e^{-i\Omega}\right)_{\ell'}{}^{\ell} \left(e^{i\Omega}\right)_{j}{}^{j'} c_{j'}^{k'\ell'}\right] \left(\Phi^\dagger e^V\right)^j \Phi_k \Phi_\ell = K_c, \tag{10.209}$$

where the last equality uses eq. (10.203) with $\omega = \Omega$. Thus K_c is supergauge invariant if c_j^{kl} is an invariant symbol of the gauge group.

One can also rewrite K_c in a slightly different form, by noting that eq. (10.203) also implies, by substituting $\omega \to -iV$, that

$$c_j^{k\ell} = \left(e^{-V}\right)_{k'}{}^{k} \left(e^{-V}\right)_{\ell'}{}^{\ell} \left(e^{V}\right)_{j}{}^{j'} c_{j'}^{k'\ell'}, \tag{10.210}$$

or, multiplying twice by the matrix e^V,

$$c_j^{k'\ell'} \left(e^V\right)_{k'}{}^{k} \left(e^V\right)_{\ell'}{}^{\ell} = \left(e^V\right)_{j}{}^{j'} c_{j'}^{k\ell}. \tag{10.211}$$

It follows that one can write the Kähler potential term eq. (10.202) in a seemingly different, but equivalent, way:

$$K_c = c_j^{k\ell}\Phi^{\dagger j} \left(e^V\Phi\right)_k \left(e^V\Phi\right)_\ell. \tag{10.212}$$

All of this readily generalizes to the case of Kähler potential terms based on general invariant symbols of the group $c_{j_1,\ldots,j_n}^{k_1,\ldots,k_m}$. To enforce supergauge invariance, one can either choose to attach the e^V factors to the Φ as in eq. (10.212), or to the $\Phi^\dagger$ fields as we did in eqs. (10.200)–(10.202).

It should be emphasized that eq. (10.200) is still not the most general nonrenormalizable supersymmetric Lagrangian, even if one restricts to chiral and gauge vector superfields. One can also include chiral, antichiral, and spacetime derivatives acting on the superfields, so that, for example, the Kähler potential can be generalized to include dependence on $D_\alpha\Phi_i$, $\overline{D}_{\dot\alpha}\Phi^{\dagger i}$, $D^2\Phi_i$, $\overline{D}^2\Phi^{\dagger i}$, etc. Such terms typically have an extra suppression at low energies compared to terms without derivatives, because of the positive mass dimension of the chiral covariant derivatives. We will not discuss these possibilities below, but will only make a remark on how supergauge invariance is maintained. The chiral covariant derivative of a chiral superfield, $D_\alpha\Phi_i$, is not gauge covariant unless Φ_i is a gauge singlet; the "covariant" in the name refers to supersymmetry transformations, not gauge transformations.

However, one can define a "gauge-covariant chiral-covariant" derivative $\mathcal{D}_\alpha$, whose action on a chiral superfield Φ is defined by

$$\mathcal{D}_\alpha \Phi \equiv e^{-V} D_\alpha (e^V \Phi), \tag{10.213}$$

where the representation indices i are suppressed. From eq. (10.174), the supergauge transformation for e^{-V} is

$$e^{-V} \to e^{-i\Omega} e^{-V} e^{i\Omega^\dagger}, \tag{10.214}$$

so that

$$e^{-V} D_\alpha (e^V \Phi) \to e^{-i\Omega} e^{-V} e^{i\Omega^\dagger} D_\alpha (e^{-i\Omega^\dagger} e^V \Phi) \;=\; e^{-i\Omega} e^{-V} D_\alpha (e^V \Phi), \tag{10.215}$$

after noting that $\Omega^\dagger$ is antichiral and thus ignored by D_α. In light of eq. (10.204), it follows that $e^{-V} D_\alpha (e^V \Phi)$ transforms like Φ under a supergauge transformation parameterized by Ω. So, using $\mathcal{D}_\alpha \Phi_i$ as a building block instead of $D_\alpha \Phi_i$, one can maintain supergauge covariance along with manifest supersymmetry. Successive applications of $\mathcal{D}$, such as

$$\mathcal{D}^2 \Phi \equiv \mathcal{D}^\alpha \mathcal{D}_\alpha \Phi = e^{-V} D^2 (e^V \Phi)\,, \tag{10.216}$$

yield quantities that also transform like Φ under a supergauge transformation.

Similarly, one can define a gauge-covariant chiral-covariant derivative $\overline{\mathcal{D}}_\alpha$, whose action on an antichiral superfield $\Phi^\dagger$ is defined by

$$\overline{\mathcal{D}}_{\dot\alpha} \Phi^\dagger \equiv \overline{D}_{\dot\alpha} (\Phi^\dagger e^V) e^{-V}\,. \tag{10.217}$$

Hence, under a supergauge transformation,

$$\overline{D}_{\dot\alpha} (\Phi^\dagger e^V) e^{-V} \to \overline{D}_{\dot\alpha} (\Phi^\dagger e^V e^{i\Omega}) e^{-i\Omega} e^{-V} e^{i\Omega^\dagger} \;=\; \overline{D}_{\dot\alpha} (\Phi^\dagger e^V) e^{-V} e^{i\Omega^\dagger}, \tag{10.218}$$

after noting that Ω is chiral and thus ignored by $\overline{D}_{\dot\alpha}$. In light of eq. (10.205), it follows that $\overline{D}_{\dot\alpha} (\Phi^\dagger e^V) e^{-V}$ transforms like $\Phi^\dagger$ under a supergauge transformation. Successive applications of $\overline{\mathcal{D}}$, such as

$$\overline{\mathcal{D}}^2 \Phi^\dagger \equiv \overline{\mathcal{D}}_{\dot\alpha}\, \overline{\mathcal{D}}^{\dot\alpha} \Phi^\dagger = \overline{D}^2 (\Phi^\dagger e^V) e^{-V}\,, \tag{10.219}$$

yield quantities that also transform like $\Phi^\dagger$ under a supergauge transformation.

Returning to the globally supersymmetric nonrenormalizable theory defined by eq. (10.200), with no extra derivatives, the part of the Lagrangian coming from the superpotential is

$$[W(\Phi_i)]_F = W^i F_i - \tfrac{1}{2} W^{ij} \psi_i \psi_j, \tag{10.220}$$

with

$$W^i = \left.\frac{\partial W}{\partial \Phi_i}\right|_{\Phi_i \to \phi_i}, \qquad W^{ij} = \left.\frac{\partial^2 W}{\partial \Phi_i \partial \Phi_j}\right|_{\Phi_i \to \phi_i}, \tag{10.221}$$

where the superfields have been replaced by their scalar components after differentiation. [Compare eqs. (8.38), (8.42), (8.46) and the surrounding discussion.] After integrating out the auxiliary fields F_i, the part of the scalar potential coming from the superpotential is

$$V = W^i W_j^\dagger (K^{-1})_i^j, \tag{10.222}$$

where K^{-1} is the inverse matrix of the Kähler metric:

$$K_j^i = \frac{\partial^2 K}{\partial \Phi_i \partial \widetilde{\Phi}^{\dagger j}}\bigg|_{\Phi_i \to \phi_i,\ \widetilde{\Phi}^{\dagger i} \to \phi^{\dagger i}}. \tag{10.223}$$

More generally, the whole component-field Lagrangian after integrating out the auxiliary fields is determined in terms of the functions W, K, and f_{ab} and their derivatives with respect to the chiral superfields, with the remaining chiral superfields replaced by their scalar components. In supergravity, there are additional contributions, some of which are discussed in Section 14.3.

10.11 *R*-symmetries

Some supersymmetric Lagrangians are also invariant under a global $U(1)_R$ symmetry. The defining feature of a continuous R-symmetry is that the anticommuting coordinates θ and $\theta^\dagger$ transform under it with charges $+1$ and -1 respectively, so

$$\theta \to e^{i\alpha}\theta, \qquad \theta^\dagger \to e^{-i\alpha}\theta^\dagger, \tag{10.224}$$

where α parameterizes the global R-transformation. It follows that

$$\widehat{Q} \to e^{-i\alpha}\widehat{Q}, \qquad \widehat{Q}^\dagger \to e^{i\alpha}\widehat{Q}^\dagger, \tag{10.225}$$

which in turn implies that the supersymmetry generators have $U(1)_R$ charges -1 and $+1$, and so do not commute with the R-symmetry charge. In particular, as noted in eqs. (9.28) and (9.29),

$$[R, Q] = -Q, \qquad [R, Q^\dagger] = Q^\dagger. \tag{10.226}$$

Thus the distinct components within a superfield always have different R charges.

If the theory is invariant under an R-symmetry, then each superfield $S(x, \theta, \theta^\dagger)$ can be assigned an R charge, denoted r_S, defined by its transformation rule,

$$S(x,\, \theta,\, \theta^\dagger) \to e^{ir_S\alpha} S(x,\, e^{-i\alpha}\theta,\, e^{i\alpha}\theta^\dagger). \tag{10.227}$$

The $U(1)_R$ charge of a product of superfields is the sum of the individual R charges.

For a chiral superfield Φ with R charge r_Φ, the ϕ, ψ, and F components transform with charges r_Φ, $r_\Phi - 1$, and $r_\Phi - 2$, respectively:

$$\phi \to e^{ir_\Phi\alpha}\phi, \qquad \psi \to e^{i(r_\Phi - 1)\alpha}\psi, \qquad F \to e^{i(r_\Phi - 2)\alpha}F. \tag{10.228}$$

The components of $\Phi^\dagger$ carry the opposite charges.

Table 10.1 $\mathrm{U}(1)_R$ charges of various objects.

	θ_α	$\theta^\dagger_{\dot\alpha}$	$d^2\theta$	$\widehat{Q}_\alpha$	D_α	$\mathcal{W}_\alpha$	A^μ	λ_α	D	W	ϕ	ψ_α	F_Φ
R	$+1$	-1	-2	-1	-1	$+1$	0	$+1$	0	$+2$	r_Φ	$r_\Phi - 1$	$r_\Phi - 2$

Gauge vector superfields will always have vanishing $\mathrm{U}(1)_R$ charge, since they are real. Hence, the nonzero components of the Wess–Zumino gauge transform as

$$A^\mu \to A^\mu, \qquad \lambda \to e^{i\alpha}\lambda, \qquad D \to D. \tag{10.229}$$

and so have $\mathrm{U}(1)_R$ charges 0, 1, and 0, respectively. Therefore, a Majorana gaugino mass term $\frac{1}{2}M_\lambda \lambda\lambda$, which will appear when supersymmetry is broken, also always breaks a continuous $\mathrm{U}(1)_R$ symmetry. The superspace integration measures $d^2\theta$ and $d^2\theta^\dagger$ and the chiral covariant derivatives D_α and $\overline{D}_{\dot\alpha}$ carry $\mathrm{U}(1)_R$ charges -2, $+2$, -1, and $+1$, respectively. It follows that the gauge field-strength superfield $\mathcal{W}_\alpha$ carries $\mathrm{U}(1)_R$ charge $+1$.

The $\mathrm{U}(1)_R$ charges of various objects are collected in Table 10.1. It is not difficult to check that all supersymmetric Lagrangian terms found above that involve gauge superfields are automatically and necessarily R-symmetric, including the couplings to chiral superfields. This is also true of the canonical Kähler potential contribution.

However, the superpotential $W(\Phi_i)$ must carry $\mathrm{U}(1)_R$ charge $+2$ in order to conserve the R-symmetry, and this is certainly not automatic, and often not true. As a simple toy example, with a single gauge-singlet superfield Φ, the allowed renormalizable terms in the superpotential are $W(\Phi) = L\Phi + \frac{1}{2}M\Phi^2 + \frac{1}{6}y\Phi^3$. If one wants to impose a continuous $\mathrm{U}(1)_R$ symmetry, then one can have at most one of these terms; L is allowed only if $r_\Phi = 2$, M is allowed only if $r_\Phi = 1$, and y is allowed only if $r_\Phi = 2/3$. The MSSM superpotential does turn out to conserve a global $\mathrm{U}(1)_R$ symmetry, but it is both anomalous and broken by Majorana gaugino masses and other supersymmetry-breaking effects.

Since continuous R-symmetries do not commute with supersymmetry, and are not conserved in the MSSM after anomalies and supersymmetry-breaking effects are included, one might wonder why they are considered at all. Perhaps the most important answer to this involves the role of $\mathrm{U}(1)_R$ symmetries in models that break global supersymmetry spontaneously, as will be discussed in Section 12.3. It is also possible to extend the particle content of the MSSM in such a way as to preserve a continuous, nonanomalous $\mathrm{U}(1)_R$ symmetry, but at the cost of introducing Dirac gaugino masses and extra Higgs fields (e.g., see Ref. [96]). It is worth pointing out that only when supersymmetry is promoted to a local symmetry, in supergravity theories, is it possible to gauge the $\mathrm{U}(1)_R$ symmetry [97, 98].

Another possibility is that a superpotential could have a discrete $\mathbb{Z}_N$ R-symmetry, which can be obtained by restricting the transformation parameter α that appears in eqs. (10.224)–(10.229) to integer multiples of $2\pi/N$. The $\mathbb{Z}_N$ R charges of all fields are then integers modulo N. However, note that the case $N = 2$ is always trivial, in the sense that any $\mathbb{Z}_2$ R-symmetry is exactly equivalent to a corresponding ordinary (i.e., non-R) $\mathbb{Z}_2$ symmetry under which all components of each supermultiplet transform the same way. This is because, when α is an integer multiple of π, both θ and $\theta^\dagger$ always just transform by changing sign, which means that fermionic fields just change sign relative to their bosonic partners. The number of fermionic fields in any Lagrangian term, in any theory, is always even, so the extra sign change for fermionic fields has no effect.

Exercises

10.1 Derive the following identities:

$$\partial_\alpha \theta_\beta = -\epsilon_{\alpha\beta}\,, \qquad \partial^{\dagger\dot{\alpha}}\theta^{\dagger\dot{\beta}} = -\epsilon^{\dot{\alpha}\dot{\beta}}\,. \tag{10.230}$$

$$\partial^\alpha \theta^\beta = -\epsilon^{\alpha\beta}\,, \qquad \partial^\dagger_{\dot{\alpha}}\theta^\dagger_{\dot{\beta}} = -\epsilon_{\dot{\alpha}\dot{\beta}}\,. \tag{10.231}$$

10.2 (a) Derive the following Fierz identities:

$$(\theta\sigma^\mu\theta^\dagger)\theta_\beta = -\tfrac{1}{2}\theta\theta(\sigma^\mu\theta^\dagger)_\beta\,, \tag{10.232}$$

$$(\theta\sigma^\mu\theta^\dagger)\theta^\dagger_{\dot{\beta}} = -\tfrac{1}{2}\theta^\dagger\theta^\dagger(\theta\sigma^\mu)_{\dot{\beta}}\,, \tag{10.233}$$

$$(\theta\sigma^\mu\theta^\dagger)(\theta\sigma^\nu\theta^\dagger) = \tfrac{1}{2}g^{\mu\nu}(\theta\theta)(\theta^\dagger\theta^\dagger). \tag{10.234}$$

(b) Prove that

$$\exp(-i\theta\sigma^\mu\theta^\dagger\,\partial_\mu) = 1 - i\theta\sigma^\mu\theta^\dagger\,\partial_\mu - \tfrac{1}{4}(\theta\theta)(\theta^\dagger\theta^\dagger)\Box, \tag{10.235}$$

where $\Box \equiv \partial_\mu\partial^\mu$.

10.3 In order to define translations in superspace, one can generalize the translation operator $\exp(ix\cdot P)$ to the supertranslation operator,

$$G(x,\theta,\theta^\dagger) = \exp(ix\cdot P + \theta Q + \theta^\dagger Q^\dagger)\,. \tag{10.236}$$

We can now extend the field operator, $\Phi(x) = \exp(ix\cdot P)\Phi(0)\exp(-ix\cdot P)$, to a *superfield* operator,

$$\Phi(x,\theta,\theta^\dagger) = G(x,\theta,\theta^\dagger)\Phi(0,0,0)G^{-1}(x,\theta,\theta^\dagger)\,. \tag{10.237}$$

In this way, we can realize a supersymmetry transformation as a translation in superspace.

(a) Using the Baker–Campbell–Hausdorff formula,

$$\exp(A)\exp(B) = \exp\big(A + B + \tfrac{1}{2}[A\,,\,B] + \cdots\big)\,, \tag{10.238}$$

show that

$$G(x',\xi,\xi^\dagger)G(x,\theta,\theta^\dagger) = G\big(x + x' + i(\xi\sigma\theta^\dagger + \xi^\dagger\overline{\sigma}\theta), \xi+\theta, \xi^\dagger+\theta^\dagger\big)\,. \tag{10.239}$$

(b) Verify that one can employ the supertranslation operator, $G(x',\xi,\xi^\dagger)$, to translate the superspace coordinates of a superfield Φ as follows:

$$G(x',\xi,\xi^\dagger)\Phi(x,\theta,\theta^\dagger)G^{-1}(x',\xi,\xi^\dagger) = \Phi\big(x+x'+i(\xi\sigma\theta^\dagger+\xi^\dagger\overline{\sigma}\theta),\xi+\theta,\xi^\dagger+\theta^\dagger\big). \tag{10.240}$$

(c) For infinitesimal x', ξ, and $\xi^\dagger$, one can approximate

$$G(x',\xi,\xi^\dagger) \simeq \mathbb{1} + i(x'\cdot P + \xi Q + \xi^\dagger Q^\dagger). \tag{10.241}$$

Expanding eq. (10.240) out to first order, show that

$$\big[\Phi\,,\,P_\mu\big] = i\,\partial_\mu\Phi\,, \tag{10.242}$$

$$\big[\Phi\,,\,\xi Q\big] = i\,\xi^\alpha\left(\partial_\alpha + i(\sigma^\mu\theta^\dagger)_\alpha\partial_\mu\right)\Phi\,, \tag{10.243}$$

$$\big[\Phi\,,\,\xi^\dagger Q^\dagger\big] = -i\left(\partial^\dagger_{\dot\alpha} + i(\theta\sigma^\mu)_{\dot\alpha}\partial_\mu\right)\xi^{\dagger\dot\alpha}\Phi\,. \tag{10.244}$$

Note that eqs. (10.242)–(10.244) are equivalent to the following three commutation relations:

$$[\Phi\,,\,P_\mu] = \widehat{P}_\mu\Phi\,, \qquad [\Phi\,,\,\xi Q] = (\xi\widehat{Q})\Phi\,, \qquad [\Phi\,,\,\xi^\dagger Q^\dagger] = (\xi^\dagger\widehat{Q}^\dagger)\Phi\,, \tag{10.245}$$

where $\widehat{P}_\mu \equiv i\partial_\mu$ and $\widehat{Q}_\alpha$, $\widehat{Q}^\dagger_{\dot\alpha}$ are defined in eqs. (10.47) and (10.48).

(d) The action of an infinitesimal supertranslation on a superfield $\Phi(x,\theta,\theta^\dagger)$ is given by

$$\begin{aligned}\delta_\xi\Phi(x,\theta,\theta^\dagger) &\equiv \Phi\big(x+x'+i(\xi\sigma\theta^\dagger+\xi^\dagger\overline{\sigma}\theta),\xi+\theta,\xi^\dagger+\theta^\dagger\big) - \Phi(x,\theta,\theta^\dagger)\\ &= i\big[\xi Q + \xi^\dagger Q^\dagger\,,\,\Phi(x)\big]\,,\end{aligned} \tag{10.246}$$

to first order in the infinitesimal Grassmann parameters ξ and $\xi^\dagger$. Using the result of part (c), show that the action of an infinitesimal supertranslation on a superfield $\Phi(x,\theta,\theta^\dagger)$ is generated by applying the differential operators $\widehat{Q}$ and $\widehat{Q}^\dagger$ as follows:

$$\delta_\xi\Phi(x,\theta,\theta^\dagger) = -i(\xi\widehat{Q} + \xi^\dagger\widehat{Q}^\dagger)\Phi(x,\theta,\theta^\dagger)\,. \tag{10.247}$$

10.4 (a) Suppose that Φ is a Grassmann-even superfield. Show that

$$(\partial_\alpha\Phi)^\dagger = -\partial^\dagger_{\dot\alpha}\Phi^\dagger\,. \tag{10.248}$$

(b) Suppose that Φ is a Grassmann-odd superfield. Show that

$$(\partial_\alpha\Phi)^\dagger = \partial^\dagger_{\dot\alpha}\Phi^\dagger\,. \tag{10.249}$$

Verify that eq. (10.72) is modified such that $\overline{D}_{\dot\alpha}\Phi^\dagger = -(D_\alpha\Phi)^\dagger$.

10.5 If Φ is a chiral superfield and $\Phi^\dagger$ is an antichiral superfield, show that

$$\int d^4x\,d^2\theta\,d^2\theta^\dagger\,\Phi(x,\theta,\theta^\dagger) = \int d^4x\,d^2\theta\,d^2\theta^\dagger\,\Phi^\dagger(x,\theta,\theta^\dagger) = 0\,. \tag{10.250}$$

10.6 (a) Given an operator $\mathcal{O}(x)$ that operates on superfields, one can define the *chiral representation* of $\mathcal{O}(x)$, denoted below by $\mathcal{O}_{\rm chiral}(x)$, as follows:

$$\mathcal{O}_{\rm chiral} = \exp(i\theta\sigma^\mu\theta^\dagger\,\partial_\mu)\mathcal{O}\exp(-i\theta\sigma^\mu\theta^\dagger\,\partial_\mu)\,, \tag{10.251}$$

where ∂_μ is a derivative with respect to x. Show that the chiral representations of D and $\overline{D}$ are given by eqs. (10.89) and (10.90) [after replacing y with x]. Likewise, show that the chiral representation of the chiral superfield,

$$\Phi_1(x) = \phi(x) + \sqrt{2}\theta\psi(x) + \theta\theta F(x)\,, \tag{10.252}$$

is related to the general expression for the chiral superfield via

$$\Phi(x,\theta,\theta^\dagger) = \exp(-i\theta\sigma^\mu\theta^\dagger\,\partial_\mu)\Phi_1(x,\theta) = \Phi_1(x - i\theta\sigma^\mu\theta^\dagger,\theta)\,. \tag{10.253}$$

(b) Given an operator $\mathcal{O}(x)$ that operates on superfields, one can define the *antichiral representation* of $\mathcal{O}(x)$, denoted below by $\mathcal{O}_{\rm antichiral}(x)$, as follows:

$$\mathcal{O}_{\rm antichiral} = \exp(-i\theta\sigma^\mu\theta^\dagger\,\partial_\mu)\mathcal{O}\exp(i\theta\sigma^\mu\theta^\dagger\,\partial_\mu)\,. \tag{10.254}$$

Show that the chiral representations of D and $\overline{D}$ are given by eqs. (10.94) and (10.95) [after replacing y with x]. Likewise, show that the antichiral representation of the antichiral superfield,

$$\Phi_2^\dagger(x) = \phi^\dagger(x) + \sqrt{2}\theta^\dagger\psi^\dagger(x) + \theta^\dagger\theta^\dagger F^\dagger(x)\,, \tag{10.255}$$

is related to the general expression for the antichiral superfield via

$$\Phi^\dagger(x,\theta,\theta^\dagger) = \exp(i\theta\sigma^\mu\theta^\dagger\,\partial_\mu)\Phi_2^\dagger(x,\theta^\dagger) = \Phi_2^\dagger(x + i\theta\sigma^\mu\theta^\dagger,\theta^\dagger)\,. \tag{10.256}$$

10.7 (a) Prove that, for any chiral superfield Φ,

$$[\Phi]_F = -\tfrac{1}{4}D^2\Phi\Big|_{\theta=\theta^\dagger=0} = \tfrac{1}{4}\partial^\alpha\partial_\alpha\Phi\Big|_{\theta=\theta^\dagger=0}\,. \tag{10.257}$$

(b) Using the chain rule, show that

$$[W(\Phi)]_F = \tfrac{1}{4}\partial^\alpha\frac{dW}{d\Phi}\partial_\alpha\Phi\Big|_{\theta=\theta^\dagger=0} = \frac{1}{4}\left\{\left(\frac{d^2W}{d\Phi^2}\partial^\alpha\Phi\partial_\alpha\Phi\right) + \frac{dW}{d\Phi}\partial^\alpha\partial_\alpha\Phi\right\}\Big|_{\theta=\theta^\dagger=0}\,. \tag{10.258}$$

Denoting $\phi = \Phi\big|_{\theta=\theta^\dagger=0}$, conclude that

$$[W(\Phi)]_F = -\frac{1}{2}\left(\frac{d^2W}{d\Phi^2}\right)_{\Phi=\phi}\psi\psi + \left(\frac{dW}{d\Phi}\right)_{\Phi=\phi}F\,. \tag{10.259}$$

(c) Prove that, for any superfield V,

$$[V]_D = \tfrac{1}{16}\overline{D}^2D^2V\Big|_{\theta=\theta^\dagger=0} = \tfrac{1}{16}(\partial^\dagger_{\dot\alpha}\partial^{\dagger\dot\alpha})(\partial^\alpha\partial_\alpha)V\Big|_{\theta=\theta^\dagger=0}\,. \tag{10.260}$$

10.8 (a) Show that

$$D\sigma^\mu \overline{D} = -\overline{D}\overline{\sigma}^\mu D + 4i\partial^\mu \,. \tag{10.261}$$

Explain why an identity of the form of eq. (1.107) is *not* satisfied in this case.

(b) Derive the following relations:

$$[D_\alpha\,,\,\overline{D}^2] = 4i\sigma^\mu_{\alpha\dot\beta}\overline{D}^{\dot\beta}\partial_\mu\,, \qquad [D^\alpha\,,\,\overline{D}^2] = -4i\overline{D}_{\dot\beta}\overline{\sigma}^{\mu\dot\beta\alpha}\partial_\mu\,, \tag{10.262}$$

$$[\overline{D}^{\dot\alpha}\,,\,D^2] = 4i\overline{\sigma}^{\mu\dot\alpha\beta}D_\beta\partial_\mu\,, \qquad [\overline{D}_{\dot\alpha}\,,\,D^2] = -4iD^\beta\sigma^\mu_{\beta\dot\alpha}\partial_\mu\,, \tag{10.263}$$

$$[D^2\,,\,\overline{D}^2] = 4i\sigma^\mu_{\alpha\dot\beta}[D^\alpha, \overline{D}^{\dot\beta}]\partial_\mu = 8i(D\sigma^\mu\overline{D})\partial_\mu + 16\Box\,, \tag{10.264}$$

$$[\overline{D}^2\,,\,D^2] = 4i\overline{\sigma}^{\mu\dot\alpha\beta}[\overline{D}_{\dot\alpha}, D_\beta]\partial_\mu = 8i(\overline{D}\overline{\sigma}^\mu D)\partial_\mu + 16\Box\,. \tag{10.265}$$

(c) Verify the following two relations:

$$D^\alpha\overline{D}^2D_\alpha = \overline{D}_{\dot\alpha}D^2\overline{D}^{\dot\alpha}\,, \tag{10.266}$$

$$(D\sigma^\mu\overline{D})(\overline{D}\overline{\sigma}^\nu D)\partial_\mu\partial_\nu = -\tfrac{1}{2}D^\alpha\overline{D}^2D_\alpha\Box\,. \tag{10.267}$$

10.9 In this exercise, we introduce the chiral, antichiral, and transverse projection operators,

$$\Pi_+ \equiv -\frac{1}{16\Box}\overline{D}^2D^2\,, \qquad \Pi_- \equiv -\frac{1}{16\Box}D^2\overline{D}^2\,, \tag{10.268}$$

$$\Pi_T \equiv \frac{1}{8\Box}D^\alpha\overline{D}^2D_\alpha = \frac{1}{8\Box}\overline{D}_{\dot\alpha}D^2\overline{D}^{\dot\alpha}\,. \tag{10.269}$$

It is also common practice to define the longitudinal projection operator,

$$\Pi_L \equiv \Pi_+ + \Pi_- = -\frac{1}{16\Box}(\overline{D}^2D^2 + D^2\overline{D}^2)\,. \tag{10.270}$$

(a) Derive the following properties (which justify calling $\Pi_\pm$ and Π_T projection operators):

$$\Pi_k\Pi_\ell = \delta_{k\ell}\Pi_k\,, \quad \text{for } k,\,\ell = +,\,-,\,T\,, \tag{10.271}$$

$$\Pi_+ + \Pi_- + \Pi_T = \mathbb{1}\,, \tag{10.272}$$

where there is no implied sum over k in eq. (10.271) and $\mathbb{1}$ is the identity operator.

(b) Suppose that $\Phi(x, \theta, \theta^\dagger)$ is an unrestricted superfield. Define

$$\Phi_\pm \equiv \Pi_\pm\Phi\,. \tag{10.273}$$

Show that Φ_+ is a chiral superfield and Φ_- is an antichiral superfield.

10.10 The operator T is defined by

$$T\Phi = -\tfrac{1}{4}\overline{D}^2\Phi^\dagger\,. \tag{10.274}$$

(a) Show that, for any chiral superfield Φ,

$$TT\Phi = -\Box\Phi\,, \tag{10.275}$$

where $\Box \equiv \partial_\mu\partial^\mu$. That is, roughly speaking, iT is the square root of $\Box$.

(b) Explain why $T\Phi$ is a chiral superfield. Using eq. (10.91), show that

$$T\Phi = F^\dagger(y) - i\sqrt{2}\theta\sigma^\mu\partial_\mu\psi^\dagger(y) - \theta\theta\Box\phi^\dagger(y)\,. \tag{10.276}$$

Compute the $\theta\theta$ component of $\Phi T\Phi$.

(c) Show that an alternative form for eq. (10.134) is

$$\mathscr{L}(x) = [\Phi_i T\Phi_i]_F + ([W(\Phi_i)]_F + \text{h.c.})\,, \tag{10.277}$$

which differs from eq. (10.134) by total derivative terms.

10.11 The (real scalar) *linear superfield*, $L(x,\theta,\bar\theta)$, is defined as a real scalar superfield that satisfies the constraint $D^2L(x,\theta,\bar\theta) = \overline{D}^2L(x,\theta,\bar\theta) = 0$.

(a) Show that $\Pi_T\, L = L$, where Π_T is the transverse projection operator defined in eq. (10.269).

(b) In the notation of eq. (10.107), show that the component fields that make up the linear superfield satisfy $b = \partial_\mu V^\mu = 0$, $D = \partial_\mu\partial^\mu a$, and $\lambda = i\sigma^\mu\partial_\mu\xi^\dagger$. Verify that the number of bosonic and the number of fermionic degrees of freedom of the linear superfield are equal.

10.12 Consider a supersymmetric model of one chiral superfield, where the Lagrangian is given by

$$\mathscr{L} = \int d^2\theta\, d^2\theta^\dagger K(\Phi,\Phi^\dagger) + \int d^2\theta\, W(\Phi) + \int d^2\theta^\dagger\, [W(\Phi)]^\dagger\,, \tag{10.278}$$

where K is an arbitrary function of the superfield Φ and its conjugate $\Phi^\dagger$.

(a) Without assuming any particular form for K or W, evaluate the Lagrangian in terms of component fields. Show that

$$\begin{aligned}
\mathscr{L} = {} & \frac{\partial^2 K}{\partial\phi\partial\phi^\dagger}\left[(\partial_\mu\phi)(\partial^\mu\phi^\dagger) + F^\dagger F + \tfrac{1}{2}i\psi^\dagger\overline{\sigma}^\mu\overset{\leftrightarrow}{\partial}_\mu\psi\right] \\
& - \frac{1}{2}\frac{\partial^3 K}{\partial\phi\partial\phi^{\dagger\,2}}\left[F\psi^\dagger\psi^\dagger + i\psi^\dagger\overline{\sigma}^\mu\psi\partial_\mu\phi^\dagger\right] \\
& - \frac{1}{2}\frac{\partial^3 K}{\partial\phi^2\partial\phi^\dagger}\left[F^\dagger\psi\psi - i\psi^\dagger\overline{\sigma}^\mu\psi\partial_\mu\phi\right] + \frac{1}{4}\frac{\partial^4 K}{\partial\phi^2\partial\phi^{\dagger\,2}}(\psi\psi)(\psi^\dagger\psi^\dagger) \\
& + F\frac{dW}{d\phi} + F^\dagger\left(\frac{dW}{d\phi}\right)^\dagger - \frac{1}{2}\left[\frac{d^2W}{d\phi^2}\psi\psi + \left(\frac{d^2W}{d\phi^2}\right)^\dagger\psi^\dagger\psi^\dagger\right],
\end{aligned} \tag{10.279}$$

where terms that can be written as total derivatives have been dropped.

HINT: Use the fact that

$$\int d^4x\, d^2\theta\, d^2\theta^\dagger\, K(\Phi, \Phi^\dagger) = \int d^4x\, \tfrac{1}{16}\, \overline{D}^2 D^2\, K(\Phi, \Phi^\dagger). \tag{10.280}$$

Then apply the chain rule to obtain expressions involving derivatives of K with respect to the component scalar field ϕ.

(b) Show that the auxiliary field F can be determined via its equation of motion, which yields

$$F = \left(\frac{\partial^2 K}{\partial\phi\partial\phi^\dagger}\right)^{-1} \left[\frac{1}{2}\frac{\partial^3 K}{\partial\phi^2\partial\phi^\dagger}\psi\psi - \left(\frac{dW}{d\phi}\right)^\dagger\right]. \tag{10.281}$$

(c) Notice that the resulting kinetic energy terms of the scalar ϕ and fermion ψ do not have the simple (canonical) form

$$\mathscr{L}_{\rm KE} = \partial_\mu\phi^\dagger\partial^\mu\phi + i\psi^\dagger\,\overline{\sigma}^\mu\partial_\mu\psi\,. \tag{10.282}$$

Rescale ϕ and ψ so that their kinetic energy terms are canonical. How does this affect the rest of the interaction Lagrangian?

(d) Check that if

$$K(\Phi, \Phi^\dagger) = \Phi^\dagger\Phi\,, \qquad W = \tfrac{1}{2}M\Phi^2 + \tfrac{1}{6}y\Phi^3\,, \tag{10.283}$$

then eq. (10.279) yields the result previously obtained in eq. (8.51) for a single chiral superfield after eliminating the auxiliary field F.

10.13 The superfields of the supersymmetric extension of QED (henceforth denoted as SUSY-QED) consist of a vector superfield V and two chiral superfields Φ_+ and Φ_- with U(1) charges $+1$ and -1, respectively. The corresponding gauge invariant superpotential is given by $W = m\Phi_+\Phi_-$. Without loss of generality, the parameter m can be taken to be real and positive by an appropriate rephasing of the chiral superfields.

(a) Construct the SUSY-QED Lagrangian in the Wess–Zumino gauge. Show that the physical states of the theory consist of a Dirac fermion (the "electron"), two complex scalar "selectrons," usually denoted by $\widetilde{e}_L$ and $\widetilde{e}_R$, a massless photon, and a massless photino.

(b) Check that the number of bosonic and the number of fermionic degrees of freedom are equal, both off-shell and on-shell.

(c) Instead of imposing the Wess–Zumino gauge condition as in part (a), explore the consequences of adding the supersymmetric gauge-fixing term

$$\mathscr{L}_{\rm GF} = -\frac{1}{8\xi}\left[(D^2V)(\overline{D}^2V)\right]_D, \tag{10.284}$$

where ξ is the gauge-fixing parameter [99–101].

10.14 If the requirement of renormalizability is dropped, then one can generalize the action of a supersymmetric Yang–Mills theory coupled to supermatter,

$$\mathscr{L} = \int d^2\theta\, d^2\theta^\dagger K(\Phi\,,\, \Phi^\dagger e^V) + \left[\int d^2\theta\, W(\Phi_i) + \text{h.c.}\right] + \left[\frac{1}{4}\int d^2\theta\, f_{ab}(\Phi)\mathcal{W}^{\alpha a}\mathcal{W}^b_\alpha + \text{h.c.}\right], \tag{10.285}$$

where K is the Kähler potential, W is the superpotential, f_{ab} is the gauge kinetic function, and $V \equiv 2g_a \boldsymbol{T}^a V^a$.

(a) Evaluate the contribution of the Kähler potential terms to the Lagrangian given in eq. (10.285) in terms of the component fields. Show that your result reduces to eq. (10.279) in the limit of $g_a \to 0$.

(b) Evaluate the contribution of the gauge kinetic function terms to the Lagrangian given in eq. (10.285) in terms of the component fields. How does your result simplify in the abelian limit?

(c) Starting from eq. (10.285), solve for the auxiliary fields F_i and D^a using their equations of motion. Using these results, determine the form of the scalar potential that generalizes the result of eq. (8.75).

10.15 The commutation relations of the $\text{U}(1)_R$ generator, R, with Q and $Q^\dagger$ are given in eq. (10.226).

(a) Show that the action of $\text{U}(1)_R$ on a superfield $S(x, \theta, \theta^\dagger)$ can be represented by a differential operator $\widehat{R}$ acting on superspace:

$$[S(x, \theta, \theta^\dagger), R] = \widehat{R}\, S(x, \theta, \theta^\dagger)\,, \tag{10.286}$$

where

$$\widehat{R} \equiv \theta^\alpha \partial_\alpha - \theta^\dagger_{\dot\alpha} \partial^{\dagger\dot\alpha} - r_S\,, \tag{10.287}$$

and r_S is a real number called the R charge of the superfield S.

(b) Under a $\text{U}(1)_R$ transformation, $S \to S + \delta_\alpha S$, where

$$\delta_\alpha S = i\alpha[R, S] = -i\alpha \widehat{R}\, S\,, \tag{10.288}$$

after employing eq. (10.286). Using eq. (10.287), verify that

$$\widehat{R}\, S(x, \theta, \theta^\dagger) = e^{i r_S \alpha} S(x, e^{-i\alpha}\theta, e^{i\alpha}\theta^\dagger)\,. \tag{10.289}$$

(c) Derive the following two identities:

$$D_\alpha \widehat{R} = (\widehat{R} + 1) D_\alpha\,, \qquad \overline{D}_{\dot\alpha} \widehat{R} = (\widehat{R} - 1)\overline{D}_{\dot\alpha}\,. \tag{10.290}$$

Show that if $S(x, \theta, \theta^\dagger)$ is a chiral [antichiral] superfield, then $\widehat{R}S(x, \theta, \theta^\dagger)$ is a chiral [antichiral] superfield.

11 Radiative Corrections in Supersymmetry

11.1 Introduction

An attractive feature of supersymmetric quantum field theories is that their ultraviolet divergences are better behaved, as compared to ordinary quantum field theories. Ref. [102] demonstrated that the loop corrections to the effective action of a supersymmetric theory of chiral superfields can be expressed as an integral over the full superspace,

$$\sum_n \int d^4x_1 \cdots d^4x_n \int d^4\theta \, g_n(x_1, \ldots, x_n) F_1(x_1, \theta, \theta^\dagger) \cdots F_n(x_n, \theta, \theta^\dagger) \,, \tag{11.1}$$

where the $F_i(x_i, \theta, \theta^\dagger)$ are local functionals of chiral and antichiral superfields and their covariant derivatives, and the g_n are translationally invariant functions on Minkowski space.

The integral over $d^4\theta$ projects out the D-term of a superfield while an integral over $d^2\theta$ projects out the F-term of a superfield. However, the distinction between integrals over the full superspace (i.e., D-terms) and integrals over half of superspace (i.e., F-terms) appears to be ambiguous, since it is possible to convert an integral over half of superspace into an integral over the full superspace. For example, for an arbitrary superfield V,

$$\int d^4x \, d^2\theta \, V(x, \theta, \theta^\dagger) = \int d^4x \, (-\tfrac{1}{4} D^2 V) \,. \tag{11.2}$$

On the left-hand side of eq. (11.2), the integration over $d^2\theta$ projects out all terms proportional to $\theta\theta$. On the right-hand side, $D^2 = -\partial^\alpha \partial_\alpha$ up to total derivative terms that can be dropped because we are integrating over d^4x. Hence, $\frac{1}{4}\partial^\alpha\partial_\alpha$ has the effect of projecting out all terms proportional to $\theta\theta$. Likewise,

$$\int d^4x \, d^2\theta^\dagger \, V(x, \theta, \theta^\dagger) = \int d^4x \, (-\tfrac{1}{4} \overline{D}^2 V) \,. \tag{11.3}$$

Hence, it follows that

$$\int d^4x \, d^2\theta \, (-\tfrac{1}{4} \overline{D}^2 V) = \int d^4x \, d^4\theta \, V(x, \theta, \theta^\dagger) \,. \tag{11.4}$$

Consider the most general supersymmetric action involving a chiral superfield Φ,

$$S = \int d^4x \, d^4\theta K(\Phi, \Phi^\dagger) + \int d^4x \, d^2\theta \, W(\Phi) + \int d^4x \, d^2\theta^\dagger \, W(\Phi^\dagger) \,, \tag{11.5}$$

where K is the Kähler potential and W is the superpotential. Suppose we attempt to convert the integral of the superpotential over half of superspace into an integral over all of superspace using eq. (11.4):

$$\int d^4x\, d^2\theta\, W(\Phi) = -4 \int d^4x\, d^4\theta\, \overline{D}^{-2} W(\Phi)\,. \tag{11.6}$$

Because of the presence of the inverse differential operator, the integrand on the right-hand side of eq. (11.6) is a nonlocal functional of chiral superfields. This provides the distinction between F-terms and D-terms. In particular, any half superspace integral that can be converted into a full superspace integral over a *local* functional of superfields will be called a D-term.

In light of eq. (11.1), D-terms are renormalized but F-terms are not renormalized. Moreover, if F-terms are absent at tree level, then they are not generated at the loop level in perturbation theory. Hence, the tree-level Kähler potential is renormalized by radiative corrections, whereas there are no loop corrections to the tree-level superpotential. This is the famous nonrenormalization theorem of $N = 1$ supersymmetry.[1] The proof of the nonrenormalization theorem in [B25] relies on the analysis of supergraphs in perturbation theory. Heuristically, this theorem is a consequence of an exact cancellation between fermion and boson loop contributions to the effective action due to supersymmetry.

Note that the nonrenormalization of the tree-level superpotential is simply a consequence of the fact that the integral of a product of chiral superfields over *all* of superspace in eq. (11.1) is zero due to eq. (10.5) [see eq. (10.250)]. Moreover, the assumption that the F_i in eq. (11.1) are *local* functionals of chiral and antichiral superfields is essential. Otherwise, one could employ eq. (11.6) and erroneously claim the existence of loop corrections to the tree-level superpotential.

We now briefly explore the consequence of the nonrenormalization of the superpotential. Consider the action of the Wess–Zumino model,

$$S_{\rm WZ} = \int d^4x \int d^4\theta\, \Phi^\dagger\Phi + \left[\int d^4x \int d^2\theta \left(\tfrac{1}{2}m\Phi^2 + \tfrac{1}{3}\lambda\Phi^3\right) + \text{h.c.}\right]. \tag{11.7}$$

The nonrenormalization theorem implies that renormalized fields and parameters are related to bare fields and parameters as follows:

$$\Phi_r = Z^{-1/2}\Phi\,, \qquad m_r = Zm\,, \qquad \lambda_r = Z^{3/2}\lambda\,, \tag{11.8}$$

where the subscript r indicates renormalized quantities and quantities with no corresponding subscript are bare quantities. Equation (11.8) is equivalent to the statement that the superpotential is unrenormalized, $W_r(\Phi_r) = W(\Phi)$. That is,

$$\tfrac{1}{2}m_r\Phi_r^2 + \tfrac{1}{3}\lambda_r\Phi_r^3 = \tfrac{1}{2}m\Phi^2 + \tfrac{1}{3}\lambda\Phi^3\,. \tag{11.9}$$

[1] The proof of the nonrenormalization theorem implicitly assumes that the function g_n in eq. (11.1) is local. However, the nonrenormalization theorem can fail if the supersymmetric theory contains massless fields due to infrared divergences [103–105]. For example, the inverse Laplacian operator $\Box^{-1}$ (from a massless propagator) can appear, resulting in a nonlocal function g_n in eq. (11.1). One can show that the nonrenormalization theorem holds for the Wilsonian effective action [106, 107], where the infrared effects are cut off [108, 109].

Note that wave function renormalization is a consequence of the renormalization of the Kähler potential (e.g., $K = \Phi^\dagger\Phi$ in the Wess–Zumino model). In Section 11.3, we shall verify these statements at the one-loop level with an explicit computation.

It is important to recognize that the nonrenormalization theorem does *not* assert that the parameters of the superpotential are not renormalized. Indeed, eq. (11.8) states that the renormalization of the parameters m and λ are governed by the wave function renormalization constant Z. Moreover, the wave function renormalization constants of the component fields of the chiral superfield are equal, i.e., $\phi_r = Z^{-1/2}\phi$ and $\psi_r = Z^{-1/2}\psi$, as a consequence of supersymmetry.

11.2 Seiberg's Proof of the Nonrenormalization Theorem

In Ref. [108], Seiberg offered a more intuitive understanding of the nonrenormalization theorem, which also forbids nonperturbative corrections to the Wilsonian effective action (see footnote 1). Seiberg's argument draws on the symmetry and holomorphy of the superpotential.[2] Consider again the example of the Wess–Zumino superpotential, $W(\Phi) = \frac{1}{2}m\Phi^2 + \frac{1}{3}\lambda\Phi^3$. Following Ref. [108], one can treat m and λ as spurion fields by viewing them as the vacuum expectation values of chiral superfields. Consequently, we shall demand that W is holomorphic in m and λ as well as in Φ. In light of Section 10.11, the superpotential carries a $\mathrm{U}(1)_R$ charge of $+2$. Hence, the theory is invariant under an enhanced $\mathrm{U}(1) \times \mathrm{U}(1)_R$ symmetry, with the charge assignments shown in Table 11.1.

Table 11.1 $\mathrm{U}(1)\times\mathrm{U}(1)_R$ charge assignments of fields and spurions.

group	Φ	$\Phi^\dagger$	m	λ
U(1)	1	−1	−2	−3
$\mathrm{U}(1)_R$	1	1	0	−1

To maintain the $\mathrm{U}(1) \times \mathrm{U}(1)_R$ symmetry and holomorphy, corrections to the Wilsonian effective superpotential must therefore be of the form

$$m\Phi^2 f\left(\frac{\lambda\Phi}{m}\right), \tag{11.10}$$

where f is an arbitrary holomorphic function. Equation (11.10) is valid for an

[2] The fact that the superpotential is a holomorphic function of chiral superfields plays a critical role in Seiberg's argument. In contrast, the renormalization of the Kähler potential is possible because the latter is a function of chiral and antichiral superfields and hence is not holomorphic.

arbitrary value of λ. Thus, we can take $|\lambda| \ll 1$, in which case perturbation theory should be valid. Expanding in powers of the coupling constant λ, the perturbative expansion should have the form

$$W_{\text{eff}} = \sum_{n=0}^{\infty} a_n \frac{\lambda^n}{m^{n-1}} \Phi^{n+2} . \tag{11.11}$$

The terms in W_{eff} must be represented diagrammatically by 1PI supergraphs constructed from propagators and three-point vertices proportional to λ. However, one cannot construct a one-loop (or higher) supergraph that behaves like $\lambda^n \Phi^{n+2}$. It is easy to show that tree-level diagrams with $n+2$ external legs, n vertices, and $n-1$ propagators would behave like $\lambda^n \Phi^{n+2}$. But the only 1PI tree-level graphs are those with either two or three external legs! Hence, we conclude that $a_0 = \frac{1}{2}$, $a_1 = \frac{1}{3}$, and $a_n = 0$ for $n \geq 2$.[3] That is, $W_{\text{eff}}(\Phi) = W_{\text{tree}}(\Phi)$, which is the statement that the superpotential is not renormalized.

11.3 Renormalization of the Wess–Zumino Model

The superpotential for the massive, interacting Wess–Zumino model is given by

$$W_{\text{WZ}} = \tfrac{1}{2} m \Phi^2 + \tfrac{1}{3} \lambda \Phi^3 . \tag{11.12}$$

Using eqs. (8.47) and (8.51), the resulting Lagrangian, expressed in terms of a complex scalar field, ϕ, and a two-component fermion, ψ, is given by

$$\begin{aligned} \mathscr{L}_{\text{WZ}} = {} & \partial_\mu \phi^\dagger \partial^\mu \phi + \tfrac{1}{2} i (\psi \sigma^\mu \partial_\mu \psi^\dagger + \psi^\dagger \overline{\sigma}^\mu \partial_\mu \psi) - \tfrac{1}{2} m \left(\psi\psi + \psi^\dagger \psi^\dagger\right) - m^2 \phi^\dagger \phi \\ & - m\lambda \left(\phi \phi^\dagger \phi^\dagger + \phi^\dagger \phi \phi\right) - \lambda^2 \phi^2 \phi^{\dagger 2} - \lambda \left(\phi \psi \psi + \phi^\dagger \psi^\dagger \psi^\dagger\right) , \end{aligned} \tag{11.13}$$

where m and λ are nonnegative real parameters (after an appropriate rephasing of the fields ϕ and ψ). In writing eq. (11.13), we have employed a symmetrical form of the fermion kinetic energy term by using eq. (1.98) to write

$$i\psi^\dagger \overline{\sigma}^\mu \partial_\mu \psi = \tfrac{1}{2} i \left[\psi \sigma^\mu \partial_\mu \psi^\dagger + \psi^\dagger \overline{\sigma}^\mu \partial_\mu \psi + \partial_\mu (\psi^\dagger \overline{\sigma}^\mu \psi)\right] , \tag{11.14}$$

and we have dropped the total derivative, which does not contribute to the action.

We shall employ the method of counterterms in the renormalization of the Wess–Zumino model. In the following, we wish to consider the renormalization of the Wess–Zumino Lagrangian. We first introduce renormalized fields and parameters (indicated with a subscript r) as follows:

$$\phi_r = Z_\phi^{-1/2} \phi , \qquad \psi_r = Z_\psi^{-1/2} \psi , \tag{11.15}$$

$$m_r = Z_m^{-1} m , \qquad \mu^\epsilon \lambda_r = Z_\lambda^{-1} \lambda , \tag{11.16}$$

where μ is the mass scale of dimensional regularization (see Appendix K.2) and the

[3] One can also conclude that $a_n = 0$ for $n \geq 2$ by noting that the Wilsonian effective action W_{eff} must have a smooth limit as $m \to 0$.

factor of μ^ϵ has been inserted so that λ_r is dimensionless in $d = 4 - 2\epsilon$ dimensions. Without loss of generality, one can rephase the renormalized fields such that Z_ϕ and Z_ψ are real. Comparing eqs. (11.15) and (11.16) with eq. (11.8), we see that the nonrenormalization of the superpotential is reproduced if

$$Z_\phi = Z_\psi = Z_m^{-1} = Z_\lambda^{-2/3} = Z \,. \tag{11.17}$$

In the following we shall show these equations hold at the one-loop quantum level.

Starting with the original Lagrangian, given in eq. (11.13) in terms of bare fields and parameters, we can rewrite it in terms of the renormalized fields as follows:

$$\begin{aligned}\mathscr{L}_{\rm WZ} = {}& \partial_\mu\phi_r^\dagger\partial^\mu\phi_r + \tfrac{1}{2}i(\psi_r\sigma^\mu\partial_\mu\psi_r^\dagger + \psi_r^\dagger\bar\sigma^\mu\partial_\mu\psi_r) - \tfrac{1}{2}m_r\left(\psi_r\psi_r + \psi_r^\dagger\psi_r^\dagger\right) - m_r^2\phi_r^\dagger\phi_r \\ & - m_r\lambda\mu^\epsilon\left(\phi_r\phi_r^\dagger\phi_r^\dagger + \phi_r^\dagger\phi_r\phi_r\right) - (\lambda_r\mu^\epsilon)^2\phi_r^2\phi_r^{\dagger 2} - \lambda_r\mu^\epsilon\left(\phi_r\psi_r\psi_r + \phi_r^\dagger\psi_r^\dagger\psi_r^\dagger\right) \\ & + \Delta\mathscr{L}_{\rm WZ}\,,\end{aligned} \tag{11.18}$$

where the counterterm Lagrangian is given by

$$\begin{aligned}\Delta\mathscr{L}_{\rm WZ} = {}& -\phi_r^\dagger\left[(Z_\phi - 1)\partial^2 + (Z_\phi Z_m^2 - 1)m_r^2\right]\phi_r \\ & + \tfrac{1}{2}i(Z_\psi - 1)\left(\psi_r\sigma^\mu\partial_\mu\psi_r^\dagger + \psi_r^\dagger\bar\sigma^\mu\partial_\mu\psi_r\right) - \tfrac{1}{2}m_r(Z_\psi Z_m - 1)\left[\psi_r\psi_r + \psi_r^\dagger\psi_r^\dagger\right] \\ & - (\lambda_r\mu^\epsilon)^2\left(Z_\phi^2 Z_\lambda^2 - 1\right)\phi_r^{\dagger 2}\phi_r^2 - (Z_\phi^{3/2}Z_m Z_\lambda - 1)m_r\lambda_r\mu^\epsilon\left[\phi_r\phi_r^{\dagger 2} + \phi_r^\dagger\phi_r^2\right] \\ & - \lambda_r\mu^\epsilon(Z_\phi^{1/2}Z_\psi Z_\lambda - 1)\left[\phi_r\psi_r\psi_r + \phi_r^\dagger\psi_r^\dagger\psi_r^\dagger\right] .\end{aligned} \tag{11.19}$$

One can read off the counterterm Feynman rules from eq. (11.19), where each counterterm is treated as a new interaction vertex.

The Feynman rules for the fermion mass counterterms are given in Fig. 11.1. Note that the structure of the counterterm diagrams in (c) and (d) are identical to the mass insertion rules of Fig. 2.4. The counterterm diagrams in (a) and (b) preserve the direction of the arrows. The rules exhibited in (a) and (b) can be summarized by a single rule, shown in Fig. 11.2, by adopting a convention where p^μ in the diagram is the momentum flowing in the direction of the fermion arrows.

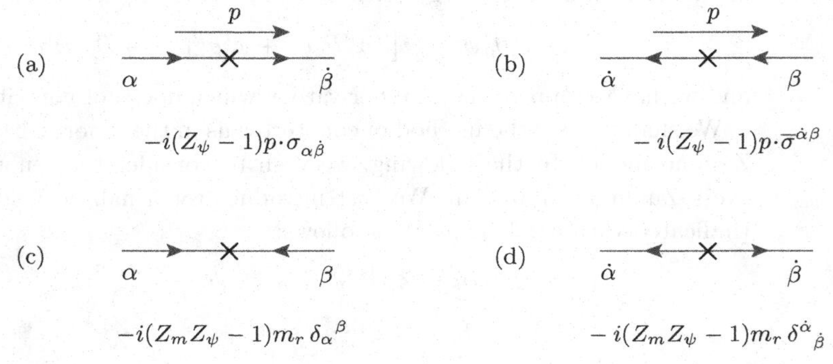

Fig. 11.1 Feynman rules for the mass counterterms of a neutral two-component spin-1/2 fermion. The location of the insertion is denoted by ×.

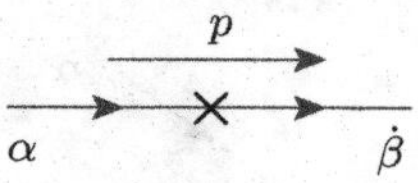

$$-i(Z_\psi - 1)p\cdot\sigma_{\alpha\dot\beta} \quad \text{or} \quad i(Z_\psi - 1)p\cdot\overline{\sigma}^{\dot\beta\alpha}$$

Fig. 11.2 This single rule summarizes the rules of Fig. 11.1(a) and (b).

$$i\left[(Z_\phi - 1)p^2 - (Z_\phi Z_m^2 - 1)\, m_r^2\right]$$

Fig. 11.3 Feynman rules for the mass counterterm of a complex scalar.

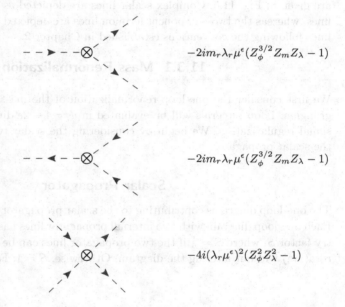

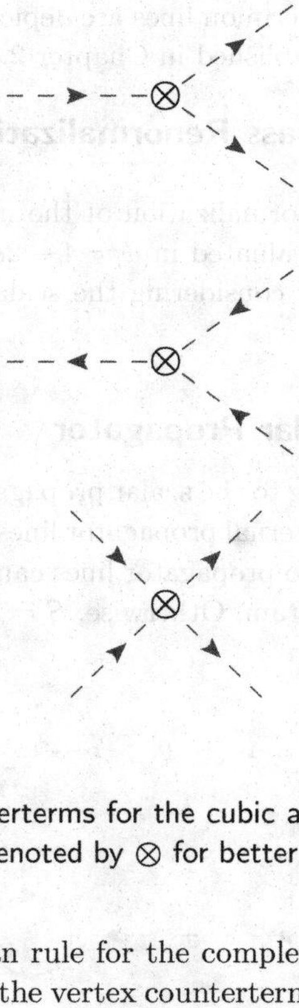

$$-2im_r\lambda_r\mu^\epsilon(Z_\phi^{3/2}Z_mZ_\lambda - 1)$$

$$-2im_r\lambda_r\mu^\epsilon(Z_\phi^{3/2}Z_mZ_\lambda - 1)$$

$$-4i(\lambda_r\mu^\epsilon)^2(Z_\phi^2Z_\lambda^2 - 1)$$

Fig. 11.4 Vertex counterterms for the cubic and quartic scalar interactions. The location of the insertion is denoted by ⊗ for better visibility.

The Feynman rule for the complex scalar mass counterterm is given in Fig. 11.3, the rules for the vertex counterterms of the cubic and quartic scalar interactions are given in Fig. 11.4, and the rules for the vertex counterterms of the Yukawa couplings

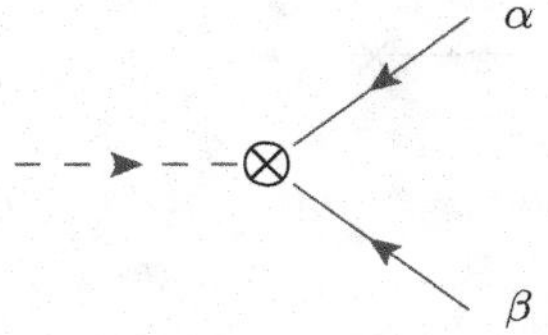

$$-2i\lambda_r\mu^\epsilon(Z_\phi^{1/2}Z_\psi Z_\lambda - 1)\delta_\alpha{}^\beta$$

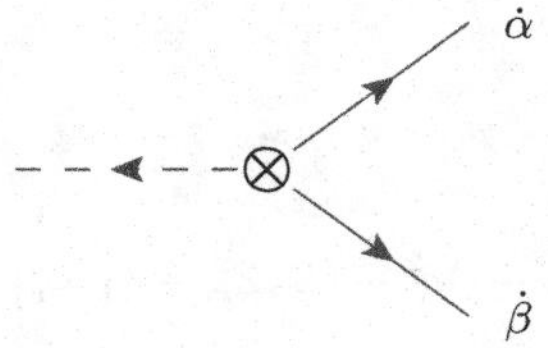

$$-2i\lambda_r\mu^\epsilon(Z_\phi^{1/2}Z_\psi Z_\lambda - 1)\delta_{\dot\alpha}{}^{\dot\beta}$$

Fig. 11.5 Vertex counterterms for the Yukawa interactions.

are given in Fig. 11.5. Complex scalar lines are depicted as arrow-directed dashed lines, whereas the two-component fermion lines are depicted as arrow directed solid lines following the conventions established in Chapter 2.

11.3.1 Mass Renormalization

We first consider the one-loop renormalization of the mass parameter in the Lagrangian. Loop integrals will be evaluated in $d = 4 - 2\epsilon$ dimensions using dimensional regularization. We begin by considering the scalar two-point function, i.e., the scalar propagator.

Scalar Propagator

The one-loop diagrams contributing to the scalar propagator are shown in Fig. 11.6. Each one-loop diagram with two internal propagator lines has an associated symmetry factor S, where $S = \frac{1}{2}$ if the two propagator lines can be interchanged (without rotation) without altering the diagram. Otherwise, $S = 1$. For example, in the two

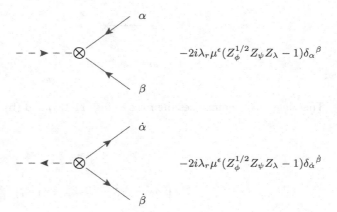

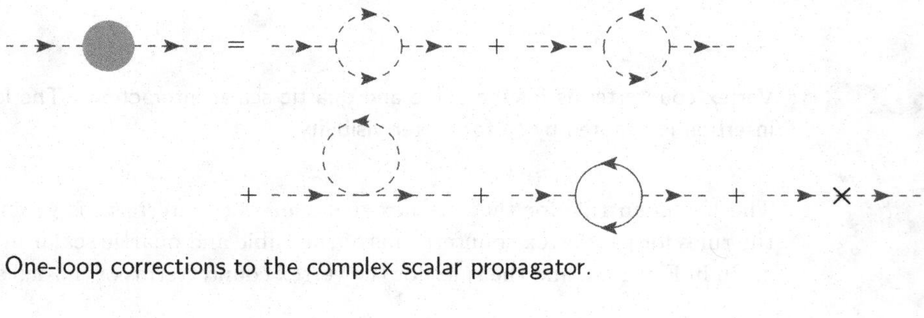

Fig. 11.6 One-loop corrections to the complex scalar propagator.

scalar loop diagrams shown in the first row of Fig. 11.6, the corresponding symmetry factors are $S = \frac{1}{2}$ and $S = 1$, respectively, whereas the fermion loop diagram in the second row has a symmetry factor of $S = \frac{1}{2}$. We evaluate the diagrams of Fig. 11.6 in $d = 4 - 2\epsilon$ dimensions in terms of the Passarino–Veltman functions defined in Appendix K. Denoting the four-momentum of the external scalar by p and the loop momentum by k (and suppressing the r subscript and $i\varepsilon$ factors in the propagator denominators), the scalar loop contributions yield

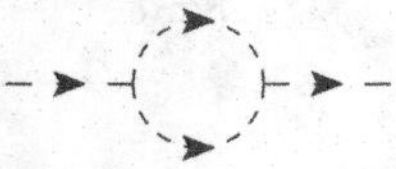

$$= \left(\tfrac{1}{2}\right)(-2im\lambda)^2\,\mu^{2\epsilon}\int\frac{d^dk}{(2\pi)^d}\left[\frac{i}{k^2-m^2}\right]\left[\frac{i}{(k-p)^2-m^2}\right]$$
$$= \frac{im^2\lambda^2}{8\pi^2}B_0(p^2;m^2,m^2)\,, \tag{11.20}$$

$$= (1)\,(-2im\lambda)^2\,\mu^{2\epsilon}\int\frac{d^dk}{(2\pi)^d}\left[\frac{i}{k^2-m^2}\right]\left[\frac{i}{(k+p)^2-m^2}\right]$$
$$= \frac{im^2\lambda^2}{4\pi^2}B_0(p^2;m^2,m^2)\,, \tag{11.21}$$

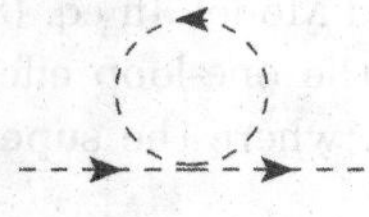

$$= (1)\,(-4i\lambda^2)\,\mu^{2\epsilon}\int\frac{d^dk}{(2\pi)^d}\,\frac{i}{k^2-m^2} = \frac{i\lambda^2}{4\pi^2}A_0(m^2)\,. \tag{11.22}$$

The fermion loop contribution to the scalar propagator is given by

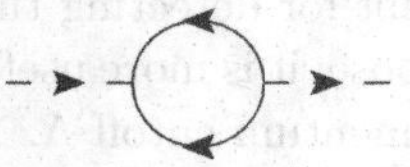

$$= -\left(\tfrac{1}{2}\right)(-2i\lambda)^2\mu^{2\epsilon}\int\frac{d^dk}{(2\pi)^d}\,\mathrm{Tr}\left[\frac{ik\cdot\overline{\sigma}}{k^2-m^2}\,\frac{i(k-p)\cdot\sigma}{[(k-p)^2-m^2]}\right]$$
$$= -4\lambda^2\mu^{2\epsilon}\int\frac{d^dk}{(2\pi)^d}\,\frac{k\cdot(k-p)}{(k^2-m^2)\left[(k-p)^2-m^2\right]}$$
$$= -4\lambda^2\mu^{2\epsilon}\int\frac{d^dk}{(2\pi)^d}\left[\frac{1}{k^2-m^2}+\frac{k\cdot p-p^2+m^2}{(k^2-m^2)\left[(k-p)^2-m^2\right]}\right]$$
$$= -\frac{i\lambda^2}{4\pi^2}\left[A_0(m^2)+(m^2-\tfrac{1}{2}p^2)B_0(p^2;m^2,m^2)\right], \tag{11.23}$$

after including the minus sign for the fermion loop and the symmetry factor of $\frac{1}{2}$. In the final step, we employed eq. (K.43) to express the Passarino–Veltman function $B_1(p^2;m^2,m^2)$ in terms of $B_0(p^2;m^2,m^2)$. Using the notation of Fig. 2.25, the sum of the contributing loop diagrams and counterterm (see Fig. 11.3) is

$$-i\Sigma(p^2) = \frac{i\lambda^2}{8\pi^2}(p^2+m^2)B_0(p^2;m^2,m^2) + i\left[(Z_\phi-1)p^2 - (Z_\phi Z_m^2-1)\,m^2\right]. \tag{11.24}$$

In the minimal subtraction renormalization scheme, the one-loop renormalization constants are chosen to be of minimal form, $Z_i - 1 = c_i/\epsilon$ (for $i = \phi$ and m), where the c_i are fixed such that all divergences, which are proportional to ϵ^{-1} at one-loop order, are exactly canceled in eq. (11.24). In light of eq. (K.21), the divergent part of B_0 is

$$B_0(p^2; m^2; m^2) \stackrel{\text{div.}}{=} \frac{1}{\epsilon}\,. \tag{11.25}$$

Hence, it follows that

$$Z_\phi = 1 - \frac{\lambda^2}{8\pi^2\epsilon} \qquad \text{and} \qquad Z_\phi Z_m^2 = 1 + \frac{\lambda^2}{8\pi^2\epsilon}\,. \tag{11.26}$$

That is, at one-loop accuracy,

$$Z_m = Z_\phi^{-1} = 1 + \frac{\lambda^2}{8\pi^2\epsilon}\,, \tag{11.27}$$

in agreement with eq. (11.17).

Cancellation of the Quadratic Divergences

In Chapter 7, the cancellation of quadratic divergences in radiative corrections in supersymmetric theories was presented as a key ingredient for employing supersymmetry to cure the hierarchy problem of the Standard Model. In eq. (8.81), we noted that the cancellation of quadratic divergences in the one-loop effective potential was a consequence of the relation $\text{STr}\, M^2(\phi) = 0$, where the supertrace is taken over a supersymmetric multiplet.

It is instructive to see how the cancellation of quadratic divergences appears in the one-loop radiative corrections to the scalar propagator of the Wess–Zumino model. Employing dimensional regularization is inconvenient for detecting the presence of quadratic sensitivity to the ultraviolet. For this purpose, it is more useful to regulate the divergence of the one-loop corrections by a momentum cutoff Λ.

The individual contributions to the scalar propagator at one-loop order obtained in eqs. (11.20)–(11.23) exhibit the following general form:

$$-i\Sigma(p^2) = c_1 A_0(m^2) + c_2 B_0(p^2; m^2, m^2)\,, \tag{11.28}$$

where c_1 and c_2 are coefficients that are determined by an explicit computation. If we regulate the ultraviolet divergences with a momentum cutoff Λ, then evaluating eqs. (K.9) and (K.10) in $d = 4$ dimensions yields (see Exercise 11.2)

$$A_0(m^2) \sim \Lambda^2\,, \qquad B_0(p^2; m^2, m^2) \sim \ln\left(\frac{\Lambda^2}{m^2}\right)\,. \tag{11.29}$$

Equations (11.22) and (11.23) show that both the scalar loop correction and the fermion loop correction to the scalar propagator are quadratically divergent due to the presence of $A_0(m^2)$. However, when these two contributions are summed, the terms proportional to $A_0(m^2)$ cancel exactly! Thus, $\Sigma(p^2)$ given in eq. (11.24) is only logarithmically divergent. Indeed, the quadratic divergences have canceled exactly due to the supersymmetric relations among the masses and couplings of the Wess–Zumino model (and the factor of -1 for the fermion loop).

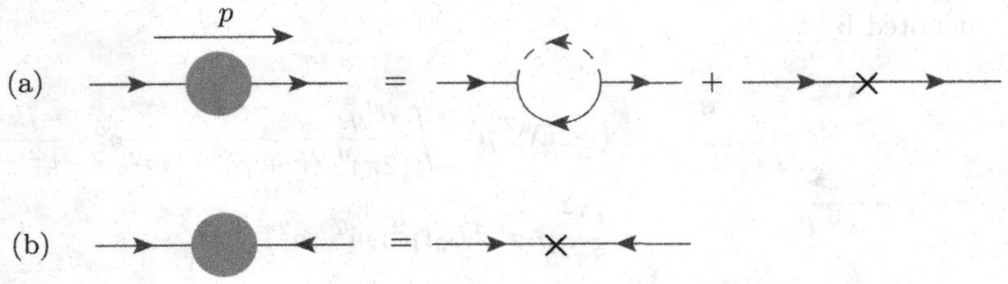

Fig. 11.7 One-loop corrections to the fermion propagators. Not shown are two additional sets of diagrams in which the directions of all arrows are reversed.

Fermion Propagator

The one-loop two-component fermion propagator diagrams are exhibited in Fig. 11.7. Denoting the external momentum of the fermion by p and the internal fermion loop momentum in the direction of its arrow by k (and suppressing the r subscript and $i\varepsilon$ factors in the propagator denominators), the scalar–fermion loop contribution of Fig. 11.7(a) yields

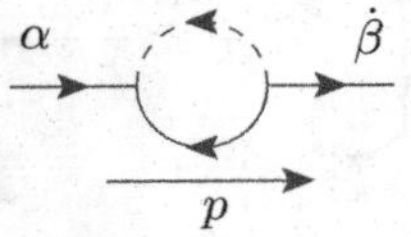

$$= (-2i\lambda)^2 \mu^{2\epsilon} \int \frac{d^d k}{(2\pi)^d} \frac{i}{(k+p)^2 - m^2} \delta_\alpha{}^\beta \frac{ik\cdot\sigma_{\beta\dot\alpha}}{k^2 - m^2} \delta^{\dot\alpha}_{\dot\beta}$$

$$= -\frac{i\lambda^2}{8\pi^2} p\cdot\sigma_{\alpha\dot\beta} B_0(p^2; m^2, m^2)\,, \qquad (11.30)$$

after employing eq. (K.43). Using the notation of Fig. 2.30 (where the direction of p is opposite to the one employed in Fig. 11.7), it follows that

$$\Xi(p^2) = -\left(\frac{\lambda^2}{8\pi^2} B_0(p^2; m^2, m^2) + Z_\psi - 1\right). \qquad (11.31)$$

In light of eq. (11.25),

$$Z_\psi = 1 - \frac{\lambda^2}{8\pi^2\epsilon}\,. \qquad (11.32)$$

Likewise, Fig. 11.7(b) yields

$$\Omega(p^2) = (Z_m Z_\psi - 1)m\,. \qquad (11.33)$$

Since there is no one-loop diagram to absorb the ϵ^{-1} poles, $Z_m Z_\psi = 1 + \mathcal{O}(\lambda^4)$. Using eq. (11.32), it follows that

$$Z_m = Z_\psi^{-1} = 1 + \frac{\lambda^2}{8\pi^2\epsilon}\,, \qquad (11.34)$$

in agreement with the results of eqs. (11.17) and (11.27).

Note that the same results are obtained if the directions of all arrows are reversed. In particular, with the external momentum p now pointing from right to left and the momentum of the internal fermion loop momentum in the direction of its arrow denoted by k,

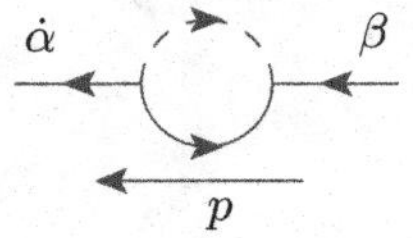

$$= (-2i\lambda)^2 \mu^{2\epsilon} \int \frac{d^d q}{(2\pi)^d} \frac{i}{(k+p)^2 - m^2} \delta^{\dot{\alpha}}{}_{\dot{\beta}} \frac{-ik\cdot\overline{\sigma}^{\dot{\beta}\alpha}}{k^2 - m^2} \delta_\alpha{}^\beta$$
$$= \frac{i\lambda^2}{8\pi^2} p\cdot\sigma^{\dot{\alpha}\beta} B_0(p^2; m^2, m^2) \,. \tag{11.35}$$

Using the notation of Fig. 2.30, it follows that

$$\Xi(p^2) = -\frac{\lambda^2}{8\pi^2} B_0(p^2; m^2, m^2) - Z_\psi + 1 \,, \tag{11.36}$$

in agreement with eq. (11.31).

11.3.2 Yukawa Coupling Renormalization

At one loop, the relevant diagrams are depicted in Fig. 11.8, where the vertex counterterm Feynman rule is given in Fig. 11.5.

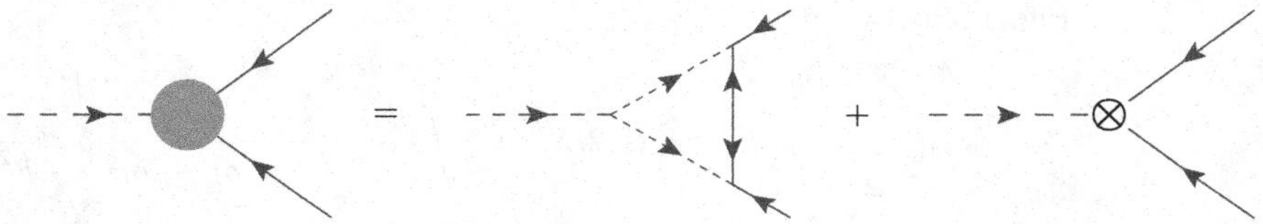

Fig. 11.8 One-loop corrections to the to the Yukawa coupling. Not shown is an additional set of diagrams in which the directions of all arrows are reversed.

It is straightforward to check that the triangle diagram is ultraviolet finite. Since there are no ϵ^{-1} poles that would otherwise be absorbed by the counterterm, it then follows that, at one-loop accuracy,

$$Z_\phi^{1/2} Z_\psi Z_\lambda = 1 \,. \tag{11.37}$$

With $Z_\phi = Z_\psi = Z$, we thus have

$$Z_\lambda = Z^{-3/2} \,, \tag{11.38}$$

in agreement with eq. (11.17).

We have thus completed our verification of eq. (11.17), thereby confirming the nonrenormalization theorem in the one-loop approximation of the Wess–Zumino model.

11.3.3 Renormalization of the Cubic and Quartic Couplings

We have yet to consider the implications of the one-loop renormalization of the cubic and quartic scalar interactions of the Wess–Zumino model. Analyzing these two cases provides independent computations of $Z_\phi^{3/2} Z_m Z_\lambda$ and $Z_\phi^2 Z_\lambda^2$, respectively, in light of the vertex counterterms exhibited in Fig. 11.4. Details of the calculation are left for the reader. It is straightforward to show that the one-loop results are consistent with eqs. (11.27), (11.34), and (11.38) obtained previously.

11.4 Regularization by Dimensional Reduction

Computations involving quantum effects in relativistic field theory are subject to ultraviolet divergences due to integrating over arbitrarily high momenta in intermediate states. In modern practice, these are dealt with by regularization followed by renormalization.

Regularization means that we modify the Lagrangian definition of the theory in such a way that all amplitudes are finite. This generally involves the introduction of one or more regularization parameters, which are arbitrary and unphysical, but which control the amplitudes in such a way that if they were removed the ultraviolet divergences would reappear. Renormalization refers to a systematic redefinition of the original ("bare") parameters of the theory in terms of new ("renormalized") parameters, in such a way that the expressions for amplitudes and physical observables written in terms of the renormalized parameters do not contain ultraviolet divergences when the regulator is removed. This is typically done order-by-order in perturbation theory; if it can be accomplished at arbitrary loop order, by a finite and fixed number of parameter redefinitions, then the theory is said to be renormalizable.

The renormalization scheme may involve trading the regularization parameters for a new renormalization mass scale, called Q here, which appears explicitly in the computation of observables, and at least implicitly in the definitions of the renormalized parameters. At each order in perturbation theory, the expressions for physical observables will depend on the choice of Q. However, because the renormalization scale choice is arbitrary and unphysical, this dependence should be formally of the same order as the lowest loop order that has not been included in the calculations. Therefore, in practice, if the renormalized couplings are not too large, the Q dependence may be under control and getting smaller as more loop effects are included. For this to hold, it is typically necessary that the renormalization scale is chosen in some range comparable to the most important physical energy scale in the problem, in order to avoid large logarithms. The implicit dependence of the renormalized parameters on Q is governed by the beta functions of the renormalization group, which allow one to change the choice of Q to fit the physical observable under consideration.

When choosing regularization and renormalization schemes, it is extremely convenient (but not strictly necessary) to keep manifest the essential symmetries of the theory. This not only allows checks on the calculations to ensure that symmetry-breaking contributions cancel as they should, but generally makes the computations much simpler. When computing radiative corrections in supersymmetric gauge theories, it is important to choose regularization and renormalization schemes that respect both gauge invariance and supersymmetry.

The most popular regularization method for treating radiative corrections within the Standard Model is dimensional regularization (DREG), in which the number of spacetime dimensions is continued to

$$d = 4 - 2\epsilon\,. \tag{11.39}$$

In d dimensions, the Lagrangian has mass dimension $[\mathscr{L}] = \mu^d$, where μ is a parameter with the dimensions of mass called the regularization scale. Consider the original bare parameters of the theory (masses and couplings), which we shall collectively call X_B. The corresponding renormalized parameters will collectively be called X. It is convenient to define X in such a way that its mass dimension corresponds to the theory in $d = 4$ dimensions. Note that in d dimensions, $[\partial_\mu] = 1$, which then fixes the mass dimension of all quantum fields (by considering the corresponding kinetic energy terms of $\mathscr{L}$). Hence, the mass dimensions of X_B and X are related as follows:

$$[X_B] = [X] + \epsilon\rho_X\,, \tag{11.40}$$

where ρ_X is equal to $N-2$ if X is an N-field coupling, so that $\rho = 2$ for scalar quartic couplings, $\rho = 1$ for gauge and Yukawa couplings and scalar cubic couplings, $\rho = 0$ for fermion masses and scalar squared masses, $\rho = -1$ for tadpoles, and $\rho = -2$ for the vacuum energy.

In the computation of loop diagrams using the Feynman rules obtained from the bare Lagrangian, the loop momentum integrals of the unregulated theory are replaced according to

$$\int d^4p \to \mu^{2\epsilon} \int d^dp. \tag{11.41}$$

This is almost always followed by renormalization by minimal subtraction, in which the regularization scale μ is replaced by an $\overline{\text{MS}}$ renormalization scale Q as follows:[4]

$$Q^2 \equiv 4\pi e^{-\gamma}\mu^2, \tag{11.42}$$

where γ is Euler's constant [see eq. (K.3)], and coupling constants are redefined in order to remove all ultraviolet divergence poles in ϵ from expressions for amplitudes and masses. This means that each Lagrangian parameter X is written as an expansion in the number of loops ℓ, containing counterterms $c^X_{\ell,n}$ with only (hence "minimal") poles in ϵ:

$$X_B = \mu^{\epsilon\rho_X}\left(X + \sum_{\ell=1}^{\infty}\sum_{n=1}^{\ell}\frac{\kappa^\ell}{\epsilon^n}c^X_{\ell,n}\right), \tag{11.43}$$

[4] The (now rarely used) minimal subtraction (MS) scheme instead just uses $Q = \mu$. The term $\overline{\text{MS}}$ refers to "modified minimal subtraction"; the modification avoids pesky $\ln(4\pi) - \gamma$ terms in relations between the $\overline{\text{MS}}$ parameters and physical quantities [see eqs. (K.18) and (K.19)].

where the X_B cannot depend on our choice of Q, the X are the corresponding $\overline{\text{MS}}$ parameters (which do depend on Q, and are finite as $\epsilon \to 0$), and

$$\kappa \equiv \frac{1}{16\pi^2}, \tag{11.44}$$

which is used as the loop-counting parameter. For the special case of a constant background scalar field parameter ϕ (for example, when computing the effective potential as in Section 11.6 below), it is traditional instead to write

$$\phi_B = \mu^{-\epsilon}\sqrt{Z_\phi}\,\phi, \tag{11.45}$$

with

$$Z_\phi = 1 + \sum_{\ell=1}^{\infty}\sum_{n=1}^{\ell}\frac{\kappa^\ell}{\epsilon^n}c^{\phi}_{\ell,n}, \tag{11.46}$$

thereby ensuring that the mass dimensions of the renormalized background scalar field parameter $[\phi] = 1$. The counterterm coefficients $c^X_{\ell,n}$ and $c^\phi_{\ell,n}$ are polynomials in the $\overline{\text{MS}}$ parameters (collectively called Y below), with no explicit dependence on Q. For example, if X is a dimensionless coupling, then $c^X_{\ell,n}$ is a polynomial function of the dimensionless $\overline{\text{MS}}$ couplings alone (including X itself). The ℓ-loop counterterms obey the homogeneous identities[5]

$$\left(-\rho_X + \sum_Y \rho_Y Y\frac{\partial}{\partial Y}\right)c^X_{\ell,n} = 2\ell c^X_{\ell,n}, \tag{11.47}$$

$$\sum_Y \rho_Y Y\frac{\partial}{\partial Y}c^\phi_{\ell,n} = 2\ell c^\phi_{\ell,n}, \tag{11.48}$$

where the sums over Y (which can include X itself) are taken over all $\overline{\text{MS}}$ parameters that appear in the polynomials $c^X_{\ell,n}$ and $c^\phi_{\ell,n}$, respectively. The counterterms are chosen in such a way that all observable quantities, when written in terms of the $\overline{\text{MS}}$ parameters, do not contain any poles in ϵ. The statement that this is possible, for a finite number of parameters X, is equivalent to the statement that the theory is renormalizable.

The renormalization group equations (RGEs) can now be derived in terms of the counterterms, as follows. Since bare quantities cannot depend on the arbitrary choice of renormalization scale Q, it follows that $Q\,dX_B/dQ = 0$. Equivalently,

$$Q\frac{dX}{dQ} + \epsilon\rho_X\left(X + \sum_{\ell=1}^{\infty}\sum_{n=1}^{\ell}\frac{\kappa^\ell}{\epsilon^n}c^X_{\ell,n}\right) + \sum_{\ell=1}^{\infty}\sum_{n=1}^{\ell}\frac{\kappa^\ell}{\epsilon^n}\sum_Y Q\frac{dY}{dQ}\frac{\partial c^X_{\ell,n}}{\partial Y} = 0. \tag{11.49}$$

Likewise, $Q\,d(\ln\phi_B)/dQ = 0$, which when applied to eq. (11.46) yields

$$\tfrac{1}{2}Q\frac{d}{dQ}\ln Z_\phi = -Q\frac{d(\ln\phi)}{dQ} + \epsilon\,. \tag{11.50}$$

[5] For a simple proof of eqs. (11.47) and (11.48) that relies only on induction and the topology of Feynman diagrams leading to the polynomial counterterms $c^X_{\ell,n}$ and $c^\phi_{\ell,n}$, see Exercise 11.5.

Making use of eqs. (11.45) and (11.46) and employing the chain rule, the end result is given by

$$\left(Q\frac{d(\ln\phi)}{dQ} - \epsilon\right)\left(1 + \sum_{\ell=1}^{\infty}\sum_{n=1}^{\ell}\frac{\kappa^\ell}{\epsilon^n}c^\phi_{\ell,n}\right) + \frac{1}{2}\sum_{\ell=1}^{\infty}\sum_{n=1}^{\ell}\frac{\kappa^\ell}{\epsilon^n}\sum_Y Q\frac{dY}{dQ}\frac{\partial c^\phi_{\ell,n}}{\partial Y} = 0. \quad (11.51)$$

We now match powers of ϵ in the expansions of eqs. (11.49) and (11.51). Since X and $\ln\phi$ are finite quantities as $\epsilon \to 0$, it follows that QdX/dQ and $Qd(\ln\phi)/dQ$ contribute only to the terms of the respective ϵ expansions of eqs. (11.49) and (11.51) with ϵ^1 and ϵ^0. Hence, we find that

$$Q\frac{dX}{dQ} = -\epsilon\rho_X X + \sum_{\ell=1}^{\infty}\kappa^\ell\left(-\rho_X + \sum_Y \rho_Y Y\frac{\partial}{\partial Y}\right)c^X_{\ell,1}, \quad (11.52)$$

$$Q\frac{d(\ln\phi)}{dQ} = \epsilon - \frac{1}{2}\sum_{\ell=1}^{\infty}\kappa^\ell\sum_Y \rho_Y Y\frac{\partial}{\partial Y}c^\phi_{\ell,1}, \quad (11.53)$$

where we have self-consistently used $QdY/dQ = -\epsilon\rho_Y Y + \cdots$ to obtain the last terms in eqs. (11.52) and (11.53).

The beta functions and scalar field anomalous dimensions are defined to be the ϵ-independent parts of eqs. (11.52) and (11.53), respectively:[6]

$$\beta_X \equiv Q\frac{dX}{dQ}\bigg|_{\epsilon=0} = Q\frac{dX}{dQ} + \epsilon\rho_X X, \quad (11.54)$$

$$\gamma_\phi \equiv -Q\frac{d\ln(\phi)}{dQ}\bigg|_{\epsilon=0} = -Q\frac{d\ln(\phi)}{dQ} + \epsilon = \frac{1}{2}Q\frac{d}{dQ}\ln Z_\phi, \quad (11.55)$$

after making use of eq. (11.50) to obtain the last equality in eq. (11.55).[7] More explicitly,

$$\beta_X = \sum_{\ell=1}^{\infty}\kappa^\ell\left(-\rho_X + \sum_Y \rho_Y Y\frac{\partial}{\partial Y}\right)c^X_{\ell,1}, \quad (11.56)$$

$$\gamma_\phi = -\frac{1}{2}\sum_{\ell=1}^{\infty}\kappa^\ell\sum_Y \rho_Y Y\frac{\partial}{\partial Y}c^\phi_{\ell,1}. \quad (11.57)$$

Consequently, we obtain the loop expansions

$$\beta_X = \sum_{\ell=1}^{\infty}\kappa^\ell\beta^{(\ell)}_X, \quad (11.58)$$

$$\gamma_\phi = \sum_{\ell=1}^{\infty}\kappa^\ell\gamma^{(\ell)}_\phi, \quad (11.59)$$

[6] The second equalities in eqs. (11.54) and (11.55) follow immediately from eqs. (11.52) and (11.53). Note that QdX/dQ, unlike β_X, crucially contains a "zero-loop" term, $-\epsilon\rho_X X$ (except when X is a fermion mass or scalar squared mass, for which $\rho_X = 0$). For a concrete example of the importance of this term, see Exercises 19.2 and 19.3(d). Similarly, $Q\,d(\ln\phi)/dQ$, unlike γ_ϕ, has a "zero-loop" term, ϵ.

[7] Indeed, some textbooks define the anomalous dimension of the field ϕ by $\gamma_\phi \equiv \frac{1}{2}Q\,d(\ln Z_\phi)/dQ$.

where, after using eqs. (11.47) and (11.48) in eqs. (11.52) and (11.53), respectively,

$$\beta_X^{(\ell)} = 2\ell c_{\ell,1}^X, \tag{11.60}$$

$$\gamma_\phi^{(\ell)} = -\ell c_{\ell,1}^\phi. \tag{11.61}$$

Thus, the RGEs at ℓ-loop order can be calculated by isolating the simple $1/\epsilon$ pole contributions to the corresponding 1PI amplitudes.

Also, from looking at the coefficients of $1/\epsilon^{n-1}$ in eqs. (11.49) and (11.51) and then matching powers of the loop-counting factor κ, we obtain (for $n > 1$)

$$2\ell c_{\ell,n}^X = \sum_{\ell'=1}^{\ell-n+1} \sum_Y \beta_Y^{(\ell')} \frac{\partial}{\partial Y} c_{\ell-\ell',n-1}^X, \tag{11.62}$$

$$\ell c_{\ell,n}^\phi = \sum_{\ell'=1}^{\ell-n+1} \left(-\gamma_\phi^{(\ell')} + \frac{1}{2} \sum_Y \beta_Y^{(\ell')} \frac{\partial}{\partial Y} \right) c_{\ell-\ell',n-1}^\phi. \tag{11.63}$$

These can be useful as checks of a multiloop counterterm calculation. Conversely, given the beta functions and anomalous dimensions, which are equivalent to knowing the simple poles in the counterterms, one can use these identities to obtain the higher-order poles. For example, the $1/\epsilon^2$ poles in the two-loop order counterterms can be obtained using only one-loop calculations.

Unfortunately, although the DREG scheme does respect gauge invariance, it violates supersymmetry explicitly because it introduces a mismatch between the numbers of vector gauge boson degrees of freedom and the gaugino degrees of freedom. This mismatch is only 2ϵ, but this can be multiplied by factors up to $1/\epsilon^\ell$ resulting from the momentum integrations in an ℓ-loop calculation. In DREG and $\overline{\text{MS}}$, supersymmetric relations between dimensionless coupling constants and related amplitudes (supersymmetric Ward identities) are therefore not respected by radiative corrections starting with the finite parts of one-loop Feynman graphs and with the divergent parts of two-loop graphs.

Instead, one may use the slightly different scheme known as regularization by dimensional reduction, or DRED, proposed by Siegel in Ref. [110]. In the DRED method, all momentum integrals are performed in $d = 4 - 2\epsilon$ dimensions as in DREG, but the vector fields still have exactly four components, at least formally, so that supersymmetry is respected. Thus, the vector index on a gauge boson field runs over all 4 dimensions, rather than only d dimensions as in DREG. It is important that to match the boson and fermion degrees of freedom, ϵ must be definitely positive in DRED, unlike in DREG.

Let us write a projector onto the d-dimensional subspace as $\hat{\delta}^\mu_\nu$, and onto the complementary 2ϵ-dimensional subspace as $\tilde{\delta}^\mu_\nu$, so that the 4-dimensional Kronecker delta tensor decomposes as

$$\delta^\mu_\nu = \hat{\delta}^\mu_\nu + \tilde{\delta}^\mu_\nu, \tag{11.64}$$

where

$$\delta^\mu_\mu = 4, \tag{11.65}$$
$$\hat{\delta}^\mu_\mu = d, \tag{11.66}$$
$$\tilde{\delta}^\mu_\mu = 4 - d, \tag{11.67}$$
$$\delta^\mu_\nu \hat{\delta}^\rho_\mu = \hat{\delta}^\rho_\nu, \tag{11.68}$$
$$\hat{\delta}^\mu_\nu \hat{\delta}^\rho_\mu = \hat{\delta}^\rho_\nu, \tag{11.69}$$
$$\delta^\mu_\nu \tilde{\delta}^\rho_\mu = \tilde{\delta}^\rho_\nu, \tag{11.70}$$
$$\hat{\delta}^\mu_\nu \tilde{\delta}^\rho_\mu = 0, \tag{11.71}$$

with corresponding metric covariant and contravariant tensors

$$\hat{g}_{\mu\nu} = \hat{\delta}^\rho_\mu g_{\rho\nu}, \qquad \hat{g}^{\mu\nu} = \hat{\delta}^\mu_\rho g^{\rho\nu}, \qquad \tilde{g}_{\mu\nu} = \tilde{\delta}^\rho_\mu g_{\rho\nu}, \qquad \tilde{g}^{\mu\nu} = \tilde{\delta}^\mu_\rho g^{\rho\nu}, \tag{11.72}$$

which satisfy the identities

$$g_{\mu\nu} = \hat{g}_{\mu\nu} + \tilde{g}_{\mu\nu}, \qquad g^{\mu\nu} = \hat{g}^{\mu\nu} + \tilde{g}^{\mu\nu}, \tag{11.73}$$
$$g^{\mu\nu} g_{\mu\nu} = 4, \tag{11.74}$$
$$\hat{g}^{\mu\nu} \hat{g}_{\mu\nu} = d, \qquad \tilde{g}^{\mu\nu} \tilde{g}_{\mu\nu} = 4 - d, \tag{11.75}$$
$$g^{\mu\nu} \hat{g}_{\mu\rho} = \hat{\delta}^\nu_\rho, \qquad g^{\mu\nu} \tilde{g}_{\mu\rho} = \tilde{\delta}^\nu_\rho, \qquad \hat{g}^{\mu\nu} \tilde{g}_{\mu\rho} = 0, \tag{11.76}$$

etc. It is often convenient to write vectors and metric tensors in components, transferring the hats and tildes to the corresponding indices:

$$g_{\mu\nu} = \begin{pmatrix} g_{\hat{\mu}\hat{\nu}} & 0 \\ 0 & g_{\tilde{\mu}\tilde{\nu}} \end{pmatrix}, \qquad g^{\mu\nu} = \begin{pmatrix} g^{\hat{\mu}\hat{\nu}} & 0 \\ 0 & g^{\tilde{\mu}\tilde{\nu}} \end{pmatrix}, \tag{11.77}$$

with each gauge vector boson in DRED split into components as

$$A_\mu = (A_{\hat{\mu}},\ E_{\tilde{\mu}}), \tag{11.78}$$

but with derivatives (and momenta) restricted to the d-dimensional subspace:

$$\partial_\mu = (\partial_{\hat{\mu}},\ 0), \qquad p^\mu = (p^{\hat{\mu}},\ 0). \tag{11.79}$$

The extra 2ϵ degrees of freedom $E_{\tilde{\mu}}$ are called "epsilon-scalars," because their multiplicity is formally 2ϵ and they transform as Lorentz scalars in the d-dimensional subspace. There are similar decompositions for the other objects that carry vector indices, including

$$\overline{\sigma}^\mu = (\overline{\sigma}^{\hat{\mu}},\ \overline{\sigma}^{\tilde{\mu}}), \qquad \sigma^\mu = (\sigma^{\hat{\mu}},\ \sigma^{\tilde{\mu}}), \tag{11.80}$$

or, in four-component spinor language,

$$\gamma^\mu = (\gamma^{\hat{\mu}},\ \gamma^{\tilde{\mu}}), \tag{11.81}$$

However, there are several issues that present special problems. These include the totally antisymmetric $\epsilon^{\mu\nu\rho\sigma}$ tensor (related to the treatment of γ_5 in four-component spinor notation), and Fierz identities that involve vector indices. These are in turn related to a more general problem: the d-dimensional, 2ϵ-dimensional, and (perhaps surprisingly) even the 4-dimensional subspace indices cannot be assigned specific values or counted in the naive way, as discussed following eq. (11.92). This is responsible for problematic aspects of DRED at high loop orders.

After regularization by DRED, the parameters are then renormalized by modified minimal subtraction, just as in eqs. (11.42)–(11.63). This is called the $\overline{\text{DR}}$ renormalization scheme. At least in low orders in perturbation theory, DRED and $\overline{\text{DR}}$ respect supersymmetry and maintain supersymmetric relations between parameters. In particular, the boundary conditions at the input scale should be applied in a supersymmetry-respecting scheme, although $\overline{\text{DR}}$ is not the only possibility. One-loop beta functions are always the same in the $\overline{\text{MS}}$ and $\overline{\text{DR}}$ schemes, but it is important to realize that the $\overline{\text{MS}}$ scheme does violate supersymmetry, so that $\overline{\text{DR}}$ is preferred.

Let us now consider how the DRED scheme Lagrangian works, in terms of the epsilon-scalars. First, the pure gauge part is $\mathscr{L}_{\text{gauge}} = -\frac{1}{4}F^a_{\mu\nu}F^{\mu\nu a}$, which can be rewritten as

$$\mathscr{L}_{\text{gauge}} = -\frac{1}{4}F^a_{\hat{\mu}\hat{\nu}}F^{\hat{\mu}\hat{\nu}a} + \frac{1}{2}D^{\hat{\mu}}E^{a\tilde{\nu}}D_{\hat{\mu}}E^a_{\tilde{\nu}} - \frac{1}{4}g^2 f^{abc}f^{ade}E^{b\tilde{\mu}}E^{c\tilde{\nu}}E^d_{\tilde{\mu}}E^e_{\tilde{\nu}}. \tag{11.82}$$

The d-dimensional field strength is $F^a_{\hat{\mu}\hat{\nu}} = \partial_{\hat{\mu}}A^a_{\hat{\nu}} - \partial_{\hat{\nu}}A^a_{\hat{\mu}} - gf^{abc}A^b_{\hat{\mu}}A^c_{\hat{\nu}}$, and

$$D_{\hat{\mu}}E^a_{\tilde{\nu}} = \partial_{\hat{\mu}}E^a_{\tilde{\nu}} - gf^{abc}A^b_{\hat{\mu}}E^c_{\tilde{\nu}} \tag{11.83}$$

is the covariant derivative of the epsilon-scalar. This shows that the epsilon-scalar transforms as an adjoint under the gauge group and as a scalar with a standard propagator under d-dimensional Lorentz transformations. Indeed, the gauge transformation parameterized by Λ^a for the 4-dimensional vector field A^a_μ decomposes into d-dimensional vector and epsilon-scalar components as

$$\delta_\Lambda A^a_{\hat{\mu}} = \partial_{\hat{\mu}}\Lambda^a - gf^{abc}A^b_{\hat{\mu}}\Lambda^c, \tag{11.84}$$

$$\delta_\Lambda E^a_{\tilde{\mu}} = -gf^{abc}E^b_{\tilde{\mu}}\Lambda^c. \tag{11.85}$$

Besides the kinetic term dictated by gauge invariance, eq. (11.82) also contains a quartic interaction for the epsilon-scalars.

Now consider the Lagrangian densities for scalar and fermion field kinetic terms, $\mathscr{L}_{\text{scalar kinetic}} = D^\mu\phi^{\dagger i}D_\mu\phi_i$ and $\mathscr{L}_{\text{fermion kinetic}} = i\psi^{\dagger I}\overline{\sigma}^\mu D_\mu\psi_I$, where for the moment we do not assume supersymmetry. These can be rewritten as

$$\mathscr{L}_{\text{scalar kinetic}} = D^{\hat{\mu}}\phi^{\dagger i}D_{\hat{\mu}}\phi_i - g^2E^{a\tilde{\mu}}E^b_{\tilde{\mu}}\phi^{\dagger j}\boldsymbol{t}^{ai}_j\boldsymbol{t}^{bk}_i\phi_k, \tag{11.86}$$

$$\mathscr{L}_{\text{fermion kinetic}} = i\psi^{\dagger I}\overline{\sigma}^{\hat{\mu}}D_{\hat{\mu}}\psi_I - gE^a_{\tilde{\mu}}\psi^{\dagger I}\overline{\sigma}^{\tilde{\mu}}\boldsymbol{T}^{aJ}_I\psi_J. \tag{11.87}$$

Here $\boldsymbol{t}^{aJ}_i$ and $\boldsymbol{T}^{aJ}_I$ are the generators for the gauge group representations carried by the scalars and fermions, respectively, with corresponding covariant derivatives $D_{\hat{\mu}}\phi_i = \partial_{\hat{\mu}}\phi_i + igA^a_{\hat{\mu}}\boldsymbol{t}^{aj}_i\phi_j$ and $D_{\hat{\mu}}\psi_I = \partial_{\hat{\mu}}\psi_I + igA^a_{\hat{\mu}}\boldsymbol{T}^{aJ}_I\psi_J$.

Thus, the use of DRED rather than DREG means that we have additional

Lagrangian interactions involving the epsilon-scalars, which can be summarized schematically as follows. Let ϵ, V, S, and F stand for epsilon-scalars, d-dimensional vectors, ordinary scalars, and fermions (including both chiral fermions and gauginos in the supersymmetric case). From eqs. (11.82)–(11.87), the additional interactions required in DRED are then of the forms

$$\epsilon\epsilon V, \qquad \epsilon\epsilon VV, \qquad \epsilon\epsilon\epsilon\epsilon, \qquad \epsilon\epsilon SS, \qquad \epsilon FF. \tag{11.88}$$

The first two of these have couplings whose structure and strength are enforced by the gauge invariance, because they come from the gauge-covariant derivative acting on the epsilon-scalar. The remaining three types of interactions are not protected by the gauge invariance, and therefore in a general field theory can be renormalized in a way that breaks the tree-level relations in terms of the gauge coupling g above. Although the list of eq. (11.88) does not contain an ϵSS coupling, this also can in principle be generated by radiative corrections. Also, there is more than one group index structure available for an $\epsilon\epsilon\epsilon\epsilon$ coupling in nonabelian gauge theories, as the product of four adjoint representations contains more than one singlet. The epsilon-scalars can also obtain squared-mass terms from radiative corrections, making them nondegenerate with their d-dimensional vector counterparts.

Couplings involving the epsilon-scalars that do not correspond exactly to the tree-level ones given in eqs. (11.82)–(11.87), but are generated by radiative corrections, are called "evanescent." The evanescent parameters are unphysical, but in the general non-supersymmetric case they are unavoidable in a consistent renormalization group analysis. In contrast, in supersymmetric theories, the renormalized epsilon-scalar interactions do remain related to their vector counterparts exactly as given in eqs. (11.82)–(11.87), because these relations are protected by supersymmetry. If supersymmetry is only softly broken, then the dimensionless epsilon-scalar couplings are still protected in this way, and only the epsilon-scalar squared masses suffer radiative corrections that split them from those of their vector counterparts.

For Feynman diagrams that do not involve vector bosons, one need not distinguish between DREG and DRED. (This is always true in any theory without gauge interactions, such as the Wess–Zumino model.) Also, note that the usual gauge fixing and ghost interactions involve vector indices only contracted with derivatives, and therefore contain only the d-dimensional vectors and not the epsilon-scalars.

In many practical situations, especially in calculations at low loop orders in supersymmetric theories, the above split of 4-dimensional vectors into d-dimensional vectors and epsilon-scalars can be evaded. Consider, for example, a helicity-preserving self-energy correction to a fermion line, as shown in Fig. 11.9. In two-component spinor notation, this will include terms with the following spinor structures:

$$\overline{\sigma}^{\hat{\mu}} \left(\frac{p \cdot \sigma}{p^2 - m^2_{A^a}} \right) \overline{\sigma}_{\hat{\mu}} = (2-d) \left(\frac{p \cdot \sigma}{p^2 - m^2_{A^a}} \right), \tag{11.89}$$

$$\overline{\sigma}^{\tilde{\mu}} \left(\frac{p \cdot \sigma}{p^2 - m^2_{E^a}} \right) \overline{\sigma}_{\tilde{\mu}} = (d-4) \left(\frac{p \cdot \sigma}{p^2 - m^2_{E^a}} \right), \tag{11.90}$$

from d-dimensional vector and epsilon-scalar exchange, respectively. The right-hand

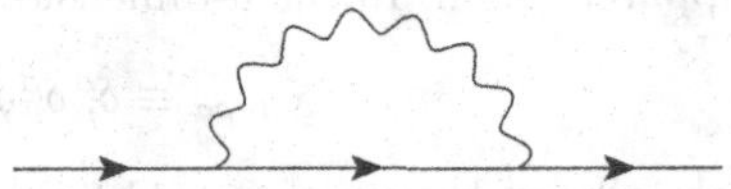

Fig. 11.9 A helicity-preserving self-energy correction for a fermion due to vector exchange.

sides of these equations can be derived using the identity of eq. (B.30), by projecting onto the relevant subspaces; the totally antisymmetric tensor does not contribute by symmetry, so problems in defining it consistently outside of exactly 4 dimensions are not relevant. Now, provided that one can treat the squared masses in the propagators as equal, these can be simply combined into

$$\overline{\sigma}^{\mu}\left(\frac{p\cdot\sigma}{p^2-m^2_{A^a}}\right)\overline{\sigma}_{\mu}=-2\left(\frac{p\cdot\overline{\sigma}}{p^2-m^2_{A^a}}\right), \tag{11.91}$$

which we see could have been obtained directly using the 4-dimensional spinor identity of eq. (B.24), avoiding the issues of epsilon-scalars and spinor identities in d dimensions. This is indeed the case if supersymmetry is exact, because the d-dimensional vector is exactly degenerate with the epsilon-scalar in that case. If supersymmetry is softly broken, then the epsilon-scalars have a squared mass that differs from the corresponding vectors, but that effect can be ascribed to radiative corrections, and so can be treated as of higher order in perturbation theory, with $m^2_{E^a}=m^2_{A^a}+\mathcal{O}(g^2/16\pi^2)$. The difference therefore can be systematically combined with, and absorbed into, the contributions of the next loop order.

In Sections 19.3 and 19.4, we provide some more detailed examples of using the $\overline{\text{DR}}$ scheme (treated in parallel with the $\overline{\text{MS}}$ scheme) to compute the one-loop self-energies and pole masses of the top quark and gluino, respectively. For more examples of the DRED scheme in action, see Ref. [111].

However, it should be recognized that the DRED scheme contains technical inconsistencies (which can be avoided at low orders in perturbation theory, but eventually cause problems), as first pointed out by its inventor Siegel in Ref. [112] and elucidated further in Refs. [113, 114]. For example, consider the totally antisymmetric four-index tensor $\epsilon_{\mu\nu\rho\sigma}$. In ordinary 4-dimensional space, by exploiting the fact that the vector indices can take on only the four values $0, 1, 2, 3$, this satisfies

$$\epsilon_{\mu_1\mu_2\mu_3\mu_4}\epsilon^{\nu_1\nu_2\nu_3\nu_4}\stackrel{?}{=}-\begin{vmatrix}\delta^{\nu_1}_{\mu_1} & \delta^{\nu_2}_{\mu_1} & \delta^{\nu_3}_{\mu_1} & \delta^{\nu_4}_{\mu_1}\\ \delta^{\nu_1}_{\mu_2} & \delta^{\nu_2}_{\mu_2} & \delta^{\nu_3}_{\mu_2} & \delta^{\nu_4}_{\mu_2}\\ \delta^{\nu_1}_{\mu_3} & \delta^{\nu_2}_{\mu_3} & \delta^{\nu_3}_{\mu_3} & \delta^{\nu_4}_{\mu_3}\\ \delta^{\nu_1}_{\mu_4} & \delta^{\nu_2}_{\mu_4} & \delta^{\nu_3}_{\mu_4} & \delta^{\nu_4}_{\mu_4}\end{vmatrix}. \tag{11.92}$$

The reason for the question mark here (and in other equations in this section) will become apparent soon. In the d-dimensional subspace, one can define

$$\hat{\epsilon}_{\mu\nu\rho\sigma} \equiv \hat{\delta}^{\alpha}_{\mu}\hat{\delta}^{\beta}_{\nu}\hat{\delta}^{\kappa}_{\rho}\hat{\delta}^{\lambda}_{\sigma}\,\epsilon_{\alpha\beta\kappa\lambda}\,, \tag{11.93}$$

from which follows, by use of eq. (11.68), (11.69), and eq. (11.92),

$$\hat{\epsilon}_{\mu_1\mu_2\mu_3\mu_4}\hat{\epsilon}^{\nu_1\nu_2\nu_3\nu_4} \stackrel{?}{=} -\begin{vmatrix} \hat{\delta}^{\nu_1}_{\mu_1} & \hat{\delta}^{\nu_2}_{\mu_1} & \hat{\delta}^{\nu_3}_{\mu_1} & \hat{\delta}^{\nu_4}_{\mu_1} \\ \hat{\delta}^{\nu_1}_{\mu_2} & \hat{\delta}^{\nu_2}_{\mu_2} & \hat{\delta}^{\nu_3}_{\mu_2} & \hat{\delta}^{\nu_4}_{\mu_2} \\ \hat{\delta}^{\nu_1}_{\mu_3} & \hat{\delta}^{\nu_2}_{\mu_3} & \hat{\delta}^{\nu_3}_{\mu_3} & \hat{\delta}^{\nu_4}_{\mu_3} \\ \hat{\delta}^{\nu_1}_{\mu_4} & \hat{\delta}^{\nu_2}_{\mu_4} & \hat{\delta}^{\nu_3}_{\mu_4} & \hat{\delta}^{\nu_4}_{\mu_4} \end{vmatrix}. \tag{11.94}$$

This leads to inconsistencies, because the evaluation of expressions can depend on the order in which this identity and the contraction identities (11.66), (11.68), and (11.69) are used. For example, on the one hand, we have directly from eq. (11.94) that $\hat{\epsilon}_{\alpha\beta\kappa\lambda}\hat{\epsilon}^{\alpha\beta\kappa\lambda} = -d(d-1)(d-2)(d-3)$, so that

$$\hat{\epsilon}^{\mu\nu\rho\sigma}\left(\hat{\epsilon}_{\alpha\beta\kappa\lambda}\hat{\epsilon}^{\alpha\beta\kappa\lambda}\right) = -d(d-1)(d-2)(d-3)\hat{\epsilon}^{\mu\nu\rho\sigma}. \tag{11.95}$$

On the other hand, evaluating the same expression again using eq. (11.94) but in a different order as indicated by the parentheses, one finds instead

$$\left(\hat{\epsilon}^{\mu\nu\rho\sigma}\hat{\epsilon}_{\alpha\beta\kappa\lambda}\right)\hat{\epsilon}^{\alpha\beta\kappa\lambda} = -24\hat{\epsilon}^{\mu\nu\rho\sigma}. \tag{11.96}$$

Equations (11.95) and (11.96) are consistent only if $d = 4$ exactly, precluding the use of eqs. (11.92) and (11.94) for DRED.

Another such identity that can arise in manipulations of σ^μ and $\overline{\sigma}^\mu$ matrices is

$$\epsilon^{\mu\nu\rho\sigma}g^{\kappa\lambda} - \epsilon^{\mu\nu\rho\kappa}g^{\sigma\lambda} - \epsilon^{\mu\nu\kappa\sigma}g^{\rho\lambda} - \epsilon^{\mu\kappa\rho\sigma}g^{\nu\lambda} - \epsilon^{\kappa\nu\rho\sigma}g^{\mu\lambda} \stackrel{?}{=} 0, \tag{11.97}$$

obtained in ordinary 4-dimensional space by antisymmetrizing over the five indices $\mu, \nu, \rho, \sigma, \kappa$ and exploiting the fact that there are only four values $0, 1, 2, 3$ available for them. However, now contracting each index with a $\hat{\delta}$-tensor, this implies

$$\hat{\epsilon}^{\mu\nu\rho\sigma}\hat{g}^{\kappa\lambda} - \hat{\epsilon}^{\mu\nu\rho\kappa}\hat{g}^{\sigma\lambda} - \hat{\epsilon}^{\mu\nu\kappa\sigma}\hat{g}^{\rho\lambda} - \hat{\epsilon}^{\mu\kappa\rho\sigma}\hat{g}^{\nu\lambda} - \hat{\epsilon}^{\kappa\nu\rho\sigma}\hat{g}^{\mu\lambda} \stackrel{?}{=} 0, \tag{11.98}$$

and now contracting with, e.g., $\hat{g}_{\kappa\lambda}$ implies

$$(d-4)\hat{\epsilon}^{\mu\nu\rho\sigma} = 0, \tag{11.99}$$

which again can hold only if $d = 4$ exactly.

In four-component spinor language, these difficulties with the antisymmetric four-index tensor can be related to the problem of defining γ_5 in DRED. If one starts with the ordinary 4-dimensional anticommutation relations

$$\{\gamma_5, \gamma^\mu\} = 0, \qquad \{\gamma^\mu, \gamma^\nu\} = 2g^{\mu\nu}, \tag{11.100}$$

then defining $\hat{\gamma}^\mu = \hat{\delta}^\mu_\nu \gamma^\nu$ results in

$$\{\gamma_5, \hat{\gamma}^\mu\} = 0, \qquad \{\hat{\gamma}^\mu, \hat{\gamma}^\nu\} = 2\hat{g}^{\mu\nu}, \tag{11.101}$$

from which it follows that

$$\hat{\gamma}^\alpha \hat{\gamma}_\alpha = d. \tag{11.102}$$

Now, using eqs. (11.101) and (11.102) with $\mathrm{Tr}[\gamma_5 \hat{\gamma}^\alpha \hat{\gamma}_\alpha]$ and $\mathrm{Tr}[\gamma_5 \hat{\gamma}^\mu \hat{\gamma}^\nu \hat{\gamma}^\alpha \hat{\gamma}_\alpha]$ and $\mathrm{Tr}[\gamma_5 \hat{\gamma}^\mu \hat{\gamma}^\nu \hat{\gamma}^\rho \hat{\gamma}^\sigma \hat{\gamma}^\alpha \hat{\gamma}_\alpha]$, and using the cyclic property of the trace, one finds

$$d\, \mathrm{Tr}[\gamma_5] = 0, \tag{11.103}$$

$$(d-2)\, \mathrm{Tr}[\gamma_5 \hat{\gamma}^\mu \hat{\gamma}^\nu] = 0, \tag{11.104}$$

$$(d-4)\, \mathrm{Tr}[\gamma_5 \hat{\gamma}^\mu \hat{\gamma}^\nu \hat{\gamma}^\rho \hat{\gamma}^\sigma] = 0, \tag{11.105}$$

forcing each of the traces to vanish. The last of these is in conflict with the identity

$$\mathrm{Tr}[\gamma_5 \hat{\gamma}^\mu \hat{\gamma}^\nu \hat{\gamma}^\rho \hat{\gamma}^\sigma] \stackrel{?}{=} -4i\hat{\epsilon}^{\mu\nu\rho\sigma}, \tag{11.106}$$

which would follow from the corresponding identity from ordinary 4-dimensional spacetime. Equation (11.105), which is required in this version of DRED with a fully anticommuting γ_5 as in eq. (11.101), is in conflict with the usual treatment of the anomaly. This is in contrast to the situation in DREG with $d > 4$, where the 't Hooft–Veltman prescription [115] for traces involving an odd number of γ_5 matrices implies the commutation relation $[\gamma_5, \gamma^{\tilde{\mu}}] = 0$ for the extra $d-4$ gamma matrices carrying index $\tilde{\mu}$, along with the anticommutation relation $\{\gamma_5, \gamma^\mu\} = 0$ for the first four indices $\mu \in \{0, 1, 2, 3\}$. This possibility is seemingly not available in DRED, where $d < 4$ gave us eq. (11.101). However, in Ref. [116] it was proposed that one can use $[\gamma_5, \gamma^{\tilde{\mu}}] = 0$ in DRED anyway, as a formal analytic continuation from $d > 4$. This was found to provide a consistent treatment of the axial anomaly at least at two-loop order.

As pointed out by Avdeev and Vladimirov in Ref. [113], this sort of DRED inconsistency is not restricted to the use of the totally antisymmetric four-index tensor (or the γ_5 matrix in four-component spinor language). Consider the following identity involving an $n \times n$ matrix determinant (note that n is definitely an integer) in d-dimensional space:

$$\hat{\delta}^{\nu_1}_{\mu_1} \hat{\delta}^{\nu_2}_{\mu_2} \cdots \hat{\delta}^{\nu_n}_{\mu_n} \begin{vmatrix} \hat{\delta}^{\mu_1}_{\nu_1} & \cdots & \hat{\delta}^{\mu_n}_{\nu_1} \\ \vdots & & \vdots \\ \hat{\delta}^{\mu_1}_{\nu_n} & \cdots & \hat{\delta}^{\mu_n}_{\nu_n} \end{vmatrix} = d(d-1)\ldots(d-n+1). \tag{11.107}$$

Now, within the determinant on the left-hand side, each $\hat{\delta}$ can be replaced with the corresponding δ, due to eqs. (11.68) and (11.69), since they are all contracted with a $\hat{\delta}$ from outside the determinant. Therefore,

$$\hat{\delta}^{\nu_1}_{\mu_1} \hat{\delta}^{\nu_2}_{\mu_2} \cdots \hat{\delta}^{\nu_n}_{\mu_n} \begin{vmatrix} \delta^{\mu_1}_{\nu_1} & \cdots & \delta^{\mu_n}_{\nu_1} \\ \vdots & & \vdots \\ \delta^{\mu_1}_{\nu_n} & \cdots & \delta^{\mu_n}_{\nu_n} \end{vmatrix} = d(d-1)\ldots(d-n+1). \tag{11.108}$$

However, in ordinary 4-dimensional space, we also have simply

$$\begin{vmatrix} \delta^{\mu_1}_{\nu_1} & \cdots & \delta^{\mu_n}_{\nu_1} \\ \vdots & & \vdots \\ \delta^{\mu_1}_{\nu_n} & \cdots & \delta^{\mu_n}_{\nu_n} \end{vmatrix} \stackrel{?}{=} 0 \qquad (\text{for } n \geq 5), \tag{11.109}$$

because the determinant is antisymmetric in all of its n raised indices. Comparing eqs. (11.108) and (11.109) yields an inconsistency for $n \geq 5$ unless $d = 4$ exactly.

The common reason for these problems is that it is actually impossible to consistently decompose an ordinary 4-dimensional space into a d-dimensional subspace and its complement, unless d is an integer. Instead, the 4-dimensional space in DRED must be interpreted as a "quasi-4-dimensional" space, sometimes known as Q4S. In Q4S, the number of dimensions is only formally equal to 4; it is really infinite-dimensional. [One way of understanding this is that the determinant in eq. (11.108) cannot vanish for any finite integer n, due to the nonvanishing of the right-hand side.] Although eqs. (11.65) and (11.74) are valid, in general vector indices cannot be assigned specific values. This means that other identities, which would seem to follow by appealing to their supposedly only being able to take on four specific values, are *not* valid. In particular, for a consistent DRED scheme, eqs. (11.92), (11.97), (11.106), and (11.109) [and identities relying on them, such as eqs. (11.94) and (11.98)] must not be used. Even though they would be valid in ordinary 4-dimensional space, they are not valid in Q4S; hence the $\stackrel{?}{=}$ notation. Similarly, spinor Fierz identities involving summed vector indices, such as

$$\overline{\sigma}^{\mu\dot{\alpha}\alpha}\overline{\sigma}^{\dot{\beta}\beta}_{\mu} \stackrel{?}{=} 2\epsilon^{\alpha\beta}\epsilon^{\dot{\alpha}\dot{\beta}}, \tag{11.110}$$

(or Fierz expressions involving $\gamma^\mu \otimes \gamma_\mu$ in four-component spinor language) are technically invalid in the decomposition of Q4S into d-dimensional and 2ϵ-dimensional subspaces, and can lead to inconsistencies if used in DRED.

The use of such Fierz identities is crucial in establishing the invariance of supersymmetric Yang–Mills gauge theories, which therefore becomes problematic in a fully consistent version of DRED. The supersymmetry variation of the Lagrangian eq. (8.57) contains a term

$$\widehat{\delta}_\epsilon \mathscr{L}_{\text{gauge}} = igf^{abc}(\epsilon^\dagger \overline{\sigma}^\mu \lambda^b)(\lambda^{\dagger a}\overline{\sigma}_\mu \lambda^c) + \text{h.c.} \tag{11.111}$$

In ordinary 4-dimensional space this becomes, by the use of the Fierz identity eq. (11.110),

$$\widehat{\delta}_\epsilon \mathscr{L}_{\text{gauge}} \stackrel{?}{=} 2igf^{abc}(\epsilon^\dagger \lambda^{\dagger a})(\lambda^b \lambda^c) + \text{h.c.}, \tag{11.112}$$

which, if true, would then vanish (as desired) due to the antisymmetry of the Lie algebra structure constants f^{abc}. Unfortunately, in the Q4S relevant for DRED, this Fierz identity cannot be applied. As concrete evidence of this, Avdeev and Vladimirov showed in Ref. [113] that amplitudes containing insertions of the term eq. (11.111) will, in principle, lead to violations of the supersymmetric Ward identities. These violations arise at lowest order in diagrams involving at least 10 γ^μ (or σ^μ, $\overline{\sigma}^\mu$) matrices, with a result proportional to the left-hand side of eq. (11.109) for $n \geq 5$. As we have seen, this expression cannot consistently be taken to vanish. If one attempts to define a consistent DRED by forbidding the use of eq. (11.109), then supersymmetric relations will be violated in such amplitudes.

Despite these problems of principle, in all known practical applications the use of DRED has been sufficient to maintain manifest supersymmetry without leading to any disasters. In many examples at low orders in perturbation theory, there are simply not enough vector indices present to trigger the inconsistency embodied in eq. (11.109), which corresponds to insertions in Feynman diagrams of the problematic Fierz identity. Furthermore, if any two of the antisymmetrized vector indices in eq. (11.109) are contracted with a metric tensor, or with two identical momenta, then the inconsistency is immediately rendered harmless. This is also quite often the case for amplitudes calculated at low orders in perturbation theory; see for example the discussion leading to eq. (11.91). Regarding the extension to arbitrary orders in perturbation theory, one point of view is that the inconsistencies can be viewed instead as ambiguities (a somewhat nicer word, at least), which can be resolved in particular examples by requiring gauge and supersymmetric invariance to be satisfied, and are therefore tolerable. For example, a calculation of a trace over spinor indices can depend on the order of contractions and other operations used to compute it, but this may correspond to the freedom to choose local finite counterterms (or alternatively, to make parameter redefinitions) in such a way as to maintain supersymmetric Ward identities. As far as we know, this plausible hope remains unproven in general, constituting a minor scandal. In other cases, the amplitude under consideration is ultraviolet finite, so that one can presumably use purely 4-dimensional vector identities with impunity. For example, we followed this course in Section 5.1.1 in computing the triangle anomaly, by doing all spacetime algebra and integrals strictly in 4 dimensions.

It is also possible to work within the $\overline{\rm MS}$ scheme, provided one is willing to forfeit manifest supersymmetry and the relations between couplings that it imposes. Indeed, it is possible in principle to completely bypass the complications involved in working with epsilon-scalars, by instead computing in $\overline{\rm MS}$ and then applying redefinitions that express all $\overline{\rm MS}$ couplings and masses in terms of their $\overline{\rm DR}$ counterparts. This procedure can even be taken as the definition of the $\overline{\rm DR}$ renormalization scheme.

For example, at one-loop order in a softly broken supersymmetric theory, the necessary redefinitions are as follows. For the gauge couplings, by demanding that scattering amplitudes computed in the two schemes are the same, one obtains

$$g_a^{\overline{\rm MS}} = g_a^{\overline{\rm DR}}\Big(1 - \frac{g_a^2}{96\pi^2}G_a\Big), \tag{11.113}$$

where G_a is the quadratic Casimir invariant of the adjoint representation for the subgroup labeled by a, normalized so that $G_a = N$ for SU(N), while $G_a = 0$ for a

U(1) gauge group. There is an important distinction between these gauge couplings and the couplings $\hat{g}_a$ which occur in the gaugino–fermion–scalar interaction

$$\mathscr{L} = -\sqrt{2}\hat{g}_a(\phi^\dagger \boldsymbol{T}^a \psi)\lambda^a. \tag{11.114}$$

Gauge invariance requires that g_a is the same everywhere it appears, but only supersymmetry guarantees that $\hat{g}_a = g_a$. Therefore, we have $\hat{g}_a = g_a$ in $\overline{\text{DR}}$ (or any other supersymmetry-respecting scheme), but we should expect that $\hat{g}_a^{\overline{\text{MS}}} \neq g_a^{\overline{\text{MS}}}$. Indeed, by computing scattering amplitudes in both schemes, one finds that, for chiral scalars and fermions in a representation r,

$$\hat{g}_a^{\overline{\text{MS}}} = g_a^{\overline{\text{DR}}}\Big(1 + \frac{g_a^2}{32\pi^2}\left[G_a - C_a(r)\right]\Big), \tag{11.115}$$

where $C_a(r)$ is the quadratic Casimir invariant, for the subgroup labeled by a, of the chiral supermultiplet representation in question (see Appendix H.3). Thus, there is a different coupling of the gaugino to each distinct irreducible representation of chiral supermultiplets in the $\overline{\text{MS}}$ scheme. Note that when working at one-loop order, we consistently neglect the distinction between $g_a^{\overline{\text{MS}}}$, $\hat{g}_a^{\overline{\text{MS}}}$, and $g_a^{\overline{\text{DR}}}$ in the one-loop correction parts, because the couplings are all the same to zeroth order.

Next, consider the Yukawa coupling y^{ijk} between a complex scalar ϕ_i and chiral fermions ψ_j and ψ_k. Again requiring that physical scattering amplitudes computed in the two schemes are the same, one finds

$$y^{ijk}_{\overline{\text{MS}}} = y^{ijk}_{\overline{\text{DR}}}\Big(1 + \frac{g_a^2}{32\pi^2}\left[C_a(j) + C_a(k) - 2C_a(i)\right]\Big). \tag{11.116}$$

The $y^{ijk}_{\overline{\text{DR}}}$ are totally symmetric in the supersymmetry-respecting $\overline{\text{DR}}$ scheme, because of the way they appear in the superpotential. Equation 11.116) then shows that the $y^{ijk}_{\overline{\text{MS}}}$ are not totally symmetric, as radiative corrections in the $\overline{\text{MS}}$ scheme do not respect supersymmetry, since the scalars and fermions are treated differently.

For supersymmetric chiral fermion masses, the result can be obtained simply by treating the index i in eq. (11.116) as belonging to a nondynamical spurion scalar field in the singlet representation, so that

$$M^{jk}_{\overline{\text{MS}}} = M^{jk}_{\overline{\text{DR}}}\Big(1 + \frac{g_a^2}{32\pi^2}\left[C_a(j) + C_a(k)\right]\Big). \tag{11.117}$$

The scalar quartic interactions similarly differ between the two schemes. For a supersymmetric theory in $\overline{\text{DR}}$, one has

$$\mathscr{L} = -\tfrac{1}{4}\lambda^{k\ell}_{ij}\phi^{\dagger i}\phi^{\dagger j}\phi_k\phi_\ell, \tag{11.118}$$

where, from eq. (8.75),

$$[\lambda_{\overline{\text{DR}}}]^{k\ell}_{ij} = y_{ijn}y^{k\ell n} + g_a^2\left(\boldsymbol{T}^{ak}_i\boldsymbol{T}^{a\ell}_j + \boldsymbol{T}^{ak}_j\boldsymbol{T}^{a\ell}_i\right). \tag{11.119}$$

The one-loop connection to the $\overline{\text{MS}}$ scheme is then found to be

$$[\lambda_{\overline{\text{MS}}}]^{k\ell}_{ij} = [\lambda_{\overline{\text{DR}}}]^{k\ell}_{ij} - \frac{g_a^2 g_b^2}{16\pi^2}\Big[\{\boldsymbol{T}^a, \boldsymbol{T}^b\}_i{}^k\,\{\boldsymbol{T}^a, \boldsymbol{T}^b\}_j{}^\ell + (i \leftrightarrow j)\Big], \tag{11.120}$$

by requiring equality of scattering amplitudes computed in the two schemes.

For each supersymmetry-breaking gaugino mass, the one-loop relation is given by

$$M_a^{\overline{\rm MS}} = M_a^{\overline{\rm DR}}\Big(1 + \frac{g_a^2}{16\pi^2} G_a\Big). \tag{11.121}$$

This can be obtained by requiring that the gaugino pole mass is the same when computed in the two schemes, as we will see in Section 19.4 for the case of the gluino.

The only other coupling of a softly broken supersymmetric theory that differs between the two schemes is the nonholomorphic scalar squared mass, which depends on the epsilon-scalar squared masses $m^2_{\epsilon,a}$, with

$$(m^2_{\overline{\rm MS}})_i^j = (m^2_{\overline{\rm DR}})_i^j - \frac{1}{16\pi^2}\Big[2g_a^2 C_a(i)\delta_i^j m^2_{\epsilon,a}\Big] \tag{11.122}$$

at one-loop order. Now, the epsilon-scalar squared mass is an arbitrary parameter coming from the renormalization procedure, with no physically observable counterpart. Therefore, it should not be surprising that one can define a more useful modified regularization scheme in which it does not appear; this is known as the $\overline{\rm DR}'$ scheme [117], where

$$(m^2_{\overline{\rm DR}'})_i^j = (m^2_{\overline{\rm MS}})_i^j, \tag{11.123}$$

but all other $\overline{\rm DR}'$ parameters are the same as their $\overline{\rm DR}$ counterparts. In other words, one can arrive at the $\overline{\rm DR}'$ scheme by starting with the $\overline{\rm MS}$ scheme and making all of the parameter redefinitions to go to the $\overline{\rm DR}$ scheme, except for nonholomorphic supersymmetry-breaking scalar squared masses. The great advantage of the $\overline{\rm DR}'$ scheme (compared to the $\overline{\rm DR}$ scheme) is that the unphysical epsilon-scalar squared masses decouple; they do not appear in the RGEs, nor in the equations relating running parameters to physical observables, such as the relations between running masses to pole masses. Note that, for the supersymmetry-preserving parameters and the holomorphic supersymmetry-breaking parameters, there is no distinction between the $\overline{\rm DR}$ and $\overline{\rm DR}'$ schemes. For more on how this works in the context of the effective potential, see Section 11.6 below.

11.5 Renormalization Group Equations

In theories with widely separated energy scales, renormalization group evolution of parameters can be used to resum potentially large logarithms of different scales. In softly broken supersymmetric theories, one therefore needs the beta functions of gauge couplings, superpotential couplings and masses, and soft supersymmetry-breaking parameters.

The beta functions for gauge couplings and for superpotential parameters are now known up to complete four-loop and three-loop orders, respectively, thanks to a series of heroic calculations culminating in Refs. [118–121] and references therein. For most applications, the one-loop or two-loop beta functions are sufficient. However, in some circumstances when extreme precision is required, one needs higher loop-order contributions, so we now review the complete known results, extended here to include theories with product groups. Consider a general supersymmetric theory with gauge components with couplings g_a, and superpotential

$$W = \tfrac{1}{6}y^{ijk}\Phi_i\Phi_j\Phi_k + \tfrac{1}{2}M^{ij}\Phi_i\Phi_j + T^i\Phi_i. \tag{11.124}$$

We will assume that there is at most one U(1) component, with the other components given by simple groups, as in the case of $SU(3)_C\times SU(2)_L\times U(1)_Y$. This avoids the complications of kinetic mixing between different U(1) components, the results of which can be inferred from the following with some effort, by consideration of the one-particle irreducible Feynman diagrams that contribute to each term.

Below, we make use of the following notation for group theory invariants (see Appendix H.3). For each Lie algebra component with coupling g_a and generators $(\mathbf{T}^a)_i{}^j$, the quadratic Casimir invariant and the dimension of the adjoint representation are denoted by G_a and d_a, respectively. The quadratic Casimir invariant of the representation corresponding to the chiral superfield index i is $C_a(i)$, where

$$(\boldsymbol{T^aT^a})_i{}^j = C_a(i)\delta_i^j. \tag{11.125}$$

The Dynkin index of a representation labeled by r is $T_a(r)$, where

$$\mathrm{Tr}_r(\boldsymbol{T^aT^b}) = T_a(r)\delta^{ab}, \tag{11.126}$$

and S_a is the sum of the $T_a(r)$ over all of the chiral supermultiplet representations, while S_{ab} is the same sum weighted by the quadratic Casimir invariants $C_b(i)$ with respect to the gauge group component b, with similar formulae for S_{abc} and S_{abcd}:

$$S_a = \sum_r T_a(r), \tag{11.127}$$

$$S_{ab} = \sum_r T_a(r)C_b(r), \tag{11.128}$$

$$S_{abc} = \sum_r T_a(r)C_b(r)C_c(r), \tag{11.129}$$

$$S_{abcd} = \sum_r T_a(r)C_b(r)C_c(r)C_d(r). \tag{11.130}$$

For example, in a supersymmetric $SU(N_c)$ gauge theory with N_f chiral superfields in each of the fundamental and antifundamental representations of $SU(N_c)$, one has $G_a = N_c$, $d_a = N_c^2 - 1$, $C_a(i) = C_F$, $S_a = N_f$, $S_{ab} = N_f C_F$, $S_{abc} = N_f C_F^2$, and $S_{abcd} = N_f C_F^3$, where $C_F = (N_c^2 - 1)/(2N_c)$. In a supersymmetric U(1) gauge theory with chiral superfields with charges q_i, one has $G_a = 0$, $d_a = 1$, $C_a(i) = q_i^2$, $S_a = \sum_i q_i^2$, $S_{ab} = \sum_i q_i^4$, $S_{abc} = \sum_i q_i^6$, and $S_{abcd} = \sum_i q_i^8$. For theories with gauge groups that are products of simple or U(1) factors, these group theory quantities will of course be more complicated.

The RGEs for the gauge couplings in the $\overline{\text{DR}}$ scheme can be written in the loop expansion form of eq. (11.58), with loop-counting parameter $\kappa = 1/16\pi^2$, as

$$\beta^{(1)}_{g_a} = g_a^3\left(S_a - 3G_a\right), \tag{11.131}$$

$$\beta^{(2)}_{g_a} = 2g_a^5 G_a\left(S_a - 3G_a\right) + 4g_a^3 g_b^2 S_{ab} - g_a^3 y^{ijk} y_{ijk} C_a(i)/d_a, \tag{11.132}$$

$$\begin{aligned}\beta^{(3)}_{g_a} = {} & g_a^7 G_a\left(S_a - 3G_a\right)\left(7G_a - S_a\right) + 8g_a^5 g_b^2 G_a S_{ab} + 6g_a^3 g_b^4 S_{ab}\left(3G_b - S_b\right)\\ & - 8g_a^3 g_b^2 g_c^2 S_{abc} - 2g_a^5 y^{ijk} y_{ijk} C_a(i) G_a/d_a\\ & + g_a^3 g_b^2 y^{ijk} y_{ijk}\left[C_b(i) - 6C_b(j)\right] C_a(i)/d_a\\ & + g_a^3 y^{ijk} y_{ijl} y_{kmn} y^{lmn}\left[3C_a(i)/2 + C_a(k)/4\right]/d_a,\end{aligned} \tag{11.133}$$

$$\begin{aligned}\beta^{(4)}_{g_a} = {} & g_a^9 G_a(S_a - 3G_a)\left(34G_a^2 - 20S_aG_a/3 - 2S_a^2/3\right)\\ & + g_a^7 g_b^2 S_{ab} G_a\left(48G_a - 32S_a/3\right) - 56 g_a^5 g_b^4 S_{ab} G_a\left(S_b - 3G_b\right)/3\\ & - 16 g_a^5 g_b^2 g_c^2 S_{abc} G_a + g_a^3 g_b^6 S_{ab}\left(84G_b^2 - \left[32 + 48\zeta(3)\right] S_b G_b + 4S_b^2/3\right)\\ & + g_a^3 g_b^4 g_c^2\left(48\left[\zeta(3) - 1\right] S_{ab} S_{bc} + 64 S_{abc}\left[S_b - 3G_b\right]/3\right) + 32 g_a^3 g_b^2 g_c^2 g_d^2 S_{abcd}\\ & + g_a^7 y^{ijk} y_{ijk} C_a(i) G_a\left(8S_a/3 - 12G_a\right)/d_a + 12 g_a^3 g_b^4 y_{ijk} y^{ijk} C_b(i) S_{ab}/d_b\\ & + 6\zeta(3) g_a^3 g_b^4 y^{ijk} y_{ijk} C_a(i)\left[C_b(i) + 2C_b(j)\right] G_b/d_a\\ & + g_a^5 g_b^2 y^{ijk} y_{ijk} C_a(i)\left[-4C_b(i)/3 - 56C_b(j)/3\right] G_a/d_a\\ & + g_a^3 g_b^4 y^{ijk} y_{ijk} C_a(i)\left[C_b(i) + 38C_b(j)/3\right]\left(S_b - 3G_b\right)/d_a\\ & + g_a^3 g_b^2 g_c^2 y^{ijk} y_{ijk} C_a(i)\Big\{[30\zeta(3) - 2] C_b(i) C_c(i) + [56/3 - 24\zeta(3)] C_b(i) C_c(j)\\ & \qquad + [68/3 + 12\zeta(3)] C_b(j) C_c(j) - [4/3 + 36\zeta(3)] C_b(j) C_c(k)\Big\}/d_a\\ & + g_a^5 y^{ijk} y_{ijl} y_{kmn} y^{lmn}\left[14C_a(i)/3 + 4C_a(k)/3\right] G_a/d_a\\ & + g_a^3 g_b^2 y^{ijk} y_{ijl} y_{kmn} y^{lmn}\Big\{-[14/3 + 6\zeta(3)] C_a(i) C_b(i) - 11 C_a(k) C_b(k)/6\\ & \qquad + [2/3 + 6\zeta(3)] C_a(i) C_b(j) + [12\zeta(3) - 5] C_a(i) C_b(k)\\ & \qquad + [38/3 - 12\zeta(3)] C_a(i) C_b(m) + 5 C_a(k) C_b(i)/3\Big\}/d_a\\ & - g_a^3 y^{ijk} y_{ijl} y_{kmn} y^{lmp} y^{nqr} y_{pqr}[38C_a(i) + 5C_a(k) + C_a(m)]/(12d_a)\\ & + g_a^3 y^{ijk} y_{iln} y_{jkm} y^{lnp} y^{mqr} y_{pqr}[C_a(i) - 2C_a(j)]/(12d_a)\\ & - 3\zeta(3) g_a^3 y^{ijk} y_{ilm} y_{jnp} y_{kqr} y^{lnq} y^{mpr} C_a(i)/d_a.\end{aligned} \tag{11.134}$$

Here $y_{ijk} = \left(y^{ijk}\right)^*$, and as usual repeated indices are always summed over, except those that appear on both sides of an equation. Thus, all indices other than a are summed over in eqs. (11.131)–(11.134).

The RGEs for the superpotential parameters can be expressed as

$$\beta_{y^{ijk}} = \gamma^i_n y^{njk} + \gamma^j_n y^{ink} + \gamma^k_n y^{ijn}, \tag{11.135}$$

$$\beta_{M^{ij}} = \gamma^i_n M^{nj} + \gamma^j_n M^{in}, \tag{11.136}$$

$$\beta_{T^i} = \gamma^i_n T^{nj}, \tag{11.137}$$

where the superfield anomalous dimension matrices are

$$\gamma_i^j = \sum_{n=1}^{\infty} \frac{1}{(16\pi^2)^n}\gamma_i^{(n)j}. \tag{11.138}$$

The one-loop, two-loop, and three-loop contributions in the $\overline{\text{DR}}$ scheme are

$$\gamma_i^{(1)j} = \tfrac{1}{2} y_{ikl} y^{jkl} - 2\delta_i^j g_a^2 C_a(i), \tag{11.139}$$

$$\begin{aligned}\gamma_i^{(2)j} &= -\tfrac{1}{2} y_{ikl} y^{jkm} y^{lnp} y_{mnp} + g_a^2 y_{ikl} y^{jkl} [2C_a(k) - C_a(i)] \\ &\quad + 2\delta_i^j g_a^2 C_a(i) \left[g_a^2 S_a + 2 g_b^2 C_b(i) - 3 g_a^2 G_a\right],\end{aligned} \tag{11.140}$$

$$\begin{aligned}\gamma_i^{(3)j} &= -\tfrac{1}{8} y_{ikl} y^{jpq} y^{kmn} y_{pmn} y^{lrs} y_{qrs} - \tfrac{1}{4} y_{ikl} y^{jkm} y^{lnp} y_{snp} y^{sqr} y_{mqr} \\ &\quad + y_{ikl} y^{jkm} y^{lnp} y_{mnq} y^{qrs} y_{prs} + \tfrac{3}{2}\zeta(3)\, y_{ikl} y^{jpq} y^{kmn} y^{lrs} y_{pmr} y_{qns} \\ &\quad + g_a^2 y_{ikl} y^{jkm} y^{lnp} y_{mnp} \Big\{ [2 + 3\zeta(3)] C_a(i) + [1 - 3\zeta(3)] C_a(k) \\ &\qquad\qquad + [4 - 6\zeta(3)][C_a(l) - C_a(p)] \Big\} \\ &\quad + g_a^2 g_b^2 y_{ikl} y^{jkl} \Big\{ [4 - 15\zeta(3)] C_a(i) C_b(i) + [12\zeta(3) - 8] C_a(i) C_b(k) \\ &\qquad\qquad - [12 + 6\zeta(3)] C_a(k) C_b(k) + [18\zeta(3) - 2] C_a(k) C_b(l) \Big\} \\ &\quad + g_a^4 y_{ikl} y^{jkl} \left[-C_a(i) - 4C_a(k)\right] (S_a - 3G_a) \\ &\quad - 3\zeta(3) g_a^4 y_{ikl} y^{jkl} \left[C_a(i) + 2C_a(k)\right] G_a \\ &\quad + \delta_i^j g_a^2 C_a(i) \Big\{ g_a^4 \left[2S_a^2 + (24\zeta(3) - 2) S_a G_a - 12 G_a^2\right] \\ &\qquad\qquad + g_a^2 g_b^2 \left[4(3G_a - S_a) C_b(i) + \big(20 - 24\zeta(3)\big) S_{ab}\right] \\ &\qquad\qquad - 16 g_b^2 g_c^2 C_b(i) C_c(i) - 5 g_a^2 y^{klm} y_{klm} C_a(k)/d_a \Big\}.\end{aligned} \tag{11.141}$$

Note that the form of the superpotential parameter beta functions in eqs. (11.135)–(11.137) is very special; to all orders in perturbation theory, one simply takes the sum of applying the anomalous dimension factor to each of the chiral supermultiplet indices $i, j, \ldots$ carried by the parameter. This is a consequence of the supersymmetric nonrenormalization theorem discussed above, which implies that the ultraviolet-divergent contributions to a given physical observable can always be written in the form of a wave function renormalization, without any vertex renormalization.

The anomalous dimensions given above can be obtained from calculation in the superfield version of theory, in which manifest supersymmetry is maintained by using an appropriate superfield gauge-fixing prescription. It is important to note that there actually *is* vertex renormalization if one instead eliminates the auxiliary fields and imposes the usual Lorentz covariant but non-supersymmetric gauge-fixing term $-(\partial_\mu A^{\mu a})^2/2\xi$. In that case, because the gauge-fixing term is not invariant under supersymmetry, the anomalous dimensions of the scalar and fermion components are not even equal to each other. At one-loop order, for example, they are given by

$$[\gamma_S^{(1)}]_i^j = \tfrac{1}{2} y_{ikl} y^{jkl} + (\xi - 1)\delta_i^j g_a^2 C_a(i), \tag{11.142}$$

$$[\gamma_F^{(1)}]_i^j = \tfrac{1}{2} y_{ikl} y^{jkl} + (\xi + 1)\delta_i^j g_a^2 C_a(i), \tag{11.143}$$

respectively, with $\xi = 0$ for Landau gauge and $\xi = 1$ for Feynman gauge. Thus, there is no covariant gauge-fixing prescription of this type in which the scalar and fermion component-field anomalous dimensions are equal to each other,[8] and neither of them is equal to the superfield anomalous dimension of eqs. (11.138)–(11.141). Nevertheless, the sum of all divergent contributions for a given process, including both wave function and vertex renormalizations, respects supersymmetry, and in totality has the form of wave function renormalization given by eqs. (11.135)–(11.137).

As is generally true in quantum field theory, beyond two-loop order, the gauge coupling beta functions depend on the choice of renormalization scheme. Another famous choice is the Novikov–Shifman–Vainshtein–Zakharov (NSVZ) scheme [122], where the beta function is found exactly in terms of the trace of the chiral superfield anomalous dimension matrix, using instanton calculus techniques. The result is

$$\beta_{g_a}^{\rm NSVZ} = \frac{g_a^3}{16\pi^2}\left(\frac{S_a - 3G_a - 2[\gamma_{\rm NSVZ}]_i^i C_a(i)/d_a}{1 - 2g_a^2 G_a/16\pi^2}\right), \tag{11.144}$$

where it is important to note that the NSVZ scheme also has a different superfield anomalous dimension matrix, $[\gamma_{\rm NSVZ}]_i^j$, than the γ_i^j of the $\overline{\rm DR}$ scheme, starting at three-loop order. Indeed, it has been shown in Refs. [119, 120] that, at least through the order pertinent to the four-loop beta functions, the NSVZ and $\overline{\rm DR}$ schemes are related by the following coupling constant redefinition:

$$g_a^{\overline{\rm DR}} = g_a^{\rm NSVZ} + \frac{1}{(16\pi^2)^2}\delta^{(2)} g_a + \frac{1}{(16\pi^2)^3}\delta^{(3)} g_a + \cdots, \tag{11.145}$$

where

$$\delta^{(2)} g_a = \tfrac{1}{2} g_a^5 G_a\,(3G_a - S_a) - g_a^3 g_b^2 S_{ab} + \tfrac{1}{4} g_a^3 y^{ijk} y_{ijk} C_a(i)/d_a, \tag{11.146}$$

$$\begin{aligned}\delta^{(3)} g_a = {}& \tfrac{1}{6} g_a^7 G_a (3G_a - S_a)(13G_a + S_a) - \tfrac{16}{3} g_a^5 g_b^2 G_a S_{ab} \\ &+ \tfrac{4}{3} g_a^3 g_b^4 S_{ab}(S_b - 3G_b) + \tfrac{10}{3} g_a^3 g_b^2 g_c^2 S_{abc} + \tfrac{4}{3} g_a^5 y^{ijk} y_{ijk} C_a(i) G_a/d_a \\ &+ g_a^3 g_b^2 y^{ijk} y_{ijk} C_a(i)\left[\tfrac{4}{3} C_b(j) - C_b(i)\right]/d_a \\ &+ g_a^3 y^{ijk} y_{ijl} y_{kmn} y^{lmn}\left[\tfrac{1}{24} C_a(k) - \tfrac{1}{3} C_a(i)\right]/d_a. \end{aligned} \tag{11.147}$$

The absence of a one-loop contribution to eq. (11.145) implies that the gauge couplings in eqs. (11.146) and (11.147) can be taken to be in either scheme; the distinction only matters at the same order as $\delta^{(4)} g_a$, which we consistently neglect. It also implies that the chiral superfield anomalous dimensions for the two schemes first differ at three-loop order, with

$$[\gamma_{\rm NSVZ}^{(3)}]_i^j = [\gamma_{\overline{\rm DR}}^{(3)}]_i^j - 4\delta_i^j C_a(i) g_a\, \delta^{(2)} g_a. \tag{11.148}$$

The result is that $\gamma_i^{(3)j}$ written in terms of NSVZ couplings can be obtained from the $\overline{\rm DR}$ formula eq. (11.141) by making the following modifications in the last three lines: $[24\zeta(3) - 2] \to 24\zeta(3)$ and $-12 \to -18$ and $[20 - 24\zeta(3)] \to [24 - 24\zeta(3)]$ and $-5 \to -6$. This can now be plugged into eq. (11.144) to obtain the four-loop NSVZ

[8] However, it is also possible to impose the noncovariant light-cone gauge-fixing scheme, after which the action contains neither auxiliary fields nor Faddeev–Popov ghosts, and is invariant under supersymmetry transformations linear in the fields. In this scheme the scalar and fermion anomalous dimensions are equal to each other and to the superfield anomalous dimensions. This comes at the cost of non-Lorentz-covariant Feynman rules. For example, see Ref. [90].

beta function for the gauge coupling. The equivalence between the four-loop order expressions for $\beta_{g_a}^{\overline{\rm DR}}$ given in eqs. (11.131)–(11.134) and $\beta_{g_a}^{\rm NSVZ}$ can then be verified directly using eqs. (11.145)–(11.147).

Another interesting choice is the holomorphic scheme [123], in which the gauge coupling is only renormalized at one-loop order. The holomorphic gauge coupling is related to the NSVZ coupling by

$$\frac{1}{(g_a^2)_{\rm holo}} = \frac{1}{(g_a^2)_{\rm NSVZ}} + \frac{1}{16\pi^2}\left[2G_a \ln[(g_a^2)_{\rm NSVZ}] - 4Z_i^i C_a(i)/d_a\right], \tag{11.149}$$

where Z_i^j is the chiral superfield wave function renormalization matrix, which is related to the NSVZ anomalous dimension by

$$Q\frac{\partial}{\partial Q} Z_i^j = [\gamma_{\rm NSVZ}]_i^j. \tag{11.150}$$

Applying eqs. (11.149) and (11.150) to eq. (11.144), one finds that the beta function for the holomorphic gauge coupling is simply given by the one-loop order result: $\beta_{(g_a)_{\rm holo}} = [(g_a)_{\rm holo}]^3(S_a - 3G_a)/16\pi^2$, exactly.

The choice of which renormalization scheme to use is dictated by what the parameters will be used for. For example, if one is calculating an observable cross section or physical mass in the $\overline{\rm DR}$ scheme, then that scheme should also be used for the beta functions, or else one should translate between schemes appropriately. The matching of a given set of theory parameters to an ultraviolet completion will also depend on the choice of scheme, but it is not clear a priori which renormalization scheme is appropriate for this.

So far in this section we have considered the case of exactly supersymmetric theories. Now suppose that we include soft breaking of supersymmetry, with the Lagrangian given in eq. (8.78). The beta functions of the supersymmetric parameters are unaffected by this. The beta functions for the soft parameters M_a, a^{ijk}, b^{ij}, t^i, and $(m^2)_i^j$ were obtained in Refs. [124–126] from the supersymmetric beta functions by a spurion method proposed in Ref. [127], by introducing an external "field" $\theta\theta$ in the superspace Lagrangian. The results for the soft parameter beta functions can be written in terms of a differential operator on the space of couplings, defined by

$$\Omega = \frac{1}{2}M_a g_a \frac{\partial}{\partial g_a} - a^{ijk}\frac{\partial}{\partial y^{ijk}}. \tag{11.151}$$

Then, at all orders in perturbation theory, one has

$$\beta_{M_a} = 2\Omega(\beta_{g_a}/g_a), \tag{11.152}$$

$$\beta_{a^{ijk}} = \left[\gamma_n^i a^{njk} - 2y^{njk}\Omega(\gamma_n^i)\right] + (i \leftrightarrow j) + (i \leftrightarrow k). \tag{11.153}$$

If there are no gauge-singlet chiral superfields, as in the case of the MSSM, then the expression for $\beta_{b^{ij}}$ is analogous to that for $\beta_{a^{ijk}}$, with the index k deleted and $a^{njk} \to b^{nj}$ and $y^{njk} \to M^{nj}$. More generally, the results for $\beta_{b^{ij}}$ and β_{t^i} in the presence of gauge singlets are given in Refs. [127, 128] at two-loop order, and a method for finding them at arbitrary loop order in terms of γ_i^j is given in Ref. [129].

For the nonholomorphic scalar squared mass, the result is more complicated. In the $\overline{\text{DR}}'$ scheme,

$$\beta_{(m^2)_i^j} = g_a^2 (\boldsymbol{T}^a)_i{}^j A_a + \Big\{ 2\Omega\Omega^* + \left(|M_a|^2 + X_a\right) g_a \frac{\partial}{\partial g_a} + \left[(m^2)_k^n y^{kpq} + (m^2)_k^p y^{nkq} + (m^2)_k^q y^{npk}\right] \frac{\partial}{\partial y^{npq}} + \left[(m^2)_n^k y_{kpq} + (m^2)_p^k y_{nkq} + (m^2)_q^k y_{npk}\right] \frac{\partial}{\partial y_{npq}} \Big\} \gamma_i^j , \tag{11.154}$$

where the conjugate of the differential operator in eq. (11.151) is

$$\Omega^* = \frac{1}{2} M_a^* g_a \frac{\partial}{\partial g_a} - a_{ijk} \frac{\partial}{\partial y_{ijk}} . \tag{11.155}$$

In eq. (11.154), there are two contributions, denoted by X_a and A_a, that do not arise in a straightforward way from the spurion method.

First, the contribution proportional to X_a comes from the effect of epsilon-scalars, which acquire soft supersymmetry-breaking squared masses through loop effects. In the $\overline{\text{DR}}$ scheme, $\beta_{(m^2)_i^j}$ contains a contribution proportional to the epsilon-scalar squared mass. In passing to the $\overline{\text{DR}}'$ scheme, the epsilon-scalar squared masses are then made to decouple by a redefinition of the ordinary scalar squared masses, which can be found order-by-order in perturbation theory; see eqs. (11.122) and (11.123) for the one-loop redefinition. This causes the appearance of X_a, which is only known in a perturbative loop expansion,

$$X_a = \frac{1}{16\pi^2} X_a^{(1)} + \frac{1}{(16\pi^2)^2} X_a^{(2)} + \cdots , \tag{11.156}$$

with the results necessary for $\beta_{(m^2)_i^j}$ at three-loop order [121, 126]:

$$X_a^{(1)} = 2 g_a^2 \left[G_a |M_a|^2 - (m^2)_k^k C_a(k)/d_a \right] , \tag{11.157}$$

$$X_a^{(2)} = g_a^4 (10 G_a - 2 S_a) G_a |M_a|^2 - 4 g_a^4 G_a (m^2)_k^k C_a(k)/d_a - 4 g_a^2 g_b^2 S_{ab} |M_b|^2 + g_a^2 y^{kpq} y_{npq} (m^2)_k^n \left[C_a(p) + \tfrac{1}{2} C_a(k) \right] / d_a + g_a^2 a^{kpq} a_{kpq} C_a(k) / 2 d_a . \tag{11.158}$$

The other exceptional contribution A_a in eq. (11.154) only occurs when the index a corresponds to an abelian subgroup, and arises in the superspace approach from D-term tadpole diagrams. Its presence can be related to the possibility of introducing a Fayet–Iliopoulos parameter ξ_a, which contributes to the scalar squared mass when the corresponding D-term is eliminated by its equation of motion. With our assumption that there is at most one such U(1) factor in the gauge group,

$$A_a = \frac{1}{16\pi^2} A_a^{(1)} + \frac{1}{(16\pi^2)^2} A_a^{(2)} + \frac{1}{(16\pi^2)^3} A_a^{(3)} + \cdots , \tag{11.159}$$

with the results in the $\overline{\text{DR}}'$ scheme obtained by Ref. [130]:

$$A_a^{(1)} = 2 (\boldsymbol{T}^a)_k{}^l (m^2)_l^k , \tag{11.160}$$

$$A_a^{(2)} = (\boldsymbol{T}^a)_k{}^l \left[8 g_b^2 C_b(k) (m^2)_l^k - 2 (m^2)_n^k y^{npq} y_{lpq} \right] , \tag{11.161}$$

$$\begin{aligned}
A_a^{(3)} = (\boldsymbol{T^a})_k{}^l \Big\{ & 3(m^2)_l^n y^{kpq} y_{npr} y^{rst} y_{qst} - \tfrac{3}{2}(m^2)_l^n y^{kpq} y_{pqr} y^{rst} y_{nst} \\
& - 4y^{knp} y_{pqr} y^{rst} y_{lst} (m^2)_n^q - 2a^{knp} a_{npq} y_{lrs} y^{qrs} - \tfrac{5}{2} a^{knp} a_{lrs} y^{qrs} y_{npq} \\
& + 16 g_b^2 C_b(k) y^{knp} y_{lnp} |M_b|^2 + (8 - 24\zeta_3) g_b^2 C_b(k) a^{knp} a_{lnp} \\
& + (12\zeta_3 - 10) g_b^2 C_b(k) \left[a^{knp} y_{lnp} M_b^* + y^{knp} a_{lnp} M_b\right] \\
& + g_b^2 (m^2)_l^n y^{kpq} y_{npq} [(10 - 24\zeta_3) C_b(k) - 12 C_b(p)] \\
& + (16 - 48\zeta_3) g_b^2 C_b(k) y^{knp} y_{lrp} (m^2)_n^r - 16 g_b^2 g_c^2 C_b(k) C_c(k) (m^2)_l^k \\
& + g_b^2 g_c^2 C_b(k) C_c(k) [(96\zeta_3 - 64)|M_b|^2 + (48\zeta_3 - 40) M_b M_c^*] \delta_l^k \\
& + 12 g_b^4 C_b(k) (3G_b - S_b)(m^2)_l^k \Big\}. \qquad (11.162)
\end{aligned}$$

The import of the above results is that knowledge of the supersymmetric parameter beta functions allows one to derive the soft supersymmetry-breaking parameter beta functions in the $\overline{\mathrm{DR}}'$ scheme through full three-loop order (four-loop order for the gaugino masses) just by taking derivatives with respect to gauge and Yukawa couplings.

For example, the one-loop contributions to the RGEs for the general soft supersymmetry-breaking Lagrangian parameters appearing in eq. (8.78) are

$$\beta_{M_a}^{(1)} = 2g_a^2 \, (S_a - 3G_a) \, M_a, \qquad (11.163)$$

$$\begin{aligned}
\beta_{(m^2)_i^j}^{(1)} = {} & \tfrac{1}{2} y_{ipq} y^{pqn} (m^2)_n^j + \tfrac{1}{2} y^{jpq} y_{pqn} (m^2)_i^n + 2 y_{ipq} y^{jpr} (m^2)_r^q \\
& + a_{ipq} a^{jpq} - 8 g_a^2 C_a(i) |M_a|^2 \delta_i^j + 2 g_a^2 (\boldsymbol{T^a})_i{}^j \mathrm{Tr}(\boldsymbol{T^a} m^2), \qquad (11.164)
\end{aligned}$$

$$\begin{aligned}
\beta_{a^{ijk}}^{(1)} = {} & \left[\tfrac{1}{2} y^{imn} y_{pmn} a^{pjk} + a^{imn} y_{pmn} y^{pjk} + g_a^2 C_a(i) (4 M_a y^{ijk} - 2a^{ijk})\right] \\
& + (i \leftrightarrow j) + (i \leftrightarrow k), \qquad (11.165)
\end{aligned}$$

$$\begin{aligned}
\beta_{b^{ij}}^{(1)} = {} & \left[\tfrac{1}{2} b^{ip} y_{pmn} y^{jmn} + \tfrac{1}{2} y^{ijp} y_{pmn} b^{mn} + M^{ip} y_{pmn} a^{mnj} \right. \\
& \left. + g_a^2 C_a(i) (4 M_a M^{ij} - 2b^{ij})\right] + (i \leftrightarrow j), \qquad (11.166)
\end{aligned}$$

$$\begin{aligned}
\beta_{t^i}^{(1)} = {} & \tfrac{1}{2} y^{imn} y_{mnp} t^p + a^{imn} y_{mnp} T^p + M^{ip} y_{pmn} b^{mn} + a^{imn} b_{mn} \\
& + 2 y^{imn} M_{mp} (m^2)_n^p. \qquad (11.167)
\end{aligned}$$

In Section 13.7, we will specialize eqs. (11.163)–(11.166) to the case of the MSSM.

11.6 Effective Potentials

The effective potential provides a convenient formalism for understanding spontaneous symmetry breaking and the structure of the possible vacuum states of a theory. In order to compute the effective potential, one writes each real scalar field $\Phi_j(x)$ in the theory (treating complex scalars as two real scalars) as a sum of a constant classical background ϕ_j and a fluctuating dynamical field $R_j(x)$:

$$\Phi_j(x) = \phi_j + R_j(x). \qquad (11.168)$$

Then the effective potential is a function of the ϕ_j, and can be evaluated as

$$V_{\rm eff}(\phi_j) = V^{(0)} + \frac{1}{16\pi^2}V^{(1)} + \frac{1}{(16\pi^2)^2}V^{(2)} + \cdots\,, \tag{11.169}$$

where $V^{(0)}$ is the usual tree-level potential obtained by setting $R_j = 0$, and the contributions $V^{(n)}$ for $n \geq 1$ are obtained from the sum of all n-loop vacuum (no external legs) Feynman diagrams, using couplings and propagator masses that depend on the background fields ϕ_j.

It is instructive to remind the reader of the origin of the last statement. We begin by expanding the effective action functional $\Gamma[\Phi]$ about a constant classical background field ϕ:

$$\Gamma[\Phi] = \sum_{n=1}^{\infty} \frac{1}{n!} \int d^4x_1 \cdots d^4x_n\, \Gamma_\phi^{(n)}(x_1,\ldots,x_n) R(x_1)\cdots R(x_n)\,, \tag{11.170}$$

where the fluctuating dynamical field $R(x)$ is defined in eq. (11.168) and flavor indices have been suppressed. The n-point 1PI Green function of the "shifted" theory is denoted by $\Gamma_\phi^{(n)}$. We can now write down a formula for $V_{\rm eff}$ by recalling the relation between the effective action functional and the effective potential,

$$\Gamma[\Phi] = \int d^4x \left[\tfrac{1}{2}Z(\Phi)\partial_\mu\Phi\partial^\mu\Phi - V_{\rm eff}(\Phi) + \text{higher derivative terms}\right]. \tag{11.171}$$

The result for $V_{\rm eff}$ takes on a particularly simple form if we make use of the momentum space Green functions,

$$\begin{aligned}\widetilde{\Gamma}_\phi^{(n)}&(p_1,\ldots,p_n)(2\pi)^4\delta^4(p_1+\cdots+p_n)\\ &= \int d^4x_1\cdots d^4x_n\, e^{i(p_1\cdot x_1+\cdots+p_n x_n)}\,\Gamma_\phi^{(n)}(x_1,\ldots,x_n)\,.\end{aligned} \tag{11.172}$$

Taking Φ to be a constant field independent of x, eq. (11.171) simplifies to

$$\Gamma[\Phi] = -V_{\rm eff}(\Phi)\int d^4x\,. \tag{11.173}$$

Setting $p_1 = \cdots = p_n = 0$ in eq. (11.172) and identifying $(2\pi)^4\delta^4(0)$ with $\int d^4x$, we end up with

$$V_{\rm eff}(\Phi) = -\sum_{n=1}^{\infty}\frac{1}{n!}\widetilde{\Gamma}_\phi^{(n)}(0)(\Phi-\phi)^n\,, \tag{11.174}$$

where the notation $\widetilde{\Gamma}_\phi^{(n)}(0)$ indicates that all the external four-momenta $p_1,\ldots,p_n$ vanish. Equation (11.174) provides the method of calculation of the effective potential described below eq. (11.169).

An alternative method for computing the effective potential can be obtained by taking the derivative of eq. (11.174) with respect to Φ and then setting $\Phi = \phi$. Only one term survives in the sum, and we obtain

$$\left.\frac{dV_{\rm eff}(\Phi)}{d\Phi}\right|_{\Phi=\phi} = -\widetilde{\Gamma}_\phi^{(1)}(0), \tag{11.175}$$

where $i\widetilde{\Gamma}^{(1)}_{\phi}(0)$ is the sum of all Feynman diagrams contributing to the tadpole (1PI one-point function) of the shifted theory. Integrating this result yields $V_{\rm eff}(\Phi)$ up to a constant of integration that is of no concern since only the field-dependent terms of $V_{\rm eff}$ are relevant. Note that if we take $\phi = v$ in eq. (11.175), where v is the loop-corrected vacuum expectation value of the scalar field where $V_{\rm eff}$ attains its minimum, then the sum of all tadpoles vanishes as expected.

By a similar calculation,

$$\left.\frac{d^2V_{\rm eff}}{d\Phi_i d\Phi_j}\right|_{\Phi_k=\phi_k} = -\widetilde{\Gamma}^{(2)}_{\phi}(0)_{ij}\,. \tag{11.176}$$

We now choose $\phi_i = v_i$ by minimizing the effective potential $V_{\rm eff}$. After diagonalizing the effective scalar squared-mass matrix,

$$(M^2_{\rm eff})_{ij} = \left.\frac{d^2V_{\rm eff}}{d\Phi_i d\Phi_j}\right|_{\Phi_k=v_k}\,, \tag{11.177}$$

one can identify

$$\widetilde{\Gamma}^{(2)}_{\phi}(0) = \boldsymbol{A^{-1}}(0) = -\boldsymbol{m^2} - \boldsymbol{\Sigma}(0)\,, \tag{11.178}$$

after using eq. (2.183), where $\boldsymbol{m^2}$ is the diagonal tree-level scalar squared-mass matrix and $i\boldsymbol{A}(p^2)$ is diagrammatically represented in Fig. 2.21. At one-loop accuracy, it follows that

$$(m^2_{\rm eff})_i = m_i^2 + \Sigma_{ii}(0)\,, \tag{11.179}$$

where the $(m^2_{\rm eff})_i$ are the eigenvalues of $M^2_{\rm eff}$. We can then employ eq. (2.192) to identify the pole masses of the scalars, Thus, we conclude that, at one-loop accuracy,

$$m^2_{p,i} = (m^2_{\rm eff})_i + \Sigma_{ii}(m_i^2) - \Sigma_{ii}(0)\,. \tag{11.180}$$

Although the effective potential itself is dependent on the gauge-fixing prescription, physically observable properties following from it, such as its value at stationary points, and observables related to the question of whether or not spontaneous symmetry breaking occurs, are gauge invariant. The radiatively corrected vacuum expectation values of the scalar fields can be determined by requiring that the effective potential is minimized:

$$\frac{\partial}{\partial\phi_j}V_{\rm eff} = 0, \tag{11.181}$$

for all j. In most cases, one can set all but one or two of the ϕ_j to 0, under the assumption that the corresponding scalar fields are stabilized at 0 by positive squared masses, and so will not obtain VEVs. It is overwhelmingly convenient to choose Landau gauge to evaluate $V_{\rm eff}$, because this avoids kinetic mixing between scalar and vector degrees of freedom that would otherwise occur for values of ϕ_j that do not minimize $V^{(0)}$. Even at one-loop order, the effective potential is quite complicated unless evaluated in Landau gauge, and the difficulties encountered in other gauge-fixing schemes are magnified in the contributions at higher loop orders.

The form of the effective potential also depends on the regularization and renormalization schemes used. The original Coleman–Weinberg calculation of the effective potential [131] used an on-shell type renormalization scheme, but in modern applications it is more appropriate to use mass-independent renormalization schemes such as $\overline{\rm MS}$ based on dimensional regularization for non-supersymmetric theories, or $\overline{\rm DR}'$ based on dimensional reduction for softly broken supersymmetric theories.

An important constraint is the RG invariance of the effective potential:

$$Q\frac{d}{dQ}V_{\rm eff} = \left(Q\frac{\partial}{\partial Q} + \sum_X \beta_X \frac{\partial}{\partial X}\right) V_{\rm eff} = 0, \qquad (11.182)$$

where X runs over all of the independent Lagrangian parameters, including the background scalar field(s). The beta function for a background scalar field ϕ is related to its anomalous dimension γ_ϕ by $\beta_\phi = -\phi\gamma_\phi$ [cf. eqs. (11.54) and (11.55) with $X = \phi$].[9] It follows that at each loop order $\ell = 1, 2, 3, \ldots$ one must have

$$Q\frac{\partial}{\partial Q}V^{(\ell)} + \sum_{n=0}^{\ell-1}\left(\sum_X \beta_X^{(\ell-n)}\frac{\partial}{\partial X}V^{(n)}\right) = 0, \qquad (11.183)$$

where we have used the expansion in eq. (11.58). This provides a useful consistency check.

For a general renormalizable field theory in the Landau gauge, one can write the ϕ_j-dependent squared-mass eigenvalues as

$$m_k^2(\phi_j), \quad m_J^2(\phi_j), \quad m_a^2(\phi_j), \qquad (11.184)$$

where the indices k, j, and a run over, respectively, the full list of real scalars, two-component fermions, and real vector bosons, after diagonalization. It is instructive to first consider the effective potential in terms of the bare quantities of the theory, working in $d = 4 - 2\epsilon$ dimensions. One finds, at one-loop order,

$$V_{\rm eff} = V_{\rm tree} + \frac{1}{16\pi^2}\Big[\sum_k h_B(m^2_{k,B}) - 2\sum_J h_B(m^2_{J,B}) + (d-1)\sum_a h_B(m^2_{a,B})\Big], \qquad (11.185)$$

where the subscript B denotes bare field-dependent squared masses, and the function from one-loop integration is

$$h_B(x) = -\tfrac{1}{2}\left(4\pi\mu^2\right)^\epsilon \Gamma(-d/2)x^{d/2} = \frac{x^2}{4}\left[-\frac{1}{\epsilon} + \ln(x/Q^2) - \frac{3}{2}\right] + \mathcal{O}(\epsilon), \qquad (11.186)$$

where Q is the renormalization scale. Now, one can apply the $\overline{\rm MS}$ scheme redefinitions as in eq. (11.43)–(11.46), and expand in ϵ. The result is the $\overline{\rm MS}$ scheme effective potential of eq. (11.169), with

$$V^{(1)}_{\overline{\rm MS}} = V_S^{(1)} + V_F^{(1)} + V_V^{(1)}, \qquad (11.187)$$

[9] Note that the γ_ϕ are the scalar component-field anomalous dimensions, as in eq. (11.142) at one-loop order with $\xi = 0$ for Landau gauge, and not the corresponding superfield anomalous dimensions.

where

$$V_S^{(1)} = \tfrac{1}{4}\sum_k (m_k^2)^2\left[\ln(m_k^2/Q^2) - \tfrac{3}{2}\right], \tag{11.188}$$

$$V_F^{(1)} = -\tfrac{1}{2}\sum_J (m_J^2)^2\left[\ln(m_J^2/Q^2) - \tfrac{3}{2}\right], \tag{11.189}$$

$$V_V^{(1)} = \tfrac{3}{4}\sum_a (m_a^2)^2\left[\ln(m_a^2/Q^2) - \tfrac{5}{6}\right]. \tag{11.190}$$

The term $-5/6$ (rather than $-3/2$) in the vector contribution arises due to the fact, reflected in the prefactor in eq. (11.185), that there are $d-1$ (rather than 3) positive-norm massive vector degrees of freedom in the dimensionally regularized theory.

For example, consider the effective potential of the (non-supersymmetric) Standard Model, as a function of the Higgs background field φ, which is equal to the VEV at the minimum. The tree-level potential is given by

$$V^{(0)} = \Lambda + \tfrac{1}{2}m^2\varphi^2 + \tfrac{1}{4}\lambda\varphi^4, \tag{11.191}$$

where the numerical values of the $\overline{\rm MS}$ running parameters are[10] $m^2 \approx -(93\ \mathrm{GeV})^2$ and $\lambda \approx 0.126$ at a renormalization scale near $Q = m_{\rm top}$. Applying the results above, the one-loop effective potential in Landau gauge and the $\overline{\rm MS}$ scheme is

$$\begin{aligned} V^{(1)} = &-3T^2\left[\ln(T/Q^2) - \tfrac{3}{2}\right] + \tfrac{3}{2}W^2\left[\ln(W/Q^2) - \tfrac{5}{6}\right] + \tfrac{3}{4}Z^2\left[\ln(Z/Q^2) - \tfrac{5}{6}\right] \\ &+ \tfrac{1}{4}H^2\left[\ln(H/Q^2) - \tfrac{3}{2}\right] + \tfrac{3}{4}G^2\left[\ln(G/Q^2) - \tfrac{3}{2}\right], \end{aligned} \tag{11.192}$$

where the field-dependent squared masses are

$$T = \tfrac{1}{2}y_t^2\varphi^2, \qquad W = \tfrac{1}{4}g^2\varphi^2, \qquad Z = \tfrac{1}{4}(g^2 + g'^2)\varphi^2, \tag{11.193}$$

$$H = 3\lambda\varphi^2 + m^2, \qquad G = \lambda\varphi^2 + m^2, \tag{11.194}$$

and the effects of all other fermions are neglected. The two-loop and three-loop contributions to $V_{\rm eff}$ have now been obtained for the Standard Model, and more generally for any renormalizable field theory in the $\overline{\rm MS}$ scheme, in Refs. [132–135].

However, for supersymmetric theories, the appearance of 5/6 rather than 3/2 in $V_V^{(1)}$ is problematic. As noted above, this is because there are only $d-1$, rather than 3, physical massive vector degrees of freedom in the dimensionally regularized theory. This implies that the contributions of a massive vector degree of freedom will not cancel with those of the corresponding massive gaugino, reflecting the fact that the $\overline{\rm MS}$ scheme does not respect supersymmetry.

Supersymmetry is respected if one uses dimensional reduction instead of dimensional regularization. Then the loop momenta are taken to be in $d = 4 - 2\epsilon$ dimensions, but there are exactly four vector degrees of freedom. It might be tempting to

[10] The parameter Λ is a field-independent vacuum energy, which does not contribute to Standard Model observables at all, except through gravitational effects as the cosmological constant. However, Λ does run with renormalization scale, and so must be included in order to ensure renormalization group invariance of the effective potential as in eqs. (11.182) and (11.183).

assume that the epsilon-scalars have the same field-dependent masses as their $4-2\epsilon$ vector counterparts. However, as pointed out in Ref. [136], in $\overline{\rm DR}$ this is actually inconsistent except in models with exact supersymmetry, unless one sticks to only one fixed value of the renormalization scale Q, because the epsilon-scalar squared mass has a beta function that gets contributions from the soft supersymmetry-breaking parameters. Therefore, in $\overline{\rm DR}$ one must allow the epsilon-scalars to have squared-mass eigenvalues $\hat{m}_a^2$ that are distinct from the m_a^2 for the ordinary vectors. To be specific, consider the explicit form of the field-dependent squared-mass matrix for the ordinary $4-2\epsilon$ vector fields:

$$m^2_{ab} = g_a g_b \phi_j \{\boldsymbol{t}^a, \boldsymbol{t}^b\}_j{}^k \phi_k, \tag{11.195}$$

where the $\boldsymbol{t}^a$ are the gauge generators acting on the scalar fields. For the epsilon-scalar squared-mass matrix, one has instead

$$\hat{m}^2_{ab} = m^2_{ab} + \delta_{ab} m^2_{\epsilon,a}, \tag{11.196}$$

where $m^2_{\epsilon,a}$ is an unphysical parameter that cannot just be set to zero, because it gets radiative corrections from the soft supersymmetry-breaking parameters in the theory. If supersymmetry is explicitly broken by soft supersymmetry-breaking couplings, the eigenvalues $\hat{m}_a^2$ of the matrix $\hat{m}^2_{ab}$ will in general differ from the eigenvalues m_a^2 of the matrix m^2_{ab}, and the corresponding couplings of the mass-eigenstate epsilon-scalars are different from the couplings of the $4-2\epsilon$ mass-eigenstate vectors.

In the $\overline{\rm DR}$ scheme, with epsilon-scalars included, one then finds

$$V^{(1)}_{\overline{\rm DR}} = V^{(1)}_S + V^{(1)}_F + V^{(1)}_V + V^{(1)}_\epsilon, \tag{11.197}$$

where $V^{(1)}_S$, $V^{(1)}_F$, $V^{(1)}_V$ are as before, and

$$V^{(1)}_\epsilon = -\frac{1}{2}\sum_a (\hat{m}_a^2)^2. \tag{11.198}$$

However, the $m^2_{\epsilon,a}$ are additional parameters with no physically observable counterpart, and so their appearance in the effective potential is quite inconvenient. The functional form of the effective potential is also not directly physically observable, so there is no contradiction; the $m^2_{\epsilon,a}$ must cancel only from observable quantities. However, clearly one would like to avoid having to include the unphysical epsilon-scalar masses in the first place. This problem was solved in the context of softly broken supersymmetric models in Ref. [117] with the introduction of the $\overline{\rm DR}'$ scheme. The point is that one can remove the dependence of the full one-loop effective potential on the $m^2_{\epsilon,a}$ by redefining the ordinary scalar squared masses and the vacuum energy term Λ (the ϕ_j-independent part of $V^{(0)}$) according to

$$(m^2_{\overline{\rm DR}'})_i^j = (m^2_{\overline{\rm DR}})_i^j - \frac{1}{16\pi^2}\left[2g_a^2 C_a(i)\delta_i^j m^2_{\epsilon,a}\right], \tag{11.199}$$

$$\Lambda_{\overline{\rm DR}'} = \Lambda_{\overline{\rm DR}} - \frac{1}{16\pi^2} d_a (m^2_{\epsilon,a})^2/2. \tag{11.200}$$

The result is the $\overline{\rm DR}'$ scheme, and the effective potential in this scheme is the one usually used for softly broken supersymmetric theories (and often slightly incorrectly referred to as the $\overline{\text{DR}}$ one):

$$V^{(1)}_{\overline{\text{DR}}'} = \sum_n (-1)^{2s_n}(2s_n+1)h(m_n^2) = \text{STr}\left[h(m_n^2)\right], \qquad (11.201)$$

where

$$h(x) = \tfrac{1}{4}x^2\left[\ln(x/Q^2) - \tfrac{3}{2}\right], \qquad (11.202)$$

and n runs over all real scalar, two-component fermion, and real vector degrees of freedom. The scalar squared masses occurring in eq. (11.201) are the ones following from the redefinition in eq. (11.199), and the vector squared masses are the eigenvalues of eq. (11.195). The $\overline{\text{DR}}'$ effective potential is independent of the unphysical parameters $m^2_{\epsilon,a}$, even when the soft terms do not vanish. It can be shown that the $m^2_{\epsilon,a}$ are also banished from the equations which relate the physical pole masses to the Q-dependent running masses in the theory, so the epsilon-scalar masses have been successfully decoupled from all practical calculations. It would be quite clumsy to use the original $\overline{\text{DR}}$ scheme in studies of realistic models like the MSSM, since in renormalization group (RG) running and evaluation of the pole masses and effective potential one would have to keep extra contributions from epsilon-scalar masses in order to avoid inconsistencies. Therefore the $\overline{\text{DR}}'$ scheme is the preferred one.

After making this painful distinction, it must be admitted that the $\overline{\text{DR}}'$ final result for the effective potential has exactly the same form that one would have obtained if one had naively set $m^2_{\epsilon,a}$ equal to zero in the first place in the $\overline{\text{DR}}$ scheme calculation. However, this naive procedure is technically inconsistent whenever RG running is involved and does not work for other calculations involving epsilon-scalars, so one should really distinguish between the two schemes as a matter of principle. The parameters appearing in the $\overline{\text{DR}}'$ effective potential obey $\overline{\text{DR}}'$ renormalization group equations, which differ from the $\overline{\text{DR}}$ ones with $m^2_{\epsilon,a}$ set equal to 0. The procedure of going from the $\overline{\text{DR}}$ scheme to the $\overline{\text{DR}}'$ scheme is similar at two loops, and is described explicitly in Ref. [133].

Note that if supersymmetry is not broken for a particular choice of background scalar field VEVs ϕ_j, then the structure of eq. (11.201) guarantees that $V^{(1)} = 0$, because the contribution of each particle cancels with that of its superpartner. This illustrates a more general fact, proved in Ref. [94], that at a minimum of the tree-level potential, the full effective potential of a supersymmetric theory must vanish at each order in perturbation theory. One application of eq. (11.201) is to the analysis of the vacuum of the supersymmetry-breaking O'Raifeartaigh model, as described in Section 12.3; see eqs. (12.28)–(12.32).

In the case of the MSSM, the most important contributions to the effective potential come from the top squarks and the top quark, because the corresponding Yukawa coupling is the largest interaction involving the Higgs fields. Taking into account the color, electric charge, and spin degrees of freedom, one has, in the approximation of a dominant top-quark Yukawa coupling,

$$V^{(1)} = 6h(m^2_{\tilde{t}_1}) + 6h(m^2_{\tilde{t}_2}) - 12h(m_t^2). \qquad (11.203)$$

However, for precise numerical work, one must take into account the remaining one-loop corrections, as well as the two-loop contributions.

Exercises

11.1 (a) The neutral fermion field of the Wess–Zumino model can be represented by a four-component fermion field Ψ_M [see eq. (3.50)], and the complex scalar field of the model, $\phi \equiv (S+iP)/\sqrt{2}$, can be expressed in terms of real neutral scalar fields S and P. Show that the Wess–Zumino Lagrangian given in eq. (11.13) can be rewritten in terms of Ψ_M, S and P as follows:

$$\begin{aligned}\mathscr{L} &= \tfrac{1}{2}(\partial_\mu S)^2 + \tfrac{1}{2}(\partial_\mu P)^2 - \tfrac{1}{2}m^2(S^2+P^2) + \tfrac{1}{2}\overline{\Psi}_M(i\gamma^\mu\partial_\mu - m)\Psi_M \\ &\quad - \frac{\lambda}{\sqrt{2}}\left[S\overline{\Psi}_M\psi_M - iP\overline{\Psi}_M\gamma_5\Psi_M\right] - \frac{m\lambda}{\sqrt{2}}S(S^2+P^2) - \tfrac{1}{4}\lambda^2(S^2+P^2)^2 .\end{aligned} \tag{11.204}$$

(b) Verify that the Wess–Zumino Lagrangian separately conserves P, T, and C. Show that S is a neutral 0^{++} scalar (using the standard $J^{\rm PC}$ notation where J is the spin of the particle) and P is a neutral 0^{-+} pseudoscalar.

(c) Introduce the renormalized fields,

$$S_r = Z_S^{-1/2}S, \qquad P_r = Z_P^{-1/2}P, \qquad \Psi_{Mr} = Z_\Psi^{-1/2}\Psi_M . \tag{11.205}$$

The renormalized parameters are defined in eq. (11.16). What are the relationships between the renormalization constants Z_S, Z_P, Z_Ψ, Z_m, and Z_λ [which replace eq. (11.17)] due to the nonrenormalization of the superpotential?

(d) Obtain the counterterm Lagrangian for the Wess–Zumino model, starting from the Lagrangian specified in eq. (11.204). Verify the Feynman rule for the four-component Majorana fermion counterterm exhibited in Fig. 11.10. Repeat the analysis of the one-loop renormalization of the Wess–Zumino model given in Section 11.3 and obtain the one-loop expressions for Z_S, Z_P, Z_Ψ, Z_m, and Z_λ. Check that the relations obtained in part (c) are satisfied.

11.2 The Passarino–Veltman loop functions $A_0(m^2)$ and $B_0(p^2; m^2, m^2)$ are defined in eqs. (K.9) and (K.10). In this exercise, we shall evaluate these loop

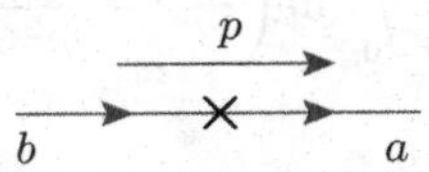

$$i(Z_\psi - 1)\not{p}_a{}^b - im_r(Z_m Z_\phi - 1)\delta_a{}^b$$

Fig. 11.10 Feynman rule for the four-component Majorana fermion counterterm, where a and b are four-component spinor indices.

integrals in $d = 4$ dimensions by using a momentum cutoff, Λ, to regulate the ultraviolet divergence.

(a) As a first step, consider the integration over q_0 in the integral expression for $A_0(m^2)$. Wick rotate the integration contour (which lies along the real axis) counterclockwise by 90° so that it now lies along the imaginary axis. Note that the value of the integral expression for $A_0(m^2)$ does not change since the integration contour, when rotated, did not cross any singularity (thanks to the $i\varepsilon$ factor in the denominator). Then, redefine the integration variable $q_0 = ik_0$ so that the integration is taken over the interval $-\infty < k_0 < \infty$. After defining $\vec{k} = \vec{q}$, convert to spherical coordinates in four Euclidean dimensions and impose an ultraviolet cutoff Λ in the k integration. Show that

$$A_0(m^2) = -2\int_0^\Lambda \frac{k^3\,dk}{k^2+m^2}\,. \tag{11.206}$$

Evaluate the integral above and show that

$$\begin{aligned} A_0(m^2) &= -\Lambda^2 + m^2\ln\left(1+\frac{\Lambda^2}{m^2}\right) \\ &= -\Lambda^2 + m^2\ln\left(\frac{\Lambda^2}{m^2}\right) + \mathcal{O}\left(\frac{m^2}{\Lambda^2}\right). \end{aligned} \tag{11.207}$$

Note that for $\Lambda \gg m^2$, eq. (11.207) exhibits both an ultraviolet quadratic divergence and a subleading ultraviolet logarithmic divergence.

(b) Repeat the analysis of part (a) to evaluate $B_0(p^2; m^2, m^2)$. In your computation, define the Euclidean four-vector $p_E = (-ip_0\,;\,\vec{p})$. Show that after introducing the Feynman parameter x [see eq. (K.6)],

$$\begin{aligned} B_0(p^2; m^2, m^2) &= 2\int_0^1 dx \int_0^\Lambda \frac{k^3\,dk}{\left[k^2+m^2+x(1-x)p_E^2\right]^2} \\ &= -2\frac{\partial}{\partial m^2}\int_0^1 dx\int_0^\Lambda \frac{k^3\,dk}{k^2+m^2+x(1-x)p_E^2}\,, \\ &= \frac{\partial}{\partial m^2}\int_0^1 dx\, A_0\big(m^2+x(1-x)p_E^2\big)\,, \end{aligned} \tag{11.208}$$

where $p_E^2 \equiv -p_0^2 + |\vec{p}|^2 = -p^2$. Assuming that $\Lambda^2 \gg p_E^2$, m^2, use the result of part (a) to obtain

$$B_0(p^2; m^2, m^2) = \ln\left(\frac{\Lambda^2}{m^2}\right) - 1 - \int_0^1 \ln\left(\frac{m^2 - p^2x(1-x)}{m^2}\right)dx + \mathcal{O}\left(\frac{m^2}{\Lambda^2}\right). \tag{11.209}$$

(c) In contrast to the analysis of parts (a) and (b), in dimensional regularization the loop integrals are evaluated in $d = 4 - 2\epsilon$ dimensions. Show that $\ln\Lambda^2$ plays the role of $1/\epsilon$ of dimensional regularization (in the limit of $\epsilon \to 0$) in identifying the ultraviolet logarithmic divergence of $A_0(m^2)$ and

$B_0(p^2; m^2, m^2)$. In particular, by comparing eqs. (11.209) and (K.21), verify the following identification:

$$\ln\left(\frac{\Lambda^2}{Q^2}\right) = \frac{1}{\epsilon} + 1\,, \tag{11.210}$$

where Q^2 is defined in eq. (11.42). Check that eq. (11.210) is also consistent with the subleading ultraviolet logarithmic divergence of $A_0(m^2)$ in light of eqs. (11.207) and (K.20).

(d) The existence of an ultraviolet quadratic divergence can be discerned by analyzing the Passarino–Veltman loop functions in the limit of $\epsilon \to 1$. Using eq. (K.92), show that one can identify

$$\Lambda^2 = \lim_{\epsilon\to 1} \frac{4\pi\mu^2}{\epsilon - 1} \tag{11.211}$$

to infer the ultraviolet quadratic divergence of $A_0(m^2)$ obtained in part (a). Note that the ultraviolet quadratic divergence is absent in $B_0(p^2; m^2, m^2)$ since this loop function is finite at $\epsilon = 1$ in light of eq. (K.93).

(e) There are other ways to define the cutoff scale Λ. For example, one could integrate from $k = 0$ to ∞ in eqs. (11.206) and (11.208) after replacing $dk \to \exp(-k^2/\Lambda^2)dk$, with $\Lambda^2 \gg m^2$. Show that with this definition of Λ,

$$A_0(m^2) = -\Lambda^2 - m^2 e^{-m^2/\Lambda^2}\mathrm{Ei}(-m^2/\Lambda^2)\,, \tag{11.212}$$

where the exponential integral function $\mathrm{Ei}(x)$ for a negative real argument is defined by

$$\mathrm{Ei}(x) = \int_{-\infty}^{x} \frac{e^t}{t}\,dt = \gamma + \ln(-x) + \sum_{k=1}^{\infty} \frac{x^k}{k\cdot k!}\,, \qquad \text{for } x < 0, \tag{11.213}$$

and γ is Euler's constant. Hence, verify that

$$A_0(m^2) = -\Lambda^2 + m^2\left[\ln\left(\frac{\Lambda^2}{m^2}\right) - \gamma\right] + \mathcal{O}\left(\frac{m^2}{\Lambda^2}\right)\,. \tag{11.214}$$

That is, the logarithmic divergences of parts (a) and (b) are modified by the replacement $\ln\Lambda^2 \to \ln\Lambda^2 - \gamma$.

(f) Modify the procedure of part (e) by replacing $dk \to \exp(-k/\Lambda')dk$, where $\Lambda' > 0$. Show that eq. (11.212) is modified as follows:

$$A_0(m^2) = -2\Lambda'^2 - 2m^2\{\mathrm{Ci}(m/\Lambda')\cos(m/\Lambda') + [\mathrm{Si}(m/\Lambda') - \tfrac{1}{2}\pi]\sin(m/\Lambda')\}\,, \tag{11.215}$$

where

$$\mathrm{Si}(x) = \int_0^x \frac{\sin t}{t}\,dt\,, \tag{11.216}$$

$$\mathrm{Ci}(x) = \gamma + \ln x + \int_0^x \frac{\cos t - 1}{t}\,dt\,, \quad \text{for } x > 0. \tag{11.217}$$

In the limit of $\Lambda' \gg m$, verify that by identifying $\Lambda = \sqrt{2}\,\Lambda'$, it follows that

$$A_0(m^2) = -\Lambda^2 + m^2\left[\ln\left(\frac{\Lambda^2}{m^2}\right) - 2\gamma - 2\ln 2\right] + \mathcal{O}\left(\frac{m}{\Lambda}\right). \quad (11.218)$$

That is, one can formally define the ultraviolet cutoff Λ such that the quadratic divergence of $A_0(m^2)$ is exactly equal to $-\Lambda^2$. With this definition of Λ, all possible cutoff procedures yield

$$A_0(m^2) = -\Lambda^2 + m^2\left[\ln\left(\frac{\Lambda^2}{m^2}\right) - c\right] + \mathcal{O}\left(\frac{m}{\Lambda}\right), \quad (11.219)$$

where only the value of c and the terms of $A_0(m^2)$ and $B_0(p^2; m^2, m^2)$ that vanish as $\Lambda \to \infty$ depend on the cutoff procedure.

11.3 In Section 11.3, we obtained explicit expressions for the renormalization constants Z_ϕ, Z_ψ, Z_m, and Z_λ of the Wess–Zumino model at one-loop accuracy and verified that the results were consistent with the prediction of the non-renormalization given in eq. (11.17). In this exercise, you will provide a cross-check of these results by computing the one-loop renormalization of the cubic scalar coupling. The relevant diagrams are given in Fig. 11.11.

(a) Draw all distinct diagrams represented by the diagrams shown below. Verify that the four diagrams corresponding to the scalar triangle diagram exhibited in Fig. 11.11 are ultraviolet finite.

(b) Show that the ultraviolet divergences of the diagrams exhibited in the

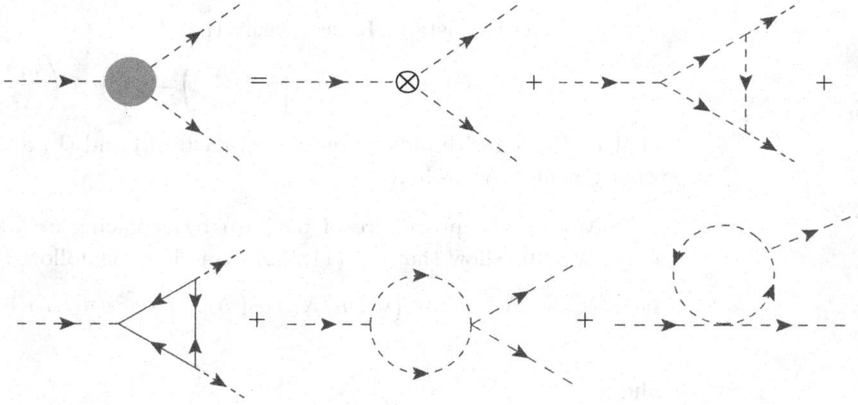

Fig. 11.11 One-loop corrections to the cubic scalar couplings. In the first row of diagrams, the scalar triangle diagram is representative of four distinct diagrams. In the second row of diagrams, the last diagram on the right is representative of two diagrams.

second row of Fig. 11.11, respectively, are

$$\frac{i\lambda^3 m}{2\pi^2\epsilon}\left[-1+\frac{1}{4}+(2)\frac{1}{2}\right]=\frac{i\lambda^3 m}{4\pi^2\epsilon}, \tag{11.220}$$

where the factor of 2 in parenthesis represents the two distinct diagrams with the topology of the last diagram in the second row of Fig. 11.11, which differ by the interchange of momenta of the two outgoing scalars.

(c) By demanding that the ϵ^{-1} pole is canceled by the cubic scalar vertex counterterm exhibited in Fig. 11.4, show that

$$Z_\phi^{3/2} Z_m Z_\lambda = 1+\frac{\lambda^2}{8\pi^2\epsilon}. \tag{11.221}$$

Verify that, at one-loop accuracy, this result is consistent with the explicit one-loop expressions for Z_ϕ, Z_m, and Z_λ obtained in Section 11.3.

11.4 Analyze the one-loop renormalization of the quartic scalar coupling of the Wess–Zumino model. Derive an explicit expression for $Z_\phi^2 Z_\lambda^2$ at one-loop accuracy, and verify that your results are consistent with eqs. (11.26) and (11.38).

11.5 The goal of this exercise is to establish eqs. (11.47) and (11.48). This can be done by considering only the topology of 1PI Feynman diagrams that lead to the polynomial counterterms $c^X_{\ell,n}$ and $c^\phi_{\ell,n}$. First, argue by induction that the quantity

$$\left(\sum_Y Y\frac{\partial}{\partial Y}-2\ell\right)c^X_{\ell,n} \tag{11.222}$$

must be the same for each counterterm $c^X_{\ell,n}$ for a given X. Do this by considering all possible ways to add a propagator line to a generic diagram with the same external lines as X, increasing the loop order ℓ by 1, and noting the effect on the coupling parameters Y that are modified or introduced as interaction vertices in the diagram by the additional line. Here, you should use the defining property that $\rho_Y = N-2$ for a parameter Y that couples N fields. Then, note that for the tree-level diagram, one can assign $c^X_{0,0}=X$ [see the first term in eq. (11.43)], so that

$$\left(\sum_Y Y\frac{\partial}{\partial Y}-2\ell\right)c^X_{0,0}=\rho_X. \tag{11.223}$$

This immediately gives eq. (11.47). The same argument works to establish eq. (11.48), by noting from the form of eq. (11.46) that at tree level one can assign $c^\phi_{0,0}=1$.

11.6 In this exercise, we will employ eq. (11.175) to compute the one-loop contribution of a neutral two-component fermion to the effective potential. For simplicity, we shall consider a theory of a real scalar field Φ and a two-component fermion field ψ with a Yukawa coupling,

$$\mathscr{L}_{\text{int}} = -\tfrac{1}{2}y\Phi(\psi\psi + \psi^\dagger\psi^\dagger)\,. \tag{11.224}$$

(a) The first step is to compute the dependence of the fermion propagator mass on the background field ϕ. In light of eq. (11.168), show that the fermion propagator mass is given by $M_F(\phi) = y\phi$. Using this notation, explain why the Feynman rule for the Yukawa coupling can be written in the form $-idM_F/d\phi$.

(b) Exhibit the tadpole diagrams that contribute to $i\widetilde{\Gamma}^{(1)}_\phi(0)$ and show that, in the one-loop approximation,

$$i\widetilde{\Gamma}^{(1)}_\phi(0) = -\frac{1}{2}\int\frac{d^dk}{(2\pi)^d}\,\frac{iM_F(\phi)}{k^2 - M_F^2(\phi) + i\varepsilon}(\delta_\alpha{}^\beta\delta_\beta{}^\alpha + \delta^{\dot\alpha}{}_{\dot\beta}\delta^{\dot\beta}{}_{\dot\alpha})\left(-i\frac{dM_F}{d\phi}\right), \tag{11.225}$$

where the symmetry factor of 1/2 for the tadpole diagrams and the factor of -1 for the fermion loop have been included.

(c) Using eq. (11.175) to compute $(dV_{\text{eff}}(\Phi)/d\Phi)_{\Phi=\phi}$, integrate to obtain

$$\frac{1}{16\pi^2}V^{(1)}(\phi) = i\mu^{2\epsilon}\int\frac{d^dk}{(2\pi)^d}\,\ln[k^2 - M_F^2(\phi) + i\varepsilon]\,, \tag{11.226}$$

where the integration constant has been dropped, and the factor of $\mu^{2\epsilon}$ has been inserted so that the Yukawa coupling is dimensionless when carrying out the integration in d spacetime dimensions.

(d) To perform the integration in $d = 4 - 2\epsilon$ dimensions using dimensional regularization, show that

$$\begin{aligned}\mu^{2\epsilon}\int\frac{d^dk}{(2\pi)^d}\,\ln(k^2 - M^2 + i\varepsilon) &= -\frac{i}{16\pi^2}\int^{M^2}A_0(m^2)\,dm^2 \\ &= -\frac{i}{32\pi^2}M^4\left[\frac{1}{\epsilon} + \frac{3}{2} - \ln\left(\frac{M^2}{Q^2}\right) + \mathcal{O}(\epsilon)\right],\end{aligned} \tag{11.227}$$

where Q^2 and $A_0(m^2)$ are defined in eqs. (11.42) and (K.9), respectively.

(e) Applying eq. (11.227) to eq. (11.226) and applying the $\overline{\text{MS}}$ scheme to eliminate Δ, show that

$$V^{(1)} = -\tfrac{1}{2}\left[M_F^2(\phi)\right]^2\left[\ln\left(\frac{M_F^2(\phi)}{Q^2}\right) - \frac{3}{2}\right], \tag{11.228}$$

thereby confirming the result of eq. (11.189).

(f) Using a similar technique as above, confirm the results of eqs. (11.188) and (11.190).

12 Spontaneous Supersymmetry Breaking

Despite the inherent beauty of supersymmetric field theories, we know that supersymmetry cannot be an exact symmetry of Nature. The observed spectrum of fundamental particles does not consist of mass-degenerate supermultiplets. Hence, supersymmetry must be broken. In this chapter, we shall discuss how supersymmetry (SUSY) breaking can arise.

12.1 General Considerations for Supersymmetry Breaking

By definition, broken supersymmetry means that the vacuum state $|0\rangle$ is not invariant under supersymmetry transformations, so $Q_\alpha|0\rangle \neq 0$ and $Q^\dagger_{\dot\alpha}|0\rangle \neq 0$. In global supersymmetry, the Hamiltonian operator H can be related to the supersymmetry generators through the algebra, as previously indicated in eq. (9.105):

$$H = P^0 = \tfrac{1}{4}(Q_1Q_1^\dagger + Q_1^\dagger Q_1 + Q_2Q_2^\dagger + Q_2^\dagger Q_2). \tag{12.1}$$

If supersymmetry is spontaneously broken in the vacuum state, then the vacuum must have positive energy,[1] since

$$\langle 0|H|0\rangle = \tfrac{1}{4}\Big(\|Q_1|0\rangle\|^2 + \|Q_1^\dagger|0\rangle\|^2 + \|Q_2|0\rangle\|^2 + \|Q_2^\dagger|0\rangle\|^2\Big) > 0\,, \tag{12.2}$$

under the assumption that the Hilbert space of physical states has positive norm. If spacetime-dependent effects and fermion condensates can be neglected, then $\langle 0|H|0\rangle = \langle 0|V|0\rangle$, where V is the scalar potential in eq. (8.75). Therefore supersymmetry will be spontaneously broken if F_i and/or D^a does not vanish in the vacuum state.

One can reach the same conclusion by considering the transformation laws of the field components of a superfield. For a chiral superfield, the component fermion field transforms according to

$$\hat\delta_\xi \psi_{\alpha i} = i\big[\xi Q + \xi^\dagger Q^\dagger\,,\, \psi_{\alpha i}\big] = -i\sqrt{2}\,(\sigma^\mu \xi^\dagger)_\alpha\, \partial_\mu \phi_i + \sqrt{2}\,\xi_\alpha F_i\,. \tag{12.3}$$

By Lorentz invariance, $\langle 0\,|\partial_\mu A_i\,|\,0\rangle = 0$. Hence,

$$\langle 0\,|\big[\xi Q + \xi^\dagger Q^\dagger\,,\, \psi_{\alpha i}\big]\,|\,0\rangle = \sqrt{2}\,\xi_\alpha\,\langle 0\,|F_i\,|\,0\rangle\,. \tag{12.4}$$

[1] These considerations do not apply in general to supergravity theories (which exhibit local supersymmetry). Indeed, there exists a theory of unbroken supergravity in anti-de Sitter space, where the vacuum energy is negative (e.g., see [B42–B44]).

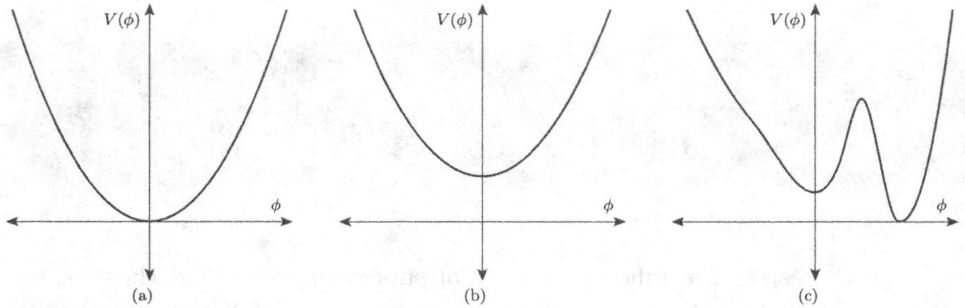

Fig. 12.1 Scalar potentials $V(\phi)$ for (a) unbroken supersymmetry, (b) spontaneously broken supersymmetry, and (c) metastable supersymmetry breaking, as a function of an order parameter ϕ.

Thus, if $Q_\alpha\,|0\rangle = 0$ and $Q^\dagger_{\dot\alpha}\,|0\rangle = 0$, then $\langle 0\,|F_i\,|\,0\rangle = 0$. Likewise, for a real vector superfield, the component gaugino field transforms according to

$$\widehat{\delta}_\xi \lambda^a_\alpha = i[\xi Q + \xi^\dagger Q^\dagger\,,\,\lambda^a_\alpha] = i\xi_\alpha D^a + \tfrac{1}{2}(\sigma^\mu\overline{\sigma}^\nu)_\alpha{}^\beta \xi_\beta F^a_{\mu\nu}\,. \tag{12.5}$$

Since $\langle 0\,|F^a_{\mu\nu}|\,0\rangle = 0$ (again, by Lorentz invariance), it follows that

$$\langle 0\,|\,[\xi Q + \xi^\dagger Q^\dagger\,,\,\lambda^a_\alpha]\,|\,0\rangle = i\xi_\alpha\,\langle 0\,|D^a|\,0\rangle\,. \tag{12.6}$$

Thus, if $Q_\alpha\,|0\rangle = 0$ and $Q^\dagger_{\dot\alpha}\,|0\rangle = 0$, then $\langle 0\,|D^a\,|\,0\rangle = 0$. If at least one of the components of the auxiliary fields F_i or D_a has a nonzero vacuum expectation value, then supersymmetry is spontaneously broken. Mechanisms of spontaneous supersymmetry breaking fall into two possible categories: F-term breaking, if $\langle 0\,|F_i\,|\,0\rangle \neq 0$ for some i; and D-term breaking, if $\langle 0\,|D^a|\,0\rangle \neq 0$ for some a.

If any state exists in which all F_i and D^a vanish, then it will have zero energy, implying that supersymmetry is not spontaneously broken in the true ground state. Conversely, one way to guarantee spontaneous supersymmetry breaking is to look for models in which the equations $F_i = 0$ and $D^a = 0$ cannot all be simultaneously satisfied for *any* values of the fields. Then the true ground state necessarily has broken supersymmetry, as does the vacuum state we live in (if it is different).

However, another possibility is that the vacuum state in which we live is not the true ground state (which may preserve supersymmetry), but is instead a higher-energy, metastable, supersymmetry-breaking state with lifetime at least of the order of the present age of the universe. Finite temperature effects can indeed cause the early universe to prefer the metastable supersymmetry-breaking local minimum of the potential over the supersymmetry-breaking global minimum. Scalar potentials for the three possibilities are illustrated qualitatively in Fig. 12.1.

Regardless of whether the vacuum state is stable or metastable, the spontaneous breaking of a global symmetry always implies a massless Goldstone mode with the same quantum numbers as the broken symmetry generator. In the case of global supersymmetry, the broken generator is the fermionic charge Q_α, so the Goldstone particle ought to be a massless neutral Weyl fermion, called the *goldstino*. To prove it, consider a general supersymmetric model with both gauge and chiral supermultiplets as in Section 8.5. The fermionic degrees of freedom consist of gauginos (λ^a) and chiral fermions (ψ_i). After some of the scalar fields in the theory obtain VEVs, the fermion mass matrix has the form

$$\boldsymbol{m}_{\mathrm{F}} = \begin{pmatrix} 0 & \sqrt{2}g_b(\langle\phi^\dagger\rangle T^b)^i \\ \sqrt{2}g_a(\langle\phi^\dagger\rangle T^a)^j & \langle W^{ji}\rangle \end{pmatrix} \tag{12.7}$$

in the (λ^a, ψ_i) basis. [The off-diagonal entries in this matrix come from the first term in the second line of eq. (8.72), and the lower right entry can be seen in eq. (8.49).] Now observe that $\boldsymbol{m}_{\mathrm{F}}$ annihilates the vector

$$\widetilde{G} = \begin{pmatrix} \langle D^a\rangle/\sqrt{2} \\ \langle F_i\rangle \end{pmatrix}. \tag{12.8}$$

The first row of $\boldsymbol{m}_{\mathrm{F}}$ annihilates $\widetilde{G}$ by virtue of the requirement eq. (8.73) that the superpotential is gauge invariant, and the second row does so because of the condition $\langle\partial V/\partial\phi_i\rangle = 0$, which must be satisfied at a local minimum of the scalar potential. Equation (12.8) is therefore proportional to the goldstino wave function; it is nontrivial if and only if at least one of the auxiliary fields has a VEV, breaking supersymmetry. So we have proved that if global supersymmetry is spontaneously broken, then there must be a massless goldstino, and that its components among the various fermions in the theory are proportional to the corresponding auxiliary field VEVs.

There is also a useful sum rule that governs the tree-level squared masses of particles in theories with spontaneously broken supersymmetry. For a general theory of the type discussed in Section 8.5, the squared masses of the real scalar degrees of freedom are the eigenvalues of the matrix

$$\boldsymbol{m}^2_{\mathrm{S}} = \begin{pmatrix} W^\dagger_{jk}W^{ik} + g_a^2(T^a\phi)_j(\phi^\dagger T^a)^i - g_aT^{ai}_j D^a & W^\dagger_{ijk}W^k + g_a^2(T^a\phi)_i(T^a\phi)_j \\ W^{ijk}W^\dagger_k + g_a^2(\phi^\dagger T^a)^i(\phi^\dagger T^a)^j & W^\dagger_{ik}W^{jk} + g_a^2(T^a\phi)_i(\phi^\dagger T^a)^j - g_aT^{aj}_i D^a \end{pmatrix}, \tag{12.9}$$

since the quadratic part of the tree-level potential is

$$V = \frac{1}{2}\begin{pmatrix} \phi^{\dagger j} & \phi_j \end{pmatrix} \boldsymbol{m}^2_{\mathrm{S}} \begin{pmatrix} \phi_i \\ \phi^{\dagger i} \end{pmatrix}. \tag{12.10}$$

Here $W^{ijk} = \delta^3 W/\delta\phi_i\delta\phi_j\delta\phi_k$, and the scalar fields on the right-hand side of eq. (12.9) are understood to be replaced by their VEVs. It follows that the sum of the real scalar squared-mass eigenvalues is

$$\mathrm{Tr}(\boldsymbol{m}^2_{\mathrm{S}}) = 2W^\dagger_{ik}W^{ik} + 2g_a^2C_a(i)\phi^{\dagger i}\phi_i - 2g_a\mathrm{Tr}(T^a)D^a, \tag{12.11}$$

with the Casimir invariants $C_a(i)$ defined by $(T^aT^a)_ij = C_a(i)\delta_i^j$. Meanwhile, the squared masses of the two-component fermions are given by the eigenvalues of

$$\boldsymbol{m}_{\rm F}^\dagger \boldsymbol{m}_{\rm F} = \begin{pmatrix} 2g_a g_b(\phi^\dagger T^a T^b \phi) & \sqrt{2} g_b (T^b\phi)_k W^{ik} \\ \sqrt{2} g_a (\phi^\dagger T^a)^k W^\dagger_{jk} & W^\dagger_{jk} W^{ik} + 2g_c^2 (T^c\phi)_j (\phi^\dagger T^c)^i \end{pmatrix}, \tag{12.12}$$

so the sum of the two-component fermion squared masses is

$$\mathrm{Tr}(\boldsymbol{m}_{\rm F}^\dagger \boldsymbol{m}_{\rm F}) = W^\dagger_{ik} W^{ik} + 4g_a^2 C_a(i) \phi^{\dagger i}\phi_i. \tag{12.13}$$

Finally, the vector squared masses are

$$\boldsymbol{m}_{\rm V}^2 = g_a^2(\phi^\dagger\{T^a, T^b\}\phi), \tag{12.14}$$

so

$$\mathrm{Tr}(\boldsymbol{m}_{\rm V}^2) = 2g_a^2 C_a(i)\phi^{\dagger i}\phi_i. \tag{12.15}$$

It follows that the supertrace of the tree-level squared-mass eigenvalues, defined in general by a weighted sum over all particles with spin j,

$$\mathrm{STr}(m^2) \equiv \sum_J (-1)^J (2J+1)\mathrm{Tr}(m_J^2), \tag{12.16}$$

satisfies the sum rule

$$\mathrm{STr}(m^2) = \mathrm{Tr}(\boldsymbol{m}_{\rm S}^2) - 2\mathrm{Tr}(\boldsymbol{m}_{\rm F}^\dagger \boldsymbol{m}_{\rm F}) + 3\mathrm{Tr}(\boldsymbol{m}_{\rm V}^2) = -2g_a \mathrm{Tr}(T^a) D^a = 0. \tag{12.17}$$

The last equality assumes that the traces of the U(1) charges over the chiral superfields are 0. This holds for $\mathrm{U}(1)_Y$ in the MSSM, and more generally for any nonanomalous gauge symmetry. The sum rule eq. (12.17) is often a useful check on models of spontaneous supersymmetry breaking.

12.2 *D*-term SUSY Breaking: The Fayet–Iliopoulos Mechanism

Supersymmetry breaking with a nonzero D-term VEV can occur through the Fayet–Iliopoulos mechanism [137]. If the gauge symmetry includes a U(1) factor, then one can introduce a Fayet–Iliopoulos (FI) term linear in the corresponding auxiliary field of the gauge supermultiplet:

$$\mathscr{L}_{\rm FI} = -\kappa D\,, \tag{12.18}$$

where κ is a constant with dimensions of [mass]^2. This term is gauge-invariant and supersymmetric by itself. [Note that for a U(1) gauge symmetry, the supersymmetry transformation δD in eq. (8.62) is a total derivative.] If we include it in the

Lagrangian, then D may be forced to get a nonzero VEV. To see this, consider the relevant part of the scalar potential from eqs. (8.57) and (8.72):

$$V = \kappa D - \tfrac{1}{2}D^2 - gD\sum_i q_i|\phi_i|^2. \tag{12.19}$$

Here the q_i are the charges of the scalar fields ϕ_i under the U(1) gauge group in question. The presence of the Fayet–Iliopoulos term modifies the equation of motion eq. (8.74) to

$$D = \kappa - g\sum_i q_i|\phi_i|^2. \tag{12.20}$$

Now suppose that the scalar fields ϕ_i that are charged under the U(1) all have nonzero superpotential masses m_i. (Gauge invariance then requires that they come in pairs with opposite charges; each of the pairs has the same squared mass $|m_i|^2$.) Then the potential will have the form

$$V = \sum_i |m_i|^2|\phi_i|^2 + \frac{1}{2}\left(\kappa - g\sum_i q_i|\phi_i|^2\right)^2. \tag{12.21}$$

Since this cannot vanish, supersymmetry must be broken; one can check that the minimum always occurs for nonzero D. Consider the simplest case in which $|m_i|^2 > gq_i\kappa$ for each i. Since the massive scalars must come in pairs with opposite q_i, this requirement can be rewritten as $|m_i|^2 > |gq_i\kappa|$. The minimum is then realized for all $\phi_i = 0$ and $D = \kappa$, with the U(1) gauge symmetry unbroken. As further evidence that supersymmetry has indeed been spontaneously broken, note that the scalars then have squared masses $|m_i|^2 - gq_i\kappa$, while their fermion partners have squared masses $|m_i|^2$. The gaugino remains massless, as can be understood from the fact that it is the goldstino, as argued on general grounds in Section 12.1.

If all charged scalars are massive but at least one has $0 < |m_i|^2 < |gq_i\kappa|$, then that scalar will have a nonzero value at the minimum of the potential, and both supersymmetry and the gauge symmetry will be spontaneously broken. Note that, if even one of the scalars is massless and has positive κ/q_i, there is an absolute minimum of the potential for the corresponding $|\phi_i|^2 = \kappa/gq_i$ with all other scalar fields vanishing, with $V = 0$. In that case, the U(1) gauge symmetry is spontaneously broken, and supersymmetry remains unbroken.

For nonabelian gauge groups, the analogue of eq. (12.18) would not be gauge-invariant and is therefore not allowed, so only U(1) D-terms can drive spontaneous symmetry breaking. In the MSSM, one might imagine that the D-term for $U(1)_Y$ has a Fayet–Iliopoulos term as the principal source of supersymmetry breaking. Unfortunately, this cannot work, because the squarks and sleptons do not have superpotential mass terms. So, at least some of them would just get nonzero VEVs in order to make eq. (12.20) vanish. That would break color and/or electromagnetism, but not supersymmetry. Therefore, a Fayet–Iliopoulos term for $U(1)_Y$ must be subdominant compared to other sources of supersymmetry breaking in the MSSM, if not absent altogether. One could instead attempt to trigger supersymmetry breaking with a Fayet–Iliopoulos term for some other U(1) gauge symmetry, which is as yet unknown because it is spontaneously broken at a very high mass scale or because it does not couple to the Standard Model particles. However, if this is the dominant source for supersymmetry breaking, it proves difficult to give appropriate masses to all of the MSSM particles, especially the gauginos. In any case, we will not discuss D-term breaking as the ultimate origin of supersymmetry violation any further (although it may not be ruled out).

12.3 F-term SUSY Breaking: The O'Raifeartaigh Mechanism

Models where spontaneous supersymmetry breaking is ultimately due to a nonzero F-term VEV, called O'Raifeartaigh models [138], have brighter phenomenological prospects. The idea is to pick a set of chiral supermultiplets $\Phi_i \supset (\phi_i, \psi_i, F_i)$ and a superpotential W in such a way that the equations

$$F_i = -\frac{\partial W^\dagger}{\partial \phi^{\dagger i}} = 0 \tag{12.22}$$

have no simultaneous solution. Then $V = \sum_i |F_i|^2$ will have to be positive at its minimum, ensuring that supersymmetry is broken. The supersymmetry-breaking minimum may be a global minimum of the potential as in Fig. 12.1(b), or only a local minimum as in Fig. 12.1(c).

The simplest example with a supersymmetry-breaking global minimum has three chiral supermultiplets $\Phi_{1,2,3}$, with superpotential

$$W = -k\Phi_1 + m\Phi_2\Phi_3 + \tfrac{1}{2}y\Phi_1\Phi_3^2. \tag{12.23}$$

Note that W contains a linear term, with k having dimensions of $[\text{mass}]^2$. Such a term is allowed if the corresponding chiral supermultiplet is a gauge singlet. In fact, a linear term is necessary to achieve F-term breaking at tree level in renormalizable theories,[2] since otherwise setting all $\phi_i = 0$ will always give a supersymmetric global minimum with all $F_i = 0$. Without loss of generality, we can choose k, m, and y to be real and positive (by a phase rotation of the fields). The scalar potential following from eq. (12.23) is

$$V = |F_1|^2 + |F_2|^2 + |F_3|^2, \tag{12.24}$$

$$F_1 = k - \tfrac{1}{2}y\phi_3^{\dagger 2}, \qquad F_2 = -m\phi_3^\dagger, \qquad F_3 = -m\phi_2^\dagger - y\phi_1^\dagger\phi_3^\dagger. \tag{12.25}$$

Clearly, $F_1 = 0$ and $F_2 = 0$ are not compatible, so supersymmetry must indeed be broken. If $m^2 > yk$ (which we assume from now on), then the absolute minimum

[2] Nonpolynomial superpotential terms, for example arising from nonperturbative effects, can avoid this requirement.

of the potential is at $\phi_2 = \phi_3 = 0$ with ϕ_1 undetermined, so $F_1 = k$ and $V = k^2$ at the minimum. The fact that ϕ_1 is undetermined is an example of a "flat direction" in the scalar potential; this is a common feature of supersymmetric models.[3]

The flat direction parameterized by ϕ_1 is an accidental feature of the classical scalar potential, and in this case it is removed ("lifted") by quantum corrections. This can be seen by computing the Coleman–Weinberg one-loop effective potential. In a loop expansion, the effective potential can be written as

$$V_{\text{eff}} = V^{(0)} + \frac{1}{16\pi^2} V^{(1)} + \cdots, \tag{12.26}$$

where the one-loop contribution is a supertrace over the scalar-field-dependent squared-mass eigenstates labeled n, with spin s_n:

$$V^{(1)} = \sum_n (-1)^{2s_n}(2s_n + 1) h(m_n^2), \tag{12.27}$$

$$h(z) \equiv \tfrac{1}{4} z^2 \left[\ln(z/Q^2) + a\right]. \tag{12.28}$$

Here Q is the renormalization scale and a is a renormalization scheme-dependent constant.[4] In the $\overline{\text{DR}}$ scheme based on dimensional reduction, $a = -3/2$. Using eqs. (12.9) and (12.12), the squared-mass eigenvalues for the six real scalar and three two-component fermion states are found to be, as a function of varying $x = |\phi_1|^2$, with $\phi_2 = \phi_3 = 0$,

$$\begin{aligned} \text{scalars:}\quad & 0,\ 0,\ m^2 + \tfrac{1}{2}y\Big(yx - k + \sqrt{4m^2x + (yx-k)^2}\Big), \\ & m^2 + \tfrac{1}{2}y\Big(yx + k - \sqrt{4m^2x + (yx+k)^2}\Big), \\ & m^2 + \tfrac{1}{2}y\Big(yx - k - \sqrt{4m^2x + (yx-k)^2}\Big), \\ & m^2 + \tfrac{1}{2}y\Big(yx + k + \sqrt{4m^2x + (yx+k)^2}\Big), \end{aligned} \tag{12.29}$$

$$\begin{aligned} \text{fermions:}\quad & 0,\ m^2 + \tfrac{1}{2}y\Big(yx + \sqrt{4m^2x + y^2x^2}\Big), \\ & m^2 + \tfrac{1}{2}y\Big(yx - \sqrt{4m^2x + y^2x^2}\Big). \end{aligned} \tag{12.30}$$

[Note that the sum rule eq. (12.17) is indeed satisfied by these squared masses.] Now, plugging these into eq. (12.28), one finds that the global minimum of the one-loop effective potential is at $x = 0$, so $\phi_1 = \phi_2 = \phi_3 = 0$. The tree-level mass spectrum of the theory at this point in field space simplifies to

$$0,\ 0,\ m^2,\ m^2,\ m^2 - yk,\ m^2 + yk \tag{12.31}$$

3 More generally, "flat directions" are noncompact lines and surfaces in the space of scalar fields along which the scalar potential vanishes. The classical scalar potential of the MSSM would have many flat directions if supersymmetry were not broken.

4 Actually, a can be different for the different spin contributions, if one chooses a renormalization scheme that does not respect supersymmetry. For example, in the $\overline{\text{MS}}$ scheme, $a = -3/2$ for the spin-0 and spin-1/2 contributions, but $a = -5/6$ for the spin-1 contribution. See Section 11.6 for a discussion and the extension to two-loop order.

for the scalars, and

$$0, \; m^2, \; m^2 \tag{12.32}$$

for the fermions. The nondegeneracy of scalars and fermions is a clear check that supersymmetry has been spontaneously broken.

The 0 eigenvalues in eqs. (12.31) and (12.32) correspond to the complex scalar ϕ_1 and its fermionic partner ψ_1. However, ϕ_1 and ψ_1 have different reasons for being massless. The masslessness of ϕ_1 corresponds to the existence of the classical flat direction, since any value of ϕ_1 gives the same energy at tree level. The one-loop potential lifts this flat direction, so that ϕ_1 gains a mass once quantum corrections are included. Expanding $V^{(1)}$ to first order in x, one finds that the complex scalar ϕ_1 receives a positive-definite squared mass equal to

$$m_{\phi_1}^2 = \frac{y^2 m^2}{16\pi^2}\left[\ln(1-r^2) - 1 + \tfrac{1}{2}\,(r+1/r)\ln\left(\frac{1+r}{1-r}\right)\right], \tag{12.33}$$

where $r = yk/m^2$. [This reduces to $m_{\phi_1}^2 = y^4k^2/48\pi^2m^2$ in the limit $yk \ll m^2$.] In contrast, the Weyl fermion ψ_1 remains exactly massless, to all orders in perturbation theory, because it is the goldstino, as predicted in Section 12.1.

The O'Raifeartaigh superpotential of eq. (12.23) provides a Lagrangian that is invariant under a $\mathrm{U}(1)_R$ symmetry (see Section 10.11) with charge assignments

$$r_{\Phi_1} = r_{\Phi_2} = 2, \qquad r_{\Phi_3} = 0. \tag{12.34}$$

This illustrates a general result, the Nelson–Seiberg Theorem [139], which says that if a theory has a scalar potential with a global minimum that breaks supersymmetry by a nonzero F-term, and the superpotential is generic (allows all terms not forbidden by symmetries), then the theory must have an exact $\mathrm{U}(1)_R$ symmetry. If the $\mathrm{U}(1)_R$ symmetry remains unbroken when supersymmetry breaks, as is the case in the O'Raifeartaigh model discussed above, then there is a problem of explaining how gauginos get masses, because nonzero gaugino mass terms have R charge of 2. On the other hand, if the $\mathrm{U}(1)_R$ symmetry is spontaneously broken, then there results a pseudo-Goldstone boson (the R-axion) which is problematic experimentally, although gravitational effects may give it a large enough mass to avoid being ruled out.

If the supersymmetry-breaking vacuum is only metastable, then one does not need an exact $\mathrm{U}(1)_R$ symmetry. This can be illustrated by adding to the O'Raifeartaigh superpotential eq. (12.23) a term ΔW that explicitly breaks the continuous R-symmetry. For example, following Ref. [140], we consider

$$\Delta W = \tfrac{1}{2}\epsilon m \Phi_2^2, \tag{12.35}$$

where ϵ is a small dimensionless parameter, so that the tree-level scalar potential is given by

$$V^{(0)} = |F_1|^2 + |F_2|^2 + |F_3|^2, \tag{12.36}$$

$$F_1 = k - \tfrac{1}{2}y\phi_3^{\dagger 2}, \qquad F_2 = -\epsilon m\phi_2^\dagger - m\phi_3^\dagger, \qquad F_3 = -m\phi_2^\dagger - y\phi_1^\dagger\phi_3^\dagger. \tag{12.37}$$

In accord with the Nelson–Seiberg Theorem, there are now (two) supersymmetric minima, with

$$\phi_1 = m/\epsilon y, \qquad \phi_2 = \pm\frac{1}{\epsilon}\sqrt{2k/y}, \qquad \phi_3 = \mp\sqrt{2k/y}. \tag{12.38}$$

However, for small enough ϵ, the local supersymmetry-breaking minimum at $\phi_1 = \phi_2 = \phi_3 = 0$ is also still present and stabilized by the one-loop effective potential, with potential barriers between it and the supersymmetric minima, so the situation is qualitatively like Fig. 12.1(c). As $\epsilon \to 0$, the supersymmetric global minima move off to infinity in field space, and there is negligible effect on the supersymmetry-breaking local minimum. One can show [140] that the lifetime of the metastable vacuum state due to quantum tunneling can be made arbitrarily large. The same effect can be realized by a variety of other perturbations to the O'Raifeartaigh model; by eliminating the continuous R-symmetry with small additional contributions to the Lagrangian, one converts the stable supersymmetry-breaking vacuum to a metastable one. (In some cases, the Lagrangian remains invariant under a discrete R-symmetry.)

The O'Raifeartaigh superpotential determines the mass scale of supersymmetry breaking $\sqrt{F_1}$ in terms of a dimensionful parameter k put in by hand. This appears somewhat artificial, since k will have to be tiny compared to M_P^2 in order to give the right order of magnitude for the MSSM soft terms. It may be more plausible to have a mechanism that can instead generate such scales naturally. This can be done in models of dynamical supersymmetry breaking, in which the small mass scales associated with supersymmetry breaking arise by dimensional transmutation. In other words, they generally feature a new asymptotically free nonabelian gauge symmetry with a gauge coupling g that is perturbative at M_P and gets strong in the infrared at some smaller scale $\Lambda \sim e^{-8\pi^2/|b|g_0^2} M_P$, where g_0 is the running gauge coupling at M_P with negative beta function $-|b|g^3/16\pi^2$. Just as in QCD, it is perfectly natural for Λ to be many orders of magnitude below the Planck scale. Supersymmetry breaking may then be best described in terms of the effective dynamics of the strongly coupled theory. Supersymmetry is still broken by the VEV of an F field, but it may be the auxiliary field of a composite chiral supermultiplet built out of fields that are charged under the new strongly coupled gauge group.

The construction of such models that break supersymmetry via strong-coupling dynamics is nontrivial if one wants a stable supersymmetry-breaking ground state. In addition to the argument from the Nelson–Seiberg Theorem that a $U(1)_R$ symmetry should be present, one can prove using the Witten index [95, 141] that any strongly coupled gauge theory with only vectorlike, massive matter cannot spontaneously break supersymmetry in its true ground state. However, things are easier if one only requires a local (metastable) minimum of the potential. Intriligator, Seiberg, and Shih showed in Ref. [142] that supersymmetric Yang–Mills theories with vectorlike matter can have metastable vacuum states with nonvanishing F-terms that break supersymmetry, and lifetimes that can be arbitrarily long. The simplest model that does this is remarkably economical; it is just supersymmetric

$SU(N_c)$ gauge theory, with N_f massive flavors of quark and antiquark supermultiplets, with $N_c+1 \leq N_f < 3N_c/2$. The recognition of the advantages of a metastable vacuum state opens up many new model building possibilities and ideas [140, 142].

The topic of known ways of breaking supersymmetry spontaneously through strongly coupled gauge theories is a big subject (e.g., see [B38]) that is in danger of becoming vast, and is beyond the scope of this book.

12.4 Effective Theory of Supersymmetry Breaking

In the absence of a fundamental theory of supersymmetry breaking, it still may be possible to account for the effects of supersymmetry breaking using an effective field theory approach. For example, suppose that Λ represents the characteristic mass scale of the fundamental dynamics responsible for supersymmetry breaking. Then, at energy scales below Λ, one can formally integrate out the degrees of freedom associated with the supersymmetry-breaking dynamics. The end result is an effective broken supersymmetric theory whose Lagrangian consists of supersymmetric terms and explicit soft supersymmetry-breaking terms.

Consider a set of light chiral superfields Φ and a set of heavy chiral superfields Ω associated with a mass scale Λ. Furthermore, assume that supersymmetry breaking is generated by an F-term that resides in the supersymmetry-breaking sector,

$$\langle F_\Omega \rangle = F \neq 0\,. \tag{12.39}$$

Note that $F^{1/2}$ identifies the energy scale of the fundamental supersymmetry breaking. One can integrate out the physics of the supersymmetry-breaking sector, as shown in the following three examples [91, 143, 144].

First, consider a holomorphic cubic multinomial of chiral superfields Φ, which we denote by $\widetilde{w}(\Phi)$. A possible term in the effective Lagrangian is

$$\frac{1}{\Lambda}\int d^2\theta\, \Omega\, \widetilde{w}(\Phi)\,, \tag{12.40}$$

since $\Omega\, \widetilde{w}(\Phi)$ is a term in the superpotential. The factor of Λ^{-1} appears on the basis of dimensional analysis. In particular, note the mass dimensions, $[\widetilde{w}] = 3$, $[\Omega]$=1, and $[\int d^2\theta] = 1$.

Since the vacuum expectation value of F_Ω, denoted by $\langle F_\Omega \rangle = F$, is nonzero, it follows that $\langle \Omega \rangle \supset \theta\theta F$. Inserting this into eq. (12.40) yields

$$\frac{1}{\Lambda}\int d^2\theta\, \theta\theta F\, \widetilde{w}(\Phi) = \frac{F}{\Lambda}\, \widetilde{w}(\phi)\,. \tag{12.41}$$

In order to achieve soft supersymmetry-breaking masses in the low-energy effective theory of order 1 TeV, one must require that $f/\Lambda \sim \mathcal{O}(1 \text{ TeV})$.

Second, we consider another possible term in the effective Lagrangian,

$$\frac{1}{\Lambda^2}\int d^4\theta\, \Phi_i^\dagger \left(e^{2gV}\right)_{ij} \Phi_j \Omega^\dagger \Omega\,, \tag{12.42}$$

which would contribute to the Kähler potential. Setting $\langle\Omega\rangle = \theta\theta F$ and evaluating the result in the Wess–Zumino gauge,

$$\frac{F^2}{\Lambda^2}\int d^4\theta\, (\theta\theta)(\theta^\dagger\theta^\dagger)\Phi_i^\dagger \left(e^{2gV}\right)_{ij} \Phi_j = \frac{F^2}{\Lambda^2}\phi^\dagger\phi\,, \tag{12.43}$$

which yields a scalar squared-mass term of order F/Λ.

Finally, a third possible term in the effective Lagrangian is

$$\frac{1}{\Lambda}\int d^2\theta\, \Omega\, \mathrm{Tr}(W^\alpha W_\alpha)\,, \tag{12.44}$$

which would contribute to the gauge kinetic function. Setting $\langle\Omega\rangle = \theta\theta F$,

$$\frac{F}{\Lambda}\int d^2\theta\, \theta\theta\, \mathrm{Tr}(W^\alpha W_\alpha) = -\frac{F}{\Lambda}\,\mathrm{Tr}(\lambda^\alpha\lambda_\alpha)\,, \tag{12.45}$$

which yields a gaugino mass term of order F/Λ.

Equations (12.41), (12.43), and (12.45) yield the catalog of soft supersymmetry-breaking terms in a supersymmetric Yang–Mills theory coupled to supermatter, previously given in eq. (8.78) and repeated here for the convenience of the reader:

$$-\mathscr{L}_{\text{soft}} = m_{ij}^2\phi_i^\dagger\phi_j + \tfrac{1}{2}\left[m_{ab}\lambda^a\lambda^b + \text{h.c.}\right] + \left[w(\phi) + \text{h.c.}\right], \tag{12.46}$$

where there is an implicit sum over repeated indices. The scalar squared-mass matrix m_{ij}^2 is hermitian and the gaugino mass matrix m_{ab} is complex symmetric. The function $w(\phi)$ is a holomorphic cubic multinomial of the scalar fields:

$$w(\phi) = c_i\phi_i + b_{ij}\phi_i\phi_j + a_{ijk}\phi_i\phi_j\phi_k\,. \tag{12.47}$$

Note that $c_i = 0$ in the absence of any gauge-singlet fields. In the literature, the b_{ij} are called the B-terms and the a_{ijk} are called the A-terms. Note the corresponding mass dimensions, $[b_{ij}] = 2$ and $[a_{ijk}] = 1$.

One notable feature of eq. (12.46) is the absence of non-supersymmetric fermion mass terms, $m_{ij}\psi_i\psi_j + \text{h.c.}$, and nonholomorphic cubic terms in the scalar fields (e.g., $\phi_i\phi_j\phi_k^\dagger$, etc.), previously given in eq. (8.79). Although such terms are technically soft in models with no gauge singlets [145–149], these terms rarely arise in actual models of fundamental supersymmetry breaking, or if present are highly suppressed [150]. Henceforth, we shall neglect them.

In general, there is no relation between $w(\phi)$ and the superpotential, which under the assumption of renormalizability has the following generic form:

$$W(\Phi) = \kappa_i\Phi_i + \mu_{ij}\Phi_i\Phi_j + \lambda_{ijk}\Phi_i\Phi_j\Phi_k\,. \tag{12.48}$$

But some models of fundamental supersymmetry breaking yield the relations

$$c_i = C\kappa_i\,, \qquad b_{ij} = B\mu_{ij}\,, \qquad a_{ijk} = A\lambda_{ijk}\,, \tag{12.49}$$

which relate the coefficients of $w(\phi)$ to the coefficients of $W(\Phi)$.

Exercises

12.1 An O'Raifeartaigh model that exhibits F-term supersymmetry breaking must involve at least three chiral superfields. Consider the O'Raifeartaigh model whose superpotential is given by eq. (12.23).

(a) Find the minimum of the scalar potential V_ϕ and show that $\langle 0|V_\phi|0\rangle > 0$. Identify the goldstino of this model.

(b) Compute the mass spectrum of the fermions and bosons, and verify that the mass sum rule, eq. (12.17), is satisfied.

12.2 Show that the mass sum rule, eq. (12.17), is valid in the limit of exact supersymmetry, i.e., when the masses of bosons and fermions are equal.

12.3 The SUSY-QED Lagrangian constructed in Exercise 10.13 did not include a Fayet–Iliopoulos term. In this exercise, we shall add the Fayet–Iliopoulos term, $\mathscr{L}_{\rm FI} = -\kappa D$, to the SUSY-QED Lagrangian [see eq. (12.18)].

(a) Show that, in the case of SUSY-QED with a Fayet–Iliopoulos term and $m^2 > e|\kappa|$, supersymmetry is broken and the goldstino can be identified as the photino (the supersymmetric partner of the photon). In the case of $m^2 < e|\kappa|$, is supersymmetry broken? Is the U(1) gauge symmetry broken?

(b) Determine the masses of the electron and its scalar partners, and the masses of the photon and photino. Consider separately the two cases where $m^2 > e|\kappa|$ and $m^2 < e|\kappa|$, respectively. Evaluate $\mathrm{STr}\, M^2$ in both cases, and compare with eq. (12.17).

12.4 Show that in a supersymmetric nonabelian gauge theory that is coupled to supermatter, only F-type supersymmetry breaking is allowed. To prove this statement, assume that a solution to $\langle F_i\rangle = 0$ exists and show that one can always find a choice of scalar fields ϕ_i that provide a solution to eq. (12.22) such that $\langle D^a\rangle = 0$ for all a.

HINT: If the ϕ_i provide a solution to eq. (12.22), then so do the corresponding gauge-transformed scalar fields, $(e^{-2ig\Lambda})_{ij}\phi_j$. The key observation is that the superpotential is a holomorphic function of the scalar fields ϕ_i. Hence, one can generate additional solutions to eq. (12.22) by taking g complex, which will modify $\langle D^a\rangle$. Conclude that there must then be a set of ϕ_i such that $\langle F_i\rangle = \langle D^a\rangle = 0$. See [B40] for further details.

PART III

REALISTIC SUPERSYMMETRIC MODELS

13 The Minimal Supersymmetric Standard Model

In Chapter 8, we have found a general recipe for constructing Lagrangians for softly broken supersymmetric theories. We are now ready to apply these general results to the MSSM. The particle content for the MSSM was described in Chapter 7. In this chapter we will complete the model by specifying the superpotential and the soft supersymmetry-breaking terms.

13.1 A Warmup Exercise: SUSY-QED

Before delving into the full structure of the MSSM, we shall briefly examine a simple subset of the MSSM that corresponds to the supersymmetric extension of QED (see Exercise 10.13) along with the corresponding soft supersymmetry-breaking terms. This model contains an electron with electric charge $Q=-e$ and mass m (in a convention where $e>0$), which is a Dirac fermion that is represented by a pair of two-component spinors, χ and η, with opposite electric charges. These two-component fermion fields are embedded into two chiral superfields [see eq. (10.91)],

$$\Phi_L(y)=\phi_L(y)+\sqrt{2}\theta\chi(y)+\theta\theta F_L(y), \qquad \Phi_R(y)=\phi_R(y)+\sqrt{2}\theta\eta(y)+\theta\theta F_R(y), \tag{13.1}$$

where ϕ_L and ϕ_R are the selectron fields. Note that the electric charges of Φ_L and Φ_R are $Q=-e$ and $Q=+e$, respectively.

The supersymmetric QED Lagrangian is obtained following the recipe given in Chapter 8. The (renormalizable) gauge-invariant superpotential is unique:

$$W=m\Phi_L\Phi_R\,. \tag{13.2}$$

The photon field A^μ and photino field λ (along with the auxiliary field D) reside in a vector supermultiplet. After using the equations of motion to eliminate the auxiliary fields F_L, F_R, and D, we obtain the softly broken SUSY-QED Lagrangian,

$$\begin{aligned}
\mathscr{L}=&-\tfrac{1}{4}F_{\mu\nu}F^{\mu\nu}-\frac{1}{2a}(\partial_\mu A^\mu)^2+i\lambda^\dagger\bar\sigma^\mu\partial_\mu\lambda+i\left(\chi^\dagger\bar\sigma^\mu\nabla_\mu\chi+\eta^\dagger\bar\sigma^\mu\nabla_\mu\eta\right)\\
&+(\nabla_\mu\phi_L)^\dagger(\nabla^\mu\phi_L)+(\nabla_\mu\phi_R)^\dagger(\nabla^\mu\phi_R)-\tfrac{1}{2}e^2\left(|\phi_L|^2-|\phi_R|^2\right)^2\\
&-m(\eta\chi+\chi^\dagger\eta^\dagger)-\sqrt{2}e\left(\phi_R^\dagger\eta\lambda+\lambda^\dagger\eta^\dagger\phi_R-\phi_L^\dagger\chi\lambda-\lambda^\dagger\chi^\dagger\phi_L\right)\\
&-m_L^2\phi_L^\dagger\phi_L-m_R^2\phi_R^\dagger\phi_R-(b\phi_L\phi_R+\text{h.c.})-\tfrac{1}{2}M(\lambda\lambda+\lambda^\dagger\lambda^\dagger)\,,
\end{aligned} \tag{13.3}$$

where we have employed the Wess–Zumino gauge, in which the vector superfield that contains the photon field A^μ is given by eq. (10.121). Note that

$$\nabla_\mu \equiv \partial_\mu + iQA_\mu\,, \tag{13.4}$$

where $Q = -e$ when acting on the fields residing in Φ_L and $Q = +e$ when acting on the fields residing in Φ_R. In addition, we have included a gauge-fixing term for the photon field. Finally, in the last line of eq. (13.3) we have added soft supersymmetry-breaking masses $m_{L,R}$ for the selectron fields, a term proportional to the complex supersymmetry-breaking b parameter [see eq. (8.78)] that leads to ϕ_L–$\phi_R^\dagger$ mixing, and a Majorana mass M for the photino.

In the treatment above, we have omitted the supersymmetric Fayet–Iliopoulos term. Including such a term would yield the following scalar potential:

$$V_{\text{scalar}} = \tfrac{1}{2}\left[e\left(|\phi_L|^2 - |\phi_R|^2\right) + \kappa\right]^2\,, \tag{13.5}$$

where κ is defined in eq. (10.155). It is a remarkable fact that if $\kappa = 0$ at tree level, then $\kappa = 0$ to all orders in perturbation theory of SUSY-QED [151].

13.2 MSSM Superpotential and Supersymmetric Interactions

The superpotential for the MSSM is given by

$$W_{\text{MSSM}} = -H_u Q \boldsymbol{y_u} \overline{u} + H_d Q \boldsymbol{y_d} \overline{d} + H_d L \boldsymbol{y_e} \overline{e} + \mu H_u H_d\,, \tag{13.6}$$

where $\boldsymbol{y_u} \equiv \widehat{\boldsymbol{y}}_{\boldsymbol{u}}^2$, $\boldsymbol{y_d} \equiv \widehat{\boldsymbol{y}}_{\boldsymbol{d}1}$, and $\boldsymbol{y_e} \equiv \widehat{\boldsymbol{y}}_{\boldsymbol{e}1}$ in the notation of eqs. (6.99) and (6.218). The chiral superfields H_u, H_d, Q, L, $\overline{u}$, $\overline{d}$, $\overline{e}$ in eq. (13.6) correspond to the chiral supermultiplets in Table 7.1. (Alternatively, they can be thought of as the corresponding scalar fields, as in Chapter 8. But, we prefer not to put the tildes on Q, L, $\overline{u}$, $\overline{d}$, $\overline{e}$ to reduce the clutter.) The dimensionless Yukawa coupling parameters $\boldsymbol{y_u}, \boldsymbol{y_d}, \boldsymbol{y_e}$ are 3×3 matrices in generation space. All gauge [SU(3)$_C$ color and SU(2)$_L$ weak isospin] and generation indices have been suppressed in eq. (13.6). More explicitly, the so-called "μ-term" can be written out as $\mu\epsilon^{ab}(H_u)_a(H_d)_b$, where ϵ^{ab} is used to tie together SU(2)$_L$ weak isospin indices $a, b \in \{1,2\}$ in a gauge-invariant way. Likewise, the term $H_u Q \boldsymbol{y_u} \overline{u}$ can be written out as $\epsilon^{ab}(H_u)_a Q_{mib}\,(\boldsymbol{y_u})^m{}_n\,\overline{u}^{ni}$, where $m, n \in \{1,2,3\}$ are generation indices, and $i \in \{1,2,3\}$ is a color index which is lowered (raised) in the $\mathbf{3}$ ($\overline{\mathbf{3}}$) representation of SU(3)$_C$. Finally, $H_d Q \boldsymbol{y_d} \overline{d}$ can be written out as $\epsilon^{ab}(H_d)_a Q_{mib}\,(\boldsymbol{y_d})^m{}_n\,\overline{d}^{ni}$, with an analogous expression for $H_d L \boldsymbol{y_e} \overline{e}$.

The μ-term in eq. (13.6) is the supersymmetric version of the Higgs boson mass in the Standard Model. It is unique, because terms $H_u^\dagger H_u$ or $H_d^\dagger H_d$ are forbidden in the superpotential, which must be holomorphic in the chiral superfields (or equivalently in the scalar fields) treated as complex variables, as shown in Section 8.2. We can also see from eq. (13.6) why both H_u and H_d are needed in order to give Yukawa couplings, and thus masses, to all of the quarks and leptons. Since the superpotential

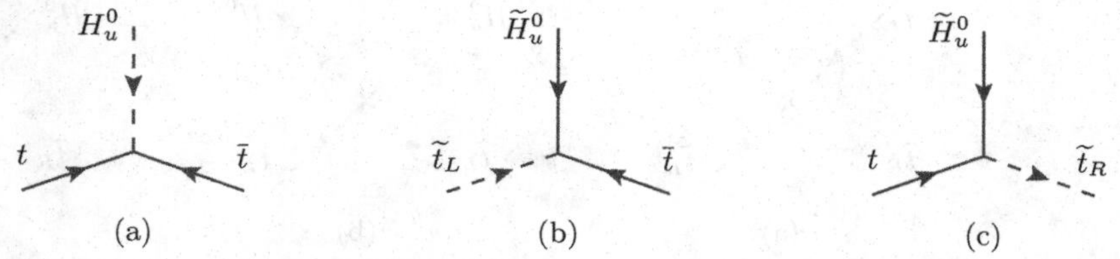

Fig. 13.1 The top-quark Yukawa coupling (a) and its supersymmetrizations (b),(c), all of coupling strength y_t.

must be holomorphic, the $H_u Q\overline{u}$ Yukawa terms cannot be replaced by something like $H_d^\dagger Q\overline{u}$. Similarly, $H_d Q\overline{d}$ and $H_d L\overline{e}$ cannot be replaced by $H_u^\dagger Q\overline{d}$ and $H_u^\dagger L\overline{e}$, respectively. The analogous Yukawa couplings would be allowed in a general non-supersymmetric two-Higgs doublet model, but are forbidden by the structure of supersymmetry. So we need both H_u and H_d, even without invoking the argument based on anomaly cancellation that was mentioned in Section 7.2.

The Yukawa matrices determine the masses and CKM mixing angles of the ordinary quarks and leptons, after the neutral scalar components of H_u and H_d get VEVs. Since the top quark, bottom quark, and tau lepton are the heaviest fermions in the Standard Model, it is often useful to make an approximation that only the $(3,3)$ generation components of each of $\boldsymbol{y_u}$, $\boldsymbol{y_d}$ and $\boldsymbol{y_e}$ are important:

$$\boldsymbol{y_u} \approx \begin{pmatrix} 0 & 0 & 0 \\ 0 & 0 & 0 \\ 0 & 0 & y_t \end{pmatrix}, \quad \boldsymbol{y_d} \approx \begin{pmatrix} 0 & 0 & 0 \\ 0 & 0 & 0 \\ 0 & 0 & y_b \end{pmatrix}, \quad \boldsymbol{y_e} \approx \begin{pmatrix} 0 & 0 & 0 \\ 0 & 0 & 0 \\ 0 & 0 & y_\tau \end{pmatrix}. \tag{13.7}$$

In this limit, it is instructive to write the superpotential in terms of the separate $\mathrm{SU}(2)_L$ weak isospin components [i.e., $Q_3 = (t\ b)$; $L_3 = (\nu_\tau\ \tau)$; $H_u = (H_u^+\ H_u^0)$; $H_d = (H_d^0\ H_d^-)$; $\overline{u}^3 = \overline{t}$; $\overline{d}^3 = \overline{b}$; $\overline{e}^3 = \overline{\tau}$]:

$$W_{\text{MSSM}} \approx y_t(H_u^0 t\overline{t} - H_u^+ b\overline{t}) - y_b(H_d^- t\overline{b} - H_d^0 b\overline{b}) - y_\tau(H_d^- \nu_\tau\overline{\tau} - H_d^0 \tau\overline{\tau}) + \mu(H_u^+ H_d^- - H_u^0 H_d^0). \tag{13.8}$$

The minus signs inside the parentheses are due to the antisymmetry of the ϵ^{ab} used to tie up the $\mathrm{SU}(2)_L$ indices. Note that the signs in eq. (13.6) were chosen so that the terms $y_t H_u^0 t\overline{t}$, $y_b H_d^0 b\overline{b}$, and $y_\tau H_d^0 \tau\overline{\tau}$, which will become the top, bottom, and tau masses when H_u^0 and H_d^0 get VEVs, have positive signs in eq. (13.8).

Since the Yukawa interactions y^{ijk} are completely symmetric under the interchange of i, j, k, we know that $\boldsymbol{y_u}$, $\boldsymbol{y_d}$, and $\boldsymbol{y_e}$ imply not only Higgs–quark–antiquark and Higgs–lepton–antilepton couplings as in the Standard Model, but also squark–higgsino–antiquark and slepton–higgsino–antilepton interactions and their conjugates. To illustrate this, we show in Fig. 13.1 some of the interactions involving the top-quark Yukawa coupling y_t. Figure 13.1(a) is the SM-like coupling of the top quark to the neutral complex scalar Higgs boson, which follows from the first

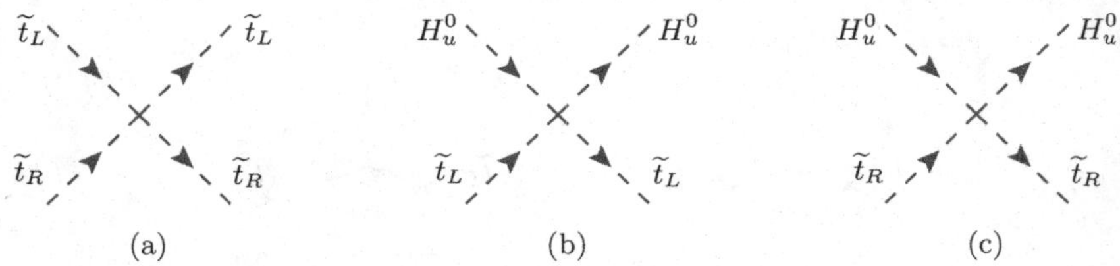

Fig. 13.2 Some of the (scalar)^4 interactions with coupling strengths proportional to y_t^2.

term in eq. (13.8). In Fig. 13.1(b), the top squark $\widetilde{t}_L$ couples to the neutral higgsino field $\widetilde{H}_u^0$ and right-handed top quark, whereas in Fig. 13.1(c) the top squark $\widetilde{t}_R$ couples to the neutral higgsino field $\widetilde{H}_u^0$ and the left-handed top quark. Note that the direction of the arrows on the top quark and top squark indicates the flow of the conserved quantum numbers (color and electric charges). For each of the three interactions, there is another with $H_u^0 \to H_u^+$ and $t \to -b$, with tildes where appropriate, corresponding to the second part of the first term in eq. (13.8). All of these interactions are required by supersymmetry to have the same strength y_t. These couplings are dimensionless and therefore can be modified by the introduction of soft supersymmetry breaking only through finite (and small) radiative corrections, so this equality of interaction strengths is also a prediction of softly broken supersymmetry. A useful mnemonic is that each of Figs. 13.1(a)–(c) can be obtained from any of the others by changing two of the particles into their superpartners.

There are also scalar quartic interactions with strength proportional to y_t^2, as can be seen from, e.g., Fig. 8.1(b) or the last term in eq. (8.50). Three of them are shown in Fig. 13.2. Using eq. (8.50) and eq. (13.8), one can see that there are five more, which can be obtained by replacing $\widetilde{t}_L \to \widetilde{b}_L$ and/or $H_u^0 \to H_u^+$ in each vertex. This illustrates the remarkable economy of supersymmetry: there are many interactions determined by only a single parameter. In a similar way, the existence of all the other quark and lepton Yukawa couplings in eq. (13.6) leads not only to Higgs–quark–quark and Higgs–lepton–lepton Lagrangian terms as in the Standard Model, but also to squark–higgsino–quark and slepton–higgsino–lepton terms, and scalar quartic couplings [(squark)^4, (slepton)^4, (squark)^2(slepton)^2, (squark)^2(Higgs)^2, and (slepton)^2(Higgs)^2]. If needed, these can all be obtained in terms of the Yukawa matrices $\boldsymbol{y_u}$, $\boldsymbol{y_d}$, and $\boldsymbol{y_e}$ as outlined above.

However, the dimensionless interactions determined by the superpotential are often not the most important ones of direct interest for phenomenology. This is because the Yukawa couplings are already known to be very small, except for those of the third generation (top, bottom, tau). Instead, production and decay processes for superpartners in the MSSM are typically dominated by the supersymmetric interactions of gauge-coupling strength, as we will explore in more detail below. The couplings of the Standard Model gauge bosons (photon, $W^\pm$, Z, and gluons) to the MSSM particles are determined completely by the gauge invariance of the kinetic terms in the Lagrangian. The gauginos also couple to (squark, quark) and (slepton, lepton) and (Higgs, higgsino) pairs as illustrated in the general case in

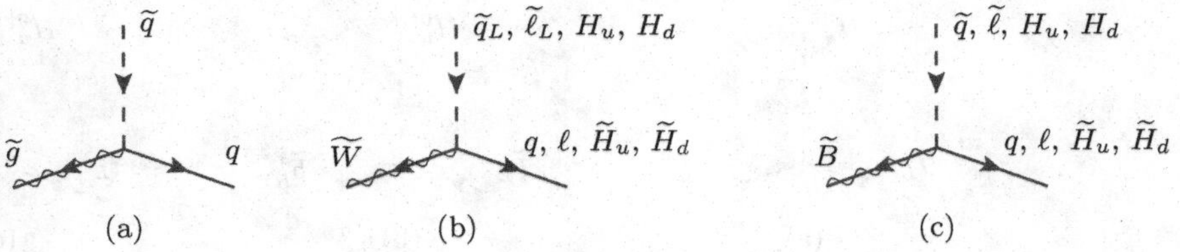

Fig. 13.3 Couplings of the gluino, wino, and bino to MSSM (scalar, fermion) pairs.

Fig. 8.3(g),(h) and the first two terms in the second line of eq. (8.72). For instance, each of the squark–quark–gluino couplings is given by $\sqrt{2}g_3(\widetilde{q}^\dagger T^a q\widetilde{g} + \text{h.c.})$, where $T^a = \lambda^a/2$ $(a = 1 \ldots 8)$ are the matrix generators for $SU(3)_C$, with λ^a the Gell-Mann matrices. The Feynman diagram for this interaction is shown in Fig. 13.3(a). In Figs. 13.3(b),(c) we show the couplings of (squark, quark), (lepton, slepton), and (Higgs, higgsino) pairs to the winos and bino, with strengths proportional to the electroweak gauge couplings g and g', respectively. The winos only couple to the left-handed squarks and sleptons, and the (lepton, slepton) and (Higgs, higgsino) pairs of course do not couple to the gluino. The bino couplings for each (scalar, fermion) pair are also proportional to the weak hypercharges Y as given in Table 7.1. The interactions shown in Fig. 13.3 provide for decays $\widetilde{q} \to q\widetilde{g}$ and $\widetilde{q} \to \widetilde{W}q'$ and $\widetilde{q} \to \widetilde{B}q$ when the final states are kinematically allowed to be on-shell. However, a complication is that the $\widetilde{W}$ and $\widetilde{B}$ states are not mass eigenstates, because of mixing due to electroweak symmetry breaking, as we will see in Section 13.9.

There are also various scalar quartic interactions in the MSSM which are uniquely determined by gauge invariance and supersymmetry, according to the last term in eq. (8.75), as illustrated in Fig. 8.3(h). Among them are (Higgs)^4 terms proportional to g^2 and g'^2 in the scalar potential. These are the direct generalization of the last term in the Standard Model Higgs potential, eq. (7.1), to the case of the MSSM. We will have occasion to identify them explicitly when we discuss the minimization of the MSSM Higgs potential in Section 13.8.

The dimensionful terms in the supersymmetric part of the MSSM Lagrangian are all dependent on μ. Using the general result of eq. (8.51), we find that μ provides for higgsino fermion mass terms

$$-\mathscr{L}_{\text{higgsino mass}} = \mu(\widetilde{H}_u^+\widetilde{H}_d^- - \widetilde{H}_u^0\widetilde{H}_d^0) + \text{h.c.}, \tag{13.9}$$

as well as Higgs squared-mass terms in the scalar potential

$$-\mathscr{L}_{\text{supersymmetric Higgs mass}} = |\mu|^2\big(|H_u^0|^2 + |H_u^+|^2 + |H_d^0|^2 + |H_d^-|^2\big). \tag{13.10}$$

Since eq. (13.10) is nonnegative definite with a minimum at $H_u^0 = H_d^0 = 0$, it is clear that we cannot understand electroweak symmetry breaking without including a negative supersymmetry-breaking squared-mass soft term for the Higgs scalars. An explicit treatment of the Higgs scalar potential will therefore have to wait until we have introduced the soft terms for the MSSM. However, we can already see a puzzle: we expect that μ should be roughly of order 10^2 or 10^3 GeV, in order

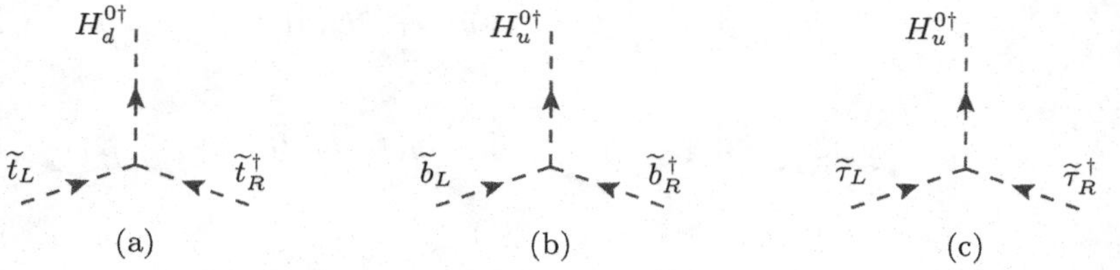

Fig. 13.4 Some of the supersymmetric (scalar)^3 couplings proportional to $\mu^* y_t$, $\mu^* y_b$, and $\mu^* y_\tau$. When H_u^0 and H_d^0 get VEVs, these contribute to (a) $\widetilde{t}_L, \widetilde{t}_R$ mixing, (b) $\widetilde{b}_L, \widetilde{b}_R$ mixing, and (c) $\widetilde{\tau}_L, \widetilde{\tau}_R$ mixing.

to allow a Higgs VEV of $v \simeq 246$ GeV without too much miraculous cancellation between $|\mu|^2$ and the negative soft squared-mass terms that we have not introduced yet. But why should $|\mu|^2$ be so small compared to, say, M_P^2, and in particular why should it be roughly of the same order as $m^2_{\rm soft}$?

The MSSM scalar potential seems to depend on two types of dimensionful parameters that are conceptually quite distinct, namely the supersymmetry-respecting mass μ and the supersymmetry-breaking soft mass terms. Yet the observed value for the electroweak breaking scale suggests that, without miraculous cancellations, both of these apparently unrelated mass scales should be within an order of magnitude or so of 100 GeV. This puzzle is called "the μ problem." Several different solutions to the μ problem have been proposed that involve extensions of the MSSM of varying intricacy. They all work in roughly the same way: the μ-term is required or assumed to be absent at tree level before symmetry breaking. It then arises from the VEV(s) of some new field(s) that are determined by minimizing a potential that depends on the soft supersymmetry-breaking terms. In this way, the value of the effective parameter μ is no longer conceptually distinct from the mechanism of supersymmetry breaking; if we can explain why $m_{\rm soft} \ll M_P$, we will also be able to understand why μ is of the same order. In Sections 14.6 and 16.1, we will study three such mechanisms: the Kim–Nilles mechanism, the Giudice–Masiero mechanism, and the Next-to-Minimal Supersymmetric Standard Model. From the point of view of the MSSM, however, we can just treat μ as an independent parameter.

In light of the second and third terms on the right-hand side of eq. (8.50), the μ-term and the Yukawa couplings in eq. (13.6) combine to yield (scalar)^3 couplings,

$$\begin{aligned}\mathscr{L}_{\text{supersymmetric (scalar)}^3} = \mu^* \Big[& H_d^{0\dagger} \widetilde{u}_L \boldsymbol{y_u} \widetilde{u}_R^\dagger + H_u^{0\dagger} \widetilde{d}_L \boldsymbol{y_d} \widetilde{d}_R^\dagger + H_u^{0\dagger} \widetilde{e}_L \boldsymbol{y_e} \widetilde{e}_R^\dagger \\ & + (H_d^-)^\dagger \widetilde{d}_L \boldsymbol{y_u} \widetilde{u}_R^\dagger + (H_u^+)^\dagger \widetilde{u}_L \boldsymbol{y_d} \widetilde{d}_R^\dagger + (H_u^+)^\dagger \widetilde{\nu}_L \boldsymbol{y_e} \widetilde{e}_R^\dagger \Big] + \text{h.c.}, \qquad (13.11)\end{aligned}$$

where the term $H_d^{0\dagger} \widetilde{u}_L \boldsymbol{y_u} \widetilde{u}_R^\dagger$ can be written out as $H_d^{0\dagger} \widetilde{u}_{Lm} (\boldsymbol{y_u})^m{}_n \widetilde{u}_R^{\dagger n}$ with analogous expressions for the other five terms in eq. (13.11). Figure 13.4 shows some of these couplings, proportional to $\mu^* y_t$, $\mu^* y_b$, and $\mu^* y_\tau$, respectively. These play an important role in determining the mixings of top squarks, bottom squarks, and tau sleptons, as we will see in Section 13.11.

13.3 R-Parity, Also Known As Matter Parity

The superpotential eq. (13.6) is minimal in the sense that it is sufficient to produce a phenomenologically viable model. However, there are other terms that one can write that are gauge-invariant and analytic in the chiral superfields, but are not included in the MSSM because they violate either baryon number (B) or total lepton number (L). The most general gauge-invariant and renormalizable superpotential would include not only eq. (13.6), but also the terms

$$W_{\Delta L=1} = \epsilon_{ab}\left(\tfrac{1}{2}\lambda^{ijk} L_i^a L_j^b \overline{e}_k + \lambda'^{ijk} L_i^a Q_j^b \overline{d}_k + \kappa^i H_u^a L_i^b\right), \qquad (13.12)$$

$$W_{\Delta B=1} = \tfrac{1}{2}\epsilon_{pqr}\lambda''^{ijk}\overline{u}_i^p \overline{d}_j^q \overline{d}_k^r, \qquad (13.13)$$

where we have included generation indices $i, j, k \in \{1, 2, 3\}$, as well as $SU(2)_L$ gauge indices, $a, b \in \{1, 2\}$, and $SU(3)_C$ gauge indices, $p, q, r \in \{1, 2, 3\}$. The chiral supermultiplets carry baryon number assignments $B = +1/3$ for Q_i; $B = -1/3$ for $\overline{u}_i, \overline{d}_i$; and $B = 0$ for all others. The total lepton number assignments are $L = +1$ for L_i; $L = -1$ for $\overline{e}_i$; and $L = 0$ for all others. Therefore, the terms in eq. (13.12) violate total lepton number by 1 unit, as well as individual lepton flavors, and those in eq. (13.13) violate baryon number by 1 unit.

The possible existence of such terms might seem rather disturbing, since corresponding B- and L-violating processes have not been seen experimentally. The most obvious experimental constraint comes from the nonobservation of proton decay, which would violate both B and L by 1 unit; see Fig. 16.4. For λ' and λ'' couplings of order unity and squark masses of order 1 TeV, the proton lifetime would be a tiny fraction of a second [see eq. (16.188)]. In contrast, the decay time of the proton into lepton–meson final states is known experimentally to be in excess of 10^{32} years. Therefore, at least one of λ'^{ijk} or λ''^{11k} for each of $i = 1, 2$; $j = 1, 2$; $k = 2, 3$ must be extremely small. Many other processes also give very strong constraints on the violation of lepton and baryon numbers; these are discussed in Section 16.5.

One could simply try to take B and L conservation as a postulate in the MSSM. However, this is clearly a step backwards from the situation in the Standard Model, where the conservation of these quantum numbers is *not* assumed, but is rather a pleasantly "accidental" consequence of the fact that there are no possible renormalizable Lagrangian terms that violate B or L. Furthermore, there is a quite general obstacle to treating B and L as fundamental symmetries of Nature, since they are known to be necessarily violated by nonperturbative electroweak effects (even though those effects are calculably negligible for experiments at ordinary energies). Therefore, in the MSSM one adds a new discrete symmetry which has the effect of eliminating the possibility of B- and L-violating terms in the renormalizable superpotential, while allowing the other terms in eq. (13.6). This new symmetry is called "R-parity" or equivalently "matter parity."

Matter parity is a multiplicatively conserved quantum number, defined as

$$P_M = (-1)^{3(B-L)} \qquad (13.14)$$

for each particle in the theory. It is easy to check that the quark and lepton supermultiplets all have $P_M = -1$, while the Higgs supermultiplets H_u and H_d have $P_M = +1$. The gauge bosons and gauginos of course do not carry baryon number or lepton number, so they are assigned matter parity $P_M = +1$. The symmetry principle to be enforced is that a term in the Lagrangian (or in the superpotential) is allowed only if the product of P_M for all of the fields in it is $+1$. It is easy to see that each of the terms in eqs. (13.12) and (13.13) is thus forbidden, while the necessary terms in eq. (13.6) are allowed. This discrete symmetry commutes with supersymmetry, as all members of a given supermultiplet have the same matter parity. The advantage of matter parity is that it can in principle be an *exact* and fundamental symmetry, which B and L themselves cannot, since they are known to be violated by nonperturbative electroweak effects.

It is often useful to recast matter parity in terms of R-parity, defined as

$$P_R = (-1)^{3(B-L)+2s} \tag{13.15}$$

for each particle of spin s. Matter parity conservation and R-parity conservation are equivalent, since the product of $(-1)^{2s}$ for the particles involved in any interaction vertex in a theory that conserves angular momentum is equal to $+1$. However, particles within the same supermultiplet do not have the same R-parity. In general, symmetries with the property that particles within the same multiplet have different charges are called R-symmetries; they do not commute with supersymmetry. Continuous U(1) R-symmetries are often encountered in the model-building literature (see Section 10.11); they should not be confused with R-parity, which is a discrete $\mathbb{Z}_2$ symmetry. In fact, the matter parity version of R-parity makes clear that there is really nothing intrinsically "R" about it; in other words, it secretly does commute with supersymmetry, so its name is somewhat suboptimal. Nevertheless, the R-parity assignment is very useful for phenomenology because all of the Standard Model particles and the Higgs bosons have even R-parity ($P_R = +1$), while all of the squarks, sleptons, gauginos, and higgsinos have odd R-parity ($P_R = -1$).

The R-parity odd particles are known as "supersymmetric particles" or "sparticles" for short, and they are distinguished by a tilde (see Tables 7.1 and 7.2). If R-parity is exactly conserved, then there can be no mixing between the sparticles and the $P_R = +1$ particles. Furthermore, every interaction vertex in the theory contains an even number of $P_R = -1$ sparticles. This has three extremely important phenomenological consequences:

- The lightest sparticle with $P_R = -1$, called the *lightest supersymmetric particle* (LSP), must be absolutely stable. If the LSP is electrically neutral, it interacts only weakly with ordinary matter, and so can make an attractive candidate for the nonbaryonic dark matter that seems to be required by cosmology.
- Each sparticle other than the LSP must eventually decay into a state that contains an odd number of LSPs (nearly always just one).
- In collider experiments, sparticles can only be produced in even numbers (nearly always two-at-a-time).

We *define* the MSSM to conserve R-parity, or equivalently matter parity, and thus the above consequences apply to the MSSM. While this decision seems to be well motivated phenomenologically by proton decay constraints and the hope that the LSP will provide a good dark matter candidate, it might appear somewhat artificial from a theoretical point of view. After all, the MSSM would not suffer any internal inconsistency if we did not impose matter parity conservation. Furthermore, it is fair to ask why matter parity should be exactly conserved, given that the discrete symmetries in the Standard Model (parity P, time reversal T, charge conjugation C, etc.) are all known to be inexact symmetries. However matter parity can be an anomaly-free discrete gauge symmetry (see Section 16.4).

A further possibility to protect the proton is if $B-L$ is a continuous U(1) gauge symmetry which is spontaneously broken at some very high energy scale. A continuous $\mathrm{U}(1)_{B-L}$ forbids the renormalizable terms that violate B and L, but this gauge symmetry must be spontaneously broken, since there is no corresponding massless vector boson. However, if gauged $\mathrm{U}(1)_{B-L}$ is only broken by scalar VEVs that carry even integer values of $3(B-L)$, then P_M will automatically survive as an exactly conserved discrete remnant subgroup. Indeed, $\mathrm{U}(1)_{B-L}$ is found in a variety of motivated extensions of the SM gauge group, such as the unification groups SO(10) and E_6, left-right symmetric models, and the unification of color and total lepton number via the Pati–Salam group $\mathrm{SU}(4)_{\mathrm{PS}}\times\mathrm{SU}(2)_L\times\mathrm{SU}(2)_R$ [152]. Consequently, there are many natural realizations of this origin for matter parity.

We discuss the consequences of R-parity violation in Section 16.5.

13.4 Soft Supersymmetry Breaking in the MSSM

To complete the description of the MSSM, we need to specify the soft supersymmetry-breaking terms. In Section 8.6, we learned how to write down the most general set of such terms in any supersymmetric theory. Applying this recipe to the MSSM,

$$\begin{aligned}\mathscr{L}_{\text{soft}}^{\text{MSSM}} = &-\tfrac{1}{2}\Big(M_3\widetilde{g}\widetilde{g}+M_2\widetilde{W}\widetilde{W}+M_1\widetilde{B}\widetilde{B}+\text{h.c.}\Big)\\ &+\Big(H_u\widetilde{Q}\,\boldsymbol{a_u}\,\widetilde{u}_R^\dagger - H_d\widetilde{Q}\,\boldsymbol{a_d}\,\widetilde{d}_R^\dagger - H_d\widetilde{L}\,\boldsymbol{a_e}\,\widetilde{e}_R^\dagger+\text{h.c.}\Big)\\ &-\widetilde{Q}^\dagger\,\boldsymbol{m_Q^2}\,\widetilde{Q}-\widetilde{L}^\dagger\,\boldsymbol{m_L^2}\,\widetilde{L}-\widetilde{u}_R^\dagger\,\boldsymbol{m_{\bar{u}}^2}\,\widetilde{u}_R-\widetilde{d}_R^\dagger\,\boldsymbol{m_{\bar{d}}^2}\,\widetilde{d}_R-\widetilde{e}_R^\dagger\,\boldsymbol{m_{\bar{e}}^2}\,\widetilde{e}_R\\ &-m_{H_u}^2H_u^\dagger H_u-m_{H_d}^2H_d^\dagger H_d-(bH_uH_d+\text{h.c.})\,,\end{aligned} \tag{13.16}$$

where M_3, M_2, and M_1 are the gluino, wino, and bino mass terms, $\widetilde{Q}\equiv(\widetilde{u}_L\,,\,\widetilde{d}_L)$ and $\widetilde{L}\equiv(\widetilde{\nu}_L\,,\,\widetilde{e}_L)$, and the SU(2)-invariant products are defined as in eq. (A.55).

The second line in eq. (13.16) contains the (scalar)^3 couplings [of the type a^{ijk} in eq. (8.78)]. Each of $\boldsymbol{a_u}$, $\boldsymbol{a_d}$, $\boldsymbol{a_e}$ is a complex 3×3 matrix in generation space, with dimensions of [mass]. They are in one-to-one correspondence with the Yukawa couplings in the superpotential given in eq. (13.6). For example, $H_u\widetilde{Q}\boldsymbol{a_u}\widetilde{u}_R^\dagger$ can be written out as $\epsilon^{ab}(H_u)_a\widetilde{Q}_{mib}\,(\boldsymbol{a_u})^m{}_n\,\widetilde{u}_R^{\dagger ni}$, with analogous expressions for $H_d\widetilde{Q}\boldsymbol{a_d}\widetilde{d}_R^\dagger$

and $H_d\widetilde{L}\boldsymbol{a_e}\widetilde{e}_R^\dagger$. Here, and from now on, we suppress the adjoint representation gauge indices on the wino and gluino fields, and the gauge indices on all of the chiral supermultiplet fields. The third line of eq. (13.16) consists of squark and slepton mass terms of the $(m^2)_i^j$ type in eq. (8.78). Each of $\boldsymbol{m^2_Q}$, $\boldsymbol{m^2_{\bar{u}}}$, $\boldsymbol{m^2_{\bar{d}}}$, $\boldsymbol{m^2_L}$, $\boldsymbol{m^2_{\bar{e}}}$ is a hermitian 3×3 matrix in generation space, which ensures that the Lagrangian is hermitian. (To avoid clutter, we do not put tildes on the $\boldsymbol{Q}$ in $\boldsymbol{m^2_Q}$, etc.) Finally, in the last line of eq. (13.16) we have supersymmetry-breaking contributions to the Higgs potential; $m^2_{H_u}$ and $m^2_{H_d}$ are squared-mass terms of the $(m^2)_i^j$ type, while b is the only squared-mass term of the type b^{ij} in eq. (8.78) that can occur in the MSSM.[1] As argued in Section 7.2, we expect

$$M_1,\ M_2,\ M_3,\ \boldsymbol{a_u},\ \boldsymbol{a_d},\ \boldsymbol{a_e} \sim m_{\text{soft}}\,, \tag{13.17}$$

$$\boldsymbol{m^2_Q},\ \boldsymbol{m^2_L},\ \boldsymbol{m^2_{\bar{u}}},\ \boldsymbol{m^2_{\bar{d}}},\ \boldsymbol{m^2_{\bar{e}}},\ m^2_{H_u},\ m^2_{H_d},\ b \sim m^2_{\text{soft}}\,, \tag{13.18}$$

with a characteristic mass scale m_{soft} that is not much larger than 10^3 GeV. Note that eq. (13.16) is the most general soft supersymmetry-breaking Lagrangian of the form eq. (8.78) that is compatible with gauge invariance and matter parity conservation.

13.5 Parameter Count of the MSSM

Unlike the supersymmetry-preserving part of the Lagrangian, $\mathscr{L}^{\text{MSSM}}_{\text{soft}}$ introduces many new parameters that were not present in the Standard Model. In this section, we demonstrate that there are 105 masses, phases, and mixing angles in the MSSM Lagrangian that cannot be rotated away by redefining the phases and flavor basis for the quark and lepton supermultiplets that have no counterpart in the Standard Model. Thus, in principle, supersymmetry (or more precisely, supersymmetry *breaking*) appears to introduce a tremendous arbitrariness in the Lagrangian.

In Section 4.8, it was shown that the Standard Model is governed by 19 independent parameters, of which 13 are associated with the flavor sector. The generalization of the parameter count to the MSSM proceeds as follows. First, the gauge sector of the MSSM consists of four Standard Model real parameters (g_3, g_2, g_1, and Θ_{QCD}) and three complex gaugino mass parameters (M_3, M_2, and M_1). The Higgs sector adds two real squared-mass parameters ($m^2_{H_d}$ and $m^2_{H_u}$) and two complex mass parameters (b and μ). In fact, two of the imaginary degrees of freedom can be removed. Consider the limit where $\mu = b = 0$, all Majorana gaugino mass parameters are zero, and all matrix $\boldsymbol{a}$-parameters are zero. The theory in this limit possesses two flavor-conserving global U(1) symmetries: a continuous R-symmetry [$\text{U}(1)_R$] and a Peccei–Quinn symmetry [$\text{U}(1)_{\text{PQ}}$]. Thus, one can make global $\text{U}(1)_R$ and $\text{U}(1)_{\text{PQ}}$ rotations on the MSSM fields to remove two unphysical degrees of freedom from among μ, b, and the three complex gaugino Majorana mass parame-

[1] The parameter we call b is often seen in the literature as $B\mu$ or m^2_{12} or m^2_3.

ters (unphysical degrees of freedom in the matrix $\boldsymbol{a}$ parameters will be addressed below). It is convenient to perform a $\mathrm{U}(1)_R$ rotation in order to make the gluino mass real and positive (i.e., $M_3 > 0$), followed by a $\mathrm{U}(1)_{\mathrm{PQ}}$ rotation to remove a complex phase from b. Since the tree-level Higgs potential depends on the Higgs mass parameters $m^2_{H_i} + |\mu|^2$ ($i = d, u$) and b, it follows that the tree-level Higgs potential is CP-conserving. Thus, three Higgs sector mass parameters can be traded in (at tree level) for two real vacuum expectation values v_d and v_u [or equivalently, $v^2 \equiv v_d^2 + v_u^2 = (246\text{ GeV})^2$ and $\tan\beta \equiv v_u/v_d$] and one Higgs mass [usually taken to be the mass of the CP-odd Higgs scalar (m_A)]. The parameters $\tan\beta$ and m_A can then be used to predict the masses of the other MSSM Higgs bosons (the CP-even Higgs states h and H and a charged Higgs pair $H^\pm$) and their couplings, as discussed in Section 13.8. Thus, among the gaugino and Higgs/higgsino mass parameters, there are seven real degrees of freedom (v, m_A, $\tan\beta$, M_3, $|M_2|$, $|M_1|$, and $|\mu|$) and three phases (arg M_2, arg M_1, and arg μ).

Next, we examine the MSSM flavor sector parameters. These include $\boldsymbol{y_u}$, $\boldsymbol{y_d}$, and $\boldsymbol{y_e}$ of the Standard Model, three arbitrary complex 3×3 matrix $\boldsymbol{a}$-parameters, $\boldsymbol{a_u}$, $\boldsymbol{a_d}$, and $\boldsymbol{a_e}$ (consisting of 27 real and 27 imaginary parameters), and five scalar hermitian 3×3 squared-mass matrices $\boldsymbol{m^2_Q}$, $\boldsymbol{m^2_{\bar{u}}}$, $\boldsymbol{m^2_{\bar{u}}}$, $\boldsymbol{m^2_L}$, $\boldsymbol{m^2_{\bar{e}}}$ (consisting of 30 real and 15 imaginary parameters). To remove the unphysical degrees of freedom, we employ global $\mathrm{U}(3)^5$ rotations on the *superfields* of the model (thereby preserving the form of the interactions of the gauginos with matter). This analysis differs from the SM analysis in that the MSSM possesses only one global lepton number L. (In particular, L_e, L_μ, and L_τ are no longer separately conserved in general, for the case of arbitrary sneutrino masses). Thus, global $\mathrm{U}(3)^5$ rotations can remove 15 real parameters and 28 phases. Hence, the flavor sector contains 69 real parameters and 41 phases. Of these, there are 9 quark and lepton masses, 3 real CKM angles, and 21 squark and slepton masses. This leaves 36 new real mixing angles to describe the squark and slepton mass eigenstates and 40 new CP-violating phases that can appear in squark and slepton interactions!

The final count gives 124 independent parameters for the MSSM, of which 110 are associated with the flavor sector. Of these 124 parameters, 18 correspond to Standard Model parameters, one corresponds to a Higgs sector parameter (the analogue of the Standard Model Higgs mass), and 105 are genuinely new parameters of the model. Indeed, the "minimal" in MSSM refers to the minimal particle content and not a minimal parameter count.

However, the MSSM in its most general form is not a phenomenologically viable theory over much of its parameter space due to the following generic features: (i) no separate conservation of L_e, L_μ, and L_τ; (ii) unsuppressed flavor-changing neutral currents (FCNCs); and (iii) electric dipole moments of the electron and neutron that are inconsistent with the experimental bounds. As a result, the 124-dimensional MSSM parameter space is severely constrained. Hence, the MSSM particle content must be supplemented by assumptions about the origin of supersymmetry breaking that lie outside the low-energy domain of the model. The implications of various models of supersymmetry breaking will be discussed in Chapter 14.

13.6 Hints of an Organizing Principle

Fortunately, there is already good experimental evidence that some sort of powerful "organizing principle" must govern the soft terms. This is because most of the new parameters in eq. (13.16) involve flavor mixing or CP violation of the types that are already severely restricted by experiment.

For example, suppose that $\boldsymbol{m}^2_{\bar{e}}$ is not diagonal in the basis $(\widetilde{e}_R, \widetilde{\mu}_R, \widetilde{\tau}_R)$ of sleptons whose superpartners are the right-handed pieces of the Standard Model mass eigenstates e, μ, τ. In that case slepton mixing occurs, and the individual lepton numbers will not be conserved. This is true even for processes that only involve the sleptons as virtual particles. A particularly strong limit on this possibility comes from the experimental constraint on $\mu \to e\gamma$, which could arise from the one-loop diagram shown in Fig. 13.5(a). The symbol "×" on the slepton line represents an insertion coming from $-(\boldsymbol{m}^2_{\bar{e}})_2{}^1 \widetilde{\mu}_R^\dagger \widetilde{e}_R$ in $\mathscr{L}^{\text{MSSM}}_{\text{soft}}$, and the slepton–bino vertices are determined by the weak hypercharge gauge coupling [see Figs. 8.3(g),(h) and eq. (8.72)]. The result of calculating this diagram gives, approximately,

$$\text{BR}(\mu \to e\gamma) = \left(\frac{|m^2_{\widetilde{\mu}_R^\dagger \widetilde{e}_R}|}{m^2_{\widetilde{\ell}_R}}\right)^2 \left(\frac{100\,\text{GeV}}{m_{\widetilde{\ell}_R}}\right)^4 10^{-6} \times \begin{cases} 15 & \text{for } m_{\widetilde{B}} \ll m_{\widetilde{\ell}_R}, \\ 5.6 & \text{for } m_{\widetilde{B}} = 0.5 m_{\widetilde{\ell}_R}, \\ 1.4 & \text{for } m_{\widetilde{B}} = m_{\widetilde{\ell}_R}, \\ 0.13 & \text{for } m_{\widetilde{B}} = 2 m_{\widetilde{\ell}_R}, \end{cases} \tag{13.19}$$

where it is assumed for simplicity that both $\widetilde{e}_R$ and $\widetilde{\mu}_R$ are nearly mass eigenstates with almost degenerate squared masses $m^2_{\widetilde{\ell}_R}$, that $m^2_{\widetilde{\mu}_R^\dagger \widetilde{e}_R} \equiv (\boldsymbol{m}^2_{\bar{e}})_2{}^1$ can be treated as a perturbation, and that the bino $\widetilde{B}$ is nearly a mass eigenstate. This result is to be compared to the experimental upper limit of $\text{BR}(\mu \to e\gamma)_{\text{exp}} < 5.7 \times 10^{-13}$ [41]. So, if the elements of $\boldsymbol{m}^2_{\bar{e}}$ were "random," with all entries of comparable size, then the prediction for $\text{BR}(\mu \to e\gamma)$ would be too large even if the sleptons and bino masses were at 1 TeV. For lighter superpartners, the constraint on $\widetilde{\mu}_R, \widetilde{e}_R$ squared-mass mixing becomes correspondingly more severe. There are also contributions to $\mu \to e\gamma$ that depend on the off-diagonal elements of $\boldsymbol{m}^2_{L}$, coming from the diagram shown in Fig. 13.5(b) involving the charged wino and the sneutrinos, as well as diagrams just like Fig. 13.5(a) but with left-handed sleptons and either $\widetilde{B}$ or $\widetilde{W}^0$ exchanged. Therefore, the slepton squared-mass matrices must not have significant mixings for $\widetilde{e}_L, \widetilde{\mu}_L$ either.

Furthermore, after the Higgs scalars acquire VEVs, the $\boldsymbol{a_e}$ matrix could imply squared-mass terms that mix $\widetilde{\ell}_L$ and $\widetilde{\ell}_R$ with different slepton flavors. For example, $-\mathscr{L}^{\text{MSSM}}_{\text{soft}}$ contains $H_d \widetilde{L} \boldsymbol{a_e} \widetilde{\bar{e}}_R^\dagger$ + h.c., which implies the presence of the terms $\langle H_d^0 \rangle (\boldsymbol{a_e})^1{}_2 \widetilde{e}_L \widetilde{\mu}_R^\dagger + \langle H_d^0 \rangle (\boldsymbol{a_e})^2{}_1 \widetilde{\mu}_L \widetilde{e}_R^\dagger$ + h.c. These terms also contribute to $\mu \to e\gamma$, as illustrated in Fig. 13.5(c). So the magnitudes of $(\boldsymbol{a_e})^1{}_2$ and $(\boldsymbol{a_e})^2{}_1$ are also con-

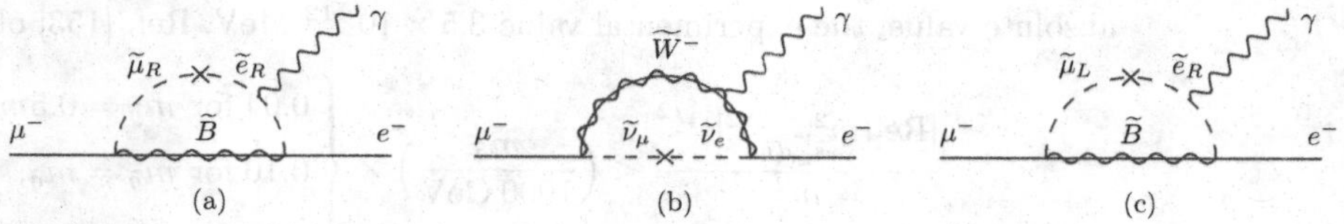

Fig. 13.5 Some of the diagrams that contribute to the process $\mu^- \to e^-\gamma$ in models with lepton flavor-violating soft supersymmetry-breaking parameters (indicated by $\times$). Diagrams (a), (b), and (c) contribute to constraints on the off-diagonal elements of $\boldsymbol{m}^2_{\bar{e}}$, $\boldsymbol{m}^2_L$, and $\boldsymbol{a}_e$, respectively.

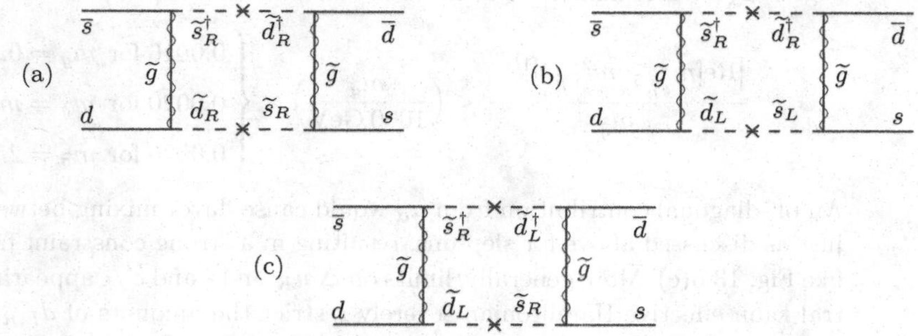

Fig. 13.6 Some of the diagrams that contribute to $K^0 \leftrightarrow \overline{K}^0$ mixing in models with strangeness-violating squark masses (indicated by $\times$). These diagrams contribute to constraints on the off-diagonal elements of (a) $\boldsymbol{m}^2_{\bar{d}}$, (b) combination of $\boldsymbol{m}^2_{\bar{d}}$ and $\boldsymbol{m}^2_Q$, and (c) $\boldsymbol{a}_d$.

strained by experiment to be small, but in a way that is more strongly dependent on other model parameters. Similarly, $(\boldsymbol{a}_e)^1{}_3, (\boldsymbol{a}_e)^3{}_1$ and $(\boldsymbol{a}_e)^2{}_3, (\boldsymbol{a}_e)^3{}_2$ are constrained, although more weakly, by the experimental limits on $\mathrm{BR}(\tau \to e\gamma)$ and $\mathrm{BR}(\tau \to \mu\gamma)$, respectively.

There are also important experimental constraints on the squark squared-mass matrices. The strongest of these come from the neutral kaon system. The effective Hamiltonian for $K^0 \leftrightarrow \overline{K}^0$ mixing gets contributions from the diagram in Fig. 13.6, among others, if $\mathscr{L}^{\mathrm{MSSM}}_{\mathrm{soft}}$ contains squared-mass terms that mix down squarks and strange squarks. The gluino–squark–quark vertices in Fig. 13.6 are all fixed by supersymmetry to be of QCD interaction strength. (There are similar diagrams in which the bino and winos are exchanged, which can be important depending on the relative sizes of the gaugino masses.) For example, suppose that there is a nonzero right-handed down-squark squared-mass mixing $(\boldsymbol{m}^2_{\bar{d}})_2{}^1$ in the basis corresponding to the quark mass eigenstates. Assuming that the supersymmetric correction to $\Delta m_K \equiv m_{K_L} - m_{K_S}$ following from Fig. 13.6(a) and others, does not exceed, in absolute value, the experimental value 3.5×10^{-12} MeV, Ref. [153] obtains

$$\frac{|\mathrm{Re}[(m^2_{\widetilde{s}^\dagger_R \widetilde{d}_R})^2]|^{1/2}}{m^2_{\widetilde{q}}} < \left(\frac{m_{\widetilde{q}}}{1000\,\mathrm{GeV}}\right) \times \begin{cases} 0.04 \text{ for } m_{\widetilde{g}} = 0.5 m_{\widetilde{q}}, \\ 0.10 \text{ for } m_{\widetilde{g}} = m_{\widetilde{q}}, \\ 0.22 \text{ for } m_{\widetilde{g}} = 2 m_{\widetilde{q}}. \end{cases} \tag{13.20}$$

Here nearly degenerate squarks with mass $m_{\widetilde{q}}$ are assumed for simplicity, with $m^2_{\widetilde{s}^\dagger_R \widetilde{d}_R} = (\boldsymbol{m^2_{\bar{d}}})_2{}^1$ treated as a perturbation. The same limit applies when $m^2_{\widetilde{s}^\dagger_R \widetilde{d}_R}$ is replaced by $m^2_{\widetilde{s}^\dagger_L \widetilde{d}_L} = (\boldsymbol{m^2_Q})_2{}^1$, in a basis corresponding to the down-type quark mass eigenstates. An even more striking limit applies to the combination of both types of flavor mixing when they are comparable in size, from diagrams including Fig. 13.6(b). The numerical constraint obtained in Ref. [153] is

$$\frac{|\mathrm{Re}[m^2_{\widetilde{s}^\dagger_R \widetilde{d}_R} m^2_{\widetilde{s}^\dagger_L \widetilde{d}_L}]|^{1/2}}{m^2_{\widetilde{q}}} < \left(\frac{m_{\widetilde{q}}}{1000\,\mathrm{GeV}}\right) \times \begin{cases} 0.0016 \text{ for } m_{\widetilde{g}} = 0.5 m_{\widetilde{q}}, \\ 0.0020 \text{ for } m_{\widetilde{g}} = m_{\widetilde{q}}, \\ 0.0026 \text{ for } m_{\widetilde{g}} = 2 m_{\widetilde{q}}. \end{cases} \tag{13.21}$$

An off-diagonal contribution from $\boldsymbol{a_d}$ would cause flavor mixing between $\widetilde{q}_L$ and $\widetilde{q}_R$ just as discussed above for sleptons, resulting in a strong constraint from diagrams like Fig. 13.6(c). More generally, limits on Δm_K and ϵ and ϵ'/ϵ appearing in the neutral kaon effective Hamiltonian severely restrict the amounts of $\widetilde{d}_{L,R}$, $\widetilde{s}_{L,R}$ squark mixings (separately and in various combinations), and associated CP-violating complex phases, that one can tolerate in the soft squared masses.

Weaker, but still interesting, constraints come from the $D^0, \overline{D}^0$ system, which limits the amounts of $\widetilde{u}, \widetilde{c}$ mixings from $\boldsymbol{m^2_{\bar{u}}}$, $\boldsymbol{m^2_Q}$, and $\boldsymbol{a_u}$. The $B^0_d, \overline{B}^0_d$ and $B^0_s, \overline{B}^0_s$ systems similarly limit the amounts of $\widetilde{d}, \widetilde{b}$ and $\widetilde{s}, \widetilde{b}$ mixings from soft supersymmetry-breaking sources. More constraints follow from rare $\Delta F = 1$ meson decays, notably those involving the parton-level processes $b \to s\gamma$ and $b \to s\ell^+\ell^-$ and $c \to u\ell^+\ell^-$ and $s \to de^+e^-$ and $s \to d\nu\overline{\nu}$, all of which can be mediated by flavor mixing in soft supersymmetry breaking.

Another challenge is presented by the electric dipole moment of the electron, for which CP-violating effects in the Standard Model ensure a nonzero, but extremely tiny, prediction:

$$|d_e^{\mathrm{SM}}| \lesssim 10^{-37}\, e \cdot \mathrm{cm}. \tag{13.22}$$

In contrast, at this writing the experimental limit (at 90% CL) from the ACME experiment [154], using measurements of the ThO molecule, is

$$|d_e^{\mathrm{exp}}| < 1.1 \times 10^{-29}\, e \cdot \mathrm{cm}. \tag{13.23}$$

This limit will likely improve in the future, but even now it significantly impacts the possibilities for CP-violating new physics near the TeV scale. Supersymmetry implies contributions to d_e from one-loop selectron–neutralino and two-loop chargino diagrams, as shown in Fig. 13.7. If the pertinent superpartner masses are

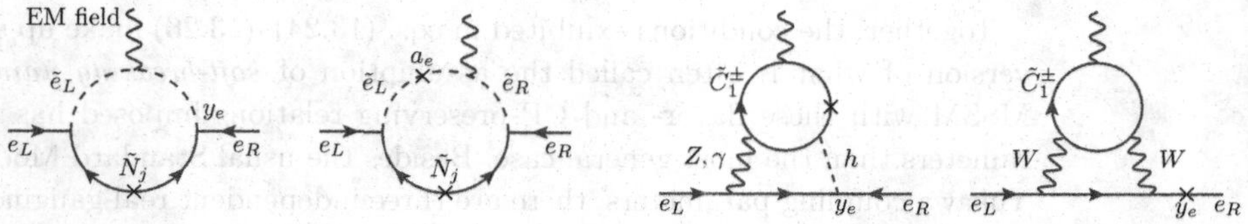

Fig. 13.7 Supersymmetric contributions to the electron's electric dipole moment from one-loop selectron–neutralino and two-loop chargino diagrams.

at the TeV scale, then the associated CP-violating phases must be less than the 1% level, with a considerable amount of model dependence. To avoid conflict with the present experimental limit, either the CP-violating phases are very small, or the charginos and either the selectrons or the neutralinos must be much heavier than 1 TeV (e.g., see Ref. [155]). There are also strict constraints on CP-violating phases in the gaugino masses and (scalar)^3 soft couplings following from limits on the electric dipole moments of the neutron. The essential lesson from experiment is that the soft supersymmetry-breaking Lagrangian cannot be arbitrary or random.

All potentially dangerous FCNC and CP-violating effects in the MSSM can be evaded if one assumes (or can explain!) that supersymmetry breaking should be suitably "universal." Consider an idealized limit in which the squark and slepton squared-mass matrices are flavor-blind, each proportional to the 3×3 identity matrix in generation space:

$$\boldsymbol{m_Q^2} = m_Q^2 \boldsymbol{I}_3; \quad \boldsymbol{m_{\bar{u}}^2} = m_{\bar{u}}^2 \boldsymbol{I}_3; \quad \boldsymbol{m_{\bar{d}}^2} = m_{\bar{d}}^2 \boldsymbol{I}_3; \quad \boldsymbol{m_L^2} = m_L^2 \boldsymbol{I}_3; \quad \boldsymbol{m_{\bar{e}}^2} = m_{\bar{e}}^2 \boldsymbol{I}_3. \tag{13.24}$$

Then all squark and slepton mixing angles are rendered trivial, because squarks and sleptons with the same electroweak quantum numbers will be degenerate in mass and can be rotated into each other at will. Supersymmetric contributions to FCNC processes will therefore be very small in such an idealized limit, up to mixing induced by $\boldsymbol{a_u}$, $\boldsymbol{a_d}$, $\boldsymbol{a_e}$. Making the further assumption that the (scalar)^3 coupling matrices are each proportional to the corresponding Yukawa coupling matrix,

$$\boldsymbol{a_u} = A_{u0}\, \boldsymbol{y_u}, \qquad \boldsymbol{a_d} = A_{d0}\, \boldsymbol{y_d}, \qquad \boldsymbol{a_e} = A_{e0}\, \boldsymbol{y_e} \tag{13.25}$$

will ensure that only the squarks and sleptons of the third generation can have large A-terms. Finally, one can avoid disastrously large CP-violating effects with the assumption that the soft parameters do not introduce new complex phases. This is automatic for $m_{H_u}^2$ and $m_{H_d}^2$, and for m_Q^2, $m_{\bar{u}}^2$, etc. if eq. (13.24) is assumed; if they were not real numbers, the Lagrangian would not be real. One can also fix μ in the superpotential and b in eq. (13.16) to be real, by an appropriate phase rotation of H_u and H_d. If one then assumes that

$$\mathrm{Im}(M_1),\ \mathrm{Im}(M_2),\ \mathrm{Im}(M_3),\ \mathrm{Im}(A_{u0}),\ \mathrm{Im}(A_{d0}),\ \mathrm{Im}(A_{e0}) = 0, \tag{13.26}$$

then the only CP-violating phase in the theory will be the ordinary CKM phase found in the ordinary Yukawa couplings.

Together, the conditions exhibited in eqs. (13.24)–(13.26) make up a rather weak version of what is often called the assumption of *soft-breaking universality*. The MSSM with these flavor- and CP-preserving relations imposed has far fewer parameters than the most general case. Besides the usual Standard Model gauge and Yukawa coupling parameters, there are three independent real gaugino masses, five real squark and slepton squared-mass parameters, three real A-term parameters, three Higgs squared-mass parameters (one of which can be traded for the known electroweak symmetry-breaking scale), and the higgsino mass parameter μ.

There are at least three other possible types of explanations for the suppression of flavor violation in the MSSM that could replace the universality hypothesis of eqs. (13.24)–(13.26). They can be referred to as the "irrelevancy," "flavor alignment," and "R-symmetry" hypotheses for the soft masses. The "irrelevancy" idea is that the sparticle masses are *extremely* heavy, so that their contributions to flavor-changing and CP-violating diagrams like Figs. 13.6(a),(b) are suppressed, as can be seen for example in eqs. (13.19)–(13.21). In practice, however, if there is no flavor-blind structure, the degree of suppression needed typically requires m_{soft} much larger than 1 TeV for at least some of the scalar masses. This seems to go directly against the motivation for supersymmetry as a cure for the hierarchy problem as discussed in Chapter 7. Nevertheless, it has been argued that this is a sensible possibility. The fact that the LHC searches conducted so far have eliminated many models with lighter squarks tends to make these models seem more attractive. Perhaps a combination of approximate flavor-blindness and heavy superpartner masses is the true explanation for the suppression of flavor-violating effects.

The "flavor alignment" idea is that the squark squared-mass matrices do not have the flavor-blindness indicated in eq. (13.24), but are arranged in flavor space to be aligned with the relevant Yukawa matrices in just such a way as to avoid large flavor-changing effects. The flavor alignment models typically require rather special flavor symmetries.

The third possibility is that the theory is (approximately) invariant under a continuous $\mathrm{U}(1)_R$ symmetry. This requires that the MSSM is supplemented by additional chiral supermultiplets in the adjoint representations of $\mathrm{SU}(3)_C$, $\mathrm{SU}(2)_L$, and $\mathrm{U}(1)_Y$, as well as an additional pair of Higgs chiral supermultiplets. The gaugino masses in this theory are purely Dirac, of the type in eq. (8.83), and the couplings $\boldsymbol{a_u}$, $\boldsymbol{a_d}$, and $\boldsymbol{a_e}$ are absent. This implies a very efficient suppression of FCNC effects even if the squark and slepton mass eigenstates are light, nondegenerate, and have large mixings in the basis determined by the Standard Model quark and lepton mass eigenstates. This can lead to unique and intriguing collider signatures. However, we will not consider these possibilities further here.

In the absence of a definite fundamental theory of supersymmetry breaking, it has been common practice to adopt an ad hoc approach to supersymmetry breaking that avoids the phenomenological problems listed at the end of Section 13.5. This approach employs the *Constrained Minimal Supersymmetric Standard Model* (CMSSM) scenario for the soft terms, which posits the following boundary conditions for the soft parameters that appear in eq. (13.16):

$$M_3 = M_2 = M_1 = m_{1/2}, \tag{13.27}$$
$$\mathbf{m_Q^2} = \mathbf{m_{\bar{u}}^2} = \mathbf{m_{\bar{d}}^2} = \mathbf{m_L^2} = \mathbf{m_{\bar{e}}^2} = m_0^2\,\mathbf{I_3}, \qquad m_{H_u}^2 = m_{H_d}^2 = m_0^2, \tag{13.28}$$
$$\mathbf{a_u} = A_0\mathbf{y_u}, \qquad \mathbf{a_d} = A_0\mathbf{y_d}, \qquad \mathbf{a_e} = A_0\mathbf{y_e}, \tag{13.29}$$
$$b = B_0\mu, \tag{13.30}$$

at some very large renormalization scale Q_0 (typically of order the Planck scale). It is a matter of some controversy whether the assumptions going into this parameterization are well motivated on purely theoretical grounds, but from a phenomenological perspective they successfully evade the most dangerous types of flavor changing and CP violation discussed above. In particular, eqs. (13.28) and (13.29) are just stronger versions of eqs. (13.24) and (13.25), respectively. If $m_{1/2}$, A_0, and B_0 all have the same complex phase, then eq. (13.26) will also be satisfied.

One can relax the universality of scalar masses by decoupling the squared-masses of H_u, H_d, and the squarks and sleptons, while preserving the flavor universality of the CMSSM approach. This leads to the nonuniversal Higgs mass models (NUHMs), thereby adding one or two new parameters to the CMSSM, depending on whether the diagonal Higgs scalar squared-mass parameters ($m_{H_d}^2$ and $m_{H_u}^2$) are set equal (NUHM1) or taken to be independent (NUHM2) at the high energy scale Q_0.

The soft-breaking universality relations eqs. (13.24)–(13.26), or stronger (more special) versions of them such as the CMSSM boundary conditions exhibited in eqs. (13.27)–(13.30), can be presumed to be the result of some specific model for the origin of supersymmetry breaking, although there is no consensus among theorists as to what the specific model should actually be. In any case, they are indicative of an assumed underlying simplicity or symmetry of the Lagrangian at some very high energy scale Q_0. If we used this Lagrangian to compute masses and cross sections and decay rates for experiments at ordinary energies near the electroweak scale, the results would involve large logarithms of order $\ln(Q_0/m_Z)$ coming from loop diagrams. As is usual in quantum field theory, the large logarithms can be conveniently resummed using renormalization group (RG) equations, by treating the couplings and masses appearing in the Lagrangian as running parameters. Therefore, eqs. (13.24)–(13.26) should be interpreted as boundary conditions on the running soft parameters at the scale Q_0, which is likely very far removed from direct experimental probes. We must then RG-evolve all of the soft parameters, the superpotential parameters, and the gauge couplings down to the electroweak scale or comparable scales where humans perform experiments.

At the electroweak scale, eqs. (13.24) and (13.25) [and eqs. (13.27)–(13.30) in the CMSSM] will no longer hold, even if they were exactly true at the input scale Q_0. However, to a good approximation, key flavor- and CP-conserving properties remain. This is because, as we will see in Section 13.7, RG corrections due to gauge interactions respect the form of eqs. (13.24) and (13.25) [and eqs. (13.28) and (13.29) of the CMSSM], while RG corrections due to Yukawa interactions are quite small except for couplings involving the top, bottom, and tau flavors. Therefore, the (scalar)^3 couplings and scalar squared-mass mixings should be quite negligible for the squarks and sleptons of the first two generations. Furthermore, RG evolution does not introduce new CP-violating phases. If universality can be arranged to hold at the input scale, supersymmetric contributions to flavor-changing and CP-violating observables can be acceptably small in comparison to present limits (although quite possibly measurable in future experiments).

One reason to be optimistic that such a program can succeed is the celebrated apparent unification of gauge couplings in the MSSM. The one-loop RGEs for the Standard Model gauge couplings g_1, g_2, g_3, previously given in eq. (6.288), are

$$\beta_{g_a} \equiv \frac{d}{dt} g_a = \frac{1}{16\pi^2} b_a g_a^3 ,$$

$$(b_1, b_2, b_3) = \begin{cases} (41/10,\ -19/6,\ -7) & \text{Standard Model,} \\ (33/5,\ 1,\ -3) & \text{MSSM,} \end{cases} \tag{13.31}$$

where $t = \ln(Q/Q_0)$, with Q the RG scale. The MSSM coefficients are larger because of the extra MSSM particles in loops. The normalization for g_1 here is chosen to agree with the canonical covariant derivative for grand unification of the gauge group $\mathrm{SU}(3)_C \times \mathrm{SU}(2)_L \times \mathrm{U}(1)_Y$ into SU(5) or SO(10). Thus in terms of the conventional electroweak gauge couplings g and g' with $e = g \sin\theta_W = g' \cos\theta_W$, one has $g_2 = g$ and $g_1 = \sqrt{5/3}\, g'$. The quantities $\alpha_a = g_a^2/4\pi$ have the nice property that their reciprocals run linearly with RG scale at one-loop order:

$$\frac{d}{dt} \alpha_a^{-1} = -\frac{b_a}{2\pi} \qquad (a = 1, 2, 3). \tag{13.32}$$

After including two-loop effects, the results are as already shown in Fig. 6.1, which compares the RG evolution of the α_a^{-1} in the Standard Model (dashed lines) and the MSSM (solid lines). From this, one can see that, unlike the Standard Model, the MSSM includes just the right particle content to ensure that the gauge couplings can unify, at a scale $M_U \sim 1.5 \times 10^{16}$ GeV. This unification is of course not perfect; α_3 tends to be slightly smaller than the common value of $\alpha_1(M_U) = \alpha_2(M_U)$ at the point where they meet, which is often taken to be the definition of M_U. However, this small difference can easily be ascribed to threshold corrections due to whatever new particles exist near M_U. Note that M_U decreases slightly as the superpartner masses are raised. While the apparent approximate unification of gauge couplings at M_U might be just an accident, it may also be taken as a strong hint in favor of a grand unified theory (GUT) or superstring models, both of which can naturally accommodate gauge coupling unification below M_P. Furthermore, if this hint is taken seriously, then we can reasonably expect to be able to apply a similar RG analysis to the other MSSM couplings and soft masses as well. Section 13.7 discusses the form of the necessary RGEs.

13.7 Renormalization Group Equations for the MSSM

In order to translate a set of predictions at an input scale into physically meaningful quantities that describe physics near the electroweak scale, it is necessary to evolve the gauge couplings, superpotential parameters, and soft terms using their renormalization group (RG) equations. This ensures that the calculations of observables will not suffer from very large logarithms. The beta functions for a general softly broken supersymmetric model were discussed in Section 11.5. In this section, we will apply these results to the special case of the MSSM, using the approximation that only the third-generation Yukawa couplings are significant, as in eq. (13.7), and considering only the leading-order approximations, for simplicity. We will use the same loop expansion notation as in eqs. (11.58) and (11.59).

For the MSSM, one has group components $SU(3)_C$, $SU(2)_L$, and $U(1)_Y$, with adjoint representation quadratic Casimir invariants and dimensions given by

$$(G_3,\, d_3) = (3,\, 8), \qquad (G_2,\, d_2) = (2,\, 3), \qquad (G_1,\, d_1) = (0,\, 1), \tag{13.33}$$

respectively. The chiral supermultiplet content quadratic Casimir invariants are

$$C_3(i) = \begin{cases} 4/3 & \text{for } \Phi_i = Q, \overline{u}, \overline{d}, \\ 0 & \text{for } \Phi_i = L, \overline{e}, H_u, H_d, \end{cases} \tag{13.34}$$

$$C_2(i) = \begin{cases} 3/4 & \text{for } \Phi_i = Q, L, H_u, H_d, \\ 0 & \text{for } \Phi_i = \overline{u}, \overline{d}, \overline{e}, \end{cases} \tag{13.35}$$

$$C_1(i) = 3Y_i^2/5 \quad \text{for each } \Phi_i \text{ with weak hypercharge } Y_i, \tag{13.36}$$

and the summed Dynkin indices are (again using the GUT normalization $g_2 = g$ and $g_1 = \sqrt{5/3}g'$ mentioned in the previous section)

$$S_3 = 6, \quad S_2 = 7, \quad S_1 = 33/5. \tag{13.37}$$

The group theory invariants defined in eq. (11.129) are

$$S_{33} = 8, \qquad S_{32} = 9/4, \qquad S_{31} = 11/20, \tag{13.38}$$

$$S_{23} = 6, \qquad S_{22} = 21/4, \qquad S_{21} = 9/20, \tag{13.39}$$

$$S_{13} = 22/5, \qquad S_{12} = 27/20, \qquad S_{11} = 199/100. \tag{13.40}$$

For the two-loop beta functions of gauge couplings in the MSSM, one therefore has

$$\beta_{g_a} = \frac{1}{16\pi^2}\beta_{g_a}^{(1)} + \frac{1}{(16\pi^2)^2}\beta_{g_a}^{(2)} + \cdots, \tag{13.41}$$

with one-loop contributions already noted above in eq. (13.31),

$$\beta_{g_3}^{(1)} = -3g_3^3, \qquad \beta_{g_2}^{(1)} = g_2^3, \qquad \beta_{g_1}^{(1)} = 33g_1^3/5, \tag{13.42}$$

and two-loop contributions

$$\beta_{g_3}^{(2)} = g_3^3 \left(14g_3^2 + 9g_2^2 + 11g_1^2/5 - 4|y_t|^2 - 4|y_b|^2\right), \tag{13.43}$$

$$\beta_{g_2}^{(2)} = g_2^3 \left(24g_3^2 + 25g_2^2 + 9g_1^2/5 - 6|y_t|^2 - 6|y_b|^2 - 2|y_\tau|^2\right), \tag{13.44}$$

$$\beta_{g_1}^{(2)} = g_1^3 \left(88g_3^2 + 27g_2^2 + 199g_1^2/5 - 26|y_t|^2 - 14|y_b|^2 - 18|y_\tau|^2\right)/5. \tag{13.45}$$

The Higgs and third-generation superfield anomalous dimensions are diagonal matrices, and from eq. (11.139) they are, at one-loop order,

$$\gamma_{H_u}^{(1)} = 3|y_t|^2 - \tfrac{3}{2}g_2^2 - \tfrac{3}{10}g_1^2, \tag{13.46}$$

$$\gamma_{H_d}^{(1)} = 3|y_b|^2 + |y_\tau|^2 - \tfrac{3}{2}g_2^2 - \tfrac{3}{10}g_1^2, \tag{13.47}$$

$$\gamma_{Q_3}^{(1)} = |y_t|^2 + |y_b|^2 - \tfrac{8}{3}g_3^2 - \tfrac{3}{2}g_2^2 - \tfrac{1}{30}g_1^2, \tag{13.48}$$

$$\gamma_{\bar{u}_3}^{(1)} = 2|y_t|^2 - \tfrac{8}{3}g_3^2 - \tfrac{8}{15}g_1^2, \tag{13.49}$$

$$\gamma_{\bar{d}_3}^{(1)} = 2|y_b|^2 - \tfrac{8}{3}g_3^2 - \tfrac{2}{15}g_1^2, \tag{13.50}$$

$$\gamma_{L_3}^{(1)} = |y_\tau|^2 - \tfrac{3}{2}g_2^2 - \tfrac{3}{10}g_1^2, \tag{13.51}$$

$$\gamma_{\bar{e}_3}^{(1)} = 2|y_\tau|^2 - \tfrac{6}{5}g_1^2. \tag{13.52}$$

The first- and second-generation anomalous dimensions in the approximation of eq. (13.7) follow by setting y_t, y_b, and y_τ to 0 in the above. Putting these into the general forms of eqs. (11.135) and (11.136) gives the running of the superpotential parameters with renormalization scale:

$$\beta_{y_t}^{(1)} = y_t\Big(6|y_t|^2 + |y_b|^2 - \tfrac{16}{3}g_3^2 - 3g_2^2 - \tfrac{13}{15}g_1^2\Big), \tag{13.53}$$

$$\beta_{y_b}^{(1)} = y_b\Big(6|y_b|^2 + |y_t|^2 + |y_\tau|^2 - \tfrac{16}{3}g_3^2 - 3g_2^2 - \tfrac{7}{15}g_1^2\Big), \tag{13.54}$$

$$\beta_{y_\tau}^{(1)} = y_\tau\Big(4|y_\tau|^2 + 3|y_b|^2 - 3g_2^2 - \tfrac{9}{5}g_1^2\Big), \tag{13.55}$$

$$\beta_{\mu}^{(1)} = \mu\Big(3|y_t|^2 + 3|y_b|^2 + |y_\tau|^2 - 3g_2^2 - \tfrac{3}{5}g_1^2\Big). \tag{13.56}$$

The presence of soft supersymmetry breaking does not affect eqs. (13.31) and (13.53)–(13.56). As a result of the supersymmetric nonrenormalization theorem, the β-functions for each supersymmetric parameter are proportional to the parameter itself. One consequence of this is that, once we have a theory that can explain why μ is of order 10^2 or 10^3 GeV at tree level, we do not have to worry about μ being made very large by radiative corrections involving the masses of some very heavy unknown particles; all such RG corrections to μ will be directly proportional to μ itself and to some combinations of dimensionless couplings.

The one-loop RGEs for the three gaugino mass parameters in the MSSM are determined by the same quantities b_a^{MSSM} that appear in the gauge coupling RG eqs. (13.31):

$$\beta_{M_a}^{(1)} = 2b_a g_a^2 M_a \qquad (b_a = 33/5,\ 1,\ -3), \tag{13.57}$$

for $a = 1, 2, 3$. It follows that the three ratios M_a/g_a^2 are each constant (RG-scale independent) up to small two-loop corrections. Since the gauge couplings are observed to unify at $Q = M_U = 1.5 \times 10^{16}$ GeV, it is a popular assumption that the gaugino masses also unify[2] near that scale, with a value called $m_{1/2}$. If so, then it follows that

$$\frac{M_1}{g_1^2} = \frac{M_2}{g_2^2} = \frac{M_3}{g_3^2} = \frac{m_{1/2}}{g_U^2} \tag{13.58}$$

at any RG scale, up to small (and known) two-loop effects and possibly much larger (and unknown) threshold effects near M_U. Here g_U is the unified gauge coupling at $Q = M_U$. The hypothesis of eq. (13.58) is particularly powerful because the gaugino mass parameters feed strongly into the RGEs for all of the other soft terms, as we are about to see.

Next we consider the one-loop RGEs for the holomorphic soft parameters $\boldsymbol{a_u}$, $\boldsymbol{a_d}$, $\boldsymbol{a_e}$. In models obeying eq. (13.25), these matrices start off proportional to the corresponding Yukawa couplings at the input scale. The RG evolution respects this property. With the approximation of eq. (13.7), one can therefore also write, at any RG scale,

$$\boldsymbol{a_u} \approx \begin{pmatrix} 0 & 0 & 0 \\ 0 & 0 & 0 \\ 0 & 0 & a_t \end{pmatrix}, \quad \boldsymbol{a_d} \approx \begin{pmatrix} 0 & 0 & 0 \\ 0 & 0 & 0 \\ 0 & 0 & a_b \end{pmatrix}, \quad \boldsymbol{a_e} \approx \begin{pmatrix} 0 & 0 & 0 \\ 0 & 0 & 0 \\ 0 & 0 & a_\tau \end{pmatrix}, \tag{13.59}$$

which defines[3] running parameters a_t, a_b, and a_τ. In this approximation, the RGEs for these parameters and b are

$$\begin{aligned} \beta_{a_t}^{(1)} &= a_t\Big(18|y_t|^2 + |y_b|^2 - \tfrac{16}{3}g_3^2 - 3g_2^2 - \tfrac{13}{15}g_1^2\Big] + 2a_b y_b^* y_t \\ &\quad + y_t\Big(\tfrac{32}{3}g_3^2 M_3 + 6g_2^2 M_2 + \tfrac{26}{15}g_1^2 M_1\Big), \end{aligned} \tag{13.60}$$

$$\begin{aligned} \beta_{a_b}^{(1)} &= a_b\Big(18|y_b|^2 + |y_t|^2 + |y_\tau|^2 - \tfrac{16}{3}g_3^2 - 3g_2^2 - \tfrac{7}{15}g_1^2\Big) + 2a_t y_t^* y_b \\ &\quad + 2a_\tau y_\tau^* y_b + y_b\Big(\tfrac{32}{3}g_3^2 M_3 + 6g_2^2 M_2 + \tfrac{14}{15}g_1^2 M_1\Big), \end{aligned} \tag{13.61}$$

$$\begin{aligned} \beta_{a_\tau}^{(1)} &= a_\tau\Big(12|y_\tau|^2 + 3|y_b|^2 - 3g_2^2 - \tfrac{9}{5}g_1^2\Big) + 6a_b y_b^* y_\tau \\ &\quad + y_\tau\Big(6g_2^2 M_2 + \tfrac{18}{5}g_1^2 M_1\Big), \end{aligned} \tag{13.62}$$

$$\begin{aligned} \beta_b^{(1)} &= b\Big(3|y_t|^2 + 3|y_b|^2 + |y_\tau|^2 - 3g_2^2 - \tfrac{3}{5}g_1^2\Big) \\ &\quad + \mu\Big(6a_t y_t^* + 6a_b y_b^* + 2a_\tau y_\tau^* + 6g_2^2 M_2 + \tfrac{6}{5}g_1^2 M_1\Big). \end{aligned} \tag{13.63}$$

[2] In GUT models, it is automatic that the gauge couplings and gaugino masses are unified at all scales $Q \geq M_U$, because in the unified theory the gauginos all live in the same representation of the unified gauge group. In many superstring models, this can also be a good approximation.

[3] Rescaled soft parameters $A_t = a_t/y_t$, $A_b = a_b/y_b$, and $A_\tau = a_\tau/y_\tau$ are often used in the literature. We do not follow this notation, because it cannot be generalized beyond the approximation of eqs. (13.7), (13.59) without introducing horrible complications such as nonpolynomial RGEs, and because a_t, a_b, and a_τ are the couplings that actually appear in the Lagrangian.

The β-function for each of these soft parameters is *not* proportional to the parameter itself, because couplings that violate supersymmetry are not protected by the supersymmetric nonrenormalization theorem. So, even if a_t, a_b, a_τ, and b vanish at the input scale, the RG corrections proportional to gaugino masses appearing in eqs. (13.60)–(13.63) ensure that they will not vanish at the electroweak scale.

Next let us consider the RGEs for the scalar squared masses in the MSSM. In the approximation of eqs. (13.7) and (13.59), the squarks and sleptons of the first two generations have only gauge interactions. This means that if the scalar squared masses satisfy a boundary condition like eq. (13.24) at an input RG scale, then when renormalized to any other RG scale, they will still be almost diagonal, with the approximate form

$$\mathbf{m_Q^2} \approx \begin{pmatrix} m^2_{Q_1} & 0 & 0 \\ 0 & m^2_{Q_1} & 0 \\ 0 & 0 & m^2_{Q_3} \end{pmatrix}, \qquad \mathbf{m_{\bar{u}}^2} \approx \begin{pmatrix} m^2_{\bar{u}_1} & 0 & 0 \\ 0 & m^2_{\bar{u}_1} & 0 \\ 0 & 0 & m^2_{\bar{u}_3} \end{pmatrix}, \tag{13.64}$$

etc. The first- and second-generation squarks and sleptons with given gauge quantum numbers remain very nearly degenerate, but the third-generation squarks and sleptons feel the effects of the larger Yukawa couplings and so their squared masses get renormalized differently. The one-loop RG equations for the first- and second-generation squark and slepton squared masses are

$$\beta^{(1)}_{m^2_{\phi_i}} = -\sum_{a=1,2,3} 8C_a(i)g_a^2|M_a|^2 + \tfrac{6}{5}Y_i g_1^2 S \tag{13.65}$$

for each scalar ϕ_i, where the $\sum_a$ is over the three gauge groups $U(1)_Y$, $SU(2)_L$, and $SU(3)_C$, with Casimir invariants $C_a(i)$ as in eqs. (13.34)–(13.36), and M_a are the corresponding running gaugino mass parameters. Also,

$$S \equiv \mathrm{Tr}[Y_j m^2_{\phi_j}] = m^2_{H_u} - m^2_{H_d} + \mathrm{Tr}[\mathbf{m_Q^2} - \mathbf{m_L^2} - 2\mathbf{m_{\bar{u}}^2} + \mathbf{m_{\bar{d}}^2} + \mathbf{m_{\bar{e}}^2}]. \tag{13.66}$$

An important feature of eq. (13.65) is that the terms on the right-hand sides proportional to gaugino squared masses are negative, so[4] the scalar squared-mass parameters grow as they are RG-evolved from the input scale down to the electroweak scale. Even if the scalars have zero or very small masses at the input scale, they can obtain large positive squared masses at the electroweak scale, thanks to the effects of the gaugino masses.

The RGEs for the squared-mass parameters of the Higgs scalars and third-generation squarks and sleptons get the same gauge contributions as in eq. (13.65), but they also have contributions due to the large Yukawa ($y_{t,b,\tau}$) and soft ($a_{t,b,\tau}$) couplings. At one-loop order, these only appear in three combinations:

$$S_t = 2|y_t|^2(m^2_{H_u} + m^2_{Q_3} + m^2_{\bar{u}_3}) + 2|a_t|^2, \tag{13.67}$$
$$S_b = 2|y_b|^2(m^2_{H_d} + m^2_{Q_3} + m^2_{\bar{d}_3}) + 2|a_b|^2, \tag{13.68}$$
$$S_\tau = 2|y_\tau|^2(m^2_{H_d} + m^2_{L_3} + m^2_{\bar{e}_3}) + 2|a_\tau|^2. \tag{13.69}$$

[4] The contributions proportional to S are relatively small in most known realistic models.

In terms of these quantities, the RGEs for the soft Higgs squared-mass parameters $m_{H_u}^2$ and $m_{H_d}^2$ are

$$\beta^{(1)}_{m^2_{H_u}} = 3S_t - 6g_2^2|M_2|^2 - \tfrac{6}{5}g_1^2|M_1|^2 + \tfrac{3}{5}g_1^2 S, \tag{13.70}$$

$$\beta^{(1)}_{m^2_{H_d}} = 3S_b + S_\tau - 6g_2^2|M_2|^2 - \tfrac{6}{5}g_1^2|M_1|^2 - \tfrac{3}{5}g_1^2 S. \tag{13.71}$$

Note that S_t, S_b, and S_τ are generally positive, so their effect is to decrease the Higgs squared masses as one evolves the RGEs down from the input scale to the electroweak scale. If y_t is the largest of the Yukawa couplings, as suggested by the experimental fact that the top quark is heavy, then S_t will typically be much larger than S_b and S_τ. This can cause the RG-evolved $m_{H_u}^2$ to run negative near the electroweak scale, helping to destabilize the point $H_u = H_d = 0$ and so provoking a Higgs VEV (for a linear combination of H_u and H_d, as we will see in Section 13.8), which is just what we want.[5] Thus a large top Yukawa coupling favors the breakdown of the electroweak symmetry breaking because it induces negative radiative corrections to the Higgs squared mass.

The third-generation squark and slepton squared-mass parameters also get contributions that depend on S_t, S_b, and S_τ. Their RGEs are given by

$$\beta^{(1)}_{m^2_{Q_3}} = S_t + S_b - \tfrac{32}{3}g_3^2|M_3|^2 - 6g_2^2|M_2|^2 - \tfrac{2}{15}g_1^2|M_1|^2 + \tfrac{1}{5}g_1^2 S, \tag{13.72}$$

$$\beta^{(1)}_{m^2_{\bar{u}_3}} = 2S_t - \tfrac{32}{3}g_3^2|M_3|^2 - \tfrac{32}{15}g_1^2|M_1|^2 - \tfrac{4}{5}g_1^2 S, \tag{13.73}$$

$$\beta^{(1)}_{m^2_{\bar{d}_3}} = 2S_b - \tfrac{32}{3}g_3^2|M_3|^2 - \tfrac{8}{15}g_1^2|M_1|^2 + \tfrac{2}{5}g_1^2 S, \tag{13.74}$$

$$\beta^{(1)}_{m^2_{L_3}} = S_\tau - 6g_2^2|M_2|^2 - \tfrac{6}{5}g_1^2|M_1|^2 - \tfrac{3}{5}g_1^2 S, \tag{13.75}$$

$$\beta^{(1)}_{m^2_{\bar{e}_3}} = 2S_\tau - \tfrac{24}{5}g_1^2|M_1|^2 + \tfrac{6}{5}g_1^2 S. \tag{13.76}$$

In eqs. (13.70)–(13.76), the terms proportional to $|M_3|^2$, $|M_2|^2$, $|M_1|^2$, and S are the same ones as in eq. (13.65). Note that the terms proportional to S_t and S_b appear with smaller numerical coefficients in the $m^2_{Q_3}$, $m^2_{\bar{u}_3}$, $m^2_{\bar{d}_3}$ RGEs than they did for the Higgs scalars, and they do not appear at all in the $m^2_{L_3}$ and $m^2_{\bar{e}_3}$ RGEs. Furthermore, the third-generation squark squared masses get a large positive contribution proportional to $|M_3|^2$ from the RG evolution, which the Higgs scalars do not get. These facts make it plausible that the Higgs scalars in the MSSM get VEVs, while the squarks and sleptons, having large positive squared mass, do not.

An examination of the RGEs (13.60)–(13.63), (13.65), and (13.70)–(13.76) reveals that if the gaugino mass parameters M_1, M_2, and M_3 are nonzero at the input scale, then all of the other soft terms will be generated too. This implies that models in which gaugino masses dominate over all other effects in the soft supersymmetry-breaking Lagrangian at the input scale can be viable. On the other hand, if the gaugino masses were to vanish at tree level, then they would not get

[5] One should think of "$m^2_{H_u}$" as a parameter unto itself, and not as the square of some mythical real number m_{H_u}. So, there is nothing strange about having $m^2_{H_u} < 0$.

any contributions to their masses at one-loop order; in that case the gauginos would be extremely light and the model would not be phenomenologically acceptable.

Viable models for the origin of supersymmetry breaking typically make predictions for the MSSM soft terms that are refinements of eqs. (13.24)–(13.26). These predictions can then be used as boundary conditions for the RGEs listed above. In Chapter 12 we will study the ideas that go into making such predictions.

13.8 Electroweak Symmetry Breaking and the Higgs Bosons

In contrast to the Standard Model with one complex Higgs doublet, the description of electroweak symmetry breaking in the MSSM is complicated by the fact that there are two complex Higgs doublets, $H_u = (H_u^+, H_u^0)$ and $H_d = (H_d^0, H_d^-)$. The classical scalar potential for the Higgs scalar fields in the MSSM is given by

$$\begin{aligned} V &= (|\mu|^2 + m_{H_u}^2)(|H_u^0|^2 + |H_u^+|^2) + (|\mu|^2 + m_{H_d}^2)(|H_d^0|^2 + |H_d^-|^2) \\ &\quad + \tfrac{1}{8}(g^2 + g'^2)(|H_u^0|^2 + |H_u^+|^2 - |H_d^0|^2 - |H_d^-|^2)^2 + \tfrac{1}{2}g^2|H_u^+ H_d^{0\dagger} + H_u^0 H_d^{-\dagger}|^2 \\ &\quad + b\,(H_u^+ H_d^- - H_u^0 H_d^0) + \text{h.c.} \end{aligned} \tag{13.77}$$

The terms proportional to $|\mu|^2$ are the F-term contributions [see eq. (13.10)], and the terms proportional to g^2 and g'^2 are the D-term contributions. Both types of terms may be derived from the general formula exhibited in eq. (8.75). Finally, the terms proportional to $m_{H_u}^2$, $m_{H_d}^2$, and b are just a rewriting of the last three terms of eq. (13.16). The full scalar potential of the theory also includes many terms involving the squark and slepton fields that we can ignore here, since they do not get VEVs because they have large positive squared mass.

We now have to demand that the minimum of this potential should break electroweak symmetry down to electromagnetism $\mathrm{SU}(2)_L \times U(1)_Y \to U(1)_{\mathrm{EM}}$, in accord with experiment. We can use the freedom to make gauge transformations to simplify this analysis. First, the freedom to make $\mathrm{SU}(2)_L$ gauge transformations allows us to rotate away a possible VEV for one of the weak isospin components of one of the scalar fields, so without loss of generality we can take $H_u^+ = 0$ at the minimum of the potential. Then one can check that a minimum of the potential satisfying $\partial V/\partial H_u^+ = 0$ must also have $H_d^- = 0$. This is good, because it means that at the minimum of the potential electromagnetism is necessarily unbroken, since the charged components of the Higgs scalars cannot get VEVs. After setting $H_u^+ = H_d^- = 0$, we are left to consider the scalar potential

$$\begin{aligned} V &= (|\mu|^2 + m_{H_u}^2)|H_u^0|^2 + (|\mu|^2 + m_{H_d}^2)|H_d^0|^2 - (b\, H_u^0 H_d^0 + \text{h.c.}) \\ &\quad + \tfrac{1}{8}(g^2 + g'^2)(|H_u^0|^2 - |H_d^0|^2)^2. \end{aligned} \tag{13.78}$$

The B-term [i.e., the term in eq. (13.78) that is proportional to b] is the only term in the potential V that depends on the phases of the fields. Therefore, a redefinition of the phase of H_u or H_d can absorb any phase in b, so we can take b to be real and positive. Then it is clear that a minimum of the potential V requires that $H_u^0 H_d^0$ is also real and positive, so $\langle H_u^0 \rangle$ and $\langle H_d^0 \rangle$ must have opposite phases. We can therefore use a $\mathrm{U}(1)_Y$ gauge transformation to make them both be real and positive without loss of generality, since H_u and H_d have opposite weak hypercharges ($\pm 1/2$). It follows that CP cannot be spontaneously broken by the Higgs scalar potential at tree level, since the VEVs and b can be simultaneously chosen real, as a convention. This implies that the Higgs scalar mass eigenstates can be assigned well-defined eigenvalues of CP, at least at tree level. (CP-violating phases in other couplings can induce loop-suppressed CP violation in the Higgs sector, but do not change the fact that $\langle H_u^0 \rangle$ and $\langle H_d^0 \rangle$ can always be chosen real and positive.)

In order for the MSSM scalar potential to be viable, we must first make sure that the potential is bounded from below for arbitrarily large values of the scalar fields, so that V will really have a minimum. (Recall from the discussion in Sections 8.2 and 8.4 that scalar potentials in purely supersymmetric theories are automatically nonnegative and so clearly bounded from below. But now that we have introduced supersymmetry breaking, we must be careful.) The scalar quartic interactions in V will stabilize the potential for almost all arbitrarily large values of H_u^0 and H_d^0. However, for the special directions in field space $|H_u^0| = |H_d^0|$, the quartic contributions to V [the second line in eq. (13.78)] are identically zero. Such directions in field space are called D-flat directions, because along them the part of the scalar potential coming from D-terms vanishes. In order for the potential to be bounded from below, we need the quadratic part of the scalar potential to be positive along the D-flat directions. This requirement yields

$$2b < 2|\mu|^2 + m_{H_u}^2 + m_{H_d}^2. \tag{13.79}$$

We shall demand that one linear combination of H_u^0 and H_d^0 has a negative squared mass near $H_u^0 = H_d^0 = 0$. It then follows that

$$b^2 > (|\mu|^2 + m_{H_u}^2)(|\mu|^2 + m_{H_d}^2). \tag{13.80}$$

If this inequality is not satisfied, then $H_u^0 = H_d^0 = 0$ will be a stable minimum of the potential (or there will be no stable minimum at all), and electroweak symmetry breaking will not occur.

Interestingly, if $m_{H_u}^2 = m_{H_d}^2$ then the constraints eqs. (13.79) and (13.80) cannot both be satisfied. In models with CMSSM boundary conditions, $m_{H_u}^2 = m_{H_d}^2$ is supposed to hold at tree level at the input scale, but the S_t contribution to the RG equation for $m_{H_u}^2$ [eq. (13.70)] naturally pushes it to negative or small values $m_{H_u}^2 < m_{H_d}^2$ at the electroweak scale. Unless this effect is significant, the parameter space in which the electroweak symmetry is broken would be quite small. So in these models electroweak symmetry breaking is actually driven by quantum corrections; this mechanism is therefore known as *radiative electroweak symmetry breaking*. Note that although a negative value for $|\mu|^2 + m_{H_u}^2$ will help eq. (13.80) to be satisfied, it is not strictly necessary. Furthermore, even if $m_{H_u}^2 < 0$, there may be no electroweak symmetry breaking if $|\mu|$ is too large or if b is too small. Still, the large negative contributions to $m^2_{H_u}$ from the RG equation are an important factor in ensuring that electroweak symmetry breaking can occur in models with simple boundary conditions for the soft terms. The realization that this works most naturally with a large top-quark Yukawa coupling provides additional motivation for these models.

Having established the conditions necessary for H_u^0 and H_d^0 to get nonzero VEVs, we can now require that they are compatible with the observed phenomenology of electroweak symmetry breaking $\mathrm{SU}(2)_L \times U(1)_Y \to U(1)_{\mathrm{EM}}$. Following eqs. (6.223) and (6.224), we write

$$\langle H_d^0 \rangle = \frac{v_d}{\sqrt{2}}\,, \qquad \langle H_u^0 \rangle = \frac{v_u}{\sqrt{2}}\,, \tag{13.81}$$

where the normalization of the VEVs is fixed by the Fermi constant [see eq. (6.38)],

$$v \equiv (v_d^2 + v_u^2)^{1/2} \simeq 246 \text{ GeV}, \tag{13.82}$$

and the ratio of the two VEVs is defined by

$$\tan\beta \equiv v_u/v_d. \tag{13.83}$$

The value of $\tan\beta$ is not fixed by present experiments, but it depends on the Lagrangian parameters of the MSSM in a calculable way. By convention, v_u and v_d are chosen to be real and positive, with $0 < \beta < \pi/2$, a requirement that will be sharpened below. One can now write down the conditions under which the minimum of the potential given in eq. (13.78) satisfies eqs. (13.82) and (13.83). To obtain the scalar potential minimum conditions, simply set $H_u^0 = v_u/\sqrt{2}$ and $H_d^0 = v_d/\sqrt{2}$ in eq. (13.78). The minimum conditions, $\partial V/\partial v_u = \partial V/\partial v_d = 0$, under the assumption that v_u, $v_d \neq 0$, yield [see eqs. (6.206), (6.207), and (6.222)]

$$|\mu|^2 + m^2_{H_d} = b\tan\beta - \tfrac{1}{2}m_Z^2\cos 2\beta\,, \tag{13.84}$$

$$|\mu|^2 + m^2_{H_u} = b\cot\beta + \tfrac{1}{2}m_Z^2\cos 2\beta\,, \tag{13.85}$$

after using $m_Z^2 = \frac{1}{4}v^2(g^2 + g'^2)$. It is easy to check that these equations satisfy the necessary conditions given by eqs. (13.79) and (13.80). They allow us to eliminate two of the Lagrangian parameters b and $|\mu|$ in favor of $\tan\beta$, but do not determine the phase of μ. Taking $|\mu|^2$, b, $m^2_{H_u}$, and $m^2_{H_d}$ as input parameters, and m_Z^2 and $\tan\beta$ as output parameters obtained by solving these two equations, one obtains

$$\sin 2\beta = \frac{2b}{m^2_{H_u} + m^2_{H_d} + 2|\mu|^2}\,, \tag{13.86}$$

$$m_Z^2 = \frac{|m^2_{H_d} - m^2_{H_u}|}{\sqrt{1 - \sin^2 2\beta}} - m^2_{H_u} - m^2_{H_d} - 2|\mu|^2. \tag{13.87}$$

Note that $\sin 2\beta$ is always positive. If $m^2_{H_u} < m^2_{H_d}$, as is usually assumed, then $\cos 2\beta$ is negative; otherwise it is positive.

As an aside, eqs. (13.86) and (13.87) highlight the "μ problem" already mentioned in Section 13.2. Without miraculous cancellations, all of the input parameters ought to be within an order of magnitude or two of m_Z^2. However, in the

MSSM, μ is a supersymmetry-respecting parameter appearing in the superpotential, while b, $m^2_{H_u}$, and $m^2_{H_d}$ are supersymmetry-breaking parameters. This has led to a widespread belief that the MSSM must be extended at very high energies to include a mechanism that relates the effective value of μ to the supersymmetry-breaking mechanism in some way; see Section 16.1 for examples.

Even if the value of μ is set by soft supersymmetry breaking, the cancellation needed by eq. (13.87) is often remarkable when evaluated in specific model frameworks, after constraints from direct searches for the Higgs bosons and superpartners are taken into account. For example, expanding for large $\tan\beta$, eq. (13.87) becomes

$$m_Z^2 = -2(m^2_{H_u} + |\mu|^2) + \frac{2}{\tan^2\beta}(m^2_{H_d} - m^2_{H_u}) + \mathcal{O}(1/\tan^4\beta). \tag{13.88}$$

Typical viable solutions for the MSSM have $-m^2_{H_u}$ and $|\mu|^2$ each much larger than m_Z^2, so that significant cancellation is needed. In particular, large top-squark squared masses, needed to avoid having the Higgs boson mass turn out too small [see the discussion surrounding eq. (13.115) below] compared to the observed value of 125 GeV, will feed into $m^2_{H_u}$. The cancellation needed in the minimal model may therefore be at the percent level, or worse. This is often referred to as the "supersymmetric little hierarchy problem." It is impossible to objectively characterize whether this should be considered worrisome, but it certainly causes subjective worry as the LHC bounds on superpartners increase, and could be taken as a hint in favor of nonminimal models.

Equations (13.85)–(13.87) are based on the tree-level potential, and involve running renormalized Lagrangian parameters, which depend on the choice of renormalization scale. In practice, one must include radiative corrections at one-loop order, at least, in order to get numerically stable results. To do this, one can compute the loop corrections ΔV to the effective potential $V_{\text{eff}}(v_u, v_d) = V + \Delta V$ as a function of the VEVs. The impact of this is that the equations governing the VEVs of the full effective potential are obtained by simply replacing

$$m^2_{H_u} \to m^2_{H_u} + \frac{1}{v_u}\frac{\partial(\Delta V)}{\partial v_u}, \qquad m^2_{H_d} \to m^2_{H_d} + \frac{1}{v_d}\frac{\partial(\Delta V)}{\partial v_d} \tag{13.89}$$

in eqs. (13.85)–(13.87), treating v_u and v_d as real variables in the differentiation. The result for ΔV has now been obtained through full two-loop order in the MSSM [133]. The most important corrections come from the one-loop diagrams involving the top squarks and top quark, and experience shows that the validity of the tree-level approximation and the convergence of perturbation theory are improved by choosing a renormalization scale roughly of order the average of the top-squark masses.

The Higgs scalar fields in the MSSM consist of two complex $SU(2)_L$-doublet, or eight real, scalar degrees of freedom. When the electroweak symmetry is broken, three of them are the would-be Goldstone bosons G^0, $G^\pm$, which become the longitudinal modes of the Z and $W^\pm$ massive vector bosons. The remaining five Higgs scalar mass eigenstates consist of two CP-even neutral scalars h and H (where

$m_h < m_H$ by convention), one CP-odd neutral scalar A^0, and a charge +1 scalar H^+ and its conjugate charge −1 scalar H^- [where $G^- \equiv (G^+)^\dagger$ and $H^- \equiv (H^+)^\dagger$]. Henceforth, we will simply omit the superscript 0 from the neutral mass eigenstate scalar fields. The gauge eigenstate fields can be expressed in terms of the mass-eigenstate fields as

$$\begin{pmatrix} H_u^0 \\ H_d^0 \end{pmatrix} = \frac{1}{\sqrt{2}} \begin{pmatrix} v_u \\ v_d \end{pmatrix} + \frac{1}{\sqrt{2}} R_\alpha \begin{pmatrix} h \\ H \end{pmatrix} + \frac{1}{\sqrt{2}} R_{\beta_0} \begin{pmatrix} G \\ A \end{pmatrix}, \tag{13.90}$$

$$\begin{pmatrix} H_u^+ \\ H_d^{-\dagger} \end{pmatrix} = R_{\beta_\pm} \begin{pmatrix} G^+ \\ H^+ \end{pmatrix}, \tag{13.91}$$

where the orthogonal rotation matrices

$$R_\alpha = \begin{pmatrix} \cos\alpha & \sin\alpha \\ -\sin\alpha & \cos\alpha \end{pmatrix} \tag{13.92}$$

and

$$R_{\beta_0} = \begin{pmatrix} \sin\beta_0 & \cos\beta_0 \\ -\cos\beta_0 & \sin\beta_0 \end{pmatrix}, \qquad R_{\beta_\pm} = \begin{pmatrix} \sin\beta_\pm & \cos\beta_\pm \\ -\cos\beta_\pm & \sin\beta_\pm \end{pmatrix}, \tag{13.93}$$

are chosen so that the quadratic part of the potential has diagonal squared masses:

$$V = \tfrac{1}{2} m_h^2 h^2 + \tfrac{1}{2} m_H^2 H^2 + \tfrac{1}{2} m_G^2 G^2 + \tfrac{1}{2} m_A^2 A^2 + m_{G^\pm}^2 |G^+|^2 + m_{H^\pm}^2 |H^+|^2 + \cdots . \tag{13.94}$$

Provided that v_u, v_d minimize the tree-level potential,[6] one finds that $\beta_0 = \beta_\pm = \beta$ and $m_{G^0}^2 = m_{G^\pm}^2 = 0$. The tree-level squared masses of the physical Higgs scalars were obtained in Section 6.2.6, and are repeated here for the reader's convenience:

$$m_A^2 = \frac{2b}{\sin 2\beta} = 2|\mu|^2 + m_{H_u}^2 + m_{H_d}^2 , \tag{13.95}$$

$$m_{H^\pm}^2 = m_A^2 + m_W^2 , \tag{13.96}$$

$$m_{H,h}^2 = \tfrac{1}{2}\left(m_A^2 + m_Z^2 \pm \sqrt{(m_A^2 - m_Z^2)^2 + 4 m_Z^2 m_A^2 \sin^2 2\beta} \right). \tag{13.97}$$

The mixing angle α is defined modulo π, and its value is conventionally chosen such that $|\alpha| \leq \frac{1}{2}\pi$. However, in light of eq. (6.217), the sign of $\sin\alpha$ is given by the sign of the off-diagonal element, $\mathcal{M}_{12}^2$, of the neutral CP-even scalar squared-mass matrix. At tree level, $\mathcal{M}_{12}^2 = -\frac{1}{2}(m_A^2 + m_Z^2)\sin 2\beta < 0$ [cf. eq. (6.230)], and we

[6] It is often more useful to expand around VEVs v_u, v_d that do not minimize the tree-level potential, for example to minimize the loop-corrected effective potential instead. In that case, the three angles β, β_0, and $\beta_\pm$ are all slightly different, and the tree-level parameters, m_G^2 and $m_{G^\pm}^2$, are nonzero and differ from each other by a small amount.

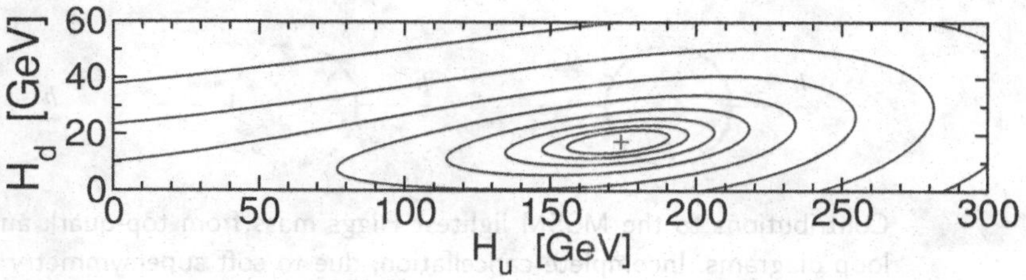

Fig. 13.8 A contour map of the MSSM Higgs potential is exhibited for a typical case with $\tan\beta \approx -\cot\alpha \approx 10$. The minimum of the potential is marked by +, and the contours are equally spaced equipotentials. Oscillations along the shallow direction, with $H_u^0/H_d^0 \approx 10$, correspond to the mass eigenstate h, while the orthogonal steeper direction corresponds to the mass eigenstate H.

conclude that the tree-level value of α lies in the range $-\frac{1}{2}\pi \leq \alpha \leq 0$. A convenient expression for α was obtained in eq. (6.238):

$$\sin\alpha = -\sqrt{\frac{m_H^2 - m_A^2 \sin^2\beta - m_Z^2 \cos^2\beta}{m_H^2 - m_h^2}}\,, \tag{13.98}$$

which uniquely determines the value of α in its allowed range.

The masses of A, H, and $H^\pm$ can in principle be arbitrarily large since they all grow with $b/\sin 2\beta$. In contrast, the mass of h is bounded from above. From eq. (13.97), one finds at tree level that

$$m_h < m_Z |\cos 2\beta|\,. \tag{13.99}$$

This corresponds to the existence of a shallow direction in the scalar potential, along the direction $(\sqrt{2}\,\mathrm{Re}\,H_u^0 - v_u\,,\ \sqrt{2}\,\mathrm{Re}\,H_d^0 - v_d) \propto (\cos\alpha, -\sin\alpha)$. The existence of this shallow direction can be traced to the supersymmetric fact that the quartic Higgs couplings are small, being given by the sum of the squares of the electroweak gauge couplings, via the D-term. A contour map of the potential, for a typical case with $\tan\beta \approx -\cot\alpha \approx 10$, is shown in Fig. 13.8. If the tree-level inequality of eq. (13.99) were robust, the lightest Higgs boson of the MSSM would have been discovered in the previous century at the CERN LEP2 e^+e^- collider, and its mass obviously could not approach the observed value of 125 GeV. However, the tree-level formula for the squared mass of h is subject to quantum corrections that are relatively drastic. The largest such contributions typically come from the top and stop loop diagrams shown[7] in Fig. 13.9.

In Section 19.8 a Feynman diagrammatic approach to computing the one-loop corrected MSSM Higgs masses is developed. Here, we shall employ a technique based on the one-loop effective potential. In addition, we shall neglect any CP-violating

[7] In general, one-loop 1-particle-reducible tadpole diagrams should also be included. However, they exactly cancel against the tree-level tadpoles, and so both can be omitted, if the VEVs v_u and v_d are taken at the minimum of the loop-corrected effective potential (see footnote 6).

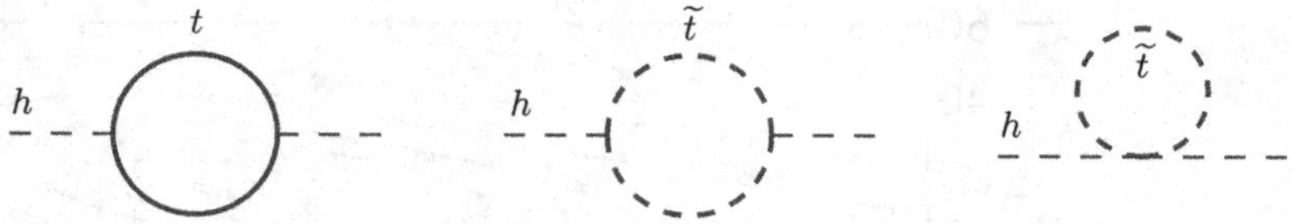

Fig. 13.9 Contributions to the MSSM lightest Higgs mass from top-quark and top-squark one-loop diagrams. Incomplete cancellation, due to soft supersymmetry breaking, leads to a large positive correction to m_h^2 in the limit of heavy top squarks.

effects that could potentially yield one-loop mixing of the neutral CP-even and CP-odd scalars. The squared-mass matrix of the CP-even scalars is obtained from the matrix of second derivatives of the one-loop effective potential:

$$(\mathcal{M}^2_{\text{eff}})_{ij} \equiv \frac{\partial^2 V_{\text{eff}}}{\partial v_i \partial v_j} = \begin{pmatrix} (\mathcal{M}^2_{\text{eff}})_{dd} & (\mathcal{M}^2_{\text{eff}})_{du} \\ (\mathcal{M}^2_{\text{eff}})_{du} & (\mathcal{M}^2_{\text{eff}})_{uu} \end{pmatrix}. \tag{13.100}$$

In particular,

$$\mathcal{M}^2_{\text{eff}} = \begin{pmatrix} M_d^2 - m_Z^2\left(\dfrac{v_u^2 - 3v_d^2}{2v^2}\right) + \dfrac{1}{16\pi^2}\dfrac{\partial^2 V^{(1)}}{(\partial v_d)^2} & -b - m_Z^2\dfrac{v_u v_d}{v^2} + \dfrac{1}{16\pi^2}\dfrac{\partial^2 V^{(1)}}{\partial v_d \partial v_u} \\ -b - m_Z^2\dfrac{v_u v_d}{v^2} + \dfrac{1}{16\pi^2}\dfrac{\partial^2 V^{(1)}}{\partial v_u \partial v_d} & M_u^2 + m_Z^2\left(\dfrac{3v_u^2 - v_d^2}{2v^2}\right) + \dfrac{1}{16\pi^2}\dfrac{\partial^2 V^{(1)}}{(\partial v_u)^2} \end{pmatrix}, \tag{13.101}$$

where $V^{(1)}$ is defined in eq. (11.169),

$$M_d^2 \equiv |\mu|^2 + m_{H_d}^2\,, \qquad M_u^2 \equiv |\mu|^2 + m_{H_u}^2\,, \tag{13.102}$$

and the minimum conditions given by eqs. (13.84) and (13.85) are modified following eq. (13.89). Using the modified minimum conditions to eliminate M_d^2 and M_u^2 and denoting $s_\beta \equiv \sin\beta = v_u/v$ and $c_\beta \equiv \cos\beta = v_d/v$, we end up with

$$\mathcal{M}^2_{\text{eff}} = \begin{pmatrix} m_A^2 s_\beta^2 + m_Z^2 c_\beta^2 & -(m_A^2 + m_Z^2)s_\beta c_\beta \\ -(m_A^2 + m_Z^2)s_\beta c_\beta & m_A^2 c_\beta^2 + m_Z^2 s_\beta^2 \end{pmatrix} + \Delta\mathcal{M}^2_{\text{eff}}, \tag{13.103}$$

where

$$\Delta\mathcal{M}^2_{\text{eff}} = \frac{1}{4\pi^2}\begin{pmatrix} v_d^2 \dfrac{\partial}{\partial v_d^2}\left(\dfrac{\partial V^{(1)}}{\partial v_d^2}\right) & v_u v_d \dfrac{\partial^2 V^{(1)}}{\partial v_u^2 \partial v_d^2} \\ v_u v_d \dfrac{\partial^2 V^{(1)}}{\partial v_u^2 \partial v_d^2} & v_u^2 \dfrac{\partial}{\partial v_u^2}\left(\dfrac{\partial V^{(1)}}{\partial v_u^2}\right) \end{pmatrix}, \tag{13.104}$$

after employing the identity

$$\frac{\partial^2 V^{(1)}}{(\partial v_i)^2} - \frac{1}{v_i}\frac{\partial V^{(1)}}{\partial v_i} = 4v_i^2 \frac{\partial}{\partial v_i^2}\left(\frac{\partial V^{(1)}}{\partial v_i^2}\right). \tag{13.105}$$

It is instructive to apply eqs. (13.103) and (13.104) to the top and stop loop contributions to the effective potential, which are given in eq. (11.203). For simplicity, we neglect top-squark mixing and the electroweak corrections to the diagonal stop squared masses. In this simple approximation [see eqs. (13.123) and (13.180)],

$$m^2_{\widetilde{t}_1} = m^2_{Q_3} + \tfrac{1}{2}y_t^2 v_u^2\,, \qquad m^2_{\widetilde{t}_2} = m^2_{\bar{u}_3} + \tfrac{1}{2}y_t^2 v_u^2\,, \qquad m_t^2 = \tfrac{1}{2}y_t^2 v_u^2\,, \tag{13.106}$$

where no mass ordering convention for $\widetilde{t}_{1,2}$ is assumed. To compute the second derivative of the effective potential, we first make use of the chain rule to obtain

$$\frac{\partial}{\partial v_u^2}\left(\frac{\partial h(x)}{\partial v_u^2}\right) = \frac{\partial^2 h}{\partial x^2}\left(\frac{\partial x}{\partial v_u^2}\right)^2 + \frac{\partial h}{\partial x}\frac{\partial}{\partial v_u^2}\left(\frac{\partial x}{\partial v_u^2}\right). \tag{13.107}$$

Then, eq. (11.202) yields

$$\frac{\partial h}{\partial x} = \frac{x}{2}\left[\ln\left(\frac{x}{Q^2}\right) - 1\right], \qquad \frac{\partial^2 h}{\partial x^2} = \frac{1}{2}\ln\left(\frac{x}{Q^2}\right). \tag{13.108}$$

In light of eq. (13.106), $\partial x/\partial v_u^2 = \frac{1}{2}y_t^2$ for $x = m^2_{\widetilde{t}_1}$, $m^2_{\widetilde{t}_2}$, and m_t^2. It follows that

$$\frac{\partial}{\partial v_u^2}\left(\frac{\partial h(x)}{\partial v_u^2}\right) = \frac{y_t^4}{8\pi^2}\ln\left(\frac{x}{Q^2}\right). \tag{13.109}$$

Hence, the leading one-loop correction to $\mathcal{M}^2_{\rm eff}$ appears as a correction to the uu matrix element of $\mathcal{M}^2_{\rm eff}$:

$$(\mathcal{M}^2_{\rm eff})_{uu} = m_A^2 c_\beta^2 + m_Z^2 s_\beta^2 + (\Delta\mathcal{M}^2_{\rm eff})_{uu}\,, \tag{13.110}$$

where

$$(\Delta\mathcal{M}^2_{\rm eff})_{uu} = \frac{3y_t^2 m_t^2}{4\pi^2}\ln\left(\frac{m_{\widetilde{t}_1}m_{\widetilde{t}_2}}{m_t^2}\right), \tag{13.111}$$

after using $m_t^2 = \frac{1}{2}y_t^2 v_u^2$. Finally, we employ eqs. (6.211) and (13.97) to obtain

$$m_h^2 = \tfrac{1}{2}\left(m_A^2 + m_Z^2 - \sqrt{(m_A^2 - m_Z^2)^2 + 4m_Z^2 m_A^2 \sin^2 2\beta}\right) + \Delta(m_h^2)\,, \tag{13.112}$$

where

$$\Delta(m_h^2) = (\Delta\mathcal{M}^2_{\rm eff})_{uu}\cos^2\alpha = \frac{3y_t^2 m_t^2 \cos^2\alpha}{4\pi^2}\ln\left(\frac{m_{\widetilde{t}_1}m_{\widetilde{t}_2}}{m_t^2}\right), \tag{13.113}$$

after using eq. (6.211), and $\cos^2\alpha$ is evaluated in terms of tree-level parameters [see eq. (6.237)]. In the Higgs decoupling limit (where $m_A \gg m_Z$), $\alpha \simeq \beta - \frac{1}{2}\pi$, which yields $\cos\alpha \simeq \sin\beta$ and $\sin\alpha \simeq -\cos\beta$. In this limit, m_h attains its upper bound as a function of β, whose leading contribution at one-loop order is given by

$$m_h^2 = m_Z^2\cos^2 2\beta + \frac{3m_t^4}{2\pi^2 v^2}\ln\left(\frac{m_{\widetilde{t}_1}m_{\widetilde{t}_2}}{m_t^2}\right), \tag{13.114}$$

after employing $m_t^2 = \frac{1}{2}v^2 y_t^2 \sin^2\beta$ [see eq. (13.106)]. The insertion of $v = 2m_W/g$ reproduces the leading logarithmic radiative correction quoted in eq. (19.316).

In the limit of top-squark masses $m_{\widetilde{t}_1}$, $m_{\widetilde{t}_2}$ much greater than the top-quark mass m_t, the logarithmic enhancement along with the parametric behavior of the coefficient of the logarithm proportional to m_t^4 provide the explanation of why the top and stop loop contributions dominate the one-loop radiatively corrected m_h^2. If top-squark mixing effects are included (see Exercise 13.2), then eq. (13.113) is modified in the Higgs decoupling limit as follows:

$$\Delta(m_h^2) = \frac{3m_t^4}{2\pi^2 v^2}\left[\ln\left(\frac{m_{\widetilde{t}_1} m_{\widetilde{t}_2}}{m_t^2}\right) + \Delta_{\rm threshold}\right], \tag{13.115}$$

where

$$\Delta_{\rm threshold} = \frac{|c_{\widetilde{t}} s_{\widetilde{t}}|^2 (m^2_{\widetilde{t}_2} - m^2_{\widetilde{t}_1})}{m_t^2}\ln\left(\frac{m^2_{\widetilde{t}_2}}{m^2_{\widetilde{t}_1}}\right) + \frac{|c_{\widetilde{t}} s_{\widetilde{t}}|^4 (m^2_{\widetilde{t}_2} - m^2_{\widetilde{t}_1})}{m_t^4}\left[m^2_{\widetilde{t}_2} - m^2_{\widetilde{t}_1} - \tfrac{1}{2}(m^2_{\widetilde{t}_2} + m^2_{\widetilde{t}_1})\ln\left(\frac{m^2_{\widetilde{t}_2}}{m^2_{\widetilde{t}_1}}\right)\right], \tag{13.116}$$

where $c_{\widetilde{t}}$ and $s_{\widetilde{t}}$ are complex elements of the top-squark mixing matrix, defined in eq. (13.181), such that

$$|c_{\widetilde{t}}|^2 + |s_{\widetilde{t}}|^2 = 1\,. \tag{13.117}$$

Explicit expressions for the elements of the top-squark mixing matrix are derived in Exercise 13.1.

One way to understand eq. (13.115) is by thinking in terms of the low-energy effective Standard Model theory obtained by integrating out the top squarks at a renormalization scale equal to the geometric mean of their masses. Then, $\Delta_{\rm threshold}$ comes from the finite threshold correction to the supersymmetric Higgs quartic coupling, via the first three diagrams shown in Fig. 13.10. The term with $\ln(m_{\widetilde{t}_1} m_{\widetilde{t}_2}/m_t^2)$ in eq. (13.115) is a consequence of the renormalization group running of the Higgs

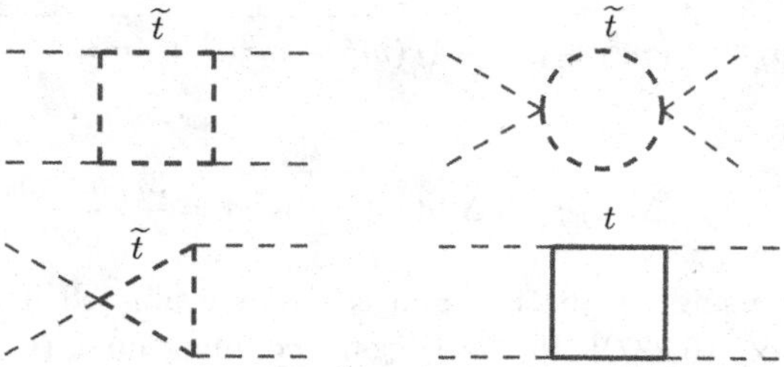

Fig. 13.10 Integrating out the top quark and top squarks yields large positive contributions to the quartic Higgs coupling in the low-energy effective theory, especially from these one-loop diagrams.

quartic coupling (due to the last diagram in Fig. 13.10) down to the top-quark mass scale, which turns out to be a good renormalization scale at which to evaluate m_h^2 within the Standard Model effective theory. For small or moderate top-squark mixing, the logarithmic running term is largest, but $\Delta_{\text{threshold}}$ can also be quite important. These corrections to the Higgs effective quartic coupling increase the steepness of the Higgs potential, thus raising m_h compared to the naive tree-level prediction.

The term proportional to $|c_{\tilde{t}}s_{\tilde{t}}|^2$ in eq. (13.116) is positive definite, while the term proportional to $|c_{\tilde{t}}s_{\tilde{t}}|^4$ is negative definite. Due to eq. (13.117), it follows that $0 \leq |c_{\tilde{t}}s_{\tilde{t}}|^2 \leq \frac{1}{4}$. Hence, the maximal value of $\Delta_{\text{threshold}}$ as a function of $|c_{\tilde{t}}s_{\tilde{t}}|^2$ is attained either when the derivative of the function vanishes or on the boundary when $|c_{\tilde{t}}s_{\tilde{t}}|^2 = \frac{1}{4}$. In either case, the maximum possible h mass occurs for rather large top-squark mixing,

$$|c_{\tilde{t}}s_{\tilde{t}}|^2 = \frac{m_t^2}{m_{\tilde{t}_2}^2 + m_{\tilde{t}_1}^2 - 2(m_{\tilde{t}_2}^2 - m_{\tilde{t}_1}^2)/\ln(m_{\tilde{t}_2}^2/m_{\tilde{t}_1}^2)}, \tag{13.118}$$

or 1/4, whichever is less. In the literature, the "maximal mixing" scenario for the MSSM Higgs sector refers to the case where eq. (13.118) is satisfied. Note that what is being maximized here is not the top-squark mixing parameter, but the value of m_h with respect to the top-squark mixing. If the maximal mixing scenario is realized, it then follows that the quantity in square brackets in eq. (13.115) is always less than $\ln(m_{\tilde{t}_2}^2/m_t^2) + 3$ (see Exercise 13.2 for further details).

Equation (13.115) already shows that m_h can easily exceed the Z boson mass, and the observed value of $m_h = 125$ GeV can in principle be accommodated. However, the above is a highly simplified account; to get reasonably accurate predictions for the Higgs scalar masses and mixings for a given set of model parameters, one must also include the remaining one-loop corrections and even the dominant two-loop and three-loop effects. Furthermore, renormalization group improvement is essential if superpartner masses (especially those of the top squarks) are significantly larger than 1 TeV. These more precise theoretical predictions of m_h in the MSSM (and beyond) along with the associated theoretical uncertainties are reviewed in Refs. [156, 157].

Including such corrections, it had been estimated long before the discovery of the 125 GeV Higgs boson that

$$m_h \lesssim 135\,\text{GeV} \tag{13.119}$$

in the MSSM. This prediction assumed that all of the sparticles that can contribute to m_h^2 in loops have masses that do not exceed a few TeV, and the bound increases logarithmically with the top-squark masses. However, in many specific model frameworks with small or moderate top-squark mixing, the bound eq. (13.119) is very far from saturated, and it turns out to be a severe challenge to accommodate values even as large as the observed $m_h = 125$ GeV, unless the top squarks are extremely heavy, or else highly mixed. Unfortunately, it is difficult to make this statement very precise, due to both the high dimensionality of the supersymmetric parameter space and the theoretical errors in the m_h prediction.

In the MSSM, the Higgs boson couplings to the quarks and leptons, which are obtained from eqs. (8.49) and (8.51), are determined by the superpotential W given by eq. (13.6). The resulting Higgs–fermion Yukawa couplings are

$$-\mathscr{L}_{\rm Y} = (\boldsymbol{y_u})^m{}_n \left[H_u^0 \widehat{u}_m \widehat{\bar{u}}^n - H_u^+ \widehat{d}_m \widehat{\bar{u}}^n\right] + (\boldsymbol{y_d})^m{}_n \left[H_d^0 \widehat{d}_m \widehat{\bar{d}}^n - H_d^- \widehat{u}_m \widehat{\bar{d}}^n\right]$$
$$+ (\boldsymbol{y_e})^m{}_n \left[H_d^0 \widehat{e}_m \widehat{\bar{e}}^n - H_d^- \widehat{\nu}_m \widehat{\bar{e}}^n\right] + \text{h.c.}\,, \tag{13.120}$$

where the hatted fields represent the interaction quark and lepton fields prior to identifying the corresponding mass eigenstates as in eq. (4.168). After identifying the quark and lepton mass eigenstates, the Higgs–fermion Yukawa interactions of the MSSM are given by [cf. eq. (6.139)]

$$-\mathscr{L}_{\rm Y} = \sum_i \Big\{ y_{ui}\left[H_u^0 u_i \bar{u}^i - H_u^+ \boldsymbol{K}_i{}^j d_j \bar{u}^i\right] + y_{di}\left[H_d^0 d_i \bar{d}^i - H_d^- (\boldsymbol{K}^\dagger)_i{}^j u_j \bar{d}^i\right]$$
$$+ y_{ei}\left[H_d^0 e_i \bar{e}^i - H_d^- \nu_i \bar{e}^i\right] + \text{h.c.}\Big\}\,, \tag{13.121}$$

where the unhatted fields are mass-eigenstate quarks and leptons, $\boldsymbol{K}$ is the CKM matrix, and

$$y_{ui} = \frac{\sqrt{2} m_{ui}}{v_u}\,, \qquad y_{di} = \frac{\sqrt{2} m_{di}}{v_d}\,, \qquad y_{ei} = \frac{\sqrt{2} m_{ei}}{v_d}\,. \tag{13.122}$$

Comparing eq. (13.120) with the GWP conditions given below eq. (6.144), we recognize the MSSM Higgs–fermion Yukawa interactions as Type-II Yukawa couplings. The Type-II structure of the MSSM Yukawa couplings derives from the form of the superpotential given in eq. (13.6) which is a consequence of supersymmetry. Indeed, in order to produce dimension-four Higgs–fermion Yukawa interaction terms that are absent from eq. (13.120), one would need to include nonholomorphic terms in the superpotential [as noted below eq. (13.6)], which would necessarily constitute a hard breaking of the supersymmetry at tree level.[8]

The MSSM Higgs–fermion interactions are given explicitly by the Type-II interaction Lagrangian exhibited in eq. (6.203). The masses and CKM mixing angles of the quarks and leptons are determined not only by the Yukawa couplings of the superpotential but also by the parameter $\tan\beta$. This is because the top, charm, and up quarks get masses proportional to $v_u = v\sin\beta$ and the bottom, strange, and down quarks and the charge leptons get masses proportional to $v_d = v\cos\beta$ [see eq. (13.122)]. At tree level,

$$m_t = \frac{y_t}{\sqrt{2}} v \sin\beta\,, \qquad m_b = \frac{y_b}{\sqrt{2}} v\cos\beta\,, \qquad m_\tau = \frac{y_\tau}{\sqrt{2}} v\cos\beta\,. \tag{13.123}$$

These relations hold for the running masses rather than the physical pole masses,

[8] An effective "wrong-Higgs" Yukawa coupling can be generated via radiative loop corrections, as shown in Section 19.9.

which are significantly larger for t, b due to radiative corrections. Including those corrections, one can relate the Yukawa couplings to $\tan\beta$ and the known fermion masses and CKM mixing angles. It is now clear why we have not neglected y_b and y_τ, even though $m_b, m_\tau \ll m_t$. At tree level, $y_b/y_t = (m_b/m_t)\tan\beta$ and $y_\tau/y_t = (m_\tau/m_t)\tan\beta$, so that y_b and y_τ cannot be neglected if $\tan\beta$ is much larger than 1. In fact, there are good theoretical motivations for considering models with large $\tan\beta$. For example, models based on the GUT gauge group SO(10) [see Section 16.3.2] can unify the running top, bottom, and tau Yukawa couplings at the unification scale; this requires $\tan\beta$ to be very roughly of order m_t/m_b.

Note that if one tries to make $\sin\beta$ too small, then y_t will become nonperturbatively large. Requiring that y_t does not blow up above the electroweak scale, one finds that $\tan\beta \gtrsim 1.2$ or so, depending on the mass of the top quark, the QCD coupling, and other details. In principle, there is also a constraint on $\cos\beta$ if one requires that y_b and y_τ do not become nonperturbatively large. This gives a rough upper bound of $\tan\beta \lesssim 65$. However, this is complicated somewhat by the fact that the bottom quark mass gets significant one-loop nonQCD corrections in the large $\tan\beta$ limit. One can obtain a stronger upper bound on $\tan\beta$ in some models where $m^2_{H_u} = m^2_{H_d}$ at the input scale, by requiring that y_b does not significantly exceed y_t. [Otherwise, S_b would be larger than S_t in eqs. (13.70) and (13.71), so one would expect $m^2_{H_d} < m^2_{H_u}$ at the electroweak scale, and the minimum of the potential would have $\langle H^0_d\rangle > \langle H^0_u\rangle$. This would be a contradiction with the supposition that $\tan\beta$ is large.] The parameter $\tan\beta$ also directly impacts the masses and mixings of the MSSM sparticles, as we will see below.

It is instructive to write the dependences on the angles β and α of the tree-level couplings of the neutral MSSM Higgs bosons. Using eqs. (6.195) and (6.196), the bosonic couplings are proportional to

$$hW^+W^-,\quad hZZ,\quad ZHA,\quad W^\pm HH^\mp \;\propto\; \sin(\beta-\alpha), \tag{13.124}$$

$$HW^+W^-,\quad HZZ,\quad ZhA,\quad W^\pm hH^\mp \;\propto\; \cos(\beta-\alpha). \tag{13.125}$$

Likewise, using eq. (6.203), the couplings to fermions are proportional to

$$hb\bar{b},\quad h\tau^+\tau^- \;\propto\; \sin(\beta-\alpha) - \tan\beta\cos(\beta-\alpha), \tag{13.126}$$

$$ht\bar{t} \;\propto\; \sin(\beta-\alpha) + \cot\beta\cos(\beta-\alpha), \tag{13.127}$$

$$Hb\bar{b},\quad H\tau^+\tau^- \;\propto\; \cos(\beta-\alpha) + \tan\beta\sin(\beta-\alpha), \tag{13.128}$$

$$Ht\bar{t} \;\propto\; \cos(\beta-\alpha) - \cot\beta\sin(\beta-\alpha), \tag{13.129}$$

$$Ab\bar{b},\quad A\tau^+\tau^- \;\propto\; \tan\beta, \tag{13.130}$$

$$At\bar{t} \;\propto\; \cot\beta. \tag{13.131}$$

The behavior of all the couplings listed above depend critically on $\cos(\beta-\alpha)$, whose tree-level value is given in eq. (6.239), which we reproduce here:

$$\cos(\beta-\alpha) = \frac{m_Z^2\sin 2\beta\cos 2\beta}{\sqrt{(m_H^2 - m_h^2)(m_H^2 - m_Z^2\cos^2 2\beta)}}\,. \tag{13.132}$$

The Higgs decoupling limit where $m_H \simeq m_A \gg m_Z$ is especially noteworthy.

In this limit, the tree-level prediction for m_h saturates its upper bound mentioned above, with $m_h^2 \approx m_Z^2 \cos^2 2\beta +$ loop corrections. The other Higgs bosons, A, H, and $H^\pm$, will be much heavier and nearly mass-degenerate, forming an isospin doublet that decouples from sufficiently low-energy processes. Moreover, eq. (13.132) implies that the angle α is very nearly $\beta - \pi/2$, with

$$\cos(\beta - \alpha) = \frac{m_Z^2 \sin 2\beta \cos 2\beta}{m_A^2} + \mathcal{O}\left(\frac{m_Z^4}{m_A^4}\right), \tag{13.133}$$

$$\sin(\beta - \alpha) = 1 - \mathcal{O}\left(\frac{m_Z^4}{m_A^4}\right). \tag{13.134}$$

As expected, in the Higgs decoupling limit, h has nearly the same couplings to quarks and leptons and electroweak gauge bosons as would the Higgs boson of the ordinary Standard Model without supersymmetry. Radiative corrections modify these tree-level predictions, but model-building experiences have shown that it is not uncommon for h to behave in a way nearly indistinguishable from a SM-like Higgs boson, even if m_A is not too huge. The measurements of the 125 GeV Higgs boson observed at the LHC are indeed consistent, so far, with the Standard Model predictions, and it is sensible to identify this particle with h. However, it should be kept in mind that the couplings of h might still turn out to deviate in measurable ways from those of a Standard Model Higgs boson. After including the effects of radiative corrections, the most significant effect for moderately large m_A is a possible enhancement of the $hb\bar{b}$ coupling compared to the value it would have in the Standard Model.

13.9 Neutralinos and Charginos

The higgsinos and electroweak gauginos mix with each other because of the effects of electroweak symmetry breaking. The neutral higgsinos ($\widetilde{H}_u^0$ and $\widetilde{H}_d^0$) and the neutral gauginos ($\widetilde{B}$, $\widetilde{W}^0$) combine to form four neutral mass eigenstates called *neutralinos*. The charged higgsinos ($\widetilde{H}_u^+$ and $\widetilde{H}_d^-$) and winos ($\widetilde{W}^+$ and $\widetilde{W}^-$) mix to form two mass eigenstates with charge ± 1 called *charginos*. We will denote[9] the neutralino and chargino mass eigenstates by $\widetilde{N}_i$ $(i = 1, 2, 3, 4)$ and $\widetilde{C}_i^\pm$ $(i = 1, 2)$. By convention, these are labeled in ascending order, so that $m_{\widetilde{N}_1} < m_{\widetilde{N}_2} < m_{\widetilde{N}_3} < m_{\widetilde{N}_4}$ and $m_{\widetilde{C}_1} < m_{\widetilde{C}_2}$. The lightest neutralino, $\widetilde{N}_1$, is usually assumed to be the LSP, unless there is a lighter gravitino or unless R-parity is not conserved, because it is the only MSSM particle that can make a good cold dark matter candidate. In this section, we will describe the mass spectrum and mixing of the neutralinos and charginos in the MSSM.

In the gauge eigenstate basis $\psi^0 = (\widetilde{B}, \widetilde{W}^0, \widetilde{H}_d^0, \widetilde{H}_u^0)$, the neutralino mass terms

[9] Other common notations use $\widetilde{\chi}_i^0$ or $\widetilde{Z}_i$ for neutralinos, and $\widetilde{\chi}_i^\pm$ or $\widetilde{W}_i^\pm$ for charginos.

in the Lagrangian are

$$\mathscr{L}_{\text{neutralino mass}} = -\tfrac{1}{2}(\psi^0)^{\mathsf{T}} \boldsymbol{M}_{\widetilde{N}} \psi^0 + \text{h.c.}, \tag{13.135}$$

where

$$\boldsymbol{M}_{\widetilde{N}} = \begin{pmatrix} M_1 & 0 & -\frac{1}{2}g'v_d & \frac{1}{2}g'v_u \\ 0 & M_2 & \frac{1}{2}gv_d & -\frac{1}{2}gv_u \\ -\frac{1}{2}g'v_d & \frac{1}{2}gv_d & 0 & -\mu \\ \frac{1}{2}g'v_u & -\frac{1}{2}gv_u & -\mu & 0 \end{pmatrix}. \tag{13.136}$$

The entries M_1 and M_2 in this matrix come directly from the MSSM soft Lagrangian [see eq. (13.16)] while the entries $-\mu$ are the supersymmetric higgsino mass terms [see eq. (13.9)]. The terms proportional to g, g' are the result of Higgs–higgsino–gaugino couplings [see eq. (8.72) and Fig. 8.3(g)], with the Higgs scalars replaced by their VEVs [eqs. (13.82), (13.83)]. This can also be written as

$$\boldsymbol{M}_{\widetilde{N}} = \begin{pmatrix} M_1 & 0 & -c_\beta\, s_W\, m_Z & s_\beta\, s_W\, m_Z \\ 0 & M_2 & c_\beta\, c_W\, m_Z & -s_\beta\, c_W\, m_Z \\ -c_\beta\, s_W\, m_Z & c_\beta\, c_W\, m_Z & 0 & -\mu \\ s_\beta\, s_W\, m_Z & -s_\beta\, c_W\, m_Z & -\mu & 0 \end{pmatrix}. \tag{13.137}$$

Here we have introduced abbreviations $s_\beta = \sin\beta$, $c_\beta = \cos\beta$, $s_W = \sin\theta_W$, and $c_W = \cos\theta_W$. One can now perform a Takagi diagonalization of the neutralino mass matrix $\boldsymbol{M}_{\widetilde{N}}$:

$$\boldsymbol{M}_{\widetilde{N}}^{\text{diag}} = \boldsymbol{N}^* \boldsymbol{M}_{\widetilde{N}} \boldsymbol{N}^{-1} = \begin{pmatrix} m_{\widetilde{N}_1} & 0 & 0 & 0 \\ 0 & m_{\widetilde{N}_2} & 0 & 0 \\ 0 & 0 & m_{\widetilde{N}_3} & 0 \\ 0 & 0 & 0 & m_{\widetilde{N}_4} \end{pmatrix}, \tag{13.138}$$

where the $m_{\widetilde{N}_i}$ $(i = 1, 2, 3, 4)$ are real and nonnegative. The unitary matrix $\boldsymbol{N}$ then determines the neutralino mass eigenstates,

$$\widetilde{N}_i = \boldsymbol{N}_{ij} \psi_j^0 \,. \tag{13.139}$$

Note that the neutralino masses can be identified as the square roots of the eigenvalues of $\boldsymbol{M}_{\widetilde{N}}^\dagger \boldsymbol{M}_{\widetilde{N}}$. The indices (i, j) on $\boldsymbol{N}_{ij}$ are (mass, gauge) eigenstate labels. The mass eigenvalues and the mixing matrix $\boldsymbol{N}_{ij}$ can be given in closed form in terms of the parameters M_1, M_2, μ, and $\tan\beta$ by solving a quartic equation, but the results are very complicated and not illuminating.

In general, the parameters M_1, M_2, and μ in the equations above can have arbitrary complex phases. The phase of μ is physical and cannot be rotated away, because we have already used up the freedom to redefine the phases of the Higgs fields, since we have picked $\langle H_u^0 \rangle$ and $\langle H_d^0 \rangle$ to be real and positive; this guarantees that the off-diagonal entries proportional to m_Z in eq. (13.137) are real. A redefinition of the phases of the gaugino fields always allows us to choose a convention in which one of the gaugino masses, say M_2, is real and positive. Within that convention, the phases of the other two gaugino masses M_1 and M_3 are physical and cannot be rotated away, because any further phase rotation of the gaugino fields would modify the relative phases of the fermion–sfermion–gaugino couplings. However, if the phases are arbitrary, then there can be potentially disastrous CP-violating effects in low-energy physics, including electric dipole moments for both the electron and the neutron. Therefore, it is common (although not strictly mandatory, because of the possibility of nontrivial cancellations in the combinations appearing in physical observables) to assume that μ, M_1, and M_3 are real in the same set of phase conventions that make M_2, b, and the tree-level VEVs $\langle H_u^0 \rangle$ and $\langle H_d^0 \rangle$ real and positive. Note that, even if the supersymmetry-breaking mechanism imposes a gaugino mass unification condition like eq. (13.27) or eq. (14.48), so that M_1, M_2, and M_3 can all be chosen real and positive, the phase of μ is still undetermined by this constraint, and even if μ is assumed to be real, its sign is undetermined.

In models which satisfy a gaugino mass unification boundary condition [see eq. (13.58)], one has the nice prediction

$$M_1 \approx \tfrac{5}{3} \tan^2 \theta_W \, M_2 \approx 0.5 M_2 \tag{13.140}$$

at the electroweak scale. If so, then the neutralino masses and mixing angles depend on only three unknown parameters. This assumption is sufficiently theoretically compelling that it has been made in almost all phenomenological studies; nevertheless it should be recognized as an assumption, to be tested someday by experiment.

There is a not-unlikely limit in which electroweak symmetry-breaking effects can be viewed as a small perturbation on the neutralino mass matrix. If

$$m_Z \ll |\mu \pm M_1|\,,\ |\mu \pm M_2|, \tag{13.141}$$

then the neutralino mass eigenstates are very nearly a "bino-like" $\widetilde{N}_1 \approx \widetilde{B}$; a "wino-like" $\widetilde{N}_2 \approx \widetilde{W}^0$; and two "higgsino-like" states $\widetilde{N}_3, \widetilde{N}_4 \approx (\widetilde{H}_u^0 \pm \widetilde{H}_d^0)/\sqrt{2}$, with mass eigenvalues

$$m_{\widetilde{N}_1} = M_1 - \frac{m_Z^2 s_W^2 (M_1 + \mu \sin 2\beta)}{\mu^2 - M_1^2} + \cdots , \tag{13.142}$$

$$m_{\widetilde{N}_2} = M_2 - \frac{m_W^2 (M_2 + \mu \sin 2\beta)}{\mu^2 - M_2^2} + \cdots , \tag{13.143}$$

$$m_{\widetilde{N}_3} = |\mu| + \frac{m_Z^2 (1 - s \sin 2\beta)(|\mu| + M_1 c_W^2 + M_2 s_W^2)}{2(|\mu| + M_1)(|\mu| + M_2)} + \cdots , \tag{13.144}$$

$$m_{\widetilde{N}_4} = |\mu| + \frac{m_Z^2 (1 + s \sin 2\beta)(|\mu| - M_1 c_W^2 - M_2 s_W^2)}{2(|\mu| - M_1)(|\mu| - M_2)} + \cdots , \tag{13.145}$$

where we have taken M_1 and M_2 real and positive by convention, and assumed μ

is real with sign $s = \pm 1$. The subscript labels of the mass eigenstates may need to be rearranged depending on the numerical values of the parameters; in particular the above labeling of $\widetilde{N}_1$ and $\widetilde{N}_2$ assumes $M_1 < M_2 \ll |\mu|$. This limit, leading to a bino-like neutralino LSP, often emerges from CMSSM boundary conditions on the soft parameters after imposing the correct electroweak symmetry breaking.

The chargino spectrum can be analyzed in a similar way. In the gauge eigenstate basis $\psi = (\widetilde{W}^+, \widetilde{H}_u^+, \widetilde{W}^-, \widetilde{H}_d^-)$, the chargino mass terms in the Lagrangian are

$$\mathscr{L}_{\text{chargino mass}} = -\tfrac{1}{2}\psi^{\mathsf{T}} \boldsymbol{M}_{\widetilde{C}}\, \psi + \text{h.c.}, \tag{13.146}$$

where, in 2×2 block form,

$$\boldsymbol{M}_{\widetilde{C}} = \begin{pmatrix} \mathbf{0} & \boldsymbol{X}^{\mathsf{T}} \\ \boldsymbol{X} & \mathbf{0} \end{pmatrix}, \tag{13.147}$$

with

$$\boldsymbol{X} = \begin{pmatrix} M_2 & gv_u/\sqrt{2} \\ gv_d/\sqrt{2} & \mu \end{pmatrix} = \begin{pmatrix} M_2 & \sqrt{2}s_\beta m_W \\ \sqrt{2}c_\beta\, m_W & \mu \end{pmatrix}. \tag{13.148}$$

The mass eigenstates are related to the gauge eigenstates by two unitary 2×2 matrices $\boldsymbol{U}$ and $\boldsymbol{V}$ according to

$$\begin{pmatrix} \widetilde{C}_1^+ \\ \widetilde{C}_2^+ \end{pmatrix} = \boldsymbol{V} \begin{pmatrix} \widetilde{W}^+ \\ \widetilde{H}_u^+ \end{pmatrix}; \qquad \begin{pmatrix} \widetilde{C}_1^- \\ \widetilde{C}_2^- \end{pmatrix} = \boldsymbol{U} \begin{pmatrix} \widetilde{W}^- \\ \widetilde{H}_d^- \end{pmatrix}. \tag{13.149}$$

Note that the mixing matrix for the positively charged left-handed fermions is different from that for the negatively charged left-handed fermions. The matrices $\boldsymbol{U}$ and $\boldsymbol{V}$ are determined by the singular value decomposition of the chargino mass matrix $\boldsymbol{M}_{\widetilde{C}}$, which yields

$$\boldsymbol{U}^* \boldsymbol{X} \boldsymbol{V}^{-1} = \begin{pmatrix} m_{\widetilde{C}_1} & 0 \\ 0 & m_{\widetilde{C}_2} \end{pmatrix}, \tag{13.150}$$

with real nonnegative entries $m_{\widetilde{C}_i}$. Because these are only 2×2 matrices, it is not hard to solve for the masses explicitly:

$$m^2_{\widetilde{C}_1}, m^2_{\widetilde{C}_2} = \tfrac{1}{2}\Big[|M_2|^2 + |\mu|^2 + 2m_W^2 \mp \sqrt{(|M_2|^2 + |\mu|^2 + 2m_W^2)^2 - 4|\mu M_2 - m_W^2 \sin 2\beta|^2}\,\Big]. \tag{13.151}$$

These are the (doubly degenerate) eigenvalues of the 4×4 matrix $\boldsymbol{M}^\dagger_{\widetilde{C}} \boldsymbol{M}_{\widetilde{C}}$, or equivalently the eigenvalues of $\boldsymbol{X}^\dagger \boldsymbol{X}$, since

$$\boldsymbol{V} \boldsymbol{X}^\dagger \boldsymbol{X} \boldsymbol{V}^{-1} = \boldsymbol{U}^* \boldsymbol{X} \boldsymbol{X}^\dagger \boldsymbol{U}^{\mathsf{T}} = \begin{pmatrix} m^2_{\widetilde{C}_1} & 0 \\ 0 & m^2_{\widetilde{C}_2} \end{pmatrix}. \tag{13.152}$$

(But they are *not* the squares of the eigenvalues of $\boldsymbol{X}$.) In the limit of eq. (13.141) with real M_2 and μ, one finds that the chargino mass eigenstates consist of a wino-like $\widetilde{C}_1^\pm$ and a higgsino-like $\widetilde{C}_2^\pm$, with masses

$$m_{\widetilde{C}_1} = M_2 - \frac{m_W^2(M_2 + \mu\sin 2\beta)}{\mu^2 - M_2^2} + \cdots \qquad (13.153)$$

$$m_{\widetilde{C}_2} = |\mu| + \frac{m_W^2(|\mu| + sM_2\sin 2\beta)}{\mu^2 - M_2^2} + \cdots . \qquad (13.154)$$

Here again the labeling assumes $M_2 < |\mu|$, and s is the sign of μ. Amusingly, the lighter chargino $\widetilde{C}_1$ is nearly degenerate with the second lightest neutralino $\widetilde{N}_2$ in this limit, but this is not an exact result. Their higgsino-like colleagues $\widetilde{N}_3$, $\widetilde{N}_4$, and $\widetilde{C}_2$ have masses of order $|\mu|$. The case of $M_1 \approx 0.5M_2 \ll |\mu|$ is not uncommonly found in viable models based on the CMSSM boundary conditions, and it has been elevated to the status of a benchmark scenario in many phenomenological studies. However, it cannot be overemphasized that such expectations are not mandatory.

The Feynman rules involving neutralinos and charginos may be inferred in terms of $\boldsymbol{N}$, $\boldsymbol{U}$, and $\boldsymbol{V}$ from the MSSM Lagrangian as discussed above; they are given in Appendix J. In practice, the masses and mixing angles for the neutralinos and charginos are best computed numerically. Note that the discussion above yields the tree-level masses. Loop corrections to these masses can be significant and must be included in any realistic phenomenological analysis of collider data.

13.10 The Gluino

The gluino is a color-octet fermion, so it cannot mix with any other particle in the MSSM, even if R-parity is violated. In this regard, it is unique among all of the MSSM sparticles. In models with universal gaugino mass parameter boundary conditions [see eq. (13.27)], the gluino mass parameter M_3 is related to the bino and wino mass parameters M_1 and M_2 by eq. (13.58), so

$$M_3 = \frac{\alpha_s}{\alpha}\sin^2\theta_W\, M_2 = \frac{3}{5}\frac{\alpha_s}{\alpha}\cos^2\theta_W\, M_1 \qquad (13.155)$$

at any RG scale, up to small two-loop corrections. This implies a rough prediction

$$M_3 : M_2 : M_1 \approx 6 : 2 : 1 \qquad (13.156)$$

near the TeV scale. It is therefore reasonable to suspect that the gluino is considerably heavier than the lighter neutralinos and charginos (even in many models where the gaugino mass unification condition is not imposed).

For more precise estimates, one must take into account the fact that M_3 is really a running mass parameter which has an implicit dependence on the RG scale Q. Because the gluino is a strongly interacting particle, M_3 runs rather quickly with Q [see eq. (13.57)]. A more useful quantity physically is the RG-scale-independent mass $M_{\widetilde{g}}$ at which the renormalized gluino propagator has a pole. Including one-loop corrections to the gluino propagator due to gluon exchange and squark–quark loops, one finds that the pole mass is given in terms of the running mass in the $\overline{\rm DR}$ scheme by

$$M_{\widetilde{g}} = M_3(Q)\Big(1 + \frac{\alpha_s}{4\pi}[15 + 6\ln(Q/M_3) + \sum A_{\widetilde{q}}]\Big), \tag{13.157}$$

where

$$A_{\widetilde{q}} = \int_0^1 dx\, x \ln\left(\frac{x m_{\widetilde{q}}^2 + (1-x)m_q^2 - x(1-x)M_3^2}{M_3^2} - i\epsilon\right). \tag{13.158}$$

The sum in eq. (13.157) is over all 12 squark–quark supermultiplets, and small effects due to squark mixing have been neglected here. [As a check, requiring the physical observable $M_{\widetilde{g}}$ to be independent of Q in eq. (13.157) reproduces the one-loop RG equation for $M_3(Q)$ in eq. (13.57).] The correction terms proportional to α_s in eq. (13.157) can be quite significant, because the gluino is strongly interacting, with a large group theory factor [the 15 in eq. (13.157)] due to its color-octet nature, and because it couples to all of the squark–quark pairs. The leading two-loop corrections to the gluino pole mass have also been found [158–160], and typically increase the prediction by another 1–2%.

13.11 Squarks and Sleptons

In principle, any scalars with the same electric charge, R-parity, and color quantum numbers can mix with each other. This means that with completely arbitrary soft terms, the mass eigenstates of the squarks and sleptons of the MSSM should be obtained by diagonalizing three 6×6 squared-mass matrices for up-type squarks ($\widetilde{u}_L$, $\widetilde{c}_L$, $\widetilde{t}_L$, $\widetilde{u}_R$, $\widetilde{c}_R$, $\widetilde{t}_R$), down-type squarks ($\widetilde{d}_L$, $\widetilde{s}_L$, $\widetilde{b}_L$, $\widetilde{d}_R$, $\widetilde{s}_R$, $\widetilde{b}_R$), and charged sleptons ($\widetilde{e}_L$, $\widetilde{\mu}_L$, $\widetilde{\tau}_L$, $\widetilde{e}_R$, $\widetilde{\mu}_R$, $\widetilde{\tau}_R$), and one 3×3 matrix for sneutrinos ($\widetilde{\nu}_e$, $\widetilde{\nu}_\mu$, $\widetilde{\nu}_\tau$). Fortunately, the general hypothesis of flavor-blind soft parameters exhibited in eqs. (13.24) and (13.25) predicts that most of these mixing angles are very small. The third-generation squarks and sleptons can have very different masses compared to their first- and second-generation counterparts, because of the effects of large Yukawa (y_t, y_b, y_τ) and soft (a_t, a_b, a_τ) couplings in the RG equations (13.72)–(13.76). Furthermore, they can have substantial mixing in pairs ($\widetilde{t}_L$, $\widetilde{t}_R$), ($\widetilde{b}_L$, $\widetilde{b}_R$), and ($\widetilde{\tau}_L$, $\widetilde{\tau}_R$). In contrast, the first- and second-generation squarks and sleptons have negligible Yukawa couplings, so they end up in seven very nearly degenerate, unmixed pairs $(\widetilde{e}_R, \widetilde{\mu}_R)$, $(\widetilde{\nu}_e, \widetilde{\nu}_\mu)$, $(\widetilde{e}_L, \widetilde{\mu}_L)$, $(\widetilde{u}_R, \widetilde{c}_R)$, $(\widetilde{d}_R, \widetilde{s}_R)$, $(\widetilde{u}_L, \widetilde{c}_L)$, $(\widetilde{d}_L, \widetilde{s}_L)$. As we have discussed in Section 13.6, this avoids the problem of disastrously large virtual sparticle contributions to FCNC processes.

Let us first consider the spectrum of first- and second-generation squarks and sleptons. In models with CMSSM boundary conditions [see eq. (13.28)], the squark and slepton running masses can be conveniently parameterized, to a good approximation, as

$$m^2_{Q_1} = m^2_{Q_2} = m_0^2 + K_3 + K_2 + \tfrac{1}{36}K_1, \tag{13.159}$$
$$m^2_{\bar{u}_1} = m^2_{\bar{u}_2} = m_0^2 + K_3 \qquad + \tfrac{4}{9}K_1, \tag{13.160}$$
$$m^2_{\bar{d}_1} = m^2_{\bar{d}_2} = m_0^2 + K_3 \qquad + \tfrac{1}{9}K_1, \tag{13.161}$$
$$m^2_{L_1} = m^2_{L_2} = m_0^2 \qquad + K_2 + \tfrac{1}{4}K_1, \tag{13.162}$$
$$m^2_{\bar{e}_1} = m^2_{\bar{e}_2} = m_0^2 \qquad + K_1. \tag{13.163}$$

A key point is that the same K_3, K_2, and K_1 appear everywhere in eqs. (13.159)–(13.163), since all of the chiral supermultiplets couple to the same gauginos with the same gauge couplings. The different coefficients in front of K_1 just correspond to the various values of weak hypercharge squared for each scalar.

In the CMSSM, m_0^2 is the same common scalar squared mass appearing in eq. (13.28). It can be very small, as in the "no-scale" limit mentioned in Section 14.3, but it could also be the dominant source of the scalar masses. The contributions K_3, K_2, and K_1 are due to the RG running[10] proportional to the gaugino masses. Explicitly, they are found at one-loop order by solving eq. (13.65):

$$K_a(Q) = \begin{Bmatrix} 3/5 \\ 3/4 \\ 4/3 \end{Bmatrix} \times \frac{1}{2\pi^2}\int_{\ln Q}^{\ln Q_0} dt\; g_a^2(t)\,|M_a(t)|^2 \qquad (a = 1,2,3). \tag{13.164}$$

Here Q_0 is the input RG scale at which the CMSSM boundary condition eq. (13.28) is applied, and Q should be taken to be evaluated near the squark and slepton mass under consideration, presumably less than about 1 TeV. The running parameters $g_a(Q)$ and $M_a(Q)$ obey eqs. (13.31) and (13.58). If the input scale is approximated by the apparent scale of gauge coupling unification $Q_0 = M_U \approx 2 \times 10^{16}$ GeV, one finds that, numerically,

$$K_1 \approx 0.15 m_{1/2}^2, \qquad K_2 \approx 0.5 m_{1/2}^2, \qquad K_3 \approx (4.5 \text{ to } 6.5) m_{1/2}^2 \tag{13.165}$$

for Q near the electroweak scale. Here $m_{1/2}$ is the common gaugino mass parameter at the unification scale. Note that $K_3 \gg K_2 \gg K_1$; this is a direct consequence of the relative sizes of the gauge couplings g_3, g_2, and g_1. Consequently, one expects the squarks to be somewhat heavier than the sleptons.

The large uncertainty in K_3 is due in part to the experimental uncertainty in the QCD coupling constant, and in part to the uncertainty in where to choose Q, since K_3 runs rather quickly below 1 TeV. If the gauge couplings and gaugino masses are unified between M_U and M_P, as would occur in a GUT model, then the effect of RG running for $M_U < Q < M_P$ can be absorbed into a redefinition of m_0^2. Otherwise, it adds a further uncertainty roughly proportional to $\ln(M_P/M_U)$, compared to the larger contributions in eq. (13.164), which go roughly like $\ln(M_U/1 \text{ TeV})$.

[10] The quantity S defined in eq. (13.66) vanishes for CMSSM boundary conditions, and remains small under RG evolution.

In addition, there is also a "hyperfine" splitting in the squark and slepton mass spectrum produced by electroweak symmetry breaking. Each squark and slepton ϕ will get a contribution Δ_ϕ to its squared mass, coming from the $SU(2)_L$ and $U(1)_Y$ D-term quartic interactions [see the last term in eq. (8.75)] of the form (squark)^2(Higgs)^2 and (slepton)^2(Higgs)^2, when the neutral Higgs scalars H_u^0 and H_d^0 get VEVs. They are model-independent for a given value of $\tan\beta$:

$$\Delta_\phi = \tfrac{1}{2}(T_{3\phi}g^2 - Y_\phi g'^2)(v_d^2 - v_u^2) = (T_{3\phi} - Q_\phi \sin^2\theta_W)\cos 2\beta\, m_Z^2, \quad (13.166)$$

where $T_{3\phi}$, Y_ϕ, and Q_ϕ are respectively the third component of weak isospin, the weak hypercharge, and the electric charge of the left-handed chiral supermultiplet to which ϕ belongs. In particular,

$$\Delta_{\tilde{u}_L} = (\tfrac{1}{2} - \tfrac{2}{3}\sin^2\theta_W)\, m_Z^2 \cos 2\beta\,, \qquad \Delta_{\tilde{u}_R} = \tfrac{2}{3}\sin^2\theta_W\, m_Z^2 \cos 2\beta\,, \quad (13.167)$$

$$\Delta_{\tilde{d}_L} = (-\tfrac{1}{2} + \tfrac{1}{3}\sin^2\theta_W)\, m_Z^2 \cos 2\beta\,, \qquad \Delta_{\tilde{d}_R} = -\tfrac{1}{3}\sin^2\theta_W\, m_Z^2 \cos 2\beta\,, \quad (13.168)$$

$$\Delta_{\tilde{e}_L} = (-\tfrac{1}{2} + \sin^2\theta_W)\, m_Z^2 \cos 2\beta\,, \qquad \Delta_{\tilde{e}_R} = -\sin^2\theta_W\, m_Z^2 \cos 2\beta\,, \quad (13.169)$$

$$\Delta_{\tilde{\nu}_L} = \tfrac{1}{2} m_Z^2 \cos 2\beta\,. \quad (13.170)$$

These D-term contributions are typically smaller than the m_0^2 and K_1, K_2, K_3 contributions, but should not be neglected. They split apart the components of the $SU(2)_L$-doublet sleptons and squarks. Including them, the first-generation squark and slepton masses are now given by

$$m_{\tilde{d}_L}^2 = m_0^2 + K_3 + K_2 + \tfrac{1}{36}K_1 + \Delta_{\tilde{d}_L}, \quad (13.171)$$

$$m_{\tilde{u}_L}^2 = m_0^2 + K_3 + K_2 + \tfrac{1}{36}K_1 + \Delta_{\tilde{u}_L}, \quad (13.172)$$

$$m_{\tilde{u}_R}^2 = m_0^2 + K_3 \quad + \tfrac{4}{9}K_1 + \Delta_{\tilde{u}_R}, \quad (13.173)$$

$$m_{\tilde{d}_R}^2 = m_0^2 + K_3 \quad + \tfrac{1}{9}K_1 + \Delta_{\tilde{d}_R}, \quad (13.174)$$

$$m_{\tilde{e}_L}^2 = m_0^2 \quad + K_2 + \tfrac{1}{4}K_1 + \Delta_{\tilde{e}_L}, \quad (13.175)$$

$$m_{\tilde{\nu}}^2 = m_0^2 \quad + K_2 + \tfrac{1}{4}K_1 + \Delta_{\tilde{\nu}}, \quad (13.176)$$

$$m_{\tilde{e}_R}^2 = m_0^2 \quad + K_1 + \Delta_{\tilde{e}_R}, \quad (13.177)$$

with identical formulae for the second-generation squarks and sleptons. The mass splittings for the left-handed squarks and sleptons are governed by model-independent sum rules,

$$m_{\tilde{e}_L}^2 - m_{\tilde{\nu}_e}^2 = m_{\tilde{d}_L}^2 - m_{\tilde{u}_L}^2 = \tfrac{1}{2}g^2(v_u^2 - v_d^2) = -\cos 2\beta\, m_W^2. \quad (13.178)$$

In the allowed range $\tan\beta > 1$, it follows that $m_{\tilde{e}_L} > m_{\tilde{\nu}_e}$ and $m_{\tilde{d}_L} > m_{\tilde{u}_L}$, with the magnitude of the splittings constrained by electroweak symmetry breaking.

Let us next consider the masses of the top squarks, for which there are several nonnegligible contributions. First, there are squared-mass terms for $\tilde{t}_L^\dagger \tilde{t}_L$ and $\tilde{t}_R^\dagger \tilde{t}_R$ that are just equal to $m_{Q_3}^2 + \Delta_{\tilde{u}_L}$ and $m_{\bar{u}_3}^2 + \Delta_{\tilde{u}_R}$, respectively, just as for the first- and second-generation squarks. Second, there are contributions equal to m_t^2 for each of $\tilde{t}_L^\dagger \tilde{t}_L$ and $\tilde{t}_R^\dagger \tilde{t}_R$. These come from F-terms in the scalar potential of the form $y_t^2 H_u^{0\dagger} H_u^0 \tilde{t}_L^\dagger \tilde{t}_L$ and $y_t^2 H_u^{0\dagger} H_u^0 \tilde{t}_R^\dagger \tilde{t}_R$ [see Figs. 13.2(b) and 13.2(c)], with the Higgs fields replaced by their VEVs. (Of course, similar contributions are present for all of the squarks and sleptons, but they are too small to worry about except in the case of the top squarks.) Third, there are contributions to the scalar potential from F-terms of the form $-\mu^* H_d^{0\dagger} y_t \widetilde{t}_R^\dagger \widetilde{t}_L$+h.c. obtained from $-\mathscr{L}$ given in eq. (13.11) and from Fig. 13.4(a). These become $-\mu^* v y_t \cos\beta\, \widetilde{t}_R^\dagger \widetilde{t}_L$+h.c. when H_d^0 is replaced by its VEV. Finally, there are contributions to the scalar potential from the soft (scalar)^3 couplings $a_t H_u^0 \widetilde{t}_R^\dagger \widetilde{Q}_3$ + h.c. [see the first term of the second line of eq. (13.16), and eq. (13.59)], which become $a_t v \sin\beta\, \widetilde{t}_R^\dagger \widetilde{t}_L$ + h.c. when H_u^0 is replaced by its VEV. Putting these all together, we have a hermitian squared-mass matrix for the top squarks, which in the gauge eigenstate basis $(\widetilde{t}_L, \widetilde{t}_R)$ is given by

$$\mathscr{L}_{\text{stops}} \supset -\begin{pmatrix} \widetilde{t}_L^\dagger & \widetilde{t}_R^\dagger \end{pmatrix} \boldsymbol{m}_{\widetilde{t}}^{\boldsymbol{2}} \begin{pmatrix} \widetilde{t}_L \\ \widetilde{t}_R \end{pmatrix}, \tag{13.179}$$

where

$$\boldsymbol{m}_{\widetilde{t}}^{\boldsymbol{2}} = \begin{pmatrix} m_{Q_3}^2 + m_t^2 + \Delta_{\widetilde{u}_L} & v(a_t^* \sin\beta - \mu y_t \cos\beta)/\sqrt{2} \\ v(a_t \sin\beta - \mu^* y_t \cos\beta)/\sqrt{2} & m_{\bar{u}_3}^2 + m_t^2 + \Delta_{\widetilde{u}_R} \end{pmatrix}. \tag{13.180}$$

This matrix can be diagonalized by a unitary matrix to give mass eigenstates [11]

$$\begin{pmatrix} \widetilde{t}_1 \\ \widetilde{t}_2 \end{pmatrix} = \begin{pmatrix} c_{\widetilde{t}} & -s_{\widetilde{t}}^* \\ s_{\widetilde{t}} & c_{\widetilde{t}}^* \end{pmatrix} \begin{pmatrix} \widetilde{t}_L \\ \widetilde{t}_R \end{pmatrix}. \tag{13.181}$$

Here $m_{\widetilde{t}_1}^2 < m_{\widetilde{t}_2}^2$ are the eigenvalues of eq. (13.180), and $|c_{\widetilde{t}}|^2 + |s_{\widetilde{t}}|^2 = 1$. Explicit formulae for the masses and mixing parameters are obtained in Exercise 13.1. If the off-diagonal elements of eq. (13.180) are real, then $c_{\widetilde{t}}$ and $s_{\widetilde{t}}$ are the cosine and sine of a stop mixing angle $\theta_{\widetilde{t}}$ that is defined modulo π.

Because of the large RG effects proportional to S_t in eq. (13.72) and eq. (13.73), at the electroweak scale one finds that $m_{\bar{u}_3}^2 < m_{Q_3}^2$, and both of these quantities are usually significantly smaller than the squark squared masses for the first two generations. The diagonal terms m_t^2 in eq. (13.180) tend to mitigate this effect somewhat, but the off-diagonal entries will typically induce a significant mixing, which always reduces the lighter top-squark squared-mass eigenvalue. Therefore, models often predict that $\widetilde{t}_1$ is the lightest of all the squarks, and that it is predominantly $\widetilde{t}_R$.

A similar analysis can be performed for the bottom squarks and charged tau sleptons, which in their respective gauge eigenstate bases $(\widetilde{b}_L, \widetilde{b}_R)$ and $(\widetilde{\tau}_L, \widetilde{\tau}_R)$ have squared-mass matrices

$$\boldsymbol{m}_{\widetilde{b}}^{\boldsymbol{2}} = \begin{pmatrix} m_{Q_3}^2 + \Delta_{\widetilde{d}_L} & v(a_b^* \cos\beta - \mu y_b \sin\beta)/\sqrt{2} \\ v(a_b \cos\beta - \mu^* y_b \sin\beta)/\sqrt{2} & m_{\bar{d}_3}^2 + \Delta_{\widetilde{d}_R} \end{pmatrix}, \tag{13.182}$$

[11] Our convention for $c_{\widetilde{t}}$, $s_{\widetilde{t}}$ has the property that $\widetilde{t}_1 = \widetilde{t}_R$ and $\widetilde{t}_2 = \widetilde{t}_L$ for zero mixing angle. Conventions found in the literature often do not have this nice property.

and

$$m_{\widetilde{\tau}}^2 = \begin{pmatrix} m_{L_3}^2 + \Delta_{\widetilde{e}_L} & v(a_\tau^* \cos\beta - \mu y_\tau \sin\beta)/\sqrt{2} \\ v(a_\tau \cos\beta - \mu^* y_\tau \sin\beta)/\sqrt{2} & m_{\overline{e}_3}^2 + \Delta_{\widetilde{e}_R} \end{pmatrix}. \tag{13.183}$$

These squared-mass matrices can be diagonalized to give mass eigenstates $\widetilde{b}_1, \widetilde{b}_2$ and $\widetilde{\tau}_1, \widetilde{\tau}_2$ [see eq. (13.181)]. Finally, as there is no $\widetilde{\nu}_R$ in the MSSM, the tau sneutrino squared mass is given by

$$m_{\widetilde{\nu}_\tau}^2 = m_{L_3}^2 + \tfrac{1}{2}\, m_Z^2 \cos 2\beta. \tag{13.184}$$

The magnitude and importance of mixing in the sbottom and stau sectors depends on how large $\tan\beta$ is. If $\tan\beta$ is not too large (in practice, this usually means less than about 10 or so, depending on the situation under study), the sbottoms and staus do not get a very large effect from the mixing terms and the RG effects due to S_b and S_τ, because $y_b, y_\tau \ll y_t$ [see eq. (13.123)]. In that case the mass eigenstates are very nearly the same as the gauge eigenstates $\widetilde{b}_L$, $\widetilde{b}_R$, $\widetilde{\tau}_L$, and $\widetilde{\tau}_R$. The latter three, and $\widetilde{\nu}_\tau$, will be nearly degenerate with their first- and second-generation counterparts with the same $\mathrm{SU}(3)_C\times\mathrm{SU}(2)_L\times\mathrm{U}(1)_Y$ quantum numbers. However, even in the case of small $\tan\beta$, $\widetilde{b}_L$ will feel the effects of the large top Yukawa coupling because it is part of the doublet $\widetilde{Q}_3$ which contains $\widetilde{t}_L$. In particular, from eq. (13.72) we see that S_t acts to decrease $m_{\widetilde{Q}_3}^2$ as it is RG-evolved down from the input scale to the electroweak scale. Therefore the mass of $\widetilde{b}_L$ can be significantly less than the masses of $\widetilde{d}_L$ and $\widetilde{s}_L$.

For larger values of $\tan\beta$, the mixing in eqs. (13.182) and (13.183) can be quite significant, because y_b, y_τ and a_b, a_τ are nonnegligible. Just as in the case of the top squarks, the lighter sbottom and stau mass eigenstates (denoted $\widetilde{b}_1$ and $\widetilde{\tau}_1$) can be significantly lighter than their first- and second-generation counterparts. Furthermore, $\widetilde{\nu}_\tau$ can be significantly lighter than the nearly degenerate $\widetilde{\nu}_e$, $\widetilde{\nu}_\mu$.

The requirement that the third-generation squarks and sleptons should all have positive squared masses implies limits on the magnitudes of $a_t^* \sin\beta - \mu y_t \cos\beta$ and $a_b^* \cos\beta - \mu y_b \sin\beta$ and $a_\tau^* \cos\beta - \mu y_\tau \sin\beta$. If they are too large, then the smaller eigenvalue of eq. (13.180), (13.182), or (13.183) will be driven negative, implying that a squark or charged slepton gets a VEV, breaking $\mathrm{SU}(3)_C$ or electromagnetism. Since this is clearly unacceptable, one can put bounds on the (scalar)^3 couplings, or equivalently on the CMSSM parameter A_0. Even if all of the squared-mass eigenvalues are positive, the presence of large (scalar)^3 couplings can yield global minima of the scalar potential, with nonzero squark and/or charged slepton VEVs, which are disconnected from the vacuum that conserves $\mathrm{SU}(3)_C$ and electromagnetism. However, it is not always immediately clear whether the mere existence of such disconnected global minima should really disqualify a set of model parameters, because the tunneling rate from our "good" vacuum to the "bad" vacua can easily be longer than the age of the universe.

13.12 Summary: The MSSM Sparticle Spectrum

In the MSSM there are 32 distinct masses corresponding to undiscovered particles, not including the gravitino. We have explained how the masses and mixing angles for these particles are computed, given an underlying model for the soft terms at some input scale. Assuming only that the mixing of first- and second-generation squarks and sleptons is negligible, the mass eigenstates of the MSSM are listed in Table 13.1.

Specific models for the soft terms typically predict the masses and the mixing angles for the MSSM in terms of far fewer parameters. For example, in the CMSSM, the only free parameters not already measured by experiment are m_0^2, $m_{1/2}$, A_0, μ, and b. In the gauge-mediated supersymmetry-breaking (GMSB) models discussed in Section 14.4, the free parameters include at least the scale Λ, the typical messenger mass scale M_{mess}, the integer number N_5 of copies of the minimal messengers, the goldstino decay constant $\langle F \rangle$, and the Higgs mass parameters μ and b.

After RG-evolving the soft terms down to the electroweak scale, one can de-

Table 13.1 The undiscovered particles (and h, which is presumably the observed Higgs boson with $m_h = 125$ GeV) in the MSSM, with sfermion mixing for the first two generations assumed to be negligible.

names	spin	P_R	mass eigenstates	gauge eigenstates
Higgs bosons	0	+1	$h\ H\ A\ H^\pm$	$H_u^0\ H_d^0\ H_u^+\ H_d^-$
squarks	0	−1	$\widetilde{u}_L\ \widetilde{u}_R\ \widetilde{d}_L\ \widetilde{d}_R$ $\widetilde{s}_L\ \widetilde{s}_R\ \widetilde{c}_L\ \widetilde{c}_R$ $\widetilde{t}_1\ \widetilde{t}_2\ \widetilde{b}_1\ \widetilde{b}_2$	$\widetilde{u}_R\ \widetilde{d}_L\ \widetilde{d}_R$ $\widetilde{s}_L\ \widetilde{s}_R\ \widetilde{c}_L\ \widetilde{c}_R$ $\widetilde{t}_L\ \widetilde{t}_R\ \widetilde{b}_L\ \widetilde{b}_R$
sleptons	0	−1	$\widetilde{e}_L\ \widetilde{e}_R\ \widetilde{\nu}_e$ $\widetilde{\mu}_L\ \widetilde{\mu}_R\ \widetilde{\nu}_\mu$ $\widetilde{\tau}_1\ \widetilde{\tau}_2\ \widetilde{\nu}_\tau$	$\widetilde{e}_L\ \widetilde{e}_R\ \widetilde{\nu}_e$ $\widetilde{\mu}_L\ \widetilde{\mu}_R\ \widetilde{\nu}_\mu$ $\widetilde{\tau}_L\ \widetilde{\tau}_R\ \widetilde{\nu}_\tau$
neutralinos	1/2	−1	$\widetilde{N}_1\ \widetilde{N}_2\ \widetilde{N}_3\ \widetilde{N}_4$	$\widetilde{B}^0\ \widetilde{W}^0\ \widetilde{H}_u^0\ \widetilde{H}_d^0$
charginos	1/2	−1	$\widetilde{C}_1^\pm\ \widetilde{C}_2^\pm$	$\widetilde{W}^\pm\ \widetilde{H}_u^+\ \widetilde{H}_d^-$
gluino	1/2	−1	$\widetilde{g}$	$\widetilde{g}$
gravitino/ goldstino	3/2	−1	$\widetilde{G}$	$\widetilde{G}$

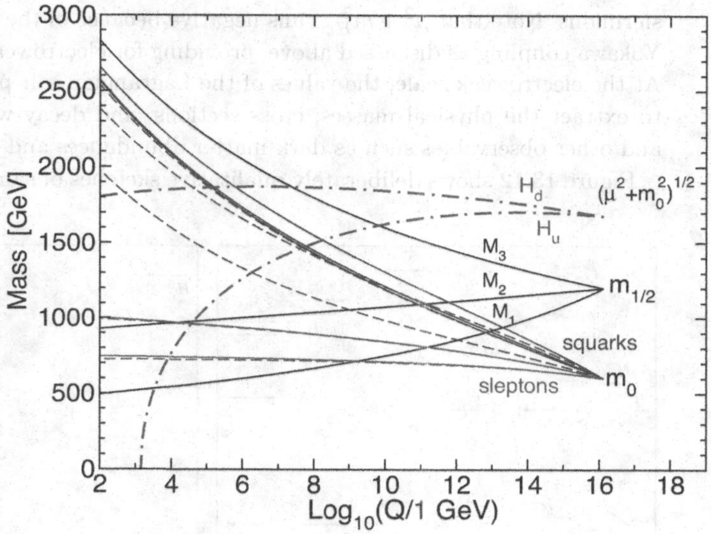

Fig. 13.11 RG evolution of scalar and gaugino mass parameters in the MSSM with typical CMSSM boundary conditions imposed at $Q_0 = 1.5 \times 10^{16}$ GeV. The parameter $\mu^2 + m_{H_u}^2$ runs negative, provoking electroweak symmetry breaking.

mand that the scalar potential gives correct electroweak symmetry breaking. This allows us to trade $|\mu|$ and b (or B_0) for one parameter $\tan\beta$, as in eqs. (13.84) and (13.85). So, to a reasonable approximation, the entire mass spectrum of the CMSSM is determined by only five unknown parameters: m_0^2, $m_{1/2}$, A_0, $\tan\beta$, and Arg(μ), while in the simplest gauge-mediated supersymmetry-breaking models one can pick parameters Λ, $M_{\rm mess}$, N_5, $\langle F \rangle$, $\tan\beta$, and Arg(μ). Both frameworks are highly predictive. Of course, it is easy to imagine that the essential physics of supersymmetry breaking is not captured by either of these two scenarios in their minimal forms. For example, the anomaly-mediated contributions could play a role, perhaps in concert with the gauge mediation or Planck-scale mediation mechanisms.

Figure 13.11 shows the RG running of scalar and gaugino masses in a typical model based on the CMSSM boundary conditions imposed at $Q_0 = 1.5 \times 10^{16}$ GeV. The parameter values employed for this illustration are $m_0 = 600$ GeV, $m_{1/2} = -A_0 = 1200$ GeV, $\tan\beta = 15$, and sign(μ)= + (although these values were chosen more for their artistic value in Fig. 13.11, and not as an attempt at realism). The goal here is to understand the qualitative trends, rather than guess the correct numerical values. The running gaugino masses are solid lines labeled by M_1, M_2, and M_3. The dot-dashed lines labeled H_u and H_d are the running values of the quantities $(\mu^2 + m_{H_u}^2)^{1/2}$ and $(\mu^2 + m_{H_d}^2)^{1/2}$, which appear in the Higgs potential. The other lines are the running squark and slepton masses, with dashed lines for the square roots of the third-generation parameters $m_{\bar{d}_3}^2$, $m_{Q_3}^2$, $m_{\bar{u}_3}^2$, $m_{L_3}^2$, and $m_{\bar{e}_3}^2$ (from top to bottom), and solid lines for the first- and second-generation sfermions. Note that $\mu^2 + m^2_{H_u}$ runs negative because of the effects of the large top Yukawa coupling as discussed above, providing for electroweak symmetry breaking. At the electroweak scale, the values of the Lagrangian soft parameters can be used to extract the physical masses, cross sections, and decay widths of the particles, and other observables such as dark matter abundances and rare process rates.

Figure 13.12 shows deliberately qualitative sketches of sample MSSM mass spec-

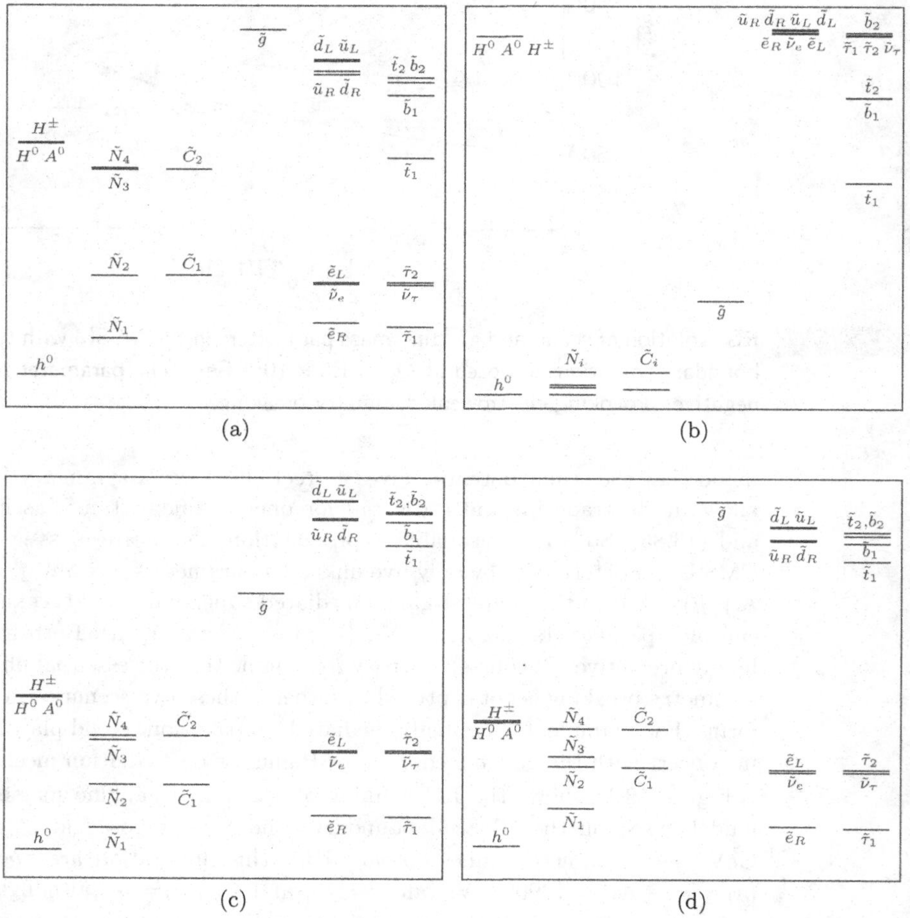

Fig. 13.12 Four sample MSSM mass spectra for the undiscovered particles, for (a) CMSSM boundary conditions with $m_0^2 \ll m_{1/2}^2$, (b) CMSSM boundary conditions with $m_0^2 \gg m_{1/2}^2$, (c) GMSB model with $N_5 = 1$, and (d) GMSB model with $N_5 = 3$. Mass scales are not equal for the four cases, and are deliberately omitted. These spectra are presented for entertainment purposes only! No warranty, expressed or implied, guarantees that they look anything like the real world.

tra obtained from three different types of models assumptions. In Fig. 13.12(a), we exhibit the output from a model with CMSSM boundary conditions with relatively low m_0^2 compared to $m_{1/2}^2$ (similar to Fig. 13.11). This model features a near-decoupling limit for the Higgs sector, and a bino-like $\widetilde{N}_1$ LSP, nearly degenerate wino-like $\widetilde{N}_2, \widetilde{C}_1$, and higgsino-like $\widetilde{N}_3, \widetilde{N}_4, \widetilde{C}_2$. The gluino is the heaviest superpartner. The squarks are all much heavier than the sleptons, and the lightest sfermion is a stau. (The second-generation squarks and sleptons are nearly degenerate with those of the first generation, and so are not shown separately.) Variations in the model parameters have important and predictable effects. For example, taking larger values of $\tan\beta$ with other model parameters held fixed will usually tend to lower $\widetilde{b}_1$ and $\widetilde{\tau}_1$ masses compared to those of the other sparticles. It is noteworthy that for sufficiently small $m_0^2/m_{1/2}^2$, the lighter stau is less massive than the lightest neutralino, and thus would be the LSP of the CMSSM. In this parameter regime, the lighter stau would be a charged stable relic that constitutes a significant fraction of the dark matter (see Section 15.4), in conflict with the bounds of Ref. [161].

Taking larger m_0^2 will tend to squeeze together the spectrum of squarks and sleptons and move them all higher compared to the neutralinos, charginos, and gluino. This is illustrated in Fig. 13.12(b), which instead has $m_0^2 \gg m_{1/2}^2$. In this model, the heaviest chargino and neutralino are wino-like.

The third sample sketch, in Fig. 13.12(c), is obtained from a typical minimal GMSB model, with $N_5 = 1$. Here we see that the hierarchy between strongly interacting sparticles and weakly interacting ones is quite large. Changing the messenger scale or Λ does not reduce the relative splitting between squark and slepton masses, because there is no analogue of the universal m_0^2 contribution here. Increasing the number of messenger fields tends to decrease the squark and slepton masses relative to the gaugino masses, but still keeps the hierarchy between squark and slepton masses intact. In the model shown, the LSP is the nearly massless gravitino and the next-to-lightest supersymmetric particle (NLSP) is a bino-like neutralino, but for larger number of messenger fields it could be either a stau, or else co-NLSPs $\widetilde{\tau}_1$, $\widetilde{e}_L$, $\widetilde{\mu}_L$, depending on $\tan\beta$.

The fourth sample sketch, in Fig. 13.12(d), is of a typical GMSB model with a nonminimal messenger sector, $N_5 = 3$. Again the LSP is the nearly massless gravitino, but this time the NLSP is the lightest stau. The heaviest superpartner is the gluino, and the heaviest chargino and neutralino are wino-like.

Additional scenarios for MSSM spectra have also been considered in the literature (e.g., see Ref. [162]). For example, in the "focus point" region of the CMSSM parameter space, the first- and second-generation squarks and leptons are significantly heavier than the gluino and the other sparticles.

It would be a mistake to rely too heavily on specific scenarios for the MSSM masses and mixings, and the above illustrations are only a tiny fraction of the available possibilities. However, it is also useful to keep in mind some general lessons that often recur in various different models. Indeed, there has emerged a sort of folklore concerning likely features of the MSSM spectrum, partly based on theoretical bias and partly on the constraints inherent in most known viable softly broken supersymmetric theories. We remark on these features mainly because they represent the prevailing prejudices among supersymmetry theorists, which is certainly a useful thing to know even if one wisely decides to remain skeptical. For example, it is perhaps not unlikely that:

- The LSP is the lightest neutralino $\widetilde{N}_1$, unless the gravitino is lighter or R-parity is not conserved. If $M_1 < M_2, |\mu|$, then $\widetilde{N}_1$ is likely to be bino-like, with a mass roughly 0.5 times the masses of $\widetilde{N}_2$ and $\widetilde{C}_1$ in many well-motivated models. If, instead, $|\mu| < M_1, M_2$, then the LSP $\widetilde{N}_1$ has a large higgsino content and $\widetilde{N}_2$ and $\widetilde{C}_1$ are not much heavier. And, if $M_2 \ll M_1, |\mu|$, then the LSP will be a wino-like neutralino, with a chargino only very slightly heavier.
- The gluino will be much heavier than the lighter neutralinos and charginos. This is certainly true in the case of the "standard" gaugino mass relation eq. (13.58); more generally, the running gluino mass parameter grows relatively quickly as it is RG-evolved into the infrared because the QCD coupling is larger than the electroweak gauge couplings. So, even if there are big corrections to the gaugino mass boundary conditions eqs. (13.27) or (14.48), the gluino mass parameter M_3 is likely to come out larger than M_1 and M_2.
- The squarks of the first and second generations are nearly degenerate and much heavier than the sleptons. This is because each squark mass gets the same large positive-definite radiative corrections from loops involving the gluino. The left-handed squarks $\widetilde{u}_L$, $\widetilde{d}_L$, $\widetilde{s}_L$, and $\widetilde{c}_L$ are likely to be heavier than their right-handed counterparts $\widetilde{u}_R$, $\widetilde{d}_R$, $\widetilde{s}_R$, and $\widetilde{c}_R$, because of the effect parameterized by K_2 in eqs. (13.171)–(13.177).
- The squarks of the first two generations cannot be lighter than about 0.8 times the mass of the gluino in models with CMSSM boundary conditions, and about 0.6 times the mass of the gluino in the simplest gauge-mediated models, as discussed in Section 14.4, if the number of messenger squark pairs is $N_5 \leq 4$. In the case of the CMSSM, this is because the gluino mass feeds into the squark masses through RG evolution; in the gauge-mediated case it is because the gluino and squark masses are tied together by eqs. (14.53) and (14.54).
- The lighter stop $\widetilde{t}_1$ and the lighter sbottom $\widetilde{b}_1$ are probably the lightest squarks. This is because stop and sbottom mixing effects and the effects of S_t and S_b in eqs. (13.72)–(13.74) both tend to decrease the lighter stop and sbottom masses.
- The lightest charged slepton is probably a stau $\widetilde{\tau}_1$. The mass difference $m_{\widetilde{e}_R} - m_{\widetilde{\tau}_1}$ is likely to be significant if $\tan\beta$ is large, because of the effects of a large tau Yukawa coupling. For smaller $\tan\beta$, $\widetilde{\tau}_1$ is predominantly $\widetilde{\tau}_R$ and it is not so much lighter than $\widetilde{e}_R$, $\widetilde{\mu}_R$.
- The left-handed charged sleptons $\widetilde{e}_L$ and $\widetilde{\mu}_L$ are likely to be heavier than their right-handed counterparts $\widetilde{e}_R$ and $\widetilde{\mu}_R$. This is because of the effect of K_2 in eq. (13.175). (Note also that $\Delta_{\widetilde{e}_L} - \Delta_{\widetilde{e}_R}$ is positive but very small because of the numerical accident of $\sin^2\theta_W \approx 1/4$.)

It should be kept in mind that each of these prejudices might be defied by the real world. The most important point is that by measuring the masses and mixing angles of the MSSM particles we will be able to gain a great deal of information that can rule out or bolster evidence for competing proposals for the origin and mediation of supersymmetry breaking.

To facilitate the exploration of MSSM phenomena in a more model-independent way while respecting the flavor and CP constraints noted in Section 13.6, one can employ the so-called *phenomenological MSSM* (pMSSM). In one version of the pMSSM, the model is governed by 19 independent real supersymmetric (low-energy effective) parameters: the three gaugino mass parameters M_1, M_2, and M_3, the Higgs sector parameters m_A and $\tan\beta$, the higgsino mass parameter μ, five sfermion squared-mass parameters for the mass-degenerate first and second generations, the five corresponding sfermion squared-mass parameters for the third generation, and three third-generation A-term parameters. The first- and second-generation A-terms are typically neglected in pMSSM studies, as their phenomenological consequences are negligible in most applications (although the second-generation slepton A-term can contribute significantly to the anomalous magnetic moment of the muon due to sparticle-mediated radiative corrections). Citations to the original literature and subsequent applications to supersymmetric particle searches at colliders and dark matter studies can be found in Refs. [26, 27].

Exercises

13.1 Define the top-squark mixing parameter by

$$X_t \equiv \frac{a_t}{y_t} - \mu^* \cot\beta\,. \tag{13.185}$$

Then, the top-squark squared-mass matrix [see eq. (13.180)] is given by

$$\boldsymbol{m}^2_{\widetilde{t}} = \begin{pmatrix} m^2_{Q_3} + m_t^2 + \Delta_{\widetilde{u}_L} & m_t X_t^* \\ m_t X_t & m^2_{\bar{u}_3} + m_t^2 + \Delta_{\widetilde{u}_R} \end{pmatrix}. \tag{13.186}$$

The top-squark mass eigenstates $\widetilde{t}_1$ and $\widetilde{t}_2$ are linear combinations of the top-squark interaction eigenstates $\widetilde{t}_L$ and $\widetilde{t}_R$:

$$\begin{pmatrix} \widetilde{t}_1 \\ \widetilde{t}_2 \end{pmatrix} = \boldsymbol{U} \begin{pmatrix} \widetilde{t}_L \\ \widetilde{t}_R \end{pmatrix}, \tag{13.187}$$

where the unitary mixing matrix $\boldsymbol{U}$ is given by

$$\boldsymbol{U} \equiv \begin{pmatrix} \cos\theta_{\widetilde{t}} & -e^{-i\chi_{\widetilde{t}}} \sin\theta_{\widetilde{t}} \\ e^{i\chi_{\widetilde{t}}} \sin\theta_{\widetilde{t}} & \cos\theta_{\widetilde{t}} \end{pmatrix}. \tag{13.188}$$

By convention, $0 \le \theta_{\widetilde{t}} \le \frac{1}{2}\pi$ and $-\pi < \chi_{\widetilde{t}} \le \pi$.

(a) Diagonalizing the top-squark squared-mass matrix yields

$$\boldsymbol{U m_{\tilde{t}}^2 U^\dagger} = \begin{pmatrix} m_{\tilde{t}_1}^2 & 0 \\ 0 & m_{\tilde{t}_2}^2 \end{pmatrix}, \tag{13.189}$$

in a convention where $m_{\tilde{t}_1}^2 < m_{\tilde{t}_2}^2$. Derive the following expressions for the masses of the top squarks $\tilde{t}_1$ and $\tilde{t}_2$:

$$m_{\tilde{t}_1,\tilde{t}_2}^2 = \frac{1}{2}\Big[m_{Q_3}^2 + m_{\bar{u}_3}^2 + 2m_t^2 + \Delta_{\tilde{u}_L} + \Delta_{\tilde{u}_R} \mp \sqrt{(m_{Q_3}^2 - m_{\bar{u}_3}^2 + \Delta_{\tilde{u}_L} - \Delta_{\tilde{u}_R})^2 + 4m_t^2|X_t|^2}\Big]. \tag{13.190}$$

(b) Using the results of Exercise G.1, show that

$$\chi_{\tilde{t}} = \arg X_t\,, \tag{13.191}$$

and derive the following expressions for the top-squark mixing angle:

$$\sin 2\theta_{\tilde{t}} = \frac{2m_t|X_t|}{m_{\tilde{t}_2}^2 - m_{\tilde{t}_1}^2}\,, \qquad \cos 2\theta_{\tilde{t}} = \frac{m_{\bar{u}_3}^2 - m_{Q_3}^2 + \Delta_{\tilde{u}_R} - \Delta_{\tilde{u}_L}}{m_{\tilde{t}_2}^2 - m_{\tilde{t}_1}^2}, \tag{13.192}$$

where $0 \leq \theta_{\tilde{t}} \leq \frac{1}{2}\pi$.

(c) If X_t is real and nonzero, then eq. (13.191) implies that $e^{i\chi_{\tilde{t}}} = \operatorname{sgn} X_t$. It is then convenient to set $\chi_{\tilde{t}} = 0$ and redefine $\theta_{\tilde{t}} \to \theta_{\tilde{t}} \operatorname{sgn} X_t$, in which case the range of $\theta_{\tilde{t}}$ is extended to $-\frac{1}{2}\pi < \theta_{\tilde{t}} \leq \frac{1}{2}\pi$, and $\boldsymbol{U}$ is a real orthogonal matrix of unit determinant. In this convention, show that

$$\sin\theta_{\tilde{t}} = \operatorname{sgn}(X_t)\left(\frac{m_{Q_3}^2 + m_t^2 + \Delta_{\tilde{u}_L} - m_{\tilde{t}_1}^2}{m_{\tilde{t}_2}^2 - m_{\tilde{t}_1}^2}\right)^{1/2}. \tag{13.193}$$

13.2 Reconsider the derivation of eq. (13.111), where the effects of top-squark mixing are taken into account. In particular, replace the expressions for the squared masses of the top squarks given in eq. (13.106) by

$$m_{\tilde{t}_{1,2}}^2 = \tfrac{1}{2}\Big[m_{Q_3}^2 + m_{\bar{u}_3}^2 + y_t^2 v_u^2 \mp \sqrt{(m_{Q_3}^2 - m_{\bar{u}_3}^2)^2 + 2y_t^2 v_u^2 |X_t|^2}\Big], \tag{13.194}$$

after employing eq. (13.190) with $m_t^2 = \frac{1}{2}y_t^2 v_u^2$ and $X_t = (a_t/y_t) - \mu^*(v_d/v_u)$, and neglecting the effects of the electroweak corrections, $\Delta_{\tilde{u}_L}$ and $\Delta_{\tilde{u}_R}$.

(a) Show that

$$(\Delta\mathcal{M}_{\rm eff}^2)_{uu} = \frac{3y_t^2 m_t^2}{4\pi^2}\bigg\{\ln\left(\frac{m_{\tilde{t}_1} m_{\tilde{t}_2}}{m_t^2}\right) + \frac{\big[\mathrm{Re}(X_t a_t^*)\big]^2}{y_t^2} f(m_{\tilde{t}_1}^2, m_{\tilde{t}_2}^2) + \frac{\mathrm{Re}(X_t a_t^*)}{y_t} g(m_{\tilde{t}_1}^2, m_{\tilde{t}_2}^2) + \frac{\mathrm{Re}(\mu a_t)\cot\beta}{4y_t m_t^2} h(m_{\tilde{t}_1}^2, m_{\tilde{t}_2}^2)\bigg\}, \tag{13.195}$$

$$(\Delta\mathcal{M}^2_{\rm eff})_{dd} = \frac{3y_t^2 m_t^2}{4\pi^2}\biggl\{\bigl[{\rm Re}(X_t\mu)\bigr]^2 f(m^2_{\tilde{t}_1}, m^2_{\tilde{t}_2}) + \frac{{\rm Re}(\mu a_t)\tan\beta}{4y_t m_t^2} h(m^2_{\tilde{t}_1}, m^2_{\tilde{t}_2})\biggr\}, \tag{13.196}$$

$$(\Delta\mathcal{M}^2_{\rm eff})_{ud} = -\frac{3y_t^2 m_t^2}{4\pi^2}\biggl\{\frac{{\rm Re}(X_t a_t^*)\,{\rm Re}(X_t\mu)}{y_t} f(m^2_{\tilde{t}_1}, m^2_{\tilde{t}_2}) + \tfrac{1}{2}\,{\rm Re}(X_t\mu) g(m^2_{\tilde{t}_1}, m^2_{\tilde{t}_2}) + \frac{{\rm Re}(\mu a_t)}{4y_t m_t^2} h(m^2_{\tilde{t}_1}, m^2_{\tilde{t}_2})\biggr\}, \tag{13.197}$$

where

$$f(a,b) \equiv \frac{1}{(a-b)^2}\left[1 - \frac{1}{2}\left(\frac{a+b}{a-b}\right)\ln\left(\frac{a}{b}\right)\right], \qquad f(a,a) = -\frac{1}{12a^2}, \tag{13.198}$$

$$g(a,b) \equiv \frac{1}{a-b}\ln\left(\frac{a}{b}\right), \qquad g(a,a) = \frac{1}{a}, \tag{13.199}$$

$$h(a,b) \equiv \frac{a\ln(a/Q^2) - b\ln(b/Q^2)}{a-b} - 1, \qquad h(a,a) = \ln(a/Q^2). \tag{13.200}$$

(b) Using eqs. (6.350) and (13.112) and the results of part (a), evaluate $\Delta(m_h^2)$ in the Higgs decoupling limit where $m_A \gg m_Z$. Show that

$$m_h^2 = m_Z^2\cos^2 2\beta + \frac{3m_t^4}{2\pi^2 v^2}\biggl\{\ln\left(\frac{m_{\tilde{t}_1} m_{\tilde{t}_2}}{m_t^2}\right) + |X_t|^4 f(m^2_{\tilde{t}_1}, m^2_{\tilde{t}_2}) + |X_t|^2 g(m^2_{\tilde{t}_1}, m^2_{\tilde{t}_2})\biggr\}. \tag{13.201}$$

Using eq. (13.192), verify the expression for $\Delta_{\rm threshold}$ given in eq. (13.116).

(c) Define $M_S^2 \equiv m_{\tilde{t}_1} m_{\tilde{t}_2}$ and assume that $m^2_{Q_3}, m^2_{\tilde{u}_3} \gg m_t^2,\ m_t|X_t|$. Conclude that $m_{\tilde{t}_2} - m_{\tilde{t}_1} \ll m_{\tilde{t}_2} + m_{\tilde{t}_1}$. After neglecting terms of order $(m^2_{\tilde{t}_2} - m^2_{\tilde{t}_1})^2$, show that eq. (13.201) yields

$$m_h^2 \simeq \frac{3m_t^4}{2\pi^2 v^2}\biggl\{\ln\left(\frac{M_S^2}{m_t^2}\right) + \frac{|X_t|^2}{M_S^2}\left(1 - \frac{|X_t|^2}{12M_S^2}\right)\biggr\}. \tag{13.202}$$

(d) The maximal mixing scenario corresponds to the case where the effect of the terms in eq. (13.201) that depend on $|X_t|$ maximizes the one-loop contribution to m_h^2. Show that this scenario is achieved when

$$|X_t|^2 = \frac{(m^2_{\tilde{t}_2} - m^2_{\tilde{t}_1})^2}{m^2_{\tilde{t}_2} + m^2_{\tilde{t}_1} - 2(m^2_{\tilde{t}_2} - m^2_{\tilde{t}_1})/\ln(m^2_{\tilde{t}_2}/m^2_{\tilde{t}_1})}. \tag{13.203}$$

Using eq. (13.192), show that the maximal mixing value of $\sin 2\theta_{\tilde{t}}$ coincides with the result obtained in eq. (13.118).

(e) Verify that the right-hand side of eq. (13.203) is positive definite for all values of $0 \leq m_{\tilde{t}_1} \leq m_{\tilde{t}_2}$. Plugging the maximal mixing value of $|X_t|^2$ back into eq. (13.201), show that

$$\Delta(m_h^2)_{\text{max.mix.}} = \frac{3m_t^4}{2\pi^2 v^2}\Bigg\{\ln\left(\frac{m_{\tilde{t}_2}^2}{m_t^2}\right)$$
$$+\frac{1}{2}\ln\left(\frac{m_{\tilde{t}_2}^2}{m_{\tilde{t}_1}^2}\right)\left[\frac{(m_{\tilde{t}_2}^2 - m_{\tilde{t}_1}^2)}{(m_{\tilde{t}_2}^2 + m_{\tilde{t}_1}^2) - 2(m_{\tilde{t}_2}^2 - m_{\tilde{t}_1}^2)/\ln(m_{\tilde{t}_2}^2/m_{\tilde{t}_1}^2)} - 1\right]\Bigg\}. \tag{13.204}$$

Check that, for $0 \leq m_{\tilde{t}_1} \leq m_{\tilde{t}_2}$,

$$\Delta(m_h^2)_{\text{max.mix.}} = \frac{3m_t^4}{2\pi^2 v^2}\left[\ln\left(\frac{m_{\tilde{t}_2}^2}{m_t^2}\right) + C\right], \tag{13.205}$$

where $1 \leq C \leq 3$ and C grows monotonically as $m_{\tilde{t}_1}$ increases from 0 to $m_{\tilde{t}_2}$.

(f) In the approximation of part (c), where $m_{\tilde{t}_2} - m_{\tilde{t}_1} \ll m_{\tilde{t}_2} + m_{\tilde{t}_1}$, show that the maximal mixing scenario is achieved when $|X_t|^2 = 6M_S^2$, after neglecting terms of order $(m_{\tilde{t}_2}^2 - m_{\tilde{t}_1}^2)^2$. Verify that the constant C, defined in eq. (13.205), is then approximately given by

$$C \simeq 3 - \frac{m_{\tilde{t}_2}^2 - m_{\tilde{t}_1}^2}{m_{\tilde{t}_2}^2 + m_{\tilde{t}_1}^2}. \tag{13.206}$$

13.3 The squared pole mass of h is given by eq. (11.180),

$$m_{p,h}^2 = m_{\text{eff},h}^2 + \Sigma_{hh}(m_h^2) - \Sigma_{hh}(0), \tag{13.207}$$

in the one-loop approximation, where $m_{\text{eff},h}^2$ is the smaller of the two eigenvalues of the matrix $\mathcal{M}_{\text{eff}}^2$ defined in eqs. (13.103) and (13.104). Show that there are no terms of $\mathcal{O}(m_t^4)$ in the one-loop approximation of the difference of self-energy functions, $\Sigma_{hh}(m_h^2) - \Sigma_{hh}(0)$. Conclude that eq. (13.201) captures the leading $\mathcal{O}(m_t^4)$ one-loop corrections to the squared pole mass of h.

13.4 The chargino masses are determined by the singular value decomposition of the matrix $\boldsymbol{X}$ [see eq. (13.148)]:

$$\boldsymbol{U}^*\boldsymbol{X}\boldsymbol{V}^{-1} = \begin{pmatrix} m_{\tilde{C}_1} & 0 \\ 0 & m_{\tilde{C}_2}\end{pmatrix}, \tag{13.208}$$

with real nonnegative entries $m_{\tilde{C}_i}$ given by eq. (13.151).

(a) Using the results of Exercise G.3, obtain explicit expressions for the matrix elements of $\boldsymbol{U}$ and $\boldsymbol{V}$ in terms of the parameters M_2, μ, m_W, and β, under the assumption that $\text{Im}(\mu M_2) \neq 0$.

(b) Show that if $\text{Im}(\mu M_2) = 0$, then it is possible to rephase the chargino gauge eigenstate fields such that μ and M_2 are real. In this case, one can choose U and V in eq. (13.208) to be real orthogonal 2×2 matrices. Using the results of Exercise G.4, obtain explicit expressions for the matrix elements of U and V in terms of the real parameters M_2, μ, m_W, and β.

13.5 (a) The neutralino mass eigenstates are given by eq. (13.139). Suppose that the elements of the neutralino mixing matrix $\boldsymbol{N}_{1k}$ satisfy

$$\boldsymbol{N}_{11} = \cos\theta_W\,, \qquad \boldsymbol{N}_{12} = \sin\theta_W\,, \qquad \boldsymbol{N}_{13} = \boldsymbol{N}_{14} = 0\,. \quad (13.209)$$

Show that the tree-level couplings of $\widetilde{N}_1$ to electrons and selectrons coincide with those of the photino of softly broken SUSY-QED.

(b) Suppose that the elements of the neutralino mixing matrix $\boldsymbol{N}_{2k}$ satisfy

$$\boldsymbol{N}_{21} = -\sin\theta_W\,, \qquad \boldsymbol{N}_{22} = \cos\theta_W\,, \qquad \boldsymbol{N}_{23} = \boldsymbol{N}_{24} = 0\,. \quad (13.210)$$

Explain why $\widetilde{N}_2$ should be called the zino, i.e., the supersymmetric partner of the Z gauge boson.

(c) Consider a neutralino mass matrix where $M_1 = M_2 \equiv M$ and $\mu = 0$. Determine the exact formulae for the neutralino mass eigenstates and their masses. Verify that the neutralino mass spectrum contains a photino of mass M and one massless higgsino state (which is a linear combination of $\widetilde{H}_u^0$ and $\widetilde{H}_d^0$). Show that the other two massive neutralinos are mixtures of the zino and the orthogonal linear combination of $\widetilde{H}_u^0$ and $\widetilde{H}_d^0$.

13.6 Consider the MSSM with one generation of quarks and leptons and their superpartners.

(a) Following the discussion of Section 13.5, determine the number of independent physical parameters (including those of the Standard Model) that govern the one-generation MSSM.

(b) The potentially complex parameters of the one-generation MSSM include μ, a_u, a_d, a_e, b, and M_i ($i = 1, 2, 3$). But some of the phases of these parameters can be removed by an appropriate rephasing of the MSSM fields. Show that for $i \neq j$, the quantities

$$\arg(\mu M_i b^*)\,, \arg(\mu a_u b^*)\,, \arg(\mu a_d b^*)\,, \arg(\mu a_e b^*)\,, \arg(M_i M_j^*) \quad (13.211)$$

are independent of field rephasings (and thus related to physical observables).

(c) Verify that one is always free to rephase the Higgs doublet fields such that b is real. Moreover, if one additionally adopts the convention where both Higgs VEVs are positive, show that $b > 0$.

(d) Suppose that all the potentially complex parameters listed in part (b) are real. Determine the number of discrete choices for the signs of these parameters that yield inequivalent models.

14 Realizations of Supersymmetry Breaking

In the MSSM, supersymmetry breaking is simply introduced explicitly. However, the soft parameters cannot be arbitrary. In order to understand how patterns like eqs. (13.24)–(13.26) can emerge, it is necessary to consider models in which supersymmetry is spontaneously broken.

Finding the ultimate cause of supersymmetry breaking is one of the most important goals for the future. For many purposes, one can simply assume that an F-term has obtained a VEV, without worrying about the specific dynamics that caused it. For understanding collider phenomenology, the most immediate concern is usually the nature of the couplings of the F-term VEV to the MSSM fields. This is the subject we turn to next.

14.1 Communication of Supersymmetry Breaking

The results of Chapter 12 imply that spontaneous supersymmetry breaking (dynamical or not) requires the MSSM to be extended. The ultimate order parameter for supersymmetry breaking cannot belong to any of the MSSM supermultiplets; a D-term VEV for $U(1)_Y$ does not lead to an acceptable spectrum, and there is no candidate gauge singlet whose F-term could develop a VEV. Therefore one must ask what effects *are* responsible for spontaneous supersymmetry breaking, and how supersymmetry breakdown is "communicated" to the MSSM particles. It is very difficult to achieve the latter in a phenomenologically viable way, working only with renormalizable interactions at tree level, even if the model is extended to involve new supermultiplets including gauge singlets. First, on general grounds it would be problematic to give masses to the MSSM gauginos, because the results of Section 8.5 inform us that renormalizable supersymmetry never has any scalar–gaugino–gaugino couplings that could turn into gaugino mass terms when the scalar gets a VEV. Second, at least some of the MSSM squarks and sleptons would have to be unacceptably light, and should have been discovered already. This can be understood from the existence of sum rules that can be obtained in the same way as eq. (12.17)) when the restrictions imposed by flavor symmetries are taken into account. For example, in the limit in which lepton flavors are conserved, the selectron mass eigenstates $\widetilde{e}_1$ and $\widetilde{e}_2$ could in general be mixtures of $\widetilde{e}_L$ and $\widetilde{e}_R$. But if they do not mix with other scalars, then part of the sum rule decouples from the

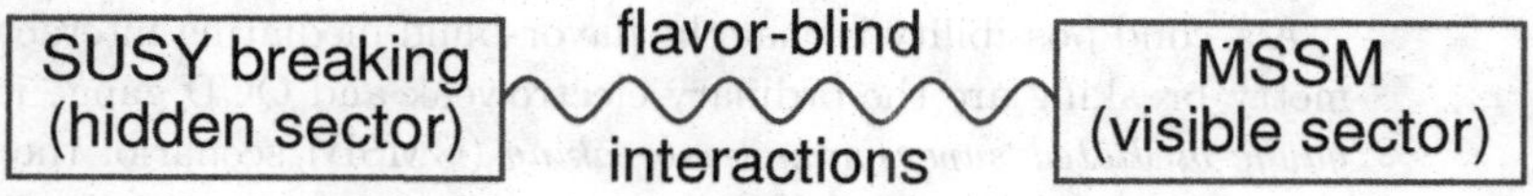

Fig. 14.1 The presumed schematic structure for supersymmetry breaking.

rest, and one obtains

$$m_{\tilde{e}_1}^2 + m_{\tilde{e}_2}^2 = 2m_e^2, \tag{14.1}$$

which is of course ruled out by experiment. Similar sum rules follow for each of the fermions of the Standard Model, at tree level and in the limits in which the corresponding flavors are conserved. In principle, the sum rules can be evaded by introducing flavor-violating mixings, but it is very difficult to see how to make a viable model in this way. Even ignoring these problems, there is no obvious reason why the resulting MSSM soft supersymmetry-breaking terms in this type of model should satisfy flavor-blindness conditions like eq. (13.24) or eq. (13.25).

For these reasons, we expect that the MSSM soft terms arise indirectly or radiatively, rather than from tree level renormalizable couplings to the supersymmetry-breaking order parameters. Supersymmetry breaking evidently occurs in a "hidden sector" of particles that have no (or only very small) direct couplings to the "visible sector" chiral supermultiplets of the MSSM. However, the two sectors do share some interactions that are responsible for mediating supersymmetry breaking from the hidden sector to the visible sector, resulting in the MSSM soft terms (see Fig. 14.1). In this scenario, the tree-level squared-mass sum rules need not hold, even approximately, for the physical masses of the visible-sector fields, so that a phenomenologically viable superpartner mass spectrum is, in principle, achievable. As a bonus, if the mediating interactions are flavor-blind, then the soft terms appearing in the MSSM will automatically obey conditions like eqs. (13.24)–(13.26).

There have been two main competing proposals for what the mediating interactions might be. The first (and historically the more popular) is that they are gravitational. More precisely, they are associated with the new physics, including gravity, that enters near the Planck scale. In this "gravity-mediated," or *Planck-scale-mediated supersymmetry-breaking* (PMSB) scenario, if supersymmetry is broken in the hidden sector by a VEV $\langle F \rangle$, then the soft terms in the visible sector should be roughly

$$m_{\text{soft}} \sim \langle F \rangle / M_P, \tag{14.2}$$

by dimensional analysis. This is because we know that m_{soft} must vanish in the limit $\langle F \rangle \to 0$ where supersymmetry is unbroken, and also in the limit $M_P \to \infty$ (corresponding to $G_N \to 0$) in which gravity becomes irrelevant. For m_{soft} of order a few hundred GeV, one would therefore expect that the scale associated with the origin of supersymmetry breaking in the hidden sector should be roughly given by $\sqrt{\langle F \rangle} \sim 10^{10}$ or 10^{11} GeV.

A second possibility is that the flavor-blind mediating interactions for supersymmetry breaking are the ordinary electroweak and QCD gauge interactions. In this *gauge-mediated supersymmetry-breaking* (GMSB) scenario, the MSSM soft terms come from loop diagrams involving some *messenger* particles. The messengers are new chiral supermultiplets that couple to a supersymmetry-breaking VEV $\langle F \rangle$, and also have $SU(3)_C \times SU(2)_L \times U(1)_Y$ interactions, which provide the necessary connection to the MSSM. Then, using dimensional analysis, one estimates, for the MSSM soft terms,

$$m_{\text{soft}} \sim \frac{\alpha_a}{4\pi} \frac{\langle F \rangle}{M_{\text{mess}}}, \tag{14.3}$$

where the $\alpha_a/4\pi$ is a loop factor for Feynman diagrams involving gauge interactions, and M_{mess} is a characteristic scale of the masses of the messenger fields. So if M_{mess} and $\sqrt{\langle F \rangle}$ are roughly comparable, then the scale of supersymmetry breaking can be as low as about $\sqrt{\langle F \rangle} \sim 10^4$ GeV (much lower than in the gravity-mediated case!) to give m_{soft} of the right order of magnitude.

14.2 The Goldstino and the Gravitino

As shown in Section 12.1, the spontaneous breaking of global supersymmetry implies the existence of a massless Weyl fermion, the goldstino. The goldstino is the fermionic component of the supermultiplet whose auxiliary field obtains a VEV.

We can derive an important property of the goldstino by considering the form of the conserved supercurrent eq. (8.76). Suppose for simplicity[1] that the only nonvanishing auxiliary field VEV is $\langle F \rangle$ with goldstino superpartner $\widetilde{G}$. Then the supercurrent conservation equation tells us that

$$0 = \partial_\mu J^\mu_\alpha = -i\langle F \rangle (\sigma^\mu \partial_\mu \widetilde{G}^\dagger)_\alpha + \partial_\mu j^\mu_\alpha + \cdots, \tag{14.4}$$

where j^μ_α is the part of the supercurrent that involves all of the other supermultiplets, and the ellipses represent other contributions of the goldstino supermultiplet to $\partial_\mu J^\mu_\alpha$, which we can ignore. [The first term in eq. (14.4) comes from the second term in eq. (8.76), using the equation of motion $F_i = -W_i^*$ for the goldstino's auxiliary field.] This equation of motion for the goldstino field allows us to write an effective Lagrangian

$$\mathscr{L}_{\text{goldstino}} = i\widetilde{G}^\dagger \overline{\sigma}^\mu \partial_\mu \widetilde{G} - \frac{1}{\langle F \rangle}(\widetilde{G}\partial_\mu j^\mu + \text{h.c.}), \tag{14.5}$$

which describes the interactions of the goldstino with all other fermion–boson pairs. In particular, since $j^\mu_\alpha = (\sigma^\nu \overline{\sigma}^\mu \psi_i)_\alpha \partial_\nu \phi^{\dagger i} + \sigma^\nu \overline{\sigma}^\rho \sigma^\mu \lambda^{\dagger a} F^a_{\nu\rho}/2\sqrt{2} + \cdots$, there are goldstino–scalar–chiral fermion and goldstino–gaugino–gauge boson vertices as

[1] More generally, if supersymmetry is spontaneously broken by VEVs for several auxiliary fields F_i and D^a, then one should make the replacement $\langle F \rangle \to \left(\sum_i |\langle F_i \rangle|^2 + \frac{1}{2}\sum_a \langle D^a \rangle^2\right)^{1/2}$ everywhere in the following.

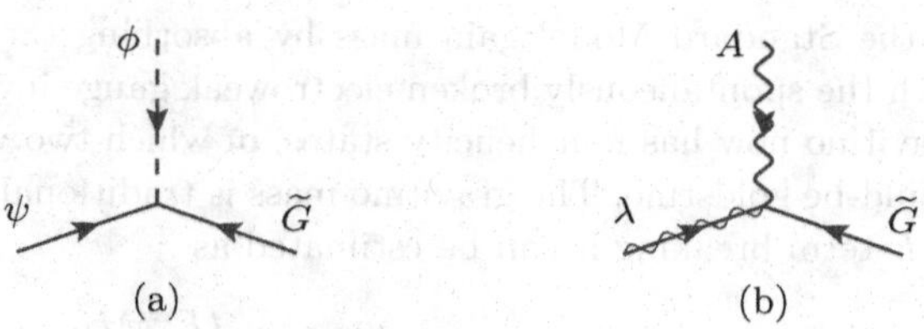

Fig. 14.2 Goldstino/gravitino $\widetilde{G}$ interactions with superpartner pairs (ϕ, ψ) and (λ, A).

shown in Fig. 14.2. Since this analysis depends only on supercurrent conservation, eq. (14.5) holds independently of the details of how supersymmetry breaking is communicated from $\langle F \rangle$ to the MSSM-sector fields (ϕ_i, ψ_i) and (λ^a, A^a). It may appear strange at first that the interaction couplings in eq. (14.5) get larger in the limit that $\langle F \rangle$ goes to zero. However, the interaction term $\widetilde{G}\partial_\mu j^\mu$ contains two derivatives, which turn out to always give a kinematic factor proportional to the squared-mass difference of the superpartners when they are on-shell, i.e., $m_\phi^2 - m_\psi^2$ and $m_\lambda^2 - m_A^2$ for Figs. 14.2(a) and 14.2(b), respectively. These can be nonzero only by virtue of supersymmetry breaking, so they must also vanish as $\langle F \rangle \to 0$, and the interaction is well defined in that limit. Nevertheless, for fixed values of $m_\phi^2 - m_\psi^2$ and $m_\lambda^2 - m_A^2$, the interaction term in eq. (14.5) can be phenomenologically important if $\langle F \rangle$ is not too large.

The preceding remarks apply to the breaking of global supersymmetry. However, taking into account gravity, supersymmetry must be promoted to a local symmetry. This means that the spinor parameter ϵ^α, which first appeared in Section 8.1, is no longer a constant, but can vary from point to point in spacetime. The resulting locally supersymmetric theory is called *supergravity*.[2] It necessarily unifies the spacetime symmetries of ordinary general relativity with local supersymmetry transformations. In supergravity, the spin-2 graviton has a spin-3/2 fermion superpartner called the gravitino, which we will denote $\widetilde{\Psi}_\mu^\alpha$. The gravitino has odd R-parity ($P_R = -1$), as can be seen from the definition in eq. (13.15). It carries both a vector index (μ) and a spinor index (α), and transforms inhomogeneously under local supersymmetry transformations:

$$\delta\widetilde{\Psi}_\mu^\alpha = \partial_\mu \epsilon^\alpha + \cdots . \tag{14.6}$$

Thus the gravitino should be thought of as the "gauge" field of local supersymmetry transformations [see eq. (8.55)]. As long as supersymmetry is unbroken, the graviton and the gravitino are both massless, each with two spin helicity states. Once supersymmetry is spontaneously broken, the gravitino acquires a mass by absorbing ("eating") the goldstino, which becomes its longitudinal (helicity $\pm 1/2$) components. This is called the *super-Higgs* mechanism, and it is analogous to the ordinary Higgs mechanism for gauge theories, by which the $W^\pm$ and Z^0 gauge bosons

[2] A comprehensive theoretical treatment of supergravity lies beyond the scope of this book but can be found in [B20, B40–B45].

in the Standard Model gain mass by absorbing the Goldstone bosons associated with the spontaneously broken electroweak gauge invariance. The massive spin-3/2 gravitino now has four helicity states, of which two were originally assigned to the would-be goldstino. The gravitino mass is traditionally called $m_{3/2}$, and in the case of F-term breaking it can be estimated as

$$m_{3/2} \sim \langle F \rangle / M_P \,. \tag{14.7}$$

This follows simply from dimensional analysis, since $m_{3/2}$ must vanish in the limits that supersymmetry is restored ($\langle F \rangle \to 0$) and that gravity is turned off ($M_P \to \infty$). Equation (14.7) implies very different expectations for the mass of the gravitino in gravity-mediated and in gauge-mediated models, because they usually make very different predictions for $\langle F \rangle$.

In the Planck-scale-mediated supersymmetry-breaking case, the gravitino mass is comparable to the masses of the MSSM sparticles [see eqs. (14.2) and (14.7)]. Therefore $m_{3/2}$ is expected to be at least of order 100 GeV or larger. Its interactions will be of gravitational strength, so the gravitino will not play any role in collider physics, but it can be important in cosmology. If it is the LSP, then it is stable and its primordial density could easily exceed the critical density, causing the universe to become matter-dominated too early. Even if it is not the LSP, the gravitino can cause problems unless its density is diluted by inflation at late times, or it decays sufficiently rapidly.

In contrast, gauge-mediated supersymmetry-breaking models predict that the gravitino is much lighter than the MSSM sparticles as long as $M_{\text{mess}} \ll M_P$. This can be seen by comparing eqs. (14.3) and (14.7). The gravitino is almost certainly the LSP in this case, and all of the MSSM sparticles will eventually decay into final states that include it. Naively, one might expect that these decays are extremely slow. However, this is not necessarily true, because the gravitino inherits the nongravitational interactions of the goldstino it has absorbed. This means that the gravitino, or more precisely its longitudinal (goldstino) components, can play an important role in collider physics experiments. The mass of the gravitino can generally be ignored for kinematic purposes, as can its transverse (helicity $\pm 3/2$) components, which really do have only gravitational interactions. Therefore in collider phenomenology discussions one may interchangeably use the same symbol $\widetilde{G}$ for the goldstino and for the gravitino of which it is the longitudinal (helicity $\pm 1/2$) part. By using the effective Lagrangian eq. (14.5), one can compute that the decay rate of any sparticle $\widetilde{X}$ into its Standard Model partner X plus a goldstino/gravitino $\widetilde{G}$ is

$$\Gamma(\widetilde{X} \to X\widetilde{G}) = \frac{m_{\widetilde{X}}^5}{16\pi \langle F \rangle^2} \left(1 - m_X^2/m_{\widetilde{X}}^2\right)^4 . \tag{14.8}$$

This corresponds to either Fig. 14.2(a) or Fig. 14.2(b), with $(\widetilde{X}, X) = (\phi, \psi)$ or (λ, A), respectively. One factor $(1 - m_X^2/m_{\widetilde{X}}^2)^2$ came from the derivatives in the interaction term in eq. (14.5) evaluated for on-shell final states, and another such factor comes from the kinematic phase space integral with $m_{3/2} \ll m_{\widetilde{X}}, m_X$.

If the supermultiplet containing the goldstino and $\langle F\rangle$ has canonically normalized kinetic terms, and the tree-level vacuum energy is required to vanish, then the estimate eq. (14.7) is sharpened to

$$m_{3/2} = \frac{\langle F\rangle}{\sqrt{3}M_P}. \tag{14.9}$$

In that case, one can rewrite eq. (14.8) as

$$\Gamma(\widetilde{X} \to X\widetilde{G}) = \frac{m_{\widetilde{X}}^5}{48\pi M_P^2 m_{3/2}^2}\left(1 - m_X^2/m_{\widetilde{X}}^2\right)^4, \tag{14.10}$$

and this is how the formula is sometimes presented, although it is less general since it assumes eq. (14.9). The decay width is larger for smaller $\langle F\rangle$, or equivalently for smaller $m_{3/2}$, if the other masses are fixed. If $\widetilde{X}$ is a mixture of superpartners of different SM particles X, then each partial width in eq. (14.8) should be multiplied by a suppression factor equal to the square of the cosine of the appropriate mixing angle. If $m_{\widetilde{X}}$ is of order 100 GeV or more, and $\sqrt{\langle F\rangle} \lesssim$ few $\times 10^6$ GeV [corresponding to $m_{3/2}$ less than roughly 1 keV according to eq. (14.9)], then the decay $\widetilde{X} \to X\widetilde{G}$ can occur quickly enough to be observed in a modern collider detector. This implies some interesting phenomenological signatures, which we will discuss further later.

We now turn to a more systematic analysis of the way in which the MSSM soft terms arise.

14.3 Planck-Scale-Mediated SUSY Breaking

Consider models in which the spontaneous supersymmetry-breaking sector connects with our MSSM sector mostly through gravitational-strength interactions, including the effects of supergravity. Let X be the chiral superfield whose F-term auxiliary field breaks supersymmetry, and consider first a globally supersymmetric effective Lagrangian, with the Planck scale suppressed effects that communicate between the two sectors included as nonrenormalizable terms of the types discussed in Section 10.10. The superpotential, the Kähler potential, and the gauge kinetic function, expanded for large M_P, are

$$W = W_{\text{MSSM}} - \frac{1}{M_P}\left(\frac{1}{6}y^{Xijk}X\Phi_i\Phi_j\Phi_k + \frac{1}{2}\mu^{Xij}X\Phi_i\Phi_j\right) + \cdots, \tag{14.11}$$

$$K = \Phi^{\dagger i}\Phi_i + \frac{1}{M_P}(n_i^j X + \overline{n}_i^j X^\dagger)\Phi^{\dagger i}\Phi_j - \frac{1}{M_P^2}k_i^j XX^\dagger\Phi^{\dagger i}\Phi_j + \cdots, \tag{14.12}$$

$$f_{ab} = \frac{\delta_{ab}}{g_a^2}\left(1 - \frac{2}{M_P}f_a X + \cdots\right). \tag{14.13}$$

Here Φ_i represent the chiral superfields of the MSSM or an extension of it, and y^{Xijk}, k_i^j, n_i^j, $\overline{n}_i^j$, and f_a are dimensionless couplings while μ^{Xij} has the dimension of mass. The leading term in the Kähler potential is chosen to give canonically normalized kinetic terms. The matrix k_i^j must be hermitian, and $\overline{n}_i^j = (n_j^i)^*$, in order for the Lagrangian to be real. To find the resulting soft supersymmetry-breaking terms in the low-energy effective theory, one can apply the superspace formalism, treating X as a spurion field by making the replacements

$$X \to \theta\theta F, \qquad X^\dagger \to \theta^\dagger\theta^\dagger F^\dagger, \tag{14.14}$$

where F denotes $\langle F_X \rangle$. The resulting supersymmetry-breaking Lagrangian, after integrating out the auxiliary fields in Φ_i, is

$$\begin{aligned}\mathcal{L}_{\text{soft}} = -\Big(&\frac{F}{2M_P} f_a \lambda^a \lambda^a + \frac{F}{6M_P} y^{Xijk} \phi_i\phi_j\phi_k + \frac{F}{2M_P}\mu^{Xij}\phi_i\phi_j \\ &+ \frac{F}{M_P} n_i^j \phi_j W^i_{\text{MSSM}} + \text{h.c.}\Big) - \frac{|F|^2}{M_P^2}(k_j^i + n_p^i \overline{n}_j^p)\phi^{*j}\phi_i,\end{aligned} \tag{14.15}$$

where ϕ_i and λ^a are the scalar and gaugino fields in the MSSM sector. Now if one assumes that $\sqrt{F} \sim 10^{10}$ or 10^{11} GeV, then eq. (14.15) has the same form as eq. (8.78), with MSSM-sector soft terms of order $m_{\text{soft}} \sim F/M_P$, perhaps of order a TeV. Actually, this discussion applies even if M_P is replaced by some other large mass scale, so in the remainder of this section we will replace M_P with a generic M, but still with the very rough estimate, $m_{\text{soft}} \sim F/M$.

In particular, if we write the visible sector superpotential as

$$W_{\text{MSSM}} = \tfrac{1}{6} y^{ijk}\Phi_i\Phi_j\Phi_k + \tfrac{1}{2}\mu^{ij}\Phi_i\Phi_j, \tag{14.16}$$

then the soft terms in that sector, in the notation of eq. (8.78), are

$$M_a = \frac{F}{M} f_a, \tag{14.17}$$

$$a^{ijk} = \frac{F}{M}(y^{Xijk} + n_p^i y^{pjk} + n_p^j y^{pik} + n_p^k y^{pij}), \tag{14.18}$$

$$b^{ij} = \frac{F}{M}(\mu^{Xij} + n_p^i \mu^{pj} + n_p^j \mu^{pi}), \tag{14.19}$$

$$(m^2)_j^i = \frac{|F|^2}{M^2}(k_j^i + n_p^i \overline{n}_j^p). \tag{14.20}$$

Note that couplings of the form $\mathcal{L}_{\text{maybe soft}}$ in eq. (8.79) do not arise from eq. (14.15). Although they actually are expected to occur, the largest possible sources for them are nonrenormalizable Kähler potential terms, which lead to

$$\mathcal{L} = -\frac{|F|^2}{M^3} x_i^{jk} \phi^{\dagger i}\phi_j\phi_k + \text{h.c.}, \tag{14.21}$$

where x_i^{jk} is dimensionless. This explains why, at least within this model framework, the couplings c_i^{jk} in eq. (8.79) are of order $|F|^2/M^3 \sim m^2_{\text{soft}}/M$, and thus negligible.

Similarly, dimensionless ("hard") supersymmetry-breaking couplings in the low-energy effective Lagrangian will arise from other nonrenormalizable operators involving the supersymmetry-breaking sector. These include three distinct types of

Table 14.1 Classification of all renormalizable supersymmetry-breaking interactions. Chiral scalars and fermions are represented by ϕ and ψ, and gauginos by λ. The last column indicates the lowest-dimension operator which can give rise to the term through spontaneous supersymmetry breaking with $\langle X\rangle = \theta\theta F$ and a high scale of suppression M. The resulting naive suppression is shown in the third column.

type	term	naive suppression	origin
soft	$\phi\phi^\dagger$	$\|F\|^2/M^2 \sim m^2_{\text{soft}}$	$\frac{1}{M^2}[XX^\dagger\Phi\Phi^\dagger]_D$
	$\phi^2+\text{h.c.}$	$\mu F/M \sim \mu m_{\text{soft}}$	$\frac{\mu}{M}[X\Phi^2]_F$
	$\phi^3+\text{h.c.}$	$F/M \sim m_{\text{soft}}$	$\frac{1}{M}[X\Phi^3]_F$
	$\lambda\lambda+\text{h.c.}$	$F/M \sim m_{\text{soft}}$	$\frac{1}{M}[XW^\alpha W_\alpha]_F$
maybe soft	$\phi^2\phi^\dagger+\text{h.c.}$	$\|F\|^2/M^3 \sim m^2_{\text{soft}}/M$	$\frac{1}{M^3}[XX^\dagger\Phi^2\Phi^\dagger]_D$
	$\psi\psi+\text{h.c.}$	$\|F\|^2/M^3 \sim m^2_{\text{soft}}/M$	$\frac{1}{M^3}[XX^\dagger D^\alpha\Phi D_\alpha\Phi]_D$
	$\psi\lambda+\text{h.c.}$	$\|F\|^2/M^3 \sim m^2_{\text{soft}}/M$	$\frac{1}{M^3}[XX^\dagger D^\alpha\Phi W_\alpha]_D$
hard	ϕ^4	$F/M^2 \sim m_{\text{soft}}/M$	$\frac{1}{M^2}[X\Phi^4]_F$
	$\phi^3\phi^\dagger$	$\|F\|^2/M^4 \sim m^2_{\text{soft}}/M^2$	$\frac{1}{M^4}[XX^\dagger\Phi^3\Phi^\dagger]_D$
	$\phi^2\phi^{\dagger 2}$	$\|F\|^2/M^4 \sim m^2_{\text{soft}}/M^2$	$\frac{1}{M^4}[XX^\dagger\Phi^2\Phi^{*2}]_D$
	$\phi\psi\psi$	$\|F\|^2/M^4 \sim m^2_{\text{soft}}/M^2$	$\frac{1}{M^4}[XX^\dagger\,\Phi D^\alpha\Phi D_\alpha\Phi]_D$
	$\phi^\dagger\psi\psi$	$\|F\|^2/M^4 \sim m^2_{\text{soft}}/M^2$	$\frac{1}{M^4}[XX^\dagger\,\Phi^\dagger D^\alpha\Phi D_\alpha\Phi]_D$
	$\phi\psi\lambda$	$\|F\|^2/M^4 \sim m^2_{\text{soft}}/M^2$	$\frac{1}{M^4}[XX^\dagger\,\Phi D^\alpha\Phi W_\alpha]_D$
	$\phi^\dagger\psi\lambda$	$\|F\|^2/M^4 \sim m^2_{\text{soft}}/M^2$	$\frac{1}{M^4}[XX^\dagger\,\Phi^\dagger D^\alpha\Phi W_\alpha]_D$
	$\phi\lambda\lambda$	$F/M^2 \sim m_{\text{soft}}/M$	$\frac{1}{M^2}[X\Phi W^\alpha W_\alpha]_F$
	$\phi^\dagger\lambda\lambda$	$\|F\|^2/M^4 \sim m^2_{\text{soft}}/M^2$	$\frac{1}{M^4}[XX^\dagger\,\Phi^\dagger W^\alpha W_\alpha]_D$

quartic scalar couplings, ϕ^4 and $\phi^3\phi^\dagger$ and $\phi^2\phi^{*2}$. There will also be various types of scalar–fermion–fermion couplings as allowed by gauge symmetries. These include nonanalytic Yukawa terms $\phi^*\psi\psi$, as well as non-supersymmetric analytic Yukawa terms with the structure $\phi_i\psi_j\psi_k$. Unlike Yukawa couplings following from the superpotential, the latter need not be symmetric under interchanges of i, j or of i, k. There are also dimensionless supersymmetry-breaking terms that arise only in extensions of the MSSM. For example, there are scalar–gaugino–gaugino $\phi\lambda^a\lambda^a$ couplings if a chiral supermultiplet transforms as a representation found in the symmetric product of the adjoint with itself. A schematic summary of all such terms and their origins is exhibited in Table 14.1, along with the estimates of their naive suppression, using $F/M = m_{\text{soft}}$. Each renormalizable supersymmetry-breaking term is listed along with the lowest-dimension nonrenormalizable operator that gives rise to it, written in superfield form in terms of either $[\cdots]_F \equiv \int d^2\theta \cdots$ or as $[\cdots]_D \equiv \int d^2\theta d^2\theta^\dagger \cdots$. Although these are hard supersymmetry-breaking terms, they do not give rise to numerically dangerous contributions to the Higgs squared mass, due to their $F/M^2 = m_{\rm soft}/M$ suppression factors. They are therefore effectively soft, despite being technically hard. They are often negligible in practice, but with some notable possible exceptions. For example, dimensionless supersymmetry-breaking operators can lift otherwise flat directions in the scalar potential, so that they can become significant at intermediate energy scales despite their suppression. It is notable that, among the dimensionless couplings, only the ϕ^4 and $\phi\lambda\lambda$ terms are suppressed by only one power of $m_{\rm soft}/M$.

Let us now return our focus to the soft supersymmetry-breaking couplings. In principle, the parameters f_a, k^i_j, n^j_i, y^{Xijk}, and μ^{Xij} ought to be determined by the fundamental underlying theory. The familiar flavor-blindness of gravity expressed in Einstein's equivalence principle does not, by itself, tell us anything about their form. Therefore, the requirement of approximate flavor-blindness in $\mathscr{L}_{\rm soft}$ is a new assumption in this framework, and is not guaranteed without further structure. Nevertheless, it has historically been popular to make a dramatic simplification by assuming a "minimal" form for the normalization of kinetic terms and gauge interactions in the nonrenormalizable Lagrangian. Specifically, it is often assumed that there is a common $f_a = f$ for the three gauginos, that $k^j_i = k\delta^j_i$ and $n^j_i = n\delta^j_i$ are the same for all scalars, with k and n real, and that the other couplings are proportional to the corresponding superpotential parameters, so that $y^{Xijk} = \alpha y^{ijk}$ and $\mu^{Xij} = \beta\mu^{ij}$ with universal real dimensionless constants α and β. Then the soft terms in $\mathscr{L}^{\rm MSSM}_{\rm soft}$ are all determined by just four parameters:

$$m_{1/2} = f\frac{\langle F\rangle}{M_P}, \qquad m_0^2 = (k+n^2)\frac{|\langle F\rangle|^2}{M_P^2}, \tag{14.22}$$

$$A_0 = (\alpha+3n)\frac{\langle F\rangle}{M_P}, \qquad B_0 = (\beta+2n)\frac{\langle F\rangle}{M_P}. \tag{14.23}$$

This corresponds to the *minimal supergravity* (mSUGRA) framework for the soft supersymmetry-breaking terms. In particular, in terms of $m_{1/2}$, m_0^2, A_0, and B_0, the parameters appearing in eq. (13.16) satisfy the CMSSM boundary conditions at a renormalization scale $Q \approx M_P$ specified in eqs. (13.27)–(13.30).[3]

Although it is unclear whether the mSUGRA framework can be well motivated on purely theoretical grounds, eqs. (13.27)–(13.30) have the virtue of being extraordinarily predictive, at least in principle.[4] As discussed in Sections 13.6–13.12, they could be applied as RG boundary conditions at the scale M_P. The RG evolution of the soft parameters down to the electroweak scale would then allow us to pre-

[3] Of course, the relation $b = B_0\mu$ [see eq. (13.30)] is content-free unless one can relate B_0 to the other parameters in some nontrivial way.

[4] The mSUGRA/CMSSM framework described above has been the subject of the bulk of phenomenological and experimental studies of supersymmetry, and has become a benchmark scenario for experimental collider search limits.

dict the entire MSSM spectrum in terms of just five parameters $m_{1/2}$, m_0^2, A_0, B_0, and μ (plus the already-measured gauge and Yukawa couplings of the MSSM). A popular approximation is to start this RG running from the unification scale $M_U \approx 2 \times 10^{16}$ GeV instead of M_P. The reason for this is more practical than principled: the apparent unification of gauge couplings gives us a strong hint that we know something about how the RG equations behave up to M_U, but unfortunately gives us little guidance about what to expect at scales between M_U and M_P. The errors made in neglecting these effects are proportional to a loop suppression factor times $\ln(M_P/M_U)$. These corrections hopefully can be partly absorbed into a redefinition of m_0^2, $m_{1/2}$, A_0, and B_0 at M_U, but in many cases could lead to other important effects that are difficult to anticipate.

Particular models of gravity-mediated supersymmetry breaking can be even more predictive, relating some of the parameters m_0^2, $m_{1/2}$, A_0, and B_0 to each other and to the mass of the gravitino $m_{3/2}$. For example, three popular kinds of models for the soft terms are:

- Dilaton-dominated: $m_0^2 = m_{3/2}^2$, $m_{1/2} = -A_0 = \sqrt{3}m_{3/2}$.
- Polonyi: $m_0^2 = m_{3/2}^2$, $A_0 = (3 - \sqrt{3})m_{3/2}$, $m_{1/2} = \mathcal{O}(m_{3/2})$.
- "No-scale": $m_{1/2} \gg m_0, A_0, m_{3/2}$.

Dilaton domination arises in a particular limit of superstring theory. While it appears to be highly predictive, it can easily be generalized in other limits. The Polonyi model has the advantage of being the simplest possible model for supersymmetry breaking in the hidden sector, but it is rather ad hoc and does not seem to have a special place in grander schemes like superstrings. The "no-scale" limit may appear in a low-energy limit of superstrings in which the gravitino mass scale is undetermined at tree level (hence the name). It implies that the gaugino masses dominate over other sources of supersymmetry breaking near M_P. RG evolution feeds the gaugino masses into the squark, slepton, and Higgs squared-mass parameters with sufficient magnitude to give acceptable phenomenology at the electroweak scale. More recent versions of the no-scale scenario, however, also can give significant A_0 and m_0^2 at the input scale. In many cases B_0 can also be predicted in terms of the other parameters, but this is quite sensitive to model assumptions. For phenomenological studies, m_0^2, $m_{1/2}$, A_0, and B_0 are usually just taken to be convenient but imperfect (and perhaps downright misleading) parameterizations of our ignorance of the supersymmetry-breaking mechanism. In a more perfect world, experimental searches might be conducted and reported using something like the larger 15-dimensional flavor-blind parameter space of eqs. (13.24)–(13.26), but such a higher-dimensional parameter space is difficult to simulate comprehensively, for practical reasons.

Let us now review in a little more detail how the soft supersymmetry-breaking terms can arise in supergravity models. The part of the scalar potential that does not depend on the gauge kinetic function can be found as follows. First, one may define the real, dimensionless *Kähler function* in terms of the Kähler potential and superpotential with the chiral superfields replaced by their scalar components:

$$G = K/M_P^2 + \ln(W/M_P^3) + \ln(W^\dagger/M_P^3). \tag{14.24}$$

Many references use units with $M_P = 1$, which simplifies the expressions but can slightly obscure the correspondence with the global supersymmetry limit of large M_P. From G, one can construct its derivatives with respect to the scalar fields and their complex conjugates: $G^i = \delta G/\delta\phi_i$; $G_i = \delta G/\delta\phi^{*i}$; and $G_i^j = \delta^2 G/\delta\phi^{*i}\delta\phi_j$. As in Section 8.2, raised (lowered) indices i correspond to derivatives with respect to ϕ_i ($\phi^{\dagger i}$). Note that $G_i^j = K_i^j/M_P^2$, which is often called the Kähler metric, does not depend on the superpotential. The inverse of this matrix is denoted $(G^{-1})_i^j$, or equivalently $M_P^2(K^{-1})_i^j$, so that $(G^{-1})_i^k G_k^j = (G^{-1})_k^j G_i^k = \delta_i^j$. In terms of these objects, the generalization of the F-term contribution to the scalar potential in ordinary renormalizable global supersymmetry turns out to be

$$V_F = M_P^4\, e^G \left[G^i (G^{-1})_i^j G_j - 3\right] \tag{14.25}$$

in supergravity. It can be rewritten as

$$V_F = K_i^j F_j F^{\dagger i} - 3e^{K/M_P^2} W W^\dagger/M_P^2, \tag{14.26}$$

where

$$F_i = -M_P^2\, e^{G/2} (G^{-1})_i^j G_j = -e^{K/2M_P^2} (K^{-1})_i^j \left(W_j^\dagger + W^\dagger K_j/M_P^2\right), \tag{14.27}$$

with $K^i = \delta K/\delta\phi_i$ and $K_j = \delta K/\delta\phi^{*j}$. The F_i are order parameters for supersymmetry breaking in supergravity (generalizing the auxiliary fields in the renormalizable global supersymmetry case). In other words, local supersymmetry will be broken if one or more of the F_i obtain a VEV. The gravitino then absorbs the would-be goldstino and obtains a squared mass

$$m_{3/2}^2 = \langle K_j^i F_i F^{*j}\rangle 3M_P^2. \tag{14.28}$$

Taking a minimal Kähler potential $K = \phi^{\dagger i}\phi_i$, one has $K_i^j = (K^{-1})_i^j = \delta_i^j$, so that expanding eqs. (14.26) and (14.27) to lowest order in $1/M_P$ just reproduces the results $F_i = -W_i^\dagger$ and $V = F_i F^{\dagger i} = W^i W_i^\dagger$, which were found in Section 8.2 [see eqs. (8.48)–(8.50)] for renormalizable global supersymmetric theories. Equation (14.28) also reproduces the expression for the gravitino mass that was quoted in eq. (14.7).

The scalar potential eq. (14.25) does not include the D-term contributions from gauge interactions, which are given by

$$V_D = \tfrac{1}{2}\,\mathrm{Re}[f_{ab}\, \widehat{D}^a \widehat{D}^b], \tag{14.29}$$

with $\widehat{D}^a = f_{ab}^{-1}\widetilde{D}^b$, where

$$\widetilde{D}^a \equiv -G^i (T^a)_i{}^j \phi_j = -\phi^{*j}(T^a)_j{}^i G_i = -K^i (T^a)_i{}^j \phi_j = -\phi^{*j}(T^a)_j{}^i K_i \tag{14.30}$$

are real order parameters of supersymmetry breaking, with the last three equalities following from the gauge invariance of W and K. Note that in the tree-level global supersymmetry case where $f_{ab} = \delta_{ab}/g_a^2$ and $K^i = \phi^{\dagger i}$, eq. (14.29) reproduces the result of Section 8.4 for the renormalizable global supersymmetry D-term scalar potential, with $\widehat{D}^a = g_a D^a$ (no sum on a). The full scalar potential, $V = V_F + V_D$, depends on W and K only through the combination G in eq. (14.24). There are many other contributions to the supergravity Lagrangian involving fermions and vectors (e.g., see [B20, B40–B45]), and these also turn out to depend only on f_{ab} and G. This allows one to consistently redefine W and K so that there are no purely holomorphic or purely antiholomorphic terms appearing in the latter.

In contrast to global supersymmetry, the scalar potential in supergravity is *not* necessarily nonnegative, because of the -3 term in eq. (14.25). Therefore, in principle, one can have supersymmetry breaking with a positive, negative, or zero vacuum energy. Recent developments in experimental cosmology imply a positive vacuum energy associated with the acceleration of the scale factor of the observable universe,

$$\rho_{\rm vac}^{\rm observed} = \frac{\Lambda}{8\pi G_N} \approx (2.3 \times 10^{-12}\ {\rm GeV})^4, \tag{14.31}$$

but this is also certainly tiny compared to the scales associated with supersymmetry breaking. Therefore, it is tempting to simply assume that the vacuum energy is 0 within the approximations pertinent for working out the supergravity effects on particle physics at collider energies. However, it is notoriously unclear *why* the terms in the scalar potential in a supersymmetry-breaking vacuum should conspire to give $\langle V \rangle \approx 0$ at the minimum. A naive estimate, without miraculous cancellations, would instead give $\langle V \rangle$ of order $|\langle F \rangle|^2$, so at least roughly $(10^{10}\ {\rm GeV})^4$ for Planck-scale-mediated supersymmetry breaking, or $(10^4\ {\rm GeV})^4$ for gauge-mediated supersymmetry breaking. Furthermore, while $\rho_{\rm vac} = \langle V \rangle$ classically, the former is a very large-distance-scale measured quantity, while the latter is associated with effective field theories at length scales comparable to and shorter than those familiar to high-energy physics. So, in the absence of a compelling explanation for the tiny value of $\rho_{\rm vac}$, it is not at all clear that $\langle V \rangle \approx 0$ is really the right condition to impose. Nevertheless, with $\langle V \rangle = 0$ imposed as a constraint, eqs. (14.26)–(14.28) tell us that $\langle K^i_j F_i F^{*j} \rangle = 3M_P^4 e^{\langle G \rangle} = 3e^{\langle K \rangle/M_P^2} |\langle W \rangle|^2/M_P^2$, and an equivalent formula for the gravitino mass is therefore $m_{3/2} = e^{\langle G \rangle/2} M_P$.

An interesting special case arises if we assume a minimal Kähler potential and divide the fields ϕ_i into a visible sector including the MSSM fields φ_i, and a hidden sector containing a field X that breaks supersymmetry for us (and other fields that we need not treat explicitly). In other words, suppose that the superpotential and the Kähler potential (expressed as functions of the scalar fields) have the forms

$$W = W_{\rm vis}(\varphi_i) + W_{\rm hid}(X), \tag{14.32}$$

$$K = \varphi^{\dagger i}\varphi_i + X^\dagger X. \tag{14.33}$$

Let us further assume that the hidden sector dynamics provides nonzero VEVs,

$$\langle X \rangle = x M_P, \qquad \langle W_{\rm hid} \rangle = w M_P^2, \qquad \langle \delta W_{\rm hid}/\delta X \rangle = w' M_P, \tag{14.34}$$

which define a dimensionless quantity x, and w, w' with dimensions of [mass].

Requiring[5] $\langle V \rangle = 0$ yields $|w' + x^* w|^2 = 3|w|^2$, and

$$m_{3/2} = \frac{|\langle F_X \rangle|}{\sqrt{3} M_P} = e^{|x|^2/2}|w|. \tag{14.35}$$

Now we suppose that it is valid to expand the scalar potential in powers of the dimensionless quantities w/M_P, w'/M_P, φ_i/M_P, etc., keeping only terms that depend on the visible-sector fields φ_i. In leading order the result is

$$\begin{aligned} V = & (W^\dagger_{\text{vis}})_i (W_{\text{vis}})^i + m^2_{3/2} \varphi^{\dagger i} \varphi_i \\ & + e^{|x|^2/2} \left[w^* \varphi_i (W_{\text{vis}})^i + (x^* w'^* + |x|^2 w^* - 3w^*) W_{\text{vis}} + \text{h.c.} \right]. \end{aligned} \tag{14.36}$$

A tricky point here is that we have rescaled the visible sector superpotential $W_{\text{vis}} \to e^{-|x|^2/2} W_{\text{vis}}$ everywhere, in order that the first term in eq. (14.36) is the usual, properly normalized, F-term contribution in global supersymmetry. The next term is a universal soft scalar squared mass of the form eq. (13.28) with

$$m_0^2 = \frac{|\langle F_X \rangle|^2}{3M_P^2} = m^2_{3/2}. \tag{14.37}$$

The second line of eq. (14.36) just gives soft (scalar)^3 and (scalar)^2 holomorphic couplings of the form exhibited in eq. (13.29), with

$$A_0 = -\frac{x^* \langle F_X \rangle}{M_P}, \qquad B_0 = \left(\frac{1}{x + w'^*/w^*} - x^* \right) \frac{\langle F_X \rangle}{M_P}, \tag{14.38}$$

since $\varphi_i (W_{\text{vis}})^i$ is equal to $3W_{\text{vis}}$ for the cubic part of W_{vis}, and to $2W_{\text{vis}}$ for the quadratic part. If the complex phases of x, w, w' can be rotated away, then eq. (14.38) implies $B_0 = A_0 - m_{3/2}$, but there are many effects that can ruin this prediction. The Polonyi model mentioned above is just the special case of this exercise in which W_{hid} is assumed to be linear in X.

However, there is no reason why W and K must have the simple form of eq. (14.32) and eq. (14.33). In general, the superpotential and Kähler potential will have terms coupling X to the MSSM fields as in eqs. (14.11) and (14.12). If one now plugs such terms into eq. (14.25), one obtains a general form like eq. (14.15) for the soft terms. It is only when special assumptions are made [like eqs. (14.32), (14.33)] that one gets the CMSSM boundary conditions [eqs. (13.27)–(13.30)]. That is, supergravity by itself does not guarantee universality or even flavor-blindness of the soft terms.

14.4 Gauge-Mediated SUSY Breaking

In gauge-mediated supersymmetry-breaking (GMSB) models, the ordinary gauge interactions, rather than gravity, are responsible for the appearance of soft supersymmetry breaking in the MSSM.[6] The basic idea is to introduce some new chiral

5 We do this only to follow popular example; as just noted, we cannot endorse this imposition.
6 For a review of GMSB models and a guide to the literature, see Ref. [13].

supermultiplets, called messengers, that couple to the ultimate source of supersymmetry breaking, and also couple indirectly to the (s)quarks and (s)leptons and Higgs(inos) of the MSSM through the ordinary $SU(3)_C\times SU(2)_L\times U(1)_Y$ gauge boson and gaugino interactions. There is still gravitational communication between the MSSM and the source of supersymmetry breaking, of course, but that effect is now relatively unimportant compared to the gauge interaction effects.

In contrast to Planck-scale mediation, GMSB can be understood entirely in terms of loop effects in a renormalizable framework. In the simplest such model, the messenger fields are a set of left-handed chiral supermultiplets q, $\overline{q}$, ℓ, $\overline{\ell}$ transforming under $SU(3)_C\times SU(2)_L\times U(1)_Y$ as

$$q \sim (\mathbf{3},\mathbf{1},-\tfrac{1}{3}),\qquad \overline{q}\sim(\overline{\mathbf{3}},\mathbf{1},\tfrac{1}{3}),\qquad \ell\sim(\mathbf{1},\mathbf{2},\tfrac{1}{2}),\qquad \overline{\ell}\sim(\mathbf{1},\mathbf{2},-\tfrac{1}{2}). \tag{14.39}$$

These supermultiplets contain messenger quarks $\psi_q, \psi_{\overline{q}}$ and scalar quarks $q, \overline{q}$ and messenger leptons $\psi_\ell, \psi_{\overline{\ell}}$ and scalar leptons $\ell, \overline{\ell}$. All of these particles must get very large masses so as not to have been discovered already. Assume they do so by coupling to a gauge-singlet chiral supermultiplet S through a superpotential:

$$W_{\text{mess}} = y_2 S\ell\overline{\ell} + y_3 S q\overline{q}. \tag{14.40}$$

The scalar component of S and its auxiliary (F-term) component are each supposed to acquire VEVs, denoted $\langle S\rangle$ and $\langle F_S\rangle$, respectively. This can be accomplished either by putting S into an O'Raifeartaigh-type model, or by a dynamical mechanism. Exactly how this happens is an interesting and important question, without a clear answer at present. Here, we will simply parameterize our ignorance of the precise mechanism of supersymmetry breaking by asserting that S participates in another part of the superpotential, call it W_{breaking}, which provides for the necessary spontaneous breaking of supersymmetry.

Let us now consider the mass spectrum of the messenger fermions and bosons. The fermionic messenger fields pair up to get mass terms:

$$\mathscr{L} = -y_2\langle S\rangle\psi_\ell\psi_{\overline{\ell}} - y_3\langle S\rangle\psi_q\psi_{\overline{q}} + \text{h.c.}, \tag{14.41}$$

as in eq. (8.51). Meanwhile, their scalar messenger partners $\ell, \overline{\ell}$ and $q, \overline{q}$ have a scalar potential given by (neglecting D-term contributions, which do not affect the following discussion)

$$V = \left|\frac{\delta W_{\text{mess}}}{\delta \ell}\right|^2 + \left|\frac{\delta W_{\text{mess}}}{\delta \overline{\ell}}\right|^2 + \left|\frac{\delta W_{\text{mess}}}{\delta q}\right|^2 + \left|\frac{\delta W_{\text{mess}}}{\delta \overline{q}}\right|^2 + \left|\frac{\delta}{\delta S}(W_{\text{mess}} + W_{\text{breaking}})\right|^2, \tag{14.42}$$

as in eq. (8.50). Now, suppose that, at the minimum of the potential, $\langle S\rangle \neq 0$ and

$$\left\langle \frac{\delta W_{\text{breaking}}}{\delta S}\right\rangle = -\langle F_S^\dagger\rangle \neq 0, \qquad \left\langle \frac{\delta W_{\text{mess}}}{\delta S}\right\rangle = 0. \tag{14.43}$$

Replacing S and F_S by their VEVs, one finds quadratic mass terms in the potential

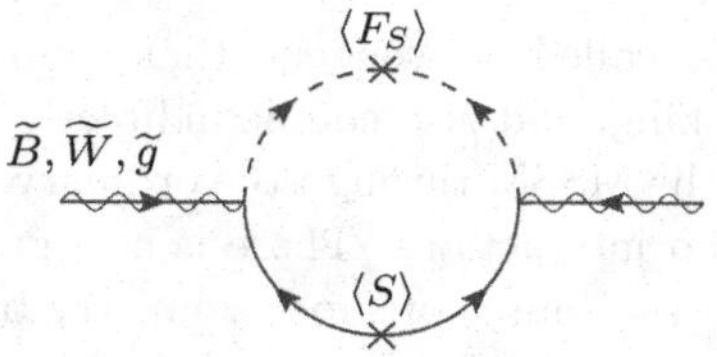

Fig. 14.3 Contributions to the MSSM gaugino masses in gauge-mediated supersymmetry-breaking models come from one-loop graphs involving virtual messenger particles.

for the messenger scalar leptons:

$$V = |y_2 \langle S \rangle|^2 (|\ell|^2 + |\bar{\ell}|^2) + |y_3 \langle S \rangle|^2 (|q|^2 + |\bar{q}|^2) \\ - (y_2 \langle F_S \rangle \ell\bar{\ell} + y_3 \langle F_S \rangle q\bar{q} + \text{h.c.}) + \text{ quartic terms.} \tag{14.44}$$

The first line in eq. (14.44) represents supersymmetric mass terms that go along with eq. (14.41), while the second line consists of soft supersymmetry-breaking masses. The complex scalar messengers $\ell, \bar{\ell}$ thus obtain a squared-mass matrix equal to

$$\begin{pmatrix} |y_2 \langle S \rangle|^2 & -y_2^* \langle F_S^\dagger \rangle \\ -y_2 \langle F_S \rangle & |y_2 \langle S \rangle|^2 \end{pmatrix}, \tag{14.45}$$

with squared-mass eigenvalues $|y_2 \langle S \rangle|^2 \pm |y_2 \langle F_S \rangle|$. In just the same way, the scalars $q, \bar{q}$ get squared masses $|y_3 \langle S \rangle|^2 \pm |y_3 \langle F_S \rangle|$.

So far, we have found that the effect of supersymmetry breaking is to split each messenger supermultiplet pair apart:

$$\ell, \bar{\ell}: \qquad m^2_{\text{fermions}} = |y_2 \langle S \rangle|^2, \qquad m^2_{\text{scalars}} = |y_2 \langle S \rangle|^2 \pm |y_2 \langle F_S \rangle|, \tag{14.46}$$

$$q, \bar{q}: \qquad m^2_{\text{fermions}} = |y_3 \langle S \rangle|^2, \qquad m^2_{\text{scalars}} = |y_3 \langle S \rangle|^2 \pm |y_3 \langle F_S \rangle|. \tag{14.47}$$

The supersymmetry violation apparent in this messenger spectrum for $\langle F_S \rangle \neq 0$ is communicated to the MSSM sparticles through radiative corrections. The MSSM gauginos obtain masses from the one-loop Feynman diagram shown in Fig. 14.3. The scalar and fermion lines in the loop are messenger fields. Recall that the interaction vertices in Fig. 14.3 are of gauge-coupling strength even though they do not involve gauge bosons; compare Fig. 8.3(g). In this way, gauge mediation provides that $q, \bar{q}$ messenger loops give masses to the gluino and the bino, and $\ell, \bar{\ell}$ messenger loops give masses to the wino and bino fields. Computing the one-loop diagrams, one finds that the resulting MSSM gaugino masses are given by

$$M_a = \frac{\alpha_a}{4\pi} \Lambda, \qquad (a = 1, 2, 3), \tag{14.48}$$

in the GUT normalization for α_a discussed in Section 13.6, where we have introduced a mass parameter

$$\Lambda \equiv \langle F_S \rangle / \langle S \rangle. \tag{14.49}$$

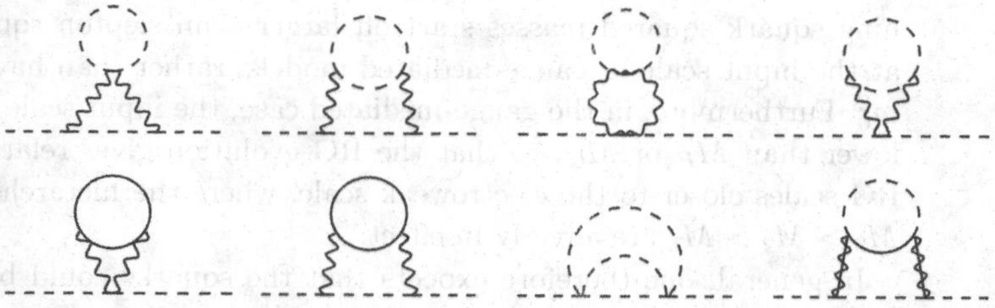

Fig. 14.4 Contributions to MSSM scalar squared masses in gauge-mediated supersymmetry-breaking models arise in leading order from these two-loop Feynman graphs.

(Note that if $\langle F_S \rangle$ were 0, then $\Lambda = 0$ and the messenger scalars would be degenerate with their fermionic superpartners and there would be no contribution to the MSSM gaugino masses.) In contrast, the corresponding MSSM gauge bosons cannot get a corresponding mass shift, since they are protected by gauge invariance. So supersymmetry breaking has been successfully communicated to the MSSM ("visible sector"). To a good approximation, eq. (14.48) holds for the running gaugino masses at an RG scale Q_0 corresponding to the average characteristic mass of the heavy messenger particles, roughly of order $M_{\text{mess}} \sim y_I \langle S \rangle$ for $I = 2, 3$. The running mass parameters can then be RG-evolved down to the electroweak scale to predict the physical masses to be measured by future experiments.

The scalars of the MSSM do not get any radiative corrections to their masses at one-loop order. The leading contribution to their masses comes from the two-loop graphs shown in Fig. 14.4, with the messenger fermions (solid lines) and messenger scalars (dashed lines) and ordinary gauge bosons and gauginos (superimposed solid and wavy lines) running around the loops. By computing these graphs, one finds that each MSSM scalar ϕ_i gets a squared mass given by

$$m_{\phi_i}^2 = 2\Lambda^2 \left[\left(\frac{\alpha_3}{4\pi}\right)^2 C_3(i) + \left(\frac{\alpha_2}{4\pi}\right)^2 C_2(i) + \left(\frac{\alpha_1}{4\pi}\right)^2 C_1(i) \right], \tag{14.50}$$

with the quadratic Casimir invariants $C_a(i)$. The squared masses in eq. (14.50) are positive (fortunately!).

Moreover, the parameterization given in eqs. (13.159)–(13.163) still holds, but m_0^2 is always 0. At the input scale Q_0, each MSSM scalar gets contributions to its squared mass that depend only on its gauge interactions, as in eq. (14.50). It is not hard to see that in general these contribute in exactly the same pattern as K_1, K_2, and K_3 in eqs. (13.159)–(13.163). The subsequent evolution of the scalar squared masses down to the electroweak scale again just yields more contributions to the K_1, K_2, and K_3 parameters. It is somewhat more difficult to give meaningful numerical estimates for these parameters in gauge-mediated models than in mSUGRA models without knowing the messenger mass scale(s) and the multiplicities of the messenger fields. However, in the gauge-mediated case one quite generally expects that the numerical values of the ratios K_3/K_2, K_3/K_1, and K_2/K_1 should be even larger than in eq. (13.165). There are two reasons for this. First, the running squark squared masses start off larger than slepton squared masses already at the input scale in gauge-mediated models, rather than having a common value m_0^2. Furthermore, in the gauge-mediated case, the input scale Q_0 is typically much lower than M_P or M_U, so that the RG evolution gives relatively more weight to RG scales closer to the electroweak scale, where the hierarchies $g_3 > g_2 > g_1$ and $M_3 > M_2 > M_1$ are already in effect.

In general, one therefore expects that the squarks could be considerably heavier than the sleptons, with the effect being more pronounced in gauge-mediated supersymmetry-breaking models than in mSUGRA models. For any specific choice of model, this effect can be easily quantified with a numerical RG analysis. The hierarchy $m_{\text{squark}} > m_{\text{slepton}}$ tends to hold even in many models that do not fit into any of the categories outlined in this chapter, because the RG contributions to squark masses from the gluino are always present and usually quite large, since QCD has a larger gauge coupling than the electroweak interactions.

The terms $\boldsymbol{a_u}$, $\boldsymbol{a_d}$, $\boldsymbol{a_e}$ arise first at two-loop order, and are suppressed by an extra factor of $\alpha_a/4\pi$ compared to the gaugino masses. So, to a very good approximation one has, at the messenger scale,

$$\boldsymbol{a_u} = \boldsymbol{a_d} = \boldsymbol{a_e} = 0, \tag{14.51}$$

a significantly stronger condition than eq. (13.25). Again, eqs. (14.50) and (14.51) should be applied at an RG scale equal to the average mass of the messenger fields running in the loops. However, evolving the RGEs down to the electroweak scale generates nonzero $\boldsymbol{a_u}$, $\boldsymbol{a_d}$, and $\boldsymbol{a_e}$ proportional to the corresponding Yukawa matrices and the nonzero gaugino masses, as indicated in Section 13.7. These will only be large for the third-generation squarks and sleptons, in the approximation of eq. (13.7). The parameter b may also be taken to vanish near the messenger scale, but this is quite model-dependent, and in any case b will be nonzero when it is RG-evolved to the electroweak scale. In practice, b can be fixed in terms of the other parameters by the requirement of correct electroweak symmetry breaking, as discussed in Section 13.8.

Because the gaugino masses arise at *one*-loop order and the scalar squared-mass contributions appear at *two*-loop order, both eq. (14.48) and eq. (14.50) correspond to the estimate eq. (14.3) for m_{soft}, with $M_{\text{mess}} \sim y_I\langle S\rangle$. Equations (14.48) and (14.50) hold in the limit of small $\langle F_S\rangle/y_I\langle S\rangle^2$, corresponding to mass splittings within each messenger supermultiplet that are small compared to the overall messenger mass scale. The subleading corrections in an expansion in $\langle F_S\rangle/y_I\langle S\rangle^2$ turn out to be quite small unless there are very large messenger mass splittings.

The model we have described so far is often called the minimal model of gauge-mediated supersymmetry breaking. Let us now generalize it to a more complicated messenger sector. Suppose that $q, \overline{q}$ and $\ell, \overline{\ell}$ are replaced by a collection of messengers (labeled by I) and denoted by $\Phi_I, \overline{\Phi}_I$, with a superpotential

$$W_{\text{mess}} = \sum_I y_I S \Phi_I \overline{\Phi}_I. \tag{14.52}$$

The bar is used to indicate that the left-handed chiral superfields $\overline{\Phi}_I$ transform as the complex conjugate representations of the left-handed chiral superfields Φ_I. Together they are said to form a "vectorlike" (self-conjugate) representation of the Standard Model gauge group. As before, the fermionic components of each pair Φ_I and $\overline{\Phi}_I$ pair up to get squared masses $|y_I\langle S\rangle|^2$ and their scalar partners mix to get squared masses $|y_I\langle S\rangle|^2 \pm |y_I\langle F_S\rangle|$. The MSSM gaugino mass parameters induced are now

$$M_a = \frac{\alpha_a}{4\pi}\Lambda \sum_I n_a(I) \qquad (a = 1, 2, 3), \tag{14.53}$$

where $n_a(I)$ is the Dynkin index for each $\Phi_I + \overline{\Phi}_I$ pair, in a normalization where $n_3 = 1$ for a $\mathbf{3} + \overline{\mathbf{3}}$ of $SU(3)_C$ and $n_2 = 1$ for a pair of doublets of $SU(2)_L$. For $U(1)_Y$, one has $n_1 = 6Y^2/5$ for each messenger pair with weak hypercharges $\pm Y$. In computing n_1 one must remember to add up the contributions for each component of an $SU(3)_C$ or $SU(2)_L$ multiplet. So, for example, $(n_1, n_2, n_3) = (2/5, 0, 1)$ for $q + \overline{q}$ and $(n_1, n_2, n_3) = (3/5, 1, 0)$ for $\ell + \overline{\ell}$. Thus the total is $\sum_I (n_1, n_2, n_3) = (1, 1, 1)$ for the minimal model, so that eq. (14.53) is in agreement with eq. (14.48). On general group-theoretic grounds, n_2 and n_3 must be integers, and n_1 is always an integer multiple of 1/5 if fractional electric charges are confined.

The MSSM scalar masses in this generalized gauge-mediation framework are now

$$m^2_{\phi_i} = 2\Lambda^2 \left[\left(\frac{\alpha_3}{4\pi}\right)^2 C_3(i) \sum_I n_3(I) + \left(\frac{\alpha_2}{4\pi}\right)^2 C_2(i) \sum_I n_2(I) + \left(\frac{\alpha_1}{4\pi}\right)^2 C_1(i) \sum_I n_1(I) \right]. \tag{14.54}$$

In writing eqs. (14.53) and (14.54) as simple sums, we have implicitly assumed that the messengers are all approximately equal in mass, with $M_{\text{mess}} \approx y_I\langle S\rangle$. Equation (14.54) is still not a bad approximation if the y_I are not very different from each other, because the dependence of the MSSM mass spectrum on the y_I is only logarithmic (due to RG running) for fixed Λ. However, if large hierarchies in the messenger masses are present, then the additive contributions to the gaugino masses and scalar squared masses from each individual messenger multiplet I should really instead be incorporated at the mass scale of that messenger multiplet. Then RG evolution is used to run these various contributions down to the electroweak or TeV scale; the individual messenger contributions to scalar and gaugino masses as indicated above can be thought of as threshold corrections to this RG running.

Messengers with masses far below the GUT scale will affect the running of gauge couplings and might therefore be expected to ruin the apparent unification shown in Fig. 6.1. However, if the messengers come in complete multiplets of the SU(5) global symmetry[7] that contains the Standard Model gauge group, and are not very different in mass, then approximate unification of gauge couplings will still occur

[7] This SU(5) may or may not be promoted to a local gauge symmetry at the GUT scale. For our present purposes, it is used only as a classification scheme, since the global SU(5) symmetry is only approximate in the effective theory at the (much lower) messenger mass scale where gauge mediation takes place.

when they are extrapolated up to the same scale M_U (but with a larger unified value for the gauge couplings at that scale). For this reason, a popular class of models is obtained by taking the messengers to consist of N_5 copies of the $\mathbf{5}+\overline{\mathbf{5}}$ of SU(5), resulting in

$$\sum_I n_1(I) = \sum_I n_2(I) = \sum_I n_3(I) = N_5\,. \tag{14.55}$$

Equations (14.53) and (14.54) then reduce to

$$M_a = \frac{\alpha_a}{4\pi}\Lambda N_5, \tag{14.56}$$

$$m^2_{\phi_i} = 2\Lambda^2 N_5 \sum_{a=1}^{3} C_a(i) \left(\frac{\alpha_a}{4\pi}\right)^2, \tag{14.57}$$

since now there are N_5 copies of the minimal messenger sector particles running around the loops. For example, the minimal model in eq. (14.39) corresponds to $N_5 = 1$. A single copy of $\mathbf{10}+\overline{\mathbf{10}}$ of SU(5) has a Dynkin index of $\sum_I n_a(I) = 3$, and thus can be substituted for three copies of $\mathbf{5}+\overline{\mathbf{5}}$. (Other combinations of messenger multiplets can also preserve the apparent unification of gauge couplings.) Note that the gaugino masses scale like N_5, while the scalar masses scale like $\sqrt{N_5}$. This means that sleptons and squarks will tend to be lighter relative to the gauginos for larger values of N_5 in nonminimal models. However, if N_5 is too large, then the running gauge couplings will diverge before they can unify at M_U. For messenger masses of order 10^6 GeV or less, for example, one needs $N_5 \leq 4$.

There are many other possible generalizations of the basic gauge-mediation scenario described above. An important general expectation in these models is that the strongly interacting sparticles (squarks, gluino) should be heavier than weakly interacting sparticles (sleptons, bino, winos), simply because of the hierarchy of gauge couplings $\alpha_3 > \alpha_2 > \alpha_1$. The common feature that makes all of these models attractive is that the masses of the squarks and sleptons depend only on their gauge quantum numbers, leading automatically to the degeneracy of squark and slepton masses needed for suppression of flavor-changing effects. But the most distinctive phenomenological prediction of gauge-mediated models may be the fact that the gravitino is the LSP. This prediction has crucial consequences for cosmology (since the gravitino could be the dark matter) and collider physics, as discussed further in Chapter 15.

14.5 Extra-Dimensional and Anomaly-Mediated SUSY Breaking

It is also possible to take the partitioning of the MSSM and supersymmetry-breaking sectors shown in Fig. 14.1 seriously as geography. This can be accomplished by assuming that there are extra spatial dimensions of the Kaluza–Klein or

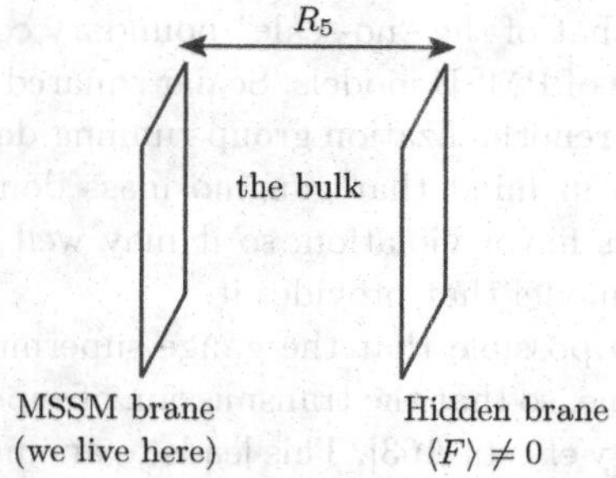

Fig. 14.5 The separation of the supersymmetry-breaking sector from the MSSM sector could take place along a hidden spatial dimension, as in the simple example shown here. The branes are four-dimensional parallel spacetime hypersurfaces in a five-dimensional spacetime.

warped type, so that a physical distance separates the visible and hidden[8] sectors. This general idea opens up numerous possibilities, which are hard to classify in a detailed way. For example, string theory suggests six such extra dimensions, with a staggeringly huge number of possible solutions.

Many of the more popular models used to explore this extra-dimensional mediated supersymmetry breaking[9] (XMSB) employ just one single hidden extra dimension with the MSSM chiral supermultiplets confined to one four-dimensional spacetime brane and the supersymmetry-breaking sector confined to a parallel brane a distance R_5 away, separated by a five-dimensional bulk, as in Fig. 14.5. Using this as an illustration, the dangerous flavor-violating terms proportional to y^{Xijk} and k^i_j in eq. (14.15) are suppressed by factors like $e^{-R_5 M_5}$, where R_5 is the size of the fifth dimension and M_5 is the five-dimensional fundamental (Planck) scale, and it is assumed that the MSSM chiral supermultiplets are confined to their brane. Therefore, it should be enough to require that $R_5 M_5 \gg 1$, in other words that the size of the fifth dimension (or, more generally, the volume of the compactified space) is relatively large in units of the fundamental length scale. Thus the suppression of flavor-violating effects does not require any fine-tuning or extreme hierarchies, because it is exponential.

One possibility is that the gauge supermultiplets of the MSSM propagate in the bulk, and so mediate supersymmetry breaking. This mediation is direct for gauginos, with

$$M_a \sim \frac{\langle F\rangle}{M_5(R_5 M_5)}, \tag{14.58}$$

but is loop-suppressed for the soft terms involving scalars. This implies that, in the simplest version of the idea, often called "gaugino mediation," soft supersymmetry breaking is dominated by the gaugino masses. The phenomenology is therefore quite

[8] The name "sequestered" is often used instead of "hidden" in this context.
[9] See, e.g., Refs. [22, 23] for reviews of various models of XMSB, with some pointers to the literature.

similar to that of the "no-scale" boundary conditions mentioned in Section 14.3 in the context of PMSB models. Scalar squared masses and the scalar cubic couplings come from renormalization group running down to the electroweak scale. It is useful to keep in mind that gaugino mass dominance is really the essential feature that defeats flavor violation, so it may well turn out to be more robust than any particular model that provides it.

It is also possible that the gauge supermultiplet fields are also confined to the MSSM brane, so that the transmission of supersymmetry breaking is due entirely to supergravity effects [163]. This leads to anomaly-mediated supersymmetry breaking (AMSB), so-named because the resulting MSSM soft terms can be understood in terms of the anomalous violation of a local superconformal invariance, an extension of scale invariance. In one formulation of supergravity, Newton's gravitational constant (or equivalently, the Planck mass scale) is set by the VEV of a scalar field ϕ that is part of a nondynamical chiral supermultiplet (called the "conformal compensator"). As a gauge fixing, this field obtains a VEV of $\langle\phi\rangle = 1$, spontaneously breaking the local superconformal invariance. Now, in the presence of spontaneous supersymmetry breaking $\langle F\rangle \neq 0$, for example on the hidden brane, the auxiliary field component also obtains a nonzero VEV, with

$$\langle F_\phi\rangle \sim \langle F\rangle/M_P \sim m_{3/2}. \tag{14.59}$$

The nondynamical conformal compensator field ϕ is taken to be dimensionless, so that F_ϕ has dimensions of [mass].

In the classical limit, there is still no supersymmetry breaking in the MSSM sector, due to the exponential suppression provided by the extra dimensions.[10] However, there is an anomalous violation of superconformal (scale) invariance manifested in the running of the couplings. This causes supersymmetry breaking to show up in the MSSM by virtue of the nonzero beta functions and anomalous dimensions of the MSSM brane couplings and fields. The resulting soft terms are (using F_ϕ to denote its VEV from now on)

$$M_a = F_\phi \beta_{g_a}/g_a, \tag{14.60}$$

$$(m^2)^i_j = \tfrac{1}{2}|F_\phi|^2 \frac{d}{dt}\gamma^i_j = \tfrac{1}{2}|F_\phi|^2 \left[\beta_{g_a}\frac{\partial}{\partial g_a} + \beta_{y^{kmn}}\frac{\partial}{\partial y^{kmn}} + \beta_{y^*_{kmn}}\frac{\partial}{\partial y^*_{kmn}}\right]\gamma^i_j, \tag{14.61}$$

$$a^{ijk} = -F_\phi \beta_{y^{ijk}}, \tag{14.62}$$

where the anomalous dimensions γ^i_j are normalized as in eqs. (11.138)–(11.141) in the general case and eqs. (13.46)–(13.52) in the MSSM. As in the GMSB scenario treated in Section 14.4, the gaugino masses arise at one-loop order, but scalar squared masses arise at two-loop order. Also, these results are approximately flavor-blind for the first two generations, because the nontrivial flavor structure derives only from the MSSM Yukawa couplings.

[10] Anomaly-mediated supersymmetry breaking can also be realized without invoking extra dimensions. See, e.g., Refs. [164, 165] for examples of viable four-dimensional AMSB models. Theoretical aspects of the AMSB mechanism in four dimensions are clarified in Ref. [166, 167].

There are several unique features of the AMSB scenario. First, there is no need to specify at which renormalization scale eqs. (14.60)–(14.62) should be applied as boundary conditions. This is because they hold at every renormalization scale, exactly, to all orders in perturbation theory. In other words, eqs. (14.60)–(14.62) are not just boundary conditions for the renormalization group equations of the soft parameters, but solutions as well. (These AMSB renormalization group trajectories can also be found from this renormalization group invariance property alone, without reference to the supergravity derivation.) In fact, even if there are heavy supermultiplets in the theory that have to be decoupled, the boundary conditions hold both above and below the arbitrary decoupling scale. This remarkable insensitivity to ultraviolet physics in AMSB ensures the absence of flavor violation in the low-energy MSSM soft terms. Another interesting prediction is that the gravitino mass $m_{3/2}$ in these models is actually much larger than the scale $m_{\rm soft}$ of the MSSM soft terms, since the latter are loop-suppressed compared to eq. (14.59).

There is only one unknown parameter, F_ϕ, among the MSSM soft terms in AMSB. Unfortunately, this exemplary falsifiability is marred by the fact that it is already falsified. The dominant contributions to the first-generation squark and slepton squared masses are

$$m^2_{\tilde{q}} = \frac{|F_\phi|^2}{(16\pi^2)^2}\left(8g_3^4 + \cdots\right), \tag{14.63}$$

$$m^2_{\tilde{e}_L} = -\frac{|F_\phi|^2}{(16\pi^2)^2}\left(\frac{3}{2}g_2^4 + \frac{99}{50}g_1^4\right), \tag{14.64}$$

$$m^2_{\tilde{e}_R} = -\frac{|F_\phi|^2}{(16\pi^2)^2}\frac{198}{25}g_1^4 . \tag{14.65}$$

The squarks have large positive squared masses, but the sleptons have negative squared masses, so the AMSB model in its simplest form is not viable. These signs come directly from those of the beta functions of the strong and electroweak gauge interactions, as can be seen from the right-hand side of eq. (14.61).

The characteristic ultraviolet insensitivity to physics at high mass scales also makes it somewhat nontrivial to modify the theory to escape this tachyonic slepton problem by deviating from the AMSB trajectory. There can be large deviations from AMSB provided by supergravity, but then in general the flavor-blindness is also forfeit. One way to modify AMSB is to introduce additional supermultiplets that contain supersymmetry-breaking mass splittings that are large compared to their average mass. Another way is to combine AMSB with gaugino mediation. Finally, there is a perhaps less motivated approach in which a common parameter m_0^2 is added to all of the scalar squared masses at some scale, and chosen large enough to allow the sleptons to have positive squared masses above LHC bounds. This allows the phenomenology to be studied in a framework conveniently parameterized by

$$F_\phi,\ m_0^2,\ \tan\beta,\ \arg(\mu), \tag{14.66}$$

with $|\mu|$ and b determined by requiring correct electroweak symmetry breaking, as described in the next section. (Some sources use $m_{3/2}$ or $M_{\rm aux}$ to denote F_ϕ.) The

MSSM gaugino masses at the leading nontrivial order are unaffected by the ad hoc addition of m_0^2:

$$M_1 = \frac{F_\phi}{16\pi^2}\frac{33}{5}g_1^2\,, \qquad M_2 = \frac{F_\phi}{16\pi^2}g_2^2\,, \qquad M_3 = -\frac{F_\phi}{16\pi^2}3g_3^2\,. \tag{14.67}$$

This implies that $|M_2| \ll |M_1| \ll |M_3|$, so the lightest neutralino is actually mostly wino, with a lightest chargino that is only of order 200 MeV heavier, depending on the values of μ and $\tan\beta$. The decay $\widetilde{C}_1^\pm \to \widetilde{N}_1\pi^\pm$ produces a very soft pion, implying unique and difficult signatures in colliders.

Another large general class of models breaks supersymmetry using the geometric or topological properties of the extra dimensions. In the Scherk–Schwarz mechanism [168], the symmetry is broken by assuming different boundary conditions for the fermion and boson fields on the compactified space. In supersymmetric models where the size of the extra dimension is parameterized by a modulus (a massless or nearly massless excitation) called a radion, the F-term component of the radion chiral supermultiplet can obtain a VEV, which becomes a source for supersymmetry breaking in the MSSM. These two ideas turn out to be often related. These mechanisms can also be combined with gaugino mediation and AMSB. It seems likely that all the possibilities are not yet fully explored.

14.6 Relating the μ-Term to the SUSY-Breaking Mechanism

The MSSM superpotential [eq. (13.6)] contains only one dimensionful parameter, μ. In order to obtain the observed electroweak symmetry-breaking scale from the minimization of the Higgs potential without a miraculous cancellation, one expects $|\mu|^2$ to be of the same order as the soft supersymmetry-breaking parameters $m^2_{H_u}$ and $m^2_{H_d}$. This presents a puzzle, because μ is a supersymmetric parameter with no connection to supersymmetry breaking in the MSSM, and so there is no reason for these parameters to be commensurate in magnitude. This strongly suggests that the MSSM should be extended in such a way that the effective value of μ is related to the supersymmetry-breaking mechanism, so that it will naturally have the same order of magnitude as the MSSM soft supersymmetry-breaking masses.

Several different solutions to this "μ problem" have been suggested where the μ-term is assumed to be absent at tree level, perhaps due to an approximate Peccei–Quinn (PQ) symmetry of the underlying theory. In such a case, μ arises in the effective theory from the VEV of some new field denoted by S. In Section 16.1, we will examine the Next-to-Minimal Supersymmetric Standard Model, whose superpotential contains the term $\lambda S H_u H_d$. Then, an effective μ parameter, $\mu_{\text{eff}} = \lambda\langle S\rangle$, is generated when S acquires a nonzero VEV. In this section, we focus instead on two particular mechanisms that involve higher-dimensional operators. These higher-dimensional operators are generated by the fundamental supersymmetry-breaking

Table 14.2 Peccei–Quinn (PQ) charges of chiral superfields. These charges are not unique, as one can add to them any linear combination of the weak hypercharge and $B-L$.

	H_u	H_d	Q	L	$\overline{u}$	$\overline{d}$	$\overline{e}$
Peccei–Quinn charge	+1	+1	−1	−1	0	0	0

dynamics, which ultimately provides the connection between the μ parameter and the effective scale of supersymmetry breaking. The first example is the Kim–Nilles mechanism [169], in which μ arises from a nonrenormalizable superpotential term in which H_uH_d couples to S^2. The second example is the Giudice–Masiero mechanism [170], in which μ arises from a nonrenormalizable Kähler potential term.

The PQ charge assignments for the chiral superfields of the MSSM appear in Table 14.2 (cf. footnote 18 of Section 6.2.4).[11] This cannot be an exact symmetry of the Lagrangian, since it has an $\mathrm{SU}(3)_C$ anomaly. However, if all other sources of PQ symmetry breaking are small, then there must result a pseudo-Goldstone boson, the axion.[12] If the scale of the breaking is too low, then the axion would be ruled out by astrophysical observations, so one must introduce an additional explicit breaking of the PQ symmetry. On the other hand, if the scale of PQ symmetry breaking is such that the axion decay constant (of order $\langle S\rangle$) is in the range

$$10^9 \text{ GeV} \lesssim f \lesssim 10^{12} \text{ GeV}, \tag{14.68}$$

then the resulting invisible axion is consistent with present astrophysical constraints.[13] (In this context, "invisible" refers to the $\mathcal{O}(\Lambda_{\mathrm{QCD}}/f)$ suppression of the axion couplings to the gauge bosons and fermions of the Standard Model.) This is an enticing possibility, since it links the solution to the strong CP problem to supersymmetry breaking. This naturally occurs in the Kim–Nilles and Giudice–Masiero mechanisms, with $f \sim \langle S\rangle \sim \sqrt{m_{\mathrm{soft}}M_P}$ in both cases.

14.6.1 The Kim–Nilles Mechanism: μ from a Nonrenormalizable Superpotential

Another way to obtain μ from the expectation value of a scalar field follows by assuming a nonrenormalizable superpotential, for example:

$$W = \frac{\lambda_\mu}{2M_P} S^2 H_u H_d, \tag{14.69}$$

[11] The PQ charges are not unique, as one can add to them any multiple of the weak hypercharge or $B-L$.

[12] For a review of axion models that provide a solution to the strong CP problem, see, e.g., Ref. [44].

[13] These bounds depend on the details of the model, and are set by constraints on anomalous stellar cooling on the low end and cosmological dark matter density on the high end.

where S is an $SU(3)_C \times SU(2)_L \times U(1)_Y$ singlet chiral superfield with PQ charge -1 (which ensures that the superpotential W respects the PQ symmetry in light of the PQ charge assignments given in Table 14.2), and λ_μ is a dimensionless coupling normalized by the Planck mass M_P. If S acquires a VEV that is parametrically of order

$$\langle S \rangle \sim \sqrt{m_{\text{soft}} M_P}, \tag{14.70}$$

then the spontaneous breaking of the PQ symmetry gives rise to an invisible axion with a decay constant f that will automatically be in the range of eq. (14.68). The low-energy effective theory will then contain the usual μ-term, with

$$\mu = \frac{\lambda_\mu}{2M_P} \langle S^2 \rangle \sim m_{\text{soft}}, \tag{14.71}$$

simultaneously solving the μ problem and the strong CP problem [169]. It is natural to also have a dimensionless, holomorphic soft supersymmetry-breaking term in the Lagrangian of the form

$$-\mathscr{L}_{\text{soft}} = \frac{a_b}{M_P} S^2 H_u H_d + \text{h.c.}, \tag{14.72}$$

where a_b is of order m_{soft}. The b-term in the MSSM will then arise as

$$b = \frac{a_b}{M_P} \langle S^2 \rangle, \tag{14.73}$$

and will be of order m^2_{soft}, as required for electroweak symmetry breaking.

To achieve the required spontaneous breaking, one can introduce an additional nonrenormalizable superpotential term, in several possible ways. For example, one could take

$$W = \frac{\lambda_S}{4M_P} S^2 \overline{S}^2, \tag{14.74}$$

where $\overline{S}$ is a chiral superfield with PQ charge $+1$. This implies a scalar potential that stabilizes S and $\overline{S}$ at large field strength:

$$V_S = |F_S|^2 + |F_{\overline{S}}|^2 = \frac{|\lambda_S|^2}{4M_P^2} |S\overline{S}|^2 (|S|^2 + |\overline{S}|^2). \tag{14.75}$$

There is also a soft supersymmetry-breaking Lagrangian:

$$-\mathscr{L}_{\text{soft}} = V_{\text{soft}} = m_S^2 |S|^2 + m_{\overline{S}}^2 |\overline{S}|^2 - \left(\frac{a_S}{4M_P} S^2 \overline{S}^2 + \text{h.c.} \right), \tag{14.76}$$

where m_S^2 and $m_{\overline{S}}^2$ are of order m^2_{soft} and a_S is of order m_{soft}. The total scalar potential $V_S + V_{\text{soft}}$ will have an appropriate VEV of order eq. (14.70) provided that m_S^2, $m_{\overline{S}}^2$ are negative or if a_S is sufficiently large. For example, with $m_S^2 = m_{\overline{S}}^2$ for simplicity, there will be a nontrivial minimum of the potential if $|a_S|^2 - 12 m_S^2 |\lambda_S|^2 > 0$, and it will be a global minimum of the potential if $|a_S|^2 - 16 m_S^2 |\lambda_S|^2 > 0$.

One pseudoscalar degree of freedom, a mixture of S and $\overline{S}$, is the axion, with a very small mass. The rest of the chiral supermultiplet from which the axion came will have masses of order m_{soft}, but couplings to the MSSM that are highly suppressed. However, if one of the fermionic members of this chiral supermultiplet (a singlino that can be properly called an "axino" $\widetilde{a}$, and which has tiny mixing with the MSSM neutralinos $\widetilde{N}_i$) is lighter than all of the MSSM odd R-parity particles, then it could be the LSP dark matter. Its relic density $\Omega_{\text{DM}}h^2$ today can be obtained from that of the would-be LSP $\widetilde{N}_1$, but suppressed by a factor of $m_{\widetilde{a}}/m_{\widetilde{N}_1}$. It is also possible that the decay of $\widetilde{N}_1$ to $\widetilde{a}$ could occur within a collider detector, rarely and with a macroscopic decay length but just often enough to provide a signal in a sufficiently large sample of superpartner pair production events.

There are several variations on the theme given above. The nonrenormalizable superpotential could be instead of the schematic form such as $S^3\overline{S} + S\overline{S}H_uH_d$, or $S^3\overline{S} + S^2H_uH_d$, or $S\overline{S}^3 + S^2H_uH_d$, each entailing a different assignment of PQ charges, but with qualitatively similar behavior. One can also introduce more than two new fields that break the PQ symmetry at the intermediate scale.

14.6.2 The Giudice–Masiero Mechanism: μ from a Nonrenormalizable Kähler Potential

The μ-term could also arise from a nonrenormalizable contribution to the superpotential in addition to the usual canonical terms for the Higgs fields:

$$K = H_u^\dagger H_u + H_d^\dagger H_d + \left(\frac{\lambda_\mu}{M_P} H_u H_d S^\dagger + \text{h.c.}\right) + \cdots . \tag{14.77}$$

Here S has PQ charge $+1$, and is a field responsible for spontaneous breaking of supersymmetry through its auxiliary F field. Giudice and Masiero [170] showed that in supergravity, the presence of such couplings in the Kähler potential will always give rise to a nonzero μ, and that the natural order of magnitude for μ is then m_{soft}. The B-term arises similarly with order of magnitude m_{soft}^2. The actual values of μ and b depend on contributions to the full superpotential and Kähler potential involving the hidden-sector fields including S (see Ref. [170] for details). These terms do not have any other direct effect on phenomenology, so without faith in a complete underlying theory it will be difficult to correlate them with future experimental results.

A different way of understanding the origin of the μ-term in this class of models is to consider a low-energy effective theory below M_P involving a nonrenormalizable Kähler potential term of the form in eq. (14.77). Even if not present in the fundamental theory, this term could arise from radiative corrections [171]. If the auxiliary field for S obtains a VEV, then one obtains

$$\mu = \frac{\lambda_\mu}{M_P}\langle F_S^\dagger\rangle. \tag{14.78}$$

This will be of the correct order of magnitude if parametrically $\langle F_S^\dagger\rangle \sim m_{\text{soft}}M_P$, which is indeed the typical size assigned to the F-terms of the hidden sector in Planck-scale-mediated models of supersymmetry breaking. The B-term in the soft supersymmetry-breaking sector at low energies could arise in this effective field theory picture from Kähler potential terms of the form

$$K = \frac{\lambda_b}{M_P^2} X^\dagger Y H_u H_d \,, \tag{14.79}$$

where $\langle F_X^\dagger \rangle \sim \langle F_Y \rangle \sim m_{\rm soft} M_P$. (One could identify both of the fields X, Y with S, but that violates the PQ symmetry.) However, this is not necessary, because with $\mu \neq 0$, the low-energy nonzero value of b will arise from threshold effects and renormalization group running.

Exercises

14.1 In the case of the scalar potential for supergravity with a minimal Kähler potential as in eq. (14.33), show that eq. (14.36) follows from the expansion mentioned in the text, requiring $\langle V \rangle = 0$, followed by the rescaling of the visible sector superpotential according to $W_{\rm vis} \to e^{-|x|^2/2} W_{\rm vis}$.

14.2 For GMSB models discussed in Section 14.4, calculate the contribution to the gaugino masses from the Feynman diagram shown in Fig. 14.3. (You may set the external momentum to zero, since the gauginos are much lighter than the messengers in the loop.) In particular:

(a) For the general GMSB model with $|\langle F_S \rangle| \ll |y_i \langle S \rangle^2|$ for each messenger supermultiplet, show that the result for the gaugino masses is eq. (14.53).

(b) Show that if $x_i \equiv |\langle F_S \rangle|/|y_i \langle S \rangle^2|$ is not small, then the result from the loop integration (still treating the external momentum as zero) generalizes to

$$M_a = \frac{\alpha_a}{4\pi} \Lambda \sum_I n_a(I) g(x_I), \tag{14.80}$$

where

$$g(x) = \frac{(1+x)\ln(1+x) + (1-x)\ln(1-x)}{x^2} \,. \tag{14.81}$$

Note that $g(0) = 1$.

14.3 In the case of AMSB models with soft terms given by eqs. (14.60)–(14.62), work out the full results for the MSSM scalar squared-mass parameters, generalizing the results of eqs. (14.63)–(14.65). In particular, include the effects of all three gauge groups for the different types of squarks, and the effects of scalar cubic couplings for stops, sbottoms, and staus. Work to leading order; note that this means one-loop order in both the anomalous dimension and the beta functions.

15 Supersymmetric Phenomenology

So far, the experimental study of supersymmetry has unfortunately been confined to setting limits. As noted in Section 13.6, there can be indirect signals for supersymmetry from processes that are rare or forbidden in the Standard Model but have contributions from sparticle loops. These include $\mu \to e\gamma$, $b \to s\gamma$, neutral meson mixing, electric dipole moments for the neutron and the electron, etc. Virtual sparticle effects can also modify the Standard Model predictions for, e.g., the fraction of Z decays with $b\bar{b}$ pairs and the anomalous magnetic moment of the muon, which exclude some models that would otherwise be viable. Some extensions of the MSSM predict proton decay and neutron–antineutron oscillations at potentially observable rates, even if R-parity is exactly conserved. However, it would be impossible to unambiguously ascribe a positive result for any of these processes to supersymmetry. There is no substitute for the direct detection of sparticles and verification of their quantum numbers and interactions. In this chapter we provide a qualitative review of some of the possible signals for direct detection of supersymmetry at colliders such as the LHC and future high-energy e^+e^- colliders. A more comprehensive treatment of supersymmetric phenomenology can be found in [B17].

15.1 Superpartner Decays

This chapter presents a qualitative overview of the decay patterns of sparticles in the MSSM, assuming that R-parity is conserved.[1] We will consider the possible decays of neutralinos, charginos, sleptons, squarks, and the gluino. All decay chains will end with the LSP in the final state, which is usually taken to be the lightest neutralino $\widetilde{N}_1$. Section 15.1.5 discusses the alternative possibility that the gravitino/goldstino $\widetilde{G}$ is the LSP. To simplify the notation, we will often not distinguish between particle and antiparticle names and labels in this chapter, with context and consistency (dictated by charge and color conservation) resolving any ambiguities.

15.1.1 Decays of Neutralinos and Charginos

Let us first consider the possible two-body decays. Each neutralino and chargino contains at least a small admixture of the electroweak gauginos $\widetilde{B}$, $\widetilde{W}^0$, or $\widetilde{W}^{\pm}$, as

[1] Modifications to MSSM phenomenology arising in models where R-parity is violated will be treated in Section 16.5.7.

we saw in Section 13.9. So $\widetilde{N}_i$ and $\widetilde{C}_i$ inherit couplings of weak interaction strength to (scalar, fermion) pairs, as shown in Figs. 13.3(b),(c). If sleptons or squarks are sufficiently light, a neutralino or chargino can therefore decay into lepton+slepton or quark+squark. To the extent that sleptons are probably lighter than squarks, the lepton+slepton final states are favored. A neutralino or chargino may also decay into any lighter neutralino or chargino plus a Higgs scalar or an electroweak gauge boson, because they inherit the gaugino–higgsino–Higgs [see Figs. 13.3(b),(c)] and $SU(2)_L$ gaugino–gaugino–vector boson [see Fig. 8.3(c)] couplings of their components. So, the possible two-body decay modes for neutralinos and charginos in the MSSM are

$$\widetilde{N}_i \to Z\widetilde{N}_j,\quad W\widetilde{C}_j,\quad h\widetilde{N}_j,\quad \ell\widetilde{\ell},\quad \nu\widetilde{\nu},\quad [A\widetilde{N}_j,\quad H\widetilde{N}_j,\quad H^{\pm}\widetilde{C}_j^{\mp},\quad q\widetilde{q}], \tag{15.1}$$

$$\widetilde{C}_i \to W\widetilde{N}_j,\quad Z\widetilde{C}_1,\quad h\widetilde{C}_1,\quad \ell\widetilde{\nu},\quad \nu\widetilde{\ell},\quad [A\widetilde{C}_1,\quad H\widetilde{C}_1,\quad H^{\pm}\widetilde{N}_j,\quad q\widetilde{q}'], \tag{15.2}$$

using a generic notation ν, ℓ, q for neutrinos, charged leptons, and quarks. The final states in brackets are the more kinematically implausible ones. (Since $m_h = 125$ GeV, it is the most likely of the Higgs scalars to appear in these decays.) For the heavier neutralinos and chargino ($\widetilde{N}_3$, $\widetilde{N}_4$, and $\widetilde{C}_2$), one or more of the two-body decays in eqs. (15.1) and (15.2) are likely to be kinematically allowed. Also, if the decays of neutralinos and charginos with a significant higgsino content into third-generation quark–squark pairs are open, they can be greatly enhanced by the top-quark Yukawa coupling, following from the interactions shown in Fig. 13.1(b),(c).

It may be that all of these two-body modes are kinematically forbidden for a given chargino or neutralino, especially for $\widetilde{C}_1$ and $\widetilde{N}_2$ decays. In that case, they have three-body decays

$$\widetilde{N}_i \to ff\widetilde{N}_j,\quad \widetilde{N}_i \to ff'\widetilde{C}_j,\quad \widetilde{C}_i \to ff'\widetilde{N}_j,\quad \text{and}\quad \widetilde{C}_2 \to ff\widetilde{C}_1, \tag{15.3}$$

through the same (but now off-shell) gauge bosons, Higgs scalars, sleptons, and squarks that appeared in the two-body decays in eqs. (15.1) and (15.2). Here f is generic notation for a lepton or quark, with f and f' distinct members of the same $SU(2)_L$ multiplet (and of course one of the f or f' in each of these decays must actually be an antifermion). Example calculations of the chargino and neutralino decay widths into various final states are given in Chapter 18.

The Feynman diagrams for the neutralino and chargino decays with $\widetilde{N}_1$ in the final state that seem most likely to be important are shown in Fig. 15.1. In many situations, the decays

$$\widetilde{C}_1^{\pm} \to \ell^{\pm}\nu\widetilde{N}_1, \qquad \widetilde{N}_2 \to \ell^+\ell^-\widetilde{N}_1 \tag{15.4}$$

can be particularly important for phenomenology, because the leptons in the final state often will result in clean signals. These decays are more likely if the intermediate sleptons are relatively light, even if they cannot be on-shell. Unfortunately, the enhanced mixing of staus, common in models, can often result in larger branching fractions for both $\widetilde{N}_2$ and $\widetilde{C}_1$ into final states with taus, rather than electrons or muons. This is one reason why tau identification may be a crucial limiting factor in attempts to discover and study supersymmetry.

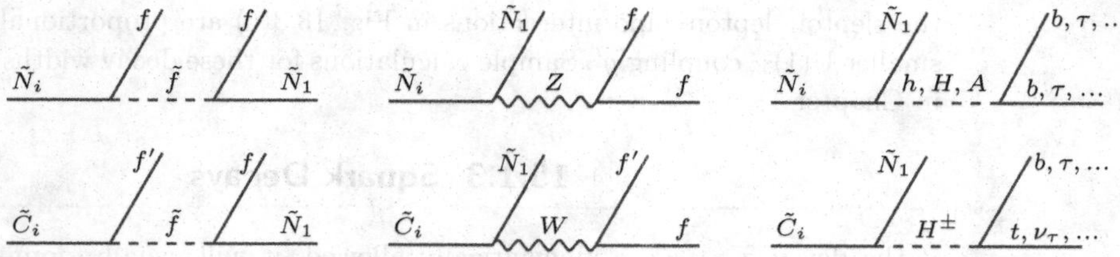

Fig. 15.1 Feynman diagrams for neutralino and chargino decays with $\widetilde{N}_1$ in the final state. The intermediate scalar or vector boson in each case can be either on-shell (so that actually there is a sequence of two-body decays) or off-shell, depending on the sparticle mass spectrum.

In other situations, decays without isolated leptons in the final state are more useful, so that one will not need to contend with background events with missing energy coming from leptonic W boson decays in Standard Model processes. Then the decays of interest are the ones with quark partons in the final state, leading to

$$\widetilde{C}_1 \to jj\widetilde{N}_1, \qquad \widetilde{N}_2 \to jj\widetilde{N}_1, \tag{15.5}$$

where j means a jet. If the second of these decays goes through an on-shell h, then these will usually be b jets.

15.1.2 Slepton Decays

Sleptons can have two-body decays into a lepton and a chargino or neutralino, because of their gaugino admixture, as may be seen directly from the couplings in Figs. 13.3(b),(c). Therefore, the two-body decays

$$\widetilde{\ell} \to \ell\widetilde{N}_i, \qquad \widetilde{\ell} \to \nu\widetilde{C}_i, \qquad \widetilde{\nu} \to \nu\widetilde{N}_i, \qquad \widetilde{\nu} \to \ell\widetilde{C}_i \tag{15.6}$$

can be of weak interaction strength. In particular, the direct decays

$$\widetilde{\ell} \to \ell\widetilde{N}_1 \quad \text{and} \quad \widetilde{\nu} \to \nu\widetilde{N}_1 \tag{15.7}$$

are (almost[2]) always kinematically allowed if $\widetilde{N}_1$ is the LSP. However, if the sleptons are sufficiently heavy, then the two-body decays

$$\widetilde{\ell} \to \nu\widetilde{C}_1, \quad \widetilde{\ell} \to \ell\widetilde{N}_2, \quad \widetilde{\nu} \to \nu\widetilde{N}_2, \quad \text{and} \quad \widetilde{\nu} \to \ell\widetilde{C}_1 \tag{15.8}$$

can be important. The right-handed sleptons do not have a coupling to the $SU(2)_L$ gauginos, so they typically prefer the direct decay $\widetilde{\ell}_R \to \ell\widetilde{N}_1$, if $\widetilde{N}_1$ is bino-like. In contrast, the left-handed sleptons may prefer to decay as in eq. (15.8) rather than the direct decays to the LSP as in eq. (15.7), if the former is kinematically open and if $\widetilde{C}_1$ and $\widetilde{N}_2$ are mostly wino. This is because the slepton–lepton–wino interactions in Fig. 13.3(b) are proportional to the $SU(2)_L$ gauge coupling g, whereas

[2] An exception occurs if the mass difference $m_{\tilde{\tau}_1} - m_{\widetilde{N}_1}$ is less than m_τ.

the slepton–lepton–bino interactions in Fig. 13.3(c) are proportional to the much smaller $U(1)_Y$ coupling g'. Sample calculations for these decay widths can be found in Chapter 18.

15.1.3 Squark Decays

If the decay $\widetilde{q} \to q\widetilde{g}$ is kinematically allowed, it will usually dominate, because the quark–squark–gluino vertex in Fig. 13.3(a) has QCD strength. Otherwise, the squarks can decay into a quark plus neutralino or chargino: $\widetilde{q} \to q\widetilde{N}_i$ or $q'\widetilde{C}_i$. The direct decay to the LSP $\widetilde{q} \to q\widetilde{N}_1$ is always kinematically favored, and for right-handed squarks it can dominate because $\widetilde{N}_1$ is mostly bino. However, the left-handed squarks may strongly prefer to decay into heavier charginos or neutralinos instead, for example $\widetilde{q} \to q\widetilde{N}_2$ or $q'\widetilde{C}_1$, because the relevant squark–quark–wino couplings are much bigger than the squark–quark–bino couplings. Squark decays to higgsino-like charginos and neutralinos are less important, except in the cases of stops and sbottoms, which have sizable Yukawa couplings. The gluino, chargino, or neutralino resulting from the squark decay will in turn decay, and so on, until a final state containing $\widetilde{N}_1$ is reached. This results in numerous and complicated decay chain possibilities called cascade decays.

It is possible that the decays $\widetilde{t}_1 \to t\widetilde{g}$ and $\widetilde{t}_1 \to t\widetilde{N}_1$ are both kinematically forbidden. If so, then the lighter top squark may decay only into charginos, by $\widetilde{t}_1 \to b\widetilde{C}_1$, or by a three-body decay $\widetilde{t}_1 \to bW\widetilde{N}_1$. If even this decay is kinematically closed, then it has only the flavor-suppressed decay to a charm quark, $\widetilde{t}_1 \to c\widetilde{N}_1$, and the four-body decay $\widetilde{t}_1 \to bff'\widetilde{N}_1$. These decays can be very slow, so that the lightest stop can be quasi-stable on the time scale relevant for collider physics, and can hadronize into bound states.

15.1.4 Gluino Decays

The decay of the gluino can only proceed through a squark, either on-shell or virtual. If two-body decays $\widetilde{g} \to q\widetilde{q}$ are open, they will dominate, again because the relevant gluino–quark–squark coupling in Fig. 13.3(a) has QCD strength. Since the top and bottom squarks can easily be much lighter than all of the other squarks, it is quite possible that $\widetilde{g} \to t\widetilde{t}_1$ and/or $\widetilde{g} \to b\widetilde{b}_1$ are the only available two-body decay mode(s) for the gluino, in which case they will dominate over all others. If instead all of the squarks are heavier than the gluino, the gluino will decay only through off-shell squarks, so $\widetilde{g} \to qq\widetilde{N}_i$ and $qq'\widetilde{C}_i$. The squarks, neutralinos, and charginos in these final states will then decay as discussed above, so there can be many competing gluino decay chains. Some of the possibilities are shown in Fig. 15.2. The cascade decays can have final-state branching fractions that are individually small and quite sensitive to the model parameters.

The simplest gluino decays, including the ones shown in Fig. 15.2, can have zero, one, or two charged leptons (in addition to two or more hadronic jets) in the final state. An important feature is that when there is exactly one charged lepton, it can

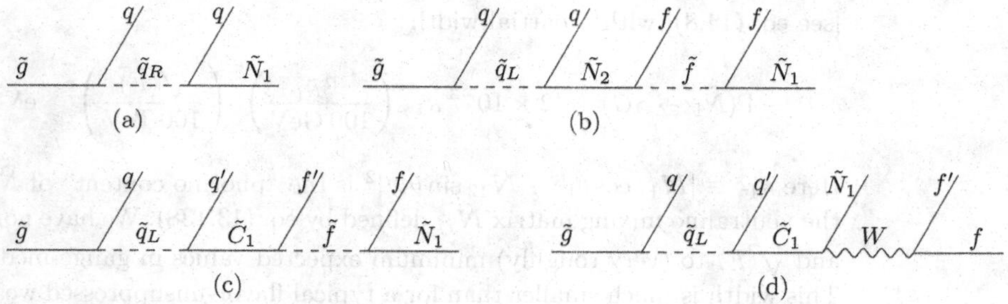

Fig. 15.2 Some of the many possible examples of gluino cascade decays ending with a neutralino LSP in the final state. The squarks appearing in these diagrams may be either on-shell or off-shell, depending on the mass spectrum of the theory.

have either charge with exactly equal probability. This follows from the fact that the gluino is a Majorana fermion, and does not "know" about electric charge; for each diagram with a given lepton charge, there is always an equal one with every particle replaced by its antiparticle.

15.1.5 Decays to the Gravitino/Goldstino

Most phenomenological studies of supersymmetry assume explicitly or implicitly that the lightest neutralino is the LSP. This is typically the case in gravity-mediated models for the soft terms. However, in gauge-mediated models (and in "no-scale" models), the LSP is instead the gravitino. As we saw in Section 14.2, a very light gravitino may be relevant for collider phenomenology, because it contains as its longitudinal component the goldstino, which has a nongravitational coupling to all sparticle–particle pairs $(\widetilde{X}, X)$. The decay rate found in eq. (14.8) for $\widetilde{X} \rightarrow X\widetilde{G}$ is usually not fast enough to compete with the other decays of sparticles $\widetilde{X}$ as mentioned above, *except* in the case that $\widetilde{X}$ is the next-to-lightest supersymmetric particle. Since the NLSP has no competing decays, it should always decay into its superpartner and the LSP gravitino.

In principle, any of the MSSM superpartners could be the NLSP in models with a light goldstino, but most models with gauge mediation of supersymmetry breaking have either a neutralino or a charged slepton playing this role. The argument for this can be seen immediately from eqs. (14.53) and (14.54); since $\alpha_1 < \alpha_2, \alpha_3$, those superpartners with only $U(1)_Y$ interactions will tend to get the smallest masses. The gauge eigenstate sparticles with this property are the bino and the right-handed sleptons $\widetilde{e}_R$, $\widetilde{\mu}_R$, $\widetilde{\tau}_R$, so the appropriate corresponding mass eigenstates should be plausible candidates for the NLSP.

First, suppose that $\widetilde{N}_1$ is the NLSP in light goldstino models. Since $\widetilde{N}_1$ contains an admixture of the photino (the linear combination of bino and neutral wino whose superpartner is the photon), it decays into photon+goldstino/gravitino [see eq. (14.8)] with a partial width

$$\Gamma(\widetilde{N}_1 \to \gamma\widetilde{G}) = 2 \times 10^{-3}\, \kappa_{1\gamma} \left(\frac{m_{\widetilde{N}_1}}{100 \text{ GeV}}\right)^5 \left(\frac{\sqrt{\langle F \rangle}}{100 \text{ TeV}}\right)^{-4} \text{eV}. \tag{15.9}$$

Here $\kappa_{1\gamma} \equiv |\boldsymbol{N}_{11} \cos\theta_W + \boldsymbol{N}_{12} \sin\theta_W|^2$ is the "photino content" of $\widetilde{N}_1$, in terms of the neutralino mixing matrix $\boldsymbol{N}_{ij}$ defined by eq. (13.138). We have normalized $m_{\widetilde{N}_1}$ and $\sqrt{\langle F \rangle}$ to (very roughly) minimum expected values in gauge-mediated models. This width is much smaller than for a typical flavor-unsuppressed weak interaction decay, but it is still large enough to allow $\widetilde{N}_1$ to decay before it has left a collider detector, if $\sqrt{\langle F \rangle}$ is less than a few thousand TeV in gauge-mediated models, or equivalently if $m_{3/2}$ is less than a keV or so when eq. (14.7) holds. In fact, from eq. (15.9), the mean decay length of an $\widetilde{N}_1$ with energy E in the lab frame is

$$d = 9.9 \times 10^{-5} \frac{1}{\kappa_{1\gamma}} \left(\frac{E^2}{m^2_{\widetilde{N}_1}} - 1\right)^{1/2} \left(\frac{m_{\widetilde{N}_1}}{100 \text{ GeV}}\right)^{-5} \left(\frac{\sqrt{\langle F \rangle}}{100 \text{ TeV}}\right)^4 \text{meters}, \tag{15.10}$$

which could be anything from submicron to multikilometer, depending on the scale of supersymmetry breaking $\sqrt{\langle F \rangle}$. (In other models with a gravitino LSP, including certain "no-scale" models, the same formulae apply with $\langle F \rangle \to \sqrt{3} m_{3/2} M_P$.)

Of course, $\widetilde{N}_1$ is not a pure photino, but contains admixtures of the superpartner of the Z boson and the neutral Higgs scalars. So, one can also have $\widetilde{N}_1 \to Z\widetilde{G}$, $h\widetilde{G}$, $A\widetilde{G}$, or $H\widetilde{G}$. Of these decays, the last two are much less likely to be kinematically allowed, and only the $\widetilde{N}_1 \to \gamma\widetilde{G}$ mode is guaranteed to be kinematically allowed for a gravitino LSP. Furthermore, even if they are open, the decays $\widetilde{N}_1 \to Z\widetilde{G}$ and $\widetilde{N}_1 \to h\widetilde{G}$ are subject to strong kinematic suppressions proportional to $(1 - m_Z^2/m^2_{\widetilde{N}_1})^4$ and $(1 - m_h^2/m^2_{\widetilde{N}_1})^4$, respectively, in view of eq. (14.8). Still, these decays may play an important role in phenomenology if $\langle F \rangle$ is not too large, $\widetilde{N}_1$ has a sizable zino or higgsino content, and $m_{\widetilde{N}_1}$ is significantly greater than m_Z or m_h.

A charged slepton makes another likely candidate for the NLSP. Actually, more than one slepton can act effectively as the NLSP, even though one of them is slightly lighter, if they are sufficiently close in mass so that each has no kinematically allowed decays except to the goldstino. In GMSB models, the squared masses obtained by $\widetilde{e}_R$, $\widetilde{\mu}_R$ and $\widetilde{\tau}_R$ are equal because of the flavor-blindness of the gauge couplings. However, this is not the whole story, because one must take into account mixing with $\widetilde{e}_L$, $\widetilde{\mu}_L$, and $\widetilde{\tau}_L$ and renormalization group running. These effects are very small for $\widetilde{e}_R$ and $\widetilde{\mu}_R$ because of the tiny electron and muon Yukawa couplings, so we can quite generally treat them as degenerate, unmixed mass eigenstates. In contrast, $\widetilde{\tau}_R$ usually has a quite significant mixing with $\widetilde{\tau}_L$, proportional to the tau Yukawa coupling. This means that the lighter stau mass eigenstate $\widetilde{\tau}_1$ is pushed lower in mass than $\widetilde{e}_R$ or $\widetilde{\mu}_R$, by an amount that depends most strongly on $\tan\beta$. If $\tan\beta$ is not too large, then the stau mixing effect leaves the slepton mass eigenstates $\widetilde{e}_R$, $\widetilde{\mu}_R$, and $\widetilde{\tau}_1$ degenerate to within less than $m_\tau \approx 1.8$ GeV, so they act effectively as co-NLSPs. In particular, this means that, even though the stau is slightly lighter, the three-body slepton decays $\widetilde{e}_R \to e\tau^\pm\widetilde{\tau}_1^\mp$ and $\widetilde{\mu}_R \to \mu\tau^\pm\widetilde{\tau}_1^\mp$ are not kinematically allowed; the only allowed decays for the three lightest sleptons are $\widetilde{e}_R \to e\widetilde{G}$ and $\widetilde{\mu}_R \to \mu\widetilde{G}$ and $\widetilde{\tau}_1 \to \tau\widetilde{G}$. This situation is called the "slepton co-NLSP" scenario.

For larger values of $\tan\beta$, the lighter stau eigenstate $\widetilde{\tau}_1$ is more than 1.8 GeV lighter than $\widetilde{e}_R$ and $\widetilde{\mu}_R$ and $\widetilde{N}_1$. Hence, the decays $\widetilde{N}_1 \to \tau\widetilde{\tau}_1$ and $\widetilde{e}_R \to e\tau\widetilde{\tau}_1$ and $\widetilde{\mu}_R \to \mu\tau\widetilde{\tau}_1$ are open. Then $\widetilde{\tau}_1$ is the sole NLSP, with all other MSSM supersymmetric particles having kinematically allowed decays into it. This is called the "stau NLSP" scenario.

In any case, a slepton NLSP can decay like $\widetilde{\ell} \to \ell\widetilde{G}$ according to eq. (14.8), with a width and decay length given by eqs. (15.9) and (15.10) after replacing $\kappa_{1\gamma} \to 1$ and $m_{\widetilde{N}_1} \to m_{\widetilde{\ell}}$. So, as for the neutralino NLSP case, the decay $\widetilde{\ell} \to \ell\widetilde{G}$ can be either fast or very slow, depending on the scale of supersymmetry breaking.

If $\sqrt{\langle F\rangle}$ is larger than roughly 10^3 TeV (or the gravitino is heavier than a keV or so), then the NLSP is so long-lived that it will usually escape a typical collider detector. If $\widetilde{N}_1$ is the NLSP, then it might as well be the LSP from the point of view of collider physics. However, the decay of $\widetilde{N}_1$ into the gravitino is still important for cosmology, since an unstable $\widetilde{N}_1$ is clearly not a good dark matter candidate, while the gravitino LSP conceivably could be. On the other hand, if the NLSP is a long-lived charged slepton, then one can see its tracks (or possibly decay kinks) inside a collider detector. The presence of a massive charged NLSP can be established by measuring an anomalously long time-of-flight or high ionization rate for a track in the detector.

15.2 Signals at Hadron Colliders

At this writing, the LHC has already excluded significant chunks of supersymmetric parameter space, based on proton–proton collisions at center-of-mass energies up to $\sqrt{s} = 13$ TeV. Further increases in LHC luminosity may allow for a discovery or even stronger limits, and proton–proton collisions at higher energies could dramatically increase the reach.

At hadron colliders, sparticles can be produced in pairs from parton collisions of electroweak strength:

$$q\overline{q} \to \widetilde{C}_i^+\widetilde{C}_j^-,\ \widetilde{N}_i\widetilde{N}_j, \qquad u\overline{d} \to \widetilde{C}_i^+\widetilde{N}_j, \qquad d\overline{u} \to \widetilde{C}_i^-\widetilde{N}_j, \tag{15.11}$$

$$q\overline{q} \to \widetilde{\ell}_i^+\widetilde{\ell}_j^-,\ \widetilde{\nu}_\ell\widetilde{\nu}_\ell^\dagger \qquad u\overline{d} \to \widetilde{\ell}_L^+\widetilde{\nu}_\ell \qquad d\overline{u} \to \widetilde{\ell}_L^-\widetilde{\nu}_\ell^\dagger, \tag{15.12}$$

as shown in Fig. 15.3, and reactions of QCD strength:

$$gg \to \widetilde{g}\widetilde{g},\ \widetilde{q}_i\widetilde{q}_j^\dagger, \tag{15.13}$$

$$gq \to \widetilde{g}\widetilde{q}_i, \tag{15.14}$$

$$q\overline{q} \to \widetilde{g}\widetilde{g},\ \widetilde{q}_i\widetilde{q}_j^\dagger, \tag{15.15}$$

$$qq \to \widetilde{q}_i\widetilde{q}_j, \tag{15.16}$$

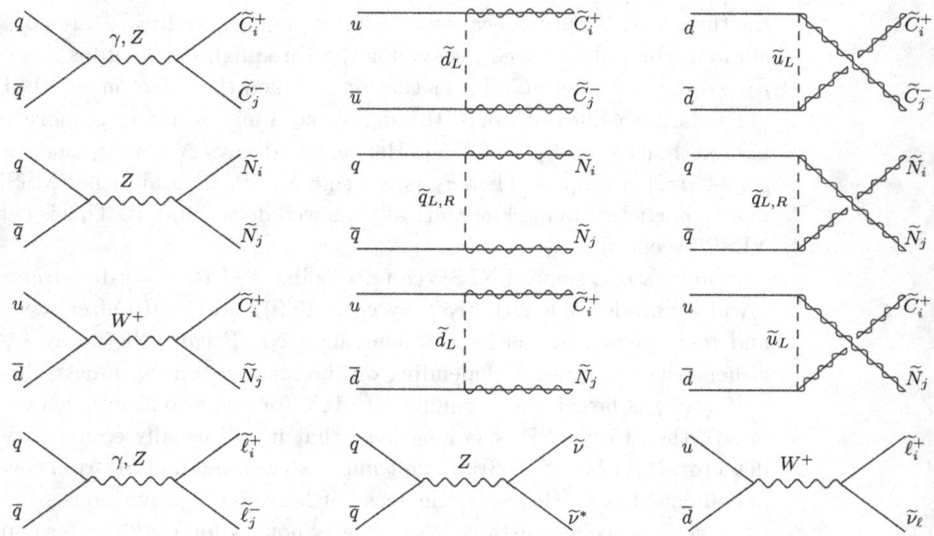

Fig. 15.3 Feynman diagrams for electroweak production of sparticles at hadron colliders from quark–antiquark annihilation. The charginos and neutralinos in the t-channel diagrams only couple because of their gaugino content, for massless initial-state quarks, and so are drawn as wavy lines superimposed on solid.

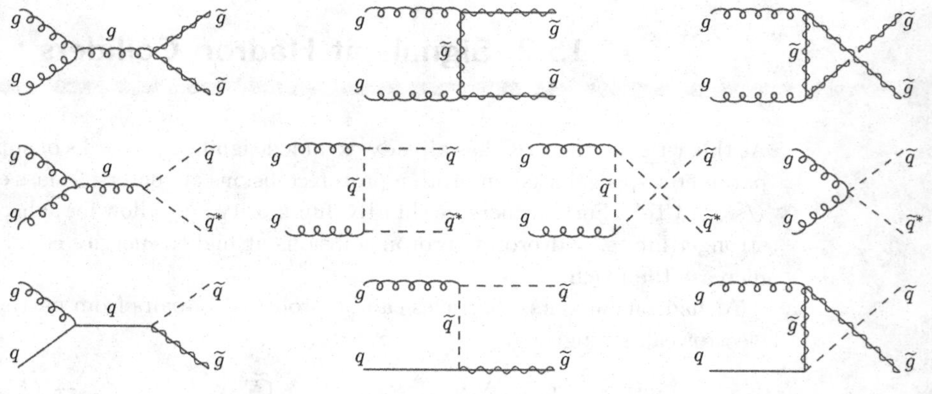

Fig. 15.4 Feynman diagrams for gluino and squark production at hadron colliders from gluon–gluon and gluon–quark fusion.

as shown in Figs. 15.4 and 15.5. The processes exhibited in eqs. (15.11) and (15.12) have contributions from electroweak vector bosons in the s-channel, and those in eq. (15.11) also have t-channel squark-exchange contributions that are of lesser importance in most models. The processes in eqs. (15.13)–(15.16) get contributions

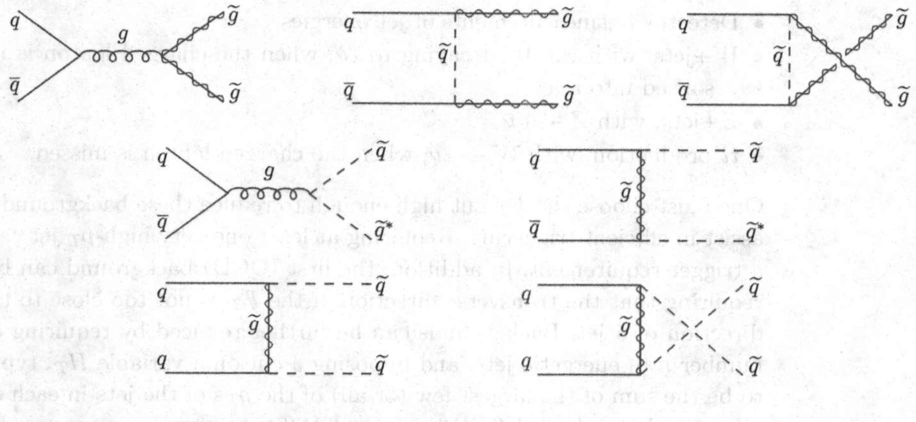

Fig. 15.5 Feynman diagrams for gluino and squark production at hadron colliders from strong quark–antiquark annihilation and quark–quark scattering.

from the t-channel exchange of an appropriate squark or gluino, and the processes in eqs. (15.13) and (15.15) also have gluon s-channel contributions. In a crude first approximation, for the hard parton collisions needed to make heavy particles, one may think of the LHC as a gluon–gluon and gluon–quark collider. However, the signals are always an inclusive combination of the results of parton collisions of all types, and generally cannot be neatly separated. Furthermore, depending on the class of model, the production of weakly interacting particles (charginos, neutralinos, and sleptons) may be so favored by the kinematics that they could be the discovery mode.

The decays of the produced sparticles result in final states with two neutralino LSPs, which escape the detector. The LSPs carry away at least $2m_{\tilde{N}_1}$ of missing energy, but at hadron colliders only the component of the missing energy that is manifest as momenta transverse to the colliding beams, usually denoted $\not{E}_T$ or E_T^{miss} (although $\vec{\not{p}}_T$ or $\vec{p}_T^{\text{miss}}$ might be more logical names), is observable. So, in general the observable signals for supersymmetry at hadron colliders will contain n leptons + m jets + $\not{E}_T$, where either n or m might be 0. There are important Standard Model backgrounds to these signals, especially from processes involving production of W and Z bosons that decay to neutrinos, which provide the $\not{E}_T$. Thus, it is important to identify specific signal region cuts for which the backgrounds can be reduced. Of course, the optimal choice of cuts depends on which sparticles are being produced and how they decay, facts that are not known in advance.

The classic $\not{E}_T$ signal for supersymmetry at hadron colliders consists of events with jets and $\not{E}_T$ but no energetic isolated leptons. The latter requirement reduces backgrounds from Standard Model processes with leptonic W decays, and is obviously most effective if the relevant sparticle decays have sizable branching fractions into channels with no leptons in the final state. The most important potential backgrounds are listed below.

- Detector mismeasurements of jet energies
- W+jets, with the W decaying to $\ell\nu$, when the charged lepton is missed or absorbed into a jet
- Z+jets, with $Z \to \nu\bar{\nu}$
- $t\bar{t}$ production, with $W \to \ell\nu$, when the charged lepton is missed.

One must choose the $\not{E}_T$ cut high enough to reduce these backgrounds, and also to assist in efficient triggering. Requiring at least one very high-p_T jet can also satisfy a trigger requirement. In addition, the first (QCD) background can be reduced by requiring that the transverse direction of the $\not{E}_T$ is not too close to the transverse direction of a jet. Backgrounds can be further reduced by requiring at least some number n of energetic jets, and imposing a cut on a variable H_T, typically defined to be the sum of the largest few (or all) of the p_Ts of the jets in each event. (There is no fixed standard definition of H_T.) Different signal regions can be defined by how many jets are required in the event, the minimum p_T cuts on those jets, how many jets are included in the definition of H_T, and other fine details. Alternatively, one can cut on $m_{\rm eff} \equiv H_T + \not{E}_T$ rather than H_T. Another cut that is often used in searches is to require a minimum value for the ratio of $\not{E}_T$ to either H_T or $m_{\rm eff}$; the backgrounds tend to have smaller values of this ratio than a supersymmetric signal would. The jets+$\not{E}_T$ signature is a favorite possibility for the first evidence for supersymmetry to be found at the LHC. It can get important contributions from every type of sparticle pair production, except slepton pair production.

Another important possibility for the LHC is the single lepton plus jets plus $\not{E}_T$ signal. It has a potentially large Standard Model background from production of $W \to \ell\nu$, either together with jets or from top decays. However, this background can be reduced by putting a cut on the transverse mass variable $m_T = \sqrt{2p_T^\ell \not{E}_T[1 - \cos(\Delta\phi)]}$, where $\Delta\phi$ is the difference in azimuthal angle between the missing transverse momentum and the lepton. For W decays, this is essentially always less than 100 GeV even after detector resolution effects, so a cut requiring $m_T > 100$ GeV nearly eliminates those background contributions at the LHC. The single lepton plus jets signal can have an extremely large rate from various sparticle production modes, and may give a good discovery or confirmation signal at the LHC.

The same-sign dilepton signal has the advantage of relatively small backgrounds. It can occur if the gluino decays with a significant branching fraction to hadrons plus a chargino, which can subsequently decay into a final state with a charged lepton, a neutrino, and $\widetilde{N}_1$. Since the gluino doesn't know anything about electric charge, the charged lepton produced from each gluino decay can have either sign with equal probability, as discussed in Section 15.1.4. This means that gluino pair production or gluino–squark production will often lead to events with two leptons with the same charge (and uncorrelated flavors) plus jets and $\not{E}_T$. This signal can also arise from squark pair production, for example if the squarks decay like $\widetilde{q} \to q\widetilde{g}$. The physics backgrounds at hadron colliders are very small, because the largest Standard Model sources for isolated lepton pairs, notably via s-channel γ/Z exchange, W^+W^-, and

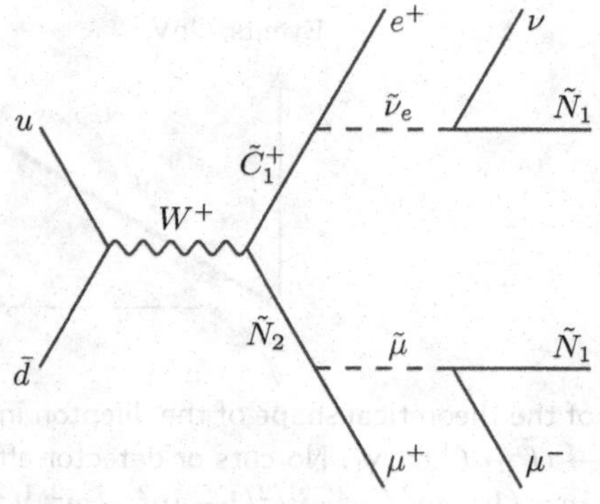

Fig. 15.6 A complete Feynman diagram for a clean (no high-p_T hadronic jets) trilepton event at a hadron collider, from production of an on-shell neutralino and a chargino, with subsequent leptonic decays, leading in this case to $\mu^+\mu^-e^+ + \not{E}_T$.

$t\bar{t}$ production, can only yield opposite-sign dileptons. Despite the backgrounds just mentioned, opposite-sign dilepton signals, for example from slepton pair production, or slepton-rich decays of heavier superpartners, with subsequent decays $\widetilde{\ell} \to \ell\widetilde{N}_1$, may also eventually give an observable signal at the LHC.

The trilepton signal is another possible discovery mode, featuring three leptons plus $\not{E}_T$, and possibly hadronic jets. This could come about from electroweak $\widetilde{C}_1\widetilde{N}_2$ production followed by the decays indicated in eq. (15.4), in which case high-p_T hadronic activity should be absent in the event. A typical Feynman diagram for such an event is shown in Fig. 15.6. It could also come from $\widetilde{g}\widetilde{g}$, $\widetilde{g}\widetilde{q}$, or $\widetilde{q}\widetilde{q}$ production, with one of the gluinos or squarks decaying through a $\widetilde{C}_1$ and the other through a $\widetilde{N}_2$. In that case, there will be very high-p_T jets from the decays, in addition to the three leptons and $\not{E}_T$. These signatures rely on the $\widetilde{N}_2$ having a significant branching fraction for the three-body decay to leptons in eq. (15.4). The competing two-body decay modes $\widetilde{N}_2 \to h\widetilde{N}_1$ and $\widetilde{N}_2 \to Z\widetilde{N}_1$ are sometimes called "spoiler" modes, since if they are kinematically allowed they can dominate, spoiling the trilepton signal. This is because if the $\widetilde{N}_2$ decay is through an on-shell h, then the final state will very likely include bottom-quark jets rather than isolated leptons, while if the decay is through an on-shell Z, then there can still be two leptons but there are Standard Model backgrounds with unfortunately similar kinematics from processes involving $Z \to \ell^+\ell^-$. Although the trilepton signal is lost, supersymmetric events with $h \to b\bar{b}$ following from $\widetilde{N}_2 \to h\widetilde{N}_1$ could eventually be useful at the LHC or future hadron colliders, given that $m_h = 125$ GeV.

One should also be aware of interesting signals that can appear for particular ranges of parameters. Final-state leptons appearing in the signals listed above might be predominantly tau, and so a significant fraction could be realized as hadronic τ jets. This is because most models predict that $\widetilde{\tau}_1$ is lighter than the selectrons and smuons. Similarly, supersymmetric events may have a preference for bottom jets,

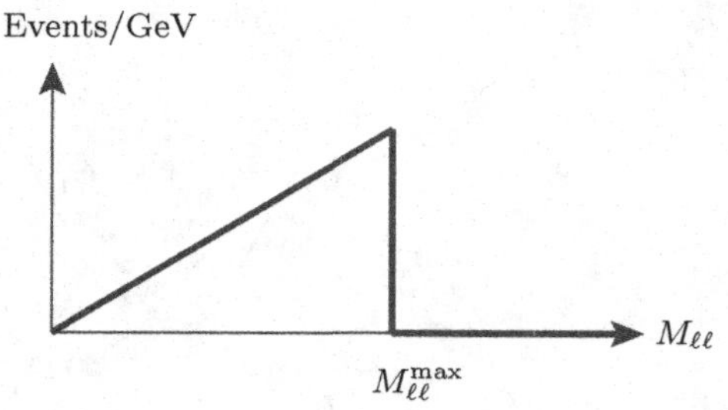

Fig. 15.7 A sketch of the theoretical shape of the dilepton invariant mass distribution from events with $\widetilde{N}_2 \to \ell\widetilde{\ell} \to \ell^+\ell^-\widetilde{N}_1$. No cuts or detector effects are included. The endpoint is at $M_{\ell\ell}^{\max} = m_{\widetilde{N}_2}(1 - m_{\widetilde{\ell}}^2/m_{\widetilde{N}_2}^2)^{1/2}(1 - m_{\widetilde{N}_1}^2/m_{\widetilde{\ell}}^2)^{1/2}$ [172].

sometimes through decays involving top quarks because $\widetilde{t}_1$ is relatively light, and sometimes because $\widetilde{b}_1$ is expected to be lighter than the squarks of the first two families, and sometimes for both reasons. In such cases, there will be at least four potentially b-taggable jets in each event. Other things being equal, the larger $\tan\beta$ is, the stronger the preference for hadronic τ and b jets will be in supersymmetric events.

After evidence for the existence of supersymmetry is acquired, the LHC or future pp collider data can be used to extract sparticle masses by analyzing the kinematics of the decays. With a neutralino LSP always escaping the detector, there are no true invariant mass peaks possible. However, various combinations of masses can be measured using kinematic edges and other reconstruction techniques. For a particularly favorable possibility, suppose the decay of the second-lightest neutralino occurs in two stages through a real slepton, $\widetilde{N}_2 \to \ell\widetilde{\ell} \to \ell^+\ell^-\widetilde{N}_1$. Then the resulting dilepton invariant mass distribution is as shown in Fig. 15.7. It features a sharp edge, allowing a precision measurement of the corresponding combination of $\widetilde{N}_2$, $\widetilde{\ell}$, and $\widetilde{N}_1$ masses (e.g., see Ref. [172]). There are significant backgrounds to this analysis, for example coming from $t\bar{t}$ production. However, the signal from $\widetilde{N}_2$ has same-flavor leptons, while the background has contributions from different flavors. Therefore the edge can be enhanced by plotting the combination $[e^+e^-] + [\mu^+\mu^-] - [\mu^+e^-] - [e^+\mu^-]$, subtracting the background.

Heavier sparticle mass combinations can also be reconstructed at the LHC using other kinematic distributions. For example, consider the gluino decay chain $\widetilde{g} \to q\widetilde{q}^\dagger \to q\bar{q}\widetilde{N}_2$ with $\widetilde{N}_2 \to \ell\widetilde{\ell}^\dagger \to \ell^+\ell^-\widetilde{N}_1$ as above. By selecting events close to the dilepton mass edge as determined in the previous paragraph, one can reconstruct a peak in the invariant mass of the $jj\ell^+\ell^-$ system, which correlates well with the gluino mass. As another example, the decay $\widetilde{q}_L \to q\widetilde{N}_2$ with $\widetilde{N}_2 \to h\widetilde{N}_1$ can be analyzed by selecting events near the peak from $h \to b\bar{b}$. There will then be a broad $jb\bar{b}$ invariant mass distribution, with a maximum value that can be related to $m_{\widetilde{N}_2}$, $m_{\widetilde{N}_1}$, and $m_{\widetilde{q}_L}$, since m_h is known. There are many other similar opportunities, depending on the specific sparticle spectrum. These techniques may determine the sparticle mass differences much more accurately than the individual masses, so that the mass of the unobserved LSP will be constrained but not precisely measured.[3]

Following the 2012 discovery of the 125 GeV Higgs boson, presumably h, the remaining Higgs scalar bosons of the MSSM are also targets of searches at the LHC. The heavier neutral Higgs scalars can be searched for in decays

$$A/H \to \tau^+\tau^-,\ \mu^+\mu^-,\ b\bar{b},\ t\bar{t}, \tag{15.17}$$

$$H \to hh, \tag{15.18}$$

$$A \to Zh \to \ell^+\ell^- b\bar{b}, \tag{15.19}$$

with prospects that vary considerably depending on the parameters of the model. The charged Higgs boson may also appear at the LHC in top-quark decays, if $m_{H^+} < m_t$. If instead $m_{H^+} > m_t$, then one can look for

$$bg \to tH^- \quad \text{or} \quad gg \to t\bar{b}H^-, \tag{15.20}$$

followed by the decay $H^- \to \tau^-\bar{\nu}_\tau$ or $H^- \to \bar{t}b$ in each case, or the charge conjugates of these processes. More details on searches for the Higgs bosons H, A, and $H^\pm$ of the MSSM at the LHC can be found in Ref. [173].

The remainder of this section briefly considers the possibility that the LSP is the goldstino/gravitino, in which case the sparticle discovery signals discussed above can be significantly improved. If the NLSP is a neutralino with a prompt decay, then $\widetilde{N}_1 \to \gamma\widetilde{G}$ will yield events with two energetic, isolated photons plus $\not{E}_T$ from the escaping gravitinos, rather than just $\not{E}_T$. So at a hadron collider the signal is $\gamma\gamma + X + \not{E}_T$ where X is any collection of leptons plus jets. The Standard Model backgrounds relevant for such events are quite small. If the $\widetilde{N}_1$ decay length is long enough, then it may be measurable because the photons will not point back to the event vertex. This would be particularly useful, as it would give an indication of the supersymmetry-breaking scale $\sqrt{\langle F\rangle}$; see eq. (14.8) and the discussion in Section 15.1.5. If the $\widetilde{N}_1$ decay occurs outside of the detector, then one just has the usual leptons + jets + $\not{E}_T$ signals as discussed above in the neutralino LSP scenario.

In the case that the NLSP is a charged slepton, then the decay $\widetilde{\ell} \to \ell\widetilde{G}$ can provide two extra leptons in each event, compared to the signals with a neutralino LSP. If the $\widetilde{\tau}_1$ is sufficiently lighter than the other charged sleptons $\widetilde{e}_R$, $\widetilde{\mu}_R$ and so is effectively the sole NLSP, then events will always have a pair of taus. If the slepton NLSP is long-lived, one can look for events with a pair of very heavy charged particle tracks or a long time-of-flight in the detector. Since slepton pair production usually has a much smaller cross section than the other processes listed in eqs. (15.11)–(15.16), this will typically be accompanied by leptons and/or jets from the same event vertex, which may be of crucial help in identifying candidate events. It is also

[3] A possible exception occurs if the lighter top squark has no kinematically allowed flavor-preserving two-body decays, which requires $m_{\widetilde{t}_1} < m_{\widetilde{N}_1} + m_t$ and $m_{\widetilde{t}_1} < m_{\widetilde{C}_1} + m_b$. Then the $\widetilde{t}_1$ will live long enough to form hadronic bound states. Scalar stoponium might then be observable at the LHC via its rare $\gamma\gamma$ decay, allowing a uniquely precise measurement of the mass through a narrow peak (limited by detector resolution) in the diphoton invariant mass spectrum.

quite possible that the decay length of $\widetilde{\ell} \to \ell\widetilde{G}$ is measurable within the detector, seen as a macroscopic kink in the charged particle track. This would again be a way to measure the scale of supersymmetry breaking through eq. (14.8).

15.3 Signals at e^+e^- Colliders

At e^+e^- colliders, all sparticles (except the gluino) can be produced at tree level:

$$e^+e^- \to \widetilde{C}_i^+\widetilde{C}_j^-,\quad \widetilde{N}_i\widetilde{N}_j,\quad \widetilde{\ell}^+\widetilde{\ell}^-,\quad \widetilde{\nu}\widetilde{\nu}^\dagger,\quad \widetilde{q}\widetilde{q}^\dagger. \tag{15.21}$$

Feynman diagrams for the production of a pair of neutral sparticles are shown in Figs. 15.8 and 15.9. The explicit computations of the scattering matrix elements for the production of $\widetilde{N}_i\widetilde{N}_j$ and $\widetilde{\nu}\widetilde{\nu}^\dagger$ are given in Section 18.8 and Exercise 18.7, respectively. Similar diagrams also govern the tree-level production of charged sparticles (see Figs. 18.16 and 18.17), which are treated in Exercises 18.8 and 18.10. The important interactions for sparticle production are the gaugino–fermion–scalar couplings shown in Figs. 13.3(b),(c) and the ordinary vector boson interactions. The cross sections are therefore determined just by the electroweak gauge couplings and the sparticle mixings.

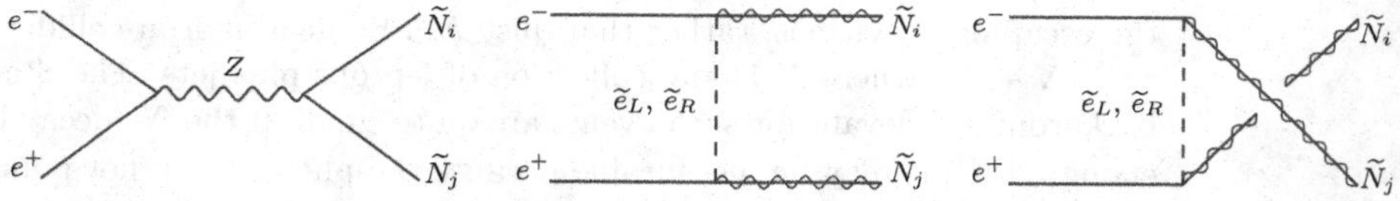

Fig. 15.8 Feynman diagrams for neutralino pair production at e^+e^- colliders.

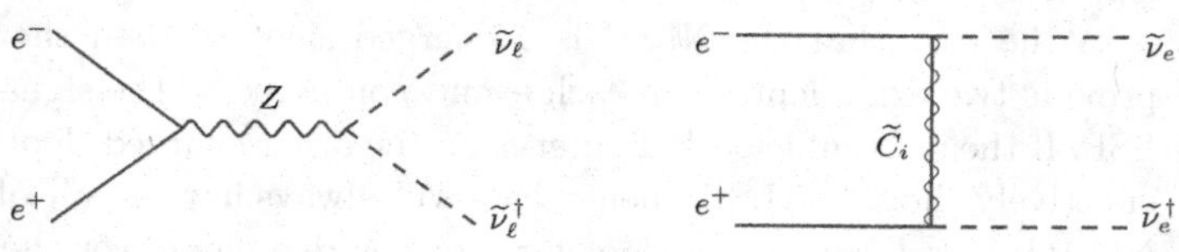

Fig. 15.9 Feynman diagrams for sneutrino pair production at e^+e^- colliders.

All of the processes in eq. (15.21) get contributions from the s-channel exchange of the Z boson and, for charged sparticle pairs, the photon. In the cases of $\widetilde{C}_i^+\widetilde{C}_j^-$, $\widetilde{N}_i\widetilde{N}_j$, $\widetilde{e}_R^+\widetilde{e}_R^-$, $\widetilde{e}_L^+\widetilde{e}_L^-$, $\widetilde{e}_L^\pm\widetilde{e}_R^\mp$, and $\widetilde{\nu}_e\widetilde{\nu}_e^\dagger$ production, there are also t-channel diagrams exchanging a virtual sneutrino, selectron, neutralino, neutralino, neutralino, and chargino, respectively. The t-channel contributions are significant if the exchanged sparticle is not too heavy. For example, the production of wino-like $\widetilde{C}_1^+\widetilde{C}_1^-$ pairs typically suffers a destructive interference between the s-channel graphs with γ, Z

exchange and the t-channel graphs with $\widetilde{\nu}_e$ exchange, if the sneutrino is not too heavy. In the case of sleptons, the pair production of smuons and staus proceeds only through s-channel diagrams, while selectron production also has a contribution from the t-channel exchanges of the neutralinos, as shown in Fig. 18.16. For this reason, selectron production may be significantly larger than smuon or stau production at e^+e^- colliders.

The pair-produced sparticles decay as discussed in Section 15.1. If the LSP is the lightest neutralino, it will always escape the detector because it has no strong or electromagnetic interactions. Every event will have two LSPs leaving the detector, so there should be at least $2m_{\widetilde{N}_1}$ of missing energy (henceforth denoted by $\not{E}$). For example, in the case of $\widetilde{C}_1^+\widetilde{C}_1^-$ production, the possible signals include a pair of acollinear leptons and $\not{E}$, or one lepton and a pair of jets plus $\not{E}$, or multiple jets plus $\not{E}$. The relative importance of these signals depends on the branching fraction of the chargino into the competing final states, $\widetilde{C}_1 \to \ell\nu\widetilde{N}_1$ and $qq'\widetilde{N}_1$. In the case of slepton pair production, the signal should be two energetic, acollinear, same-flavor leptons plus $\not{E}$. There is a potentially large Standard Model background for the acollinear leptons plus $\not{E}$ and the lepton plus jets plus $\not{E}$ signals, coming from W^+W^- production with one or both of the W bosons decaying leptonically. However, these and other Standard Model backgrounds can be kept under control with angular cuts, and beam polarization if available. It is not difficult to construct the other possible signatures for sparticle pairs, which can become quite complicated for the heavier charginos, neutralinos, and squarks.

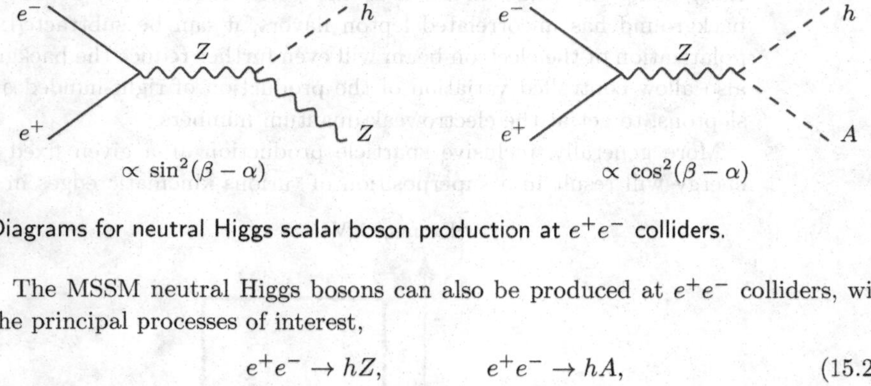

Fig. 15.10 Diagrams for neutral Higgs scalar boson production at e^+e^- colliders.

The MSSM neutral Higgs bosons can also be produced at e^+e^- colliders, with the principal processes of interest,

$$e^+e^- \to hZ, \qquad e^+e^- \to hA, \tag{15.22}$$

shown in Fig. 15.10. At tree level, the first of these has a cross section given by the corresponding Standard Model cross section multiplied by a factor of $\sin^2(\beta-\alpha)$, which approaches 1 in the Higgs decoupling limit of $m_A \gg m_Z$ discussed in Section 13.8. The other process is complementary, since (up to kinematic factors) its cross section is the same but multiplied by $\cos^2(\beta-\alpha)$, which is significant if m_A is not large.

If $\sqrt{s}$ is high enough [note the mass relation eq. (13.96)], one can also have

$$e^+e^- \to H^+H^-, \tag{15.23}$$

with a cross section that is fixed, at tree level, in terms of $m_{H^\pm}$, and also

$$e^+e^- \to HZ, \qquad e^+e^- \to HA, \tag{15.24}$$

with cross sections proportional to $\cos^2(\beta-\alpha)$ and $\sin^2(\beta-\alpha)$, respectively. At sufficiently high $\sqrt{s}$, the process

$$e^+e^- \to \nu_e\bar{\nu}_e h \tag{15.25}$$

following from W^+W^- fusion provides the best way to study the Higgs boson decays, which can differ substantially from those in the Standard Model.

At a future e^+e^- collider, the processes in eq. (15.21) can all be probed close to their kinematic limits, given sufficient integrated luminosity. (In the case of sneutrino pair production, this assumes that some of the decays are visible, rather than just $\widetilde{\nu} \to \nu\widetilde{N}_1$.) Establishing the couplings and masses of the particles can be done by making use of polarized beams and the relatively clean e^+e^- collider environment. For example, consider the production and decay of sleptons in $e^+e^- \to \widetilde{\ell}^+\widetilde{\ell}^-$ with $\widetilde{\ell} \to \ell\widetilde{N}_1$. The resulting leptons will have (up to significant but calculable effects of initial-state radiation, beam energy spread, kinematic cuts, and detector efficiencies and resolutions) a flat energy distribution as shown in Fig. 15.11. By measuring the endpoints of this distribution, one can precisely and uniquely determine both $m_{\widetilde{\ell}_R}$ and $m_{\widetilde{N}_1}$ (e.g., see Ref. [174]). There is a large $W^+W^- \to \ell^+\ell'^-\nu_\ell\bar{\nu}_{\ell'}$ background, but this can be brought under control using angular cuts, since the positively (negatively) charged leptons from the background tend to go preferentially along the same direction as the positron (electron) beam. Also, since the background has uncorrelated lepton flavors, it can be subtracted. Changing the polarization of the electron beam will even further reduce the background, and will also allow controlled variation of the production of right-handed and left-handed sleptons, to get at the electroweak quantum numbers.

More generally, inclusive sparticle production at a given fixed e^+e^- collision energy will result in a superposition of various kinematic edges in lepton and jet

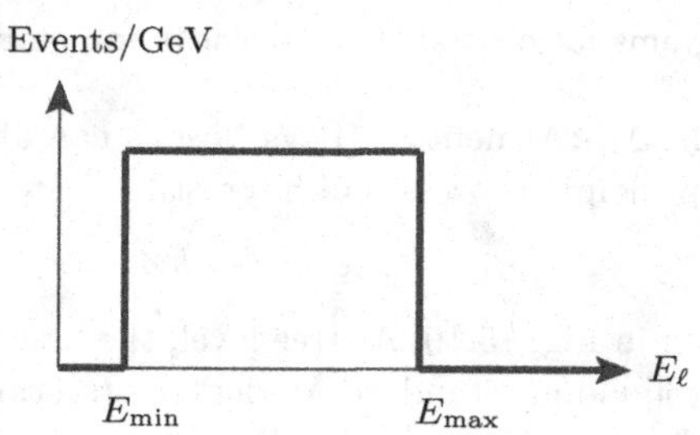

Fig. 15.11 A sketch of the theoretical shape of the lepton energy distribution from events with $e^+e^- \to \widetilde{\ell}^+\widetilde{\ell}^- \to \ell^+\ell^-\widetilde{N}_1\widetilde{N}_1$ at a future e^+e^- linear collider. No cuts or initial-state radiation or beam energy spread or detector effects are included. The endpoints are given by $E_{\rm max,min} = \frac{1}{4}\sqrt{s}(1 - m^2_{\widetilde{N}_1}/m^2_{\widetilde{\ell}})[1 \pm (1 - 4m^2_{\widetilde{\ell}}/s)^{1/2}]$ [174], allowing precision reconstruction of both $\widetilde{\ell}$ and $\widetilde{N}_1$ masses.

energies, and distinctive distributions in dilepton and dijet energies and invariant masses. By varying the beam polarization and changing the beam energy, these observables give information about the couplings and masses of the sparticles. For example, in the ideal limit of a right-handed polarized electron beam, the reaction

$$e^+e^-_R \to \widetilde{C}_1^+\widetilde{C}_1^- \tag{15.26}$$

is suppressed if $\widetilde{C}_1$ is pure wino, because in the first diagram of Fig. 18.17 the right-handed electron only couples to the $\mathrm{U}(1)_Y$ gauge boson linear combination of γ, Z while the wino only couples to the orthogonal $\mathrm{SU}(2)_L$ gauge boson linear combination, and in the second diagram the electron–sneutrino–chargino coupling involves purely left-handed electrons. Therefore, the polarized beam cross section can be used to determine the charged wino mixing with the charged higgsino. Even more precise information about the sparticle masses can be obtained by varying the beam energy in small discrete steps very close to thresholds, an option unavailable at hadron colliders. The rise of the production cross section above threshold provides information about the spin and "handedness," because the production cross sections for $\widetilde{\ell}_R^+\widetilde{\ell}_R^-$ and $\widetilde{\ell}_L^+\widetilde{\ell}_L^-$ are p-wave and therefore rise like β^3 above threshold, where β is the velocity of one of the produced sparticles. In contrast, the rates for $\widetilde{e}_L^\pm\widetilde{e}_R^\mp$ and for chargino and neutralino pair production are s-wave, and therefore should rise like β just above threshold. By measuring the angular distributions of the final-state leptons and jets with respect to the beam axis, the spins of the sparticles can be inferred. These will provide crucial tests that the new physics that has been discovered is indeed supersymmetry.

In general, a future e^+e^- collider will provide an excellent way of testing softly broken supersymmetry and measuring the model parameters, if it has enough energy. Furthermore, the processes $e^+e^- \to hZ$, hA, HZ, HA, H^+H^-, and $h\nu_e\bar{\nu}_e$ should be able to definitively test the Higgs sector of supersymmetry at a linear collider.

The situation may be qualitatively better if the gravitino is the LSP as in gauge-mediated models, because of the decays mentioned in Section 15.1.5. If the lightest neutralino is the NLSP and the decay $\widetilde{N}_1 \to \gamma\widetilde{G}$ occurs within the detector, then even the process $e^+e^- \to \widetilde{N}_1\widetilde{N}_1$ leads to a dramatic signal of two energetic photons plus missing energy. There are significant backgrounds to the $\gamma\gamma\not{E}$ signal, but they are easily removed by cuts. Each of the other sparticle pair-production modes of eq. (15.21) will lead to the same signals as in the neutralino LSP case, but now with two additional energetic photons, which should make the experimentalists' tasks quite easy. If the decay length for $\widetilde{N}_1 \to \gamma\widetilde{G}$ is much larger than the size of a detector, then the signals revert to those found in the neutralino LSP scenario. In an intermediate regime for the $\widetilde{N}_1 \to \gamma\widetilde{G}$ decay length, one may see events with one or both photons displaced from the event vertex by a macroscopic distance.

If the NLSP is a charged slepton, $\widetilde{\ell}$, then $e^+e^- \to \widetilde{\ell}^+\widetilde{\ell}^-$ followed by prompt decays $\widetilde{\ell} \to \ell\widetilde{G}$ will yield two energetic same-flavor leptons in every event, with a different energy distribution than the acollinear leptons arising from either $\widetilde{C}_1^+\widetilde{C}_1^-$ or $\widetilde{\ell}^+\widetilde{\ell}^-$ production in the neutralino LSP scenario. Pair production of non-NLSP sparticles will yield unmistakable signals, which are the same as those found in the neutralino NLSP case but with two additional energetic leptons (not necessarily of the same flavor). An even more striking possibility is that the NLSP is a slepton that decays very slowly. If the slepton NLSP is so long-lived that it decays outside the detector, then slepton pair production will lead to events featuring a pair of charged particle tracks with high ionization rates that betray their very large mass. If the sleptons decay within the detector, then one can look for large-angle kinks in the charged particle tracks, or a macroscopic impact parameter. The pair production of any of the other heavy charged sparticles will also yield heavy charged particle tracks or decay kinks, plus leptons and/or jets, but no $\not{E}$ unless the decay chains happen to include neutrinos. It may also be possible to identify the presence of a heavy charged NLSP by measuring its anomalously long time-of-flight through the detector.

In both the neutralino and slepton NLSP scenarios, a measurement of the decay length to $\widetilde{G}$ would provide a great opportunity to measure the supersymmetry-breaking scale $\sqrt{\langle F \rangle}$, as discussed in Section 15.1.5.

15.4 The Lightest Supersymmetric Particle and Dark Matter

Evidence from experimental cosmology has now solidified to the point that, with some plausible assumptions, the dark matter density is known to be [41]

$$\Omega_{\rm DM} h^2 \simeq 0.120\,, \tag{15.27}$$

with statistical errors of order 1%, and systematic errors that are less clear. Here $\Omega_{\rm DM}$ is the average energy density in nonbaryonic dark matter divided by the total critical density that would lead to a spatially flat homogeneous universe, and h is the Hubble constant in units of 100 km sec^{-1} Mpc^{-1}, observed to be $h \approx 0.7$. Equation (15.27) yields matter density

$$\rho_{\rm DM} \simeq 1.26 \times 10^{-6}\ {\rm GeV\,cm^{-3}}, \tag{15.28}$$

averaged over the universe as a whole.

One of the nice features of supersymmetry with exact R-parity conservation is that a stable electrically neutral LSP might be this cold dark matter.[4] In the MSSM, there are three obvious candidates: the lightest sneutrino, the gravitino, and the lightest neutralino (e.g., see Ref. [176]). The possibility of a sneutrino LSP making up the dark matter with a cosmologically interesting density has been largely ruled out by direct searches. The gravitino was the first proposed supersymmetric

[4] One can exclude the possibility that a significant fraction of the dark matter consists of a stable relic of unit charge [175], for masses below about 10^{11} GeV [161]. Thus, if the LSP considered in this section is charged, then the SUSY model is ruled out unless there is a lighter singlet fermion that mixes with the neutralinos, such as in the models discussed in Sections 14.6.1 and 16.1, or if R-parity-violating effects are present (see Section 16.5.7).

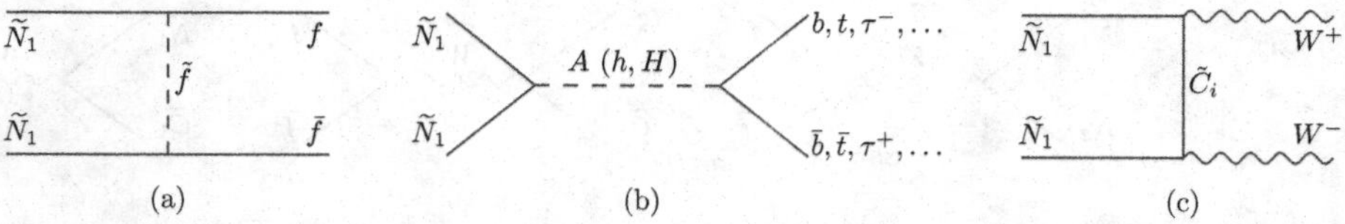

Fig. 15.12 The contributions to the annihilation cross section for neutralino dark matter LSPs from (a) t-channel slepton and squark exchange, (b) near-resonant annihilation through a Higgs boson (s-wave for A, and p-wave for h, H), and (c) t-channel chargino exchange.

dark matter candidate [177]. In particular, if the gravitino is the LSP, as in many gauge-mediated supersymmetry-breaking models, then gravitinos from reheating after inflation or from other sparticle decays could be the dark matter. However, such gravitinos would be impossible to detect directly, even if they have the right cosmological density today, as they interact too weakly. Consequently, the most attractive prospects for direct detection of supersymmetric dark matter are based on the idea that the lightest neutralino $\widetilde{N}_1$ is the LSP.

In the early universe, sparticles existed in thermal equilibrium with the ordinary Standard Model particles. As the universe cooled and expanded, the heavier sparticles could no longer be produced, and they eventually annihilated or decayed into neutralino LSPs. Some of the remaining LSPs pair-annihilated into SM final states. If there are other sparticles that are only slightly heavier than the LSP, then they existed in thermal equilibrium in comparable numbers to the LSP, and their co-annihilations are also important in determining the resulting dark matter density. Eventually, as the density decreased, the pair-annihilation and pair-creation processes go out of equilibrium, causing chemical decoupling. Thus, the $\widetilde{N}_1$ experienced "freeze out," with a density today determined by the small pair-annihilation rate and the subsequent dilution due to the expansion of the universe.

To obtain the observed dark matter density today, the thermal-averaged effective annihilation cross section times the relative speed v of the LSPs should be about

$$\langle \sigma v \rangle \sim 1\,\text{pb} \simeq \alpha^2/(144\,\text{GeV})^2, \tag{15.29}$$

where $\alpha \simeq 1/137$ [see eq. (D.104)]. That is, a neutralino LSP roughly has the correct (electroweak) interaction strength and mass. More detailed and precise estimates can be obtained with publicly available computer programs, so that the predictions of specific candidate models of supersymmetry breaking can be compared to eq. (15.27). Some of the diagrams that are typically important for neutralino LSP pair annihilation are shown in Fig. 15.12. Depending on the mass of $\widetilde{N}_1$, various other processes, including $\widetilde{N}_1\widetilde{N}_1 \to ZZ$, Zh, hh or even $W^{\pm}H^{\mp}$, ZA, hA, hH, HA, HH, AA, or H^+H^-, may also have been important. Some of the diagrams that can lead to co-annihilation of the LSPs with slightly heavier sparticles are shown in Figs. 15.13 and 15.14.

If $\widetilde{N}_1$ is mostly higgsino or mostly wino, then the annihilation diagram Fig. 15.12(c) and the co-annihilation mechanisms provided by Fig. 15.13 are typically much too

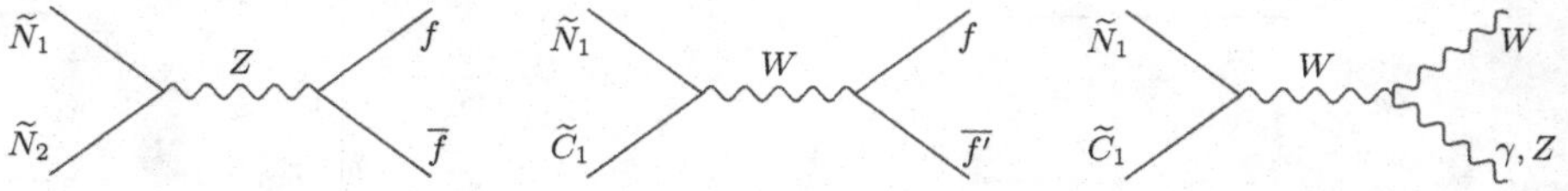

Fig. 15.13 Some contributions to the co-annihilation of dark matter $\widetilde{N}_1$ LSPs with slightly heavier $\widetilde{N}_2$ and $\widetilde{C}_1$. All three diagrams are particularly important if the LSP is higgsino-like, and the last two diagrams are important if the LSP is wino-like.

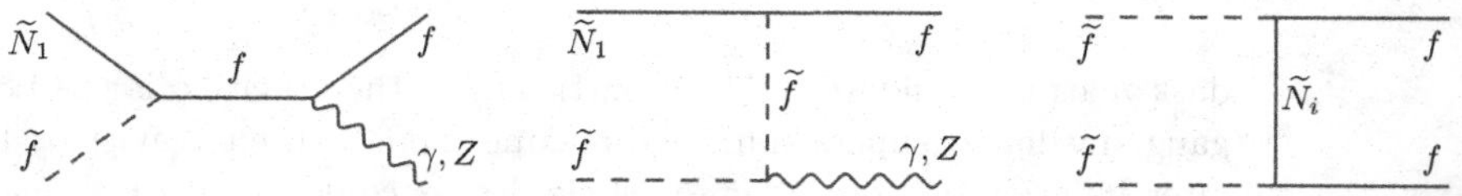

Fig. 15.14 Some contributions to the co-annihilation of dark matter $\widetilde{N}_1$ LSPs with slightly heavier sfermions, which in popular models are most plausibly staus (or perhaps top squarks).

efficient to account for the total dark matter density, unless the LSP is very heavy (e.g., of order 1 TeV for pure higgsinos and 2–3 TeV for winos). This is often considered to be somewhat at odds with the idea that supersymmetry is the solution to the hierarchy problem. However, for lighter higgsino-like or wino-like LSPs, nonthermal mechanisms can be invoked to provide the right dark matter abundance.

A recurring feature of many models of supersymmetry breaking is that the lightest neutralino is mostly bino; see the discussions in Sections 13.9 and 13.12. It turns out that in much of the parameter space not already ruled out by other collider constraints, the predicted relic density of a bino-like $\widetilde{N}_1$, is too high, because either the LSP couplings are too small or the sparticles are too heavy, or both, leading to an annihilation cross section that is too low. To avoid this, there must be significant contributions to $\langle \sigma v \rangle$. The possibilities can be classified qualitatively in terms of the diagrams that contribute most strongly to the annihilation.

First, if at least one sfermion is not too heavy, then the diagram of Fig. 15.12(a) is effective in reducing the dark matter density. In models with a bino-like $\widetilde{N}_1$, the most important contribution usually comes from $\widetilde{e}_R$, $\widetilde{\mu}_R$, and $\widetilde{\tau}_1$ slepton exchange. The parameter regime where this works out right is often referred to by the jargon "bulk region," because it corresponded to the main allowed region with dark matter density less than the critical density, before $\Omega_{\rm DM}h^2$ was accurately known and prior to the LEP and early LHC searches. However, the diagram of Fig. 15.12(a) is subject to a p-wave suppression. Hence, in many models, the sleptons that are light enough to reduce the relic density sufficiently are also light enough to be excluded already, or correspond to light Higgs bosons that are excluded, or have difficulties with other indirect constraints. In the mSUGRA/CMSSM framework described in Section 14.3, the viable bulk region remaining after LEP usually takes m_0 and $m_{1/2}$ to be less than about 100 GeV and 250 GeV, respectively, depending on other parameters.

Within the CMSSM, this part of parameter space has now been excluded by the LHC. If the final state of neutralino pair annihilation is instead $t\bar{t}$, then there is no p-wave suppression. This typically requires a top squark that is less than about 150 GeV heavier than the LSP, which in turn has $m_{\widetilde{N}_1}$ between about m_t and $m_t + 100$ GeV. This parameter regime does not occur in the CMSSM framework, but can be natural if the ratio of gluino and wino mass parameters, M_3/M_2, is smaller than the unification prediction of eq. (13.155) by a factor of a few.

A second way of annihilating excess bino-like LSPs to the correct density is obtained if $2m_{\widetilde{N}_1} \approx m_A$, or m_h, or m_H, as shown in Fig. 15.12(b), so that the cross section is near a resonance pole. An A resonance annihilation will be s-wave, and so more efficient than a p-wave h or H resonance. Therefore, the most commonly found realization involves annihilation through A. Because the $Ab\bar{b}$ coupling is proportional to $m_b \tan\beta$, this usually entails large values of $\tan\beta$. (Annihilation through h is also possible.) The region of parameter space where this happens is often called the "A-funnel" or "Higgs funnel" or "Higgs resonance region."

A third effective annihilation mechanism is obtained if $\widetilde{N}_1$ mixes to obtain a significant higgsino or wino admixture. Then both Fig. 15.12(c) and the co-annihilation diagrams of Fig. 15.13 can be important. In the focus point region of parameter space (e.g., see Ref. [162]), where $|\mu|$ is not too large, an LSP with a significant higgsino content can yield the correct relic abundance even for very heavy squarks and sleptons. This is motivated by focusing properties of the renormalization group equations, which allow $|\mu|^2 \ll m_0^2$ in the CMSSM framework. In particular, squarks are required to be very heavy, typically several TeV. This possibility is attractive, given the recent LHC results that exclude many models with squarks lighter than 1 TeV. It is also possible to arrange for just enough wino content in the LSP to do the job, by choosing M_1/M_2 appropriately.

A fourth possibility, the "sfermion co-annihilation region" of parameter space, is obtained if there is a sfermion that happens to be less than a few GeV heavier than the LSP. In many model frameworks, this is most naturally the lightest stau, but it could also be the lightest top squark A significant density of this sfermion will then coexist with the LSP around the freeze-out time, and so annihilations involving the sfermion with itself or with the LSP, including those of the type shown in Fig. 15.14, will further dilute the number of sparticles and so the eventual dark matter density.

It is important to keep in mind that a set of MSSM Lagrangian parameters that "fails" to predict the correct relic dark matter abundance by the standard thermal mechanisms is *not* ruled out as a model for collider physics. This is because simple extensions can completely change the relic abundance prediction without changing the predictions for colliders much or at all. For example, if the model predicts a neutralino dark matter abundance that is too small, one need only assume another sector (even a completely disconnected one) with a stable neutral particle, or that the dark matter is supplied by some nonthermal mechanism such as out-of-equilibrium decays of heavy particles. If the predicted neutralino dark matter abundance appears to be too large, one can assume that R-parity is slightly broken, so that the offending LSP decays before nucleosynthesis; this would require some other unspecified dark matter candidate. Or, the dark matter LSP might be some particle that the lightest neutralino decays into. One possibility is a gravitino LSP. Another example is obtained by extending the model to solve the strong CP problem with an invisible axion, which can allow the LSP to be a very weakly interacting axino (the fermionic supersymmetric partner of the axion). In such cases, the dark matter density after the lightest neutralino decays would be reduced compared to its naively predicted value by a factor of $m_{\text{LSP}}/m_{\widetilde{N}_1}$, provided that other sources for the LSP relic density are absent. A correct density for neutralino LSPs can also be obtained by assuming that they are produced nonthermally in reheating of the universe after neutralino freeze-out but before nucleosynthesis. Another possibility is entropy dilution of the LSP relic abundance, for example by the post-freeze-out decay of a heavier species. Finally, in the absence of a compelling explanation for the apparent cosmological constant, it seems possible that the standard model of cosmology will still need to be modified in ways not yet imagined.

If neutralino LSPs really make up the cold dark matter, then their local mass density in our neighborhood ought to be of order 0.3 $\text{GeV}\,\text{cm}^{-3}$ [much larger than the average density given by eq. (15.28)] in order to explain the dynamics of our own galaxy. LSP neutralinos could be detectable directly through their weak interactions with ordinary matter, or indirectly by their ongoing annihilations. However, the dark matter halo is subject to significant uncertainties in overall size, velocity, and clumpiness, so even if the Lagrangian parameters were known exactly, the signal rates would be quite indefinite, possibly even by orders of magnitude.

The direct detection of $\widetilde{N}_1$ depends on their elastic scattering off of heavy nuclei in a detector. At the parton level, $\widetilde{N}_1$ can interact with a quark by virtual exchange of squarks in the s-channel, or Higgs scalars or a Z boson in the t-channel. It can also scatter off of gluons through one-loop diagrams. The scattering mediated by neutral Higgs scalars is suppressed by tiny Yukawa couplings, but is coherent for the quarks and so can actually be the dominant contribution for nuclei with larger atomic weights, if the squarks are heavy. The strange quark has a larger Yukawa coupling than up and down, leading to important uncertainties from the nucleon strangeness content. The energy transferred to the nucleus in these elastic collisions is typically of order $\frac{1}{2}m_{\widetilde{N}_1}v^2$, where $v \sim 220$ $\text{km}\,\text{s}^{-1}$ is the typical dark matter velocity, or roughly 27 keV $(m_{\widetilde{N}_1}/100\ \text{GeV})$ per event. There are important backgrounds from natural radioactivity and cosmic rays, which can be reduced by shielding and pulse-shape analysis. A wide variety of current or future experiments are sensitive to some, but not all, of the parameter space of the MSSM that predicts a dark matter abundance in the range of eq. (15.27).

Another, more indirect, way to detect neutralino LSPs is through ongoing annihilations. This can occur in regions of space where the density is greatly enhanced. If the LSPs lose energy by repeated elastic scattering with ordinary matter, they can eventually become concentrated inside massive astronomical bodies like the Sun or the Earth. In that case, the annihilation of neutralino pairs into final states leading to neutrinos is the most important process, since no other particles can escape from the center of the object where the annihilation is going on. In particular, muon neutrinos and antineutrinos from $\widetilde{N}_1\widetilde{N}_1 \to W^+W^-$ or ZZ (or possibly $\widetilde{N}_1\widetilde{N}_1 \to \tau^+\tau^-$ or $\nu\overline{\nu}$, although these are p-wave suppressed) will travel large distances, and can be detected in neutrino telescopes. The neutrinos undergo a charged-current weak interaction in the earth, water, or ice under or within the detector, leading to energetic upward-going muons pointing back to the center of the Sun or Earth.

Another possibility is that neutralino LSP annihilation in the galactic center (or the halo) could result in high-energy photons from cascade decays of the heavy Standard Model particles that are produced. An even more striking signal could be provided by loop-induced annihilations $\widetilde{N}_1\widetilde{N}_1 \to \gamma\gamma$ or $\widetilde{N}_1\widetilde{N}_1 \to \gamma Z$, which would both produce essentially monochromatic photons with energy $E = m_{\widetilde{N}_1}$ (see Exercise 15.3). The final states γh, γH, and γA would also lead to monochromatic photons. These photons could be detected in air Cerenkov telescopes or in space-based detectors. There are also interesting possible signatures from neutralino LSP annihilation in the galactic halo producing detectable quantities of high-energy positrons, antiprotons, antideuterium, and/or anti-^{3}He.

Exercises

15.1 If supersymmetry has not been discovered when you read this, find the latest collider exclusion limits on: gluinos, top squarks, other squarks, sleptons, charginos, neutralinos, and the heavier Higgs bosons predicted by the MSSM [27]. In each case, what assumptions regarding production and decay modes are made in the experimental search results? Pay special attention to any such assumptions that are made for convenience or simplicity, but may not be valid in reality. In the "compressed supersymmetry" limit, the mass difference between the produced superpartners and the LSP becomes small. How do the exclusions vary as one approaches this limit?

15.2 One possibility for a good cold dark matter candidate, with a predicted thermal relic abundance in accord with observation, is a nearly pure higgsino-like neutralino LSP of mass near $\mu = 1.1$ TeV. Smaller μ is also possible, and would imply the predicted thermal abundance of dark matter is much less, so either nonthermal sources are needed or something else (for example, axions) could be the dark matter. Larger μ is also allowed, if there is a nonthermal source of late entropy production in the form of Standard Model particles only, or if the neutral higgsino is unstable due to R-parity violation. In any case, the higgsino-like neutralino $\widetilde{N}_1$ is accompanied by a slightly heavier chargino $\widetilde{C}_1^\pm$. At colliders, the latter can be produced in association with the LSP as $\widetilde{C}_1^\pm\widetilde{N}_1$ or in pairs $\widetilde{C}_1^+\widetilde{C}_1^-$. The dominant decay of the chargino is to a very soft charged pion, $\widetilde{C}_1^\pm \to \pi^\pm\widetilde{N}_1$.

(a) If all other superpartners are much heavier, then the mass difference between the charged and neutral higgsinos is determined by a one-loop self-energy computation involving virtual weak bosons:

$$\Delta m \equiv M_{\widetilde{C}_1} - M_{\widetilde{N}_1} = \tfrac{1}{2}\alpha m_Z f(\mu^2/m_Z^2), \tag{15.30}$$

where the loop integral function is

$$f(r) = \frac{\sqrt{r}}{\pi}\int_0^1 dx\,(2-x)\ln[1 + x/r(1-x)^2]. \tag{15.31}$$

How does $f(r)$ behave for large r? Evaluate Δm numerically for $\mu = 100$ GeV, 200 GeV, 500 GeV, 1000 GeV, and ∞.

(b) The charged higgsinos decay width can be computed as

$$\Gamma(\widetilde{C}_1^{\pm} \to \pi^{\pm}\widetilde{N}_1) = \frac{G_F}{\pi^2}\cos^2\theta_C\, f_\pi^2 (\Delta m)^3 \sqrt{1 - m_\pi^2/(\Delta m)^2}, \tag{15.32}$$

where the Cabibbo angle factor is $\cos^2\theta_C \approx 0.95$ and $f_\pi = 130$ MeV is the pion decay constant. For the values of μ given in part (a), what are the momentum of the charged pion in the $\widetilde{C}_1$ rest frame, and the proper decay length $c\tau = c/\Gamma$, in millimeters? One way of searching for $\widetilde{C}_1$ production is by looking for "disappearing tracks," which end where $\widetilde{C}_1$ decays, since the pion is too soft to register. Another is by observing an anomalous rate of energy deposition dE/dx along the $\widetilde{C}_1$ track, due to its large mass.

NOTE: This exercise uses material taken from Ref. [178].

15.3 Consider the indirect detection of neutralino dark matter LSPs through the one-loop decays $\widetilde{N}_1\widetilde{N}_1 \to \gamma\gamma$ and $\widetilde{N}_1\widetilde{N}_1 \to \gamma Z$. In the limit of large $m_{\widetilde{N}_1}$, and taking the dark matter to be nearly at rest, what is the difference in monochromatic photon energies for these two kinds of annihilations? Evaluate your result numerically for the masses that give a thermally produced relic abundance in agreement with observations, $m_{\widetilde{N}_1} \approx 1.1$ TeV and 2.8 TeV for nearly pure higgsinos and winos, respectively.

15.4 Suppose that the gluino is the lightest among all the colored superpartners. Some of the possible gluino decay modes are exhibited in Fig. 15.2.

(a) Which gluino decay modes should have the largest branching ratios?

(b) If the supersymmetric model contains a massless goldstino or a very light gravitino, then the decay $\widetilde{g} \to g\widetilde{G}$ is possible. Estimate its partial decay width and determine whether this decay mode dominates or is negligible compared with the decay mode considered in part (a).

(c) If both the gravitino and squarks are heavier than the gluino, then a two-body decay mode of the gluino still exists, $\widetilde{g} \to g\widetilde{N}_1$. Exhibit the one-loop diagrams that yield the $g\widetilde{N}_1$ final state and estimate its partial decay width. Under what circumstances does this decay mode compete with or dominate the decay modes considered in part (a)?

16 Beyond the MSSM

There are a number of reasons to consider realistic models of supersymmetry that go beyond the MSSM. The first reason is connected to the μ problem, which was introduced in Section 14.6. Naively, one would expect the dimensionful parameter μ to be of the same order of magnitude as the largest possible mass scale. Instead, a phenomenologically reliable theory requires μ to be of order the supersymmetry-breaking scale. Since μ is the coefficient of a supersymmetric term of the Lagrangian, the association of μ with a scale that arises from the supersymmetry-breaking mechanism is puzzling.

Perhaps a more severe deficiency of the MSSM is that it inherits one of the fundamental problems of the Standard Model, namely that there are no right-handed neutrinos (and their superpartners) in the MSSM and consequently neutrino masses are exactly zero. However, given the overwhelming evidence for neutrino masses and mixing, any viable model of fundamental particles must provide a mechanism for generating neutrino masses. In supersymmetric extensions of the Standard Model, various mechanisms exist for producing massive neutrinos. Although one can devise models for generating massive Dirac neutrinos, the most common approaches for incorporating neutrino masses are based on L-violating extensions of the MSSM, which generate massive Majorana neutrinos. There are two examples of L-violating supersymmetric models that are of particular interest. In the first example, we construct the seesaw-extended MSSM, which generalizes the seesaw extension of the Standard Model. In the second example, we relax the assumption of R-parity conservation in the MSSM [16, 179].

R-parity-violating (RPV) supersymmetry allows for the possibility of both baryon and lepton number violation. If present, the lepton number violation can provide a mechanism for generating neutrino masses without introducing right-handed neutrinos (e.g., see [B142]). The possibility of baryon number violation also introduces new phenomena that are absent in the MSSM. Of course, RPV supersymmetry models are also highly constrained, since in an unconstrained RPV model, the proton decay rate would be many orders of magnitude larger than the present experimental bound [180].

In this chapter, we shall investigate three particular extensions of the MSSM.[1] We first consider the next-to-minimal supersymmetric extension of the Standard Model, which adds a new chiral superfield S that couples to $H_u H_d$ in a renormalizable superpotential term, while employing a discrete symmetry to eliminate the $H_u H_d$

[1] A textbook treatment of a variety of extensions of the MSSM can be found in [B27].

term of the superpotential. If the scalar field that resides in S gains a VEV, then an effective μ-term will be generated proportional to that VEV, thereby providing a mechanism for understanding the origin of the μ-term. Second, we construct the seesaw-extended MSSM while preserving R-parity but providing a mechanism for neutrino masses. Finally, we relax the assumption of R-parity in the MSSM and explore some of the consequences.

16.1 The Next-to-Minimal Supersymmetric Standard Model

The simplest possible extension of the particle content of the MSSM is obtained by adding a new gauge-singlet chiral supermultiplet with even matter parity. The resulting model is called the Next-to-Minimal Supersymmetric Standard Model (NMSSM). Reviews of the NMSSM can be found in Refs. [181, 182]. The most general renormalizable superpotential for this field content is

$$W_{\text{NMSSM}} = W_{\text{MSSM}} + \lambda S H_u H_d + \tfrac{1}{3}\kappa S^3 + \tfrac{1}{2}\mu_S S^2, \tag{16.1}$$

where S stands for both the new chiral supermultiplet and its scalar component and

$$H_u H_d \equiv \epsilon_{ij} H_u^i H_d^j = H_u^+ H_d^- - H_u^0 H_d^0 \,. \tag{16.2}$$

The λ term establishes that S should have PQ charge -2 from Table 14.2, but then the presence of either nonzero κ or μ_S explicitly breaks the PQ symmetry, avoiding a dangerous pseudo-Goldstone boson. The soft supersymmetry-breaking Lagrangian is

$$\mathscr{L}_{\text{soft}}^{\text{NMSSM}} = \mathscr{L}_{\text{soft}}^{\text{MSSM}} - \left(a_\lambda S H_u H_d + \tfrac{1}{3} a_\kappa S^3 + \tfrac{1}{2} b_S S^2 + \text{h.c.}\right) - m_S^2 |S|^2 \,. \tag{16.3}$$

It should be noted that one can also include linear (tadpole) terms, ξS in the superpotential, and tS in $\mathscr{L}_{\text{soft}}^{\text{NMSSM}}$. In global supersymmetry, the supersymmetric tadpole parameter ξ can always be removed by redefining S by a constant shift, but it will then reappear in the parameter t, where it cannot be removed in general. This could be problematic, since radiative corrections can generate the tadpole couplings with magnitudes of order

$$\xi \sim \Lambda m_{\text{soft}} \,, \qquad t \sim \Lambda m_{\text{soft}}^2, \tag{16.4}$$

where m_{soft} is the soft supersymmetry-breaking scale, and Λ could be of order M_P when supergravity effects are included. Such tadpoles can arise whenever the theory contains a singlet chiral superfield. When the singlet superfield couples to the doublet Higgs superfield, as here, they are dangerous unless they are suppressed for some reason. This is because they would give rise to very large VEVs for the scalar component of S, which would in turn provoke large VEVs for the scalar components of H_u and H_d. However, these tadpoles can be suppressed or forbidden if S is actually charged under some additional hidden symmetry (for example, a gauged discrete symmetry) at very high energies.

One of the virtues of the NMSSM is that it can provide a solution to the μ problem discussed in Section 14.6. To understand this, suppose we set $\mu_S = \mu = 0$ and $\xi = 0$ so that there are no mass terms or dimensionful parameters in the superpotential at all, and also set the corresponding terms $b_S = b = 0$ and $t = 0$ in the supersymmetry-breaking Lagrangian. If λ, κ, a_λ, and a_κ are chosen auspiciously, then phenomenologically acceptable VEVs will be induced for S, H_u^0, and H_d^0, each parametrically of order m_{soft}. The absence of dimensionful terms in W_{NMSSM}, and the corresponding terms in $V_{\text{soft}}^{\text{NMSSM}}$, can be enforced by introducing a new symmetry. The simplest way is to notice that the new superpotential and Lagrangian will be invariant under a $\mathbb{Z}_3$ discrete symmetry, under which every field in a chiral supermultiplet transforms as $\Phi \to e^{2\pi i/3}\Phi$, and all gauge and gaugino fields are inert. Imposing this symmetry indeed eliminates μ, μ_S, b, b_S, and t.

However, if this $\mathbb{Z}_3$ symmetry were exact, then because it must be spontaneously broken by the VEVs of S, H_u, and H_d, domain walls are expected to be produced in the electroweak symmetry-breaking phase transition in the early universe. This occurs because there are degenerate minima of the potential (related by the $\mathbb{Z}_3$ symmetry), and causally disconnected regions of the universe would fall randomly into these three minima. The domain walls between the regions with different minima would come to dominate the cosmological energy density, causing unobserved anisotropies in the microwave background radiation and/or ruining the successful predictions of primordial nucleosynthesis. Several ways of avoiding this problem have been proposed, including the assumption that inflation takes place after the domain walls are formed. One can also avoid the domains by embedding the discrete symmetry into a continuous gauged symmetry at very high energies. Finally, the $\mathbb{Z}_3$ degeneracy of the vacuum can be eliminated by allowing either higher-dimensional (nonrenormalizable) terms in the Lagrangian, or a highly suppressed μ-term, to explicitly break the discrete symmetry.

Henceforth, we shall adopt the $\mathbb{Z}_3$ discrete symmetry and forbid dimensionful parameters in the superpotential and in the holomorphic soft SUSY-breaking terms. Focusing first on the Higgs sector of the $\mathbb{Z}_3$-invariant NMSSM, the Higgs scalar potential is given by

$$\mathcal{V} = \mathcal{V}_{\text{SUSY}} + \mathcal{V}_{\text{soft}}\,, \tag{16.5}$$

where the supersymmetric contributions to the Higgs scalar potential consist of an F-term contribution and a D-term contribution [see eq. (8.75)],

$$\begin{aligned}\mathcal{V}_{\text{SUSY}} = {}& \tfrac{1}{8}(g^2+g'^2)(H_u^\dagger H_u - H_d^\dagger H_d)^2 + \tfrac{1}{2}g^2|H_d^\dagger H_u|^2 + |\lambda|^2|H_uH_d|^2 \\ &+ |\lambda|^2 S^\dagger S(H_u^\dagger H_u + H_d^\dagger H_d) + |\kappa|^2(S^\dagger S)^2 + [\lambda\kappa^*(S^\dagger)^2 H_uH_d + \text{h.c.}],\end{aligned} \tag{16.6}$$

and the soft supersymmetry-breaking terms contributing to the Higgs scalar potential are given by

$$\mathcal{V}_{\text{soft}} = m_S^2 S^\dagger S + m_{H_u}^2 H_u^\dagger H_u + m_{H_d}^2 H_d^\dagger H_d + \left(a_\lambda S H_u H_d + \tfrac{1}{3}a_\kappa S^3 + \text{h.c.}\right), \tag{16.7}$$

where the parameters λ, κ, a_λ, and a_κ are generically complex, whereas $m^2_{H_u}$, $m^2_{H_d}$, and m^2_S are real.

We shall minimize the scalar potential given in eq. (16.5) by assuming that only the neutral Higgs fields acquire a nonvanishing VEV:[2]

$$\langle H_u^0 \rangle = \frac{v_u}{\sqrt{2}}, \qquad \langle H_d^0 \rangle = \frac{v_d}{\sqrt{2}}, \qquad \langle S \rangle = \frac{v_s}{\sqrt{2}} \equiv s, \tag{16.8}$$

with $v \equiv (v_u^2 + v_d^2)^{1/2} \simeq 246$ GeV. It is convenient to perform field redefinitions of the Higgs fields such that v_u, v_d, and s are real and nonnegative. In this convention, we define

$$s_\beta \equiv \sin\beta = \frac{v_u}{v}, \qquad c_\beta \equiv \cos\beta = \frac{v_d}{v}, \tag{16.9}$$

where $0 \leq \beta \leq \frac{1}{2}\pi$, in which case the Higgs fields of the NMSSM can be written as

$$H_u = \begin{pmatrix} H_u^+ \\ \frac{1}{\sqrt{2}}(v s_\beta + h_u + i a_u) \end{pmatrix}, \qquad H_d = \begin{pmatrix} \frac{1}{\sqrt{2}}(v c_\beta + h_d + i a_d) \\ H_d^- \end{pmatrix},$$

$$S = \frac{1}{\sqrt{2}}(v_s + h_s + i a_s). \tag{16.10}$$

The scalar potential minimum conditions are easily obtained by inserting eq. (16.10) into eq. (16.5) and then setting the coefficients of the terms that are linear in h_u, a_u, h_d, a_d, h_s, and a_s, respectively, to zero, which yields six real equations. Under our assumption that v_d, v_u, and s are nonzero, the two equations obtained by setting the coefficients of the terms linear in a_d and a_u to zero yield one nontrivial relation. Hence, we are left with five independent conditions:

$$m^2_{H_d} = s\,\mathrm{Re}(a_\lambda + \lambda\kappa^* s)\tan\beta - \tfrac{1}{8}(g^2 + g'^2)v^2 c_{2\beta} - \tfrac{1}{2}|\lambda|^2(v^2 s_\beta^2 + 2s^2), \tag{16.11}$$

$$m^2_{H_u} = s\,\mathrm{Re}(a_\lambda + \lambda\kappa^* s)\cot\beta + \tfrac{1}{8}(g^2 + g'^2)v^2 c_{2\beta} - \tfrac{1}{2}|\lambda|^2(v^2 c_\beta^2 + 2s^2), \tag{16.12}$$

$$m^2_S = \frac{v^2 s_{2\beta}}{4s}\mathrm{Re}(a_\lambda + 2\lambda\kappa^* s) - s\,\mathrm{Re}\,a_\kappa - \tfrac{1}{2}|\lambda|^2 v^2 - 2|\kappa|^2 s^2 = 0, \tag{16.13}$$

$$\mathrm{Im}(a_\lambda + \lambda\kappa^* s) = 0, \tag{16.14}$$

$$\tfrac{1}{2}v^2 s_{2\beta}\,\mathrm{Im}(a_\lambda + 2\lambda^*\kappa s) - 2s^2\,\mathrm{Im}\,a_\kappa = 0, \tag{16.15}$$

where $s_{2\beta} \equiv \sin 2\beta$, $c_{2\beta} = \cos 2\beta$. Note that eqs. (16.14) and (16.15) yield[3]

$$\mathrm{Im}\,a_\lambda = -s\,\mathrm{Im}(\lambda\kappa^*), \qquad \mathrm{Im}\,a_\kappa = \frac{3v^2}{4s}s_{2\beta}\,\mathrm{Im}(\lambda^*\kappa), \tag{16.16}$$

[2] We define $s \equiv v_s/\sqrt{2}$ to eliminate factors of $\sqrt{2}$ in eqs. (16.11)–(16.16).

[3] In obtaining eq. (16.16), it is convenient to employ $\mathrm{Im}(\lambda\kappa^*)$ in the equation for $\mathrm{Im}\,a_\lambda$ and $\mathrm{Im}(\lambda^*\kappa) = -\mathrm{Im}(\lambda\kappa^*)$ in the equation for $\mathrm{Im}\,a_\kappa$. One motivation for this choice is that the beta functions of a_κ and a_λ are proportional to κ (not κ^*) and λ (not λ^*), respectively. Hence, it is more natural to write expressions in terms of combinations where a_κ and κ appear together and likewise where a_λ and λ appear together.

which implies that the quantities

$$b \equiv s(a_\lambda + \lambda\kappa^* s)\,, \tag{16.17}$$

$$b' \equiv \tfrac{3}{2}\lambda^*\kappa v_u v_d - s a_\kappa\,, \tag{16.18}$$

are real in a convention where v_u, v_d, and s are real and nonnegative. Moreover, the only independent physical complex phase of the NMSSM Higgs sector is the phase of $\lambda\kappa^*$.

With S acquiring a VEV, an effective μ-term for $H_u H_d$ will arise from eq. (16.1), with

$$\mu \equiv \lambda s. \tag{16.19}$$

Likewise, an effective μ_S-term for S^2 will arise from eq. (16.1), with $\mu_S \equiv 2\kappa s$. Both μ and μ_S are determined by dimensionless couplings and soft SUSY-breaking terms of order $m_{\rm soft}$, instead of being free parameters that are conceptually independent of SUSY breaking. With the conventions of real nonnegative VEVS as chosen here, the complex phase of μ is the same as that of λ, and the complex phase of μ_S is the same as that of κ.

There is also an effective B-term that can be identified with b given by eq. (16.17), which is a real quantity [in light of eq. (16.14)] that is naturally of order $m^2_{\rm soft}$. Noting that $m_Z^2 = \frac{1}{4}(g^2 + g'^2)v^2$, it follows that eqs. (16.11) and (16.12) can be rewritten as

$$|\mu|^2 + m^2_{H_d} = b\tan\beta - \tfrac{1}{2}m_Z^2 c_{2\beta} - \tfrac{1}{2}|\lambda|^2 v^2 s^2_\beta\,, \tag{16.20}$$

$$|\mu|^2 + m^2_{H_u} = b\cot\beta + \tfrac{1}{2}m_Z^2 c_{2\beta} - \tfrac{1}{2}|\lambda|^2 v^2 c^2_\beta\,. \tag{16.21}$$

The MSSM limit of the NMSSM Higgs sector can be defined to be a limit in which $\lambda \sim \kappa \to 0$, $a_\lambda \sim a_\kappa \to 0$, and $s \to \infty$, while keeping λs, κs, $a_\lambda s$, and $a_\kappa s$ fixed (which implies that μ, b, and b' are also kept fixed in the MSSM limit). Indeed, by comparing with eqs. (13.84) and (13.85), we see that we recover the MSSM results for the 2HDM scalar potential minimum conditions by setting $\lambda \to 0$ in eqs. (16.20) and (16.21).

Finally, we note that when S acquires a VEV, an effective b_S-term is generated that is given by

$$b_S = 2sa_\kappa - v_u v_d \lambda^*\kappa = -\tfrac{2}{3}(b' - 2sa_\kappa)\,, \tag{16.22}$$

which is naturally of order $m^2_{\rm soft}$. The complex phase of b_S is related to the complex phase of a_κ, since b' is real in the convention of positive scalar field VEVs.

The charged scalar mass matrix can be obtained by inserting eq. (16.10) into eq. (16.5) and examining the coefficients of the terms that are quadratic in the scalar fields H_u^+, H_d^- and their complex conjugates after employing eqs. (16.11)–(16.13). The resulting 2×2 charged scalar squared-mass matrix possesses two orthonormal eigenstates corresponding to the charged Goldstone field,

$$G^+ = -c_\beta (H_d^-)^\dagger + s_\beta H_u^+\,, \tag{16.23}$$

and the charged Higgs field,

$$H^+ = s_\beta (H_d^-)^\dagger + c_\beta H_u^+ \,. \tag{16.24}$$

Taking the trace of the charged scalar squared-mass matrix, one obtains

$$m_{H^\pm}^2 = m_W^2 - \tfrac{1}{2}|\lambda|^2 v^2 + \frac{b}{s_\beta c_\beta} \,. \tag{16.25}$$

This should be compared with the tree-level expression for the mass of the charged Higgs boson in the MSSM, which can be obtained from eq. (16.25) by setting $\lambda = 0$ and identifying $\overline{M}_A$ with the mass of the CP-odd Higgs scalar of the MSSM [see eq. (6.234)]. In particular, the tree-level MSSM inequality $m_{H^\pm} \geq m_W$ can be violated in the NMSSM, although this observation is less interesting in light of the current experimental mass bounds for charged Higgs bosons that have excluded charged Higgs boson masses below m_W.

Likewise, the 6×6 neutral scalar squared-mass matrix can be obtained by inserting eq. (16.10) into eq. (16.5) and examining the coefficients of the terms that are quadratic in the scalar fields, a_u, h_d, a_d, h_s, and a_s after employing eqs. (16.11)–(16.13) to eliminate $m_{H_u}^2$, $m_{H_d}^2$, and m_S^2. To analyze the neutral scalar squared-mass matrix, one first identifies the massless neutral Goldstone boson eigenstate,

$$G^0 = -c_\beta a_d + s_\beta a_u \,. \tag{16.26}$$

The CP-odd scalar interaction eigenstate fields orthogonal to G^0 are a_s and

$$a \equiv s_\beta a_d + c_\beta a_u \,. \tag{16.27}$$

Then, by a suitable orthogonal similarity transformation, one can transform the 6×6 squared-mass matrix into a block-diagonal form whose first row and column consist entirely of zeros (corresponding to the massless Goldstone boson) and a 5×5 block which has the form

$$\mathcal{M}_N^2 = \begin{pmatrix} M_{HH}^2 & M_{HA}^2 \\ (M_{HA}^2)^{\mathsf{T}} & M_{AA}^2 \end{pmatrix} , \tag{16.28}$$

where the 3×3 matrix M_{HH}^2, with respect to the basis of CP-even interaction eigenstate scalar fields, $\{h_d, h_u, h_s\}$, is given by

$$M_{HH}^2 = \begin{pmatrix} m_Z^2 c_\beta^2 + \overline{M}_A^2 c_\beta^2 & (|\lambda|^2 v^2 - m_Z^2 - \overline{M}_A^2) s_\beta c_\beta & \frac{v}{s\sqrt{2}} s_\beta \left[2|\mu|^2 \cot\beta - b - s\,\mathrm{Re}(\mu^*\kappa)\right] \\ (|\lambda|^2 v^2 - m_Z^2 - \overline{M}_A^2) s_\beta c_\beta & m_Z^2 s_\beta^2 + \overline{M}_A^2 c_\beta^2 & \frac{v}{s\sqrt{2}} c_\beta \left(2|\mu|^2 \tan\beta - b - s\,\mathrm{Re}(\mu^*\kappa)\right] \\ \frac{v}{s\sqrt{2}} s_\beta \left[2|\mu|^2 \cot\beta - b - s\,\mathrm{Re}(\mu^*\kappa)\right] & \frac{v}{s\sqrt{2}} c_\beta \left[2|\mu|^2 \tan\beta - b - s\,\mathrm{Re}(\mu^*\kappa)\right] & 4|\kappa|^2 s^2 + \frac{1}{2} v^2 b s_\beta c_\beta / s^2 + \frac{1}{2}\,\mathrm{Re}\, b_S \end{pmatrix} , \tag{16.29}$$

the 2×2 matrix $\mathcal{M}^2_{AA}$, with respect to the basis of CP-odd interaction eigenstate scalar fields, $\{a, a_s\}$, is given by

$$\mathcal{M}^2_{AA} = \begin{pmatrix} \overline{M}^2_A & \frac{v}{s\sqrt{2}}[b - 3s\,\mathrm{Re}(\mu^*\kappa)] \\ \frac{v}{s\sqrt{2}}[b - 3s\,\mathrm{Re}(\mu^*\kappa)] & \frac{1}{2}v^2 b s_\beta c_\beta/s^2 - \frac{3}{2}\mathrm{Re}\,b_S \end{pmatrix}, \tag{16.30}$$

where

$$\overline{M}^2_A \equiv \frac{b}{s_\beta c_\beta}, \tag{16.31}$$

and the 3×2 matrix $\mathcal{M}^2_{HA}$, which mixes the CP-even and CP-odd scalar interaction eigenstate fields, is given by

$$\mathcal{M}^2_{HA} = \frac{v}{\sqrt{2}}\,\mathrm{Im}(\lambda\kappa^*) \begin{pmatrix} 0 & -3ss_\beta \\ 0 & -3sc_\beta \\ s & 2\sqrt{2}vs_\beta c_\beta \end{pmatrix}. \tag{16.32}$$

In the MSSM limit of the NMSSM Higgs sector introduced below eq. (16.21), the upper 2×2 matrix block of $\mathcal{M}^2_{HH}$ coincides with the 2×2 squared-mass matrix for the CP-even Higgs scalars of the MSSM, and $\overline{M}_A$ is the mass of the CP-odd Higgs scalar of the MSSM. Moreover, $\mathcal{M}^2_{HA} \to 0$ and the mixing of the CP-even and CP-odd MSSM-like scalars with the CP-even and CP-odd singlet-like scalars, respectively, vanish in the MSSM limit.

It is convenient to introduce the Higgs basis as follows. We first define two hypercharge $Y = \frac{1}{2}$, weak isospin doublet scalar fields, Φ_d and Φ_u [cf. eq. (6.218)]:

$$\Phi^j_d \equiv \epsilon_{ij} H^{\dagger i}_d, \qquad \Phi^j_u = H^j_u. \tag{16.33}$$

Then, the Higgs basis fields are defined by

$$H_1 = \begin{pmatrix} H_1^+ \\ H_1^0 \end{pmatrix} \equiv c_\beta \Phi_d + s_\beta \Phi_u = \begin{pmatrix} G^+ \\ \frac{1}{\sqrt{2}}\left(v + c_\beta h_d + s_\beta h_u + iG^0\right) \end{pmatrix}, \tag{16.34}$$

$$H_2 = \begin{pmatrix} H_2^+ \\ H_2^0 \end{pmatrix} \equiv -s_\beta \Phi_d + c_\beta \Phi_u = \begin{pmatrix} H^+ \\ \frac{1}{\sqrt{2}}\left(-s_\beta h_d + c_\beta h_u + ia\right) \end{pmatrix}, \tag{16.35}$$

which have the property that $\langle H_1^0 \rangle = v/\sqrt{2}$ and $\langle H_2^0 \rangle = 0$. Note that the definition of S is not modified when defining the Higgs basis of the NMSSM.

In Exercise 16.4, the squared-mass matrices $\mathcal{M}^2_{HH}$, $\mathcal{M}^2_{AA}$, and $\mathcal{M}^2_{HA}$ are evaluated when expressed with respect to the neutral Higgs basis fields. Here, we simply note that the upper diagonal 2×2 block of the transformed $\mathcal{M}^2_{HH}$ is given by

$$\begin{pmatrix} m_Z^2 c_{2\beta}^2 + \frac{1}{2}|\lambda|^2 v^2 s_{2\beta}^2 & -(m_Z^2 - \frac{1}{2}|\lambda|^2 v^2) s_{2\beta} c_{2\beta} \\ -(m_Z^2 - \frac{1}{2}|\lambda|^2 v^2) s_{2\beta} c_{2\beta} & \overline{M}^2_A + (m_Z^2 - \frac{1}{2}|\lambda|^2 v^2) s_{2\beta}^2 \end{pmatrix}. \tag{16.36}$$

In light of eq. (6.166), it follows that if h is the lightest neutral scalar, then

$$m_h^2 \leq m_Z^2 c_{2\beta}^2 + \tfrac{1}{2}|\lambda|^2 v^2 s_{2\beta}^2 \,. \tag{16.37}$$

This tree-level inequality should be compared with the corresponding result of the MSSM Higgs sector, which is attained in the limit of $\lambda = 0$ [see eq. (6.226)]. This result shows that it is possible for the tree-level value of the Higgs mass to be as large as its observed value of 125 GeV if λ and β are chosen appropriately. Of course, a more accurate computation of the scalar masses requires one to include radiative corrections. The effects of radiative corrections $\Delta V(v_u, v_d, s)$ to the effective potential are included by replacing $m_S^2 \to m_S^2 + [\partial(\Delta V)/\partial s]/2s$, in addition to eq. (13.89).

In our analysis above, we have assumed that the scalar fields VEVs do not violate $\mathrm{U(1)_{EM}}$. This assumption is consistent if $m_{H^\pm}^2 > 0$. Using eq. (16.25), it follows that

$$\tfrac{1}{2}|\lambda|^2 v^2 < m_W^2 + \overline{M}_A^2 \,. \tag{16.38}$$

Likewise, the stability of the electroweak symmetry-breaking vacuum requires that all the principal minors of $\mathcal{M}_N^2$ should be positive (after removing the neutral Goldstone boson state). This implies that M_{HH}^2 and M_{AA}^2 are positive definite matrices, or equivalently all the principal minors of M_{HH}^2 and M_{AA}^2 are positive. For example, since M_{AA}^2 is positive definite it follows that $\overline{M}_A^2 > 0$ and $\det M_{AA}^2 > 0$. In particular,

$$\det M_{AA}^2 = \overline{M}_A^2 \left[\tfrac{3}{2} v^2\, \mathrm{Re}(\lambda^* \kappa) s_\beta c_\beta - s\, \mathrm{Re}\, a_\kappa\right] - \tfrac{3}{2} v^2 \left[\mathrm{Re}(\mu^* \kappa)\right]^2 > 0 \,. \tag{16.39}$$

Hence, it follows that the coefficient of $\overline{M}_A^2$ in eq. (16.39) is positive. Using this result along with eqs. (16.9) and (16.31), we can conclude that[4]

$$b > 0\,, \qquad b' > 0\,. \tag{16.40}$$

In addition, by employing eqs. (16.17) and (16.31), we can simplify the inequality given in eq. (16.39) to obtain the following form:

$$\tfrac{3}{2} v_u v_d\, \mathrm{Re}\, a_\lambda\, \mathrm{Re}(\mu^* \kappa) - bs\, \mathrm{Re}\, a_\kappa > 0 \,. \tag{16.41}$$

Next, consider the conditions that arise from the positivity of the principal minors of M_{HH}^2. The determinant of eq. (16.36) yields

$$\overline{M}_A^2 m_Z^2 c_{2\beta}^2 + \tfrac{1}{2}|\lambda|^2 v^2 s_{2\beta}^2 (\overline{M}_A^2 + m_Z^2 - \tfrac{1}{2}|\lambda|^2 v^2) \,, \tag{16.42}$$

which is positive in light of eq. (16.38). In addition, we require the 33 element of M_{HH}^2 to be positive, which can be rewritten as

$$4|\kappa|^2 s^2 + s\, \mathrm{Re}\, a_\kappa + \frac{v_u v_d}{2s} \mathrm{Re}\, a_\lambda > 0 \,. \tag{16.43}$$

Combining eqs. (16.40) and (16.43) shows that $\mathrm{Re}\, a_\kappa$ is bounded both from above

[4] Note that eqs. (16.16)–(16.18) imply that, in a convention where the VEVs v_u, v_d, and s are real and positive, the quantities b and b' are real. By imposing the stability conditions for the electroweak symmetry-breaking vacuum, we have learned that these quantities are in fact positive.

and from below. Studies of NMSSM parameter scans reveal that negative values of a_κ are typically favored. For example, in the MSSM limit of the NMSSM Higgs sector, eqs. (16.41) and (16.43) yield

$$-4|\kappa|^2 s^2 < s\,\mathrm{Re}\,a_\kappa < 0\,, \qquad \text{in the MSSM limit.} \tag{16.44}$$

Finally, under the assumption that the electroweak symmetry-breaking vacuum is stable (and not metastable with a finite lifetime), it follows that the energy density of the electroweak symmetry-breaking vacuum, denoted by $\mathcal{V}_{\rm vac}$, lies below the electroweak-conserving vacuum energy density (where all scalar field VEVs vanish), the latter being equal to zero. Hence, we shall demand that $\mathcal{V}_{\rm vac} < 0$ in the electroweak breaking vacuum, where

$$\begin{aligned}\mathcal{V}_{\rm vac} &= \tfrac{1}{2}m^2_{H_d}v_d^2 + \tfrac{1}{2}m^2_{H_u}v_u^2 + m_S^2 s^2 - s^2 v_u v_d\,\mathrm{Re}(\lambda^*\kappa) - s v_u v_d\,\mathrm{Re}\,a_\lambda + \tfrac{2}{3}s^3\,\mathrm{Re}\,a_\kappa \\ &\quad + \tfrac{1}{32}(g^2+g'^2)(v_u^2 - v_d^2)^2 + \tfrac{1}{4}|\lambda|^2\{v_u^2 v_d^2 + 2s^2(v_u^2+v_d^2)\} + |\kappa|^2 s^4 \\ &= -\tfrac{1}{2}|\lambda|^2 v^2 s^2 - |\kappa|^2 s^4 - \tfrac{1}{8}m_Z^2 v^2 c^2_{2\beta} + \tfrac{1}{3}b' s^2 + \tfrac{1}{4}v^2 s_{2\beta}(b - \tfrac{1}{4}|\lambda|^2 v^2 s_{2\beta})\,,\end{aligned} \tag{16.45}$$

after making use of eqs. (16.11)–(16.13) to eliminate $m^2_{H_d}$, $m^2_{H_u}$, and m_S^2 and employing eqs. (16.17) and (16.22). Hence, we can rewrite the condition $\mathcal{V}_{\rm vac} < 0$ in the following form:

$$\tfrac{1}{2}|\lambda|^2 v^2 s^2 + |\kappa|^2 s^4 + \tfrac{1}{8}v^2(m_Z^2 c^2_{2\beta} + \tfrac{1}{2}|\lambda|^2 v^2 s^2_{2\beta}) - \tfrac{1}{3}b' s^2 - \tfrac{1}{2}v_u v_d b > 0. \tag{16.46}$$

In contrast to the 2HDM (and the MSSM Higgs sector), where $V_{\rm min} < 0$ is automatically satisfied for a stable electroweak breaking vacuum [see eq. (6.54)], the condition that the energy density of the NMSSM electroweak breaking vacuum lies below that of the electroweak conserving vacuum imposes two additional constraints on the NMSSM parameters due to the presence of the last two terms that provide negative contributions to eq. (16.46) in light of eq. (16.40). First, one obtains an upper bound on b,

$$b < \frac{1}{s_{2\beta}}\left\{2|\mu|^2 + \tfrac{1}{2}(m_Z^2 c^2_{2\beta} + \tfrac{1}{2}|\lambda|^2 v^2 s^2_{2\beta}) + \frac{4s^2}{v^2}(|\kappa|^2 s^2 - \tfrac{1}{3}b')\right\}. \tag{16.47}$$

Second, given that $b > 0$, it follows that the right-hand side of the above inequality must be positive. This requirement yields an upper bound on b',

$$b' < 3|\kappa|^2 s^2 + \frac{3v^2}{8s^2}\left[m_Z^2 c^2_{2\beta} + \tfrac{1}{2}|\lambda|^2 v^2 s^2_{2\beta} + 4|\mu|^2\right]. \tag{16.48}$$

Note that in the MSSM limit of the NMSSM Higgs sector, eq. (16.48) implies that $s\,\mathrm{Re}\,a_\kappa > -3|\kappa|^2 s^2$, which is compatible with eq. (16.44). Consequently, in the MSSM limit eq. (16.47) yields $b < \infty$, implying no upper bound for the mass of the CP-odd scalar, $\overline{M}_A$, as expected. Likewise, eqs. (16.25) and (16.47) yield an upper bound on $m^2_{H^\pm}$,

$$s^2_{2\beta}m^2_{H^\pm} < 4|\mu|^2 + m_Z^2 c^2_{2\beta} + m_W^2 s^2_{2\beta} + \frac{8s^2}{v^2}(|\kappa|^2 s^2 - \tfrac{1}{3}b')\,, \tag{16.49}$$

which is unbounded in the MSSM limit.

As we have noted below eq. (16.16), CP violation in the scalar sector of the NMSSM is controlled by $\mathrm{Im}(\lambda^*\kappa)$, which also governs the mixing of the CP-even and CP-odd scalar interaction eigenstates. However, the presence of this new source of CP violation could generate effects that are phenomenologically problematical (e.g., electric dipole moments for the electron and neutron that are larger than the present-day experimental bounds). Consequently, it is often assumed that λ, κ, a_λ, and a_κ are all real in the same convention where s, v_u, and v_d are real and positive. This assumption is natural if the mediation mechanism for supersymmetry breaking does not introduce new CP-violating phases. Moreover, in light of eqs. (16.40) and (16.41), a sufficient (although not necessary) condition to achieve the stability of the electroweak vacuum is to assume that $\lambda\kappa > 0$, $a_\lambda > 0$, and $a_\kappa < 0$ (in the convention where all VEVs are positive), with $|a_\kappa|$ bounded as dictated by eq. (16.43).

Henceforth, we shall assume that all NMSSM Higgs sector parameters are real (subject to the stability conditions discussed above). In this case, $M^2_{HA} = 0$, and the neutral scalar spectrum consists of three CP-even scalars, denoted by h, H, and H_s, where $m_h < m_H$, and h, H are the MSSM-like CP-even scalars and H_S is the CP-even scalar field that is mostly composed of the singlet component h_s. Likewise, there are two CP-odd scalars, A and A_s, where A is MSSM-like and A_s is the CP-odd scalar field that is mostly composed of the singlet component a_s.

The NMSSM contains, besides the particles of the MSSM, a neutral $P_R = +1$ CP-even scalar, a neutral $P_R = +1$ CP-odd scalar, and a $P_R = -1$ Weyl fermion "singlino." These fields have no gauge couplings of their own, so they can only interact with Standard Model particles by mixing with the neutral MSSM fields with the same spin and charge. The neutral CP-even scalar mixes with the MSSM particles h and H, and the neutral CP-odd scalar mixes with A.

One of the effects of replacing the μ-term of the MSSM by the dynamical field S is that the squared mass of the lightest Higgs boson is raised, by an amount bounded at tree level by

$$\Delta(m_h^2) \leq \tfrac{1}{2}\lambda^2 v^2 \sin^2 2\beta\,, \tag{16.50}$$

as previously noted in eq. (16.37). This extra contribution comes from the $|F_S|^2$ contribution to the scalar potential. The neutral Higgs scalars have reduced couplings to the electroweak gauge bosons, compared to those in the Standard Model, because of the mixing with the singlets. Because the observed Higgs boson (of mass 125 GeV) at the LHC appears to have properties like those of a SM Higgs boson, it is unlikely to have a large admixture of the singlet field S. Moreover, there could be a yet-undiscovered neutral Higgs scalar that is mostly electroweak singlet and even lighter than 125 GeV.

The effect of eq. (16.50) to increase m_h is limited, for two reasons. First, it vanishes in the large $\tan\beta$ limit; note that for, e.g., $\tan\beta \geq 5$, the factor $\sin^2 2\beta$ is less than 0.15, and for $\tan\beta \geq 10$ it is less than 0.04. Second, there is an upper bound on λ if one requires that it remains perturbative at scales well above the TeV scale. To see this, consider the one-loop renormalization group equations for the dimensionless couplings λ and κ:

$$\frac{d}{dt}\lambda = \frac{\lambda}{16\pi^2}\left(4\lambda^2 + 2\kappa^2 + 3y_t^2 + 3y_b^2 + y_\tau^2 - 3g_2^2 - \tfrac{3}{5}g_1^2\right), \tag{16.51}$$

$$\frac{d}{dt}\kappa = \frac{\kappa}{16\pi^2}\left(6\kappa^2 + 6\lambda^2\right), \tag{16.52}$$

where $t = \ln(Q/Q_0)$, $g_2 = g$, and $g_1 = (5/3)^{1/2}g'$ [see eq. (6.287)]. As one evolves the renormalization scale Q, the dominant effects on the running of λ are positive beta-function contributions from y_t^2, λ^2, and possibly κ^2 and y_b^2, which overwhelm the negative electroweak gauge coupling contributions. This implies that λ cannot be too large near the TeV scale, or it will become nonperturbatively large in the ultraviolet. If one puts in the numerical values for y_t and y_b, one obtains a bound of $|\lambda| \lesssim 0.8$ at the TeV scale. A practical way to obtain this bound is to start with a large value, say $\lambda = 3$, at the GUT scale, and evolve down; the resulting bound at the TeV scale is not very sensitive to the chosen large value at the GUT scale. Note that a nonzero value of κ can only tighten the upper bound on λ. It is possible to suppose that λ does become nonperturbative at some scale Q lower than the GUT scale, but this would endanger the apparent unification of the gauge couplings, since λ contributes nontrivially to the beta functions of g_2 and g_1 at two-loop order.

The odd R-parity singlino $\widetilde{S}$ mixes with the four MSSM neutralinos, which yields five neutralinos in the NMSSM. The singlino could be the LSP, depending on the parameters of the model, and so it could be the dark matter. The neutralino mass matrix in the $\psi^0 = (\widetilde{B}, \widetilde{W}^0, \widetilde{H}_d^0, \widetilde{H}_u^0, \widetilde{S})$ gauge eigenstate basis is [cf. eq. (13.136)]

$$\boldsymbol{M}_{\widetilde{N}} = \begin{pmatrix} M_1 & 0 & -\frac{1}{2}g'v_d & \frac{1}{2}g'v_u & 0 \\ 0 & M_2 & \frac{1}{2}gv_d & -\frac{1}{2}gv_u & 0 \\ -\frac{1}{2}g'v_d & \frac{1}{2}gv_d & 0 & -\mu & -\frac{1}{\sqrt{2}}\lambda v_u \\ \frac{1}{2}g'v_u & -\frac{1}{2}gv_u & -\mu & 0 & -\frac{1}{\sqrt{2}}\lambda v_d \\ 0 & 0 & -\frac{1}{\sqrt{2}}\lambda v_u & -\frac{1}{\sqrt{2}}\lambda v_d & 2\kappa s \end{pmatrix}. \tag{16.53}$$

The singlino mixes directly with the higgsinos, but only indirectly with the gauginos via terms suppressed by the Higgs VEVs and electroweak couplings. This implies that the physical states will include two mostly gaugino-like neutralinos. In the MSSM limit of the NMSSM Higgs sector, the quantities v/s and $\lambda v/\kappa s$ are small (with $\mu \equiv \lambda s$ and κs held fixed), which implies that mixing effects of the singlet Higgs and singlino are also small, and they nearly decouple. In that case, the phenomenology of the NMSSM is almost indistinguishable from that of the MSSM. However, a potentially striking effect is that if the lightest neutralino is mostly singlino with small mixing, then the NLSP could be long-lived, with a macroscopic decay length when produced in collider experiments For larger λ, the singlino mixing is important and the experimental signals for sparticles and the Higgs scalars can be altered in important ways due to reduced interaction strengths.

16.2 The Supersymmetric Seesaw

Neutrino masses can be incorporated into the Standard Model by introducing SU(3)×SU(2)×U(1) singlet right-handed neutrinos (ν_R) whose mass parameters are very large, typically near the grand unification scale. In addition, one must also include standard Yukawa couplings between the lepton doublets, the Higgs doublet, and ν_R. The Higgs vacuum expectation value then induces an off-diagonal ν_L–ν_R mass on the order of the electroweak scale. Diagonalizing the neutrino mass matrix (in the three-generation model) yields three superheavy neutrino states, and three very light neutrino states that are identified with the light neutrinos observed in Nature. This is the seesaw mechanism, which has been exhibited in detail in Section 6.1.

It is straightforward to construct a supersymmetric generalization of the seesaw model of neutrino masses by promoting the right-handed neutrino field to a superfield. In particular, by expanding the chiral superfield content of the MSSM shown in Table 7.1, we obtain the chiral superfields of the seesaw-extended MSSM (henceforth denoted as SE-MSSM), as shown in Table 16.1. Note that the hypercharge Y is normalized as in eq. (4.139).

Table 16.1 The leptonic and Higgs chiral supermultiplets in the seesaw-extended MSSM. The spin-0 fields are complex scalars, and the spin-1/2 fields are left-handed two-component Weyl fermions.

names		spin-0	spin-1/2	$SU(3)_C,\ SU(2)_L,\ U(1)_Y$
sleptons, leptons	L	$(\widetilde{\nu}_L\ \widetilde{e}_L)$	$(\nu\ e)$	$(\mathbf{1},\ \mathbf{2},\ -\frac{1}{2})$
(×3 generations)	$\overline{\nu}$	$\widetilde{\nu}_R^\dagger$	$\overline{\nu}$	$(\mathbf{1},\ \mathbf{0},\ 0)$
	$\overline{e}$	$\widetilde{e}_R^\dagger$	$\overline{e}$	$(\mathbf{1},\ \mathbf{1},\ 1)$
Higgs, higgsinos	H_u	$(H_u^+\ H_u^0)$	$(\widetilde{H}_u^+\ \widetilde{H}_u^0)$	$(\mathbf{1},\ \mathbf{2},\ +\frac{1}{2})$
	H_d	$(H_d^0\ H_d^-)$	$(\widetilde{H}_d^0\ \widetilde{H}_d^-)$	$(\mathbf{1},\ \mathbf{2},\ -\frac{1}{2})$

In this section, the seesaw-extended MSSM is constructed following Ref. [183]. Under the assumption of R-parity conservation, two additional terms can appear in the leptonic part of the MSSM superpotential [cf. eq. (13.6)],

$$W_{\text{SE-MSSM}} = \overline{\nu}\boldsymbol{y_\nu} L H_u - \overline{e}\boldsymbol{y_e} L H_d + \tfrac{1}{2}\overline{\nu}\boldsymbol{M}_{\overline{\nu}}\overline{\nu} + \mu H_u H_d\,, \tag{16.54}$$

where $\boldsymbol{M}_{\overline{\nu}}$ is a complex symmetric 3×3 matrix. It is convenient to perform field redefinitions of the (charged and neutral) lepton superfields:

$$L \to V_L L\,, \qquad \overline{e} \to V_R \overline{e}\,, \qquad \overline{\nu} \to V_N \overline{\nu}\,, \tag{16.55}$$

where V_L, V_R, and V_N are 3×3 unitary matrices. Note that the kinetic energy terms (and the couplings of the lepton superfields to the gauge fields) are invariant under the above unitary transformations. However, the coefficients of the terms of the superpotential [eq. (16.54)] are modified:

$$\boldsymbol{y_e} \to V_L^T \boldsymbol{y_e} V_R\,, \qquad \boldsymbol{y_\nu} \to V_L^T \boldsymbol{y_\nu} V_N\,, \qquad \boldsymbol{M_{\bar{\nu}}} \to V_N^T \boldsymbol{M_{\bar{\nu}}} V_N\,. \tag{16.56}$$

We shall choose V_L, V_R, and V_N such that

$$V_L^T \boldsymbol{y_e} V_R = \mathrm{diag}(y_e\,,\, y_\mu\,,\, y_\tau)\,, \tag{16.57}$$

$$V_N^T \boldsymbol{M_{\bar{\nu}}} V_N = \mathrm{diag}(M_1\,,\, M_2\,,\, M_3)\,, \tag{16.58}$$

where the elements of the two diagonal matrices above are real and nonnegative. As shown in Appendix G, it is always possible to find unitary matrices V_L and V_R such that eq. (16.57) is satisfied: this is the singular value decomposition of an arbitrary complex matrix. Likewise, it is always possible to find a unitary matrix V_N such that eq. (16.58) holds: this is the Takagi diagonalization of an arbitrary complex symmetric matrix. Thus, the redefinition of the lepton superfields [eq. (16.55)] implies that one can assume from the beginning without loss of generality that $\boldsymbol{y_e}$ and $\boldsymbol{M_{\bar{\nu}}}$ in eq. (13.6) are real nonnegative diagonal matrices.[5] The (transformed) $\boldsymbol{y_\nu}$ is in general an arbitrary complex 3×3 matrix, although we shall assume that the (dimensionless) eigenvalues of $\boldsymbol{y_\nu}$ are at most $\mathcal{O}(1)$ [as in the case of $\boldsymbol{y_e}$].

To complete the description of the SE-MSSM, we need to specify the soft supersymmetry-breaking terms. In Section 8.6, we learned how to write down the most general set of such terms in any supersymmetric theory. The soft supersymmetry-breaking terms of the MSSM were given in eq. (13.16). Under the assumption of R-parity conservation, three additional terms can appear among the soft supersymmetry-breaking terms of the SE-MSSM involving the slepton, sneutrino, and Higgs fields:

$$\begin{aligned}\mathscr{L}_{\text{soft}}^{\text{SE}-\text{MSSM}} = &-\left(\widetilde{\bar{\nu}}\,\boldsymbol{a_\nu}\,\widetilde{L}H_u - \widetilde{\bar{e}}\,\boldsymbol{a_e}\,\widetilde{L}H_d + \text{h.c.}\right) - \widetilde{L}^\dagger\,\boldsymbol{m_L^2}\,\widetilde{L} - \widetilde{\bar{\nu}}\,\boldsymbol{m_{\bar{\nu}}^2}\,\widetilde{\bar{\nu}}^\dagger - \widetilde{\bar{e}}\,\boldsymbol{m_{\bar{e}}^2}\,\widetilde{\bar{e}}^\dagger \\ &- m_{H_u}^2 H_u^\dagger H_u - m_{H_d}^2 H_d^\dagger H_d - \left(\widetilde{\bar{\nu}}\boldsymbol{b_\nu}\widetilde{\bar{\nu}} + bH_uH_d + \text{h.c.}\right).\end{aligned} \tag{16.59}$$

In eq. (16.59), each of $\boldsymbol{a_\nu}$ and $\boldsymbol{a_e}$ is a complex 3×3 matrix in generation space, with dimensions of [mass]. They are in one-to-one correspondence with the Yukawa couplings in the superpotential. Each of $\boldsymbol{m_L^2}$, $\boldsymbol{m_{\bar{\nu}}^2}$, $\boldsymbol{m_{\bar{e}}^2}$ is a 3×3 matrix in generation space that can have complex entries, but they must be hermitian so that the Lagrangian is real. Finally, in addition to supersymmetry-breaking squared-mass parameters that contribute to the Higgs potential, $m_{H_u}^2$, $m_{H_d}^2$, and b, we have a new squared-mass term of the type b^{ij} in eq. (8.78), $\boldsymbol{b_\nu}$, which is a 3×3 complex symmetric matrix in generation space. In general the 3×3 soft supersymmetry-breaking parameters in eq. (16.59) do not take a simplified form in the basis defined by eqs. (16.57) and (16.58). Finally, we remind the reader that after electroweak

[5] After electroweak symmetry breaking, eq. (1.175) corresponds to working in a basis in which the charged lepton mass matrices are (real) nonnegative and diagonal.

symmetry breaking, the neutral Higgs fields acquire vacuum expectation values

$$\langle H_u^0 \rangle = \frac{v_u}{\sqrt{2}}\,, \qquad \langle H_d^0 \rangle \equiv \frac{v_d}{\sqrt{2}}\,, \tag{16.60}$$

where $v^2 \equiv |v_u|^2 + |v_d|^2 \simeq (246\text{ GeV})^2$.

We shall make the standard seesaw assumption that the magnitudes of the eigenvalues of $\boldsymbol{M}_{\bar{\nu}}$ are significantly larger that the scale of electroweak symmetry breaking. That is, in light of eq. (6.11),

$$\|\boldsymbol{M}_{\bar{\nu}}\| \sim \mathcal{O}(M_N) \gg v\,, \tag{16.61}$$

where we have introduced a mass scale M_N that characterizes the size of the eigenvalues of $\boldsymbol{M}_{\bar{\nu}}$ (for simplicity we take the eigenvalues of $\boldsymbol{M}_{\bar{\nu}}$ to be of similar order of magnitude). We expect that the nonsinglet soft supersymmetry-breaking squared-mass parameters, $\boldsymbol{m}^2_{\boldsymbol{L}}$ and $\boldsymbol{m}^2_{\bar{\boldsymbol{e}}}$, are of order the square of the supersymmetry-breaking scale, $M_{\rm SUSY}$, that governs the effective supersymmetry-breaking scale of the MSSM. We will also assume that the μ parameter is related to $M_{\rm SUSY}$ by one of the mechanisms discussed in Section 14.6. That is,

$$\|\boldsymbol{m}^2_{\boldsymbol{L}}\| \sim \|\boldsymbol{m}^2_{\bar{\boldsymbol{e}}}\| \sim M^2_{\rm SUSY}\,, \qquad |\mu| \sim M_{\rm SUSY}\,. \tag{16.62}$$

The parameters $\boldsymbol{a_e}$ and $|b|^{1/2}$ are likewise expected to be at most of $\mathcal{O}(M_{\rm SUSY})$. In contrast, the parameters $\boldsymbol{a_\nu}$ and $\boldsymbol{b_\nu}$ are unconnected to electroweak symmetry breaking at tree level. Nevertheless, they do contribute via loop corrections to neutrino mass splittings. Consequently, we shall assume that

$$\|\boldsymbol{a_\nu}\| \lesssim M_{\rm SUSY}\,, \qquad \|\boldsymbol{b_\nu}\| \lesssim M_{\rm SUSY} M_N\,. \tag{16.63}$$

Finally, the singlet soft supersymmetry-breaking parameter $\boldsymbol{m}^2_{\bar{\boldsymbol{\nu}}}$ is also unconnected to electroweak symmetry breaking at tree level. However, the one-loop corrections to the Higgs mass parameters depend quadratically on $\boldsymbol{m}^2_{\bar{\boldsymbol{\nu}}}$, so to avoid unnatural fine-tuning of the electroweak symmetry-breaking scale, one expects that $\boldsymbol{m}^2_{\bar{\boldsymbol{\nu}}}$ cannot be much larger than $M^2_{\rm SUSY}$:

$$\|\boldsymbol{m}^2_{\bar{\boldsymbol{\nu}}}\| \sim M^2_{\rm SUSY}\,. \tag{16.64}$$

Employing these assumptions, one can perturbatively diagonalize the neutrino mass matrix and sneutrino squared-mass matrix to identify the mass eigenstates and the corresponding physical masses.

It is now straightforward to derive all the interactions of the SE-MSSM following the procedures outlined in Chapters 8 and 13. In a three-generation model, the neutrino mass matrix is a 6×6 complex symmetric matrix, which can be written in block (partitioned) form in terms of 3×3 matrix blocks. The sneutrino squared-mass matrix is a 12×12 hermitian matrix, which can be written in block (partitioned) form in terms of 6×6 matrix blocks. Each of these 6×6 matrices can be further partitioned in terms of 3×3 matrix blocks. One can easily check that the neutrino mass spectrum and mixing matrices are precisely those of the seesaw-extended

Standard Model given in Section 6.1, after identifying [see eq. (6.21)]

$$\boldsymbol{M_D} \equiv \frac{v_u}{\sqrt{2}}\boldsymbol{y_\nu}\,, \tag{16.65}$$

where v_u is defined in eq. (16.60).

Of particular interest is the squared-mass matrix of the sneutrinos. Contributing to this squared-mass matrix are terms arising from the F- and D-terms of the scalar potential (computed in the usual way from the superpotential given in eq. (16.54) and the soft supersymmetry-breaking terms specified in eq. (16.59). This calculation yields the following 12×12 hermitian sneutrino squared-mass matrix in block form:

$$-\mathscr{L}_{\rm mass} = \tfrac{1}{2}\begin{pmatrix}\phi_L^\dagger & \phi_N^\dagger\end{pmatrix}\begin{pmatrix}\mathcal{M}^2_{LL} & \mathcal{M}^2_{LN} \\ (\mathcal{M}^2_{LN})^\dagger & \mathcal{M}^2_{NN}\end{pmatrix}\begin{pmatrix}\phi_L \\ \phi_N\end{pmatrix}, \tag{16.66}$$

where $\phi_L \equiv (\widetilde{L}_1\,, \widetilde{L}_1^*)^{\mathsf T}$ and $\phi_N \equiv (\widetilde{\overline{\nu}}\,, \widetilde{\overline{\nu}}^*)^{\mathsf T}$ are six-dimensional vectors. The 6×6 hermitian matrices $\mathcal{M}^2_{LL}$, $\mathcal{M}^2_{NN}$ and the 6×6 complex matrix $\mathcal{M}^2_{LN}$ can be written in block partitioned form as

$$\mathcal{M}^2_{AB} \equiv \begin{pmatrix} M^2_{A^\dagger B} & M^{2\,*}_{A^{\mathsf T} B} \\ M^2_{A^{\mathsf T} B} & M^{2\,*}_{A^\dagger B}\end{pmatrix}, \tag{16.67}$$

where the subscripts A and B can take on possible values L and N [this labeling allows one to keep track of the origin of the various matrix blocks]. The $M^2_{A^\dagger A}$ are 3×3 hermitian matrices and the $M^2_{A^{\mathsf T} A}$ are 3×3 complex symmetric matrices, for $A = L\,, N$. There are no restrictions on the 3×3 complex matrices $M^2_{A^\dagger B}$ and $M^2_{A^{\mathsf T} B}$ for $A \neq B$. Explicitly, the 3×3 blocks in eq. (16.67) are given by

$$M^2_{L^\dagger L} = \boldsymbol{m^2_L} + \tfrac{1}{2}m_Z^2\cos 2\beta\,\mathbb{1}_{3\times 3} + \boldsymbol{M^*_D M^{\mathsf T}_D}\,, \tag{16.68}$$

$$M^2_{N^\dagger N} = \boldsymbol{M^2_{\overline{\nu}}} + \boldsymbol{m^2_{\overline{\nu}}} + \boldsymbol{M^\dagger_D M_D}\,, \tag{16.69}$$

$$M^2_{L^\dagger N} = \boldsymbol{M^*_D M_{\overline{\nu}}}\,, \tag{16.70}$$

$$M^2_{L^{\mathsf T} N} = -\boldsymbol{x_\nu}\,, \tag{16.71}$$

$$M^2_{N^{\mathsf T} N} = -2\boldsymbol{b_\nu}\,, \tag{16.72}$$

$$M^2_{L^{\mathsf T} L} = 0\,, \tag{16.73}$$

where we have introduced the complex 3×3 matrix parameter $\boldsymbol{x_\nu}$ as follows:

$$\boldsymbol{x_\nu} \equiv \frac{1}{\sqrt{2}}(v_u\boldsymbol{a_\nu} + \mu^* v_d\boldsymbol{y_\nu})\,. \tag{16.74}$$

Under the assumptions of eqs. (16.61)–(16.64), the 12×12 sneutrino mass matrix, written in terms of 6×6 matrix blocks with estimated magnitudes,

$$\boldsymbol{M^2_{\tilde\nu}} \equiv \begin{pmatrix}\mathcal{M}^2_{LL} & \mathcal{M}^2_{LN} \\ (\mathcal{M}^2_{LN})^\dagger & \mathcal{M}^2_{NN}\end{pmatrix} = \begin{pmatrix}\mathcal{O}(v^2) & \mathcal{O}(vM_N) \\ \mathcal{O}(vM_N) & \mathcal{O}(M_N^2)\end{pmatrix}, \tag{16.75}$$

also exhibits a seesaw-type behavior, analogous to the seesaw-type mass matrix [eq. (6.20)] of the neutrino sector. Following the standard procedure for diagonalizing a hermitian matrix, we introduce a 12 × 12 unitary matrix:

$$\boldsymbol{V} = \begin{pmatrix} \mathcal{I} - \frac{1}{2}\mathcal{M}^2_{LN}\mathcal{M}^{-4}_{NN}(\mathcal{M}^2_{LN})^\dagger & \mathcal{M}^2_{LN}\mathcal{M}^{-2}_{NN} \\ -\mathcal{M}^{-2}_{NN}(\mathcal{M}^2_{LN})^\dagger & \mathcal{I} - \frac{1}{2}\mathcal{M}^{-2}_{NN}(\mathcal{M}^2_{LN})^\dagger\mathcal{M}^2_{LN}\mathcal{M}^{-2}_{NN} \end{pmatrix}, \tag{16.76}$$

where $\mathcal{I}$ is the 6 × 6 identity matrix. One can easily compute

$$\boldsymbol{V}^\dagger \boldsymbol{M}^2_{\widetilde{\nu}}\,\boldsymbol{V} = \begin{pmatrix} \mathcal{M}^2_{LL} - \mathcal{M}^2_{LN}\mathcal{M}^{-2}_{NN}(\mathcal{M}^2_{LN})^\dagger + \mathcal{O}(v^4 M_N^{-2}) & \mathcal{O}(v^3 M_N^{-1}) \\ \mathcal{O}(v^3 M_N^{-1}) & \mathcal{M}^2_{NN} + \mathcal{O}(v^2) \end{pmatrix}. \tag{16.77}$$

Hence, the effective 6 × 6 hermitian squared-mass matrix for the light sneutrinos reads

$$\mathcal{M}^2_{\widetilde{\nu}_\ell} \equiv \mathcal{M}^2_{LL} - \mathcal{M}^2_{LN}\mathcal{M}^{-2}_{NN}\left(\mathcal{M}^2_{LN}\right)^\dagger + \mathcal{O}(v^4 M_N^{-2}), \tag{16.78}$$

analogous to the light effective neutrino mass matrix of eq. (6.26). Likewise, the effective 6 × 6 hermitian squared-mass matrix for the superheavy sneutrinos reads

$$\mathcal{M}^2_{\widetilde{\nu}_h} \equiv \mathcal{M}^2_{NN} + \tfrac{1}{2}\left[\mathcal{M}^{-2}_{NN}(\mathcal{M}^2_{LN})^\dagger\mathcal{M}^2_{LN} + (\mathcal{M}^2_{LN})^\dagger\mathcal{M}^2_{LN}\mathcal{M}^{-2}_{NN}\right] + \mathcal{O}(v^4 M_N^{-2}), \tag{16.79}$$

where we have exhibited the $\mathcal{O}(v^2)$ corrections to the leading term. As expected, the masses of half of the sneutrino eigenstates are of order the electroweak symmetry-breaking scale, whereas the other half are superheavy, of order M_N.

Following the notation of Table 16.1, the (complex) sneutrino interaction eigenstates are denoted by $\widetilde{\nu}_L \equiv \widetilde{L}_1$ and $\widetilde{\nu}_R \equiv \widetilde{\overline{\nu}}^\dagger$. The latter convention reflects the fact that in the lepton-number-conserving limit of $\boldsymbol{M}_{\bar{\nu}} = \boldsymbol{b}_\nu = 0$, the lepton numbers of $\widetilde{\nu}_L$ and $\widetilde{\nu}_R$ are identical. In analogy to ν_ℓ and ν_h defined in eq. (6.24), we define transformed (light and heavy) sneutrino states $\widetilde{\nu}_\ell$ and $\widetilde{\nu}_h$ by

$$\begin{pmatrix} \phi_L \\ \phi_N \end{pmatrix} = \boldsymbol{V}\begin{pmatrix} \phi_\ell \\ \phi_h \end{pmatrix}, \tag{16.80}$$

where $\phi_\ell \equiv (\widetilde{\nu}_\ell\,,\,\widetilde{\nu}_\ell^\dagger)^T$ and $\phi_h \equiv (\widetilde{\nu}_h^\dagger\,,\,\widetilde{\nu}_h)^T$ are six-dimensional vectors. Sneutrino–antisneutrino oscillations are a consequence of the $\Delta L = 2$ elements in the light and heavy sneutrino squared-mass matrices $\mathcal{M}^2_{\widetilde{\nu}_\ell}$ and $\mathcal{M}^2_{\widetilde{\nu}_h}$, and are governed by $M^2_{N^TN}$ and $M^2_{L^TN}$.

Using the form of $\mathcal{M}^2_{AB}$ (A, $B = L$ or N) given by eq. (16.67) with the M^2_{AB} given in eqs. (16.68)–(16.73), the effective 6 × 6 hermitian squared-mass matrix for the light sneutrinos [eq. (16.78)] is given by

$$\mathcal{M}^2_{\widetilde{\nu}_\ell} \equiv \begin{pmatrix} M^2_{LC} & (M^2_{LV})^* \\ M^2_{LV} & (M^2_{LC})^* \end{pmatrix}, \tag{16.81}$$

where the lepton-number-conserving (LC) and lepton-number-violating (LV) matrix elements are given by

$$M^2_{LC} \equiv \boldsymbol{m}^2_L + \tfrac{1}{2}m_Z^2\cos 2\beta\,\mathbb{1}_{3\times 3} + \mathcal{O}(v^4 M_N^{-2})\,, \tag{16.82}$$

$$M^2_{LV} \equiv \boldsymbol{M}_D \boldsymbol{M}_{\bar{\nu}}^{-1}\boldsymbol{x}_\nu^{\mathsf{T}} + \boldsymbol{x}_\nu \boldsymbol{M}_{\bar{\nu}}^{-1}\boldsymbol{M}_D^{\mathsf{T}} - 2\boldsymbol{M}_D\boldsymbol{M}_{\bar{\nu}}^{-1}\boldsymbol{b}_\nu \boldsymbol{M}_{\bar{\nu}}^{-1}\boldsymbol{M}_D^{\mathsf{T}} + \mathcal{O}(v^5 M_N^{-3})\,, \tag{16.83}$$

where M^2_{LC} is a 3×3 hermitian matrix, and M^2_{LV} is a 3×3 complex symmetric matrix. The $M_N \to \infty$ limit of eqs. (16.82) and (16.83) is noteworthy. In this limit, $M^2_{LV} = 0$ and the lepton-number-violating effects completely decouple, as expected. In addition, M^2_{LC} reproduces the well-known tree-level light sneutrino 3×3 squared-mass matrix of the MSSM [see eq. (13.166)].

The physical light sneutrino states can be identified by diagonalizing $\mathcal{M}^2_{\tilde{\nu}_\ell}$. Note that if $M^2_{LV} = 0$, then the eigenvalues[6] of $\mathcal{M}^2_{\tilde{\nu}_\ell}$ are doubly degenerate, corresponding to the fact that the conserved lepton number implies that the six light sneutrino states are comprised of three sneutrino–antisneutrino pairs. If $M^2_{LV} \neq 0$, then lepton number is violated and the sneutrinos and antisneutrinos can mix. This mixing splits the degenerate pairs and yields (in general) six nondegenerate light sneutrinos. In particular, the resulting sneutrino mass eigenstates are self-conjugate real fields.

For simplicity, we briefly examine the case of one generation. In this case, M^2_{LC} is a real number and M^2_{LV} is a complex number. Note that if $M^2_{LV} \neq 0$, then the sneutrino mass eigenstates are no longer eigenstates of lepton number L. Thus, the would-be sneutrino and antisneutrino fields mix, and we denote the resulting sneutrino mass-eigenstate fields by S_1 and S_2. The sneutrino squared-mass matrix, $\mathcal{M}^2_{\tilde{\nu}_\ell}$, is a 2×2 hermitian matrix, with eigenvectors[7]

$$S_1 = \frac{1}{\sqrt{2}}\left(e^{i\theta/2}\tilde{\nu}_\ell + e^{-i\theta/2}\tilde{\nu}_\ell^\dagger\right), \tag{16.84}$$

$$S_2 = \frac{1}{i\sqrt{2}}\left(e^{i\theta/2}\tilde{\nu}_\ell - e^{-i\theta/2}\tilde{\nu}_\ell^\dagger\right), \tag{16.85}$$

where $\theta \equiv \arg M^2_{LV}$, and the fields $\tilde{\nu}_\ell$ and $\tilde{\nu}_\ell^*$ are eigenstates of lepton number, with $L = \pm 1$, respectively. The squared-mass diagonalization matrix has been chosen such that S_1 and S_2 are self-conjugate real fields. The corresponding mass eigenvalues are

$$m^2_{S_1,S_2} = M^2_{LC} \pm |M^2_{LV}| = m_L^2 + \tfrac{1}{2}m_Z^2\cos 2\beta \pm \frac{2|M_D|}{M_{\bar{\nu}}}\left|x_\nu - \frac{bM_D}{M_{\bar{\nu}}}\right|, \tag{16.86}$$

where the corresponding one-generation quantities are written without boldface.

[6] Under the assumption that R-parity is not spontaneously broken, the (real) eigenvalues of the hermitian matrix M^2_{LC} are nonnegative.

[7] In light of eqs. (16.90) and (16.91), if $\theta = 0 \pmod{2\pi}$ then S_1 is a CP-even scalar and S_2 is a CP-odd scalar, whereas if $\theta = \pi \pmod{2\pi}$ then S_1 is CP-odd and S_2 is CP-even. For any other choice of θ, S_1 and S_2 are scalar states of indefinite CP.

The corresponding sneutrino mass splitting, $\Delta m_{\widetilde{\nu}_\ell} \equiv |m_{S_2} - m_{S_1}|$, is given by

$$\frac{\Delta m_{\widetilde{\nu}_\ell}}{m_{\nu_\ell}} = \frac{2}{m_{\widetilde{\nu}_\ell}} \left| \frac{x_\nu}{M_D} - \frac{b}{M_{\bar{\nu}}} \right| , \tag{16.87}$$

where $m_{\nu_\ell} \equiv |m_D|^2/M_{\bar{\nu}}$ is the mass of the light neutrino and $m_{\widetilde{\nu}_\ell} \equiv \frac{1}{2}(m_{S_1} + m_{S_2})$ is the average light sneutrino mass. In light of eq. (16.63), it follows that both terms on the right-hand side of eq. (16.87) are of the same order, which implies that $\Delta m_{\widetilde{\nu}_\ell} \sim \mathcal{O}(m_{\nu_\ell})$.

A similar analysis employed for the light sneutrinos can be employed to obtain the effective 6×6 hermitian squared-mass matrix for the heavy sneutrinos [eq. (16.79)]:

$$\mathcal{M}^2_{\widetilde{\nu}_h} \equiv \begin{pmatrix} \boldsymbol{M^2_H} & -2\boldsymbol{b^*_\nu} \\ -2\boldsymbol{b_\nu} & \boldsymbol{M^{2*}_H} \end{pmatrix} + \mathcal{O}(v^4 M_N^{-2}) , \tag{16.88}$$

where the 3×3 hermitian matrix $\boldsymbol{M^2_H}$ is defined by

$$\boldsymbol{M^2_H} \equiv \boldsymbol{M^2_{\widetilde{\bar{\nu}}}} + \boldsymbol{M^\dagger_D M_D} + \tfrac{1}{2}\boldsymbol{M^{-1}_{\bar{\nu}} M^{\mathsf{T}}_D M^*_D M_{\bar{\nu}}} + \tfrac{1}{2}\boldsymbol{M_{\bar{\nu}} M^{\mathsf{T}}_D M^*_D M^{-1}_{\bar{\nu}}} . \tag{16.89}$$

The heavy sneutrino mass eigenstates are determined by diagonalizing $\mathcal{M}^2_{\widetilde{\nu}_h}$. At leading order, the mass eigenstates are mass-degenerate sneutrino–antisneutrino pairs. The lepton-number-violating off-block-diagonal matrix $\boldsymbol{b_\nu}$ is responsible for sneutrino–antisneutrino mixing, and generates mass splittings between nearly degenerate heavy sneutrino pairs of order $\Delta m_{\widetilde{\nu}_h} \sim \mathcal{O}(M_{\rm SUSY})$.

The complex elements of the sneutrino squared-mass matrix govern CP-violating sneutrino phenomena, due to the nondegeneracy of masses of the real and imaginary parts of the sneutrino fields. It is convenient to define a new basis of sneutrino interaction eigenstates of definite CP. Under a CP transformation, sneutrino states are transformed into antisneutrino states, $\mathrm{CP}\,|\widetilde{\nu}\rangle = |\widetilde{\nu}^\dagger\rangle$ and $\mathrm{CP}\,|\widetilde{\nu}^\dagger\rangle = |\widetilde{\nu}\rangle$. The eigenstates of CP are given by

$$|\widetilde{\nu}^{(+)}\rangle \equiv \frac{1}{\sqrt{2}}\left(|\widetilde{\nu}\rangle + |\widetilde{\nu}^\dagger\rangle\right) , \qquad |\widetilde{\nu}^{(-)}\rangle \equiv \frac{1}{i\sqrt{2}}\left(|\widetilde{\nu}\rangle - |\widetilde{\nu}^\dagger\rangle\right) , \tag{16.90}$$

with corresponding eigenvalues

$$\mathrm{CP}|\widetilde{\nu}^{(+)}\rangle = +|\widetilde{\nu}^{(+)}\rangle , \qquad \mathrm{CP}|\widetilde{\nu}^{(-)}\rangle = -|\widetilde{\nu}^{(-)}\rangle . \tag{16.91}$$

One can now decompose the light and heavy sneutrino mass eigenstates into states of definite CP,

$$\widetilde{\nu}_\ell = \frac{1}{\sqrt{2}}\left[\widetilde{\nu}^{(+)}_\ell + i\,\widetilde{\nu}^{(-)}_\ell\right] , \tag{16.92}$$

$$\widetilde{\nu}_h = \frac{1}{\sqrt{2}}\left[\widetilde{\nu}^{(+)}_h + i\,\widetilde{\nu}^{(-)}_h\right] . \tag{16.93}$$

With respect to the CP basis,

$$-\mathscr{L}_{\rm mass} = \tfrac{1}{2}(\widetilde{\nu}_\ell^{(+)T}\,,\,\widetilde{\nu}_\ell^{(-)T})\mathcal{P}^\dagger \mathcal{M}^2_{\widetilde{\nu}_\ell}\mathcal{P}\begin{pmatrix}\widetilde{\nu}_\ell^{(+)}\\ \widetilde{\nu}_\ell^{(-)}\end{pmatrix} + \tfrac{1}{2}(\widetilde{\nu}_h^{(+)T}\,,\,\widetilde{\nu}_h^{(-)T})\mathcal{P}^T \mathcal{M}^2_{\widetilde{\nu}_h}\mathcal{P}^*\begin{pmatrix}\widetilde{\nu}_h^{(+)}\\ \widetilde{\nu}_h^{(-)}\end{pmatrix}, \tag{16.94}$$

where $\mathcal{P}$ is the 6×6 unitary matrix,

$$\mathcal{P} \equiv \frac{1}{\sqrt{2}}\begin{pmatrix}\mathbb{1}_{3\times3} & i\mathbb{1}_{3\times3}\\ \mathbb{1}_{3\times3} & -i\mathbb{1}_{3\times3}\end{pmatrix}. \tag{16.95}$$

For example, with respect to the CP basis, the effective squared-mass matrix for the light sneutrinos is given by the 6×6 real symmetric matrix

$$\overline{\mathcal{M}}^2_{\widetilde{\nu}_\ell} \equiv \mathcal{P}^\dagger \mathcal{M}^2_{\widetilde{\nu}_\ell}\mathcal{P} = \begin{pmatrix}\mathrm{Re}(M^2_{LC}+M^2_{LV}) & -\mathrm{Im}(M^2_{LC}+M^2_{LV})\\ \mathrm{Im}(M^2_{LC}-M^2_{LV}) & \mathrm{Re}(M^2_{LC}-M^2_{LV})\end{pmatrix}. \tag{16.96}$$

If $\mathrm{Im}\, M^2_{LC} = \mathrm{Im}\, M^2_{LV} = 0$, then the sneutrino mass eigenstates are also eigenstates of CP. If in addition $\mathrm{Re}\, M^2_{LV} \neq 0$, then the would-be sneutrino–antisneutrino pairs are organized into CP-even/CP-odd pairs of nearly mass-degenerate sneutrinos.

16.3 Supersymmetric Grand Unified Models

In Section 6.3, we reviewed grand unified extensions of the Standard Model. In models based on a simple unifying gauge group, there exists a single gauge coupling constant above the unification scale, $M_{\rm GUT}$. Below $M_{\rm GUT}$, the grand unified gauge group is partially broken. The running of the gauge couplings corresponding to the unbroken gauge groups is governed by renormalization group equations. Ultimately, one ends up with the SM gauge group $\mathrm{SU}(3)_C\times\mathrm{SU}(2)_L\times\mathrm{U}(1)_Y$ at the electroweak scale. The calculations exhibited in Section 6.3.5 yield a relation among the SU(3), SU(2), and U(1) gauge couplings (g_s, g, and g') evaluated at m_Z. In the simplest SU(5) and SO(10) grand unified models in which the unifying gauge group breaks down to $\mathrm{SU}(3)_C\times\mathrm{SU}(2)_L\times\mathrm{U}(1)_Y$ at $M_{\rm GUT}$, the predicted relation among g_s, g and g' is not respected by the experimental data. In contrast, in a supersymmetric extension of the Standard Model, the gauge coupling running is modified, and the predicted relation among g_s, g, and g' is approximately satisfied by the data. The unification of couplings in the simplest supersymmetric GUT models (SUSYGUTs) motivates us to explore some of the consequences of SU(5) and SO(10) SUSYGUT models. In this section, we briefly highlight the theory and phenomenology of some simple SUSYGUT models. A more comprehensive treatment and a guide to the literature can be found in [B32, B36].

16.3.1 Supersymmetric SU(5) Grand Unification

The MSSM was constructed by taking the 2HDM extension of the Standard Model and adding the corresponding superpartners. The supersymmetric SU(5) grand unified model is constructed in the same way. The SU(5) gauge fields lie in a vector supermultiplet that transforms as the 24-dimensional adjoint representation of SU(5). Each generation of quarks and leptons lies in chiral supermultiplets, $\overline{F}$ and T, that transform as a reducible $\overline{\mathbf{5}}\oplus\mathbf{10}$ representation of SU(5). In order to allow for nonzero neutrino masses, one can add an SU(5) singlet chiral supermultiplet to the model.

Following the results of Section 6.3, we introduce an SU(5) adjoint chiral supermultiplet, Σ^a, whose scalar component will also be called Σ^a, and is responsible for breaking the SU(5) gauge symmetry down to SU(3)×SU(2)×U(1). As in eq. (6.249), it is convenient to introduce a 5×5 traceless complex matrix of adjoint superfields,

$$\Sigma=\Sigma^a \boldsymbol{t^a}\,, \tag{16.97}$$

where the $\boldsymbol{t^a}$ are the generators of SU(5) in the fundamental representation. To obtain the scalar potential of the Higgs fields that reside in this supermultiplet, consider the most general gauge invariant cubic superpotential,

$$W=\tfrac{1}{2}m\,\mathrm{Tr}\,\Sigma^2+\tfrac{1}{3}\lambda\,\mathrm{Tr}\,\Sigma^3\,. \tag{16.98}$$

Employing eq. (8.75), the scalar potential is given by[8]

$$V=F_a^\dagger F_a+\tfrac{1}{2}\sum_c D^c D^c=\left(\frac{\partial W}{\partial\Sigma^a}\right)\left(\frac{\partial W^\dagger}{\partial\Sigma^{\dagger a}}\right)+\tfrac{1}{2}\sum_c g^2(\Sigma^{\dagger a}(T^c)_{ab}\Sigma^b)^2\,. \tag{16.99}$$

We search for supersymmetry-preserving scalar field solutions that minimize the scalar potential. First, consider the scalar field that satisfies

$$D^c=\Sigma^{\dagger a}(T^c)_{ab}\Sigma_b=0\,, \tag{16.100}$$

where the generator T^a in the adjoint representation is given by $T^c_{ab}=-if^c_{ab}$, where the f^c_{ab} are the structure constants of the SU(5) Lie algebra. That is,

$$f^c_{ab}\Sigma^{\dagger a}\Sigma^b=0\,. \tag{16.101}$$

Following eq. (16.97), we construct a 5×5 traceless complex matrix of scalar fields, $\Sigma\equiv\Sigma^a\boldsymbol{t^a}$, which can be decomposed as follows:

$$\Sigma=\Sigma_R+i\Sigma_I=(\Sigma^a_R+i\Sigma^a_I)\boldsymbol{t^a}\,, \tag{16.102}$$

where Σ_R and Σ_I are 5×5 traceless hermitian matrices. It follows from eq. (16.101) that Σ_R and Σ_I commute. This can be seen by first inserting $\Sigma^a=\Sigma^a_R+i\Sigma^a_I$ into eq. (16.101), which yields

$$f^c_{ab}\Sigma^a_R\Sigma^b_I=0\,, \tag{16.103}$$

[8] For adjoint fields, there is no significance to the height of the adjoint index.

due to the antisymmetry properties of the structure constants. It then follows that

$$[\Sigma_R\,,\,\Sigma_I] = \Sigma_R^a\Sigma_I^b[\boldsymbol{t^a},\boldsymbol{t^b}] = if_{ab}^c\Sigma_R^a\Sigma_I^b\boldsymbol{t^c} = 0\,. \tag{16.104}$$

Hence, one can simultaneously diagonalize the hermitian Σ_R and Σ_I matrices. That is, there exists an SU(5) matrix U such that $U\Sigma_R U^\dagger$ and $U\Sigma_I U^\dagger$ are traceless real diagonal matrices. Thus, without loss of generality, we assume that the vacuum expectation value $\langle\Sigma\rangle$ is a complex traceless diagonal matrix. That is, $\langle\Sigma\rangle$ is an element of the Cartan subalgebra of the complexification of $\mathfrak{su}(5)$, which means that $\langle\Sigma\rangle$ is a complex linear combination of $\boldsymbol{t^3},\boldsymbol{t^8},\boldsymbol{t^{23}}$, and $\boldsymbol{t^{24}}$. Using $\mathrm{Tr}(\boldsymbol{t^a}\boldsymbol{t^b}) = \frac{1}{2}\delta^{ab}$, it follows that $\Sigma^a = 2\,\mathrm{Tr}(\boldsymbol{t^a}\Sigma)$, which implies that the components of $\langle\Sigma^a\rangle$ are nonzero only when a = 3, 8, 23, and 24, corresponding to the generators of the Cartan subalgebra. We conclude that $f_{ab}^c = 0$ if $\langle\Sigma^a\rangle$ and $\langle\Sigma^b\rangle$ are nonzero, and therefore

$$\langle D^c\rangle = -if_{ab}^c\langle\Sigma^a\rangle^\dagger\langle\Sigma^b\rangle = 0\,. \tag{16.105}$$

Next, consider scalar field solutions to the equation

$$F_a^\dagger = -\frac{\partial W}{\partial\Sigma^a} = 0\,, \tag{16.106}$$

where the superpotential W is specified in eq. (16.98). Using the chain rule,

$$\frac{\partial}{\partial\Sigma^a}\,\mathrm{Tr}\,\Sigma^n = \frac{\partial\Sigma_{ij}}{\partial\Sigma^a}\frac{\partial}{\partial\Sigma_{ij}}\,\mathrm{Tr}\,\Sigma^n = n\,\mathrm{Tr}(\boldsymbol{t^a}\Sigma^{n-1})\,, \tag{16.107}$$

where the indices $i,j \in \{1,2,\ldots,5\}$. Hence, eq. (16.106) yields

$$m\,\mathrm{Tr}(\boldsymbol{t^a}\Sigma) + \lambda\,\mathrm{Tr}(\boldsymbol{t^a}\Sigma^2) = 0\,. \tag{16.108}$$

It is convenient to rewrite eq. (16.108) as

$$\mathrm{Tr}\big(\boldsymbol{t^a}\{m\Sigma + \lambda[\Sigma^2 - \tfrac{1}{5}\,\mathrm{Tr}(\Sigma^2)\mathbb{1}_{5\times5}]\}\big) = 0\,, \tag{16.109}$$

since the extra term that we have added inside the square brackets does not contribute because $\mathrm{Tr}\,\boldsymbol{t^a} = 0$. Indeed, the term inside the square brackets in eq. (16.109) is traceless and thus can be written as a linear combination of the SU(5) generators. Using $\mathrm{Tr}(\boldsymbol{t^a}\boldsymbol{t^b}) = \frac{1}{2}\delta^{ab}$, it follows that for any 5×5 traceless complex matrix A, the equation $\mathrm{Tr}(\boldsymbol{t^a}A) = 0$ is satisfied if and only if $A = 0$. Thus, we can conclude that

$$m\Sigma + \lambda[\Sigma^2 - \tfrac{1}{5}\,\mathrm{Tr}(\Sigma^2)\mathbb{1}_{5\times5}] = 0\,. \tag{16.110}$$

Having performed a gauge transformation such that Σ is a complex traceless diagonal 5×5 matrix, one can find all possible traceless diagonal solutions to eq. (16.110). Three classes of solutions emerge:

$$(i) \qquad \langle\Sigma\rangle = 0\,, \tag{16.111}$$

$$(ii) \qquad \langle\Sigma\rangle = \frac{m}{3\lambda}\,\mathrm{diag}(1,1,1,1,-4)\,, \tag{16.112}$$

$$(iii) \qquad \langle\Sigma\rangle = \frac{m}{\lambda}\,\mathrm{diag}(2,2,2,-3,-3)\,, \tag{16.113}$$

up to gauge transformations that permute the order of the diagonal elements of $\langle\Sigma\rangle$. In case (1), SU(5) is unbroken; in case (ii), SU(5) is broken down to SU(4)×U(1); and in case (iii) SU(5) is broken down to SU(3)×SU(2)×U(1). In all three cases, eqs. (16.100) and (16.106) are satisfied, implying that supersymmetry is unbroken. Thus at this stage of the analysis, cases (i), (ii), and (iii) are completely degenerate. Indeed, this degeneracy is not lifted to all finite orders in perturbation theory.

In a realistic model, soft supersymmetry-breaking terms are introduced to lift the degeneracy such that case (iii) corresponds to the minimum energy solution. In this case, SU(5) is broken down to the Standard Model gauge group. As in the case of the non-supersymmetric SU(5) GUT examined in Section 6.3, we can break $\mathrm{SU}(3)_C\times\mathrm{SU}(2)_L\times\mathrm{U}(1)_Y$ down to $\mathrm{SU}(3)\times\mathrm{U}(1)_{\mathrm{EM}}$ by introducing an additional chiral supermultiplet H that transforms as the fundamental representation $\mathbf{5}$ of $\mathfrak{su}(5)$. This supermultiplet will contain a Higgs doublet field that transforms as $(\mathbf{1},\mathbf{2},\frac{1}{2})$ under $\mathrm{SU}(3)_C\times\mathrm{SU}(2)_L\times\mathrm{U}(1)_Y$. However, the Higgs sector of the MSSM contains two Higgs doublets that reside in chiral superfields of opposite sign hypercharge. Thus, in the SU(5) SUSYGUT, we shall add a second chiral supermultiplet $\overline{H}$ that transforms as a $\overline{\mathbf{5}}$ of $\mathfrak{su}(5)$, which contains a second Higgs doublet field that transforms as $(\mathbf{1},\mathbf{2},-\frac{1}{2})$ under $\mathrm{SU}(3)_C\times\mathrm{SU}(2)_L\times\mathrm{U}(1)_Y$.

The presence of both H and $\overline{H}$ allows us to incorporate the quark and lepton Yukawa couplings into the SUSYGUT superpotential. Thus, the full superpotential of the SU(5) SUSYGUT is given by

$$\begin{aligned} W = {}&\tfrac{1}{2}m\,\mathrm{Tr}\,\Sigma^2 + \tfrac{1}{3}\lambda\,\mathrm{Tr}\,\Sigma^3 + \mu H_i\overline{H}^i + \beta\overline{H}^i\Sigma_i{}^jH_j \\ &+ y_1\overline{F}^iT_{ij}\overline{H}^j + y_2\epsilon^{ijk\ell m}T_{ij}T_{k\ell}H_m, \end{aligned} \tag{16.114}$$

where $\overline{F}$ and T are the chiral supermultiplets containing the quarks, leptons, and their supersymmetric partners, and generation indices have been suppressed. In specifying the superpotential above, we have imposed a matter parity symmetry, $\overline{F}\to-\overline{F}, T\to-T$, which removes any potential R-parity-violating interactions, as the symmetry distinguishes between $\overline{F}$ and $\overline{H}$, which otherwise transform the same way under SU(5). Prior to including soft SUSY-breaking terms, none of the scalar fields that reside in H, $\overline{H}$, $\overline{F}$, and T acquire nonzero vacuum expectation values. Thus, our previous analysis remains valid, and three degenerate supersymmetry-conserving minima exist, as previously noted. Employing eq. (16.113), in which SU(5) breaks down to $\mathrm{SU}(3)_C\times\mathrm{SU}(2)_L\times\mathrm{U}(1)_Y$, one can use eq. (16.114) to compute the masses of the fields that reside in Σ, H, and $\overline{H}$.

Here, we focus on the masses of the fields that reside in H and $\overline{H}$, which we write more explicitly in the form

$$H = \begin{pmatrix} H_3 \\ H_u \end{pmatrix}, \qquad \overline{H} = \begin{pmatrix} \overline{H}_3 \\ H_d \end{pmatrix}, \tag{16.115}$$

where H_3 is a color triplet with hypercharge $-\frac{1}{3}$, $\overline{H}_3$ is a color antitriplet with hypercharge $\frac{1}{3}$, and H_u and H_d are the SU(2) doublet Higgs superfields of the

MSSM with hypercharges $\frac{1}{2}$ and $-\frac{1}{2}$, respectively. After inserting eq. (16.113) into eq. (16.114), the resulting mass terms involving H and $\overline{H}$ are

$$W_{\text{mass}} \supset \left(\mu + \frac{2m\beta}{\lambda}\right) \overline{H}_3 H_3 + \left(\mu - \frac{3m\beta}{\lambda}\right) H_u H_d \,. \tag{16.116}$$

At this stage of the analysis, there is no electroweak symmetry breaking. Thus, we must demand that the colored fields residing in H_3 and $\overline{H}_3$ are superheavy, or order the GUT scale, whereas the fields residing in H_u and H_d are massless. Hence, one must set

$$\mu = \frac{3m\beta}{\lambda} \,. \tag{16.117}$$

Having done so, eq. (16.116) reduces to $W_{\text{mass}} = (5m\beta/\lambda)\overline{H}_3 H_3$. The dimensionless couplings β, $\lambda \sim \mathcal{O}(1)$, whereas m is of order the GUT scale. Thus, having performed the fine-tuning of the parameters in eq. (16.117), one obtains superheavy color Higgs triplets and massless Higgs doublets. We noted in Section 6.3 a similar fine-tuning of the parameters in the context of the so-called doublet–triplet splitting problem. In the SU(5) SUSYGUT, the fine-tuning of the parameters is still required, but this fine-tuning is technically natural since the parameters of the superpotential are protected from radiative corrections by the no-renormalization theorems, discussed in Chapter 11. Once we add soft supersymmetry-breaking terms, one can now generate the spontaneous breaking of $\mathrm{SU}(3)_C\times\mathrm{SU}(2)_L\times\mathrm{U}(1)_Y$ down to $\mathrm{SU}(3)\times\mathrm{U}(1)_{\mathrm{EM}}$, which will provide masses to the Higgs bosons residing in H_u and H_d of order the electroweak scale. The fine-tuning condition given in eq. (16.117) will receive corrections at the loop level of order the supersymmetry-breaking scale multiplied by the relevant coupling constant and loop-factor suppression. In this sense, the doublet–triplet splitting is stable once eq. (16.117) is implemented.

It is also possible to try to solve the doublet–triplet splitting without any fine-tuning at all. One class of models uses a "sliding singlet" chiral superfield S. To see how this works, suppose we replace the term $\mu H_i \overline{H}^i$ in eq. (16.114) by a term $\lambda_\mu S H_i \overline{H}^i$. Then, once the scalar components of the electroweak doublets H_u and H_d obtain VEVs v_u and v_d that break the electroweak symmetry, the minimization conditions following from the supersymmetric potential would imply

$$v_u(\lambda_\mu\langle S\rangle - 3\beta\langle\Sigma_{jj}\rangle) = v_d(\lambda_\mu\langle S\rangle - 3\beta\langle\Sigma_{jj}\rangle) = 0 \,, \tag{16.118}$$

for each of $j = 4, 5$, so that we automatically get

$$\mu = \lambda_\mu\langle S\rangle = 3\beta m/\lambda \,, \tag{16.119}$$

as required for the triplets to be much heavier than the doublets. Unfortunately, the simplest attempts to implement this fail once one includes soft supersymmetry-breaking effects, which will generically shift the VEV of S away from the result given in eq. (16.118), eliminating the hierarchy between the doublet and triplet masses. However, by enlarging the model in various ways, it is possible in principle for the sliding singlet to accomplish the goal of making the color triplet superheavy while keeping the Higgs doublets light.

Another model-building strategy to avoid light color triplet scalars in supersymmetric GUTS is the "missing doublet" mechanism. Here, the idea is to enlarge the Higgs sector of the theory to include larger representations in which there are color triplets but no corresponding electroweak doublets. The color triplets in these large representations can pair up with the color triplets in the $\overline{\mathbf{5}}$ and $\mathbf{5}$, giving them both large masses, leaving the electroweak doublets behind as light fields. The simplest version of this in SU(5) uses Higgs scalar multiplets in the $\mathbf{50}$, the $\overline{\mathbf{50}}$, and the $\mathbf{75}$ representations, which together form a real representation, and so are free of anomalies. From the branching rules in Table H.9, we see that the $\mathbf{50} \oplus \overline{\mathbf{50}}$ has a charge $\pm 1/3$ color triplet–antitriplet pair (and many other components) but no electroweak doublets. From the results in Table H.8, we note that the following terms are allowed in the superpotential:

$$W = M(\mathbf{50})(\overline{\mathbf{50}}) + y\overline{H}(\mathbf{50})(\mathbf{75}) + y'H(\overline{\mathbf{50}})(\mathbf{75}). \tag{16.120}$$

Then, as long as we arrange for the $\mathbf{75}$ field to obtain a VEV of order the GUT scale M_{GUT} along its $\text{SU}(3)_C \times \text{SU}(2)_L \times \text{U}(1)_Y$-singlet direction, all of the components of the $\mathbf{50} \oplus \overline{\mathbf{50}}$ together with the color triplet components of H and $\overline{H}$ will naturally get mass of that same order, leaving the electroweak doublet components of H and $\overline{H}$ light, as desired.

As we noted in Section 6.3, the unification of the SU(3), SU(2), and U(1) gauge couplings is significantly improved in the supersymmetric version of the GUT model because of the modification of the running of the couplings due to the superpartners of the Standard Model particles. The relations among the Yukawa couplings are more model-dependent. In the minimal non-supersymmetric SU(5) GUT, we found an approximate unification of the third-generation down-type Yukawa couplings, whereas the predicted unification for the first- and second-generation down-type Yukawa couplings was strongly violated by the experimental data. This suggests that we need $H + \overline{H}$ to include an admixture of $\mathbf{45} \oplus \overline{\mathbf{45}}$ Higgs representations in addition to the $\mathbf{5} \oplus \overline{\mathbf{5}}$. In the corresponding supersymmetric GUT, the impact on Yukawa unification is not as dramatic as in the case of gauge coupling unification. For example, the successful unification of third-generation down-type Yukawa couplings, assuming they come about entirely from coupling to the $\mathbf{5} \oplus \overline{\mathbf{5}}$ Higgs fields, is modified by roughly 10% due to the impact of the superpartners on the running of couplings from the GUT scale to the electroweak scale.

In supersymmetric SU(5), we have the additional feature that there are relations between squark and slepton masses that will automatically hold, generation by generation, due to the way in which they are embedded into unified representations. In particular, even if no other organizing principle governs supersymmetry breaking, we must have

$$m^2_{\widetilde{Q}} = m^2_{\widetilde{\bar{u}}} = m^2_{\widetilde{\bar{e}}}\,, \qquad m^2_{\widetilde{\bar{d}}} = m^2_{\widetilde{L}}\,, \tag{16.121}$$

at the GUT breaking scale. In the simplest versions of supersymmetric SU(5) unification, the gaugino masses are also unified at the GUT breaking scale:

$$M_1 = M_2 = M_3. \tag{16.122}$$

However, this assumes that the F-term that effectively communicates supersymmetry breaking to the gauginos is in a singlet of SU(5). The other possibilities are that this F-term lives in any symmetric product of two adjoint representations that includes a Standard Model singlet component. Here, the relevant SU(5) tensor product is, as seen in Table H.8,

$$(\mathbf{24} \otimes \mathbf{24})_S = \mathbf{1} \oplus \mathbf{24} \oplus \mathbf{75} \oplus \mathbf{200}. \tag{16.123}$$

An F-term VEV in each of these representations will contribute to the gaugino masses at the GUT scale in the ratios for (M_1, M_2, M_3):

$$\begin{aligned} &\mathbf{1} : (1,1,1), &&\mathbf{24} : (-1,-3,2), \\ &\mathbf{75} : (-5,3,1), &&\mathbf{200} : (10,2,1), \end{aligned} \tag{16.124}$$

which can be obtained by a computation outlined in Ref. [184]. By taking linear combinations of F-term VEVs, one can obtain any desired ratio of gaugino masses.

16.3.2 Supersymmetric Grand Unification beyond SU(5)

The gauge group SO(10) provides a more ambitious and predictive unification scheme, since it contains SU(5) as a subgroup and puts together all Standard Model fermions (and their superpartners) for each generation into a single **16** representation, and it combines both Higgs doublets H_u and H_d into a single **10** representation. As we have seen, an SU(5) GUT requires large representations in order to accommodate realistic fermion masses. The group SO(10) has an even richer set of ways to include large representations, and the possibilities are too many to review here. We will therefore make only a few remarks that distinguish the SO(10) case from the SU(5) case.

First, there are several distinct ways that SO(10) can be broken down to the Standard Model gauge group with Higgs field representations. Note that SO(10) has two maximal subgroups that contain the Standard Model as sub-subgroups; there is a reduction to SU(5)$\times$U(1)$_X$ [see eq. (6.274)], and an alternative reduction to the Pati–Salam group SU(4)$_{\text{PS}}\times$SU(2)$_L\times$SU(2)$_R$. Here SU(4)$_{\text{PS}}$ unifies color and a gauged U(1)$_{B-L}$ symmetry, and was historically the first proposal to unify the QCD gauge group with a part of the electroweak gauge sector. Either of these maximal subgroups could be realized at some intermediate scale. From Table H.12 we see that the **45** contains an SU(5)$\times$U(1)$_X$ singlet component while the **54** does not, and conversely from Table H.13 we see that the **54** contains a singlet under SU(4)$_{\text{PS}}\times$SU(2)$_L\times$SU(2)$_R$ while the **45** does not. These tables also show that the **210** contains singlets of both maximal subgroups. Each **45**, **54**, and **210** multiplet contains components in the adjoint representation of SU(5) that could play the role of the Σ field in the previous subsection. The VEVs of these three multiplets can play a role in obtaining the SM gauge group, but none of these will break the $U(1)_{B-L}$ subgroup of SO(10). This can be done with either a $\mathbf{16} \oplus \overline{\mathbf{16}}$ or a $\mathbf{126} \oplus \overline{\mathbf{126}}$. Note that the **45**, **54**, and **210** are real representations of SO(10), which implies that the fermionic fields of these multiplets can obtain masses of order the GUT breaking scale. The **16** and the **126** are complex representations of SO(10). Hence, apart from the three **16**s that include the SM quarks and leptons, each additional **16** and the **126**s must be accompanied by their conjugates when included in the Higgs sector; otherwise there would be exotic light chiral fermions that could not obtain a superheavy mass.

Both the $\mathbf{16} \oplus \overline{\mathbf{16}}$ and the $\mathbf{126} \oplus \overline{\mathbf{126}}$ options for the Higgs fields will ensure the breaking of the $U(1)_{B-L}$ subgroup of SO(10) if they get VEVs. Both of them can also provide neutrino masses via the seesaw mechanism, through superpotential terms of the schematic forms

$$W = \frac{1}{M}\langle\overline{\mathbf{16}}_H\rangle\langle\overline{\mathbf{16}}_H\rangle(\mathbf{16}_{\text{matter}})(\mathbf{16}_{\text{matter}}) \qquad (16.125)$$

or

$$W = \langle\overline{\mathbf{126}}_H\rangle(\mathbf{16}_{\text{matter}})(\mathbf{16}_{\text{matter}})\,. \qquad (16.126)$$

However, there is a crucial difference between these possibilities. From either Table H.12 or the combination of Tables H.13 and H.4, we see that the Standard Model singlet components of the **16** and the **126** have $U(1)_{B-L}$ charges equal to -1 and -2, respectively. This means that a VEV for $\mathbf{16} \oplus \overline{\mathbf{16}}$ will break $U(1)_{B-L}$ completely, but the $\mathbf{126} \oplus \overline{\mathbf{126}}$ option will preserve a residual $\mathbb{Z}_2$ subgroup of $U(1)_{B-L}$, under which each field transforms by a multiplicative factor of $(-1)^{3(B-L)}$. Comparing with eq. (13.14), we see that this residual $\mathbb{Z}_2$ symmetry is nothing other than the matter parity symmetry, which is equivalent to R-parity. Therefore, if SO(10) is broken by the $\mathbf{126} \oplus \overline{\mathbf{126}}$ combination, R-parity will automatically be a symmetry of the low-energy theory. On the other hand, if the $\mathbf{16} \oplus \overline{\mathbf{16}}$ combination is used to break SO(10), then this $\mathbb{Z}_2$ subgroup of SO(10) is spontaneously broken, and if R-parity is present as an unbroken symmetry, it has to be included separately. Thus two (of many) possible patterns of symmetry breaking in SO(10) can be summarized as

$$\mathrm{SO}(10) \xrightarrow{\langle\mathbf{45}\rangle,\,\langle\mathbf{16}\rangle} \mathrm{SU}(3)_C \times \mathrm{SU}(2)_L \times \mathrm{U}(1)_Y, \qquad (16.127)$$

$$\mathrm{SO}(10) \xrightarrow{\langle\mathbf{45}\rangle,\,\langle\mathbf{126}\rangle} \mathrm{SU}(3)_C \times \mathrm{SU}(2)_L \times \mathrm{U}(1)_Y \times \mathbb{Z}_2, \qquad (16.128)$$

where the $\mathbb{Z}_2$ is matter parity. In addition to Higgs representations in the **45**, **54**, or **210**, and either $\mathbf{126} \oplus \overline{\mathbf{126}}$ or $\mathbf{16} \oplus \overline{\mathbf{16}}$ to break the SO(10) symmetry down to that of the Standard Model, one needs Higgs representations to break the electroweak symmetry and provide for ordinary fermion masses. Due to the group theory fact that $\mathbf{16} \otimes \mathbf{16} = \mathbf{10}_S \oplus \mathbf{120}_A \oplus \mathbf{126}_S$ (see Table H.11), at least the **10** or **126** representations are necessary for this purpose, since the Standard Model Yukawa couplings are not antisymmetric.

More generally, automatic R-parity conservation will occur in any theory with gauged $U(1)_{B-L}$ provided that it is broken only by Higgs fields with even integer values of $3(B-L)$. In the case of SO(10), this means that the Higgs VEVs can be in any representations that have even SO(10) quadrality (see Table H.10), which include $\mathbf{10}, \mathbf{45}$, $\mathbf{54}$, $\mathbf{120}$, $\mathbf{126}$, $\overline{\mathbf{126}}$, $\mathbf{210}$, $\mathbf{210}'$,..., but not $\mathbf{16}, \overline{\mathbf{16}}, \mathbf{144}$, or $\overline{\mathbf{144}}$, if one wants R-parity to be an automatic symmetry. In terms of the Pati–Salam subgroup, the Higgs representations whose VEVs preserve R-parity automatically are those that have even SU(4) quadrality (see Table H.4), which include the $\mathbf{1}$, $\mathbf{10}$, $\mathbf{15}$, $\mathbf{35}$, while the $\mathbf{4}$, $\mathbf{20}''$, and $\mathbf{36}$ will break this $\mathbb{Z}_2$. Note that supersymmetric SU(5) does not have this possibility of an automatic R-parity conservation, because it does not contain $U(1)_{B-L}$ as a subgroup. For SU(5), R-parity is either absent or must be imposed as a separate, ad hoc symmetry.

In order to solve the doublet–triplet splitting problem, the missing doublet mechanism employed in SU(5) [see eq. (16.120)] can also be realized in SO(10). The most straightforward way to do this is to simply promote each of the relevant SU(5) representations to the smallest SO(10) representation that contains it. From Tables H.12 and H.9, we see that

$$\mathrm{SU}(5) \quad \rightarrow \quad \mathrm{SO}(10) \tag{16.129}$$

$$(\mathbf{50},\ \overline{\mathbf{50}},\ \mathbf{75}) \rightarrow (\overline{\mathbf{126}},\ \mathbf{126},\ \mathbf{210}) \tag{16.130}$$

will work for this purpose, where the $\mathbf{210}$ should acquire a VEV to give masses to the $\mathbf{126} \oplus \overline{\mathbf{126}}$ fields along with the triplets in $H + \overline{H} = \mathbf{10}_H$. This can be accomplished with a superpotential schematically of the form

$$W = M(\mathbf{126})(\overline{\mathbf{126}}) + y(\mathbf{10}_H)(\mathbf{126})(\mathbf{210}) + y'(\mathbf{10}_H)(\overline{\mathbf{126}})(\mathbf{210}), \tag{16.131}$$

which are allowed interactions, as one can see from Table H.11.

However, there is another way of solving the doublet–triplet splitting problem known as the "missing VEV" mechanism, which has no analogue in SU(5). Here the idea is that the $\mathbf{45}$ of SO(10) contains a component $(\mathbf{15}, \mathbf{1}, \mathbf{1})$ of the $\mathrm{SU}(4)_{\mathrm{PS}} \times \mathrm{SU}(2)_L \times \mathrm{SU}(2)_R$ maximal subgroup, which can give mass to color triplets but not to singlets when it gets a VEV. The relevant tensor product for $\mathrm{SU}(4)_{\mathrm{PS}}$ from Table H.5 is $(\mathbf{6} \otimes \mathbf{6})_A = \mathbf{15}$, which lives inside $(\mathbf{10} \otimes \mathbf{10})_A = \mathbf{45}$ of SO(10) as given in Table H.11. The crucial fact is that the $\mathbf{6}$ of $\mathrm{SU}(4)_{\mathrm{PS}}$ contains only a color triplet and antitriplet, as shown in Table H.6. Since this is an antisymmetric tensor product, one must start with two distinct Higgs fields in the $\mathbf{10}$ of SO(10), with a superpotential coupling of the form

$$W = y(\mathbf{10}_H)(\mathbf{10}_{H'})(\mathbf{45}) + M(\mathbf{10}_{H'})(\mathbf{10}_{H'}). \tag{16.132}$$

The purpose of the last mass term is to avoid having two H_u and two H_d Higgs multiplets in the low-energy theory, as opposed to one each as in the MSSM. This general structure can be augmented by other Higgs representations in a variety of ways.

In SO(10), the minimal way to include the ordinary Yukawa couplings of the Standard Model is a superpotential of the form

$$W = y(\mathbf{10}_H)(\mathbf{16}_{\mathrm{matter}})(\mathbf{16}_{\mathrm{matter}}), \tag{16.133}$$

which predicts a unification of Yukawa couplings at the GUT scale:

$$y_t = y_b = y_\tau = y_{\nu_\tau}, \tag{16.134}$$

a stronger relation than in SU(5). For this to be valid in the MSSM requires large $\tan\beta$ of roughly 50 or so, so that the equality of y_t and y_b is counteracted by a large ratio v_u/v_d to provide for $m_t \gg m_b$. For the first- and second-generation quarks and leptons, eq. (16.134) would then be badly violated by the observed masses, so one needs the MSSM Higgs doublets to contain at least a small admixture of the **120** and/or $\overline{\mathbf{126}}$ representations. It is the latter that have Yukawa couplings to the lighter Standard Model quarks and leptons. For the soft supersymmetry-breaking terms, we also have, for each generation,

$$m^2_{\widetilde{Q}} = m^2_{\widetilde{\bar{u}}} = m^2_{\widetilde{\bar{d}}} = m^2_{\widetilde{L}} = m^2_{\widetilde{\bar{e}}}, \tag{16.135}$$

and in the MSSM Higgs sector,

$$m^2_{H_u} = m^2_{H_d}, \tag{16.136}$$

to the extent that one can neglect the **120** and/or $\overline{\mathbf{126}}$ contamination just mentioned. For the gaugino masses, the most straightforward possibility occurs if the effective F-term is a singlet under SO(10), which would imply unification $M_1 = M_2 = M_3$ at the GUT scale. However, one can also have F-terms contributing to gaugino masses in any component of the symmetric product of the adjoint representation with itself:

$$(\mathbf{45} \otimes \mathbf{45})_S = \mathbf{1} \oplus \mathbf{54} \oplus \mathbf{210} \oplus \mathbf{770}. \tag{16.137}$$

As in the case of SU(5) [see eq. (16.124)], this is enough to give any desired ratios of M_1, M_2, and M_3 at the GUT scale. In fact, if the F-term contains a **54** of SO(10), then the corresponding contributions to gaugino masses (M_1, M_2, M_3) are in the same ratio $(-1, -3, 2)$ as predicted by the **24** of SU(5), which it contains.

Another intriguing scenario is the so-called flipped SU(5) model that arises by considering an embedding SU(5)×U(1)$_X$ inside SO(10) such that the U(1)$_Y$ generator is a linear combination of the U(1)$_X$ generator and the $\boldsymbol{T^{24}}$ generator of SU(5). One generation of fermions resides in the $\overline{\mathbf{5}} \oplus \mathbf{10} \oplus \mathbf{1}$ consisting of $\{\bar{u}, e, \nu\}$, $\{u, d, \bar{d}, \bar{\nu}\}$, and $\{\bar{e}\}$, respectively (see Exercise 16.8). Some of the benefits of supersymmetric flipped SU(5) models were first given in Ref. [185].

An even more ambitious unification scheme uses the exceptional group E$_6$, which includes SO(10) as a subgroup, and can include Higgs and quark and lepton superfields all within its **27** fundamental representation. Like SO(10), all representations of E$_6$ are anomaly-free. Of all the exceptional Lie groups, only E$_6$ has complex representations, and therefore the possibility of chiral quarks and leptons in a straightforward symmetry-breaking scheme without introducing extra space dimensions. The Yukawa interactions are implemented schematically by symmetric superpotential couplings of the form (see Table H.15)

$$W = y(\mathbf{27})(\mathbf{27})(\mathbf{27}). \tag{16.138}$$

Note that there are no allowed mass terms at tree level of the form $W = M(\mathbf{27})(\mathbf{27})$ because the **27** is a complex representation. This could be a way of explaining the suppression of the μ-term within the MSSM. The adjoint representation of E_6 is the **78**, which of course contains the adjoint representations of the SO(10) and SU(5) subgroups, but there is another copy of the SO(10) adjoint within the **351** of E_6, as shown in Table H.16. From the same table, we see that the $\mathbf{351'}$ is the smallest E_6 representation that contains the $\overline{\mathbf{126}}$ of SO(10) necessary for the seesaw mechanism with automatic R-parity conservation, through superpotential couplings of the form

$$W = y(\mathbf{27})(\mathbf{27})(\mathbf{351'}) \,. \tag{16.139}$$

Couplings of this form, and/or of the form $(\mathbf{27})(\mathbf{27})(\mathbf{351})$, are also necessary in order to avoid generation-by-generation equality of the MSSM quark and lepton Yukawa couplings. As shown in Table H.14, there are no other smaller representations of E_6; the next smallest is the **650**.

There are three maximal subgroups of E_6 that are compatible with the Standard Model fermion embedding structure, namely SO(10)×U(1), SU(6)×SU(2), and $SU(3)_C \times SU(3)_L \times SU(3)_R$, with branching rules shown in Tables H.16, H.17, and H.18. Of these, the $SU(3)_C \times SU(3)_L \times SU(3)_R$ path, called "trinification," can itself be viewed as a GUT, since it features symmetries under which the three SU(3)s are exchanged if one sets the corresponding gauge couplings equal to each other. This is an attractive possibility for unification, since it does not have a doublet–triplet splitting problem at all, because the color triplets live in completely different representations than the MSSM Higgs bosons.

16.3.3 Proton Decay

In Section 6.3.6, we noted that by imposing $SU(3)_C \times SU(2)_L \times U(1)_Y$ invariance and Lorentz invariance, all possible operators of dimension four or less involving the fields of the Standard Model conserve baryon number (B) and lepton number (L). This is no longer true if one considers operators greater than four. Nevertheless, this observation provides a reason why B and L are conserved by the Standard Model Lagrangian. At dimension five, there exists one L-violating operator, which was exhibited in eq. (6.1), and provides a Majorana mass term for the would-be massless neutrinos of the Standard Model. The first B-violating operator arises at dimension six and can mediate proton decay. Indeed, grand unified theories provide an ultraviolet realization of this mechanism. The lifetime of the proton in a typical (non-supersymmetric) GUT is expected to be roughly

$$\tau_p \sim \left(\frac{M_{\rm GUT}}{10^{16}~\text{GeV}}\right)^4 10^{35}~\text{years} \,. \tag{16.140}$$

This mass dependence follows dimensional analysis. Indeed, if proton decay is mediated by an operator of dimension $d > 4$, then the operator coefficient scales as M^{4-d}, where M is the scale at which the operator is generated by the corresponding ultraviolet theory. The proton lifetime is inversely proportional to the squared decay matrix element, and therefore scales as $M^{2(d-4)}$. This explains the $M_{\rm GUT}^4$ behavior shown in eq. (16.140), which is a consequence of an effective $d = 6$ B-violating operator.

Given eq. (16.140) and the current experimental limit of

$$\Gamma^{-1}(p \to e^+\pi^0) \gtrsim 10^{34} \text{ years}, \tag{16.141}$$

it follows that the GUT scale cannot be significantly less than about 10^{16} GeV. In Section 6.3.5, we showed that in SUSYGUT models, gauge coupling unification is successful and the corresponding GUT scale where the couplings unify is roughly given by 2×10^{16} GeV. Hence, it naively appears that SUSYGUT models are consistent with the experimental nonobservation of proton decay.

Our naive intuition fails in the case of the supersymmetric extension of the Standard Model and its ultraviolet completion by a SUSYGUT model. Because of the existence of colored scalar-quark fields, one can construct dimension-four Lorentz-invariant, $\mathrm{SU}(3)_C\times\mathrm{SU}(2)_L\times\mathrm{U}(1)_Y$-invariant operators that are B-violating. An example of such an operator is $\varepsilon_{ijk}\widetilde{u}^i_R d^j_R s^k_R$, where i, j, k are color indices. This operator is a singlet under $\mathrm{SU}(3)_C\times\mathrm{SU}(2)_L\times\mathrm{U}(1)_Y$ and yet carries net baryon number. Thus, in supersymmetric models, the previous natural explanation for the near absence of baryon number violation in Nature is lost.

In the MSSM, such dimension-four operators are absent due to the imposition of a matter parity symmetry, in which the MSSM Lagrangian is assumed to be invariant under the transformations $q \to -q$ and $\widetilde{q} \to -\widetilde{q}$ for quark and squark fields. Likewise, the matter parity symmetry is also imposed on lepton and slepton fields, whereas the Higgs fields, gauge fields, and their fermionic partners do not change under the symmetry. Thus, requiring $\mathrm{SU}(3)_C\times\mathrm{SU}(2)_L\times\mathrm{U}(1)_Y$ invariance along with the matter parity symmetry just described leads to automatic baryon number and lepton number conservation for all supersymmetric operators composed of MSSM fields of dimension four or less.[9]

It should be noted that it is possible to construct supersymmetry-breaking, dimension-four operators, consistent with $\mathrm{SU}(3)_C\times\mathrm{SU}(2)_L\times\mathrm{U}(1)_Y$ gauge invariance and the matter parity symmetry, that violate B and/or L. An example is

$$\mathcal{O}_4 = \frac{m_{\rm soft}}{M_P}\epsilon^{ijk}\epsilon^{\alpha\beta}\epsilon^{\rho\gamma}\widetilde{q}_{i\alpha p}\widetilde{q}_{j\beta q}\widetilde{q}_{k\rho r}\widetilde{\ell}_{\gamma s}\,, \tag{16.142}$$

where i, j, and k are color SU(3) indices; α, β, ρ, and γ are SU(2) indices; and p, q, r, and s are quark and lepton generation indices.[10] We have included the expected suppression due to a superheavy mass scale for a supersymmetry-breaking term, as discussed in Section 14.3. As we shall see below, this is more suppressed than the supersymmetric dimension-five terms, which we first discuss in some detail. We return to the implications of $\mathcal{O}_4$ below.

[9] Of course, we recognize the impact of the matter parity symmetry to be the same as that of R-parity. The possibility of extending the MSSM Lagrangian to include RPV terms is the subject of Section 16.5.

[10] Looking ahead to eq. (16.143), one simply replaces the superfields by their scalar components to obtain B- and/or L-violating dimension-four terms.

We first analyze operators with dimension greater than four. We again impose Lorentz invariance, $\mathrm{SU}(3)_C\times\mathrm{SU}(2)_L\times\mathrm{U}(1)_Y$, and the matter parity symmetry. Consider possible supersymmetric operators of dimension five that violate B and/or L. The complete list is quite short, consisting of three possible terms in the superpotential:

$$W \supset \frac{c^{pq}}{M_1}\epsilon^{\alpha\beta}\epsilon^{\rho\gamma}L_{\alpha p}(H_u)_\beta L_q(H_u)_\gamma + \frac{g^{pqrs}}{M_2}\epsilon^{ijk}\epsilon^{\alpha\beta}\epsilon^{\rho\gamma}Q_{i\alpha p}Q_{j\beta q}Q_{k\rho r}L_{\gamma s}$$
$$+\frac{h_{pqrs}}{M_3}\epsilon_{ijk}\bar{u}^{ip}\bar{u}^{jq}\bar{d}^{kr}\bar{e}^{s}\,, \qquad (16.143)$$

where i, j, and k are $\mathrm{SU}(3)_C$ indices; α, β, ρ, and γ are $\mathrm{SU}(2)_L$ indices; p, q, r, and s are quark and lepton generation indices; c, g, and h are dimensionless constants that depend on the generation labels; and the M_i are mass scales at which the corresponding higher-dimensional operators are generated. We recognize the term proportional to c as the L-violating operator responsible for the supersymmetric seesaw discussed in Section 16.2. The operators proportional to g and h violate B and L while conserving $B-L$, and are present in the MSSM, although they have no counterpart in the Standard Model.

Since the terms in eq. (16.143) are suppressed only by M^{-1}, one might fear that the proton would decay much too rapidly. A careful analysis shows that this conclusion is incorrect. If we consider how such operators would mediate proton decay in a supersymmetric grand unified model, we are led to diagrams such as in Fig. 16.1. Proton decay via these dimension-five operators must occur via a loop diagram which leads to extra powers of the coupling constant compared to tree-level exchange of a superheavy baryon-number-violating gauge boson. Furthermore, one can estimate the short-distance QCD corrections. Whereas these corrections yield an enhancement of the dimension-six operators in the ordinary grand unified theories, in the supersymmetric theory they lead to a suppression of the dimension-five operators. The end result is that the contribution to the proton decay rate due to diagrams such as in Fig. 16.1 is not significantly larger than the contributions from dimension-six operators that result in the estimate given in eq. (16.140). Nevertheless, one can expect the contribution of the dimension-five operators to provide the leading contribution to the proton decay amplitude in many SUSYGUT models. In such cases, the dominant decay final states of the proton are markedly different from $p\to e^+\pi^0$, which is the dominant decay mode arising from the tree-level exchange of superheavy gauge bosons, for reasons we explain below.

First, we note that the dimension-five operators that contribute to proton decay must involve quark superfields of at least two different generations. A closer examination of eq. (16.143) shows that $g_{ppps}=h_{ppps}=0$ due to the presence of the ϵ^{ijk} factor, in light of the fact that the SU(2) indices run over two possible values (corresponding to the up-type and the down-type quark superfield, respectively). As an example, if we extract the F component of the second term in the superpotential given in eq. (16.143) to obtain terms in the effective Lagrangian, one of the

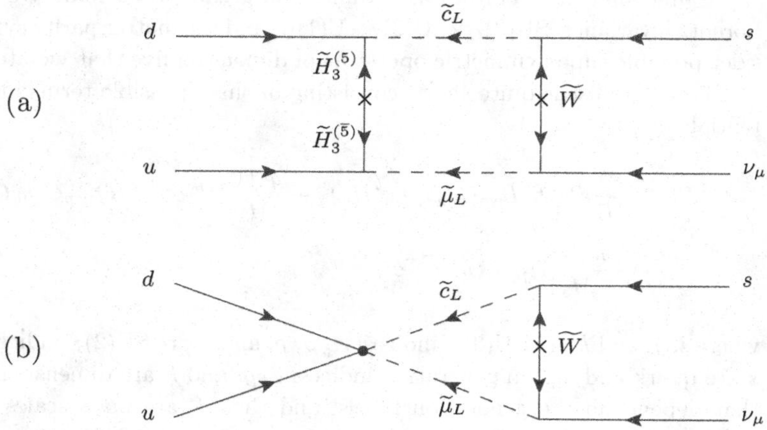

Fig. 16.1 Proton decay mediated by supersymmetric particles. The loop diagram (a), which is mediated by the exchange of one light and one superheavy supersymmetric fermion, can be as important as the tree-level diagrams involving exchange of one superheavy grand unified gauge boson. The superheavy fermion line $\widetilde{H}_3$ can be shrunk to a point as in diagram (b), resulting in an effective dimension-five vertex. Adding in the spectator u quark (and noting that the incoming s and ν_μ can be reinterpreted as an outgoing $\bar{s}$ and $\bar{\nu}_\mu$, respectively) then yields a mechanism for the decay $p \to K^+\bar{\nu}_\mu$.

contributing terms is given by

$$\begin{aligned}\mathscr{L}_{\rm int} &= \frac{g^{pqrs}}{M_2}\epsilon^{ijk}\epsilon^{\alpha\beta}\epsilon^{\rho\gamma}\widetilde{q}_{i\alpha p}\widetilde{q}_{j\beta q}q_{kpr}\ell_{\gamma s} \\ &= \frac{g^{pqrs}}{M_2}\epsilon^{ijk}\left(\widetilde{u}_{ip}\widetilde{d}_{jq}u_{kr}e_s - \widetilde{d}_{ip}\widetilde{u}_{jq}u_{kr}e_s + \widetilde{d}_{ip}\widetilde{u}_{jq}d_{kr}\nu_s - \widetilde{u}_{ip}\widetilde{d}_{jq}d_{kr}\nu_s\right) .\end{aligned} \tag{16.144}$$

The four-point vertex appearing in Fig. 16.1(b) arises from the interaction Lagrangian exhibited in eq. (16.144). Indeed, the quarks and squarks that comprise the four-point vertex shown in Fig. 16.1(b) cannot be all of the same generation because of the relation $g_{ppps} = 0$ noted above. This explains why final-state proton decay products containing a kaon are preferred.

There is a second reason which favors the participation of higher generations in the context of SU(5) and similar models of grand unification. In the SU(5) SUSYGUT model, $\widetilde{H}_3^{(5)}$ and $\widetilde{H}_3^{(\bar{5})}$ are color triplet higgsino components of the chiral supermultiplets that transform under SU(5) as $\mathbf{5}$ and $\overline{\mathbf{5}}$, respectively. A gauge invariant mass term $\widetilde{H}_3^{(5)}\widetilde{H}_3^{(\bar{5})}$ exists which is represented by the cross in Fig. 16.1(a). Because the color triplet higgsinos couple to fermions and their scalar partners with strength proportional to the corresponding fermion mass, it follows that the contributions from diagrams involving members of the higher generations are enhanced. Thus, a diagram similar to Fig. 16.1(a) with an intermediate state of $\widetilde{t\tau}$ will also be important (depressed only by a mixing angle to convert t to s in the final state). The conclusion is that if dimension-five supersymmetric operators mediate proton decay, the dominant decay final states will be $K^+\bar{\nu}_\tau$, $K^+\bar{\nu}_\mu$, and μ^+K. As previously noted, this differs significantly from the non-supersymmetric SU(5) GUT model, which favors a $\pi^0 e^+$ final state and where the production of second-generation fermions is suppressed.

One may be concerned that the dimension-five operators may still result in a proton decay rate which contradicts the experimental limits. The most recent results from proton decay experiments yield $\tau_p/\mathrm{BR}(p \to \mu^+ K) > 1.6 \times 10^{33}$ years at 90% CL [41]. For example, one would expect diagrams similar to the one shown in Fig. 16.1, where the $\widetilde{W}$ exchange is replaced by neutral gaugino exchange. It seems plausible that gluino exchange could dominate. Because of the strong coupling constant which would appear, the result could be a proton decay rate larger than anticipated previously. In some models no such problem exists. For example, it has been shown that if one assumes that all squark and slepton masses are separately degenerate, then the sum of all Feynman diagrams involving the dimension-five operators and gluino exchange exactly vanishes. The reason is connected to the flavor independence of gluino interactions. Models that do not exhibit such mass degeneracies will have nonvanishing gluino exchanges. Furthermore, note the appearance of the helicity flip which must occur on the gaugino line in Fig. 16.1. Thus, in principle, one could use the present limits on proton decay to constrain the values of the gaugino masses.

In conclusion, dimension-five operators in supersymmetric theories lead to new expectations for the proton decay branching ratios. One would expect final states involving kaons and either muons or neutrinos to be the dominant decay channels, while the standard SU(5) proton decay modes such as $e^+\pi^0$ mediated by dimension-six operators would be subdominant. One always has the option of inventing a discrete symmetry to ban the baryon-number-violating dimension-five operators (in the same manner that baryon-number-violating dimension-four operators were eliminated with a matter parity symmetry). In such a model, proton decay would be mediated via dimension-six operators in the same way as it is in non-supersymmetric grand unified theories. Thus, we would now expect the branching ratio predictions to be similar to that of a non-supersymmetric SU(5) GUT model. Of course, if proton decay is discovered, it will have significant implications for supersymmetric extensions of the Standard Model.

We return to a discussion of the dimension-four operator $\mathcal{O}_4$ in eq. (16.142). Like the dimension-five operator in eq. (16.144), it is suppressed by one power of the very high mass scale, here M_P. Unlike the operator in eq. (16.144), now all external states are supersymmetric scalar fields. Compared to the Feynman diagram in Fig. 16.1(b), the proton decay Feynman diagram for $\mathcal{O}_4$ involves two loops, each with the exchange of a virtual gaugino. The issues of first- and second-generation (s)quark fields remain. Assuming all supersymmetric masses are equal, then on dimensional grounds the contribution of the two operators to the proton decay amplitude must have the same mass dependence. The proton decay amplitude due to $\mathcal{O}_4$ will however be suppressed by an additional loop factor of $g^2/(16\pi^2)$, where g denotes the relevant gauge coupling in the second loop. We have assumed the arising logarithms are of order 1. Thus the contribution of $\mathcal{O}_4$ is relatively suppressed by about three orders of magnitude in the amplitude.

16.4 Discrete Gauge Symmetries in Supersymmetry

In this section we extend the discussion of Section 5.3 to a supersymmetric model with the field content of the MSSM (see Table 7.1), following Refs. [186, 187]. Such an extension introduces a larger set of possible renormalizable and nonrenormalizable operators that violate baryon and/or lepton number [e.g., see eqs. (13.12) and (13.13)]. In order to constrain these operators, one must also consider the effect of the low-energy discrete symmetries. We first introduce some extra notation that is helpful in the analysis.

We restrict ourselves to a generation blind discrete $\mathbb{Z}_N$ symmetry acting on the fermions of the MSSM chiral superfields

$$\psi_i \to e^{2\pi i X_i/N}\psi_i\,, \qquad \boldsymbol{X} \equiv (X_Q, X_{\bar{u}}, X_{\bar{d}}, X_L, X_{\bar{e}}, X_{\widetilde{H}_d}, X_{\widetilde{H}_u})\,. \tag{16.145}$$

The index i of ψ_i and the components of the charge vector $\boldsymbol{X}$ run over the fermions of the chiral supermultiplets of the MSSM. The low-energy discrete symmetry should originate from a gauged anomaly-free $\mathrm{U}(1)_X$ symmetry, and we discuss the resulting constraints further below. Note that by an appropriate rescaling of the $\mathrm{U}(1)_X$ generator, the X charges of the MSSM fields can be taken to be integers. Likewise, we shall rescale the hypercharge $\mathrm{U}(1)_Y$ generator [as we did in eq. (5.118) in the non-supersymmetric case] such that the charges of the MSSM fields are all integers. In particular,

$$\boldsymbol{Y} \equiv (Y_Q, Y_{\bar{u}}, Y_{\bar{d}}, Y_L, Y_{\bar{e}}, Y_{\widetilde{H}_d}, Y_{\widetilde{H}_u}) = (1,\ -4,\ \ 2,\ -3,\ \ 6,\ -3,\ \ 3)\,. \tag{16.146}$$

From the start we impose the $\mathrm{U}(1)_X$ invariance of the minimal set of trilinear superpotential terms in the MSSM, which give masses to the quarks and leptons [see eq. (13.6)]. This results in the constraints on the discrete charges

$$X_Q + X_{\bar{d}} = -X_{\widetilde{H}_d}\,, \qquad X_Q + X_{\bar{u}} = -X_{\widetilde{H}_u}\,, \qquad X_L + X_{\bar{e}} = -X_{\widetilde{H}_d}\,. \tag{16.147}$$

We purposefully do not impose the constraint arising from the μ-term $H_u H_d$, which could have another origin (see Section 14.6). Thus, of the original seven charges shown in eq. (16.145), only four are independent.

As in Section 5.3, we consider a gauged $\mathrm{U}(1)_X$ at the high energy scale. If a given $\mathrm{U}(1)_X$ is anomaly-free, then an alternative $\mathrm{U}(1)'_X$ where all the X charges are shifted by a multiple of $\boldsymbol{Y}$ is also anomaly-free. We do not regard such a $\mathrm{U}(1)'_X$

Table 16.2 The charges $\boldsymbol{X}_R$, $\boldsymbol{X}_A$, and $\boldsymbol{X}_L$ of the corresponding three independent $\mathbb{Z}_N$ generators R_N, A_N, and L_N, respectively, are defined as in eq. (16.145). Y denotes the rescaled hypercharge needed to identify the complete set of equivalent $\mathbb{Z}_N$ symmetries via eq. (16.148).

	Q	$\bar{u}$	$\bar{d}$	L	$\bar{e}$	$\widetilde{H}_d$	$\widetilde{H}_u$
$\boldsymbol{X}_R$	0	−1	1	0	1	−1	1
$\boldsymbol{X}_A$	0	0	−1	−1	0	1	0
$\boldsymbol{X}_L$	0	0	0	−1	1	0	0
$\boldsymbol{Y}$	1	−4	2	−3	6	−3	3

as an independent symmetry, and correspondingly the resultant $\mathbb{Z}_N$ is not regarded as an independent discrete gauge symmetry. Consequently, discrete symmetries that are related under a shift of the charges,

$$\boldsymbol{X} \to \boldsymbol{X} + (1,\ -4,\quad 2,\ -3,\quad 6,\ -3,\quad 3) \mod N\,, \tag{16.148}$$

are equivalent. We use this equivalence to choose the charge

$$X_Q = 0\,. \tag{16.149}$$

Thus, any generation-independent $\mathbb{Z}_N$ can be written in terms of three independent discrete generators, which we choose as R_N, A_N, and L_N, defined as in eq. (16.145) with the corresponding charges $\boldsymbol{X}_{R,A,L}$ given in Table 16.2. Note that the symmetries generated by R_N, A_N, and L_N are independent. One may classify all possible $\mathbb{Z}_N$ symmetries by three integers $\{m, n, p\}$ such that the corresponding discrete generator T_N is given by

$$T_N = R_N^m \times A_N^n \times L_N^p\,, \qquad m, n, p = 0, 1, \ldots, N-1\,. \tag{16.150}$$

The corresponding $\mathbb{Z}_N$ charges are given by

$$\boldsymbol{X}_T = m\boldsymbol{X}_R + n\boldsymbol{X}_A + p\boldsymbol{X}_L\,. \tag{16.151}$$

The $\boldsymbol{X}_T$ charges obtained from Table 16.2 are

$$\begin{aligned} X_Q &= 0, & X_{\bar{u}} &= -m, & X_{\bar{d}} &= m-n,\\ X_L &= -n-p, & X_{\bar{e}} &= m+p, & &\\ X_{\widetilde{H}_d} &= -m+n, & X_{\widetilde{H}_u} &= m. & & \end{aligned} \tag{16.152}$$

In Sections 16.4.1 and 16.4.2, we express the anomaly cancellation conditions in terms of m, n, and p.

16.4.1 Three Classes of Discrete Gauge Symmetries

If a Lagrangian is invariant under a given $\mathbb{Z}_N$ discrete symmetry generated by T_N, then it is also invariant under the discrete symmetry generated by T_N^k, with $k = 2, 3, \ldots, N-1$. For example, acting on a given operator $\mathcal{O}_i$, T_N^2 operates as $T_N(T_N\mathcal{O}_i)$. The reverse need not be true, except when $k = N-1$, since

$$(T_N^{N-1})^{N-1} = T_N^{N^2-2N+1} = T_N\,, \tag{16.153}$$

after making use of the fact that $T_N^N = \mathbb{1}$. Thus any Lagrangian invariant under a discrete symmetry generated by T_N is also invariant under a discrete symmetry generated by T_N^{N-1}, and vice versa. The two symmetries are equivalent and are not listed separately in what follows. We now provide a classification of the discrete gauge symmetries of a supersymmetric model with the field content of the MSSM.

(a) *Generalized matter parities (GMP).* These symmetries forbid both the renormalizable baryon- and lepton-number-violating terms in the superpotential for the supersymmetric SM [see eqs. (13.12) and (13.13)]. The discrete charges must satisfy

$$X_{\bar{u}} + 2X_{\bar{d}} \neq 0 \bmod N\,, \tag{16.154}$$

$$2X_L + X_{\bar{e}} \neq 0 \bmod N\,. \tag{16.155}$$

We consider a basis where the LH_u terms have been rotated away (see Section 16.5.4), and thus restrict ourselves to dimension-four operators. Since eq. (16.147) implies that $2X_L + X_{\bar{e}} = X_L + X_Q + X_{\bar{d}}$, only one constraint is required for the lepton-number-violating operators. Using eq. (16.152) we can rewrite eqs. (16.154) and (16.155) as

$$m - 2n \neq 0 \bmod N\,, \tag{16.156}$$

$$m - 2n - p \neq 0 \bmod N\,. \tag{16.157}$$

In light of eq. (16.150), a generic GMP discrete generator can be written as

$$\begin{aligned} T_N &= R_N^m A_N^n L_N^p \\ &= (R_N^2 A_N)^n R_N^{(m-2n)} L_N^p\,, \qquad m-2n \neq 0,\; m-2n-p \neq 0\,. \end{aligned} \tag{16.158}$$

For $N = 2$ or $N = 3$ the possible solutions are

$$N = 2: \quad R_2, R_2A_2\,, \tag{16.159}$$

$$N = 3: \quad R_3, A_3, R_3A_3, R_3A_3L_3, R_3^2L_3, A_3^2L_3\,. \tag{16.160}$$

We have dropped pair-wise equivalent symmetries, e.g., $(A_3^2L_3)^2 = A_3L_3^2$. The symmetry R_2 is matter parity, $P_{\rm M}$ [see eq. (13.14)].

(b) *Generalized lepton parities (GLP).* These symmetries allow for the baryon-number-violating operator $\bar{u}\bar{d}\bar{d}$, but not the lepton-number-violating operators $LQ\bar{d}$, $LL\bar{e}$, and LH_u, with charge constraints

$$m - 2n = 0 \bmod N\,, \qquad m - 2n - p \neq 0 \bmod N\,. \tag{16.161}$$

Using these constraints, a generic GLP discrete generator can be written as

$$T_N = (R_N^2 A_N)^n L_N^p\,, \qquad p \neq 0. \tag{16.162}$$

The possible solutions for $N = 2, 3$ are

$$N = 2: \quad L_2, A_2 L_2\,, \tag{16.163}$$

$$N = 3: \quad L_3, R_3^2 A_3 L_3, R_3 A_3^2 L_3\,. \tag{16.164}$$

(c) *Generalized baryon parities (GBP).* These symmetries allow for dimension-four lepton number violation, but not baryon number violation. The conditions on m, n, p are

$$m - 2n \neq 0 \mod N\,, \tag{16.165}$$

$$m - 2n - p = 0 \mod N\,. \tag{16.166}$$

The generic GBP discrete generator can be written as

$$T_N = R_N^m A_N^n L_N^p = (R_N^2 A_N)^n (R_N L_N)^p, \qquad p \neq 0\,. \tag{16.167}$$

The possible solutions for $N = 2, 3$ are

$$N = 2: \quad R_2 L_2, R_2 A_2 L_2\,, \tag{16.168}$$

$$N = 3: \quad R_3 L_3, A_3 L_3, R_3 A_3 L_3^2\,. \tag{16.169}$$

16.4.2 Anomaly Cancellation

In light of the charge assignments in eq. (16.152), the linear anomaly constraints, eqs. (5.113) and (5.116), for the MSSM can be rewritten as

$$3n = 0 \mod N\,, \tag{16.170}$$

$$3(n+p) - n = 0 \mod N\,, \tag{16.171}$$

$$3(5n + p - m) - 2n = \begin{cases} 0 \mod N, & \text{for } N \text{ odd,} \\ 0 \mod N/2, & \text{for } N \text{ even,} \end{cases} \tag{16.172}$$

where the factor of 3 corresponds to the three generations of quark and lepton superfields. Using $3n = 0 \mod N$, one can rewrite eq. (16.171) as $3p - n = 0 \mod N$. One can use this latter result along with eq. (16.170) to further simplify eq. (16.172). We then obtain the following equivalent version of the linear anomaly constraints:

$$3n = 0 \mod N\,, \tag{16.173}$$

$$3p - n = 0 \mod N\,, \tag{16.174}$$

$$3m + n = \begin{cases} 0 \mod N, & \text{for } N \text{ odd,} \\ 0 \mod N/2, & \text{for } N \text{ even.} \end{cases} \tag{16.175}$$

It is now straightforward to check the 18 $\mathbb{Z}_2$ and $\mathbb{Z}_3$ symmetries enumerated in Section 16.4.1. For example, the $\mathbb{Z}_3$ GLP $R_3 A_3^2 L_3$ has $m = 1, n = 2$, and $p = 1$,

Table 16.3 The dimension-three, -four and -five supersymmetric operators which violate lepton and/or baryon number and are allowed by the discrete gauge anomaly-free $\mathbb{Z}_2$ or $\mathbb{Z}_3$ symmetries. The subscripts F, D refer to the F- and D-terms, respectively.

	R_2	R_3	L_3	$\mathbb{Z}_3^B \equiv R_3L_3$	$R_3^2L_3$
$[LL\bar{e}]_F, [LQ\bar{d}]_F, [LH_u]_F$				✓	
$[\bar{u}\bar{d}\bar{d}]_F$			✓		
$[QQQL]_F$	✓	✓			
$[\bar{u}\bar{u}\bar{d}\bar{e}]_F$	✓	✓			
$[QQQH_d]_F$			✓		
$[Q\bar{u}\bar{e}H_d]_F$				✓	
$[LLH_uH_u]_F$	✓			✓	
$[LH_dH_uH_u]_F$				✓	
$[\bar{u}\bar{d}^\dagger \bar{e}]_D$				✓	
$[H_u^\dagger H_d\bar{e}]_D$				✓	
$[Q\bar{u}L^\dagger]_D$				✓	
$[QQ\bar{d}^\dagger]_D$			✓		

which violates eq. (16.174) and is thus not discrete gauge anomaly-free. The only symmetries that satisfy the anomaly-free constraints are

$$N = 2: \quad \text{GMP}: R_2\,, \tag{16.176}$$

$$N = 3: \quad \text{GLP}: R_3, L_3, \tag{16.177}$$

$$\text{GLP}: R_3L_3, R_3^2L_3\,. \tag{16.178}$$

The symmetry R_3L_3 is often called baryon triality and denoted $\mathbb{Z}_3^B$. Including a $\mathrm{U}(1)_Y$ hypercharge shift, one can redefine $\mathbb{Z}_3^B$ as

$$\mathbb{Z}_3^B = \exp\left(2\pi i[B - 2\mathrm{Y}]/3\right). \tag{16.179}$$

Table 16.3 shows how the five symmetries in eqs. (16.176)–(16.178) act on the dimension-three, -four and -five lepton- and/or baryon-number-violating operators. As discussed in Section 13.3, matter parity, R_2, does not forbid baryon and/or lepton number violation at the nonrenormalizable level. Indeed, matter parity allows the potentially dangerous dimension-five operators $[QQQL]_F$, $[\bar{u}\bar{d}\bar{d}\bar{e}]_F$, and $[LLH_uH_u]_F$. In contrast, baryon triality forbids any dimension-four or dimension-five baryon-number-violating operator. In light of eq. (16.179), only operators that violate baryon number by multiples of three units are allowed by the $\mathbb{Z}_3^B$ symmetry.

16.5 R-Parity Violation

The MSSM can be extended to allow for subsets of the superpotential terms given in eqs. (13.12) and (13.13), i.e., to allow for *explicit* R-parity violation (RPV).[11] In this section, our emphasis is on the most prominent differences with the R-parity-conserving MSSM, including proton decay, sneutrino mixing, neutrino masses, neutralino decay (and thus no neutralino dark matter), and a possible light neutralino.

16.5.1 The R-Parity-Violating Supersymmetric Lagrangian

Employing the superfields of the MSSM, $Q, \overline{u}, \overline{d}\, L, \overline{e}, H_d, H_u$ (see Table 7.1), the most general renormalizable superpotential is given by

$$W_{\rm RPV} = W_{\rm MSSM} + W_{\Delta L=1} + W_{\Delta B=1}\,, \tag{16.180}$$

where $W_{\Delta L=1}$ and $W_{\Delta B=1}$ were given in eqs. (13.12) and (13.13) and are repeated below with all flavor indices in the lowered position for typographical convenience:[12]

$$W_{\Delta L=1} = \epsilon_{ab}\big(\tfrac{1}{2}\lambda_{ijk}L_i^a L_j^b \bar{e}_k + \lambda'_{ijk}L_i^a Q_j^b \bar{d}_k + \kappa_i H_u^a L_i^b\big)\,, \tag{16.181}$$

$$W_{\Delta B=1} = \tfrac{1}{2}\epsilon_{pqr}\lambda''_{ijk}\overline{u}_i^p \overline{d}_j^q \overline{d}_k^r\,. \tag{16.182}$$

We can write the resulting Yukawa couplings in the Lagrangian in terms of the components of the superfields:

$$\mathscr{L}_{LL\bar{e}} = -\tfrac{1}{2}\lambda_{ijk}\left(\widetilde{\ell}^\dagger_{Rk}\nu_i\ell_j + \widetilde{\nu}_i\ell_j\bar{\ell}_k + \widetilde{\ell}_{Lj}\bar{\ell}_k\nu_i - \widetilde{\ell}^\dagger_{Rk}\ell_i\nu_j - \widetilde{\nu}_j\bar{\ell}_k\ell_i - \widetilde{\ell}_{Li}\nu_j\bar{\ell}_k\right) + \text{h.c.}, \tag{16.183}$$

$$\mathscr{L}_{LQ\bar{d}} = -\lambda'_{ijk}\left(\widetilde{d}^\dagger_{Rk}\nu_i d_j + \widetilde{\nu}_i d_j\bar{d}_k + \widetilde{d}_{Lj}\bar{d}_k\nu_i - \widetilde{d}^\dagger_{Rk}\ell_i u_j - \widetilde{u}_{Lj}\bar{d}_k\ell_i - \widetilde{\ell}_{Li}u_j\bar{d}_k\right) + \text{h.c.}, \tag{16.184}$$

$$\mathscr{L}_{LH_u} = -\kappa_i(\nu_i\widetilde{H}_u^0 - \ell_i\widetilde{H}_u^+) + \text{h.c.}, \tag{16.185}$$

[11] In the absence of explicit RPV, if the sneutrino fields acquire nonzero VEVs, then R parity is *spontaneously* broken. We shall not consider this possibility further in this section. A detailed treatment of spontaneous R-parity violation can be found in [B142].

[12] In this section, we shall not follow the flavor-index conventions of Appendix A.2. Instead, all flavor indices appear as subscripts and the dagger (†) denotes conjugation.

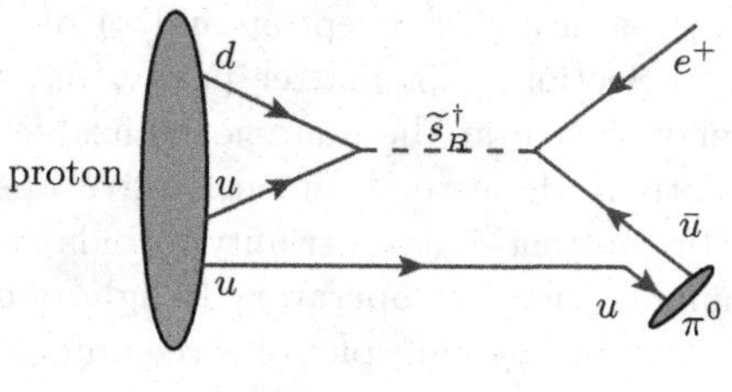

Fig. 16.2 Feynman diagram for proton decay $p \to e^+\pi^0$ via the operators $\lambda''_{112}\overline{u}_1\overline{d}_1\overline{d}_2$ and $\lambda'_{112}L_1Q_1\overline{d}_2$ mediated by a virtual SU(2) singlet strange squark.

$$\mathscr{L}_{\bar{u}\bar{d}\bar{d}} = -\tfrac{1}{2}\lambda''_{ijk}\epsilon_{pqr}\left(\widetilde{u}^{\dagger p}_{Ri}\bar{d}^q_j\bar{d}^r_k + \widetilde{d}^{\dagger q}_{Rj}\bar{u}^p_i\bar{d}^r_k + \widetilde{d}^{\dagger r}_{Rk}\bar{u}^p_i\bar{d}^q_j\right) + \text{h.c.} \qquad (16.186)$$

The corresponding Feynman rules for the trilinear terms are given in Appendix J (see Figs. J.13–J.15). The bilinear term in eq. (16.181) induces mixing between neutrinos and higgsinos, as well as between charged leptons and charginos. The former leads at tree level to one massive Majorana neutrino, as we discuss below.

The soft supersymmetry-breaking Lagrangian of the MSSM is given in eq. (13.16). In the RPV extension of the MSSM, this is supplemented by the bilinear and trilinear scalar interactions

$$\mathscr{L}^{\text{RPV}}_{\text{soft}} = -\tfrac{1}{2}h_{ijk}\widetilde{L}_i\widetilde{L}_j\widetilde{e}^\dagger_{Rk} - h'_{ijk}\widetilde{L}_i\widetilde{Q}_j\widetilde{d}^\dagger_{Rk} + b_i\widetilde{L}_iH_u - \tfrac{1}{2}h''_{ijk}\widetilde{u}^\dagger_{Ri}\widetilde{d}^\dagger_{Rj}\widetilde{d}^\dagger_{Rk} + \text{h.c.}, \qquad (16.187)$$

where $SU(3)_C$ color indices have been suppressed. The first three sets of terms on the right-hand side of eq. (16.187) violate lepton number and flavor, whereas the last set of terms violates baryon number. The bilinear $\widetilde{L}_iH_u$ term mixes the sneutrino and charged slepton fields with the scalar Higgs fields. The couplings $h_{ijk}, h'_{ijk}, h''_{ijk}$ have mass dimension one and the b_i have mass dimension two.

16.5.2 Proton Decay

The terms in eq. (16.181) violate lepton number by one unit; those in eq. (16.182) violate baryon number by one unit. Combining operators we can thus obtain dimension-six interactions which violate both baryon and lepton number, leading to the possibility of proton decay [180]. A candidate Feynman diagram is shown in Fig. 16.2. On dimensional grounds, we expect, for the resulting decay rate,

$$\Gamma(p \to e^+\pi^0) \sim \frac{\alpha_\lambda\alpha_{\lambda'}m_p^5}{\widetilde{m}_{sR}^4}, \qquad (16.188)$$

where m_p is the mass of the proton, $\widetilde{m}_{sR}$ is the mass of the virtual strange squark, and $\alpha_{\lambda^{(\prime)}} = \lambda^{(\prime)2}/(4\pi)$. A full calculation involves the hadronic matrix element, which can be estimated using lattice gauge theory techniques. Experimentally the

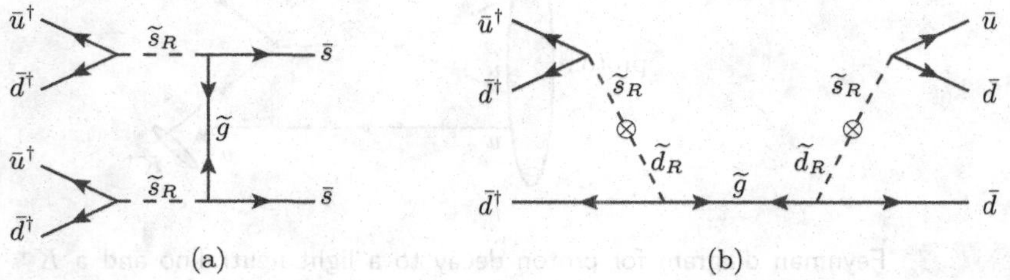

Fig. 16.3 In (a), we exhibit a Feynman diagram for di-nucleon decay $pp \to K^+K^+$, where the two accompanying spectator up quarks have been omitted. In (b), we show a possible diagram for neutron–antineutron oscillation. The $\otimes$ indicates flavor mixing of the down-type squarks. In both diagrams we have omitted the ovals for the bound states.

upper bound on the partial mean lifetime is [41]

$$\frac{\tau_p}{\mathrm{BR}(p \to e^+\pi^0)} \gtrsim 1.6 \times 10^{34}\,\text{years}, \quad \text{at 90\% CL}. \tag{16.189}$$

Using eq. (16.188), we obtain the bound on the product of couplings,

$$\lambda'_{112}\lambda''_{112} \lesssim 10^{-25}\left(\frac{m_{\widetilde{s}R}}{1\,\mathrm{TeV}}\right)^2, \tag{16.190}$$

which is very restrictive. A natural explanation is that at least one of the two couplings vanishes, for example due to a symmetry (see Section 16.4). Replacing the intermediate $\widetilde{s}_R$ by a $\widetilde{b}_R$ gives the analogous bound on the product $\lambda'_{113}\lambda''_{113}$. Other possible final states via the same type of s-channel diagram include $\bar{\nu}_i K^+$ $(\lambda''_{11k}\lambda'_{i2k})$, and $\mu^+\pi^0$ $(\lambda''_{11k}\lambda'_{21k})$, $k = 2, 3$. The decays e^+K^0, and μ^+K^0 via the couplings $\lambda''_{12k}\lambda'_{11k,21k}$, $k = 1, 3$, can be obtained by tree-level t-channel squark-exchange diagrams. Allowing for left–right squark mixing, one can also obtain the decay modes $\nu_i\pi^+$ and $\nu_i K^+$ via t-channel diagrams.

Using just $\bar{u}\bar{d}\bar{d}$ operators, processes that violate baryon number by two units are possible. In Fig. 16.3 we show two examples: (a) the decay of a heavy nucleus containing at least two protons, $pp \to K^+K^+$, and (b) neutron–antineutron oscillations, $n \leftrightarrow \bar{n}$. Both involve a virtual Majorana fermion and a gluino (which can also be replaced by a neutralino). These processes yield bounds on a single RPV coupling: $\lambda''_{112} \lesssim \mathcal{O}(10^{-6})$.

If the lightest neutralino is lighter than the proton such that $m_{\widetilde{N}_1} < m_p - m_{K^+}$, then a direct decay, $p \to K^+\widetilde{N}_1$, is possible using only one nonzero λ''_{112} [180], as shown in Fig. 16.4. The possibility of a light neutralino in RPV supersymmetry will be considered further in Section 16.5.8.

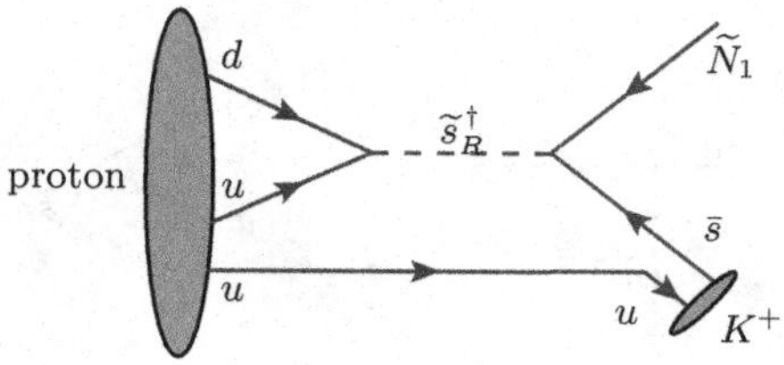

Fig. 16.4 Feynman diagram for proton decay to a light neutralino and a K^+ via the operator $\lambda''_{112}\bar{u}_1\bar{d}_1\bar{d}_2$.

16.5.3 Motivation, Discrete Symmetries, and GUTs

In light of eq. (16.190), we expect either $\lambda'_{112}\lambda''_{112} = 0$ by some symmetry, or at least one of the couplings is extremely small and the theory is highly fine-tuned. We have discussed a set of discrete symmetries in Section 16.4, including the anomaly cancellation conditions. A standard solution is to impose a discrete multiplicative symmetry such as R-parity, $P_{\rm R} = (-1)^{3(B-L)+2s}$ [see eq. (13.15)], or matter parity, $P_{\rm M} = (-1)^{3(B-L)}$ [see eq. (13.14)]. Matter parity, which is the symmetry R_2 identified in Section 16.4, prohibits all the terms in eqs. (16.181) and (16.182), but not the dangerous operators in eq. (16.143) (see Table 16.3). The bound in eq. (16.190) can be equally satisfied with a stable proton, if all the lepton-number-violating operators, or all the baryon-number-violating operators (modulo a light neutralino) are prohibited. For example, under lepton triality $Z_3^L \equiv L_3$ (see Table 16.3) $LL\bar{e}$, $LQ\bar{d}$, and LH_u are prohibited, but $\bar{u}\bar{d}\bar{d}$ is still allowed. Similarly, under baryon triality $Z_3^B = R_3L_3$ [see eq. (16.179)], $\bar{u}\bar{d}\bar{d}$ is prohibited, but $LL\bar{e}$, $LQ\bar{d}$, and LH_u are allowed. A further special symmetry is proton hexality $Z_6^P \equiv R_6^5L_6^2$ [188], which also protects the proton and is discussed in more detail in Exercise 16.9.

We briefly discuss the connections between RPV and SU(5) grand unified theories. The RPV superpotential given in eqs. (16.181) and (16.181) consists of terms that are either trilinear in the superfields (denoted below by $W^{(3)}_{\rm RPV}$) and bilinear in the superfields (denoted below by $W^{(2)}_{\rm RPV}$). The trilinear RPV operators involve only the matter fields. In SU(5) these are contained in the $\overline{\mathbf{5}}$ and $\mathbf{10}$ representations, denoted ψ_i and ψ^{ij}, respectively, in eq. (6.260). We denote the corresponding superfields as Ψ_i and Ψ^{ij}. The only SU(5) gauge invariant product of three fields is the singlet contained in the product $\overline{\mathbf{5}} \otimes \overline{\mathbf{5}} \otimes \mathbf{10}$,

$$[\Psi_i]_\ell[\Psi_j]_m[\Psi^{ij}]_n \supset W^{(3)}_{\rm RPV}\,, \tag{16.191}$$

where i, j are SU(5) indices and ℓ, m, n are generation indices. Since all the trilinear terms of the RPV superpotential exhibited in eqs. (16.181) and (16.182) arise from the same SU(5) superpotential term, it follows that $\frac{1}{2}\lambda_{ijk} = \lambda'_{ijk} = \frac{1}{2}\lambda''_{ijk} = \lambda_{\rm SU(5)}$ at $M_{\rm GUT}$. Then, by eq. (16.190), we have $\lambda_{\rm SU(5)} \lesssim 2.7 \times 10^{-13}$, modulo renormalization group factors of order 1, making RPV interactions irrelevant for the LHC, and most other physics applications.

However, given the two Higgs multiplets in SU(5), $H, \overline{H}$ [see eq. (16.114)], bilinear RPV operators are also possible:

$$[\Psi_i]_\ell H^i \supset W^{(2)}_{\rm RPV}\,, \tag{16.192}$$

where i is an SU(5) index and ℓ a generation index. Besides the κ-term in eq. (16.181), this also leads to a dangerous mixing between the colored Higgs triplet in H^i and the down-type (s)quarks in Ψ_i. As we shall see in Section 16.5.5, these terms also lead to neutrino masses, and thus the associated coupling must be very small.

16.5.4 The R-Parity-Violating Spectrum

We discuss in some detail the RPV spectrum, as an example of how the MSSM can be modified. Here, we follow the treatment of Ref. [189].

Lepton–Higgs Mixing: Generalized Notation

The superfields L_i and H_d have the same $\mathrm{SU}(2)_L\times\mathrm{U}(1)_Y$ gauge quantum numbers. They are only distinguished, if at all, by the discrete symmetries discussed in Section 16.5.3. For baryon parity or baryon triality, they remain indistinguishable. Thus within RPV theories it is often convenient to combine them into a set of four superfields,

$$\mathcal{L}^a_\alpha = \{H^a_d,\, L^a_i\}\,, \qquad \alpha\in\{0,1,2,3\},\quad i\in\{1,2,3\}; \tag{16.193}$$

where $a\in\{1,2\}$ is an $\mathrm{SU}(2)_L$ index. We can rewrite the relevant superpotential in terms of the chiral superfields $\mathcal{L}_\alpha$:

$$W_{\rm RPV}(\mathcal{L}_\alpha) = \epsilon_{ab}\big(\tfrac{1}{2}\lambda_{\alpha\beta k}\mathcal{L}^a_\alpha\mathcal{L}^b_\beta\bar{e}_k + \lambda'_{\alpha jk}\mathcal{L}^a_\alpha Q^b_j\bar{d}_k + \mu_\alpha H^a_u\mathcal{L}^b_\alpha\big)\,. \tag{16.194}$$

Here $\lambda_{i0k}\equiv(\boldsymbol{y}_e)_{ik}$, $\lambda'_{i0k}\equiv(\boldsymbol{y}_d)_{ik}$, $\mu_\alpha\equiv(\mu,\kappa_i)$, and $\lambda_{\alpha\beta k}=-\lambda_{\beta\alpha k}$.

We can also rewrite the relevant scalar soft supersymmetry-breaking terms,

$$\mathscr{L}^{\rm RPV}_{\rm soft} = -\widetilde{\mathcal{L}}^\dagger_\alpha(\boldsymbol{m}^2_{\widetilde{\mathcal{L}}})_{\alpha\beta}\widetilde{\mathcal{L}}_\beta - \Big(\tfrac{1}{2}h_{\alpha\beta k}\widetilde{\mathcal{L}}_\alpha\widetilde{\mathcal{L}}_\beta\widetilde{e}^\dagger_{Rk} + \bar{h}'_{\alpha jk}\widetilde{\mathcal{L}}_\alpha\widetilde{Q}_j\widetilde{d}^\dagger_{Rk} - b_\alpha\widetilde{\mathcal{L}}_\alpha H_u + \text{h.c.}\Big)\,, \tag{16.195}$$

where the scalar components of the superfields $\mathcal{L}^a_\alpha$ are denoted by

$$\widetilde{\mathcal{L}}^a_\alpha \equiv \begin{pmatrix}\widetilde{\nu}_\alpha\\ \widetilde{\ell}^-_\alpha\end{pmatrix}, \qquad \alpha\in\{0,1,2,3\}, \tag{16.196}$$

with $\widetilde{\nu}_0\equiv H^0_d$ and $\widetilde{\ell}^-_0\equiv H^-_d$ in the notation of Section 13.8. Note that, in eq. (16.195), $(\boldsymbol{m}^2_{\widetilde{\mathcal{L}}})_{\alpha\beta}$ are the elements of the soft supersymmetry-breaking (hermitian) squared-mass matrix of the Higgs and slepton scalar field, $h_{\alpha\beta k}$ and $\bar{h}'_{\alpha jk}$ are coefficients of the generalized trilinear terms [see eq. (13.16)], the b_α are the bilinear soft-breaking terms, with $b_0\equiv b$ of eq. (13.16), and b_i of eq. (16.187).

Modified Higgs Potential

It is not particularly illuminating to give the full expression for the scalar potential. We list here only the tree-level neutral scalar potential [cf. eq. (13.78)], which necessarily includes the sneutrinos. Employing the notation of eq. (16.195),

$$V_{\rm neutral} = (m^2_{H_u} + |\bar{\mu}|^2)|H^0_u|^2 + [(\boldsymbol{m}^2_{\widetilde{L}})_{\alpha\beta} + \mu_\alpha\mu^*_\beta]\widetilde{\nu}_\alpha\widetilde{\nu}^\dagger_\beta - (b_\alpha\widetilde{\nu}_\alpha H^0_u + b^*_\alpha\widetilde{\nu}^\dagger_\alpha H^{0\dagger}_u) + \tfrac{1}{8}(g^2+g'^2)\left[|H^0_u|^2 - |\widetilde{\nu}_\alpha|^2\right]^2, \tag{16.197}$$

where

$$|\widetilde{\nu}_\alpha|^2 \equiv |H^0_d|^2 + \widetilde{\nu}_i\widetilde{\nu}^i\,, \qquad |\bar{\mu}|^2 \equiv \mu^*_\alpha\mu_\alpha\,. \tag{16.198}$$

In minimizing the full scalar potential, we assume that only the neutral scalar fields acquire VEVs. In addition, we shall choose a scalar field basis (by rephasing the sneutrino and Higgs fields) such that the VEVs are real: $\langle H^0_u\rangle \equiv v_u/\sqrt{2}$, and $\langle\widetilde{\nu}_\alpha\rangle \equiv v_\alpha/\sqrt{2}$. The latter spontaneously breaks lepton number. It is convenient to introduce the notation

$$|\bar{b}|^2 \equiv b^*_\alpha b_\alpha\,, \qquad \bar{v}^2_d \equiv \sum_\alpha v^2_\alpha\,. \tag{16.199}$$

Because of the normalization of the VEVs, it follows that $v^2 = v^2_u + \bar{v}^2_d = (246\,\text{GeV})^2$.

We now have five minimization conditions for $V_{\rm neutral}$:

$$\left.\frac{\partial V_{\rm neutral}}{\partial H^0_u}\right|_{\rm min} = (m^2_{H_u} + |\bar{\mu}|^2)v_u - b_\alpha v_\alpha + \tfrac{1}{8}(g^2+g'^2)(v^2_u - \bar{v}^2_d)v_u = 0\,, \tag{16.200}$$

$$\left.\frac{\partial V_{\rm neutral}}{\partial \widetilde{\nu}_\alpha}\right|_{\rm min} = [(\boldsymbol{m}^2_{\widetilde{L}})_{\alpha\beta} + \mu_\alpha\mu^*_\beta]v_\beta - b_\alpha v_u - \tfrac{1}{8}(g^2+g'^2)(v^2_u - \bar{v}^2_d)v_\alpha = 0. \tag{16.201}$$

Equations (16.200) and (16.201) can be solved iteratively to obtain the VEVs in terms of the superpotential and soft parameters. As in the MSSM [see eqs. (13.86) and (13.87)], $|\bar{\mu}|^2$ is fixed by electroweak symmetry breaking.

To simplify the subsequent analysis, we shall henceforth assume that the parameters b_α and μ_α (and hence the κ_i) are real in a scalar field basis where the VEVs are real. Under these assumptions, the sneutrino/Higgs sector is CP-conserving and the sneutrino and Higgs mass eigenstates are eigenstates of CP.

CP-even Higgs and Sneutrino Masses

After electroweak symmetry breaking, the sneutrino fields $\widetilde{\nu}_i$ mix with the neutral Higgs fields, H^0_u and $H^0_d = \widetilde{\nu}_0$. Since CP is conserved under the assumptions noted above, the physical neutral scalars will be either CP-even or CP-odd. The CP-even and CP-odd scalar squared-mass matrices are most easily derived by inserting

$$H^0_u \equiv \frac{1}{\sqrt{2}}(v_u + r_u + ia_u), \qquad \widetilde{\nu}_\alpha \equiv \frac{1}{\sqrt{2}}(v_\alpha + r_\alpha + ia_\alpha) \tag{16.202}$$

into eq. (16.197), where $r_{u,\alpha}$, $a_{u,\alpha}$ are real scalar fields that vanish at the minimum of the scalar potential given in eq. (16.197). As in the MSSM, the neutral Higgs mass eigenstates consist of the CP-even scalars h and H (with $m_h < m_H$) and the CP-odd scalar A. We shall also denote the CP-even and CP-odd sneutrino mass eigenstates by $(\widetilde{\nu}_+)_i$ and $(\widetilde{\nu}_-)_i$, respectively. Inserting eq. (16.202) into eq. (16.197), it follows that $V_{\text{neutral}} = V_{\text{even}} + V_{\text{odd}}$, where V_{even} is identified by setting $r_{u,\alpha} = a_{u,\alpha} = 0$. Likewise, one can identify $V_{\text{odd}} = V_{\text{neutral}} - V_{\text{even}}$, with $r_{u,\alpha} = 0$. In particular,

$$\begin{aligned} V_{\text{even}} &= \tfrac{1}{2}(m^2_{H_u} + |\bar{\mu}|^2)v_u^2 + \tfrac{1}{2}[(\boldsymbol{m}^2_{\widetilde{L}})_{\alpha\beta} + \mu_\alpha\mu_\beta]v_\alpha v_\beta - b_\alpha v_u v_\alpha \\ &\quad + \tfrac{1}{32}(g^2 + g'^2)(v_u^2 - \bar{v}_d^2)^2 \,, \end{aligned} \tag{16.203}$$

$$\begin{aligned} V_{\text{odd}} &= \tfrac{1}{2}(m^2_{H_u} + |\bar{\mu}|^2)a_u^2 + \tfrac{1}{2}[(\boldsymbol{m}^2_{\widetilde{L}})_{\alpha\beta} + \mu_\alpha\mu_\beta]a_\alpha a_\beta + b_\alpha a_u a_\alpha \\ &\quad + \tfrac{1}{32}(g^2 + g'^2)[(a_u^2 - a_\alpha^2)^2 + 2(a_u^2 - a_\alpha^2)(v_u^2 - \bar{v}_d^2)] \,. \end{aligned} \tag{16.204}$$

The squared-mass matrix of the CP-even neutral scalars is obtained via

$$(\mathcal{M}^2_{\text{CP+}})_{pq} = \frac{\partial^2 V_{\text{even}}}{\partial v_p \partial v_q} \,, \tag{16.205}$$

where $p, q \in \{u, 0, 1, 2, 3\}$. Using eqs. (16.200) and (16.201), we end up with

$$\mathcal{M}^2_{\text{CP+}} = \begin{pmatrix} b_\alpha v_\alpha / v_u + \tfrac{1}{4}(g^2 + g'^2)v_u^2 & -b_\gamma - \tfrac{1}{4}(g^2 + g'^2)v_u v_\gamma \\ -b_\beta - \tfrac{1}{4}(g^2 + g'^2)v_u v_\beta & (\boldsymbol{m}^2_{\widetilde{\nu}\widetilde{\nu}^*})_{\beta\gamma} + \tfrac{1}{4}(g^2 + g'^2)v_\beta v_\gamma \end{pmatrix} \,, \tag{16.206}$$

with

$$(\boldsymbol{m}^2_{\widetilde{\nu}\widetilde{\nu}^*})_{\alpha\beta} \equiv (\boldsymbol{m}^2_{\widetilde{L}})_{\alpha\beta} + \mu_\alpha\mu_\beta - \tfrac{1}{8}(g^2 + g'^2)(v_u^2 - \bar{v}_d^2)\delta_{\alpha\beta} \,. \tag{16.207}$$

The CP-even scalar mass eigenstates are denoted by $h, H, (\widetilde{\nu}_+)_i$. The case when R-parity-violating effects are small ($|\bar{b}|^{1/2}$, $v_i \ll v$) will be considered below.

CP-odd Neutral Scalar Masses

The CP-odd neutral scalar masses are determined from the mass matrix given by

$$(\mathcal{M}^2_{\text{CP}-})_{pq} = \left.\frac{\partial^2 V_{\text{odd}}}{\partial a_p \partial a_q}\right|_{a_p=0} = \begin{pmatrix} b_\alpha v_\alpha / v_u & b_\gamma \\ b_\beta & (\boldsymbol{m}^2_{\widetilde{\nu}\widetilde{\nu}^*})_{\beta\gamma} \end{pmatrix} \,, \tag{16.208}$$

which is a 5×5 matrix. Using the definition of $(\boldsymbol{m}^2_{\widetilde{\nu}\widetilde{\nu}^*})_{\alpha\beta}$ given in eq. (16.207), we can rewrite the minimization condition eq. (16.201) such that

$$(\boldsymbol{m}^2_{\widetilde{\nu}\widetilde{\nu}^*})_{\alpha\beta} v_\beta = v_u b_\alpha \,. \tag{16.209}$$

The vector $(-v_u, v_\gamma)$ is an eigenvector of $\mathcal{M}^2_{\text{CP}-}$, with eigenvalue 0. The four orthogonal states correspond to the four physical massive states A, $(\widetilde{\nu}_-)_i$. We discuss below the identification of the states $(\widetilde{\nu}_-)_i$ in the case of a small perturbation due to the RPV parameters b_i, v_i.

Basis Choice and RPV Alignment

It is important to consider the alignment between various RPV terms. To discuss the case of μ_α and b_α, we define the angle

$$\cos\zeta \equiv \frac{\mu_\alpha b_\alpha}{\bar{\mu}\,\bar{b}}\,, \tag{16.210}$$

where $\bar{\mu} \equiv (|\bar{\mu}|^2)^{1/2}$ and $\bar{b} \equiv (|\bar{b}|^2)^{1/2}$. For $\cos\zeta = 1$, the bilinear superpotential and soft-breaking terms are said to be *aligned.* This has direct consequences for the basis choice. The μ_i-terms in the superpotential can be rotated away by an orthogonal transformation, $\mathcal{L}_\alpha = \mathcal{O}_{\alpha\beta}\mathcal{L}'_\beta$. More explicitly,

$$\begin{pmatrix} H_d \\ L_1 \\ L_2 \\ L_3 \end{pmatrix} = \begin{pmatrix} c_3 & -s_3 & 0 & 0 \\ c_2 s_3 & c_2 c_3 & -s_2 & 0 \\ c_1 s_2 s_3 & c_1 s_2 c_3 & c_1 c_2 & -s_1 \\ s_1 s_2 s_3 & s_1 s_2 c_3 & s_1 c_2 & c_1 \end{pmatrix} \begin{pmatrix} H'_d \\ L'_1 \\ L'_2 \\ L'_3 \end{pmatrix}, \tag{16.211}$$

where $c_i \equiv \cos\theta_i$ and $s_i \equiv \sin\theta_i$ and

$$c_1 = \frac{\kappa_2}{\sqrt{\kappa_2^2+\kappa_3^2}}\,, \qquad c_2 = \frac{\kappa_1}{\bar{\kappa}}\,, \qquad c_3 = \frac{\mu_0}{\bar{\mu}}\,, \tag{16.212}$$

$$s_1 = \frac{\kappa_3}{\sqrt{\kappa_2^2+\kappa_3^2}}\,, \qquad s_2 = \frac{\sqrt{\kappa_2^2+\kappa_3^2}}{\bar{\kappa}}\,, \qquad s_3 = \frac{\bar{\kappa}}{\bar{\mu}}\,, \tag{16.213}$$

where $\bar{\kappa} \equiv (\kappa_1^2+\kappa_2^2+\kappa_3^2)^{1/2}$. The sole bilinear term in the superpotential is then $\bar{\mu} H_u H'_d$. However, since eq. (16.211) transforms the superfields $\mathcal{L}_\alpha$, the soft supersymmetry-breaking terms exhibited in eq. (16.195) will also be affected. For $\cos\zeta = 1$ the orthogonal transformation above simultaneously rotates away the bilinear soft terms $b_i \widetilde{\mathcal{L}}_i H_u$. For $\cos\zeta \neq 1$ this is not possible, and there will always be L_i–H_d mixing in the model.

For a basis-independent analysis it is convenient to introduce a second angle

$$\cos\xi \equiv \frac{\sum_\alpha v_\alpha \mu_\alpha}{\bar{v}_d\,\bar{\mu}}\,, \tag{16.214}$$

quantifying the alignment between v_α and μ_α. We can use eq. (16.209) to derive a necessary and sufficient condition for $\cos\xi = 1$. If there exists a c number with

$$(\boldsymbol{m}^2_{\widetilde{\nu}\widetilde{\nu}^*})_{\alpha\beta}\mu_\beta = c\, b_\alpha\,, \tag{16.215}$$

it follows that μ_α and v_α are aligned. This result can be established by combining eqs. (16.209) and (16.215) to eliminate b_α. Since $\boldsymbol{m}^2_{\widetilde{\nu}\widetilde{\nu}^*}$ must be invertible to guarantee that eq. (16.209) yields a unique solution for v_α, it follows that v_α and μ_α are proportional and hence aligned as claimed. With a field redefinition analogous to eq. (16.211) one can also rotate the direction of the VEV such that $v_i = 0$, for $i = 1, 2, 3$. Many calculations simplify for this basis choice, as we discuss below.

Sneutrino Mass Difference

An important feature in RPV models is that a mass difference between $(\widetilde{\nu}_+)_i$ and $(\widetilde{\nu}_-)_i$ can be generated. This can be seen by considering the sum rule

$$\mathrm{Tr}(\mathcal{M}^2_{\mathrm{CP}+}) = m_Z^2 + \mathrm{Tr}(\mathcal{M}^2_{\mathrm{CP}-})\,, \tag{16.216}$$

which is verified by inserting eqs. (16.206) and (16.208) above, and making use of eq. (16.286). This is the generalization of the tree-level MSSM sum rule obtained from eq. (13.97), i.e., $m_h^2+m_H^2 = m_Z^2+m_A^2$. To compute the sneutrino–antisneutrino mass splitting, one must evaluate the nonzero eigenvalues of the CP-even and CP-odd scalar squared-mass matrices, $\mathcal{M}^2_{\mathrm{CP}+}$ and $\mathcal{M}^2_{\mathrm{CP}-}$, and identify the sneutrino mass eigenstates. It is convenient to choose a basis where $v_i = 0$. To simplify the analysis, we consider the case of one lepton generation and drop the index i. Then,

$$\mathcal{M}^2_{\mathrm{CP}+} = \begin{pmatrix} b_0\cot\beta + \frac{1}{4}(g^2+g'^2)v_u^2 & -b_0 - \frac{1}{4}(g^2+g'^2)v_u\bar{v}_d & -b_1 \\ -b_0 - \frac{1}{4}(g^2+g'^2)v_u\bar{v}_d & b_0\tan\beta + \frac{1}{4}(g^2+g'^2)\bar{v}_d^2 & b_1\tan\beta \\ -b_1 & -b_1\tan\beta & m^2_{\widetilde{\nu}\widetilde{\nu}^*} \end{pmatrix}, \tag{16.217}$$

$$\mathcal{M}^2_{\mathrm{CP}-} = \begin{pmatrix} b_0\cot\beta & b_0 & b_1 \\ b_0 & b_0\tan\beta & b_1\tan\beta \\ b_1 & b_1\tan\beta & m^2_{\widetilde{\nu}\widetilde{\nu}^*} \end{pmatrix}, \tag{16.218}$$

where we have introduced the single-generation parameter

$$m^2_{\widetilde{\nu}\widetilde{\nu}^*} \equiv (\boldsymbol{m}^2_{\widetilde{\nu}\widetilde{\nu}^*})_{11} = (\boldsymbol{m}^2_{\widetilde{\mathcal{L}}})_{11} + \mu_1^2 - \tfrac{1}{8}(g^2+g'^2)(v_u^2 - \bar{v}_d^2)\,, \tag{16.219}$$

and $\tan\beta = v_u/\bar{v}_d$. We now assume $b_1 \ll b_0, v^2$ and treat the entries with b_1 as a perturbation. The unperturbed CP-even scalar squared-mass matrix is given by

$$\left[\mathcal{M}^2_{\mathrm{CP}+}\right]^{(0)} = \begin{pmatrix} b_0\cot\beta + \frac{1}{4}(g^2+g'^2)v_u^2 & -b_0 - \frac{1}{4}(g^2+g'^2)v_u\bar{v}_d & 0 \\ -b_0 - \frac{1}{4}(g^2+g'^2)v_u\bar{v}_d & b_0\tan\beta + \frac{1}{4}(g^2+g'^2)\bar{v}_d^2 & 0 \\ 0 & 0 & m^2_{\widetilde{\nu}\widetilde{\nu}^*} \end{pmatrix}. \tag{16.220}$$

The upper left 2×2 matrix coincides with that of the MSSM and the unperturbed sneutrino squared mass is $m^2_{\widetilde{\nu}_+} = m^2_{\widetilde{\nu}\widetilde{\nu}^*}$. The perturbation is an off-diagonal matrix,

$$\left[\mathcal{M}^2_{\mathrm{CP}+}\right]_{\mathrm{pert.}} = \begin{pmatrix} 0 & 0 & -b_1 \\ 0 & 0 & b_1\tan\beta \\ -b_1 & -b_1\tan\beta & 0 \end{pmatrix}. \tag{16.221}$$

Hence, the first-order perturbative correction to the sneutrino mass eigenvalues is zero. The second-order correction to the sneutrino squared mass $m^2_{\widetilde{\nu}_+}$ is given by

$$(m^2_{\widetilde{\nu}_+})^{(2)} = \sum_{i+} \frac{|\langle \widetilde{\nu}_+ | \left[\mathcal{M}^2_{\rm CP+}\right]_{\rm pert.} |i+\rangle|^2}{m^2_{\widetilde{\nu}\widetilde{\nu}^*} - m^2_{i+}} . \tag{16.222}$$

The index $i+$ runs over the two orthogonal unperturbed mass eigenvalues and eigenstates. The resulting squared-mass eigenvalue is

$$m^2_{\widetilde{\nu}_+} = m^2_{\widetilde{\nu}\widetilde{\nu}^*} + \frac{b_1^2}{\cos^2\beta}\left[\frac{\sin^2(\beta-\alpha)}{m^2_{\widetilde{\nu}\widetilde{\nu}^*} - m^2_H} + \frac{\cos^2(\beta-\alpha)}{m^2_{\widetilde{\nu}\widetilde{\nu}^*} - m^2_h}\right] , \tag{16.223}$$

where α is the CP-even Higgs mixing angle of the MSSM defined in eqs. (6.231) and (6.232). Similarly, one obtains for the CP-odd sneutrino squared mass

$$m^2_{\widetilde{\nu}_-} = m^2_{\widetilde{\nu}\widetilde{\nu}^*} + \frac{b_1^2}{(m^2_{\widetilde{\nu}\widetilde{\nu}^*} - m^2_A)\cos^2\beta} . \tag{16.224}$$

Using eqs. (6.239) and (6.240), the sneutrino squared-mass difference is

$$\Delta m^2_{\widetilde{\nu}} \equiv m^2_{\widetilde{\nu}_+} - m^2_{\widetilde{\nu}_-} = \frac{4b_1^2 m_Z^2 m^2_{\widetilde{\nu}\widetilde{\nu}^*}\sin^2\beta}{(m^2_{\widetilde{\nu}\widetilde{\nu}^*} - m^2_H)(m^2_{\widetilde{\nu}\widetilde{\nu}^*} - m^2_h)(m^2_{\widetilde{\nu}\widetilde{\nu}^*} - m^2_A)} . \tag{16.225}$$

In RPV models, sneutrino–antisneutrino mixing exists [190], with phenomenological consequences analogous to those of K^0–$\bar{K}^0$ and B^0–$\bar{B}^0$ mixing.

RPV Alignment at Different Energy Scales

The RPV alignment condition, $\cos\zeta = 1$, is not stable under the renormalization group equations. To see this we consider the RGEs for the bilinear superpotential parameters κ_i and the corresponding soft parameters b_i (see Section 11.5),

$$\frac{d}{dt}\kappa_i = \kappa_i\gamma^{H_u}_{H_u} + \kappa_j\gamma^{L_i}_{L_j} + \mu\gamma^{L_i}_{H_u} , \tag{16.226}$$

$$\frac{d}{dt}b_i = b_i\gamma^{H_u}_{H_u} + b_j\gamma^{L_i}_{L_j} + b_0\gamma^{L_i}_{H_d} - 2\mu\,(\gamma_1)^{L_i}_{H_d} - 2[\kappa_i\,(\gamma_1)^{H_u}_{H_u} + \kappa_k\,(\gamma_1)^{L_i}_{L_k}], \tag{16.227}$$

with $t = \ln(Q)$. The γ^Y_X are the anomalous dimensions defined in eq. (11.138). Moreover, $(\gamma_1)^X_Y \equiv \mathcal{O}\gamma^X_Y$, where the operator $\mathcal{O}$ on the space of couplings is defined in eq. (11.151). Note that the $\gamma^{L_i}_{L_p}$ are modified from their MSSM values due to nonzero λ_{ijk}, λ'_{ijk}. Likewise, $\gamma^{L_i}_{H_u} \neq 0$ due to the mixing of H_d and the L_i. We list the one-loop RPV expressions, for the case where the SM Yukawa couplings and the MSSM trilinear soft-breaking terms are dominated by the third generation:

$$\gamma^{(1)L_j}_{L_i} = \left(|y_\tau|^2 - \tfrac{3}{2}g'^2 - \tfrac{3}{10}g_1^2\right)\delta_{i3}\delta_{j3} + \lambda_{ikl}\lambda^*_{jkl} + 3\lambda'_{ikl}\lambda'^*_{jkl} , \tag{16.228}$$

$$\gamma^{(1)L_i}_{H_d} = -y_\tau\lambda^*_{i33} - 3y_b\lambda'^*_{i33} , \tag{16.229}$$

$$(\gamma_1)^{(1)L_j}_{L_i} = -y^*_\tau a_\tau - h_{ikq}\lambda^*_{jkq} - 3h'_{ikq}\lambda'^*_{jkq} - \delta^j_i\left(\tfrac{3}{10}M_1 g_1^2 + \tfrac{3}{2}M_2 g'^2\right) , \tag{16.230}$$

$$(\gamma_1)^{(1)L_3}_{H_d} = 3a_b\lambda'^*_{i33} + 3a_\tau\lambda^*_{i33} . \tag{16.231}$$

Even if $\kappa_i = 0$ and $b_i = 0$ at a given scale, we see from eq. (16.226) that if $\mu \neq 0$ (which is required to obtain viable chargino masses) and λ_{i33} or $\lambda'_{i33} \neq 0$, then $\kappa_i \neq 0$ and $b_i \neq 0$ are regenerated through RG evolution. In addition, the RGEs for κ_i and b_i are clearly distinct. Thus even if $\cos\zeta = 1$ at one scale this will not be true for all scales.

16.5.5 Neutralino–Neutrino Mixing and Neutrino Masses

The neutral higgsinos and neutrinos mix due to the superpotential term $\mu_\alpha \epsilon_{ab} H_u^a \mathcal{L}_\alpha^b$. Neutralino–neutrino mixing is also generated by electroweak symmetry breaking, $v_i \neq 0$, due to the neutralino–neutrino–sneutrino interactions. With respect to the basis $\{\widetilde{B}, \widetilde{W}^3, \widetilde{H}_u, \nu_\alpha\}$, $\alpha = 0, 1, 2, 3$, the 7×7 neutralino mass matrix is given by[13]

$$M_{\widetilde{N}_7} = \begin{pmatrix} M_1 & 0 & -m_Z s_W v_u/v & m_Z s_W v_\beta/v \\ 0 & M_2 & m_Z c_W v_u/v & -m_Z c_W v_\beta/v \\ -m_Z s_W v_u/v & m_Z c_W v_u/v & 0 & -\mu_\beta \\ m_Z s_W v_\alpha/v & -m_Z c_W v_\alpha/v & -\mu_\alpha & 0_{\alpha\beta} \end{pmatrix}. \tag{16.232}$$

Note that $\det M_{\widetilde{N}_7} = 0$, since the last four column vectors span only a two-dimensional space. Thus we have five nonzero masses: the four neutralinos and one massive neutrino. Since $v_i/v, \mu_i/\bar{\mu} \ll 1$, the neutrino is light and the four heavy states are dominated by the original neutralino components; we retain the names neutralino and neutrino for these states. Since we have two massless eigenstates at tree level, we can write, for the characteristic polynomial,

$$\det(M_{\widetilde{N}_7} - \lambda I_7) = -\lambda^2 \sum_{n=0}^{5} c_n \lambda^n . \tag{16.233}$$

We denote the first polynomial coefficient $c_0 \equiv \det' M_{\widetilde{N}_7}$, which equals the product of the five nonzero masses. A straightforward calculation yields

$$\det{}' M_{\widetilde{N}_7} = m_Z^2 \bar{\mu}^2 M_{\widetilde{\gamma}} \cos^2\beta \sin^2\xi , \tag{16.234}$$

where $M_{\widetilde{\gamma}} \equiv \cos^2\theta_W M_1 + \sin^2\theta_W M_2$ denotes the would-be photino mass (if the photino were a mass eigenstate), and ξ is the alignment angle defined in eq. (16.214). To first order in the neutrino mass, the neutralino masses are unchanged by the small parameters v_i, μ_i. The product of the four neutralino masses, denoted by $\det M_{\widetilde{N}_4}$, is given by the determinant of the upper 4×4 mass matrix in eq. (16.232),

$$\det M_{\widetilde{N}_4} = \bar{\mu}(m_Z^2 M_{\widetilde{\gamma}} \sin 2\beta - M_1 M_2 \bar{\mu}) . \tag{16.235}$$

[13] The upper left 4×4 block of $M_{\widetilde{N}_7}$ corresponds to $M_{\widetilde{N}}$ of the MSSM given in eq. (13.137), after replacing μ_0, v_0 by μ, v_d and reordering the basis by interchanging $\widetilde{H}_u \leftrightarrow \widetilde{H}_d$.

The tree-level nonzero neutrino mass is then given by the ratio

$$m_\nu = \frac{\det' \boldsymbol{M}_{\widetilde{N}_7}}{\det \boldsymbol{M}_{\widetilde{N}_4}} = \frac{m_Z^2 \bar{\mu} M_{\tilde{\gamma}} \cos^2\beta \sin^2\xi}{m_Z^2 M_{\tilde{\gamma}} \sin 2\beta - M_1 M_2 \bar{\mu}} \approx \frac{m_Z^2}{M_2} \cos^2\beta \sin^2\xi \,. \tag{16.236}$$

The last step holds for the approximation $M_1 \approx M_{\tilde{\gamma}}$ and $M_2\bar{\mu} \gg m_Z^2 \sin 2\beta$. We see that a necessary condition for $m_\nu \neq 0$ is $\sin\xi \neq 0$. Thus v_α and μ_α must be misaligned [which implies that eq. (16.215) is not satisfied].

One can obtain a rough bound on the superpotential parameters $\kappa_i = \mu_i$ by employing the cosmological bound on the sum of neutrino masses, $\sum m_\nu \lesssim m_\nu^{\rm cosmo}$. If we fix the basis such that $v_i = 0$, then using eq. (16.214) we obtain

$$\sin^2\xi = 1 - \left(\frac{v_\alpha \mu_\alpha}{\bar{v}_d \bar{\mu}}\right)^2 = 1 - \frac{\mu_0^2}{\mu_0^2 + \bar{\kappa}^2} \approx \frac{\bar{\kappa}^2}{2\mu_0^2}, \tag{16.237}$$

for $\bar{\kappa}^2 \ll \mu_0^2$ [where $\bar{\kappa}$ is defined below eq. (16.213)]. It then follows that

$$m_\nu \approx \frac{\cos^2\beta\, m_Z^2 \bar{\kappa}^2}{2\mu_0^2 M_2} < m_\nu^{\rm cosmo} \,. \tag{16.238}$$

As an example, for $\tan\beta = 10$ ($\cos^2\beta \approx 0.01$) and $\mu_0 = M_2 = 500\,{\rm GeV}$, we obtain $\bar{\kappa} \lesssim 10^8 \cdot m_\nu^{\rm cosmo} \sim \mathcal{O}(10\,{\rm MeV})$ after setting $m_\nu^{\rm cosmo} \sim 0.1\,{\rm eV}$. Thus $\mu_i = \kappa_i \ll \mu_0$, which justifies the approximations above. Given that $\sin\xi \lesssim 10^{-5}$, and $\cos^2\beta$ is also small, eq. (16.236) is related to the standard seesaw formula $m_\nu \sim m_D^2/M$ [cf. eq. (6.5)] where $m_Z \cos\beta \sin\xi \sim 100\,{\rm keV}$ corresponds to the (Dirac) mass m_D.

Assuming a normal hierarchical spectrum of neutrino masses, $m_{\nu_1} \ll m_{\nu_2} < m_{\nu_3}$, a global fit to the current neutrino data yields [41],

$$m_{\nu_2} \simeq 8.6 \times 10^{-3}\ {\rm eV}\,, \qquad m_{\nu_3} \simeq 0.05\ {\rm eV}. \tag{16.239}$$

As previously noted, the tree-level neutrino spectrum obtained above consists of two massless neutrinos and one massive neutrino. For example, one can numerically realize $m_{\nu_3} \simeq 0.05$ eV for $\tan\beta = 10$, $\mu_0, M_2 = 500\,{\rm GeV}$, and $\bar{\kappa} = \mathcal{O}(10\,{\rm MeV})$ in light of eqs. (16.236) and (16.237).

16.5.6 Radiative Neutrino Masses

We have seen that only one neutrino attains mass via tree-level neutralino–neutrino mixing in the RPV SUSY model considered above, whereas at least two massive neutrinos must exist in light of the observed neutrino mixing data. In the case of a normal hierarchical spectrum of neutrino masses, it follows that the mass of ν_2 exhibited in eq. (16.239) must then be generated through radiative corrections.

v eq. (2.352). Since the corresponding tree-level neutrino mass vanishes, we obtain in the one-loop approximation

$$m_{\nu_2} = \tfrac{1}{2}\,{\rm Re}\big[\boldsymbol{\Omega}(0)^{22} + \overline{\boldsymbol{\Omega}}(0)_{22}\big]\,, \tag{16.240}$$

where $\boldsymbol{\Omega}$ and $\overline{\boldsymbol{\Omega}}$ are the Majorana neutrino self-energy functions shown in Fig. 2.30.

The one-loop contributions to $\boldsymbol{\Omega}(0)^{22}$ and $\overline{\boldsymbol{\Omega}}(0)_{22}$ that involve the $LQ\bar{d}$ operator

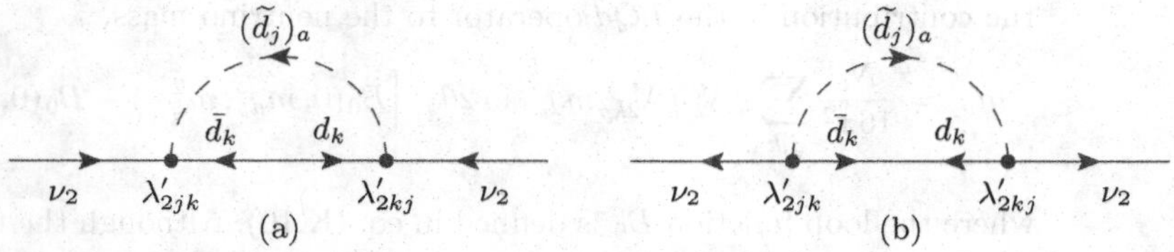

Fig. 16.5 One-loop radiative neutrino mass Feynman diagrams contributing to the Majorana neutrino self-energy functions, (a) $\mathbf{\Omega}(0)^{22}$ and (b) $\overline{\mathbf{\Omega}}(0)_{22}$, which are defined more generally in Fig. 2.30. Each diagram represents two separate diagrams corresponding to the exchange of the two down-type squark mass eigenstates $a = 1, 2$ defined in eq. (16.241). The relevant vertex Feynman rules are given in Fig. J.14.

are shown in Fig. 16.5. The arrow on the scalar line indicates the direction of flow of electric charge. Using the Feynman rules exhibited in Fig. J.14, we see that the vertex labeled by λ'_{2jk} involves the interaction eigenstate $\widetilde{d}_{Lj}$, whereas the vertex labeled by λ'_{2kj} involves the interaction eigenstate $\widetilde{d}_{Rj}$. Hence, the contributions of the diagrams shown in Fig. 16.5 vanish unless there is $\widetilde{d}_{Lj}$–$\widetilde{d}_{Rj}$ mixing. That is, the mass-eigenstate squarks which appear in the loop are admixtures of $\widetilde{d}_{Lj}$ and $\widetilde{d}_{Rj}$.

For simplicity, we shall again neglect CP-violating effects by taking λ'_{2jk}, λ'_{2kj}, and the relevant squark mixing parameters to be real. In addition, we shall neglect intergenerational mixing among the squarks. Following the notation of eq. (13.181), the two down-squark mass eigenstates can be written as

$$(\widetilde{d}_j)_a = V_{a1}\widetilde{d}_{Lj} + V_{a2}\widetilde{d}_{Rj}\,, \qquad \text{for } a = 1, 2, \tag{16.241}$$

where

$$V = \begin{pmatrix} \cos\theta_{\widetilde{d}_j} & -\sin\theta_{\widetilde{d}_j} \\ \sin\theta_{\widetilde{d}_j} & \cos\theta_{\widetilde{d}_j} \end{pmatrix}. \tag{16.242}$$

The contributions of the diagrams shown in Fig. 16.5 yield

$$\begin{aligned}-i\mathbf{\Omega}(0)^{22} = -i\overline{\mathbf{\Omega}}(0)_{22} = N_c \sum_{j,k}\sum_{a=1,2}\int \frac{d^dq}{(2\pi)^d}\Bigg[(-i\lambda'_{2kj})V_{a1}\frac{im_{d_k}}{q^2 - m^2_{d_k} + i\varepsilon}(-i\lambda'_{2jk}) \\ \times V_{a2}\frac{i}{q^2 - m^2_{\widetilde{d}_{ja}} + i\varepsilon}\Bigg]\,, \end{aligned}\tag{16.243}$$

evaluated in $d = 4 - 2\epsilon$ dimensions, where $N_c = 3$ is the color factor for the quark/squark loop, $j, k \in \{1, 2, 3\}$ runs over the three generations, $a \in \{1, 2\}$ runs over the two mass eigenstates of eq. (16.241), and m_{d_k} is the mass of the down-type quark of generation k. Writing the squark mixing matrix elements in terms of the squark mixing angle $\theta_{\widetilde{d}_j}$ [see eq. (16.242)], we can now employ eq. (16.240) to obtain the contribution of the $LQ\bar{d}$ operator to the neutrino mass,

$$m_{\nu_2} = \frac{N_c}{16\pi^2} \sum_{j,k} \left\{ \lambda'_{2jk} \lambda'_{2kj} m_{d_k} \sin 2\theta_{\widetilde{d}_j} \left[B_0(0, m_{d_k}, m^2_{\widetilde{d}_{j1}}) - B_0(0, m_{d_k}, m^2_{\widetilde{d}_{j2}}) \right] \right\}, \tag{16.244}$$

where the loop function B_0 is defined in eq. (K.10). Although the two B_0 functions above are divergent (in the limit of $\epsilon \to 0$), the divergences cancel in the difference, resulting in a finite contribution to the neutrino mass. Using eqs. (K.44) and (K.48), it follows that, in the limit of $m_{\widetilde{d}_{j1}}, m_{\widetilde{d}_{j2}} \gg m_{d_k}$,[14]

$$m_{\nu_2} \simeq \frac{N_c}{16\pi^2} \sum_{j,k} \lambda'_{2kj} \lambda'_{2jk} m_{d_k} \sin 2\theta_{\widetilde{d}_j} \ln \left(\frac{m^2_{\widetilde{d}_{j2}}}{m^2_{\widetilde{d}_{j1}}} \right). \tag{16.245}$$

A similar computation can be carried out to obtain the contribution of the $LL\bar{e}$ operator to the neutrino mass (see Exercise 16.20).

The general neutrino mass matrix includes the entries from eqs. (16.236) and (16.245). Here we determine the order of magnitude of a single trilinear coupling λ'_{233} ($j = k = 3$) in eq. (16.245), assuming it is fixed by the value of m_{ν_2} given in eq. (16.239). In this case $m_{d_k} = m_b = 4.2\,\text{GeV}$. For example, if $m_{\widetilde{d}_{32}} = 2\,\text{TeV}$, $m_{\widetilde{d}_{31}} = 1.5\,\text{TeV}$, and $\theta_{\widetilde{d}_3} = \frac{1}{8}\pi$ (the sensitivity to the squark masses is only logarithmic), we obtain $\lambda'_{233} \approx 2 \times 10^{-5}$.

16.5.7 *R*-Parity Violating Phenomenology

The RPV interactions in eqs. (16.181), (16.182), and (16.187) lead to significant differences in the supersymmetric phenomenology [191], e.g., for potential searches at colliders, and also in astrophysics and cosmology. The four main changes are:

1. Lepton and/or baryon number, as well as lepton flavor (for hierarchical couplings), are violated.
2. The LSP is unstable with a lifetime that could be short enough to be observed as a displaced track in a collider detector.
3. As a corollary, the LSP need not be the lightest neutralino.
4. Supersymmetric particles can be singly produced, e.g., as an s-channel resonance.

We expand on these points in a bit more detail below.

1. As discussed above, the proton can decay (violating baryon number) and Majorana neutrino masses can be generated (violating lepton number). If the RPV couplings are hierarchical, then the Yukawa couplings also violate lepton flavor. This can lead, for example, to the decays $\mu^- \to e^-\gamma$ or $\tau^- \to (e^-/\mu^-)\gamma$ as exhibited in Fig. 16.6. An explicit bound is obtained in Exercise 16.16.

[14] If the right-hand side of eq. (16.245) is negative, then a field redefinition, $\nu_2 \to i\nu_2$, is sufficient to yield a positive mass. More precisely, the physical neutrino mass is equal to the absolute value of the expressions that appear on the right-hand side of eqs. (16.240), (16.244), and (16.245).

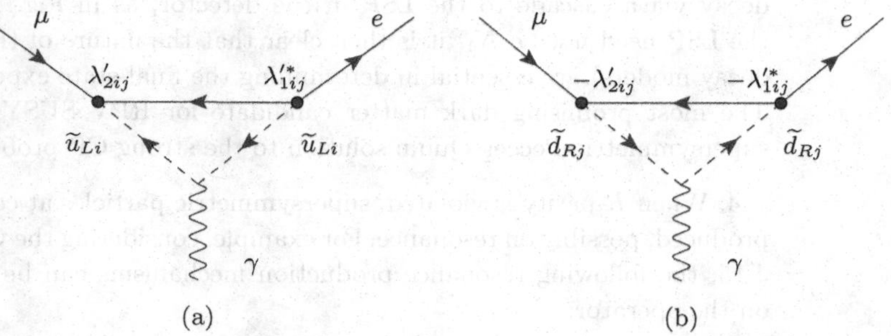

Fig. 16.6 Two graphs contributing to the decay $\mu \to e\gamma$ via the operators $L_{\mu,e}Q_i\bar{d}_j$.

2. Since R-parity is violated, the LSP is no longer protected by a symmetry and can decay. For example, for a neutralino LSP and the operator $L_2Q_1\bar{d}_1$ we have at tree level the three diagrams shown in Fig. 18.7. For a pure photino neutralino, the decay rate is given by [see eq. (18.62)]

$$\Gamma_{\widetilde{\gamma}} = \frac{3\alpha\lambda'^2_{211}}{128\pi^2}\frac{m^5_{\widetilde{\gamma}}}{m^4_{\widetilde{f}}}, \tag{16.246}$$

under the assumption that $m_{\widetilde{\mu}_L} = m_{\widetilde{u}_L} = m_{\widetilde{d}_R} = m_{\widetilde{f}} \gg m_{\widetilde{\gamma}}$.

At a hadron collider like the LHC, the decay occurs in the detector if the boosted decay length $\beta_{\widetilde{\gamma}}c\gamma_L\tau_{\widetilde{\gamma}} \lesssim 5$ meters, or

$$\lambda'_{211} \gtrsim 2.5\cdot 10^{-8}\sqrt{\beta_{\widetilde{\gamma}}\gamma_L}\left(\frac{\widetilde{m}}{1\,\text{TeV}}\right)^2\left(\frac{100\,\text{GeV}}{m_{\widetilde{\gamma}}}\right)^{5/2}. \tag{16.247}$$

Here $\beta_{\widetilde{\gamma}}$ is the relativistic speed of the photino and γ_L is the Lorentz boost. This is well below existing bounds, allowing for ample new signatures with little or no missing transverse momentum $\not{p}_T$. Other LSPs will decay differently; for example, a stau LSP can decay via the operator $L_1L_2\bar{e}_3$ as $\widetilde{\tau}^-_R \to (e^-\nu_\mu/\mu^-\nu_e)$, i.e., via a two-body mode.

3. In the MSSM, the LSP is stable and can have a nonvanishing cosmological relic density. Since a charged stable relic is excluded, as noted in footnote 4 of Section 15.4, it follows that the LSP of the MSSM is electrically neutral. If R-parity is violated, the LSP is unstable and thus is not a candidate for the dark matter (unless the lifetime is very long), in which case the constraint on the electric charge of the LSP can be dropped. For example, the region of CMSSM parameter space in which the lightest stau is the LSP (see Section 13.12) is now viable if RPV interactions are present. More generally, in RPV SUSY, any supersymmetric particle can be the LSP:

$$\text{LSP} \in \{\widetilde{N}_1,\, \widetilde{C}_1,\, \widetilde{g},\, \widetilde{q},\, \widetilde{t},\, \widetilde{\ell}^{\pm},\, \widetilde{\tau}^{\pm},\, \widetilde{\nu}\}, \tag{16.248}$$

where we have restricted ourselves to the MSSM particle content, neglecting the gravitino. We would expect produced squarks or gluinos at colliders to promptly decay via a cascade to the LSP in the detector, as in Fig. 15.2. However, since the LSP need not be $\widetilde{N}_1$, it is then clear that the nature of the LSP as well as its decay mode(s) are essential in determining the final-state experimental signatures. The most promising dark matter candidate for RPV SUSY is the axino of the supersymmetric Peccei–Quinn solution to the strong CP problem.

4. When R-parity is violated, supersymmetric particles at colliders can be singly produced, possibly on resonance. For example, considering the vertices in Figs. J.13–J.15, the following resonance production mechanisms can be relevant, depending on the operator:

$$e^+ + e^- \to \widetilde{\nu}_{Lj}\,, \qquad [L_1 L_j \bar{e}_1]\,, \tag{16.249}$$

$$e^- + u_j \to \widetilde{d}_{Rk}\,, \qquad [L_1 Q_j \bar{d}_k]\,, \tag{16.250}$$

$$\bar{u}_i + \bar{d}_j \to \widetilde{d}_{Rk}\,, \qquad [\bar{u}_i \bar{d}_j \bar{d}_k]\,. \tag{16.251}$$

These processes can be realized at e^+e^-–colliders, at ep–colliders, and at hadron colliders. The kinematic reach is larger than for pair production; however, the production cross section is proportional to the square of a potentially small Yukawa coupling. Particles can also be produced singly via t-channel processes.

16.5.8 Light Neutralino

A stable neutralino contributes to the dark matter of the universe. For a mass

$$0.7\,\text{eV} \lesssim m_{\widetilde{N}_1} \lesssim 10\,\text{GeV}\,, \tag{16.252}$$

the relic density is too large. The lower bound is due to Cowsik and McLelland [192], and the upper bound is due to Lee and Weinberg [193]. Here, we consider neutralinos with mass below the upper bound: the neutralino can decay via an RPV interaction, making the bound void. For simplicity, in the discussion below we shall assume that the parameters μ, M_1, and M_2 are real (in a basis where the VEVs are real). The case of complex parameters is treated in Exercise 16.21.

Experimental searches set a limit on the lightest chargino mass $m_{\widetilde{C}_1} > m_{\widetilde{C}}^{\text{exp}}$. Scanning over the parameters μ, M_2, $\tan\beta$ of the chargino mass matrix, $\boldsymbol{M}_{\widetilde{C}}$ yields

$$|\mu|,\, M_2 \gtrsim m_{\widetilde{C}}^{\text{exp}}\,. \tag{16.253}$$

Consider models with the gaugino mass unification boundary condition, which relates M_1 and M_2 [see eq. (13.140)]:

$$M_1 = \tfrac{5}{3}\tan^2\theta_{\text{W}} M_2 \approx \tfrac{1}{2} M_2\,. \tag{16.254}$$

Using eq. (16.253) we find

$$M_1 \gtrsim \tfrac{1}{2} m_{\widetilde{C}}^{\text{exp}}\,. \tag{16.255}$$

Combining eqs. (16.253) and (16.255) and scanning over the neutralino mass matrix parameters, we find the lower neutralino mass bound

$$m_{\widetilde{N}_1} \gtrsim \tfrac{1}{2} m_{\widetilde{C}}^{\rm exp}\,, \tag{16.256}$$

which excludes the light neutralino region. However, if we drop the assumption of gaugino mass unification, then M_1 and M_2 are independent parameters. We have additional freedom in determining the lightest neutralino mass. In fact it is straightforward to show that there can always be a zero mass neutralino. Setting $\det \boldsymbol{M}_{\widetilde{N}} = 0$ and using the explicit form for $\boldsymbol{M}_{\widetilde{N}}$ given in eq. (13.137), we obtain

$$\mu\left[M_2 m_Z^2 s_W^2 \sin 2\beta + M_1(-M_2\mu + m_Z^2 c_W^2 \sin 2\beta)\right] = 0\,. \tag{16.257}$$

The solution $\mu = 0$ is excluded by the chargino mass bound. Solving for M_1 yields

$$M_1 = \frac{M_2 m_Z^2 s_W^2 \sin 2\beta}{\mu M_2 - m_Z^2 c_W^2 \sin 2\beta}\,. \tag{16.258}$$

For every choice of $\mu, M_2, \tan\beta$, there is an M_1 such that there is a massless neutralino. Note that M_1 is typically small. For example, for $M_2, \mu \gg m_Z$ one can neglect the second term in the denominator. In that case

$$M_1 \simeq \frac{m_Z^2 s_W^2 \sin 2\beta}{\mu} \simeq 280\,\text{MeV}\,, \tag{16.259}$$

where in the last step we have used $\tan\beta = 5$ and $\mu = 500$ GeV as an example. Since $M_1 \ll M_2$, $|\mu|$, and m_Z, it follows that the lightest neutralino $\widetilde{N}_1$ is dominantly bino, with $m_{\widetilde{N}_1} \approx M_1$ [see eq. (13.142)].

The question arises whether such a light or even massless neutralino is consistent with present experiments and observations. It turns out that the answer is yes [194]. As an example, we consider the decay $Z \to \widetilde{N}_1\widetilde{N}_1$, which contributes to the invisible width of the Z boson. The relevant Feynman rule is given in Fig. J.3 and eq. (J.18). The tree-level decay width is

$$\Gamma(Z \to \widetilde{N}_1\widetilde{N}_1) = \frac{g^2}{4\pi}\frac{(|N_{14}|^2 - |N_{13}|^2)^2 m_Z}{24 c_W^2}\left[1 - \left(\frac{2m_{\widetilde{N}_1}}{m_Z}\right)^2\right]^{3/2}\,. \tag{16.260}$$

Only the higgsino components of the lightest neutralino contribute. For light neutralinos, the term in square brackets can be set to 1 and we obtain a simple constraint on $(|N_{14}|^2 - |N_{13}|^2)$, which is easily satisfied for a light bino-like neutralino.

Exercises

16.1 The scalar potential of the NMSSM Higgs sector is given by eqs. (16.5)–(16.7). Assume that the minimum of the scalar potential is realized by the following VEVs: $\langle H_u^0\rangle = v_u/\sqrt{2}$, $\langle H_d\rangle = v_d/\sqrt{2}$, and $\langle S\rangle = s$.

(a) Minimizing the scalar potential with respect to the complex fields H_u,

H_d, and S, show that the minimization conditions are given, in a convention where v_u, v_d, and s are real and nonnegative, by the following three complex equations:

$$v_d m_{H_d}^2 + |\lambda|^2 v_d(s^2 + \tfrac{1}{2}v_u^2) + \tfrac{1}{8}(g^2 + g'^2)v_d(v_d^2 - v_u^2) - (a_\lambda + \lambda\kappa^* s)sv_u = 0\,, \tag{16.261}$$

$$v_u m_{H_u}^2 + |\lambda|^2 v_u(s^2 + \tfrac{1}{2}v_d^2) + \tfrac{1}{8}(g^2 + g'^2)v_u(v_u^2 - v_d^2) - (a_\lambda + \lambda\kappa^* s)sv_d = 0\,, \tag{16.262}$$

$$sm_S^2 + \tfrac{1}{2}|\lambda|^2 s(v_u^2 + v_d^2) + 2|\kappa|^2 s^3 + s(sa_\kappa - \lambda^*\kappa v_u v_d) - \tfrac{1}{2}a_\lambda v_u v_d = 0\,. \tag{16.263}$$

(b) Under the assumption that v_u, v_d, and s are nonvanishing, show that the results of part (a) are equivalent to the five real equations given in eqs. (16.11)–(16.15). Explain why a potential sixth real equation is redundant.

16.2 Suppose that the parameters λ, κ, a_λ, and a_κ appearing in eqs. (16.6) and (16.7) are real. In this case, the NMSSM scalar potential is explicitly CP-conserving. Does the possibility of spontaneous CP violation exist in this case? That is, does there exist a CP-violating extremum where the complex phases of H_u, H_d, and S cannot be removed in a real scalar field basis (i.e., in a basis in which all the scalar potential parameters are real)? If such an extremum exists, find the conditions on the NMSSM Higgs sector parameters such that the CP-violating extremum is a global minimum.

16.3 The 6×6 neutral scalar squared-mass matrix of the NMSSM Higgs sector with respect to the neutral scalar interaction eigenstate fields $\{h_d, h_u, h_s, a_d, a_u, a_s\}$ has the following form:

$$\mathcal{N}^2 = \begin{pmatrix} M_{HH}^2 & M_{HA'}^2 \\ (M_{HA'}^2)^{\mathsf{T}} & M_{A'A'}^2 \end{pmatrix}, \tag{16.264}$$

where the 3×3 matrix M_{HH}^2 is given by eq. (16.29) in a convention that the scalar field VEVs are all real and positive.

(a) Show that the 3×3 matrix $M_{A'A'}^2$, with respect to the basis of CP-odd eigenstate scalar fields $\{a_d, a_u, a_s\}$, is given by

$$M_{A'A'}^2 = \begin{pmatrix} b\tan\beta & b & as_\beta \\ b & b\cot\beta & ac_\beta \\ as_\beta & ac_\beta & a's_\beta c_\beta - 3s\,\mathrm{Re}\,a_\kappa \end{pmatrix}, \tag{16.265}$$

where b is defined in eq. (16.17) and

$$a \equiv \frac{v}{\sqrt{2}}\,\mathrm{Re}(a_\lambda - 2\lambda^*\kappa s)\,, \qquad a' \equiv \frac{v^2}{2s}\,\mathrm{Re}(a_\lambda + 4\lambda^*\kappa s)\,. \tag{16.266}$$

(b) Show that the 3×3 matrix $M^2_{HA'}$ is given by

$$M^2_{HA'} = \frac{v}{\sqrt{2}}\,\mathrm{Im}(\lambda\kappa^*)\begin{pmatrix} 0 & 0 & -3ss_\beta \\ 0 & 0 & -3sc_\beta \\ ss_\beta & sc_\beta & 2\sqrt{2}vs_\beta c_\beta \end{pmatrix}. \tag{16.267}$$

(c) Verify that $M^2_{A'A'}$ has a zero eigenvalue and find the corresponding eigenvector. Using this eigenvector, perform an orthogonal similarity transformation, $R^{\mathsf{T}}M^2_{A'A'}R$, to produce a matrix whose first row and first column consist entirely of zeros. Then show that the product $M^2_{HA'}R$ is a matrix whose first column consists entirely of zeros. Finally, by deleting the first row and column of zeros in $R^{\mathsf{T}}M^2_{A'A'}R$, show that one obtains the 2×2 matrix M^2_{AA} given in eq. (16.30). Likewise, by deleting the first column of zeros in $M^2_{HA'}R$, show that one obtains the 3×2 matrix M^2_{HA} given in eq. (16.32).

(d) By a similar analysis, derive the corresponding charged scalar squared-mass matrix and identify its zero eigenvalue and corresponding eigenvector. Then, perform the analogous orthogonal similarity transformation employed in part (c) to identify the squared mass of the charged Higgs boson.

16.4 The Higgs basis of scalar fields of the NMSSM is defined in eqs. (16.34) and (16.35). It is convenient to re-express the tree-level neutral scalar mass matrices M^2_{HH}, M^2_{AA}, and M^2_{HA}, given in eqs. (16.29), (16.30), and (16.32), with respect to the neutral Higgs basis fields (e.g., see Ref. [195]).

(a) Transforming to the Higgs basis, show that the squared-mass matrices M^2_{HH} and M^2_{HA} are transformed,

$$M^2_{HH} \longrightarrow \begin{pmatrix} c_\beta & s_\beta & 0 \\ -s_\beta & c_\beta & 0 \\ 0 & 0 & 1 \end{pmatrix} M^2_{HH} \begin{pmatrix} c_\beta & -s_\beta & 0 \\ s_\beta & c_\beta & 0 \\ 0 & 0 & 1 \end{pmatrix}, \tag{16.268}$$

$$M^2_{HA} \longrightarrow \begin{pmatrix} c_\beta & s_\beta & 0 \\ -s_\beta & c_\beta & 0 \\ 0 & 0 & 1 \end{pmatrix} M^2_{HA}\,, \tag{16.269}$$

whereas M^2_{AA} is unchanged.

(b) Show that the squared-mass matrix M^2_{HH}, when expressed with respect to the neutral Higgs basis fields, is given by

$$M^2_{HH} = \begin{pmatrix} \overline{M}^2_Z c^2_{2\beta} + \frac{1}{2}|\lambda|^2 v^2 & -\overline{M}^2_Z s_{2\beta} c_{2\beta} & \frac{v}{s\sqrt{2}}\left\{2|\mu|^2 - \left[b + s\,\mathrm{Re}(\mu^*\kappa)\right] s_{2\beta}\right\} \\ -\overline{M}^2_Z s_{2\beta} c_{2\beta} & \overline{M}^2_A + \overline{M}^2_Z s^2_{2\beta} & -\frac{v}{s\sqrt{2}} c_{2\beta}\left[b + s\,\mathrm{Re}(\mu^*\kappa)\right] \\ \frac{v}{s\sqrt{2}}\left\{2|\mu|^2 - \left[b + s\,\mathrm{Re}(\mu^*\kappa)\right] s_{2\beta}\right\} & -\frac{v}{s\sqrt{2}} c_{2\beta}\left[b + s\,\mathrm{Re}(\mu^*\kappa)\right] & 4|\kappa|^2 s^2 + \frac{1}{4}v^2 b s_{2\beta}/s^2 + \frac{1}{2}\,\mathrm{Re}\, b_S \end{pmatrix}, \tag{16.270}$$

where $\overline{M}^2_A \equiv 2b/s_{2\beta}$ [see eq. (16.31)], and we have introduced the squared-mass parameter

$$\overline{M}^2_Z \equiv m^2_Z - \tfrac{1}{2}|\lambda|^2 v^2 . \tag{16.271}$$

In particular, check that the 11 and 12 elements of M^2_{HH} are related as follows:

$$(M^2_{HH})_{12} = \left[(M^2_{HH})_{11} - m^2_Z c_{2\beta} - |\lambda|^2 v^2 s^2_\beta\right] \cot\beta . \tag{16.272}$$

Likewise, show that the squared-mass matrix M^2_{HA}, when expressed with respect to the neutral Higgs basis fields, is given by

$$M^2_{HA} = \frac{v}{\sqrt{2}}\,\mathrm{Im}(\lambda^*\kappa) \begin{pmatrix} 0 & -3s s_{2\beta} \\ 0 & -3s c_{2\beta} \\ s & \sqrt{2} v s_{2\beta} \end{pmatrix} . \tag{16.273}$$

(c) Consider the limit of the NMSSM Higgs sector in which $\kappa \sim \mathcal{O}(1)$, $a_\kappa \sim \mathcal{O}(v)$, and $s \gg v$ while holding fixed $|\mu| \equiv |\lambda| s \sim \mathcal{O}(v)$ and $s a_\lambda \sim \mathcal{O}(v^2)$. In this limit, show that the mixing of the SM-like scalar mass eigenstate with the other MSSM-like [singlet-like] neutral scalar interaction eigenstates is suppressed by a factor of $\mathcal{O}(v/s)$ [$\mathcal{O}(v^2/s^2)$]. Verify that the lightest neutral Higgs scalar, h, is an approximate CP-even scalar mass eigenstate whose tree-level squared mass is given by

$$m^2_h = m^2_Z c^2_{2\beta} + \mathcal{O}(v^3/s) , \tag{16.274}$$

and whose couplings to gauge bosons, quarks, leptons, and the hhh and $hhhh$ couplings approximately coincide with those of the SM Higgs boson up to corrections of $\mathcal{O}(v/s)$. Moreover, show that the masses of all the other scalars (both charged and neutral) of the NMSSM Higgs sector are all parametrically much larger than v. We recognize this scenario as the decoupling limit of the NMSSM Higgs sector (in analogy with the decoupling limit of the 2HDM treated in Section 6.2.3). Obtain approximate expressions for the squared masses of the heavy Higgs bosons of the NMSSM in the decoupling limit. Finally, explain why the decoupling limit of the NMSSM can never be achieved if $s \sim \mathcal{O}(v)$.

NOTE: The decoupling limit is one method of realizing an approximate Higgs alignment limit (see Section 6.2.3). The possibility of Higgs alignment without requiring a decoupling scenario is examined in part (d).

(d) In the exact Higgs alignment limit, the mixing between the neutral scalar interaction eigenstate, $h \equiv \sqrt{2}\,\mathrm{Re}\,H_1^0 - v$, and the other four neutral scalar interaction eigenstates, $\sqrt{2}\,\mathrm{Re}\,H_2^0$, $a \equiv \sqrt{2}\,\mathrm{Im}\,H_2^0$, $h_s \equiv \sqrt{2}\,\mathrm{Re}\,S$ and $a_s \equiv \sqrt{2}\,\mathrm{Im}\,S$, vanishes exactly. In this case h can be identified as a scalar mass eigenstate whose tree-level properties coincide with those of the SM Higgs boson. Using eq. (16.272), and assuming that radiative corrections to $\mathcal{M}_{HH}^2$ are small and can be neglected, show that a necessary condition for exact alignment at tree level is given by

$$|\lambda|^2 = \frac{m_h^2 - m_Z^2 c_{2\beta}}{v^2 s_\beta}\,. \tag{16.275}$$

What are the other two conditions on the NMSSM Higgs sector parameters that are needed to guarantee that the tree-level Higgs alignment is exact? *HINT*: Note that, in the exact Higgs alignment limit, $m_h^2 = (\mathcal{M}_{HH}^2)_{11}$. (Why?)

(e) Show that it is possible to find numerical values of the NMSSM Higgs sector parameters that yield a CP-even Higgs state with the properties of the SM Higgs boson and $m_h = 125$ GeV. In such a scenario, is the value of $|\lambda|$ perturbative all the way up to the GUT scale?

16.5 Consider the superpotential of the seesaw-extended MSSM given in eq. (16.54), and assume that the smallest eigenvalue of the matrix $\boldsymbol{M_{\bar{\nu}}}$ has a characteristic mass scale of $M_N \gg v$, where $v = 246$ GeV is the scale of electroweak symmetry breaking.

(a) Show that if one integrates out the heavy right-handed neutrino supermultiplet, one obtains the superpotential for the MSSM supplemented by a new higher-dimensional term of the form

$$W_{\text{seesaw}} = \frac{f}{M_N}(LH_u)(LH_u), \tag{16.276}$$

where f is a dimensionless constant.

(b) Using eq. (16.276), show that one recovers the results of the seesaw-extended Standard Model given in eqs. (6.1)–(6.9).

(c) Investigate the corresponding supersymmetric phenomena involving the sneutrinos that arises from eq. (16.276).

16.6 Determine all possible nonzero 5×5 complex traceless matrices A that satisfy

$$A^2 + cA - \tfrac{1}{5}\,\mathrm{Tr}(A^2)\mathbb{1}_{5\times 5} = 0\,, \tag{16.277}$$

where c is a complex parameter. Excluding the trivial solution $A = 0$, show that at least three of the diagonal elements of A are equal, and exactly two of the diagonal elements are distinct.

16.7 Following eq. (6.264), denote the SU(5) Higgs supermultiplet Σ in terms of their $\mathrm{SU}(3)_C\times\mathrm{SU}(2)_L\times\mathrm{U}(1)_Y$ component superfields by

$$\Sigma = \left(\begin{array}{ccc|cc} & & & \overline{H}^1_X & \overline{H}^1_Y \\ & \mathcal{H} - \dfrac{2H_B}{\sqrt{30}}\mathbb{1}_{3\times 3} & & \overline{H}^2_X & \overline{H}^2_Y \\ & & & \overline{H}^3_X & \overline{H}^3_Y \\ \hline H_{X1} & H_{X2} & H_{X3} & \dfrac{H_{W3}}{\sqrt{2}} + \dfrac{3H_B}{\sqrt{30}} & H^+_W \\ H_{Y1} & H_{Y2} & H_{Y3} & H^-_W & -\dfrac{H_{W3}}{\sqrt{2}} + \dfrac{3H_B}{\sqrt{30}} \end{array}\right), \tag{16.278}$$

where, e.g., $\mathcal{H}$ is a 3×3 traceless complex matrix scalar field that transforms under the adjoint eight-dimensional representation of color SU(3), etc.

(a) Starting from the superpotential given in eq. (16.98), and assuming that the vacuum expectation value of the scalar field component of Σ is given by eq. (16.113), evaluate the masses of the scalar fields. Show that

$$W_{\rm mass} = m\big(\tfrac{5}{2}\,\mathrm{Tr}\,\mathcal{H}^2 - 5H^+_W H^-_W - \tfrac{5}{2}H_{W3}H_{W3} - \tfrac{1}{2}H_B H_B\big)\,. \tag{16.279}$$

That is, the masses of scalars and fermions that reside in the superfields $\mathcal{H}$, $H^\pm_W$, H_{W3}, and H_B are all of $\mathcal{O}(m)$, which is assumed to be the scale of grand unification ($\sim 10^{16}$ GeV).

(b) Show that no mass term for H_{Xi} H_{Yi}, $\overline{H}^i_X$, and $\overline{H}^i_Y$ ($i = 1,2,3$) appear in eq. (16.279). Identify the CP-odd scalar component of these superfields as the Goldstone bosons that provide superheavy masses for the X, Y, $\overline{X}$, and $\overline{Y}$ gauge bosons of SU(5).

(c) Since supersymmetry is not broken (because the F- and D-terms vanish in the $\mathrm{SU}(3)_C\times\mathrm{SU}(2)_L\times\mathrm{U}(1)_Y$ vacuum), show that the superfields H_{Xi}, H_{Yi}, $\overline{H}^i_X$, and $\overline{H}^i_Y$ ($i = 1,2,3$) can be absorbed into the corresponding vector superfields that contain the X, Y, $\overline{X}$, and $\overline{Y}$ gauge bosons of SU(5), thereby producing massive vector supermultiplets. Identify all the physical degrees of freedom of these massive vector supermultiplets and check that the number of bosonic and the number of fermionic degrees of freedom are equal. In addition, verify that all component fields of each massive vector supermultiplet have the same mass.

16.8 In the flipped SU(5) model, one generation of fermions resides in the reducible representation $\overline{\mathbf{5}} \oplus \mathbf{10} \oplus \mathbf{1}$. Compared to the standard embedding of SU(5) inside SO(10), we interchange the fermion fields $\bar{u} \leftrightarrow \bar{d}$ and $\bar{e} \leftrightarrow \bar{\nu}$ (e.g., see [B36]). Defining X to be the $\mathrm{U}(1)_X$ generator [see eq. (6.268)] and denoting the U(1) generator that is embedded inside SU(5) by $\boldsymbol{T^{24}}$ (as in Section 6.3.1), show that the SM hypercharge U(1) generator is

$$Y = -\frac{1}{5}\left[\sqrt{5/3}\,\boldsymbol{T^{24}} + X\right]. \tag{16.280}$$

16.9 Proton hexality is a $\mathbb{Z}_6$ symmetry defined by $P_6 \equiv R_6^5 L_6^2$ [188].

(a) Identify the charges of the MSSM chiral superfields under this symmetry.

(b) Is proton hexality a GMP, a GLP, or a GBP?

(c) Show that P_6 is discrete gauge anomaly-free. In particular, show that P_6 satisfies eqs. (16.173), (16.174), and (16.175).

(d) Which operators, of those listed in Table 16.3, are allowed by P_6? How does this list of allowed operators compare with the operators allowed by $R_p \equiv R_2$? Among the operators that mediate proton decay, which ones are allowed or forbidden by P_6 and R_2, respectively?

(e) Show that P_6 is equivalent to the direct product of matter parity R_2 and baryon triality $R_2 L_2$.

16.10 Extending the MSSM to N_g generations and N_D pairs of Higgs doublets, show that the generalization of eqs. (16.173)–(16.175) is given by

$$nN_g = 0 \mod N\,, \tag{16.281}$$

$$pN_g - nN_D = 0 \mod N\,, \tag{16.282}$$

$$mN_g + nN_D = \begin{cases} 0 \mod N, & \text{for } N = \text{odd}\,, \\ 0 \mod N/2, & \text{for } N = \text{even}\,. \end{cases} \tag{16.283}$$

Show that if $N_g = 1$ then the only solution is $n = p = 0$, $N = 2$, and $m = 1$, corresponding to R_2 or matter parity.

16.11 How do the five discrete symmetries listed in eqs. (16.176)–(16.178) act on the superpotential term $\mu H_u H_d$?

16.12 Consider the charges X_i specified in eq. (5.107),

$$X_i = q_i + m_i N, \qquad q_i, m_i \in \mathbb{Z}\,, \tag{16.284}$$

where the m_i are arbitrary unknown integers. For the MSSM fields the q_i

are given in eq. (16.152). Consider the $U(1)^3_X$ cubic anomaly. Show that the condition $D_{XXX} = 0$, or $\sum_i X_i^3 = 0$, can be rewritten as

$$n(13n^2 + 18np - 21nm + 18p^2 + 21m^2) + p(3p^2 - 9pm - 9m^2) - 3m^3 = 3N \sum_i \left(q_i^2 m_i + q_i m_i^2 N + \frac{1}{3} N^2 m_i^3 \right) . \quad (16.285)$$

Thus for $N = 3$, the right-hand side of eq. (16.284) is an arbitrary integer multiple of 3. Show that of the four $\mathbb{Z}_3$ symmetries in eqs. (16.177) and (16.178) only B_3 can satisfy this equation. Thus, in terms of the low-energy MSSM particle content of the $\mathbb{Z}_{2,3}$ discrete symmetries, only R_2 (see Exercise 16.10) and B_3 are discrete gauge anomaly-free.

16.13 In Table 16.3 we listed the operators of dimension three, four, and five composed of MSSM chiral superfields, which are $\mathrm{SU}(3)_C \times \mathrm{SU}(2)_L \times \mathrm{U}(1)_Y$ gauge invariant and which violate baryon and/or lepton number. Here we shall determine the $\mathrm{SU}(3)_C \times \mathrm{SU}(2)_L \times \mathrm{U}(1)_Y$ gauge invariant, dimension-four hard supersymmetry-breaking terms (see Table 14.1) that violate baryon number and/or lepton number, considering just the scalar components of the MSSM chiral superfields.

(a) Consider the $\phi^\dagger \phi\phi\phi$ hard supersymmetry-breaking operators. There are two terms for which $|\Delta B| = 1, |\Delta L| = 1$, namely $Q^\dagger L \bar{d}\,\bar{d}$, $e^\dagger \bar{d}\,\bar{d}\,\bar{d}$. Show that there are three operators with $\Delta B = 1$, $\Delta L = 0$; two operators with $\Delta B = 0$, $\Delta L = 2$; and ten operators with $\Delta B = 0$, $\Delta L = 1$.

(b) Consider the $\phi^\dagger \phi^\dagger \phi\phi$ hard supersymmetry-breaking operators. Show that there are two operators with $\Delta B = 1$, $\Delta L = 1$; two operators with $\Delta B = 1$, $\Delta L = 0$; one operator with $\Delta B = 0$, $\Delta L = 2$; and nine operators with $\Delta B = 0$, $\Delta L = 1$.

16.14 In Section 16.5.4, we obtained the R-parity-violating spectrum in a supersymmetric model governed by the Lagrangian given in Section 16.5.1, under the assumption that the RPV parameters were real in a scalar field basis where the VEVs are real. If this assumption is relaxed, then new sources of CP violation appear. You may assume that R-parity-violating effects are small and can be treated perturbatively.

(a) Write out the squared-mass matrices for the neutral color singlet scalars of the model. Determine the masses of the neutral Higgs bosons and the sneutrinos. In the one-generation model, evaluate the resulting sneutrino squared-mass difference.

(b) How many neutrinos acquire mass via tree-level neutralino–neutrino mixing? Determine the masses of the corresponding massive neutrinos.

16.15 Consider the RPV SUSY model defined in Section 16.5.1.

(a) Show that the $W^\pm$ gauge boson squared mass is now given by

$$m_W^2 = \tfrac{1}{4}g^2(v_u^2 + \bar{v}_d^2)\,, \tag{16.286}$$

where $\bar{v}_d$ is defined in eq. (16.199). What is the expression for m_Z^2? Do the photon and gluon remain massless?

(b) Show that the down-type quark and lepton mass matrices are given by

$$(\boldsymbol{m}_d)_{ij} = \frac{1}{\sqrt{2}}\lambda'_{\alpha ij}v_\alpha\,, \qquad (\boldsymbol{m}_\ell)_{\alpha k} = \frac{1}{\sqrt{2}}\lambda_{\beta\alpha k}v_\beta\,, \tag{16.287}$$

whereas the up-type quark mass matrix is unchanged.

(c) In the MSSM, the bottom-squark squared-mass matrix $\boldsymbol{m}^2_{\widetilde{b}}$ is given by eq. (13.182). Show that, in the RPV SUSY model, the off-diagonal element of this matrix is given by

$$(m^2_{\widetilde{b}})_{LR} = \frac{1}{\sqrt{2}}(\bar{h}'^{*}_{\alpha 33}v_\alpha - \lambda'^{*}_{\alpha 33}\mu_\alpha v_u)\,, \tag{16.288}$$

where there is an implicit sum over the repeated index $\alpha \in \{0,1,2,3\}$. How are the diagonal elements of $\boldsymbol{m}^2_{\widetilde{b}}$ modified with respect to the MSSM? How are the top-squark squared-mass matrix $\boldsymbol{m}^2_{\widetilde{t}}$ and tau-slepton squared-mass matrix $\boldsymbol{m}^2_{\widetilde{\tau}}$ modified with respect to their MSSM expressions given in eqs. (13.180) and (13.183), respectively?

(d) After electroweak symmetry breaking, the charginos and charged leptons mix. The resulting mass matrix is obtained from the Lagrangian

$$\mathscr{L} = -\frac{1}{2}(\psi^+, \psi^-)\begin{pmatrix} 0 & X^{\mathsf{T}} \\ X & 0 \end{pmatrix}\begin{pmatrix} \psi^+ \\ \psi^- \end{pmatrix}, \tag{16.289}$$

with respect to the basis

$$\psi^+ = (-i\widetilde{W}^+\,,\, \widetilde{H}_u^+\,,\, \bar{e}_k), \qquad \psi^+ = (-i\widetilde{W}^+\,,\, \ell^\dagger_\alpha), \tag{16.290}$$

where $k = 1,2,3$ and $\alpha = 0,1,2,3$. Show that the 5×5 matrix X is given by

$$X = \begin{pmatrix} M_2 & gv_u/\sqrt{2} & 0_k \\ gv_\alpha/\sqrt{2} & \mu_\alpha & (\boldsymbol{m}_\ell)_{\alpha k} \end{pmatrix}. \tag{16.291}$$

16.16 The trilinear couplings λ_{ijk}, λ'_{ijk} violate lepton number, which has not been observed. Here we derive example bounds from tau decay on λ_{13k} and λ_{23k}.

(a) Show that, in addition to the Standard Model (SM) contribution to tau decay, there is a second tree-level Feynman diagram involving $\widetilde{e}_{Rk}$ exchange due to the RPV $\lambda_{13k}L_1L_3\bar{e}_k$ interaction.

(b) Assuming that the masses of the virtual particles that appear in the Feynman diagrams of part (a) are significantly larger than the masses of the external fermions, show that the resulting amplitudes are given by

$$\mathcal{M}_{\rm SM} \simeq \frac{-8G_F}{\sqrt{2}}(x^\dagger_{\nu_\tau}\bar\sigma^\mu x_\tau)(x^\dagger_e\bar\sigma_\mu y_{\nu_e})\,, \tag{16.292}$$

$$\mathcal{M}_{\rm RPV} \simeq \frac{-|\lambda_{13k}|^2}{8m^2_{\tilde e_{Rk}}}(x^\dagger_{\nu_\tau}\bar\sigma^\mu x_\tau)(x^\dagger_e\bar\sigma_\mu y_{\nu_e}), \tag{16.293}$$

where G_F is the Fermi constant. To obtain eq. (16.293), employ an appropriate Fierz identity.

(c) The two amplitudes in part (b) have the identical current structure. Show that when added together, the result is a shift in the Fermi constant as measured in tau decay,

$$\frac{G_{F\tau}}{\sqrt{2}} = \frac{g'^2}{8m_W^2}\left[1 + r_{13k}(\tilde e_{Rk})\right]\,, \quad r_{ijk}(\tilde f) \equiv \frac{|\lambda_{ijk}|^2 m_W^2}{g'^2\, m^2_{\tilde f}} = \frac{\sqrt{2}|\lambda_{ijk}|^2}{8G_F\, m^2_{\tilde f}}, \tag{16.294}$$

which is observable only when compared to other tau decay measurements.

(d) Show that the decay $\tau^- \to \mu^-\nu_\tau\bar\nu_\mu$ yields an alternative Fermi constant

$$\frac{G'_{F\tau}}{\sqrt{2}} = \frac{g'^2}{8m_W^2}\left[1 + r_{23k}(\tilde e_{Rk})\right]\,. \tag{16.295}$$

(e) Show that, for small $r_{ijk}(\tilde f)$,

$$R_\tau \equiv \frac{\Gamma(\tau^- \to \mu^-\nu_\tau\bar\nu_\mu)}{\Gamma(\tau^- \to e^-\nu_\tau\bar\nu_e)} = R_\tau({\rm SM})\left\{1 + 2\left[r_{13k}(\tilde e_{Rk}) - r_{23k}(\tilde e_{Rk})\right]\right\}, \tag{16.296}$$

where R_τ(SM) is the pure SM theory prediction.

(f) Find the latest theory prediction R_τ(SM), as well as the experimentally measured value R_τ(expt), the latter including the $2\,\sigma$ error. Then assume that only one $r_{ijk} \neq 0$ at a time. This is called the single coupling dominance hypothesis, which need not hold. Use the latest value for the Fermi constant to obtain upper bounds on $|r_{13k}|$, $|r_{23k}|$, at $2\,\sigma$. Translate these into individual bounds on $|\lambda_{13k}|$, $|\lambda_{23k}|$, as a function of $m_{\tilde e_{Rk}}$. Dropping the single coupling dominance hypothesis, there can be cancellations between two RPV contributions, thus weakening or annulling any bound.

16.17 In Chapter 11 the general renormalization group equations in supersymmetric theories were presented. In this exercise, we shall derive some of the RGEs of the RPV extended MSSM. In eq. (11.136) the general beta function is given for the bilinear parameters in the superpotential:

$$\beta_{M^{ij}} = \gamma^i_n M^{nj} + \gamma^j_n M^{in}\,. \tag{16.297}$$

At one loop the γ_i^j are given in eq. (11.139) in terms of the gauge couplings and the trilinear parameters of the superpotential:

$$\gamma_i^{(1)j} = \tfrac{1}{2} y_{ikl} y^{jkl} - 2\delta_i^j g_a^2 C_a(i) \,. \tag{16.298}$$

Furthermore, as noted below eq. (11.153), in theories with no gauge-singlet chiral superfields, the beta function for b^{ij} is given by

$$\beta_{b^{ij}} = [\gamma_n^i b^{nj} - 2M^{nj}\Omega(\gamma_n^i)] + (i \leftrightarrow j) \,. \tag{16.299}$$

The operator Ω on the space of couplings is defined in eq. (11.151). These functions are explicitly calculated for the MSSM in Section 13.7. We would like to compute $\beta_{M^{ij}}$ and $\beta_{b^{ij}}$ in the RPV extended MSSM.

(a) Show that $\gamma_{H_u}^{(1)H_u}$ given in eq. (13.46) is unchanged in the RPV extended MSSM.

(b) Compute the modifications to $\gamma_{L_i}^{(1)L_j}$ due to RPV. The special case of dominant third-generation SM Yukawa couplings is given in eq. (16.228).

(c) Compute at one loop the anomalous dimension $\gamma_{H_d}^{(1)L_j}$, which is specific to RPV SUSY. The fields L_i and H_d have identical quantum numbers. The special case of dominant third-generation SM Yukawa couplings is given in eq. (16.229).

(d) Compute $\Omega(\gamma_{L_i}^{(1)L_j})$ and $\Omega(\gamma_{H_d}^{(1)L_j})$.

(e) In light of parts (a)–(d), compute $d\kappa_i/dt$ and db_i/dt. Show that the alignment of κ_i and b_i is not stable under RG evolution.

16.18 Starting from eq. (16.218) for the CP-odd scalar squared-mass matrix $\mathcal{M}^2_{\rm CP-}$ and following the analogous perturbative procedure leading to the squared-mass eigenvalue $m^2_{\tilde{\nu}+}$ given in eq. (16.223), derive the squared-mass eigenvalue $m^2_{\tilde{\nu}-}$ given in eq. (16.224).

16.19 Consider the RPV extension of the MSSM governed by the superpotential exhibited in eq. (16.180). Under the following three different assumptions, count the number of parameters that govern the corresponding RPV model (including the 19 parameters of the Standard Model), following the technique outlined in Section 13.5.

(a) $W_{\Delta L} = 0$ and $W_{\Delta B} \neq 0$.

(b) $W_{\Delta L} \neq 0$ and $W_{\Delta B} = 0$.

(c) $W_{\Delta L} \neq 0$ and $W_{\Delta B} \neq 0$.

Identify how many parameters are associated with the gauge sector and how many parameters are associated with the flavor sector in each of the three cases above. Determine the circumstances under which the vacuum angle Θ_2 associated with the SU(2) gauge group (which is not a physical parameter of the SM or MSSM) is promoted to a physical parameter of the RPV model.

16.20 A calculation similar to the one presented in Section 16.5.6 can be performed to obtain the contribution of the $LL\bar{e}$ operator to the one-loop neutrino mass. If the contribution of the $LQ\bar{d}$ operator is negligible, show that

$$m_{\nu_2} \simeq \frac{1}{16\pi^2} \sum_{j,k=1,3} \lambda_{2kj}\lambda_{2jk} m_{\ell_k} \sin 2\theta_{\widetilde{\ell}_j} \ln\left(\frac{m^2_{\widetilde{\ell}_{j2}}}{m^2_{\widetilde{\ell}_{j1}}}\right), \tag{16.300}$$

in the limit of $m_{\widetilde{\ell}_{j1}}, m_{\widetilde{\ell}_{j2}} \gg m_{\ell_k}$, where m_{ℓ_k} is the mass of the charged lepton of generation k and $\widetilde{\ell}_j$ is the corresponding charged slepton.

16.21 Consider the case of a complex 4×4 neutralino mass matrix $\boldsymbol{M}_{\widetilde{N}}$ that possesses a light neutralino mass eigenstate. For simplicity, choose a basis such that $M_2 \in \mathbb{R}$ and $M_1, \mu \in \mathbb{C}$ (see the discussion in Section 13.5) with

$$M_1 = |M_1|e^{i\phi_1}, \qquad \mu = |\mu|e^{i\phi_\mu}. \tag{16.301}$$

(a) The complex condition $\det \boldsymbol{M}_{\widetilde{N}} = 0$ is equivalent to two real conditions,

$$\operatorname{Re}\det \boldsymbol{M}_{\widetilde{N}} = 0, \qquad \operatorname{Im}\det \boldsymbol{M}_{\widetilde{N}} = 0. \tag{16.302}$$

Using the explicit form of $\boldsymbol{M}_{\widetilde{N}}$ [see eq. (13.137)], show that eq. (16.302) yields

$$m_Z^2 c_W^2 \sin 2\beta \sin\phi_1 - |\mu| M_2 \sin(\phi_1 + \phi_\mu) = 0, \tag{16.303}$$

$$M_2 m_Z^2 s_W^2 \sin 2\beta + |M_1|\left[m_Z^2 c_W^2 \sin 2\beta \cos\phi_1 - M_2 \mu \cos(\phi_1 + \phi_\mu)\right] = 0. \tag{16.304}$$

Show that in the CP-conserving limit, where $\phi_1 = \phi_\mu = 0$, eq. (16.304) reproduces the condition given in eq. (16.258).

(b) Verify that the solutions to eqs. (16.303) and (16.304) are given by

$$|\mu| = \frac{m_Z^2 c_W^2 \sin 2\beta \sin\phi_1}{M_2 \sin(\phi_1 + \phi_\mu)}, \qquad |M_1| = -M_2 \tan^2\theta_W \frac{\sin(\phi_1 + \phi_\mu)}{\sin\phi_\mu}. \tag{16.305}$$

(c) Alternatively, verify that one can also solve eqs. (16.303) and (16.304) for M_2 and $|M_1|$:

$$M_2 = \frac{m_Z^2 c_W^2 \sin 2\beta \sin\phi_1}{|\mu| \sin(\phi_1 + \phi_\mu)}, \qquad |M_1| = -\frac{m_Z^2 s_W^2 \sin 2\beta \sin\phi_1}{|\mu| \sin\phi_\mu}. \tag{16.306}$$

(d) Show that eqs. (16.303) and (16.304) can *not* always be solved for arbitrary choices of the complex phases ϕ_1, ϕ_μ.

PART IV

SAMPLE CALCULATIONS IN THE STANDARD MODEL AND ITS SUPERSYMMETRIC EXTENSION

17 Practical Calculations Involving Two-Component Fermions

One occasionally encounters the misconception that two-component spinor notation is somehow inherently ill-suited or unwieldy for practical use. Perhaps this is due in part to a lack of examples of calculations using two-component language in the pedagogical literature. In this chapter, we seek to dispel this idea by presenting Feynman rules for external fermions using two-component spinor notation, intended for practical calculations of cross sections, decays, and radiative corrections. As a warmup exercise, we shall first apply the technology developed in Chapter 2 to some Standard Model processes. In two subsequent chapters, we present detailed computations of a number of tree-level supersymmetric decay and scattering processes and a number of one-loop computations relevant in both the Standard Model and its supersymmetric extension.

Before jumping into the explicit evaluation of Feynman graphs, we should adopt a particular labeling convention for the fermions and antifermions. There is an option of labeling fermion lines in Feynman diagrams by particle names or by field names; each choice has advantages and disadvantages. In the sample calculations that follow, we have chosen to label fermion lines in graphs with the two-component $(\frac{1}{2},0)$ and $(0,\frac{1}{2})$ fields that correspond to the physical particle (see Table 17.1). Details of this convention are explained in Section 17.1.

17.1 Conventions for Fermion and Antifermion Names and Fields

In this section, we establish conventions for labeling Feynman diagrams that contain two-component fermion fields of the Standard Model (SM) and its minimal supersymmetric extension (MSSM). In the case of Majorana fermions, there is a one-to-one correspondence between the particle names and the undaggered $(\frac{1}{2},0)$ [left-handed] fields. In contrast, for Dirac fermions there are always two distinct two-component fields that correspond to each particle name. For a quark or lepton generically denoted by f, we employ the two-component undaggered $(\frac{1}{2},0)$ [left-handed] fields f and $\bar{f}$ (where the bar is part of the field name and does *not* refer to complex conjugation of any kind). This is illustrated in Table 17.1, which lists the SM and MSSM fermion particle names together with the corresponding two-component fields. For each particle, we list the two-component field with the same

Table 17.1 Fermion and antifermion names and the corresponding two-component fields of the Standard Model and the MSSM. In the listing of two-component fields, the first is an undaggered $(\frac{1}{2},0)$ [left-handed] field and the second is a daggered $(0,\frac{1}{2})$ [right-handed] field. The bars on the two-component (antifermion) fields are part of their names, and do not denote any form of complex conjugation. (The neutrinos are considered to be exactly massless and the left-handed antineutrino $\bar{\nu}$ is absent from the spectrum.)

fermion name	two-component fields
ℓ^- (lepton)	$\ell\,,\,\bar{\ell}^\dagger$
ℓ^+ (antilepton)	$\bar{\ell}\,,\,\ell^\dagger$
ν (neutrino)	$\nu\,,\,-$
$\bar{\nu}$ (antineutrino)	$-\,,\,\nu^\dagger$
q (quark)	$q\,,\,\bar{q}^\dagger$
$\bar{q}$ (antiquark)	$\bar{q}\,,\,q^\dagger$
f (quark or lepton)	$f\,,\,\bar{f}^\dagger$
$\bar{f}$ (antiquark or antilepton)	$\bar{f}\,,\,f^\dagger$
$\widetilde{N}_i$ (neutralino)	$\chi_i^0\,,\,\chi_i^{0\dagger}$
$\widetilde{C}_i^+$ (chargino)	$\chi_i^+\,,\,\chi_i^{-\dagger}$
$\widetilde{C}_i^-$ (antichargino)	$\chi_i^-\,,\,\chi_i^{+\dagger}$
$\widetilde{g}$ (gluino)	$\widetilde{g}\,,\,\widetilde{g}^\dagger$

quantum numbers, i.e., the field that contains the annihilation operator for that one-particle state (which creates the one-particle state when acting to the *left* on the vacuum $\langle 0|$). In the explicit calculations presented in this and subsequent chapters, we always label fermion lines with two-component fields (rather than particle names), and adopt the following conventions:

• In the Feynman rules for interaction vertices, the external lines are always labeled by the undaggered $(\frac{1}{2},0)$ [left-handed] field, regardless of whether the corresponding arrow is pointed in or out of the vertex. Two-component fermion lines with arrows pointing away from the vertex correspond to dotted indices, and two-component fermion lines with arrows pointing toward the vertex always correspond to undotted indices. This also applies to Feynman diagrams where the roles of the initial state and the final state are ambiguous (such as self-energy diagrams).

• Internal fermion lines in Feynman diagrams are also always labeled by the undaggered $(\frac{1}{2}, 0)$ [left-handed] field(s). Internal fermion lines containing a propagator with opposing arrows can carry two labels (e.g., see Fig. 2.17).

• Initial-state external fermion lines (which always have physical three-momenta pointing into the vertex) in Feynman diagrams for complete processes are labeled by the corresponding undaggered $(\frac{1}{2}, 0)$ [left-handed] field if the arrow is into the vertex, and by the daggered $(0, \frac{1}{2})$ [right-handed] field if the arrow is away from the vertex.

• Final-state external fermion lines (which always have physical three-momenta pointing out of the vertex) in Feynman diagrams for complete processes are labeled by the corresponding daggered $(0, \frac{1}{2})$ [right-handed] field if the arrow is into the vertex, and by the undaggered $(\frac{1}{2}, 0)$ [left-handed] field if the arrow is away from the vertex.

The application of our labeling conventions to processes involving Majorana fermions is completely straightforward. For example, the conventions for employing the neutralino states as external particles are summarized in Fig. 17.1. The corresponding rules for labeling external Dirac fermions are summarized in Fig. 17.2. The labeling conventions employed in Fig. 17.2 differ slightly from the ones used in Section 2.8, where *all* internal and external initial-state and final-state fermion lines were labeled by the corresponding *undaggered* $(\frac{1}{2}, 0)$ left-handed fields. In this latter convention, the conserved quantities (charges, lepton numbers, baryon numbers, etc.) of the labeled fields follow the direction of the arrow that adorns the corresponding fermion line in the diagram. In contrast, in the convention of Fig. 17.2,

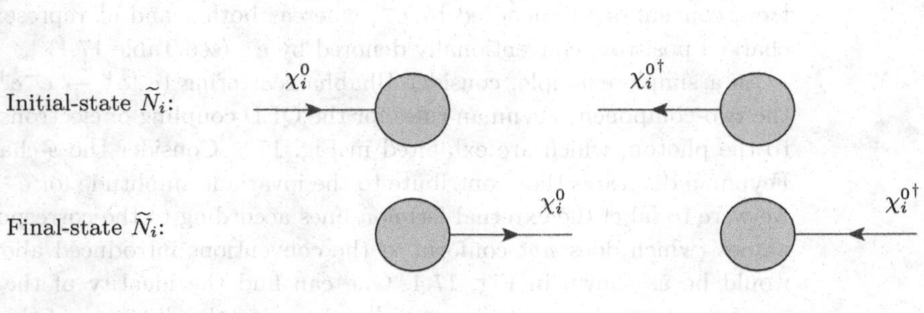

Fig. 17.1 The two-component field labeling conventions for external Majorana fermion lines in a Feynman diagram for a physical process. The top row corresponds to an initial-state neutralino, and the second row to a final-state neutralino. The labels above each line are the two-component field names. (The neutralino is its own antiparticle.)

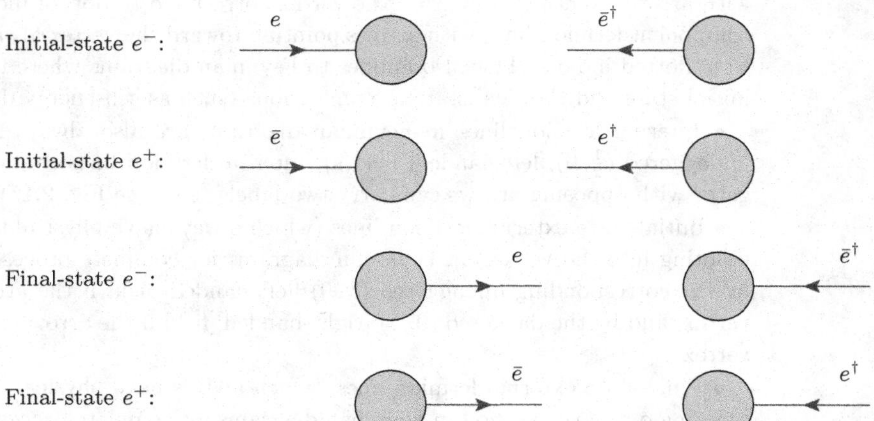

Fig. 17.2 The two-component field labeling conventions for external Dirac fermion lines in a Feynman diagram for a physical process. The top row corresponds to an initial-state electron, the second row to an initial-state positron, the third row to a final-state electron, and the fourth row to a final-state positron. The labels above each line are the two-component field names. The corresponding conventions for a massless neutrino are obtained by deleting the diagrams with $\bar{e}$ or $\bar{e}^\dagger$, and changing e and $e^\dagger$ to ν and $\nu^\dagger$, respectively.

the field labels used for external fermion lines always correspond to the physical particle, and the corresponding conserved quantities of the labeled fields follow the direction of the particle three-momentum. As an example, for either initial or final states, the two-component fields e and $\bar{e}^\dagger$ both represent a negatively charged electron, conventionally denoted by e^-, whereas both $\bar{e}$ and $e^\dagger$ represent a positively charged positron, conventionally denoted by e^+ (see Table 17.1).

As a simple example, consider Bhabha scattering ($e^-e^+ \to e^-e^+$). We require the two-component Feynman rules for the QED coupling of electrons and positrons to the photon, which are exhibited in Fig. 17.3. Consider the s-channel tree-level Feynman diagrams that contribute to the invariant amplitude for $e^-e^+ \to e^-e^+$. If we were to label the external fermion lines according to the corresponding particle names (which does *not* conform to the conventions introduced above), the result would be as shown in Fig. 17.4. One can find the identity of the external two-component fermion fields by carefully observing the direction of the arrow of each fermion line. In contrast, the same diagrams, relabeled with two-component fields following the conventions established in this section (see Fig. 17.2), are shown in Fig. 17.5. An explicit computation of the invariant amplitude for Bhabha scattering is given in Section 17.3.

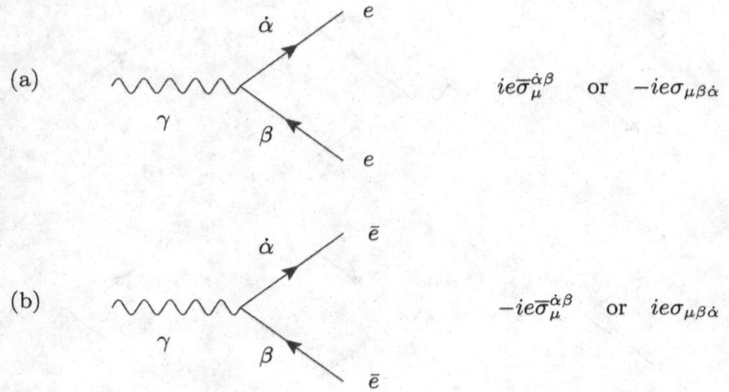

Fig. 17.3 The two-component Feynman rules for the QED vertex. Following the conventions outlined in this section, we label these rules with the $(\frac{1}{2}, 0)$ [left-handed] fields e and $\bar{e}$, which comprise the Dirac electron. Note that $Q_e = -1$, and the electromagnetic coupling constant e (not to be confused with the two-component electron field that is denoted by the same letter) is conventionally defined such that $e > 0$ (see Fig. I.2).

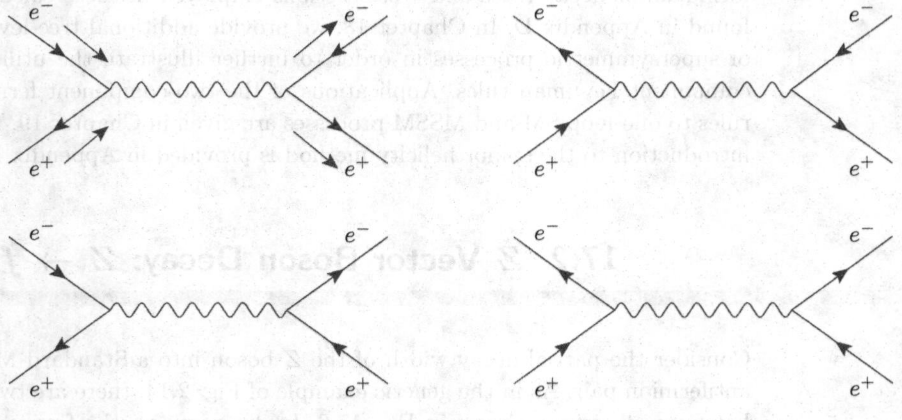

Fig. 17.4 Tree-level s-channel Feynman diagrams for $e^+e^- \to e^+e^-$, with the external lines labeled according to the particle names. The initial state is on the left, and the final state is on the right. Thus, the physical momentum flow of the external particles, as well as the flow of the labeled charges, are indicated by the arrows adjacent to the corresponding fermion lines in the upper left diagram.

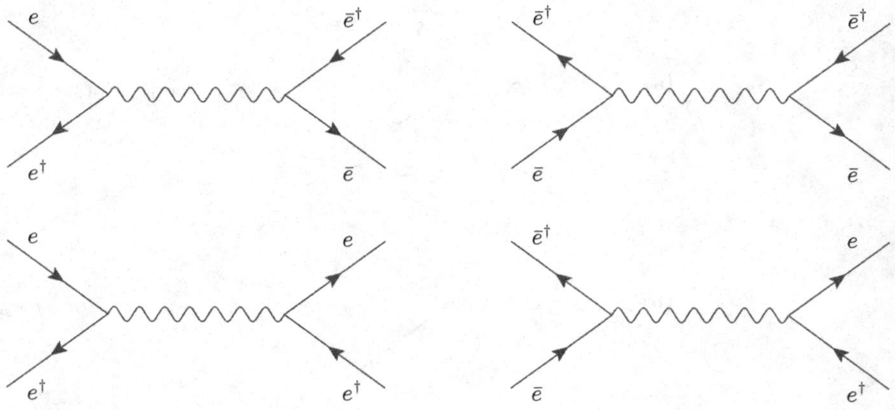

Fig. 17.5 Tree-level s-channel Feynman diagrams for $e^+e^- \to e^+e^-$. These diagrams are the same as in Fig. 17.4, but with the external lines relabeled by the two-component fermion fields according to the conventions of Fig. 17.2.

In order to get some practice with applying the two-component fermion Feynman rules while employing the convention for labeling external fermion lines introduced above, we now proceed to exhibit four explicit calculations involving Standard Model particles and interactions. A summary of relativistic kinematics and the formulae for decay rates and cross sections employed in these calculations can be found in Appendix D. In Chapter 18, we provide additional tree-level calculations of supersymmetric processes in order to further illustrate the utility of the two-component Feynman rules. Applications of the two-component fermion Feynman rules to one-loop SM and MSSM processes are given in Chapter 19. Finally, a brief introduction to the spinor helicity method is provided in Appendix F.

17.2 Z Vector Boson Decay: $Z \to f\bar{f}$

Consider the partial decay width of the Z boson into a Standard Model fermion–antifermion pair. As in the generic example of Fig. 2.14, there are two contributing Feynman diagrams, shown in Fig. 17.6. In diagram (a), the fermion particle f in the final state is created by a two-component field f in the Feynman rule, and the antifermion particle $\bar{f}$ by a two-component field $f^\dagger$. In diagram (b), the fermion particle f in the final state is created by a two-component field $\bar{f}$, and the antifermion particle $\bar{f}$ by a two-component field $\bar{f}^\dagger$. Denote the Z four-momentum and helicity (p, λ_Z) and the final-state fermion (f) and antifermion $(\bar{f})$ momentum and helicities (k_f, λ_f) and $(k_{\bar{f}}, \lambda_{\bar{f}})$, respectively, where $k_f^2 = k_{\bar{f}}^2 = m_f^2$ and $p^2 = m_Z^2$.

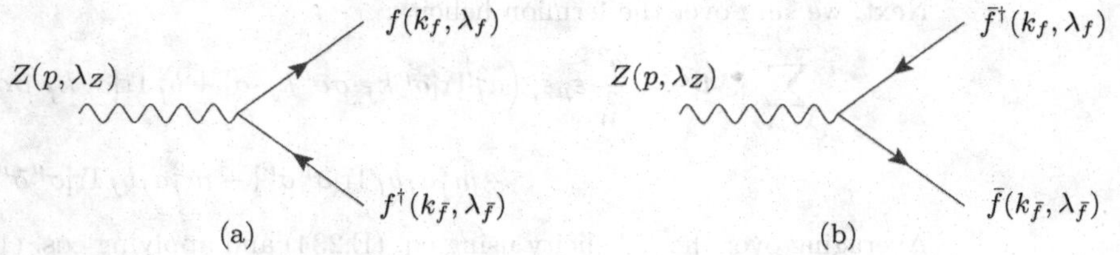

Fig. 17.6 The Feynman diagrams for Z decay into a fermion–antifermion pair.

The dot products are fixed by the kinematics:

$$k_f \cdot k_{\bar{f}} = \frac{1}{2} m_Z^2 - m_f^2 \,, \qquad p \cdot k_f = p \cdot k_{\bar{f}} = \tfrac{1}{2} m_Z^2 \,. \tag{17.1}$$

According to the rules of Fig. I.2, the matrix elements for the two Feynman graphs are

$$i\mathcal{M}_a = -i \frac{g}{c_W} (T_3^f - s_W^2 Q_f)\, \varepsilon_\mu x_f^\dagger \overline{\sigma}^\mu y_{\bar{f}} \,, \tag{17.2}$$

$$i\mathcal{M}_b = ig \frac{s_W^2}{c_W} Q_f\, \varepsilon_\mu y_f \sigma^\mu x_{\bar{f}}^\dagger \,, \tag{17.3}$$

where $x_i \equiv x(\vec{\boldsymbol{k}}_i, \lambda_i)$ and $y_i \equiv y(\vec{\boldsymbol{k}}_i, \lambda_i)$, for $i = f, \bar{f}$, and $\varepsilon_\mu \equiv \varepsilon_\mu(p, \lambda_Z)$.

Using the Bouchiat–Michel formulae developed in Appendix E.5, one can explicitly evaluate $\mathcal{M}_a$ and $\mathcal{M}_b$ as functions of the final-state fermion helicities. The result of this computation is given in eqs. (E.263) and (E.264). If the final-state helicities are not measured, then it is simpler to square the amplitude and sum over the final-state helicities.

It is convenient to define

$$a_f \equiv T_3^f - Q_f s_W^2 \,, \qquad b_f \equiv -Q_f s_W^2 \,. \tag{17.4}$$

Using eqs. (1.105) and (1.106), the squared matrix element for the decay is given by

$$|\mathcal{M}|^2 = \frac{g^2}{c_W^2} \varepsilon_\mu \varepsilon_\nu^* \Big(a_f x_f^\dagger \overline{\sigma}^\mu y_{\bar{f}} + b_f y_f \sigma^\mu x_{\bar{f}}^\dagger \Big) \Big(a_f y_{\bar{f}}^\dagger \overline{\sigma}^\nu x_f + b_f x_{\bar{f}} \sigma^\nu y_f^\dagger \Big) \,. \tag{17.5}$$

Summing over the antifermion helicity using eqs. (2.63)–(2.66) gives

$$\sum_{\lambda_{\bar{f}}} |\mathcal{M}|^2 = \frac{g^2}{c_W^2} \varepsilon_\mu \varepsilon_\nu^* \Big(a_f^2 x_f^\dagger \overline{\sigma}^\mu k_{\bar{f}} \cdot \sigma \overline{\sigma}^\nu x_f + b_f^2 y_f \sigma^\mu k_{\bar{f}} \cdot \overline{\sigma} \sigma^\nu y_f^\dagger$$
$$- m_f a_f b_f x_f^\dagger \overline{\sigma}^\mu \sigma^\nu y_f^\dagger - m_f a_f b_f y_f \sigma^\mu \overline{\sigma}^\nu x_f \Big) \,. \tag{17.6}$$

Next, we sum over the fermion helicity:

$$\sum_{\lambda_f,\lambda_{\bar{f}}} |\mathcal{M}|^2 = \frac{g^2}{c_W^2}\varepsilon_\mu\varepsilon_\nu^*\Big(a_f^2\,\mathrm{Tr}[\bar\sigma^\mu k_{\bar f}\!\cdot\!\sigma\bar\sigma^\nu k_f\!\cdot\!\sigma] + b_f^2\,\mathrm{Tr}[\sigma^\mu k_{\bar f}\!\cdot\!\bar\sigma\sigma^\nu k_f\!\cdot\!\bar\sigma]$$

$$- m_f^2 a_f b_f\,\mathrm{Tr}[\bar\sigma^\mu\sigma^\nu] - m_f^2 a_f b_f\,\mathrm{Tr}[\sigma^\mu\bar\sigma^\nu]\Big). \tag{17.7}$$

Averaging over the Z helicity using eq. (E.234) and applying eqs. (1.92)–(1.94), we obtain

$$\frac{1}{3}\sum_{\lambda_f,\lambda_{\bar{f}}} |\mathcal{M}|^2 = \frac{g^2}{3c_W^2}\left[(a_f^2+b_f^2)\left(2k_f\!\cdot\!k_{\bar f} + 4\,k_f\!\cdot\!p\,k_{\bar f}\!\cdot\!p/m_Z^2\right) + 12a_f b_f m_f^2\right]$$

$$= \frac{2g^2}{3c_W^2}\left[(a_f^2+b_f^2)(m_Z^2 - m_f^2) + 6a_f b_f m_f^2\right], \tag{17.8}$$

where we have used eq. (17.1). The well-known result for the partial width of the Z is obtained after employing eq. (D.90):

$$\Gamma(Z\to f\bar f) = \frac{N_c^f}{16\pi m_Z}\left(1-\frac{4m_f^2}{m_Z^2}\right)^{1/2}\left(\frac{1}{3}\sum_{\lambda_f,\lambda_{\bar f}}|\mathcal{M}|^2\right)$$

$$= \frac{N_c^f g^2 m_Z}{24\pi c_W^2}\left(1-\frac{4m_f^2}{m_Z^2}\right)^{1/2}\left[(a_f^2+b_f^2)\left(1-\frac{m_f^2}{m_Z^2}\right) + 6a_f b_f\frac{m_f^2}{m_Z^2}\right]. \tag{17.9}$$

Here we have also included a factor of N_c^f (equal to 1 for leptons and 3 for quarks) for the sum over colors.

17.3 Bhabha Scattering: $e^+e^- \to e^+e^-$

In our next example, we consider the computation of Bhabha scattering in QED (that is, we consider photon exchange but neglect Z exchange) We denote the initial-state electron and positron momenta and helicities by (p_1,λ_1) and (p_2,λ_2) and the final-state electron and positron momenta and helicities by (p_3,λ_3) and (p_4,λ_4), respectively. In this calculation, we shall neglect the electron mass by setting $p_i^2=0$ for $i=1,\ldots,4$. Then, the dot products of momenta can be expressed in terms of the Mandelstam variables using eqs. (D.28)–(D.30):

$$p_1\!\cdot\!p_2 = p_3\!\cdot\!p_4 \equiv \tfrac{1}{2}s\,, \tag{17.10}$$

$$p_1\!\cdot\!p_3 = p_2\!\cdot\!p_4 \equiv -\tfrac{1}{2}t\,, \tag{17.11}$$

$$p_1\!\cdot\!p_4 = p_2\!\cdot\!p_3 \equiv -\tfrac{1}{2}u\,. \tag{17.12}$$

There are eight distinct Feynman diagrams. First, there are four s-channel diagrams, as shown in Fig. 17.5, with amplitudes that follow from the Feynman rules

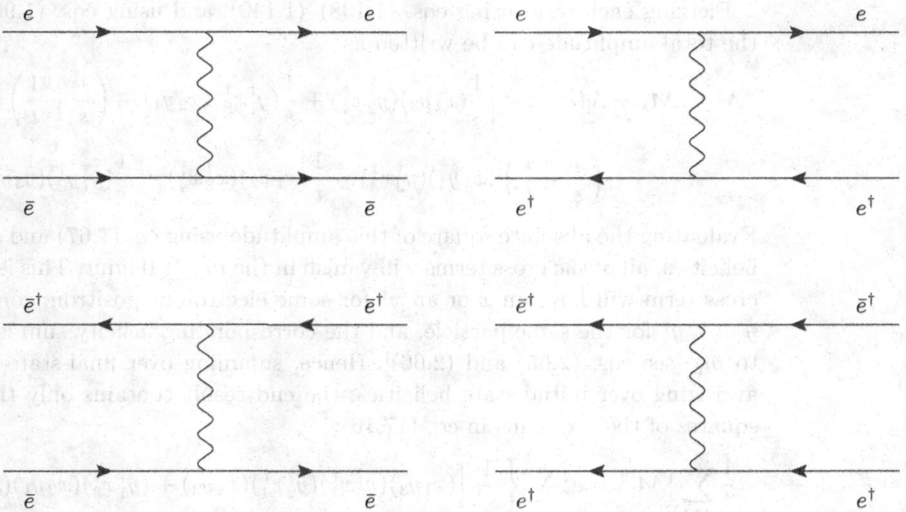

Fig. 17.7 Tree-level t-channel Feynman diagrams for $e^+e^- \to e^+e^-$. The momentum flow of the external particles is from left to right.

of Fig. 17.3 (more generally, see Fig. I.2):

$$i\mathcal{M}_s = \left(\frac{-ig^{\mu\nu}}{s}\right)\Big[(-ie\,x_1\sigma_\mu y_2^\dagger)(ie\,y_3\sigma_\nu x_4^\dagger) + (-ie\,y_1^\dagger\overline{\sigma}_\mu x_2)(ie\,y_3\sigma_\nu x_4^\dagger)$$
$$+(-ie\,x_1\sigma_\mu y_2^\dagger)(ie\,x_3^\dagger\overline{\sigma}_\nu y_4) + (-ie\,y_1^\dagger\overline{\sigma}_\mu x_2)(ie\,x_3^\dagger\overline{\sigma}_\nu y_4)\Big], \quad (17.13)$$

where $x_i \equiv x(\vec{\boldsymbol{p}}_i, \lambda_i)$ and $y_i \equiv y(\vec{\boldsymbol{p}}_i, \lambda_i)$, for $i = 1, 4$. In light of Fig. 2.6, the photon propagator in Feynman gauge is $-ig^{\mu\nu}/s$. Here, we have chosen to write the external fermion spinors in the order 1, 2, 3, 4. This dictates in each term the use of either the $\overline{\sigma}$ or σ forms of the Feynman rules of Fig. 17.3. One can group the terms of eq. (17.13) together more compactly:

$$i\mathcal{M}_s = e^2\left(\frac{-ig^{\mu\nu}}{s}\right)\left(x_1\sigma_\mu y_2^\dagger + y_1^\dagger\overline{\sigma}_\mu x_2\right)\left(y_3\sigma_\nu x_4^\dagger + x_3^\dagger\overline{\sigma}_\nu y_4\right). \quad (17.14)$$

There are also four t-channel diagrams, as shown in Fig. 17.7. The corresponding amplitudes for these four diagrams can be written

$$i\mathcal{M}_t = (-1)e^2\left(\frac{-ig^{\mu\nu}}{t}\right)\left(x_1\sigma_\mu x_3^\dagger + y_1^\dagger\overline{\sigma}_\mu y_3\right)\left(x_2\sigma_\nu x_4^\dagger + y_2^\dagger\overline{\sigma}_\nu y_4\right). \quad (17.15)$$

Here, the overall factor of (-1) comes from Fermi–Dirac statistics, since the external fermion wave functions are written in an odd permutation $(1, 3, 2, 4)$ of the original order $(1, 2, 3, 4)$ established by the first term in eq. (17.13).

Fierzing each term using eqs. (1.138)–(1.140), and using eqs. (1.96) and (1.97), the total amplitude can be written as

$$\mathcal{M} = \mathcal{M}_s + \mathcal{M}_t = 2e^2\left[\frac{1}{s}(x_1y_3)(y_2^\dagger x_4^\dagger) + \frac{1}{s}(y_1^\dagger x_3^\dagger)(x_2y_4) + \left(\frac{1}{s}+\frac{1}{t}\right)(y_1^\dagger x_4^\dagger)(x_2y_3)\right.$$
$$\left. + \left(\frac{1}{s}+\frac{1}{t}\right)(x_1y_4)(y_2^\dagger x_3^\dagger) - \frac{1}{t}(x_1x_2)(x_3^\dagger x_4^\dagger) - \frac{1}{t}(y_1^\dagger y_2^\dagger)(y_3y_4)\right]. \quad (17.16)$$

Evaluating the absolute square of this amplitude using eq. (1.67) and summing over helicities, all of the cross terms will vanish in the $m_e \to 0$ limit. This is because each cross term will have an x or an $x^\dagger$ for some electron or positron, combined with a y or a $y^\dagger$ for the same particle, and the corresponding helicity sum is proportional to m_e [see eqs. (2.65) and (2.66)]. Hence, summing over final-state helicities and averaging over initial-state helicities, the end result contains only the sum of the squares of the six terms in eq. (17.16):

$$\frac{1}{4}\sum_{\{\lambda\}}|\mathcal{M}|^2 = e^4\sum_{\{\lambda\}}\left\{\frac{1}{s^2}\left[(x_1y_3)(y_3^\dagger x_1^\dagger)(y_2^\dagger x_4^\dagger)(x_4y_2) + (y_1^\dagger x_3^\dagger)(x_3y_1)(x_2y_4)(y_4^\dagger x_2^\dagger)\right]\right.$$
$$+ \left(\frac{1}{s}+\frac{1}{t}\right)^2\left[(y_1^\dagger x_4^\dagger)(x_4y_1)(x_2y_3)(y_3^\dagger x_2^\dagger) + (x_1y_4)(y_4^\dagger x_1^\dagger)(y_2^\dagger x_3^\dagger)(x_3y_2)\right]$$
$$\left. + \frac{1}{t^2}\left[(x_1x_2)(x_2^\dagger x_1^\dagger)(x_3^\dagger x_4^\dagger)(x_4x_3) + (y_1^\dagger y_2^\dagger)(y_2y_1)(y_3y_4)(y_4^\dagger y_3^\dagger)\right]\right\}, \quad (17.17)$$

where $\Sigma_{\{\lambda\}}$ stands for the fourfold sum over the helicities $\lambda_1, \lambda_2, \lambda_3, \lambda_4$. Performing these helicity sums using eqs. (2.63) and (2.64) and using the trace identities eq. (B.113),

$$\frac{1}{4}\sum_{\{\lambda\}}|\mathcal{M}|^2 = 8e^4\left[\frac{p_2\cdot p_4\, p_1\cdot p_3}{s^2} + \frac{p_1\cdot p_2\, p_3\cdot p_4}{t^2} + \left(\frac{1}{s}+\frac{1}{t}\right)^2 p_1\cdot p_4\, p_2\cdot p_3\right]$$
$$= 2e^4\left[\frac{t^2}{s^2} + \frac{s^2}{t^2} + \left(\frac{u}{s}+\frac{u}{t}\right)^2\right]. \quad (17.18)$$

Thus, the differential cross section for Bhabha scattering is given by

$$\frac{d\sigma}{dt} = \frac{1}{16\pi s^2}\left(\frac{1}{4}\sum_{\{\lambda\}}|\mathcal{M}|^2\right) = \frac{2\pi\alpha^2}{s^2}\left[\frac{t^2}{s^2} + \frac{s^2}{t^2} + \left(\frac{u}{s}+\frac{u}{t}\right)^2\right]. \quad (17.19)$$

17.4 Polarized Muon Decay

So far we have only treated cases where the initial-state fermion helicities or spins are averaged and the final-state helicities or spins are summed. In the case of the polarized decay of a particle or polarized scattering we must project out the appropriate polarization of the particles in the helicity or spin sums. This is achieved

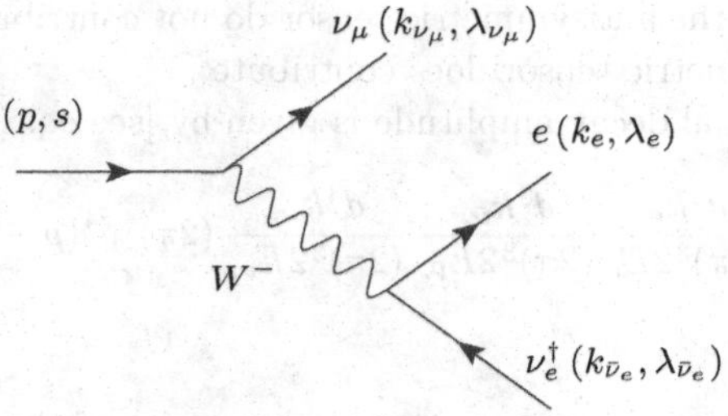

Fig. 17.8 Feynman diagram for electroweak muon decay.

by replacing the helicity/spin sums given in eqs. (2.63)–(2.66) by the relevant projection operators for definite helicity or spin exhibited in eqs. (2.51)–(2.62). As an example, we consider the decay of a polarized muon. The leading-order Feynman diagram for muon decay is shown in Fig. 17.8 (and the relevant four-momenta, the initial-state muon spin, and the final-state helicities are indicated).

In our computation, the mass of the muon is denoted by m_μ, the neutrinos are assumed to be massless, and the electron mass is neglected. The spin of the muon is measured in its rest frame with respect to a fixed z-axis. Assume that the muon at rest is polarized such that its spin component along the $\hat{\boldsymbol{z}}$-direction is $s = +1/2$. Then, the decay amplitude is given by[1]

$$i\mathcal{M} = \left(\frac{-ig}{\sqrt{2}}\right)^2 \left(x^\dagger_{\nu_\mu}\overline{\sigma}_\rho x_\mu\right)\left(x^\dagger_e\overline{\sigma}_\tau y_{\bar{\nu}_e}\right)\left(\frac{-ig^{\rho\tau}}{D_W}\right), \tag{17.20}$$

where $D_W = (p - k_{\nu_\mu})^2 - m_W^2$ is the denominator of the W-boson propagator. In eq. (17.20), $x_\mu \equiv x(\vec{\boldsymbol{p}}, s = \frac{1}{2})$ for the spin-polarized initial-state muon, and $x^\dagger_{\nu_\mu} \equiv x(\vec{\boldsymbol{k}}_{\nu_\mu}, \lambda_{\nu_\mu})$, $x^\dagger_e \equiv x^\dagger(\vec{\boldsymbol{k}}_e, \lambda_e)$, and $y_{\bar{\nu}_e} \equiv y(\vec{\boldsymbol{k}}_{\bar{\nu}_e}, \lambda_{\bar{\nu}_e})$. Evaluating the absolute square of this amplitude using eq. (1.106), we obtain

$$|\mathcal{M}|^2 = \frac{g^4}{4D_W^2}\left(x^\dagger_{\nu_\mu}\overline{\sigma}^\rho x_\mu\right)\left(x^\dagger_\mu\overline{\sigma}^\tau x_{\nu_\mu}\right)\left(x^\dagger_e\overline{\sigma}_\rho y_{\bar{\nu}_e}\right)\left(y^\dagger_{\bar{\nu}_e}\overline{\sigma}_\tau x_e\right). \tag{17.21}$$

Summing over the neutrino and electron helicities using eqs. (2.63)–(2.64), and using eq. (2.51) for the muon spin (with $s = 1/2$) yields

$$\begin{aligned}\sum_{\{\lambda\}}|\mathcal{M}|^2 &= \frac{g^4}{8D_W^2}\mathrm{Tr}[k_{\nu_\mu}\!\cdot\sigma\,\overline{\sigma}^\rho(p\cdot\sigma - m_\mu S\cdot\sigma)\overline{\sigma}^\tau]\,\mathrm{Tr}[k_e\!\cdot\sigma\,\overline{\sigma}_\rho k_{\bar{\nu}_e}\!\cdot\sigma\,\overline{\sigma}_\tau] \\ &= \frac{2g^4}{D_W^2}\,k_e\!\cdot k_{\nu_\mu}\,k_{\bar{\nu}_e}\!\cdot(p - m_\mu S)\,,\end{aligned} \tag{17.22}$$

where S^μ (in an arbitrary reference frame) is given by eq. (2.18) [with $\hat{\boldsymbol{s}} = \hat{\boldsymbol{z}}$] and $\Sigma_{\{\lambda\}}$ indicates a three-fold sum over the helicities λ_e, λ_{ν_μ}, and $\lambda_{\bar{\nu}_e}$. To obtain the second line of eq. (17.22) we have used eq. (1.93) twice. Note that the resulting

[1] Throughout this section μ and ν are particle labels. Hence, we employ ρ and τ as Lorentz vector indices.

terms linear in the antisymmetric tensor do not contribute, but the term quadratic in the antisymmetric tensor does contribute.

The differential decay amplitude is given by [see eq. (D.87)]

$$d\Gamma = \frac{1}{2m_\mu}\frac{d^3\boldsymbol{k_e}}{(2\pi)^3 2E_e}\frac{d^3\boldsymbol{k_{\bar{\nu}_e}}}{(2\pi)^3 2E_{\bar{\nu}_e}}\frac{d^3\boldsymbol{k_{\nu_\mu}}}{(2\pi)^3 2E_{\nu_\mu}}(2\pi)^4\delta^4(p-k_e-k_{\bar{\nu}_e}-k_{\nu_\mu})\sum_{\{\lambda\}}|\mathcal{M}|^2\,, \tag{17.23}$$

where the E_i, (for $i = e, \bar{\nu}_e, \nu_\mu$) are the energies of the final-state particles in the muon rest frame. Note that, in contrast with eq. (D.87), we do not average over the initial-state muon spins, since it is specified. In light of $m_W \gg m_\mu$, it is a very good approximation to take $D_W^2 \simeq m_W^4$. We can now use eq. (D.132) (with the neutrino masses set to zero) to integrate over the neutrino momenta:

$$\int \frac{d^3\boldsymbol{k_{\bar{\nu}_e}}}{(2\pi)^3 2E_{\bar{\nu}_e}}\frac{d^3\boldsymbol{k_{\nu_\mu}}}{(2\pi)^3 2E_{\nu_\mu}}(2\pi)^4\delta^4(q-k_{\bar{\nu}_e}-k_{\nu_\mu})k^\rho_{\bar{\nu}_e}k^\tau_{\nu_\mu} = \frac{1}{96\pi}(q^2 g^{\rho\tau}+2q^\rho q^\tau)\,, \tag{17.24}$$

where $q = p - k_e$. It then follows that

$$d\Gamma = \frac{g^4}{1536\pi^4 m_\mu m_W^4}\left[q^2\, k_e\cdot(p-m_\mu S) + 2q\cdot k_e\, q\cdot(p-m_\mu S)\right]\frac{d^3\boldsymbol{k_e}}{E_e}\,. \tag{17.25}$$

In the muon rest frame, $k_e = E_e(1;\sin\theta\cos\phi,\sin\theta\sin\phi,\cos\theta)$ and $S = (0;0,0,1)$, so that

$$\begin{aligned} q^2 &= m_\mu^2 - 2E_e m_\mu\,, & k_e\cdot(p-m_\mu S) &= m_\mu E_e(1+\cos\theta)\,,\\ q\cdot k_e &= m_\mu E_e\,, & q\cdot(p-m_\mu S) &= m_\mu(m_\mu - E_e - E_e\cos\theta). \end{aligned} \tag{17.26}$$

Noting that the maximum energy of the electron is $m_\mu/2$ (when the neutrino and antineutrino both recoil in the direction opposite to that of the electron), we obtain

$$\begin{aligned} \frac{d\Gamma}{d\cos\theta} &= \frac{g^4 m_\mu^2}{768\pi^3 m_W^4}\int_0^{m_\mu/2} dE_e E_e^2\left[3-\frac{4E_e}{m_\mu}+\left(1-\frac{4E_e}{m_\mu}\right)\cos\theta\right]\\ &= \frac{g^4 m_\mu^5}{3\cdot 2^{12}\pi^3 m_W^4}\left(1-\tfrac{1}{3}\cos\theta\right)\,. \end{aligned} \tag{17.27}$$

In terms of the Fermi constant [see eq. 4.132], we can rewrite eq. (17.27) as

$$\frac{d\Gamma}{d\cos\theta} = \frac{G_F^2 m_\mu^5}{384\pi^3}\left(1-\tfrac{1}{3}\cos\theta\right)\,. \tag{17.28}$$

Integrating over $\cos\theta$ reproduces the well-known formula for the total muon decay width,

$$\Gamma = \frac{G_F^2 m_\mu^5}{192\pi^3}\,. \tag{17.29}$$

17.5 Top-Quark Condensation in a Nambu–Jona–Lasinio Model

The previous examples have involved renormalizable field theories. However, there are cases in which it is preferable to use effective four-fermion interactions. The obvious historical example is the four-fermion Fermi theory of weak decays. This has been superseded by a more complete and accurate theory of the weak interactions but is still useful for leading-order calculations of low-energy processes. Another case of some interest is the use of strong coupling four-fermion interactions to drive symmetry breaking via a Nambu–Jona–Lasinio model (see, e.g., Chapter 6 of [B145]), as in the top-quark condensate approach to electroweak symmetry breaking [196].

Consider an effective four-fermion Lagrangian involving the top quark, written in two-component fermion form as

$$\mathscr{L} = it^\dagger\bar{\sigma}^\mu\partial_\mu t + i\bar{t}^\dagger\bar{\sigma}^\mu\partial_\mu\bar{t} + \frac{G}{\Lambda^2}(t\bar{t})(t^\dagger\bar{t}^\dagger). \tag{17.30}$$

Here the Standard Model gauge interactions have been suppressed; the quantities within parentheses are color singlets. Note also that there is no top-quark Yukawa coupling to a Higgs scalar boson, nor a top-quark mass term, which would normally appear in the form $-m_t(t\bar{t}+t^\dagger\bar{t}^\dagger)$. Instead, the effective top-quark mass is supposed to be driven by a nonperturbatively large and positive dimensionless coupling G, with Λ the cutoff scale at which G arises from some more fundamental physics.

The Feynman rule for the four-fermion interaction is given in Fig. 17.9. The resulting gap equation for the dynamically generated top-quark mass is shown in Fig. 17.10. Evaluating this using the Feynman rules of Figs. 2.4 and 2.5, one finds

$$-im_t\delta_i^j\delta_\alpha^\beta = (-1)\int^\Lambda \frac{d^4k}{(2\pi)^4}\left(i\frac{G}{\Lambda^2}\delta_i^j\delta_n^k\delta_\alpha^\beta\delta_{\dot{\alpha}}^{\dot{\beta}}\right)\left(\delta_k^n\delta_{\dot{\beta}}^{\dot{\alpha}}\frac{im_t}{k^2-m_t^2+i\epsilon}\right). \tag{17.31}$$

Here i, j, k, n are color indices of the fundamental representation of $SU(3)$, and $\alpha, \beta, \dot{\alpha}, \dot{\beta}$ are two-component spinor indices. The factor of (-1) on the right-hand side is due to the presence of a fermion loop.

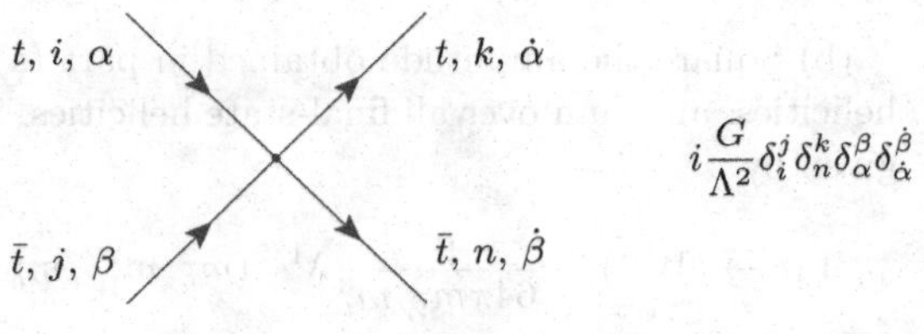

Fig. 17.9 Feynman rule for the four-fermion interaction in the top-quark condensate model. The indices $i, j, k, n \in \{1, 2, 3\}$ are for color in the fundamental representation of $SU(3)$, and the indices $\alpha, \beta, \dot{\alpha}, \dot{\beta}$ are two-component spinor indices.

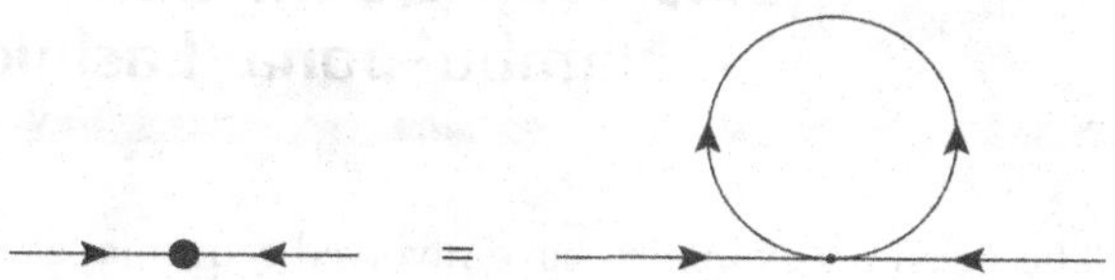

Fig. 17.10 The Nambu–Jona–Lasinio gap equation for a possible dynamically generated top-quark mass m_t.

Euclideanizing the loop integration over k^μ by $k^2 \to -k_E^2$ and $\int d^4k \to i \int d^4k_E$, and then rewriting the integration in terms of $x = k_E^2$, this amounts to

$$m_t = \frac{N_c G m_t}{16\pi^2\Lambda^2}\delta^{\dot{\alpha}}_{\dot{\beta}}\delta^{\dot{\alpha}}_{\dot{\beta}}\int_0^{\Lambda^2}\frac{x\,dx}{x+m_t^2} = \frac{3Gm_t}{8\pi^2}\left[1 - \frac{m_t^2}{\Lambda^2}\ln\left(\frac{\Lambda^2}{m_t^2}\right) + \cdots\right], \tag{17.32}$$

after summing over the respective dotted spinor and color indices (with $N_c = 3$).

For small or negative G, only the trivial solution $m_t = 0$ is possible. However, for $G \geq G_{\text{critical}} = 8\pi^2/3 \approx 26$, a positive solution for m_t^2/Λ^2 exists. Although this minimal version of the model cannot explain the top-quark mass and the observed features of electroweak symmetry breaking, extensions of this model may be viable [196].

Exercises

17.1 Consider the decay width of a top quark into a bottom quark and W^+ vector boson. For simplicity, treat this as a one-generation exercise and ignore the CKM mixing among the three quark generations.

(a) Using two-component fermion Feynman rules, show that the decay amplitude for $t \to bW^+$ is given by

$$i\mathcal{M} = -i\frac{g}{\sqrt{2}}\varepsilon^*_\mu x_b^\dagger \bar{\sigma}^\mu x_t\,, \tag{17.33}$$

where ε_μ is the polarization vector of the W^+, and $x_b^\dagger$ and x_t are the external-state spinor wave functions of the bottom and top quark, respectively.

(b) Square the amplitude obtained in part (a), average over the top-quark helicities, and sum over all final-state helicities. Show that the decay width is given by

$$\Gamma(t \to bW^+) = \frac{g^2}{64\pi m_W^2 m_t^3}\lambda^{1/2}(m_t^2, m_W^2, m_b^2)\Big[\lambda(m_t^2, m_W^2, m_b^2) + 3m_W^2(m_t^2 + m_b^2 - m_W^2)\Big], \tag{17.34}$$

where the kinematic triangle function λ is defined in eq. (D.1).

(c) Check that, in the limit of $m_b = 0$, one obtains the well-known result

$$\Gamma(t \to bW^+) = \frac{g^2 m_t}{64\pi}\left(2 + \frac{m_t^2}{m_W^2}\right)\left(1 - \frac{m_W^2}{m_t^2}\right)^2 , \qquad (17.35)$$

which exhibits the Nambu–Goldstone enhancement factor (m_t^2/m_W^2) for the longitudinal W contribution compared to the two transverse W contributions.

17.2 Equation (17.27) can also be derived by making use of eq. (D.67).

(a) Label the momenta of final-state particles e, ν_μ, and $\bar{\nu}_e$ by 1, 2, and 3, respectively. Show that if all final-state particle masses are neglected,

$$\Gamma = \frac{1}{512\pi^5 m}\int_0^{\frac{1}{2}m} dE_1 \int_{\frac{1}{2}m - E_1}^{\frac{1}{2}m} dE_3 \int d\Omega_1 \int_0^{2\pi} d\phi_{13} \sum_{\{\lambda\}} |\mathcal{M}|^2 , \qquad (17.36)$$

where the E_i are the energies of the final-state particles in the muon rest frame and $m \equiv m_\mu$. Note that $\Omega_1 = (\theta_1, \phi_1)$ is the solid angle of $\vec{p}_1$ with respect to the fixed z-axis and (θ_{13}, ϕ_{13}) are the polar and azimuthal angles of $\vec{p}_3$ with respect to an axis parallel to $\vec{p}_1$.

(b) After approximating $D_W^2 \simeq m_W^4$, verify that in the muon rest frame where $S = (0; 0, 0, 1)$ defines the fixed z-axis, eq. (17.22) yields

$$\sum_{\{\lambda\}} |\mathcal{M}|^2 = \frac{g^4 m^2 E_3 (m - 2E_3)(1 + \cos\theta_3)}{m_W^4} , \qquad (17.37)$$

where $\cos\theta_3 = \hat{\boldsymbol{p}}_3 \cdot \hat{\boldsymbol{z}}$.

(c) In light of eq. (D.65), show that

$$\cos\theta_{13} = 1 - \frac{m(E_1 + E_3 - \frac{1}{2}m)}{E_1 E_3} . \qquad (17.38)$$

Define the z'-axis to lie along $\vec{p}_1$ with the z-axis lying in the x'–z' plane. Then (θ_{13}, ϕ_{13}) are the polar and azimuthal angles of $\vec{p}_3$ with respect to the z'-axis and θ_1 is the polar angle of the z-axis with respect to the z'-axis. Derive the following equation that relates $\cos\theta_3$ to θ_1 and (θ_{13}, ϕ_{13}):

$$\cos\theta_3 = \cos\theta_1 \cos\theta_{13} + \sin\theta_1 \sin\theta_{13} \cos\phi_{13} . \qquad (17.39)$$

(d) Using the results of parts (a)–(c), show that

$$\frac{d\Gamma}{d\Omega_1} = \frac{g^4 m}{256\pi^4 m_W^4}\int_0^{\frac{1}{2}m} \frac{dE_1}{E_1}\int_{\frac{1}{2}m - E_1}^{\frac{1}{2}m} dE_3\, (m - 2E_3)$$
$$\times \left[E_1 E_3 + \left(E_1 E_3 - m(E_1 + E_3) + \tfrac{1}{2}m^2\right)\cos\theta_1\right] . \qquad (17.40)$$

Carry out the integration over E_3 and ϕ_1 and confirm the result obtained in eq. (17.27).

18 Tree-Level Supersymmetric Processes

In this chapter, we present example Feynman diagrammatic calculations of supersymmetric decay and scattering processes, employing the two-component fermion techniques developed in Chapter 2. We present the first calculations in some detail to get the reader acquainted with the technical details. In all cases, the fermion lines in Feynman diagrams are labeled by two-component field names, rather than the particle names, as explained in Section 17.1. We shall denote the incoming momentum of particle i by p_i and outgoing momenta of particle f by k_f. The spin-1/2 helicities are labeled λ_i and λ_f, respectively.

All the computations presented in this chapter can also be carried out using four-component fermion techniques. In cases where Majorana fermions appear, the corresponding calculations can be performed unambiguously by employing the Feynman rules for four-component Majorana fermions developed in Chapter 3. The reader can check that these results reproduce the ones exhibited in this chapter.

18.1 Sneutrino Decay: $\widetilde{\nu}_e \to \widetilde{C}_i^+ e^-$

Consider the two-body decay of an electron-type sneutrino to a chargino and an electron: $\widetilde{\nu}_e \to \widetilde{C}_i^+ e^-$. Because only the left-handed electron can couple to the chargino and sneutrino (with the excellent approximation that the electron Yukawa coupling can be neglected), there is just one Feynman diagram, shown in Fig. 18.1. The external wave functions of the electron and of the chargino are denoted as $x_e \equiv x(\vec{\boldsymbol{k}}_e, \lambda_e)$ and $x_{\widetilde{C}} \equiv x(\vec{\boldsymbol{k}}_{\widetilde{C}}, \lambda_{\widetilde{C}})$, respectively. From the corresponding Feynman rule given in Fig. J.6, the amplitude is

$$i\mathcal{M} = -igV_{i1}\, x^\dagger_{\widetilde{C}} x^\dagger_e, \tag{18.1}$$

where V is one of the two matrices used in the diagonalization of the chargino mass matrix [see eq. (J.19)]. Computing the absolute square of the amplitude yields

$$|\mathcal{M}|^2 = g^2|V_{i1}|^2\,(x^\dagger_{\widetilde{C}} x^\dagger_e)(x_e x_{\widetilde{C}})\,. \tag{18.2}$$

Summing over the electron and chargino spin polarizations using eq. (2.63) and contracting all the indices, we end up with

$$\sum_{\lambda_e,\lambda_{\widetilde{C}}} |\mathcal{M}|^2 = g^2|V_{i1}|^2 \mathrm{Tr}[k_e\cdot\overline{\sigma}\, k_{\widetilde{C}}\cdot\sigma] = 2g^2|V_{i1}|^2\, k_e\cdot k_{\widetilde{C}} = g^2|V_{i1}|^2(m^2_{\widetilde{\nu}_e} - m^2_{\widetilde{C}_i}). \tag{18.3}$$

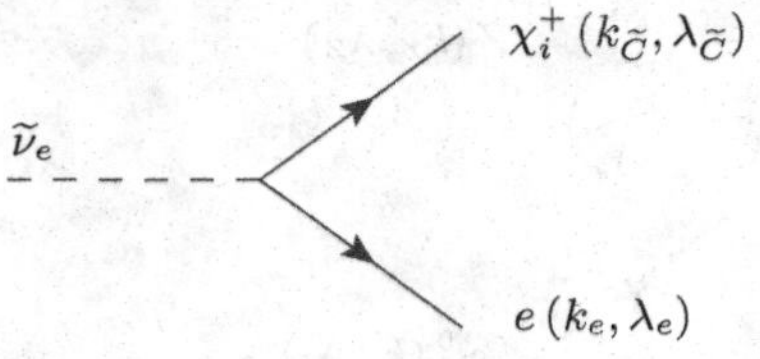

Fig. 18.1 The Feynman diagram for the decay $\widetilde{\nu}_e \to \widetilde{C}_i^+ e^-$ in the MSSM.

After employing eq. (D.90), the sneutrino decay width is given by

$$\Gamma(\widetilde{\nu}_e \to \widetilde{C}_i^+ e^-) = \frac{1}{16\pi m_{\widetilde{\nu}_e}} \left(1 - \frac{m^2_{\widetilde{C}_i}}{m^2_{\widetilde{\nu}_e}}\right) \sum_{\lambda_e, \lambda_{\widetilde{C}}} |\mathcal{M}|^2 = \frac{g^2}{16\pi} |V_{i1}|^2 m_{\widetilde{\nu}_e} \left(1 - \frac{m^2_{\widetilde{C}_i}}{m^2_{\widetilde{\nu}_e}}\right)^2 . \tag{18.4}$$

18.2 $\widetilde{N}_i \to Z\widetilde{N}_j$

For this two-body decay we assume that $m_{\widetilde{N}_i} > m_{\widetilde{N}_j} + m_Z$. There are two tree-level Feynman diagrams, shown in Fig. 18.2 with the definitions of the helicities and the momenta. Using the Feynman rules of Fig. J.3, the two amplitudes are given by[1]

$$i\mathcal{M}_1 = -i\frac{g}{c_W} \mathcal{O}''^L_{ji} x_i \sigma^\mu x_j^\dagger \varepsilon^*_\mu , \tag{18.5}$$

$$i\mathcal{M}_2 = i\frac{g}{c_W} \mathcal{O}''^L_{ij} y_i^\dagger \overline{\sigma}^\mu y_j \varepsilon^*_\mu , \tag{18.6}$$

where the fermion external wave functions are $x_i = x(\vec{\boldsymbol{p}}_i, \lambda_i)$, $y_i^\dagger = y^\dagger(\vec{\boldsymbol{p}}_i, \lambda_i)$, $x_j^\dagger = x^\dagger(\vec{\boldsymbol{k}}_j, \lambda_j)$, $y_j = y(\vec{\boldsymbol{k}}_j, \lambda_j)$; and the Z-boson wave function is $\varepsilon^*_\mu = \varepsilon_\mu(\vec{\boldsymbol{k}}_Z, \lambda_Z)^*$. Noting that $\mathcal{O}''^L_{ji} = \mathcal{O}''^{L*}_{ij}$ [see eq. (J.18)], and applying eqs. (1.105) and (1.106), we find that the squared matrix element is

$$|\mathcal{M}|^2 = \frac{g^2}{c_W^2} \varepsilon^*_\mu \varepsilon_\nu \Big[|\mathcal{O}''^L_{ij}|^2 (x_i \sigma^\mu x_j^\dagger x_j \sigma^\nu x_i^\dagger + y_i^\dagger \overline{\sigma}^\mu y_j y_j^\dagger \overline{\sigma}^\nu y_i) - \left(\mathcal{O}''^L_{ij}\right)^2 y_i^\dagger \overline{\sigma}^\mu y_j x_j \sigma^\nu x_i^\dagger - \left(\mathcal{O}''^{L*}_{ij}\right)^2 x_i \sigma^\mu x_j^\dagger y_j^\dagger \overline{\sigma}^\nu y_i \Big] . \tag{18.7}$$

Summing over the final-state neutralino spin using eqs. (2.63)–(2.66) yields

$$\sum_{\lambda_j} |\mathcal{M}|^2 = \frac{g^2}{c_W^2} \varepsilon^*_\mu \varepsilon_\nu \Big[|\mathcal{O}''^L_{ij}|^2 (x_i \sigma^\mu k_j \cdot \overline{\sigma} \sigma^\nu x_i^\dagger + y_i^\dagger \overline{\sigma}^\mu k_j \cdot \sigma \overline{\sigma}^\nu y_i) + \left(\mathcal{O}''^L_{ij}\right)^2 m_{\widetilde{N}_j} y_i^\dagger \overline{\sigma}^\mu \sigma^\nu x_i^\dagger + \left(\mathcal{O}''^{L*}_{ij}\right)^2 m_{\widetilde{N}_j} x_i \sigma^\mu \overline{\sigma}^\nu y_i \Big] . \tag{18.8}$$

[1] When comparing with the four-component Feynman rule of Ref. [10], note that $\mathcal{O}''^L_{ij} = -\mathcal{O}''^{R*}_{ij}$ [see eq. (J.18)].

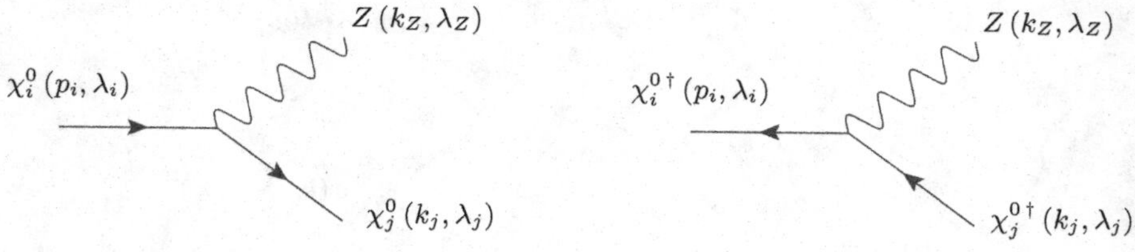

Fig. 18.2 The Feynman diagrams for $\widetilde{N}_i \to \widetilde{N}_j Z$ in the MSSM.

Averaging over the initial-state neutralino spins in the same way gives

$$\begin{aligned}\frac{1}{2}\sum_{\lambda_i,\lambda_j} |\mathcal{M}|^2 &= \frac{g^2}{2c_W^2}\varepsilon_\mu^*\varepsilon_\nu \Big[|\mathcal{O}_{ij}^{\prime\prime L}|^2 \Big(\mathrm{Tr}[\sigma^\mu k_j\!\cdot\!\overline{\sigma}\sigma^\nu p_i\!\cdot\!\overline{\sigma}] + \mathrm{Tr}[\overline{\sigma}^\mu k_j\!\cdot\!\sigma\overline{\sigma}^\nu p_i\!\cdot\!\sigma]\Big) \\ &\qquad - \big(\mathcal{O}_{ij}^{\prime\prime L}\big)^2 m_{\widetilde{N}_i} m_{\widetilde{N}_j}\mathrm{Tr}[\overline{\sigma}^\mu\sigma^\nu] - \big(\mathcal{O}_{ij}^{\prime\prime L*}\big)^2 m_{\widetilde{N}_i} m_{\widetilde{N}_j}\mathrm{Tr}[\sigma^\mu\overline{\sigma}^\nu]\Big] \\ &= \frac{2g^2}{c_W^2}\varepsilon_\mu^*\varepsilon_\nu \Big\{|\mathcal{O}_{ij}^{\prime\prime L}|^2\,(k_j^\mu p_i^\nu + p_i^\mu k_j^\nu - p_i\!\cdot\! k_j g^{\mu\nu}) \\ &\qquad - \mathrm{Re}\Big[\big(\mathcal{O}_{ij}^{\prime\prime L}\big)^2\Big] m_{\widetilde{N}_i} m_{\widetilde{N}_j} g^{\mu\nu}\Big\}, \end{aligned} \tag{18.9}$$

where in the last equality we have applied eqs. (1.92)–(1.94). Using eq. (E.234) we obtain

$$\begin{aligned}\frac{1}{2}\sum_{\lambda_i,\lambda_j,\lambda_Z} |\mathcal{M}|^2 = \frac{2g^2}{c_W^2}\Big\{&|\mathcal{O}_{ij}^{\prime\prime L}|^2\,(p_i\!\cdot\! k_j + 2p_i\!\cdot\! k_Z k_j\!\cdot\! k_Z/m_Z^2) \\ &+ 3m_{\widetilde{N}_i} m_{\widetilde{N}_j}\,\mathrm{Re}\big[\big(\mathcal{O}_{ij}^{\prime\prime L}\big)^2\big]\Big\}. \end{aligned} \tag{18.10}$$

Employing eqs. (D.3)–(D.5), we obtain the total decay width,

$$\begin{aligned}\Gamma(\widetilde{N}_i \to Z\widetilde{N}_j) &= \frac{1}{16\pi m_{\widetilde{N}_i}^3}\lambda^{1/2}(m_{\widetilde{N}_i}^2, m_Z^2, m_{\widetilde{N}_j}^2)\left(\frac{1}{2}\sum_{\lambda_i,\lambda_j,\lambda_Z}|\mathcal{M}|^2\right) \\ &= \frac{g^2 m_{\widetilde{N}_i}}{16\pi c_W^2}\lambda^{1/2}(1, r_Z, r_j)\Big[|\mathcal{O}_{ij}^{\prime\prime L}|^2\,(1 + r_j - 2r_Z + (1-r_j)^2/r_Z) \\ &\qquad + 6\,\mathrm{Re}\big[\big(\mathcal{O}_{ij}^{\prime\prime L}\big)^2\big]\sqrt{r_j}\Big], \end{aligned} \tag{18.11}$$

where

$$r_j \equiv m_{\widetilde{N}_j}^2/m_{\widetilde{N}_i}^2\,, \qquad r_Z \equiv m_Z^2/m_{\widetilde{N}_i}^2\,, \tag{18.12}$$

and the kinematic triangle function $\lambda^{1/2}$ is defined in eq. (D.1).

18.3 $\widetilde{N}_i \to \widetilde{N}_j\widetilde{N}_k\widetilde{N}_\ell$

Next we consider the decay of a neutralino $\widetilde{N}_i$ to three lighter neutralinos: $\widetilde{N}_j, \widetilde{N}_k, \widetilde{N}_\ell$. This decay is not likely to be phenomenologically relevant, because a variety of two-body decay modes will always be available. Furthermore, the calculation itself is quite complicated because of the large number of Feynman diagrams involved. Therefore, we consider this only as a matter-of-principle example of a process with four external-state Majorana fermions, and will restrict ourselves to writing down the contributing matrix element amplitudes.

At tree level, the decay can proceed via a virtual Z boson; the Feynman graphs are shown in Fig. 18.3. In addition, it can proceed via the exchange of any of the neutral scalar Higgs bosons of the MSSM, $\phi = h, H, A$, as shown in Fig. 18.4. Since any of the final-state neutralinos can directly couple to the initial-state neutralino there are two more diagrams for each one shown in Figs. 18.3 and 18.4, for a total of 48 tree-level diagrams (counting each intermediate Higgs boson state as distinct). In all cases, the four-momenta of the neutralinos $\widetilde{N}_i$, $\widetilde{N}_j$, $\widetilde{N}_k$, $\widetilde{N}_\ell$ are denoted p_i, k_j, k_k, k_ℓ, respectively.

We obtain the sum of the four diagrams in Fig. 18.3 by implementing the rules

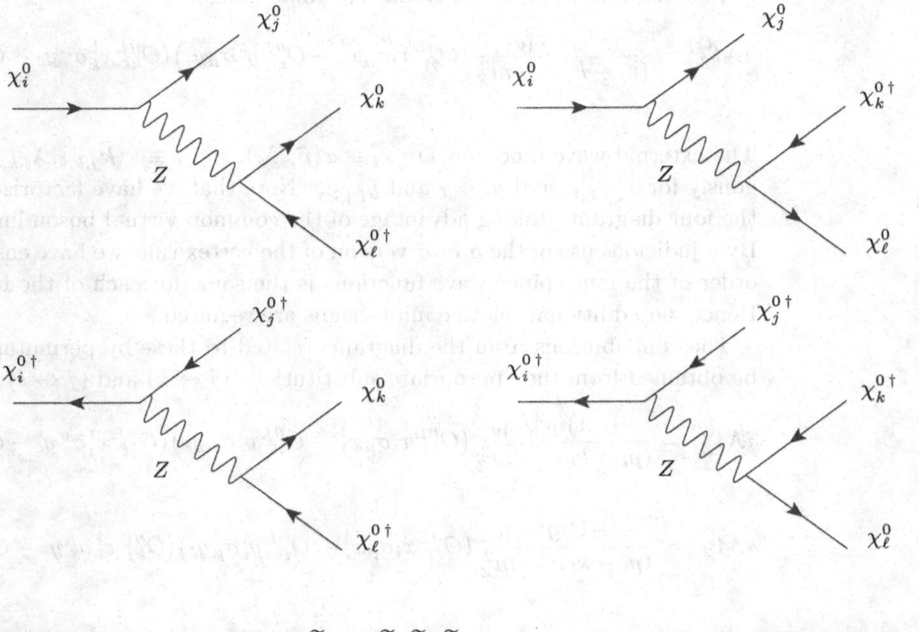

Fig. 18.3 Four Feynman diagrams for $\widetilde{N}_i \to \widetilde{N}_j\widetilde{N}_k\widetilde{N}_\ell$ in the MSSM via Z exchange. There are four more where $\widetilde{N}_j \leftrightarrow \widetilde{N}_k$ and another four where $\widetilde{N}_j \leftrightarrow \widetilde{N}_\ell$.

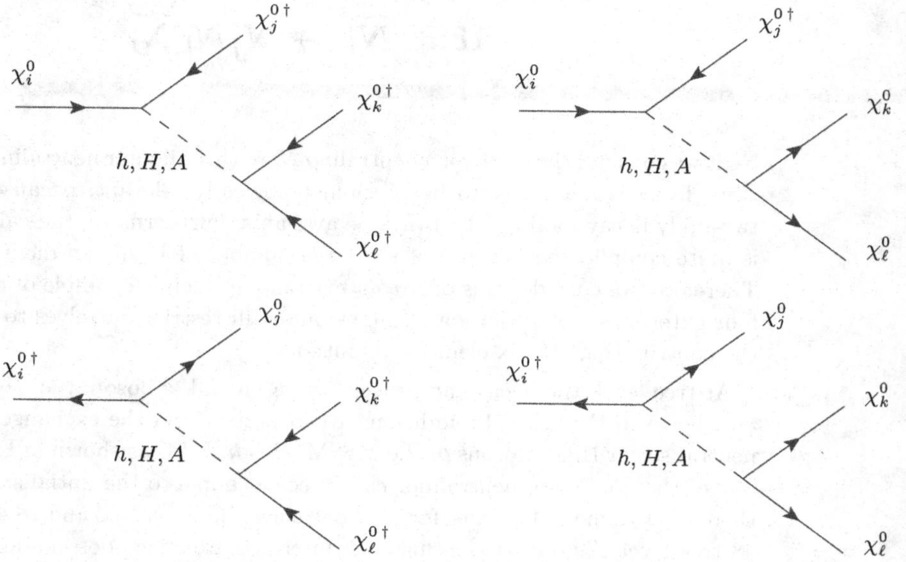

Fig. 18.4 Four Feynman diagrams for $\widetilde{N}_i \to \widetilde{N}_j \widetilde{N}_k \widetilde{N}_\ell$ in the MSSM via $\phi = h,\, H,\, A$ exchange. There are four more where $\widetilde{N}_j \leftrightarrow \widetilde{N}_k$ and another four where $\widetilde{N}_j \leftrightarrow \widetilde{N}_\ell$.

of Fig. J.3, and using the 't Hooft–Feynman gauge:

$$i\mathcal{M}_Z^{(1)} = \frac{-ig^2/c_W^2}{(p_i - k_j)^2 - m_Z^2} \left(O''^L_{ji} x_i \sigma_\mu x_j^\dagger - O''^L_{ij} y_i^\dagger \overline{\sigma}_\mu y_j\right)\left(O''^L_{k\ell} x_k^\dagger \overline{\sigma}^\mu y_\ell - O''^L_{\ell k} y_k \sigma^\mu x_\ell^\dagger\right). \tag{18.13}$$

The external wave functions are $x_i \equiv x(\vec{\boldsymbol{p}}_i, \lambda_i)$, $x_{j,k,\ell} \equiv x(\vec{\boldsymbol{k}}_{j,k,\ell}, \lambda_{j,k,\ell})$, and analogously for $x^\dagger_{i,j,k,\ell}$, and $y_{i,j,k,\ell}$ and $y^\dagger_{i,j,k,\ell}$. Note that we have factorized the sum of the four diagrams, taking advantage of the common virtual boson line propagator. By a judicious use of the σ or $\overline{\sigma}$ version of the vertex rule, we have ensured that the order of the four spinor wave functions is the same for each of the four diagrams. Hence, no additional relative minus signs are required.

The contributions from the diagrams related to these by permutations can now be obtained from the appropriate substitutions $(j \leftrightarrow k)$ and $(j \leftrightarrow \ell)$:

$$i\mathcal{M}_Z^{(2)} = \frac{-(-1)ig^2/c_W^2}{(p_i - k_k)^2 - m_Z^2} \left(O''^L_{ki} x_i \sigma_\mu x_k^\dagger - O''^L_{ik} y_i^\dagger \overline{\sigma}_\mu y_k\right)\left(O''^L_{j\ell} x_j^\dagger \overline{\sigma}^\mu y_\ell - O''^L_{\ell j} y_j \sigma^\mu x_\ell^\dagger\right), \tag{18.14}$$

$$i\mathcal{M}_Z^{(3)} = \frac{-(-1)ig^2/c_W^2}{(p_i - k_\ell)^2 - m_Z^2} \left(O''^L_{\ell i} x_i \sigma_\mu x_\ell^\dagger - O''^L_{i\ell} y_i^\dagger \overline{\sigma}_\mu y_\ell\right)\left(O''^L_{kj} x_k^\dagger \overline{\sigma}^\mu y_j - O''^L_{jk} y_k \sigma^\mu x_j^\dagger\right). \tag{18.15}$$

The extra factors of (-1) in $i\mathcal{M}_Z^{(2)}$ and $i\mathcal{M}_Z^{(3)}$ are present because in both cases the order of the spinors is an odd permutation of the canonical order set by $i\mathcal{M}_Z^{(1)}$.

Note that if we were to proceed to a computation of the decay rate, the very first step would be to apply the Fierz relations of eqs. (1.138)–(1.140) to eliminate all of the σ and $\overline{\sigma}$ matrices in the above amplitudes.

The diagrams in Fig. 18.4 combine to give a contribution

$$i\mathcal{M}_\phi^{(1)} = \frac{-i}{(p_i-k_j)^2 - m_\phi^2}(Y^{ij}x_i y_j + Y_{ij}y_i^\dagger x_j^\dagger)(Y^{k\ell}y_k y_\ell + Y_{k\ell}x_k^\dagger x_\ell^\dagger)\,, \quad (18.16)$$

where we have used the Feynman rules of Fig. J.5, and adopted the shorthand notation $Y^{ij} = (Y_{ij})^* = Y^{\phi\chi_i^0\chi_j^0}$. Again we have factored the amplitude using the common virtual boson propagator. As in the Z-exchange diagrams, the other contributions can be obtained by the appropriate substitutions:

$$i\mathcal{M}_\phi^{(2)} = (-1)\frac{-i}{(p_i-k_k)^2 - m_\phi^2}(Y^{ik}x_i y_k + Y_{ik}y_i^\dagger x_k^\dagger)(Y^{j\ell}y_j y_\ell + Y_{j\ell}x_j^\dagger x_\ell^\dagger)\,, \quad (18.17)$$

$$i\mathcal{M}_\phi^{(3)} = (-1)\frac{-i}{(p_i-k_\ell)^2 - m_\phi^2}(Y^{i\ell}x_i y_\ell + Y_{i\ell}y_i^\dagger x_\ell^\dagger)(Y^{kj}y_k y_j + Y_{kj}x_k^\dagger x_j^\dagger)\,. \quad (18.18)$$

The extra factors of (-1) in $i\mathcal{M}_\phi^{(2)}$ and $i\mathcal{M}_\phi^{(3)}$ are due to the order of the spinors, which in both cases is an odd permutation of the canonical order set by $i\mathcal{M}_Z^{(1)}$.

The total matrix element is obtained by adding all the contributing diagrams:

$$\mathcal{M} = \sum_{n=1}^{3}\mathcal{M}_Z^{(n)} + \sum_\phi \sum_{n=1}^{3}\mathcal{M}_\phi^{(n)}\,. \quad (18.19)$$

Employing the relevant kinematic relations of the three-body decay and either eq. (D.97) or eq. (D.99) yields the total decay rate. Note that final states differing by the interchange of identical particles must be considered as a single state, counted once. Given an N-body final state made up of ν_r particles of type r (where $r \leq N$), we define a statistical factor

$$S = \prod_r \nu_r!\,, \qquad \text{where} \qquad \sum_r \nu_r = N\,. \quad (18.20)$$

In computing the total decay rate, the integration over phase space must be divided by S to avoid overcounting. In the present example, $N = 3$ with $S = 2$ [$S = 6$] in the case of two [three] identical neutralinos in the final state, respectively.

18.4 Three-Body Slepton Decays: $\widetilde{\ell}_R^- \to \ell^-\tau^\pm\widetilde{\tau}_1^\mp$

We next consider the three-body decays of sleptons, $\widetilde{\ell}_R^- \to \ell^-\tau^\pm\widetilde{\tau}_1^\mp$ for $\ell = e, \mu$, mediated by a virtual neutralino. The usual assumption in supersymmetric phenomenology is that these decays will have a very small branching fraction, because a two-body decay to a lighter neutralino and lepton is always open. However, in GMSB models with a nonminimal messenger sector, the sleptons can be lighter than the lightest neutralino. In that case, the mostly right-handed smuon and selectron, $\widetilde{\mu}_R$ and $\widetilde{e}_R$, will decay by $\widetilde{\ell}_R^- \to \ell^- \tau^\pm \widetilde{\tau}_1^\mp$. The lightest stau mass eigenstate, $\widetilde{\tau}_1^\pm$, is a mixture of the weak eigenstates, $\widetilde{\tau}_L^\pm$ and $\widetilde{\tau}_R^\pm$, as described in Appendix J.4:

$$\widetilde{\tau}_1^- = R^*_{\widetilde{\tau}_1} \widetilde{\tau}_R^- + L^*_{\widetilde{\tau}_1} \widetilde{\tau}_L^- , \tag{18.21}$$

and $\widetilde{\tau}_1^+ = (\widetilde{\tau}_1^-)^*$, while the $\widetilde{\mu}_R$ and $\widetilde{e}_R$ are taken to be unmixed.

First consider the decay $\widetilde{\ell}_R^- \to \ell^- \tau^+ \widetilde{\tau}_1^-$, which proceeds by the diagrams in the top row of Fig. 18.5. The momenta and polarizations of the particles are also indicated on the diagram. Using the Feynman rules of Fig. J.9, we find that the amplitudes of these two diagrams, for each neutralino $\widetilde{N}_j$ exchanged, are

$$i\mathcal{M}_1 = (-ia_j^{\widetilde{\ell}*})(-ia_j^{\widetilde{\tau}})\, y_1 \left[\frac{-i(p-k_1)\cdot\sigma}{(p-k_1)^2 - m^2_{\widetilde{N}_j}} \right] x_2^\dagger , \tag{18.22}$$

$$i\mathcal{M}_2 = (-ia_j^{\widetilde{\ell}*})(-ib_j^{\widetilde{\tau}})\, y_1 \left[\frac{i m_{\widetilde{N}_j}}{(p-k_1)^2 - m^2_{\widetilde{N}_j}} \right] y_2 , \tag{18.23}$$

where

$$a_j^{\widetilde{\ell}} = \sqrt{2} g' N_{j1} , \qquad a_j^{\widetilde{\tau}} = Y_\tau N_{j3} L^*_{\widetilde{\tau}_1} + \sqrt{2} g' N_{j1} R^*_{\widetilde{\tau}_1} , \tag{18.24}$$

$$b_j^{\widetilde{\tau}} = Y_\tau N^*_{j3} R^*_{\widetilde{\tau}_1} - \frac{1}{\sqrt{2}} (g N^*_{j2} + g' N^*_{j1}) L^*_{\widetilde{\tau}_1} . \tag{18.25}$$

The spinor wave functions are $y_1 = y(\vec{\boldsymbol{k}}_1, \lambda_1) =, y_2 = y(\vec{\boldsymbol{k}}_2, \lambda_2)$, and $x_2^\dagger = x^\dagger(\vec{\boldsymbol{k}}_2, \lambda_2)$.

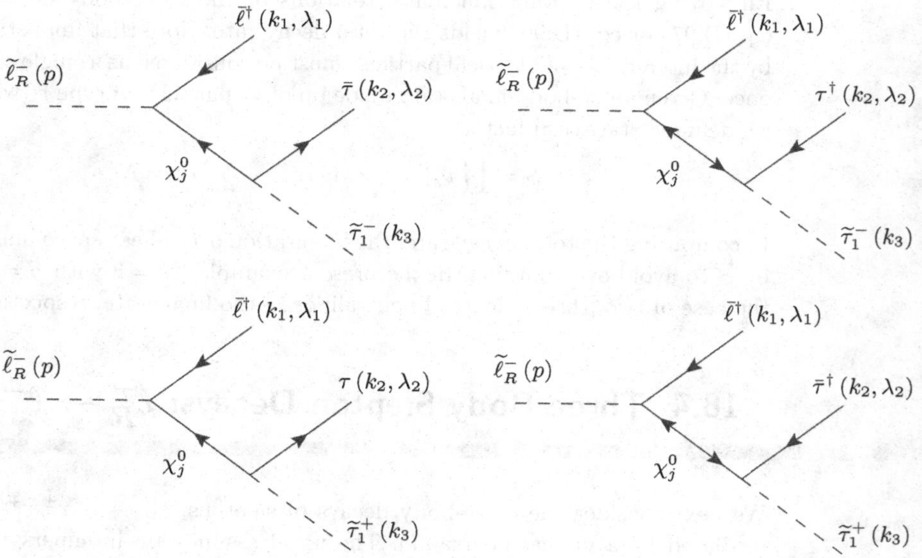

Fig. 18.5 Feynman diagrams for the three-body slepton decays $\widetilde{\ell}_R^- \to \ell^- \tau^+ \widetilde{\tau}_1^-$ (top row) and $\widetilde{\ell}_R^- \to \ell^- \tau^- \widetilde{\tau}_1^+$ (bottom row) in the MSSM.

It is convenient to introduce dimensionless variables following eq. (D.98):

$$z_\ell \equiv \frac{2E_\ell}{m_{\widetilde{\ell}_R}} = \frac{2p\cdot k_1}{m^2_{\widetilde{\ell}_R}}, \qquad z_\tau \equiv \frac{2E_\tau}{m_{\widetilde{\ell}_R}} = \frac{2p\cdot k_2}{m^2_{\widetilde{\ell}_R}}, \tag{18.26}$$

where E_ℓ and E_τ are the energies of ℓ and τ in the rest frame of the decaying slepton. In addition, we also define the mass ratios

$$r_{\widetilde{N}_j} \equiv m_{\widetilde{N}_j}/m_{\widetilde{\ell}_R}, \qquad r_{\widetilde{\tau}} \equiv m_{\widetilde{\tau}_1}/m_{\widetilde{\ell}_R}, \tag{18.27}$$

$$r_\tau \equiv m_\tau/m_{\widetilde{\ell}_R}, \qquad r_\ell \equiv m_\ell/m_{\widetilde{\ell}_R}. \tag{18.28}$$

The total amplitude then can be written as

$$\mathcal{M} = \sum_{j=1}^{4}\left[c_j y_1 (p-k_1)\cdot\sigma x_2^\dagger + d_j y_1 y_2\right], \tag{18.29}$$

where

$$c_j = \frac{-a_j^{\widetilde{\ell}*} a_j^{\widetilde{\tau}}}{m^2_{\widetilde{\ell}_R}(r^2_{\widetilde{N}_j} - 1 + z_\ell)}, \qquad d_j = \frac{a_j^{\widetilde{\ell}*} b_j^{\widetilde{\tau}} m_{\widetilde{N}_j}}{m^2_{\widetilde{\ell}_R}(r^2_{\widetilde{N}_j} - 1 + z_\ell)}. \tag{18.30}$$

We consistently neglect the electron and muon masses and Yukawa couplings (so $r_\ell = 0$) in the matrix elements, but not in the kinematic integration over phase space, where the muon mass effects can be important.

Using eqs. (1.67) and (1.105), we find

$$|\mathcal{M}|^2 = \sum_{j,k}\Big[c_j c_k^*\, y_1(p-k_1)\cdot\sigma x_2^\dagger\, x_2(p-k_1)\cdot\sigma y_1^\dagger + d_j d_k^* y_1 y_2\, y_2^\dagger y_1^\dagger$$
$$+ c_j d_k^* y_1(p-k_1)\cdot\sigma x_2^\dagger\, y_2^\dagger y_1^\dagger + c_j^* d_k x_2(p-k_1)\cdot\sigma y_1^\dagger\, y_1 y_2\Big]. \tag{18.31}$$

Summing over the lepton helicities using eqs. (2.63)–(2.66) gives

$$\sum_{\lambda_1,\lambda_2}|\mathcal{M}|^2 = \sum_{j,k}\Big[c_j c_k^* \mathrm{Tr}[(p-k_1)\cdot\sigma k_2\cdot\overline{\sigma}(p-k_1)\cdot\sigma k_1\cdot\overline{\sigma}] + d_j d_k^* \mathrm{Tr}[k_2\cdot\sigma k_1\cdot\overline{\sigma}]$$
$$- c_j d_k^* m_\tau \mathrm{Tr}[(p-k_1)\cdot\sigma k_1\cdot\overline{\sigma}] - c_j^* d_k m_\tau \mathrm{Tr}[(p-k_1)\cdot\sigma k_1\cdot\overline{\sigma}]\Big]. \tag{18.32}$$

Taking the traces using eqs. (1.92) and (1.93) yields

$$\sum_{\lambda_1,\lambda_2}|\mathcal{M}|^2 = \sum_{j,k}\Big\{c_j c_k^*[4k_1\cdot(p-k_1)k_2\cdot(p-k_1) - 2k_1\cdot k_2(p-k_1)^2] + 2d_j d_k^* k_1\cdot k_2$$
$$- 4\,\mathrm{Re}[c_j d_k^*] m_\tau k_1\cdot(p-k_1)\Big\}$$
$$= \sum_{j,k}\Big\{c_j c_k^* m^4_{\widetilde{\ell}_R}[(1-z_\ell)(1-z_\tau) - r^2_{\widetilde{\tau}} + r^2_\tau]$$
$$+ d_j d_k^* m^2_{\widetilde{\ell}_R}(z_\ell + z_\tau - 1 + r^2_{\widetilde{\tau}} - r^2_\tau) - 2\,\mathrm{Re}[c_j d_k^*] m_\tau m^2_{\widetilde{\ell}_R} z_\ell\Big\}. \tag{18.33}$$

The slepton decay rate is then obtained using eq. (D.99).

We now turn to the competing decay $\widetilde{\ell}_R^- \to \ell^- \tau^- \widetilde{\tau}_1^+$, with diagrams appearing in the bottom row of Fig. 18.5. By appealing again to the Feynman rules of Fig. J.8, we find that the amplitude has exactly the same form as in eqs. (18.22) and (18.23), except now with $a_j^{\widetilde{\tau}} \leftrightarrow b_j^{\widetilde{\tau}*}$. Therefore, the entire previous calculation goes through precisely as before, but now with

$$c_j = \frac{-a_j^{\widetilde{\ell}*} b_j^{\widetilde{\tau}*}}{m_{\widetilde{\ell}_R}^2 (r_{\widetilde{N}_j}^2 - 1 + z_\ell)}, \qquad d_j = \frac{a_j^{\widetilde{\ell}*} a_j^{\widetilde{\tau}*} m_{\widetilde{N}_j}}{m_{\widetilde{\ell}_R}^2 (r_{\widetilde{N}_j}^2 - 1 + z_\ell)}. \tag{18.34}$$

If $m_{\widetilde{\ell}_R} - m_{\widetilde{\tau}_1} - m_\tau$ is not too large, the resulting decays can have a macroscopic length in a detector, and the ratio of the two decay modes can provide an interesting probe of the supersymmetric Lagrangian.

18.5 Neutralino Decay to Photon and Goldstino: $\widetilde{N}_i \to \gamma \widetilde{G}$

The goldstino $\widetilde{G}$ is a massless Weyl fermion that couples to the neutralino and photon fields according to the nonrenormalizable Lagrangian term

$$\mathscr{L} = -\frac{a_i}{2} (\chi_i^0 \sigma^\mu \overline{\sigma}^\rho \sigma^\nu \partial_\mu \widetilde{G}^\dagger) (\partial_\nu A_\rho - \partial_\rho A_\nu) + \text{h.c.} \tag{18.35}$$

Here χ_i^0 is the left-handed two-component fermion field that corresponds to the neutralino $\widetilde{N}_i$ particle, $\widetilde{G}$ is the two-component fermion field corresponding to the (nearly) massless goldstino, and the effective coupling is

$$a_i \equiv \frac{1}{\sqrt{2}\langle F \rangle} (N_{i1}^* \cos\theta_W + N_{i2}^* \sin\theta_W), \tag{18.36}$$

where N_{ij} is the mixing matrix for the neutralinos [see eq. (J.21)] and $\langle F \rangle$ is the F-term expectation value associated with supersymmetry breaking. Therefore $\widetilde{N}_i$ can decay to γ plus $\widetilde{G}$ through the diagrams shown in Fig. 18.6, with amplitudes

$$i\mathcal{M}_1 = i\frac{a_i}{2}\, x_{\widetilde{N}} k_{\widetilde{G}} \cdot \sigma \,(\varepsilon^* \cdot \overline{\sigma}\, k_\gamma \cdot \sigma - k_\gamma \cdot \overline{\sigma}\, \varepsilon^* \cdot \sigma)\, x_{\widetilde{G}}^\dagger, \tag{18.37}$$

$$i\mathcal{M}_2 = -i\frac{a_i^*}{2}\, y_{\widetilde{N}}^\dagger k_{\widetilde{G}} \cdot \overline{\sigma} \,(\varepsilon^* \cdot \sigma\, k_\gamma \cdot \overline{\sigma} - k_\gamma \cdot \sigma\, \varepsilon^* \cdot \overline{\sigma})\, y_{\widetilde{G}}. \tag{18.38}$$

Here $x_{\widetilde{N}} \equiv x(\vec{\boldsymbol{p}}, \lambda_{\widetilde{N}})$, $y_{\widetilde{N}}^\dagger \equiv y^\dagger(\vec{\boldsymbol{p}}, \lambda_{\widetilde{N}})$, and $x_{\widetilde{G}}^\dagger \equiv x^\dagger(\vec{\boldsymbol{k}}_{\widetilde{G}}, \lambda_{\widetilde{G}})$, $y_{\widetilde{G}} \equiv y(\vec{\boldsymbol{k}}_{\widetilde{G}}, \lambda_{\widetilde{G}})$, and $\varepsilon^* = \varepsilon^*(\vec{\boldsymbol{k}}_\gamma, \lambda_\gamma)$ are the external wave function factors for the neutralino, goldstino, and photon, respectively. Using the on-shell condition $k_\gamma \cdot \varepsilon^* = 0$, we have $k_\gamma \cdot \sigma \varepsilon^* \cdot \overline{\sigma} = -\varepsilon^* \cdot \sigma k_\gamma \cdot \overline{\sigma}$ and $k_\gamma \cdot \overline{\sigma} \varepsilon^* \cdot \sigma = -\varepsilon^* \cdot \overline{\sigma} k_\gamma \cdot \sigma$ from eqs. (1.88) and (1.89). Hence, we can rewrite the total amplitude as

$$\mathcal{M} = \mathcal{M}_1 + \mathcal{M}_2 = x_{\widetilde{N}} A x_{\widetilde{G}}^\dagger + y_{\widetilde{N}}^\dagger B y_{\widetilde{G}}, \tag{18.39}$$

where

$$A = a_i\, k_{\widetilde{G}} \cdot \sigma\, \varepsilon^* \cdot \overline{\sigma}\, k_\gamma \cdot \sigma, \qquad B = -a_i^*\, k_{\widetilde{G}} \cdot \overline{\sigma}\, \varepsilon^* \cdot \sigma\, k_\gamma \cdot \overline{\sigma}. \tag{18.40}$$

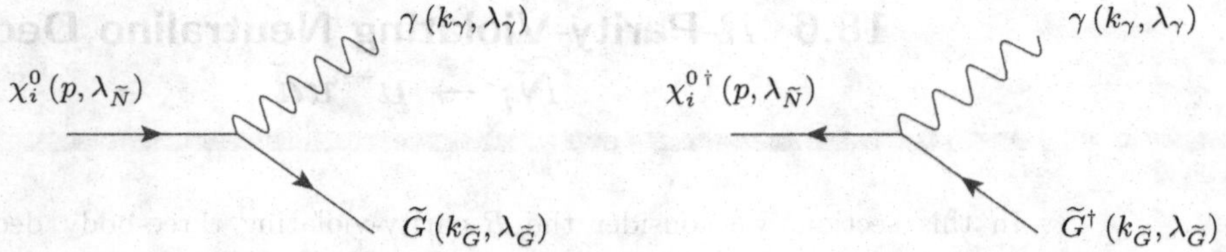

Fig. 18.6 The two Feynman diagrams for $\widetilde{N}_i \to \gamma\widetilde{G}$ in supersymmetric models with a light goldstino.

The complex square of the matrix element is therefore

$$|\mathcal{M}|^2 = x_{\widetilde{N}} A x^\dagger_{\widetilde{G}} x_{\widetilde{G}} \hat{A} x^\dagger_{\widetilde{N}} + y^\dagger_{\widetilde{N}} B y_{\widetilde{G}} y^\dagger_{\widetilde{G}} \hat{B} y_{\widetilde{N}} + x_{\widetilde{N}} A x^\dagger_{\widetilde{G}} y^\dagger_{\widetilde{G}} \hat{B} y_{\widetilde{N}} + y^\dagger_{\widetilde{N}} B y_{\widetilde{G}} x_{\widetilde{G}} \hat{A} x^\dagger_{\widetilde{N}}, \tag{18.41}$$

where $\hat{A}$ and $\hat{B}$ are obtained from A and B by reversing the order of the σ and $\overline{\sigma}$ matrices and taking the complex conjugates of a_i and ε [see eq. (2.148)].

Summing over the goldstino helicities using eqs. (2.63)–(2.66) now yields

$$\sum_{\lambda_{\widetilde{G}}} |\mathcal{M}|^2 = x_{\widetilde{N}} A k_{\widetilde{G}}\!\cdot\!\overline{\sigma} \hat{A} x^\dagger_{\widetilde{N}} + y^\dagger_{\widetilde{N}} B k_{\widetilde{G}}\!\cdot\!\sigma \hat{B} y_{\widetilde{N}}. \tag{18.42}$$

(The $A, \hat{B}$ and $\hat{A}, B$ cross terms vanish because of $m_{\widetilde{G}} = 0$.) Averaging over the neutralino helicities using eqs. (2.63) and (2.64), we find

$$\begin{aligned} \frac{1}{2} \sum_{\lambda_{\widetilde{N}}, \lambda_{\widetilde{G}}} |\mathcal{M}|^2 &= \frac{1}{2} \mathrm{Tr}[A k_{\widetilde{G}}\!\cdot\!\overline{\sigma} \hat{A} p\!\cdot\!\overline{\sigma}] + \frac{1}{2} \mathrm{Tr}[B k_{\widetilde{G}}\!\cdot\!\sigma \hat{B} p\!\cdot\!\sigma] \\ &= \frac{1}{2} |a_i|^2 \mathrm{Tr}[\varepsilon^*\!\cdot\!\overline{\sigma}\, k_\gamma\!\cdot\!\sigma\, k_{\widetilde{G}}\!\cdot\!\overline{\sigma}\, k_\gamma\!\cdot\!\sigma\, \varepsilon\!\cdot\!\overline{\sigma}\, k_{\widetilde{G}}\!\cdot\!\sigma\, p\!\cdot\!\overline{\sigma}\, k_{\widetilde{G}}\!\cdot\!\sigma] + (\sigma \leftrightarrow \overline{\sigma}). \end{aligned} \tag{18.43}$$

We now use

$$k_\gamma\!\cdot\!\sigma\, k_{\widetilde{G}}\!\cdot\!\overline{\sigma}\, k_\gamma\!\cdot\!\sigma = 2k_{\widetilde{G}}\!\cdot\!k_\gamma\, k_\gamma\!\cdot\!\sigma, \tag{18.44}$$

$$k_{\widetilde{G}}\!\cdot\!\sigma\, p\!\cdot\!\overline{\sigma}\, k_{\widetilde{G}}\!\cdot\!\sigma = 2k_{\widetilde{G}}\!\cdot\!p\, k_{\widetilde{G}}\!\cdot\!\sigma, \tag{18.45}$$

which follow from eq. (1.90), and the corresponding identities with $\sigma \leftrightarrow \overline{\sigma}$, to obtain

$$\frac{1}{2} \sum_{\lambda_{\widetilde{N}}, \lambda_{\widetilde{G}}} |\mathcal{M}|^2 = 2|a_i|^2 (k_{\widetilde{G}}\!\cdot\!k_\gamma)(k_{\widetilde{G}}\!\cdot\!p) \mathrm{Tr}[\varepsilon^*\!\cdot\!\overline{\sigma}\, k_\gamma\!\cdot\!\sigma\, \varepsilon\!\cdot\!\overline{\sigma}\, k_{\widetilde{G}}\!\cdot\!\sigma] + (\sigma \leftrightarrow \overline{\sigma}). \tag{18.46}$$

Applying eq. (E.250) for the sum over photon helicities and the trace identities, eqs. (1.93) and (1.94), we obtain

$$\frac{1}{2} \sum_{\lambda_\gamma, \lambda_{\widetilde{N}}, \lambda_{\widetilde{G}}} |\mathcal{M}|^2 = 16|a_i|^2 (k_{\widetilde{G}}\!\cdot\!k_\gamma)^2 (k_{\widetilde{G}}\!\cdot\!p) = 2|a_i|^2 m^6_{\widetilde{N}_i}\,. \tag{18.47}$$

Hence, in light of eq. (D.90), the decay rate is given by

$$\Gamma(\widetilde{N}_i \to \gamma\widetilde{G}) = |N_{i1} \cos\theta_W + N_{i2} \sin\theta_W|^2 \frac{m^5_{\widetilde{N}_i}}{16\pi |\langle F \rangle|^2}. \tag{18.48}$$

18.6 R-Parity-Violating Neutralino Decay: $\widetilde{N}_i \to \mu^- u\bar{d}$

In this section, we consider the R-parity-violating three-body decay of a neutralino, $\widetilde{N}_i \to \mu^- u\bar{d}$, which arises due to the L-violating $LQ\bar{d}$ coupling governed by eq. (16.184). This is of particular interest when the neutralino is the LSP, since it determines the final-state signatures. The three Feynman diagrams are shown in Fig. 18.7, including the definitions of the momenta and helicities. We have neglected sfermion mixing, i.e., we assume $\widetilde{\mu}_L$, $\widetilde{u}_L$, and $\widetilde{d}_R$ are mass eigenstates. Using the Feynman rules given in Figs. J.14 and J.7 (or J.9), we obtain the corresponding contributions to the decay amplitude,

$$i\mathcal{M}_1 = (i\lambda'^*)\left[\frac{i}{\sqrt{2}}(gN_{i2} + g'N_{i1})\right]\left[\frac{i}{(p_i - k_\mu)^2 - m^2_{\widetilde{\mu}_L}}\right] y_i^\dagger x_\mu^\dagger x_u^\dagger x_d^\dagger\,, \qquad (18.49)$$

$$i\mathcal{M}_2 = (i\lambda'^*)\left[-\frac{i\sqrt{2}}{3}g'N_{i1}\right]\left[\frac{i}{(p_i - k_d)^2 - m^2_{\widetilde{d}_R}}\right] y_i^\dagger x_d^\dagger x_\mu^\dagger x_u^\dagger\,, \qquad (18.50)$$

$$i\mathcal{M}_3 = (i\lambda'^*)\left[-\frac{i}{\sqrt{2}}(gN_{i2} + \tfrac{1}{3}g'N_{i1})\right]\left[\frac{i}{(p_i - k_u)^2 - m^2_{\widetilde{u}_L}}\right] y_i^\dagger x_u^\dagger x_d^\dagger x_\mu^\dagger\,. \qquad (18.51)$$

Here we have defined $\lambda' \equiv \lambda'_{211}$, and the external wave functions are denoted by $y_i^\dagger \equiv y^\dagger(\vec{p}_i, \lambda_i)$, $x_\mu^\dagger \equiv x^\dagger(\vec{k}_\mu, \lambda_\mu)$, $x_u^\dagger \equiv x^\dagger(\vec{k}_u, \lambda_u)$, and $x_d^\dagger \equiv x^\dagger(\vec{k}_d, \lambda_d)$, respectively.

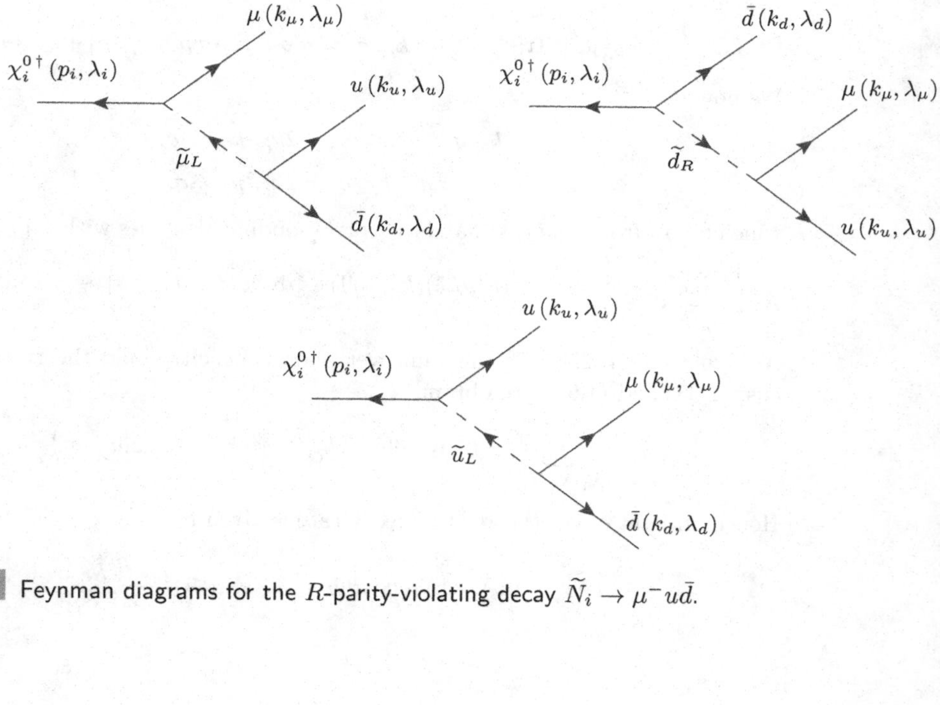

Fig. 18.7 Feynman diagrams for the R-parity-violating decay $\widetilde{N}_i \to \mu^- u\bar{d}$.

In what follows, we will neglect all of the final-state fermion masses. It is convenient to introduce dimensionless variables

$$z_\mu \equiv \frac{2p_i\cdot k_\mu}{m^2_{\widetilde{N}_i}} = \frac{2E_\mu}{m_{\widetilde{N}_i}}, \qquad z_q \equiv \frac{2p_i\cdot k_q}{m^2_{\widetilde{N}_i}} = \frac{2E_q}{m_{\widetilde{N}_i}}, \quad \text{for } q = d, u, \tag{18.52}$$

which satisfy $z_\mu + z_d + z_u = 2$. Then we can rewrite the total matrix element as

$$\mathcal{M} = c_1 y_i^\dagger x_\mu^\dagger x_u^\dagger x_d^\dagger + c_2 y_i^\dagger x_d^\dagger x_\mu^\dagger x_u^\dagger + c_3 y_i^\dagger x_u^\dagger x_d^\dagger x_\mu^\dagger, \tag{18.53}$$

where

$$c_1 \equiv \frac{\lambda'^*(gN_{i2} + g'N_{i1})}{\sqrt{2}[m^2_{\tilde{\mu}_L} - m^2_{\widetilde{N}_i}(1 - z_\mu)]}, \qquad c_2 \equiv -\frac{\sqrt{2}\lambda'^* g' N_{i1}}{3[m^2_{\tilde{d}_R} - m^2_{\widetilde{N}_i}(1 - z_d)]}, \tag{18.54}$$

$$c_3 \equiv -\frac{\lambda'^*(gN_{i2} + \frac{1}{3}g'N_{i1})}{\sqrt{2}[m^2_{\tilde{u}_L} - m^2_{\widetilde{N}_i}(1 - z_u)]}. \tag{18.55}$$

Before squaring the amplitude, it is convenient to use the Fierz identity given in eq. (1.137) to reduce the number of terms:

$$\mathcal{M} = (c_1 - c_3) y_i^\dagger x_\mu^\dagger x_u^\dagger x_d^\dagger + (c_2 - c_3) y_i^\dagger x_d^\dagger x_\mu^\dagger x_u^\dagger. \tag{18.56}$$

Using eq. (1.67), we obtain

$$|\mathcal{M}|^2 = |c_1 - c_3|^2 y_i^\dagger x_\mu^\dagger x_\mu y_i x_u^\dagger x_d^\dagger x_d x_u + |c_2 - c_3|^2 y_i^\dagger x_d^\dagger x_d y_i x_\mu^\dagger x_u^\dagger x_u x_\mu$$
$$-2\,\mathrm{Re}[(c_1 - c_3)(c_2^* - c_3^*) y_i^\dagger x_\mu^\dagger x_\mu x_u x_u^\dagger x_d^\dagger x_d y_i], \tag{18.57}$$

where eq. (1.96) was used on the last term. Summing over the fermion helicities using eqs. (2.63)–(2.66), we obtain

$$\sum_{\{\lambda\}} |\mathcal{M}|^2 = |c_1 - c_3|^2 \mathrm{Tr}[k_\mu\cdot\overline{\sigma} p_i\cdot\sigma]\mathrm{Tr}[k_d\cdot\overline{\sigma} k_u\cdot\sigma] + |c_2 - c_3|^2 \mathrm{Tr}[k_d\cdot\overline{\sigma} p_i\cdot\sigma]\mathrm{Tr}[k_u\cdot\overline{\sigma} k_\mu\cdot\sigma]$$
$$-2\,\mathrm{Re}\big[(c_1 - c_3)(c_2^* - c_3^*)\mathrm{Tr}[k_\mu\cdot\overline{\sigma} k_u\cdot\sigma k_d\cdot\overline{\sigma} p_i\cdot\sigma]\big]. \tag{18.58}$$

Applying the trace formulae, eqs. (1.92) and (1.94), we obtain

$$\begin{aligned}\sum_{\{\lambda\}} |\mathcal{M}|^2 &= 4|c_1 - c_3|^2 p_i\cdot k_\mu\, k_d\cdot k_u + 4|c_2 - c_3|^2 p_i\cdot k_d\, k_\mu\cdot k_u \\ &\quad - 4\,\mathrm{Re}\big[(c_1 - c_3)(c_2^* - c_3^*)\big](k_\mu\cdot k_u\, p_i\cdot k_d + p_i\cdot k_\mu\, k_d\cdot k_u - k_\mu\cdot k_d\, p_i\cdot k_u) \\ &= m^4_{\widetilde{N}_i}\Big[|c_1|^2 z_\mu(1 - z_\mu) + |c_2|^2 z_d(1 - z_d) + |c_3|^2 z_u(1 - z_u) \\ &\quad - 2\,\mathrm{Re}[c_1c_2^*](1 - z_\mu)(1 - z_d) - 2\,\mathrm{Re}[c_1c_3^*](1 - z_\mu)(1 - z_u) \\ &\quad - 2\,\mathrm{Re}[c_2c_3^*](1 - z_d)(1 - z_u)\Big],\end{aligned} \tag{18.59}$$

where in the last equality we have used eq. (18.52) and

$$2k_\mu\cdot k_d = (1 - z_u)m^2_{\widetilde{N}_i}, \qquad 2k_\mu\cdot k_u = (1 - z_d)m^2_{\widetilde{N}_i}, \qquad 2k_d\cdot k_u = (1 - z_\mu)m^2_{\widetilde{N}_i}. \tag{18.60}$$

Using eq. (D.99), the decay rate is given by

$$\Gamma = \frac{N_c m_{\widetilde{N}_i}}{256\pi^3} \int_0^1 dz_\mu \int_{1-z_\mu}^1 dz_d \left(\frac{1}{2} \sum_{\{\lambda\}} |\mathcal{M}|^2 \right), \tag{18.61}$$

where a factor of $N_c = 3$ has been included for the sum over colors. In the limit of heavy sfermions, the integrations over z_d and z_μ can be easily carried out:

$$\Gamma = \frac{N_c m^5_{\widetilde{N}_i}}{2^{11} \cdot 3\pi^3} \left(|c'_1|^2 + |c'_2|^2 + |c'_3|^2 - \mathrm{Re}[c'_1 c'^*_2 + c'_1 c'^*_3 + c'_2 c'^*_3] \right), \tag{18.62}$$

where the c'_i are obtained from c_i of eqs. (18.54) and (18.55) by neglecting the terms proportional to $m^2_{\widetilde{N}_i}$ in the denominators.

18.7 $e^- e^- \to \widetilde{e}^-_L \widetilde{e}^-_R$

Consider the scattering of two electrons into a pair of nonidentical selectrons. There are two Feynman graphs (neglecting the electron mass, Yukawa couplings, and selectron mixing), shown in Fig. 18.8, with momenta and helicities as indicated. Note that these two graphs are related by interchange of the identical initial-state electrons. It is convenient to introduce the Mandelstam invariant variables s, t, and u defined in eqs. (D.23)–(D.25).

Using the Feynman rules of Fig. J.7, the scattering amplitude for the first graph, for each neutralino $\widetilde{N}_i$ exchanged in the t channel, is

$$i\mathcal{M}_t = \left[i\frac{g}{\sqrt{2}} \left(N^*_{i2} + \frac{s_W}{c_W} N^*_{i1} \right) \right] \left[-i\sqrt{2} g \frac{s_W}{c_W} N_{i1} \right] x_1 \left[\frac{i(k_1 - p_1)\cdot\sigma}{t - m^2_{\widetilde{N}_i}} \right] y^\dagger_2. \tag{18.63}$$

We employ the notation for the external wave functions $x_i = (\vec{\boldsymbol{p}}_i, \lambda_i)$, $i = 1, 2$, and analogously for $y_i, x^\dagger_i, y^\dagger_i$. The scattering amplitude for the second (u-channel)

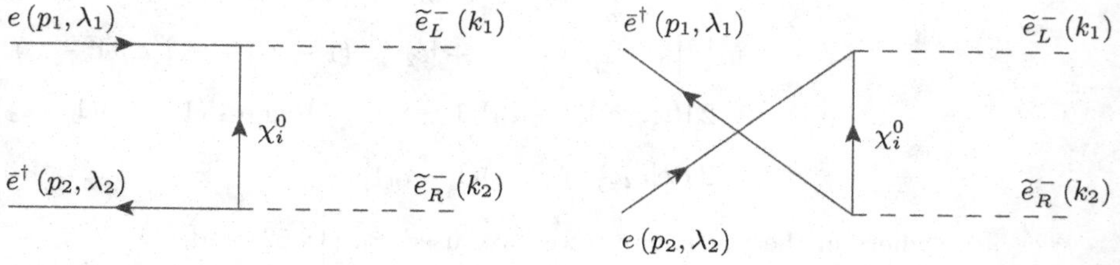

Fig. 18.8 Feynman diagrams for $e^- e^- \to \widetilde{e}^-_L \widetilde{e}^-_R$.

graph is the same with the two incoming electrons exchanged, $e_1 \leftrightarrow e_2$:

$$i\mathcal{M}_u = (-1)\left[i\frac{g}{\sqrt{2}}\left(N_{i2}^* + \frac{s_W}{c_W}N_{i1}^*\right)\right]\left[-i\sqrt{2}g\frac{s_W}{c_W}N_{i1}\right] x_2 \left[\frac{i(k_1-p_2)\cdot\sigma}{u-m_{\widetilde{N}_i}^2}\right] y_1^\dagger\,. \tag{18.64}$$

Since we have written the fermion wave function spinors in the opposite order in $\mathcal{M}_2$ as compared to $\mathcal{M}_1$, there is a factor (-1) for Fermi–Dirac statistics. Alternatively, starting at the electron with momentum p_1 and using the Feynman rules as above, we can directly write

$$i\mathcal{M}_u = \left[i\frac{g}{\sqrt{2}}\left(N_{i2}^* + \frac{s_W}{c_W}N_{i1}^*\right)\right]\left[-i\sqrt{2}g\frac{s_W}{c_W}N_{i1}\right] y_1^\dagger \left[\frac{-i(k_1-p_2)\cdot\overline{\sigma}}{u-m_{\widetilde{N}_i}^2}\right] x_2\,. \tag{18.65}$$

In this case, the wave function spinors are written in the same order as in $\mathcal{M}_t$, so there is no extra factor of (-1) as in eq. (18.65). However, now the Feynman rule for the propagator has an extra minus sign, as can be seen in Fig. 2.3. We can also obtain eq. (18.65) from eq. (18.64) by using eq. (1.98).

Hence, the total amplitude is given by

$$\mathcal{M} = \mathcal{M}_t + \mathcal{M}_u = x_1 a\cdot\sigma y_2^\dagger + y_1^\dagger b\cdot\overline{\sigma} x_2\,, \tag{18.66}$$

where

$$a^\mu \equiv \frac{g^2 s_W}{c_W}(k_1^\mu - p_1^\mu)\sum_{i=1}^4 N_{i1}(N_{i2}^* + \frac{s_W}{c_W}N_{i1}^*)\frac{1}{t-m_{\widetilde{N}_i}^2}\,, \tag{18.67}$$

$$b^\mu \equiv -\frac{g^2 s_W}{c_W}(k_1^\mu - p_2^\mu)\sum_{i=1}^4 N_{i1}(N_{i2}^* + \frac{s_W}{c_W}N_{i1}^*)\frac{1}{u-m_{\widetilde{N}_i}^2}\,. \tag{18.68}$$

Using eqs. (1.105) and (1.106) it follows that

$$|\mathcal{M}|^2 = \left(x_1 a\cdot\sigma y_2^\dagger\right)\left(y_2 a^*\cdot\sigma x_1^\dagger\right) + \left(y_1^\dagger b\cdot\overline{\sigma} x_2\right)\left(x_2^\dagger b^*\cdot\overline{\sigma} y_1\right) + \left(x_1 a\cdot\sigma y_2^\dagger\right)\left(x_2^\dagger b^*\cdot\overline{\sigma} y_1\right) + \left(y_1^\dagger b\cdot\overline{\sigma} x_2\right)\left(y_2 a^*\cdot\sigma x_1^\dagger\right)\,. \tag{18.69}$$

Averaging over the initial-state electron helicities using eqs. (2.63)–(2.66), the a, b^* and a^*, b cross terms are proportional to m_e and can thus be neglected in our approximation. We get

$$\frac{1}{4}\sum_{\lambda_1,\lambda_2}|\mathcal{M}|^2 = \tfrac{1}{4}\,\mathrm{Tr}[a\cdot\sigma\, p_2\cdot\overline{\sigma}\, a^*\cdot\sigma\, p_1\cdot\overline{\sigma}] + \tfrac{1}{4}\,\mathrm{Tr}[b\cdot\overline{\sigma}\, p_2\cdot\sigma\, b^*\cdot\overline{\sigma}\, p_1\cdot\sigma]\,. \tag{18.70}$$

In light of eqs. (1.93) and (1.94), these terms can be simplified using the identities

$$\mathrm{Tr}[(k_1-p_1)\cdot\sigma\, p_2\cdot\overline{\sigma}\,(k_1-p_1)\cdot\sigma\, p_1\cdot\overline{\sigma}] = \mathrm{Tr}[(k_1-p_2)\cdot\overline{\sigma}\, p_2\cdot\sigma\,(k_1-p_2)\cdot\overline{\sigma}\, p_1\cdot\sigma] = tu - m_{\widetilde{e}_L}^2 m_{\widetilde{e}_R}^2\,. \tag{18.71}$$

Thus, after employing eq. (D.110), we obtain the following differential cross section:

$$\frac{d\sigma}{dt} = \frac{\pi\alpha^2}{4s_W^2 c_W^2}\left(\frac{tu - m_{\tilde{e}_L}^2 m_{\tilde{e}_R}^2}{s^2}\right)\sum_{i,j=1}^{4} N_{j1}N_{i1}^*\left(N_{j2}^* + \frac{s_W}{c_W}N_{j1}^*\right)\left(N_{i2} + \frac{s_W}{c_W}N_{i1}\right)$$
$$\times\left[\frac{1}{(t-m_{\widetilde{N}_i}^2)(t-m_{\widetilde{N}_j}^2)} + \frac{1}{(u-m_{\widetilde{N}_i}^2)(u-m_{\widetilde{N}_j}^2)}\right]. \qquad (18.72)$$

18.8 $e^+e^- \to \widetilde{N}_i\widetilde{N}_j$

Consider the pair production of neutralinos via e^+e^- annihilation. There are four Feynman graphs for s-channel Z exchange, shown in Fig. 18.9, and four for t/u-channel selectron exchange, shown in Fig. 18.10. The momenta and helicities are as labeled in the graphs (where the momenta of the electron and positron are p_1 and p_2, respectively). We denote the neutralino masses by $m_{\widetilde{N}_i}$ and $m_{\widetilde{N}_j}$ and the selectron masses by $m_{\tilde{e}_L}$ and $m_{\tilde{e}_R}$. The electron mass will again be neglected.

By applying the Feynman rules of Figs. I.2 and J.3, we obtain, for the sum of the s-channel diagrams,[2]

$$i\mathcal{M}_Z = \frac{-ig^{\mu\nu}}{D_Z}\left[\frac{ig(s_W^2-\frac{1}{2})}{c_W}x_1\sigma_\mu y_2^\dagger + \frac{igs_W^2}{c_W}y_1^\dagger\bar{\sigma}_\mu x_2\right]$$
$$\times\left[\frac{ig}{c_W}O_{ij}^{\prime\prime L}x_i^\dagger\bar{\sigma}_\nu y_j - \frac{ig}{c_W}O_{ji}^{\prime\prime L}y_i\sigma_\nu x_j^\dagger\right], \qquad (18.73)$$

where O''_{ij} is given in eq. (J.18), the external fermion wave functions are denoted by $x_1 \equiv x(\vec{\boldsymbol{p}}_1,\lambda_1)$, $y_2^\dagger \equiv y^\dagger(\vec{\boldsymbol{p}}_2,\lambda_2)$, $x_i^\dagger \equiv x^\dagger(\vec{\boldsymbol{k}}_i,\lambda_i)$, $y_j \equiv y(\vec{\boldsymbol{k}}_j,\lambda_j)$, and

$$D_Z \equiv s - m_Z^2 + i\Gamma_Z m_Z. \qquad (18.74)$$

For the four t/u-channel diagrams, by applying the rules of Fig. J.7, we obtain

$$i\mathcal{M}_{\tilde{e}_L}^{(t)} = (-1)\left[\frac{i}{t-m_{\tilde{e}_L}^2}\right]\left[\frac{ig}{\sqrt{2}}\left(N_{i2}^* + \frac{s_W}{c_W}N_{i1}^*\right)\right]\left[\frac{ig}{\sqrt{2}}\left(N_{j2} + \frac{s_W}{c_W}N_{j1}\right)\right]x_1 y_i y_2^\dagger x_j^\dagger, \qquad (18.75)$$

$$i\mathcal{M}_{\tilde{e}_L}^{(u)} = \left[\frac{i}{u-m_{\tilde{e}_L}^2}\right]\left[\frac{ig}{\sqrt{2}}\left(N_{j2}^* + \frac{s_W}{c_W}N_{j1}^*\right)\right]\left[\frac{ig}{\sqrt{2}}\left(N_{i2} + \frac{s_W}{c_W}N_{i1}\right)\right]x_1 y_j y_2^\dagger x_i^\dagger, \qquad (18.76)$$

$$i\mathcal{M}_{\tilde{e}_R}^{(t)} = (-1)\left[\frac{i}{t-m_{\tilde{e}_R}^2}\right]\left(-i\sqrt{2}g\frac{s_W}{c_W}N_{i1}\right)\left(-i\sqrt{2}g\frac{s_W}{c_W}N_{j1}^*\right)y_1^\dagger x_i^\dagger x_2 y_j, \qquad (18.77)$$

$$i\mathcal{M}_{\tilde{e}_R}^{(u)} = \left[\frac{i}{u-m_{\tilde{e}_R}^2}\right]\left(-i\sqrt{2}g\frac{s_W}{c_W}N_{j1}\right)\left(-i\sqrt{2}g\frac{s_W}{c_W}N_{i1}^*\right)y_1^\dagger x_j^\dagger x_2 y_i. \qquad (18.78)$$

[2] Because we neglect the electron mass, we may drop the $Q^\mu Q^\nu$ term of the Z propagator, where $Q \equiv p_1 + p_2$ is the propagating four-momentum in the s-channel (see Fig. 2.6).

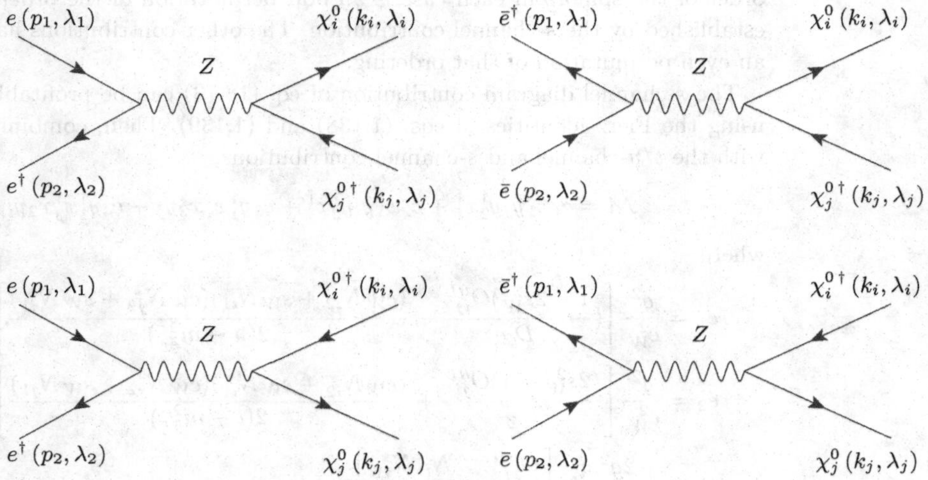

Fig. 18.9 The four Feynman diagrams for $e^+e^- \to \widetilde{N}_i\widetilde{N}_j$ via s-channel Z exchange.

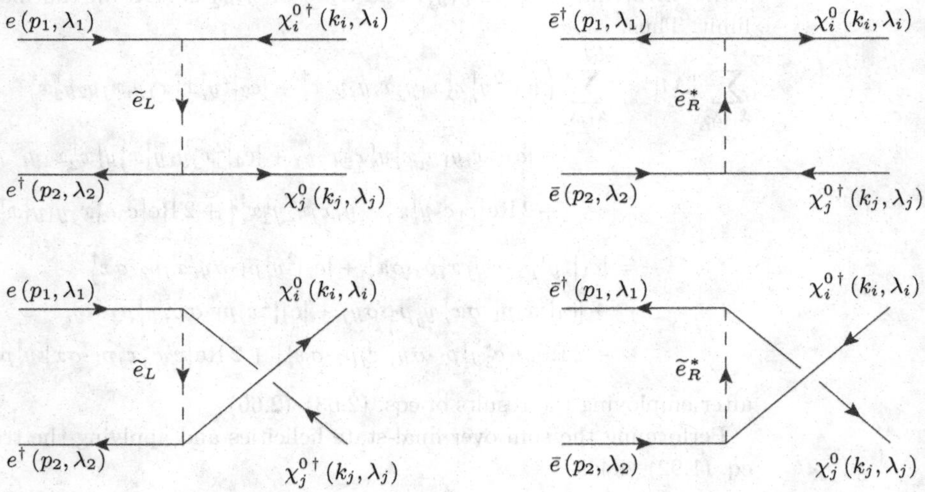

Fig. 18.10 The four Feynman diagrams for $e^+e^- \to \widetilde{N}_i\widetilde{N}_j$ via t/u-channel selectron exchange.

The extra factors of (-1) in each of eqs. (18.75) and (18.77) are present because the order of the spinors in each case is an odd permutation of the ordering $(1, 2, i, j)$ established by the s-channel contribution. The other contributions have spinors in an even permutation of that ordering.

The s-channel diagram contribution of eq. (18.73) can be profitably rearranged using the Fierz identities of eqs. (1.138) and (1.139). Then, combining the result with the t/u-channel and s-channel contributions,

$$\mathcal{M} = c_1 x_1 y_j y_2^\dagger x_i^\dagger + c_2 x_1 y_i y_2^\dagger x_j^\dagger + c_3 y_1^\dagger x_i^\dagger x_2 y_j + c_4 y_1^\dagger x_j^\dagger x_2 y_i, \tag{18.79}$$

where

$$c_1 = \frac{g^2}{c_W^2}\left[\frac{(1-2s_W^2)O''^L_{ij}}{D_Z} - \frac{(c_W N_{i2} + s_W N_{i1})(c_W N_{j2}^* + s_W N_{j1}^*)}{2(u - m_{\tilde{e}_L}^2)}\right], \tag{18.80}$$

$$c_2 = \frac{g^2}{c_W^2}\left[\frac{(2s_W^2 - 1)O''^L_{ji}}{D_Z} + \frac{(c_W N_{i2}^* + s_W N_{i1}^*)(c_W N_{j2} + s_W N_{j1})}{2(t - m_{\tilde{e}_L}^2)}\right], \tag{18.81}$$

$$c_3 = -\frac{2g^2 s_W^2}{c_W^2}\left[\frac{O''^L_{ij}}{D_Z} - \frac{N_{i1} N_{j1}^*}{t - m_{\tilde{e}_R}^2}\right], \tag{18.82}$$

$$c_4 = \frac{2g^2 s_W^2}{c_W^2}\left[\frac{O''^L_{ji}}{D_Z} - \frac{N_{i1}^* N_{j1}}{u - m_{\tilde{e}_R}^2}\right]. \tag{18.83}$$

Squaring the amplitude and averaging over electron and positron helicities, only terms involving $x_1 x_1^\dagger$ or $y_1 y_1^\dagger$, and $x_2 x_2^\dagger$ or $y_2 y_2^\dagger$ survive in the massless electron limit. Thus,

$$\begin{aligned}
\sum_{\lambda_1,\lambda_2} |\mathcal{M}|^2 = \sum_{\lambda_1,\lambda_2} \Big(& |c_1|^2 y_j^\dagger x_1^\dagger x_1 y_j x_i y_2 y_2^\dagger x_i^\dagger + |c_2|^2 y_i^\dagger x_1^\dagger x_1 y_i x_j y_2 y_2^\dagger x_j^\dagger \\
& + |c_3|^2 x_i y_1 y_1^\dagger x_i^\dagger y_j^\dagger x_2^\dagger x_2 y_j + |c_4|^2 x_j y_1 y_1^\dagger x_j^\dagger y_i^\dagger x_2^\dagger x_2 y_i \\
& + 2\,\mathrm{Re}\big[c_1 c_2^* y_i^\dagger x_1^\dagger x_1 y_j x_j y_2 y_2^\dagger x_i^\dagger\big] + 2\,\mathrm{Re}\big[c_3 c_4^* x_j y_1 y_1^\dagger x_i^\dagger y_i^\dagger x_2^\dagger x_2 y_j\big]\Big) \\
= {} & |c_1|^2 y_j^\dagger p_1\!\cdot\!\overline{\sigma} y_j\, x_i p_2\!\cdot\!\sigma x_i^\dagger + |c_2|^2 y_i^\dagger p_1\!\cdot\!\overline{\sigma} y_i\, x_j p_2\!\cdot\!\sigma x_j^\dagger \\
& + |c_3|^2 x_i p_1\!\cdot\!\sigma x_i^\dagger\, y_j^\dagger p_2\!\cdot\!\overline{\sigma} y_j + |c_4|^2 x_j p_1\!\cdot\!\sigma x_j^\dagger\, y_i^\dagger p_2\!\cdot\!\overline{\sigma} y_i \\
& + 2\,\mathrm{Re}\big[c_1 c_2^* y_i^\dagger p_1\!\cdot\!\overline{\sigma} y_j\, x_j p_2\!\cdot\!\sigma x_i^\dagger\big] + 2\,\mathrm{Re}\big[c_3 c_4^* x_j p_1\!\cdot\!\sigma x_i^\dagger\, y_i^\dagger p_2\!\cdot\!\overline{\sigma} y_j\big],
\end{aligned} \tag{18.84}$$

after employing the results of eqs. (2.63)–(2.66).

Performing the sum over final-state helicities and applying the trace identity of eq. (1.92) yields

$$\begin{aligned}
\sum_{\{\lambda\}} |\mathcal{M}|^2 = {} & (|c_1|^2 + |c_4|^2) 4 p_1\!\cdot\!k_j\, p_2\!\cdot\!k_i + (|c_2|^2 + |c_3|^2) 4 p_1\!\cdot\!k_i\, p_2\!\cdot\!k_j \\
& + 4\,\mathrm{Re}[c_1 c_2^* + c_3 c_4^*] m_{\widetilde{N}_i} m_{\widetilde{N}_j} p_1\!\cdot\!p_2 \\
= {} & (|c_1|^2 + |c_4|^2)(u - m_{\widetilde{N}_i}^2)(u - m_{\widetilde{N}_j}^2) + (|c_2|^2 + |c_3|^2)(t - m_{\widetilde{N}_i}^2)(t - m_{\widetilde{N}_j}^2) \\
& + 2\,\mathrm{Re}[c_1 c_2^* + c_3 c_4^*] m_{\widetilde{N}_i} m_{\widetilde{N}_j} s.
\end{aligned} \tag{18.85}$$

The differential cross section $d\sigma/dt$ can be obtained by employing eq. (D.110). Accounting for the identical fermions in the final state when $i = j$, the total cross section is given by [see eq. (D.111)]

$$\sigma = \frac{1}{1+\delta_{ij}} \int_{t^-}^{t^+} \frac{d\sigma}{dt} dt \,, \tag{18.86}$$

where $t^\pm$ are given in eq. (D.37).

18.9 $\widetilde{N}_1\widetilde{N}_1 \to f\bar{f}$

Consider the annihilation rate for $\widetilde{N}_1\widetilde{N}_1 \to f\bar{f}$, where f is any kinematically allowed quark, charged lepton, or neutrino. The case of $f = e^-$ is the reversed reaction of the process examined in Section 18.8 (with $i = j = 1$). In R-parity-conserving supersymmetric models in which $\widetilde{N}_1$ is the lightest supersymmetric particle (and hence is stable), the $\widetilde{N}_1\widetilde{N}_1$ annihilation process is relevant for the computation of the neutralino relic density. In particular, $\widetilde{N}_1\widetilde{N}_1 \to f\bar{f}$ can be an important contribution to cold dark matter annihilation.

The annihilation of $\widetilde{N}_1\widetilde{N}_1$ into heavy quarks (c, b, and t), followed by the decay of the heavy quarks, can yield observable signatures suitable for indirect dark matter detection. For example, the annihilation of neutralinos in the galaxy provides a possible source of indirect dark matter detection via the observation of positrons in cosmic rays. Neutralino dark matter can also be captured in the Sun. The neutrinos that arise (either directly or indirectly) from the neutralino annihilation in the Sun can be detected on Earth.

In the computation of the relic density, one computes $v_{\rm rel}\sigma_{\rm ann}$, where $\sigma_{\rm ann}$ is the $\widetilde{N}_1\widetilde{N}_1$ annihilation cross section and $v_{\rm rel}$ is the relative velocity of the two neutralinos in the center-of-momentum frame. The square of the relative velocity is taken to be its thermal average, $v^2_{\rm rel} \simeq 6k_BT/m_{\widetilde{N}_1}$, which is typically nonrelativistic when the temperature is of order the freeze-out temperature (where the neutralino falls out of thermal equilibrium). Hence, it is sufficient to compute the annihilation cross section for $\widetilde{N}_1\widetilde{N}_1 \to f\bar{f}$ in the nonrelativistic limit.

As in Section 18.8, there are four Feynman graphs for s-channel Z exchange, shown in Fig. 18.11. In addition, there are s-channel neutral Higgs exchange graphs, shown in Fig. 18.12, that yield contributions to the annihilation amplitude proportional to the fermion mass, m_f.[3] Likewise, there are four Feynman graphs for t/u-channel $\widetilde{f}_L$ and $\widetilde{f}_R$ exchange, shown in Fig. 18.13. However, because we do not set m_f to zero, four additional t/u-channel graphs contribute, shown in Fig. 18.14, that are sensitive to the higgsino components of the neutralino.

[3] In regions of parameter space where $m_{\widetilde{N}_1} \simeq \frac{1}{2}m_Z$ or $m_{\widetilde{N}_1} \simeq \frac{1}{2}m_\phi$ (where $\phi = h$, H, or A), the resonant $2 \to 1$ annihilation $\widetilde{N}_1\widetilde{N}_1 \to Z$ or $\widetilde{N}_1\widetilde{N}_1 \to \phi$ dominates the $2 \to 2$ annihilation processes considered here.

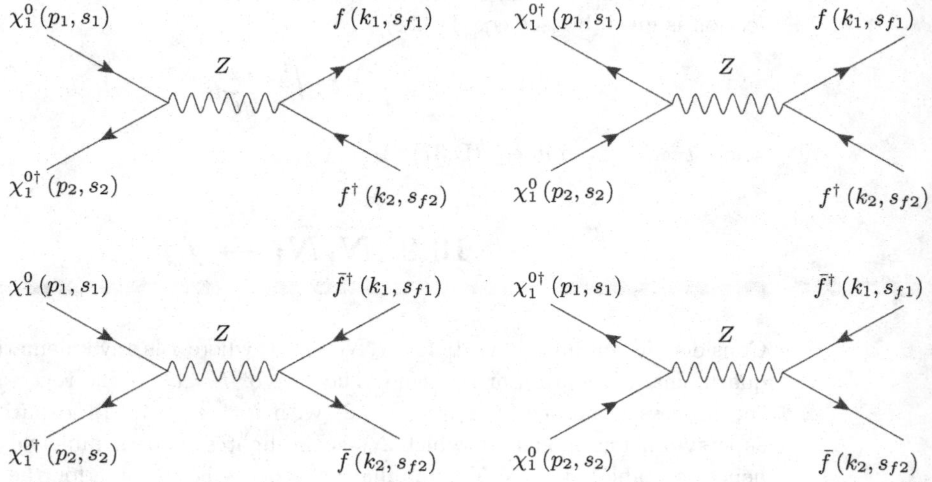

Fig. 18.11 The four Feynman diagrams for $\widetilde{N}_1\widetilde{N}_1 \to f\bar{f}$ via s–channel Z exchange, where f is a quark or lepton.

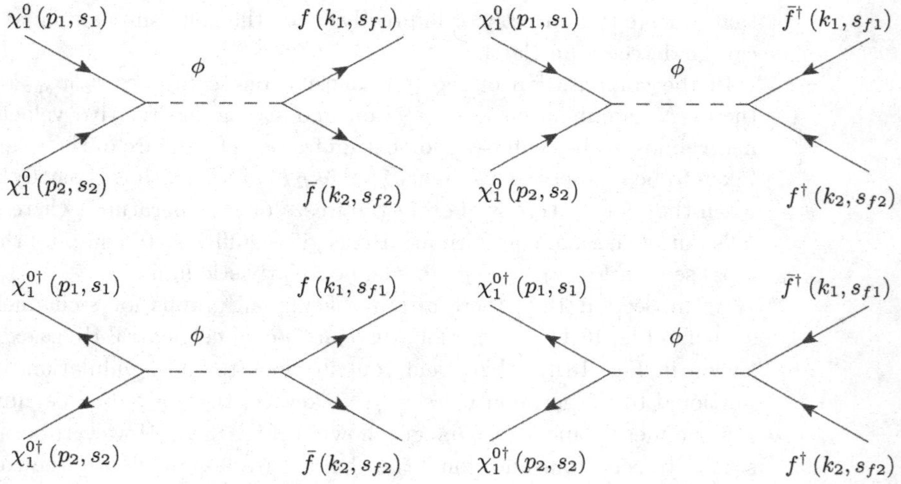

Fig. 18.12 Feynman diagrams for $\widetilde{N}_1\widetilde{N}_1 \to f\bar{f}$ via s–channel Higgs exchange. There are four diagrams for each possible neutral Higgs state $\phi = h,\, H$, and A.

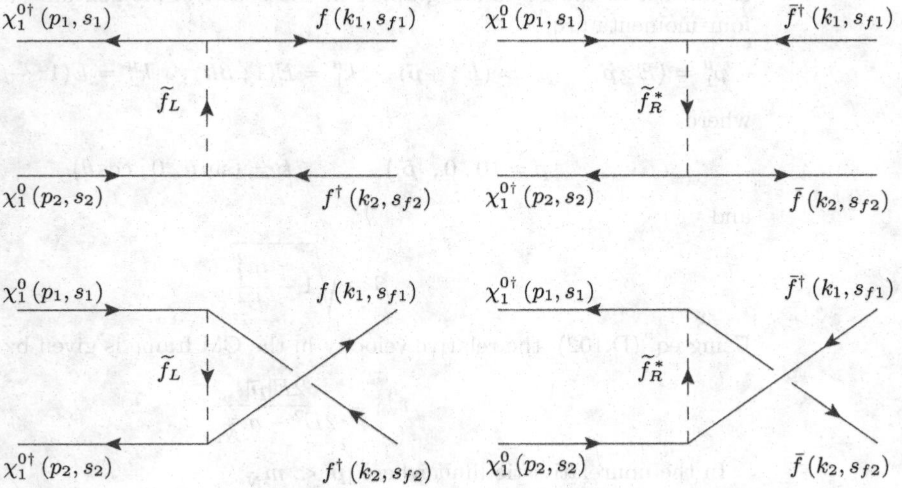

Fig. 18.13 The four Feynman diagrams for $\widetilde{N}_i\widetilde{N}_j \to f\bar{f}$ via t/u–channel $\widetilde{f}_L$ and $\widetilde{f}_R$ exchange, where $\widetilde{f}_L$ and $\widetilde{f}_R$ couple to the gaugino components of the neutralino.

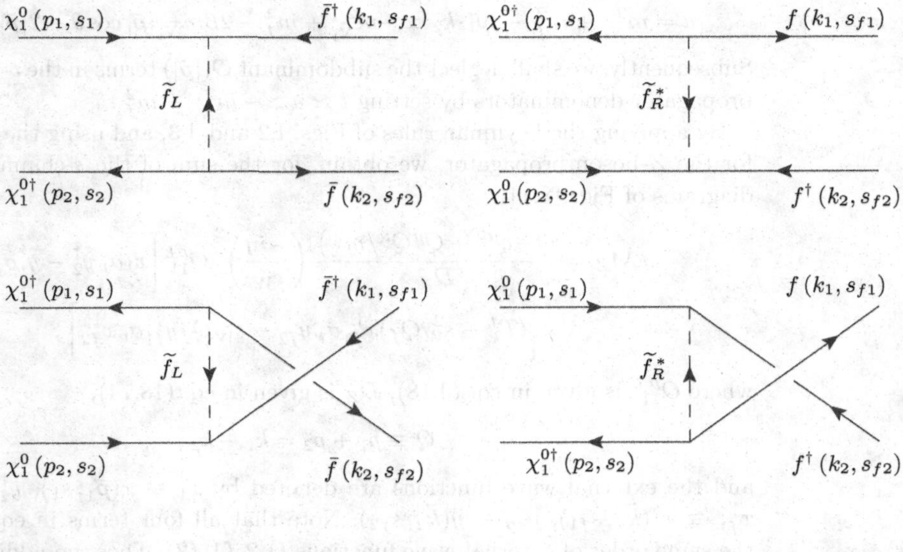

Fig. 18.14 The four Feynman diagrams for $\widetilde{N}_i\widetilde{N}_j \to f\bar{f}$ via t/u–channel $\widetilde{f}_L$ and $\widetilde{f}_R$ exchange, where $\widetilde{f}_L$ and $\widetilde{f}_R$ couple to the higgsino components of the neutralino.

The neutralino and the final-state fermion four-momenta and polarizations are as labeled in the Feynman graphs. In the center-of-momentum (CM) frame, the four-momenta are

$$p_1^\mu = (E\,;\vec{\boldsymbol{p}})\,,\quad p_2^\mu = (E\,;-\vec{\boldsymbol{p}})\,,\quad k_1^\mu = E(1\,;\beta\hat{\boldsymbol{k}})\,,\quad k_2^\mu = E(1\,;-\beta\hat{\boldsymbol{k}})\,, \tag{18.87}$$

where

$$\vec{\boldsymbol{p}} = \bigl(0\,,0\,,|\vec{\boldsymbol{p}}|\bigr)\,,\qquad \hat{\boldsymbol{k}} = \bigl(\sin\theta\,,0\,,\cos\theta\bigr)\,, \tag{18.88}$$

and

$$\beta \equiv \sqrt{1-\frac{m_f^2}{E^2}}\,. \tag{18.89}$$

Using eq. (D.102), the relative velocity in the CM frame is given by

$$v_{\rm rel} = \frac{2E|\vec{\boldsymbol{p}}|}{2E^2 - m^2_{\widetilde{N}_1}}\,. \tag{18.90}$$

In the nonrelativistic limit where $|\vec{\boldsymbol{p}}| \ll m_{\widetilde{N}_1}$,

$$E \simeq m_{\widetilde{N}_1} + \frac{|\vec{\boldsymbol{p}}|^2}{2m_{\widetilde{N}_1}}\,,\qquad v_{\rm rel} \simeq \frac{2|\vec{\boldsymbol{p}}|}{m_{\widetilde{N}_1}}\,, \tag{18.91}$$

and the Mandelstam invariants are given by

$$s = 4E^2 = 4m^2_{\widetilde{N}_1} + 4|\vec{\boldsymbol{p}}|^2\,, \tag{18.92}$$

$$t = m^2_{\widetilde{N}_1} + m_f^2 - 2p_1\!\cdot\! k_1 \simeq -m^2_{\widetilde{N}_1} + m_f^2 + 2\beta m_{\widetilde{N}_1}|\vec{\boldsymbol{p}}|\cos\theta - 2|\vec{\boldsymbol{p}}|^2\,, \tag{18.93}$$

$$u = m^2_{\widetilde{N}_1} + m_f^2 - 2p_1\!\cdot\! k_2 \simeq -m^2_{\widetilde{N}_1} + m_f^2 - 2\beta m_{\widetilde{N}_1}|\vec{\boldsymbol{p}}|\cos\theta - 2|\vec{\boldsymbol{p}}|^2\,. \tag{18.94}$$

Subsequently, we shall neglect the subdominant $\mathcal{O}(|\vec{\boldsymbol{p}}|)$ terms in the t- and u-channel propagator denominators by setting $t \simeq u \simeq -m^2_{\widetilde{N}_1} + m_f^2$.

By applying the Feynman rules of Figs. I.2 and J.3, and using the unitary gauge for the Z-boson propagator, we obtain, for the sum of the s-channel Z-exchange diagrams of Fig. 18.11,

$$i\mathcal{M}_Z = \frac{i\left(-g^{\mu\nu} + Q^\mu Q^\nu/m_Z^2\right)}{D_Z}\left(\frac{-ig}{c_W}\right)^2 O''^{L}_{11}\left[x_1\sigma_\mu y_2^\dagger - y_1^\dagger\overline{\sigma}_\mu x_2\right]$$
$$\times\left[(T_3^f - s_W^2 Q_f)x_{f1}^\dagger\overline{\sigma}_\nu y_{f2} - s_W^2 Q_f y_{f1}\sigma_\nu x_{f2}^\dagger\right]\,, \tag{18.95}$$

where O''^{L}_{11} is given in eq. (J.18), D_Z is given in eq. (18.74),

$$Q \equiv p_1 + p_2 = k_1 + k_2\,, \tag{18.96}$$

and the external wave functions are denoted by $x_1 \equiv x(\vec{\boldsymbol{p}}_1, s_1)$, $y_2^\dagger \equiv y^\dagger(\vec{\boldsymbol{p}}_2, s_2)$, $x_{f1}^\dagger \equiv x^\dagger(\vec{\boldsymbol{k}}_1, s_{f1})$, $y_{f2} \equiv y(\vec{\boldsymbol{k}}_2, s_{f2})$. Note that all four terms in eq. (18.95) have the same order of external wave functions (1,2,f1,f2). Thus, no additional relative signs arise (beyond the sign associated with the choice of the σ or $\overline{\sigma}$ version of the vertex Feynman rules). One can simplify the terms that originate from the $Q^\mu Q^\nu$

part of the Z-boson propagator by writing $Q^\mu = (p_1+p_2)^\mu$ and $Q^\nu = (k_1+k_2)^\nu$. Contracting the μ and ν indices with the help of eqs. (2.10)–(2.13) yields

$$(p_1+p_2)^\mu \left[x_1\sigma_\mu y_2^\dagger - y_1^\dagger\overline{\sigma}_\mu x_2\right] = 2m_{\widetilde{N}_1}(x_1x_2 - y_1^\dagger y_2^\dagger)\,, \tag{18.97}$$

and

$$(k_1+k_2)^\nu \left[(T_3^f - s_W^2 Q_f)x_{f1}^\dagger\overline{\sigma}_\nu y_{f2} - s_W^2 Q_f y_{f1}\sigma_\nu x_{f2}^\dagger\right] = T_3^f m_f(y_{f1}y_{f2} - x_{f1}^\dagger x_{f2}^\dagger). \tag{18.98}$$

Hence, we shall write

$$\mathcal{M}_Z \equiv \mathcal{M}_Z^{(1)} + \mathcal{M}_Z^{(2)}\,, \tag{18.99}$$

where

$$\begin{aligned} i\mathcal{M}_Z^{(1)} &= \frac{-ig^{\mu\nu}}{D_Z}\left(\frac{-ig}{c_W}\right)^2 O_{11}^{\prime\prime L}\left[x_1\sigma_\mu y_2^\dagger - y_1^\dagger\overline{\sigma}_\mu x_2\right] \\ &\quad\times\left[(T_3^f - s_W^2 Q_f)x_{f1}^\dagger\overline{\sigma}_\nu y_{f2} - s_W^2 Q_f y_{f1}\sigma_\nu x_{f2}^\dagger\right], \end{aligned} \tag{18.100}$$

$$i\mathcal{M}_Z^{(2)} = \frac{im_f m_{\widetilde{N}_1}}{m_Z^2 D_Z}\left(\frac{-ig}{c_W}\right)^2 O_{11}^{\prime\prime L}(2T_3^f)\left(x_1x_2 - y_1^\dagger y_2^\dagger\right)\left(y_{f1}y_{f2} - x_{f1}^\dagger x_{f2}^\dagger\right). \tag{18.101}$$

Next, we apply the Feynman rules of Figs. J.1 and J.5 to obtain the sum of the four s-channel Higgs exchange diagrams (for $\phi = h$, H, and A) of Fig. 18.12,

$$\begin{aligned} i\mathcal{M}_H = \sum_{\phi=h,H,A} \frac{i}{D_\phi}\left(\frac{-m_f}{\sqrt{2}\,v_f}\right)&\left[(Y^{\phi\chi_1^0\chi_1^0})x_1x_2 + (Y^{\phi\chi_1^0\chi_1^0})^* y_1^\dagger y_2^\dagger\right] \\ &\times\left[k_{f\phi}y_{f1}y_{f2} + k_{f\phi}^* x_{f1}^\dagger x_{f2}^\dagger\right], \end{aligned} \tag{18.102}$$

where $Y^{\phi\chi_1^0\chi_1^0}$ is given by eq. (J.24), and $D_\phi \equiv s - m_\phi^2 + i\Gamma_\phi m_\phi$. In addition, we have introduced the following notation:

$$k_{f\phi} \equiv \begin{cases} k_{d\phi}\,, & \text{for } f = d, e^-\,, \\ k_{u\phi}\,, & \text{for } f = u\,, \\ 0, & \text{for } f = \nu\,, \end{cases} \qquad v_f \equiv \begin{cases} v_d\,, & \text{for } f = d, e^-\,, \\ v_u\,, & \text{for } f = u\,, \nu\,, \end{cases} \tag{18.103}$$

where v_u, v_d are the neutral Higgs vacuum expectation values [see eq. (J.2)] and $k_{u\phi}$ and $k_{d\phi}$ are defined in eqs. (J.8) and (J.9). As the order of the spinor wave functions is $(1, 2, f1, f2)$ for all four terms of $\mathcal{M}_H$, no extra minus signs appear.

We next evaluate the t/u-channel exchange diagrams shown in Figs. 18.13 and 18.14. We neglect $\widetilde{f}_L$–$\widetilde{f}_R$ mixing. Eight Feynman graphs contribute, and we denote the total invariant amplitude by

$$\mathcal{M}_{\widetilde{f}} = \sum_{j=1}^{2}(\mathcal{M}_{\widetilde{f}_L}^{(tj)} + \mathcal{M}_{\widetilde{f}_L}^{(uj)} + \mathcal{M}_{\widetilde{f}_R}^{(tj)} + \mathcal{M}_{\widetilde{f}_R}^{(uj)})\,, \tag{18.104}$$

where $j = 1, 2$ labels the contributions of Figs. 18.13 and 18.14, respectively, and the other superscripts (t or u) and subscripts ($\widetilde{f}_L$ or $\widetilde{f}_R$) indicate the exchange channel and the exchanged particle, respectively. These matrix elements are evaluated by applying the rules of Fig. J.7.

The graphs of Fig. 18.13 are sensitive to the gaugino components of $\widetilde{N}_1$, and yield

$$i\mathcal{M}^{(t1)}_{\widetilde{f}_L} = (-1)\left(-ig\sqrt{2}\right)^2\left(\frac{i}{t-m^2_{\widetilde{f}_L}}\right)\left|T^f_3 N_{12}+\frac{s_W}{c_W}(Q_f - T^f_3)N_{11}\right|^2(y^\dagger_1 x^\dagger_{f1})(x_2 y_{f2}), \tag{18.105}$$

$$i\mathcal{M}^{(u1)}_{\widetilde{f}_L} = \left(-ig\sqrt{2}\right)^2\left(\frac{i}{u-m^2_{\widetilde{f}_L}}\right)\left|T^f_3 N_{12}+\frac{s_W}{c_W}(Q_f - T^f_3)N_{11}\right|^2(x_1 y_{f2})(y^\dagger_2 x^\dagger_{f1}), \tag{18.106}$$

$$i\mathcal{M}^{(t1)}_{\widetilde{f}_R} = (-1)\left(i\sqrt{2}g\frac{s_W}{c_W}Q_f\right)^2\left(\frac{i}{t-m^2_{\widetilde{f}_R}}\right)|N_{11}|^2\,(x_1 y_{f1})(y^\dagger_2 x^\dagger_{f2})\,, \tag{18.107}$$

$$i\mathcal{M}^{(u1)}_{\widetilde{f}_R} = \left(i\sqrt{2}g\frac{s_W}{c_W}Q_f\right)^2\left(\frac{i}{u-m^2_{\widetilde{f}_R}}\right)|N_{11}|^2\,(y^\dagger_1 x^\dagger_{f2})(x_2 y_{f1})\,. \tag{18.108}$$

The extra factors of (-1) in eqs. (18.105) and (18.107) are present because the order of the external wave functions in these cases is an odd permutation of the ordering $(1, 2, f1, f2)$ established in the computation of the s-channel amplitudes.

The graphs of Fig. 18.14 are sensitive to the higgsino components of $\widetilde{N}_1$ and yield

$$i\mathcal{M}^{(t2)}_{\widetilde{f}_L} = (-1)\left(\frac{-im_f}{v_f}\right)^2\left(\frac{i}{t-m^2_{\widetilde{f}_L}}\right)|N_{1f}|^2\,(x_1 y_{f1})(y^\dagger_2 x^\dagger_{f2})\,, \tag{18.109}$$

$$i\mathcal{M}^{(u2)}_{\widetilde{f}_L} = \left(\frac{-im_f}{v_f}\right)^2\left(\frac{i}{u-m^2_{\widetilde{f}_L}}\right)|N_{1f}|^2\,(y^\dagger_1 x^\dagger_{f2})(x_2 y_{f1})\,, \tag{18.110}$$

$$i\mathcal{M}^{(t2)}_{\widetilde{f}_R} = (-1)\left(\frac{-im_f}{v_f}\right)^2\left(\frac{i}{t-m^2_{\widetilde{f}_R}}\right)|N_{1f}|^2\,(y^\dagger_1 x^\dagger_{f1})(x_2 y_{f2})\,, \tag{18.111}$$

$$i\mathcal{M}^{(u2)}_{\widetilde{f}_R} = \left(\frac{-im_f}{v_f}\right)^2\left(\frac{i}{u-m^2_{\widetilde{f}_R}}\right)|N_{1f}|^2\,(x_1 y_{f2})(y^\dagger_2 x^\dagger_{f1})\,, \tag{18.112}$$

where v_f is defined in eq. (18.103), and

$$N_{1f} \equiv \begin{cases} N_{13}\,, & \text{for } f = d,\, e^-\,, \\ N_{14}\,, & \text{for } f = u\,, \\ 0\,, & \text{for } f = \nu\,. \end{cases} \tag{18.113}$$

As before, the explicit factors of (-1) are due to the ordering of the external wave functions.

It is convenient to write the total matrix element for $\widetilde{N}_1\widetilde{N}_1 \to f\bar{f}$ as the sum of products of separate neutralino and final-state fermionic currents. The contributions of the s-channel diagrams are already in this form. The contributions of the t- and

u-channel diagrams given in eqs. (18.105)–(18.112) can be rearranged using the Fierz identities of eqs. (1.138)–(1.140):

$$y_1^\dagger x_{f1}^\dagger x_2 y_{f2} = -\tfrac{1}{2}(y_1^\dagger \overline{\sigma}^\mu x_2)(x_{f1}^\dagger \overline{\sigma}_\mu y_{f2}), \tag{18.114}$$

$$x_1 y_{f2} y_2^\dagger x_{f1}^\dagger = -\tfrac{1}{2}(x_1 \sigma^\mu y_2^\dagger)(x_{f1}^\dagger \overline{\sigma}_\mu y_{f2}), \tag{18.115}$$

$$x_1 y_{f1} y_2^\dagger x_{f2}^\dagger = -\tfrac{1}{2}(x_1 \sigma^\mu y_2^\dagger)(y_{f1} \sigma_\mu x_{f2}^\dagger), \tag{18.116}$$

$$y_1^\dagger x_{f2}^\dagger x_2 y_{f1} = -\tfrac{1}{2}(y_1^\dagger \overline{\sigma}^\mu x_2)(y_{f1} \sigma_\mu x_{f2}^\dagger). \tag{18.117}$$

Combining the results of the s-, t-, and u-channel contributions, we have, for the total amplitude,

$$\begin{aligned}\mathcal{M} = {}& \frac{m_f m_{\widetilde{N}_1}}{m_Z^2} c_0 \left(x_1 x_2 - y_1^\dagger y_2^\dagger\right)\left(y_{f1} y_{f2} - x_{f1}^\dagger x_{f2}^\dagger\right) + c_1 (y_1^\dagger \overline{\sigma}^\mu x_2)(x_{f1}^\dagger \overline{\sigma}_\mu y_{f2}) \\ & + c_2 (x_1 \sigma^\mu y_2^\dagger)(x_{f1}^\dagger \overline{\sigma}_\mu y_{f2}) + c_3 (x_1 \sigma^\mu y_2^\dagger)(y_{f1} \sigma_\mu x_{f2}^\dagger) + c_4 (y_1^\dagger \overline{\sigma}^\mu x_2)(y_{f1} \sigma_\mu x_{f2}^\dagger) \\ & + m_f \Big[c_5 (x_1 x_2)(y_{f1} y_{f2}) + c_6 (x_1 x_2)(x_{f1}^\dagger x_{f2}^\dagger) + c_7 (y_1^\dagger y_2^\dagger)(y_{f1} y_{f2}) \\ & \qquad + c_8 (y_1^\dagger y_2^\dagger)(x_{f1}^\dagger x_{f2}^\dagger) \Big], \end{aligned} \tag{18.118}$$

where the coefficients $c_0, c_1, \ldots, c_4$ are given by

$$c_0 = -g^2 \frac{2 T_3^f O_{11}''^L}{c_W^2 D_Z}, \tag{18.119}$$

$$c_1 = -g^2 \left[\frac{(T_3^f - s_W^2 Q_f) O_{11}''^L}{c_W^2 D_Z} + \frac{|T_3^f N_{12} + \frac{s_W}{c_W}(Q_f - T_3^f) N_{11}|^2}{t - m_{\tilde{f}_L}^2} \right] - \frac{m_f^2}{2 v_f^2} \left(\frac{|N_{1f}|^2}{t - m_{\tilde{f}_R}^2} \right), \tag{18.120}$$

$$c_2 = g^2 \left[\frac{(T_3^f - s_W^2 Q_f) O_{11}''^L}{c_W^2 D_Z} + \frac{|T_3^f N_{12} + \frac{s_W}{c_W}(Q_f - T_3^f) N_{11}|^2}{u - m_{\tilde{f}_L}^2} \right] + \frac{m_f^2}{2 v_f^2} \left(\frac{|N_{1f}|^2}{u - m_{\tilde{f}_R}^2} \right), \tag{18.121}$$

$$c_3 = -g^2 \frac{s_W^2}{c_W^2} Q_f \left[\frac{O_{11}''^L}{D_Z} + \frac{Q_f |N_{11}|^2}{t - m_{\tilde{f}_R}^2} \right] - \frac{m_f^2}{2 v_f^2} \left(\frac{|N_{1f}|^2}{t - m_{\tilde{f}_L}^2} \right), \tag{18.122}$$

$$c_4 = g^2 \frac{s_W^2}{c_W^2} Q_f \left[\frac{O_{11}''^L}{D_Z} + \frac{Q_f |N_{11}|^2}{u - m_{\tilde{f}_R}^2} \right] + \frac{m_f^2}{2 v_f^2} \left(\frac{|N_{1f}|^2}{u - m_{\tilde{f}_L}^2} \right). \tag{18.123}$$

The coefficients $c_5, \ldots, c_8$ are obtained from eq. (18.102) and represent the s-channel Higgs exchange contributions to the annihilation matrix element.

In the nonrelativistic limit, $|\vec{p}| \ll m_{\widetilde{N}_1}$. Then $t \simeq u \simeq -m_{\widetilde{N}_1}^2 + m_f^2$, and we can approximate[4] $c_1 = -c_2$ and $c_3 = -c_4$. Hence, the total amplitude, eq. (18.118), can

[4] In particular, we assume that $\tilde{f}_L$ and $\tilde{f}_R$ are significantly heavier than all other particles in the annihilation process. Consequently, we can ignore all $\mathcal{O}(|\vec{p}|/m_{\tilde{f}_{L,R}})$ terms in $c_1 + c_2$ and $c_3 + c_4$.

be written as

$$\mathcal{M} = \frac{m_f m_{\widetilde{N}_1}}{m_Z^2} c_0 \left(x_1 x_2 - y_1^\dagger y_2^\dagger\right)\left(y_{f1} y_{f2} - x_{f1}^\dagger x_{f2}^\dagger\right)$$
$$+ \left[y_1^\dagger \overline{\sigma}^\mu x_2 - x_1 \sigma^\mu y_2^\dagger\right]\left[c_1 (x_{f1}^\dagger \overline{\sigma}_\mu y_{f2}) - c_3 (y_{f1} \sigma_\mu x_{f2}^\dagger)\right] + \mathcal{M}_H, \qquad (18.124)$$

where the s-channel Higgs exchange contributions, $\mathcal{M}_H$, will be neglected for simplicity in the subsequent analysis. The spin-averaged squared matrix element for $\widetilde{N}_1 \widetilde{N}_1 \to f\bar{f}$ then takes the following form:

$$\frac{1}{4}\sum_{\{s\}} |\mathcal{M}_Z + \mathcal{M}_{\widetilde{f}}|^2 = N_{\mu\nu}\left[|c_1|^2 F_1^{\mu\nu} + |c_3|^2 F_2^{\mu\nu} - 2\,\mathrm{Re}(c_1 c_3^*) F_{12}^{\mu\nu}\right] \qquad (18.125)$$
$$+ \frac{m_f^2 m_{\widetilde{N}_1}^2}{m_Z^4} |c_0|^2 N F + \frac{2 m_f m_{\widetilde{N}_1}}{m_Z^2} \mathrm{Re}[c_0^*(c_1 + c_3)] N_\mu F^\mu, \qquad (18.126)$$

where $N_{\mu\nu}$, N_μ, and N are spin-averaged tensor, vector, and scalar quantities that depend on the initial-state neutralino kinematics; $F_1^{\mu\nu}$, $F_2^{\mu\nu}$, $F_{12}^{\mu\nu}$, F^μ, and F are spin-summed tensor, vector, and scalar quantities that depend on the final-state fermion kinematics; and $\{s\} \equiv \{s_1, s_2, s_{f1}, s_{f2}\}$. These quantities are easily computed using the projection operators of eqs. (2.63)–(2.66) and the standard trace techniques to perform the spin averages and sums. Explicitly, the spin-averaged neutralino quantities are

$$N \equiv \frac{1}{4}\sum_{s_1, s_2} (x_1 x_2 - y_1^\dagger y_2^\dagger)(x_2^\dagger x_1^\dagger - y_2 y_1) = p_1 \cdot p_2 + m_{\widetilde{N}_1}^2 = 2E^2, \qquad (18.127)$$

$$N^\mu \equiv \frac{1}{4}\sum_{s_1, s_2} (y_1^\dagger \overline{\sigma}^\mu x_2 - x_1 \sigma^\mu y_2^\dagger)(x_2^\dagger x_1^\dagger - y_2 y_1)$$
$$= -m_{\widetilde{N}_1}(p_1 + p_2)^\mu = \begin{cases} -2 m_{\widetilde{N}_1} E\,, & \mu = 0\,, \\ 0\,, & \mu = i\,, \end{cases} \qquad (18.128)$$

and a symmetric second-rank tensor,

$$N^{\mu\nu} \equiv \frac{1}{4}\sum_{s_1, s_2} (y_1^\dagger \overline{\sigma}^\mu x_2 - x_1 \sigma^\mu y_2^\dagger)(x_2^\dagger \overline{\sigma}^\nu y_1 - y_2 \sigma^\nu x_1^\dagger)$$
$$= p_1^\mu p_2^\nu + p_2^\mu p_1^\nu - g^{\mu\nu}(p_1 \cdot p_2 - m_{\widetilde{N}_1}^2)$$
$$= \begin{cases} 2 m_{\widetilde{N}_1}^2\,, & \mu = \nu = 0\,, \\ 0\,, & \mu = 0\,, \nu = j \text{ or } \mu = i\,, \nu = 0\,, \\ 2\left[|\vec{\boldsymbol{p}}|^2 \delta^{ij} - p^i p^j\right]\,, & \mu = i,\, \nu = j\,, \end{cases} \qquad (18.129)$$

where the final results given in eqs. (18.127)–(18.129) have been evaluated in the

CM frame. Similarly, the spin-summed final-state fermion quantities are

$$F \equiv \sum_{s_{f1},s_{f2}} (y_{f1}y_{f2} - x^\dagger_{f1}x^\dagger_{f2})(y^\dagger_{f2}y^\dagger_{f1} - x_{f2}x_{f1}) = 4(k_1\cdot k_2 + m_f^2) = 8E^2\,, \quad (18.130)$$

$$\begin{aligned} F^\mu &\equiv \sum_{s_{f1},s_{f2}} (x^\dagger_{f1}\overline{\sigma}^\mu y_{f2})(y^\dagger_{f2}y^\dagger_{f1} - x_{f2}x_{f1}) = -\sum_{s_{f1},s_{f2}} (y_{f1}\sigma^\mu x^\dagger_{f2})(y^\dagger_{f2}y^\dagger_{f1} - x_{f2}x_{f1}) \\ &= 2m_f(k_1+k_2)^\mu = \begin{cases} 4m_f E\,, & \mu = 0\,, \\ 0\,, & \mu = i\,, \end{cases} \end{aligned} \quad (18.131)$$

after evaluating the above quantities in the CM frame, and

$$F_1^{\mu\nu} \equiv \sum_{s_{f1},s_{f2}} (x^\dagger_{f1}\overline{\sigma}^\mu y_{f2})(y^\dagger_{f2}\overline{\sigma}^\nu x_{f1}) = k_{1\rho}k_{2\lambda}\,\mathrm{Tr}(\sigma^\rho\overline{\sigma}^\mu\sigma^\lambda\overline{\sigma}^\nu)\,, \quad (18.132)$$

$$F_2^{\mu\nu} \equiv \sum_{s_{f1},s_{f2}} (y_{f1}\sigma^\mu x^\dagger_{f2})(x_{f2}\sigma^\nu y^\dagger_{f1}) = k_{1\rho}k_{2\lambda}\,\mathrm{Tr}(\overline{\sigma}^\rho\sigma^\mu\overline{\sigma}^\lambda\sigma^\nu)\,, \quad (18.133)$$

$$\begin{aligned} F_{12}^{\mu\nu} &\equiv \sum_{s_{f1},s_{f2}} (y_{f1}\sigma^\mu x^\dagger_{f2})(y^\dagger_{f2}\overline{\sigma}^\nu x_{f1}) = \sum_{s_{f1},s_{f2}} (x^\dagger_{f1}\overline{\sigma}^\mu y_{f2})(x_{f2}\sigma^\nu y^\dagger_{f1}) \\ &= -m_f^2\,\mathrm{Tr}(\sigma^\mu\overline{\sigma}^\nu) = -2m_f^2 g^{\mu\nu} \end{aligned} \quad (18.134)$$

Since $N^{\mu\nu}$ is symmetric, the antisymmetric parts of $F_1^{\mu\nu}$ and $F_2^{\mu\nu}$ do not contribute in eq. (18.126). The symmetric parts of $F_1^{\mu\nu}$ and $F_2^{\mu\nu}$ are equal and given by

$$\begin{aligned} [F_1^{\mu\nu}]_{\rm symm} &= [F_2^{\mu\nu}]_{\rm symm} = 2(k_1^\mu k_2^\nu + k_1^\nu k_2^\mu - k_1\cdot k_2 g^{\mu\nu}) \\ &= \begin{cases} 2m_f^2\,, & \mu=\nu=0, \\ 0, & \mu=0,\ \nu=j \text{ or } \mu=i,\ \nu=0, \\ F^{ij} & \mu=i \text{ and } \nu=j, \end{cases} \end{aligned} \quad (18.135)$$

where

$$F^{ij} \equiv 2m_f^2(2\hat{\boldsymbol{k}}^i\hat{\boldsymbol{k}}^j - \delta^{ij}) - 4E^2(\hat{\boldsymbol{k}}^i\hat{\boldsymbol{k}}^j - \delta^{ij}). \quad (18.136)$$

The spin-averaged squared matrix element for $\widetilde{N}_1\widetilde{N}_1 \to f\bar{f}$ given by eq. (18.126) can now be fully evaluated, resulting in

$$\begin{aligned} \frac{1}{4}\sum_{\{s\}} |\mathcal{M}_Z + \mathcal{M}_{\tilde{f}}|^2 &= 4(|c_1|^2 + |c_3|^2)\left[m_{\widetilde{N}_1}^2 m_f^2 + 2|\vec{\boldsymbol{p}}|^2(E^2(1+\cos^2\theta) - m_f^2\cos^2\theta)\right] \\ &+ 8m_f^2\,\mathrm{Re}(c_1c_3^*)\left[m_{\widetilde{N}_1}^2 - 2|\vec{\boldsymbol{p}}|^2\right] + \frac{16m_f^2 m_{\widetilde{N}_1}^2}{m_Z^4}E^2\Big[E^2|c_0|^2 - m_Z^2\,\mathrm{Re}[c_0^*(c_1+c_3)]\Big]. \end{aligned} \quad (18.137)$$

In the nonrelativistic limit, we use eq. (18.91) and drop terms of $\mathcal{O}(|\vec{\boldsymbol{p}}|^4)$.

To compute $v_{\rm rel}\sigma_{\rm ann}$, we first make use of eqs. (18.90) and (D.106) to obtain the differential annihilation cross section in the CM frame:

$$v_{\rm rel}\frac{d\sigma_{\rm ann}}{d\Omega} = \frac{1}{64\pi^2(s-2m^2_{\widetilde{N}_1})}\left(1-\frac{4m_f^2}{s}\right)^{1/2}\frac{1}{4}\sum_{\{s\}}|\mathcal{M}|^2 . \tag{18.138}$$

Inserting the squared matrix element obtained above into eq. (18.138) and integrating over solid angles, we end up with

$$\begin{aligned} v_{\rm rel}\sigma_{\rm ann} &= \frac{1}{8\pi(2E^2-m^2_{\widetilde{N}_1})}\left(1-\frac{m_f^2}{E^2}\right)^{1/2} \\ &\times\Bigg\{\left(|c_1|^2+|c_3|^2\right)\left[m^2_{\widetilde{N}_1}m_f^2+\frac{2|\vec{p}|^2}{3}\left(4m^2_{\widetilde{N}_1}-m_f^2\right)\right] \\ &+\frac{4m_f^2m^2_{\widetilde{N}_1}}{m_Z^4}\left[m^2_{\widetilde{N}_1}(m^2_{\widetilde{N}_1}+2|\vec{p}|^2)|c_0|^2-m_Z^2(m^2_{\widetilde{N}_1}+|\vec{p}|^2)\,{\rm Re}[c_0^*(c_1+c_3)]\right] \\ &+2m_f^2\,{\rm Re}(c_1c_3^*)\left[m^2_{\widetilde{N}_1}-2|\vec{p}|^2\right]+\mathcal{O}(|\vec{p}|^4)\Bigg\}, \end{aligned} \tag{18.139}$$

where the effects of the s-channel Higgs boson exchanges have been omitted.

The momentum dependence of eq. (18.139) reflects the famous p-wave suppression of the annihilation cross section in the $m_f = 0$ limit. In general, the annihilation cross section in the nonrelativistic limit behaves as $v_{\rm rel}\,\sigma_{\rm ann} \propto |\vec{p}|^{2\ell}$. Applying this result to eq. (18.139) in the $m_f = 0$ limit implies that $\ell = 1$. This is a consequence of the Majorana nature of the neutralino. In particular, in the limit of $m_f = 0$, the $f\bar{f}$ pair is in a $J = 1$ angular momentum state. However, Fermi–Dirac statistics dictates that at threshold, a pair of identical Majorana fermions in a $J = 1$ state must have relative orbital angular momentum $\ell = 1$ (corresponding to p-wave annihilation). The s-wave annihilation (corresponding to the Majorana fermion pair in a $J = 0$ state) is suppressed by a factor of m_f^2, as is evident from eq. (18.139).

18.10 Gluino Pair Production from Gluon Fusion: $gg \to \widetilde{g}\widetilde{g}$

In this section we will compute the cross section for the process $gg \to \widetilde{g}\widetilde{g}$. The relevant Feynman diagrams are shown in Fig. 18.15. The initial-state gluons have $SU(3)_C$ adjoint representation indices a and b, with momenta p_1 and p_2 and polarization vectors $\varepsilon_1^\mu = \varepsilon^\mu(\vec{p}_1,\lambda_1)$ and $\varepsilon_2^\mu = \varepsilon^\mu(\vec{p}_2,\lambda_2)$, respectively. The final-state gluinos carry adjoint representation indices c and d, with momenta k_1 and k_2 and wave function spinors $x_1^\dagger = x^\dagger(\vec{k}_1,\lambda_1')$ or $y_1 = y(\vec{k}_1,\lambda_1')$ and $x_2^\dagger = x^\dagger(\vec{k}_2,\lambda_2')$ or $y_2 = y(\vec{k}_2,\lambda_2')$, respectively.

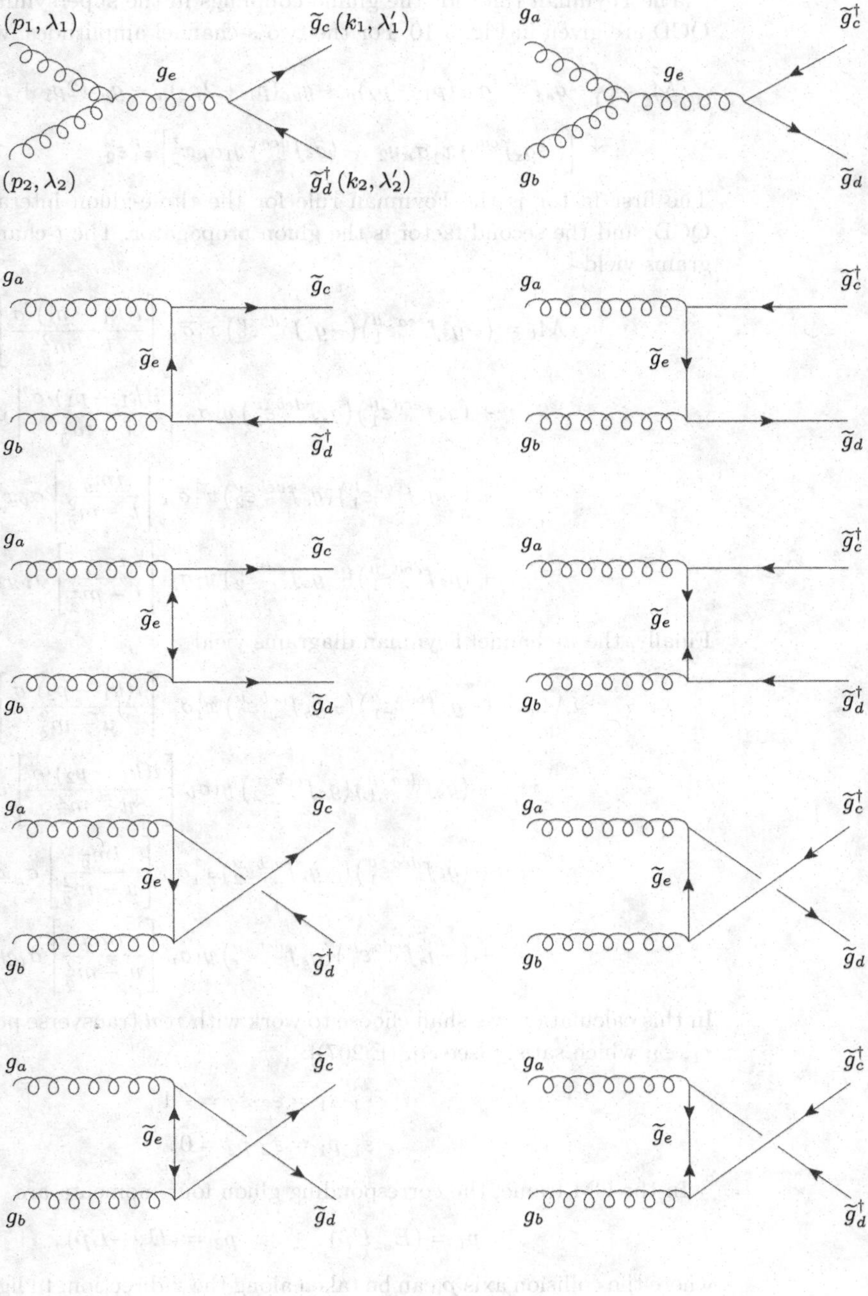

Fig. 18.15 The 10 Feynman diagrams for $gg \to \widetilde{g}\widetilde{g}$. The momentum and spin polarization assignments are indicated on the first diagram.

The Feynman rules for the gluino couplings in the supersymmetric extension of QCD are given in Fig. J.10. For the two s-channel amplitudes, we obtain

$$i\mathcal{M}_s = \left\{-g_s f^{abe}[g_{\mu\nu}(p_1 - p_2)_\rho + g_{\nu\rho}(p_1 + 2p_2)_\mu - g_{\mu\rho}(2p_1 + p_2)_\nu]\right\}\left(\frac{-ig^{\rho\kappa}}{s}\right)$$
$$\times \left[(-g_s f^{cde})\, x_1^\dagger \overline{\sigma}_\kappa y_2 + (g_s f^{dce})\, y_1 \sigma_\kappa x_2^\dagger\right] \varepsilon_1^\mu \varepsilon_2^\nu. \qquad (18.140)$$

The first factor is the Feynman rule for the three-gluon interaction of standard QCD, and the second factor is the gluon propagator. The t-channel Feynman diagrams yield

$$i\mathcal{M}_t = (-g_s f^{cea}\varepsilon_1^\mu)(-g_s f^{edb}\varepsilon_2^\nu)\, x_1^\dagger \overline{\sigma}_\mu \left[\frac{i(k_1 - p_1)\cdot\sigma}{t - m_{\tilde{g}}^2}\right] \overline{\sigma}_\nu y_2$$
$$+ (g_s f^{eca}\varepsilon_1^\mu)(g_s f^{deb}\varepsilon_2^\nu)\, y_1 \sigma_\mu \left[\frac{i(k_1 - p_1)\cdot\overline{\sigma}}{t - m_{\tilde{g}}^2}\right] \sigma_\nu x_2^\dagger$$
$$+ (-g_s f^{cea}\varepsilon_1^\mu)(g_s f^{deb}\varepsilon_2^\nu)\, x_1^\dagger \overline{\sigma}_\mu \left[\frac{i m_{\tilde{g}}}{t - m_{\tilde{g}}^2}\right] \sigma_\nu x_2^\dagger$$
$$+ (g_s f^{eca}\varepsilon_1^\mu)(-g_s f^{edb}\varepsilon_2^\nu)\, y_1 \sigma_\mu \left[\frac{i m_{\tilde{g}}}{t - m_{\tilde{g}}^2}\right] \overline{\sigma}_\nu y_2. \qquad (18.141)$$

Finally, the u-channel Feynman diagrams yield

$$i\mathcal{M}_u = (-g_s f^{eda}\varepsilon_1^\mu)(-g_s f^{ceb}\varepsilon_2^\nu)\, x_1^\dagger \overline{\sigma}_\nu \left[\frac{i(k_1 - p_2)\cdot\sigma}{u - m_{\tilde{g}}^2}\right] \overline{\sigma}_\mu y_2$$
$$+ (g_s f^{dea}\varepsilon_1^\mu)(g_s f^{ecb}\varepsilon_2^\nu)\, y_1 \sigma_\nu \left[\frac{i(k_1 - p_2)\cdot\overline{\sigma}}{u - m_{\tilde{g}}^2}\right] \sigma_\mu x_2^\dagger$$
$$+ (g_s f^{dea}\varepsilon_1^\mu)(-g_s f^{ceb}\varepsilon_2^\nu)\, x_1^\dagger \overline{\sigma}_\nu \left[\frac{i m_{\tilde{g}}}{u - m_{\tilde{g}}^2}\right] \sigma_\mu x_2^\dagger$$
$$+ (-g_s f^{eda}\varepsilon_1^\mu)(g_s f^{ecb}\varepsilon_2^\nu)\, y_1 \sigma_\nu \left[\frac{i m_{\tilde{g}}}{u - m_{\tilde{g}}^2}\right] \overline{\sigma}_\mu y_2. \qquad (18.142)$$

In this calculation, we shall choose to work with *real* transverse polarization vectors ε_1, ε_2, which satisfy [see eq. (E.207)]:

$$\varepsilon_1\cdot\varepsilon_1 = \varepsilon_2\cdot\varepsilon_2 = -1\,, \qquad (18.143)$$
$$\varepsilon_1\cdot p_1 = \varepsilon_2\cdot p_2 = 0\,. \qquad (18.144)$$

In the CM frame, the corresponding gluon four-momenta are

$$p_1 = (E\,;\, E\hat{\boldsymbol{p}})\,, \qquad p_2 = (E\,;\, -E\hat{\boldsymbol{p}})\,, \qquad (18.145)$$

where the collision axis $\hat{\boldsymbol{p}}$ can be taken along the z-direction. In light of eq. (18.144), it follows that, in the CM frame,

$$\varepsilon_1\cdot p_2 = \varepsilon_2\cdot p_1 = \varepsilon_1\cdot(k_1 + k_2) = \varepsilon_2\cdot(k_1 + k_2) = 0\,, \qquad (18.146)$$

for all choices of $\lambda_1, \lambda_2 = \pm 1$.

The sums over gluon polarizations will be performed using eq. (E.247). In particular, in the CM frame we recognize that $\widetilde{p}_1 = p_2$ and $\widetilde{p}_2 = p_1$. It then follows that $2\widetilde{p}_1 \cdot \widetilde{p}_2 = s$. Hence, for real polarization vectors,

$$\sum_{\lambda_1} \varepsilon_1^\mu \varepsilon_1^\nu = \sum_{\lambda_2} \varepsilon_2^\mu \varepsilon_2^\nu = -g^{\mu\nu} + \frac{2\,(p_1^\mu p_2^\nu + p_2^\mu p_1^\nu)}{s}. \tag{18.147}$$

Before squaring the amplitude, it is convenient to rewrite the last two terms in each of eqs. (18.141) and (18.142) by using the identities [see eq. (2.13)]

$$m_{\widetilde{g}} x_1^\dagger = y_1(k_1 \cdot \sigma)\,, \qquad m_{\widetilde{g}} y_1 = x_1^\dagger (k_1 \cdot \overline{\sigma})\,. \tag{18.148}$$

Using eqs. (1.90) and (1.91), the resulting matrix element is then reduced to a sum of terms that each contain exactly one σ or $\overline{\sigma}$ matrix. We define convenient factors

$$G_s \equiv \frac{g_s^2 f^{abe} f^{cde}}{s}, \tag{18.149}$$

$$G_t \equiv \frac{g_s^2 f^{ace} f^{bde}}{t - m_{\widetilde{g}}^2}, \tag{18.150}$$

$$G_u \equiv \frac{g_s^2 f^{ade} f^{bce}}{u - m_{\widetilde{g}}^2}. \tag{18.151}$$

Then the total amplitude is given by

$$\mathcal{M} = \mathcal{M}_s + \mathcal{M}_t + \mathcal{M}_u = x_1^\dagger a \cdot \overline{\sigma} y_2 + y_1 a^* \cdot \sigma x_2^\dagger, \tag{18.152}$$

where

$$\begin{aligned} a^\mu \equiv & -(G_t + G_s)\varepsilon_1 \cdot \varepsilon_2\, p_1^\mu - (G_u - G_s)\varepsilon_1 \cdot \varepsilon_2\, p_2^\mu - 2G_t k_1 \cdot \varepsilon_1\, \varepsilon_2^\mu - 2G_u k_1 \cdot \varepsilon_2\, \varepsilon_1^\mu \\ & - i\epsilon^{\mu\nu\rho\kappa} \varepsilon_{1\nu} \varepsilon_{2\rho} (G_t p_1 - G_u p_2)_\kappa . \end{aligned} \tag{18.153}$$

Squaring the amplitude using eqs. (1.105) and (1.106), we get

$$\begin{aligned} |\mathcal{M}|^2 = & \, x_1^\dagger a \cdot \overline{\sigma} y_2 y_2^\dagger a^* \cdot \overline{\sigma} x_1 + y_1 a^* \cdot \sigma x_2^\dagger x_2 a \cdot \sigma y_1^\dagger + x_1^\dagger a \cdot \overline{\sigma} y_2 x_2 a \cdot \sigma y_1^\dagger \\ & + y_1 a^* \cdot \sigma x_2^\dagger y_2^\dagger a^* \cdot \overline{\sigma} x_1 . \end{aligned} \tag{18.154}$$

Summing over the gluino helicities using eqs. (2.63)–(2.66), and performing the resulting traces using eqs. (1.92)–(1.94), we find

$$\sum_{\lambda_1', \lambda_2'} |\mathcal{M}|^2 = 8\, \mathrm{Re}[a \cdot k_1 a^* \cdot k_2] - 4 a \cdot a^*\, k_1 \cdot k_2 - 4i\epsilon^{\mu\nu\rho\kappa} k_{1\mu} k_{2\nu} a_\rho a_\kappa^* - 4 m_{\widetilde{g}}^2\, \mathrm{Re}[a^2]. \tag{18.155}$$

Inserting the explicit form for a^μ given in eq. (18.153) then yields

$$\begin{aligned} \sum_{\lambda_1', \lambda_2'} |\mathcal{M}|^2 = & \, 2(t - m_{\widetilde{g}}^2)(u - m_{\widetilde{g}}^2)[(G_t + G_u)^2 + 4(G_s + G_t)(G_s - G_u)(\varepsilon_1 \cdot \varepsilon_2)^2] \\ & + 16(G_t + G_u)[G_s(t - u) + G_t(t - m_{\widetilde{g}}^2) + G_u(u - m_{\widetilde{g}}^2)](\varepsilon_1 \cdot \varepsilon_2)(k_1 \cdot \varepsilon_1)(k_1 \cdot \varepsilon_2) \\ & - 32(G_t + G_u)^2 (k_1 \cdot \varepsilon_1)^2 (k_1 \cdot \varepsilon_2)^2 . \end{aligned} \tag{18.156}$$

Using eq. (18.147) to perform the sums over gluon polarizations,

$$\sum_{\lambda_1,\lambda_2} 1 = 4, \qquad \sum_{\lambda_1,\lambda_2} (\varepsilon_1 \cdot \varepsilon_2)^2 = 2, \tag{18.157}$$

$$\sum_{\lambda_1,\lambda_2} (\varepsilon_1 \cdot \varepsilon_2)(k_1 \cdot \varepsilon_1)(k_1 \cdot \varepsilon_2) = m_{\widetilde{g}}^2 - \frac{(t-m_{\widetilde{g}}^2)(u-m_{\widetilde{g}}^2)}{s}, \tag{18.158}$$

$$\sum_{\lambda_1,\lambda_2} (k_1 \cdot \varepsilon_1)^2(k_1 \cdot \varepsilon_2)^2 = \left[m_{\widetilde{g}}^2 - \frac{(t-m_{\widetilde{g}}^2)(u-m_{\widetilde{g}}^2)}{s}\right]^2. \tag{18.159}$$

We perform the sum over colors using Ref. [197]:

$$f^{abe}f^{cde}f^{abe'}f^{cde'} = 2f^{abe}f^{cde}f^{ace'}f^{bde'} = N_c^2(N_c^2-1) = 72\,, \tag{18.160}$$

which yields

$$\sum_{\rm colors} G_s^2 = \frac{72g_s^4}{s^2}, \qquad \sum_{\rm colors} G_t^2 = \frac{72g_s^4}{(t-m_{\widetilde{g}}^2)^2}, \tag{18.161}$$

$$\sum_{\rm colors} G_u^2 = \frac{72g_s^4}{(u-m_{\widetilde{g}}^2)^2}, \qquad \sum_{\rm colors} G_sG_t = \frac{36g_s^4}{s(t-m_{\widetilde{g}}^2)}, \tag{18.162}$$

$$\sum_{\rm colors} G_sG_u = -\frac{36g_s^4}{s(u-m_{\widetilde{g}}^2)}, \qquad \sum_{\rm colors} G_tG_u = \frac{36g_s^4}{(t-m_{\widetilde{g}}^2)(u-m_{\widetilde{g}}^2)}. \tag{18.163}$$

Putting all the pieces together and averaging over the initial-state colors and polarizations, we end up with

$$\frac{d\sigma}{dt} = \frac{9\pi\alpha_s^2}{4s^4}\left[2(t-m_{\widetilde{g}}^2)(u-m_{\widetilde{g}}^2) - 3s^2 - 4m_{\widetilde{g}}^2 s + \frac{s^2(s+2m_{\widetilde{g}}^2)^2}{(t-m_{\widetilde{g}}^2)(u-m_{\widetilde{g}}^2)} - \frac{4m_{\widetilde{g}}^4 s^4}{(t-m_{\widetilde{g}}^2)^2(u-m_{\widetilde{g}}^2)^2}\right]. \tag{18.164}$$

Since the final state has identical particles, the total cross section is given by

$$\sigma = \frac{1}{2}\int_{t^-}^{t^+} \frac{d\sigma}{dt}dt\,, \tag{18.165}$$

where $t^{\pm}$ is defined in eq. (D.37).

Exercises

18.1 Consider the tree-level decay of the chargino, $\widetilde{C}_i^+ \to \widetilde{\nu}_e e^+$. The external wave functions for the chargino and the positron are denoted by $x_{\widetilde{C}}$ and y_e, respectively.

(a) Show that the tree-level decay amplitude for $\widetilde{C}_i^+ \to \widetilde{\nu}_e e^+$ is given by

$$\mathcal{M} = -igV_{i1}^*\, x_{\widetilde{C}}\, y_e\,, \tag{18.166}$$

where V_{i1} is an element of the chargino mixing matrix V [see eq. (13.149)].

(b) Summing over the electron helicity, averaging over the chargino helicity, and neglecting the electron mass, derive the partial decay width,

$$\Gamma(\widetilde{C}_i^+ \to \widetilde{\nu}e^+) = \frac{g^2}{32\pi}|V_{i1}|^2 m_{\widetilde{C}_i}\left(1 - \frac{m_{\widetilde{\nu}_e}^2}{m_{\widetilde{C}_i}^2}\right)^2. \qquad (18.167)$$

18.2 Consider the decay of a neutralino to a lighter neutralino and neutral Higgs boson $\phi = h$, H, or A. Denote the masses of the two neutralinos and the Higgs boson by $m_{\widetilde{N}_i}$, $m_{\widetilde{N}_j}$, and m_ϕ, respectively, and assume that $m_{\widetilde{N}_i} > m_{\widetilde{N}_j} + m_\phi$.

(a) Using the two-component fermion Feynman rules, draw the two tree-level diagrams that contribute to the decay amplitude, and show that

$$i\mathcal{M} = -i\left[Y\, x_i y_j + Y^*\, y_i^\dagger x_j^\dagger\right], \qquad (18.168)$$

where the coupling $Y \equiv Y^{\phi\chi_i^0\chi_j^0}$ is defined in eq. (J.24).

(b) After squaring the decay amplitude, average over the initial-state spins and sum over the final-state spins. Show that the decay rate is given by

$$\Gamma(\widetilde{N}_i \to \phi\widetilde{N}_j) = \frac{m_{\widetilde{N}_i}}{16\pi}\lambda^{1/2}(1, r_\phi, r_j)\Big[|Y^{\phi\chi_i^0\chi_j^0}|^2(1 + r_j - r_\phi) + 2\,\mathrm{Re}\big[(Y^{\phi\chi_i^0\chi_j^0})^2\big]\sqrt{r_j}\Big], \qquad (18.169)$$

where $r_j \equiv m_{\widetilde{N}_j}^2/m_{\widetilde{N}_i}^2$ and $r_\phi \equiv m_\phi^2/m_{\widetilde{N}_i}^2$. The results for $\phi = h$, H, and A can now be obtained by using eqs. (J.8) and (J.9) in eq. (J.24).

18.3 Consider the decays of the neutral Higgs scalar bosons $\phi = h$, H, and A of the MSSM into Standard Model fermion–antifermion pairs.

(a) Using the two-component fermion Feynman rules, draw the two tree-level diagrams that contribute to the decay amplitude, and show that

$$i\mathcal{M} = -\frac{iY_f}{\sqrt{2}}\left[k_{f\phi}^*\, x_1^\dagger x_2^\dagger + k_{f\phi}\, y_1 y_2\right], \qquad (18.170)$$

where Y_f is the Yukawa coupling of the fermion and $k_{f\phi}$ is the Higgs mixing parameter from eq. (J.9).

(b) Derive the tree-level decay rate,

$$\Gamma(\phi \to f\bar{f}) = \frac{N_c^f|Y_f|^2}{16\pi m_\phi}\left(1 - \frac{4m_f^2}{m_\phi^2}\right)^{1/2}\left\{|k_{f\phi}|^2(m_\phi^2 - 2m_f^2) - 2\,\mathrm{Re}[(k_{f\phi})^2]m_f^2\right\}, \qquad (18.171)$$

where $N_c^f = 3$ for quarks and 1 for leptons.

(c) Show that the decay rates of the CP-even scalars to $f\bar{f}$ is proportional to $p_{\rm CM}^3$, whereas the decay rates of the CP-odd scalar to $f\bar{f}$ is proportional to $p_{\rm CM}$, where $p_{\rm CM}$ is the magnitude of the CM three-momentum [see eq. (D.7)].

18.4 Consider the decay of a neutralino $\widetilde{N}_i$ to three lighter neutralinos: $\widetilde{N}_j, \widetilde{N}_k, \widetilde{N}_\ell$. In Section 18.3, the calculation of the decay rate was performed using two-component fermion Feynman rules. Repeat this calculation using the four-component fermion Feynman rules for Majorana fermions developed in Chapter 3.

18.5 Consider the MSSM tree-level scattering process $e^-e^- \to \widetilde{e}^-\widetilde{e}^-$. You may neglect the mass of the electron and the mixing of $\widetilde{e}_L$ and $\widetilde{e}_R$ in this exercise.

(a) Using the two-component fermion Feynman rules, draw the two tree-level diagrams that contribute to the scattering amplitude of $e^-e^- \to \widetilde{e}_R^-\widetilde{e}_R^-$, and show that

$$i\mathcal{M} = \left(-i\sqrt{2}g\frac{s_W}{c_W}\right)^2 \sum_{i=1}^{4} m_{\widetilde{N}_i}(N_{i1})^2 \left[\frac{i}{t-m_{\widetilde{N}_i}^2} + \frac{i}{u-m_{\widetilde{N}_i}^2}\right] y_1^\dagger y_2^\dagger\,. \qquad (18.172)$$

(b) Show that the spin-averaged differential cross section is given by

$$\frac{d\sigma}{dt} = \frac{\pi\alpha^2}{c_W^4 s}\left|\sum_{i=1}^{4} m_{\widetilde{N}_i}(N_{i1})^2 \left(\frac{1}{t-m_{\widetilde{N}_i}^2} + \frac{1}{u-m_{\widetilde{N}_i}^2}\right)\right|^2\,. \qquad (18.173)$$

Integrate eq. (18.173) over t to obtain the total cross section. Do not forget to include a factor of 1/2 to account for the identical particles in the final state [see eq. (D.111)].

(c) Repeat the computation of the spin-averaged differential cross section for the scattering amplitude of $e^-e^- \to \widetilde{e}_L^-\widetilde{e}_L^-$. How is eq. (18.173) modified?

(d) In softly broken supersymmetric QED, the scattering of $e^-e^- \to \widetilde{e}^-\widetilde{e}^-$ is mediated by the exchange of a photino. Repeat the computations of parts (b) and (c) for this model.

18.6 In Section 18.7 and in Exercise 18.5, two-component fermion Feynman rules were employed in the computation of the cross sections for $e^-e^- \to \widetilde{e}_L^-\widetilde{e}_R^-$, $\widetilde{e}_R^-\widetilde{e}_R^-$, and $\widetilde{e}_L^-\widetilde{e}_L^-$. Repeat these calculations by employing the four-component fermion Feynman rules developed in Chapter 3.

18.7 Consider the pair production of sneutrinos in electron–positron collisions. Using two-component fermion Feynman rules, there are two tree-level graphs featuring the s-channel exchange of the Z. Neglecting the electron mass and Yukawa coupling, there is only one graph involving the t-channel exchange of the charginos.

(a) Draw the three tree-level diagrams that contribute to the scattering amplitude of $e^+e^- \to \widetilde{\nu}\widetilde{\nu}^*$, and show that

$$\mathcal{M} = c_1 x_1(k_1 - k_2)\cdot\sigma y_2^\dagger + c_2 y_1^\dagger(k_1 - k_2)\cdot\overline{\sigma} x_2 + c_3 x_1(k_1 - p_1)\cdot\sigma y_2^\dagger\,, \quad (18.174)$$

where

$$c_1 \equiv \frac{g^2(1 - 2s_W^2)}{4c_W^2 D_Z}\,, \qquad c_2 \equiv -\frac{g^2 s_W^2}{2c_W^2 D_Z}\,, \qquad c_3 \equiv g^2 \sum_{i=1}^{2} \frac{|V_{i1}|^2}{m_{\widetilde{C}_j}^2 - t}\,, \quad (18.175)$$

where $D_Z \equiv s - m_Z^2 + i\Gamma_Z m_Z$ is the denominator of the Z boson propagator.

(b) Show that the spin-averaged differential cross section is given by

$$\frac{d\sigma}{d\Omega} = \frac{s\beta^3}{256\pi^2}\sin^2\theta\big(|c_1|^2 + |c_2|^2 + \tfrac{1}{4}c_3^2 + \text{Re}[c_1]c_3\big)\,, \quad (18.176)$$

where

$$t = m_{\widetilde{\nu}}^2 - \tfrac{1}{2}(1 - \beta\cos\theta)s\,, \qquad \beta \equiv \left(1 - \frac{4m_{\widetilde{\nu}}^2}{s}\right)^{1/2}\,, \quad (18.177)$$

and θ is the angle between the initial-state electron and the final-state sneutrino in the CM frame.

(c) Integrate the result obtained in eq. (18.176) to obtain the total cross section σ.

18.8 Consider the tree-level pair production of charged sfermions in e^+e^- collisions represented by the diagrams of Fig. 18.16. Consider separately the cases of squark pair production, selectron pair production, and charged slepton ($\widetilde{\ell} \neq \widetilde{e}$) pair production.

(a) Using two-component fermion Feynman rules, draw all the tree-level diagrams that contribute to each of the sfermion pair production processes represented in Fig. 18.16.

(b) Modify the computation performed in Exercise 18.7 to obtain the scattering amplitude, the spin-averaged differential cross section, and the total cross section for each of the sfermion pair production processes of Fig. 18.16.

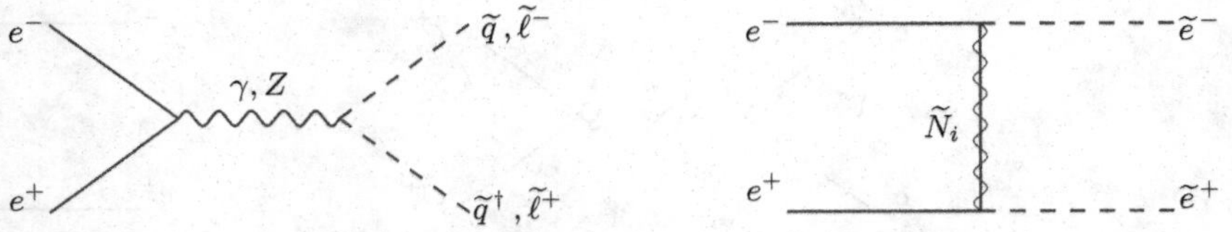

Fig. 18.16 Diagrams for squark and charged slepton pair production at e^+e^- colliders.

18.9 Consider the associated production of a positively charged chargino and a neutralino in the scattering of a u-quark and $\bar{d}$-antiquark. Using two-component fermion Feynman rules, there are two tree-level graphs featuring the s-channel exchange of the W^+. In addition, there is a t-channel graph mediated by $\widetilde{d}_L$ and a u-channel graph mediated by $\widetilde{u}_L$.

(a) Draw the four tree-level diagrams that contribute to the scattering amplitude for $u\bar{d} \to \widetilde{C}_i^+ \widetilde{N}_j$.

(b) Show that the tree-level amplitude has the form of eq. (18.79) with

$$c_1 = -\sqrt{2}g^2 \left[\frac{O_{ji}^{L*}}{s - m_W^2} + \left(\frac{1}{2} N_{j2}^* + \frac{s_W}{6c_W} N_{j1}^* \right) \frac{V_{i1}}{u - m_{\widetilde{u}_L}} \right], \tag{18.178}$$

$$c_2 = -\sqrt{2}g^2 \left[\frac{O_{ji}^{R*}}{s - m_W^2} + \left(\frac{1}{2} N_{j2}^* - \frac{s_W}{6c_W} N_{j1}^* \right) \frac{U_{i1}^*}{t - m_{\widetilde{d}_L}} \right], \tag{18.179}$$

$$c_3 = c_4 = 0, \tag{18.180}$$

after an appropriate application of the Fierz relations given in eqs. (1.138) and (1.140) to rewrite the s-channel amplitude in a form without σ or $\overline{\sigma}$ matrices.

(c) Compute the spin- and color-averaged differential cross section for $u\bar{d} \to \widetilde{C}_i^+ \widetilde{N}_j$, where the masses of the quarks are neglected and the possible final-state spins are summed over. Show that

$$\frac{d\sigma}{dt} = \frac{1}{192\pi s^2} \Big[|c_1|^2 (u - m_{\widetilde{C}_i}^2)(u - m_{\widetilde{N}_j}^2) + |c_2|^2 (t - m_{\widetilde{C}_i}^2)(t - m_{\widetilde{N}_j}^2) + 2\,\mathrm{Re}[c_1 c_2^*] m_{\widetilde{C}_i} m_{\widetilde{N}_j} s \Big]. \tag{18.181}$$

18.10 Consider the tree-level pair production of charginos via e^+e^- annihilation represented by the diagrams shown in Fig. 18.17.

(a) Using two-component fermion Feynman rules, show that the diagrams represented in Fig. 18.17 correspond to four tree-level Feynman graphs for s-channel Z exchange, four graphs for s-channel photon exchange, and one graph for t/u-channel sneutrino exchange.

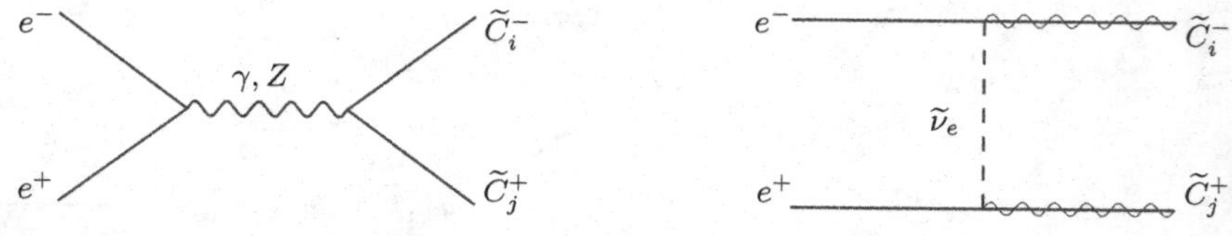

Fig. 18.17 Diagrams for chargino pair production at e^+e^- colliders.

(b) Modify the computations in Section 18.8 and obtain the differential cross section for $e^+e^- \to \widetilde{C}_i^+\widetilde{C}_j^-$.

(c) Integrate the result of part (a) to obtain the total cross section for chargino production.

18.11 The calculation of the s-channel Z-exchange diagrams that contribute to $\widetilde{N}_i\widetilde{N}_j \to f\bar{f}$ annihilation was performed using the unitary gauge Z boson propagator. As a check of the results exhibited in eqs. (18.99)–(18.101), repeat the calculation using the Z boson propagator in the 't Hooft–Feynman gauge (with gauge parameter $\xi = 1$).

(a) Show that in the 't Hooft–Feynman gauge, eq. (18.99) is replaced by

$$\mathcal{M}_Z = \mathcal{M}_Z^{(1)} + \mathcal{M}_G , \tag{18.182}$$

where $\mathcal{M}_Z^{(1)}$ is given by eq. (18.100) and $\mathcal{M}_G$ arises from diagrams of Fig. 18.12 with $\phi = G^0$ (the neutral Goldstone boson).

(b) In the 't Hooft–Feynman gauge, the Goldstone boson propagator was exhibited in Section 4.5.3. Moreover, using eqs. (J.8) and (J.9),

$$\frac{k_{fG^0}}{v_f} = \frac{2iT_f}{v} . \tag{18.183}$$

Hence, using eq. (18.102) with $\phi = G^0$, show that

$$i\mathcal{M}_G = \frac{m_f}{\sqrt{2}\,vD_Z}(2T_f)Y^{G^0\chi_1^0\chi_1^0}\left(x_1x_2 - y_1^\dagger y_2^\dagger\right)\left(y_{f1}y_{f2} - x_{f1}^\dagger x_{f2}^\dagger\right) , \tag{18.184}$$

where $iY^{G^0\chi_1^0\chi_1^0}$ is real in light of eq. (J.37).

(c) Verify that $\mathcal{M}_G = \mathcal{M}_Z^{(2)}$, where the latter is given by eq. (18.101). Conclude that eq. (18.182) reproduces the corresponding result obtained in the unitary gauge, as expected from gauge invariance.

18.12 (a) In softly broken supersymmetric QED, compute the annihilation rate for photinos into fermion–antifermion pairs in the nonrelativistic limit, where sfermion mixing effects are neglected. Show that $v_{\rm rel}\sigma_{\rm ann} \propto |\vec{p}|^2$. Compare with the result obtained in Ref. [198] and verify that the published result should be multiplied by a factor of 2.

(b) In eq. (18.139), a formula for $v_{\rm rel}\sigma_{\rm ann}$ is presented that neglects the effects of s-channel Higgs boson exchanges. Complete the computation of the annihilation cross section by including these terms, along with the effects of interference between the neglected contributions and the ones already taken into account.

18.13 Reconstruct the computation of the cross section for the process $gg \to \widetilde{g}\widetilde{g}$ exhibited in Section 18.10.

(a) The scattering amplitude for $gg \to \widetilde{g}\widetilde{g}$ obtained in eqs. (18.140)–(18.142) has the following structure:

$$\mathcal{M}_s + \mathcal{M}_t + \mathcal{M}_u = \mathcal{M}_{\mu\nu}\varepsilon_1^\mu \varepsilon_2^\nu\,. \tag{18.185}$$

Without choosing any specific frame of reference, show that the replacement of ϵ_1 with p_1 or ϵ_2 with p_2 yields zero. That is,

$$p_{1\mu}\varepsilon_{2\nu}\mathcal{M}^{\mu\nu} = p_{2\nu}\varepsilon_{1\mu}\mathcal{M}^{\mu\nu} = 0\,. \tag{18.186}$$

(b) Using the properties of the gluon polarization vector, show that the expression for $i\mathcal{M}_s$ in eq. (18.140) can be replaced with

$$i\widetilde{\mathcal{M}}_s = \left\{-g_s f^{abe}[g_{\mu\nu}(p_1 - p_2)_\rho + 2g_{\nu\rho}p_{2\mu} - 2g_{\mu\rho}p_{1\nu}]\right\}\left(\frac{-ig^{\rho\kappa}}{s}\right)$$
$$\times\left[(-g_s f^{cde})\, x_1^\dagger \overline{\sigma}_\kappa y_2 + (g_s f^{dce})\, y_1 \sigma_\kappa x_2^\dagger\right]\varepsilon_1^\mu \varepsilon_2^\nu. \tag{18.187}$$

(c) Repeat the calculation of part (a), where eq. (18.185) for $\mathcal{M}^{\mu\nu}$ is modified by replacing $\mathcal{M}_s$ with $\widetilde{\mathcal{M}}_s$. That is,

$$\widetilde{\mathcal{M}}_s + \mathcal{M}_t + \mathcal{M}_u = \widetilde{\mathcal{M}}_{\mu\nu}\varepsilon_1^\mu \varepsilon_2^\nu\,. \tag{18.188}$$

By judicious use of the Dirac equations [see eqs. (2.10)–(2.13)], show that

$$p_{1\mu}\widetilde{\mathcal{M}}^{\mu\nu} = -g_s^2(f^{abe}f^{cde} + f^{ace}f^{dbe} + f^{ade}f^{bce})(x_1^\dagger\overline{\sigma}^\nu y_2 + y_1\sigma^\nu x_2^\dagger) = 0\,, \tag{18.189}$$

$$p_{2\nu}\widetilde{\mathcal{M}}^{\mu\nu} = g_s^2(f^{abe}f^{cde} + f^{ace}f^{dbe} + f^{ade}f^{bce})(x_1^\dagger\overline{\sigma}^\mu y_2 + y_1\sigma^\mu x_2^\dagger) = 0\,, \tag{18.190}$$

after employing the Jacobi identity [see eq. (H.3)].

(d) In contrast to eqs. (18.189) and (18.190), verify that

$$p_{1\mu}\mathcal{M}^{\mu\nu} \neq 0\,, \qquad p_{2\nu}\mathcal{M}^{\mu\nu} \neq 0\,. \tag{18.191}$$

Check that the right-hand sides of these two equations have the forms indicated by eq. (E.253).

(e) Evaluate $\frac{1}{4}\sum_{\{\lambda\}}|\widetilde{\mathcal{M}}|^2$, where $\{\lambda\} \equiv \{\lambda_1, \lambda_2, \lambda_1', \lambda_2'\}$ and $\frac{1}{4}\sum_{\{\lambda\}}$ indicates the average over initial gluon polarizations and the sum over final-state gluino helicities. Because eq. (E.249) is now satisfied in light of part (c), you may safely employ the simpler polarization sum formula given in eq. (E.250) for *both* initial gluons when averaging over the initial gluon polarizations. Perform the gluon polarization sums and show that the resulting expression for $d\sigma/dt$ (after including the appropriate color factors) reproduces eq. (18.164).

19 One-Loop Calculations

In this chapter, we present example one-loop Feynman diagrammatic calculations in the Standard Model and MSSM, employing the two-component fermion techniques developed in Chapter 2.

19.1 Wave Function Renormalization in Softly Broken SUSY-QED

In this section, we will evaluate the wave function renormalization constants of the electron in softly broken SUSY-QED that is governed by the Lagrangian given in Section 13.1. For simplicity, we shall ignore the contribution of eq. (19.358), which mixes $\phi_L \equiv \widetilde{e}_L$ and $\phi_R^\dagger \equiv \widetilde{e}_R$, by setting $b = 0$. In addition, we will set the coefficient of the Fayet–Iliopoulos term to zero.

The relevant Feynman rules for softly broken SUSY-QED are shown in Figs. 17.3, 19.1, and 19.2. Note that the QED coupling of electrons to photons is exhibited in Fig. 17.3, whereas the corresponding supersymmetric couplings of selectrons to photons in the absence of $\widetilde{e}_L$–$\widetilde{e}_R$ mixing (i.e., with $b = 0$) are shown in Fig. 19.1. Generalizing the latter to include $\widetilde{e}_L$–$\widetilde{e}_R$ mixing is left as an exercise for the reader.

We begin by evaluating the one-loop self-energy functions of the electron mediated by an electron–photon loop in QED. We shall take the electron to be on mass shell. There are four relevant diagrams exhibited in Fig. 19.3. Two regularization procedures will be employed simultaneously: dimensional regularization (DREG) and dimensional reduction (DRED). The integration over the loop momentum will

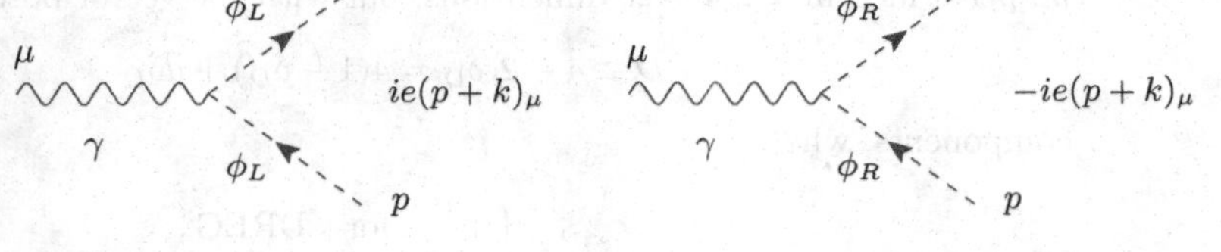

Fig. 19.1 The Feynman rules for the photon interaction of the selectrons with four-momenta p and k, respectively, and Lorentz index μ. The arrows on the scalar lines indicate the flow of positive charge for the $\phi_R \equiv \widetilde{e}_R^\dagger$ line and negative charge for the $\phi_L \equiv \widetilde{e}_L$ line.

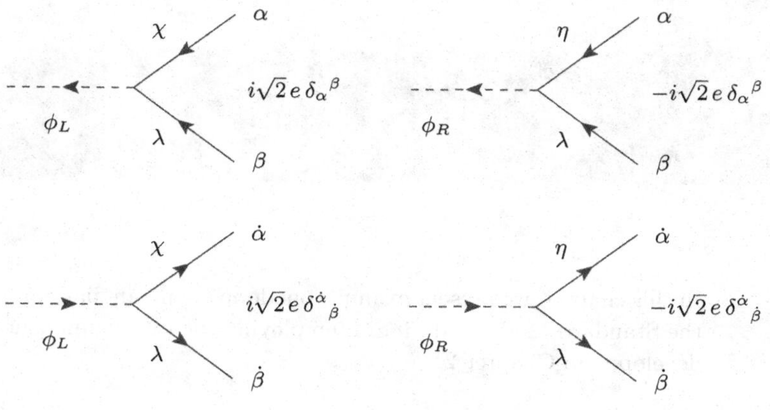

Fig. 19.2 Feynman rules for the interaction of the photino λ with the electron and selectron. The corresponding two-component spinor labels α, β and $\dot\alpha$, $\dot\beta$ are specified by employing one of the two possible choices exhibited in Fig. 2.7.

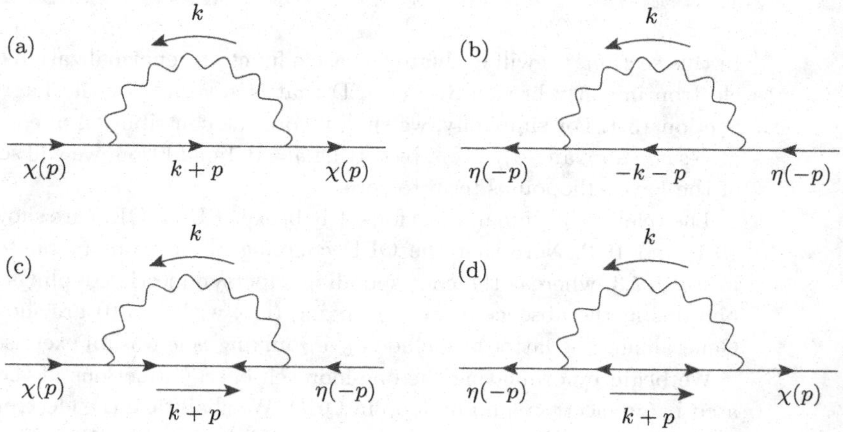

Fig. 19.3 Electron–photon loop contributions to the one-loop electron self-energy in QED. The directions of momentum flow are indicated by the arrows.

be performed in $d \equiv 4 - 2\epsilon$ dimensions, but with the vector bosons possessing

$$D \equiv 4 - 2\epsilon\delta_{\rm D} = 4(1 - \delta_D) + d\delta_D \tag{19.1}$$

components, where

$$\delta_D \equiv \begin{cases} 1 & \text{for DREG}, \\ 0 & \text{for DRED}. \end{cases} \tag{19.2}$$

In other words, the metric $g^{\mu\nu}$ appearing explicitly in the vector propagator is treated as four-dimensional in DRED, but as d-dimensional in DREG.

Comparing with the diagrams shown in Fig. 2.32, where the flow of the momentum p is opposite in direction to that in Fig. 19.3, we first evaluate diagram (a) of Fig. 19.3:

$$\begin{aligned}-ip\cdot\overline{\sigma}\Sigma_L(p^2)_{\rm QED} &= \int\frac{d^dk}{(2\pi)^d}\,(ie\mu^\epsilon\overline{\sigma}^\mu)\frac{i(p+k)\cdot\sigma}{(k+p)^2-m^2+i\varepsilon}(ie\mu^\epsilon\overline{\sigma}^\nu)\left(\frac{-ig_{\mu\nu}}{k^2+i\varepsilon}\right)\\ &= (D-2)e^2\mu^{2\epsilon}\overline{\sigma}^\mu\int\frac{d^dk}{(2\pi)^d}\,\frac{(p+k)_\mu}{(k^2+i\varepsilon)\left[(k+p)^2-m^2+i\varepsilon\right]}\,, \qquad (19.3)\end{aligned}$$

where D is defined in eq. (19.1), m is the electron mass, and we have employed the photon propagator in the Feynman gauge. Note that we have replaced $e\to e\mu^\epsilon$ (where μ is an arbitrary mass scale) so that e remains dimensionless when evaluating the integral in $d=4-2\epsilon$ dimensions. The last integral in eq. (19.3) can be expressed in terms of the Passarino–Veltman loop functions B_0 and B_1 [see eqs. (K.10) and (K.14)]. The end result is

$$\Sigma_L(p^2)_{\rm QED} = (2-D)\frac{\alpha}{4\pi}\left[B_0(p^2;0,m^2)+B_1(p^2;0,m^2)\right], \qquad (19.4)$$

where $\alpha\equiv e^2/(4\pi)$.

Next, we evaluate diagram (b) of Fig. 19.3:

$$\begin{aligned}-ip\cdot\sigma\Sigma_R(p^2)_{\rm QED} &= \int\frac{d^dk}{(2\pi)^d}\,(ie\mu^\epsilon\sigma^\mu)\frac{i(p+k)\cdot\overline{\sigma}}{(k+p)^2-m^2+i\varepsilon}(ie\mu^\epsilon\sigma^\nu)\left(\frac{-ig_{\mu\nu}}{k^2+i\varepsilon}\right)\\ &= (D-2)e^2\mu^{2\epsilon}\sigma^\mu\int\frac{d^dk}{(2\pi)^d}\,\frac{(p+k)_\mu}{(k^2+i\varepsilon)\left[(k+p)^2-m^2+i\varepsilon\right]}, \qquad (19.5)\end{aligned}$$

which yields

$$\Sigma_R(p^2)_{\rm QED} = (2-D)\frac{\alpha}{4\pi}\left[B_0(p^2;0,m^2)+B_1(p^2;0,m^2)\right]. \qquad (19.6)$$

In particular, $\Sigma_L(p^2)=\Sigma_R(p^2)$ in QED, as expected for a vectorlike (parity-conserving) theory.

The computation of diagram (c) of Fig. 19.3 yields

$$\begin{aligned}-i\Sigma_D(p^2)_{\rm QED} &= \int\frac{d^dk}{(2\pi)^d}\,(ie\mu^\epsilon\overline{\sigma}^\mu)\frac{im}{(k+p)^2-m^2+i\varepsilon}(ie\mu^\epsilon\sigma^\nu)\left(\frac{-ig_{\mu\nu}}{k^2+i\varepsilon}\right)\\ &= -De^2\mu^{2\epsilon}m\int\frac{d^dk}{(2\pi)^d}\,\frac{1}{(k^2+i\varepsilon)\left[(k+p)^2-m^2+i\varepsilon\right]}\,. \qquad (19.7)\end{aligned}$$

After employing eqs. (K.10) and (K.38), we obtain

$$\Sigma_D(p^2)_{\rm QED} = D\frac{\alpha m}{4\pi}B_0(p^2;0,m^2)\,. \qquad (19.8)$$

Finally, diagram (d) of Fig. 19.3 produces

$$-i\overline{\Sigma}_D(p^2)_{\rm QED} = \int\frac{d^dk}{(2\pi)^d}\,(ie\mu^\epsilon\sigma^\mu)\frac{im}{(k+p)^2-m^2+i\varepsilon}(ie\mu^\epsilon\overline{\sigma}^\nu)\left(\frac{-ig_{\mu\nu}}{k^2+i\varepsilon}\right), \qquad (19.9)$$

which yields

$$\overline{\Sigma}_D(p^2)_{\rm QED} = D\frac{\alpha m}{4\pi}B_0(p^2;0,m^2)\,. \tag{19.10}$$

In particular, $\Sigma_D(p^2) = \overline{\Sigma}_D(p^2)$ in QED, as expected for a vectorlike (parity-conserving) theory.

Using the results obtained above, we can compute the QED contributions to the electron wave function renormalization constants. In light of eqs. (2.310), (2.324), and (2.325), it follows that $\delta Z_{\rm QED} \equiv (\delta Z_L)_{\rm QED} = (\delta Z_R)_{\rm QED}$. More explicitly,[1]

$$\delta Z_{\rm QED} = \Sigma_L(m^2) + 2m^2\Sigma'_L(m^2) + 2m\Sigma'_D(m^2)\,, \tag{19.11}$$

where

$$\Sigma'_L(m^2) \equiv \left.\frac{d\Sigma'_L(p^2)}{dp^2}\right|_{p^2=m^2}, \qquad \Sigma'_D(m^2) \equiv \left.\frac{d\Sigma'_D(p^2)}{dp^2}\right|_{p^2=m^2}. \tag{19.12}$$

Hence, we end up with

$$\delta Z_{\rm QED} = -\frac{\alpha}{2\pi}\Big\{(1-\epsilon\delta_D)\big[B_0(m^2;0,m^2) + B_1(m^2;0,m^2) + 2m^2B'_1(m^2;0,m^2)\big] \\ -2m^2B'_0(m^2;0,m^2)\Big\}, \tag{19.13}$$

where δ_D is defined in eq. (19.2) and

$$B'_i(m^2;m^2,0) \equiv \left.\frac{\partial B_i(p^2;m^2,0)}{\partial p^2}\right|_{p^2=m^2}, \qquad \text{for } i=0,\,1. \tag{19.14}$$

The functions $B_0(m^2;m^2,0)$ and $B_1(m^2;m^2,0)$ are easily evaluated. Starting from the integral representations given by eqs. (K.93) and (K.94), it follows that

$$B_0(m^2;0,m^2) = (4\pi)^\epsilon\,\Gamma(\epsilon)\left(\frac{m^2}{\mu^2}\right)^{-\epsilon}\int_0^1 x^{-2\epsilon}dx = \frac{(4\pi)^\epsilon\,\Gamma(\epsilon)}{1-2\epsilon}\left(\frac{m^2}{\mu^2}\right)^{-\epsilon}, \tag{19.15}$$

$$B_1(m^2;0,m^2) = -(4\pi)^\epsilon\,\Gamma(\epsilon)\left(\frac{m^2}{\mu^2}\right)^{-\epsilon}\int_0^1 x^{1-2\epsilon}dx = -\,\frac{(4\pi)^\epsilon\,\Gamma(\epsilon)}{2(1-\epsilon)}\left(\frac{m^2}{\mu^2}\right)^{-\epsilon}. \tag{19.16}$$

The singularity at $\epsilon = 0$ corresponds to an ultraviolet divergence.

In contrast, the loop functions $B'_0(m^2;0,m^2)$ and $B'_1(m^2;0,m^2)$ are ultraviolet finite since the ultraviolet singularity (which is independent of p^2) disappears when taking the derivative. However, $B'_0(m^2;0,m^2)$ is infrared divergent due to the massless photon that appears in the loop (corresponding to the kinematic region where the internal photon line goes on-shell). This infrared divergence is also regulated

[1] Most textbooks denote the electron wave function renormalization constant by Z_2, in which case $\delta Z_{\rm QED} \equiv Z_2 - 1$.

by performing the integration over the Feynman parameter x prior to taking the limit of $\epsilon \to 0$. Taking the derivative of eq. (K.93) with respect to p^2 yields

$$B_0'(p^2; m_a^2, m_b^2) = \frac{(4\pi)^\epsilon\,\Gamma(1+\epsilon)}{\mu^2} \times \int_0^1 \left(\frac{p^2x^2 - (p^2 + m_a^2 - m_b^2)x + m_a^2 - i\varepsilon}{\mu^2}\right)^{-1-\epsilon} x(1-x)dx\,, \tag{19.17}$$

which is ultraviolet finite after taking the $\epsilon \to 0$ limit. It then follows that

$$\begin{aligned} m^2 B_0'(m^2; 0, m^2) &= (4\pi)^\epsilon\,\Gamma(1+\epsilon)\left(\frac{m^2}{\mu^2}\right)^{-\epsilon}\int_0^1 x^{-1-2\epsilon}(1-x)dx \\ &= (4\pi)^\epsilon\,\Gamma(1+\epsilon)\left(\frac{m^2}{\mu^2}\right)^{-\epsilon} B(-2\epsilon, 2)\,, \end{aligned} \tag{19.18}$$

where Euler's beta function is defined by

$$B(x,y) = \frac{\Gamma(x)\Gamma(y)}{\Gamma(x+y)} = \int_0^1 t^{x-1}(1-t)^{y-1}\,dt\,. \tag{19.19}$$

Using the well-known properties of the gamma function, $\Gamma(x+1) = x\Gamma(x)$ and $\Gamma(1+n) = n!$ for any nonnegative integer n, it follows that

$$B(-2\epsilon, 2) = \frac{\Gamma(-2\epsilon)\Gamma(2)}{\Gamma(2-2\epsilon)} = -\frac{1}{2\epsilon(1-2\epsilon)}\,. \tag{19.20}$$

Hence, we end up with

$$m^2 B_0'(m^2; 0, m^2) = -\frac{(4\pi)^\epsilon\,\Gamma(1+\epsilon)}{2\epsilon(1-2\epsilon)}\left(\frac{m^2}{\mu^2}\right)^{-\epsilon}\,, \tag{19.21}$$

where the infrared singularity is due to the behavior of the integrand in eq. (19.18) at $x = 0$ (in the $\epsilon \to 0$ limit).

In contrast, $B_1'(m^2; 0, m^2)$ is infrared finite. Taking the derivative of eq. (K.94) with respect to p^2 yields

$$B_1'(p^2; m_a^2, m_b^2) = -\frac{(4\pi)^\epsilon\,\Gamma(1+\epsilon)}{\mu^2} \times \int_0^1 \left(\frac{p^2x^2 - (p^2 + m_a^2 - m_b^2)x + m_a^2 - i\varepsilon}{\mu^2}\right)^{-1-\epsilon} x^2(1-x)dx\,, \tag{19.22}$$

which is ultraviolet finite after taking the $\epsilon \to 0$ limit. It then follows that

$$\begin{aligned} m^2 B_1'(m^2; 0, m^2) &= -(4\pi)^\epsilon\,\Gamma(1+\epsilon)\left(\frac{m^2}{\mu^2}\right)^{-\epsilon}\int_0^1 x^{-2\epsilon}(1-x)dx \\ &= -\frac{(4\pi)^\epsilon\,\Gamma(1+\epsilon)}{2(1-\epsilon)(1-2\epsilon)}\left(\frac{m^2}{\mu^2}\right)^{-\epsilon}\,, \end{aligned} \tag{19.23}$$

which is manifestly finite in the limit of $\epsilon \to 0$.

Hence, eq. (19.13) yields

$$\delta Z_{\rm QED} = -\frac{\alpha}{2\pi}(4\pi)^\epsilon \left(\frac{m^2}{\mu^2}\right)^{-\epsilon} \left[\frac{(1-\epsilon\delta_D)\Gamma(\epsilon)}{2(1-\epsilon)} + \frac{\Gamma(1+\epsilon)}{\epsilon(1-2\epsilon)}\right], \tag{19.24}$$

where the first term in the bracketed expression contains the ultraviolet divergence and the second term contains the infrared divergence. Some authors will distinguish the two roles played by ϵ by employing the notation, $\epsilon_{\rm UV}$ and $\epsilon_{\rm IR}$, which allows one to track the different sources of divergences throughout the calculation [see eq. (K.5)]. The infrared divergence will ultimately cancel when computing any infrared safe observable. The ultraviolet divergence will cancel in the computation of S-matrix elements once the bare mass and coupling parameters are replaced by the corresponding renormalized parameters.

The wave function renormalization constant of the electron is a gauge-dependent quantity. The result exhibited in eq. (19.24) was obtained for the Feynman gauge, corresponding to the choice of gauge parameter $\xi = 1$ [see eq. (19.331)]. It is straightforward to obtain a more general form for $\delta Z_{\rm QED}$, where the gauge parameter ξ is arbitrary. This is left as an exercise for the reader (see Exercise 19.1); the resulting expression is given in eq. (19.339).

In softly broken SUSY-QED, there are additional contributions to δZ_L and δZ_R in the one-loop approximation due to selectron–photino loops, as shown in Fig. 19.4.

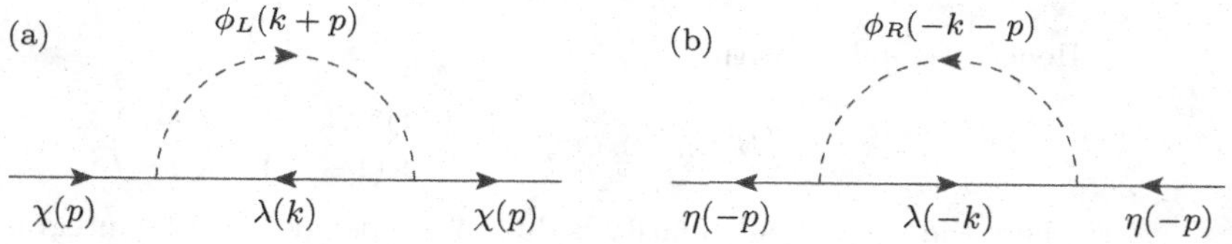

Fig. 19.4 Selectron–photino loop contributions to the one-loop electron self-energy in softly broken SUSY-QED. The directions of momentum flow are indicated by the arrows.

Note the absence of χ–η self-energy graphs due to the structure of the electron–selectron–photino vertex Feynman rules exhibited in Fig. 19.2. This immediately implies that, in the one-loop approximation,

$$\Sigma_D(p^2)_{\rm SQED} = \overline{\Sigma}_D(p^2)_{\rm SQED} = 0\,, \tag{19.25}$$

where SQED indicates the superpartner contributions of SUSY-QED. Diagram (a) of Fig. 19.4 yields

$$\begin{aligned} -ip\!\cdot\!\overline{\sigma}\Sigma_L(p^2)_{\rm SQED} &= \int\frac{d^dk}{(2\pi)^d}\,(i\sqrt{2}\,e\mu^\epsilon)\frac{-ik\!\cdot\!\overline{\sigma}}{k^2-M^2+i\varepsilon}(i\sqrt{2}\,e\mu^\epsilon)\frac{i}{(k+p)^2-m_{\widetilde{L}}^2+i\varepsilon} \\ &= -2e^2\mu^{2\epsilon}\overline{\sigma}^\mu\int\frac{d^dk}{(2\pi)^d}\,\frac{k_\mu}{(k^2-M^2+i\varepsilon)\big[(k+p)^2-m_{\widetilde{L}}^2+i\varepsilon\big]}, \end{aligned} \tag{19.26}$$

where M is the photino mass and m_L is the mass of $\widetilde{e}_L$. It then follows that

$$\Sigma_L(p^2)_{\rm SQED} = \frac{\alpha}{2\pi} B_1(p^2, M^2, m_L^2) \,. \tag{19.27}$$

Diagram (b) of Fig. 19.4 yields

$$-ip\cdot\sigma\Sigma_R(p^2)_{\rm SQED} = \int \frac{d^dk}{(2\pi)^d}(-i\sqrt{2}\,e\mu^\epsilon)\frac{-ik\cdot\sigma}{k^2-M^2+i\varepsilon}(-i\sqrt{2}\,e\mu^\epsilon)\frac{i}{(k+p)^2-m_R^2+i\varepsilon}$$

$$= -2e^2\mu^{2\epsilon}\sigma^\mu \int \frac{d^dk}{(2\pi)^d}\frac{k_\mu}{(k^2-M^2+i\varepsilon)\left[(k+p)^2-m_R^2+i\varepsilon\right]}\,, \tag{19.28}$$

where m_R is the mass of $\widetilde{e}_R$. It then follows that

$$\Sigma_R(p^2)_{\rm SQED} = \frac{\alpha}{2\pi} B_1(p^2, M^2, m_R^2) \,. \tag{19.29}$$

Using the results obtained above, we can compute the SQED contributions to the electron wave function renormalization constants. In light of eqs. (2.310), (2.324), and (2.325), and making use of eq. (19.25), it follows that

$$(\delta Z_L)_{\rm SQED} = \Sigma_L(m^2) + \mathcal{D}(m^2)\,,$$

$$(\delta Z_R)_{\rm SQED} = \Sigma_R(m^2) + \mathcal{D}(m^2)\,, \tag{19.30}$$

where

$$\mathcal{D}(m^2) \equiv m^2\left[\Sigma_L'(m^2) + \Sigma_R'(m^2)\right]. \tag{19.31}$$

Note that if $m_L \neq m_R$, then parity is violated. Consequently, it follows that $\Sigma_L(m^2) \neq \Sigma_R(m^2)$ and $\delta Z_L \neq \delta Z_R$, as expected for a parity-nonconserving theory.

Using the results obtained in eqs. (19.27) and (19.29), it follows that

$$(\delta Z_L)_{\rm SQED} = \frac{\alpha}{2\pi}\left\{ B_1(m^2; M^2, m_L^2) + m^2\left[B_1'(m^2; M^2, m_L^2) + B_1'(m^2; M^2, m_R^2)\right]\right\}, \tag{19.32}$$

$$(\delta Z_R)_{\rm SQED} = \frac{\alpha}{2\pi}\left\{ B_1(m^2; M^2, m_R^2) + m^2\left[B_1'(m^2; M^2, m_L^2) + B_1'(m^2; M^2, m_R^2)\right]\right\}. \tag{19.33}$$

As a final exercise, let us examine the supersymmetric limit where $m_L = m_R = m$ and $M = 0$. In this case, $\delta Z_{\rm SQED} \equiv (\delta Z_L)_{\rm SQED} = (\delta Z_R)_{\rm SQED}$, where

$$\delta Z_{\rm SQED} = \frac{\alpha}{2\pi}\left[B_1(m^2; 0, m^2) + 2m^2 B_1'(m^2; 0, m^2)\right]. \tag{19.34}$$

It is noteworthy that $\delta Z_{\rm SQED}$ is not infrared divergent in the supersymmetric limit. Adding the QED and SQED contributions given in eqs. (19.13) and (19.34), the end result (in the Feynman gauge) is

$$\delta Z = \frac{\alpha}{2\pi}\left\{ \epsilon\delta_D\left[B_0(m^2; 0, m^2) + B_1(m^2; 0, m^2) + 2m^2 B_1'(m^2; 0, m^2)\right] - B_0(m^2; 0, m^2) + 2m^2 B_0'(m^2; 0, m^2)\right\}. \tag{19.35}$$

Note that, in the limit of $\epsilon \to 0$,

$$\epsilon\big[B_0(m^2;0,m^2)+B_1(m^2;0,m^2)+2m^2B_1'(m^2;0,m^2)\big] = (4\pi)^\epsilon \frac{\Gamma(1+\epsilon)}{2(1-\epsilon)(1-2\epsilon)}\left(\frac{m^2}{\mu^2}\right)^{-\epsilon} = \frac{1}{2}+\mathcal{O}(\epsilon)\,. \tag{19.36}$$

Hence, eq. (19.35) yields the following supersymmetric limit:

$$\delta Z = \frac{\alpha}{2\pi}\left[\frac{1}{2}\delta_D - B_0(m^2;0,m^2)+2m^2B_0'(m^2;0,m^2)\right]+\mathcal{O}(\epsilon)\,. \tag{19.37}$$

In order to preserve supersymmetric Ward identities, we shall employ the DRED procedure by taking $\delta_D = 0$. One can then employ eqs. (19.15) and (19.21) to obtain

$$\delta Z = -\frac{\alpha}{2\pi}(4\pi)^\epsilon\left(\frac{m^2}{\mu^2}\right)^{-\epsilon}\left[\frac{\Gamma(\epsilon)}{1-2\epsilon}+\frac{\Gamma(1+\epsilon)}{\epsilon(1-2\epsilon)}\right]. \tag{19.38}$$

As in eq. (19.24), the term proportional to $\Gamma(\epsilon)$ exhibits the ultraviolet divergence, whereas ϵ^{-1} corresponds to the infrared divergence.

19.2 Electroweak Vector Boson Self-Energies from Fermion Loops

In this section, we consider the contributions to the self-energy functions of the Standard Model electroweak vector bosons coming from quark and lepton loops. The independent self-energies are given by $\Pi^{WW}_{\mu\nu}$, $\Pi^{ZZ}_{\mu\nu}$, $\Pi^{\gamma Z}_{\mu\nu} = \Pi^{Z\gamma}_{\mu\nu}$, and $\Pi^{\gamma\gamma}_{\mu\nu}$, as shown in Figs. 19.5 and 19.6. In each case, $i\Pi_{\mu\nu}$ is equal to the sum of Feynman diagrams for two-point functions with amputated external legs, and is implicitly a function of the external momentum p^μ.

First consider the self-energy function for the W boson, shown in Fig. 19.5. The W boson only couples to left-handed fermions, so there is only one Feynman diagram for each Standard Model weak isodoublet. Taking the external momentum flowing from left to right to be p, and the loop momentum flowing counterclockwise in the

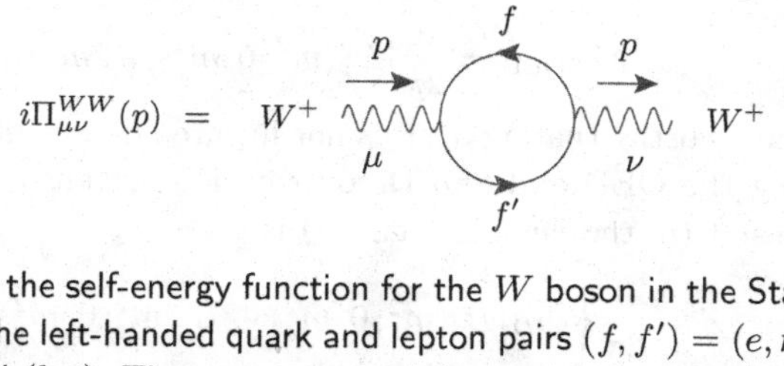

Fig. 19.5 Contributions to the self-energy function for the W boson in the Standard Model, from loops involving the left-handed quark and lepton pairs $(f, f') = (e, \nu_e)$, (μ, ν_μ), (τ, ν_τ), (d, u), (s, c), and (b, t). The momentum of the positively charged W^+ flows from left to right.

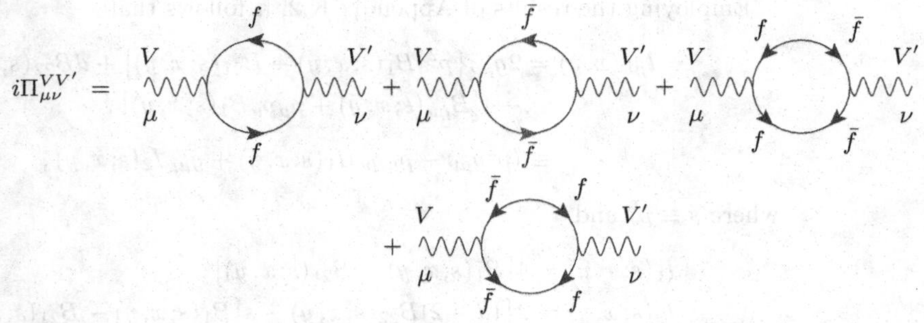

Fig. 19.6 Contributions to the diagonal and off-diagonal self-energy functions for the neutral vector bosons $V, V' = \gamma, Z$ in the Standard Model, from loops involving the three families of leptons and quarks: $f = e, \nu_e, \mu, \nu_\mu, \tau, \nu_\tau, d, u, s, c, b, t$.

upper fermion line (f) to be k, we have, from the Feynman rules of Fig. I.2,

$$i\Pi_{\mu\nu}^{WW}(p) = (-1)\int \frac{d^dk}{(2\pi)^d} \sum_{(f,f')} N_c^f \\ \times \text{Tr}\left[\left(-i\frac{g}{\sqrt{2}}\mu^\epsilon \overline{\sigma}_\mu\right)\left(\frac{ik\cdot\sigma}{k^2-m_f^2}\right)\left(-i\frac{g}{\sqrt{2}}\mu^\epsilon \overline{\sigma}_\nu\right)\left(\frac{i(k+p)\cdot\sigma}{(k+p)^2-m_{f'}^2}\right)\right], \tag{19.39}$$

where we have replaced $g \to g\mu^\epsilon$ (where μ is an arbitrary mass scale) so that g remains dimensionless when evaluating the integral in $d = 4 - 2\epsilon$ dimensions. The sum in eq. (19.39) is over the six isodoublet pairs $(f, f') = (e, \nu_e)$, (μ, ν_μ), (τ, ν_τ), (d, u), (s, c), and (b, t) with CKM mixing neglected, and

$$N_c^f = \begin{cases} 3, & f = \text{quarks}, \\ 1, & f = \text{leptons}. \end{cases} \tag{19.40}$$

The first factor of (-1) in eq. (19.39) is due to the presence of a closed fermion loop. The trace is taken over the two-component dotted spinor indices. Using eq. (B.20), it follows that

$$\Pi_{\mu\nu}^{WW}(p) = \frac{g^2}{32\pi^2}\sum_f N_c^f I_{\mu\nu}(m_f^2, m_{f'}^2), \tag{19.41}$$

where we have defined

$$I_{\mu\nu}(x,y) = 16\pi^2 i\,\mu^{2\epsilon}\int \frac{d^dk}{(2\pi)^d}\,\frac{4k_\mu k_\nu + 2k_\mu p_\nu + 2k_\nu p_\mu - 2k\cdot(k+p)\,g_{\mu\nu}}{(k^2-x)[(k+p)^2-y]}. \tag{19.42}$$

We do not explicitly exhibit here the term proportional to $\epsilon_{\mu\nu\alpha\beta}$, as it integrates to zero.

Employing the results of Appendix K.2, it follows that

$$
\begin{aligned}
I_{\mu\nu}(x,y) &= 2g_{\mu\nu}\{p^2\left[B_1(s;x,y)+B_{21}(s;x,y)\right]+dB_{22}(s;x,y)\} \\
&\quad -4\left[B_{\mu\nu}(s;x,y)+p_\mu p_\nu B_1(s;x,y)\right], \\
&\equiv (p^2 g_{\mu\nu}-p_\mu p_\nu)I_1(s;x,y)+g_{\mu\nu}I_2(s;x,y), \qquad (19.43)
\end{aligned}
$$

where $s \equiv p^2$ and

$$
\begin{aligned}
I_1(s;x,y) &= 4\left[B_1(s;x,y)+B_{21}(s;x,y)\right], \qquad (19.44)\\
I_2(s;x,y) &= 2\left\{(d-2)B_{22}(s;x,y)-s\left[B_1(s;x,y)+B_{21}(s;x,y)\right]\right\}. \qquad (19.45)
\end{aligned}
$$

Using eqs. (K.28) and (K.36), it follows that

$$
\begin{aligned}
B_1(s;x,y)+B_{21}(s;x,y) &= \frac{1}{6s^2}\Big\{\left[s-2(x-y)\right]A_0(x)+\left[s+2(x-y)\right]A_0(y) \\
&\quad -\left[s^2+s(x+y)-2(x-y)^2\right]B_0(s;x,y)+\tfrac{1}{3}s^2-s(x+y)\Big\}. \qquad (19.46)
\end{aligned}
$$

Likewise, using eqs. (K.28), (K.34), and (K.37),

$$
\begin{aligned}
&(d-2)B_{22}(s;x,y)-s\left[B_1(s;x,y)+B_{21}(s;x,y)\right] \\
&\quad = \frac{1}{2s}\Big\{(x-y)\left[A_0(x)-A_0(y)\right]+\left[s(x+y)-(x-y)^2\right]B_0(s;x,y)\Big\}. \qquad (19.47)
\end{aligned}
$$

It is convenient to isolate the ultraviolet divergence explicitly by defining the finite functions $A(x)$ and $B(s;x,y)$ such that

$$
A_0(x) \equiv \frac{x}{\epsilon}+A(x), \qquad B_0(s;x,y) \equiv \frac{1}{\epsilon}+B(s;x,y). \qquad (19.48)
$$

In light of eqs. (K.18)–(K.21),

$$
\begin{aligned}
A(x) &= x\left[1-\ln(x/Q^2)\right], \qquad (19.49)\\
B(s;x,y) = B(s;y,x) &= -\int_0^1 \ln\left(\frac{zx+(1-z)y-z(1-z)s-i\varepsilon}{Q^2}\right)dz, \qquad (19.50)
\end{aligned}
$$

where the $\overline{\rm MS}$ renormalization scale Q is defined in eq. (K.19). Explicit forms for $B(s;x,y)$ in various kinematic regions are given in eqs. (K.45)–(K.49).

After dropping terms of $\mathcal{O}(\epsilon)$ we end up with

$$
\begin{aligned}
I_1(s;x,y) &= -\frac{2}{3\epsilon}+\frac{2}{3s^2}\Big\{(s-2x+2y)A(x)+(s+2x-2y)A(y) \\
&\quad -\left[s^2+s(x+y)-2(x-y)^2\right]B(s;x,y)-s(x+y)+\tfrac{1}{3}s^2\Big\}, \qquad (19.51)
\end{aligned}
$$

$$
I_2(s;x,y) = \frac{x+y}{\epsilon}+\left(\frac{x-y}{s}\right)\left[A(x)-A(y)\right]+\left[x+y-\frac{(x-y)^2}{s}\right]B(s;x,y). \qquad (19.52)
$$

The photon and Z boson have mixed self-energy functions, defined in Fig. 19.6. Applying the pertinent Feynman rules from Fig. I.2, we obtain

$$
\begin{aligned}
i\Pi_{\mu\nu}^{VV'}(p) = (-1)\mu^{2\epsilon}\int\frac{d^dk}{(2\pi)^d}\sum_f N_c^f \\
\times\mathrm{Tr}\bigg\{&\left(-iG_V^f\overline{\sigma}_\mu\right)\left(\frac{ik\cdot\sigma}{k^2-m_f^2}\right)\left(-iG_{V'}^f\overline{\sigma}_\nu\right)\left(\frac{i(k+p)\cdot\sigma}{(k+p)^2-m_f^2}\right) \\
&+\left(-iG_V^{\bar f}\overline{\sigma}_\mu\right)\left(\frac{ik\cdot\sigma}{k^2-m_f^2}\right)\left(-iG_{V'}^{\bar f}\overline{\sigma}_\nu\right)\left(\frac{i(k+p)\cdot\sigma}{(k+p)^2-m_f^2}\right) \\
&+\left(-iG_V^{f}\overline{\sigma}_\mu\right)\left(\frac{im_f}{k^2-m_f^2}\right)\left(iG_{V'}^{\bar f}\sigma_\nu\right)\left(\frac{im_f}{(k+p)^2-m_f^2}\right) \\
&+\left(-iG_V^{\bar f}\overline{\sigma}_\mu\right)\left(\frac{im_f}{k^2-m_f^2}\right)\left(iG_{V'}^{f}\sigma_\nu\right)\left(\frac{im_f}{(k+p)^2-m_f^2}\right)\bigg\}, \qquad (19.53)
\end{aligned}
$$

where V and V' can each be either γ or Z, and $\sum_f$ is taken over the 12 Standard Model fermions. The corresponding Vff and $V\bar{f}\bar{f}$ couplings are[2]

$$G_\gamma^f = -G_\gamma^{\bar f} = eQ_f\,, \qquad (19.54)$$

$$G_Z^f = \frac{g}{c_W}(T_3^f - s_W^2 Q_f)\,, \qquad G_Z^{\bar f} = \frac{g}{c_W}s_W^2 Q_f\,, \qquad (19.55)$$

where $s_W \equiv \sin\theta_W$. The four terms in eq. (19.53) correspond to the four diagrams in Fig. 19.6, in the same order.

The first two terms in eq. (19.53) are computed exactly as for $\Pi_{\mu\nu}^{WW}$, while in the last two terms we use eq. (B.113) to compute the trace. It follows that the neutral electroweak vector boson self-energy function matrix, after dropping terms that vanish as $\epsilon \to 0$, is given by

$$
\begin{aligned}
\Pi_{\mu\nu}^{VV'}(p) = \frac{1}{16\pi^2}\sum_f N_c^f\Big[&(G_V^f G_{V'}^f + G_V^{\bar f}G_{V'}^{\bar f})I_{\mu\nu}(m_f^2, m_f^2) \\
&+ g_{\mu\nu}(G_V^f G_{V'}^{\bar f} + G_V^{\bar f}G_{V'}^{f})m_f^2 I_3(m_f^2, m_f^2)\Big], \qquad (19.56)
\end{aligned}
$$

where $I_{\mu\nu}(x,y)$ was defined in eqs. (19.43), (19.51), and (19.52), and

$$I_3(x,y) = -2i(16\pi^2)\,\mu^{2\epsilon}\int\frac{d^dk}{(2\pi)^d}\,\frac{1}{(k^2-x)[(k+p)^2-y]} = 2B_0(s;x,y)\,. \qquad (19.57)$$

The photon self-energy function is a simple special case of eq. (19.56):

$$\Pi_{\mu\nu}^{\gamma\gamma}(p) = \frac{1}{16\pi^2}\sum_f 2N_c^f(eQ_f)^2\left[I_{\mu\nu}(m_f^2, m_f^2) - g_{\mu\nu}m_f^2 I_3(m_f^2, m_f^2)\right]. \qquad (19.58)$$

Employing the explicit results of $I_{\mu\nu}$ obtained above, and making use of eqs. (19.48)

[2] Note that there is no contribution from the left-handed two-component antineutrino fields, $\bar\nu_e$, $\bar\nu_\mu$, $\bar\nu_\tau$, which do not exist in the Standard Model.

and (19.50) to evaluate I_3 yields

$$\Pi^{\gamma\gamma}_{\mu\nu}(p) = -\frac{\alpha}{3\pi}\sum_f N_c^f Q_f^2\,(p^2 g_{\mu\nu} - p_\mu p_\nu)\left\{\frac{1}{\epsilon} - \frac{1}{3} - \frac{2}{p^2}\left[A(m_f^2) - m_f^2\right] + \left(1 + \frac{2m_f^2}{p^2}\right)B(p^2; m_f^2, m_f^2)\right\}. \quad (19.59)$$

Equation (19.59) satisfies

$$p^\mu \Pi^{\gamma\gamma}_{\mu\nu}(p) = p^\nu \Pi^{\gamma\gamma}_{\mu\nu}(p) = 0\,, \quad (19.60)$$

as required by the Ward identity of QED, and is regular in the limit $p^2 \to 0$.

In each of eqs. (19.41), (19.56), and (19.58), there are $1/\epsilon$ poles contained in the loop integral functions. In the $\overline{\text{MS}}$ renormalization scheme, these poles are simply removed by counterterms (which have no other effect at one-loop order), and the regularization scale μ has been replaced by the $\overline{\text{MS}}$ renormalization scale Q as a consequence of eq. (K.19).

In eqs. (19.39) and (19.53), we chose to write a $\overline{\sigma}_\mu$ for the left vertex in the Feynman diagram in each case. This is an arbitrary choice; we could have chosen instead to use $-\sigma_\mu$ for the left vertex in any given diagram, as mentioned in the caption for Fig. I.2. This would have dictated the replacements $\overline{\sigma} \leftrightarrow -\sigma$ throughout the expression for the diagram, including for the fermion propagators, as was indicated in Fig. 2.5. It is not hard to check that the result, after computing the spinor index traces, is unaffected. Note that the contribution proportional to $\epsilon_{\mu\nu\rho\kappa}$ from eq. (B.19) or eq. (B.20) vanishes; this is clear because the self-energy function is symmetric under interchange of vector indices, and there is only one independent momentum in the problem.

19.3 Self-Energy and Pole Mass of the Top Quark

We next consider the one-loop calculation of the self-energy and the pole mass of the top quark in the Standard Model, including the effects of the gauge interactions and the top-quark and bottom-quark Yukawa couplings. We treat this as a one-generation problem, neglecting CKM mixing. Consequently, the corresponding Yukawa couplings Y_t and Y_b are real and positive (by a suitable phase redefinition of the Higgs field[3]). Using the formalism of Section 2.10.1 for Dirac fermions, the independent 1PI self-energy functions are given by[4] Σ_{Lt}, Σ_{Rt}, and Σ_{Dt} (defined in Fig. 2.32) as shown in Fig. 19.7.

[3] As shown in Section 1.6, after the fermion mass matrix diagonalization procedure, the tree-level fermion masses are real and nonnegative. If CKM mixing is neglected, it follows from eq. (4.172) that the corresponding diagonal Yukawa couplings are real and positive if the phase of the Higgs field is chosen such that the neutral Higgs vacuum expectation value $v > 0$.

[4] Since the Yukawa couplings can be chosen real (in the one-generation model), $\overline{\Sigma}_{Lt} = \Sigma_{Lt}$. Note that after suppressing the color degrees of freedom, Σ_{Lt}, Σ_{Rt}, and Σ_{Dt} are one-dimensional matrices, so we do not employ boldface letters in this case.

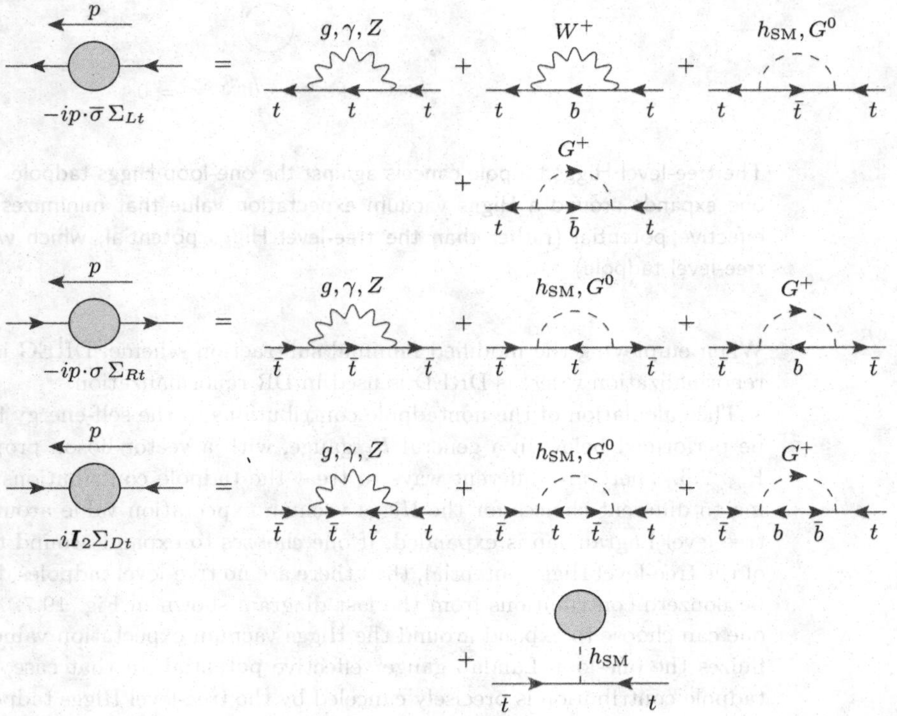

Fig. 19.7 One-loop contributions to the 1PI self-energy functions for the top quark in the Standard Model. The external momentum of the physical top quark, p^μ, flows from right to left. The loop momentum k^μ in the text is taken to flow clockwise. Spinor and color indices are suppressed, and I_2 is the 2×2 identity matrix. The external legs are amputated. The last diagram contains one-loop tadpole contributions.

Note that in these diagrams the physical top quark moves from right to left, carrying momentum p^μ. Then, according to the general formula obtained in eq. (2.283), the complex pole squared mass of the top quark is given by

$$M_t^2 - i\Gamma_t M_t = \frac{(m_t + \Sigma_{Dt})^2}{(1-\Sigma_{Lt})(1-\Sigma_{Rt})}, \tag{19.61}$$

where m_t is the tree-level mass. Working consistently to one-loop order,

$$M_t^2 - i\Gamma_t M_t = \left[m_t^2(1+\Sigma_{Lt}+\Sigma_{Rt}) + 2m_t\Sigma_{Dt}\right]\Big|_{s=m_t^2+i\varepsilon}. \tag{19.62}$$

It remains to calculate the self-energy functions Σ_{Lt}, Σ_{Rt} and Σ_{Dt}. As in Section 19.1, two regularization procedures will be used simultaneously, DREG and DRED, which depend on the choice of the parameter δ_D defined in eq. (19.2).

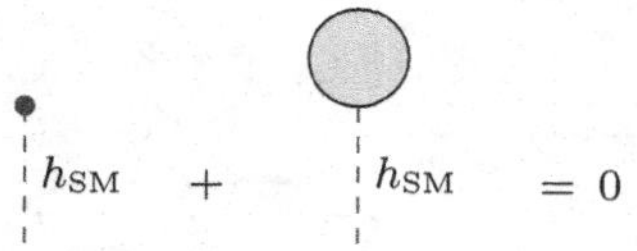

Fig. 19.8 The tree-level Higgs tadpole cancels against the one-loop Higgs tadpole, provided that one expands around a Higgs vacuum expectation value that minimizes the one-loop effective potential (rather than the tree-level Higgs potential, which would yield no tree-level tadpole).

When employing the modified minimal subtraction scheme, DREG is used in $\overline{\rm MS}$ renormalization whereas DRED is used in $\overline{\rm DR}$ renormalization.

The calculation of the nontadpole contributions to the self-energy functions will be performed below in a general R_ξ gauge, with a vector boson propagator as in Fig. 2.6. There are different ways to treat the tadpole contributions, corresponding to different choices for the Higgs vacuum expectation value around which the tree-level Lagrangian is expanded. If one chooses to expand around the minimum of the tree-level Higgs potential, then there are no tree-level tadpoles, but there will be nonzero contributions from the last diagram shown in Fig. 19.7. Alternatively, one can choose to expand around the Higgs vacuum expectation value v that minimizes the one-loop Landau gauge[5] effective potential. In that case, the one-loop tadpole contribution is precisely canceled by the tree-level Higgs tadpole, as shown in Fig. 19.8. Here, we have in mind the latter prescription; the calculation for the pole mass is therefore complete without tadpole contributions, provided that the tree-level top-quark mass is taken to be

$$m_t = Y_t v, \tag{19.63}$$

where Y_t is the $\overline{\rm MS}$ or $\overline{\rm DR}$ Yukawa coupling, and v is the Higgs vacuum expectation value at the minimum of the one-loop effective potential in Landau gauge. To be consistent with this choice, $\xi = 0$ should be taken in all formulae below that involve electroweak gauge bosons or Goldstone bosons. (The gluon contribution is naturally independent of ξ because the gauge symmetry is unbroken, providing a check of gauge-fixing invariance.) Nevertheless, for the sake of generality we will keep the dependence on ξ in the computation of the individual nontadpole self-energy diagrams below.

Consider the one-loop calculation of the self-energy Σ_{Lt}, which is the sum of individual diagram contributions:

$$\Sigma_{Lt} = [\Sigma_{Lt}]_g + [\Sigma_{Lt}]_\gamma + [\Sigma_{Lt}]_Z + [\Sigma_{Lt}]_W + [\Sigma_{Lt}]_{h_{\rm SM}} + [\Sigma_{Lt}]_{G^0} + [\Sigma_{Lt}]_{G^+} \,. \tag{19.64}$$

[5] This procedure is considerably more involved outside of Landau gauge, because the propagators mix the longitudinal components of the vector boson with the Goldstone bosons for $\xi \neq 0$ if one expands around a Higgs vacuum expectation value that does not minimize the tree-level potential. This is the same reason the effective potential is traditionally calculated specifically in Landau gauge.

First, consider the diagrams involving exchanges of the scalars $\phi = h_{\rm SM}, G^0, G^\pm$. These contributions all have the same form,

$$-ip\cdot\overline{\sigma}\,[\Sigma_{Lt}]_\phi = \mu^{2\epsilon}\int\frac{d^dk}{(2\pi)^d}\,(-iY^*)\left(\frac{i(k+p)\cdot\overline{\sigma}}{(k+p)^2-m_f^2}\right)(-iY)\left(\frac{i}{k^2-m_\phi^2}\right), \quad (19.65)$$

where the loop momentum k^μ flows clockwise, and the couplings and propagator masses are, using the Feynman rules of Figs. I.3 and I.4,

$$\text{for } \phi = h_{\rm SM}: \qquad Y = Y_t/\sqrt{2}, \qquad m_\phi^2 = m_{h_{\rm SM}}^2, \qquad m_f = m_t, \quad (19.66)$$

$$\text{for } \phi = G^0: \qquad Y = iY_t/\sqrt{2}, \qquad m_\phi^2 = \xi m_Z^2, \qquad m_f = m_t, \quad (19.67)$$

$$\text{for } \phi = G^\pm: \qquad Y = Y_b, \qquad m_\phi^2 = \xi m_W^2, \qquad m_f = m_b. \quad (19.68)$$

Once again, the factor of $\mu^{2\epsilon}$ is present so that the Yukawa coupling Y is dimensionless when carrying out the integration in d spacetime dimensions. Multiplying both sides of eq. (19.65) by $p\cdot\sigma$ and taking the trace over spinor indices using eq. (B.113), one finds

$$[\Sigma_{Lt}]_\phi = i|Y|^2\frac{\mu^{2\epsilon}}{p^2}\int\frac{d^dk}{(2\pi)^d}\frac{p\cdot(k+p)}{(k^2-m_\phi^2)[(k+p)^2-m_f^2]}\,. \quad (19.69)$$

Employing the results of Appendix K.2, it follows that

$$[\Sigma_{Lt}]_\phi \equiv -\frac{|Y|^2}{16\pi^2}I_{SF}(s;m_\phi^2,m_f^2)\,, \quad (19.70)$$

where $s \equiv p^2$ and

$$I_{SF}(s;x,y) = B_0(s;x,y) + B_1(s;x,y)\,. \quad (19.71)$$

Hence, in light of eqs. (K.28) and (19.48),

$$\begin{aligned} I_{SF}(s;x,y) &= \frac{1}{2s}\left[A_0(x) - A_0(y) + (s-x+y)B_0(s;x,y)\right] \\ &= \frac{1}{2\epsilon} + \frac{1}{2s}\left[A(x) - A(y) + (s-x+y)B(s;x,y)\right]\,, \end{aligned} \quad (19.72)$$

where the finite functions A and B are defined in eqs. (19.49) and (19.50). Note that $I_{SF}(s;x,y)$ has a smooth limit as $s \to 0$ in light of eqs. (K.97), (K.104), and (K.103).

Next, consider the contributions to Σ_{Lt} involving the vector bosons $V = g, \gamma, Z, W$. These have the common form

$$\begin{aligned} -ip\cdot\overline{\sigma}[\Sigma_{Lt}]_V = \mu^{2\epsilon}&\int\frac{d^dk}{(2\pi)^d}\,(-iG\,\overline{\sigma}_\mu)\left(\frac{i(k+p)\cdot\sigma}{(k+p)^2-m_f^2}\right)(-iG\,\overline{\sigma}_\nu) \\ &\times\left(\frac{-i}{k^2-m_V^2}\right)\left(g^{\mu\nu} - \frac{(1-\xi)k^\mu k^\nu}{k^2-\xi m_V^2}\right), \end{aligned} \quad (19.73)$$

where again the loop momentum k flows clockwise, and, using the rules of Figs. I.2

and J.10,

$$\begin{aligned}
&\text{for } V = g: && G = g_s \boldsymbol{T^a}\,, && m_f = m_t, && (19.74)\\
&\text{for } V = \gamma: && G = eQ_t\,, && m_f = m_t, && (19.75)\\
&\text{for } V = Z: && G = g(T_3^t - s_W^2 Q_t)/c_W\,, && m_f = m_t, && (19.76)\\
&\text{for } V = W: && G = g/\sqrt{2}\,, && m_f = m_b. && (19.77)
\end{aligned}$$

In the case of gluon exchange ($V = g$), the T^a are the $SU(3)_C$ generators (with color indices suppressed). The adjoint representation index a is summed over, producing a factor of the Casimir invariant $(T^aT^a)_{ij} = C_F\delta_{ij} = \frac{4}{3}\delta_{ij}$. We now use eq. (B.118) to obtain $\overline{\sigma}_\mu\,\sigma_\rho\,\overline{\sigma}_\nu\,g^{\mu\nu} = -(D-2)\,\overline{\sigma}_\rho$, where D is defined in eq. (19.1); note that this introduces a difference between the $\overline{\text{MS}}$ and $\overline{\text{DR}}$ schemes. In addition, we used $k\cdot\overline{\sigma}(k+p)\cdot\sigma k\cdot\overline{\sigma} = (k^2 + 2k\cdot p)k\cdot\overline{\sigma} - k^2 p\cdot\overline{\sigma}$, which follows from eq. (1.91). After multiplying by $p\cdot\sigma$ and taking the trace over spinor indices, we obtain

$$\begin{aligned}
[\Sigma_{Lt}]_V &= -i\,G^2\frac{\mu^{2\epsilon}}{p^2}\int\frac{d^dk}{(2\pi)^d}\,\frac{1}{(k^2-m_V^2)[(k+p)^2-m_f^2]}\bigg[(2-D)p\cdot(k+p)\\
&\qquad - \left(k^2 k\cdot p + 2(k\cdot p)^2 - k^2p^2\right)\frac{1-\xi}{k^2-\xi m_V^2}\bigg]\\
&= -i\,G^2\frac{\mu^{2\epsilon}}{p^2}\int\frac{d^dk}{(2\pi)^d}\,\frac{1}{(k^2-m_V^2)[(k+p)^2-m_f^2]}\bigg[(2-D)p\cdot(k+p)\\
&\qquad -(1-\xi)p\cdot(k-p) - \Big[\xi m_V^2 p\cdot(k-p) + 2(k\cdot p)^2\Big]\left(\frac{1-\xi}{k^2-\xi m_V^2}\right)\bigg].
\end{aligned} \tag{19.78}$$

Employing the identity

$$\frac{1-\xi}{(k^2-m_V^2)(k^2-\xi m_V^2)} = \frac{1}{m_V^2}\left[\frac{1}{k^2-m_V^2} - \frac{1}{k^2-\xi m_V^2}\right], \tag{19.79}$$

we then obtain

$$\begin{aligned}
[\Sigma_{Lt}]_V &= -i\,G^2\frac{\mu^{2\epsilon}}{p^2}\\
&\times\Bigg\{\int\frac{d^dk}{(2\pi)^d}\,\frac{1}{(k^2-m_V^2)[(k+p)^2-m_f^2]}\bigg[(1-D+\xi)p\cdot k + (3-D-\xi)p^2\\
&\qquad -\xi\,p\cdot(k-p) - \frac{2(k\cdot p)^2}{m_V^2}\bigg]\\
&+\int\frac{d^dk}{(2\pi)^d}\,\frac{1}{(k^2-\xi m_V^2)[(k+p)^2-m_f^2]}\left[\xi\,p\cdot(k-p) + \frac{2(k\cdot p)^2}{m_V^2}\right]\Bigg\}.
\end{aligned} \tag{19.80}$$

Using the results of Appendix K.2, it then follows that

$$[\Sigma_{Lt}]_V \equiv -\frac{1}{16\pi^2}G^2 I_{VF}(s; m_V^2, m_f^2), \tag{19.81}$$

where $s \equiv p^2$ and

$$\begin{aligned}I_{VF}(s;x,y) &= (D-3)B_0(s;x,y) + (D-1)B_1(s;y,x) \\ &\quad + \frac{2}{x}\left[sB_{21}(s;x,y) + B_{22}(s;x,y)\right] + \xi\big[B_0(s;\xi x,y) - B_1(s;\xi x,y)\big] \\ &\quad - \frac{2}{x}\left[sB_{21}(s;\xi x,y) + B_{22}(s;\xi x,y)\right]. \end{aligned} \tag{19.82}$$

Simplifying the above result with the help of eq. (K.30), we obtain

$$\begin{aligned}I_{VF}(s;x,y) &= (D-3)B_0(s;x,y) + (D-2)B_1(s;x,y) + \xi\, B_0(s;\xi x,y) \\ &\quad - \left(\frac{s-y}{x}\right)\big[B_1(s;x,y) - B_1(s;\xi x,y)\big]. \end{aligned} \tag{19.83}$$

Finally, we employ eqs. (K.28) and (K.38) to obtain

$$\begin{aligned}I_{VF}(s;x,y) &= -\left(\frac{D-2}{2s}\right)A_0(y) + \left(\frac{(D-2)x-s+y}{2sx}\right)A_0(x) + \left(\frac{s-y}{2sx}\right)A_0(\xi x) \\ &\quad + \left(\frac{(D-2)(s-x+y)}{2s} + \frac{(s-y)^2 - x(s+y)}{2sx}\right)B_0(s;x,y) \\ &\quad - \left(\frac{(s-y)^2 - \xi x(s+y)}{2sx}\right)B_0(s;\xi x,y). \end{aligned} \tag{19.84}$$

Using eqs. (19.1) and (19.48), we end up with

$$\begin{aligned}I_{VF}(s;x,y) &= \frac{\xi}{\epsilon} - \delta_D + \frac{A(x)-A(y)}{s} - \frac{(s-y)\big[A(x)-A(\xi x)\big]}{2sx} \\ &\quad + \frac{1}{s}\left(s-x+y+\frac{(s-y)^2 - x(s+y)}{2x}\right)B(s;x,y) \\ &\quad - \left(\frac{(s-y)^2 - \xi x(s+y)}{2sx}\right)B(s;\xi x,y), \end{aligned} \tag{19.85}$$

after dropping terms that vanish as $\epsilon \to 0$. Combining the results of eqs. (19.70) and (19.81)) and making the substitution $s = m_t^2$, we obtain

$$\begin{aligned}\Sigma_{Lt} = -\frac{1}{16\pi^2}\Big[&\left(g_s^2 C_F + e^2 Q_t^2\right)I_{VF}(m_t^2;0,m_t^2) \\ &+ [g(T_3^t - s_W^2 Q_t)/c_W]^2 I_{VF}(m_t^2;m_Z^2,m_t^2) + \tfrac{1}{2}g^2 I_{VF}(m_t^2;m_W^2,m_b^2) \\ &+ \tfrac{1}{2}Y_t^2 I_{SF}(m_t^2;m_{h_{\rm SM}}^2,m_t^2) + \tfrac{1}{2}Y_t^2 I_{SF}(m_t^2;\xi m_Z^2,m_t^2) + Y_b^2 I_{SF}(m_t^2;\xi m_W^2,m_b^2)\Big]. \end{aligned} \tag{19.86}$$

It is useful to note that, for massless gauge bosons,

$$I_{VF}(y;0,y) = \xi\left[\frac{1}{\epsilon} - \ln\left(\frac{y}{Q^2}\right) + 2\right] + 1 - \delta_D, \tag{19.87}$$

after making use of the identity

$$B(y;0,y) = 1 + \frac{A(y)}{y} + \mathcal{O}(\epsilon) = 2 - \ln\left(\frac{y}{Q^2}\right) + \mathcal{O}(\epsilon). \tag{19.88}$$

One can check that for $\xi = 1$, eqs. (19.81) and (19.87) reproduce the result previously obtained in eq. (19.4).

The contributions to

$$\Sigma_{Rt} = [\Sigma_{Rt}]_g + [\Sigma_{Rt}]_\gamma + [\Sigma_{Rt}]_Z + [\Sigma_{Rt}]_{h_{\rm SM}} + [\Sigma_{Rt}]_{G^0} + [\Sigma_{Rt}]_{G^\pm} \qquad (19.89)$$

are obtained similarly. Note that there is no W boson contribution, since the right-handed top quark is an $SU(2)_L$ singlet. For the scalar exchange diagrams with $\phi = h_{\rm SM}, G^0, G^\pm$, the general form is

$$-ip\cdot\sigma[\Sigma_{Rt}]_\phi = \mu^{2\epsilon}\int\frac{d^dk}{(2\pi)^d}(-iY)\left(\frac{i(k+p)\cdot\sigma}{(k+p)^2-m_f^2}\right)(-iY^*)\left(\frac{i}{k^2-m_\phi^2}\right), \qquad (19.90)$$

which yields

$$[\Sigma_{Rt}]_\phi = -\frac{1}{16\pi^2}|Y|^2 I_{SF}(s; m_\phi^2, m_f^2). \qquad (19.91)$$

Here the couplings and propagator masses for $h_{\rm SM}$ and G^0 are the same as in eqs. (19.66) and (19.67), but now, instead of eq. (19.68),

$$\text{for } \phi = G^\pm: \qquad Y = -Y_t\,, \qquad m_\phi^2 = \xi m_W^2\,, \qquad m_f = m_b\,, \qquad (19.92)$$

from Fig. I.4. For the contributions due to exchanges of vectors $v = g, \gamma, Z$, the general form is given by

$$\begin{aligned}-ip\cdot\sigma[\Sigma_{Rt}]_V = \mu^{2\epsilon}&\int\frac{d^dk}{(2\pi)^d}(iG\,\sigma_\mu)\left(\frac{i(k+p)\cdot\overline{\sigma}}{(k+p)^2-m_f^2}\right)(iG\,\sigma_\nu)\\ &\times\left(\frac{-i}{k^2-m_V^2}\right)\left(g^{\mu\nu}-\frac{(1-\xi)k^\mu k^\nu}{k^2-\xi m_V^2}\right), \qquad (19.93)\end{aligned}$$

where

$$\text{for } V = g: \qquad G = -g_s\boldsymbol{T^a}\,, \qquad (19.94)$$

$$\text{for } V = \gamma: \qquad G = -eQ_t\,, \qquad (19.95)$$

$$\text{for } V = Z: \qquad G = gs_W^2 Q_t/c_W\,, \qquad (19.96)$$

after using the rules of Figs. I.2 and J.10 with $m_f = m_t$ in each case. We then make use of $\sigma_\mu\,\overline{\sigma}_\rho\,\sigma_\nu\,g^{\mu\nu} = -(D-2)\,\sigma_\rho$ [see eq. (B.117)] and

$$k\cdot\sigma(k+p)\cdot\overline{\sigma}k\cdot\sigma = (k^2+2k\cdot p)k\cdot\sigma - k^2p\cdot\sigma\,, \qquad (19.97)$$

in light of eq. (1.90). After multiplying by $p\cdot\overline{\sigma}$ and taking the trace over spinor indices [using eq. (B.113)], we obtain

$$[\Sigma_{Rt}]_V = -\frac{1}{16\pi^2}G^2 I_{VF}(s; m_V^2, m_t^2)\,, \qquad (19.98)$$

in terms of the same function appearing in eqs. (19.85) and (19.87). Adding up the contributions from eqs. (19.91) and (19.98) and taking $s = m_t^2$ yields

$$\Sigma_{Rt} = -\frac{1}{16\pi^2}\Big[\Big(g_s^2 C_F + e^2 Q_t^2\Big) I_{VF}(m_t^2; 0, m_t^2) + (g^2 Q_t^2 s_W^4/c_W^2) I_{VF}(m_t^2; m_Z^2, m_t^2) + \tfrac{1}{2}Y_t^2 I_{SF}(m_t^2; m_{h_{\rm SM}}^2, m_t^2) + \tfrac{1}{2}Y_t^2 I_{SF}(m_t^2; \xi m_Z^2, m_t^2) + Y_t^2 I_{SF}(m_t^2; \xi m_W^2, m_b^2)\Big]. \tag{19.99}$$

Next, consider the contributions to

$$\Sigma_{Dt} = [\Sigma_{Dt}]_g + [\Sigma_{Dt}]_\gamma + [\Sigma_{Dt}]_Z + [\Sigma_{Dt}]_{h_{\rm SM}} + [\Sigma_{Dt}]_{G^0} + [\Sigma_{Dt}]_{G^\pm}, \tag{19.100}$$

ignoring the tadpole contribution for now. The diagrams involving the exchange of scalars $\phi = h_{\rm SM}, G^0, G^\pm$ yield

$$\begin{aligned} -i[\Sigma_{Dt}]_\phi &= \mu^{2\epsilon}\int \frac{d^dk}{(2\pi)^d}(-iY_1)\left(\frac{im_f}{(k+p)^2 - m_f^2}\right)(-iY_2)\left(\frac{i}{k^2 - m_\phi^2}\right), \\ &= im_f Y_1 Y_2 \mu^{2\epsilon}\int \frac{d^dk}{(2\pi)^d}\frac{1}{(k^2 - m_\phi^2)[(k+p)^2 - m_f^2]}. \end{aligned} \tag{19.101}$$

In light of eq. (K.38), it follows that

$$[\Sigma_{Dt}]_\phi \equiv -\frac{1}{16\pi^2} m_f Y_1 Y_2 I_{S\overline{F}}(s; m_\phi^2, m_f^2), \tag{19.102}$$

where $s \equiv p^2$ and

$$I_{S\overline{F}}(s; x, y) = B_0(s; x, y) = \frac{1}{\epsilon} + B(s; x, y), \tag{19.103}$$

after dropping terms that vanish as $\epsilon \to 0$. The relevant couplings and masses are, from Figs. I.3 and I.4,

$$\text{for } \phi = h_{\rm SM}: \quad Y_1 = Y_2 = Y_t/\sqrt{2}, \quad m_\phi^2 = m_{h_{\rm SM}}^2, \quad m_f = m_t, \tag{19.104}$$
$$\text{for } \phi = G^0: \quad Y_1 = Y_2 = iY_t/\sqrt{2}, \quad m_\phi^2 = \xi m_Z^2, \quad m_f = m_t, \tag{19.105}$$
$$\text{for } \phi = G^\pm: \quad Y_1 = Y_b, \quad Y_2 = -Y_t, \quad m_\phi^2 = \xi m_W^2, \quad m_f = m_b. \tag{19.106}$$

The contributions from vector boson exchanges are of the form

$$\begin{aligned} -i\boldsymbol{I}_2[\Sigma_{Dt}]_V &= \mu^{2\epsilon}\int \frac{d^dk}{(2\pi)^d}(iG_1\sigma_\mu)\left(\frac{im_f}{(k+p)^2 - m_f^2}\right)(-iG_2\overline{\sigma}_\nu) \\ &\quad \times \left(\frac{-i}{k^2 - m_V^2}\right)\left(g^{\mu\nu} - \frac{(1-\xi)k^\mu k^\nu}{k^2 - \xi m_V^2}\right). \end{aligned} \tag{19.107}$$

Using $\sigma_\mu \overline{\sigma}_\nu g^{\mu\nu} = D\boldsymbol{I}_2$ [see eq. (B.116)], where D is defined in eq. (19.1), and $k\cdot\sigma\, k\cdot\overline{\sigma} = k^2\boldsymbol{I}_2$ [using eq. (1.88)], we end up with

$$[\Sigma_{Dt}]_V = im_f G_1 G_2 \mu^{2\epsilon}\int \frac{d^dk}{(2\pi)^d}\frac{1}{(k^2 - m_V^2)[(k+p)^2 - m_f^2]}\left[D - \frac{(1-\xi)k^2}{k^2 - \xi m_V^2}\right]. \tag{19.108}$$

Employing the identity

$$D - \frac{(1-\xi)k^2}{k^2 - \xi m_V^2} = D - 1 + \frac{\xi(k^2 - m_V^2)}{k^2 - \xi m_V^2}\,, \tag{19.109}$$

we obtain

$$[\Sigma_{Dt}]_V = i m_f G_1 G_2 \mu^{2\epsilon}\Bigg\{(D-1)\int\frac{d^dk}{(2\pi)^d}\,\frac{1}{(k^2-m_V^2)[(k+p)^2-m_f^2]}$$
$$+\xi\int\frac{d^dk}{(2\pi)^d}\,\frac{1}{(k^2-\xi m_V^2)[(k+p)^2-m_f^2]}\Bigg\}. \tag{19.110}$$

In light of eq. (K.38), it follows that

$$[\Sigma_{Dt}]_V \equiv -\frac{1}{16\pi^2} m_f G_1 G_2 I_{V\overline{F}}(s; m_V^2, m_f^2)\,, \tag{19.111}$$

where $s \equiv p^2$ and

$$I_{V\overline{F}}(s;x,y) = (D-1)B_0(s;x,y) + \xi B_0(s;\xi x, y)$$
$$= \frac{3+\xi}{\epsilon} - 2\delta_D + 3B(s;x,y) + \xi B(s;\xi x,y)\,, \tag{19.112}$$

after dropping terms that vanish as $\epsilon \to 0$. It is useful to note that, for massless gauge bosons,

$$I_{V\overline{F}}(y;0,y) = (3+\xi)\left[\frac{1}{\epsilon} + 2 - \ln\left(\frac{y}{Q^2}\right)\right] - 2\delta_D\,. \tag{19.113}$$

The relevant couplings are obtained from the rules of Figs. I.2 and J.10:

$$\text{for } V = g: \qquad G_1 = -G_2 = g_s \boldsymbol{T^a}, \tag{19.114}$$
$$\text{for } V = \gamma: \qquad G_1 = -G_2 = eQ_t, \tag{19.115}$$
$$\text{for } V = Z: \qquad G_1 = g(T_3^t - s_W^2 Q_t)/c_W, \qquad G_2 = g s_W^2 Q_t/c_W, \tag{19.116}$$

and $m_f = m_t$ in each case. Adding up these contributions and taking $s = m_t^2$, we end up with

$$\Sigma_{Dt} = -\frac{m_t}{16\pi^2}\Big\{g^2\left[(T_3^t - s_W^2 Q_t)s_W^2 Q_t/c_W^2\right] I_{V\overline{F}}(m_t^2; m_Z^2, m_t^2)$$
$$-(g_s^2 C_F + e^2 Q_t^2) I_{V\overline{F}}(m_t^2; 0, m_t^2) + \tfrac{1}{2}Y_t^2 I_{S\overline{F}}(m_t^2; m_{h_{\rm SM}}^2, m_t^2)$$
$$-\tfrac{1}{2}Y_t^2 I_{S\overline{F}}(m_t^2; \xi m_Z^2, m_t^2) - Y_b^2 I_{S\overline{F}}(m_t^2; \xi m_W^2, m_b^2)\Big\}, \tag{19.117}$$

where $Y_t = m_t Y_b / m_b$ was used in the last term.

In each of the self-energy functions above, there are poles in $1/\epsilon$ contained within the functions I_{VF}, I_{SF}, $I_{V\overline{F}}$, and $I_{S\overline{F}}$. In the $\overline{\rm MS}$ or $\overline{\rm DR}$ schemes, these poles are simply canceled by counterterms (which have no other effect at one-loop order). The one-loop top-quark pole mass can now be obtained by plugging eqs. (19.86), (19.99), and (19.117) into eq. (19.62) with $\xi = 0$, as discussed earlier. It is not hard to check that the terms from massless Goldstone boson exchange just cancel against the terms from the vector exchange diagrams that came from ξm_W^2 and ξm_Z^2.

As a simple example, consider the one-loop pole mass with only QCD effects included. Then the result of eq. (19.62) has no imaginary part. Taking the square root (and dropping a two-loop order part) yields

$$\begin{aligned} M_{t,\text{pole}} &= m_t\left(1 + \tfrac{1}{2}\Sigma_{Lt} + \tfrac{1}{2}\Sigma_{Rt}\right) + \Sigma_{Dt} \\ &= m_t\left\{1 - \frac{C_F g_s^2}{16\pi^2}\left[I_{VF}(m_t^2; 0, m_t^2) - I_{V\overline{F}}(m_t^2; 0, m_t^2)\right]\right\} \\ &= m_t\left\{1 + \frac{\alpha_s}{4\pi}C_F\left[5 - \delta_D - 3\ln\left(\frac{m_t^2}{Q^2}\right)\right]\right\}, \end{aligned} \tag{19.118}$$

after removing the poles in $1/\epsilon$ as noted below eq. (19.117). As another check, consider the imaginary part of the squared pole mass of the top quark. At leading order, eq. (19.62) implies

$$\begin{aligned} \Gamma_t &= -\text{Im}[m_t(\Sigma_{Lt} + \Sigma_{Rt}) + 2\Sigma_{Dt}] \\ &= \frac{m_t}{16\pi^2}\,\text{Im}\Big[\tfrac{1}{2}g^2 I_{VF}(m_t^2; m_W^2, m_b^2) + (Y_t^2 + Y_b^2)I_{SF}(m_t^2; \xi m_W^2, m_b^2,) \\ &\qquad - 2Y_b^2 I_{S\overline{F}}(m_t^2; \xi m_W^2, m_b^2,)\Big] \\ &= \frac{1}{32\pi^2 m_t}\left\{(g^2 + Y_t^2 + Y_b^2)(m_t^2 + m_b^2 - m_W^2) - 4Y_b^2 m_t^2\right\}\text{Im}[B(m_t^2; m_W^2, m_b^2)]. \end{aligned} \tag{19.119}$$

The cancellation of the ξ dependence is a successful check of gauge invariance, since the tadpole diagram in Fig. 19.7 does not contribute to the absorptive part of the self-energy. In light of eqs. (K.45)–(K.47),

$$\text{Im}[B(s; x, y)] = \begin{cases} 0 & \text{for} \quad s \leq (\sqrt{x} + \sqrt{y})^2, \\ \pi\lambda^{1/2}(s, x, y)/s & \text{for} \quad s > (\sqrt{x} + \sqrt{y})^2. \end{cases} \tag{19.120}$$

Equation (19.119) reproduces eq. (17.34) for the top-quark width at leading order.

19.4 Self-Energy and Pole Mass of the Gluino

The Feynman diagrams for the gluino self-energy are shown in Fig. 19.9. Since the gluino is a Majorana fermion, we can use the general formalism of Section 2.10.1. We will compute the self-energy functions $\Xi_{\tilde{g}} \equiv \Xi_{\tilde{g}}{}^{\tilde{g}}$ and $\Omega_{\tilde{g}} \equiv \Omega^{\tilde{g}\tilde{g}}$ defined in Fig. 2.30, and infer $\overline{\Omega}_{\tilde{g}} \equiv \overline{\Omega}_{\tilde{g}\tilde{g}}$ from the latter by replacing all Lagrangian parameters by their complex conjugates.[6] At one-loop order, it follows from the general result

[6] Suppressing the color degrees of freedom, Ξ, Ω, and $\overline{\Omega}$ are one-dimensional matrices, so we do not employ boldface letters in this case.

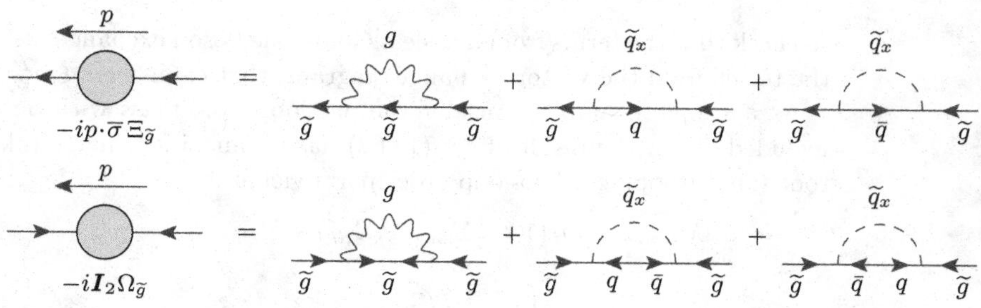

Fig. 19.9 Self-energy functions for the gluino in supersymmetry. The external momentum p^μ flows from right to left. The loop momentum k^μ in the text is taken to flow clockwise. Spinor and color indices are suppressed, and $\boldsymbol{I}_2$ is the 2×2 identity matrix. The index $x = 1, 2$ labels the two squark mass eigenstates of a given flavor $q = u, d, s, c, b, t$. Both x and q must be summed over. The external legs are amputated.

of eq. (2.267) that the complex pole squared mass of the gluino is related to the tree-level mass $m_{\tilde{g}}$ by

$$M_{\tilde{g}}^2 - iM_{\tilde{g}}\Gamma_{\tilde{g}} = \left[m_{\tilde{g}}^2(1+2\Xi_{\tilde{g}}) + m_{\tilde{g}}(\Omega_{\tilde{g}} + \overline{\Omega}_{\tilde{g}})\right]\Big|_{s=m_{\tilde{g}}^2+i\varepsilon}. \tag{19.121}$$

It is convenient to split the self-energy functions into gluon/gluino loop and squark/quark loop contributions, as

$$\Xi_{\tilde{g}} = [\Xi_{\tilde{g}}]_g + \sum_q \sum_{i=1,2} [\Xi_{\tilde{g}}]_{\tilde{q}_i}\,, \quad \text{and} \quad \Omega_{\tilde{g}} = [\Omega_{\tilde{g}}]_g + \sum_q \sum_{i=1,2} [\Omega_{\tilde{g}}]_{\tilde{q}_i}\,, \tag{19.122}$$

where the sum over q runs over the six squark flavors u, d, s, c, b, and t, and $i = 1, 2$ corresponds to the two squark mass eigenstates [i.e., the two appropriate linear combinations (for fixed squark flavor) of $\tilde{q}_L$ and $\tilde{q}_R$]. The gluon exchange contributions, following from the Feynman rules of Fig. J.10, are

$$-ip\cdot\overline{\sigma}\,[\Xi_{\tilde{g}}]_g\,\delta^{ab} = \mu^{2\epsilon}\int\frac{d^dk}{(2\pi)^d}\,(-g_s f^{aec}\overline{\sigma}_\mu)\left(\frac{i(k+p)\cdot\sigma}{(k+p)^2-m_{\tilde{g}}^2}\right)(-g_s f^{ebc}\overline{\sigma}_\nu)$$
$$\times\left(\frac{-i}{k^2}\right)\left(g^{\mu\nu}-(1-\xi)\frac{k^\mu k^\nu}{k^2}\right), \tag{19.123}$$

$$-i\boldsymbol{I}_2[\Omega_{\tilde{g}}]_g\,\delta^{ab} = \mu^{2\epsilon}\int\frac{d^dk}{(2\pi)^d}\,(g_s f^{eac}\sigma_\mu)\left(\frac{im_{\tilde{g}}}{(k+p)^2-m_{\tilde{g}}^2}\right)(-g_s f^{ebc}\overline{\sigma}_\nu)$$
$$\times\left(\frac{-i}{k^2}\right)\left(g^{\mu\nu}-(1-\xi)\frac{k^\mu k^\nu}{k^2}\right). \tag{19.124}$$

The internal gluon and gluino lines carry $SU(3)_C$ adjoint representation indices c and e, respectively, while the external gluinos on the left and right carry indices a and b, respectively. The gluino external momentum p^μ flows from right to left, and the loop momentum k^μ flows clockwise. We can immediately evaluate the integrals of eqs. (19.123) and (19.124) by comparing with the derivations of eqs. (19.81) and (19.111) in Section 19.3, and by using $-f^{aec}f^{ebc} = f^{eac}f^{ebc} = \delta^{ab}C_A$ [with $C_A = 3$ for $SU(3)_C$]. Hence, it follows that

$$[\Xi_{\widetilde{g}}]_g = -\frac{\alpha_s}{4\pi}C_A I_{VF}(s;0,m^2_{\widetilde{g}}), \tag{19.125}$$

$$[\Omega_{\widetilde{g}}]_g = \frac{\alpha_s}{4\pi}C_A m_{\widetilde{g}} I_{V\overline{F}}(s;0,m^2_{\widetilde{g}}), \tag{19.126}$$

where the loop integral functions I_{VF} and $I_{V\overline{F}}$ were defined in eqs. (19.85) and (19.112).

Next consider the virtual squark exchange diagrams contributing to $\Xi_{\widetilde{g}}$. Labeling the quark and squark with color indices j and k, respectively, we have, for each squark mass eigenstate labeled by i,

$$\begin{aligned}
-ip\cdot\overline{\sigma}\,[\Xi_{\widetilde{g}}]_{\widetilde{q}_i}\,\delta^{ab} &= \mu^{2\epsilon}\int\frac{d^dk}{(2\pi)^d}\left[-i\sqrt{2}g_s(\boldsymbol{T}^a)_j{}^k L_{\widetilde{q}_i}\right]\left(\frac{i(k+p)\cdot\overline{\sigma}}{(k+p)^2-m_q^2}\right)\\
&\quad\times\left[-i\sqrt{2}g_s(\boldsymbol{T}^b)_k{}^j L^*_{\widetilde{q}_i}\right]\left(\frac{i}{k^2-m^2_{\widetilde{q}_i}}\right)\\
&\quad+\mu^{2\epsilon}\int\frac{d^dk}{(2\pi)^d}\left[i\sqrt{2}g_s(\boldsymbol{T}^a)_k{}^j R^*_{\widetilde{q}_i}\right]\left(\frac{i(k+p)\cdot\overline{\sigma}}{(k+p)^2-m_q^2}\right)\\
&\quad\times\left[i\sqrt{2}g_s(\boldsymbol{T}^b)_j{}^k R_{\widetilde{q}_i}\right]\left(\frac{i}{k^2-m^2_{\widetilde{q}_i}}\right),
\end{aligned} \tag{19.127}$$

after employing the Feynman rules shown in Fig. J.12, which are presented in terms of the squark mixing parameters $L_{\widetilde{q}_i}$ and $R_{\widetilde{q}_i}$ defined in eq. (J.39). Comparing with the derivation of eq. (19.70) in Section 19.3, and making use of $\mathrm{Tr}(\boldsymbol{T}^a\boldsymbol{T}^b) = \frac{1}{2}\delta^{ab}$ and $|L_{\widetilde{q}_i}|^2 + |R_{\widetilde{q}_i}|^2 = 1$, we obtain

$$[\Xi_{\widetilde{g}}]_{\widetilde{q}_i} = -\frac{\alpha_s}{4\pi}I_{SF}(s;m^2_{\widetilde{q}_i},m_q^2). \tag{19.128}$$

Similarly, for the last two diagrams of Fig. 19.9, we again use the Feynman rules shown in Fig. J.12 to obtain

$$\begin{aligned}
-i[\Omega_{\widetilde{g}}]_{\widetilde{q}_i}\,\delta^{ab} &= \mu^{2\epsilon}\int\frac{d^dk}{(2\pi)^d}\left[-i\sqrt{2}g_s(\boldsymbol{T}^a)_k{}^j L^*_{\widetilde{q}_i}\right]\left(\frac{im_q}{(k+p)^2-m_q^2}\right)\\
&\quad\times\left[i\sqrt{2}g_s(\boldsymbol{T}^b)_j{}^k R_{\widetilde{q}_i}\right]\left(\frac{i}{k^2-m^2_{\widetilde{q}_i}}\right)\\
&\quad+\mu^{2\epsilon}\int\frac{d^dk}{(2\pi)^d}\left[i\sqrt{2}g_s(\boldsymbol{T}^a)_j{}^k R_{\widetilde{q}_i}\right]\left(\frac{im_q}{(k+p)^2-m_q^2}\right)\\
&\quad\times\left[-i\sqrt{2}g_s(\boldsymbol{T}^b)_k{}^j L^*_{\widetilde{q}_i}\right]\left(\frac{i}{k^2-m^2_{\widetilde{q}_i}}\right).
\end{aligned} \tag{19.129}$$

As before, j and k are the color indices for the quark and the squark, respectively. Comparing to the derivation of eq. (19.102) in Section 19.3, we obtain

$$[\Omega_{\widetilde{g}}]_{\widetilde{q}_i} = \frac{\alpha_s}{2\pi}L^*_{\widetilde{q}_i}R_{\widetilde{q}_i}m_q I_{S\overline{F}}(s;m^2_{\widetilde{q}_i},m_q^2)\,. \tag{19.130}$$

Summing up the results obtained above, and taking $s = m_{\tilde{g}}^2$, we have

$$\Xi_{\tilde{g}} = -\frac{\alpha_s}{4\pi}\left[C_A I_{VF}(m_{\tilde{g}}^2; 0, m_{\tilde{g}}^2) + \sum_q \sum_{i=1,2} I_{SF}(m_{\tilde{g}}^2; m_{\tilde{q}_i}^2, m_q^2)\right], \tag{19.131}$$

$$\Omega_{\tilde{g}} = \frac{\alpha_s}{4\pi}\left[C_A m_{\tilde{g}} I_{V\overline{F}}(m_{\tilde{g}}^2; 0, m_{\tilde{g}}^2) + 2\sum_q \sum_{i=1,2} L_{\tilde{q}_i}^* R_{\tilde{q}_i} m_q I_{S\overline{F}}(m_{\tilde{g}}^2; m_{\tilde{q}_i}^2, m_q^2)\right]. \tag{19.132}$$

As previously noted, we can now write down $\overline{\Omega}_{\tilde{g}}$ by replacing the Lagrangian parameters of eq. (19.132) by their complex conjugates:

$$\overline{\Omega}_{\tilde{g}} = \frac{\alpha_s}{4\pi}\left[C_A m_{\tilde{g}} I_{V\overline{F}}(m_{\tilde{g}}^2; 0, m_{\tilde{g}}^2) + 2\sum_q \sum_{i=1,2} L_{\tilde{q}_i} R_{\tilde{q}_i}^* m_q I_{S\overline{F}}(m_{\tilde{g}}^2; m_{\tilde{q}_i}^2, m_q^2)\right]. \tag{19.133}$$

In each of the self-energy functions above, there are poles in $1/\epsilon$ contained within the functions I_{VF}, I_{SF}, $I_{V\overline{F}}$, and $I_{S\overline{F}}$. In the $\overline{\text{MS}}$ or $\overline{\text{DR}}$ schemes, these poles are simply canceled by counterterms (which have no other effect at one-loop order). Inserting the results of eqs. (19.131)–(19.133) into eq. (19.121), one obtains

$$M_{\tilde{g}}^2 - iM_{\tilde{g}}\Gamma_{\tilde{g}} = m_{\tilde{g}}^2\left[1 + \frac{\alpha_s}{2\pi}\left\{C_A\left[5 - \delta_D - 3\ln\left(\frac{m_{\tilde{g}}^2}{Q^2}\right)\right]\right.\right.$$
$$\left.\left. - \sum_q \sum_{i=1,2}\left[I_{SF}(m_{\tilde{g}}^2; m_{\tilde{q}_i}^2, m_q^2) - 2\,\mathrm{Re}[L_{\tilde{q}_i}^* R_{\tilde{q}_i}]\frac{m_q}{m_{\tilde{g}}} I_{S\overline{F}}(m_{\tilde{g}}^2; m_{\tilde{q}_i}^2, m_q^2)\right]\right\}\right], \tag{19.134}$$

with δ_D defined in eq. (19.2), where the $1/\epsilon$ poles are understood to have been removed from I_{SF} and $I_{S\overline{F}}$ by counterterms as noted below eq. (19.133).

19.5 The Anomalous Magnetic Moment of the Muon

In this section, we shall employ the two-component fermion technology in the computation of the anomalous magnetic moment of the muon in softly broken SUSY-QED. In addition to the standard one-loop QED contribution first obtained by Schwinger, we shall include the contribution of loops involving the smuons and photino. We shall make use of the results of Section 13.1 where the electron is replaced by a muon, which is a Dirac fermion that is represented by a pair of two-component spinors $\chi \equiv \mu$ and $\eta \equiv \bar{\mu}$ with opposite electric charges, and the corresponding supersymmetric partners are the smuon fields, $\phi_L = \tilde{\mu}_L$ and $\phi_R = \tilde{\mu}_R^*$. The muon μ has electric charge $-e$ and mass m, whereas the corresponding smuons have mass m_L and m_R. The photino is a Majorana fermion of mass M. As in Section 19.1, we shall ignore $\widetilde{\mu}_L$–$\widetilde{\mu}_R$ mixing by setting $b = 0$. In addition, we will set the coefficient of the Fayet–Iliopoulos term to zero.

19.5.1 Electromagnetic Vertex Structure

Consider the scattering of a negatively charged muon (with electric charge $Q = -e$) off an external static A^μ field. If we employ four-component spinor notation, then the first-order S-matrix amplitude is given by

$$\langle p', s' |\, S^{(1)}\, | p, s\rangle = ie\bar{u}(\vec{\boldsymbol{p}}', s')\Gamma^\mu(p, p')u(\vec{\boldsymbol{p}}, s)\widetilde{A}_\mu(q)\,. \tag{19.135}$$

Here, $\widetilde{A}_\mu(q)$ is the four-dimensional Fourier transform of $A_\mu(x)$,

$$\widetilde{A}_\mu(q) \equiv \int d^4x\, A_\mu(x)\, e^{iq\cdot x}\,, \tag{19.136}$$

and the effective electromagnetic vertex function is

$$\Gamma^\mu(p, p') = F_1(q^2)\gamma^\mu + \frac{i}{2m}\Sigma^{\mu\nu}q_\nu F_2(q^2) + \frac{1}{2m}\gamma_5\Sigma^{\mu\nu}q_\nu F_3(q^2) + \frac{1}{4m^2}(q^\mu q^\nu - q^2 g^{\mu\nu})\gamma_\nu\gamma_5 F_4(q^2)\,, \tag{19.137}$$

following the Dirac gamma matrix conventions of eqs. (A.90) and (A.91), where $q \equiv p' - p$ and $p^2 = p'^2 = m^2$ is the squared mass of the muon. Note that if we employ the Dirac equation [see eqs. (3.172) and (3.173)], we can rewrite the term proportional to $F_4(q^2)$ as

$$\frac{1}{2m}\left(q^\mu - \frac{q^2}{2m}\gamma^\mu\right)\gamma_5 F_4(q^2)\,. \tag{19.138}$$

The form of $\Gamma^\mu(p, p')$ satisfies

$$q_\mu\Gamma^\mu(p, p') = F_1(q^2)\not{q}\,, \tag{19.139}$$

which (after invoking the Dirac equation) yields

$$q_\mu\bar{u}(\vec{\boldsymbol{p}}', s')\Gamma^\mu(p, p')u(\vec{\boldsymbol{p}}, s) = F_1(q^2)\bar{u}(\vec{\boldsymbol{p}}', s')(\not{p}' - \not{p})u(\vec{\boldsymbol{p}}, s) = 0\,, \tag{19.140}$$

as required by current conservation (or equivalently by gauge invariance). Hermiticity implies that the form factors $F_i(q^2)$ [$i = 1, 2, 3, 4$] are real-valued functions of the Lorentz-invariant scalar q^2. Note that eq. (19.137) is the most general Lorentz-covariant form that could have been written, subject to eq. (19.140), that makes use of all the relevant four-vectors and gamma matrices. Any other such terms can be reduced to the ones shown in eq. (19.137) by using the Dirac equation and the Gordon identities listed in eqs. (3.326)–(3.329).

The terms appearing in the effective vertex $u(\vec{\boldsymbol{p}}')\Gamma^\mu(p, p')u(\vec{\boldsymbol{p}})\widetilde{A}_\mu(q)$ have well-defined properties under parity (P), time reversal (T), and charge conjugation (C). In particular, the behavior of the terms with corresponding coefficients $F_i(q^2)$ under

Table 19.1 P, T, and C properties of the terms of the effective vertex, $\bar{u}(\vec{p}\,')\,\Gamma^\mu(p,p')u(\vec{p})\widetilde{A}_\mu(q)$. The chirality flip properties are also displayed.

form factor	name	P	T	C	chirality flip
F_1	electric charge	+	+	+	no
F_2	anomalous magnetic dipole	+	+	+	yes
F_3	electric dipole	−	−	+	yes
F_4	anapole	−	+	−	no

P, T, and C are given in Table 19.1. This table also indicates the chirality properties of each term, which depend on whether the corresponding effective operator is chirality conserving (i.e., $P_L\Gamma^\mu P_L = P_R\Gamma^\mu P_R = 0$) or chirality flipping (i.e., $P_L\Gamma^\mu P_R = P_R\Gamma^\mu P_L = 0$), where $P_{R,L} = \frac{1}{2}(1 \pm \gamma_5)$ [see eq. (F.69)].

Since the A^μ field is static (i.e., time-independent), it then follows that

$$\widetilde{A}_\mu(q) = \int_{-\infty}^{\infty} dt\, e^{-i(E-E')t} \int d^3x\, A_\mu(\vec{x})e^{i(\vec{p}-\vec{p}\,')\cdot\vec{x}} = 2\pi\delta(E'-E)\widetilde{A}_\mu(\vec{q})\,, \tag{19.141}$$

where $\vec{q} \equiv \vec{p}\,' - \vec{p}$ and $\widetilde{A}_\mu(\vec{q})$ is the three-dimensional Fourier transform of $A_\mu(\vec{x})$,

$$\widetilde{A}_\mu(\vec{q}) \equiv \int d^3x\, A_\mu(\vec{x})e^{-i\vec{q}\cdot\vec{x}}\,. \tag{19.142}$$

In nonrelativistic scattering theory, the S-matrix element is related to the interaction potential $V(\vec{x})$ via[7]

$$\langle\vec{p}\,', s'|\, S\,|\vec{p}, s\rangle = (2\pi)^3 2E\delta^3(\vec{p}\,' - \vec{p})\delta_{ss'} - 2\pi i\delta(E'-E)\,\langle\vec{p}\,', s'|\, V(\vec{x})|\vec{p}^{\,(+)}, s\rangle\,, \tag{19.143}$$

where $|\vec{p}^{\,(+)}\rangle$ is given by the Lippmann–Schwinger equation,

$$|\vec{p}^{\,(+)}, s\rangle = |\vec{p}, s\rangle + \lim_{\varepsilon\to 0^+} \frac{1}{E - H_0 + i\varepsilon} V|\vec{p}^{\,(+)}, s\rangle\,, \tag{19.144}$$

and the Hamiltonian is given by $H = H_0 + V$. Thus, to leading order in the perturbation V, it follows that

$$\langle\vec{p}\,', s'|\, S^{(1)}\,|\vec{p}, s\rangle = -2\pi i\delta(E'-E)\,\langle\vec{p}\,', s'|\, V\,|\vec{p}, s\rangle\,, \tag{19.145}$$

which yields an expression for the Fourier transform of the interaction potential V,

$$\langle\vec{p}\,', s'|\, V(\vec{x})\,|\vec{p}, s\rangle = -e\bar{u}(\vec{p}\,', s')\Gamma^\mu(p,p')u(\vec{p}, s)\widetilde{A}_\mu(\vec{q})\,. \tag{19.146}$$

The presence of the energy-conserving delta function in eq. (19.145) implies that $q^0 = E' - E = 0$, in which case $q^2 = -|\vec{q}|^2$.

[7] Here, we employ a covariant normalization of one-particle momentum states, as in eq. (2.7).

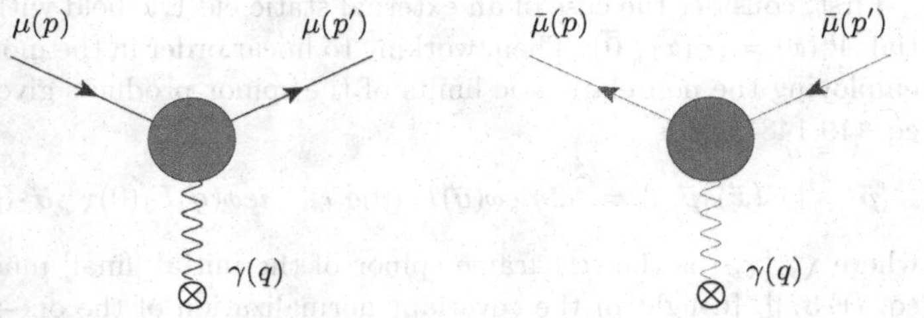

Fig. 19.10 Generic graphs for the photon muon interaction. The gray circles can include all possible higher-order corrections. The momenta p and p' flow from left to right. The $\otimes$ symbol indicates the presence of an external static localized classical A^μ field, and $q = p' - p$.

Since this analysis based on four-component spinor notation is already fairly well known, we shall convert to two-component spinor notation to take advantage of the formalism that has been provided in this book. Thus, we can rewrite the results of eqs. (19.135) and (19.137) as

$$\begin{aligned}\langle \vec{p}\,', s'|\, V(\vec{x})\, |\vec{p}, s\rangle = -e\widetilde{A}_\mu(\vec{q})\Big\{ & F_1(-|\vec{q}|^2)\left[y(\vec{p}\,', s')\sigma^\mu y^\dagger(\vec{p}, s) + x^\dagger(\vec{p}\,', s')\overline{\sigma}^\mu x(\vec{p}, s)\right] \\ & + \frac{i}{m}F_2(-|\vec{q}|^2)\left[y(\vec{p}\,', s')\sigma^{\mu\nu} x(\vec{p}, s) + x^\dagger(\vec{p}\,', s')\overline{\sigma}^{\mu\nu} y^\dagger(\vec{p}, s)\right] q_\nu \\ & - \frac{1}{m}F_3(-|\vec{q}|^2)\left[y(\vec{p}\,', s')\sigma^{\mu\nu} x(\vec{p}, s) - x^\dagger(\vec{p}\,', s')\overline{\sigma}^{\mu\nu} y^\dagger(\vec{p}, s)\right] q_\nu \\ & + \frac{1}{4m^2}F_4(-|\vec{q}|^2)\left[y(\vec{p}\,', s')\sigma_\nu y^\dagger(\vec{p}, s) - x^\dagger(\vec{p}\,', s')\overline{\sigma}_\nu x(\vec{p}, s)\right] \\ & \times\left(q^\mu q^\nu - q^2 g^{\mu\nu}\right)\Big\}. \end{aligned} \tag{19.147}$$

It is convenient to make use of the Gordon identities (see Exercise 2.1) to eliminate $\sigma^{\mu\nu}$ and $\overline{\sigma}^{\mu\nu}$ from eq. (19.147). We then obtain

$$\begin{aligned}\langle \vec{p}\,', s'|\, V(\vec{x})\, |\vec{p}, s\rangle = -e\widetilde{A}_\mu(\vec{q})\Big\{ & \left(F_1(-|\vec{q}|^2) + F_2(-|\vec{q}|^2)\right) \\ & \times\left[y(\vec{p}\,', s')\sigma^\mu y^\dagger(\vec{p}, s) + x^\dagger(\vec{p}\,', s')\overline{\sigma}^\mu x(\vec{p}, s)\right] \\ & - \frac{1}{2m}F_2(-|\vec{q}|^2)\left[y(\vec{p}\,', s')x(\vec{p}, s) + x^\dagger(\vec{p}\,', s')y^\dagger(\vec{p}, s)\right](p+p')^\mu \\ & - \frac{i}{2m}F_3(-|\vec{q}|^2)\left[y(\vec{p}\,', s')x(\vec{p}, s) - x^\dagger(\vec{p}\,', s')y^\dagger(\vec{p}, s)\right](p+p')^\mu \\ & + \frac{1}{4m^2}F_4(-|\vec{q}|^2)\left[y(\vec{p}\,', s')\sigma_\nu y^\dagger(\vec{p}, s) - x^\dagger(\vec{p}\,', s')\overline{\sigma}_\nu x(\vec{p}, s)\right] \\ & \times\left(q^\mu q^\nu - q^2 g^{\mu\nu}\right)\Big\}. \end{aligned} \tag{19.148}$$

The form factors F_i ($i = 1, 2, 3, 4$) can be computed in perturbation theory by evaluating Feynman graphs that generically are of the form shown in Fig. 19.10.

First, consider the case of an external static electric field with four-vector potential $A^\mu(x) = \big(\phi(x)\,;\,\vec{\boldsymbol{0}}\big)$. Then, working to linear order in the momenta $\vec{\boldsymbol{p}}$ and $\vec{\boldsymbol{p}}'$ and employing the nonrelativistic limits of the spinor products given in Appendix E.2, eq. (19.148) yields

$$\langle \vec{\boldsymbol{p}}', s'|\, V(\vec{\boldsymbol{x}})\, |\vec{\boldsymbol{p}}, s\rangle = -2me\widetilde{\phi}(\vec{\boldsymbol{q}})F_1(0)\delta_{s's} + ie\widetilde{\phi}(\vec{\boldsymbol{q}})F_3(0)\chi^\dagger_{s'}\vec{\boldsymbol{\sigma}}\boldsymbol{\cdot}(\vec{\boldsymbol{p}}' - \vec{\boldsymbol{p}})\chi_s\,, \tag{19.149}$$

where χ_s $[\chi_{s'}]$ is the rest-frame spinor of the initial [final] muon [after employing eq. (E.37)]. In light of the covariant normalization of the one-particle momentum states [see eq. (2.7)],[8] it follows that $\langle\vec{\boldsymbol{x}}|\vec{\boldsymbol{p}}\rangle = \sqrt{2E_{\vec{\boldsymbol{p}}}}\, e^{i\vec{\boldsymbol{p}}\cdot\vec{\boldsymbol{x}}}$. Since $E_{\vec{\boldsymbol{p}}} \to m$ in the nonrelativistic limit, we end up with

$$\langle \vec{\boldsymbol{p}}', s'|\, V(\vec{\boldsymbol{x}})\, |\vec{\boldsymbol{p}}, s\rangle = -eF_1(0)\,\langle \vec{\boldsymbol{p}}', s'|\, \phi(\vec{\boldsymbol{x}})\, |\vec{\boldsymbol{p}}, s\rangle - \frac{eF_3(0)}{2m}\langle \vec{\boldsymbol{p}}', s'|\boldsymbol{\sigma}\boldsymbol{\cdot}\vec{\boldsymbol{E}}(\vec{\boldsymbol{x}})\, |\vec{\boldsymbol{p}}, s\rangle, \tag{19.150}$$

after an integration by parts and using $\vec{\boldsymbol{E}} = -\vec{\boldsymbol{\nabla}}\phi$. That is,

$$V(\vec{\boldsymbol{x}}) = -eF_1(0)\phi(\vec{\boldsymbol{x}}) - \frac{eF_3(0)}{2m}\vec{\boldsymbol{\sigma}}\boldsymbol{\cdot}\vec{\boldsymbol{E}}(\vec{\boldsymbol{x}})\,. \tag{19.151}$$

Since the potential energy of a negatively charged muon with electric dipole moment $\vec{\boldsymbol{d}}$ in an external static electric field is given by

$$V(\vec{\boldsymbol{x}}) = -e\phi(\vec{\boldsymbol{x}}) - \vec{\boldsymbol{d}}\boldsymbol{\cdot}\vec{\boldsymbol{E}}(\vec{\boldsymbol{x}})\,, \tag{19.152}$$

it follows that $F_1(0) = 1$, which is equivalent to the statement that the charge Q of the muon is

$$Q = -eF_1(0)\,, \tag{19.153}$$

and

$$\vec{\boldsymbol{d}} = \frac{e}{m}F_3(0)\vec{\boldsymbol{S}}\,, \tag{19.154}$$

where $\vec{\boldsymbol{S}} = \frac{1}{2}\vec{\boldsymbol{\sigma}}$ is the nonrelativistic spin operator.

Second, consider the case of an external static magnetic field with four-vector potential $A^\mu(x) = \big(0\,;\,\vec{\boldsymbol{A}}(\vec{\boldsymbol{x}})\big)$. We again work to linear order in the momenta $\vec{\boldsymbol{p}}$ and $\vec{\boldsymbol{p}}'$ and employ the nonrelativistic limits of the spinor products given in Appendix E.2. Equation (19.148) yields

$$\langle \vec{\boldsymbol{p}}', s'|\, V(\vec{\boldsymbol{x}})\, |\vec{\boldsymbol{p}}, s\rangle = eF_1(0)\widetilde{\vec{\boldsymbol{A}}}\boldsymbol{\cdot}(\vec{\boldsymbol{p}}+\vec{\boldsymbol{p}}')\delta_{s's} - ie\big[F_1(0) + F_2(0)\big]\,\chi^\dagger_{s'}\vec{\boldsymbol{\sigma}}\boldsymbol{\cdot}(\widetilde{\vec{\boldsymbol{A}}}\times\vec{\boldsymbol{q}})\,\chi_s\,, \tag{19.155}$$

after employing eqs. (E.59) and (E.64). The Fourier transform of the magnetic field, $\vec{\boldsymbol{B}} = \vec{\boldsymbol{\nabla}} \times \vec{\boldsymbol{A}}$, is given by

$$\widetilde{\vec{\boldsymbol{B}}}(\vec{\boldsymbol{q}}) = \int (\vec{\boldsymbol{\nabla}} \times \vec{\boldsymbol{A}})e^{-i\vec{\boldsymbol{q}}\cdot\vec{\boldsymbol{x}}}\, d^3x = i\vec{\boldsymbol{q}} \times \widetilde{\vec{\boldsymbol{A}}}\,, \tag{19.156}$$

[8] In contrast, the position eigenstates are normalized as usual: $\langle\vec{\boldsymbol{x}}|\,\vec{\boldsymbol{x}}'\rangle = \delta^3(\vec{\boldsymbol{x}} - \vec{\boldsymbol{x}}')$.

after an integration by parts. Hence, eq. (19.155) yields

$$\langle \vec{p}\,', s'|\, V(\vec{x})\,|\vec{p}, s\rangle = eF_1(0)\widetilde{\vec{A}}\cdot(\vec{p}+\vec{p}\,')\delta_{s's} + e\big[F_1(0)+F_2(0)\big]\chi^{\dagger}_{s'}\vec{\sigma}\cdot\widetilde{\vec{B}}(\vec{q})\chi_s\,. \tag{19.157}$$

Hence, in the nonrelativistic limit, we end up with

$$\begin{aligned}\langle \vec{p}\,', s'|\, V(\vec{x})\,|\vec{p}, s\rangle &= \frac{e}{2m}F_1(0)\,\langle \vec{p}\,', s'|\, \vec{P}\cdot\vec{A} + \vec{A}\cdot\vec{P}\,|\vec{p}, s\rangle \\ &\quad + \frac{e}{2m}\left[F_1(0)+F_2(0)\right]\langle \vec{p}\,', s'|\boldsymbol{\sigma}\cdot\vec{B}(\vec{x})\,|\vec{p}, s\rangle,\end{aligned} \tag{19.158}$$

where $\vec{P}$ is the momentum operator.

Using $F_1(0) = 1$, we conclude that

$$V(\vec{x}) = \frac{e}{2m}(\vec{P}\cdot\vec{A} + \vec{A}\cdot\vec{P}) + \frac{e}{2m}\left[1+F_2(0)\right]\boldsymbol{\sigma}\cdot\vec{B}(\vec{x})\,. \tag{19.159}$$

We recognize the first term on the right-hand side of eq. (19.159) as the part of the electromagnetic Hamiltonian of a particle of charge $-e$ that is linear in the vector potential. Comparing the second term on the right-hand side of eq. (19.159) with the potential $V = -\vec{m}\cdot\vec{B}(\vec{x})$ of a particle of charge $-e$ and magnetic dipole moment $\vec{m}$ moving in a static magnetic field, we can identify

$$\vec{m} = -\frac{e}{m}\left[1+F_2(0)\right]\vec{S} = -\frac{eg}{2m}\vec{S}\,, \tag{19.160}$$

where g is the so-called g-factor of the particle. We conclude that

$$F_2(0) = a \equiv \frac{g-2}{2}\,. \tag{19.161}$$

That is, we can identify the anomalous magnetic moment of the muon, $a = F_2(0)$.

What remains is the question of the significance of the form factor F_4. Suppose that we probe a particle of charge $-e$ with external static electric and magnetic fields. The nonrelativistic limit of the term on the right-hand side of eq. (19.148) that is proportional to F_4 is given by

$$\begin{aligned}\langle \vec{p}\,', s'|\, V(\vec{x})\,|\vec{p}, s\rangle &\supset -\frac{e}{2m}F_4(0)\widetilde{A}^i(\vec{q})\,(q_iq_j - |\vec{q}|^2\delta_{ij})\chi^{\dagger}_{s'}\sigma^j\chi_s \\ &= -\frac{e}{2m}F_4(0)\left[\widetilde{\vec{A}}\cdot\vec{q}\,\chi^{\dagger}_{s'}\vec{\sigma}\cdot\vec{q}\,\chi_s - |\vec{q}|^2\,\chi^{\dagger}_{s'}\vec{\sigma}\cdot\widetilde{\vec{A}}\,\chi_s\right] \\ &= \frac{ie}{2m}F_4(0)\left[\chi^{\dagger}_{s'}\,\vec{\sigma}\cdot(\vec{q}\times\widetilde{\vec{B}})\,\chi_s\right] \\ &= \frac{e}{4m^2}F_4(0)\,\langle \vec{p}\,', s'|\,\vec{\sigma}\cdot(\vec{\nabla}\times\vec{B}(\vec{x}))\,|\vec{p}, s\rangle\,,\end{aligned} \tag{19.162}$$

after using the nonrelativistic limit of the spinor products given in Appendix E.2 and using eq. (19.156) in the penultimate step. Note that there is no term in eq. (19.162) proportional to the electric field, $\vec{E}(\vec{x})$, since $y(\vec{p}\,')\sigma_0 y^{\dagger}(\vec{p}) - x^{\dagger}(\vec{p}\,')\overline{\sigma}_0 x(\vec{p})$ vanishes in the limit of $\vec{p}$, $\vec{p}\,' \to 0$, and $q^0 = 0$ due to energy conservation.

The static anapole moment $\vec{a}$ is defined as

$$\vec{a} = -\frac{e}{2m^2}F_4(0)\vec{S}\,. \tag{19.163}$$

Using eq. (19.162), the P-violating interaction potential of a particle of charge $-e$ and anapole moment $\vec{\boldsymbol{a}}$ moving in a static magnetic field is given by

$$V(\vec{\boldsymbol{x}}) = \frac{e}{4m^2} F_4(0) \vec{\boldsymbol{\sigma}} \cdot (\vec{\boldsymbol{\nabla}} \times \vec{\boldsymbol{B}}(\vec{\boldsymbol{x}})) = -\vec{\boldsymbol{a}} \cdot \vec{\boldsymbol{J}}(\vec{\boldsymbol{x}}) \,, \tag{19.164}$$

where $\vec{\boldsymbol{J}}(\vec{\boldsymbol{x}}) = \vec{\boldsymbol{\nabla}} \times \vec{\boldsymbol{B}}(\vec{\boldsymbol{x}})$ is the external current that produces the static magnetic field. Consequently, the interaction potential vanishes unless the source of the magnetic field is nonzero. As noted in Ref. [199], this implies that the coupling of the anapole moment to an external electromagnetic field is of relevance only in matter. Initial claims in the literature asserted that the static anapole moment in the Standard Model is gauge-dependent and hence unphysical. However, it was subsequently shown in Ref. [200] how to properly define a gauge invariant anapole moment of a free pointlike fermion, which is observable in principle.

19.5.2 QED Contribution to the Anomalous Magnetic Moment

In this section, we examine the one-loop QED contribution to the anomalous magnetic moment of the muon. Since the explicit computation using the two-component fermion formalism does not appear in the literature, we shall provide the details here. In particular, we need to evaluate the relevant one-loop QED diagrams and identify the term in the effective electromagnetic vertex Γ^μ that is proportional to F_2. This can be done by determining an appropriate set of projection operators, as follows. Starting from the expression for Γ^μ given in eq. (19.137) and converting to two-component spinor notation, we seek two functions $g_1(q^2)$ and $g_2(q^2)$ that satisfy the following identity:[9]

$$\begin{aligned} F_2(q^2) = \sum_{s,s'} \Big\{ & g_1(q^2) \left[y(\vec{\boldsymbol{p}}, s) \sigma_\mu y^\dagger(\vec{\boldsymbol{p}}', s') + x^\dagger(\vec{\boldsymbol{p}}, s) \overline{\sigma}_\mu x(\vec{\boldsymbol{p}}', s') \right] \\ & + g_2(q^2) \frac{(p+p')_\mu}{2m} \left[y(\vec{\boldsymbol{p}}, s) x(\vec{\boldsymbol{p}}', s') + x^\dagger(\vec{\boldsymbol{p}}, s) y^\dagger(\vec{\boldsymbol{p}}', s') \right] \Big\} z(\vec{\boldsymbol{p}}', s') \Gamma^\mu z(\vec{\boldsymbol{p}}, s) \,, \end{aligned} \tag{19.165}$$

where

$$\begin{aligned} z(\vec{\boldsymbol{p}}', s') \Gamma^\mu z(\vec{\boldsymbol{p}}, s) \equiv & \left[F_1(q^2) + F_2(q^2) \right] \left[y(\vec{\boldsymbol{p}}', s') \sigma^\mu y^\dagger(\vec{\boldsymbol{p}}, s) + x^\dagger(\vec{\boldsymbol{p}}', s') \overline{\sigma}^\mu x(\vec{\boldsymbol{p}}, s) \right] \\ & - \frac{1}{2m} F_2(q^2) (p+p')^\mu \left[y(\vec{\boldsymbol{p}}', s') x(\vec{\boldsymbol{p}}, s) + x^\dagger(\vec{\boldsymbol{p}}', s') y^\dagger(\vec{\boldsymbol{p}}, s) \right] \\ & - \frac{i}{2m} F_3(q^2) (p+p')^\mu \left[y(\vec{\boldsymbol{p}}', s') x(\vec{\boldsymbol{p}}, s) - x^\dagger(\vec{\boldsymbol{p}}', s') y^\dagger(\vec{\boldsymbol{p}}, s) \right] \\ & + \frac{1}{4m^2} F_4(q^2) (q^\mu q^\nu - q^2 g^{\mu\nu}) \big[y(\vec{\boldsymbol{p}}', s') \sigma_\nu y^\dagger(\vec{\boldsymbol{p}}, s) \\ & \qquad - x^\dagger(\vec{\boldsymbol{p}}', s') \overline{\sigma}_\nu x(\vec{\boldsymbol{p}}, s) \big] \,. \end{aligned} \tag{19.166}$$

Multiplying out the terms in eq. (19.165), one can perform one simplification

[9] Note carefully the ordering of the unprimed and primed momentum and spin variables appearing as arguments of the x-spinor and y-spinor wave functions in eq. (19.165).

by employing the Dirac equations for the x and y spinors given in eqs. (2.10)–(2.13). The sum over spins of the spinor products is then evaluated with the help of eqs. (2.63)–(2.66). The end result (in d spacetime dimensions) is

$$F_2(q^2) = F_1(q^2)\big[4\big[2m^2 + \tfrac{1}{2}q^2(d-2)\big]g_1(q^2) + 2(4m^2 - q^2)g_2(q^2)\big] + F_2(q^2)\big[2(d-1)q^2 g_1(q^2) + \frac{q^2(4m^2-q^2)}{2m^2}\,g_2(q^2)\big]\,. \tag{19.167}$$

The terms proportional to F_3 and F_4 appearing on the right-hand side of eq. (19.165) vanish exactly. This is not surprising in light of the P, T, and C transformation properties exhibited in Table 19.1. To give one example of the calculations involved,

$$\begin{aligned}
&\sum_{s,s'}\big[y(\vec{\boldsymbol{p}},s)\sigma_\mu y^\dagger(\vec{\boldsymbol{p}}',s') + x^\dagger(\vec{\boldsymbol{p}},s)\overline{\sigma}_\mu x(\vec{\boldsymbol{p}}',s')\big]\big[y(\vec{\boldsymbol{p}}',s')\sigma^\mu y^\dagger(\vec{\boldsymbol{p}},s) + x^\dagger(\vec{\boldsymbol{p}}',s')\overline{\sigma}^\mu x(\vec{\boldsymbol{p}},s)\big]\\
&= \left(\sum_s y^{\dagger\dot\sigma}(\vec{\boldsymbol{p}},s)y^\alpha(\vec{\boldsymbol{p}},s)\right)\left(\sum_{s'} y^{\dagger\dot\beta}(\vec{\boldsymbol{p}}',s')y^\rho(\vec{\boldsymbol{p}}',s')\right)\sigma_{\mu\alpha\dot\beta}\sigma^\mu_{\rho\dot\sigma}\\
&\quad+\left(\sum_s x_\sigma(\vec{\boldsymbol{p}},s)y^\alpha(\vec{\boldsymbol{p}},s)\right)\left(\sum_{s'} y^{\dagger\dot\beta}(\vec{\boldsymbol{p}}',s')x^\dagger_{\dot\rho}(\vec{\boldsymbol{p}}',s')\right)\sigma_{\mu\alpha\dot\beta}\overline{\sigma}^{\mu\dot\rho\sigma}\\
&\quad+\left(\sum_s y^{\dagger\dot\sigma}(\vec{\boldsymbol{p}},s)x^\dagger_{\dot\alpha}(\vec{\boldsymbol{p}},s)\right)\left(\sum_{s'} x_\beta(\vec{\boldsymbol{p}}',s')y^\rho(\vec{\boldsymbol{p}}',s')\right)\overline{\sigma}^{\dot\alpha\beta}_\mu\sigma^\mu_{\rho\dot\sigma}\\
&\quad+\left(\sum_s x_\sigma(\vec{\boldsymbol{p}},s)x^\dagger_{\dot\alpha}(\vec{\boldsymbol{p}},s)\right)\left(\sum_{s'} x_\beta(\vec{\boldsymbol{p}}',s')x^\dagger_{\dot\rho}(\vec{\boldsymbol{p}}',s')\right)\overline{\sigma}^{\dot\alpha\beta}_\mu\overline{\sigma}^{\mu\dot\rho\sigma}\\
&= \mathrm{Tr}(\sigma_\mu p'\!\cdot\!\overline{\sigma}\sigma^\mu p\!\cdot\!\overline{\sigma}) + 2m^2\,\mathrm{Tr}(\sigma_\mu\overline{\sigma}^\mu) + \mathrm{Tr}(\overline{\sigma}_\mu p'\!\cdot\!\sigma\overline{\sigma}^\mu p\!\cdot\!\sigma)\\
&= 4dm^2 - 4(d-2)p\!\cdot\!p' = 4\big[2m^2 + (\tfrac{1}{2}d-1)q^2\big]\,.
\end{aligned}\tag{19.168}$$

In obtaining eq. (19.168), we made use of the muon mass-shell conditions and the conservation of four-momentum ($q = p' - p$), which can be used to express any dot product in terms of q^2 and m^2:

$$p^2 = p'^2 = m^2\,, \qquad p\!\cdot\!p' = m^2 - \tfrac{1}{2}q^2\,, \qquad q\!\cdot\!p' = -q\!\cdot\!p = \tfrac{1}{2}q^2\,. \tag{19.169}$$

We have evaluated the traces above in d spacetime dimensions using the results of Appendix B.4, since we anticipate that g_1 and g_2 will eventually multiply integrals that are potentially divergent in the intermediate steps of our calculation. However, the final answer for the anomalous magnetic moment is finite. Thus, in our initial analysis, we shall keep $d \neq 4$ where necessary, and we will set $d = 4$ once all potentially divergent integrals have canceled.

Equation (19.167) must be satisfied for all values of q^2. Hence, on the right-hand side of eq. (19.167) the coefficient of F_1 must vanish and the coefficient of F_2 must be equal to 1. These conditions determine that

$$g_1(q^2) = \frac{m^2}{(\tfrac{1}{2}d-1)q^2(4m^2-q^2)}\,, \qquad g_2(q^2) = -\frac{2m^2\big[2m^2 + (\tfrac{1}{2}d-1)q^2\big]}{(\tfrac{1}{2}d-1)q^2(4m^2-q^2)^2}\,. \tag{19.170}$$

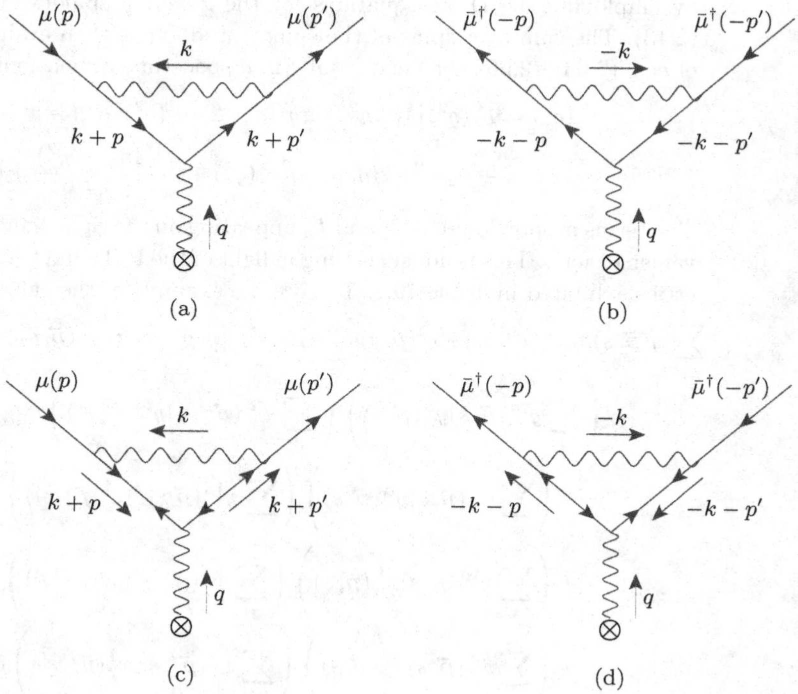

Fig. 19.11 One-loop QED contributions to the anomalous magnetic moment of the muon. The first set of four contributing diagrams is exhibited above. The internal loop consists of a photon of four-momentum k and a pair of muons following the two-component fermion field labeling conventions specified in Fig. 17.2. The four-momenta point in the directions indicated by the arrows. In particular, the direction of flow of each of the four-momenta is the same in all four diagrams. The $\otimes$ symbol indicates the presence of an external static classical four-vector potential A^μ field, and $q = p' - p$.

Eight diagrams, shown in Figs. 19.11 and 19.12, contribute at one-loop order to the QED anomalous magnetic moment of the muon. Figure 19.11(a) yields

$$\begin{aligned}\langle p', s'|\, S_a^{(1)}\, |p, s\rangle &= \widetilde{A}_\mu(q) \int \frac{d^d k}{(2\pi)^d}\, x^\dagger(p', s') ie\overline{\sigma}^\alpha \left(\frac{i(p'+k)\cdot\sigma}{(p'+k)^2 - m^2 + i\varepsilon}\right) ie\overline{\sigma}^\mu \\ &\qquad \times \left(\frac{i(p+k)\cdot\sigma}{(p+k)^2 - m^2 + i\varepsilon}\right) ie\overline{\sigma}^\beta x(p, s) \left(\frac{-ig_{\alpha\beta}}{k^2 + i\varepsilon}\right) \\ &= -2e^3 \widetilde{A}_\mu(q) \int \frac{d^d k}{(2\pi)^d}\, \frac{1}{D} \Big\{ [x^\dagger(\vec{\boldsymbol{p}}', s') \big[(p+k)\cdot\overline{\sigma}\, \sigma^\mu\, (p'+k)\cdot\overline{\sigma} \\ &\qquad - \tfrac{1}{2}(4-d) k\cdot\overline{\sigma}\, \sigma^\mu\, k\cdot\overline{\sigma}\big] x(\vec{\boldsymbol{p}}, s) \Big\}, \qquad (19.171)\end{aligned}$$

after making use of eq. (B.122) to simplify the numerator, where

$$D \equiv \left[(k+p)^2 - m^2 + i\varepsilon\right]\left[(k+p')^2 - m^2 + i\varepsilon\right]\left[k^2 + i\varepsilon\right], \qquad (19.172)$$

and we have employed the photon propagator in the Feynman gauge [corresponding to choosing $a = 1$ in eq. (13.3)]. Note that we only need to keep numerator factors that are quadratic in the loop momentum k in the term proportional to $d-4$, since the corresponding integral is potentially logarithmically divergent.

Similarly, Fig. 19.11(b) yields

$$\begin{aligned}\langle p', s'|\, S_b^{(1)}\, |p, s\rangle &= \widetilde{A}_\mu(q) \int \frac{d^d k}{(2\pi)^d}\, y(p', s') ie\sigma^\alpha \left(\frac{i(p'+k)\cdot\overline{\sigma}}{(p'+k)^2 - m^2 + i\varepsilon}\right) ie\sigma^\mu \\ &\quad \times \left(\frac{i(p+k)\cdot\overline{\sigma}}{(p+k)^2 - m^2 + i\varepsilon}\right) ie\sigma^\beta y^\dagger(p, s) \left(\frac{-ig_{\alpha\beta}}{k^2 + i\varepsilon}\right) \\ &= -2e^3 \widetilde{A}_\mu(q) \int \frac{d^d k}{(2\pi)^d}\, \frac{1}{D} \Big\{ y(\vec{\boldsymbol{p}}', s')\big[(p+k)\cdot\sigma\, \overline{\sigma}^\mu\, (p'+k)\cdot\sigma \\ &\qquad - \tfrac{1}{2}(4-d) k\cdot\sigma\, \overline{\sigma}^\mu\, k\cdot\sigma\big] y^\dagger(\vec{\boldsymbol{p}}, s) \Big\}, \end{aligned} \qquad (19.173)$$

after making use of eq. (B.121) to simplify the numerator.

The amplitudes represented by the diagrams in Fig. 19.11(c) and (d) are ultraviolet finite and can be evaluated by setting $d = 4$:

$$\begin{aligned}\langle p', s'|\, S_c^{(1)}\, |p, s\rangle &= \widetilde{A}_\mu(q) \int \frac{d^4 k}{(2\pi)^4}\, x^\dagger(p', s') ie\overline{\sigma}^\alpha \left(\frac{im}{(p'+k)^2 - m^2 + i\varepsilon}\right) ie\sigma^\mu \\ &\quad \times \left(\frac{im}{(p+k)^2 - m^2 + i\varepsilon}\right) ie\overline{\sigma}^\beta x(p, s) \left(\frac{-ig_{\alpha\beta}}{k^2 + i\varepsilon}\right), \\ \langle p', s'|\, S_d^{(1)}\, |p, s\rangle &= \widetilde{A}_\mu(q) \int \frac{d^4 k}{(2\pi)^4}\, y(p', s') ie\sigma^\alpha \left(\frac{im}{(p'+k)^2 - m^2 + i\varepsilon}\right) ie\overline{\sigma}^\mu \\ &\quad \times \left(\frac{im}{(p+k)^2 - m^2 + i\varepsilon}\right) ie\sigma^\beta y^\dagger(p, s) \left(\frac{-ig_{\alpha\beta}}{k^2 + i\varepsilon}\right). \end{aligned} \qquad (19.174)$$

Simplifying the numerators with the help of eqs. (B.24) and (B.25), we end up with

$$\langle p', s'|\, S_{c+d}^{(1)}\, |p, s\rangle = -2e^3 m^2 \widetilde{A}_\mu(q) \int \frac{d^4 k}{(2\pi)^4}\, \frac{x^\dagger(\vec{\boldsymbol{p}}', s')\overline{\sigma}^\mu x(\vec{\boldsymbol{p}}, s) + y(\vec{\boldsymbol{p}}', s')\sigma^\mu y^\dagger(\vec{\boldsymbol{p}}, s)}{D}. \qquad (19.175)$$

The amplitudes represented by the diagrams in Fig. 19.12(e)–(h) are ultraviolet

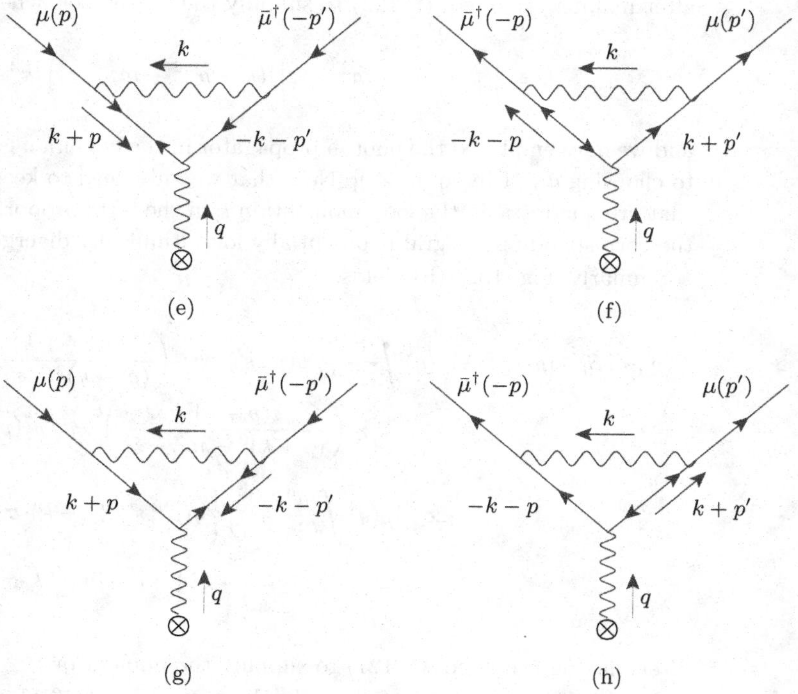

Fig. 19.12 One-loop QED contributions to the anomalous magnetic moment of the muon. The second set of four contributing diagrams is exhibited above. The internal loop consists of a photon of four-momentum k and a pair of muons following the two-component fermion field labeling conventions specified in Fig. 17.2. The four-momenta point in the directions indicated by the arrows. In particular, the direction of flow of each of the four-momenta is the same in all four diagrams. The $\otimes$ symbol indicates the presence of an external static classical four-vector potential A^μ field, and $q = p' - p$.

finite and can be evaluated by setting $d = 4$:

$$\langle p', s'| S_e^{(1)} |p, s\rangle = \widetilde{A}_\mu(q) \int \frac{d^4k}{(2\pi)^4}\, y(p', s') ie\sigma^\alpha \left(\frac{i(p'+k)\cdot\overline{\sigma}}{(p'+k)^2 - m^2 + i\varepsilon}\right) ie\sigma^\mu \times \left(\frac{im}{(p+k)^2 - m^2 + i\varepsilon}\right) ie\overline{\sigma}^\beta x(p, s) \left(\frac{-ig_{\alpha\beta}}{k^2 + i\varepsilon}\right), \tag{19.176}$$

$$\langle p', s'| S_f^{(1)} |p, s\rangle = \widetilde{A}_\mu(q) \int \frac{d^4k}{(2\pi)^4}\, x^\dagger(p', s') ie\overline{\sigma}^\alpha \left(\frac{i(p'+k)\cdot\sigma}{(p'+k)^2 - m^2 + i\varepsilon}\right) ie\overline{\sigma}^\mu \times \left(\frac{im}{(p+k)^2 - m^2 + i\varepsilon}\right) ie\sigma^\beta y^\dagger(p, s) \left(\frac{-ig_{\alpha\beta}}{k^2 + i\varepsilon}\right), \tag{19.177}$$

$$\langle p', s'|\, S_g^{(1)}\, |p, s\rangle = \widetilde{A}_\mu(q) \int \frac{d^4k}{(2\pi)^4}\, y(p', s') ie\sigma^\alpha \left(\frac{im}{(p'+k)^2 - m^2 + i\varepsilon} \right) ie\overline{\sigma}^\mu \times \left(\frac{i(p+k)\cdot\sigma}{(p+k)^2 - m^2 + i\varepsilon} \right) ie\overline{\sigma}^\beta x(p, s) \left(\frac{-ig_{\alpha\beta}}{k^2 + i\varepsilon} \right), \tag{19.178}$$

$$\langle p', s'|\, S_h^{(1)}\, |p, s\rangle = \widetilde{A}_\mu(q) \int \frac{d^4k}{(2\pi)^4}\, x^\dagger(p', s') ie\overline{\sigma}^\alpha \left(\frac{im}{(p'+k)^2 - m^2 + i\varepsilon} \right) ie\sigma^\mu \times \left(\frac{i(p+k)\cdot\overline{\sigma}}{(p+k)^2 - m^2 + i\varepsilon} \right) ie\sigma^\beta y^\dagger(p, s) \left(\frac{-ig_{\alpha\beta}}{k^2 + i\varepsilon} \right). \tag{19.179}$$

Adding the four amplitudes above yields

$$\langle p', s'|\, S_{e+f+g+h}^{(1)}\, |p, s\rangle = 4e^3 m \widetilde{A}_\mu(q) \times \int \frac{d^4k}{(2\pi)^4}\, (p + p' + 2k)^\mu \frac{y(\vec{\boldsymbol{p}}', s')x(\vec{\boldsymbol{p}}, s) + x^\dagger(\vec{\boldsymbol{p}}', s')y^\dagger(\vec{\boldsymbol{p}}, s)}{D}. \tag{19.180}$$

When computing the first-order S-matrix element for the scattering of an on-shell muon off a classical electromagnetic field, one should also include the effects of wave function renormalization of the external on-shell muons. However, these contributions are proportional to the sum of tree-level amplitudes multiplied by δZ_L or δZ_R. Thus, they do not contribute to $F_2(q^2)$, as these contributions will vanish when applying the projection operators specified in eq. (19.165).

In light of eq. (19.165), the contributions to $F_2(q^2)$ from Fig. 19.11(a)–(d) and Fig. 19.12(e)–(h), denoted below by $F_2(q^2)_{\rm QED}$, are given by

$$\begin{aligned} ieF_2(q^2)_{\rm QED} = \sum_{s,s'} \Big\{ & g_1(q^2)\big[y(\vec{\boldsymbol{p}}, s)\sigma_\mu y^\dagger(\vec{\boldsymbol{p}}', s') + x^\dagger(\vec{\boldsymbol{p}}, s)\overline{\sigma}_\mu x(\vec{\boldsymbol{p}}', s')\big] \\ & + g_2(q^2)\frac{(p+p')_\mu}{2m}\big[y(\vec{\boldsymbol{p}}, s)x(\vec{\boldsymbol{p}}', s') + x^\dagger(\vec{\boldsymbol{p}}, s)y^\dagger(\vec{\boldsymbol{p}}', s')\big]\Big\} \\ & \times \int \frac{d^dk}{(2\pi)^d}\frac{1}{D}\Big\{ x^\dagger(\vec{\boldsymbol{p}}', s')Q^\mu x(\vec{\boldsymbol{p}}, s) + y(\vec{\boldsymbol{p}}', s')R^\mu y^\dagger(\vec{\boldsymbol{p}}, s) \\ & + y(\vec{\boldsymbol{p}}', s')S^\mu x(\vec{\boldsymbol{p}}, s) + x^\dagger(\vec{\boldsymbol{p}}', s')T^\mu y^\dagger(\vec{\boldsymbol{p}}, s)\Big\}, \end{aligned} \tag{19.181}$$

where

$$Q^\mu \equiv -2e^3\big[(p+k)\cdot\overline{\sigma}\,\sigma^\mu\,(p'+k)\cdot\overline{\sigma} + m^2\overline{\sigma}^\mu - \tfrac{1}{2}(4-d)k\cdot\overline{\sigma}\,\sigma^\mu\,k\cdot\overline{\sigma}\big], \tag{19.182}$$

$$R^\mu \equiv -2e^3\big[(p+k)\cdot\sigma\,\overline{\sigma}^\mu\,(p'+k)\cdot\sigma + m^2\sigma^\mu - \tfrac{1}{2}(4-d)k\cdot\sigma\,\overline{\sigma}^\mu\,k\cdot\sigma\big], \tag{19.183}$$

$$S^\mu = T^\mu \equiv 4e^3 m(p+p'+2k)^\mu. \tag{19.184}$$

One can now perform the sum over spins [see eq. (19.168)]:

$$
\begin{aligned}
ieF_2(q^2)_{\rm QED} = \int \frac{d^dk}{(2\pi)^d}\frac{1}{D}\bigg\{ & g_1\big[\mathrm{Tr}\big(m^2(Q\cdot\sigma+R\cdot\overline{\sigma})+\overline{\sigma}_\mu p'\cdot\sigma Q^\mu p\cdot\sigma+\sigma_\mu p'\cdot\overline{\sigma}R^\mu p\cdot\overline{\sigma}\big) \\
& +m\,\mathrm{Tr}(S\cdot\sigma\, p'\cdot\overline{\sigma}+S\cdot\overline{\sigma}\,p\cdot\sigma+T\cdot\sigma\,p\cdot\overline{\sigma}+T\cdot\overline{\sigma}\,p'\cdot\sigma)\big] \\
& +\frac{g_2(p+p')_\mu}{2m}\big[\mathrm{Tr}\big(m^2(S^\mu+T^\mu)+S^\mu p\cdot\overline{\sigma}\,p'\cdot\sigma+T^\mu p\cdot\sigma\, p'\cdot\overline{\sigma}\big) \\
& +m\,\mathrm{Tr}\big(Q^\mu(p+p')\cdot\sigma+R^\mu(p+p')\cdot\overline{\sigma}\big)\big]\bigg\}, \qquad (19.185)
\end{aligned}
$$

where g_1 and g_2 are given by eq. (19.170). When performing a trace over a product of four $\sigma/\overline{\sigma}$ matrices, the term proportional to the Levi-Civita epsilon tensor will always vanish since there are only two independent four-momenta in this problem. Consequently, the traces obtained in evaluating eq. (19.185) are unchanged under an interchange of $\sigma\leftrightarrow\overline{\sigma}$. This implies that the traces in eq. (19.185) involving R^μ and S^μ give the same results as those involving Q^μ and T^μ, respectively. Hence,

$$
\begin{aligned}
ieF_2(q^2)_{\rm QED} = 2\int \frac{d^dk}{(2\pi)^d}\frac{1}{D}\bigg\{ & g_1\big[\mathrm{Tr}\big(m^2 Q\cdot\sigma+\overline{\sigma}_\mu p'\cdot\sigma Q^\mu p\cdot\sigma\big) \\
& +m\,\mathrm{Tr}(S\cdot\sigma\,p'\cdot\overline{\sigma}+S\cdot\overline{\sigma}\,p\cdot\sigma)\big] \\
& +\frac{g_2(p+p')_\mu}{2m}\big[\mathrm{Tr}\big(S^\mu(p\cdot\overline{\sigma}+m^2)\big)+m\,\mathrm{Tr}\big(Q^\mu(p+p')\cdot\sigma\big)\big]\bigg\}.
\end{aligned}
\qquad (19.186)
$$

Employing the results of Appendix B.4 to evaluate the traces and using the mass-shell conditions $p^2=p'^2=m^2$, we end up with

$$
\begin{aligned}
F_2(q^2)_{\rm QED} = 4ie^2\mu^{2\epsilon}\int \frac{d^dk}{(2\pi)^d}\frac{1}{D}\bigg\{ & g_1\Big[8p\cdot p'(p\cdot p'-2m^2)+4(2\,p\cdot p'-3m^2)k\cdot(p+p') \\
& +4(d-2)k\cdot p\,k\cdot p'+(d-2)(d-4)k^2p\cdot p'-m^2(d-2)^2k^2\Big] \\
& +g_2\Big[-4p\cdot p'(m^2+p\cdot p')-2(m^2+p\cdot p')k\cdot(p+p') \\
& +(d-2)\big([k\cdot(p+p')]^2-k^2(m^2+p\cdot p')\big)\Big]\bigg\}, \qquad (19.187)
\end{aligned}
$$

after replacing $e\to e\mu^\epsilon$ so that e remains dimensionless when evaluating the integral in $d=4-2\epsilon$ dimensions. We have so far been careful not to set $d=4$ in any numerator coefficient that multiplies a term that is quadratic in k, since the corresponding integral is potentially divergent in the limit of $d=4$.

We perform the integration over k using the results of Appendix K.2 to express the result in terms of the Passarino–Veltman loop functions and note that the potentially divergent integral is proportional to C_{24}. To obtain the corresponding coefficient, we simply replace

$$
-16\pi^2 i\mu^{2\epsilon}\int\frac{d^dk}{(2\pi)^d}\frac{k^\mu k^\nu}{D} \quad\longrightarrow\quad g^{\mu\nu}C_{24}(p^2,q^2,p'^2;0,m^2,m^2)\,, \qquad (19.188)
$$

$$
-16\pi^2 i\mu^{2\epsilon}\int\frac{d^dk}{(2\pi)^d}\frac{k^2}{D} \quad\longrightarrow\quad dC_{24}(p^2,q^2,p'^2;0,m^2,m^2)\,, \qquad (19.189)
$$

where we have made use of eq. (K.1) following the rules of dimensional regularization. Thus, the coefficient of C_{24} is proportional to

$$(d-2)^2\{g_1[(d-2)p\cdot p' - dm^2] - g_2(m^2+p\cdot p')\}$$
$$= -4(d-2)^2\{[2m^2 + (\tfrac{1}{2}d-1)q^2]g_1 + \tfrac{1}{2}(4m^2-q^2)g_2\} = 0\,, \quad (19.190)$$

after employing eqs. (19.169) and (19.170). Thus, C_{24} does not appear in the expression for $F_2(q^2)_{\rm QED}$ given by eq. (19.187), which implies that $F_2(q^2)_{\rm QED}$ is finite for $d=4$. Moreover, even prior to setting $d=4$, there is no term of the form $(d-4)C_{24}$ which is finite but nonzero in the limit of $d\to 4$. This observation implies that we would have been justified in setting $d=4$ in eqs. (19.170) and (19.187) from the beginning.

Setting $d=4$ in eqs. (19.170) and (19.187) and making use of eq. (19.169) yields

$$F_2(q^2)_{\rm QED} = -4ie^2\int\frac{d^dk}{(2\pi)^d}\frac{1}{D}\left\{\frac{2m^2}{4m^2-q^2}[k\cdot(p+p')-k^2]\right.$$
$$\left. - 8g_1 k\cdot p\, k\cdot p' - 2g_2[k\cdot(p+p')]^2\right\}, \quad (19.191)$$

where we have employed the $d=4$ forms for g_1 and g_2 in obtaining the first line of eq. (19.191). In deriving the above result, the sum of the terms in the numerator of the integrand of eq. (19.187) that are independent of k vanishes exactly, since

$$8g_1 p\cdot p'(p\cdot p' - 2m^2) - 4g_2 p\cdot p'(m^2+p\cdot p')$$
$$= (q^2-2m^2)[2(q^2+2m^2)g_1 + (4m^2-q^2)g_2] = 0\,. \quad (19.192)$$

The integration over k is now easily performed using the results of Appendix K.2. We have already shown in eq. (19.190) that terms proportional to C_{24} exactly cancel. In addition, there are no terms proportional to C_{12} in light of the identity $q\cdot(p+p') = -(p-p')\cdot(p+p') = 0$, for on-shell muons. We then obtain

$$F_2(q^2)_{\rm QED} = \frac{e^2}{4\pi^2}\left\{m^2C_{11} - \frac{2m^2}{4m^2-q^2}[m^2C_{21} + q^2(C_{22}-C_{23})]\right.$$
$$\left. - 2g_1[2m^2(2m^2-q^2)C_{21} - q^4(C_{22}-C_{23})] - \tfrac{1}{2}g_2(4m^2-q^2)^2C_{21}\right\}, \quad (19.193)$$

where the suppressed arguments of the C_{ij} are $(p^2, q^2, p'^2; 0, m^2, m^2)$, subject to the mass-shell conditions $p^2 = p'^2 = m^2$. Plugging in the expressions for g_1 and g_2, the coefficient of $C_{22}-C_{23}$ exactly cancels. The end result is remarkably simple:

$$F_2(q^2)_{\rm QED} = \frac{\alpha m^2}{\pi}[C_{11}(m^2,q^2,m^2;0,m^2,m^2) + C_{21}(m^2,q^2,m^2;0,m^2,m^2)], \quad (19.194)$$

where $\alpha\equiv e^2/(4\pi)$.

The QED contribution to the anomalous magnetic moment of the muon is now obtained by setting $q^2 = 0$. We can evaluate C_{11} and C_{12} immediately by employing eqs. (K.70), (K.72), and (K.76):

$$C_{11}(m^2, 0, m^2; 0, m^2, m^2) = \frac{1}{m^2}, \tag{19.195}$$

$$C_{21}(m^2, 0, m^2; 0, m^2, m^2) = -\frac{1}{2m^2}. \tag{19.196}$$

Inserting these results back into eq. (19.194) yields the well-known Schwinger result,

$$F_2(0)_{\rm QED} = \frac{\alpha}{2\pi}. \tag{19.197}$$

19.5.3 Supersymmetric Particle Contributions to the Anomalous Magnetic Moment

In softly broken SUSY-QED, the one-loop anomalous magnetic moment of the muon consists of a sum of eq. (19.197) and the contributions from one-loop diagrams involving a loop consisting of a photino and a smuon pair, shown in Fig. 19.13.

The amplitude represented by Fig. 19.13(a) is given by

$$\begin{aligned}\langle p', s'|\, S_a^{(1)}\, |p, s\rangle &= ie(ie\sqrt{2})^2 \widetilde{A}_\mu(q) \int \frac{d^dk}{(2\pi)^d} (p+p'+2k)^\mu \frac{i}{(k+p)^2 - m_L^2 + i\varepsilon} \\ &\quad \times \frac{i}{(k+p')^2 - m_L^2 + i\varepsilon} \left(\frac{x^\dagger(\vec{\boldsymbol{p}}', s')\left[-ik\cdot\overline{\sigma}\right]x(\vec{\boldsymbol{p}}, s)}{k^2 - M^2 + i\varepsilon} \right) \\ &= 2e^3 \widetilde{A}_\mu(q) \int \frac{d^dk}{(2\pi)^d} \frac{(p+p'+2k)^\mu\, x^\dagger(\vec{\boldsymbol{p}}', s')\, k\cdot\overline{\sigma}\, x(\vec{\boldsymbol{p}}, s)}{\left[(k+p)^2 - m_L^2 + i\varepsilon\right]\left[(k+p')^2 - m_L^2 + i\varepsilon\right]\left[k^2 - M^2 + i\varepsilon\right]}, \end{aligned} \tag{19.198}$$

where m_L is the mass of the smuon $\widetilde{\mu}_L$ and M is the mass of the photino.

Likewise, the amplitude represented by Fig. 19.13(b) is given by

$$\begin{aligned}\langle p', s'|\, S_b^{(1)}\, |p, s\rangle &= ie(-ie\sqrt{2})^2 \widetilde{A}_\mu(q) \int \frac{d^dk}{(2\pi)^d} (p+p'+2k)^\mu \frac{i}{(k+p)^2 - m_R^2 + i\varepsilon} \\ &\quad \times \frac{i}{(k+p')^2 - m_R^2 + i\varepsilon} \left(\frac{y(\vec{\boldsymbol{p}}', s')\left[-ik\cdot\sigma\right]y^\dagger(\vec{\boldsymbol{p}}, s)}{k^2 - M^2 + i\varepsilon} \right) \\ &= 2e^3 \widetilde{A}_\mu(q) \int \frac{d^dk}{(2\pi)^d} \frac{(p+p'+2k)^\mu\, y(\vec{\boldsymbol{p}}', s')\, k\cdot\sigma\, y^\dagger(\vec{\boldsymbol{p}}, s)}{\left[(k+p)^2 - m_R^2 + i\varepsilon\right]\left[(k+p')^2 - m_R^2 + i\varepsilon\right]\left[k^2 - M^2 + i\varepsilon\right]}, \end{aligned} \tag{19.199}$$

where m_R is the mass of the smuon $\widetilde{\mu}_R$. Note that, in contrast to diagram (a), the sign of the photino–smuon–muon vertex has flipped, as indicated by eq. (13.3) and by the Feynman rules exhibited in Fig. 19.2. We have also made use of the rule for the photino propagator given in Fig. 2.3, where the choice of σ or $\overline{\sigma}$ is dictated by

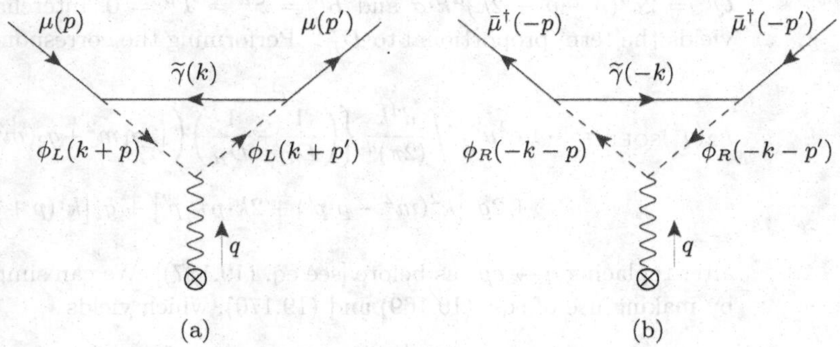

Fig. 19.13 One-loop (softly broken) SUSY-QED contributions to the anomalous magnetic moment of the muon. The internal loop consists of a photino of four-momentum k and a pair of smuons: $\phi_L \equiv \widetilde{\mu}_L$ in diagram (a) and $\phi_R \equiv \widetilde{\mu}_R$ in diagram (b). The four-momenta are shown in parentheses and all four-momenta point in the directions of the arrows. In particular, the direction of flow of each of the four-momenta is the same in diagrams (a) and (b). The $\otimes$ symbol indicates the presence of an external static classical four-vector potential A^μ field, and $q = p' - p$.

the spinor wave functions that appear in the amplitude. The integrals above will be computed in d dimensions since some of the terms exhibit a logarithmic divergence in the ultraviolet. However, as we shall demonstrate below, the expression that we eventually obtain for the anomalous magnetic moment is a finite quantity, so all potentially divergent quantities must cancel out in obtaining our final result.

It is convenient to introduce a notation for the denominators of eqs. (19.198) and (19.199):

$$D_i \equiv \big[(k+p)^2 - m_i^2 + i\varepsilon\big]\big[(k+p')^2 - m_i^2 + i\varepsilon\big]\big[k^2 - M^2 + i\varepsilon\big]\,, \quad \text{for } i = L,\, R. \tag{19.200}$$

Then, it follows from eq. (19.165) that the contributions to $F_2(q^2)$ from Fig. 19.13(a) and (b), denoted below by $F(q^2)_{\rm SQED}$, are given by

$$\begin{aligned} ieF_2(q^2)_{\rm SQED} = \sum_{s,s'} \Big\{ & g_1(q^2)\big[y(\vec{\boldsymbol{p}},s)\sigma_\mu y^\dagger(\vec{\boldsymbol{p}}\,',s') + x^\dagger(\vec{\boldsymbol{p}},s)\overline{\sigma}_\mu x(\vec{\boldsymbol{p}}\,',s')\big] \\ & + g_2(q^2)\frac{(p+p')_\mu}{2m}\big[y(\vec{\boldsymbol{p}},s)x(\vec{\boldsymbol{p}}\,',s') + x^\dagger(\vec{\boldsymbol{p}},s)y^\dagger(\vec{\boldsymbol{p}}\,',s')\big]\Big\} \\ & \times 2e^3 \int \frac{d^dk}{(2\pi)^d}\,(p+p'+2k)^\mu \left\{\frac{x^\dagger(\vec{\boldsymbol{p}}\,',s')\,k\!\cdot\!\overline{\sigma}\,x(\vec{\boldsymbol{p}},s)}{D_L} + \frac{y(\vec{\boldsymbol{p}}\,',s')\,k\!\cdot\!\sigma\,y^\dagger(\vec{\boldsymbol{p}},s)}{D_R}\right\}, \end{aligned} \tag{19.201}$$

where g_1 and g_2 are given by eq. (19.170). To compute the sum over spins, the term proportional to D_L^{-1} can be obtained by employing eqs. (19.181) and (19.185), where

$Q^\mu = 2e^3(p+p'+2k)^\mu k\cdot\overline{\sigma}$ and $R^\mu = S^\mu = T^\mu = 0$. Interchanging $\sigma \leftrightarrow \overline{\sigma}$ then yields the term proportional to D_R^{-1}. Performing the corresponding traces yields

$$F_2(q^2)_{\rm SQED} = -4ie^2\mu^{2\epsilon}\int\frac{d^dk}{(2\pi)^d}\left\{\left(\frac{1}{D_L}+\frac{1}{D_R}\right)\left([2g_1m^2+g_2(m^2+p\cdot p')]k\cdot(p+p')\right.\right.$$
$$\left.\left.+2g_1\left[k^2(m^2-p\cdot p')+2k\cdot p\,k\cdot p'\right]+g_2\left[k\cdot(p+p')\right]^2\right)\right\} \tag{19.202}$$

after replacing $e\to e\mu^\epsilon$ as before [see eq. (19.187)]. We can simplify this expression by making use of eqs. (19.169) and (19.170), which yields

$$F_2(q^2)_{\rm SQED} = 4ie^2\mu^{2\epsilon}\int\frac{d^dk}{(2\pi)^d}\left\{\left(\frac{1}{D_L}+\frac{1}{D_R}\right)\left(\frac{m^2}{4m^2-q^2}\left[k\cdot(p+p')-\frac{k^2}{\frac{1}{2}d-1}\right]\right.\right.$$
$$\left.\left.-4g_1k\cdot p\,k\cdot p'-g_2\left[k\cdot(p+p')\right]^2\right)\right\}, \tag{19.203}$$

where we have set $d=4$ in the terms of the integrand that are proportional to k, but we have (temporarily) retained the factors of d in the terms of the integrand that are quadratic in k.

We can now perform the integration over k using the results of Appendix K.2 to express the result in terms of the Passarino–Veltman loop functions. The potentially divergent integral is proportional to C_{24}. To obtain the corresponding coefficient, we make the replacements indicated by eqs. (19.188) and (19.189). Thus, the coefficient of C_{24} is proportional to

$$4g_1p\cdot p'+g_2(p+p')^2+\frac{m^2d}{(\frac{1}{2}d-1)(4m^2-q^2)}$$
$$=2g_1(2m^2-q^2)+g_2(4m^2-q^2)+\frac{m^2d}{(\frac{1}{2}d-1)(4m^2-q^2)}=0\,, \tag{19.204}$$

after employing eqs. (19.169) and (19.170). As in the computation of the QED contribution to the anomalous magnetic moment, the coefficient of C_{24} vanishes for all values of d. Thus, we would have been justified in setting $d=4$ in eqs. (19.170) and (19.203) from the beginning.

Setting $d=4$ in eq. (19.203) and comparing the result to eq. (19.191), we see that the two results are the same up to an overall factor of $-\frac{1}{2}$ and the different propagator factors. Thus, we can immediately make use of eq. (19.194) to obtain our final result,

$$F_2(q^2)_{\rm SQED} = -\frac{\alpha m^2}{2\pi}\Big[C_{11}(m^2,q^2,m^2;M^2,m_{\widetilde{L}}^2,m_{\widetilde{L}}^2)+C_{21}(m^2,q^2,m^2;M^2,m_{\widetilde{L}}^2,m_{\widetilde{L}}^2)$$
$$+C_{11}(m^2,q^2,m^2;M^2,m_{\widetilde{R}}^2,m_{\widetilde{R}}^2)+C_{21}(m^2,q^2,m^2;M^2,m_{\widetilde{R}}^2,m_{\widetilde{R}}^2)\Big]\,, \tag{19.205}$$

after including contributions from both $\widetilde{\mu}_L$ and $\widetilde{\mu}_R$ in the loop.

The supersymmetric contribution to the anomalous magnetic moment of the muon is obtained by setting $q^2 = 0$. Using eqs. (K.70), (K.72), and (K.76),

$$F_2(0)_{\rm SQED} = -\frac{\alpha m^2}{2\pi}\left\{\int_0^1 \frac{x^2(1-x)\,dx}{m^2x^2 + (m_L^2 - m^2 - M^2)x + M^2} + (m_L^2 \to m_R^2)\right\}. \tag{19.206}$$

This integral is elementary and can be expressed in terms of a logarithm. Rather then providing the exact expression, we shall examine two limiting cases in the next subsection.

19.5.4 Limiting Cases

If smuons and photinos exist in Nature, then M, m_L, $m_R \gg m$. Thus, to a very good approximation, we can set $m = 0$ in the integrand of eq. (19.206). If we define the dimensionless variables

$$z_L \equiv \frac{m_L^2}{M^2}, \qquad z_R \equiv \frac{m_R^2}{M^2}, \tag{19.207}$$

then eq. (19.206) yields

$$F_2(0)_{\rm SQED} = \frac{\alpha m^2}{12\pi M^2}\left\{\frac{z_L^3 - 6z_L^2 + 3z_L + 2 + 6z_L \ln z_L}{(z_L-1)^4} + (m_L^2 \to m_R^2)\right\}. \tag{19.208}$$

In the limit of $M = m_L = m_R \gg m$, eq. (19.208) simplifies to

$$F_2(0)_{\rm SQED} = -\frac{\alpha m^2}{12\pi M^2}. \tag{19.209}$$

We now examine the supersymmetric limit which, although not realized in Nature, is of academic interest. In the supersymmetric limit, $m_L = m_R = m$ and $M = 0$. Then eq. (19.205) yields

$$F_2(q^2)_{\rm SQED} = -\frac{\alpha m^2}{\pi}\left[C_{11}(m^2, q^2, m^2; 0, m^2, m^2) + C_{21}(m^2, q^2, m^2; 0, m^2, m^2)\right]. \tag{19.210}$$

Adding this result to that of eq. (19.194) yields the remarkable result,

$$F_2(q^2) = F_2(q^2)_{\rm QED} + F_2(q^2)_{\rm SQED} = 0\,. \tag{19.211}$$

Thus, in unbroken supersymmetric QED, the one-loop anomalous magnetic moment of the muon vanishes.

Such a striking result calls out for an explanation. Indeed, the result $F(q^2) = 0$ persists to all orders in the loop expansion. An explanation was provided by Ferrara and Remiddi in Ref. [201], where they argued that there is no way to write the magnetic dipole operator,

$$\mathscr{L}_{\rm dipole} = \frac{1}{m}F_{\mu\nu}(\eta\sigma^{\mu\nu}\chi + \chi^\dagger\overline{\sigma}^{\mu\nu}\eta^\dagger)\,, \tag{19.212}$$

in a supersymmetric invariant way. To demonstrate their claim, let us attempt to construct a term in the effective superspace Lagrangian that contains the magnetic dipole operator specified in eq. (19.212). Written in terms of superfields, it must be expressed as an integral either over $d^2\theta$ (corresponding to an F-term) or over $d^2\theta d^2\theta^\dagger$ (corresponding to a D-term), as discussed in Section 10.6. The integrand must be a combination of the superfields Φ_L, Φ_R and the SUSY-QED abelian field-strength superfield $\mathcal{W}^\alpha$ defined in eq. (10.144), which we rewrite in a more convenient form using eq. (1.118):

$$\mathcal{W}_\alpha(y,\theta,\theta^\dagger) = \lambda_\alpha + \theta_\alpha D - (\sigma^{\mu\nu}\theta)_\alpha F_{\mu\nu} + i\theta\theta(\sigma^\mu\partial_\mu\lambda^\dagger)_\alpha\,, \tag{19.213}$$

where $y \equiv x - i\theta\sigma^\mu\theta^\dagger$ [see eq. (10.88)] and $\partial_\mu \equiv \partial/\partial y^\mu$. Moreover, the chiral covariant derivative D_α must be employed so that the integrand possesses no free spinor index. This last requirement implies that no F-term of the desired form exists since applying D_α to a chiral superfield yields a superfield that is neither chiral nor antichiral. To construct a D-term would require us to find an operator whose $\theta\theta\theta^\dagger\theta^\dagger$ coefficient consists precisely of the fields $F_{\mu\nu}$, χ, and η (with no extra derivatives). In light of eqs. (13.1) and (19.213), this is not possible if supersymmetry is exact.

On the other hand, if supersymmetry is broken, then we can introduce a spurion chiral superfield, $X = \theta\theta F$, as in eq. (14.14). The form of the superspace Lagrangian that produces eq. (19.212) as one of its terms is unique (up to an overall coefficient C):

$$\mathscr{L}_{\rm eff} = \frac{C}{m^2 M_P}\int d^2\theta\, d^2\theta^\dagger\, \tfrac{1}{2}\left[X^\dagger\Phi_L \overleftrightarrow{D}_\alpha \Phi_R \mathcal{W}^\alpha + {\rm h.c.}\right], \tag{19.214}$$

where $\Phi_L\overleftrightarrow{D}_\alpha\Phi_R \equiv \Phi_L(D_\alpha\Phi_R) - (D_\alpha\Phi_L)\Phi_R$, and the insertion of the spurion superfield corresponds to Planck-scale-mediated supersymmetry breaking, treated in Section 14.3. We have also included two powers of the muon mass, m, in the denominator to ensure that the mass dimension of the effective Lagrangian is four. As in eq. (14.2), we will identify the effective soft supersymmetry-breaking scale by $m_{\rm soft} \sim F/M_P$.

Plugging the θ expansions of X, Φ_L, Φ_R, and $\mathcal{W}$ into eq. (19.214), we focus on the term proportional to $F_{\mu\nu}$ (with no additional derivatives with respect to y). When D_α acts on Φ_R, we obtain

$$\begin{aligned} -2\theta^\beta\chi_\gamma\delta_\beta{}^\gamma\eta_\alpha\epsilon^{\alpha\rho}(\sigma^{\mu\nu})_\rho{}^\tau\theta_\tau F_{\mu\nu} &= -F_{\mu\nu}\theta^\beta\chi_\gamma\eta^\rho\theta_\tau\Big\{(\sigma^{\mu\nu})_\beta{}^\tau\delta_\rho{}^\gamma + \delta_\beta{}^\tau(\sigma^{\mu\nu})_\rho{}^\gamma \\ &\qquad - ig_{\rho\kappa}\left[(\sigma^{\mu\kappa})_\beta{}^\tau(\sigma^{\nu\rho})_\rho{}^\gamma - (\sigma^{\nu\kappa})_\beta{}^\tau(\sigma^{\mu\rho})_\rho{}^\gamma\right]\Big\} \\ &= (\theta\theta)\eta\sigma^{\mu\nu}\chi F_{\mu\nu}\,, \end{aligned} \tag{19.215}$$

after employing the Fierz identity given in eq. (B.86) and using $\theta\sigma^{\mu\nu}\theta = 0$. When D_α acts on Φ_L we obtain the same result after using eq. (1.133). Hence, we end up with

$$\mathscr{L}_{\rm eff} = C\frac{m_{\rm soft}}{m^2}F_{\mu\nu}(\eta\sigma^{\mu\nu}\chi + \chi^\dagger\overline{\sigma}^{\mu\nu}\eta^\dagger) + \cdots\,, \tag{19.216}$$

after identifying $m_{\rm soft} = F/M_P$, where $\cdots$ represents additional interactions (e.g., $i\phi_R\chi\sigma^\mu\partial_\mu\lambda^\dagger - i\phi_L\eta\sigma^\mu\partial_\mu\lambda^\dagger$) which are not pertinent to the present discussion. In the absence of supersymmetry breaking, i.e., if $m_{\rm soft} = 0$, then the anomalous magnetic moment vanishes. Moreover, by comparing eqs. (19.216) and (19.212), it follows that the first-order supersymmetric correction yields $F_2(0) = \mathcal{O}(m_{\rm soft}/m)$.

We can verify this last statement by evaluating eq. (19.206) under the assumption that the deviation from exact supersymmetry is small. For example, suppose we take $m_L^2 = m_R^2 = m^2 + M^2$, under the assumption that the photino mass is much smaller than the muon mass, i.e., $M \ll m$. In this example, we are identifying $m_{\rm soft} = M$. Then, adding the results of eqs. (19.197) and (19.206),

$$F_2(0) = \frac{\alpha}{2\pi}\left\{1 - 2\int_0^1 \frac{x^2(1-x)}{x^2+z^2}\,dx\right\} = -\frac{\alpha z}{2\pi}\left[\arctan\left(\frac{1}{z}\right) - z\ln\left(\frac{z^2}{z^2+1}\right)\right], \tag{19.217}$$

where $z \equiv M/m$. That is, $F_2(0) \simeq -\alpha M/(4m)$, to leading order in M/m, as anticipated by the discussion following eq. (19.216).

19.6 Anapole Moment of the Muon

In Sections 19.5.2 and 19.5.3, we evaluated the first-order S-matrix element for the scattering of an on-shell muon off a classical electromagnetic field and then projected out the relevant form factor of the electromagnetic vertex, $F_2(q^2)$, to determine the anomalous magnetic moment $F_2(0) = \frac{1}{2}(g-2)$. Likewise, by projecting out the form factor $F_4(q^2)$, one may determine the static anapole moment, which is proportional to $F_4(0)$ [see eq. (19.163)]. In this section, we provide some details of the computation of the anapole moment, since one new feature arises.

The form factor $F_4(q^2)$ is given by

$$F_4(q^2) = g_4(q^2)\sum_{s,s'}\left[y(\vec{p},s)\sigma_\mu y^\dagger(\vec{p}\,',s') - x^\dagger(\vec{p},s)\overline{\sigma}_\mu x(\vec{p}\,',s')\right]z(\vec{p}\,',s')\Gamma^\mu z(\vec{p},s). \tag{19.218}$$

Plugging in the expression for $z(\vec{p}\,',s')\Gamma^\mu z(\vec{p},s)$ given in eq. (19.166), one can easily determine $g_4(q^2)$. In particular,

$$\begin{aligned}\frac{4m^2}{g_4(q^2)} &= (q^\mu q^\nu - q^2 g^{\mu\nu})\sum_{s,s'}\left[y(\vec{p},s)\sigma_\mu y^\dagger(\vec{p}\,',s') - x^\dagger(\vec{p},s)\overline{\sigma}_\mu x(\vec{p}\,',s')\right]\\ &\qquad\times\left[y(\vec{p}\,',s')\sigma_\nu y^\dagger(\vec{p},s) - x^\dagger(\vec{p}\,',s')\overline{\sigma}_\nu x(\vec{p},s)\right]\\ &= (q^\mu q^\nu - q^2 g^{\mu\nu})\Big\{\mathrm{Tr}(\sigma_\mu p'\!\cdot\!\overline{\sigma}\sigma_\nu p\!\cdot\!\overline{\sigma}) + \mathrm{Tr}(\overline{\sigma}_\mu p'\!\cdot\!\sigma\overline{\sigma}_\nu p\!\cdot\!\sigma) - m^2\,\mathrm{Tr}(\sigma_\mu\overline{\sigma}_\nu + \overline{\sigma}_\mu\sigma_\nu)\Big\}\\ &= 4q^2(4m^2 - q^2)\,,\end{aligned} \tag{19.219}$$

after evaluating the traces in $d = 4$ dimensions.[10] That is,

$$g_4(q^2) = \frac{m^2}{q^2(4m^2 - q^2)}\,. \tag{19.220}$$

Since the anapole moment is parity odd, the only nonzero contributions to the anapole moment of the muon in softly broken SUSY-QED arise from one-loop diagrams involving the smuons, assuming that $m_L \neq m_R$, as the latter is the source of parity violation. Thus, we can immediately exploit the results of Section 19.5.3. In light of eq. (19.201), it follows that the vertex diagrams exhibited in Fig. 19.13 yield

$$\begin{aligned} ieF_4(q^2)_{\text{vertex}} &= \frac{m^2}{q^2(4m^2-q^2)} \sum_{s,s'} \left[y(\vec{\boldsymbol{p}}, s)\sigma_\mu y^\dagger(\vec{\boldsymbol{p}}{}', s') - x^\dagger(\vec{\boldsymbol{p}}, s)\overline{\sigma}_\mu x(\vec{\boldsymbol{p}}{}', s') \right] \\ &\times 2e^3 \int \frac{d^d k}{(2\pi)^d} (p + p' + 2k)^\mu \left\{ \frac{x^\dagger(\vec{\boldsymbol{p}}{}', s')\, k\cdot\overline{\sigma}\, x(\vec{\boldsymbol{p}}, s)}{D_L} + \frac{y(\vec{\boldsymbol{p}}{}', s')\, k\cdot\sigma\, y^\dagger(\vec{\boldsymbol{p}}, s)}{D_R} \right\}, \end{aligned} \tag{19.221}$$

where the denominator factors D_L and D_R are defined in eq. (19.200). Converting the spin sums to traces and replacing $e \to e\mu^\epsilon$ as before [see eq. (19.187)],

$$\begin{aligned} F_4(q^2)_{\text{vertex}} &= \frac{-2ie^2\mu^{2\epsilon}m^2}{q^2(4m^2-q^2)} \int \frac{d^d k}{(2\pi)^d} \left\{ \frac{1}{D_L} \Big[m^2\, \mathrm{Tr}\big[(p+p'+2k)\cdot\sigma\, k\cdot\overline{\sigma}\big] \right. \\ &\qquad - \mathrm{Tr}\big[(p+p'+2k)\cdot\overline{\sigma}\, p'\cdot\sigma\, k\cdot\overline{\sigma}\, p\cdot\sigma\big] \Big] \\ &\quad \left. - \frac{1}{D_R} \Big[m^2\, \mathrm{Tr}\big[(p+p'+2k)\cdot\overline{\sigma}\, k\cdot\sigma\big] - \mathrm{Tr}\big[(p+p'+2k)\cdot\sigma\, p'\cdot\overline{\sigma}\, k\cdot\sigma\, p\cdot\overline{\sigma}\big] \Big] \right\} \\ &= \frac{-4ie^2\mu^{2\epsilon}m^2}{q^2(4m^2-q^2)} \int \frac{d^d k}{(2\pi)^d} \left[k^2(4m^2-q^2) - 4p'\cdot k\, p\cdot k \right] \left(\frac{1}{D_L} - \frac{1}{D_R} \right). \end{aligned} \tag{19.222}$$

As expected, $F_4(q^2)_{\text{vertex}} = 0$ when $m_L \neq m_R$. Carrying out the loop integration, we can express $F_4(q^2)_{\text{vertex}}$ in terms of the Passarino–Veltman C-functions. Using the results of Appendix K.2,

$$\begin{aligned} F_4(q^2)_{\text{vertex}} &= \frac{\alpha m^2}{\pi q^2(4m^2-q^2)} \Big\{ m^2 q^2 \left[C_{21} + 4C_{22} - 4C_{23} \right] \\ &\qquad + \left[4m^2(d-1) - q^2(d-2) \right] C_{24} - (L \to R) \Big\}, \end{aligned} \tag{19.223}$$

where the arguments of the C-functions above are $(p^2, q^2, p'^2; M^2, m_L^2, m_L^2)$. In the terms denoted by "$(L \to R)$", one simply makes the replacement $m_L^2 \to m_R^2$.

Equation (19.223) cannot be the complete one-loop result for $F_4(q^2)$. Although the divergence cancels when the $(L \to R)$ terms are included, one would have

[10] Since $g_4(q^2)$ multiplies an expression that will end up being finite in the limit of $d \to 4$, there is no need to perform the trace calculations in $d \neq 4$ spacetime dimensions in obtaining eq. (19.220).

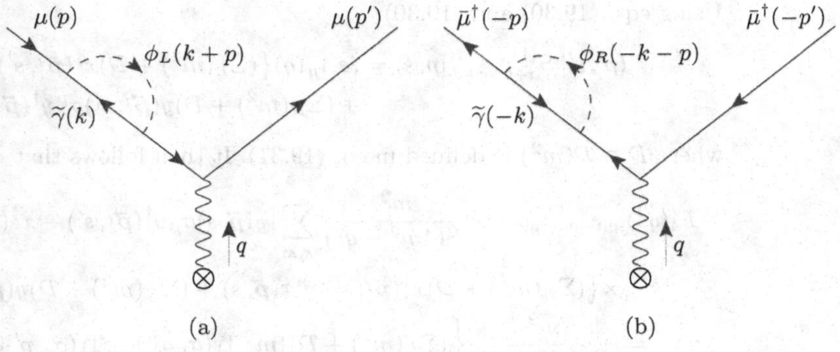

Fig. 19.14 One-loop (softly broken) SUSY-QED contributions to the anapole moment of the muon consist of two classes of diagrams. The vertex corrections are shown in Fig. 19.13. In addition, one must consider wave function renormalization on the external on-shell muon lines. In diagrams (a) and (b), the self-energy insertions appear on the incoming muon line. In diagrams (c) and (d), not explicitly exhibited, the self-energy insertions appear on the outgoing muon line. The four-momenta are shown in parentheses and all four-momenta point in the directions of the arrows. The ⊗ symbol indicates the presence of an external static classical four-vector potential A^μ field, and $q = p' - p$.

expected the divergences to cancel if the loop involving $\widetilde{\mu}_R$ had been omitted. Moreover, $F_4(q^2)_{\text{vertex}}$ does not have a finite static limit (when $q^2 = 0$), and there is a spurious singularity at $q^2 = 4m^2$. It is not difficult to see what is missing. When computing the first-order S-matrix element for the scattering of an on-shell muon off a classical electromagnetic field, one must also include the parity-violating effects of wave function renormalization of the external on-shell muons. In contrast to the computation of $F_2(q^2)$, the effects of wave function renormalization in the calculation of $F_4(q^2)$ do not cancel out when the projection operator in eq. (19.218) is applied if $m_L^2 \neq m_R^2$.

In Fig. 19.14, we exhibit the one-loop Feynman graphs where the one-loop self-energy corrections appear on the incoming muon line. Likewise, there are two diagrams not shown, in which the one-loop self-energy corrections appear on the outgoing muon line. These diagrams should be interpreted in the context of the LSZ reduction formulae discussed in Section 2.11. Thus, the one-loop contributions to the first-order S-matrix elements represented by diagrams (a) and (b) of Fig. 19.14 and the corresponding diagrams (c) and (d) [not shown] where the self-energy corrections appear on the outgoing muon line are given by

$$\langle p', s'|\, S^{(1)}_{a+c}\, |p, s\rangle = \widetilde{A}_\mu(q)\tfrac{1}{2}(\delta Z_L + \delta Z_L^*)x^\dagger(\vec{\boldsymbol{p}}\,', s')ie\overline{\sigma}^\mu x(\vec{\boldsymbol{p}}, s)\,, \qquad (19.224)$$

$$\langle p', s'|\, S^{(1)}_{b+d}\, |p, s\rangle = \widetilde{A}_\mu(q)\tfrac{1}{2}(\delta Z_R + \delta Z_R^*)y(\vec{\boldsymbol{p}}\,', s')ie\sigma^\mu y^\dagger(\vec{\boldsymbol{p}}, s)\,. \qquad (19.225)$$

Using eqs. (19.30) and (19.30),

$$\langle p', s'|\, S^{(1)}_{a+b+c+d}\, |p, s\rangle = ie\widetilde{A}_\mu(q)\{(\Sigma_L(m^2)+\mathcal{D})x^\dagger(\vec{\boldsymbol p}', s')\overline{\sigma}^\mu x(\vec{\boldsymbol p}, s) + (\Sigma_R(m^2)+\mathcal{D})y(\vec{\boldsymbol p}', s')\sigma^\mu y^\dagger(\vec{\boldsymbol p}, s)\}\,, \qquad (19.226)$$

where $\mathcal{D} \equiv \mathcal{D}(m^2)$ is defined in eq. (19.31). It then follows that

$$\begin{aligned} F_4(q^2)_{\text{self-energies}} &= \frac{m^2}{q^2(4m^2-q^2)}\sum_{s,s'}\big[y(\vec{\boldsymbol p}, s)\sigma_\mu y^\dagger(\vec{\boldsymbol p}', s') - x^\dagger(\vec{\boldsymbol p}, s)\overline{\sigma}_\mu x(\vec{\boldsymbol p}', s')\big] \\ &\quad \times\{(\Sigma_L(m^2)+\mathcal{D})x^\dagger(\vec{\boldsymbol p}', s')\overline{\sigma}^\mu x(\vec{\boldsymbol p}, s) + (\Sigma_R(m^2)+\mathcal{D})y(\vec{\boldsymbol p}', s')\sigma^\mu y^\dagger(\vec{\boldsymbol p}, s)\} \\ &= \frac{m^2}{q^2(4m^2-q^2)}\Big\{(\Sigma_L(m^2)+\mathcal{D})\big[m^2\,\mathrm{Tr}(\sigma_\mu\overline{\sigma}^\mu) - \mathrm{Tr}(\overline{\sigma}_\mu\, p'\!\cdot\!\sigma\,\overline{\sigma}^\mu\, p\!\cdot\!\sigma)\big] \\ &\qquad - (\Sigma_R(m^2)+\mathcal{D})\big[m^2\,\mathrm{Tr}(\overline{\sigma}_\mu\sigma^\mu) - \mathrm{Tr}(\sigma_\mu\, p'\!\cdot\!\overline{\sigma}\,\sigma^\mu\, p\!\cdot\!\overline{\sigma})\big]\Big\} \\ &= \frac{m^2}{q^2(4m^2-q^2)}\big[4m^2(d-1) - q^2(d-2)\big]\big(\Sigma_L(m^2) - \Sigma_R(m^2)\big)\,. \end{aligned} \qquad (19.227)$$

The evaluation of the self-energy functions is provided in Section 19.1. Using eqs. (19.27) and (19.29), it follows that

$$\Sigma_L(m^2) - \Sigma_R(m^2) = \frac{\alpha}{2\pi}\big[B_1(m^2; M^2, m_{\tilde L}^2) - B_1(m^2; M^2, m_{\tilde R}^2)\big]\,. \qquad (19.228)$$

Hence we end up with

$$\begin{aligned} F_4(q^2)_{\text{self-energies}} &= \frac{\alpha m^2}{2\pi q^2(4m^2-q^2)}\big[4m^2(d-1) - q^2(d-2)\big] \\ &\quad \times\big[B_1(m^2; M^2, m_{\tilde L}^2) - B_1(m^2; M^2, m_{\tilde R}^2)\big]\,. \end{aligned} \qquad (19.229)$$

Adding the contributions of eqs. (19.223) and (19.229), we arrive at

$$\begin{aligned} F_4(q^2) = \frac{\alpha m^2}{\pi q^2(4m^2-q^2)}\Big\{ & m^2q^2\big[C_{21} + 4C_{22} - 4C_{23}\big] \\ & + \big[4m^2(d-1) - q^2(d-2)\big]\big[C_{24} + \tfrac{1}{2}B_1(m^2; M^2, m_{\tilde L}^2)\big] - (L\to R)\Big\}, \end{aligned} \qquad (19.230)$$

where the arguments of the C-functions are $(m^2, q^2, m^2; M^2, m_{\tilde L}^2, m_{\tilde L}^2)$ after making use of the on-shell conditions, $p^2 = p'^2 = m^2$. The above result simplifies significantly using the two identities

$$C_{21}(m^2, q^2, m^2; M^2, m_{\tilde L}^2, m_{\tilde L}^2) = 2C_{23}(m^2, q^2, m^2; M^2, m_{\tilde L}^2, m_{\tilde L}^2)\,, \qquad (19.231)$$

$$\begin{aligned} 2C_{24}(m^2, q^2, m^2; M^2, m_{\tilde L}^2, m_{\tilde L}^2) &= -B_1(m^2; M^2, m_{\tilde L}^2) \\ &\quad + q^2\big[C_{23}(m^2, q^2, m^2; M^2, m_{\tilde L}^2, m_{\tilde L}^2) - 2C_{22}(m^2, q^2, m^2; M^2, m_{\tilde L}^2, m_{\tilde L}^2)\big], \end{aligned} \qquad (19.232)$$

which are derived in Exercise K.10. Indeed, the apparent singularities at $q^2 = 0$

and $q^2 = 4m^2$ are canceled. Moreover, the end result is manifestly finite, so we may set $d = 4$ in our final expression:

$$F_4(q^2) = \frac{\alpha m^2}{\pi}\Big\{C_{23}(m^2, q^2, m^2; M^2, m_{\tilde{L}}^2, m_{\tilde{L}}^2) - 2C_{22}(m^2, q^2, m^2; M^2, m_{\tilde{L}}^2, m_{\tilde{L}}^2) - (L \to R)\Big\}. \tag{19.233}$$

Using the integral representations of the C-functions given in eqs. (K.73) and (K.74) and assuming that $m \ll m_{L,R}$, it follows that

$$F_4(0) = \frac{\alpha m^2}{6\pi}\int_0^1 \left(\frac{1}{(m_{\tilde{L}}^2 - M^2)x + M^2} - \frac{1}{(m_{\tilde{R}}^2 - M^2)x + M^2}\right) x^3\, dx\,. \tag{19.234}$$

The static anapole moment of the muon is given by eq. (19.163). As an example, in the limit of $M = 0$,

$$F_4(0) = \frac{\alpha m^2}{18\pi}\left(\frac{1}{m_{\tilde{L}}^2} - \frac{1}{m_{\tilde{R}}^2}\right). \tag{19.235}$$

19.7 One-Loop MSSM Contributions to $g-2$ of the Muon

In the Standard Model, the theoretical predication for the anomalous magnetic moment of the muon, a_μ [see eq. (19.161)], is known to extraordinary precision. Thus, the experimentally measured value provides a very sensitive test of the Standard Model. In light of a long-standing discrepancy between the SM prediction and the experimentally determined value, it is tempting to invoke new physics beyond the Standard Model to account for the observed deviation. Even if the discrepancy fades away with more sensitive experiments, it is noteworthy that unexcluded regions of the MSSM parameter space exist that yield deviations in a_μ from its Standard Model value that can be detected in future experimental studies.

As of this writing, there exists an apparent 4.2σ discrepancy [202] between the Standard Model prediction and the average of the measurements of the anomalous magnetic moment of the muon from experiments at Brookhaven and Fermilab. Unlike many other such discrepancies, this one has lasted a long time and has so far successfully resisted attempts to explain it in conventional ways. It is possible that explaining it will require new physics beyond the Standard Model; even if this proves not to be true, the measurement can be used to put nontrivial constraints on supersymmetric models that are complementary to other bounds.

Therefore, in this section we review the one-loop supersymmetric contributions to a_μ in the MSSM [203]. These come about from diagrams with loops with a chargino and a muon sneutrino, and those with loops with a neutralino and a smuon. Summations are performed over all of the chargino, neutralino, and smuon mass eigenstates. The one-loop superpartner contributions to a_μ, including all effects of mixing and possible complex phases, are

$$\Delta a_\mu^{\widetilde{C}} = \frac{m_\mu}{16\pi^2}\sum_i\Big[\frac{m_\mu}{12m^2_{\widetilde{\nu}_\mu}}(|c_i^L|^2+|c_i^R|^2)F_1^C(x_i)+\frac{2m_{\widetilde{C}_i}}{3m^2_{\widetilde{\nu}_\mu}}\,\mathrm{Re}[c_i^L c_i^R]F_2^C(x_i)\Big], \tag{19.236}$$

$$\Delta a_\mu^{\widetilde{N}} = \frac{m_\mu}{16\pi^2}\sum_{j,k}\Big[-\frac{m_\mu}{12m^2_{\widetilde{\mu}_k}}(|n_{jk}^L|^2+|n_{jk}^R|^2)F_1^N(x_{jk})+\frac{m_{\widetilde{N}_j}}{3m^2_{\widetilde{\mu}_k}}\,\mathrm{Re}[n_{jk}^L n_{jk}^R]F_2^N(x_{jk})\Big], \tag{19.237}$$

where $i=1,2$ and $j=1,2,3,4$ and $k=1,2$ are chargino, neutralino, and smuon mass-eigenstate labels, respectively. The kinematic loop functions depend on the variables $x_i = m^2_{\widetilde{C}_i}/m^2_{\widetilde{\nu}_\mu}$ and $x_{jk} = m^2_{\widetilde{N}_j}/m^2_{\widetilde{\mu}_k}$, and are given by

$$F_1^C(x) = \frac{2}{(1-x)^4}\left[2+3x-6x^2+x^3+6x\ln x\right], \tag{19.238}$$

$$F_2^C(x) = -\frac{3}{2(1-x)^3}\left[3-4x+x^2+2\ln x\right], \tag{19.239}$$

$$F_1^N(x) = \frac{2}{(1-x)^4}\left[1-6x+3x^2+2x^3-6x^2\ln x\right], \tag{19.240}$$

$$F_2^N(x) = \frac{3}{(1-x)^3}\left[1-x^2+2x\ln x\right]. \tag{19.241}$$

They are normalized so that, for degenerate superpartners, $F_1^C(1) = F_2^C(1) = 1$ and $F_1^N(1) = F_2^N(1) = 1$. Also, the pertinent chargino and neutralino couplings, which can be obtained by specializing the results of Figs. J.6 and J.7, are

$$c_i^R = y_\mu \mathbf{U}_{i2}, \tag{19.242}$$

$$c_i^L = -g\mathbf{V}_{i1}, \tag{19.243}$$

$$n_{jk}^R = \sqrt{2}g'\mathbf{N}_{j1}\mathbf{X}_{k2} + y_\mu \mathbf{N}_{j3}\mathbf{X}_{k1}, \tag{19.244}$$

$$n_{jk}^L = \frac{1}{\sqrt{2}}(g\mathbf{N}_{j2}+g'\mathbf{N}_{j1})\mathbf{X}^*_{k1} - y_\mu\mathbf{N}_{j3}\mathbf{X}^*_{k2}. \tag{19.245}$$

Here, $y_\mu = gm_\mu/\sqrt{2}m_W\cos\beta$ is the muon Yukawa coupling, and $\mathbf{N}$ and $\mathbf{U}$ and $\mathbf{V}$ are the unitary neutralino and chargino mixing matrices defined in eqs. (13.138) and (13.150). The unitary smuon-mixing matrix $\mathbf{X}$ is defined by

$$\mathbf{X}M^2_{\widetilde{\mu}}\mathbf{X}^\dagger = \mathrm{diag}\,(m^2_{\widetilde{\mu}_1}, m^2_{\widetilde{\mu}_2}), \tag{19.246}$$

where the smuon squared-mass matrix in the gauge-eigenstate basis $(\widetilde{\mu}_L, \widetilde{\mu}_R)$ is given by [see eq. (13.183)][11]

$$M^2_{\widetilde{\mu}} = \begin{pmatrix} m_L^2 + (s_W^2 - \frac{1}{2})m_Z^2\cos 2\beta & m_\mu(A_\mu^* - \mu\tan\beta) \\ m_\mu(A_\mu - \mu^*\tan\beta) & m_R^2 - s_W^2\, m_Z^2\cos 2\beta \end{pmatrix}. \tag{19.247}$$

[11] In eq. (19.247), we employ the notation $m_\mu A_\mu$ in place of a_μ (see footnote 5 of Section 8.6), since in this section we have denoted $a_\mu = \frac{1}{2}(g-2)$ of the muon.

The muon sneutrino mass is given by $m^2_{\tilde{\nu}_\mu} = m^2_{\tilde{L}} + \frac{1}{2}m^2_Z \cos 2\beta$ [see eq. (13.184)].

The MSSM correction $\Delta a^{\rm SUSY}_\mu = \Delta a^{\widetilde{C}}_\mu + \Delta a^{\widetilde{N}}_\mu$ has an enhancement for large $\tan\beta$, as can be seen by numerical evaluation. One can also see this analytically by expanding all of the chargino, neutralino, and smuon mass eigenvalues and mixing matrices above in small m_W, m_Z compared to the superpartner masses; a particularly simple analytic approximation can be obtained by taking the further limit $|\mu|^2, |M_1|^2, |M_2|^2, m^2_{\tilde{L}}, m^2_{\tilde{R}} = M^2_{\rm SUSY}$ so that the superpartners are nearly degenerate. This leads to

$$\Delta a^{\rm SUSY}_\mu = \frac{\tan\beta}{192\pi^2}\frac{m^2_\mu}{M^2_{\rm SUSY}}(5g^2 + g'^2) = 14\tan\beta\left(\frac{100\ {\rm GeV}}{M_{\rm SUSY}}\right)^2 10^{-10} \qquad (19.248)$$

in the large $\tan\beta$ limit, with the chargino diagram contribution positive and dominating the negative neutralino diagram contribution; they are respectively proportional to $6g^2$ and to $g'^2 - g^2$. The large $\tan\beta$ scaling can be understood as follows: a_μ requires a muon chirality flip, which costs a suppression of m_μ if it comes from a muon line. However, the chirality flip can be performed instead by the internal chargino or neutralino line, with the result proportional to the muon Yukawa coupling y_μ to the higgsino component of the chargino or neutralino, as seen in eq. (19.242) for example, leading to an enhancement $y_\mu \propto m_\mu \tan\beta$ at large $\tan\beta$. One can check that the contributions proportional to $\tan\beta$ come from the terms proportional to $m_{\widetilde{C}_i}\,{\rm Re}[c^L_i c^R_i]$ and $m_{\widetilde{N}_j}\,{\rm Re}[n^L_{jk} n^R_{jk}]$ in eqs. (19.236) and (19.237).

Another interesting limit occurs if $M_1 \ll M_2, \mu$, so that only loops containing a light bino-like neutralino and the smuons are important. In that limit,

$$\Delta a^{\rm light\ bino}_\mu = \frac{g'^2}{48\pi^2}\frac{m^2_\mu M_1\,{\rm Re}[\mu\tan\beta - A^*_\mu]}{m^2_{\tilde{\mu}_2} - m^2_{\tilde{\mu}_1}}\left[\frac{F^N_2(x_{11})}{m^2_{\tilde{\mu}_1}} - \frac{F^N_2(x_{12})}{m^2_{\tilde{\mu}_2}}\right], \qquad (19.249)$$

where $x_{1k} = M_1^2/m^2_{\tilde{\mu}_k}$ for $k = 1, 2$. Note that eq. (19.249) has a smooth limit as the sleptons become degenerate. This can yield a significant contribution in the case where all neutralinos and charginos except the light bino become heavy, if $\mu\tan\beta$ is large. For example, if $m_{\tilde{\mu}_1} \approx m_{\tilde{\mu}_2} = 2M_1$, then eq. (19.249) yields

$$\Delta a^{\rm light\ bino}_\mu \simeq 18\tan\beta\left(\frac{100\ {\rm GeV}}{m_{\tilde{\mu}}}\right)^3\left(\frac{\mu - A_\mu\cot\beta}{1000\ {\rm GeV}}\right)10^{-10}. \qquad (19.250)$$

This formula will eventually fail to be accurate for extremely large $\mu\tan\beta$, in accord with decoupling, since then $m_{\tilde{\mu}_1} \approx m_{\tilde{\mu}_2}$ must fail badly.

The two-loop leading log contribution yields a suppression given by

$$\Delta a^{\rm SUSY}_{\mu,\,2\text{-loop}} = \Delta a^{\rm SUSY}_{\mu,\,1\text{-loop}}\left(1 - \frac{4\alpha}{\pi}\ln\frac{M_{\rm SUSY}}{m_\mu}\right), \qquad (19.251)$$

where $M_{\rm SUSY}$ is a typical superpartner mass. The logarithmic suppression factor varies between about 7% and 9% when the superpartner masses are small enough to give contributions large enough to worry about.

19.8 One-Loop Corrected MSSM Higgs Masses

In this section we employ Feynman diagrammatic techniques to derive formulae for the one-loop radiatively corrected Higgs boson masses of the MSSM.[12]

We begin by considering the tree-level scalar potential for the neutral Higgs scalar fields in the MSSM given in eq. (13.78). The neutral components of the scalar Higgs fields are expanded around their VEVs:

$$H^0_{d,u} \equiv \frac{h_{d,u} + i a_{d,u} + v_{d,u}}{\sqrt{2}}\,, \tag{19.252}$$

where $v^2 \equiv v_d^2 + v_u^2 \simeq (246~\text{GeV})^2$. Here v_d [v_u] are the VEVs of the neutral Higgs fields that couple exclusively to the down-type [up-type] quark and lepton fields, respectively. Without loss of generality, one can assume that v_d and v_u are real and nonnegative (after rephasing the fields H_d and H_u as necessary). Plugging eq. (19.252) into eq. (13.78) yields

$$V = V_0 + t_d h_d + t_u h_u + \tfrac{1}{2}(\mathcal{M}_e^2)_{ij} h_i h_j + \tfrac{1}{2}(\mathcal{M}_o^2)_{ij} a_i a_j + \cdots\,, \tag{19.253}$$

where repeated indices $i, j = d, u$ are summed over, and terms that are cubic or quartic in the scalar fields are not explicitly shown.

Explicitly, the linear (tadpole) terms in the scalar potential are given by

$$t_d \equiv \left.\frac{\partial V}{\partial h_d}\right|_{h=a=0} = v_d\left(M_d^2 + \tfrac{1}{8}G^2(v_d^2 - v_u^2) - b\frac{v_u}{v_d}\right), \tag{19.254}$$

$$t_u \equiv \left.\frac{\partial V}{\partial h_u}\right|_{h=a=0} = v_u\left(M_u^2 + \tfrac{1}{8}G^2(v_u^2 - v_d^2) - b\frac{v_d}{v_u}\right), \tag{19.255}$$

where

$$G^2 \equiv g^2 + g'^2\,, \qquad M_d^2 \equiv m^2_{H_d} + |\mu|^2\,, \qquad M_u^2 \equiv m^2_{H_u} + |\mu|^2\,. \tag{19.256}$$

Likewise, the quadratic terms in the scalar fields yield 2×2 CP-even and CP-odd scalar squared-mass matrices [in the (h_d, h_u) basis]:

$$\mathcal{M}_e^2 \equiv \left.\frac{\partial^2 V}{\partial h_i \partial h_j}\right|_{h=a=0} = \begin{pmatrix} M_d^2 + \frac{1}{8}G^2(3v_d^2 - v_u^2) & -\frac{1}{4}G^2 v_u v_d - b \\ -\frac{1}{4}G^2 v_u v_d - b & M_u^2 + \frac{1}{8}G^2(3v_u^2 - v_d^2) \end{pmatrix}, \tag{19.257}$$

$$\mathcal{M}_o^2 \equiv \left.\frac{\partial^2 V}{\partial a_i \partial a_j}\right|_{h=a=0} = \begin{pmatrix} M_d^2 + \frac{1}{8}G^2(v_d^2 - v_u^2) & b \\ b & M_u^2 + \frac{1}{8}G^2(v_u^2 - v_d^2) \end{pmatrix}. \tag{19.258}$$

All parameters appearing in the above formulae should be interpreted as bare (unrenormalized) parameters.

[12] In practice, two-loop (and even some three-loop) contributions to the MSSM Higgs boson masses are numerically significant. Moreover, if the supersymmetry-breaking scale is large, then renormalization group improvement is required to obtain accurate numerical results [156, 157].

Fig. 19.15 The sum of all one-loop tadpole graphs at zero external momentum contributing to the one-point 1PI Green function is denoted by $-iT_\phi$.

We ensure that $v_{u,d}$ are stationary points of the full one-loop effective potential by enforcing the tadpole cancellation conditions exhibited in Fig. 19.8,

$$-i(t_{d,u} + T_{d,u}) = 0\,, \tag{19.259}$$

where $t_{u,d}$ are functions of the bare parameters given in eqs. (19.254) and (19.255) and $-iT_{d,u}$ consist of the sum of all Feynman diagrams contributing to the one-point 1PI Green functions of h_d and h_u, respectively (see Fig. 19.15). For simplicity, we take the gaugino mass parameters, the μ parameter, and the A-terms to be real, thus neglecting potential CP-violating effects that could arise from CP-violating parameters in the sparticle sector. Under this assumption, there is no mixing at one loop between CP-even and CP-odd Higgs scalar eigenstates, and we can treat the analysis of the CP-even and CP-odd scalar squared-mass matrices separately.

Using eq. (19.259), the CP-odd scalar squared-mass matrix simplifies to

$$\mathcal{M}_o^2 = \begin{pmatrix} b\dfrac{v_u}{v_d} - \dfrac{T_d}{v_d} & b \\ b & b\dfrac{v_d}{v_u} - \dfrac{T_u}{v_u} \end{pmatrix}. \tag{19.260}$$

Diagonalizing this matrix and expanding to leading order in $T_{u,d}$, the bare masses for the CP-odd scalar A and the Goldstone boson G are found:

$$m_A^2 = \frac{v^2}{v_u v_d}b - \frac{v_u^2}{v^2}\frac{T_d}{v_d} - \frac{v_d^2}{v^2}\frac{T_u}{v_u}\,, \qquad m_G^2 = -\frac{1}{v^2}\left(T_d v_d + T_u v_u\right). \tag{19.261}$$

Solving for b, M_d^2 and M_u^2 and making use of eq. (19.259) yields

$$b = \left(\frac{v_u v_d}{v^2}\right) m_A^2 + \left(\frac{v_u}{v}\right)^4 \frac{T_d}{v_u} + \left(\frac{v_d}{v}\right)^4 \frac{T_u}{v_d}\,, \tag{19.262}$$

$$M_d^2 = \left(\frac{v_u}{v}\right)^2 m_A^2 + \left[\left(\frac{v_u}{v}\right)^4 - 1\right]\frac{T_d}{v_d} + \left(\frac{v_d v_u}{v^2}\right)^2 \frac{T_u}{v_u} + \tfrac{1}{8}G^2(v_u^2 - v_d^2)\,, \tag{19.263}$$

$$M_u^2 = \left(\frac{v_d}{v}\right)^2 m_A^2 + \left(\frac{v_u v_d}{v^2}\right)^2 \frac{T_d}{v_d} + \left[\left(\frac{v_d}{v}\right)^4 - 1\right]\frac{T_u}{v_u} - \tfrac{1}{8}G^2(v_u^2 - v_d^2)\,. \tag{19.264}$$

Inserting these results into $\mathcal{M}_e^2$, we obtain

$$\mathcal{M}_e^2 = \begin{pmatrix} \mathcal{M}_{dd}^2 & \mathcal{M}_{du}^2 \\ \mathcal{M}_{du}^2 & \mathcal{M}_{uu}^2 \end{pmatrix}, \tag{19.265}$$

where

$$\mathcal{M}^2_{dd} = m_A^2 s_\beta^2 + m_Z^2 c_\beta^2 + \frac{T_d}{v_d}(s_\beta^4 - 1) + \frac{T_u}{v_u} s_\beta^2 c_\beta^2 \,, \tag{19.266}$$

$$\mathcal{M}^2_{uu} = m_A^2 c_\beta^2 + m_Z^2 s_\beta^2 + \frac{T_d}{v_d} s_\beta^2 c_\beta^2 + \frac{T_u}{v_u}(c_\beta^4 - 1) \,, \tag{19.267}$$

$$\mathcal{M}^2_{du} = -(m_A^2 + m_Z^2) s_\beta c_\beta - \frac{T_u}{v_u} c_\beta^3 s_\beta - \frac{T_d}{v_d} s_\beta^3 c_\beta \,, \tag{19.268}$$

with $m_Z^2 \equiv \frac{1}{4} G^2 v^2$ and

$$s_\beta \equiv \sin\beta = \frac{v_u}{v}, \qquad c_\beta \equiv \cos\beta = \frac{v_d}{v} \,. \tag{19.269}$$

The eigenvalues of the squared-mass matrix $\mathcal{M}^2_e$ correspond to the bare squared masses, m_H^2 and m_h^2, for the heavier and lighter of the two neutral CP-even Higgs bosons H and h, respectively, where

$$m^2_{H,h} = \frac{1}{2}\left(\mathcal{M}^2_{dd} + \mathcal{M}^2_{uu} \pm \sqrt{(\mathcal{M}^2_{dd} - \mathcal{M}^2_{uu})^2 + 4\left[\mathcal{M}^2_{du}\right]^2}\right) . \tag{19.270}$$

It is noteworthy that the tree-level sum rule,

$$\operatorname{Tr} \mathcal{M}^2_e = m_Z^2 + \operatorname{Tr} \mathcal{M}^2_o \,, \tag{19.271}$$

still holds when $v_{u,d}$ are stationary points of the full one-loop effective potential. In particular, one can check that

$$m_h^2 + m_H^2 = m_Z^2 + m_A^2 + m_G^2 \,, \tag{19.272}$$

where $m_h^2 + m_H^2 = \mathcal{M}^2_{dd} + \mathcal{M}^2_{uu}$ and m_G^2 is given by eq. (19.261).

We can extend the above analysis to include the charged Higgs boson and Goldstone boson fields. Starting from eq. (13.77), one can identify the terms of the scalar potential that are quadratic in the charged scalar fields by replacing $H^0_{d,u}$ with their vacuum expectation values, $\langle H^0_{d,u}\rangle = v_{d,u}/\sqrt{2}$:

$$V \supset (\mathcal{M}^2_\pm)_{ij} H_i^+ H_j^- \,, \tag{19.273}$$

where repeated indices $i, j = d, u$ are summed over and

$$\mathcal{M}^2_\pm = \begin{pmatrix} M_d^2 + \frac{1}{4} g^2 v_u^2 + \frac{1}{8} G^2 (v_d^2 - v_u^2) & b + \frac{1}{4} g^2 v_u v_d \\ b + \frac{1}{4} g^2 v_u v_d & M_u^2 + \frac{1}{4} g^2 v_d^2 + \frac{1}{8} G^2 (v_u^2 - v_d^2) \end{pmatrix} . \tag{19.274}$$

We can eliminate M_d^2 and M_u^2 via eqs. (19.254) and (19.255). After employing eq. (19.259), we end up with

$$\mathcal{M}^2_\pm = \begin{pmatrix} \left(b + \frac{1}{4} g^2 v_u v_d\right) \dfrac{v_u}{v_d} - \dfrac{T_d}{v_d} & b + \frac{1}{4} g^2 v_u v_d \\ b + \frac{1}{4} g^2 v_u v_d & \left(b + \frac{1}{4} g^2 v_u v_d\right) \dfrac{v_d}{v_u} - \dfrac{T_u}{v_u} \end{pmatrix} . \tag{19.275}$$

Comparing with eq. (19.261), it immediately follows that

$$m^2_{H^\pm} = m_A^2 + m_W^2 \,, \qquad m^2_{G^\pm} = m_G^2 \,, \tag{19.276}$$

after using $m_W^2 = \frac{1}{4}g^2v^2$.

It is convenient to replace the bare masses (denoted by a lower case m) by physical masses (denoted by an upper case M) in the one-loop approximation:

$$m_\phi^2 = M_\phi^2 - \mathrm{Re}\,\Sigma_{\phi\phi}(M_\phi^2)\,, \quad \text{for } \phi = h, H, A, H^\pm\,, \tag{19.277}$$

$$m_V^2 = M_V^2 - \mathrm{Re}\,A_{VV}(M_V^2)\,, \quad \text{for } V = W^\pm, Z\,, \tag{19.278}$$

following the notation and conventions discussed in Section 2.9.1.

The physical Higgs masses are gauge invariant quantities. Nevertheless, if we work in the R_ξ gauge in evaluating self-energy functions that involve loops containing gauge bosons and/or Goldstone bosons, the gauge parameter ξ will appear at intermediate stages of the calculation. At one-loop accuracy, all factors of ξ will eventually cancel in the final expression for the physical squared masses of the Higgs bosons. It is convenient to work in the Landau gauge where $\xi = 0$ and the Goldstone boson pole masses are zero. Thus, applying eq. (19.277) to $\phi = G, G^\pm$ with $M_G = M_{G^\pm} = 0$, it follows that[13]

$$m_G^2 = M_G^2 - \Sigma_{GG}(0) = -\Sigma_{GG}(0)\,, \tag{19.279}$$

$$m_{G^\pm}^2 = M_{G^\pm}^2 - \Sigma_{G^+G^-}(0) = -\Sigma_{G^+G^-}(0)\,. \tag{19.280}$$

In light of eqs. (19.261), (19.269), and (19.276),

$$\Sigma_{GG}(0) = \Sigma_{G^+G^-}(0) = \frac{T_d c_\beta + T_u s_\beta}{v}\,. \tag{19.281}$$

The replacements of eqs. (19.277) and (19.278) sidestep the need to introduce squared-mass counterterms in the calculation of the physical Higgs masses. Employing eqs. (19.277) and (19.278) in eq. (19.276) and working to one-loop accuracy,

$$M_{H^\pm}^2 = M_W^2 + M_A^2 + \mathrm{Re}\,\Sigma_{H^+H^-}(M_W^2 + M_A^2) - \mathrm{Re}\,A_{WW}(M_W^2) - \mathrm{Re}\,\Sigma_{AA}(M_A^2)\,, \tag{19.282}$$

since $\Sigma_{H^+H^-}(M_W^2 + M_A^2)$ differs from $\Sigma_{H^+H^-}(M_{H^\pm}^2)$ by terms of two-loop order in perturbation theory. To complete the computation, one must explicitly evaluate the contributions of the MSSM particle spectrum to the three one-loop self-energy functions that appear in eq. (19.282).

We now turn to the computation of the radiatively corrected squared masses of the CP-even neutral Higgs bosons. In contrast to eq. (19.276), the tree-level expressions for the squared masses of the CP-even neutral Higgs bosons depend on $\tan\beta$. Consequently, the counterterms associated with the parameters v_u and v_d, which are divergent because they have been fixed to the expectation values of the bare fields $H_{u,d}$, are now relevant.

In light of eq. (2.286), the renormalized VEVs are given by

$$v_{d,r} = Z_{H_d}^{-1/2} v_d = v_d\big(1 - \tfrac{1}{2}\delta Z_{H_d}\big)\,, \qquad v_{u,r} = Z_{H_u}^{-1/2} v_u = v_u\big(1 - \tfrac{1}{2}\delta Z_{H_u}\big)\,, \tag{19.283}$$

[13] Note that the absorptive parts of $\Sigma_{GG}(0)$ and $\Sigma_{G^+G^-}(0)$ are zero. Thus in the CP-conserving limit, $\Sigma_{GG}(0)$ and $\Sigma_{G^+G^-}(0)$ are both real quantities.

at one-loop accuracy. The counterterms for the VEVs are defined by

$$\delta v_d \equiv v_{d,r} - v_d = -\tfrac{1}{2} v_d \delta Z_{H_d}\,, \qquad \delta v_u \equiv v_{u,r} - v_u = -\tfrac{1}{2} v_u \delta Z_{H_u}\,. \quad (19.284)$$

The neutral Higgs masses depend on the bare parameter $\tan\beta \equiv v_u/v_d$, which can be replaced by a renormalized parameter and a counterterm,

$$\tan\beta \to \tan\beta - \delta\tan\beta\,, \quad (19.285)$$

where

$$\frac{\delta\tan\beta}{\tan\beta} = \frac{\delta v_u}{v_u} - \frac{\delta v_d}{v_d} = \tfrac{1}{2}\left(\delta Z_{H_d} - \delta Z_{H_u}\right). \quad (19.286)$$

Likewise, we can express the shifts of the parameters s_β and c_β in terms of $\delta\tan\beta$:

$$s_\beta \to s_\beta - \delta s_\beta = s_\beta - c_\beta^3 \delta\tan\beta\,, \quad (19.287)$$

$$c_\beta \to c_\beta - \delta c_\beta = c_\beta + c_\beta^2 s_\beta \delta\tan\beta\,. \quad (19.288)$$

Making the substitutions of eqs. (19.277), (19.278), (19.287), (19.288),

$$\mathcal{M}^2_{dd} = M_A^2 s_\beta^2 + M_Z^2 c_\beta^2 + \delta\mathcal{M}^2_{dd}\,, \quad (19.289)$$

$$\mathcal{M}^2_{uu} = M_A^2 c_\beta^2 + M_Z^2 s_\beta^2 + \delta\mathcal{M}^2_{uu}\,, \quad (19.290)$$

$$\mathcal{M}^2_{du} = -(M_A^2 + M_Z^2) s_\beta c_\beta + \delta\mathcal{M}^2_{du}\,, \quad (19.291)$$

where β is the one-loop renormalized parameter and

$$\delta\mathcal{M}^2_{dd} = -\mathrm{Re}\,\Sigma_{AA}(M_A^2) s_\beta^2 - \mathrm{Re}\,A_{ZZ}(M_Z^2) c_\beta^2 + \frac{T_d}{v_d}(s_\beta^4 - 1) + \frac{T_u}{v_u} s_\beta^2 c_\beta^2 - 2 s_\beta c_\beta^3 (M_A^2 - M_Z^2)\delta\tan\beta\,, \quad (19.292)$$

$$\delta\mathcal{M}^2_{uu} = -\mathrm{Re}\,\Sigma_{AA}(M_A^2) c_\beta^2 - \mathrm{Re}\,A_{ZZ}(M_Z^2) s_\beta^2 + \frac{T_d}{v_d} s_\beta^2 c_\beta^2 + \frac{T_u}{v_u}(c_\beta^4 - 1) + 2 s_\beta c_\beta^3 (M_A^2 - M_Z^2)\delta\tan\beta\,, \quad (19.293)$$

$$\delta\mathcal{M}^2_{du} = \left[\mathrm{Re}\,\Sigma_{AA}(M_A^2) + \mathrm{Re}\,A_{ZZ}(M_Z^2)\right] s_\beta c_\beta - \frac{T_d}{v_d} s_\beta^3 c_\beta - \frac{T_u}{v_u} c_\beta^3 s_\beta + (M_A^2 + M_Z^2) c_\beta^2 c_{2\beta}\delta\tan\beta\,. \quad (19.294)$$

Using eqs. (19.289)–(19.291) we can now perturbatively expand eq. (19.270) at one-loop accuracy and rewrite the bare squared-mass parameters in terms of physical (renormalized) parameters, following eq. (19.277); i.e., $m_H^2 = M_H^2 - \mathrm{Re}\,\Sigma_{HH}(M_H^2)$ and $m_h^2 = M_h^2 - \mathrm{Re}\,\Sigma_{hh}(M_h^2)$, respectively. In particular,

$$M_H^2 - \mathrm{Re}\,\Sigma_{HH}(\widehat{M}_H^2) = \widehat{M}_H^2 + \tfrac{1}{2}\left(\delta\mathcal{M}^2_{dd} + \delta\mathcal{M}^2_{uu}\right) + \frac{(M_Z^2 - M_A^2)c_{2\beta}(\delta\mathcal{M}^2_{dd} - \delta\mathcal{M}^2_{uu}) - 2(M_Z^2 + M_A^2)s_{2\beta}\delta\mathcal{M}^2_{du}}{2(\widehat{M}_H^2 - \widehat{M}_h^2)}\,, \quad (19.295)$$

$$M_h^2 - \mathrm{Re}\,\Sigma_{hh}(\widehat{M}_h^2) = \widehat{M}_h^2 + \tfrac{1}{2}\left(\delta\mathcal{M}^2_{dd} + \delta\mathcal{M}^2_{uu}\right) - \frac{(M_Z^2 - M_A^2)c_{2\beta}(\delta\mathcal{M}^2_{dd} - \delta\mathcal{M}^2_{uu}) - 2(M_Z^2 + M_A^2)s_{2\beta}\delta\mathcal{M}^2_{du}}{2(\widehat{M}_H^2 - \widehat{M}_h^2)}\,, \quad (19.296)$$

where $s_{2\beta} \equiv \sin 2\beta$, $c_{2\beta} \equiv \cos 2\beta$, and

$$\widehat{M}^2_{H,h} \equiv \frac{1}{2}\left(M_Z^2 + M_A^2 \pm \sqrt{(M_A^2 - M_Z^2)^2 + 4M_A^2 M_Z^2 s_{2\beta}^2}\right). \qquad (19.297)$$

We recognize $\widehat{M}^2_{H,h}$ as the eigenvalues of the tree-level CP-even Higgs boson squared-mass matrix with the bare parameters m_A, m_Z, and β replaced by the corresponding physical (renormalized) masses M_A and M_Z and the one-loop renormalized parameter β. One can also employ this squared-mass matrix to define the mixing angle α, which can be expressed in terms of M_A^2, M_Z^2, and the renormalized parameter β. In light of eqs. (6.235), (6.237), and (6.238), it follows that

$$\cos 2\alpha = \frac{(M_Z^2 - M_A^2)c_{2\beta}}{\widehat{M}^2_H - \widehat{M}^2_h}, \qquad \sin 2\alpha = \frac{-(M_Z^2 + M_A^2)s_{2\beta}}{\widehat{M}^2_H - \widehat{M}^2_h}. \qquad (19.298)$$

Using eq. (19.298), one can immediately derive the following useful identity:

$$M_A^2 \sin[2(\beta - \alpha)] = -M_Z^2 \sin[2(\beta + \alpha)]. \qquad (19.299)$$

One can now rewrite eqs. (19.295) and (19.296) as

$$M_H^2 = \widehat{M}^2_H + \operatorname{Re}\Sigma_{HH}(\widehat{M}^2_H) + \delta\mathcal{M}^2_{dd}\cos^2\alpha + \delta\mathcal{M}^2_{uu}\sin^2\alpha + \delta\mathcal{M}^2_{du}\sin 2\alpha\,, \qquad (19.300)$$

$$M_h^2 = \widehat{M}^2_h + \operatorname{Re}\Sigma_{hh}(\widehat{M}^2_h) + \delta\mathcal{M}^2_{dd}\sin^2\alpha + \delta\mathcal{M}^2_{uu}\cos^2\alpha - \delta\mathcal{M}^2_{du}\sin 2\alpha\,. \qquad (19.301)$$

Plugging eqs. (19.292)–(19.294) into the above equations, we obtain

$$\begin{aligned} M_H^2 = \widehat{M}^2_H &+ \operatorname{Re}\Sigma_{HH}(\widehat{M}^2_H) - \cos^2(\beta+\alpha)\operatorname{Re}A_{ZZ}(M_Z^2) - s^2_{\beta-\alpha}\operatorname{Re}\Sigma_{AA}(M_A^2) \\ &+ \frac{T_d}{v_d}[s_\beta^2 s^2_{\beta-\alpha} - \cos^2\alpha] + \frac{T_u}{v_u}[c_\beta^2 s^2_{\beta-\alpha} - \sin^2\alpha] + 2m_Z^2 c_\beta^2 \sin[2(\beta+\alpha)]\delta\tan\beta\,, \end{aligned} \qquad (19.302)$$

$$\begin{aligned} M_h^2 = \widehat{M}^2_h &+ \operatorname{Re}\Sigma_{hh}(\widehat{M}^2_h) - \sin^2(\beta+\alpha)\operatorname{Re}A_{ZZ}(M_Z^2) - c^2_{\beta-\alpha}\operatorname{Re}\Sigma_{AA}(M_A^2) \\ &+ \frac{T_d}{v_d}[s_\beta^2 c^2_{\beta-\alpha} - \sin^2\alpha] + \frac{T_u}{v_u}[c_\beta^2 c^2_{\beta-\alpha} - \cos^2\alpha] - 2m_Z^2 c_\beta^2 \sin[2(\beta+\alpha)]\delta\tan\beta\,, \end{aligned} \qquad (19.303)$$

after making use of eq. (19.299), where $s_{\beta-\alpha} \equiv \sin(\beta-\alpha)$ and $c_{\beta-\alpha} \equiv \cos(\beta-\alpha)$. To obtain $M^2_{H,h}$ one must explicitly evaluate the hh, HH, AA, and ZZ one-loop self-energy functions and the one-loop tadpole functions, T_d and T_u. It is convenient to evaluate the one-loop tadpole functions with respect to the neutral CP-even Higgs boson mass basis:[14]

$$T_H \equiv T_u \sin\alpha + T_d \cos\alpha\,, \qquad T_h \equiv T_u \cos\alpha - T_d \sin\alpha\,. \qquad (19.304)$$

[14] Since T_u and T_d are one-loop quantities, it is consistent to define T_h and T_H at one-loop accuracy by employing the mixing angle α defined in eq. (19.298) that is based on tree-level relations.

One can then rewrite eqs. (19.302) and (19.303) in a more useful form,

$$M_H^2 = \widehat{M}_H^2 + \mathrm{Re}\,\Sigma_{HH}(\widehat{M}_H^2) - \cos^2(\beta+\alpha)\,\mathrm{Re}\,A_{ZZ}(M_Z^2) - s^2_{\beta-\alpha}\,\mathrm{Re}\,\Sigma_{AA}(M_A^2)$$
$$+ c^2_{\beta-\alpha}\Sigma_{GG}(0) - 2c_{\beta-\alpha}\frac{T_H}{v} + 2m_Z^2 c_\beta^2 \sin[2(\beta+\alpha)]\delta\tan\beta\,, \qquad (19.305)$$

$$M_h^2 = \widehat{M}_h^2 + \mathrm{Re}\,\Sigma_{hh}(\widehat{M}_h^2) - \sin^2(\beta+\alpha)\,\mathrm{Re}\,A_{ZZ}(M_Z^2) - c^2_{\beta-\alpha}\,\mathrm{Re}\,\Sigma_{AA}(M_A^2)$$
$$+ s^2_{\beta-\alpha}\Sigma_{GG}(0) - 2s_{\beta-\alpha}\frac{T_h}{v} - 2m_Z^2 c_\beta^2 \sin[2(\beta+\alpha)]\delta\tan\beta\,, \qquad (19.306)$$

where

$$\Sigma_{GG}(0) = \frac{1}{v}\left[T_H c_{\beta-\alpha} + T_h s_{\beta-\alpha}\right], \qquad (19.307)$$

after making use of eqs. (19.281) and (19.304). The following relations among one-loop tadpole functions are noteworthy:

$$\frac{T_H}{v} = c_{\beta-\alpha}\Sigma_{GG}(0) + s_\beta c_\beta s_{\beta-\alpha}\left(\frac{T_d}{v_d} - \frac{T_u}{v_u}\right), \qquad (19.308)$$

$$\frac{T_h}{v} = s_{\beta-\alpha}\Sigma_{GG}(0) - s_\beta c_\beta s_{\beta-\alpha}\left(\frac{T_d}{v_d} - \frac{T_u}{v_u}\right). \qquad (19.309)$$

Adding eqs. (19.305) and (19.306) yields the one-loop correction to the tree-level squared-mass sum rule of the MSSM Higgs sector,

$$M_h^2 + M_H^2 = M_A^2 + M_Z^2 + \mathrm{Re}\,\Sigma_{hh}(\widehat{M}_h^2) + \mathrm{Re}\,\Sigma_{HH}(\widehat{M}_H^2) - \mathrm{Re}\,\Sigma_{AA}(M_A^2)$$
$$- \mathrm{Re}\,A_{ZZ}(M_Z^2) - \Sigma_{GG}(0)\,, \qquad (19.310)$$

where we have employed eq. (19.307). Note that one may derive eq. (19.310) directly from eq. (19.272) after rewriting the bare squared-mass parameters in terms of the renormalized squared masses using eqs. (19.277), (19.278), and (19.281).

A notable prediction of the MSSM is that the tree-level mass of the lightest CP-even Higgs boson is bounded from above, and its maximal value is achieved in the case of $\beta = \frac{1}{2}\pi$ and $M_A > M_Z$. In this limit, $v_d = 0$ and $v_u = v$, in which case $t_d = T_d = 0$ and there is no mixing of h_u and h_d (i.e., $\alpha = 0$). It then follows that $\widehat{M}_h = M_Z$ and $\widehat{M}_H = M_A$, and eqs. (19.305) and (19.306) simplify to

$$M_h^2 = M_Z^2 + \mathrm{Re}\,\Sigma_{hh}(M_Z^2) - \mathrm{Re}\,A_{ZZ}(M_Z^2) - \frac{T_h}{v}\,, \qquad (19.311)$$

$$M_H^2 = M_A^2 + \mathrm{Re}\,\Sigma_{HH}(M_A^2) - \mathrm{Re}\,\Sigma_{AA}(M_A^2)\,, \qquad (19.312)$$

which reproduces the result for the upper bound of M_h in the one-loop approximation obtained in Ref. [204]. Note that $\delta\tan\beta$ does not appear in eq. (19.311), which means that the one-loop upper bound of M_h is not sensitive to the one-loop definition of $\tan\beta$. That is, any $\tan\beta$ dependence appearing in eq. (19.311) appears for the first time in one-loop quantities, which means that one is free to employ the tree-level definition of $\tan\beta$ in eqs. (19.311) and (19.312).

In the Higgs decoupling limit where $M_A \gg M_Z$, it follows that $c_{\beta-\alpha} = 0$ and

$s_{\beta-\alpha} = 1$. In this limit, employing eqs. (19.305) and (19.306) at one-loop accuracy yields[15]

$$M_h^2 = c_{2\beta}^2\left[M_Z^2 - \operatorname{Re} A_{ZZ}(M_Z^2)\right] + \operatorname{Re}\Sigma_{hh}(M_Z^2 c_{2\beta}^2) - \frac{T_h}{v} + 4M_Z^2 c_\beta^2 s_{2\beta} c_{2\beta}\delta\tan\beta\,, \tag{19.313}$$

$$\begin{aligned} M_H^2 = M_A^2 + s_{2\beta}^2\left[M_Z^2 - \operatorname{Re} A_{ZZ}(M_Z^2)\right] + \operatorname{Re}\Sigma_{HH}(M_A^2) - \operatorname{Re}\Sigma_{AA}(M_A^2) \\ - 4M_Z^2 c_\beta^2 s_{2\beta} c_{2\beta}\delta\tan\beta\,, \end{aligned} \tag{19.314}$$

after making use of eq. (19.307). It is instructive to look at the leading contributions to the one-loop radiatively corrected mass of the light CP-even Higgs boson of the MSSM. Numerically, the leading effect is due to the loop contributions of the top quarks and the supersymmetric top-quark partners. Because of the dependence on the couplings of the top quark and top squarks that depend on the Higgs–top-quark Yukawa coupling y_t [see eq. (13.123)], it is sufficient to evaluate the leading m_t^4 behavior of the self-energy functions that appear in eq. (19.313). One can check that there are no terms that behave like m_t^4 in eq. (19.286). Hence, we can neglect the term in eq. (19.313) proportional to $\delta\tan\beta$. Likewise, there are no terms that behave like m_t^4 in $A_{ZZ}(M_Z^2)$. Hence, we are left with extracting the leading m_t^4 behavior of

$$M_h^2 = M_Z^2 c_{2\beta}^2 + \operatorname{Re}\Sigma_{hh}(M_Z^2 c_{2\beta}^2) - \frac{T_h}{v}\,, \tag{19.315}$$

due to loops of top quarks and their supersymmetric scalar partners. Evaluating eq. (19.315) at one-loop order in the limit of $M_Z \ll M_t \ll M_A, M_S$, where M_S is the geometric mean of the two top-squark squared masses, $M_S^2 \equiv m_{\tilde t_1} m_{\tilde t_2}$,

$$M_h^2 \simeq M_Z^2 c_{2\beta}^2 + \frac{3g^2 m_t^4}{8\pi^2 m_W^2}\left[\ln\left(\frac{M_S^2}{m_t^2}\right) + \frac{X_t^2}{M_S^2}\left(1 - \frac{X_t^2}{12M_S^2}\right)\right], \tag{19.316}$$

which reproduces the result of eq. (13.202), where $m_t X_t \equiv v(a_t s_\beta - \mu y_t c_\beta)/\sqrt{2}$ is the off-diagonal entry of the top-squark squared-mass matrix [see. eqs. (13.180) and (13.185)], and a_t and μ have been assumed to be real (for simplicity).

An alternative technique for deriving the radiatively corrected neutral CP-even Higgs masses of the MSSM employs the effective potential, as discussed in Section 13.8 and in Exercise 13.2. In fact, the leading logarithmic behavior of eq. (19.316) can be understood quite easily using the renormalization group equations of the Standard Model. In the decoupling limit, there exists a scale M_S below which the effective field theory of the MSSM coincides with that of the Standard Model. At the scale M_S, we can employ the MSSM relation $M_h^2 = M_Z^2 c_{2\beta}^2$. Using eq. (4.159), it follows that

$$\lambda(M_S) = \tfrac{1}{8}(g^2 + g'^2)c_{2\beta}^2\,, \tag{19.317}$$

[15] At one-loop accuracy, one may replace $m_Z^2\,\delta\tan\beta$ with $M_Z^2\,\delta\tan\beta$, since $\delta\tan\beta$ is a one-loop quantity.

which serves as a boundary condition of the renormalization group equation for λ,

$$\frac{d\lambda}{dt} = \beta_\lambda \,, \qquad \text{where } t \equiv \ln\mu. \tag{19.318}$$

In first approximation, we can take the right-hand side of eq. (19.318) to be independent of t, in which case

$$\lambda(m_t) = \lambda(M_S) - \tfrac{1}{2}\beta_\lambda \ln\left(\frac{M_S^2}{m_t^2}\right). \tag{19.319}$$

The one-loop beta function for λ in the Standard Model (SM) is given by

$$16\pi^2\beta_\lambda = 24\lambda^2 + \tfrac{3}{8}\left[2g^4 + (g^2+g'^2)^2\right] - 2\sum_i N_{c_i} y_i^4 - \lambda\Big(9g^2 + 3g'^2 - 4\sum_i N_{c_i} y_i^2\Big), \tag{19.320}$$

where the sum is taken over all SM fermions, f_i, that couple to the SM Higgs boson, with $y_i = g m_{f_i}/(\sqrt{2}m_W)$ and $N_{ci} = 3$ [$N_{ci} = 1$] for quarks [charged leptons]. To obtain the leading logarithmic behavior of the radiatively corrected Higgs mass, it suffices to retain the term in eq. (19.320) that is proportional to y_t^4:

$$\beta_\lambda = -\frac{3y_t^4}{8\pi^2} = -\frac{3g^4 m_t^4}{32\pi^2 m_W^4}. \tag{19.321}$$

Finally, in light of eqs. (4.159) and (19.319), we identify

$$M_h^2 = 2\lambda(m_t^2)v^2 = M_Z^2 c_{2\beta}^2 + \frac{3g^2 m_t^4}{8\pi^2 m_W^2}\ln\left(\frac{M_S^2}{m_t^2}\right), \tag{19.322}$$

in agreement with the leading logarithmic behavior of eq. (19.316) and the result of eq. (13.114) obtained using the one-loop effective potential.

19.9 The MSSM Wrong-Higgs Yukawa Couplings

The tree-level MSSM Lagrangian consists of supersymmetry-conserving mass and interaction terms, supplemented by soft supersymmetry-breaking operators. In particular, all tree-level dimension-four gauge invariant interactions must respect supersymmetry. When supersymmetry is broken, in principle all supersymmetry-breaking operators consistent with gauge invariance can be generated in the effective low-energy theory below the scale of supersymmetry breaking. The MSSM Higgs sector provides an especially illuminating example of this phenomenon.

The most general 2HDM Yukawa Lagrangian, consistent with gauge invariance, is given by eq. (6.139). For simplicity, we will focus on the Higgs couplings to the third generation of quarks (neglecting the generation indices and the couplings to leptons). Using the more familiar MSSM Higgs field notation of eq. (6.218),

$$\begin{aligned}\mathscr{L}_Y = &-y_t(H_u^0 t\bar{t} - H_u^+ b\bar{t}) - w_t(H_d^{0\dagger} t\bar{t} + H_d^+ b\bar{t}) \\ &- y_b(H_d^0 b\bar{b} - H_d^- t\bar{b}) - w_b(H_u^{0\dagger} b\bar{b} + H_u^- t\bar{b}) + \text{h.c.}\end{aligned} \tag{19.323}$$

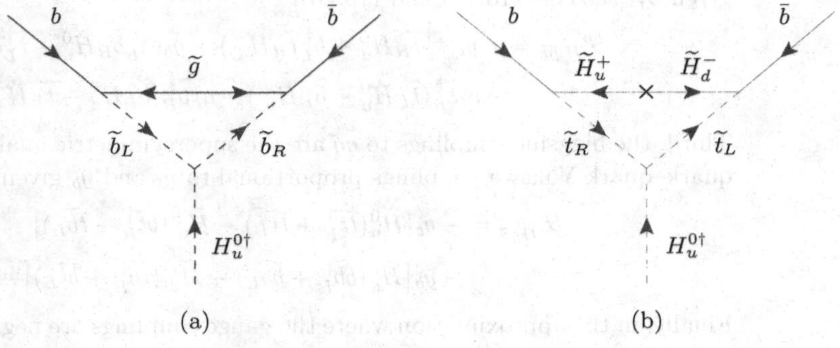

Fig. 19.16 One-loop MSSM contributions to the wrong-Higgs Yukawa couplings to $b\bar{b}$. In diagram (b), the $\times$ serves as a reminder that the exchanged charged higgsino is a Dirac fermion that is comprised of a pair of two-component fermions, $\widetilde{H}_u^+$ and $\widetilde{H}_d^-$ [see eq. (19.326)].

Imposing supersymmetry on the form of the Yukawa Lagrangian implies that we must eliminate the nonholomorphic couplings by setting $w_t = w_b = 0$, which yields the Type-II Yukawa interactions in agreement with eq. (13.121).

However, consider the MSSM under the assumption that all squark, gluino, chargino, and neutralino masses (characterized by a mass scale M_S) are significantly heavier than the heaviest scalar of the Higgs sector [whose mass is of $\mathcal{O}(m_{H^\pm})$ in light of eqs. (6.234) and (6.243)]. In this case, one can formally integrate out all the supersymmetric particles below the scale M_S. The resulting low-energy effective theory is the non-supersymmetric two-Higgs doublet extension of the Standard Model. In this effective theory, the so-called wrong-Higgs Yukawa couplings, w_t and w_b, are nonzero. In this section, we shall compute w_b by performing a one-loop computation under the assumption that $M_S \gg m_{H^\pm}$.

From eq. (19.323), the Feynman rule for the $H_u^{0\dagger} b\bar{b}$ vertex is $-iw_b$. The dominant contributions to this quantity are generated at one-loop order due to the two Feynman diagrams exhibited in Fig. 19.16.[16] We shall simplify the analysis by ignoring squark mixing, although a more complete calculation must take this into account since we will be assuming that μ, a_b, and a_t are nonzero [see eqs. (13.180) and (13.182)]. Finally, we shall ignore CP-violating effects by taking μ, a_b, and a_t and M_3 to be real parameters. In what follows, we shall first assume that μ and M_3 are positive real parameters (a condition we shall later relax).

To evaluate the diagrams displayed in Fig. 19.16, we employ Feynman rules obtained from the following interaction Lagrangians. First, the gluino couplings to quarks and squarks are given by eq. (J.41), and the corresponding Feynman rules are shown in Fig. J.11. Second, the couplings of Higgs bosons to squarks are

[16] We shall neglect subdominant corrections to w_b/y_b that are proportional to y_b, g^2, and g'^2.

given by [see eqs. (13.11) and (13.16)]

$$\mathscr{L}_{H\widetilde{q}\widetilde{q}} = \mu\big[y_t(\widetilde{t}_L^\dagger \widetilde{t}_R H_d^0 + \widetilde{b}_L^\dagger \widetilde{t}_R H_d^-) + y_b(\widetilde{b}_L^\dagger \widetilde{b}_R H_u^0 + \widetilde{t}_L^\dagger \widetilde{b}_R H_u^+)\big]$$
$$- a_t \widetilde{t}_R^\dagger(\widetilde{t}_L H_u^0 - \widetilde{b}_L H_u^+) - a_b \widetilde{b}_R^\dagger(\widetilde{b}_L H_d^0 - \widetilde{t}_L H_d^-) + \text{h.c.} \qquad (19.324)$$

Third, the higgsino couplings to $q\widetilde{q}$ are the supersymmetric analogues of the Higgs–quark–quark Yukawa couplings proportional to y_t and y_b given in eq. (19.323):

$$\mathscr{L}_{\widetilde{H}q\widetilde{q}} = -\, y_t\big[\widetilde{H}_u^0(t\widetilde{t}_R^\dagger + t\widetilde{t}_L) - \widetilde{H}_u^+(b\widetilde{t}_R^\dagger + t\widetilde{b}_L)\big]$$
$$- y_b\big[\widetilde{H}_d^0(b\widetilde{b}_R^\dagger + b\widetilde{b}_L) - \widetilde{H}_d^-(t\widetilde{b}_R^\dagger + b\widetilde{t}_L)\big] + \text{h.c.} \qquad (19.325)$$

Finally, in the approximation where the gauge couplings are neglected, the chargino masses [see eqs. (13.146)–(13.148)] and the gluino mass are obtained from

$$\mathscr{L}_{\text{mass}} = -\tfrac{1}{2}M_3\widetilde{g}\widetilde{g} - M_2\widetilde{W}^+\widetilde{W}^- - \mu\widetilde{H}_u^+\widetilde{H}_d^- + \text{h.c.}, \qquad (19.326)$$

where the mixing of gauginos and higgsinos (proportional to g) is neglected. The gluino of mass $M_{\widetilde{g}} = M_3$ is a Majorana fermion, and the charged Dirac fermion of mass $M_{\widetilde{H}^\pm} = \mu$ comprises the pair of two-component higgsino fields, $\widetilde{H}_u^+$ and $\widetilde{H}_d^-$.

Under the assumption that $M_S \gg m_{H^\pm}$, one can compute the leading contribution to the diagrams of Fig. 19.16 by setting all external four-momenta equal to zero. Performing the integration over the loop momentum then yields the Passarino–Veltman function $C_0(0,0,0;m_a^2,m_b^2,m_c^2)$, where the arguments of C_0 are the squared masses of the particles appearing in the loop. The vertex factors are obtained from $i\mathscr{L}_{\text{int}}$ given by eqs. (19.324), (19.325), and (J.41). Thus, we obtain

$$-iw_b\delta_{jk} = (i\mu y_b)2g_s^2\left(\boldsymbol{T}^a\boldsymbol{T}^a\right)_{jk} i^3 M_3 \frac{i}{16\pi^2}C_0(0,0,0;,M_3^2,m^2_{\widetilde{b}_L},m^2_{\widetilde{b}_R})$$
$$+(-ia_t)(-iy_t)(-iy_b)\delta_{jk}\, i^3\mu\, \frac{i}{16\pi^2}C_0(0,0,0;,\mu^2,m^2_{\widetilde{t}_L},m^2_{\widetilde{t}_R}), \qquad (19.327)$$

where j, k are color indices and the factor of i^3 derives from the numerators of the three propagators in the loop. The gluino and charged higgsino propagators have been obtained by employing the rules exhibited in Figs. 2.2(d) and 2.5(d), which contain positive masses in their respective numerators.

Equation (19.327) is usually expressed in terms of the function [see eq. (K.78)]:

$$\mathcal{I}(m_a,m_b,m_c) \equiv -C_0(0,0,0;m_a^2,m_b^2,m_c^2) \qquad (19.328)$$
$$= \frac{m_a^2 m_b^2\ln(m_a^2/m_b^2) + m_b^2 m_c^2\ln(m_b^2/m_c^2) + m_c^2 m_a^2\ln(m_c^2/m_a^2)}{(m_a^2-m_b^2)(m_b^2-m_c^2)(m_a^2-m_c^2)},$$

where $\mathcal{I}(m,m,m) = 1/(2m^2)$. Hence, eq. (19.327) yields

$$w_b = y_b\left[\frac{C_F\alpha_s\mu M_3}{2\pi}\mathcal{I}(M_{\widetilde{g}},m_{\widetilde{b}_L},m_{\widetilde{b}_R}) + \frac{\mu a_t y_t}{16\pi^2}\mathcal{I}(M_{\widetilde{H}^\pm},m_{\widetilde{t}_L},m_{\widetilde{t}_R})\right], \qquad (19.329)$$

where $(\boldsymbol{T}^a\boldsymbol{T}^a)_{jk} = C_F\delta_{jk}$, with $C_F = 4/3$, is the Casimir operator in the fundamental representation of SU(3)$_C$. Equation (19.329) was derived under the assumption that M_3 and μ are positive. However, in light of Exercise 13.6, it follows that eq. (19.329) remains valid if M_3 and μ are real quantities of either sign.

A remarkable feature of the above result is that, in the limit of $M_S \gg m_{H^\pm}$, the right-hand side of eq. (19.329) does not decouple if μ, M_3, $a_t \sim \mathcal{O}(M_S)$. That is, apart from the one-loop suppression factor, the contribution of w_b to the Yukawa interactions of the effective low-energy 2HDM theory can yield significant deviations from the Type-II Yukawa interactions of the tree-level MSSM Higgs sector. For example, setting $\langle H_u^0\rangle = v_u/\sqrt{2}$ and $\langle H_d^0\rangle = v_d/\sqrt{2}$ in eq. (19.323) yields

$$m_b = \frac{y_b v}{\sqrt{2}}\cos\beta\left(1+\frac{w_b\tan\beta}{y_b}\right) \equiv \frac{y_b v}{\sqrt{2}}\cos\beta(1+\Delta_b)\,, \tag{19.330}$$

which defines the quantity Δ_b. The dominant contributions to Δ_b are $\tan\beta$-enhanced, with $\Delta_b \simeq (w_b/y_b)\tan\beta$. Thus, the tree-level relation between the b-quark mass and the b-quark Yukawa coupling receives a significant radiative correction if $\tan\beta$ is large. This can significantly modify the tree-level predictions for the couplings of $b\bar{b}$ to the physical Higgs bosons of the MSSM (see Exercise 19.16).

Exercises

19.1 In the calculation of the wave function renormalization constant of the electron presented in Section 19.1, the Feynman gauge was used, corresponding to $\xi = 1$ (see Fig. 2.6). Repeat the calculation of $\delta Z_{\rm QED}$ (using either the two-component fermion or four-component fermion formalism) by taking $\xi \neq 1$ and using the Feynman rule for the photon propagator in the Feynman gauge,

$$\frac{-i}{k^2+i\varepsilon}\left(g^{\mu\nu}-(1-\xi)\frac{k^\mu k^\nu}{k^2}\right). \tag{19.331}$$

(a) Using DREG in d spacetime dimensions, show that

$$\begin{aligned}\Sigma(p^2) &\equiv \Sigma_L(p^2)_{\rm QED} = \Sigma_R(p^2)_{\rm QED}\\ &= -\frac{\alpha}{2\pi}\Big\{\left[\tfrac{1}{2}(1+\xi)-\epsilon\right]B_0(p^2;0,m^2)+\left[\tfrac{1}{2}(3-\xi)-\epsilon\right]B_1(p^2;0,m^2)\\ &\qquad +(1-\xi)\left[p^2C_{22}(0,p^2,p^2;0,0,m^2)+C_{24}(0,p^2,p^2;0,0,m^2)\right]\Big\}\,,\end{aligned} \tag{19.332}$$

$$\Sigma_D(p^2)_{\rm QED} = \overline{\Sigma}_D(p^2)_{\rm QED} = \frac{\alpha m}{2\pi}\left[\tfrac{1}{2}(3+\xi)-\epsilon\right]B_0(p^2;0,m^2)\,, \tag{19.333}$$

where $\epsilon \equiv 2-\frac{1}{2}d$ and the Passarino–Veltman loop functions B_0, B_1, C_{22}, and C_{24} are defined in Appendix K.2.

(b) Using eqs. (K.120) and (K.121), show that one can rewrite eq. (19.332) in the following form:

$$\begin{aligned}\Sigma(p^2) = -\frac{\alpha}{2\pi}\Big\{&\left[\tfrac{1}{2}(1+\xi)-\epsilon\right]B_0(p^2;0,m^2)+(1-\epsilon)B_1(p^2;0,m^2)\\ &-\tfrac{1}{2}(1-\xi)(p^2-m^2)B_0'(p^2;0,m^2)\Big\}\,.\end{aligned} \tag{19.334}$$

Using this result, derive the following two expressions:

$$\Sigma(m^2) = -\frac{\alpha}{2\pi}\left\{\left[\tfrac{1}{2}(1+\xi)-\epsilon\right]B_0(m^2;0,m^2)+(1-\epsilon)B_1(m^2;0,m^2)\right\}, \tag{19.335}$$

$$\Sigma'(m^2) = -\frac{\alpha}{2\pi}\left[(\xi-\epsilon)B_0'(m^2;0,m^2)+(1-\epsilon)B_1'(m^2;0,m^2)\right], \tag{19.336}$$

where the prime indicates differentiation with respect to p^2:

$$B_i'(m^2;0,m^2) \equiv \frac{\partial}{\partial p^2}B_i(p^2;0,m^2)\bigg|_{p^2=m^2}. \tag{19.337}$$

The step in which terms multiplied by $p^2 - m^2$ are dropped in the limit of $p^2 \to m^2$ must be justified (see Exercise K.6).

(c) Using eq. (19.11) and the results of part (a) and (b), derive the following expression for $\delta Z_{\rm QED}$:

$$\delta Z_{\rm QED} = -\frac{\alpha}{2\pi}\left\{\left[\tfrac{1}{2}(1+\xi)-\epsilon\right]B_0(m^2;0,m^2)+(1-\epsilon)B_1(m^2;0,m^2)\right.$$
$$\left.+(\xi-3)m^2B_0'(m^2;0,m^2)+2m^2(1-\epsilon)B_1'(m^2;0,m^2)\right\}. \tag{19.338}$$

Note that the infrared divergence resides entirely in $B_0'(m^2;0,m^2)$. Thus, the infrared divergence at one loop is absent in the Yennie gauge, where $\xi = 3$.

(d) Using eqs. (19.15), (19.16), (19.21), and (19.23), evaluate eq. (19.338). After some simplification, show that

$$\delta Z_{\rm QED} = -\frac{\alpha}{4\pi}(4\pi)^\epsilon\left(\frac{m^2}{\mu^2}\right)^{-\epsilon}\left\{\left(\frac{\xi-2\epsilon}{1-2\epsilon}\right)\Gamma(\epsilon)+(3-\xi)\frac{\Gamma(1+\epsilon)}{\epsilon(1-2\epsilon)}\right\}. \tag{19.339}$$

As expected, the wave function renormalization constant of the electron in QED is gauge-dependent. The term on the right-hand side of eq. (19.339) that is proportional to $\Gamma(\epsilon)$ contains the ultraviolet divergence, whereas the term that is proportional to $\Gamma(1+\epsilon)/\epsilon$ corresponds to the infrared divergence. Setting $\xi = 1$ reproduces the result given in eq. (19.24).

(e) How is the result of eq. (19.339) modified if the computations of this exercise are repeated using DRED?

19.2 In QED, the electron mass counterterm is defined by $\delta m \equiv m_p - m$, where $p^2 = m_p^2$ is the pole of the loop-corrected electron propagator and m is the unrenormalized bare mass of the electron.

(a) Using eq. (2.297) and the results of parts (a) and (b) of Exercise 19.1, show that the expression for the one-loop mass shift of the electron, δm, is gauge invariant (i.e., independent of ξ), and verify that

$$\delta m = m\Sigma(m^2) + \Sigma_D(m^2) = \frac{\alpha m}{2\pi}\Big[B_0(m^2;0,m^2) - (1-\epsilon)B_1(m^2;0,m^2)\Big]$$
$$= \frac{3\alpha m}{4\pi}\left[\frac{1}{\epsilon} - \ln\left(\frac{m^2}{Q^2}\right) + \frac{4}{3} + \mathcal{O}(\epsilon)\right], \tag{19.340}$$

where the $\overline{\text{MS}}$ renormalization scale Q is defined in eq. (11.42).

(b) Argue that δm must be independent of Q. Noting that

$$Q\frac{de}{dQ} = \beta_e - \epsilon e\,, \tag{19.341}$$

in light of eq. (11.54), and working consistently at $\mathcal{O}(e^2)$, verify that the right-hand side of eq. (19.340) is independent of Q in the limit of $\epsilon \to 0$.

HINT: Note that the one-loop QED beta function is given by $\beta_e = e^3/(12\pi^2)$.

19.3 In QED, the unrenormalized photon self-energy function given by eq. (19.59) with $N_c^f = 1$ and $Q_f = -1$ (for the electron, $f = e$) has the form

$$\Pi^{\gamma\gamma}_{\mu\nu}(p) = (p_\mu p_\nu - p^2 g_{\mu\nu})\Pi(p^2)\,. \tag{19.342}$$

(a) Show that $\Pi(p^2)$ can be written in the following form:

$$\Pi(p^2) = -\frac{2\alpha}{\pi}\left[B_1(p^2;m^2,m^2) + B_{21}(p^2;m^2,m^2)\right]$$
$$= \frac{2\alpha}{\pi}\left\{\frac{1}{6\epsilon} - \int_0^1 x(1-x)\ln\left(\frac{m^2 - p^2 x(1-x)}{Q^2}\right)dx + \mathcal{O}(\epsilon)\right\}, \tag{19.343}$$

where m is the electron mass and Q^2 is defined in eq. (11.42). Note that when written in the form of eq. (19.343), it is straightforward to evaluate the limit $\lim_{p^2\to 0}\Pi(p^2)$.

(b) The quantities $\Pi(p^2)$, α, and m should be understood to be bare quantities (prior to renormalization). For convenience, we shall affix the subscript B to the bare quantities. The renormalized fields and parameters in some renormalization scheme (denoted by the subscript R) are given by

$$A^\mu_B = Z_3^{1/2} A^\mu_R\,, \qquad \alpha_B = Z_\alpha \mu^{2\epsilon}\alpha_R\,, \qquad m_p = m_B + \delta m\,, \tag{19.344}$$

where m_p is the physical (pole) mass of the electron. We also define Z_1 and Z_2 as the renormalization constants of the $ee\gamma$ vertex and the electron field, respectively. Since $Z_1 \equiv Z_2 Z_3^{1/2} Z_\alpha^{1/2}$ (e.g., see Ref. [B64]), the Ward identity of QED, $Z_1 = Z_2$, implies that $Z_\alpha = Z_3^{-1}$. Using this result, prove that

$$\frac{\mu^{-2\epsilon}\alpha_B}{1+\Pi_B(p^2)} = \frac{\alpha_R}{1+\Pi_R(p^2)}\,. \tag{19.345}$$

HINT: Consider the bare and renormalized versions of eqs. (2.221) and (4.55).

Two different choices for the renormalization scheme are considered in part (c) below.

(c) After renormalization, the function $\Pi(p^2)$ given by eq. (19.343) is rendered finite in the limit of $\epsilon \to 0$. In the $\overline{\text{MS}}$ renormalization scheme at one loop, one simply subtracts off the term proportional to ϵ^{-1}. That is,

$$\Pi_{\overline{\text{MS}}}(p^2) \equiv \Pi(p^2) - \frac{\alpha}{3\pi\epsilon} = -\frac{2\alpha}{\pi}\int_0^1 x(1-x)\ln\left(\frac{m^2 - p^2x(1-x)}{Q^2}\right) dx + \mathcal{O}(\epsilon)\,. \tag{19.346}$$

Alternatively, in the on-shell (OS) renormalization scheme where $\Pi_{\text{OS}}(0) = 0$,

$$\Pi_{\text{OS}}(p^2) \equiv \Pi(p^2) - \Pi(0) = -\frac{2\alpha}{\pi}\int_0^1 x(1-x)\ln\left(\frac{m^2 - p^2x(1-x)}{m^2}\right) dx + \mathcal{O}(\epsilon)\,. \tag{19.347}$$

In this renormalization scheme, α_{OS} can be identified with the electromagnetic coupling measured in the Thomson limit, $\alpha_{\text{OS}} \simeq 1/137$. In particular, neither α_{OS} nor $\Pi_{\text{OS}}(p^2)$ depends on the $\overline{\text{MS}}$ renormalization scale Q.

Because eq. (19.345) holds for any renormalization scheme, show that, in the one-loop approximation,

$$\alpha_{\overline{\text{MS}}}(Q)\left[1 - \Pi_{\overline{\text{MS}}}(p^2)\right] = \alpha_{\text{OS}}\left[1 - \Pi_{\text{OS}}(p^2)\right]. \tag{19.348}$$

Employing eqs. (19.346)–(19.348), derive the one-loop relation,

$$\alpha_{\overline{\text{MS}}}(Q) = \alpha_{\text{OS}}\left\{1 - \frac{\alpha_{\text{OS}}}{3\pi}\ln\left(\frac{m^2}{Q^2}\right) + \mathcal{O}(\alpha_{\text{OS}}^2)\right\}. \tag{19.349}$$

Note that, in the one-loop approximation, $\alpha_{\text{OS}} = \alpha_{\overline{\text{MS}}}(Q = m)$.

(d) Using eq. (19.341) and the QED one-loop beta function, show that the electromagnetic coupling in the $\overline{\text{MS}}$ scheme satisfies

$$Q\frac{d\alpha}{dQ} = -2\epsilon\alpha + \frac{2\alpha^2}{3\pi} + \mathcal{O}(\alpha^3)\,. \tag{19.350}$$

Solving this equation to one-loop accuracy and making use of the results of part (c), verify that

$$\alpha_{\overline{\text{MS}}}(Q) = \alpha_{\text{OS}}\left\{1 - \epsilon\ln\left(\frac{Q^2}{m^2}\right) + \mathcal{O}(\epsilon^2)\right\} + \mathcal{O}(\alpha_{\text{OS}}^2)\,, \tag{19.351}$$

thereby confirming that an $\mathcal{O}(\epsilon)$ term has been omitted in the derivation of eq. (19.349). Inserting the above result into eq. (19.340), conclude that, at one-loop accuracy,

$$\delta m = \frac{3m\alpha_{\text{OS}}}{4\pi}\left[\frac{1}{\epsilon} + \frac{4}{3} + \mathcal{O}(\epsilon)\right]. \tag{19.352}$$

That is, in terms of on-shell parameters, δm is explicitly independent of the $\overline{\text{MS}}$ renormalization scale Q, as expected.

19.4 Examine the limit as $x \to 0$ of $I_{VF}(s;x,y)$ and $I_{V\overline{F}}(s;x,y)$.

(a) Starting from eqs. (19.84) and (19.112), show that

$$I_{VF}(s;0,y) = \left[\epsilon(1-\delta_D)+\xi(1-\epsilon)\right]\left[\left(1+\frac{y}{s}\right)B_0(s;0,y)-\frac{A_0(y)}{s}\right], \quad (19.353)$$

$$I_{V\overline{F}}(s;0,y) = (3-2\epsilon\delta_D+\xi)B_0(s;0,y), \quad (19.354)$$

in $d = 4-2\epsilon$ spacetime dimensions, where δ_D is defined in eq. (19.2).

HINT: For example, taking the $x \to 0$ limit of eq. (19.83) yields

$$I_{VF}(s;0,y) = (D-3+\xi)B_0(s;0,y)+(D-2)B_1(s;0,y) - (1-\xi)(s-y)\left(\frac{\partial B_1(s;x,y)}{\partial x}\right)_{x=0}. \quad (19.355)$$

Then make use of eqs. (K.28), (K.61), and (K.122) to derive eq. (19.353).

(b) In light of eqs. (19.86), (19.99), and (19.117), it follows that the self-energy functions of QED are given by

$$\Sigma_L(p^2)_{\rm QED} = \Sigma_R(p^2)_{\rm QED} = -\frac{\alpha}{4\pi}I_{VF}(p^2;0,m^2), \quad (19.356)$$

$$\Sigma_D(p^2)_{\rm QED} = \overline{\Sigma}_D(p^2)_{\rm QED} = \frac{\alpha m}{4\pi}I_{V\overline{F}}(p^2;0,m^2). \quad (19.357)$$

Verify that these expressions are equivalent to the results given in eqs. (19.333) and (19.334) when DREG is employed.

HINT: Compare eqs. (19.334) and (19.355) in light of eq. (K.61).

19.5 Consider softly broken SUSY-QED that includes the mixing of $\widetilde{e}_L$ and $\widetilde{e}_R$ due to the presence of a mixing term in the SUSY-QED Lagrangian,

$$\mathscr{L}_{\rm mix} = -(b\phi_L\phi_R + {\rm h.c.}). \quad (19.358)$$

Show that two new diagrams contribute to the electron self-energy at one loop that supplement those exhibited in Fig. 19.4. Compute the contributions of the selectron–photino loop to the one-loop computation of the electron wave function renormalization constants $(\delta Z_L)_{\rm SQED}$ and $(\delta Z_R)_{\rm SQED}$, thereby generalizing the results obtained in eqs. (19.32) and (19.33).

19.6 (a) Using the Gordon identities obtained in Exercise 3.9, show that the electromagnetic vertex function given in eq. (19.137) can be rewritten as

$$\bar{u}(p')\Gamma^\mu(p,p')u(p) = \bar{u}(p')\widetilde{\Gamma}^\mu(p,p')u(p), \quad (19.359)$$

where

$$\widetilde{\Gamma}^\mu(p,p') = \gamma^\mu[F_1(q^2)+F_2(q^2)] - \frac{(p+p')^\mu}{2m}[F_2(q^2)-iF_3(q^2)\gamma_5] + \frac{1}{4m^2}(q^\mu q^\nu - q^2 g^{\mu\nu})\gamma_\nu\gamma_5 F_4(q^2). \quad (19.360)$$

(b) Show that eq. (19.165), when written in terms of four-component spinor notation, takes the following form:

$$F_2(q^2) = \sum_{s,s'} \bar{u}(p)\left(g_1(q^2)\gamma_\mu + \frac{g_2(q^2)}{2m}(p+p')_\mu\right) u(p')\bar{u}(p')\widetilde{\Gamma}^\mu(p,p')u(p)$$
$$= \mathrm{Tr}\left\{\left(g_1(q^2)\gamma_\mu + \frac{g_2(q^2)}{2m}(p+p')_\mu\right)(\not{p}' + m)\widetilde{\Gamma}^\mu(p,p')(\not{p}+m)\right\}, \tag{19.361}$$

where $g_1(q^2)$ and $g_2(q^2)$ are given by eq. (19.170) when the traces are evaluated in d spacetime dimensions.

(c) In four-component spinor notation, there is only one QED diagram that contributes at one-loop order to the anomalous magnetic moment of the muon. Show that, in the Feynman gauge, the contribution of the QED diagram to $\widetilde{\Gamma}^\mu$ is given by

$$\widetilde{\Gamma}^\mu(p,p') = 2ie^2 \int \frac{d^dk}{(2\pi)^d}\frac{1}{D}\Big[(\not{k}+\not{p})\gamma^\mu(\not{k}+\not{p}') + m^2\gamma^\mu - 2m(p+p'+2k)^\mu$$
$$+ \tfrac{1}{2}(d-4)\not{k}\gamma^\mu\not{k}\Big], \tag{19.362}$$

where D is given by eq. (19.172), after applying $d = 4$ in any coefficient of the numerator that is not quadratic in k. Insert eq. (19.362) into eq. (19.361), evaluate the trace, and check that your result reproduces eq. (19.187).

19.7 In the calculation of the anomalous magnetic moment of the muon, in Section 19.5.2 (and in part (c) of Exercise 19.6), the Feynman gauge was used, corresponding to $\xi = 1$. Repeat the calculation by taking $\xi \neq 1$ and employing the Feynman rule for the photon propagator given in eq. (19.331). Show that the anomalous magnetic moment is independent of the gauge parameter ξ and thus is gauge invariant.

19.8 (a) Using eqs. (10.89) and (19.213), show that

$$\tfrac{1}{2}D_\alpha W^\alpha = -D + i\theta\sigma^\mu\partial_\mu\lambda^\dagger + i\theta^\dagger\bar{\sigma}^\mu\partial_\mu(\lambda - i\sigma^{\beta\rho}F_{\beta\rho})$$
$$+ i\theta^\dagger\bar{\sigma}^\mu\theta\partial_\mu D + \theta\theta\,\theta^\dagger\Box\lambda^\dagger, \tag{19.363}$$

where all fields are a function of $y \equiv x - i\theta\sigma^\mu\theta^\dagger$ and $\partial_\mu \equiv \partial/\partial y^\mu$.

(b) In light of eq. (19.363), explain why the effective superspace Lagrangian at order M_P^{-1} given in eq. (19.214) is the unique expression (up to an overall constant) that contains the magnetic dipole operator exhibited in eq. (19.212).

19.9 In eq. (19.165), the form factor $F_2(q^2)$ was obtained by constructing a projection operator which when applied to the electromagnetic vertex yields $F_2(q^2)$.

(a) Show that the corresponding projection operators that yield the form factors $F_3(q^2)$ and $F_4(q^2)$ are given by

$$F_3(q^2) = \frac{ig_3(q^2)}{2m}(p+p')_\mu \sum_{s,s'} \left[y(\vec{p},s)x(\vec{p}\,',s') - x^\dagger(\vec{p},s)y^\dagger(\vec{p}\,',s')\right] \times z(\vec{p}\,',s')\Gamma^\mu z(\vec{p},s), \tag{19.364}$$

$$F_4(q^2) = g_4(q^2)\sum_{s,s'} \left[y(\vec{p},s)\sigma_\mu y^\dagger(\vec{p}\,',s') - x^\dagger(\vec{p},s)\overline{\sigma}_\mu x(\vec{p}\,',s')\right] z(\vec{p}\,',s')\Gamma^\mu z(\vec{p},s), \tag{19.365}$$

where $z(\vec{p}\,',s')\Gamma^\mu z(\vec{p},s)$ is defined in eq. (19.166) and

$$g_3(q^2) = \frac{2m^2}{q^2(4m^2-q^2)}, \qquad g_4(q^2) = \frac{m^2}{(\frac{1}{2}d-1)q^2(4m^2-q^2)}, \tag{19.366}$$

after evaluating the traces in d dimensions [see eq. (19.168)].

(b) In light of the results of Exercise 19.6, show that in the four-component spinor formalism, eqs. (19.364) and (19.365) yield

$$F_3(q^2) = -\frac{ig_3(q^2)}{2m}(p+p')_\mu \operatorname{Tr}\{\gamma_5(\not{p}'+m)\widetilde{\Gamma}^\mu(p,p')(\not{p}+m)\}, \tag{19.367}$$

$$F_4(q^2) = g_4(q^2)\operatorname{Tr}\{\gamma_\mu\gamma_5(\not{p}'+m)\widetilde{\Gamma}^\mu(p,p')(\not{p}+m)\}, \tag{19.368}$$

where $\widetilde{\Gamma}(p,p')$ is defined in eq. (19.360), and $g_3(q^2)$ and $g_4(q^2)$ were obtained in part (a).

19.10 Consider the softly broken QED Lagrangian including the $\widetilde{\mu}_L$–$\widetilde{\mu}_R$ mixing term specified in eq. (19.358).

(a) Using the two-component fermion formalism, enumerate all the Feynman diagrams that contribute at one-loop order to $F_2(q^2)_{\rm SQED}$. Evaluate the contributions to $\langle p', s' | S^{(1)} | p, s\rangle$ from the diagrams that you have identified in an approximation where eq. (19.358) is treated perturbatively. That is, eq. (19.358) is regarded as an interaction, which yields a Feynman rule corresponding to the insertion of b in analogy with the Feynman rules for fermion mass insertions exhibited in Fig. 2.4.

(b) What are the modifications to the results of Section 19.5.3? Obtain the modification to eq. (19.205) to first order in b. Derive an integral expression for $F_2(0)_{\rm SQED}$ under the assumption that m_L, $m_R \gg |b| \gg m$, where m is the muon mass.

(c) Repeat part (b) where the $\widetilde{\mu}_L$–$\widetilde{\mu}_R$ mixing is now treated exactly.

19.11 Consider the softly broken SUSY-QED Lagrangian including the $\widetilde{\mu}_L$–$\widetilde{\mu}_R$ mixing term specified in eq. (19.358), under the assumption that b is a complex mixing parameter. Such a term violates P and T, which implies that the muon can possess an electric dipole moment.

(a) Show that $F_3(q^2)_{\rm QED} = 0$.

(b) Using the two-component fermion formalism, evaluate the contributions to $\langle p', s'|\, S^{(1)}\, |p, s\rangle$ from the diagrams enumerated in part (a) of Exercise 19.10, in the approximation where the $\widetilde{\mu}_L$–$\widetilde{\mu}_R$ mixing is treated to first order in b.

(c) Using eq. (19.364), compute $F_3(q^2)_{\rm SQED}$ and show that it is finite and nonzero. Then, set $q^2 = 0$ and obtain an integral expression for the electric dipole moment of the muon. Your result should be proportional to $\operatorname{Im} b$ (why?). Find an explicit formula for the electric dipole moment under the assumption that $m_L, m_R \gg m$, where m is the muon mass. Consider separately the cases of mass-degenerate and mass-nondegenerate smuons.

19.12 Obtain a general expression for the one-loop contribution to the electric dipole moment of the electron in the MSSM. Explore the results of the limiting case of $M_1 \ll M_2, |\mu|$, where M_1 and M_2 are gaugino mass parameters and μ is the coefficient of $H_u H_d$ in the MSSM superpotential. Consider separately the cases of mass-degenerate and mass-nondegenerate selectrons.

19.13 In the MSSM, the masses of the two neutral CP-even Higgs bosons can be determined via eqs. (19.305) and (19.306) in terms of $\tan\beta$ and other MSSM mass parameters (such as m_A) and the top-squark mass and mixing parameters). Since eqs. (19.305) and (19.306) depend on $\delta\tan\beta$, one must specify a renormalization scheme to define $\tan\beta$. One choice is to employ the $\overline{\rm DR}$ scheme to fix the wave function renormalization constants δZ_{H_u} and δZ_{H_d} [see eq. (19.286)]. Another choice is to define $\tan\beta$ via a physical process, such as the decay rate $\Gamma(A \to \tau\tau)$. In this exercise, a third proposal is put forward.

(a) In the "HiggsMass" (HM) scheme of Ref. [205], M_H is taken to be an input parameter (determined by experiment). The counterterm $\delta\tan\beta$ is obtained by setting $M_H^2 = \widehat{M}_H^2$, where $\widehat{M}_H^2$ is defined in eq. (19.297). In the convention where $0 \leq \beta \leq \frac{1}{2}\pi$, show that β is determined (up to a two-fold ambiguity) by

$$s_{2\beta} = \frac{\left[(M_H^2 - M_Z^2)(M_H^2 - M_A^2)\right]^{1/2}}{M_A M_Z}\,. \tag{19.369}$$

(b) Obtain an explicit expression for $\delta\tan\beta$ in the HM scheme. Inserting the result for $\delta\tan\beta$ into eq. (19.306), show that

$$\begin{aligned} M_h^2 = \widehat{M}_h^2 &+ \operatorname{Re}\Sigma_{hh}(\widehat{M}_h^2) + \operatorname{Re}\Sigma_{HH}(M_H^2) - \operatorname{Re}\Sigma_{AA}(m_A^2) \\ &- \operatorname{Re} A_{ZZ}(m_Z^2) - \Sigma_{GG}(0)\,, \end{aligned} \tag{19.370}$$

where $\widehat{M}_h^2$ is defined in eq. (19.297).

19.14 Evaluate the one-loop contributions of the top quark and top squarks to $\mathrm{Re}\,A_{ZZ}(p^2)$, $\mathrm{Re}\,\Sigma_{hh}(p^2)$, $\mathrm{Re}\,\Sigma_{HH}(p^2)$, $\mathrm{Re}\,\Sigma_{AA}(p^2)$, $\Sigma_{GG}(0)$, and the one-loop tadpole functions T_H and T_h. You may neglect the effects of top-squark mixing by setting the MSSM parameters μ and a_t to zero. For simplicity, you may take the masses of $\widetilde{t}_L$ and $\widetilde{t}_R$ to be equal to M_S.

(a) Check that eq. (19.307) is satisfied.

(b) Evaluate the leading term of the self-energy functions and the one-loop tadpole functions in the limit of $p^2 \ll m_t^2 \ll M_A^2, M_S^2$. Identify the terms that are proportional to the fourth power of the top-quark mass.

(c) Compute the one-loop corrected squared mass M_h^2 in the Higgs decoupling limit. Verify that the leading terms, which are proportional to the fourth power of the top-quark mass and enhanced by a factor of $\ln M_{\widetilde{t}}^2$, can be obtained from eq. (19.315). Show that all ultraviolet divergences cancel, and confirm the $X_t = 0$ limit of eq. (19.316).

(d) Repeat the calculations performed above for nonzero values of the top-squark mixing parameters μ and a_t. Show that the leading stop mixing contribution to the one-loop radiatively corrected squared mass M_h^2 in the Higgs decoupling limit depends only on M_S and X_t, and verify the result of eq. (19.316).

19.15 Evaluate the one-loop contributions of the third-generation quarks and squarks to $\mathrm{Re}\,A_{WW}(p^2)$, $\mathrm{Re}\,\Sigma_{H^+H^-}(p^2)$, and $\mathrm{Re}\,\Sigma_{AA}(p^2)$. You may neglect the effects of top- and bottom-squark mixing by setting the MSSM parameters μ, a_t, and a_b to zero. For simplicity, you may take the masses of $\widetilde{t}_{L,R}$ and $\widetilde{b}_{L,R}$ to be equal to M_S.

(a) Use eq. (19.282) to compute the third-generation quark and squark contributions to the one-loop radiatively corrected squared mass $M_{H^\pm}^2$ in the limit of $M_W \ll m_t \ll M_{H^\pm} \ll M_S^2$. Identify the leading logarithmic term proportional to $\ln(M_S^2/m_t^2)$.

(b) In the CP-conserving 2HDM, the tree-level value of the squared mass of the charged Higgs boson is given by

$$m_{H^\pm}^2 = m_A^2 + \tfrac{1}{2}(\lambda_4 + \lambda_5)v^2\,, \qquad (19.371)$$

where λ_4 and λ_5 are parameters of the scalar potential in the Φ-basis. In light of eq. (6.222), $\lambda_4 = -\frac{1}{2}g^2$ and $\lambda_5 = 0$, and we recover the expected tree-level result for $m_{H^\pm}$ given by eq. (19.276). One can use renormalization group methods to obtain the leading logarithmic contribution at one loop to the squared mass of the charged Higgs boson in the MSSM, following the technique described at the end of Section 19.8. Here, we need to employ the one-loop beta function for λ_4 in the CP-conserving 2HDM (with $\lambda_5 = \lambda_6 = \lambda_7 = 0$):

$$16\pi^2\beta_{\lambda_4} = 2\lambda_4(\lambda_1 + \lambda_2 + 4\lambda_3 + 2\lambda_4) - 3\lambda_4(3g^2 + g'^2) + 3g^2g'^2 + 6\lambda_4(y_b^2 + y_t^2) - 12y_b^2y_t^2\,, \tag{19.372}$$

where the Yukawa couplings y_t and y_b are related to the corresponding top and bottom quark masses as follows:

$$y_t = \frac{gm_t}{\sqrt{2}\,m_W s_\beta}\,, \qquad y_b = \frac{gm_b}{\sqrt{2}\,m_W c_\beta}\,. \tag{19.373}$$

Obtain an approximate solution to the renormalization group equation for λ_4 with the boundary condition $\lambda_4(M_S) = -\frac{1}{2}g^2(M_S)$ [see eq. (19.319)], where only terms on the right-hand side of eq. (19.372) involving y_t^2 and y_b^2 are retained (admittedly not a particularly good approximation). Under these assumptions, $g^2(M_S) = g^2(m_t)$ since the one-loop beta function of g is independent of y_t and y_b. Using the expression obtained for $\lambda_4(M_S)$, conclude that

$$\begin{aligned} m_{H^\pm}^2 &= m_A^2 + \tfrac{1}{2}\lambda_4(m_t)v^2 \\ &= m_A^2 + m_W^2 + \frac{3g^2}{32\pi^2 m_W^2}\left[\frac{2m_t^2m_b^2}{s_\beta^2c_\beta^2} - m_W^2\left(\frac{m_t^2}{s_\beta^2} + \frac{m_b^2}{c_\beta^2}\right)\right]\ln\left(\frac{M_S^2}{m_t^2}\right). \end{aligned} \tag{19.374}$$

Check that eq. (19.374) reproduces the results obtained in part (a).

19.16 In light of eq. (19.330), derive the following expression for the $hb\bar{b}$ coupling:

$$g_{hb\bar{b}} = -\frac{m_b \sin\alpha}{v\cos\beta}\left[1 - \left(\frac{\Delta_b}{1+\Delta_b}\right)(1 + \cot\alpha\cot\beta)\right]. \tag{19.375}$$

Show that $g_{hb\bar{b}}$ reduces to its SM value when $m_A \gg m_Z$ [see eq. (6.352)]. Obtain the corresponding expressions for $g_{Hb\bar{b}}$, $g_{Ab\bar{b}}$, and $g_{H^+b\bar{t}}$.

19.17 (a) Show how the result of eq. (19.329) is modified by accounting for $\widetilde{t}_L$–$\widetilde{t}_R$ mixing and $\widetilde{b}_L$–$\widetilde{b}_R$ mixing, which must be present due to the nonzero values of a_t, a_b, and μ.

(b) Improve the result of eq. (19.329) under the assumption that $y_b \sim y_t$.

(c) Extend the result of eq. (19.329) under the assumption that the parameters M_3, μ, and a_t are complex.

(d) Derive the corresponding formulae for w_t, making the same assumptions as used in the derivation of eq. (19.329). Feel free to improve your result by applying similar considerations addressed in parts (a)–(c) of this exercise.

PART V

THE APPENDICES

Appendix A Notations and Conventions

A.1 Matrix Notation and the Summation Convention

If A is a matrix, then

$$\begin{aligned}&A^{\mathsf{T}} \text{ is the transpose of } A\,,\\&A^* \text{ is the complex conjugate of } A\,,\\&A^\dagger \text{ is the hermitian conjugate of } A\,,\\&\boldsymbol{I}_n \text{ (or } \mathbb{1}_{n\times n}) \text{ is the } n\times n \text{ identity matrix}\,.\end{aligned}$$

The summation convention can be illustrated by the rule for matrix multiplication: $(AB)_{ij} = A_{ik}B_{kj}$, where the sum over k is implicit. That is, the summation convention states that all repeated pairs of *dummy* indices are summed over. Note that a diagonal matrix can be written as

$$A_{ij} = a_i\delta_{ij} = \mathrm{diag}(a_1, a_2, \ldots)\,, \tag{A.1}$$

where the Kronecker delta $\delta_{ij} = 1$ if $i = j$ and $\delta_{ij} = 0$ if $i \neq j$. Note that there is no implicit sum over the repeated index i in eq. (A.1) since i and j are *fixed* indices, which assume the same (fixed) values on both sides of the equation.

A.2 Conjugation and the Flavor Index

In theories with a collection of complex fields, $\zeta_i(x)$, the flavor index i labels the individual field components. It is convenient to designate the conjugated fields[1] $[\zeta_i(x)]^\dagger$ by raising the flavor index i:

$$\zeta^{\dagger i}(x) \equiv [\zeta_i(x)]^\dagger\,. \tag{A.2}$$

When applied to neutral two-component anticommuting fermion fields, the $(\frac{1}{2}, 0)$ fermion fields, $\xi_{\alpha i}$, always have lowered flavor indices and the $(0, \frac{1}{2})$ fermion fields, $\xi^{\dagger\dot{\alpha} i}$, always have raised indices. When applied to two-component anticommuting fields $\chi(x)$ and $\eta(x)$ that comprise a Dirac fermion, it is more convenient to employ

[1] The term *conjugation* (denoted by †) means complex conjugation when applied to a classical field, or hermitian conjugation when applied to a quantum field operator.

opposite flavor-index heights, $\chi_i(x)$ and $\eta^i(x)$, to reflect the fact that these fields have opposite U(1) charge. In this case, we define $\eta_i^\dagger(x) \equiv [\eta^i(x)]^\dagger$.

The above conventions can be generalized to objects that possess more than one flavor index. If M^{ij} are the elements of a complex symmetric matrix M, then we can treat the multiplet ζ_i as a vector $\boldsymbol{\zeta}$ and consider the following quantity:

$$\boldsymbol{\zeta}^{\mathsf{T}} M \boldsymbol{\zeta} \equiv M^{ij}\zeta_i\zeta_j \,, \tag{A.3}$$

where $\zeta_i\zeta_j = \zeta_j\zeta_i$ and the summation convention has been employed such that two repeated flavor indices are summed over by contracting raised indices with lowered indices (or vice versa).[2] It then follows from eq. (A.2) that

$$[\boldsymbol{\zeta}^{\mathsf{T}} M \boldsymbol{\zeta}]^\dagger \equiv M_{ij}\zeta^{\dagger i}\zeta^{\dagger j} \,, \tag{A.4}$$

after introducing the following notation for the complex conjugate of M^{ij}:

$$M_{ij} \equiv (M^{ij})^* \,. \tag{A.5}$$

Likewise, if $M^i{}_j$ are the elements of a complex matrix M, then

$$\left[M^i{}_j \chi_i \eta^j\right]^\dagger = M_i{}^j \chi^{\dagger i} \eta_j^\dagger \,, \tag{A.6}$$

where we have denoted the complex conjugate of $M^i{}_j$ by

$$M_i{}^j \equiv (M^i{}_j)^* = (M^\dagger)^j{}_i \,. \tag{A.7}$$

In the special case where M is a hermitian matrix, the matrix elements of M satisfy $(M^i{}_j)^* = M^j{}_i$, or equivalently, $M_i{}^j = M^j{}_i$, in the notation of eq. (A.7).

Quantities with more than two indices are similarly treated. For example, the Yukawa couplings arise in the interactions of a scalar boson ϕ_I with pairs of two-component fermions $\psi_j\psi_k$. The complex conjugate of the Yukawa couplings y^{Ijk} are then denoted by $y_{Ijk} \equiv (y^{Ijk})^*$.

A.3 Conventional Units for Particle Physics

In particle physics, it is common practice to introduce so-called natural units in which $\hbar = c = 1$. The electric charge of the electron is negative and is denoted by $-e$, where $e > 0$. The units used for electromagnetic quantities (often called rationalized Gaussian units) follow the Heaviside–Lorentz system, in which e in Gaussian units is replaced by $e/\sqrt{4\pi}$. Hence, the fine structure constant is given by

$$\alpha = \frac{e^2}{4\pi\hbar c} = \frac{e^2}{4\pi} \simeq \frac{1}{137.036} \,, \tag{A.8}$$

using natural units. Moreover, one can relate units of time and length to GeV^{-1} [41]:

$$1\ \text{GeV}^{-1} = 6.582119569 \times 10^{-25}\ \text{s} = 1.973269804 \times 10^{-14}\ \text{cm}. \tag{A.9}$$

[2] When applied to anticommuting two-component fermion fields, $\zeta_i\zeta_j = \zeta_j\zeta_i$ and $\zeta^{\dagger i}\zeta^{\dagger j} = \zeta^{\dagger j}\zeta^{\dagger i}$ in light of eqs. (B.58) and (B.59).

A.4 Spacetime Notation

Contravariant four-vectors are defined with raised indices. For example,

$$a^\mu \equiv (a^0\,;\,\vec{\boldsymbol{a}}) \equiv (a^0\,;\,a^i)\,. \tag{A.10}$$

Here, we use Greek indices such as μ, ν, ρ, and τ to be spacetime indices that run over $0,1,2,3$, while Latin indices such as i, j, k, and ℓ are space indices that run over $1,2,3$. Examples of contravariant four-vectors include positions, momenta, gauge fields, and currents:

$$x^\mu = (t\,;\,\vec{\boldsymbol{x}})\,, \tag{A.11}$$

$$p^\mu = (E\,;\,\vec{\boldsymbol{p}})\,, \tag{A.12}$$

$$A^\mu(x) = (\Phi(\vec{\boldsymbol{x}},t)\,;\,\vec{\boldsymbol{A}}(\vec{\boldsymbol{x}},t))\,, \tag{A.13}$$

$$J^\mu(x) = (\rho(\vec{\boldsymbol{x}},t)\,;\,\vec{\boldsymbol{J}}(\vec{\boldsymbol{x}},t))\,, \tag{A.14}$$

in units with $c = 1$. Covariant four-vectors (e.g., derivatives) are defined with lowered indices. For example,

$$\partial_\mu \equiv \frac{\partial}{\partial x^\mu} = (\partial/\partial t\,;\,\vec{\boldsymbol{\nabla}})\,. \tag{A.15}$$

The following shorthand notation is also useful:

$$A(x)\overleftrightarrow{\partial}_\mu B(x) \equiv A(x)\big[\partial_\mu B(x)\big] - \big[\partial_\mu A(x)\big]B(x)\,. \tag{A.16}$$

The spacetime metric is taken to be

$$g_{\mu\nu} = \mathrm{diag}(1,-1,-1,-1)\,, \tag{A.17}$$

corresponding to the *mostly minus* convention. The metric can be used to lower indices and convert contravariant four-vectors into covariant four-vectors,

$$a_\mu \equiv g_{\mu\nu}a^\nu = (a_0\,;\,a_i) = (a^0\,;\,-\vec{\boldsymbol{a}})\,. \tag{A.18}$$

The summation convention introduced in Appendix A.1 also applies to repeated spacetime (or space) indices. Note that $a_0 = a^0$ and $a_i = -a^i$.

To convert covariant indices into contravariant indices we use the inverse metric $g^{\mu\nu} \equiv (g^{-1})_{\mu\nu}$, which is numerically equal to $g_{\mu\nu}$ in flat Minkowski spacetime:

$$g^{\mu\rho}g_{\rho\nu} = g^\mu_\nu = \delta^\mu_\nu = \begin{cases} 1\,, & \mu = \nu\,, \\ 0\,, & \mu \neq \nu\,, \end{cases} \tag{A.19}$$

where δ^μ_ν is the Kronecker delta function.

The totally anticommuting four-dimensional Levi-Civita tensor $\epsilon^{\mu\nu\rho\tau}$ is equal to

$$\epsilon^{\mu\nu\rho\tau} = \begin{cases} +1\,, & \mu\nu\rho\tau \text{ an even permutation of } 0123\,, \\ -1\,, & \mu\nu\rho\tau \text{ an odd permutation of } 0123\,, \\ 0 & \text{otherwise}\,. \end{cases} \tag{A.20}$$

In particular, $\epsilon^{0123} = +1$. Lowering the four indices to get $\epsilon_{\mu\nu\rho\tau}$, we see that $\epsilon_{0123} = -1$. We shall also define the three-dimensional Levi-Civita tensor

$$\epsilon^{ijk} \equiv \epsilon^{0ijk} . \tag{A.21}$$

Thus, $\epsilon^{123} = +1$. Equations that involve only tensors with *space* indices will always be displayed in this book with *all* space indices in the raised position.[3]

Finally, we note that if the gauge field strength tensor is $F^{\mu\nu}$, then the dual gauge field strength tensor is defined by

$$\widetilde{F}_{\alpha\beta} \equiv \tfrac{1}{2}\epsilon_{\alpha\beta\mu\nu}F^{\mu\nu} . \tag{A.22}$$

A.5 The Pauli Matrices

The sigma matrices are defined with a raised (contravariant) index:

$$\sigma^\mu = (\mathbb{1}_{2\times 2}\,;\, \vec{\boldsymbol{\sigma}}) , \qquad \overline{\sigma}^\mu = (\mathbb{1}_{2\times 2}\,;\, -\vec{\boldsymbol{\sigma}}) , \tag{A.23}$$

where the three-vector of Pauli matrices is given by $\vec{\boldsymbol{\sigma}} \equiv (\sigma^1\,, \sigma^2\,, \sigma^3)$ [see eq. (A.43)] and $\mathbb{1}_{2\times 2}$ is the 2×2 identity matrix. The corresponding quantities with lower (covariant) index are

$$\sigma_\mu = g_{\mu\nu}\sigma^\nu = (\mathbb{1}_{2\times 2}\,;\, -\vec{\boldsymbol{\sigma}}) , \qquad \overline{\sigma}_\mu = g_{\mu\nu}\overline{\sigma}^\nu = (\mathbb{1}_{2\times 2}\,;\, \vec{\boldsymbol{\sigma}}) . \tag{A.24}$$

Various identities involving products of sigma matrices are given in Appendix B. The generators of the $(\frac{1}{2}, 0)$ and $(0, \frac{1}{2})$ representations of the Lorentz group are, respectively, given by[4]

$$\sigma^{\mu\nu} \equiv \frac{i}{4}(\sigma^\mu\overline{\sigma}^\nu - \sigma^\nu\overline{\sigma}^\mu) , \qquad \overline{\sigma}^{\mu\nu} \equiv \frac{i}{4}(\overline{\sigma}^\mu\sigma^\nu - \overline{\sigma}^\nu\sigma^\mu) . \tag{A.25}$$

The components of $\sigma^{\mu\nu}$ and $\overline{\sigma}^{\mu\nu}$ are given by

$$\sigma^{ij} = \overline{\sigma}^{ij} = \tfrac{1}{2}\epsilon^{ijk}\sigma^k , \qquad \sigma^{i0} = -\sigma^{0i} = -\overline{\sigma}^{i0} = \overline{\sigma}^{0i} = \tfrac{1}{2}i\sigma^i . \tag{A.26}$$

In adopting the above definition of the sigma matrices, we differ from the corresponding conventions of Wess and Bagger [B40]. The Wess–Bagger (WB) definition of the sigma matrices can be written (with lowered index μ) as[5]

$$\sigma_\mu^{\rm WB} = (\mathbb{1}_{2\times 2}\,;\, \vec{\boldsymbol{\sigma}}) , \qquad \overline{\sigma}_\mu^{\rm WB} = (\mathbb{1}_{2\times 2}\,;\, -\vec{\boldsymbol{\sigma}}) . \tag{A.27}$$

In contrast to eq. (3.10), the WB definition of σ and $\overline{\sigma}$ yields $\gamma_5 = {\rm diag}(\mathbb{1}_{2\times 2}\,, -\mathbb{1}_{2\times 2})$ in the chiral representation. This convention associates a lowered undotted [raised dotted] two-component spinor with a right-handed [left-handed] four-component spinor. Indeed, this was the common convention in the older literature (for ex-

[3] As an example, $\theta^{ij} = \epsilon^{ijk}\theta^k$ with an implicit sum over $k \in \{1, 2, 3\}$, which relates θ^k defined below eq. (1.4) to θ^{ij}.

[4] Some authors define $\sigma^{\mu\nu}$ and $\overline{\sigma}^{\mu\nu}$ without the factor of i.

[5] Note that the spinor structure of the σ and $\overline{\sigma}$ matrices [see eq. (A.44)] and the definitions of the various (two-index and four-index) epsilon tensors [see eqs. (A.20), (A.37), and (A.38)] are identical both in the WB conventions and in our conventions.

ample, see [B67, B75, B80, B133, B136]). However, in the modern formulation of electroweak theory in terms of left-handed fermions, it is now more common to associate a lower undotted [raised dotted] two-component spinor with a left-handed [right-handed] four-component spinor as in eq. (3.13). This is the motivation for our conventions for the sigma matrices given in eqs. (A.23) and (A.24).

A.6 How to Translate between Metric Signature Conventions

In the mostly minus convention used in this book, the metric tensor is defined as in eq. (A.17), where $g_{00} = +1$. In contrast, in the mostly plus convention, the signs of the diagonal elements of $g_{\mu\nu}$ defined in eq. (A.17) are reversed so that $g_{00} = -1$.

In order to facilitate the comparison of the mostly minus and mostly plus conventions, we provide the key ingredients needed for translating between Minkowski metrics of opposite signature. In our conventions established in eqs. (A.11)–(A.24), each of the following objects (with the Lorentz index heights as shown) is defined *independently* of the metric signature:

$$\text{no sign change:}\quad x^\mu\,,\, p^\mu\,,\, J^{\mu\nu}\,,\, J_{\mu\nu}\,,\, \partial_\mu\,,\, \sigma^\mu\,,\, \overline{\sigma}^\mu\,,\, S^\mu\,,\, J^\mu\,,\, A^\mu\,,\, \nabla_\mu\,,\, G^\mu{}_\nu\,,\\ \gamma^\mu\,,\, \gamma_5\,,\, \delta^\mu_\nu\,,\, \epsilon^{\mu\nu\rho\sigma}\,,\, \epsilon_{\mu\nu\rho\sigma}\,,\, \theta^\mu{}_\nu\,, \tag{A.28}$$

whereas the following change sign when the Minkowski metric signature is reversed:

$$\text{sign change:}\quad g_{\mu\nu}\,,\, g^{\mu\nu}\,,\, x_\mu\,,\, p_\mu\,,\, \partial^\mu\,,\, \sigma_\mu\,,\, \overline{\sigma}_\mu\,,\, S_\mu\,,\, J_\mu\,,\, A_\mu\,,\, \nabla^\mu\,,\, G^{\mu\nu}\,,\, G_{\mu\nu}\,,\\ \gamma_\mu\,,\, \theta^{\mu\nu}\,,\, \theta_{\mu\nu}\,. \tag{A.29}$$

Here, $J^{\mu\nu}$ is the angular momentum tensor, the spin four-vector S^μ is defined in eq. (2.18), J^μ is any conserved current, A^μ is any gauge vector potential, and ∇_μ and $G_{\mu\nu}$ are the corresponding covariant derivative and antisymmetric tensor field strength, respectively. The Dirac gamma matrices are defined in eq. (3.7), and the tensor $\theta^{\mu\nu}$ parameterizes Lorentz transformations [see eqs. (1.2), (A.65), and (A.67)–(A.69)]. The list of eq. (A.29) can be deduced from eq. (A.28) by using the metric tensor and its inverse to lower and raise Lorentz indices, simply because each metric or inverse metric changes sign when the metric signature is reversed. Given any other object not included in eqs. (A.28) and (A.29), it is straightforward to make the appropriate assignment by considering how the object is defined. For example, we must assign $\sigma_{\mu\nu}$, $\overline{\sigma}_{\mu\nu}$, $\sigma^{\mu\nu}$, and $\overline{\sigma}^{\mu\nu}$ to the list of eq. (A.28), based on the definitions given in eqs. (1.115) and (1.116). In general, objects that do not carry Lorentz vector indices (including all fermion spinor fields and spinor wave functions) are defined to be the same in the two metric signatures, with the obvious exception of scalar quantities formed from an odd number of objects from the list of eq. (A.29). For example, the dot product of two four-vectors may or may not change sign when the Minkowski metric signature is reversed. By writing out the dot product explicitly using the metric tensor to contract the indices, one can use eqs. (A.28) and (A.29) to determine the behavior of a dot product under the reversal of the metric signature. In particular, $p\cdot A$ changes sign whereas $\sigma\cdot\partial$ and $\overline{\sigma}\cdot\partial$ do not change sign, when the Minkowski metric signature is reversed.

The translation between Minkowski metrics of opposite signatures is now straightforward. Given any relativistic covariant quantity or equation in the convention where $g_{00}=+1$, one need only employ eqs. (A.28) and (A.29) to obtain the same quantity or equation in the convention where $g_{00}=-1$, and vice versa. Note that for any relativistic covariant term appearing additively in a valid equation, the *relative* sign that results from changing between Minkowski metrics of opposite signature is simply given by $\mathcal{S}=(-1)^{\mathcal{N}}$, where $\mathcal{N}\equiv N_m+N_d+N_G+\cdots$. Here N_m is the number of metric tensors appearing either explicitly or implicitly through contracted upper and lower indices, N_d is the number of spacetime and/or covariant derivatives, N_G is the number of gauge field strength tensors, and the ellipsis $(\cdots)$ accounts for any additional quantities whose contravariant forms (with all Lorentz indices raised) appear in the list of eq. (A.29).

As an example, let us verify that under the reversal of the Minkowski metric signature the gauge covariant derivative ∇_μ does not change sign and the gauge field strength tensor $G^{\mu\nu}$ changes sign. In the metric signature with $g_{00}=+1$, we define

$$\nabla_\mu \equiv \boldsymbol{I}_{d_R}\partial_\mu + igA_\mu\,, \qquad (g_{00}=+1)\,, \tag{A.30}$$

where $A_\mu \equiv A_\mu^a \boldsymbol{T^a}$ is the matrix gauge field for a representation R of dimension d_R, and $\boldsymbol{I}_{d_R}$ is the $d_R\times d_R$ identity matrix. Since under the reversal of the metric signature, ∂_μ does not change sign [according to eq. (A.28)] whereas A_μ changes sign [according to eq. (A.29)], it follows that the quantity defined in the metric signature where $g_{00}=-1$,

$$\nabla_\mu \equiv \boldsymbol{I}_{d_R}\partial_\mu - igA_\mu\,, \qquad (g_{00}=-1) \tag{A.31}$$

has the same overall sign as eq. (A.30). It follows that when the metric signature is reversed, ∇_μ does not change sign whereas $\nabla_\mu \equiv g^{\mu\nu}\nabla_\nu$ does change sign, as indicated in eqs. (A.28) and (A.29). Next, consider the matrix gauge field strength tensor $G_{\mu\nu}\equiv G^a_{\mu\nu}\boldsymbol{T^a}$, defined by

$$G_{\mu\nu} \equiv \frac{-i}{g}[\nabla_\mu\,,\,\nabla_\nu] = \partial_\mu A_\nu - \partial_\nu A_\mu + ig[A_\mu\,,\,A_\nu]\,, \qquad (g_{00}=+1)\,, \tag{A.32}$$

where the commutator $[\nabla_\mu\,,\,\nabla_\nu]$ is an operator that acts on fields that transform with respect to an arbitrary representation R. In the metric signature with $g_{00}=-1$, we define the gauge field strength tensor as a commutator of covariant derivatives with the *opposite* overall sign:

$$G_{\mu\nu} \equiv \frac{i}{g}[\nabla_\mu\,,\,\nabla_\nu] = \partial_\mu A_\nu - \partial_\nu A_\mu - ig[A_\mu\,,\,A_\nu]\,, \qquad (g_{00}=-1)\,, \tag{A.33}$$

where ∇_μ is now defined as in eq. (A.31). Hence, both $G_{\mu\nu}$ and $G^{\mu\nu}=g^{\mu\rho}g^{\nu\sigma}G_{\rho\sigma}$ change sign when the metric signature is reversed, as stated in eq. (A.29).

As another simple illustration, consider the σ-matrix identity

$$\overline{\sigma}^\mu \sigma^\nu \overline{\sigma}^\rho = g^{\mu\nu}\overline{\sigma}^\rho - g^{\mu\rho}\overline{\sigma}^\nu + g^{\nu\rho}\overline{\sigma}^\mu - i\epsilon^{\mu\nu\rho\kappa}\overline{\sigma}_\kappa\,, \qquad (g_{00}=+1)\,. \qquad (A.34)$$

In the opposite metric signature with $g_{00} = -1$, we apply the results of eqs. (A.28) and (A.29) and then multiply both sides of the equation by -1 to obtain

$$\overline{\sigma}^\mu \sigma^\nu \overline{\sigma}^\rho = -g^{\mu\nu}\overline{\sigma}^\rho + g^{\mu\rho}\overline{\sigma}^\nu - g^{\nu\rho}\overline{\sigma}^\mu + i\epsilon^{\mu\nu\rho\kappa}\overline{\sigma}_\kappa\,, \qquad (g_{00}=-1)\,. \qquad (A.35)$$

Finally, in the sigma matrix conventions of Wess and Bagger [B40], both eqs. (A.34) and (A.35) are modified by changing the overall sign of $i\epsilon^{\mu\nu\rho\kappa}\overline{\sigma}_\kappa$. In general, to convert the identities of Appendix B to the conventions of WB, one must first employ the mostly plus metric signature, and then interchange $\sigma \leftrightarrow \overline{\sigma}$ [see eq. (A.27)].

A.7 Two-Component Spinor Notation

As discussed in Chapter 1, two-component spinors come in two varieties: undotted and dotted. The undotted spinor is denoted by χ_α and the dotted spinor is denoted by $\chi^{\dagger\dot{\alpha}}$. We use Greek letters such as α, β, and γ to denote undotted spinor indices and $\dot{\alpha}$, $\dot{\beta}$, and $\dot{\gamma}$ to denote dotted spinor indices. Both types of indices take on one of two values: 1 or 2. Undotted [dotted] spinor indices can be raised and lowered with a two-dimensional undotted [dotted] epsilon tensor:

$$\chi_\alpha = \epsilon_{\alpha\beta}\chi^\beta\,, \qquad \chi^\alpha = \epsilon^{\alpha\beta}\chi_\beta\,, \qquad \chi^\dagger_{\dot{\alpha}} = \epsilon_{\dot{\alpha}\dot{\beta}}\chi^{\dagger\dot{\beta}}\,, \qquad \chi^{\dagger\dot{\alpha}} = \epsilon^{\dot{\alpha}\dot{\beta}}\chi^\dagger_{\dot{\beta}}\,. \qquad (A.36)$$

Explicitly, the undotted ϵ-tensor is given by

$$\epsilon^{\alpha\beta} = \begin{pmatrix} 0 & 1 \\ -1 & 0 \end{pmatrix}, \qquad \epsilon_{\alpha\beta} = \begin{pmatrix} 0 & -1 \\ 1 & 0 \end{pmatrix}, \qquad (A.37)$$

so that $\epsilon^{12} = -\epsilon^{21} = \epsilon_{21} = -\epsilon_{12} = 1$; and

$$\epsilon^{\dot{\alpha}\dot{\beta}} = \begin{pmatrix} 0 & 1 \\ -1 & 0 \end{pmatrix}, \qquad \epsilon_{\dot{\alpha}\dot{\beta}} = \begin{pmatrix} 0 & -1 \\ 1 & 0 \end{pmatrix}, \qquad (A.38)$$

so that $\epsilon^{\dot{1}\dot{2}} = -\epsilon^{\dot{2}\dot{1}} = \epsilon_{\dot{2}\dot{1}} = -\epsilon_{\dot{1}\dot{2}} = 1$. That is, ϵ symbols with raised dotted and undotted indices are numerically equal (and likewise for ϵ with lowered indices).

We also introduce the two-index symmetric Kronecker delta symbol,

$$\delta_1^1 = \delta_2^2 = 1\,, \qquad \delta_2^1 = \delta_1^2 = 0\,, \qquad (A.39)$$

and $\delta_{\dot{\alpha}}^{\dot{\beta}} \equiv (\delta_\alpha^\beta)^*$. Equation (1.41) implies that the numerical values of the undotted and dotted Kronecker delta symbols coincide. The epsilon symbols with undotted and with dotted indices, respectively, satisfy

$$\epsilon_{\alpha\beta}\epsilon^{\gamma\delta} = -\delta_\alpha^\gamma\delta_\beta^\delta + \delta_\alpha^\delta\delta_\beta^\gamma, \qquad \epsilon_{\dot{\alpha}\dot{\beta}}\epsilon^{\dot{\gamma}\dot{\delta}} = -\delta_{\dot{\alpha}}^{\dot{\gamma}}\delta_{\dot{\beta}}^{\dot{\delta}} + \delta_{\dot{\alpha}}^{\dot{\delta}}\delta_{\dot{\beta}}^{\dot{\gamma}}\,, \qquad (A.40)$$

from which it follows that

$$\epsilon_{\alpha\beta}\epsilon^{\beta\gamma} = \epsilon^{\gamma\beta}\epsilon_{\beta\alpha} = \delta^{\gamma}_{\alpha}, \qquad \epsilon_{\dot{\alpha}\dot{\beta}}\epsilon^{\dot{\beta}\dot{\gamma}} = \epsilon^{\dot{\gamma}\dot{\beta}}\epsilon_{\dot{\beta}\dot{\alpha}} = \delta^{\dot{\gamma}}_{\dot{\alpha}}\,, \tag{A.41}$$

$$\epsilon_{\alpha\beta}\epsilon_{\gamma\delta} + \epsilon_{\alpha\gamma}\epsilon_{\delta\beta} + \epsilon_{\alpha\delta}\epsilon_{\beta\gamma} = 0\,, \qquad \epsilon_{\dot{\alpha}\dot{\beta}}\epsilon_{\dot{\gamma}\dot{\delta}} + \epsilon_{\dot{\alpha}\dot{\gamma}}\epsilon_{\dot{\delta}\dot{\beta}} + \epsilon_{\dot{\alpha}\dot{\delta}}\epsilon_{\dot{\beta}\dot{\gamma}} = 0\,. \tag{A.42}$$

Equation (A.42) is often referred to as the Schouten identities.

The explicit forms for the Pauli sigma matrices defined in eq. (A.23) are

$$\sigma^0 = \overline{\sigma}^0 = \begin{pmatrix} 1 & 0 \\ 0 & 1 \end{pmatrix}, \qquad \sigma^1 = -\overline{\sigma}^1 = \begin{pmatrix} 0 & 1 \\ 1 & 0 \end{pmatrix},$$

$$\sigma^2 = -\overline{\sigma}^2 = \begin{pmatrix} 0 & -i \\ i & 0 \end{pmatrix}, \qquad \sigma^3 = -\overline{\sigma}^3 = \begin{pmatrix} 1 & 0 \\ 0 & -1 \end{pmatrix}. \tag{A.43}$$

We associate the following two-component index structure to the sigma matrices defined above: $\sigma^{\mu}_{\alpha\dot{\beta}}$ and $\overline{\sigma}^{\dot{\alpha}\beta}_{\mu}$. The relations between σ^{μ} and $\overline{\sigma}^{\mu}$ are

$$\sigma^{\mu}_{\alpha\dot{\alpha}} = \epsilon_{\alpha\beta}\epsilon_{\dot{\alpha}\dot{\beta}}\overline{\sigma}^{\mu\,\dot{\beta}\beta}\,, \qquad \overline{\sigma}^{\mu\,\dot{\alpha}\alpha} = \epsilon^{\alpha\beta}\epsilon^{\dot{\alpha}\dot{\beta}}\sigma^{\mu}_{\beta\dot{\beta}}\,, \tag{A.44}$$

$$\epsilon^{\alpha\beta}\sigma^{\mu}_{\beta\dot{\alpha}} = \epsilon_{\dot{\alpha}\dot{\beta}}\overline{\sigma}^{\mu\dot{\beta}\alpha}\,, \qquad \epsilon^{\dot{\alpha}\dot{\beta}}\sigma^{\mu}_{\alpha\dot{\beta}} = \epsilon_{\alpha\beta}\overline{\sigma}^{\mu\dot{\alpha}\beta}\,. \tag{A.45}$$

From the sigma matrices, one can construct [see eq. (A.25)],

$$\sigma^{\mu\nu}{}_{\alpha}{}^{\beta} = \frac{i}{4}(\sigma^{\mu}_{\alpha\dot{\rho}}\overline{\sigma}^{\nu\dot{\rho}\beta} - \sigma^{\nu}_{\alpha\dot{\rho}}\overline{\sigma}^{\mu\dot{\rho}\beta})\,, \tag{A.46}$$

$$\overline{\sigma}^{\mu\nu\dot{\alpha}}{}_{\dot{\beta}} = \frac{i}{4}(\overline{\sigma}^{\mu\dot{\alpha}\rho}\sigma^{\nu}_{\rho\dot{\beta}} - \overline{\sigma}^{\nu\dot{\alpha}\rho}\sigma^{\mu}_{\rho\dot{\beta}})\,. \tag{A.47}$$

The components of $\sigma^{\mu\nu}$ and $\overline{\sigma}^{\mu\nu}$ are explicitly given in eq. (A.26).

The matrices $\sigma^{\mu\nu}$ and $\overline{\sigma}^{\mu\nu}$ satisfy a hermiticity relation, $(\sigma^{\mu\nu})^{\dagger} = \overline{\sigma}^{\mu\nu}$, and the self-duality relations,

$$\sigma^{\mu\nu} = -\tfrac{1}{2}i\epsilon^{\mu\nu\rho\kappa}\sigma_{\rho\kappa}\,, \qquad \overline{\sigma}^{\mu\nu} = \tfrac{1}{2}i\epsilon^{\mu\nu\rho\kappa}\overline{\sigma}_{\rho\kappa}\,. \tag{A.48}$$

The self-duality relations can be used to obtain the following two identities:

$$g^{\kappa\rho}\sigma^{\mu\nu} - g^{\nu\rho}\sigma^{\mu\kappa} + g^{\mu\rho}\sigma^{\nu\kappa} - i\epsilon^{\mu\nu\kappa}{}_{\lambda}\sigma^{\lambda\rho} = 0\,, \tag{A.49}$$

$$g^{\kappa\rho}\overline{\sigma}^{\mu\nu} - g^{\nu\rho}\overline{\sigma}^{\mu\kappa} + g^{\mu\rho}\overline{\sigma}^{\nu\kappa} + i\epsilon^{\mu\nu\kappa}{}_{\lambda}\overline{\sigma}^{\lambda\rho} = 0\,. \tag{A.50}$$

A number of useful properties and identities involving $\sigma^{\mu\nu}$ and $\overline{\sigma}^{\mu\nu}$ can be derived. For example, eqs. (A.40) and (B.21) imply that

$$(\sigma^{\mu\nu})_{\alpha}{}^{\beta} = \epsilon_{\alpha\tau}\epsilon^{\beta\gamma}(\sigma^{\mu\nu})_{\gamma}{}^{\tau}\,, \qquad (\overline{\sigma}^{\mu\nu})^{\dot{\alpha}}{}_{\dot{\beta}} = \epsilon^{\dot{\alpha}\dot{\tau}}\epsilon_{\dot{\beta}\dot{\gamma}}(\overline{\sigma}^{\mu\nu})^{\dot{\gamma}}{}_{\dot{\tau}}\,, \tag{A.51}$$

$$\epsilon^{\tau\alpha}(\sigma^{\mu\nu})_{\alpha}{}^{\beta} = \epsilon^{\beta\gamma}(\sigma^{\mu\nu})_{\gamma}{}^{\tau}\,, \qquad \epsilon_{\dot{\tau}\dot{\alpha}}(\overline{\sigma}^{\mu\nu})^{\dot{\alpha}}{}_{\dot{\beta}} = \epsilon_{\dot{\beta}\dot{\gamma}}(\overline{\sigma}^{\mu\nu})^{\dot{\gamma}}{}_{\dot{\tau}}\,, \tag{A.52}$$

$$\epsilon_{\gamma\beta}(\sigma^{\mu\nu})_{\alpha}{}^{\beta} = \epsilon_{\alpha\tau}(\sigma^{\mu\nu})_{\gamma}{}^{\tau}\,, \qquad \epsilon^{\dot{\gamma}\dot{\beta}}(\overline{\sigma}^{\mu\nu})^{\dot{\alpha}}{}_{\dot{\beta}} = \epsilon^{\dot{\alpha}\dot{\tau}}(\overline{\sigma}^{\mu\nu})^{\dot{\gamma}}{}_{\dot{\tau}}\,. \tag{A.53}$$

A.8 Group Indices and the ϵ Symbol

In contrast to eqs. (A.37) and (A.38), when considering the invariant tensors of SU(2) and U(2), the conventions for the epsilon symbols with group indices are

$$\epsilon^{ij} = -\epsilon^{ji} = \epsilon_{ij} = -\epsilon_{ji}\,, \qquad \text{with } \epsilon^{12} = \epsilon_{12} = +1\,. \tag{A.54}$$

Given two SU(2) doublet fields, e.g., $(H_u)_i \equiv (H_u^+\,,\, H_u^0)$ and $(H_d)_i \equiv (H_d^0\,,\, H_d^-)$, it is conventional to define the SU(2) invariant product $H_u H_d$ by

$$H_u H_d \equiv \epsilon^{ij}(H_u)_i(H_d)_j = H_u^+ H_d^- - H_u^0 H_d^0\,. \tag{A.55}$$

The rank-n Levi-Civita epsilon symbol is the totally antisymmetric invariant tensor of SU(n) that generalizes eq. (A.54) such that

$$\epsilon^{i_1 i_2 \cdots i_n} = \epsilon_{i_1 i_2 \cdots i_n}\,, \qquad \text{with } \epsilon^{12\cdots n} = \epsilon_{12\cdots n} = +1\,. \tag{A.56}$$

A.9 Lorentz Transformations

The 4×4 Lorentz transformation matrix, $\Lambda^\mu{}_\nu$ governs the transformation of contravariant four-vectors:

$$V'^\mu = \Lambda^\mu{}_\nu V^\nu\,. \tag{A.57}$$

The invariance of $g_{\mu\nu}V^\mu V^\nu$ under Lorentz transformations implies that

$$\Lambda^\mu{}_\nu g_{\mu\rho}\Lambda^\rho{}_\lambda = g_{\lambda\nu}\,. \tag{A.58}$$

Likewise, we may introduce $\Lambda_\mu{}^\nu$, which governs the transformation of covariant four-vectors:

$$V'_\mu = \Lambda_\mu{}^\nu V_\nu\,. \tag{A.59}$$

One then easily derives a relation between $\Lambda^\mu{}_\nu$ and $\Lambda_\mu{}^\nu$:

$$\Lambda_\mu{}^\nu = g_{\mu\alpha}g^{\nu\beta}\Lambda^\alpha{}_\beta\,. \tag{A.60}$$

Using eqs. (A.58) and (A.60), it follows that $\Lambda_\mu{}^\nu\Lambda^\mu{}_\beta = g^\nu{}_\beta = \delta^\nu_\beta$. This relation allows one to identify

$$\Lambda_\mu{}^\nu = (\Lambda^{-1})^\nu{}_\mu\,, \qquad \Lambda^\mu{}_\nu = (\Lambda^{-1})_\nu{}^\mu\,. \tag{A.61}$$

Using eq. (A.61), one can invert eqs. (A.57) and (A.59):

$$V^\nu = V'^\mu\Lambda_\mu{}^\nu\,, \qquad V_\nu = V'_\mu\Lambda^\mu{}_\nu\,. \tag{A.62}$$

With respect to the Lorentz transformation Λ, a general n-component field Φ transforms according to a representation R of the Lorentz group as

$$\Phi(x^\mu) \to \Phi'(x'^\mu) = M_R(\Lambda)\,\Phi(x^\mu)\,, \tag{A.63}$$

where $M_R(\Lambda)$ is the corresponding (finite) d_R-dimensional matrix representation. Equivalently, the functional form of the transformed field Φ obeys

$$\Phi'(x^\mu) = M_R(\Lambda)\Phi(x^\nu \Lambda_\nu{}^\mu)\,, \tag{A.64}$$

after using eqs. (A.61) and (A.62). For proper orthochronous Lorentz transformations (which are continuously connected to the identity),

$$M_R = \exp\left(-\tfrac{1}{2}i\theta_{\mu\nu}S^{\mu\nu}\right) \simeq \mathbb{1}_{d_R\times d_R} - \tfrac{1}{2}i\theta_{\mu\nu}S^{\mu\nu}\,, \tag{A.65}$$

where $\mathbb{1}_{d_R\times d_R}$ is the $d_R \times d_R$ identity matrix and $\theta_{\mu\nu}$ parameterizes the Lorentz transformation Λ [eq. (1.2)]. The six independent components of the matrix-valued antisymmetric tensor $S^{\mu\nu}$ are the d_R-dimensional generators of the Lorentz group and satisfy the commutation relations

$$[S^{\mu\nu}\,, S^{\lambda\kappa}] = i(g^{\mu\kappa}\,S^{\nu\lambda} + g^{\nu\lambda}\,S^{\mu\kappa} - g^{\mu\lambda}\,S^{\nu\kappa} - g^{\nu\kappa}\,S^{\mu\lambda})\,. \tag{A.66}$$

The $\sigma^{\mu\nu}$ and $\overline{\sigma}^{\mu\nu}$ satisfy the commutation relations of the $S^{\mu\nu}$, and thus can be identified as the generators of the Lorentz group in the $(\frac{1}{2},0)$ and $(0,\frac{1}{2})$ representations, respectively. That is, for the $(\frac{1}{2},0)$ representation with a lowered undotted index (e.g., ψ_α), $S^{\mu\nu} = \sigma^{\mu\nu}$, while for the $(0,\frac{1}{2})$ representation with a raised dotted index (e.g., $\psi^{\dagger\dot{\alpha}}$), $S^{\mu\nu} = \overline{\sigma}^{\mu\nu}$. In particular, the infinitesimal forms for the 4×4 Lorentz transformation matrix Λ and the corresponding matrices M and $(M^{-1})^\dagger$ that transform the $(\frac{1}{2},0)$ and $(0,\frac{1}{2})$ spinors, respectively, are given by

$$\Lambda^\mu{}_\nu \simeq \delta^\mu_\nu + \tfrac{1}{2}\left(\theta_{\rho\nu}g^{\rho\mu} - \theta_{\nu\lambda}g^{\lambda\mu}\right)\,, \tag{A.67}$$

$$M \simeq \mathbb{1}_{2\times 2} - \tfrac{1}{2}i\theta_{\mu\nu}\sigma^{\mu\nu}\,, \tag{A.68}$$

$$(M^{-1})^\dagger \simeq \mathbb{1}_{2\times 2} - \tfrac{1}{2}i\theta_{\mu\nu}\overline{\sigma}^{\mu\nu}. \tag{A.69}$$

The inverses of these quantities are obtained (to first order in θ) by replacing $\theta \to -\theta$ in the above formulae. Using eqs. (1.123), (A.68), and (A.69), we obtain

$$(M^{-1})_\gamma{}^\tau = \epsilon^{\tau\alpha}M_\alpha{}^\beta\epsilon_{\beta\gamma}\,, \qquad (M^{-1\,\dagger})^{\dot{\gamma}}{}_{\dot{\tau}} = \epsilon_{\dot{\tau}\dot{\alpha}}\,(M^\dagger)^{\dot{\alpha}}{}_{\dot{\beta}}\,\epsilon^{\dot{\beta}\dot{\gamma}}\,. \tag{A.70}$$

The infinitesimal forms given by eqs. (A.67)–(A.69) can also be used [with the assistance of eqs. (B.34)–(B.37)] to establish the following two results:

$$M^\dagger\overline{\sigma}^\mu M = \Lambda^\mu{}_\nu\,\overline{\sigma}^\nu\,, \qquad M^{-1}\sigma^\mu(M^{-1})^\dagger = \Lambda^\mu{}_\nu\,\sigma^\nu\,. \tag{A.71}$$

As an example, consider a pure boost from the rest frame to a frame where $p^\mu = (E_{\boldsymbol{p}}\,, \vec{\boldsymbol{p}})$, which corresponds to $\theta_{ij} = 0$ and $\zeta^i = \theta^{i0} = -\theta^{0i}$. We assume that the mass-shell condition is satisfied, i.e., $p^0 = E_{\vec{\boldsymbol{p}}} \equiv (|\vec{\boldsymbol{p}}|^2 + m^2)^{1/2}$. The matrices $M_\alpha{}^\beta$ and $[(M^{-1})^\dagger]^{\dot{\alpha}}{}_{\dot{\beta}}$ that govern the Lorentz transformations of spinor fields with a lowered undotted index and spinor fields with a raised dotted index, respectively, are given by

$$\exp\left(-\frac{i}{2}\theta_{\mu\nu}S^{\mu\nu}\right) = \begin{cases} M = \exp\left(-\tfrac{1}{2}\vec{\boldsymbol{\zeta}}\cdot\vec{\boldsymbol{\sigma}}\right) = \sqrt{\dfrac{p\cdot\sigma}{m}}\,, & \text{for } (\tfrac{1}{2},0)\,, \\ (M^{-1})^\dagger = \exp\left(\tfrac{1}{2}\vec{\boldsymbol{\zeta}}\cdot\vec{\boldsymbol{\sigma}}\right) = \sqrt{\dfrac{p\cdot\overline{\sigma}}{m}}\,, & \text{for } (0,\tfrac{1}{2})\,, \end{cases} \tag{A.72}$$

where

$$\sqrt{p\cdot\sigma} \equiv \frac{(E_{\boldsymbol{p}}+m)\,\mathbb{1}_{2\times 2} - \vec{\boldsymbol{\sigma}}\cdot\vec{\boldsymbol{p}}}{\sqrt{2(E_{\boldsymbol{p}}+m)}}\,, \qquad \sqrt{p\cdot\overline{\sigma}} \equiv \frac{(E_{\boldsymbol{p}}+m)\,\mathbb{1}_{2\times 2} + \vec{\boldsymbol{\sigma}}\cdot\vec{\boldsymbol{p}}}{\sqrt{2(E_{\boldsymbol{p}}+m)}}\,. \tag{A.73}$$

These matrix square roots are defined to be the unique nonnegative definite hermitian matrices (i.e., with nonnegative eigenvalues) whose squares are equal to the nonnegative definite hermitian matrices $p\cdot\sigma$ and $p\cdot\overline{\sigma}$, respectively.[6]

In some applications, the complex conjugate of the two-component spinor wave function is required. One can then employ the identity $\sigma^2\vec{\boldsymbol{\sigma}}\,\sigma^2 = -\vec{\boldsymbol{\sigma}}^*$ to obtain

$$\sqrt{p\cdot\sigma^*} = \sigma^2\sqrt{p\cdot\overline{\sigma}}\,\sigma^2\,, \qquad \sqrt{p\cdot\overline{\sigma}^*} = \sigma^2\sqrt{p\cdot\sigma}\,\sigma^2\,, \tag{A.74}$$

where $\sigma^* = (\mathbb{1}_{2\times 2}\,;\,\vec{\boldsymbol{\sigma}}^*)$ and $\overline{\sigma}^* = (\mathbb{1}_{2\times 2}\,;\,-\vec{\boldsymbol{\sigma}}^*)$.

According to eq. (A.72), the spinor index structure of $\sqrt{p\cdot\sigma}$ and $\sqrt{p\cdot\overline{\sigma}}$ corresponds to that of $M_\alpha{}^\beta$ and $[(M^{-1})^\dagger]^{\dot\alpha}{}_{\dot\beta}$, respectively. Hence, we can rewrite eq. (A.73) as

$$\left[\sqrt{p\cdot\sigma}\,\right]_\alpha{}^\beta \equiv \left[\sqrt{p\cdot\sigma\,\overline{\sigma}^0}\,\right]_\alpha{}^\beta = \frac{(p\cdot\sigma_{\alpha\dot\alpha})\overline{\sigma}^{0\,\dot\alpha\beta} + m\delta_\alpha^\beta}{\sqrt{2(E_{\boldsymbol{p}}+m)}}\,, \tag{A.75}$$

$$\left[\sqrt{p\cdot\overline{\sigma}}\,\right]^{\dot\alpha}{}_{\dot\beta} \equiv \left[\sqrt{p\cdot\overline{\sigma}\,\sigma^0}\,\right]^{\dot\alpha}{}_{\dot\beta} = \frac{(p\cdot\overline{\sigma}^{\dot\alpha\alpha})\sigma^0_{\alpha\dot\beta} + m\delta^{\dot\alpha}_{\dot\beta}}{\sqrt{2(E_{\boldsymbol{p}}+m)}}\,, \tag{A.76}$$

since $\sigma^0 = \overline{\sigma}^0 = \mathbb{1}_{2\times 2}$. Using eqs. (1.90) and (1.91), one can easily verify that

$$\left[\sqrt{p\cdot\sigma}\,\right]_\alpha{}^\gamma\left[\sqrt{p\cdot\sigma}\,\right]_\gamma{}^\beta = (p\cdot\sigma\,\overline{\sigma}^0)_\alpha{}^\beta\,, \tag{A.77}$$

$$\left[\sqrt{p\cdot\overline{\sigma}}\,\right]^{\dot\alpha}{}_{\dot\gamma}\left[\sqrt{p\cdot\overline{\sigma}}\,\right]^{\dot\gamma}{}_{\dot\beta} = (p\cdot\overline{\sigma}\,\sigma^0)^{\dot\alpha}{}_{\dot\beta}\,, \tag{A.78}$$

where implicit factors of $\overline{\sigma}^0$ and σ^0 inside the square roots of eq. (A.77) have been suppressed.

Because $p\cdot\sigma$ and $p\cdot\overline{\sigma}$ are hermitian, we could have defined their hermitian matrix square roots by the hermitian conjugate of eq. (A.72). In this case, the spinor index structure of $\sqrt{p\cdot\sigma}$ and $\sqrt{p\cdot\overline{\sigma}}$ would correspond to that of $[(M^\dagger]^{\dot\alpha}{}_{\dot\beta}$ and $[M^{-1}]_\alpha{}^\beta$, respectively. That is, instead of eqs. (A.75) and (A.76), we would now rewrite eq. (A.73) in the following form:

$$\left[\sqrt{p\cdot\sigma}\,\right]^{\dot\alpha}{}_{\dot\beta} \equiv \left[\sqrt{\overline{\sigma}^0\,p\cdot\sigma}\,\right]^{\dot\alpha}{}_{\dot\beta} = \frac{\overline{\sigma}^{0\,\dot\alpha\beta}(p\cdot\sigma_{\beta\dot\beta}) + m\delta^{\dot\alpha}_{\dot\beta}}{\sqrt{2(E_{\boldsymbol{p}}+m)}}\,, \tag{A.79}$$

$$\left[\sqrt{p\cdot\overline{\sigma}}\,\right]_\alpha{}^\beta \equiv \left[\sqrt{\sigma^0\,p\cdot\overline{\sigma}}\,\right]_\alpha{}^\beta = \frac{\sigma^0_{\alpha\dot\beta}(p\cdot\overline{\sigma}^{\dot\beta\beta}) + m\delta_\alpha^\beta}{\sqrt{2(E_{\boldsymbol{p}}+m)}}\,. \tag{A.80}$$

Using eqs. (1.90) and (1.91), one can again confirm that

$$\left[\sqrt{p\cdot\sigma}\,\right]^{\dot\alpha}{}_{\dot\gamma}\left[\sqrt{p\cdot\sigma}\,\right]^{\dot\gamma}{}_{\dot\beta} = (\overline{\sigma}^0\,p\cdot\sigma)^{\dot\alpha}{}_{\dot\beta}\,, \tag{A.81}$$

$$\left[\sqrt{p\cdot\overline{\sigma}}\,\right]_\alpha{}^\gamma\left[\sqrt{p\cdot\overline{\sigma}}\,\right]_\gamma{}^\beta = (\sigma^0\,p\cdot\overline{\sigma})_\alpha{}^\beta\,, \tag{A.82}$$

[6] Note that $p\cdot\sigma$ and $p\cdot\overline{\sigma}$ are nonnegative matrices when p^μ satisfies the mass-shell condition, $g_{\mu\nu}p^\mu p^\nu = m^2$.

where implicit factors of $\overline{\sigma}^0$ and σ^0 inside the square roots of eq. (A.81) have been suppressed.

The proper choice of the spinor index structure for $\sqrt{p\cdot\sigma}$ and $\sqrt{p\cdot\overline{\sigma}}$ can always be determined for any covariant expression. That is, if we employ the spinor index-free notation (and suppress the factors of σ^0 and $\overline{\sigma}^0$), it will always be clear from the context which spinor index structure for $\sqrt{p\cdot\sigma}$ and $\sqrt{p\cdot\overline{\sigma}}$ is implicit.

As an example that we will exploit in Appendix E, consider an arbitrary four-vector S^μ, defined in a reference frame where $p^\mu = (E\,;\,\vec{p})$, whose rest-frame value is S^μ_R, i.e.,

$$S^\mu = \Lambda^\mu{}_\nu S^\nu_R\,, \qquad \text{with} \qquad \Lambda = \begin{pmatrix} E/m & p^j/m \\ p^i/m & \delta_{ij} + \dfrac{p^i p^j}{m(E+m)} \end{pmatrix}. \tag{A.83}$$

Then, using eqs. (A.71) and (A.72), it follows that

$$\sqrt{p\cdot\sigma}\, S\cdot\overline{\sigma}\, \sqrt{p\cdot\sigma} = mS_R\cdot\overline{\sigma}\,, \tag{A.84}$$

$$\sqrt{p\cdot\overline{\sigma}}\, S\cdot\sigma\, \sqrt{p\cdot\overline{\sigma}} = mS_R\cdot\sigma\,. \tag{A.85}$$

The spinor index structure of eqs. (A.84) and (A.85) is easily established:

$$\left[\sqrt{p\cdot\sigma}\right]^{\dot\beta}{}_{\dot\gamma}\, S\cdot\overline{\sigma}^{\dot\gamma\alpha}\left[\sqrt{p\cdot\sigma}\right]_\alpha{}^\beta = mS_R\cdot\overline{\sigma}^{\dot\beta\beta}\,, \tag{A.86}$$

$$\left[\sqrt{p\cdot\overline{\sigma}}\right]_\beta{}^\gamma\, S\cdot\sigma_{\gamma\dot\alpha}\left[\sqrt{p\cdot\overline{\sigma}}\right]^{\dot\alpha}{}_{\dot\beta} = mS_R\cdot\sigma_{\beta\dot\beta}\,. \tag{A.87}$$

Using eqs. (A.75)–(A.83) and (B.30)–(B.31), one can directly verify the above results.

A.10 Relating Higher-Rank Spinors and Lorentz Tensors

Consider a spinor of rank (n, n) denoted by $S_{\alpha_1\alpha_2\ldots\alpha_n\dot\beta_1\dot\beta_2\ldots\dot\beta_n}$. The object obtained by multiplying S by $\overline{\sigma}^{\mu_1\,\dot\beta_1\alpha_1}\cdots\overline{\sigma}^{\mu_n\,\dot\beta_n\alpha_n}$ has the transformation properties of a rank-n contravariant Lorentz tensor. For example, there is a one-to-one correspondence between each bispinor, $V_{\alpha\dot\beta}$, and the associated Lorentz four-vector V^μ:

$$V^\mu \equiv \tfrac{1}{2}\overline{\sigma}^{\mu\dot\beta\alpha}V_{\alpha\dot\beta}\,, \qquad\qquad V_{\alpha\dot\beta} = V^\mu\sigma_{\mu\alpha\dot\beta}\,. \tag{A.88}$$

In particular, if V^μ is a real four-vector then $V_{\alpha\dot\beta}$ is hermitian (and vice versa). To clarify this last remark, consider the bispinor $V_{\alpha\dot\beta}$ regarded as a 2×2 matrix. Then,

$$(V^{\mathsf{T}})_{\alpha\dot\beta} \equiv V_{\beta\dot\alpha}\,, \qquad (V^*)_{\alpha\dot\beta} \equiv (V_{\alpha\dot\beta})^*\,, \qquad (V^\dagger)_{\alpha\dot\beta} \equiv (V_{\beta\dot\alpha})^*. \tag{A.89}$$

In taking the transpose of $V_{\alpha\dot\beta}$, the rows and columns of V are interchanged without altering the fact that the first spinor index is undotted and the second spinor index is dotted. A hermitian bispinor satisfies $V = V^\dagger$, or equivalently, $V_{\alpha\dot\beta} = (V_{\beta\dot\alpha})^*$.

A.11 Four-Component Spinors and the Dirac Matrices

We begin by introducing the Dirac gamma matrices, which are 4×4 matrices that satisfy the anticommutation relations

$$\{\gamma^\mu, \gamma^\nu\} \equiv \gamma^\mu\gamma^\nu + \gamma^\nu\gamma^\mu = 2g^{\mu\nu}\boldsymbol{I}_4 . \tag{A.90}$$

We also define the matrices

$$\gamma_5 \equiv i\gamma^0\gamma^1\gamma^2\gamma^3 , \qquad \Sigma^{\mu\nu} \equiv \frac{i}{2}[\gamma^\mu, \gamma^\nu] \equiv \frac{i}{2}\left(\gamma^\mu\gamma^\nu - \gamma^\nu\gamma^\mu\right) , \tag{A.91}$$

and note the following duality relation:

$$\gamma_5\Sigma^{\mu\nu} = \tfrac{1}{2}i\epsilon^{\mu\nu\rho\kappa}\Sigma_{\rho\kappa} . \tag{A.92}$$

Finally, we introduce the chiral projection operators

$$P_L \equiv \tfrac{1}{2}(\boldsymbol{I}_4 - \gamma_5) , \qquad P_R \equiv \tfrac{1}{2}(\boldsymbol{I}_4 + \gamma_5) . \tag{A.93}$$

Any 4×4 matrix can be expressed as a complex linear combination of 16 linearly independent 4×4 matrices. A convenient basis to use consists of the following 16 matrices: $\boldsymbol{I}_4$, γ_5, γ^μ, $\gamma^\mu\gamma_5$, and $\Sigma^{\mu\nu}$.

Dirac matrices act on a four-component Dirac spinor field, $\Psi(x)$. Given a four-component spinor $\Psi(x)$, we define the Dirac adjoint field $\overline{\Psi}$, the space-reflected field Ψ^P, the time-reversed field Ψ^t, and the charge-conjugated field Ψ^c as follows:

$$\overline{\Psi}(x) \equiv \psi^\dagger A , \tag{A.94}$$

$$\Psi^P(x) \equiv i\gamma^0\Psi(x) , \tag{A.95}$$

$$\Psi^t(x) \equiv -\gamma^0 B^{-1}\overline{\Psi}^{\mathsf{T}}(x) = -\gamma^0 B^{-1}A^{\mathsf{T}}\Psi^*(x) , \tag{A.96}$$

$$\Psi^C(x) \equiv C\overline{\Psi}^{\mathsf{T}}(x) = D\Psi^*(x) , \tag{A.97}$$

where

$$D \equiv CA^{\mathsf{T}} . \tag{A.98}$$

The Dirac conjugation matrix A, the time-reversal matrix B, the charge-conjugation matrix C, and the matrix D satisfy [B135]

$$A\gamma^\mu A^{-1} = \gamma^{\mu\dagger} , \tag{A.99}$$

$$B\gamma^\mu B^{-1} = \gamma^{\mu\mathsf{T}} , \tag{A.100}$$

$$C^{-1}\gamma^\mu C = -\gamma^{\mu\mathsf{T}} , \tag{A.101}$$

$$D^{-1}\gamma^\mu D = -\gamma^{\mu *} . \tag{A.102}$$

From these definitions and the defining properties of the gamma matrices [eq. (A.90)], one can prove that B and C are antisymmetric matrices in all representations of the gamma matrices. However, eqs. (A.99)–(A.101) do not fix the overall scale of the matrices A, B, and C. Thus, there is some freedom in the definition of these matrices (independently of the chosen gamma matrix representation). We shall remove some of this freedom by imposing the following additional relations:[7]

$$(\Psi^C)^C = \Psi\,, \qquad (\Psi^t)^t = -\Psi\,. \tag{A.103}$$

After imposing eq. (A.103), one can prove that the following results are satisfied in all gamma matrix representations:

$$BA^{-1} = -(AB^{-1})^*\,, \qquad (AC)^{-1} = (AC)^*\,, \tag{A.104}$$

$$CB = -\gamma_5\,, \qquad A^\dagger A^{-1} = e^{i\theta_A}\boldsymbol{I}_4\,, \tag{A.105}$$

$$D^*D = DD^* = \boldsymbol{I}_4\,. \tag{A.106}$$

where the value of θ_A is conventional. The standard choice, $\theta_A = 0$, corresponds to the condition that $\overline{\Psi}(x)\Psi(x)$ is a hermitian quantity in all gamma matrix representations. In this convention, $A^\dagger = A$.

In all the expressions above, the four-component spinor indices have been suppressed. We can assign a lowered four-component spinor index a to a Dirac spinor field, Ψ_a where $a \in \{1,2,3,4\}$. The conjugate spinor $\overline{\Psi}^a$ is assigned a raised spinor index. It is also convenient to introduce barred four-component spinor indices for hermitian-conjugated four-component spinors:

$$\Psi^\dagger_{\bar{a}} \equiv (\Psi_a)^\dagger\,, \qquad \overline{\Psi}^{\dagger\bar{a}} \equiv (\overline{\Psi}^a)^\dagger\,. \tag{A.107}$$

The spinor index structure of the Dirac conjugation matrix A is then fixed by noting that the Dirac conjugate spinor, $\overline{\Psi}^b \equiv \Psi^\dagger_{\bar{a}} A^{\bar{a}b}$, has a raised unbarred spinor index, whereas the hermitian-conjugated spinor has a lowered barred spinor index.

The charge-conjugation matrix C can be used to raise and lower four-component spinor indices, which we shall employ in defining the spinors Ψ^a, $\Psi^{\dagger\bar{a}}$, $\overline{\Psi}_a$, and $\overline{\Psi}^\dagger_{\bar{a}}$:

$$\Psi_a = C_{ab}\Psi^b\,, \qquad \Psi^a = (C^{-1})^{ab}\Psi_b\,, \tag{A.108}$$

$$\Psi^\dagger_{\bar{a}} = C_{\bar{a}\bar{b}}\Psi^{\dagger\bar{b}}\,, \qquad \Psi^{\dagger\bar{a}} = (C^{-1})^{\bar{a}\bar{b}}\Psi^\dagger_{\bar{b}}\,, \tag{A.109}$$

where

$$C_{\bar{a}\bar{b}} \equiv (C_{ab})^*\,, \qquad (C^{-1})^{\bar{a}\bar{b}} \equiv [(C^{-1})^{ab}]^*\,. \tag{A.110}$$

Equations (A.108) and (A.109) also apply to $\overline{\Psi}^a$, $\overline{\Psi}_a$, and their hermitian conjugates. In particular, one can identify the Dirac conjugate spinor with a lowered spinor index ($\overline{\Psi}_a$) as the charge-conjugated spinor, $\Psi^C \equiv C\overline{\Psi}^{\mathsf{T}}$, and the Dirac spinor with a raised spinor index (Ψ^a) as the Dirac conjugate of the charge-conjugated spinor, $\overline{\Psi^C} = -\Psi^{\mathsf{T}}C^{-1}$. That is,

$$\Psi^C_a \equiv \overline{\Psi}_a = C_{ab}\overline{\Psi}^b\,, \qquad \overline{\Psi^C}^{\,a} = \Psi^a = (C^{-1})^{ab}\Psi_b\,. \tag{A.111}$$

Using eqs. (A.108) and (A.109) along with the properties of the matrices A and

[7] The condition $(\Psi^P)^P = -\Psi$ is automatically satisfied and provides no further constraint.

C, it then follows that

$$\overline{\Psi}_a = (A^{-1})_{a\bar{b}}\Psi^{\dagger\bar{b}}\,, \qquad \overline{\Psi}^{\,a} = \Psi^{\dagger}_{\bar{b}}A^{\bar{b}a}\,, \tag{A.112}$$

$$\Psi^{\dagger}_{\bar{a}} = \overline{\Psi}^{\,b}(A^{-1})_{b\bar{a}}\,, \qquad \Psi^{\dagger\bar{a}} = A^{\bar{a}b}\,\overline{\Psi}_b\,. \tag{A.113}$$

The charge-conjugated spinor can also be written as $\Psi^C_a \equiv D_a{}^{\bar{c}}\Psi^{\dagger}_{\bar{c}}$ [see eq. (A.97)]. The spinor index structure of D (and its inverse) derives from

$$D_a{}^{\bar{c}} \equiv C_{ab}(A^{\mathsf{T}})^{b\bar{c}} = C_{ab}A^{\bar{c}b}\,, \qquad (D^{-1})_{\bar{a}}{}^{c} \equiv (C^*A)_{\bar{a}}{}^{c} = C_{\bar{a}\bar{b}}A^{\bar{b}c}\,, \tag{A.114}$$

where we have used $D^{-1} = D^*$. Combining the results of eqs. (A.108), (A.109), (A.112), and (A.113) then yields

$$\overline{\Psi}_a = D_a{}^{\bar{c}}\,\Psi^{\dagger}_{\bar{c}}\,, \qquad \Psi^{\dagger}_{\bar{a}} = (D^{-1})_{\bar{a}}{}^{c}\,\overline{\Psi}_c\,, \tag{A.115}$$

$$\overline{\Psi}^{\,a} = -\Psi^{\dagger\bar{c}}(D^{-1})_{\bar{c}}{}^{a}\,, \qquad \Psi^{\dagger\bar{a}} = -\overline{\Psi}^{\,c}\,D_c{}^{\bar{a}}\,. \tag{A.116}$$

In summary, a four-component spinor Ψ_a and its charge-conjugated spinor Ψ^C_a possess a lowered unbarred spinor index, whereas the corresponding Dirac conjugates, $\overline{\Psi}^{\,a}$ and $\overline{\Psi^C}^{\,a}$, possess a raised unbarred spinor index. The corresponding hermitian-conjugated spinors exhibit barred spinor indices (with the height of each spinor index unchanged). Following eqs. (A.108) and (A.109), one can also lower or raise a four-component unbarred or barred spinor index by multiplying by the appropriate matrix C, C^{-1}, C^*, or $(C^{-1})^*$, respectively. If we identify the identity matrix, the Dirac gamma matrices, and their products collectively by Γ, then the four-component spinor index structures of these matrices and their inverses are given by $\Gamma_a{}^b$ and $(\Gamma^{-1})_a{}^b$. For the matrices A, B, C, D, and their inverses, the four-component spinor index structure is given by

$$A^{\bar{a}b}\,,\ (A^{-1})_{a\bar{b}}\,,\ B^{ab}\,,\ (B^{-1})_{ab}\,, C_{ab}\,,\ (C^{-1})^{ab}\,, D_a{}^{\bar{b}}\,,\ (D^{-1})_{\bar{a}}{}^{b}\,. \tag{A.117}$$

The corresponding complex-conjugated matrices exhibit the analogous spinor index structure with unbarred spinor indices changed to barred spinor indices and vice versa, whereas matrix transposition simply interchanges row and column indices.

In Section 3.1, we introduced the chiral (or high-energy) representation for the gamma matrices, which is defined by

$$\gamma^0_C = \begin{pmatrix} 0 & \boldsymbol{I}_2 \\ \boldsymbol{I}_2 & 0 \end{pmatrix}, \qquad \gamma^i_C = \begin{pmatrix} 0 & \sigma^i \\ -\sigma^i & 0 \end{pmatrix}. \tag{A.118}$$

This representation follows naturally from the two-component form of the Dirac Lagrangian. But, given any representation of the gamma matrices, γ^μ [and corresponding four-component Dirac spinor $\Psi(x)$], a new representation $\widetilde{\gamma}^\mu \equiv \mathcal{S}\gamma^\mu\mathcal{S}^{-1}$ also satisfies eq. (A.90), where $\mathcal{S}$ is any nonsingular matrix. The Dirac spinor with respect to the new representation is given by $\widetilde{\Psi}(x) = \mathcal{S}\Psi(x)$. The matrices A, B, C, and D are defined according to eqs. (A.99)–(A.98) with respect to the gamma matrices appropriate to the given representation. In addition, we impose eq. (A.103) and $\widetilde{A}^\dagger = \widetilde{A}$. This is still not quite sufficient to fix $\widetilde{A}$, $\widetilde{B}$, and $\widetilde{C}$ uniquely, and we find

$$\widetilde{A} = \pm|z|(\mathcal{S}^{-1})^\dagger A\mathcal{S}^{-1}\,, \quad \widetilde{B} = z(\mathcal{S}^{-1})^{\mathsf{T}} B\mathcal{S}^{-1}\,, \quad \widetilde{C} = z^{-1}\mathcal{S}C\mathcal{S}^{\mathsf{T}}\,, \tag{A.119}$$

where z is an arbitrary complex number. Note that there is only a phase ambiguity in the definition of $\widetilde{D} = \widetilde{C}\widetilde{A}^{\mathsf{T}}$. In particular,

$$\widetilde{D} = e^{i\varsigma}\mathcal{S}D\mathcal{S}^{*-1}\,. \tag{A.120}$$

Most modern textbooks impose one additional constraint by restricting to a class of γ-matrices such that $\gamma^{0\dagger} = \gamma^0$ and $\gamma^{i\dagger} = -\gamma^i$. This constraint is satisfied by the chiral representation and the two representations (Dirac and Majorana) introduced below. Representations in this class are related by a unitary similarity transformation, $\widetilde{\gamma}^\mu = \mathcal{S}\gamma^\mu\mathcal{S}^{-1}$, with $\mathcal{S}^\dagger\mathcal{S} = \boldsymbol{I}_4$. One can then show that it is possible to choose A, B, and C to be unitary matrices in all the gamma matrix representations of this class. Moreover, in this convention, $|z| = 1$, so that in any gamma matrix representation of this class one is left with a sign ambiguity in the definition of A and a phase ambiguity in the definitions of B and C.

Another common representation for the gamma matrices is the Dirac (or low-energy) representation:

$$\gamma_D^\mu = \mathcal{S}\gamma_C^\mu\mathcal{S}^{-1}\,, \qquad \mathcal{S} = \frac{1}{\sqrt{2}}\begin{pmatrix} \boldsymbol{I}_2 & \boldsymbol{I}_2 \\ -\boldsymbol{I}_2 & \boldsymbol{I}_2 \end{pmatrix}. \tag{A.121}$$

Note that in this case, $\mathcal{S}^{-1} = \mathcal{S}^{\mathsf{T}} = \mathcal{S}^\dagger$. As a result, the representations of A, B, C, and D in terms of gamma matrices is the same for both the chiral and the Dirac representations. A convenient choice for A, B, C, and D in terms of the gamma matrices is given by

$$A = \gamma^0\,, \qquad B = \gamma^1\gamma^3\,, \qquad C = i\gamma^0\gamma^2\,, \qquad D = -i\gamma^2\,. \tag{A.122}$$

Finally, we mention one other representation for the Dirac matrices in which all four gamma matrices are purely imaginary, *i.e.*, $\gamma^{\mu *} = -\gamma^\mu$. This is the Majorana representation:

$$\gamma_M^\mu = \mathcal{S}\gamma_D^\mu\mathcal{S}^{-1}\,, \qquad \mathcal{S} = \frac{1}{\sqrt{2}}\begin{pmatrix} \boldsymbol{I}_2 & \sigma^2 \\ \sigma^2 & -\boldsymbol{I}_2 \end{pmatrix}. \tag{A.123}$$

Here, one finds that $\mathcal{S} = \mathcal{S}^{-1} = \mathcal{S}^\dagger$. A convenient choice for A, B, and C in terms of the gamma matrices is given by

$$A = \gamma^0\,, \qquad B = \gamma^0\gamma_5\,, \qquad C = -\gamma^0\,. \tag{A.124}$$

In particular, in this convention $D = CA^{\mathsf{T}} = \boldsymbol{I}_4$, corresponding to the choice of $e^{i\varsigma} = -i$ in eq. (A.120). That is, in the Majorana representation with the conventions above, $\Psi^C = \Psi^*$. Hence, in the Majorana representation, the four-component self-conjugate Majorana fermion (which satisfies $\Psi^C = \Psi = \Psi^*$) is a real field.

Appendix B Compendium of Sigma Matrix and Fierz Identities

B.1 Sigma Matrix Identities

Various sigma matrices – σ^μ, $\overline{\sigma}^\mu$, $\sigma^{\mu\nu}$ and $\overline{\sigma}^{\mu\nu}$ – were defined in eqs. (A.43), (A.46), and (A.47). The following identities involving them are particularly useful.

$$\sigma^\mu_{\alpha\dot{\alpha}}\overline{\sigma}^{\dot{\beta}\beta}_\mu = 2\delta_\alpha{}^\beta\delta^{\dot{\beta}}{}_{\dot{\alpha}}, \tag{B.1}$$

$$\sigma^\mu_{\alpha\dot{\alpha}}\sigma_{\mu\beta\dot{\beta}} = 2\epsilon_{\alpha\beta}\epsilon_{\dot{\alpha}\dot{\beta}}, \tag{B.2}$$

$$\overline{\sigma}^{\mu\dot{\alpha}\alpha}\overline{\sigma}^{\dot{\beta}\beta}_\mu = 2\epsilon^{\alpha\beta}\epsilon^{\dot{\alpha}\dot{\beta}}, \tag{B.3}$$

$$(\sigma^{\mu\nu})_\alpha{}^\beta(\sigma_{\mu\nu})_\rho{}^\kappa = 2\delta_\alpha{}^\kappa\delta_\rho{}^\beta - \delta_\alpha{}^\beta\delta_\rho{}^\kappa, \tag{B.4}$$

$$(\overline{\sigma}^{\mu\nu})^{\dot{\alpha}}{}_{\dot{\beta}}(\overline{\sigma}_{\mu\nu})^{\dot{\rho}}{}_{\dot{\kappa}} = 2\delta^{\dot{\alpha}}{}_{\dot{\kappa}}\delta^{\dot{\rho}}{}_{\dot{\beta}} - \delta^{\dot{\alpha}}{}_{\dot{\beta}}\delta^{\dot{\rho}}{}_{\dot{\kappa}}, \tag{B.5}$$

$$(\sigma^{\mu\nu})_\alpha{}^\beta(\overline{\sigma}_{\mu\nu})^{\dot{\rho}}{}_{\dot{\kappa}} = 0, \tag{B.6}$$

$$[\sigma^\mu\overline{\sigma}_\mu]_\alpha{}^\beta = 4\delta_\alpha{}^\beta, \tag{B.7}$$

$$[\overline{\sigma}^\mu\sigma_\mu]^{\dot{\alpha}}{}_{\dot{\beta}} = 4\delta^{\dot{\alpha}}{}_{\dot{\beta}}, \tag{B.8}$$

$$[\sigma^\mu\overline{\sigma}^\nu + \sigma^\nu\overline{\sigma}^\mu]_\alpha{}^\beta = 2g^{\mu\nu}\delta_\alpha{}^\beta, \tag{B.9}$$

$$[\overline{\sigma}^\mu\sigma^\nu + \overline{\sigma}^\nu\sigma^\mu]^{\dot{\alpha}}{}_{\dot{\beta}} = 2g^{\mu\nu}\delta^{\dot{\alpha}}{}_{\dot{\beta}}, \tag{B.10}$$

$$[\sigma^\mu\overline{\sigma}^\nu]_\alpha{}^\beta = -2i(\sigma^{\mu\nu})_\alpha{}^\beta + g^{\mu\nu}\delta_\alpha{}^\beta, \tag{B.11}$$

$$[\overline{\sigma}^\mu\sigma^\nu]^{\dot{\alpha}}{}_{\dot{\beta}} = -2i(\overline{\sigma}^{\mu\nu})^{\dot{\alpha}}{}_{\dot{\beta}} + g^{\mu\nu}\delta^{\dot{\alpha}}{}_{\dot{\beta}}, \tag{B.12}$$

$$[\sigma^{\mu\nu}\sigma_{\mu\nu}]_\alpha{}^\beta = 3\delta_\alpha{}^\beta, \tag{B.13}$$

$$[\overline{\sigma}^{\mu\nu}\overline{\sigma}_{\mu\nu}]^{\dot{\alpha}}{}_{\dot{\beta}} = 3\delta^{\dot{\alpha}}{}_{\dot{\beta}}\,. \tag{B.14}$$

Three identities that generalize the results of eqs. (B.4)–(B.6) are

$$g_{\rho\kappa}(\sigma^{\mu\rho})_\alpha{}^\beta(\sigma^{\nu\kappa})_\gamma{}^\delta = \tfrac{1}{4}g^{\mu\nu}(2\delta_\alpha{}^\delta\delta_\gamma{}^\beta - \delta_\alpha{}^\beta\delta_\gamma{}^\delta) - \tfrac{1}{2}i[\delta_\gamma{}^\beta(\sigma^{\mu\nu})_\alpha{}^\delta - \delta_\alpha{}^\delta(\sigma^{\mu\nu})_\gamma{}^\beta], \tag{B.15}$$

$$g_{\rho\kappa}(\overline{\sigma}^{\mu\rho})^{\dot{\alpha}}{}_{\dot{\beta}}(\overline{\sigma}^{\nu\kappa})^{\dot{\gamma}}{}_{\dot{\delta}} = \tfrac{1}{4}g^{\mu\nu}(2\delta^{\dot{\alpha}}{}_{\dot{\delta}}\delta^{\dot{\gamma}}{}_{\dot{\beta}} - \delta^{\dot{\alpha}}{}_{\dot{\beta}}\delta^{\dot{\gamma}}{}_{\dot{\delta}}) - \tfrac{1}{2}i[\delta^{\dot{\gamma}}{}_{\dot{\beta}}(\overline{\sigma}^{\mu\nu})^{\dot{\alpha}}{}_{\dot{\delta}} - \delta^{\dot{\alpha}}{}_{\dot{\delta}}(\overline{\sigma}^{\mu\nu})^{\dot{\gamma}}{}_{\dot{\beta}}], \tag{B.16}$$

$$g_{\rho\kappa}(\sigma^{\mu\rho})_\alpha{}^\beta(\overline{\sigma}^{\nu\kappa})^{\dot{\beta}}{}_{\dot{\alpha}} = -\tfrac{1}{8}(2\delta_\alpha{}^\gamma\delta_\delta{}^\beta - \delta_\alpha{}^\beta\delta_\delta{}^\gamma)(\sigma^\mu_{\gamma\dot{\alpha}}\overline{\sigma}^{\nu\dot{\beta}\delta} + \sigma^\nu_{\gamma\dot{\alpha}}\overline{\sigma}^{\mu\dot{\beta}\delta})\,. \tag{B.17}$$

Equations (B.1)–(B.6) and eqs. (B.15)–(B.17) are examples of Fierz identities. A comprehensive set of Fierz identities is provided in Appendix B.3.

Additional useful identities are displayed here for completeness. Henceforth, we suppress the spinor indices.

$$\mathrm{Tr}[\sigma^\mu\overline{\sigma}^\nu] = \mathrm{Tr}[\overline{\sigma}^\mu\sigma^\nu] = 2g^{\mu\nu}\,, \tag{B.18}$$

$$\mathrm{Tr}[\sigma^\mu\overline{\sigma}^\nu\sigma^\rho\overline{\sigma}^\kappa] = 2\left(g^{\mu\nu}g^{\rho\kappa} - g^{\mu\rho}g^{\nu\kappa} + g^{\mu\kappa}g^{\nu\rho} + i\epsilon^{\mu\nu\rho\kappa}\right), \tag{B.19}$$

$$\mathrm{Tr}[\overline{\sigma}^\mu\sigma^\nu\overline{\sigma}^\rho\sigma^\kappa] = 2\left(g^{\mu\nu}g^{\rho\kappa} - g^{\mu\rho}g^{\nu\kappa} + g^{\mu\kappa}g^{\nu\rho} - i\epsilon^{\mu\nu\rho\kappa}\right), \tag{B.20}$$

$$\mathrm{Tr}\;\sigma^{\mu\nu} = \mathrm{Tr}\;\overline{\sigma}^{\mu\nu} = 0\,, \tag{B.21}$$

$$\mathrm{Tr}[\sigma^{\mu\nu}\sigma^{\rho\kappa}] = \tfrac{1}{2}\left[g^{\mu\rho}g^{\nu\kappa} - g^{\mu\kappa}g^{\nu\rho} - i\epsilon^{\mu\nu\rho\kappa}\right], \tag{B.22}$$

$$\mathrm{Tr}[\overline{\sigma}^{\mu\nu}\overline{\sigma}^{\rho\kappa}] = \tfrac{1}{2}\left[g^{\mu\rho}g^{\nu\kappa} - g^{\mu\kappa}g^{\nu\rho} + i\epsilon^{\mu\nu\rho\kappa}\right], \tag{B.23}$$

$$\overline{\sigma}^\mu\sigma^\nu\overline{\sigma}_\mu = -2\overline{\sigma}^\nu, \tag{B.24}$$

$$\sigma^\mu\overline{\sigma}^\nu\sigma_\mu = -2\sigma^\nu, \tag{B.25}$$

$$\sigma^\mu\overline{\sigma}^\nu\sigma^\rho\overline{\sigma}_\mu = 4g^{\nu\rho}, \tag{B.26}$$

$$\overline{\sigma}^\mu\sigma^\nu\overline{\sigma}^\rho\sigma_\mu = 4g^{\nu\rho}, \tag{B.27}$$

$$\overline{\sigma}^\mu\sigma^\nu\overline{\sigma}^\rho\sigma^\kappa\overline{\sigma}_\mu = -2\overline{\sigma}^\kappa\sigma^\rho\overline{\sigma}^\nu, \tag{B.28}$$

$$\sigma^\mu\overline{\sigma}^\nu\sigma^\rho\overline{\sigma}^\kappa\sigma_\mu = -2\sigma^\kappa\overline{\sigma}^\rho\sigma^\nu, \tag{B.29}$$

$$\overline{\sigma}^\mu\sigma^\nu\overline{\sigma}^\rho = g^{\mu\nu}\overline{\sigma}^\rho - g^{\mu\rho}\overline{\sigma}^\nu + g^{\nu\rho}\overline{\sigma}^\mu - i\epsilon^{\mu\nu\rho\kappa}\overline{\sigma}_\kappa, \tag{B.30}$$

$$\sigma^\mu\overline{\sigma}^\nu\sigma^\rho = g^{\mu\nu}\sigma^\rho - g^{\mu\rho}\sigma^\nu + g^{\nu\rho}\sigma^\mu + i\epsilon^{\mu\nu\rho\kappa}\sigma_\kappa, \tag{B.31}$$

$$\epsilon^{\mu\nu\kappa}{}_\lambda\sigma^{\lambda\rho} = -i\left(g^{\kappa\rho}\sigma^{\mu\nu} - g^{\nu\rho}\sigma^{\mu\kappa} + g^{\mu\rho}\sigma^{\nu\kappa}\right), \tag{B.32}$$

$$\epsilon^{\mu\nu\kappa}{}_\lambda\overline{\sigma}^{\lambda\rho} = i\left(g^{\kappa\rho}\overline{\sigma}^{\mu\nu} - g^{\nu\rho}\overline{\sigma}^{\mu\kappa} + g^{\mu\rho}\overline{\sigma}^{\nu\kappa}\right), \tag{B.33}$$

$$\sigma^{\mu\nu}\sigma^\rho = \tfrac{1}{2}i(g^{\nu\rho}\sigma^\mu - g^{\mu\rho}\sigma^\nu + i\epsilon^{\mu\nu\rho\kappa}\sigma_\kappa), \tag{B.34}$$

$$\overline{\sigma}^{\mu\nu}\overline{\sigma}^\rho = \tfrac{1}{2}i(g^{\nu\rho}\overline{\sigma}^\mu - g^{\mu\rho}\overline{\sigma}^\nu - i\epsilon^{\mu\nu\rho\kappa}\overline{\sigma}_\kappa), \tag{B.35}$$

$$\overline{\sigma}^\mu\sigma^{\nu\rho} = \tfrac{1}{2}i(g^{\mu\nu}\overline{\sigma}^\rho - g^{\mu\rho}\overline{\sigma}^\nu - i\epsilon^{\mu\nu\rho\kappa}\overline{\sigma}_\kappa), \tag{B.36}$$

$$\sigma^\mu\overline{\sigma}^{\nu\rho} = \tfrac{1}{2}i(g^{\mu\nu}\sigma^\rho - g^{\mu\rho}\sigma^\nu + i\epsilon^{\mu\nu\rho\kappa}\sigma_\kappa), \tag{B.37}$$

$$\begin{aligned}\sigma^{\mu\nu}\sigma^{\rho\kappa} &= -\tfrac{1}{4}(g^{\nu\rho}g^{\mu\kappa} - g^{\mu\rho}g^{\nu\kappa} + i\epsilon^{\mu\nu\rho\kappa}) \\ &\quad + \tfrac{1}{2}i(g^{\nu\rho}\sigma^{\mu\kappa} + g^{\mu\kappa}\sigma^{\nu\rho} - g^{\mu\rho}\sigma^{\nu\kappa} - g^{\nu\kappa}\sigma^{\mu\rho}),\end{aligned} \tag{B.38}$$

$$\begin{aligned}\overline{\sigma}^{\mu\nu}\overline{\sigma}^{\rho\kappa} &= -\tfrac{1}{4}(g^{\nu\rho}g^{\mu\kappa} - g^{\mu\rho}g^{\nu\kappa} - i\epsilon^{\mu\nu\rho\kappa}) \\ &\quad + \tfrac{1}{2}i(g^{\nu\rho}\overline{\sigma}^{\mu\kappa} + g^{\mu\kappa}\overline{\sigma}^{\nu\rho} - g^{\mu\rho}\overline{\sigma}^{\nu\kappa} - g^{\nu\kappa}\overline{\sigma}^{\mu\rho}).\end{aligned} \tag{B.39}$$

One can recursively derive trace formulae for products of six or more $\sigma/\overline{\sigma}$ matrices by using the results of eqs. (B.30) and (B.31) to reduce the number of $\sigma/\overline{\sigma}$ matrices by two. For example,

$$\begin{aligned}\mathrm{Tr}[\sigma^\mu\overline{\sigma}^\nu\sigma^\rho\overline{\sigma}^\kappa\sigma^\lambda\overline{\sigma}^\delta] &= g^{\mu\nu}\mathrm{Tr}[\sigma^\rho\overline{\sigma}^\kappa\sigma^\lambda\overline{\sigma}^\delta] - g^{\mu\rho}\mathrm{Tr}[\sigma^\nu\overline{\sigma}^\kappa\sigma^\lambda\overline{\sigma}^\delta] + g^{\nu\rho}\mathrm{Tr}[\sigma^\mu\overline{\sigma}^\kappa\sigma^\lambda\overline{\sigma}^\delta] \\ &\quad + i\epsilon^{\mu\nu\rho\epsilon}\mathrm{Tr}[\sigma_\epsilon\overline{\sigma}^\kappa\sigma^\lambda\overline{\sigma}^\delta]\,,\end{aligned} \tag{B.40}$$

$$\begin{aligned}\mathrm{Tr}[\overline{\sigma}^\mu\sigma^\nu\overline{\sigma}^\rho\sigma^\kappa\overline{\sigma}^\lambda\sigma^\delta] &= g^{\mu\nu}\mathrm{Tr}[\overline{\sigma}^\rho\sigma^\kappa\overline{\sigma}^\lambda\sigma^\delta] - g^{\mu\rho}\mathrm{Tr}[\overline{\sigma}^\nu\sigma^\kappa\overline{\sigma}^\lambda\sigma^\delta] + g^{\nu\rho}\mathrm{Tr}[\overline{\sigma}^\mu\sigma^\kappa\overline{\sigma}^\lambda\sigma^\delta] \\ &\quad - i\epsilon^{\mu\nu\rho\epsilon}\mathrm{Tr}[\overline{\sigma}_\epsilon\sigma^\kappa\overline{\sigma}^\lambda\sigma^\delta]\,.\end{aligned} \tag{B.41}$$

We then use eqs. (B.19) and (B.20) to evaluate the remaining traces over four $\sigma/\overline{\sigma}$ matrices.

B.2 Two-Component Spinor Product Identities

Next, we consider two-component spinor identities that arise when manipulating products of two or four spinors. The heights of indices must be consistent in the sense that lowered indices must always be contracted with raised indices. As a convention, *descending* contracted undotted indices ("from northwest to southeast") and *ascending* contracted dotted indices ("from southwest to northeast"),

$$^{\alpha}{}_{\alpha} \qquad \text{and} \qquad {}_{\dot\alpha}{}^{\dot\alpha}\,, \tag{B.42}$$

can be suppressed. In all spinor products given in this book, contracted indices always have heights that conform to eq. (B.42)). For example, in an index-free notation, we define

$$\chi\eta \equiv \chi^{\alpha}\eta_{\alpha}, \tag{B.43}$$

$$\chi^{\dagger}\eta^{\dagger} \equiv \chi^{\dagger}_{\dot\alpha}\eta^{\dagger\dot\alpha}, \tag{B.44}$$

$$\chi^{\dagger}\overline{\sigma}^{\mu}\eta \equiv \chi^{\dagger}_{\dot\alpha}\overline{\sigma}^{\mu\dot\alpha\beta}\eta_{\beta}, \tag{B.45}$$

$$\chi\sigma^{\mu}\eta^{\dagger} \equiv \chi^{\alpha}\sigma^{\mu}_{\alpha\dot\beta}\eta^{\dagger\dot\beta}, \tag{B.46}$$

$$\chi\sigma^{\mu\nu}\eta \equiv \chi^{\alpha}(\sigma^{\mu\nu})_{\alpha}{}^{\beta}\eta_{\beta}, \tag{B.47}$$

$$\chi^{\dagger}\overline{\sigma}^{\mu\nu}\eta^{\dagger} \equiv \chi^{\dagger}_{\dot\alpha}(\overline{\sigma}^{\mu\nu})^{\dot\alpha}{}_{\dot\beta}\eta^{\dagger\dot\beta}. \tag{B.48}$$

All the spinor-index-contracted products above have natural interpretations as products of matrices and vectors by regarding η_{α} and $\eta^{\dagger\dot\alpha}$ as column vectors and $\chi^{\dagger}_{\dot\alpha}$ and χ^{α} as row vectors of the two-dimensional spinor space. However, the reader is cautioned that, in the index-free notation (with undotted and dotted indices suppressed), the undaggered and daggered spinors cannot be uniquely identified as column or row vectors until their locations within the spinor product are specified. Nevertheless, the proper identifications are straightforward, as any spinor on the left end of a spinor product can be identified as a row vector and any spinor on the right end of a spinor product can be identified as a column vector.

For an *anticommuting* two-component spinor ψ, the product $\psi^{\alpha}\psi^{\beta}$ is antisymmetric with respect to the interchange of the spinor indices α and β. Hence, this product of spinors must be proportional to $\epsilon^{\alpha\beta}$. Similar conclusions hold for the corresponding spinor products with raised undotted indices and with lowered and raised dotted indices, respectively. Thus,

$$\psi^{\alpha}\psi^{\beta} = -\tfrac{1}{2}\epsilon^{\alpha\beta}\psi\psi\,, \qquad \psi_{\alpha}\psi_{\beta} = \tfrac{1}{2}\epsilon_{\alpha\beta}\psi\psi\,, \tag{B.49}$$

$$\psi^{\dagger\dot\alpha}\psi^{\dagger\dot\beta} = \tfrac{1}{2}\epsilon^{\dot\alpha\dot\beta}\psi^{\dagger}\psi^{\dagger}\,, \qquad \psi^{\dagger}_{\dot\alpha}\psi^{\dagger}_{\dot\beta} = -\tfrac{1}{2}\epsilon_{\dot\alpha\dot\beta}\psi^{\dagger}\psi^{\dagger}\,, \tag{B.50}$$

where $\psi\psi \equiv \psi^{\alpha}\psi_{\alpha}$ and $\psi^{\dagger}\psi^{\dagger} \equiv \psi^{\dagger}_{\dot\alpha}\psi^{\dagger\dot\alpha}$ as in eqs. (B.43) and (B.44). Note that the minus signs above can be understood to be a consequence of the extra minus sign that arises when the indices of the epsilon symbol are lowered or raised [see eqs. (1.55) and (1.56)].

The behavior of the spinor products under hermitian conjugation (for quantum field operators) or complex conjugation (for classical fields) is as follows:

$$(\chi\eta)^\dagger = \eta^\dagger\chi^\dagger, \tag{B.51}$$

$$(\chi\sigma^\mu\eta^\dagger)^\dagger = \eta\sigma^\mu\chi^\dagger, \tag{B.52}$$

$$(\chi^\dagger\overline{\sigma}^\mu\eta)^\dagger = \eta^\dagger\overline{\sigma}^\mu\chi, \tag{B.53}$$

$$(\chi\sigma^\mu\overline{\sigma}^\nu\eta)^\dagger = \eta^\dagger\overline{\sigma}^\nu\sigma^\mu\chi^\dagger, \tag{B.54}$$

$$(\chi\sigma^{\mu\nu}\eta)^\dagger \equiv \eta^\dagger\overline{\sigma}^{\mu\nu}\chi^\dagger, \tag{B.55}$$

where we have used the hermiticity properties $(\sigma^\mu)^\dagger = \sigma^\mu$, $(\overline{\sigma}^\mu)^\dagger = \overline{\sigma}^\mu$, and $(\sigma^{\mu\nu})^\dagger = \overline{\sigma}^{\mu\nu}$. More generally,

$$(\chi\Sigma\eta)^\dagger = \eta^\dagger\Sigma_r\chi^\dagger, \qquad (\chi\Sigma\eta^\dagger)^\dagger = \eta\Sigma_r\chi^\dagger, \qquad (\chi^\dagger\Sigma\eta)^\dagger = \eta^\dagger\Sigma_r\chi, \tag{B.56}$$

where in each case Σ stands for any sequence of alternating σ and $\overline{\sigma}$ matrices, and Σ_r is obtained from Σ by reversing the order of all of the σ and $\overline{\sigma}$ matrices. Note that eqs. (B.51)–(B.56) apply to both anticommuting and commuting spinors.

In addition to manipulating expressions containing anticommuting fermion fields, we often must deal with products of *commuting* spinor wave functions that arise when evaluating the Feynman rules. In the following expressions we denote the generic spinor by z_i. In the various identities listed below, an extra minus sign arises when manipulating a product of anticommuting fermion fields. Thus, we employ the notation

$$(-1)^A \equiv \begin{cases} +1, & \text{commuting spinors,} \\ -1, & \text{anticommuting spinors.} \end{cases} \tag{B.57}$$

The following identities hold for the z_i:

$$z_1 z_2 = -(-1)^A z_2 z_1, \tag{B.58}$$

$$\overline{z}_1\overline{z}_2 = -(-1)^A \overline{z}_2\overline{z}_1, \tag{B.59}$$

$$z_1\sigma^\mu\overline{z}_2 = (-1)^A \overline{z}_2\overline{\sigma}^\mu z_1, \tag{B.60}$$

$$z_1\sigma^\mu\overline{\sigma}^\nu z_2 = -(-1)^A z_2\sigma^\nu\overline{\sigma}^\mu z_1, \tag{B.61}$$

$$\overline{z}_1\overline{\sigma}^\mu\sigma^\nu\overline{z}_2 = -(-1)^A \overline{z}_2\overline{\sigma}^\nu\sigma^\mu\overline{z}_1, \tag{B.62}$$

$$\overline{z}_1\overline{\sigma}^\mu\sigma^\rho\overline{\sigma}^\nu z_2 = (-1)^A z_2\sigma^\nu\overline{\sigma}^\rho\sigma^\mu\overline{z}_1, \tag{B.63}$$

$$z_1\sigma^{\mu\nu} z_2 = (-1)^A z_2\sigma^{\mu\nu} z_1, \tag{B.64}$$

$$\overline{z}_1\overline{\sigma}^{\mu\nu}\overline{z}_2 = (-1)^A \overline{z}_2\overline{\sigma}^{\mu\nu}\overline{z}_1. \tag{B.65}$$

Finally, we exhibit a number of identities for two-component spinor products. From eq. (A.40), one easily obtains the following identities:

$$(z_1 z_2)(z_3 z_4) = -(z_1 z_3)(z_4 z_2) - (z_1 z_4)(z_2 z_3), \tag{B.66}$$

$$(\overline{z}_1\overline{z}_2)(\overline{z}_3\overline{z}_4) = -(\overline{z}_1\overline{z}_3)(\overline{z}_4\overline{z}_2) - (\overline{z}_1\overline{z}_4)(\overline{z}_2\overline{z}_3), \tag{B.67}$$

where we have used eqs. (B.58) and (B.59) to cancel out any residual factors of

$(-1)^A$. Similarly, eqs. (B.1)–(B.3) can be used to derive

$$(z_1\sigma^\mu \bar{z}_2)(\bar{z}_3\bar{\sigma}_\mu z_4) = -2(z_1 z_4)(\bar{z}_2\bar{z}_3)\,, \tag{B.68}$$

$$(\bar{z}_1\bar{\sigma}^\mu z_2)(\bar{z}_3\bar{\sigma}_\mu z_4) = 2(\bar{z}_1\bar{z}_3)(z_4 z_2)\,, \tag{B.69}$$

$$(z_1\sigma^\mu \bar{z}_2)(z_3\sigma_\mu \bar{z}_4) = 2(z_1 z_3)(\bar{z}_4\bar{z}_2)\,, \tag{B.70}$$

and eqs. (B.4)–(B.6) can be used to derive

$$(z_1\sigma^{\mu\nu} z_2)(z_3\sigma_{\mu\nu} z_4) = -2(z_1 z_4)(z_2 z_3) - (z_1 z_2)(z_3 z_4)\,, \tag{B.71}$$

$$(z_1^\dagger\bar{\sigma}^{\mu\nu} z_2^\dagger)(z_3^\dagger\bar{\sigma}_{\mu\nu} z_4^\dagger) = -2(z_1^\dagger z_4^\dagger)(z_2^\dagger z_3^\dagger) - (z_1^\dagger z_2^\dagger)(z_3^\dagger z_4^\dagger)\,, \tag{B.72}$$

$$(z_1\sigma^{\mu\nu} z_2)(z_3^\dagger\bar{\sigma}_{\mu\nu} z_4^\dagger) = 0\,, \tag{B.73}$$

where we have again employed eqs. (B.58) and (B.59) to eliminate any residual factors of $(-1)^A$. Thus, eqs. (B.66)–(B.73) hold for both commuting and anticommuting spinors.

A more extensive list of Fierz identities can be found in Appendix B.3.

B.3 Fierz Identities

B.3.1 Two-Component Spinor Fierz Identities

We begin with the basic identity for 2×2 matrices,

$$\delta_{ab}\delta_{cd} = \tfrac{1}{2}\left[\delta_{ad}\delta_{cb} + \sigma^i_{ad}\sigma^i_{cb}\right]\,, \tag{B.74}$$

where there is an implicit sum over the repeated superscript $i \in \{1,2,3\}$. Equation (B.74) is a consequence of the completeness of $\{\mathbb{1}_{2\times 2}\,,\, \sigma^i\}$ in the four-dimensional vector space of 2×2 matrices. Applying these considerations to matrices that possess two indices, where each of the two indices is either undotted or dotted, one can establish four isomorphic four-dimensional vector spaces, each of which is spanned by four linearly independent hermitian matrices:

$$\mathcal{V} = \{\delta_\alpha{}^\beta\,,\, i(\sigma^{0i})_\alpha{}^\beta\}\,, \quad \overline{\mathcal{V}} = \{\delta^{\dot\alpha}{}_{\dot\beta}\,,\, i(\bar{\sigma}^{0i})^{\dot\alpha}{}_{\dot\beta}\}\,, \quad \mathcal{V}' = \{\sigma^\mu_{\alpha\dot\beta}\}\,, \quad \overline{\mathcal{V}}' = \{\bar{\sigma}^{\mu\,\dot\alpha\beta}\}\,. \tag{B.75}$$

Note that the σ^{jk} and $\bar{\sigma}^{jk}$ are completely determined by the six matrices σ^{0i} and $\bar{\sigma}^{0i}$ (where $i,j,k \in \{1,2,3\}$) due to the self-duality relations given by eq. (A.48).

It is therefore convenient to consider the set of matrices

$$\Gamma \equiv \left\{\delta_\alpha{}^\beta\,,\, \sigma^\mu_{\alpha\dot\beta}\,,\, \sigma^{\mu\nu}{}_\alpha{}^\beta\,,\, \delta^{\dot\alpha}{}_{\dot\beta}\,,\, \sigma^{\mu\,\dot\alpha\beta}\,,\, \bar{\sigma}^{\mu\nu\,\dot\alpha}{}_{\dot\beta}\right\}\,. \tag{B.76}$$

Elements of Γ will be denoted by $\Gamma^{(n)}$ $(n = 1,2,\ldots,6)$. Starting from eq. (B.74), one can establish a set of 21 identities of the following form:

$$(\Gamma^{(k)})^I_{AB}(\Gamma^{(n)})^J_{CD} = \sum_{p,q,K,L} (C^{kn}_{pq})^{IJ}_{KL}\,(\Gamma^{(p)})^K_{AD}(\Gamma^{(q)})^L_{CB}\,, \tag{B.77}$$

where each label I, J, K, and L can represent zero, one, or two Lorentz spacetime indices, and A, B, C, and D represent two-component spinor indices, each of which may be undotted or dotted and in the lowered or raised position as appropriate. The sum in eq. (B.77) is taken over the matrices specified in eq. (B.76), and the C^{kn}_{pq} are numerical coefficients [see eqs. (B.79)–(B.99)].

Let us multiply eq. (B.77) by four (commuting or anticommuting) two-component spinors $Z_{1A}Z_{2B}Z_{3C}Z_{4D}$, where the Z_i are either the undaggered or daggered spinor z_i or $z_i^\dagger$, depending on whether the corresponding spinor index is undotted or dotted. This procedure yields generalized Fierz identities of the form (e.g., see [B18])

$$(Z_1\Gamma^{(k)I}Z_2)(Z_3\Gamma^{(n)J}Z_4) = (-1)^A \sum_{p,q,K,L} (C^{kn}_{pq})^{IJ}_{KL}(Z_1\Gamma^{(p)K}Z_4)(Z_3\Gamma^{(q)L}Z_2)\,, \tag{B.78}$$

where $(-1)^A = +1$ [-1] for commuting [anticommuting] spinors.[1] The explicit expressions for the 21 identities represented by eq. (B.77) are as follows:

$$\delta_\alpha{}^\beta\delta^{\dot\beta}{}_{\dot\alpha} = \tfrac{1}{2}\sigma^\mu_{\alpha\dot\alpha}\overline\sigma^{\dot\beta\beta}_\mu\,, \tag{B.79}$$

$$\delta_\alpha{}^\beta\delta_\gamma{}^\tau = \tfrac{1}{2}\left[\delta_\alpha{}^\tau\delta_\gamma{}^\beta + (\sigma^{\mu\nu})_\alpha{}^\tau(\sigma_{\mu\nu})_\gamma{}^\beta\right], \tag{B.80}$$

$$\delta^{\dot\alpha}{}_{\dot\beta}\delta^{\dot\gamma}{}_{\dot\tau} = \tfrac{1}{2}\left[\delta^{\dot\alpha}{}_{\dot\tau}\delta^{\dot\gamma}{}_{\dot\beta} + (\overline\sigma^{\mu\nu})^{\dot\alpha}{}_{\dot\tau}(\overline\sigma_{\mu\nu})^{\dot\gamma}{}_{\dot\beta}\right], \tag{B.81}$$

$$\delta_\alpha{}^\beta\sigma^\mu_{\gamma\dot\alpha} = \tfrac{1}{2}\sigma^\mu_{\alpha\dot\alpha}\delta_\gamma{}^\beta - i\sigma_{\nu\,\alpha\dot\alpha}(\sigma^{\mu\nu})_\gamma{}^\beta\,, \tag{B.82}$$

$$\delta_\alpha{}^\beta\overline\sigma^{\mu\,\dot\beta\gamma} = \tfrac{1}{2}\delta_\alpha{}^\gamma\overline\sigma^{\mu\,\dot\beta\beta} + i(\sigma^{\mu\nu})_\alpha{}^\gamma\overline\sigma^{\dot\beta\beta}_\nu\,, \tag{B.83}$$

$$\delta^{\dot\alpha}{}_{\dot\beta}\sigma^\mu_{\beta\dot\gamma} = \tfrac{1}{2}\delta^{\dot\alpha}{}_{\dot\gamma}\sigma^\mu_{\beta\dot\beta} + i(\overline\sigma^{\mu\nu})^{\dot\alpha}{}_{\dot\gamma}\sigma_{\nu\,\beta\dot\beta}\,, \tag{B.84}$$

$$\delta^{\dot\alpha}{}_{\dot\beta}\overline\sigma^{\mu\,\dot\gamma\alpha} = \tfrac{1}{2}\overline\sigma^{\mu\,\dot\alpha\alpha}\delta^{\dot\gamma}{}_{\dot\beta} - i\overline\sigma^{\dot\alpha\alpha}_\nu(\overline\sigma^{\mu\nu})^{\dot\gamma}{}_{\dot\beta}\,, \tag{B.85}$$

$$\begin{aligned}\delta_\alpha{}^\beta(\sigma^{\mu\nu})_\gamma{}^\tau = \tfrac{1}{2}\Big\{&(\sigma^{\mu\nu})_\alpha{}^\tau\delta_\gamma{}^\beta + \delta_\alpha{}^\tau(\sigma^{\mu\nu})_\gamma{}^\beta \\ &- ig_{\rho\kappa}\left[(\sigma^{\mu\kappa})_\alpha{}^\tau(\sigma^{\nu\rho})_\gamma{}^\beta - (\sigma^{\nu\kappa})_\alpha{}^\tau(\sigma^{\mu\rho})_\gamma{}^\beta\right]\Big\},\end{aligned} \tag{B.86}$$

$$\delta_\alpha{}^\beta(\overline\sigma^{\mu\nu})^{\dot\beta}{}_{\dot\alpha} = -\tfrac{1}{4}i\left[\sigma^\mu_{\alpha\dot\alpha}\overline\sigma^{\nu\,\dot\beta\beta} - \sigma^\nu_{\alpha\dot\alpha}\overline\sigma^{\mu\,\dot\beta\beta} + i\epsilon^{\mu\nu\rho\kappa}\sigma_{\rho\,\alpha\dot\alpha}\overline\sigma^{\dot\beta\beta}_\kappa\right], \tag{B.87}$$

$$\delta^{\dot\alpha}{}_{\dot\beta}(\sigma^{\mu\nu})_\beta{}^\alpha = -\tfrac{1}{4}i\left[\overline\sigma^{\mu\,\dot\alpha\alpha}\sigma^\nu_{\beta\dot\beta} - \overline\sigma^{\nu\,\dot\alpha\alpha}\sigma^\mu_{\beta\dot\beta} - i\epsilon^{\mu\nu\rho\kappa}\overline\sigma^{\dot\alpha\alpha}_\rho\sigma_{\kappa\,\beta\dot\beta}\right], \tag{B.88}$$

$$\begin{aligned}\delta^{\dot\alpha}{}_{\dot\beta}(\overline\sigma^{\mu\nu})^{\dot\gamma}{}_{\dot\tau} = \tfrac{1}{2}\Big\{&(\overline\sigma^{\mu\nu})^{\dot\alpha}{}_{\dot\tau}\delta^{\dot\gamma}{}_{\dot\beta} + \delta^{\dot\alpha}{}_{\dot\tau}(\overline\sigma^{\mu\nu})^{\dot\gamma}{}_{\dot\beta} \\ &- ig_{\rho\kappa}\left[(\overline\sigma^{\mu\kappa})^{\dot\alpha}{}_{\dot\tau}(\overline\sigma^{\nu\rho})^{\dot\gamma}{}_{\dot\beta} - (\overline\sigma^{\nu\kappa})^{\dot\alpha}{}_{\dot\tau}(\overline\sigma^{\mu\rho})^{\dot\gamma}{}_{\dot\beta}\right]\Big\},\end{aligned} \tag{B.89}$$

$$\sigma^\mu_{\alpha\dot\alpha}\sigma^\nu_{\beta\dot\beta} = \tfrac{1}{2}\left[\sigma^\mu_{\alpha\dot\beta}\sigma^\nu_{\beta\dot\alpha} + \sigma^\nu_{\alpha\dot\beta}\sigma^\mu_{\beta\dot\alpha} - g^{\mu\nu}\sigma^\lambda_{\alpha\dot\beta}\sigma_{\lambda\,\beta\dot\alpha} + i\epsilon^{\mu\nu\rho\kappa}\sigma_{\rho\,\alpha\dot\beta}\,\sigma_{\kappa\,\beta\dot\alpha}\right], \tag{B.90}$$

$$\overline\sigma^{\mu\dot\alpha\alpha}\overline\sigma^{\nu\dot\beta\beta} = \tfrac{1}{2}\left[\overline\sigma^{\mu\dot\alpha\beta}\overline\sigma^{\nu\dot\beta\alpha} + \overline\sigma^{\nu\dot\alpha\beta}\overline\sigma^{\mu\dot\beta\alpha} - g^{\mu\nu}\overline\sigma^{\lambda\,\dot\alpha\beta}\overline\sigma^{\dot\beta\alpha}_\lambda - i\epsilon^{\mu\nu\rho\kappa}\overline\sigma^{\dot\alpha\beta}_\rho\,\overline\sigma^{\dot\beta\alpha}_\kappa\right], \tag{B.91}$$

[1] It is often convenient to reverse the order of the spinors Z_2 and Z_3 on the right-hand side of eq. (B.78) by using eqs. (B.58)–(B.64) to eliminate the factor of $(-1)^A$.

$$\begin{aligned}
\sigma^{\mu}_{\alpha\dot{\alpha}}\overline{\sigma}^{\nu\dot{\beta}\beta} &= \tfrac{1}{2}\big[g^{\mu\nu}\delta_{\alpha}{}^{\beta}\delta^{\dot{\beta}}{}_{\dot{\alpha}} - 2i(\sigma^{\mu\nu})_{\alpha}{}^{\beta}\delta^{\dot{\beta}}{}_{\dot{\alpha}} + 2i\delta_{\alpha}{}^{\beta}(\overline{\sigma}^{\mu\nu})^{\dot{\beta}}{}_{\dot{\alpha}} \\
&\qquad - 4g_{\rho\kappa}(\sigma^{\mu\kappa})_{\alpha}{}^{\beta}(\overline{\sigma}^{\nu\rho})^{\dot{\beta}}{}_{\dot{\alpha}}\big]\,, && \text{(B.92)}
\end{aligned}$$

$$\begin{aligned}
(\sigma^{\mu\nu})_{\alpha}{}^{\beta}\sigma^{\rho}_{\gamma\dot{\alpha}} &= \tfrac{1}{2}\left[\sigma^{\nu}_{\alpha\dot{\alpha}}(\sigma^{\mu\rho})_{\gamma}{}^{\beta} - \sigma^{\mu}_{\alpha\dot{\alpha}}(\sigma^{\nu\rho})_{\gamma}{}^{\beta} + i\epsilon^{\mu\nu\kappa}{}_{\lambda}\sigma_{\kappa\,\alpha\dot{\alpha}}(\sigma^{\lambda\rho})_{\gamma}{}^{\beta}\right. \\
&\qquad \left. - \tfrac{1}{2}i\left(g^{\mu\rho}\sigma^{\nu}_{\alpha\dot{\alpha}} - g^{\nu\rho}\sigma^{\mu}_{\alpha\dot{\alpha}} - i\epsilon^{\mu\nu\rho\kappa}\sigma_{\kappa\,\alpha\dot{\alpha}}\right)\delta_{\gamma}{}^{\beta}\right], && \text{(B.93)}
\end{aligned}$$

$$\begin{aligned}
\overline{\sigma}^{\rho\,\dot{\alpha}\beta}(\sigma^{\mu\nu})_{\gamma}{}^{\alpha} &= \tfrac{1}{2}\left[\overline{\sigma}^{\nu\,\dot{\alpha}\alpha}(\sigma^{\mu\rho})_{\gamma}{}^{\beta} - \overline{\sigma}^{\mu\,\dot{\alpha}\alpha}(\sigma^{\nu\rho})_{\gamma}{}^{\beta} + i\epsilon^{\mu\nu\kappa}{}_{\lambda}\overline{\sigma}^{\dot{\alpha}\alpha}_{\kappa}(\sigma^{\lambda\rho})_{\gamma}{}^{\beta}\right. \\
&\qquad \left. + \tfrac{1}{2}i\left(g^{\mu\rho}\overline{\sigma}^{\nu\,\dot{\alpha}\alpha} - g^{\nu\rho}\overline{\sigma}^{\mu\,\dot{\alpha}\alpha} - i\epsilon^{\mu\nu\rho\kappa}\overline{\sigma}^{\dot{\alpha}\alpha}_{\kappa}\right)\delta_{\gamma}{}^{\beta}\right], && \text{(B.94)}
\end{aligned}$$

$$\begin{aligned}
\sigma^{\rho}_{\alpha\dot{\beta}}(\overline{\sigma}^{\mu\nu})^{\dot{\gamma}}{}_{\dot{\alpha}} &= \tfrac{1}{2}\left[\sigma^{\nu}_{\alpha\dot{\alpha}}(\overline{\sigma}^{\mu\rho})^{\dot{\gamma}}{}_{\dot{\beta}} - \sigma^{\mu}_{\alpha\dot{\alpha}}(\overline{\sigma}^{\nu\rho})^{\dot{\gamma}}{}_{\dot{\beta}} - i\epsilon^{\mu\nu\kappa}{}_{\lambda}\sigma_{\kappa\,\alpha\dot{\alpha}}(\overline{\sigma}^{\lambda\rho})^{\dot{\gamma}}{}_{\dot{\beta}}\right. \\
&\qquad \left. + \tfrac{1}{2}i\left(g^{\mu\rho}\sigma^{\nu}_{\alpha\dot{\alpha}} - g^{\nu\rho}\sigma^{\mu}_{\alpha\dot{\alpha}} + i\epsilon^{\mu\nu\rho\kappa}\sigma_{\kappa\,\alpha\dot{\alpha}}\right)\delta^{\dot{\gamma}}{}_{\dot{\beta}}\right], && \text{(B.95)}
\end{aligned}$$

$$\begin{aligned}
(\overline{\sigma}^{\mu\nu})^{\dot{\alpha}}{}_{\dot{\beta}}\overline{\sigma}^{\rho\,\dot{\gamma}\alpha} &= \tfrac{1}{2}\left[\overline{\sigma}^{\nu\,\dot{\alpha}\alpha}(\overline{\sigma}^{\mu\rho})^{\dot{\gamma}}{}_{\dot{\beta}} - \overline{\sigma}^{\mu\,\dot{\alpha}\alpha}(\overline{\sigma}^{\nu\rho})^{\dot{\gamma}}{}_{\dot{\beta}} - i\epsilon^{\mu\nu\kappa}{}_{\lambda}\overline{\sigma}^{\dot{\alpha}\alpha}_{\kappa}(\overline{\sigma}^{\lambda\rho})^{\dot{\gamma}}{}_{\dot{\beta}}\right. \\
&\qquad \left. - \tfrac{1}{2}i\left(g^{\mu\rho}\overline{\sigma}^{\nu\,\dot{\alpha}\alpha} - g^{\nu\rho}\overline{\sigma}^{\mu\,\dot{\alpha}\alpha} + i\epsilon^{\mu\nu\rho\kappa}\overline{\sigma}^{\dot{\alpha}\alpha}_{\kappa}\right)\delta^{\dot{\gamma}}{}_{\dot{\beta}}\right], && \text{(B.96)}
\end{aligned}$$

$$\begin{aligned}
(\sigma^{\mu\nu})_{\alpha}{}^{\beta}(\sigma^{\rho\kappa})_{\gamma}{}^{\tau} &= \tfrac{1}{2}(\sigma^{\mu\nu})_{\alpha}{}^{\tau}(\sigma^{\rho\kappa})_{\gamma}{}^{\beta} + \tfrac{1}{8}\delta_{\alpha}{}^{\tau}\delta_{\gamma}{}^{\beta}(g^{\mu\rho}g^{\nu\kappa} - g^{\mu\kappa}g^{\nu\rho} - i\epsilon^{\mu\nu\rho\kappa}) \\
&\quad + \tfrac{1}{4}i\delta_{\alpha}{}^{\tau}\left(g^{\mu\rho}\sigma^{\nu\kappa} + g^{\nu\kappa}\sigma^{\mu\rho} - g^{\nu\rho}\sigma^{\mu\kappa} - g^{\mu\kappa}\sigma^{\nu\rho}\right)_{\gamma}{}^{\beta} \\
&\quad - \tfrac{1}{4}i\delta_{\gamma}{}^{\beta}\left(g^{\mu\rho}\sigma^{\nu\kappa} + g^{\nu\kappa}\sigma^{\mu\rho} - g^{\nu\rho}\sigma^{\mu\kappa} - g^{\mu\kappa}\sigma^{\nu\rho}\right)_{\alpha}{}^{\tau} \\
&\quad + \tfrac{1}{4}\left[(\sigma^{\mu\rho})_{\alpha}{}^{\tau}(\sigma^{\nu\kappa})_{\gamma}{}^{\beta} + (\sigma^{\nu\kappa})_{\alpha}{}^{\tau}(\sigma^{\mu\rho})_{\gamma}{}^{\beta}\right. \\
&\qquad \left. - (\sigma^{\nu\rho})_{\alpha}{}^{\tau}(\sigma^{\mu\kappa})_{\gamma}{}^{\beta} - (\sigma^{\mu\kappa})_{\alpha}{}^{\tau}(\sigma^{\nu\rho})_{\gamma}{}^{\beta}\right] \\
&\quad + \tfrac{1}{4}g_{\lambda\sigma}\left[g^{\mu\kappa}(\sigma^{\rho\sigma})_{\alpha}{}^{\tau}(\sigma^{\nu\lambda})_{\gamma}{}^{\beta} + g^{\nu\rho}(\sigma^{\kappa\sigma})_{\alpha}{}^{\tau}(\sigma^{\mu\lambda})_{\gamma}{}^{\beta}\right. \\
&\qquad \left. - g^{\nu\kappa}(\sigma^{\rho\sigma})_{\alpha}{}^{\tau}(\sigma^{\mu\lambda})_{\gamma}{}^{\beta} - g^{\mu\rho}(\sigma^{\kappa\sigma})_{\alpha}{}^{\tau}(\sigma^{\nu\lambda})_{\gamma}{}^{\beta}\right], && \text{(B.97)}
\end{aligned}$$

$$\begin{aligned}
(\overline{\sigma}^{\mu\nu})^{\dot{\alpha}}{}_{\dot{\beta}}(\overline{\sigma}^{\rho\kappa})^{\dot{\gamma}}{}_{\dot{\tau}} &= \tfrac{1}{2}(\overline{\sigma}^{\mu\nu})^{\dot{\alpha}}{}_{\dot{\tau}}(\overline{\sigma}^{\rho\kappa})^{\dot{\gamma}}{}_{\dot{\beta}} + \tfrac{1}{8}\delta^{\dot{\alpha}}{}_{\dot{\tau}}\delta^{\dot{\gamma}}{}_{\dot{\beta}}(g^{\mu\rho}g^{\nu\kappa} - g^{\mu\kappa}g^{\nu\rho} + i\epsilon^{\mu\nu\rho\kappa}) \\
&\quad + \tfrac{1}{4}i\delta^{\dot{\alpha}}{}_{\dot{\tau}}\left(g^{\mu\rho}\overline{\sigma}^{\nu\kappa} + g^{\nu\kappa}\overline{\sigma}^{\mu\rho} - g^{\nu\rho}\overline{\sigma}^{\mu\kappa} - g^{\mu\kappa}\overline{\sigma}^{\nu\rho}\right)^{\dot{\gamma}}{}_{\dot{\beta}} \\
&\quad - \tfrac{1}{4}i\delta^{\dot{\gamma}}{}_{\dot{\beta}}\left(g^{\mu\rho}\overline{\sigma}^{\nu\kappa} + g^{\nu\kappa}\overline{\sigma}^{\mu\rho} - g^{\nu\rho}\overline{\sigma}^{\mu\kappa} - g^{\mu\kappa}\overline{\sigma}^{\nu\rho}\right)^{\dot{\alpha}}{}_{\dot{\tau}} \\
&\quad + \tfrac{1}{4}\left[(\overline{\sigma}^{\mu\rho})^{\dot{\alpha}}{}_{\dot{\tau}}(\overline{\sigma}^{\nu\kappa})^{\dot{\gamma}}{}_{\dot{\beta}} + (\overline{\sigma}^{\nu\kappa})^{\dot{\alpha}}{}_{\dot{\tau}}(\overline{\sigma}^{\mu\rho})^{\dot{\gamma}}{}_{\dot{\beta}}\right. \\
&\qquad \left. - (\overline{\sigma}^{\nu\rho})^{\dot{\alpha}}{}_{\dot{\tau}}(\overline{\sigma}^{\mu\kappa})^{\dot{\gamma}}{}_{\dot{\beta}} - (\overline{\sigma}^{\mu\kappa})^{\dot{\alpha}}{}_{\dot{\tau}}(\overline{\sigma}^{\nu\rho})^{\dot{\gamma}}{}_{\dot{\beta}}\right] \\
&\quad + \tfrac{1}{4}g_{\lambda\sigma}\left[g^{\mu\kappa}(\overline{\sigma}^{\rho\sigma})^{\dot{\alpha}}{}_{\dot{\tau}}(\overline{\sigma}^{\nu\lambda})^{\dot{\gamma}}{}_{\dot{\beta}} + g^{\nu\rho}(\overline{\sigma}^{\kappa\sigma})^{\dot{\alpha}}{}_{\dot{\tau}}(\overline{\sigma}^{\mu\lambda})^{\dot{\gamma}}{}_{\dot{\beta}}\right. \\
&\qquad \left. - g^{\nu\kappa}(\overline{\sigma}^{\rho\sigma})^{\dot{\alpha}}{}_{\dot{\tau}}(\overline{\sigma}^{\mu\lambda})^{\dot{\gamma}}{}_{\dot{\beta}} - g^{\mu\rho}(\overline{\sigma}^{\kappa\sigma})^{\dot{\alpha}}{}_{\dot{\tau}}(\overline{\sigma}^{\nu\lambda})^{\dot{\gamma}}{}_{\dot{\beta}}\right], && \text{(B.98)}
\end{aligned}$$

$$\begin{aligned}
(\sigma^{\mu\nu})_{\alpha}{}^{\beta}(\overline{\sigma}^{\rho\kappa})^{\dot{\beta}}{}_{\dot{\alpha}} &= \tfrac{1}{8}\left[(g^{\mu\rho}g^{\nu\kappa} - g^{\mu\kappa}g^{\nu\rho})\sigma^{\lambda}_{\alpha\dot{\alpha}}\overline{\sigma}^{\dot{\beta}\beta}_{\lambda} + i\epsilon^{\mu\nu\rho\lambda}\sigma_{\lambda\,\alpha\dot{\alpha}}\overline{\sigma}^{\kappa\,\dot{\beta}\beta}\right. \\
&\qquad - i\epsilon^{\mu\nu\kappa\lambda}\sigma_{\lambda\,\alpha\dot{\alpha}}\overline{\sigma}^{\rho\,\dot{\beta}\beta} - i\epsilon^{\mu\rho\kappa\lambda}\sigma^{\nu}_{\alpha\dot{\alpha}}\overline{\sigma}^{\dot{\beta}\beta}_{\lambda} + i\epsilon^{\nu\rho\kappa\lambda}\sigma^{\mu}_{\alpha\dot{\alpha}}\overline{\sigma}^{\dot{\beta}\beta}_{\lambda} \\
&\qquad - g^{\mu\rho}(\sigma^{\kappa}_{\alpha\dot{\alpha}}\overline{\sigma}^{\nu\,\dot{\beta}\beta} + \sigma^{\nu}_{\alpha\dot{\alpha}}\overline{\sigma}^{\kappa\,\dot{\beta}\beta}) + g^{\nu\rho}(\sigma^{\kappa}_{\alpha\dot{\alpha}}\overline{\sigma}^{\mu\,\dot{\beta}\beta} + \sigma^{\mu}_{\alpha\dot{\alpha}}\overline{\sigma}^{\kappa\,\dot{\beta}\beta}) \\
&\qquad \left. + g^{\mu\kappa}(\sigma^{\rho}_{\alpha\dot{\alpha}}\overline{\sigma}^{\nu\,\dot{\beta}\beta} + \sigma^{\nu}_{\alpha\dot{\alpha}}\overline{\sigma}^{\rho\,\dot{\beta}\beta}) - g^{\nu\kappa}(\sigma^{\rho}_{\alpha\dot{\alpha}}\overline{\sigma}^{\mu\,\dot{\beta}\beta} + \sigma^{\mu}_{\alpha\dot{\alpha}}\overline{\sigma}^{\rho\,\dot{\beta}\beta})\right].
\end{aligned}$$

(B.99)

Of the above Fierz identities [in the form given by eq. (B.78)], 11 also appear in Appendix A of [B18].[2]

The derivation of the 21 identities listed above is straightforward. Equations (B.79)–(B.81) are equivalent to the completeness relation of eq. (B.74). The next eight identities [eqs. (B.82)–(B.89)] are easily derived starting from eqs. (B.79)–(B.81). As a simple example, using the results of eqs. (B.80) and (1.112), it follows that

$$\begin{aligned}\delta_\alpha{}^\beta \sigma^\mu_{\gamma\dot\alpha} &= \delta_\alpha{}^\beta \delta_\gamma{}^\tau \sigma^\mu_{\tau\dot\alpha} = \tfrac{1}{2}\left[\delta_\alpha{}^\tau \delta_\gamma{}^\beta + (\sigma^{\rho\kappa})_\alpha{}^\tau (\sigma_{\rho\kappa})_\gamma{}^\beta\right]\sigma^\mu_{\tau\dot\alpha} \\ &= \tfrac{1}{2}\left[\sigma^\mu_{\alpha\dot\alpha}\delta_\gamma{}^\beta + (\sigma^{\rho\kappa}\sigma^\mu)_{\alpha\dot\alpha}(\sigma_{\rho\kappa})_\gamma{}^\beta\right] \\ &= \tfrac{1}{2}\left[\sigma^\mu_{\alpha\dot\alpha}\delta_\gamma{}^\beta + \tfrac{1}{2}i(g^{\kappa\mu}\sigma^\rho - g^{\rho\mu}\sigma^\kappa + i\epsilon^{\rho\kappa\mu\nu}\sigma_\nu)_{\alpha\dot\alpha}(\sigma_{\rho\kappa})_\gamma{}^\beta\right] \\ &= \tfrac{1}{2}\delta_\alpha{}^\gamma \overline{\sigma}^{\mu\,\dot\beta\beta} + i(\sigma^{\mu\nu})_\alpha{}^\gamma \overline{\sigma}_\nu^{\dot\beta\beta}\,, \end{aligned} \tag{B.100}$$

where eq. (A.48) was employed in the final step. We can now use eqs. (B.82)–(B.85) to derive eqs. (B.90)–(B.96). Finally, starting from eqs. (B.86)–(B.89) we may employ the same technique once more to derive eqs. (B.97)–(B.99).[3]

Equations (B.79)–(B.81) can also be used to derive four additional identities, which yield Fierz identities of a different form. Simply multiply each of these equations by two ϵ symbols (with appropriately chosen undotted and/or dotted spinor indices), and use eqs. (A.44) and (A.51). Two of the resulting identities coincide with eqs. (B.2) and (B.3), while the other two are[4]

$$\epsilon_{\alpha\beta}\epsilon^{\gamma\tau} = -\tfrac{1}{2}\left[\delta_\alpha{}^\gamma \delta_\beta{}^\tau - (\sigma^{\mu\nu})_\alpha{}^\gamma (\sigma_{\mu\nu})_\beta{}^\tau\right], \tag{B.101}$$

$$\epsilon^{\dot\alpha\dot\beta}\epsilon_{\dot\gamma\dot\tau} = -\tfrac{1}{2}\left[\delta^{\dot\alpha}{}_{\dot\gamma}\delta^{\dot\beta}{}_{\dot\tau} - (\overline{\sigma}^{\mu\nu})^{\dot\alpha}{}_{\dot\gamma}(\overline{\sigma}_{\mu\nu})^{\dot\beta}{}_{\dot\tau}\right]. \tag{B.102}$$

Multiplying eqs. (B.2), (B.3), (B.101), and (B.102) by four (commuting or anticommuting) two-component spinors $Z_{1A}Z_{2B}Z_{3C}Z_{4D}$ yields the corresponding Fierz identities of the form

$$(Z_1\Gamma^{(k)I}Z_2)(Z_3\Gamma^{(n)J}Z_4) = (-1)^A \sum_{p,q,K,L} (C^{kn}_{pq})^{IJ}_{KL}(Z_1\Gamma^{(p)K}Z_3)(Z_2\Gamma^{(q)L}Z_4)\,, \tag{B.103}$$

which differs from eq. (B.78) in the ordering of the spinors on the right-hand side.

Finally, we note that if one multiplies the Schouten identities [eq. (A.42)] by four (commuting or anticommuting) two-component spinors, one obtains the Fierz identities given in eqs. (B.66) and (B.67). These two identifies do not assume the simple forms of either eq. (B.78) or eq. (B.103).

[2] Note that in [B18], $\epsilon^{\mu\nu\rho\kappa}$ has the opposite sign with respect to our conventions, and $\sigma^{\mu\nu}$ is defined without an overall factor of i. Taking these differences into account, we have confirmed that the results of Appendix A of [B18] match the corresponding results obtained here.

[3] In particular, the identities given in eqs. (B.32) and (B.33) are especially useful in the evaluation of eqs. (B.93)–(B.98).

[4] Note that eqs. (B.101) and (B.102) are equivalent to the previously obtained eqs. (B.4) and (B.5).

B.3.2 Four-Component Spinor Fierz Identities

In the four-component spinor formalism, the Fierz identities consist of relations among products of two bilinear covariants, in which the fermion fields appear in two different orders. The corresponding two-component spinor Fierz identities have been treated in detail in Appendix B.3.1. In principle, the latter can be converted into four-component spinor Fierz identities using the techniques developed in this appendix. However, it is easier to derive the four-component spinor Fierz identities directly using the properties of the gamma matrix algebra.

Instead of eqs. (B.79)–(B.81), the equivalent identity relevant for four-component spinors is

$$\delta_a^b\delta_c^d = \tfrac{1}{4}\big[\delta_a^d\delta_c^b + (\gamma_5)_a{}^d(\gamma_5)_c{}^b + (\gamma^\mu)_a{}^d(\gamma_\mu)_c{}^b - (\gamma^\mu\gamma_5)_a{}^d(\gamma_\mu\gamma_5)_c{}^b + \tfrac{1}{2}(\Sigma^{\mu\nu})_a{}^d(\Sigma_{\mu\nu})_c{}^b\big]\,. \tag{B.104}$$

This is the fundamental identity from which many additional identities can be derived. One of many possible Fierz identities can be obtained by multiplying eq. (B.104) by $\overline{\Psi}{}_1^a\Psi_{2b}\overline{\Psi}{}_3^c\Psi_{4d} = (-1)^A\,\overline{\Psi}{}_1^a\Psi_{4d}\overline{\Psi}{}_3^c\Psi_{2b}$, where $(-1)^A = +1$ [-1] for commuting [anticommuting] Dirac or Majorana spinors. More generally [206, 207],

$$(\overline{\Psi}_1\Gamma^{(k)I}\Psi_2)(\overline{\Psi}_3\Gamma_I^{(k)}\Psi_4) = (-1)^A\sum_{n=1}^{5}F^k{}_n\,(\overline{\Psi}_1\Gamma^{(n)J}\Psi_4)(\overline{\Psi}_3\Gamma_J^{(n)}\Psi_2)\,, \tag{B.105}$$

where the sum is taken over the 4×4 matrices, $\Gamma^{(n)}\in\Gamma$, which have been ordered as follows:[5]

$$\Gamma = \{\Gamma^I\} = \{\mathbb{1}\,,\,\gamma^\mu\,,\,\Sigma^{\mu\nu}\ (\mu<\nu)\,,\,\gamma^\mu\gamma_5\,,\,\gamma_5\}\,, \tag{B.106}$$

with I (and J) representing zero, one, or two spacetime indices (sums over repeated indices I and J are implied), and

$$F = \frac{1}{4}\begin{pmatrix} 1 & 1 & \frac{1}{2} & -1 & 1 \\ 4 & -2 & 0 & -2 & -4 \\ 12 & 0 & -2 & 0 & 12 \\ -4 & -2 & 0 & -2 & 4 \\ 1 & -1 & \frac{1}{2} & 1 & 1 \end{pmatrix}\,. \tag{B.107}$$

For example, taking $k=1$ in eq. (B.105) yields a result equivalent to eq. (B.104):

$$\begin{aligned}(\overline{\Psi}_1\Psi_2)(\overline{\Psi}_3\Psi_4) = \tfrac{1}{4}(-1)^A\big[&(\overline{\Psi}_1\Psi_4)(\overline{\Psi}_3\Psi_2) + (\overline{\Psi}_1\gamma_5\Psi_4)(\overline{\Psi}_3\gamma_5\Psi_2) \\ &+ (\overline{\Psi}_1\gamma^\mu\Psi_4)(\overline{\Psi}_3\gamma_\mu\Psi_2) - (\overline{\Psi}_1\gamma^\mu\gamma_5\Psi_4)(\overline{\Psi}_3\gamma_\mu\gamma_5\Psi_2) \\ &+ \tfrac{1}{2}(\overline{\Psi}_1\Sigma^{\mu\nu}\Psi_4)(\overline{\Psi}_3\Sigma_{\mu\nu}\Psi_2)\big]\,.\end{aligned} \tag{B.108}$$

[5] The 16 matrices of Γ constitute a complete set that spans the 16-dimensional vector space of 4×4 matrices.

For a comprehensive treatment of all possible four-component spinor Fierz identities, see Ref. [208]. Simple derivations of generalized Fierz identities have also been given in Refs. [207, 209].

B.4 Sigma Matrix Identities in $d \neq 4$ Dimensions

In Appendix B.1, we derived a number of identities involving σ^μ, $\overline{\sigma}^\mu$, $\sigma^{\mu\nu}$, and $\overline{\sigma}^{\mu\nu}$. In a theory regularized by dimensional continuation, one must give meaning to the sigma matrices and their respective identities in $d \neq 4$ dimensions. In many cases, it is possible to reinterpret the sigma matrix identities for $d \neq 4$. However, the Fierz identities, which depend on the completeness of $\{\mathbb{1}_{2\times 2}\,, \sigma^i\}$ in the vector space of 2×2 matrices, do not have a consistent, unambiguous meaning outside of four dimensions (e.g., see Refs. [112, 113, 210] and references therein). In this section, we examine the class of sigma matrix identities that can unambiguously be extended to $d \neq 4$ dimension and can thus be employed in the context of dimensional regularization.

When considering a theory regularized by dimensional continuation, one must be careful in treating cases with contracted spacetime vector indices $\mu, \nu, \kappa, \rho, \ldots$. Instead of taking on four possible values, these vector indices formally run over d values, where d is infinitesimally different from 4. This means that some identities that would hold in unregularized four-dimensional theories are inconsistent and must not be used; other identities remain valid if d replaces 4 in the appropriate spots; and still other identities hold without modification.

Two important identities that do hold in $d \neq 4$ dimensions are

$$[\sigma^\mu\overline{\sigma}^\nu + \sigma^\nu\overline{\sigma}^\mu]_\alpha{}^\beta = 2g^{\mu\nu}\delta_\alpha{}^\beta\,, \tag{B.109}$$

$$[\overline{\sigma}^\mu\sigma^\nu + \overline{\sigma}^\nu\sigma^\mu]^{\dot{\alpha}}{}_{\dot{\beta}} = 2g^{\mu\nu}\delta^{\dot{\alpha}}{}_{\dot{\beta}}\,. \tag{B.110}$$

Equivalently,

$$(\sigma^\mu\overline{\sigma}^\nu)_\alpha{}^\beta = g^{\mu\nu}\delta_\alpha{}^\beta - 2i(\sigma^{\mu\nu})_\alpha{}^\beta\,, \tag{B.111}$$

$$(\overline{\sigma}^\mu\sigma^\nu)^{\dot{\alpha}}{}_{\dot{\beta}} = g^{\mu\nu}\delta^{\dot{\alpha}}{}_{\dot{\beta}} - 2i(\overline{\sigma}^{\mu\nu})^{\dot{\alpha}}{}_{\dot{\beta}}\,. \tag{B.112}$$

The trace identities,

$$\mathrm{Tr}[\sigma^\mu\overline{\sigma}^\nu] = \mathrm{Tr}[\overline{\sigma}^\mu\sigma^\nu] = 2g^{\mu\nu}\,, \tag{B.113}$$

$$\mathrm{Tr}\,\sigma^{\mu\nu} = \mathrm{Tr}\,\overline{\sigma}^{\mu\nu} = 0\,, \tag{B.114}$$

then follow. We also note that the spinor index trace identity,

$$\mathrm{Tr}[\mathbb{1}_{2\times 2}] = \delta^\alpha_\alpha = \delta^{\dot{\alpha}}_{\dot{\alpha}} = 2\,, \tag{B.115}$$

continues to hold in $d \neq 4$ dimensions since we do not modify the dimension of the matrices $\{\mathbb{1}_{2\times 2}\,, \sigma^i\}$. However, as noted above, the Fierz identities given in

Appendix B.3 do not have a consistent, unambiguous meaning outside of four dimensions. Nevertheless, the following identities that are implied by eq. (B.79) do consistently generalize to $d \neq 4$ spacetime dimensions:

$$[\sigma^\mu \overline{\sigma}_\mu]_\alpha{}^\beta = d\delta_\alpha^\beta \,, \qquad [\overline{\sigma}^\mu \sigma_\mu]^{\dot\alpha}{}_{\dot\beta} = d\delta_{\dot\beta}^{\dot\alpha} \,. \tag{B.116}$$

Using eq. (B.116) along with the repeated use of eqs. (B.109) and (B.110) then yields

$$[\sigma^\mu \overline{\sigma}^\nu \sigma_\mu]_{\alpha\dot\beta} = -(d-2)\sigma^\nu_{\alpha\dot\beta} \,, \tag{B.117}$$

$$[\overline{\sigma}^\mu \sigma_\nu \overline{\sigma}_\mu]^{\dot\alpha\beta} = -(d-2)\overline{\sigma}_\nu^{\dot\alpha\beta} \,, \tag{B.118}$$

$$[\sigma^\mu \overline{\sigma}^\nu \sigma^\rho \overline{\sigma}_\mu]_\alpha{}^\beta = 4g^{\nu\rho}\delta_\alpha^\beta - (4-d)[\sigma^\nu \overline{\sigma}^\rho]_\alpha{}^\beta \,, \tag{B.119}$$

$$[\overline{\sigma}^\mu \sigma^\nu \overline{\sigma}^\rho \sigma_\mu]^{\dot\alpha}{}_{\dot\beta} = 4g^{\nu\rho}\delta_{\dot\beta}^{\dot\alpha} - (4-d)[\overline{\sigma}^\nu \sigma^\rho]^{\dot\alpha}{}_{\dot\beta} \,, \tag{B.120}$$

$$[\sigma^\mu \overline{\sigma}^\nu \sigma^\rho \overline{\sigma}^\kappa \sigma_\mu]_{\alpha\dot\beta} = -2[\sigma^\kappa \overline{\sigma}^\rho \sigma^\nu]_{\alpha\dot\beta} + (4-d)[\sigma^\nu \overline{\sigma}^\rho \sigma^\kappa]_{\alpha\dot\beta} \,, \tag{B.121}$$

$$[\overline{\sigma}^\mu \sigma^\nu \overline{\sigma}^\rho \sigma^\kappa \overline{\sigma}_\mu]^{\dot\alpha\beta} = -2[\overline{\sigma}^\kappa \sigma^\rho \overline{\sigma}^\nu]^{\dot\alpha\beta} + (4-d)[\overline{\sigma}^\nu \sigma^\rho \overline{\sigma}^\kappa]^{\dot\alpha\beta} \,. \tag{B.122}$$

Identities that involve explicitly and inextricably the four-dimensional $\epsilon^{\mu\nu\rho\kappa}$ symbol, such as those given in eqs. (B.30)–(B.41), are also only meaningful in exactly four spacetime dimensions. This applies as well to the trace identities that follow from them, such as those given in eqs. (B.19) and (B.20). For example, the identity

$$\mathrm{Tr}\left[\overline{\sigma}^\mu \sigma^\nu \overline{\sigma}^\rho \sigma^\kappa - \sigma^\mu \overline{\sigma}^\nu \sigma^\rho \overline{\sigma}^\kappa\right] = -4i\epsilon^{\mu\nu\rho\kappa} \tag{B.123}$$

has no unique definition for $d \neq 4$. Consequently, the appearance of the $\epsilon^{\mu\nu\rho\kappa}$ symbol in the evaluation of integrals that arise in loop calculations, where it is necessary to perform the computation in $d \neq 4$ dimensions (until the end of the calculation where the limit $d \to 4$ is taken), can lead to ambiguities. However, in practice one typically finds that the above expressions appear multiplied by the metric and/or other external tensors (such as four-momenta appropriate to the problem at hand). In almost all such cases, two of the indices appearing in eqs. (B.19) and (B.20) are symmetrized, which eliminates the $\epsilon^{\mu\nu\rho\kappa}$ term, rendering the resulting expressions unambiguous. Similarly, the sum of the trace identities given in eqs. (B.19) and (B.20) can be assigned an unambiguous meaning in $d \neq 4$ dimensions:

$$\mathrm{Tr}[\sigma^\mu \overline{\sigma}^\nu \sigma^\rho \overline{\sigma}^\kappa] + \mathrm{Tr}[\overline{\sigma}^\mu \sigma^\nu \overline{\sigma}^\rho \sigma^\kappa] = 4\left(g^{\mu\nu}g^{\rho\kappa} - g^{\mu\rho}g^{\nu\kappa} + g^{\mu\kappa}g^{\nu\rho}\right). \tag{B.124}$$

Similar ambiguities arise when employing four-component spinor notation. Indeed, the $\epsilon^{\mu\nu\rho\kappa}$ symbol can arise when using identities involving the Dirac gamma matrices as well as in trace identities that follow from them. In all such cases, the $\epsilon^{\mu\nu\rho\kappa}$ symbol can be associated with γ_5. For example, eq. (B.123) is equivalent to

$$\mathrm{Tr}(\gamma_5 \gamma^\mu \gamma^\nu \gamma^\rho \gamma^\kappa) = -4i\epsilon^{\mu\nu\rho\kappa} \,. \tag{B.125}$$

In the literature various schemes have been proposed for defining the properties of γ_5 in $d \neq 4$ dimensions.[6] In two-component notation, this would translate into a procedure for dealing with general traces involving four or more $\sigma/\overline{\sigma}$ matrices.

[6] See the discussion following eq. (11.106). For a review and further references to the literature, see, e.g., Ref. [211].

Exercises

B.1 Prove that

$$g_{\rho\kappa}(\sigma^{\mu\rho})_\alpha{}^\beta(\overline{\sigma}^{\nu\kappa})^{\dot\beta}{}_{\dot\alpha} = g_{\rho\kappa}(\sigma^{\nu\kappa})_\alpha{}^\beta(\overline{\sigma}^{\mu\rho})^{\dot\beta}{}_{\dot\alpha}\,. \tag{B.126}$$

B.2 A check of eqs. (B.97)–(B.99) can be carried out by multiplying these results by $g_{\mu\rho}g_{\nu\kappa}$ and summing over the two repeated Lorentz index pairs. In this exercise, you will show that the end result is equivalent to eqs. (B.4)–(B.6)].

(a) Show that the above procedure yields

$$(\sigma^{\mu\nu})_\alpha{}^\beta(\sigma_{\mu\nu})_\gamma{}^\tau = -\tfrac{1}{2}(\sigma^{\mu\nu})_\alpha{}^\tau(\sigma_{\mu\nu})_\gamma{}^\beta + \tfrac{3}{2}\delta_\alpha{}^\tau\delta_\gamma{}^\beta\,, \tag{B.127}$$

$$(\overline{\sigma}^{\mu\nu})^{\dot\alpha}{}_{\dot\beta}(\overline{\sigma}_{\mu\nu})^{\dot\gamma}{}_{\dot\tau} = -\tfrac{1}{2}(\overline{\sigma}^{\mu\nu})^{\dot\alpha}{}_{\dot\tau}(\overline{\sigma}_{\mu\nu})^{\dot\gamma}{}_{\dot\beta} + \tfrac{3}{2}\delta^{\dot\alpha}{}_{\dot\tau}\delta^{\dot\gamma}{}_{\dot\beta}\,, \tag{B.128}$$

$$(\sigma^{\mu\nu})_\alpha{}^\beta(\overline{\sigma}_{\mu\nu})^{\dot\gamma}{}_{\dot\tau} = 0\,. \tag{B.129}$$

(b) Verify that eqs. (B.127)–(B.128) are equivalent to eqs. (B.4)–(B.5).

B.3 A check of eqs. (B.90)–(B.92) can be carried out by multiplying these results by $g_{\mu\nu}$ and summing over the repeated Lorentz index pair. In this exercise, you will show that the end result is equivalent to eqs. (B.1)–(B.3).

(a) Show that the above procedure yields

$$\sigma^\mu_{\alpha\dot\alpha}\sigma_{\mu\,\beta\dot\beta} = -\sigma^\mu_{\alpha\dot\beta}\sigma_{\mu\,\beta\dot\alpha}\,, \tag{B.130}$$

$$\overline{\sigma}^{\mu\,\dot\alpha\alpha}\overline{\sigma}_\mu^{\dot\beta\beta} = -\overline{\sigma}^{\mu\,\dot\alpha\beta}\overline{\sigma}_\mu^{\dot\beta\alpha}\,, \tag{B.131}$$

$$\sigma^\mu_{\alpha\dot\alpha}\overline{\sigma}_\mu^{\dot\beta\beta} = 2\,\delta_\alpha{}^\beta\delta^{\dot\beta}{}_{\dot\alpha}\,. \tag{B.132}$$

(b) Verify that eqs. (B.130) and (B.131) imply that

$$\sigma^\mu_{\alpha\dot\alpha}\sigma_{\mu\,\beta\dot\beta} = 2\,\epsilon_{\alpha\beta}\epsilon_{\dot\alpha\dot\beta}\,, \tag{B.133}$$

$$\overline{\sigma}^{\mu\,\dot\alpha\alpha}\overline{\sigma}_\mu^{\dot\beta\beta} = 2\,\epsilon^{\alpha\beta}\epsilon^{\dot\alpha\dot\beta}\,, \tag{B.134}$$

where the coefficients on the right-hand side of eqs. (B.133) and (B.134) can be determined, e.g., by substituting $\alpha=\dot\alpha=1$ and $\beta=\dot\beta=2$.

B.4 (a) Show that eqs. (B.116)–(B.122) are equivalent to the first four gamma matrix identities in $d\neq 4$ spacetime dimensions shown below:

$$\gamma_\mu\gamma^\mu = d\,, \tag{B.135}$$

$$\gamma_\mu\gamma^\alpha\gamma^\mu = (2-d)\gamma^\alpha\,, \tag{B.136}$$

$$\gamma_\mu\gamma^\alpha\gamma^\beta\gamma^\mu = 4g^{\alpha\beta} + (d-4)\gamma^\alpha\gamma^\beta\,, \tag{B.137}$$

$$\gamma_\mu\gamma^\alpha\gamma^\beta\gamma^\rho\gamma^\mu = -2\gamma^\rho\gamma^\beta\gamma^\alpha + (4-d)\gamma^\alpha\gamma^\beta\gamma^\rho\,, \tag{B.138}$$

$$\gamma_\mu\gamma_\alpha\gamma_\beta\gamma_\rho\gamma_\sigma\gamma^\mu = 2(\gamma_\alpha\gamma_\sigma\gamma_\rho\gamma_\beta + \gamma_\beta\gamma_\rho\gamma_\sigma\gamma_\alpha) + (d-4)\gamma_\alpha\gamma_\beta\gamma_\rho\gamma_\sigma\,. \tag{B.139}$$

(b) Using eq. (B.139), obtain the corresponding sigma matrix identities.

C

Appendix C Behavior of Fermion Bilinears under P, T, C

C.1 Two-Component Fermion Field Bilinear Covariants

In Chapter 1, we examined the behavior of two-component spinor fields under P, T, and C. With these results, one can easily compute the behavior of fermion bilinears under the discrete symmetries. First we examine fermion bilinears constructed from anticommuting two-component neutral fermion fields. We summarize the fermion field transformation laws:

$$\mathcal{P}\,\xi_\alpha(x)\mathcal{P}^{-1} = \eta_P i\sigma^0_{\alpha\dot\beta}\xi^{\dagger\dot\beta}(x_P)\,, \tag{C.1}$$

$$\mathcal{T}\,\xi_\alpha(x)\mathcal{T}^{-1} = \eta_T\sigma^0_{\beta\dot\alpha}\xi^{\beta}(x_T)\,, \tag{C.2}$$

$$\mathcal{C}\,\xi_\alpha(x)\mathcal{C}^{-1} = \eta_C\,\xi_\alpha(x)\,, \tag{C.3}$$

where $x_P \equiv (t\,;-\vec{x})$ and $x_T \equiv (-t\,;\vec{x})$. The phases η_P, η_T, and η_C are restricted to be real (either ± 1).

From the above results, one easily derives

$$\mathcal{CP}\,\xi_\alpha(x)(\mathcal{CP})^{-1} = \eta_{CP} i\sigma^0_{\alpha\dot\beta}\xi^{\dagger\dot\beta}(x_P)\,, \tag{C.4}$$

$$\mathcal{CPT}\,\xi_\alpha(x)(\mathcal{CPT})^{-1} = -i\eta_{CPT}\,\xi^\dagger_{\dot\alpha}(-x)\,, \tag{C.5}$$

where $\eta_{CP} \equiv \eta_C\eta_P$, $\eta_{CPT} \equiv \eta_{CP}\eta_T$, and $-x \equiv (-t\,;-\vec{x})$. The CPT invariance of Lorentz-invariant local quantum field theory is consistent with a convention such that

$$\eta_{CPT} \equiv \eta_C\eta_P\eta_T = +1\,. \tag{C.6}$$

Using the above results, the behavior of the fermion bilinears under the discrete symmetries is given in Table C.1.

Next, we examine the case of a charged fermion field, which is specified by a pair of anticommuting two-component fermion fields χ and η of opposite charge. The transformation laws for χ_α are given by

$$\mathcal{P}\,\chi_\alpha(x)\mathcal{P}^{-1} = \eta_P i\sigma^0_{\alpha\dot\beta}\eta^{\dagger\dot\beta}(x_P)\,, \tag{C.7}$$

$$\mathcal{T}\,\chi_\alpha(x)\mathcal{T}^{-1} = \eta_T\sigma^0_{\beta\dot\alpha}\chi^{\beta}(x_T)\,, \tag{C.8}$$

$$\mathcal{C}\chi_\alpha(x)\mathcal{C}^{-1} = \eta_C\eta_\alpha(x)\,, \tag{C.9}$$

with no restriction initially on the phases η_P, η_T, and η_C. The corresponding transformation laws for η_α are given by

$$\mathcal{P}\,\eta_\alpha(x)\mathcal{P}^{-1} \equiv \eta^P_\alpha(x) = \eta^*_P i\sigma^0_{\alpha\dot\beta}\xi^{\dagger\dot\beta}\,, \tag{C.10}$$

$$\mathcal{T}\,\eta_\alpha(x)\mathcal{T}^{-1} \equiv \eta^*_T\sigma^0_{\beta\dot\alpha}\eta^\beta(x_T)\,, \tag{C.11}$$

$$\mathcal{C}\,\eta_\alpha(x)\mathcal{C}^{-1} \equiv \eta^*_C\chi_\alpha(x)\,. \tag{C.12}$$

From the above results, one easily derives

$$\mathcal{CP}\,\chi_\alpha(x)(\mathcal{CP})^{-1} = \eta_{CP} i\sigma^0_{\alpha\dot\beta}\chi^{\dagger\dot\beta}(x_P)\,, \tag{C.13}$$

$$\mathcal{CP}\,\eta_\alpha(x)(\mathcal{CP})^{-1} = \eta^*_{CP} i\sigma^0_{\alpha\dot\beta}\eta^{\dagger\dot\beta}(x_P)\,. \tag{C.14}$$

Likewise,

$$\mathcal{CPT}\chi_\alpha(x)(\mathcal{CPT})^{-1} = -i\eta_{CPT}\,\chi^\dagger_{\dot\alpha}(-x)\,, \tag{C.15}$$

$$\mathcal{CPT}\,\eta_\alpha(x)(\mathcal{CPT})^{-1} = -i\eta^*_{CPT}\eta^\dagger_{\dot\alpha}(-x)\,. \tag{C.16}$$

As in eq. (C.6), we fix $\eta_{CPT} = +1$. Although no further restrictions on the phases are required, it is always possible to rotate η_P and η_C by an arbitrary phase by redefining the parity and charge-conjugation operators (by multiplying by an appropriate gauge transformation). Thus, one is free to establish a convention in which all phases are real (either ±1), as in the case of the neutral self-conjugate fermion. In this convention, $\mathcal{CP} = \mathcal{PC}$, $\mathcal{TC} = \mathcal{CT}$, and $\mathcal{TP} = -\mathcal{PT}$.

Using the above results, the behavior of the bilinear covariants made up of anti-commuting two-component charged fermion fields is easily determined, and is exhibited in Table C.2. In order to be general, we have not imposed any reality conditions on the phases η_P, η_T, and η_C. In Table C.2, we display the behavior of certain linear combinations of fermion bilinears, denoted generically by $B_{12}(x)$, under P, T, and C transformations. The indices 1 and 2 refer to two different charged fermion fields. The interchange of the particle labels 1 and 2 by C, CP, and CPT introduces an extra minus sign due to the anticommutativity property of the fermion fields.

Here, we shall deviate from the flavor-index conventions employed in Chapters 1 and 2 by using lowered flavor indices for all two-component spinors and their conjugates. Note that we have introduced a factor of i in defining the bilinear covariants, $i(\chi_1^\dagger\eta_2^\dagger - \eta_1\chi_2)$ and $i(\chi_1^\dagger\bar\sigma^{\mu\nu}\eta_2^\dagger - \eta_1\bar\sigma^{\mu\nu}\chi_2)$. It then follows that $B^\dagger_{12} = B_{21}$. Moreover, in the case of a single Dirac fermion (where $\chi_1 = \chi_2 \equiv \chi$ and $\eta_1 = \eta_2 \equiv \eta$), all the corresponding bilinear covariants listed in Table C.2 are hermitian.

Since T and CPT are antiunitary operators, any explicit factor of i that appears in the bilinear covariants will change sign under these antiunitary transformations. Hence, we end up with the expected result that

$$\mathcal{CPT}\,B_{12}(x)\,(\mathcal{CPT})^{-1} = B_{21}(-x)\,, \tag{C.17}$$

$$\mathcal{CPT}\,B^\mu_{12}(x)\,(\mathcal{CPT})^{-1} = -B^\mu_{21}(-x)\,, \tag{C.18}$$

$$\mathcal{CPT}\,B^{\mu\nu}_{12}(x)\,(\mathcal{CPT})^{-1} = B^{\mu\nu}_{21}(-x)\,. \tag{C.19}$$

The overall sign on the right-hand side above is equal to $(-1)^s$ for a Lorentz tensor of rank s.

Table C.1 Transformation properties of the bilinear covariants made up of anticommuting two-component fermion fields under the discrete symmetries. The phases η_P, η_T and η_C are restricted to be real (either ± 1), subject to the convention that the phase $\eta_{CPT} \equiv \eta_C \eta_P \eta_T = 1$, independently of the particle species. The following notation is employed: $\Lambda_P \equiv \mathrm{diag}(1, -1, -1, -1)$, $\Lambda_T \equiv \mathrm{diag}(-1, 1, 1, 1)$, $x = (t\,; \vec{x})$, $x_P = (t\,; -\vec{x})$, and $x_T = (-t\,; \vec{x})$. For the product of two phase factors, we denote $\eta_{X12} \equiv \eta_{X1}\eta_{X2}$ for $X = P, T, C$ and CP, respectively.

	P	**T**	**C**	**CP**	**CPT**
$\xi_1\xi_2(x)$	$\eta_{P12}\xi_1^\dagger\xi_2^\dagger(x_P)$	$\eta_{T12}\xi_1\xi_2(x_T)$	$\eta_{C12}\xi_1\xi_2(x)$	$\eta_{CP12}\xi_1^\dagger\xi_2^\dagger(x_P)$	$\xi_1^\dagger\xi_2^\dagger(-x)$
$\xi_1^\dagger\xi_2^\dagger(x)$	$\eta_{P12}\xi_1\xi_2(x_P)$	$\eta_{T12}\xi_1^\dagger\xi_2^\dagger(x_T)$	$\eta_{C12}\xi_1^\dagger\xi_2^\dagger(x)$	$\eta_{CP12}\xi_1\xi_2(x_P)$	$\xi_1\xi_2(-x)$
$\xi_1^\dagger\overline{\sigma}^\mu\xi_2(x)$	$\eta_{P12}(\Lambda_P)^\mu{}_\nu\xi_1\sigma^\nu\xi_2^\dagger(x_P)$	$-\eta_{T12}(\Lambda_T)^\mu{}_\nu\xi_1^\dagger\overline{\sigma}^\nu\xi_2(x_T)$	$\eta_{C12}\xi_1^\dagger\overline{\sigma}^\mu\xi_2(x)$	$\eta_{CP12}(\Lambda_P)^\mu{}_\nu\xi_1\sigma^\nu\xi_2^\dagger(x_P)$	$\xi_1\sigma^\mu\xi_2^\dagger(-x)$
$\xi_1\sigma^\mu\xi_2^\dagger(x)$	$\eta_{P12}(\Lambda_P)^\mu{}_\nu\xi_1^\dagger\overline{\sigma}^\nu\xi_2(x_P)$	$-\eta_{T12}(\Lambda_T)^\mu{}_\nu\xi_1\sigma^\nu\xi_2^\dagger(x_T)$	$\eta_{C12}\xi_1\sigma^\mu\xi_2^\dagger(x)$	$\eta_{CP12}(\Lambda_P)^\mu{}_\nu\xi_1^\dagger\overline{\sigma}^\nu\xi_2(x_P)$	$\xi_1^\dagger\overline{\sigma}^\mu\xi_2(-x)$
$\xi_1\sigma^{\mu\nu}\xi_2(x)$	$\eta_{P12}(\Lambda_P)^\mu{}_\rho(\Lambda_P)^\nu{}_\tau\xi_1^\dagger\overline{\sigma}^{\rho\tau}\xi_2^\dagger(x_P)$	$-\eta_{T12}(\Lambda_T)^\mu{}_\rho(\Lambda_T)^\nu{}_\tau\xi_1\sigma^{\rho\tau}\xi_2(x_T)$	$\eta_{C12}\xi_1\sigma^{\mu\nu}\xi_2(x)$	$\eta_{CP12}(\Lambda_P)^\mu{}_\rho(\Lambda_P)^\nu{}_\tau\xi_1^\dagger\overline{\sigma}^{\rho\tau}\xi_2^\dagger(x_P)$	$-\xi_1^\dagger\overline{\sigma}^{\mu\nu}\xi_2^\dagger(-x)$
$\xi_1^\dagger\overline{\sigma}^{\mu\nu}\xi_2^\dagger(x)$	$\eta_{P12}(\Lambda_P)^\mu{}_\rho(\Lambda_P)^\nu{}_\tau\xi_1\sigma^{\rho\tau}\xi_2(x_P)$	$-\eta_{T12}(\Lambda_T)^\mu{}_\rho(\Lambda_T)^\nu{}_\tau\xi_1^\dagger\overline{\sigma}^{\rho\tau}\xi_2^\dagger(x_T)$	$\eta_{C12}\xi_1^\dagger\overline{\sigma}^{\mu\nu}\xi_2^\dagger(x)$	$\eta_{CP12}(\Lambda_P)^\mu{}_\rho(\Lambda_P)^\nu{}_\tau\xi_1\sigma^{\rho\tau}\xi_2(x_P)$	$-\xi_1\sigma^{\mu\nu}\xi_2(-x)$

Table C.2 Transformation properties of the bilinear covariants made up of anticommuting two-component charged fermion fields [denoted generically by $B_{12}(x)$] under the discrete symmetries. Note that $B_{21} = B_{12}^\dagger$, which implies that B_{11} is hermitian when fermions 1 and 2 are the same. Any explicit factor of i appearing in the bilinear covariants changes sign under the antiunitary T and CPT transformations. The phases η_P, η_T, and η_C are arbitrary, subject to the convention that the phase $\eta_{CPT} \equiv \eta_C\eta_P\eta_T = 1$, independently of the particle species. The following notation is employed: $\Lambda_P \equiv \mathrm{diag}(1, -1, -1, -1)$, $\Lambda_T \equiv \mathrm{diag}(-1, 1, 1, 1)$, $x = (t\,; \vec{x})$, $x_P = (t\,; -\vec{x})$, and $x_T = (-t\,; \vec{x})$. For the product of two phase factors, we denote $\overline{\eta}_{X12} \equiv \eta_{X1}^*\eta_{X2}$ for $X = P, T, C$ and CP, respectively.

$B_{12}(x)$	**P**	**T**	**C**	**CP**	**CPT**
$(\eta_1\chi_2 + \chi_1^\dagger\eta_2^\dagger)(x)$	$\overline{\eta}_{P12}B_{12}(x_P)$	$\overline{\eta}_{T12}B_{12}(x_T)$	$\overline{\eta}_{C12}B_{21}(x)$	$\overline{\eta}_{CP12}B_{21}(x_P)$	$B_{21}(-x)$
$i(\chi_1^\dagger\eta_2^\dagger - \eta_1\chi_2)(x)$	$-\overline{\eta}_{P12}B_{12}(x_P)$	$-\overline{\eta}_{T12}B_{12}(x_T)$	$\overline{\eta}_{C12}B_{21}(x)$	$-\overline{\eta}_{CP12}B_{21}(x_P)$	$B_{21}(-x)$
$(\chi_1^\dagger\overline{\sigma}^\mu\chi_2 + \eta_1\sigma^\mu\eta_2^\dagger)(x)$	$\overline{\eta}_{P12}(\Lambda_P)^\mu{}_\nu B_{12}^\nu(x_P)$	$-\overline{\eta}_{T12}(\Lambda_T)^\mu{}_\nu B_{12}^\nu(x_T)$	$-\overline{\eta}_{C12}B_{21}^\mu(x)$	$-\overline{\eta}_{CP12}(\Lambda_P)^\mu{}_\nu B_{21}^\nu(x_P)$	$-B_{21}^\mu(-x)$
$(\eta_1\sigma^\mu\eta_2^\dagger - \chi_1^\dagger\overline{\sigma}^\mu\chi_2)(x)$	$-\overline{\eta}_{P12}(\Lambda_P)^\mu{}_\nu B_{12}^\nu(x_P)$	$-\overline{\eta}_{T12}(\Lambda_T)^\mu{}_\nu B_{12}^\nu(x_T)$	$\overline{\eta}_{C12}B_{21}^\mu(x)$	$-\overline{\eta}_{CP12}(\Lambda_P)^\mu{}_\nu B_{21}^\nu(x_P)$	$-B_{21}^\mu(-x)$
$(\eta_1\sigma^{\mu\nu}\chi_2 + \chi_1^\dagger\overline{\sigma}^{\mu\nu}\eta_2^\dagger)(x)$	$\overline{\eta}_{P12}(\Lambda_P)^\mu{}_\rho(\Lambda_P)^\nu{}_\tau B_{12}^{\rho\tau}(x_P)$	$-\overline{\eta}_{T12}(\Lambda_T)^\mu{}_\rho(\Lambda_T)^\nu{}_\tau B_{12}^{\rho\tau}(x_T)$	$-\overline{\eta}_{C12}B_{21}^{\mu\nu}(x)$	$-\overline{\eta}_{CP12}(\Lambda_P)^\mu{}_\rho(\Lambda_P)^\nu{}_\tau B_{21}^{\rho\tau}(x_P)$	$B_{21}^{\mu\nu}(-x)$
$i(\chi_1^\dagger\overline{\sigma}^{\mu\nu}\eta_2^\dagger - \eta_1\sigma^{\mu\nu}\chi_2)(x)$	$-\overline{\eta}_{P12}(\Lambda_P)^\mu{}_\rho(\Lambda_P)^\nu{}_\tau B_{12}^{\rho\tau}(x_P)$	$\overline{\eta}_{T12}(\Lambda_T)^\mu{}_\rho(\Lambda_T)^\nu{}_\tau B_{12}^{\rho\tau}(x_T)$	$-\overline{\eta}_{C12}B_{21}^{\mu\nu}(x)$	$\overline{\eta}_{CP12}(\Lambda_P)^\mu{}_\rho(\Lambda_P)^\nu{}_\tau B_{21}^{\rho\tau}(x_P)$	$B_{21}^{\mu\nu}(-x)$

C.2 Four-Component Fermion Field Bilinear Covariants

In Chapter 3, we examined the behavior of four-component spinor fields under P, T, and C. We first summarize the corresponding transformation laws for a four-component Dirac fermion field $\Psi(x)$:

$$\mathcal{P}\,\Psi(x)\mathcal{P}^{-1} = \eta_P i\gamma^0\Psi(x_P) = \eta_P i\Psi^p(x_P)\,, \tag{C.20}$$

$$\mathcal{T}\,\Psi(x)\mathcal{T}^{-1} = \eta_T(\gamma^0 A^{-1})^{\mathsf{T}}B\Psi(x_T) = \eta_T\Psi^{t*}(x_T)\,, \tag{C.21}$$

$$\mathcal{C}\,\Psi(x)\mathcal{C}^{-1} = \eta_C C\overline{\Psi}^{\mathsf{T}}(x) = \eta_C D\Psi^*(x) = \eta_C\Psi^C(x)\,, \tag{C.22}$$

where $x_P \equiv (t\,;\,-\vec{x})$ and $x_T \equiv (-t\,;\,\vec{x})$ and the phases η_P, η_T, and η_C are complex numbers of unit modulus. For convenience, we also list the corresponding transformation laws for the Dirac adjoint field:

$$\mathcal{P}\,\overline{\Psi}(x)\mathcal{P}^{-1} = -i\eta_P^*\overline{\Psi}(x_p)\gamma^0\,, \tag{C.23}$$

$$\mathcal{T}\,\overline{\Psi}(x)\mathcal{T}^{-1} = \eta_T^*\overline{\Psi}(x_T)B^{-1}(A^\dagger\gamma^0)^{\mathsf{T}}\,, \tag{C.24}$$

$$\mathcal{C}\,\overline{\Psi}(x)\mathcal{C}^{-1} = -\eta_C^*\Psi^{\mathsf{T}}(x)C^{-1}\,. \tag{C.25}$$

From the above results, it follows that

$$\mathcal{CP}\,\Psi(x)(\mathcal{CP})^{-1} = i\eta_{CP}\gamma^0 CA^{\mathsf{T}}\Psi^*(x_P)\,, \tag{C.26}$$

$$\mathcal{CPT}\,\Psi(x)(\mathcal{CPT})^{-1} = i\eta_{CPT}[\gamma_5\Psi(-x)]^*\,, \tag{C.27}$$

where $\eta_{CP} \equiv \eta_C\eta_P$ and $\eta_{CPT} \equiv \eta_{CP}\eta_T$, and

$$\mathcal{CP}\,\overline{\Psi}(x)(\mathcal{CP})^{-1} = i\eta_{CP}^*\Psi^{\mathsf{T}}(x_P)C^{-1}\gamma^0\,, \tag{C.28}$$

$$\mathcal{CPT}\,\overline{\Psi}(x)(\mathcal{CPT})^{-1} = -i\eta_{CPT}^*[A^\dagger\gamma_5\Psi(-x)]^{\mathsf{T}}\,. \tag{C.29}$$

Following eq. (C.6), we adopt a convention where $\eta_{CPT} = +1$.

Using the above results, the behavior of the bilinear covariants composed of anticommuting four-component charged fermion fields is exhibited in Table C.3. In order to be general, we have not imposed any reality conditions on the phases η_P, η_T, and η_C. In Table C.3, we have displayed the behavior the four-component fermion bilinear covariants, denoted generically by $D_{12}(x)$, under P, T, and C transformations, where the indices 1 and 2 refer to two different charged fermion fields. Here, we shall deviate from the flavor-index conventions employed in Chapter 3 by using lowered flavor indices for both Ψ and $\overline{\Psi}$. More explicitly, we consider

$$D_{12}(x) \equiv \overline{\Psi}_1\Gamma\Psi_2\,, \quad \text{for } \Gamma = \mathbb{1}\,,\, i\gamma_5\,,\, \gamma^\mu, \gamma^\mu\gamma_5\,,\, \Sigma^{\mu\nu}\,,\, i\Sigma^{\mu\nu}\gamma_5. \tag{C.30}$$

For all cases, Γ given in eq. (C.30) satisfies the condition [see eq. (3.97)]

$$\Gamma^\dagger = A\Gamma A^{-1}\,. \tag{C.31}$$

By choosing Γ to satisfy eq. (C.31), the bilinear covariants satisfy $D_{12}^\dagger = D_{21}$ in the standard convention where $A^\dagger = A$. In particular, for a single species of Dirac fermion fields, the bilinear covariants $D_{11}(x)$ are hermitian.

In the derivation of the P and T transformed bilinear covariants, we have made use of the identity

$$\gamma^0\gamma^\mu\gamma_0 = (\Lambda_P)^\mu{}_\nu\gamma^\nu\,, \tag{C.32}$$

where $(\Lambda_P)^\mu{}_\nu = \text{diag}(1,-1,-1,-1)$. It then follows that

$$\gamma^0\Gamma\gamma^0 = \eta^0_\Gamma\widetilde{\Gamma}\,, \qquad \eta^0_\Gamma = \begin{cases} +1\,, & \text{for } \Gamma = \mathbb{1}\,,\, \gamma^\mu\,,\, \Sigma^{\mu\nu}\,, \\ -1\,, & \text{for } \Gamma = \gamma_5\,,\, \gamma^\mu\gamma_5\,,\, \Sigma^{\mu\nu}\gamma_5\,, \end{cases} \tag{C.33}$$

where we have defined $\widetilde{\Gamma} = \Gamma$ if Γ is a scalar or pseudoscalar quantity, and

$$\widetilde{\Gamma}^\mu \equiv (\Lambda_P)^\mu{}_\nu\Gamma^\nu\,, \tag{C.34}$$

$$\widetilde{\Gamma}^{\mu_1\mu_2} \equiv (\Lambda_P)^{\mu_1}{}_{\nu_1}(\Lambda_P)^{\mu_2}{}_{\nu_2}\Gamma^{\nu_1\nu_2}\,, \tag{C.35}$$

if Γ is a (pseudo)vector or second-rank (pseudo)tensor quantity, respectively. In the case of the T transformed bilinear covariants, it is convenient to express the results in terms of $(\Lambda_T)^\mu{}_\nu = -(\Lambda_P)^\mu{}_\nu = \text{diag}(-1,1,1,1)$.

The transformation laws for a four-component Majorana fermion field $\Psi_M(x)$ are also given by eqs. (C.20)–(C.25). However, due to the Majorana condition

$$\Psi_M(x) = C\overline{\Psi}{}^{\mathsf{T}}_M(x)\,, \tag{C.36}$$

where C is the charge-conjugation matrix [eq. (A.97)], it follows that the phases η_P, η_T, and η_C are restricted to be real (either ± 1). In addition, $\eta_{CPT} = +1$, as indicated in eq. (C.6). In Table C.4, we have displayed the behavior the four-component Majorana fermion bilinear covariants, denoted generically by $M_{12}(x)$, under P, T, and C transformations. More explicitly, we consider

$$M_{12}(x) \equiv \overline{\Psi}_1\Gamma\Psi_2\,, \quad \text{for } \Gamma = \mathbb{1}\,,\, i\gamma_5, i\gamma^\mu\,,\, \gamma^\mu\gamma_5\,,\, i\Sigma^{\mu\nu}\,,\, \Sigma^{\mu\nu}\gamma_5. \tag{C.37}$$

For all cases, Γ given in eq. (C.37) satisfies the condition [see eq. (3.105)],

$$\Gamma^\dagger = AC\,\Gamma^{\mathsf{T}}(AC)^{-1}\,. \tag{C.38}$$

By choosing Γ to satisfy eq. (C.38), it follows that the Majorana fermion bilinear covariants $M_{12}(x)$ are hermitian.

In obtaining the C, CP, and CPT transformed bilinear covariants in Table C.4, we have re-expressed $M_{21}(x)$ in terms of $M_{12}(x)$ by making use of the relations given in eqs. (3.98)–(3.100):

$$\overline{\Psi}_{M2}\Gamma\Psi_{M1} = \eta^C_\Gamma\overline{\Psi}_{M1}\Gamma\Psi_{M2}\,, \tag{C.39}$$

where

$$\eta^C_\Gamma = \begin{cases} +1\,, & \text{for } \Gamma = \mathbb{1}\,,\, i\gamma_5\,,\, \gamma^\mu\gamma_5\,, \\ -1\,, & \text{for } \Gamma = i\gamma^\mu\,,\, i\Sigma^{\mu\nu}\,,\, \Sigma^{\mu\nu}\gamma_5\,. \end{cases} \tag{C.40}$$

One consequence of eqs. (C.39) and (C.40) is that $M_{11}(x) = 0$ for $\Gamma = i\gamma^\mu$, $i\Sigma^{\mu\nu}$, and $\Sigma^{\mu\nu}\gamma_5$. That is,

$$\overline{\Psi}_M\gamma^\mu\Psi_M = \overline{\Psi}_M\Sigma^{\mu\nu}\Psi_M = \overline{\Psi}_M\Sigma^{\mu\nu}\gamma_5\Psi_M = 0\,. \tag{C.41}$$

Table C.3 Transformation properties of the bilinear covariants made up of anticommuting four-component Dirac fermion fields [denoted generically by $D_{12}(x)$] under the discrete symmetries. Note that $D_{21} = D_{12}^\dagger$, which implies that D_{11} is hermitian when fermions 1 and 2 are the same. Any explicit factor of i appearing in the bilinear covariants changes sign under the antiunitary T and CPT transformations. The phases η_P, η_C, and η_T are arbitrary, subject to the convention that the phase $\eta_{CPT} \equiv \eta_C\eta_P\eta_T = 1$, independently of the particle species. The following notation is employed: $\Lambda_P \equiv \mathrm{diag}(1,-1,-1,-1)$, $\Lambda_T \equiv \mathrm{diag}(-1,1,1,1)$, $x = (t\,;\vec{x})$, $x_P = (t\,;-\vec{x})$, and $x_T = (-t\,;\vec{x})$. For the product of two phase factors, we denote $\eta_{X12} \equiv \eta^*_{X1}\eta_{X2}$ for $X = P, T, C$ and CP, respectively.

$D_{12}(x)$	P	T	C	CP	CPT
$\overline{\Psi}_1(x)\Psi_2(x)$	$\bar{\eta}_{P12}D_{12}(x_P)$	$\bar{\eta}_{T12}D_{12}(x_T)$	$\bar{\eta}_{C12}D_{21}(x)$	$\bar{\eta}_{CP12}D_{21}(x_P)$	$D_{21}(-x)$
$i\overline{\Psi}_1(x)\gamma_5\Psi_2(x)$	$-\bar{\eta}_{P12}D_{12}(x_P)$	$-\bar{\eta}_{T12}D_{12}(x_T)$	$\bar{\eta}_{C12}D_{21}(x)$	$-\bar{\eta}_{CP12}D_{21}(x_P)$	$D_{21}(-x)$
$\overline{\Psi}_1(x)\gamma^\mu\Psi_2(x)$	$\bar{\eta}_{P12}(\Lambda_P)^\mu{}_\nu D^\nu_{12}(x_P)$	$-\bar{\eta}_{T12}(\Lambda_T)^\mu{}_\nu D^\nu_{12}(x_T)$	$-\bar{\eta}_{C12}D^\mu_{21}(x)$	$-\bar{\eta}_{CP12}(\Lambda_P)^\mu{}_\nu D^\nu_{21}(x_P)$	$-D^\mu_{21}(-x)$
$\overline{\Psi}_1(x)\gamma^\mu\gamma_5\Psi_2(x)$	$-\bar{\eta}_{P12}(\Lambda_P)^\mu{}_\nu D^\nu_{12}(x_P)$	$-\bar{\eta}_{T12}(\Lambda_T)^\mu{}_\nu D^\nu_{12}(x_T)$	$\bar{\eta}_{C12}D^\mu_{21}(x)$	$-\bar{\eta}_{CP12}(\Lambda_P)^\mu{}_\nu D^\nu_{21}(x_P)$	$-D^\mu_{21}(-x)$
$\overline{\Psi}_1(x)\Sigma^{\mu\nu}\Psi_2(x)$	$\bar{\eta}_{P12}(\Lambda_P)^\mu{}_\rho(\Lambda_P)^\nu{}_\tau D^{\rho\tau}_{12}(x_P)$	$-\bar{\eta}_{T12}(\Lambda_T)^\mu{}_\rho(\Lambda_T)^\nu{}_\tau D^{\rho\tau}_{12}(x_T)$	$-\bar{\eta}_{C12}D^{\mu\nu}_{21}(x)$	$-\bar{\eta}_{CP12}(\Lambda_P)^\mu{}_\rho(\Lambda_P)^\nu{}_\tau D^{\rho\tau}_{21}(x_P)$	$D^{\mu\nu}_{21}(-x)$
$i\overline{\Psi}_1(x)\Sigma^{\mu\nu}\gamma_5\Psi_2(x)$	$-\bar{\eta}_{P12}(\Lambda_P)^\mu{}_\rho(\Lambda_P)^\nu{}_\tau D^{\rho\tau}_{12}(x_P)$	$\bar{\eta}_{T12}(\Lambda_T)^\mu{}_\rho(\Lambda_T)^\nu{}_\tau D^{\rho\tau}_{12}(x_T)$	$-\bar{\eta}_{C12}D^{\mu\nu}_{21}(x)$	$\bar{\eta}_{CP12}(\Lambda_P)^\mu{}_\rho(\Lambda_P)^\nu{}_\tau D^{\rho\tau}_{21}(x_P)$	$D^{\mu\nu}_{21}(-x)$

Table C.4 Transformation properties of the hermitian bilinear covariants made up of anticommuting four-component Majorana fermion fields [denoted generically by $M_{12}(x)$] under the discrete symmetries. Any explicit factor of i appearing in M_{12} changes sign under the antiunitary T and CPT transformations. The phases η_P, η_T and η_C are restricted to be real (either ±1), subject to the convention that the phase $\eta_{CPT} \equiv \eta_C\eta_P\eta_T = 1$, independently of the particle species. The following notation is employed: $\Lambda_P \equiv \mathrm{diag}(1,-1,-1,-1)$, $\Lambda_T \equiv \mathrm{diag}(-1,1,1,1)$, $x = (t\,;\vec{x})$, $x_P = (t\,;-\vec{x})$, and $x_T = (-t\,;\vec{x})$. For the product of two phase factors, we denote $\eta_{X12} \equiv \eta_{X1}\eta_{X2}$ for $X = P, T, C$ and CP, respectively.

$M_{12}(x)$	P	T	C	CP	CPT
$\overline{\Psi}_{M1}(x)\Psi_{M2}(x)$	$\eta_{P12}M_{12}(x_P)$	$\eta_{T12}M_{12}(x_T)$	$\eta_{C12}M_{12}(x)$	$\eta_{CP12}M_{12}(x_P)$	$M_{12}(-x)$
$i\overline{\Psi}_{M1}(x)\gamma_5\Psi_{M2}(x)$	$-\eta_{P12}M_{12}(x_P)$	$-\eta_{T12}M_{12}(x_T)$	$\eta_{C12}M_{12}(x)$	$-\eta_{CP12}M_{12}(x_P)$	$M_{12}(-x)$
$i\overline{\Psi}_{M1}(x)\gamma^\mu\Psi_{M2}(x)$	$\eta_{P12}(\Lambda_P)^\mu{}_\nu M^\nu_{12}(x_P)$	$\eta_{T12}(\Lambda_T)^\mu{}_\nu M^\nu_{12}(x_T)$	$\eta_{C12}M^\mu_{12}(x)$	$\eta_{CP12}(\Lambda_P)^\mu{}_\nu M^\nu_{12}(x_P)$	$-M^\mu_{12}(-x)$
$\overline{\Psi}_{M1}(x)\gamma^\mu\gamma_5\Psi_{M2}(x)$	$-\eta_{P12}(\Lambda_P)^\mu{}_\nu M^\nu_{12}(x_P)$	$-\eta_{T12}M(\Lambda_T)^\mu{}_\nu M^\nu_{12}(x_T)$	$\eta_{C12}M^\mu_{12}(x)$	$-\eta_{CP12}(\Lambda_P)^\mu{}_\nu M^\nu_{12}(x_P)$	$-M^\mu_{12}(-x)$
$i\overline{\Psi}_{M1}(x)\Sigma^{\mu\nu}\Psi_{M2}(x)$	$\eta_{P12}(\Lambda_P)^\mu{}_\rho(\Lambda_P)^\nu{}_\tau M^{\rho\tau}_{12}(x_P)$	$\eta_{T12}(\Lambda_T)^\mu{}_\rho(\Lambda_T)^\nu{}_\tau M^{\rho\tau}_{12}(x_T)$	$\eta_{C12}M^{\mu\nu}_{12}(x)$	$\eta_{CP12}(\Lambda_P)^\mu{}_\rho(\Lambda_P)^\nu{}_\tau M^{\rho\tau}_{12}(x_P)$	$M^{\mu\nu}_{12}(-x)$
$\overline{\Psi}_{M1}(x)\Sigma^{\mu\nu}\gamma_5\Psi_{M2}(x)$	$-\eta_{P12}(\Lambda_P)^\mu{}_\rho(\Lambda_P)^\nu{}_\tau M^{\rho\tau}_{12}(x_P)$	$-\eta_{T12}(\Lambda_T)^\mu{}_\rho(\Lambda_T)^\nu{}_\tau M^{\rho\tau}_{12}(x_T)$	$\eta_{C12}M^{\mu\nu}_{12}(x)$	$-\eta_{CP12}(\Lambda_P)^\mu{}_\rho(\Lambda_P)^\nu{}_\tau M^{\rho\tau}_{12}(x_P)$	$M^{\mu\nu}_{12}(-x)$

D

Appendix D Kinematics and Phase Space

In this appendix, we summarize a number of important results of relativistic particle kinematics, in units where the speed of light $c = 1$. For more details, see [B146].

D.1 Relativistic Kinematics

We begin by defining the ubiquitous kinematic triangle function,

$$\begin{aligned}\lambda(x^2, y^2, z^2) &\equiv x^4 + y^4 + z^4 - 2x^2y^2 - 2x^2z^2 - 2y^2z^2 \\ &= (x^2 - y^2 - z^2)^2 - 4y^2z^2 \\ &= \left[x^2 - (y+z)^2\right]\left[x^2 - (y-z)^2\right].\end{aligned} \tag{D.1}$$

If p, p_1, and p_2 are four-momenta that satisfy $p = p_1 + p_2$, then the kinematic triangle function $\lambda(p^2, p_1^2, p_2^2)$ is proportional to the symmetric Gram determinant:

$$\lambda((p_1+p_2)^2, p_1^2, p_2^2) = -4\det\begin{pmatrix} p_1^2 & p_1{\cdot}p_2 \\ p_1{\cdot}p_2 & p_2^2\end{pmatrix} = 4\left[(p_1{\cdot}p_2)^2 - p_1^2p_2^2\right]. \tag{D.2}$$

Consider the two-body decay $a \to 1 + 2$, where the four-momenta and masses of particles a, 1, and 2 are denoted (p, m), (p_1, m_1), and (p_2, m_2), respectively, where $p^2 = m^2$ and $p_i^2 = m_i^2$ (for $i = 1, 2$). Using momentum conservation, $p = p_1 + p_2$, all dot products of momenta can be expressed in terms of the particle masses:

$$2p{\cdot}p_1 = m^2 + m_1^2 - m_2^2\,, \tag{D.3}$$

$$2p{\cdot}p_2 = m^2 - m_1^2 + m_2^2\,, \tag{D.4}$$

$$2p_1{\cdot}p_2 = m^2 - m_1^2 - m_2^2\,. \tag{D.5}$$

It is often convenient to work in the center-of-momentum (CM) reference frame, where $p = (m\,;\,\vec{0})$. In the CM frame, the energies of particles 1 and 2 are given by

$$E_1 = \frac{m^2 + m_1^2 - m_2^2}{2m}\,, \qquad E_2 = \frac{m^2 + m_2^2 - m_1^2}{2m}\,, \tag{D.6}$$

and the magnitudes of the corresponding CM three-momenta are

$$p_{\rm CM} \equiv |\vec{\boldsymbol{p}}_{\boldsymbol{1}}| = |\vec{\boldsymbol{p}}_{\boldsymbol{2}}| = \frac{\lambda^{1/2}(m^2, m_1^2, m_2^2)}{2m}\,, \tag{D.7}$$

where $\lambda^{1/2}$ is the square root of the kinematic triangle function defined in eq. (D.1).

Next, consider the three-body decay $a \to 1+2+3$, where the four-momenta and masses of particles a, 1, 2, and 3 are denoted by (p, m), (p_1, m_1), (p_2, m_2), and (p_3, m_3), respectively, where $p^2 = m^2$ and $p_i^2 = m_i^2$ (for $i = 1, 2, 3$). Using momentum conservation, $p = p_1 + p_2 + p_3$, one can define the Lorentz-invariant quantities

$$s_1 \equiv s_{12} = (p_1 + p_2)^2 = (p - p_3)^2\,, \tag{D.8}$$
$$s_2 \equiv s_{23} = (p_2 + p_3)^2 = (p - p_1)^2\,, \tag{D.9}$$
$$s_3 \equiv s_{31} = (p_3 + p_1)^2 = (p - p_2)^2\,. \tag{D.10}$$

It immediately follows that s_1, s_2, and s_3 must satisfy

$$(m_1 + m_2)^2 \le s_1 \le (m - m_3)^2\,, \tag{D.11}$$
$$(m_2 + m_3)^2 \le s_2 \le (m - m_1)^2\,, \tag{D.12}$$
$$(m_1 + m_3)^2 \le s_3 \le (m - m_2)^2\,. \tag{D.13}$$

Only two of the three variables s_1, s_2, and s_3 are independent due to the relation

$$s_1 + s_2 + s_3 = m^2 + m_1^2 + m_2^2 + m_3^2\,. \tag{D.14}$$

The above relations can be used to express all dot products in terms of s_1, s_2, s_3, and the squared masses:

$$2p\cdot p_1 = m^2 + m_1^2 - s_2\,, \qquad 2p_1\cdot p_2 = s_1 - m_1^2 - m_2^2\,, \tag{D.15}$$
$$2p\cdot p_2 = m^2 + m_2^2 - s_3\,, \qquad 2p_1\cdot p_3 = s_3 - m_1^2 - m_3^2\,, \tag{D.16}$$
$$2p\cdot p_3 = m^2 + m_3^2 - s_1\,, \qquad 2p_2\cdot p_3 = s_2 - m_2^2 - m_3^2\,. \tag{D.17}$$

In the CM frame where $p = (m\,;\,\vec{\mathbf{0}})$, the energies of particles 1, 2, 3 are given by

$$E_1 = \frac{m^2 + m_1^2 - s_2}{2m}\,, \qquad E_2 = \frac{m^2 + m_2^2 - s_3}{2m}\,, \qquad E_3 = \frac{m^2 + m_3^2 - s_1}{2m}\,, \tag{D.18}$$

and the magnitudes of the corresponding CM three-momenta are

$$|\vec{\boldsymbol{p}}_1| = \frac{\lambda^{1/2}(m^2, m_1^2, s_2)}{2m}\,, \qquad |\vec{\boldsymbol{p}}_2| = \frac{\lambda^{1/2}(m^2, m_2^2, s_3)}{2m}\,, \qquad |\vec{\boldsymbol{p}}_3| = \frac{\lambda^{1/2}(m^2, m_3^2, s_1)}{2m}\,. \tag{D.19}$$

Suppose that p, p_1, p_2, and p_3 are four-momenta that satisfy $p = p_1 + p_2 + p_3$. Generalizing eq. (D.2) to the case of three independent four-momenta, we define the basic four-particle kinematic function, which is proportional to the symmetric Gram determinant:

$$G\big((p_1+p_2)^2, (p_1+p_3)^2, (p_1+p_2+p_3)^2, p_1^2, p_2^2, p_3^2\big)$$
$$= -4\det\begin{pmatrix} p_1^2 & p_1\cdot p_2 & p_1\cdot p_3 \\ p_1\cdot p_2 & p_2^2 & p_2\cdot p_3 \\ p_1\cdot p_3 & p_2\cdot p_3 & p_3^2 \end{pmatrix}. \tag{D.20}$$

Note that G is a function of six independent variables: three dot products and three squared four-momenta. An explicit algebraic expression for G is given by[1]

$$\begin{aligned} G(x,y,z,u,v,w) &= xy(x+y)+zu(z+u)+vw(v+w)+x(zw+uv)+y(zv+uw) \\ &\quad -xy(z+u+v+w)-zu(x+y+v+w)-vw(x+y+z+u)\,. \end{aligned} \tag{D.21}$$

The physical region of the three-body decay in the s_1–s_2 plane must satisfy the inequality $G(s_1, s_2, m^2, m_2^2, m_1^2, m_3^2) \leq 0$. It then follows that $s_1^- \leq s_1 \leq s_1^+$, where

$$\begin{aligned} s_1^{\pm} = m_1^2 + m_2^2 - \frac{1}{2s_2}\Big[&(s_2 - s + m_1^2)(s_2 + m_2^2 - m_3^2) \\ &\mp \lambda^{1/2}(s_2, s, m_1^2)\lambda^{1/2}(s_2, m_2^2, m_3^2)\Big]\,, \end{aligned} \tag{D.22}$$

where $(m_2 + m_3)^2 \leq s_2 \leq (m - m_1)^2$ in light of eq. (D.12).

The $2 \to 2$ particle scattering process $a + b \to 1 + 2$ is also characterized by three Lorentz-invariant quantities: the Mandelstam variables, s, t, and u:

$$s \equiv (p_a + p_b)^2 = (p_1 + p_2)^2\,, \tag{D.23}$$

$$t \equiv (p_a - p_1)^2 = (p_b - p_2)^2\,, \tag{D.24}$$

$$u \equiv (p_a - p_2)^2 = (p_b - p_1)^2\,, \tag{D.25}$$

where

$$s \geq \max\{(m_a + m_b)^2\,,\,(m_1 + m_2)^2\}\,. \tag{D.26}$$

The four-momenta and masses of particles a, b, 1, and 2 are denoted by (p_a, m_a), (p_b, m_b), (p_1, m_1), and (p_2, m_2), respectively, where $p_i^2 = m_i^2$ (for $i = a, b, 1, 2$) and $p_a + p_b = p_1 + p_2$ as a consequence of momentum conservation. Only two of the three variables s, t, and u are independent due to the relation

$$s + t + u = m_a^2 + m_b^2 + m_1^2 + m_2^2\,. \tag{D.27}$$

The above relations can be used to express all dot products in terms of s, t, u, and the squared masses:

$$2p_a{\cdot}p_b = s - m_a^2 - m_b^2\,, \qquad 2p_1{\cdot}p_2 = s - m_1^2 - m_2^2\,, \tag{D.28}$$

$$2p_a{\cdot}p_1 = m_a^2 + m_1^2 - t\,, \qquad 2p_b{\cdot}p_1 = m_b^2 + m_1^2 - u\,, \tag{D.29}$$

$$2p_a{\cdot}p_2 = m_a^2 + m_2^2 - u\,, \qquad 2p_b{\cdot}p_2 = m_b^2 + m_2^2 - t\,. \tag{D.30}$$

In the CM frame where $p_a + p_b = (\sqrt{s}\,;\,\vec{\boldsymbol{0}})$, the energies of particles a, b, 1, and 2 are given by

$$E_a = \frac{s + m_a^2 - m_b^2}{2\sqrt{s}}\,, \qquad E_b = \frac{s + m_b^2 - m_a^2}{2\sqrt{s}}\,, \tag{D.31}$$

$$E_1 = \frac{s + m_1^2 - m_2^2}{2\sqrt{s}}\,, \qquad E_2 = \frac{s + m_2^2 - m_1^2}{2\sqrt{s}}\,, \tag{D.32}$$

[1] Equation (D.21), which was first defined in Ref. [212], is also given in eq. IV-5.23 of [B146]. We have noted a typographical error in the latter; in the second line of eq. IV-5.23, the first term yzw should read yzv.

and the magnitudes of the corresponding CM three-momenta are

$$p_i \equiv |\vec{p}_a| = |\vec{p}_b| = \frac{\lambda^{1/2}(s, m_a^2, m_b^2)}{2\sqrt{s}}\,, \tag{D.33}$$

$$p_f \equiv |\vec{p}_1| = |\vec{p}_2| = \frac{\lambda^{1/2}(s, m_1^2, m_2^2)}{2\sqrt{s}}\,. \tag{D.34}$$

The CM scattering angle is defined as the angle θ between incoming particle a and outgoing particle 1 in the CM frame. In terms of invariant quantities,

$$\cos\theta = \frac{s(t-u) + (m_a^2 - m_b^2)(m_1^2 - m_2^2)}{\lambda^{1/2}(s, m_a^2, m_b^2)\lambda^{1/2}(s, m_1^2, m_2^2)}\,. \tag{D.35}$$

Starting from eq. (D.35), a laborious computation yields

$$\sin^2\theta = -4s\,\frac{G(s, t, m_2^2, m_a^2, m_b^2, m_1^2)}{\lambda(s, m_a^2, m_b^2)\lambda(s, m_1^2, m_2^2)}\,, \tag{D.36}$$

where G is the basic four-particle kinematic function defined in eq. (D.21). This result implies that the physical region of the scattering process $a + b \to 1 + 2$ must satisfy $G(s, t, m_2^2, m_a^2, m_b^2, m_1^2) \leq 0$. It then follows that, in the s–t plane at fixed values of s [subject to eq. (D.26)], t must lie in the interval $t^- \leq t \leq t^+$, where

$$t^\pm = m_a^2 + m_b^2 - \frac{1}{2s_2}\Big[(s + m_a^2 - m_b^2)(s_2 + m_1^2 - m_2^2) \mp \lambda^{1/2}(s, m_a^2, m_b^2)\lambda^{1/2}(s, m_1^2, m_2^2)\Big]\,. \tag{D.37}$$

For completeness, we briefly examine the $2 \to 3$ scattering process $a + b \to 1 + 2 + 3$, which depends on five independent Lorentz-invariant quantities that are conventionally chosen to be

$$s_1 \equiv s_{12} = (p_1 + p_2)^2 = (p_a + p_b - p_3)^2\,, \tag{D.38}$$

$$s_2 \equiv s_{23} = (p_2 + p_3)^2 = (p_a + p_b - p_1)^2\,, \tag{D.39}$$

$$t_1 \equiv t_{a1} = (p_a - p_1)^2 = (p_2 + p_3 - p_b)^2\,, \tag{D.40}$$

$$t_2 \equiv t_{b3} = (p_b - p_3)^2 = (p_1 + p_2 - p_a)^2\,, \tag{D.41}$$

$$s \equiv s_{ab} = (p_a + p_b)^2 = (p_1 + p_2 + p_3)^2\,, \tag{D.42}$$

where we have extended the notation employed in the $2 \to 2$ scattering process above for the four-momenta and masses. The 10 possible dot products can be expressed in terms of these invariants as follows:

$$2p_a\cdot p_b = s - m_a^2 - m_b^2\,, \qquad 2p_b\cdot p_2 = s_2 + t_2 - t_1 - m_3^2\,, \tag{D.43}$$

$$2p_a\cdot p_1 = m_a^2 + m_1^2 - t_1\,, \qquad 2p_b\cdot p_3 = m_b^2 + m_3^2 - t_2\,, \tag{D.44}$$

$$2p_a\cdot p_2 = s_1 + t_1 - t_2 - m_1^2\,, \qquad 2p_1\cdot p_2 = s_1 - m_1^2 - m_2^2\,, \tag{D.45}$$

$$2p_a\cdot p_3 = s - s_1 + t_2 - m_b^2\,, \qquad 2p_1\cdot p_3 = s - s_1 - s_2 + m_2^2\,, \tag{D.46}$$

$$2p_b\cdot p_1 = s - s_2 + t_1 - m_a^2\,, \qquad 2p_2\cdot p_3 = s_2 - m_2^2 - m_3^2\,. \tag{D.47}$$

D.2 Lorentz-Invariant Phase Space

The n-particle Lorentz-invariant phase space (Lips) element is defined by

$$d\,\mathrm{Lips}(p_1, p_2, \ldots, p_n) \equiv \prod_{i=1}^{n} \frac{d^3\boldsymbol{p_i}}{(2\pi)^3\, 2E_i}\,, \tag{D.48}$$

where $p_i = (E_i\,;\, \vec{\boldsymbol{p}}_i)$ and $E_i = (|\vec{\boldsymbol{p}}_i|^2 + m_i^2)^{1/2}$. The corresponding n-particle phase space integral is given by

$$R_n(s) \equiv (2\pi)^{3n} \int d\,\mathrm{Lips}(p_1, p_2, \ldots, p_n)\, \delta^4\left(p - \sum_{i=k}^{n} p_k\right), \tag{D.49}$$

where s is the Lorentz-invariant quantity, $s \equiv p^2$. In Appendices D.4 and D.5, the integrand above will be multiplied by a squared matrix element which can be expressed as a function of dot products of momentum and spin four-vectors.

The case of $n = 2$ is easily analyzed. In this case, we must evaluate

$$R_2(s) = \int \frac{d^3\boldsymbol{p_1}}{2E_1} \frac{d^3\boldsymbol{p_2}}{2E_2}\, \delta^4(p - p_1 - p_2)\,. \tag{D.50}$$

The first step is to make use of the identity

$$\frac{1}{2E}\delta(p_0 - E) = \delta(p^2 - m^2)\, \Theta(p_0)\,, \tag{D.51}$$

where $E \equiv (|\vec{\boldsymbol{p}}|^2 + m^2)^{1/2}$, $p^2 = p_0^2 - |\vec{\boldsymbol{p}}|^2$, and $\Theta(x)$ is the step function,

$$\Theta(x) \equiv \begin{cases} 1, & \text{for } x > 0\,, \\ 0, & \text{for } x < 0\,. \end{cases} \tag{D.52}$$

It then follows that

$$\int \frac{d^3\boldsymbol{p}}{2E} = \int d^4p \frac{1}{2E}\delta(p_0 - E) = \int d^4p\, \delta(p^2 - m^2)\Theta(p_0)\,, \tag{D.53}$$

after writing $d^4p = d^3p\, dp_0$. Using eq. (D.53), we can rewrite eq. (D.50) as

$$\begin{aligned} R_2(s) &= \int \frac{d^3\boldsymbol{p_1}}{2E_1}\, d^4p_2\, \delta(p_2^2 - m_2^2)\Theta(p_{20})\delta^4(p - p_1 - p_2) \\ &= \int \frac{d^3\boldsymbol{p_1}}{2E_1}\, \delta((p - p_1)^2 - m_2^2)\,\Theta((p - p_1)_0)\,, \end{aligned} \tag{D.54}$$

after using the four-dimensional delta function to integrate over p_2.

We shall evaluate the Lorentz-invariant quantity $R_2(s)$ in the CM frame,

$$p_1 = (E_1\,;\, \vec{\boldsymbol{p}}_{\mathrm{CM}})\,, \qquad p_2 = (E_2\,;\, -\vec{\boldsymbol{p}}_{\mathrm{CM}})\,, \qquad p = (\sqrt{s}\,;\, \vec{\boldsymbol{0}})\,, \tag{D.55}$$

where [see eq. (D.6)]

$$E_1 = \frac{s + m_1^2 - m_2^2}{2\sqrt{s}}\,, \qquad E_2 = \frac{s + m_2^2 - m_1^2}{2\sqrt{s}}\,, \tag{D.56}$$

and the magnitude of $\vec{p}_{\rm CM}$ is given by

$$p_{\rm CM} \equiv |\vec{p}_{\rm CM}| = (E_1^2 - m_1^2)^{1/2} = \frac{\lambda^{1/2}(s, m_1^2, m_2^2)}{2\sqrt{s}}\,. \tag{D.57}$$

Note that in the CM frame,

$$(p - p_1)^2 - m_2^2 = p^2 - 2p{\cdot}p_1 + p_1^2 - m_2^2 = s - 2\sqrt{s}E_1 + m_1^2 - m_2^2\,, \tag{D.58}$$

after using $p_1^2 = m_1^2$ and $p{\cdot}p_1 = \sqrt{s}E_1$. Inserting the result of eq. (D.58) back into eq. (D.54) yields

$$R_2(s) = \int \frac{d^3\boldsymbol{p_1}}{2E_1}\,\delta(s - 2\sqrt{s}E_1 + m_1^2 - m_2^2)\Theta(\sqrt{s} - E_1)\,. \tag{D.59}$$

In light of eq. (D.57), it follows that $E_1 dE_1 = p_{\rm CM} dp_{\rm CM}$. Thus in the CM frame,

$$d^3\boldsymbol{p_1} = p_{\rm CM}^2 dp_{\rm CM} d\Omega = p_{\rm CM} E_1 dE_1 d\Omega\,. \tag{D.60}$$

It follows that

$$\begin{aligned} R_2(s) &= \frac{1}{2}\int p_{\rm CM} dE_1 d\Omega\,\delta(s - 2\sqrt{s}E_1 + m_1^2 - m_2^2)\Theta(\sqrt{s} - E_1) \\ &= \frac{p_{\rm CM}}{4\sqrt{s}}\int d\Omega = \frac{\lambda^{1/2}(s, m_1^2, m_2^2)}{8s}\int d\Omega\,, \end{aligned} \tag{D.61}$$

after using the delta function to integrate over E_1. Note that the step function condition is automatically satisfied since four-momentum conservation implies that $\sqrt{s} = E_1 + E_2$ [see eq. (D.55)], and the energies E_1 and E_2 are nonnegative.

In eq. (D.61), the solid angle Ω corresponds to the orientation of the three-momentum $\vec{p}_1$ in the CM frame. In many applications, this orientation is arbitrary and thus we may integrate over solid angles to obtain

$$R_2(s) = \frac{\pi\lambda^{1/2}(s, m_1^2, m_2^2)}{2s}\,. \tag{D.62}$$

The case of $n = 3$ is more complicated. Here, we must evaluate

$$R_3(s) = \int \frac{d^3\boldsymbol{p_1}}{2E_1}\frac{d^3\boldsymbol{p_2}}{2E_2}\frac{d^3\boldsymbol{p_3}}{2E_3}\,\delta^3(\vec{p} - \vec{p}_1 - \vec{p}_2 - \vec{p}_3)\,\delta(\sqrt{s} - E_1 - E_2 - E_3)\,. \tag{D.63}$$

We first use the three-dimensional delta function to integrate over $\vec{p}_2$, which yields

$$R_3(s) = \frac{1}{8}\int \frac{d^3\boldsymbol{p_1}\,d^3\boldsymbol{p_3}}{E_1 E_2 E_3}\,\delta(\sqrt{s} - E_1 - E_2 - E_3)\,. \tag{D.64}$$

In the CM frame, $\vec{p} = \vec{0}$, which implies that

$$E_2^2 = |\vec{p}_1 + \vec{p}_3|^2 + m_2^2 = |\vec{p}_1|^2 + |\vec{p}_3|^2 + 2|\vec{p}_1|\,|\vec{p}_3|\cos\theta_{13} + m_2^2\,, \tag{D.65}$$

where θ_{13} and ϕ_{13} are the polar and azimuthal angles of $\vec{p}_3$ with respect to an axis parallel to $\vec{p}_1$. In spherical coordinates,

$$d^3\boldsymbol{p_1}\,d^3\boldsymbol{p_3} = |\vec{p}_1| E_1 dE_1\, d\Omega_1\, |\vec{p}_3| E_3\, dE_3\, d\cos\theta_{13}\, d\phi_{13}\,, \tag{D.66}$$

where $\Omega_1 = (\theta_1, \phi_1)$ is the solid angle of $\vec{p}_1$ with respect to a fixed z-axis.

We can now make use of eq. (D.65) to write $|\vec{p}_1|\,|\vec{p}_3|\,d\cos\theta_{13} = E_2\,dE_2$. Integration over E_2 is then trivial due to the delta function in eq. (D.64). It follows that

$$R_3(s) = \frac{1}{8}\int dE_1\,dE_3\,d\Omega_1\,d\phi_{13}\,\Theta(1-\cos^2\theta_{13})$$

$$= \frac{1}{8}\int_{m_1}^{E_1^{\max}} dE_1 \int_{E_3^-}^{E_3^+} dE_3 \int d\Omega_1\,d\phi_{13}\,, \tag{D.67}$$

where eqs. (D.12) and (D.18) yield

$$E_1^{\max} = \frac{s + m_1^2 - (m_2+m_3)^2}{2\sqrt{s}}\,, \tag{D.68}$$

and the step function condition in eq. (D.67) determines the integration limits,

$$E_3^{\pm} = \frac{1}{2(s+m_1^2-2\sqrt{s}E_1)}\biggl[(\sqrt{s}-E_1)(s+m_1^2-2\sqrt{s}E_1-m_2^2+m_3^2)$$

$$\pm(E_1^2-m_1^2)^{1/2}\lambda^{1/2}(s+m_1^2-2\sqrt{s}E_1, m_2^2, m_3^2)\biggr]\,. \tag{D.69}$$

The choice of fixed z-axis is arbitrary in applications to decays into three-body final states where the initial-state and final-state spins are not measured, in which case we may integrate over Ω_1 to obtain 4π. Similarly, one is also free to integrate over ϕ_{13} to obtain 2π. Hence, the end result is given by

$$R_3(s) = \pi^2 \int dE_1\,dE_3\,\Theta(1-\cos^2\theta_{13})\,. \tag{D.70}$$

An alternative form of eq. (D.70) is obtained by using eq. (D.18) to change integration variables from E_1, E_3 to the invariant quantities s_1, s_2:

$$R_3(s) = \frac{\pi^2}{4s}\int ds_1\,ds_2\,\Theta(-G(s_1,s_2,s,m_2^2,m_1^2,m_3^2))$$

$$= \frac{\pi^2}{4s}\int_{(m_2+m_3)^2}^{(\sqrt{s}-m_1)^2} ds_2 \int_{s_1^-}^{s_1^+} ds_1\,, \tag{D.71}$$

where G is the basic four-particle kinematic function defined in eq. (D.21) and the integration limits, $s_1^{\pm}$, are given in eq. (D.22).

The results obtained above for three-body phase space will be applied to decays into three-body final states in eq. (D.97). The application to $2 \to 3$ scattering processes is more involved, and we refer the reader to [B146] for further details.

We end this section by quoting the total volume of n-body phase space in the extreme relativistic limit (where all final-state particle masses can be neglected):

$$R_n(s) = \left(\frac{\pi}{2}\right)^{n-1}\frac{s^{n-2}}{(n-1)!\,(n-2)!}\,, \qquad \text{for } m_i = 0\ (i=1,2,\ldots,n). \tag{D.72}$$

D.3 Dimensionally Regularized Phase Space

If one or more final-state particle masses are zero, then the corresponding decay width or cross section will exhibit an infrared divergence. In practice, only infrared safe observables are experimentally accessible. In the computation of such observables, the infrared divergences, which appear at intermediate stages of the calculation, will ultimately cancel. However, to realize such a cancellation, one must regularize the infrared divergences to perform a well-defined computation. It is often convenient to employ dimensional regularization (which is already used in regularizing the ultraviolet divergences, as discussed in Appendix K.1) to regularize the infrared divergences. To accomplish this, one must extend the analysis of Appendix D.2 from $3+1$ to $(d-1)+1$ spacetime dimensions.

In this section, we reconsider the three-particle phase space integral in $(d-1)+1$ spacetime dimensions,

$$R_3(s) = \int \frac{d^{d-1}\boldsymbol{p_1}}{2E_1} \int \frac{d^{d-1}\boldsymbol{p_2}}{2E_2} \int \frac{d^{d-1}\boldsymbol{p_3}}{2E_3}\, \delta^d(p - p_1 - p_2 - p_3)\,, \tag{D.73}$$

where $s = p^2$. We shall examine the case where particles 2 and 3 have masses m_2 and m_3, respectively, and particle 1 is massless.

In $(d-1)+1$ spacetime dimensions, the volume element in spherical coordinates is given by

$$d^{d-1}\boldsymbol{p_3} = |\vec{\boldsymbol{p}}_3|^{n-2} d|\vec{\boldsymbol{p}}_3| d\Omega_1^{(d-2)} = |\vec{\boldsymbol{p}}_3|^{d-3} E_3 dE_3 d\Omega_3^{(d-2)}\,, \tag{D.74}$$

after imposing the mass-shell condition, $E_3^2 = |\vec{\boldsymbol{p}}_3|^2 + m_3^2$. In eq. (D.74), the solid angle $\Omega_3^{(n-2)}$ corresponds to the orientation of $\vec{\boldsymbol{p}}_3$ in the CM frame. The integral over the solid angle yields the $(d-2)$-dimensional surface area of the boundary of a $(d-1)$-dimensional solid ball, which is well known:

$$\int d\Omega_3^{(d-2)} = \frac{2\pi^{\frac{1}{2}d-1}}{\Gamma(\frac{1}{2}d-1)} \int_{-1}^{1} (\sin\theta_{13})^{d-4}\, d\cos\theta_{13}\,, \tag{D.75}$$

where θ_{13} is the polar angle of $\vec{\boldsymbol{p}}_3$ with respect to a fixed z-axis that lies along $\vec{\boldsymbol{p}}_1$.

Following the corresponding derivation of Appendix D.2, we work in the reference frame where $\vec{\boldsymbol{p}} = \vec{\boldsymbol{p}}_1 + \vec{\boldsymbol{p}}_2 + \vec{\boldsymbol{p}}_3 = \boldsymbol{0}$. It follows that

$$R_3(s) = \frac{1}{8} \int \frac{E_1^{d-3}\, dE_1\, d\Omega_1^{(d-2)}\, (E_3^2 - m_3^2)^{(d-3)/2}\, dE_3}{E_2} \times \frac{2\pi^{\frac{1}{2}d-1}}{\Gamma(\frac{1}{2}d-1)} \int \frac{(\sin\theta_{13})^{d-4}}{E_2}\, d\cos\theta_{13}\, \delta(\sqrt{s} - E_1 - E_2 - E_3), \tag{D.76}$$

where E_2 is given by eq. (D.65), with $m_1 = 0$ and $E_1 = |\vec{\boldsymbol{p}}_1|$. We can integrate over the orientation of $\vec{p}_1$ for free. Hence, it follows that

$$R_3(s) = \frac{(4\pi)^{\frac{1}{2}d-1}\Gamma(\frac{1}{2}d-1)}{8\Gamma(d-2)} \int E_1^{d-3}\, dE_1\, (E_3^2 - m_3^2)^{(d-3)/2}\, dE_3$$
$$\times \frac{2\pi^{\frac{1}{2}d-1}}{\Gamma(\frac{1}{2}d-1)} \int \frac{(\sin\theta_{13})^{d-4}}{E_2}\, d\cos\theta_{13}\, \delta(\sqrt{s} - E_1 - E_2 - E_3). \tag{D.77}$$

In light of eq. (D.65),

$$\frac{1}{E_2}\delta(\sqrt{s} - E_1 - E_2 - E_3) = \frac{1}{E_1(E_3^2 - m_3^2)^{1/2}}\, \delta(\cos\theta_{13} - \cos\overline{\theta}_{13}), \tag{D.78}$$

where

$$\cos\overline{\theta}_{13} = \frac{s - 2\sqrt{s}(E_1 - E_3) + 2E_1E_3}{2E_1(E_3^2 - m_3^2)^{1/2}}. \tag{D.79}$$

We can now perform the integration over $\cos\theta_{13}$ in eq. (D.77), which yields

$$R_3(s) = \frac{(2\pi)^{d-2}}{4\,\Gamma(d-2)} \int_0^{E_1^{\max}} E_1^{d-4}\, dE_1 \int_{E_3^-}^{E_3^+} (E_3^2 - m_3^2)^{(d-4)/2}(1 - \cos^2\overline{\theta}_{13})^{(d-4)/2}\, dE_3. \tag{D.80}$$

The integration limits are obtained by setting $m_1 = 0$ in eqs. (D.68) and (D.69):

$$E_1^{\max} = \frac{s - (m_2 + m_3)^2}{2\sqrt{s}}, \tag{D.81}$$

$$E_3^{\pm} = \frac{(\sqrt{s} - E_1)(s - 2\sqrt{s}E_1 - m_2^2 + m_3^2) \pm E_1\lambda^{1/2}(s - 2\sqrt{s}E_1, m_2^2, m_3^2)}{2(s - 2\sqrt{s}E_1)}. \tag{D.82}$$

Inserting the expression for $\cos\overline{\theta}_{13}$ [eq. (D.79)] into eq. (D.80) and writing $d = 4-2\epsilon$, we end up with

$$R_3(s) = \frac{\pi^{2-2\epsilon}}{\Gamma(2-2\epsilon)} \int_0^{E_1^{\max}} dE_1 \int_{E_3^-}^{E_3^+} \Big\{(s - 2\sqrt{s}E_1)(\sqrt{s} - 2E_3)(2E_1 + 2E_3 - \sqrt{s})$$
$$- 4m_3^2E_1^2 - (m_2^2 - m_3^2)^2 + 2(m_2^2 - m_3^2)\big[s - 2\sqrt{s}(E_1 + E_3) + 2E_1E_3\big]\Big\}^{-\epsilon} dE_3. \tag{D.83}$$

It is convenient to introduce dimensionless variables

$$x \equiv \frac{2\sqrt{s}E_1}{s - (m_2 + m_3)^2}, \qquad y \equiv \frac{E_3 - E_3^-}{E_3^+ - E_3^-}, \qquad r \equiv \frac{m_2 + m_3}{\sqrt{s}}, \qquad \delta \equiv \frac{m_2 - m_3}{\sqrt{s}}. \tag{D.84}$$

It then follows that

$$R_3(s) = \frac{\pi^{2-2\epsilon}s^{1-2\epsilon}(1 - r^2)^{\frac{5}{2}-3\epsilon}}{4\Gamma(2-2\epsilon)} \int_0^1 \frac{x^{1-2\epsilon}(1-x)^{\frac{1}{2}-\epsilon}}{\sqrt{1 - x(1-r^2)}} \left(\frac{1 - \delta^2 - x(1-r^2)}{1 - x(1-r^2)}\right)^{\frac{1}{2}-\epsilon} dx$$
$$\times \int_0^1 y^{-\epsilon}(1-y)^{-\epsilon}\, dy\,. \tag{D.85}$$

When evaluating a decay or scattering amplitude for a process with final-state particles 1, 2, and 3, where particle 1 is massless, an infrared divergence will arise due to the soft emission of particle 1 in the limit of $E_1 \to 0$ (or equivalently $x \to 0$). For example, the propagator of particle j (where $j = 2$ or 3) just prior to the soft emission will contribute to the amplitude a factor proportional to

$$\frac{1}{(p_1 + p_j)^2 - m_j^2} = \frac{1}{2p_1 \cdot p_j} = \frac{1}{2E_1(E_j - |\vec{p}_j| \cos\theta_{1j})} . \tag{D.86}$$

Thus, squaring the resulting amplitude yields a factor proportional to x^{-2}, which when inserted into the phase space integral yields an integrand that behaves like $x^{-1-2\epsilon}$ as $x \to 0$. Indeed, the infrared divergence, $\int_0^1 x^{-1} dx$, is regulated when x^{-1} is replaced by $x^{-1-2\epsilon}$ (assuming $\epsilon < 0$).

D.4 Decay Rate

The total decay rate for an unstable particle of mass m and spin J is

$$\Gamma = \frac{1}{2m} \frac{1}{2J+1} \sum_{i,f} \int d\,\mathrm{Lips}(p_1, p_2, \ldots, p_n)(2\pi)^4 \delta^4 \left(p - \sum_{k=1}^{n} p_k \right) |\mathcal{M}_{fi}|^2, \tag{D.87}$$

where f collectively refers to the final-state particles, $(2J+1)^{-1} \sum_i$ averages over the initial spins of the decaying particle, and $\mathcal{M}_{fi}$ is the Lorentz-invariant matrix element for the decay process. The lifetime of the particle is then given by $\tau = \hbar/\Gamma$.

Consider the decay of a particle of mass m and spin J into a two-body final state. Then, eqs. (D.49), (D.61), and (D.87) yield

$$\frac{d\Gamma}{d\Omega} = \frac{p_{\mathrm{CM}}}{32\pi^2 m^2} \frac{1}{2J+1} \sum_{i,f} |\mathcal{M}_{fi}|^2 . \tag{D.88}$$

Note that Ω refers to the solid angle of emitted particle 1 with respect to a fixed z-axis in the rest frame of the decaying particle. After summing over final-state spins, we are free to choose the z-axis arbitrarily. Thus, the integration over Ω is trivial and yields 4π. Hence the decay rate of the unstable particle is given by

$$\Gamma = \frac{p_{\mathrm{CM}}}{8\pi m^2} \frac{1}{2J+1} \sum_{f} |\mathcal{M}_{fi}|^2 , \tag{D.89}$$

where the sum over f is a sum over final-state spins (and any other internal degrees of freedom if present). Finally, one can rewrite this result using eq. (D.7) to obtain

$$\Gamma = \frac{\lambda^{1/2}(m^2, m_1^2, m_2^2)}{16\pi m^3} \frac{1}{2J+1} \sum_{i,f} |\mathcal{M}_{fi}|^2. \tag{D.90}$$

Note that in deriving eq. (D.89), we integrated over the full 4π steradians. In the case where the two decay products are identical particles, the integration over

4π steradians is double counting, since one cannot distinguish whether particle 1 or particle 2 has been emitted at a solid angle $\Omega = (\theta, \phi)$ with respect to a fixed z-axis. To avoid double counting, we simply integrate over 2π steradians. That is, the width computed in eq. (D.89) must be reduced by a factor of 2. We can write this more explicitly as

$$\Gamma = \left(\frac{1}{1+\delta_{12}}\right) \frac{\lambda^{1/2}(M^2, m_1^2, m_2^2)}{16\pi M^3} \frac{1}{2J+1} \sum_{i,f} |\mathcal{M}_{fi}|^2 \,, \tag{D.91}$$

where $\delta_{12} = 1$ if the two particles in the final state are identical; otherwise, $\delta_{12} = 0$.

If the spin states of the initial and final particles are known, then one can obtain an equation for the decay angular distribution of the final-state particles. In the CM frame with a suitably chosen z-axis, suppose that the decaying particle is in the state $|J, M\rangle$, which is an eigenstate of J_z with eigenvalue M. We shall also suppose that the two final states consist of particles of spin s_i and helicity λ_i (for $i = 1, 2$). Then, the decay angular distribution is given by

$$\frac{d\Gamma}{d\Omega} = \frac{p_{\rm CM}}{32\pi^2 m^2} \left|\mathcal{M}^{JM}_{\lambda_1\lambda_2}(\theta,\phi)\right|^2 , \tag{D.92}$$

where[2]

$$\mathcal{M}^{JM}_{\lambda_1\lambda_2}(\theta,\phi) = \sqrt{\frac{2J+1}{4\pi}} D^J_{M\lambda}(\phi,\theta,0)^* \mathcal{M}^J_{\lambda_1\lambda_2} . \tag{D.93}$$

In the above formula, $\lambda \equiv \lambda_1 - \lambda_2$, and $\mathcal{M}^J_{\lambda_1\lambda_2}$ is a reduced decay amplitude which is a function of J, the outgoing helicities, and the particle masses, but is independent of the angles θ and ϕ.

If the decay process is mediated by parity invariant interactions, then there are nontrivial constraints on the reduced decay amplitude. Further restrictions are obtained if the final-state particles are identical, as summarized below.

1. Parity: $$\mathcal{M}^J_{\lambda_1\lambda_2} = \eta\eta_1\eta_2(-1)^{s_1+s_2-J}\mathcal{M}^J_{-\lambda_1,-\lambda_2} \,, \tag{D.94}$$

2. Identical particles: $$\mathcal{M}^J_{\lambda_1\lambda_2} = (-1)^J \mathcal{M}^J_{\lambda_2\lambda_1} \,, \tag{D.95}$$

where η is the intrinsic parity of the decaying spin J particle and η_i is the intrinsic parity of final-state particle i. Note that, in the case of identical particles, J must be an integer.

In Feynman diagrammatic computations of decay processes, one can compute the helicity amplitudes by employing explicit representations for the spin wave functions, which are presented in Appendix E. Then, the reduced decay amplitudes can be identified using eq. (D.93).

We end this section with a brief treatment of the decay of a particle of mass m and spin J into a three-body final state. In this case, eqs. (D.49), (D.71), and (D.87)

[2] In this appendix, we adopt the convention of [B147] where the Euler angle $\gamma = 0$ appears in the spin J rotation matrix $D^J_{M\lambda}(\phi, \theta, \gamma)$.

can be employed to obtain

$$\frac{d^2\Gamma}{ds_1 ds_2} = \frac{1}{256\pi^3 m^3} \frac{1}{2J+1} \sum_{i,f} |\mathcal{M}_{fi}|^2 . \tag{D.96}$$

The total width is obtained by employing the kinematic constraints on s_1 and s_2 given in eqs. (D.11) and (D.22):

$$\Gamma = \frac{1}{256\pi^3 m^3} \frac{1}{2J+1} \int_{(m_2+m_3)^2}^{(m-m_1)^2} ds_2 \int_{s_1^-}^{s_1^+} ds_1 \sum_{i,f} |\mathcal{M}_{fi}|^2 . \tag{D.97}$$

It is convenient to introduce dimensionless variables [see eq. (D.18)]

$$z_1 \equiv \frac{2E_1}{m} = 1 + r_1 - \frac{s_2}{m^2} , \qquad z_3 \equiv \frac{2E_3}{m} = 1 + r_3 - \frac{s_1}{m^2} , \tag{D.98}$$

where $r_i \equiv m_i/m$ (for $i = 1, 2, 3$). Then, eq. (D.97) can be re-expressed as

$$\Gamma = \frac{m}{256\pi^3} \frac{1}{2J+1} \int_{2r_1}^{1+r_1^2-(r_2+r_3)^2} dz_1 \int_{z_3^-}^{z_3^+} dz_3 \sum_{i,f} |\mathcal{M}_{fi}|^2 , \tag{D.99}$$

where eq. (D.69) sets the limits on the integration over z_3:

$$z_3^{\pm} = \frac{1}{2(1-z_1+r_1^2)} \Big[(2-z_1)(1+r_1^2-r_2^2+r_3^2-z_1) \pm (z_1^2-4r_1^2)^{1/2} \lambda^{1/2}(1+r_1^2-z_1, r_2^2, r_3^2) \Big] . \tag{D.100}$$

D.5 Cross Section

Next, we consider the cross section for the scattering process $a+b \to 1+2+\cdots+n$. The total cross section is given by

$$\sigma = \frac{1}{4\sqrt{(p_a \cdot p_b)^2 - m_a^2 m_b^2}} \int d\,\mathrm{Lips}\,(2\pi)^4 \delta^4 \left(p_a + p_b - \sum_{k=1}^{n} p_k \right) \sum_f |\mathcal{M}_{fi}|^2, \tag{D.101}$$

where $\mathcal{M}_{fi}$ is the Lorentz-invariant scattering amplitude, and all unobserved internal spin degrees of freedom of the final-state particles are summed over. The relativistic invariant flux factor appearing in the denominator of eq. (D.101) is $F = 4(p_a \cdot p_b) v_{\rm rel}$, where the relative velocity of the incoming particles a and b is a relativistic invariant given by

$$v_{\rm rel} = \frac{\sqrt{(p_a \cdot p_b)^2 - m_a^2 m_b^2}}{p_a \cdot p_b} = \frac{\lambda^{1/2}(s, m_a^2, m_b^2)}{s - m_a^2 - m_b^2} , \tag{D.102}$$

after employing eqs. (D.2) and (D.5) (see Exercise D.3). Note that $0 \le v_{\rm rel} \le 1$ (since physical velocities cannot exceed the speed of light). The flux factor in the

CM frame is given by

$$F = 4\sqrt{(p_a \cdot p_b)^2 - m_a^2 m_b^2} = 4p_i\sqrt{s}\,, \tag{D.103}$$

in light of eq. (D.33).

The integration of the squared matrix element over Lorentz-invariant phase space yields a dimensionless quantity. It follows that the cross section given by eq. (D.101) has units of GeV^{-2} (see Appendix A.3). To obtain a cross section in units of picobarns (1 pb $= 10^{-12}$ barns $= 10^{-36}$ cm^2), one should multiply eq. (D.101) by the following conversion factor:

$$1\ \text{GeV}^{-2} \simeq 3.8938 \times 10^8\ \text{pb}. \tag{D.104}$$

If the spins s_a and s_b of the initial-state particles are not measured, then one must average over the initial spin states. The spin-averaged cross section is obtained by summing the squared scattering amplitude over all possible initial spin states and dividing by the total number of spin states, $(2s_a+1)(2s_b+1)$. In this case, we make the following replacement in eq. (D.101):

$$\sum_f \longrightarrow \frac{1}{(2s_a+1)(2s_b+1)}\sum_{i,f}\,. \tag{D.105}$$

As an example, consider the $2 \to 2$ particle scattering process $a + b \to 1 + 2$, with corresponding momentum four-vectors p_a, p_b, p_1, and p_2. Using the explicit expression obtained for $R_2(s)$ in eqs. (D.49) and (D.61), the differential cross section (assuming no spins are measured) in the CM frame is given by

$$\frac{d\sigma}{d\Omega} = \frac{1}{64\pi^2 s}\,\frac{p_f}{p_i}\,\frac{1}{(2s_a+1)(2s_b+1)}\sum_{i,f}|\mathcal{M}_{fi}|^2\,, \tag{D.106}$$

after employing eq. (D.103).

Note that $d\sigma/d\Omega$ is not Lorentz invariant since it depends on angles defined in the CM frame. It is possible to define a Lorentz-invariant differential cross section, since the variable t can be used in place of the CM scattering angle. In particular,

$$\begin{aligned} t = (p_a - p_1)^2 &= m_a^2 + m_1^2 - 2p_a \cdot p_1 = m_a^2 + m_1^2 - 2E_aE_1 + 2p_ip_f\cos\theta \\ &= m_a^2 + m_1^2 - \frac{(s - m_a^2 + m_b^2)(s - m_1^2 + m_2^2)}{2s} \\ &\quad + \frac{1}{2s}\,\lambda^{1/2}(s, m_a^2, m_b^2)\,\lambda^{1/2}(s, m_1^2, m_2^2)\cos\theta\,, \end{aligned} \tag{D.107}$$

after employing eqs. (D.31)–(D.34). It follows that

$$dt = \frac{1}{2s}\lambda^{1/2}(s, m_a^2, m_b^2)\lambda^{1/2}(s, m_1^2, m_2^2)\,d\cos\theta\,. \tag{D.108}$$

The azimuthal angle ϕ of particle 1 in the CM frame corresponds to rotations around the z-axis. The integral over this angle is trivial. Hence,

$$d\Omega = 2\pi\,d\cos\theta = 4\pi s\,\lambda^{-1/2}(s, m_a^2, m_b^2)\,\lambda^{-1/2}(s, m_1^2, m_2^2)\,dt = \frac{\pi}{p_ip_f}\,dt\,. \tag{D.109}$$

Inserting this result into eq. (D.106) and employing eq. (D.33) yields the Lorentz-invariant expression

$$\frac{d\sigma}{dt} = \frac{1}{16\pi\lambda(s, m_a^2, m_b^2)} \frac{1}{(2s_a+1)(2s_b+1)} \sum_{i,f} |\mathcal{M}_{fi}|^2 . \tag{D.110}$$

Note that $t^- \leq t \leq t^+$, where the $t^\pm$ [whose explicit form is given in eq. (D.37)] can be deduced from eq. (D.107) by imposing the requirement that $|\cos\theta| \leq 1$.

Finally, the total cross section can be obtained from eq. (D.106) by integrating over the solid angle. In light of the discussion following eq. (D.90) in the case of identical final-state particles,

$$\sigma = \left(\frac{1}{1+\delta_{12}}\right) \int \frac{d\sigma}{d\Omega}\, d\Omega\,, \tag{D.111}$$

where the integration is taken over the full 4π steradians. In particular, $\delta_{12} = 1$ if the two particles in the final state are identical; otherwise, $\delta_{12} = 0$. No extra factor is needed in the case of identical initial-state particles, since the issue of double counting does not arise in this case.

If the spin states of the initial and final particles are known, then one can obtain an equation for the polarized cross section. Denote the initial-state helicities by λ_a and λ_b and the final-state helicities by λ_1 and λ_2. The differential cross section for the scattering of particles of definite helicities in the CM frame is given by

$$\frac{d\sigma}{d\Omega} = \frac{1}{64\pi^2 s} \frac{p_f}{p_i} |\mathcal{M}_{\lambda_1,\lambda_2;\lambda_a,\lambda_b}(s,\theta,\phi)|^2 . \tag{D.112}$$

The helicity amplitudes for the scattering process, $\mathcal{M}_{\lambda_1,\lambda_2;\lambda_a,\lambda_b}(s,\theta,\phi)$, are given by the Jacob–Wick partial wave expansion [213], using the conventions of [B147]:

$$\mathcal{M}_{\lambda_1\lambda_2;\lambda_a\lambda_b}(s,\theta,\phi) = \frac{8\pi\sqrt{s}}{(p_i p_f)^{1/2}} \sum_{J=J_0}^{\infty} (2J+1) d^J_{\lambda_i\lambda_f}(\theta) e^{i\lambda_i\phi} \mathcal{M}^J_{\lambda_1\lambda_2;\lambda_a\lambda_b}(s), \tag{D.113}$$

summed over integers (half-integers) for integer (half-integer) λ_i and λ_f, where

$$J_0 \equiv \max\{|\lambda_i|, |\lambda_f|\}\,, \qquad \lambda_i \equiv \lambda_a - \lambda_b, \qquad \lambda_f \equiv \lambda_1 - \lambda_2, \tag{D.114}$$

and p_i, p_f are given by eqs. (D.33) and (D.34). In eq. (D.113), $\mathcal{M}^J_{\lambda_1\lambda_2;\lambda_a\lambda_b}(s)$ is a reduced scattering amplitude that is independent of the scattering angles, and $d^j_{m'm}(\theta) \equiv \langle j\ m'|\exp(-i\theta J_y)|j\ m\rangle$ are matrix elements of Wigner's little d matrix.

The normalization of $\mathcal{M}^J(s)$ differs from that of [B147] by a factor of 2 (see eq. 10-8 of [B148]) and has been chosen such that the partial wave unitarity condition is given by

$$|\mathcal{M}^J_{\lambda_1\lambda_2;\lambda_a\lambda_b}(s)| \leq 1\,. \tag{D.115}$$

Consequently, the cross section for the Jth partial wave is bounded by

$$\sigma_J \leq \frac{4\pi(2J+1)}{(2s_a+1)(2s_b+1)p_i^2}\,. \tag{D.116}$$

If the scattering process is mediated by parity- and/or time-reversal-invariant interactions, then there are nontrivial constraints on the reduced scattering amplitude. Additional restrictions are obtained if the initial-state and/or final-state particles are identical. These constraints are summarized below [213].

1. Parity:

$$\mathcal{M}^J_{\lambda_1\lambda_2;\lambda_a\lambda_b}(s) = (\eta_1\eta_2/\eta_a\eta_b)\,(-1)^{s_1+s_2-s_a-s_b}\mathcal{M}^J_{-\lambda_1-\lambda_2;-\lambda_a-\lambda_b}(s)\,, \quad \text{(D.117)}$$

where s_i and η_i are the spin and the intrinsic parity of particle i, respectively.

2. Time reversal:

$$\mathcal{M}^J_{\lambda_1\lambda_2;\lambda_a\lambda_b}(s) = \mathcal{M}^J_{\lambda_a\lambda_b;\lambda_1\lambda_2}(s). \quad \text{(D.118)}$$

3. Identical particles:

$$\mathcal{M}^J_{\lambda_1\lambda_2;\lambda_a\lambda_b}(s) = \begin{cases} (-1)^J\mathcal{M}^J_{\lambda_1\lambda_2;\lambda_b\lambda_a}(s), & a,\, b \text{ identical},\\ (-1)^J\mathcal{M}^J_{\lambda_2\lambda_1;\lambda_a\lambda_b}(s), & 1,\, 2 \text{ identical}. \end{cases} \quad \text{(D.119)}$$

As previously noted, J must be an integer in the case of identical particles.

Exercises

D.1 Consider the decay of a particle of mass M into three massless particles.

(a) Show that $s_1 \geq 0$, $s_2 \geq 0$, and $s_1 + s_2 \leq M^2$.

(b) Verify that $G(s_1, s_2, s, 0, 0, 0) = s_1 s_2(s - s_1 - s_2)$.

(c) Explicitly evaluate $R_3(s)$ by performing the integral in eq. (D.71), and check that your result confirms eq. (D.72) in the case of $n = 3$.

D.2 Consider particles a and b in the center-of-momentum frame, with four-vector momenta $p_a = (E_a\,;\,\vec{p})$ and $p_b = (E_b\,;\,-\vec{p})$, with corresponding squared masses, $m_j^2 = E_j^2 - |\vec{p}|$ for $j = a, b$.

(a) Show that in the nonrelativistic limit where $|\vec{p}| \ll m_a, m_b$, the kinematic triangle function is given by

$$\lambda(s, m_a^2, m_b^2) = 4(m_a + m_b)^2|\vec{p}|^2 + \mathcal{O}(|\vec{p}|^4). \quad \text{(D.120)}$$

(b) Show that in the nonrelativistic limit, eq. (D.102) for the relative velocity of particles a and b yields the expected result,

$$v_{\rm rel} = |\vec{v}_a - \vec{v}_b| = \frac{(m_a + m_b)|\vec{p}|}{m_a m_b}. \quad \text{(D.121)}$$

D.3 Consider the scattering of particles a and b, with corresponding four-momenta p_a and p_b. Denote the corresponding velocity vectors in a generic inertial reference frame by $\vec{\boldsymbol{v}}_{\boldsymbol{a}}$ and $\vec{\boldsymbol{v}}_{\boldsymbol{b}}$ and the corresponding energies by E_a and E_b.

(a) In a generic reference frame, verify that

$$1 - \vec{\boldsymbol{v}}_{\boldsymbol{a}}\cdot\vec{\boldsymbol{v}}_{\boldsymbol{b}} = \frac{p_a\cdot p_b}{E_a E_b}. \tag{D.122}$$

(b) Show that the relative velocity of particles a and b in a generic inertial reference frame is given by

$$v_{\rm rel} = \frac{\sqrt{|\vec{\boldsymbol{v}}_{\boldsymbol{a}} - \vec{\boldsymbol{v}}_{\boldsymbol{b}}|^2 - |\vec{\boldsymbol{v}}_{\boldsymbol{a}}\times\vec{\boldsymbol{v}}_{\boldsymbol{b}}|^2}}{1-\vec{\boldsymbol{v}}_{\boldsymbol{a}}\cdot\vec{\boldsymbol{v}}_{\boldsymbol{b}}}, \tag{D.123}$$

in units of $c = 1$. Restore the factors of c and show that in the nonrelativistic limit, $v_{\rm rel} = |\vec{\boldsymbol{v}}_{\boldsymbol{a}} - \vec{\boldsymbol{v}}_{\boldsymbol{b}}|$, independently of the directions of vectors $\vec{\boldsymbol{v}}_{\boldsymbol{a}}$ and $\vec{\boldsymbol{v}}_{\boldsymbol{b}}$.

(c) Define the four-vector current $J^\mu = (n\,;\,n\vec{\boldsymbol{v}})$, where $n \equiv J^0$ is the particle number density in a generic inertial reference frame. Using the covariant normalization of the one-particle states as in eqs. (2.6) and (2.7), it follows that $n = 2E$ for the number density corresponding to a one-particle state of energy E. The Lorentz-invariant incident flux F for the scattering of particles a and b is defined by $F \equiv (J_a\cdot J_b)v_{\rm rel}$. In light of eq. (D.102), show that

$$F = 4(p_a\cdot p_b)v_{\rm rel} = 4\sqrt{(p_a\cdot p_b)^2 - m_a^2 m_b^2}\,. \tag{D.124}$$

NOTE: Equation (D.124) is sometimes written as $F = n_a n_b v_M$, where the Lorentz-noninvariant Møller velocity is defined as

$$v_M \equiv \sqrt{|\vec{\boldsymbol{v}}_{\boldsymbol{a}} - \vec{\boldsymbol{v}}_{\boldsymbol{b}}|^2 - |\vec{\boldsymbol{v}}_{\boldsymbol{a}}\times\vec{\boldsymbol{v}}_{\boldsymbol{b}}|^2} = \frac{\sqrt{(p_a\cdot p_b)^2 - m_a^2 m_b^2}}{E_a E_b}\,. \tag{D.125}$$

D.4 Show that the three-body phase space integral (with unit squared matrix element) is given by

$$\frac{dR_3(s)}{d\Omega} = \frac{\pi}{16s}\int_{(m_2+m_3)^2}^{(\sqrt{s}-m_1)^2}\frac{ds_2}{s_2}\,\lambda^{1/2}(s,s_2,m_1^2)\,\lambda^{1/2}(s_2,m_2^2,m_3^2)\,. \tag{D.126}$$

Verify that eq. (D.126) coincides with the result obtained by integrating over s_1 in eq. (D.71) using the integration limits $s_1^\pm$ given in eq. (D.22).

D.5 Consider two particles with four-momenta $p_1 = (E_1; \vec{\boldsymbol{p}}_{\boldsymbol{1}})$ and $p_2 = (E_2; \vec{\boldsymbol{p}}_{\boldsymbol{2}})$ and squared masses $m_i^2 = E_i^2 - |\vec{\boldsymbol{p}}_{\boldsymbol{i}}|^2$ for $i = 1, 2$, respectively.

(a) Consider the two-particle phase space integral in $(d-1)+1$ spacetime dimensions,

$$R_2(s) = \int \frac{d^{d-1}\boldsymbol{p_1}}{2E_1}\,\frac{d^{d-1}\boldsymbol{p_2}}{2E_2}\,\delta^d(p-p_1-p_2)\,, \tag{D.127}$$

where $s \equiv p^2$. Working in the CM frame and following the analysis of Appendix D.2, show that

$$R_2(s) = \frac{p_{\rm CM}^{d-3}}{4\sqrt{s}} \int d\Omega_1^{(d-2)}, \tag{D.128}$$

where $p_{\rm CM}$ is given in eq. (D.57). After employing eq. (D.75) and introducing $d = 4 - 2\epsilon$, show that one can rewrite eq. (D.128) as

$$\begin{aligned} R_2(s) &= \left(\frac{\pi}{4}\right)^{1-\epsilon} \frac{[\lambda(s, m_1^2, m_2^2)]^{\frac{1}{2}-\epsilon}}{s^{1-\epsilon}\Gamma(1-\epsilon)} \int_{-1}^{1} (\sin\theta_1)^{-2\epsilon}\, d\cos\theta_1 \\ &= \left(\frac{\pi}{s}\right)^{1-\epsilon} \frac{[\lambda(s, m_1^2, m_2^2)]^{\frac{1}{2}-\epsilon}}{2\Gamma(1-\epsilon)} \int_0^1 y^{-\epsilon}(1-y)^{-\epsilon}\, dy\,, \end{aligned} \tag{D.129}$$

where $y \equiv \cos^2(\theta_1/2)$ and θ_1 is the polar angle of $\vec{p}_1$ with respect to a fixed z-axis.

(b) Verify the results for the following two-body phase space integrals in $(d-1)+1$ spacetime dimensions in terms of $R_2(q^2)$ given in eq. (D.129), where q is a four-vector that does not depend on p_1 and p_2.

$$\int \frac{d^{d-1}\boldsymbol{p_1}}{2E_1} \frac{d^{d-1}\boldsymbol{p_2}}{2E_2} \delta^4(q-p_1-p_2)\, p_{1\mu} = \frac{q^2+m_1^2-m_2^2}{2q^2} q_\mu R_2(q^2)\,, \tag{D.130}$$

$$\begin{aligned} &\int \frac{d^{d-1}\boldsymbol{p_1}}{2E_1} \frac{d^{d-1}\boldsymbol{p_2}}{2E_2} \delta^4(q-p_1-p_2)\, p_{1\mu}p_{1\nu} \\ &\quad = -\frac{1}{4(d-1)q^4} \Big[\lambda q^2 g_{\mu\nu} - [d\lambda + 4(d-1)m_1^2 q^2]\, q_\mu q_\nu\Big] R_2(q^2), \end{aligned} \tag{D.131}$$

$$\begin{aligned} &\int \frac{d^{d-1}\boldsymbol{p_1}}{2E_1} \frac{d^{d-1}\boldsymbol{p_2}}{2E_2} \delta^4(q-p_1-p_2)\, p_{1\mu}p_{2\nu} \\ &\quad = \frac{1}{4(d-1)q^4} \Big\{\lambda q^2 g_{\mu\nu} - [d\lambda - 2(d-1)q^2(q^2-m_1^2-m_2^2)]\, q_\mu q_\nu\Big\} R_2(q^2), \end{aligned} \tag{D.132}$$

where $\lambda \equiv \lambda(q^2, m_1^2, m_2^2)$ is the kinematic triangle function.

HINT: Using the Lorentz covariance of the above expressions, it follows that

$$\int \frac{d^{d-1}\boldsymbol{p_1}}{2E_1} \frac{d^{d-1}\boldsymbol{p_2}}{2E_2} \delta^4(q-p_1-p_2)\, p_{1\mu} = A q_\mu\,, \tag{D.133}$$

$$\int \frac{d^{d-1}\boldsymbol{p_1}}{2E_1} \frac{d^{d-1}\boldsymbol{p_2}}{2E_2} \delta^4(q-p_1-p_2)\, p_{1\mu}p_{1\nu} = B q^2 g_{\mu\nu} + C q_\mu q_\nu\,, \tag{D.134}$$

for $k = 1, 2$. Multiplying eq. (D.133) by q^μ yields an equation for A that can be easily solved. Multiplying eq. (D.134) first by $g^{\mu\nu}$ and then by $q^\mu q^\nu$ yields two equations for B and C, which again can be easily solved. Finally, set $p_{2\nu} = q_\nu - p_{1\nu}$ on the left-hand side of eq. (D.132) and use eqs. (D.130) and (D.131) to obtain the final result exhibited in eq. (D.132).

Appendix E The Spin-1/2 and Spin-1 Wave Functions

In this appendix, we first construct the explicit forms for the eigenstates of the spin operator $\frac{1}{2}\vec{\sigma}\cdot\hat{s}$, and examine their properties. For massive fermions, it is possible to transform to the rest frame, and quantize the spin along a fixed axis in space. The corresponding two-component spinor wave functions will be called fixed-axis spinors. For both massive or massive fermions, one can quantize the spin along the direction of momentum. The corresponding two-component spinor wave functions are helicity spinors. Helicity spinor wave functions are most conveniently applied to massless fermions or fermions in the high-energy relativistic limit, $E \gg m$. Fixed-axis spinors are most conveniently applied to massive fermions in the non-relativistic limit. We then exhibit the relation between the two-component spinor wave functions and the more traditional four-component spinor wave functions that are treated in most textbooks of quantum field theory.

Finally, we present the explicit forms for the spin-1 wave functions. For massless spin-1 particles, only two transverse polarization states exist, corresponding to helicity ± 1. For massive spin-1 particles, one must also include a longitudinal polarization state, corresponding to helicity 0.

E.1 Fixed-Axis Spinor Wave Functions

Consider a spin-1/2 fermion in its rest frame and quantize the spin along a fixed axis specified by the unit vector, $\hat{s} \equiv (\sin\theta\cos\phi\,, \sin\theta\sin\phi\,, \cos\theta)$, with polar angle θ and azimuthal angle ϕ with respect to a fixed z-axis. The relevant basis of two-component fixed-axis spinors χ_s are eigenstates of $\frac{1}{2}\vec{\sigma}\cdot\hat{s}$, i.e.,

$$\tfrac{1}{2}\vec{\sigma}\cdot\hat{s}\,\chi_s = s\chi_s\,, \qquad s = \pm\tfrac{1}{2}\,. \tag{E.1}$$

The formulae for the corresponding two-component spinor wave functions x and y were given in eqs. (2.24)–(2.27). We summarize the results here:

$$x_\alpha(\vec{p}, s) = \sqrt{p\cdot\sigma}\,\chi_s\,, \qquad x^\alpha(\vec{p}, s) = -2s\chi^\dagger_{-s}\sqrt{p\cdot\overline{\sigma}}\,, \tag{E.2}$$

$$y_\alpha(\vec{p}, s) = 2s\sqrt{p\cdot\sigma}\,\chi_{-s}\,, \qquad y^\alpha(\vec{p}, s) = \chi^\dagger_s\sqrt{p\cdot\overline{\sigma}}\,, \tag{E.3}$$

$$x^{\dagger\dot{\alpha}}(\vec{p}, s) = -2s\sqrt{p\cdot\overline{\sigma}}\,\chi_{-s}\,, \qquad x^\dagger_{\dot{\alpha}}(\vec{p}, s) = \chi^\dagger_s\sqrt{p\cdot\sigma}\,, \tag{E.4}$$

$$y^{\dagger\dot{\alpha}}(\vec{p}, s) = \sqrt{p\cdot\overline{\sigma}}\,\chi_s\,, \qquad y^\dagger_{\dot{\alpha}}(\vec{p}, s) = 2s\chi^\dagger_{-s}\sqrt{p\cdot\sigma}\,, \tag{E.5}$$

where $\sqrt{p\cdot\sigma}$ and $\sqrt{p\cdot\overline{\sigma}}$ are defined in eq. (A.73). Note that eqs. (E.2)–(E.5) imply that the x and y spinors are related:

$$y(\vec{p}, s) = 2sx(\vec{p}, -s)\,, \qquad y^\dagger(\vec{p}, s) = 2sx^\dagger(\vec{p}, -s)\,. \tag{E.6}$$

In order to explicitly construct the eigenstates of $\frac{1}{2}\vec{\sigma}\cdot\hat{s}$, we first consider the case where $\hat{s} = \hat{z}$. In this case, we define the eigenstates of $\frac{1}{2}\sigma^3$ to be

$$\chi_{1/2}(\hat{z}) = \begin{pmatrix} 1 \\ 0 \end{pmatrix}, \qquad \chi_{-1/2}(\hat{z}) = \begin{pmatrix} 0 \\ 1 \end{pmatrix}. \tag{E.7}$$

By convention, we have set an arbitrary overall multiplicative phase factor for each spinor of eq. (E.7) to unity. We then determine $\chi_s(\hat{s})$ from $\chi_s(\hat{z})$ by employing the spin-1/2 rotation operator that corresponds to a rotation from $\hat{z}$ to $\hat{s}$. This rotation is represented by a 3×3 matrix $\mathcal{R}$ such that $\hat{s} = \mathcal{R}\hat{z}$. However, this rotation operator is not unique. In its most general form, the rotation operator can be parameterized in terms of three Euler angles (e.g., see [B122, B134]):

$$\mathcal{R}(\phi\,,\theta\,,\gamma) \equiv R(\hat{z}\,,\phi)\,R(\hat{y}\,,\theta)\,R(\hat{z}\,,\gamma)\,. \tag{E.8}$$

The Euler angles can be chosen to lie in the range $0 \leq \theta \leq \pi$ and $0 \leq \phi\,,\gamma < 2\pi$. In general, $R(\hat{n}\,,\theta)$ is a 3×3 real orthogonal matrix with unit determinant that represents a rotation by an angle θ about a fixed axis $\hat{n}$:

$$R^{ij}(\hat{n}\,,\theta) = \exp(-i\theta\hat{n}\cdot\vec{\mathcal{S}}) = n^i n^j + (\delta^{ij} - n^i n^j)\cos\theta - \epsilon^{ijk}n^k\sin\theta\,, \tag{E.9}$$

where the $\vec{\mathcal{S}} = (\mathcal{S}^1\,,\mathcal{S}^2\,,\mathcal{S}^3)$ are three 3×3 matrices whose matrix elements are given by $(\mathcal{S}^i)^{jk} = -i\epsilon^{ijk}$ [see eq. (1.4)].

However, the angle γ is arbitrary, since $R(\hat{z}\,,\gamma)\hat{z} = \hat{z}$. Thus,

$$\hat{s} = \mathcal{R}\hat{z} = (\sin\theta\cos\phi\,,\sin\theta\sin\phi\,,\cos\theta)\,, \tag{E.10}$$

independently of the choice of γ. For $\theta = 0$ or $\theta = \pi$, where $\hat{s}$ is parallel to the z-axis, the azimuthal angle ϕ is undefined. Since $\hat{s} \to -\hat{s}$ corresponds in general to $\theta \to \pi - \theta$ and $\phi \to \phi + \pi$ (mod 2π), we shall adopt a convention whereby

$$\phi = \begin{cases} 0\,, & \text{for } \hat{s} = \hat{z}\,, \quad (\theta = 0)\,, \\ \pi\,, & \text{for } \hat{s} = -\hat{z}\,, \quad (\theta = \pi)\,. \end{cases} \tag{E.11}$$

Using the spin-1/2 rotation operator corresponding to $\mathcal{R}(\phi\,,\theta\,,\gamma)$, one can compute $\chi_s(\hat{s})$,

$$\chi_s(\hat{s}) = \mathcal{D}(\phi\,,\theta\,,\gamma)\,\chi_s(\hat{z})\,, \tag{E.12}$$

where $\mathcal{D}$ is the spin-1/2 unitary representation matrix

$$\mathcal{D}(\phi\,,\theta\,,\gamma) \equiv D(\hat{z},\phi)\,D(\hat{y},\theta)\,D(\hat{z},\gamma)\,, \tag{E.13}$$

and D is the 2×2 unitary matrix

$$D(\hat{n},\theta) \equiv \exp\left(-i\theta\hat{n}\cdot\vec{\sigma}/2\right) = \cos\frac{\theta}{2} - i\hat{n}\cdot\vec{\sigma}\sin\frac{\theta}{2}\,. \tag{E.14}$$

Equation (E.12) yields explicit forms for the eigenstates of $\frac{1}{2}\vec{\boldsymbol{\sigma}}\boldsymbol{\cdot}\hat{\boldsymbol{s}}$:

$$\chi_{1/2}(\hat{\boldsymbol{s}}) = e^{-i\gamma/2}\begin{pmatrix} e^{-i\phi/2}\cos\dfrac{\theta}{2} \\ e^{i\phi/2}\sin\dfrac{\theta}{2}\end{pmatrix}, \qquad \chi_{-1/2}(\hat{\boldsymbol{s}}) = e^{i\gamma/2}\begin{pmatrix} -e^{-i\phi/2}\sin\dfrac{\theta}{2} \\ e^{i\phi/2}\cos\dfrac{\theta}{2}\end{pmatrix}. \tag{E.15}$$

The well-known two-to-one mapping between SU(2) and SO(3) associates two spin-1/2 rotation matrices $\mathcal{D}$ for each rotation matrix $\mathcal{R}$. Moreover,

$$\mathcal{D}(\phi + 2\pi\,,\,\theta\,,\,\gamma) = -\mathcal{D}(\phi\,,\,\theta\,,\,\gamma)\,, \tag{E.16}$$

which implies that a rotation of a spinor by 2π yields an overall change of sign in the spinor wave function (an effect that can be observed in quantum interference experiments!). Strictly speaking, we should take the range of the Euler angles to be $0 \leq \phi < 4\pi$, $0 \leq \theta \leq \pi$, and $0 \leq \gamma < 2\pi$. However, when constructing the spinor wave function of a spin-1/2 particle whose spin quantization axis is given by eq. (E.10), we will fix the overall sign of the spinor wave function by convention.

Since the overall phase of the spinor wave function in eq. (E.15) is unphysical, the value of γ is a matter of convention. In this book, we shall employ the conventional choice of $\gamma = 0$, as advocated in [B147].[1] The corresponding rotation matrix is then given by

$$\mathcal{R}(\phi\,,\,\theta\,,\,0) = R(\hat{\boldsymbol{z}}\,,\,\phi)\,R(\hat{\boldsymbol{y}}\,,\,\theta) = \begin{pmatrix} \cos\theta\cos\phi & -\sin\phi & \sin\theta\cos\phi \\ \cos\theta\sin\phi & \cos\phi & \sin\theta\sin\phi \\ -\sin\theta & 0 & \cos\theta \end{pmatrix}. \tag{E.17}$$

Employing the corresponding spin-1/2 rotation operator $\mathcal{D}(\phi\,,\,\theta\,,\,0)$ in eq. (E.12) yields the spinor wave functions given in eq. (E.15) with $\gamma = 0$. In particular, note that

$$\chi_s(-\hat{\boldsymbol{z}}) = i\chi_{-s}(\hat{\boldsymbol{z}})\,, \qquad s = \pm\tfrac{1}{2}\,, \tag{E.18}$$

after making use of eq. (E.11).

The spinor wave functions χ_s defined by eq. (E.12) are normalized such that

$$\chi_s^\dagger(\hat{\boldsymbol{s}})\chi_{s'}(\hat{\boldsymbol{s}}) = \delta_{ss'}\,, \tag{E.19}$$

and satisfy the following completeness relation:

$$\sum_s \chi_s(\hat{\boldsymbol{s}})\chi_s^\dagger(\hat{\boldsymbol{s}}) = \begin{pmatrix} 1 & 0 \\ 0 & 1 \end{pmatrix}. \tag{E.20}$$

[1] An alternative convention that assigns $\gamma = -\phi$ is adopted in the older particle physics literature (e.g., see [B133, B148] and Ref. [213]), with the good feature that $\mathcal{R}(\phi\,,\,0\,,\,-\phi) = \mathbb{1}_{3\times 3}$ independently of the angle ϕ (which is undefined when $\theta = 0$). See Appendix C of Ref. [1] for the appropriate modifications to the results of this appendix in the $\gamma = -\phi$ convention.

Moreover, the spinor wave functions $\chi_s(\hat{\boldsymbol{s}})$ and $\chi_{-s}(\hat{\boldsymbol{s}})$ are connected by the following relation:

$$\chi_{-s}(\hat{\boldsymbol{s}}) = -2si\sigma^2\,\chi^*_s(\hat{\boldsymbol{s}})\,. \tag{E.21}$$

Consider a spin-1/2 fermion with four-momentum $p^\mu = (E\,,\,\vec{\boldsymbol{p}})$, with relativistic energy $E = (|\vec{\boldsymbol{p}}|^2 + m^2)^{1/2}$, and the direction of $\vec{\boldsymbol{p}}$ given by

$$\hat{\boldsymbol{p}} = (\sin\theta_p\cos\phi_p\,,\,\sin\theta_p\sin\phi_p\,,\,\cos\theta_p)\,. \tag{E.22}$$

Using eq. (A.73), one can employ eqs. (E.2)–(E.5) to obtain explicit expressions for the two-component spinor wave functions $x(\vec{\boldsymbol{p}}, s)$, $y(\vec{\boldsymbol{p}}, s)$, $x^\dagger(\vec{\boldsymbol{p}}, s)$, and $y^\dagger(\vec{\boldsymbol{p}}, s)$.

Additional properties of the χ_s can be derived by introducing an orthonormal set of unit three-vectors $\hat{\boldsymbol{s}}^a$ that provide a basis for a right-handed coordinate system. Explicitly,

$$\hat{\boldsymbol{s}}^a \boldsymbol{\cdot} \hat{\boldsymbol{s}}^b = \delta^{ab}\,, \tag{E.23}$$

$$\hat{\boldsymbol{s}}^a \times \hat{\boldsymbol{s}}^b = \epsilon^{abc}\,\hat{\boldsymbol{s}}^c\,. \tag{E.24}$$

We shall identify

$$\hat{\boldsymbol{s}}^3 \equiv \hat{\boldsymbol{s}} \tag{E.25}$$

as the quantization axis used in defining the third component of the spin of the fermion in its rest frame. The unit vectors $\hat{\boldsymbol{s}}^1$ and $\hat{\boldsymbol{s}}^2$ are then chosen such that eqs. (E.23) and (E.24) are satisfied. To explicitly construct the $\hat{\boldsymbol{s}}^a$, we begin with the orthonormal set $\{\hat{\boldsymbol{x}}\,,\,\hat{\boldsymbol{y}}\,,\,\hat{\boldsymbol{z}}\}$, and employ the *same* rotation operator $\mathcal{R}$ used to define $\chi_s(\hat{\boldsymbol{s}})$. That is,

$$(\hat{\boldsymbol{s}}^1\,,\,\hat{\boldsymbol{s}}^2\,,\,\hat{\boldsymbol{s}}^3) = (\mathcal{R}\hat{\boldsymbol{x}},\,\mathcal{R}\hat{\boldsymbol{y}}\,,\,\mathcal{R}\hat{\boldsymbol{z}})\,, \qquad \text{where} \quad \mathcal{R} \equiv \mathcal{R}(\phi\,,\,\theta\,,\,0)\,, \tag{E.26}$$

and ϕ, θ, and $\gamma = 0$ are the Euler angles used to define the spinor wave function in eq. (E.12). From eq. (E.26), one can immediately derive the completeness relation (as a consequence of $\mathcal{R}\mathcal{R}^\mathsf{T} = \mathbb{1}$),

$$\sum_a (\hat{\boldsymbol{s}}^a)^i(\hat{\boldsymbol{s}}^a)^j = \delta^{ij}\,, \tag{E.27}$$

where i and j label the space components of the three-vector $\hat{\boldsymbol{s}}^a$.

The explicit forms for the $\hat{\boldsymbol{s}}^a$ can be obtained by making use of eqs. (E.17) and (E.26), which yield

$$\begin{aligned}\hat{\boldsymbol{s}}^1 &= (\cos\theta\cos\phi,\,\cos\theta\sin\phi,\,-\sin\theta)\,,\\ \hat{\boldsymbol{s}}^2 &= (-\sin\phi,\,\cos\phi,\,0)\,,\\ \hat{\boldsymbol{s}}^3 &= (\sin\theta\cos\phi,\,\sin\theta\sin\phi,\,\cos\theta)\,.\end{aligned} \tag{E.28}$$

We can use the $\hat{\boldsymbol{s}}^a$ to extend the defining equation of χ_s [eq. (E.1)]:

$$\tfrac{1}{2}\vec{\boldsymbol{\sigma}}\boldsymbol{\cdot}\hat{\boldsymbol{s}}^a\chi_{s'}(\hat{\boldsymbol{s}}) = \tfrac{1}{2}\tau^a_{ss'}\chi_s(\hat{\boldsymbol{s}})\,, \tag{E.29}$$

where the $\tau^a_{ss'}$ are the matrix elements of the Pauli matrices. Here, we use the symbol τ rather than σ to emphasize that the indices of the Pauli matrices τ^a are spin labels s, s' and *not* spinor indices α, $\dot\alpha$. The first [second] row and column of the τ-matrices correspond to $s = 1/2$ [$-1/2$]. For example, $\tau^3_{ss'} = 2s\delta_{ss'}$ (no sum over s). Note that $\frac{1}{2}\vec{\boldsymbol{\sigma}}\cdot(\boldsymbol{s^1} \pm i\boldsymbol{s^2})$ serve as ladder operators that connect the spinor wave functions $\chi_{1/2}$ and $\chi_{-1/2}$. Using eq. (E.19), it follows that eq. (E.29) is equivalent to

$$\chi^\dagger_s(\hat{\boldsymbol{s}})\,\vec{\boldsymbol{\sigma}}\boldsymbol{\cdot}\hat{\boldsymbol{s}}^{\boldsymbol{a}}\chi_{s'}(\hat{\boldsymbol{s}}) = \tau^a_{ss'}\,. \tag{E.30}$$

It is instructive to prove eq. (E.30) directly. Employing eq. (E.12) and using the fact that $\mathcal{D}$ is a unitary matrix,

$$\chi^\dagger_s(\hat{\boldsymbol{s}})\,\vec{\boldsymbol{\sigma}}\boldsymbol{\cdot}\hat{\boldsymbol{s}}^{\boldsymbol{a}}\chi_{s'}(\hat{\boldsymbol{s}}) = \chi^\dagger_s(\hat{\boldsymbol{z}})\,[\mathcal{D}(\phi\,,\theta\,,\,0)]^{-1}\vec{\boldsymbol{\sigma}}\boldsymbol{\cdot}\hat{\boldsymbol{s}}^{\boldsymbol{a}}\,\mathcal{D}(\phi\,,\theta\,,\,0)\chi_{s'}(\hat{\boldsymbol{z}})\,. \tag{E.31}$$

The above result can be simplified by a repeated use of the identity

$$e^{i\theta\hat{\boldsymbol{n}}\cdot\vec{\boldsymbol{\sigma}}/2}\,\sigma^j\,e^{-i\theta\hat{\boldsymbol{n}}\cdot\vec{\boldsymbol{\sigma}}/2} = R^{jk}(\hat{\boldsymbol{n}}\,,\,\theta)\sigma^k\,, \tag{E.32}$$

which is valid for any fixed axis $\hat{\boldsymbol{n}}$, where $R(\hat{\boldsymbol{n}}\,,\,\theta)$ is the rotation matrix defined in eq. (E.9). It follows that

$$[\mathcal{D}(\phi\,,\theta\,,\,0)]^{-1}\,\sigma^j\,\mathcal{D}(\phi\,,\theta\,,\,0) = \mathcal{R}^{jk}(\phi\,,\theta\,,\,0)\,\sigma^k\,, \tag{E.33}$$

where $\mathcal{R}(\phi\,,\theta\,,\,0)$ is defined in eq. (E.8). Since $\mathcal{R}^{\mathsf{T}} = \mathcal{R}^{-1}$,

$$\chi^\dagger_s(\hat{\boldsymbol{s}})\,\vec{\boldsymbol{\sigma}}\boldsymbol{\cdot}\hat{\boldsymbol{s}}^{\boldsymbol{a}}\chi_{s'}(\hat{\boldsymbol{s}}) = \chi^\dagger_s(\hat{\boldsymbol{z}})\,\vec{\boldsymbol{\sigma}}\boldsymbol{\cdot}\left[\mathcal{R}^{-1}\hat{\boldsymbol{s}}^{\boldsymbol{a}}\right]\,\chi_{s'}(\hat{\boldsymbol{z}})\,. \tag{E.34}$$

Equation (E.26) implies that $(\mathcal{R}^{-1}\hat{\boldsymbol{s}}^{\boldsymbol{1}}\,,\,\mathcal{R}^{-1}\hat{\boldsymbol{s}}^{\boldsymbol{2}}\,,\,\mathcal{R}^{-1}\hat{\boldsymbol{s}}^{\boldsymbol{3}}) = (\hat{\boldsymbol{x}}\,,\,\hat{\boldsymbol{y}}\,,\,\hat{\boldsymbol{z}})$, and it follows that

$$\vec{\boldsymbol{\sigma}}\boldsymbol{\cdot}\left[\mathcal{R}^{-1}\hat{\boldsymbol{s}}^{\boldsymbol{a}}\right] = \sigma^a\,. \tag{E.35}$$

Consequently, we end up with

$$\chi^\dagger_s(\hat{\boldsymbol{s}})\,\vec{\boldsymbol{\sigma}}\boldsymbol{\cdot}\hat{\boldsymbol{s}}^{\boldsymbol{a}}\chi_{s'}(\hat{\boldsymbol{s}}) = \chi^\dagger_s(\hat{\boldsymbol{z}})\sigma^a\chi_{s'}(\hat{\boldsymbol{z}}) \equiv \tau^a_{ss'}\,, \tag{E.36}$$

which defines the matrix elements of the Pauli matrices, and our proof of eq. (E.30) is complete.

Using the completeness relation given by eq. (E.27), we can rewrite eq. (E.30) as

$$\chi^\dagger_s(\hat{\boldsymbol{s}})\,\sigma^i\chi_{s'}(\hat{\boldsymbol{s}}) = \tau^a_{ss'}\,\hat{\boldsymbol{s}}^{\boldsymbol{ai}}\,. \tag{E.37}$$

Taking the hermitian conjugate of eq. (E.37) is equivalent to interchanging $s \leftrightarrow s'$, since the σ^i are hermitian matrices and $(\tau^a_{ss'})^* = \tau^a_{s's}$. To evaluate expressions similar to eq. (E.37) that contain products of σ-matrices, it is sufficient to use the relation $\sigma^i\sigma^j = \delta^{ij}\mathbb{1} + i\epsilon^{ijk}\sigma^k$ as many times as needed to reduce the final expression to terms containing at most one σ-matrix. For example, using eqs. (E.19) and (E.37), it follows that

$$\chi^\dagger_s(\hat{\boldsymbol{s}})\,\sigma^i\sigma^j\chi_{s'}(\hat{\boldsymbol{s}}) = \delta_{ss'}\delta^{ij} + i\epsilon^{ijk}\tau^a_{ss'}\,\hat{\boldsymbol{s}}^{\boldsymbol{ak}}\,. \tag{E.38}$$

Finally, we provide a list of relations among the two-component spinor wave functions that are particularly useful when analyzing the properties of the two-component spinor fields under space inversion and time reversal. The first set of relations is given by

$$x_\alpha(-\vec{\boldsymbol{p}}, s) = \sigma^0_{\alpha\dot{\beta}} y^{\dagger\dot{\beta}}(\vec{\boldsymbol{p}}, s)\,, \qquad x^\alpha(-\vec{\boldsymbol{p}}, s) = -y^\dagger_{\dot{\beta}}(\vec{\boldsymbol{p}}, s)\,\overline{\sigma}^{0\dot{\beta}\alpha}\,, \tag{E.39}$$

$$y_\alpha(-\vec{\boldsymbol{p}}, s) = -\sigma^0_{\alpha\dot{\beta}} x^{\dagger\dot{\beta}}(\vec{\boldsymbol{p}}, s)\,, \qquad y^\alpha(-\vec{\boldsymbol{p}}, s) = x^\dagger_{\dot{\beta}}(\vec{\boldsymbol{p}}, s)\,\overline{\sigma}^{0\dot{\beta}\alpha}\,, \tag{E.40}$$

$$x^{\dagger\dot{\alpha}}(-\vec{\boldsymbol{p}}, s) = -\overline{\sigma}^{0\dot{\alpha}\beta} y_\beta(\vec{\boldsymbol{p}}, s)\,, \qquad x^\dagger_{\dot{\alpha}}(-\vec{\boldsymbol{p}}, s) = y^\beta(\vec{\boldsymbol{p}}, s)\sigma^0_{\beta\dot{\alpha}}\,, \tag{E.41}$$

$$y^{\dagger\dot{\alpha}}(-\vec{\boldsymbol{p}}, s) = \overline{\sigma}^{0\dot{\alpha}\beta} x_\beta(\vec{\boldsymbol{p}}, s)\,, \qquad y^\dagger_{\dot{\alpha}}(-\vec{\boldsymbol{p}}, s) = -x^\beta(\vec{\boldsymbol{p}}, s)\sigma^0_{\beta\dot{\alpha}}\,. \tag{E.42}$$

The above results follow from eqs. (E.2)–(E.5) after noting that $\sqrt{p\cdot\sigma} \longleftrightarrow \sqrt{p\cdot\overline{\sigma}}$ under the interchange of $\vec{\boldsymbol{p}} \longleftrightarrow -\vec{\boldsymbol{p}}$, in light of eq. (A.73). The second set of relations is related to the first set by eq. (E.6):

$$x_\alpha(-\vec{\boldsymbol{p}}, -s) = -2s\sigma^0_{\alpha\dot{\beta}} x^{\dagger\dot{\beta}}(\vec{\boldsymbol{p}}, s)\,, \qquad x^\alpha(-\vec{\boldsymbol{p}}, -s) = 2sx^\dagger_{\dot{\beta}}(\vec{\boldsymbol{p}}, s)\,\overline{\sigma}^{0\dot{\beta}\alpha}\,, \tag{E.43}$$

$$y_\alpha(-\vec{\boldsymbol{p}}, -s) = -2s\sigma^0_{\alpha\dot{\beta}} y^{\dagger\dot{\beta}}(\vec{\boldsymbol{p}}, s)\,, \qquad y^\alpha(-\vec{\boldsymbol{p}}, -s) = 2sy^\dagger_{\dot{\beta}}(\vec{\boldsymbol{p}}, s)\,\overline{\sigma}^{0\dot{\beta}\alpha}\,, \tag{E.44}$$

$$x^{\dagger\dot{\alpha}}(-\vec{\boldsymbol{p}}, -s) = 2s\overline{\sigma}^{0\dot{\alpha}\beta} x_\beta(\vec{\boldsymbol{p}}, s)\,, \qquad x^\dagger_{\dot{\alpha}}(-\vec{\boldsymbol{p}}, -s) = -2sx^\beta(\vec{\boldsymbol{p}}, s)\sigma^0_{\beta\dot{\alpha}}\,, \tag{E.45}$$

$$y^{\dagger\dot{\alpha}}(-\vec{\boldsymbol{p}}, -s) = 2s\overline{\sigma}^{0\dot{\alpha}\beta} y_\beta(\vec{\boldsymbol{p}}, s)\,, \qquad y^\dagger_{\dot{\alpha}}(-\vec{\boldsymbol{p}}, -s) = -2sy^\beta(\vec{\boldsymbol{p}}, s)\sigma^0_{\beta\dot{\alpha}}\,. \tag{E.46}$$

E.2 Fixed-Axis Spinors in the Nonrelativistic Limit

Consider an on-shell massive fermion of three-momentum $\vec{\boldsymbol{p}}$, mass m, and spin quantum number s, where $s = \pm\frac{1}{2}$ are the possible projections of the spin vector (in units of $\hbar$) along the fixed $\hat{\boldsymbol{s}}$ direction [see eq. (E.1)]. The spinor wave functions x, y, and their hermitian conjugates are given by eqs. (E.2)–(E.5). In the nonrelativistic limit,

$$\sqrt{p\cdot\sigma} \simeq \sqrt{m}\left(\mathbb{1} - \frac{\vec{\boldsymbol{\sigma}}\cdot\vec{\boldsymbol{p}}}{2m}\right), \qquad \sqrt{p\cdot\overline{\sigma}} \simeq \sqrt{m}\left(\mathbb{1} + \frac{\vec{\boldsymbol{\sigma}}\cdot\vec{\boldsymbol{p}}}{2m}\right), \tag{E.47}$$

where we keep terms only up to $\mathcal{O}(|\vec{\boldsymbol{p}}|/m)$. Inserting these results into eqs. (E.2)–(E.5) yields

$$x_\alpha(\vec{p}, s) \simeq \sqrt{m}\left(\mathbb{1} - \frac{\vec{\boldsymbol{\sigma}}\cdot\vec{\boldsymbol{p}}}{2m}\right)\chi_s(\hat{\boldsymbol{s}})\,, \tag{E.48}$$

$$x^\alpha(\vec{p}, s) \simeq -2s\sqrt{m}\,\chi^\dagger_{-s}(\hat{\boldsymbol{s}})\left(\mathbb{1} + \frac{\vec{\boldsymbol{\sigma}}\cdot\vec{\boldsymbol{p}}}{2m}\right), \tag{E.49}$$

$$y_\alpha(\vec{p}, s) \simeq 2s\sqrt{m}\left(\mathbb{1} - \frac{\vec{\boldsymbol{\sigma}}\cdot\vec{\boldsymbol{p}}}{2m}\right)\chi_{-s}(\hat{\boldsymbol{s}})\,, \tag{E.50}$$

$$y^\alpha(\vec{p}, s) \simeq \sqrt{m}\,\chi^\dagger_s(\hat{\boldsymbol{s}})\left(\mathbb{1} + \frac{\vec{\boldsymbol{\sigma}}\cdot\vec{\boldsymbol{p}}}{2m}\right), \tag{E.51}$$

for the undotted spinor wave functions, and

$$x^{\dagger\dot\alpha}(\vec{p},s) \simeq -2s\sqrt{m}\left(\mathbb{1} + \frac{\vec{\boldsymbol\sigma}\boldsymbol{\cdot}\vec{\boldsymbol p}}{2m}\right)\chi_{-s}(\hat{\boldsymbol s}), \tag{E.52}$$

$$x^{\dagger}_{\dot\alpha}(\vec{p},s) \simeq \sqrt{m}\,\chi^{\dagger}_{s}(\hat{\boldsymbol s})\left(\mathbb{1} - \frac{\vec{\boldsymbol\sigma}\boldsymbol{\cdot}\vec{\boldsymbol p}}{2m}\right), \tag{E.53}$$

$$y^{\dagger\dot\alpha}(\vec{p},s) \simeq \sqrt{m}\left(\mathbb{1} + \frac{\vec{\boldsymbol\sigma}\boldsymbol{\cdot}\vec{\boldsymbol p}}{2m}\right)\chi_{s}(\hat{\boldsymbol s}), \tag{E.54}$$

$$y^{\dagger}_{\dot\alpha}(\vec{p},s) \simeq 2s\sqrt{m}\,\chi^{\dagger}_{-s}(\hat{\boldsymbol s})\left(\mathbb{1} - \frac{\vec{\boldsymbol\sigma}\boldsymbol{\cdot}\vec{\boldsymbol p}}{2m}\right), \tag{E.55}$$

for the dotted spinor wave functions.

In the computation of the S-matrix amplitudes for scattering and decay processes, one typically must evaluate a bilinear product of spinors, i.e., quantities of the form

$$z_1(\vec{\boldsymbol p}_1, s_1)\,\Gamma\, z_2(\vec{\boldsymbol p}_2, s_2), \tag{E.56}$$

where z_1 and z_2 represent one of the two-component spinor wave functions x, y, $x^\dagger$, or $y^\dagger$, and Γ is a 2×2 matrix (in spinor space) that is either the identity matrix or is made up of alternating products of σ and $\overline{\sigma}$. In the nonrelativistic limit, these bilinears take on rather simple forms. In what follows, we work to first order in $|\vec{\boldsymbol p}_i|/m_i$. For example,

$$\begin{aligned} y^\alpha(\vec{\boldsymbol p}_1, s_1)x_\alpha(\vec{\boldsymbol p}_2, s_2) &\simeq \sqrt{m_1 m_2}\,\chi^{\dagger}_{s_1}(\hat{\boldsymbol s})\left(\mathbb{1} + \frac{\vec{\boldsymbol\sigma}\boldsymbol{\cdot}\vec{\boldsymbol p}}{2m_1} - \frac{\vec{\boldsymbol\sigma}\boldsymbol{\cdot}\vec{\boldsymbol p}}{2m_2}\right)\chi_{s_2}(\hat{\boldsymbol s}) \\ &\simeq \sqrt{m_1 m_2}\left[\delta_{s_1,s_2} + \left(\frac{\vec{\boldsymbol p}_1}{2m_1} - \frac{\vec{\boldsymbol p}_2}{2m_2}\right)\boldsymbol{\cdot}\hat{\boldsymbol s}^a\tau^a_{s_1,s_2}\right], \end{aligned} \tag{E.57}$$

where we have used the results of eqs. (E.19) and (E.37). Similarly,

$$\begin{aligned} y^\alpha(\boldsymbol p_1, s_1)\sigma^\mu_{\alpha\dot\beta}y^{\dagger\dot\beta}(\boldsymbol p_2, s_2) &\simeq \sqrt{m_1 m_2}\,\chi^{\dagger}_{s_1}(\hat{\boldsymbol s})\left[\sigma^\mu + \frac{\vec{\boldsymbol\sigma}\boldsymbol{\cdot}\vec{\boldsymbol p}_1}{2m_1}\sigma^\mu + \sigma^\mu\frac{\vec{\boldsymbol\sigma}\boldsymbol{\cdot}\vec{\boldsymbol p}_2}{2m_2}\right]\chi_{s_2}(\hat{\boldsymbol s}) \\ &\simeq \sqrt{m_1 m_2}\,Z^\mu_{s_1,s_2}(\vec{\boldsymbol p}_1, \vec{\boldsymbol p}_2), \end{aligned} \tag{E.58}$$

where

$$Z^\mu_{ss'}(\vec{\boldsymbol p}_1, \vec{\boldsymbol p}_2) \equiv \begin{cases} \delta_{ss'} + \left(\dfrac{\vec{\boldsymbol p}_1}{2m_1} + \dfrac{\vec{\boldsymbol p}_2}{2m_2}\right)\boldsymbol{\cdot}\hat{\boldsymbol s}^a\tau^a_{ss'}, & \text{for } \mu = 0, \\ \hat{\boldsymbol s}^{ai}\tau^a_{ss'} + \left(\dfrac{p_1^i}{2m_1} + \dfrac{p_2^i}{2m_2}\right)\delta_{ss'} + \left(\dfrac{p_2^j}{2m_2} - \dfrac{p_1^j}{2m_1}\right) i\epsilon^{ijk}\hat{\boldsymbol s}^{ak}\tau^a_{ss'}, & \\ \qquad\qquad \text{for } \mu = i \in \{1,2,3\}, \end{cases} \tag{E.59}$$

is obtained after using the results of eqs. (E.37) and (E.38).

In summary, we list the nonrelativistic forms of the spinor bilinears. Referring to eq. (E.56), if $\Gamma = \mathbb{1}$, then

$$x^\alpha(\vec{p}_1, s_1)x_\alpha(\vec{p}_2, s_2) \simeq 2s_2\sqrt{m_1 m_2}\left[\delta_{-s_2,s_1} + \left(\frac{\vec{p}_1}{2m_1} - \frac{\vec{p}_2}{2m_2}\right)\cdot\hat{s}^a\tau^a_{-s_2,s_1}\right], \tag{E.60}$$

$$y^\alpha(\vec{p}_1, s_1)y_\alpha(\vec{p}_2, s_2) \simeq 2s_2\sqrt{m_1 m_2}\left[\delta_{s_1,-s_2} + \left(\frac{\vec{p}_1}{2m_1} - \frac{\vec{p}_2}{2m_2}\right)\cdot\hat{s}^a\tau^a_{s_1,-s_2}\right], \tag{E.61}$$

$$x^\alpha(\vec{p}_1, s_1)y_\alpha(\vec{p}_2, s_2) \simeq \sqrt{m_1 m_2}\left[-\delta_{s_2,s_1} + \left(\frac{\vec{p}_1}{2m_1} - \frac{\vec{p}_2}{2m_2}\right)\cdot\hat{s}^a\tau^a_{s_2,s_1}\right], \tag{E.62}$$

$$y^\alpha(\vec{p}_1, s_1)x_\alpha(\vec{p}_2, s_2) \simeq \sqrt{m_1 m_2}\left[\delta_{s_1,s_2} + \left(\frac{\vec{p}_1}{2m_1} - \frac{\vec{p}_2}{2m_2}\right)\cdot\hat{s}^a\tau^a_{s_1,s_2}\right], \tag{E.63}$$

where we have used

$$\tau^a_{s's} = -4ss'\tau^a_{-s,-s'}, \qquad s, s' = \pm\tfrac{1}{2} \tag{E.64}$$

to arrive at the final forms given in eqs. (E.60) and (E.62). However, in using the above results, one must now pay close attention to the ordering of the subscript indices of the τ^a. The corresponding formulae for dotted spinor wave function bilinears are obtained by taking the hermitian conjugates of eqs. (E.60)–(E.63), which complex-conjugates the τ^a that appear on the right-hand side of these equations. After employing $(\tau^a_{ss'})^* = \tau^a_{s's}$, we obtain

$$x^\dagger_{\dot\alpha}(\vec{p}_1, s_1)x^{\dagger\dot\alpha}(\vec{p}_2, s_2) \simeq 2s_1\sqrt{m_1 m_2}\left[\delta_{s_2,-s_1} - \left(\frac{\vec{p}_1}{2m_1} - \frac{\vec{p}_2}{2m_2}\right)\cdot\hat{s}^a\tau^a_{s_2,-s_1}\right], \tag{E.65}$$

$$y^\dagger_{\dot\alpha}(\vec{p}_1, s_1)y^{\dagger\dot\alpha}(\vec{p}_2, s_2) \simeq 2s_1\sqrt{m_1 m_2}\left[\delta_{-s_1,s_2} - \left(\frac{\vec{p}_1}{2m_1} - \frac{\vec{p}_2}{2m_2}\right)\cdot\hat{s}^a\tau^a_{-s_1,s_2}\right], \tag{E.66}$$

$$y^\dagger_{\dot\alpha}(\vec{p}_1, s_1)x^{\dagger\dot\alpha}(\vec{p}_2, s_2) \simeq -\sqrt{m_1 m_2}\left[\delta_{s_2,s_1} + \left(\frac{\vec{p}_1}{2m_1} - \frac{\vec{p}_2}{2m_2}\right)\cdot\hat{s}^a\tau^a_{s_2,s_1}\right], \tag{E.67}$$

$$x^\dagger_{\dot\alpha}(\vec{p}_1, s_1)y^{\dagger\dot\alpha}(\vec{p}_2, s_2) \simeq \sqrt{m_1 m_2}\left[\delta_{s_1,s_2} - \left(\frac{\vec{p}_1}{2m_1} - \frac{\vec{p}_2}{2m_2}\right)\cdot\hat{s}^a\tau^a_{s_1,s_2}\right]. \tag{E.68}$$

Likewise, if $\Gamma = \sigma^\mu$, then

$$x^\alpha(\boldsymbol{p_1}, s_1)\sigma^\mu_{\alpha\dot\beta}x^{\dagger\dot\beta}(\boldsymbol{p_2}, s_2) \simeq 4s_1 s_2\sqrt{m_1 m_2}\, Z^\mu_{-s_1,-s_2}(\vec{p}_1, \vec{p}_2), \tag{E.69}$$

$$y^\alpha(\boldsymbol{p_1}, s_1)\sigma^\mu_{\alpha\dot\beta}y^{\dagger\dot\beta}(\boldsymbol{p_2}, s_2) \simeq \sqrt{m_1 m_2}\, Z^\mu_{s_1,s_2}(\vec{p}_1, \vec{p}_2), \tag{E.70}$$

$$x^\alpha(\boldsymbol{p_1}, s_1)\sigma^\mu_{\alpha\dot\beta}y^{\dagger\dot\beta}(\boldsymbol{p_2}, s_2) \simeq -2s_1\sqrt{m_1 m_2}\, Z^\mu_{-s_1,s_2}(\vec{p}_1, \vec{p}_2), \tag{E.71}$$

$$y^\alpha(\boldsymbol{p_1}, s_1)\sigma^\mu_{\alpha\dot\beta}x^{\dagger\dot\beta}(\boldsymbol{p_2}, s_2) \simeq -2s_2\sqrt{m_1 m_2}\, Z^\mu_{s_1,-s_2}(\vec{p}_1, \vec{p}_2), \tag{E.72}$$

where $Z^\mu_{ss'}(\vec{p}_1, \vec{p}_2)$ is defined in eq. (E.59). If $\Gamma = \overline{\sigma}^\mu$, one can use $z_1\sigma^\mu z_2^\dagger = z_2^\dagger\overline{\sigma}^\mu z_1$ [i.e., eq. (1.98) for commuting spinors] to obtain the corresponding formulae for the spinor wave function bilinears:

$$x^\dagger_{\dot\alpha}(\boldsymbol{p_1},s_1)\overline{\sigma}^{\mu\dot\alpha\beta}x_\beta(\boldsymbol{p_2},s_2) \simeq 4s_1s_2\sqrt{m_1m_2}\,Z^\mu_{-s_2,-s_1}(\vec{\boldsymbol{p}}_2,\vec{\boldsymbol{p}}_1)\,, \tag{E.73}$$

$$y^\dagger_{\dot\alpha}(\boldsymbol{p_1},s_1)\overline{\sigma}^{\mu\dot\alpha\beta}y_\beta(\boldsymbol{p_2},s_2) \simeq \sqrt{m_1m_2}\,Z^\mu_{s_2,s_1}(\vec{\boldsymbol{p}}_2,\vec{\boldsymbol{p}}_1)\,, \tag{E.74}$$

$$y^\dagger_{\dot\alpha}(\boldsymbol{p_1},s_1)\overline{\sigma}^{\mu\dot\alpha\beta}x_\beta(\boldsymbol{p_2},s_2) \simeq -2s_2\sqrt{m_1m_2}\,Z^\mu_{-s_2,s_1}(\vec{\boldsymbol{p}}_2,\vec{\boldsymbol{p}}_1)\,, \tag{E.75}$$

$$x^\dagger_{\dot\alpha}(\boldsymbol{p_1},s_1)\overline{\sigma}^{\mu\dot\alpha\beta}y_\beta(\boldsymbol{p_2},s_2) \simeq -2s_1\sqrt{m_1m_2}\,Z^\mu_{s_2,-s_1}(\vec{\boldsymbol{p}}_2,\vec{\boldsymbol{p}}_1)\,, \tag{E.76}$$

where $Z^\mu_{s_2s_1}(\vec{\boldsymbol{p}}_2,\vec{\boldsymbol{p}}_1)$ is obtained from the expression given by eq. (E.59) by interchanging $\{s_1\,,\,\vec{\boldsymbol{p}}_1\,,\,m_1\}$ and $\{s_2\,,\,\vec{\boldsymbol{p}}_2\,,\,m_2\}$. The above results can also be derived directly from eqs. (E.48)–(E.55), after employing eq. (E.64).

It is straightforward to evaluate the spinor wave function bilinears when Γ is a product of two or more $\sigma/\overline{\sigma}$ matrices. As the corresponding expressions are considerably more complicated, we shall not write them out explicitly here.

E.3 Helicity Spinor Wave Functions

All the results of Appendix E.1 also apply to the helicity spinors χ_λ, which are defined to be eigenstates of $\frac{1}{2}\vec{\boldsymbol{\sigma}}\cdot\hat{\boldsymbol{p}}$, i.e.,

$$\tfrac{1}{2}\vec{\boldsymbol{\sigma}}\cdot\hat{\boldsymbol{p}}\,\chi_\lambda(\hat{\boldsymbol{p}}) = \lambda\chi_\lambda(\hat{\boldsymbol{p}})\,, \qquad \lambda = \pm\tfrac{1}{2}\,, \tag{E.77}$$

where $\hat{\boldsymbol{p}} = (\sin\theta_p\cos\phi_p\,,\,\sin\theta_p\sin\phi_p\,,\,\cos\theta_p)$. It follows that

$$\sqrt{p\cdot\sigma}\,\chi_\lambda(\hat{\boldsymbol{p}}) = \omega_{-\lambda}(\vec{\boldsymbol{p}})\,\chi_\lambda(\hat{\boldsymbol{p}})\,, \qquad \sqrt{p\cdot\overline{\sigma}}\,\chi_\lambda(\hat{\boldsymbol{p}}) = \omega_\lambda(\vec{\boldsymbol{p}})\,\chi_\lambda(\hat{\boldsymbol{p}})\,, \tag{E.78}$$

where $\omega_\lambda(\vec{\boldsymbol{p}}) \equiv (E+2\lambda|\vec{\boldsymbol{p}}|\,)^{1/2}$ and $E=\sqrt{|\vec{\boldsymbol{p}}|^2+m^2}$. As a result, the explicit forms for the two-component spinor wave functions given in eqs. (E.2)–(E.5) simplify:

$$x_\alpha(\vec{\boldsymbol{p}},\lambda) = \omega_{-\lambda}\,\chi_\lambda(\hat{\boldsymbol{p}})\,, \qquad x^\alpha(\vec{\boldsymbol{p}},\lambda) = -2\lambda\,\omega_{-\lambda}\,\chi^\dagger_{-\lambda}(\hat{\boldsymbol{p}})\,, \tag{E.79}$$

$$y_\alpha(\vec{\boldsymbol{p}},\lambda) = 2\lambda\,\omega_\lambda\,\chi_{-\lambda}(\hat{\boldsymbol{p}})\,, \qquad y^\alpha(\vec{\boldsymbol{p}},\lambda) = \omega_\lambda\,\chi^\dagger_\lambda(\hat{\boldsymbol{p}})\,, \tag{E.80}$$

$$x^{\dagger\dot\alpha}(\vec{\boldsymbol{p}},\lambda) = -2\lambda\,\omega_{-\lambda}\,\chi_{-\lambda}(\hat{\boldsymbol{p}})\,, \qquad x^\dagger_{\dot\alpha}(\vec{\boldsymbol{p}},\lambda) = \omega_{-\lambda}\,\chi^\dagger_\lambda(\hat{\boldsymbol{p}})\,, \tag{E.81}$$

$$y^{\dagger\dot\alpha}(\vec{\boldsymbol{p}},\lambda) = \omega_\lambda\,\chi_\lambda(\hat{\boldsymbol{p}})\,, \qquad y^\dagger_{\dot\alpha}(\vec{\boldsymbol{p}},\lambda) = 2\lambda\,\omega_\lambda\,\chi^\dagger_{-\lambda}(\hat{\boldsymbol{p}})\,, \tag{E.82}$$

where $\omega_{\pm\lambda} \equiv \omega_{\pm\lambda}(\vec{\boldsymbol{p}})$. Note that eqs. (E.79)–(E.82) imply that the x and y spinors are related:

$$y(\vec{\boldsymbol{p}},\lambda) = 2\lambda x(\vec{\boldsymbol{p}},-\lambda)\,, \qquad y^\dagger(\vec{\boldsymbol{p}},\lambda) = 2\lambda x^\dagger(\vec{\boldsymbol{p}},-\lambda)\,. \tag{E.83}$$

In analogy with the $\hat{\boldsymbol{s}}^a$, it is convenient to introduce an orthonormal set of unit three-vectors $\hat{\boldsymbol{p}}^a$ such that $\hat{\boldsymbol{p}}^3 = \hat{\boldsymbol{p}}$. Then, eqs. (E.23)–(E.30) apply as well to the two-component helicity spinors, after taking $\hat{\boldsymbol{s}}^a = \hat{\boldsymbol{p}}^a$.

In scattering processes, it is often convenient to work in the rest frame of the incoming particles, in which the corresponding incoming fermion three-momenta are denoted by $\vec{p}$ and $-\vec{p}$, respectively. The helicity spinor wave function of the second fermion is defined to be the spinor wave function obtained from $\chi_\lambda(\hat{z})$ via a rotation by a polar angle $\pi - \theta_p$ and an azimuthal angle $\phi_p + \pi$ with respect to the $\hat{z}$-direction:

$$\chi_\lambda(-\hat{p}) = \mathcal{D}(\phi_p + \pi\,,\, \pi - \theta_p\,,\, 0)\, \chi_\lambda(\hat{z})\,. \tag{E.84}$$

Using the properties of the spin-1/2 rotation matrices, one can derive

$$\mathcal{D}(\phi_p + \pi\,,\, \pi - \theta_p\,,\, 0) = -\mathcal{D}(\phi_p\,,\, \theta_p\,,\, 0)\, D(\hat{z}, 0)\, D(\hat{x}, \pi)\,. \tag{E.85}$$

After employing the relation $D(\hat{x}, \pi)\chi_\lambda(\hat{z}) = -i\sigma^1\chi_\lambda(\hat{z}) = -i\chi_{-\lambda}(\hat{z})$, eqs. (E.84) and (E.85) yield

$$\chi_\lambda(-\hat{p}) = i\chi_{-\lambda}(\hat{p})\,, \tag{E.86}$$

which generalizes the result of eq. (E.18). Note that $\chi_\lambda(\hat{p})$ possesses the peculiar property that

$$\chi_\lambda(-(-\hat{p})) = -\chi_\lambda(\hat{p})\,. \tag{E.87}$$

This is a consequence of the fact that the result of two successive inversions is equivalent to $\phi_p \to \phi_p + 2\pi$, which yields an overall change of sign of a spinor wave function [see eq. (E.16)].

Suppose that the two fermions considered above have equal mass. In the center-of-mass frame, if the four-momentum of one of the fermions is $p^\mu = (E\,;\, \vec{p})$, then the four-momentum of the other fermion is

$$\widetilde{p}^\mu \equiv (E\,;\, -\vec{p})\,. \tag{E.88}$$

The following *numerical* identities are then satisfied: $\sigma{\cdot}\widetilde{p} = \overline{\sigma}{\cdot}p$ and $\overline{\sigma}{\cdot}\widetilde{p} = \sigma{\cdot}p$. However, in order to maintain covariance with respect to the undotted and dotted spinor indices, we shall write these identities as

$$\widetilde{p}{\cdot}\sigma_{\alpha\dot{\beta}} = \sigma^0_{\alpha\dot{\alpha}}(p{\cdot}\overline{\sigma}^{\dot{\alpha}\beta})\,\sigma^0_{\beta\dot{\beta}}\,, \qquad \widetilde{p}{\cdot}\overline{\sigma}^{\dot{\alpha}\beta} = \overline{\sigma}^{0\dot{\alpha}\alpha}(p{\cdot}\sigma_{\alpha\dot{\beta}})\,\overline{\sigma}^{0\dot{\beta}\beta}\,. \tag{E.89}$$

Taking the matrix square root on both sides of the two equations above removes one of the factors of σ^0 and $\overline{\sigma}^0$, respectively [see eqs. (A.75)–(A.81)]. For example, using eqs. (E.2) and (E.86),

$$x_\alpha(-\vec{p}, -\lambda) = \sqrt{\widetilde{p}{\cdot}\sigma}\,\chi_{-\lambda}(-\hat{p}) = i\sigma^0\sqrt{p{\cdot}\overline{\sigma}}\;\chi_\lambda(\vec{p}) = i\sigma^0_{\alpha\dot{\beta}}\, y^{\dagger\dot{\beta}}(\vec{p}, \lambda). \tag{E.90}$$

Similar manipulations yield the following helicity spinor wave function relations:[2]

$$x_\alpha(-\vec{p}, -\lambda) = i\sigma^0_{\alpha\dot{\beta}}\, y^{\dagger\dot{\beta}}(\vec{p}, \lambda)\,, \qquad x^\alpha(-\vec{p}, -\lambda) = -iy^\dagger_{\dot{\beta}}(\vec{p}, \lambda)\,\overline{\sigma}^{0\dot{\beta}\alpha}\,, \tag{E.91}$$

$$y_\alpha(-\vec{p}, -\lambda) = i\sigma^0_{\alpha\dot{\beta}}\, x^{\dagger\dot{\beta}}(\vec{p}, \lambda)\,, \qquad y^\alpha(-\vec{p}, -\lambda) = -ix^\dagger_{\dot{\beta}}(\vec{p}, \lambda)\,\overline{\sigma}^{0\dot{\beta}\alpha}\,, \tag{E.92}$$

$$x^{\dagger\dot{\alpha}}(-\vec{p}, -\lambda) = i\overline{\sigma}^{0\dot{\alpha}\beta}\, y_\beta(\vec{p}, \lambda)\,, \qquad x^\dagger_{\dot{\alpha}}(-\vec{p}, -\lambda) = -iy^\beta(\vec{p}, \lambda)\,\sigma^0_{\beta\dot{\alpha}}\,, \tag{E.93}$$

$$y^{\dagger\dot{\alpha}}(-\vec{p}, -\lambda) = i\overline{\sigma}^{0\dot{\alpha}\beta}\, x_\beta(\vec{p}, \lambda)\,, \qquad y^\dagger_{\dot{\alpha}}(-\vec{p}, -\lambda) = -ix^\beta(\vec{p}, \lambda)\,\sigma^0_{\beta\dot{\alpha}}\,. \tag{E.94}$$

[2] Note that eqs. (E.91)–(E.94) can also be obtained directly from eqs. (E.79)–(E.82).

One subtlety in using the results of eqs. (E.91)–(E.94) is worth mentioning. An example will illustrate the relevant issue. Suppose we wish to express $x_\alpha(\vec{\boldsymbol{p}},\lambda)$ in terms of $y^{\dagger\dot{\beta}}(-\vec{\boldsymbol{p}},-\lambda)$. One method for achieving this is to multiply both sides of the first relation given in eq. (E.94) by $-i\sigma^0_{\alpha\dot{\alpha}}$. It then follows that

$$x_\alpha(\vec{\boldsymbol{p}},\lambda) = -i\sigma^0_{\alpha\dot{\beta}}\, y^{\dagger\dot{\beta}}(-\vec{\boldsymbol{p}},-\lambda)\,. \tag{E.95}$$

One can derive eq. (E.95) more quickly by substituting $\vec{\boldsymbol{p}} = -\vec{\boldsymbol{p}}'$ and $\lambda = -\lambda'$ in the first relation given in eq. (E.91). However, in order to confirm the overall sign in eq. (E.95), one must also observe that

$$x(-(-\vec{\boldsymbol{p}}),\lambda) = -x(\vec{\boldsymbol{p}},\lambda)\,, \qquad y(-(-\vec{\boldsymbol{p}}),\lambda) = -y(\vec{\boldsymbol{p}},\lambda)\,, \tag{E.96}$$

as a consequence of eq. (E.87). Likewise, suppose we compute $x_\alpha(-\vec{\boldsymbol{p}},-\lambda)$ by substituting $\vec{\boldsymbol{p}} = -\vec{\boldsymbol{p}}'$ and $\lambda = -\lambda'$ in eq. (E.95). Once again, the end result differs from eq. (E.91) by an overall minus sign unless one properly takes eq. (E.96) into account. Indeed, a quick comparison between the results of eqs. (E.91)–(E.94) and of eqs. (2.377)–(2.380) will confirm the importance of eq. (E.96) in the derivation of the latter results from the former ones.

A second set of relations among the helicity spinor wave functions can be obtained by applying eq. (E.83) to eqs. (E.91)–(E.94):

$$x_\alpha(-\vec{\boldsymbol{p}},\lambda) = -2\lambda i\sigma^0_{\alpha\dot{\beta}}\, x^{\dagger\dot{\beta}}(\vec{\boldsymbol{p}},\lambda)\,, \qquad x^\alpha(-\vec{\boldsymbol{p}},\lambda) = 2\lambda i x^\dagger_{\dot{\beta}}(\vec{\boldsymbol{p}},\lambda)\,\overline{\sigma}^{0\dot{\beta}\alpha}\,, \tag{E.97}$$

$$y_\alpha(-\vec{\boldsymbol{p}},\lambda) = 2\lambda i\sigma^0_{\alpha\dot{\beta}}\, y^{\dagger\dot{\beta}}(\vec{\boldsymbol{p}},\lambda)\,, \qquad y^\alpha(-\vec{\boldsymbol{p}},\lambda) = -2\lambda i y^\dagger_{\dot{\beta}}(\vec{\boldsymbol{p}},\lambda)\,\overline{\sigma}^{0\dot{\beta}\alpha}\,, \tag{E.98}$$

$$x^{\dagger\dot{\alpha}}(-\vec{\boldsymbol{p}},\lambda) = -2\lambda i\overline{\sigma}^{0\dot{\alpha}\beta}\,, x_\beta(\vec{\boldsymbol{p}},\lambda)\,, \qquad x^\dagger_{\dot{\alpha}}(-\vec{\boldsymbol{p}},\lambda) = 2\lambda i x^\beta(\vec{\boldsymbol{p}},\lambda)\,\sigma^0_{\beta\dot{\alpha}}\,, \tag{E.99}$$

$$y^{\dagger\dot{\alpha}}(-\vec{\boldsymbol{p}},\lambda) = 2\lambda i\overline{\sigma}^{0\dot{\alpha}\beta}\, y_\beta(\vec{\boldsymbol{p}},\lambda)\,, \qquad y^\dagger_{\dot{\alpha}}(-\vec{\boldsymbol{p}},\lambda) = -2\lambda i y^\beta(\vec{\boldsymbol{p}},\lambda)\,\sigma^0_{\beta\dot{\alpha}}\,. \tag{E.100}$$

E.4 Covariant Spin Operators for a Spin-1/2 Fermion

Consider a massive spin-1/2 fermion of mass m and four-momentum p. We define a set of three four-vectors $S^{a\mu}$ ($a \in \{1,2,3\}$) such that the $S^{a\mu}$ and p^μ/m form an orthonormal set of four-vectors. In the rest frame of the fermion, where $p^\mu = (m\,;\,\vec{\boldsymbol{0}})$, we can define

$$S^{a\mu} \equiv (0\,;\,\hat{\boldsymbol{s}}^a)\,, \qquad a \in \{1,2,3\}\,, \tag{E.101}$$

where the $\hat{\boldsymbol{s}}^a$ are a mutually orthonormal set of unit three-vectors that form a basis for a right-handed coordinate system. Explicit forms for the $\hat{\boldsymbol{s}}^a$ are given in eq. (E.28). Using eq. (A.83), the three four-vectors $S^{a\mu}$, in a reference frame in which the four-momentum of the fermion is $p^\mu = (E\,;\,\vec{\boldsymbol{p}})$, are given by

$$S^{a\mu} = \left(\frac{\vec{\boldsymbol{p}}\boldsymbol{\cdot}\hat{\boldsymbol{s}}^a}{m}\,;\,\hat{\boldsymbol{s}}^a + \frac{(\vec{\boldsymbol{p}}\boldsymbol{\cdot}\hat{\boldsymbol{s}}^a)\,\vec{\boldsymbol{p}}}{m(E+m)}\right)\,, \qquad a \in \{1,2,3\}\,. \tag{E.102}$$

We identify $\hat{\boldsymbol{s}} = \hat{\boldsymbol{s}}^3$ as the quantization axis used in defining the third component of the spin of the fermion in its rest frame in eq. (E.1). It then follows that the spin four-vector, previously introduced in eq. (2.18), is given by $S^\mu = S^{3\mu}$.

The orthonormal set of four four-vectors p^μ/m and the $S^{a\mu}$ satisfy the following Lorentz-covariant relations:

$$p\cdot S^a = 0\,, \tag{E.103}$$

$$S^a\cdot S^b = -\delta^{ab}\,, \tag{E.104}$$

$$\epsilon^{\mu\nu\lambda\sigma} p_\mu S^1_\nu S^2_\lambda S^3_\sigma = -m\,, \tag{E.105}$$

$$S^a_\mu S^b_\nu - S^a_\nu S^b_\mu = \epsilon^{abc}\epsilon_{\mu\nu\rho\sigma} S^{c\rho}\frac{p^\sigma}{m}\,, \tag{E.106}$$

$$S^a_\mu\, S^a_\nu = -g_{\mu\nu} + \frac{p_\mu p_\nu}{m^2}\,, \tag{E.107}$$

where the sum over the repeated indices is implicit. The following two relations are noteworthy:

$$[(\sigma\cdot S^a)\,(\overline{\sigma}\cdot S^b)]_\alpha{}^\beta = -\delta^{ab}\delta_\alpha{}^\beta - \frac{i\epsilon^{abc}}{m}[(\sigma\cdot p)\,(\overline{\sigma}\cdot S^c)]_\alpha{}^\beta\,, \tag{E.108}$$

$$[(\overline{\sigma}\cdot S^a)\,(\sigma\cdot S^b)]^{\dot\alpha}{}_{\dot\beta} = -\delta^{ab}\delta^{\dot\alpha}{}_{\dot\beta} + \frac{i\epsilon^{abc}}{m}[(\overline{\sigma}\cdot p)\,(\sigma\cdot S^c)]^{\dot\alpha}{}_{\dot\beta}\,. \tag{E.109}$$

It is convenient to define a matrix-valued spin four-vector $\mathscr{S}^\mu$, whose matrix elements are given by

$$\mathscr{S}^\mu_{ss'} \equiv S^{a\mu}\tau^a_{ss'}\,, \qquad s, s' = \pm\tfrac{1}{2}\,, \tag{E.110}$$

where $\tau^a_{ss'}$ are the matrix elements of the Pauli matrices [see eq. (E.29)]. Then, we can rewrite eqs. (E.104) and (E.106) as

$$\tfrac{1}{3}\, g_{\mu\nu}\mathscr{S}^\mu\mathscr{S}^\nu = -\mathbb{1}_{2\times 2}\,, \tag{E.111}$$

$$\mathscr{S}^\mu\mathscr{S}^\nu - \mathscr{S}^\nu\mathscr{S}^\mu = \frac{2i}{m}\epsilon^{\mu\nu\rho\sigma}\mathscr{S}_\rho p_\sigma\,, \tag{E.112}$$

where the product $\mathscr{S}^\mu\mathscr{S}^\nu$ corresponds to ordinary 2×2 matrix multiplication. The $\mathscr{S}^\mu$ serve as covariant spin operators for a spin-1/2 fermion. In particular, in the rest frame, the $\frac{1}{2}\mathscr{S}^i$ satisfy the usual SU(2) commutation relations, with $(\frac{1}{2}\vec{\mathscr{S}})^2 = \frac{3}{4}$ as expected for a spin-1/2 particle. For future reference, we also define the following two quantities (see Exercise E.2):

$$S^\mu_- \equiv \tfrac{1}{2}S^{a\mu}\tau^a_{\frac{1}{2},-\frac{1}{2}} = \tfrac{1}{2}(S^{1\mu} - iS^{2\mu})\,, \qquad S^\mu_+ \equiv \tfrac{1}{2}S^{a\mu}\tau^a_{-\frac{1}{2},\frac{1}{2}} = \tfrac{1}{2}(S^{1\mu} + iS^{2\mu})\,. \tag{E.113}$$

It is often desirable to work with helicity states. In this case, we choose

$$\hat{\boldsymbol{s}}^a = \hat{\boldsymbol{p}}^a\,, \tag{E.114}$$

where the $\hat{\boldsymbol{p}}^a$ are an orthonormal triad of unit three-vectors with $\hat{\boldsymbol{p}}^3 \equiv \hat{\boldsymbol{p}}$. Moreover, since $\hat{\boldsymbol{p}}^{\boldsymbol{a}}\cdot\hat{\boldsymbol{p}} = 0$ for $a \neq 3$, it follows that $S^{a\mu} = (0\,;\,\hat{\boldsymbol{p}}^{\boldsymbol{a}})$ for $a = 1, 2$ in all reference frames obtained from the rest frame by a boost in the $\hat{\boldsymbol{p}}$ direction. In a reference frame where $p^\mu = (E\,;\,\vec{\boldsymbol{p}})$, eq. (E.102) yields

$$S^{1\mu} = (0\,;\,\hat{\boldsymbol{p}}^{\boldsymbol{1}}) = (0\,;\,\cos\theta\cos\phi,\ \cos\theta\sin\phi,\ -\sin\theta)\,, \tag{E.115}$$

$$S^{2\mu} = (0\,;\,\hat{\boldsymbol{p}}^{\boldsymbol{2}}) = (0\,;\,-\sin\phi,\ \cos\phi,\ 0)\,, \tag{E.116}$$

$$S^{3\mu} = \left(\frac{|\vec{\boldsymbol{p}}|}{m}\,;\,\frac{E}{m}\hat{\boldsymbol{p}}\right), \tag{E.117}$$

in a coordinate system where $\hat{\boldsymbol{p}} = (\sin\theta\cos\phi,\ \sin\theta\sin\phi,\ \cos\theta)$. One can check that eqs. (E.101)–(E.107) are also satisfied by the $S^{a\mu}$ defined in eqs. (E.115)–(E.117). As expected, $S^{3\mu}$ is the spin four-vector for helicity states obtained in eq. (2.19). In the high-energy limit ($E \gg m$),

$$mS^{a\mu} = p^\mu\,\delta^{a3} + \mathcal{O}(m)\,. \tag{E.118}$$

In eqs. (E.2)–(E.5), we presented explicit forms for the two-component spinor wave functions. The spinor wave functions x and y are related by the multiplication of either $S\!\cdot\!\sigma$ or $S\!\cdot\!\overline{\sigma}$ as exhibited in eqs. (2.29)–(2.32) [where $S \equiv S^3$]. These latter equations can be generalized as follows:

$$(S^a\!\cdot\!\overline{\sigma})^{\dot{\alpha}\beta}x_\beta(\vec{\boldsymbol{p}}, s') = \tau^a_{ss'}\,y^{\dagger\dot{\alpha}}(\vec{\boldsymbol{p}}, s), \qquad (S^a\!\cdot\!\sigma)_{\alpha\dot{\beta}}y^{\dagger\dot{\beta}}(\vec{\boldsymbol{p}}, s') = -\tau^a_{ss'}x_\alpha(\vec{\boldsymbol{p}}, s), \tag{E.119}$$

$$(S^a\!\cdot\!\sigma)_{\alpha\dot{\beta}}x^{\dagger\dot{\beta}}(\vec{\boldsymbol{p}}, s') = -\tau^a_{s's}y_\alpha(\vec{\boldsymbol{p}}, s), \qquad (S^a\!\cdot\!\overline{\sigma})^{\dot{\alpha}\beta}y_\beta(\vec{\boldsymbol{p}}, s') = \tau^a_{s's}x^{\dagger\dot{\alpha}}(\vec{\boldsymbol{p}}, s), \tag{E.120}$$

$$x^\alpha(\vec{\boldsymbol{p}}, s')(S^a\!\cdot\!\sigma)_{\alpha\dot{\beta}} = -\tau^a_{s's}y^\dagger_{\dot{\beta}}(\vec{\boldsymbol{p}}, s), \qquad y^\dagger_{\dot{\alpha}}(\vec{\boldsymbol{p}}, s')(S^a\!\cdot\!\overline{\sigma})^{\dot{\alpha}\beta} = \tau^a_{s's}x^\beta(\vec{\boldsymbol{p}}, s), \tag{E.121}$$

$$x^\dagger_{\dot{\alpha}}(\vec{\boldsymbol{p}}, s')(S^a\!\cdot\!\overline{\sigma})^{\dot{\alpha}\beta} = \tau^a_{ss'}y^\beta(\vec{\boldsymbol{p}}, s), \qquad y^\alpha(\vec{\boldsymbol{p}}, s')(S^a\!\cdot\!\sigma)_{\alpha\dot{\beta}} = -\tau^a_{ss'}x^\dagger_{\dot{\beta}}(\vec{\boldsymbol{p}}, s), \tag{E.122}$$

where there are implicit sums over the repeated labels $s = \pm\frac{1}{2}$. As expected, the case of $a = 3$ simply reproduces the results of eqs. (2.29)–(2.32). The above equations also apply to helicity wave functions $x(\vec{\boldsymbol{p}}, \lambda)$ and $y(\vec{\boldsymbol{p}}, \lambda)$ by replacing s, s' with λ, λ' and defining the $S^{a\mu}$ by eqs. (E.115)–(E.117).

The derivation of eqs. (E.119)–(E.122) for arbitrary a closely follows the corresponding derivation of eqs. (2.29)–(2.32) given in Section 2.2. In the present case, we need to employ eqs. (A.84) and (A.85) to obtain

$$\sqrt{p\!\cdot\!\sigma}\,S^a\!\cdot\!\overline{\sigma}\,\sqrt{p\!\cdot\!\sigma} = m\vec{\boldsymbol{\sigma}}\boldsymbol{\cdot}\hat{\boldsymbol{s}}^{\boldsymbol{a}}\,, \tag{E.123}$$

$$\sqrt{p\!\cdot\!\overline{\sigma}}\,S^a\!\cdot\!\sigma\,\sqrt{p\!\cdot\!\overline{\sigma}} = -m\vec{\boldsymbol{\sigma}}\boldsymbol{\cdot}\hat{\boldsymbol{s}}^{\boldsymbol{a}}\,, \tag{E.124}$$

which generalizes eqs. (2.22) and (2.23). For example, using eqs. (E.123) and (E.124) and the definitions for $x_\alpha(\vec{\boldsymbol{p}}, s)$ and $y^{\dagger\dot{\alpha}}(\vec{\boldsymbol{p}}, s)$, we find (suppressing spinor indices)

$$\sqrt{p\!\cdot\!\sigma}\,S^a\!\cdot\!\overline{\sigma}\,x(\vec{\boldsymbol{p}}, s') = \sqrt{p\!\cdot\!\sigma}\,S^a\!\cdot\!\overline{\sigma}\,\sqrt{p\!\cdot\!\sigma}\,\chi_{s'} = m\vec{\boldsymbol{\sigma}}\boldsymbol{\cdot}\hat{\boldsymbol{s}}^{\boldsymbol{a}}\,\chi_{s'} = m\tau^a_{ss'}\,\chi_s\,, \tag{E.125}$$

after using eq. (E.29). Multiplying both sides of eq. (E.125) by $\sqrt{p\!\cdot\!\overline{\sigma}}$, we end up with

$$S^a\cdot\overline{\sigma}\, x(\vec{\boldsymbol{p}}, s') = \tau^a_{ss'}\sqrt{p\cdot\overline{\sigma}}\,\chi_s = \tau^a_{ss'}\, y^\dagger(\vec{\boldsymbol{p}}, s)\,. \tag{E.126}$$

Similarly,

$$S^a\cdot\sigma x^\dagger(\vec{\boldsymbol{p}}, s') = 2s'\tau^a_{-s,-s'}\sqrt{p\cdot\sigma}\,\chi_{-s} = -\tau^a_{s's}\, y(\vec{\boldsymbol{p}}, s)\,, \tag{E.127}$$

where we have used

$$4ss'\tau^a_{-s,-s'} = -\tau^a_{s's}\,, \qquad \text{for} \quad s, s' = \pm 1/2\,. \tag{E.128}$$

All the results of eqs. (E.119)–(E.122) can be derived in this manner.

E.5 Two-Component Bouchiat–Michel formulae

Bouchiat and Michel derived a useful set of formulae [214, 215] that generalize the spin projection operators employed in four-component spinor computations. Here, we shall obtain the corresponding formulae using two-component spinor notation. We first consider the case of a massive fermion ($m \neq 0$). To establish the two-component Bouchiat–Michel formulae, we begin with the following identity:

$$\tfrac{1}{2}(\delta_{ss'} + \vec{\boldsymbol{\sigma}}\cdot\hat{\boldsymbol{s}}^a\,\tau^a_{ss'}) \sum_{t=\pm 1/2} \chi_t\chi_t^\dagger = \chi_{s'}\chi_s^\dagger\,. \tag{E.129}$$

To verify eq. (E.129), we first make use of eq. (E.29) to write $\vec{\boldsymbol{\sigma}}\cdot\hat{\boldsymbol{s}}^a\chi_t = \tau^a_{t't}\chi_{t'}$ and evaluate the product of two Pauli matrices:

$$\tau^a_{ss'}\tau^a_{t't} = 2\,\delta_{st}\delta_{s't'} - \delta_{ss'}\delta_{tt'}\,. \tag{E.130}$$

We then use eq. (E.123) and the completeness relation given in eq. (E.20) to rewrite eq. (E.129) in terms of $\mathscr{S}^\mu_{ss'}$ defined in eq. (E.110):

$$\chi_{s'}\chi_s^\dagger = \tfrac{1}{2}\left(\delta_{ss'} + \frac{1}{m}\sqrt{p\cdot\sigma}\,\mathscr{S}_{ss'}\cdot\overline{\sigma}\sqrt{p\cdot\sigma}\right). \tag{E.131}$$

Hence, with both spinor indices in the lowered position,

$$\begin{aligned}
x(\vec{\boldsymbol{p}}, s')x^\dagger(\vec{\boldsymbol{p}}, s) &= \sqrt{p\cdot\sigma}\,\chi_{s'}\chi_s^\dagger\sqrt{p\cdot\sigma}\\
&= \tfrac{1}{2}\sqrt{p\cdot\sigma}\left[\delta_{ss'} + \frac{1}{m}\sqrt{p\cdot\sigma}\,\mathscr{S}_{ss'}\cdot\overline{\sigma}\sqrt{p\cdot\sigma}\right]\sqrt{p\cdot\sigma}\\
&= \tfrac{1}{2}\left[p\cdot\sigma\delta_{ss'} + \frac{1}{m}\,p\cdot\sigma\,\mathscr{S}_{ss'}\cdot\overline{\sigma}\,p\cdot\sigma\right]\\
&= \tfrac{1}{2}\left(p\cdot\sigma\delta_{ss'} - m\mathscr{S}_{ss'}\cdot\sigma\right)\,.
\end{aligned} \tag{E.132}$$

In the final step of eq. (E.132), we simplified the product of three dot products by noting that $p\cdot S^a = 0$ implies that $\mathscr{S}_{ss'}\cdot\overline{\sigma}\;p\cdot\sigma = -p\cdot\overline{\sigma}\;\mathscr{S}_{ss'}\cdot\sigma$. Equation (E.132) is the two-component version of one of the Bouchiat–Michel formulae.

We list below a complete set of Bouchiat–Michel formulae:

$$x_\alpha(\vec{p}, s')x^\dagger_{\dot{\beta}}(\vec{p}, s) = \tfrac{1}{2}(p\,\delta_{ss'} - m\mathscr{S}_{ss'})\cdot\sigma_{\alpha\dot{\beta}}\,, \tag{E.133}$$

$$y^{\dagger\dot{\alpha}}(\vec{p}, s')y^\beta(\vec{p}, s) = \tfrac{1}{2}(p\,\delta_{ss'} + m\mathscr{S}_{ss'})\cdot\overline{\sigma}^{\dot{\alpha}\beta}\,, \tag{E.134}$$

$$x_\alpha(\vec{p}, s')y^\beta(\vec{p}, s) = \tfrac{1}{2}\left[m\delta_{ss'}\delta_\alpha{}^\beta - [(\sigma\cdot\mathscr{S}_{ss'})\,(\overline{\sigma}\cdot p)]_\alpha{}^\beta\right], \tag{E.135}$$

$$y^{\dagger\dot{\alpha}}(\vec{p}, s')x^\dagger_{\dot{\beta}}(\vec{p}, s) = \tfrac{1}{2}\left[m\delta_{ss'}\delta^{\dot{\alpha}}{}_{\dot{\beta}} + [(\overline{\sigma}\cdot\mathscr{S}_{ss'})\,(\sigma\cdot p)]^{\dot{\alpha}}{}_{\dot{\beta}}\right]. \tag{E.136}$$

If we set $s = s'$, we recover eqs. (2.51)–(2.54) as expected. The Bouchiat–Michel formulae can also be verified directly by using the explicit forms for the two-component spinor wave functions [eq. (2.404)] and the $\mathscr{S}^\mu_{ss'}$ [defined in eq. (E.110)]. The latter depends on the explicit form of the $\hat{\boldsymbol{s}}^a$ via eq. (E.102).

An equivalent set of Bouchiat–Michel formulae can be obtained by raising and/or lowering the appropriate free spinor indices using eqs. (1.83) and (1.123):

$$x^{\dagger\dot{\alpha}}(\vec{p}, s')x^\beta(\vec{p}, s) = \tfrac{1}{2}(p\,\delta_{s's} - m\mathscr{S}_{s's})\cdot\overline{\sigma}^{\dot{\alpha}\beta}\,, \tag{E.137}$$

$$y_\alpha(\vec{p}, s')y^\dagger_{\dot{\beta}}(\vec{p}, s) = \tfrac{1}{2}(p\,\delta_{s's} + m\mathscr{S}_{s's})\cdot\sigma_{\alpha\dot{\beta}}\,, \tag{E.138}$$

$$y_\alpha(\vec{p}, s')x^\beta(\vec{p}, s) = -\tfrac{1}{2}\left[m\delta_{s's}\delta_\alpha{}^\beta + [(\sigma\cdot\mathscr{S}_{s's})\,(\overline{\sigma}\cdot p)]_\alpha{}^\beta\right], \tag{E.139}$$

$$x^{\dagger\dot{\alpha}}(\vec{p}, s')y^\dagger_{\dot{\beta}}(\vec{p}, s) = -\tfrac{1}{2}\left[m\delta_{s's}\delta^{\dot{\alpha}}{}_{\dot{\beta}} - [(\overline{\sigma}\cdot\mathscr{S}_{s's})\,(\sigma\cdot p)]^{\dot{\alpha}}{}_{\dot{\beta}}\right]. \tag{E.140}$$

In this derivation, the spin labels in eqs. (E.137)–(E.140) are reversed relative to those in eqs. (E.133)–(E.136) due to eq. (E.128). Eight additional relations of the Bouchiat–Michel type can be obtained by replacing one x spinor with a y spinor (or vice versa). Recalling that the x and y spinors are related by [see eq. (E.6)],

$$y(\vec{p}, s) = 2s x(\vec{p}, -s)\,, \qquad y^\dagger(\vec{p}, s) = 2s x^\dagger(\vec{p}, -s)\,, \tag{E.141}$$

all possible spinor bilinears can be obtained from eqs. (E.133)–(E.140).

Equations (E.133)–(E.140) also apply to helicity spinor wave functions $x(\vec{p}, \lambda)$ and $y(\vec{p}, \lambda)$ after replacing s, s' with λ, λ' and using the $S^{a\mu}$ as defined in eqs. (E.115)–(E.117). Strictly speaking, all results involving the spinor wave functions obtained up to this point apply in the case of a massive spin-1/2 fermion. If we take the massless limit, then the four-vector $S^{3\mu}$ does not exist, as its definition depends on the existence of a rest frame. (In contrast, the four-vectors $S^{1\mu}$ and $S^{2\mu}$ do exist in the massless limit.) Nevertheless, massless helicity spinor wave functions are well defined; explicit forms can be found in eqs. (2.42)–(2.45). Using these forms, one can derive the Bouchiat–Michel formulae for a massless spin-1/2 fermion:

$$x_\alpha(\vec{p}, \lambda')x^\dagger_{\dot{\beta}}(\vec{p}, \lambda) = (\tfrac{1}{2} - \lambda)\,\delta_{\lambda\lambda'}\,p\cdot\sigma_{\alpha\dot{\beta}}\,, \tag{E.142}$$

$$y^{\dagger\dot{\alpha}}(\vec{p}, \lambda')y^\beta(\vec{p}, \lambda) = (\tfrac{1}{2} + \lambda)\,\delta_{\lambda\lambda'}\,p\cdot\overline{\sigma}^{\dot{\alpha}\beta}\,, \tag{E.143}$$

$$x_\alpha(\vec{p}, \lambda')y^\beta(\vec{p}, \lambda) = -(\tfrac{1}{2} - \lambda')(\tfrac{1}{2} + \lambda)\,[(\sigma\cdot S_-)(\overline{\sigma}\cdot p)]_\alpha{}^\beta\,, \tag{E.144}$$

$$y^{\dagger\dot{\alpha}}(\vec{p}, \lambda')x^\dagger_{\dot{\beta}}(\vec{p}, \lambda) = (\tfrac{1}{2} + \lambda')(\tfrac{1}{2} - \lambda)\,[(\overline{\sigma}\cdot S_+)(\sigma\cdot p)]^{\dot{\alpha}}{}_{\dot{\beta}}\,, \tag{E.145}$$

where S_-^μ and S_+^μ are defined in eq. (E.113), and explicit forms for $\sigma\cdot S_-$ and $\overline{\sigma}\cdot S_+$ are given in eq. (E.262).

The equivalent set of Bouchiat–Michel formulae, obtained by raising and/or lowering the appropriate free spinor indices, is given by

$$x^{\dagger\dot{\alpha}}(\vec{\boldsymbol{p}},\lambda')x^{\beta}(\vec{\boldsymbol{p}},\lambda) = (\tfrac{1}{2}-\lambda)\,\delta_{\lambda\lambda'}\,p\cdot\overline{\sigma}^{\dot{\alpha}\beta}\,, \tag{E.146}$$

$$y_{\alpha}(\vec{\boldsymbol{p}},\lambda')y^{\dagger}_{\dot{\beta}}(\vec{\boldsymbol{p}},\lambda) = (\tfrac{1}{2}+\lambda)\,\delta_{\lambda\lambda'}\,p\cdot\sigma_{\alpha\dot{\beta}}\,, \tag{E.147}$$

$$y_{\alpha}(\vec{\boldsymbol{p}},\lambda')x^{\beta}(\vec{\boldsymbol{p}},\lambda) = -(\tfrac{1}{2}+\lambda')(\tfrac{1}{2}-\lambda)\,[(\sigma\cdot S_-)(\overline{\sigma}\cdot p)]_{\alpha}{}^{\beta}\,, \tag{E.148}$$

$$x^{\dagger\dot{\alpha}}(\vec{\boldsymbol{p}},\lambda')y^{\dagger}_{\dot{\beta}}(\vec{\boldsymbol{p}},\lambda) = (\tfrac{1}{2}-\lambda')(\tfrac{1}{2}+\lambda)\,[(\overline{\sigma}\cdot S_+)(\sigma\cdot p)]^{\dot{\alpha}}{}_{\dot{\beta}}\,. \tag{E.149}$$

Eight additional relations of the Bouchiat–Michel type can be obtained by replacing one x spinor with a y spinor (or vice versa), using the results of eq. (E.141). As a check, one can verify that the above results follow from eqs. (E.133)–(E.140) by replacing s with λ, setting $mS^{a\mu} = p^{\mu}\,\delta^{a3}$, applying the mass-shell condition ($p^2 = m^2$), and taking the $m\to 0$ limit at the end of the computation.

We now demonstrate how to use the Bouchiat–Michel formulae to evaluate helicity amplitudes involving two equal mass spin-1/2 fermions. A typical amplitude involving a fermion–antifermion pair, evaluated in the center-of-mass frame of the pair, has the generic structure

$$z(\vec{\boldsymbol{p}},\lambda)\,\Gamma\,z'(-\vec{\boldsymbol{p}},\lambda')\,, \tag{E.150}$$

where z is one of the two-component spinor wave functions x, $x^{\dagger}$, y, or $y^{\dagger}$, and Γ is a 2×2 matrix (in spinor space) that is either the identity matrix, or is made up of alternating products of σ and $\overline{\sigma}$. As an illustration, we evaluate

$$\begin{aligned} x^{\dagger}_{\dot{\alpha}}(\vec{\boldsymbol{p}},\lambda)\,\Gamma^{\dot{\alpha}\beta}\,y_{\beta}(-\vec{\boldsymbol{p}},\lambda') &= 2\lambda'\,\Gamma^{\dot{\alpha}\beta}\,x_{\beta}(-\vec{\boldsymbol{p}},-\lambda')x^{\dagger}_{\dot{\alpha}}(\vec{\boldsymbol{p}},\lambda) \\ &= 2i\lambda'\,\Gamma^{\dot{\alpha}\beta}\sigma^{0}_{\beta\dot{\beta}}\,y^{\dagger\dot{\beta}}(\vec{\boldsymbol{p}},\lambda')x^{\dagger}_{\dot{\alpha}}(\vec{\boldsymbol{p}},\lambda)\,, \end{aligned} \tag{E.151}$$

where we have used eqs. (E.91) and (E.141). We can now employ eq. (E.145) to convert the right hand side of eq. (E.151) into a trace. By a similar computation, all expressions of the form of eq. (E.150) can be expressed as traces

$$x^{\dagger}_{\dot{\alpha}}(\vec{\boldsymbol{p}},\lambda)\,\Gamma^{\dot{\alpha}\beta}\,y_{\beta}(-\vec{\boldsymbol{p}},\lambda') = i\lambda'\,\mathrm{Tr}\left[\Gamma\,\sigma^{0}(m\delta_{\lambda\lambda'}+\overline{\sigma}\cdot\mathscr{S}_{\lambda\lambda'}\,\sigma\cdot p)\right]\,, \tag{E.152}$$

$$y^{\alpha}(\vec{\boldsymbol{p}},\lambda)\,\Gamma_{\alpha\dot{\beta}}\,x^{\dagger\dot{\beta}}(-\vec{\boldsymbol{p}},\lambda') = -i\lambda'\,\mathrm{Tr}\left[\Gamma\,\overline{\sigma}^{0}(m\delta_{\lambda\lambda'}-\sigma\cdot\mathscr{S}_{\lambda\lambda'}\,\overline{\sigma}\cdot p)\right]\,, \tag{E.153}$$

$$y^{\alpha}(\vec{\boldsymbol{p}},\lambda)\,\Gamma_{\alpha}{}^{\beta}\,y_{\beta}(-\vec{\boldsymbol{p}},\lambda') = i\lambda'\,\mathrm{Tr}\left[\Gamma\,\sigma^{0}(\overline{\sigma}\cdot p\,\delta_{\lambda\lambda'}+m\overline{\sigma}\cdot\mathscr{S}_{\lambda\lambda'})\right]\,, \tag{E.154}$$

$$x^{\dagger}_{\dot{\alpha}}(\vec{\boldsymbol{p}},\lambda)\,\Gamma^{\dot{\alpha}}{}_{\dot{\beta}}\,x^{\dagger\dot{\beta}}(-\vec{\boldsymbol{p}},\lambda') = -i\lambda'\,\mathrm{Tr}\left[\Gamma\,\overline{\sigma}^{0}(\sigma\cdot p\,\delta_{\lambda\lambda'}-m\sigma\cdot\mathscr{S}_{\lambda\lambda'})\right]\,, \tag{E.155}$$

after making use of eqs. (E.133) and (E.136). Similarly, there are four additional results that make use of eqs. (E.137) and (E.140):

$$y^{\dagger}_{\dot{\alpha}}(\vec{\boldsymbol{p}},\lambda)\,\Gamma^{\dot{\alpha}\beta}\,x_{\beta}(-\vec{\boldsymbol{p}},\lambda') = i\lambda'\,\mathrm{Tr}\left[\Gamma\,\sigma^{0}(m\delta_{\lambda'\lambda}-\overline{\sigma}\cdot\mathscr{S}_{\lambda'\lambda}\,\sigma\cdot p)\right]\,, \tag{E.156}$$

$$x^{\alpha}(\vec{\boldsymbol{p}},\lambda)\,\Gamma_{\alpha\dot{\beta}}\,y^{\dagger\dot{\beta}}(-\vec{\boldsymbol{p}},\lambda') = -i\lambda'\,\mathrm{Tr}\left[\Gamma\,\overline{\sigma}^{0}(m\delta_{\lambda'\lambda}+\sigma\cdot\mathscr{S}_{\lambda'\lambda}\,\overline{\sigma}\cdot p)\right]\,, \tag{E.157}$$

$$x^{\alpha}(\vec{\boldsymbol{p}},\lambda)\,\Gamma_{\alpha}{}^{\beta}\,x_{\beta}(-\vec{\boldsymbol{p}},\lambda') = -i\lambda'\,\mathrm{Tr}\left[\Gamma\,\sigma^{0}(\overline{\sigma}\cdot p\,\delta_{\lambda'\lambda}-m\overline{\sigma}\cdot\mathscr{S}_{\lambda'\lambda})\right]\,, \tag{E.158}$$

$$y^{\dagger}_{\dot{\alpha}}(\vec{\boldsymbol{p}},\lambda)\,\Gamma^{\dot{\alpha}}{}_{\dot{\beta}}\,y^{\dagger\dot{\beta}}(-\vec{\boldsymbol{p}},\lambda') = i\lambda'\,\mathrm{Tr}\left[\Gamma\,\overline{\sigma}^{0}(\sigma\cdot p\,\delta_{\lambda'\lambda}+m\sigma\cdot\mathscr{S}_{\lambda'\lambda})\right]\,. \tag{E.159}$$

For amplitudes involving equal mass fermions (or equal mass antifermions), other combinations of spinor bilinears appear in which one x spinor above is replaced by a y spinor or vice versa. These amplitudes can be reduced to one of the eight listed above by using eq. (E.141).

In the massless limit, one can again put $mS^{a\mu} = p^\mu \delta^{a3}$, set $p^2 = m^2$, and take $m \to 0$ at the end of the computation. Alternatively, one can repeat the derivation of eqs. (E.152)–(E.159) using the results of eqs. (E.142) and (E.149). For completeness, we record the end results here:

$$x^\dagger_{\dot\alpha}(\vec{p},\lambda)\,\Gamma^{\dot\alpha\beta}\,y_\beta(-\vec{p},\lambda') = i(\tfrac{1}{2}+\lambda')(\tfrac{1}{2}-\lambda)\ \mathrm{Tr}(\Gamma\,\sigma^0\overline{\sigma}\cdot S_-\,\sigma\cdot p)\,, \tag{E.160}$$

$$y^\alpha(\vec{p},\lambda)\,\Gamma_{\alpha\dot\beta}\,x^{\dagger\dot\beta}(-\vec{p},\lambda') = -i(\tfrac{1}{2}-\lambda')(\tfrac{1}{2}+\lambda)\ \mathrm{Tr}(\Gamma\,\overline{\sigma}^0\sigma\cdot S_-\,\overline{\sigma}\cdot p)\,, \tag{E.161}$$

$$y^\alpha(\vec{p},\lambda)\,\Gamma_\alpha{}^\beta\,y_\beta(-\vec{p},\lambda') = i(\tfrac{1}{2}+\lambda)\,\delta_{\lambda\lambda'}\ \mathrm{Tr}(\Gamma\,\sigma^0\,\overline{\sigma}\cdot p)\,, \tag{E.162}$$

$$x^\dagger_{\dot\alpha}(\vec{p},\lambda)\,\Gamma^{\dot\alpha}{}_{\dot\beta}\,x^{\dagger\dot\beta}(-\vec{p},\lambda') = i(\tfrac{1}{2}-\lambda)\,\delta_{\lambda\lambda'}\ \mathrm{Tr}(\Gamma\,\overline{\sigma}^0\,\sigma\cdot p)\,. \tag{E.163}$$

The equivalent set of formulae, obtained by raising and/or lowering the appropriate free spinor indices as before, is given by

$$y^\dagger_{\dot\alpha}(\vec{p},\lambda)\,\Gamma^{\dot\alpha\beta}\,x_\beta(-\vec{p},\lambda') = i(\tfrac{1}{2}-\lambda')(\tfrac{1}{2}+\lambda)\ \mathrm{Tr}(\Gamma\,\sigma^0\overline{\sigma}\cdot S_+\,\sigma\cdot p)\,, \tag{E.164}$$

$$x^\alpha(\vec{p},\lambda)\,\Gamma_{\alpha\dot\beta}\,y^{\dagger\dot\beta}(-\vec{p},\lambda') = -i(\tfrac{1}{2}+\lambda')(\tfrac{1}{2}-\lambda)\ \mathrm{Tr}(\Gamma\,\overline{\sigma}^0\sigma\cdot S_+\,\overline{\sigma}\cdot p)\,, \tag{E.165}$$

$$x^\alpha(\vec{p},\lambda)\,\Gamma_\alpha{}^\beta\,x_\beta(-\vec{p},\lambda') = i(\tfrac{1}{2}-\lambda)\,\delta_{\lambda\lambda'}\ \mathrm{Tr}(\Gamma\,\sigma^0\,\overline{\sigma}\cdot p)\,, \tag{E.166}$$

$$y^\dagger_{\dot\alpha}(\vec{p},\lambda)\,\Gamma^{\dot\alpha}{}_{\dot\beta}\,y^{\dagger\dot\beta}(-\vec{p},\lambda') = i(\tfrac{1}{2}+\lambda)\,\delta_{\lambda\lambda'}\ \mathrm{Tr}(\Gamma\,\overline{\sigma}^0\,\sigma\cdot p)\,. \tag{E.167}$$

E.6 Four-Component Spinor Wave Functions

In Chapter 3, we showed that the traditional four-component spinor wave functions, $u(\vec{p},s)$ and $v(\vec{p},s)$, can be expressed in terms of the two-component spinor wave functions as

$$u(\vec{p},s) = \begin{pmatrix} x_\alpha(\vec{p},s) \\ y^{\dagger\dot\alpha}(\vec{p},s) \end{pmatrix}, \qquad \bar{u}(\vec{p},s) = \left(y^\alpha(\vec{p},s),\ x^\dagger_{\dot\alpha}(\vec{p},s)\right), \tag{E.168}$$

$$v(\vec{p},s) = \begin{pmatrix} y_\alpha(\vec{p},s) \\ x^{\dagger\dot\alpha}(\vec{p},s) \end{pmatrix}, \qquad \bar{v}(\vec{p},s) = \left(x^\alpha(\vec{p},s),\ y^\dagger_{\dot\alpha}(\vec{p},s)\right), \tag{E.169}$$

where the u and v spinors are related by

$$v(\vec{p},s) = C\bar{u}(\vec{p},s)^{\mathsf{T}}\,, \qquad u(\vec{p},s) = C\bar{v}(\vec{p},s)^{\mathsf{T}}\,, \tag{E.170}$$

$$\bar{v}(\vec{p},s) = -u(\vec{p},s)^{\mathsf{T}}C^{-1}\,, \qquad \bar{u}(\vec{p},s) = -v(\vec{p},s)^{\mathsf{T}}C^{-1}\,. \tag{E.171}$$

The spin quantum number takes on values $s = \pm\frac{1}{2}$, and refers either to the component of the spin as measured in the rest frame with respect to a fixed axis, or to the helicity. Note that the u and v spinors also satisfy

$$v(\vec{p}, s) = -2s\gamma_5 u(\vec{p}, -s)\,, \qquad u(\vec{p}, s) = 2s\gamma_5 v(\vec{p}, -s)\,, \tag{E.172}$$

which follows from eq. (E.141).

Explicit forms for the four-component spinor wave functions in the chiral representation can be obtained using eqs. (E.2)–(E.5), where $\chi_s(\hat{s})$ is given in eq. (E.15). For example,

$$u(\vec{p}, s) = \begin{pmatrix} \sqrt{p\cdot\sigma}\,\chi_s \\ \sqrt{p\cdot\overline{\sigma}}\,\chi_s \end{pmatrix}, \qquad v(\vec{p}, s) = \begin{pmatrix} 2s\sqrt{p\cdot\sigma}\,\chi_{-s} \\ -2s\sqrt{p\cdot\overline{\sigma}}\,\chi_{-s} \end{pmatrix}. \tag{E.173}$$

For helicity spinors, further simplifications result by employing eqs. (E.79)–(E.82), which yields

$$u(\vec{p}, \lambda) = \begin{pmatrix} \omega_{-\lambda}\,\chi_\lambda(\hat{p}) \\ \omega_\lambda\,\chi_\lambda(\hat{p}) \end{pmatrix}, \qquad v(\vec{p}, \lambda) = \begin{pmatrix} 2\lambda\,\omega_\lambda\,\chi_{-\lambda}(\hat{p}) \\ -2\lambda\,\omega_{-\lambda}\,\chi_{-\lambda}(\hat{p}) \end{pmatrix}, \tag{E.174}$$

where $\omega_{\pm\lambda} = \omega_{\pm\lambda}(\vec{p}) \equiv (E + 2\lambda|\vec{p}|\,)^{1/2}$ and $E = \sqrt{|\vec{p}|^2 + m^2}$.

Additional relations among the four-component spinor wave functions arise in the study of the properties of the fermion fields under space inversion and time reversal. For the convenience of the reader, we repeat the behavior of the four-component fermion fields given in eqs. (3.106) and (3.107):

$$\mathcal{P}\Psi(x)\mathcal{P}^{-1} = \eta_P iP\Psi(x_P)\,, \tag{E.175}$$

$$\mathcal{T}\Psi(x)\mathcal{T}^{-1} = \eta_T T\Psi(x_T)\,, \tag{E.176}$$

where the Dirac parity operator P and time-reversal operator T are given by

$$P \equiv \gamma^0\,, \qquad T \equiv (\gamma^0 A^{-1})^{\mathsf{T}} B\,. \tag{E.177}$$

In all common representations of the Dirac gamma matrices, $A = \gamma^0$, in which case $T = B$. However, by employing the definition of T given in eq. (E.177), one can verify that the four-component index structure of the Dirac time-reversal operator is $T_{\bar{a}}{}^{b}$, in light of the results of Appendix A.11. This implies that $T\Psi$ behaves as the complex conjugate of a four-component field.

The Dirac parity and time-reversal operators can be employed to derive a number of useful relations among the four-component fermion wave functions. First, employing the Dirac parity operator yields

$$u(\vec{p}, s) = Pu(-\vec{p}, s)\,, \tag{E.178}$$

$$v(\vec{p}, s) = -Pv(-\vec{p}, s)\,. \tag{E.179}$$

As noted in the derivation of eqs. (E.39)–(E.42), the above results are immediately obtained after noting that $\sqrt{p\cdot\sigma} \longleftrightarrow \sqrt{p\cdot\overline{\sigma}}$ under the interchange $\vec{p} \longleftrightarrow -\vec{p}$.

In contrast to the spin quantum number s, which is defined by quantizing the spin along a fixed axis $\hat{\boldsymbol{s}}$ independently of the space inversion, the helicity quantum number λ changes sign under a space inversion. After employing eqs. (E.91)–(E.94), the corresponding results for the helicity spinor wave functions are

$$u(\vec{\boldsymbol{p}}, \lambda) = -i\, Pu(-\vec{\boldsymbol{p}}, -\lambda)\,, \tag{E.180}$$

$$v(\vec{\boldsymbol{p}}, \lambda) = -i\, Pv(-\vec{\boldsymbol{p}}, -\lambda)\,. \tag{E.181}$$

Second, employing the Dirac time-reversal operator yields[3]

$$u^*(\vec{\boldsymbol{p}}, s) = 2sTu(-\vec{\boldsymbol{p}}, -s)\,, \tag{E.182}$$

$$v^*(\vec{\boldsymbol{p}}, s) = 2sTv(-\vec{\boldsymbol{p}}, -s)\,. \tag{E.183}$$

These results are equivalent to the corresponding results for the two-component spinor wave functions obtained in eqs. (E.43)–(E.46). One can also derive eqs. (E.182) and (E.183) directly from the explicit forms for the u and v spinors given in eq. (E.173). For example, the proof of eq. (E.182) proceeds as follows. Working in the chiral representation where $A = \gamma^0$, it follows that $T = B = \begin{pmatrix} i\sigma^2 & 0 \\ 0 & i\sigma^2 \end{pmatrix}$. Making use of eqs. (A.74), (E.21), and (E.173), it follows that

$$\begin{aligned}
Tu(-\vec{\boldsymbol{p}}, -s) &= \begin{pmatrix} i\sigma^2 & 0 \\ 0 & i\sigma^2 \end{pmatrix} \begin{pmatrix} \sqrt{p\cdot\overline{\sigma}}\,\chi_{-s} \\ \sqrt{p\cdot\sigma}\,\chi_{-s} \end{pmatrix} = \begin{pmatrix} i\sigma^2\sqrt{p\cdot\overline{\sigma}}\, i\sigma^2\big[-i\sigma^2\chi_{-s}\big] \\ i\sigma^2\sqrt{p\cdot\sigma}\, i\sigma^2\big[-i\sigma^2\chi_{-s}\big] \end{pmatrix} \\
&= \begin{pmatrix} -\sqrt{p\cdot\sigma^*}(-i\sigma^2)(-2si\sigma^2\chi_s^*) \\ -\sqrt{p\cdot\overline{\sigma}^*}(-i\sigma^2)(-2si\sigma^2\chi_s^*) \end{pmatrix} = 2s \begin{pmatrix} \sqrt{p\cdot\sigma^*}\,\chi_s^* \\ \sqrt{p\cdot\overline{\sigma}^*}\,\chi_s^* \end{pmatrix} \\
&= 2su^*(\vec{\boldsymbol{p}}, s)\,.
\end{aligned} \tag{E.184}$$

Note that the spin quantum number s changes sign under a time-reversal transformation. In contrast, the helicity quantum number λ is fixed under time reversal. Employing eqs. (E.97)–(E.100) in eqs. (E.168) and (E.169), one can easily derive

$$u^*(\vec{\boldsymbol{p}}, \lambda) = -2i\lambda\, Tu(-\vec{\boldsymbol{p}}, \lambda)\,, \tag{E.185}$$

$$v^*(\vec{\boldsymbol{p}}, \lambda) = 2i\lambda\, Tv(-\vec{\boldsymbol{p}}, \lambda)\,. \tag{E.186}$$

Finally, we can combine the above results to obtain one additional set of useful relations. In light of eq. (E.170), we can write

$$v(\vec{\boldsymbol{p}}, s) = CA^{\mathsf{T}}u^*(\vec{\boldsymbol{p}}, s) = 2sCA^{\mathsf{T}}TP^{-1}u(\vec{\boldsymbol{p}}, -s)\,, \tag{E.187}$$

$$u(\vec{\boldsymbol{p}}, s) = CA^{\mathsf{T}}v^*(\vec{\boldsymbol{p}}, s) = -2sCA^{\mathsf{T}}TP^{-1}u(\vec{\boldsymbol{p}}, -s)\,, \tag{E.188}$$

[3] As expected from the discussion below eq. (E.177), multiplying a four-component spinor wave function by T yields a complex-conjugated spinor wave function.

after applying the Dirac time-reversal and space-inversion operators in succession. However, it is a simple matter to compute

$$CA^{\mathsf{T}}TP^{-1} = CA^{\mathsf{T}}(\gamma^0 A^{-1})^{\mathsf{T}} B\gamma^0 = C\gamma^{0\mathsf{T}} B\gamma^0 = \gamma^0\gamma_5\gamma^0 = -\gamma_5\,, \qquad \text{(E.189)}$$

after employing the relation $CB = \gamma_5$, and using the properties of the charge-conjugation matrix. Hence, it follows that

$$v(\vec{\boldsymbol{p}}, s) = -2s\gamma_5 u(\vec{\boldsymbol{p}}, -s)\,, \qquad u(\vec{\boldsymbol{p}}, s) = 2s\gamma_5 v(\vec{\boldsymbol{p}}, -s)\,. \qquad \text{(E.190)}$$

One can check that the results of eqs. (E.187), (E.188), and (E.190) are also valid for the helicity spinor wave functions after replacing s with λ:

$$v(\vec{\boldsymbol{p}}, \lambda) = -2\lambda\gamma_5 u(\vec{\boldsymbol{p}}, -\lambda)\,, \qquad u(\vec{\boldsymbol{p}}, \lambda) = 2\lambda\gamma_5 v(\vec{\boldsymbol{p}}, -\lambda)\,. \qquad \text{(E.191)}$$

E.7 Four-Component Bouchiat–Michel Formulae

The formulae originally obtained by Bouchiat and Michel employed four-component spinor notation. Using the results of Appendices E.5 and E.6, one can recover these results quite easily. Consider first the case of a massive spin-1/2 fermion. Converting eqs. (E.119)–(E.122) to four-component spinor notation, we obtain

$$\gamma_5 \not{S}^a\, u(\vec{\boldsymbol{p}}, s') = \tau^a_{ss'}\, u(\vec{\boldsymbol{p}}, s)\,, \qquad \gamma_5 \not{S}^a\, v(\vec{\boldsymbol{p}}, s') = \tau^a_{s's}\, v(\vec{\boldsymbol{p}}, s)\,, \qquad \text{(E.192)}$$

$$\bar{u}(\vec{\boldsymbol{p}}, s')\, \gamma_5 \not{S}^a = \tau^a_{ss'}\, \bar{u}(\vec{\boldsymbol{p}}, s)\,, \qquad \bar{v}(\vec{\boldsymbol{p}}, s')\, \gamma_5 \not{S}^a = \tau^a_{s's}\, \bar{v}(\vec{\boldsymbol{p}}, s)\,. \qquad \text{(E.193)}$$

In the case of $a = 3$, eqs. (E.192) and (E.193) reduce to eqs. (3.174) and (3.175).

The four-component Bouchiat–Michel formulae [214, 215] can be obtained from eqs. (E.133)–(E.140):

$$u(\vec{\boldsymbol{p}}, s')\bar{u}(\vec{\boldsymbol{p}}, s) = \tfrac{1}{2}\left[\delta_{ss'} + \gamma_5\gamma_\mu \mathscr{S}^\mu_{ss'}\right](\not{p} + m)\,, \qquad \text{(E.194)}$$

$$v(\vec{\boldsymbol{p}}, s')\bar{v}(\vec{\boldsymbol{p}}, s) = \tfrac{1}{2}\left[\delta_{s's} + \gamma_5\gamma_\mu \mathscr{S}^\mu_{s's}\right](\not{p} - m)\,, \qquad \text{(E.195)}$$

where $\mathscr{S}^\mu_{ss'} \equiv S^{a\mu}\tau^a_{ss'}$. As expected, the above results for $s = s'$ correspond to the spin projection operators given in eqs. (3.176) and (3.177). Related formulae involving products of u and v spinors can be obtained using [see eq. (3.171)]

$$v(\vec{\boldsymbol{p}}, s) = -2s\gamma_5 u(\vec{\boldsymbol{p}}, -s)\,, \qquad u(\vec{\boldsymbol{p}}, s) = 2s\gamma_5 v(\vec{\boldsymbol{p}}, -s)\,. \qquad \text{(E.196)}$$

Equations (E.192)–(E.195) also apply to helicity u and v spinors, after replacing s, s' with λ, λ' and using the S^a as defined in eq. (E.117). The four-component versions of eqs. (E.91)–(E.94) yield

$$u(-\boldsymbol{p}, -\lambda) = i\gamma^0\, u(\boldsymbol{p}, \lambda)\,, \qquad v(-\boldsymbol{p}, -\lambda) = i\gamma^0\, v(\boldsymbol{p}, \lambda)\,, \qquad \text{(E.197)}$$

$$\bar{u}(-\boldsymbol{p}, -\lambda) = -\bar{u}(\boldsymbol{p}, \lambda)\, i\gamma^0\,, \qquad \bar{v}(-\boldsymbol{p}, -\lambda) = -\bar{v}(\boldsymbol{p}, \lambda)\, i\gamma^0\,. \qquad \text{(E.198)}$$

In order to consider the massless limit, one must employ helicity spinors, as discussed in Appendix E.5. For $a = 1, 2$, eqs. (E.192) and (E.193) apply in the $m \to 0$ limit as written. The corresponding massless limit for the case of $a = 3$ is smooth and results in eq. (3.178). Similarly, the massless limit of the Bouchiat–Michel formulae for helicity spinors can be obtained by setting $mS^{a\mu} = p^\mu \delta^{a3}$, applying the mass-shell condition ($p^2 = m^2$), and taking the $m \to 0$ limit at the end of the computation. The end result is

$$u(\vec{p}, \lambda')\bar{u}(\vec{p}, \lambda) = \tfrac{1}{2}(1 + 2\lambda\gamma_5)\, \not{p}\, \delta_{\lambda\lambda'} + \tfrac{1}{2}\gamma_5[\not{S}^1 \tau^1_{\lambda\lambda'} + \not{S}^2 \tau^2_{\lambda\lambda'}]\, \not{p}\,, \tag{E.199}$$

$$v(\vec{p}, \lambda')\bar{v}(\vec{p}, \lambda) = \tfrac{1}{2}(1 - 2\lambda\gamma_5)\, \not{p}\, \delta_{\lambda'\lambda} + \tfrac{1}{2}\gamma_5[\not{S}^1 \tau^1_{\lambda'\lambda} + \not{S}^2 \tau^2_{\lambda'\lambda}]\, \not{p}\,. \tag{E.200}$$

As expected, when $\lambda = \lambda'$, we recover the helicity projection operators for massless spin-1/2 particles given in eqs. (3.180) and (3.181).

As before, we can use the Bouchiat–Michel formulae to evaluate helicity amplitudes involving two equal mass spin-1/2 fermions. A typical amplitude involving a fermion–antifermion pair, evaluated in the center-of-mass frame of the pair, has the generic structure

$$\overline{w}(\vec{p}, \lambda)\, \Gamma\, w'(-\vec{p}, \lambda')\,, \tag{E.201}$$

where w is either a u or v spinor, w' is respectively either a v or u spinor, and Γ is a product of Dirac gamma matrices. For example,

$$\begin{aligned} \bar{u}(\vec{p}, \lambda)\, \Gamma\, v(-\vec{p}, \lambda') &= -2\lambda' \bar{u}(\vec{p}, \lambda)\, \Gamma\, \gamma_5\, u(-\vec{p}, -\lambda') \\ &= -2i\lambda'\, \bar{u}(\vec{p}, \lambda)\, \Gamma\, \gamma_5\, \gamma^0\, u(\vec{p}, \lambda')\,, \end{aligned} \tag{E.202}$$

where we have used the results of eqs. (E.196) and (E.197). One can now use the Bouchiat–Michel formula given in eq. (E.194) to convert eq. (E.202) into a trace. By a similar computation, one can employ eq. (E.195) to convert $\bar{v}(\vec{p}, \lambda)\, \Gamma\, u(-\vec{p}, \lambda')$ into a trace. The two computations yield

$$\bar{u}(\vec{p}, \lambda)\, \Gamma\, v(-\vec{p}, \lambda') = -i\lambda' \operatorname{Tr}\Big[\Gamma\gamma_5\gamma^0(\delta_{\lambda\lambda'} + \gamma_5\gamma_\mu \mathcal{S}^\mu_{\lambda\lambda'})(\not{p} + m)\Big]\,, \tag{E.203}$$

$$\bar{v}(\vec{p}, \lambda)\, \Gamma\, u(-\vec{p}, \lambda') = i\lambda' \operatorname{Tr}\Big[\Gamma\gamma_5\gamma^0(\delta_{\lambda'\lambda} + \gamma_5\gamma_\mu \mathcal{S}^\mu_{\lambda'\lambda})(\not{p} - m)\Big]\,. \tag{E.204}$$

These results are the four-component spinor versions of eqs. (E.152)–(E.155) and eqs. (E.156)–(E.159), respectively. For amplitudes that involve a pair of equal mass fermions [or equal mass antifermions], w and w' in eq. (E.201) are both u spinors [or both v spinors]. Using eq. (3.171), these amplitudes can then be evaluated using the results of eqs. (E.203) and (E.204).

In the massless limit, one can again put $mS^{a\mu} = p^\mu \delta^{a3}$, set $p^2 = m^2$, and take $m \to 0$ at the end of the computation. Alternatively, one can repeat the derivation of eqs. (E.203)–(E.204) using the results of eqs. (E.199) and (E.200). For completeness, we record the end results here:

$$\bar{u}(\vec{p},\lambda)\,\Gamma\,v(-\vec{p},\lambda') = \tfrac{1}{2}i\delta_{\lambda\lambda'}\,\mathrm{Tr}\big[\Gamma\gamma^0(1+2\lambda\gamma_5)\not{p}\big] + i\lambda'\,\mathrm{Tr}\big[\Gamma\gamma^0(\not{s}^1\tau^1_{\lambda\lambda'}+\not{s}^2\tau^2_{\lambda\lambda'})\,\not{p}\big], \tag{E.205}$$

$$\bar{v}(\vec{p},\lambda)\,\Gamma\,u(-\vec{p},\lambda') = \tfrac{1}{2}i\delta_{\lambda'\lambda}\,\mathrm{Tr}\big[\Gamma\gamma^0(1-2\lambda\gamma_5)\not{p}\big] - i\lambda'\,\mathrm{Tr}\big[\Gamma\gamma^0(\not{s}^1\tau^1_{\lambda'\lambda}+\not{s}^2\tau^2_{\lambda'\lambda})\,\not{p}\big]. \tag{E.206}$$

E.8 Polarization Vectors for Spin-1 Bosons

Spin-1 helicity wave functions (also called polarization vectors) are four-vectors denoted by $\varepsilon^\mu(k,\lambda)$, where $k^\mu=(E_k\,;\,\vec{k})$ is the four-momentum of the particle and $E_k=\big(|\vec{k}|^2+m^2\big)^{1/2}$. The helicity λ can take on three possible values ($\lambda=-1,0,1$) if the particle is massive, and two possible values ($\lambda=\pm1$) if the particle is massless. The spin-1 polarization four-vectors satisfy[4]

$$k\cdot\varepsilon(\vec{k}\,,\,\lambda)=0\,,\qquad\qquad \varepsilon(\vec{k}\,,\,\lambda)\cdot\varepsilon(\vec{k}\,,\,\lambda')^* = -\delta_{\lambda\lambda'}\,. \tag{E.207}$$

We first consider a massless spin-1 particle moving in the z-direction with four-momentum $k^\mu=E(1\,;\,0\,,\,0\,,\,1)$. The textbook expression for the helicity ±1 polarization four-vectors of a massless spin-1 boson is given by [B72, B147]

$$\varepsilon^\mu(\hat{\boldsymbol{z}},\pm1)=\big(0\,;\,\hat{\boldsymbol{\varepsilon}}(\hat{\boldsymbol{z}},\pm1)\big)=\frac{1}{\sqrt{2}}\,(0\,;\mp1\,,\,-i\,,\,0)\,, \tag{E.208}$$

where $\hat{\boldsymbol{\varepsilon}}$ is the polarization three-vector.[5] Note that the $\varepsilon^\mu(\hat{\boldsymbol{z}},\lambda)$ are normalized eigenvectors of the spin-1 operator $\vec{\mathcal{S}}\cdot\hat{\boldsymbol{z}}$:

$$(\vec{\mathcal{S}}\cdot\hat{\boldsymbol{z}})^\mu{}_\nu\,\varepsilon^\nu(\hat{\boldsymbol{z}},\lambda)=\lambda\varepsilon^\mu(\hat{\boldsymbol{z}},\lambda)\,, \tag{E.209}$$

where $\mathcal{S}^i\equiv\frac{1}{2}\epsilon^{ijk}\mathcal{S}_{jk}$, and the matrix elements of the 4×4 matrices $\mathcal{S}_{jk}$ are given by eq. (1.4).

If we transform $\varepsilon^\mu(\hat{\boldsymbol{z}},\lambda)$ by employing a three-dimensional rotation $\mathcal{R}$ such that $\hat{\boldsymbol{k}}=\mathcal{R}\,\hat{\boldsymbol{z}}$, then we can obtain the polarization vector for a massless spin-1 boson of energy E moving in the direction $\hat{\boldsymbol{k}}=(\sin\theta\cos\phi\,,\,\sin\theta\sin\phi\,,\,\cos\theta)$. That is,

$$\varepsilon^\mu(\hat{\boldsymbol{k}},\lambda)=\Lambda^\mu{}_\nu(\phi\,,\theta\,,\,\gamma)\,\varepsilon^\nu(\hat{\boldsymbol{z}},\lambda)\,, \tag{E.210}$$

[4] Since the four-vector k of an on-shell spin-1 boson is determined by $\vec{k}$, it is often convenient to denote the polarization vector by $\varepsilon(\vec{k}\,,\,\lambda)$.

[5] Note that $\hat{\boldsymbol{\varepsilon}}(\hat{\boldsymbol{z}},\pm1)=\mp2^{-1/2}(1,\pm i,0)$. Some authors define the polarization three-vector with the $\mp$ sign omitted [which would also remove the $(-1)^\lambda$ in eq. (E.215)]. Our motivation for including the $\mp$ sign is to maintain consistency with the Condon–Shortley phase convention for the eigenfunctions of the spin-1 angular momentum operators $\vec{S}^2$ and S_z (e.g., see [B122]). In particular, using the radial unit vector $\hat{\boldsymbol{r}}=(\sin\theta\cos\phi,\sin\theta\sin\phi,\cos\theta)$, one obtains the relation $\hat{\boldsymbol{r}}\cdot\hat{\boldsymbol{\varepsilon}}(\hat{\boldsymbol{z}},\pm1)=(4\pi/3)^{1/2}Y_{1,\pm1}(\theta,\phi)$ between the polarization three-vector defined above and the $\ell=1$ spherical harmonics.

where

$$\Lambda^0{}_0 = 1\,, \qquad \Lambda^i{}_0 = \Lambda^0{}_i = 0\,, \quad \text{and} \quad \Lambda^i{}_j = \mathcal{R}^{ij}(\phi\,,\theta\,,\gamma)\,, \tag{E.211}$$

and $\mathcal{R}(\phi\,,\theta\,,\gamma)$ is the rotation matrix introduced in eq. (E.8). A simple computation yields

$$\varepsilon^\mu(\hat{\boldsymbol{k}},\pm 1) = \frac{1}{\sqrt{2}}\,e^{\mp i\gamma}\,(0\,;\,\mp\cos\theta\cos\phi + i\sin\phi\,,\,\mp\cos\theta\sin\phi - i\cos\phi\,,\,\pm\sin\theta)\,. \tag{E.212}$$

Note that $\varepsilon^\mu(\hat{\boldsymbol{k}},\pm 1)$ depends only on the direction of $\vec{\boldsymbol{k}}$ and not on its magnitude $E = |\vec{\boldsymbol{k}}|$. One can easily check that the $\varepsilon^\mu(\hat{\boldsymbol{k}},\pm 1)$ are normalized eigenstates of $\vec{\boldsymbol{S}}\cdot\hat{\boldsymbol{k}}$ with corresponding eigenvalues ± 1. Similar to the corresponding discussion in Appendix E.1 for the spin-1/2 spinor wave functions, the choice of the Euler angle γ is arbitrary. Following [B147], we again adopt the convention where $\gamma = 0$ (see footnote 1).

The expressions given by eqs. (E.208) and (E.212) also apply in the case of a massive spin-1 particle. In addition, there is a helicity $\lambda = 0$ polarization vector which depends on the magnitude of the momentum as well as its direction:

$$\varepsilon^\mu(|\vec{\boldsymbol{k}}|\hat{\boldsymbol{z}}\,,\,0) = \frac{1}{m}(|\vec{\boldsymbol{k}}|\,;\,0\,,\,0\,,\,E)\,, \tag{E.213}$$

where $E = (|\vec{\boldsymbol{k}}|^2 + m^2)^{1/2}$. One can use eq. (E.210) to obtain the helicity zero polarization vector for a massive spin-1 particle moving in an arbitrary direction $\hat{\boldsymbol{k}}$,

$$\varepsilon^\mu(\vec{\boldsymbol{k}},0) = \frac{1}{m}\left(|\vec{\boldsymbol{k}}|\,;\,E\hat{\boldsymbol{k}}\right) = \frac{1}{m}\left(|\vec{\boldsymbol{k}}|\,;\,E\sin\theta\cos\phi\,,\,E\sin\theta\sin\phi\,,\,E\cos\theta\right)\,. \tag{E.214}$$

Note that both the massless and massive spin-1 polarization vectors satisfy

$$\varepsilon^\mu(\vec{\boldsymbol{k}}\,,\,\lambda)^* = (-1)^\lambda\varepsilon^\mu(\vec{\boldsymbol{k}}\,,\,-\lambda)\,. \tag{E.215}$$

One can check that the $\varepsilon^\mu(\vec{\boldsymbol{k}},\lambda)$ also satisfy the conditions for a spin-1 polarization four-vector given in eq. (E.207).

If the spin-1 boson three-momentum is $-\vec{\boldsymbol{k}}$, then its polarization vector can be obtained from eqs. (E.212) and (E.214) by taking $\theta \to \pi - \theta$ and $\phi \to \phi + \pi$. It can also be derived from eqs. (E.210) and (E.211) by making use of the spin-1 analogue of eq. (E.85),

$$\mathcal{R}(\phi + \pi\,,\,\pi - \theta\,,\,0) = \mathcal{R}(\phi\,,\,\theta\,,\,0)\,R(\hat{\boldsymbol{z}},0)\,R(\hat{\boldsymbol{x}},\pi)\,, \tag{E.216}$$

where R is the rotation matrix given by eq. (E.9). Applying the above rotation matrices to the polarization three-vectors defined in eq. (E.208), we obtain

$$R(\hat{\boldsymbol{x}},\pi)\,\hat{\boldsymbol{\varepsilon}}(\hat{\boldsymbol{z}},\lambda) = -\hat{\boldsymbol{\varepsilon}}(\hat{\boldsymbol{z}},-\lambda)\,, \tag{E.217}$$

$$R(\hat{\boldsymbol{z}},\beta)\,\hat{\boldsymbol{\varepsilon}}(\hat{\boldsymbol{z}},\lambda) = e^{-i\lambda\beta}\,\hat{\boldsymbol{\varepsilon}}(\hat{\boldsymbol{z}},\lambda)\,. \tag{E.218}$$

It then follows that

$$\varepsilon^\mu(-\vec{\boldsymbol{k}}\,,\,\lambda) = (\Lambda_P)^\mu{}_\nu\,\varepsilon^\nu(\vec{\boldsymbol{k}}\,,\,-\lambda)\,, \qquad \lambda = 0\,,\pm 1\,, \tag{E.219}$$

where $(\Lambda_P)^\mu{}_\nu = \text{diag}(1,-1,-1,-1)$. We can rewrite eq. (E.219) in the form

$$\varepsilon^\mu(\vec{\boldsymbol{k}}\,,\,\lambda) = (\Lambda_P)^\mu{}_\nu\,\varepsilon^\nu(-\vec{\boldsymbol{k}}\,,\,-\lambda)\,, \tag{E.220}$$

to emphasize the similarity of this result with that of eq. (E.180). Indeed, under space inversion, both $\vec{\boldsymbol{k}}$ and λ flip signs. If we now combine the results of eqs. (E.215) and (E.220), we end up with

$$\varepsilon^\mu(\vec{\boldsymbol{k}}\,,\,\lambda)^* = -(-1)^\lambda(\Lambda_T)^\mu{}_\nu\,\varepsilon^\nu(-\vec{\boldsymbol{k}}\,,\,\lambda)\,, \tag{E.221}$$

where $(\Lambda_T)^\mu{}_\nu = \text{diag}(-1,1,1,1)$. The form of eq. (E.221) emphasizes the similarity of this result with that of eq. (E.185). Indeed, under time reversal, $\vec{\boldsymbol{k}}$ flips sign whereas λ is unchanged.

In the evaluation of spin-averaged cross sections of processes involving one or more external spin-1 bosons, one must evaluate polarization sums of the form

$$\mathbb{P}_{\mu\nu} \equiv \sum_\lambda \varepsilon_\mu(\vec{\boldsymbol{k}},\lambda)\varepsilon_\nu(\vec{\boldsymbol{k}},\lambda)^*\,. \tag{E.222}$$

Consider first the computation of $\mathbb{P}_{\mu\nu}$ in case of $m\neq 0$, in which case the sum over polarizations runs over $\lambda=-1,0,+1$. We define the following four four-vectors:

$$e^{(0)} \equiv \frac{k}{m}\,,\qquad e^{(1)}\equiv\varepsilon(\vec{\boldsymbol{k}},1)\,,\qquad e^{(2)}\equiv\varepsilon(\vec{\boldsymbol{k}},0)\,,\qquad e^{(3)}\equiv\varepsilon(\vec{\boldsymbol{k}},-1)\,, \tag{E.223}$$

where $k=(E\,;\,\vec{\boldsymbol{k}})$ and $E=(|\vec{\boldsymbol{k}}|^2+m^2)^{1/2}$. In light of the fact that $k^2=m^2$, it follows that $e^{(0)}\cdot e^{(0)\,*}=1$. Combining this result with eq. (E.207), it follows that

$$e^{(\alpha)}\cdot e^{(\beta)\,*} = g^{\alpha\beta}\,, \tag{E.224}$$

where $g^{\alpha\beta} = \text{diag}(1\,,\,-1\,,\,-1\,,\,-1)$. It is convenient to introduce a bra and ket notation, where $|\alpha\rangle\equiv e^{(\alpha)}$ and $\langle\alpha|\equiv e^{(\alpha)\,*}$. Then eq. (E.224) can be rewritten as

$$\langle\beta|\alpha\rangle = e^{(\alpha)}\cdot e^{(\beta)\,*} = g^{\alpha\beta}\,. \tag{E.225}$$

That is, we can interpret the ket $|\alpha\rangle$ as a contravariant four-vector (with an upper index) and the bra $\langle\beta|$ as a covariant four-vector (with a lower index).

Since $|\alpha\rangle$ is a suitable basis set for four-vectors in Minkowski space, one can deduce a completeness relation. Any four-vector $|\zeta\rangle$ can be expanded in the basis spanned by the $|\alpha\rangle$:

$$|\zeta\rangle = \sum_\alpha c_\alpha\,|\alpha\rangle\,. \tag{E.226}$$

Thus,

$$\langle\beta|\zeta\rangle = \sum_\alpha c_\alpha\,\langle\beta|\alpha\rangle = \sum_\alpha c_\alpha g^{\alpha\beta} = c_\beta g^{\beta\beta}\,, \tag{E.227}$$

where there is no implicit sum over the thrice-repeated index β in eq. (E.227). Indeed, only one term, corresponding to $\alpha=\beta$, survives in the sum over α. It follows that

$$c_\beta = g^{\beta\beta}\,\langle\beta|\zeta\rangle \qquad \text{(no sum over } \beta\text{)}, \tag{E.228}$$

where we have used $(g^{\beta\beta})^2 = 1$ (for any fixed value of $\beta = 0$, 1, 2, or 3). Plugging this result back into eq. (E.226) yields

$$|\zeta\rangle = \sum_\alpha g^{\alpha\alpha} |\alpha\rangle \langle\alpha|\zeta\rangle \,, \tag{E.229}$$

which is equivalent to

$$\left(\mathbb{1} - \sum_\alpha g^{\alpha\alpha} |\alpha\rangle \langle\alpha|\right) |\zeta\rangle = 0 \,, \tag{E.230}$$

where $\mathbb{1}$ is the unit operator. That is,

$$\sum_\alpha g^{\alpha\alpha} |\alpha\rangle \langle\alpha| = \mathbb{1} \,, \tag{E.231}$$

which is the desired completeness relation.

Making the Lorentz indices explicit, the completeness relation [eq. (E.231)] can be written more explicitly as

$$\sum_\alpha g^{\alpha\alpha} |\alpha\rangle^\mu \langle\alpha|_\nu = \delta^\mu_\nu \,. \tag{E.232}$$

As before, we identify $|\alpha\rangle^\mu = e^{(\alpha)\,\mu}$ and $\langle\alpha|_\nu = e^{(\alpha)}_\nu$, for $\alpha = 0$, 1, 2, and 3. Using eq. (E.223), we see that eq. (E.232) is equivalent to

$$\frac{k^\mu k_\nu}{m^2} - \sum_{\lambda=-1,0,1} \varepsilon^\mu(\vec{\boldsymbol{k}},\lambda)\varepsilon_\nu(\vec{\boldsymbol{k}},\lambda)^* = \delta^\mu_\nu \,. \tag{E.233}$$

Multiplying by the metric tensor to lower the index μ, we end up with

$$\mathbb{P}_{\mu\nu} \equiv \sum_{\lambda=-1,0,1} \varepsilon_\mu(k,\lambda)\varepsilon_\nu(k,\lambda)^* = -g_{\mu\nu} + \frac{k_\mu k_\nu}{m^2} \,. \tag{E.234}$$

In the case of $m = 0$, the sum over polarizations runs only over the transverse polarization states, $\lambda = -1, +1$, since there is no longitudinal polarization state for a physical massless spin-1 boson. Moreover, the transversality condition depends on the reference frame. Consequently, the expression for $\mathbb{P}_{\mu\nu}$ cannot be a Lorentz covariant tensor, in contrast to the Lorentz covariant expression given in eq. (E.234) in the case of $m \neq 0$. Nevertheless, we may compute $\mathbb{P}_{\mu\nu}$ using a similar technique to the computation above. In this case, we define the following four-vectors:

$$e^{(0)} \equiv n = (1\,;\,\vec{\boldsymbol{0}})\,, \quad e^{(1)} \equiv \varepsilon(\vec{\boldsymbol{k}},\lambda=1)\,, \quad e^{(2)} \equiv (0\,;\,\hat{\boldsymbol{k}})\,, \quad e^{(3)} \equiv \varepsilon(\vec{\boldsymbol{k}},\lambda=-1). \tag{E.235}$$

Note that n^ν is a fixed four-vector which is independent of the reference frame that defines the four-momentum k. Consequently, we shall refer to n^μ as a fake four-vector, since n does not transform as a Lorentz four-vector under Lorentz transformations. The physical transverse polarization vectors $\varepsilon(\vec{\boldsymbol{k}},\lambda=\pm1)$ satisfy

$$n \cdot \varepsilon(\vec{\boldsymbol{k}},\lambda=\pm1) = 0 \,, \tag{E.236}$$

which is the mathematical statement that the transversality condition depends on the reference frame.

One can rewrite $e^{(2)}$ in a more useful way as follows. Note that the four-momentum of the massless spin-1 boson satisfies

$$k = (E\,;\vec{\boldsymbol{k}}) = E(1\,;\hat{\boldsymbol{k}}) = E\big[n + e^{(2)}\big] = k\cdot n\big[n + e^{(2)}\big]\,, \tag{E.237}$$

where we have used the definition of n to write $E = k\cdot n$. It then follows that

$$e^{(2)} = \frac{k}{k\cdot n} - n\,. \tag{E.238}$$

It is straightforward to check that [see eq. (E.224)]

$$e^{(\alpha)}\cdot e^{(\beta)\,*} = g^{\alpha\beta}\,. \tag{E.239}$$

Hence, we can again use the bra and ket notation to derive the completeness relation [see eq. (E.232)],

$$\sum_\alpha g^{\alpha\alpha}\,|\alpha)^\mu\,(\alpha|_\nu = \delta^\mu_\nu\,. \tag{E.240}$$

Using eq. (E.235), the completeness relation is equivalent to

$$n_\mu n_\nu - e^{(2)}_\mu e^{(2)}_\nu - \sum_{\lambda=\pm 1}\varepsilon_\mu(k,\lambda)\varepsilon_\nu(k,\lambda)^* = g_{\mu\nu}\,, \tag{E.241}$$

after lowering the index μ with the metric tensor. Using eq. (E.238),

$$e^{(2)}_\mu e^{(2)}_\nu = \left(\frac{k_\mu}{k\cdot n} - n_\mu\right)\left(\frac{k_\nu}{k\cdot n} - n_\nu\right) = \frac{k_\mu k_\nu}{(k\cdot n)^2} - \frac{k_\mu n_\nu + k_\nu n_\mu}{k\cdot n} + n_\mu n_\nu\,. \tag{E.242}$$

Plugging the above result for $e^{(2)}_\mu e^{(2)}_\nu$ into eq. (E.241), we arrive at our final result for the polarization sum for a massless spin-1 boson,

$$\mathbb{P}_{\mu\nu} \equiv \sum_{\lambda=\pm 1}\varepsilon_\mu(k,\lambda)\varepsilon_\nu(k,\lambda)^* = -g_{\mu\nu} + \frac{k_\mu n_\nu + k_\nu n_\mu}{k\cdot n} - \frac{k_\mu k_\nu}{(k\cdot n)^2}\,. \tag{E.243}$$

There is an equivalent form for $\mathbb{P}_{\mu\nu}$ that is a little more convenient. First, we rewrite eq. (E.243) in the following equivalent form:

$$\mathbb{P}_{\mu\nu} = -g_{\mu\nu} - \frac{1}{2(k\cdot n)^2}\big[k_\mu\big(k_\nu - 2(k\cdot n)n_\nu\big) + k_\nu\big(k_\mu - 2(k\cdot n)n_\mu\big)\big]\,. \tag{E.244}$$

Following eq. (E.88), we define the following fake four-vector (which depends on the fake four-vector n^μ):

$$\widetilde{k}^\mu \equiv 2(k\cdot n)n^\mu - k^\mu = (E\,;-\vec{\boldsymbol{k}})\,. \tag{E.245}$$

It then follows that $\widetilde{k}\cdot\widetilde{k} = 0$ and

$$k\cdot\widetilde{k} = E^2 + |\vec{\boldsymbol{k}}|^2 = 2E^2 = 2(k\cdot n)^2\,. \tag{E.246}$$

Thus, we can write $\mathbb{P}_{\mu\nu}$ as

$$\mathbb{P}_{\mu\nu} \equiv \sum_{\lambda=\pm 1} \varepsilon_\mu(k,\lambda)\varepsilon_\nu(k,\lambda)^* = -g_{\mu\nu} + \frac{k_\mu \widetilde{k}_\nu + k_\nu \widetilde{k}_\mu}{k\cdot\widetilde{k}} . \tag{E.247}$$

As advertised, the expression for $\mathbb{P}_{\mu\nu}$ [eq. (E.243) or (E.247)] is not a Lorentz covariant tensor, due to the presence of the fake four-vectors n or $\widetilde{k}$, which do not transform like Lorentz four-vectors under Lorentz transformations. Nevertheless, in any process that involves massless spin-1 particles, the vectors n or $\widetilde{k}$ resulting from the evaluation of a polarization sum must ultimately cancel out in the computation of a physical observable. This cancellation is guaranteed by the gauge invariance and the Lorentz invariance of the theory.

For example, consider the scattering (or decay) amplitude in an abelian gauge theory (e.g., QED) for a process with n external massless gauge bosons, which is of the form

$$\mathcal{M} = \mathcal{M}^{\mu_1\mu_2\cdots\mu_n}\varepsilon_{\mu_1}(k_1,\lambda_1)\varepsilon_{\mu_2}(k_2,\lambda_2)\cdots\varepsilon_{\mu_n}(k_n,\lambda_n) . \tag{E.248}$$

Then, gauge invariance implies the following Ward identity:

$$(k_i)_{\mu_i}\mathcal{M}^{\mu_1\mu_2\cdots\mu_n} = 0 , \qquad \text{for } i = 1,2,\ldots,n. \tag{E.249}$$

After squaring the amplitude and summing over the gauge boson polarizations using either eq. (E.243) or eq. (E.247), it follows that the terms in $\mathbb{P}_{\mu\nu}$ that are proportional to k_μ and k_ν will not contribute, due to the Ward identity, eq. (E.249). That is, when summing over the abelian gauge boson polarizations one obtains the correct result by employing

$$\mathbb{P}_{\mu\nu} \equiv \sum_{\lambda=\pm 1} \varepsilon_\mu(k,\lambda)\varepsilon_\nu(k,\lambda)^* \longrightarrow -g_{\mu\nu} . \tag{E.250}$$

In contrast, consider the scattering (or decay) amplitude in a nonabelian gauge theory (e.g., QCD) for a process with n external massless gauge bosons, which again is of the form given in eq. (E.248). In this case, the corresponding Ward identity imposed by gauge invariance is

$$(k_j)_{\mu_j}\mathcal{M}^{\mu_1\cdots\mu_j\cdots\mu_n}\prod_{\substack{m=1\\ m\neq j}}^{n}\varepsilon_{\mu_m}(k_m,\lambda_m) = 0 , \tag{E.251}$$

for any choice of $j = 1,2,\ldots,n$, which is a weaker constraint than eq. (E.249). That is, given the scattering amplitude of the form given by eq. (E.248), replacing one polarization vector by its four-momentum yields zero. Equation (E.251) is a consequence of the property that $k\cdot\varepsilon(k,\lambda) = 0$, which implies that employing either $\varepsilon_\mu(k,\lambda)$ or $\varepsilon_\mu(k,\lambda) + Ck_\mu$ (for any constant C) as the gauge boson polarization vector must yield the same result for any physical process.

Indeed, eq. (E.249) is not generally satisfied in nonabelian gauge theories for scattering and decay processes with two or more external nonabelian gauge bosons. Hence, one can employ eq. (E.250) for only one of the gauge boson polarization sums. For the polarization sums of the other gauge bosons, one must use either eq. (E.243) or eq. (E.247). Nevertheless, after all polarization sums have been performed, the gauge invariance and Lorentz invariance of the theory will ensure that the final result is independent of the four-vector n^μ.

To give an explicit example, consider a scattering process that involves two incoming nonabelian gauge bosons with four-momenta k_1 and k_2, respectively. The scattering amplitude is of the form

$$\mathcal{M} = \mathcal{M}^{\mu\nu}\varepsilon_\mu(k_1, \lambda_1)\varepsilon_\nu(k_2, \lambda_2)\,. \tag{E.252}$$

Then, eq. (E.251) implies that

$$k_{1\mu}\mathcal{M}^{\mu\nu} = A_1 k_2^\nu\,, \qquad k_{2\nu}\mathcal{M}^{\mu\nu} = A_2 k_1^\mu\,, \tag{E.253}$$

for some functions A_1 and A_2 of the kinematic variables. In particular, if A_1 and/or A_2 are nonzero, then eq. (E.249) is not satisfied. In contrast, eq. (E.251) is satisfied for $j = 1, 2$ after employing eq. (E.207).

In practice, it is often possible to manipulate the form of the amplitude for a process with n external massless nonabelian gauge bosons such that eq. (E.249) is satisfied. Using the example above, one can rewrite eq. (E.252) as

$$\mathcal{M} = \widetilde{\mathcal{M}}^{\mu\nu}\varepsilon_\mu(k_1, \lambda_1)\varepsilon_\nu(k_2, \lambda_2)\,, \tag{E.254}$$

where

$$\widetilde{\mathcal{M}}^{\mu\nu} = \mathcal{M}^{\mu\nu} - B^\nu k_1^\mu - C^\mu k_2^\nu\,, \tag{E.255}$$

due to eq. (E.207) which ensures that the extra terms appearing in eq. (E.255) cancel when multiplied by $\varepsilon_\mu(k_1, \lambda_1)\varepsilon_\nu(k_2, \lambda_2)$. In light of eq. (E.253) [and using $k_1^2 = k_2^2 = 0$], it follows that

$$k_{1\mu}\widetilde{\mathcal{M}}^{\mu\nu} = k_{2\nu}\widetilde{\mathcal{M}}^{\mu\nu} = 0\,, \tag{E.256}$$

if B^ν and C^μ are chosen such that $A_1 = k_1 \cdot C$ and $A_2 = k_2 \cdot B$. Consequently, by employing eq. (E.254) in the evaluation of $\sum_{\lambda_1,\lambda_2} |\mathcal{M}|^2$, one can safely apply eq. (E.250) to both gauge boson polarization sums.

Exercises

E.1 Define the spinor product

$$T(\hat{\boldsymbol{p}}_a\,, \hat{\boldsymbol{p}}_b)_{\tau_a,\tau_b} \equiv \chi^\dagger_{\tau_a/2}(\hat{\boldsymbol{p}}_a)\chi_{\tau_b/2}(\hat{\boldsymbol{p}}_b)\,, \tag{E.257}$$

where τ_a, $\tau_b = \pm 1$. Explicit forms for χ_λ can be found in eq. (E.15).

(a) With the help of eq. (E.12), obtain the following explicit expression in the convention where $\gamma = 0$:

$$\begin{aligned} T(\hat{\boldsymbol{a}}, \hat{\boldsymbol{b}})_{\tau_a\,\tau_b} = \chi^\dagger_{\tau_a/2}(\hat{\boldsymbol{z}}) \exp(i\theta_a\sigma^2/2) \exp(i\phi_a\sigma^3/2) \exp(-i\phi_b\sigma^3/2) \\ \times \exp(-i\theta_b\sigma^2/2)\, \chi_{\tau_b/2}(\hat{\boldsymbol{z}})\,. \end{aligned} \tag{E.258}$$

(b) Verify the following two properties:

$$T(\hat{\boldsymbol{a}},\hat{\boldsymbol{b}})_{\tau_a\,\tau_b} = T(\hat{\boldsymbol{b}},\hat{\boldsymbol{a}})^*_{\tau_b\,\tau_a}\,, \qquad T(\hat{\boldsymbol{a}},\hat{\boldsymbol{b}})_{-\tau_a,-\tau_b} = \tau_a\,\tau_b\,T(\hat{\boldsymbol{a}},\hat{\boldsymbol{b}})^*_{\tau_a\,\tau_b}\,. \tag{E.259}$$

(c) In light of the results of part (b), it is sufficient to consider explicit forms for only two of the spinor products. Use the result of part (a) to obtain

$$T(\hat{\boldsymbol{p}}_{\boldsymbol{a}},\hat{\boldsymbol{p}}_{\boldsymbol{b}})_{++} = e^{i(\phi_a-\phi_b)/2}\cos\frac{\theta_a}{2}\cos\frac{\theta_b}{2} + e^{-i(\phi_a-\phi_b)/2}\sin\frac{\theta_a}{2}\sin\frac{\theta_b}{2}\,, \tag{E.260}$$

$$T(\hat{\boldsymbol{p}}_{\boldsymbol{a}},\hat{\boldsymbol{p}}_{\boldsymbol{b}})_{-+} = e^{-i(\phi_a-\phi_b)/2}\cos\frac{\theta_a}{2}\sin\frac{\theta_b}{2} - e^{i(\phi_a-\phi_b)/2}\sin\frac{\theta_a}{2}\cos\frac{\theta_b}{2}\,, \tag{E.261}$$

where (θ_a,ϕ_a) and (θ_b,ϕ_b) are the polar and azimuthal angles of $\hat{\boldsymbol{p}}_{\boldsymbol{a}}$ and $\hat{\boldsymbol{p}}_{\boldsymbol{b}}$, respectively.[6]

E.2 Consider the two quantities, $S^\mu_- \equiv \frac{1}{2}S^{a\mu}\tau^a_{\frac12,-\frac12}$ and $S^\mu_+ \equiv \frac{1}{2}S^{a\mu}\tau^a_{-\frac12,\frac12}$ introduced in eq. (E.113). Using eqs. (E.114)–(E.117), obtain the following explicit expressions in the convention where $\gamma = 0$:

$$\sigma\!\cdot\! S_- = \begin{pmatrix} \frac12\sin\theta & e^{-i\phi}\sin^2\dfrac{\theta}{2} \\ -e^{i\phi}\cos^2\dfrac{\theta}{2} & -\frac12\sin\theta \end{pmatrix}, \qquad \overline{\sigma}\!\cdot\! S_+ = \begin{pmatrix} \frac12\sin\theta & -e^{-i\phi}\cos^2\dfrac{\theta}{2} \\ e^{i\phi}\sin^2\dfrac{\theta}{2} & -\frac12\sin\theta \end{pmatrix}. \tag{E.262}$$

E.3 Derive the following two identities:

$$\begin{aligned} x^\dagger(\vec{p},\lambda)\overline{\sigma}^\mu y(-\vec{p},\lambda') = 2i\lambda'\big[&mg^{\mu0}\delta_{\lambda\lambda'} + p^\mu\mathscr{S}^0_{\lambda\lambda'} - p^0\mathscr{S}^\mu_{\lambda\lambda'} \\ &- 2m(\mathscr{S}^\mu\mathscr{S}^0 - \mathscr{S}^0\mathscr{S}^\mu)_{\lambda\lambda'}\big]\,, \end{aligned} \tag{E.263}$$

$$\begin{aligned} y(\vec{p},\lambda)\sigma^\mu x^\dagger(-\vec{p},\lambda') = 2i\lambda'\big[&-mg^{\mu0}\delta_{\lambda\lambda'} + p^\mu\mathscr{S}^0_{\lambda\lambda'} - p^0\mathscr{S}^\mu_{\lambda\lambda'} \\ &+ 2m(\mathscr{S}^\mu\mathscr{S}^0 - \mathscr{S}^0\mathscr{S}^\mu)_{\lambda\lambda'}\big]\,, \end{aligned} \tag{E.264}$$

where the covariant spin operators $\mathscr{S}^\mu$ are defined in eq. (E.110).

Note that eqs. (E.263) and (E.264) provide explicit forms for the $Z^0 \to f\bar{f}$ decay helicity amplitudes defined in eqs. (17.2) and (17.3).

E.4 Derive the following two identities, which are the four-component spinor analogues of eqs. (E.263) and (E.264):

$$\begin{aligned} \bar{u}(\vec{p},\lambda)\,\tfrac12\gamma^\mu(1-\gamma_5)\,v(-\vec{p},\lambda') = 2i\lambda'\big[&mg^{\mu0}\delta_{\lambda\lambda'} + p^\mu\mathscr{S}^0_{\lambda\lambda'} - p^0\mathscr{S}^\mu_{\lambda\lambda'} \\ &+ i\epsilon^{0\mu\nu\rho}(\mathscr{S}_{\lambda\lambda'})_\nu p_\rho\big]\,, \end{aligned} \tag{E.265}$$

$$\begin{aligned} \bar{u}(\vec{p},\lambda)\,\tfrac12\gamma^\mu(1+\gamma_5)\,v(-\vec{p},\lambda') = 2i\lambda'\big[&-mg^{\mu0}\delta_{\lambda\lambda'} + p^\mu\mathscr{S}^0_{\lambda\lambda'} - p^0\mathscr{S}^\mu_{\lambda\lambda'} \\ &- i\epsilon^{0\mu\nu\rho}(\mathscr{S}_{\lambda\lambda'})_\nu p_\rho\big]\,. \end{aligned} \tag{E.266}$$

[6] In the case where $\hat{\boldsymbol{p}}_{\boldsymbol{a}}$ and/or $\hat{\boldsymbol{p}}_{\boldsymbol{b}}$ are parallel to the negative z-axis, one should employ the convention of eq. (E.11) and choose the corresponding azimuthal angle equal to π.

E.5 (a) Using eqs. (E.79)–(E.82), derive the following four identities:

$$x^{\dagger}_{\dot\alpha}(\vec{p},\lambda)\,\Gamma^{\dot\alpha\beta}\,y_\beta(\vec{p}\,',-\lambda') = -2\lambda'\omega_{-\lambda}(\vec{p})\omega_{-\lambda'}(\vec{p}\,')\chi^\dagger\lambda(\hat{p})\,\Gamma\chi_{\lambda'}(\hat{p}\,')\,, \quad \text{(E.267)}$$

$$y^\alpha(\vec{p},\lambda)\,\Gamma_{\alpha\dot\beta}\,x^{\dagger\dot\beta}(\vec{p}\,',-\lambda') = 2\lambda'\omega_{\lambda}(\vec{p})\omega_{\lambda'}(\vec{p}\,')\chi^\dagger\lambda(\hat{p})\,\Gamma\chi_{\lambda'}(\hat{p}\,')\,, \quad \text{(E.268)}$$

$$y^\alpha(\vec{p},\lambda)\,\Gamma_{\alpha}{}^{\beta}\,y_\beta(\vec{p}\,',-\lambda') = -2\lambda'\omega_{\lambda}(\vec{p})\omega_{-\lambda'}(\vec{p}\,')\chi^\dagger\lambda(\hat{p})\,\Gamma\chi_{\lambda'}(\hat{p}\,')\,, \quad \text{(E.269)}$$

$$x^{\dagger}_{\dot\alpha}(\vec{p},\lambda)\,\Gamma^{\dot\alpha}{}_{\dot\beta}\,x^{\dagger\dot\beta}(\vec{p}\,',-\lambda') = 2\lambda'\omega_{-\lambda}(\vec{p})\omega_{\lambda'}(\vec{p}\,')\chi^\dagger\lambda(\hat{p})\,\Gamma\chi_{\lambda'}(\hat{p}\,')\,, \quad \text{(E.270)}$$

where Γ is a product of alternating σ and $\overline{\sigma}$ matrices. The spinor index structure determines the identity of the first and last matrix (e.g., $\Gamma^{\dot\alpha}{}_{\dot\beta}$ indicates a string of matrices that begins with a $\overline{\sigma}$ and ends with a σ, etc.).

(b) Check that the results of part (a) are consistent with eqs. (E.152)–(E.155).

E.6 (a) Evaluate explicitly the rotation matrix $\mathcal{R}(\phi\,,\,\theta\,,\,-\phi)$ [see eq. (E.8)].

(b) How would the results in Appendix E be modified if one employed the convention where $\gamma = -\phi$ in eq. (E.15)?

E.7 (a) Consider a spin-1/2 fermion with four-momentum p, helicity λ, and mass m. Derive the identity

$$\bar{v}(p,\lambda)\gamma^\mu u(p,\lambda) = -4\lambda m S^{a\mu}\tau^a_{-\lambda,\lambda}\,. \quad \text{(E.271)}$$

Show that eq. (E.271) is equivalent to

$$\bar{v}(p,\pm\tfrac{1}{2})\gamma^\mu u(p,\pm\tfrac{1}{2}) = \mp m(\hat{\boldsymbol{p}}^{\boldsymbol{1}} \pm i\hat{\boldsymbol{p}}^{\boldsymbol{2}})\,, \quad \text{(E.272)}$$

where eq. (E.272) should be read as two separate equations corresponding to the upper and lower set of signs, respectively. In eq. (E.272), the $\hat{\boldsymbol{p}}^{\boldsymbol{a}}$ (for $a = 1,2,3$) are an orthonormal triad of unit three-vectors with $\hat{\boldsymbol{p}}^{\boldsymbol{3}} = \hat{\boldsymbol{p}}$. Explicit forms for the $\hat{\boldsymbol{s}}^{\boldsymbol{a}} = \hat{\boldsymbol{p}}^{\boldsymbol{a}}$ are given in eq. (E.28).

(b) Employing the results of part (a) with $m \neq 0$, show that the polarization vector for the helicity ±1 states of a vector boson can be expressed in terms of four-component helicity spinor wave functions,

$$\varepsilon^\mu(\hat{\boldsymbol{k}},2\lambda) = \frac{1}{m\sqrt{2}}\bar{v}(\vec{k},\lambda)\gamma^\mu u(\vec{k},\lambda)\,, \quad \text{for } \lambda = \pm\tfrac{1}{2}. \quad \text{(E.273)}$$

Comparing with the explicit form of $\varepsilon^\mu(\hat{\boldsymbol{k}},\pm1)$ given in eq. (E.212), check the validity of eq. (E.273) by using eq. (E.272) or the explicit forms for the four-component helicity spinor wave functions in eq. (E.174).

(c) Using eqs. (E.180) and (E.181) and the results of part (b), provide an independent derivation of eq. (E.220) for $\lambda = \pm1$. Similarly, using eqs. (E.185) and (E.186), provide an independent derivation of eq. (E.221) for $\lambda = \pm 1$. The case of $\lambda = 0$ can be addressed directly by employing eq. (E.214).

E.8 Here is a slick way of computing the polarization sum $\mathbb{P}_{\mu\nu}$ for a spin-1 boson.

(a) In light of eq. (E.215), show that

$$\mathbb{P}_{\mu\nu} = \mathbb{P}^*_{\mu\nu} = \mathbb{P}_{\nu\mu} \,. \tag{E.274}$$

(b) For a massive spin-1 boson, the expression for $\mathbb{P}_{\mu\nu}$ is a Lorentz-covariant real second-rank symmetric tensor. Since $g^{\mu\nu}$ and $k^\mu k^\nu$ are the only second-rank tensors that can be constructed using the four-vector k, it follows that

$$\mathbb{P}_{\mu\nu} \equiv \sum_{\lambda=-1,0,+1} \varepsilon_\mu(k,\lambda)\varepsilon_\nu(k,\lambda)^* = Ag_{\mu\nu} + Bk_\mu k_\nu \,, \tag{E.275}$$

where A and B are real Lorentz-invariant functions of k^2. For a spin-1 particle of four-momentum k and mass m, A and B are functions of $k^2 = m^2$. By multiplying $\mathbb{P}_{\mu\nu}$ respectively by $g^{\mu\nu}$ and by $k^\mu k^\nu$, derive two equations for A and B. By solving these equations you should be able to reproduce eq. (E.234).

(c) For a massless spin-1 boson, the method employed in part (b) must be modified, since the expression for $\mathbb{P}_{\mu\nu}$ is not a Lorentz-covariant tensor. Generalize the technique used in part (b) by noting that the sum over transverse polarizations can depend on the fixed four-vector $n = (1\,;\,\vec{\mathbf{0}})$ as well as the other relevant Lorentz tensors, $g_{\mu\nu}$, k^μ, and $\epsilon_{\mu\nu\alpha\beta}$. Using eq. (E.274), show that one can eliminate $\epsilon_{\mu\nu\alpha\beta}$ from consideration and obtain the following general form for $\mathbb{P}_{\mu\nu}$:

$$\mathbb{P}_{\mu\nu} \equiv \sum_{\lambda=-1,+1} \varepsilon_\mu(k,\lambda)\varepsilon_\nu(k,\lambda)^* = Ag_{\mu\nu} + Bk_\mu k_\nu + Cn_\mu n_\nu + D(n_\mu k_\nu + n_\nu k_\mu), \tag{E.276}$$

where the coefficients A, B, C, and D are real functions of $k \cdot n$. You should then be able to deduce four equations for the four unknowns A, B, C, and D. Solving these equations, verify the result exhibited in eq. (E.243).

(d) Derive the following identity for the transverse polarization vectors of a massless spin-1 boson (for λ, $\lambda' = \pm 1$):

$$\epsilon^{\alpha\beta\mu\nu}\varepsilon_\mu(\vec{\boldsymbol{k}},\lambda)\varepsilon_\nu(\vec{\boldsymbol{k}},\lambda')^* = -i\lambda\left(\frac{n^\alpha k^\beta - n^\beta k^\alpha}{n\cdot k}\right)\delta_{\lambda\lambda'} \,. \tag{E.277}$$

Appendix F The Spinor Helicity Method

In many practical calculations (e.g., QED or QCD at high energies), the masses of the fermions can be neglected. In such cases, the computation of multiparticle helicity amplitudes simplifies considerably. The spinor helicity method is a powerful technique for computing helicity amplitudes for multiparticle processes involving massless spin-1/2 and spin-1 particles.

The spinor helicity method also plays a central role in the modern scattering amplitudes program. In a scattering process, the traditional Feynman diagram techniques become more and more impractical as the number of final-state particles increases and/or as the number of loops of perturbation theory increases. Nevertheless, the end result of such computations can in some cases produce a remarkably simple result. As an example, we quote the famous Parke–Taylor formula for the n-point tree color-ordered amplitude for gluon–gluon scattering [216]:[1]

$$A_n[1^+, 2^+, \ldots, i^-, \ldots, j^-, \ldots, n^+] = \frac{\langle ij \rangle^4}{\langle 12 \rangle \langle 23 \rangle \cdots \langle n1 \rangle}\,, \tag{F.1}$$

where the angle brackets will be defined in the next section. The + and − superscripts refer to helicity ± 1 gluons in a convention where all gluons are considered outgoing (with the understanding that an outgoing helicity $\lambda = \pm 1$ gluon corresponds by crossing symmetry to an incoming helicity $\lambda = \mp 1$ gluon). Thus, one can use eq. (F.1) to evaluate the amplitude for the scattering of two negative helicity gluons (1 and 2) into two negative helicity gluons (i and j) plus $n - 4$ positive helicity gluons. One can show that if either i or j (or both) were assigned positive helicities, the resulting amplitude would vanish. Thus, eq. (F.1) corresponds to the maximally helicity-violating (MHV) nonvanishing n-gluon amplitude.

The Parke–Taylor formula was initially stated as a conjecture based on considerations based on supersymmetric gauge theories. Indeed, many of the insights of the modern amplitudes program have been drawn from computations in supersymmetry. A textbook treatment of the spinor helicity formalism can be found in [B149], which goes far beyond the introductory material presented in this appendix.

Generalizations of the spinor helicity methods that incorporate massive spin-1/2 and spin-1 particles also exist, although they tend to be cumbersome. In this appendix, we restrict our presentation to the massless case.

[1] The full tree-level amplitude for gluon–gluon scattering is given by a product of A_n with the appropriate color factors and powers of the strong coupling constant, and then a sum over all possible permutations.

F.1 Massless Spinors: the Bracket Notation

The spinor helicity technique employs a very useful bra and ket notation for massless four-component spinors [217]:

$$|p\pm\rangle \equiv u(p, \pm\tfrac{1}{2}) = v(p, \mp\tfrac{1}{2}), \tag{F.2}$$
$$\langle p\pm| \equiv \bar{u}(p, \pm\tfrac{1}{2}) = \bar{v}(p, \mp\tfrac{1}{2}). \tag{F.3}$$

The $\pm$ notation specified by the bra and ket indicates the chirality (i.e., the eigenvalue of γ_5) of the corresponding four-component spinor [see eq. (3.178)]. One can then define two Lorentz-invariant massless spinor products,[2]

$$\langle p\, q\rangle \equiv \langle p-|q+\rangle = \bar{u}(p, -\tfrac{1}{2})\, u(q, +\tfrac{1}{2}), \tag{F.4}$$
$$[p\, q] \equiv \langle p+|q-\rangle = \bar{u}(p, +\tfrac{1}{2})\, u(q, -\tfrac{1}{2}). \tag{F.5}$$

It is noteworthy that $\not{p} \equiv \gamma^\mu p_\mu$ can be expressed in terms of the bras and kets defined in eqs. (F.2) and (F.3) by employing the following identity:

$$\not{p} = |p+\rangle\, \langle p+| \;+\; |p-\rangle\, \langle p-|\,. \tag{F.6}$$

The two-component spinor formalism is especially economical in the case of massless spin-1/2 fermions. Hence, it is instructive to reformulate eqs. (F.2)–(F.5) using two-component spinor notation. First, we consider the explicit forms for the two-component helicity spinor wave functions [given by eqs. (2.42)–(2.45)] in the massless limit:

$$x_\alpha(\vec{\boldsymbol{p}}, -\tfrac{1}{2}) = y_\alpha(\vec{\boldsymbol{p}}, \tfrac{1}{2}) = (2E)^{1/2}\, \chi_{-1/2}(\hat{\boldsymbol{p}}), \tag{F.7}$$
$$x^\alpha(\vec{\boldsymbol{p}}, -\tfrac{1}{2}) = y^\alpha(\vec{\boldsymbol{p}}, \tfrac{1}{2}) = (2E)^{1/2}\, \chi^\dagger_{1/2}(\hat{\boldsymbol{p}}), \tag{F.8}$$
$$x^{\dagger\dot{\alpha}}(\vec{\boldsymbol{p}}, -\tfrac{1}{2}) = y^{\dagger\dot{\alpha}}(\vec{\boldsymbol{p}}, \tfrac{1}{2}) = (2E)^{1/2}\, \chi_{1/2}(\hat{\boldsymbol{p}}), \tag{F.9}$$
$$x^\dagger_{\dot{\alpha}}(\vec{\boldsymbol{p}}, -\tfrac{1}{2}) = y^\dagger_{\dot{\alpha}}(\vec{\boldsymbol{p}}, \tfrac{1}{2}) = (2E)^{1/2}\, \chi^\dagger_{-1/2}(\hat{\boldsymbol{p}}), \tag{F.10}$$

where $E = |\vec{\boldsymbol{p}}|$. For all other choices of helicities, the corresponding helicity spinor wave functions vanish. Hence, we can redefine the bras and kets of eqs. (F.2) and (F.3) as two-component spinor objects,

$$|p+\rangle = y^{\dagger\dot{\alpha}}(\vec{\boldsymbol{p}}, \tfrac{1}{2}) = x^{\dagger\dot{\alpha}}(\vec{\boldsymbol{p}}, -\tfrac{1}{2}), \qquad \langle p+| = y^{\alpha}(\vec{\boldsymbol{p}}, \tfrac{1}{2}) = x^{\alpha}(\vec{\boldsymbol{p}}, -\tfrac{1}{2}), \tag{F.11}$$
$$|p-\rangle = x_{\alpha}(\vec{\boldsymbol{p}}, -\tfrac{1}{2}) = y_{\alpha}(\vec{\boldsymbol{p}}, \tfrac{1}{2}), \qquad \langle p-| = x^\dagger_{\dot{\alpha}}(\vec{\boldsymbol{p}}, -\tfrac{1}{2}) = y^\dagger_{\dot{\alpha}}(\vec{\boldsymbol{p}}, \tfrac{1}{2}). \tag{F.12}$$

The association of undotted and dotted indices in eqs. (F.11) and (F.12) is a consequence of our convention for the Dirac gamma matrices [see eq. (3.7)]. Note that in this convention, the left-handed [right-handed] projection operator P_L [P_R] projects

[2] Note that for massless spinors, $P_R u(p,\lambda) = (\tfrac{1}{2}+\lambda)u(p,\lambda)$ and $P_L u(p,\lambda) = (\tfrac{1}{2}-\lambda)u(p,\lambda)$, where $P_{R,L} \equiv \tfrac{1}{2}(1 \pm \gamma_5)$. It then follows that $\langle p-|q-\rangle = \langle p+|q+\rangle = 0$ due to $P_L P_R = P_R P_L = 0$.

out the lowered undotted [raised dotted] index components of the four-component spinor [see eq. (3.5)].[3]

It is convenient to introduce an alternative notation for the sigma matrices:

$$\sigma_+^\mu \equiv \sigma^\mu\,, \qquad \sigma_-^\mu \equiv \overline{\sigma}^\mu\,. \tag{F.13}$$

Then, the $|p\pm\rangle$ and $\langle p\pm|$ defined in eqs. (F.11) and (F.12) satisfy the massless Dirac equation [see eqs. (2.10)–(2.13)]

$$p\cdot\sigma_\pm\,|p\pm\rangle = 0, \qquad \langle p\pm|\,p\cdot\sigma_\pm = 0\,. \tag{F.14}$$

Lorentz-invariant spinor products can be formed from the bras and kets in the usual way and satisfy

$$\langle p\pm|q\mp\rangle^* = \langle q\mp|p\pm\rangle\,, \tag{F.15}$$

$$\langle p\pm|\sigma_\pm^\mu|q\pm\rangle^* = \langle q\pm|\sigma_\pm^\mu|p\pm\rangle\,, \tag{F.16}$$

where we have used the fact that the $\sigma_\pm^\mu$ are hermitian. Moreover, the following properties are noteworthy:[4]

$$|p\pm\rangle\,\langle p\pm| = p\cdot\sigma_\mp\,, \tag{F.17}$$

$$\langle p\pm|\,\sigma_\pm^\mu\,|p\pm\rangle = 2p^\mu\,, \tag{F.18}$$

$$\langle p\pm|q\mp\rangle = -\,\langle q\pm|p\mp\rangle\,, \tag{F.19}$$

$$\langle p+|\,\sigma_+^\mu\,|q+\rangle = \langle q-|\,\sigma_-^\mu\,|p-\rangle\,, \tag{F.20}$$

$$\langle p\pm|\,\sigma_\pm^\mu\sigma_\mp^\nu\,|q\mp\rangle = -\,\langle q\pm|\,\sigma_\pm^\nu\sigma_\mp^\mu\,|p\mp\rangle\,. \tag{F.21}$$

Each of eqs. (F.14)–(F.21) should be read as two separate equations corresponding to the upper and lower set of signs, respectively. Moreover, eq. (F.17) is simply a rewriting of eq. (F.6) using two-component spinor notation.

The proofs of the properties listed above are straightforward. Equations (F.17) and (F.18) follow from eqs. (2.59) and (2.60). For example,

$$\langle p+|\,\sigma_+^\mu\,|p+\rangle = y^\alpha(\vec{\boldsymbol p},\tfrac12)\sigma^\mu_{\alpha\dot\beta}\,y^{\dagger\dot\beta}(\vec{\boldsymbol p},\tfrac12) = \sigma^\mu_{\alpha\dot\beta}\,y^{\dagger\dot\beta}(\vec{\boldsymbol p},\tfrac12)y^\alpha(\vec{\boldsymbol p},\tfrac12) = \mathrm{Tr}\,(\sigma^\mu p\cdot\overline{\sigma}) = 2p^\mu, \tag{F.22}$$

and similarly for $\langle p-|\,\sigma_-^\mu\,|p-\rangle$. Note that eqs. (F.19)–(F.21) are consequences of eqs. (1.96)–(1.100). Equations (F.20) and (F.21) generalize easily to the case of a product of an even and odd number of $\sigma/\overline{\sigma}$ matrices. For any positive integer n,

$$\langle p+|\,\sigma_+^{\mu_1}\sigma_-^{\mu_2}\cdots\sigma_+^{\mu_{2n-1}}\,|q+\rangle = \langle q-|\,\sigma_-^{\mu_{2n-1}}\cdots\sigma_+^{\mu_2}\sigma_-^{\mu_1}\,|p-\rangle\,, \tag{F.23}$$

$$\langle p\pm|\,\sigma_\pm^{\mu_1}\sigma_\mp^{\mu_2}\cdots\sigma_\mp^{\mu_{2n}}\,|q\mp\rangle = -\,\langle q\pm|\,\sigma_\pm^{\mu_{2n}}\cdots\sigma_\pm^{\mu_2}\sigma_\mp^{\mu_1}\,|p\mp\rangle\,. \tag{F.24}$$

3 In the literature on the spinor helicity method, one often finds $|p+\rangle$ associated with a lowered undotted index and $|p-\rangle$ associated with an upper dotted index. This is due to a different convention for the sigma matrices, such as the WB definition given in eq. (A.27). Numerically, this is equivalent to a convention for the Dirac gamma matrices in which σ^μ and $\overline{\sigma}^\mu$ are interchanged in eq. (3.7), resulting in an overall change of sign in the matrix representation of γ_5. As a result, in this latter convention (not adopted in this book) the lowered undotted [raised dotted] index components are associated with positive [negative] chirality.

4 One can check that eqs. (F.14)–(F.21) have the correct two-component spinor structure if the suppressed undotted and dotted spinor indices are restored.

The covariance with respect to the undotted and dotted spinors allows only two possible Lorentz-invariant spinor products,

$$\langle p\, q\rangle \equiv \langle p-|q+\rangle = x^\dagger(\vec{p}, -\tfrac{1}{2})\, y^\dagger(\vec{q}, \tfrac{1}{2})\,, \tag{F.25}$$

$$[p\, q] \equiv \langle p+|q-\rangle = y(\vec{p}, \tfrac{1}{2})\, x(\vec{q}, -\tfrac{1}{2})\,. \tag{F.26}$$

In particular, $\langle p\, q\rangle$ is a sum over dotted indices and $[p\, q]$ is a sum over undotted indices.[5] In terms of the spinor products $T(\hat{\boldsymbol{p}}\,,\,\hat{\boldsymbol{q}})_{\tau,\tau'} \equiv \chi^\dagger_{\tau/2}(\hat{\boldsymbol{p}})\chi_{\tau'/2}(\hat{\boldsymbol{q}})$ defined in eq. (E.257),

$$\langle p\, q\rangle = (2E_p)^{1/2}(2E_q)^{1/2}T(\hat{\boldsymbol{p}}\,,\,\hat{\boldsymbol{q}})_{-+}\,, \tag{F.27}$$

$$[p\, q] = (2E_p)^{1/2}(2E_q)^{1/2}T(\hat{\boldsymbol{p}}\,,\,\hat{\boldsymbol{q}})_{+-}\,, \tag{F.28}$$

where $E_p = |\vec{p}|$ and $E_q = |\vec{q}|$. Using eq. (E.261), one can obtain explicit forms for $T(\hat{\boldsymbol{p}}\,,\,\hat{\boldsymbol{q}})_{-+}$ and $T(\hat{\boldsymbol{p}}\,,\,\hat{\boldsymbol{q}})_{+-} = -T(\hat{\boldsymbol{p}}\,,\,\hat{\boldsymbol{q}})^*_{-+}$.

Using eqs. (F.19) and (F.15), the spinor products satisfy the following relations:

$$\langle p\, q\rangle = -\langle q\, p\rangle\,, \tag{F.29}$$

$$[p\, q] = -[q\, p]\,, \tag{F.30}$$

$$\langle p\, q\rangle^* = -[p\, q]\,. \tag{F.31}$$

One immediate consequence of eqs. (F.29) and (F.30) is

$$\langle p\, p\rangle = \langle p-|p+\rangle = 0\,, \qquad\qquad [p\, p] = \langle p+|p-\rangle = 0\,. \tag{F.32}$$

We next compute the absolute square of the spinor product:

$$\begin{aligned} |\langle p\, q\rangle|^2 &= x^\dagger_{\dot\alpha}(\vec{p}, -\tfrac{1}{2})\, y^{\dagger\dot\alpha}(\vec{q}, \tfrac{1}{2})\, x_\alpha(\vec{p}, -\tfrac{1}{2})\, y^\alpha(\vec{q}, \tfrac{1}{2}) \\ &= x_\alpha(\vec{p}, -\tfrac{1}{2})\, x^\dagger_{\dot\alpha}(\vec{p}, -\tfrac{1}{2})\, y^{\dagger\dot\alpha}(\vec{q}, \tfrac{1}{2})\, y^\alpha(\vec{q}, \tfrac{1}{2}) \\ &= p\cdot\sigma_{\alpha\dot\alpha}\, q\cdot\overline{\sigma}^{\dot\alpha\alpha} = p_\mu q_\nu\, \mathrm{Tr}(\sigma^\mu\overline{\sigma}^\nu) = 2p\cdot q\,. \end{aligned} \tag{F.33}$$

Using this result and eq. (F.31) yields

$$\left|\langle p\, q\rangle\right|^2 = \left|[p\, q]\right|^2 = 2p\cdot q\,, \tag{F.34}$$

which indicates that the spinor products are roughly the square roots of the corresponding dot products. One other noteworthy relation is

$$\begin{aligned} \langle p_1\, p_2\rangle\, [p_2\, p_3]\, \langle p_3\, p_4\rangle\, [p_4\, p_1] &= \mathrm{Tr}\,(\sigma\cdot p_1\,\overline{\sigma}\cdot p_2\,\sigma\cdot p_3\,\overline{\sigma}\cdot p_4) \\ &= 2(g_{\mu\nu}g_{\rho\kappa} - g_{\mu\rho}g_{\nu\kappa} + g_{\mu\kappa}g_{\nu\rho} + i\epsilon_{\mu\nu\rho\kappa})p_1^\mu p_2^\nu p_3^\rho p_4^\kappa\,, \end{aligned} \tag{F.35}$$

where the trace has been evaluated using eq. (B.19). The first line of eq. (F.35) immediately follows from eqs. (2.59) and (2.60) after plugging in the definition of the spinor products.

When applied to massless spinors, eq. (F.34) indicates that the square of the helicity amplitude of a multifermion scattering process can be expressed in terms

[5] This is to be contrasted with the alternative convention used in the two-component spinor formalism that is often found in the literature (see footnote 3), in which $\langle p\, q\rangle$ is written as a sum over undotted indices and $[p\, q]$ is written as a sum over dotted indices.

of products of dot products of pairs of fermion momenta. If more than one diagram contributes to a helicity amplitude, then it is often possible to combine the contributions after a rearrangement of momenta via the Fierz identities. Using eqs. (B.66)–(B.70), it follows that

$$\langle p_1\, p_2\rangle\,\langle p_3\, p_4\rangle = \langle p_1\, p_3\rangle\,\langle p_2\, p_4\rangle + \langle p_1\, p_4\rangle\,\langle p_3\, p_2\rangle\,, \tag{F.36}$$

$$[p_1\, p_2]\,[p_3\, p_4] = [p_1\, p_3]\,[p_2\, p_4] + [p_1\, p_4]\,[p_3\, p_2]\,, \tag{F.37}$$

$$\langle p_1 + |\sigma^\mu_+|p_2+\rangle\,\langle p_3 + |\sigma_{+\mu}|p_4+\rangle = 2\,[p_1\, p_3]\,\langle p_4\, p_2\rangle\,, \tag{F.38}$$

$$\langle p_1 - |\sigma^\mu_-|\, p_2-\rangle\,\langle p_3 - |\sigma_{-\mu}|p_4-\rangle = 2\,\langle p_1\, p_3\rangle\,[p_4\, p_2]\,, \tag{F.39}$$

$$\langle p_1 + |\sigma^\mu_+|p_2+\rangle\,\langle p_3 - |\sigma_{-\mu}|p_4-\rangle = 2\,[p_1\, p_4]\,\langle p_3\, p_2\rangle\,. \tag{F.40}$$

Equations (F.36) and (F.37) are often called the Schouten identities, as they follow from eq. (A.42).

F.2 Including Massless Vector Bosons

We now extend the spinor helicity formalism to multiparticle processes involving massless fermions and massless spin-1 bosons. This can be accomplished by expressing the massless spin-1 polarization vector in terms of products of massless spin-1/2 spinor wave functions, as suggested by eq. (A.88).

The properties of spin-1 polarization vectors in the helicity basis are given in Appendix E.8. For a massless spin-1 boson with four-momentum $k^\mu = E(1\,;\,\hat{\boldsymbol{k}})$, it is straightforward to verify that[6]

$$\varepsilon^\mu(\hat{\boldsymbol{k}}\,,\pm 1) = \frac{1}{\sqrt{2}}\,\frac{\langle k \mp |\sigma^\mu_\mp|\,\widetilde{k}\mp\rangle}{\langle k \pm |\widetilde{k}\mp\rangle} \tag{F.41}$$

precisely reproduces the result of eq. (E.212), where $\widetilde{k}^\mu \equiv E(1\,;\,-\hat{\boldsymbol{k}})$ and the massless spinor wave functions are given in eq. (E.15). Equation (F.41) is inconvenient to use since the four-vector $\widetilde{k}$ cannot be covariantly defined in terms of k [see eq. (E.245)]. It is therefore more convenient to employ a slightly different strategy by introducing a "reference" four-vector p (in practical computations, p is taken to be another four-momentum vector in the scattering process of interest), with the properties that $p^2 = 0$ and $p\cdot k \neq 0$. Then, one can replace eq. (F.41) with the following alternative expression for the spin-1 polarization vectors:

$$\varepsilon^\mu(k\,,p\,,\pm 1) = \frac{1}{\sqrt{2}}\,\frac{\langle k\mp|\,\sigma^\mu_\mp\,|p\mp\rangle}{\langle k \pm|p\mp\rangle}\,, \tag{F.42}$$

where the reference four-momentum p is included as an argument of ε^μ for the

[6] In the literature on the spinor helicity method (e.g., see Ref. [217]), the spin-1 polarization vector ε is often employed in Feynman diagram computations for an *outgoing* final-state vector boson. In contrast, we follow the more standard textbook convention where ε [ε^*] is used in the Feynman rule for an incoming [outgoing] vector boson, as indicated at the end of Section 2.4.

sake of clarity. One can verify that eq. (F.42) satisfies the conditions for a valid polarization four-vector $\varepsilon^\mu(\vec{\boldsymbol{k}},\lambda)$ given in eq. (E.207) in the phase convention of eq. (E.215). In addition, $\varepsilon^\mu(k\,,p\,,\pm 1)$ is orthogonal to the reference four-momentum p^μ in light of eq. (F.14):

$$p\cdot\varepsilon(k\,,p\,,\lambda)=0\,. \tag{F.43}$$

The significance of the reference four-vector p can be discerned from the property that if a different reference momentum is chosen, then ε^μ is shifted by a term proportional to k^μ. Explicitly, if $\varepsilon^\mu(k\,,\,p\,,\,\lambda)$ is a polarization vector with reference momentum p, then[7]

$$\varepsilon^\mu(k\,,\,q\,,\,\pm 1)=\varepsilon^\mu(k\,,\,p\,,\,\pm 1)+\frac{\sqrt{2}\,\langle q\pm|p\mp\rangle}{\langle k\pm|q\mp\rangle\,\langle k\pm|p\mp\rangle}\,k^\mu\,. \tag{F.44}$$

In particular, if we choose $q=\widetilde{k}$, we see that the difference of the two spin-1 polarization vectors given by eqs. (E.212) and (F.42) is proportional to k^μ. This shift of the reference momentum from p to q in eq. (F.42) does not affect eq. (E.207) since $k^2=0$ for a massless spin-1 particle. Moreover, this shift does not affect the final result for any observable (in particular, the sum of amplitudes of any gauge invariant set of Feynman diagrams remains unchanged). Thus, the presence of the arbitrary four-vector p just reflects the gauge invariance of a theory of massless spin-1 particles.

One can also verify that $\varepsilon^\mu(k\,,\,p\,,\,\lambda)$ defined in eq. (F.42) behaves as expected under rotations. Using eq. (E.12), massless spinors transform as

$$|k\pm\rangle\longrightarrow\mathcal{D}(\phi,\theta,\gamma)\,|k\pm\rangle\,,\qquad\qquad \langle k\pm|\longrightarrow\langle k\pm|\,[\mathcal{D}(\phi,\theta,\gamma)]^{-1}\,, \tag{F.45}$$

under a rotation specified by the Euler angles ϕ, θ, and γ. We shall rotate the spin-1 polarization vectors by rotating both $\vec{\boldsymbol{k}}$ and the reference momentum $\vec{\boldsymbol{p}}$ simultaneously (since one is always free to shift the reference vector with no physical consequences). Using eq. (E.33), it follows that

$$[\mathcal{D}(\phi,\theta,\gamma)]^{-1}\,\sigma^\mu_\pm\,\mathcal{D}(\phi,\theta,\gamma)=\Lambda^\mu{}_\nu\sigma^\nu_\pm\,, \tag{F.46}$$

where $\Lambda^\mu{}_\nu$ is specified by eq. (E.211). Indeed, if we simultaneously rotate both k and p via $k^\mu\to\Lambda^\mu{}_\nu k^\nu$ and $p^\mu\to\Lambda^\mu{}_\nu p^\nu$, then

$$\varepsilon^\mu(k\,,\,p\,,\,\lambda)\longrightarrow\Lambda^\mu{}_\nu\,\varepsilon^\nu(k\,,\,p\,,\,\lambda)\,, \tag{F.47}$$

as expected. Henceforth, we adopt the convention of Appendix E by setting $\gamma=0$. Performing a similar computation as the one above, one can check that, under $\vec{\boldsymbol{k}}\to-\vec{\boldsymbol{k}}$ and $\vec{\boldsymbol{p}}\to-\vec{\boldsymbol{p}}$, eq. (E.219) is satisfied.[8]

[7] To derive eq. (F.44), evaluate $\varepsilon^\mu(k\,,\,q\,,\,\lambda)-\varepsilon^\mu(k\,,\,p\,,\,\lambda)$, and simplify the resulting expression using eqs. (F.17), (F.20), and (F.21).

[8] Here, we have used eqs. (B.30) and (B.31) to write $\sigma^0_\pm\sigma^\mu_\mp\sigma^0_\pm=-\sigma^\mu_\pm+2g^{\mu 0}\sigma^0_\pm=g^{\mu\mu}\sigma^\mu_\pm$ (no sum over μ).

The following additional property of $\varepsilon^\mu(k\,,\,p\,,\,\lambda)$ defined in eq. (F.42) is noteworthy [see eq. (E.247)]:

$$\sum_{\lambda=\pm 1} \varepsilon_\mu(k\,,\,p\,,\,\lambda)\varepsilon_\nu(k\,,\,p\,,\,\lambda)^* = -g_{\mu\nu} + \frac{p_\mu k_\nu + p_\nu k_\mu}{p\cdot k}\,. \tag{F.48}$$

This result can be proven by using eqs. (F.15) and (F.16) and manipulating the resulting expression with the help of eqs. (F.17) and (F.23). The end result is

$$\sum_{\lambda=\pm 1} \varepsilon_\mu(k\,,\,p\,,\,\lambda)\varepsilon_\nu(k\,,\,p\,,\,\lambda)^* = \frac{\langle k+|\,(\sigma_\mu p\cdot\overline{\sigma}\sigma_\nu + \sigma_\nu p\cdot\overline{\sigma}\sigma_\mu)\,|k+\rangle}{2\,\langle k+|\,p\cdot\sigma\,|k+\rangle}\,. \tag{F.49}$$

Using eq. (B.31) to simplify the product of three $\sigma/\overline{\sigma}$ matrices, and employing eq. (F.18), then yields eq. (F.48).

Finally, using the Fierz identities given in eqs. (B.79)–(B.81), one derives from eq. (F.42) that

$$\sigma_\pm\cdot\varepsilon(k\,,\,\pm 1) = \frac{\sqrt{2}\,|p\mp\rangle\,\langle k\mp|}{\langle k\pm|p\mp\rangle}\,,\qquad \sigma_\pm\cdot\varepsilon(k\,,\,\pm 1)^* = \frac{\sqrt{2}\,|k\mp\rangle\,\langle p\mp|}{\langle p\mp|k\pm\rangle}\,, \tag{F.50}$$

$$\sigma_\mp\cdot\varepsilon(k\,,\,\pm 1) = \frac{\sqrt{2}\,|k\pm\rangle\,\langle p\pm|}{\langle k\pm|p\mp\rangle}\,,\qquad \sigma_\mp\cdot\varepsilon(k\,,\,\pm 1)^* = \frac{\sqrt{2}\,|p\pm\rangle\,\langle k\pm|}{\langle p\mp|k\pm\rangle}\,. \tag{F.51}$$

Note that each equation in eqs. (F.50) and (F.51) represents two separate expressions, corresponding to the upper and lower signs in each equation, respectively. In four-component spinor notation, where the bras and kets are defined as in eqs. (F.2) and (F.3), one can rewrite eqs. (F.50) and (F.51) as

$$\not{\varepsilon}(k,+1) = \frac{\sqrt{2}}{[k\,p]}\Big\{|p-\rangle\,\langle k-| \,+\, |k+\rangle\,\langle p+|\Big\}\,, \tag{F.52}$$

$$\not{\varepsilon}(k,-1) = \frac{\sqrt{2}}{\langle k\,p\rangle}\Big\{|k-\rangle\,\langle p-| \,+\, |p+\rangle\,\langle k+|\Big\}\,, \tag{F.53}$$

where $\not{\varepsilon} \equiv \gamma^\mu\varepsilon_\mu$. The corresponding expressions for $\not{\varepsilon}(k,\pm 1)^*$ can be similarly obtained and are consistent with eq. (E.215) in light of eqs. (F.29) and (F.30). The above results are reminiscent of the expression for $\not{p}$ in eq. (F.6).

F.3 Simple Application of the Spinor Helicity Method

The following simple example demonstrates the spinor helicity technique. Consider Compton scattering in QED, $e^-(\vec{p}_1,\lambda_1)\gamma(\vec{k}_1,\lambda_1') \to e^-(\vec{p}_2,\lambda_2)\gamma(\vec{k}_2,\lambda_2')$, in the limit of massless electrons. The amplitude for this process is given by eq. (2.166) with $m=0$ and $G_L = G_R = -e$. Writing out the "crossed" term explicitly, and noting that $s\equiv(p_1+k_1)^2 = 2p_1\!\cdot\! k_1$ and $u\equiv(p_1-k_2)^2 = -2p_1\!\cdot\! k_2$ for massless particles,

$$i\mathcal{M} = \frac{-ie^2}{2p_1\cdot k_1}\Big\{x^\dagger(\vec{p}_2,\lambda_2)\,\overline{\sigma}\cdot\varepsilon_2^*\,\sigma\cdot(p_1+k_1)\,\overline{\sigma}\cdot\varepsilon_1\,x(\vec{p}_1,\lambda_1)$$
$$+\,y(\vec{p}_2,\lambda_2)\,\sigma\cdot\varepsilon_2^*\,\overline{\sigma}\cdot(p_1+k_1)\,\sigma\cdot\varepsilon_1 y^\dagger(\vec{p}_1,\lambda_1)\Big\}$$
$$+\frac{ie^2}{2p_1\cdot k_2}\Big\{x^\dagger(\vec{p}_2,\lambda_2)\,\overline{\sigma}\cdot\varepsilon_1\,\sigma\cdot(p_1-k_2)\,\overline{\sigma}\cdot\varepsilon_2^*\,x(\vec{p}_1,\lambda_1)$$
$$+\,y(\vec{p}_2,\lambda_2)\,\sigma\cdot\varepsilon_1\,\overline{\sigma}\cdot(p_1-k_2)\,\sigma\cdot\varepsilon_2^* y^\dagger(\vec{p}_1,\lambda_1)\Big\}. \qquad \text{(F.54)}$$

The results of eqs. (2.42)–(2.45) imply that the helicity amplitudes with $\lambda_1 \neq \lambda_2$ vanish. Using eqs. (F.11) and (F.12), we identify

$$i\mathcal{M}(\lambda_1=\lambda_2=\pm\tfrac{1}{2}) = \frac{-ie^2}{2p_1\cdot k_1}\langle p_2\pm|\,\sigma_\pm\cdot\varepsilon_2^*\,\sigma_\mp\cdot(p_1+k_1)\,\sigma_\pm\cdot\varepsilon_1\,|\,p_1\pm\rangle$$
$$+\frac{ie^2}{2p_1\cdot k_2}\langle p_2\pm|\,\sigma_\pm\cdot\varepsilon_1\,\sigma_\mp\cdot(p_1-k_2)\,\sigma_\pm\cdot\varepsilon_2^*\,|\,p_1\pm\rangle\,. \qquad \text{(F.55)}$$

Further simplification ensues when we apply the results of eqs. (F.50) and (F.51). To use these results, we must select a reference momentum p, which can be any lightlike four-vector that is not parallel to the corresponding photon polarization vector. One is free to choose a different reference momentum for each photon polarization vector. Moreover, when computing two different helicity amplitudes (each of which is a gauge invariant quantity), one may select a different reference momentum for the *same* photon polarization vector in the two computations. The decision of which reference momenta to choose is somewhat of an art; experience will teach you which choices lead to the most simplification in a given calculation.

We shall denote the reference momenta for ε_1 and ε_2 by $p^{(1)}$ and $p^{(2)}$, respectively. Two different choices are considered below: either $p^{(1)}=p_1$ and $p^{(2)}=p_2$ or $p^{(1)}=p_2$ and $p^{(2)}=p_1$. The amplitudes, $\mathcal{M}(\lambda_1=\lambda_2=\pm\frac{1}{2})$, vanish for either of the two choices for the reference momenta unless the photon helicities are equal, i.e., $\lambda_1'=\lambda_2'$. This leaves only four possible nonvanishing helicity amplitudes.

For the case of $\lambda_1=\lambda_2=\pm\frac{1}{2}$ and $\lambda_1'=\lambda_2'=\pm1$ (i.e., $\lambda_1\lambda_1'>0$), we choose reference momenta $p^{(1)}=p_2$ and $p^{(2)}=p_1$. Then, the second term vanishes on the right-hand side of eq. (F.55). Making use of eqs. (F.17), (F.50), and (F.51), we find

$$i\mathcal{M}(\lambda_1=\lambda_2=\tfrac{1}{2},\,\lambda_1'=\lambda_2'=1) = \frac{-ie^2}{p_1\cdot k_1}\,\frac{\langle p_1\,k_1\rangle\,\langle k_1\,p_1\rangle\;[p_2\,k_2]}{\langle p_1\,k_2\rangle}\,. \qquad \text{(F.56)}$$

Using eq. (F.34) to write the dot product in terms of spinor products, we obtain

$$i\mathcal{M}(\lambda_1=\lambda_2=\tfrac{1}{2},\,\lambda_1'=\lambda_2'=1) = 2ie^2\frac{\langle p_1\,k_1\rangle}{\langle p_1\,k_1\rangle^*}\,\frac{[p_2\,k_2]}{\langle p_1\,k_2\rangle}\,, \qquad \text{(F.57)}$$

after making use of eq. (F.29). A similar computation yields

$$i\mathcal{M}(\lambda_1=\lambda_2=-\tfrac{1}{2},\,\lambda_1'=\lambda_2'=-1) = 2ie^2\frac{[p_1\,k_1]}{[p_1\,k_1]^*}\,\frac{\langle p_2\,k_2\rangle}{[p_1\,k_2]}\,. \qquad \text{(F.58)}$$

For the case of $\lambda_1 = \lambda_2 = \pm\frac{1}{2}$ and $\lambda'_1 = \lambda'_2 = \mp 1$ (i.e., $\lambda_1\lambda'_1 < 0$), we choose reference momenta $p^{(1)} = p_1$ and $p^{(2)} = p_2$. Then, the first term vanishes on the right-hand side of eq. (F.55). A similar calculation to the one given above yields

$$i\mathcal{M}(\lambda_1 = \lambda_2 = -\tfrac{1}{2}\,,\,\lambda'_1 = \lambda'_2 = 1) = 2ie^2 \frac{[p_1\,k_2]}{[p_1\,k_2]^*}\frac{\langle p_2\,k_1\rangle}{[p_1\,k_1]}\,, \tag{F.59}$$

$$i\mathcal{M}(\lambda_1 = \lambda_2 = \tfrac{1}{2}\,,\,\lambda'_1 = \lambda'_2 = -1) = 2ie^2 \frac{\langle p_1\,k_2\rangle}{\langle p_1\,k_2\rangle^*}\frac{[p_2\,k_1]}{\langle p_1\,k_1\rangle}\,. \tag{F.60}$$

Note that each pair of helicity amplitudes above is simply related:[9]

$$\left[\mathcal{M}_{\lambda_2\,,\,\lambda'_2\,;\,\lambda_1\,,\,\lambda'_1}(s,\theta,\phi)\right]^* = \mathcal{M}_{-\lambda_2\,,\,-\lambda'_2\,;\,-\lambda_1\,,\,-\lambda'_1}(s,\theta,\phi)\,. \tag{F.61}$$

Thus, in this calculation we only need to evaluate two nonzero helicity amplitudes. It is clear that we have simplified the computation enormously by our choice of reference momenta. With a less judicious choice, the calculation is significantly more tedious, although gauge invariance guarantees that one must arrive at the same result for the helicity amplitudes quoted above.

One can easily evaluate the spinor products above in the CM reference frame. Writing $p_1^\mu = E(1\,;\,\hat{\boldsymbol{z}})$, $k_1^\mu = E(1\,;\,-\hat{\boldsymbol{z}})$, $p_2^\mu = E(1\,;\,\hat{\boldsymbol{p}}_{\rm CM})$, and $k_2^\mu = E(1\,;\,-\hat{\boldsymbol{p}}_{\rm CM})$, and using the results of eqs. (E.261) and (F.27), we obtain

$$\langle p_1\,k_1\rangle = 2iE\,, \qquad \langle p_1\,k_2\rangle = 2iEe^{i\phi/2}\cos(\theta/2), \tag{F.62}$$

$$\langle p_2\,k_2\rangle = 2iE\,, \qquad \langle p_2\,k_1\rangle = 2iEe^{-i\phi/2}\cos(\theta/2), \tag{F.63}$$

where θ and ϕ are the polar and azimuthal angles of $\hat{\boldsymbol{p}}_{\rm CM}$. All other relevant spinor products can be found using eqs. (F.29)–(F.31).

Exercises

F.1 It is always possible to define the plane of the scattering process to be the x–z plane, in which case $\phi = 0$ and all the spinor products in eqs. (F.62) and (F.63) are manifestly real. Nevertheless, by keeping the explicit ϕ dependence, one maintains a useful check of the calculation. For example, in light of the Jacob–Wick partial wave expansion [see eq. (D.113)], the tree-level amplitude for Compton scattering satisfies

$$\mathcal{M}_{\lambda_2\,,\,\lambda'_2\,;\,\lambda_1\,,\,\lambda'_1}(s,\theta,\phi) = e^{i(\lambda_1-\lambda'_1)\phi}\mathcal{M}_{\lambda_2\,,\,\lambda'_2\,;\,\lambda_1\,,\,\lambda'_1}(s,\theta)\,, \tag{F.64}$$

where $\mathcal{M}(s,\theta) \equiv \mathcal{M}(s,\theta,\phi=0)$. Inserting the explicit forms for the spinor products into eqs. (F.57)–(F.60), verify that the ϕ dependence of the helicity amplitudes satisfies eq. (F.64).

[9] Although eq. (F.61) is not a general result that is valid for arbitrary parity-conserving $2 \to 2$ scattering amplitudes, one can verify that it is satisfied for the tree-level helicity amplitudes for the Compton scattering of massless electrons, in light of eqs. (D.113) and (D.117).

F.2 To compute the unpolarized cross section for Compton scattering, one must sum the absolute squares of the helicity amplitudes obtained in Appendix F.3 and divide by 4 to average over the initial helicities.

(a) Show that

$$\begin{aligned}|\mathcal{M}(\lambda_1 = \lambda_2 = \tfrac{1}{2}\,,\, \lambda_1' = \lambda_2' = 1)|^2 &= |\mathcal{M}(\lambda_1 = \lambda_2 = -\tfrac{1}{2}\,,\, \lambda_1' = \lambda_2' = -1)|^2 \\ &= 4e^4 \frac{p_1 \cdot k_1}{p_1 \cdot k_2}\,, \end{aligned} \tag{F.65}$$

$$\begin{aligned}|\mathcal{M}(\lambda_1 = \lambda_2 = -\tfrac{1}{2}\,,\, \lambda_1' = \lambda_2' = 1)|^2 &= |\mathcal{M}(\lambda_1 = \lambda_2 = \tfrac{1}{2}\,,\, \lambda_1' = \lambda_2' = -1)|^2 \\ &= 4e^4 \frac{p_1 \cdot k_2}{p_1 \cdot k_1}\,. \end{aligned} \tag{F.66}$$

(b) Verify the well-known textbook result in the zero electron mass limit,

$$\tfrac{1}{4}\sum_{\{\lambda\}} |\mathcal{M}|^2 = 2e^4 \left(\frac{p_1 \cdot k_1}{p_1 \cdot k_2} + \frac{p_1 \cdot k_2}{p_1 \cdot k_1} \right)\,. \tag{F.67}$$

(c) Evaluate the unpolarized differential cross section, $d\sigma/d\Omega$, for Compton scattering in the CM reference frame.

F.3 Using the spinor helicity method, compute the tree-level scattering amplitude for the process $gg \to gg$ in QCD, where the two incoming gluons and the two outgoing gluons all have helicity $\lambda = -1$.

(a) Show that the scattering amplitude can be written as (e.g., see [B149])

$$\mathcal{M} = g_s^2 \{ A_4[1^+, 2^+, 3^-, 4^-]\, \mathrm{Tr}(\boldsymbol{T}^{a_1}\boldsymbol{T}^{a_2}\boldsymbol{T}^{a_3}\boldsymbol{T}^{a_4}) + \text{perms of (234)} \}, \tag{F.68}$$

using the notation of eq. (F.1), where g_s is the coupling constant of $\mathrm{SU}(3)_C$ and "perms of (234)" indicates the five additional terms obtained by permuting the indices 2, 3, and 4.

(b) Verify the result of eq. (F.1) for the case of $n = 4$.

(c) Suppose that the two incoming gluons and the two outgoing gluons all have helicity $\lambda = +1$. How would the results of part (a) and (b) be modified?

F.4 (a) In light of footnote 2, show that, for massless spinors,

$$\bar{u}(p, \lambda')\, \Gamma\, u(q, \lambda) \propto \begin{cases} \delta_{\lambda\lambda'}\,, & \text{for } \Gamma = \gamma^\mu\,, \gamma^\mu\gamma_5, \\ \delta_{\lambda,-\lambda'}\,, & \text{for } \Gamma = 1\,,\, \gamma_5\,,\, \Sigma^{\mu\nu}\,,\, \Sigma^{\mu\nu}\gamma_5. \end{cases} \tag{F.69}$$

(b) In part (a), set $q = p$ and evaluate the spinor products $\bar{u}(p, \lambda')\, \Gamma\, u(p, \lambda)$ explicitly using the four-component Bouchiat–Michel formulae given in Appendix E.7. Check that eq. (F.69) is satisfied.

Appendix G Matrix Decompositions for Fermion Mass Diagonalization

In scalar field theory, the diagonalization of the tree-level squared-mass matrix M^2 is straightforward. For a theory of n complex scalar fields, M^2 is a hermitian $n \times n$ matrix, which can be diagonalized by a unitary matrix W:[1]

$$W^\dagger M^2 W = m^2 = \mathrm{diag}(m_1^2, m_2^2, \ldots, m_n^2)\,. \tag{G.1}$$

For a theory of n real scalar fields, M^2 is a real symmetric $n \times n$ matrix, which can be diagonalized by an orthogonal matrix Q:

$$Q^\mathsf{T} M^2 Q = m^2 = \mathrm{diag}(m_1^2, m_2^2, \ldots, m_n^2)\,. \tag{G.2}$$

In both cases, the eigenvalues m_k^2 of M^2 are real. These are the standard matrix diagonalization problems treated in all elementary textbooks on linear and/or matrix algebra (e.g., see [B128]).

In spin-1/2 fermion field theory, the most general fermion mass matrix, obtained from the Lagrangian, written in terms of two-component spinors, is complex and symmetric. If the Lagrangian exhibits a U(1) symmetry, then a basis can be found such that fields that are charged under the U(1) pair up into Dirac fermions. The fermion mass matrix then decomposes into the direct sum of a complex Dirac fermion mass matrix and a complex symmetric neutral fermion mass matrix. In this appendix, we review the linear algebra theory relevant for the matrix decompositions associated with the general charged and neutral spin-1/2 fermion mass matrix diagonalizations. The diagonalization of the Dirac fermion mass matrix is governed by the singular value decomposition (SVD) of a complex matrix, as shown in Appendix G.1. In contrast, the diagonalization of a neutral fermion mass matrix is governed by the Autonne–Takagi factorization of a complex symmetric matrix, which is treated in Appendix G.2.[2] These two techniques are compared and contrasted in Appendix G.3. Dirac fermions can also arise in the case of a pseudoreal representation of fermion fields. This latter case requires the reduction of a complex antisymmetric fermion mass matrix to real normal form. The relevant theorem and its proof are given in Appendix G.4.

[1] More generally, a complex $n \times n$ matrix A is unitarily diagonalizable if and only if A is *normal* (i.e., A commutes with its hermitian adjoint $A^\dagger$). However, in contrast to eq. (G.1), a nonhermitian normal matrix possesses at least one (nonreal) complex eigenvalue.

[2] One may choose not to work in a basis where the fermion fields are eigenstates of the U(1) charge operator. In this case, all fermions are governed by a complex symmetric mass matrix, which can be Takagi diagonalized according to the procedure described in Appendix G.2.

G.1 Singular Value Decomposition (SVD)

The diagonalization of the charged (Dirac) fermion mass matrix requires the singular value decomposition of an arbitrary complex matrix M.

Theorem G.1. *For any complex [or real] $n \times n$ matrix M, unitary [or real orthogonal] matrices L and R exist such that*

$$L^{\mathsf{T}} M R = M_D = \mathrm{diag}(m_1, m_2, \ldots, m_n), \tag{G.3}$$

where the m_k are real and nonnegative. This is called the singular value decomposition of the matrix M (e.g., see [B128]).

In general, the m_k are *not* the eigenvalues of M. Rather, the m_k are the *singular values* of the general complex matrix M, which are defined to be the nonnegative square roots of the eigenvalues of $M^\dagger M$ (or equivalently of $MM^\dagger$). An equivalent definition of the singular values can be established as follows. Since $M^\dagger M$ is a hermitian nonnegative matrix, its eigenvalues are real and nonnegative and its eigenvectors, v_k, defined by $M^\dagger M v_k = m_k^2 v_k$, can be chosen to be orthonormal.[3] Consider first the eigenvectors corresponding to the nonzero eigenvalues of $M^\dagger M$. Then, we define the vectors w_k such that $M v_k = m_k w_k^*$. It follows that $m_k^2 v_k = M^\dagger M v_k = m_k M^\dagger w_k^*$, which yields $M^\dagger w_k^* = m_k v_k$. Note that these equations also imply that $MM^\dagger w_k^* = m_k^2 w_k^*$. The orthonormality of the v_k implies the orthonormality of the w_k, and vice versa. For example,

$$\delta_{jk} = \langle v_j | v_k \rangle = \frac{1}{m_j m_k} \langle M^\dagger w_j^* | M^\dagger w_k^* \rangle = \frac{1}{m_j m_k} \langle w_j | M M^\dagger w_k^* \rangle = \frac{m_k}{m_j} \langle w_j^* | w_k^* \rangle \,, \tag{G.4}$$

which yields $\langle w_k | w_j \rangle = \delta_{jk}$. If M is a real matrix, then the eigenvectors v_k can be chosen to be real, in which case the corresponding w_k are also real.

If v_i is an eigenvector of $M^\dagger M$ with zero eigenvalue, then $0 = v_i^\dagger M^\dagger M v_i = \langle M v_i | M v_i \rangle$, which implies that $M v_i = 0$. Likewise, if w_i^* is an eigenvector of $MM^\dagger$ with zero eigenvalue, then $0 = w_i^{\mathsf{T}} M M^\dagger w_i^* = \langle M^{\mathsf{T}} w_i | M^{\mathsf{T}} w_i \rangle^*$, which implies that $M^{\mathsf{T}} w_i = 0$. Because the eigenvectors of $M^\dagger M$ [$MM^\dagger$] can be chosen orthonormal, the eigenvectors corresponding to the zero eigenvalues of M [$M^\dagger$] can be taken to be orthonormal.[4] Finally, these eigenvectors are also orthogonal to the eigenvectors corresponding to the nonzero eigenvalues of $M^\dagger M$ [$MM^\dagger$]. That is, if the indices i and j run over the eigenvectors corresponding to the zero and nonzero eigenvalues of $M^\dagger M$ [$MM^\dagger$], respectively, then

$$\langle v_j | v_i \rangle = \frac{1}{m_j} \langle M^\dagger w_j^* | v_i \rangle = \frac{1}{m_j} \langle w_j^* | M v_i \rangle = 0 \,, \tag{G.5}$$

[3] We define the inner product of two vectors to be $\langle v | w \rangle \equiv v^\dagger w$. Then, v and w are orthonormal if $\langle v | w \rangle = 0$. The norm of a vector is defined by $\| v \| = \langle v | v \rangle^{1/2}$.

[4] This analysis shows that the number of linearly independent eigenvectors of $M^\dagger M$ [$MM^\dagger$] with zero eigenvalue coincides with the number of linearly independent eigenvectors of M [$M^\dagger$] with zero eigenvalue.

and similarly $\langle w_j | w_i \rangle = 0$.

Thus, we can define the singular values of a general complex $n \times n$ matrix M to be the simultaneous solutions (with real nonnegative m_k) of[5]

$$M v_k = m_k w_k^* \,, \qquad w_k^{\mathsf{T}} M = m_k v_k^\dagger \,. \tag{G.6}$$

The corresponding v_k (w_k), normalized to have unit norm, are called the right (left) singular vectors of M. In particular, the number of linearly independent v_k coincides with the number of linearly independent w_k and is equal to n.

Proof of the singular value decomposition theorem In light of eqs. (G.4) and (G.5), the right [left] singular vectors can be chosen to be orthonormal. Hence, the unitary matrix R [L] can be constructed such that its kth column is given by the right [left] singular vector v_k [w_k]. It then follows from eq. (G.6) that

$$w_k^{\mathsf{T}} M v_\ell = m_k \delta_{k\ell} \qquad \text{(no sum over } k\text{)}. \tag{G.7}$$

In matrix form, eq. (G.7) coincides with eq. (G.3), and the singular value decomposition is established. If M is real, then the right and left singular vectors, v_k and w_k, can be chosen to be real, in which case eq. (G.3) holds for real orthogonal matrices L and R.

The singular values of a complex matrix M are unique (up to ordering), as they correspond to the eigenvalues of $M^\dagger M$ (or equivalently the eigenvalues of $MM^\dagger$). The unitary matrices L and R are not unique. The matrix R must satisfy

$$R^\dagger M^\dagger M R = M_D^2 \,, \tag{G.8}$$

which follows directly from eq. (G.3) by computing $M_D^\dagger M_D = M_D^2$. That is, R is a unitary matrix that diagonalizes the nonnegative definite matrix $M^\dagger M$. Since the eigenvectors of $M^\dagger M$ are orthonormal, each v_k corresponding to a nondegenerate eigenvalue of $M^\dagger M$ can be multiplied by an arbitrary phase $e^{i\theta_k}$. For the case of degenerate eigenvalues, any orthonormal linear combination of the corresponding eigenvectors is also an eigenvector of $M^\dagger M$. It follows that, within the subspace spanned by the eigenvectors corresponding to nondegenerate eigenvalues, R is uniquely determined up to multiplication on the right by an arbitrary diagonal unitary matrix. Within the subspace spanned by the eigenvectors of $M^\dagger M$ corresponding to a degenerate eigenvalue, R is determined up to multiplication on the right by an arbitrary unitary matrix.

Once R is fixed, L is obtained from eq. (G.3):

$$L = (M^{\mathsf{T}})^{-1} R^* M_D \,. \tag{G.9}$$

However, if some of the diagonal elements of M_D are zero, then L is not uniquely defined. Writing M_D in 2×2 block form such that the upper left block is a diagonal

[5] One can always find a solution to eq. (G.6) such that the m_k are real and nonnegative. Given a solution where m_k is complex, we simply write $m_k = |m_k| e^{i\theta}$ and redefine $w_k \to w_k e^{i\theta}$ to remove the phase θ.

matrix with positive diagonal elements and the other three blocks are equal to the zero matrix of the appropriate dimensions, it follows that $M_D = M_D W$, where

$$W = \left(\begin{array}{c|c} \mathbb{1} & \mathbb{O} \\ \hline \mathbb{O} & W_0 \end{array}\right). \tag{G.10}$$

W_0 is an arbitrary unitary matrix whose dimension is equal to the number of zeros that appear in the diagonal elements of M_D, and $\mathbb{1}$ and $\mathbb{O}$ are respectively the identity matrix and zero matrix of the appropriate size. Hence, we can multiply both sides of eq. (G.9) on the right by W, which means that L is only determined up to multiplication on the right by an arbitrary unitary matrix whose form is given by eq. (G.10).[6] □

If M is a real matrix, then the singular value decomposition of M is given by eq. (G.3), where L and R are real orthogonal matrices. This result is easily established by replacing "phase" with "sign" and replacing "unitary" by "real orthogonal" in the above proof.

G.2 Takagi Diagonalization

The mass matrix of neutral Majorana fermions (or a system of two-component fermions in a generic basis) is complex and symmetric. This mass matrix must be diagonalized in order to identify the physical fermion mass eigenstates and to compute their masses. However, the fermion mass matrix is *not* diagonalized by the standard unitary similarity transformation. Instead a different diagonalization equation is employed that was discovered independently by Autonne [218] and Takagi [219] and then rediscovered many times since [B128].

Theorem G.2. *For any complex symmetric $n \times n$ matrix M, there exists a unitary matrix Ω such that*

$$\Omega^{\mathsf{T}} M \Omega = M_D = \mathrm{diag}(m_1, m_2, \ldots, m_n), \tag{G.11}$$

where the m_k are real and nonnegative. This is the Takagi diagonalization[7] *of the complex symmetric matrix M.*

[6] Of course, one can reverse the above procedure by first determining the unitary matrix L. Equation (G.3) implies that $L^{\mathsf{T}} M M^\dagger L^* = M_D^2$, in which case L is determined up to multiplication on the right by an arbitrary [diagonal] unitary matrix within the subspace spanned by the eigenvectors corresponding to the degenerate [nondegenerate] eigenvalues of $MM^\dagger$. Having fixed L, one can obtain $R = M^{-1} L^* M_D$ from eq. (G.3). As above, R is only determined up to multiplication on the right by a unitary matrix whose form is given by eq. (G.10).

[7] In [B128], eq. (G.11) is called the Autonne–Takagi factorization of a complex symmetric matrix, although in the literature this is sometimes referred to as Takagi factorization. In this book, we shall refer to eq. (G.11) as the Takagi *diagonalization* of a complex symmetric matrix to emphasize and contrast this with the more standard diagonalization of normal matrices by a unitary similarity transformation (see footnote 1 and Exercise G.2).

In general, the m_k are *not* the eigenvalues of M. Rather, the m_k are the singular values of the symmetric matrix M. From eq. (G.11) it follows that

$$\Omega^\dagger M^\dagger M \Omega = M_D^2 = \mathrm{diag}(m_1^2, m_2^2, \ldots, m_n^2)\,. \tag{G.12}$$

If all of the singular values m_k are nondegenerate, then one can find a solution to eq. (G.11) for Ω from eq. (G.12). This is no longer true if some of the singular values are degenerate. For example, if $M = \left(\begin{smallmatrix} 0 & m \\ m & 0 \end{smallmatrix}\right)$, then the singular value $|m|$ is doubly degenerate, but eq. (G.12) yields $\Omega^\dagger\Omega = \mathbb{1}_{2\times 2}$, which does not specify Ω. That is, in the degenerate case, the physical fermion states *cannot* be determined by the diagonalization of $M^\dagger M$. Instead, one must make direct use of eq. (G.11). Below, we shall present a constructive method for determining Ω that is applicable in both the nondegenerate and the degenerate cases.

Equation (G.11) can be rewritten as $M\Omega = \Omega^* M_D$, where the columns of Ω are orthonormal. If we denote the kth column of Ω by v_k, then

$$M v_k = m_k v_k^*\,, \tag{G.13}$$

where the m_k are the singular values of M and the vectors v_k are normalized to have unit norm. The v_k are called the *Takagi vectors* of the complex symmetric $n \times n$ matrix M. The Takagi vectors corresponding to nondegenerate nonzero [zero] singular values are unique up to an overall sign [phase]. Any orthogonal [unitary] linear combination of Takagi vectors corresponding to a set of degenerate nonzero [zero] singular values is also a Takagi vector corresponding to the same singular value.

Using the above results, one can determine the degree of nonuniqueness of the matrix Ω. For definiteness, we fix an ordering of the diagonal elements of M_D.[8] If the singular values of M are distinct, then the matrix Ω is uniquely determined up to multiplication by a diagonal matrix whose entries are either ± 1 (i.e., a diagonal orthogonal matrix). If there are degeneracies corresponding to nonzero singular values, then within the degenerate subspace, Ω is unique up to multiplication on the right by an arbitrary orthogonal matrix. Finally, in the subspace corresponding to zero singular values, Ω is unique up to multiplication on the right by an arbitrary unitary matrix.

For a real symmetric matrix M, the Takagi diagonalization [eq. (G.11)] still holds for a unitary matrix Ω, which is easily determined as follows. Any real symmetric matrix M can be diagonalized by a real orthogonal matrix Z:

$$Z^{\mathsf{T}} M Z = \mathrm{diag}(\varepsilon_1 m_1\,,\, \varepsilon_2 m_2\,,\, \ldots\,,\, \varepsilon_n m_n)\,, \tag{G.14}$$

where the m_k are real and nonnegative and the $\varepsilon_k m_k$ are the real eigenvalues of M with corresponding signs $\varepsilon_k = \pm 1$.[9] Then, the Takagi diagonalization of M is achieved by taking $\Omega_{jk} = \varepsilon_k^{1/2} Z_{jk}$ (no sum over k).

8 Permuting the order of the singular values is equivalent to permuting the order of the columns of Ω.

9 In the case of $m_k = 0$, we conventionally choose the corresponding $\varepsilon_k = +1$.

Proof of the Takagi diagonalization theorem To prove the existence of the Takagi diagonalization of a complex symmetric matrix, it is sufficient to provide an algorithm for constructing the orthonormal Takagi vectors v_k that make up the columns of Ω. This is achieved by rewriting the $n \times n$ complex matrix equation $Mv = mv^*$ [with m real and nonnegative] as a $2n \times 2n$ real matrix equation:

$$M_R \begin{pmatrix} \operatorname{Re} v \\ \operatorname{Im} v \end{pmatrix} \equiv \begin{pmatrix} \operatorname{Re} M & -\operatorname{Im} M \\ -\operatorname{Im} M & -\operatorname{Re} M \end{pmatrix} \begin{pmatrix} \operatorname{Re} v \\ \operatorname{Im} v \end{pmatrix} = m \begin{pmatrix} \operatorname{Re} v \\ \operatorname{Im} v \end{pmatrix}, \quad \text{where } m \geq 0 . \tag{G.15}$$

Since $M = M^{\mathsf{T}}$, the $2n \times 2n$ matrix $M_R \equiv \left(\begin{smallmatrix} \operatorname{Re} M & -\operatorname{Im} M \\ -\operatorname{Im} M & -\operatorname{Re} M \end{smallmatrix}\right)$ is a real symmetric matrix.[10] In particular, M_R is diagonalizable by a real orthogonal similarity transformation, and its eigenvalues are real. Moreover, if m is an eigenvalue of M_R with eigenvector $(\operatorname{Re} v\,,\, \operatorname{Im} v)$, then $-m$ is an eigenvalue of M_R with (orthogonal) eigenvector $(-\operatorname{Im} v\,,\, \operatorname{Re} v)$. This observation implies that M_R has an equal number of positive and negative eigenvalues and an even number of zero eigenvalues.[11] Thus, eq. (G.13) has been converted into an ordinary eigenvalue problem for a real symmetric matrix. Since $m \geq 0$, we solve the eigenvalue problem $M_R u = mu$ for the real eigenvectors $u \equiv (\operatorname{Re} v\,,\, \operatorname{Im} v)$ corresponding to the nonnegative eigenvalues of M_R,[12] which then immediately yields the complex Takagi vectors, v. It is straightforward to prove that the total number of linearly independent Takagi vectors is equal to n. Simply note that the orthogonality of $(\operatorname{Re} v_1\,,\, \operatorname{Im} v_1)$ and $(-\operatorname{Im} v_1\,,\, \operatorname{Re} v_1)$ with $(\operatorname{Re} v_2\,,\, \operatorname{Im} v_2)$ implies that $v_1^\dagger v_2 = 0$.

Thus, we have derived a constructive method for obtaining the Takagi vectors, v_k. If there are degeneracies, one can always choose the v_k in the degenerate subspace to be orthonormal. The Takagi vectors then make up the columns of the matrix Ω in eq. (G.11). □

G.3 Relating Takagi Diagonalization and the SVD

The Takagi diagonalization is a special case of the singular value decomposition. If the complex matrix M in eq. (G.3) is symmetric, $M = M^{\mathsf{T}}$, then the Takagi diagonalization corresponds to $\Omega = L = R$. In this case, the right and left singular vectors coincide ($v_k = w_k$) and are identified with the Takagi vectors defined in eq. (G.13). However as previously noted, the matrix Ω cannot be determined from

[10] The $2n \times 2n$ matrix M_R is a real representation of the $n \times n$ complex matrix M.
[11] Note that $(-\operatorname{Im} v\,,\, \operatorname{Re} v)$ corresponds to replacing v_k in eq. (G.13) by iv_k. However, for $m < 0$ these solutions are not relevant for Takagi diagonalization (where the m_k are by definition nonnegative). The case of $m = 0$ is considered in footnote 12.
[12] For $m = 0$, the corresponding vectors $(\operatorname{Re} v\,,\, \operatorname{Im} v)$ and $(-\operatorname{Im} v\,,\, \operatorname{Re} v)$ are two linearly independent eigenvectors of M_R; but these yield only one independent Takagi vector v (since v and iv are linearly dependent).

eq. (G.12) in cases where there is a degeneracy among the singular values.[13] For example, one possible singular value decomposition of the matrix $M = \left(\begin{smallmatrix} 0 & m \\ m & 0 \end{smallmatrix}\right)$ [with m assumed real and positive] can be obtained by choosing $R = \left(\begin{smallmatrix} 1 & 0 \\ 0 & 1 \end{smallmatrix}\right)$ and $L = \left(\begin{smallmatrix} 0 & 1 \\ 1 & 0 \end{smallmatrix}\right)$, in which case $L^{\mathsf{T}} M R = \left(\begin{smallmatrix} m & 0 \\ 0 & m \end{smallmatrix}\right) = M_D$. Of course, this is not a Takagi diagonalization because $L \neq R$. Since R is only defined modulo the multiplication on the right by an arbitrary 2×2 unitary matrix $\mathcal{O}$, then at least one singular value decomposition exists that is also a Takagi diagonalization. For the example under consideration, it is not difficult to deduce the Takagi diagonalization: $\Omega^{\mathsf{T}} M \Omega = M_D$, where

$$\Omega = \frac{1}{\sqrt{2}} \begin{pmatrix} 1 & i \\ 1 & -i \end{pmatrix} \mathcal{O}, \tag{G.16}$$

and $\mathcal{O}$ is any 2×2 orthogonal matrix. Details can be found in Exercise G.5.

Since the Takagi diagonalization is a special case of the singular value decomposition, it seems plausible that one could prove the former from the latter. This turns out to be correct; for completeness, we provide the proof below. Our second proof depends on the following lemma:

Lemma G.3. *For any symmetric unitary matrix V, there exists a unitary matrix U such that $V = U^{\mathsf{T}} U$.*

Proof of the lemma For any $n \times n$ unitary matrix V, there exists a hermitian matrix H such that $V = \exp(iH)$ (this is the polar decomposition of V). If $V = V^{\mathsf{T}}$ then $H = H^{\mathsf{T}} = H^*$ (since H is hermitian); therefore H is real symmetric. But any real symmetric matrix can be diagonalized by an orthogonal similarity transformation. It follows that V can also be diagonalized by an orthogonal similarity transformation. Since the eigenvalues of any unitary matrix are pure phases, there exists a real orthogonal matrix Q such that $Q^{\mathsf{T}} V Q = \mathrm{diag}\,(\mathrm{e}^{i\theta_1}, \mathrm{e}^{i\theta_2}, \ldots, \mathrm{e}^{i\theta_n})$. Thus, the unitary matrix

$$U = \mathrm{diag}\,(e^{i\theta_1/2}, e^{i\theta_2/2}, \ldots, e^{i\theta_n/2})\, Q^{\mathsf{T}} \tag{G.17}$$

satisfies $V = U^{\mathsf{T}} U$ and the lemma is proved. Note that U is unique modulo multiplication on the left by an arbitrary real orthogonal matrix. □

Second Proof of the Takagi diagonalization Starting from the singular value decomposition of M, there exist unitary matrices L and R such that $M = L^* M_D R^\dagger$, where M_D is the diagonal matrix of singular values. Since $M = M^{\mathsf{T}} = R^* M_D L^\dagger$, we have two different singular value decompositions for M. However, as noted below eq. (G.8), R is unique modulo multiplication on the right by an arbitrary [diagonal] unitary matrix, V, within the [non-]degenerate subspace. Thus, it follows that a [diagonal] unitary matrix V exists such that $L = RV$. Moreover, $V = V^{\mathsf{T}}$.

[13] This is in contrast to the singular value decomposition, where R can be determined from eq. (G.8) modulo right multiplication by a [diagonal] unitary matrix in the [non-]degenerate subspace and L is then determined by eq. (G.9) modulo multiplication on the right by eq. (G.10).

This is manifestly true within the nondegenerate subspace where V is diagonal. Within the degenerate subspace, M_D is proportional to the identity matrix so that $L^*R^\dagger = R^*L^\dagger$. Inserting $L = RV$ then yields $V^\mathsf{T} = V$. Using the lemma proved above, there exists a unitary matrix U such that $V = U^\mathsf{T}U$. That is,

$$L = RU^\mathsf{T}U\,, \tag{G.18}$$

for some unitary matrix U. Moreover, it is now straightforward to show that

$$M_D U^* = U^* M_D\,. \tag{G.19}$$

To see this, note that within the degenerate subspace, eq. (G.19) is trivially true since M_D is proportional to the identity matrix. Within the nondegenerate subspace, V is diagonal; hence we may choose $U = U^\mathsf{T} = V^{1/2}$, so that eq. (G.19) is true since diagonal matrices commute. Using eqs. (G.18) and (G.19), we can write the singular value decomposition of M as follows:

$$M = L^*M_DR^\dagger = R^*U^\dagger U^*M_DR^\dagger = (RU^\mathsf{T})^*M_DU^*R^\dagger = \Omega^*M_D\Omega^\dagger\,, \tag{G.20}$$

where $\Omega \equiv RU^\mathsf{T}$ is a unitary matrix. Thus the existence of the Takagi diagonalization of an arbitrary complex symmetric matrix [eq. (G.11)] is once again proved. □

In the diagonalization of the two-component fermion mass matrix, M, the eigenvalues of $M^\dagger M$ typically fall into two classes: nondegenerate eigenvalues corresponding to neutral (Majorana) fermion mass eigenstates, and degenerate pairs corresponding to charged (Dirac) mass eigenstates. In this case, the sector of the neutral fermions corresponds to a nondegenerate subspace of the space of fermion fields. Hence, in order to identify the neutral fermion mass eigenstates, it is sufficient to diagonalize $M^\dagger M$ with a unitary matrix R [as in eq. (G.8)], and then adjust the overall phase of each column of R so that the resulting matrix Ω satisfies $\Omega^\mathsf{T}M\Omega = M_D$, where M_D is a diagonal matrix of the nonnegative fermion masses. This last result is a consequence of eqs. (G.18)–(G.20), where $\Omega = RV^{1/2}$ and V is a diagonal matrix of phases.

G.4 Real Normal Form of a Complex Antisymmetric Matrix

In the case of two-component fermions that transform under a pseudoreal representation of a compact Lie group [see eq. (1.179)], the corresponding mass matrix is in general complex and antisymmetric. In this case, one must employ the antisymmetric analogue of the Takagi diagonalization of a complex symmetric matrix [B128], where the complex antisymmetric matrix is transformed into a 2×2 block-diagonal form.

Theorem G.4. *For any complex [or real] antisymmetric $n \times n$ matrix M, there exists a unitary [or real orthogonal] matrix U such that*

$$U^{\mathsf{T}} M U = N \equiv \operatorname{diag}\left\{ \begin{pmatrix} 0 & m_1 \\ -m_1 & 0 \end{pmatrix}, \begin{pmatrix} 0 & m_2 \\ -m_2 & 0 \end{pmatrix}, \ldots, \begin{pmatrix} 0 & m_p \\ -m_p & 0 \end{pmatrix}, \mathbb{O}_{n-2p} \right\}, \tag{G.21}$$

where N is written in block-diagonal form with 2×2 matrices appearing along the diagonal, followed by an $(n-2p) \times (n-2p)$ block of zeros (denoted by $\mathbb{O}_{n-2p}$), and the m_j are real and positive. N is called the real normal form of an antisymmetric matrix.

Proof of the reduction of an antisymmetric matrix to real normal form
A number of proofs can be found in the literature [220–222] and in textbooks (e.g., see [B127]). Here we provide a proof inspired by Ref. [221]. Following Appendix G.3, we first consider the eigenvalue equation for $M^\dagger M$:

$$M^\dagger M v_k = m_k^2 v_k\,, \qquad m_k > 0\,, \qquad \text{and} \qquad M^\dagger M u_k = 0\,, \tag{G.22}$$

where we have distinguished the eigenvectors corresponding to positive eigenvalues and zero eigenvalues, respectively. The quantities m_k are the positive singular values of M. Noting that $u_k^\dagger M^\dagger M u_k = \langle M u_k \,|\, M u_k \rangle = 0$, it follows that

$$M u_k = 0\,, \tag{G.23}$$

so that the u_k are the eigenvectors corresponding to the zero eigenvalues of M. For each eigenvector of $M^\dagger M$ with $m_k \neq 0$, we define a new vector

$$w_k \equiv \frac{1}{m_k} M^* v_k^*\,. \tag{G.24}$$

It follows that $m_k^2 v_k = M^\dagger M v_k = m_k M^\dagger w_k^*$, which yields $M^\dagger w_k^* = m_k v_k$. Comparing with eq. (G.6), we identify v_k and w_k as the right and left singular vectors, respectively, corresponding to the nonzero singular values of M. For any antisymmetric matrix, $M^\dagger = -M^*$. Hence,

$$M v_k = m_k w_k^*\,, \qquad M w_k = -m_k v_k^*\,, \tag{G.25}$$

and

$$M^\dagger M w_k = -m_k M^\dagger v_k^* = m_k M^* v_k^* = m_k^2 w_k\,, \qquad m_k > 0\,. \tag{G.26}$$

That is, the w_k are also eigenvectors of $M^\dagger M$.

The key observation is that for fixed k the vectors v_k and w_k are orthogonal, since eq. (G.25) implies that

$$\langle w_k | v_k \rangle = \langle v_k | w_k \rangle^* = -\frac{1}{m_k^2} \langle M w_k | M v_k \rangle = -\frac{1}{m_k^2} \langle w_k | M^\dagger M v_k \rangle = -\langle w_k | v_k \rangle, \tag{G.27}$$

which yields $\langle w_k | v_k \rangle = 0$. Thus, if all the m_k are distinct, it follows that m_k^2 is a doubly degenerate eigenvalue of $M^\dagger M$, with corresponding linearly independent eigenvectors v_k and w_k, where $k \in \{1, 2, \ldots, p\}$ (and $p \leq \frac{1}{2}n$). The remaining zero eigenvalues are $(n-2p)$-fold degenerate, with corresponding eigenvectors u_k (for $k \in \{1, 2, \ldots, n-2p\}$). If some of the m_k are degenerate, these conclusions still apply. For example, suppose that $m_j = m_k$ for $j \neq k$, which means that m_k^2 is at least a three-fold degenerate eigenvalue of $M^\dagger M$. Then, there must exist an eigenvector v_j that is orthogonal to v_k and w_k, such that $M^\dagger M v_j = m_k^2 v_j$. We now construct $w_j \equiv M^* v_j^* / m_k$ according to eq. (G.24). According to eq. (G.27), w_j is orthogonal to v_j. However, we still must show that w_j is also orthogonal to v_k and w_k. But this is straightforward:

$$\langle w_j | w_k \rangle = \langle w_k | w_j \rangle^* = \frac{1}{m_k^2} \langle M v_k | M v_j \rangle = \frac{1}{m_k^2} \langle v_k | M^\dagger M v_j \rangle = \langle v_k | v_j \rangle = 0 \,, \quad \text{(G.28)}$$

$$\langle w_j | v_k \rangle = \langle v_k | w_j \rangle^* = -\frac{1}{m_k^2} \langle M w_k | M v_j \rangle = -\frac{1}{m_k^2} \langle w_k | M^\dagger M v_j \rangle = -\langle w_k | v_j \rangle = 0 \,, \quad \text{(G.29)}$$

where we have used the assumed orthogonality of v_j with v_k and w_k, respectively. It follows that v_j, w_j, v_k, and w_k are linearly independent eigenvectors corresponding to a fourfold degenerate eigenvalue m_k^2 of $M^\dagger M$. Additional degeneracies are treated in the same way.

Thus, the number of nonzero eigenvalues of $M^\dagger M$ must be an even number, denoted by $2p$ above. Moreover, one can always choose the complete set of eigenvectors $\{u_k \,, v_k \,, w_k\}$ of $M^\dagger M$ to be orthonormal. These orthonormal vectors can be used to construct a unitary matrix U with matrix elements:

$$\begin{aligned} &U_{\ell\,,2k-1} = (w_k)_\ell \,, \qquad U_{\ell\,,2k} = (v_k)_\ell \,, && k \in \{1\,,2\,,\ldots\,,p\}\,, \\ &U_{\ell\,,k+2p} = (u_k)_\ell \,, && k \in \{1\,,2\,,\ldots\,,n-2p\}\,, \end{aligned} \quad \text{(G.30)}$$

for $\ell \in \{1\,,2\,,\ldots\,,n\}$, where, e.g., $(v_k)_\ell$ is the ℓth component of the vector v_k with respect to the standard orthonormal basis. The orthonormality of $\{u_k \,, v_k \,, w_k\}$ implies that $(U^\dagger U)_{\ell k} = \delta_{\ell k}$ as required. Equations (G.23) and (G.25) are thus equivalent to the matrix equation $MU = U^* N$, which immediately yields eq. (G.21), and the theorem is proven. If M is a real antisymmetric matrix, then all the eigenvectors of $M^\dagger M$ can be chosen to be real, in which case U is a real orthogonal matrix.

Finally, we address the nonuniqueness of the matrix U. For definiteness, we fix an ordering of the 2×2 blocks containing the m_k in the matrix N. In the subspace corresponding to a nonzero singular value of degeneracy d, the matrix U is unique up to multiplication on the right by a $2d \times 2d$ unitary matrix S that satisfies

$$S^{\mathsf{T}} J S = J \,, \quad \text{(G.31)}$$

where the $2d \times 2d$ matrix J, defined by

$$J = \text{diag}\left\{ \begin{pmatrix} 0 & 1 \\ -1 & 0 \end{pmatrix} , \begin{pmatrix} 0 & 1 \\ -1 & 0 \end{pmatrix} , \ldots , \begin{pmatrix} 0 & 1 \\ -1 & 0 \end{pmatrix} \right\} , \quad \text{(G.32)}$$

is a block-diagonal matrix with d blocks of 2×2 matrices. A unitary matrix S that satisfies eq. (G.31) is an element of the unitary symplectic group, Sp(d). If there are no degeneracies among the m_k, then $d = 1$. Identifying Sp(1) ≅ SU(2), it follows that within the subspace corresponding to a nondegenerate singular value, the matrix U is unique up to multiplication on the right by an arbitrary SU(2) matrix. Finally, in the subspace corresponding to the zero eigenvalues of M, the matrix U is unique up to multiplication on the right by an arbitrary unitary matrix. □

Exercises

G.1 The diagonalization of a general 2×2 hermitian matrix can be performed analytically. Consider a general 2×2 hermitian nondiagonal matrix,

$$A = \begin{pmatrix} a & c \\ c^* & b \end{pmatrix} \tag{G.33}$$

where a and b are real numbers and the complex number $c \neq 0$ expressed in polar exponential form is $c = |c|e^{i\phi}$, with $0 \leq \phi < 2\pi$. A hermitian matrix can be diagonalized by a unitary similarity transformation,

$$U^{-1}AU = D \equiv \begin{pmatrix} \lambda_1 & 0 \\ 0 & \lambda_2 \end{pmatrix}, \tag{G.34}$$

where λ_1 and λ_2 are the eigenvalues of A. By convention, we shall take $\lambda_1 \leq \lambda_2$.

(a) Show that the eigenvalues of A are given by

$$\lambda_{1,2} = \tfrac{1}{2}\left[a + b \mp \sqrt{(a-b)^2 + 4|c|^2}\right]. \tag{G.35}$$

(b) Parameterize the unitary matrix U as follows:

$$U = \begin{pmatrix} \cos\theta & e^{i\chi}\sin\theta \\ -e^{-i\chi}\sin\theta & \cos\theta \end{pmatrix}, \tag{G.36}$$

where $0 \leq \theta \leq \frac{1}{2}\pi$ and $0 \leq \chi < 2\pi$. Derive the following expressions that fix the form of U:

$$\sin 2\theta = \frac{2|c|}{\sqrt{(b-a)^2 + 4|c|^2}}, \qquad \cos 2\theta = \frac{b-a}{\sqrt{(b-a)^2 + 4|c|^2}}, \tag{G.37}$$

and $\chi = \phi = \arg c$. Note that the sign of $b - a$ determines uniquely the value of θ in its defined range.

(c) How would the results of part (b) change if the order of the diagonal elements of D in eq. (G.34) were reversed?

(d) A real symmetric matrix can be diagonalized by a real orthogonal similarity transformation. Suppose that c is real and nonzero in eq. (G.33). We can then use the results of parts (a) and (b) to diagonalize the 2×2 real symmetric matrix A. The eigenvalues of A are still given by eq. (G.35), where $c^2 = |c|^2$. If one redefines $\theta \to \theta\,\mathrm{sgn}\,c$ in eq. (G.36), show that the range of θ can be taken to be $-\frac{1}{2}\pi < \theta \leq \frac{1}{2}\pi$ without loss of generality. As a result, show that the diagonalizing matrix U is real and orthogonal, with $e^{i\chi} = \mathrm{sgn}\,c$ and

$$U = \begin{pmatrix} \cos\theta & \sin\theta \\ -\sin\theta & \cos\theta \end{pmatrix}, \qquad \text{where} \quad \begin{cases} c > 0 & \Longrightarrow \qquad 0 < \theta < \frac{1}{2}\pi\,, \\ c < 0 & \Longrightarrow \qquad -\frac{1}{2}\pi < \theta < 0\,, \end{cases} \tag{G.38}$$

where the angle θ is determined by

$$\sin 2\theta = \frac{2c}{\sqrt{(b-a)^2 + 4c^2}}\,, \qquad \cos 2\theta = \frac{b-a}{\sqrt{(b-a)^2 + 4c^2}}\,. \tag{G.39}$$

The signs of c and $b-a$ determine uniquely the value of θ in its defined range.

(e) Assume a fixed ordering of the diagonal elements of D as in eq. (G.34). How unique is the choice of U? Consider separately the two cases treated above: (i) A is a hermitian matrix and (ii) A is a real symmetric matrix.

G.2 A normal matrix B is defined as a matrix that commutes with its hermitian conjugate, i.e., $BB^\dagger = B^\dagger B$. A matrix is normal if and only if it can be diagonalized by a unitary similarity transformation. Consider the most general 2×2 complex normal matrix. In this exercise, you will show that the diagonalization of B can be reduced to the problem of diagonalizing a hermitian matrix (in which case, you can make use of the results of Exercise G.1).

(a) Show that the most general 2×2 complex normal matrix is of the form

$$B = \begin{pmatrix} a & |b|e^{i\alpha} \\ |b|e^{i\beta} & d \end{pmatrix}, \tag{G.40}$$

where a, b, and d are complex numbers and

$$\mathrm{Im}\big[(d-a)e^{-i(\alpha+\beta)/2}\big] = 0\,. \tag{G.41}$$

(b) Show that if eq. (G.41) is satisfied, then the matrix

$$A = e^{-i(\alpha+\beta)/2}(B - a\mathbb{1}_{2\times 2}) \tag{G.42}$$

is hermitian, where $\mathbb{1}_{2\times 2}$ is the 2×2 identity matrix.

(c) Using the results of Exercise G.1, diagonalize A and determine the diagonalizing matrix U [see eq. (G.36)]. Show that U also diagonalizes B. Determine the eigenvalues of B to complete the diagonalization procedure.

G.3 The singular value decomposition of a general 2×2 complex matrix can be performed analytically [223]. Consider a general complex nondiagonal matrix

$$M = \begin{pmatrix} a & c \\ \tilde{c} & b \end{pmatrix} . \tag{G.43}$$

Denote the singular values of M by m_1 and m_2, where $0 \leq m_1 \leq m_2$.

(a) Show that the two singular values are the nonnegative square roots of m_1^2 and m_2^2, where

$$m_{1,2}^2 = \tfrac{1}{2}\left[|a|^2 + |b|^2 + |c|^2 + |\tilde{c}|^2 \mp \Delta\right] , \tag{G.44}$$

and

$$\begin{aligned} \Delta &\equiv \left[(|a|^2 - |b|^2 + |\tilde{c}|^2 - |c|^2)^2 + 4|ac^* + b^*\tilde{c}|^2\right]^{1/2} \\ &= \left[|a|^2 + |b|^2 + |c|^2 + |\tilde{c}|^2)^2 - 4|ab - c\tilde{c}|^2\right]^{1/2} . \end{aligned} \tag{G.45}$$

Note that $m_1 = m_2$ if and only if $|a| = |b|$, $|c| = |\tilde{c}|$, and $ac^* + b^*\tilde{c} = 0$ are satisfied. Furthermore, $m_1 = 0$ when $\det M = ab - c\tilde{c} = 0$.

(b) Assume that $m_1 \neq m_2$. The singular value decomposition of M is

$$L^{\mathsf{T}} M R = \begin{pmatrix} m_1 & 0 \\ 0 & m_2 \end{pmatrix} . \tag{G.46}$$

Parameterize the unitary matrices L and R as follows:

$$L = \begin{pmatrix} e^{-i\alpha_L} \cos\theta_L & e^{i(\phi_L - \beta_L)} \sin\theta_L \\ -e^{-i(\phi_L + \alpha_L)} \sin\theta_L & e^{-i\beta_L} \cos\theta_L \end{pmatrix} , \tag{G.47}$$

$$R = \begin{pmatrix} e^{-i\alpha_R} \cos\theta_R & e^{i(\phi_R - \beta_R)} \sin\theta_R \\ -e^{-i(\phi_R + \alpha_R)} \sin\theta_R & e^{-i\beta_R} \cos\theta_R \end{pmatrix} , \tag{G.48}$$

where $0 \leq \theta_{L,R} \leq \frac{1}{2}\pi$ and $0 \leq \alpha_{L,R}, \beta_{L,R}, \phi_{L,R} < 2\pi$.

Using the results of Exercise G.1, obtain the following expressions for $\theta_{L,R}$ and $e^{i\phi_{L,R}}$ by diagonalizing $M^\dagger M$ and $M^* M^{\mathsf{T}}$ with a diagonalizing matrix R and L, respectively:

$$\cos\theta_{L,R} = \sqrt{\frac{\Delta + |b|^2 - |a|^2 \mp |c|^2 \pm |\tilde{c}|^2}{2\Delta}} , \tag{G.49}$$

$$\sin\theta_{L,R} = \sqrt{\frac{\Delta - |b|^2 + |a|^2 \pm |c|^2 \mp |\tilde{c}|^2}{2\Delta}} , \tag{G.50}$$

where Δ is given by eq. (G.45) and

$$e^{i\phi_L} = \frac{a^*\tilde{c} + bc^*}{|a^*\tilde{c} + bc^*|}\,, \qquad e^{i\phi_R} = \frac{a^*c + b\tilde{c}^*}{|a^*c + b\tilde{c}^*|}\,. \tag{G.51}$$

(c) Derive the following alternative expressions for θ_L and θ_R:

$$\tan\theta_L = \frac{|a|^2 + |c|^2 - m_1^2}{|a^*\tilde{c} + bc^*|}\,, \qquad \tan\theta_R = \frac{|a^*c + b\tilde{c}^*|}{m_2^2 - a^2 - |\tilde{c}|^2}\,. \tag{G.52}$$

Verify that the following relation is satisfied:

$$\frac{\tan\theta_L}{\tan\theta_R} = \left|\frac{c^*(ab - c\tilde{c}) + \tilde{c}m_1^2}{\tilde{c}^*(ab - c\tilde{c}) + cm_1^2}\right|\,. \tag{G.53}$$

(d) Show that eq. (G.46) determines $\alpha_L + \alpha_R$ and $\beta_L + \beta_R$. Hence, without loss of generality, one can choose $\alpha \equiv \alpha_L = \alpha_R$ and $\beta \equiv \beta_L = \beta_R$. Having adopted this convention, verify the following results:

$$\alpha = \tfrac{1}{2}\arg\{a(|b|^2 - m_1^2) - b^*c\tilde{c}\}\,, \tag{G.54}$$

$$\beta = \tfrac{1}{2}\arg\{b(m_2^2 - |a|^2) + a^*c\tilde{c}\}\,, \tag{G.55}$$

where $m_{1,2}^2$ are given in eq. (G.44). Show that α is indeterminate if $m_1 = 0$.

(e) Consider separately the case of degenerate nonzero singular values, i.e., $m \equiv m_1 = m_2 \neq 0$. In part (a), we saw that the degenerate case arises when $|a| = |b|$, $|c| = |\tilde{c}|$, and $ac^* = -b^*\tilde{c}$. Show that in this case one can re-express b in terms of a, c, and $\tilde{c}$ and rewrite M in the form

$$M = D_1 \begin{pmatrix} |a| & |c| \\ |c| & -|a| \end{pmatrix} D_2\,, \tag{G.56}$$

where D_1 and D_2 are diagonal matrices whose diagonal elements are pure phases. Hence, in the singular decomposition of M, one can absorb D_1 and D_2 into the unitary matrices L and R. Derive the following form for the singular decomposition of M:

$$\begin{pmatrix} i\cos\theta & -i\sin\theta \\ -\sin\theta & \cos\theta \end{pmatrix}\begin{pmatrix} |a| & |c| \\ |c| & -|a| \end{pmatrix}\begin{pmatrix} i\cos\theta & \sin\theta \\ -i\sin\theta & \cos\theta \end{pmatrix} = \begin{pmatrix} m & 0 \\ 0 & m \end{pmatrix}, \tag{G.57}$$

where $m = \sqrt{|a|^2 + |c|^2}$ and θ is determined by

$$\cos\theta = \sqrt{\frac{1 - |a|/m}{2}}\,, \qquad \sin\theta = \sqrt{\frac{1 + |a|/m}{2}}\,. \tag{G.58}$$

(f) Show that the singular value decomposition of a 2×2 complex matrix M with degenerate nonzero singular values, $L^{\mathsf{T}}MR = m\mathbb{1}_{2\times 2}$, is satisfied for $L = \mathbb{1}_{2\times 2}$ and $R = m^{-1}M^\dagger$. Discuss the uniqueness of the singular value decomposition of M by considering separately the two cases treated above: (i) $m_1 \neq m_2$ and (ii) $m_1 = m_2 \neq 0$.

G.4 The singular value decomposition of a general 2×2 real matrix can be performed analytically. The results of Exercise G.3 can be used with a few minor modifications.

(a) Parameterize the unitary matrices L and R as follows:

$$L = \begin{pmatrix} \cos\theta_L & \varepsilon_L \sin\theta_L \\ -\sin\theta_L & \varepsilon_L \cos\theta_L \end{pmatrix}, \qquad R = \begin{pmatrix} \cos\theta_R & \varepsilon_R \sin\theta_R \\ -\sin\theta_R & \varepsilon_R \cos\theta_R \end{pmatrix}, \tag{G.59}$$

where $-\frac{1}{2}\pi < \theta_{L,R} \leq \frac{1}{2}\pi$, and $\varepsilon_{L,R} = \pm 1$. Notice that the ranges of the angles θ_L and θ_R have been extended in comparison with the conventions employed in Exercise G.3. Show that the product $\varepsilon_L \varepsilon_R$ is fixed by the condition

$$\varepsilon_L \varepsilon_R = \operatorname{sgn}(\det M), \tag{G.60}$$

but one is free to set either ε_L or ε_R to unity without loss of generalization. Derive the following expressions that uniquely determine θ_L and θ_R:

$$\tan\theta_L = \frac{a^2 + c^2 - m_1^2}{a\tilde{c} + bc}, \qquad \tan\theta_R = \frac{ac + b\tilde{c}}{m_2^2 - a^2 - \tilde{c}^2}. \tag{G.61}$$

Show that the following relation is satisfied:

$$\frac{\tan\theta_L}{\tan\theta_R} = \frac{cm_2 + \varepsilon_L \varepsilon_R \tilde{c} m_1}{\tilde{c} m_2 + \varepsilon_L \varepsilon_R c m_1}. \tag{G.62}$$

(b) Consider separately the case of degenerate nonzero singular values, i.e., $m \equiv m_1 = m_2 \neq 0$. Verify that the degenerate case arises when $a = \pm b$ and $c = \mp\tilde{c}$, which yields $m = (a^2 + c^2)^{1/2}$. Since eq. (G.46) implies that $MR = mL$, one can take R to be an arbitrary 2×2 real orthogonal matrix. Show that θ_L is determined as follows:

$$\cos\theta_L = \frac{a\cos\theta_R - c\sin\theta_R}{\sqrt{a^2 + c^2}}, \qquad \sin\theta_L = \pm\left(\frac{c\cos\theta_R + a\sin\theta_R}{\sqrt{a^2 + c^2}}\right), \tag{G.63}$$

subject to the constraint that $\operatorname{sgn}(\det M) = \varepsilon_L \varepsilon_R = \pm 1$.

G.5 The Takagi diagonalization of a general 2×2 complex symmetric matrix can be performed analytically [223]. Consider a general 2×2 complex symmetric nondiagonal matrix,

$$M = \begin{pmatrix} a & c \\ c & b \end{pmatrix}, \tag{G.64}$$

where $c \neq 0$. The Takagi diagonalization of M is

$$U^{\mathsf{T}} M U = D = \begin{pmatrix} m_1 & 0 \\ 0 & m_2 \end{pmatrix}, \tag{G.65}$$

where U is a unitary matrix and the singular values of M, m_1, and m_2 are nonnegative. Assume that $m_1 \neq m_2$ (the degenerate case will be treated in Exercise G.6). Henceforth, we shall adopt a convention where $m_1 < m_2$. We can parameterize the 2×2 unitary matrix in eq. (G.65) by $U \equiv VP$, where $P = \text{diag}(e^{-i\alpha}, e^{-i\beta})$ is a diagonal matrix of phases and

$$V = \begin{pmatrix} \cos\theta & e^{i\phi}\sin\theta \\ -e^{-i\phi}\sin\theta & \cos\theta \end{pmatrix}. \tag{G.66}$$

(a) Show that one may restrict the ranges of the four angles θ, ϕ, α, and β as follows: $0 \leq \theta \leq \frac{1}{2}\pi$ and $0 \leq \alpha\,,\, \beta < \pi$ and $0 \leq \phi < 2\pi$.

(b) Show that eq. (G.65) is equivalent to

$$MV = V^* \begin{pmatrix} \sigma_1 & 0 \\ 0 & \sigma_2 \end{pmatrix}, \tag{G.67}$$

where $|\sigma_1| < |\sigma_2|$ and

$$\sigma_1 \equiv m_1 e^{2i\alpha} = a - c\,e^{-i\phi}\tan\theta = b\,e^{-2i\phi} - c\,e^{-i\phi}\cot\theta\,, \tag{G.68}$$

$$\sigma_2 \equiv m_2 e^{2i\beta} = b + c\,e^{i\phi}\tan\theta \;\; = a\,e^{2i\phi} + c\,e^{i\phi}\cot\theta\,. \tag{G.69}$$

(c) Using part (b), obtain an expression for $\tan 2\theta$ in terms of ϕ and the elements of M. Since $\tan 2\phi$ is real, conclude that $\text{Im}(bc^*\,e^{-i\phi} - ac^*\,e^{i\phi}) = 0$. Derive an expression for $e^{i\phi}$ up to a possible ambiguity of an overall sign. Resolve this ambiguity by imposing $|\sigma_1| < |\sigma_2|$ and show that

$$e^{i\phi} = \frac{a^*c + bc^*}{|a^*c + bc^*|}\,, \qquad \tan 2\theta = \frac{2|a^*c + bc^*|}{|b|^2 - |a|^2}\,. \tag{G.70}$$

Using the latter result, derive an expression for $\tan\theta$.

(d) The singular values of M can be derived by taking the nonnegative square roots of the eigenvalues of $M^\dagger M$. Show that

$$m_{1,2}^2 = |\sigma_{1,2}|^2 = \tfrac{1}{2}\left[|a|^2 + |b|^2 + 2|c|^2 \mp \sqrt{(|b|^2 - |a|^2)^2 + 4|bc^* + a^*c|^2}\right]. \tag{G.71}$$

(e) The results of part (d) can be independently checked by using eqs. (G.68) and (G.69) to first prove that

$$|\sigma_2|^2 - |\sigma_1|^2 = |a^*c + bc^*|^2(\tan\theta + \cot\theta)\,, \tag{G.72}$$

after employing the result for $e^{i\phi}$ given in eq. (G.70). Then, one can make use of $\text{Tr}(M^\dagger M) = |\sigma_1|^2 + |\sigma_2|^2 = |a|^2 + |b|^2 + 2|c|^2$ to isolate separate expressions for $|\sigma_1|^2$ and $|\sigma_2|^2$.

(f) Finally, using the results of parts (b) and (c), deduce the following explicit expressions for α and β:

$$\alpha = \tfrac{1}{2}\arg\left[a(|b|^2 - |\sigma_1|^2) - b^* c^2\right], \tag{G.73}$$

$$\beta = \tfrac{1}{2}\arg\left[b(|\sigma_2|^2 - |a|^2) + a^* c^2\right]. \tag{G.74}$$

(g) If $\det M = ab - c^2 = 0$, then $m_1 = 0$. Show that $m_2 = |a| + |b|$,

$$\tan\theta = |a/b|^{1/2}, \qquad \phi = \arg(b/c) = \arg(c/a), \qquad \beta = \tfrac{1}{2}\arg b, \tag{G.75}$$

and the angle α is indeterminate. Explain why α is arbitrary.

(h) How unique is the Takagi diagonalization of M? Consider separately the cases of (i) $0 < m_1 < m_2$ and (ii) $m_1 = 0$ and $m_2 > 0$.

G.6 In Exercise G.5, we considered the Takagi diagonalization of a 2×2 complex symmetric (nondiagonal) matrix M with nondegenerate singular values. In this exercise, the case of degenerate singular values will be addressed.

(a) Starting from eqs. (G.64) and (G.65) under the assumption that $c \neq 0$, show that $m_1 = m_2$ if and only if

$$a^* c + b c^* = 0. \tag{G.76}$$

Verify that if eq. (G.76) is satisfied, then $|a| = |b|$ and

$$m = m_1 = m_2 = \sqrt{|b|^2 + |c|^2}. \tag{G.77}$$

(b) Suppose eq. (G.76) is satisfied. Note that ϕ and θ are indeterminate in light of eq. (G.70). Nevertheless, these two indeterminate angles are related if $a,\ b \neq 0$. Using eqs. (G.68), (G.69), and (G.76), derive the following two results:

$$\tan 2\theta = \left[\mathrm{Re}(b/c)c_\phi + \mathrm{Im}(b/c)s_\phi\right]^{-1}, \tag{G.78}$$

$$c\sigma_1^* + c^*\sigma_2 = 0, \tag{G.79}$$

where $c_\phi \equiv \cos\phi$ and $s_\phi \equiv \sin\phi$. If $a = b = 0$, note that eq. (G.68) yields $\theta = \frac{1}{4}\pi$, independently of the value of the indeterminate angle ϕ. Thus, in this special case, θ is uniquely determined in the range $0 \leq \theta \leq \frac{1}{2}\pi$.

(c) In the degenerate case under consideration, eqs. (G.73) and (G.74) are no longer valid, as their derivation relies on the results of eq. (G.70). Employing eqs. (G.68), (G.69), and (G.76) and the results of part (b), show that

$$\sigma_1 = me^{2i\alpha} = -ce^{-i\phi}\left[(1 + A^2)^{1/2} + iB\right], \tag{G.80}$$

$$\sigma_2 = me^{2i\beta} = ce^{i\phi}\left[(1 + A^2)^{1/2} - iB\right], \tag{G.81}$$

where $m = (|b|^2 + |c|^2)^{1/2}$ and

$$A \equiv \mathrm{Re}(b/c)c_\phi + \mathrm{Im}(b/c)s_\phi, \qquad B \equiv \mathrm{Re}(b/c)s_\phi - \mathrm{Im}(b/c)c_\phi. \tag{G.82}$$

Thus, the angles α and β are separately determined in terms of the indeterminate angle ϕ. Nevertheless, show that the sum $\alpha+\beta$ is independent of ϕ:

$$e^{2i(\alpha+\beta)} = -\frac{c}{c^*}\,. \tag{G.83}$$

Thus, the matrix U in eq. (G.65) is now fixed in terms of the quantity $\alpha+\beta$ and the indeterminate angle ϕ.

(d) How unique is the Takagi diagonalization of a 2×2 complex symmetric nondiagonal matrix (assuming a, $b \neq 0$) in the case of degenerate singular values?

(e) Apply the results of parts (b) and (c) to the Takagi diagonalization of[14]

$$M = \begin{pmatrix} 0 & c \\ c & 0 \end{pmatrix}, \tag{G.84}$$

where c is a nonzero complex number. Write out explicitly the Takagi diagonalization of M and show that it can be written as $U^{\mathsf{T}}MU = |c|\mathbb{1}_{2\times 2}$, where

$$U = \frac{1}{\sqrt{2}}\begin{pmatrix} i & 1 \\ -i & 1 \end{pmatrix}\begin{pmatrix} \pm\cos(\phi/2) & \sin(\phi/2) \\ \mp\sin(\phi/2) & \cos(\phi/2) \end{pmatrix}. \tag{G.85}$$

Since ϕ is indeterminate, we see that the Takagi diagonalization of M in this special case is unique only up to multiplication on the right by an arbitrary orthogonal matrix. Show that eqs. (G.16) and (G.85) are equivalent.

G.7 Real symmetric matrices are always diagonalizable by a real orthogonal similarity transformation. It is noteworthy that not all *complex* symmetric matrices are diagonalizable by a similarity transformation. In contrast, complex symmetric matrices are *always* Takagi diagonalizable. For example, consider the symmetric matrix

$$M = \begin{pmatrix} 1 & i \\ i & -1 \end{pmatrix}. \tag{G.86}$$

(a) Show that M is *not* diagonalizable by a similarity transformation.

(b) Construct the Takagi diagonalization of the matrix M.

[14] Note that $M^\dagger M = |c|^2\mathbb{1}_{2\times 2}$. Hence, in this case, the Takagi diagonalization of M cannot be deduced by diagonalizing $M^\dagger M$.

Appendix H Lie Group and Algebra Techniques for Gauge Theories

H.1 Lie Groups, Lie Algebras, and their Representations

Consider a compact connected Lie group G, which can always be expressed as a direct product of compact connected simple groups and U(1) groups. If no U(1) factors are present, then G is semisimple. For any $U \in G$,

$$U = \exp(-i\theta^a \boldsymbol{T^a}) , \tag{H.1}$$

where the $\boldsymbol{T^a}$ are the generators of G and the θ^a are real numbers that parameterize the elements of G. The corresponding real Lie algebra $\mathfrak{g}$ consists of arbitrary real linear combinations of the generators, $\theta^a \boldsymbol{T^a}$. The Lie group generators $\boldsymbol{T^a}$, which serve as a basis for the Lie algebra $\mathfrak{g}$, satisfy the commutation relations

$$[\boldsymbol{T^a}, \boldsymbol{T^b}] = i f_c^{ab} \boldsymbol{T^c} , \tag{H.2}$$

where the structure constants f_c^{ab} are real numbers that satisfy $f_c^{ab} = -f_c^{ba}$ and the Jacobi identity,

$$f_c^{ab} f_e^{cd} + f_c^{da} f_e^{cb} + f_c^{bd} f_e^{ca} = 0 . \tag{H.3}$$

The generator indices run over $a, b, c \in \{1, 2, \ldots, d_G\}$, where d_G is the dimension of the Lie algebra. If $\mathfrak{g}$ is a semisimple compact Lie algebra, then a basis for $\mathfrak{g}$ exists such that $\mathrm{Tr}(\boldsymbol{T^a T^b}) \propto \delta^{ab}$ (where the proportionality constant is a positive real number). With respect to this new basis, the structure constants $f^{abc} \equiv \delta^{cd} f_d^{ab}$ are totally antisymmetric with respect to the interchange of the indices a, b, c.

The elements of the compact Lie group G act on a multiplet of fields. This group action is specified by a d_R-dimensional matrix representation R of G acting on a *complex* d_R-dimensional vector space. All representations of a compact group are equivalent (via a similarity transformation) to a unitary representation. Hence, the group elements $U \in G$ can be represented by $d_R \times d_R$ unitary matrices, $D_R(U) = \exp(-i\theta^a \boldsymbol{T^a_R})$, where the $\boldsymbol{T^a_R}$ are $d_R \times d_R$ hermitian matrices that satisfy eq. (H.2) and thus provide a representation of the Lie group generators. For any representation R of a semisimple group, $\mathrm{Tr}\, \boldsymbol{T^a_R} = 0$ for all a. A representation R' is unitarily equivalent to R if there exists a fixed unitary matrix S such that $D_{R'}(U) = S^{-1} D_R(U) S$ for all $U \in G$. Similarly, the corresponding generators satisfy $\boldsymbol{T^a_{R'}} = S^{-1} \boldsymbol{T^a_R} S$ for all $a \in \{1, 2, \ldots, d_G\}$.

For semisimple compact Lie groups, two representations are noteworthy. If G is one of the classical groups, SU(n) [for $n \geq 2$], SO(n) [for $n \geq 3$], or Sp($n/2$) [for even integer values of $n \geq 2$ in light of eqs. (G.31) and (G.32)], then the $n \times n$ matrices that define these groups comprise the *fundamental* (or *defining*) *representation* F, with $d_F = n$. For example, the fundamental representation of SU(n) consists of $n \times n$ unitary matrices with determinant equal to 1, and the corresponding generators comprise a suitably chosen basis for the $n \times n$ traceless hermitian matrices, as shown in Table 4.1. Every Lie group G also possesses an *adjoint representation* A, with $d_A = d_G$. The matrix elements of the generators in the adjoint representation are

$$(\boldsymbol{T}^a_{\boldsymbol{A}})^b{}_c = -if^{ab}_c\,. \tag{H.4}$$

Since the f^{ab}_c are real, $i\boldsymbol{T}^a_{\boldsymbol{A}}$ (for $a \in \{1, 2, \ldots, d_G\}$) are real matrices.

Given the unitary matrices $D_R(U)$ of the representation R of G, the matrices $[D_R(U)]^*$ constitute the *conjugate* representation $\overline{R}$. Equivalently, if the $\boldsymbol{T}^a_{\boldsymbol{R}}$ define a representation R of the Lie algebra $\mathfrak{g}$, then the $-(\boldsymbol{T}^a_{\boldsymbol{R}})^* = -(\boldsymbol{T}^a_{\boldsymbol{R}})^{\mathsf{T}}$ define a representation $\overline{R}$ of the same dimension d_R. If $\overline{R}$ and R are unitarily equivalent, then the representation R is *self-conjugate.* Otherwise, the representation R is *complex* (also called "strictly complex" in the language of [B119]). For example, the fundamental representation F is complex for $G =$ SU(n) [$n \geq 3$], SO($4k+2$) [$k \geq 1$], or E_6, in which case the conjugate $\overline{F}$ is called the *antifundamental representation.*

However, the representation matrices $D_R(U)$ of a self-conjugate representation can also be complex. If R and $\overline{R}$ are unitarily equivalent to a representation R' that satisfies the reality property $[D_{R'}(U)]^* = D_{R'}(U)$ for all $U \in G$ (equivalently, the matrices $i\boldsymbol{T}^a_{\boldsymbol{R'}}$ are real for all a), then the representation R is *real* (also called "strictly real" in the language of [B119]). If R and $\overline{R}$ are unitarily equivalent representations, but neither is unitarily equivalent to a representation that satisfies the reality property above, then R is said to be *pseudoreal.*

Self-conjugate representations are either real or pseudoreal. Moreover, for a self-conjugate representation, there exists a constant unitary matrix W such that [B119]

$$[D_R(U)]^* = WD_R(U)W^{-1}\,, \quad \text{or equivalently,} \quad (i\boldsymbol{T}^a_{\boldsymbol{R}})^* = W(i\boldsymbol{T}^a_{\boldsymbol{R}})W^{-1}\,, \tag{H.5}$$

where

$$WW^* = \mathbb{1}\,, \qquad W^{\mathsf{T}} = W\,, \qquad \text{for real representations}\,, \tag{H.6}$$

$$WW^* = -\mathbb{1}\,, \qquad W^{\mathsf{T}} = -W\,, \qquad \text{for pseudoreal representations}\,, \tag{H.7}$$

and $\mathbb{1}$ is the $d_R \times d_R$ identity matrix. Taking the determinant of eq. (H.7), and using the fact that W is unitary (and hence invertible), it follows that $1 = (-1)^{d_R}$. Therefore, a pseudoreal representation must be even-dimensional. If one redefines the basis for the Lie group generators by $\boldsymbol{T}^a_{\boldsymbol{R}} \to V^{-1}\boldsymbol{T}^a_{\boldsymbol{R}}V$, where V is unitary, then $W \to V^{\mathsf{T}}WV$. We can make use of this change of basis to transform W to a canonical form. Since W is unitary, its singular values (i.e., the positive square roots of the eigenvalues of $W^\dagger W$) are all equal to 1. Hence, in the two cases corresponding to $W^{\mathsf{T}} = \pm W$, respectively, eqs. (G.11) and (G.21) yield the following canonical forms (for an appropriately chosen V):

$$W = \mathbb{1}\,, \qquad \text{for a real representation } R\,, \tag{H.8}$$

$$W = J\,, \qquad \text{for a pseudoreal representation } R\,, \tag{H.9}$$

where $J \equiv \text{diag}\left\{\left(\begin{smallmatrix} 0 & 1 \\ -1 & 0 \end{smallmatrix}\right), \left(\begin{smallmatrix} 0 & 1 \\ -1 & 0 \end{smallmatrix}\right), \ldots, \left(\begin{smallmatrix} 0 & 1 \\ -1 & 0 \end{smallmatrix}\right)\right\}$ is a $d_R \times d_R$ matrix.

There are many examples of complex, real, and pseudoreal representations. For example, the fundamental representation of SU(n) is complex for $n \geq 3$. The adjoint representation of any Lie group is real in light of eqs. (H.1) and (H.4). The simplest example of a pseudoreal representation is the two-dimensional representation of SU(2), where $\boldsymbol{T^a} = \frac{1}{2}\tau^a$ (and the τ^a are the usual Pauli matrices).[1] More generally, the generators of a pseudoreal representation must satisfy

$$(i\boldsymbol{T^a_R})^* = C^{-1}(i\boldsymbol{T^a_R})C\,, \tag{H.10}$$

for some fixed unitary antisymmetric matrix C [previously denoted by W^{-1} in eqs. (H.5) and (H.7)]. For the two-dimensional representation of SU(2) given above, $C^{ab} = (i\tau^2)^{ab} \equiv \epsilon^{ab}$ is the familiar SU(2)-invariant tensor (see Appendix A.8).

Finally, we note that for U(1), all irreducible representations are one-dimensional. The structure constants vanish and any d-dimensional representation of the U(1) generator is given by the $d \times d$ identity matrix multiplied by the corresponding U(1) charge. For a Lie group that is a direct product of U(1) groups and a semisimple group, $\text{Tr}\,\boldsymbol{T^a_R}$ is nonzero when a corresponds to one of the U(1) generators, unless the sum of the U(1) charges of the states of the representation R vanishes.

H.2 Matrix Exponentials

In light of eq. (H.1), we present a number of theorems involving the matrix exponential [B117]. These theorems employ the adjoint operator ad_A defined by

$$\text{ad}_A(B) = [A, B] \equiv AB - BA\,, \tag{H.11}$$

which is a linear operator acting on the vector space of $n \times n$ matrices. Note that

$$(\text{ad}_A)^n(B) = \underbrace{[A, [\cdots [A, [A, B]] \cdots]]}_{n \text{ times}} \tag{H.12}$$

involves n nested commutators. We also introduce two auxiliary functions, $f(z)$ and $g(z)$, that are defined by their power series,

$$f(z) \equiv \frac{1 - e^{-z}}{z} = \sum_{n=0}^{\infty} \frac{(-1)^n}{(n+1)!} z^n\,, \qquad |z| < \infty\,, \tag{H.13}$$

[1] No unitary matrix W exists such that the $Wi\tau^a W^{-1}$ are real for all $a = 1, 2, 3$. Thus, the two-dimensional representation of SU(2) is not real. However, $(i\tau^a)^* = (i\tau^2)(i\tau^a)(i\tau^2)^{-1}$ for $a = 1, 2, 3$, which proves that the two-dimensional representation of SU(2) is pseudoreal.

$$g(z) \equiv \frac{\ln z}{z-1} = \sum_{n=0}^{\infty} \frac{(1-z)^n}{n+1}\,, \qquad |1-z| < 1\,. \tag{H.14}$$

Theorem H.1.

$$e^A B e^{-A} = \exp(\mathrm{ad}_A)(B) \equiv \sum_{n=0}^{\infty} \frac{1}{n!}(\mathrm{ad}_A)^n(B) = B + [A,B] + \tfrac{1}{2}[A,[A,B]] + \cdots\,. \tag{H.15}$$

Proof Define $B(t) \equiv e^{tA} B e^{-tA}$ and compute the Taylor series of $B(t)$ around the point $t = 0$. A simple computation yields $B(0) = B$ and

$$\frac{dB(t)}{dt} = Ae^{tA}Be^{-tA} - e^{tA}Be^{-tA}A = [A, B(t)] = \mathrm{ad}_A(B(t))\,. \tag{H.16}$$

Higher derivatives can also be computed. It is a simple exercise to show that

$$\frac{d^n B(t)}{dt^n} = (\mathrm{ad}_A)^n(B(t))\,. \tag{H.17}$$

Putting $t = 1$ in the resulting Taylor series expansion of $B(t)$ completes the proof. □

Theorem H.2.

$$e^{-A(t)}\frac{d}{dt}e^{A(t)} = f(\mathrm{ad}_A)\left(\frac{dA}{dt}\right) = \frac{dA}{dt} - \frac{1}{2!}\left[A, \frac{dA}{dt}\right] + \frac{1}{3!}\left[A, \left[A, \frac{dA}{dt}\right]\right] - \cdots\,, \tag{H.18}$$

where $f(z)$ is defined via its Taylor series in eq. (H.13). In general $A(t)$ does not commute with dA/dt.

Proof Define

$$B(s,t) \equiv e^{-sA(t)}\frac{d}{dt}e^{sA(t)}\,, \tag{H.19}$$

and compute the Taylor series of $B(s,t)$ around the point $s = 0$. It is straightforward to verify that $B(0,t) = 0$ and

$$\left.\frac{d^n B(s,t)}{ds^n}\right|_{s=0} = (-\,\mathrm{ad}_{A(t)})^{n-1}\left(\frac{dA}{dt}\right)\,, \tag{H.20}$$

for all positive integers n. Assembling the Taylor series for $B(s,t)$ and inserting $s = 1$ completes the proof. □

Theorem H.3.

$$\left(\frac{d}{dt}e^{A+tB}\right)_{t=0} = e^A f(\mathrm{ad}_A)(B)\,, \tag{H.21}$$

where $f(z)$ is defined in eq. (H.13) and A and B are independent of t.

Proof Define $A(t) \equiv A + tB$, and use Theorem H.2. □

Corollary.

$$\exp(A+\epsilon B) = e^A\left[1+\epsilon\, f(\mathrm{ad}_A)(B) + \mathcal{O}(\epsilon^2)\right] . \tag{H.22}$$

Proof Starting from Theorem H.3, let us denote the right-hand side of eq. (H.21) by $F(A,B) \equiv e^A\, f(\mathrm{ad}_A)(B)$. Then, using the definition of the derivative, it follows that

$$\left(\frac{d}{dt}e^{A+tB}\right)_{t=0} = \left(\lim_{\epsilon\to 0}\frac{e^{A+(t+\epsilon)B}-e^{A+tB}}{\epsilon}\right)_{t=0} = \lim_{\epsilon\to 0}\frac{e^{A+\epsilon B}-e^A}{\epsilon} = F(A,B) . \tag{H.23}$$

In particular, eq. (H.23) implies that

$$e^{A+\epsilon B} = e^A + \epsilon F(A,B) + \mathcal{O}(\epsilon^2) . \tag{H.24}$$

Employing the definition of $F(A,B)$ yields eq. (H.22). □

Theorem H.4. The Baker–Campbell–Hausdorff (BCH) formula

$$e^A e^B = \exp\left\{B + \int_0^1 g\left[\exp(t\ \mathrm{ad}_A)\exp(\mathrm{ad}_B)\right](A)\, dt\right\} , \tag{H.25}$$

where $g(z)$ is defined via its Taylor series in eq. (H.14).

Since $g(z)$ is defined for $|1-z|<1$, it follows that the BCH formula for $\ln\left(e^A e^B\right)$ converges, provided that $\|e^A e^B - \boldsymbol{I}\| < 1$, where $\boldsymbol{I}$ is the identity matrix and $\|\cdots\|$ is a suitably defined matrix norm. Expanding the BCH formula, using the series definition of $g(z)$, yields

$$e^A e^B = \exp\left(A + B + \tfrac{1}{2}[A,B] + \tfrac{1}{12}[A,[A,B]] + \tfrac{1}{12}[B,[B,A]] + \cdots\right) , \tag{H.26}$$

assuming that the resulting series is convergent. Indeed, if G is a Lie group and its corresponding Lie algebra is $\mathfrak{g}$, then the BCH formula (if convergent) implies that if $A, B \in \mathfrak{g}$ then the product of the Lie Group elements e^A, $e^B \in$ G yields an element of G that is also the exponential of some element of the Lie algebra $\mathfrak{g}$.

Proof See, e.g., pp. 161–162 of [B117] or Section 5.5 of [B110]. □

The Zassenhaus formula for matrix exponentials is sometimes referred to as the dual of the BCH formula. It provides an expression for $\exp(A+B)$ as an infinite produce of matrix exponentials. It is convenient to insert a parameter t into the argument of the exponential.

Theorem H.5. The Zassenhaus formula

$$e^{t(A+B)} = e^{tA}e^{tB}\exp\{-\tfrac{1}{2}t^2[A,B]\}\exp\{\tfrac{1}{6}t^3(2[B,[A,B]] + [A,[A,B]])\}\cdots , \tag{H.27}$$

or more explicitly,

$$e^{t(A+B)} = e^{tA}e^{tB}e^{t^2C_2}e^{t^3C_3}\cdots , \tag{H.28}$$

where the C_n are defined recursively as

$$C_2 = \frac{1}{2}\left[\frac{\partial^2}{\partial t^2}\left(e^{-tB}e^{-tA}e^{t(A+B)}\right)\right]_{t=0} = -\tfrac{1}{2}[A,B]\,, \tag{H.29}$$

$$C_3 = \frac{1}{3!}\left[\frac{\partial^3}{\partial t^3}\left(e^{-t^2C_2}e^{-tB}e^{-tA}e^{t(A+B)}\right)\right]_{t=0} = -\tfrac{1}{3}[A+2B,C_2]\,, \tag{H.30}$$

and in general

$$C_n = \frac{1}{n!}\left[\frac{\partial^n}{\partial t^n}\left(e^{-t^{n-1}C_{n-1}}\dots e^{-t^2C_2}e^{-tB}e^{-tA}e^{t(A+B)}\right)\right]_{t=0}. \tag{H.31}$$

Proof See, e.g., Ref. [224]. □

H.3 Dynkin Index and Casimir Operator

In this section, we define the Dynkin index [225] and Casimir operator of a representation of a semisimple compact Lie algebra $\mathfrak{g}$.

The *second-order Dynkin index* $I_2(R)$ of the representation R is defined by [B116]

$$\mathrm{Tr}(\boldsymbol{T}^a_{\boldsymbol{R}}\boldsymbol{T}^b_{\boldsymbol{R}}) = I_2(R)\delta^{ab}\,, \tag{H.32}$$

where $I_2(R)$ is a positive real number that depends on R. Once $I_2(R)$ is defined for one representation, its value is uniquely fixed for any other representation. Note that an alternative unboldfaced notation (not to be confused with the boldfaced $\boldsymbol{T}^a_{\boldsymbol{R}}$),

$$T_R \equiv I_2(R)\,, \tag{H.33}$$

is often used in the literature (as well as in other parts of this book).

If the representation R is reducible, it can be decomposed into the direct sum of irreducible representations, $R = \sum_k R_k$. In this case, $I_2(R)$ is given by

$$I_2(R) = \sum_k I_2(R_k)\,. \tag{H.34}$$

The second-order Dynkin index of a tensor product of R_1 and R_2 is given by

$$I_2(R_1\otimes R_2) = d_{R_1}I_2(R_2) + d_{R_2}I_2(R_1)\,. \tag{H.35}$$

Moreover, if $R_1 = R_2$, one can derive a result for the second-order Dynkin index of the symmetric and antisymmetric parts of the tensor product $R\otimes R$ [B121]:

$$I_2\big[(R\otimes R)_S\big] = (d_R+2)I_2(R)\,, \qquad I_2\big[(R\otimes R)_A\big] = (d_R-2)I_2(R)\,. \tag{H.36}$$

Finally, we note that if $\overline{R}$ is the complex conjugate of the representation R, then

$$I_2(\overline{R}) = I_2(R)\,. \tag{H.37}$$

A Casimir operator of a Lie algebra $\mathfrak{g}$ is an operator that commutes with all the generators $\boldsymbol{T}^a$. If the representation R of the generators (acting on a complex vector space) is *irreducible*, then Schur's lemma implies that the Casimir operator is an (R-dependent) multiple of the identity (e.g., see Section 4.7.2 of [B112]). The quadratic Casimir operator of an irreducible representation R is given by

$$(\boldsymbol{T}^2_{\boldsymbol{R}})_i{}^j \equiv (\boldsymbol{T}^a_{\boldsymbol{R}})_i{}^k(\boldsymbol{T}^a_{\boldsymbol{R}})_k{}^j = C_R\delta_i{}^j\,, \tag{H.38}$$

for i, j, $k \in \{1, 2, \ldots, d_R\}$ (with an implicit sum over k). Setting $i = j$ and summing over j then yields the eigenvalue, C_R, of the quadratic Casimir operator:

$$C_R = \frac{I_2(R)d_G}{d_R}\,. \tag{H.39}$$

For a simple Lie algebra the adjoint representation (with dimension $d_A = d_G$) is irreducible, and it immediately follows that $C_A = I_2(A)$. For a reducible representation, $\boldsymbol{T}^2_{\boldsymbol{R}}$ is a block-diagonal matrix consisting of $d_{R_k} \times d_{R_k}$ blocks given by $C_{R_k}\mathbb{1}$ for each irreducible component R_k of R.

The example of the simple Lie algebra $\mathfrak{su}(n)$ is well known. The standard convention in the physics literature employs antisymmetric structure constants f^{abc} and normalizes the generators of the fundamental representation F of $\mathfrak{su}(n)$ such that

$$\mathrm{Tr}(\boldsymbol{T}^a_{\boldsymbol{F}}\boldsymbol{T}^b_{\boldsymbol{F}}) = \tfrac{1}{2}\delta^{ab}\,. \tag{H.40}$$

The dimension of this Lie algebra (equal to the number of generators) is given by $n^2 - 1$. As previously noted, $d_F = n$ and $I_2(F) = \frac{1}{2}$ in light of eq. (H.40). It then follows that $C_F = (n^2 - 1)/(2n)$. One can also check that $C_A = I_2(A) = n$.

The Lie algebras $\mathfrak{su}(n)$ [$n \geq 3$] are the only simple Lie algebras that possess a cubic Casimir operator [226]. First, we define the symmetrized trace of three generators [227–230]:

$$D^{abc} \equiv \mathrm{STr}(\boldsymbol{T}^a\boldsymbol{T}^b\boldsymbol{T}^c) = \tfrac{1}{6}\,\mathrm{Tr}(\boldsymbol{T}^a\boldsymbol{T}^b\boldsymbol{T}^c + \mathrm{perm}) = \tfrac{1}{2}\,\mathrm{Tr}\left[\{\boldsymbol{T}^a, \boldsymbol{T}^b\}\boldsymbol{T}^c\right], \tag{H.41}$$

where "perm" indicates the five additional terms obtained by permuting the indices a, b, and c, and $\{\boldsymbol{T}^a, \boldsymbol{T}^b\} \equiv \boldsymbol{T}^a\boldsymbol{T}^b + \boldsymbol{T}^b\boldsymbol{T}^a$. That is, for a given representation R,

$$D^{abc}(R) = \tfrac{1}{2}\,\mathrm{Tr}\left[\{\boldsymbol{T}^a_{\boldsymbol{R}}, \boldsymbol{T}^b_{\boldsymbol{R}}\}\boldsymbol{T}^c_{\boldsymbol{R}}\right]. \tag{H.42}$$

By convention, for the n-dimensional fundamental representation of $\mathfrak{su}(n)$ we define

$$d^{abc} \equiv 2\,\mathrm{Tr}\left[\{\boldsymbol{T}^a_{\boldsymbol{F}}, \boldsymbol{T}^b_{\boldsymbol{F}}\}\boldsymbol{T}^c_{\boldsymbol{F}}\right]. \tag{H.43}$$

One important property of the d^{abc} is [197, 226]

$$d^{abc}d^{abc} = \frac{(n^2-1)(n^2-4)}{n}\,. \tag{H.44}$$

In general, $D^{abc}(R)$ is proportional to d^{abc}. In particular, the *cubic index* $I_3(R)$ of a representation R is defined such that

$$D^{abc}(R) = I_3(R)d^{abc}\,. \tag{H.45}$$

It then follows that, for any irreducible representation R of $\mathfrak{su}(n)$,

$$\mathrm{Tr}(\boldsymbol{T}^a_{\boldsymbol{R}}\boldsymbol{T}^b_{\boldsymbol{R}}\boldsymbol{T}^c_{\boldsymbol{R}}) = I_3(R)d^{abc} + \tfrac{1}{2}iI_2(R)f^{abc}\,. \tag{H.46}$$

In particular, $I_3(F) = \frac{1}{4}$ in light of eqs. (H.42) and (H.43). Having fixed $I_3(F)$, the cubic index is then uniquely determined for all irreducible representations of $\mathfrak{su}(n)$. If the hermitian generators of the representation R are $\boldsymbol{T}^{\boldsymbol{a}}_{\boldsymbol{R}}$, then the generators of the complex conjugate representation $\overline{R}$ are $-\boldsymbol{T}^{\boldsymbol{a}\,\mathsf{T}}_{\boldsymbol{R}}$. It then follows that

$$I_3(\overline{R}) = -I_3(R)\,. \tag{H.47}$$

Hence, the cubic index of a self-conjugate representation vanishes. Note that the index of a complex representation of $\mathfrak{su}(n)$ can also vanish in special cases [231].

For a reducible representation $R = \sum_k R_k$,

$$I_3(R) = \sum_k I_3(R_k)\,. \tag{H.48}$$

The cubic index of a tensor product of R_1 and R_2 is given by

$$I_3(R_1 \otimes R_2) = d_{R_1} I_3(R_2) + d_{R_2} I_3(R_1)\,. \tag{H.49}$$

Moreover, if $R_1 = R_2$, one can derive a result for the cubic index of the symmetric and antisymmetric part of the tensor product $R \otimes R$ (e.g., see [B121] or Ref. [232]):

$$I_3\big[(R\otimes R)_S\big] = (d_R + 4) I_3(R)\,, \qquad I_3\big[(R\otimes R)_A\big] = (d_R - 4) I_3(R)\,. \tag{H.50}$$

For a non-semisimple Lie group (i.e., a Lie group that is a direct product of simple Lie groups and at least one U(1) group), D^{abc} is generally nonvanishing. For example, suppose that the $\boldsymbol{T}^{\boldsymbol{a}}_{\boldsymbol{R}}$ constitute an irreducible representation of the generators of $G \otimes \mathrm{U}(1)$, where G is a semisimple Lie group. Then the U(1) generator (which we denote by setting the index $\boldsymbol{a} = \boldsymbol{Q}$) is $\boldsymbol{T}^{\boldsymbol{Q}}_{\boldsymbol{R}} \equiv q\mathbb{1}$, where q is the corresponding U(1) charge. It then follows that $D^{Qab} = qI_2(R)\delta^{ab}$. More generally, for a compact non-semisimple Lie group, D^{abc} can be nonzero when either one or three of its indices corresponds to a U(1) generator.

In the computation of the anomaly, the quantity $\mathrm{Tr}(\boldsymbol{T}^{\boldsymbol{a}}_{\boldsymbol{R}}\boldsymbol{T}^{\boldsymbol{b}}_{\boldsymbol{R}}\boldsymbol{T}^{\boldsymbol{c}}_{\boldsymbol{R}})$ appears (see Section 5.1). We can evaluate this trace using eqs. (H.2) and (H.45):

$$\mathrm{Tr}(\boldsymbol{T}^{\boldsymbol{a}}_{\boldsymbol{R}}\boldsymbol{T}^{\boldsymbol{b}}_{\boldsymbol{R}}\boldsymbol{T}^{\boldsymbol{c}}_{\boldsymbol{R}}) = I_3(R) d^{abc} + \tfrac{1}{2} i I_2(R) f^{abc}\,. \tag{H.51}$$

The cubic Casimir operator of an *irreducible* representation R is given by

$$(\boldsymbol{T}^3_{\boldsymbol{R}})_i{}^j \equiv d^{abc}(\boldsymbol{T}^{\boldsymbol{a}}_{\boldsymbol{R}}\boldsymbol{T}^{\boldsymbol{b}}_{\boldsymbol{R}}\boldsymbol{T}^{\boldsymbol{c}}_{\boldsymbol{R}})_i{}^j = C_{3R}\delta_i{}^j\,. \tag{H.52}$$

Using eqs. (H.44) and (H.45), we obtain a relation between the eigenvalue of the cubic Casimir operator, C_{3R}, and the cubic index:

$$C_{3R} = \frac{(n^2-1)(n^2-4) I_3(R)}{n d_R}\,. \tag{H.53}$$

Again, we provide two examples. For the fundamental representation of $\mathfrak{su}(n)$, $I_3(F) = \frac{1}{4}$ and $C_{3F} = (n^2-1)(n^2-4)/(4n^2)$. For the adjoint representation, $I_3(A) = C_{3A} = 0$, since the adjoint representation is self-conjugate. A general formula for the eigenvalue of the cubic Casimir operator in an arbitrary irreducible representation of $\mathfrak{su}(n)$ [or equivalently the cubic index $I_3(R)$, which is related to C_{3R} by eq. (H.53)] can be found in Refs. [226, 231–233].

H.4 The Techniques of Cartan and Dynkin

The techniques introduced by Cartan and further developed by Dynkin are especially useful for treating simple and semisimple Lie algebras (see, e.g., [B97–B102, B104, B106–B121, B123, B124]). It is convenient to introduce $\boldsymbol{X}^a \equiv -i\boldsymbol{T}^a$ to remove the factor of i in eq. (H.2):

$$[\boldsymbol{X}^a, \boldsymbol{X}^b] = f_c^{ab}\boldsymbol{X}^c . \tag{H.54}$$

A real Lie algebra $\mathfrak{g}$ is a vector space that consists of real linear combinations of the basis vectors $\{\boldsymbol{X}^a\}$. The commutator given in eq. (H.54) is then used to define a vector product of any two vectors in $\mathfrak{g}$. One can generalize this construction to a complex Lie algebra where the vector space $\mathfrak{g}$ consists of complex linear combinations of the basis vectors. If $\mathfrak{g}$ is a complex semisimple Lie algebra, then it is always possible to choose a basis such that the structure constants f_c^{ab} are real. Thus, without loss of generality, we can assume that the f_c^{ab} are real numbers.

The Killing form is defined in terms of a real symmetric metric tensor:

$$g^{ab} = \frac{1}{\eta} f_d^{ac} f_c^{bd} , \tag{H.55}$$

where η is a real positive normalization factor that will be chosen in Appendix H.4.4. The inverse of g^{ab} will be denoted by g_{ab}; or equivalently, $g^{ab}g_{bc} = \delta_c^a$. We shall use the negative of the metric tensor to raise and lower indices. In particular,

$$f^{abc} \equiv -g^{cd} f_d^{ab} \tag{H.56}$$

is a completely antisymmetric third-rank tensor as a consequence of the antisymmetry of f_d^{ab} under the interchange $a \leftrightarrow b$ and the Jacobi identity [eq. (H.3)]. If $\mathfrak{g}$ is a semisimple compact real Lie algebra, then the metric tensor is negative definite, and one can always find a basis $\{\boldsymbol{X}^a\}$ (via Sylvester's law of inertia [B128]) such that $g^{cd} = -\delta^{cd}$. In this convention (adopted elsewhere in this book), $f^{abc} = f_c^{ab}$ [which motivates the minus sign employed in eq. (H.56)].

The adjoint representation consists of $d_G \times d_G$ matrices that represent the $\boldsymbol{X}^a$. These matrices, which we denote henceforth by $\boldsymbol{F}^a$, are defined by [see eq. (H.4)]

$$(\boldsymbol{F}^a)^b{}_c = -f_c^{ab} , \tag{H.57}$$

where b and c label the rows and columns of the $\boldsymbol{F}^a$. Equation (H.55) can be rewritten as

$$g^{ab} = \frac{1}{\eta}\,\mathrm{Tr}(\boldsymbol{F}^a\boldsymbol{F}^b) . \tag{H.58}$$

The quadratic Casimir operator, C_2, is defined by

$$C_2 \equiv g_{ab}\boldsymbol{X}^a\boldsymbol{X}^b . \tag{H.59}$$

Using eq. (H.54) and the total antisymmetry of f^{abc} [eq. (H.56)], it follows that

$$[C_2\,, \boldsymbol{X}^a] = 0 , \qquad a = 1, 2, \ldots, d_G . \tag{H.60}$$

For a given representation of the simple Lie algebra $\mathfrak{g}$, the generators $\boldsymbol{X}^a$ are represented by $d_R \times d_R$ matrices $\boldsymbol{R}^a$. As noted above eq. (H.38), if R is an irreducible representation of $\mathfrak{g}$ (acting on a complex d_R-dimensional vector space), then any operator that commutes with all the $\boldsymbol{R}^a$ $(a = 1, 2, \ldots, d_G)$ must be a multiple of the identity operator. Thus, we shall write

$$C_2(R) = g_{ab}\boldsymbol{R}^a\boldsymbol{R}^b = C_R\mathbb{1}\,, \tag{H.61}$$

where $\mathbb{1}$ is the $d_R \times d_R$ identity matrix, and C_R is a number that depends only on the representation R. One can immediately prove the following theorem.

Theorem H.6. *For the adjoint representation (denoted by $R = A$) of a simple Lie algebra, $C_A = \eta$.*

Proof The adjoint representation of a simple Lie algebra is irreducible. Using the explicit form for the adjoint representation generators given in eq. (H.57) and the definition of C_R (for $R = A$) given in eq. (H.61),

$$C_2(A)^c{}_e \equiv g_{ab}(\boldsymbol{F}^a)^c{}_d(\boldsymbol{F}^b)^d{}_e = g_{ab}f_d^{ac}\,f_e^{bd} = C_A\delta_e^c\,. \tag{H.62}$$

Multiplying both sides of eq. (H.62) by δ_c^e, summing over c and e, and employing eq. (H.55),

$$d_G C_A = g_{ab}f_d^{ac}\,f_c^{bd} = \eta g_{ab}g^{ab} = \eta d_G\,, \tag{H.63}$$

and we immediately obtain $C_A = \eta$. □

As indicated following eq. (H.56), in the case of a semisimple compact real Lie algebra, it is common practice to choose a basis $\{\boldsymbol{X}^a\}$ of generators in eq. (H.54) such that the metric tensor defined in eq. (H.55) is given by $g^{ab} = -\delta^{ab}$. With this basis choice, the expression for the quadratic Casimir operator defined in eq. (H.61) reduces to the one given in eq. (H.38), after identifying $\boldsymbol{R}^a \equiv -i\boldsymbol{T}^a_{\boldsymbol{R}}$.

H.4.1 Root Vectors

We shall work in the Cartan–Weyl basis of a complex semisimple Lie algebra $\mathfrak{g}$, where the generators consist of $\{H^j\,,\,E^{\boldsymbol{\alpha}}\}$, which satisfy

$$[H^j\,,\,H^k] = 0\,, \tag{H.64}$$

$$[H^j\,,\,E^{\boldsymbol{\alpha}}] = \alpha^j E^{\boldsymbol{\alpha}}\,, \tag{H.65}$$

$$[E^{\boldsymbol{\alpha}}\,,\,E^{-\boldsymbol{\alpha}}] = g_{jk}\alpha^k H^j\,, \tag{H.66}$$

$$[E^{\boldsymbol{\alpha}}\,,\,E^{\boldsymbol{\beta}}] = \begin{cases} N_{\boldsymbol{\alpha\beta}}E^{\boldsymbol{\alpha}+\boldsymbol{\beta}}\,, & \text{if } \boldsymbol{\alpha}+\boldsymbol{\beta} \text{ is a root and } \boldsymbol{\alpha}+\boldsymbol{\beta}\neq 0\,,\\ 0\,, & \text{if } \boldsymbol{\alpha}+\boldsymbol{\beta} \text{ is not a root}\,, \end{cases} \tag{H.67}$$

where $H^{j\dagger} = H^j$ and $E^{\boldsymbol{\alpha}\dagger} = E^{-\boldsymbol{\alpha}}$. Here, $j \in \{1, 2, \ldots, \ell\}$ defines the *rank* ℓ of the Lie algebra $\mathfrak{g}$ (and corresponds to the maximal number of commuting generators), and the root vectors are real ℓ-dimensional vectors, $\boldsymbol{\alpha} = (\alpha^1\,,\,\alpha^2\,,\,\ldots\,,\,\alpha^\ell) \neq \boldsymbol{0}$, whose components are defined by eq. (H.65). Moreover, the phases of the E^{α} can be chosen such that the constants $N_{\alpha\beta}$ are real. The set of root vectors is denoted by Δ. One can easily prove that $\boldsymbol{\alpha} \in \Delta$ implies that $-\boldsymbol{\alpha} \in \Delta$ and $k\boldsymbol{\alpha} \notin \Delta$ if $k \neq \pm 1$.

Comparing eqs. (H.64) and (H.65) with eq. (H.54), it follows that

$$f_{\alpha}^{ij} = f_j^{i\alpha} = f_k^{ij} = 0\,, \qquad f_{\beta}^{i\alpha} = \delta_{\beta}^{\alpha}\alpha^i\,, \tag{H.68}$$

where the indices $i, j, k \in \{1, 2, \ldots, \ell\}$ and the "indices" $\boldsymbol{\alpha}$ and $\boldsymbol{\beta}$ run over all the roots. As a result, the upper $\ell \times \ell$ block of the metric tensor is given by

$$g^{ij} = \frac{1}{\eta} f_d^{ic} f_c^{jd} = \frac{1}{\eta} \sum_{\alpha \in \Delta} \alpha^i \alpha^j, \tag{H.69}$$

and the off-diagonal blocks by $g^{\alpha,j} = g^{j,\alpha} = 0$. In particular, since the H^j are hermitian, it follows that g^{ij} is positive definite.[2] The inverse metric tensor g_{ij}, which satisfies $g_{ik}g^{kj} = \delta_i^j$, appears explicitly in eq. (H.66). One can use g_{ij} to define a Euclidean inner product of vectors that live in the ℓ-dimensional root vector space:

$$(\boldsymbol{\alpha}\,, \boldsymbol{\beta}) = g_{jk}\alpha^j \beta^k\,. \tag{H.70}$$

The length of the root $\boldsymbol{\alpha}$ is then defined by $(\boldsymbol{\alpha}, \boldsymbol{\alpha})^{1/2}$, and the angle $\phi_{\alpha\beta}$ between roots $\boldsymbol{\alpha}$ and $\boldsymbol{\beta}$ is given by

$$\cos\phi_{\alpha\beta} = \frac{(\boldsymbol{\alpha}, \boldsymbol{\beta})}{(\boldsymbol{\alpha}, \boldsymbol{\alpha})^{1/2}(\boldsymbol{\beta}, \boldsymbol{\beta})^{1/2}}\,, \qquad \text{where } 0 \leq \phi_{\alpha\beta} \leq \pi. \tag{H.71}$$

Using eqs. (H.69) and (H.70), it follows that[3]

$$(\boldsymbol{\beta}, \boldsymbol{\gamma}) = \frac{1}{\eta} \sum_{\alpha \in \Delta} (\boldsymbol{\alpha}, \boldsymbol{\beta})(\boldsymbol{\alpha}, \boldsymbol{\gamma})\,, \qquad \text{for } \boldsymbol{\beta}, \boldsymbol{\gamma} \in \Delta. \tag{H.72}$$

Theorem H.7. *Suppose that $\boldsymbol{\alpha}, \boldsymbol{\beta} \in \Delta$ and $\boldsymbol{\beta} \neq \pm\boldsymbol{\alpha}$. Let p and q be the largest nonnegative integers for which $\boldsymbol{\beta} - p\boldsymbol{\alpha} \in \Delta$ and $\boldsymbol{\beta} + q\boldsymbol{\alpha} \in \Delta$, respectively. Then, $\boldsymbol{\beta} + k\boldsymbol{\alpha} \in \Delta$ for all integer values of k that satisfy $-p \leq k \leq q$. Moreover, $p + q \leq 3$ and*

$$p - q = \frac{2(\boldsymbol{\alpha}, \boldsymbol{\beta})}{(\boldsymbol{\alpha}, \boldsymbol{\alpha})} \quad \textit{is an integer.} \tag{H.73}$$

As a consequence of Schwarz's inequality, $0 \leq (\boldsymbol{\alpha}, \boldsymbol{\beta})^2 \leq (\boldsymbol{\alpha}, \boldsymbol{\alpha})(\boldsymbol{\beta}, \boldsymbol{\beta})$. Assuming that $(\boldsymbol{\alpha}, \boldsymbol{\alpha}) \leq (\boldsymbol{\beta}, \boldsymbol{\beta})$, it then follows that, for a simple Lie algebra $\mathfrak{g}$, the only possible values of $(\boldsymbol{\beta}, \boldsymbol{\beta})/(\boldsymbol{\alpha}, \boldsymbol{\alpha})$ are 1, 2, and 3. Moreover, at most two different root lengths can occur. A proof of this theorem can be found in many of the books cited at the beginning of Appendix H.4.

[2] Indeed, this implies that one can define a new basis of commuting hermitian generators such that $g^{ij} = \delta^{ij}$, although we shall not do so explicitly here.

[3] The factor $1/\eta$ on the right-hand side of eq. (H.72), which is typically absent in the literature, appears here due to the normalization chosen in defining the metric tensor g^{ij} and our definition of the inner product.

In eq. (H.66), we have chosen to normalize of the generators E_α of the Cartan–Weyl basis such that[4]

$$g^{\alpha,-\alpha} = 1\,. \tag{H.74}$$

In this convention, one can show that

$$N^2_{\boldsymbol{\alpha\beta}} = \tfrac{1}{2}(\boldsymbol{\alpha}\,,\,\boldsymbol{\alpha})q(p+1)\,, \qquad\qquad N_{-\boldsymbol{\alpha},-\boldsymbol{\beta}} = -N_{\boldsymbol{\alpha},\boldsymbol{\beta}}\,, \tag{H.75}$$

where p and q are determined as specified in Theorem H.7 above. Conventionally, the $N_{\boldsymbol{\alpha\beta}}$ are taken to be real, as previously noted.

One can therefore introduce an ordering of the root vectors by defining $\boldsymbol{\alpha} > \boldsymbol{\beta}$ if the first nonzero component of $\boldsymbol{\alpha} - \boldsymbol{\beta}$ with respect to some fixed basis is positive. The roots can be divided into two sets: the set of positive roots, denoted by Δ_+, and the set of negative roots, denoted by Δ_-. Note that the quadratic Casimir operator can be written in terms of the Cartan–Weyl basis as

$$C_2 = \sum_{j=1}^{\ell} g_{ij} H^i H^j + \sum_{\alpha\in\Delta_+} \left(E^{\alpha}E^{-\alpha} + E^{-\alpha}E^{\alpha}\right). \tag{H.76}$$

Finally, we define a *simple* root to be a positive root that cannot be expressed as a sum of two other positive roots. One can prove that there are precisely ℓ positive roots in a semisimple Lie algebra of rank ℓ. The set of simple roots is denoted by Π.

Theorem H.8. *If $\boldsymbol{\alpha}$, $\boldsymbol{\beta} \in \Pi$ and $\boldsymbol{\alpha} \neq \boldsymbol{\beta}$, then $\boldsymbol{\alpha} - \boldsymbol{\beta}$ is not a root, and*

$$(\boldsymbol{\alpha}\,,\,\boldsymbol{\beta}) \leq 0\,. \tag{H.77}$$

Proof If $\boldsymbol{\alpha} - \boldsymbol{\beta} \in \Delta_+$, then $\boldsymbol{\alpha} = (\boldsymbol{\alpha} - \boldsymbol{\beta}) + \boldsymbol{\beta}$ shows that $\boldsymbol{\alpha}$ is the sum of two positive roots, which is impossible since $\boldsymbol{\alpha} \in \Pi$. Likewise, if $\boldsymbol{\beta} - \boldsymbol{\alpha} \in \Delta_+$, then $\boldsymbol{\beta} = (\boldsymbol{\beta} - \boldsymbol{\alpha}) + \boldsymbol{\alpha}$ shows that $\boldsymbol{\beta}$ is the sum of two positive roots, which is impossible since $\boldsymbol{\beta} \in \Pi$. Hence $\boldsymbol{\alpha} - \boldsymbol{\beta}$ is not a root, which implies that $p = 0$ in eq. (H.73), and it follows that $(\boldsymbol{\alpha}\,,\,\boldsymbol{\beta}) \leq 0$. □

As a consequence of Theorems H.7 and H.8, if $\boldsymbol{\alpha}$, $\boldsymbol{\beta} \in \Pi$ and $(\boldsymbol{\alpha}, \boldsymbol{\alpha}) \leq (\boldsymbol{\beta}, \boldsymbol{\beta})$, then the possible values of $\phi_{\boldsymbol{\alpha\beta}}$ are 120° if $(\boldsymbol{\beta}, \boldsymbol{\beta})/(\boldsymbol{\alpha}, \boldsymbol{\alpha}) = 1$; 135° if $(\boldsymbol{\beta}, \boldsymbol{\beta})/(\boldsymbol{\alpha}, \boldsymbol{\alpha}) = 2$; or 150° if $(\boldsymbol{\beta}, \boldsymbol{\beta})/(\boldsymbol{\alpha}, \boldsymbol{\alpha}) = 3$.

It is convenient to introduce the $\ell \times \ell$ Cartan matrix A, whose matrix elements are given by[5]

$$A_{ij} \equiv \frac{2(\boldsymbol{\alpha}_i\,,\,\boldsymbol{\alpha}_j)}{(\boldsymbol{\alpha}_i\,,\,\boldsymbol{\alpha}_i)}\,, \tag{H.78}$$

where i, j label the ℓ simple roots. Note that all elements of the Cartan matrix are integers such that $A_{ii} = 2$ and $A_{ij} \leq 0$ for $i \neq j$. As a result of the constraints

[4] More generally, eq. (H.66) is given by $[E^{\alpha}\,,\,E^{-\alpha}] = g^{\alpha,-\alpha}\,g_{ij}\alpha^j H^i$. Since the normalization of E^{α} is not fixed by eq. (H.65), we are free to rescale the E^{α} and $E^{-\alpha}$ such that $g^{\alpha,-\alpha} = 1$, as noted in eq. (H.74).

[5] Some books define the expression given in eq. (H.78) to be the transpose of the Cartan matrix.

of Theorems H.7 and H.8, the possible Cartan matrices corresponding to possible complex simple Lie algebras are limited. After fixing a convention for the labeling of the ℓ simple roots $\boldsymbol{\alpha_i}$, one can completely classify the complex simple Lie algebras of rank ℓ. The list consists of four infinite series of Lie algebras, $A_\ell \equiv \mathfrak{sl}(\ell+1,\mathbb{C})$ $[\ell \geq 1]$, $B_\ell \equiv \mathfrak{so}(2\ell+1,\mathbb{C})$ $[\ell \geq 3]$, $C_\ell \equiv \mathfrak{sp}(\ell,\mathbb{C})$ $[\ell \geq 2]$,[6] and $D_\ell \equiv \mathfrak{so}(2\ell,\mathbb{C})$ $[\ell \geq 4]$,[7] and five exceptional Lie algebras, denoted by G_2, F_4, E_6, E_7, and E_8. There is a one-to-one correspondence between the possible Cartan matrices of the complex simple Lie algebras and the celebrated Dynkin diagrams. Details can be found in the books on Lie algebras previously cited.

We next introduce the Weyl reflection, which acts on a root vector as follows:

$$S_i(\boldsymbol{\alpha}) \equiv \boldsymbol{\alpha} - \frac{2(\boldsymbol{\alpha}\,,\,\boldsymbol{\alpha_i})}{(\boldsymbol{\alpha_i}\,,\,\boldsymbol{\alpha_i})}\boldsymbol{\alpha_i}\,, \qquad \boldsymbol{\alpha} \in \Delta \quad \text{and} \quad \boldsymbol{\alpha_i} \in \Pi\,. \tag{H.79}$$

One can prove that $S_i(\boldsymbol{\alpha}) \in \Delta$. Three additional properties of S_i are

$$S_i(\boldsymbol{\alpha_i}) = -\boldsymbol{\alpha_i}\,, \tag{H.80}$$

$$(S_i(\boldsymbol{\alpha})\,,\,\boldsymbol{\beta}) = (\boldsymbol{\alpha}\,,\,S_i(\boldsymbol{\beta}))\,, \tag{H.81}$$

$$(S_i(\boldsymbol{\alpha})\,,\,S_i(\boldsymbol{\alpha})) = (\boldsymbol{\alpha},\boldsymbol{\alpha})\,. \tag{H.82}$$

Additional properties of the Weyl reflection are summarized by the following theorem.

Theorem H.9. *If $\boldsymbol{\alpha} \in \Delta_+$ and $\boldsymbol{\alpha} \neq \boldsymbol{\alpha_i}$ (where $\boldsymbol{\alpha_i} \in \Pi$ is one of the simple roots), then $S_i(\boldsymbol{\alpha}) > 0$. Moreover, if $S_i(\boldsymbol{\alpha}) = S_i(\boldsymbol{\beta})$, then $\boldsymbol{\alpha} = \boldsymbol{\beta}$.*

Proof Any positive root $\boldsymbol{\alpha} \in \Delta_+$ can be written as

$$\boldsymbol{\alpha} = k_i\boldsymbol{\alpha_i} + \sum_{j\neq i} k_j\boldsymbol{\alpha_j} \qquad \text{(no sum over } i\text{)}, \tag{H.83}$$

where k_i and k_j are nonnegative integers. Then,

$$S_i(\boldsymbol{\alpha}) = \boldsymbol{\alpha} - \frac{2(\boldsymbol{\alpha}\,,\,\boldsymbol{\alpha_i})}{(\boldsymbol{\alpha_i}\,,\,\boldsymbol{\alpha_i})}\boldsymbol{\alpha_i} = -\boldsymbol{\alpha_i}\Big[k_i + \sum_{j\neq i} k_j A_{ij}\Big] + \sum_{j\neq i} k_j\boldsymbol{\alpha_j}\,. \tag{H.84}$$

Since $S_i(\boldsymbol{\alpha}) \in \Delta$, it follows that $S_i(\boldsymbol{\alpha})$ is either positive or negative. If $S_i(\boldsymbol{\alpha}) < 0$, then we must have $k_j = 0$ for $j \neq i$, in which case $\boldsymbol{\alpha} = \boldsymbol{\alpha_i}$ (i.e., $k_i = 1$) and we recover eq. (H.80). Hence if $\boldsymbol{\alpha} \neq \boldsymbol{\alpha_i}$, it then follows that $S_i(\boldsymbol{\alpha}) > 0$. If $S_i(\boldsymbol{\alpha}) = S_i(\boldsymbol{\beta})$, then $\boldsymbol{\alpha} - \boldsymbol{\beta} = \kappa\boldsymbol{\alpha_i}$, where

$$\kappa = \frac{2(\boldsymbol{\alpha} - \boldsymbol{\beta}\,,\,\boldsymbol{\alpha_i})}{(\boldsymbol{\alpha_i}\,,\,\boldsymbol{\alpha_i})}\,. \tag{H.85}$$

Inserting $\boldsymbol{\alpha} - \boldsymbol{\beta} = \kappa\boldsymbol{\alpha_i}$ into the expression above yields $\kappa\boldsymbol{\alpha_i} = 2\kappa\boldsymbol{\alpha_i}$, and we conclude that $\kappa = 0$, which implies that $\boldsymbol{\alpha} = \boldsymbol{\beta}$. □

6 The reader is warned that what we call $\mathfrak{sp}(\ell,\mathbb{C})$ is often called $\mathfrak{sp}(2\ell,\mathbb{C})$ in the literature.

7 Our choices of allowed values of ℓ have been made to avoid duplication in the list of simple Lie algebras, since $\mathfrak{so}(3,\mathbb{C}) \cong \mathfrak{sp}(1,\mathbb{C}) \cong \mathfrak{sl}(2,\mathbb{C})$, $\mathfrak{sp}(2,\mathbb{C}) \cong \mathfrak{so}(5,\mathbb{C})$, and $\mathfrak{so}(6,\mathbb{C}) \cong \mathfrak{su}(4,\mathbb{C})$. In addition, we have not included $\mathfrak{so}(4,\mathbb{C}) \cong \mathfrak{sl}(2,\mathbb{C}) \oplus \mathfrak{sl}(2,\mathbb{C})$ as this is a semisimple Lie algebra.

One consequence of Theorem H.9 is that S_i maps the set of positive roots excluding $\boldsymbol{\alpha}_i$ onto itself, where the map is bijective. Thus, if we define the Weyl vector $\boldsymbol{\delta}$ to be half the sum of the positive roots,

$$\boldsymbol{\delta} \equiv \tfrac{1}{2} \sum_{\boldsymbol{\alpha}\in\Delta_+} \boldsymbol{\alpha}\,, \tag{H.86}$$

then, using eq. (H.80),

$$S_i(\boldsymbol{\delta}) = \tfrac{1}{2} S_i \left(\boldsymbol{\alpha}_i + \sum_{j\neq i} \boldsymbol{\alpha}_j \right) = \tfrac{1}{2} \left(-\boldsymbol{\alpha}_i + \sum_{j\neq i} \boldsymbol{\alpha}_j \right) = \boldsymbol{\delta} - \boldsymbol{\alpha}_i\,. \tag{H.87}$$

Hence, eq. (H.81) yields

$$(S_i(\boldsymbol{\delta})\,,\, \boldsymbol{\alpha}_i) = (\boldsymbol{\delta}\,,\, S_i(\boldsymbol{\alpha}_i))\,. \tag{H.88}$$

Using eqs. (H.80) and (H.87), it follows that

$$(\boldsymbol{\delta} - \boldsymbol{\alpha}_i\,,\, \boldsymbol{\alpha}_i) = -(\boldsymbol{\delta}\,,\, \boldsymbol{\alpha}_i)\,. \tag{H.89}$$

Rearranging the above result then yields

$$\frac{2(\boldsymbol{\delta}\,,\, \boldsymbol{\alpha}_i)}{(\boldsymbol{\alpha}_i\,,\, \boldsymbol{\alpha}_i)} = 1\,, \qquad \text{for } \boldsymbol{\alpha}_i \in \Pi\,. \tag{H.90}$$

Finally, we introduce the dual root (or co-root) of $\boldsymbol{\alpha} \in \Delta$,

$$\boldsymbol{\alpha}^\vee \equiv \frac{2\boldsymbol{\alpha}}{(\boldsymbol{\alpha}\,,\, \boldsymbol{\alpha})}\,. \tag{H.91}$$

In terms of the dual root, the Cartan matrix can be defined as

$$A_{ij} = (\boldsymbol{\alpha}_i^\vee\,,\, \boldsymbol{\alpha}_j)\,, \tag{H.92}$$

and the Weyl reflection acts on a root vector as follows:

$$S_i(\boldsymbol{\alpha}) = \boldsymbol{\alpha} - (\boldsymbol{\alpha}\,,\, \boldsymbol{\alpha}_i^\vee)\boldsymbol{\alpha}_i\,. \tag{H.93}$$

Equation (H.90) then can be rewritten as

$$(\boldsymbol{\delta}\,,\, \boldsymbol{\alpha}_i^\vee) = 1\,, \qquad \text{for } \boldsymbol{\alpha}_i \in \Pi\,. \tag{H.94}$$

For completeness we record a formula for the length of the Weyl vector $\boldsymbol{\delta}$. In the normalization convention adopted in eq. (H.55), which fixes the scalar product defined in eq. (H.70),

$$(\boldsymbol{\delta}, \boldsymbol{\delta}) = \tfrac{1}{24}\eta d_G\,, \tag{H.95}$$

where d_G is the dimension of the Lie algebra and η will be explicitly evaluated in Appendix H.4.4. Elementary proofs of eq. (H.95), known as the *strange formula*, can be found in Refs. [234–236].

In this book, we are primarily interested in real Lie algebras of compact Lie groups. Remarkably, there is a one-to-one correspondence between the semisimple complex Lie algebras and the semisimple compact real Lie algebras. This can be achieved by rewriting the commutation relations of the Cartan–Weyl basis given in eqs. (H.64)–(H.67) in terms of H^j, $Y^{\boldsymbol{\alpha}}$, $Z^{\boldsymbol{\alpha}}$, where

$$Y^{\boldsymbol{\alpha}} \equiv E^{\boldsymbol{\alpha}} + E^{-\boldsymbol{\alpha}}\,, \qquad Z^{\boldsymbol{\alpha}} \equiv -i(E^{\boldsymbol{\alpha}} - E^{-\boldsymbol{\alpha}})\,. \tag{H.96}$$

Note that H^j, $Y^{\boldsymbol{\alpha}}$, and $Z^{\boldsymbol{\alpha}}$ are all hermitian generators. Using the properties of the $N_{\boldsymbol{\alpha\beta}}$ given in eq. (H.75), it follows that

$$[H^j\,,\,H^k] = 0\,, \tag{H.97}$$
$$[H^j\,,\,Y^{\boldsymbol{\alpha}}] = i\alpha^j Z^{\boldsymbol{\alpha}}\,, \tag{H.98}$$
$$[H^j\,,\,Z^{\boldsymbol{\alpha}}] = -i\alpha^j Y^{\boldsymbol{\alpha}}\,, \tag{H.99}$$
$$[Y^{\boldsymbol{\alpha}}\,,\,Y^{\boldsymbol{\beta}}] = iN_{\boldsymbol{\alpha},-\boldsymbol{\beta}}Z^{\boldsymbol{\alpha}-\boldsymbol{\beta}} + iN_{\boldsymbol{\alpha\beta}}Z^{\boldsymbol{\alpha}+\boldsymbol{\beta}}\,, \tag{H.100}$$
$$[Z^{\boldsymbol{\alpha}}\,,\,Z^{\boldsymbol{\beta}}] = iN_{\boldsymbol{\alpha},-\boldsymbol{\beta}}Z^{\boldsymbol{\alpha}-\boldsymbol{\beta}} - iN_{\boldsymbol{\alpha\beta}}Z^{\boldsymbol{\alpha}+\boldsymbol{\beta}}\,, \tag{H.101}$$
$$[Y^{\boldsymbol{\alpha}}\,,\,Z^{\boldsymbol{\beta}}] = \begin{cases} 2ig_{ij}\alpha^i H^j\,, & \text{if } \boldsymbol{\alpha}-\boldsymbol{\beta}=0 \text{ or } \boldsymbol{\alpha}+\boldsymbol{\beta}=0\,, \\ iN_{\boldsymbol{\alpha},-\boldsymbol{\beta}}Y^{\boldsymbol{\alpha}-\boldsymbol{\beta}} - iN_{\boldsymbol{\alpha\beta}}Y^{\boldsymbol{\alpha}+\boldsymbol{\beta}}\,, & \text{if } \boldsymbol{\alpha}-\boldsymbol{\beta}\neq 0 \text{ and } \boldsymbol{\alpha}+\boldsymbol{\beta}\neq 0\,, \end{cases} \tag{H.102}$$

where

$$N_{\boldsymbol{\alpha\beta}} = 0 \text{ if } \boldsymbol{\alpha}+\boldsymbol{\beta} \text{ is not a root, and } N_{\boldsymbol{\alpha},-\boldsymbol{\beta}} = 0 \text{ if } \boldsymbol{\alpha}-\boldsymbol{\beta} \text{ is not a root.}$$

The elements of the complex semisimple Lie algebra $\mathfrak{g}$ are arbitrary complex linear combinations of the generators $\{H^j, Y^{\boldsymbol{\alpha}}, Z^{\boldsymbol{\alpha}}\}$, where $j = 1, 2, \ldots, \ell = \text{rank}\,\mathfrak{g}$ and $\boldsymbol{\alpha} \in \Delta$. However, if we restrict the elements of the Lie algebra to real linear combinations of these generators, the resulting Lie algebra (henceforth denoted by $\mathfrak{g}_R$), is a compact real form of $\mathfrak{g}$.[8] Indeed, the commutation relations of the generators of $\mathfrak{g}_R$ are of the form specified in eq. (H.2), with hermitian generators and real structure constants (after extracting the factor of i). The compact real forms of the four infinite series of Lie algebras identified above are $\mathfrak{su}(\ell+1)$ $[\ell \geq 1]$, $\mathfrak{so}(2\ell+1, \mathbb{R})$ $[\ell \geq 3]$, $\mathfrak{sp}(\ell)$ $[\ell \geq 2]$,[9] and $\mathfrak{so}(2\ell, \mathbb{R})$ $[\ell \geq 4]$. Henceforth, we shall simply write $\mathfrak{so}(n)$ for $\mathfrak{so}(n, \mathbb{R})$, as is the usual convention.[10] The compact real Lie algebras $\mathfrak{sp}(1)$, $\mathfrak{so}(3)$, $\mathfrak{so}(5)$, and $\mathfrak{so}(6)$ are related to the Lie algebras previously listed according to the isomorphisms (see footnote 7)

$$\mathfrak{su}(2) \cong \mathfrak{so}(3) \cong \mathfrak{sp}(1)\,, \qquad \mathfrak{so}(5) \cong \mathfrak{sp}(2)\,, \qquad \mathfrak{so}(6) \cong \mathfrak{su}(4)\,. \tag{H.103}$$

The remaining results presented in this appendix will apply both to the complex semisimple Lie algebra $\mathfrak{g}$ and to its corresponding compact real form $\mathfrak{g}_R$.

[8] With respect to the basis $\{iH^j, iY^{\boldsymbol{\alpha}}, iZ^{\boldsymbol{\alpha}}\}$, the metric tensor g^{ab} is negative definite, which implies that the semisimple real Lie algebra $\mathfrak{g}_R$ is compact. See, e.g., [B101] for a discussion of the construction of real forms of $\mathfrak{g}$ that are not compact.

[9] What we call $\mathfrak{sp}(\ell)$ is often called $\mathfrak{usp}(2\ell)$ in the literature, as it is the Lie algebra of the unitary symplectic group of $2\ell \times 2\ell$ matrices.

[10] To be precise, we should introduce a separate notation for the compact real forms of the exceptional Lie algebras. However, for simplicity of notation, we will continue to employ the same symbols for the exceptional Lie groups and their corresponding Lie algebras (both complex and their corresponding compact real forms).

H.4.2 Irreducible Representations and Weights

Consider a simple Lie algebra $\mathfrak{g}$ of rank ℓ. To construct the irreducible representations (commonly called *irreps*) of $\mathfrak{g}$, one determines the basis vectors of the representation space, denoted collectively by $|\boldsymbol{m}\rangle$. These vectors are chosen to be the simultaneous eigenvectors of the commuting hermitian generators H^j:

$$H^j\,|\boldsymbol{m}\rangle = m^j\,|\boldsymbol{m}\rangle\,. \tag{H.104}$$

The components of the ℓ-dimensional vector $\boldsymbol{m} = (m^1\,,\,m^2\,,\,\ldots\,,\,m^\ell)$ are the corresponding eigenvalues of H^j. The ℓ-dimensional vector space in which the $\boldsymbol{m}$ reside is called the vector space of weight vectors. We can formulate an ordering of vectors of the weight space by introducing the rule that $\boldsymbol{m} > \boldsymbol{n}$ if the first nonzero component of $\boldsymbol{m} - \boldsymbol{n}$ is positive. An important theorem in Lie algebra representation theory states that, for a given irrep, the *highest* weight $|\boldsymbol{M}\rangle$ is nondegenerate and uniquely fixes the representation. Moreover,

$$E^{\boldsymbol{\alpha}}\,|\boldsymbol{M}\rangle = 0\,, \qquad \text{for all } \boldsymbol{\alpha}\in\Delta_+\,. \tag{H.105}$$

Given a conventional ordered list, $\{\boldsymbol{\alpha}_1\,,\,\boldsymbol{\alpha}_2\,,\,\ldots\,,\,\boldsymbol{\alpha}_\ell\}$, of the simple roots of $\mathfrak{g}$, one can prove that the following quantities are nonnegative integers:

$$a_i \equiv \frac{2(\boldsymbol{M}\,,\,\boldsymbol{\alpha}_i)}{(\boldsymbol{\alpha}_i\,,\,\boldsymbol{\alpha}_i)} = (\boldsymbol{M}\,,\,\boldsymbol{\alpha}_i^\vee)\,, \qquad i\in\{1,2,\ldots,\ell\}\,. \tag{H.106}$$

Thus, an irrep can be identified by $\boldsymbol{a} \equiv (a_1\ a_2\ \cdots\ a_\ell)$, where the a_i are called the *Dynkin labels* of the irrep.[11]

Since $\boldsymbol{M}$ is a vector that lives in an ℓ-dimensional space, it can be expanded in terms of the root vectors with rational coefficients:

$$\boldsymbol{M} = \sum_{k=1}^{\ell} q_k \boldsymbol{\alpha}_k\,. \tag{H.107}$$

Inserting this expansion into eq. (H.106) and using eq. (H.78) yields $a_j = \sum_{k=1}^{\ell} A_{jk}q_k$. Inverting this result, we obtain

$$q_k = \sum_{j=1}^{\ell}(A^{-1})_{kj}a_j\,. \tag{H.108}$$

The dimension of the irrep with maximal weight $\boldsymbol{M}$ is given by the Weyl dimension formula,

$$d_R = \prod_{\boldsymbol{\alpha}\in\Delta_+}\frac{(\boldsymbol{M}+\boldsymbol{\delta},\boldsymbol{\alpha})}{(\boldsymbol{\delta},\boldsymbol{\alpha})} = \prod_{\boldsymbol{\alpha}\in\Delta_+}\frac{\sum_{j=1}^{\ell} k_j^{\alpha}(a_j+1)(\boldsymbol{\alpha}_j,\boldsymbol{\alpha}_j)}{\sum_{j=1}^{\ell} k_j^{\alpha}(\boldsymbol{\alpha}_j,\boldsymbol{\alpha}_j)}\,, \tag{H.109}$$

[11] In the case of $\mathfrak{g} = \mathfrak{su}(\ell+1)$ where ℓ is a positive integer, the irrep $(a_1\ a_2\ \cdots\ a_\ell)$ can be represented by a Young tableau such that a_k equals the number of columns of the Young tableau with k boxes. Note that the jth row of the Young tableau consists of p_j boxes such that $p_1 \geq p_2 \geq \cdots \geq p_\ell \geq p_{\ell+1} \equiv 0$, and $a_j = p_j - p_{j+1}$, for $j = 1, 2, \ldots, \ell$. The Young tableaux represent the permutation symmetries of the indices of tensors that transform irreducibly under the SU$(\ell+1)$ group.

Table H.1 Pseudoreal irreducible representations of the simple Lie algebras.

simple Lie algebra	Dynkin label conditions
$\mathfrak{su}(\ell+1)$ $[\ell = 1 \pmod 4]$	$(a_1 a_2 \cdots a_\ell) = (a_\ell a_{\ell-1} \cdots a_1)$, for $a_{(\ell+1)/2} = 1 \pmod 2$
$\mathfrak{so}(2\ell+1)$ $[\ell = 1, 2 \pmod 4]$	$a_\ell = 1 \pmod 2$
$\mathfrak{so}(2\ell)$ $[\ell = 2 \pmod 4]$	$a_{\ell-1} + a_\ell = 1 \pmod 2$
$\mathfrak{sp}(\ell)$	$\begin{cases} a_1 + a_3 + \cdots + a_\ell = 1 \pmod 2, & \text{if } \ell \text{ is odd} \\ a_1 + a_3 + \cdots + a_{\ell-1} = 1 \pmod 2, & \text{if } \ell \text{ is even} \end{cases}$
E_7	$a_4 + a_6 + a_7 = 1 \pmod 2$

where the expansion of the positive root $\boldsymbol{\alpha}$ in terms of the simple root is given by $\boldsymbol{\alpha} = \sum_{j=1}^{\ell} k_j^\alpha \boldsymbol{\alpha}_j$ and the k_j^α are nonnegative integers. Starting from eq. (H.109), expressions for d_R can be derived that depend only on the Dynkin labels of the corresponding representation R. In the cases of the four infinite series of simple Lie algebras, explicit formulae for d_R can be found in Section 3.3 of Ref. [69].

An irrep of a simple Lie algebra $\mathfrak{g}$ of rank ℓ (and its compact real form) is uniquely specified by the Dynkin labels a_i [see eq. (H.106)], which can be assembled into a vector with ℓ nonnegative integer components, $\boldsymbol{a} \equiv (a_1 a_2 \cdots a_\ell)$. By considering all possible choices for the a_i, one obtains all finite-dimensional irreps of $\mathfrak{g}$.

Consider the irreps of $\mathfrak{su}(\ell+1)$ [for $\ell \geq 1$], $\mathfrak{so}(2\ell+1)$ [for $\ell \geq 3$], $\mathfrak{so}(2\ell)$ [for $\ell \geq 4$], and $\mathfrak{sp}(\ell)$ [for $\ell \geq 2$]. Note that this list of simple compact Lie algebras excludes $\mathfrak{so}(3)$, $\mathfrak{so}(5)$, $\mathfrak{so}(6)$, and $\mathfrak{sp}(1)$ in light of the isomorphisms listed in eq. (H.103). However, the conventions employed in defining the labeling of the ℓ simple roots $\boldsymbol{\alpha}_i$ of A_ℓ, B_ℓ, C_ℓ, and D_ℓ imply the following correspondences:

- irreps (a_1) of $\mathfrak{so}(3)$ and (a_1) of $\mathfrak{sp}(1)$ coincide with irreps (a_1) of $\mathfrak{su}(2)$,
- irreps $(a_1 a_2)$ of $\mathfrak{so}(5)$ coincide with irreps $(a_2 a_1)$ of $\mathfrak{sp}(2)$,
- irreps $(a_1 a_2 a_3)$ of $\mathfrak{so}(6)$ coincide with irreps $(a_2 a_1 a_3)$ of $\mathfrak{su}(4)$. (H.110)

We noted in Appendix H.1 that irreps of $\mathfrak{g}$ are either complex, real, or pseudoreal. If $\mathfrak{g}$ is a simple compact Lie algebra that admits complex representations, then $\mathfrak{g}$ must be one of the following Lie algebras: $\mathfrak{su}(\ell+1)$ $[\ell \geq 2]$, $\mathfrak{so}(4k+2)$ $[k \geq 1]$, or E_6. In terms of the Dynkin labels,[12]

$$\mathfrak{su}(\ell+1): \quad (a_1 a_2 \cdots a_{\ell-1} a_\ell)^* = (a_\ell a_{\ell-1} \cdots a_2 a_1), \tag{H.111}$$

$$\mathfrak{so}(4k+2): \quad (a_1 \cdots a_{2k-1} a_{2k} a_{2k+1})^* = (a_1 \cdots a_{2k-1} a_{2k+1} a_{2k}), \tag{H.112}$$

$$E_6: \quad (a_1 a_2 a_3 a_4 a_5 a_6)^* = (a_5 a_4 a_3 a_2 a_1 a_6). \tag{H.113}$$

A complex irrep is a representation whose Dynkin labels differ from those of its conjugate representation (denoted by the star above). A noncomplex irrep is either real or pseudoreal. The conditions on the Dynkin labels that identify a pseudoreal irrep (thereby distinguishing it from a real irrep) are shown in Table H.1 [69].

12 Note that eq. (H.111) with $\ell = 3$ and eq. (H.112) with $k = 1$ are consistent in light of eq. (H.110).

Some of the irreps of the Lie algebras $\mathfrak{su}(n)$, $\mathfrak{so}(n)$, and $\mathfrak{sp}(n)$ merit special attention. The fundamental representations of SU(n), SO(n), and Sp(n) defined in Appendix H.1, when expanded around the identity matrix, yield the fundamental n-dimensional matrix representations of $\mathfrak{su}(n)$ and $\mathfrak{so}(n)$ and the fundamental $2n$-dimensional representation of $\mathfrak{sp}(n)$, respectively. In terms of the Dynkin labels, the fundamental irreps are

$$\boldsymbol{a} = (1\,0\,0\,\cdots\,0\,0\,0)\,, \tag{H.114}$$

for $\mathfrak{su}(\ell+1)$ [for $\ell \geq 1$], $\mathfrak{so}(2\ell+1)$ [for $\ell \geq 2$], $\mathfrak{so}(2\ell)$ [for $\ell \geq 3$], $\mathfrak{sp}(\ell)$ [for $\ell \geq 1$], and $\boldsymbol{a} = (2)$ is the fundamental irrep of $\mathfrak{so}(3)$.

In light of eq. (H.111), the antifundamental irrep of $\mathfrak{su}(\ell+1)$ [for $\ell \geq 2$] is the representation that is the conjugate of the fundamental irrep with Dynkin labels given by

$$\boldsymbol{a} = (0\,0\,0\cdots\,0\,0\,1)\,. \tag{H.115}$$

The adjoint representation of a simple Lie algebra $\mathfrak{g}$ is irreducible. Moreover, the nonzero weights of the adjoint representation, $\boldsymbol{m_i} = \boldsymbol{\alpha_i}$, coincide with the root vectors of $\mathfrak{g}$. The Dynkin labels, $\boldsymbol{a} = (a_1, a_2, \cdots, a_\ell)$, of the adjoint representation of the rank ℓ simple (nonexceptional) Lie algebras are given by

$$\mathfrak{su}(\ell+1)\,,\quad \ell \geq 2: \quad \boldsymbol{a} = (1\,0\,0\,\cdots\,0\,0\,1)\,, \tag{H.116}$$

$$\mathfrak{so}(2\ell+1)\,,\quad \ell \geq 3: \quad \boldsymbol{a} = (0\,1\,0\,0\,\cdots\,0\,0)\,, \tag{H.117}$$

$$\mathfrak{so}(2\ell)\,,\quad \ell \geq 4: \quad \boldsymbol{a} = (0\,1\,0\,0\,\cdots\,0\,0)\,, \tag{H.118}$$

$$\mathfrak{sp}(\ell)\,,\quad \ell \geq 1: \quad \boldsymbol{a} = (2\,0\,0\,0\,\cdots\,0\,0)\,. \tag{H.119}$$

In light of eq. (H.110), the Dynkin labels of the adjoint representations of $\mathfrak{su}(2)$, $\mathfrak{so}(3)$, $\mathfrak{so}(5)$, and $\mathfrak{so}(6)$ are (2), (2), (02), and (011), respectively. It follows that $\mathfrak{so}(3)$ is the special case whose fundamental and adjoint irreps coincide.

It is noteworthy that the $\mathfrak{so}(n)$ Lie algebras [$n \geq 3$] possess irreducible spinor representations that cannot be obtained by decomposing the tensor product of fundamental representations. There is one elementary irreducible spinor representation of $\mathfrak{so}(2\ell+1)$, of dimension 2^ℓ, given by $\boldsymbol{a} = (0\,0\,\cdots\,0\,1)$ [for $\ell \geq 3$]. For $\mathfrak{so}(2\ell)$, there are two elementary irreducible spinor representations, each of dimension $2^{\ell-1}$, given by $\boldsymbol{a} = (0\,0\,\cdots\,0\,1\,0)$ and $\boldsymbol{a} = (0\,0\,\cdots\,0\,0\,1)$ [for $\ell \geq 4$], respectively.[13]

H.4.3 A Formula for C_R and the Second-Order Index $I_2(R)$

In an irreducible representation R, the highest weight $|\boldsymbol{M}\rangle$ is an eigenvector of the Casimir operator with eigenvalue C_R:

$$C_2(R)\,|\boldsymbol{M}\rangle = C_R\,|\boldsymbol{M}\rangle\,, \tag{H.120}$$

[13] The elementary spinor representations of $\mathfrak{so}(3)$ and $\mathfrak{so}(5)$ correspond to the fundamental representations of $\mathfrak{su}(2)$ and $\mathfrak{sp}(2)$, respectively. The two elementary spinor representations of $\mathfrak{so}(6)$ correspond to the fundamental and antifundamental representations of $\mathfrak{su}(4)$.

where eq. (H.76) is applied in the irrep R. After employing eqs. (H.66), (H.104), and (H.105) and using the inner product defined in eq. (H.70), we obtain

$$\begin{aligned} C_2(R)\,|\boldsymbol{M}\rangle &= (\boldsymbol{M},\boldsymbol{M})\,|\boldsymbol{M}\rangle + \sum_{\boldsymbol{\alpha}\in\Delta_+} (E^{\boldsymbol{\alpha}}E^{-\boldsymbol{\alpha}} + E^{-\boldsymbol{\alpha}}E^{\boldsymbol{\alpha}})\,|\boldsymbol{M}\rangle \\ &= (\boldsymbol{M},\boldsymbol{M})\,|\boldsymbol{M}\rangle + \sum_{\boldsymbol{\alpha}\in\Delta_+} \left[E^{\boldsymbol{\alpha}}\,,\,E^{-\boldsymbol{\alpha}}\right]|\boldsymbol{M}\rangle \\ &= (\boldsymbol{M},\boldsymbol{M})\,|\boldsymbol{M}\rangle + \sum_{\boldsymbol{\alpha}\in\Delta_+} (\boldsymbol{\alpha},\boldsymbol{M})\,|\boldsymbol{M}\rangle\,. \end{aligned} \tag{H.121}$$

In terms of $\boldsymbol{\delta} \equiv \frac{1}{2}\sum_{\boldsymbol{\alpha}\in\Delta_+}\boldsymbol{\alpha}$ [see eq. (H.86)], one can rewrite eq. (H.121) as follows:

$$C_2(R)\,|\boldsymbol{M}\rangle = (\boldsymbol{M}\,,\,\boldsymbol{M}+2\boldsymbol{\delta})\,|\boldsymbol{M}\rangle\,. \tag{H.122}$$

In light of eq. (H.120), we can therefore identify

$$C_R = (\boldsymbol{M}\,,\,\boldsymbol{M}+2\boldsymbol{\delta})\,. \tag{H.123}$$

Using eqs. (H.90), (H.106), and (H.107), it follows that

$$(\boldsymbol{M}\,,\,\boldsymbol{M}+2\boldsymbol{\delta}) = \left(\sum_{k=1}^{\ell} q_k\boldsymbol{\alpha}_k\,,\,\boldsymbol{M}+2\boldsymbol{\delta}\right) = \tfrac{1}{2}\sum_{k=1}^{\ell}(\boldsymbol{\alpha}_k\,,\,\boldsymbol{\alpha}_k)\,(a_k+2)q_k\,. \tag{H.124}$$

After inserting eq. (H.108) for the q_k, we end up with

$$C_R = \tfrac{1}{2}\sum_{j=1}^{\ell}\sum_{k=1}^{\ell}(\boldsymbol{\alpha}_k\,,\,\boldsymbol{\alpha}_k)(a_k+2)(A^{-1})_{kj}a_j\,. \tag{H.125}$$

Equation (H.125) is sometimes rewritten in terms of the symmetrized Cartan matrix, which is defined by

$$G_{ij} \equiv \frac{2}{(\boldsymbol{\alpha}_j\,,\,\boldsymbol{\alpha}_j)}A_{ij} = \frac{4(\boldsymbol{\alpha}_i\,,\,\boldsymbol{\alpha}_j)}{(\boldsymbol{\alpha}_i\,,\,\boldsymbol{\alpha}_i)(\boldsymbol{\alpha}_j\,,\,\boldsymbol{\alpha}_j)} = (\boldsymbol{\alpha}_i^\vee\,,\,\boldsymbol{\alpha}_j^\vee)\,. \tag{H.126}$$

The inverse of the symmetrized Cartan matrix, which we shall denote by G^{ij}, is therefore given by

$$G^{ij} = \tfrac{1}{2}(\boldsymbol{\alpha}_i\,,\,\boldsymbol{\alpha}_i)A^{-1}_{ij}\,. \tag{H.127}$$

One can immediately check that $G_{ij}G^{jk} = \delta_i^k$ as required. Hence, eq. (H.125) can be rewritten as

$$C_R = \sum_{j=1}^{\ell}\sum_{k=1}^{\ell}(a_k+2)G^{kj}a_j\,. \tag{H.128}$$

For any irreducible representation,

$$\mathrm{Tr}(\boldsymbol{R}^a\boldsymbol{R}^b) = I_2(R)g^{ab}\,, \tag{H.129}$$

where $I_2(R)$ is the second-order Dynkin index of the representation R. As noted below eq. (H.63), in the case of a semisimple compact real Lie algebra, it is common practice to choose a basis $\{\boldsymbol{X}^a\}$ of generators in eq. (H.54) such that the metric tensor defined in eq. (H.55) is given by $g^{ab} = -\delta^{ab}$. With this basis choice, eq. (H.129) reduces to the one given in eq. (H.32), after identifying $\boldsymbol{R^a} \equiv -i\boldsymbol{T^a_R}$.

By virtue of eq. (H.58), the second-order Dynkin index of the adjoint representation is $I_2(A) = \eta$. For an arbitrary irreducible representation R of dimension d_R, taking the trace of eq. (H.61) yields

$$C_R = \frac{I_2(R)d_G}{d_R} \tag{H.130}$$

[previously obtained in eq. (H.39)], where d_G is the dimension of the Lie algebra:

$$\mathfrak{su}(n): \quad d_G = n^2 - 1\,, \tag{H.131}$$

$$\mathfrak{so}(n): \quad d_G = \tfrac{1}{2}n(n-1)\,, \tag{H.132}$$

$$\mathfrak{sp}(n): \quad d_G = n(2n+1)\,. \tag{H.133}$$

For the adjoint representation ($R = A$), we have $d_R = d_G$, in which case eq. (H.130) yields the expected result,

$$C_A = I_2(A) = \eta\,. \tag{H.134}$$

H.4.4 The Dual Coxeter Number

The time has now come to identify η, which was employed in the normalization of the metric tensor g^{ab} in eq. (H.55). We begin by introducing the maximal weight of the adjoint representation, denoted by $\boldsymbol{\theta}$, which coincides with the highest positive root. For a semisimple Lie algebra, there are at most two roots of different length, called long roots and short roots.[14] Furthermore, one can prove that $(\boldsymbol{\theta},\boldsymbol{\theta}) \geq (\boldsymbol{\alpha},\boldsymbol{\alpha})$ for all $\boldsymbol{\alpha} \in \Delta$, which implies that $\boldsymbol{\theta}$ must be a long root.

Following eq. (H.91), we define the dual of $\boldsymbol{\theta}$ via

$$\boldsymbol{\theta}^\vee \equiv \frac{2\boldsymbol{\theta}}{(\boldsymbol{\theta},\boldsymbol{\theta})}\,. \tag{H.135}$$

One can expand $\boldsymbol{\theta}^\vee$ in terms of the dual roots with positive integer coefficients,

$$\boldsymbol{\theta}^\vee = \sum_{k=1}^{\ell} c_k^\vee \boldsymbol{\alpha}_k^\vee\,. \tag{H.136}$$

The *dual Coxeter number* is then defined as

$$h^\vee \equiv 1 + \sum_{k=1}^{\ell} c_k^\vee\,. \tag{H.137}$$

We now choose η in eq. (H.55) to be the dual Coxeter number, $\eta \equiv h^\vee$, which fixes the inner products of the roots and the root lengths. In particular, we may compute the length of the highest positive root $\boldsymbol{\theta}$ as follows. Taking the inner product of the

[14] Roots are conventionally called long in cases where all roots are of the same length.

Weyl vector with $\boldsymbol{\theta}^\vee$ using eqs. (H.86) and (H.136), and employing eq. (H.94), it follows that

$$(\boldsymbol{\delta},\boldsymbol{\theta}^\vee) = \sum_{k=1}^{\ell} c_k^\vee (\boldsymbol{\delta},\boldsymbol{\alpha}_k^\vee) = \sum_{k=1}^{\ell} c_k^\vee \,. \tag{H.138}$$

Since $\boldsymbol{\theta}$ is the maximal weight of the adjoint representation, eqs. (H.137) and (H.138) yield

$$h^\vee = 1 + (\boldsymbol{\delta},\boldsymbol{\theta}^\vee) = 1 + \frac{(\boldsymbol{\delta},2\boldsymbol{\theta})}{(\boldsymbol{\theta},\boldsymbol{\theta})} = \frac{(\boldsymbol{\theta}+2\boldsymbol{\delta},\boldsymbol{\theta})}{(\boldsymbol{\theta},\boldsymbol{\theta})} = \frac{C_A}{(\boldsymbol{\theta},\boldsymbol{\theta})}\,, \tag{H.139}$$

after using eq. (H.122) and the symmetry property of the inner product. Note that $C_A/(\boldsymbol{\theta},\boldsymbol{\theta})$ and the dual Coxeter number $h^\vee$ are independent of an overall rescaling of the metric tensor. Since $C_A = \eta$ [see eq. (H.134)], we conclude that, in the convention where the metric tensor is defined by eq. (H.55) with $\eta = h^\vee$,[15]

$$(\boldsymbol{\theta},\boldsymbol{\theta}) = 1\,. \tag{H.140}$$

In order to provide an explicit evaluation of $h^\vee$, we begin by multiplying eq. (H.69) by g_{ij} and summing over i and j, which yields

$$\sum_{\boldsymbol{\alpha}\in\Delta} (\boldsymbol{\alpha}\,,\boldsymbol{\alpha}) = \eta\ell\,, \tag{H.141}$$

where ℓ is the rank of the group. This result can be used to compute $(\boldsymbol{\theta}\,,\boldsymbol{\theta})$ as follows. First, we examine the simply laced Lie algebras, which are defined as the semisimple Lie algebras whose roots are all of equal length. Since there are $d_G-\ell$ nonzero roots, it follows from eq. (H.141) that $\eta\ell = (d_G-\ell)(\boldsymbol{\theta}\,,\boldsymbol{\theta})$, or

$$(\boldsymbol{\theta}\,,\boldsymbol{\theta}) = \frac{\eta\ell}{d_G-\ell}\,, \quad \text{for } \mathfrak{g}=\mathfrak{su}(\ell+1)\ [\ell\geq 1] \text{ and } \mathfrak{so}(2\ell)\ [\ell\geq 2]\,. \tag{H.142}$$

Next, we examine the semisimple Lie algebras, $\mathfrak{so}(2\ell+1)$ $[\ell\geq 2]$ and $\mathfrak{sp}(\ell)$ $[\ell\geq 1]$, that possess both short and long roots (which we denote generically by $\boldsymbol{\alpha}_S$ and $\boldsymbol{\alpha}_L$, respectively). In both cases, one can show that

$$(\boldsymbol{\theta},\boldsymbol{\theta}) = (\boldsymbol{\alpha}_L,\boldsymbol{\alpha}_L) = 2(\boldsymbol{\alpha}_S,\boldsymbol{\alpha}_S)\,, \tag{H.143}$$

since $\boldsymbol{\theta}$ is necessarily one of the long roots as previously noted. For $\mathfrak{so}(2\ell+1)$ $[\ell\geq 2]$, there are $\ell-1$ long roots and one short root. Using Weyl reflections to generate the remaining roots yields $(\ell-1)(d_G-\ell)/\ell$ long roots and $(d_G-\ell)/\ell$ short roots. For $\mathfrak{sp}(\ell)$ $[\ell\geq 1]$, there is one long root and $\ell-1$ short roots. Using Weyl reflections to generate the remaining roots yields $(d_G-\ell)/\ell$ long roots and $(\ell-1)(d_G-\ell)/\ell$ short roots. Hence, for $\mathfrak{so}(2\ell+1)$ $[\ell\geq 2]$, eq. (H.141) gives

$$\eta\ell = \left[\frac{(\ell-1)(d_G-\ell)}{\ell} + \frac{d_G-\ell}{2\ell}\right](\boldsymbol{\theta}\,,\boldsymbol{\theta}) = \frac{(2\ell-1)(d_G-\ell)}{2\ell}(\boldsymbol{\theta}\,,\boldsymbol{\theta})\,, \tag{H.144}$$

15 Beware: in the mathematics literature, the common convention is to take $(\boldsymbol{\theta},\boldsymbol{\theta})=2$, which is equivalent to replacing η with 2η in eq. (H.55) and in Theorem H.6, whereas the value of the dual Coxeter number given by eq. (H.139) is unchanged.

and for $\mathfrak{sp}(\ell)$ [$\ell \geq 1$], eq. (H.141) gives

$$\eta\ell = \left[\frac{d_G - \ell}{\ell} + \frac{(\ell-1)(d_G-\ell)}{2\ell}\right](\boldsymbol{\theta}\,,\boldsymbol{\theta}) = \frac{(\ell+1)(d_G-\ell)}{2\ell}(\boldsymbol{\theta}\,,\boldsymbol{\theta})\,. \quad \text{(H.145)}$$

Therefore,

$$(\boldsymbol{\theta}\,,\boldsymbol{\theta}) = \frac{2\eta\ell^2}{d_G-\ell}\times\begin{cases}\dfrac{1}{2\ell-1}\,, & \text{for } \mathfrak{g}=\mathfrak{so}(2\ell+1) \qquad (\ell\geq 2)\,,\\[2ex] \dfrac{1}{\ell+1}\,, & \text{for } \mathfrak{g}=\mathfrak{sp}(\ell) \qquad (\ell \geq 1)\,.\end{cases} \quad \text{(H.146)}$$

Since $(\boldsymbol{\theta},\boldsymbol{\theta}) = 1$ in the convention where $\eta = h^\vee$, we can employ eqs. (H.131)–(H.133) in eqs. (H.142)–(H.146) and solve for $h^\vee$. The end result is given by

$$\eta = h^\vee = \begin{cases} n\,, & \text{for } \mathfrak{su}(n) \qquad (n\geq 2)\,,\\ 2\,, & \text{for } \mathfrak{so}(3)\cong\mathfrak{su}(2)\,,\\ n-2\,, & \text{for } \mathfrak{so}(n) \qquad (n\geq 5)\,,\\ n+1\,, & \text{for } \mathfrak{sp}(n) \qquad (n\geq 1)\,.\end{cases} \quad \text{(H.147)}$$

H.4.5 The Quadratic Casimir Operator and Second-Order Index for the Irreps of $\mathfrak{su}(n)$, $\mathfrak{so}(n)$, and $\mathfrak{sp}(n)$

Explicit forms for the Cartan matrices (and their inverses) corresponding to the simple Lie algebras can be found in many books (see, e.g., [B115]). In order to compute the quadratic Casimir operator, we shall employ eq. (H.125), which depends on the length of the simple roots and the elements of the inverse Cartan matrix.

Given the Cartan matrix, one can compute the inner product of any two simple roots as follows. First we note the following result obtained in [B102]:

$$(\boldsymbol{\alpha_i}\,,\boldsymbol{\alpha_i}) = \eta\left[\tfrac{1}{2}\sum_{\beta\in\Delta+}\left\{\sum_{j=1}^{\ell}k_j^\beta A_{ij}\right\}^2\right]^{-1}, \qquad \boldsymbol{\alpha_i}\in\Pi\,, \quad \text{(H.148)}$$

$$(\boldsymbol{\alpha_i}\,,\boldsymbol{\alpha_j}) = \tfrac{1}{2}A_{ij}(\boldsymbol{\alpha_i}\,,\boldsymbol{\alpha_i})\,, \quad \text{for } i\neq j, \quad \text{(H.149)}$$

where A_{ij} are the matrix elements of the Cartan matrix [defined in eq. (H.78)], the expansion of the positive root $\boldsymbol{\beta}$ in terms of simple roots is given by $\boldsymbol{\beta} = \sum_{j=1}^{\ell}k_j^\beta\boldsymbol{\alpha_j}$, and the k_j^β are nonnegative integers. In light of eq. (H.147), the final results are given by [B102]

$$\mathfrak{su}(\ell+1): \quad (\boldsymbol{\alpha_k}\,,\boldsymbol{\alpha_k}) = 1\,, \qquad k=1,2,\ldots,\ell\,, \quad \text{(H.150)}$$

$$\mathfrak{so}(2\ell+1):\quad (\boldsymbol{\alpha}_k\,,\,\boldsymbol{\alpha}_k)=\begin{cases}1\,, & \text{for } k=1,2,\ldots,\ell-1\,,\\[6pt] \tfrac{1}{2}\,, & \text{for } k=\ell\quad(\ell\geq 2)\,,\end{cases} \tag{H.151}$$

$$\mathfrak{so}(2\ell):\quad (\boldsymbol{\alpha}_k\,,\,\boldsymbol{\alpha}_k)=1\,,\qquad \text{for } k=1,2,\ldots,\ell\quad(\ell\geq 3)\,, \tag{H.152}$$

$$\mathfrak{sp}(\ell):\quad (\boldsymbol{\alpha}_k\,,\,\boldsymbol{\alpha}_k)=\begin{cases}\tfrac{1}{2}\,, & \text{for } k=1,2,\ldots,\ell-1\,,\\[6pt] 1\,, & \text{for } k=\ell\,.\end{cases} \tag{H.153}$$

Next, we exhibit below the inverse Cartan matrices for $\mathfrak{su}(\ell+1)$, $\mathfrak{so}(2\ell)$, $\mathfrak{so}(2\ell+1)$ and $\mathfrak{sp}(\ell)$, where ℓ is the rank of the corresponding Lie algebras. Since $A_{ii}=2$, it follows that $A^{-1}=\frac{1}{2}$ for the rank $\ell=1$ simple Lie algebras, $\mathfrak{su}(2)$, $\mathfrak{so}(3)$, and $\mathfrak{sp}(1)$. For the higher-rank Lie algebras,

$\mathfrak{su}(\ell+1)\ \ [\ell\geq 2]:$

$$A^{-1}=\frac{1}{\ell+1}\begin{pmatrix} \ell & \ell-1 & \ell-2 & \cdots & 3 & 2 & 1\\ \ell-1 & 2(\ell-1) & 2(\ell-2) & \cdots & 6 & 4 & 2\\ \ell-2 & 2(\ell-2) & 3(\ell-2) & \cdots & 9 & 6 & 3\\ \vdots & \vdots & \vdots & \ddots & \vdots & \vdots & \vdots\\ 3 & 6 & 9 & \cdots & 3(\ell-2) & 2(\ell-2) & \ell-2\\ 2 & 4 & 6 & \cdots & 2(\ell-2) & 2(\ell-1) & \ell-1\\ 1 & 2 & 3 & \cdots & \ell-2 & \ell-1 & \ell \end{pmatrix}, \tag{H.154}$$

$$\mathfrak{so}(2\ell+1)\ \ [\ell\geq 2]:\qquad A^{-1}=\begin{pmatrix} 1 & 1 & 1 & \cdots & 1 & 1 & \frac{1}{2}\\ 1 & 2 & 2 & \cdots & 2 & 2 & 1\\ 1 & 2 & 3 & \cdots & 3 & 3 & \frac{3}{2}\\ \vdots & \vdots & \vdots & \ddots & \vdots & \vdots & \vdots\\ 1 & 2 & 3 & \cdots & \ell-2 & \ell-2 & \frac{1}{2}(\ell-2)\\ 1 & 2 & 3 & \cdots & \ell-2 & \ell-1 & \frac{1}{2}(\ell-1)\\ 1 & 2 & 3 & \cdots & \ell-2 & \ell-1 & \frac{1}{2}\ell \end{pmatrix}, \tag{H.155}$$

$$\mathfrak{so}(2\ell)\ [\ell \geq 3]: \qquad A^{-1} = \begin{pmatrix} 1 & 1 & 1 & \cdots & 1 & \frac{1}{2} & \frac{1}{2} \\ 1 & 2 & 2 & \cdots & 2 & 1 & 1 \\ 1 & 2 & 3 & \cdots & 3 & \frac{3}{2} & \frac{3}{2} \\ \vdots & \vdots & \vdots & \ddots & \vdots & \vdots & \vdots \\ 1 & 2 & 3 & \cdots & \ell-2 & \frac{1}{2}(\ell-2) & \frac{1}{2}(\ell-2) \\ \frac{1}{2} & 1 & \frac{3}{2} & \cdots & \frac{1}{2}(\ell-2) & \frac{1}{4}\ell & \frac{1}{4}(\ell-2) \\ \frac{1}{2} & 1 & \frac{3}{2} & \cdots & \frac{1}{2}(\ell-2) & \frac{1}{4}(\ell-2) & \frac{1}{4}\ell \end{pmatrix}, \tag{H.156}$$

$$\mathfrak{sp}(\ell)\ [\ell \geq 2]: \qquad A^{-1} = \begin{pmatrix} 1 & 1 & 1 & \cdots & 1 & 1 & 1 \\ 1 & 2 & 2 & \cdots & 2 & 2 & 2 \\ 1 & 2 & 3 & \cdots & 3 & 3 & 3 \\ \vdots & \vdots & \vdots & \ddots & \vdots & \vdots & \vdots \\ 1 & 2 & 3 & \cdots & \ell-2 & \ell-2 & \ell-2 \\ 1 & 2 & 3 & \cdots & \ell-2 & \ell-1 & \ell-1 \\ \frac{1}{2} & 1 & \frac{3}{2} & \cdots & \frac{1}{2}(\ell-2) & \frac{1}{2}(\ell-1) & \frac{1}{2}\ell \end{pmatrix}. \tag{H.157}$$

Two isomorphic simple Lie algebras, $\mathfrak{g}_1 \cong \mathfrak{g}_2$, can have different Cartan matrices due to a difference in the ordering conventions of the ℓ simple roots of $\mathfrak{g}_1$ and $\mathfrak{g}_2$. For example, $\mathfrak{so}(6) \cong \mathfrak{su}(4)$ [see eq. (H.103)], whereas A^{-1} for $\mathfrak{so}(6)$ and $\mathfrak{su}(4)$ are

$$\mathfrak{so}(6): \ A^{-1} = \begin{pmatrix} 1 & \frac{1}{2} & \frac{1}{2} \\ \frac{1}{2} & \frac{3}{4} & \frac{1}{4} \\ \frac{1}{2} & \frac{1}{4} & \frac{3}{4} \end{pmatrix}, \qquad \mathfrak{su}(4): \ A^{-1} = \begin{pmatrix} \frac{3}{4} & \frac{1}{2} & \frac{1}{4} \\ \frac{1}{2} & 1 & \frac{1}{2} \\ \frac{1}{4} & \frac{1}{2} & \frac{3}{4} \end{pmatrix}. \tag{H.158}$$

In particular, consider A^{-1} for $\mathfrak{so}(6)$.[16] If the first and second columns are interchanged and then the first and second rows are interchanged (corresponding to an interchange $\boldsymbol{\alpha_1} \leftrightarrow \boldsymbol{\alpha_2}$ of the simple roots), one ends up with A^{-1} for $\mathfrak{su}(4)$.

[16] The last two rows and the last two columns of A^{-1} for $\mathfrak{so}(2\ell)$ do not follow the same pattern as the preceding rows and columns. Hence, the inverse Cartan matrix of $\mathfrak{so}(6)$ is obtained by setting $\ell = 3$ in eq. (H.156) and keeping only the last three rows and three columns of A^{-1}.

One can now employ eq. (H.125) to evaluate of the eigenvalue of the quadratic Casimir operator, C_R, for an irrep R of $\mathfrak{g}$. Consider first the fundamental irreps of $\mathfrak{su}(n)$, $\mathfrak{so}(n)$, and $\mathfrak{sp}(n)$. Using the results provided above, a straightforward calculation yields

$$\mathfrak{su}(\ell+1)\,,\quad [\ell\geq 1]:\quad C_F=\tfrac{1}{2}(A^{-1})_{11}+\sum_{k=1}^{\ell}(A^{-1})_{k1}=\frac{\ell(\ell+2)}{2(\ell+1)}\,, \tag{H.159}$$

$$\mathfrak{so}(2\ell+1)\,,\quad [\ell\geq 2]:\quad C_F=\tfrac{1}{2}(A^{-1})_{11}+\sum_{k=1}^{\ell-1}(A^{-1})_{k1}+\tfrac{1}{2}(A^{-1})_{\ell 1}=\ell\,, \tag{H.160}$$

$$\mathfrak{so}(2\ell)\,,\quad [\ell\geq 3]:\quad C_F=\tfrac{1}{2}(A^{-1})_{11}+\sum_{k=1}^{\ell}(A^{-1})_{k1}=\tfrac{1}{2}(2\ell-1)\,, \tag{H.161}$$

$$\mathfrak{sp}(\ell)\,,\quad [\ell\geq 1]:\quad C_F=\tfrac{1}{4}(A^{-1})_{11}+\tfrac{1}{2}\sum_{k=1}^{\ell-1}(A^{-1})_{k1}+(A^{-1})_{\ell 1}=\tfrac{1}{4}(2\ell+1)\,. \tag{H.162}$$

The case of $\mathfrak{so}(3)$ must be treated separately by employing $(\boldsymbol{\alpha},\boldsymbol{\alpha})=1$, $A^{-1}=\frac{1}{2}$ and $a_1=2$ in eq. (H.125). The end results are

$$\mathfrak{su}(n):\quad C_F=\frac{n^2-1}{2n}\qquad (n\geq 2)\,, \tag{H.163}$$

$$\mathfrak{so}(3):\quad C_F=2\,, \tag{H.164}$$

$$\mathfrak{so}(n):\quad C_F=\tfrac{1}{2}(n-1)\qquad (n\geq 5)\,, \tag{H.165}$$

$$\mathfrak{sp}(n):\quad C_F=\tfrac{1}{4}(2n+1)\qquad (n\geq 1)\,. \tag{H.166}$$

Given the dimensions of the fundamental irreps (d_F),

$$\mathfrak{su}(n):\quad d_F=n\,, \tag{H.167}$$

$$\mathfrak{so}(n):\quad d_F=n\,, \tag{H.168}$$

$$\mathfrak{sp}(n):\quad d_F=2n\,, \tag{H.169}$$

and making use of eq. (H.39), one obtains the second-order Dynkin index of the fundamental tensor representation exhibited below:

$$\mathfrak{su}(n):\quad I_2(F)=\tfrac{1}{2}\qquad (n\geq 2)\,, \tag{H.170}$$

$$\mathfrak{so}(3):\quad I_2(F)=2\,, \tag{H.171}$$

$$\mathfrak{so}(n):\quad I_2(F)=1\qquad (n\geq 5)\,, \tag{H.172}$$

$$\mathfrak{sp}(n):\quad I_2(F)=\tfrac{1}{2}\qquad (n\geq 1)\,. \tag{H.173}$$

Moreover, for any representation R, the quantity $I_2(R)/I_2(F)$ is a positive integer that is independent of the choice of normalization of the metric tensor.

One can repeat the above analysis in the case of the adjoint representation. The Dynkin labels for the adjoint representation have been previously given in eqs. (H.116)–(H.119). The corresponding eigenvalues of the quadratic Casimir operator can be evaluating using eq. (H.125). For example,

$$\mathfrak{su}(\ell+1)\,,\quad \ell\geq 2: \quad C_A = \tfrac{1}{2}\Bigg[(A^{-1})_{11}+(A^{-1})_{\ell 1}+(A^{-1})_{1\ell}+(A^{-1})_{\ell\ell} + 2\sum_{k=1}^{\ell}[(A^{-1})_{k1}+(A^{-1})_{k\ell}]\Bigg] = \ell+1\,. \qquad \text{(H.174)}$$

That is, $C_A = \ell + 1 = \eta$ in light of eq. (H.147), in agreement with eq. (H.134). Similarly, the reader can check that $C_A = \eta$ for the other simple Lie algebras.

H.5 Tables of Dimensions, Indices, and Branching Rules

The tables of dimensions, indices, and branching rules for representations of simple Lie algebras presented in Section H.5.3 are particularly useful for grand unified theory model building. We focus on the irreducible representations (irreps) of the following Lie algebras: $\mathfrak{su}(3)$, $\mathfrak{su}(4)$, $\mathfrak{su}5)$, $\mathfrak{so}(10)$, and the exceptional Lie algebra E_6. More extensive tables can be found in [B116] and in Refs. [68–70].

H.5.1 Introduction to the Tables

An irrep of $\mathfrak{su}(n)$ is specified by its Dynkin labels, $(a_1a_2\cdots a_{n-1})$. The n-ality of this irrep [e.g., triality for $n=3$, quadrality for $n=4$, quintality for $n=5$, etc.] is defined by $a_1+2a_2+\cdots+(n-1)a_{n-1} \pmod n$. In light of eq. (H.111), each complex irrep $(a_1a_2\cdots a_{n-1})$ with n-ality k is conjugate to $(a_{n-1}\cdots a_2a_1)$ with n-ality $n-k$, for $k=0,1,2,\ldots,n$ (with n-ality n equivalent to n-ality zero). Real irreps *must* have n-ality 0 (or equivalently n), although not all irreps with n-ality 0 or n are real. The notation for complex irreps follows that of Ref. [68]. Conjugate irreps are not explicitly listed in Tables H.2, H.4, and H.7.[17]

An irrep of $\mathfrak{so}(10)$ is denoted by its Dynkin labels, $(a_1a_2a_3a_4a_5)$. The quadrality of this irrep is defined by $2a_1+2a_3-a_4+a_5 \pmod 4$. Spinor irreps, which satisfy $a_4+a_5=1 \pmod 2$, have quadrality 1 and the corresponding conjugate spinor irreps have quadrality 3. The irreps of quadrality 0 or 2 may be real or complex (if complex, then the corresponding conjugate irrep is also of quadrality 0 or 2, respectively). Complex irreps of $\mathfrak{so}(10)$ possess Dynkin indices $(a_1a_2a_3a_4a_5)$ with $a_4\neq a_5$ and conjugation simply interchanges $a_4\leftrightarrow a_5$. Conjugate irreps are not explicitly listed in Table H.10.[18]

An irrep of E_6 is denoted by its Dynkin labels, $(a_1a_2a_3a_4a_5a_6)$. The triality of this irrep is defined by $a_1-a_2+a_4-a_5 \pmod 3$. A triality k irrep, $(a_1a_2a_3a_4a_5a_6)$

[17] We caution the reader that what we designate as the irrep **6** of $\mathfrak{su}(3)$ is called $\overline{\mathbf{6}}$ in Ref. [69].
[18] We caution the reader that what we designate (following Ref. [68]) as the irrep **126** of $\mathfrak{so}(10)$ is called $\overline{\mathbf{126}}$ in Refs. [69, 70].

is conjugate to $(a_5a_4a_3a_2a_1a_6)$ with triality $3-k$ (for $k=0,1,2$ or 3 and triality 3 equivalent to triality 0). Real irreps must have triality 0 (or 3), although the converse is not true in general. Conjugate irreps are not explicitly listed in Table H.14.

The tables provide information on the dimension of the representation R, the second-order Dynkin index $I_2(R)$ (denoted in the tables by "index"), the eigenvalue of the quadratic Casimir operator (denoted in the tables by "Casimir"), and the anomaly coefficient (denoted in the tables by "anomaly"), which is defined by

$$A(R)=\frac{2I_3(R)}{I_2(F)}\,, \tag{H.175}$$

where $I_3(R)$ is the cubic index, in a convention where $I_2(F)=\frac{1}{2}$. In particular,

$$\tfrac{1}{2}\,\mathrm{Tr}\,[\{\boldsymbol{T}^a_{\boldsymbol{R}},\boldsymbol{T}^b_{\boldsymbol{R}}\}\boldsymbol{T}^c_{\boldsymbol{R}}]=\tfrac{1}{4}A(R)d^{abc}\,, \tag{H.176}$$

in light of eqs. (H.42) and (H.45). If $\mathfrak{g}$ is a simple Lie algebra, then $A(R)\neq 0$ only in the case of $\mathfrak{g}=\mathfrak{su}(n)$ for $n\geq 3$. In our normalization convention, $A(F)=1$ in the fundamental n-dimensional irrep of $\mathfrak{su}(n)$.

For completeness, an explicit formula for $A(R)$ will now be presented. Given an irrep R of $\mathfrak{su}(n)$ $[n\geq 3]$ specified by its Dynkin labels, $(a_1a_2\cdots a_{n-1})$, we first introduce n nonnegative integers, $p_1\geq p_2\geq\cdots\geq p_{n-1}\geq 0$ and $p_n\equiv 0$, such that $a_j=p_j-p_{j+1}$, for $j=1,2,\ldots,n-1$. The p_k correspond to the number of boxes in the kth row of the Young tableau that represents R (see footnote 11). In terms of the p_j, we define

$$\sigma_j\equiv p_j-j+\tfrac{1}{2}(n+1)-\frac{1}{n}\sum_{k=1}^{n}p_k\,,\qquad \text{for } j=1,2,\ldots,n, \tag{H.177}$$

which satisfy $\sum_{j=1}^n\sigma_j=0$. Then, for a representation R of $\mathfrak{su}(n)$ $[n\geq 3]$ of dimension d_R, the following formulae can be derived:

$$A(R)=\frac{2nd_R}{(n^2-1)(n^2-4)}\sum_{j=1}^{n}\sigma_j^3\,, \tag{H.178}$$

where

$$d_R=\frac{\prod_{j<k}^{n}(\sigma_j-\sigma_k)}{1!2!\cdots(n-1)!}\,. \tag{H.179}$$

Given an irrep R of a simple Lie algebra and its conjugate irrep $\overline{R}\equiv\overline{R}$, we have

$$I_2(\overline{R})=I_2(R)\,,\qquad C_{\overline{R}}=C_R\,,\qquad A(\overline{R})=-A(R)\,. \tag{H.180}$$

The conjugacy class of the irrep is indicated by specifying the quadrality of SU(4) and SO(10) irreps, the quintality of SU(5) irreps, and the triality of E_6 irreps.

One can check the results of the tables of tensor products by employing

$$d_{R_1\otimes R_2}=d_{R_1}d_{R_2}\,, \tag{H.181}$$

$$I_2(R_1\otimes R_2)=d_{R_1}I_2(R_2)+d_{R_2}I_2(R_1)\,, \tag{H.182}$$

$$A(R_1\otimes R_2)=d_{R_1}A(R_2)+d_{R_2}A(R_1)\,, \tag{H.183}$$

where the second-order Dynkin index and anomaly coefficient of a direct sum of irreps are given by

$$I_2(R_1 \oplus R_2) = I_2(R_1) + I_2(R_2)\,, \tag{H.184}$$

$$A(R_1 \oplus R_2) = A(R_1) + A(R_2)\,. \tag{H.185}$$

In addition, all irreps contained in $R_1 \otimes R_2$ have the same n-ality, which is equal to the sum of the n-alities of R_1 and R_2.

H.5.2 Subalgebras, Branching Rules, and Index Sum Rules

Consider a simple Lie algebra $\mathfrak{g}$ and a proper subalgebra, $\mathfrak{h} \subset \mathfrak{g}$, which may be simple, semisimple, or a direct sum of simple and $\mathfrak{u}(1)$ Lie algebras. Then, $\mathfrak{h}$ is a maximal subalgebra of $\mathfrak{g}$ if no subalgebra $\mathfrak{p} \subset \mathfrak{g}$ exists such that $\mathfrak{h} \subset \mathfrak{p} \subset \mathfrak{g}$. If the root vectors of $\mathfrak{h}$ are also root vectors of $\mathfrak{g}$ and the commuting generators of $\mathfrak{h}$ are linear combinations of the commuting generators of $\mathfrak{g}$, then $\mathfrak{h}$ is called a regular subalgebra of $\mathfrak{g}$. If $\mathfrak{h}$ is a maximal regular subalgebra of $\mathfrak{g}$, then rank $\mathfrak{h}$ = rank $\mathfrak{g}$. A subalgebra $\mathfrak{h}$ is called a special subalgebra of $\mathfrak{g}$ if the root vectors of $\mathfrak{h}$ are *not* a subset of the root vectors of $\mathfrak{g}$. Moreover, if a special subalgebra $\mathfrak{h}$ is maximal, then it must be either simple or semisimple (i.e., with no $\mathfrak{u}(1)$ factors present) and rank $\mathfrak{h}$ < rank $\mathfrak{g}$. Further details can be found in [B101, B106, B120] and in Ref. [225].

Branching rules exhibit how irreps of $\mathfrak{g}$ decompose into a direct sum of irreps of $\mathfrak{h}$. The branching rules will depend on how $\mathfrak{h}$ is embedded inside $\mathfrak{g}$. Given a branching rule $R \longrightarrow \sum_j R_j$, the second-order Dynkin index of an irrep R of $\mathfrak{g}$ can be related to the second-order Dynkin indices of the irreps R_i of $\mathfrak{h}$.

First, we suppose that $\mathfrak{h} = \mathfrak{h}_1 \oplus \mathfrak{p}$, where $\mathfrak{h}_1$ is simple and $\mathfrak{p}$ is simple, semisimple, or a direct sum of simple and $\mathfrak{u}(1)$ Lie algebras. Given the branching rule

$$R \longrightarrow \sum_j (R_1^{(j)}, R_{\mathfrak{p}}^{(j)})\,, \tag{H.186}$$

where R_1 is an irrep of $\mathfrak{h}_1$ and $R_{\mathfrak{p}}$ is an irrep of $\mathfrak{p}$ of dimension $d(R_{\mathfrak{p}})$, one obtains an index sum rule,

$$I_2(R) j_2(\mathfrak{h}_1 \subset \mathfrak{g}) = \sum_j d(R^{(j)}(\mathfrak{p})) I_2(R_1^{(j)})\,, \tag{H.187}$$

where $j_2(\mathfrak{h}_1 \subset \mathfrak{g})$ is called the embedding index [B106, 225] (using the notation of Ref. [237]) and depends on how $\mathfrak{h}_1$ is embedded inside $\mathfrak{g}$. The embedding index does not depend on the choice of irrep R of $\mathfrak{g}$. Thus, one can compute $j_2(\mathfrak{h}_1 \subset \mathfrak{g})$ by comparing both sides of eq. (H.187) when applied to the branching rule for any choice of R. Typically, it is most convenient to choose R to be the fundamental representation of $\mathfrak{g}$. Once $j_2(\mathfrak{h}_1 \subset \mathfrak{g})$ is known, eq. (H.187) can provide useful checks of the results of the tables of branching rules. If the maximal subalgebra of $\mathfrak{g}$ is a direct sum of m simple subalgebras ($\mathfrak{h}_j$) and n $\mathfrak{u}(1)$ subalgebras (for $m \geq 1$, $n \geq 0$), then any one of the $\mathfrak{h}_j$ subalgebras could have been chosen to be $\mathfrak{h}_1$. Hence, eq. (H.187) actually consists of m separate identities. To make use of all of these identities, one would have to first evaluate $j_2(\mathfrak{h}_j \subset \mathfrak{g})$ for each $j = 1, 2, \ldots, m$ as indicated above.

It can be shown that for simple Lie algebras $\mathfrak{h}_1 \subset \mathfrak{g}$, the embedding index is always a positive integer [225, 236, 238]. Moreover, for a regular simple subalgebra of $\mathfrak{g}$, the embedding index is given by the following explicit formula [B106]:

$$j_2(\mathfrak{h}_1 \subset \mathfrak{g}) = \frac{(\boldsymbol{\theta}(\mathfrak{g}), \boldsymbol{\theta}(\mathfrak{g}))}{(\boldsymbol{\theta}(\mathfrak{h}_1), \boldsymbol{\theta}(\mathfrak{h}_1))}, \tag{H.188}$$

where $\boldsymbol{\theta}(\mathfrak{g})$ $[\boldsymbol{\theta}(\mathfrak{h}_1)]$ is the highest positive root of $\mathfrak{g}$ $[\mathfrak{h}_1]$. Since $\mathfrak{h}_1$ is a regular subalgebra of $\mathfrak{g}$, the highest positive root $\boldsymbol{\theta}(\mathfrak{h}_1)$ is also a root of $\mathfrak{g}$. It then follows that

$$j_2(\mathfrak{h}_1 \subset \mathfrak{g}) = 1 \text{ for a simply laced Lie algebra } \mathfrak{g} \tag{H.189}$$

whose roots are all of the same length. Indeed, for the branching rules displayed in the tables below, all the Lie algebras exhibited are simply laced and their corresponding maximal subalgebras are regular, in which case eq. (H.187) applies with $j_2(\mathfrak{h}_1 \subset \mathfrak{g}) = 1$. In contrast, the possible values of $j_2(\mathfrak{h}_1 \subset \mathfrak{g})$ for $\mathfrak{so}(2n+1)$ $[n \geq 2]$ and $\mathfrak{sp}(n)$ $[n \geq 3]$ are 1 or 2. For example, if $\mathfrak{g} = \mathfrak{so}(9)$ and $\mathfrak{h} = \mathfrak{su}(4) \oplus \mathfrak{su}(2)$, then the branching rule of the fundamental representation, $\mathbf{9} \longrightarrow (\mathbf{6}, \mathbf{1}) + (\mathbf{1}, \mathbf{3})$, yields $j_2(\mathfrak{su}(4) \subset \mathfrak{so}(9)) = 2$.

One can also extend eq. (H.187) to the case where $\mathfrak{h}_1 = \mathfrak{u}(1)$. Since the normalization of the $\mathfrak{u}(1)$ charge (denoted generically by q) is arbitrary, the embedding index $j_2(\mathfrak{u}(1) \subset \mathfrak{g})$ is no longer required to be an integer. In particular, one can simply insert $I_2(R_1^{(j)}) = q_j^2$ in eq. (H.187), while absorbing any factors associated with the normalization of the $\mathfrak{u}(1)$ charge into the definition of $j_2(\mathfrak{u}(1) \subset \mathfrak{g})$. The embedding index then can be evaluated by applying eq. (H.187) to the branching ratio for the fundamental representation of $\mathfrak{g}$. Having obtained $j_2(\mathfrak{u}(1) \subset \mathfrak{g})$, then for any other representation R,

$$I_2(R) j_2(\mathfrak{u}(1) \subset \mathfrak{g}) = \sum_j q_j^2 d(R^{(j)}(\mathfrak{p}))\,. \tag{H.190}$$

As an illustration of this result, consider the branching rule for irreps of $\mathfrak{su}(5)$ to $\mathfrak{su}(3) \oplus \mathfrak{su}(2) \oplus \mathfrak{u}(1)_Y$. Then, given $\mathbf{5} \longrightarrow (\mathbf{3}, \mathbf{1}, -\tfrac{1}{3}) + (\mathbf{1}, \mathbf{2}, \tfrac{1}{2})$, one can apply eq. (H.187) with $\mathfrak{u}(1)_Y$ playing the role of $\mathfrak{h}_1$ and $A(\mathbf{5}) = \tfrac{1}{2}$ to obtain

$$I_2(\mathbf{5}) j_2(\mathfrak{u}(1) \subset \mathfrak{su}(5)) = 3(-\tfrac{1}{3})^2 + 2(\tfrac{1}{2})^2 = \tfrac{5}{6}\,, \tag{H.191}$$

where $I_2(\mathbf{5}) = \tfrac{1}{2}$. Hence, it follows that

$$j_2(\mathfrak{u}(1) \subset \mathfrak{su}(5)) = \tfrac{5}{3}\,. \tag{H.192}$$

One can now verify the result of eq. (H.190) in the case of $R = \mathbf{10}$, by employing the $\mathfrak{u}(1)_Y$ charges and the value of $j_2(\mathfrak{u}(1) \subset \mathfrak{su}(5))$ obtained in eq. (H.192). Using the branching rule $\mathbf{10} \longrightarrow (\mathbf{3}, \mathbf{2}, \tfrac{1}{6}) + (\overline{\mathbf{3}}, \mathbf{1}, -\tfrac{2}{3}) + (\mathbf{1}, \mathbf{1}, 1)$, eq. (H.190) yields the expected result,

$$I_2(\mathbf{10}) = \tfrac{3}{5}\left[6(\tfrac{1}{6})^2 + 3(-\tfrac{2}{3})^2 + 1\right] = \tfrac{3}{2}\,. \tag{H.193}$$

It is instructive to see how the computation of $I_2(R)$ is modified if one employs a different normalization of the $\mathfrak{u}(1)$ charges. Let us denote the $\mathfrak{u}(1)$ generator inside of $\mathfrak{su}(5)$ by $\mathcal{Q} \equiv \boldsymbol{t^{24}}$, which is given explicitly by eq. (6.246). The diagonal elements of $\mathcal{Q}$ correspond to the properly normalized $\mathfrak{u}(1)$ charges, which are related to the $\mathfrak{u}(1)_Y$ charges by $q_1 = -r/3$ and $q_2 = r/2$, where $r^2 = 3/5$. Consequently, if one employs the properly normalized $\mathfrak{u}(1)$ charges and the known value of $I_2(\mathbf{5})$ in the index sum rule [eq. (H.190)], then

$$I_2(\mathbf{5}) j_2\big(\mathfrak{u}(1) \subset \mathfrak{su}(5)\big) = \tfrac{1}{2} j_2\big(\mathfrak{u}(1) \subset \mathfrak{su}(5)\big)$$
$$= 3(-\tfrac{1}{3}r)^2 + 2(\tfrac{1}{2}r)^2 = \tfrac{5}{6}r^2 = \tfrac{1}{2}\,, \qquad \text{(H.194)}$$

which yields $j_2\big(\mathfrak{u}(1) \subset \mathfrak{su}(5)\big) = 1$, in contrast to eq. (H.192). It then follows that one can obtain $I_2(R)$ for any representation R by employing the properly renormalized $\mathfrak{u}(1)$ charges and $j_2\big(\mathfrak{u}(1) \subset \mathfrak{su}(5)\big) = 1$ in eq. (H.190). Equivalently, by rescaling the $\mathfrak{u}(1)_Y$ charges by a factor r and employing $j_2\big(\mathfrak{u}(1) \subset \mathfrak{su}(5)\big) = 1$, one can use the index sum rule to determine the value of r^2. This method was implicitly used in obtaining eq. (6.287).

The index sum rule obtained above for the second-order Dynkin index can be generalized in the case of the anomaly coefficient (which is proportional to the third-order index). In this case, one must introduce a new embedding index, $j_3(\mathfrak{h}_1 \subset \mathfrak{g})$. The relevant index sum rule analogous to eq. (H.187) is

$$A(R) j_3(\mathfrak{h}_1 \subset \mathfrak{g}) = \sum_j d(R^{(j)}(\mathfrak{p}))A(R_1^{(j)})\,, \qquad \text{(H.195)}$$

as a consequence of the $(H_1)^3$ triangle anomaly, where H_1 is the Lie group associated with $\mathfrak{h}_1$. In eq. (H.195), we have introduced the embedding index, $j_3(\mathfrak{h}_1 \subset \mathfrak{g})$, which depends on how $\mathfrak{h}_1$ is embedded inside $\mathfrak{g}$. Recall that the only simple Lie algebras with nonvanishing $A(R)$ are $\mathfrak{su}(n)$ [$n \geq 3$]. One can show that, for $n > m \geq 3$, if $\mathfrak{g} = \mathfrak{su}(n)$ and $\mathfrak{h}_1 = \mathfrak{su}(m)$ is a regular subalgebra of $\mathfrak{g}$, then $j_3(\mathfrak{h}_1 \subset \mathfrak{g}) = 1$.[19] On the other hand, for any other choice of the simple Lie subalgebra $\mathfrak{h}_1$ that has a vanishing anomaly coefficient, the right-hand side of eq. (H.195) is zero, which implies that $j_3(\mathfrak{h}_1 \subset \mathfrak{g}) = 0$, in which case eq. (H.195) is trivial. One can also extend eq. (H.195) to the case where $\mathfrak{h}_1 = \mathfrak{u}(1)$ by using $A(R_1^{(j)}) = q_j^3$ in eq. (H.195), while absorbing any additional factors associated with the normalization of the $\mathfrak{u}(1)$ charge into the definition of the embedding index $j_3(\mathfrak{u}(1) \subset \mathfrak{g})$. The embedding index can then be evaluated by applying eq. (H.195) to the branching ratio for the fundamental representation of $\mathfrak{g}$.

If R is the representation of a simple Lie algebra with vanishing anomaly coefficient, then eq. (H.195) simplifies to

$$\sum_j d(R^{(j)}(\mathfrak{p}))A(R_1^{(j)}) = 0\,. \qquad \text{(H.196)}$$

19 If $\mathfrak{h}_1$ is a special subalgebra of $\mathfrak{g}$ then $j_3(\mathfrak{h}_1 \subset \mathfrak{g})$ is an integer that can be determined by evaluating eq. (H.195) in the case where R is the fundamental representation of $\mathfrak{g}$. For example, if $\mathfrak{g} = \mathfrak{su}(6)$ and $\mathfrak{h} = \mathfrak{su}(3) \oplus \mathfrak{su}(2)$, then the branching rule of the fundamental representation, $\mathbf{6} \longrightarrow (\mathbf{3}, \mathbf{2})$, yields $j_3\big(\mathfrak{su}(3) \subset \mathfrak{su}(6)\big) = 2$.

This result can be applied for all possible choices of $\mathfrak{h}_j$, corresponding to the vanishing of the $(H_j)^3$ triangle anomalies for $j = 1, 2, \ldots, r$. Note that eq. (H.196) also applies to the cancellation of the U(1)^3 triangle anomaly by taking $A(R_1^{(j)}) = q_j^3$, independently of the choice of the normalization of the U(1) charge. For example, consider the case of $\mathfrak{g} = \mathfrak{so}(10)$ and $\mathfrak{h} = \mathfrak{su}(5) \oplus \mathfrak{u}(1)$. Choosing $R = \mathbf{16}$ and $\mathfrak{h}_1 = \mathfrak{u}(1)$, and employing the branching rule $\mathbf{16} \longrightarrow (\mathbf{10}, -1) + (\overline{\mathbf{5}}, 3) + (\mathbf{1}, -5)$, eq. (H.196) yields the expected result,

$$(10)(-1)^3 + (5)(3)^3 + (-5)^3 = 0 . \tag{H.197}$$

Additional checks of the branching rules can be made by employing the mixed anomaly. Suppose the maximal subalgebra of $\mathfrak{g}$ is of the form $\mathfrak{h} = \mathfrak{h}_1 \oplus \mathfrak{u}(1) \oplus \mathfrak{p}'$, where $\mathfrak{h}_1$ is a simple Lie algebra and $\mathfrak{p}'$ is simple, semisimple, or a direct sum of simple and $\mathfrak{u}(1)$ Lie algebras. Writing the branching rule in the form

$$R \longrightarrow \sum_j (R_1^{(j)}, R_{\mathfrak{p}'}^{(j)}, q_j) , \tag{H.198}$$

where R_1 is an irrep of $\mathfrak{h}_1$, $R_{\mathfrak{p}'}$ is an irrep of $\mathfrak{p}'$ of dimension $d(R_{\mathfrak{p}'})$, and q is the $\mathfrak{u}(1)$ charge (arbitrarily normalized), it follows that, as a consequence of the $(H_1)^2$U(1) triangle anomaly,

$$A(R) j_3(\mathfrak{h}_1 \oplus \mathfrak{u}(1) \subset \mathfrak{g}) = \sum_j q_j d(R^{(j)}(\mathfrak{p}')) I_2(R_1^{(j)}) , \tag{H.199}$$

where $j_3(\mathfrak{h}_1 \oplus \mathfrak{u}(1) \subset \mathfrak{g})$ is the embedding index, which depends on how $\mathfrak{h}_1 \oplus \mathfrak{u}(1)$ is embedded inside $\mathfrak{g}$. As in eqs. (H.187) and (H.195), one can absorb factors associated with the normalization of the $\mathfrak{u}(1)$ charge into the definition of the embedding index $j_3(\mathfrak{h}_1 \oplus \mathfrak{u}(1) \subset \mathfrak{g})$. The embedding index can then be evaluated by applying eq. (H.199) to the branching ratio for the fundamental representation of $\mathfrak{g}$.

In order to illustrate eqs. (H.195) and (H.199), consider the case of $\mathfrak{g} = \mathfrak{su}(5)$ and the maximal regular subalgebra $\mathfrak{h} = \mathfrak{su}(3) \oplus \mathfrak{su}(2) \oplus \mathfrak{u}(1)_Y$. We can check that $j_3(\mathfrak{su}(3) \subset \mathfrak{su}(5)) = 1$ by applying the branching rule $\mathbf{5} \longrightarrow (\mathbf{3}, \mathbf{1}, -\tfrac{1}{3}) + (\mathbf{1}, \mathbf{2}, \tfrac{1}{2})$ in eq. (H.195). All other branching rules can be checked using eq. (H.195). For example, in light of the branching rule $\mathbf{10} \longrightarrow (\mathbf{3}, \mathbf{2}, \tfrac{1}{6}) + (\overline{\mathbf{3}}, \mathbf{1}, -\tfrac{2}{3}) + (\mathbf{1}, \mathbf{1}, 1)$, eq. (H.195) yields $A(\mathbf{10}) = 1 = (2)(1) - 1 + 0$, as expected.

Suppose that $\mathfrak{h}_1 = \mathfrak{u}(1)$. Applying the branching rule for $\mathbf{5}$, eq. (H.195) yields

$$A(\mathbf{5}) j_3(\mathfrak{u}(1) \subset \mathfrak{su}(5)) = \left[3(-\tfrac{1}{3})^3 + 2(\tfrac{1}{2})^3\right] = \tfrac{5}{36} , \tag{H.200}$$

where $A(\mathbf{5}) = 1$. One can now check the result of eq. (H.195) in the case of $R = \mathbf{10}$, by employing the $\mathfrak{u}(1)_Y$ charges and the value of $j_3(\mathfrak{u}(1) \subset \mathfrak{su}(5))$ obtained in eq. (H.200). We again obtain the correct value,

$$A(\mathbf{10}) = \tfrac{36}{5}\left[6(\tfrac{1}{6})^3 + 3(-\tfrac{2}{3})^3 + 1\right] = 1 . \tag{H.201}$$

As previously noted, the normalization of the $\mathfrak{u}(1)$ charges is not significant, since any normalization factors can be absorbed into the definition of $j_3(\mathfrak{u}(1) \subset \mathfrak{su}(5))$. However, one could have chosen to employ the properly normalized $\mathfrak{u}(1)$ charges, which are denoted below by q_j. When doing so, in order to use $A(R_1^{(j)}) = q_j^3$ in eq. (H.195), one must take into account the factor of $\frac{1}{4}d^{abc}$ [see eq. (H.176)] for values of the indices a, b, and c corresponding to the $\mathfrak{u}(1)$ charge direction inside $\mathfrak{g}$. Consequently, one must replace eq. (H.195) with

$$\tfrac{1}{4}d^{QQQ}A(R)j_3\big(\mathfrak{u}(1)\subset\mathfrak{su}(5)\big) = \sum_j q_j^3 d(R^{(j)}(\mathfrak{p}))\,, \qquad \text{(H.202)}$$

where Q indicates the direction of the $\mathfrak{u}(1)$ generator inside $\mathfrak{g}$. Using eq. (H.202) to define the embedding index, we anticipate that

$$j_3\big(\mathfrak{u}(1)\subset\mathfrak{su}(5)\big) = 1\,. \qquad \text{(H.203)}$$

We can check this via the following computation:

$$\tfrac{1}{4}A(\mathbf{5})d^{QQQ}j_3\big(\mathfrak{u}(1)\subset\mathfrak{su}(5)\big) = 3(-\tfrac{1}{3}r)^3 + 2(\tfrac{1}{2}r)^3 = \frac{5r^3}{36} = \frac{5}{36}\left(\frac{3}{5}\right)^{3/2}, \qquad \text{(H.204)}$$

after rescaling the $\mathfrak{u}(1)_Y$ charges by $r = (3/5)^{1/2}$. Using eq. (H.176) and making use of the explicit form of $Q = \boldsymbol{t^{24}}$,

$$\operatorname{Tr} Q^3 = \tfrac{1}{4}d^{QQQ} = \frac{5}{36}\left(\frac{3}{5}\right)^{3/2}, \qquad \text{(H.205)}$$

it follows that eq. (H.203) is confirmed. One can now repeat the computation of $A(\mathbf{10})$ using eq. (H.202) with $j_3\big(\mathfrak{u}(1)\subset\mathfrak{su}(5)\big) = 1$ and properly normalized $\mathfrak{u}(1)$ charges to verify that $A(\mathbf{10}) = 1$.

To check the validity of eq. (H.199), one can consider either the SU(3)^{2}U(1) or SU(2)^{2}U(1) mixed triangle anomalies. In the case of $\mathfrak{h}_1 = \mathfrak{su}(3)$, eq. (H.199) yields

$$j_3\big(\mathfrak{su}(3)\oplus\mathfrak{u}(1)\subset\mathfrak{su}(5)\big) = \tfrac{1}{2}\cdot 1(-\tfrac{1}{3}) = -\tfrac{1}{6}\,. \qquad \text{(H.206)}$$

Applying this result to the case of $R = \mathbf{10}$, eq. (H.199) yields

$$A(\mathbf{10}) = -6\cdot\tfrac{1}{2}\cdot 2(-\tfrac{1}{6}) = 1\,, \qquad \text{(H.207)}$$

which is correct. The same result is obtained In the case of $\mathfrak{h}_1 = \mathfrak{su}(2)$ where eq. (H.199) is used to obtain

$$j_3\big(\mathfrak{su}(2)\oplus\mathfrak{u}(1)\subset\mathfrak{su}(5)\big) = \tfrac{1}{2}\cdot 1\cdot\tfrac{1}{2} = \tfrac{1}{4}\,. \qquad \text{(H.208)}$$

Applying this result to the case of $R = \mathbf{10}$, eq. (H.199) yields

$$A(\mathbf{10}) = 4\cdot 3\cdot\tfrac{1}{2}\cdot\tfrac{1}{6} = 1\,. \qquad \text{(H.209)}$$

Note that the above results can be reproduced using properly normalized $\mathfrak{u}(1)$ charges and $j_3\big(\mathfrak{u}(1)\subset\mathfrak{su}(5)\big) = 1$, by including the appropriate factor of $\frac{1}{4}d^{abc}$ on the left-hand side of eq. (H.199) and making use of

$$\operatorname{Tr}\big[Q(\boldsymbol{t^3})^2\big] = -\frac{1}{6}\sqrt{\frac{3}{5}}\,, \qquad \operatorname{Tr}\big[Q(\boldsymbol{t^{23}})^2\big] = \frac{1}{4}\sqrt{\frac{3}{5}}\,, \qquad \text{(H.210)}$$

after using the explicit forms of $\boldsymbol{t}^3$ and $\boldsymbol{t}^{23}$ given in eqs. (6.244) and (6.245).

H.5.3 Tables of Lie Algebra Representations Used in GUT Models

In Tables H.2–H.18, we present selected irreps, tensor products, and branching rules that are most often used in constructing models of grand unification. The irreps are specified by their Dynkin labels as indicated in Appendix H.5.1.

In the tables of irreps presented below, real and complex irreps are distinguished. (In light of Table H.1, no pseudoreal irreps appear.) The irreps that are conjugate to the ones exhibited in Tables H.2, H.4, H.7, H.10, and H.14 can be determined from eqs. (H.111)–(H.113) and thus are not explicitly shown. The second-order Dynkin index and the Casimir and anomaly coefficients of the conjugated representations can be obtained by employing eq. (H.180). The corresponding n-alities and Dynkin labels are indicated at the beginning of Appendix H.5.1. For a tensor product of the form $R \otimes R$, we indicate the irreps that appear in the symmetric and antisymmetric parts of the tensor product with the subscripts S and A, respectively.

$\mathfrak{su}(3)$ Representations and Tensor Products

Table H.2 Smallest irreps of $\mathfrak{su}(3)$.

dimension	real?	index	Casimir	anomaly	triality	Dynkin label
1	yes	0	0	0	0	(00)
3	no	1/2	4/3	1	1	(10)
6	no	5/2	10/3	7	2	(20)
8	yes	3	3	0	0	(11)
10	no	15/2	6	27	0	(30)
15	no	10	16/3	14	1	(21)
15′	no	35/2	28/3	77	1	(40)
21	no	35	40/3	−182	1	(05)
24	no	25	25/3	−64	1	(13)
27	yes	27	8	0	0	(22)

Table H.3 Selected tensor products for $\mathfrak{su}(3)$ irreps.

$$\mathbf{3} \otimes \mathbf{3} = \overline{\mathbf{3}}_A \oplus \mathbf{6}_S$$
$$\mathbf{3} \otimes \overline{\mathbf{3}} = \mathbf{1} \oplus \mathbf{8}$$
$$\mathbf{3} \otimes \mathbf{6} = \mathbf{8} \oplus \mathbf{10}$$
$$\mathbf{3} \otimes \overline{\mathbf{6}} = \overline{\mathbf{3}} \oplus \mathbf{15}$$
$$\mathbf{3} \otimes \mathbf{8} = \mathbf{3} \oplus \overline{\mathbf{6}} \oplus \mathbf{15}$$
$$\mathbf{6} \otimes \mathbf{6} = \overline{\mathbf{6}}_S \oplus \mathbf{15}_A \oplus \mathbf{15}'_S$$
$$\mathbf{6} \otimes \overline{\mathbf{6}} = \mathbf{1} \oplus \mathbf{8} \oplus \mathbf{27}$$
$$\mathbf{6} \otimes \mathbf{8} = \overline{\mathbf{3}} \oplus \mathbf{6} \oplus \overline{\mathbf{15}} \oplus \overline{\mathbf{24}}$$
$$\mathbf{8} \otimes \mathbf{8} = \mathbf{1}_S \oplus \mathbf{8}_S \oplus \mathbf{8}_A \oplus \mathbf{10}_A \oplus \overline{\mathbf{10}}_A \oplus \mathbf{27}_S$$

$\mathfrak{su}(4)$ Representations, Tensor Products, and Branching Rules

Table H.4 Smallest irreps of $\mathfrak{su}(4)$.

dimension	real?	index	Casimir	anomaly	quadrality	Dynkin label
1	yes	0	0	0	0	(000)
4	no	1/2	15/8	1	1	(100)
6	yes	1	5/2	0	2	(010)
10	no	3	9/2	8	2	(200)
15	yes	4	4	0	0	(101)
20	no	13/2	39/8	−7	1	(011)
20′	yes	8	6	0	0	(020)
20″	no	21/2	63/8	−35	1	(003)
35	no	28	12	112	0	(400)
36	no	33/2	55/8	21	1	(201)
45	no	24	8	48	0	(210)
50	yes	35	21/2	0	2	(030)

Table H.5 Selected tensor products for $\mathfrak{su}(4)$ irreps.

$$\mathbf{4}\otimes\mathbf{4} = \mathbf{6}_A\oplus\mathbf{10}_S$$
$$\mathbf{4}\otimes\overline{\mathbf{4}} = \mathbf{1}\oplus\mathbf{15}$$
$$\mathbf{4}\otimes\mathbf{6} = \overline{\mathbf{4}}\oplus\overline{\mathbf{20}}$$
$$\mathbf{4}\otimes\mathbf{10} = \overline{\mathbf{20}}\oplus\overline{\mathbf{20''}}$$
$$\mathbf{4}\otimes\overline{\mathbf{10}} = \overline{\mathbf{4}}\oplus\mathbf{36}$$
$$\mathbf{4}\otimes\mathbf{15} = \mathbf{4}\oplus\mathbf{20}\oplus\mathbf{36}$$
$$\mathbf{6}\otimes\mathbf{6} = \mathbf{1}_S\oplus\mathbf{15}_A\oplus\mathbf{20}'_S$$
$$\mathbf{10}\otimes\mathbf{10} = \mathbf{20}'_S\oplus\mathbf{35}_S\oplus\mathbf{45}_A$$
$$\mathbf{15}\otimes\mathbf{15} = \mathbf{1}_S\oplus\mathbf{15}_S\oplus\mathbf{15}_A\oplus\mathbf{20}'_S\oplus\mathbf{45}_A\oplus\overline{\mathbf{45}}_A\oplus\mathbf{84}_S$$

Table H.6 Branching rules for $\mathfrak{su}(4)$ irreps to $\mathfrak{su}(3)_C\oplus\mathfrak{u}(1)_{B-L}$.

1	$(\mathbf{1},0)$
4	$(\mathbf{3},1/3)+(\mathbf{1},-1)$
6	$(\mathbf{3},-2/3)+(\overline{\mathbf{3}},2/3)$
10	$(\mathbf{6},2/3)+(\mathbf{3},-2/3)+(\mathbf{1},-2)$
15	$(\mathbf{8},0)+(\mathbf{3},4/3)+(\overline{\mathbf{3}},-4/3)+(\mathbf{1},0)$
20	$(\mathbf{8},-1)+(\overline{\mathbf{6}},1/3)+(\overline{\mathbf{3}},5/3)+(\mathbf{3},1/3)$
20′	$(\mathbf{8},0)+(\mathbf{6},-4/3)+(\overline{\mathbf{6}},4/3)$
20″	$(\overline{\mathbf{10}},-1)+(\overline{\mathbf{6}},1/3)+(\overline{\mathbf{3}},5/3)+(\mathbf{1},3)$
35	$(\mathbf{15}',4/3)+(\mathbf{10},0)+(\mathbf{6},-4/3)+(\mathbf{3},-8/3)+(\mathbf{1},-4)$
36	$(\mathbf{15},1/3)+(\mathbf{8},-1)+(\mathbf{6},5/3)+(\mathbf{3},1/3)+(\overline{\mathbf{3}},-7/3)+(\mathbf{1},-1)$

$\mathfrak{su}(5)$ Representations, Tensor Products, and Branching Rules

Table H.7 Smallest irreps of $\mathfrak{su}(5)$.

dimension	real?	index	Casimir	anomaly	quintality	Dynkin label
1	yes	0	0	0	0	(0000)
5	no	1/2	12/5	1	1	(1000)
10	no	3/2	18/5	1	2	(0100)
15	no	7/2	28/5	9	2	(2000)
24	yes	5	5	0	0	(1001)
35	no	14	48/5	−44	2	(0003)
40	no	11	33/5	−16	2	(0011)
45	no	12	32/5	−6	1	(0101)
50	no	35/2	42/5	−15	1	(0020)
70	no	49/2	42/5	29	1	(2001)
70′	no	42	72/5	−156	1	(0004)
75	yes	25	8	0	0	(0110)

Table H.8 Selected tensor products for $\mathfrak{su}(5)$ irreps.

$$\mathbf{5}\otimes\mathbf{5} = \mathbf{10}_A \oplus \mathbf{15}_S$$
$$\mathbf{5}\otimes\overline{\mathbf{5}} = \mathbf{1}\oplus\mathbf{24}$$
$$\mathbf{5}\otimes\overline{\mathbf{10}} = \overline{\mathbf{5}}\oplus\overline{\mathbf{45}}$$
$$\mathbf{5}\otimes\mathbf{10} = \overline{\mathbf{10}}\oplus\overline{\mathbf{40}}$$
$$\mathbf{5}\otimes\overline{\mathbf{50}} = \mathbf{75}\oplus\mathbf{175}'$$
$$\mathbf{10}\otimes\mathbf{10} = \overline{\mathbf{5}}_S\oplus\overline{\mathbf{45}}_A\oplus\overline{\mathbf{50}}_S$$
$$\mathbf{10}\otimes\overline{\mathbf{10}} = \mathbf{1}\oplus\mathbf{24}\oplus\mathbf{75}$$
$$\mathbf{24}\otimes\mathbf{24} = \mathbf{1}_S\oplus\mathbf{24}_S\oplus\mathbf{24}_A\oplus\mathbf{75}_S\oplus\mathbf{126}_A\oplus\overline{\mathbf{126}}_A\oplus\mathbf{200}_S$$

Table H.9 Branching rules for $\mathfrak{su}(5)$ irreps to $\mathfrak{su}(3)_C\oplus\mathfrak{su}(2)_L\oplus\mathfrak{u}(1)_Y$.

1	$(\mathbf{1},\mathbf{1},0)$
5	$(\mathbf{3},\mathbf{1},-1/3)+(\mathbf{1},\mathbf{2},1/2)$
10	$(\mathbf{3},\mathbf{2},1/6)+(\overline{\mathbf{3}},\mathbf{1},-2/3)+(\mathbf{1},\mathbf{1},1)$
15	$(\mathbf{6},\mathbf{1},-2/3)+(\mathbf{3},\mathbf{2},1/6)+(\mathbf{1},\mathbf{3},1)$
24	$(\mathbf{8},\mathbf{1},0)+(\mathbf{1},\mathbf{3},0)+(\mathbf{1},\mathbf{1},0)+(\mathbf{3},\mathbf{2},-5/6)+(\overline{\mathbf{3}},\mathbf{2},5/6)$
35	$(\overline{\mathbf{10}},\mathbf{1},1)+(\overline{\mathbf{6}},\mathbf{2},1/6)+(\overline{\mathbf{3}},\mathbf{3},-2/3)+(\mathbf{1},\mathbf{4},-3/2)$
40	$(\overline{\mathbf{6}},\mathbf{2},1/6)+(\mathbf{8},\mathbf{1},1)+(\overline{\mathbf{3}},\mathbf{3},-2/3)+(\overline{\mathbf{3}},\mathbf{1},-2/3)+(\mathbf{3},\mathbf{2},1/6)+(\mathbf{1},\mathbf{2},-3/2)$
45	$(\mathbf{8},\mathbf{2},1/2)+(\overline{\mathbf{6}},\mathbf{1},-1/3)+(\overline{\mathbf{3}},\mathbf{2},-7/6)+(\overline{\mathbf{3}},\mathbf{1},4/3)+(\mathbf{3},\mathbf{3},-1/3)+$ $(\mathbf{3},\mathbf{1},-1/3)+(\mathbf{1},\mathbf{2},1/2)$
50	$(\mathbf{8},\mathbf{2},1/2)+(\mathbf{6},\mathbf{1},4/3)+(\overline{\mathbf{6}},\mathbf{3},-1/3)+(\overline{\mathbf{3}},\mathbf{2},-7/6)+(\mathbf{3},\mathbf{1},-1/3)+(\mathbf{1},\mathbf{1},-2)$
75	$(\mathbf{8},\mathbf{3},0)+(\mathbf{8},\mathbf{1},0)+(\mathbf{1},\mathbf{1},0)+(\mathbf{6},\mathbf{2},5/6)+(\overline{\mathbf{6}},\mathbf{2},-5/6)+(\mathbf{3},\mathbf{2},-5/6)+$ $(\overline{\mathbf{3}},\mathbf{2},5/6)+(\mathbf{3},\mathbf{1},5/3)+(\overline{\mathbf{3}},\mathbf{1},-5/3)$

$\mathfrak{so}(10)$ Representations, Tensor Products, and Branching Rules

Table H.10 Smallest irreps of $\mathfrak{so}(10)$.

dimension	real?	index	Casimir	quadrality	Dynkin label
1	yes	0	0	0	(00000)
10	yes	1	9/2	2	(10000)
16	no	2	45/8	1	(00001)
45	yes	8	8	0	(01000)
54	yes	12	10	0	(20000)
120	yes	28	21/2	2	(00100)
126	no	35	25/2	2	(00002)
144	no	34	85/8	1	(10010)
210	yes	56	12	0	(00011)
210′	yes	77	33/2	2	(30000)

Table H.11 Selected tensor products for $\mathfrak{so}(10)$ irreps.

$$\mathbf{10} \otimes \mathbf{10} = \mathbf{1}_S \oplus \mathbf{45}_A \oplus \mathbf{54}_S$$
$$\mathbf{16} \otimes \mathbf{10} = \overline{\mathbf{16}} \oplus \overline{\mathbf{144}}$$
$$\mathbf{16} \otimes \mathbf{16} = \mathbf{10}_S \oplus \mathbf{120}_A \oplus \mathbf{126}_S$$
$$\mathbf{16} \otimes \overline{\mathbf{16}} = \mathbf{1} \oplus \mathbf{45} \oplus \mathbf{210}$$
$$\mathbf{45} \otimes \mathbf{10} = \mathbf{10} \oplus \mathbf{120} \oplus \mathbf{320}$$
$$\mathbf{45} \otimes \mathbf{16} = \mathbf{16} \oplus \mathbf{144} \oplus \mathbf{560}$$
$$\mathbf{45} \otimes \mathbf{45} = \mathbf{1}_S \oplus \mathbf{45}_A \oplus \mathbf{54}_S \oplus \mathbf{210}_S \oplus \mathbf{770}_S \oplus \mathbf{945}_A$$
$$\mathbf{54} \otimes \mathbf{10} = \mathbf{10} \oplus \mathbf{210'} \oplus \mathbf{320}$$
$$\mathbf{120} \otimes \mathbf{10} = \mathbf{45} \oplus \mathbf{210} \oplus \mathbf{945}$$
$$\mathbf{126} \otimes \mathbf{10} = \mathbf{210} \oplus \mathbf{1050}$$
$$\mathbf{210} \otimes \mathbf{10} = \mathbf{120} \oplus \mathbf{126} \oplus \overline{\mathbf{126}} \oplus \mathbf{1728}$$

Table H.12 Branching rules for $\mathfrak{so}(10)$ irreps to $\mathfrak{su}(5) \oplus \mathfrak{u}(1)$.

1	$(\mathbf{1}, 0)$
10	$(\mathbf{5}, 2) + (\overline{\mathbf{5}}, -2)$
16	$(\mathbf{10}, -1) + (\overline{\mathbf{5}}, 3) + (\mathbf{1}, -5)$
45	$(\mathbf{24}, 0) + (\mathbf{10}, 4) + (\overline{\mathbf{10}}, -4) + (\mathbf{1}, 0)$
54	$(\mathbf{24}, 0) + (\mathbf{15}, 4) + (\overline{\mathbf{15}}, -4)$
120	$(\mathbf{45}, 2) + (\overline{\mathbf{45}}, -2) + (\mathbf{10}, -6) + (\overline{\mathbf{10}}, 6) + (\mathbf{5}, 2) + (\overline{\mathbf{5}}, -2)$
126	$(\overline{\mathbf{50}}, -2) + (\mathbf{45}, 2) + (\overline{\mathbf{15}}, 6) + (\mathbf{10}, -6) + (\overline{\mathbf{5}}, -2) + (\mathbf{1}, -10)$
144	$(\overline{\mathbf{45}}, 3) + (\mathbf{40}, -1) + (\mathbf{24}, -5) + (\mathbf{15}, -1) + (\mathbf{10}, -1) + (\mathbf{5}, 7) + (\overline{\mathbf{5}}, 3)$
210	$(\mathbf{75}, 0) + (\mathbf{40}, -4) + (\overline{\mathbf{40}}, 4) + (\mathbf{24}, 0) + (\mathbf{10}, 4) + (\overline{\mathbf{10}}, -4) +$ $(\mathbf{5}, -8) + (\overline{\mathbf{5}}, 8) + (\mathbf{1}, 0)$
210′	$(\mathbf{70}, 2) + (\overline{\mathbf{70}}, -2) + (\mathbf{35}, -6) + (\overline{\mathbf{35}}, 6)$

Table H.13 Branching rules for $\mathfrak{so}(10)$ irreps to $\mathfrak{su}(4)_{PS} \oplus \mathfrak{su}(2)_L \oplus \mathfrak{su}(2)_R$.

1	$(\mathbf{1}, 0, 0)$
10	$(\mathbf{6}, \mathbf{1}, \mathbf{1}) + (\mathbf{1}, \mathbf{2}, \mathbf{2})$
16	$(\mathbf{4}, \mathbf{2}, \mathbf{1}) + (\overline{\mathbf{4}}, \mathbf{1}, \mathbf{2})$
45	$(\mathbf{15}, \mathbf{1}, \mathbf{1}) + (\mathbf{6}, \mathbf{2}, \mathbf{2}) + (\mathbf{1}, \mathbf{3}, \mathbf{1}) + (\mathbf{1}, \mathbf{1}, \mathbf{3})$
54	$(\mathbf{20'}, \mathbf{1}, \mathbf{1}) + (\mathbf{6}, \mathbf{2}, \mathbf{2}) + (\mathbf{1}, \mathbf{3}, \mathbf{3}) + (\mathbf{1}, \mathbf{1}, \mathbf{1})$
120	$(\mathbf{15}, \mathbf{2}, \mathbf{2}) + (\mathbf{6}, \mathbf{3}, \mathbf{1}) + (\mathbf{6}, \mathbf{1}, \mathbf{3}) + (\mathbf{10}, \mathbf{1}, \mathbf{1}) + (\overline{\mathbf{10}}, \mathbf{1}, \mathbf{1}) + (\mathbf{1}, \mathbf{2}, \mathbf{2})$
126	$(\mathbf{15}, \mathbf{2}, \mathbf{2}) + (\mathbf{10}, \mathbf{1}, \mathbf{3}) + (\overline{\mathbf{10}}, \mathbf{3}, \mathbf{1}) + (\mathbf{6}, \mathbf{1}, \mathbf{1})$
144	$(\mathbf{20}, \mathbf{2}, \mathbf{1}) + (\overline{\mathbf{20}}, \mathbf{1}, \mathbf{2}) + (\mathbf{4}, \mathbf{2}, \mathbf{3}) + (\overline{\mathbf{4}}, \mathbf{3}, \mathbf{2}) + (\mathbf{4}, \mathbf{2}, \mathbf{1}) + (\overline{\mathbf{4}}, \mathbf{1}, \mathbf{2})$
210	$(\mathbf{10}, \mathbf{2}, \mathbf{2}) + (\overline{\mathbf{10}}, \mathbf{2}, \mathbf{2}) + (\mathbf{15}, \mathbf{3}, \mathbf{1}) + (\mathbf{15}, \mathbf{1}, \mathbf{3}) + (\mathbf{15}, \mathbf{1}, \mathbf{1}) + (\mathbf{6}, \mathbf{2}, \mathbf{2}) + (\mathbf{1}, \mathbf{1}, \mathbf{1})$
210′	$(\mathbf{50}, \mathbf{1}, \mathbf{1}) + (\mathbf{20'}, \mathbf{2}, \mathbf{2}) + (\mathbf{6}, \mathbf{3}, \mathbf{3}) + (\mathbf{1}, \mathbf{4}, \mathbf{4}) + (\mathbf{6}, \mathbf{1}, \mathbf{1}) + (\mathbf{1}, \mathbf{2}, \mathbf{2})$

E_6 Representations, Tensor Products, and Branching Rules

Table H.14 Smallest irreps of E_6.

dimension	real?	index	Casimir	triality	Dynkin label
1	yes	0	0	0	(000000)
27	no	3	26/3	1	(100000)
78	yes	12	12	0	(000001)
351	no	75	50/3	1	(000100)
351′	no	84	56/3	1	(000020)
650	yes	150	18	0	(100010)

Table H.15 Selected tensor products for E_6 irreps.

$$\mathbf{27} \otimes \mathbf{27} = \overline{\mathbf{27}}_S \oplus \overline{\mathbf{351}}_A \oplus \overline{\mathbf{351}}'_S$$
$$\mathbf{27} \otimes \overline{\mathbf{27}} = \mathbf{1} \oplus \mathbf{78} \oplus \mathbf{650}$$
$$\mathbf{78} \otimes \mathbf{27} = \mathbf{27} \oplus \mathbf{351} \oplus \mathbf{1728}$$
$$\mathbf{78} \otimes \mathbf{78} = \mathbf{1}_S \oplus \mathbf{78}_A \oplus \mathbf{650}_S \oplus \mathbf{2430}_S \oplus \mathbf{2925}_A$$

Table H.16 Branching rules for E_6 irreps to $\mathfrak{so}(10) \oplus \mathfrak{u}(1)$.

1	$(\mathbf{1}, 0)$
27	$(\mathbf{16}, 1) + (\mathbf{10}, -2) + (\mathbf{1}, 4)$
78	$(\mathbf{45}, 0) + (\mathbf{1}, 0) + (\mathbf{16}, -3) + (\overline{\mathbf{16}}, 3)$
351	$(\mathbf{144}, 1) + (\mathbf{120}, -2) + (\mathbf{45}, 4) + (\mathbf{16}, 1) + (\overline{\mathbf{16}}, -5) + (\mathbf{10}, -2)$
351′	$(\mathbf{144}, 1) + (\overline{\mathbf{126}}, -2) + (\mathbf{54}, 4) + (\overline{\mathbf{16}}, -5) + (\mathbf{10}, -2) + (\mathbf{1}, -8)$
650	$(\mathbf{210}, 0) + (\mathbf{144}, -3) + (\overline{\mathbf{144}}, 3) + (\mathbf{54}, 0) + (\mathbf{45}, 0) + (\mathbf{16}, -3) + (\overline{\mathbf{16}}, 3) + (\mathbf{10}, 6) + (\mathbf{10}, -6) + (\mathbf{1}, 0)$

Table H.17 Branching rules for E_6 irreps to $\mathfrak{su}(6) \oplus \mathfrak{su}(2)$.

1	$(\mathbf{1},\mathbf{1})$
27	$(\mathbf{15},\mathbf{1})+(\overline{\mathbf{6}},\mathbf{2})$
78	$(\mathbf{35},\mathbf{1})+(\mathbf{20},\mathbf{2})+(\mathbf{1},\mathbf{3})$
351	$(\mathbf{105},\mathbf{1})+(\overline{\mathbf{84}},\mathbf{2})+(\mathbf{21},\mathbf{1})+(\mathbf{15},\mathbf{3})+(\overline{\mathbf{6}},\mathbf{2})$
351$'$	$(\mathbf{105}',\mathbf{1})+(\overline{\mathbf{84}},\mathbf{2})+(\mathbf{21},\mathbf{3})+(\mathbf{15},\mathbf{1})$
650	$(\mathbf{189},\mathbf{1})+(\mathbf{70},\mathbf{2})+(\overline{\mathbf{70}},\mathbf{2})+(\mathbf{35},\mathbf{3})+(\mathbf{35},\mathbf{1})+(\mathbf{20},\mathbf{2})+(\mathbf{1},\mathbf{1})$

Table H.18 Branching rules for E_6 irreps to $\mathfrak{su}(3)_C \oplus \mathfrak{su}(3)_L \oplus \mathfrak{su}(3)_R$.

1	$(\mathbf{1},\mathbf{1},\mathbf{1})$
27	$(\mathbf{3},\mathbf{3},\mathbf{1})+(\overline{\mathbf{3}},\mathbf{1},\overline{\mathbf{3}})+(\mathbf{1},\overline{\mathbf{3}},\mathbf{3})$
78	$(\mathbf{8},\mathbf{1},\mathbf{1})+(\mathbf{1},\mathbf{8},\mathbf{1})+(\mathbf{1},\mathbf{1},\mathbf{8})+(\mathbf{3},\overline{\mathbf{3}},\overline{\mathbf{3}})+(\overline{\mathbf{3}},\mathbf{3},\mathbf{3})$
351	$(\mathbf{8},\overline{\mathbf{3}},\mathbf{3})+(\overline{\mathbf{3}},\mathbf{8},\overline{\mathbf{3}})+(\mathbf{3},\mathbf{3},\mathbf{8})+(\mathbf{6},\mathbf{1},\overline{\mathbf{3}})+(\overline{\mathbf{6}},\mathbf{3},\mathbf{1})+(\overline{\mathbf{3}},\mathbf{1},\mathbf{6})+(\mathbf{3},\overline{\mathbf{6}},\mathbf{1})+$ $(\mathbf{1},\mathbf{6},\mathbf{3})+(\mathbf{1},\overline{\mathbf{3}},\overline{\mathbf{6}})+(\mathbf{3},\mathbf{3},\mathbf{1})+(\overline{\mathbf{3}},\mathbf{1},\overline{\mathbf{3}})+(\mathbf{1},\overline{\mathbf{3}},\mathbf{3})$
351$'$	$(\mathbf{8},\overline{\mathbf{3}},\mathbf{3})+(\overline{\mathbf{3}},\mathbf{8},\overline{\mathbf{3}})+(\mathbf{3},\mathbf{3},\mathbf{8})+(\mathbf{6},\mathbf{1},\mathbf{6})+(\overline{\mathbf{6}},\overline{\mathbf{6}},\mathbf{1})+(\mathbf{1},\mathbf{6},\overline{\mathbf{6}})+(\mathbf{3},\mathbf{3},\mathbf{1})+$ $(\overline{\mathbf{3}},\mathbf{1},\overline{\mathbf{3}})+(\mathbf{1},\overline{\mathbf{3}},\mathbf{3})$
650	$(\mathbf{8},\mathbf{8},\mathbf{1})+(\mathbf{8},\mathbf{1},\mathbf{8})+(\mathbf{1},\mathbf{8},\mathbf{8})+(\mathbf{8},\mathbf{1},\mathbf{1})+(\mathbf{1},\mathbf{8},\mathbf{1})+(\mathbf{1},\mathbf{1},\mathbf{8})+$ $(\mathbf{6},\mathbf{3},\mathbf{3})+(\overline{\mathbf{6}},\overline{\mathbf{3}},\overline{\mathbf{3}})+(\overline{\mathbf{3}},\mathbf{3},\overline{\mathbf{6}})+(\mathbf{3},\overline{\mathbf{3}},\mathbf{6})+(\overline{\mathbf{3}},\overline{\mathbf{6}},\mathbf{3})+(\mathbf{3},\mathbf{6},\overline{\mathbf{3}})+$ $2\times(\mathbf{3},\overline{\mathbf{3}},\overline{\mathbf{3}})+2\times(\overline{\mathbf{3}},\mathbf{3},\mathbf{3})+2\times(\mathbf{1},\mathbf{1},\mathbf{1})$

Exercises

H.1 This exercise provides an explicit construction to show that $\mathfrak{u}(n)$ is a Lie subalgebra of the Lie algebra $\mathfrak{so}(2n)$. First, a $2n \times 2n$ real antisymmetric matrix is introduced that satisfies

$$\Sigma_0^{\mathsf{T}}\Sigma_0 = \Sigma_0\Sigma_0^{\mathsf{T}} = c^2 \mathbb{1}_{2n\times 2n}\,, \tag{H.211}$$

where c is a real number. The Lie algebra of $\mathfrak{so}(2n)$ in the fundamental (defining) $2n$-dimensional representation consists of the set of real antisymmetric $2n \times 2n$ matrices (see Table 4.1). The corresponding $n(2n-1)$ generators of $\mathfrak{so}(2n)$ are denoted by $\{\boldsymbol{T}^a\,,\,\boldsymbol{S}^b\}$, where the $i\boldsymbol{T}^a$ and $i\boldsymbol{S}^b$ are real antisymmetric $2n \times 2n$ matrices that satisfy

$$\boldsymbol{T}^a\Sigma_0 + \Sigma_0\boldsymbol{T}^{a\mathsf{T}} = 0\,, \tag{H.212}$$

$$\boldsymbol{S}^b\Sigma_0 - \Sigma_0\boldsymbol{S}^{b\mathsf{T}} = 0\,. \tag{H.213}$$

It then follows that the $\boldsymbol{T}^a$ span a $\mathfrak{u}(n)$ Lie subalgebra of $\mathfrak{so}(2n)$ and the $\boldsymbol{T}^a$ and $\boldsymbol{S}^b$ satisfy $\mathrm{Tr}(\boldsymbol{T}^a\boldsymbol{S}^b) = 0$.

(a) Using eqs. (H.211)–(H.213), show that

$$c^2\, \boldsymbol{T}^{a\mathsf{T}} = -\Sigma_0^{\mathsf{T}} \boldsymbol{T}^a \Sigma_0\,, \qquad c^2\, \boldsymbol{S}^{b\mathsf{T}} = \Sigma_0^{\mathsf{T}} \boldsymbol{S}^b \Sigma_0\,. \tag{H.214}$$

(b) In light of eq. (G.21), show that there exists a real orthogonal matrix $\mathcal{O}$ such that

$$\mathcal{O}\,\Sigma_0\,\mathcal{O}^{\mathsf{T}} = cJ\,, \quad \text{where } J \equiv \begin{pmatrix} \mathbf{0} & \mathbb{1}_{n\times n} \\ -\mathbb{1}_{n\times n} & \mathbf{0} \end{pmatrix}, \tag{H.215}$$

where $\mathbf{0}$ is the $n \times n$ zero matrix and $\mathbb{1}_{n\times n}$ is the $n \times n$ identity matrix. Using the matrix $\mathcal{O}$, one can define

$$\widetilde{\boldsymbol{T}}^a \equiv \mathcal{O}\boldsymbol{T}^a\mathcal{O}^{\mathsf{T}}, \qquad \widetilde{\boldsymbol{S}}^b \equiv \mathcal{O}\boldsymbol{S}^b\mathcal{O}^{\mathsf{T}}. \tag{H.216}$$

Then, show that

$$\widetilde{\boldsymbol{T}}^a = -J\widetilde{\boldsymbol{T}}^a J, \qquad \widetilde{\boldsymbol{S}}^a = J\widetilde{\boldsymbol{S}}^a J. \tag{H.217}$$

In particular, $i\widetilde{\boldsymbol{T}}^a$ and $i\widetilde{\boldsymbol{S}}^b$ are real antisymmetric $2n \times 2n$ matrices.

(c) Using eq. (H.217), show that $i\widetilde{\boldsymbol{T}}^a$ and $i\widetilde{\boldsymbol{S}}^b$ take the following block form:

$$i\,\widetilde{\boldsymbol{T}}^a = \begin{pmatrix} A & -B \\ B & A \end{pmatrix}, \qquad i\widetilde{\boldsymbol{S}}^b = \begin{pmatrix} C & D \\ D & -C \end{pmatrix}, \tag{H.218}$$

where A, C, and D are real $n\times n$ antisymmetric matrices, and B is a real $n\times n$ symmetric matrix. Check that the commutator of two matrices of the type $i\widetilde{\boldsymbol{T}}^a$ yields a matrix of the same form. This implies that the set of matrices of the form $i\widetilde{\boldsymbol{T}}^a$ constitutes a subalgebra of $\mathfrak{so}(2n)$.

(d) Since the $\boldsymbol{T}^a$ and $\widetilde{\boldsymbol{T}}^a$ are related by a similarity transformation, they span the same subalgebra of $\mathfrak{so}(2n)$. Moreover, eq. (H.218) establishes a mapping, f, from the $2n \times 2n$ real antisymmetric matrices, $i\widetilde{\boldsymbol{T}}^a$, to the $n \times n$ antihermitian matrices, $A + iB$, that preserves matrix multiplication [i.e., the mapping f satisfies $f(S_1S_2) = f(S_1)f(S_2)$, where $S_i = \left(\begin{smallmatrix} A_i & -B_i \\ B_i & A_i \end{smallmatrix}\right)$ and $f(S_i) = A_i + iB_i$ if A_i and B_i $(i = 1, 2)$ are real matrices]. By counting the number of degrees of freedom necessary to specify a real antisymmetric matrix A and a real symmetric matrix B, show that $A+iB$ provides n^2 antihermitian generators that span the Lie algebra $\mathfrak{u}(n)$. Finally, verify that $\text{Tr}(\boldsymbol{T}^a\boldsymbol{S}^b) = 0$.

(e) In light of eq. (H.218), count the number of degrees of freedom needed to specify $i\widetilde{\boldsymbol{S}}^b$. Verify that the number of $\mathfrak{so}(2n)$ generators, $\{\boldsymbol{T}^a\,, \boldsymbol{S}^b\}$, is $n(2n-1)$, as expected.

(f) Any element $\mathcal{R}_{2n} \in \text{SO}(2n)$ can be written as the exponential of some linear combination of the generators $\{i\widetilde{T}^a\,,\,i\widetilde{S}^b\}$ given in eq. (H.218). Show that, for any $\mathcal{R}_{2n}\in\mathrm{SO}(2n)$, it is also possible to write $\mathcal{R}_{2n}=\mathcal{R}\,\mathcal{Q}_U$, with

$$\mathcal{R}\equiv\exp\begin{pmatrix} C & D \\ D & -C\end{pmatrix}\,,\qquad \mathcal{Q}_U\equiv\exp\begin{pmatrix} A & -B \\ B & A\end{pmatrix}\,,\tag{H.219}$$

where A, C, and D are real $n\times n$ antisymmetric matrices and B is a real $n\times n$ symmetric matrix. In particular, $\mathcal{Q}_U$ is the $2n$-dimensional real representation of the group $\mathrm{U}(n)$. That is, given an $n\times n$ unitary matrix $U=\exp(A+iB)$, show that

$$\mathcal{Q}_U\equiv\begin{pmatrix} \mathrm{Re}\,U & -\,\mathrm{Im}\,U \\ \mathrm{Im}\,U & \mathrm{Re}\,U\end{pmatrix}\,,\tag{H.220}$$

which we recognize [see Exercise 4.3(d)] as the $2n\times 2n$ real orthogonal matrix that provides the explicit form for the embedding of $\mathrm{U}(n)$ inside $\mathrm{SO}(2n)$.

HINT: In obtaining eq. (H.220), consider making use of the fact that the mapping f defined in part (d) preserves matrix multiplication.

H.2 This exercise provides an explicit construction to show that $\mathfrak{u}(n)$ is a Lie subalgebra of the Lie algebra $\mathfrak{so}(2n+1)$. First, a $(2n+1)\times(2n+1)$ real antisymmetric matrix is introduced that satisfies

$$\Sigma_0^T\Sigma_0=\Sigma_0\Sigma_0^T=c^2\begin{pmatrix}\mathbb{1}_{2n\times 2n} & \mathbf{0} \\ \mathbf{0}^{\mathsf{T}} & 0\end{pmatrix}\,,\tag{H.221}$$

where c is a real number and $\mathbf{0}$ $[\mathbf{0}^{\mathsf{T}}]$ is a $2n$-dimensional zero column [row] vector. The corresponding $n(2n+1)$ generators of $\mathfrak{so}(2n)$ are denoted by $\{\boldsymbol{T}^a\,,\,\boldsymbol{S}^b\,,\,\boldsymbol{V}^b\}$, where the $i\boldsymbol{T}^a$, $i\boldsymbol{S}^b$, and $i\boldsymbol{V}^b$ are real antisymmetric $2n\times 2n$ matrices that satisfy

$$\boldsymbol{T}^a\Sigma_0+\Sigma_0\boldsymbol{T}^{a\mathsf{T}}=0\,,\tag{H.222}$$

$$\boldsymbol{S}^b\Sigma_0-\Sigma_0\boldsymbol{S}^{b\mathsf{T}}=0\,,\tag{H.223}$$

$$\Sigma_0^{\mathsf{T}}\boldsymbol{V}^b\Sigma_0=0\,.\tag{H.224}$$

(a) Modify the analysis used to solve Exercise H.1 to verify that the $\boldsymbol{T}^a$ span a $\mathfrak{u}(n)$ Lie subalgebra of $\mathfrak{so}(2n+1)$.

(b) Show that $i\boldsymbol{T}^a$, $i\boldsymbol{S}^b$, and $i\boldsymbol{V}^b$ are related by similarity transformations to $i\widetilde{\boldsymbol{T}}^a$, $i\widetilde{\boldsymbol{S}}^b$, and $\widetilde{\boldsymbol{V}}^c$, respectively, where $i\widetilde{\boldsymbol{T}}^a$ and $i\widetilde{\boldsymbol{S}}^b$ take the following block form:

$$i\widetilde{\boldsymbol{T}}^a=\begin{pmatrix} A & -B & 0 \\ B & A & 0 \\ 0 & 0 & 0\end{pmatrix}\,,\qquad i\widetilde{\boldsymbol{S}}^b=\begin{pmatrix} C & D & 0 \\ D & -C & 0 \\ 0 & 0 & 0\end{pmatrix}\,,\tag{H.225}$$

and $i\widetilde{V}^c$ is a real $(2n+1)\times(2n+1)$ antisymmetric matrix with the following block structure:

$$i\widetilde{V}^b = \begin{pmatrix} 0 & v \\ -v^\mathsf{T} & 0 \end{pmatrix}, \tag{H.226}$$

where A, C, and D are real $n\times n$ antisymmetric matrices, B is a real $n\times n$ symmetric matrix, v is a real $2n$-dimensional column vector, and the zero entries fill up the remaining elements of the $(2n+1)\times(2n+1)$ matrices exhibited in eqs. (H.225) and (H.226).

(c) Verify that $\mathrm{Tr}(\boldsymbol{T}^a\boldsymbol{S}^b) = \mathrm{Tr}(\boldsymbol{T}^a\boldsymbol{V}^b) = \mathrm{Tr}(\boldsymbol{S}^b\boldsymbol{V}^b) = 0$. In light of eqs. (H.225) and (H.226), count the number of degrees of freedom needed to specify $i\widetilde{\boldsymbol{T}}^a$, $i\widetilde{\boldsymbol{S}}^b$, and $i\widetilde{\boldsymbol{V}}^c$. Verify that the number of $\mathfrak{so}(2n+1)$ generators, $\{\boldsymbol{T}^a, \boldsymbol{S}^b, \boldsymbol{V}^b\}$, is $n(2n+1)$, as expected.

(d) How are the results of part (f) of Exercise H.1 modified for describing the embedding of $\mathrm{U}(n)$ inside $\mathrm{SO}(2n+1)$?

H.3 Starting from the Zassenhaus formula [eqs. (H.28)–(H.31)] with $A\to A/t$, show that, for $|t|\ll 1$,

$$\exp(A+tB) = e^A\left[1 + t\, f(\mathrm{ad}_A)(B) + \mathcal{O}(t^2)\right], \tag{H.227}$$

in agreement with eq. (H.22), where the function f is defined in eq. (H.13).

H.4 Consider an irreducible representation R of a Lie group G of dimension d_R. The matrix representation of an element of G is given by the $d_R\times d_R$ matrix $U = \exp(-i\theta^a\boldsymbol{T}^a)$. One can construct a tensor product representation, $R\otimes R$, where the corresponding matrix representation of $U\in G$ is given by the Kronecker product of matrices, $U\otimes U$, where

$$(U\otimes U)^{ij}{}_{k\ell} \equiv U^i{}_k U^j{}_\ell\,. \tag{H.228}$$

The generators of G in the direct product representation are obtained via

$$U\otimes U = \exp(-i\theta^a\mathcal{T}^a)\,, \tag{H.229}$$

where $\mathcal{T}^a$ is the generator in the direct product representation $R\otimes R$.

(a) Show that

$$(\mathcal{T}^a)^{ij}{}_{k\ell} = (\boldsymbol{T}^a)^i{}_k\delta^j_\ell + (\boldsymbol{T}^a)^j{}_\ell\delta^i_k\,, \tag{H.230}$$

where $i,j,k,\ell\in\{1,2,\ldots,d_R\}$. Note that eq. (H.230) is equivalent to

$$\mathcal{T}^a = \boldsymbol{T}^a\otimes\boldsymbol{I} + \boldsymbol{I}\otimes\boldsymbol{T}^a\,, \tag{H.231}$$

where $\boldsymbol{I}$ is the $d_R\times d_R$ identity matrix. Verify that $\boldsymbol{T}^a$ and $\mathcal{T}^a$ satisfy the same commutation relations [see eq. (H.2)].

(b) Generalize the results of part (a) to obtain an expression for the generator $\boldsymbol{T}^a$ in the direct product representation $R_1 \otimes R_2$ of G . Show that

$$\boldsymbol{T}^a_{R_1\otimes R_2} = \boldsymbol{T}^a_{R_1} \otimes \boldsymbol{I}_{R_2} + \boldsymbol{I}_{R_1} \otimes \boldsymbol{T}^a_{R_2}\,, \tag{H.232}$$

where $\boldsymbol{I}_{R_k}$ is the $d_{R_k} \times d_{R_k}$ identity matrix (for $k = 1, 2$). Using the result of eq. (H.232), verify the results quoted in eqs. (H.35) and (H.49).

HINT: Given an $m \times n$ matrix A, a $q \times r$ matrix B, an $n \times p$ matrix C, and an $r \times s$ matrix D, then $(A \otimes B)(C \otimes D) = AC \otimes BD$.

(c) Following the results of part (b), construct the generators, $\boldsymbol{T}^a_{\pm}$, of the symmetric (upper sign) and antisymmetric (lower sign) parts of the representation $R \otimes R$, denoted by $(R \otimes R)_S$ and $R \otimes R)_A$, respectively. Show that

$$(\boldsymbol{T}^a_{\pm})^{ij}{}_{k\ell} = \tfrac{1}{2}\left[(\boldsymbol{T}^a)^i{}_k\delta^j_\ell + (\boldsymbol{T}^a)^j{}_\ell\delta^i_k \pm (\boldsymbol{T}^a)^i{}_\ell\delta^j_k \pm (\boldsymbol{T}^a)^j{}_k\delta^i_\ell\right]. \tag{H.233}$$

Note that one cannot express $\boldsymbol{T}^a_{\pm}$ as a sum of tensor products.

(d) Using eq. (H.233), derive the following expression:

$$\begin{aligned}(\boldsymbol{T}^a_{\pm}\boldsymbol{T}^b_{\pm})^{ij}{}_{k\ell} &= \tfrac{1}{2}\left[(\boldsymbol{T}^a\boldsymbol{T}^b)^i{}_k\delta^j_\ell + [(\boldsymbol{T}^a\boldsymbol{T}^b)^j{}_\ell\delta^i_k \pm (\boldsymbol{T}^a\boldsymbol{T}^b)^i{}_\ell\delta^j_k \pm (\boldsymbol{T}^a\boldsymbol{T}^b)^j{}_k\delta^i_\ell\right] \\ &+\tfrac{1}{2}\left[(\boldsymbol{T}^a)^i{}_k(\boldsymbol{T}^b)^j{}_\ell + (\boldsymbol{T}^a)^j{}_\ell(\boldsymbol{T}^b)^i{}_k \pm (\boldsymbol{T}^a)^i{}_\ell(\boldsymbol{T}^b)^j{}_k \pm (\boldsymbol{T}^a)^j{}_k(\boldsymbol{T}^b)^i{}_\ell\right],\end{aligned} \tag{H.234}$$

where the upper [lower] signs on the left are taken with the upper [lower] signs on the right. Using the above result, verify that both $\boldsymbol{T}^a_+$ and $\boldsymbol{T}^a_-$ satisfy the commutation relations given in eq. (H.2).

(e) The trace of $A^{ij}{}_{k\ell}$ is given by $\operatorname{Tr} A = A^{ij}{}_{ij}$ (summed over the repeated indices). Using the results of part (d) and noting that $\delta^i_i = d_R$, show that

$$\operatorname{Tr}(\boldsymbol{T}^a_{\pm}\boldsymbol{T}^b_{\pm}) = (d_R \pm 2)\operatorname{Tr}(\boldsymbol{T}^a\boldsymbol{T}^b)\,, \tag{H.235}$$

which yields eq. (H.36). Use a similar technique to verify the result of eq. (H.50).

H.5 The Dynkin labels for the adjoint representation of $\mathfrak{su}(n)$, $\mathfrak{so}(n)$, and $\mathfrak{sp}(n)$ are given in eqs. (H.116)–(H.119), with the special case of $\mathfrak{su}(2)$ given below eq. (H.119). Equation (H.174) exhibits a computation of C_A in the case of $\mathfrak{su}(n)$ for $n \geq 3$. Using eq. (H.125), evaluate C_A for $\mathfrak{su}(2)$, $\mathfrak{so}(n)$, and $\mathfrak{sp}(n)$. Check that your results are consistent with eqs. (H.134) and (H.147).

H.6 The exceptional Lie algebras E_6, E_7, and E_8 are simply laced, which indicates that the lengths of all simple roots are equal. These root lengths are equal to 1 in the normalization convention adopted in this appendix.

(a) By computing η in a convention where $(\boldsymbol{\theta}, \boldsymbol{\theta}) = 1$, show that the dual Coxeter numbers for E_6, E_7, and E_8 are given by $h^\vee = (d_G - \ell)/\ell$, where the ranks ℓ of E_6, E_7, and E_8 are given by 6, 7, and 8, respectively, and the corresponding dimensions of these Lie algebras are 78, 133, and 248. Evaluate the numerical value of $h^\vee$ for the three cases.

(b) The inverse of the Cartan matrix for E_6 is

$$A^{-1} = \frac{1}{3}\begin{pmatrix} 4 & 5 & 6 & 4 & 2 & 3 \\ 5 & 10 & 12 & 8 & 4 & 6 \\ 6 & 12 & 18 & 12 & 6 & 9 \\ 4 & 8 & 12 & 10 & 5 & 6 \\ 2 & 4 & 6 & 5 & 4 & 3 \\ 3 & 6 & 9 & 6 & 3 & 6 \end{pmatrix}. \tag{H.236}$$

The Dynkin labels of the fundamental representation of E_6 are (100000) and its dimension is 27. Using eq. (H.125), show that the eigenvalue of the Casimir operator in the fundamental representation of E_6 is given by

$$C_F = \tfrac{1}{2}(A^{-1})_{11} + \sum_{k=1}^{\ell}(A^{-1})_{k1} = \tfrac{26}{3}\,, \tag{H.237}$$

and the second-order Dynkin index is given $I_2(F) = 3$.

(c) The Dynkin labels of the adjoint representation of E_6 are (000001) and its dimension is 78. Using eq. (H.125), show that the eigenvalue of the Casimir operator in the fundamental representation of E_6 is given by

$$C_A = \tfrac{1}{2}(A^{-1})_{66} + \sum_{k=1}^{\ell}(A^{-1})_{k6} = 12\,, \tag{H.238}$$

and check the consistency with the result of part (a).

(d) Obtain the expressions for the inverse Cartan matrices and other relevant data for E_7 and E_8 (e.g., using [B102]), and repeat the computations of part (b) and (c).

H.7 To complete the exploration of exceptional Lie algebras, one must consider F_4, which possesses two long simple roots and two short simple roots that satisfy eq. (H.143), and G_2, which possesses one long simple root and one short simple root that satisfy $(\boldsymbol{\alpha}_L, \boldsymbol{\alpha}_L)/(\boldsymbol{\alpha}_S, \boldsymbol{\alpha}_S) = 3$. The rank and dimension of F_4 are 4 and 52, respectively, whereas the rank and dimension of G_2 are 2 and 14, respectively.

(a) By computing η in a convention where $(\boldsymbol{\theta}, \boldsymbol{\theta}) = 1$, evaluate the dual Coxeter numbers of F_4 and G_2.

(b) Obtain the expressions for the inverse Cartan matrices and other relevant data for F_4 and G_2 (e.g., using [B102]), and repeat the computations of part (b) and (c) of Exercise H.6.

H.8 Consider an irreducible representation $R = (a_1 a_2 \cdots a_{n-1})$ of $\mathfrak{su}(n)$ $[n \geq 3]$ of dimension d_R, with anomaly coefficient $A(R)$, where the a_i are the Dynkin labels that specify the representation. One can also associate the representation R with a Young tableau consisting of p_k boxes in row k (where $1 \leq k \leq n$) such that $p_1 \geq p_2 \geq \cdots \geq p_{n-1} \geq p_n \equiv 0$.

(a) Show that $p_{n-k} = a_{n-1} + a_{n-2} + \cdots + a_{n-k}$, for $k = 1, 2, \ldots, n-1$.

(b) Formulae for $A(R)$ and d_R in terms of the p_k were given in eqs. (H.178) and (H.179), respectively. Derive the following alternative expressions for $A(R)$ and d_R in terms of the Dynkin indices:

$$A(R) = \frac{2d_R(n-3)!}{(n+2)!} \sum_{i=1}^{n-1}\sum_{j=1}^{n-1}\sum_{k=1}^{n-1} c_{ijk}(a_i+1)(a_j+1)(a_k+1)\,, \quad \text{(H.239)}$$

where

$$d_R = \prod_{j=1}^{n-1}\left[\frac{1}{j!}\prod_{k=j}^{n-1}\left(j + \sum_{i=k-j+1}^{k} a_i\right)\right] \quad \text{(H.240)}$$

$$= \prod_{j=1}^{n-1}\frac{1}{j!}(j+a_1+\cdots+a_j)(j+a_2+\cdots+a_{j+1})\cdots(j+a_{n-j}+\cdots+a_{n-1})\,,$$

and the coefficients c_{ijk} are totally symmetric under the permutation of its indices such that

$$c_{ijk} = i(n-2j)(n-k)\,, \quad \text{for } i \leq j \leq k. \quad \text{(H.241)}$$

H.9 Consider the tensor product of two fundamental representations of the Lie algebra $\mathfrak{su}(n)$, denoted by $\mathbf{n} \otimes \mathbf{n}$.

(a) Show that $\mathbf{n} \otimes \mathbf{n}$ is a reducible representation that decomposes into the sum of two irreducible representations that consist of the symmetric and antisymmetric combinations, $(\mathbf{n} \otimes \mathbf{n})_S$ and $(\mathbf{n} \otimes \mathbf{n})_A$, respectively, with corresponding dimensions $d_S = \frac{1}{2}n(n+1)$ and $d_A = \frac{1}{2}n(n-1)$.

(b) Show that the Dynkin indices for the irreducible representations obtained in part (a) are $(\mathbf{n} \otimes \mathbf{n})_S = (200\cdots0)$ and $(\mathbf{n} \otimes \mathbf{n})_A = (010\cdots0)$, where $\cdots$ represent a string of $n-5$ zeros. (For the cases of $n = 3$, 4, and 5, simply remove the requisite number of trailing zeros.)

HINT: Use the relation between the Dynkin indices and the Young tableaux that represent the second-rank symmetric and antisymmetric tensors which transform irreducibly under SU(n), respectively (see footnote 11).

(c) Using eq. (H.125), show that the eigenvalues of the quadratic Casimir operator for the $\mathfrak{su}(n)$ representations $(\mathbf{n}\otimes\mathbf{n})_S$ and $(\mathbf{n}\otimes\mathbf{n})_A$ are

$$C_{(\mathbf{n}\otimes\mathbf{n})_S} = \frac{(n-1)(n+2)}{n}\,, \qquad C_{(\mathbf{n}\otimes\mathbf{n})_A} = \frac{(n+1)(n-2)}{n}\,. \tag{H.242}$$

(d) Show that the second-order indices for the $\mathfrak{su}(n)$ representations $(\mathbf{n}\otimes\mathbf{n})_S$ and $(\mathbf{n}\otimes\mathbf{n})_A$ are given by

$$I_2\big((\mathbf{n}\otimes\mathbf{n})_S\big) = \tfrac{1}{2}(n+2)\,, \qquad I_2\big((\mathbf{n}\otimes\mathbf{n})_A\big) = \tfrac{1}{2}(n-2)\,. \tag{H.243}$$

H.10 Consider the fundamental representation $R = F$ of a Lie algebra $\mathfrak{g}$. Then Exercise H.9 shows that for $\mathfrak{g} = \mathfrak{su}(n)$, both $(F\otimes F)_S$ and $(F\otimes F)_A$ are irreducible representations. In contrast, show that:

(a) For $\mathfrak{g} = \mathfrak{so}(n)$, $n \geq 3$, $n \neq 4$, $(F\otimes F)_S$ is reducible and $(F\otimes F)_A$ is irreducible. Resolve $(F\otimes F)_S$ into a direct sum of irreps.

(b) For $\mathfrak{g} = \mathfrak{sp}(n)$, $n \geq 2$, $(F\otimes F)_S$ is irreducible and $(F\otimes F)_A$ is reducible. Resolve $(F\otimes F)_A$ into a direct sum of irreps.

H.11 Given a simple Lie algebra $\mathfrak{g}$ and four irreps, R_i, R_j, R_k, and R_ℓ (not necessarily distinct), consider the tensor product

$$R \equiv R_i \otimes R_j \otimes R_k \otimes R_\ell = \bigoplus_r n_r S_r\,, \tag{H.244}$$

where the reducible representation R decomposes into a direct sum of irreps of $\mathfrak{g}$, denoted by S_r with corresponding multiplicity n_r. Let n_s be the number of singlets (i.e., trivial one-dimensional irreps) that appear in the direct sum.

Consider the decomposition of the following tensor products into irreps:

$$R_i \otimes R_j = \bigoplus_r n_r^{(ij)} S_r^{(ij)}\,, \qquad \overline{R}_k \otimes \overline{R}_\ell = \bigoplus_r \overline{n}_r^{(k\ell)}\overline{S}_r^{(k\ell)}\,, \tag{H.245}$$

where $n_r^{(ij)}$ and $\overline{n}_r^{(k\ell)}$ are the corresponding multiplicity factors. Denote the set of irreps that are common to $\{S_r^{(ij)}\}$ and $\{\overline{S}_r^{(k\ell)}\}$ by $\{R_m^{(ijk\ell)}\}$. Show that

$$n_s = \sum_m n_m^{(ij)}\overline{n}_m^{(k\ell)}\,. \tag{H.246}$$

H.12 Given the four irreps R_i, R_j, R_k, and R_ℓ of Exercise H.11 and the decomposition of the tensor products specified in eqs. (H.244) and (H.245), the following formula has been conjectured in Ref. [239] to hold:

$$n_s(C_{R_i} + C_{R_j} + C_{R_k} + C_{R_\ell}) = \sum_m n_m^{(ij)}\overline{n}_m^{(k\ell)} C_{R_m^{(ijk\ell)}} + (j \leftrightarrow k) + (j \leftrightarrow \ell)\,, \tag{H.247}$$

where C_R is the eigenvalue of the quadratic Casimir operator in irrep R, the sum over m is taken over the irreps $\{R_m^{(ijk\ell)}\}$ specified above eq. (H.246), and n_s is given by eq. (H.246). Since $R_i \otimes R_j = R_j \otimes R_i$, etc., the formula given in eq. (H.247) is completely symmetric under the interchange of the indices i, j, k, and ℓ. Note that if $n_s = 0$, then no terms appear on the right-hand side of eq. (H.247), in which case the formula is trivial.

(a) If R is a complex irrep of $\mathfrak{g}$, consider the special case where $R_i = R_j = R$ and $R_k = R_\ell = \overline{R}$. Given the tensor product decompositions $R\otimes R = \bigoplus_r n_r S_r$ and $R \otimes \overline{R} = \bigoplus_r n'_r S'_r$ into irreps, show that

$$4n_s C_R = \sum_r n_r^2 C_{S_r} + 2\sum_r n_r'^2 C_{S'_r}, \tag{H.248}$$

where $n_s = \sum_r n_r^2 = \sum_r n_r'^2$. If R is a real irrep, then $R = \overline{R}$, in which case eq. (H.248) still applies with $C_{S'_r} = C_{S_r}$ and $n'_r = n_r$.

(b) Equations (H.247) and (H.248) provide additional checks of the tables in Appendix H.5.3. Consider the following example: $\mathfrak{g} = \mathfrak{su}(n)$ with $R_i = R_j = \mathbf{n}$ and $R_k = R_\ell = \overline{\mathbf{n}}$. Show that the decomposition of $\mathbf{n} \otimes \mathbf{n} \otimes \overline{\mathbf{n}} \otimes \overline{\mathbf{n}}$ into irreps contains $n_s = 2$ singlets. Recalling that $\mathbf{n} \otimes \overline{\mathbf{n}} = \mathbf{n^2-1} \oplus \mathbf{1}$, where $\mathbf{n^2-1}$ is the adjoint representation of $\mathfrak{su}(n)$, use eq. (H.248) to obtain

$$8C_F = C_{(\mathbf{n}\otimes\mathbf{n})_S} + C_{(\mathbf{n}\otimes\mathbf{n})_A} + 2C_A\,. \tag{H.249}$$

Employing the results of Exercise H.9, check that eq. (H.249) is satisfied.

(c) In light of eq. (H.130) and the known values of $I_2(F) = \frac{1}{2}$ and $I_2(A) = n$ for $\mathfrak{su}(n)$, show that eq. (H.249) is equivalent to the following relation:

$$(n-1)I_2\big((\mathbf{n}\otimes\mathbf{n})_S\big) + (n+1)I_2\big((\mathbf{n}\otimes\mathbf{n})_A\big) = n^2 - 2\,. \tag{H.250}$$

In addition, show that eqs. (H.182) and (H.184) yield

$$I_2\big((\mathbf{n}\otimes\mathbf{n})_S\big) + I_2\big((\mathbf{n}\otimes\mathbf{n})_A\big) = n\,. \tag{H.251}$$

Verify that eqs. (H.250) and (H.251) yield the results given in eq. (H.243).

(d) Suppose that $\mathfrak{g} = \mathfrak{su}(3)$ and $R_i = R_j = R_k = R_\ell = \mathbf{8}$. Show that $n_s = 8$ and verify that eq. (H.248) yields $20C_\mathbf{8} = 6C_\mathbf{10} + 3C_\mathbf{27}$. Check that this result is consistent with Table H.2.

(e) Suppose that $\mathfrak{g} = \mathfrak{su}(4)$ and $R_i = R_j = R_k = R_\ell = \mathbf{4}$. Show that $n_s = 1$ and verify that eq. (H.247) implies that $4C_\mathbf{4} = 3C_\mathbf{6}$. Check that this result is consistent with Table H.4.

(f) Suppose that $\mathfrak{g} = \mathfrak{so}(10)$ and $R_i = R_j = \mathbf{10}$ and $R_k = R_\ell = \mathbf{54}$. Show that $n_s = 3$ and eq. (H.247) is satisfied, using the tables of Ref. [69].

(g) Suppose that $\mathfrak{g} = \mathrm{E}_6$ with $R_i = \mathbf{27}$, $R_j = \mathbf{78}$, $R_k = \overline{\mathbf{351}}$, and $R_\ell = \mathbf{650}$. Show that $n_s = 5$ and eq. (H.247) is satisfied, using the tables of Ref. [69].

Appendix I Interaction Vertices of the SM and Its Seesaw Extension

In this appendix, we provide the Feynman rules for vertices of the Standard Model (see Section 4.7) and its seesaw extension (see Section 6.1) involving a pair of fermions. To complete the tabulation of all SM Feynman rules, one must include the rules for the purely bosonic interactions of the SM. These can be found in [B75].

I.1 Standard Model Fermion Interaction Vertices

We begin with the Feynman rules for the QCD color interactions of the quarks, which are given in Fig. I.1. Next, the Feynman rules for the electroweak interactions of the quarks and leptons with the charged and neutral gauge bosons are exhibited in Fig. I.2. For each of the rules of Fig. I.2, we have chosen to employ $\overline{\sigma}_\mu^{\dot\alpha\beta}$. If the indices are lowered, then one should take $\overline{\sigma}_\mu^{\dot\alpha\beta} \to -\sigma_{\mu\beta\dot\alpha}$. Finally, the Feynman rules for the interactions of the quark and lepton mass eigenstates with the Higgs fields are shown in Fig. I.3.

When covariant gauge fixing is employed (e.g., the 't Hooft–Feynman gauge or Landau gauge), the Goldstone bosons appear explicitly in internal lines of Feynman diagrams. The Feynman rules for fermion interactions with the neutral Goldstone boson G are flavor-diagonal, whereas the corresponding rules for $G^\pm$ exhibit flavor-changing interactions that depend on the CKM matrix elements, as shown in Fig. I.4. In the derivation of the couplings of the Goldstone bosons to the fermion

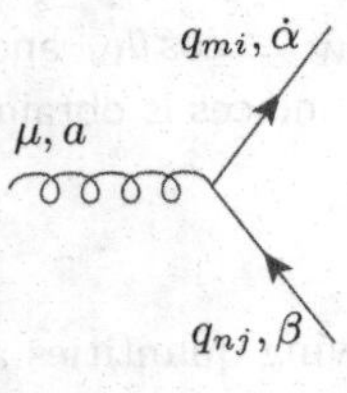

$-ig_s\delta_i^j(\boldsymbol{T}^a)_m{}^n\,\overline{\sigma}_\mu^{\dot\alpha\beta}$

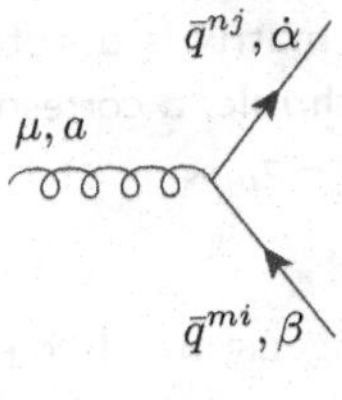

$ig_s\delta_i^j(\boldsymbol{T}^a)_m{}^n\,\overline{\sigma}_\mu^{\dot\alpha\beta}$

Fig. I.1 Fermionic Feynman rules for QCD that involve the gluon, with $q = u, d, c, s, t, b$. Lowered (raised) indices m, n correspond to the fundamental (antifundamental) representation of SU(3)$_C$. The gluon interactions are flavor-diagonal (where i, j are flavor indices). For each rule shown above, a corresponding rule with lowered spinor indices is obtained by replacing $\overline{\sigma}_\mu^{\dot\alpha\beta} \to -\sigma_{\mu\beta\dot\alpha}$.

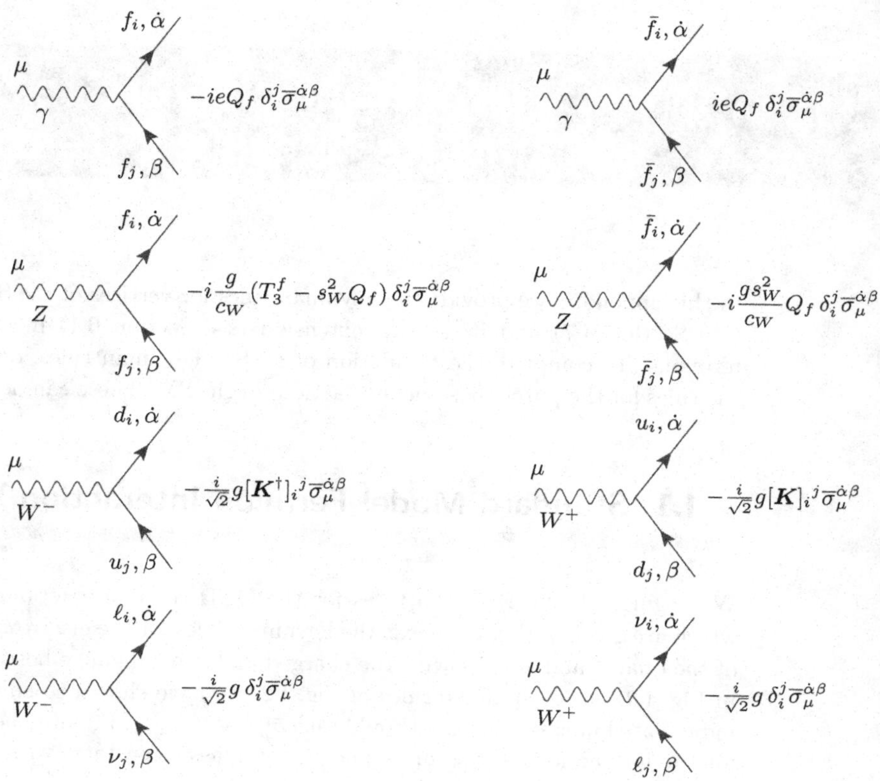

Fig. I.2 Feynman rules for the two-component fermion interactions with electroweak gauge bosons in the Standard Model. The couplings of the fermions to γ and Z are flavor-diagonal. In all couplings, i and j label the fermion families; an upper [lowered] flavor index in the corresponding Feynman rule is associated with a fermion line that points into [out from] the vertex. For the $W^\pm$ bosons, the charge indicated is flowing into the vertex. The electric charge is denoted by Q_f (in units of $e > 0$), with $Q_e = -1$ for the electron. $T_3^f = 1/2$ for $f = u$, ν, and $T_3^f = -1/2$ for $f = d$, ℓ. The CKM mixing matrix is denoted by $\boldsymbol{K}$, and $s_W \equiv \sin\theta_W$, $c_W \equiv \cos\theta_W$ and $e \equiv g\sin\theta_W$. For each rule, a corresponding one with lowered spinor indices is obtained by replacing $\overline{\sigma}_\mu^{\dot\alpha\beta} \to -\sigma_{\mu\beta\dot\alpha}$.

mass eigenstates [see eqs. (4.169)–(4.171)], the following quantities appear:

$$(L_d)_k{}^j(\boldsymbol{y}_u)^k{}_m(R_u)^m{}_i = y_{ui}(L_d)_k{}^j(L_u^\dagger)_i{}^k = y_{ui}(L_u^\dagger L_d)_i{}^j = [\boldsymbol{K}]_i{}^j y_{ui}\,, \tag{I.1}$$

$$(L_u)_k{}^j(\boldsymbol{y}_d)^k{}_m(R_d)^m{}_i = y_{di}(L_u)_k{}^j(L_d^\dagger)_i{}^k = y_{di}(L_d^\dagger L_u)_i{}^j = [\boldsymbol{K}^\dagger]_i{}^j y_{di}\,, \tag{I.2}$$

with no sum over the repeated index i. The CKM matrix, $\boldsymbol{K}$, appears by virtue of eqs. (4.179) and (4.183). Hence, the interaction Lagrangian for the coupling of the

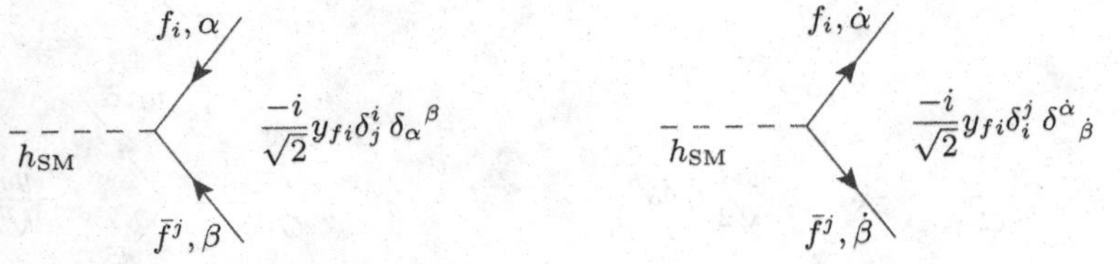

Fig. I.3 Feynman rules for the Standard Model Higgs boson interactions with fermions, where $y_{fi} \equiv \sqrt{2} m_{fi}/v$, and i, j label the families.

Goldstone bosons to the fermion mass eigenstates is given by

$$\begin{aligned} \mathscr{L}_{\rm int} &= y_{ui}[\boldsymbol{K}]_i{}^j d_j \bar{u}^i G^+ - y_{di}[\boldsymbol{K}^\dagger]_i{}^j u_j \bar{d}^i G^- - y_{\ell i} \nu_i \bar{\ell}^i G^- \\ &\quad + \frac{i}{\sqrt{2}} \left[y_{di} d_i \bar{d}^i - y_{ui} u_i \bar{u}^i + y_{\ell i} \ell_i \bar{\ell}^i \right] G + \text{h.c.}, \end{aligned} \tag{I.3}$$

which yields the diagrammatic Feynman rules shown in Fig. I.4.

I.2 Interaction Vertices of the Seesaw-Extended Standard Model

To accommodate massive neutrinos, we employ the seesaw mechanism, which is treated in Section 6.1. The charged-current neutrino interactions are given in terms of the neutrino mass-eigenstate fields by eq. (6.8), which we reproduce below:

$$\begin{aligned} \mathscr{L}_{\rm int} &= -\frac{g}{\sqrt{2}} \left[\check{\nu}^{\dagger i} \bar{\sigma}^\mu \ell_i W_\mu^+ + \ell^{\dagger i} \bar{\sigma}^\mu \check{\nu}_i W_\mu^- \right] \\ &= -\frac{g}{\sqrt{2}} \left[(\boldsymbol{U}^\dagger)_j{}^i \nu_\ell^{\dagger j} \bar{\sigma}^\mu \ell_i W_\mu^+ + \boldsymbol{U}_i{}^j \ell^{\dagger i} \bar{\sigma}^\mu \nu_{\ell j} W_\mu^- \right], \end{aligned} \tag{I.4}$$

where $\boldsymbol{U}$ is the PMNS mixing matrix [see eq. (6.6)] The neutral current neutrino interactions are flavor-diagonal (which follows from the unitarity of U), and are thus equivalent to those of the Standard Model. Finally, the couplings of the neutrinos to the Higgs and Goldstone fields arise from eq. (6.3) and from the term in eq. (4.169) proportional to $\boldsymbol{y}_\ell$. Neglecting terms of $\mathcal{O}(m_\nu^2/v^2)$, one obtains eq. (6.9), which we reproduce below:

$$\begin{aligned} \mathscr{L}_{\rm int} &= \frac{\sqrt{2}}{v} \sum_{i,j} \left[(m_{\nu_\ell})_j (\boldsymbol{U}^\dagger)_j{}^i (\nu_\ell)_j \, \ell_i \, G^+ - (m_\ell)_i \boldsymbol{U}_i{}^j (\nu_\ell)_j \, \bar{\ell}^i \, G^- + \text{h.c.} \right] \\ &\quad - \frac{1}{v} \sum_j (m_{\nu_\ell})_j \left[(\nu_\ell)_j (\nu_\ell)_j (h_{\rm SM} + iG) + \text{h.c.} \right]. \end{aligned} \tag{I.5}$$

The Feynman rules for the interactions of the neutrino with the electroweak gauge bosons, the Higgs boson, and the Goldstone bosons are exhibited in Fig. I.5.

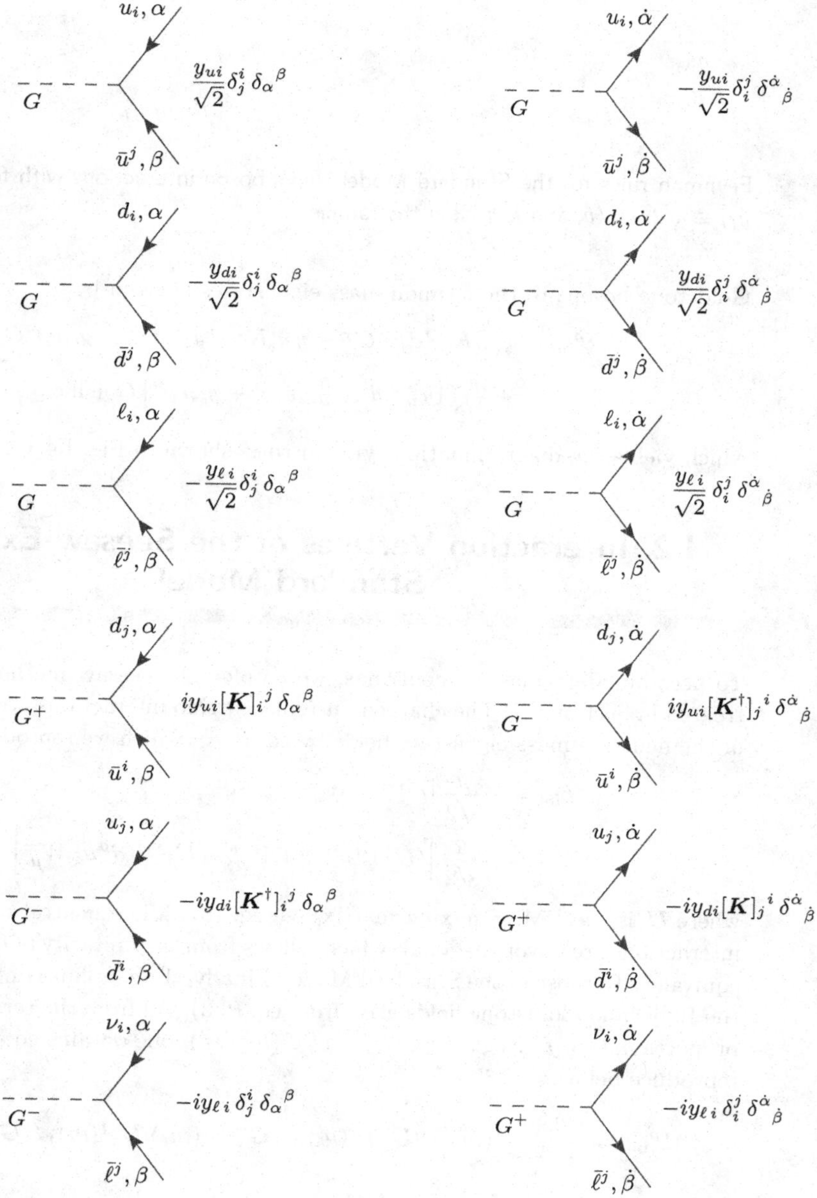

Fig. I.4 Feynman rules for the Standard Model Goldstone boson interactions with quarks and leptons, where $y_{fi} \equiv \sqrt{2}m_{fi}/v$, and i, j label the families.

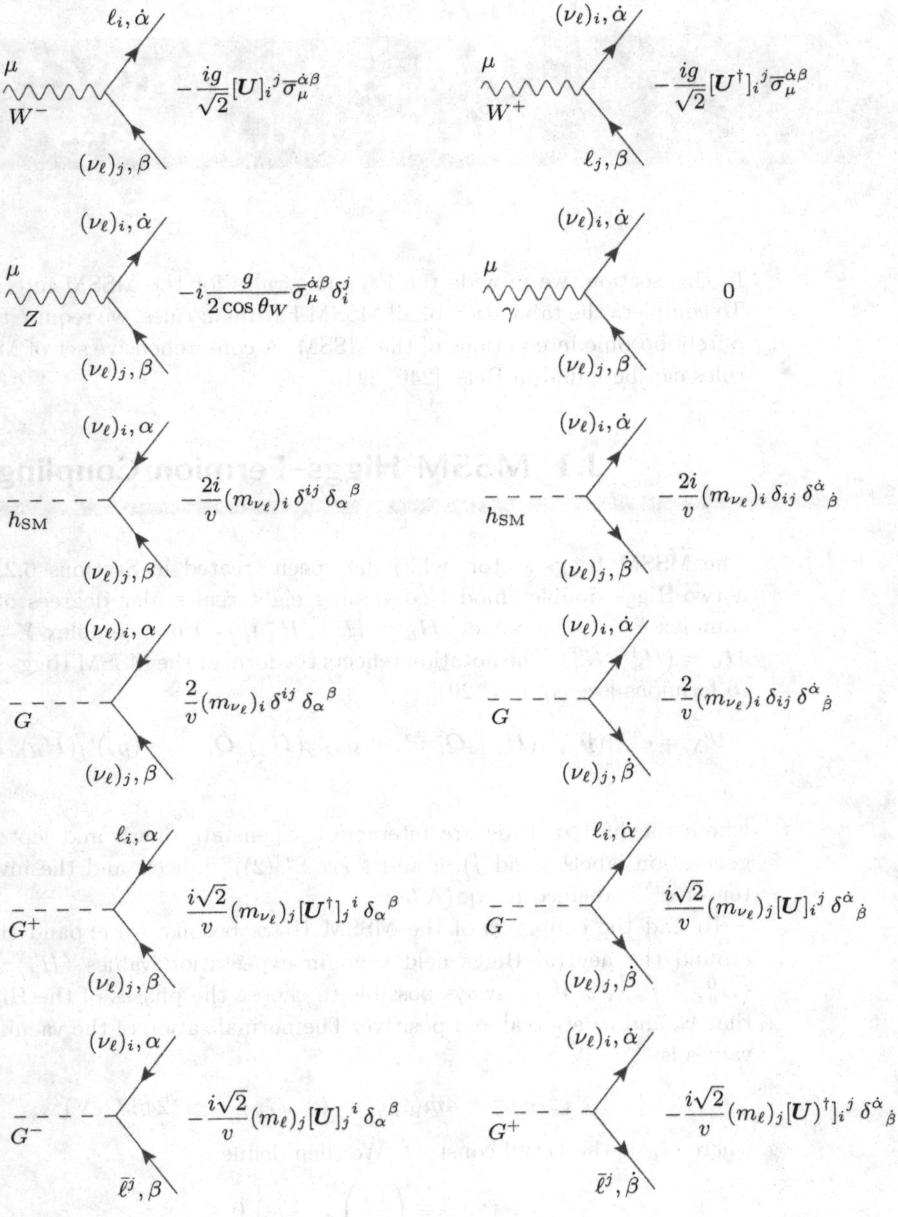

Fig. I.5 Feynman rules for the interactions of the two-component light neutrino (ν_ℓ) with electroweak gauge bosons, the Standard Model Higgs boson, and the Goldstone bosons, where i, j label the family and $v \simeq 246$ GeV. For the $W^\pm$ bosons and $G^\pm$ scalars, the charge indicated is flowing into the vertex. The PMNS mixing matrix is denoted by $\boldsymbol{U}$. For the rules involving $W^\pm$ and Z bosons, a corresponding rule with lowered spinor indices is obtained by replacing $\overline{\sigma}_\mu^{\dot{\alpha}\beta} \to -\sigma_{\mu\beta\dot{\alpha}}$. In the h_{SM} and G interactions, a factor of 2 is included to account for the identical neutrinos.

Appendix J MSSM and RPV Fermion Interaction Vertices

In this section, we provide the Feynman rules for the MSSM interaction vertices. To complete the tabulation of all MSSM Feynman rules, we require the rules for the purely bosonic interactions of the MSSM. A comprehensive set of MSSM Feynman rules can be found in Refs. [240, 241].

J.1 MSSM Higgs–Fermion Couplings

The MSSM Higgs sector, which has been treated in Sections 6.2.6 and 13.8, is a two-Higgs doublet model containing eight real scalar degrees of freedom: one complex $Y = -\frac{1}{2}$ doublet, $\boldsymbol{H_d} = (H_d^0\,,\, H_d^-)$, and one complex $Y = +\frac{1}{2}$ doublet, $\boldsymbol{H_u} = (H_u^+\,,\, H_u^0)$. The notation reflects the form of the MSSM Higgs sector coupling to fermions [see eq. (13.120)]:

$$\mathscr{L}_{\rm Y} = \epsilon^{ab}\left[(\boldsymbol{y}_u)^i{}_j(\boldsymbol{H_u})_a\hat{\boldsymbol{Q}}_{bi}\hat{\bar{u}}^j - (\boldsymbol{y}_d)^i{}_j(\boldsymbol{H_d})_a\hat{\boldsymbol{Q}}_{bi}\hat{\bar{d}}^j - (\boldsymbol{y}_\ell)^i{}_j(\boldsymbol{H_d})_a\hat{\boldsymbol{L}}_{bi}\hat{\bar{\ell}}^j\right] + \text{h.c.}\,, \tag{J.1}$$

where the hatted fields are interaction eigenstate quark and lepton fields (with generation labels i and j), a and b are $\mathrm{SU}(2)_L$ indices, and the invariant $\mathrm{SU}(2)_L$ tensor ϵ^{ab} is defined in eq. (A.54).

To find the couplings of the MSSM Higgs bosons, we expand the Higgs fields around the neutral Higgs field vacuum expectation values $\langle H_d^0\rangle \equiv v_d/\sqrt{2}$ and $\langle H_u^0\rangle \equiv v_u/\sqrt{2}$. It is always possible to choose the phases of the Higgs fields such that v_u and v_d are real and positive. The normalization of the vacuum expectation values is

$$v_d^2 + v_u^2 = 4m_W^2/g^2 = (\sqrt{2}G_F)^{-1} \simeq (246\ \text{GeV})^2\,, \tag{J.2}$$

where G_F is the Fermi constant. We then define

$$\tan\beta \equiv \left(\frac{v_u}{v_d}\right)\,, \qquad 0 \leq \beta \leq \frac{\pi}{2}\,. \tag{J.3}$$

The tree-level MSSM Higgs sector conserves CP, which implies that the neutral Higgs mass eigenstates possess definite CP quantum numbers.[1] Spontaneous elec-

[1] When one-loop corrections are taken into account, new MSSM phases can enter in the loops that cannot be removed. In this case, the physical neutral Higgs states can be mixtures of CP-even and CP-odd scalar states.

troweak symmetry breaking results in three Goldstone bosons $G^{\pm}$, G (the neutral Goldstone boson is a CP-odd scalar field), which are absorbed and become the longitudinal components of the $W^{\pm}$ and Z. The remaining five physical Higgs particles consist of a charged Higgs pair $H^{\pm}$, one neutral CP-odd scalar A, and two neutral CP-even scalars h and H.

It is convenient to define $\phi^- \equiv (\phi^+)^\dagger$, $H_u^- \equiv (H_u^+)^\dagger$, and $H_d^+ \equiv (H_d^-)^\dagger$. One can then parameterize the mixing angles between Higgs gauge eigenstates and mass eigenstates by

$$H_u^0 = \frac{1}{\sqrt{2}}\left(v_u + \sum_{\phi^0} k_{u\phi^0}\phi^0\right), \qquad H_u^{\pm} = \sum_{\phi^{\pm}} k_{u\phi^{\pm}}\phi^{\pm}\,, \tag{J.4}$$

$$H_d^0 = \frac{1}{\sqrt{2}}\left(v_d + \sum_{\phi^0} k_{d\phi^0}\phi^0\right), \qquad H_d^{\mp} = \sum_{\phi^{\pm}} k_{d\phi^{\pm}}\phi^{\pm}\,. \tag{J.5}$$

For $\phi^{\pm} = (H^{\pm},\, G^{\pm})$,

$$k_{u\phi^{\pm}} = (\cos\beta,\; \sin\beta)\,, \tag{J.6}$$

$$k_{d\phi^{\pm}} = (\sin\beta,\; -\cos\beta)\,, \tag{J.7}$$

and for $\phi^0 = (h,\, H,\, A,\, G)$,

$$k_{u\phi^0} = (\cos\alpha,\; \sin\alpha,\; i\cos\beta,\; i\sin\beta)\,, \tag{J.8}$$

$$k_{d\phi^0} = (-\sin\alpha,\; \cos\alpha,\; i\sin\beta,\; -i\cos\beta)\,, \tag{J.9}$$

where the mixing angle α parameterizes the orthogonal matrix that diagonalizes the tree-level 2×2 CP-even Higgs squared-mass matrix, which is given explicitly in eq. (6.230).

The Higgs–fermion Yukawa couplings in the gauge-interaction basis are given by eq. (J.1). We use eqs. (J.4) and (J.5) to express the interaction eigenstate Higgs fields in terms of the physical Higgs fields and Goldstone fields. We can identify the quark and lepton mass matrices simply by setting $H_u^0 = v_u/\sqrt{2}$, $H_d^0 = v_d/\sqrt{2}$, and $H_u^+ = H_d^- = 0$ in eq. (J.1):

$$(\boldsymbol{M}_u)^i{}_j = v_u(\boldsymbol{y}_u)^i{}_j\,, \qquad (\boldsymbol{M}_d)^i{}_j = v_d(\boldsymbol{y}_d)^i{}_j\,, \qquad (\boldsymbol{M}_\ell)^i{}_j = v_d(\boldsymbol{y}_\ell)^i{}_j\,. \tag{J.10}$$

We then use eqs. (4.173) and (4.174) to express the interaction eigenstate quark and lepton fields in terms of the corresponding mass-eigenstate fields. Equations (4.175)-(4.177) ensure that the fermion mass matrices are diagonal (with real nonnegative elements) in the fermion mass-eigenstate basis. In this basis, the resulting neutral Higgs–fermion interactions are diagonal. The diagonalized Yukawa couplings are related to the corresponding fermion masses by eq. (13.122), which we reproduce here for the reader's convenience:

$$y_{ui} = \frac{\sqrt{2}m_{ui}}{v_u}\,, \qquad y_{di} = \frac{\sqrt{2}m_{di}}{v_d}\,, \qquad y_{ei} = \frac{\sqrt{2}m_{ei}}{v_d}\,. \tag{J.11}$$

We have used the same symbol for the Yukawa couplings in the MSSM as we did

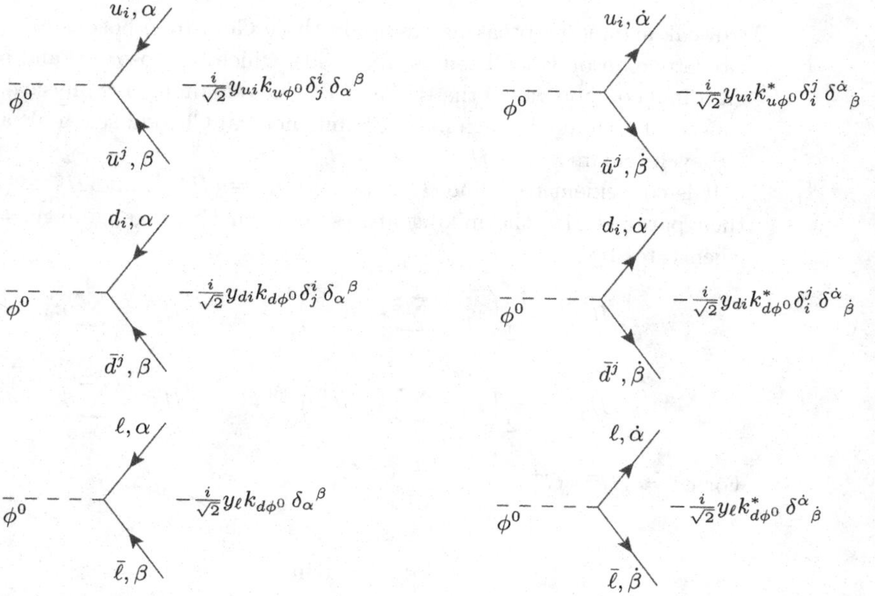

Fig. J.1 Feynman rules for the interactions of neutral Higgs bosons $\phi^0 = (h, H, A, G)$ with fermion–antifermion pairs in the MSSM. The repeated index i is not summed.

for the Standard Model Yukawa couplings in Appendix I.1. However, it is important to note that the MSSM Yukawa couplings are normalized differently because of the presence of two neutral Higgs field vacuum expectation values. Using a superscript SM to denote the Standard Model Yukawa couplings of Appendix I.1, the MSSM Yukawa couplings defined here are related by

$$y_{ui} = \frac{y_{ui}^{\rm SM}}{\sin\beta}\,, \qquad y_{di} = \frac{y_{di}^{\rm SM}}{\cos\beta}\,, \qquad y_{\ell i} = \frac{y_{\ell i}^{\rm SM}}{\cos\beta}\,. \tag{J.12}$$

The interactions of the neutral Higgs and Goldstone scalars $\phi^0 = (h, H, A, G)$ with Standard Model fermions are given in Fig. J.1. Note that the rules involving undotted spinor indices are proportional to either couplings $k_{d\phi^0}$ or $k_{u\phi^0}$, whereas the rules involving dotted spinor indices are proportional to the corresponding complex-conjugated couplings. For the CP-even scalars, h and H, the corresponding couplings are real. Hence, starting with the rule for the coupling of the CP-even neutral scalars to fermions with undotted indices, one obtains the corresponding rule for the coupling to fermions with dotted indices (with the direction of the arrows reversed) by replacing $\delta_\alpha{}^\beta \to \delta^{\dot{\alpha}}{}_{\dot{\beta}}$.

In contrast, for the CP-odd scalars, A and G, the corresponding couplings $k_{d\phi^0}$ and $k_{u\phi^0}$ are purely imaginary. Therefore, starting with the rule for the coupling of the CP-odd neutral scalars to fermions with undotted indices, one obtains the corresponding rule for the coupling to fermions with dotted indices (with the direc-

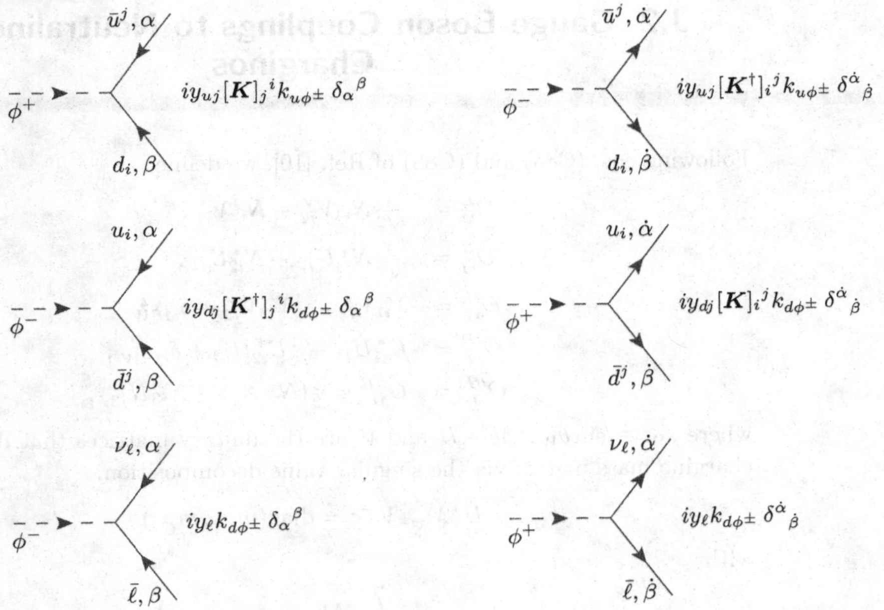

Fig. J.2 Feynman rules for the interactions of charged Higgs bosons $\phi^\pm = (H^\pm, G^\pm)$ with fermion–antifermion pairs in the MSSM. The repeated index j is not summed.

tion of the arrows reversed) by replacing $\delta_\alpha{}^\beta \to -\delta^{\dot{\alpha}}{}_{\dot{\beta}}$. The latter minus sign is a signal that A and G are CP-odd scalars. In particular, because the Feynman rules for A and G arise from a term in $\mathscr{L}_{\rm int}$ proportional to $i\,{\rm Im}\, H^0$, the latter i flips sign when the rule is conjugated, resulting in the extra minus sign noted above. As an additional consequence, since the Feynman rules are obtained from $i\mathscr{L}_{\rm int}$, the overall A and G rules are real.

The couplings of the charged Higgs and Goldstone bosons to quark–antiquark pairs are not flavor-diagonal and involve the CKM matrix $\boldsymbol{K}$. Starting with eq. (J.1) and transforming to the mass-eigenstate basis, we make use of eqs. (I.1) and (I.2) to obtain

$$\begin{aligned}\mathscr{L}_{\rm int} = {}& y_{ui}[\boldsymbol{K}]_i{}^j d_j \bar{u}^i H^+ \cos\beta + y_{ui}[\boldsymbol{K}]_i{}^j d_j \bar{u}^i G^+ \sin\beta \\ &+ y_{di}[\boldsymbol{K}^\dagger]_i{}^j u_j \bar{d}^i H^- \sin\beta - y_{di}[\boldsymbol{K}^\dagger]_i{}^j u_j \bar{d}^i G^- \cos\beta \\ &+ y_{\ell i}\nu_i \bar{\ell}^i H^- \sin\beta - y_{\ell i}\nu_i \bar{\ell}^i G^- \cos\beta + \text{h.c.}\end{aligned} \tag{J.13}$$

The resulting charged scalar Feynman rules of the MSSM are given in Fig. J.2. Note that when eq. (J.12) is taken into account, the fermion couplings to the neutral and charged Goldstone bosons are equivalent to those of the Standard Model [see eq. (I.3)].

J.2 Gauge Boson Couplings to Neutralinos and Charginos

Following eqs. (C83) and (C88) of Ref. [10], we define

$$O^L_{ij} = -\tfrac{1}{\sqrt{2}} N_{i4} V^*_{j2} + N_{i2} V^*_{j1}\,, \tag{J.14}$$

$$O^R_{ij} = \tfrac{1}{\sqrt{2}} N^*_{i3} U_{j2} + N^*_{i2} U_{j1}\,, \tag{J.15}$$

$$O'^L_{ij} = -V_{i1} V^*_{j1} - \tfrac{1}{2} V_{i2} V^*_{j2} + \delta_{ij} s^2_W\,, \tag{J.16}$$

$$O'^R_{ij} = -U^*_{i1} U_{j1} - \tfrac{1}{2} U^*_{i2} U_{j2} + \delta_{ij} s^2_W\,, \tag{J.17}$$

$$O''^L_{ij} = -O''^R_{ji} = \tfrac{1}{2}(N_{i4} N^*_{j4} - N_{i3} N^*_{j3})\,, \tag{J.18}$$

where $s_W \equiv \sin\theta_W$. Here U and V are the unitary matrices that diagonalize the chargino mass matrix via the singular value decomposition,

$$U^* M_{\psi^\pm} V^{-1} = \mathrm{diag}(m_{\widetilde{C}_1}, m_{\widetilde{C}_2})\,, \tag{J.19}$$

with

$$M_{\psi^\pm} = \begin{pmatrix} M_2 & g v_u/\sqrt{2} \\ g v_d/\sqrt{2} & \mu \end{pmatrix}\,, \tag{J.20}$$

and N is a unitary matrix that Takagi diagonalizes the neutralino mass matrix,

$$N^* M_{\psi^0} N^{-1} = \mathrm{diag}(m_{\widetilde{N}_1}, m_{\widetilde{N}_2}, m_{\widetilde{N}_3}, m_{\widetilde{N}_4})\,, \tag{J.21}$$

with

$$M_{\psi^0} = \begin{pmatrix} M_1 & 0 & -\tfrac{1}{2} g' v_d & \tfrac{1}{2} g' v_u \\ 0 & M_2 & \tfrac{1}{2} g v_d & -\tfrac{1}{2} g v_u \\ -\tfrac{1}{2} g' v_d & \tfrac{1}{2} g v_d & 0 & -\mu \\ \tfrac{1}{2} g' v_u & -\tfrac{1}{2} g v_u & -\mu & 0 \end{pmatrix}\,, \tag{J.22}$$

in a convention in which v_u and v_d are real and positive. The gaugino mass parameters M_1, M_2 and the higgsino mass parameter μ are potentially complex.

The gauge boson interactions with the neutralinos and charginos in the form of Feynman rules can be obtained by using Figs. 2.8—2.10. The Feynman rules for the Z and γ interactions with charginos and neutralinos are given in Fig. J.3 and the corresponding rules for the $W^\pm$ interactions are given in Fig. J.4. For each of these rules, one has a version with lowered spinor indices by replacing $\overline{\sigma}^{\dot{\alpha}\beta}_\mu \to -\sigma_{\mu\beta\dot{\alpha}}$. We label fermion lines with the symbols of the two-component fermion fields as given in Table 17.1. The $Z\widetilde{N}_i\widetilde{N}_j$ interaction vertex also subsumes the O''^R_{ij} interaction found in four-component Majorana Feynman rules as in Ref. [10], due to the result of eq. (3.101) and the relation $O''^R_{ij} = -O''^L_{ji}$ of eq. (J.18).

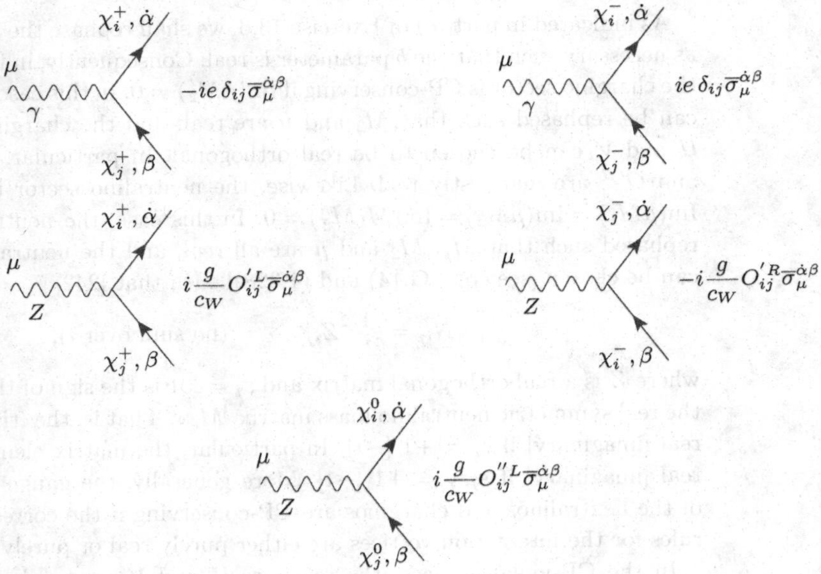

Fig. J.3 Feynman rules for chargino and neutralino interactions with neutral gauge bosons. The coupling matrices are defined in eqs. (J.16)–(J.18) and $c_W \equiv \cos\theta_W$. For each rule, a corresponding rule with lowered spinor indices is obtained by replacing $\overline{\sigma}_\mu^{\dot\alpha\beta} \to -\sigma_{\mu\beta\dot\alpha}$.

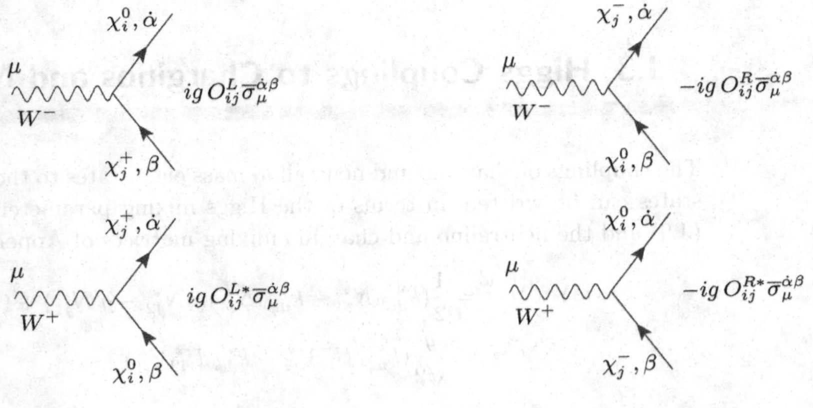

Fig. J.4 Feynman rules for chargino and neutralino interactions with $W^\pm$ gauge bosons. The charge indicated on the W boson is flowing into the vertex in each case. The coupling matrices are defined in eqs. (J.14) and (J.15). For each rule shown above, a corresponding rule with lowered spinor indices is obtained by replacing $\overline{\sigma}_\mu^{\dot\alpha\beta} \to -\sigma_{\mu\beta\dot\alpha}$.

As indicated in part (c) of Exercise 13.6, we shall rephase the Higgs doublet fields as necessary such that the b parameter is real. Consequently, in light of eq. (13.211), the chargino sector is CP-conserving if $\text{Im}(\mu M_2) = 0$. In this case, the chargino fields can be rephased such that M_2 and μ are real, and the chargino mixing matrices U and V can be chosen to be real orthogonal. In particular, the couplings O'^L and O'^R are manifestly real. Likewise, the neutralino sector is CP-conserving if $\text{Im}(\mu M_1) = \text{Im}(\mu M_2) = \text{Im}(M_1 M_2^*) = 0$. In this case, the neutralino fields can be rephased such that M_1, M_2 and μ are all real, and the neutralino mixing matrix can be chosen [see eqs. (G.14) and (1.328)] such that [242]

$$N_{ij} = \varepsilon_i^{1/2} Z_{ij} \qquad \text{(no sum over } i\text{)}, \tag{J.23}$$

where Z is a real orthogonal matrix and $\varepsilon_i = \pm 1$ is the sign of the ith eigenvalue of the real symmetric neutralino mass matrix, M_{ψ^0}. That is, the ith row of N is purely real [imaginary] if $\varepsilon_i = +1$ [-1]. In particular, the matrix element O''^L_{ij} is purely real [imaginary] if $\varepsilon_i \varepsilon_j = +1$ [-1]. More generally, the gauge boson interactions of the neutralinos and charginos are CP-conserving if the corresponding Feynman rules for the interaction vertices are either purely real or purely imaginary.

In the CP-violating case, the matrices U and V cannot be chosen to be real orthogonal, and N cannot be written in the form of eq. (J.23). Nevertheless, the diagonal couplings O'^L_{ii}, O'^R_{ii}, and O''^L_{ii} are manifestly real. This indicates that the diagonal $Z^0 \widetilde{C}_i^+ \widetilde{C}_i^-$ and $Z^0 \widetilde{N}_i \widetilde{N}_i$ couplings are CP-conserving at tree level, even in the presence of a CP-violating chargino and neutralino sector. Similarly, the diagonal $\gamma \widetilde{C}_i^+ \widetilde{C}_i^-$ couplings are CP-conserving, whereas the off-diagonal $\gamma \widetilde{C}_i^\pm \widetilde{C}_j^\mp$ couplings ($i \neq j$) vanish at tree level, as expected from gauge invariance.

J.3 Higgs Couplings to Charginos and Neutralinos

The couplings of chargino and neutralino mass eigenstates to the Higgs mass eigenstates can be written, in terms of the Higgs mixing parameters of eqs. (J.8) and (J.9) and the neutralino and chargino mixing matrices of Appendix J.2, as [242]

$$Y^{\phi^0 \chi_i^0 \chi_j^0} = \frac{1}{2}(k^*_{d\phi^0} N^*_{i3} - k^*_{u\phi^0} N^*_{i4})(g N^*_{j2} - g' N^*_{j1}) + (i \leftrightarrow j)\,, \tag{J.24}$$

$$Y^{\phi^0 \chi_i^- \chi_j^+} = \frac{g}{\sqrt{2}}(k^*_{u\phi^0} U^*_{i1} V^*_{j2} + k^*_{d\phi^0} U^*_{i2} V^*_{j1})\,, \tag{J.25}$$

$$Y^{\phi^+ \chi_i^0 \chi_j^-} = k_{d\phi^\pm}\left[g(N^*_{i3} U^*_{j1} - \frac{1}{\sqrt{2}} N^*_{i2} U^*_{j2}) - \frac{g'}{\sqrt{2}} N^*_{i1} U^*_{j2}\right], \tag{J.26}$$

$$Y^{\phi^- \chi_i^0 \chi_j^+} = k_{u\phi^\pm}\left[g(N^*_{i4} V^*_{j1} + \frac{1}{\sqrt{2}} N^*_{i2} V^*_{j2}) + \frac{g'}{\sqrt{2}} N^*_{i1} V^*_{j2}\right], \tag{J.27}$$

for $\phi^0 = h, H, A, G$ and $\phi^\pm = H^\pm, G^\pm$. We exhibit the Higgs boson and Goldstone boson interactions with the neutralinos and charginos in Fig. J.5. For each of the

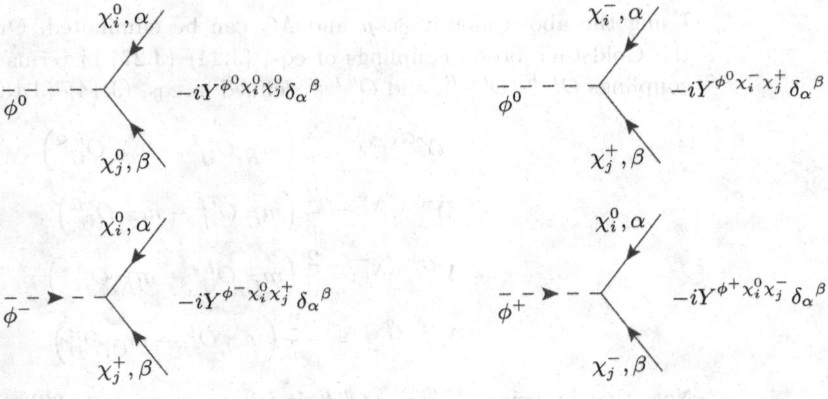

Fig. J.5 Feynman rules for the interactions of neutral Higgs bosons $\phi^0 = (h, H, A, G)$ with neutralino pairs and chargino pairs, respectively, and the interaction of charged Higgs bosons $\phi^\pm = (H^\pm, G^\pm)$ with chargino–neutralino pairs. For each rule shown above, there is a corresponding rule with all arrows reversed, undotted indices changed to dotted indices with the opposite height, and the Y coupling (without the explicit $-i$) replaced by its complex conjugate.

Feynman rules in Fig. J.5, one can reverse all arrows by taking $\delta_\alpha{}^\beta \to \delta^{\dot\alpha}{}_{\dot\beta}$ and complex conjugating the corresponding coupling (but not the overall factor of $-i$).

Goldstone bosons may appear as internal lines in Feynman graphs that are evaluated in the 't Hooft–Feynman gauge. The propagation of a Goldstone boson yields a result that is identical to the propagation of the corresponding longitudinal gauge boson in the unitary gauge. It is thus convenient to express the Goldstone boson couplings to the neutralinos and charginos in terms of the corresponding gauge boson couplings. To accomplish this, we record a number of identities among the neutralino and chargino mixing matrices. First, eqs. (J.19) and (J.20) yield

$$M_2 U_{i1}^* + \frac{g v_d}{\sqrt{2}} U_{i2}^* = m_{\widetilde{C}_i} V_{i1}\,, \qquad \frac{g v_u}{\sqrt{2}} U_{i1}^* + \mu U_{i2}^* = m_{\widetilde{C}_i} V_{i2}\,, \tag{J.28}$$

$$M_2 V_{i1}^* + \frac{g v_u}{\sqrt{2}} V_{i2}^* = m_{\widetilde{C}_i} U_{i1}\,, \qquad \frac{g v_d}{\sqrt{2}} V_{i1}^* + \mu V_{i2}^* = m_{\widetilde{C}_i} U_{i2}\,. \tag{J.29}$$

Next, we make use of eqs. (J.21) and (J.22) to derive

$$m_{\widetilde{N}_i} N_{i4} = \sum_{j=1}^{4} N_{ij}^* (M_{\psi^0})_{j4} = \tfrac{1}{2} v_u \left(g' N_{i1}^* - g N_{i2}^*\right) - \mu N_{i3}^*\,, \tag{J.30}$$

$$m_{\widetilde{N}_i} N_{i3} = \sum_{j=1}^{4} N_{ij}^* (M_{\psi^0})_{j3} = -\tfrac{1}{2} v_d \left(g' N_{i1}^* - g N_{i2}^*\right) - \mu N_{i4}^*\,, \tag{J.31}$$

$$m_{\widetilde{N}_i} N_{i2} = \sum_{j=1}^{4} N_{ij}^* (M_{\psi^0})_{j2} = N_{i2}^* M_2 + \tfrac{1}{2} g \left(v_d N_{i3}^* - v_u N_{i4}^*\right)\,. \tag{J.32}$$

Using the above identities, μ and M_2 can be eliminated. One can then rewrite the Goldstone boson couplings of eqs. (J.24)–(J.27) in terms of the gauge boson couplings $O^{L,R}$, $O'^{L,R}$, and $O''^{L,R}$ defined in eqs. (J.14)–(J.18). It follows that

$$iY^{G\chi_i^0\chi_j^0} = \frac{2}{v}\left(m_{\widetilde{N}_i} O''^{L}_{ij} - m_{\widetilde{N}_j} O''^{R}_{ij}\right) , \tag{J.33}$$

$$iY^{G\chi_i^-\chi_j^+} = \frac{2}{v}\left(m_{\widetilde{C}_i} O'^{L}_{ij} - m_{\widetilde{C}_j} O'^{R}_{ij}\right) , \tag{J.34}$$

$$Y^{G^+\chi_i^0\chi_j^-} = \frac{2}{v}\left(m_{\widetilde{C}_j} O^{L*}_{ij} - m_{\widetilde{N}_i} O^{R*}_{ij}\right) , \tag{J.35}$$

$$Y^{G^-\chi_i^0\chi_j^+} = -\frac{2}{v}\left(m_{\widetilde{N}_i} O^{L}_{ij} - m_{\widetilde{C}_j} O^{R}_{ij}\right) . \tag{J.36}$$

Note that by using $O''^{R}_{ij} = -O''^{L}_{ji}$, it follows from eq. (J.33) that $iY^{G\chi_i^0\chi_j^0}$ is symmetric under the interchange of i and j, as expected.

In general, for a CP-violating chargino and neutralino sector, the couplings $Y^{\phi^0\chi_i^0\chi_i^0}$ and $Y^{\phi^0\chi_i^+\chi_i^-}$ for $\phi^0 = h, H, A$ are neither purely real nor purely imaginary. That is, the diagonal neutralino and chargino couplings to the physical neutral Higgs bosons are generically CP-violating. However, for $\phi^0 = G$, the diagonal neutralino and chargino couplings to the neutral Goldstone boson (when multiplied by i) are manifestly real. In particular, eqs. (J.33) and (J.34) yield

$$iY^{G\chi_i^0\chi_i^0} = \frac{4m_{\widetilde{N}_i}}{v} O''^{L}_{ii} = \frac{2m_{\widetilde{N}_i}}{v}\left[|N_{i4}|^2 - |N_{i3}|^2\right] , \tag{J.37}$$

$$iY^{G\chi_i^-\chi_i^+} = \frac{2m_{\widetilde{C}_i}}{v}(O'^{L}_{ii} - O'^{R}_{ii}) = \frac{m_{\widetilde{C}_i}}{v}\left[|V_{i2}|^2 - |U_{i2}|^2\right] , \tag{J.38}$$

where the unitarity of U and V has been used to obtain the final expression in eq. (J.38). It follows that the diagonal neutralino and chargino couplings to the neutral Goldstone boson are CP-conserving. This result is not surprising, as the corresponding diagonal tree-level couplings of the (longitudinal) Z^0 boson are always CP-conserving, as noted at the end of Appendix J.2.

J.4 Chargino and Neutralino Couplings to Fermions and Sfermions

In the MSSM, the scalar partners of the two-component fields q and $\bar{q}^\dagger$ are the squarks, denoted by $\widetilde{q}_L$ and $\widetilde{q}_R$, respectively. In our notation, $\widetilde{q}_L^\dagger$ and $\widetilde{q}_R^\dagger$ denote both the complex conjugate fields and the names of the corresponding antisquarks. Thus u, $\widetilde{u}_L$, and $\widetilde{u}_R$ all have electric charges $+2/3$, whereas $\bar{u}$, $\widetilde{u}_L^\dagger$, and $\widetilde{u}_R^\dagger$ all have electric charges $-2/3$. Likewise, the scalar partners of the two-component fields ℓ and $\bar{\ell}^\dagger$ are the charged sleptons, denoted by $\widetilde{\ell}_L$ and $\widetilde{\ell}_R$, respectively, with $\ell = e, \mu, \tau$. The sneutrino, $\widetilde{\nu}$, is the superpartner of the neutrino. There is no $\widetilde{\nu}_R$, since there is no $\bar{\nu}$ in the theory.

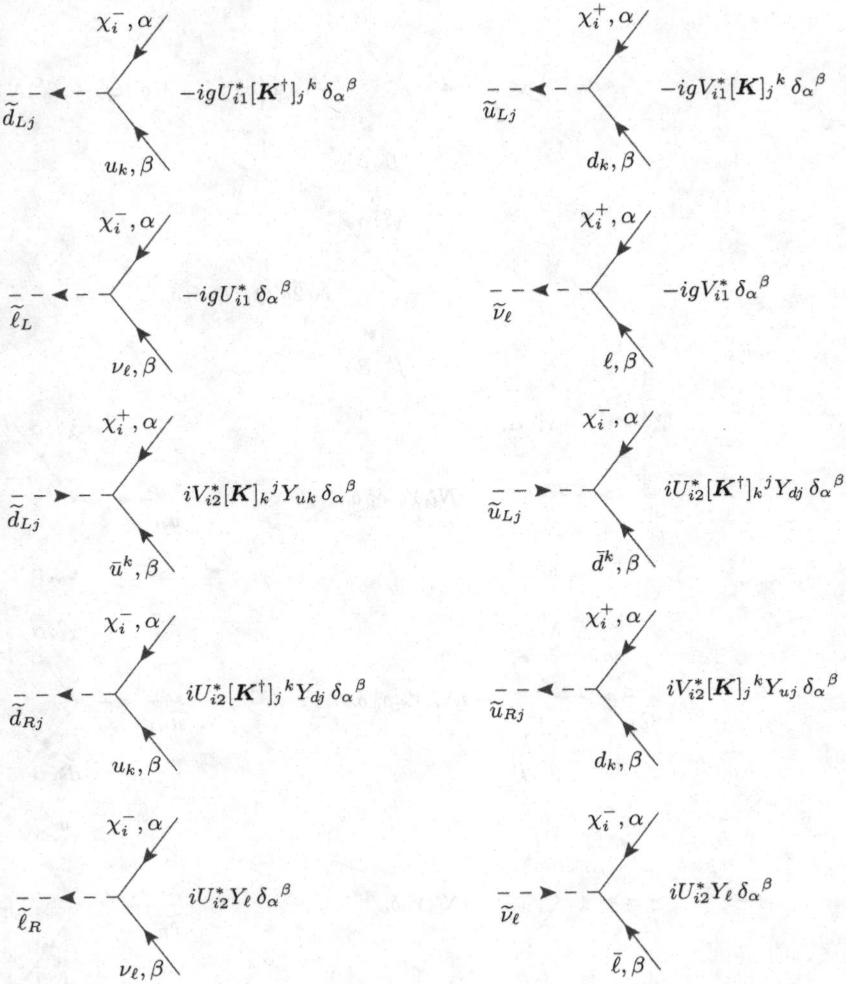

Fig. J.6 Feynman rules for the interactions of charginos with fermion/sfermion pairs in the MSSM. The fermions (quarks, charged leptons, and neutrinos) are mass eigenstate fields. For each rule shown above, there is a corresponding rule with all arrows reversed, undotted indices changed to dotted indices with the opposite height, and the coupling (without the explicit i) replaced by its complex conjugate. Note that chargino interaction vertices involving $\bar{u}\widetilde{d}_R$ and $\bar{d}\widetilde{u}_R$ do not occur in the MSSM.

The Feynman rules for the chargino–quark–squark interactions are exhibited in Fig. J.6, and the Feynman rules for the neutralino–quark–squark interactions are exhibited in Fig. J.7. Here we have taken the quark and lepton two-component fields to be in a mass-eigenstate basis, and the squark and slepton field basis consists

$$\tilde{f}_{Lj} \quad\blacktriangleleft\quad \overset{\chi_i^0,\,\alpha}{\diagup}\quad f_k,\beta \qquad -i\sqrt{2}\left[gT_3^f N_{i2}^* + g'(Q_f - T_3^f)N_{i1}^*\right]\delta_j^k\,\delta_\alpha{}^\beta$$

$$\tilde{f}_{Rj} \quad\blacktriangleright\quad \overset{\chi_i^0,\,\alpha}{\diagup}\quad \bar{f}^k,\beta \qquad i\sqrt{2}\,g'Q_f N_{i1}^*\,\delta_k^j\,\delta_\alpha{}^\beta$$

$$\tilde{u}_{Lj} \quad\blacktriangleright\quad \overset{\chi_i^0,\,\alpha}{\diagup}\quad \bar{u}^k,\beta \qquad -iN_{i4}^* Y_{uj}\delta_k^j\,\delta_\alpha{}^\beta$$

$$\tilde{u}_{Rj} \quad\blacktriangleleft\quad \overset{\chi_i^0,\,\alpha}{\diagup}\quad u_k,\beta \qquad -iN_{i4}^* Y_{uj}\delta_j^k\,\delta_\alpha{}^\beta$$

$$\tilde{d}_{Lj} \quad\blacktriangleright\quad \overset{\chi_i^0,\,\alpha}{\diagup}\quad \bar{d}^k,\beta \qquad -iN_{i3}^* Y_{dj}\delta_k^j\,\delta_\alpha{}^\beta$$

$$\tilde{d}_{Rj} \quad\blacktriangleleft\quad \overset{\chi_i^0,\,\alpha}{\diagup}\quad d_k,\beta \qquad -iN_{i3}^* Y_{dj}\delta_j^k\,\delta_\alpha{}^\beta$$

$$\tilde{\ell}_L \quad\blacktriangleright\quad \overset{\chi_i^0,\,\alpha}{\diagup}\quad \bar{\ell},\beta \qquad -iN_{i3}^* Y_\ell\,\delta_\alpha{}^\beta$$

$$\tilde{\ell}_R \quad\blacktriangleleft\quad \overset{\chi_i^0,\,\alpha}{\diagup}\quad \ell,\beta \qquad -iN_{i3}^* Y_\ell\,\delta_\alpha{}^\beta$$

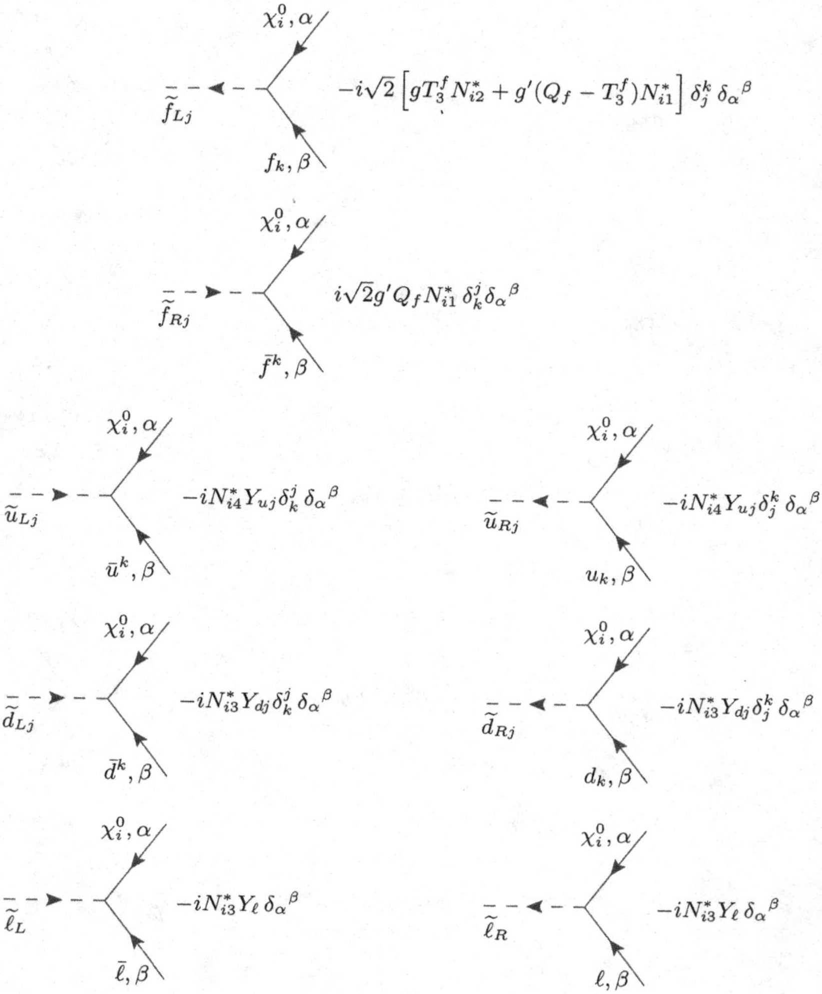

Fig. J.7 Feynman rules for the interactions of neutralinos with fermion/sfermion pairs in the MSSM. The fermions (quarks, charged leptons, and neutrinos) are mass eigenstate fields. For each rule shown above, there is a corresponding rule with all arrows reversed, undotted indices changed to dotted indices with the opposite height, and the coupling (without the explicit i) replaced by its complex conjugate.

of the superpartners of these fields, as described above. Therefore, in practical applications, one must include unitary rotation matrix elements relating the squarks and sleptons as given to the mass eigenstates, which can be different.

Under the assumption that only the sfermions of the third generation (with a given electric charge) exhibit significant mixing, the relation between the gauge

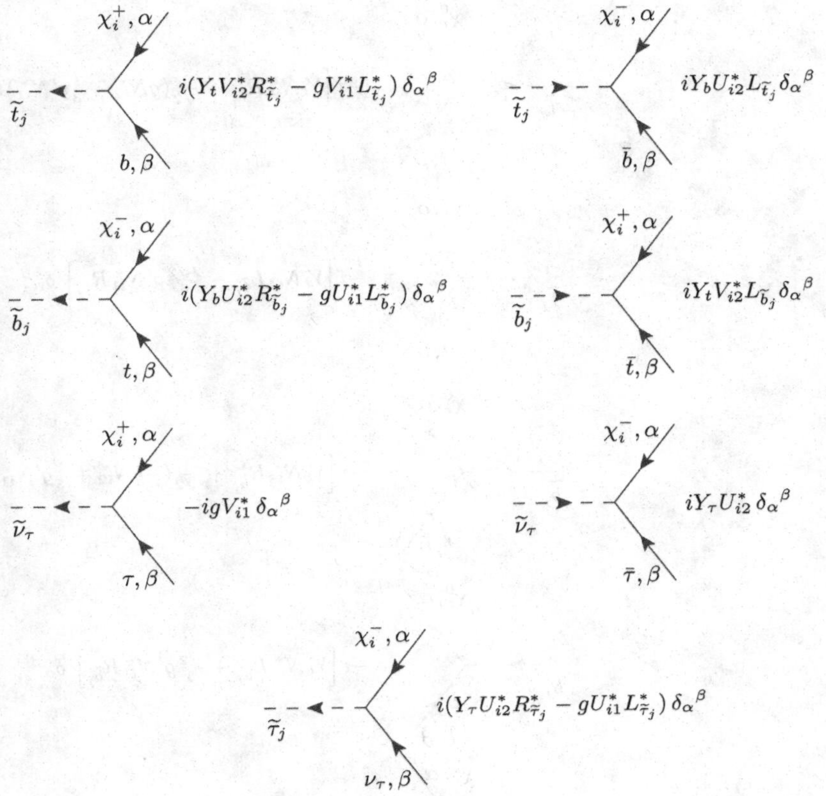

Fig. J.8 Feynman rules for the interactions of charginos with third-family fermion/sfermion pairs in the MSSM. The fermions (quarks, charged leptons, and neutrinos) are mass-eigenstate fields. CKM mixing is neglected, and the sfermions are assumed to mix only within the third family. The corresponding rules for the first and second families, with the approximation of no mixing and vanishing fermion masses, can be obtained from these by setting $Y_f = 0$ and $L_{\widetilde{f}_2} = R_{\widetilde{f}_1} = 1$ and $L_{\widetilde{f}_1} = R_{\widetilde{f}_2} = 0$ (so that $\widetilde{f}_1 = \widetilde{f}_R$ and $\widetilde{f}_2 = \widetilde{f}_L$). For each rule shown above, there is a corresponding rule with all arrows reversed, undotted indices changed to dotted indices with the opposite height, and the coupling (without the explicit i) replaced by its complex conjugate.

eigenstates $\widetilde{f}_L$, $\widetilde{f}_R$ and the mass eigenstates $\widetilde{f}_1$, $\widetilde{f}_2$ (for $f = t, b, \tau$) is given by

$$\begin{pmatrix} \widetilde{f}_R \\ \widetilde{f}_L \end{pmatrix} = X_{\widetilde{f}} \begin{pmatrix} \widetilde{f}_1 \\ \widetilde{f}_2 \end{pmatrix}, \qquad X_{\widetilde{f}} \equiv \begin{pmatrix} R_{\widetilde{f}_1} & R_{\widetilde{f}_2} \\ L_{\widetilde{f}_1} & L_{\widetilde{f}_2} \end{pmatrix}, \tag{J.39}$$

where $X_{\widetilde{f}}$ is a 2×2 unitary matrix [cf. Exercise 13.1]. The Feynman rules for squarks and sleptons that mix within each generation are shown in Figs. J.8 and J.9.

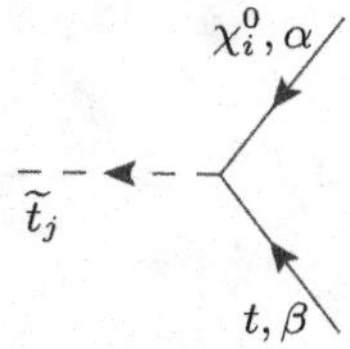

$$-i\left[Y_t N_{i4}^* R_{\widetilde{t}_j}^* + \frac{1}{\sqrt{2}}(gN_{i2}^* + \tfrac{1}{3}g' N_{i1}^*)L_{\widetilde{t}_j}^*\right]\delta_\alpha{}^\beta$$

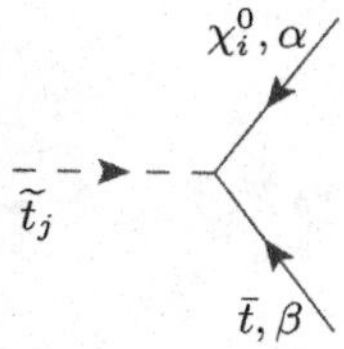

$$-i\left[Y_t N_{i4}^* L_{\widetilde{t}_j} - \tfrac{2\sqrt{2}}{3}g' N_{i1}^* R_{\widetilde{t}_j}\right]\delta_\alpha{}^\beta$$

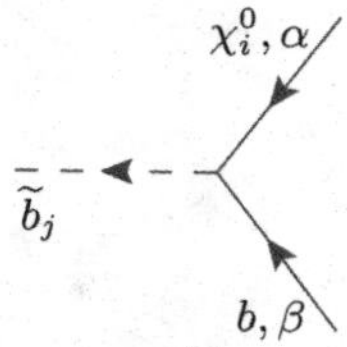

$$-i\left[Y_b N_{i3}^* R_{\widetilde{b}_j}^* + \frac{1}{\sqrt{2}}(-gN_{i2}^* + \tfrac{1}{3}g' N_{i1}^*)L_{\widetilde{b}_j}^*\right]\delta_\alpha{}^\beta$$

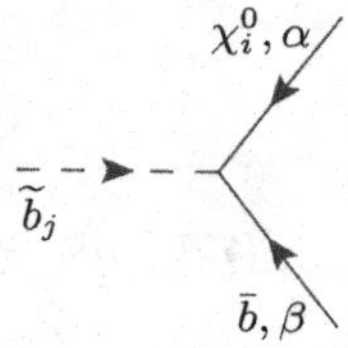

$$-i\left[Y_b N_{i3}^* L_{\widetilde{b}_j} + \tfrac{\sqrt{2}}{3}g' N_{i1}^* R_{\widetilde{b}_j}\right]\delta_\alpha{}^\beta$$

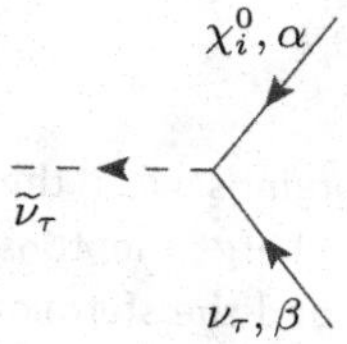

$$-\frac{i}{\sqrt{2}}(gN_{i2}^* - g' N_{i1}^*)\,\delta_\alpha{}^\beta$$

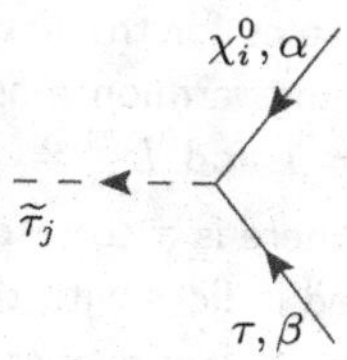

$$-i\left[Y_\tau N_{i3}^* R_{\widetilde{\tau}_j}^* - \frac{1}{\sqrt{2}}(gN_{i2}^* + g' N_{i1}^*)L_{\widetilde{\tau}_j}^*\right]\delta_\alpha{}^\beta$$

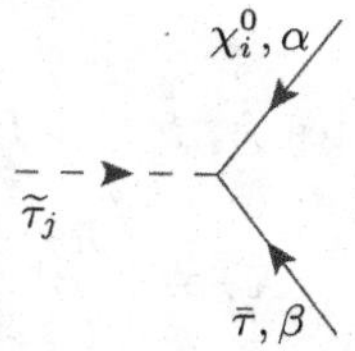

$$-i\left[Y_\tau N_{i3}^* L_{\widetilde{\tau}_j} + \sqrt{2}g' N_{i1}^* R_{\widetilde{\tau}_j}\right]\delta_\alpha{}^\beta$$

Fig. J.9 Feynman rules for the interactions of neutralinos with third-family fermion/sfermion pairs in the MSSM. The comments of the caption of Fig. J.8 also apply here.

For each Feynman rule in Figs. J.6—J.9, one can reverse all arrows by taking $\delta_\alpha{}^\beta \to \delta^{\dot\alpha}{}_{\dot\beta}$ and complex conjugating the corresponding coupling (but not the overall factor of $-i$).

J.5 SUSY-QCD Feynman Rules

In supersymmetric (SUSY) QCD, the Lagrangian governing the gluon interactions with colored fermions (gluinos and quarks) in two-component spinor notation, which derives from the covariant derivatives in the kinetic terms, is given by

$$\mathscr{L}_{\rm int} = ig_s f^{abd}(\tilde g_a^\dagger \,\overline\sigma_\mu \,\tilde g_b)A_d^\mu - g_s(\boldsymbol{T^a})_j{}^k \sum_q \left[q^{\dagger j}\overline\sigma_\mu q_k - \bar q_k^\dagger \overline\sigma_\mu \bar q^j\right] A_a^\mu\,. \quad \text{(J.40)}$$

Here g_s is the strong coupling constant, $a,b,d \in \{1,2,\ldots,8\}$ are $\mathrm{SU}(3)_C$ adjoint representation indices, and f^{abd} are the SU(3) structure constants. Raised (lowered) indices $j,k \in \{1,2,3\}$ are color indices in the fundamental (antifundamental) representation. We have denoted the two-component gluino field by $\tilde g_a$ as in Table 17.1, and the gluon field by A_a^μ. The sum $\sum_q$ is over the six flavors $q = u,d,s,c,b,t$ (in either the mass-eigenstate or electroweak gauge-eigenstate basis). The corresponding Feynman rules are shown in Fig. J.10. The gluino–squark–quark Lagrangian is given by

$$\mathscr{L}_{\rm int} = -\sqrt{2}g_s(\boldsymbol{T^a})_j{}^k \sum_q \left[\tilde g_a q_k\, \tilde q_L^{\dagger j} + \tilde g_a^\dagger q^{\dagger j}\,\tilde q_{Lk} - \tilde g_a \bar q^j\,\tilde q_{Rk} - \tilde g_a^\dagger \bar q_k^\dagger\,\tilde q_R^{\dagger j}\right], \quad \text{(J.41)}$$

where the squark fields are taken to be in the same basis as the quarks. The Feynman rules resulting from these Lagrangian terms are shown in Fig. J.11.

For practical applications, one typically takes the quark fields as the familiar mass eigenstates, and then performs a unitary rotation on the squarks in the corresponding basis to obtain their mass-eigenstate basis. In the approximation described at the end of Appendix J.4 [see eq. (J.39) and the accompanying text], one obtains the Feynman rules of Fig. J.12, as an alternative to those of Fig. J.11.

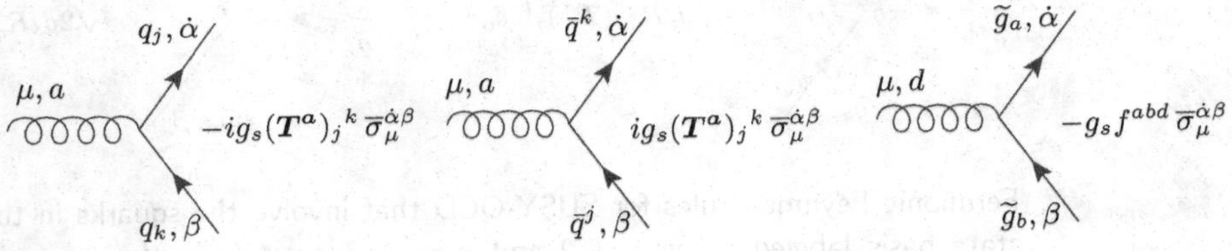

Fig. J.10 Fermionic Feynman rules for SUSY-QCD that involve the gluon, with $q = u,d,c,s,t,b$. Lowered (raised) indices j,k correspond to the fundamental (antifundamental) representation of $\mathrm{SU}(3)_C$. For each rule shown above, a corresponding rule with lowered spinor indices is obtained by replacing $\overline\sigma_\mu^{\dot\alpha\beta} \to -\sigma_{\mu\beta\dot\alpha}$.

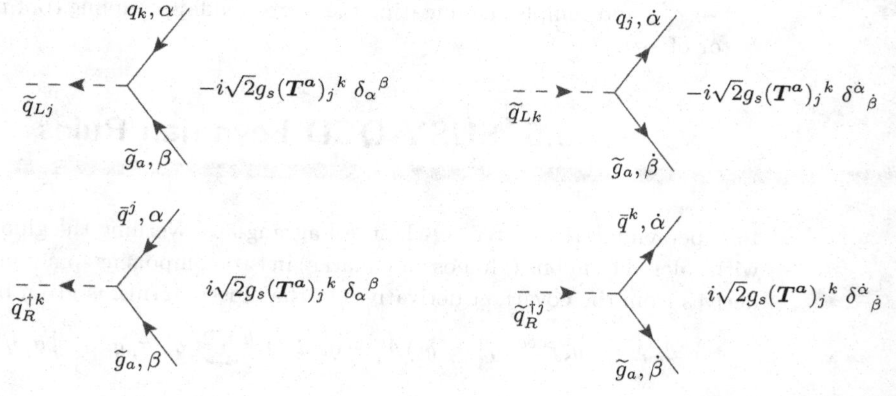

Fig. J.11 Fermionic Feynman rules for SUSY-QCD that involve the squarks, in a basis corresponding to the quark mass eigenstates $q = u, d, c, s, t, b$. Lowered (raised) indices j, k correspond to the fundamental (antifundamental) representation of SU(3)$_C$, and the index a labels the adjoint representation carried by the gluino. The spinor index heights can be exchanged in each case by replacing $\delta_\alpha{}^\beta \to \delta_\beta{}^\alpha$ or $\delta^{\dot{\alpha}}{}_{\dot{\beta}} \to \delta^{\dot{\beta}}{}_{\dot{\alpha}}$. For an alternative set of rules, incorporating $\widetilde{q}_L$–$\widetilde{q}_R$ mixing, see Fig. J.12.

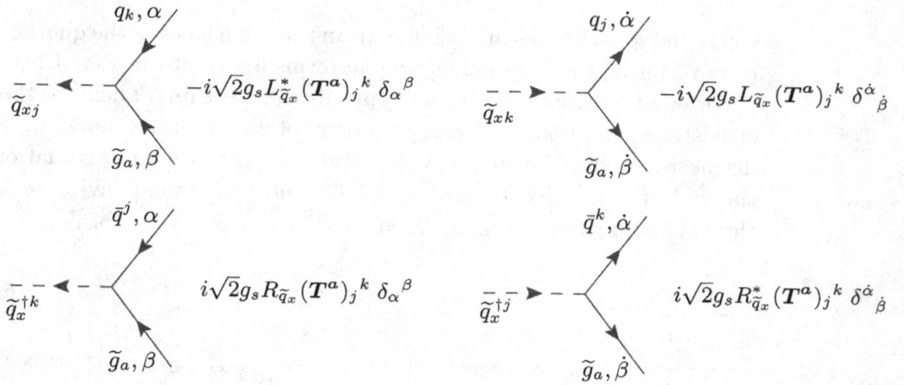

Fig. J.12 Fermionic Feynman rules for SUSY-QCD that involve the squarks in the mass-eigenstate basis labeled by $x = 1, 2$ and $q = u, d, c, s, t, b$, in the approximation where mixing is allowed only within a given flavor, as in eq. (J.39). Lowered (raised) indices j, k correspond to the fundamental (antifundamental) representation of SU(3)$_C$, and the index a labels the adjoint representation carried by the gluino.

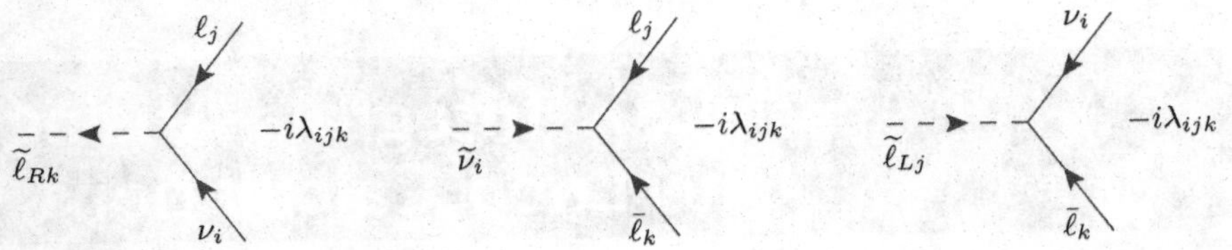

Fig. J.13 Feynman rules for the Yukawa couplings of two-component fermions due to the supersymmetric, R-parity-violating Yukawa Lagrangian $\mathscr{L}_{LL\bar{e}}$ [see eq. (16.183)].

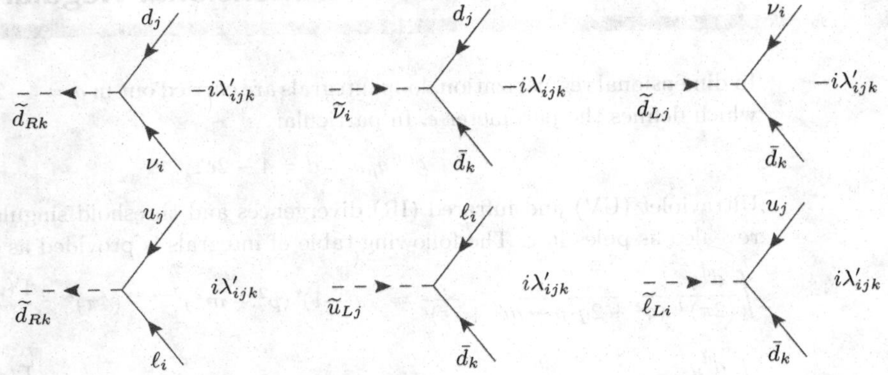

Fig. J.14 Feynman rules for the Yukawa couplings of two-component fermions for the supersymmetric, R-parity-violating Yukawa Lagrangian $\mathscr{L}_{LQ\bar{d}}$ [see eq. (16.184)].

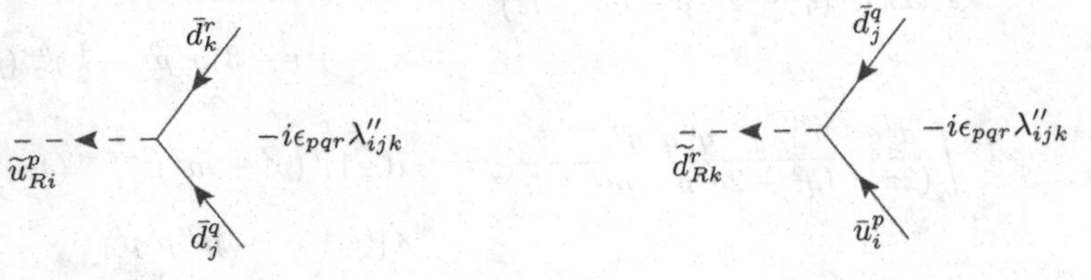

Fig. J.15 Feynman rules for the Yukawa couplings of two-component fermions due to the supersymmetric, R-parity-violating Yukawa Lagrangian $\mathscr{L}_{\bar{u}\bar{d}\bar{d}}$ [see eq. (16.186)].

J.6 Trilinear R-Parity-Violating Yukawa Couplings

In Figs. J.13, J.14, and J.15, we present the Feynman rules for the trilinear R-parity-violating Lagrangian interactions given in eqs. (16.183), (16.184), and (16.186). For each diagram there is another diagram with all arrows reversed and $\lambda_{ijk} \to \lambda^*_{ijk}$, $\lambda'_{ijk} \to \lambda'^*_{ijk}$, and $\lambda''_{ijk} \to \lambda''^*_{ijk}$, respectively.[2]

[2] Note that in this section we do not follow the flavor-index conventions of Appendix A.2. Instead, all flavor indices appear as subscripts and a superscript star (*) denotes complex conjugation.

Appendix K Integrals Arising in One-Loop Calculations

K.1 The Formulae of Dimensional Regularization

In dimensional regularization, loop integrals are carried out in $d = 4-2\epsilon$ dimensions, which defines the parameter ϵ. In particular,

$$g^{\mu\nu} g_{\mu\nu} = d = 4 - 2\epsilon \,. \tag{K.1}$$

Ultraviolet (UV) and infrared (IR) divergences and threshold singularities will be revealed as poles in ϵ. The following table of integrals is provided as a reference:

$$\int \frac{d^d q}{(2\pi)^d} \frac{1}{(q^2 + 2q{\cdot}p - m^2 + i\varepsilon)^r} = \quad i(-1)^r (p^2 + m^2)^{2-\epsilon-r} (4\pi)^{\epsilon-2} \frac{\Gamma(\epsilon + r - 2)}{\Gamma(r)} ,$$

$$\int \frac{d^d q}{(2\pi)^d} \frac{q^\mu}{(q^2 + 2q{\cdot}p - m^2 + i\varepsilon)^r} = -i(-1)^r (p^2 + m^2)^{2-\epsilon-r} (4\pi)^{\epsilon-2} \frac{\Gamma(\epsilon + r - 2)}{\Gamma(r)} p^\mu ,$$

$$\begin{aligned}\int \frac{d^d q}{(2\pi)^d} \frac{q^\mu q^\nu}{(q^2 + 2q{\cdot}p - m^2 + i\varepsilon)^r} = &\quad i(-1)^r (p^2 + m^2)^{2-\epsilon-r} (4\pi)^{\epsilon-2} \frac{\Gamma(\epsilon + r - 3)}{\Gamma(r)} \\ &\times \left[(\epsilon + r - 3) p^\mu p^\nu - \tfrac{1}{2} g^{\mu\nu} (p^2 + m^2) \right] ,\end{aligned}$$

$$\begin{aligned}\int \frac{d^d q}{(2\pi)^d} \frac{q^\mu q^\nu q^\alpha}{(q^2 + 2q{\cdot}p - m^2 + i\varepsilon)^r} = &-i(-1)^r (p^2 + m^2)^{2-\epsilon-r} (4\pi)^{\epsilon-2} \frac{\Gamma(\epsilon + r - 3)}{\Gamma(r)} \\ &\times \left[(\epsilon + r - 3) p^\mu p^\nu p^\alpha \right. \\ &\left. - \tfrac{1}{2} (g^{\mu\nu} p^\alpha + g^{\mu\alpha} p^\nu + g^{\nu\alpha} p^\mu)(p^2 + m^2) \right] ,\end{aligned}$$

$$\begin{aligned}\int \frac{d^d q}{(2\pi)^d} \frac{q^\mu q^\nu q^\alpha q^\beta}{(q^2 + 2q{\cdot}p - m^2 + i\varepsilon)^r} = &\quad i(-1)^r (p^2 + m^2)^{2-\epsilon-r} (4\pi)^{\epsilon-2} \frac{\Gamma(\epsilon + r - 4)}{\Gamma(r)} \\ &\times \Big\{ (\epsilon + r - 4) \Big[(\epsilon + r - 3) p^\mu p^\nu p^\alpha p^\beta \\ - \tfrac{1}{2} (g^{\mu\nu} p^\alpha p^\beta + g^{\mu\alpha} p^\nu p^\beta + g^{\mu\beta} p^\nu p^\alpha &+ g^{\nu\alpha} p^\mu p^\beta + g^{\nu\beta} p^\mu p^\alpha + g^{\alpha\beta} p^\mu p^\nu)(p^2 + m^2) \Big] \\ &+ \tfrac{1}{4} (g^{\mu\nu} g^{\alpha\beta} + g^{\mu\alpha} g^{\nu\beta} + g^{\mu\beta} g^{\nu\alpha})(p^2 + m^2)^2 \Big\} ,\end{aligned}$$

where ε (not to be confused with ϵ) is a positive infinitesimal constant. One can expand about $\epsilon = 0$ by using

$$\Gamma(-N + \epsilon) = \frac{(-1)^N}{N!} \left[\frac{1}{\epsilon} + \psi(N + 1) + \mathcal{O}(\epsilon) \right] , \tag{K.2}$$

where N is a nonnegative integer, $\psi(x) \equiv \Gamma'(x)/\Gamma(x)$ with $\Gamma'(x) \equiv d\Gamma(x)/dx$,

$$\psi(1) = -\gamma, \qquad \psi(N+1) = -\gamma + \sum_{k=1}^{N} \frac{1}{k}, \tag{K.3}$$

and $\gamma = -\Gamma'(1) = 0.5772156649\cdots$ is Euler's constant.

In addition, all scaleless integrals are *defined* by the dimensionless regularization procedure to be zero. For example, for *all* values of r,

$$\int \frac{d^d q}{(2\pi)^d} \frac{1}{(q^2)^r} \equiv 0. \tag{K.4}$$

In the special case of $r = 2$, eq. (K.4) is both UV and IR divergent when $d = 4$. Following Appendix C.4 of [B65], one can isolate the UV and IR divergent parts of eq. (K.4) with $r = 2$ by writing (see Exercise K.4)

$$\int \frac{d^d q}{(2\pi)^d} \frac{1}{(q^2)^2} = \int \frac{d^d q}{(2\pi)^d} \left\{ \frac{1}{q^2(q^2 - m^2)} - \frac{m^2}{(q^2)^2(q^2 - m^2)} \right\} = \frac{i}{16\pi^2} \left(\frac{1}{\epsilon_{\rm UV}} - \frac{1}{\epsilon_{\rm IR}} \right), \tag{K.5}$$

where $d = 4 - 2\epsilon_{\rm UV} = 4 - 2\epsilon_{\rm IR}$ when evaluating the above two integrals, respectively, and performing the expansion around $d = 4$. This properly keeps track of the UV divergences for the purposes of identifying counterterms and beta functions.

Finally, we record some of the Feynman parameter formulae:

$$\frac{1}{A^\alpha B^\beta} = \frac{\Gamma(\alpha+\beta)}{\Gamma(\alpha)\Gamma(\beta)} \int_0^1 dx \, \frac{x^{\alpha-1}(1-x)^{\beta-1}}{\left[xA + (1-x)B\right]^{\alpha+\beta}}, \tag{K.6}$$

$$\frac{1}{A^\alpha B^\beta C^\delta} = \frac{\Gamma(\alpha+\beta+\delta)}{\Gamma(\alpha)\Gamma(\beta)\Gamma(\delta)} \int_0^1 x \, dx \int_0^1 dy \, \frac{x^{\alpha+\beta-2}\, y^{\alpha-1}(1-x)^{\delta-1}(1-y)^{\beta-1}}{\left[xyA + x(1-y)B + (1-x)C\right]^{\alpha+\beta+\delta}}, \tag{K.7}$$

and more generally,

$$\frac{1}{A_1^{\alpha_1} A_2^{\alpha_2} \cdots A_N^{\alpha_N}} = \frac{\Gamma(\alpha_1 + \alpha_2 + \cdots + \alpha_N)}{\Gamma(\alpha_1)\Gamma(\alpha_2)\cdots\Gamma(\alpha_N)} \int_0^1 dx_1 \cdots \int_0^1 dx_N \, \delta\left(\sum_{j=1}^{N} x_j - 1 \right) \times \frac{x_1^{\alpha_1 - 1} x_2^{\alpha_2 - 1} \cdots x_N^{\alpha_N - 1}}{(x_1 A_1 + x_2 A_2 + \cdots + x_N A_N)^{\alpha_1 + \alpha_2 + \cdots + \alpha_N}}. \tag{K.8}$$

K.2 The Passarino–Veltman Loop Functions

We collect here the relevant integrals that arise in one-loop computations of one-point, two-point, and three-point Green functions, based on work that first appeared in Ref. [243, 244]. However, in contrast to the original presentations, the metric convention specified in eq. (A.17) is employed and a different normalization constant is chosen in defining the loop functions, following the conventions of Ref. [245].

The Passarino–Veltman loop functions, which arise when evaluating the one-loop contributions to one-point, two-point, and three-point Green functions (or the corresponding amplitudes of a physical process), are defined as follows:[1]

$$A_0(m^2) = -16\pi^2 i\mu^{2\epsilon} \int \frac{d^d q}{(2\pi)^d} \frac{1}{q^2 - m^2 + i\varepsilon}, \tag{K.9}$$

$$B_0(p^2; m_a^2, m_b^2); B^\mu; B^{\mu\nu} = -16\pi^2 i\mu^{2\epsilon} \int \frac{d^d q}{(2\pi)^d} \frac{1; q^\mu; q^\mu q^\nu}{D_B}, \tag{K.10}$$

$$C_0(p_1^2, p_2^2, p^2; m_a^2, m_b^2, m_c^2); C^\mu; C^{\mu\nu} = -16\pi^2 i\mu^{2\epsilon} \int \frac{d^d q}{(2\pi)^d} \frac{1; q^\mu; q^\mu q^\nu}{D_C}, \tag{K.11}$$

where the integrals are evaluated in $d = 4 - 2\epsilon$ dimensions,

$$D_B \equiv (q^2 - m_a^2 + i\varepsilon)[(q+p)^2 - m_b^2 + i\varepsilon], \tag{K.12}$$

$$D_C \equiv (q^2 - m_a^2 + i\varepsilon)[(q+p_1)^2 - m_b^2 + i\varepsilon][(q+p_1+p_2)^2 - m_c^2 + i\varepsilon], \tag{K.13}$$

and $p = -(p_1 + p_2)$. In eqs. (K.9)–(K.11) all external momenta are flowing into the diagrammatic representation of the Green function. By including the $\mu^{2\epsilon}$ factor in eqs. (K.9)–(K.11), where μ is the mass scale of dimensional regularization, one ensures that the arguments of all logarithms that arise in the evaluation of the above integrals are dimensionless.

The loop functions with explicit Lorentz indices can be decomposed in terms of Lorentz scalar functions as follows:

$$B^\mu = B_1 p^\mu, \tag{K.14}$$

$$B^{\mu\nu} = B_{21} p^\mu p^\nu + B_{22} g^{\mu\nu}, \tag{K.15}$$

$$C^\mu = C_{11} p_1^\mu + C_{12} p_2^\mu, \tag{K.16}$$

$$C^{\mu\nu} = C_{21} p_1^\mu p_1^\nu + C_{22} p_2^\mu p_2^\nu + C_{23}(p_1^\mu p_2^\nu + p_2^\mu p_1^\nu) + C_{24} g^{\mu\nu}, \tag{K.17}$$

where the suppressed arguments of B_1, the B_{ij}, and the C_{ij} are the same as the arguments of B_0 and C_0 appearing in eqs. (K.10) and (K.11), respectively. The derivatives of B-type integrals are also of interest and will be analyzed below.[2]

Among the integrals listed above, A_0, B_0, B_1, B_{21}, B_{22}, and C_{24} are divergent as $\epsilon \to 0$. The integrals C_0 and C_{ij} for $ij \neq 24$ are ultraviolet convergent and can be evaluated by setting $\epsilon = 0$ (assuming that no infrared divergences are present). It is convenient to introduce the quantity

$$\Delta \equiv (4\pi)^\epsilon\, \Gamma(\epsilon) = \frac{1}{\epsilon} - \gamma + \ln(4\pi) + \mathcal{O}(\epsilon), \tag{K.18}$$

after using eqs. (K.2) and (K.3). Using the results of Appendix K.1, one can evaluate A_0 and the B-type and C-type loop functions explicitly in $d = 4 - 2\epsilon$ dimensions.

[1] All squared masses, m^2, m_a^2, m_b^2, and m_c^2 are nonnegative real parameters. The four-momenta p_1, p_2, and p correspond to either on-shell or off-shell particles, depending on the application.

[2] One can also consider B-type and C-type tensor integrals with more than two Lorentz indices, the derivatives of C-type integrals, and the D-type and E-type Passarino–Veltman loop functions that arise when evaluating the respective one-loop contributions to four-point and five-point Green functions. These loop integrals will not be treated here; for further details and references to the literature, the reader may consult [B82, B150] and Refs. [33, 244, 246].

Since the product $\mu^{2\epsilon}\Delta$ always appears when evaluating the divergent one-loop integrals, it is convenient to introduce the $\overline{\text{MS}}$ renormalization scale Q:

$$\mu^{2\epsilon}\Delta = \frac{1}{\epsilon} + \ln Q^2 + \mathcal{O}(\epsilon)\,, \quad \text{where } Q^2 \equiv 4\pi e^{-\gamma}\mu^2 . \tag{K.19}$$

The integral expressions for A_0 and the B-type loop functions are then obtained:

$$A_0(m^2) = m^2\left[\frac{1}{\epsilon} + 1 - \ln\left(\frac{m^2}{Q^2}\right)\right], \tag{K.20}$$

$$B_0(p^2; m_a^2, m_b^2) = \frac{1}{\epsilon} - \int_0^1 \ln\left(\frac{p^2x^2 - (p^2 + m_a^2 - m_b^2)x + m_a^2 - i\varepsilon}{Q^2}\right) dx\,, \tag{K.21}$$

$$B_1(p^2; m_a^2, m_b^2) = -\frac{1}{2\epsilon} + \int_0^1 \ln\left(\frac{p^2x^2 - (p^2 + m_a^2 - m_b^2)x + m_a^2 - i\varepsilon}{Q^2}\right) x\, dx\,, \tag{K.22}$$

$$B_{21}(p^2; m_a^2, m_b^2) = \frac{1}{3\epsilon} - \int_0^1 \ln\left(\frac{p^2x^2 - (p^2 + m_a^2 - m_b^2)x + m_a^2 - i\varepsilon}{Q^2}\right) x^2\, dx\,, \tag{K.23}$$

$$\begin{aligned} B_{22}(p^2; m_a^2, m_b^2) &= \frac{1}{4}\left(\frac{1}{\epsilon} + 1\right)\left(m_a^2 + m_b^2 - \tfrac{1}{3}p^2\right) \\ &\quad - \frac{1}{2}\int_0^1 \left[p^2x^2 - (p^2 + m_a^2 - m_b^2)x + m_a^2\right] \\ &\qquad \times \ln\left(\frac{p^2x^2 - (p^2 + m_a^2 - m_b^2)x + m_a^2 - i\varepsilon}{Q^2}\right) dx\,, \end{aligned} \tag{K.24}$$

where terms of $\mathcal{O}(\epsilon)$ have been dropped. Starting from eq. (K.9), it follows that $A_0(0) = 0$ in light of eq. (K.4). Limiting cases of the B-type loop integrals are given in Exercise K.2.

It is possible to evaluate B_1 in terms of A_0 and B_0 by noting that

$$p^2 B_1(p^2; m_a^2, m_b^2) = p_\mu B^\mu(p^2; m_a^2, m_b^2) = -16\pi^2 i\,\mu^{2\epsilon}\int \frac{d^dq}{(2\pi)^d}\frac{p\cdot q}{D_B}\,, \tag{K.25}$$

where D_B is given in eq. (K.12). To simplify this result, we shall employ the method of partial fractions by making use of following algebraic identity:

$$p\cdot q = \tfrac{1}{2}\left[(q+p)^2 - q^2 - p^2\right] = \tfrac{1}{2}\left[(q+p)^2 - m_b^2 - (q^2 - m_a^2) - p^2 + m_b^2 - m_a^2\right]. \tag{K.26}$$

Plugging this result into eq. (K.25) yields

$$\begin{aligned} p^2 B_1(p^2; m_a^2, m_b^2) = 8\pi^2 i\,\mu^{2\epsilon}\Bigg\{&\int \frac{d^dq}{(2\pi)^d}\left(\frac{1}{(q+p)^2 - m_b^2 + i\varepsilon} - \frac{1}{q^2 - m_a^2 + i\varepsilon}\right) \\ &+ (p^2 + m_a^2 - m_b^2)\mu^{2\epsilon}\int \frac{d^dq}{(2\pi)^d}\frac{1}{D_B}\Bigg\}. \end{aligned} \tag{K.27}$$

The end result is

$$p^2 B_1(p^2;m_a^2,m_b^2) = \tfrac{1}{2}\Big[A_0(m_a^2) - A_0(m_b^2) - (p^2+m_a^2-m_b^2)B_0(p^2;m_a^2,m_b^2)\Big]\,. \tag{K.28}$$

The case of $p^2=0$ is treated in Exercise K.2.

Similarly, B_{21} and B_{22} can be expressed in terms of A_0, B_0, and B_1. Starting from eq. (K.15), it follows that

$$p^\mu B_{\mu\nu} = p_\nu(p^2 B_{21} + B_{22})\,, \qquad g^{\mu\nu}B_{\mu\nu} = p^2 B_{21} + dB_{22}\,, \tag{K.29}$$

where the arguments of the B-type loop functions are $(p^2;m_a^2,m_b^2)$. Following the steps used in the derivation of eq. (K.28), we obtain two equations,

$$p^2 B_{21} + B_{22} = \tfrac{1}{2}A_0(m_b^2) - \tfrac{1}{2}(p^2+m_a^2-m_b^2)B_1\,, \tag{K.30}$$

$$p^2 B_{21} + dB_{22} = A_0(m_b^2) + m_a^2 B_0\,. \tag{K.31}$$

Solving for B_{21} and B_{22} yields

$$(d-1)p^2 B_{21} = \left(\tfrac{1}{2}d-1\right)A_0(m_b^2) - \tfrac{1}{2}d(p^2+m_a^2-m_b^2)B_1 - m_a^2 B_0\,, \tag{K.32}$$

$$(d-1)B_{22} = \tfrac{1}{2}A_0(m_b^2) + m_a^2 B_0 + \tfrac{1}{2}(p^2+m_a^2-m_b^2)B_1\,. \tag{K.33}$$

After expanding about $\epsilon=0$ and dropping terms of $\mathcal{O}(\epsilon)$, it follows that

$$dB_{21} = 4B_{21} - \tfrac{2}{3}\,, \qquad dB_{22} = 4B_{22} - \tfrac{1}{2}\left(m_a^2+m_b^2-\tfrac{1}{3}p^2\right), \tag{K.34}$$

$$dB_1 = 4B_1 + 1\,, \qquad \tfrac{1}{2}dA_0(m_b^2) = 2A_0(m_b^2) - m_b^2\,. \tag{K.35}$$

We end up with

$$p^2 B_{21}(p^2;m_a^2,m_b^2) = \frac{1}{3}\bigg\{A_0(m_b^2) - m_a^2 B_0(p^2;m_a^2,m_b^2) - \tfrac{1}{2}(m_a^2+m_b^2) + \frac{p^2}{6} - 2(p^2+m_a^2-m_b^2)B_1(p^2;m_a^2,m_b^2)\bigg\}, \tag{K.36}$$

$$B_{22}(p^2;m_a^2,m_b^2) = \frac{1}{6}\bigg\{A_0(m_b^2) + 2m_a^2 B_0(p^2;m_a^2,m_b^2) + m_a^2 + m_b^2 - \frac{p^2}{3} + (p^2+m_a^2-m_b^2)B_1(p^2;m_a^2,m_b^2)\bigg\}. \tag{K.37}$$

The symmetry properties of B_0, B_1, B_{21}, and B_{22} under an interchange of the arguments $m_a^2 \leftrightarrow m_b^2$ are noteworthy:

$$B_0(p^2;m_a^2,m_b^2) = B_0(p^2;m_b^2,m_a^2)\,, \tag{K.38}$$

$$B_1(p^2;m_a^2,m_b^2) = -B_1(p^2;m_b^2,m_a^2) - B_0(p^2;m_b^2,m_a^2)\,, \tag{K.39}$$

$$B_1(p^2;m_a^2,m_b^2) = B_{21}(p^2;m_b^2,m_a^2) + 2B_1(p^2;m_b^2,m_a^2) + B_0(p^2;m_b^2,m_a^2)\,, \tag{K.40}$$

$$B_{22}(p^2;m_a^2,m_b^2) = B_{22}(p^2;m_b^2,m_a^2)\,. \tag{K.41}$$

Note that eqs. (K.39) and (K.40) imply that

$$B_1(p^2;m_a^2,m_b^2) + B_{21}(p^2;m_a^2,m_b^2) = B_1(p^2;m_b^2,m_a^2) + B_{21}(p^2;m_b^2,m_a^2)\,. \tag{K.42}$$

In the limit of $m_a^2 = m_b^2$, eq. (K.39) yields

$$B_1(p^2; m^2, m^2) = -\tfrac{1}{2} B_0(p^2; m^2, m^2)\,. \tag{K.43}$$

One can now perform the integration in eq. (K.21). The end result is

$$B_0(p^2; m_a^2, m_b^2) = \frac{1}{\epsilon} + B(p^2; m_a^2, m_b^2) + \mathcal{O}(\epsilon)\,, \tag{K.44}$$

where the function B is explicitly evaluated below in five distinct cases. The resulting expressions employ the kinematic triangle function $\lambda(p^2, p_1^2, p_2^2)$ defined in eq. (D.1).

<u>Case 1</u>: $p^2 > (m_a + m_b)^2$

$$B(p^2; m_a^2, m_b^2) = \ln\left(\frac{Q^2}{m_b^2}\right) + 2 + \left(\frac{p^2 + m_a^2 - m_b^2}{2p^2}\right)\ln\left(\frac{m_b^2}{m_a^2}\right) \tag{K.45}$$

$$-\frac{\lambda^{1/2}(p^2, m_a^2, m_b^2)}{p^2}\left[\ln\left(\frac{\left[p^2-(m_a-m_b)^2\right]^{1/2} + \left[p^2-(m_a+m_b)^2\right]^{1/2}}{\left[p^2-(m_a-m_b)^2\right]^{1/2} - \left[p^2-(m_a+m_b)^2\right]^{1/2}}\right) - i\pi\right].$$

<u>Case 2</u>: $p^2 < (m_a - m_b)^2$ and $p^2 \neq 0$

$$B(p^2; m_a^2, m_b^2) = \ln\left(\frac{Q^2}{m_b^2}\right) + 2 + \left(\frac{p^2 + m_a^2 - m_b^2}{2p^2}\right)\ln\left(\frac{m_b^2}{m_a^2}\right) \tag{K.46}$$

$$-\frac{\lambda^{1/2}(p^2, m_a^2, m_b^2)}{p^2}\ln\left(\frac{\left[(m_a+m_b)^2-p^2\right]^{1/2} + \left[(m_a-m_b)^2-p^2\right]^{1/2}}{\left[(m_a+m_b)^2-p^2\right]^{1/2} - \left[(m_a-m_b)^2-p^2\right]^{1/2}}\right).$$

<u>Case 3</u>: $(m_a - m_b)^2 < p^2 < (m_a + m_b)^2$

$$B(p^2; m_a^2, m_b^2) = \ln\left(\frac{Q^2}{m_b^2}\right) + 2 + \left(\frac{p^2 + m_a^2 - m_b^2}{2p^2}\right)\ln\left(\frac{m_b^2}{m_a^2}\right)$$

$$-\frac{2\left[-\lambda(p^2, m_a^2, m_b^2)\right]^{1/2}}{p^2}\arctan\left(\frac{\sqrt{p^2-(m_a-m_b)^2}}{\sqrt{(m_a+m_b)^2-p^2}}\right), \tag{K.47}$$

where the principal value of the real arctangent function satisfies $|\arctan x| \leq \tfrac{1}{2}\pi$.

<u>Case 4</u>: $p^2 = 0$ and $m_a \neq m_b$

$$B(0; m_a^2, m_b^2) = \frac{1}{m_a^2 - m_b^2}\left[m_a^2 \ln\left(\frac{Q^2}{m_a^2}\right) - m_b^2 \ln\left(\frac{Q^2}{m_b^2}\right)\right] + 1\,. \tag{K.48}$$

<u>Case 5</u>: $p^2 = 0$ and $m \equiv m_a = m_b$

$$B(0; m^2, m^2) = \ln\left(\frac{Q^2}{m^2}\right)\,. \tag{K.49}$$

Sometimes a loop integral arises in which one or more of the propagator denominators are raised to a power. For example, taking the derivative with respect to m^2

of eq. (K.9) yields

$$\frac{\partial A_0(m^2)}{\partial m^2} = -16\pi^2 i\mu^{2\epsilon}\int \frac{d^dq}{(2\pi)^d}\frac{1}{(q^2-m_a^2+i\varepsilon)^2}$$
$$= (4\pi)^\epsilon\Gamma(\epsilon)\left(\frac{m^2}{\mu^2}\right)^{-\epsilon} = \frac{(1-\epsilon)A_0(m^2)}{m^2} = \frac{A_0(m^2)}{m^2} - 1 + \mathcal{O}(\epsilon). \quad \text{(K.50)}$$

For a B-type loop integral, consider

$$\frac{\partial B_0}{\partial m_a^2}(p^2;m_a^2,m_b^2) = -16\pi^2 i\mu^{2\epsilon}\int \frac{d^dq}{(2\pi)^d}\frac{1}{(q^2-m_a^2+i\varepsilon)^2[(q+p)^2-m_b^2+i\varepsilon]}$$
$$= -\frac{(4\pi)^\epsilon\,\Gamma(1+\epsilon)}{\mu^2}\int_0^1\left(\frac{p^2x^2-(p^2+m_a^2-m_b^2)x+m_a^2-i\varepsilon}{\mu^2}\right)^{-1-\epsilon}(1-x)dx$$
$$= -\int_0^1 \frac{(1-x)dx}{p^2x^2-(p^2+m_a^2-m_b^2)x+m_a^2-i\varepsilon} + \mathcal{O}(\epsilon)\,. \quad \text{(K.51)}$$

The presence of the factor $\Gamma(1+\epsilon)$ indicates that the above integral is ultraviolet finite. However, in some cases the integration over the Feynman parameter x will expose the presence of an infrared divergence or a threshold singularity. In particular, the integral on the last line of eq. (K.51) diverges if $m_a^2 = 0$ [infrared divergence] or if $p^2 = (m_a + m_b)^2$ [threshold singularity]. In both cases, one must take one step back and carry out the integration using the second line of eq. (K.51). After the integration over x, the existence of the infrared divergence or the threshold singularity is then revealed by the ϵ^{-1} behavior of the result as $\epsilon \to 0$.

Derivatives of B_1, B_{21}, and B_{22} with respect to m_a^2 can be similarly defined. It is noteworthy that the derivative of a B-type loop integral with respect to m_a^2 is a special type of C-type loop integral. In particular, eqs. (K.10) and (K.11) yield

$$\frac{\partial B_0}{\partial m_a^2}(p^2;m_a^2,m_b^2) = C_0(0,p^2,p^2;m_a^2,m_a^2,m_b^2)\,, \quad \text{(K.52)}$$
$$\frac{\partial B_1}{\partial m_a^2}(p^2;m_a^2,m_b^2) = C_{11}(0,p^2,p^2;m_a^2,m_a^2,m_b^2)\,, \quad \text{(K.53)}$$
$$\frac{\partial B_{21}}{\partial m_a^2}(p^2;m_a^2,m_b^2) = C_{22}(0,p^2,p^2;m_a^2,m_a^2,m_b^2)\,, \quad \text{(K.54)}$$
$$\frac{\partial B_{22}}{\partial m_a^2}(p^2;m_a^2,m_b^2) = C_{24}(0,p^2,p^2;m_a^2,m_a^2,m_b^2)\,. \quad \text{(K.55)}$$

In the computation of wave function renormalization in the one-loop approximation, one encounters derivatives of B-type loop integrals with respect to p^2. For example, taking the derivative with respect to p^2 of eq. (K.93) yields

$$B_0'(p^2;m_a^2,m_b^2) \equiv \frac{\partial}{\partial p^2}B_0(p^2;m_a^2,m_b^2)$$
$$= \frac{(4\pi)^\epsilon\,\Gamma(1+\epsilon)}{\mu^2}\int_0^1\left(\frac{p^2x^2-(p^2+m_a^2-m_b^2)x+m_a^2-i\varepsilon}{\mu^2}\right)^{-1-\epsilon}x(1-x)dx$$
$$= \int_0^1 \frac{x(1-x)dx}{p^2x^2-(p^2+m_a^2-m_b^2)x+m_a^2-i\varepsilon} + \mathcal{O}(\epsilon)\,, \quad \text{(K.56)}$$

where B' indicates differentiation with respect to the first argument of the B-type loop function. The integral in eq. (K.56) is ultraviolet finite. If $p = (m_a+m_b)^2$ where $m_a,\ m_b \neq 0$, then the integral appearing in the last line of eq. (K.56) diverges [due to the vanishing of the denominator at $x = m_a/(m_a + m_b)$], which indicates the presence of a threshold singularity. As shown in Exercise K.3, one can employ the middle line of eq. (K.56) to regularize the divergence.

The symmetry properties given by eqs. (K.38)–(K.42) are also respected by the derivatives of the B-type loop functions with respect to p^2. For example,

$$B_0'(p^2; m_a^2, m_b^2) = B_0'(p^2; m_b^2, m_a^2)\,. \tag{K.57}$$

Derivatives of B_1, B_{21}, and B_{22} with respect to p^2 can be similarly obtained:

$$-p^2 B_1'(p^2; m_a^2, m_b^2) = B_1 + \tfrac{1}{2}B_0 + \tfrac{1}{2}(p^2 + m_a^2 - m_b^2)B_0'\,, \tag{K.58}$$

$$-p^2 B_{21}'(p^2; m_a^2, m_b^2) = \tfrac{1}{2}(p^2 + m_a^2 - m_b^2)B_1' - B_1 - \tfrac{3}{2}B_{21}\,, \tag{K.59}$$

$$B_{22}'(p^2; m_a^2, m_b^2) = \tfrac{1}{2}(B_1 + B_{21})\,, \tag{K.60}$$

where the suppressed arguments of the loop functions above are $(p^2; m_a^2, m_b^2)$. Equation (K.58) is derived by taking the derivative with respect to p^2 of eq. (K.28). The integral representations given in eqs. (K.94)–(K.96) were employed in obtaining eq. (K.60). Finally, eq. (K.59) is derived by taking the derivative with respect to p^2 of eq. (K.30) and then employing eq. (K.60) to eliminate B_{22}'.

We can relate the derivatives of the B-type loop integrals with respect to m_a^2 to B_0, B_1, B_0', and B_1' as follows. First, in light of the integral representations given in eqs. (K.93)–(K.96), the following three relations are easily obtained:

$$\frac{\partial B_1}{\partial m_a^2}(p^2; m_a^2, m_b^2) = B_0'(p^2; m_a^2, m_b^2)\,, \tag{K.61}$$

$$\frac{\partial B_{21}}{\partial m_a^2}(p^2; m_a^2, m_b^2) = B_1'(p^2; m_a^2, m_b^2)\,, \tag{K.62}$$

$$\frac{\partial B_{22}}{\partial m_a^2}(p^2; m_a^2, m_b^2) = \tfrac{1}{2}\left[B_0(p^2; m_a^2, m_b^2) + B_1(p^2; m_a^2, m_b^2)\right]. \tag{K.63}$$

We now take the derivative of eqs. (K.28), (K.30), and (K.31) with respect to m_a^2 and employ eqs. (K.61) and (K.63) to eliminate the derivatives of the B-type loop integrals with respect to m_a^2. One of the three equations thusly obtained reproduces eq. (K.58), whereas the other two equations are

$$p^2 B_0'(p^2; m_a^2, m_b^2) = \frac{1}{2}\left[\frac{\partial A_0(m_a^2)}{\partial m_a^2} - B_0 - (p^2 + m_a^2 - m_b^2)\frac{\partial B_0}{\partial m_a^2}\right], \tag{K.64}$$

$$p^2 B_1'(p^2; m_a^2, m_b^2) = (1 - \tfrac{1}{2}d)B_0 - \tfrac{1}{2}dB_1 + m_a^2\frac{\partial B_0}{\partial m_a^2}\,, \tag{K.65}$$

where the suppressed arguments of the loop functions above are $(p^2; m_a^2, m_b^2)$. Adding eqs. (K.58) and (K.65) yields

$$2m_a^2\frac{\partial B_0}{\partial m_a^2}(p^2; m_a^2, m_b^2) = (d-3)B_0 + (d-2)B_1 - \tfrac{1}{2}(p^2 + m_a^2 - m_b^2)B_0'\,. \tag{K.66}$$

Using eqs. (K.64) and (K.66) to solve for B_0' and $\partial B_0/\partial m_a^2$, the end result is

$$\lambda(p^2,m_a^2,m_b^2)B_0'(p^2;m_a^2,m_b^2) = (p^2+m_a^2-m_b^2)\big[(d-3)B_0+(d-2)B_1\big] - 2m_a^2\left(\frac{\partial A_0(m_a^2)}{\partial m_a^2} - B_0\right), \tag{K.67}$$

$$\lambda(p^2,m_a^2,m_b^2)\frac{\partial B_0}{\partial m_a^2}(p^2;m_a^2,m_b^2) = (p^2+m_a^2-m_b^2)\left(\frac{\partial A_0(m_a^2)}{\partial m_a^2} - B_0\right) - 2p^2\big[(d-3)B_0+(d-2)B_1\big]. \tag{K.68}$$

Next, we present integral expressions for C_0 and the C_{ij}.

$$C_0(p_1^2,p_2^2,p^2;m_a^2,m_b^2,m_c^2) = -\int_0^1 dx\int_0^x \frac{dy}{D-i\varepsilon}, \tag{K.69}$$

$$C_{11}(p_1^2,p_2^2,p^2;m_a^2,m_b^2,m_c^2) = \int_0^1 x\,dx\int_0^x \frac{dy}{D-i\varepsilon}, \tag{K.70}$$

$$C_{12}(p_1^2,p_2^2,p^2;m_a^2,m_b^2,m_c^2) = \int_0^1 dx\int_0^x \frac{(x-y)dy}{D-i\varepsilon}, \tag{K.71}$$

$$C_{21}(p_1^2,p_2^2,p^2;m_a^2,m_b^2,m_c^2) = -\int_0^1 x^2\,dx\int_0^x \frac{dy}{D-i\varepsilon}, \tag{K.72}$$

$$C_{22}(p_1^2,p_2^2,p^2;m_a^2,m_b^2,m_c^2) = -\int_0^1 dx\int_0^x \frac{(x-y)^2dy}{D-i\varepsilon}, \tag{K.73}$$

$$C_{23}(p_1^2,p_2^2,p^2;m_a^2,m_b^2,m_c^2) = -\int_0^1 x\,dx\int_0^x \frac{(x-y)dy}{D-i\varepsilon}, \tag{K.74}$$

$$C_{24}(p_1^2,p_2^2,p^2;m_a^2,m_b^2,m_c^2) = \frac{1}{4\epsilon} - \frac{1}{2}\int_0^1 dx\int_0^x dy\,\ln\left(\frac{D-i\varepsilon}{Q^2}\right), \tag{K.75}$$

after dropping terms of $\mathcal{O}(\epsilon)$, where

$$D \equiv p^2x^2 + p_2^2y^2 + (p_1^2-p_2^2-p^2)xy + (m_c^2-m_a^2-p^2)x + (m_b^2-m_c^2+p^2-p_1^2)y + m_a^2. \tag{K.76}$$

In addition, the following expression for C_{24} is useful and worthy of display:

$$C_{24}(p_1^2,p_2^2,p^2;m_a^2,m_b^2,m_c^2) = \tfrac{1}{4}\big[B_0(p_2^2;m_b^2,m_c^2) + (p_1^2+m_a^2-m_b^2)C_{11} + (p^2-p_1^2+m_b^2-m_c^2)C_{12} + 2m_a^2C_0 + 1\big], \tag{K.77}$$

where the suppressed arguments of C_0, C_{11}, and C_{12} are the same as those of C_{24}. Indeed, it is possible to express all the C_{ij} (and their derivatives) in terms of A_0, B_0, and C_0 following the same partial fractioning strategy that was used to obtain B_1 in terms of B_0 and A_0. The explicit relations for the C_{ij} are given in Exercise K.7.

The integral given by eq. (K.69) can be explicitly evaluated. The resulting expression, which involves logarithms and dilogarithms, is given in Refs. [33, 243], although it is not particularly illuminating in the most general case. However, one can derive a useful set of expressions in the limit of $p_1^2 = p_2^2 = p^2 = 0$. For example,

$$C_0(0,0,0;m_a^2,m_b^2,m_c^2) = \frac{1}{m_b^2-m_c^2}\left[B_0(0;m_a^2,m_b^2) - B_0(0;m_a^2,m_c^2)\right]$$
$$= \frac{m_a^2 m_b^2 \ln\left(\frac{m_a^2}{m_b^2}\right) + m_b^2 m_c^2 \ln\left(\frac{m_b^2}{m_c^2}\right) + m_c^2 m_a^2 \ln\left(\frac{m_c^2}{m_a^2}\right)}{(m_a^2-m_b^2)(m_b^2-m_c^2)(m_c^2-m_a^2)}\,, \tag{K.78}$$

$$C_{24}(0,0,0;m_a^2,m_b^2,m_c^2) = \tfrac{1}{4}\left[B_0(0;m_b^2,m_c^2) + m_a^2 C_0(0,0,0;m_a^2,m_b^2,m_c^2) + \tfrac{1}{2}\right]. \tag{K.79}$$

If two or three of the masses are degenerate, then it follows that

$$C_0(0,0,0;m^2,m^2,m_c^2) = -\frac{1}{m^2-m_c^2}\left[1 + \frac{m_c^2}{m^2-m_c^2}\ln\left(\frac{m_c^2}{m^2}\right)\right], \tag{K.80}$$

$$C_0(0,0,0;m^2,m^2,m^2) = -\frac{1}{2m^2}\,. \tag{K.81}$$

One other limiting case is noteworthy. The following loop function arises in the calculation of the one-loop amplitude for the decay of a neutral Higgs boson to $Z\gamma$:

$$C_0(p_1^2,0,p^2;m^2,m^2,m^2) = \frac{1}{p^2-p_1^2}\int_0^1 \frac{dx}{x}\ln\left(\frac{m^2-x(1-x)p^2-i\varepsilon}{m^2-x(1-x)p_1^2-i\varepsilon}\right)$$
$$= \frac{1}{p^2-p_1^2}\left[G(p^2/m^2) - G(p_1^2/m^2)\right]. \tag{K.82}$$

The function $G(z)$ can be explicitly evaluated:

$$G(z) \equiv \int_0^1 \frac{dx}{x}\ln\left[1-zx(1-x)-i\varepsilon\right]$$
$$= \begin{cases} 2\ln^2\left(\frac{\sqrt{-z}}{2}+\sqrt{1-\frac{z}{4}}\right), & \text{for } z<0\,,\\ -2\left[\arcsin\left(\tfrac{1}{2}\sqrt{z}\right)\right]^2, & \text{for } 0\le z\le 4\,,\\ -2\left[\frac{\pi}{2}+i\ln\left(\frac{\sqrt{z}}{2}+\sqrt{\frac{z}{4}-1}\right)\right]^2, & \text{for } z>4\,, \end{cases} \tag{K.83}$$

where $0 \le \arcsin(\frac{1}{2}\sqrt{z}) \le \frac{1}{2}\pi$ (for $0 \le z \le 4$) employs the principal range of the real arcsine function. In the limit of $p_1^2 = p^2$, one may apply eq. (K.52) to obtain

$$C_0(p^2,0,p^2;m^2,m^2,m^2) = \frac{\partial B_0}{\partial m_a^2}(p^2;m_a^2,m_b^2)\bigg|_{m_b^2=m_a^2} = \frac{1}{m^2}\left(\frac{\partial G}{\partial z}\right)_{z=p^2/m^2}$$
$$= \frac{1}{p^2-4m^2}\left[\frac{A_0(m^2)}{m^2} - B_0(p^2;m^2,m^2)+1\right], \tag{K.84}$$

after using eqs. (K.43) and (K.68) and employing the following substitutions that are valid when terms of $\mathcal{O}(\epsilon)$ are neglected:

$$dB_0 = 4B_0 - 2\,, \qquad dB_1 = 4B_1 + 1\,, \qquad \frac{\partial A_0(m_a^2)}{\partial m_a^2} = \frac{A_0(m_a^2)}{m_a^2} - 1\,. \tag{K.85}$$

In the limit of $p^2 = p_1^2 = 0$, we recover eq. (K.81).

Finally, we record a number of useful symmetry properties of the C-functions when their arguments are permuted. First,

$$\begin{aligned} C_0\left(p_1^2, p_2^2, p^2; m_a^2, m_b^2, m_c^2\right) &= C_0\left(p_2^2, p_1^2, p^2; m_c^2, m_b^2, m_a^2\right) \\ &= C_0\left(p^2, p_1^2, p_2^2; m_c^2, m_a^2, m_b^2\right) = C_0\left(p_1^2, p^2, p_2^2; m_b^2, m_a^2, m_c^2\right) \\ &= C_0\left(p_2^2, p^2, p_1^2; m_b^2, m_c^2, m_a^2\right) = C_0\left(p^2, p_2^2, p_1^2; m_a^2, m_c^2, m_b^2\right). \end{aligned} \tag{K.86}$$

The same symmetry properties are also satisfied by C_{24}. For completeness, we list below the symmetry properties of the other C_{ij} functions.

$$\begin{aligned} C_{11}\left(p_1^2, p_2^2, p^2; m_a^2, m_b^2, m_c^2\right) &= \left[-C_{12} - C_0\right]\left(p_2^2, p_1^2, p^2; m_c^2, m_b^2, m_a^2\right) \\ &= \left[-C_{11} + C_{12} - C_0\right]\left(p^2, p_1^2, p_2^2; m_c^2, m_a^2, m_b^2\right) \\ &= \left[-C_{11} + C_{12} - C_0\right]\left(p_1^2, p^2, p_2^2; m_b^2, m_a^2, m_c^2\right) \\ &= \left[-C_{12} - C_0\right]\left(p_2^2, p^2, p_1^2; m_b^2, m_c^2, m_a^2\right) \\ &= C_{11}\left(p^2, p_2^2, p_1^2; m_a^2, m_c^2, m_b^2\right). \end{aligned} \tag{K.87}$$

$$\begin{aligned} C_{12}\left(p_1^2, p_2^2, p^2; m_a^2, m_b^2, m_c^2\right) &= \left[-C_{11} - C_0\right]\left(p_2^2, p_1^2, p^2; m_c^2, m_b^2, m_a^2\right) \\ &= \left[-C_{11} - C_0\right]\left(p^2, p_1^2, p_2^2; m_c^2, m_a^2, m_b^2\right) \\ &= C_{12}\left(p_1^2, p^2, p_2^2; m_b^2, m_a^2, m_c^2\right) \\ &= \left[C_{11} - C_{12}\right]\left(p_2^2, p^2, p_1^2; m_b^2, m_c^2, m_a^2\right) \\ &= \left[C_{11} - C_{12}\right]\left(p^2, p_2^2, p_1^2; m_a^2, m_c^2, m_b^2\right). \end{aligned} \tag{K.88}$$

$$\begin{aligned} C_{21}\left(p_1^2, p_2^2, p^2; m_a^2, m_b^2, m_c^2\right) &= \left[C_{22} + 2C_{12} + C_0\right]\left(p_2^2, p_1^2, p^2; m_c^2, m_b^2, m_a^2\right) \\ &= \left[C_{21} + C_{22} - 2C_{23} + 2C_{11} - 2C_{12} + C_0\right]\left(p^2, p_1^2, p_2^2; m_c^2, m_a^2, m_b^2\right) \\ &= \left[C_{21} + C_{22} - 2C_{23} + 2C_{11} - 2C_{12} + C_0\right]\left(p_1^2, p^2, p_2^2; m_b^2, m_a^2, m_c^2\right) \\ &= \left[C_{22} + 2C_{12} + C_0\right]\left(p_2^2, p^2, p_1^2; m_b^2, m_c^2, m_a^2\right) \\ &= C_{21}\left(p^2, p_2^2, p_1^2; m_a^2, m_c^2, m_b^2\right). \end{aligned} \tag{K.89}$$

$$\begin{aligned} C_{22}\left(p_1^2, p_2^2, p^2; m_a^2, m_b^2, m_c^2\right) &= \left[C_{21} + 2C_{11} + C_0\right]\left(p_2^2, p_1^2, p^2; m_c^2, m_b^2, m_a^2\right) \\ &= \left[C_{21} + 2C_{11} + C_0\right]\left(p^2, p_1^2, p_2^2; m_c^2, m_a^2, m_b^2\right) \\ &= C_{22}\left(p_1^2, p^2, p_2^2; m_b^2, m_a^2, m_c^2\right) \\ &= \left[C_{21} + C_{22} - 2C_{23}\right]\left(p_2^2, p^2, p_1^2; m_b^2, m_c^2, m_a^2\right) \\ &= \left[C_{21} + C_{22} - 2C_{23}\right]\left(p^2, p_2^2, p_1^2; m_a^2, m_c^2, m_b^2\right). \end{aligned} \tag{K.90}$$

$$\begin{aligned} C_{23}\left(p_1^2, p_2^2, p^2; m_a^2, m_b^2, m_c^2\right) &= \left[C_{23} + C_{12} + C_{11} + C_0\right]\left(p_2^2, p_1^2, p^2; m_c^2, m_b^2, m_a^2\right) \\ &= \left[C_{21} - C_{23} + 2C_{11} - C_{12} + C_0\right]\left(p^2, p_1^2, p_2^2; m_c^2, m_a^2, m_b^2\right) \\ &= \left[C_{22} - C_{23} - C_{12}\right]\left(p_1^2, p^2, p_2^2; m_b^2, m_a^2, m_c^2\right) \\ &= \left[C_{22} - C_{23} - C_{11} + C_{12}\right]\left(p_2^2, p^2, p_1^2; m_b^2, m_c^2, m_a^2\right) \\ &= \left[C_{21} - C_{23}\right]\left(p^2, p_2^2, p_1^2; m_a^2, m_c^2, m_b^2\right). \end{aligned} \tag{K.91}$$

Exercises

K.1 Derive the following exact expressions for A-type and B-type loop integrals:

$$A_0(m^2) = \frac{(4\pi)^\epsilon\,\Gamma(\epsilon) m^2}{1-\epsilon}\left(\frac{m^2}{\mu^2}\right)^{-\epsilon}, \tag{K.92}$$

$$B_0(p^2; m_a^2, m_b^2) = (4\pi)^\epsilon\,\Gamma(\epsilon)\int_0^1 \left(\frac{p^2x^2 - (p^2+m_a^2-m_b^2)x + m_a^2 - i\varepsilon}{\mu^2}\right)^{-\epsilon} dx\,, \tag{K.93}$$

$$B_1(p^2; m_a^2, m_b^2) = -(4\pi)^\epsilon\,\Gamma(\epsilon)\int_0^1 \left(\frac{p^2x^2 - (p^2+m_a^2-m_b^2)x + m_a^2 - i\varepsilon}{\mu^2}\right)^{-\epsilon} x\,dx\,, \tag{K.94}$$

$$B_{21}(p^2; m_a^2, m_b^2) = (4\pi)^\epsilon\,\Gamma(\epsilon)\int_0^1 \left(\frac{p^2x^2 - (p^2+m_a^2-m_b^2)x + m_a^2 - i\varepsilon}{\mu^2}\right)^{-\epsilon} x^2\,dx\,, \tag{K.95}$$

$$B_{22}(p^2; m_a^2, m_b^2) = \frac{(4\pi)^\epsilon\,\Gamma(\epsilon)}{2(1-\epsilon)}\mu^2\int_0^1 \left(\frac{p^2x^2 - (p^2+m_a^2-m_b^2)x + m_a^2 - i\varepsilon}{\mu^2}\right)^{1-\epsilon} dx\,. \tag{K.96}$$

Check that in the limit of $\epsilon \to 0$, you recover the results of eqs. (K.20)–(K.24).

K.2 (a) Derive the following limiting cases starting from eq. (K.10) and using the method of partial fractions:

$$B_0(0; m_a^2, m_b^2) = \frac{A_0(m_a^2) - A_0(m_b^2)}{m_a^2 - m_b^2}, \tag{K.97}$$

$$B_0(0; m^2, 0) = \frac{A_0(m^2)}{m^2}, \tag{K.98}$$

$$B_0(0; m^2, m^2) = \frac{\partial A_0(m^2)}{\partial m^2}, \tag{K.99}$$

$$\frac{\partial B_0}{\partial m_a^2}(0; m_a^2, m_b^2) = \frac{1}{m_a^2 - m_b^2}\left[\frac{\partial A_0(m_a^2)}{\partial m_a^2} - B_0(0; m_a^2, m_b^2)\right], \tag{K.100}$$

$$\left.\frac{\partial B_0}{\partial m_a^2}(0; m_a^2, m_b^2)\right|_{m_a=m_b=m} = -\frac{\epsilon}{2m^2}\frac{\partial A_0(m^2)}{\partial m^2} = -\frac{1}{2m^2} + \mathcal{O}(\epsilon), \tag{K.101}$$

$$\left.\frac{\partial B_0}{\partial m_a^2}(0; m_a^2, m_b^2)\right|_{m_a=0} = \frac{A_0(m_b^2)}{m_b^4}, \tag{K.102}$$

where $\partial A_0(m^2)/\partial m^2$ is given in eq. (K.50).

(b) Derive the following exact relation:

$$B_1(0; m_a^2, m_b^2) = -\tfrac{1}{2}\left[B_0(0; m_a^2, m_b^2) + (m_a^2 - m_b^2)B_0'(0; m_a^2, m_b^2)\right], \tag{K.103}$$

where $B_0'(0; m_a^2, m_b^2) \equiv \left(\partial B_0(p^2; m_a^2, m_b^2)/\partial p^2\right)|_{p^2=0}$.

(c) Starting from eq. (K.56), derive the following additional limiting cases after dropping terms of $\mathcal{O}(\epsilon)$:

$$B_0'(0; m_a^2, m_b^2) = \frac{1}{(m_a^2 - m_b^2)^3}\left[\tfrac{1}{2}(m_a^4 - m_b^4) + m_b^2 A_0(m_a^2) - m_a^2 A_0(m_b^2)\right], \tag{K.104}$$

$$B_0'(0; m^2, m^2) = \frac{1}{6m^2}, \qquad B_0'(0; m^2, 0) = \frac{1}{2m^2}. \tag{K.105}$$

(d) If $\lambda(p^2, m_a^2, m_b^2) = 0$, then it follows that either $p^2 = (m_a + m_b)^2$ or $p^2 = (m_b - m_a)^2$. The case of $p^2 = (m_a + m_b)^2$ is treated in Exercise K.3. Show that if $p^2 = (m_b - m_a)^2$, then B_0' and $\partial B_0/\partial m_a^2$ are given by

$$B_0'((m_b - m_a)^2; m_a^2, m_b^2) = \frac{m_a + m_b}{2(m_b - m_a)^3}\left(\frac{A_0(m_a^2)}{m_a^2} - \frac{A_0(m_b^2)}{m_b^2}\right) - \frac{2}{(m_b - m_a)^2}, \tag{K.106}$$

$$\frac{\partial B_0}{\partial m_a^2}((m_b - m_a)^2; m_a^2, m_b^2) = \frac{1}{2(m_b - m_a)^2}\left[\frac{A_0(m_a^2)}{m_a^2} - \frac{A_0(m_b^2)}{m_b^2} - \frac{2(m_b - m_a)}{m_a}\right], \tag{K.107}$$

assuming that $m_a \neq m_b$. Verify that one recovers eqs. (K.105) and (K.101) by taking the limit of eqs. (K.106) and (K.107), respectively, as $m_a \to m_b$.

K.3 When $p^2 = (m_a + m_b)^2$, the ultraviolet finite loop functions $B_0'(p^2; m_a^2, m_b^2)$ and $\partial B_0(p^2, m_a^2, m_b^2)/\partial m_a^2$ exhibit a threshold singularity in the limit $\epsilon \to 0$.

(a) Using the integral expression for B_0' given by eq. (K.56) prior to performing the expansion around $\epsilon = 0$, show that, for $p^2 = (m_a + m_b)^2$,

$$B_0'((m_a + m_b)^2; m_a^2, m_b^2) = (4\pi\mu^2)^\epsilon\, \Gamma(1+\epsilon) \times \int_0^1 dx\, x(1-x)\left(\left[m_a - (m_a + m_b)x\right]^2\right)^{-1-\epsilon}. \tag{K.108}$$

This integral is well defined if $\operatorname{Re}\epsilon < -\frac{1}{2}$. After integration, we shall analytically continue the result in the complex ϵ-plane to the region near $\epsilon = 0$.

(b) Evaluate the integral in eq. (K.108) as follows. Break up the integration range into two intervals, $0 \leq x \leq m_a/(m_a + m_b)$ and $m_a/(m_a + m_b) \leq x \leq 1$. In the first interval, define a new integration variable by $y = x(m_a + m_b)/m_a$, and in the second interval define $y = (1 - x)(m_a + m_b)/m_b$. Evaluate both integrals under the assumption that $\operatorname{Re}\epsilon < -\frac{1}{2}$, and show that

$$B_0'((m_a + m_b)^2; m_a^2, m_b^2) = \frac{(4\pi)^\epsilon \Gamma(\epsilon)}{2(1 - 4\epsilon^2)(m_a + m_b)^2} \times \left\{\left(\frac{m_a^2}{\mu^2}\right)^{-\epsilon}\left[\frac{m_b - m_a}{m_a + m_b} - 2\epsilon\right] + \left(\frac{m_b^2}{\mu^2}\right)^{-\epsilon}\left[\frac{m_a - m_b}{m_a + m_b} - 2\epsilon\right]\right\}, \tag{K.109}$$

which reveals the threshold singularity as $\epsilon \to 0$. Expand about $\epsilon = 0$ and determine the leading $\mathcal{O}(\epsilon^{-1})$ and the $\mathcal{O}(1)$ terms of the expansion.

(c) Evaluate $\partial B_0(p^2, m_a^2, m_b^2)/\partial m_a^2$ at $p^2 = (m_a + m_b)^2$ following the strategy employed in deriving eq. (K.109). Show that for $m_a \neq 0$ and $\mathrm{Re}\,\epsilon < -\frac{1}{2}$,

$$\frac{\partial B_0}{\partial m_a^2}\big((m_a+m_b)^2; m_a^2, m_b^2\big) = \frac{(4\pi)^\epsilon\Gamma(\epsilon)}{2(m_a+m_b)^2}\left\{\left(\frac{m_a^2}{\mu^2}\right)^{-\epsilon}\left[1+\frac{m_b}{m_a}\right] + \frac{1}{1+2\epsilon}\left[\left(\frac{m_a^2}{\mu^2}\right)^{-\epsilon} - \left(\frac{m_b^2}{\mu^2}\right)^{-\epsilon}\right]\right\}. \quad \text{(K.110)}$$

(d) If one of the two masses vanishes, show that

$$2m^2 B_0'(m^2; m^2, 0) = 2m^2 B_0'(m^2; 0, m^2) = -\frac{(4\pi)^\epsilon\,\Gamma(\epsilon)}{1-2\epsilon}\left(\frac{m^2}{\mu^2}\right)^{-\epsilon}, \quad \text{(K.111)}$$

$$2m^2\frac{\partial B_0}{\partial m_a^2}(m^2; m_a^2, m^2)\bigg|_{m_a^2=0} = -\frac{(4\pi)^\epsilon\,\Gamma(\epsilon)}{1+2\epsilon}\left(\frac{m^2}{\mu^2}\right)^{-\epsilon}. \quad \text{(K.112)}$$

K.4 (a) Show that the dimensional regularization procedure implies that

$$\frac{\partial^r A_0(m^2)}{\partial(m^2)^r}\bigg|_{m=0} = 0\,, \quad \text{(K.113)}$$

for all nonnegative integers r.

(b) Verify that eq. (K.5) is equivalent to the $m_a \to 0$ limit of eq. (K.100), after identifying the poles in ϵ that characterize the UV and IR divergences.

K.5 (a) Using the integral representations of B_0 and B_1 obtained in Exercise K.1, derive the following three identities:

$$B_0(m^2; m^2, 0) = B_0(m^2; 0, m^2) = (4\pi)^\epsilon\Gamma(\epsilon)\left(\frac{m^2}{\mu^2}\right)^{-\epsilon}\frac{1}{1-2\epsilon}\,, \quad \text{(K.114)}$$

$$B_1(m^2; m^2, 0) = -\tfrac{1}{2}(4\pi)^\epsilon\Gamma(\epsilon)\left(\frac{m^2}{\mu^2}\right)^{-\epsilon}\frac{1}{(1-\epsilon)(1-2\epsilon)}\,, \quad \text{(K.115)}$$

$$B_1(m^2; 0, m^2) = -\frac{A_0(m^2)}{2m^2} = -\tfrac{1}{2}(4\pi)^\epsilon\Gamma(\epsilon)\left(\frac{m^2}{\mu^2}\right)^{-\epsilon}\frac{1}{1-\epsilon}\,. \quad \text{(K.116)}$$

(b) Using the integral representation of B_1 obtained in Exercise K.1 and the results of part (a), derive the following identities, assuming that $\mathrm{Re}\,\epsilon < 0$:

$$2m^2 B_1'(m^2; 0, m^2) = B_1(m^2; m^2, 0) - B_1(m^2; 0, m^2)\,, \quad \text{(K.117)}$$

$$2m^2 B_1'(m^2; m^2, 0) = -2B_1(m^2; m^2, 0)\,, \quad \text{(K.118)}$$

$$2m^2\frac{\partial B_1}{\partial m_a^2}(m^2; m_a^2, m^2)\bigg|_{m_a^2=0} = 2m^2 B_0'(m^2; 0, m^2) = -B_0(m^2; 0, m^2). \quad \text{(K.119)}$$

That is, $B_1'(m^2; 0, m^2)$ is finite as $\epsilon \to 0$, whereas the other two derivatives above exhibit an infrared divergence. Note that eqs. (K.118) and (K.117) can also be derived from eq. (K.58).

K.6 The following quantity appears in the computation of the one-loop self-energy function of the electron in QED with arbitrary gauge parameter ξ:

$$\mathcal{C}(p^2, m^2) \equiv p^2 C_{22}(0, p^2, p^2; 0, 0, m^2) + C_{24}(0, p^2, p^2; 0, 0, m^2) . \qquad \text{(K.120)}$$

The following identity is a consequence of eqs. (K.54) and (K.55):

$$\begin{aligned}\mathcal{C}(p^2, m^2) &= \frac{\partial}{\partial m_a^2}\left[p^2 B_{21}(p^2; m_a^2, m^2) + B_{22}(p^2; m_a^2, m^2)\right]\Bigg|_{m_a^2=0} \\ &= -\tfrac{1}{2}B_1(p^2; 0, m^2) - \tfrac{1}{2}(p^2 - m^2)B_0'(p^2; 0, m^2) , \qquad \text{(K.121)}\end{aligned}$$

after employing eqs. (K.62) and (K.63) and making use of eq. (K.58) to simplify the resulting expression.

(a) In light of eq. (K.67), show that if $p^2 \neq m^2$, then

$$B_0'(p^2; 0, m^2) = \frac{(1-2\epsilon)B_0(p^2; 0, m^2) + 2(1-\epsilon)B_1(p^2; 0, m^2)}{p^2 - m^2} , \qquad \text{(K.122)}$$

$$B_0''(p^2; 0, m^2) = \frac{-2\epsilon B_0'(p^2; 0, m^2) + 2(1-\epsilon)B_1'(p^2; 0, m^2)}{p^2 - m^2} , \qquad \text{(K.123)}$$

in $d = 4 - 2\epsilon$ spacetime dimensions, where the prime superscripts indicate differentiation with respect to p^2.

(b) Verify the following limits:

$$\lim_{p^2 \to m^2}(p^2 - m^2)B_0'(p^2; 0, m^2) = 0 , \qquad \text{(K.124)}$$

$$\lim_{p^2 \to m^2}(p^2 - m^2)B_0''(p^2; 0, m^2) = 0 . \qquad \text{(K.125)}$$

(c) Using the results of part (b), show that $\mathcal{C}(m^2, m^2) = -\tfrac{1}{2}B_1(m^2; 0, m^2)$ and

$$\frac{\partial \mathcal{C}(p^2, m^2)}{\partial p^2}\Bigg|_{p^2=m^2} = -\tfrac{1}{2}\left[B_0'(m^2; 0, m^2) + B_1'(m^2; 0, m^2)\right] . \qquad \text{(K.126)}$$

K.7 Show that the C_{ij} functions can be expressed in terms of A_0, B_0, B_1, and C_0 via the following relations [245]:

$$C_{11} = \frac{4(p_1 \cdot p_2 R_2 - p_2^2 R_1)}{\lambda(p^2, p_1^2, p_2^2)} , \qquad C_{12} = \frac{4(p_1 \cdot p_2 R_1 - p_1^2 R_2)}{\lambda(p^2, p_1^2, p_2^2)} , \qquad \text{(K.127)}$$

$$C_{21} = \frac{4(p_1 \cdot p_2 R_5 - p_2^2 R_3)}{\lambda(p^2, p_1^2, p_2^2)} , \qquad C_{22} = \frac{4(p_1 \cdot p_2 R_4 - p_1^2 R_6)}{\lambda(p^2, p_1^2, p_2^2)} , \qquad \text{(K.128)}$$

$$C_{23} = \frac{4(p_1 \cdot p_2 R_3 - p_1^2 R_5)}{\lambda(p^2, p_1^2, p_2^2)} = \frac{4(p_1 \cdot p_2 R_6 - p_2^2 R_4)}{\lambda(p^2, p_1^2, p_2^2)} , \qquad \text{(K.129)}$$

where $\lambda(p^2, p_1^2, p_2^2) \equiv 4\left[(p_1\cdot p_2)^2 - p_1^2 p_2^2\right]$, with $p = -p_1 - p_2$, is the kinematic triangle function [see eq. (D.2)]. The R_i are defined as follows:

$$
\begin{aligned}
R_1 &\equiv \tfrac{1}{2}\left[B_0\left(p^2; m_a^2, m_c^2\right) - B_0\left(p_2^2; m_b^2, m_c^2\right) - \left(p_1^2 + m_a^2 - m_b^2\right) C_0\right], \\
R_2 &\equiv \tfrac{1}{2}\left[B_0\left(p_1^2; m_a^2, m_b^2\right) - B_0\left(p^2; m_a^2, m_c^2\right) + \left(p_1^2 - p^2 - m_b^2 + m_c^2\right) C_0\right], \\
R_3 &\equiv -C_{24} - \tfrac{1}{2}\left[(p_1^2 + m_a^2 - m_b^2)C_{11} - B_1\left(p^2; m_a^2, m_c^2\right) - B_0\left(p_2^2; m_b^2, m_c^2\right)\right], \\
R_4 &\equiv -\tfrac{1}{2}\left[(p_1^2 + m_a^2 - m_b^2)C_{12} - B_1\left(p^2; m_a^2, m_c^2\right) + B_1\left(p_2^2; m_b^2, m_c^2\right)\right], \\
R_5 &\equiv -\tfrac{1}{2}\left[(p^2 - p_1^2 + m_b^2 - m_c^2)C_{11} - B_1\left(p_1^2; m_a^2, m_b^2\right) + B_1\left(p^2; m_a^2, m_c^2\right)\right], \\
R_6 &\equiv -C_{24} - \tfrac{1}{2}\left[(p^2 - p_1^2 + m_b^2 - m_c^2)C_{12} + B_1\left(p^2; m_a^2, m_c^2\right)\right], \qquad \text{(K.130)}
\end{aligned}
$$

where the expression for C_{24} given by eq. (K.135) is derived in Exercise K.8 and the suppressed arguments of C_0 and the C_{ij} are $(p_1^2, p_2^2, p^2; m_a^2, m_b^2, m_c^2)$.

K.8 (a) Derive the following two identities:

$$g_{\mu\nu}B^{\mu\nu} \equiv -16\pi^2 i\mu^{2\epsilon}\int \frac{d^dq}{(2\pi)^d}\frac{q^2}{D_B} = A_0(m_b^2) + m_a^2 B_0(p^2; m_a^2, m_b^2), \qquad \text{(K.131)}$$

$$
\begin{aligned}
g_{\mu\nu}C^{\mu\nu} &\equiv -16\pi^2 i\mu^{2\epsilon}\int \frac{d^dq}{(2\pi)^d}\frac{q^2}{D_C} \\
&= B_0(p_2^2; m_b^2, m_c^2) + m_a^2 C_0(p_1^2, p_2^2, p^2; m_a^2, m_b^2, m_c^2), \qquad \text{(K.132)}
\end{aligned}
$$

where D_B and D_C are defined in eqs. (K.12) and (K.13), respectively.

(b) In light of eq. (K.17),

$$g_{\mu\nu}C^{\mu\nu} = p_1^2 C_{21} + p_2^2 C_{22} + 2p_1\cdot p_2 C_{23} + dC_{24}, \qquad \text{(K.133)}$$

where the arguments of all the C-type loop functions are $(p_1^2, p_2^2, p^2; m_a^2, m_b^2, m_c^2)$. Using the results of Exercise K.7, show that

$$
\begin{aligned}
p_1^2 C_{21} + p_2^2 C_{22} + 2p_1\cdot p_2 C_{23} &= R_3 + R_6 \\
&= \tfrac{1}{2}B_0(p_2^2; m_b^2, m_c^2) - 2C_{24} - \tfrac{1}{2}(p_1^2 + m_a^2 - m_b^2)C_{11} \\
&\quad - \tfrac{1}{2}(p^2 - p_1^2 + m_b^2 - m_c^2)C_{12}. \qquad \text{(K.134)}
\end{aligned}
$$

(c) Using eq. (K.132) and the results of part (b), derive the following expression for C_{24}:

$$
\begin{aligned}
(d-2)C_{24}(p_1^2, p_2^2, p^2; m_a^2, m_b^2, m_c^2) &= \tfrac{1}{2}\left[B_0(p_2^2; m_b^2, m_c^2) + (p_1^2 + m_a^2 - m_b^2)C_{11}\right. \\
&\quad \left. + (p^2 - p_1^2 + m_b^2 - m_c^2)C_{12} + 2m_a^2 C_0\right], \\
&\qquad \text{(K.135)}
\end{aligned}
$$

where $d \equiv 4 - 2\epsilon$ and the arguments of C_0, C_{11}, and C_{12} are the same as those of C_{24}. Show that eq. (K.135) yields the identity given in eq. (K.77) after dropping terms of $\mathcal{O}(\epsilon)$.

K.9 (a) Starting from the integral representations of the C-type loop integrals, derive the following identities:

$$C_0(p^2,0,p^2;m_a^2,m_b^2,m_c^2) = \frac{1}{m_b^2-m_c^2}\left[B_0(p^2;m_a^2,m_b^2) - B_0(p^2;m_a^2,m_c^2)\right], \tag{K.136}$$

$$C_{11}(p^2,0,p^2;m_a^2,m_b^2,m_c^2) = \frac{1}{m_b^2-m_c^2}\left[B_1(p^2;m_a^2,m_b^2) - B_1(p^2;m_a^2,m_c^2)\right], \tag{K.137}$$

$$C_{21}(p^2,0,p^2;m_a^2,m_b^2,m_c^2) = \frac{1}{m_b^2-m_c^2}\left[B_{21}(p^2;m_a^2,m_b^2) - B_{21}(p^2;m_a^2,m_c^2)\right]. \tag{K.138}$$

(b) Show that the limit of eq. (K.136) as $m_b^2 \to m_c^2$ yields eq. (K.52).

K.10 Suppose that p and p' are four-momenta of particles of mass m. Furthermore, assume that the on-shell conditions, $p^2 = p'^2 = m^2$, are satisfied.

(a) Using eq. (K.43) and the results of Exercise K.7, derive the following three identities:

$$C_{11}(p^2,q^2,p^2;M^2,m^2,m^2) = 2C_{12}(p^2,q^2,p^2;M^2,m^2,m^2)\,, \tag{K.139}$$

$$C_{21}(p^2,q^2,p^2;M^2,m^2,m^2) = 2C_{23}(p^2,q^2,p^2;M^2,m^2,m^2)\,, \tag{K.140}$$

$$\begin{aligned}2C_{24}(p^2,q^2,p^2;M^2,m^2,m^2) &= -B_1(p^2;M^2,m^2)\\ &\quad + q^2\left[C_{23}(p^2,q^2,p^2;M^2,m^2,m^2) - 2C_{22}(p^2,q^2,p^2;M^2,m^2,m^2)\right].\end{aligned} \tag{K.141}$$

Note that eqs. (K.139) and (K.140) can also be obtained by employing the last equality of eqs. (K.88) and (K.91), respectively.

(b) Setting $q^2 = 0$ in eq. (K.140) yields

$$2C_{24}(p^2,0,p^2;M^2,m^2,m^2) + B_1(p^2;M^2,m^2) = 0\,. \tag{K.142}$$

Derive this result directly from the integral representations given in eqs. (K.22) and (K.75).

(c) Using eqs. (K.73)–(K.75) and the results of eqs. (K.141) and (K.142), show that

$$\begin{aligned}\left(\frac{\partial C_{24}(p^2,q^2,p^2;M^2,m^2,m^2)}{\partial q^2}\right)_{q^2=0} &= \frac{1}{12}\int_0^1 \frac{x^3\,dx}{p^2x^2+(m^2-M^2-p^2)x+M^2}\\ &= \tfrac{1}{2}C_{23}(p^2,0,p^2;M^2,m^2,m^2) - C_{22}(p^2,0,p^2;M^2,m^2,m^2).\end{aligned} \tag{K.143}$$

Bibliography

[B1] I. M. Benn and R. W. Tucker, *An Introduction to Spinors and Geometry with Applications in Physics* (Adam Hilger, Bristol, UK, 1987)

[B2] P. Budinich and A. Trautman, *The Spinorial Chessboard* (Springer-Verlag, Berlin, 1988)

[B3] M. Carmeli and S. Malin, *Theory of Spinors: An Introduction* (World Scientific, Singapore, 2000)

[B4] E. Cartan, *The Theory of Spinors* (Dover Publictions, New York, 1981)

[B5] G. Coddens, *From Spinors to Quantum Mechanics* (Imperial College Press, London, 2015)

[B6] E. M. Corson, *An Introduction to Tensors, Spinors, and Relativistic Wave Equations* (Chelsea Publishing Company, New York, 1953)

[B7] F. R. Harvey, *Spinors and Calibrations* (Academic Press, San Diego, 1990)

[B8] J. Hladik, *Spinors in Physics* (Springer-Verlag, New York, 1999)

[B9] D. J. Hurley and M. A. Vandyck, *Geometry, Spinors and Applications* (Praxis Publishing, Chichester, UK, 2000)

[B10] P. Lounesto, *Clifford Algebras and Spinors*, 2nd edition (Cambridge University Press, Cambridge, UK, 2001)

[B11] P. O'Donnell, *Introduction to 2-Spinors in General Relativity* (World Scientific, Singapore, 2003)

[B12] R. Penrose and W. Rindler, *Spinors and Space-Time, Vol. 1: Two-Spinor Calculus and Relativistic Fields* (Cambridge University Press, Cambridge, UK, 1984)

[B13] G. F. Torres del Castillo, *Spinors in Four-Dimensional Spaces* (Birkhäuser, New York, 1984)

[B14] J. Vaz, Jr. and R. Da Rocha, Jr., *An Introduction to Clifford Algebras and Spinors* (Oxford University Press, Oxford, UK, 2016)

[B15] V. A. Zhelnorovich, *Theory of Spinors and its Application in Physics and Mechanics* (Springer Nature, Cham, Switzerland, 2019)

[B16] I. J. R. Aitchison, *Supersymmetry in Particle Physics: An Elementary Introduction* (Cambridge University Press, Cambridge, UK, 2007)

[B17] H. Baer and X. Tata, *Weak Scale Sypersymmetry: From Superfields to Scattering Events* (Cambridge University Press, Cambridge, UK, 2006)

[B18] D. Bailin and A. Love, *Supersymmetric Gauge Field Theory and String Theory* (Institute of Physics Publishing, Bristol, UK, 1994)

[B19] P. Binétruy, *Supersymmetry: Theory, Experiment and Cosmology* (Oxford University Press, Oxford, UK, 2006)

[B20] I. L. Buchbinder and S. M. Kuzenko, *Ideas and Methods of Supersymmetry and Supergravity or a Walk through Superspace* (Institute of Physics Publishing, Bristol, UK, 1995)

[B21] M. Dine, *Supersymmetry and String Theory*, 2nd edition (Cambridge University Press, Cambridge, UK, 2015)

[B22] M. Drees, R. M. Godbole, and P. Roy, *Theory and Phenomenology of Sparticles* (World Scientific, Singapore, 2004)

[B23] P. G. O. Freund, *Introduction to Supersymmetry* (Cambridge University Press, Cambridge, UK, 1986)

[B24] A. S. Galperin, E. A. Ivanov, V. I. Ogievetsky, and E. S. Sokatchev, *Harmonic Superspace* (Cambridge University Press, Cambridge, UK, 2001)

[B25] J. Gates, M. T. Grisaru, M. Roček, and W. Siegel, *Superspace or One Thousand and One Lessons in Supersymmetry* (Benjamin/Cummings Publishing Company, Reading, MA, 1983)

[B26] E. B. Manoukian, *Quantum Field Theory II: Introductions to Quantum Gravity, Supersymetry and String Theory* (Springer International Publishing, Cham, Switzerland, 2016)

[B27] S. Khalil and S. Moretti, *Supersymmetry Beyond Minimality: From Theory to Experiment* (CRC Press, Boca Raton, FL, 2018)

[B28] P. Labelle, *Supersymmetry Demystified* (McGraw Hill, New York, 2017)

[B29] J. Łopuszański, *An Introduction to Symmetry and Supersymmetry in Quantum Field Theory* (World Scientific, Singapore, 1991)

[B30] R. N. Mohapatra, *Unification and Supersymmetry: The Frontiers of Quark-Lepton Physics*, 3rd edition (Springer-Verlag, New York, 2003)

[B31] H. J. W. Müller-Kirsten and A. Wiedemann, *Introduction to Supersymmetry*, 2nd edition (World Scientific, Singapore, 2010)

[B32] P. Nath, *Supersymmetry, Supergravity, and Unification* (Cambridge University Press, Cambridge, UK, 2017)

[B33] A. Petrov, *Quantum Superfield Supersymmetry* (Springer Nature, Cham, Switzerland, 2021)

[B34] O. Piguet and K. Sibold, *Renormalized Supersymmetry* (Birkhäuser, Boston, MA, 1986)

[B35] N. Polonsky, *Supersymmetry: Structure and Phenomena* (Springer-Verlag, Berlin, 2001)

[B36] S. Raby, *Supersymmetric Grand Unified Theories* (Springer International Publishing, Cham, Switzerland, 2017)

[B37] P. P. Srivastava, *Supersymmetry, Superfields and Supergravity: An Introduction* (Adam Hilger, Bristol, UK, 1986)

[B38] J. Terning, *Modern Supersymmetry* (Oxford Science Publications, Oxford, UK, 2006)

[B39] S. Weinberg, *The Quantum Theory of Fields, Vol. 3: Supersymmetry* (Cambridge University Press, Cambridge, UK, 2000)

[B40] J. Wess and J. Bagger, *Supersymmetry and Supergravity* (Princeton University Press, Princeton, NJ, 1992)

[B41] P. C. West, *Introduction to Supersymmetry and Supergravity*, 2nd edition (World Scientific, Singapore, 1990)

[B42] G. Dall'Agata and M. Zagermann, *Supergravity: From First Principles to Modern Applications* (Springer-Verlag, Berlin, 2021)

[B43] D. Z. Freedman and A. Van Proeyen, *Supergravity* (Cambridge University Press, Cambridge, UK, 2012)

[B44] Y. Tanii, *Introduction to Supergravity* (Springer, Tokyo, 2014)

[B45] M. Rausch de Traubenberg and M. Valenzuela, *A Supergravity Primer: From Geometrical Principles to the Final Lagrangian* (World Scientific, Singapore, 2020)

[B46] G. Kane and M. Shifman, editors, *The Supersymmetric World: The Beginnings of the Theory* (World Scientific, Singapore, 2000)

[B47] A. I. Akhiezer and S. V. Peletminsky, *Fields and Fundamental Interactions* (Taylor & Francis, London, 2002)

[B48] T. Banks, *Modern Quantum Field Theory: A Concise Introduction* (Cambridge University Press, Cambridge, UK, 2008)

[B49] L. Baulieu, J. Iliopoulos, and R. Sénéor, *From Classical to Quantum Fields* (Oxford University Press, Oxford, UK, 2014)

[B50] R. A. Bertlmann, *Anomalies in Quantum Field Theory* (Oxford University Press, Oxford, UK, 2000)

[B51] C. P. Burgess, *Introduction to Effective Field Theory* (Cambridge University Press, Cambridge, UK, 2021)

[B52] A. Das, *Lectures on Quantum Field Theory*, 2nd edition (World Scientific, Singapore, 2021)

[B53] E. Fradkin, *Quantum Field Theory: An Integrated Approach* (Princeton University Press, Princeton, NJ, 2021)

[B54] F. Gelis, *Quantum Field Theory: From Basics to Modern Topics* (Cambridge University Press, Cambridge, UK, 2019)

[B55] W. Greiner and J. Reinhardt, *Field Quantization* (Springer, Berlin, 1996)

[B56] C. Itzykson and J.-B. Zuber, *Quantum Field Theory* (McGraw-Hill, New York, 1980)

[B57] R. Kleiss, *Quantum Field Theory: A Diagrammatic Approach* (Cambridge University Press, Cambridge, UK, 2021)

[B58] E. B. Manoukian, *Quantum Field Theory I: Foundations and Abelian and Non-Abelian Gauge Theories* (Springer International Publishing, Cham, Switzerland, 2016)

[B59] M. Maggiore, *A Modern Introduction to Quantum Field Theory* (Oxford University Press, Oxford, UK, 2005)

[B60] U.-G. Meissner and A. Rusetsky, *Effective Field Theories* (Cambridge University Press, Cambridge, UK, 2022)

[B61] V. P. Nair, *Quantum Field Theory: A Modern Perspective* (Springer Science, New York, 2005)

[B62] H. Năstase, *Introduction to Quantum Field Theory* (Cambridge University Press, Cambridge, UK, 2020)

[B63] T. Padmanabhan, *Quantum Field Theory: The Why, What and How* (Springer International Publishing, Cham, Switzerland, 2016)
[B64] M. Peskin and D. Schroeder, *Introduction to Quantum Field Theory* (Westview Press, Boulder, CO, 1995)
[B65] A. A. Petrov and A. E. Blechman, *Effective Field Theories* (World Scientific, Singapore, 2016)
[B66] P. Roman, *Introduction to Quantum Field Theory* (John Wiley & Sons, New York, 1969)
[B67] L. H. Ryder, *Quantum Field Theory*, 2nd edition (Cambridge University Press, Cambridge, UK, 1996)
[B68] M. D. Schwartz, *Quantum Field Theory and the Standard Model* (Cambridge University Press, Cambridge, UK, 2014)
[B69] M. Shifman, *Advanced Topics in Quantum Field Theory*, 2nd edition (Cambridge University Press, Cambridge, UK, 2022)
[B70] M. Srednicki, *Quantum Field Theory* (Cambridge University Press, Cambridge, UK, 2007)
[B71] R. Ticciati, *Quantum Field Theory for Mathematicians* (Cambridge University Press, Cambridge, UK, 1999)
[B72] S. Weinberg, *The Quantum Theory of Fields, Vol. 1: Foundations* (Cambridge University Press, Cambridge, UK, 1995)
[B73] S. Weinberg, *The Quantum Theory of Fields, Vol. 2: Modern Applications* (Cambridge University Press, Cambridge, UK, 1996)
[B74] A. Zee, *Quantum Field Theory in a Nutshell*, 2nd edition (Princeton University Press, Princeton, NJ, 2010)
[B75] D. Bailin and A. Love, *Introduction to Gauge Field Theory*, revised edition (Institute of Physics Publishing, Bristol, UK, 1993)
[B76] M. Böhm, A. Denner, and H. Joos, *Gauge Theories of the Strong and Electromagnetic Interaction* (B.G. Teubner, Stuttgart, Germany, 2001)
[B77] T.-P. Cheng and L.-F. Li, *Gauge Theory of Elementary Particle Physics* (Oxford University Press, Oxford, UK, 1984)
[B78] M. J. D. Hamilton, *Mathematical Gauge Theory* (Springer International Publishing, Cham, Switzerland, 2017)
[B79] S. Pokorski, *Gauge Field Theories*, 2nd edition (Cambridge University Press, Cambridge, UK, 2000)
[B80] F. Scheck, *Electroweak and Strong Interactions: An Introduction to Theoretical Particle Physics*, 2nd edition (Springer-Verlag, Berlin, 1996)
[B81] J. C. Taylor, *Gauge Theories of Weak Interactions* (Cambridge University Press, Cambridge , UK, 1976)
[B82] D. Yu. Bardin and G. Passarino, *The Standard Model in the Making* (Oxford University Press, Oxford, UK, 1999)
[B83] I. I. Bigi and A. I. Sanda, *CP Violation*, 2nd edition (Cambridge University Press, Cambridge, UK, 2009)
[B84] G. C. Branco, L. Lavoura, and J. P. Silva, *CP Violation* (Oxford University Press, Oxford, UK, 1999)

[B85] A.J. Buras, *Gauge Theory of Weak Decays* (Cambridge University Press, Cambridge, UK, 2020)

[B86] C. Burgess and G. Moore, *The Standard Model: A Primer* (Cambridge University Press, Cambridge, UK, 2015)

[B87] J. F. Donoghue, E. Golowich, and B. R. Holstein, *Dynamics of the Standard Model*, 2nd edition (Cambridge University Press, Cambridge, UK, 2014)

[B88] T. Hübsch, *Advanced Concepts in Particle and Field Theory* (Cambridge University Press, Cambridge, UK, 2015)

[B89] P. Langacker, *The Standard Model and Beyond*, 2nd edition (CRC Press, Boca Raton, FL, 2017)

[B90] Y. Nagashima, *Elementary Particle Physics, Vol. 1: Quantum Field Theory and Particles* (Wiley-VCH, Weinheim, Germany, 2010)

[B91] Y. Nagashima, *Elementary Particle Physics, Vol. 2: Foundations of the Standard Model* (Wiley-VCH, Weinheim, Germany, 2013)

[B92] Y. Nagashima, *Beyond the Standard Model of Elementary Particle Physics* (Wiley-VCH, Weinheim, Germany, 2014)

[B93] S. Raby, *Introduction to the Standard Model and Beyond* (Cambridge University Press, Cambridge, UK, 2021)

[B94] P. Ramond, *Journeys Beyond the Standard Model* (Westview Press, Boulder, CO, 1999)

[B95] G. G. Ross, *Grand Unified Theories* (Westview Press, Boulder, CO, 1985)

[B96] M. S. Sozzi, *Discrete Symmetries and CP Violation: From Experiment to Theory* (Oxford University Press, Oxford, UK, 2008)

[B97] H. Bacry, *Lectures on Group Theory and Particle Theory* (Gordon and Breach Science Publishers, New York, 1977)

[B98] G. G. A. Bäuerle and E. A. de Kerf, *Lie Algebras Part 1: Finite and Infinite Dimensional Lie Algebras and Applications in Physics* (North-Holland Elsevier Science Publishers, Amsterdam, 1990)

[B99] A. M. Bincer, *Lie Groups and Lie Algebras: A Physicist's Perspective* (Oxford University Press, Oxford, UK, 2013)

[B100] R. N. Cahn, *Semi-Simple Lie Algebras and Their Representations* (Dover Publications, Mineola, NY, 2006)

[B101] R. Campoamor-Stursberg and M. Rausch de Traubenberg, *Group Theory in Physics: A Practitioner's Guide* (World Scientific, Singapore, 2019)

[B102] J. F. Cornwell, *Group Theory in Physics, Vol. 2* (Academic Press, London, 1985)

[B103] P. Cvitanović, *Group Theory: Birdtracks, Lie's, and Exceptional Groups* (Princeton University Press, Princeton, NJ, 2008)

[B104] A. Das and S. Okubo, *Lie Groups and Lie Algebras for Physicists* (World Scientific, Singapore, 2014)

[B105] P. G. Fré and A. Fedotov, *Groups and Manifods: Lectures for Physicists with Examples in Mathematica* (Walter de Gruyter GmbH, Berlin, 2018)

[B106] J. Fuchs and C. Schweigert, *Symmetries, Lie Algebra and Representations* (Cambridge University Press, Cambridge, UK, 1997)

[B107] H. Georgi, *Lie Algebras in Particle Physics*, 2nd edition (Westview Press, Boulder, CO, 1999)

[B108] R. Gilmore, *Lie Groups, Lie Algebras, and Some of Their Applications* (Dover Publications, Mineola, NY, 2002)

[B109] M. Guidry and Y. Sun, *Symmetry, Broken Symmetry,and Topology in Modern Physics* (Cambridge University Press, Cambridge, UK, 2022)

[B110] B. Hall, *Lie Groups, Lie Algebras, and Representations: An Elementary Introduction*, 2nd edition (Springer International, Cham, Switzerland, 2015)

[B111] F. Iachello, *Lie Algebras and Applications*, 2nd edition (Springer-Verlag, Berlin, 2015)

[B112] A. P. Isaev and V. A. Rubakov, *Theory of Groups and Symmetries: Finite Groups, Lie Groups and Lie Algebras* (World Scientific, Singapore, 2018)

[B113] A. P. Isaev and V. A. Rubakov, *Theory of Groups and Symmetries: Representations of Groups and Lie Algebras, Applications* (World Scientific, Singapore, 2021)

[B114] H. F. Jones, *Groups, Representations and Physics*, 2nd edition (Institute of Physics Publishing, Bristol, UK, 1998)

[B115] Z.-Q. Ma, *Group Theory for Physicists*, 2nd edition (World Scientific, Singapore, 2019)

[B116] W. G. McKay and J. Patera, *Tables of Dimensions, Indices, and Branching Rules for Representations of Simple Lie Algebras* (Marcel Dekker, New York, 1981)

[B117] W. Miller, Jr., *Symmetry Groups and Their Applications* (Academic Press, New York, 1972)

[B118] N. Mukunda and S. Chaturvedi, *Continuous Groups for Physicists* (Cambridge University Press, Cambridge, UK, 2022)

[B119] L. O'Raifeartaigh, *Group Structure of Gauge Theories* (Cambridge University Press, Cambridge, UK, 1986)

[B120] P. B. Pal, *A Physicist's Introduction to Algebraic Structures* (Cambridge University Press, Cambridge, UK, 2019)

[B121] P. Ramond, *Group Theory: A Physicist's Survey* (Cambridge University Press, Cambridge, UK, 2010)

[B122] W.-K. Tung, *Group Theory in Physics* (World Scientific, Singapore, 1985)

[B123] J. D. Vergados, *Group and Representation Theory* (World Scientific, Singapore, 2017)

[B124] B. G. Wybourne, *Classical Groups for Physicists* (John Wiley & Sons, New York, 1974)

[B125] A. Zee, *Group Theory in a Nutshell for Physicists* (Princeton University Press, Princeton, NJ, 2016)

[B126] D. Bernstein, *Scalar, Vector, and Matrix Mathematics: Theory, Facts, and Formulas*, 3rd edition (Princeton University Press, Princeton, NJ, 2018)

[B127] W.H. Greub, *Linear Algebra*, 4th edition (Springer-Verlag, New York, 1975)

[B128] R. A. Horn and C. R. Johnson, *Matrix Analysis*, 2nd edition (Cambridge University Press, Cambridge, UK, 2013)

[B129] R. A. Horn and C. R. Johnson, *Topics in Matrix Analysis* (Cambridge University Press, Cambridge, UK, 1991)

[B130] R. U. Sexl and H. K. Urbantke, *Relativity, Groups, Particles: Special Relativity and Relativistic Symmetry in Field and Particle Physics* (Springer-Verlag, Vienna, 2001)

[B131] K. Sundermeyer, *Symmetries in Fundamental Physics*, 2nd edition (Springer International Publishing, Cham, Switzerland, 2014)

[B132] M. Markoutsakis, *Geometry, Symmetries, and Classical Physics: A Mosaic* (CRC Press, Boca Raton, FL, 2022)

[B133] P. A. Carruthers, *Spin and Isospin in Particle Physics* (Gordon and Breach Science Publishers, New York, 1971)

[B134] M. Chaichian and R. Hagedorn, *Symmetries in Quantum Mechanics: From Angular Momentum to Supersymmetry* (IoP Publishing, Bristol, UK, 1998)

[B135] D. Bailin, *Weak Interactions*, 2nd edition (Adam Hilger, Bristol, UK, 1982)

[B136] Yu. V. Novozhilov, *Introduction to Elementary Particle Theory* (Pergamon Press, Oxford, UK, 1975)

[B137] S. Profumo, *An Introduction to Particle Dark Matter* (World Scientific, Singapore, 2017)

[B138] M. Fukugita and T. Yanagida, *Physics of Neutrinos and Applications to Astrophysics* (Springer-Verlag, Berlin, 2003)

[B139] C. Giunti and C. W. Kim, *Fundamentals of Neutrino Physics and Astrophysics* (Oxford University Press, Oxford, UK, 2007)

[B140] R. N. Mohapatra and P. B. Pal, *Massive Neutrinos in Physics and Astrophysics*, 3rd edition (World Scientific, Singapore, 2004)

[B141] Z.-Z. Xing and S. Zhou, *Neutrinos in Particle Physics, Astronomy and Cosmology* (Springer-Verlag, Berlin, 2011)

[B142] J. W. F. Valle and J. C. Romão, *Neutrinos in High Energy and Astroparticle Physics* (Wiley-VCH, Weinheim, Germany, 2015)

[B143] J. F. Gunion, H. E. Haber, G. Kane, and S. Dawson, *The Higgs Hunter's Guide* (Westview Press, Boulder, CO, 2000)

[B144] A. I. Miller, *Early Quantum Electrodynamics: A Source Book* (Cambridge University Press, Cambridge, UK, 1994)

[B145] V. A. Miransky, *Dynamical Symmetry Breaking in Quantum Field Theory* (World Scientific, Singapore, 1993)

[B146] K. Byckling and K. Kajantie, *Particle Kinematics* (John Wiley & Sons, London, 1973)

[B147] E. Leader, *Spin in Particle Physics* (Cambridge University Press, Cambridge, UK, 2001)

[B148] M. L. Perl, *High Energy Hadron Physics* (John Wiley & Sons, New York, 1974)

[B149] H. Elvang and Y.-T. Huang, *Scattering Amplitudes in Gauge Theory and Gravity* (Cambridge University Press, Cambridge, UK, 2015)

[B150] S. Weinzierl, *Feynman Integrals: A Comprehensive Treatment for Students and Researchers* (Springer Nature, Cham, Switzerland, 2022)

References

[1] H. K. Dreiner, H. E. Haber, and S. P. Martin, Phys. Rept. **494**, 1 (2010)

[2] S. P. Martin, *A Supersymmetry Primer*, Adv. Ser. Direct. High Energy Phys. **18**, 1 (1998); **21**, 1 (2010); expanded version in arXiv:hep-ph/9709356.

[3] H. E. Haber and L. Stephenson Haskins, in *Anticipating the Next Discoveries in Particle Physics*, Proceedings of TASI-2016, eds. R. Essig and I. Low (World Scientific, 2018), pp. 355–499.

[4] L. Corwin, Y. Ne'eman, and S. Sternberg, Rev. Mod. Phys. **47**, 573 (1975)

[5] P. Fayet and S. Ferrara, Phys. Rept. **32**, 249 (1977)

[6] A. Salam and J. A. Strathdee, Fortsch. Phys. **26**, 57 (1978)

[7] P. van Nieuwenhuizen, Phys. Rept. **68**, 189 (1981)

[8] L. Susskind, Phys. Rept. **104**, 181 (1984)

[9] H. P. Nilles, Phys. Rept. **110**, 1 (1984)

[10] H. E. Haber and G. L. Kane, Phys. Rept. **117**, 75 (1985)

[11] M. F. Sohnius, Phys. Rept. **128**, 39 (1985)

[12] A. B. Lahanas and D. V. Nanopoulos, Phys. Rept. **145**, 1 (1987)

[13] G. F. Giudice and R. Rattazzi, Phys. Rept. **322**, 419 (1999)

[14] M. Carena and H. E. Haber, Prog. Part. Nucl. Phys. **50**, 63 (2003)

[15] D. J. H. Chung, L. L. Everett, G. L. Kane, S. F. King, J. D. Lykken, and L.-T. Wang, Phys. Rept. **407**, 1 (2005)

[16] R. Barbier et al., Phys. Rept. **420**, 1 (2005)

[17] A. Djouadi, Phys. Rept. **459**, 1 (2008)

[18] T. Goto, *Formulae for Supersymmetry: MSSM and More*, unpublished

[19] J. D. Lykken, in *Fields, Strings and Duality*, Proceedings of TASI-1996, eds. C. Efthimiou and B. Greene (World Scientific, 1997), pp. 85–153.

[20] A. Bilal, *Introduction to Supersymmetry*, arXiv:hep-th/0101055

[21] J. M. Figueroa-O'Farrill, *BUSSTEPP Lectures on Supersymmetry*, arXiv: hep-th/0109172

[22] M. Quirós, in *Particle Physics and Cosmology*, Proceedings of TASI-2002, eds. H. E. Haber and A. E. Nelson (World Scientific, 2004), pp. 549–601.

[23] M. A. Luty, in *Physics in* $D \geq 4$, Proceedings of TASI-2004, eds. J. Terning, C. E. M. Wagner, and D. Zeppenfeld (World Scientific, 2006), pp. 495–582.

[24] Y. Shirman, in *The Dawn of the LHC Era*, Proceedings of TASI-2008, ed. T. Han (World Scientific, 2010), pp. 359–422.

[25] D. Bertolini, J. Thaler, and Z. Thomas, in *Searching for New Physics at Small and Large Scales*, Proceedings of TASI-2012, eds. M. Schmaltz and E. Pierpaoli (World Scientific, 2013), pp. 421–496.

[26] B. C. Allanach and H. E. Haber, in Ref. [41], pp. 1000–1018.
[27] M. D'Onofrio and F. Moortgat, in Ref. [41], pp. 1019–1037.
[28] E. Massa, Nuovo Cim. B **9**, 41 (1972)
[29] E. Leader and C. Lorcé, Phys. Rept. **541**, 163 (2014) [Erratum: ibid. **802**, 23 (2019)].
[30] G. Luders, Annals Phys. **2**, 1 (1957)
[31] D. Andrica and R.-A. Rohan, Carpathian J. Math. **30**, 23 (2014)
[32] B. A. Kniehl and A. Pilaftsis, Nucl. Phys. B **474**, 286 (1996)
[33] A. Denner, Fortsch. Phys. **41**, 307 (1993)
[34] D. Espriu, J. Manzano, and P. Talavera, Phys. Rev. D **66**, 076002 (2002)
[35] V. Hnizdo, Eur. J. Phys. **32**, 287 (2011)
[36] B. Kayser, Phys. Rev. D **30**, 1023 (1984)
[37] A. Denner, H. Eck, O. Hahn, and J. Kublbeck, Nucl. Phys. B **387**, 467 (1992)
[38] A. Pilaftsis, Phys. Rev. D **65**, 115013 (2002)
[39] W. Grimus and M. Löschner, Int. J. Mod. Phys. A **31**, 1630038 (2017) [Erratum: ibid. **32**, 1792001 (2017)].
[40] E. S. Abers and B. W. Lee, Phys. Rept. **9**, 1 (1973)
[41] R. L. Workman et al. (Particle Data Group), PTEP **2022**, 083C01 (2022)
[42] S. Dimopoulos and D. W. Sutter, Nucl. Phys. B **452**, 496 (1995)
[43] A. Hook, PoS **TASI2018**, 004 (2019) [arXiv:1812.02669].
[44] L. Di Luzio, M. Giannotti, E. Nardi, and L. Visinelli, Phys. Rept. **870**, 1 (2020)
[45] R. Jackiw, in *Current Algebra and Anomalies* (Princeton University Press, Princeton, NJ, 1985), pp. 81–210.
[46] A. Bilal, *Lectures on Anomalies*, arXiv:hep-th/0101055
[47] S. L. Adler, Phys. Rev. **177**, 2426 (1969)
[48] L. Alvarez-Gaume and E. Witten, Nucl. Phys. B **234**, 269 (1984)
[49] P. Batra, B. A. Dobrescu, and D. Spivak, J. Math. Phys. **47**, 082301 (2006)
[50] D. Costa, B. Dobrescu, and P. Fox, Phys. Rev. Lett. **123**, 151601 (2019)
[51] B. C. Allanach, B. Gripaios, and J. Tooby-Smith, JHEP **05**, 065 (2020)
[52] L. M. Krauss and F. Wilczek, Phys. Rev. Lett. **62**, 1221 (1989)
[53] T. Banks and M. Dine, Phys. Rev. D **45**, 1424 (1992)
[54] G. A. Christos, Phys. Rept. **116**, 251 (1984)
[55] T. Kimura, Prog. Theor. Phys. **42**, 1191 (1969)
[56] A. Pawl, JHEP **03**, 034 (2005)
[57] G. C. Branco, P. M. Ferreira, L. Lavoura, M. N. Rebelo, M. Sher, and J. P. Silva, Phys. Rept. **516**, 1 (2012)
[58] H. E. Haber and D. O'Neil, Phys. Rev. D **74**, 015018 (2006) [Erratum: ibid. **74**, 059905 (2006)].
[59] R. Boto, T. V. Fernandes, H. E. Haber, J. C. Romão, and J. P. Silva, Phys. Rev. D **101**, 055023 (2020)
[60] H. E. Haber and D. O'Neil, Phys. Rev. D **83**, 055017 (2011)
[61] S. L. Glashow and S. Weinberg, Phys. Rev. D **15**, 1958 (1977)

[62] E. A. Paschos, Phys. Rev. D **15**, 1966 (1977)
[63] L. J. Hall and M. B. Wise, Nucl. Phys. B **187**, 397 (1981)
[64] M. Aoki, S. Kanemura, K. Tsumura, and K. Yagyu, Phys. Rev. D **80**, 015017 (2009)
[65] R. D. Peccei and H. R. Quinn, Phys. Rev. D **16**, 1791 (1977)
[66] A. Arbey, F. Mahmoudi, O. Stal, and T. Stefaniak, Eur. Phys. J. C **78**, 182 (2018)
[67] P. Langacker, Phys. Rept. **72**, 185 (1981)
[68] R. Slansky, Phys. Rept. **79**, 1 (1981)
[69] N. Yamatsu, arXiv:1511.08771v2
[70] R. Feger, T. W. Kephart, and R. J. Saskowski, Comput. Phys. Commun. **257**, 107490 (2020)
[71] H. Georgi and S. L. Glashow, Phys. Rev. Lett. **32**, 438 (1974)
[72] H. Georgi, H. R. Quinn, and S. Weinberg, Phys. Rev. Lett. **33**, 451 (1974)
[73] H. Georgi and C. Jarlskog, Phys. Lett. B **86**, 297 (1979)
[74] J. A. Casas and A. Ibarra, Nucl. Phys. B **618**, 171 (2001)
[75] S. Pascoli, S. T. Petcov, and W. Rodejohann, Phys. Rev. D **68**, 093007 (2003)
[76] J. A. Casas, A. Ibarra, and F. Jimenez-Alburquerque, JHEP **04**, 064 (2007)
[77] L.-F. Li, Phys. Rev. D **9**, 1723 (1974)
[78] V. Elias, S. Eliezer, and A. R. Swift, Phys. Rev. D **12**, 3356 (1975)
[79] F. Buccella, H. Ruegg, and C. A. Savoy, Nucl. Phys. B **169**, 68 (1980)
[80] N. Craig, *Naturalness: A Snowmass White Paper*, arXiv:2205.05708
[81] D. A. Eliezer and R. P. Woodard, Nucl. Phys. B **325**, 389 (1989)
[82] C. Csaki, C. Grojean, and J. Terning, Rev. Mod. Phys. **88**, 045001 (2016)
[83] T. Cohen, PoS **TASI2018**, 011 (2019) [arXiv:1903.03622].
[84] V. F. Weisskopf, Phys. Rev. **56**, 72 (1939)
[85] V. Weisskopf, Z. Phys. **89**, 27 (1934) [Erratum: ibid. **90**, 817 (1934)].
[86] H. Murayama, ICTP Ser. Theor. Phys. **16**, 296 (2000)
[87] J. Wess and B. Zumino, Phys. Lett. B **49**, 52 (1974)
[88] J. Wess and B. Zumino, Nucl. Phys. B **70**, 39 (1974)
[89] J. Wess and B. Zumino, Nucl. Phys. B **78**, 1 (1974)
[90] D. M. Capper and D. R. T. Jones, Phys. Rev. D **31**, 3295 (1985)
[91] L. Girardello and M. T. Grisaru, Nucl. Phys. B **194**, 65 (1982)
[92] S. R. Coleman and J. Mandula, Phys. Rev. **159**, 1251 (1967)
[93] R. Haag, J. T. Łopuszański, and M. Sohnius, Nucl. Phys. B **88**, 257 (1975)
[94] B. Zumino, Nucl. Phys. B **89**, 535 (1975)
[95] E. Witten, Nucl. Phys. B **202**, 253 (1982)
[96] G. D. Kribs, E. Poppitz, and N. Weiner, Phys. Rev. D **78**, 055010 (2008)
[97] D. Z. Freedman, Phys. Rev. D **15**, 1173 (1977)
[98] A. H. Chamseddine and H. K. Dreiner, Nucl. Phys. B **458**, 65 (1996)
[99] M. Dine, P. Draper, H. E. Haber, and L. Stephenson Haskins, Phys. Rev. D **94**, 095003 (2016)
[100] B. A. Ovrut and J. Wess, Phys. Rev. D **25**, 409 (1982)

[101] R. D. C. Miller, Phys. Lett. B **129**, 72 (1983)
[102] M. T. Grisaru, W. Siegel, and M. Rocek, Nucl. Phys. B **159**, 429 (1979)
[103] P. C. West, Phys. Lett. B **258**, 375 (1991)
[104] I. Jack, D. R. T. Jones, and P. C. West, Phys. Lett. B **258**, 382 (1991)
[105] D. C. Dunbar, I. Jack, and D. R. T. Jones, Phys. Lett. B **261**, 62 (1991)
[106] M. A. Shifman and A. I. Vainshtein, Nucl. Phys. B **277**, 456 (1986)
[107] M. A. Shifman and A. I. Vainshtein, Nucl. Phys. B **359**, 571 (1991)
[108] N. Seiberg, Phys. Lett. B **318**, 469 (1993)
[109] E. Poppitz and L. Randall, Phys. Lett. B **389**, 280 (1996)
[110] W. Siegel, Phys. Lett. B **84**, 193 (1979)
[111] D. M. Capper, D. R. T. Jones, and P. van Nieuwenhuizen, Nucl. Phys. B **167**, 479 (1980)
[112] W. Siegel, Phys. Lett. B **94**, 37 (1980)
[113] L. V. Avdeev and A. A. Vladimirov, Nucl. Phys. B **219**, 262 (1983)
[114] D. Stockinger, JHEP **03**, 076 (2005)
[115] G. 't Hooft and M. J. G. Veltman, Nucl. Phys. B **44**, 189 (1972)
[116] D. R. T. Jones and J. P. Leveille, Nucl. Phys. B **206**, 473 (1982) [Erratum: ibid. **222**, 517 (1983)].
[117] I. Jack, D. R. T. Jones, S. P. Martin, M. T. Vaughn, and Y. Yamada, Phys. Rev. D **50**, R5481 (1994)
[118] I. Jack, D. R. T. Jones, and C. G. North, Nucl. Phys. B **473**, 308 (1996)
[119] I. Jack, D. R. T. Jones, and C. G. North, Phys. Lett. B **386**, 138 (1996)
[120] I. Jack, D. R. T. Jones, and C. G. North, Nucl. Phys. B **486**, 479 (1997)
[121] I. Jack, D. R. T. Jones, and A. Pickering, Phys. Lett. B **435**, 61 (1998)
[122] V. A. Novikov, M. A. Shifman, A. I. Vainshtein, and V. I. Zakharov, Phys. Lett. B **166**, 329 (1986)
[123] N. Arkani-Hamed and H. Murayama, JHEP **06**, 030 (2000)
[124] I. Jack and D. R. T. Jones, Phys. Lett. B **415**, 383 (1997)
[125] I. Jack, D. R. T. Jones, and A. Pickering, Phys. Lett. B **426**, 73 (1998)
[126] I. Jack, D. R. T. Jones, and A. Pickering, Phys. Lett. B **432**, 114 (1998)
[127] Y. Yamada, Phys. Rev. D **50**, 3537 (1994)
[128] S. P. Martin and M. T. Vaughn, Phys. Rev. D **50**, 2282 (1994) [Erratum: ibid. **78**, 039903 (2008)].
[129] I. Jack, D. R. T. Jones, and R. Wild, Phys. Lett. B **509**, 131 (2001)
[130] I. Jack and D. R. T. Jones, Phys. Rev. D **63**, 075010 (2001)
[131] S. R. Coleman and E. J. Weinberg, Phys. Rev. D **7**, 1888 (1973)
[132] C. Ford, I. Jack, and D. R. T. Jones, Nucl. Phys. B **387**, 373 (1992) [Erratum: ibid. **504**, 551 (1997)].
[133] S. P. Martin, Phys. Rev. D **65**, 116003 (2002)
[134] S. P. Martin, Phys. Rev. D **89**, 013003 (2014)
[135] S. P. Martin, Phys. Rev. D **96**, 096005 (2017)
[136] I. Jack and D. R. T. Jones, Phys. Lett. B **333**, 372 (1994)
[137] P. Fayet and J. Iliopoulos, Phys. Lett. B **51**, 461 (1974)
[138] L. O'Raifeartaigh, Nucl. Phys. B **96**, 331 (1975)

[139] A. E. Nelson and N. Seiberg, Nucl. Phys. B **416**, 46 (1994)
[140] K. A. Intriligator, N. Seiberg, and D. Shih, JHEP **07**, 017 (2007)
[141] I. Affleck, M. Dine, and N. Seiberg, Nucl. Phys. B **241**, 493 (1984)
[142] K. A. Intriligator, N. Seiberg, and D. Shih, JHEP **04**, 021 (2006)
[143] A. Pomarol and S. Dimopoulos, Nucl. Phys. B **453**, 83 (1995)
[144] R. Rattazzi, Phys. Lett. B **375**, 181 (1996)
[145] L. J. Hall and L. Randall, Phys. Rev. Lett. **65**, 2939 (1990)
[146] I. Jack and D. R. T. Jones, Phys. Lett. B **457**, 101 (1999)
[147] C. S. Ün, Ş. H. Tanyıldızı, S. Kerman, and L. Solmaz, Phys. Rev. D **91**, 105033 (2015)
[148] U. Chattopadhyay and A. Dey, JHEP **10**, 027 (2016)
[149] G. Ross, K. Schmidt-Hoberg, and F. Staub, Phys. Lett. B **759**, 110 (2016)
[150] S. P. Martin, Phys. Rev. D **61**, 035004 (2000)
[151] W. Fischler, H. P. Nilles, J. Polchinski, S. Raby, and L. Susskind, Phys. Rev. Lett. **47**, 757 (1981)
[152] J. C. Pati and A. Salam, Phys. Rev. D **10**, 275 (1974) [Erratum: ibid. **11**, 703 (1975)].
[153] M. Ciuchini, V. Lubicz, L. Conti, et al., JHEP **10**, 008 (1998)
[154] V. Andreev et al. (ACME Collaboration), Nature **562**, 355 (2018)
[155] C. Cesarotti, Q. Lu, Y. Nakai, A. Parikh and M. Reece, JHEP **05**, 059 (2019)
[156] P. Draper and H. Rzehak, Phys. Rept. **619**, 1 (2016)
[157] P. Slavich, S. Heinemeyer, et al., Eur. Phys. J. C **81**, 450 (2021)
[158] Y. Yamada, Phys. Lett. B **623**, 104 (2005)
[159] S. P. Martin, Phys. Rev. D **72**, 096008 (2005)
[160] S. P. Martin, Phys. Rev. D **74**, 075009 (2006) [Erratum: ibid. **98**, 119901 (2018)].
[161] D. Dunsky, L. J. Hall, and K. Harigaya, JCAP **07**, 015 (2019)
[162] J. L. Feng, Ann. Rev. Nucl. Part. Sci. **63**, 351 (2013)
[163] L. Randall and R. Sundrum, Nucl. Phys. B **557**, 79 (1999)
[164] M. Luty and R. Sundrum, Phys. Rev. D **67**, 045007 (2003)
[165] R. Harnik, H. Murayama, and A. Pierce, JHEP **08**, 034 (2002)
[166] F. D'Eramo, J. Thaler, and Z. Thomas, JHEP **06**, 151 (2012)
[167] F. D'Eramo, J. Thaler, and Z. Thomas, JHEP **09**, 125 (2013)
[168] J. Scherk and J. H. Schwarz, Nucl. Phys. B **153**, 61 (1979)
[169] J. E. Kim and H. P. Nilles, Phys. Lett. B **138**, 150 (1984)
[170] G. F. Giudice and A. Masiero, Phys. Lett. B **206**, 480 (1988)
[171] L. J. Hall, J. D. Lykken, and S. Weinberg, Phys. Rev. D **27**, 2359 (1983)
[172] C. G. Lester, M. A. Parker, and M. J. White, JHEP **01**, 080 (2006)
[173] M. Carena, C. Grojean, M. Kado, and V. Sharma, in Ref. [41], pp. 201–260.
[174] J. A. Conley, H. K. Dreiner, and P. Wienemann, Phys. Rev. D **83**, 055018 (2011)
[175] J. R. Ellis, J. S. Hagelin, D. V. Nanopoulos, K. A. Olive, and M. Srednicki, Nucl. Phys. B **238**, 453 (1984)

[176] G. Jungman, M. Kamionkowski, and K. Griest, Phys. Rept. **267**, 195 (1996)

[177] H. Pagels and J. R. Primack, Phys. Rev. Lett. **48**, 223 (1982)

[178] S. D. Thomas and J. D. Wells, Phys. Rev. Lett. **81**, 34 (1998)

[179] H. K. Dreiner, *An Introduction to Explicit R-parity Violation*, Adv. Ser. Direct. High Energy Phys. **18**, 462 (1998); **21**, 565 (2010).

[180] N. Chamoun, F. Domingo and H.K. Dreiner, Phys. Rev. D **104**, 015020 (2021)

[181] M. Maniatis, Int. J. Mod. Phys. A **25**, 3505 (2010)

[182] U. Ellwanger, C. Hugonie, and A. M. Teixeira, Phys. Rept. **496**, 1 (2010)

[183] A. Dedes, H. E. Haber, and J. Rosiek, JHEP **11**, 059 (2007)

[184] S. P. Martin, Phys. Rev. D **79**, 095019 (2009)

[185] I. Antoniadis, J. R. Ellis, J. S. Hagelin, and D. V. Nanopoulos, Phys. Lett. B **194**, 231 (1987)

[186] L. E. Ibáñez and G. G. Ross, Nucl. Phys. B **368**, 3 (1992)

[187] C. Luhn, Ph.D. thesis, University of Bonn (2006)

[188] H. K. Dreiner, C. Luhn, and M. Thormeier, Phys. Rev. D **73**, 075007 (2006)

[189] B. C. Allanach, A. Dedes, and H. K. Dreiner, Phys. Rev. D **69**, 115002 (2004) [Erratum: ibid. **72**, 079902 (2005)].

[190] Y. Grossman and H. E. Haber, Phys. Rev. D **59**, 093008 (1999)

[191] D. Dercks, H. K. Dreiner, M. E. Krauss, T. Opferkuch, and A. Reinert, Eur. Phys. J. C **77**, 856 (2017)

[192] R. Cowsik and J. McClelland, Phys. Rev. Lett. **29**, 669 (1972)

[193] B.W. Lee and S. Weinberg, Phys. Rev. Lett. **39**, 165 (1977)

[194] H. K. Dreiner, S. Heinemeyer, O. Kittel, U. Langenfeld, A. M. Weber, and G. Weiglein, Eur. Phys. J. C **62**, 547 (2009)

[195] M. Carena, H. E. Haber, I. Low, N. R. Shah, and C. E. M. Wagner, Phys. Rev. D **93**, 035013 (2016)

[196] C. T. Hill and E. H. Simmons, Phys. Rept. **381**, 235 (2003) [Erratum: ibid. **390**, 553 (2004)].

[197] H. E. Haber, SciPost Phys. Lect. Notes **21**, 1 (2021)

[198] H. Goldberg, Phys. Rev. Lett. **50**, 1419 (1983) [Erratum: ibid. **103**, 099905 (2009)].

[199] M. Nowakowski, E. Paschos, and J. Rodriguez, Eur. J. Phys. **26**, 545 (2005)

[200] A. Gongora and R. G. Stuart, Z. Phys. C **55**, 101 (1992)

[201] S. Ferrara and E. Remiddi, Phys. Lett. B **53**, 347 (1974)

[202] B. Abi et al. (Muon $g-2$ Collaboration), Phys. Rev. Lett. **126**, 141801 (2021)

[203] T. Moroi, Phys. Rev. D **53**, 6565 (1996) [Erratum: ibid. **56**, 4424 (1997)].

[204] H. E. Haber and R. Hempfling, Phys. Rev. Lett. **66**, 1815 (1991)

[205] P. Draper and H. E. Haber, Eur. Phys. J. C **73**, 2522 (2013) [Erratum: ibid. **81**, 593 (2021)].

[206] R. H. Good, Rev. Mod. Phys. **27**, 187 (1955)

[207] J. F. Nieves and P. B. Pal, Am. J. Phys. **72**, 1100 (2004)

[208] Y. Takahashi, J. Math. Phys. **24**, 1783 (1982)
[209] C. C. Nishi, Am. J. Phys. **73**, 1160 (2005)
[210] M. Blatter, Helv. Phys. Acta **65**, 1011 (1992)
[211] F. Jegerlehner, Eur. Phys. J. C **18**, 673 (2001)
[212] P. Nyborg, H. Song, W. Kernan, and R. Good, Phys. Rev. **140**, B914 (1965)
[213] M. Jacob and G. C. Wick, Annals Phys. **7**, 404 (1959)
[214] C. Bouchiat and L. Michel, Nucl. Phys. **5**, 416 (1958)
[215] L. Michel, Nuovo Cim. **14**, 95 (1959)
[216] S. J. Parke and T. R. Taylor, Phys. Rev. Lett. **56**, 2459 (1986)
[217] Z. Xu, D.-H. Zhang, and L. Chang, Nucl. Phys. B **291**, 392 (1987)
[218] L. Autonne, Ann. Univ. Lyon, Nouvelle Série I, Fasc. **38**, 1 (1915)
[219] T. Takagi, Japan J. Math. **1**, 83 (1925)
[220] B. Zumino, J. Math. Phys. **3**, 1055 (1962)
[221] H. G. Becker, Lett. Nuovo Cim. **8**, 185 (1973)
[222] O. Napoly, Phys. Lett. B **106**, 125 (1981)
[223] H. E. Haber, Int. J. Mod. Phys. A **36**, 2130003 (2021)
[224] M. Suzuki, Commun. Math. Phys. **57**, 193 (1977)
[225] E. Dynkin, Am. Math. Soc. Transl. Series 2 **6**, 111 (1957)
[226] S. Okubo, Phys. Rev. D **16**, 3528 (1977)
[227] T. van Ritbergen, A. N. Schellekens, and J. A. M. Vermaseren, Int. J. Mod. Phys. A **14**, 41 (1999)
[228] A. Mountain, J. Math. Phys. **39**, 5601 (1998)
[229] J. A. de Azcarraga, A. J. Macfarlane, A. J. Mountain, and J. C. Perez Bueno, Nucl. Phys. B **510**, 657 (1998)
[230] A. J. Macfarlane and H. Pfeiffer, J. Math. Phys. **41**, 3192 (2000)
[231] A. B. Balantekin, J. Math. Phys. **23**, 486 (1982)
[232] J. Patera and R. T. Sharp, J. Math. Phys. **22**, 2352 (1981)
[233] J. Banks and H. Georgi, Phys. Rev. D **14**, 1159 (1976)
[234] H. Fegan and B. Steer, Math. Proc. Cambridge Phil. Soc. **105**, 249 (1989)
[235] J. Burns, Q. J. Math. **51**, 295 (2000)
[236] H. Braden, J. London Math. Soc. (2) **43**, 313 (1991)
[237] K. R. Dienes and J. March-Russell, Nucl. Phys. B **479**, 113 (1996)
[238] M. Esole and M. J. Kang, arXiv:2012.13401
[239] S.P. Martin, Nucl. Phys. B **338**, 244 (1990)
[240] J. Rosiek, Phys. Rev. D **41**, 3464 (1990) [Erratum: arXiv:hep-ph/9511250].
[241] M. Kuroda, arXiv:hep-ph/9902340
[242] J. F. Gunion and H. E. Haber, Nucl. Phys. B **272**, 1 (1986) [Erratum: ibid. **402**, 567 (1993)].
[243] G. 't Hooft and M. J. G. Veltman, Nucl. Phys. B **153**, 365 (1979)
[244] G. Passarino and M. J. G. Veltman, Nucl. Phys. B **160**, 151 (1979)
[245] W. Hollik, Adv. Ser. Direct. High Energy Phys. **14**, 37 (1995)
[246] R. K. Ellis, Z. Kunszt, K. Melnikov, and G. Zanderighi, Phys. Rept. **518**, 141 (2012)

Index

Printed in the United States
by Baker & Taylor Publisher Services